W9-CSU-499

third • edition

Life

third • edition

Ricki Lewis

the university at albany

WCB McGraw-Hill

Boston, Massachusetts Burr Ridge, Illinois Dubuque, Iowa
Madison, Wisconsin New York, New York San Francisco, California St. Louis, Missouri

WCB/McGraw-Hill

A Division of The **McGraw-Hill** *Companies*

LEWIS: LIFE

This book is printed on recycled paper containing 10% postconsumer waste.

1 2 3 4 5 6 7 8 9 0 VNH/VNH 9 0 9 8 7

ISBN 0-697-28563-4

Editorial Director: *Kevin Kane*
Publisher: *Michael Lange*
Sr. Developmental Editor: *Elizabeth M. Sievers*
Marketing Manager: *Julie Keck*
Project Manager: *Donna Nemmers*
Production Supervisor: *Cheryl Horch*
Designer: *K. Wayne Harms*
Cover photo: *AP/Wide World Photos*
Photo research coordinator: *Lori Hancock*
Art Editor: *Joyce Watters*
Compositor: *York Graphic Services, Inc.*
Typeface: *10/12 Times*
Printer: *Von Hoffmann Press, Inc.*

Library of Congress Cataloging-in-Publication Data

Lewis, Ricki
Life / Ricki Lewis. — 3rd ed.
p. cm.
Includes index.
ISBN 0-697-28563-4
1. Biology. 2. Human biology. I. Title.
QH308.2.L485 1997
570 — dc21 97–15290
CIP

INTERNATIONAL EDITION

1 2 3 4 5 6 7 8 9 0 VNH/VNH 9 0 9 8 7

When ordering this title, use ISBN 0-07-115374-8.

www.mhhe.com

Dedicated to Larry,

Heather, Sarah, & Carly

Publisher's Note to the Instructor

Full-Color Customization

In this third edition of Life, *your options are unlimited!* You can now create the ideal text for your course. Take a look at the contents of *Life.* Are there some chapters you don't address in your course? If so, select just those chapters that you do cover. Then let WCB/McGraw-Hill professionally print and bind the *full-color* general biology text that is right for your classroom.

Of course, you can still select the standard, full-color text. It's your choice. Contact your WCB/McGraw-Hill sales representative.

Recycled Paper

Life is printed on recycled paper stock. All of its ancillaries, as well as all advertising pieces for *Life,* will also be printed on recycled paper. In addition, when possible, the inks used in printing are soy based.

Our goal in using these materials for *Life* and its ancillary package is to minimize the environmental impact of our products.

Brief Contents

Unit 1
Overview of Biology 1
1 What Is Life? 3
2 Thinking Scientifically About Life 17
3 The Chemistry of Life 31

Unit 2
Cell Biology 57
4 Cells 59
5 Cellular Architecture 83
6 The Energy of Life 109
7 How Cells Release Energy 125
8 Photosynthesis 145
9 Cell Division and Death 167

Unit 3
Reproduction and Development 189
10 The Making of Gametes 191
11 Animal Development 209
12 Human Reproductive and Developmental Concerns 237

Unit 4
Genetics 259
13 Transmission Genetics 261
14 Chromosomes 285
15 DNA Structure and Replication 307
16 Gene Function 325

Unit 5
Evolution 349
17 The Origin and Diversification of Life 351
18 Darwinian Evolution 373
19 Evidence for Evolution 389
20 The Forces of Evolutionary Change 403

Unit 6
The Diversity of Life 423
21 Classifying Life 425
22 Monera 439
23 Protista 457
24 Fungi 475
25 Plantae 491
26 Animalia 507

Unit 7
Plant Life 531
27 Plant Form and Function 533
28 Plant Life Cycles 555
29 Plant Responses to Stimuli 575

Unit 8
Animal Life 591
30 Animal Tissues 593
31 The Nervous System 609
32 The Senses 639
33 The Endocrine System 659
34 The Musculoskeletal System 681
35 The Circulatory System 707
36 The Respiratory System 731
37 Digestion and Nutrition 749
38 Regulation of Temperature and Body Fluids 769
39 The Immune System 789

Unit 9
Behavior and Ecology 809
40 Behavior 811
41 Populations 835
42 Communities and Ecosystems 853
43 Biomes 871
44 Environmental Concerns 889

Contents

Preface xxi
Multimedia Correlation Chart xxiv

Unit 1
Overview of Biology 1

Chapter 1 ~ What Is Life? 3

Chapter Outline* 3
Distinguishing Life from Death 4
The Characteristics of Life 4
Organization 5
Bioethics Connections 1.1: What is Bioethics? 6
Energy Use and Metabolism 8
Maintenance of Internal Constancy 9
Reproduction, Growth, and Development 9
Irritability and Adaptation 10
Biodiversity 12
What's in a Name? 12
Biology in Action 1.1: Some Like It Hot: Thermophilic Bacteria 13
The Five Kingdoms 14
Evolution Is the Backbone of Biology 15
Summary* 15
To Review* 15
To Think About* 16
To Learn More* 16

Chapter 2 ~ Thinking Scientifically About Life 17

DES Daughters and Apopka Alligators 18
Theories Attempt to Explain Nature 18
Scientific Inquiry 18
Observations 19
Background Information 20
Formulating a Hypothesis 21
Devising an Experiment 21
Biology in Action 2.1: Elephants Calling—The Powers of Observation 22
Scientific Inquiry Continues 23
Designing Experiments 24
Experimental Controls 24
Types of Experiments 25
Applying the Scientific Method: Challenges and Limitations 28
Interpreting Results 28
Expecting the Unexpected 28
Scientific Discovery by "Accident" 28

Chapter 3 ~ The Chemistry of Life 31

Why Study Chemistry? 32
Chemistry Basics 32
Elements 32
Atoms 32
Molecules 34
Chemical Bonds 35
Biology in Action 3.1: NO and CO—Tiny Biological Messengers 36
Water—The Matrix of Life 40
Solutions 40
Acids and Bases 40
Water and Temperature 42
The Organic Molecules of Life 43
Carbohydrates 43
Lipids 44
Proteins 46
Nucleic Acids 50

Unit 2
Cell Biology 57

Chapter 4 ~ Cells 59

Cells—The Units of Life 60
The First Microscopes 60
The Cell Theory Emerges 61
Technology Improves 62
Cell Complexity 63
Cell Size 64
Prokaryotic Cells 64
Eukaryotic Cells 66
Archaean Cells 67
Organelles 69
Making Milk 69

*These elements appear in every chapter.

Other Organelles and Structures 71
Biology in Action 4.1: Lysosomal Storage Diseases 72
Simpler Than Cells 75
Viruses 75
Viroids 77
Prions 77
Bioethics Connections 4.1: Should We Destroy the Last Smallpox Virus? 78
The Origin of Eukaryotic Cells 79

Chapter 5 ~ Cellular Architecture 83

The Cell Surface and "Self" 84
Cellular Name Tags 84
The Cell Membrane—Cellular Gates 85
The Protein-Phospholipid Bilayer 85
Membrane Proteins 85
Movement Across Membranes 87
Diffusion 88
A Special Case of Diffusion—The Movement of Water (Osmosis) 88
Facilitated Diffusion 92
Active Transport 92
Large-Scale Membrane Transport 93
Exocytosis and Endocytosis 93
Cell Connections 96
Animal Cell Junctions 96
Plant Cell Junctions 96
The Cytoskeleton—Cellular Support 96
Biology in Action 5.1: Breaching the Blood-Brain Barrier 97
Microtubules 97
Microfilaments 100
Intermediate Filaments 100
The Cytoskeleton Molds Specialized Cells 100
Cell-to-Cell Interactions 103
Signal Transduction 103
Cell Adhesion 103
Bioethics Connections 5.1: Nicotine Addiction and Cell Biology 104

Chapter 6 ~ The Energy of Life 109

Life Requires Energy 110
What Is Energy? 110
Sources of Energy 110
Potential and Kinetic Energy 112
The Laws of Thermodynamics 112
The First Law of Thermodynamics 113
The Second Law of Thermodynamics 113
Metabolism 114
Anabolism (Synthesis) 114
Catabolism (Degradation) 114
Energy in Chemical Reactions 115
Chemical Equilibrium 116
Oxidation, Reduction, and Energy Content 117
ATP Cellular Energy Currency 117
High-Energy Phosphate Bonds 117
Biology in Action 6.1: Firefly Bioluminescence 118
Coupled Reactions 119
Other Compounds Involved in Energy Metabolism 119
Enzymes and Energy 121
Control of Metabolism 122
The Major Energy Transformations—Respiration and Photosynthesis 123

Chapter 7 ~ How Cells Release Energy 125

"Breathing" Respiration and Cellular Respiration 126
Ways That Cells Release Energy 127
Mitochondria 127
Energy Transfer from Nutrients to ATP 128
Routes to ATP Formation 129
Substrate-Level Phosphorylation 129
Oxidative Phosphorylation 129
Glycolysis—Glucose Breaks Down to Pyruvic Acid 129
The First Half of Glycolysis—Glucose Activation 129
The Second Half of Glycolysis—Energy Extraction 131
Acetyl CoA Formation Bridges Glycolysis and the Krebs Cycle 131
The Krebs Cycle 132
Oxidative Phosphorylation and Electron Transport 133

The Respiratory Chain 133
ATP Synthesis 133
Biology in Action 7.1: Disrupting the Respiratory Chain 135
ATP Accounting—An Estimate 136
Control of Cellular Respiration 136
Energy Retrieval in the Absence of Oxygen 137
Fermentation Pathways 137
Biology in Action 7.2: Whole-Body Metabolism—Energy on an Organismal Level 138
Anaerobic Electron Transport 140
How Proteins and Lipids Enter the Energy Pathways 140

Chapter 8 ~ Photosynthesis 145

The Importance of Photosynthesis 146
Photosynthesis: From Carbon Dioxide to Carbohydrate 146
The Beginnings of Photosynthesis 146
The Nature of Light 147
Probing Light 147
Types of Light 147
Biology in Action 8.1: Early Thoughts on Photosynthesis 148
Pigment Molecules 149
Chlorophyll 149
Accessory Pigments 150
Autumn Leaves and More 150
Chloroplasts 151
Photosystems and Reaction Centers in Chloroplasts 151
The Fate of Absorbed Light Energy 153
The Light-Dependent Reactions 153
Photosystem II Starts Photosynthesis 154
Photosystem I 155
Why Two Photosystems Increase Energy Extraction 156
Biology in Action 8.2: How Photosynthetic Bacteria Capture Light Energy 156
The Light-Independent Reactions 159
The Calvin Cycle 159
How Experiments Revealed the Steps of the Calvin Cycle 159
The Efficiency of Photosynthesis 160
Photorespiration 160
C_4 Photosynthesis 160
CAM Photosynthesis 162
The Big Picture of Biological Energy Use 162
Possible Origins of the Energy Pathways 163

Chapter 9 ~ Cell Division and Death 167

The Balance Between Cell Division and Cell Death 168
The Cell Cycle 168
Interphase 168
Participants in Mitosis 170
The Stages of Mitosis 171
Cytokinesis 173
Control of Cell Division 175
A Cellular Clock—Telomeres 175
Cell Division Signals 176
Apoptosis 177
Cell Division in a Tissue 178
Stem Cells 178
Cell Populations 179
Cancer—When the Cell Cycle Goes Awry 180
Characteristics of Cancer Cells 180
The Causes of Cancer 183
Genes That Cause Cancer 183
Environmental Influences 184
Bioethics Connections 9.1: How Much Should a Physician Tell a Cancer Patient? 185

Unit 3

Reproduction and Development 189

Chapter 10 ~ The Making of Gametes 191

Why Reproduce? 192
Reproductive Diversity 192
Conjugation—Sex Without Reproduction 192

Biology in Action 10.1: The Complicated Sex Life of the Aphid 194
Cell Fusion—Sex with Reproduction 194
Gamete Formation in the Human Body 196
The Human Male Reproductive System 196
The Human Female Reproductive System 197
Meiosis 198
Gamete Maturation 201
Sperm Development 203
Ovum Development 204

Chapter 11 ~ Animal Development 209

A Descriptive Science Becomes Experimental 210
Preformation Versus Epigenesis 210
Early Experiments with Embryos 210
Differentiation and Selective Gene Activation 211
Two Beginnings: Early Development in a Worm and a Fly 213
A Worm—*Caenorhabditis elegans* 213
A Fly—*Drosophila melanogaster* 215
The Stages of Human Prenatal Development 216
The Preembryonic Stage 217
The Embryonic Stage 223
Bioethics Connections 11.1: Preembryos as Research Subjects 224
Biology in Action 11.1: A Heart in the Wrong Place 227
The Fetal Period 229
Labor and Birth 229
Postnatal Growth and Development 230
Aging 232
What Triggers Aging? 233
The Human Life Span 234
Causes of Death 234
Is Longevity Inherited? 234

Chapter 12 ~ Human Reproductive and Developmental Concerns 237

Two Louises: Research and Reproductive Technology 238
Infertility 239
Infertility in the Male 239
Infertility in the Female 239
Infertility Tests 240
Assisted Reproductive Technologies 241
Donated Sperm—Artificial Insemination 241
A Donated Uterus—Surrogate Motherhood 242
Bioethics Connections 12.1: Surrogates and the Definition of Motherhood 242
In Vitro Fertilization 242
Gamete Intrafallopian Transfer 243
Oocyte Banking and Donation 244
Preimplantation Genetic Diagnosis 244
Problems in Pregnancy 245
Spontaneous Abortion 245
Born Too Soon—Prematurity 245
Birth Defects 246
Neonatology 249
Birth Control 250
Available Methods of Birth Control 250
Experimental Methods of Birth Control 250
Terminating a Pregnancy 252
Sexually Transmitted Diseases 253
Herpes Simplex 253
Chlamydia 253
Acquired Immune Deficiency Syndrome (AIDS) 253
Homosexuality 255
Psychological Aspects of Homosexuality 255
Biological Studies—Of Genes and Brains 255

Unit 4

Genetics 259

Chapter 13 ~ Transmission Genetics 261

The History of Genetics 262
Mendel's First Law (Segregation) Traces One Gene 262
How Biologists Describe Single Genes 265
Mendel's Laws and Meiosis 265
Mendelian Traits in Humans 267
Modes of Inheritance 267
Pedigrees 268
Mendel's Second Law (Independent Assortment) Traces Two Genes 269
When Gene Expression Appears to Alter Mendelian Ratios 272
Lethal Allele Combinations 272
Multiple Alleles 272
Different Dominance Relationships 273
Epistasis—Gene Masking at *General Hospital* 274
Penetrance and Expressivity 275
Environmental Influences 275
Pleiotropy 276
Phenocopies—When It's Not Really in the Genes 276
Complex Traits 276

Measuring Multifactorial Traits 277
Studying Multifactorial Traits 278
Bioethics Connections 13.1: Blaming Genes 281

Chapter 14 ~ Chromosomes 285

Studying Chromosomes 286
Linked Genes Do Not Follow Mendel's Laws 286
Peas and Flies Reveal Linkage 286
Linkage Maps 288
Sex-Linkage and Sex Chromosomes 290
Sex Determination in Humans 290
Biology in Action 14.1: Sex Determination—Male or Female? 290
Sex-Linked Recessive Inheritance 291
Sex-Linked Dominant Inheritance 292
Y Linkage 293
X Inactivation—Equalizing the Sexes 293
Chromosome Structure 296
Size and Shape 296
Stains and Probes 296
Evolution of the Karyotype 296
Abnormal Chromosome Numbers 297
Extra Chromosome Sets—Polyploidy 297
Extra and Missing Chromosomes—Aneuploidy 298
Smaller-Scale Chromosome Abnormalities 301
Deletions and Duplications 301
Chromosome Rearrangements 301
Chromosomes and Health 302
Bioethics Connections 14.1: Using Human Genome Information 302

Chapter 15 ~ DNA Structure and Replication 307

A Multifunctional Molecule 308
Identifying and Describing the Genetic Material 308
DNA Transmits Traits 309
The Hereditary Material—DNA, Not Protein 310
Deciphering the Structure of DNA 311
Gene and Protein—An Important Partnership 312
DNA Structure Encodes Information 312
DNA Is Highly Coiled 313
DNA Replication—Maintaining Genetic Information 313
Semiconservative Replication 313
Biology in Action 15.1: DNA Fingerprinting: An Application of Understanding DNA Structure 314
Steps and Participants in DNA Replication 317
DNA Repair 319
Biology in Action 15.2: PCR: An Application of Understanding DNA Replication 320

Chapter 16 ~ Gene Function 325

Transcription—The Genetic Crossroads 326
The Lac Operon Reveals Transcriptional Complexity 326
Transcription Factors 327
Steps of Transcription 327
Types of RNA 329
Messenger RNA (mRNA) Carries the Genetic Information 329
Ribosomal and Transfer RNAs Help Translate the Message 329
RNA Processing 330
Translation Expresses Genetic Information 331
The Genetic Code—From Genetic Message to Protein Product 331
Building a Protein 333
Biology in Action 16.1: From Genetics to Biotechnology 334
Protein Folding 336
Computers Coax Meaning from Genes 336
Mutation—Genetic Misinformation 338
Understanding Sickle Cell Disease 338
Causes of Mutation 339
Types of Mutations 341
Biology in Action 16.2: Chernobyl: Radiation Raises Mutation Rates 342
Natural Protection Against Mutation 344

Unit 5
Evolution 349

Chapter 17 ~ The Origin and Diversification of Life 351

Setting the Scene 352

Where Did Life Come from? 352
Spontaneous Generation 353
Life from Space 353
Chemical Evolution 353
The RNA World 355
The Origin of Metabolism 357
Biology in Action 17.1: Handedness in Biomolecules 358
The History of Life on Earth 359
Precambrian Time 359
The Paleozoic Era 361
The Mesozoic Era 365
The Cenozoic Era 366
The Road to Humanity 367
Bioethics Connections 17.1: Does Creationism Belong in a Biology Classroom? 370

Chapter 18 ~ Darwinian Evolution 373

A Tale of Three Islands 374
Mauritius 374
The Hawaiian Islands 375
The Galápagos Islands 375
Macroevolution and Microevolution 375
The Influence of Geology—Clues to Evolution in Rock Layers 376
Biology in Action 18.1: Evolutionary Thought Through History 378
The Voyage of the HMS *Beagle* 379
Darwin's Observations on the Voyage 379
After the Voyage 380
Natural Selection—The Mechanism Behind Darwinian Evolution 382
Sexual Selection 382
The Direction of Natural Selection Can Change 382
Evolution and Epidemiology 384
Emerging Infectious Diseases 384
Biology in Action 18.2: Crowd Diseases 385
An Evolving Infectious Illness—Ebola 386
The Rise of Antibiotic Resistance 387

Chapter 19 ~ Evidence for Evolution 389

The Stories of Life 390
Fossils 390
How Fossils Form 390
Determining the Age of a Fossil 392
Comparing Structures in Modern Species 393
Comparative Anatomy 393
Vestigial Organs 395
Comparative Embryology 395
Molecular Evolution 395
Comparing Chromosomes 395
Comparing Protein Sequences 396
Comparing DNA Sequences 397
Molecular Trees 397
Molecular Clocks 398
Reconciling Fossil, Biogeographical, and Molecular Evidence 400
Biology in Action 19.1: What Exactly Is a Velvet Worm? 400

Chapter 20 ~ The Forces of Evolutionary Change 403

Bullwinkle's Story—Sticking to One's Own Species 404
Evolution After Darwin—The Genetics of Populations 404
When Gene Frequencies Stay Constant—Hardy-Weinberg Equilibrium 405
When Gene Frequencies Change—Evolution 406
Nonrandom Mating 406
Biology in Action 20.1: Dogs and Cats—Products of Artificial Selection 407
Migration 408
Genetic Drift 408
Mutation 410
Natural Selection 410
How Species Arise 413
Premating Reproductive Isolation 413
Postmating Reproductive Isolation 413
Biology in Action 20.2: Sunflower Speciation in the Laboratory 415
The Pace of Evolution—Slow or Fast? 415
How Species Become Extinct 416
What Causes Mass Extinctions? 417
Mass Extinctions Through Geologic Time 418

Unit 6

The Diversity of Life 423

Chapter 21 ~ Classifying Life 425

On the Meaning of Names 426
Taxonomy Then and Now 427
Biology in Action 21.1: Taxonomic Finds 428
Characteristics to Consider 428
The Evolution of Taxonomy 429
The History of the Kingdom Concept 429
The Introduction of Domains 430
More Restrictive Taxonomic Levels 433
Challenges of Taxonomy 433
Biology in Action 21.2: Symbiosis 434
How Many Species Live on Earth? 435
Estimating Biodiversity 436

Chapter 22 ~ Monera 439

The Monera—A Biological Success 440
Bacteria and Archaea 440
Microbial Ecology 441
Studying Moneran Diversity 442
Laboratory Culture 442
Molecular Methods 442
Moneran Characteristics 443
Classifying Monera 443
Distinguishing Bacteria from Archaea 444
Energy Acquisition 445
Biology in Action 22.1: Life in a Sponge 446
A Closer Look at the Bacteria 446
Distinguishing Characteristics 446
Reproduction 448
Bacteria as Human Pathogens 449
The Archaea—Life in the Extremes 450
Classification 451
Monera—Crucial to Life on Earth 452
Industrial Microbiology 452

Chapter 23 ~ Protista 457

Protista—The Simplest Eukaryotes 458
Protozoa 458
Protozoan Classification 459
Types of Protozoa 460
Protozoa and Human Disease 461
Biology in Action 23.1: Malaria 462
Algae 463
Algal Classification 463
Types of Algae 465
Chytrids, Water Molds, and Slime Molds 469
Chytrids 470
Water Molds 470
Slime Molds 470

Chapter 24 ~ Fungi 475

A Killer Fungus 476
Characteristics of Fungi 476
Differences and Similarities from Plants and Animals 476
The Fungal Body 477
Reproduction 478
Fungal Diversity 480
Fungal Divisions 480
Zygomycetes 480
Ascomycetes 481
Bioethics Connections 24.1: LSD 482
Basidiomycetes 482
Deuteromycetes 483
Lichens 484
The Impact of Fungi on the Living World 484
Evolution 485
Ecology 485
Diseases Fungi Cause 485
Biology in Action 24.1: Yeast—A "Stripped-Down Human Cell" 488

Chapter 25 ~ Plantae 491

The Value of Diversity 492
Basic Plant Characteristics 492
Bryophytes 492
Bryophyte Diversity 493
Bryophyte Ecology 495
Biology in Action 25.1: Buried in the Bog 495
Uses of Bryophytes 495
Seedless Vascular Plants 496
Seedless Vascular Plant Characteristics 496
Seedless Vascular Plant Diversity 497
Uses of Seedless Vascular Plants 498
Seed-Producing Vascular Plants 498
Gymnosperms—Naked Seed Plants 499
Angiosperms—Flowering Plants 501
Biology in Action 25.2: How Does a Plant Become a Fossil? 502
Nature's Botanical Medicine Cabinet 504
Preserving Plant Diversity 504

Chapter 26 ~ Animalia 507

What Is an Animal? 508
Blueprint of an Animal 508
Bioethics Connections 26.1: Animal Research: From Science to People 509
Levels of Organization 509
Development 509
Body Cavity 511
Symmetry 511
Patterns of Animal Development 512
Animal Phyla 514
Phylum Porifera: Sponges 515
Phylum Cnidaria: Hydras, Jellyfish, Sea Anemones, Corals, and Others 516
Phylum Platyhelminthes: Flatworms, Including Planarians, Flukes, and Tapeworms 517
Phylum Nematoda: Roundworms 518
Phylum Mollusca: Clams, Snails, Octopuses, Squid, and Others 519
Phylum Annelida: Segmented Worms, Including Oligochaetes, Polychaetes, and Leeches 521
Phylum Arthropoda: Insects, Crustaceans, Arachnids, Myriapods, and Others 521
Phylum Echinodermata: Sea Stars, Brittle Stars, Sea Urchins, Sand Dollars, Sea Lilies, Sea Cucumbers, and Others 523
Phylum Chordata: Lancelets, Tunicates, and Vertebrates 524

Unit 7

Plant Life 531

Chapter 27 ~ Plant Form and Function 533

A Garden of Plant Cells and Tissues 534
Primary Growth 534
Meristems 534
Ground Tissue 535
Dermal Tissue 536
Vascular Tissues 537
Biology in Action 27.1: How Water Reaches Treetops 540
Parts of the Plant Body 540
Stems 541
Leaves 541
Roots 546
Secondary Growth 550
Vascular Cambium 550
Cork Cambium 552

Chapter 28 ~ Plant Life Cycles 555

Plant Reproduction and the Wandering Wasp 556
Asexual Reproduction 556
Alternation of Generations 557
Biology in Action 28.1: New Routes to Plant Reproduction 558
The Pine Life Cycle 560
Flowering Plant Life Cycle 563
Flower Structure 563
Gamete Formation 563
Biology in Action 28.2: Unraveling the Genetic Controls of Flower Formation 564
Pollination 567
Fertilization 568
Seed Development 568
Seed Dormancy 569
Fruit Formation 569
Fruit and Seed Dispersal 570
Seed Germination 571
Plant Development 572
A Comparison of Gymnosperm and Angiosperm Life Cycles 572

Chapter 29 ~ Plant Responses to Stimuli 575

The Sundew Plant's Response to a Stimulus 576
Plant Hormones 576
Auxins 577
Gibberellins 577
Cytokinins 577
Ethylene 577
Abscisic Acid 578
Hormonal Interactions 578
Tropisms—Growth Movements 578
Phototropism—Response to Unidirectional Light 578
Gravitropism—Response to Gravity 579
Thigmotropism—Response to Touch 579
Biology in Action 29.1: Plants in Space 580

Nastic Movements—Nongrowth Movements 581
Thigmonasty 581
Photonasty 582
Thigmomorphogenesis 582
Seasonal Responses of Plants to the Environment 582
Flowering—Response to Photoperiod 582
Do Plants Measure Day or Night? 584
Phytochrome—A Pigment Controlling Photoperiodism 584
Other Responses Influenced by Photoperiod and Phytochrome 586
Senescence 587
Dormancy 587
Circadian Rhythms 587
Biology in Action 29.2: Plant Communication from Within 588

Unit 8

Animal Life 591

Chapter 30 ~ Animal Tissues 593

Specialized Cells 594
Epithelium 594
Connective Tissues 594
Nervous Tissue 596
Muscle Tissue 597
A Sample Organ System—The Integument 598
A Closer Look at Human Skin 598
Specializations of the Integument 599
Replacing Damaged Tissues 603
Transplanted Tissues and Organs 604
Tissue Engineering 606
Bioethics Connections 30.1: From Death Row to the Internet 606

Chapter 31 ~ The Nervous System 609

The Nervous System—Always in Control 610
Evolutionary Trends in Nervous Systems 610
Neurons—Functional Units of the Nervous System 611
A Neuron's Message 611
The Resting Potential 612
The Action Potential 614
The Myelin Sheath and Saltatory Conduction 614
Sympathetic Transmission 616
Disposal of Neurotransmitters 617
Synaptic Integration—A Neuron's Response 617
Types of Neurotransmitters 618
A Comparative View of Nervous Systems 619
Invertebrate Nervous Systems 619
Vertebrate Nervous Systems 620
The Human Central Nervous System 622
The Spinal Cord 622
The Brain 624
Biology in Action 31.1: Spinal Cord Injuries 624
The Cerebrum 629
The Cerebral Cortex 629
Specializations of the Cerebral Hemispheres 629
Memory 630
The Peripheral Nervous System 630
The Somatic Nervous System 630
The Autonomic Nervous System 630
The Unhealthy Nervous System 631
Molecular Causes 631
Biology in Action 31.2: Addiction! 634
Environmental Causes 635
Protection of the Nervous System 635

Chapter 32 ~ The Senses 639

The Senses Paint Pictures of the World 640
Sensory Systems and Adaptation 640
General Principles of Sensory Reception 641
Sensory Receptors—Detecting, Transducing, and Amplifying Stimuli 641
Biology in Action 32.1: Snake Tongues, Butterfly Rears, and Hippo Jaws 642
Sensory Adaptation 644
Sensory System Diversity 644
Invertebrate Senses 644
Vertebrate Sensory Systems 645
Human Sensory Systems 645
Chemoreception 646
Biology in Action 32.2: Mixed-up Senses—Synesthesia 646
Photoreception 648
Mechanoreception 652

Chapter 33 ~ The Endocrine System 659

The Human Metamorphosis—Puberty 660
Hormones—Chemical Messengers and Regulators 660
The Endocrine and Nervous Systems Maintain Homeostasis 661
Pheromones 662
How Hormones Exert Their Effects 662
- Water-Soluble Hormones—Activating Second Messengers 662
- Fat-Soluble Hormones—Directing DNA Action 663
- Controlling Hormone Levels 663

Diversity of Hormones and Glands 664
- Hormones in Invertebrates 664
- Hormones and Glands in Vertebrates 665

Pituitary Hormones 667
- Anterior Pituitary Hormones 667
- Posterior Pituitary Hormones 667
- Intermediate Region Hormones 669

Endocrine Control of Metabolism 669
- The Thyroid Gland 669
- The Parathyroid Glands 670
- The Adrenal Glands 670
- The Pancreas 672

Endocrine Control of Reproduction 674
- The Menstrual Cycle 674
- The Testes 675

The Pineal Gland 675
- *Biology in Action 33.1: Steroids and Athletes—An Unhealthy Combination 676*

Prostaglandins 676
- *Bioethics Connections 33.1: Misrepresenting Melatonin 677*

Neuroendocrine-Immune Interactions 677

Chapter 34 ~ The Musculoskeletal System 681

The Musculoskeletal System and Movement 682
Responding to the Environment 682
- Musculoskeletal Plasticity 682

Skeletal System Diversity 683
- Hydrostatic Skeletons 683
- Exoskeletons 683
- Endoskeletons 684

Vertebrate Musculoskeletal Systems—A Closer Look 684
- Functions 684
- Structure 685

Cells and Tissues of the Vertebrate Skeleton 687
- Cartilage 687
- Bone 688

Biological Movement 689

- Muscle Cell Types 689
- *Biology in Action 34.1: The Human Skeleton over a Lifetime 690*
- Amoeboid Movement 692
- Invertebrate Muscles 692
- Vertebrate Skeletal Muscle 692

Muscle and Bone Working Together 700
- Joints—Where Bone Meets Bone 700
- Lever Systems 700
- Muscle Tone 701
- Effects of Exercise on Muscle 701
- *Biology in Action 34.2: Inborn Athletic Ability and Muscle Fiber Types 703*

Chapter 35 ~ The Circulatory System 707

Circulatory Systems 708
- Types of Circulatory Systems 708
- Respiratory Pigments 708
- Vertebrate Circulatory Systems 710

The Human Circulatory System 710
- Blood Composition 711
- *Biology in Action 35.1: The Unhealthy Circulatory System 712*
- *Biology in Action 35.2: The Return of the Medicinal Leech 718*

The Human Heart 718
- Blood Vessels 722
- Blood Vessels and Circulation 725
- Blood Pressure 725
- Control of Circulatory Functions 726
- Exercise and the Circulatory System 727
- The Lymphatic System 727

Chapter 36 ~ The Respiratory System 731

Breathing and Respiration 732
Evolution and Diversity Among Gas Exchange Systems 732
- Body Surface 733

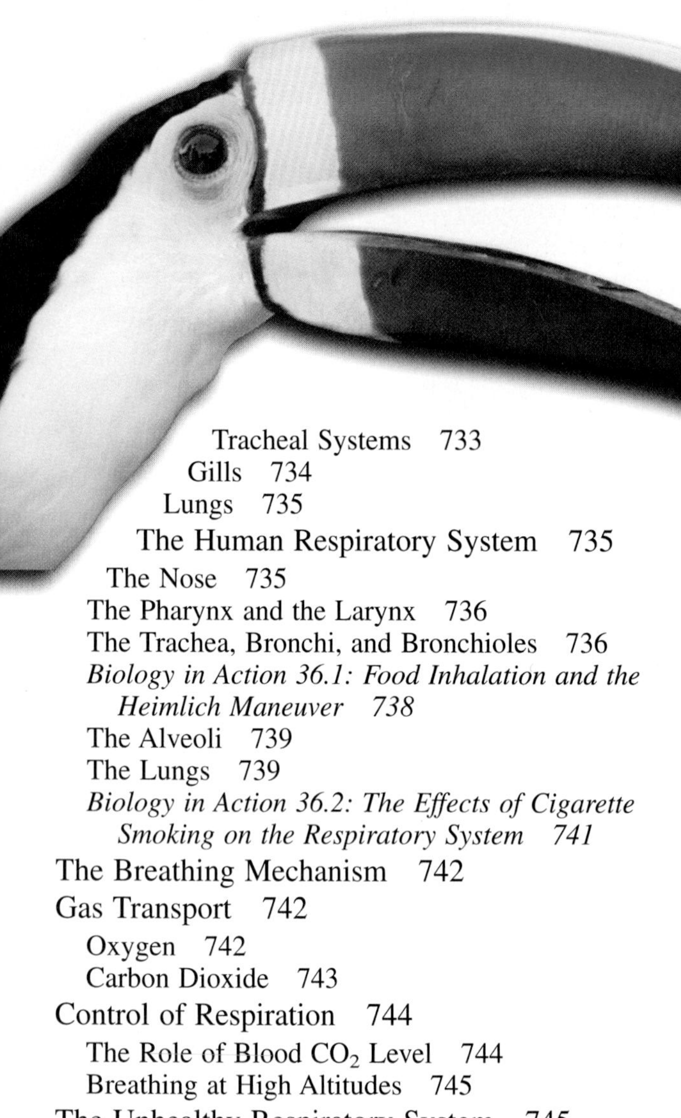

Tracheal Systems 733
Gills 734
Lungs 735
The Human Respiratory System 735
The Nose 735
The Pharynx and the Larynx 736
The Trachea, Bronchi, and Bronchioles 736
Biology in Action 36.1: Food Inhalation and the Heimlich Maneuver 738
The Alveoli 739
The Lungs 739
Biology in Action 36.2: The Effects of Cigarette Smoking on the Respiratory System 741
The Breathing Mechanism 742
Gas Transport 742
Oxygen 742
Carbon Dioxide 743
Control of Respiration 744
The Role of Blood CO_2 Level 744
Breathing at High Altitudes 745
The Unhealthy Respiratory System 745

Chapter 37 ~ Digestion and Nutrition 749

Why Animals Eat 750
The Evolution of Eating 751
Digestive Diversity 752
Types of Digestive Systems 752
Specific Digestive System Adaptations 754
Human Nutrition 755
Nutrient Deficiencies 757
An Overview of the Human Digestive System 757
Biology in Action 37.1: Starvation 758
Food's Journey Through the Human Body 760
The Mouth and the Esophagus 760
The Stomach 760
The Small Intestine 761
The Large Intestine 763
The Pancreas, Liver, and Gallbladder 764
What We Can't—and Can—Control 765
Biology in Action 37.2: Obesity 766

Chapter 38 ~ Regulation of Temperature and Body Fluids 769

The Camel—Playing It Cool 770
Temperature Regulation 771
Source of Body Heat—External or Internal? 771
Body Temperature—Constant or Variable? 771
Coping with Cold 771
Biology in Action 38.1: A Day in the Life of a Marine Iguana 772
Beating the Heat 772
Osmoregulation 776
Nitrogenous Waste Removal 776
Nephridia 778
The Human Urinary System—An Overview 780
The Kidney 780
The Nephron 781
The Glomerular Capsule 781
The Proximal Convoluted Tubule 783
The Nephron Loop 783
The Distal Convoluted Tubule 784
The Collecting Duct 784
Control of Kidney Function 784
The Unhealthy Urinary System 785
Bedwetting 785
Urinary Tract Infections 785
Biology in Action 38.2: Urinalysis—Clues to Health 785
Kidney Stones 786
Kidney Failure 786

Chapter 39 ~ The Immune System 789

The Immune System Protects 790
Nonspecific Defenses 790
Phagocytosis 790
Inflammation 791
Antimicrobial Substances 791
Fever 792
Specific Defenses 792
Immune System Characteristics 792
Steps in the Immune Response 793
Development of the Human Immune System 798
Immunity Breakdown 798
AIDS 798
Severe Combined Immune Deficiency 799
Chronic Fatigue Syndrome 800
Autoimmunity and Allergy 800
Autoimmunity—Attacking Self 800
Allergies—Attacking the Innocuous 800
Biology in Action 39.1: The Origin of AIDS 801

Altering Immune Function 802
Vaccines—Augmenting Immunity 803
Transplants—Suppressing Immunity 804
Immunotherapy 804
Biology in Action 39.2: Two Vaccine Success Stories 805

Unit 9

Behavior and Ecology 809

Chapter 40 ~ Behavior 811

Studying Behavior 812
Genes and Experience Shape Behavior 812
Innate Behavior 813
Fixed Action Patterns 813
Releasers 814
Reaction Chains 814
Learning 814
Habituation 815
Classical Conditioning 815
Operant Conditioning 816
Imprinting 816
Insight Learning 816
Latent Learning 816
Orientation and Navigation 817
The Compass Sense 817
Homing 817
Biology in Action 40.1: DNA Testing Solves the Mystery of Green Turtle Migration 818
Biological Societies 819
Biology in Action 40.2: Three Animal Societies 820
Advantages and Disadvantages of Group Living 821
More Favorable Surroundings 821

Better Defense 822
Enhanced Reproductive Success 822
Improved Foraging Efficiency 822
Improved Learning 823
Disadvantages of Group Living 823
Social Structure and the Situation 823
Communication and Group Cohesion 824
Chemical Messages 824
Auditory Messages 824
Tactile Messages 824
Visual Messages 824
Altruism 826
Aggression 827
Territoriality 827
Dominance Hierarchies 828
Cannibalism 829
Mating Behavior 829
Mating Systems 830
Two Views of Mating Behavior—Spiders and Sparrows 830
Primate Mating Behavior 831

Chapter 41 ~ Populations 835

Effects of Overpopulation 836
Ecology: An Organism's Place in the World 836
Habitat and Niche 836
Characteristics of Populations 838
Biology in Action 41.1: Conducting a Wildlife Census 839
Potential Population Growth 840
Intrinsic Rate of Increase and Exponential Growth 840
Survivorship, Fecundity, and Age Structure 842
Regulation of Population Size 842
Environmental Resistance and the Carrying Capacity 842
Density-Independent and Density-Dependent Factors 842
Logistic Growth Curve 843
"Boom and Bust" Cycles 844
Biology in Action 41.2: Booming and Busting Sheep 844
K-selection Versus r-selection 844
Interspecific Interactions 845
Interspecific Competition 845
Predator-Prey and Parasite-Host Interactions 845
Human Population Growth 848
Bioethics Connections 41.1: Human Population Control in China—Did It Go Too Far? 849

Chapter 42 ~ Communities and Ecosystems 853

From Krill to Whale—The Antarctic Ecosystem 854

Communities and Their Characteristics 854
Biotic Succession 856
- Primary and Secondary Succession 856
- Mechanisms of Succession 858
- Disturbance and Disequilibrium 858

Ecosystems and Their Characteristics 858
- Energy Flow Through an Ecosystem 859
- *Biology in Action 42.1: A Sampling of Ecosystems 860*
- Stable Isotope Tracing—Deciphering Food Webs 862
- Ecological Pyramids 863

Biogeochemical Cycles 864
- The Water Cycle 865
- The Carbon Cycle 865
- The Nitrogen Cycle 865
- The Phosphorus Cycle 866

Bioaccumulation and Biomagnification 866
- DDT 866
- Mercury 868

Chapter 43 ~ Biomes 871

Terrestrial Biomes 872
- Tropical Rain Forest 872
- Temperate Deciduous Forest 873
- Temperate Coniferous Forest 873
- Taiga 876
- Grasslands and Savannas 877
- Tundra 878
- Desert 878

Freshwater Biomes 879
- Lakes and Ponds 879
- Rivers and Streams 881

Marine Biomes 881
- The Coast 881
- *Biology in Action 43.1: Mangrove Swamps 882*
- The Ocean 883

El Niño Affects Ecosystems and Biomes 884

Chapter 44 ~ Environmental Concerns 889

The *Exxon Valdez* Oil Spill 890
Environmentalism 891
- The 1960s and the 1990s 891
- Environmental Insults May Be Subtle 892

The Air 892
- Air Pollution 892
- *Biology in Action 44.1: A Legacy of Cold War 893*
- Acid Precipitation 894
- Thinning of the Ozone Layer 894
- Global Warming 896

The Waters 897
- Chemical Pollution 897
- Altered Rivers 899
- Lake Eutrophication 899
- Endangered Estuaries 899
- Polluted Oceans 899
- *Biology in Action 44.2: The Rhine—A River in Recovery 900*
- *Biology in Action 44.3: Adding Iron to the Ocean 903*

The Land 904
- Deforestation 904
- Desertification 905

Biodiversity and Resilience 906
- Biodiversity and Ecosystem Stability 906
- The Resiliency of Life 907

Appendix A *Units of Measurement* 909

Appendix B *Metric Conversion* 910

Appendix C *Periodic Table of Elements* 911

Appendix D *Tree of Life* 912

Appendix E *Taxonomy* 913

Glossary 917

Credits 939

Index 947

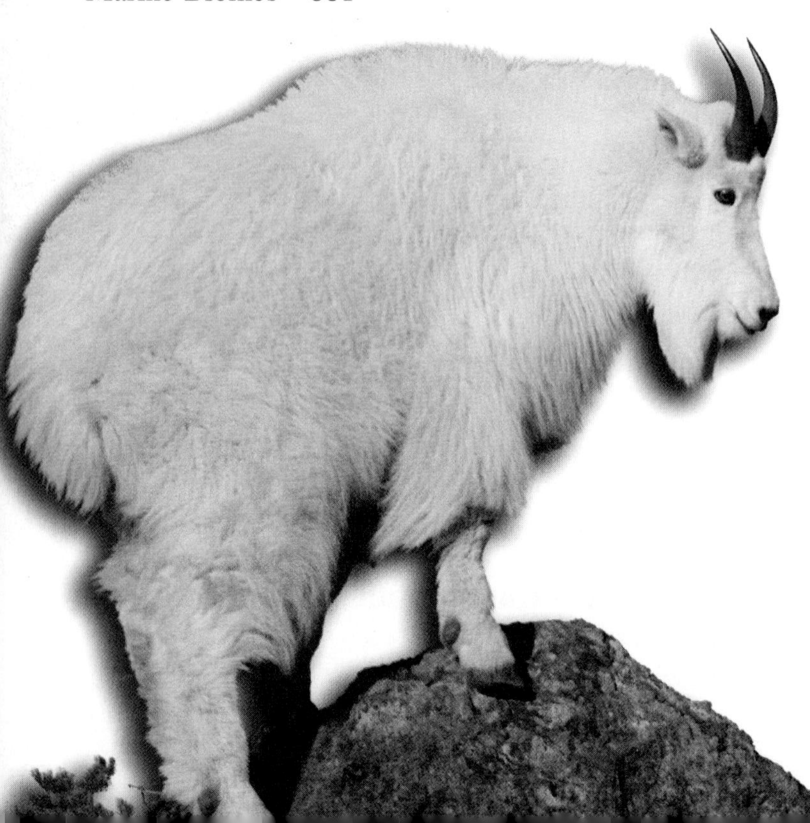

Preface

It's a great time to study *Life!* The late 1990s will be remembered as the time when the press proclaimed "the biology textbooks will have to be rewritten."

The call for textbook authors to get back to work referred to the Archaea, the cells that represent a third form of life on earth. Although biologists have known about these organisms for years and have long puzzled about their place in our organizational scheme of life, it was exciting to see the "news" burst into the public consciousness. Ironically, the announcement of the genome sequencing of an archaeon was dwarfed by the discovery of chemical smudges on Mars that resembled, some researchers hypothesized, microbial life.

Bracketing the archaean/Martian discoveries were others with the same theme—discovering new life. Researchers found microorganisms thriving in rocks, and an immense heated lake beneath the ice of Antarctica that might also house microscopic life. Chemical evidence of metabolic processes in ancient Greenland sediments set back the first signs of life to nearly 4 billion years ago. The message is clear: we still have a lot to learn about life.

To today's biology students, the scientific method is no longer just a dull list of steps to be memorized and spit out on an exam—it is a powerful way of thinking that can explain natural phenomena. To today's textbook authors, writing a new edition no longer means finding a better way to depict mitosis or the carbon cycle, but creating a knowledge base that students can use to put into perspective the exciting discoveries they read about on a daily basis.

The goal of *Life* is to convey both the knowledge and excitement, with a unique flavor and readability. "Science literacy" is a current educational buzzword—but it has always been *Life*'s raison d'être (reason for being). In many places, the book pairs classical topics with news or technology. For example, in chapter 18, Darwinian evolution segues smoothly into discussions of emerging infectious diseases and antibiotic resistance, topics students will certainly have heard of, but not necessarily connected to evolution. Similarly, the new chapter on tissues (chapter 30) leads into a discussion of ways to replace damaged tissue (transplants and tissue engineering).

What's New to the Third Edition

Organizational Changes

If you compare this new table of contents with that of the second edition, you will immediately recognize the changes in emphasis in this third edition. New chapters and material expand the scope and coverage of Lewis' *Life.*

- **New Diversity Unit**
  Unit 6, "The Diversity of Life," is new to the third edition and presents an overview of diversity in Chapter 21, "Classifying Life" and five chapters that survey the five kingdoms (chapters 22–26). The survey chapters provide representative but interesting organisms without going overboard in detailed information.
- **Reorganized Coverage of Evolution**
  A new chapter, Chapter 17 entitled "The Origin and Diversification of Life," appears in Unit 5, "Evolution." Previously, this information was presented earlier in the book in the context of the chemical nature of life. By moving this material to Unit 5, it provides a wonderful introduction to the study of evolution but also reinforces the "chemistry" in biology that is usually lost on students after they pass chapter 3. The information on the origins of life and the primordial soup experiment has been updated. New and interesting information has also been added to expand the discussion about the diversification of life on earth.
- **Expanded Coverage of Animal Biology**
  A new chapter, Chapter 30 entitled "Animal Tissues," opens Unit 8, "Animal Life," and provides fundamental information about tissue types before discussing body systems. Consideration of tissue types leads into a look at two new medical technologies, transplants and tissue engineering.
- **Expanded Coverage of Ecology**
  A separate chapter on Biomes, Chapter 43, expands coverage of diverse biomes. Chapter 42, "Communities and Ecosystems," has also been expanded and updated.
- **Streamlined and Integrated Genetics Coverage**
  The second edition contained a separate chapter on Genetic Technology, chapter 17, which has been integrated into related topics in the third edition. The discussions of genetic technologies now accompany explanations of the processes that gave rise to them. For example, PCR and DNA fingerprinting are discussed in chapter 15 with DNA Structure and Replication; recombinant DNA, gene therapy, transgenics, antisense, and knockouts are discussed in chapter 16 with Gene Function.

Pedagogical Changes

The third edition has many more summary charts, new examples, more questions, updated references, and updated material. To help students learn, several new features are in this edition:

- **Mastering Concepts**
  Short lists of questions follow each major text section and help the student understand what was just covered. These replace the Learning Objectives and Key Concepts from the

Running on Empty

Michael P. was noticeably weak from birth. He didn't move much, had "floppy" muscles, had difficulty breathing, and grew exhausted merely from the effort of eating. At the age of two-and-a-half months, he suffered his first seizure, staring and jerking his limbs for several frightening minutes. Despite medication, his seizures continued more and more frequently.

The doctors were puzzled because most of Michael's many medical tests showed normal results—with one notable exception. His cerebrospinal fluid (the fluid that bathes the brain and spinal cord) was unusually low in glucose and a small organic chemical called lactic acid. How could these deficits explain the baby's disturbing symptoms?

to be continued . . .

Running on Empty

continued from page 126.

The lack of glucose and pyruvic acid in Michael P.'s cerebrospinal fluid told physicians that his cells were not performing glycolysis or fermentation. Hypothesizing that a profound lack of ATP was causing the boy's symptoms, medical researchers decided to intervene beyond the block in the boy's metabolic pathway and take a detour to energy production. When Michael was seven-and-a-half months old, he was placed on a diet rich in certain fatty acids. Within 4 days, he appeared to be healthy for the very first time in his life!

The diet reactivated aerobic respiration at the point of acetyl CoA formation by supplying an alternative to glucose. Other children with similar symptoms have since enjoyed spectacular recoveries thanks to similar dietary interventions, but we do not yet know the long-term effects of the therapy. This medical success story, however, illustrates the importance of the energy pathways—and how valuable our understanding of them can be.

second edition but are more valuable because they require the student to be more active in the learning process. They accomplish more in less space, and don't "give" the students easy answers.

- **Split Vignettes**
  True tales open each chapter but stop at a precipice which launches the student into the chapter material. When the cliffhanger continues at the chapter's end, the content level is slightly higher, and includes chapter concepts. With the split vignettes, students need never ask, "Why do we have to know this?" The "split vignette" from chapter 7 is reproduced here.
- **"Bioethics Connections"**
  Because of biology's intimate ties to human health care, students will have to make many life decisions that will call upon their knowledge of how the human body works. Bioethics Connections prepare them for coming changes in medicine. These boxes consider such "hot topics" as nicotine addiction, melatonin hype, embryo and cloning research, and how to use human genome project information. The "Bioethics Connections" box 11.1 is reproduced here.

11.1 bioethics connections

Preembryos as Research Subjects

The needle piercing a secondary oocyte in figure 11.A will deliver a sperm cell into the oocyte in an effort to help the people who donated these cells become parents. This procedure, discussed in chapter 12, would not have been possible without experiments that followed the fate of fertilized ova, for a few cell divisions, in laboratory glassware.

Biology or Politics?

Basic research on human prenatal development provides new infertility treatments and contraceptives. However, many people object to using prenatal humans in the experiments that make reproductive technologies possible. As a result of the controversy over the ethics of this type of experimentation, the U.S. government, at the time of this writing, does not fund such research.

In 1993, when President Clinton lifted a ban on government-funded human embryo research begun in 1980, the National Institutes of Health (NIH) established a Human Embryo Research Panel, consisting of scientists, ethicists, and attorneys. The panel outlined the goals

- It defined when
  include when th
  not occur any ot
  fewest possible
- It determined th
  abdominal surge
  informed consen
- It recommended
  14 days of devel

The NIH recom
servative outcome
stituted the ban on
ceived solely for re
(R-Calif.) led the c
a letter that more th
"The issue of huma
Given the new reali
will presumably fe
to use every legisla
being spent on grot

- **"Chemistry Explains Biology"**
  These are short boxes unique to the chemistry chapter that ease understanding and introduce relevancy of the material presented. This chapter was evaluated by students, who compared it point-by-point to the competition and gave it a grade of "A".
- **New Figures**
  Several new figures connect concepts within chapters. For example, figure 20.7 clarifies at a glance the effects of migration, mutation, genetic drift, and natural selection on populations—and *shows* clearly that Hardy-Weinberg equilibrium is more theoretical than realistic. Figures 32.2 and 32.3 do what other books do not—show in one place, up front, what the human senses have in common at the cellular and molecular levels. Many new summary charts in the genetics unit help students learn difficult concepts and master terminology.
- **Scientific Method/Experiments**
  Chapter 2 introduces the scientific method in terms of the multilevel problem of environmental estrogens. By the chapter's end, students will see how women with certain cancers, eagles with misshapen beaks, and alligators with stunted penises are related—while learning how scientists work.

Woven throughout the narrative, in the split vignettes, To Think About questions, Biology in Action boxes, and Bioethics Connections, abound diverse examples of researchers in action, conducting individual laboratory experiments to large-scale epidemiological investigations. Figure 18.3, for example, depicts an elegant experiment that demonstrates evolution happening right now. The new Biology in Action 44.3, "Adding Iron to the Ocean," describes a large-scale, intriguing experiment.

By touching on the scientific method throughout the book—and not just in one chapter—students become comfortable with this approach, see how they use it to think in their everyday lives, and glimpse the bigger picture of biology that experiments taken together reveal.

- **Summary**
  The end of chapter summary is now in list form which makes it easier for students to identify and review key concepts.
- **References and Resources**
  As always, suggested readings are eclectic and interesting and are often the source material for a chapter's examples. Many are written by the author and are available on the *Life* home page. These selected readings are indicated with an icon in the "References and Resources" lists.

denotes articles by author Ricki Lewis that may be accessed via *Life*'s home page /http://www.mhhe.com/sciencemath/biology/life/.

Supplemental Material

Instructor's Manual and Test Item File

The Instructor's Manual, written by Roberta Meehan and Joe Coelho, has been completely rewritten and includes for each chapter: an overview of chapter objectives, list of key terms, an annotated outline, teaching tips, answers to end of chapter review questions ("To Review" and "To Think About"), and additional topics for discussion. The test item file has been completely updated and revised.

Microtest Computerized Testing Program

The printed test bank is available in a computerized format for Windows and Macintosh platforms.

Student Study Guide

The study guide, written by Donald Breakwell and Allan Stevens, contains a list of key terms, student activities to help reinforce chapter concepts, multiple choice questions for review, and answers to all activities and review questions.

Transparencies

A set of 300 transparencies of key figures in the text is available to adopters. A list of transparencies is available in the Instructor's Manual.

Laboratory Manual

The lab manual, written by Alice Jacklet, has been updated and may be customized in full color. It follows the new organization of *Life,* 3d edition and so is a valuable supplement to that text but it can be used in conjunction with any introductory biology textbook.

Laboratory Resource Guide

A separate laboratory resource guide is now available with the laboratory manual and assists the professor in setting up and performing the laboratory exercises.

Technology

WCB/McGraw-Hill offers a wide range of technology products to assist you in teaching and can improve student learning. A correlation table provided on the next page indicates where related material for each chapter can be found on various WCB/McGraw-Hill Multimedia CD-ROM products. Also provided are descriptions of these and other WCB/McGraw-Hill technology products for your introductory general biology course.

Exploring the Internet

http://www.mhhe.com/sciencemath/biology/life/

The *Life* Home Page allows students and teachers from all over the world to communicate. By visiting this site, students can access additional study aids, explore links to other relevant biology sites, catch up on current information including articles written by Ricki Lewis, and pursue other activities. The site includes frequently-updated suggested readings listed by chapter and study questions based on popular movies and television programs. The inside back cover of the text provides directions for accessing the *Life* Home Page.

The Internet Primer by Fritz J. Erickson and John A. Vonk

This short, concise primer shows students and instructors how to access and use the Internet. The guide provides enough information to get started by describing the most critical elements of using the Internet.

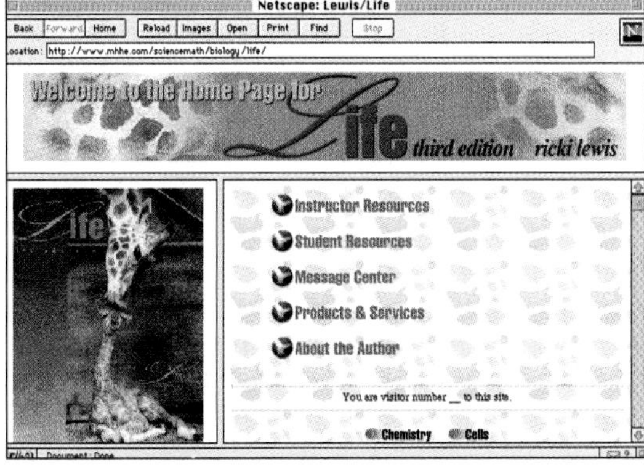

The third edition of *Life* has three technology learning tools that are correlated to the chapters. The guide on pg. xxiv indicates where the multimedia programs can supplement the material in the text. The Correlation Guide refers to *Life Science Animations* videotape series, *Explorations in Human Biology* CD-ROM, *Explorations in Cell Biology and Genetics* CD-ROM, and *Dynamic Human* CD-ROM.

Technology Correlation

Lewis Chapter	Explorations in Cell Biology & Genetics	Explorations in Human Biology	The Dynamic Human	Life Science Animations
3	Module 1			tape 1, concept 1
4	Module 2			tape 1, concepts 2,4
5	Modules 3, 4	Module 2		tape 1, concepts 3, 4
6	Module 6			tape 1, concept 11
7	Module 8			tape 1, concepts 5, 6, 7
8	Module 9			tape 1, concepts 8, 9, 10
9	Module 5	Module 6		tape 2, concept 12
10	Module 10		Reproductive System/ Anatomy, Explorations	tape 2, concepts 13, 14, 19, 20
11		Module 3		tape 2, concept 21
12			Reproductive System/ Clinical Concepts	
13	Modules 11, 13	Module 14		tape 4, concept 40
14	Module 12	Module 15		
15	Module 14			tape 2, concept 15
16	Module 16	Module 1		tape 2, concepts 16, 17, 18
20				tape 4, concept 45; tape 5, concept 53
27				tape 5, concepts 46, 47, 48
28				tape 5, concept 50
29				tape 5, concept 49
31		Modules 8, 9, 10	Nervous System/ Anatomy, Explorations	tape 3, concepts 22, 23, 24, 25
32			Nervous System/Explorations	tape 3, concepts 26, 27
33		Module 11	Reproductive System/ Explorations & Endocrine System/Clinical Concepts, Explorations	tape 3, concept 28
34		Module 4	Skeletal & Muscular Systems/ Explorations	tape 3, concepts 29, 30, 31
35		Module 5	Cardiovascular System/ Explorations; Lymphatic System/Anatomy	tape 3, concept 32; tape 4, concepts 37, 38, 39
36			Respiratory System/ Anatomy, Explorations	
37			Digestive System/ Explorations	tape 3, concepts 33, 34, 35; tape 4, concept 36
38			Urinary System/ Anatomy, Explorations	
39		Modules 12, 13	Lymphatic System/Clinical Concepts, Explorations	tape 5, concepts 41, 42, 43, 44
42				tape 5, concepts, 51, 52
44		Module 16		

The Dynamic Human CD-ROM

This guide to anatomy and physiology interactively illustrates the complex relationships between anatomical structures and their functions in the human body. Realistic, three-dimensional visuals are the premier feature of this exciting learning tool. The program covers each body system, demonstrating to the viewer the anatomy, physiology, histology, and clinical applications of each system. The correlation chart in this edition of *Life* indicates where material in the text can be supported by this interactive program.

Explorations in Human Biology CD-ROM; Explorations in Cell Biology and Genetics CD-ROM

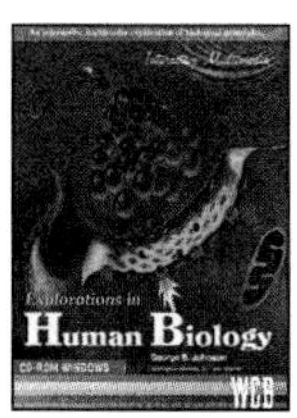

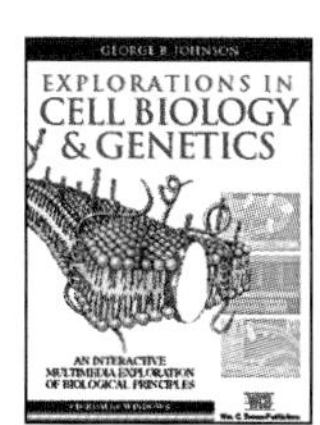

These interactive CDs, by George Johnson, feature fascinating topics in biology. *Explorations in Human Biology* and *Explorations in Cell Biology and Genetics* have 33 different modules that allow students to study a high-interest biological topic in an interactive way. The correlation chart in this edition of *Life* indicates where material in the text can be supported by this interactive program.

Life Science Animations Videotape

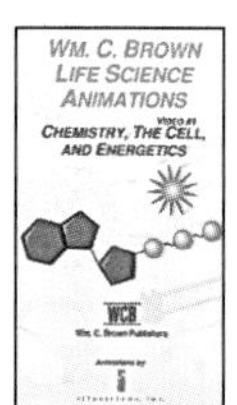

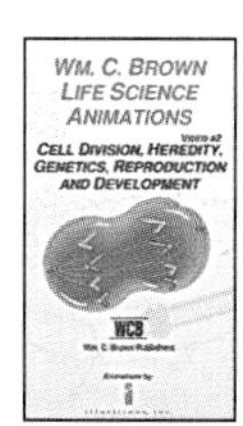

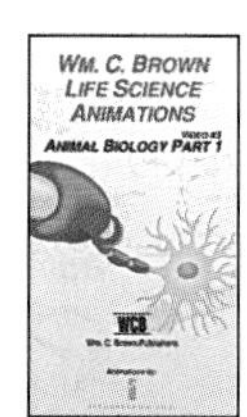

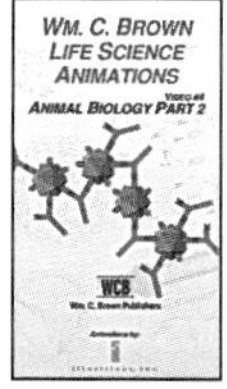

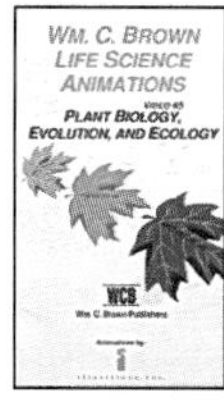

Fifty-three animations of key physiological processes are available on videotapes. The animations bring visual movement to biological processes that are difficult to understand based on the static image in the text. The correlation chart in this edition of *Life* indicates where material in the text can be supported by the animations in this video series.

The WCB/McGraw-Hill Visual Resource Library

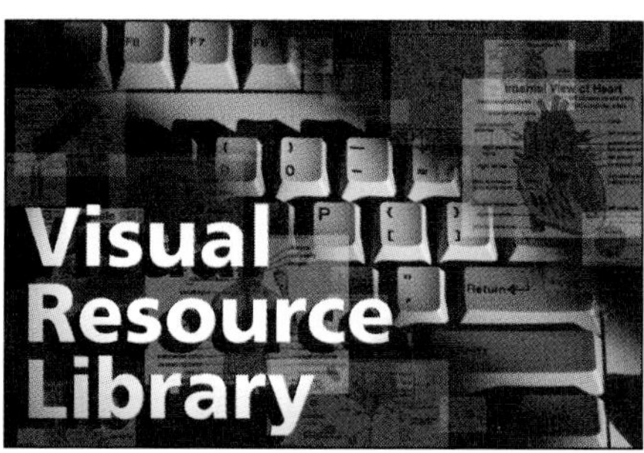

Nearly all text illustrations will be available on CD-ROM for use in the classroom. The user-friendly program allows you to easily pull the art files into PowerPoint or other presentation tools in order to prepare your own, customized classroom presentations. Several QuickTime movies are available summarizing key biological processes to make your lecture even more dynamic.

Life Science Living Lexicon CD-ROM

A *Life Science Living Lexicon* CD-ROM, by William Marchuk contains a comprehensive collection of life science terms, including definitions of their roots, prefixes, and suffixes as well as audio pronunciations and illustrations. The *Lexicon* is student-interactive, providing quizzing and note-taking capabilities. It contains 4,500 terms, which can be broken down for study into the following categories: anatomy and physiology, botany, cell and molecular biology, genetics, ecology and evolution, and zoology.

Biosource Videodisc

BioSource Videodisc, by WCB/McGraw-Hill and Sandpiper Multimedia, Inc., features 20 minutes of animations and nearly 10,000 full-color illustrations and photos, many from leading WCB/McGraw-Hill biology textbooks.

Bioethics Forums Videodisc

Bioethics Forums is an interactive program that explores societal dilemmas arising from recent breakthroughs in biology, genetics, and biomedical technology. The scenarios are fictional, but the underlying science and social issues are real. *Bioethics Forums* encourages students to explore the science behind decisions as well as the processes of ethical reasoning and decision making. This would enhance the "Bioethics Connections" in this edition of *Life.*

The Secret of Life Video Modules WGBH, Boston and BBC-TV

WGBH has produced eight 15-minute video modules that illuminate the biological universe with unique stories and animation. Each module concludes with a series of stimulating questions for class discussion.

The Secret of Life Videodisc WGBH, Boston

A two-sided videodisc will be available as a companion to *Life,* third edition. Topic coverage includes biotechnology, human reproduction, portraits of modern science and research, and human genetics.

Virtual Biology Laboratory CD-ROM by John Beneski and Jack Waber, West Chester University

This CD-ROM is designed primarily for nonscience major students. The exercises (10 modules) are designed to expose students to the types of tools used by biologists, allow students to perform experiments without the use of wet lab setups, and support and illustrate topics and concepts from a traditional biology course.

HealthQuest CD-ROM

This is an interactive CD-ROM designed to help students address the behavioral aspects of personal health and wellness. **HealthQuest** allows users to assess their current health and wellness status, determine their health risks and relative life expectancy, explore options and make decisions to improve the behaviors that impact their health.

Virtual Physiology Laboratory CD-ROM

Virtual Physiology Lab CD-ROM features ten simulations of the most common and important animal-based experiments ordinarily performed in the physiology component of your laboratory. This revolutionary program allows students to repeat laboratory experiments until they adeptly master the principles involved. The program contains video, audio, and text to clarify complex physiological functions.

Our CD-ROM products may be packaged with the text at a cost savings. Contact your WCB/McGraw-Hill sales representative for details.

From WCB/McGraw-Hill

How to Study Science, 2nd Edition by Fred Drewes, Suffolk County Community College

This excellent new workbook offers students helpful suggestions for meeting the considerable challenges of a college science course. It offers tips on how to take notes, how to get the most out of laboratories, and how to overcome science anxiety. The book's unique design helps students develop critical thinking skills while facilitating careful note taking. (ISBN 0–697–15905–1)

A Life Science Living Lexicon by William N. Marchuk, Red Deer College

This portable, inexpensive reference helps introductory-level students quickly master the vocabulary of the life sciences. Not a dictionary, it carefully explains the rules of word construction and derivation, in addition to giving complete definitions of all important terms. (ISBN 0–697–12133–X)

Biology Study Cards by Kent Van De Graaff, R. Ward Rhees, and Christopher H. Creek, Brigham Young University

This boxed set of 300 two-sided study cards provides a quick yet thorough visual synopsis of all key biological terms and concepts in the general biology curriculum. Each card features a masterful illustration, pronunciation guide, definition, and description in context. (ISBN 0–697–03069–5)

Critical Thinking Case Study Workbook

Written by Robert Allen, this ancillary includes 34 critical thinking case studies that are designed to immerse students in the "process of science" and challenge them to solve problems in the same way biologists do. The case studies are divided into three levels of difficulty (introductory, intermediate, and advanced) to afford instructors greater choice and flexibility. An answer key accompanies this workbook. (ISBN 0–697–34250–6)

The AIDS Booklet
Frank D. Cox

This booklet describes how AIDS is commonly spread so that readers can protect themselves and their friends against this debilitating and deadly disease. This booklet is updated quarterly to give readers the most current information. Also visit the Mader Home Page for additional AIDS material. (ISBN 0–697–26261–8)

Chemistry for Biology
Carolyn Chapman

This workbook is a self-paced introduction or review of the basic principles of chemistry that are most useful in other areas of science. (ISBN 0–697–24121–1)

Acknowledgments

Several people helped create *Life* 3/e. Many thanks for superb editing skills and moral support to Michael Lange and Laura Beaudoin. Special thanks to Kennie Harris, developmental editor extraordinaire, to Roberta Meehan, reviewer extraordinaire, and to Toni Michaels, the very best photo editor. Thanks also to the artists and reviewers, and especially to contributors George Cox and David Shier. Finally, thanks to my wonderful family.

Ricki Lewis, 1997

Reviewers

Many instructors have contributed to the making of this text and its teaching package. We gratefully acknowledge the assistance of the many reviewers whose constructive criticism proved invaluable to the development of this text.

Donald P. French
Oklahoma State University

Eric Juterbock
Ohio State University

Linda K. Butler
University of Texas at Austin

Theodosia R. Wade
Oxford College of Emory University

Sheila Schneider
Cuyahoga Community College

George W. Cox
Biosphere & Biosurvival

Heather Hall

Linda Kolbus
Cuyahoga Community College

R. Dean Decker
University of Richmond

Lori Garrett
Danville Area Community College

Loretta M. Parsons
Skidmore College

Andrew H. Lapinski
Reading Area Community College

Jennifer McGehee Marsh
University of Louisville

Pamela Tabery
Northampton County Community College

James Horwitz
Palm Beach Community College

Kenneth J. Curry
University of Southern Mississippi

Robert T. Grammar
Belmont University

William H. Chrouser
Warner Southern College

Lonnie Guralnick
Western Oregon State College

Joe Coelho
Western Illinois University

John Hoagland
Danville Area Community College

Lauren D. Howard
Norwich University

John C. Harley
Eastern Kentucky University

Thomas Justice
McLennon Community College

Michelle A. Briggs
Lycoming College

Marirose Ethington
Genesee Community College

Jon R. Maki
Eastern Kentucky University

Allan P. Russell
Mt. Wachusett Community College

Craig T. Johnson
University of Texas at San Antonio

Rita A. Hoots
Yuba College

Life in action. A lion bellows as a cub nips his rear.

Unit 1

Overview of Biology

Abundant and diverse life on a Kenyan plain.

1 What Is Life?

chapter outline

Distinguishing Life from Death 4

The Characteristics of Life 4

Organization
Energy Use and Metabolism
Maintenance of Internal Constancy
Reproduction, Growth, and Development
Irritability and Adaptation

Biodiversity 12

What's in a Name?
The Five Kingdoms
Evolution Is the Backbone of Biology

Alive or Dead?

Sometime during her midday nap, three-month-old Baby F. stopped breathing. When the young mother discovered this, she called for help and began cardiopulmonary resuscitation (CPR). Paramedics arrived five minutes later. They transported Baby F., while continuing CPR, to the nearest hospital. There, her heart started within minutes, and soon she began to breathe again. Doctors determined that Baby F. was in a coma—she was neither awake nor aware. Nurses kept her warm and gave her fluids. The child's brain was active, her heart beat, and she breathed unassisted, but she remained unconscious.

A day later, Baby F.'s condition worsened. Her eyes failed to respond to light, a sign that her nervous system was shutting down. A respirator had to force air into and out of her lungs so that she could breathe. Still, a brain scan showed apparently normal structure and electrical brain-wave activity.

By the third day, Baby F. couldn't see, feel, cough, gag, suck, or breathe unassisted. Finally, her brain-wave pattern went from regular peaks to a flat line. Two physicians, after observing the "flat line" for 30 minutes, declared Baby F. "brain dead." She would be considered officially dead 48 hours later, if a physical exam and brain-wave pattern again showed no responses.

Two days later, Baby F. remained brain dead, but her physicians kept her on a respirator because her brain still showed no evidence of injury, bleeding, or swelling, the usual causes of brain death.

On the 19th day, something amazing happened—a brain scan showed blood flowed in part of Baby F.'s brain. When physicians injected Baby F. with chemicals to stimulate brain function, she responded. Baby F. slept on, but a small part of her brain was seemingly awake.

Then, suddenly, 43 days after she stopped breathing, Baby F. began to breathe again on her own.

to be continued . . .

Distinguishing Life from Death

The true story of Baby F. illustrates how difficult it can be to distinguish life from death. The definition of death once seemed clear—life ceased when the heart stopped beating and the lungs stopped inflating and deflating. But the development of heart transplants in the late 1960s made this definition obsolete. How could a heart donor be considered dead if his or her heart continued to beat in another body? A few years later, invention of the respirator further blurred the line between life and death by restoring breathing.

At present, death in humans is defined as the cessation of brain activity (table 1.1). Yet criteria for brain death can come into question, just as previous criteria for death once did. Was Baby F. alive when only part of her brain was active?

The definition of death as cessation of brain activity is suitable for humans in a hospital setting—but is it applicable to the at least 30 million other types of organisms? When, for example, is a mushroom dead? A moss? Some organisms can even enter deathlike states when the environment is harsh. In a laboratory at Cornell University, for example, researchers are observing eggs hatch that were laid more than 350 years ago by a small crustacean called a copepod (fig. 1.1).

The copepod eggs in the Cornell lab were laid in 1639 and then covered rapidly with mud, which blocked the temperature and light cues that would usually stimulate hatching. Instead, the eggs entered a deathlike state called **diapause,** a type of suspended animation. Only when researchers collected samples from a pond bottom and exposed them to light in a laboratory did the eggs "awaken."

A three-century sleep is nothing compared to the dormancy of some other organisms. Botanists recently grew shoots from a 1,288-year-old lotus seed from China, and experts in ancient DNA analysis have brought back to life, so to speak, bacteria discovered in a Dominican bee that was preserved in tree sap some 25 million years ago!

The definition of death is a key part of **bioethics,** a field of philosophy that examines principles of morality, individual rights, justice, and values as they apply to medical technologies and research activities. Because bioethics bridges the study of life and everyday experience, it is a recurring theme in this book (Bioethics Connections 1.1).

Table 1.1 Criteria for Brain Death

1. Coma (unconscious, cannot be aroused)
2. Cannot breathe without respirator
3. Cannot move; muscles are flaccid
4. Eyes do not move or pupils change size in response to changes in light intensity
5. No heartbeat
6. Cannot swallow, yawn, or make sounds
7. Flat brain waves
8. Above signs are not due to explainable circumstances, such as a drug overdose or very low body temperature
9. Above signs must be observed by at least two medical professionals, at least twice during a certain time period, determined by the age of the individual

The Characteristics of Life

The stories of Baby F., the over 350-year-old copepods, and the ancient seeds and bacteria illustrate that it isn't easy to establish which biological functions must cease to cause death, because to do so is really to ask the broader question, What is life? What combination of activities distinguishes the living from the nonliving?

We all have an intuitive sense of what life is: if we see a rabbit on a rock, we "know" that the rabbit is alive and the rock is not. But it's difficult to state just what it is that makes the rabbit alive and the rock not.

(a)

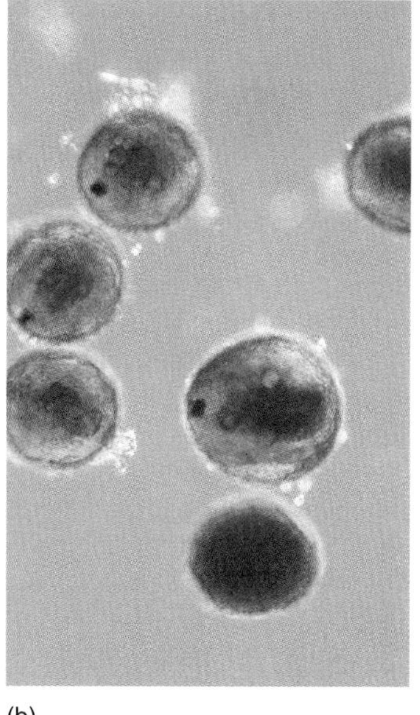
(b)

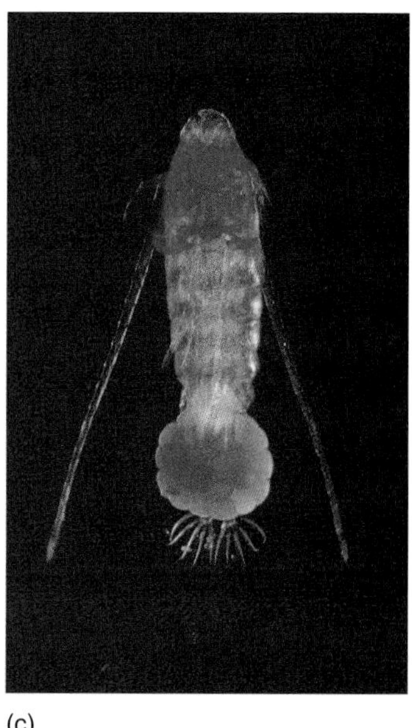
(c)

FIGURE 1.1

More than 350 years ago, a copepod, *Diaptomus sanguineus,* deposited eggs in a pond in Rhode Island. A storm buried the eggs before they could hatch, and they entered a state of suspended animation called diapause. (*a*) A graduate student from Cornell University sampled some of the 6 billion eggs on the pond's bottom. (*b*) Given light in a laboratory, the centuries-old eggs are hatching. (*c*) The adult female copepod is 1 millimeter long; the eggs are about one-tenth this size.

Scientists define life in terms of qualities that the living uniquely share. But in the second before a newborn baby takes its first breath—or after a person dies—we may ask, as the nineteenth-century thinkers called vitalists did, "Is there some indefinable 'essence' that renders a particular collection of structures or processes living?" This question lies at the heart of near-death experiences, as well as at the core of such headline-grabbing controversies as abortion rights and the right to die.

Throughout history, the question, "What is life?" has stymied thinkers in many fields. Eighteenth-century French physician Marie Francois Xavier Bichat poetically but imprecisely defined life as "the ensemble of functions that resist death." Others less eloquent and still no more precise have hypothesized that life is a kind of "black box" that endows a group of associated biochemicals with the qualities of life. Still others have tried to break life down into the smallest parts that still exhibit the characteristics of life and then to identify what makes those units different from their nonliving components.

Five qualities in combination constitute life (table 1.2). An **organism** is a collection of structures that function coordinately and exhibit these qualities. These characteristics alone may also occur in inanimate objects. A computer responds to stimuli, as does a self-flushing toilet, but neither is alive. A fork placed in a pot of boiling water absorbs heat energy and passes it to a hand grabbing it, just as a living organism receives and transfers energy, but this does not make the fork alive. A supermarket is highly organized, but it is obviously not alive!

Table 1.2 Characteristics of Life
1. Organization
2. Energy use and metabolism
3. Maintenance of internal constancy
4. Reproduction, growth, and development
5. Irritability and adaptation

Organization

Living matter consists of structures organized in a particular three-dimensional relationship, often following a pattern of structures within structures within structures. Bichat was the first to notice this pattern in the human body. During the bloody French revolution, as he performed autopsies, Bichat noticed that the body's largest structures, or **organs,** were sometimes linked to form collections of organs, or **organ systems,** but were also composed of simpler structures, which he named **tissues** (from the French word for "very thin"). Had he used a microscope, Bichat would have seen that tissues themselves are comprised of even smaller units called **cells.** All life consists of cells. Within complex cells such as those that make up plants and animals are structures called **organelles** that carry out specific functions. Organelles, and ultimately all living structures, are composed of chemicals (fig. 1.2).

The chemicals that make up organisms are called **biochemicals.** An important biochemical is **deoxyribonucleic acid,** or DNA. The sequences of DNA's four types of building blocks are like a language

1.1 bioethics connections

What Is Bioethics?

In 1975, 21-year-old Karen Ann Quinlan drank alcohol after taking a prescription sedative, and her heart and lungs stopped working. When she was found, Karen had no pulse, was not breathing, had dilated pupils, and was unresponsive. CPR restored her pulse. At the hospital, a respirator enabled her to breathe. Within 12 hours, some functions returned—Karen's pupils constricted, and she moved, gagged, grimaced, and even opened her eyes. Within a few months, she could breathe unaided for short periods.

Because Karen's responses were random and not purposeful, and she seemed unaware of herself and her environment, she was said to be in a "persistent vegetative state." This differs from a coma in that at times the eyes open and the person appears to be awake. Karen's basic life functions were intact, but she had to be fed and given water intravenously. Fourteen months after Karen took the fateful pills and alcohol, her parents asked that life support stop. Doctors removed Karen's respirator, and she lived for nine more years at a nursing home, where she died of an infection, never regaining awareness.

Removing Karen Ann Quinlan's life support, addressing the difficult issue of the right to die, was just one concern in the 1970s that led to renewed interest in bioethics (table 1.A). The 1970s was also the time of abortion reform, raising questions of when life begins, and the beginning of heart transplants, raising questions of when life ends. At that time the public also became aware of unethical biomedical research. Most horrifying was the Tuskegee syphilis study, conducted in the 1940s. Four hundred poor black men with the sexually transmitted disease syphilis were not treated so that researchers could examine the course of this curable illness.

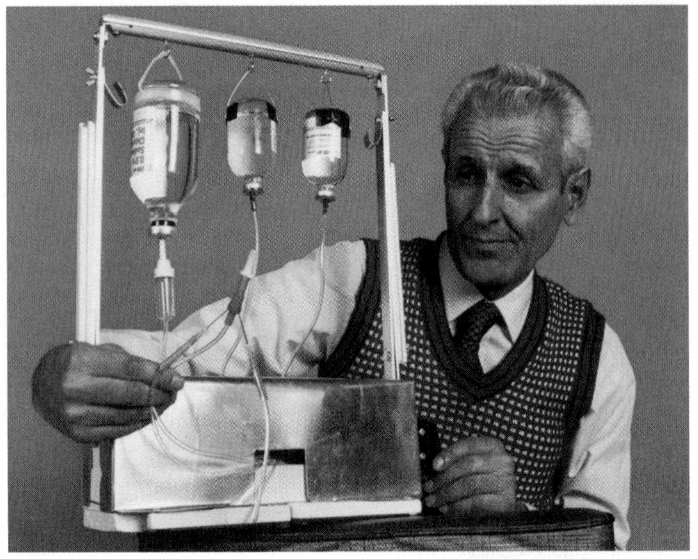

FIGURE 1.A
Dr. Jack Kevorkian, pictured here with a suicide device, is a controversial physician who helps very ill people commit suicide. Many people feel that Dr. Kevorkian provides a useful service, but others question whether some of the people he has assisted are truly near death from their illnesses.

Several aspects of the doctor-patient relationship became and remain a major focus of bioethics. Foremost among them is paternalism, a system under which a physician decides a course of treatment without honoring the patient's autonomy (the ability to choose rationally). Paternalism often entails not fully informing a patient of all options and outcomes. Although paternalism is usually meant to serve the "patient's own good," it can ignore a patient's right to participate in treatment decisions.

Another aspect of the doctor-patient relationship is confidentiality. When is it ethical for a physician to discuss a patient's case with, or reveal the results of medical tests to, someone other than the patient? This question arises in psychiatry, when an ill person may harm others, and in determining when a diagnosis of an illness should be given to insurers or employers.

Today, our aging population, intense interest in health care delivery, and the avalanche of new medical tests and treatments made possible by biotechnology and the Human Genome Project (a worldwide effort to identify all human genes) make bioethical questions perhaps more compelling than ever (fig. 1.A and table 1.B).

Table 1.A General Areas of Bioethics
Doctor-patient relationships
Informed consent in experiments and medical procedures
Right to die and euthanasia
Use of animals in research
Environmental ethics
Reproductive technologies
Genetic screening and prenatal diagnosis
The beginning of life
Access to limited medical resources
Abortion, fetal rights, father's rights

Table 1.B Specific Bioethics Issues
1. Under what circumstances should a physician assist, or fail to prevent, a person's dying?
2. Should humans in very early stages of prenatal development be used for research or to correct medical conditions in adults?
3. On what criteria should we ration organs available for transplant?
4. If a person is diagnosed with a devastating inherited illness years before symptoms arise, should the physician inform the person's employer and health insurer?
5. Under what circumstances should a physician not tell a patient that he or she is dying?

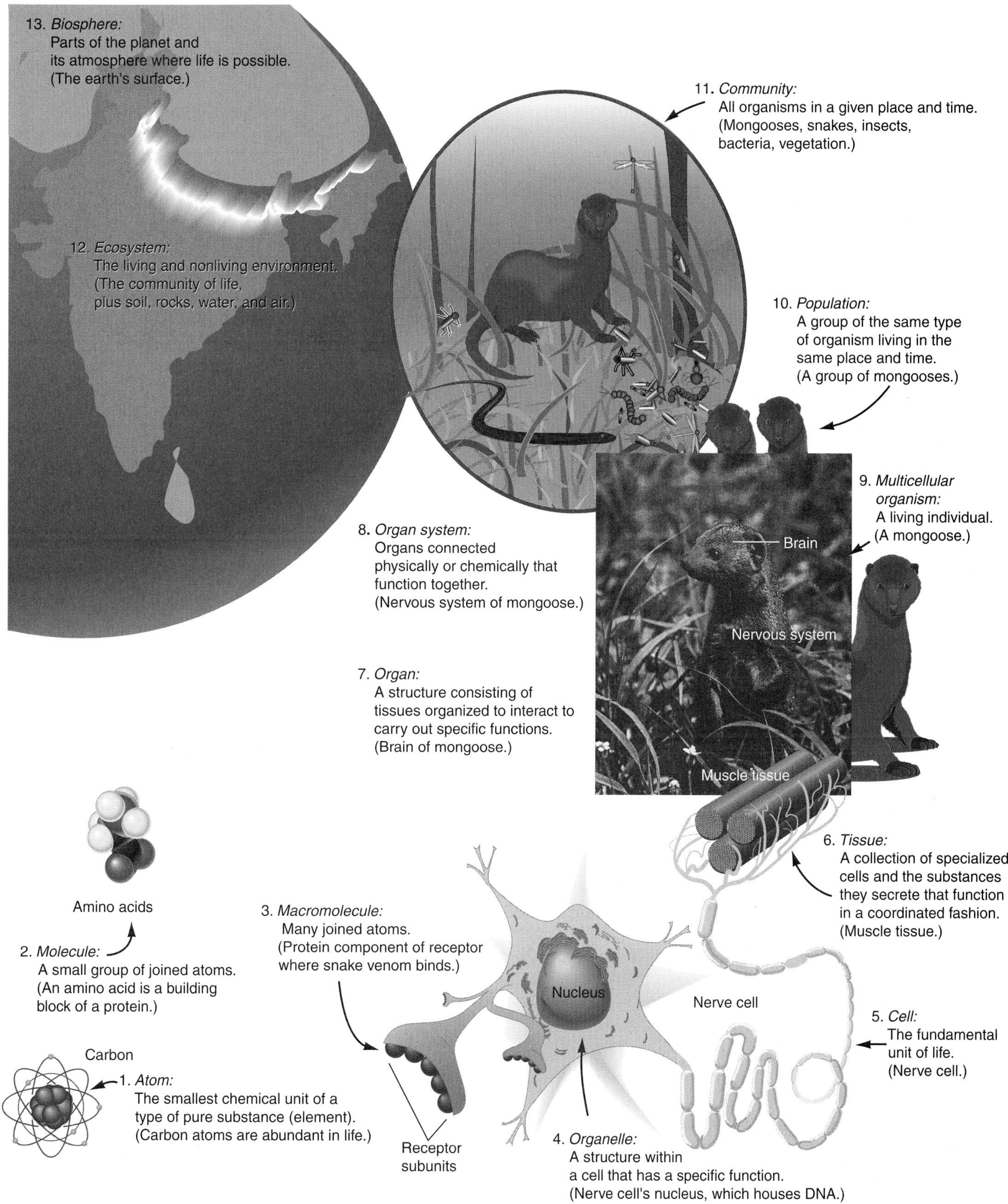

13. *Biosphere:*
Parts of the planet and
its atmosphere where life is possible.
(The earth's surface.)
11. *Community:*
All organisms in a given place and time.
(Mongooses, snakes, insects,
bacteria, vegetation.)
12. *Ecosystem:*
The living and nonliving environment.
(The community of life,
plus soil, rocks, water, and air.)
10. *Population:*
A group of the same type
of organism living in the
same place and time.
(A group of mongooses.)
9. *Multicellular organism:*
A living individual.
(A mongoose.)
8. *Organ system:*
Organs connected
physically or chemically that
function together.
(Nervous system of mongoose.)
Brain
Nervous system
7. *Organ:*
A structure consisting of
tissues organized to interact to
carry out specific functions.
(Brain of mongoose.)
Muscle tissue
6. *Tissue:*
A collection of specialized
cells and the substances
they secrete that function
in a coordinated fashion.
(Muscle tissue.)
Amino acids
3. *Macromolecule:*
Many joined atoms.
(Protein component of receptor
where snake venom binds.)
2. *Molecule:*
A small group of joined atoms.
(An amino acid is a building
block of a protein.)
Nucleus
Nerve cell
5. *Cell:*
The fundamental
unit of life.
(Nerve cell.)
Carbon
1. *Atom:*
The smallest chemical unit of a
type of pure substance (element).
(Carbon atoms are abundant in life.)
Receptor
subunits
4. *Organelle:*
A structure within
a cell that has a specific function.
(Nerve cell's nucleus, which houses DNA.)

FIGURE 1.2

Levels of biological organization.

of life, encoding information that tells cells how to construct particular proteins. Cells' responses to genetic instructions—DNA—enables them to specialize and to function in tissues, organs, and organ systems. DNA is responsible for all inherited traits and is passed from generation to generation. DNA's capacity to change, or **mutate,** makes possible biological change over time, or **evolution.**

Understanding how an organism functions at the biochemical level can sometimes explain its more obvious characteristics. Consider the mongoose, a small mammal made famous in Rudyard Kipling's *The Jungle Book* for its snake-killing behavior. Mongooses not only stand up to poisonous snakes, but if bitten, they can also survive. How can the mongoose withstand a cobra's deadly venom? Biologists once credited the mongoose's swiftness—it darts at a snake's head, cracks the snake's skull with a well-placed bite, and thereby avoids being bitten. But how do mongooses that are bitten survive? Biochemistry provides an answer.

Snake venom kills by binding to structures, called receptors, on certain nerve cells. The venom blocks the nerve cells' ability to send messages to nearby muscle cells and paralyzes the victim. A mongoose's nerve cell receptors are slightly different from those of nearly all other animals except snakes, and the difference prevents venom from binding. So a mongoose's bravado in the face of a poisonous snake may be because it learns that the bite, although painful, isn't deadly.

Organs, tissues, cells, and organelles are highly organized and carry out specific functions. At all levels, and in all organisms, structural organization is closely tied to function. Disrupt the structure, and function ceases. Shaking a fertilized hen's egg, for example, stops the embryo within from developing. If function is disrupted, a structure will eventually break down. Unused muscles begin to atrophy (waste away)—a phenomenon familiar to anyone who has ever had a broken limb immobilized for weeks in a cast. Biological function and form, then, are interdependent.

Biological organization is apparent in all life. Though they are outwardly very different, humans, eels, and evergreens are all nonetheless organized into specialized organ systems, organs, tissues, and cells. Bacteria, although less complex than animal or plant cells, are also composed of highly organized structures.

An organism, however, is more than just a collection of smaller parts. The organization of different levels of life imparts distinct characteristics. For example, individual chemicals cannot harness energy from sunlight. But groups of chemicals embedded in the membranes of an organelle within a plant cell use solar energy to synthesize nutrients. Qualities such as this, which occur as biological complexity grows, are called **emergent properties** (fig. 1.3). These characteristics are not magical; they arise from properties common to all matter.

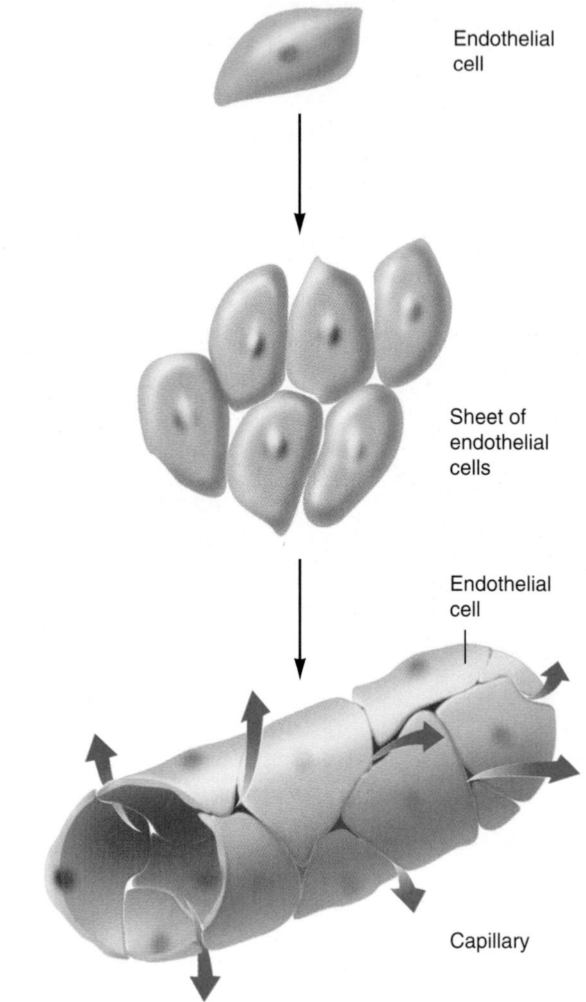

FIGURE 1.3

An emergent property—from tiles to tubes. Endothelial tissue consists of cells that look like tiles adhering to one another to form a sheet. This sheet folds to form a tiny tube called a capillary, which is the smallest type of blood vessel. In the brain, the endothelial cells of capillaries are so tightly packed that substances (including many drugs and poisons) cannot leak from the bloodstream to the brain.

Energy Use and Metabolism

The organization of life seems contrary to the natural tendency of matter to be random or disordered, a state called **entropy.** Energy is required to maintain organization or order. A house cleaned on Sunday that grows progressively messier each day of the week is analogous to entropy. Just as a house does not regain its organization without some energy input (cleaning!), organisms must expend energy to maintain their highly organized state.

An organism must be able to acquire and use energy to build new structures, repair or break down old ones, and reproduce. These functions occur at the whole-body or organismal level, as well as at the organ, tissue, and cellular levels. The term **metabolism** refers to the chemical reactions within cells that direct this acquisition and use of energy. Metabolic reactions both build up (synthesize) and break down (degrade) biochemicals.

Organisms obtain energy to carry out life processes from the environment. **Producers** are organisms that extract energy from the nonliving environment, such as light energy from the sun or chemical energy in rocks. Plants are producers, as are the bacteria described at the beginning of chapter 3. They appear to take nourishment from nothing at all but actually obtain it from chemicals in rocks. **Consumers,** in contrast, obtain energy by eating nutrients that make up other organisms. **Decomposers** obtain nutrients from dead organisms. Fungi, such as mushrooms, are decomposers.

(a)

(b)

(c)

FIGURE 1.4

Reproduction. (*a*) Bacteria called *Escherichia coli* live in the intestines of many mammals, including humans. They usually cause no harm, but certain strains may produce a deadly toxin. These microorganisms reproduce asexually every 20 to 30 minutes. (*b*) Many more oak seedlings, such as this one, sprout than survive to grow into mature trees. (*c*) A bear gives birth to two or three cubs during the winter. She awakens briefly from her hibernation to deliver the cubs, then goes back to sleep for two more months, while they nurse.

Energy requirements link life into chains and webs of "who eats whom," beginning with producers that capture energy from the nonliving environment and continuing through several levels of consumers and decomposers. This interdependency of life suggests other ways that we can view organization in the living world, also shown in figure 1.2.

A **population** includes two or more members of the same type of organism, or **species,** living in the same place at the same time. A **community** includes the populations of different species in a particular region, and an **ecosystem** includes both the living and nonliving components of an area. Finally, the **biosphere** refers to the parts of the planet that can support life.

Maintenance of Internal Constancy

A living organism is composed of the same basic chemical elements as nonliving matter, yet it is obviously quite different. The chemical environment inside cells must remain within a constant range, even in the face of drastic changes in the outside environment, to stay alive. This ability to maintain constancy is called **homeostasis.** A cell must maintain a certain temperature and water balance. It must take in nutrients, rid itself of wastes, and regulate its many chemical reactions to prevent deficiencies or excesses of specific chemicals. An important characteristic of life, then, is the ability to sense and react to environmental change, thus keeping conditions within cells constantly compatible with being alive.

Reproduction, Growth, and Development

Organisms reproduce—that is, they make other individuals like themselves. These new organisms then grow, increasing in size, and develop, adding anatomical detail and taking on specialized functions. Reproduction and development can be as simple as a bacterium dividing in two or as complex as the germination and growth of a tree seedling or the conception, gestation, and birth of a mammal (fig. 1.4). On the organismal level, reproduction transmits genetic instructions (DNA) for developing the defining characteristics of the organism from one generation to the next. Reproduction is vital if a population of organisms is to survive for more than one generation, and on a biosphere level, it is essential for life itself to continue.

Reproduction occurs in two basic ways. In **asexual reproduction** in single-celled or **unicellular** organisms, cellular contents double; then the cell divides to form two new individuals genetically identical to the original parent cell. Some many-celled or **multicellular** organisms reproduce asexually. For example, a potato growing on an underground stem can sprout leaves and roots that constitute an entirely new plant. Asexual reproduction also occurs in some simple animals, such as sponges and sea anemones, when a fragment of the parent animal detaches and develops into a new individual. Asexual reproduction yields **clones,** which are cells or organisms that are genetically identical to each other and to the previous generation.

In **sexual reproduction,** genetic material from two individuals combines to form a third individual, which has a new combination of inherited traits. By mixing trait combinations at each generation, sexual reproduction produces greater diversity than does asexual reproduction. Humans reproduce sexually. However, identical twins are clones, because they derive from a single fertilized egg that has split in two—but they arise from sexual reproduction.

Irritability and Adaptation

Living organisms sense and respond to certain environmental stimuli while ignoring others. The tendency to respond immediately to stimuli, **irritability,** can be essential for survival. A woman touching a thorn quickly pulls her hand away; a dog lifts its ears in response to a whistle; a plant grows toward the sun.

Whereas irritability is an immediate response to a stimulus, an **adaptation** is an inherited characteristic or behavior that enables an organism to successfully reproduce in a given environment. The copepod eggs preserved in time through diapause are adapted to survive a harsh environment, essentially "waiting it out" until conditions improve.

Adaptation is a response developed over time, and because of this, it is usually considered more complex and specific than irritability. Adaptations are likely to differ from one species to another and within the same species in different environments. Both a human and a dog will pull an extremity away from fire (irritability). But on a hot summer day, the human sweats, while the dog pants (adaptations).

Adaptations can be diverse and very striking. The color patterns of many organisms enable them to literally fade into the background, such as the snake in figure 1.5. Such a camouflaged organism can hide from predators or await prey unnoticed.

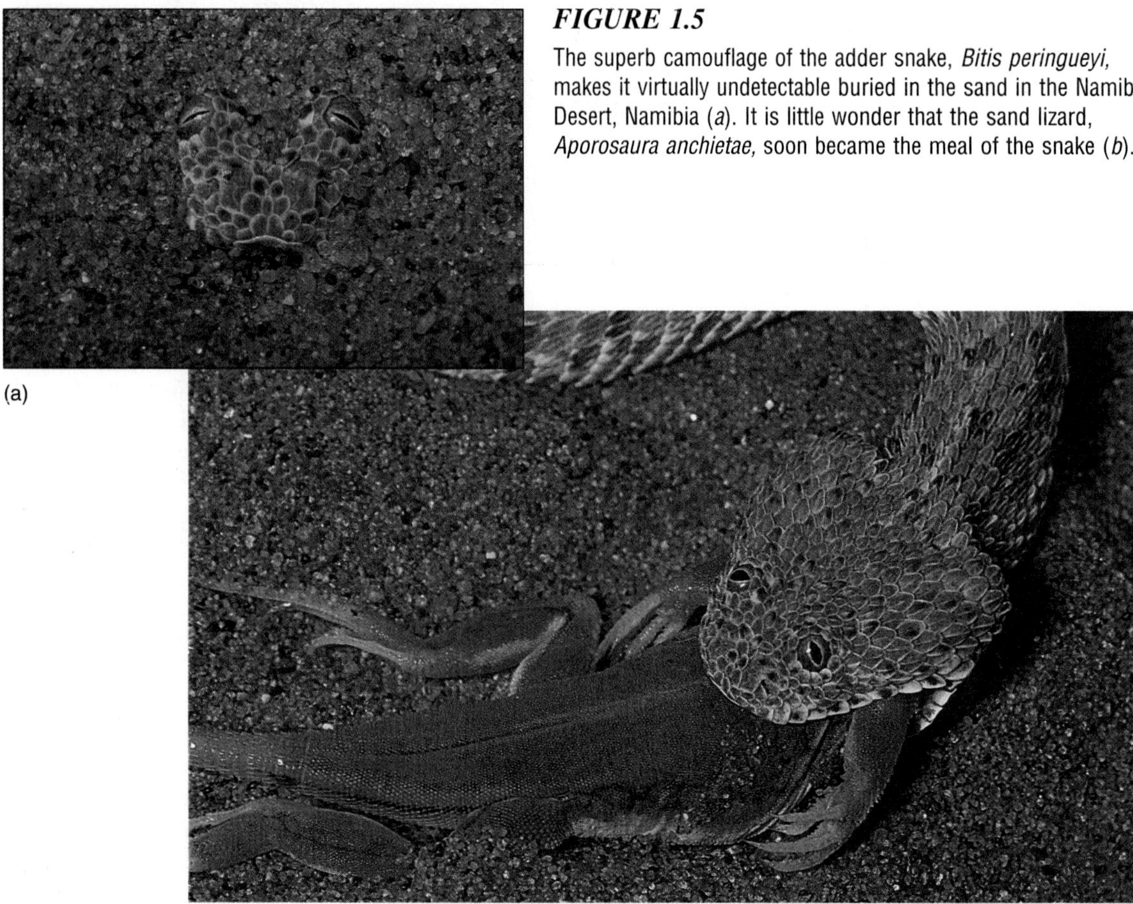

(a)

(b)

FIGURE 1.5

The superb camouflage of the adder snake, *Bitis peringueyi,* makes it virtually undetectable buried in the sand in the Namib Desert, Namibia (*a*). It is little wonder that the sand lizard, *Aporosaura anchietae,* soon became the meal of the snake (*b*).

Plants have interesting adaptations. Most trees, for example, can survive strong winds because of their wide and sturdy trunks, flexible branches that sway without snapping, and strong, well-spread roots. Figure 1.6 shows another adaptation to wind not readily apparent during the violence of a storm. A researcher who noticed that the fronds of certain algae close into cones and cylinder shapes in turbulent streams studied several types of trees during simulated "wind tunnel storms" to see if similar reactions might occur. Indeed, he found that the leaves of some types of trees do close into tighter shapes to minimize wind damage.

Adaptations accumulate in a population of organisms when individuals with certain inherited traits are more likely than others to survive in a particular environment. Trees with leaves that curl to resist wind are more likely to survive for the 20 or more years it takes for most trees to reproduce—even in the face of severe windstorms that strike, on average, every 5 years. Because trees with curling leaves survive more often than trees without this adaptation, more of them pass the advantageous trait to future generations. Trees with the adaptation eventually predominate in the population—unless environmental conditions change and protection against the wind is no longer important.

Over time, adaptation can mold the characteristics of a population by enabling individuals with certain combinations of genes to preferentially survive. **Natural selection** is the elimination of certain individuals from a population whose inherited characteristics make survival and reproduction in a given environment difficult. As individuals that have inherited particularly adaptive traits contribute more offspring, they come to make up more of the population. As the population changes, evolution is occurring. Evolution may be as subtle as the increasing prevalence of a particular trait or as profound as the extinction of a species. Largely because of natural selection, fewer than 1% of the species that have ever existed on earth are alive today.

Mastering Concepts

1. Why is it difficult to define death?
2. What are some of the ways people have tried to define life?
3. What combination of characteristics distinguishes the living from the nonliving?
4. How does life change over time?

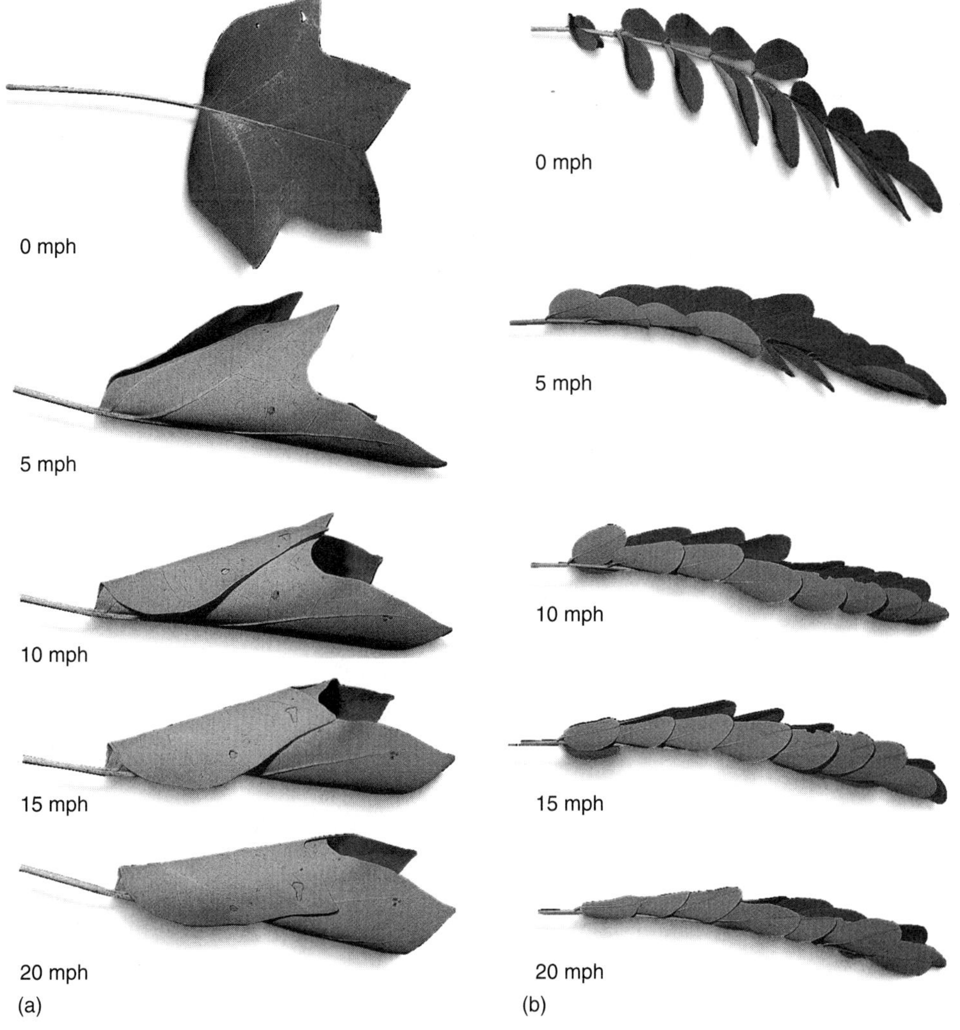

FIGURE 1.6

Leaves of many trees are adapted to minimize wind damage. Researcher Steven Vogel subjected various leaves to a wind tunnel and snapped these photos, which reveal a previously unrealized adaptation to survive a storm. Leaves characteristically fold into a more compact form—a cone for a tulip tree (*a*) and a cylinder for a black locust tree (*b*).

Biodiversity

Adaptation and natural selection explain the many variations on the theme of life—species—that have populated earth. Life is united in its basic defining characteristics, yet it is also diverse, especially in the adaptive strategies of organisms to maintain those characteristics. **Biodiversity** refers to the many different types of organisms on earth.

We are all aware of biodiversity, and we are limited in our study of it by what we can see or otherwise sense. A child explores the residents of a creek bottom; a biologist searches the treetops of a tropical rain forest to count and categorize insects; a paleontologist wipes away dust and debris to uncover a remnant of past life (fig. 1.7). Biology in Action 1.1 follows the harrowing experience of one researcher exploring an area once thought to be too forbidding (by human standards) for life—hot springs in Yellowstone National Park. Researchers have discovered previously unknown microscopic life-forms in these springs, and they estimate that we have identified only 0.1% of the microbial species living there. The hot springs may resemble conditions on the very early earth, perhaps affording us a glimpse into the living past.

What's in a Name?

We humans love to classify things. We organize large stores by departments, grouping similar items so it is easy to quickly find a sweater, spark plugs, or cat litter. We sort soiled laundry by color and fabric types. We assign the hundreds of children in a school to grades and classes, based on their ages, abilities, and interests.

Similarly, biologists organize life into groups called **taxa** (singular: taxon). The biological science of **taxonomy** classifies life based on what we know about the evolutionary relationships of organisms—that is, how recently one type of organism shared a common ancestor with another type of organism. The more recently such an event occurred, the closer related the two types of organisms are presumed to be. Researchers can guess at how closely related types of organisms are by their similarities, both anatomical and biochemical.

Several levels of taxa with increasingly restrictive criteria are used to describe organisms. Just as a student is assigned to a particular school district, school, grade, and class, an organism is assigned to a specific **kingdom, division** or **phylum, class, order, family, genus,** and species. With the recent discovery of unicellular organisms called Archaea, taxonomists have added another taxon, the **domain,** to the scheme as the broadest category. Life has three domains.

(a)

(b)

FIGURE 1.7

Amateur as well as professional scientists can discover and enjoy biodiversity. (*a*) Marshes and ponds are homes to abundant and diverse species. (*b*) This paleontologist searches for signs of past life preserved in a wall of rock.

1.1 biology in action

Some Like It Hot: Thermophilic Bacteria

"I still have nightmares about being boiled alive," says Francis Barany, a biochemist at Cornell University Medical College in New York City. Barany fell into a pool of boiling mud in Yellowstone Park in 1991 while searching for bacteria. His story illustrates vividly the excitement of discovery—although his fall was frightening, Barany gained the opportunity to explore biodiversity, clues to the origin of life, and a brand new biotechnology.

Yellowstone National Park has long fascinated biologists, who flock to its pools of steaming water. Hot springs, topped with colored mats of algae in which a few types of thermophilic (heat-loving) bacteria thrive (fig. 1.B), are found today only near volcanoes and lakes. But long ago, as life began, such hot environments were the norm. Biologists hope that by studying how organisms live in pockets of today's environment that resemble the conditions of early earth, they might discover clues to how life itself began. This has already happened. In some Yellowstone hot springs live microorganisms, called Archaea, that may be the most ancient types of organisms on earth.

Researchers study thermophilic bacteria for another reason—the microbes' biochemicals can withstand great heat. Of special interest are enzymes, which are proteins that control the rates of specific biochemical reactions. Heat dismantles most enzymes, so those from the hot springs bacteria are valuable in the laboratory because they remain active when heated.

The first such "thermostable" enzyme used in the lab—to mass-produce genetic material—came from a bacterium called *Thermus aquaticus,* discovered in 1965 by Indiana University microbiologist Thomas Brock. In Yellowstone's hot springs, Brock dipped glass slides deep into the water and pulled them up a few days later. The slides were covered with bacteria. It wasn't until 1987 that other researchers who pioneered a method to mass-produce DNA discovered the value of using the microbe's DNA-copying enzyme in automating their procedure. The technique required a heating step to separate strands of DNA, and the heat inactivated the enzyme needed to copy the DNA. Researchers had to repeatedly add new enzyme. The thermostable enzyme, however, works over and over, untouched by the heat.

Francis Barany ventured into one of the park's 10,000 hot springs in search of an enzyme for a new gene-copying method. Here is his story:

It was evening, and we had been collecting so well that we thought we might as well finish, and the next day relax. We ended the day at Artist's Paint Pot, a hot springs and mud hole where the water had dried out, leaving hot mud. Hot springs are just beautiful, deep blue in the interior, the surface yellow-brown, where the bacteria are. It was constantly steaming, with bubbling water.

While collecting my thirteenth sample, the ground simply gave way, collapsed under me, and my left leg sank into the hot mud. It was 70°C; that's 158°F! It blistered the skin right off. I jumped out, and hopped onto land, screaming. Someone poured cold water on my leg, ripped my jeans off. I remember screaming that I had a permit to collect, and to save that thirteenth sample!

FIGURE 1.B
The water in this hot spring in Yellowstone Park is boiling—but certain microorganisms thrive in it.

The members of a domain or kingdom share many characteristics, yet they are quite diverse—just as students in a school district are alike in that they live in the same area but are nonetheless individuals. A human, snake, and mosquito, for example, are all members of the animal kingdom, but they are clearly very different from each other and do not share the other more specific taxa (phylum to species). A human, muskrat, and giraffe are more closely related—all belong to the same kingdom, phylum, and class (Mammalia). A human, orangutan, and chimpanzee are even more closely related, sharing the same kingdom, phylum, class, and order (Primates).

Taxonomists give each type of organism an abbreviated name, called its **binomial name,** consisting of its genus and species. A human is *Homo sapiens.* Our full taxonomic name is Animalia Chordata Mammalia Primates Hominidae *Homo sapiens.* Another familiar organism is Plantae Angiospermophyta Monocotyledoneae Liliales Liliaceae *Allium sativum*—otherwise known as garlic. The most restrictive taxon, species, designates organisms that can breed only among themselves or distinctive types of asexually reproducing organisms.

The Five Kingdoms

Organisms within a kingdom demonstrate the same general strategies for staying alive, but they differ from each other in the details of how they do so. Two criteria that biologists consider to assign organisms to kingdoms are cellular complexity and mode of nutrition. An organism may be unicellular or multicellular. If it is unicellular, its cells may be a simple type, called **prokaryotic,** or a complex type with organelles, called **eukaryotic.** The cells of a multicellular organism are always eukaryotic. An organism that is a prokaryotic cell is called a prokaryote. An organism with eukaryotic cells is called a eukaryote.

Kingdom **Monera** includes the bacteria, which are unicellular prokaryotes. One type, the cyanobacteria, can extract energy from sunlight through a series of chemical reactions called photosynthesis. At first taxonomists classified Archaea in Monera because they are unicellular, but now we know that they are so different from bacteria that they should probably be classified differently. However, some biologists consider Archaea to belong to a sixth kingdom.

Kingdom **Protista** includes simple eukaryotes such as the unicellular protozoa, algae, and certain molds. Some are producers and some are consumers.

Kingdom **Plantae** includes multicellular eukaryotes that obtain energy from sunlight—plants. A green pigment called chlorophyll enables certain plant cells to harness solar energy. Plants are **autotrophs,** which means that they make their own food. Plant cells are cuboidal or brick-shaped, because of rigid outer structures called **cell walls,** which are composed of a biochemical called cellulose. Plants have specialized tissues and organs, but they do not have nervous and muscular systems.

Members of kingdom **Fungi** are eukaryotes that obtain nutrients from decomposing organisms. They have cell walls that consist of a substance called chitin. Yeasts are unicellular fungi. Familiar multicellular fungi are morels and mushrooms.

Kingdom **Animalia** (animals) includes multicellular eukaryotes that have nervous and muscular systems. They are **heterotrophs,** deriving energy from the chemical bonds of nutrients that make up other organisms. Animal cells do not have cell walls.

Figure 1.8 compares representative members of each kingdom.

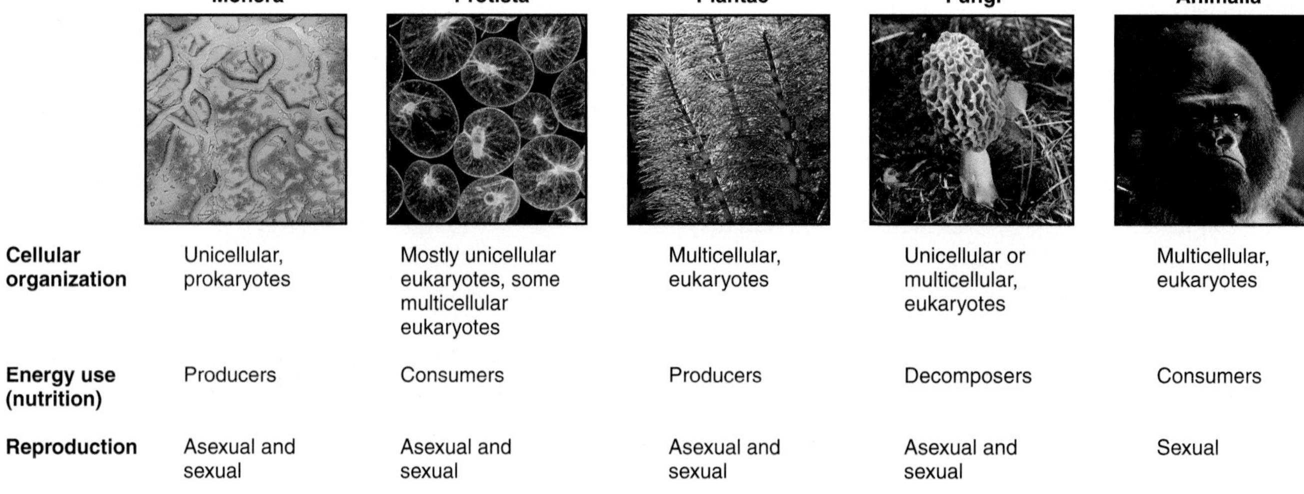

	Monera	Protista	Plantae	Fungi	Animalia
Cellular organization	Unicellular, prokaryotes	Mostly unicellular eukaryotes, some multicellular eukaryotes	Multicellular, eukaryotes	Unicellular or multicellular, eukaryotes	Multicellular, eukaryotes
Energy use (nutrition)	Producers	Consumers	Producers	Decomposers	Consumers
Reproduction	Asexual and sexual	Asexual and sexual	Asexual and sexual	Asexual and sexual	Sexual

FIGURE 1.8

Meeting the challenges of being alive. Members of the different kingdoms of life satisfy some life requirements in very diverse ways—such as irritability responses and adaptations—yet also in some strikingly similar ways—such as the biochemical reactions of metabolism. This chart compares three other basic characteristics of life—cellular organization, energy use, and reproduction.

Alive or Dead? ~

continued from page 4.

When Baby F. began to breathe on the 43rd day after she had mysteriously stopped, was she alive or dead? While the baby's parents and physicians wrestled with the difficult question, nature resolved the issue. Baby F. died on day 71 of pneumonia—her lungs had filled with fluid to the point that even a respirator could not assist her breathing.

An autopsy did not solve the puzzle of Baby F.'s curious state between life and death. Her brain had not "liquefied," as the brains of people on respirators for a long time are described. The "thinking" regions were decomposed, but parts of the structures controlling vital functions were intact.

Baby F.'s physicians used her case to argue for a change to the brain death criteria, allowing greater leeway for infants, especially if cause of death is not obvious. However, other physicians responded to their published report in support of the brain death criteria, concluding that in Baby F.'s case, a few brain stem cells and a local blood supply had somehow persisted as the surrounding vital tissue died. Her spare breaths, these physicians claimed, were merely a remnant of her former capacity to breathe, not sufficient to support life and therefore not a sign of life.

The lessons to be learned from the tragic case of Baby F. extend beyond a child's struggle to live. They go straight to the heart of the question of what combination of abilities distinguishes the living from the nonliving. They also call into question our ability to fashion definitions of life and death. More philosophically, this true tale illustrates that scientists and physicians do not always have an answer or explanation for a natural phenomenon.

Evolution Is the Backbone of Biology

Evolution is a continuing tale that weaves together the disparate threads of diverse adaptations and shared ancestries, connecting life of past ages with that of today. Evolution forms the very basis of the study of life, because all earth's inhabitants are, in a sense, related: a tuna and a trout obviously so, because they share a fishlike ancestor. But also related are the tuna, a human, and a frog, in that they share a common vertebrate ancestor—some animal with a backbone that gave rise to all three modern vertebrates. The tuna and an oak tree are related even farther back, because they each descend from a mass of cells that was the first type of multicellular organism. All multicellular organisms, in turn, ultimately can trace their lineages to ancient organisms that were single cells. The specialized tuna, trout, human, frog, and oak, and even the single-celled amoeba, evolved as genetic changes accumulated in populations of their distant ancestors. The organisms that populate the planet today are but a small sampling of all those that have lived here.

Mastering Concepts

1. How can living organisms be very similar, yet diverse?
2. What do taxonomic names mean?
3. What are the five kingdoms of life?
4. Why is evolution so important in biology?

SUMMARY

1. A living **organism** is distinguished from an inanimate object by the presence of a combination of characteristics.

2. An organism is organized as structures of increasing size and complexity, from **biochemicals,** to **cells,** to **tissues, organs,** and **organ systems,** to individuals, **populations, communities, ecosystems,** and the **biosphere.**

3. Emergent properties arise as the level of organization of life increases and are the consequence of physical and chemical laws.

4. Life requires energy to maintain its organization and functions. **Metabolism** directs the acquisition and use of energy.

5. Organisms must maintain an internal constancy in the face of changing environmental conditions, a state called **homeostasis.**

6. Organisms develop, grow, and reproduce.

7. Organisms respond to the environment, through **irritability** in the short term and by **natural selection** over generations. Natural selection eliminates inherited traits that decrease chance of survival in a certain environment by differential survival and reproduction.

8. Biologists classify organisms with a series of names that reflect probable evolutionary relationships.

9. The five kingdoms of life are **Monera, Protista, Plantae, Fungi,** and **Animalia.** Cell complexity, mode of nutrition, and other factors distinguish their members.

10. Evolution through natural selection explains why organisms are alike yet diverse and how common ancestry unites all species.

11. Bioethics is a discipline of philosophy concerned with issues of rights, morality, justice, and values that arise from biomedical research and practice.

TO REVIEW . . .

1. How did technology alter the definition of death based on ability to breathe and presence of a heartbeat?

2. How is death defined today?

3. List the characteristics of life.

4. Define the following terms:
  a. metabolism
  b. homeostasis
  c. irritability
  d. adaptation

5. Place the following terms in size order, from smallest to largest: ecosystem, cell, organelle, organ system, tissue, atom, community, molecule, organ, biosphere, macromolecule.

6. Cite two ways that asexual and sexual reproduction differ.

7. Give three examples of adaptations, from this chapter or from your own observations of life.

8. How does natural selection act on adaptations in a way that causes evolution to occur?

9. Why do biologists assign taxonomic designations to organisms? What do these names reflect?

10. What are the five kingdoms of life? How are their members distinguished?

11. List a way that organisms in the following pairs of kingdoms differ from each other:

- **a.** plants and fungi
- **b.** monera and protista
- **c.** animals and plants
- **d.** monera and animals

12. What is bioethics?

TO THINK ABOUT . . .

1. An 86-year-old woman was found on the floor of her apartment, with no signs of breathing or a heartbeat. Paramedics and a coroner declared her dead at the scene. A few hours later, the woman lay under a sheet on a table in a morgue. An attendant noticed the sheet rhythmically rising and falling. The woman was breathing! What might have been done to better determine whether the woman was dead or alive?

2. In an episode of the television series *Star Trek,* three members of the crew of the starship USS *Enterprise,* exploring a planet, meet an enemy who dehydrates each one into a small box. He then rearranges the boxes and crushes one of them. He later attempts to restore the people to life, but only two of them reappear. What biological principle does this action illustrate?

3. What general principle discussed in the chapter does the function of aggregated endothelium in keeping dangerous substances out of the brain illustrate?

4. Should national parks receive a portion of revenue from biotechnology companies that use thermophilic bacteria from natural hot springs in their products?

5. An android is a being that inhabits many science fiction stories. It is engineered from hardware and operates on software with neural networks that impressively mimic the human nervous system, giving the android intelligence, expanded memory, and self-awareness. Is an android a person or a nonliving device? Which characteristics of life does it have? Which does it lack?

6. How has research conducted on the hot springs in Yellowstone National Park altered our view of the biosphere?

7. Some people believe that human life begins with the union of sperm and egg, while others believe that it begins at birth. Some even think that human life has no start or finish—they argue that it is a continuum, because we all possess cells that have the potential to be alive when joined with another cell of the appropriate sort. What do you think? When does life begin, and why?

TO LEARN MORE . . .

References and Resources*

Belkin, Lisa. 1993. *First, do no harm.* New York: Fawcett Crest. A compelling look at several unforgettable young patients on the brink of death.

Fischman, Joshua. May 19, 1995. Have 25-million-year-old bacteria returned to life? *Science,* vol. 268. Can an organism remain in suspended animation for millions of years?

Morell, Virginia. August 23, 1996. Life's last domain. *Science,* vol. 273. We are beginning to understand characteristics of the Archaea.

Okamoto, Ken and Tsuyoshi Sugimoto. September 1995. Return of spontaneous respiration in an infant who fulfilled current criteria to determine brain death. *Pediatrics,* vol. 96. The medical journal report of Baby F.

Orstan, Aydin. Spring 1990. How to define life: A hierarchical approach. *Perspectives in Biology and Medicine.* A historical account of how different thinkers have defined life.

Slome, L.R., et. al. February 6, 1997. Physician-assisted suicide and patients with human immunodeficiency virus disease. *The New England Journal of Medicine*, vol. 336. Do physicians have the right to assist patients who wish to die?

Vogel, Steven. September 1993. When leaves save the tree. *Natural History.* Vogel placed leaves in wind tunnels and observed an adaptation.

**Life*'s website features frequently updated references.

2 Thinking Scientifically About Life

Wildlife biologists study organisms in their natural habitats. Scientific experiments may take place in diverse settings, but they fit into the process of scientific thinking in much the same way.

chapter outline

***DES Daughters and Apopka Alligators* 18**

***Theories Attempt to Explain Nature* 18**

***Scientific Inquiry* 18**
Observations
Background Information
Formulating a Hypothesis
Devising an Experiment
Scientific Inquiry Continues

***Designing Experiments* 24**
Experimental Controls
Types of Experiments

***Applying the Scientific Method: Challenges and Limitations* 28**
Interpreting Results
Expecting the Unexpected
Scientific Discovery by "Accident"

A Tale of Two Drugs

In late December 1973, Anne Needham developed a facial rash, a foul vaginal discharge, and fatigue. When severe abdominal pain and vaginal bleeding began in early 1974, Anne became alarmed. So did her gynecologist, who tested a growth he found in Anne's reproductive tract. A week later, Anne learned that she had a rare vaginal cancer that only strikes women over age 50. But Anne was only 20—and she was not the only case. In 1966, physicians at a Boston hospital treated a 15-year-old girl for the same rare cancer. By 1969, five more teens with the problem appeared at Boston hospitals. What was happening?

When a sudden cluster of cases of a very rare illness appears in uncharacteristic types of people, it attracts the attention of epidemiologists, researchers who track the causes of illnesses. Within a few years, as the cancer affected more young women, epidemiologists identified a common factor among them—their mothers had taken a drug, diethylstilbestrol (DES), while pregnant with them. The drug, a synthetic form of the hormone estrogen, supposedly prevented miscarriage. Investigators later found that DES did no such thing—but could cause the rare cancer and a less serious condition called adenosis in many "DES daughters." "DES sons" have some reproductive anomalies related to exposure to the drug before birth.

From 1949 to 1971, nearly 2 million pregnant women in the United States took DES, often being told that they were taking a vitamin. Some of their children suffered because of widespread use of a drug before doctors fully understood its effects.

In the early 1990s, 11,000 healthy women in the United States who have relatives with breast cancer began taking a drug called tamoxifen. It isn't a synthetic estrogen, like DES, but it interferes with estrogen's action. Unlike the women in the DES trial, those in the "tamoxifen prevention trial" knew why they were taking the drug—to see if it would prevent breast cancer in people supposedly genetically predisposed to develop it. However well planned, the trial of tamoxifen ended early when the premises of the investigation were found to be incorrect. Tamoxifen, like DES, might actually harm some women.

to be continued . . .

DES Daughters and Apopka Alligators

In 1980, a large amount of a pesticide, DDT, spilled into Lake Apopka, Florida. Years later, researchers studying wildlife in the area noted oddities among the resident alligator population. The eggs didn't look normal, and male hatchlings had stunted reproductive organs. Baby male alligators' blood revealed a hormonal imbalance. The animals had too much estrogen, a hormone associated with development of female structures, and too little testosterone, a hormone associated with development of male structures. Female hatchlings also had excess estrogen (fig. 2.1).

Since 1980, the eggs of Apopka alligators had absorbed a chemical called DDE, which is a breakdown product of DDT. The pesticide had long been linked with harming wildlife, and as a result, the U.S. government had banned its use in 1972. DDE thinned birds' eggshells, so they shattered and the developing young fell to their deaths. Many populations, such as those of bald eagles, plummeted. Others had increased incidences of birth defects. Among seagulls, males developed female reproductive organs, and females were born with two sets of reproductive organs.

What do young women with a rare cancer caused by exposure to the drug DES (see opening vignette) have in common with hormonally unbalanced alligators? Both the women and the alligators had reproductive problems and unusual exposures to estrogen-like chemicals. The DES daughters and Apopka alligators provide evidence for an emerging idea, called the estrogen mimic theory. It proposes that certain drugs, industrial chemicals, and environmental pollutants act as "estrogen mimics," either directly or indirectly shifting hormone levels so that forms of estrogen that promote cell division predominate, thus setting the stage for reproductive problems and cancer.

Theories Attempt to Explain Nature

A **theory** is a systematically organized body of knowledge that applies to a variety of situations. The estrogen mimic theory has implications for life at all levels. The phenomenon can be studied at the level of molecules, cells, tissues, organs, populations, ecosystems, and, because it involves widespread pollution, even the entire living planet, the biosphere.

Theories attempt to logically explain natural phenomena, but we cannot know something scientific with absolute certainty. Theories change, as we learn more. The history of science is full of long-established ideas changing as we learned more about nature, often thanks to new technology. Figure 2.2 illustrates a recent example of this—a microorganism that drastically altered our view of what causes a common illness, ulcers. A large body of evidence had pointed to emotional stress as making a person more prone to develop a peptic (stomach) ulcer. When two young Australian physicians proposed in 1983 that a bacterial infection causes ulcers, the medical community scoffed. How could bacteria survive in the high-acid environment of the stomach lining? So one doctor performed a daring experiment—he drank some of the bacteria! The resulting upset stomach proved the point. These bacteria, called *Helicobacter pylori,* thrive in the digestive tract lining. His discovery changed the way doctors treat ulcers.

Scientific Inquiry

Theories build on interpretations of observations, evidence, and the results of experiments. Those interpretations arise from application of the **scientific method,** which is a general way of thinking and of organizing an investigation. The scientific method is

(a) (b)

(c)

FIGURE 2.1

The estrogen mimic theory proposes that estrogen-like chemicals in pesticides cause reproductive problems and breast cancer. (*a*) DDT was sprayed directly on people! (*b*) Bird populations exposed to DDE and other environmental estrogens experience many problems, such as this eagle with a malformed beak. (*c*) Nests with too many eggs, such as this one, can indicate that females are sharing nests. This can happen for a variety of reasons, one of which is that the number of males is unusually small. Enlarged egg clutches that coincide with exposure to certain toxins may indicate that the pollution has altered the sex ratio, harming males.

not a strict list of rules, like a recipe for a cheesecake, but rather a framework in which to consider ideas and evidence. Because scientists can take several approaches to answering a particular question, there is no "one" scientific method. An ecologist, a chemist, a geneticist, and a physicist conduct very different types of investigations. Instead, we use the term "scientific inquiry" to refer to the sequence of logical steps scientists follow to answer a question about nature.

Scientific inquiry consists of everyday activities—observing, questioning, reasoning, predicting, testing, interpreting, and concluding (fig. 2.3). It includes thinking, detective work, and seeing connections between seemingly unrelated events—such as the reproductive problems of DES daughters and Apopka alligators. We can use the effects of estrogen and chemicals that mimic it as a way to explore the method of scientific inquiry and the development of a scientific theory.

Observations

The scientific method begins with observations:

> Humans exposed to the synthetic estrogen DES are at increased risk of developing certain reproductive problems.
>
> Alligators exposed to DDE have a higher risk of developing certain reproductive problems.

Other observations link estrogen exposure to reproductive problems:

> Estrogen applied to human breast or uterine cells growing in culture stimulates them to divide.
>
> Women are more likely to develop breast cancer if they begin menstruating early and cease menstruating late—factors that expose them to more estrogen over their lifetimes than women who begin menstruating later and end earlier.

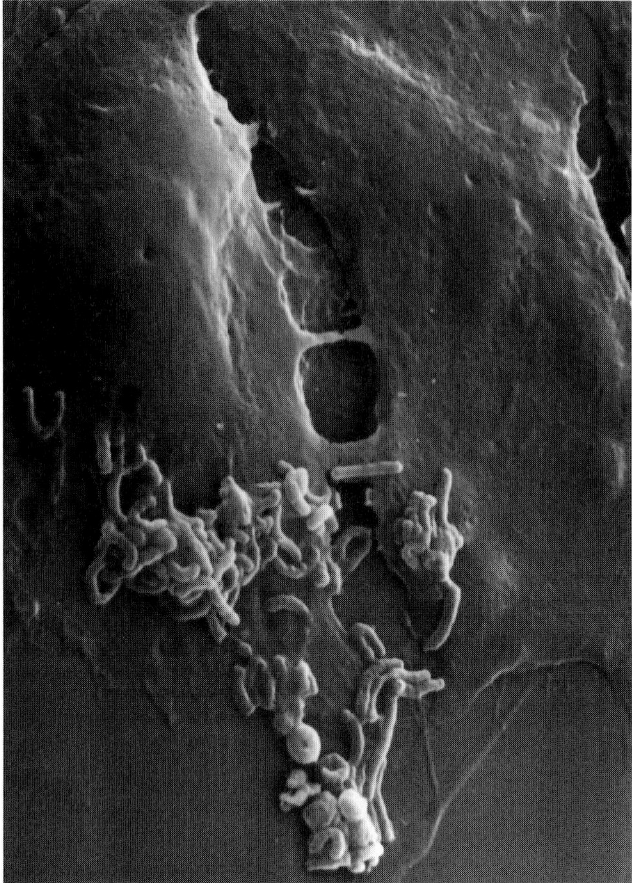

FIGURE 2.2

In 1982, two young Australian physicians, Barry Marshall and J. Robin Warren, isolated bacteria from beneath the protective mucous linings in the stomachs of patients with inflammation or ulcers. When the medical community, long convinced that ulcers stem from stress, did not at first accept the idea of a bacterial cause, Marshall swallowed the bacteria! His inflamed digestive tract painfully, but temporarily, proved the point.

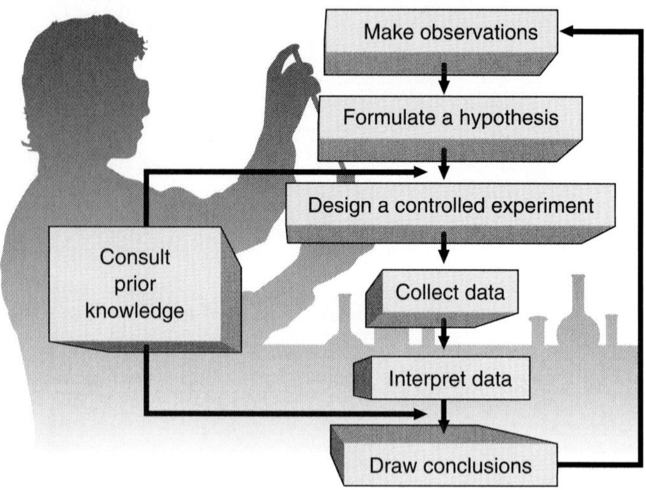

FIGURE 2.3

Steps of the scientific method.

The estrogen mimic idea may explain several other observations:

Human sperm counts worldwide dropped by 50% between 1938 and 1991. Use of pesticides containing estrogen-like chemicals has increased since 1940.

Since the 1940s, the incidence of human testicular cancer has increased more than 30% and of human breast cancer 25 to 30%.

The incidence of birth defects affecting the human male reproductive system has risen sharply. These include undescended testicles and malformed penises. Abnormal sperm have become more common (fig. 2.4).

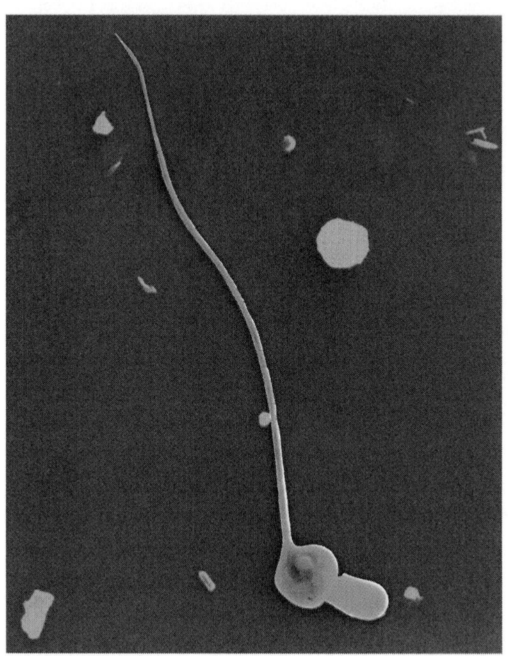

FIGURE 2.4

This sperm cell has an abnormal outgrowth emanating from its head, which may have been caused by exposure to estrogen-like chemicals in the environment. (×1900)

Background Information

Considering existing knowledge is important in scientific inquiry. For example, to understand what happens when estrogen is in excess or exerts an abnormal effect, it is important to understand its normal role.

The estrogens are a group of similar biochemicals. They are hormones, which are biochemicals made in one part of an organism that travel to another site, where they exert an effect. Estrogen molecules enter certain cells and bind to proteins called receptors (fig. 2.5), which fit them in much the same way that a catcher's mitt fits an incoming baseball. The binding occurs in the nucleus, the part of the cell that houses the genetic material. A cell that is sensitive to estrogen's stimulation may have many thousands of such receptor proteins. Estrogen molecules bound to receptors then turn on certain genes, whose actions produce the cell's hormone-associated effects, such as dividing.

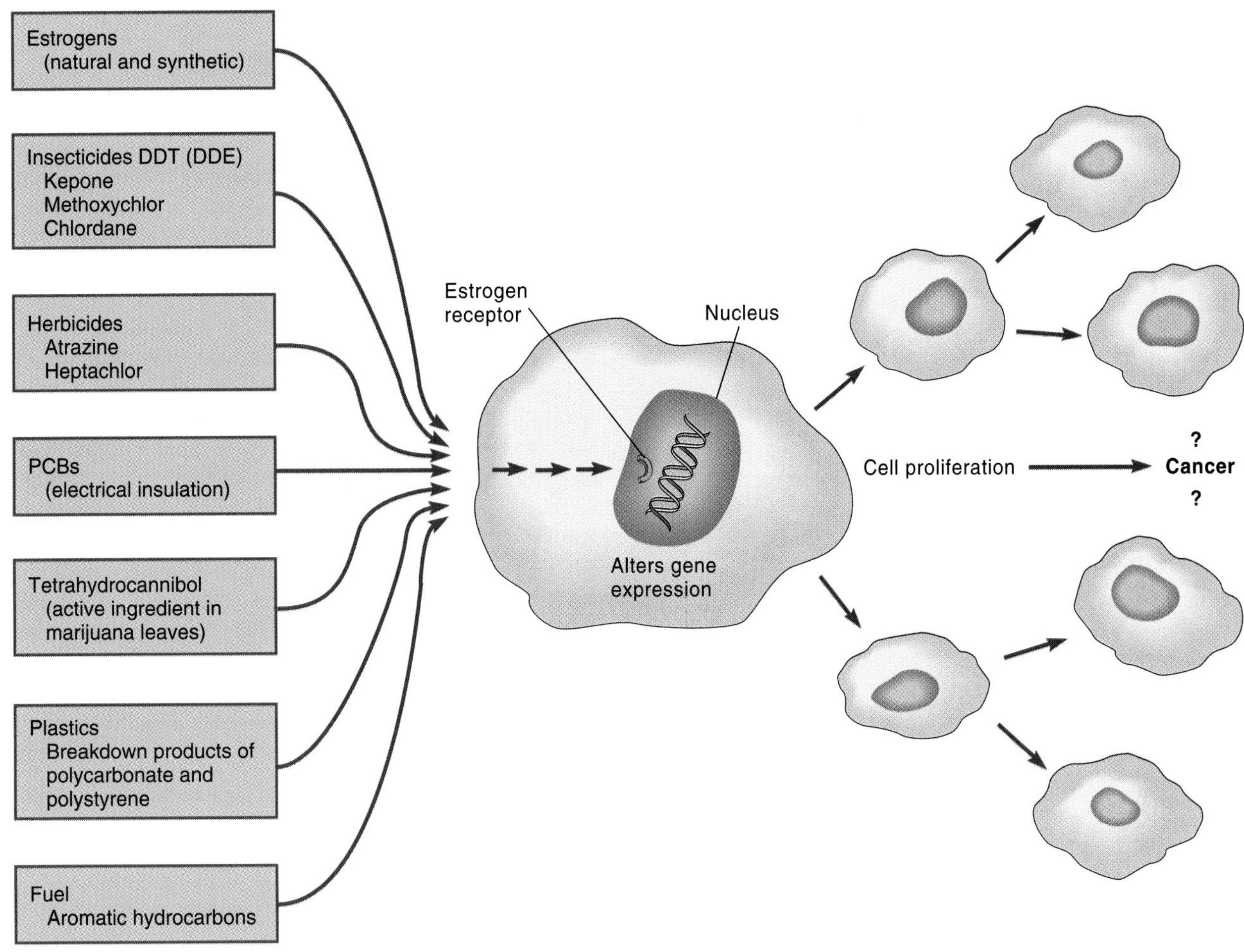

FIGURE 2.5

Several types of commonly used or once-used chemicals can enter cells and bind estrogen receptors, activating genes that stimulate the cell to divide—possibly causing cancer or other problems.

Some estrogens promote cell division, and some do not. The proportion of cell-division-promoting to non-cell-division-promoting estrogens, as well as the ratio of estrogens to other sex hormones, determines whether the reproductive system develops normally. Chemicals acting as estrogen mimics may disrupt this crucial hormonal balance in several ways. DDE, for example, shifts the balance of estrogens to those that stimulate cell division and blocks the function of testosterone, a male sex hormone. DDE may therefore make inappropriate forms of estrogen predominate.

Formulating a Hypothesis

The observations that begin the scientific method lead first to a general question. In the case of investigating estrogen mimics, a general question might be:

Does excess estrogen exposure cause reproductive problems in animals, including humans?

The next step in the scientific method is formulating a statement to explain an observation, based on previous knowledge. This educated guess, or **hypothesis,** is specific and testable and should examine only one changeable factor, or **variable.** A hypothesis is more than just a question or a hunch; it is based on already known facts.

Often a hypothesis is posed as an "if . . . then" proposition. This phrasing is specific, attempts to explain observations, and makes a prediction that can be tested:

If exposure to large amounts of estrogen causes reproductive problems in animals, then animal eggs intentionally exposed to estrogen or similar substances should not hatch, or they will hatch to yield abnormal offspring.

Devising an Experiment

To support or disprove the hypothesis, the researcher collects information. This often entails devising an **experiment,** which is a test that yields information. An experiment can disprove the hypothesis but can never prove it, because there is always the possibility of discovering additional information.

2.1 biology in action

Elephants Calling—The Powers of Observation

If you think only humans possess extensive language, consider an elephant's extended family. An elephant clan, led by an elder female, or matriarch, and consisting of females and young males, has quite a vocabulary. Elephant "language" includes calls in infrasound, or sounds too low for human ears to detect.

But one curious pair of human ears heard the low-pitched language of the great beasts in 1984. Cornell University researcher Katy Payne was standing near caged elephants at a zoo when she felt a "throbbing" in the air. The sensation brought back memories of singing in the church choir as a child; standing in front of the church organ pipes, she felt a similar throbbing. "In the zoo I felt the same thing again, without any sound. I guessed the elephants might be making powerful sounds like an organ's notes, but even lower in pitch," she writes in her children's book, *Elephants Calling*.

To test her hypothesis, Payne and two friends used equipment that could detect infrasound to record elephants in a circus and a zoo. Infrasound was there!

Payne, already well known for having discovered the songlike language of whales, published her results, which caught the attention of elephant researchers Joyce Poole and Cynthia Moss. They invited her to join them in the wild. So Payne moved her study to the Amboseli plain, a salty, dusty stretch of land at the foot of Mount Kilimanjaro in Tanzania. Her lab was a truck sitting among the elephants, who grew so used to its presence that they regarded it as a natural part of the scenery.

Payne's already highly developed powers of observation were further sharpened living among the elephants. Just as we might watch people converse in a foreign language and then discern the meanings of words from accompanying actions and expressions, Payne and her fellow elephant watchers soon became attuned to the subtle communications between mother and calf; the urges of a male ready to mate; and messages to move to find food or water. Writes Payne, "It is amazing how much you can learn about animals if you watch for a long time without disturbing them. They do odd things, which at first you don't understand. Then gradually your mind opens to what it would be like to have different eyes, different ears, and different taste; different needs, different fears, and different knowledge from ours."

One day, when Payne, Poole, and the "R" elephant family (all of which were given names starting with R) were mesmerized by two bulls fighting for dominance, the youngest family member, baby Raoul, slipped away (fig. 2.A). Finding an intriguing hole, Raoul stuck his inquisitive trunk inside—then leapt back, bellowing, as a very surprised warthog bounded out of his invaded burrow. Raoul's mother, Renata, responded with a roar. Payne described the scene:

> *Elephants in all directions answer Renata, and they answer each other with roars, screams, bellows, trumpets, and rumbles. Male and female elephants of all sizes and ages charge past each other and us with eyes wide, foreheads high, trunks, tails, and ears swinging wildly. The air throbs with infrasound made not only by elephants' voices but also by their thundering feet. Running legs and swaying bodies loom toward and above us and veer away at the last second.*

To determine whether DDE in the environment caused the reproductive abnormalities in the Lake Apopka alligators, experiments re-created conditions that exposed the eggs to estrogen-like chemicals (fig. 2.6). Investigators collected alligator eggs from a clean lake and incubated them at high temperatures, which in alligators ensures that the hatchlings are male. The researchers painted some eggs with estrogen, some with DDE, and some with an inert substance, which would determine whether painting an egg can alter development.

Among the estrogen-treated eggs, only 20% produced male hatchlings—a drastic change from the expected 100%. Only 40% of the eggs painted with DDE hatched males, which had excess estrogen and deficient testosterone in the same proportions as did male alligators from Lake Apopka. Another 40% of the DDE-treated eggs hatched intersexes, which had both male and female reproductive structures. Intersexes are not seen among alligators in clean lakes. The remaining 20% of the DDE-treated eggs produced female hatchlings. All of the eggs painted with the inert substance hatched normal male alligators.

These experimental results support the hypothesis and lead to the conclusion that DDE alters reproductive structures in alligators. But for a conclusion to become widely accepted, the results must be repeatable—other scientists must perform the same experiments and observe the same results.

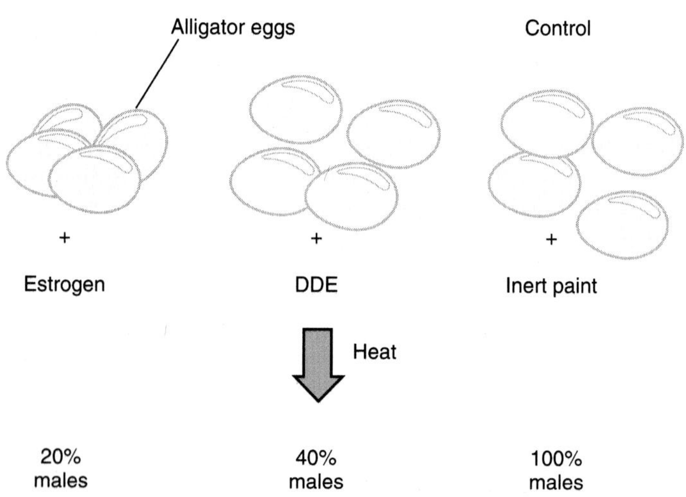

FIGURE 2.6

Normally, alligator eggs incubated at high temperature hatch males. Painting eggs with estrogen or DDE drastically lowers the percentage of male hatchlings. The experiment was set up so that the researchers did not know which eggs had received which treatment until after results had been tallied. This experimental design prevented unintentional bias, such as looking more carefully for abnormalities in animals that a researcher knows has received estrogen or DDE.

FIGURE 2.A
Raoul rarely ventures far from his mother, Renata.

The R family reunited, but all the animals were clearly shaken. Unable to resist comparing the pachyderms to people, Payne writes, "Renata does not seem angry at Raoul. Perhaps elephants don't ask for explanations."

The kind of observation that leads to new knowledge or understanding, Payne says, happens only rarely. "You have to be alone and undistracted. You have to be concentrating on what's there, as if it were the only thing in the world and you were a tiny child again. The observation comes the way a dream—or a poem—comes. Being ready is what brings it to you."

Observation: The air near the elephant's cage at the zoo seems to vibrate.

Background knowledge: Organ pipes make similar vibrations when playing very low notes.

Hypothesis: Elephants ccmmunicate by infrasound, making sounds that are too low for the human ear to hear.

Experiment: Record the elephants with equipment that can detect infrasound.

Results: Elephants make infrasound.

Further observations: Elephants emit infrasound only in certain situations involving communication.

Conclusion: Elephants communicate with infrasound.

Further question: Do different patterns of sounds communicate different messages?

Scientific Inquiry Continues

Scientific inquiry does not end with a conclusion, because each discovery leads to further questions. Once experiments in painting alligator eggs provided evidence that DDE and estrogen cause reproductive abnormalities, a new question arose:

How do estrogen mimics exert their effects?

This new question suggests a new hypothesis:

If estrogen mimics produce the same or similar effects as estrogen, then the mimics should bind to estrogen receptors.

Chemical experiments support this hypothesis—estrogen receptors bind estrogens and DDE. Figure 2.5 lists some other chemicals that estrogen receptors bind. Additional experiments revealed that DDE also binds to testosterone receptors, thereby indirectly increasing the proportion of estrogen. The estrogen overload may have caused the stunted sex organs of male Apopka alligators.

The cycle of scientific inquiry continues, as puzzle pieces further connect to reveal a fuller picture of the estrogen mimic story. Because estrogen exposure correlates to increased risk of reproductive abnormalities and cancer, and pesticides introduce estrogen-like compounds into the environment, we can connect these facts to pose yet another hypothesis:

Are pesticide residues responsible for the increase in breast cancer and reproductive problems in humans?

The experiments and observations discussed so far do not directly address this new hypothesis. The simultaneous increases in incidence of certain reproductive problems and pesticide use do not demonstrate cause-and-effect but rather a correlation—events occurring at the same time that may or may not be causally related. It will take further experiments to address directly the idea that pesticide exposure causes breast cancer and other reproductive health problems.

Biology in Action 2.1 relates another example of how a scientist's powers of observation can lead to new knowledge through the process of scientific inquiry. We now take a closer look at experimental design, the backbone of science.

Mastering Concepts

1. What is a scientific theory?
2. What are the steps of scientific inquiry?
3. How is scientific inquiry a continuous process?

Table 2.1 Animal Research and Medical Progress

Time	Type of Animal	Research Results
Before 1900	Chickens	Treatment for vitamin B deficiency disease beriberi
	Sheep	Treatment for anthrax
	Dogs, rabbits	Treatment for rabies
1900–20	Dogs, rabbits	Cardiac catheterization technique to view heart's blood vessels
		Dialysis to treat kidney failure
	Dogs	Treatment for vitamin D deficiency disease rickets
	Dogs, monkeys	Treatment for niacin deficiency disease pellagra
1920–40	Dogs	Discovery of insulin's action
		Development of general anesthesia
	Cats	Development of anticlotting drugs
1940s	Monkeys	Discovery of Rh factor (blood type)
	Horses	Diphtheria vaccine
	Many species	Discovery and development of several antibiotic drugs
1950s	Rabbits, monkeys, rodents	Polio vaccine, cancer chemotherapy
	Dogs	Open heart surgery and cardiac pacemakers
	Rats, mice	Discovery that DNA is the genetic material
1960s	Monkeys	Rubella (German measles) vaccine
	Rats, mice	Development of antidepressant drugs
	Leeches	Rediscovery of chemical in leech saliva that prevents blood clotting, useful in reattaching severed human body parts
1970s	Monkeys, armadillos	Treatment for leprosy
	Dogs	Heart transplants, coronary bypass surgery
	Rabbits, rats	Development of drugs to treat ulcers
1980s	Mice, rabbits	Development of monoclonal antibodies, proteins used to diagnose and treat a variety of disorders
	Dogs, cows, pigs, sheep	Surgical transplant techniques
		Antirejection drugs for transplant recipients
1990s	Mice, rats, dogs	Models of human genetic diseases
	Pigs, baboons	Organs for transplantation into humans
	Chimps, macaques, cats, mice	AIDS vaccine development
	Chimps	Effect of salt intake on blood pressure
	Goats, sheep	Produce drugs in milk
	Rats	Grafts to treat Parkinson's disease in humans
	Mice	Arthritis treatment

Information from the Department of Health & Human Services.

Designing Experiments

Scientists use several approaches when they design experiments in order to make data as valid as possible. A useful experiment would examine dozens of alligator eggs, or hundreds of cells, because a large **sample size** helps ensure meaningful results. For example, consider the task of estimating the academic ability of the average student in a particular school that has 600 students. If the average is obtained from the grades of only six students, by chance the examiner could choose six *A* students, or six *F* students. This result is not representative of the school as a whole. However, a sample of 60 students reduces the chance that all of them will be exceptionally good or bad students. The same situation is true for the alligator egg experiment. Sampling only intersexes among the DDE-treated eggs might seemingly support the conclusion that DDE exposure causes alligators to always develop as intersexes, when larger samples reveal that they do so only 40% of the time.

Obtaining a large sample is not always practical or possible. When scientists study very rare disorders, only a few patients may be available. Although scientists prefer to work with a large sample, small-scale (called pilot) experiments are nevertheless valuable because they may indicate whether continuing research is likely to yield valid, meaningful results. For example, physicians saw only a few cases of the DES-related cancer, but when the link to the drug was established, they knew to examine all exposed offspring to see if they had other health problems.

Experimental Controls

Logically, the unusual can only be distinguished by comparison to the usual. This is why understanding the effects of estrogen mimics requires understanding estrogen's normal functions. Well-designed scientific experiments compare a group of "normal" individuals or components to a group undergoing experimental treat-

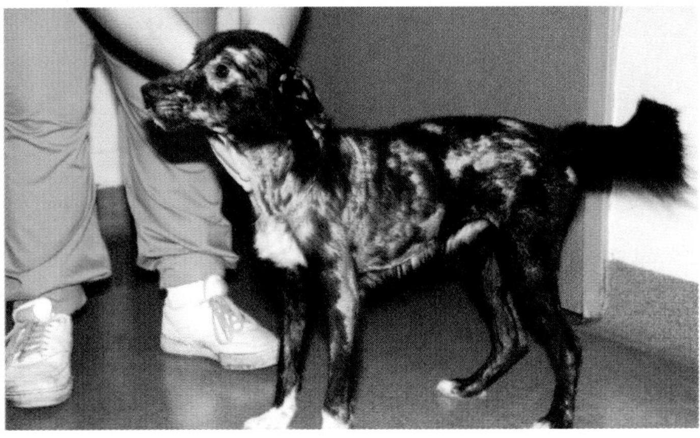

(a)

(b)

FIGURE 2.7

Two views of animal research. (*a*) The animal rights movement is threatening biomedical research that uses nonhuman animals. This photo is typical of those in the literature of PETA (People for the Ethical Treatment of Animals). This dog was intentionally infected with scabies mites to study the infection process. (*b*) This photo depicts a happier laboratory animal.

ment. Ideally, the only difference between the normal group and the experimental group is the one factor being tested. This normal group constitutes an **experimental control** and provides a basis of comparison. Designing experiments to include controls helps ensure that a single factor, or variable, truly causes an observed effect.

Experimental controls may take several forms. A **placebo** is a type of control that presents the control group with the same experience as the experimental group. In the alligator egg experiment (fig. 2.6), the eggs painted with an inert substance received a placebo. In medical research, a placebo is often a stand-in for a drug being tested—a sugar pill or a treatment already known to be effective.

Another safeguard used often in medical experiments is a **double-blind** design, in which neither the researchers nor the participants know who received the substance being evaluated and who received the placebo. The "code" of who received which treatment is broken only after the data are tabulated. The alligator egg study used a double-blind protocol. Researchers didn't know which eggs had received which of the three treatments until they compiled the results. This avoided bias, such as looking more carefully for reproductive problems in alligators known to have been exposed to estrogen or DDE.

Types of Experiments

Experimental designs vary. An experiment may use a microscope to take a biologist within a cell, observe leaves in a wind tunnel to demonstrate an adaptation, or explore a remote habitat.

Animal Studies

Researchers use nonhuman animals in experiments for a number of reasons. Animals such as fruit flies and rodents have short life spans, thus enabling a researcher to observe effects of a treatment over many generations. Plus, it is simple to breed large numbers of these animals and provide adequate sample sizes. If animals are raised in laboratories, researchers can control conditions such as nutrition, light and dark cycles, and temperature. Table 2.1 lists some of the medical advances that experiments using animals have made possible. Bioethics Connection 26.1 discusses how experiments using rats answered a long-standing question about the amount of alcohol that can alter functioning of a fetus's brain cells.

Many studies expose nonhuman organisms to treatments and situations that might raise ethical objections if they were conducted on humans, although animal studies sometimes elicit objections, too (fig. 2.7). The results of animal studies are extrapolated to humans or serve as the basis for further investigation. This is why the public responded with great excitement when researchers identified a biochemical that induces weight loss in mice that has a counterpart in humans (fig. 2.8).

More than 90% of animal experiments use rodents, and fewer than 1% use monkeys. The remainder include either species that are easy to work with, such as the fruit fly *Drosophila melanogaster* or the roundworm *Caenorhabditis elegans,* or particular animals that suffer from a disease that humans have too. This is the case for armadillos, which develop a form of leprosy. It is common practice to breed mice that have human disease-causing genes, providing models on which researchers can study the disease process and try new treatments. Nude mice, named for their hairless appearance, lack immune systems and are studied as models for human immune deficiency conditions (fig. 2.9). Organizations that represent people with disorders studied in mice sometimes underwrite the cost of developing the mice.

Many experiments use cells or tissues instead of whole animals (fig. 2.10). It is certainly more humane to observe the inflammatory effects of a cosmetic ingredient on cells growing in a dish than on the eyes of rabbits, as researchers once did.

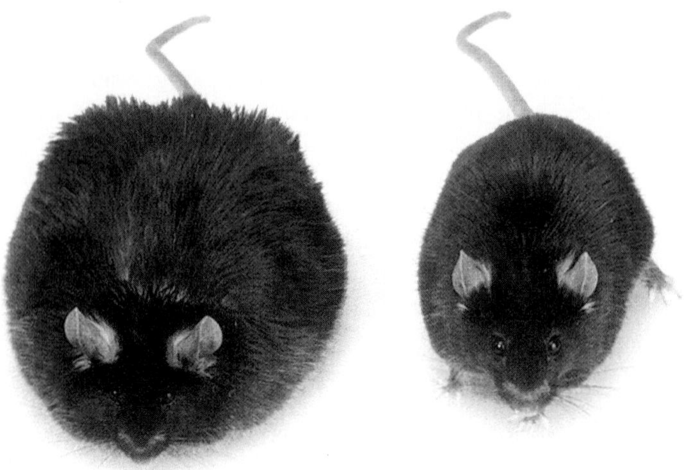

FIGURE 2.8

A gene called "ob" (for obese) in mice normally instructs cells to produce a protein, called leptin, that limits fat storage, thus causing mice to maintain a healthy body weight. Obese mice given leptin lose weight. When researchers at Rockefeller University announced that they had identified a human version of the leptin-encoding gene, the news made headlines everywhere. Understanding how leptin works may lead to development of new weight loss aids for humans.

FIGURE 2.9

A nude mouse lacks an immune system and is used in biomedical research on immunity.

Computer Experimentation

Many biological phenomena are informational, such as the building block sequences of proteins and genes, the pattern of electrical activity of a beating heart or network of nerve cells, and the comings and goings of organisms in ecosystems. This quantitative nature of life makes many aspects of biology amenable to computer analysis.

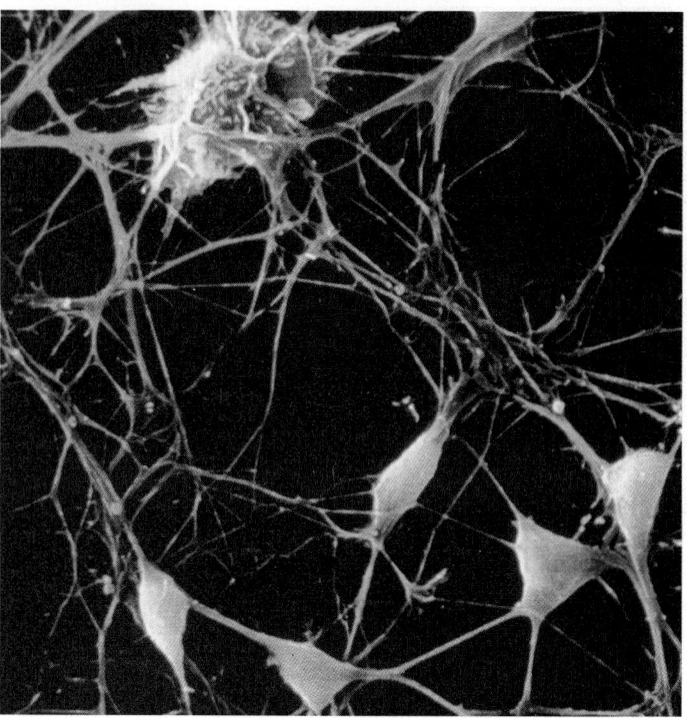

FIGURE 2.10

Researchers can perform some experiments on cells or tissues growing in culture, rather than on animals. This is a culture of human Schwann cells, which in the body insulate nerve cells. (x295)

A computer can help a researcher "see" how a string of amino acids folds into the three-dimensional form necessary for a protein to have a biological function (fig. 2.11*a*). A researcher who has just deciphered a gene's sequence scans a computer database to find similar genes in other species, which may provide hints to the gene's function. Doctors-in-training learn anatomy and physiology using computer simulations based on real imaging data (fig. 2.11*b*), and biology students not wishing to kill a frog can dissect a virtual frog.

Computers can solve environmental problems, too. When a wild area of southern California was slated for development, ecologists entered into a computer all of the species known to inhabit the area. Consulting other data, the computer then predicted the species most vulnerable to habitat destruction, so that scientists could alter plans and spare these species.

Epidemiological Studies

Epidemiology is the study of disease-related data from real-life situations. This can range from one-person anecdotal accounts to large-scale studies involving thousands of individuals. Epidemiological studies may be retrospective and evaluate the health effects of past exposure to a pesticide or drug, for example, or they may be prospective and set up a situation and then observe what happens.

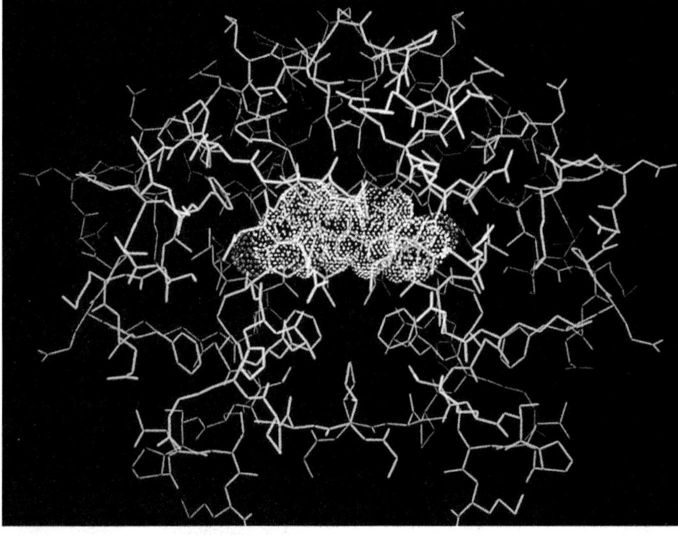

(a)

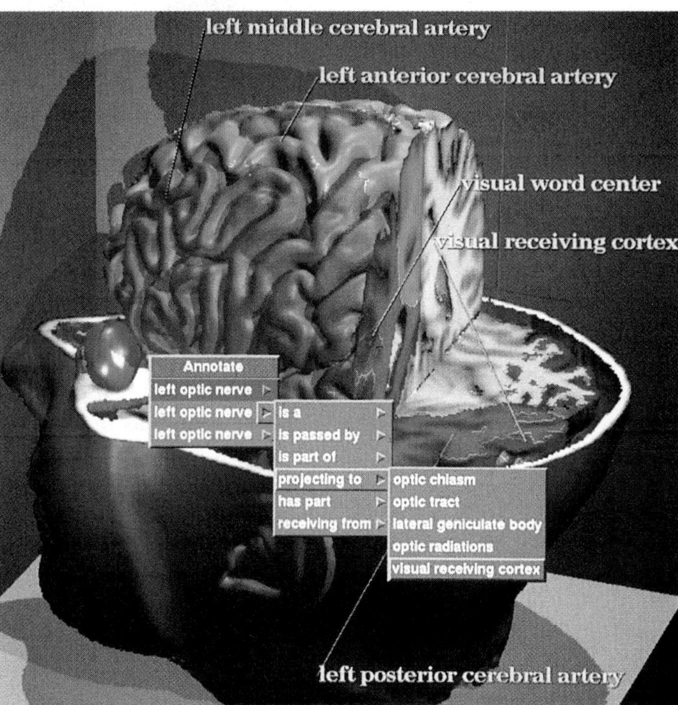

(b)

FIGURE 2.11

Computers aid life science research. (*a*) Computer modeling allows researchers to predict the three-dimensional structure of a protein. The computer program applies "rules" of attractive and repulsive forces between the amino acid building blocks of proteins. A protein's three-dimensional structure is essential for its function. This protein is uteroglobin, found in a rabbit's uterus. (*b*) Medical students can remove tissue layers from a simulated human head, using information from real scans.

One type of epidemiological investigation compares two groups of individuals differing in one variable, such as exposure to a particular substance. Ideally, each person from one group matches a person in the other group in as many characteristics as possible, such as sex, race, age, diet, and occupation. Such a one-to-one comparison is called a **case-controlled study.**

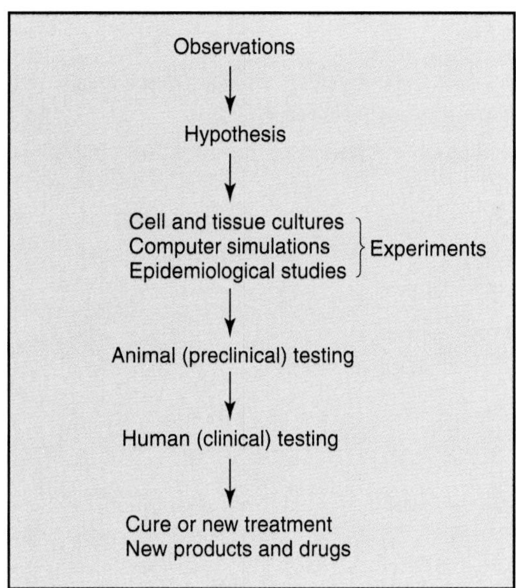

FIGURE 2.12

The research process.

Researchers took this approach to examine a possible link between DDE exposure and increased risk of breast cancer. One study compared DDE levels in blood sampled in the 1960s from 50 white women, 50 African American women, and 50 Asian women who subsequently developed breast cancer, to DDE blood levels in 50 white, 50 African American, and 50 Asian women of the same ages who did not develop breast cancer. The hypothesis was:

If DDE exposure increases breast cancer risk, then the women in the group with breast cancer should have higher DDE levels than the women in the group that did not develop breast cancer.

Results were surprising. A high correlation between DDE in the blood and elevated breast cancer risk was seen among African American women, and a lower correlation was seen among white women. However, among Asian American women, higher levels of blood DDE correlated to *lower* risk of developing breast cancer. What further hypothesis might researchers test to explain these results?

Meta-Analysis

Scientific research includes many experimental approaches using organisms, cells growing in culture, computer simulations, and epidemiological studies (fig. 2.12). Each approach has limitations. An experiment may address only one aspect of a question; a pilot study may involve only a few people; an epidemiological investigation may lack controls; or too many variables may confound results and their interpretations. A **meta-analysis** attempts to compensate for the problems with individual studies by combining results from many studies. This approach can increase sample sizes and more clearly reveal trends and correlations. Meta-analyses will be important in amassing information on the causes of breast cancer.

Mastering Concepts

1. What factors determine whether an experiment will provide meaningful information?
2. Why are nonhuman animals used in scientific research?
3. What are some alternatives to using live animals in experimentation?
4. What is epidemiology?
5. What is a meta-analysis?

Applying the Scientific Method: Challenges and Limitations

The generalized scientific method is flexible enough so that scientists in different fields can follow it. All scientists observe, form and test hypotheses, record results, and draw conclusions, but the ways in which they do so can vary. Yet as helpful as this way of thinking is, it has limitations.

Interpreting Results

The scientific method is neither foolproof nor always easy to implement. Experimental evidence may lead to multiple interpretations or unexpected conclusions, and even the most carefully designed experiment can fail to provide a definitive answer. Consider the observation that animals fed large doses of vitamin E live significantly longer than similar animals who do not ingest the vitamin. Does vitamin E slow aging? Possibly, but excess vitamin E causes weight loss, and other experiments associate weight loss with longevity. Does vitamin E extend life, or does the weight loss? The experiment of feeding animals large doses of vitamin E does not distinguish between these possibilities. Can you think of further experiments to clarify whether vitamin E or weight loss extends life?

Another limitation of implementing the scientific method is that researchers may misinterpret observations or experimental results. For example, scientists once "proved" that life spontaneously arises from nonliving matter by exposing decaying meat to the air and observing the maggots that soon appeared on it. There was another explanation.

We simply cannot study certain natural phenomena. Many experiments have attempted to re-create the sequence of chemical reactions that might, on the early earth, have formed the chemicals that led to life. Although the experiments produced interesting results and inspired much later work, we cannot really know if they accurately re-created conditions at the beginning of life.

Expecting the Unexpected

An investigator should try to keep an open mind towards observations, not allowing biases or expectations to cloud interpretation of the results. To do so, scientists must expect the unexpected. But it is human nature to be cautious in accepting an observation that does not fit existing knowledge. The careful demonstration that life does not arise from decaying meat surprised many people who believed that mice were created from mud, flies from rotted beef, and beetles from cow dung. Similarly, earlier this century, researchers were so enamored of the much-studied proteins that they were hesitant to follow up suggestions that a less-understood biochemical—the nucleic acid DNA—was really the genetic material.

Perhaps no one knew better the frustration of an unexpected scientific discovery than Barbara McClintock. In the 1940s, she studied inheritance of kernel color in corn. McClintock noticed that some kernels of her meticulously bred plants had an odd pattern of spots. After many experiments, she concluded that the spots indicated that genes were moving, jumping into other genes and disrupting their functions. This idea seemed as preposterous at the time as the thought that the earth orbited the sun once did. Genes couldn't move!

Despite McClintock's evidence and the repeated publication of her findings, many scientists did not believe her—possibly because she was a woman in a field then dominated by men. But in the 1970s, researchers began finding evidence of "jumping genes" in other species, including our own. McClintock was finally recognized as the perceptive researcher she always was and won a Nobel prize.

Scientific Discovery by "Accident"

Scientific discovery is not always a well-planned sequence of observing, hypothesizing, and experimenting. Sometimes knowledge comes simply from being in the right place at the right time or from being particularly sensitive to the unusual. For example, the observation that a drug to treat high blood pressure causes hair

Table 2.2 From Discovery to Technology

Discovery	Observation
Ether is an anesthetic.	Used as a recreational drug, ether renders people temporarily unable to feel pain.
Aspartame is a sweetener.	A by-product of a chemical synthesis of an ulcer drug tasted sweet when a researcher accidentally put some in his mouth.
Cellulase whitens laundry.	An enzyme originally studied for use in degrading human waste before flush toilets were invented, when added to soap accidentally, removed stains on fabrics.
A platinum-containing chemical stops cancer cells from dividing.	An electromagnet stops bacteria from dividing. A biophysicist found that platinum in the electrodes stopped the cell division. Similarly, platinum as the basis of a cancer drug stops cell division in humans.
Penicillin is an antibiotic.	A mold that accidentally fell on a dish of bacteria stopped the bacteria from growing.

FIGURE 2.13

Edward Jenner was at first ridiculed for his idea of using infection with one virus (cowpox) to prevent infection by another (smallpox). Many people at the time were afraid of receiving a smallpox vaccine because they believed it would cause them to grow cow parts.

growth as a side effect in some people led to development of a baldness treatment. Table 2.2 lists other examples of astute observations leading to practical applications. Often such an unexpected, or serendipitous, finding can be linked with other findings or produce new information.

Scientific discoveries very much depend upon our willingness to accept unusual ideas. When Edward Jenner noticed in 1796 that young girls who worked with cows did not develop the devastating illness smallpox, he used the observation to invent a smallpox vaccine. He intentionally infected young human volunteers with the cowpox virus and hoped that this would alert their immune systems to fight the related smallpox virus. Jenner had to test his vaccine in private, because people tried to lynch him for the very idea of intentionally infecting people (fig. 2.13). Today, the fact that infection with one strain of HIV may protect against a more deadly strain echoes Jenner's observation of two centuries ago and may lead to development of an AIDS vaccine.

Scientific research seeks to understand nature. Because humans are part of nature, we sometimes tend to view scientific research, and particularly biological research, as aimed at improving the human condition. How often do newspaper or magazine articles validate research results by reporting that the work "could prevent birth defects or cure cancer"? But knowledge for knowledge's sake, without any immediate application or payoff, is valuable in and of itself—because we can never know when information will prove useful.

Mastering Concepts

1. What are some limitations of scientific inquiry and experimentation?
2. How can our biases interfere with scientific discovery?
3. What is a serendipitous discovery?

SUMMARY

1. Scientific **theories** are ideas based on observations and evidence that attempt to explain natural phenomena.

2. Scientific inquiry, often called the **scientific method,** is a way of thinking that involves observing, questioning, reasoning,

A Tale of Two Drugs~

continued from page 18.

The tale of tamoxifen vividly illustrates the ongoing nature of the scientific method.

Tamoxifen, first synthesized in 1966, binds to estrogen receptors in cells in the breast and elsewhere. The drug blocks estrogen's action yet exerts actions of its own. Clinical trials conducted over two decades showed that tamoxifen prevents recurrence of certain types of breast cancer and also increases bone mineral density, lowers blood cholesterol, and lowers risk of breast cancer in the other breast. Would tamoxifen prevent cancer in healthy women with strong family histories of the disease?

Researchers calculated that at least 16,000 women would have to participate in the tamoxifen prevention trial to show any significant protective effect. Half of the women would receive tamoxifen and half placebo, and neither participants nor physicians would know who received what until the study's end. Treatment would continue for 5 years, and researchers would follow the participants for at least 7 years. Note the specificity of the hypothesis:

> *If among the 8,000 treated women, the number of breast cancers is at least 62 fewer than the number among the 8,000 women given placebo, then this is evidence that tamoxifen can prevent breast cancer.*

Some physicians, recalling the danger of widespread use of DES before we fully understood its actions, voiced concern about giving 8,000 healthy women tamoxifen. Still, 11,000 healthy women with relatives who had breast cancer signed up for the trial. But in 1994 and 1995, several developments supported the doubting doctors' concerns:

Uterine cancer occurred in some women taking tamoxifen after they had breast cancer, at three times the rate among women who had breast cancer and didn't take tamoxifen.

Discoveries of several genes that cause breast cancer indicated a flaw in the design of the tamoxifen prevention trial. The participants did not all have the exact same illness.

Breast cancer survivors who took tamoxifen were found not to experience any health benefit from the drug, compared to survivors who did not take the drug, after 5 years. Physicians were notified to take patients off tamoxifen after 5 years.

The trial was halted. Thanks to rigorous application of the scientific method, a drug trial was prevented from possibly leading to a medical disaster, as did DES a generation ago.

predicting, testing, interpreting, concluding, and posing further questions. Scientific inquiry builds theories.

3. Scientific inquiry begins when a scientist makes an observation, raises questions about it, and reasons to construct an explanation, or **hypothesis.**

4. Experiments test the validity of the hypothesis, and conclusions are based on data analysis. **Experimental controls** ensure that only one **variable** is examined at a time.

5. **Placebo**-controlled, **double-blind** experiments minimize bias.

6. Other scientists must repeat an experiment to verify the results.

7. Experiments take many forms, including observations of natural events, tests on living organisms, computer simulations, and **epidemiological** evidence.

8. A **meta-analysis** combines results of many studies to reach a consensus.

9. The scientific method does not always yield a complete answer or explanation or may produce ambiguous results. Discoveries may be unusual, unexpected, or serendipitous.

TO REVIEW . . .

1. What is a theory?

2. How does scientific inquiry (the scientific method) help to build theories?

3. List the steps in scientific inquiry.

4. What is a hypothesis?

5. What is an experiment?

6. Define the following terms and indicate how each is part of designing an experiment:

- **a.** variable
- **b.** sample size
- **c.** experimental control
- **d.** placebo
- **e.** double-blind design
- **f.** case-controlled study

7. How is a hypothesis more than just a question?

8. How does an epidemiological study differ from a study using laboratory animals?

9. What did the physician who swallowed ulcer-causing bacteria have in common with Barbara McClintock and Edward Jenner?

10. How do the evidence and observations supporting the estrogen mimic theory examine life at the molecular, cellular, whole organism, and population levels?

11. Give an example of how computers can provide scientific information.

12. What are the benefits and limitations of a meta-analysis?

TO THINK ABOUT . . .

1. In a 1994 study, researchers asked 845 women with breast cancer and 961 women who did not have breast cancer questions concerning their sexual history and behavior. The breast cancer patients were significantly more likely to have had an abortion than the healthy women. The researchers concluded that having had an abortion raises risk of developing breast cancer by 50%.

- **a.** How was this study controlled?
- **b.** In what way is the design of the study flawed?
- **c.** Does this study reveal a cause-and-effect relationship or a correlation?
- **d.** What further questions might be asked, information obtained, and experiments designed to continue to examine the relationship, if there is one, between having had an abortion and increased breast cancer risk?

2. A chemical manufacturing plant closes because of a leak of a toxic substance. To examine the possible health effects of the contamination, researchers ask workers and residents to fill out surveys and list their health problems and record the extent of their exposure to the toxins. What can be learned from this type of study? What are its limitations?

3. Cauliflower, broccoli, and soy products contain chemicals that reduce the effects of estrogen.

- **a.** What experimental results discussed in the chapter might be influenced by the effect of these foods in the diet?
- **b.** Suggest a type of epidemiological study to address the question of the effect of these plant-based biochemicals on breast cancer risk. Follow the steps of the scientific method.

TO LEARN MORE . . .

References and Resources*

Blaser, Martin J. February 1996. The bacteria behind ulcers. *Scientific American.* The tale of a daring young physician who swallowed ulcer-causing bacteria.

Botting, Jack H. and Adrian R. Morrison. February 1997. Animal research is vital to medicine. *Scientific American.* Many medical procedures and medicines would not exist without animal research.

Dold, Catherine. September 1996. Hormone hell. *Discover.* Are estrogen mimics in the environment causing reproductive problems in wildlife?

Garrett, Laurie. 1994. *The Coming Plague.* New York: Penguin Books. Vivid descriptions of many epidemiological investigations.

Kanigel, Robert. January/February 1987. Specimen no. 1913. *The Sciences.* A rat's brief life in the service of science—from its point of view. An informative account of the use of animals in experiments.

Madhusree, Mukerjee. February 1997. Trends in animal research. *Scientific American.* Increased use of cell and tissue culture and other techniques improve the lives of animals used in research.

Stix, Gary, Sasah Nemecek and Philip Yam. January 1997. Science versus antiscience? *Scientific American.* Media refutations of scientific ideas often reveal a misunderstanding of the scientific method.

**Life*'s website features frequently updated references.

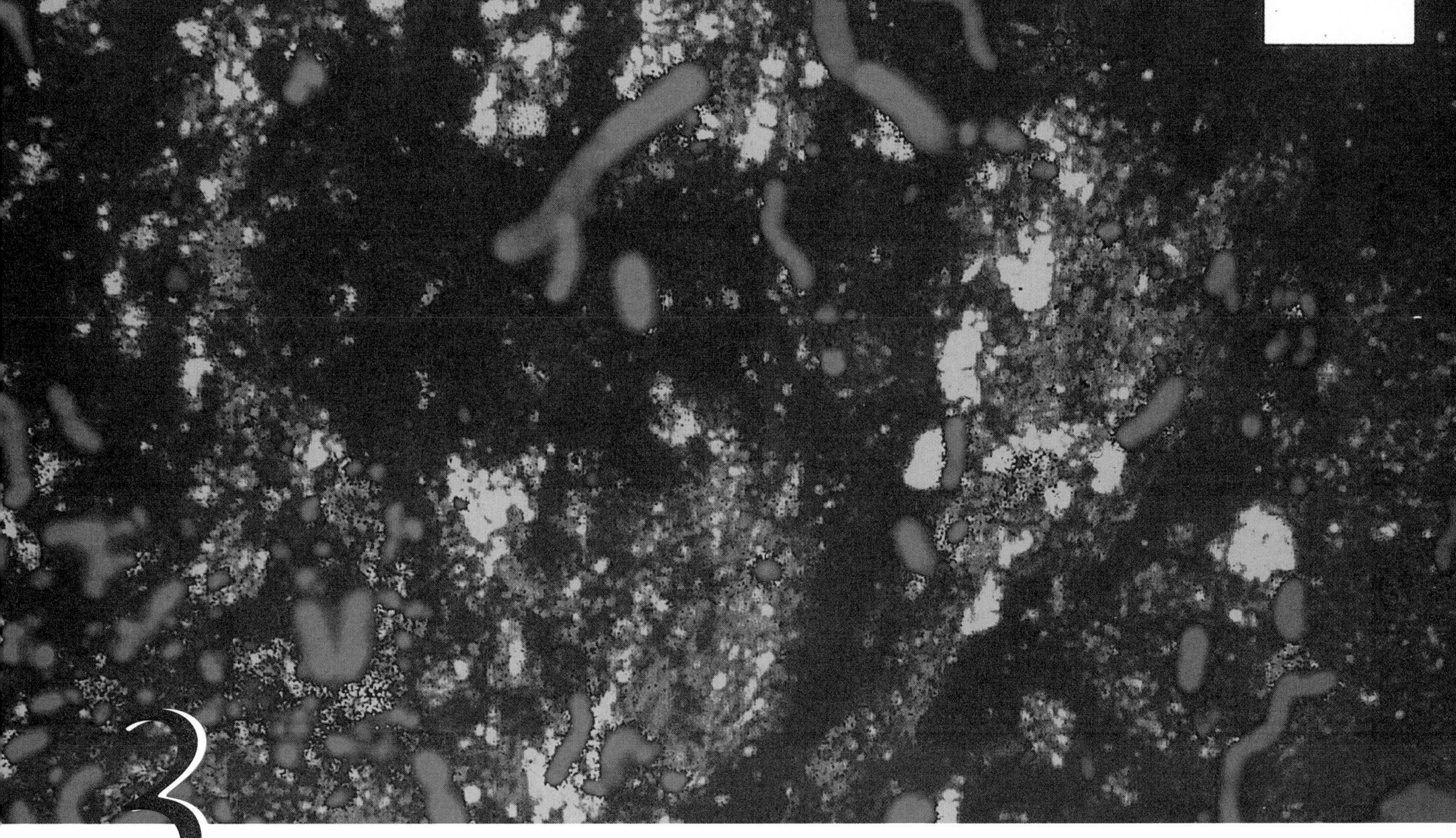

3 The Chemistry of Life

These bacteria (red) require the volcanic basalt rock on which they live (green) to survive. The microorganisms use hydrogen released from the rock to extract energy from carbon-containing chemicals in the environment. Scalebar = 5 micrometers.

chapter outline

***Why Study Chemistry?* 32**

***Chemistry Basics* 32**
Elements
Atoms
Molecules
Chemical Bonds

***Water—The Matrix of Life* 40**
Solutions
Acids and Bases
Water and Temperature

***The Organic Molecules of Life* 43**
Carbohydrates
Lipids
Proteins
Nucleic Acids

The Mystery of the Rock Eaters

Microbiologist Todd Stevens and geochemist James McKinley were perplexed. They drilled 1,000 meters down into volcanic rock near the Columbia River in Richland, Washington, to study movement of pollutants in groundwater and discovered abundant bacteria that seemed to be living on nothing but rock. The researchers named the underground find of bacteria SLiME, which stands for "subsurface lithoautotrophic microbial ecosystem"—meaning microorganisms that eat rocks underground, and their surroundings.

Although bacteria have been known to inhabit some unusual (to us) parts of the planet, including deep underground, usually the energy source that keeps them alive and reproducing is obvious. The energy source is often the sun but may also be heat from the earth's core or chemicals from other organisms. None of these explanations accounted for the SLiME bacteria. How were these microbes extracting the raw materials and energy required to survive?

to be continued . . .

Why Study Chemistry?

A college student has a blood test as part of a routine physical examination. When the results are in, the physician tells the student that her LDL cholesterol is high and her HDL low and suggests that she avoid foods high in saturated fats. What does all of this information mean?

A man roams the supermarket shampoo section, befuddled. All he wants to do is clean his hair, but the ingredients lists on the shampoo bottles are endless—collagen, elastin, chlorophyll, RNA, jojoba oil, and many unpronounceable others. Will any of these actually clean hair? And, should he care whether the product is pH balanced, organic, and enzyme-activated?

A truck crashes into a tree near a residential neighborhood. The newscaster on the scene worriedly reports that the truck has released nitrogen and helium gases. Is the public's health threatened?

Reasons to study chemistry are all around us (fig. 3.1). Tests to diagnose illness may detect a telltale biochemical, and drugs to treat disease are chemicals, whether they come from a chemist's flask or an organism's body. Health food store shelves are stocked with chemicals, which despite the healthful surroundings may actually endanger a consumer who does not know a lipid from a lipase or a free radical from a fatty acid.

We eat chemicals, drink chemicals, and inhale chemicals and, most important, we, and all other organisms, *are* chemicals. Chemistry plays a role in nearly all aspects of life science.

Chemistry Basics

Organisms consist of matter (material that takes up space) and energy (the ability to do work). Chapter 6 discusses the energy of life in detail; this chapter concentrates on the chemicals that make up living matter.

Organisms can be viewed as subsets of chemicals that include instructions to reproduce and require energy to maintain their organization. Organisms must capture energy and convert it to a usable form. They then channel energy to a specific process, such as bringing a substance into a cell or duplicating genetic material.

Elements

All matter can be broken down into pure substances called **elements.** There are 92 known, naturally occurring elements and at least 17 synthetic ones. The elements are arranged into a chart, called the **periodic table,** in columns according to their properties.

It is interesting to look at the periodic table positions of the elements that make up organisms. These elements of life appear across the table, indicating that they represent a diverse sampling of all chemicals (fig. 3.2).

Twenty-five elements are essential to life. Elements required in large amounts—such as carbon, hydrogen, oxygen, nitrogen, sulfur, and phosphorus—are termed **bulk elements;** those required in small amounts are called **trace elements.** Many trace elements are important in ensuring that vital chemical reactions occur fast enough to sustain life. Some elements that are toxic in large amounts, such as arsenic, may be vital in very small amounts; these are called **ultratrace elements.**

Atoms

An **atom** is the smallest possible "piece" of an element that retains the characteristics of the element. An atom is composed of subatomic particles. The three major types of subatomic particles are **protons** and **neutrons,** which form a centralized mass called the **nucleus,** and **electrons,** which surround the nucleus. Atoms of different elements have characteristic numbers of protons and may vary greatly in size. Hydrogen, the simplest type of atom, has only 1 proton and 1 electron; in contrast, an atom of uranium has 92 protons, 146 neutrons, and 92 electrons.

Protons carry a positive charge, electrons carry a negative charge, and neutrons are electrically neutral. Within an atom, when the number of protons equals the number of electrons, the atom is electrically neutral. That is, it does not have a charge. An electron has about 1/2,000 the mass of a proton or neutron. If the nucleus of an atom were the size of a meatball, then the electrons would be about 0.62 mile (1 kilometer) away from it! Table 3.1 summarizes characteristics of the three types of subatomic particles.

The periodic table depicts in shorthand the structures of the atoms of each element. Each element has a symbol, which can come

* Good Source of Fiber
* Low in Sodium

Nutrition Information per serving

Serving size: 1.25 oz. (about 1/3 cup) (35.4 g)
Servings per package: 11

	1.25 oz. (35.4 g) cereal	with 1/2 cup vitamin A & D skim milk
Calories	140	180*
Protein	3 g	7 g
Carbohydrate	27 g	33 g
Fat	3 g	3 g*
Cholesterol	8	8*
Sodium	70 mg	130 mg
Potassium	140 mg	350 mg

Percentage of U.S. recommended daily allowances (U.S. RDA)

Protein	4%	10%
Vitamin A	30%	35%
Vitamin C	**	2%
Thiamine	30%	35%
Riboflavin	30%	40%
Niacin	30%	30%
Calcium	**	15%
Iron	20%	20%
Vitamin D	15%	30%
Vitamin B6	30%	30%
Folic acid	30%	30%
Vitamin B12	30%	40%
Phosphorus	10%	20%
Magnesium	10%	15%
Zinc	8%	10%
Copper	10%	10%

*Whole milk supplies an additional 30 calories, 4 g fat, and 15 mg cholesterol.
**Contains less than 2% U.S. RDA of these nutrients.

Ingredients

Natural whole wheat, rolled oats, dried fruit (raisins and dates), brown sugar, barley, pecans, sugar, partially hydrogenated sunflower oil, salt, and BHT (added to preserve freshness).

Vitamins and Minerals

Vitamin A Palmitate, Niacinamide, Iron, Zinc oxide, (source of Zinc), Vitamin B6, Riboflavin (Vitamin B2), Thiamine, Mononitrate (Vitamin B1), Vitamin B12, Folic acid, and Vitamin D.

Carbohydrate Information

	1.25 oz. cereal	with 1/2 cup skim milk
Dietary fiber	3 g	3 g
Complex Carbohydrate	15 g	15 g
Natural sugar from fruit	3 g	3 g
Sucrose and other sugars	6 g	12 g
Total Carbohydrate	27 g	33 g

FIGURE 3.1

The label on this breakfast cereal reveals one reason why it is helpful to understand some chemical terms. Do you know what you're eating?

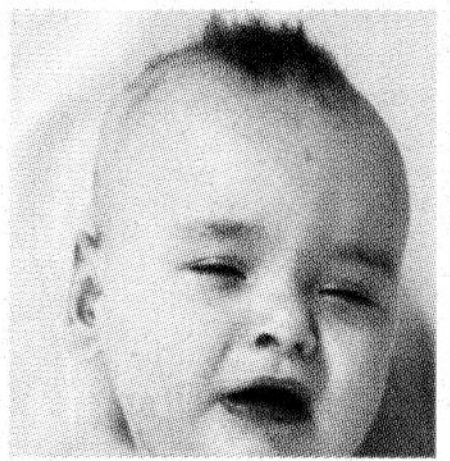

Chemistry Explains Biology

Copper Woes

The human body requires certain amounts of trace elements. Inherited disorders called "inborn errors of metabolism" can cause a trace element to be overabundant or lacking—often with severe effects on health. Consider copper.

In Wilson disease, the digestive tract absorbs too much copper from food, causing stomachaches and headaches, changes in one's voice and coordination, and an inflamed liver. Greenish rings around the irises are the key clue to a diagnosis of the disease. A drug, penicillamine, rids the body of excess copper, which turns the urine a coppery color. Symptoms of Wilson disease usually appear during adolescence. The treatment cannot reverse symptoms but prevents the disease from worsening.

In the opposite biochemical situation, Menkes disease, the intestines cannot release copper from food into the bloodstream. Lack of copper causes symptoms that begin shortly after birth. They include yellowish skin, low body temperature, extreme lethargy, seizures, mental retardation, impaired growth, and white, kinky, stubbly hair that gives the disease its other name, "kinky hair syndrome." The white, stubby hair of a child with Menkes disease provided a clue to the cause. It reminded an astute Australian researcher of the peculiar wool of sheep that graze in his native land on copper-deficient soil. This observation led to the discovery that an inherited copper deficiency causes Menkes disease. Menkes disease cannot be treated, and death occurs in childhood.

from the English word for that element (*He* for *helium,* for example) or from the word in another language (*Na* for *sodium,* which is *natrium* in Latin). The **atomic number** above the element symbol and name shows the number of protons in the atom, which establishes the identity of the atom. The elements are arranged sequentially in the periodic table by atomic number.

The **atomic weight,** beneath the element symbol, reflects the total number of protons and neutrons in the nucleus of the atom (fig. 3.3). For biological purposes, we can approximate the weight (mass in the presence of gravity) of a proton and that of a neutron to be one. By comparison, the contribution of an electron to an atom's weight or mass is negligible. Therefore, we add the number of protons and neutrons to approximate the atomic weight. Subtracting the atomic number (the number of protons) from the atomic weight (the number of protons and neutrons) yields the number of neutrons.

The atoms of an element can exist in different forms, called **isotopes,** that have different numbers of neutrons. Isotopes of an element all have the same number of protons but different weights (masses) because they have different numbers of neutrons. Often one isotope of an element is very abundant and others rare.

Some isotopes are unstable, breaking down to a more stable form at a known rate and releasing radioactive energy in the process. Such radioactive isotopes have a variety of uses in biomedical and life science research, and we will encounter them in context in later chapters.

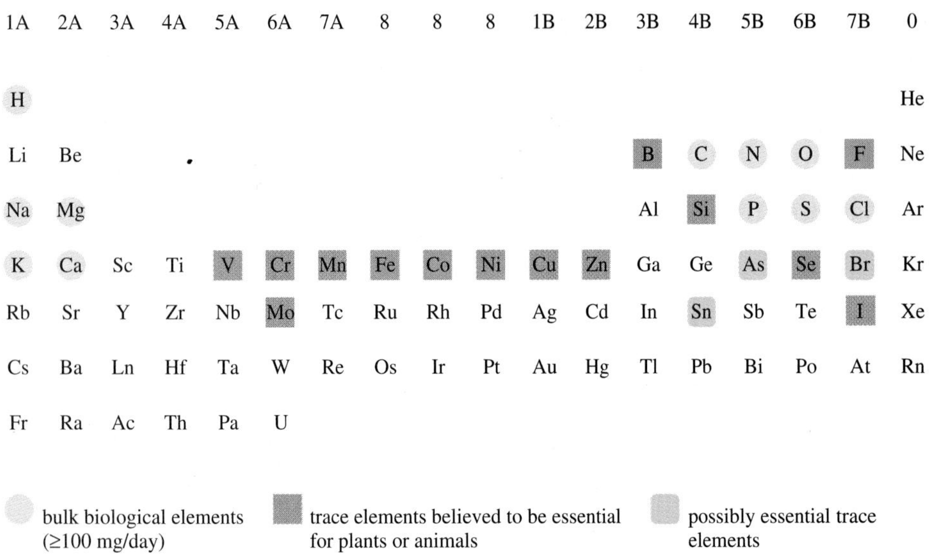

1A	2A	3A	4A	5A	6A	7A	8	8	8	1B	2B	3B	4B	5B	6B	7B	0
H																	He
Li	Be											B	C	N	O	F	Ne
Na	Mg											Al	Si	P	S	Cl	Ar
K	Ca	Sc	Ti	V	Cr	Mn	Fe	Co	Ni	Cu	Zn	Ga	Ge	As	Se	Br	Kr
Rb	Sr	Y	Zr	Nb	Mo	Tc	Ru	Rh	Pd	Ag	Cd	In	Sn	Sb	Te	I	Xe
Cs	Ba	Ln	Hf	Ta	W	Re	Os	Ir	Pt	Au	Hg	Tl	Pb	Bi	Po	At	Rn
Fr	Ra	Ac	Th	Pa	U												

bulk biological elements (≥100 mg/day)

trace elements believed to be essential for plants or animals

possibly essential trace elements

FIGURE 3.2

Elements of life. Robert J. P. Williams of the inorganic chemistry laboratory, Oxford University, writes: "The true chemistry of living systems is a fascinating combination of many elements from the Periodic Table used in some optimum fashion but adjusted according to species which have become stabilized during evolutionary development." The periodic table organizes the elements by their similarities in chemical reactivity. Even without knowing much chemistry, it is easy to see that the elements of life represent a diverse sampling from the chemical world.

Table 3.1 Subatomic Particles

Particle	Charge	Mass	Function	Symbol	Location
Electron	–	0	Bonding	e^-	Orbitals
Neutron	0	1	Isotope	n	Nucleus
Proton	+	1	Identity	p	Nucleus

Table 3.2 On the Matter of Matter

Designation	Definition
Element	A pure chemical substance
Atom	The smallest piece of an element that retains the characteristics of that element
Compound	A pure substance formed when atoms of different elements bond
Molecule	The smallest piece of a compound that retains the characteristics of that compound

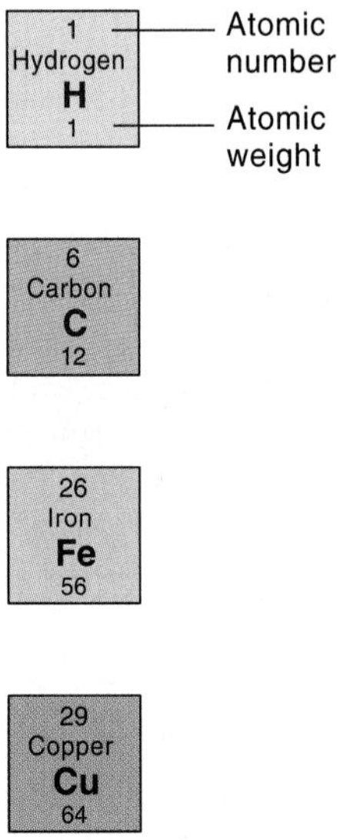

1
Hydrogen
H
1

Atomic number

Atomic weight

6
Carbon
C
12

26
Iron
Fe
56

29
Copper
Cu
64

FIGURE 3.3

Four elements, as represented in the periodic table.

Molecules

Atoms of two or more elements joined form a **compound,** which is a chemical substance with properties distinct from those of its constituent elements. A **molecule** is the smallest unit of a compound, just as an atom is the smallest unit of an element. A molecule usually consists of atoms of different elements, but a few exceptions are "diatomic" (two-atom) hydrogen, oxygen, and nitrogen gases. Table 3.2 reviews these designations of matter.

A compound's characteristics can differ strikingly from those of its constitutive elements. Consider table salt, which is the compound sodium chloride. A molecule of salt contains one atom of sodium (Na) and one atom of chlorine (Cl). Sodium is a silvery, highly reactive solid metal, while chlorine is a yellow, corrosive gas. But when equal numbers of these two types of atoms combine, the resulting compound is a white crystalline solid—table salt. The same is true for the swamp gas methane. Its components are carbon, a black sooty solid, and hydrogen, a light, combustible gas.

Scientists describe molecules by writing the symbols of their constituent elements and indicating the numbers of atoms of each element in one molecule as subscripts. For example, methane is written CH_4, which denotes 1 carbon atom bonded (joined) to 4 hydrogen atoms. A molecule of the sugar glucose, $C_6H_{12}O_6$, has 6 atoms of carbon (C), 12 of hydrogen (H), and 6 of oxygen (O). A coefficient indicates the number of molecules—6 molecules of glucose is written $6C_6H_{12}O_6$.

In **chemical reactions,** two molecules interact with each other to yield different molecules. Chemical reactions are depicted as equations with the starting materials, or **reactants,** on the left and the end products on the right. The total number of atoms of each element is the same on either side of the equation. Consider the equation for the formation of glucose in a plant cell:

$$6CO_2 + 6H_2O \longrightarrow C_6H_{12}O_6 + 6O_2$$

In words, this means "six molecules of carbon dioxide and six molecules of water react to produce one molecule of glucose plus six molecules of diatomic oxygen." Note that each side of the equation indicates 6 total atoms of carbon, 18 of oxygen, and 12 of hydrogen.

Organisms are composed mostly of water and carbon-containing molecules, especially **organic molecules,** which contain carbon and hydrogen. Many organic molecules also include oxygen, nitrogen, phosphorus, and sulfur. Some organic molecules of life are so large that they are termed **macromolecules.** The plant pigment chlorophyll, for example, contains 55 carbon atoms, 68 hydrogens, 5 oxygens, 4 nitrogens, and 1 magnesium. **Molecular weight** measures a molecule's size. It is calculated by adding the atomic weights of the constituent atoms.

Carbon monoxide (CO) and carbon dioxide (CO_2) are not considered organic because of their lack of hydrogen and simple structure. However, even small molecules can have profound effects on life (Biology in Action 3.1).

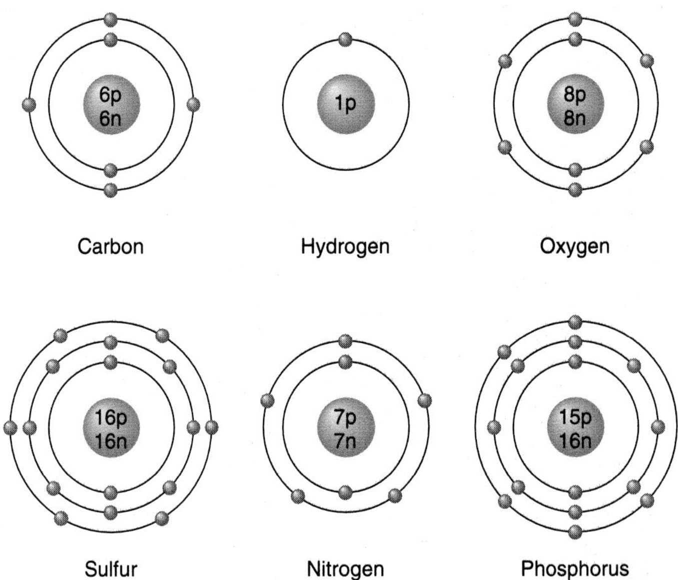

FIGURE 3.4

Structures of the atoms prevalent in life. Shown here are Bohr models of the six most common atoms that make up organisms.

Chemical Bonds

Atoms join, forming molecules, through attractive forces called **chemical bonds.** There are several types of chemical bonds. The number of electrons in an atom's outermost shell, or **valence electrons,** determines which type of chemical bond can form with the least amount of energy input and is therefore most likely to occur. The number of valence electrons also determines the atom's chemical properties. Elements whose atoms have the same number of valence electrons have similar characteristics, and they appear in the same column in the periodic table. Understanding how electrons interact explains how chemical bonds form.

Electrons Determine Reactivity

Electrons move about the nucleus near the speed of light; it is impossible to represent their positions accurately in a static diagram. Often electrons are illustrated as dots moving in concentric circles, much as planets orbit the sun (fig. 3.4). Although these depictions, called **Bohr models,** show interactions between atoms, they do not realistically capture the cloudlike distribution of electrons moving constantly about the nucleus.

A more accurate way to describe electron behavior is to envision areas around the nucleus where the probability of finding an electron is the greatest. An **electron orbital** is the volume of space where a particular electron is found 90% of the time. An orbital is not a physical entity but a way of imagining or representing electron organization.

Groups of electron orbitals form levels of energy called **energy shells,** which are organized a little like a series of boxes, one within another. Electrons tend to occupy the lowest possible energy shell; that is, the shell closest to the nucleus.

Atoms are least likely to combine with other atoms when their outermost energy shells are filled. Atoms have a natural tendency to fill their outermost shells with eight electrons, and thus this is called the **octet rule.** Atoms fulfill the octet rule, and achieve stability, by sharing, donating, or receiving electrons in their outermost shells until they have eight. The two types of strong chemical bonds are covalent and ionic bonds.

Covalent Bonds—Electron Sharing

Covalent bonds hold together many of the molecules of life. These bonds form when two atoms share valence electrons, and they tend to form among atoms that have three, four, or five valence electrons. Sharing electrons enables these atoms to fill their outermost shells with eight electrons.

Carbon has four electrons in its outermost shell, and therefore it requires four more electrons to satisfy the octet rule. A

3.1 biology in action

NO and CO—Tiny Biological Messengers

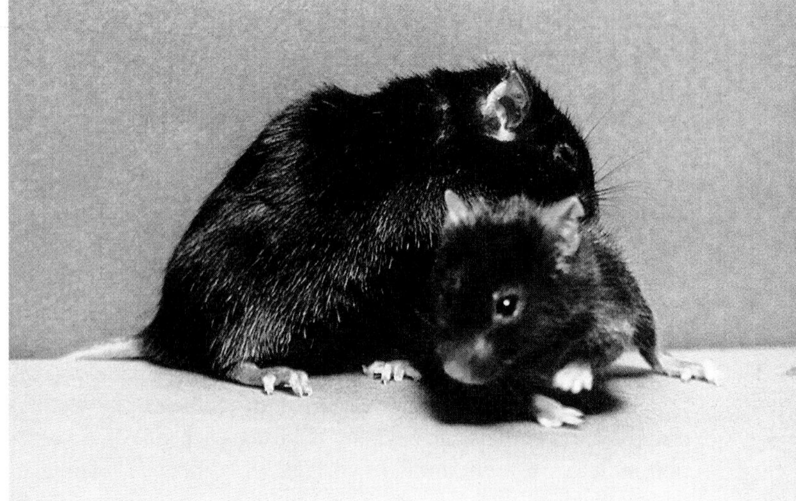

FIGURE 3.A
Male mice unable to make NO become highly aggressive, fighting other males to the death and forcing intercourse on females.

Before 1992, nitric oxide (NO) was known for its presence in smog, cigarettes, and acid rain. But by late 1992, researchers had implicated the small molecule in several roles in physiology, including digestion, memory, immunity, respiration, and circulation. Scientists offered one possible explanation for NO's varied functions in life—it has one unpaired electron, making it highly reactive.

In the male human body, NO enables the penis to become erect by dilating blood vessels in the organ, thus engorging it with blood so that it stiffens and rises. Elsewhere in the body, cells lining certain blood vessels release NO, which signals nearby muscles to relax. This gives the vessels room to expand, thus lowering blood pressure. NO also helps the immune system prevent several types of infections.

NO was the first gas found to be a neurotransmitter, a messenger that enables nerve cells to communicate with each other. Most known neurotransmitters are large molecules that travel slowly between cells. Evidence of NO's role in nervous system function is vividly seen in male mice genetically altered so that they cannot produce NO. They become so violent that they literally tear each other to pieces (fig. 3.A).

If NO, a small, gaseous molecule, is a neurotransmitter, do other small molecules transmit nerve cell messages too? Solomon Snyder, the Johns Hopkins University neuroscientist who has discovered and described many neurotransmitters and who bred the murderous mice, suspected that carbon monoxide (CO) might also be a neurotransmitter.

Like NO, CO had a nasty reputation—as a colorless, odorless, and lethal gas. CO released from faulty gas heaters kills hundreds of people each year and makes many others ill. CO binds to the protein hemoglobin (Hb) and prevents it from carrying oxygen in the blood. As the proportion of hemoglobin that binds CO rises, symptoms worsen (table 3.A).

CO also participates in normal physiology. It is in the spleen, an organ where old red blood cells are dismantled. Laboratory studies show that CO binds to another type of biological communicator called a second messenger. This indicates that CO might function as a neurotransmitter. Snyder hypothesized that if CO acts as a neurotransmitter, the enzyme responsible for synthesizing it would probably be found in specific brain regions, as the enzymes for other neurotransmitters are. Experiments support his hypothesis:

CO, its enzyme, and the second messenger it binds are found in the same parts of the brain—the memory center, smell center, and lower parts controlling vital functions.

Nerve cells growing in culture display second-messenger activity if CO is present, but not if it is absent.

In rats, blocked CO production impairs memory.

Researchers who study the brain now have an entirely new class of neurotransmitters to work with, headed by two tiny chemicals that were right under their noses. More important, the functions of CO and NO have reaffirmed the statement made in chapter 2: When implementing the scientific method, one must always expect the unexpected.

Table 3.A Rising Carbon Monoxide Levels

Situation	% Hb Bearing CO	Symptoms
Nonsmoker	2%	None
Smoker	5–9%	Cough
Moderate CO exposure	10–20%	Headache, dizziness, nausea and vomiting, fatigue, confusion
Severe CO exposure	20–60%	Increased breathing and heart rate, seizures, coma, death

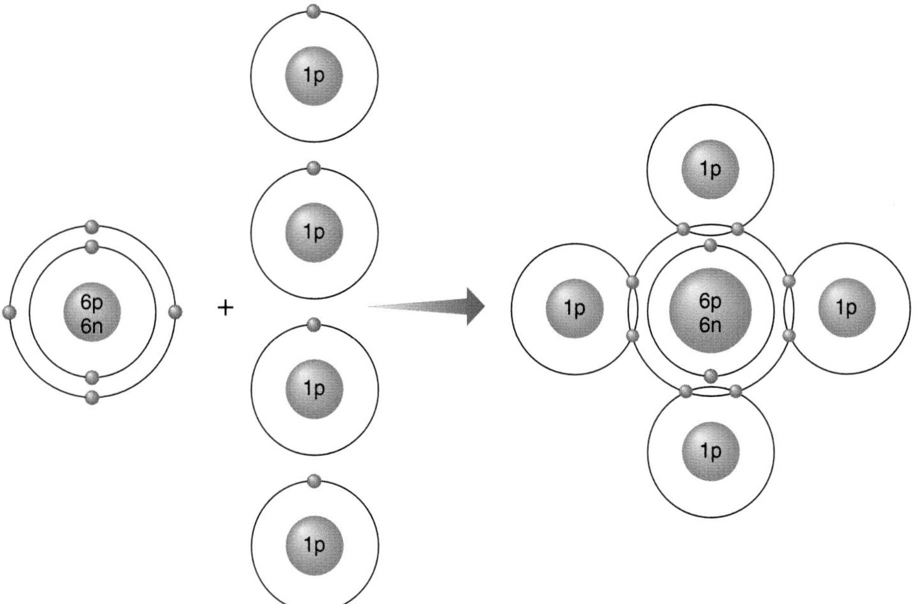

FIGURE 3.5

A covalently bonded molecule—methane (CH_4). One carbon and four hydrogen atoms complete their outermost shells by sharing electrons. Note that the first electron shell is complete with two electrons.

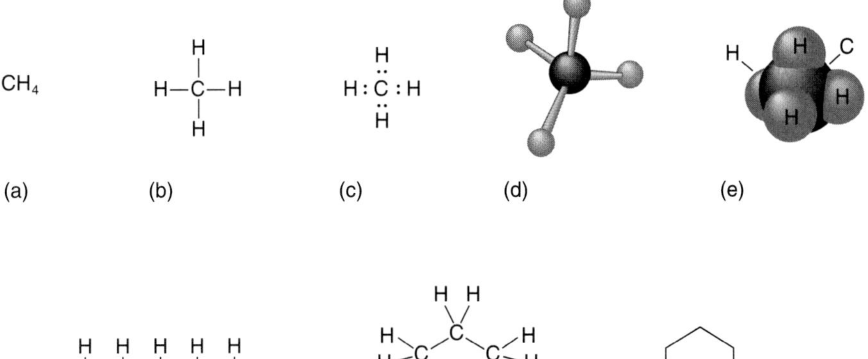

FIGURE 3.6

Different types of diagrams used to represent molecules. Consider the swamp gas methane. (*a*) The molecular formula CH_4 indicates that methane consists of one carbon atom bonded to four hydrogen atoms. (*b*) The structural formula shows those bonds as single. (*c*) The electron dot diagram shows the number and arrangement of shared electron pairs. (*d*) A ball-and-stick model gives further information—the angles of the bonds between the hydrogens and the carbon. (*e*) A space-filling model shows bond relationships as well as the overall shape of a molecule.

The organic compounds of life are built of carbon chains (*f*) and rings (*g*), with attached hydrogens, oxygens, and sometimes other elements such as nitrogen, phosphorus, and sulfur. (*h*) The carbons of ring-shaped molecules are sometimes indicated only by the vertices of the bond lines. The hydrogen atoms in such structures are usually not explicitly shown.

carbon atom can attain the stable eight-electron configuration in its outer shell by sharing electrons with four hydrogen atoms, each of which has one electron in its only shell (fig. 3.5). The resulting molecule is the swamp gas methane (CH_4). Figure 3.6 shows several ways to represent its chemical structure.

Covalent bonds are most simply depicted with lines between the interacting atoms. Each bond contains two electrons, one from each atom. Each atom shares the electrons in the bond to help fill its outer octet of electrons. By sharing four electrons with four hydrogen atoms, the carbon in methane fulfills the octet rule.

Methane forms from the sharing of single electron pairs in covalent bonds. Carbon atoms can also bond with each other and share two or three electron pairs, forming double and triple covalent bonds, respectively (fig. 3.7). The fact that a carbon atom may form four covalent bonds, which satisfies the octet rule, allows this element to assemble into long chains, intricate branches, and

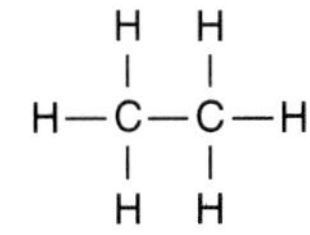

(a) Ethane

H H
C=C
H H

(b) Ethylene

H—C≡C—H

(c) Acetylene

FIGURE 3.7

Carbon atoms form four covalent bonds. Two carbon atoms can bond, forming single, double, or triple bonds. Note that as the number of bonds between carbon atoms increases, the number of hydrogens bonded to decreases. (*a*) Ethane, a component of natural gas, is a hydrocarbon built around two singly bonded carbon atoms. (*b*) Ethylene consists of two carbon atoms linked by a double bond. Ethylene is a plant hormone, triggering flowers to drop and fruit to ripen. (*c*) Acetylene, consisting of two carbons held by a triple bond, is a flammable gas used in torches.

Table 3.3 Effects of Sodium and Potassium Deficiency and Excess

Abnormality	Causes	Symptoms
Na^+ deficiency (hyponatremia)	Prolonged sweating, vomiting, or diarrhea Kidney disease Adrenal gland abnormality	Seizures, possibly coma
Na^+ excess (hypernatremia)	High fever Diabetes insipidus (inability of kidneys to reabsorb water)	Confusion, stupor, coma
K^+ deficiency (hypokalemia)	Adrenal gland abnormality Diuretic drugs Kidney disease Prolonged vomiting or diarrhea	Muscle weakness, paralysis Heartbeat irregularities Difficulty breathing
K^+ excess (hyperkalemia)	Kidney disease Adrenal gland abnormality Drugs that cause kidneys to retain K^+	Paralysis Heart attack Heartbeat irregularities

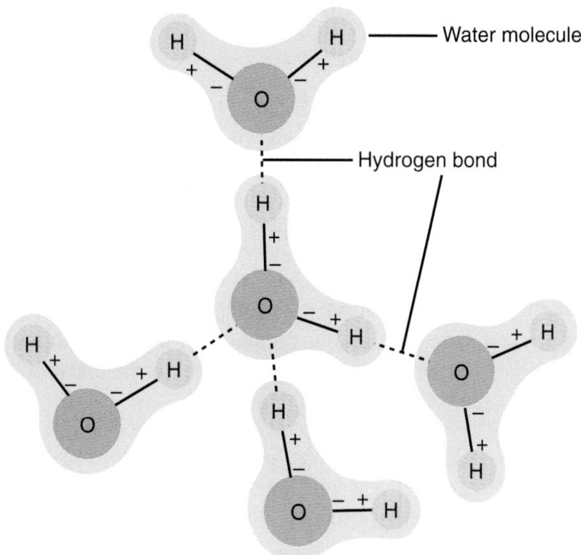

FIGURE 3.8

Polar covalent bonds hold together the two hydrogen atoms and one oxygen atom of water (H_2O). Because the oxygen attracts the negatively charged hydrogen electrons more strongly than the hydrogen nuclei do, the oxygen atom bears a partial negative charge and the hydrogens carry a partial positive charge.

rings. Carbon chains and rings bonded to hydrogens are called **hydrocarbons.** Carbon bonded to other chemical groups yields a great variety of biological molecules.

Methane is held together by **nonpolar covalent bonds,** in which the atoms share all of the electrons in the covalent bond equally. In contrast, in a **polar covalent bond,** electrons draw more towards one atom's nucleus than the other. Water (H_2O), which consists of two hydrogen atoms bonded to an oxygen atom, illustrates a molecule with polar covalent bonds (fig. 3.8). The nucleus of each oxygen atom attracts the electrons on the hydrogen atoms more than the hydrogen nuclei do. As a result, the area near the oxygen carries a partial negative charge (from attracting the negatively charged electrons), and the area near the hydrogens is slightly positively charged (as the electrons draw away).

The tendency of an atom to attract electrons is termed **electronegativity.** Oxygen is highly electronegative. Its ability to accept electrons is crucial in helping organisms extract energy from nutrients.

Ionic Bonds

In a covalent bond, atoms share electrons. In an **ionic bond,** the electronegativity of one atom is so great that the atom takes an electron from the other atom. An atom with one, two, or three electrons in the outermost shell loses them to an atom with, correspondingly, seven, six, or five valence electrons. A sodium (Na) atom, for example, has one valence electron. When it donates this electron to an atom of chlorine (Cl), which has seven electrons in its outer shell, the two atoms bond ionically to form NaCl (fig. 3.9*a*).

Once an atom loses or gains electrons, it has an electric charge and is called an **ion.** Atoms that lose electrons lose negative charges and thus carry a positive charge. Atoms that gain electrons become negatively charged. The attraction between oppositely charged ions forms the ionic bond. In NaCl, the oppositely charged ions Na^+ and Cl^- attract in such an ordered manner that a crystal results (fig. 3.9*b*). A large lattice built of oppositely charged ions, like NaCl, is called a **salt.** A salt can consist of any number of ions and is described as a ratio of one type of ion to another. Ionic bonds are strong in a chemical sense, but they weaken in water. Because many of the molecules of life are dissolved in water, ionic bonds in organisms tend to be weak.

Most biochemicals are negatively charged ions. Many small, positively charged ions are abundant in organisms, such as sodium (Na^+) and potassium (K^+). These ions balance the larger, negatively charged molecules. In the human body, fluids within cells tend to have high concentrations of K^+, whereas fluids outside cells tend to have high concentrations of Na^+. Disruptions of this distribution can cause severe symptoms (table 3. 3).

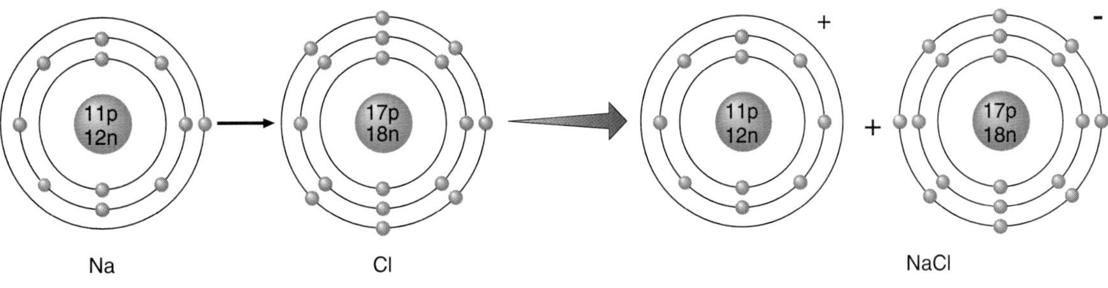

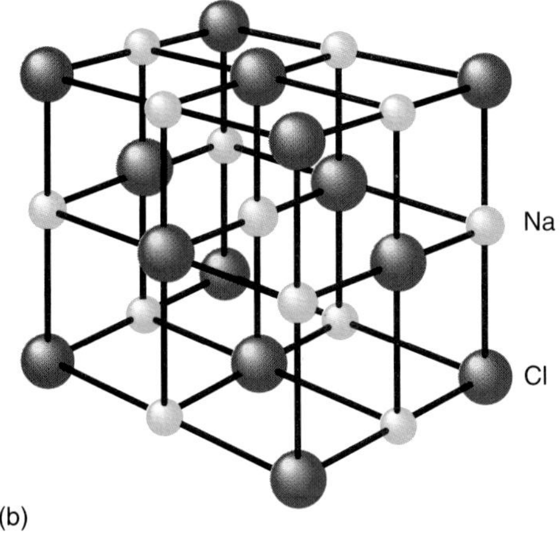

FIGURE 3.9

An ionically bonded molecule. (*a*) A sodium atom (Na) can donate the one electron in its valence shell to a chlorine atom (Cl), which has seven electrons in its outermost shell. This satisfies the octet rule. The resulting ions (Na^+ and Cl^-) bond to form the compound sodium chloride (NaCl), better known as table salt. (*b*) The ions that constitute NaCl occur in a repeating pattern.

Ions are important in many biological functions. Nerve transmission depends upon passage of Na^+ and K^+ in and out of nerve cells. Muscle contraction depends upon movement of calcium (Ca^{2+}) ions into and out of muscle cells. Even the start of a new life is ionically controlled. In many types of organisms, egg cells release calcium ions, which clear a path for the sperm's entry.

Weak Chemical Bonds

Weak chemical bonds are also important in the molecules of life. **Hydrogen bonds** form when hydrogen atoms covalently bonded to one atom are attracted to an atom that is part of another molecule (see fig. 3.8). A hydrogen bond occurs because of the hydrogen atom's partial positive charge, which results from the atom's participation in a polar covalent bond. Hydrogen bonds are abundant in water. They help provide the great strength of DNA, the genetic material. DNA consists of ladderlike chains held together at the "rungs" by hydrogen bonds. Any one hydrogen bond is not particularly strong, but the many that hold together a long DNA molecule impart a great overall strength, like the teeth of a zipper.

Different parts of the large and intricately shaped chemicals of life may be temporarily charged because their electrons are always in motion. Dynamic attractions between molecules or within molecules that occur when oppositely charged regions approach one another are called **van der Waals attractions.** They help shape the molecules of life.

Interaction with water is another type of chemical attraction that helps determine the spatial arrangement of atoms comprising molecules. **Hydrophilic** (water-loving) parts of molecules are attracted to water; **hydrophobic** (water-fearing) parts are repelled by water. Large molecules such as proteins contort so that their hydrophilic regions touch water and their hydrophobic regions shun it. Interactions with water help foster the molecular surfaces of cells. Table 3.4 summarizes the chemical bonds important in biology.

Mastering Concepts

1. Of what types of chemicals do living organisms consist?
2. What is the relationship between an element and an atom?
3. Which chemical elements do organisms require in large amounts?
4. What types of particles make up an atom?
5. What do an element's atomic number and atomic weight indicate?
6. What is an isotope?
7. What is the relationship between a compound and a molecule?
8. How do the number and positions of electrons determine an atom's ability to form bonds?
9. Distinguish between nonpolar and polar covalent bonds, ionic bonds, and hydrogen bonds.
10. How do van der Waals attractions and interactions with water help molecules assume their three-dimensional forms?

Table 3.4 How Atoms Attract One Another

Type	Chemical Basis	Strength	Example
Covalent bonds	Atoms share electron pairs	Strong	Hydrocarbons
Hydrogen bonds	Atoms with partial negative charges attract hydrogen atoms with positive charges (due to their participation in a polar covalent bond)	Weak	Water
Ionic bonds	One atom donates an electron to another	Weak in water	Sodium chloride
van der Waals attractions	Oppositely charged regions between or within molecules attract one another	Weak	Protein structure
Water interactions	Hydrophobic parts of molecules repel water; hydrophilic parts are attracted to water	Strong	DNA structure

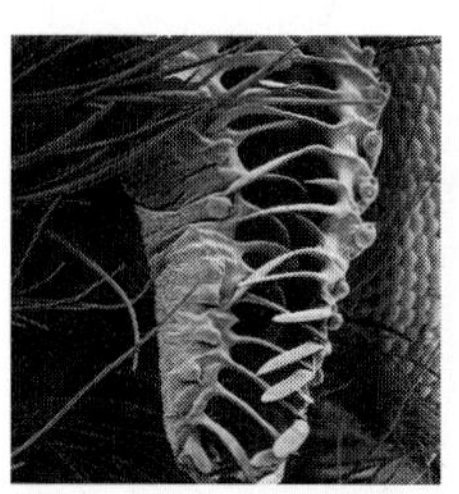

Chemistry Explains Biology

Seeking Salt

Imagine drinking 600 times your weight in just a few hours and excreting most of it as you did so! This is exactly what male moths of species *Gluphisia septentrionis* do to obtain sufficient sodium ions in a salt-poor environment. The moths do this by a behavior called "puddling"—they sip mud puddles. The animals' mouths have sifterlike structures to admit salty water while keeping muck out.

Researchers Scott Smedley and Thomas Eisner discovered the salt-acquiring talent of the male moths by measuring the sodium ion content of puddle water and comparing it to that of the copious urine that spurts from the insects' rectums every few seconds, for hours.

The moths collect and concentrate the vital Na^+ in their reproductive organs and then transfer the ions to females during a 5-hour mating act. The female uses some of the ions herself and then passes some to her eggs. Says Smedley, "It's like dad giving the kids a one-a-day vitamin plus minerals."

Water—The Matrix of Life

Water has been called the "mater and matrix" of life. "Mater" means mother, and indeed life as we know it could not have begun were it not for this unusual substance. Water is also the matrix, or medium, of life, because it is vital in most biochemical reactions. The discovery of evidence of water on Mars suggests that life may once have been present there.

Solutions

In one second, the hydrogen bonds between a single water molecule and its nearest neighbors form and re-form some 500 billion times! This action of hydrogen bonding makes water fluid. Fluidity is the ability of a substance to flow. The strong attractions between water molecules, which account for their constant rebonding, is called **cohesion.** Water readily forms hydrogen bonds to many other compounds, a property called **adhesion,** which is important in many biological processes. Adhesion enables water to form body fluids that carry vital, dissolved electrically charged chemicals called **electrolytes.**

Cohesion and adhesion are important in plant physiology. Movement of water from a plant's roots to its highest leaves depends upon cohesion of water within the plant's water-conducting tubes. Water entering roots is drawn up as water evaporates from leaf cells. Water movement up trees also depends upon the adhesion of water to the walls of the conducting tubes. Water's adhesiveness accounts for **imbibition,** the tendency of substances to absorb water and swell. Rapidly imbibed water swells a seed so that it bursts through the seed coat, releasing the embryo within (fig. 3.10).

Biologists use several terms to describe water's ability to carry other chemicals. A **solvent** is a chemical in which other chemicals, called **solutes,** dissolve. A **solution** consists of one or more chemicals dissolved in a solvent. In an **aqueous solution,** water is the solvent. Water molecules flow between the positively and negatively charged ions of a salt, and they also separate molecules held together by polar covalent bonds. Recall that the hydrogen atoms in a water molecule carry a partial positive charge, which causes them to attract other atoms and form new hydrogen bonds. In this sense, water molecules "pull apart" molecules held by polar covalent bonds.

Nonpolar covalently bonded molecules, such as fats and oils, do not carry a charge, are not attracted to hydrogen atoms, and thus do not usually dissolve in water. This is why oil on the surface of a cup of coffee forms a visible drop and a drop of water on an oily surface beads up.

Acids and Bases

In an aqueous solution, hydrogens from one water molecule are attracted to another, transforming two molecules of water ($2H_2O$) into a **hydronium ion** (H_3O^+) and a **hydroxide ion** (OH^-). (The H_3O^+, H_2O plus H^+, can be considered the chemical equivalent of H^+.) In pure water, the numbers of free H^+s and OH^-s are equal. However, another chemical dissolved in water can alter the balance of positive and negative charges. Substances that add more H^+ to a solution are **acids.** Common acids include hydrochloric acid (HCl), sulfuric acid (H_2SO_4), phosphoric acid (H_3PO_4), and nitric acid (HNO_3). Substances that decrease the number of H^+s are **bases** and are alkaline (nonacid). In water, bases typically dissociate (come apart) to yield OH^-, which binds with some of the

(a)

(b)

FIGURE 3.10

Cohesion and imbibition. (*a*) Water's cohesion and adhesion enable it to be drawn up more than 328 feet (100 meters) to the tops of tall trees. (*b*) Seeds imbibe water, eventually bursting and germinating. This is a germinating coconut.

H^+ present in the dissociated water to re-form H_2O. A common base is sodium hydroxide (NaOH), or lye. Shampoo advertised as nonalkaline is actually acidic!

The pH Scale

Scientists use a system of measurement called the **pH scale** to gauge how acidic or basic a solution is in terms of its H^+ (hydrogen ion) concentration. The pH scale ranges from 0 to 14, with 0 representing strong acidity (high H^+ concentration) and 14 strong basicity or alkalinity (low H^+ concentration). A neutral solution, such as pure water, has a pH of 7 (fig. 3.11).

Many fluids in the human body function within a narrow pH range. Illness results when pH changes beyond this range. The normal pH of blood, for example, is 7.35 to 7.45. Blood pH of 7.5 to 7.8, a condition called alkalosis, makes one feel agitated and dizzy. Alkalosis can be caused by breathing rapidly at high altitudes, taking too many antacids, enduring high fever, or feeling extreme anxiety. Acidosis, in which blood pH falls to 7.0 to 7.3, makes one feel disoriented and fatigued and may impair breathing. This condition may result from severe vomiting, diabetes, brain damage, or lung or kidney disease.

Buffer Systems Regulate pH in Organisms

Organisms have pairs of weak acids and bases, called **buffer systems,** that maintain the pH of body fluids in a comfortable range by gaining or losing hydrogens to neutralize strong acids and bases. Carbonic acid (H_2CO_3) and sodium bicarbonate ($NaHCO_3$) form one such buffer system. It is found inside cells and in some of the fluids outside cells.

A strong acid, such as hydrochloric acid (HCl), reacts with sodium bicarbonate ($NaHCO_3$) and produces the weaker carbonic acid (H_2CO_3) and sodium chloride (NaCl). The acidity of the fluid decreases:

$$\underset{\text{strong acid}}{HCl + NaHCO_3} \longrightarrow \underset{\text{weak acid}}{H_2CO_3} + NaCl$$

A strong base, such as sodium hydroxide (NaOH), reacts with carbonic acid (H_2CO_3) and produces the weaker base sodium bicarbonate ($NaHCO_3$) and water:

$$\underset{\text{strong base}}{NaOH + H_2CO_3} \longrightarrow \underset{\text{weak base}}{NaHCO_3} + H_2O$$

Other buffer systems maintain the pH of urine and blood within certain ranges.

Acid solutions affect life on a global scale. Sulfur- and nitrogen-based pollutants react with water in the atmosphere and return to earth as acid precipitation. Just slightly lowering the pH of a body of water can drastically alter the types of organisms that can survive there. Some organisms, however, are adapted to low-pH environments, such as the bacteria that cause peptic ulcers.

Chemistry Explains Biology

Living in Acid

In sulfur-rich sediments and streams live bacteria, algae, protozoa, and insects that have adapted to acidic conditions as low as pH 2. Acid environments arise when sulfur-containing compounds, called sulfides, are exposed to oxygen and react with it to form sulfuric acid (H_2SO_4). Sulfides are common in swamps, bogs, and volcanoes and are exposed in mining operations. In the photograph, yellowish, sulfur-containing compounds coat the rocks near Mount Saint Helens, a volcano in Washington State.

Organisms that live in acidic surroundings are "extremophiles," species that thrive in extreme habitats. Other extremophiles "prefer" very high pH, high or low temperature, or high or low salinity. These organisms fascinate biologists because they exist in an environment that normally dismantles DNA, protein, and biochemicals that extract or store chemical energy. The molecules that extremophiles have are important to us in a variety of industrial processes.

Water and Temperature

Several temperature-related characteristics of water affect life. Water controls temperature in organisms because a great deal of heat is required to raise its temperature—a characteristic called high **heat capacity.** Because of water's high heat capacity, an organism may be exposed to considerable heat before its aqueous body fluids become dangerously warm. Water's high **heat of vaporization** means that a lot of heat is required to evaporate water. This is why evaporating sweat draws heat out from the body—another important factor in regulating body temperature.

Water's unusual tendency to expand upon freezing also affects life. The bonds of ice are farther apart than those of liquid water. Therefore, ice is less dense and floats on the surface of liquid water. This characteristic benefits water-inhabiting (aquatic) organisms. When the air temperature drops sufficiently, a small amount of water freezes at the surface of a body of water, forming a solid cap of ice that retains heat in the water below. Many organisms can then survive in the depths of such shielded lakes and ponds (fig. 3.12). If water were to become denser upon freezing, ice would sink to the bottom, and the body of water would gradually turn to ice from the bottom up, entrapping the organisms that live there.

On the other hand, organisms do freeze and die because water inside cells expands when it freezes and breaks apart the cells. Many types of organisms have adaptations that enable them to survive freezing, such as producing biochemicals that act as antifreeze, thus keeping body fluids liquid at temperatures low enough to freeze pure water. Another strategy is to temporarily dehydrate, such as do the tardigrades in figure 3.13. The larvae of flies called midges exhibit yet another adaptation to freezing—ice forms between their cells, not inside them.

Table 3.5 lists some of the important properties of water.

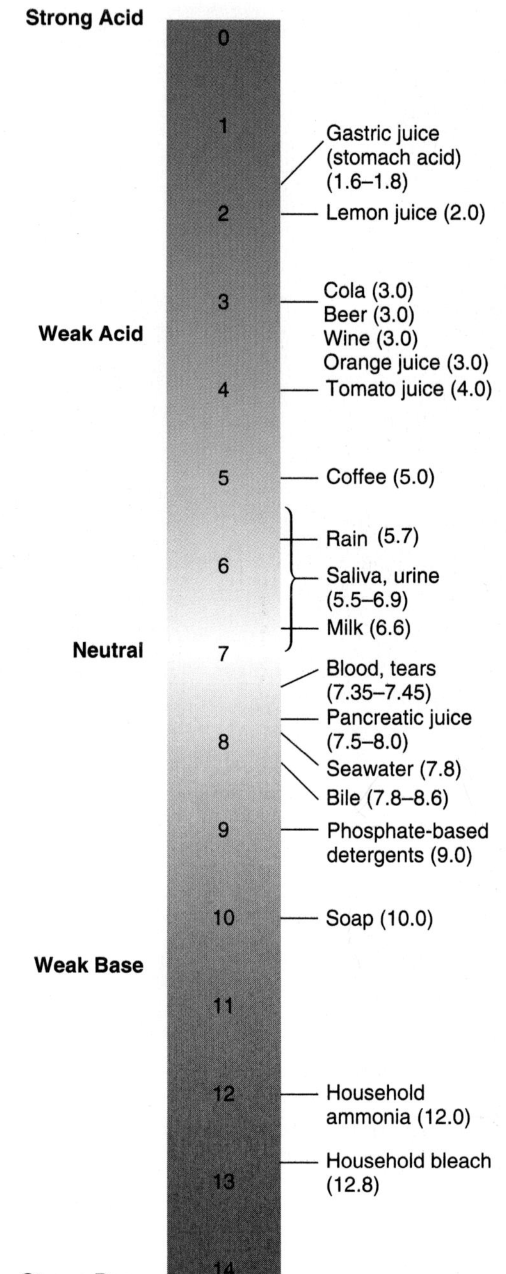

FIGURE 3.11

The pH scale is commonly used to measure the strength of acids and bases.

Mastering Concepts

1. List ways that water is essential to life.
2. What are acids and bases?
3. How does pH measure acidity and basicity?
4. What is the function of buffer systems?
5. What properties of water enable it to regulate body temperature?
6. How does the fact that water expands upon freezing affect life?

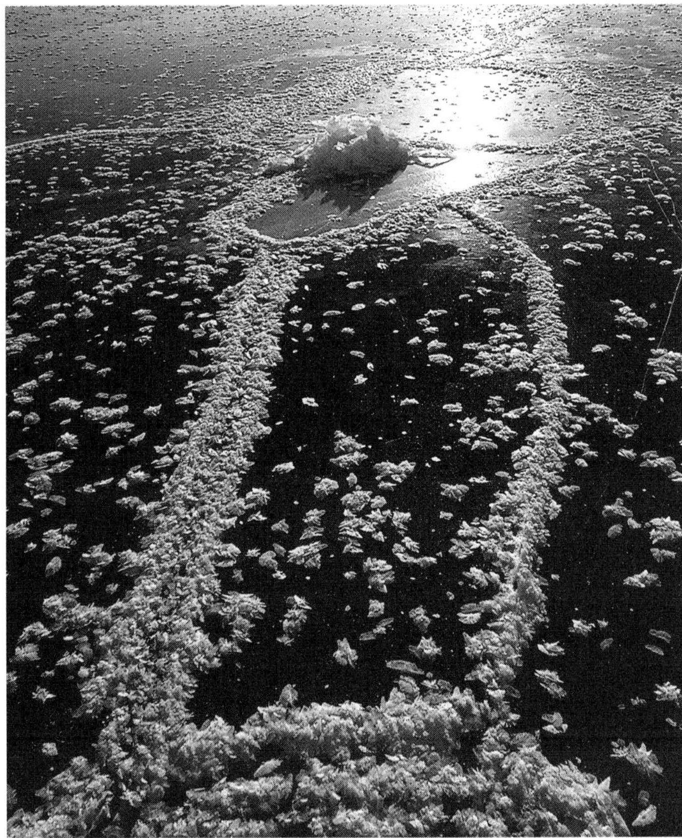

FIGURE 3.12
When ponds freeze, aquatic life survives on the bottom, which does not freeze.

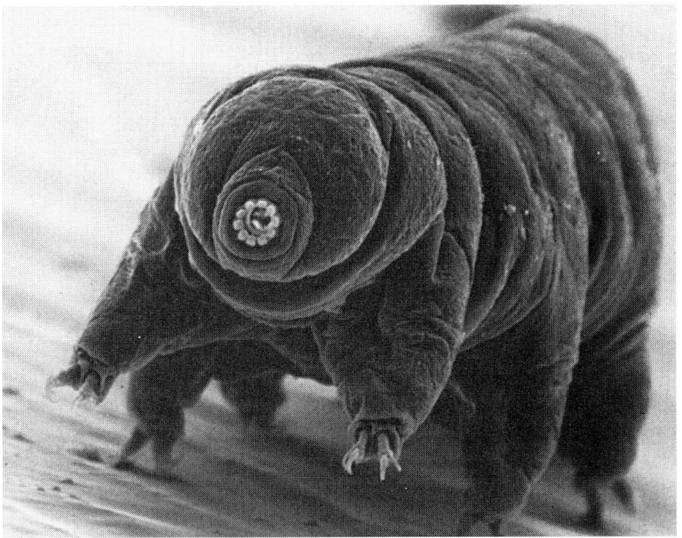

FIGURE 3.13
Tardigrades are called waterbears. They are small (less than a millimeter long) and live on aquatic vegetation. Under environmental stress, including freezing, a tardigrade gradually dries up, its body composition changing from 85% water to about 3% water. This creates a state of suspended animation called cryptobiosis that can extend a tardigrade's life span from 1 to 60 years.

Table 3.5 Properties of Water

Property	Example of Importance to Life
Cohesion	Water flows, forming habitats and a matrix for biological molecules
Imbibition	Embryo within seed absorbs water, swelling and bursting through seed coat
High heat capacity	Water regulates constant internal body temperature
High heat of vaporization	Evaporation (sweating) cools

The Organic Molecules of Life

Organisms are composed of, and must take in large amounts of, four types of organic compounds: carbohydrates, lipids, proteins, and nucleic acids. Vitamins are another group of biologically important organic molecules, but they are required in smaller amounts.

Carbohydrates

Carbohydrates consist of carbon, hydrogen, and oxygen, in the proportion 1:2:1. That is, a carbohydrate has twice as many hydrogen atoms as either carbon or oxygen atoms. Familiar carbohydrates are sugars and starches. Carbohydrates store energy, which is released when their bonds break. Some carbohydrates physically support cells and tissues.

Carbohydrates can be a single, small unit, generally called a **monomer,** or hundreds or even thousands of monomers linked to form long molecules called **polymers.** A **monosaccharide** is the smallest type of carbohydrate, and it contains three to seven carbons. Monosaccharides can differ from each other by how their atoms are bonded. For example, three six-carbon monosaccharides with the same molecular formula ($C_6H_{12}O_6$) but different chemical structures are glucose (blood sugar), galactose, and fructose (fruit sugar). Three-carbon monosaccharides, called trioses, form as cells break down glucose to release its energy, the subject of chapter 7.

Carbohydrates of intermediate length, consisting of 2 to 100 monomers, are called **oligosaccharides.** The smallest oligosaccharide is a **disaccharide,** which forms when two monosaccharides link by releasing a molecule of water (H_2O). Figure 3.14 shows how the disaccharide sucrose (table sugar) forms as a molecule of glucose and a molecule of fructose lose a water and join. This type of chemical reaction is called **dehydration synthesis** ("made by losing water"). In the opposite reaction, **hydrolysis** ("breaking with water"), a disaccharide and water react to form two monosaccharides.

Many plants contain sucrose, including sugarcane and beets (fig. 3.15). Maltose, a disaccharide formed from two glucose molecules, provides energy in sprouting seeds and is used to make

Chemistry Explains Biology

Frozen Turtles

When a biologist's little girl left her pet painted turtle in a wading pool during an early frost, the family thought it was time to find a new pet. But once indoors, the animal thawed, in apparent perfect health!

The biologist gradually froze painted turtles in his laboratory to study their survival mechanism. The animals froze from the skin inwards, with only the brain and heart spared. Water inside cells remained liquid, but extracellular fluids—urine, blood plasma, and fluid in the abdomen—froze solid. About half of the turtle's body fluid froze!

beer. Lactose, or milk sugar, is a disaccharide formed from glucose and galactose. Sugars, which are monosaccharides and disaccharides, are sometimes called simple carbohydrates.

Oligosaccharides are often anchored to proteins in cell membranes, emanating from the cell's surface and creating a topography, like the surface of the earth (fig. 3.16). Carbohydrates on cell surfaces are important in immunity, which is based on the distinctiveness of cell membranes in individuals. Oligosaccharides are also important in enabling proteins called antibodies to assume their characteristic three-dimensional shapes, which is essential to their function in protecting an animal's body from infection.

Polysaccharides ("many sugars") are long polymers of monosaccharides (sugars) linked by dehydration synthesis (fig. 3.17). They are also called complex carbohydrates. The most common polysaccharides, glycogen, starch, and cellulose, are all long chains of glucose, but they differ from each other by their branching patterns.

Cellulose and chitin are familiar complex carbohydrates that are glucose polymers (fig. 3.18). Cellulose forms wood and parts of plant cell walls, and chitin forms the outer coverings of insects, crabs, and lobsters and supports cell walls of fungi. Cellulose is the most common organic compound in nature, and chitin is the second most common.

Lipids

Lipids contain the same elements as carbohydrates but with proportionately less oxygen. The lipids are diverse molecules that dissolve in organic solvents but not in water. The most familiar lipids are fats and oils.

Despite their bad press, lipids are vital to life. They are necessary for growth and for the utilization of some vitamins. Fats slow digestion, thereby delaying hunger. Some lipids make leaves, fur, and feathers water-repellent, and others cushion organs and help to retain body heat by providing insulation. Fat is also an excellent energy source, providing more than twice as much energy as equal weights of carbohydrate or protein.

Fat cells aggregate as **adipose tissue.** White adipose tissue forms most of the fat in human adults. Another type, brown adipose tissue, releases energy as heat and keeps organisms warm, particularly mammals that hibernate. Brown adipose tissue is rare in adult humans, but in newborns it is layered around the neck and shoulders and along the spine.

Lipids enclose cells. In our own bodies, lipids ensheath nerve cells and speed nerve transmission. Human milk is rich in lipids, partly to suit the rapid growth of the brain in the first 2 years of life. A human infant is essentially a "fathead."

Glucose + Fructose = Sucrose ($-H_2O$)

FIGURE 3.14

Glucose + fructose − H_2O = sucrose. Dehydration synthesis between two monosaccharides yields a disaccharide.

FIGURE 3.15

Sugarcane is a rich source of sucrose, a disaccharide.

Chemistry Explains Biology

How Sweet It Is!

Different sugars produce different degrees of sweetness on the human tongue. When anthropologists introduced sucrose—in the form of candy—to native peoples in the Amazon basin accustomed to eating the fructose in wild fruits, the natives preferred the sweeter sucrose (a disaccharide) and asked for more! It's healthier to satisfy a sweet tooth with fructose (a monosaccharide), because it brings along fiber in the fruit. Candy brings fat.

This chapter discusses the triglycerides, fatty acids, sterols, and waxes. Two other types of lipids, phospholipids and glycolipids, are considered in the next two chapters in the context of their functions in cell membranes. A phospholipid is a lipid bonded to a phosphate group (a phosphorus atom bonded to four oxygen atoms). Glycolipids are lipids bonded to sugars.

Triglycerides and Fatty Acids

One type of fat, a **triglyceride,** consists of a three-carbon alcohol (an organic group with an OH) called **glycerol,** from which three hydrocarbon chains called **fatty acids** extend (fig. 3.19). Single covalent bonds link the carbons of fatty acids and form tail-like structures of up to 36 carbon atoms.

We describe fatty acids by their degree of saturation, which is a measure of their hydrogen content. A **saturated** fatty acid contains all the hydrogens it possibly can, which occurs when single bonds connect all the carbons. A fatty acid is **unsaturated** if it has at least one double bond and **polyunsaturated** if it has more than one double bond. A **monounsaturated** fatty acid has just one double bond. Olive oil is a monounsaturated fat. Lipids in plants are less saturated than those in animals.

The sites of unsaturation (double bonds) in the fatty acids cause them to form kinks and spread their "tails." This gives the fat an oily consistency at room temperature, while more saturated animal fats tend to be solid. A food-processing technique called hydrogenation, used to produce margarine, adds hydrogen to an oil to solidify it—in essence, saturating a formerly unsaturated fat.

The site of the double bond in a dietary fat or oil can dramatically affect our health. Unsaturated fatty acids are designated omega-3 and omega-6, depending on the location of the first double bond. In naming fatty acids, the omega carbon is the carbon at the end bearing a methyl group (CH_3). (Omega is the last letter of the Greek alphabet.) An omega-3 fatty acid has a double bond between the third and fourth carbons from the end. In an omega-6 fatty acid, the double bond is between the sixth and seventh carbons (fig. 3.19). Most fats in red meat and dairy products contain omega-6 unsaturated fatty acids, which may contribute to or cause heart disease (fig. 3.20). Greenland Eskimos and Dutch men, who have very low incidences of heart disease but bleed easily, eat mostly omega-3 fatty acids in their fish-rich diets.

Sterols

Sterols are lipids that have four carbon rings. Vitamin D and cortisone are sterols. A very familiar sterol, cholesterol, is part of cell membranes (fig. 3.21). Cells use cholesterol as a starting material to synthesize other lipids, including the sex hormones testosterone and estrogen. Liver cells manufacture cholesterol when they break down saturated fats. Eating cholesterol adds to this, and the excess cholesterol collects on the inner linings of blood vessels and impedes blood flow. Because the liver essentially converts saturated fat into cholesterol, it is important to limit dietary intake of saturated fats as well as cholesterol.

Waxes

Waxes are combinations of fatty acids with either alcohols or other hydrocarbons that usually form a hard, water-repellent covering.

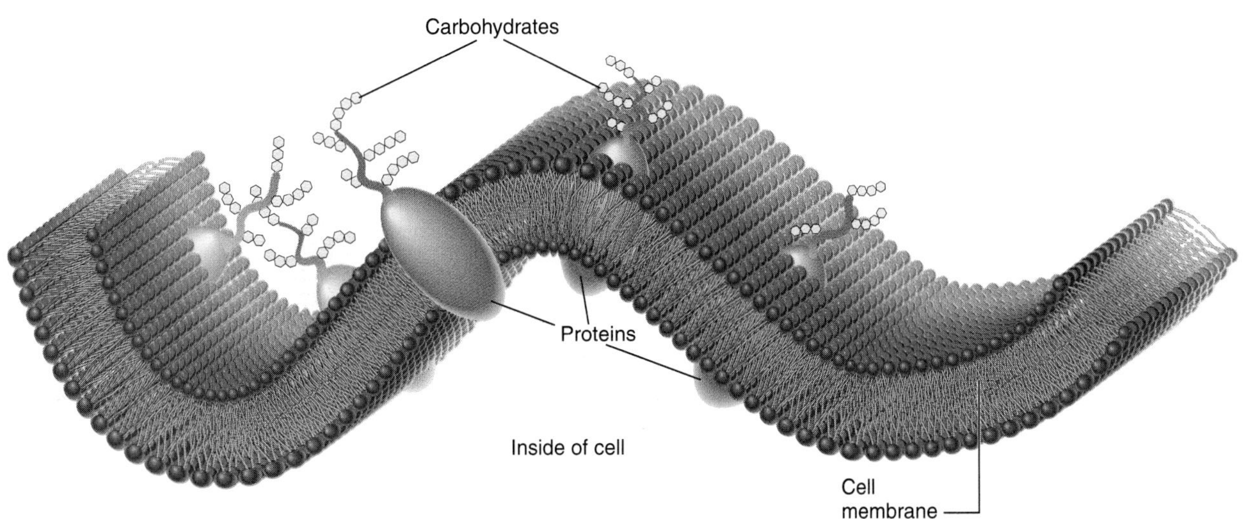

FIGURE 3.16

Oligosaccharides (medium-length carbohydrates) extend from the outside surfaces of cells.

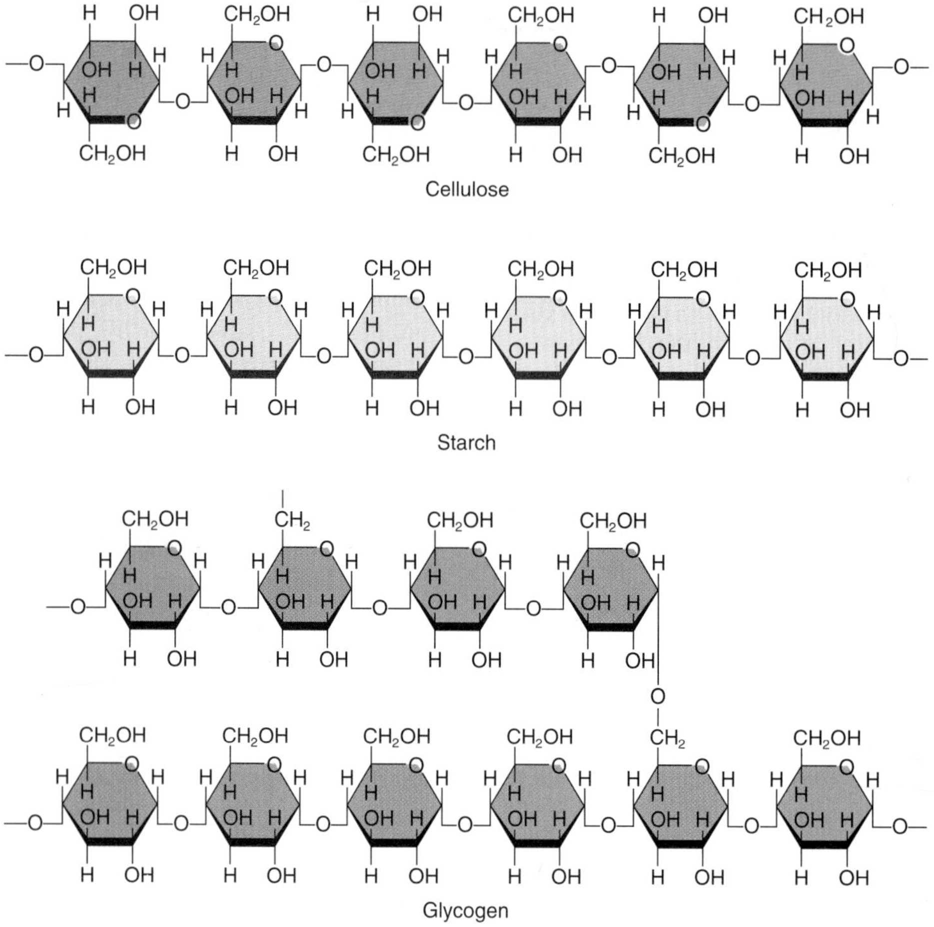

Cellulose

Starch

Glycogen

FIGURE 3.17

The complex carbohydrates (polysaccharides) cellulose, starch, and glycogen are all chains of glucose subunits, but they differ in structure. The glucose subunits of cellulose (found in plants) link together differently than the glucose subunits of starch (also found in plants). Glycogen, a glucose polymer found in animals, is more highly branched than either cellulose or starch.

These lipids help waterproof fur, feathers, leaves, fruits, and some stems (fig. 3.22). Jojoba oil, used in cosmetics and shampoos, is unusual in that it is a liquid wax.

Proteins

Proteins consist of monomers of **amino acids** linked to form **polypeptide chains**. A protein consists of one or more polypeptide chains. Amino acids in organisms are of 20 types, even though chemically many others exist. These 20 different types make possible tens of millions of different amino acid sequences. In contrast, a polysaccharide is a polymer of one or a few types of monomers.

Proteins have diverse biological functions. They enable blood to clot, muscles to contract, oxygen to reach tissues, and nutrients to break down to release energy. Proteins called **enzymes** allow biochemical reactions to proceed fast enough to sustain life. Proteins mark our cell surfaces as distinctly ours. They enable certain cells to migrate and others to attach to one another and build tissues. Table 3.6 lists some of the more than 50,000 kinds of proteins in the human body.

FIGURE 3.18

The strong jaws of insects are constructed of a complex carbohydrate, chitin. They can crunch through the complex carbohydrate of plants, called cellulose. Although chitin and cellulose are both polysaccharides, the repeating single sugars of each are different.

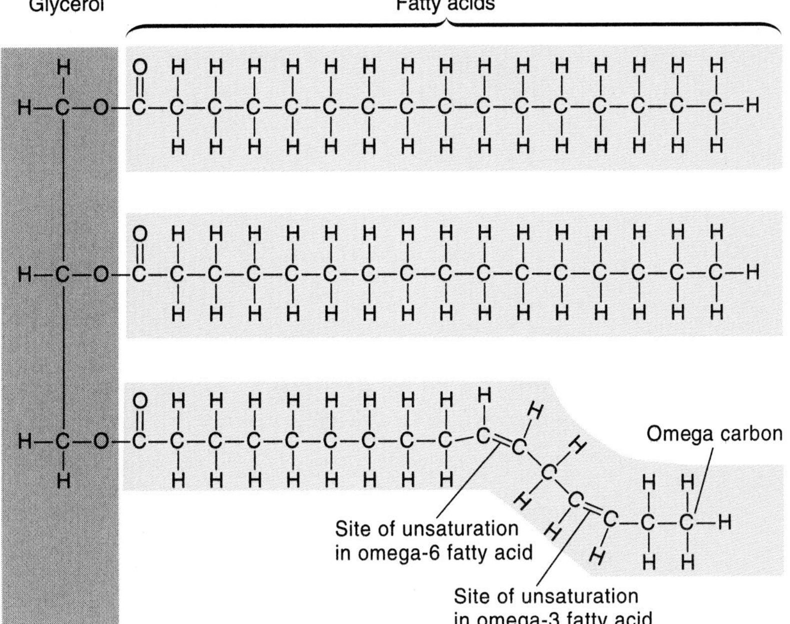

FIGURE 3.19

This lipid is an unsaturated derivative of tripalmitin, a triglyceride. Dietary fats rich in omega-3 fatty acids are healthier than those rich in omega-6 fatty acids. The double bonds bend the fatty acid tails, making the lipid oilier. The bend is shown schematically here.

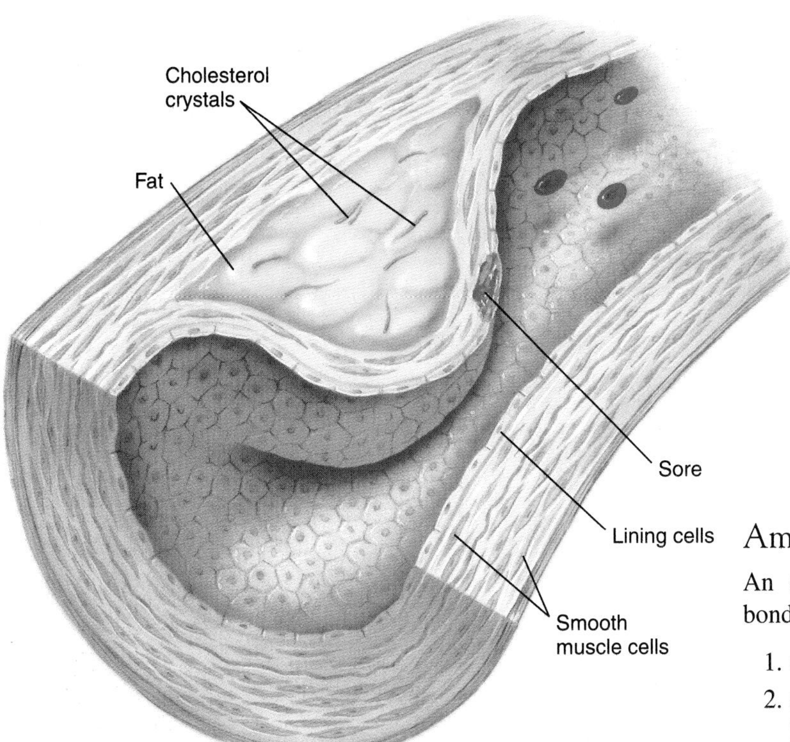

FIGURE 3.20

Excess dietary fat can be deposited as cholesterol-rich fatty plaques on the inner linings of blood vessels and impede blood flow.

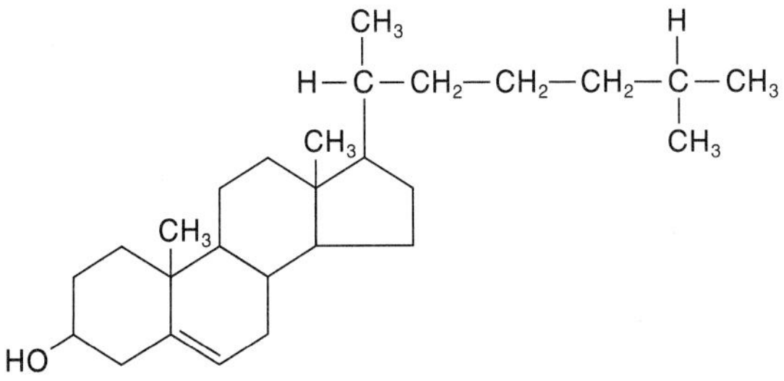

FIGURE 3.21

Cholesterol is the starting material for the synthesis of several types of steroid hormones.

Amino Acid and Protein Structure

An amino acid (fig. 3.23*a*) contains a central carbon atom bonded to

1. a hydrogen atom;
2. a **carboxyl group** (COO^-), which is a carbon atom double-bonded to one oxygen and single-bonded to another bearing a negative charge;
3. an **amino group,** which is a nitrogen atom single-bonded to three hydrogen atoms, one with a positive charge (NH_3^+);
4. a side chain, or **R group,** which can be any of several chemical groups.

The nature of the R group distinguishes the 20 types of biological amino acids. An R group may be as simple as the lone hydrogen atom in glycine or as complex as the two organic rings of

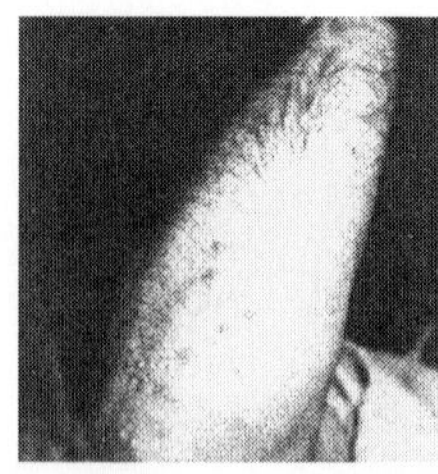

Chemistry Explains Biology

Excess Cholesterol

Lipoproteins carry cholesterol in the bloodstream. As their name suggests, lipoproteins consist of lipid and protein. Low-density lipoprotein (LDL) particles carry cholesterol to the arteries. Excess LDL cholesterol that does not enter cells accumulates on the inner linings of blood vessels, eventually impeding blood flow. High-density lipoproteins (HDL), in contrast, carry cholesterol to the liver, where it is removed from the bloodstream. High levels of LDL cholesterol increase risk of heart disease, whereas high levels of HDL cholesterol promote heart health.

In the severe form of an inherited illness called familial hypercholesterolemia, cells lack the receptors that admit LDL cholesterol. As a result, blood serum cholesterol levels skyrocket, and the fatty substance not only accumulates in blood vessels but also appears as yellowish deposits called xanthomas visible as bulges beneath the skin.

FIGURE 3.22

Waxes waterproof the coat of the otter and the cuticles of the grasses growing in the background.

tryptophan. Figure 3.23*b* shows three amino acids. The R groups of two amino acids, cysteine and methionine, contain sulfur. Note that proteins contain large amounts of nitrogen, another way that they differ from carbohydrates and lipids.

Two adjacent amino acids join by dehydration synthesis, just as two monosaccharides shed a water molecule to yield a disaccharide. When two amino acids bond, the acid-group carbon of one amino acid bonds to the nitrogen of the other and forms a **peptide bond** (fig. 3.24). Two joined amino acids are a dipeptide,

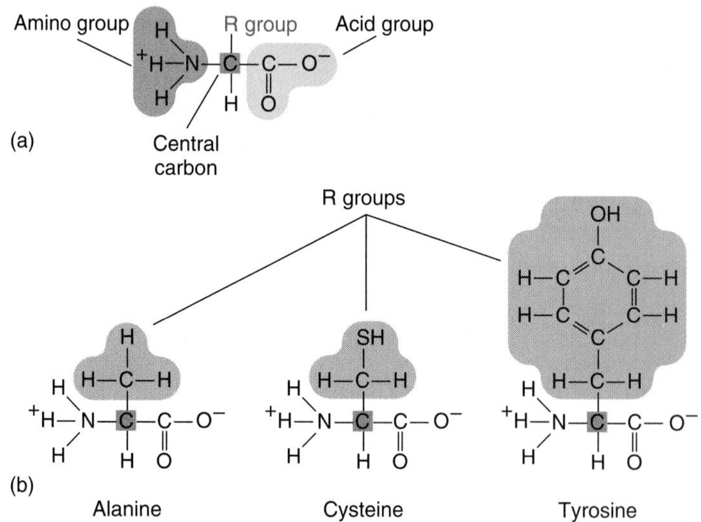

FIGURE 3.23

Amino acids, the building blocks of protein. (*a*) An amino acid consists of a central carbon atom covalently bonded to an amino group, an acid (carboxyl) group, a hydrogen, and a side group designated "R." (*b*) Different amino acids are distinguished by their R groups. Twenty types of amino acids are important in life; three of them are shown here.

Table 3.6 Protein Diversity in the Human Body

Proteins	Function
Actin, myosin, dystrophin	Muscle contraction
Antibodies, antigens, cytokines	Immunity
Carbohydrases, lipases, proteases	Digestive enzymes
Casein	Milk protein
Collagen, elastin	Connective tissue
Colony stimulating factors	Blood cell formation
DNA polymerase	DNA replication
Ferritin	Iron transport
Fibrin, thrombin	Blood clotting
Growth factors	Cell division
Hemoglobin, myoglobin	Oxygen transport
Insulin, glucagon	Control of blood glucose level
Integrins, laminins	Cell adhesion
Keratin	Hair structure
Tubulin	Cell movements
Tumor suppressors	Prevent cancer

Chemistry Explains Biology

Spider Silk

Spider silk is the strongest natural fiber known. Spiders use silk to build webs to capture prey, to store prey, and to protect their eggs. The eight-legged animals rely on their silk-making ability for survival.

Silks are proteins, manufactured in sets of glands. A spider usually has several sets of glands, each of which produces a different type of silk. In addition to producing webs, a spider continually secretes a strand of silk called a dragline. Should danger threaten, the spider can temporarily escape the web on its dragline Tarzan-style.

Dragline silk is best studied in the golden orb weaver spider, *Nephila clavipes.* This especially strong and elastic silk consists of two types of proteins that are dry and practically indestructible once outside the animal's body. Dragline silk proteins include a few types of amino acids that repeat in short sequences. This imparts a conformation of coiled sheets, like the steps of a spiral staircase. This shape is similar to that of elastin, a connective tissue protein in the human body also known for great strength and resiliency. Spider silk is an elegant illustration of how a protein's structure determines its functions.

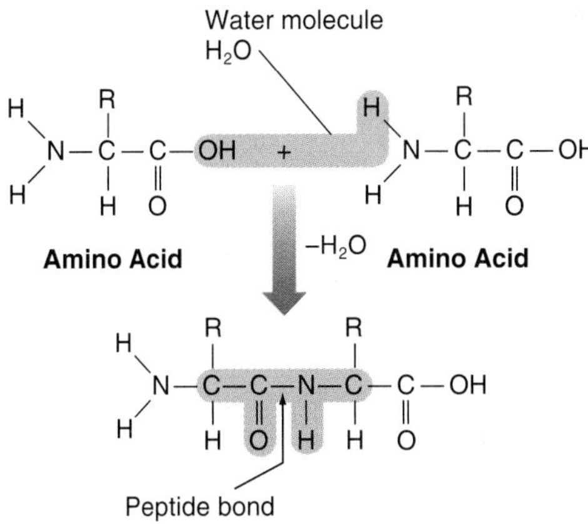

FIGURE 3.24

A peptide bond forms by dehydration synthesis. This reaction takes place when an OH from a carboxyl group of one amino acid combines with a hydrogen from the amino group of another amino acid, creating a water molecule and linking the carboxyl carbon of the first amino acid to the nitrogen of the other. (The amino acids shown here are in a non-ionized form.)

three a tripeptide; larger chains with fewer than 100 amino acids are peptides, and finally, those with 100 or more amino acids are polypeptides. Hydrolysis (adding water) breaks a protein into its constituent amino acids.

Protein Folding

As a protein is synthesized in a cell, it folds into a three-dimensional structure, or **conformation.** A protein takes its unique shape from interactions with other proteins and water molecules and from bonds that form between its atoms. A class of proteins called chaperones help guide the protein into its characteristic conformation. Thousands of water molecules typically surround a growing peptide, contorting it as its hydrophilic R groups move toward the water and its hydrophobic R groups move away from it. Hydrogen bonds between R groups also mold a protein's conformation.

Disulfide bonds that form between sulfur atoms bridge sulfur-containing parts of amino acid chains. Disulfide bonds are abundant in keratin, the protein that forms various protective coverings in vertebrates (animals with backbones), including hair, scales, beaks, wool, feathers, nails, claws, quills, horns, hooves, and tortoise shells (figs. 3.25 and 3.26). Figure 3.27 shows how four disulfide bonds tie a polypeptide into a more compact shape.

The conformation of a protein may be described at four levels (fig. 3.28). The amino acid sequence of a polypeptide chain determines its **primary (1°) structure.** Chemical attractions between amino acids that are close together in the 1° structure fold the polypeptide chain into its **secondary (2°) structure.** Common secondary structures form a few distinctive shapes, called motifs, that include barrels, loops, spirals, coils, sheets, and corkscrews. A single polypeptide may fold into several motifs. Figure 3.29 shows two common ones, alpha helices and beta-pleated sheets. The keratins depicted in figures 3.25 and 3.26 are alpha helices.

Secondary structures wind into **tertiary (3°) structures** as R groups and water interact. Finally, proteins consisting of more than one polypeptide have **quaternary (4°) structure.** For example, hemoglobin, the blood protein that carries oxygen, has four polypeptide chains. The liver protein ferritin has 20 identical polypeptides of 200 amino acids each. In contrast, the muscle protein myoglobin is a single polypeptide chain wound into a globular shape.

A protein's conformation (structure) determines its function. A digestive enzyme holds a large nutrient molecule, immobilizing it so that it can be broken into pieces. An antibody grabs a bacterium by a surface protein and displays it to the immune system. Muscle proteins form long, aligned fibers that slide past one another. This shortens their combined length, which, on a cellular level, produces contraction.

Enzymes

Among the most important of all biological molecules are enzymes. Enzymes are proteins that speed the rates of specific chemical reactions without being consumed in the process, a phenomenon called **catalysis.** Enzymes bind reactants and bring them in contact with each other, so that less energy is required for the

(a) (b) (c) (d)

FIGURE 3.25
A gallery of keratin-based structures. Alpha-keratin forms (*a*) the hair of a hare, (*b*) the beak of a bird, (*c*) the scales of a snake, and (*d*) the horn of a ram.

reaction to proceed. Without enzymes, many biochemical reactions would proceed far too slowly to support life; some enzymes increase reaction rates a billion times. Enzymes function under specific pH and temperature conditions.

The key to an enzyme's specificity lies in its **active site,** a region to which the reactants, or **substrates,** bind. A substrate fits into the active site of an enzyme as the enzyme contorts slightly around it, forming a short-lived **enzyme-substrate complex** (fig. 3.30). An enzyme can hold two substrate molecules that react to form one product molecule, or it can hold a single substrate that splits to yield two products. Then the complex breaks down to release the products (or product) of the reaction. The enzyme is unchanged, and its active site is empty and ready to pick up more substrate.

Nucleic Acids

Synthesizing a protein is a more complex task than synthesizing a carbohydrate or fat because of the great variability of amino acid sequences. How does an organism "know" how to synthesize a particular protein? A protein's amino acid sequence is encoded in a sequence of chemical units of another type of biochemical, a **nucleic acid.**

(a)

(b)

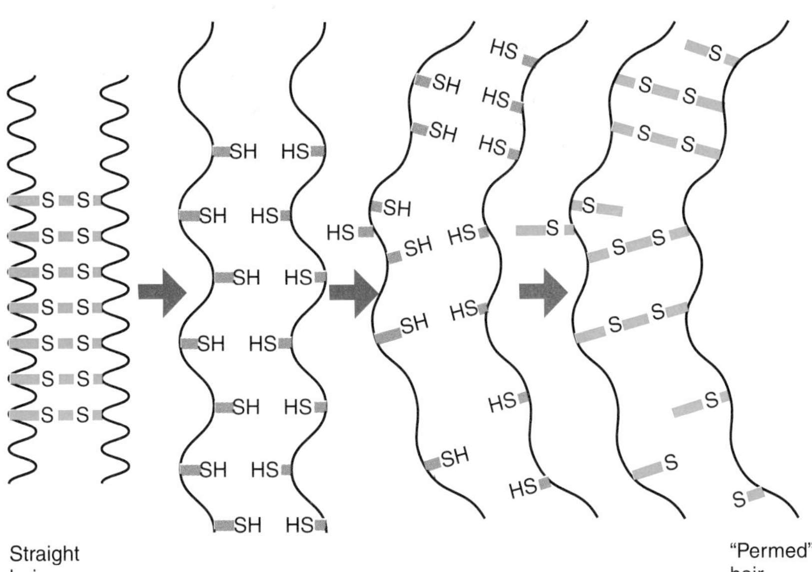

FIGURE 3.26

Hair consists of triplets of helices, each a keratin polypeptide. Disulfide bonds hold the helices together. In a permanent wave, applying certain chemicals blocks the sulfur atoms from interacting. Then the hair is wrapped around a mold, such as a plastic tube. The chemical is replaced with another type, and disulfide bonds re-form—but between different sulfur-bearing amino acids than before. The new, offset-bonding twists the hair. A curl is born. Actress Nicole Kidman with (*a*) straight hair; (*b*) after perm, with curly hair.

The nucleic acids **deoxyribonucleic acid** (DNA) and **ribonucleic acid** (RNA) are polymers of monomers called **nucleotides.** A nucleotide consists of a five-carbon sugar (deoxyribose in DNA and ribose in RNA), a phosphate group (PO_4), and one of five types of nitrogen-containing compounds called **nitrogenous bases.** The nitrogenous bases are **adenine** (A), **guanine** (G), **thymine** (T), **cytosine** (C), and **uracil** (U). DNA contains A, C, G, and T; RNA contains A, C, G, and U. The DNA polymer is a double helix and resembles a spiral staircase in which alternating sugars and phosphates form the rails and nitrogenous bases form the rungs.

Long sequences of DNA nucleotides provide information that RNA molecules use to guide assembly of amino acids into polypeptide chains. Each group of three DNA bases in a row specifies a particular amino acid, in a correspondence called the **genetic code.** A sequence of genetic code specifying a polypeptide is a **gene.** DNA, therefore, is the genetic material.

FIGURE 3.27

Proteins are composed of polypeptides, which are pɔlymers of amino acids. In a cow enzyme, bovine ribonuclease, part of the conformation derives from disulfide bonds that fold the amino acid chain into a compact shape. Disulfide bonds form between the sulfur-containing amino acid cysteine.

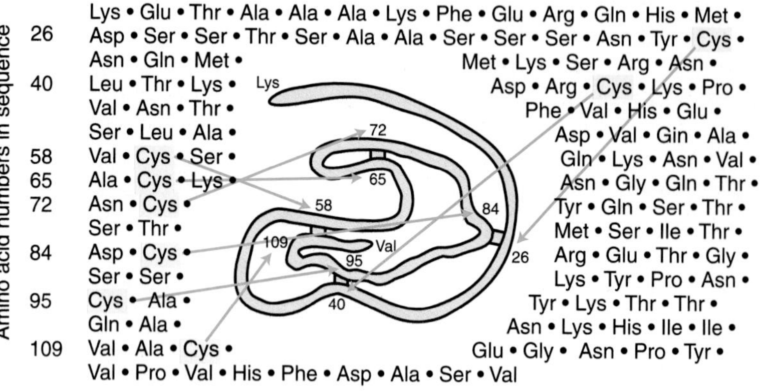

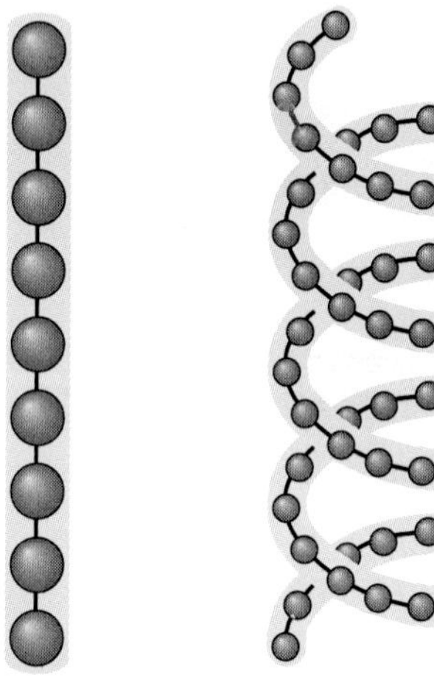

Primary structure

Secondary structure

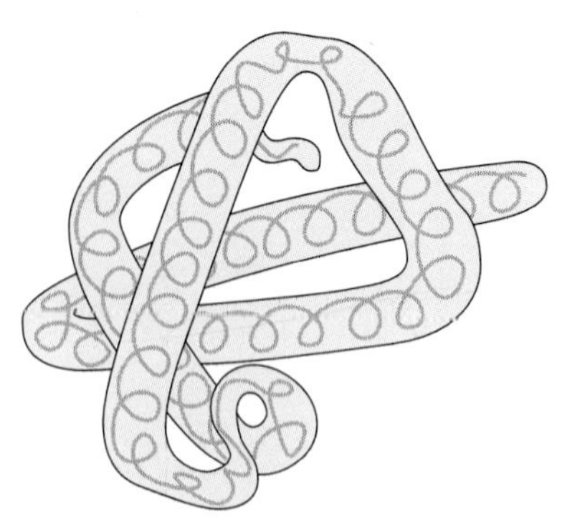

Tertiary structure

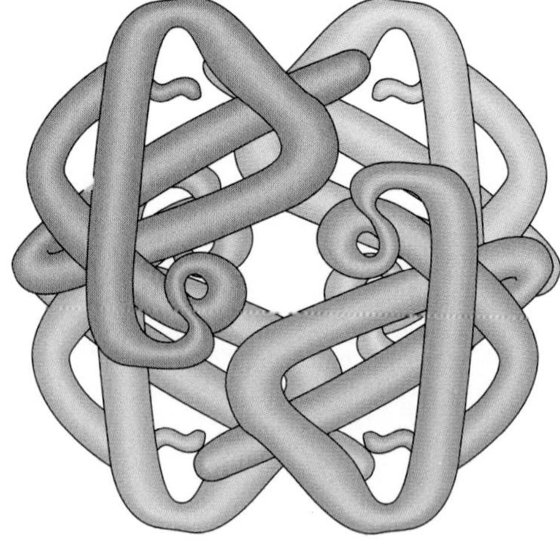

Quaternary structure

FIGURE 3.28

Protein conformation. A protein's primary structure is the amino acid sequences of its polypeptides. Secondary structures are barrels, loops, spirals, coils, sheets, or corkscrews created when amino acids close to each other in the primary structure attract. Attractions between amino acids farther apart in the primary structure fold the secondary structures into tertiary structures. A protein consisting of more t! an one polypeptide chain has a quaternary structure.

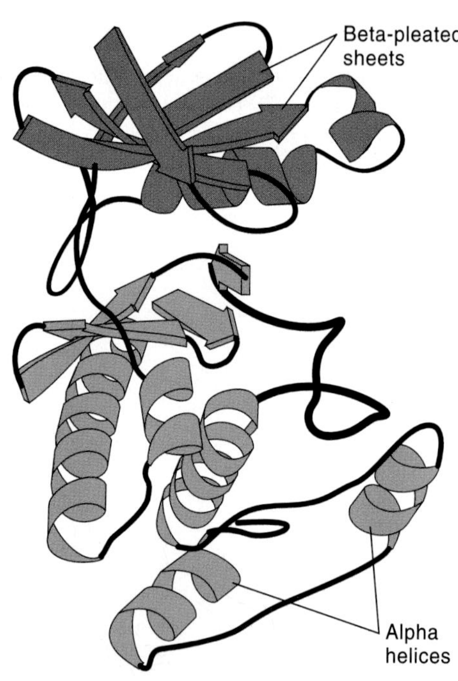

FIGURE 3.29

A depiction of an enzyme whose secondary structure includes beta-pleated sheets (purple) and alpha helices (pink).

Reprinted with permission from Daniel R. Knighton, et al, "Crystal structure of the catalytic subunit of cyclic adenosine monophosphate-dependent protein kinase," *Science* 153:407–414, 26 July 1991. Copyright 1991 American Association for the Advancement of Science.

DNA differs from other macromolecules of life in that it can duplicate itself exactly. In the rungs of the molecule, base pairs form only between A and T and between G and C. To replicate, the double helix splits down the middle and unwinds, separating each base pair. Each half of the pair then chemically attracts a free nucleotide with the base of the proper, or complementary, type. The final result is two new DNA molecules, each identical to the original (fig. 3.31). The ability of DNA to replicate provides a means for biological information, and life itself, to perpetuate.

Table 3.7 reviews the characteristics of the major types of organic molecules that make up organisms.

Mastering Concepts

1. What are the chemical compositions and different types of carbohydrates, lipids, proteins, and nucleic acids?
2. What are the functions of carbohydrates, lipids, proteins, and nucleic acids?
3. Why are proteins extremely varied in organisms?
4. What is the significance of a protein's conformation?
5. How do nucleic acids instruct cells to manufacture proteins?

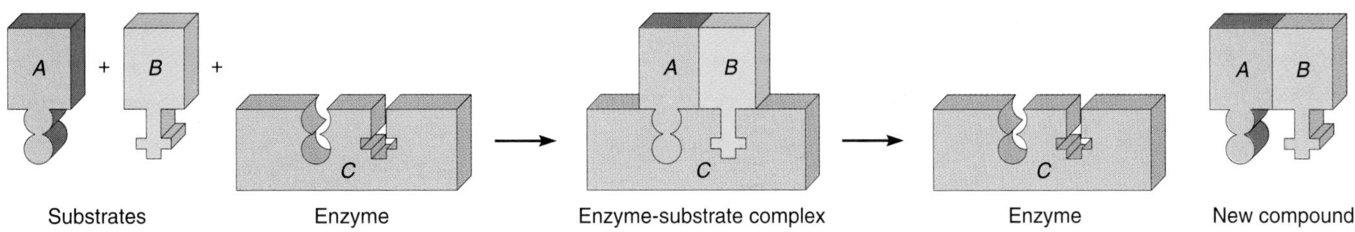

FIGURE 3.30

Enzyme action. In this highly schematic depiction, substrate molecules A and B fit into the active site of enzyme C. An enzyme-substrate complex forms as the active site moves slightly to accommodate its occupants. A new compound, AB, is released, and the enzyme is recycled. Enzyme-catalyzed reactions can break down as well as build up substrate molecules.

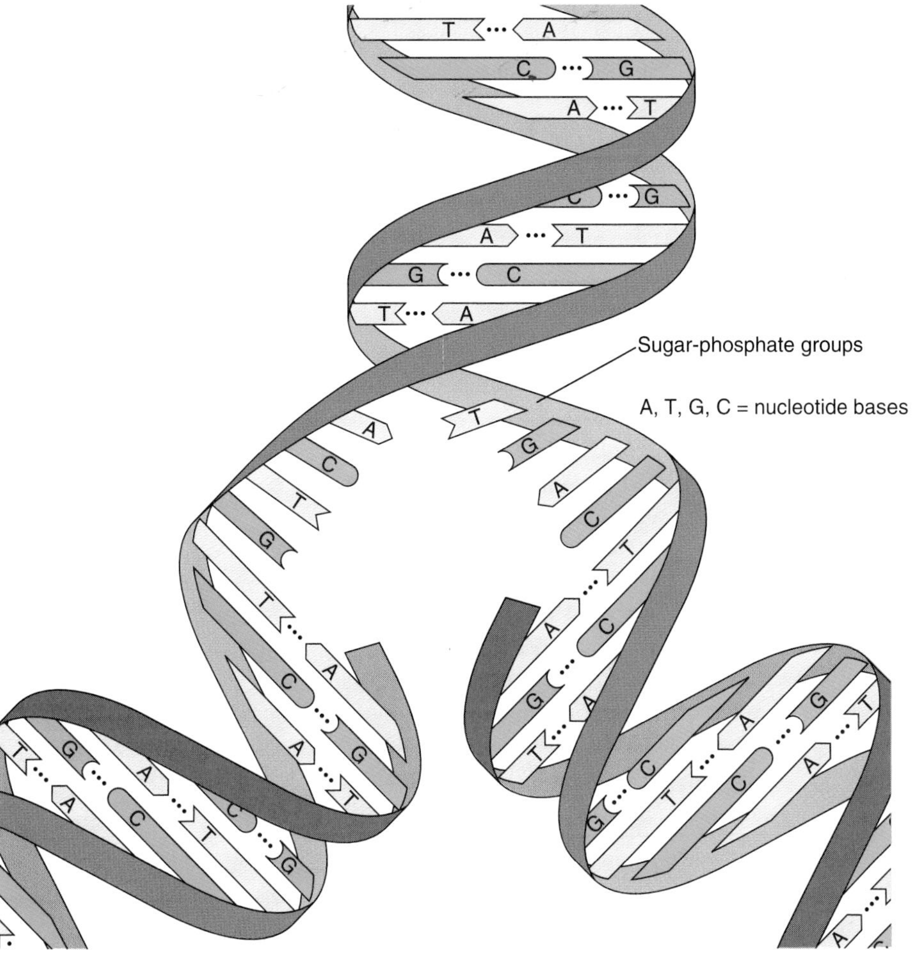

FIGURE 3.31

A DNA molecule resembles a spiral staircase, where the steps are specific pairs of nucleotide bases and the sides are chains of sugar-phosphate groups. The molecule twists into a double helix because the "rungs" are specific (complementary) base pairs: A–T (or T–A) and G–C (or C–G). The molecule replicates by locally unwinding so that the base pairs part. Each exposed base then chemically attracts a free complementary nucleotide, forming two identical DNA double helices from the original one. Hydrogen bonds hold the base pairs together.

The Mystery of the Rock Eaters~

continued from page 32

Clues to the SLiME bacteria's way of life came from the chemicals near them—notably methane, a swamp gas (CH_4). Bacteria called methanogens are known to use energy from hydrogen obtained from other organisms to use the carbon in carbon dioxide (CO_2) to build the organic molecules of life. Methanogens release methane. Deep in the volcanic rock was methane and abundant hydrogen gas (H_2)—but far too much hydrogen to be coming from organisms.

Just as the researchers were pondering the source of the hydrogen, an explosion gave them the answer. Near to where they were exploring, a lit welder's torch had touched a piece of basalt and set off an explosion that released hydrogen—from the rock! A chemical reaction between iron-containing compounds in the rock and groundwater, in the presence of the flame, released hydrogen gas. Then, the bacteria used the energy in the hydrogen atoms to use the carbon present in the environment. But this explanation was just a hypothesis.

Stevens and McKinley decided to test the hypothesis by re-creating the underground scenario to see if hydrogen from rocks could support life. First, they showed that crushed basalt mixed with water releases hydrogen gas. Next, they mixed water, basalt, and some of the bacteria and sealed them in a container for a year. After that time, they opened the container and found that the bacteria had multiplied! The combination of rocks and water could indeed provide chemical energy for life.

Chemistry explains many aspects of life—even a seeming impossibility, life subsisting on rocks.

Table 3.7 The Macromolecules of Life

Type of Molecule	Structure	Functions	Examples
Lipids	Hydrocarbon chains plus oxygen	Membranes Energy storage	Fats, oils, waxes
Carbohydrates	Sugars and polymers of sugars	Energy storage Support	Sugars, starches, chitin, glycogen
Proteins	Polymers of amino acids	Transport Muscle action Blood clotting Support Immunity	Hemoglobin Myosin, actin Thrombin, fibrin Collagen, elastin Antibodies
Nucleic acids	Polymers of nucleotides	Transfer of genetic information	DNA, RNA

SUMMARY

1. All life consists of chemicals. Chemicals affect our lives in many ways.

2. Organisms obtain energy from the environment and convert it to a usable form.

3. Matter can be broken down into pure substances called **elements.** Elements are organized in the **periodic table** according to their number of subatomic particles. **Bulk elements,** those essential to life in large quantities, include carbon, hydrogen, oxygen, nitrogen, sulfur, and phosphorus.

4. An **atom** is the smallest unit of an element. Subatomic particles include the positively charged **protons** and neutral **neutrons** that form the nucleus and the negatively charged, much smaller **electrons** that circle the nucleus.

5. Atomic number is an element's characteristic number of protons, and the **atomic weight** is the weight of its protons and neutrons. **Isotopes** of an element differ by the number of neutrons.

6. Compounds are built of bonded atoms of different elements, in a consistent ratio. A **molecule** is the smallest unit of a compound that retains the characteristics of that compound. A compound's characteristics differ from those of its constituent elements.

7. Chemical shorthand indicates the numbers of atoms and molecules in a compound. In a **chemical reaction,** different compounds are broken down and form, but the total number of atoms of each element remains the same.

8. Electrons move constantly; they occupy volumes of space called **electron orbitals,** which form levels of energy called **energy shells.**

9. An atom's tendency to fill its outermost shell with **valence electrons** drives atoms to bond and form molecules.

10. Covalent bonds share electrons and form between atoms that have three, four, or five valence electrons. Carbon atoms form up to four covalent bonds and build **hydrocarbons.** A **nonpolar covalent bond** shares all electrons equally. In a **polar covalent bond,** the nucleus of one type of atom attracts electrons more than the nucleus of another type of atom, resulting in opposite charges in different parts of the molecule.

11. In an **ionic bond,** an atom with one, two, or three valence electrons loses them to an atom with seven, six, or five valence electrons, respectively; the resulting bonds form a lattice of positively and negatively charged ions. Ionic bonds in organisms are weak.

12. Hydrogen bonds form when a hydrogen in one molecule is drawn to part of a neighboring molecule because of unequal electrical charge distribution.

13. van der Waals attractions occur between parts of molecules that are temporarily oppositely charged.

14. Most biochemical reactions occur in an aqueous environment. Water is **cohesive** and **adhesive,** enabling many substances to dissolve in it. Water helps regulate temperature in organisms because of its high **heat capacity** and high **heat of vaporization. pH** is a measure of H^+ concentration, or how acidic or basic a solution is. Water has a pH of 7, which means that the numbers of H^+ and OH^- in water are equal.

15. An **acid** adds H^+ to a solution, lowering the pH below 7. A **base** adds OH^-, raising pH to between 7 and 14. **Buffer systems** consisting of weak acid-and-base pairs maintain the pH ranges of body fluids.

16. The hydrogen bonds that keep water fluid are constantly forming and breaking. Water is a solvent for salts and for polar covalent bonds. Freezing water expands and becomes less dense.

17. Carbohydrates provide energy and support. They consist of carbon, hydrogen, and oxygen in the ratio 1:2:1. **Monosaccharides** contain three to seven carbons and differ from each other by how their atoms are bonded. **Oligosaccharides** consist of 2 to 100 monomers and include the **disaccharides.** Two monosaccharides combine to form a disaccharide by **dehydration synthesis. Polysaccharides** are long polymers of monosaccharides (sugars) linked by dehydration synthesis.

18. Lipids are a diverse group of organic compounds that provide energy, slow digestion, waterproof the outsides of organisms, cushion organs, and preserve body heat. Lipids include fats and oils, do not dissolve in water, and contain carbon, hydrogen, and oxygen but have less oxygen than carbohydrates. **Triglycerides** consist of **glycerol** and three **fatty acids,** which may be **saturated** (no double bonds), **unsaturated** (at least one double bond), or **polyunsaturated** (more than one double bond). Double bonds make a lipid oily at room temperature, whereas saturated fats are more solid. Unsaturated fats and omega-3 fatty acids are more healthful to eat than are saturated fats and omega-6 fatty acids. **Sterols** are lipids containing four carbon rings.

19. Proteins have many functions and a great diversity of structures. They consist of 20 types of **amino acids,** each of which includes a hydrogen, an **amino group,** an acid or **carboxyl group,** and an **R group** bonded to a central carbon. Amino acids join by forming **peptide bonds** through dehydration synthesis. A protein's **conformation,** or three-dimensional shape, is vital to its function and is determined by the amino acid sequence **(primary structure)** and interactions between amino acids close together **(secondary structure)** and far apart **(tertiary structure)** in the sequence. Water molecules and chaperone proteins also contribute to conformation. A protein with more than one polypeptide has a quaternary structure.

20. Nucleic acid sequences encode amino acid sequences. DNA and RNA are polymers consisting of a sugar-phosphate backbone and sequences of **nitrogenous bases.** DNA includes deoxyribose and the bases **adenine, cytosine, guanine,** and **thymine.** RNA contains ribose and has uracil instead of thymine. A **nucleotide,** a nucleic acid monomer, consists of a phosphate, a base, and a sugar. DNA can replicate. DNA carries genetic information. RNA translates the information to enable the cell to synthesize proteins.

TO REVIEW . . .

1. Define the following terms:
- **a.** atom
- **b.** element
- **c.** molecule
- **d.** compound
- **e.** isotope
- **f.** ion

2. Arrange the terms in the following lists by size, from smallest to largest:
- **a.** proton, electron, molecule, atom
- **b.** fructose, cellulose, sucrose
- **c.** protein, amino acid, amino group, dipeptide
- **d.** hydrogen, zinc, carbon, oxygen, sulfur, nitrogen

3. The vitamin biotin contains 10 atoms of carbon, 16 of hydrogen, 3 of oxygen, 2 of nitrogen, and 1 of sulfur. What is its molecular formula?

4. Describe how each of these chemical attractions forms:
- **a.** covalent bond
- **b.** ionic bond
- **c.** hydrogen bond
- **d.** disulfide bond
- **e.** van der Waals attraction
- **f.** peptide bond

5. Cite three illustrations of the concept of emergent properties (review chapter 1 if necessary) that appear in this chapter.

6. List four characteristics of water that make it essential to life.

7. The following diagram shows the chemical structure of the amino acid threonine. Label the amino group, the carboxyl group, and the R group.

```
             CH3
              |
          H — C — OH
              |
      +H3N — C — C — O-
              |   ||
              H   O
```

8. What do a protein and a nucleic acid have in common? What are two ways that a protein differs from a complex carbohydrate and a lipid? What do all three types of macromolecules have in common?

9. Diagram the way a dipeptide bond would form between alanine and cysteine.

10. Outline the steps of the scientific method used to implicate NO and CO as biological messenger molecules.

TO THINK ABOUT . . .

1. If a shampoo is labeled "nonalkaline," would it more likely have a pH of 3, 7, or 12?

2. A topping for ice cream contains fructose, hydrogenated soybean oil, salt, and cellulose. What types of chemicals are in it?

3. An advertisement for a supposedly healthful cookie claims that it contains an "organic carbohydrate." Why is this statement silly?

4. At a restaurant, a server suggests a sparkling carbonated beverage, claiming that it contains no carbohydrates. The label on the product lists water and fructose as ingredients. Is the server correct?

5. A Horta is a fictional animal whose biochemistry is based on the element silicon. Consult the periodic table. Is silicon a likely substitute for carbon in a life-form? Cite a reason for your answer.

6. If water is left in a covered outdoor swimming pool over a very cold winter, the center of the cover bulges up. What characteristic of water makes this happen?

7. Amyotrophic lateral sclerosis, also known as ALS or Lou Gehrig's disease, paralyzes muscles. An inherited form of the illness is caused by a gene (sequence of DNA) that encodes an abnormal enzyme that contains zinc and copper. The abnormal enzyme fails to rid the body of a toxic form of oxygen. Which of the molecules mentioned in this description is a

- **a.** protein?
- **b.** nucleic acid?
- **c.** bulk element?
- **d.** trace element?

8. A man on a very low-fat diet proclaims to his friend, "I'm going to get my cholesterol down to zero!" Why is this an impossible (and undesirable) goal?

TO LEARN MORE . . .

References and Resources

Fredrickson, James K., and Tullis C. Onstott. October 1996. Microbes deep inside the earth. *Scientific American.* Microbes live inside rocks!

Lewis, Ricki. October 1996. Unraveling the weave of spider silk. *BioScience.* Spider silk has a number of uses—with a little help from genetic engineering.

Rose, George D. January/February 1996. No assembly required. *The Sciences.* A protein must fold properly to function.

Snyder, Solomon H. September 1996. No NO prevents parkinsonism. *Nature Medicine,* vol. 2. Excess NO is linked to Parkinson disease.

Williams, Robert J. P. 1991. The chemical elements of life. *Journal of Chemical Society, Dalton Transactions.* A detailed examination of why and how 24 elements are part of life.

Denotes articles by author Ricki Lewis that may be accessed via *Life*'s home page /http://www.mhhe.com/sciencemath/biology/life/.

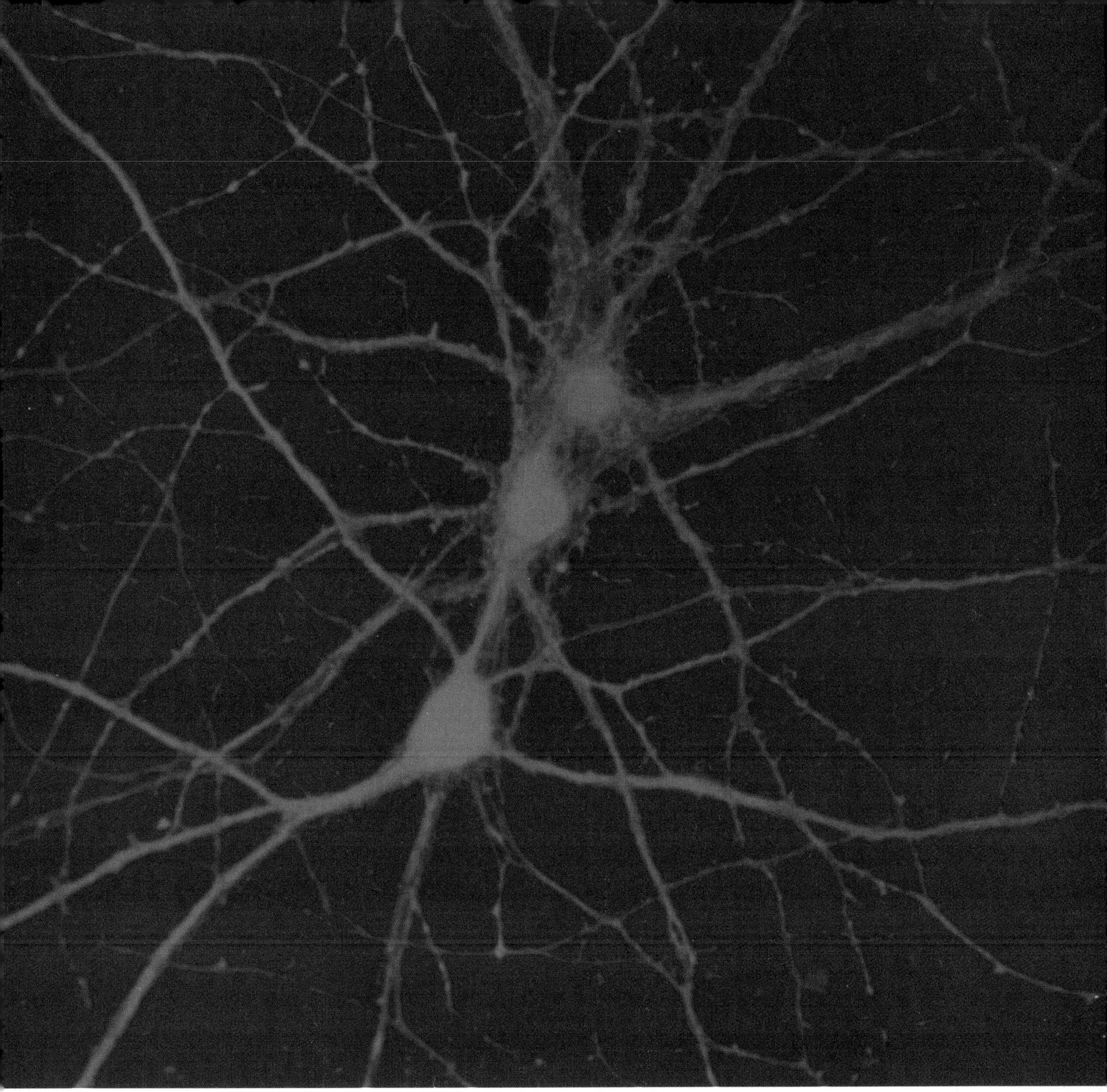

Unit 2

Cell Biology

Cells are worlds within themselves, exhibiting the characteristics of life. Also vitally important to a multicellular organism are interactions between cells, such as these communicating rat nerve cells.

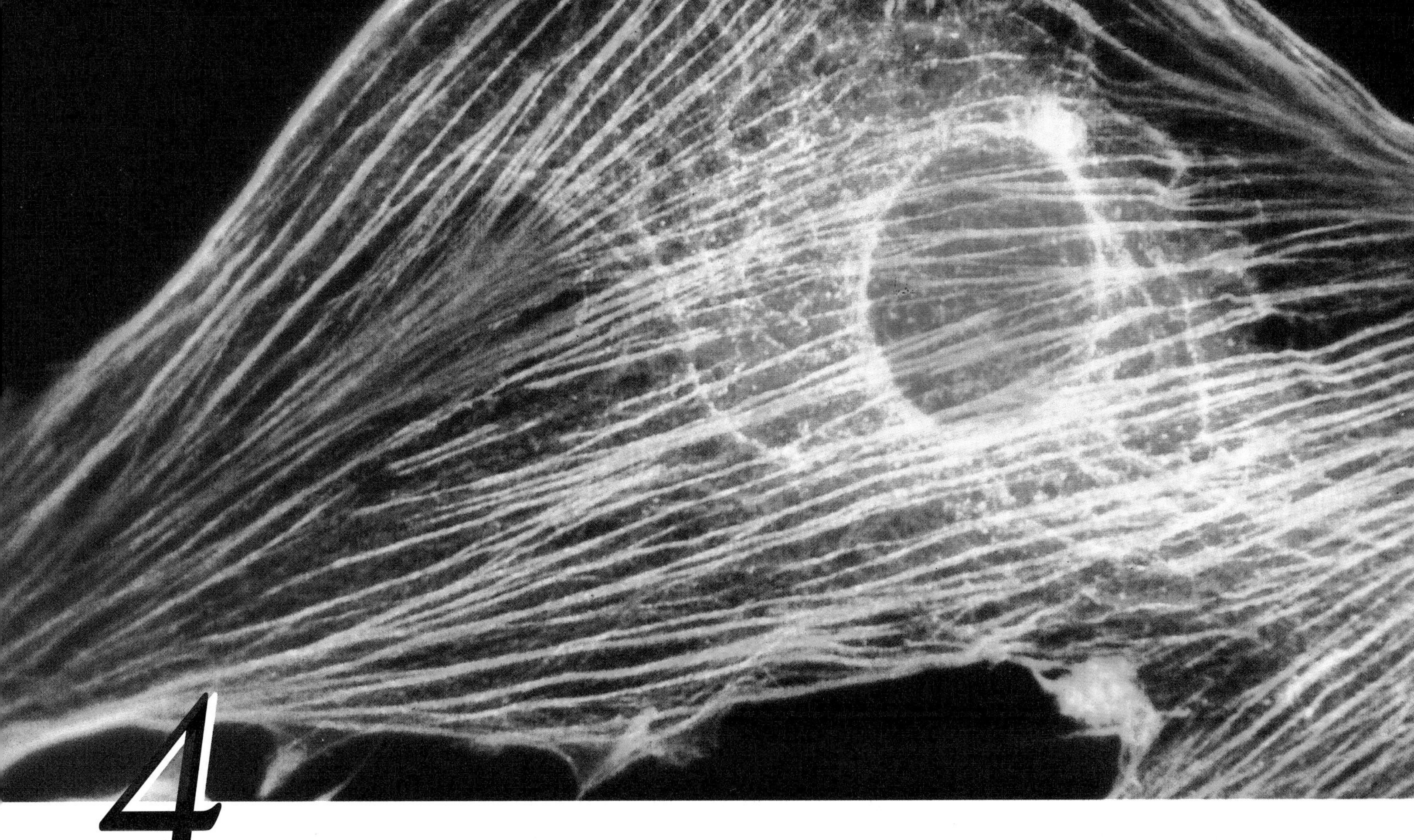

4 Cells

Cell biologists use a variety of techniques to highlight particular structures within cells, which otherwise are transparent. This is a mouse fibroblast, a type of connective tissue cell. The blue oval is the nucleus, which contains DNA. The threads are proteins (tubulin and actin) that make up part of the cell's inner framework, or cytoskeleton.

chapter outline

Cells—The Units of Life 60

The First Microscopes

The Cell Theory Emerges

Technology Improves

Cell Complexity 63

Cell Size

Prokaryotic Cells

Eukaryotic Cells

Archaean Cells

Organelles 69

Making Milk

Other Organelles and Structures

Simpler Than Cells 75

Viruses

Viroids

Prions

The Origin of Eukaryotic Cells 79

Cancer Cells— A Personal Reflection

I never thought I would care very much about the cells composing my thyroid gland. All that changed on August 4, 1993. On that day, a physician-friend, looking at me from across a room, said, "What's that lump in your neck?" And so began my medical journey.

I wasn't very worried, even when a specialist stuck seven thin needles into my neck to sample thyroid cells for testing. Each doctor I'd seen assured me that thyroid tumors are benign (noncancerous) 99% of the time. Still, I consulted my anatomy books to ponder my thyroid. I learned that cancer affecting the outermost cells of the butterfly-shaped gland was nearly always treatable, but cancer affecting cells in the gland's interior, near the blood supply, was far more dangerous.

The doctor's phone call came early on a Monday morning. Knowing that doctors don't like to deliver bad news on a Friday, I was panic struck. I knew I had become a statistic.

to be continued . . .

Cells—The Units of Life

A human, a hyacinth, a mushroom, and a bacterium appear to have little in common other than being alive. However, on a microscopic level, these organisms are astoundingly similar. All organisms consist of microscopic structures called **cells.** Within cells, highly coordinated biochemical activities carry on the basic functions of life, as well as, in some cases, specialized functions (fig. 4.1). The next two chapters introduce the cell, and the chapters that follow delve into specific cellular events.

Cells, as the basic units of life, exhibit all of the characteristics of life. A cell requires energy, genetic information to direct biochemical activities, and structures to carry out these activities. Movement occurs within living cells, and some cells, such as the sperm cell, can move about in the environment. All cells have some structures in common that allow them to reproduce, grow, respond to stimuli, and obtain energy and convert it to a usable form. In short, the functions of cells are similar to the functions of whole organisms.

A cell consists of living matter, or **protoplasm,** bounded by an outer membrane. **Unicellular** organisms such as bacteria consist of a single cell. **Multicellular** organisms, such as ourselves, consist of many cells. Structures called **organelles** (little organs) within the cells of multicellular organisms and within some of the more complex unicellular organisms carry out specific functions. In multicellular organisms, different numbers of particular types of organelles endow some cells with specialized functions, such as support, contraction, or message transmission. For example, an active muscle cell contains many more organelles that enable it to use energy than does an adipose cell, which is little more than a blob of fat.

Our knowledge of the structures inside cells depends upon technology, because most cells are too small for the unaided human eye to see. Today, cell biologists use sophisticated microscopes to greatly magnify cell contents and many types of stains to color otherwise transparent cellular structures.

The First Microscopes

The ability to make objects appear larger probably dates back to ancient times, when people noticed that pieces of glass or very smooth, clear pebbles could magnify small or distant objects. By the thirteenth century, the ability of such "lenses" to aid people with poor vision was widely recognized in the Western world.

Three centuries later, people began using paired lenses. Many sources trace the origin of a double-lens compound microscope to Dutch spectacle makers Johann and Zacharius Janssen. Reports claim that their children were unwittingly responsible for this important discovery. One day in 1590, a Janssen youngster was playing with two lenses, stacking them and looking through them at distant objects. Suddenly he screamed—the church spire looked as if it was coming toward him! The elder Janssens quickly looked through both pieces of glass, and the faraway spire indeed looked as if it were approaching. One lens had magnified the spire, and the other lens had further enlarged the magnified image. This observation led the Janssens to invent the first compound optical device, a telescope. Soon, similar double-lens systems were constructed to focus on objects too small for the unaided human eye to see. The compound microscope was born.

The study of cells—cell biology—began in 1660, when English physicist Robert Hooke melted strands of spun glass to create lenses that he focused on bee stingers, fish scales, fly legs, feathers, and any type of insect he could hold still. When he looked at cork, which is bark from a type of oak tree, it appeared to be divided into little boxes, which were left by cells that were once alive. Hooke called these units "cells," because they looked like the cubicles (cellae) where monks studied and prayed. Although Hooke did not realize the significance of his observation, he was the first person to see the outlines of cells.

In 1673, lenses were improved again, at the hands of Antonie van Leeuwenhoek of Holland. Leeuwenhoek used only a single lens, but it was more effective at magnifying and produced a clearer image than most two-lens microscopes then available. One of his first objects of study was tartar scraped from his own teeth, and his words best describe what he saw there:

> *To my great surprise, I found that it contained many very small animalcules, the motions of which were very pleasing to behold. The motion of these little creatures, one among another, may be likened to that of a great number of gnats or flies disporting in the air.*

Leeuwenhoek opened up a vast new world to the human eye and mind (fig. 4.2). He viewed bacteria and protozoa that people hadn't known existed. However, he failed to see the single-celled

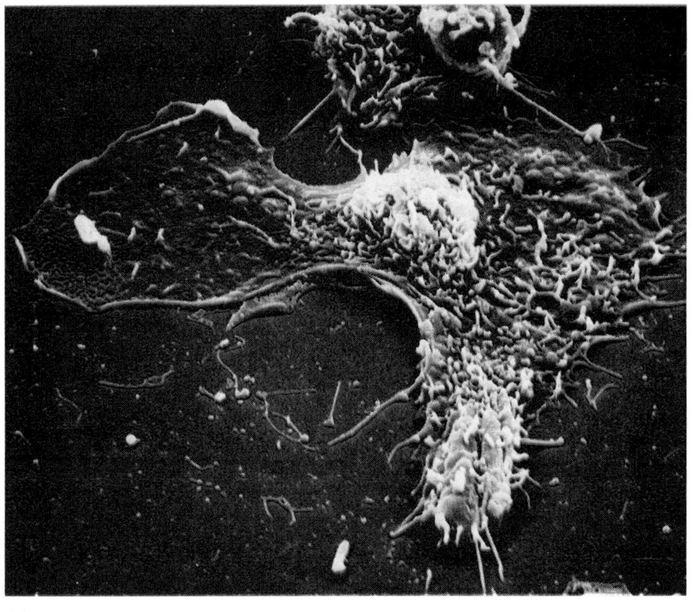

(a)

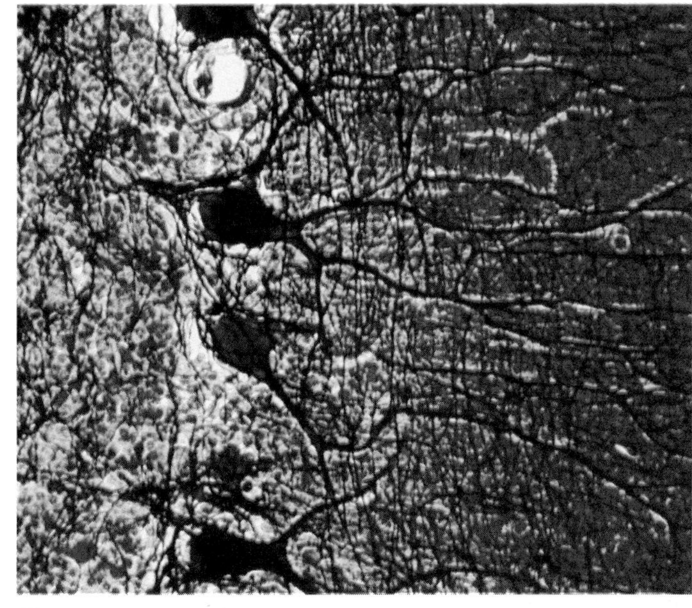

(b)

FIGURE 4.1

Specialized cells. (*a*) A macrophage is a giant cell of the immune system that travels through the human body detecting "foreign" substances and microorganisms. It sends out extensions called pseudopods that capture and engulf its target, which is then drawn into the cell and destroyed by enzymes. (×2,000) (*b*) Billions of nerve cells like these interconnect within the human brain. Note the roundish cell bodies and long extensions. (×400) (*c*) The green discs in these leaf cells of the common water plant *Elodea* are chloroplasts. Chloroplasts contain structures and biochemicals that permit the cell to utilize energy from sunlight to synthesize nutrient molecules. (×250)

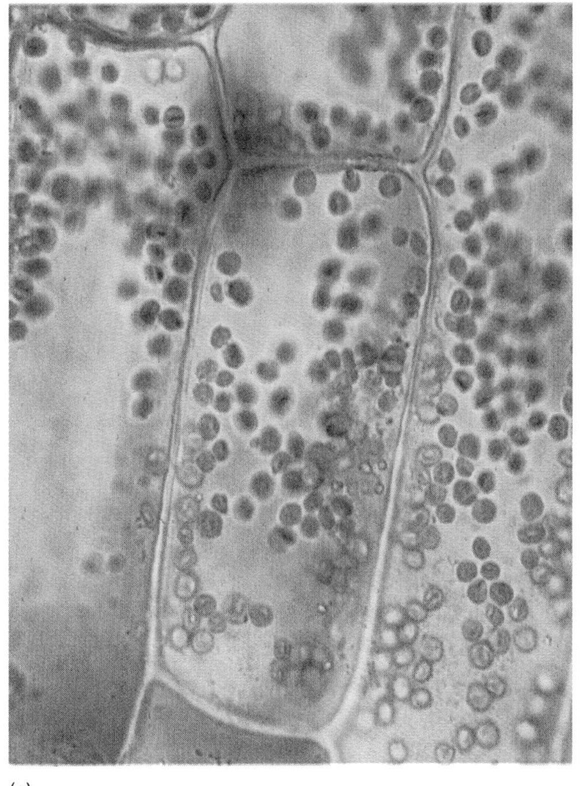

(c)

"animalcules" reproduce, and therefore he perpetuated the popular idea at the time that life arises from the nonliving or from nothing. Nevertheless, Leeuwenhoek did describe with remarkable accuracy microorganisms and microscopic parts of larger organisms.

The Cell Theory Emerges

Despite the accumulation of microscopists' drawings of cells during the seventeenth and eighteenth centuries, the **cell theory**—the idea that the cell is the fundamental unit of all life—did not emerge until the nineteenth century. Historians attribute this delay in the development of the cell theory to poor technology, including crude microscopes and lack of procedures to preserve and study living cells without damaging them. Neither the evidence itself nor early interpretations of it suggested that all organisms were composed of cells. Hooke had not observed actual cells but rather their absence. Leeuwenhoek made important observations, but he did not systematically describe or categorize the structures that cells had in common.

In the nineteenth century more powerful microscopes, with better magnification and illumination, revealed details of life at the subcellular level. In the early and mid-1830s, observers saw a darkened area, the **nucleus,** in plant cells and then in animal cells and the translucent, moving **cytoplasm,** which made up the rest of the cell.

In 1839, German biologists Matthias J. Schleiden and Theodor Schwann put together many observations made with microscopes to develop the cell theory. Schleiden first noted that cells were the basic units of plants, and then Schwann compared animal cells to plant cells. After observing many different plant and animal cells, Schleiden and Schwann concluded that cells were "elementary particles of organisms, the unit of structure and function." They described the components of the cell as a cell body and nucleus contained within a surrounding membrane. Schleiden called a cell

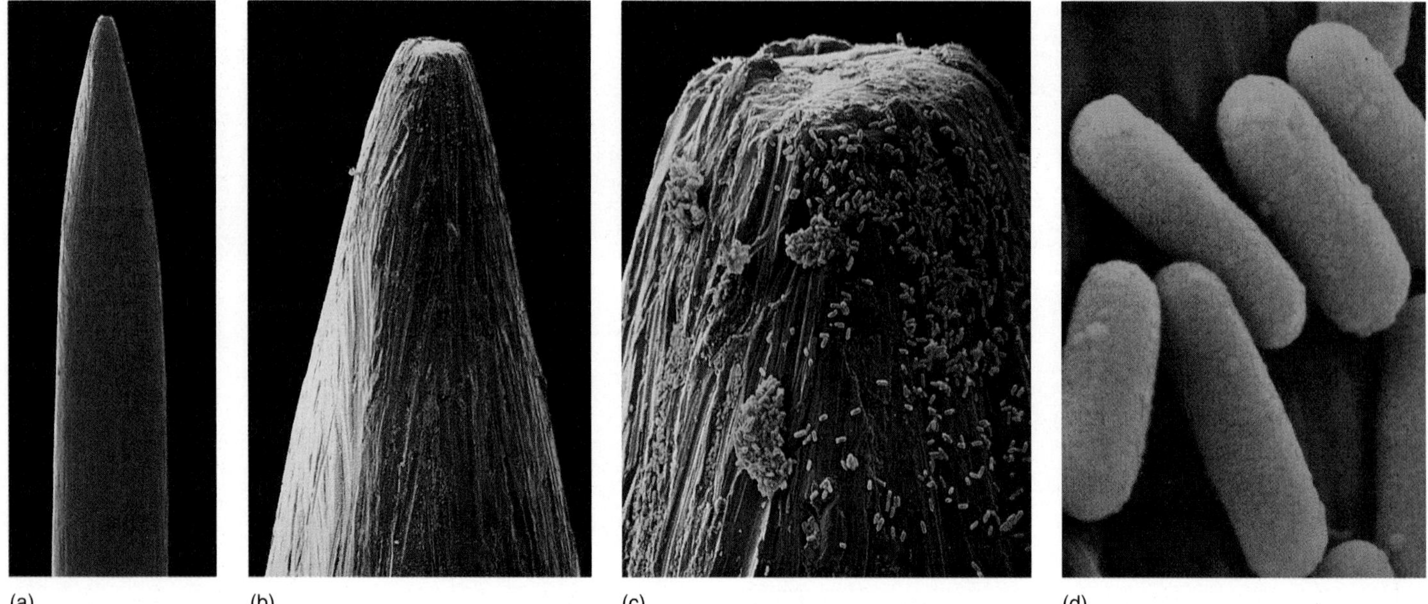

FIGURE 4.2

The living world beyond our vision. (*a*) To the naked eye, this pin does not appear to be a likely site for bacterial growth. However, when the pin is examined under the scanning electron microscope at increasing magnifications (*b, c,* and *d*), rod-shaped bacteria are apparent. (Magnifications: (*a*) ×7; (*b*) ×35; (*c*) ×178; (*d*) ×4,375.)

a "peculiar little organism" and realized that a cell can be a living entity on its own; but the new theory also recognized that in plants and animals, cells are part of a larger living organism.

Many cell biologists extended Schleiden and Schwann's observations and ideas. German physiologist Rudolph Virchow added the important corollary in 1855 that all cells come from preexisting cells, contradicting the still-popular theory of life arising from the nonliving or from nothingness. Virchow's statement also challenged another popular concept, that cells develop on their own from the inside out, the nucleus somehow forming a cell body around itself, and then the cell body growing a cell membrane, like a maturing fruit. Virchow's observation set the stage for descriptions of cell division in the 1870s and 1880s (chapter 9).

Virchow's thinking was ahead of his time, because he hypothesized the idea that diseases affecting the whole body result from abnormal cells. Today, many new treatments for diverse disorders are based on knowledge of the disease process at the cellular level.

Technology Improves

During the second half of the nineteenth century, heightened activity in the field of microscopy revealed many more details of the structures within cells. At this time, researchers used a **light microscope,** which sends visible light through a sample and focuses it through a glass lens. By the twentieth century, subcellular function and the microscopic aspects of disease were increasingly under study. It became clear that researchers needed a device that could visualize structures even smaller than those seen under the light microscope. In the 1940s, the invention of the **electron microscope** met this need.

An electron microscope sends a beam of electrons through a specimen and uses a magnetic field to focus the beam. An electron microscope offers greater resolution (the ability to distinguish two objects as being separate), greater magnification, and better depth than an image in a light microscope. A transmission electron microscope provides flat images, whereas a scanning electron microscope's images have a three-dimensional quality.

The 1980s brought the **scanning probe microscope,** which reveals three-dimensional views of cell surfaces and biological molecules such as DNA. This type of microscope passes a probe over a sample, which registers the hills and valleys of the surface. A computer converts this information into an image. A scanning probe microscope is unique in that it provides images at the atomic level, magnifying objects up to 100 million times. It can image specimens in water, which is similar to the environment in an organism. Figure 4.3 depicts the range of sizes of objects each of the three major types of microscopes can detect. Figure 4.4 shows how differently blood cells appear using light, electron, and scanning probe microscopes.

The existence of cells is an undisputed fact, despite the term "cell theory," because we can easily observe them. Yet the cell theory is still evolving. Until just a few years ago, scientists viewed a complex cell as a structure containing a nucleus and jellylike cytoplasm, in which organelles were suspended in no particular organization. Researchers are now learning that organelles have precise locations in cells. Rather than drifting about randomly, organelles are transported along tracks in the cell like cars along a roller coaster track, powered by 20 or so different types of "motor molecules." An organelle is likely to be located near other structures with which it interacts. So although the cell theory is still considered to be a product of the nineteenth century, we are

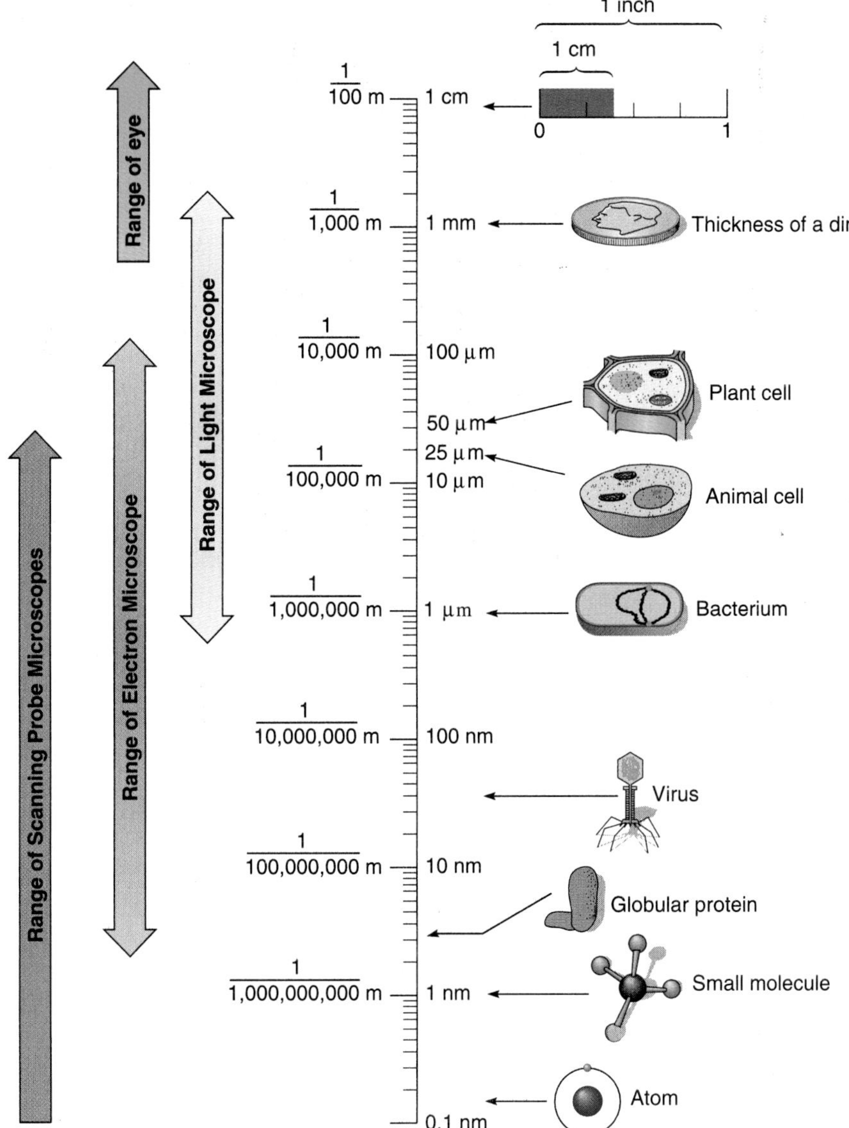

FIGURE 4.3

Ranges of the light, electron, and scanning probe microscopes. Biologists use the metric system to measure size. The basic unit of length is the meter (m), which equals 39.37 inches (slightly more than a yard). Smaller metric units measure many chemical and biological structures. A centimeter (cm) is 0.01 meter (about 2/5 of an inch); a millimeter (mm) is 0.001 meter; a micrometer (μm) is 0.000001 meter; a nanometer (nm) is 0.000000001 meter; an angstrom unit (Å) is 1/10 of a nanometer. Although each segment of the scale is depicted equally here, each segment actually represents only 1/10 of the length of the segment beneath it. The sizes of some chemical and biological structures are indicated next to the scale.

constantly learning about the organization within cells and about how cells interact to build larger-scale biological organization. We still have much to learn about the organization and functions within cells.

Mastering Concepts

1. What is a cell?
2. What types of structures are within complex cells?
3. Distinguish between unicellular and multicellular organisms.
4. How were microscopes developed?
5. How did Robert Hooke and Antonie van Leeuwenhoek contribute to the beginnings of cell biology?
6. What is the cell theory?
7. Distinguish between the light, electron, and scanning probe microscopes.
8. How is the cell theory still developing?

Cell Complexity

Cells were long thought to be of two fundamental types. **Prokaryotic cells** lack the nuclei and other complex organelles of **eukaryotic cells.** Organisms consisting of prokaryotic cells are called prokaryotes, and organisms consisting of eukaryotic cells are called eukaryotes. The Greek word *karyon* means "kernel," or a little less literally, "nucleus." *Eukaryote* means "true nucleus," and *prokaryote* means "before a nucleus existed." Although simpler, prokaryotes have obviously met the challenges of life. Prokaryotes were the earliest cells to have left fossil evidence, and today they make up the majority of living cells on earth.

The distinctions between prokaryotes and eukaryotes are not as clear-cut as cell biologists once thought. Cells called **archaea** also exist. These unicellular organisms are neither prokaryotic nor eukaryotic but share characteristics with each. Our knowledge of the archaea lags far behind that of the other two types of cells.

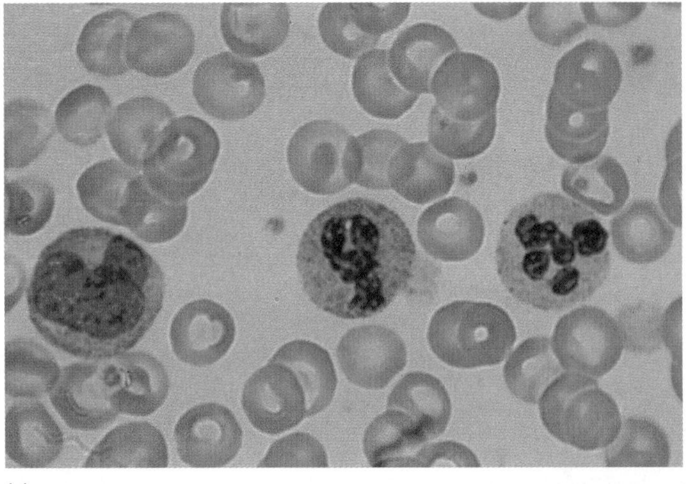

(a)

(b)

FIGURE 4.4

Three views of red blood cells. (*a*) A "smear" of blood visualized under a light microscope appears flat. (*b*) The disc-shaped red blood cells appear with much greater depth with the scanning electron microscope, but material must be killed and fixed in place before viewing, thus limiting applications. (*c*) The scanning probe microscope captures images of cells and molecules in their natural state, as parts of living organisms.

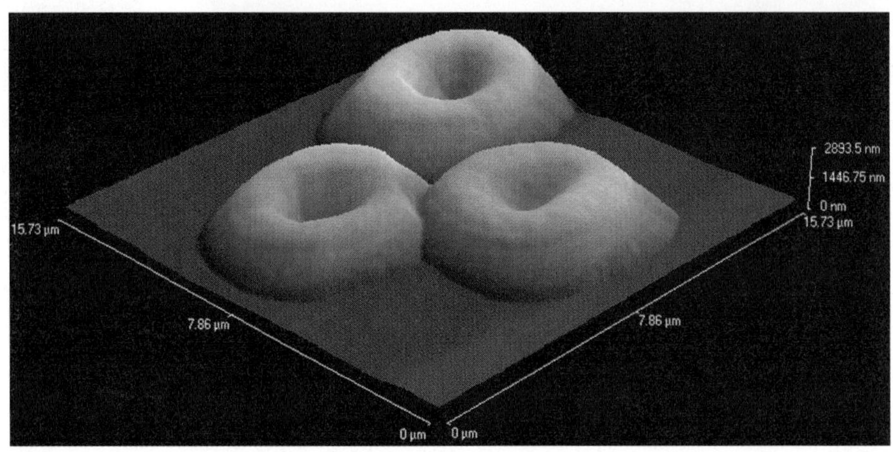

(c)

Cell Size

Cells require relatively large surface areas through which they interact with the environment. Nutrients, water, oxygen, carbon dioxide, and waste products must enter or leave a cell through its surfaces. As a cell grows, its volume increases at a faster rate than its surface area, a phenomenon you can observe with simple calculations (fig. 4.5). Put another way, much of the interior of a large cell is far away from the cell's surface.

The ultimate inability of a cell's surface area to keep pace with its volume can limit a cell's size, and this may be why prokaryotic cells are usually small. Another adaptation to maximize surface area is for the cell membrane to fold, just as inlets and capes extend the perimeter of a shoreline.

Most eukaryotic cells are 100 to 1,000 times the volume of typical prokaryotic cells (fig. 4.6); however, there are a few notable exceptions. A tiny single-celled eukaryote, *Nanochlorum eukaryotum,* is smaller than some bacteria, which are prokaryotes. A giant among bacteria, *Epulopiscium fishelsoni,* is more than half a millimeter long, which is a million times larger than the commonly studied bacterium *Escherichia coli.* It is nearly visible to the human eye! The giant bacterium lives inside a type of brown surgeonfish in the Red Sea. Researchers at first thought it was a eukaryote because of its size, but the electron microscope revealed the hallmarks of a prokaryote—no nucleus or complex inner structures.

Prokaryotic Cells

All prokaryotes lack true nuclei. These organisms include the familiar bacteria, as well as the **cyanobacteria,** organisms once known as blue-green algae because of their characteristic pigments and their similarity to the true (eukaryotic) algae. Figure 4.7 shows the major features of prokaryotic cells, and figure 4.8 shows a few prokaryotes. Bacteria cause many illnesses, but they are also very valuable in food and beverage processing and pharmaceutical production.

Most prokaryotic cells are surrounded by rigid **cell walls** that consist of peptidoglycans, which are molecules consisting of amino acid chains and carbohydrates. Polysaccharides on the cell wall protect the cell or enable it to attach to specific types of surfaces. Many antibiotic drugs halt bacterial infection by interfering with the microorganism's ability to build its cell wall.

Microbiologists classify prokaryotes by shape, cell wall structure, and biochemical characteristics. Prokaryotic cells may be

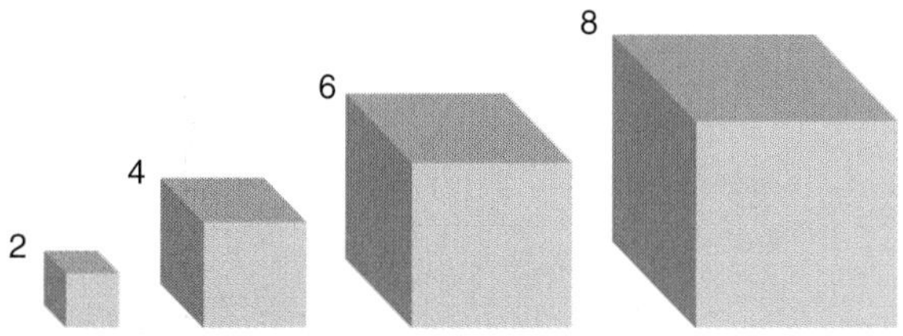

	2	4	6	8
Surface area (square inches)	24	96	216	384
Volume (cubic inches)	8	64	216	512
Surface area/volume	3.0	1.5	1.0	0.75

FIGURE 4.5

The important relationship between surface area and volume. As a cell enlarges, the amount of material inside it (its volume) increases faster than the area of its surface. The surface of a cell is vital to its functioning because communication and exchange of molecules and ions between the cell's interior and the outer environment takes place through the surface. The chemical reactions of life occur more readily when surface area is maximized. Imagine that a cell is a simple cube. Compare the surface area and volume of four increasingly larger cubes. (Surface area equals the area of each face multiplied by the number of faces; volume equals the length of a side cubed.) Can you see that volume increases faster than surface area?

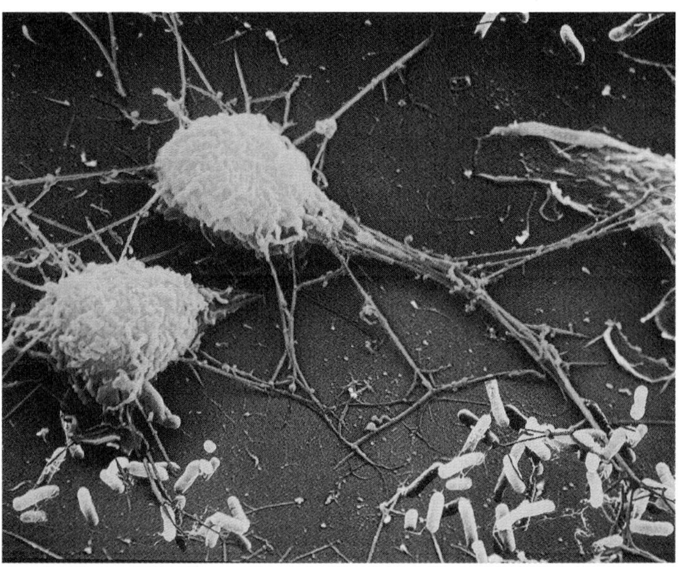

FIGURE 4.6

A giant among cells, this human macrophage extends parts of itself towards nearby bacteria. (×3,200) The scene of this eukaryote-prokaryote encounter: a catheter removed from a patient's abdomen due to bacterial infection.

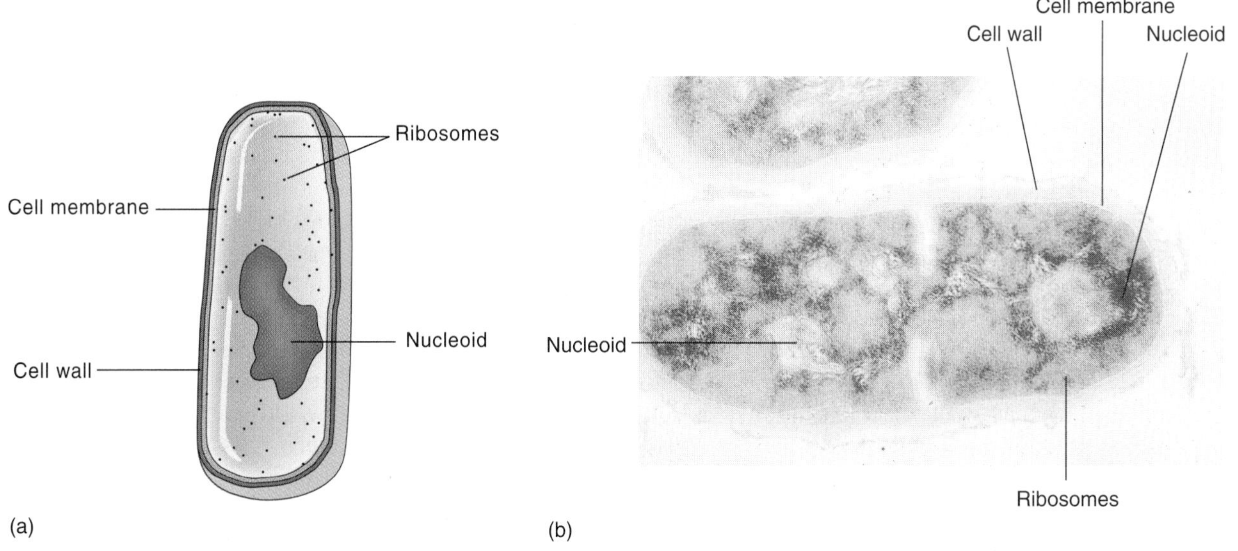

FIGURE 4.7

A generalized prokaryotic cell. The single DNA molecule of a prokaryotic cell resides in a seemingly dense region of the cytoplasm called the nucleoid, but there is no true membrane-bounded nucleus to house the DNA, as exists in more complex cells. The prokaryotic cell contains many ribosomes, as well as fats, proteins, carbohydrates, and pigments. A prokaryotic cell is surrounded by a cell membrane and in many cases by a cell wall as well. Part (*a*) is a schematic view of the micrograph of a bacterium shown in (*b*).

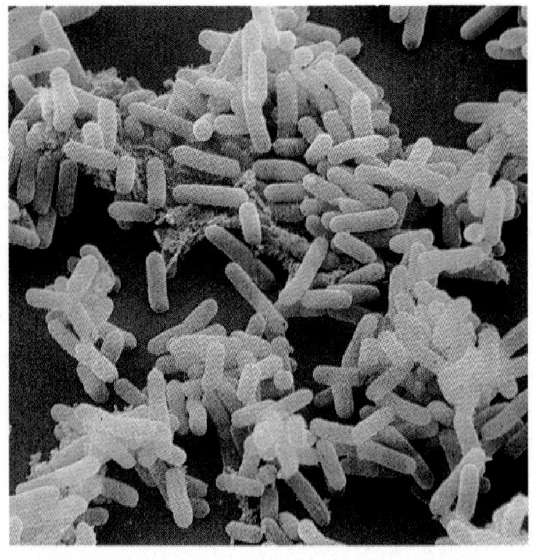

(a)

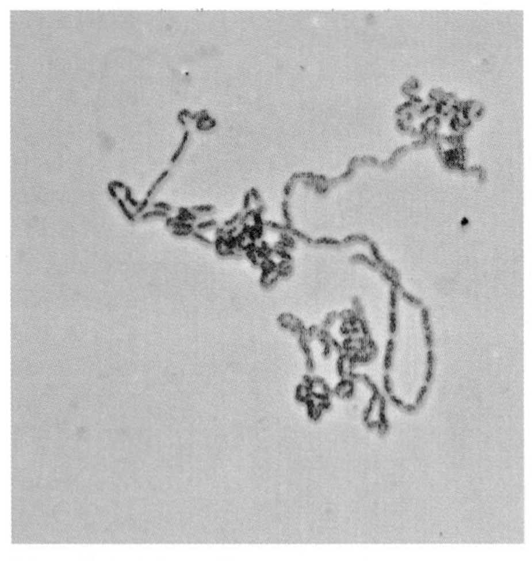

(b)

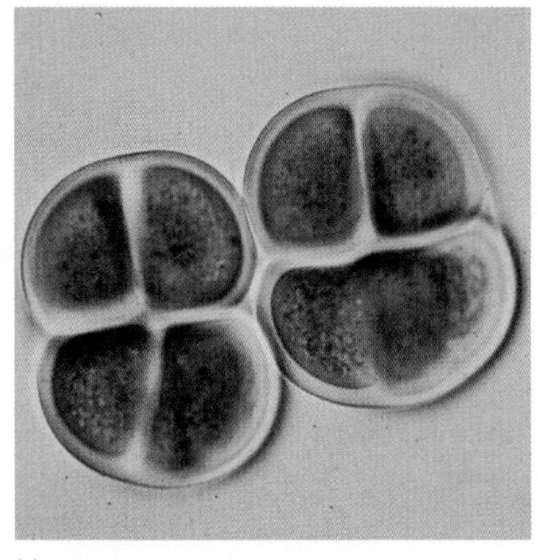

(c)

FIGURE 4.8

A trio of prokaryotes. (*a*) *Escherichia coli* (× 35,000) made headlines in the early 1990s when a strain that manufactures a toxin caused deadly food poisoning in several people who ate contaminated hamburgers and other foods. But genetic researchers have used this prokaryote, which normally inhabits the human digestive tract, for decades. Recombinant DNA technology, described in Biology in Action 16.1, was founded using this organism. Today, thousands of people with diabetes can credit *E. coli* as the source of their daily insulin injections. (*b*) *Lactobacillus bulgaris* (×600) is a rod-shaped bacterium that lives in a slightly acidic environment of pH 4.5 to 6.4. It is cultured with another bacterium, *Streptococcus thermophilus,* to manufacture yogurt from milk. *S. thermophilus* removes oxygen from the milk culture and lowers the pH, creating conditions for *L. bulgaricus* to further acidify the milk and flavor the resulting yogurt. (*c*) Cyanobacteria are prokaryotes capable of photosynthesis (converting solar energy to chemical energy). This is *Chroococcus furgidus.* (×600)

round (cocci), rod-shaped (bacilli), spiral (spirilla), comma-shaped (vibrios), or spindle-shaped (fusiform). Bacteria termed "gram-positive" have cell walls that are very thick layers of peptidoglycan, which stains them dark blue or purple in the presence of a dye called crystal violet. Other "gram-negative" bacteria have thinner peptidoglycan layers plus an outer layer that makes the cell stain pink when exposed to crystal violet.

Beneath the prokaryote's cell wall is a **cell membrane.** The cell membrane pinches inward in places, which may indicate sites where the cell can divide in two. Embedded in the cell membrane and in some species in internal arrays of membranes are enzymes that enable the cell to obtain and use energy.

Unlike bacteria, cyanobacteria contain internal membranes that are continuous with the cell membrane. These membranes, however, are not extensive enough to subdivide the cell into compartments, as membranes do in more complex cells. The cyanobacterium's membranes are studded with pigment molecules that absorb and extract energy from sunlight. In some prokaryotes, tail-like appendages called **flagella,** which enable the cell to move, are anchored in the cell wall and underlying cell membrane. (Chapter 5 covers membrane structure and function in detail.)

The genetic material of a prokaryote is a single circle of DNA. A prokaryote's DNA associates with proteins that are different than those in more complex cells. The part of a prokaryotic cell where the DNA is located is called the **nucleoid** (nucleus-like), and it sometimes appears fibrous under a microscope. Nearby are molecules of RNA and **ribosomes,** which are spherical organelles consisting of RNA and protein. Ribosomes are structural supports for protein synthesis, and the RNA portions of them may help catalyze protein synthesis. Because the DNA, RNA, and ribosomes in prokaryotic cells are in close contact, protein synthesis is rapid compared to the process in more complex cells, whose cellular components are separated.

Eukaryotic Cells

Plants, animals, fungi, and protista have eukaryotic cells. Organelles create specialized compartments where specific biochemical reactions can occur. Saclike organelles (lysosomes, peroxisomes, and vacuoles) contain biochemicals that might harm other cellular contents. Some organelles consist of membranes studded with enzymes that allow certain chemical reactions to occur on their surfaces. On some membranes, different enzymes are physically laid out in the order in which they participate in biochemical reactions. In general, organelles keep related biochemicals and structures sufficiently close together to make them function more efficiently. The compartmentalization organelles provide also makes it unnecessary for the entire cell to maintain a high concentration of a particular biochemical.

A eukaryotic cell is organized like many small plastic bags within a large plastic bag (fig. 4.9). The most prominent organelle is the nucleus, which contains the genetic material (DNA) organized with protein into rod-shaped chromosomes. The remainder of the cell consists of other organelles and the cytoplasm. About half of the volume of an animal cell consists of organelles; in contrast, a plant cell contains about 90% water. The cytoplasm and

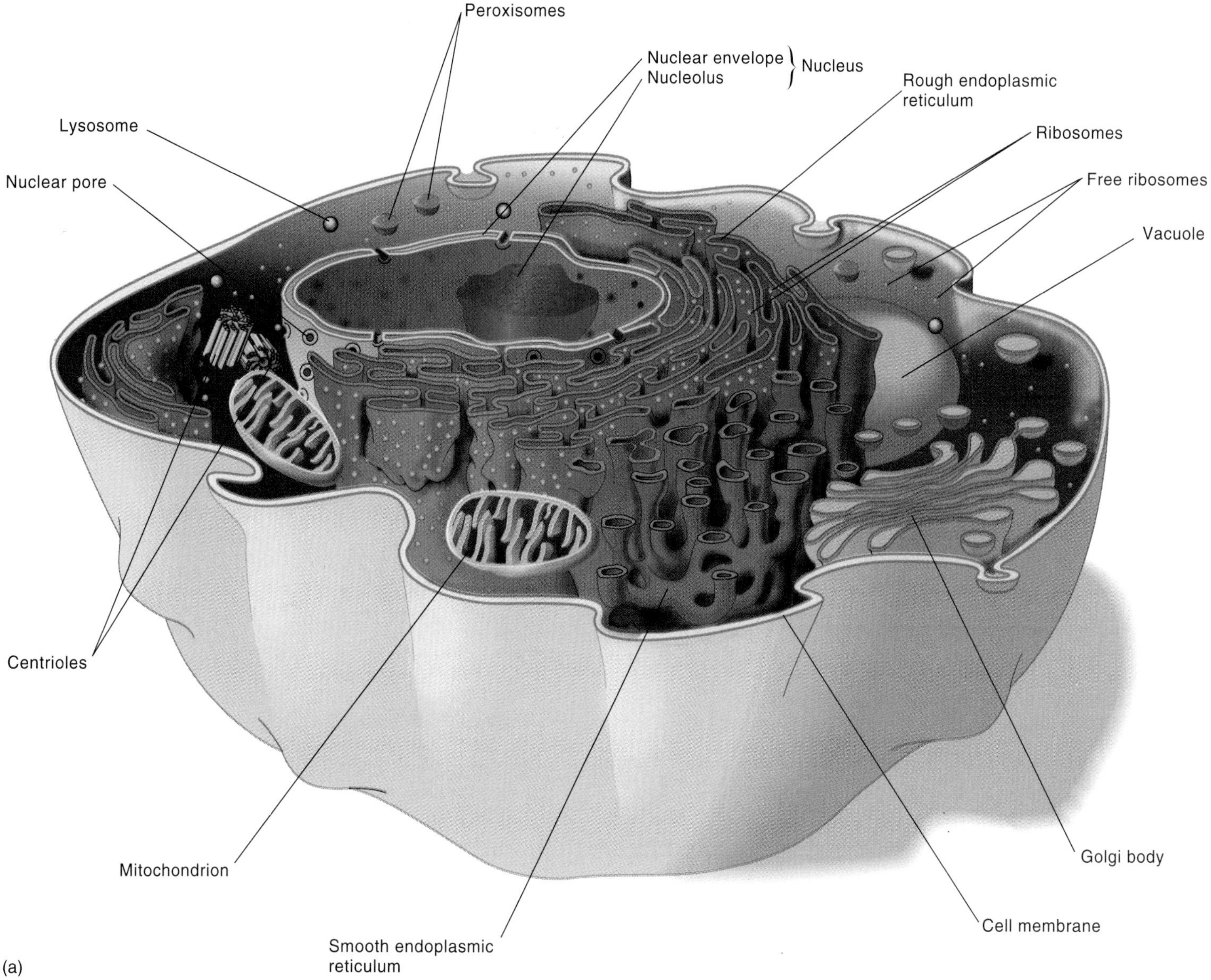

(a)

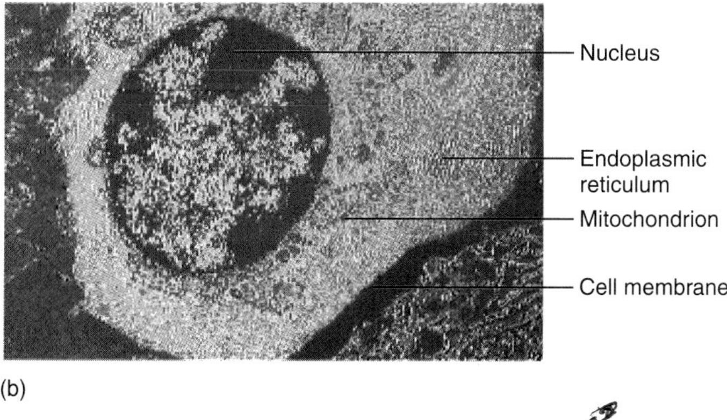

(b)

FIGURE 4.9

Animal and plant cells are eukaryotic. (*a*) A generalized animal cell, with a photograph of a plasma cell, which is a type of white blood cell, that reveals several cell parts. (*b*) A generalized plant cell, which has a cell wall, chloroplasts, and a large vacuole that animal cells lack. A plant cell lacks the lysosomes an animal cell has. The drawing and photograph of the *Coleus* leaf cell display various organelles and structures.

organelles, including the nucleus, are considered the living parts of the cell. Nonliving cellular components include stored nutrients, minerals, and pigment molecules. Arrays of protein rods and tubules within plant and animal cells form the **cytoskeleton,** which helps to give the cell its shape. Protein rods and tubules also form cellular appendages that enable certain cells to move, and they form structures involved in cell division. (Chapter 5 examines the cytoskeleton in depth.) Table 4.1 summarizes the differences between prokaryotic and eukaryotic cells.

Archaean Cells

In the late 1970s, researchers discovered a type of microorganism not seen before that had characteristics of prokaryotes and eukaryotes. How did this "third form of life" fit into the story of life on earth at the cellular level? We knew nothing of the archaea for

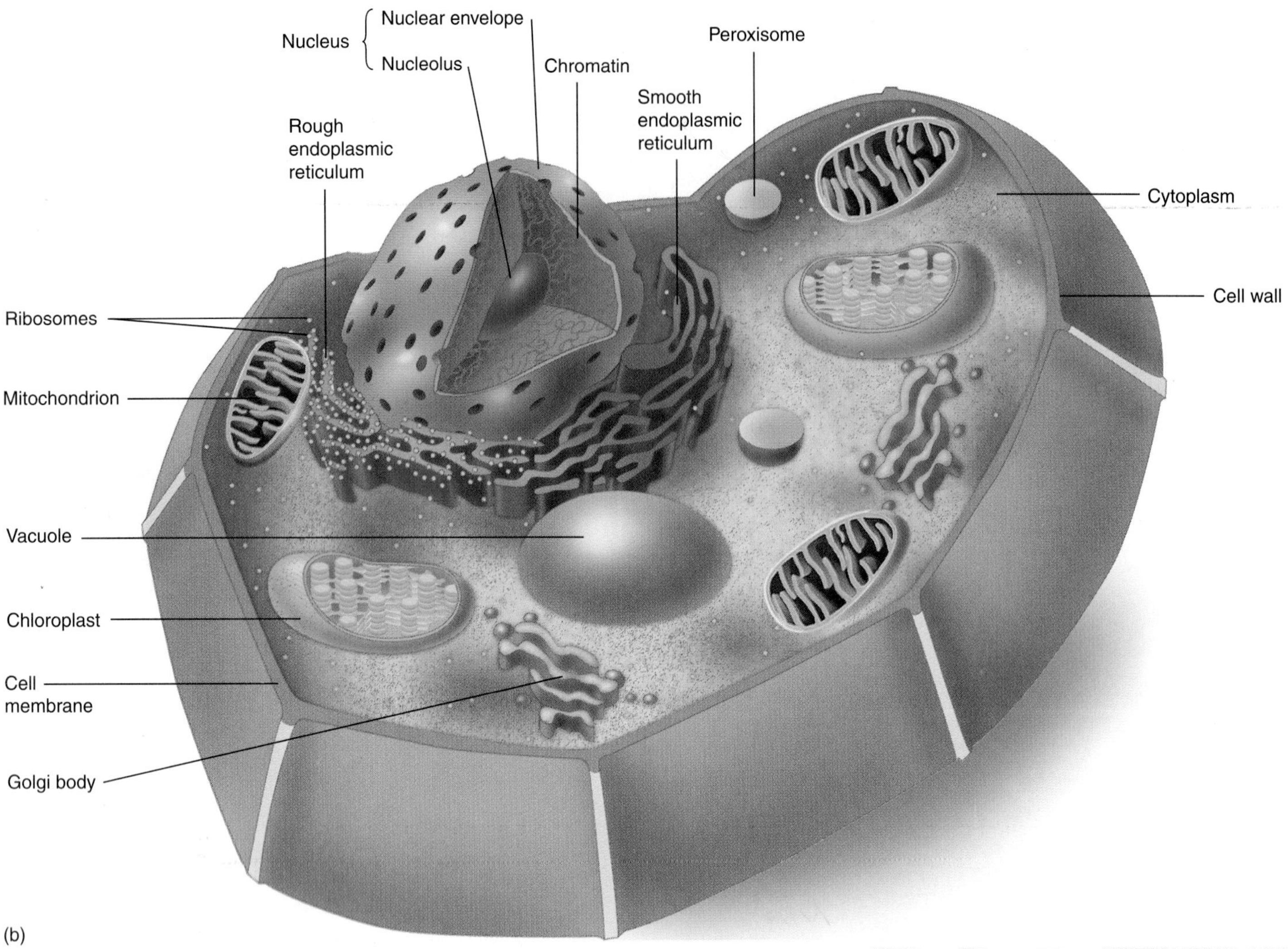

Nucleus
Nuclear envelope
Nucleolus
Chromatin
Peroxisome
Smooth endoplasmic reticulum
Rough endoplasmic reticulum
Cytoplasm
Cell wall
Ribosomes
Mitochondrion
Vacuole
Chloroplast
Cell membrane
Golgi body

(b)

Chloroplast
Cell wall
Cell membrane
Central vacuole
Ribosomes
Nucleus
Mitochondrion
Peroxisome

a long time, because many of them live in very harsh environments that are difficult to simulate in a laboratory. More recently, archaea have been found in less extreme habitats, such as the surface layers of oceans. These organisms may live in places where we have simply not thought to look!

The archaea resemble prokaryotes in that they lack nuclei, but proteins wrap around their DNA as they do in eukaryotes. Their outer coverings (cell membranes) are distinct (fig. 4.10). The structures and building block sequences of certain molecules in the archaea are as different from those in prokaryotes as they are from those in eukaryotes. When researchers deciphered the genetic makeup of an archaeon in 1996, they discovered that more than half of its genes had no known counterparts among prokaryotes or eukaryotes! Although we clearly know very little about these organisms, some scientists hypothesize that they may represent a life-form that could exist on another planet, because of the archaea's ability to survive at low temperatures and under high pressures.

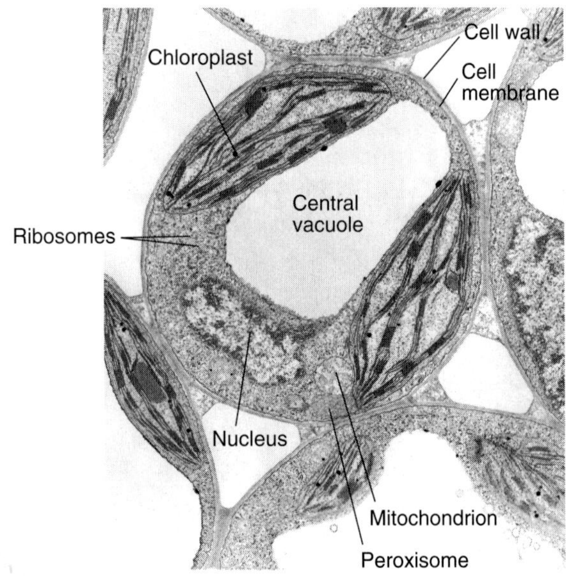

The archaea create a dilemma for taxonomists, because these microorganisms seem to straddle two of the five kingdoms—the Monera (prokaryotes) and the Protista (including single-celled eukaryotes). To account for the hard-to-place archaea, researchers developed a system of three new taxonomic groups, called

Table 4.1 Comparison of Prokaryotic and Eukaryotic Cells

Characteristic	Prokaryotic Cells	Eukaryotic Cells
Organisms	Bacteria (including cyanobacteria)	Protista, fungi, plants, animals
Cell size	Usually 1–10 μm across	Usually 10–100 μm across
Oxygen required	By some	By many
Membrane-bound organelles	No	Yes
Ribosomes	Yes	Yes
DNA form	Circular	Coiled linear strands, complexed with protein
DNA location	In nucleoid	In nucleus
DNA length	Short	Long
Protein synthesis	RNA and protein synthesis not spatially separated	RNA and protein synthesis are spatially separated
Membranes	Some	Many
Cytoskeleton	No	Yes
Cellular organization	Single cells or colonies	Some single-celled, most multicellular with specialized cells

Source: From Bruce Alberts et al., *Molecular Biology of the Cell,* 1983 Garland Publishing, Inc. Used by permission.

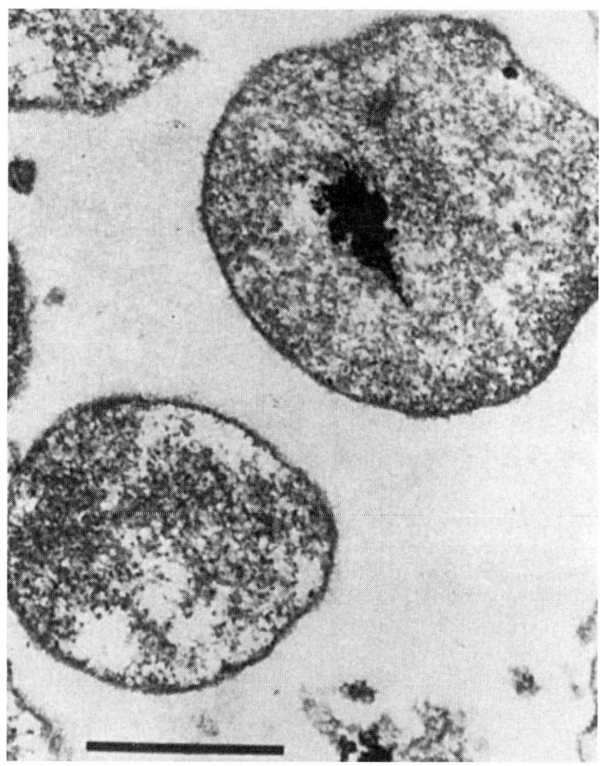

FIGURE 4.10

Thermoplasma are archaea that combine characteristics of prokaryotic and eukaryotic cells. The DNA wraps around proteins much as it does in eukaryotic cells, but there are no nuclei. *Thermoplasma* thrive in the high heat and acidic conditions of smoldering coal deposits. When researcher Gary Darland first discovered these cells, he was so startled by the habitat and mix of features that he named it the "wonder organism." Bar = 0.5 μm.

domains, to supersede kingdoms. The three domains are the bacteria, archaea, and eukarya, and members are distinguished, so far, by cell complexity, nucleic acid sequences, and the types and organizations of lipids in cell membranes. New molecular criteria are being added all the time. The following discussions focus on prokaryotes and eukaryotes simply because we do not yet know very much about the archaea.

Organelles

Organelles divide the activities of life at the cellular level, like departments in a large store group items that are used together. Organelles also interact with each other, providing basic life functions as well as the distinguishing characteristics of a particular cell type. Examining a coordinated function, milk secretion, illustrates how organelles interact to produce, package, and release from the cell a complex mixture of biochemicals (fig. 4.11).

Making Milk

Special cells in the mammary glands of female mammals produce milk. Dormant most of the time, these cells increase their metabolic activity during pregnancy and then undergo a burst of productivity shortly after the female gives birth. Organelles form a secretory network that enables individual cells to manufacture milk, a complex mixture of immune system cells and proteins, fats, and carbohydrates in a proportion ideal for development of a particular species. Human milk is rich in lipids, which the rapidly growing newborn's brain requires. In contrast, cow milk contains a higher proportion of protein, better suited to a calf's rapid muscle growth. We follow here secretion of human milk.

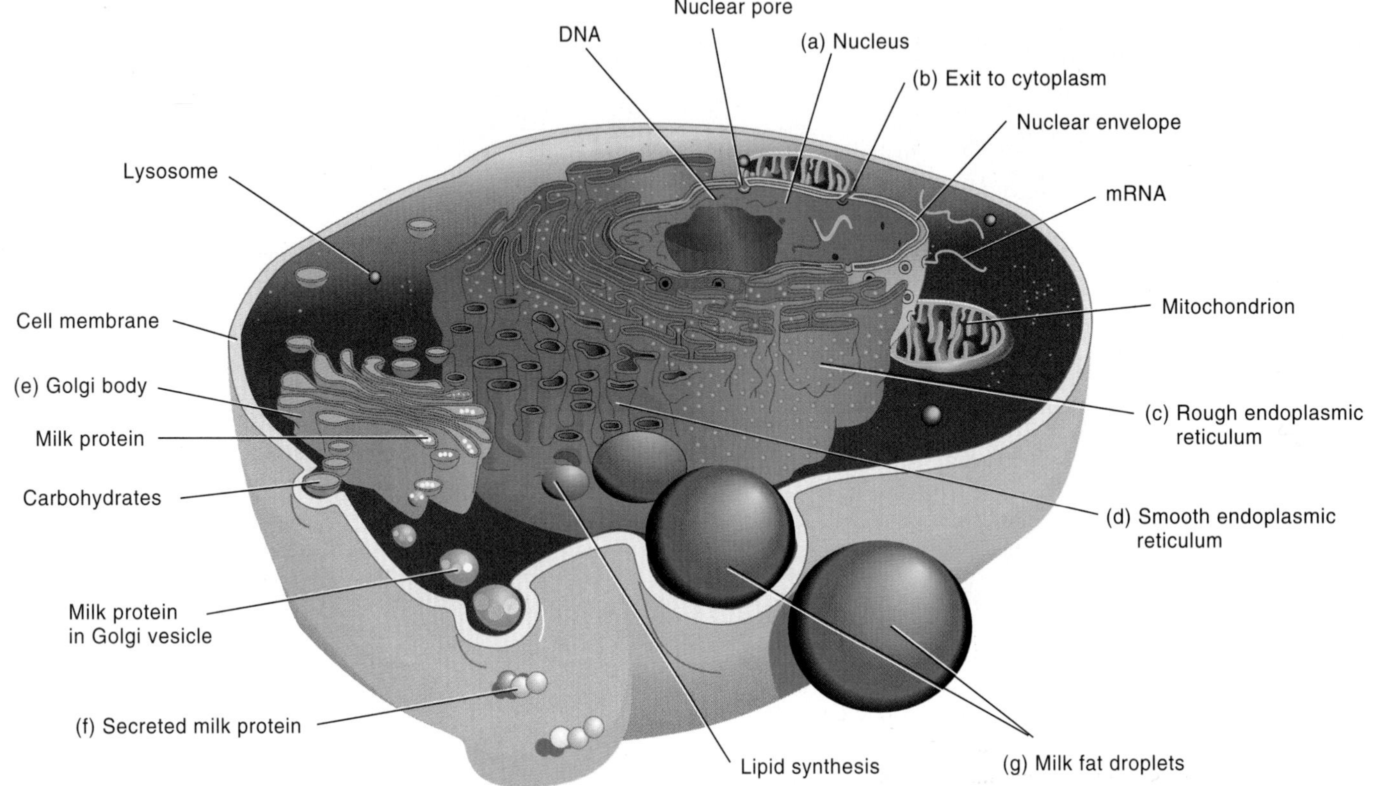

FIGURE 4.11

The process of milk secretion illustrates how organelles interact to synthesize, transport, store, and export biochemicals. Secretion begins in the nucleus (*a*), where messenger RNA (mRNA) bearing the genetic information for milk protein production is synthesized. The mRNA then exits through a nuclear pore to the cytoplasm (*b*). Many proteins are synthesized on the membranes of the rough endoplasmic reticulum, or ER (*c*), using amino acids in the cytoplasm. Lipids are synthesized in the smooth ER (*d*), and sugars are synthesized, assembled, and stored in the Golgi body (*e*). In an active cell in a mammary gland, milk proteins (*f*) are released from vesicles that bud from the Golgi body. Fat droplets (*g*) pick up a layer of lipid from the cell membrane as they exit the cell. When the baby suckles (photo), he or she receives a chemically complex secretion—human milk.

The Nucleus

Secretion begins in the nucleus, where certain genes are copied into another nucleic acid, RNA. These genes include those encoding milk protein, as well as those specifying enzymes required to synthesize the carbohydrates and lipids in milk. The RNA moves from the interior of the nucleus towards the cytoplasm and then exits the nucleus through holes, called **nuclear pores,** in the two-layered **nuclear envelope** that separates the nucleus from the cytoplasm. Nuclear pores are not merely perforations but channels composed of more than 100 different proteins.

The Cytoplasm

In the cytoplasm, the RNA encounters, but does not actually enter, a maze of enzyme-studded, interconnected membranous tubules and sacs that winds from the nuclear envelope to the cell membrane. This labyrinth is the **endoplasmic reticulum** (ER). (*Endoplasmic* means "within the plasm" and *reticulum* means "network.") The portion of this membranous system nearest the nucleus is flattened and studded with ribosomes.

Ribosomes consist of proteins and RNA (a different type of RNA than the type that carries a gene's information). The parts

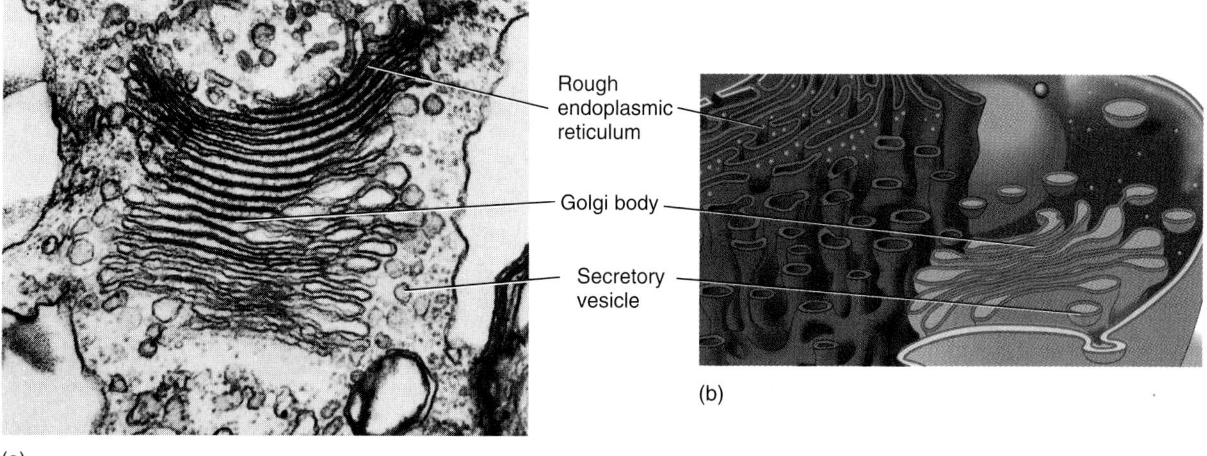

FIGURE 4.12

Golgi bodies carry out secretion. (*a*) A transmission electron micrograph of a Golgi body. (×36,000) (*b*) A Golgi body consists of membranous sacs that continually receive vesicles from the ER and produce vesicles, which enclose components of secretions.

of ribosomes are assembled in a region within the nucleus called the **nucleolus.** The ribosome-studded part of the ER is called **rough ER** because of its fuzzy appearance under the electron microscope. The RNA carrying the gene's message associates with ribosomes on the rough ER, where it strings amino acids together, forming proteins. These proteins may either exit the cell, as does the milk protein casein, or be incorporated into the cell membrane. Proteins synthesized on ribosomes not associated with ER are released into the cytoplasm, where they serve specific functions.

The ER acts as a quality control center for the cell. Its chemical environment enables proteins to fold into the conformations necessary for their functions. Misfolded proteins are pulled out of the production line and degraded in the ER.

As the rough ER winds from near the nucleus outwards toward the cell membrane, the ribosomes become fewer and the diameters of the tubules widen, forming a section called **smooth ER.** Here, lipids are synthesized and added to the proteins arriving from the rough ER. The lipids and proteins move along the smooth ER as its tubules narrow and end. Then the proteins and lipids exit the ER in membrane-bounded sacs, called **vesicles,** that pinch off of the tubular endings of the ER membrane.

A loaded vesicle takes its contents to the next stop in the secretory production line, a structure called a **Golgi body** (fig. 4.12). This organelle is a stack of flat, membrane-enclosed sacs that functions as a processing center. Here, simple carbohydrates are synthesized and linked to form starches, or they attach to proteins to form glycoproteins or to lipids to form glycolipids. Proteins fold into their final forms in the Golgi body. The components of complex secretions, such as milk, are temporarily stored here.

Vesicles budding off of the Golgi body contain milk proteins. The vesicles move toward the cell membrane, and then they fleetingly become part of the cell membrane and open out facing the exterior of the cell and release the proteins. Fat droplets retain a layer of surrounding membrane when they leave the cell.

When a baby suckles, hormones (chemical messengers) released in the mother's system stimulate muscle cells surrounding balls of glandular cells to contract and squeeze milk from them. The milk is released into ducts that lead to the nipple.

Other Organelles and Structures

Mitochondria

The activities of secretion, as well as the many chemical reactions taking place in the cytoplasm, require a steady supply of energy. In eukaryotic cells, energy is extracted from nutrient molecules in organelles called **mitochondria** (fig. 4.13). The number of mitochondria in a cell can vary from a few to tens of thousands. A typical liver cell has about 1,700 mitochondria; cells with high energy requirements, such as muscle cells, may have many more.

A mitochondrion has an outer membrane similar to those of the ER and Golgi body and an intricately folded inner membrane. The folds of the inner membrane, called **cristae,** contain enzymes that catalyze the biochemical reactions that acquire energy. This organelle is especially interesting because it contains genetic material, a point we will return to at the chapter's end.

Another unique characteristic of mitochondria is that they are inherited from the mother only. This is because mitochondria are found in the middle regions of sperm cells but not in the head region, which is the portion that enters the egg to fertilize it. A class of inherited diseases whose symptoms result from abnormal mitochondria are always passed from mother to offspring. These mitochondrial illnesses usually produce extreme muscle weakness, because muscle is a highly active tissue dependent upon the functioning of many mitochondria.

Lysosomes

Eukaryotic cells break down molecules and structures as well as produce them. Organelles called **lysosomes** are the cell's "garbage disposals"; they chemically dismantle captured bacteria, worn-out organelles and debris, and some nutrients (fats, proteins, and

4.1 biology in action

Lysosomal Storage Diseases

The key to the functioning of a lysosome can be summed up in one word—balance. The lysosomes in a particular type of cell contain a balanced mix of enzymes appropriate to the function of that cell. Disrupt that balance, and the cell's function changes, sometimes drastically.

The absence or malfunction of just one lysosomal enzyme can devastate health by creating a lysosomal storage disease. In these inherited disorders, the molecule that is normally degraded by a missing or abnormal lysosomal enzyme accumulates in the lysosome. The lysosome swells with the excess waste, crowding organelles and interfering with the cell's biochemical activities. Usually only the cells of a particular tissue are affected. Defects in different enzymes are associated with particular symptoms, depending upon the type of tissue containing the affected cells.

Tay-Sachs disease results from a missing lysosomal enzyme that normally breaks down lipids in cells surrounding nerve cells. Without the enzyme, cells of the nervous system gradually drown in lipid. Symptoms are usually noted at about 6 months of age, when an infant begins to lag behind in the acquisition of motor skills. On a cellular level, however, signs of Tay-Sachs disease are evident even earlier as enlarged lysosomes. Children who have inherited Tay-Sachs disease soon lose their vision and hearing and later are paralyzed. All affected children die before they are 4 years old.

Pompe disease is another childhood lysosomal storage disease. The absent enzyme normally breaks down the complex carbohydrate glycogen into simple sugars. As a result, glycogen builds up in muscle and liver cells. The young patients usually die of heart failure, because the muscle cells of the heart swell so greatly that they can no longer function properly. Another lysosomal storage disease, Hurler disease, causes bone deformities. Affected bone cells contain huge lysosomes swollen with mucuslike substances called mucopolysaccharides.

carbohydrates) (fig. 4.14). White blood cells have many lysosomes because these cells engulf debris and bacteria. Liver cells, which break down toxins, also require many lysosomes.

A lysosome originates as a sac that buds off of the ER or a Golgi body. In humans, each lysosome contains more than 40 different types of digestive enzymes. Vesicles bring material to lysosomes, either from within the cell or from the cell membrane's budding inward.

Lysosomal enzymes can function only in a very acidic environment (pH 5). The lysosome maintains a highly acidic, protected environment for the enzymes without harming other cellular constituents. Lysosomes are sometimes called "suicide sacs," because if they rupture and release their enzymes, the entire cell is digested from within and dies. Lysosomes play a role in aging by destroying cells in this way. The correct balance of enzymes within a lysosome is important to human health (Biology in Action 4.1).

Plant cells do not have lysosomes. Instead, enzymes in the **central vacuole,** a membrane-bounded organelle that temporarily stores a substance or transports it within a cell, provides a similar "disposal" function.

Peroxisomes

Peroxisomes are single-membrane-bounded sacs present in all eukaryotic cells that contain several types of enzymes. Peroxisomes reproduce by enlarging and then splitting in two. In some species, exposure to environmental toxins triggers an explosive proliferation of peroxisomes, which helps cells to survive the insult.

Peroxisomes help the cell use oxygen. Specifically, the reactions that some peroxisomal enzymes catalyze produce hydrogen peroxide (H_2O_2). This compound introduces oxygen-free radicals, which are oxygen atoms with unpaired electrons that can damage the cell. To counteract the free-radical buildup, peroxisomes contain abundant catalase, an enzyme that removes an oxygen from hydrogen peroxide. Catalase then combines the oxygen from H_2O_2 with hydrogen atoms removed from various organic molecules, producing harmless water molecules. Liver and kidney cells contain many peroxisomes to help dismantle toxins from the blood (fig. 4.15).

The 40 types of peroxisomal enzymes catalyze a variety of biochemical reactions, including:

- synthesizing bile acids, which are used in fat digestion;
- breaking down lipids called very-long-chain fatty acids;
- degrading rare biochemicals;
- metabolizing potentially toxic compounds that form as a result of oxygen exposure.

Like the lysosomal storage disorders described in Biology in Action 4.1, abnormal peroxisomal enzymes also harm health. A defect in a receptor on the peroxisome's membrane may affect several of the enzymes and cause a variety of symptoms. Some peroxisome disorders affect only a single enzyme type. One such disorder was the subject of a 1992 film called *Lorenzo's Oil.*

Six-year-old Lorenzo Odone suffered from adrenoleukodystrophy (ALD). His peroxisomes lacked one of the two major proteins in the outer membrane. Normally, this protein transports an enzyme into the peroxisome, where it catalyzes a reaction that helps break down a type of very-long-chain fatty acid. Without the enzyme transporter protein, the fatty acid builds up in the cells of the brain and spinal cord, eventually stripping these cells of fatty coverings necessary for nerve

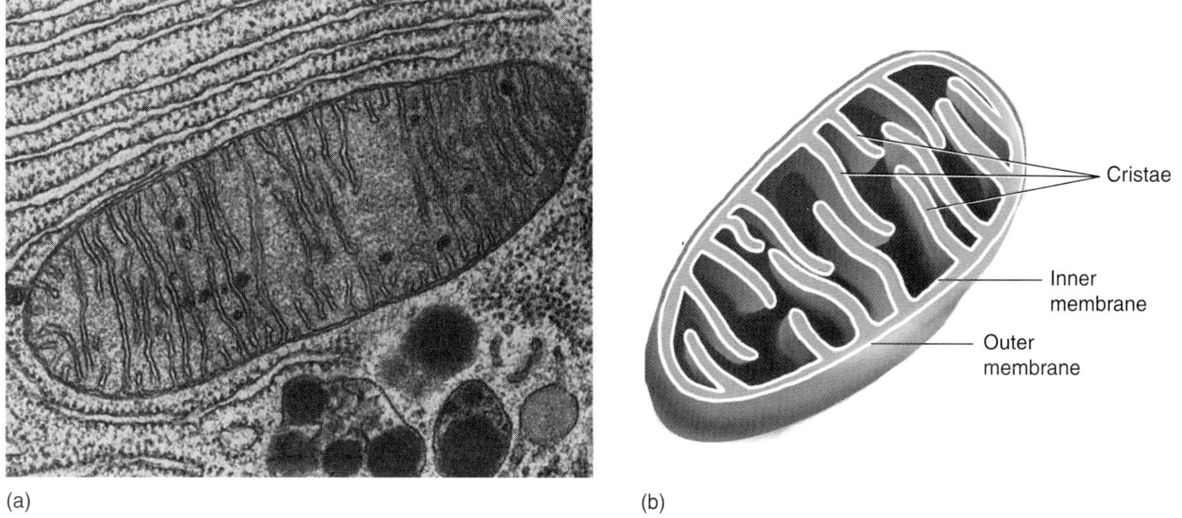

FIGURE 4.13

Mitochondria are the sites of energy reactions. (*a*) A transmission electron micrograph of a mitochondrion. (×79,000) (*b*) Infoldings of the inner membrane, called cristae, form partitions within this saclike organelle

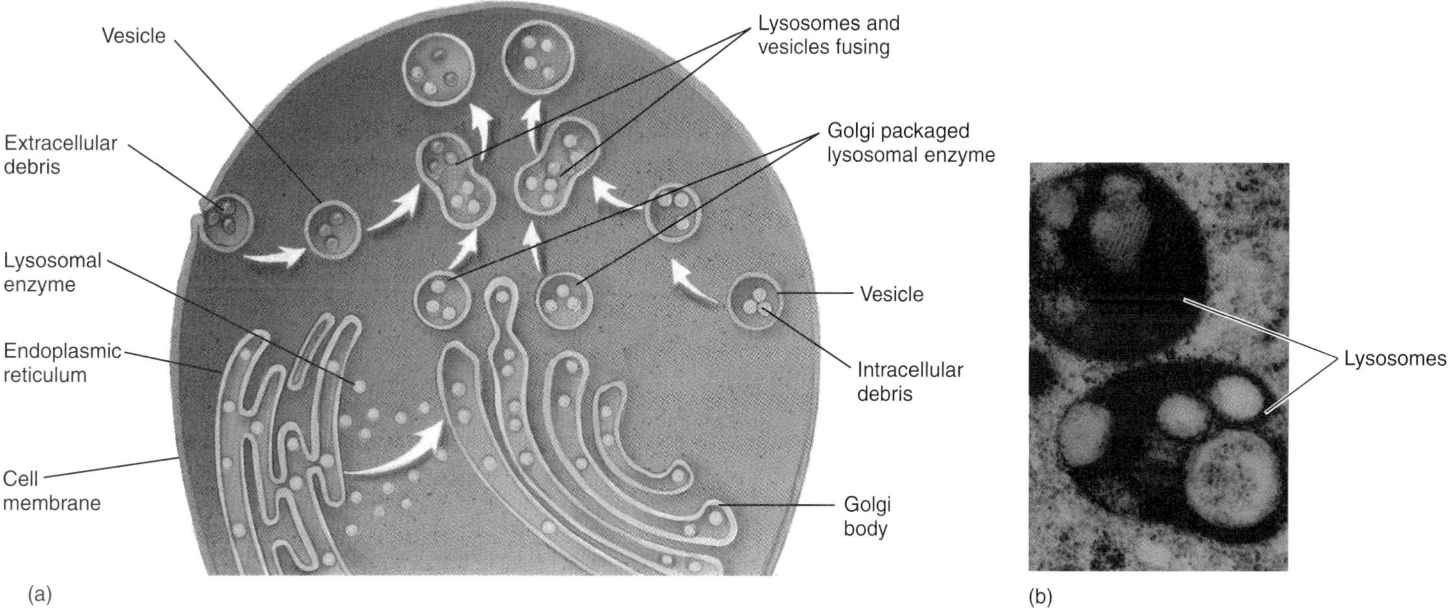

FIGURE 4.14

Lysosomes contain enzymes. (*a*) Lysosomal enzymes are synthesized on the rough ER and are transported to the Golgi body. The Golgi body detects and separates the specific enzymes destined for lysosomes by recognizing a particular sugar attached to these enzymes. The lysosome-specific enzymes are then packaged into vesicles that eventually become lysosomes. Lysosomes fuse with vesicles carrying debris from the outside or from within the cell, and the lysosomal enzymes then degrade the debris. (*b*) In this transmission electron micrograph, lysosomes appear as membranous sacs. (×14,100)

conduction. Table 4.2 lists the symptoms of ALD and other peroxisomal disorders. Ingesting a triglyceride from rapeseed oil prevents the buildup of the very-long-chain fatty acids for a few years and stalls symptoms. However, the treatment eventually impairs blood clotting and other vital functions and does not halt progression of the illness.

In plants, leaf cells have many peroxisomes. In these cells, a dense area where catalase concentrates often appears as a crystal in electron micrographs (fig. 4.15*b*). Peroxisomes in plants help to break down organic molecules synthesized in **photosynthesis,** the process by which plants harness solar energy to synthesize simple carbohydrates.

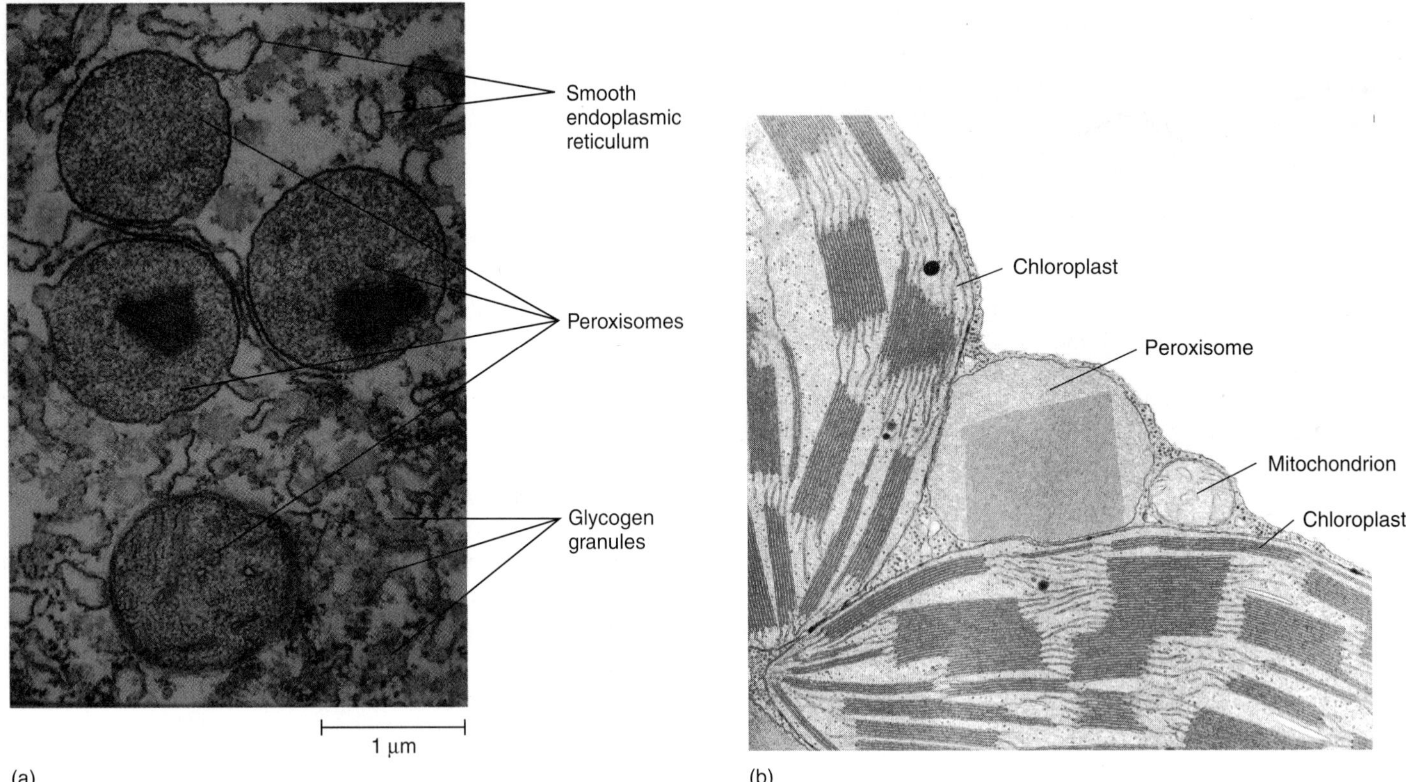

FIGURE 4.15

Peroxisomes assist in detoxification. (*a*) In animals, peroxisomes are abundant in liver cells. (*b*) This plant peroxisome is sandwiched between two chloroplasts in a leaf cell, where it participates in a process related to photosynthesis. The square within the peroxisome is a large crystal of catalase, an enzyme found in all peroxisomes.

Table 4.2 Peroxisomal Disorders

Disorder	Defect
Infantile Refsums disease	Accumulation of very-long-chain fatty acids; too few peroxisomes in liver and skin cells. Symptoms include mental retardation, abnormal face, impaired vision and hearing, enlarged liver, weak bones.
Primary hyperoxaluria type I	A peroxisomal enzyme enters mitochondria instead of peroxisomes, where it cannot catalyze its reaction. Toxins accumulate.
X-linked adrenoleukodystrophy	Accumulation of very-long-chain fatty acids, which depletes fatty sheaths of brain cells. Symptoms include weakness, dizziness, low blood sugar, darkening skin, behavioral complications, loss of muscular control.
Zellweger syndrome	Peroxisomes are absent because their parts cannot be assembled. Only 35 of the 40 enzyme types can function free in the cytoplasm. Symptoms include abnormal face, hands, and feet; kidney cysts; malformed liver.

Organelles Restricted to Certain Kingdoms

Most of the organelles and structures discussed so far, including the nucleus, ribosomes, ER, Golgi bodies, mitochondria, and peroxisomes, are present in nearly all eukaryotic cells. A few structures, such as lysosomes, are peculiar to specific kingdoms. Here are a few others.

Centrioles are oblong structures built of protein rods called **microtubules.** They are present in pairs in animal cells, oriented at right angles to one another near the nucleus. Centrioles organize other microtubules to pull replicated chromosomes into two groups during cell division. In plants, a cylindrical structure also built of microtubules, called a **phragmoplast,** divides one cell into two.

An organelle unique to plants and green algae, the **chloroplast,** imparts their color. Chloroplasts house the chemical reactions of photosynthesis. The inner membrane of a chloroplast is studded with enzymes and pigments necessary for photosynthesis. The membrane is organized into stacks, called **grana,** of flattened membranous disks called **thylakoids.** Like mitochondria, chloroplasts contain genetic material. The green pigment **chlorophyll** within the chloroplasts harnesses the solar energy. An actively photosynthesizing plant cell, such as a leaf cell facing the sun, may contain more than 50 chloroplasts (see fig. 4.1*c*). Chloroplasts are the most abundant form of a general class of pigment-containing organelles, called plastids, found in plant cells.

Mature plant cells, unlike other eukaryotic cells, usually have a large, centrally located sac or **vacuole.** Water stored in a plant cell's vacuole can amount to 90% of the cell's total volume.

A rigid cell wall surrounds and imparts shape and strength to the cells of many organisms, including plants, fungi, and most microorganisms. Animal cells are notable in lacking cell walls. The composition of the cell wall varies among members of the different kingdoms. Plant cell walls consist mostly of the carbohydrate cellulose, and fungal cell walls consist of chitin.

Table 4.3 summarizes organelle structures and functions.

Mastering Concepts

1. What are the three major types of cells?
2. What are the major differences between prokaryotic and eukaryotic cells?
3. Name and describe the structures that make up a prokaryotic cell.
4. Which organelles interact to produce and secrete a complex substance?
5. What is the function of the nucleus and its contents?
6. Which organelle houses the reactions that extract chemical energy from nutrient molecules?
7. Which two organelles are membrane-bounded sacs containing a variety of enzymes?
8. Which organelles are specific to particular kingdoms?

Simpler Than Cells

The simplest form of life is probably a prokaryote, such as a bacterium. However, several types of "infectious agents" appear to be living while they are infecting cells but otherwise seem to be nonliving, complex collections of chemicals. These entities include the familiar viruses and several not-so-familiar structures that straddle the border between the living and the nonliving.

Viruses

Viruses frequent the headlines—and our bodies. Colds, flu, herpes, and measles are all viral infections. Some viral scourges are more dramatic. Hantavirus outbreaks cause death from respiratory failure in days. Hemorrhagic fever caused by the tiny Ebola virus erupts seemingly out of nowhere in an African village, dissolving infected human bodies within 2 weeks (fig. 4.16*a*). The human immunodeficiency virus (HIV), which causes AIDS, is a slower killer (fig. 4.16*b*).

Because viruses can devastate the human body, one might expect them to be rather complex in structure. Yet compared to any cell, viruses are extremely simple in structure.

A virus consists of two major components:

1. A nucleic acid core. This may be RNA or DNA, and it may be in a single strand or a double strand.
2. A protein coat, called a **capsid.** Protein coats may form cylinders, icosahedrons (20-faced geometric structures), or combinations of these two basic shapes.

Some viruses have an envelope surrounding the capsid, which houses the nucleic acid core. The envelope may consist of carbohydrates and lipids that the virus takes from the infected cell's membrane, plus proteins, which sometimes extend like spikes from the virus. The viral genetic material encodes the proteins. The viruses that cause rabies, influenza, AIDS, and herpes infections have envelopes.

Table 4.3 Structures and Functions of Organelles

Organelle	Structure	Function
Chloroplast	Stacks of flattened sacs containing pigments	Photosynthesis
Endoplasmic reticulum	Membrane network; rough ER has ribosomes, smooth ER does not	Site of protein synthesis and folding; lipid synthesis
Golgi body	Stacks of membrane-enclosed sacs.	Sugars made and linked to form starches or joined to lipids or proteins; proteins finish folding; secretions stored
Lysosome	Sac containing digestive enzymes	Debris degraded; cell contents recycled
Mitochondrion	Two membranes; inner one enzyme-studded.	Releases energy from nutrients
Nucleus	Perforated sac containing DNA	Separates DNA from rest of cell
Peroxisome	Sac containing enzymes	Oxygen use, other reactions
Ribosome	Two associated globular subunits of RNA and protein	Scaffold for protein synthesis; RNA may help catalyze protein synthesis
Vesicle	Membrane-bounded sac	Temporarily stores or transports substances

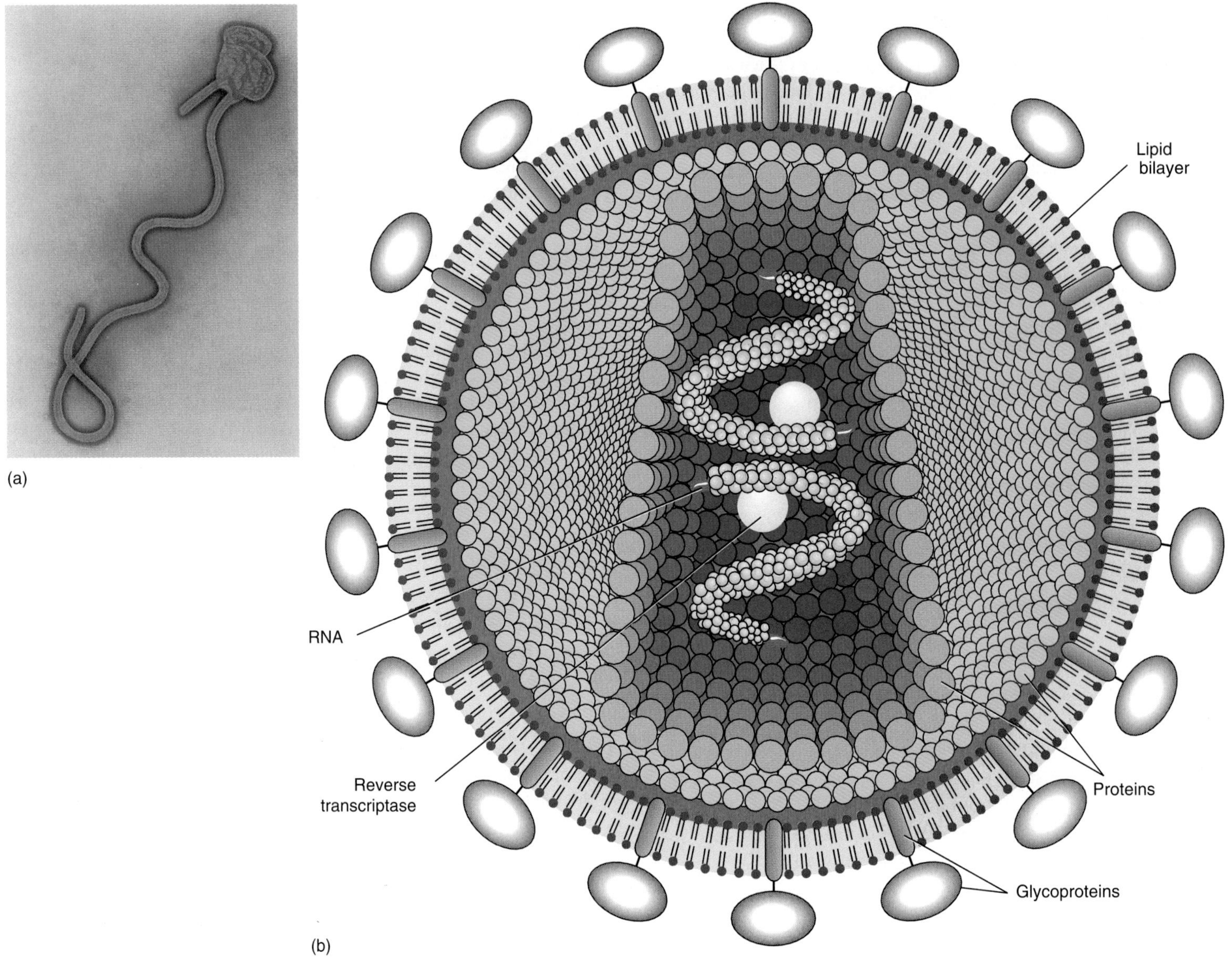

FIGURE 4.16

Viruses are not cells. (*a*) Ebola virus is a single strand of RNA and just seven proteins. Virologists do not know the species where Ebola survives between outbreaks among humans. People become infected by touching body fluids of those who have died of the infection. Symptoms progress rapidly, from headache and fever to vomiting of blood, to tearing apart of the internal organs, until the person "bleeds out" and dies. (*b*) HIV consists of two identical strands of RNA surrounded by several layers of proteins, plus an envelope of lipids and glycoproteins. Once inside a human cell (usually a T cell, part of the immune system), the virus uses an enzyme, reverse transcriptase, to copy its RNA into DNA. The virus then inserts this copy into the host DNA. HIV damages the body's ability to protect itself against infection and cancer by killing T cells and by using these cells to reproduce itself. The amount of genetic material in HIV is 1/100,000 that of a human cell, yet among viruses, it is complex.

A virus's symmetrical structure enables it to carry a minimum of genetic material. The capsid often consists of many copies of the same protein, like a greenhouse built of identical panes of glass. Only one or a few viral genes are necessary to instruct the host cell to produce the proteins, which then join ("self-assemble") and form the capsid.

Viruses infect nearly all types of organisms, including prokaryotes, fungi, algae, plants, and animals. The first virus discovered infects tobacco cells. Any particular type of virus, however, infects only certain species. The same virus may have different effects in different species. For example, Borna virus causes adult rats to become extremely agitated and aggressive, but it merely disturbs social behavior in tree shrews. Researchers are currently investigating whether Borna virus infection causes mental illness in humans.

Viruses can exist outside cells but must enter cells to reproduce. Here, they commandeer the DNA replication and protein synthetic machinery of the host (the nucleus, ribosomes, and ER) and use it to produce more viruses, which may burst from the cell. This is what happens in HIV infection. Or, viruses in a cell may slowly destroy it without bursting it. A virus infecting a cell may have no effect at all. Our own cells contain DNA sequences that were once, perhaps even in our distant ancestors, viral nucleic acid sequences. About 1% of our DNA is of such viral origin. It is apparently harmless.

Biologists use several criteria to classify viruses, including type of nucleic acid and whether it is single- or double-stranded, capsid symmetry, envelope structure, the types and range of organisms the viruses infect, and the diseases they cause. Viruses can "evolve" in the sense that their DNA or RNA sequence changes rapidly. This is because viruses replicate very often and cannot "repair" errors in the sequences, as cells can. The changeable nature of viruses is why we cannot develop a vaccine for the common cold, why flu vaccines must be developed new each season, and why a vaccine has yet to be perfected that can prevent HIV infection. (A vaccine is a weakened form of or part of a virus that stimulates the body to mount an immune response.)

Bioethics Connections 4.1 considers another virus that has greatly affected human populations, *Variola.* It causes smallpox.

Viroids

Viroids are even more streamlined and simpler than viruses. They consist only of highly wound single strands of RNA, which, untwisted and stretched out, would be about 3 feet (1 meter) long. Viroids cause several exotic-sounding plant diseases, including "avocado sun blotch," "potato spindle-tuber disease," "coconut cadang cadang," and "tomato bunchy top" (fig. 4.17). They nearly destroyed the chrysanthemum industry in the United States in the early 1950s.

Viroids pass from plant to plant mechanically, as parts of one plant touch another, or they can be transmitted from generation to generation. Once inside a host cell, the viroid RNA accumulates, building to 200 to 10,000 copies, in the nucleolus. The viroids are somehow mistaken for DNA and replicated. It isn't known how viroids cause symptoms.

FIGURE 4.17

Plants can fall victim to an infectious particle called a viroid. Shown here are a tomato with "tomato bunchy top," caused by a viroid (*left*), and a healthy tomato plant (*right*).

Table 4.4 compares the amount of genetic material in cells, viruses, and viroids, which gives an idea of their relative complexities.

Prions

Composed only of glycoprotein, **prions** are 1/100 to 1/1,000 the size of the smallest known virus. The name prion derives from "protein infectious agent." Prions were first described in 1966 as proteins that replicate to cause a disease called scrapie in sheep. British researchers originated the idea of prions, and in 1982, University of California at San Francisco biologist S. B. Prusiner further described prions and implicated them in several disorders.

Prions are an abnormal form of a glycoprotein present in the cells of many mammals. The abnormal prion protein forms a gummy mass that causes different symptoms depending on its location. The buildup usually occurs in the brain, causing a characteristic spongy appearance. An epidemic of "mad cow disease" (bovine spongiform encephalopathy) in the United Kingdom beginning in 1985 brought a prion illness to public attention (fig. 4.18*a*). Cows were infected by eating a scrapie agent in food made from sheep by-products. Infected beef caused a new variant of a human condition, called Creutzfeldt-Jakob syndrome. It is rare, and the outbreak subsided after cattle raisers stopped feeding their animals sheep products.

Table 4.4 Genetic Comparisons of Cells, Viruses, and Viroids

Organism or Infectious Agent	Number of DNA or RNA Nucleotides
Eukaryotes	
Human (*Homo sapiens*)	3 billion
Mouse (*Mus musculus*)	3 billion
Fruit fly (*Drosophila melanogaster*)	165 million
Roundworm (*Caenorhabditis elegans*)	100 million
Prokaryotes	
Escherichia coli	4.8 million
Haemophilus influenzae	1.8 million
Viruses	
Variola	200,000
Herpes simplex	160,000
Parvovirus	4,800
Viroid	
Potato spindle-tuber disease agent	359

4.1 bioethics connections

Should We Destroy the Last Smallpox Virus?

In two freezers, one at the Centers for Disease Control and Prevention in Atlanta and another at the Research Institute for Viral Preparations in Moscow, lie the last smallpox viruses on the planet. Infectious disease experts and virologists are debating whether or not to destroy the remaining smallpox virus, called *Variola.* Both sides offer compelling arguments, but a look at history places them in perspective.

History of a Scourge

Smallpox is the deadliest infectious disease known to strike humans, in terms of numbers of people killed. Smallpox ravaged the Roman Empire and enabled the Spaniards to defeat the Aztecs in Mexico, whose immune systems could not handle the foreign virus. Survivors of smallpox often are left with severe scars. Luckiest were those few individuals who had mild cases and were left with relatively smooth skin plus immunity against reinfection.

English physician Edward Jenner's invention of the first vaccine in 1796, against *Variola,* was the beginning of the end of the scourge (Biology in Action 39.2). It took many decades before the vaccine was improved and distributed widely enough to impact upon the disease's prevalence. By 1967, when the World Health Organization (WHO) began its eradication campaign, some 10 million people in 40 nations still contracted smallpox each year. The last case in the United States occurred in 1950.

The WHO campaign was remarkably successful, and in October 1977, the last victim in the general population, a Somali man, died of smallpox. A year later, though, a shocking case prompted public health officials and scientists to question the wisdom of maintaining samples of the virus. A photographer, Janet Parker, acquired smallpox while visiting a laboratory in England that kept the virus, obviously not sufficiently contained. Parker developed smallpox and died. The head of the laboratory, overcome with guilt, killed himself.

To Destroy or Not to Destroy

In 1979, the world was declared free of smallpox. The Global Commission for the Certification of Smallpox Eradication formed and requested that all *Variola* samples be destroyed or sent to facilities in the United States, United Kingdom, South Africa, or Russia. The WHO agreed in 1980. In 1983, the African and Soviet virus samples were sent to the other two sites. In 1986, WHO raised the idea of destroying all samples, if the scientific and public health communities approved. In 1990, scientists requested that the decision be postponed until they could learn the virus's DNA sequence, so that they could continue to study it. Researchers in the United States and the former Soviet Union collaborated and sequenced all 200,000 DNA bases that constitute variola virus by 1993.

People debated the fate of the smallpox virus against a backdrop of emerging viral illnesses, such as AIDS. The remaining *Variola* samples are tentatively scheduled to be destroyed in the year 2000, although the debate over their fate continues. The "destroy" arguments tend to be political and practical; the "do not destroy" arguments are scientific in tone. The reasoning is as follows:

Destroy!

1. A terrorist could use the stored virus for biological warfare.
2. Damage to the freezers storing the virus—such as from a bomb, earthquake, or other disaster—could unleash a deadly smallpox epidemic.
3. Knowledge of the DNA sequence of *Variola* will enable researchers to continue studying the virus, without needing the actual virus.

Do Not Destroy!

1. Learning how the smallpox virus evades the human immune system will provide clues on how to combat HIV. Both *Variola* and HIV seem to infect only humans, and in many cases they defeat the human immune system.
2. Knowing the DNA sequence of a virus is not sufficient to understand how it causes symptoms and evades the immune system.
3. Viral infections reemerge. Should a new virus evolve to fit the niche that *Variola* occupied, or an existing virus (such as monkeypox) expand and change to affect humans, the lack of supplies of smallpox virus could hamper research.
4. Smallpox could reappear even if we destroy the frozen samples. *Variola* may have been stored in other laboratories, and smallpox victims buried in Soviet permafrost could some day thaw and release active virus.

A final note is more philosophical. Even though a virus is not technically alive, do we have the right to destroy it? In the words of one researcher, "It's taken millions of years for nature to make the *Variola* virus, and why should 10 guys sitting around a table say, 'Let's destroy it'?"

(a)

(b)

FIGURE 4.18

(*a*) Cows afflicted with bovine spongiform encephalopathy—better known as "mad cow disease"—must be removed to stop the spread of infection. (*b*) This child can no longer walk, stand, sit, or talk because of nervous system degeneration caused by kuru, a prion disease. The child died a few months after this photo was taken. Kuru is known only among the Fore people of eastern New Guinea. Before the 1980s, women and children became infected when preparing the brains of war dead, which were eaten. Today, the Fore people no longer practice cannibalism, and kuru has ceased.

Another human prion disorder causes a form of insomnia that is fatal (table 4.5). Some prion diseases are inherited as a mutation in a gene encoding prion protein. Other prion illnesses are contracted from contaminated tissue, such as in cannibalism (kuru) (fig. 4.18*b*), and from receiving infected tissue, such as in a transplant or by using blood products pooled from donors. Abnormal prion protein arises from the action of an abnormal gene, but once the protein forms, it can be transmitted from individual to individual, like an infectious agent.

Table 4.5 Prion Diseases

Disease	Symptoms
Creutzfeldt-Jakob syndrome	Memory loss, nervousness, unsteady gait, jerky motions, loss of facial expression; death within 2 years
Fatal familial insomnia	Onset in middle age of inability to sleep; tremors, a dreamlike state, coma, and death within 10 months
Gerstmann-Straussler-Scheinker syndrome	Onset in 50s; loss of coordination, dementia, lack of leg reflexes, deposits throughout central nervous system; death in 2 to 10 years
Kuru	Loss of ability to walk, stand, sit, talk; death within months.

Mastering Concepts

1. How does the structure of a virus differ from the structure of a cell?
2. Describe the possible effects of a virus on a host cell.
3. How do viroids and prions differ from cells?

The Origin of Eukaryotic Cells

How did eukaryotic cells arise? The **endosymbiont theory** proposes that eukaryotic cells formed as large prokaryotic cells incorporated smaller and simpler prokaryotic cells (an endosymbiont is an organism that can live only inside another organism). The compelling evidence supporting the endosymbiont theory is that mitochondria and chloroplasts, present only in eukaryotic cells, greatly resemble prokaryotic cells.

In the 1920s, investigators noted the resemblances between eukaryotic cell components and free-living prokaryotes. These early investigators proposed that eukaryotes evolved by incorporating prokaryotes, but their ideas were ridiculed at the time because there was little proof to support them. In the 1960s, studies with the electron microscope dramatically confirmed the similarities proposed four decades earlier between eukaryotic organelles and the prokaryotic bacteria and cyanobacteria. Then genetic studies revealed DNA in mitochondria and chloroplasts, further suggesting that these structures may have descended from once free-living organisms. These converging ideas and observations led biologist Lynn Margulis at the University of Massachusetts at Amherst to propose the endosymbiont theory in the late 1960s.

Mitochondria and chloroplasts resemble bacteria in size, shape, and membrane structure. Each organelle contains its own DNA and reproduces by splitting in two. In addition, the DNA, messenger RNA, and ribosomes within chloroplasts and mitochondria function in close association with each other, just as they do in prokaryotes during protein synthesis. In contrast, DNA in a eukaryotic cell's nucleus is physically separated from the RNA and ribosomes.

Prokaryotic cells and the organelles of eukaryotic cells are similar in other ways. Pigments in the chloroplasts of eukaryotic red algae are similar to pigments cyanobacteria use to carry out photosynthesis. Sperm tails and centrioles may descend from ancient spiral-shaped bacteria. Mitochondria most closely resemble the aerobic (oxygen-using) purple nonsulfur bacteria, which can photosynthesize. Accordingly, the endosymbiont theory proposes that mitochondria descended from aerobic bacteria, that the chloroplasts of red algae descended from cyanobacteria, and that the chloroplasts of green plants came from yet another type of photosynthetic microorganism. How might this hypothesized merger of organisms have taken place?

Picture a mat of bacteria and cyanobacteria, thriving in a pond some 2.5 billion years ago. Over many millions of years, the flourishing cyanobacteria pumped oxygen into the atmosphere as a byproduct of photosynthesis. Eventually, only those organisms that could tolerate free oxygen survived.

Free oxygen reacts readily with molecules in organisms, producing oxides that can no longer carry out biological functions. One way for a large cell to survive in an oxygen-rich environment would be to engulf an aerobic bacterium in an inward-budding vesicle of its cell membrane. The aerobic endosymbiont would contribute biochemical reactions to detoxify free oxygen. Eventually, the membrane of the enveloping vesicle became the outer membrane of the mitochondrion. The outer membrane of the engulfed aerobic bacterium became the inner membrane system of the mi-

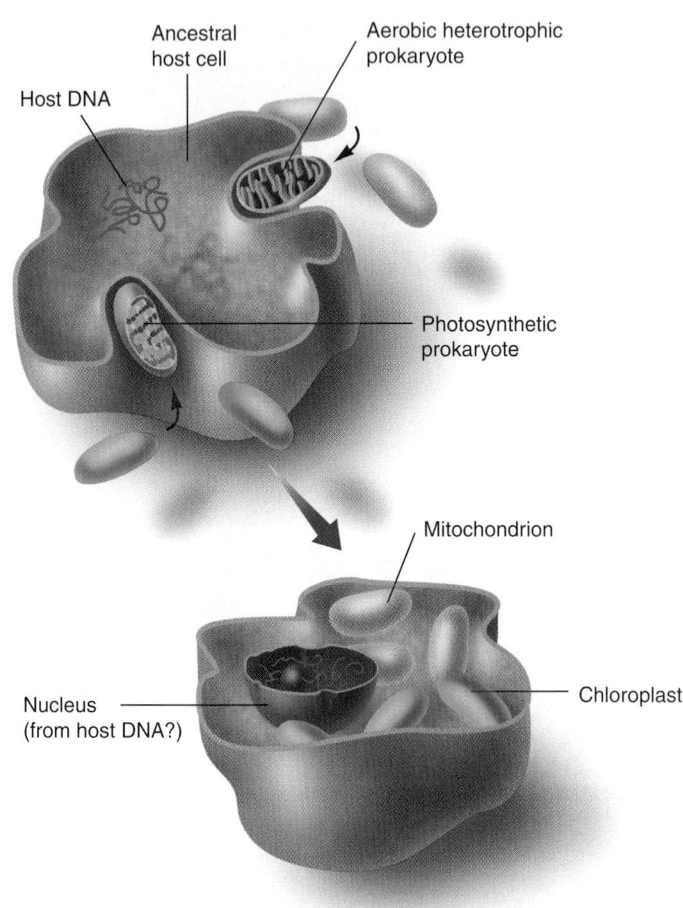

FIGURE 4.19

The endosymbiont theory proposes that mitochondria and chloroplasts descend from once free-living microorganisms.

Cancer Cells— A Personal Reflection~

continued from page 60.

I was the 1 in 100 whose thyroid lump defied the odds. Through my tears, my brain registered the words "papillary carcinoma," and I at least knew I had the "good" type of thyroid cancer. Surgery and radiation soon followed, and I was, and am, fine. But I will never forget the terror of discovering that I had cancer.

Preventing and conquering cancer are compelling reasons to study cell biology. By describing and understanding what is normal, we can begin to battle what is not. And cancer cells are far from normal.

A healthy cell has a characteristic shape, with a boundary that allows entry to some substances, yet blocks others. Not so the misshapen cancer cell, with its fluid surface and less discriminating boundaries. The cancer cell breaches the controls that hold other cells in place, squeezing into spaces where other cells do not, secreting biochemicals that blast pathways through healthy tissue. The cancer even creates its own personal blood supply. The renegade cell's genetic controls differ from those of healthy cells and it transmits these differences when it divides. Cancer cells disregard the "rules" of normal cell division that enable the body to develop and maintain distinct organs. To defy so many biological traditions, the cancer cell uses up tremendous amounts of energy, causing further disruptions.

Understanding the biology of cancer cells is our most powerful weapon against them. New combinations of drugs are targeting cancer cells at vulnerable points—altering their surfaces to attract the immune system; dismantling the fibers that enable them to divide often; bolstering the immune system's cancer-fighting biochemical arsenal. Detecting cancer cells' heightened energy use or altered surfaces permits earlier diagnoses, when treatments are more effective.

We hear so often of those who have died of cancer. Yet many people living perfectly normal, productive lives either have cancer or have had it. Many times we can defeat this most frightening of illnesses. Our ability to do so rests solidly on our knowledge of the biology of cells, the units of life.

tochondrion, complete with respiratory enzymes (fig. 4.19). The smaller bacterium found a new home in the larger cell; in return, the host cell could survive in the newly oxygenated atmosphere.

Similarly, large cells that picked up cyanobacteria or other small cells capable of photosynthesis obtained the forerunners of chloroplasts and thus became the early ancestors of red algae or of green plants. Once such ancient cells had acquired their endosymbiont organelles, genetic changes impaired the ability of the captured prokaryotes to live on their own outside the host cell. The larger cells and the captured prokaryotes came to depend on one another for survival. The result of this biological interdependency, according to the endosymbiont theory, is the compartmentalized cells of modern eukaryotes, including our own.

Although none of us was present 2.5 billion years ago to witness the appearance of complex cells, each of the steps to the endosymbiont hypothesis are seen in present-day organisms. In the absence of a time machine, the elegant endosymbiont theory may be the best window we have to look into the critical juncture in cell history when cells became more complex. As we learn more about the mysterious archaea, it will be interesting to see how the theory evolves.

Mastering Concepts

1. What is the evidence that certain complex organelles descend from simpler cells engulfed long ago?
2. How are the hypothesized evolution of mitochondria and chloroplasts related?

SUMMARY

1. **Cells,** the basic units of life, are the microscopic components of all living organisms. Cells exhibit the characteristics of life.

2. Even the simplest cells are highly organized; complex cells may carry out very specialized functions.

3. Organisms may be **unicellular** (one cell) or **multicellular** (many cells). Cells in multicellular organisms specialize by expressing subsets of genes.

4. Cells were observed first in the late seventeenth century, when Robert Hooke viewed cork with a crude lens. Antonie van Leeuwenhoek viewed many cells under the light microscope he invented. Stains revealed subcellular details. Modern tools include the **light microscope, electron microscope,** and the **scanning probe microscope.**

5. The **cell theory** states that all life is composed of cells, that cells are the functional units of life, and that all cells come from preexisting cells. Many illnesses result from abnormalities at the subcellular level.

6. **Prokaryotic** organisms are unicellular and include bacteria and **cyanobacteria.** A prokaryotic cell lacks a nucleus and has DNA, **ribosomes** (structures that help manufacture proteins), enzymes for obtaining energy, and various other biochemicals. A **cell membrane** and usually a rigid **cell wall** enclose the cell contents. Prokaryotes are abundant and ancient.

7. The more complex **eukaryotic** cells sequester certain biochemical activities in **organelles.** A eukaryotic cell houses DNA in a membrane-bounded **nucleus;** synthesizes, stores, transports, and releases molecules along a network of organelles **(endoplasmic reticulum, Golgi bodies, vesicles);** degrades wastes in **lysosomes;** processes toxins and oxygen in **peroxisomes;** extracts energy from nutrients in **mitochondria;** and in plants, extracts solar energy in **chloroplasts.** A **cell membrane** surrounds eukaryotic cells. **Cell walls** protect and support cells of many organisms, except animals.

8. **Archaea** are a third form of life that share some characteristics with prokaryotes and eukaryotes but also have unique structures and biochemistry.

9. Simpler than cells are **viruses, viroids,** and **prions.** Viruses consist of nucleic acid cores, protein **capsids,** and in some cases, envelopes. Viruses require a host cell to reproduce and have varied effects on specific species. Viroids are infectious RNA, and prions are infectious glycoproteins.

10. The **endosymbiont theory** proposes that chloroplasts and mitochondria evolved from once free-living prokaryotes engulfed by larger prokaryotes—chloroplasts descend from cyanobacteria, and mitochondria descend from aerobic bacteria. Evidence for the endosymbiont theory is that mitochondria and chloroplasts resemble small aerobic bacteria in size, shape, and membrane structure and in the ways their DNA, RNA, and ribosomes interact to manufacture proteins.

TO REVIEW . . .

1. Describe the structures and functions of each of the following organelles:

a. nucleus
b. ribosome
c. lysosome
d. smooth ER
e. rough ER
f. chloroplast
g. mitochondrion
h. peroxisome
i. Golgi body

2. Identify the cell parts indicated in this diagram:

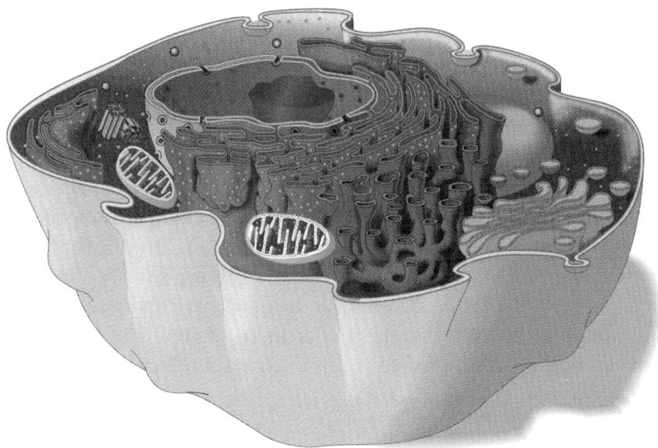

3. How do the structures of a virus, bacterium, and human cell differ?

4. Until very recently, life scientists thought that there were only two forms of life. How has that view changed?

5. A nerve cell is long, with many branches. A muscle cell is spindle-shaped. A white blood cell can move and engulf particles. Each of these cell types has the same types of organelles, yet each also has a characteristic structure and function. How can these cells contain the same components, yet be so different from one another?

6. As a cube increases in size from 3 centimeters to 5 centimeters to 7 centimeters on a side, how does the surface area to volume ratio change?

7. Name three structures or activities in both prokaryotic and eukaryotic cells. List eight differences between the two cell types.

8. What are the differences between viruses, viroids, and prions?

9. How can a virus have an intricate capsid, with 20 faces, yet have very few genes encoding that protein?

10. State the cell theory and the endosymbiont theory. What evidence supports each?

TO THINK ABOUT . . .

1. A liver cell has a volume of 5,000 μm^3. Its total membrane area, including the inner membranes lining organelles as well as the outer cell membrane, fills an area of 110,000 μm^2. A cell in the pancreas that manufactures digestive enzymes has a volume of 1,000 μm^3 and a total membrane area of 13,000 μm^2. Which cell is probably more efficient in carrying out activities that require extensive membrane surfaces? State the reason for your answer.

2. What advantages does compartmentalization confer on a large cell?

3. In an inherited condition called glycogen cardiomyopathy, teenagers develop muscle weakness, including heart muscle. Affected muscle cells contain huge lysosomes, swollen with glycogen. What is the biochemical basis of the glycogen buildup?

4. Why does a muscle cell contain many mitochondria and a white blood cell contain many lysosomes?

5. Cells of the green alga *Chlamydomonas reinhardtii* are grown for a few hours in a medium that contains amino acids "labeled" with radioactive hydrogen, called tritium. The *Chlamydomonas* cells are then returned to a nonradioactive medium. At various times, the cells are applied to photographic film (which radiation from the tritium exposes, producing dark silver grains). The film is then examined under an electron microscope. After 3 minutes, the film shows radioactivity in the rough ER; after 20 minutes, in the smooth ER; after 45 minutes, in the Golgi body; after 90 minutes, in vesicles near one end of each of the cells. After 2 hours, no radioactive label is evident. What cellular process has this experiment traced? How might a similar technique be used to follow a lysosome's activity?

6. The amoeba *Pelomyxa palustris* is a single-celled eukaryote with no mitochondria, but it contains symbiotic bacteria that can live in the presence of oxygen. How does this observation support or argue against the endosymbiont theory of the origin of eukaryotic cells?

7. Some of the DNA in the nuclei of certain species can jump from one chromosome to another. Do you think that movable DNA could account for the presence of DNA in mitochondria and chloroplasts? If so, what subcellular structures might transfer DNA from the nucleus to the mitochondria and chloroplasts?

8. Why is it biologically incorrect to call a virus an organism?

9. Cite two ways in which the cell theory is still evolving.

10. Give two examples of organisms that we did not know about until recently either because they cannot survive in a laboratory or because their natural habitat is extreme (to us).

TO LEARN MORE . . .

References and Resources

Culliton, Barbara J. January 1997. The microbial plague continues. *Nature Medicine,* vol. 3. Despite advances in medical knowledge and technology, we continue to be assaulted by bacteria, viruses, and prions—some apparently new.

Dickson, David. December 1996. A sane response to mad cow link. *Nature Medicine,* vol. 2. A new variant of Creutzfeldt-Jakob disease is linked to cows in England fed parts of infected sheep. The sheep had a prion disease.

de Duve, Christian. April 1996. The birth of complex cells. *Scientific American.* The endosymbiont theory proposes that complex cells arose by engulfing simpler complex cells.

Gray, Michael W. September 26, 1996. The third form of life. *Nature,* vol. 383. Cells come in three basic varieties.

Lewis, Ricki. February 19, 1996. Telomere findings may yield tips for treating cancer, geriatric disorders. *The Scientist.* Chromosome ends normally shorten.

Lewis, Ricki. February 15, 1996. Telomerase level may aid in decisions on cancer therapy. *Genetic Engineering News.* The length of the telomeres of cancer cell chromosomes predicts severity of illness.

Lewis, Ricki. January/February 1996. Probing the complex world of the cell. *Biophotonics International.* Fluorescence microscopy highlights cell functions, including apoptosis and signal transduction.

Margulis, Lynn, and Michael F. Dolan. January/February 1997. Swimming against the current. *The Sciences.* Reconstructing endosymbiosis as it led to the evolution of the cell.

Oliwenstein, Lori. January 1996. Bacterium tells all, human tells a lot. *Discover.* Prokaryotes and eukaryotes differ greatly in genetic complexity.

Preston, Richard. 1994. *The hot zone.* New York: Doubleday. The true and riveting tale of Ebola virus loose in a laboratory animal facility in the United States.

denotes articles by author Ricki Lewis that may be accessed via *Life*'s home page /http://www.mhh.com/science math/biology/life.

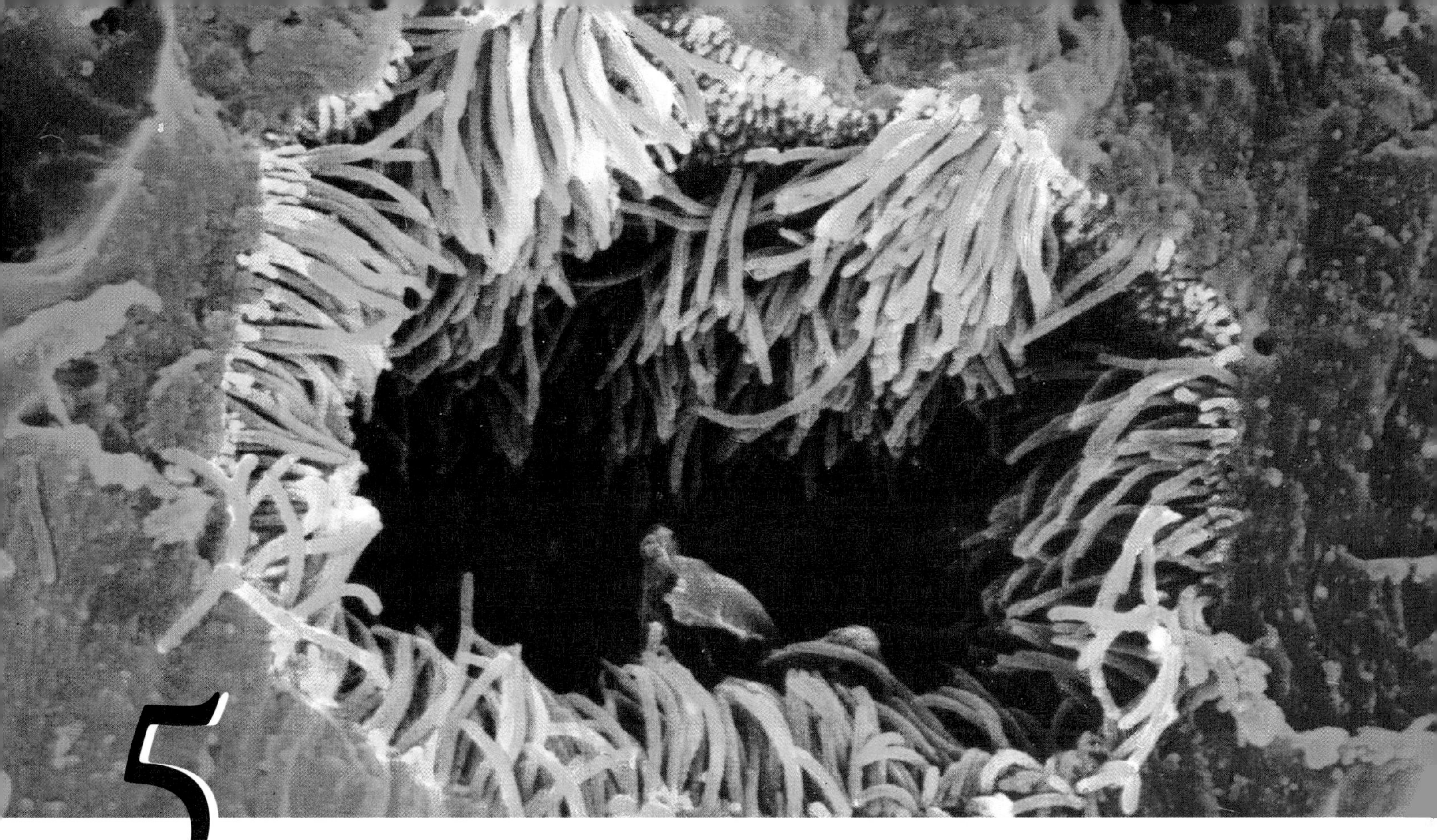

5 Cellular Architecture

These cells lining a rat's respiratory passageway have very distinctive shapes and functions. The extensive fringes, consisting of waving outgrowths called cilia (yellow) and extensions of the cell membrane called microvilli (green), greatly increase the cell's surface area—an adaptation to its function of absorbing digested nutrients.

chapter outline

The Cell Surface and "Self" ***84***
Cellular Name Tags

The Cell Membrane—Cellular Gates ***85***
The Protein-Phospholipid Bilayer
Membrane Proteins

Movement Across Membranes ***87***
Diffusion
A Special Case of Diffusion—The Movement of Water (Osmosis)
Facilitated Diffusion
Active Transport

Large-Scale Membrane Transport ***93***
Exocytosis and Endocytosis

Cell Connections ***96***
Animal Cell Junctions
Plant Cell Junctions

The Cytoskeleton—Cellular Support ***96***
Microtubules
Microfilaments
Intermediate Filaments
The Cytoskeleton Molds Specialized Cells

Cell-to-Cell Interactions ***103***
Signal Transduction
Cell Adhesion

Sisters or Brothers?

The sisters seemed quite ordinary, until they reached adolescence and did not mature sexually. No one questioned that the young women were female, until a physician performed internal examinations on the girls. Although from the outside the sisters indeed looked female, their insides told a different developmental story.

Each girl lacked internal female organs and structures (uterus, fallopian tubes, cervix, and upper one-third of the vagina), but because the lower two-thirds of the vagina appeared normal, they had been raised as girls. However, an ultrasound scan showed in each girl's abdomen small testes, which are male structures that produce sperm cells. Doctors decided, with the girls' and their parents' permission, to probe further.

The next step was a chromosome test. Two of our chromosomes, the X and the Y, indicate sex, XX a female and XY a male. The chromosome test showed that the sisters were XY—they were in this sense really brothers. Their outsides and feelings did not match their insides and chromosomes. The reason for the confusion lies in cell biology—certain of the girls' cells had failed to communicate with one another.

to be continued . . .

The Cell Surface and "Self"

The 2-year-old's health appeared to be returning mere hours after the transplant. Her new liver, which replaced a damaged one, came from a child just killed in a car accident. The patient's skin was losing its sickly yellow pallor, and she was much more alert. Yet, just as the new organ began taking over the jobs the old one had abandoned, immune system cells detected the new organ and, interpreting it as "foreign," began to produce chemicals to attack it. Even though the donor's liver was carefully "matched" to the little girl—the pattern of molecules on the surfaces of its cells was very similar to that on the cells of her own liver cells—the match was not perfect. A rejection reaction was in progress, and the transplanted organ would soon cease to function. The little girl would need another transplant to survive.

The rejection of a transplanted organ illustrates the importance of cell surfaces in the coordinated functioning of a multicellular organism. The cell surface is one component of the cellular architecture—structures that give a cell its particular three-dimensional shape and topography, help determine the locations and movements of organelles and biochemicals within the cell, and participate in the cell's interactions with other cells and the extracellular environment.

The cellular architecture consists of surface molecules embedded in the cell membrane, which is the outer covering of a cell. Just beneath the cell membrane protein fibers form part of the cell's interior scaffolding, or cytoskeleton. Together, the cell surface and cytoskeleton form a dynamic structural framework that helps to distinguish one cell from another (fig. 5.1). The various components of the cellular architecture must communicate and interact for the cell to carry out the processes of life.

Cellular Name Tags

At a conference where most participants do not know one another, name tags are often used to identify people. Cells also have name tags in the form of carbohydrates, lipids, and proteins that protrude from their surfaces. These cellular name tags are found on all cells, from the single cell of a bacterium to the trillions of cells of the human body. Some surface molecules distinguish cells of different species, like company affiliations on name tags. Other surface structures distinguish individuals within a species. A multicellular organism must be able to determine which cells are part of the body and which are not. Within the human body, the immune system performs this function (see chapter 39). The immune system is a huge collection of white blood cells and the biochemicals they produce. These cells recognize the surfaces of a person's cells as "self" and all other cell surfaces as "nonself." When the immune system encounters the nonself surfaces of some bacteria or other foreign agents that enter the body, it attacks. Unfortunately for the child who received the liver transplant, her immune system recognized the cells of her transplanted liver as nonself.

Surface structures also distinctively mark cells of different tissues in an individual, so that a bone cell's surface is different from the surface of a nerve cell or a muscle cell. Surface differences between cell types are particularly important during the development of an embryo, when different cells sort and grow into specific tissues and organs. Cell surfaces also change over time; the number of surface molecules declines as a cell ages.

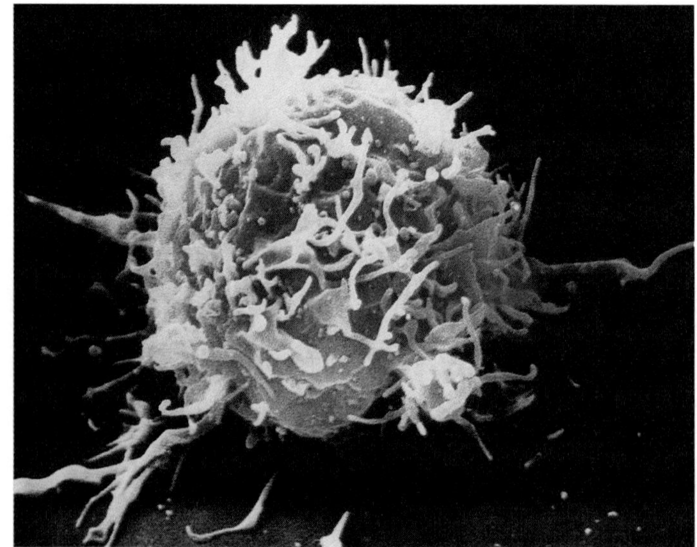

FIGURE 5.1

Cellular architecture. A white blood cell's inner skeleton and surface features enable it to move in the body and to recognize "foreign" cell surfaces—such as those of transplanted tissue. This T lymphocyte, also known as a T cell, carries out the tissue rejection reaction.

Individual patterns of cell surface features may be used to predict the likelihood that a person will develop certain disorders. The **human leukocyte antigens** (HLA) are cell surface molecules that appear in different patterns in different people. The name comes from the fact that these cell surface molecules, or **antigens,** are found on leukocytes, the white blood cells. A person who has certain HLA cell surface molecules may have a much greater chance of developing a certain disease than a person with a different HLA makeup. Researchers do not usually know the reason for the correlation between an HLA type and a disorder.

One HLA-linked disease is ankylosing spondylitis, which inflames and deforms the vertebrae. A person with an HLA antigen called B27 is 100 times more likely to develop the condition than a person who lacks the B27 antigen. This HLA prediction is based on the observation that more than 90% of people who suffer from this disorder have B27, although the antigen is present in only 5% of the general population. However, predictions of disease risk based on HLA tests should be approached cautiously: 10% of people who have ankylosing spondylitis do *not* have the B27 antigen, and some people who have the antigen never develop the disease. Other HLA-associated disorders include psoriasis, rheumatoid arthritis, and juvenile-onset diabetes mellitus.

Mastering Concepts

1. What are components of the cellular architecture?
2. What is the basis of immune system function?
3. What are antigens?

The Cell Membrane—Cellular Gates

Just as the character of a community is molded by the people who enter and leave it, the special characteristics of different cell types are shaped in part by the substances that enter and leave them. The cell membrane, a selective barrier that completely surrounds the cell, monitors the movements of molecules in and out of the cell. The chemical characteristics and the pattern of the molecules forming a membrane determine which substances can cross it. In eukaryotes, membranes are also found within cells as parts of organelles.

The Protein-Phospholipid Bilayer

The structure of a biological membrane is possible because of a chemical property of the phospholipid molecules that compose it. One end of a phospholipid is attracted to water, while the other end is repelled by it (fig. 5.2). **Phospholipids** are lipid molecules with attached phosphate groups (PO_4, a phosphorus atom bonded to four oxygen atoms).

The phosphate end of a phospholipid molecule, which is attracted to water, is hydrophilic (water-loving); the other end, consisting of two fatty acid chains, moves away from water because it is hydrophobic (water-fearing). Because of these water preferences, phospholipid molecules in water spontaneously arrange into a **phospholipid bilayer.** In this two-layered, sandwichlike structure, the hydrophilic surfaces are on the outsides of the "sandwich," exposed to the watery medium outside and inside the cell. The hydrophobic surfaces face each other on the inside of the "sandwich," away from water.

The phospholipid bilayer forms the structural backbone of a biological membrane. The pharmaceutical industry uses phospholipid bilayers to construct microscopic bubbles called liposomes, which are used to encapsulate a variety of drugs. Liposomes deliver anticancer drugs, antiinflammatory drugs, antibiotics, and even a baldness remedy and artificial tears.

In a cell membrane, the phospholipid bilayer's hydrophobic interior forms a barrier to most substances dissolved in water. However, some proteins embedded in the bilayer create passageways through which water-soluble molecules and ions pass. Other proteins are carriers, transporting substances across the membrane. The membranes of living cells, then, consist of phospholipid bilayers and the proteins embedded in them (fig. 5.3).

Membrane Proteins

Proteins within the oily lipid bilayer can move about, sometimes at remarkable speed. Because of this movement, the protein-phospholipid bilayer is often called a fluid mosaic. A pigment protein in cells of the retina of a human eye, for example, moves to different depths within the phospholipid bilayer of the cell membrane depending upon the intensity of the incoming light. The lipid molecules can also move laterally.

One way to classify membrane proteins is by their location in the phospholipid bilayer. Membrane proteins may lie completely within the phospholipid bilayer or traverse the membrane to extend out of one or both sides (fig. 5.4). In animal cells, membrane proteins often attach to branchlike carbohydrate molecules to form glycoproteins, which protrude from the membrane's outer surface. The proteins and glycoproteins that jut from the cell membrane create the surface characteristics that are so important to a cell's interactions with other cells.

The functions of membrane proteins are related to their locations within the phospholipid bilayer. We saw in the case of transplant rejection that one function of cell membrane proteins is to establish a cell's surface as "self." Some membrane proteins exposed on the outer face of the membrane are receptors, binding outside molecules and triggering cascades of chemical reactions in the cell that lead to a specific response. This process is called **signal transduction.** Another important class of membrane proteins are **cellular adhesion molecules** (CAMs). These proteins enable specific cell types to stick to each other, making cell-to-cell interactions possible. Signal transduction and cell adhesion are discussed at the end of the chapter. Table 5.1 reviews functions of membrane proteins.

Table 5.1 Types of Membrane Proteins

Protein Type	Function
Carrier proteins	Transport substances across membranes
Cell surface proteins	Establish self
Cellular adhesion molecules	Enable cells to stick to each other
Receptor proteins	Mediate signal transduction

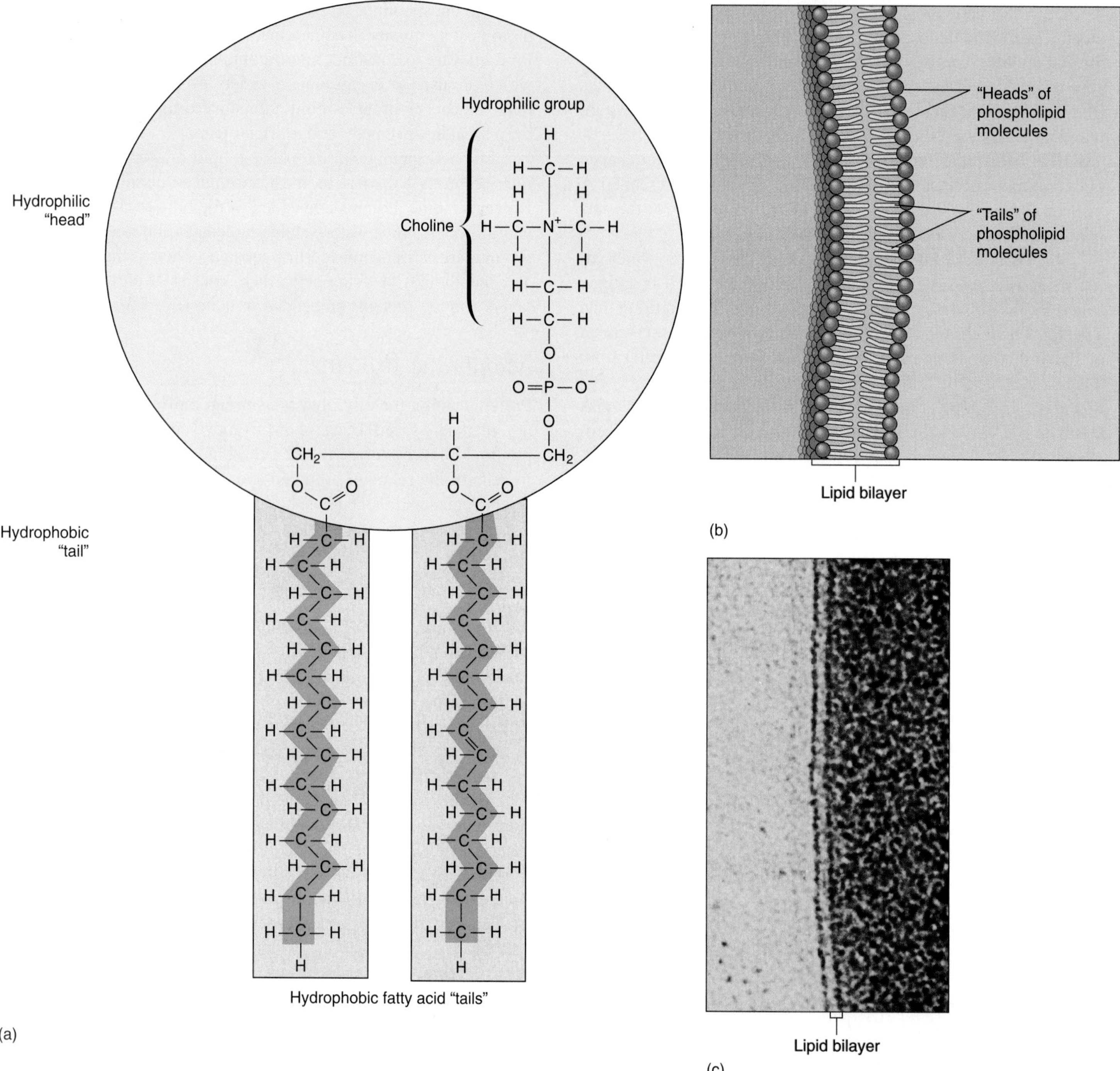

FIGURE 5.2

The two faces of membrane lipids. (*a*) A phospholipid is literally a two-faced molecule, with one end attracted to water (hydrophilic, or water-loving) and the other repelled by it (hydrophobic, or water-fearing). Membrane lipids are often depicted as a circle with two tails. (*b*) and (*c*) A depiction and an electron micrograph of a phospholipid bilayer.

Disruptions in the protein/lipid ratio of a membrane can affect health. Consider the cell membranes of certain types of cells that support nerve cells. About three-quarters of their membranes consist of a lipid called myelin. The cells enfold nerve cells in tight layers, wrapping their fatty cell membranes into a sheath that provides insulation which speeds nerve impulse transmission. In multiple sclerosis, the cells coating the nerves that lead to certain muscles lack myelin. The resulting blocked neural messages to muscles impair vision and movement and cause numbness and tremor. In Tay-Sachs disease, the reverse happens. The cell membranes accumulate too much lipid, and brain cells beneath them cannot transmit messages to each other and to muscle cells. An affected child gradually loses the ability to see, hear, and move and dies by age four.

Outside the Cell
Extracellular fluid
Glycolipid
Sugar molecules
Proteins
Glycoprotein
Lipid bilayer
Protein fibers
Cytoplasm
Inside the Cell

FIGURE 5.3

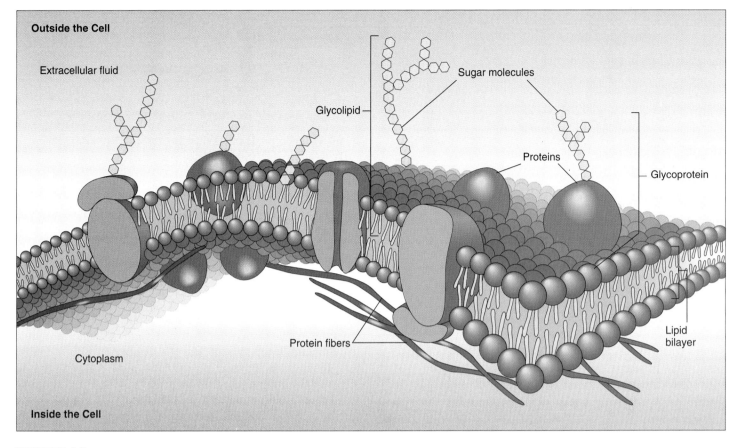

In a cell membrane, mobile proteins embed throughout a phospholipid bilayer, which is also mobile, to produce a somewhat fluid structure. An underlying mesh of protein fibers supports the cell membrane. Jutting from the membrane's outer face are carbohydrate molecules linked to proteins (glycoproteins) and lipids (glycolipids). Because a membrane includes several types of molecules, and proteins move within the fluid phospholipid bilayer, the structural organization of a membrane is described as a fluid mosaic.

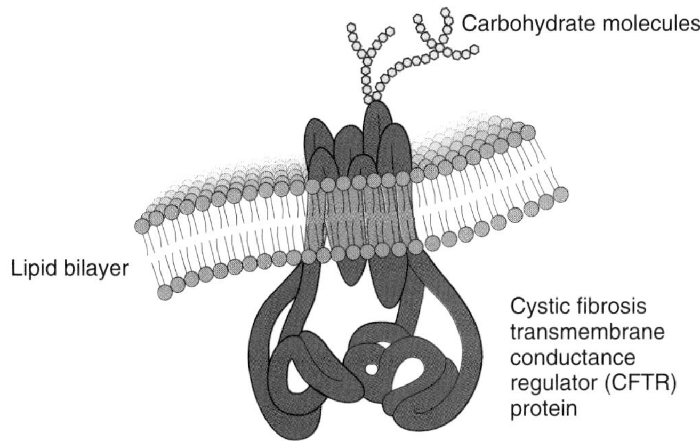

FIGURE 5.4

The protein that, when abnormal, causes cystic fibrosis traverses the cell membrane, emerging from both sides. It functions as a channel for chloride ions in several tissues. When abnormally shaped, the channel traps chloride ions inside cells, which draws fluids inward. As a result, in the respiratory passages and certain digestive organs, secretions dry out, causing symptoms.
Source: M.C. Iannuzi and F.S. Collins, "Reverse Genetics and Cystic Fibrosis" in American Journal of Respiratory Cellular and Molecular Biology 2:309–316, 1990 American Lung Association.

Mastering Concepts

1. How is the cell surface important in health?
2. What molecules make up a cell membrane?
3. How are the components of a cell membrane organized?
4. How are the locations of membrane proteins related to their functions?
5. How do membranes differ in different cell types?

Movement Across Membranes

A muscle cell can contract and a nerve cell can conduct a message only if certain molecules and ions are maintained at certain levels inside and outside the cell. In all cells, the cell membrane oversees these vital concentration differences. Before considering how cells control which substances enter and leave, it is helpful to define some terms.

An **aqueous solution** is a homogeneous mixture of a substance (the **solute**) dissolved in water (the **solvent**). Lemonade made from a powdered mix illustrates the relationship between solute and solvent: the solvent is water, the solute is the powdered

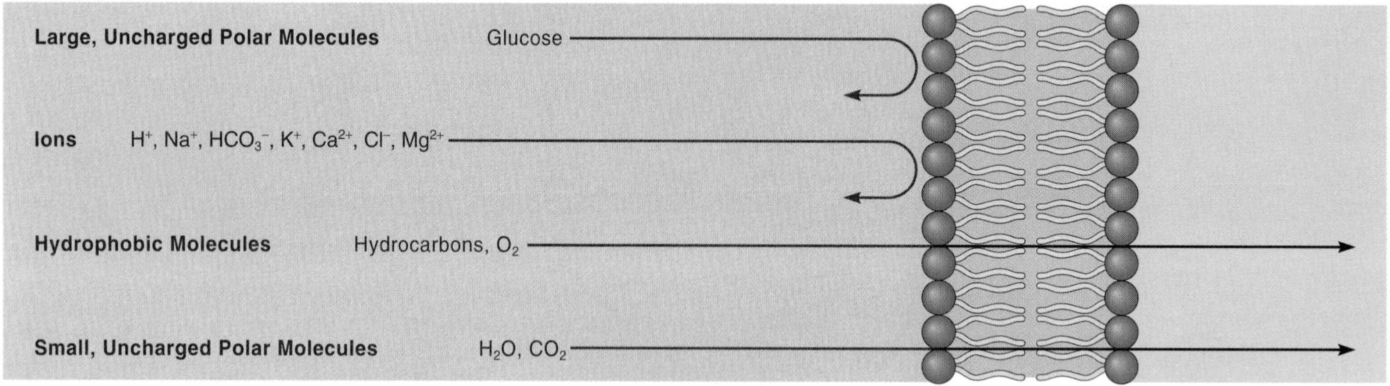

FIGURE 5.5

A phospholipid bilayer is selectively permeable, allowing in hydrophobic molecules such as hydrocarbons, oxygen, and small, uncharged polar molecules but blocking ions and large uncharged polar molecules.

mix, and the solution is lemonade. Concentration refers to the relative number of one kind of molecule compared to the total number of molecules present, and it is usually given in terms of the solute. When solute concentration is high, the proportion of solvent (water) present is low, and the solution is concentrated. When solute concentration is low, solvent is proportionately high, and the solution is dilute.

Diffusion

The cell membrane is selectively permeable—that is, some molecules are able to pass freely through the membrane (either between the molecules of the phospholipid bilayer or through protein-lined channels), while others are not. Figure 5.5 depicts the permeability of a phospholipid bilayer to certain molecules. For example, oxygen (O_2), carbon dioxide (CO_2), and water (H_2O) freely cross biological membranes. They do so by a process called **diffusion.**

Diffusion is the movement of a substance from a region where it is more concentrated to a region where it is less concentrated without using energy. Diffusion occurs because molecules are in constant motion, and they move so that two regions of differing concentration tend towards evening out. An easy way to observe diffusion is to place a tea bag in a cup of hot water. Compounds in the tea leaves dissolve gradually and diffuse throughout the cup. The tea is at first concentrated near the bag, but the brownish color eventually spreads to create a uniform brew.

The natural tendency of a substance to move from where it is highly concentrated to where it is less so is called "moving down" or "following" its **concentration gradient.** A gradient is a general term that refers to a difference in some quality between two neighboring regions. Gradients may be created by concentration, electrical, and pressure differences.

Molecules crossing a membrane because of this natural tendency to travel from higher to lower concentration is called simple diffusion because it does not require energy or a carrier molecule. Simple diffusion eventually reaches a point where the concentration of the substance is the same on both sides of the membrane. After this, molecules of the substance continue to flow randomly back and forth across the membrane at the same rate, so that the concentration remains the same on both sides. This point of equal movement back and forth is called **dynamic equilibrium.**

To envision dynamic equilibrium, picture a party taking place in two rooms. The festivities are in full swing. Everyone has arrived, and no one has yet left. People walk between the rooms in a way that maintains the same number of partiers in each room, but the specific occupants change. The party is in dynamic equilibrium.

A Special Case of Diffusion—The Movement of Water (Osmosis)

The fluids that continually bathe our cells consist of molecules dissolved in water. Because cells are constantly exposed to water, it is important to understand how a cell regulates water entry. If too much water enters a cell, it swells; if too much water leaves, it shrinks. Either response may affect a cell's functioning. Water moves across biological membranes by a form of simple diffusion called **osmosis,** which is influenced by the concentration of dissolved substances inside and outside the cell (fig. 5.6).

Tonicity

In osmosis, water is driven to move because the membrane is impermeable to the solute, and the solute concentrations differ on each side of the membrane. Water moves across the membrane in the direction that dilutes the solute on the side where it is more concentrated.

Variants of the word **tonicity** are used to describe osmosis in relative terms. Tonicity refers to the differences in solute concentration in two compartments separated by an impermeable membrane. Most cells are **isotonic** to the surrounding fluid, which means that solute concentrations are the same within and outside the cell. In this situation, there is no net flow of water, and a cell's shape does not change.

Disrupting a cell's isotonic state changes its shape as water rushes in or leaks out. If a cell is placed in a solution in which

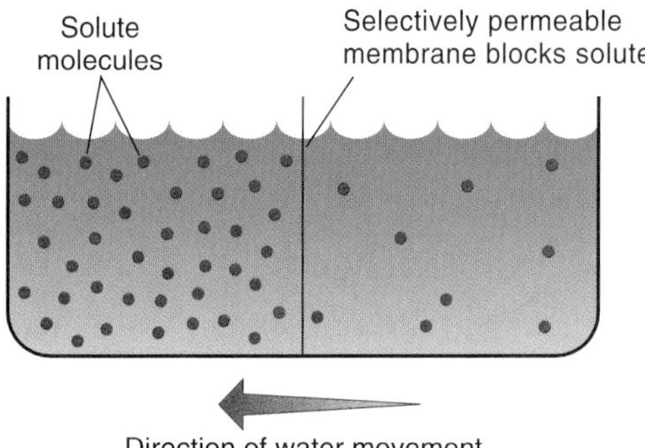

FIGURE 5.6

Osmosis—a biologically important type of simple diffusion. Either a synthetic nonliving material (called an artificial membrane) or a biological membrane can be used to demonstrate osmosis, which is the movement of water down its concentration gradient. The membrane does not allow solute molecules to pass through. The differing concentrations of solute on either side of the membrane drive the movement of water molecules from a region where their concentration is high to where it is low; that is, water moves through the membrane to the side where the solute is more concentrated and the concentration of water is consequently lower.

the concentration of solute is lower than it is inside the cell, water enters the cell to dilute the higher solute concentration there. In this situation, the solution outside the cell is **hypotonic** to the inside of the cell. The cell swells. In the opposite situation, if a cell is placed in a solution in which the solute concentration is higher than it is inside the cell, water leaves the cell to dilute the higher solute concentration outside. In this case, the outside is **hypertonic** to the inside. This cell shrinks.

Hypotonic and hypertonic are relative terms and can refer to the surrounding solution or to the solution inside the cell. It may help to remember that *hyper* means "over," *hypo* means "under," and *iso* means "the same." A solution in one region is hypotonic or hypertonic to a solution in another region.

Osmosis and Cell Shape

The effects of immersing a cell in a hypertonic or hypotonic solution can be demonstrated with a human red blood cell, which is normally suspended in an isotonic solution called plasma (fig. 5.7). In this state, the cell is doughnut-shaped, with a central indentation. Placing a red blood cell in a hypertonic solution forces water out of the cell, and it shrinks. Placing the cell in a hypotonic solution has the opposite effect. To dilute its insides, the cell admits water, and it swells.

Because shrinking and swelling cells may not function normally, unicellular organisms can regulate osmosis to maintain their shapes. Many cells alter membrane transport activities, changing the concentrations of different solutes on either side of the cell membrane, in a way that drives osmosis in a direction that maintains the cell's shape. This enables some single-celled inhabitants of the ocean to remain isotonic to their salty environment, so their shapes persist. In contrast, the paramecium, a single-celled organism that lives in ponds, must

Water in = water out

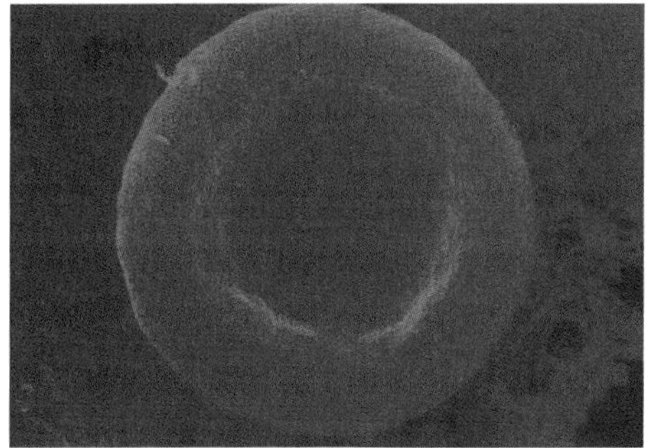

Isotonic (no net change in water movement)

(a)

Water out > water in

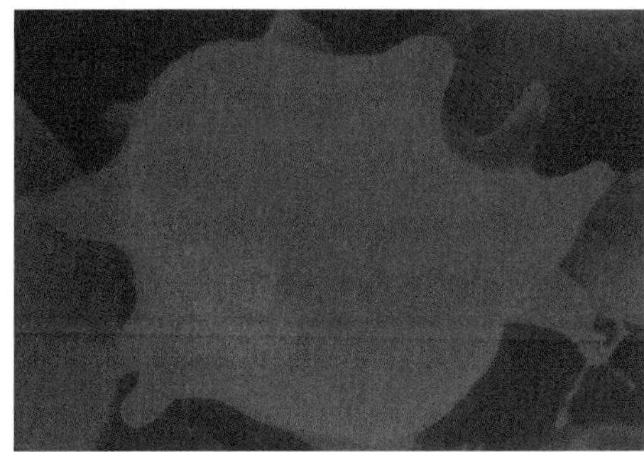

Blood cells in hypertonic solution
(water diffuses outward)

(b)

Water in > water out

Blood cells in hypotonic solution
(water diffuses inward)

(c)

FIGURE 5.7

A red blood cell changes shape in response to changing plasma solute concentrations. (*a*) A human red blood cell is normally shaped like a doughnut without a hole. When the surrounding blood plasma is isotonic to the cell's interior, water enters and leaves the cell at the same rate and the cell maintains its characteristic shape. (*b*) When the salt concentration of the plasma increases, so that the surrounding fluid is hypertonic to the cell, water leaves the cells to dilute the outside solute faster than water enters the cell. The cell shrinks. (*c*) When the salt concentration of the plasma decreases relative to the salt concentration inside the cell, so that the surrounding fluid is hypotonic to the cell, water flows into the cell faster than it leaves. The cell swells and may even burst.

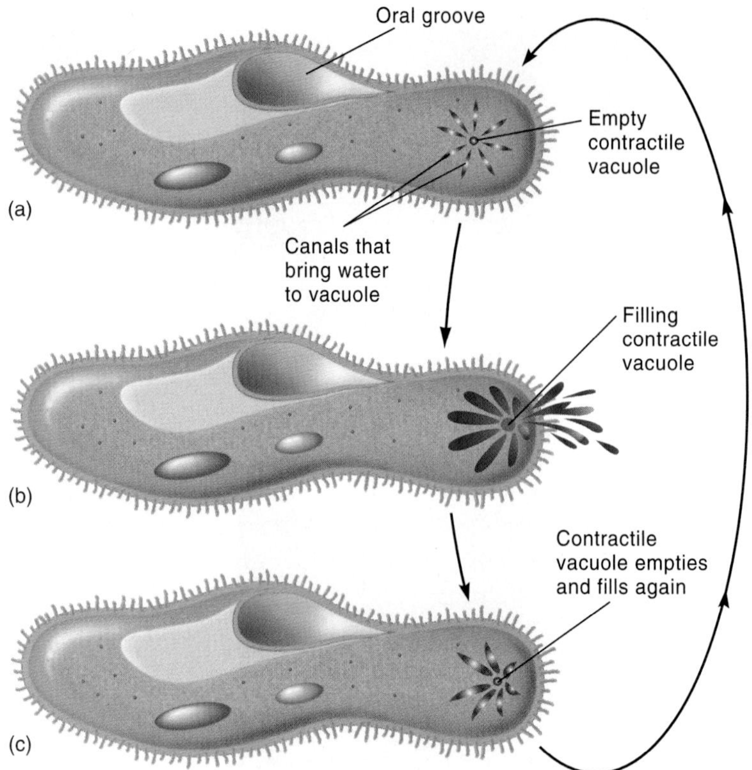

FIGURE 5.8

How paramecia keep their shapes. Because solutes within paramecia are more concentrated than they are in the surrounding water (making the inner solution hypertonic to the outside), water tends to enter these unicellular organisms faster than it leaves. Contractile vacuoles fill and then pump excess water out of the cells across the cell membrane. The contractile vacuole resides in the cell's interior (*a*). It moves near the cell membrane as it fills with water (*b*), and then releases the water to the outside. The organelle then resumes its empty shape (*c*) and moves away from the cell membrane back to the interior of the cell.

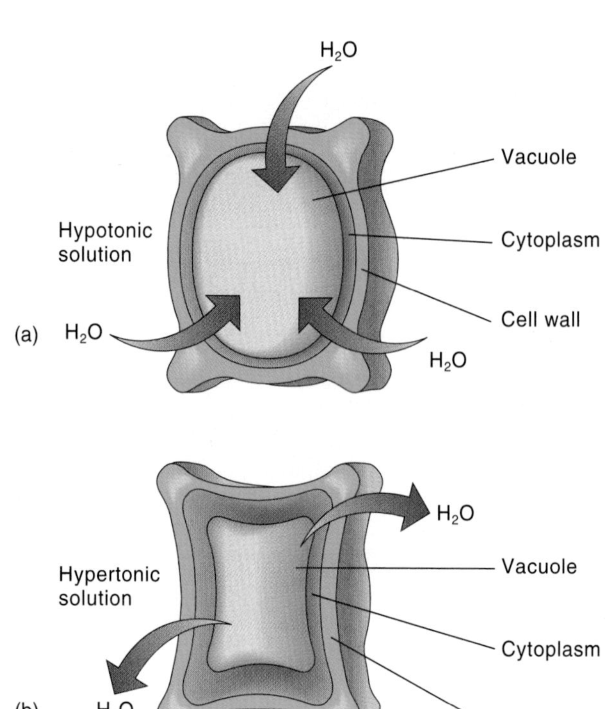

FIGURE 5.9

How plant cells keep their shapes. Like paramecia, plant cells usually contain more concentrated solutes than their surroundings, causing water to flow into the cell. (*a*) In a hypotonic solution, water enters the cell and collects in vacuoles. The cell swells against its rigid, restraining cell wall, generating turgor pressure. (*b*) When a plant cell is placed in a hypertonic environment (so that solutes are more concentrated outside the cells), water flows out of the vacuole, and the cell shrinks. Turgor pressure is low, and the plant wilts.

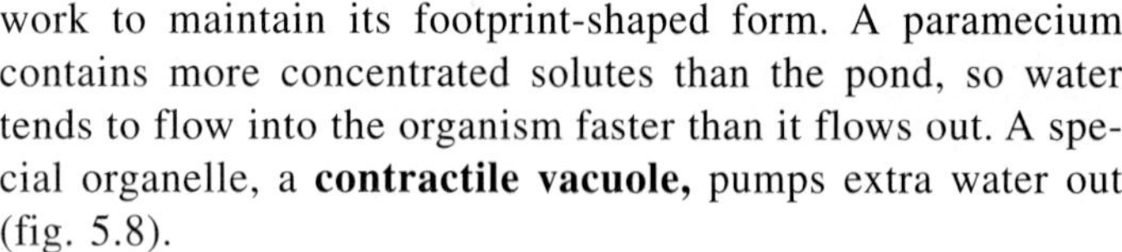

work to maintain its footprint-shaped form. A paramecium contains more concentrated solutes than the pond, so water tends to flow into the organism faster than it flows out. A special organelle, a **contractile vacuole,** pumps extra water out (fig. 5.8).

Plant cells also face the challenge of maintaining their shapes even with a concentrated interior. Instead of expelling the extra water that rushes in, as the paramecium does, plant cells expand until their cell walls restrain their cell membranes. The resulting rigidity, caused by the force of water against the cell wall, is called **turgor pressure** (fig. 5.9). A piece of wilted lettuce demonstrates the effect of losing turgor pressure. When placed in water, the leaf becomes crisp, as the individual cells expand like inflated balloons.

Osmosis and the Human Kidneys

Osmosis influences the concentration of our urine (fig. 5.10). Cells in the brain called osmoreceptors shrink when body fluids are too concentrated, indicating water loss. The shrinking osmoreceptors signal the pituitary gland in the brain to release antidiuretic hormone (ADH). The bloodstream transports ADH to cells lining the kidney tubules, where it alters their permeabilities so that water exits the tubules and enters capillaries (microscopic blood vessels) that entwine about the tubules. Without ADH's action, this water would remain in the kidney tubules and leave the body in dilute urine. Instead, it returns to the bloodstream, precisely where it's needed.

In the opposite situation, when a person drinks too much water, osmoreceptors swell with excess water. This inhibits ADH release, and the kidney cells remain impermeable to water. As a result, urine is very dilute and copious, ridding the body of the excess. Drinking alcohol causes abundant, dilute urine by inhibiting release of ADH.

Proteins in the membranes of certain kidney cells, called aquaporins, control water reabsorption. Abnormal aquaporin is one cause of a condition called diabetes insipidus, in which the kidney cells cannot adequately reabsorb water. People with this disorder urinate profusely, up to 7 gallons (26 liters) a day!

Dehydrated Person

Waterlogged Person

Osmoreceptors in brain shrink, triggering pituitary gland to release ADH.

Osmoreceptors in brain swell, preventing pituitary gland from releasing ADH.

Kidneys

Kidney tubule

H_2O

Urine

Dark yellow

Pale yellow

Capillary

FIGURE 5.10

To counteract dehydration, osmoreceptors in the brain shrink, allowing the pituitary gland to release ADH. This hormone travels in the bloodstream to the kidneys. Here, it alters the permeability of certain cells in the kidney tubules, so that water enters the capillaries near the tubules, returning to the bloodstream instead of leaving as urine. The reverse occurs when a person drinks too much—excess water leaves in the urine. Kidney tubule permeability alters so that water doesn't return to the already waterlogged bloodstream.

FIGURE 5.11

Simple diffusion, facilitated diffusion, and active transport. In passive transport, molecules cross cell membranes in the direction dictated by their concentration gradient: they move from the side of the membrane where they are more highly concentrated to the side where they are less highly concentrated. This movement, which does not require energy, occurs by either simple or facilitated diffusion. (*a*) In simple diffusion, molecules move through spaces between the lipid and protein components of the membrane. (*b*) In facilitated diffusion, molecules move across the membrane with the aid of a carrier protein embedded within the membrane. (*c*) In active transport, a molecule crosses a membrane against its concentration gradient, using membrane protein pumps and energy from ATP or a gradient of protons or other ions across the membrane.

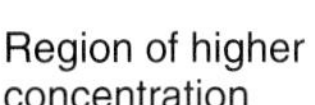

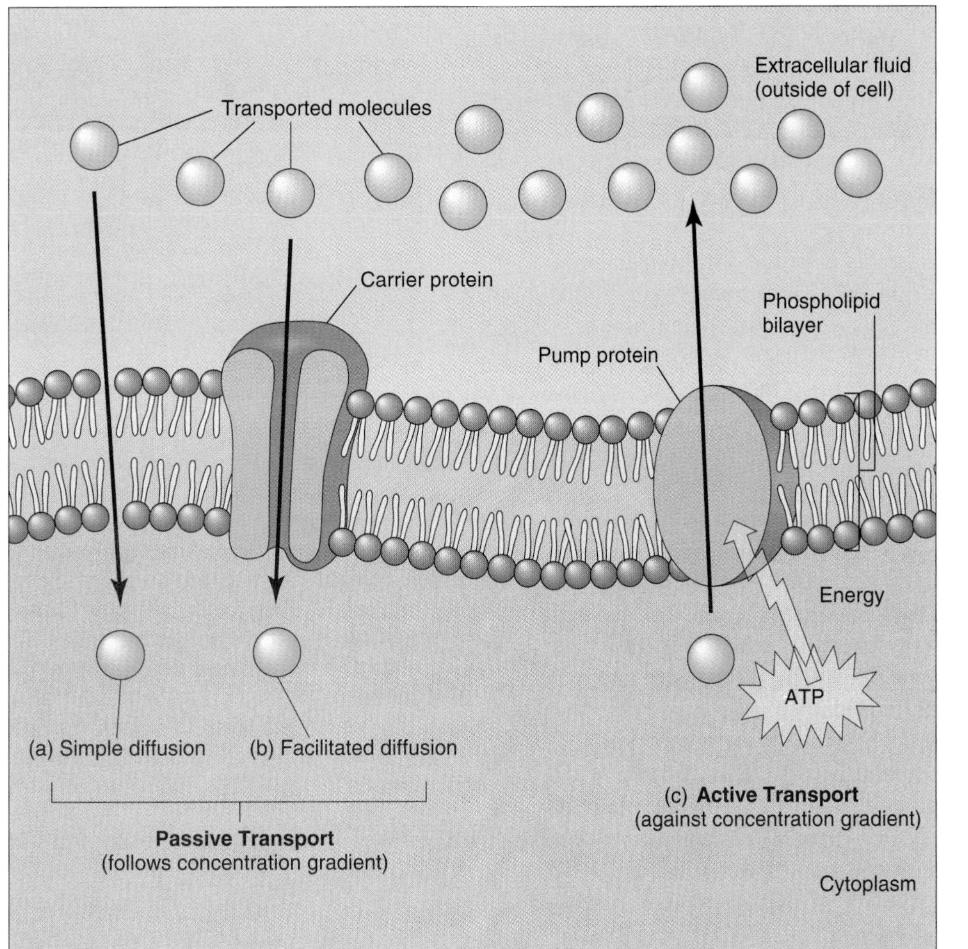

Facilitated Diffusion

Many polar or ionic biochemicals that move in and out of cells cannot simply diffuse across the cell membrane because the hydrophobic portion repels them. One way for these molecules to cross the cell membrane is by moving through channels within carrier proteins. Whether the transported molecule moves into or out of the cell depends upon its concentration on both sides of the membrane. The process of crossing a membrane down a concentration gradient with the aid of a carrier protein is termed **facilitated diffusion.** This route transports glucose, amino acids, and iron from the blood to the brain. Both facilitated diffusion and simple diffusion are considered forms of passive transport because they do not require an input of energy.

Active Transport

Life is, in a sense, a matter of partitioning and maintaining certain concentrations of specific biochemicals within cells. Often passive transport alone cannot maintain these concentrations. How does a cell import more of an essential substance—perhaps a nutrient—that is already more concentrated inside the cell than outside? How would a cell export a metabolic by-product that is already more concentrated outside the cell?

Movement of substances against their concentration gradients is possible, but it requires energy. The molecule that often provides biological energy is called **adenosine triphosphate** (ATP), which is discussed in the next chapter. When a phosphate group splits from ATP, energy is released; that energy can then be harnessed to help drive a cellular function. The movement of a substance against a concentration gradient with the aid of a carrier protein and energy is called **active transport.** Figure 5.11 compares simple diffusion, facilitated diffusion, and active transport.

The first active transport system discovered was a membrane carrier protein called the **sodium-potassium pump** found in the cell membranes of most animal cells. Cells must contain high concentrations of potassium ions (K^+) and low concentrations of sodium ions (Na^+) to perform such basic functions as maintaining their volume and synthesizing protein, as well as to conduct more specific activities, such as transmitting nerve impulses and enabling the lungs and kidneys to function.

The sodium-potassium pump first binds Na^+ from inside the cell (fig. 5.12). When ATP splits, the freed phosphate also binds to the pump. This changes the shape of the pump, allowing Na^+ to move out of the cell and K^+ to move into the cell. The sodium-potassium pump thus uses energy from ATP to help keep the proper concentrations of Na^+, K^+, and other substances inside and outside of the cell. For every three sodium ions pumped out, two potassium ions are pumped in.

Another example of active transport is the functioning of the drug fluoxetine (Prozac), used to treat depression. Researchers attribute the inexplicable sadness of clinical depression to a deficiency of a biochemical called serotonin between certain nerve cells in the brain. Serotonin is a type of nervous system messenger molecule called a neurotransmitter. Fluoxetine blocks a

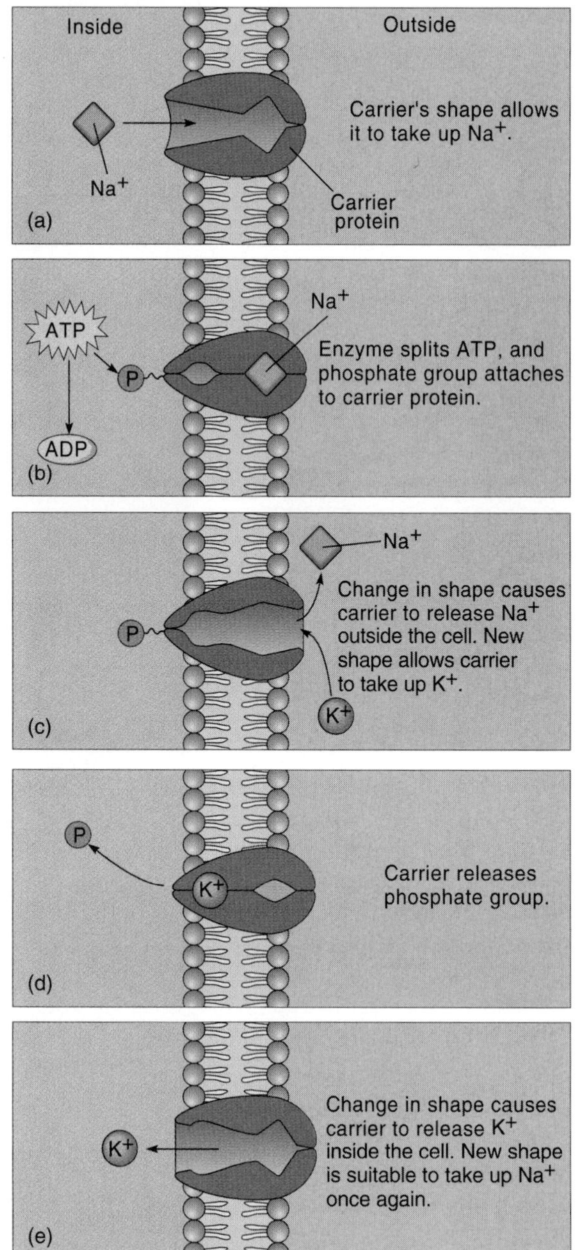

FIGURE 5.12

The sodium-potassium pump. This "pump," actually a carrier protein embedded in the cell membrane, uses energy to move potassium ions (K^+) into the cell and sodium ions (Na^+) out of the cell. The pump first binds Na^+ on the inside face of the membrane (*a*). ATP is split to yield ADP and a phosphate group, and the liberated phosphate binds to the carrier or pump (*b*). This binding alters the conformation of the pump and causes it to release the Na^+ to the outside. The altered pump can now take up K^+ from outside the cell (*c*). Next, the pump releases the bound phosphate (*d*), which alters the conformation of the carrier protein once again. This time, the change in shape releases the K^+ to the cell's interior (*e*). The pump is also back in the proper shape to bind intracellular Na^+.

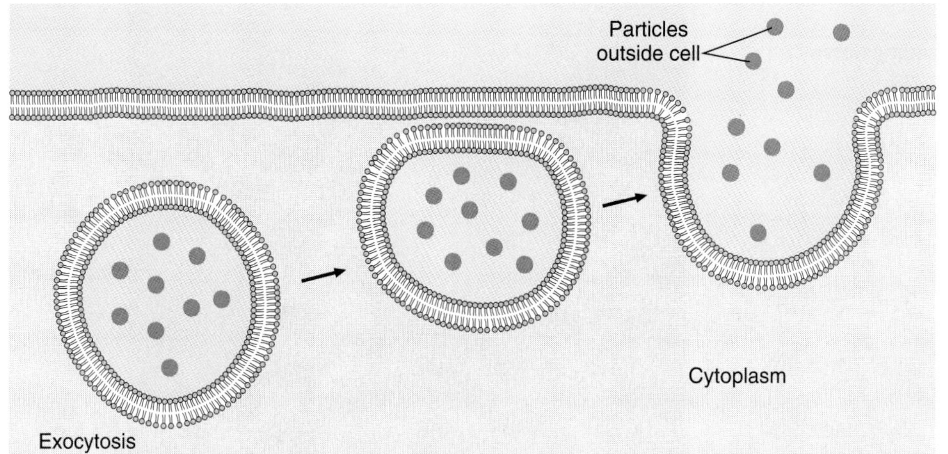

(a)

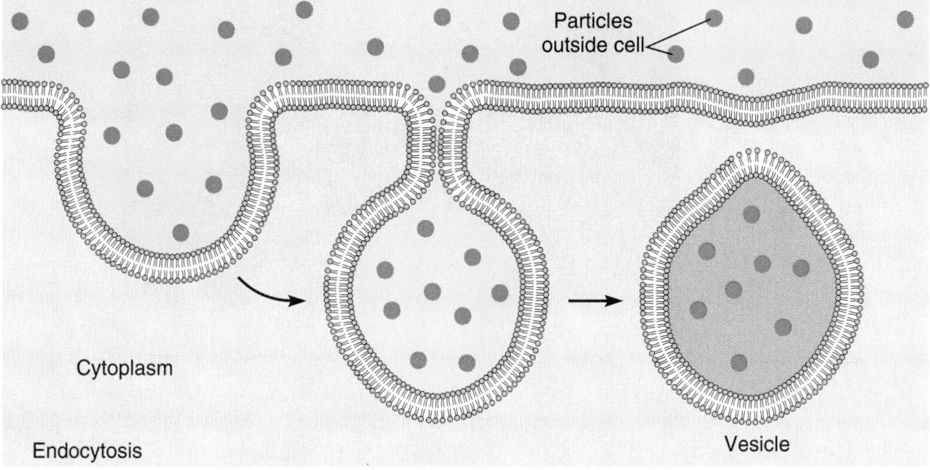

(b)

FIGURE 5.13

Exocytosis and endocytosis. (*a*) Biochemicals or particles exit cells by exocytosis. Phospholipid bilayer spheres within the cell surround or take up the structures to be exported. These structures or particles travel to the cell membrane and merge with it. The particles are then released to the outside. (*b*) Endocytosis transports substances into the cell. Large particles and even whole bacteria can enter a cell through endocytosis. A small portion of the cell membrane buds inward, entrapping the particles, and a vesicle forms, which brings the substances into the cell.

membrane protein that would otherwise actively transport serotonin into the nerve cells. Serotonin remains between the cells longer, and the sending cells release more of it, actions that alleviate depression in some people.

Mastering Concepts

1. What is diffusion?
2. How do differing concentrations of solutes in neighboring aqueous solutions drive the movement of water molecules?
3. How do facilitated diffusion and active transport differ from each other and from diffusion?

Large-Scale Membrane Transport

Most molecules dissolved in water are small, and they can cross cell membranes by simple diffusion, facilitated diffusion, or active transport. Large molecules (and even bacteria) can also enter and leave cells with the help of cell membranes.

Exocytosis and Endocytosis

Exocytosis transports large particles out of cells, such as some components of milk (see fig. 4.11). Inside a cell, a vesicle made of a phospholipid bilayer surrounds or takes up the substances to be sent out of the cell. The vesicle moves to the cell membrane and joins with it, releasing the substance outside the membrane (fig. 5.13*a*). For example, exocytosis in the front tip of a sperm cell releases enzymes that enable the tiny cell to penetrate the much larger egg cell.

Endocytosis is the reverse of exocytosis. A vesicle that has released its contents via exocytosis merges with the cell membrane (fig. 5.13*b*). Once inside, the vesicle is recycled by joining an intermediate stage vesicle, called an **endosome.** Endocytosis may also bring a substance into a cell. Some vesicles that enter by endocytosis move to and fuse with lysosomes, where enzymes dismantle the cargo. Certain substances that enter cells by endocytosis are used rather than destroyed. This is the case for some neurotransmitters.

The coordination of exocytosis and endocytosis is well studied in the human nervous system (fig. 5.14). The cycle of neurotransmitter function begins as active transport loads vesicles in a nerve cell sending a message. The filled vesicles move to the nerve

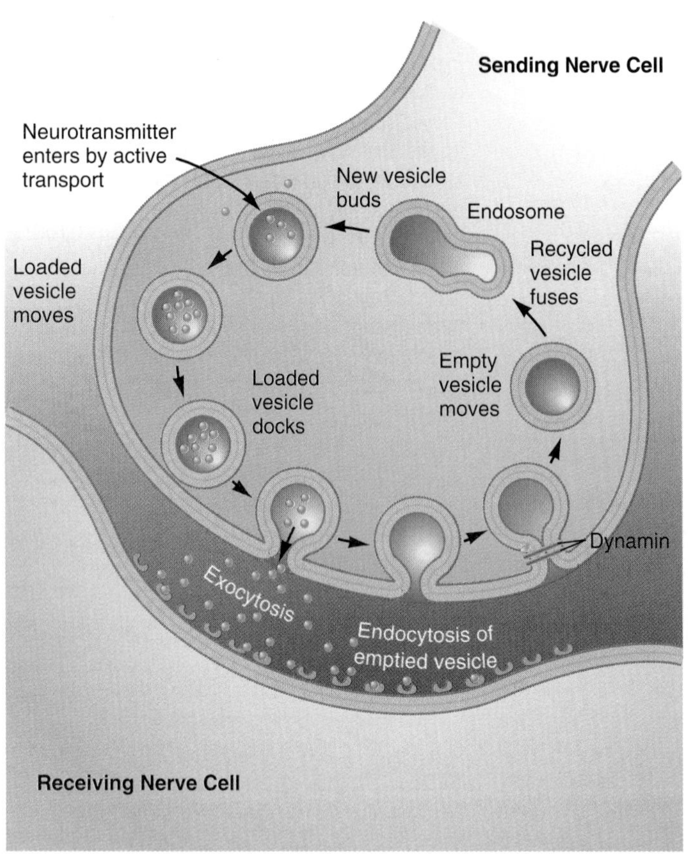

(a)

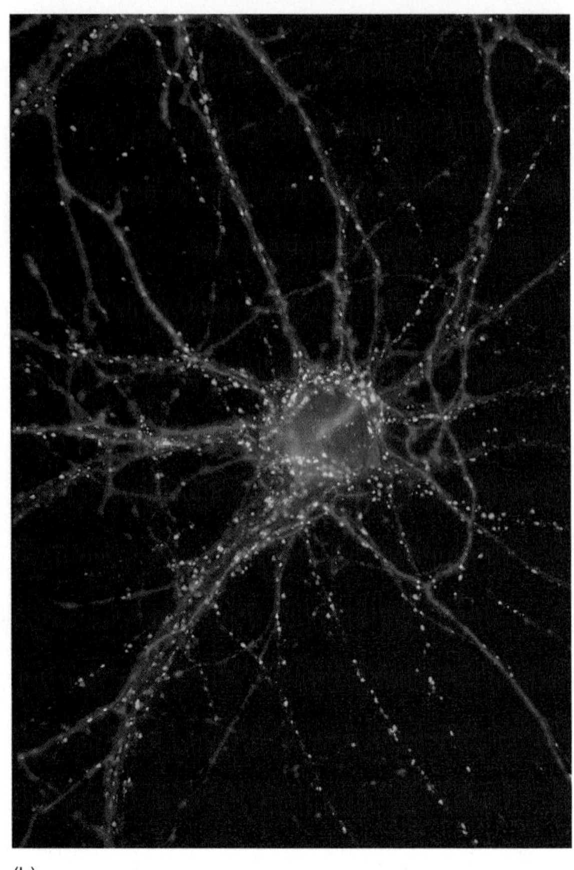

(b)

FIGURE 5.14

Exocytosis, endocytosis and nerve cell function. (*a*) A human nerve cell tip rapidly sends out neurotransmitter molecules by exocytosis and reclaims the vesicle by endocytosis. (*b*) In this brain neuron, the green dots represent vesicles loaded with neurotransmitter molecules about to be released from nerve cell endings by exocytosis.

ending, where they merge with the cell membrane and release the contents, all in less than a microsecond. This is exocytosis. Almost immediately, the emptied vesicles return to the cell membrane, by endocytosis. The engulfed vesicles then move to endosomes, from which new vesicles bud, and the cycle begins anew. Interference with this process is deadly. Botulinum and tetanus toxins block the cycle of exocytosis and endocytosis in nerve cells and cause fatal paralysis.

Receptor-Mediated Endocytosis

When biologists first viewed endocytosis in white blood cells in the 1930s, they thought it might be a general route of entry into a cell. They believed a cell would gulp in anything at its surface, just as if a theater manager let anyone in rather than just ticket holders. We now recognize a more specific form of endocytosis called **receptor-mediated endocytosis.** In this process, a receptor protein on a cell's surface binds a particular biochemical, called a **ligand;** the cell membrane then indents, embracing the ligand and drawing it into the cell (fig. 5.15). Receptor-mediated endocytosis enables liver cells to take in dietary cholesterol from the bloodstream.

Lipoprotein particles carry cholesterol. One type of cholesterol carrier is called a **low-density lipoprotein,** or LDL. The protein portions of LDL particles, called **apolipoproteins,** bind to LDL receptors clustered in protein-lined pits in the surfaces of liver cells. The cell membranes envelop the LDL particles, forming loaded vesicles and bringing them into the cell, where they move towards lysosomes. Within lysosomes, enzymes liberate the cholesterol from the LDL carriers. The cholesterol then enters the endoplasmic reticulum, and the receptor recycles to the cell surface, where it can bind more cholesterol-laden LDL particles.

Vesicle Trafficking

How does a cholesterol-containing vesicle "know" how to find a lysosome, or any other organelle? Vesicles move about the cell following specific routes in a process called **vesicle trafficking.** As is often the case in organisms, proteins provide the specificity required in a generalized process.

A vesicle "docks" at its target membrane guided by a series of proteins and then anchors to proteins called receptors. Some guiding proteins are located within the phospholipid bilayers of the organelle and the approaching vesicle, while others are initially free in the cytoplasm. Certain combinations of proteins join and form a complex that helps draw the vesicle to its destination (fig. 5.16).

Table 5.2 summarizes the transport mechanisms that move substances across membranes.

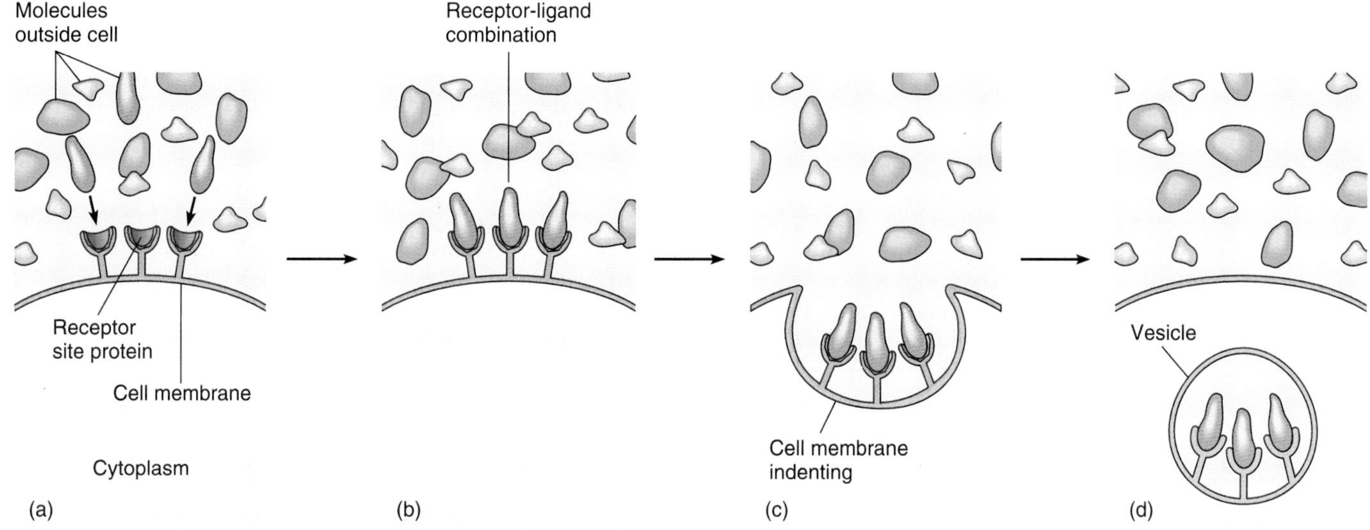

FIGURE 5.15

Receptor-mediated endocytosis. A specific molecule binds to a receptor site protein (*a*). The combination of the molecule binding its receptor (*b*) stimulates the cell membrane to indent (*c*), forming a vesicle (*d*) that transports the molecule into the cytoplasm.

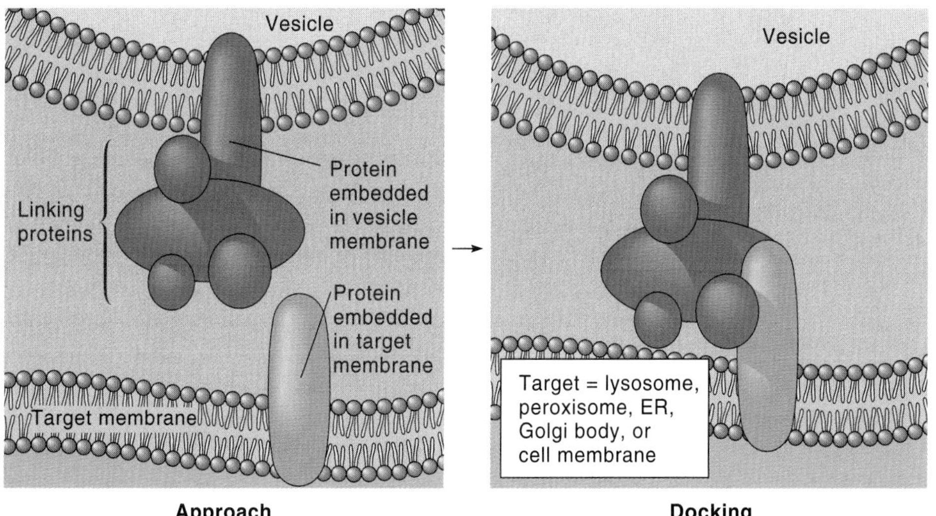

FIGURE 5.16

Vesicles recognize and dock at the appropriate organelle by means of a series of linking proteins.

Table 5.2 Movement Across Membranes

Mechanism	Characteristics	Example
Diffusion	Follows concentration gradient	Oxygen diffuses from lung into capillaries
Osmosis	Diffusion of water	Water reabsorbed from kidney tubules
Facilitated diffusion	Follows concentration gradient, assisted by carrier protein	Glucose diffuses into red blood cells
Active transport	Moves against concentration gradient, assisted by carrier protein and energy, usually ATP	Salts reabsorbed from kidney tubules
Endocytosis	Membrane engulfs substance and draws it into cell in membrane-bounded vesicle	White blood cells ingest bacteria
Exocytosis	Membrane-bounded vesicle fuses with cell membrane, releasing its contents outside of cell	Nerve cells release neurotransmitters

Mastering Concepts

1. How do exocytosis and endocytosis transport large particles across cell membranes?
2. How can endocytosis become specialized?
3. What structures guide vesicles inside a cell?

Cell Connections

Many cells are separated from each other. Blood cells are suspended in fluid, cartilage cells lie in a flexible background, and bone cells are in a hard, mineralized matrix. Yet other cells are tightly packed, with specializations of their cell membranes, called **intercellular junctions,** knitting them together.

Animal Cell Junctions

Animal cells have several types of intercellular junctions (fig. 5.17). In a **tight junction,** the cell membranes of adjacent cells fuse at a localized point. The area of fusion surrounds the two cells like a belt, closing the space between the cells. This region is also called a "kiss site." Tight junctions join cells into sheets, such as those that line the inside of the human digestive tract. In serious diarrheal infections, bacteria produce toxins that loosen the tight junctions in intestinal lining cells, which causes fluid to rapidly exit.

Extensive tight junctions knit together the cells of microscopic blood vessels (capillaries) in the human brain. This anatomical organization prevents chemical fluctuations from damaging delicate brain tissue. Biology in Action 5.1 explains how researchers are applying knowledge of membrane transport to manipulate this "blood-brain barrier" to treat certain disorders.

Another type of intercellular junction in animal cells, called a **desmosome,** rivets or "spot welds" adjacent cells. Desmosomes hold skin cells in place. A third type of junction, a **gap junction,** joins muscle cells forming the heart and muscle cells lining the digestive tract. Gap junctions link the cytoplasm of adjacent cells but allows them to communicate by permitting exchange of ions, nutrients, and other small molecules.

Plant Cell Junctions

Sturdy cell walls, consisting primarily of the carbohydrate cellulose, surround plant cell membranes. Adjacent plant cells join by a layer of a sticky carbohydrate, pectin, which is used in thickening fruit preserves (fig. 5.18*a*).

Tiny channels, called **plasmodesmata,** link plant cells. Plasmodesmata are clustered in weakened, thin parts of the cell wall (fig. 5.18*b*). Because the cell membrane continues along a plasmodesma from one plant cell to another, biochemicals can move between cells. Plasmodesmata are particularly plentiful in parts of plants that conduct water or nutrients and in cells that secrete substances such as oils and nectar.

Table 5.3 summarizes intercellular junctions.

Mastering Concepts

1. What types of structures link cells in animals?
2. What types of structures link cells in plants?

The Cytoskeleton—Cellular Support

The cytoskeleton, a meshwork of tiny protein rods and tubes, molds the distinctive structures of cells, positioning organelles and providing overall shape. The chapter 4 opening photo vividly highlights two major cytoskeletal proteins (tubulin and actin) in a mouse cell. The protein girders of the cytoskeleton are dynamic structures that constantly break down and rebuild as the cell performs specific activities. Some cytoskeletal elements function as rails, forming con-

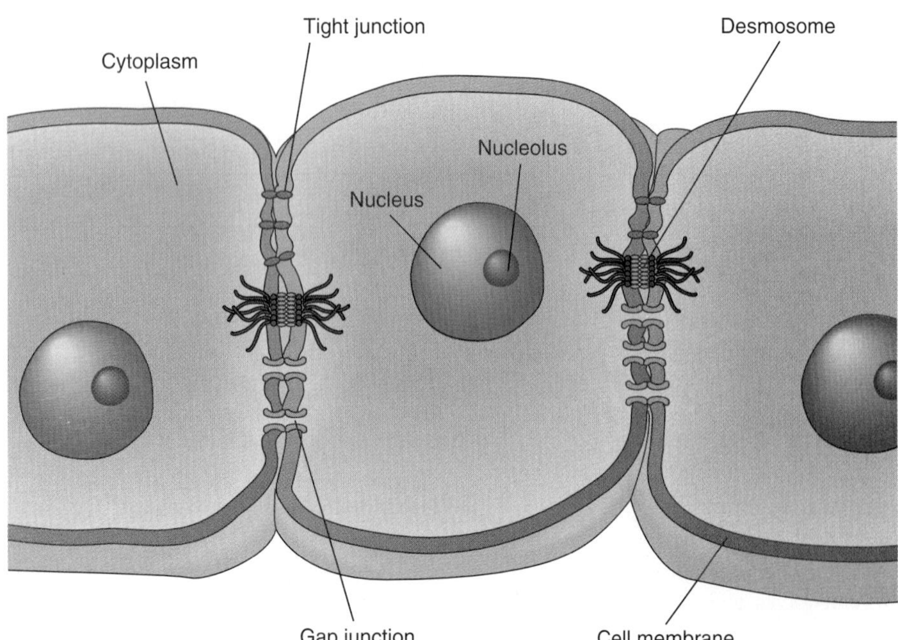

FIGURE 5.17

Tight junctions fuse neighboring membranes, desmosomes form "spot welds," and gap junctions allow small molecules to move between the cytoplasm of adjacent cells.

5.1 biology in action

Breaching the Blood-Brain Barrier

A 400-mile (644-kilometer) network of blood vessels winds through the human brain. Elsewhere in the body, capillaries adhere at a few tight junctions, which leaves holes. Brain capillaries, in contrast, touch as smoothly as bricks cemented together. The resulting blood-brain barrier keeps out many useful drugs.

A Selective Barrier

As researchers learn more about how substances cross the blood-brain barrier, they are developing new ways to deliver useful drugs into the brain. The barrier readily admits lipid-soluble drugs, because the cell membranes of the lining (endothelial) cells that form the capillaries are lipid-rich. Heroin, valium, nicotine, cocaine, and alcohol are lipid soluble and can cross the barrier, which is why these substances act rapidly. Oxygen also enters the brain by directly crossing through endothelial cell membranes.

Water-soluble molecules must take other routes across the blood-brain barrier. Insulin, for example, enters the blood side of an endothelial cell by endocytosis, moves through the cell, and arrives on the brain side by exocytosis. Glucose, amino acids, and iron cross the barrier with the aid of carrier molecules. The brain requires a constant supply of these nutrients.

Delivering Drugs Across the Barrier

Until recently, penetrating the blood-brain barrier required drilling into the skull and delivering a needed drug in a catheter (tube). A less drastic approach is to inject a substance called mannitol into the carotid artery in the neck. Mannitol dehydrates the endothelium, loosening the tight junctions for 15 seconds, enough time for a drug injected into a vein to enter. Mannitol infusion can deliver drugs to battle brain tumors when other treatments fail, but one drawback is that other molecules may also rush in and cause seizures.

An experimental drug called RMP-7 also loosens the tight junctions of the blood-brain barrier. It is used to deliver the drug amphotericin-B to treat a fungal brain infection in AIDS patients. Delivering this drug directly to the brain enables patients to receive lower doses, thus easing the severe nausea that earns the drug its nickname, "amphoterrible."

Mannitol and RMP-7 admit whatever chemicals are near the barrier. A more targeted approach uses receptor-mediated endocytosis and a receptor for a protein called transferrin. This receptor normally ferries iron from the blood to the brain. Each cell membrane forming the blood-brain barrier has up to 100,000 transferrin receptors! A transferrin molecule carrying iron binds to the receptor, moves into the endothelial cell, migrates through the cytoplasm, and departs on the brain side of the cell, where it releases the iron. Researchers hope that a drug linked to the transferrin receptor will follow the same path, without admitting unwanted molecules.

Table 5.3 Intercellular Junctions

Type	Function	Location
Tight junctions	Close spaces between animal cells by fusing cell membranes	Inside lining of small intestine
Desmosomes	Spot weld adjacent animal cell membranes	Outer skin layer
Gap junctions	Form channels between animal cells, allowing exchange of substances	Muscle cells in heart and digestive tract
Plasmodesmata	Allow substances to move between plant cells	Weakened areas of cell walls

duits for cellular contents on the move; other components of the cytoskeleton, the **motor molecules,** power the movements of organelles along these rails. **Microtubules** and **microfilaments** are the most abundant cytoskeletal elements (fig. 5.19).

The motor molecules, fibers, and hollow tubules that make up the cytoskeleton carry out a variety of cellular tasks. Specifically, the cytoskeleton

moves chromosomes apart during cell division;
controls vesicle trafficking;
builds organelles by helping to move components;
enables cellular appendages and cells themselves to move;
builds plants' cell walls;
secretes and takes up neurotransmitters in nerve cells.

Let's take a closer look at the components of the cytoskeleton.

Microtubules

All eukaryotic cells contain long, hollow microtubules that provide many cellular movements (fig. 5.20). A microtubule is only

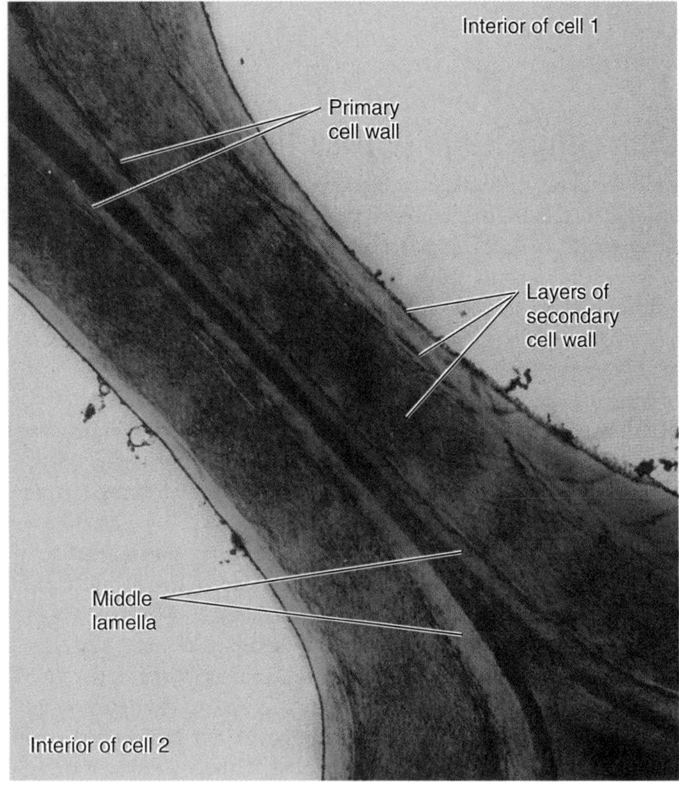

(a)

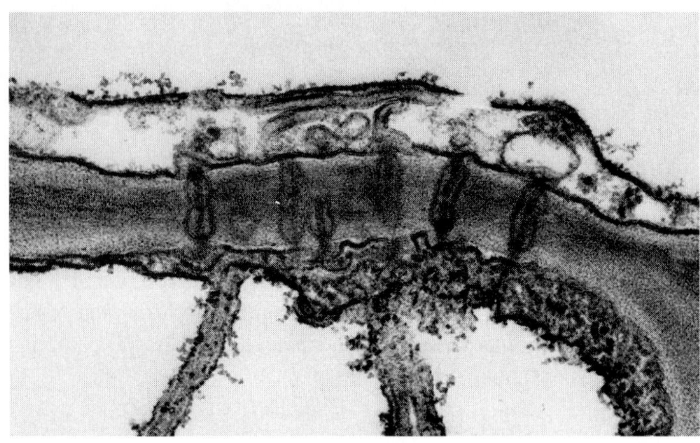

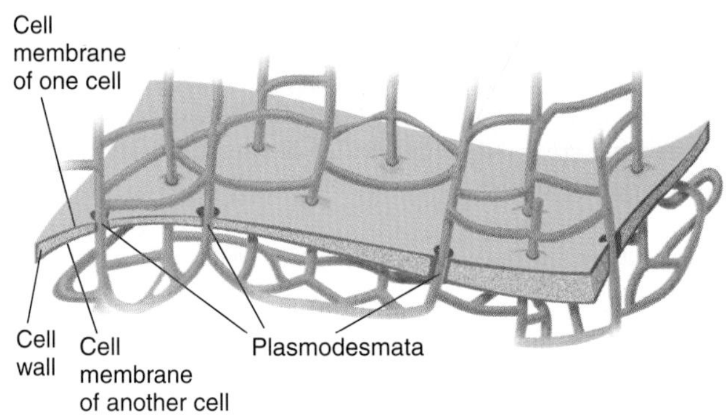

(b)

FIGURE 5.18

Plant cell connections. (*a*) Plant cells adhere to one another by the sticky polysaccharide pectin, which is laid down in a layer called the middle lamella. (×3,000) (*b*) Plant cells exchange substances through continuations of their cell membranes called plasmodesmata.

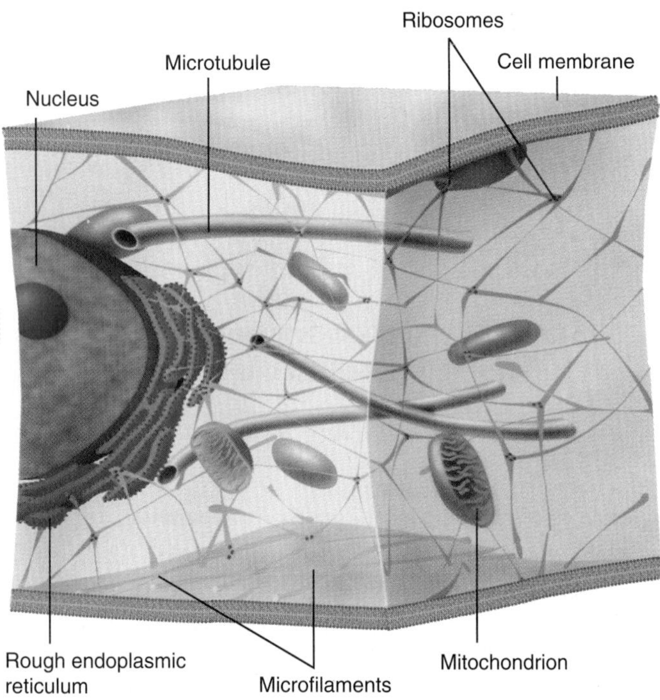

FIGURE 5.19

The cytoskeleton is a scaffolding of protein tubes (microtubules) and rods (microfilaments) that support the cell and help determine the locations and movements of organelles.

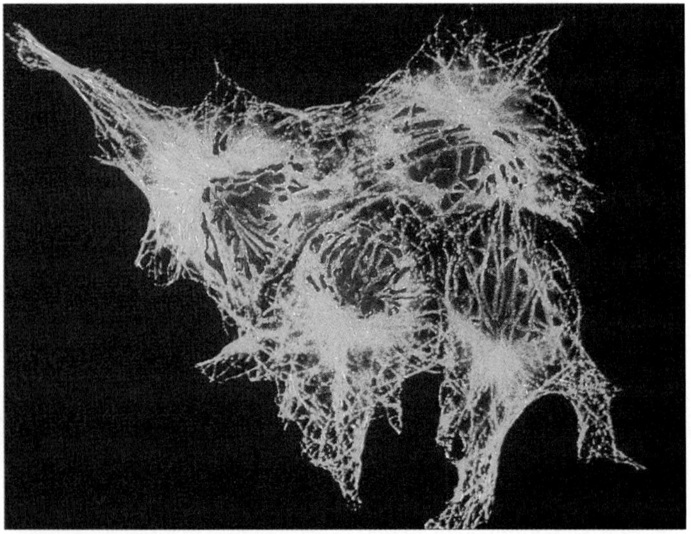

FIGURE 5.20

Microtubules in a mammalian cell.

FIGURE 5.21

Microtubules self-assemble. Microtubules assemble when they gain tubulin dimers at one end and tear down when they lose tubulin dimers from the other end.

about a millionth of an inch (about 25 nanometers) in diameter. Its walls consist of pairs, or dimers, of the protein **tubulin.**

Cells contain both formed microtubules and individual tubulin molecules. When the cell requires microtubules to carry out a specific function—dividing, for example—the free tubulin dimers self-assemble into more tubules. After the cell divides, some of the microtubules dissociate into individual tubulin dimers. This replenishes the cell's supply of building blocks (fig. 5.21). Cells are in a perpetual state of flux, building up and breaking down microtubules.

Some drugs used to fight cancer dismantle microtubules that pull a cell's duplicated chromosomes apart, halting cell division. These drugs damage cancer cells because these cells divide too frequently, but they also harm healthy cells that divide often, thus causing side effects. Many older cancer drugs work by preventing tubulin from assembling into microtubules. The new cancer drug Taxol works by preventing microtubules from breaking down into free tubulin dimers, which also arrests cell division.

Microtubules form locomotor organelles, which move or enable cells to move. The two types of locomoter organelles are **cilia** and **flagella** (fig. 5.22). Cilia are hairlike structures that move in a coordinated fashion, producing a wavelike motion (fig. 5.23).

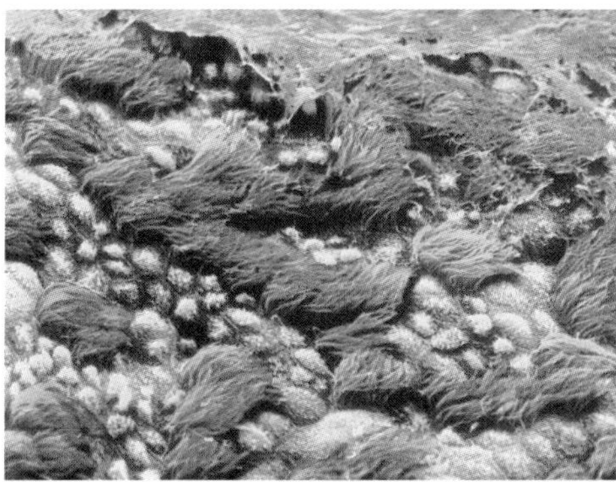

(a)

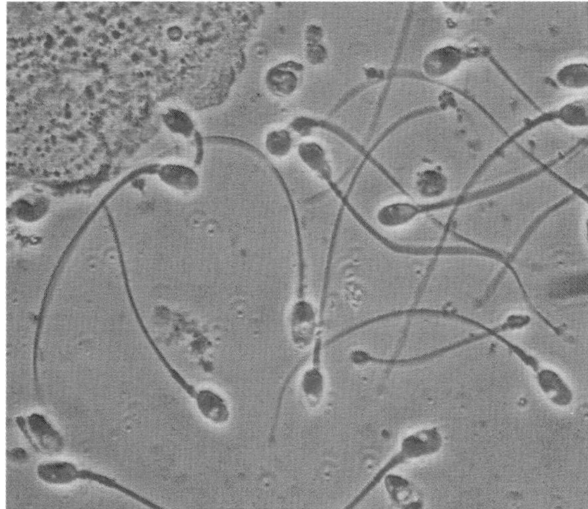

(b)

FIGURE 5.22

Cilia and flagella. The cilia in (*a*) line the human respiratory tract, where their coordinated movements propel dust particles upward so the person can exhale them. The flagella on the human sperm cells in (*b*) enable them to swim. (×1,200)

An individual cilium is constructed of nine microtubule pairs that surround a central, separated pair and form a pattern described as "9 + 2" (fig. 5.24). A type of motor molecule called **dynein** connects the outer microtubule pairs and also links them to the central pair. Dynein molecules use energy from ATP to shift in a way that slides adjacent microtubules against each other. This movement bends the cilium (or flagellum). Coordinated movement of these cellular extensions sets up a wave that moves the cell or propels substances along its surface.

Cilia perform many vital functions in animal cells, including beating particles up and out of respiratory tubules and enabling egg cells to move in the female reproductive tract towards sperm cells. A single-celled organism may have thousands of individual cilia, enabling it to "swim" through water.

Cilia tend to be short and numerous, like a fringe. A flagellum, also built of a 9 + 2 microtubule array, is larger. Flagella are more like tails, and their whiplike movement enables cells to move. Sperm cells in many species have prominent flagella (see fig. 5.22*b*). A human sperm cell has only one flagellum, but a sperm cell of a cycad (a type of tree) has thousands of flagella.

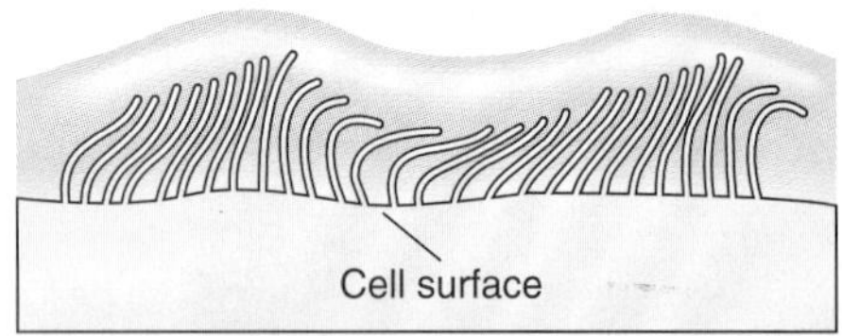

FIGURE 5.23

Sliding microtubules, powered by ATP-driven dynein movements, bend cilia. Coordinated movements of adjacent cilia produce a wave pattern, much like people do with their arms at athletic events or rock concerts.

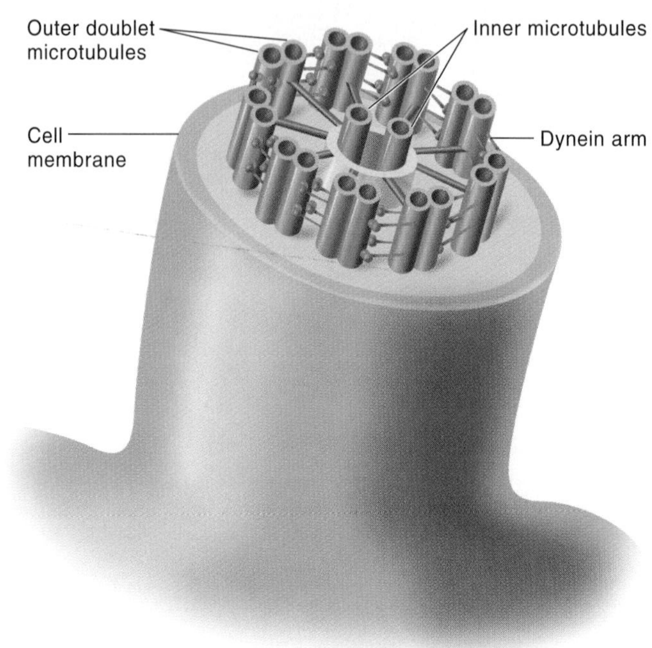

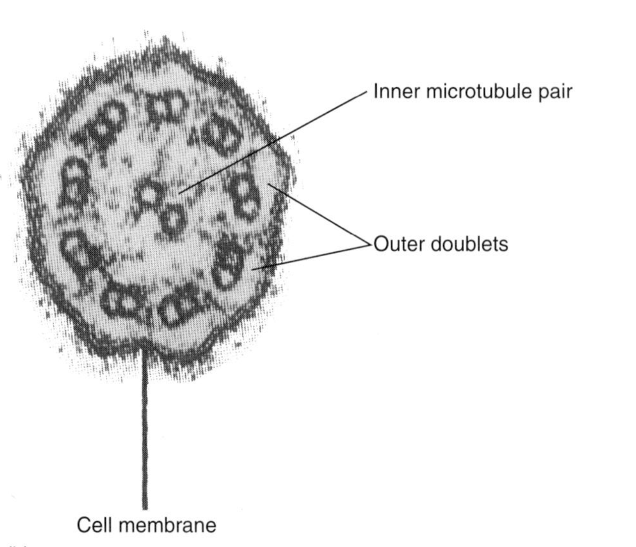

FIGURE 5.24

The microtubules that form cilia and flagella have a characteristic "9 + 2" organization. (*a*) Dynein joins the outer microtubule doublets to each other and to the central pair of microtubules. (*b*) A transmission electron micrograph showing a cross section of a flagellum in the green alga *Ulvaria.*

Microfilaments

Another building block of the cytoskeleton is the microfilament, a tiny rod of the protein **actin** (fig. 5.25). Microfilaments are about one-third as thick as microtubules, and they are solid rather than hollow. In muscles, actin microfilaments are interspersed with larger filaments of a protein called **myosin.** When the actin microfilaments and larger myosin rods slide past one another, muscle cells contract. Microfilaments also enable white blood cells to bring particles into the cell by endocytosis. The microfilaments align beneath the cell membrane and propel portions of the membrane outward to entrap particles. You can actually see microfilaments in action by watching the blood clot that forms on a scraped knee. Microfilaments and other proteins retract the clot as the wounded area fills in new skin.

Intermediate Filaments

When cells are treated with harsh substances such as detergents, microtubules and microfilaments break down; but another type of filament, intermediate in size between the other two, remains. These remaining filaments, called **intermediate filaments** because of their size, may provide scaffolding in the cell because they are seen in parts of the cell under structural tension. Intermediate filaments may, for example, hold the nucleus in place. Different cell types have different types of intermediate filaments.

The Cytoskeleton Molds Specialized Cells

A cell in a multicellular organism, as a part of a whole, faces stresses that a unicellular organism rarely, if ever, encounters. This is especially true for red blood cells, which travel through an endless network of conduits, and muscle cells, which are part of a densely packed, highly active tissue. The cytoskeletons of these cells make their specialized functions possible.

Red Blood Cells

Much of what we know about the cohesion between the cell membrane and the cytoskeleton comes from studies of human red blood cells. The doughnut shapes of these cells enable them to roll through the narrowest blood vessels on their 300-mile (483-kilometer), 120-day journey through the circulatory system (fig. 5.26*a*).

FIGURE 5.25

Microfilaments in a mammalian cell.

FIGURE 5.26

The red cell membrane. (*a*) Red blood cells must withstand great turbulent force in the circulation. The cytoskeleton beneath the cell membrane enables these cells to retain their shapes, which are adapted to movement and providing surface area for the oxygen-carrying molecule hemoglobin to function. (*b*) Proteins called ankyrins bind spectrin from the cytoskeleton to the interior face of the cell membrane. On its other end, ankyrin binds a large glycoprotein that helps transport ions across the cell membrane. Abnormal ankyrin causes the cell membrane to collapse—a problem for a cell whose function depends upon its shape.

(a)

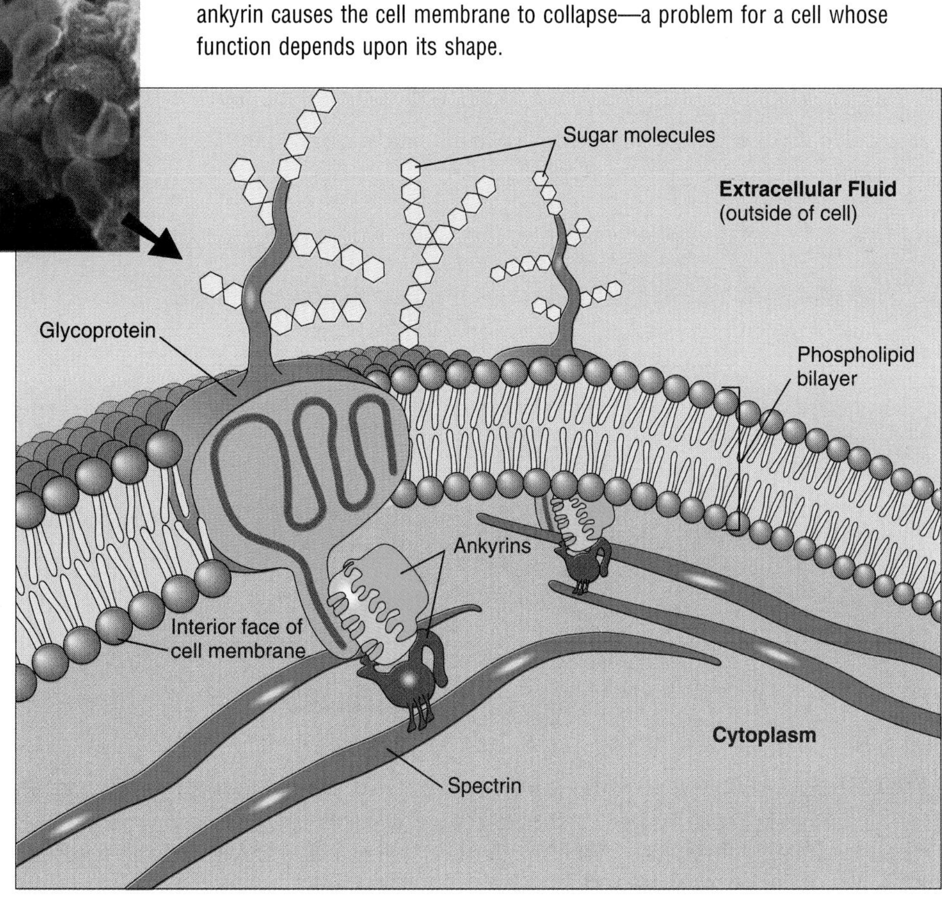

(b)

A red blood cell's strength derives from rodlike proteins called **spectrin** that form a meshwork beneath the cell membrane (fig. 5.26*b*). Proteins called **ankyrins** attach the spectrin rods to the membrane. Spectrin molecules are like the steel girders of the red blood cell architecture, and the ankyrins are like the nuts and bolts. If either is absent, the cell's structural support collapses. Abnormal ankyrin in red blood cells causes anemia (too few red blood cells) because the cells balloon out and block circulation and then die. Researchers expect to find defective ankyrin behind problems in other cell types, because it seems to be a key protein in establishing contact between the components of the cell's architecture.

Mastering Concepts

1. What are some functions of the cytoskeleton?
2. What is the structure of a microtubule?
3. How are microtubules organized to form cilia and flagella?
4. What type of protein composes microfilaments?
5. What is the role of intermediate filaments?
6. How do the cytoskeletons of red blood cells and muscle cells enable them to withstand physical forces?

Muscle Cells

Red blood cells and muscle cells are similar in that each must maintain its integrity and shape in the face of great physical force. The red blood cell can dissipate some of the force because it moves within a fluid. Muscle cells, in contrast, are held within a larger structure—a muscle—where they must withstand powerful forces of contraction, as well as rapid shape changes.

A muscle cell is filled with tiny filaments of actin and myosin, the proteins that slide past one another, providing contraction (fig. 5.27*a*). A far less abundant protein in muscle cells, **dystrophin,** is also very important. Dystrophin is the linchpin in muscle cell structure. It physically links actin in the cytoskeleton to glycoproteins that are part of the cell membrane (fig. 5.27*b*). Some of the glycoproteins, in turn, touch **laminin,** a cross-shaped protein that is anchored in noncellular material surrounding the muscle cell, called extracellular matrix. By holding together the cytoskeleton, cell membrane, and extracellular matrix, dystrophin and the associated glycoproteins greatly strengthen the muscle cell, which enables it to maintain its structure and function during repeated rounds of contraction. The muscle weakness disorders called muscular dystrophies result from missing or abnormal dystrophin or the glycoproteins it touches.

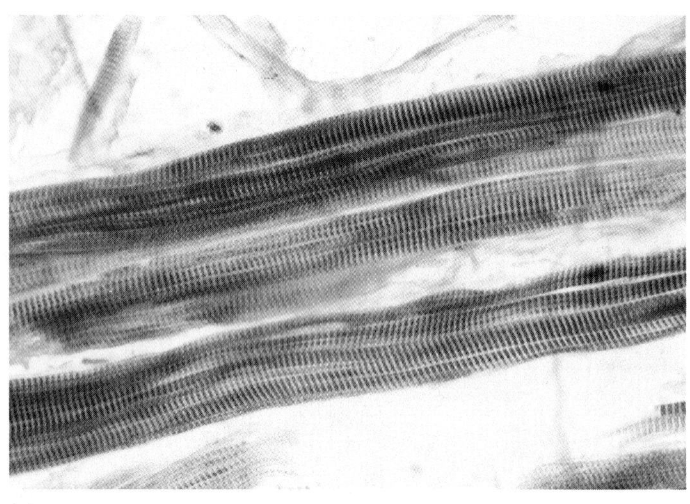

(a)

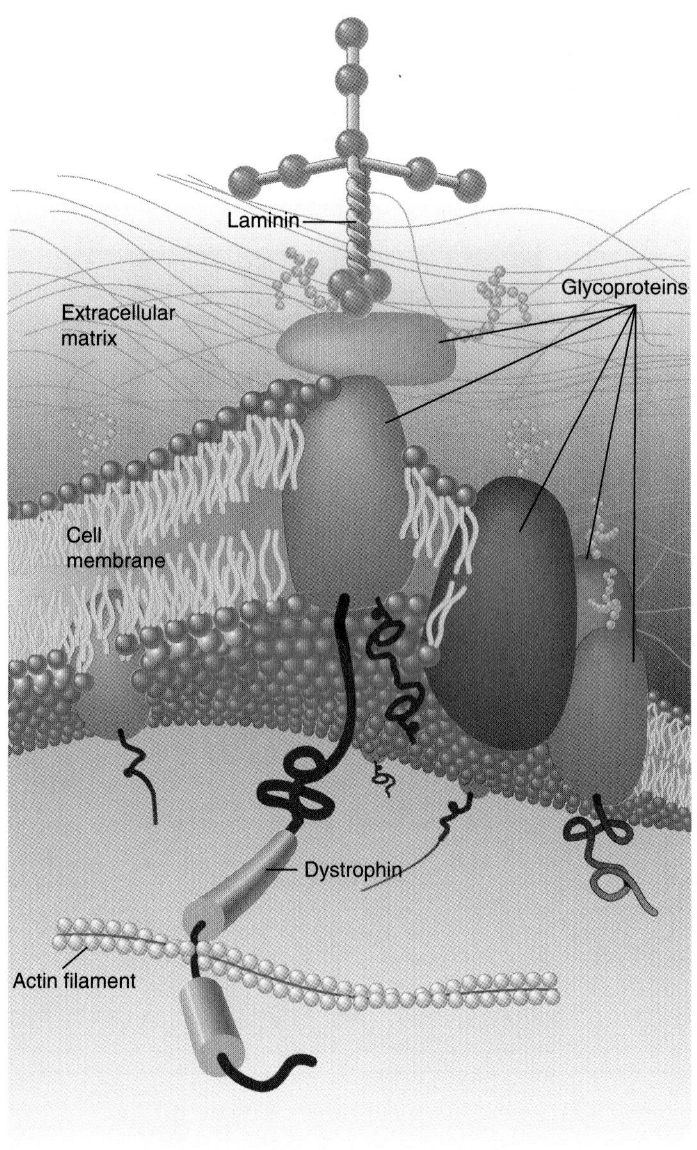

(b)

FIGURE 5.27

(*a*) Skeletal muscle appears striped, or striated, due to the orderly organization of the contractile proteins actin and myosin. (*b*) Dystrophin is a membrane protein that stabilizes muscle cell structure by linking the cytoskeleton to the basal lamina. (Actin is present in the cytoskeleton of all cells and is also one of the two major contractile proteins of muscle.)

Cell-to-Cell Interactions

In prokaryotes and single-celled eukaryotes, the cell is the entire organism. In multicellular organisms, cells "talk" to each other, and contact each other, in thousands of ways. Not surprisingly, researchers are finding that the versatile proteins play a major role in cell-to-cell interactions. We look now at two broad types of interactions between cells—signal transduction and cell adhesion.

Signal Transduction

A cell's organelles, cell membrane, and cytoskeleton are dynamic structures that must communicate with each other and with the external environment to maintain life. A key participant in this biological communication network is the cell membrane, which helps assess and transmit incoming messages in a process called signal transduction. A cell must discriminate between those signals it must respond to to survive and those it must ignore, such as an inappropriate signal to divide.

Signal transduction is a cascade of protein action. It begins at a cell's surface, where protruding receptor molecules bind incoming molecules called **first messengers.** These may be hormones (chemicals carried in the blood to a specific site, where they act) or growth factors (which act more locally). When a first messenger docks at a cell surface and binds its receptor, the receptor contorts, touching other nearby molecules and causing them to change shape. This triggers changes in other molecules inside the cell, which activates linked sequences of chemical reactions. The culmination of all this signaling is a specific cellular response, such as secretion, cell division, contraction, or energy release. The molecules that cause the cell's response to the first messenger are called **second messengers.** Figure 5.28 shows the role of one important second messenger, called cyclic AMP or cAMP for short. It derives from ATP, the biological energy molecule.

Because genes encode proteins, and proteins carry out signal transduction, abnormalities in this vital process can be inherited. The two sisters described in the opening vignette, whose physical appearance did not match their chromosomal sex, had such a block in signal transduction.

Cell Adhesion

Cells touch each other through adhesion, which is a precise sequence of interactions between proteins that bring cells into contact. One well-studied process involving cell adhesion is inflammation—the painful, red swelling at a site of injury or infection. At the cellular level, inflammation is the movement of white blood cells in the circulation to an endangered body part, where they stop and squeeze between the cells of the blood vessel walls to reach the site of injury or infection.

Cellular adhesion molecules (CAMs) help guide white blood cells. Different types of CAMs act in sequence (fig. 5.29). First, CAMs called **selectins** coat the white blood cell's tiny projections. Selectins provide traction, slowing the white blood cell they cling to by also binding to carbohydrates on the capillary wall. Next, clotting blood, bacteria, or decaying tissue release chemicals that signal the white blood cell to stop and also activate a second type of CAM, called an **integrin.** An integrin acts as a link, latching onto the white blood cell at one end and a third type of CAM, called an **adhesion receptor protein,** that extends from the capillary wall at the injury site.

In the next step, the adhesion receptor protein touches the cytoskeleton within the capillary cell. The integrin and adhesion receptor protein then pull the white blood cell between the lining cells to the other side, where the damage is. A simple event—a white blood cell crossing from a blood vessel to nearby tissue—is indeed complex, like the individual steps of a precisely choreographed dance.

As we saw with signal transduction, defects in cell adhesion can affect health. Consider the plight of a young woman named Brooke Blanton. She first experienced the effects of faulty cell adhesion as an infant, when her teething sores did not heal. These

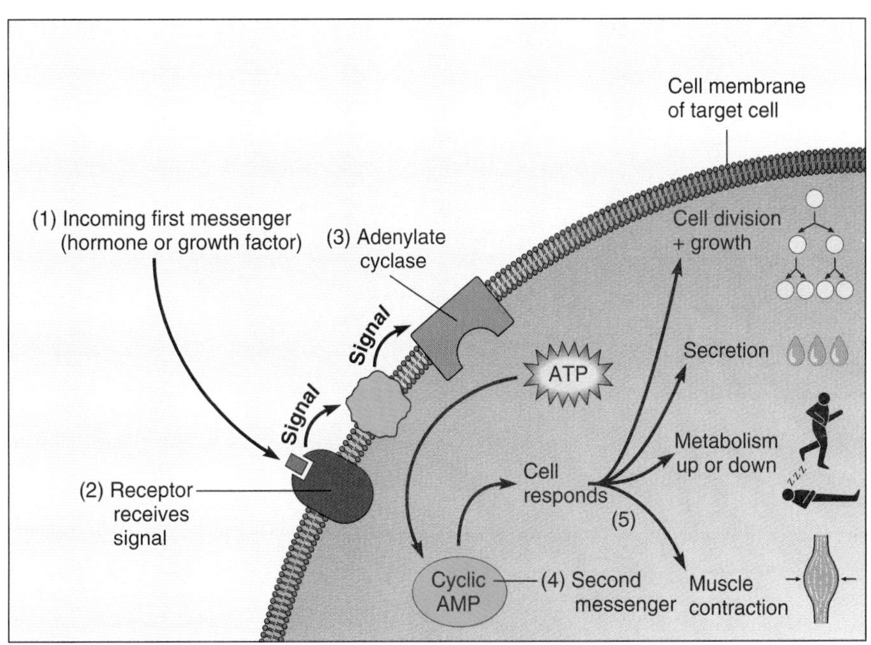

FIGURE 5.28

Signal transduction. A first messenger (*1*) binds a receptor (*2*), triggering a cascade of biochemical activity at the cell's surface. An enzyme, adenylate cyclase (*3*), catalyzes a reaction inside the cell. In this reaction, ATP loses phosphates, becoming cyclic AMP, the second messenger (*4*). cAMP stimulates various responses, such as cell division, secretion, metabolic changes, and muscle contraction. (*5*) The splitting of ATP also provides energy.

5.1 bioethics connections

Nicotine Addiction and Cell Biology

Why are so many people addicted to the nicotine in cigarettes (fig. 5.A)? Cell biology may explain how an enjoyable habit becomes a physical dependency.

Is Smoking Addictive?

Tobacco has been linked to cancer since the mid-1700s. Recognition of smoking tobacco as an addictive behavior is more recent. A 1964 Surgeon General's report distinguished between "addicting" and "habituating" drugs and defined "addiction" as causing intoxication. On this basis, cigarette smoking was considered merely habituating, a definition that tobacco companies still promote (fig. 5.B). More recent Surgeon General's reports reflect accumulating scientific evidence. The 1988 report stated clearly that nicotine causes addiction, and the 1994 report calls smoking a "totally preventable public health tragedy." It was also in 1994 that the Food and Drug Administration (FDA) begin investigating the possibility of regulating tobacco as a drug, because of its ability to addict.

According to the Diagnostic & Statistical Manual of Mental Disorders, a person addicted to tobacco

1. must smoke more to attain the same effects (tolerance) over time;

FIGURE 5.A
Celebrities helped glamorize smoking.

SURGEON GENERAL'S WARNING: Smoking Causes Lung Cancer, Heart Disease, Emphysema, And May Complicate Pregnancy.

FIGURE 5.B
This photograph, published by the American Medical Association in 1944, shows a man using a prosthetic hand to light a cigarette—indicating that even the medical community accepted smoking as a routine part of life. Today physicians feel quite differently, as the Surgeon General's warning on cigarette advertisements indicates.

2. experiences withdrawal symptoms when smoking stops, including weight gain, difficulty concentrating, insomnia, restlessness, anxiety, depression, slowed metabolism, and lowered heart rate;
3. smokes more often and for longer than intended;
4. spends considerable time obtaining cigarettes;
5. devotes less time to other activities;
6. continues to smoke despite knowing it is unhealthy;
7. wants to stop, but cannot easily do so.

A Biological Explanation

Nicotine causes the addiction, and the addiction supplies enough of the other chemicals in cigarette smoke to destroy health. The site of nicotine's activity is the nerve cell (neuron). Following nicotine's effects reviews several concepts discussed in this chapter.

An activated form of nicotine binds protein receptors, called nicotinic receptors, that are parts of cell membranes of certain brain neurons. These receptors normally receive the neurotransmitter acetylcholine. When sufficient nicotine binds, a channel within the receptor opens, allowing positive ions to enter the neuron (fig. 5.C). When a certain number of positive ions enter, the neuron is stimulated to release (by exocytosis) the neurotransmitter dopamine from its other end. The dopamine provides the pleasurable feelings associated with smoking. Addiction stems from two sources, researchers hypothesize—seeking the good feelings of sending off all that dopamine, and avoiding painful withdrawal symptoms.

Binding nicotinic receptors isn't the only effect of nicotine on the brain. When a smoker increases the number of cigarettes smoked, the number of nicotinic receptors on the brain cells increases. This happens because the way that the nicotine binds impairs the recycling of receptor proteins by endocytosis, so receptors are produced faster than they are taken apart. However, after a period of steady nicotine exposure, many of the receptors malfunction and no longer admit the positive ions that trigger nerve transmission. This may be why as time goes on it takes more nicotine to produce the same effects.

Many questions remain concerning the biological effects of tobacco smoking. Why don't lab animals experience withdrawal? Why do people who have successfully stopped smoking often start again 6 months later, even though withdrawal eases within 2 weeks of quitting? Why do some people become addicted easily, yet others smoke only a few cigarettes a day and can stop anytime?

While scientists try to answer these questions, society must deal with questions of rights and responsibilities that cigarette smoking causes. The following bioethical and practical questions concern cigarette smoking:

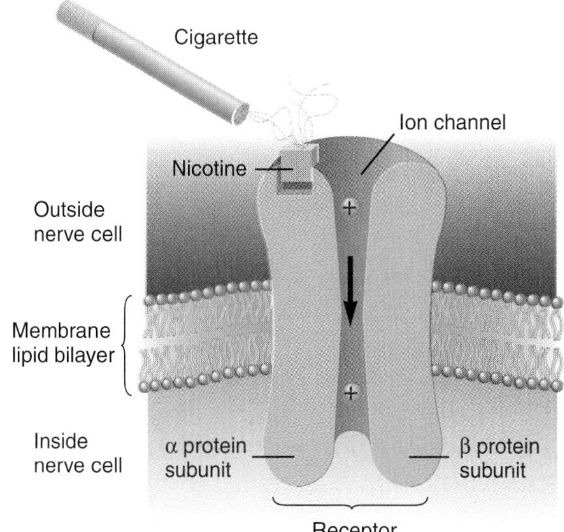

FIGURE 5.C
The seat of tobacco addiction lies in nicotine's binding to nerve cell surface receptors that normally bind the neurotransmitter acetylcholine. Not only does nicotine alter the receptor so that positive ions enter the cell, triggering dopamine release, but the chemical's repeated presence in a heavy smoker stimulates excess receptors to accumulate—but they soon become nonfunctional. Nicotine's effects are complex.

1. Under the Federal Food, Drug & Cosmetic Act, the FDA can regulate nicotine as a drug if it can show that manufacturers intend the product to be used as a drug that changes the body in some way. What would the consequences be of treating cigarettes as a drug?
2. How can we accommodate the rights of smokers to smoke and of nonsmokers to breathe nonsmoky air?
3. Should people whose heart or lungs are failing due to smoking-related illnesses have the same access to donor organs for transplants as nonsmokers?
4. If research funds are limited, should more support go towards better understanding of how nicotine causes addiction on a cellular level or to identifying ways to help people fight or prevent nicotine addiction?
5. Smoking is on the rise among young people. Social scientists believe that the increase reflects aggressive advertising by cigarette manufacturers and a rising popularity of smoking as smokers unite to object to restrictions. Should we, and how can we, prevent young people from starting to smoke?

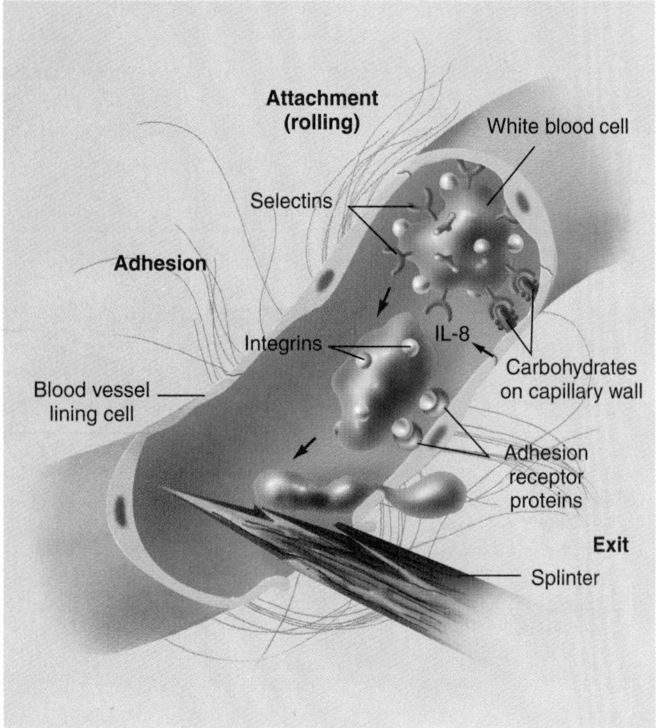

FIGURE 5.29

Cellular adhesion molecules (CAMs) direct white blood cells to injury sites, such as this splinter. Selectin proteins latch onto a rolling white blood cell and bind carbohydrates on the inner blood vessel wall at the same time, slowing the cell from moving at 2,500 micrometers per second to a more leisurely 50 micrometers per second. Then integrin proteins anchor the white blood cell to the blood vessel wall. Finally, the white blood cell squeezes between lining cells at the injury site and exits the bloodstream.

and other small wounds never accumulated pus, which consists of bacteria, cellular debris, and white blood cells and is a sign of the body fighting infection. Doctors eventually diagnosed Brooke with a newly recognized disorder called leukocyte-adhesion deficiency. Her body lacks the CAMs that enable white blood cells to stick to blood vessel walls. As a result, her blood cells move right past wounds. Brooke must avoid injury and infection, and she receives antiinfective treatments for even the slightest wound.

More common disorders may reflect abnormal cell adhesion. Lack of cell adhesion eases the journey of cancer cells from one part of the body to another. Arthritis may occur when white blood cells are reined in by the wrong adhesion molecules and inflame a joint where there isn't an injury. Cell adhesion is critical to many other functions. CAMs guide cells surrounding an embryo to grow toward maternal cells and form the placenta, the supportive organ linking a pregnant woman to the fetus. Sequences of CAMs also help establish connections between nerve cells that underlie learning and memory.

This chapter and the last offer many illustrations of how understanding cell biology can help explain medical problems. Bioethics Connection 5.1 explores another health-related phenomenon that results from disturbances at the cellular level—nicotine addiction.

Mastering Concepts

1. What is signal transduction?
2. What are first and second messengers?
3. How do cell adhesion molecules orchestrate a white blood cell's migration to an injury site?
4. In what ways is cell adhesion important in the human body?

SUMMARY

1. The cell surface is a selective interface between the cell and the outside environment. It receives and transmits incoming messages, controls which substances enter and leave the cell, and mediates attachments to and interactions with other cells and extracellular material.

2. The features of a cell's surface identify it as belonging to a particular species, to a particular individual, and to a particular tissue within that individual. The surface consists of molecules embedded in and extending from the cell membrane. Cell surfaces guide embryonic cells to form tissues and organs. The **human leukocyte antigen** cell surface markers establish self and are associated with increased risk of developing certain disorders.

3. A biological membrane consists of a **phospholipid bilayer** embedded with movable proteins, glycoproteins, and glycolipids, like a fluid mosaic. The percentage and distribution of membrane proteins varies in different cell types. Membrane proteins include receptors, carriers, adhesion molecules, and cell surface proteins establishing self.

4. Substances cross cell membranes in several ways. In **diffusion,** a molecule passes through openings in a membrane following its **concentration gradient. Osmosis** is the simple diffusion of water across the cell membrane or other semipermeable membrane. Terms describing **tonicity (isotonic, hypotonic, hypertonic)** predict whether cells will swell or shrink when the surroundings change. Several adaptations maintain cell shape in the face of solute concentration changes.

5. In **facilitated diffusion,** molecules cross a membrane following the **concentration gradient** with the aid of a carrier protein. In **active transport,** molecules cross a membrane against the concentration gradient and require a carrier protein and energy from ATP.

6. In **exocytosis,** vesicles inside the cell carry substances to the cell membrane, where they fuse with the membrane and release the cargo outside. In **endocytosis,** molecules are brought into the cell by a vesicle in the cell membrane. In some processes, vesicles shuttle between exocytosis and endocytosis, including intermediate **endosomes.** Endocytosis is more specific when a receptor in the vesicle binds a specific molecule. Within cells, proteins guide vesicles to particular organelles.

7. **Intercellular junctions** join cells. In animals they include **tight junctions, desmosomes,** and **gap junctions.** In plants they include a pectin layer and **plasmodesmata.**

8. The cytoskeleton is a network of rods and tubes that provides cells with form, support, and the ability to move. **Microtubules** self-assemble from hollow **tubulin** subunits to become **cilia, flagella,** and the spindle fibers that separate one

Sisters or Brothers?

continued from page 84.

The sisters who looked and felt like females were actually males at a genetic level. The reason—a block in signal transduction.

The girls' development began to veer from normal months before they were born. The miscommunication started in a fetal structure similar to an adult male's testes that gives rise to the reproductive system. Two cell types are involved—interstitial cells (once called Leydig cells) and sustentacular cells (once called Sertoli cells). If a person is genetically destined to develop as a male, sustentacular cells produce a hormone that inhibits development of internal female structures. Meanwhile, interstitial cells produce testosterone, the male hormone that stimulates a penis to form. To steer development towards maleness, the interstitial cells must receive a message, in the form of luteinizing hormone (LH), from the pituitary gland in the brain.

The sisters had underdeveloped interstitial cells, a condition called Leydig cell hypoplasia. As a result, their rudimentary reproductive tracts did not receive enough testosterone as fetuses to form penises. The genitalia instead developed ambiguously, resembling female organs. On the inside, the girls were more malelike, because their normal sustentacular cells had suppressed development of female internal organs. But because at birth they looked like girls, that is how they were raised.

What, precisely, was wrong with the girls' interstitial cells? An examination of cells from the girls' rudimentary testes pinpointed the block in signal transduction. The interstitial cells lacked surface receptors that normally bind LH. Unable to receive LH's message, the cells never received the signal to produce testosterone. So as fetuses, the siblings had turned off female development but hadn't completely turned on male development. Clues to their underlying maleness persisted in their chromosomes and two tiny patches of testeslike tissue deep within their bodies. But in their hearts and minds, the sisters felt like members of the gender that they had always appeared to be, and they continued to live happily as women.

cell into two during cell division. Microtubules have a characteristic 9 + 2 configuration. **Dynein** causes adjacent microtubules to slide, which moves the overall structure. **Microfilaments** are solid and smaller than microtubules. They are composed of the protein **actin** and provide contractile motion. **Intermediate filaments** provide scaffolding.

9. Red blood cells and muscle cells derive their strength from cell membrane proteins that connect cytoskeletal elements to extracellular materials.

10. In **signal transduction,** receptors in the cell membrane receive input from **first messengers** (hormones and growth factors) and transmit the messages through a series of membrane proteins. Eventually this signaling activates a **second messenger,** which stimulates the cell to carry out a specific function.

11. Cellular adhesion molecules enable cells to contact each other in precise steps that carry out a particular function.

TO REVIEW . . .

1. Why don't all human cells have the same pattern of surface features?

2. Why are some substances able to cross a cell membrane easily, and some not?

3. Which components of the cell membrane or cytoskeleton must be targeted in order to
- **a.** breach the blood-brain barrier?
- **b.** help the body to accept transplanted tissue?
- **c.** correct the defect that causes muscular dystrophy?
- **d.** treat diarrhea caused by a bacterial toxin?

4. Cite an illustration of how the cell membrane is a dynamic rather than a static structure.

5. Explain the differences between diffusion, facilitated diffusion, and active transport.

6. How does tonicity affect cell size? List three ways that cells maintain their shapes in the face of changing extracellular solute concentrations.

7. Endocytosis provides a general route for a substance to enter a cell. How can this process be more selective?

8. Distinguish between a desmosome and a plasmodesma.

9. List five functions of the cytoskeleton.

10. Describe the roles that receptors play in
- **a.** transporting substances across a cell membrane.
- **b.** signal transduction.
- **c.** vesicle trafficking.
- **d.** cell adhesion in inflammation.

11. What are the functions of the following proteins?
- **a.** tubulin
- **b.** dynein
- **c.** actin
- **d.** ankyrin
- **e.** spectrin
- **f.** CAMs
- **g.** dystrophin
- **h.** aquaporins

12. How does each of the following processes illustrate the interaction of components of the cellular architecture?
- **a.** maintaining the integrity of the red blood cell membrane
- **b.** the ability of muscle cells to withstand the force of contraction
- **c.** signal transduction
- **d.** cell adhesion in leukocyte trafficking

TO THINK ABOUT . . .

1. Breakdown in a component of the cellular architecture (cell surface, cell membrane, cytoskeleton) can have drastic effects on human health. For the following medical conditions discussed in the chapter, describe how the defect at the molecular and/or cellular level explains the symptoms:
- **a.** leukocyte-adhesion deficiency
- **b.** a form of muscular dystrophy in which muscle cells lack dystrophin
- **c.** Leydig cell hypoplasia
- **d.** diabetes insipidus
- **e.** nicotine addiction
- **f.** multiple sclerosis
- **g.** paralysis due to tetanus or botulism toxins

2. For the following two disorders not described in the chapter, state the cellular process that is abnormal:

a. In the mild form of familial hypercholesterolemia (FH), liver cells have half the normal number of LDL receptors. Individuals with this condition usually suffer a first heart attack in young adulthood. People with the severe form of FH have liver cells that completely lack LDL receptors. They develop cholesterol deposits under their skin, have extremely high serum cholesterol levels, and usually die of a heart attack before childhood ends.

b. In neurofibromatosis, tumors grow in nervous tissue beneath the skin. Normally, a protein called neurofibromin blocks a growth factor docked at a cell surface receptor from activating cell division. In neurofibromatosis, affected cells do not receive the message to stop cell division, and a tumor forms.

3. Many drugs used to treat cancer stop working because of a protein in the cell membranes of some cancer cells, called P-gp. This protein pumps many types of drugs out of the cell membrane before they can enter the cytoplasm, causing a phenomenon called "multidrug resistance." P-gp requires ATP to function, and it continually cycles from its site of synthesis in the cytoplasm to the cell membrane, carried within vesicles. P-gp is recycled when vesicles bud inward from the cell membrane.

a. Is the pumping action of P-gp an active or a passive process?

b. What term describes the role of vesicles in producing, transporting, and recycling P-gp?

c. How might you discover whether P-gp contacts chemotherapy drugs by endocytosis or receptor-mediated endocytosis?

d. How might researchers alter P-gp to make cancer drug treatment more effective?

4. Describe how dynein and dystrophin are vital for the functioning of certain cells, even though they are not very abundant.

5. A toxin produced by *Shigella* bacteria causes profuse diarrhea by opening tight junctions in small intestinal lining cells. What is another way that water can leave a cell?

6. How would you test the hypothesis that a particular medical condition is associated with a particular HLA type?

7. In many species, sperm cells have surface proteins that bind to receptor molecules on the egg cell's surface. Use this information to suggest two possible methods of birth control.

8. Liver cells are packed with glucose. What mechanism could be used to transport more glucose into a liver cell? Why would only this mode of transport work?

9. A drop of a 5% salt ($NaCl$) solution is added to a leaf of the aquatic plant *Elodea.* When the leaf is viewed under a microscope, colorless regions appear at the edges of each cell as the cell membranes shrink from the cell walls. What is happening to these cells?

10. Would a substance that destroys the integrity of the blood-brain barrier be dangerous? Why or why not?

TO LEARN MORE . . .

References and Resources

Brown, Michael S., and Joseph L. Goldstein. April 6, 1986. A receptor-mediated pathway for cholesterol homeostasis. *Science,* vol. 232. A classic paper tracing how liver cells handle dietary cholesterol.

Campbell, Kevin P. and Rachelle H. Crosbie. November 28, 1996. Utrophin to the rescue. *Nature*, vol. 384. A protein called utrophin may stand in for dystrophin to treat muscular dystrophy.

Ember, Lois R. November 28, 1994. The nicotine connection. *Chemistry & Engineering News,* vol. 72. Nicotine addiction results from activity at cell surfaces.

Hooper, Celia. June 1990. Ankyrins aweigh! *The Journal of NIH Research,* vol. 2. As its name and this article title imply, ankyrin helps anchor proteins of the cytoskeleton to the cell membrane.

Lewis, Ricki. January/February 1996. Probing the complex world of the cell. *Biophotonics International.* Fluorescence microscopy highlights cell functions, including apoptosis and signal transduction.

Lewis, Ricki. March 1994. Gateway to the brain. *BioScience,* vol. 44. Understanding the function of the blood-brain barrier leads to development of new drug delivery systems.

McManus, Michael, et. al. November 9, 1995. Regulation of cell volume in health and disease. *The New England Journal of Medicine,* vol. 333. Recovery after a severe head injury depends upon how quickly cell swelling is controlled.

Sansom, Mark S. P. January 30, 1997. Structure of a molecular hole-punch. *Nature*, vol. 385. Bacterial toxins harm humans by punching holes in our cells.

Sudhof, Thomas C. June 22, 1995. The synaptic vesicle cycle: a cascade of protein-protein interactions. *Nature,* vol. 375. Neurotransmitter release demonstrates vesicle trafficking.

Thompson, Dick. September 28, 1992. The glue of life. *Time.* People missing adhesion molecules live dangerously.

denotes articles by author Ricki Lewis that may be accessed via *Life*'s home page /http://www.mhh.com/science math/biology/life.

6 The Energy of Life

Swallowing and digesting a meal almost as big as oneself requires a huge energy investment. This African rock python is consuming a Thomson's gazelle.

chapter outline

Life Requires Energy ***110***

What Is Energy? ***110***

Sources of Energy

Potential and Kinetic Energy

The Laws of Thermodynamics ***112***

The First Law of Thermodynamics

The Second Law of Thermodynamics

Metabolism ***114***

Anabolism (Synthesis)

Catabolism (Degradation)

Energy in Chemical Reactions

Chemical Equilibrium

Oxidation, Reduction, and Energy Content

ATP: Cellular Energy Currency ***117***

High-Energy Phosphate Bonds

Coupled Reactions

Other Compounds Involved in Energy Metabolism

Enzymes and Energy ***121***

Control of Metabolism

The Major Energy Transformations—Respiration and Photosynthesis ***123***

Needing Energy to Get Energy

The African rock python lay in wait for the lone gazelle. When the gazelle came close, oblivious to the danger, the snake moved suddenly, positioning the victim's head and holding it in place while it swiftly entwined its 30-foot-long body snugly around the mammal. Each time the gazelle exhaled, the snake wound tighter, shutting down the victim's heart and lungs in less than a minute. Then the swallowing began.

How can a 200-pound snake eat a 130-pound meal? Thanks to adaptations of the reptile's digestive system, the snake can indeed swallow its meal, and thanks to remarkable control over its energy use, what seems impossible to us is a way of life for the snake.

The snake begins its meal by opening its jaws at an angle of 130 degrees (compared with 30 degrees for the most gluttonous human) and positions its mouth over the gazelle's head, using strong muscles to gradually envelop and push along the carcass. Saliva coats the prey, easing its journey to the snake's stomach. After several hours, the huge meal arrives at the stomach, and the remainder of the digestive tract readies itself for several weeks of dismantling the gazelle. When digestion completes, only a few chunks of hair will remain to be excreted.

Eating a tremendous meal once every few months places great energy demands on the snake. How does the reptile have enough energy left over from its meal to carry on the activities of life?

to be continued . . .

Life Requires Energy

A tortoiseshell butterfly expends a great deal of energy to stay alive. Not only does it fly from flower to flower, but it migrates seasonally over long distances. A less obvious way the insect uses energy is to power the many biochemical reactions of its metabolism. The butterfly obtains this energy from nectar a flower produces (fig. 6.1); the flowering plant gets its energy from the sun. All animals ultimately extract life-powering energy from plants and from certain microorganisms that capture the energy in sunlight and convert it into chemical energy.

Evidence of energy pervades our lives. We experience a burst of energy soon after eating a candy bar. Thunderstorms, volcanoes, and earthquakes violently display energy. A vibrating guitar string, a sunbeam, and a flying hockey puck each demonstrates energy. All aspects of life require energy, from replicated chromosomes dividing and distributing into two cells, to cells migrating in an early embryo to form tissues, to a flower blossoming. We saw in the preceding two chapters that energy is required to actively transport substances across membranes against their concentration gradients and to move secreted products through various organelles as they make their way out of the cell. Because energy is so basic to life, we need to understand just what it is and how organisms use it.

FIGURE 6.1

A butterfly expends great amounts of energy flying from flower to flower and migrating sometimes thousands of miles. This tortoiseshell butterfly is feeding on a flower.

What Is Energy?

The term *energy* was coined about two centuries ago, when the Industrial Revolution redefined familiar ideas of energy from the power behind horse-drawn carriages and falling water to the power of the internal combustion engine. As people began to think more about harnessing energy, biologists realized that understanding energy could not only improve the quality of life but reveal how life itself is possible. **Bioenergetics** is the study of how living organisms use energy to perform the activities of life.

Energy is the ability to do work—that is, to change or move matter against an opposing force, such as gravity or friction. Because energy is an ability, it is not as tangible as matter, which has mass and takes up space. One way to measure energy is in calories. A **calorie** (cal) is the amount of energy required to raise the temperature of 1 gram of water from 14.5°C to 15.5°C. The most common unit for measuring the energy content of food and the heat output of organisms is the Calorie (Cal), or kilocalorie (kcal), which is the energy required to raise the temperature of a kilogram of water 1°C. The Calorie or kilocalorie equals 1,000 calories.

Sources of Energy

Energy on earth comes from various sources (fig. 6.2):

— the sun;
— the wind;
— **geothermal energy** from the earth's core;
— the pull of the moon (tidal power);
— fossil fuels such as coal, oil, and gas;
— organisms (biomass).

FIGURE 6.2
Sources of energy: (*a*) the sun, (*b*) the wind, (*c*) the earth, (*d*) moving water, (*e*) fossil fuels, and (*f*) organisms.

Most organisms obtain energy from the sun, either directly or indirectly (through other organisms). In the preceding list, the energy in fossil fuels and in organisms originates as solar energy.

Deep within the sun, temperatures of 10,000,000°C fuse hydrogen atoms, forming helium and releasing gamma rays. These rays then emit electrons and **photons,** which are particles of light energy that travel in waves. Life depends upon the ability to transform this solar energy into chemical energy before it is converted to heat—the eventual fate of all energy.

The sun's total energy output is about 3.8 sextillion megawatts of electricity. Earth intercepts only about two-billionths of this. Although this is the equivalent of burning about 200 trillion tons of coal a year, most of this energy doesn't reach life. Nearly a third of the sun's incoming energy reflects back to space, and another half is absorbed by the planet, converted to heat, and returned to space. Another 19% of incoming solar radiation powers wind and the tides and drives photosynthesis, the process by which plants

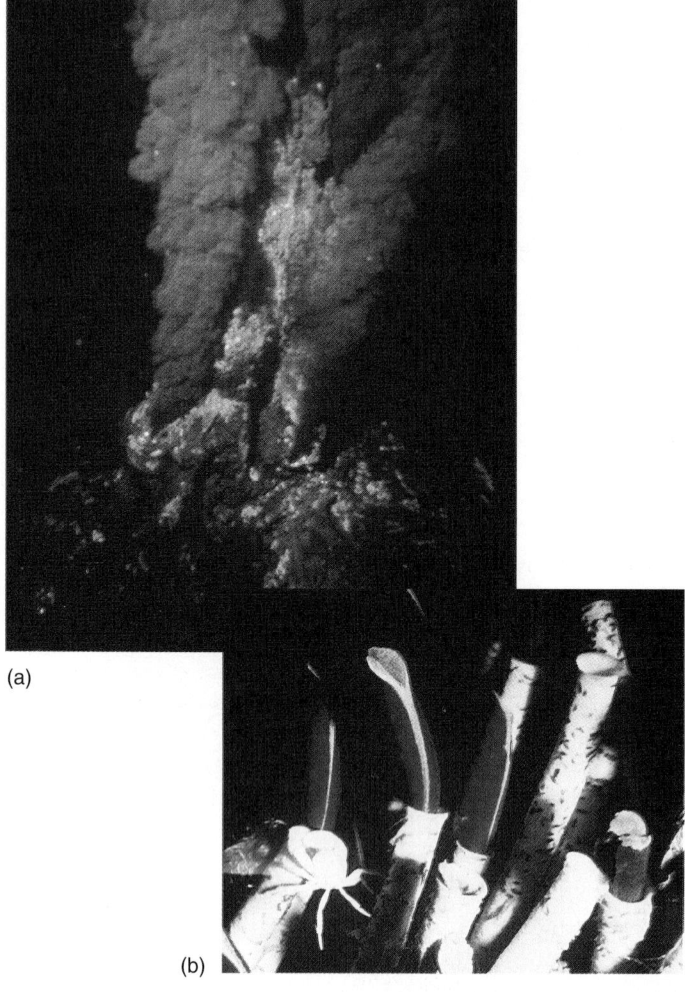

FIGURE 6.3

Geothermal energy supports life. (*a*) Chemoautotrophic bacteria support life in deep-sea hydrothermal vents. The bacteria derive energy from inorganic chemicals such as hydrogen sulfide. (*b*) Bacteria, in turn, support other organisms, such as crabs, clams, and the tube worms shown here.

harness solar energy. Of this 19%, only 0.05 to 1.5% is incorporated into plant material—and only about a tenth of that makes its way into the bodies of animals that eat plants. Far less solar energy reaches animals that consume plant-eating animals. Obviously, life on earth functions using just a tiny portion of solar energy.

Most organisms depend, directly or indirectly, on organisms that photosynthesize to acquire solar energy, such as plants. However, a few species can obtain and use geothermal energy. Bacteria thrive around dark cracks in the ocean floor where hot molten rock seeps through, creating an intensely hot environment (fig. 6.3). These areas are called deep-sea hydrothermal vents. Bacteria there extract energy from inorganic chemicals surrounding them, specifically from the chemical bonds of hydrogen sulfide (H_2S), a gas present at the vents. The bacteria then add the freed hydrogens to carbon dioxide to synthesize organic compounds that are nutrients for them and the organisms that consume them. Organisms that use inorganic chemicals to manufacture nutrient molecules are called **chemoautotrophs.** This term comes from the more general term **autotroph,** which refers to organisms that synthesize their own nutrients.

Potential and Kinetic Energy

To stay alive, an organism must continually take in energy through eating or from the sun or the earth and convert it into a usable form. Every aspect of life centers on converting energy from one form to another. Two basic types of energy constantly interconvert: potential energy and kinetic energy.

Potential energy is stored energy available to do work. When matter assumes certain positions or arrangements, it contains potential energy. A teaspoon of sugar, a snake about to pounce on its prey, a child at the top of a slide, and a baseball player about to throw a ball illustrate potential energy. In organisms, potential energy is stored in the chemical bonds of nutrient molecules, such as carbohydrates, lipids, and proteins.

Kinetic energy is energy being used to do work. Burning sugar, the snake pouncing, the child riding down the slide, and the soaring baseball all demonstrate kinetic energy. A chameleon shooting out its sticky tongue to capture a butterfly uses kinetic energy, as does an elephant trumpeting and a Venus's flytrap closing its leaves around an insect. The adder snake in figure 1.5 demonstrates potential energy as it awaits the approach of a sand lizard. Somewhere between figures 1.5*a* and 1.5*b,* the snake strikes using kinetic energy to grab its lizard meal. Figures 6.4 and 6.5 illustrate potential and kinetic energy.

Kinetic energy transfers motion to matter. The movement in the pitcher's arm transfers energy to the ball, which then takes flight. Similarly, flowing water turns a turbine, and a growing root can push aside concrete to break through a sidewalk. Put another way, kinetic energy moves objects. Heat and sound are types of kinetic energy because they result from the movement of molecules.

Potential energy is translated into the kinetic energy of motion, and that burst of energy can be used to do work. Under most conditions, potential and kinetic energy are readily interconvertible, although not at 100% efficiency.

Mastering Concepts

1. What is energy?
2. What are some sources of energy?
3. How does solar energy support life?
4. Distinguish between potential and kinetic energy.

The Laws of Thermodynamics

Rules called the **laws of thermodynamics** regulate the energy conversions vital for life, as well as those that occur in the nonliving world. These laws operate on a system and its surroundings; the system is the collection of matter under consideration, and the surroundings are the rest of the universe. In figure 6.6, the elephant is the system and its watery background the surroundings. An open system exchanges energy with its surroundings, while a closed system is isolated from its surroundings—it does not exchange energy with anything outside the system. Thus, the open system that is the elephant loses heat to its surroundings as it splashes about.

A thermos bottle and its contents illustrate closed and open systems. The uncapped thermos, an open system, loses steam from the heated food to the surrounding air, while a closed thermos prevents such an exchange of energy.

(a)

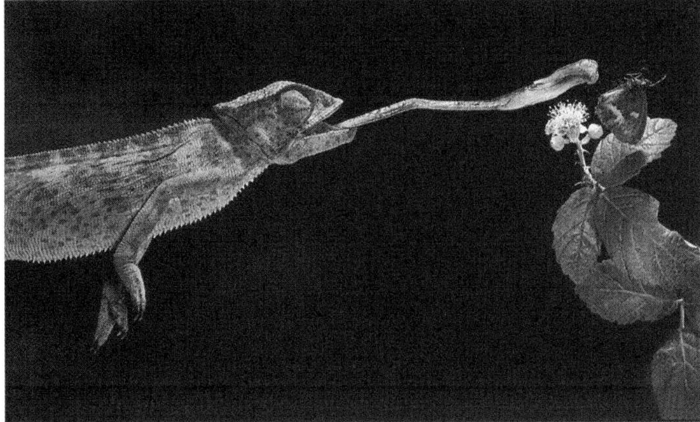

(b)

FIGURE 6.4

Potential and kinetic energy. (*a*) The ball this pitcher is about to throw has potential energy. (*b*) The living world abounds with illustrations of kinetic energy, the energy of action. Here, a chameleon's tongue whips out to ensnare a butterfly.

FIGURE 6.5

The difference between potential energy and kinetic energy is clear to anyone who has ever pedaled a bicycle to the top of a steep hill, and then enjoyed the burst of kinetic energy on the ride down.

FIGURE 6.6

A system and its surroundings. The elephant obtains energy from the chemical bonds of the nutrient molecules in its food. When the animal cavorts in the water, it dissipates some energy as heat and motion into the surroundings.

The laws of thermodynamics are unbreakable and apply to all energy transformations—gasoline combustion in a car's engine, a burning chunk of wood, or a cell breaking down glucose.

The First Law of Thermodynamics

The laws of thermodynamics follow common sense. The first law of thermodynamics is the law of energy conservation. It states that energy cannot be created or destroyed but only converted to other forms. This means that the total amount of energy in a system and its surroundings remains constant; thus, on a grander scale, the amount of energy in the universe is constant.

In a practical sense, the first law of thermodynamics explains why we can't get something for nothing. The energy released when a baseball hurtles towards the outfield doesn't appear out of nowhere—it comes from a batter's muscles. Likewise, green plants do not manufacture energy from nothingness; they trap the energy in sunlight and convert it into chemical bonds. The energy in sunlight, in turn, comes from chemical reactions in the sun's matter.

According to the first law of thermodynamics, the amount of energy an organism uses cannot exceed the amount of energy it takes in through the chemical bonds contained in the nutrient molecules of food. Even when starving, an organism cannot use more energy than its tissues already contain. Similarly, the amount of chemical energy that a plant's leaves produce during photosynthesis cannot exceed the amount of energy in the light it has absorbed. No system can use or release more energy than it takes in.

The Second Law of Thermodynamics

The second law of thermodynamics concerns the concept of **entropy,** which is a tendency towards randomness or disorder (fig. 6.7). This law states that all energy transformations are inefficient because every reaction results in increased entropy and loses some usable energy to the surroundings as heat. Unlike other forms of energy, heat energy results from random molecule movements.

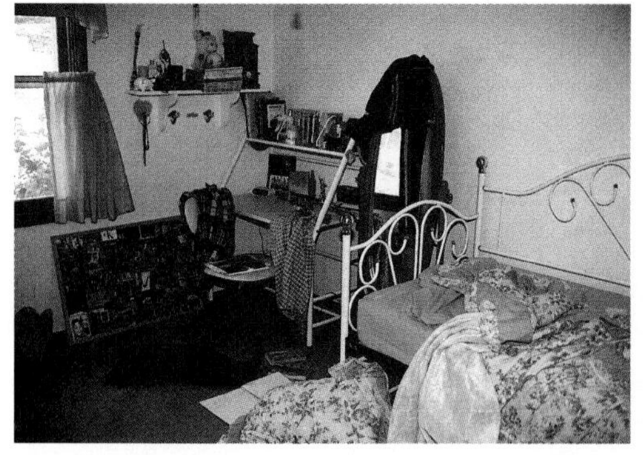

(a)

(b)

FIGURE 6.7

A messy room symbolizes entropy—extreme disorder. (*a*) Cleaning the room (*b*) requires an input of energy.

Any other form of energy can be converted completely to heat, but heat cannot be completely converted to any other form of energy. Because all energy eventually becomes heat, and heat is disordered, all energy transformations head towards increasing disorder (entropy). In general, the more disordered a system is, the higher its entropy.

Because of the second law of thermodynamics, events tend to be irreversible (proceed in one direction) unless energy is added to reverse them. Processes that occur without an energy input are termed spontaneous. A house becomes messier—not neater—as the week progresses unless someone expends energy to clean. In natural processes, irreversibility results from the loss of usable energy as heat during energy transformation. It is impossible to reorder the molecules that have dispersed as heat.

The second law of thermodynamics governs cell energetics. Cells derive energy from nutrient molecules and use it to perform such activities as growth, repair, and reproduction. The chemical reactions that free this energy are (as the law predicts) inefficient and release much heat. The cells of most organisms are able to extract and use only about half of the energy in their food. Although organisms can transform energy—storing it in tissues or using it to repair a wound, for example—ultimately, much of the energy is dissipated as heat.

Because organisms are highly organized, they may seem to defy the second law of thermodynamics—but only when they are considered alone, as closed systems. Organisms remain organized because they are *not* closed systems. They use incoming energy and matter, from sources such as sunlight and food, in a constant effort to maintain their organization and stay alive. Although the entropy of one system, such as an organism or a cell, may decrease as the system becomes more organized, the organization is temporary; eventually the system (organism) will die. The entropy of the universe as a whole is always increasing.

We're familiar with energy transformations that release large amounts of energy at once—explosions, lightning, or a plane taking off. Cellular energy transformations, by contrast, release energy in tiny increments. Cells extract energy from glucose in small amounts at a time via the biochemical pathways of **cellular respiration,** the subject of the next chapter. If all the energy in the glucose chemical bonds was released at once, it would be converted mostly to heat and produce deadly high temperatures. Instead, cells extract energy from glucose and other molecules by slowly stripping electrons from them in a series of chemical reactions called **electron transport systems.** Each reaction lowers the potential energy of the reactant molecule. Some of this energy is lost as heat, but much of it is trapped in the chemical bonds of other molecules in the cell. Then the energy is either stored or used to power the activities of life.

Mastering Concepts

1. On what do the laws of thermodynamics act?
2. What is the first law of thermodynamics?
3. What is the second law of thermodynamics?

Metabolism

Metabolism, Greek for "change," consists of the chemical reactions that change or transform energy in cells. The reactions of metabolism are organized into step-by-step sequences called **metabolic pathways,** in which the product of one reaction becomes the starting point, or substrate, of another (fig. 6.8). Enzymes enable metabolic reactions to proceed fast enough to sustain life, a point we return to later in the chapter. Metabolism in its entirety is an enormously complex network of interrelated biochemical reactions organized into chains and cycles.

Metabolism on a familiar, whole-body level, refers to how rapidly an organism "burns" food. On a cellular level, metabolism includes building and breaking down molecules within cells.

Anabolism (Synthesis)

Anabolism refers to metabolic pathways that construct large molecules from small ones. Just as erecting a building requires energy (fig. 6.9), anabolic reactions require energy, because molecules are being built. In anabolism, a small number of chemical subunits join in different ways to produce many different types of large molecules. The pathways of anabolism diverge (branch out) because a few types of precursor molecules combine to yield many different types of products. For example, the 20 different types of amino acids link together in varied sequences to form thousands of types of proteins.

Catabolism (Degradation)

Catabolism refers to metabolic pathways that break down large molecules into smaller ones. These pathways release energy, just

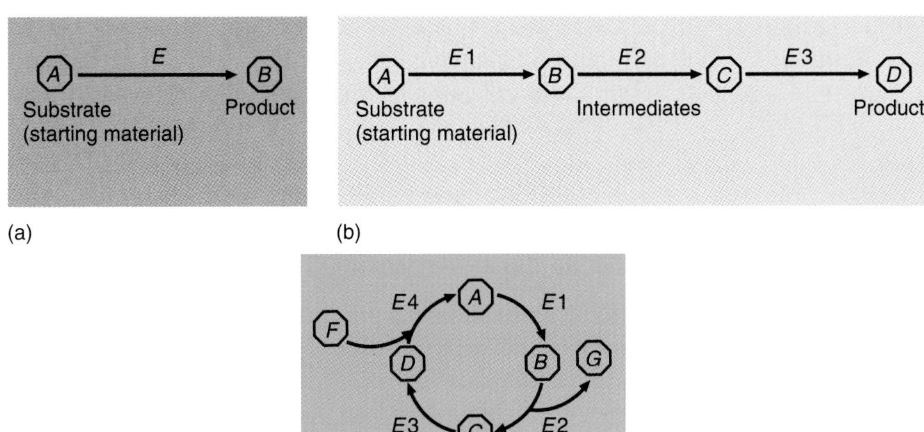

FIGURE 6.8

The chemical reactions of life form chains and cycles of metabolic pathways. Depicted are (*a*) an enzyme-catalyzed chemical reaction, (*b*) a biochemical pathway, and (*c*) a biochemical cycle. In each case, *E* represents the enzyme that catalyzes a specific reaction. Note that in a cycle, the product of the last reaction is also the starting material of the first. Cycles release by-products that are important in other biochemical pathways, such as compound *G* in (*c*).

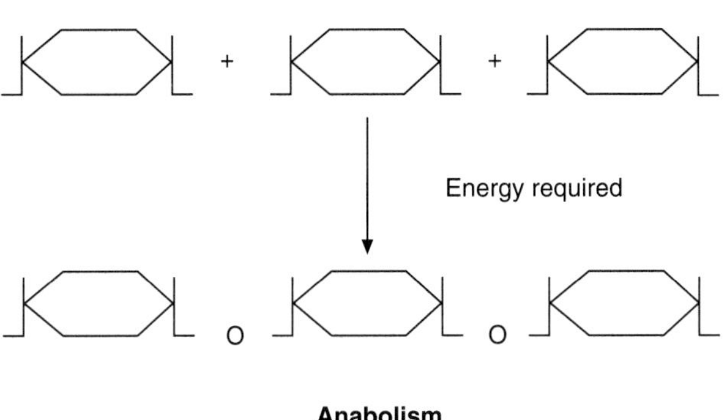

FIGURE 6.9

Anabolism—building something—requires energy, whether it is stringing sugar units to form a polysaccharide or assembling blocks to build a pyramid.

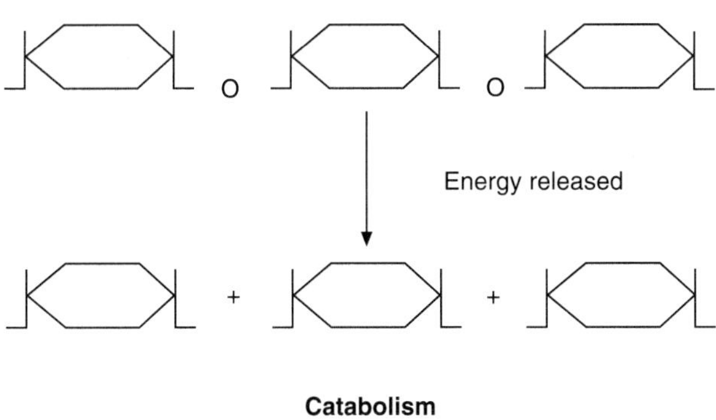

FIGURE 6.10

Catabolism—tearing something down—releases energy, whether it is a polysaccharide digested to its component sugars or a building blown up in seconds.

as a collapsing building releases energy in the form of sound, heat, and motion (fig. 6.10). Catabolic pathways converge (come together) because many different types of large molecules degrade to yield fewer types of small molecules. Thousands of distinct proteins, for example, break down to yield the same 20 types of amino acids. Of the varied foods that organisms consume, all of the nutrients are ultimately broken down or converted to just a few types of building block compounds—glucose, glycerol and fatty acids, and amino acids.

Energy in Chemical Reactions

Each reaction in a metabolic pathway rearranges atoms into new compounds, and each reaction either absorbs or releases energy. According to the first law of thermodynamics, it takes the same amount of energy to break a particular kind of bond as it does to form that same bond—energy is not created or destroyed. The amount of energy stored in a chemical bond (potential energy) is called its **bond energy.** Some chemical bonds are stronger than others—that is, they have greater bond energies. When a stronger bond breaks, it releases more energy. Conversely, forming a stronger bond requires a greater input of energy.

The potential energy of a compound, then, is contained in its chemical bonds. When these bonds break, some of the energy released can be used to do work—for example, to form other bonds. The amount of stored energy potentially available to form new bonds is called the **free energy** of the molecule. Chemical reactions change the amount of free energy stored in a bond and potentially available to do work. In these energetic terms, two types of chemical reactions occur—endergonic and exergonic reactions.

Endergonic Reactions

In an **endergonic reaction** (energy inward), the products contain more energy than the reactants. Endergonic reactions are not spontaneous—that is, they do not proceed unless there is an input of energy. Entropy decreases in endergonic reactions.

An example of an endergonic reaction is combining the monosaccharides glucose and fructose to form water and sucrose (table sugar). This synthesis requires energy, and the products of the reaction store more energy in their chemical bonds than do the reactants. Because energy cannot be created or destroyed, this reaction requires an absorption of energy from the surroundings in order for the products to have gained energy.

Exergonic Reactions

In an **exergonic reaction** (energy outward), the products contain less energy than the reactants. Energy is released, and the reaction is spontaneous. Entropy increases in exergonic reactions.

The breakdown of glucose ($C_6H_{12}O_6$) to carbon dioxide (CO_2) and water (H_2O) is spontaneous and exergonic. The carbon dioxide and water products store less energy than glucose; some of the energy released can do work, and the rest is lost as heat.

The reactions of anabolism (synthesis) are endergonic (require energy), and the reactions of catabolism (degradation) are exergonic (release energy). Figure 6.11 compares and contrasts endergonic and exergonic reactions.

Chemical Equilibrium

Most chemical reactions can proceed in both directions—that is, when enough product forms, some of it converts back to reactants. Reversible reactions are indicated by arrows between reactants and products going in both directions, such as:

$$\underset{\text{(carbon dioxide)}}{CO_2} + \underset{\text{(water)}}{H_2O} \rightleftharpoons \underset{\text{(carbonic acid)}}{H_2CO_3}$$

Such reactions that proceed in both directions reach a point called **chemical equilibrium,** where the reaction goes in both directions at the same rate. However, there may be different amounts of products and reactants at equilibrium—it is their rate of formation that equalizes. At chemical equilibrium, energy is not being gained or lost. When a reaction departs from equilibrium (that is, when either reactants or products accumulate), energy is lost or gained.

Because all activities of life require energy, cells must remain far from equilibrium to obtain energy to stay alive. They do this by continually preventing the accumulation of any reactants in metabolic pathways. For example, the large difference in free energy between glucose and its breakdown products, carbon dioxide and water, propels cellular metabolism strongly in one direction—as soon as glucose enters the cell, it is quickly broken down to release energy. This energy allows the cell to stave off equilibrium by continuously forming new products required for life.

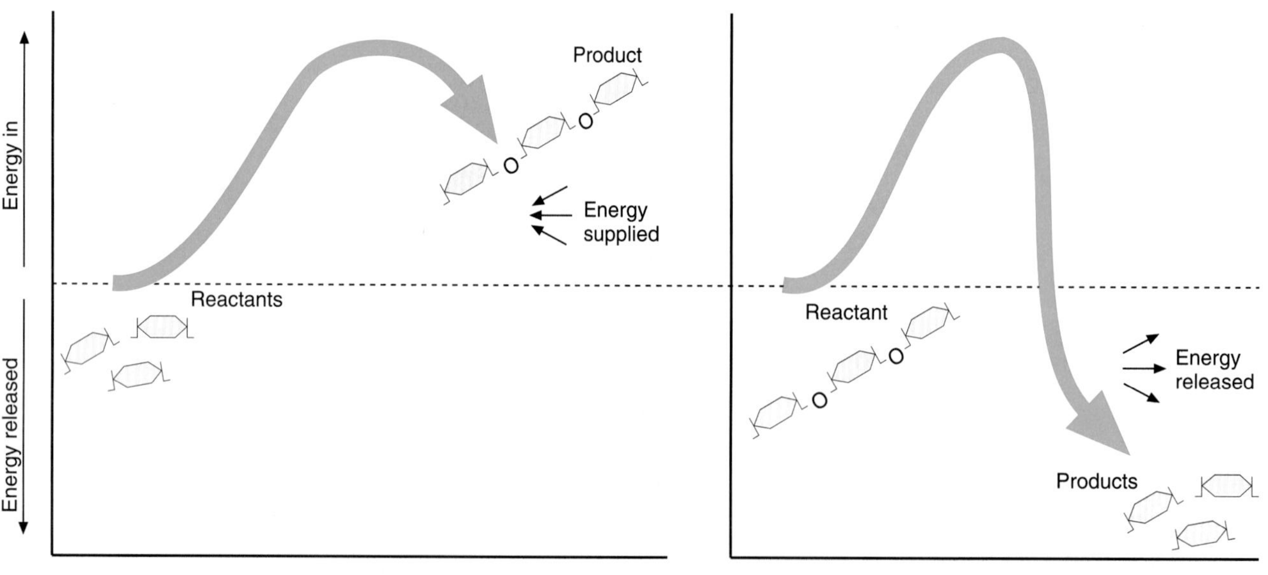

FIGURE 6.11

Endergonic and exergonic reactions contrasted.

Oxidation, Reduction, and Energy Content

Most energy transformations in organisms occur in chemical reactions called oxidations and reductions. In **oxidation,** a molecule loses electrons. The name comes from the observation that many reactions in which molecules lose electrons involve oxygen. Oxidation is the equivalent of adding oxygen because oxygen strongly attracts electrons away from the original atom. Oxidation reactions, such as the breakdown of glucose to carbon dioxide and water, are catabolic. That is, they degrade molecules into simpler products as they release energy.

In **reduction,** a molecule gains electrons. Reduction changes the chemical properties of a molecule. Reduction reactions, such as the formation of lipids, are usually anabolic. They require a net input of energy.

Oxidations and reductions tend to be linked, occurring simultaneously, because electrons removed from one molecule during oxidation join another molecule and reduce it (fig. 6.12). That is, if one molecule is reduced (gains electrons), then another must be oxidized (loses electrons).

Many energy transformations in living systems involve carbon oxidations and reductions. Reduced carbon contains more energy than oxidized carbon. This is why reduced molecules such as methane (CH_4) are explosive, while oxidized molecules such as carbon dioxide (CO_2) are not. The same principle applies to other compounds: The more reduced they are, the more energy they contain. Anyone who has ever dieted is at least intuitively familiar with this concept of energy content. Saturated fats are highly reduced, and they contain more than twice as many kilocalories by weight as proteins or carbohydrates. Thus, a fat-laden meal of a bacon double cheeseburger with fries and a shake may contain the same number of kcal (and thus the same amount of energy) as a bathtub full of celery sticks.

Mastering Concepts

1. What does metabolism mean in a cellular sense?
2. How are anabolism and catabolism opposites?
3. Distinguish between endergonic and exergonic reactions.
4. What distinguishes a reaction that has reached chemical equilibrium?
5. What are oxidation and reduction reactions, and why are they linked?

ATP: Cellular Energy Currency

Organisms store potential energy in the chemical bonds of nutrient molecules. It takes energy to make these bonds, and energy is released when bonds of these molecules break. Much of the released energy of life is stored temporarily in the covalent bonds of **adenosine triphosphate,** a compound more commonly known as **ATP.** Chapter 5 introduced two examples of how a cell uses ATP energy—to power active transport and to move cilia and flagella.

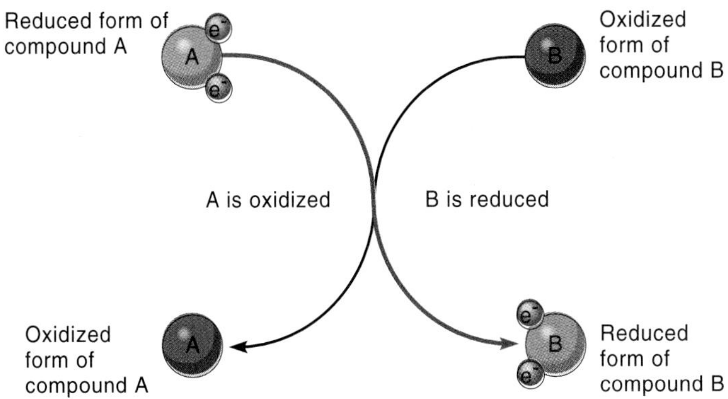

FIGURE 6.12

Oxidation and reduction reactions are paired. Here, compound A is initially in the reduced (electron-rich) form. It becomes oxidized in the reaction shown, losing a pair of electrons to compound B, which is reduced in the process.

High-Energy Phosphate Bonds

ATP is composed of the nitrogen-containing base adenine, the five-carbon sugar ribose, and three phosphate groups (each group includes a phosphorus atom bonded to four oxygen atoms and is indicated as PO_4) (fig. 6.13). Recall from chapter 3 that the building blocks of the ATP molecule—specifically, the adenine, ribose, and one phosphate—form an RNA nucleotide.

The ATP molecule has two additional phosphate groups, and this is where the energy-storing capacity comes in. The covalent bonds joining these two extra phosphate groups can break, releasing a large amount of energy. These phosphate bonds release so much energy that we illustrate them with squiggly lines. However, ATP's "high energy" is not contained completely in its phosphate bonds but in the complex interaction of atoms that make up the entire molecule.

When the endmost phosphate group of an ATP molecule detaches, energy is released and the molecule becomes adenosine diphosphate (ADP, which has two phosphate groups rather than three). Still another phosphate bond can break to yield adenosine monophosphate (AMP) and another release of energy. ATP thus provides an energy currency for the cell. When a cell requires energy for an activity, it "spends" ATP by converting it to ADP, inorganic phosphate (symbolized P_i), and energy (fig. 6.14). This reaction is represented as:

$$ATP + H_2O \rightarrow ADP + P_i + \text{energy}$$

In the reverse situation, energy can be temporarily stored by adding a phosphate to ADP, forming ATP. This energy comes from molecules broken down in other reactions that occur at the same time and in the same place as ATP synthesis.

Biology in Action 6.1 addresses the role of ATP in bioluminescence, the reaction that gives fireflies a "glow."

ATP is an effective biological energy currency for several reasons. First, converting ATP to ADP + P_i releases about twice the amount of energy required to drive most reactions in cells. The extra energy dissipates as heat. Second, ATP is readily available. The large amounts of energy in the bonds of fats and starches are not as easy to access—they must first be converted to ATP

6.1 biology in action

Firefly Bioluminescence

Photosynthesis transforms light energy into chemical energy. In **bioluminescence,** the reverse can happen—chemical energy is converted into light energy.

We see bioluminescence during summer in the glow of a firefly's abdomen (fig. 6.A). More than 1,900 species of firefly are known, and each uses a distinctive repertoire of light signals to attract a mate. Typically, flying males emit a pattern of flashes. Wingless females, called glowworms, usually are on leaves, where they glow in response to the male. In one species, *Photuris versicolor,* the female emits the mating signal of another species and then eats the tricked male who approaches her. Some frogs consume so many fireflies that they glow!

Johns Hopkins University researchers William McElroy and Marlene DeLuca asked Baltimore schoolchildren to bring them jars of fireflies and, in the 1960s, used the insects to decipher the firefly bioluminescence reaction. McElroy and DeLuca found that light is emitted when an organic acid called *luciferin* reacts with ATP, yielding the intermediate compound *luciferyl adenylate* (fig. 6.B). The enzyme *luciferase* then catalyzes reaction of this intermediate with molecular oxygen (O_2) to yield *oxyluciferin*—and a flash of light. Oxyluciferin is then reduced to luciferin, and the cycle starts over.

Chemical companies sell luciferin and luciferase, which researchers use to detect ATP. When ATP appears in a sample of any substance, it indicates contamination by an organism. For example, the manufacturers of Coca-Cola use firefly luciferin and luciferase to detect bacteria in syrups used to produce the beverages. Contaminated syrups glow in the presence of luciferin and luciferase because the ATP in the bacteria sets the bioluminescence reaction into motion. Firefly luciferin and luciferase were also aboard the *Viking* spacecraft sent to Mars. Scientists sent the compounds to detect possible life—a method that would only succeed if Martians use ATP. The opening text in chapter 29 describes an experiment that used luciferin to detect patterns of photosynthesis activity in plants.

Although we understand the biochemistry of the firefly's glow, the ways animals use their bioluminescence are still very much a mystery. This is particularly true for the bioluminescent synchrony seen in trees. When night falls, first one firefly, then another, then more, begin flashing from the tree. Soon the tree twinkles like a Christmas tree. But then, order slowly descends. In small parts of the tree, the lights begin to blink on and off together. The synchrony spreads. A half hour later, the entire tree seems to blink on and off every second. Biologists studying animal behavior have joined mathematicians studying order to try to figure out just what the fireflies are doing—or saying—when they synchronize their glow.

FIGURE 6.A
Bioluminescence from fireflies lights the air around this Iowa farmhouse.

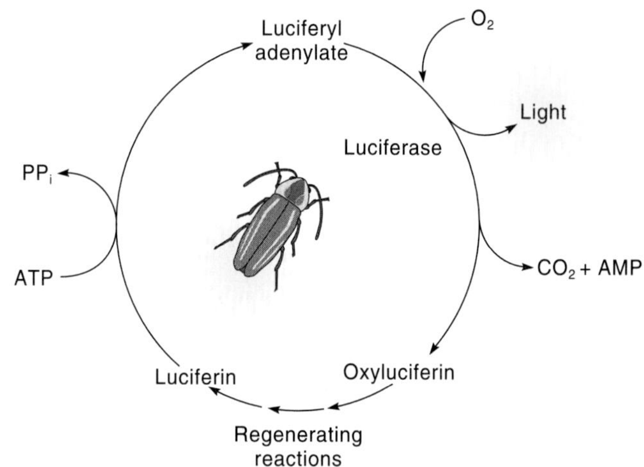

FIGURE 6.B
Luciferin reacts with ATP to produce luciferyl adenylate and two phosphate groups (designated PP_i). Luciferyl adenylate then reacts with oxygen in a luciferase-catalyzed reaction to release CO_2, AMP (adenosine monophosphate), and light. A series of reactions that reduces oxyluciferin regenerates luciferin.

before the cell can use them. Finally, ATP's terminal (end) phosphate bond, unlike the covalent bonds between carbon and hydrogen in organic molecules, is unstable—so it breaks and releases energy easily.

All cells use ATP for energy transformations. Just as you can use currency to purchase a great variety of different products, cells use ATP in all kinds of different chemical reactions to do different kinds of work. If you ran out of ATP, you would die instantly. Life requires huge amounts of ATP. A typical adult uses the equivalent of 2 billion ATP molecules a minute just to stay alive, or about 440 pounds (200 kilograms) of ATP a day. However, ATP is used so rapidly that only a few grams are available at any given instant. Organisms recycle ATP at a furious pace, adding phosphate groups to ADP to reconstitute ATP, using the

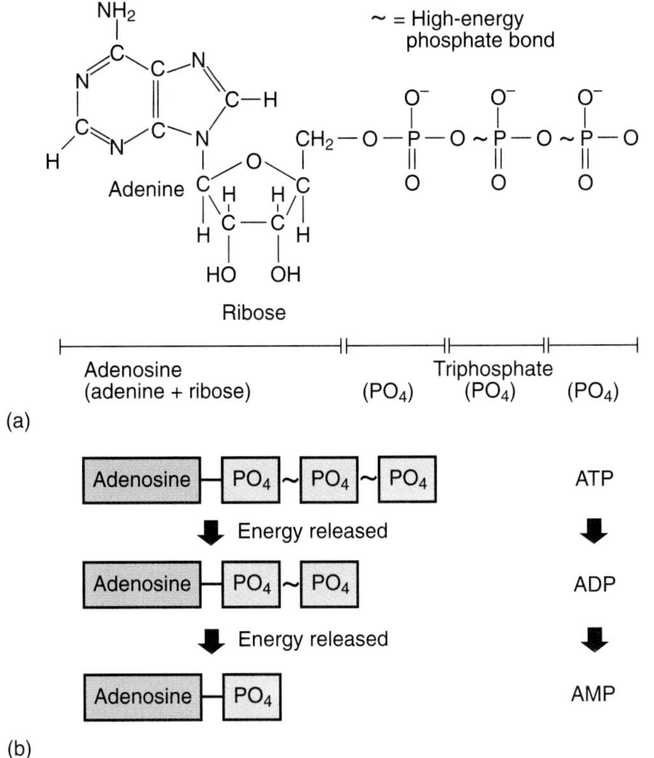

FIGURE 6.13

ATP—energy currency of the cell. (*a*) Adenosine triphosphate consists of an adenine and a ribose, which together form the "adenosine" part of the molecule, plus three phosphate (PO_4) groups. (*b*) Energy is required to form bonds between negatively charged phosphate groups. The squiggly lines that represent these bonds indicate that a great deal of energy is released when they break.

ATP to drive reactions, and turning over the entire supply every minute or so. The reaction that converts ADP to ATP is represented as:

$$ADP + P_i + \text{energy} \rightarrow ATP + \text{heat}$$

Coupled Reactions

Cells couple the breakdown of nutrients to ATP production, and they couple the breakdown of ATP to other reactions that occur at the same time and place in the cell. **Coupled reactions,** as their name implies, are reactions that occur in pairs. One reaction drives the other, which does work or synthesizes new molecules. Consider, for example, the formation of sucrose and water from glucose and fructose. This reaction is not spontaneous; it requires a net input of energy. When ATP provides additional energy, the reactants now have more energy available than the products, and the reaction proceeds. ATP breakdown is coupled to sucrose synthesis.

ATP accomplishes most of its work by transferring its terminal (end) phosphate group to another molecule in a process called **phosphorylation.** In muscle cells, for example, ATP transfers phosphate groups to contractile proteins. Similarly, plants use ATP energy to build cell walls. Phosphorylation energizes the molecules receiving the phosphate groups so that they can be used in later reactions. The original energy "cost" of this phosphorylation is thus returned in subsequent reactions.

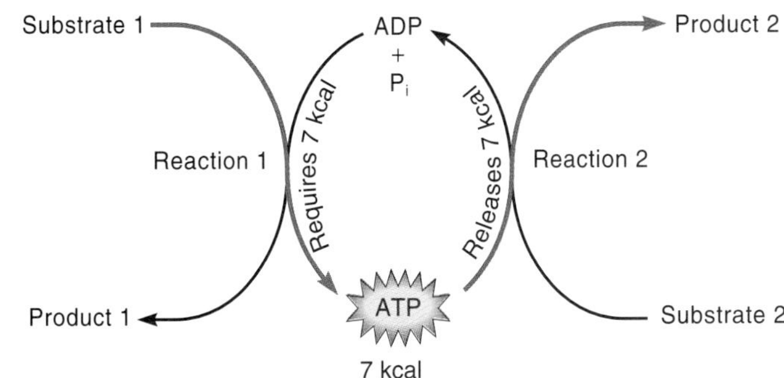

FIGURE 6.14

ATP breakdown and formation link biochemical reactions. Here, reaction 1 releases energy, which is used to phosphorylate ADP to ATP. Reaction 2 requires energy, which is released when ATP splits to yield ADP.

Table 6.1 Some Molecules Involved in Cellular Energy Transformations

Molecule	Mechanism
ATP	Compound that phosphorylates molecules, energizing them to participate in other reactions; conversions between ATP and ADP link many biochemical reactions
Cofactors	Substances, usually ions, that transfer chemical groups such as phosphates between molecules
Coenzymes	Vitamin-derived cofactors that transfer electrons or protons
NAD^+	Coenzyme that transfers electrons; used to synthesize ATP
$NADP^+$	Coenzyme that supplies hydrogen to reduce CO_2 in photosynthesis
FAD	Coenzyme that transfers electrons; used to synthesize ATP in cellular respiration
Cytochromes	Iron-containing molecules that transfer electrons in metabolic pathways

Other Compounds Involved in Energy Metabolism

Several other compounds besides ATP participate in the cell's energy transformations (table 6.1). Nonprotein helpers called **cofactors** assist other chemicals in enabling certain reactions to proceed. Cofactors are often ions. Mg^{2+}, for example, is a cofactor required to transfer phosphate groups between molecules.

Organic cofactors, which are called **coenzymes,** usually carry protons or electrons. Coenzymes are often nucleotides, as is ATP. But a coenzyme's energy content, unlike ATP's, depends on its ability to donate electrons or protons, not on the presence or absence of a particular phosphate bond. Vitamins act as coenzymes

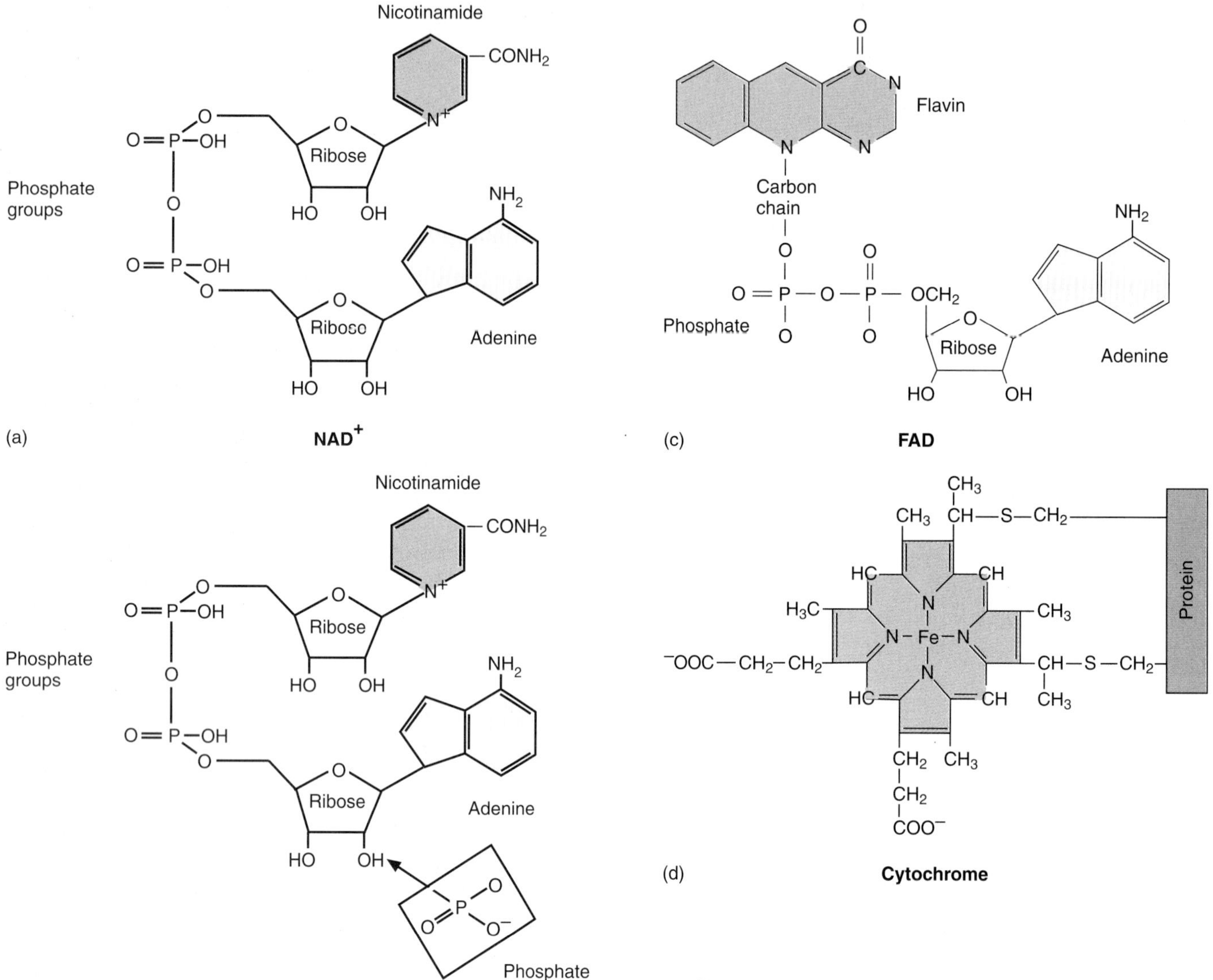

(a) **NAD^+**

(b) **$NADP^+$**

(c) **FAD**

(d) **Cytochrome**

FIGURE 6.15
Other molecules important in cellular energy reactions.

in cells and help to drive metabolic reactions. Vitamins, therefore, do not directly supply energy, as many people believe. Rather, vitamins make possible the reactions that extract energy from food.

NAD^+, $NADP^+$, and FAD are other important molecules that function as coenzymes. We introduce them here and return to them in the next two chapters (fig. 6.15).

Nicotinamide Adenine Dinucleotide (NAD^+)

Nicotinamide adenine dinucleotide (NAD^+), like ATP, consists of adenine, ribose, and phosphate groups; but NAD^+ also has a nitrogen-containing ring, called nicotinamide, which is derived from niacin (vitamin B_3). The nicotinamide is the active part of the molecule. NAD^+ is reduced when it accepts two electrons and two protons from a substrate. Both electrons and one proton actually join the NAD^+, leaving a proton (H^+). This reaction is written as:

$$NAD^+ + 2H^+ + 2e^- \rightarrow NADH + H^+$$

$NADH + H^+$ is fully reduced and is therefore packed with potential energy. The cell uses it to synthesize ATP and to reduce other compounds. Other biochemical pathways also tap the energy in $NADH + H^+$ to produce ATP.

Nicotinamide Adenine Dinucleotide Phosphate ($NADP^+$)

The structure of **nicotinamide adenine dinucleotide phosphate ($NADP^+$)** is similar to that of NAD^+ but with an added phosphate group. $NADP^+$ supplies the hydrogen that reduces carbon dioxide to carbohydrate during photosynthesis. This process "fixes" atmospheric carbon into organic molecules that then serve as nutrients. The energy that drives this process comes from ATP.

Flavin Adenine Dinucleotide (FAD)

Flavin adenine dinucleotide (FAD) is derived from riboflavin (vitamin B_2). FAD, like NAD, carries two electrons. However, it also accepts both protons to become $FADH_2$. In cells, FAD helps break glucose down to carbon dioxide and water.

Cytochromes

The **cytochromes** are iron-containing molecules that transfer electrons in metabolic pathways. When oxidized, the iron in cytochromes is in the Fe^{3+} form. When the iron accepts an electron, it is reduced to Fe^{2+}. There are several types of cytochromes, all of which carry electrons in cells. In metabolic pathways, several cytochromes align to form electron transport systems, with each molecule accepting an electron from the molecule before it and passing an electron to the next. Figure 6.16 shows such a chain; we will see more of them in the next two chapters. Small amounts of energy are released at each step of an electron transport chain, and the cell uses this energy in other reactions. Cytochromes take part in all energy transformations in life, suggesting that the cellular strategy for energy transformation is quite ancient.

Mastering Concepts

1. How does ATP supply energy for cellular functions?
2. What are the functions of cofactors and coenzymes in cellular metabolism?
3. What are some other compounds that participate in cellular energy reactions?

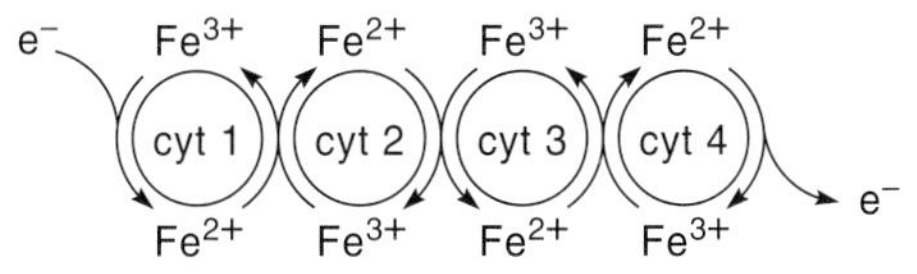

FIGURE 6.16

Different cytochrome proteins align, forming electron transport systems. Electrons pass down the chain in a series of oxidation-reduction reactions, releasing energy in small, usable increments. Electrons are passed between iron molecules attached to the cytochromes. Iron in ferric form (Fe^{3+}) gains an electron, becoming reduced to ferrous (Fe^{2+}) form. The cytochromes are embedded in membranes.

Enzymes and Energy

Many chemical reactions require an initial boost of energy, called the **energy of activation.** Sometimes this energy comes from heat in the environment. For example, heat provides the energy of activation to ignite a piece of wood. The amount of heat required to activate most metabolic reactions in cells would swiftly be lethal were it not for enzymes, which lower the required energy of activation (fig. 6.17).

Recall from chapter 3 that an enzyme is a protein that catalyzes (speeds) specific chemical reactions without taking part in them or being altered by them. An enzyme catalyzes a chemical reaction by decreasing the energy of activation. An enzyme binds the substrate in a way that enables the reaction to proceed. The conformation of the enzyme is such that the substrate fits into its active site, but not as precisely as a key fits a lock. Rather, the active site contorts slightly, as if it is hugging the substrate (fig. 6.18). Without enzymes, many biochemical reactions would not occur fast enough to support life.

Consider the activity of the enzyme carbonic anhydrase, necessary for carbon dioxide transport in red blood cells. This enzyme catalyzes formation of carbonic acid (H_2CO_3) from carbon dioxide (CO_2) and water (H_2O), mentioned earlier to illustrate a reversible reaction:

$$\underset{\text{(carbon dioxide)}}{CO_2} + \underset{\text{(water)}}{H_2O} \rightarrow \underset{\text{(carbonic acid)}}{H_2CO_3}$$

Without carbonic anhydrase, only about one molecule of carbonic acid forms per second—not fast enough to help organisms rid themselves of metabolic waste. However, with the enzyme present, carbonic acid forms at a rate of about 600,000 molecules per second.

Temperature affects enzyme function. Higher temperatures increase enzyme activity—to a point. The activity of an enzyme doubles for every increase of 10°C, but beyond about 60°C, entropy grows too great, and the enzyme unwinds from its characteristic three-dimensional form. At this point, the reaction halts. Although the enzymes of a few organisms—such as the hot springs bacteria

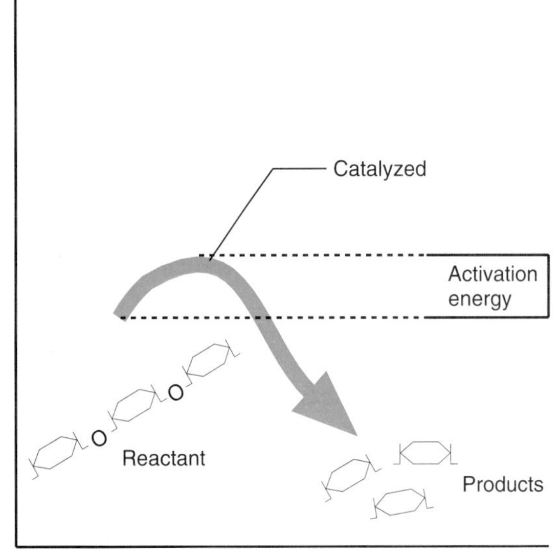

FIGURE 6.17

Enzymes lower the activation energies of specific reactions, without actually participating in the reactions. This speeds reaction rates.

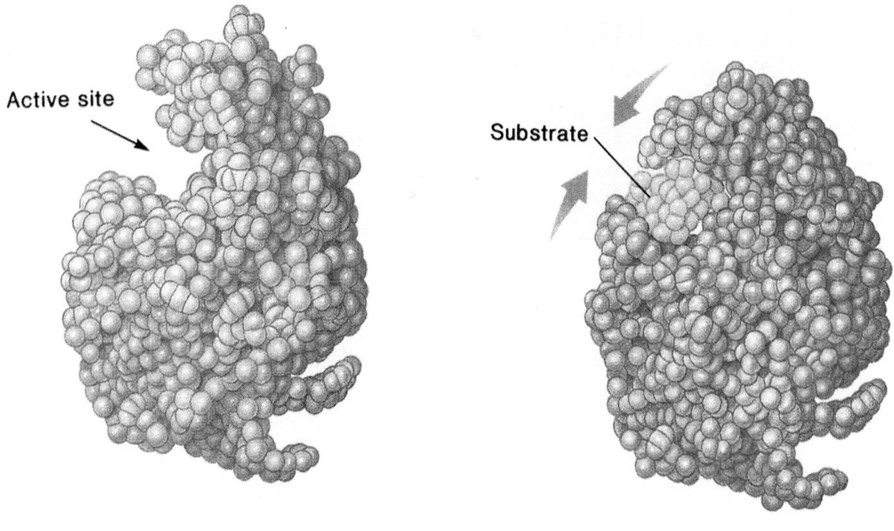

FIGURE 6.18

Lysozyme is an enzyme found in human tears and in chicken egg whites. It catalyzes a reaction that breaks down oligosaccharides. The oligosaccharide substrate fits into the enzyme's active site, but not as precisely as a key fits a lock. When the substrate nestles into the active site's cavity, the surrounding part of the enzyme contorts slightly to accommodate it, forming an "induced fit."

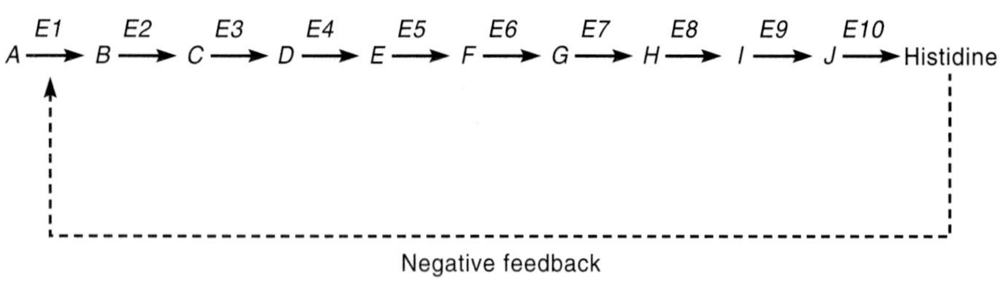

FIGURE 6.19

Negative feedback (feedback inhibition). Like other organisms, *E. coli* requires the amino acid histidine to synthesize proteins. This diagram isolates the 10 enzyme-catalyzed, linked reactions leading to histidine production. When sufficient histidine accumulates, the histidine inhibits the activity of E1, the first enzyme in the pathway, temporarily halting further production of the amino acid.

discussed in chapter 1—can tolerate high temperatures, most enzymes work best at much lower temperatures. Most of the enzymes in the human body function optimally near body temperature.

Control of Metabolism

A cell is in a constant state of flux, as it rapidly dismantles some molecules and synthesizes others. Disturbances in the balance between these two actions threaten the life of the cell. Several biological mechanisms regulate cellular metabolism and preserve this delicate balance.

Certain enzymes control metabolism by functioning as pacesetters at important junctures in biochemical pathways. The enzyme whose reaction precedes the slowest reaction sets the pace for the pathway's productivity, just as the slowest runner on a relay team limits the overall pace for the whole team. This is because each subsequent reaction in the metabolic pathway (like each subsequent relay runner) requires the product of the preceding reaction to continue. The reaction catalyzed by this enzyme is called the rate-limiting step; the enzyme is a **regulatory enzyme** because it regulates the pathway's pace and productivity.

Inhibition of a Metabolic Pathway

Often a regulatory enzyme is highly sensitive to a chemical cue. When this cue is the end product of the pathway the enzyme takes part in, the enzyme may "turn off" for a time, and the regulation is called **negative feedback,** or feedback inhibition. Negative feedback prevents too much of one substance from accumulating—an excess of a particular biochemical effectively shuts down its own synthesis until its levels fall. At that point, the pathway resumes its activity. Negative feedback is somewhat like a thermostat—when the temperature in a building reaches a certain level, the thermostat shuts the heat off for a while. Falling temperature cues the thermostat to ignite the furnace again.

A product molecule can inhibit its own synthesis in two general ways. It may bind to the regulatory enzyme's active site, preventing it from binding substrate and temporarily shutting down the pathway. This is sometimes called competitive inhibition because a substance other than the substrate competes to occupy the active site. Alternatively, product molecules may bind to the regulatory enzyme at a site other than the active site, but in a way that alters the shape of the enzyme so that it can no longer bind substrate. This indirect approach is sometimes called noncompetitive inhibition because the inhibitor does not directly compete to occupy the active site. Both competitive and noncompetitive inhibition are forms of negative feedback.

Figure 6.19 illustrates negative feedback. It shows the 10-step pathway that certain bacteria use to synthesize a nutrient, the amino acid histidine. When excess histidine accumulates, it binds (noncompetitively) to the junction between two subunits of the enzyme that regulates the pathway. This temporarily destabilizes the enzyme and impairs catalysis. For a time, histidine synthesis ceases. When levels of histidine in the bacterium fall, the block on the regulatory enzyme lifts, and the cell can once again synthesize the amino acid.

Substances from outside the body, such as drugs and poisons, can also inhibit enzyme function. A foreign chemical similar in conformation to the substrate may block the enzyme's active site and prevent the substrate from binding. The drug sulfanilamide, for example, competitively inhibits certain enzymes in bacteria and is therefore useful to fight certain infections. Sometimes, increasing the concentration of the substrate can overcome such a competitive inhibitor.

Poisons that bind noncompetitively, to a part of the enzyme other than the active site, also inhibit enzyme function. Some nerve gases, for example, inhibit the activity of the enzyme acetylcholinesterase, which normally recycles the neurotransmitter acetylcholine back into nerve cells after it sends its message. When the nerve gas binds to acetylcholinesterase, the enzyme cannot bind its substrate, and nerve transmission ceases. Penicillin, a drug, acts in a similar way, binding irreversibly to an enzyme that bacteria require to build cell walls. Cell walls fall apart, the bacteria die, and the infection halts.

Activation of a Metabolic Pathway

Much rarer than negative feedback in living systems is **positive feedback,** in which a molecule activates the pathway leading to its production. Blood clotting, for example, begins when a biochemical pathway synthesizes fibrin, a threadlike protein. The products of the later reactions in the clotting pathway stimulate the enzymes that activate earlier reactions. As a result, fibrin accumulates faster and faster, until there is enough to stem the blood flow. When the clot forms and the blood flow stops, the clotting pathway shuts down—an example of negative feedback.

Mastering Concepts

1. How does an enzyme speed a chemical reaction?
2. How does temperature affect enzyme activity?
3. How can an enzyme control the productivity of a biochemical pathway?
4. In what two ways can the product of a biochemical pathway turn off its own synthesis?
5. Can the product of a biochemical pathway increase its own synthesis?

The Major Energy Transformations—Respiration and Photosynthesis

The energy transformations that sustain life are similar in all organisms. The most important of these pathways are respiration and photosynthesis, the topics of the next two chapters. As we shall see, these pathways are intimately related.

The *energy-requiring* stage of biological energy acquisition and utilization is photosynthesis. During this process, chloroplasts in leaves absorb light energy. They use this energy to release oxygen and reduce carbon dioxide (a low-energy compound) to carbohydrate (a high-energy compound). Carbohydrate, in turn, fuels the activities of the plant and ultimately all other organisms. During cellular respiration, the *energy-releasing* stage of the

Needing Energy to Get Energy

continued from page 110.

The snake pays dearly for its meal of a 130-pound gazelle. While most organisms that eat frequently or continuously invest 10 to 23% of a meal's energy in digesting it and assimilating its nutrients, most snakes invest a whopping 32% in energy acquisition. The African rock python tops that—it expends an equivalent of half the energy in the chemical bonds of its meal just to digest it. Eating a megameal is an energetically expensive proposition.

Two particular activities handle the python's meal. First, abundant hydrochloric acid (HCl) must be present in the stomach to lower the pH sufficiently for digestive enzymes to function. This isn't the spurt of HCl we require to tackle an occasional hamburger but a constant supply lasting weeks. To continually produce HCl, the snake must maintain an energy level equivalent to that of an active greyhound or racehorse but far beyond the few minutes it takes these animals to run a race. To do this, as soon as the snake's jaws fit around the gazelle's head, oxygen consumption increases some 36-fold. As we'll see in the next chapter, oxygen is critical for extracting the maximal amount of energy from food.

The second digestive/energy adaptation of the massive rock python affects the intestines, where digested nutrients are absorbed into the bloodstream. After a feast, literally overnight, this organ doubles or triples in weight! Stimulation for rapid cell division comes from the expanding stomach, which sends hormonal and nervous signals to the awaiting intestines. The blooming intestines can produce 60 times more digestive enzymes than they do during the many months between meals.

Enormous energy is required to support this sudden burst of intestinal activity. To study how the digestive tract handles such a load, researchers grew patches of small intestine from a Burmese python and a sidewinder rattlesnake in the laboratory and applied amino acids tagged with radioactive amino acids so that their fate could be traced. The investigators found that within a day, the small intestine's ability to absorb the amino acids increased 6-fold, and that after 3 days, it had increased 16-fold.

Producing buckets of HCl and supporting rapidly expanding intestines greatly taps the snake's energy reserves. The reptile can do this, though, by shutting down its digestive tract between meals. HCl level plummets, and the intestinal lining shrinks to nearly nothing.

biological energy process, energy-rich carbohydrate molecules are oxidized to carbon dioxide and water. Cellular respiration liberates the energy necessary to power life.

SUMMARY

1. Energy is the ability to do work. Energy comes from the sun, the wind, moving water, the earth, and the tides and is held in the bonds of molecules of organisms.

2. Potential energy is stored energy potentially available to do work, while **kinetic energy** actually does work.

3. The **laws of thermodynamics** govern the energy transformations of life. The first law concerns the conservation of energy and states that energy cannot be created or destroyed but only converted to other forms. The second law states that all energy transformations are inefficient, because every reaction results in increased **entropy** (disorder) and the loss of usable energy as heat.

4. Metabolism is the sum of the energy and matter conversions in a cell. It consists of enzyme-catalyzed reactions organized into often interconnected pathways and cycles. **Anabolism** includes reactions that synthesize molecules. These reactions diverge from a few reactants to many products and require energy. **Catabolism** includes reactions that break down molecules, converge from many reactants to a few products, and release energy.

5. Each reaction of a **metabolic pathway** rearranges atoms into new compounds, which changes the amount of free energy available in chemical bonds to do work. In **endergonic reactions,** products have more energy than reactants, and entropy decreases. These reactions are not spontaneous—they require energy input. In **exergonic reactions,** products have less energy than reactants, entropy increases, and energy is released.

6. Many energy transformations in organisms occur via **oxidation** and **reduction** reactions. Oxidation is the loss of electrons from a molecule; reduction is the gain of electrons. Oxidation and reduction reactions occur simultaneously, in pairs.

7. ATP, the energy currency of cells, stores energy in its high-energy phosphate bonds. Many energy transformations involve coupled reactions, in which the cell uses the energy released by ATP to drive another reaction. Other compounds that take part in cellular metabolism include **cofactors, coenzymes, NAD^+, $NADP^+$, FAD,** and **cytochromes.**

8. Enzymes are proteins that speed spontaneous biochemical reactions to a biologically useful rate by lowering the **energy of activation.** In **negative feedback,** a pathway's product interacts with a regulatory enzyme for the pathway, temporarily shutting down its own synthesis when its levels rise. In **positive feedback,** a product stimulates its own production.

TO REVIEW . . .

1. What units are used to measure energy?

2. What are some sources of energy?

3. Give one example of potential energy and one of kinetic energy.

4. Cite everyday illustrations of the first and second laws of thermodynamics. How do these principles underlie every organism's ability to function?

5. State the differences between endergonic and exergonic reactions.

6. Give an example of entropy.

7. Why isn't heat usable energy?

8. State three differences between anabolic and catabolic reactions.

9. What is chemical equilibrium?

10. Why are oxidation and reduction reactions linked?

11. Cite three reasons why ATP is an excellent source of biological energy.

12. What do ATP, NAD^+, $NADP^+$, and FAD have in common chemically?

13. How does an enzyme speed a chemical reaction?

14. Cite three ways that the product of a biochemical pathway can control its own rate of synthesis.

TO THINK ABOUT . . .

1. Some people claim that life's high degree of organization defies the laws of thermodynamics. How is the organization of life actually consistent with the principles of thermodynamics?

2. How might you explain the fact that even species as diverse as humans and yeast use the same biochemical pathways to extract energy from nutrient molecules?

3. Cytochrome c is an electron carrier that is nearly identical in all species. No known disorders affecting cytochrome c exist. What does this suggest about the importance of this molecule?

4. When a person eats a fatty diet and excess cholesterol accumulates in the bloodstream, cells temporarily turn off their synthesis of cholesterol. What phenomenon described in the chapter does this control of cholesterol illustrate?

5. In 1991, a large volcanic eruption in the Philippines threw great dust clouds into the atmosphere and lowered temperatures around the world for many months. How might this event have affected energy transformations in organisms?

TO LEARN MORE . . .

References and Resources*

Baum, Rudy M. February 22, 1993. Views on biological, long-range electron transfer stir debate. *Chemical and Engineering News,* vol. 73. Chemists are isolating parts of the chain of electron carriers in cellular respiration.

Cossins, Andrew R., and Neil Roberts. January 4, 1996. The gut in feast and famine. *Nature,* vol. 379. Energy adaptations enable a snake to digest a meal that is 65% of its body dry weight.

Diamond, Jared. April 1994. Dining with the snakes. *Discover.* A blow-by-blow description of how a snake eats.

Herring, Peter J. May 13, 1993. Light genes will out. *Nature,* vol. 363. The bioluminescent bacteria that live in anglerfish and flashlight fish are very unusual.

Storey, Richard D. March 1992. Textbook errors and misconceptions in biology: Cell energetics. *The American Biology Teacher.* Understanding the laws of thermodynamics helps us understand the energy pathways in a cell.

**Life*'s website features frequently updated references.

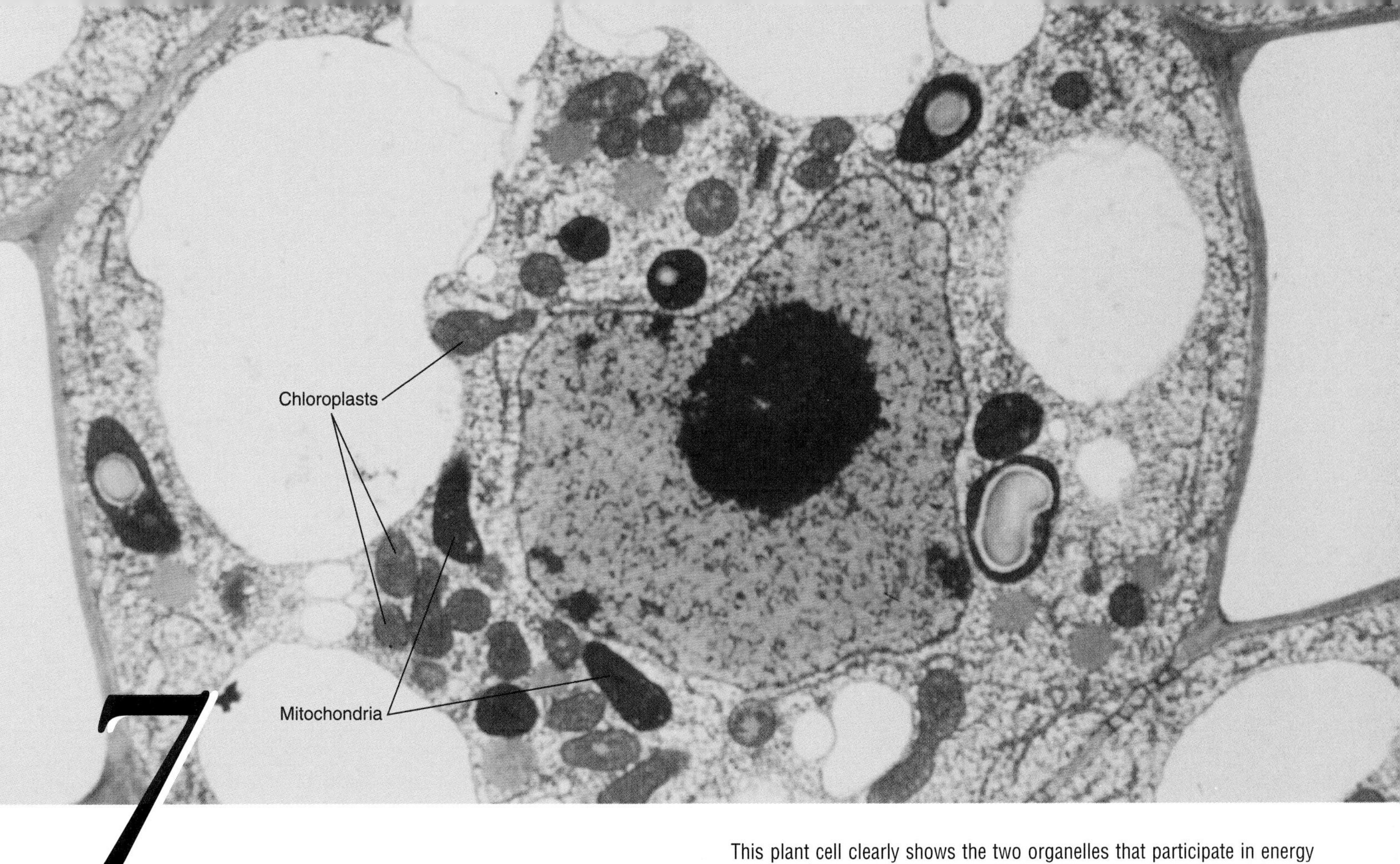

This plant cell clearly shows the two organelles that participate in energy procurement (the chloroplast) and energy release (the mitochondrion).

7 How Cells Release Energy

chapter outline

"Breathing" Respiration and Cellular Respiration 126

Ways That Cells Release Energy 127
Mitochondria
Energy Transfer from Nutrients to ATP

Routes to ATP Formation 129
Substrate-Level Phosphorylation
Oxidative Phosphorylation

Glycolysis—Glucose Breaks Down to Pyruvic Acid 129
The First Half of Glycolysis—Glucose Activation
The Second Half of Glycolysis—Energy Extraction

Acetyl CoA Formation Bridges Glycolysis and the Krebs Cycle 131

The Krebs Cycle 132

Oxidative Phosphorylation and Electron Transport 133
The Respiratory Chain
ATP Synthesis

ATP Accounting—An Estimate 136

Control of Cellular Respiration 136

Energy Retrieval in the Absence of Oxygen 137
Fermentation Pathways
Anaerobic Electron Transport

How Proteins and Lipids Enter the Energy Pathways 140

Running on Empty

Michael P. was noticeably weak from birth. He didn't move much, had "floppy" muscles, had difficulty breathing, and grew exhausted merely from the effort of eating. At the age of two-and-a-half months, he suffered his first seizure, staring and jerking his limbs for several frightening minutes. Despite medication, his seizures continued more and more frequently.

The doctors were puzzled because most of Michael's many medical tests showed normal results—with one notable exception. His cerebrospinal fluid (the fluid that bathes the brain and spinal cord) was unusually low in glucose and a small organic chemical called lactic acid. How could these deficits explain the baby's disturbing symptoms?

to be continued . . .

"Breathing" Respiration and Cellular Respiration

The word "respiration" to most people means "breathing"—as in "artificial respiration" or "the respiratory system." Respiration at the cellular level, though, has a different meaning. **Cellular respiration** refers to the biochemical pathways that extract energy from the bonds of nutrient molecules, in the presence of oxygen. Yet cellular respiration and the familiar, whole-body respiration are very much related. Both processes take in oxygen and release carbon dioxide (CO_2), which is called gas exchange (fig. 7.1).

The function of a respiratory system is to filter and convey air to cells, where gas exchange can occur. Specifically, oxygen gas in inhaled air enters red blood cells in the bloodstream, and carbon dioxide leaves the bloodstream and is exhaled.

In cellular respiration, energy is slowly liberated from food molecules by stripping off electrons from intermediate compounds (also called intermediates) and channeling them through a series of electron carriers. The released energy is then used to phosphorylate ADP to ATP, the biological energy store. Oxygen is the final acceptor of the moving electrons. Without oxygen, much energy would remain locked in nutrient molecules. In the reactions that remove electrons from the compounds that result when nutrients break down, carbon to carbon bonds are cleaved and each carbon is combined with oxygen, forming CO_2, which is a metabolic waste. Cellular respiration, then, explains why our respiratory systems obtain oxygen and get rid of CO_2. Cellular respiration is one of several approaches that organisms use to extract energy from food.

Cellular respiration begins where digestion leaves off. Consider the nut-munching chipmunk in figure 7.2. The nut passes through the rodent's digestive system and breaks into clumps of cells. As the nut's cells break apart, they release proteins, carbohydrates, and lipids. After the chipmunk digests these macromolecules into their component amino acids, monosaccharides, and fatty acids and glycerol, they are small enough to enter the blood and lymphatic systems and be transported to the body's tissues.

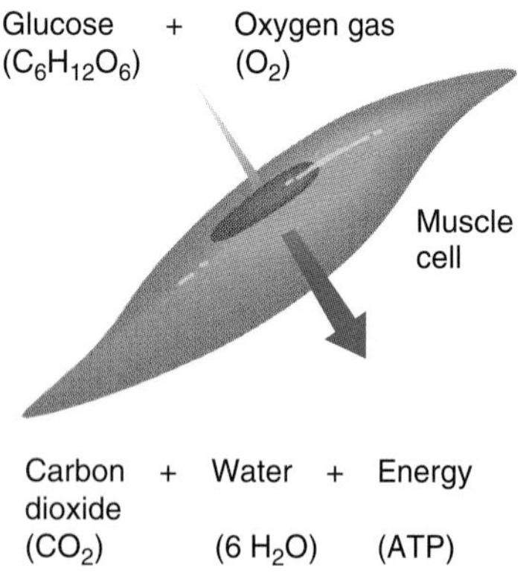

Respiration (breathing + gas exchange)

Cellular Respiration (extracting energy from nutrients)

FIGURE 7.1

"Breathing" respiration and cellular respiration are linked. The athlete breathes in oxygen and exhales carbon dioxide (CO_2), a metabolic waste. Oxygen enters the bloodstream in the lungs and is distributed to all cells. There, in mitochondria, oxygen enables the reactions of cellular respiration to occur, extracting energy from glucose and other nutrient molecules. This energy powers the athlete's muscles to contract.

FIGURE 7.2

Eating and digestion are the first steps in procuring biological energy. Nutrient molecules must be absorbed into cells lining the intestines and then transferred to the bloodstream, where they are transported to cells. Once in cells, nutrients are broken down in biochemical reactions, pathways, and cycles that release energy in their chemical bonds.

When these smaller nutrient molecules enter the animal's cells, they break down further, and some may be converted to glucose. When glucose and other nutrients break down, energy is released and temporarily stored in the high-energy phosphate bonds of ATP.

This chapter describes how cells extract energy from nutrients. The journey from food to energy entails several biochemical pathways, with many chemical names, and may appear overwhelming. If we consider energy release in major stages, and also one step at a time, the logic becomes clear and, like much of life science, it makes sense.

Mastering Concepts

1. How does cellular respiration differ from whole-body respiration?
2. What is gas exchange?
3. What is the relationship between digestion of foods and cellular respiration?
4. What is the role of oxygen in cellular respiration?

Ways That Cells Release Energy

The cells of all organisms extract energy from nutrients. The process always begins with a pathway called **glycolysis.** Other biochemical cycles or pathways follow glycolysis, the specific type depending upon the species and whether or not oxygen is in the environment.

Eukaryotes and some prokaryotes capture energy as ATP using the pathways of cellular respiration, also called aerobic respiration because it occurs in the presence of oxygen. In aerobic respiration, glucose breaks down, releasing carbon dioxide (CO_2), water (H_2O), and energy. Organisms that carry out aerobic respiration are called **aerobes.** Organisms called **anaerobes,** and cells temporarily lacking oxygen, use other pathways that produce little or no ATP but lose heat energy or transfer energy into organic molecules other than ATP. Fermentation pathways and anaerobic electron transport are energy-releasing pathways that do not use oxygen and are discussed later in the chapter (fig. 7.3).

If glucose were burned to release energy outside of an organism, it would create a small fire lasting several minutes. In a cell, aerobic respiration controls this process, so that the heat does not destroy living tissue and usable energy is released gradually (fig. 7.4). In organisms, part of the energy from glucose becomes trapped in ATP. The rest of the released energy—the heat lost during respiration—never causes an organism to combust because its release is spread over many biochemical reactions. The carbon liberated from stored chemicals is recycled into other organic molecules.

Cellular respiration harvests energy in stages (fig. 7.5). In the first stage, glycolysis, glucose splits into smaller molecules in the cytoplasm. A short second stage, **acetyl CoA formation,** takes the products of glycolysis into mitochondria, where the third stage, the **Krebs cycle,** occurs. The Krebs cycle completely breaks down the products of glycolysis and captures some of the energy. The reactants in these first three stages lose electrons and protons, which are ferried on coenzymes to the fourth stage, an electron transport system called the **respiratory chain.** The chain is a series of proteins embedded in the inner mitochondrial membrane. These proteins pass electrons in a series of oxidation and reduction reactions, capturing their energy to phosphorylate ADP. Oxygen is the final electron acceptor.

The electrons and protons essential to generating ATP come from hydrogen ions released from reactants as the various reactions occur. Figure 7.6 reviews a Bohr model of a hydrogen ion, which is just one proton and one electron.

Mitochondria

Mitochondria, which house the reactions of the Krebs cycle and the respiratory chain, are particularly plentiful in cells that require a lot of energy, such as muscle cells. Recall from chapter 4 that a mitochondrion consists of an outer membrane and

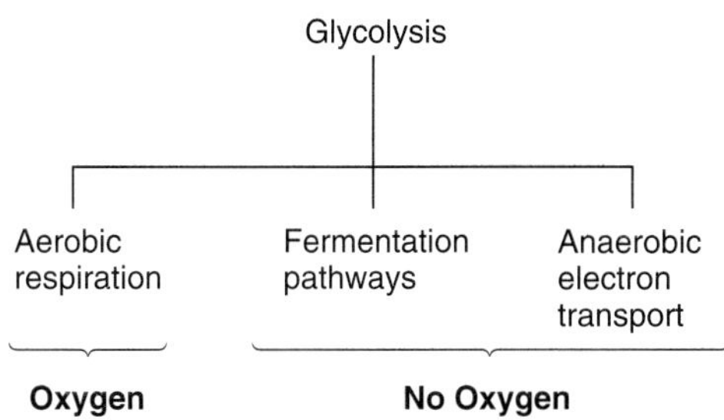

FIGURE 7.3

All cells undergo glycolysis. Subsequent pathways continue the release of energy from nutrient molecules. These are three common pathways.

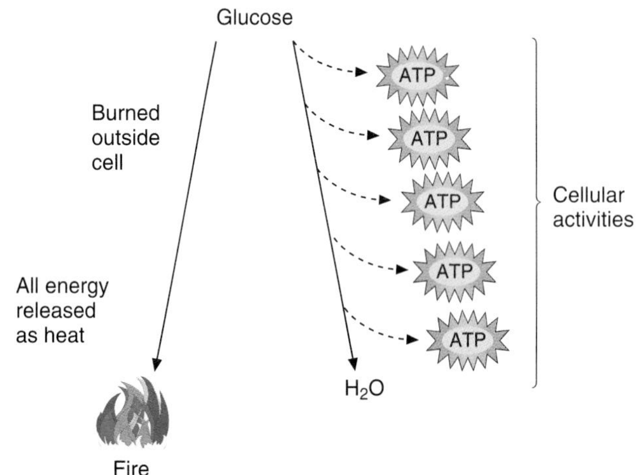

FIGURE 7.4

Release of energy from the bonds holding together glucose molecules can occur all at once, as it does outside of organisms when ignited, or can break down gradually, releasing energy in stages, as it does in organisms. Hydrogens split from intermediate compounds, providing electrons whose energy is used to phosphorylate ADP to ATP, the biological energy currency. Oxygen is the final electron acceptor.

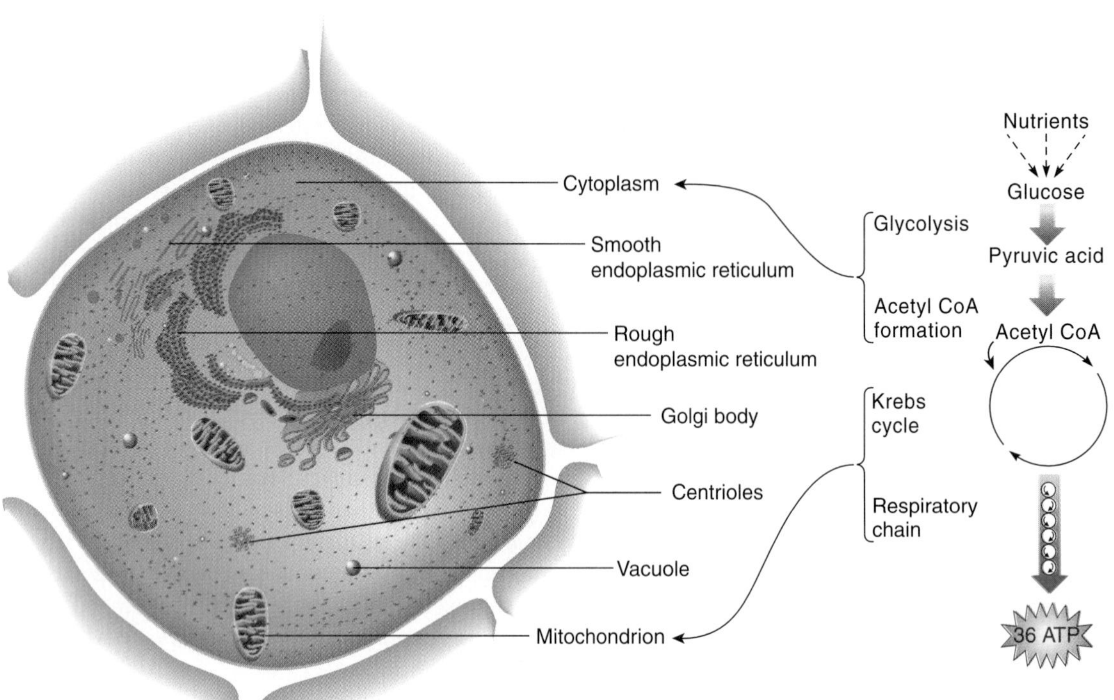

FIGURE 7.5

An overview of cellular respiration. This highly schematic overview shows the stages of energy procurement that constitute cellular respiration. As the chapter progresses, the illustrations of biochemical pathways will become more detailed. Look to the insets that repeat this diagram with different sections highlighted to follow the part of the overall pathway under discussion. Note that glycolysis and acetyl CoA formation occur in the cytoplasm, and the Krebs cycle and respiratory chain occur in mitochondria.

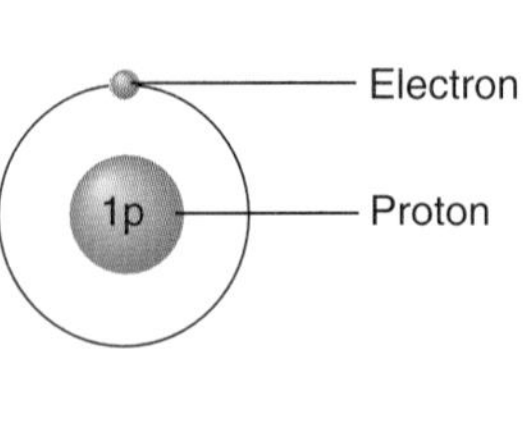

FIGURE 7.6

A hydrogen atom is just a proton and an electron.

FIGURE 7.7

A mitochondrion's interior is a highly folded membrane studded with enzymes and electron carriers important in cellular respiration. An outer membrane surrounds the organelle. ATP synthase is an enzyme that catalyzes phosphorylation of ADP to ATP.

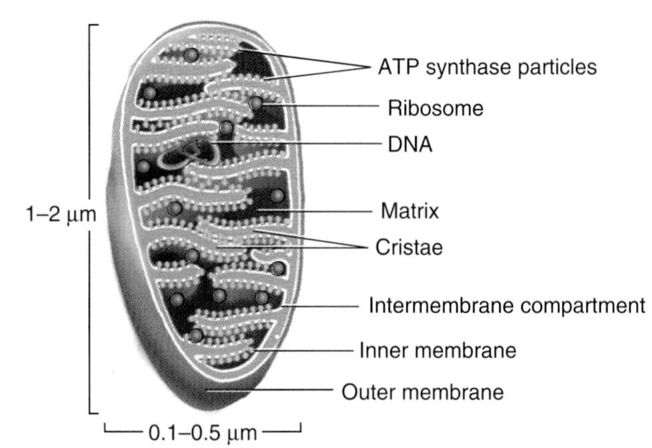

a highly folded inner membrane, an organization that creates two compartments (fig. 7.7). The innermost compartment, called the **matrix,** is where the Krebs cycle occurs. A second area between the two membranes is called the **intermembrane compartment.**

The inner mitochondrial membrane is folded into numerous projections called **cristae,** which are studded with enzymes and electron carrier molecules. This extensive folding greatly increases the surface area on which the reactions of the respiratory chain can occur. In aerobic prokaryotes, which lack mitochondria, respiratory enzymes are embedded in the cell membrane.

Energy Transfer from Nutrients to ATP

The overall result of the energy-releasing pathways is to use energy stored in organic molecules to phosphorylate ADP to ATP. Cellular respiration is quite efficient, with one glucose molecule theoretically yielding 36 ATPs in eukaryotes and 38 in prokaryotes. Just as we use ordinary currency, the cell can use its energy currency—ATP—in several ways (table 7.1).

Table 7.1 Some Functions in the Human Body Using ATP

Some Functions in the Human Body Using ATP
Cilia and flagella movement
Muscle contraction (all voluntary and involuntary movements)
Return of ions to the bloodstream in kidney tubules
Delivery of neurotransmitters from one nerve cell to another
Absorption of monosaccharides into bloodstream in the small intestine
Movement of chromosomes in cell division
Transport of NADH into mitochondria

In animals, the reactions of cellular respiration begin with glucose. But before this, other carbohydrates in food, such as sucrose (table sugar) or fructose (fruit sugar), must be digested or converted to glucose. Plants begin cellular respiration with glucose that may derive from other sources, too. In leaves, for example,

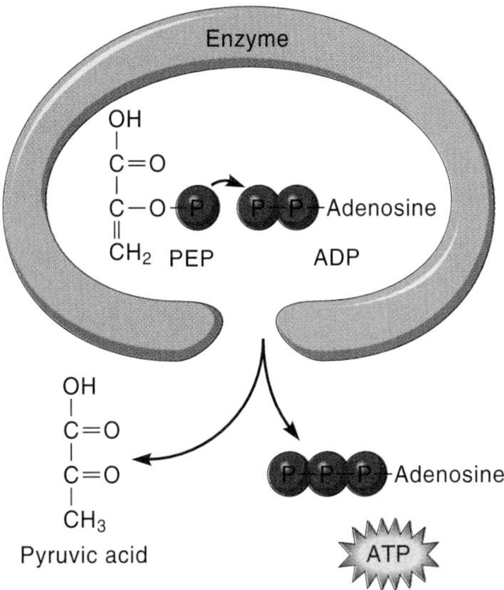

FIGURE 7.8

Substrate-level phosphorylation. A phosphate group on an organic substrate is transferred to ADP, phosphorylating it to ATP. The reaction is enzyme-catalyzed. This particular reaction is the final step in glycolysis. A molecule of phosphoenolpyruvate (PEP) loses a phosphate to ADP, forming pyruvic acid and ATP.

cellular respiration begins with sucrose, but in other parts of the same plant, such as roots and stems, it may begin with starch. Later in the chapter, we will see how the other major nutrients—proteins and lipids—enter the energy pathways. But for now, we will focus on glucose.

Mastering Concepts

1. What are the stages of cellular respiration?
2. Where in the cell does cellular respiration begin?
3. How is the energy from nutrient molecules temporarily stored?
4. What is the starting material for cellular respiration?

Routes to ATP Formation

The general equation for glucose respiration is:

glucose	+	oxygen	⟶	carbon dioxide	+	water	+	energy
$C_6H_{12}O_6$	+	$6O_2$		$6CO_2$	+	$6H_2O$	+	ATP

In English, this equation means: The energy held in the bonds of the nutrient molecule glucose ends up in the products carbon dioxide and water and in ATP, or it is lost as heat.

Substrate-Level Phosphorylation

ATP synthesis occurs in two ways. In the first, **substrate-level phosphorylation,** a phosphate (PO_4) transfers from organic compounds (substrates) to ADP, forming ATP. (Recall that phosphorylation is the process of taking up a phosphorus-containing group, which energizes the molecule receiving it.) An enzyme that binds both the substrate and ADP catalyzes the transfer (fig. 7.8). The energy of substrate-level phosphorylation comes from the phosphate bond of the substrate. Some energy escapes as heat when the phosphate group transfers.

Oxidative Phosphorylation

The second way respiration generates ATP is through **oxidative phosphorylation,** so named because it depends on oxidation-reduction reactions. Electrons from NADH pass along a chain of electron carriers through a series of oxidation-reduction reactions. The energy from this electron movement sets up differing proton concentrations on each side of the inner mitochondrial membrane and establishes a gradient of protons. When protons move down their concentration gradient (traveling from where they are highly concentrated on one side of the membrane to the other side, where they are not highly concentrated), they contact an enzyme, **ATP synthase,** at one site, which triggers phosphorylation of ADP to ATP at a different site on the same enzyme.

Substrate-level phosphorylation is the simpler and more direct mechanism for producing ATP, but it accounts for only a small percentage of the ATP produced in cellular respiration—specifically, that from glycolysis and the Krebs cycle. In the respiratory chain that follows these reactions, ATP forms through oxidative phosphorylation.

Mastering Concepts

1. What is the overall equation that describes cellular respiration?
2. What does the equation mean?
3. What is substrate-level phosphorylation, and in what stages of cellular respiration does it generate ATP?
4. What is oxidative phosphorylation, and in what stages of cellular respiration does it generate ATP?

Glycolysis—Glucose Breaks Down to Pyruvic Acid

Glucose contains considerable bond energy, but cells recover only a small portion of it during glycolysis. In this pathway, glucose splits into two three-carbon compounds. The entire process requires 10 steps, all of which occur in the cytoplasm (fig. 7.9). The first half of the pathway activates glucose so that energy can be extracted from its bonds. The second half of the pathway actually extracts the energy.

The First Half of Glycolysis—Glucose Activation

The first step of glycolysis uses one molecule of ATP to phosphorylate glucose. Phosphorylation activates glucose so that the appropriate enzyme can carry out the next step. Because phosphate is negatively charged, and since charged compounds cannot easily cross the cell membrane, phosphorylation also traps glucose in the cell.

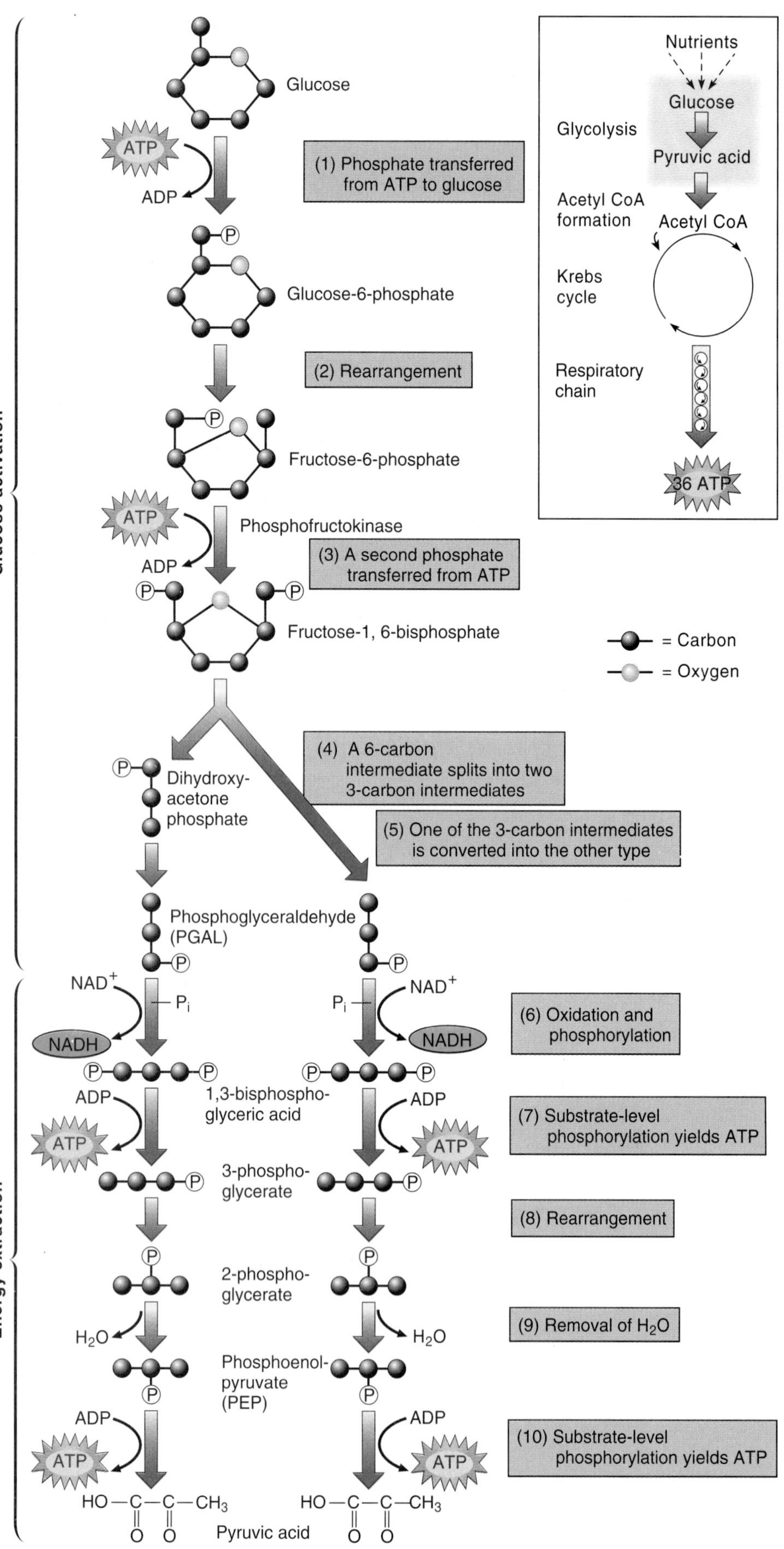

FIGURE 7.9

In the glycolysis reactions, glucose is rearranged and split into two three-carbon intermediates, each of which is rearranged further to eventually yield two molecules of pyruvic acid. Along the way, four ATPs and two NADHs are produced. Two ATPs are consumed in activating glucose, so the net ATP yield is two ATP molecules per molecule of glucose.

In step 2, the atoms of phosphorylated glucose are rearranged; in step 3 the new molecule (fructose-6-phosphate) is phosphorylated again by a second ATP, forming fructose-1,6-bisphosphate. This compound splits into two three-carbon compounds in steps 4 and 5, and each of the three-carbon products has one phosphate. One of the products, phosphoglyceraldehyde (PGAL), is further metabolized in glycolysis. The other product, dihydroxyacetone phosphate, is converted to PGAL and then is catabolized along with the other PGAL. Formation of the two PGALs derived from each glucose molecule marks the halfway point of glycolysis (steps 1 through 5 in fig. 7.9). So far, energy in the form of ATP has been invested, but no ATP has been produced.

The Second Half of Glycolysis—Energy Extraction

The first energy-obtaining step of the glycolysis pathway occurs in the second half when NAD^+ is reduced to NADH through the coupled oxidation of PGAL (step 6). After this step, some of the energy from glucose is stored in the energy-rich electrons of NADH. This oxidation also releases enough energy to add a second phosphate group to PGAL, making it 1,3-bisphosphoglyceric acid. Finally, the cell is ready to make ATP.

Substrate-level phosphorylation occurs when one of the phosphates of 1,3-bisphosphoglyceric acid is transferred to ADP (step 7). The three-carbon molecule that remains, 3-phosphoglycerate, is then rearranged to form 2-phosphoglycerate (step 8). This compound then loses water and becomes phosphoenolpyruvate (PEP) (step 9). PEP then becomes **pyruvic acid** when it donates its phosphate to a second ADP (step 10). Each PGAL from the first half of glycolysis progresses through this pathway to make two ATPs and one pyruvic acid in the second half. ("Pyruvate" refers to the ionized form of pyruvic acid.)

Because one molecule of glucose yields two molecules of PGAL, and each molecule of PGAL yields two molecules of ATP and one of pyruvic acid, each glucose produces four ATPs and two pyruvic acids. However, because the first half of glycolysis requires two ATPs, the net gain of glycolysis is two ATPs per molecule of glucose.

At the end of glycolysis, a small amount of the chemical energy that started out in glucose ends up in ATP and NADH. However, most of the energy of glucose remains in the bonds of pyruvic acid. The energy in the bonds of pyruvic acid is tapped to synthesize ATP in the mitochondrion.

Mastering Concepts

1. What are the starting materials and products of glycolysis?
2. Describe the reactions that activate glucose.
3. Describe the reactions that extract energy from the two PGAL molecules derived from glucose.
4. What is the net gain of ATP for each glucose molecule undergoing glycolysis?

Acetyl CoA Formation Bridges Glycolysis and the Krebs Cycle

The pyruvic acid transported into the mitochondrial matrix is not directly used in the Krebs cycle. First it loses a molecule of carbon dioxide as NAD^+ is reduced to NADH. The remaining molecule, called an acetyl group, attaches to a coenzyme to form **acetyl coenzyme A,** abbreviated **acetyl CoA** (fig. 7.10).

The conversion of pyruvic acid to acetyl CoA links glycolysis and the Krebs cycle. Pyruvic acid is the final product of glycolysis, and acetyl CoA is the compound that enters the Krebs cycle.

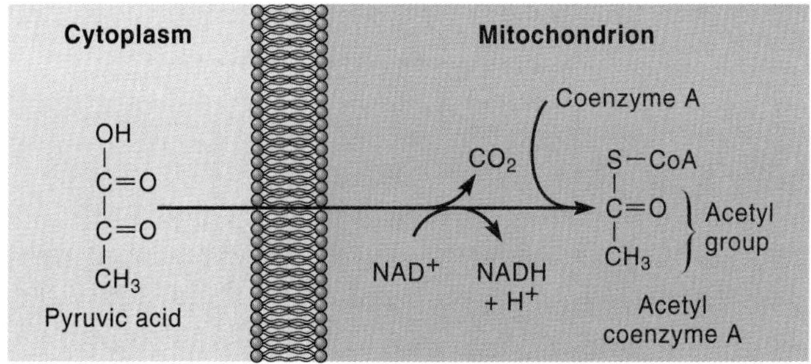

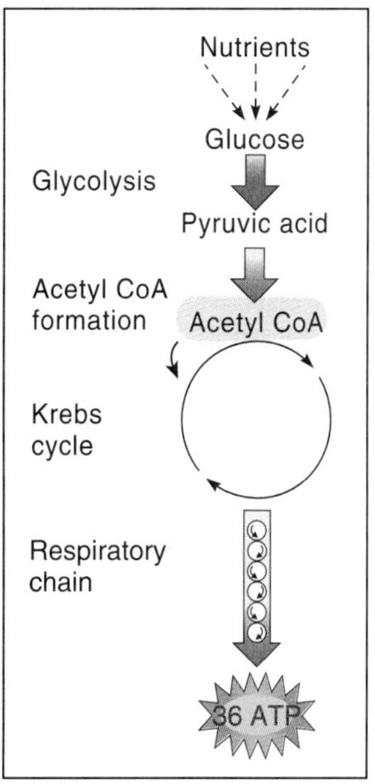

FIGURE 7.10

Acetyl CoA formation bridges glycolysis and the Krebs cycle. After pyruvic acid enters the mitochondrion, crossing both membranes, it loses CO_2, reduces NAD^+ to NADH, and combines with coenzyme A to yield acetyl CoA.

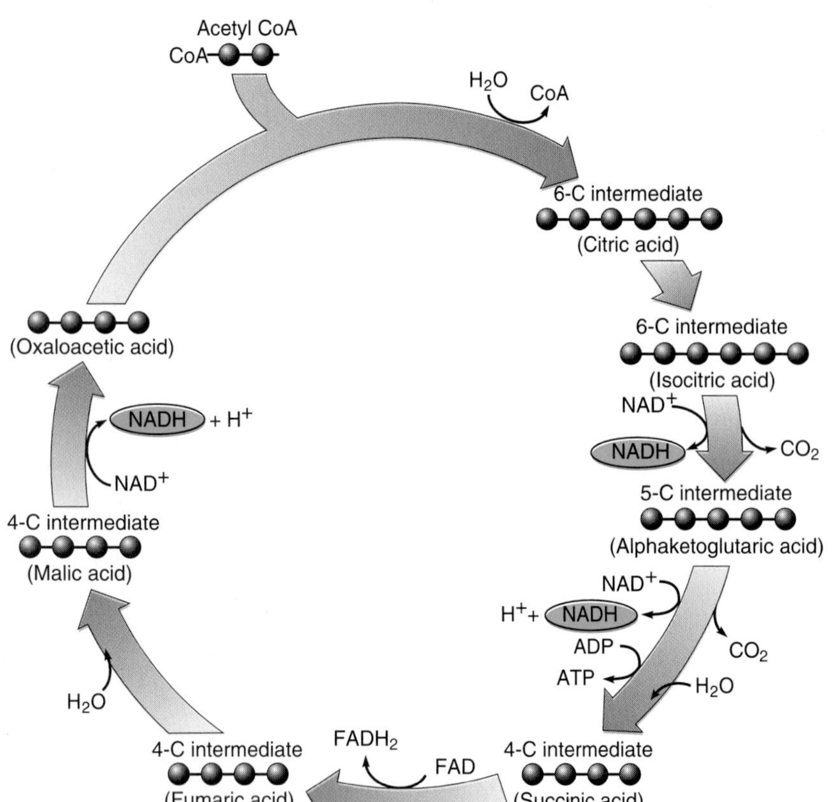

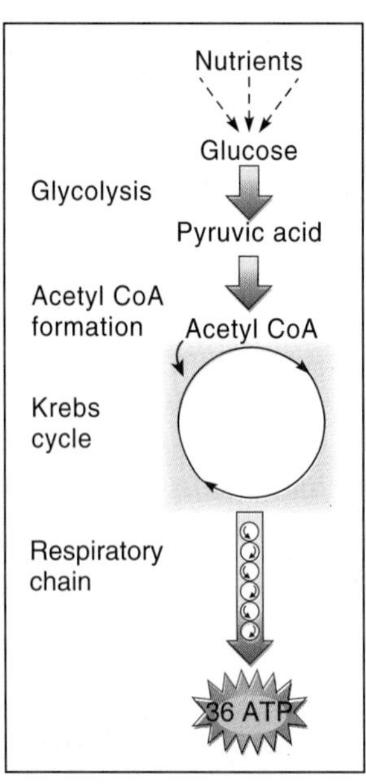

FIGURE 7.11

Each turn of the Krebs cycle generates one molecule of ATP, three molecules of NADH, one molecule of $FADH_2$, and two molecules of CO_2. One glucose molecule yields two molecules of acetyl CoA. Therefore, one glucose molecule is associated with two turns of the Krebs cycle.

The Krebs Cycle

The Krebs cycle is a cycle because the last step regenerates the reactants of the first step (fig. 7.11). Seven of the eight steps occur in the inner mitochondrial membrane. In addition to continuing the catabolism of glucose, the Krebs cycle forms intermediate compounds. The cell then uses the carbon skeletons of these compounds to manufacture other organic molecules, such as amino acids. Organisms release enormous quantities of CO_2 generated in the Krebs cycle to the environment (fig. 7.12).

In the first step of the Krebs cycle, coenzyme A is cleaved from acetyl CoA, and the acetyl group attaches to four-carbon oxaloacetic acid. The resulting six-carbon compound is citric acid. (The Krebs cycle is also called the citric acid cycle.) Figure 7.13 shows an organism that was useful in identifying this first step of the Krebs cycle.

In the next step, citric acid rearranges to isocitric acid, which becomes the substrate for two oxidation steps that together remove two molecules of carbon dioxide. These steps also reduce two molecules of NAD^+ to NADH.

In the first oxidation, CO_2 removal yields alphaketoglutaric acid. In the second oxidation, CO_2 is removed from the alphaketoglutaric acid, producing succinic acid. This reaction provides energy for the substrate-level phosphorylation of ADP to ATP and reduces a molecule of NAD^+ to NADH.

Three more oxidations occur after the formation of succinic acid: succinic acid is oxidized to fumaric acid, fumaric acid is oxidized to malic acid, and malic acid is oxidized to oxaloacetic acid. The oxidation of succinic acid also reduces FAD to $FADH_2$. Finally, oxidation of malic acid to oxaloacetic acid reduces a third NAD^+ to NADH.

FIGURE 7.12

The microbes, insects, worms, fungi, plant roots, and other occupants of soil continually release tremendous amounts of carbon dioxide (CO_2) from cellular respiration. In forests, a third of the CO_2 released to the atmosphere from soil comes from the top layer of leaf litter. Organisms here metabolize nutrient molecules, releasing CO_2. These nutrients, in turn, derive ultimately from photosynthesis. Chapter 42 further explores this cycling of carbon in the environment.

FIGURE 7.13

Deciphering the steps of a biochemical cycle is very difficult, requiring many observations and experiments. Clues to identifying the initial step of the Krebs cycle came in 1919, when two biochemists suggested that the fungus *Aspergillus niger* produces citric acid by combining oxaloacetic acid with a two-carbon compound. Years later, Krebs included this step in the cycle that would bear his name. It took much work to determine that the two-carbon compound that combines with oxaloacetic acid in the cycle is acetyl CoA. Krebs conducted many of the experiments that revealed the steps of the cycle on pigeon tissues. ($\times$ 220)

To summarize, each molecule of glucose sends two molecules of acetyl CoA into the Krebs cycle, which produces only one ATP through substrate-level phosphorylation per acetyl CoA. Most of the energy derived from the oxidative steps of the Krebs cycle is stored in the high-energy electrons of NADH and $FADH_2$. The cell harvests this energy along the respiratory chain.

Mastering Concepts

1. What happens to the pyruvic acid generated in glycolysis before it enters the Krebs cycle?
2. Where does the Krebs cycle occur?
3. How does the Krebs cycle generate CO_2, ATP, NADH, and $FADH_2$?
4. What is the fate of oxaloacetic acid?

Oxidative Phosphorylation and Electron Transport

Most of the ATP generated in cellular respiration comes from oxidative phosphorylation along the respiratory chain. However, the cell does not directly use the high-energy electrons of NADH and $FADH_2$, derived from the Krebs cycle, to synthesize ATP. Rather, these electrons initiate a series of oxidation-reduction reactions that move electrons through several carrier molecules and form an electron transport system, the respiratory chain (fig. 7.14).

The Respiratory Chain

The respiratory chain is like a series of tiny, successively stronger magnets; each carrier has a greater ability to accept and donate electrons than the previous molecule. Each carrier thus pulls electrons from a weaker neighbor and gives them up to a stronger one. The strongest and final carrier in the chain is oxygen. Once reduced, oxygen combines with protons (H^+) in the mitochondrial matrix to yield water. Without oxygen, the electron flow would halt, and ATP would not be generated. If NADH and $FADH_2$ donated electrons directly to oxygen, the single-step reaction would probably release a damaging amount of heat. Passing electrons in a stepwise fashion enables the energy to be harnessed to drive ATP synthesis. The respiratory chain is so important that blocking it with a poison is deadly (Biology in Action 7.1).

ATP Synthesis

Today we know that energy from the electrons passing down the respiratory chain fuels the formation of the proton gradient (the differing H^+ concentrations on each side of the membrane), which in turn drives ATP synthesis. But figuring out the sequence of events was quite a challenge (table 7.2). Many scientists hypothesized that the respiratory chain forms high-energy compounds directly that fuel ATP synthesis. They searched for these compounds but never found them.

Taking another approach, British researcher Peter Mitchell focused on *where* ATP synthesis occurs rather than on *what* powers it. He looked at the inner mitochondrial membrane, where he knew the carrier molecules of the respiratory chain are located. Did the membrane itself, or perhaps a molecule embedded in it, play a role in ATP synthesis?

To test this idea, Mitchell re-created events that occur in mitochondria. He isolated the inner membranes of mitochondria and made vesicles of them. When he added oxygen and NADH to the medium containing the vesicles, the respiratory chain functioned as it does in intact cells! But Mitchell noticed something else. The pH of the medium surrounding the vesicles fell, indicating an increase in protons (H^+). (Recall that pH measures hydrogen ion concentration.) Mitchell formulated another hypothesis: Protons pumped out of the vesicles during electron transport form a gradient, which provides the energy to drive ATP synthesis.

Further studies using the electron microscope revealed knobby structures protruding from one face of each of Mitchell's experimental vesicles (fig. 7.15). When researchers isolated these structures, they discovered that they are complexes of ATP synthase. This multisubunit enzyme spans the inner mitochondrial membrane, and it catalyzes the phosphorylation of ADP to ATP. The enzyme also forms a channel in the membrane through which protons can flow. The portion of the enzyme outside the membrane has sites where ADP and inorganic phosphate (P_i) bond to form

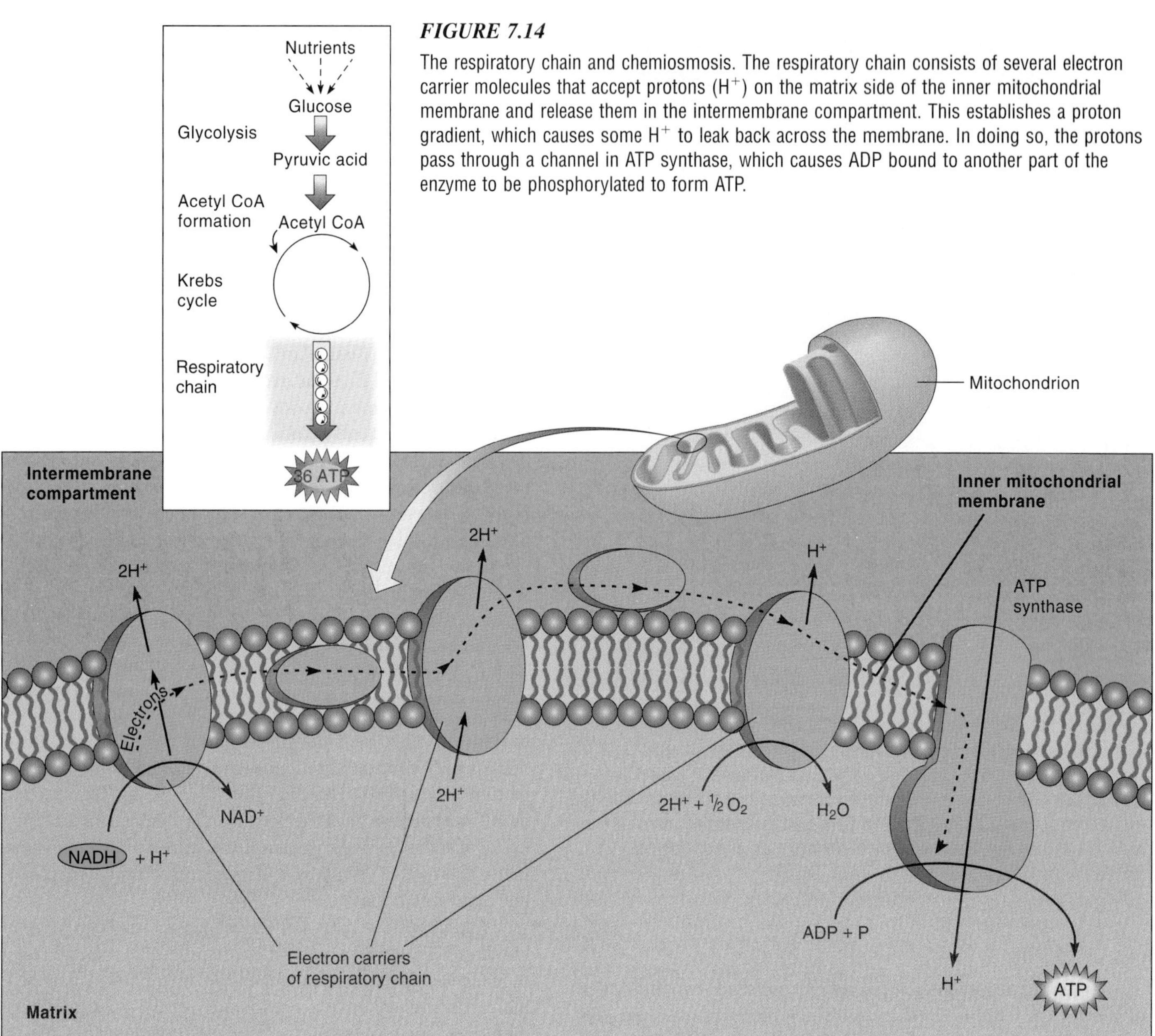

FIGURE 7.14

The respiratory chain and chemiosmosis. The respiratory chain consists of several electron carrier molecules that accept protons (H^+) on the matrix side of the inner mitochondrial membrane and release them in the intermembrane compartment. This establishes a proton gradient, which causes some H^+ to leak back across the membrane. In doing so, the protons pass through a channel in ATP synthase, which causes ADP bound to another part of the enzyme to be phosphorylated to form ATP.

Table 7.2 Demonstrating Chemiosmosis—And the Scientific Method

Observation:	Mitochondria are the sites of ATP formation.
Hypothesis:	If the inner mitochondrial membrane is the site of ATP synthesis, then pieces of membrane cultured in a laboratory and given O_2 and NADH should demonstrate ATP formation.
Experiment:	Create vesicles from inner membranes of mitochondria, and add O_2 and NADH to simulate conditions in a cell. See what happens as cellular respiration proceeds.
Results:	ATP is produced; pH drops in medium containing vesicles.
Conclusion:	A change in concentration of H^+ occurs on the two sides of the inner mitochondrial membrane as ATP forms from ADP.
New hypothesis:	If formation of a proton (H^+) gradient causes ATP formation, then a structure or mechanism must physically link the two activities.
Further experiments:	Electron microscopy of experimental vesicles reveals clusters of ATP synthase. Further study of enzyme shows that it both provides a channel in the membrane for H^+ and is the site of phosphorylation of ADP to ATP.
Chemiosmotic theory:	A proton gradient across the inner mitochondrial membrane drives ATP synthesis.

7.1 biology in action

Disrupting the Respiratory Chain

Blocking the transfer of electrons through the respiratory chain can be deadly—or helpful—depending upon the type of organism and the circumstances. A number of substances can disrupt the respiratory chain in certain types of cells.

Cyanide

Cyanide (CN^-) kills animal cells by binding to the electron carrier that passes electrons to oxygen. However, cyanide doesn't harm certain plants, fungi, and bacteria that have a cyanide-resistant respiration pathway. This is a short, alternative electron transport system that branches off at succinic acid formation in the Krebs cycle. This alternative respiratory pathway also ends with oxygen, but it is coupled to little or no oxidative phosphorylation. Therefore, it produces heat rather than much ATP.

Cyanide-resistant respiration occurs in some carbohydrate-rich plant cells where glycolysis and the Krebs cycle occur very rapidly. Perhaps the cyanide-resistant pathway in such cells handles the overflow of electrons. In the eastern skunk cabbage, *Symplocarpus foetidus,* the heat generated in this metabolic pathway is part of the organism's ecology, enabling it to live in its own private sauna even in winter in the northeastern United States. Parts of the plant maintain temperatures that are 20°C (68°F) above the air temperature (fig. 7.A). In the skunk cabbage's tropical relatives, which include philodendrons, voodoo lilies, and dumbcanes, the heat from metabolism vaporizes foul-smelling biochemicals with such descriptive names as putrescine, cadaverine, and skatole. The stench, like that of rotting flesh, attracts pollinating insects.

Uncouplers

Another respiratory poison is 2,4-dinitrophenol (DNP), used in dye manufacture, in wood preservation, and as an insecticide. This compound is an uncoupler, which means that it halts aerobic respiration by making the inner mitochondrial membrane permeable to protons (H^+). This prevents the cell from forming the proton gradient necessary to drive ATP synthesis. Electrons move from carrier to carrier, but ATP doesn't form. The energy from the electrons instead dissipates as heat. When a person breathes in DNP, metabolic rate races as electrons pass through the respiratory chain with no energy payoff. On a whole-body level, temperature rises, and the person sweats profusely, suffers extreme nausea and vomiting, and then collapses and dies.

A natural uncoupler enables some animals to hibernate to survive very cold temperatures. Brown adipose (fat) cells, which get their color from abundant mitochondria, carry a large protein called thermogenin in their inner mitochondrial membranes. Thermogenin forms an alternate channel for protons, disrupting the formation of the proton gradient and uncoupling proton passage from ATP formation. The energy that normally would synthesize ATP is instead released as heat—just what the hibernating animal needs. Rerouting ATP formation to release heat is called thermogenesis. Adult humans do not have very much brown fat, but infants have deposits around their necks and shoulders, where thermogenesis occurs and helps maintain body temperature.

FIGURE 7.A
Skunk cabbage diverts some of its metabolic energy to heat, which melts snow. This enables shoots to grow in winter.

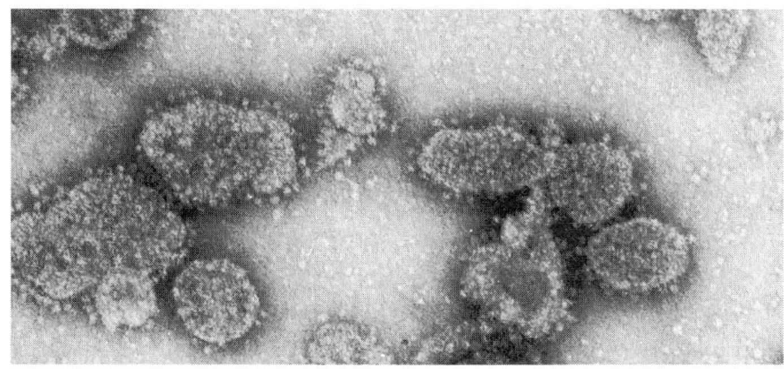

FIGURE 7.15
ATP synthase molecules protrude from an isolated mitochondrial inner membrane on the matrix side, resembling lollipops.

ATP. ATP is released only when protons move through the ATP synthase channel. This is how the proton movement triggers ATP formation—by contacting the enzyme on which ADP is phosphorylated to ATP (see fig. 7.14). ATP synthase is also found in chloroplast membranes and in the respiratory membranes in bacteria. It appears to be common to nearly all, if not all, organisms.

Mitchell called this theory of ATP synthesis **chemiosmosis.** The name derives from the fact that the process entails chemical reactions as well as transport across membranes (osmosis). The theory states that the cell uses the energy released from the flow of electrons through the respiratory chain to actively pump protons into the intermembrane compartment. The membrane is not very permeable to protons, so they do not readily leak back across it. Continued electron transport pumps more and more protons into the intermembrane compartment and sets up a proton gradient. As protons move through ATP synthase channels from the intermembrane space to the matrix, ADP is phosphorylated to ATP. Thus, the interaction between electron transport and ATP synthesis is indirect—the electron transport chain produces a proton gradient, and that gradient drives ATP synthesis.

Mastering Concepts

1. What happens to the electrons captured in NADH and $FADH_2$ that the Krebs cycle generates?
2. How do oxidation and reduction reactions constitute an electron transport system?
3. What is the role of oxygen in the respiratory chain?
4. How does a proton (H^+) gradient link the electrons from the Krebs cycle to ATP formation?

ATP Accounting—An Estimate

In following the energy pathways, it is easy to lose track of the overall function of the process—converting the energy in a molecule of glucose into a form the cell can easily use. How productive is cellular respiration? That is, how many ATPs can one molecule of glucose generate? To estimate the yield of ATPs, we can add the presumed maximum net number of ATPs generated from glycolysis, the Krebs cycle, and oxidative phosphorylation (table 7.3).

Substrate-level phosphorylation yields two ATPs from glycolysis and two ATPs from the Krebs cycle (one ATP each from two turns of the cycle). These are the only steps that produce ATP directly. Most of the ATP generated from respiration comes from oxidative phosphorylation. But when we try to determine exactly how much ATP this step produces, some uncertainty enters the picture.

To estimate the number of ATPs resulting from oxidative phosphorylation, we can assume that each pair of actively transported protons produces one ATP. Each NADH would produce three ATPs, but each $FADH_2$ would produce only two, because this molecule enters the respiratory chain a step later. One molecule of glucose yields two NADH molecules from glycolysis, two NADHs from converting two molecules of pyruvic acid to acetyl CoA, and six NADHs and two $FADH_2$s from two turns of the Krebs cycle. This totals 10 NADHs, which would yield 30 ATPs, and two $FADH_2$s, which would yield four more ATPs. Add the four ATPs from substrate-level phosphorylation and the total is 38 ATPs. However, NADH from glycolysis must be actively transported into the mitochondrion, at a cost of as many as two ATPs for each NADH. This reduces the net production of ATPs from a molecule of glucose to 36.

Table 7.3 One Glucose Can Yield 36 ATPs

Pathways	Coenzymes Reduced	ATP Yield
Glycolysis		
Substrate-level phosphorylation:		2 ATP
Reduction of NAD^+:	2 NADH	
Pyruvic Acid ⟶ Acetyl CoA (×2)		
Reduction of NAD^+:	2 NADH	
Krebs Cycle (×2)		
Substrate-level phosphorylation:		2 ATP
Reduction of NAD^+:	6 NADH	
Reduction of FAD:	2 $FADH_2$	
Respiratory Chain		
Oxidation of 10 NADH × 3 ATP/NADH		30 ATP
Oxidation of 2 $FADH_2$ × 2 ATP/$FADH_2$		4 ATP
		38 ATP
Energy expended to actively transport NADH from glycolysis into mitochondrion		−2 ATP
	Total	36 ATP

Source: Randy Moore, et. al., *Botany*, copyright 1995 The McGraw-Hill Companies, Inc.

The number of kilocalories stored in 36 ATPs is about 38% of the total kilocalories stored in the glucose bonds—and this is a theoretical maximum. To put this energy yield into perspective, an automobile uses about 20 to 25% of the energy contained in gasoline's chemical bonds.

The estimate that aerobic respiration yields 36 net ATPs is just that—an estimate. This is because the NADHs generated by oxidative phosphorylation do not always produce exactly three ATPs each. Respiration efficiency may vary in different cell types; a highly active muscle cell might generate more ATPs than a relatively inactive adipose cell. In addition, the theoretical maximum number of ATPs assumes that all of the protons produced in the respiratory chain contact ATP synthase. This is not true—the proton gradient plays a part in other processes too, so not all protons are routed through ATP synthase.

Mastering Concepts

1. How efficiently does ATP capture the energy in the bonds of a glucose molecule?
2. Calculate the theoretical estimate of 36 ATPs generated per glucose.
3. In what ways is 36 ATPs an approximation?

Control of Cellular Respiration

A cell is hardly the site of neat pathways and cycles, as diagrams may suggest. It is more like a complex soup, brimming with thou-

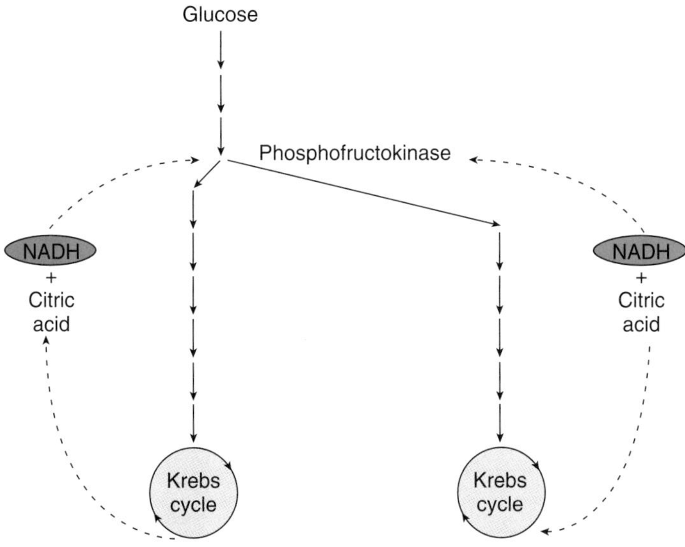

FIGURE 7.16

Negative feedback regulates aerobic respiration. NADH and citric acid from the Krebs cycle, as well as accumulating ATP, bind to the regulatory enzyme phosphofructokinase in a way that temporarily halts its ability to catalyze a key reaction in glycolysis. Accumulating ADP restores the enzyme's activity.

sands of copies of each intermediate and fewer copies of enzymes and carrier molecules. Cellular metabolism consists of many interconnecting pathways occurring simultaneously, subject to feedback controls. The pathways are coordinated so that substances neither build up nor become depleted.

The cell regulates glycolysis and aerobic respiration to prevent glycolysis from proceeding so rapidly that it overwhelms the Krebs cycle and the respiratory chain and to ensure that some ATP is always available. Cells use some of the mechanisms discussed in chapter 6 to regulate the energy pathways.

One important regulatory enzyme is phosphofructokinase, which catalyzes the third step in glycolysis (fig. 7.16). This reaction yields fructose-1,6-bisphosphate, which is the six-carbon intermediate compound that splits into two three-carbon compounds. A negative feedback loop inhibits phosphofructokinase when NADH or citric acid (an intermediate product of the Krebs cycle) accumulates. NADH and citric acid alter the shape of phosphofructokinase so that it can temporarily no longer catalyze its reaction in glycolysis. ATP buildup also inhibits this enzyme, and excess ADP activates it. All of these interactions keep the pace of glycolysis in step with that of subsequent pathways.

Biology in Action 7.2 considers energy use on the whole-body level.

Mastering Concepts

1. Why does a cell regulate the rates of energy-releasing reactions?
2. What mechanism can temporarily halt glycolysis?

Table 7.4 Types of Organisms That Use Fermentation Pathways

Alcoholic Fermentation	Lactic Acid Fermentation
Algae	Algae
Bacteria	Animal cells (muscle)
Fungi (yeast)	Bacteria
Plants	Fungi (yeast)
Protozoa	Protozoa
	Water molds

Energy Retrieval in the Absence of Oxygen

Cellular respiration via the Krebs cycle and the respiratory chain requires oxygen as the terminal electron acceptor. However, anaerobes live in environments lacking oxygen, and some cells in aerobic organisms are occasionally without oxygen. When oxygen is not available, cells use other pathways to extract energy from nutrients.

Fermentation Pathways

Some anaerobes use **fermentation** pathways that occur in the cytoplasm. There are many types of fermentation pathways, but all oxidize NADH to NAD^+, which is recycled to glycolysis. However, fermentation is far less efficient than cellular (aerobic) respiration for energy procurement because it dismantles fewer glucose bonds and does not yield ATP (besides the ATP generated in glycolysis).

A fermentation pathway includes glycolysis plus one or two reactions in which pyruvic acid is converted to other biochemicals, most commonly by lactic acid or the alcohol ethanol and carbon dioxide (table 7.4). Fermentation also occurs in parts of an organism that are in an anaerobic environment, such as in a plant partially submerged in a pond or in cells deep within a multicellular organism that lack direct access to oxygen.

Alcoholic Fermentation

In **alcoholic fermentation,** yeast cells convert pyruvic acid to ethanol and carbon dioxide, oxidizing NADH to NAD^+ (fig. 7.17*a*). Fermentation by brewer's yeast is used industrially to help manufacture baked goods and alcoholic beverages (fig. 7.17*b*). Depending upon the substance being fermented, the variety of yeast used, and whether carbon dioxide is allowed to escape during the process, yeast fermentation may be used to produce wine or champagne from grapes, the syrupy drink called mead from honey, and cider from apples. Beer is brewed by fermenting grain—barley, rice, or corn.

Using alcoholic fermentation in an industrial process inadvertently led to discovery of lactic acid fermentation. In 1856, a

7.2 biology in action

Whole-Body Metabolism—Energy on an Organismal Level

Metabolic reactions supply cells with the energy to survive, as well as to accomplish specialized functions in multicellular organisms. However, we are most familiar with biological energy on a larger scale—the energy that moves muscles, keeps blood flowing, and powers other bodily functions essential to whole-body metabolism.

Energy enters an animal's body as food molecules and exits as heat or activity, either in movement or in the many internal energy-requiring functions that are part of life. Usually, energy intake and output are balanced. When they are not, the result can be underweight or obesity.

The energy a vertebrate requires to stay alive is described as its basal metabolic rate (BMR). In a person, BMR measures the kilocalories required for heartbeat; breathing; nerve, kidney, and gland function; and the maintenance of body temperature when the subject is awake, physically and mentally relaxed, and has not eaten for 12 hours. BMR does not include the energy needed for digestion or physical activity. For an adult human male, the average BMR is 1,750 kilocalories in 24 hours; for a female, 1,450 kilocalories. But because most people do not usually spend 24 hours a day completely at rest and fasting, the number of kilocalories needed to get through a day generally exceeds the basal requirement.

Several factors influence BMR, including age, sex, weight, body proportion, and regular activity level. A sedentary student's BMR may be 1,700 kilocalories in 24 hours. His roommate, who is on the swim team and runs 5 miles each morning, may burn 5,000 kilocalories per day and have a BMR of 2,500. When the students sleep, their metabolic rates are likely to fall below basal level; but the swimmer's rate will still be higher because regular exercise elevates metabolism, even during rest.

BMR rises from birth to about age five and then declines until adolescence, when it peaks again. As we age, BMR drops as our energy needs decline. It also falls during starvation as the body attempts to conserve energy. BMR increases during some illnesses, when body temperature rises, and during pregnancy and breast-feeding.

BMR is also related to body composition. For two individuals of the same weight, age, and sex, the one with a greater proportion of lean tissue (muscle, nerve, liver, and kidney) will have a higher BMR because lean tissue consumes more energy than relatively inactive fat tissue.

The thyroid gland in the neck affects BMR. It manufactures the hormone thyroxine, which increases the body's energy expenditure. Too much thyroxine can double the BMR, resulting in excessive consumption of kilocalories and weight loss. Too little thyroxine slows the BMR. The body burns fewer kilocalories, and weight increases.

In general, the smaller the organism, the higher its BMR. A rat has a higher metabolic rate than a cat, which has a higher rate than a dog, which has a higher rate than a human. This is partly because smaller organisms have higher surface-to-volume ratios and therefore lose more heat to the environment. Consequently, they must burn relatively more fuel to maintain a nearly constant body temperature. Even within a species, smaller individuals have higher metabolic rates.

Whole-body metabolic rate is related to life span in some insects. In houseflies, a measurement called the metabolic potential equals the amount of oxygen the fly consumes in a lifetime. This value is relatively constant from fly to fly, but individuals differ in the rate at which aerobic respiration uses the oxygen. Experiments that compared groups of houseflies with different activity levels showed that the less active the fly, the longer it lives, although all flies consume roughly the same amount of oxygen over their life spans (fig. 7.B).

Does this fact about housefly metabolism suggest that people should foresake the gym and take up residence on the sofa in front of the television? No, because there are differences between human and fly muscles. Researchers who study aging hypothesize that flies succumb to eventual death from excess exercise because toxic byproducts of oxygen metabolism destroy their tissues. In humans, exercise strengthens our muscles and cardiovascular systems. Because of differences in whole-body metabolism, what's good for the fly is harmful to the human.

FIGURE 7.B
This fly, caught by the camera while dining on sugar, will live longer if it flies less and faces cool temperatures. An active insect living in a very warm climate takes in the same amount of oxygen as a more sedentary fly but lives a shorter time.

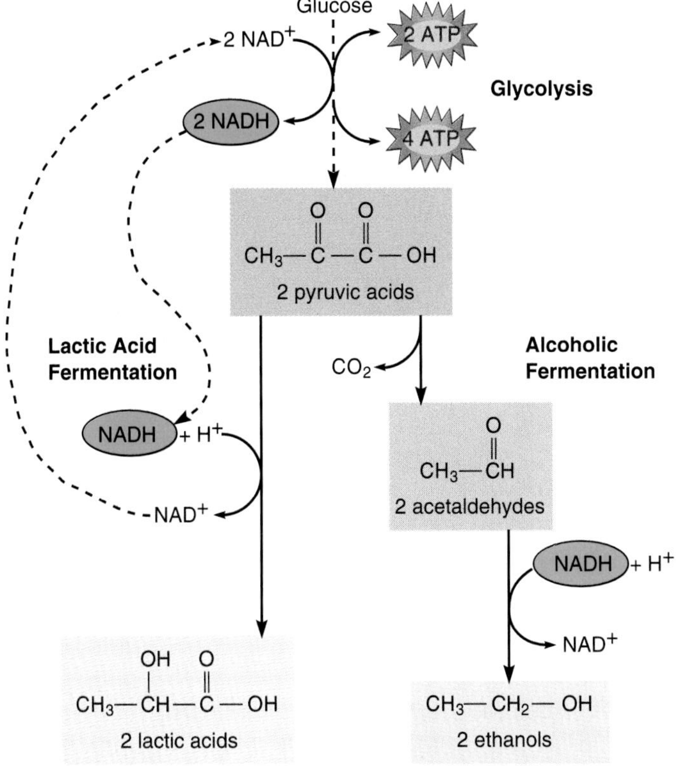

(a)

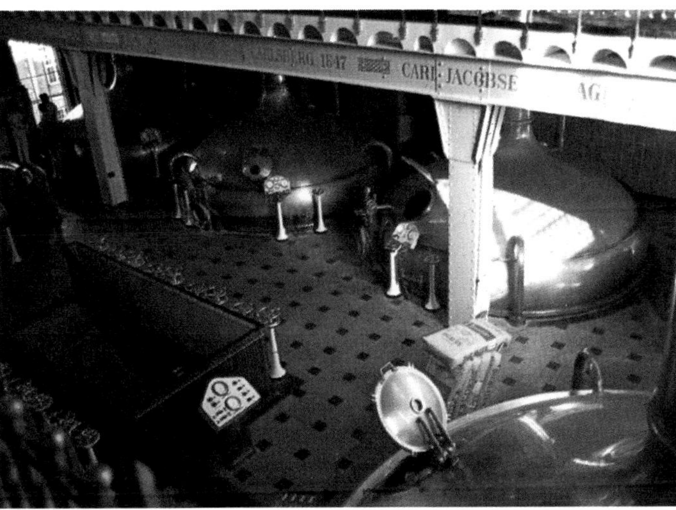

(b)

FIGURE 7.17

Fermentation. (*a*) The two major fermentation pathways produce lactic acid or ethanol. The fermentation pathways recycle NAD^+ to glycolysis (shown only for lactic acid fermentation) but do not generate additional ATP. (*b*) Ale and beer begin as germinating barley grain. After activated enzymes and a long soak in water break down complex carbohydrates and proteins, and other grains are added, the resulting mash is heated with dried flowers from a vine, called hops. The heat arrests the enzyme activity and yeast is added. Yeast fermentation adds alcohol to the brew. Yeast strains that ferment the bottom of the brew produce beer; yeast that ferments the top lowers the pH further, producing ale.

French manufacturer of industrial chemicals who had produced ethanol from fermenting sugar beets noticed that his brew had turned sour-tasting and didn't contain much alcohol. He asked for help from Louis Pasteur, who determined that the type of microorganisms fermenting the beets had changed, and they produced lactic acid instead of alcohol.

Lactic Acid Fermentation

Some anaerobic bacteria and other single-celled organisms, and some animal cells temporarily deprived of oxygen, convert pyruvic acid to the three-carbon compound lactic acid in a single step, oxidizing NADH to NAD^+. **Lactic acid fermentation** occurs in human muscle cells that are working so strenuously that their production of pyruvic acid exceeds the oxygen supply. In this "oxygen-debt" condition, the muscle cells revert to fermentation to extract energy. If enough lactic acid accumulates, the muscle fatigues and cramps (fig. 7.18). When oxygen is once again present, lactic acid is converted back to pyruvic acid in the liver. From this pyruvic acid, the body can extract more energy.

Athletic coaches sometimes measure lactic acid levels in the blood to assess the physical condition of swimmers and sprinters. One study reported that lactic acid accumulates to triple the normal levels in the bloodstreams of children who cry vigorously as they are prepared for surgery but not in calm children in the same situation. This suggests that stress may trigger lactic acid fermentation.

Lactic acid fermentation also occurs in cancer cells deep within tumors far from oxygen-delivering blood. These cells survive by metabolizing glucose through glycolysis and lactic acid fermentation. The lactic acid forms a "halo" around the cancer cells, which inactivates immune system biochemicals that would otherwise destroy them.

Some baking and brewing processes require ethanol *and* lactic acid and therefore use both yeast and bacteria. Genetic researchers have made these processes easier by growing yeast cells that contain a bacterial gene enabling the cells to undergo lactic acid fermentation as well as their normal alcoholic fermentation. Cider manufacturers who use the engineered yeast no longer need to use only high-acid apples to produce cider. Figure 7.19 summaries the position of the fermentation reactions in the energy pathways.

Other Fermentation Pathways

Certain bacteria use fermentation pathways other than alcoholic or lactic acid fermentation. These organisms catabolize a variety of organic compounds and release a variety of gases, including carbon dioxide, hydrogen (H_2), hydrogen sulfide (H_2S), and ammonia (NH_3). Some types of *Clostridium* bacteria ferment nearly anything organic—except (unfortunately, from an ecological point of view) plastics!

One less common fermentation pathway proved very useful during World War II. The Allies needed an organic compound called acetone to produce a type of gunpowder. Chemical manu-

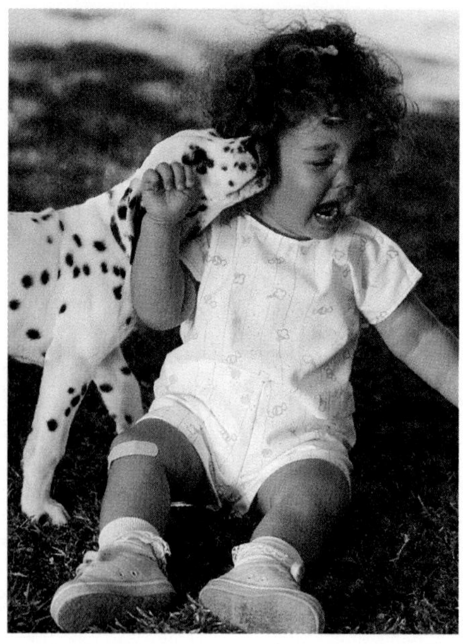

(a) (b)

FIGURE 7.18

Feeling lactic acid production. (*a*) If President Bill Clinton and Vice President Al Gore pick up their paces, their strained muscle cells will begin to respire anaerobically, producing lactic acid that might cause cramps. Seasoned runners know to keep walking after a sprint to dissipate the lactic acid. (*b*) Lactic acid levels in youngsters crying vigorously are triple those of calmer kids. Could stress temporarily switch metabolism to lactic acid fermentation?

facturers at the time extracted acetone from wood, and supplies were rapidly running out. English chemist Chaim Weizmann solved the problem by using a fermentation pathway of the anaerobic bacterium *Clostridium acetobutylicum.* The bacteria converted grain or molasses into the needed acetone, producing a second valuable chemical, butanol, used to manufacture synthetic rubber.

Anaerobic Electron Transport

Electron acceptors in energy reactions can be organic molecules (lactic acid or alcoholic fermentation) or oxygen at the end of the respiratory chain. Inorganic molecules can also accept electrons in pathways called **anaerobic respiration.** Three common inorganic electron acceptors are nitrate (NO_3^-), sulfate (SO_4^{2-}) and carbon dioxide. Anaerobic respiration pathways are more energy efficient than fermentation but less so than aerobic respiration.

Some *Pseudomonas* and *Bacillus* bacteria use anaerobic respiration when oxygen gas is not present in their environment. Such organisms that can follow aerobic or anaerobic energy-releasing pathways are called **facultative anaerobes.** These bacteria combine nitrate with electrons and protons to yield nitrite (NO_2^-) and water. However, because nitrite is toxic, an additional reaction converts nitrite to harmless nitrogen gas (N_2). Another example of anaerobic respiration is the reduction of CO_2 to methane gas (CH_4) by full-time anaerobic bacteria called, appropriately, methanogens.

Mastering Concepts

1. What do different fermentation pathways have in common?
2. What are the products of alcoholic fermentation?
3. What are the products of lactic acid fermentation?
4. Describe an alternative type of pathway to aerobic respiration and fermentation.

How Proteins and Lipids Enter the Energy Pathways

Figure 7.20 shows how the other major nutrients—proteins and lipids—fit into the energy pathways. Amino acids liberated from dietary protein are usually used to manufacture more protein. However, when an organism exhausts immediate carbohydrate supplies, cells may use protein as an energy source. When this happens, dietary amino acids are rearranged and broken down. They enter the energy pathways as either pyruvic acid, acetyl CoA, or an intermediate of the Krebs cycle, depending upon the type of amino acid broken down. Ammonia is stripped from the amino groups of the amino acids and eventually excreted.

The fat in food—such as that in the cheese on the meat in figure 7.20—is digested into glycerol and fatty acids, which enter the lymphatic system and ultimately the blood. Glycerol

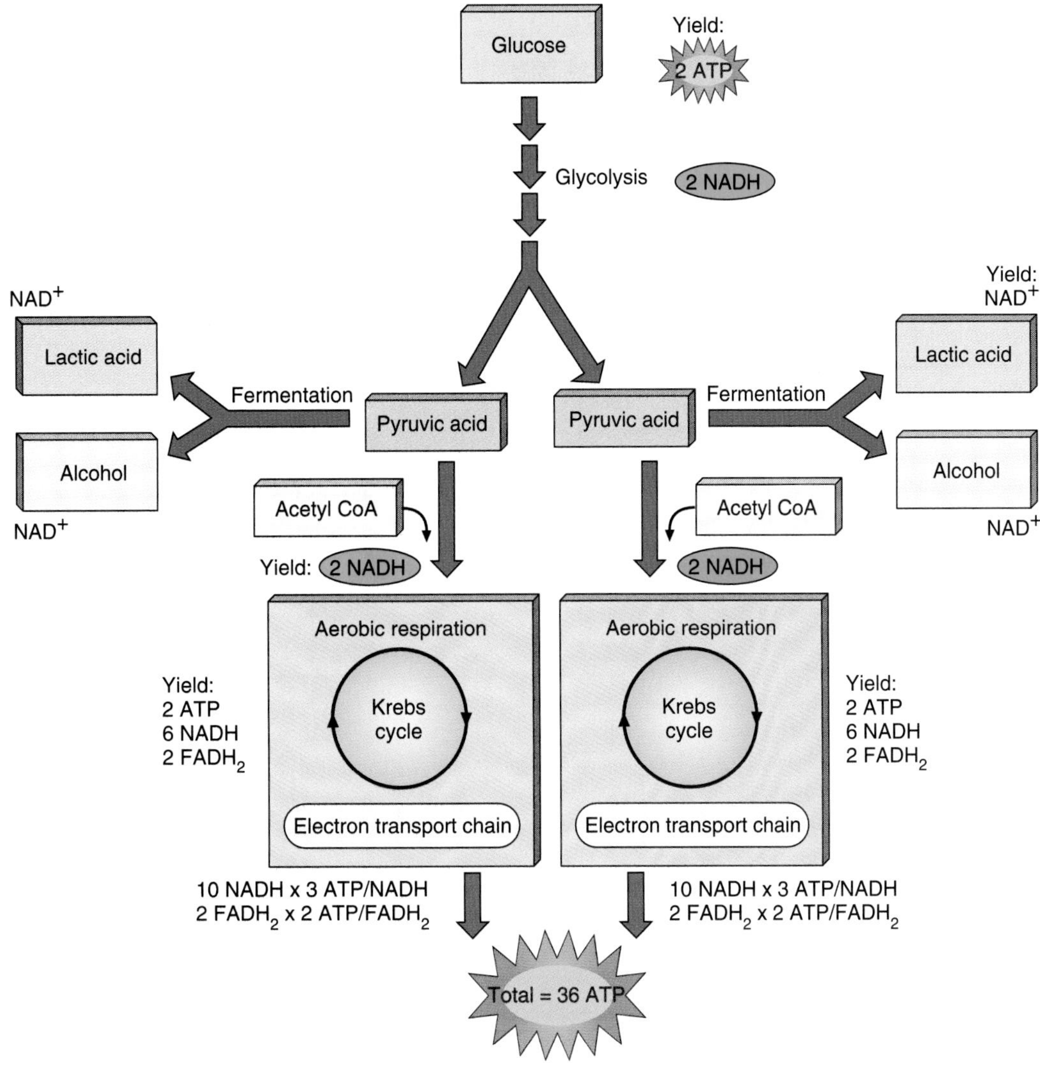

FIGURE 7.19

Extracting energy from glucose.

is converted to pyruvic acid and continues from there through acetyl CoA formation, the Krebs cycle, and the respiratory chain. Fatty acids enter cells and are transported into mitochondria, where they are catabolized to acetyl CoA. From here, the pathways continue as they would for glucose catabolism.

Plants use lipids to fuel activities such as seed germination. In oily seeds, triglycerides break down into glycerol and fatty acids. The fatty acids are then cut into two-carbon pieces released as acetyl CoA. This reaction repeats for every pair of carbons until all of the fatty acids are converted into molecules of acetyl CoA.

Mastering Concepts

1. At what points in the energy pathways can amino acids from digested proteins enter?
2. At what points in the energy pathways can glycerol and fatty acids from digested fats enter?

SUMMARY

1. The respiratory system provides the oxygen that **cellular respiration** requires as a final electron acceptor. The overall reaction for cellular respiration is:

$$C_6H_{12}O_6 + 6O_2 \longrightarrow 6CO_2 + 6H_2O + ATP.$$

2. All cells begin energy release from nutrients with **glycolysis** and then may follow any of several pathways.

3. Cellular (aerobic) respiration harvests energy gradually. **Acetyl CoA formation,** the **Krebs cycle,** and the **respiratory chain** follow glycolysis. Glycolysis takes place in the cytoplasm; acetyl CoA formation, the Krebs cycle, and the respiratory chain occur in mitochondria.

4. ATP synthesis occurs by **substrate-level phosphorylation** (phosphate transfer between organic compounds) or by **oxidative phosphorylation** (passage of electrons along carrier molecules through oxidation-reduction reactions, setting up a proton gradient that powers phosphorylation of ADP to ATP).

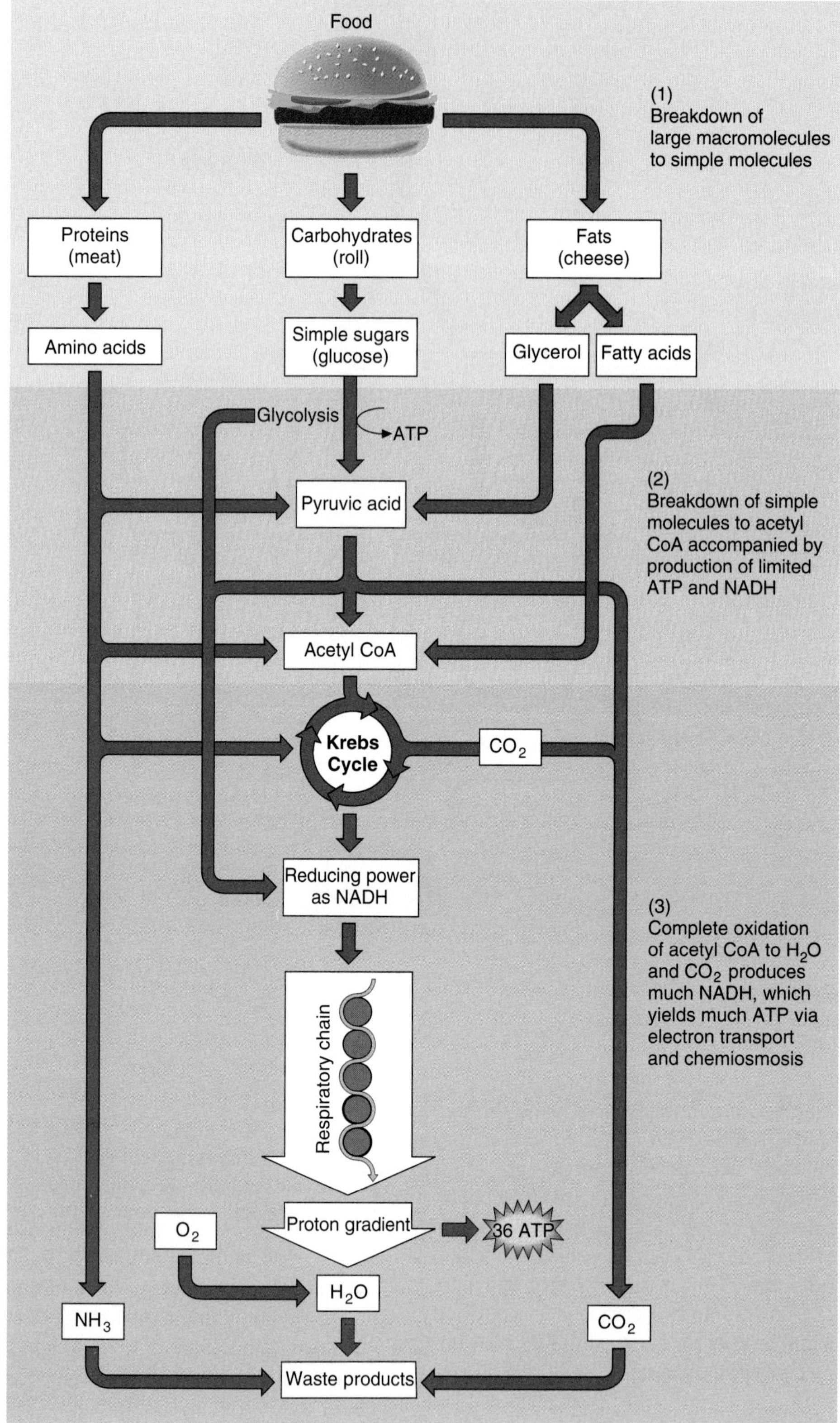

FIGURE 7.20

How nutrients enter the energy pathways. Most cells use carbohydrate as an initial source of energy. Amino acids enter the energy pathways as pyruvic acid, acetyl CoA, or intermediates of the Krebs cycle. Fats are digested to glycerol and fatty acids, which are catabolized to acetyl CoA.

Running on Empty

continued from page 126.

The lack of glucose and pyruvic acid in Michael P.'s cerebrospinal fluid told physicians that his cells were not performing glycolysis or fermentation. Hypothesizing that a profound lack of ATP was causing the boy's symptoms, medical researchers decided to intervene beyond the block in the boy's metabolic pathway and take a detour to energy production. When Michael was seven-and-a-half months old, he was placed on a diet rich in certain fatty acids. Within 4 days, he appeared to be healthy for the very first time in his life!

The diet reactivated aerobic respiration at the point of acetyl CoA formation by supplying an alternative to glucose. Other children with similar symptoms have since enjoyed spectacular recoveries thanks to similar dietary interventions, but we do not yet know the long-term effects of the therapy. This medical success story, however, illustrates the importance of the energy pathways—and how valuable our understanding of them can be.

5. In the first half of glycolysis, glucose breaks down into two molecules of the three-carbon compound PGAL. In the second half of glycolysis, the PGALs are oxidized as NAD^+s are reduced to NADHs, contribute phosphate groups to form two ATPs, and react and are rearranged to form two molecules of **pyruvic acid.**

6. In the mitochondria, pyruvic acid is converted into **acetyl CoA** in a coupled reaction that also reduces NAD^+ to NADH.

7. Acetyl CoA enters the Krebs cycle. This cycle is a series of oxidation-reduction reactions that produces ATP, NADH, $FADH_2$, and CO_2. Substrate-level phosphorylation produces ATP in the Krebs cycle.

8. Energy-rich electrons from NADH and $FADH_2$ fuel the respiratory chain. Electrons move through a series of carriers that release energy at each step. The terminal electron acceptor, oxygen, is reduced to form water.

9. Electron transport energy establishes a proton gradient that pumps protons from the mitochondrial **matrix** into the **intermembrane compartment.** As protons diffuse back into the matrix through channels in ATP synthase, their energy drives phosphorylation of ADP to ATP.

10. Adding the ATPs produced in substrate-level phosphorylation and oxidative phosphorylation and subtracting those used to enter the mitochondrion predicts a theoretical maximum energy of 36 ATPs per glucose.

11. Negative feedback coordinates the rates of glycolysis, acetyl CoA formation, and the Krebs cycle.

12. In the absence of oxygen, alcoholic, lactic acid, or other **fermentation** pathways may run. Fermentation does not produce ATP but oxidizes NADH to NAD^+, which is recycled to glycolysis. **Alcoholic fermentation** reduces pyruvic acid to ethanol and loses carbon dioxide. **Lactic acid fermentation** reduces pyruvic acid to lactic acid.

13. Anaerobic electron transport uses inorganic electron carriers and is less efficient than aerobic respiration but more so than fermentation.

14. Amino acids enter the energy pathways as pyruvic acid, acetyl CoA, or an intermediate of the Krebs cycle. Fatty acids and glycerol enter as acetyl CoA.

TO REVIEW . . .

1. How are breathing and cellular respiration connected? How are they different?

2. Why is aerobic respiration a more efficient energy-extracting pathway than glycolysis alone?

3. How does substrate-level phosphorylation differ from oxidative phosphorylation?

4. Describe how and at which point each of the following compounds fits into or assists in the energy pathways:

- **a.** coenzyme A
- **b.** $FADH_2$
- **c.** pyruvic acid
- **d.** ATP synthase
- **e.** fructose-1,6-bisphosphate

5. Cite a reaction or pathway that occurs in each of the following locations:

- **a.** cytoplasm
- **b.** mitochondrial matrix
- **c.** inner mitochondrial membrane
- **d.** intermembrane space

6. How is the structure of a mitochondrion adapted to its function?

7. What is the immediate source of the six carbons in citric acid in the Krebs cycle?

8. Why don't NADH and $FADH_2$ in the respiratory chain directly donate electrons to oxygen?

9. Cite four controls over levels of phosphofructokinase, an enzyme that functions in glycolysis.

10. At what point does oxygen (O_2) enter the energy pathways? What is its role?

11. Explain how a eukaryote's metabolism of a single glucose molecule might yield 36 ATP molecules. Why is the total number of 36 theoretical?

12. Distinguish between an aerobe, an anaerobe, and a facultative anaerobe.

13. Why would lack of oxygen completely inhibit the Krebs cycle and the respiratory chain but not glycolysis?

14. Under what circumstances would uncoupling electron transport from ATP synthesis be adaptive?

15. Which energy pathway is used to manufacture beer?

TO THINK ABOUT . . .

1. The amino acid sequence of the respiratory chain carrier molecule cytochrome c is almost identical in all organisms. Disorders caused by abnormal cytochrome c are unknown. What do these two observations suggest about the importance of this molecule?

2. A student regularly runs 3 miles (1.9 kilometers) each afternoon at a slow, leisurely pace. One day, she runs a mile as fast as she can. Afterwards she is winded and feels pain in her chest and leg muscles. She thought she was in great shape! What, in terms of energy metabolism, has she experienced?

3. Fructose intolerance is an inherited disorder in which a missing enzyme makes a person unable to use fructose, a simple sugar abundant in fruit. Infants with the condition have very low mental and motor function. Older children are very lethargic and mildly mentally disabled. By adulthood, the nervous system of a person with fructose intolerance deteriorates, eventually causing mental illness and death.

Molecules derived from fructose are intermediates in the first few reactions of glycolysis. The missing enzyme would normally catalyze these reactions. Considering this information about the whole-body and biochemical effects of fructose intolerance, suggest what might be happening on a cellular level to these people.

4. Two men weigh 230 pounds each. One is 6 feet tall and a bodybuilder, with tight, bulging muscles. The other is 5 feet, 4 inches tall and obese. Which man probably has the higher basal metabolic rate? Why?

TO LEARN MORE . . .

References and Resources*

Allen, Robert G. February 1992. The long and short of it. *Natural History.* Houseflies seem to be allotted a certain amount of oxygen that limits their life spans.

Baum, Rudy M. February 22, 1993. Views on biological, long-range electron transfer stir debate. *Chemical and Engineering News,* vol. 73. Chemists are isolating parts of the respiratory chain and deciphering precisely how the electron carriers align.

Bentley, Ronald. Spring 1994. A history of the reaction between oxaloacetate and acetate for citrate biosynthesis: An unsung contribution to the tricarboxylic acid cycle. *Perspectives in Biology and Medicine,* vol. 37. Don't let the title scare you—this is a great description of experiments used to decipher a single step of the Krebs cycle.

Blake, Colin. January 16, 1997. Phosphotransfer hinges in PGK. *Nature,* vol. 385. A look at an enzyme that is crucial to both cellular respiration and photosynthesis.

Collins, Steve. August 1995. The limit of human adaptation to starvation. *Nature Medicine,* vol. 1. This study examines the basal metabolic rate and adaptations to starvation among 573 people who nearly starved to death in Somalia in the famine of 1992–93.

Levi, Primo. October 1984. Travels with C. *The Sciences.* An eloquent travelogue of how a single atom of carbon circulates through the living world.

Lewis, Ricki. October 1989. Mitochondria—eclectic organelles. *Biology Digest.* Energy reactions occur in the mitochondria—but these organelles are fascinating for a number of reasons.

Wivagg, D. 1987. Research reviews: How many ATPs per glucose molecule? *The American Biology Teacher,* vol. 49. Many approximations are used to calculate respiratory efficiency.

**Life*'s website features frequently updated references.

Without the sun, life would eventually cease.

8 Photosynthesis

chapter outline

The Importance of Photosynthesis 146
Photosynthesis: From Carbon Dioxide to Carbohydrate
The Beginnings of Photosynthesis

The Nature of Light 147
Probing Light
Types of Light

Pigment Molecules 149
Chlorophyll
Accessory Pigments
Autumn Leaves and More

Chloroplasts 151
Photosystems and Reaction Centers in Chloroplasts
The Fate of Absorbed Light Energy

The Light-Dependent Reactions 153
Photosystem II Starts Photosynthesis
Photosystem I
Why Two Photosystems Increase Energy Extraction

The Light-Independent Reactions 159
The Calvin Cycle
How Experiments Revealed the Steps of the Calvin Cycle

The Efficiency of Photosynthesis 160
Photorespiration

C_4 Photosynthesis 160

CAM Photosynthesis 162

The Big Picture of Biological Energy Use 162
Possible Origins of the Energy Pathways

Life Without Photosynthesis?

Imagine the world without sunlight, the consequence of an explosive event that throws dust and debris into the atmosphere and blocks the sun for many months. It might be eruption of an enormous volcano, a meteor crashing to earth, or a nuclear holocaust.

Scientists aptly term the aftermath of a nuclear blast *nuclear winter.* In a first nuclear winter, light would be about a tenth of its normal intensity. Plants would die as they break down the scarce carbohydrates for energy faster than they can manufacture them using sunlight. As the plants die, animals cannot find enough to eat. Perhaps in a year or two, life-giving light would again trickle through the hazy atmosphere, but by then it would be too late. The lethal chain reaction would already be in motion, destroying food webs at their bases.

In these days of relative peace, nuclear winter scenarios do not appear in science journals with the frequency that they did in the 1980s, when people feared nuclear weapons. Yet something close to a limited nuclear winter occurred in the 1960s, in Vietnam, and not from a nuclear weapon. The target was leaves.

to be continued . . .

The Importance of Photosynthesis

If asked to designate the most important biochemical pathway, most biologists would cite **photosynthesis,** the process by which plants harness solar energy into chemical energy. All life on this planet depends upon it. Disrupting photosynthesis is devastating (fig. 8.1).

Photosynthesis: From Carbon Dioxide to Carbohydrate

In green plants and algae, carbon dioxide (CO_2) and water (H_2O) react in the presence of sunlight and pigment molecules to produce oxygen (O_2) and the sugar glucose ($C_6H_{12}O_6$). Overall, these reactions convert six carbon dioxide molecules and 12 molecules of water into one glucose molecule, six oxygen molecules, and six water molecules. As we saw in the last chapter, glucose is an excellent energy storage molecule in all organisms. The reactions of photosynthesis can be summarized as follows:

$$6CO_2 + 12H_2O \xrightarrow[\text{pigments}]{\text{light +}} C_6H_{12}O_6 + 6O_2 + 6H_2O$$

The oxygen released in photosynthesis comes from the H_2O and not from the CO_2. This can be demonstrated by exposing a plant to water containing a heavy oxygen isotope designated ^{18}O. The "labeled" isotope appears in the oxygen gas released in photosynthesis, proving that the oxygen came from the water.

Certain bacteria photosynthesize using hydrogen sulfide (H_2S) in place of H_2O. These microorganisms release sulfur (S_2) rather than oxygen (O_2), along with an energy-rich organic product.

FIGURE 8.1

Bombs and sprayed herbicides destroyed more than half of Vietnam's mangrove forests. Loss of leaves devastated communities of many types of organisms. Leaves, as the major conduits linking solar energy to biological energy, are essential for life.

The Beginnings of Photosynthesis

Cells are open systems that can absorb but not create energy. Before photosynthesis evolved, organisms must have extracted energy from organic (carbon-containing) compounds in their environment. These first organisms were **heterotrophs,** meaning that they absorbed energy-rich organic compounds and manufactured other organic compounds from them. As these organisms oxidized the carbon compounds from their surroundings, they released CO_2 into the environment. This lifestyle was adequate for awhile, but eventually, environmental energy stores probably began to run out. Because the earliest organisms couldn't use the carbon in CO_2, they were doomed to extinction as soon as they depleted the organic compounds in their habitats. Many biologists accept this theory—but of course, we cannot prove what happened so long ago.

The evolution of photosynthesis 3 billion years ago gave life a new energy source. Photosynthetic organisms, rather than relying on the dwindling supply of energy-rich compounds in the environment, began to use the energy in sunlight. These were the first photosynthetic **autotrophs,** organisms that make organic compounds from inorganic compounds such as water and CO_2. Their ability to transduce (convert) light energy into chemical energy soon supported most other forms of life on the planet.

The rise of plants radically altered the earth. It decreased the concentration of CO_2 in the atmosphere; lowered global temperature, adding to the polar ice caps and lowering sea level; and filled the atmosphere with a waste product that many other organisms eventually would find essential for life: oxygen. The proportion of oxygen in the atmosphere gradually rose from a tiny fraction of a percent to 21% today. Oxygen allowed the evolution of aerobic respiration and more diverse life-forms. All the oxygen in the air we breathe has cycled through plants in photosynthesis.

Our evolutionary look at photosynthesis continues at the end of the chapter. Biology in Action 8.1 describes some early ideas and experiments that led to our understanding of this most important process.

Mastering Concepts

1. What is photosynthesis? Describe the process in words and in chemical symbols.
2. How did the origin of photosynthesis alter life on earth?

The Nature of Light

Today, only about a tenth of the earth's known species can photosynthesize, but their importance is profound. Photosynthesis begins with reactions that require light.

Each minute, the sun converts more than 120 million tons of matter to radiant energy, releasing much of it outward as waves of radiation. After an 8-minute journey, about two-billionths of this energy reaches the earth's upper atmosphere. Only about 1% of this light is used for photosynthesis—yet this tiny fraction of the sun's power produces some 3 quadrillion pounds of carbohydrates a year!

Probing Light

Our understanding of light began about 300 years ago, when Sir Isaac Newton showed that white light passing through a prism, water droplet, or soap bubble separates into a band of colors. Newton also demonstrated that separated light, passed through another prism, can recombine to form white light. Based on these discoveries, Newton proposed that white light is actually a combined spectrum of colors ranging from violet to red. Two centuries later, in the 1860s, Scottish mathematician James Maxwell showed that visible light is, in turn, a small sliver of a much larger spectrum of radiation: the **electromagnetic spectrum** (fig. 8.2).

In 1905, Albert Einstein extended the ideas of Newton and Maxwell and proposed that light consists of packets of energy called **photons.** The intensity (brightness) of light depends on the number of photons (the amount of energy) absorbed per unit of time. Each photon carries a fixed amount of energy determined by how the photon vibrates. The slower the vibration, the less energy the photon carries. The distance a photon moves during a complete vibration is its **wavelength** (l) (fig. 8.3). The wavelength of visible light is measured in nanometers (nm), or billionths of a meter, and ranges from 390 to 760 nanometers. A photon's energy is inversely proportional to the wavelength of the light. That is, the longer the wavelength (the distance traveled during a vibration), the less energy per photon.

Types of Light

Sunlight consists of about 4% ultraviolet (UV) radiation, 53% infrared (IR) radiation, and 44% visible light. Each of these kinds of light has different energy characteristics and effects on organisms.

UV radiation contains high-energy photons that drive electrons from molecules, forming ions, which is why UV is called **ionizing radiation.** UV breaks weak chemical bonds and causes sunburn and

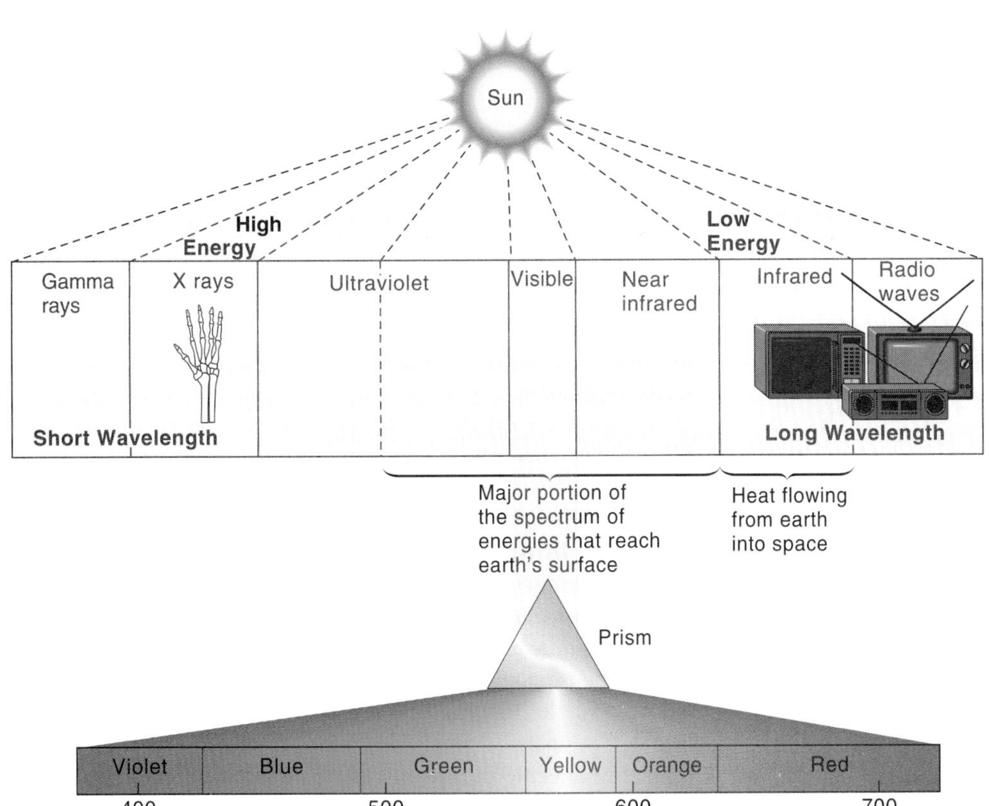

FIGURE 8.2

Visible light is a part of a continuous spectrum of radiation termed the electromagnetic spectrum.

8.1 biology in action

Early Thoughts on Photosynthesis

Today researchers are identifying and describing in molecular detail the various pigments and proteins that interact to capture solar energy and transduce it to chemical energy. However, people have investigated photosynthesis for centuries, with cruder tools, but with no less compelling questions.

An early discussion of photosynthesis appears in Jonathan Swift's eighteenth-century novel *Gulliver's Travels.* In chapter 5 of part 3, "A Voyage to Laputa," Swift writes, "He had been eight years upon a project for extracting sun-beams out of cucumbers, which were to be put into vials hermetically sealed, and let out to warm the air in raw inclement summers." Because the idea of extracting the sun's warmth from a vegetable seemed preposterous, the expression "getting sunbeams from cucumbers" became synonymous with seeking to do the impossible. Historians disagree as to whether Swift was indulging his famed sarcasm or demonstrating a rather deep understanding of photosynthesis, considering how little was known about the process when the novel was published in 1726.

One of the first investigators to explore plant metabolism was Flemish physician and alchemist Jan van Helmont. In the early 1600s, he grew willow trees in weighed amounts of soil, applied known amounts of water, and noted that the trees gained more than 100 pounds (45 kilograms), while the soil lost only a few ounces (fig. 8.A). Because he had applied large amounts of water, van Helmont concluded (incorrectly) that plants grew solely by absorbing water. He also noted that burning wood exudes carbon dioxide, but he did not connect this and plant growth. In the 1720s, English botanist and chemist Stephen Hales hypothesized that some substance in the air, rather than water, nourishes plants. Swift may have taken his "sunbeams from cucumbers" imagery from Hales's work.

Unitarian minister and chemist Joseph Priestley studied photosynthesis more systematically in the summer of 1771. Priestley knew

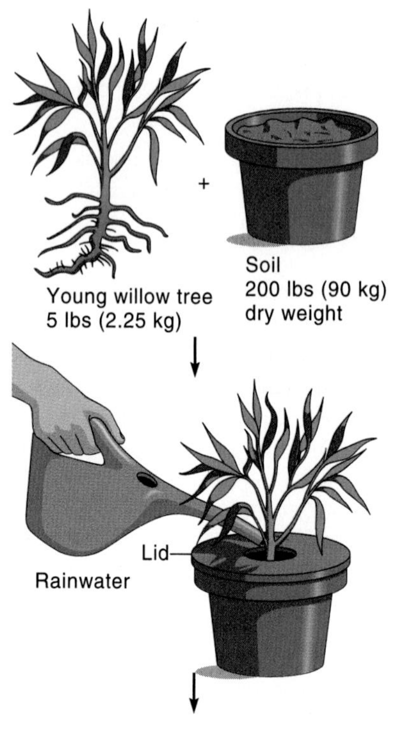

FIGURE 8.A
Jan van Helmont was one of the first scientists to study plant growth. He grew willow trees, carefully recording the amounts of soil and water he used and the weight of the trees after 5 years. He erroneously concluded that the increase in plant matter was due directly to the plentiful water.

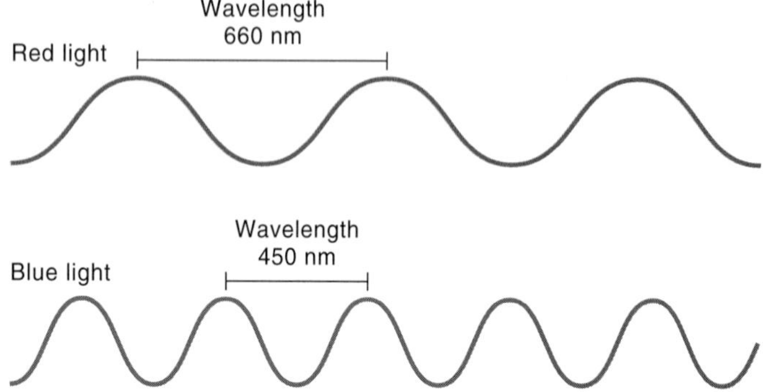

FIGURE 8.3
The wavelike nature of light. Different colors have different wavelengths. The longer the wavelength of light, the less energy it has.

skin cancer. Because glass absorbs UV, a window blocks UV radiation and prevents it from burning your skin. The ozone (O_3) in the atmosphere also absorbs UV radiation. This is one reason why people are concerned about the thinning of the ozone layer.

IR radiation doesn't contain enough energy per photon to be useful to organisms. Most of its energy is converted immediately to heat. Unlike UV, IR penetrates glass, heating a window on a sunny day.

Visible light provides the right amount of energy for biochemical reactions. Red and blue light are the most effective for photosynthesis. Shorter wavelength radiation, such as X rays and UV, contains enough energy to break chemical bonds and release energy. Visible light, in contrast, contains just enough energy to "excite" or energize molecules. When a photon impinges upon a pigment molecule in a plant cell, the pigment molecule absorbs it, causing certain electrons to jump to higher energy levels or even out of the atom completely. Photon-boosted electrons are packed

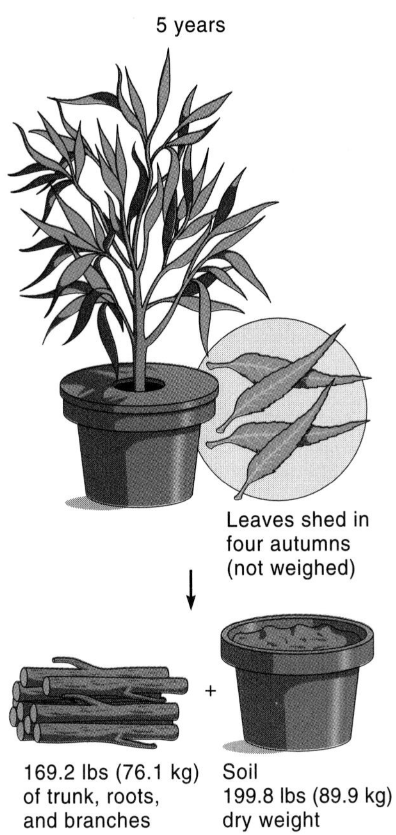

that if he placed a mouse in an enclosed container with a lit candle, the mouse would die. Hypothesizing that burning somehow "injured" the air so that the mouse couldn't breathe, Priestley looked for a substance that could "purify" the air. A mint plant worked. If a closed chamber held both a burning candle and a mint leaf, a second candle could still burn several days later. Priestley wrote, "I have been so happy as by accident to hit upon a method of restoring air which has been injured by the burning of candles, and to have discovered at least one of the restoratives which nature employs for this purpose. It is vegetation." He also discovered that a mouse could live in a container as long as a plant was present. Priestley concluded that plants purify air, allowing animals to breathe. However, he couldn't always repeat his experimental results—a key part of the scientific method. Part of the problem may have been that Priestley moved his experiments inside to a dark area, where the plant could not photosynthesize and produce oxygen.

Eight summers after Priestley conducted his experiments, Dutch physician Jan Ingenhousz extended knowledge about photosynthesis in his book, *Experiments upon Vegetables, Discovering Their Great Power of Purifying the Common Air in the Sunshine, and of Injuring it in the Shade or at Night.* Ingenhousz conducted more than 500 experiments to identify the components of photosynthesis.

Ingenhousz hypothesized that the sun's warmth powered photosynthesis. He found that when leaves were placed in water in an inverted jar near a fire but shielded from light, "bad" air formed. But leaves from the same plant in a jar exposed to air and sunlight produced "good" air. Ingenhousz called this *dephlogisticated air* (from *phlogiston,* which refers to burning matter). He wrote, "No dephlogisticated air is obtained in a warm room, if the sun does not shine upon the jar containing the leaves. . . . The sun by itself has no power to mend air without the concurrence of plants." He concluded that leaves are the site of "dephlogistication" and that they operate differently in the light than in the dark—two very important contributions to our modern knowledge of photosynthesis. But Ingenhousz also recommended that people place houseplants outdoors at night—otherwise, he asserted, the plants might poison the air!

with potential energy, which they release when they fall back to their original positions nearer the nucleus or when they move to an electron-accepting molecule.

Mastering Concepts

1. What are some different types of light, and what is the basis of their differences?
2. What is a unit of light called?
3. What is the relationship between a photon's energy and its wavelength?

Pigment Molecules

Light striking an object is either reflected, transmitted, or absorbed. Only absorbed light can cause an effect. Organisms absorb light through pigment molecules. Pigments are colored because they absorb some wavelengths of light and transmit others. Black pigment absorbs all wavelengths, whereas white pigment absorbs none.

Chlorophyll

Photosynthetic organisms use a variety of pigment molecules to capture photon energy. **Chlorophyll a** is the primary photosynthetic pigment in all photosynthetic organisms except bacteria. It absorbs wavelengths corresponding to red and orange (600–700 nm) and blue and violet (400–500 nm) and reflects and transmits green wavelengths. Chlorophyll, therefore, is responsible for the green color of plants.

Chlorophyll a is a huge molecule—its chemical formula is $C_{55}H_{22}O_5N_4Mg$ (fig. 8.4). Chlorophyll's long hydrocarbon tail is

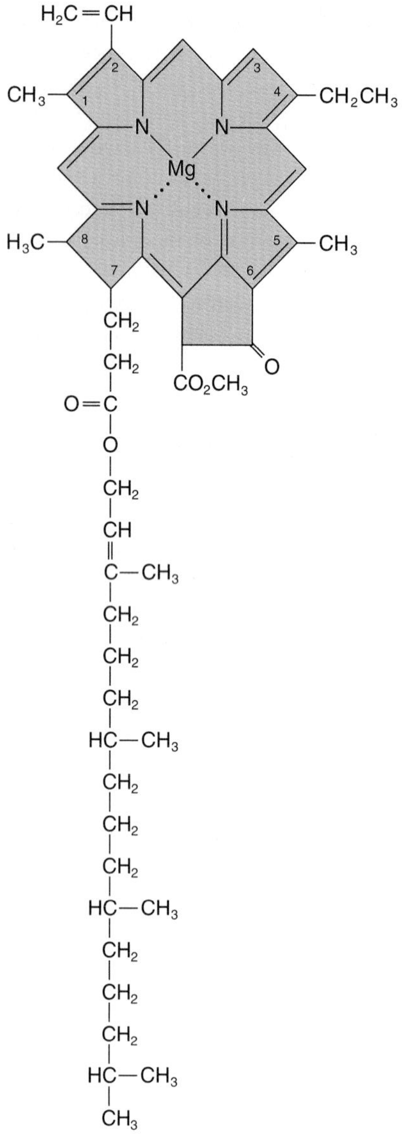

FIGURE 8.4

The green pigment chlorophyll a is the dominant pigment in the photosynthetic cells of plants. Each of the large molecules of this pigment consists of a central magnesium (Mg) atom surrounded by four nitrogens (N) and various organic rings. A long, hydrophobic (water-hating) tail anchors the molecule into the chloroplast membranes.

FIGURE 8.5

Fall leaves. Senescent leaves lose their chlorophyll and then other pigments become visible.

Table 8.1 Accessory Pigments

Pigment	Organisms	Color
Bacteriorhodopsin	Purple bacteria	Purple
Carotenoids	Plants, algae	Red, orange, yellow
Chlorophyll b	Plants, green algae, some prokaryotes	Bluish green
Fucoxanthin	Brown algae	Brown
Phycocyanin	Red algae, cyanobacteria	Blue
Phycoerythrin	Red algae	Red
Xanthophylls (oxidized carotenoids)	Plants, algae, bacteria	Red, yellow

hydrophobic, and this is why the molecule is anchored in lipids within the chloroplasts of plant cells, rather than dissolved in the watery part of the cell surrounding them.

Accessory Pigments

Plants and algae contain a rainbow of pigments. Some pigments are present all year long, and others make seasonal appearances. Having different types of pigments extends the range of light wavelengths useful in photosynthesis. For example, some pigments absorb wavelengths that chlorophyll a cannot absorb and pass that energy to chlorophyll a. These pigments that expand the accessible wavelengths are called **accessory pigments** (table 8.1).

Autumn Leaves and More

Most plants regularly destroy and resynthesize chlorophyll. In a single year, the 300 million tons of chlorophyll on earth turn over about three times. When plants break down chlorophyll faster than they resynthesize it, the green pigment no longer masks the others; this produces the spectacular colors of autumn leaves. Fall foliage watching is a billion-dollar industry in Vermont and New Hampshire, thanks to chlorophyll breakdown (fig. 8.5).

Many hues of autumn leaves come from accessory pigments called **carotenoids,** which include carotenes and their oxidized derivatives, xanthophylls. Carotenoid pigments absorb wavelengths between 460 and 550 nanometers, producing yellow, orange, and red colors. Carotenoids provide the distinctive colors of carrots, tomatoes, bananas, and squashes. Animals do not manufacture carotenoids, but they eat them in plants and then use the pigments to color various structures. Colorful frogs, fishes, and corals, a squid's ink, flamingo feathers, and egg yolks all come from carotenoids, as does the red pigment released when a lobster boils.

Carotenoids also protect plants from damage from oxygen, which tends to form high-energy free radicals that can destroy cellular structures. Many herbicides kill plants by blocking carotenoid synthesis, removing protection from oxygen-free radical damage.

Mastering Concepts

1. What can happen to light striking a surface?
2. What is the general structure of chlorophyll?
3. How are accessory pigments adaptive?
4. What are two functions of carotenoids?

Chloroplasts

Leaves derive their green color from oblong organelles called **chloroplasts.** A chloroplast is a type of **plastid,** which is an organelle that synthesizes or stores nutrients. Chloroplasts are the sites of photosynthesis in plants and algae (fig. 8.6). Most photosynthetic cells contain 40 to 200 chloroplasts, adding up to half a million per square millimeter of leaf.

A chloroplast consists of two membranes that enclose a gelatinous matrix called the **stroma.** The stroma contains ribosomes, DNA, and enzymes used to synthesize carbohydrates. Suspended in the stroma are folded sacs of **thylakoid membranes,** which enclose an area called the **thylakoid space.** These membranes are unique to chloroplasts. Ten to 20 thylakoids may stack into structures called **grana.** Thylakoids and grana contain chlorophyll. Cells that contain chlorophyll (and thus are green) tend to be located on the parts of plants that face the sun.

Photosynthesis in diatoms and dinoflagellates, which are unicellular marine-dwelling eukaryotes, occurs on thylakoid membranes encapsulated in other types of plastids (fig. 8.7*a* and *b*). In contrast to pigments in plants, which are eukaryotes, chlorophylls and accessory pigments in prokaryotes are embedded in membranes that are not encapsulated (fig. 8.7*c* and *d*).

Photosystems and Reaction Centers in Chloroplasts

A single chlorophyll a molecule can absorb only a small amount of photon energy. Several chlorophylls located near each other capture far more energy because they can pass the energy on to other molecules, freeing themselves to absorb other photons as they strike. If 200 chlorophyll molecules aggregate and pass a photon along, the chance that a second photon will quickly follow is

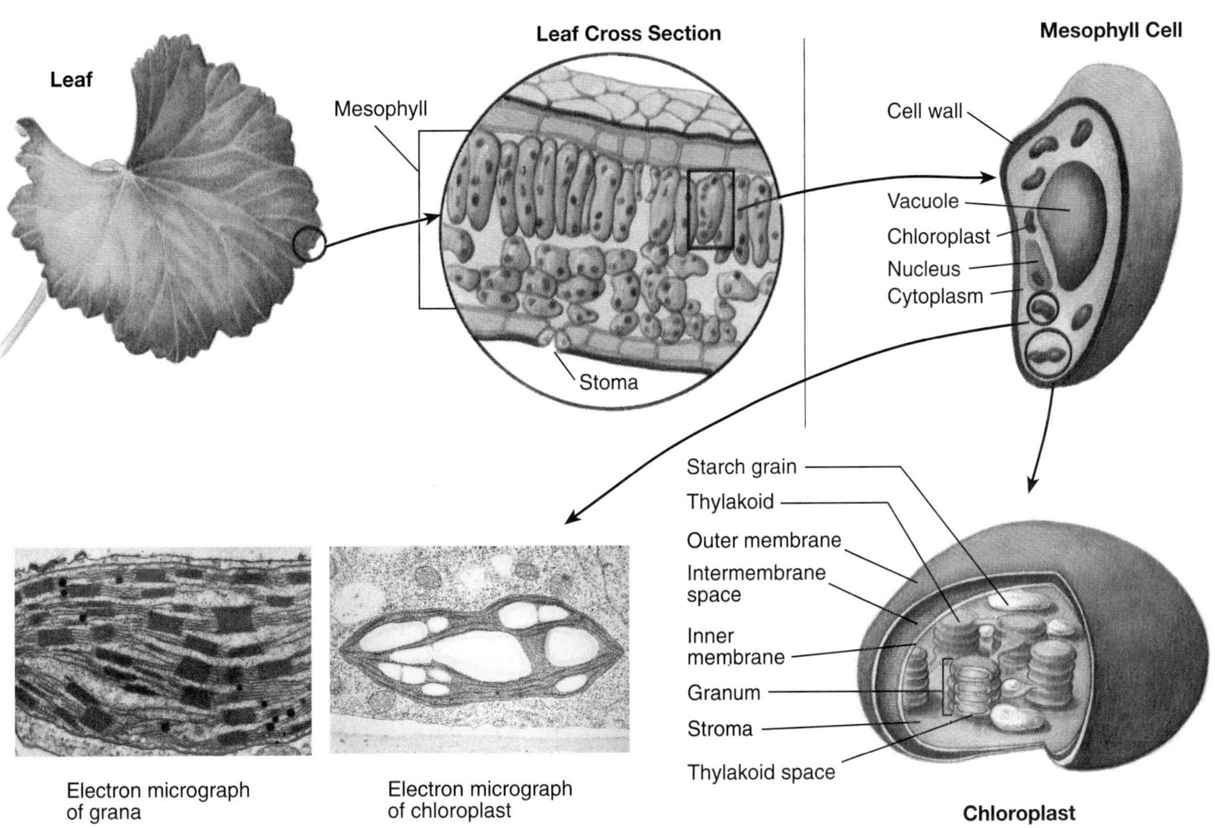

FIGURE 8.6

Chloroplasts. A chloroplast in cross section. Leaf mesophyll tissue consists of cells with many chloroplasts. The chloroplast is the site of photosynthesis. This double-membraned organelle contains flat, interconnected sacs called thylakoids, which are organized into stacks called grana. Light is absorbed and converted to chemical energy in the grana. The chemical energy is then used in the stroma to manufacture carbohydrates.

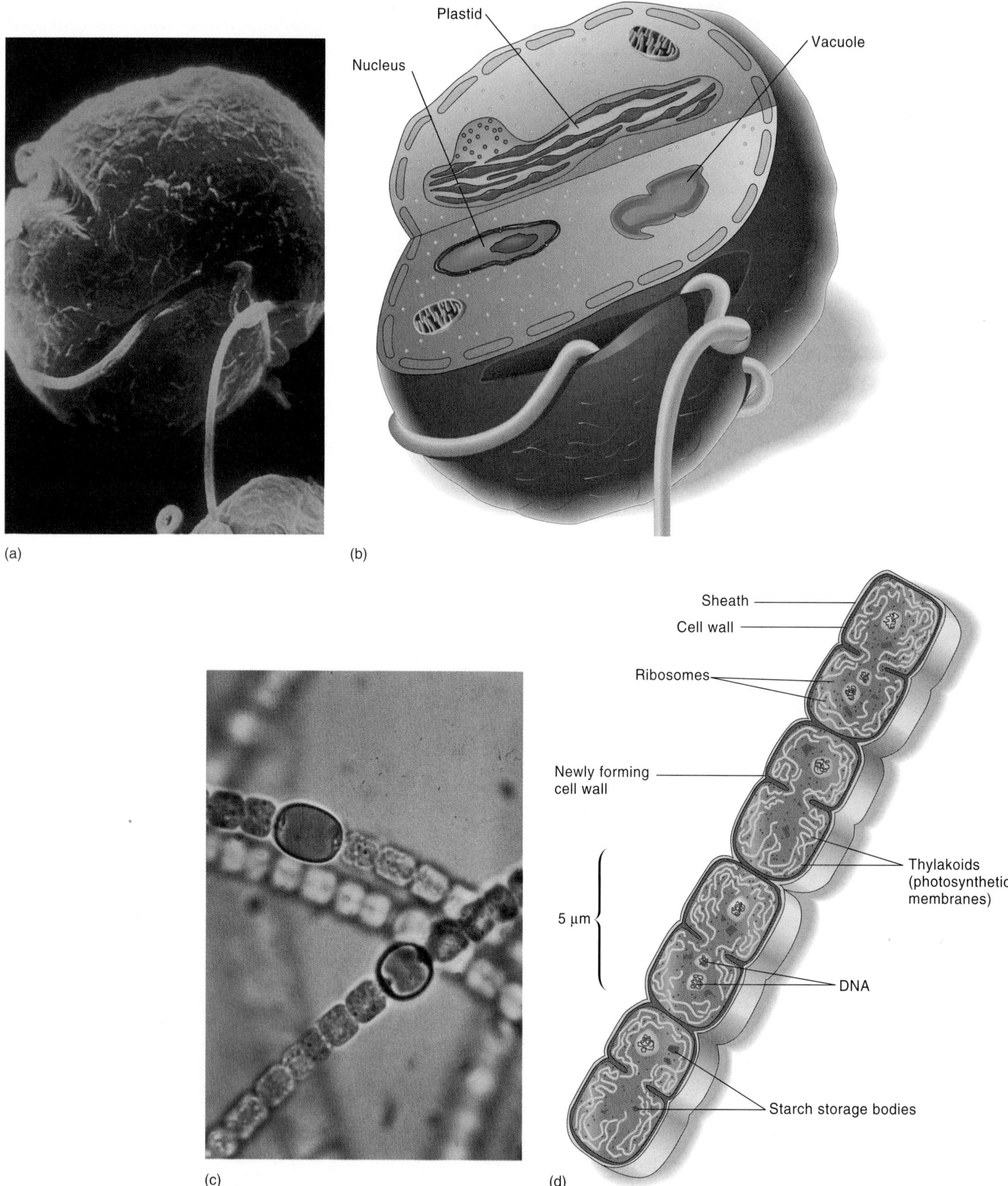

FIGURE 8.7

Photosynthetic diversity. A dinoflagellate such as *Gonyaulax tamarensis* (*a*) has photosynthetic membranes within a plastid (*b*). When populous in oceans, these dinoflagellates tinge the water a deep red because of other photosynthetic pigments, creating a condition called a "red tide." Cyanobacteria such as *Anabaena* (*c*) photosynthesize using thylakoid membranes (*d*). *Anabaena* lives in freshwater lakes and ponds. Dinoflagellates are eukaryotes, and cyanobacteria are prokaryotes.

40,000 times greater than if the first photon hits a single chlorophyll. To effectively accomplish the tremendous job of photosynthesis, then, chlorophyll molecules congregate with other pigment molecules and electron carrier molecules to form a **photosystem.** This consists of a photon-harvesting pigment molecule complex and an associated electron transport system. Photosynthesis in algae and higher plants occurs along two linked photosystems; one photosystem passes electrons to the other.

Photosynthesis begins within the pigment complex of a photosystem. The pigment complex contains a collection of molecules called an antenna complex (fig. 8.8). It includes about 300 chlorophyll molecules, 50 accessory pigments, and proteins, all anchored in a protein matrix. Several chlorophylls within the antenna complex act as antennae, capturing photon energy and passing it along to other chlorophyll molecules. Within microseconds after the antenna complex absorbs photon energy from sunlight, the energy passes to the **reaction center,** a special pair of energy-gathering molecules of chlorophyll a coupled with associated proteins. Each antenna complex contains one reaction center. The electrons of the reaction center's two chlorophyll molecules are less excited than electrons in other chlorophylls. When the photon strikes a reactive (reaction center) chlorophyll, one of its electrons is so energized that it is boosted to a higher energy orbital.

Eukaryotes have two kinds of reaction center chlorophylls because they use two photosystems. The reactive chlorophyll of photosystem I absorbs light energy mostly at 700 nanometers and is therefore called P700 (P stands for pigment). The reactive chlorophyll of photosystem II is called P680 and absorbs energy of 680 nanometers. Although reaction centers comprise less than 1% of the chlorophyll in plants, they are vitally important in photosynthesis because they first capture the light energy.

The Fate of Absorbed Light Energy

Energy from boosted chlorophyll electrons can meet any of several fates. Their energy can escape as heat or as an afterglow of light called **fluorescence.** Fluorescent light has a longer wavelength, and therefore less energy, than the light that originally excited the pigment. Chlorophyll fluoresces deep red. Alternatively, photon energy can pass to neighboring molecules. This is what happens in photosynthesis.

Mastering Concepts

1. How do pigments participate in photosynthesis?
2. What is the structure of a chloroplast?
3. How are pigment and protein molecules organized in chloroplasts?
4. Why are reaction centers vital to photosynthesis?

The Light-Dependent Reactions

Photosynthesis occurs in two stages—a **light-dependent stage** and a **light-independent stage.** The light-dependent stage requires light, but the light-independent stage can occur in light or dark.

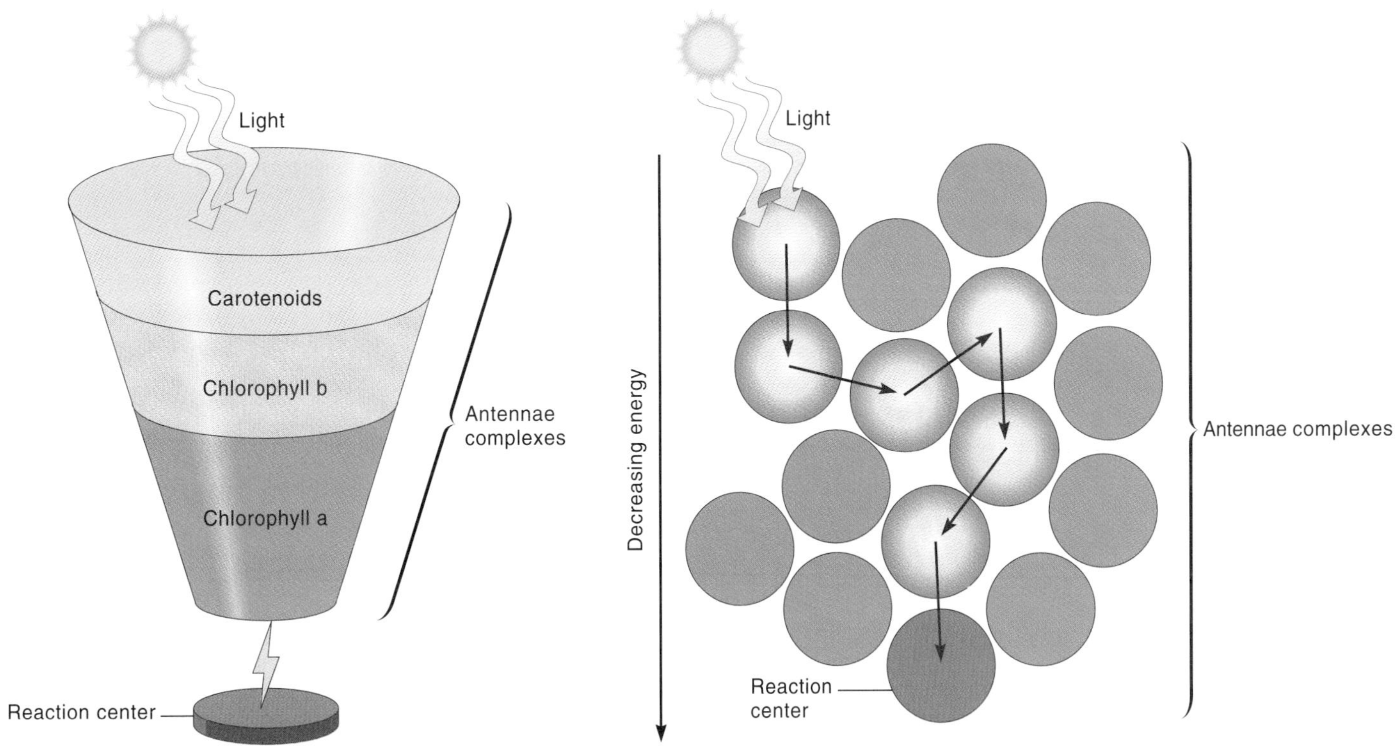

FIGURE 8.8

The light-harvesting system in chloroplasts functions like a funnel; it collects photons and passes their energy to the reaction center.

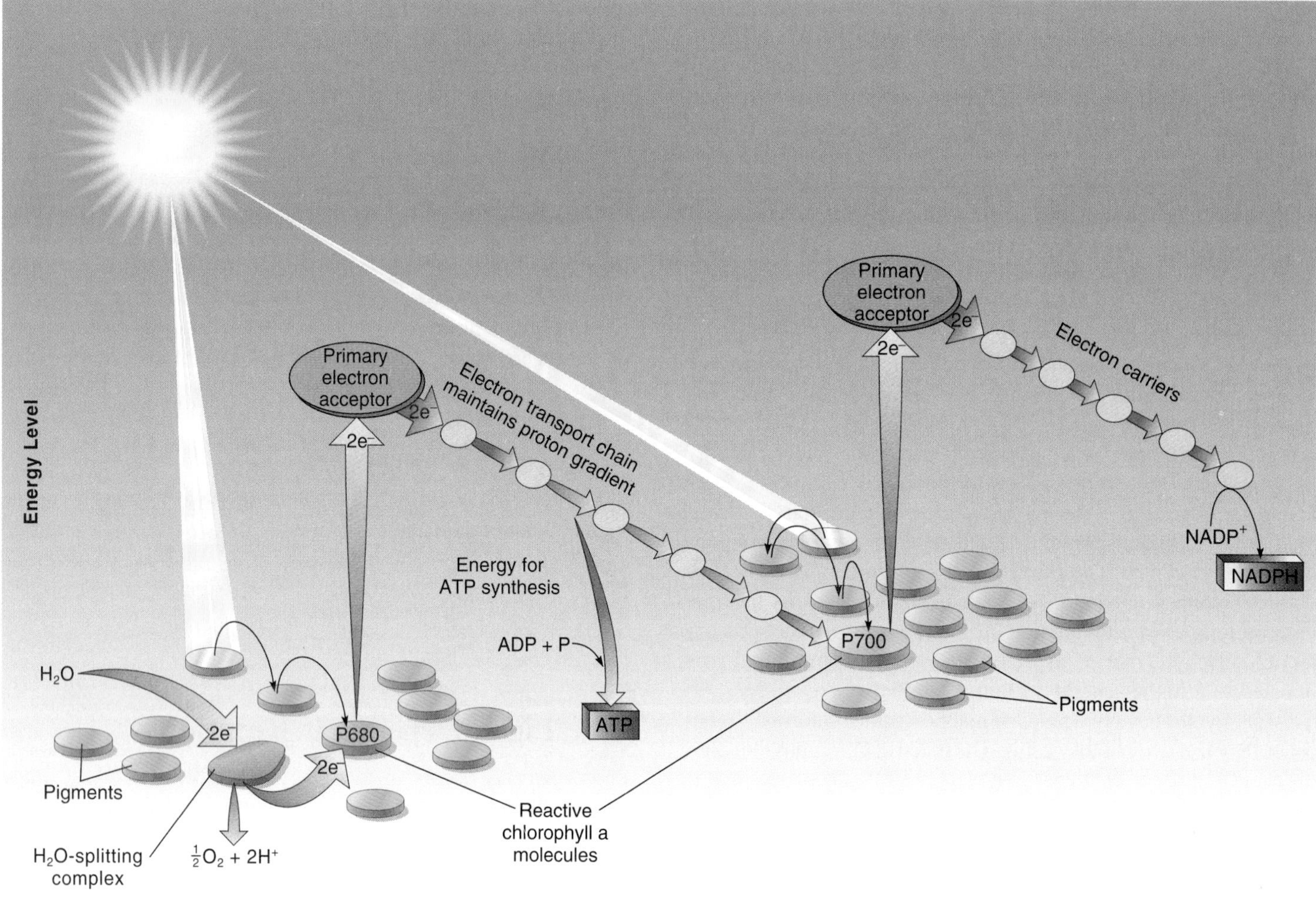

FIGURE 8.9

The light-dependent reactions of photosynthesis. The sun's energy propels electrons from reactive molecules of chlorophyll a to primary electron acceptors. From the acceptors, the electrons flow through a series of electron carrier molecules. Electrons flow continuously from water to $NADP^+$, reducing $NADP^+$ to NADPH. Energy from the electron transport system that links photosystems I and II is used to synthesize ATP. The products of the light-dependent reactions—ATP and NADPH—are the starting materials for the light-independent reactions.

Table 8.2 Requirements for the Light-Dependent Reactions of Photosynthesis

1. Light-harvesting pigment molecules and electron transfer systems must be in close physical proximity.
2. Absorbed light must be in the visible part of the electromagnetic spectrum.
3. Energy must move rapidly from pigments to the reaction center to the electron transport system. Proton gradient formation and ATP synthesis must also be fast.
4. Electron transport must be unidirectional, establishing a proton gradient.
5. Pigment molecules and electron transport molecules must be used over and over.

The light-dependent reactions of photosynthesis use photon energy to oxidize water, release oxygen, produce ATP, and reduce $NADP^+$ to NADPH. The light-independent reactions later use the ATP and NADPH to reduce CO_2 to carbohydrate. Table 8.2 summarizes the conditions necessary for the light-dependent reactions to occur.

The light-dependent reactions take place in two **photosystems:** photosystem I and photosystem II.

Photosystem II Starts Photosynthesis

Photosynthesis begins in the cluster of pigment molecules of photosystem II. This may seem illogical, but the two photosystems were named as they were discovered; photosystem II was discovered after photosystem I, but it functions first in the overall process.

Pigment molecules in photosystem II absorb the energy from incoming photons in sunlight. The energy is transferred from one pigment molecule to another until it reaches a chlorophyll a reaction center, where it excites an electron. The excited electron is ejected from this chlorophyll a molecule and is grabbed by the first electron carrier molecule of the electron transport system that links the two photosystems (fig. 8.9). Photosystems II and I are often called a "Z-scheme" because the electron transport system connecting them, when diagrammed, resembles the letter Z. The reactive chlorophyll a molecules replace their lost electrons when light energy is used to split water (H_2O) into oxygen gas (O_2) and protons (H^+), releasing electrons. As electrons pass from carrier to carrier, the energy they lose drives the active transport of protons from the stroma to the thylakoid space (fig. 8.10). The protons accumulate inside the thylakoids and establish a pH difference and therefore a proton gradient between the inside and outside of the membrane. Then, protons diffuse down their concentration gradient back into the stroma, exiting the thylakoid membrane through ATP synthase channels. This movement triggers phosphorylation of ADP to ATP, which is released from the ATP synthase. (ATP synthase is also the enzyme that couples proton passage to ATP formation in mitochondria.)

Photosystem II is remarkably fast. It removes electrons from water, releases oxygen gas, replaces the ejected electrons, and synthesizes ATP—in under a billionth of a second! This light-driven production of ATP through an electron transport chain is called **photophosphorylation.** It is very similar to oxidative phosphorylation in mitochondria.

Photosystem I

Photosystem I begins much as photosystem II does. Photon energy strikes energy-absorbing molecules of chlorophyll a, which pass the energy to reactive chlorophyll a. The reactive chlorophyll molecules eject electrons to the first electron carrier molecule in a second electron transport system. The boosted electrons in photosystem I are then replaced with electrons passing down the first electron transport system from photosystem II. This sequence of electron movement continues until finally, the transported electrons of photosystem I reduce a molecule of $NADP^+$ to NADPH. This NADPH, plus the ATP that photosystem II generates, become the energy sources that power the light-independent reactions to follow.

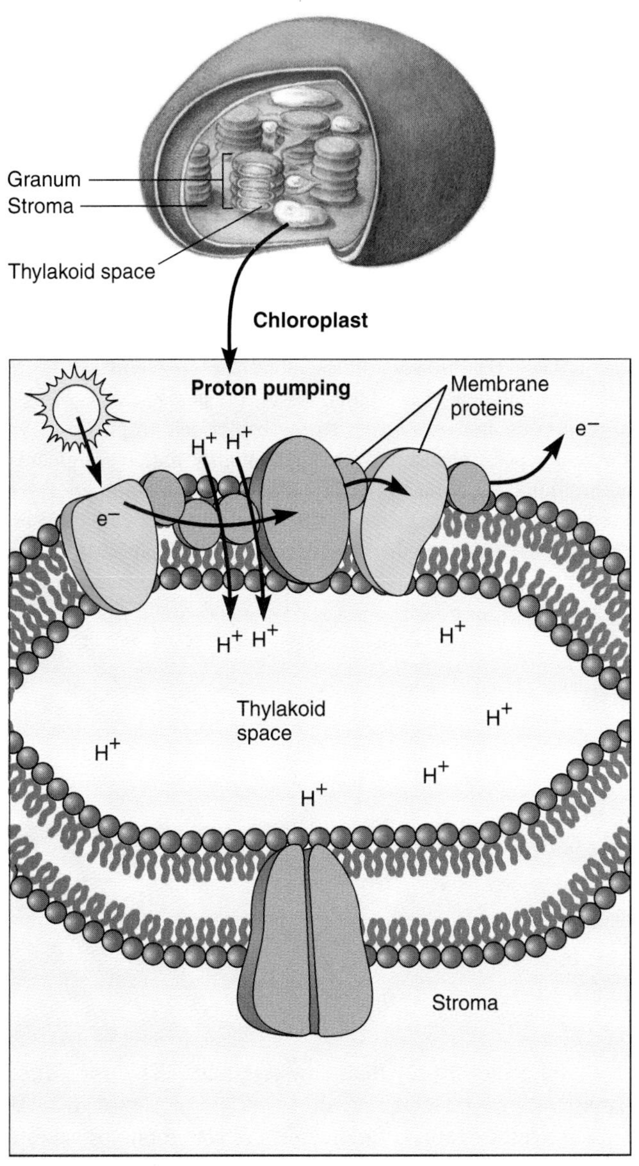

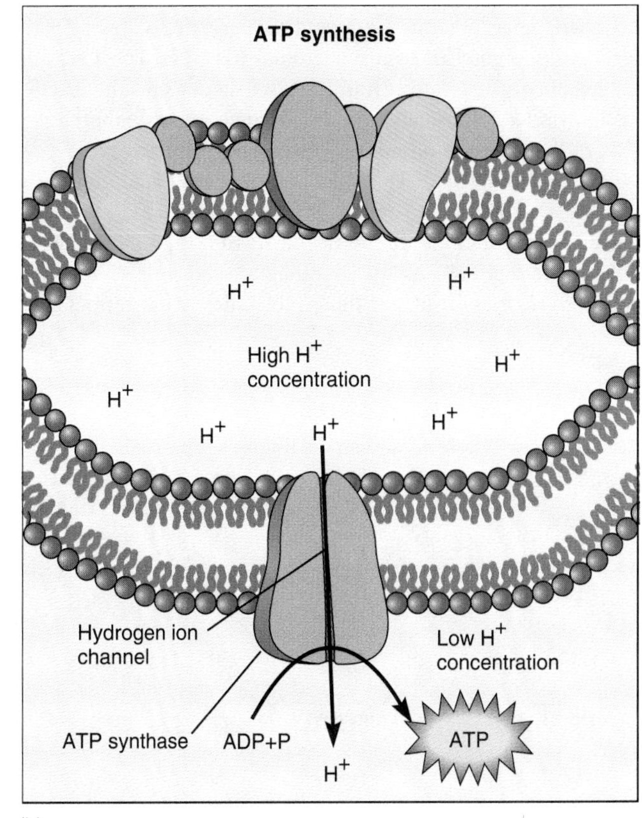

FIGURE 8.10

Chemiosmosis couples a proton gradient to ATP synthesis in the chloroplast. As electrons pass between carrier molecules in the electron transport system that links photosystems I and II, protons are actively transported into the space within the thylakoid membranes. The energy of the proton gradient phosphorylates ADP to ATP. Proton pumping (*a*) and ATP synthesis (*b*) occur simultaneously and continuously in the light.

biology in 8.2 action

How Photosynthetic Bacteria Capture Light Energy

Photosynthesis in bacteria vividly illustrates how interacting chemicals provide a characteristic of life. The bacterial photosynthetic machinery is a cluster of molecular assemblies, resembling doughnuts, that are organized in a highly symmetric fashion that enables them to capture and transfer solar energy in mere trillionths of a second. Bacterial photosystems are like those of plants in that clusters of pigment molecules entrap photon energy and transfer it. Yet the bacterial molecules and their organization in complexes differ.

Structure

Photosynthesis in prokaryotes is well studied in *Rhodopseudomonas acidophila,* a bacterium that lives beneath algae in murky ponds and captures stray light rays that penetrate the surface (figure 8.B*1*). The organism's cell membrane houses light-harvesting complexes (LHs), which are doughnut-shaped groups of associated polypeptides and pigments. Because the cell membrane wraps around the cell several times, it creates layers of LHs (*2*). It is the precise organization of these LHs that makes possible the remarkable speed of photosynthesis.

Bacterial LHs come in two varieties—LH2, whose structure researchers determined in 1995, and the less well known LH1. In *R. acidophila,* eight LH2s cluster around LH1 (*3*). The LH2s capture photon energy and deliver it to LH1, which contains the reactive pigments (bacteriochlorophylls) that continue photosynthesis. A closer look at the structure of an LH2 reveals how it works.

An LH2 consists of 18 coiled polypeptides (*4*), organized in two circles. The nine polypeptide coils on the outside slant, whereas those forming a circle on the inside stand upright. The coils on the outside are more loosely packed than are the coils on the inside. In addition, each coil touches three molecules of bacteriochlorophyll, and 14 of the coils also touch a carotenoid molecule. Because the bacteriochlorophylls are in two different chemical environments (outside and inside of the structure), they absorb different wavelengths of light. Specifically, the bottom bacteriochlorophylls absorb light of 800 nanometers wavelength, whereas those above absorb light of 850 nanometers.

Function

Photosynthesis begins as photons impinge upon an LH2 and bacteriochlorophylls absorb them. Because of the spaces separating the outside pigments, instead of passing the absorbed energy to each other, they pass it to the bacteriochlorophylls in the inner circle. The energy transfer takes 2 to 3 ten-trillionths of a second. The eight LH2s and the sole LH1 join at the exact same level in space, like eight plates arranged on a table about a central ninth plate. Because of this proximity, photon energy jumps from LH2 to LH2 and to the all-important LH1.

The LH1 contains a region called the reaction center, which houses two highly reactive bacteriochlorophyll molecules. These molecules absorb energy of wavelength 875 nanometers. Recall that the longer the wavelength of light, the lower the energy. Therefore, the light coming from the LH2s has higher energy than the energy that LH1 can absorb. This keeps the energy flowing in one direction, from the LH2s to the LH1. The LH2s are sometimes called "outer antennas" and the LH1 an "inner antenna." LH1 functions as an energy sink.

The photon energy transferred to the bacteriochlorophylls in the reaction center pulls electrons from these highly reactive pigments. This is the important point in photosynthesis when photon energy is converted to chemical energy. The released electrons then reduce acceptor molecules, which are the starting point for the light-independent reactions of photosynthesis, where carbon dioxide is reduced to yield carbohydrates.

Why Two Photosystems Increase Energy Extraction

Why does eukaryotic photosynthesis require two photosystems? The answer is that two photosystems are more efficient than one—they increase the amount of energy extracted from sunlight.

Photosynthetic prokaryotes use only one photosystem, different from those in plants, and recycle boosted electrons back to the same reaction center from which they originated. This short-form photosynthesis generates chemical energy, but not enough to allow cells to synthesize many molecules. A single photosystem is hypothesized to have been the sole strategy for harnessing solar energy for the first billion years of life. It would have yielded no NADPH or oxygen. Splitting oxygen to replace the light-boosted electrons probably required more energy than these early bacterial systems could muster. Biology in Action 8.2 describes bacterial

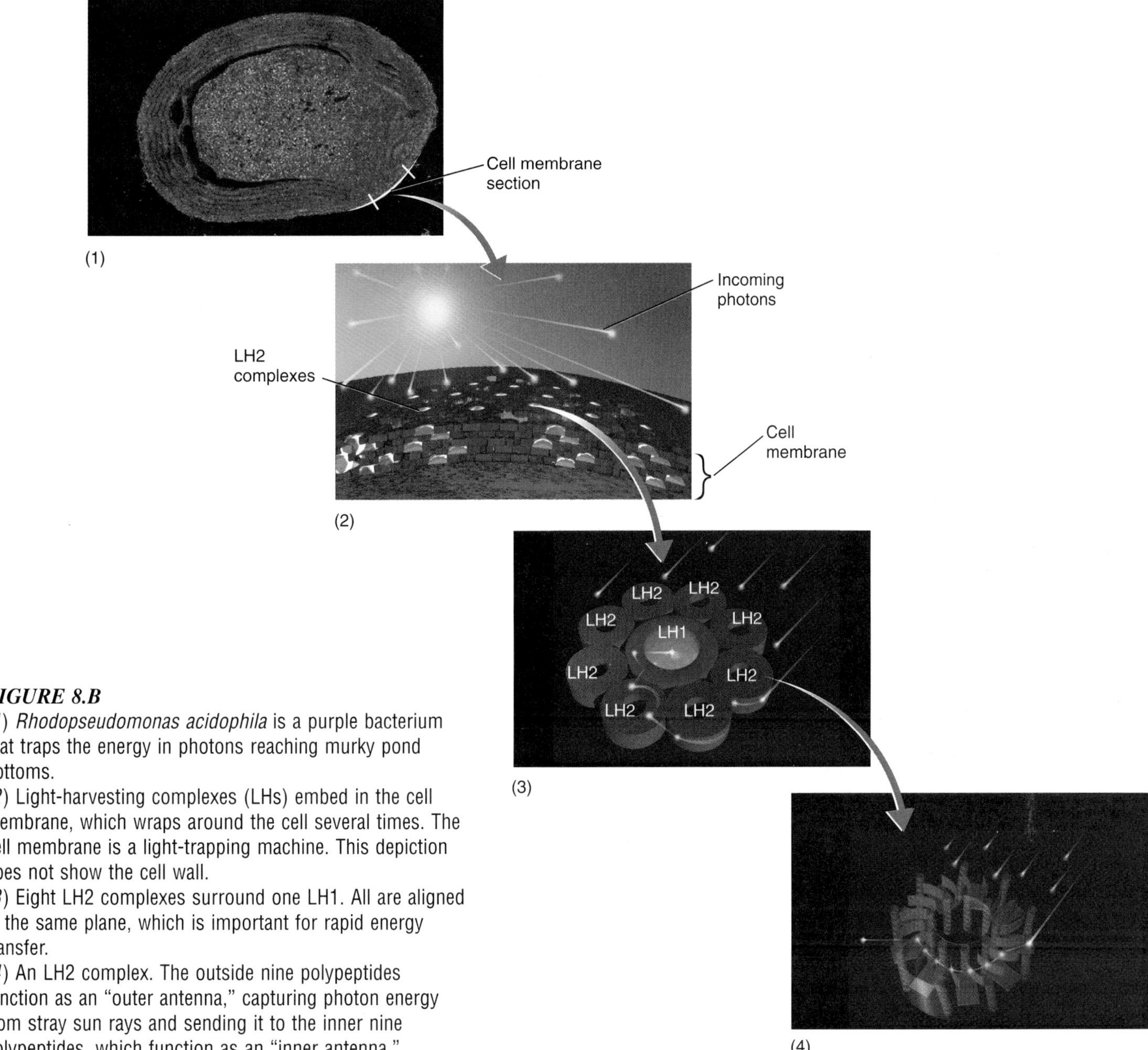

FIGURE 8.B
(*1*) *Rhodopseudomonas acidophila* is a purple bacterium that traps the energy in photons reaching murky pond bottoms.
(*2*) Light-harvesting complexes (LHs) embed in the cell membrane, which wraps around the cell several times. The cell membrane is a light-trapping machine. This depiction does not show the cell wall.
(*3*) Eight LH2 complexes surround one LH1. All are aligned in the same plane, which is important for rapid energy transfer.
(*4*) An LH2 complex. The outside nine polypeptides function as an "outer antenna," capturing photon energy from stray sun rays and sending it to the inner nine polypeptides, which function as an "inner antenna."

photosynthesis. The process is like a simplified version of plant photosynthesis but with its own distinctive pigments, proteins, and molecular complexes.

Grafting a second, more powerful photosystem onto a version of the single bacterial system would have enabled plants to extract more energy from sunlight. This second photosystem, called photosystem II, acts first. It passes electrons in a straight line rather than cycling them back to the original reaction center, as photosystem I does. The two photosystems, connected in a series, produce more energy—much as two batteries increase the voltage in a flashlight. In addition, the newer photosystem II has a unique arrangement of pigments that allows it to harvest shorter wavelengths of light.

The evolution of photosystem II forever changed life on earth. No longer did organisms have to rely directly on compounds in the environment as energy sources; they could now harness solar energy. Moreover, the oxygen pumped into the atmosphere by photosynthesis allowed the evolution of aerobic respiration.

Mastering Concepts

1. What are the products of the light-dependent reactions of photosynthesis?
2. What conditions must exist for the light-dependent reactions to proceed?
3. What events capture and transfer photon energy in photosystem II?
4. How are electrons passed from photosystem II to photosystem I?
5. How are the boosted electrons from photosystem II replaced?
6. How is ATP produced in photosystem II?
7. What happens in photosystem I?
8. How do two photosystems increase the efficiency of photosynthesis?

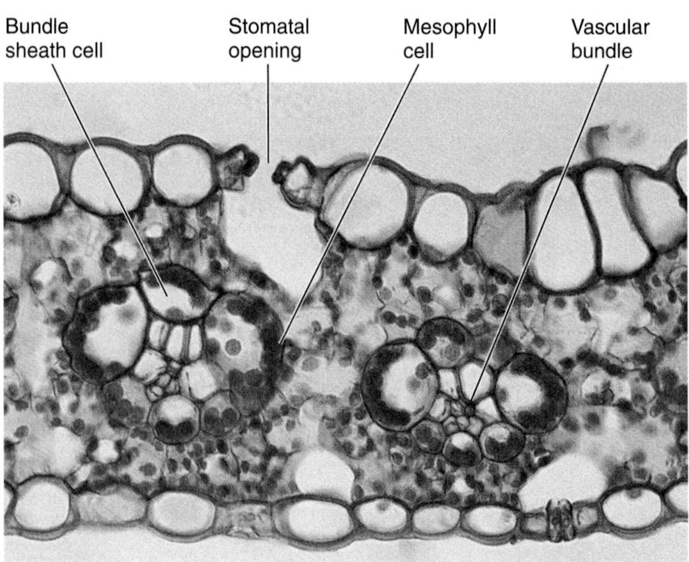

FIGURE 8.11

A stoma is an opening on the underside of a leaf that allows CO_2 to enter leaf cells. The stoma can open and close to regulate evaporation of water from the leaf.

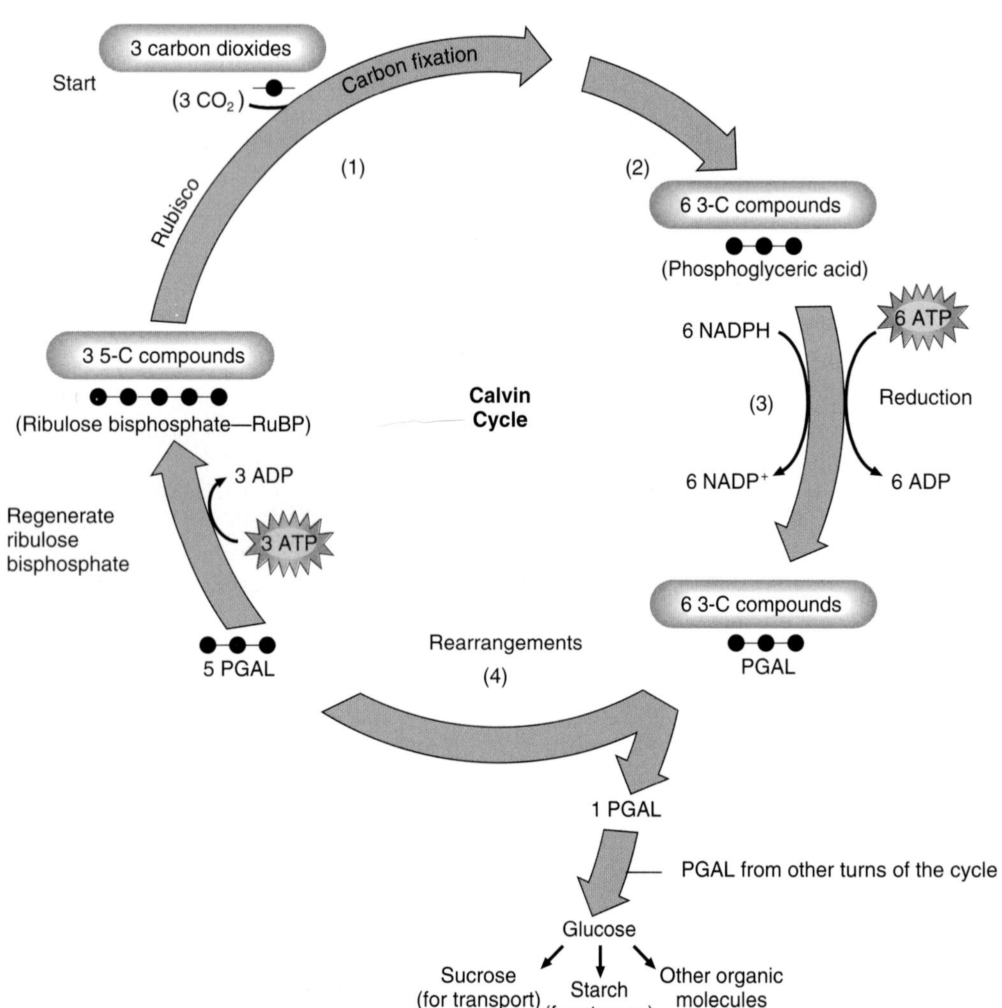

FIGURE 8.12

The products of the light-dependent reactions—ATP and NADPH—drive the Calvin cycle, shown here in simplified form. The cycle "fixes" carbon from CO_2 into organic molecules. Rubisco catalyzes the first step, the reaction between RuBP and CO_2. The Calvin cycle generates a three-carbon molecule, PGAL, which reacts to produce glucose and other organic molecules and also regenerates RuBP. This pathway operates in most photosynthesizing plants, but some species have significantly modified cycles.

The Light-Independent Reactions

The light-independent reactions of photosynthesis incorporate, or fix, carbon from CO_2 into organic compounds that cells use for energy storage and to build tissues. Where does the CO_2 in biological systems originate? Photosynthetic algae and bacteria and some aquatic plants absorb dissolved CO_2 from the water that surrounds them. In plants, atmospheric CO_2 enters through tiny leaf and stem openings called **stomata** (singular: stoma) (fig. 8.11).

The Calvin Cycle

The light-independent reactions occur in both darkness and light. These reactions form a metabolic cycle known as the Calvin cycle, named for American biochemist Melvin Calvin. It is also known as the C_3 cycle because a three-carbon compound, **phosphoglyceric acid** (PGA), is the first stable compound in the pathway. Plants that use only the Calvin cycle to fix carbon from CO_2 are called **C_3 plants.** They include cereals, peanuts, tobacco, spinach, sugar beets, soybeans, most trees, and lawn grasses.

If ATP and NADPH (the products of the light-dependent reactions) are plentiful, the Calvin cycle continually fixes the carbon from CO_2 into small organic molecules. The plant cell can then use these smaller molecules to synthesize more complex nutrient molecules. The Calvin cycle begins when CO_2 reacts with **ribulose bisphosphate** (RuBP), a five-carbon sugar with two phosphate groups, to yield an unstable six-carbon molecule. This molecule immediately breaks down to two molecules of PGA (fig. 8.12). The enzyme that catalyzes the reaction between RuBP and CO_2 is **RuBP carboxylase/oxygenase,** also known as **rubisco.** This is one of the most important and abundant proteins in the living world. Next, energy from ATP and NADPH convert phosphoglyceric acid to **phosphoglyceraldehyde** (PGAL), which is the carbohydrate product of the Calvin cycle. PGAL reacts further to produce glucose and other nutrient molecules, as well as regenerating RuBP, and perpetuates the cycle.

How Experiments Revealed the Steps of the Calvin Cycle

A series of clever experiments identified the intermediates of the Calvin cycle using ^{14}C, or radioactive carbon. ^{14}C is identical chemically to the common isotope ^{12}C, but it is radioactive because it contains more neutrons.

In the early 1950s, Calvin and his colleagues exposed a dense culture of the green alga *Chlorella* to CO_2 whose carbon was the ^{14}C radioactive isotope. Then Calvin identified the chemicals that progressively incorporated the radioactive carbon "label" and traced the path of the ^{14}C as it moved from CO_2 to carbohydrate. When Calvin exposed the alga to radioactively labeled CO_2 for 60 seconds, he found that ^{14}C labeled many compounds in the alga. But when he repeated the experiment with a 7-second exposure, most of the ^{14}C appeared in phosphoglyceric acid. This suggested that the molecule that initially joins CO_2 is a two-carbon compound, because phosphoglyceric acid has three carbons. Calvin searched for this two-carbon acceptor, but he couldn't find it. He then began looking for another molecule that, combined with CO_2, would yield a three-carbon product. He found the elusive acceptor in RuBP. Calvin used a similar approach to decipher the other biochemical reactions of photosynthesis. Figure 8.13 depicts the relationship between the light-dependent and light-independent reactions of photosynthesis.

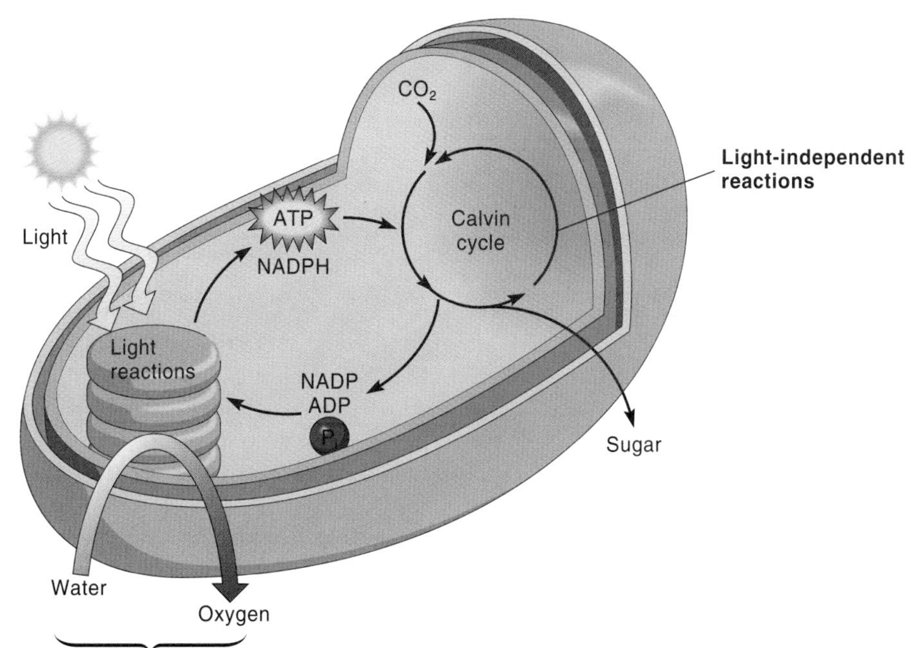

FIGURE 8.13

Summary of photosynthesis. The light-dependent reactions yield ATP and NADPH, which provide chemical energy and reducing power to the Calvin cycle and other light-independent reactions. The light-dependent reactions occur in the grana, and the light-independent reactions take place in the stroma.

Mastering Concepts

1. What is the function of the Calvin cycle?
2. What is the role of ribulose bisphosphate in the Calvin cycle?
3. Why is rubisco an important molecule?
4. What does the energy from ATP and NADPH power in the Calvin cycle?
5. How did radioactive labels help identify intermediates of the Calvin cycle?

The Efficiency of Photosynthesis

The atmosphere is only 0.035% CO_2, yet each year, plants use 200 billion tons of carbon from CO_2 to manufacture organic compounds. That sounds like a large output, but the proportion of photon energy that plants actually harness in the carbohydrate products of photosynthesis, collectively called **photosynthate,** is less impressive. The process of energy acquisition is quite inefficient. If each photosystem absorbs the maximum possible number of photons, theoretical calculations yield an efficiency rate of 30%. In reality, this rate is much lower.

On cloudy days, field measurements of the photosynthetic efficiency of individual plants average about 0.1%, while measurements for intensively cultivated plants average about 3%. Some meticulously cared for laboratory-grown plants can reach 25% efficiency. The plant with the greatest natural efficiency—8%—is *Oenothera claviformis,* the annual winter evening primrose grown in Death Valley, California. Sugarcane follows at 7%.

Why do plants waste so much solar energy? One contributing factor is a process that counters photosynthesis called **photorespiration.**

Photorespiration

In photorespiration, the Calvin cycle "fixes" the oxygen in CO_2 rather than the carbon. Increased amounts of oxygen thus inhibit photosynthesis in C_3 plants; the plants fix oxygen and release CO_2 rather than the other way around. But how can a plant's exposure to oxygen increase, if the proportion of CO_2 and oxygen (O_2) in the atmosphere stays relatively constant? The answer is that the plant itself influences the CO_2 and O_2 proportions it contacts as well as how much carbon is available for fixation.

A plant can capture atmospheric CO_2 only when its stomata are open. As CO_2 diffuses into the leaf, water diffuses out through the stomata. As long as the plant has plenty of water, stomata stay open; but on hot, dry days, stomata close to conserve water. When this happens, the Calvin cycle's continued fixation of carbon depletes the CO_2 supply inside the leaf, and O_2 accumulates. No additional CO_2 can enter because of the closed stomata. The greatly increasing relative amount of O_2 inside the plant cells has a drastic effect—rubisco stops fixing carbon from CO_2 and starts fixing O_2.

The products of this photorespiration reaction are a two-carbon compound, phosphoglycolic acid, and the three-carbon phosphoglyceric acid (PGA). The PGA stays in the Calvin cycle, but the phosphoglycolic acid reacts further, eventually using ATP energy to release CO_2 and to be converted to PGA, which can then enter the Calvin cycle. However, the plant often loses this CO_2, rather than fixing its carbon—which lowers photosynthetic efficiency.

The reactions initiated when rubisco fixes oxygen rather than carbon from CO_2 are called photorespiration because they occur only in light (photo) and consume oxygen and release CO_2 (respiration). It is not the same as cellular respiration in the mitochondria, which also consumes oxygen. Photorespiration is wasteful because it uses ATP and NADPH and releases CO_2 that must be fixed again. As much as half of the carbon fixed in the Calvin cycle is released by photorespiration during hot, dry days. Thus, photorespiration counters photosynthesis.

Why would a seemingly wasteful metabolic process like photorespiration evolve? One hypothesis is that the key enzyme, rubisco, cannot discriminate perfectly between CO_2 and O_2. Or photorespiration may be a no longer useful evolutionary holdover, a metabolic relic of the time when the atmosphere lacked free oxygen and competition for CO_2 didn't matter. Still another possibility is that the buildup of oxygen from photosynthesis led to photorespiration after millions of years.

Mastering Concepts

1. How efficient is photosynthesis?
2. Under what environmental conditions does photorespiration occur?
3. How does photorespiration counter photosynthesis?

C_4 Photosynthesis

Photorespiration peaks in hot, dry conditions, when plants begin to close their stomata to conserve water. In this situation, the plants lose carbon. Any plant that can avoid photorespiration has a significant advantage. An adaptation called **C_4 photosynthesis** enables certain plants to avoid photorespiration. This route is called the C_4 pathway, and plants that use it are called C_4 plants, because the first stable compound contains four carbons.

When biologists repeated Calvin's study labeling carbon from CO_2 in different types of plants, their results usually confirmed Calvin's. But in 1965, a study of sugarcane produced strikingly different results. When sugarcane was exposed to radioactive CO_2 for 1 second, 80% of the radioactivity eventually appeared not in a three-carbon compound, as in the Calvin cycle, but in four-carbon malic acid. Most of this malic acid surfaced in thin-walled **mesophyll cells** (fig. 8.14), which lack most of the enzymes of the Calvin cycle. After about 10 seconds, the radioactive label appeared in PGA in the thick-walled **bundle-sheath cells** that surround veins. If CO_2 was being fixed into malic acid, was the Calvin cycle occurring?

Further studies of sugarcane leaves revealed that the plant does fix carbon from CO_2 via the Calvin cycle but only after fixing it with another set of reactions that happen first. A more complex anatomy in C_4 plants accommodates this extra biochemical pathway.

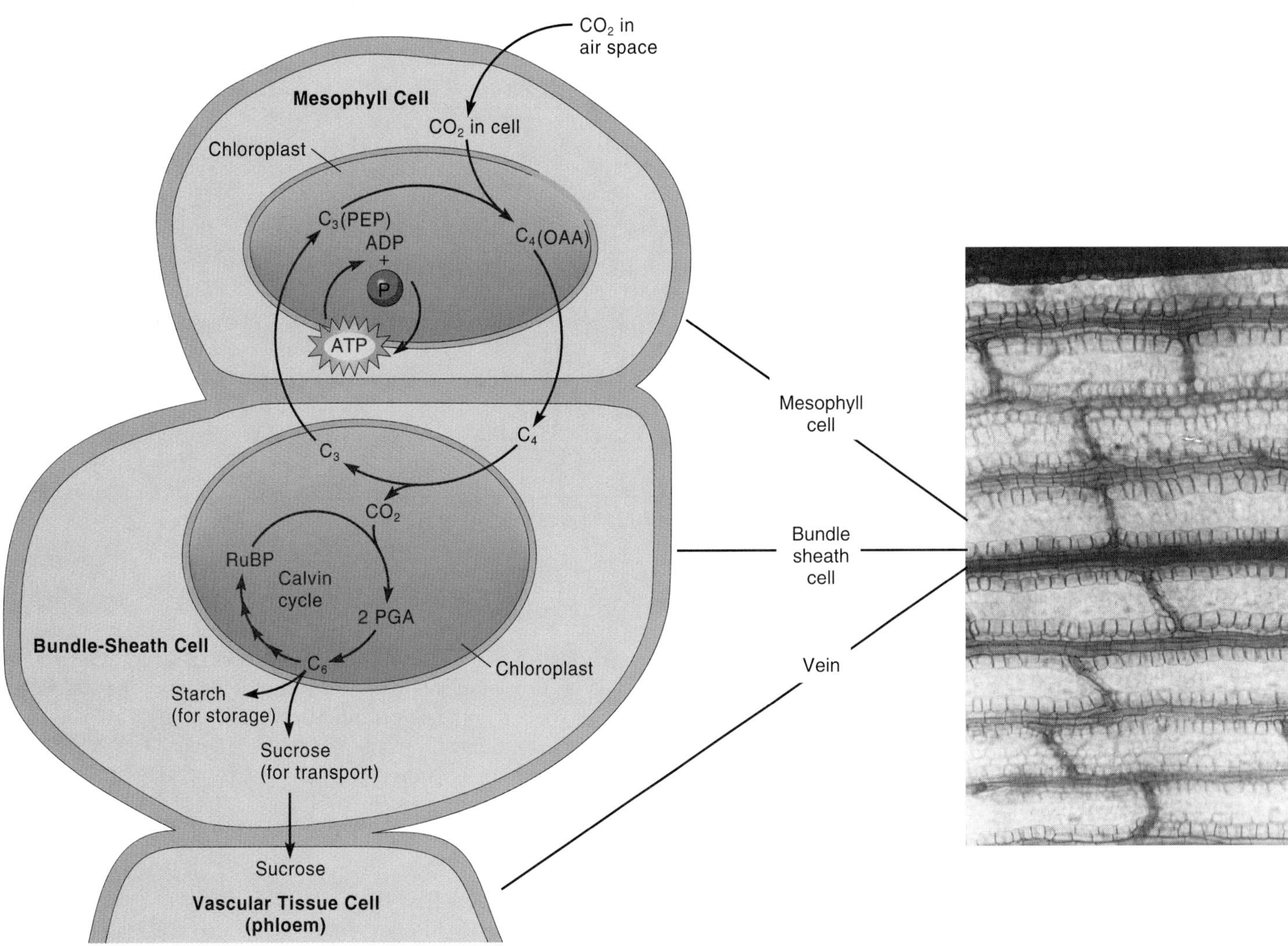

FIGURE 8.14

C_4 photosynthesis. C_4 plants such as corn have a characteristic cellular organization. A bundle-sheath cell is sandwiched between a photosynthetic mesophyll cell and a vascular tissue cell, which transports photosynthate. CO_2 enters and is fixed in mesophyll cells, combining with phosphoenolpyruvate (PEP), a three-carbon compound, to yield the four-carbon compound oxaloacetic acid (OAA), which is converted to malic acid and aspartic acid. These compounds are then carried to adjacent bundle-sheath cells, where they are split into CO_2 and a three-carbon compound. The Calvin cycle fixes this CO_2, and the three-carbon compound returns to the mesophyll cell, where it is converted to PEP and picks up another CO_2.

In C_4 photosynthesis, CO_2 diffuses into the leaf through stomata and fixes carbon in mesophyll cells. CO_2 combines with a three-carbon compound, phosphoenolpyruvate (PEP), to form the four-carbon oxaloacetic acid (OAA). OAA then is converted to malic acid and aspartic acid, both of which are transported to adjacent bundle-sheath cells. There, the acids split into CO_2 and three-carbon compounds. Malic acid and aspartic acid therefore temporarily store CO_2.

Pumping CO_2 into bundle-sheath cells keeps the internal concentration of CO_2 20 to 120 times greater than normal. This prevents photorespiration and makes C_4 plants more efficient photosynthesizers than C_3 plants in hot, dry, sunny weather. The Calvin cycle fixes the CO_2 released in the bundle-sheath cells. Meanwhile, the three-carbon compound returns to the mesophyll cell, where it is converted to PEP, the initial CO_2 acceptor of C_4 photosynthesis. C_4 can fix CO_2 even when the stomata begin to close. As a result, C_4 plants need only about half as much water as C_3 plants need for photosynthesis.

C_4 photosynthesis evolved in the hot, dry tropics. It probably has persisted because atmospheric CO_2 concentrations have decreased during the last 50 to 60 million years. Today, C_4 plants are all flowering plants growing in hot, open environments. They represent diverse species, suggesting that C_4 photosynthesis may have evolved independently several times. C_4 plants include economically important crops such as corn, millet, and sorghum.

Even though C_4 plants dominate in hot and dry ecosystems because they avoid the photorespiration that affects C_3 plants, in other habitats they are not as abundant. C_4 plants are at an energetic disadvantage in these other environments because they must invest two ATPs for every carbon that goes from the mesophyll cell to the bundle-sheath cell. Still other plants combine aspects of C_3 and C_4 photosynthesis.

Table 8.3 Photosynthetic Characteristics of C_3, C_4, and CAM Plants

Characteristic	C_3 Plants	C_4 Plants	CAM Plants
Leaf anatomy	Bundle-sheath cells have sparse chloroplasts	Bundle-sheath cells have dense chloroplasts	Large vacuoles
Water (grams) to produce 1 gram of dry photosynthate	450–950	250–350	50–55
CO_2 concentration in atmosphere	0.03–0.07%	0–0.01%	0–0.005%
Photorespiration	Yes	In bundle-sheath cells	No
Best temperature for photosynthesis	15–25°C	30–40°C	About 35°C
Tons of dry photosynthate per hectare per year	20–25	35–40	Low and variable

Source: Randy Moore, et al., *Botany*. Copyright © 1995 The McGraw-Hill Companies, Inc.

Mastering Concepts

1. In what environment is C_4 photosynthesis adaptive?
2. Describe the additional biochemical pathway of C_4 photosynthesis.
3. What types of plant cells are involved in C_4 photosynthesis?

CAM Photosynthesis

Just as farmers store a bountiful harvest for use over the long winter, some plants that live in very dry environments take in CO_2 at night, but they do not fix it in the Calvin cycle until daylight. This strategy was discovered in plants of genus *Crassulacea,* and so it is called crassulacean acid metabolism (CAM). Plants with **CAM photosynthesis** include many cacti, pineapple, Spanish moss, orchids, some ferns, and the wax plant. This form of photosynthesis is an adaptation to the day and night temperature and humidity differences in the desert. At night, when temperature drops and humidity rises, stomata open and CO_2 enters the plant. During the hot and dry day, stomata close to conserve water—but the plant already has CO_2 from its nighttime activity.

Botanists first observed CAM photosynthesis, indirectly, in the nineteenth century, when they noted that certain plants are acidic at night but become more alkaline during the day. It wasn't until 1958 that botanists discovered the reason for the fluctuating pH—at night, plants open their stomata, take in CO_2, and combine it with the three-carbon compound phosphoenolpyruvate (PEP) to form malic acid. Unlike C_4 metabolism, which occurs in two different cell types, in CAM photosynthesis, malic acid forms and is stored in large vacuoles in the same cells that have chloroplasts. In the daytime, malic acid enters the chloroplast, where it releases CO_2 and sends a three-carbon compound back to the vacuole. The CO_2 is fixed in the Calvin cycle in the chloroplast of the same cell.

The different variations on the photosynthesis theme enable plants to capture and use solar energy in a variety of environments. About 85% of plants are C_3. Only 0.4% are C_4 plants, but we are very aware of this group because it includes many crop plants. About 10% of plants use CAM photosynthesis. The remaining plant species combine certain reactions of C_3, C_4, or CAM photosynthesis. Some C_3 plants, for example, switch to CAM photosynthesis in times of drought, and some plants use different pathways at different stages in development. Table 8.3 compares C_3, C_4, and CAM plants.

Mastering Concepts

1. How is CAM photosynthesis an adaptation to the desert environment?
2. How is CAM photosynthesis like C_4 metabolism, and how is it different?

The Big Picture of Biological Energy Use

The products of photosynthesis provide energy—directly to plants and indirectly to nearly all other organisms. Plants use about half of all photosynthate as fuel for cellular respiration and photorespiration. They also use photosynthate to manufacture a variety of compounds, including amino acids, starch, sucrose, cellulose, glucose, rubber, quinine, and spices.

Photosynthesis, glycolysis, and cellular respiration are intimately related (fig. 8.15). The organic products of photosynthesis supply the starting materials for glycolysis. The oxygen produced by photosynthesis becomes the final electron acceptor in aerobic respiration. CO_2 generated in the Krebs cycle enters the Calvin cycle. Finally, water produced by aerobic respiration is split in photosynthesis, releasing electrons that replace the electrons boosted out of reactive chlorophyll a molecules when they absorb photon energy.

Together, these biological energy reactions sustain life. Table 8.4 summarizes the factors that affect photosynthesis.

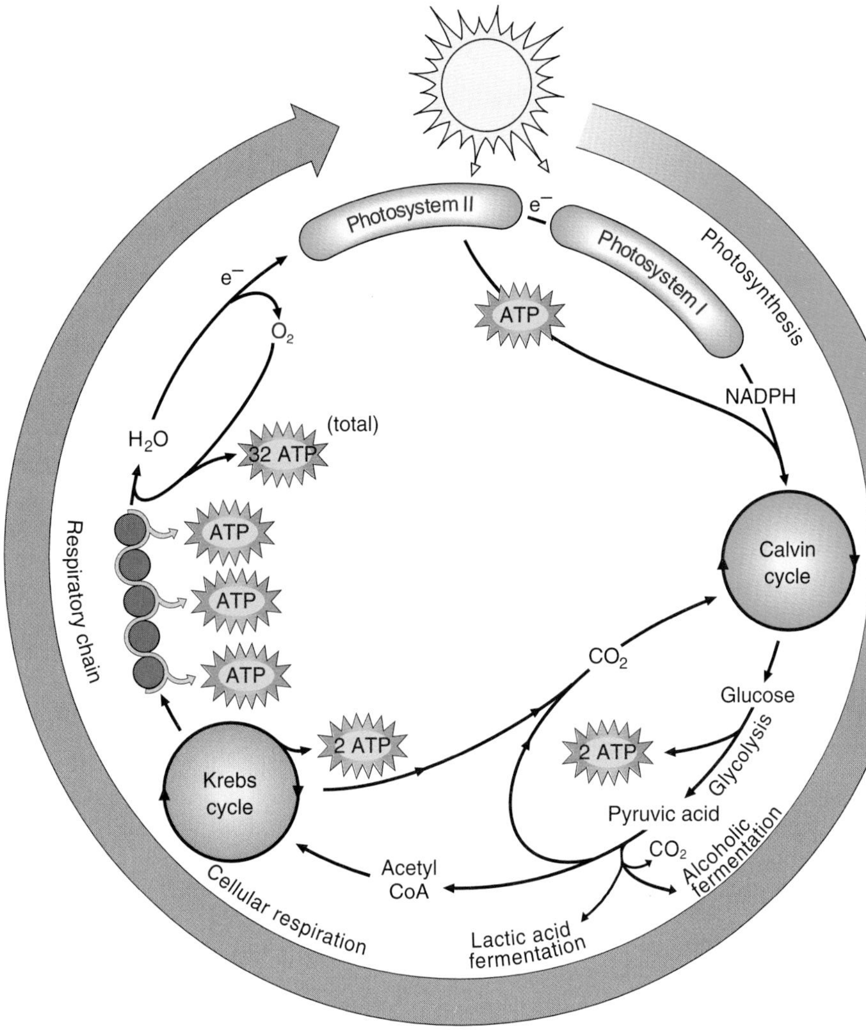

FIGURE 8.15

An overview of energy metabolism. The biological energy reactions are interrelated.

Table 8.4 Factors Controlling Photosynthesis

Factor	Effect
CO_2 concentration	Increasing CO_2 concentration increases the rate of photosynthesis in C_3 plants but not in C_4 plants
Light intensity	The maximum rate of photosynthesis occurs when light is brightest; there is no photosynthesis in the dark
Stage of plant development	Plants in active stages, such as seedlings, require more energy and therefore have higher rates of photosynthesis
Storage or use of photosynthate	The photosynthetic rate increases when injured plants use carbohydrates for energy; it decreases when carbohydrate use diminishes
Temperature	Photosynthetic rate increases at higher temperatures
Water availability	Lack of water closes stomata, inhibiting photosynthesis

Possible Origins of the Energy Pathways

Glycolysis is probably the most ancient of the energy pathways because it is common to nearly all cells. The other pathways are more specialized—photosynthesis occurs only in green plants, algae, and some prokaryotes; fermentation and aerobic respiration take place only under certain environmental conditions.

Glycolysis evolved when the atmosphere lacked or had very little oxygen. These reactions enabled the earliest organisms to extract energy from simple organic compounds in the nonliving environment. Photosynthesis may have evolved from glycolysis because some of the reactions of the Calvin cycle are the reverse of some of those of glycolysis. Broadly speaking, glycolysis converts a six-carbon compound into two three-carbon compounds; some of the reactions of the Calvin cycle do the opposite.

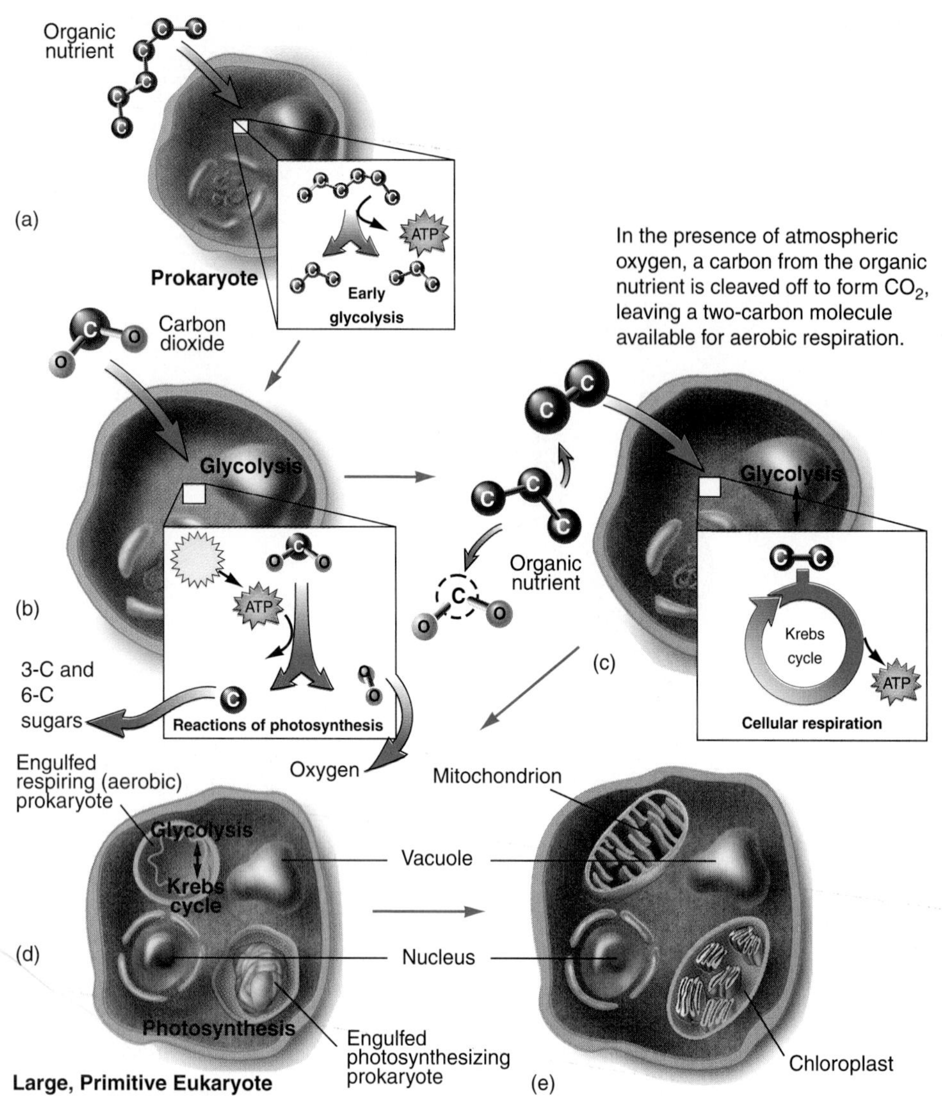

FIGURE 8.16

A scenario for the evolution of energy pathways. (*a*) The first prokaryotes used a pathway similar to glycolysis to feed on organic compounds of nonbiological origin in the oxygen-poor environment. (*b*) Somehow, perhaps by reversing some of the reactions of glycolysis, photosynthesis arose in some of these organisms on internal membranes. (*c*) A by-product, oxygen, was to profoundly change the atmosphere and life on earth. Oxygen in the air led to a more energy-efficient extension of glycolysis—respiration. (*d*) Eventually, a large cell engulfed both a photosynthesizing prokaryote and a respiring one. (*e*) The result, according to the endosymbiont theory (chapter 4), was the first photosynthesizing eukaryote, which may have been the ancestor of modern plants. The photosynthesizing prokaryote (*b*) gave rise to the chloroplast; the respiring one (*c*) gave rise to the mitochondrion.

The reactions of photosynthesis arose through a still-unknown chemical rearrangement that botanists hypothesize occurred from 3.5 to 3.8 billion years ago. Photosynthesis altered life on earth forever. No longer did organisms have to depend on organic compounds in their surroundings for energy; they now had a way to produce nutrients constantly, from sunlight. Photosynthesis, over time, released oxygen into the primitive atmosphere, paving the way for an explosion of new life-forms capable of using this new atmospheric component. In addition, electrical storms split water in the atmosphere, releasing single oxygen atoms that joined with diatomic oxygen to produce ozone (O_3). As ozone accumulated high in the atmosphere, it blocked harmful ultraviolet radiation from reaching the planet's surface, which prevented some genetic damage and allowed new life-forms to arise.

Photosynthesis could not have debuted in a plant cell, because such complex organisms were not present on the early earth. The first photosynthetic organisms may have been prokaryotes, such as anaerobic bacteria that used hydrogen sulfide (H_2S) in photosynthesis instead of the water plants use. These first photosynthetic bacteria would have released sulfur, rather than oxygen, into the environment. Eventually, evolutionary changes in pigment molecules enabled some bacteria to use water instead of hydrogen sulfide. If a large cell engulfed an ancient bacterium of this sort, it may have become a eukaryotic-like cell that may have been the ancestor of modern plants. (Recall the endosymbiont theory discussed in chapter 4.) Aerobic respiration might have evolved in a similar way, when larger cells engulfed bacteria capable of using oxygen (fig. 8.16).

After endosymbiosis, different types of complex cells probably diverged, leading to evolution of a great variety of eukaryotic life-forms. Today, the interrelationships of the biological reactions of photosynthesis, glycolysis, and aerobic respiration, and the great similarities of these reactions in diverse species, demonstrate a unifying theme of biology: All types of organisms are related at the biochemical level.

Mastering Concepts

1. Which energy pathway is probably the most ancient?
2. Why were the first energy pathways anaerobic?
3. What is the evidence that photosynthesis may have evolved from glycolysis?
4. How did photosynthesis forever alter life on earth?

Life Without Photosynthesis?

continued from page 146.

To plants and the organisms that eat them, leaves are the specialized organs that capture the sun's energy and convert it to usable chemical energy. To U.S. military strategists during the Vietnam war, leaves were a nuisance that provided a cover for the enemy. Their solution was called "area denial"—bombing millions of trees out of existence or spraying them with herbicides that mimicked plant growth hormones and caused them to drop their leaves.

From 1961 to 1971, aircraft sprayed 20 million gallons of herbicide. From 1965 through 1973, other planes dropped a total of 2.8 million tons of bombs, in near-daily doses. The landscape, once lush and green, became gray and lifeless.

The principal target of the defoliation and destruction was the mangrove forest. Mangrove trees have distinctive aboveground roots that screen salt from saltwater and enable these plants to survive where sea meets land. Fallen leaves spread nutrients on the wet land, supporting vast communities of snails, crabs, and insects. Dozens of bird species lived in the treetops.

In 1969, ecologists toured the destroyed region, called the Rung Sat Special Zone, and cried when they saw the damage. They knew that a similar habitat in Costa Rica would house uncountable numbers of animal species; residents of the bombed and sprayed region in southeast Asia were rare. It will be more than a century, ecologists predict, before the mangrove community recovers, if it ever does. Meanwhile, crab populations have grown as these animals consume the mangrove seedlings struggling to survive, cutting off forest regrowth. Peasants have grown rice on the denuded areas, creating breeding grounds for the mosquitoes that carry the malaria parasite—a disease not naturally found in healthy mangrove forests. The once-lush region seems destined to become a wasteland.

SUMMARY

1. Photosynthesis is the light-driven conversion of water and CO_2 to carbohydrate. All life depends on photosynthesis; photosynthesis depends on light.

2. Photons move in waves. The longer the **wavelength,** the less energy per photon. Light occurs in a spectrum of colors representing different wavelengths. Red and blue light are most effective for photosynthesis.

3. Only absorbed light can affect life, and pigment molecules absorb light. **Chlorophyll a** is the primary photosynthetic pigment and occurs in all photosynthetic organisms except bacteria, which use bacteriochlorophyll. **Accessory pigments** absorb wavelengths of light that chlorophyll a cannot absorb, extending the range of wavelengths useful for photosynthesis. They may also protect.

4. Some protists photosynthesize using **thylakoid membranes** encapsulated in **plastids.** Photosynthetic bacteria use unencapsulated membranes. Plants and algae use specialized plastids, the **chloroplasts.**

5. A chloroplast consists of a gelatinous matrix called the **stroma** that contains stacks of thylakoid membranes called **grana.** Pigments aggregate in the grana and absorb light, funneling the energy to reactive chlorophylls that aggregate with proteins to form **reaction centers.**

6. In reaction centers, light energizes electrons, which move along an electron transport system, releasing energy. The energy pumps protons from the stroma into the thylakoid space. The resulting proton gradient powers ATP synthesis as protons flow back into the stroma.

7. In plants, the **light-dependent reactions** oxidize water, produce ATP, and reduce $NADP^+$ to NADPH. This happens along two **photosystems,** each consisting of a photon-harvesting pigment molecule complex and an electron transport system. The two photosystems are linked, and electrons flow from one to the other.

8. The **light-independent reactions** use ATP and NADPH to fix carbon into organic compounds that can form nutrients. In the Calvin cycle, **rubisco** catalyzes the reaction of CO_2 with **ribulose bisphosphate** (RuBP) to yield two molecules of PGA. ATP and NADPH power the conversion of PGA to PGAL, the immediate carbohydrate product (**photosynthate**) of photosynthesis. This is C_3 photosynthesis.

9. Photon energy produces very little carbohydrate. **Photorespiration** contributes to this inefficiency by fixing oxygen instead of carbon from CO_2. This happens when **stomata** close to conserve water, cutting off the supply of atmospheric CO_2. Photorespiration wastes CO_2.

10. C_4 plants use **C_4 photosynthesis** to resist photorespiration. An additional pathway recycles CO_2 to the Calvin cycle and enables these plants to continue photosynthesis even when little atmospheric CO_2 enters their cells.

11. In **CAM photosynthesis,** certain plants that live in hot, dry habitats open their stomata and take in CO_2 at night and store it as malic acid in vacuoles. During the day, they split off CO_2 and fix it in chloroplasts in the same cells storing the malic acid.

12. The energy pathways are interrelated, with common intermediates and some reactions that mirror the reactions of the others. Glycolysis may be the oldest energy pathway because it is more prevalent—the other pathways are more specialized. The pathways may have arisen when larger cells engulfed prokaryotes that were forerunners to mitochondria and chloroplasts.

TO REVIEW . . .

1. What roles do the following molecules play in photosynthesis in plants?

- **a.** ATP
- **b.** NADPH
- **c.** CO_2
- **d.** H_2O
- **e.** ribulose bisphosphate (RuBP)
- **f.** PGAL
- **g.** glucose
- **h.** chlorophyll a

2. Define these terms and arrange them in size order, from smallest to largest:

- **a.** thylakoid membrane
- **b.** chloroplast
- **c.** reaction center
- **d.** photosystem
- **e.** electron transport system

3. List the reactants and products of the light-dependent and light-independent reactions of photosynthesis.

4. What was a probable CO_2 source for the earliest photosynthesizing organisms?

5. What evidence indicates that white light consists of several colors?

6. Why are plants green? Answer this question both in terms of light and in terms of the relevant organelle.

7. Accessory pigments vary in color and chemical structure, yet they share a common function in photosynthesis. What is it?

8. List three similarities between photophosphorylation and oxidative phosphorylation.

9. Why are two photosystems better than one?

10. How were isotopes used to show
   a. that the oxygen produced from photosynthesis comes from water?
   b. the intermediates of the Calvin cycle?
   c. the steps of C_4 photosynthesis?
   d. the steps of CAM photosynthesis?

11. How does photorespiration counter photosynthesis?

12. Describe how Priestley's experiment consisting of a jar, a leaf, and a mouse followed the scientific method.

13. What did Ingenhousz's experiments demonstrate about photosynthesis? How did he employ the scientific method?

14. The light-independent reactions of photosynthesis were once called the dark reactions. Why is the older term not quite accurate?

15. How is CAM photosynthesis adaptive to survival in a particular environment?

TO THINK ABOUT . . .

1. The National Aeronautics and Space Administration (NASA) operates a program to establish "controlled ecological life-support systems," also known as space farms, to be used aboard spacecraft. These closed life-support systems are based on plants, using light as the only energy input. Use your knowledge of photosynthesis to explain how a space farm might operate.

2. Photosynthesis takes place in plants, algae, and some microbes. How does it affect a meat-eating animal?

3. What color would plants be if they absorbed all wavelengths of visible light? Why?

4. The jellyfish *Aequorea victoria* undergoes a bioluminescence reaction that causes it to glow green. However, the protein that causes the color emission absorbs blue light. What property of light does this type of bioluminescence illustrate?

5. How might you measure the efficiency of photosynthesis? Why is it more difficult to measure photorespiration?

6. A plant is exposed to an environment where the proportion of oxygen is increased from the normal value (21%) to 42%. This appears to have no effect on the plant's photosynthetic rate. What kind of plant is this, in terms of the type of photosynthesis that it conducts? How do you know?

7. Suppose rubisco is genetically altered so that it loses its ability to fix oxygen. What effect might this have on corn as compared to rice?

8. A mutant strain of corn lacks carotenoids. Seedlings must be grown in the absence of oxygen, and as photosynthesis begins, the oxygen it generates must be removed for the plants to live. Why must the oxygen be removed?

9. Why must electron transport in photosynthesis be unidirectional?

10. When vegetables and flowers are grown in greenhouses in the winter, their growth rate greatly increases if the CO_2 level is raised to two or three times the level in the natural environment. What is the biological basis for the increased rate of growth?

11. Some photosynthetic bacteria use hydrogen sulfide (H_2S) instead of water in photosynthesis. They produce sulfur instead of oxygen. How do these bacteria demonstrate that the oxygen source in photosynthesis is water rather than CO_2?

12. An alternative scenario for the origin of photosynthesis proposes that the pathway began in archaea living near deep-sea hydrothermal vents. These organisms used H_2S from the vents as a primary energy source but, according to this hypothesis, also had a molecule similar to modern-day bacteriochlorophyll that enabled them to harness the energy in infrared (ir) wavelengths that provide heat at the thermal vents. This second energy source would have given these organisms a survival advantage, and the bacteriochlorophyll persisted, eventually altering to become the chlorophyll seen in plants today.
   a. What evidence discussed in the chapter argues against this hypothesis?
   b. What evidence might you look for today that would support this hypothesis?

13. How would you design an experiment to investigate the types of organisms in which photosynthesis may have originated?

TO LEARN MORE . . .

References and Resources

Barber, Jim. August 3, 1995. Short-circuiting the Z-scheme. *Nature,* vol. 376. Photosynthesis using only photosystem II is possible, as an algal mutant lacking photosystem I demonstrates.

Caldwell, Mark. December 1995. The amazing all-natural light machine. *Discover.* How molecules join in bacteria to harness solar energy.

Gest, Howard. Winter 1991. Sunbeams, cucumbers, and purple bacteria: The discovery of photosynthesis revisited. *Perspectives in Biology and Medicine,* vol. 34, no. 2. The details of photosynthesis are not obvious; some early thinkers developed interesting views of the process.

Lewis, Ricki. September 1996. Photosynthesis the plankton way. *Photonics Spectra.* A precise organization of polypeptides and pigments forms bacterial photosystems.

Nisbet, Elian, et al. February 9, 1995. Origins of photosynthesis. *Nature,* vol. 373. Several scenarios could account for the origin of photosynthesis.

Pfeiffer, E. W. November 1990. Degreening Vietnam. *Natural History.* Destroying leaves ultimately destroys life.

denotes articles by author Ricki Lewis that may be accessed via *Life*'s home page /http://www.mhhe.com/sciencemath/biology/life/.

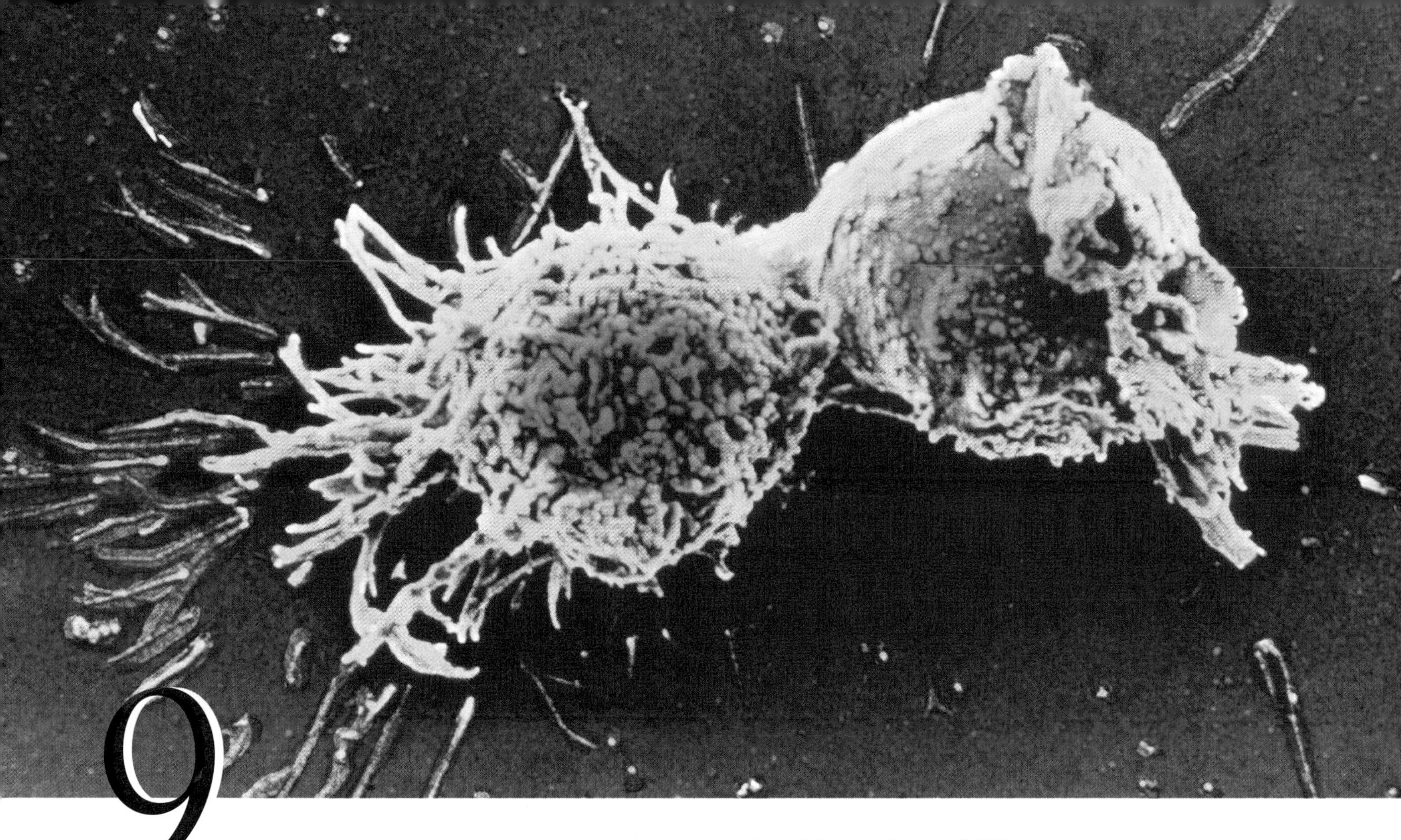

One cell becomes two. (× 2,000)

9 Cell Division and Death

chapter outline

The Balance Between Cell Division and Cell Death 168

The Cell Cycle 168
Interphase
Participants in Mitosis
The Stages of Mitosis
Cytokinesis

Control of Cell Division 175
A Cellular Clock—Telomeres
Cell Division Signals

Apoptosis 177

Cell Division in a Tissue 178
Stem Cells
Cell Populations

Cancer—When the Cell Cycle Goes Awry 180
Characteristics of Cancer Cells

The Causes of Cancer 182
Genes That Cause Cancer
Environmental Influences

The Anatomy of a Sunburn

A generation ago, many people were sun worshippers. Come summer, teens would slather on baby oil, grab a reflecting device to concentrate the rays, and lie out in the midday sun for hours at a time. Our love affair with the sun began with the end of World War II, when people once again had time to be out in the sun.

Although the effects of the sun are more obvious on light-complected individuals, even people with very dark skin can suffer the pain of a sunburn. But sunburn isn't the most dangerous consequence of too much sun. Ultraviolet wavelengths in sunlight can inactivate genes that normally limit the number of times cells divide, setting the stage for the out-of-control cell division that is cancer.

Today, many a former tanned teen lives with the legacy of too much sun, collectively called photoaging—wrinkles, spots, and, for thousands of people each year, skin cancer. But for those of us whose skin was damaged years ago, it may help to know that the human body has a protein that provides two-pronged, built-in protection against sun damage—with the unassuming name p53.

to be continued . . .

The Balance Between Cell Division and Cell Death

Imagine a sculptor creating the form of a pig from clay. The object begins as a mere lump, but the sculptor's hands soon work the pliable material into a head and body region. Then, the sculptor adds clay to some regions, and removes it from others, gradually altering the overall shape so that distinctly piglike features emerge. To add detail or to enlarge the pig, the artist continues to add and subtract clay, maintaining and refining its characteristic shape.

Manipulating clay to sculpt a form is much like the ebb and flow of the number of cells in regions of an animal's body as it grows and develops. The process of cell division adds somatic (nonsex) cells to an animal's body, while a form of genetically programmed cell death, called **apoptosis,** removes cells. Cell division occurs in other multicellular organisms (plants and fungi), but so far apoptosis has only been studied in animals.

Cell division and death must occur in a coordinated fashion to maintain an animal's form (fig. 9.1). Both processes are particularly vital in the embryo and fetus, the two major stages of prenatal (before birth) development, discussed in chapter 11. At these times, some tissues normally overgrow, and then certain cells die. For example, in the fingers and toes of vertebrates (animals with backbones), including humans, apoptosis whittles down the initially weblike hands and feet to distinct, separate fingers and toes (fig. 9.2). After birth, cell division and death protect. Cell division fills in new skin to heal a skinned knee; apoptosis peels away sunburnt skin cells that might otherwise lead to cancer.

Because cell division and death are ongoing, they must maintain a balance between overgrowth and loss of tissue. In this way, a liver in a child is the same shape, although of a different size, as a liver in an adult. Cancer is one consequence of disruption of this balance. Cancer occurs when cell division is too frequent or occurs too many times or when apoptosis is too infrequent. This chapter examines the opposing forces of cell division and death.

The Cell Cycle

A cell can be in the process of dividing; not dividing but alive, well, and specialized; or in the throes of death. A series of events called the **cell cycle** describes whether a cell is actively dividing or not (fig. 9.3). The cycle consists of a sequence of activities when the cell synthesizes or duplicates certain molecules, assembles them into larger structures such as membranes and chromosomes, and apportions its contents into two cells. When one cell divides into two, the resulting progeny cells must receive complete genetic instructions, as well as molecules and organelles to sustain life and possibly also to specialize.

So important is the cell cycle to the form and function of a multicellular organism that several "checkpoints" control it. Checkpoints are actually interacting proteins that ensure that the proper sequence of events unfolds. For example, a cell cannot divide until its genetic material is duplicated, yet it must not actually copy its DNA more than once or the progeny cells will have too much DNA. Checkpoints also ensure that the cell cycle can pause briefly so that errors in the sequence of newly formed DNA molecules can be repaired before they are replicated and therefore perpetuated.

The cell cycle includes three major stages: **mitosis,** or **karyokinesis,** when the nucleus is actively dividing; **cytokinesis,** when other cell contents are distributed into the progeny cells; and **interphase,** when the cell is not dividing. Karyokinesis and cytokinesis can overlap in time, depending upon the species. The next chapter considers another form of cell division that results in formation of sex cells.

Interphase

Biologists in the 1950s and 1960s mistakenly described interphase as a time when the cell is at rest, perhaps because the traditional stains used in microscopy do not bind to the highly unwound interphase DNA. However, interphase is actually a very active time in the life of a cell. Not only do the basic biochemical life functions continue, but the cell also replicates its genetic material, a process discussed in detail in chapter 15. Interphase is divided into phases of gap (designated "G" and sometimes called "growth") and synthesis ("S").

G_1 Phase

During the first gap phase, or **G_1 phase,** the cell carries out the basic functions of life and performs specialized activities. At the same time, G_1 is a critical checkpoint that determines a cell's fate. This checkpoint, called the restriction point, "decides" whether a cell goes on to divide, remains in G_1 as a specialized cell or perhaps to repair damaged DNA, or dies.

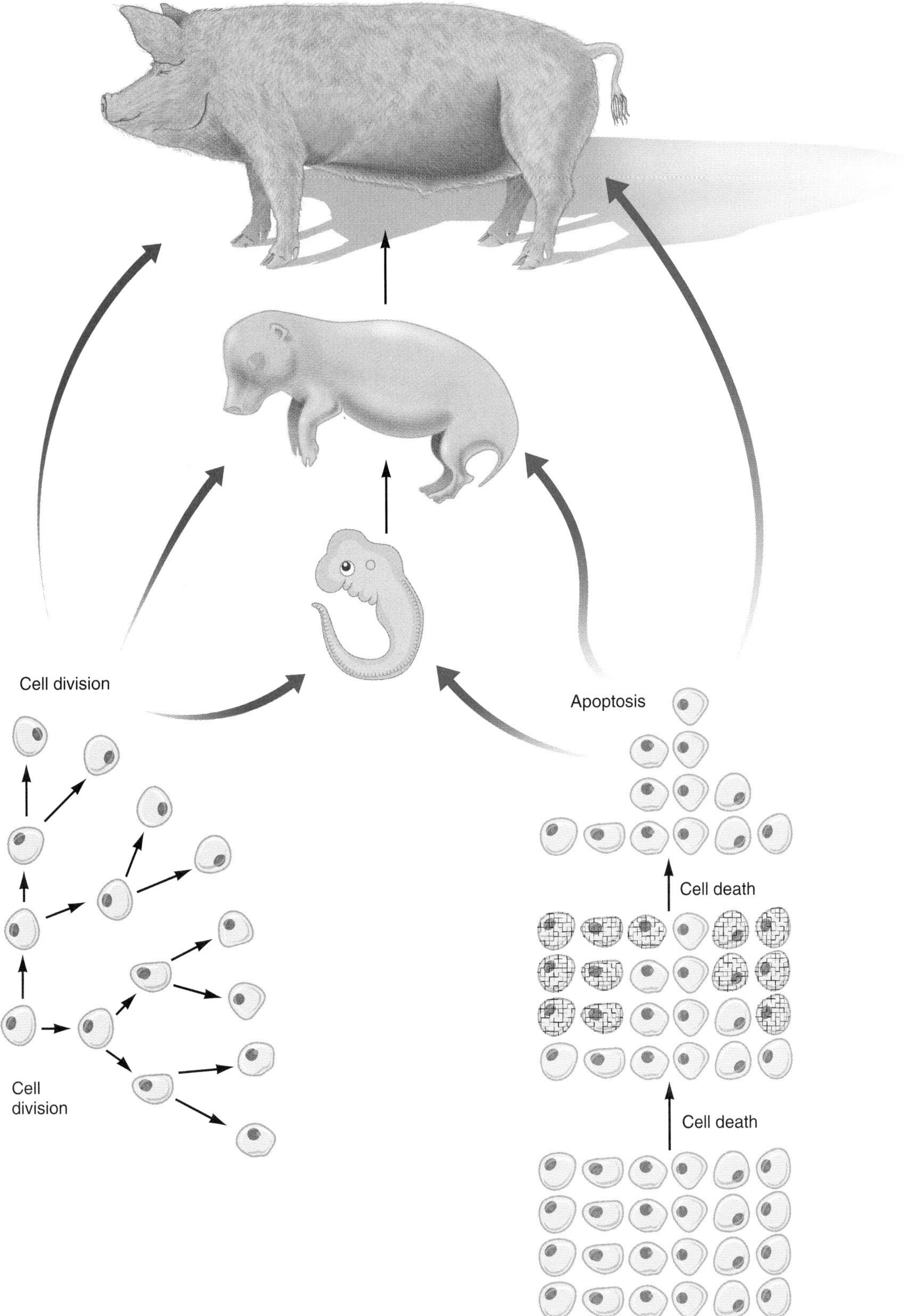

FIGURE 9.1

Biological structures in animal bodies enlarge—and therefore the organisms can grow—because of increases in cell number from cell division. Decreases in cell number from apoptosis, a form of programmed cell death, help form the distinctive shapes of body parts.

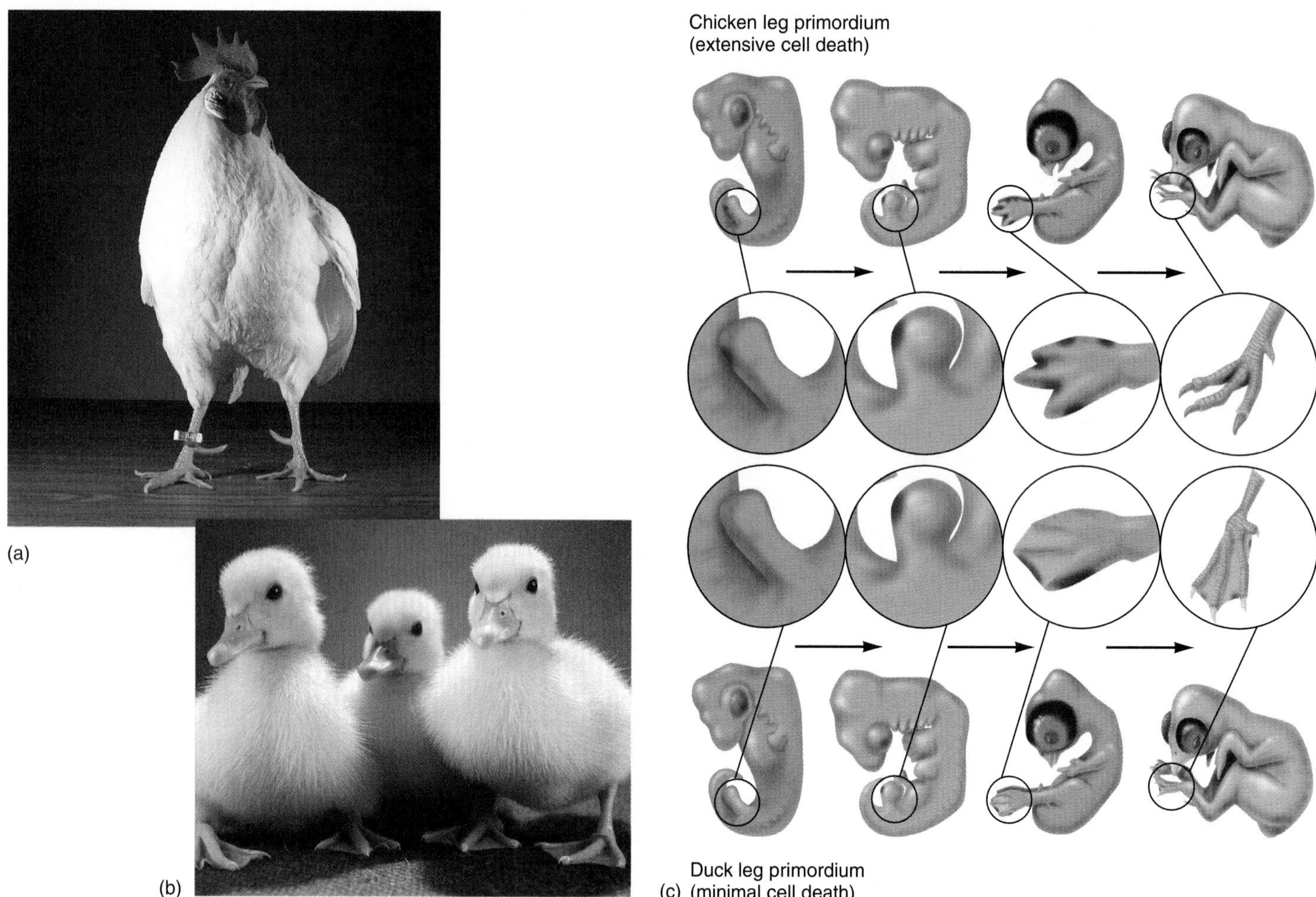

FIGURE 9.2

Programmed cell death carves digits from webbed precursor structures. (*a*) In the chicken, much cell death occurs, and as a result "chicken fingers" are distinctive. (*b*) The webbed foot of the duck reflects less extensive cell death. (*c*) Comparing chicken and duck embryos reveals how different amounts of cell death account for the differences in their feet.

Because G_1 is a decision point in the life of a cell, it is the cell cycle period whose duration varies the most among different types of specialized cells. Slow-growing cells, such as human liver cells, may remain in G_1 for years. Yet human bone marrow cells speed through G_1 in 16 to 24 hours, and early embryonic cells may skip G_1 entirely. Interestingly, the total time spent in the other cell cycle stages is nearly the same in all cells.

G_1 is also a time of preparation for possible impending division. The cell synthesizes proteins, lipids, and carbohydrates. If it divides, the cell will require these molecular supplies to surround the two cells that form from the original cell.

S Phase

The next period of interphase, the **S phase,** is a time of great synthetic activity as the cell initiates the immense job of replicating the genetic material. In most human cells, assembling billions of DNA nucleotides during S phase takes from 8 to 10 hours. Many proteins are manufactured during the S phase, including those that are part of the chromosomes and proteins that coordinate the many events taking place in the nucleus and cytoplasm. The cell also synthesizes more microtubule components (tubulin proteins) during the S phase (see fig. 5.21). During mitosis, microtubules will assemble into a structure called the **mitotic spindle,** which separates the two sets of chromosomes.

G_2 Phase

In the second gap phase, the **G_2 phase,** the cell synthesizes more proteins. Membrane material is assembled and stored as small empty vesicles beneath the cell membrane. The extra membrane will be used to provide enough material to enclose two cells rather than one. The DNA winds more tightly around its associated proteins, and this start of chromosome condensation signals impending mitosis. Interphase has ended.

Participants in Mitosis

During mitosis, genetic material that replicated in the previous S phase divides. The important structures in the process are the replicated chromosomes and the mitotic spindle that separates them into two sets.

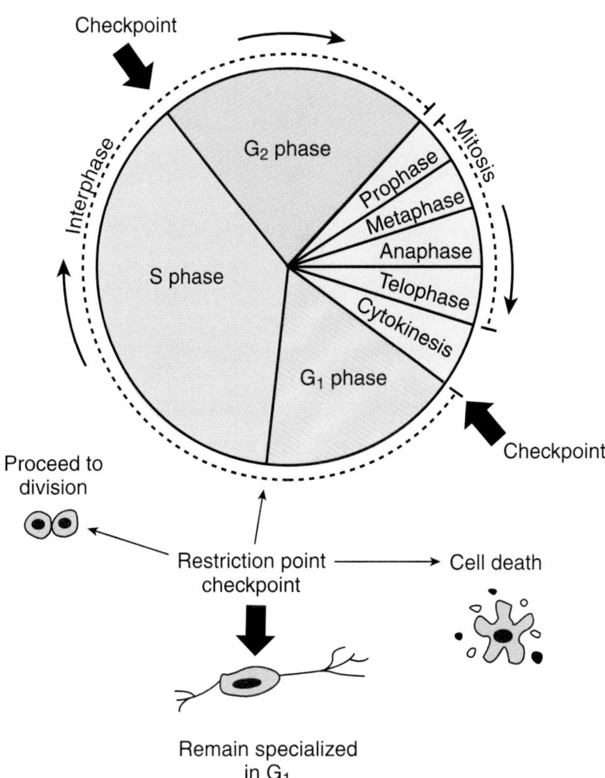

FIGURE 9.3

The cell cycle is divided into interphase, when cellular components replicate, and cell division (karyokinesis or mitosis, and cytokinesis), when the cell splits in two, distributing its contents into two progeny cells. Interphase is divided into two gap phases (G_1 and G_2), when specific molecules and structures duplicate, and a synthesis phase (S), when the genetic material replicates. Mitosis can be described as consisting of stages—prophase, metaphase, anaphase, and telophase. Several checkpoints control the cell cycle. Of particular importance is the checkpoint called the restriction point, at the end of G_1. It determines a cell's fate—whether it will continue in the cell cycle and divide, stay in G_1 as a specialized cell, or die.

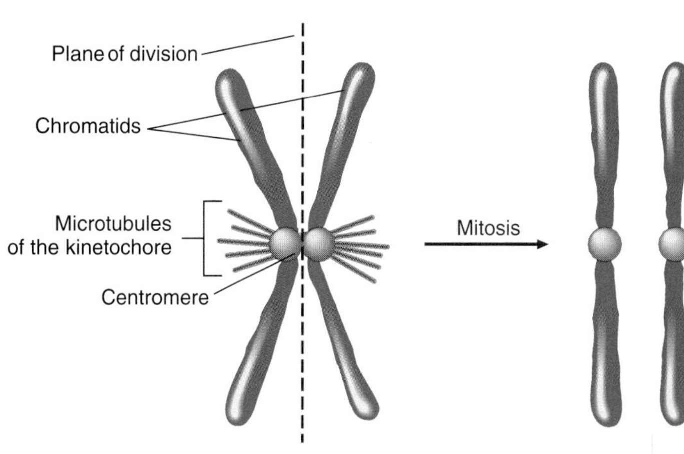

(a) One replicated chromosome consisting of two chromatids

(b) Two unreplicated chromosomes consisting of one chromatid each (progeny chromosomes)

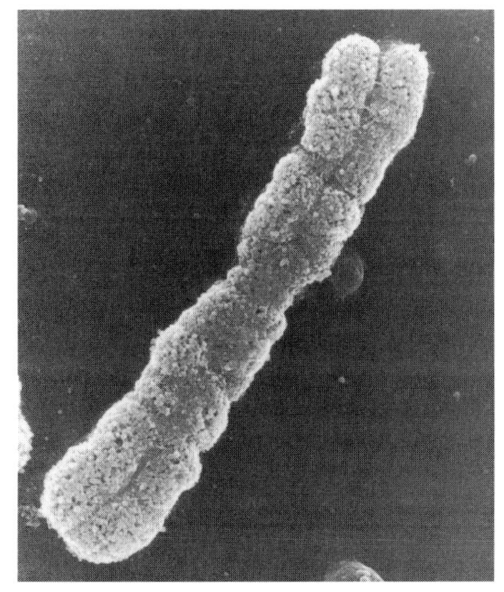

(c)

FIGURE 9.4

Replicated and unreplicated chromosomes. (*a*) Chromosomes are replicated during S phase, before mitosis begins. The two genetically identical chromatids of a replicated chromosome attach at the centromere. (*b*) In anaphase, the centromeres part, and each chromosome consists of only one chromatid. (*c*) A human chromosome is in the midst of forming two chromatids from one. A longitudinal furrow extends from the chromosome tips inward.

Chromosomes and Chromatids

A replicated chromosome consists of two strands of identical chromosomal material, called **chromatids,** joined at a small section of the DNA sequence called the **centromere** (fig. 9.4). The centromere in a replicated chromosome is duplicated too, but it may appear as a constriction. At a certain time during mitosis, the chromatids separate. Thus, chromosomes may consist of one or two chromatids, depending upon the stage of mitosis they are in. The number of chromosomes remains the same in a cell before mitosis and in each of its progeny cells after mitosis.

The Mitotic Spindle

Microtubules carry out the movements of mitosis, attaching chromosomes to the mitotic spindle and pulling apart the chromosome sets. The mitotic spindle forms rapidly from trios of microtubules fused along their length. Nine such trios aggregate and form a structure called a **centriole** (fig. 9.5). A centriole, in turn, is packaged into a structure called a **centrosome.** Short microtubules extending from centrioles are called **astral rays.** On the chromosomes, structures called **kinetochores,** also built of microtubules, attach to the mitotic spindle. Plant cells lack centrioles but have different microtubule-based structures to separate replicated chromosomes.

Table 9.1 reviews some terms introduced so far.

The Stages of Mitosis

Mitosis **(M phase)** is a continuous process but is considered in stages for ease of understanding. The stages are: **prophase, metaphase, anaphase,** and **telophase.** The longer the duration of a particular stage, the more cells are in that stage at any given time, as figure 9.6 shows.

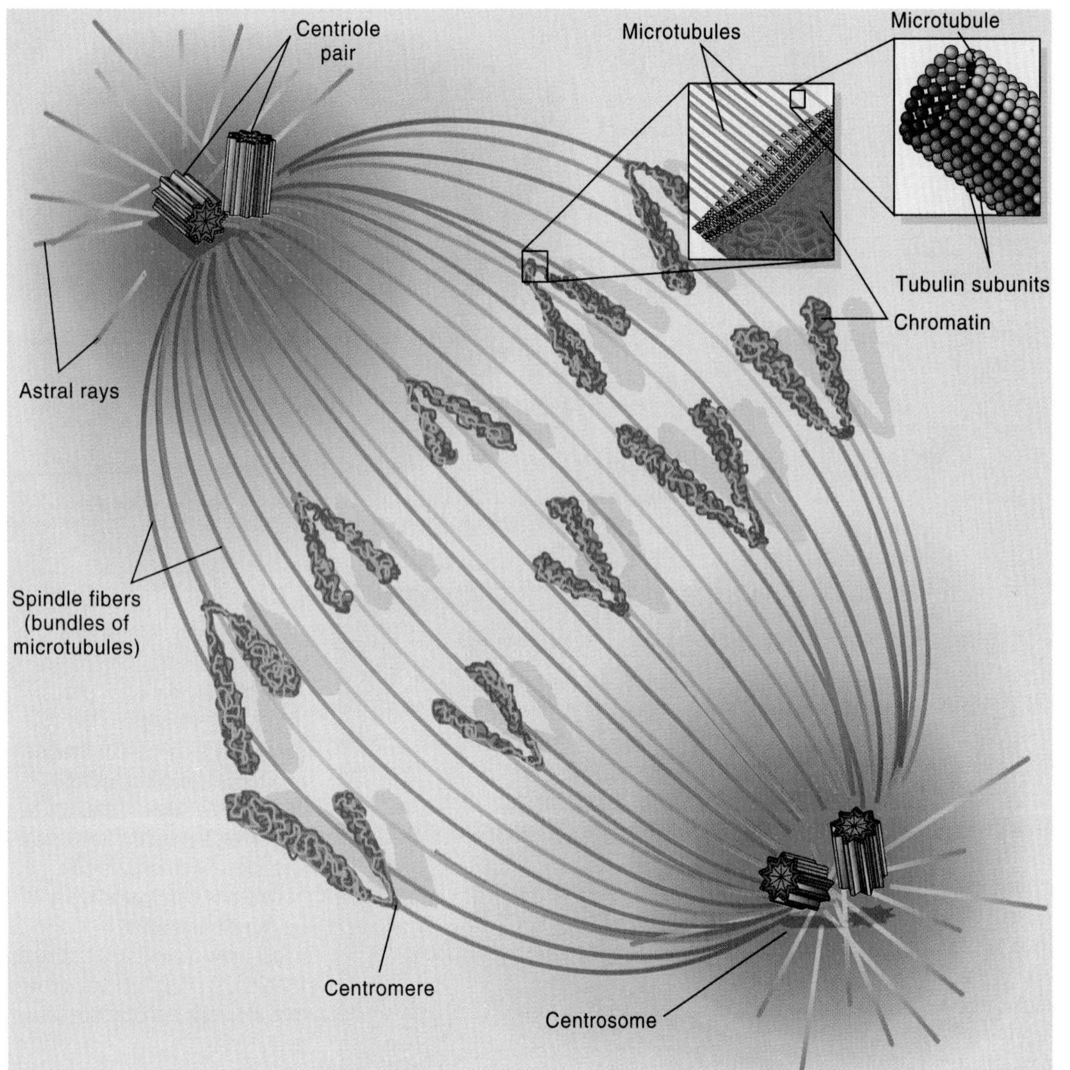

FIGURE 9.5

The mitotic spindle consists of highly organized microtubules that form the fibers that grow outwards from two structures called centrioles. The centrioles, also built of microtubules, occupy opposite ends of the cell. Short microtubules extending from the centrioles are called astral rays. Kinetochores are complexes of microtubules on chromosomes where the chromosomes attach to spindle fibers.

Table 9.1 Miniglossary of Cell Division Terms

Term	Definition
Astral rays	Short microtubules extending from centrioles important in cytokinesis.
Centriole	A structure consisting of nine trios of microtubules that gives rise to the mitotic spindle.
Centromere	A part of a chromosome that joins two chromatids.
Centrosome	A structure that houses a centriole.
Chromatid	A single strand of a chromosome. A replicated chromosome consists of two chromatids; an unreplicated chromosome consists of one chromatid.
Cytokinesis	Distribution of molecules, cytoplasm, and organelles to progeny cells following division of the genetic material.
Karyokinesis	Distribution of replicated genetic material to two progeny cells.
Kinetochore	An aggregation of microtubules on a chromosome where it attaches to the mitotic spindle.

Prophase

During prophase, the first stage of mitosis, DNA coils very tightly around chromosomal proteins, shortening and thickening the chromosomes (fig. 9.7). They are now visible when stained and viewed under a microscope. Chromosomes can separate into two groups more easily when they are condensed.

As prophase begins, microtubules throughout the cell disassemble. At the same time, in animal cells, the centrosome replicates by forming a second centriole at a 90-degree angle to the first. The two centrosomes then migrate towards opposite ends, or poles, of the cell. Microtubules extend from the centrosomes, then back away, and then extend again, like fingers groping for something to hold. When a microtubule touches a centromere, it stops elongating. This is the mitotic spindle forming.

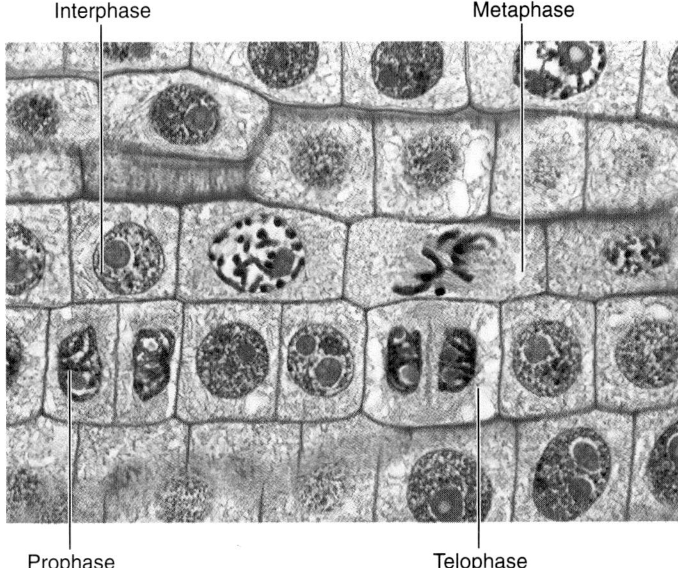

FIGURE 9.6

The cells of an onion root tip reveal stages of the cell cycle. The number of cells in a particular stage reflects the duration of that stage. Interphase takes the longest, and therefore most cells are in interphase.

Table 9.2 The Stages of Mitosis

Stage	Events
Prophase	Chromosomes condense
	Microtubules disassemble, re-form, and approach centromeres
	Mitotic spindle forms
	Nucleolus may become visible
	Nuclear membrane breaks down
Metaphase	Chromosomes attached to mitotic spindle align down equator
Anaphase	Centromeres part, sending one set of chromosomes to each pole
	Cytokinesis begins in some cells
Telophase	Mitotic spindle disassembles
	Nuclear membranes form around two nuclei
	Nucleoli may become invisible
	Cytokinesis distributes molecules and organelles to progeny cells
	Progeny cells separate

During middle prophase in some cells, the nucleolus becomes visible as a darkened area in the nucleus. The nucleolus is the site of ribosome assembly. In other cells the nucleolus is visible at all times. Toward the end of prophase, the nuclear membrane breaks down, and the nucleolus is no longer visible.

Metaphase

As metaphase begins, the mitotic spindle aligns the chromosomes down the center, or equator, of the cell. Because of this alignment, each progeny cell will receive a complete set of genetic information when the centromeres eventually part. That is, each resulting cell will contain one chromatid from each chromatid pair. Metaphase chromosomes are under a great deal of tension. They appear motionless because they are pulled with equal force from both sides, somewhat like the rope in a game of tug-of-war.

Anaphase

In the next stage of mitosis, anaphase, the centromeres, one per chromatid, move apart, relieving the tension and sending one chromatid from each pair to opposite poles of the cell. Anaphase is similar to the tug-of-war rope breaking in the middle and the participants falling into two groups. As the chromatids separate, some microtubules in the spindle shorten and some lengthen in a way that actually moves the poles farther apart, stretching the dividing cell.

Telophase

In telophase, the final stage of mitosis, an animal cell looks like a dumbbell with a set of chromosomes at each end. The mitotic spindle disassembles, and nucleoli and nuclear membranes re-form at each end of the stretched-out cell. Division of the genetic material is now complete. During or just after this karyokinesis, cytokinesis completes as organelles and macromolecules distribute into the two forming progeny cells. Finally, the progeny cells physically separate. They are usually about the same size.

Table 9.2 summarizes the stages of mitosis.

Cytokinesis

Cytokinesis begins during anaphase or telophase. In an animal cell, the first sign of cytokinesis is a slight indentation around the cell at the equator. This indentation, called a **cleavage furrow,** results from a band of microfilaments forming beneath the cell membrane that will contract like a drawstring to separate the progeny cells.

The location of groups of astral rays, called asters, determines the location of the cleavage furrow. Cells with abnormal numbers of asters vividly illustrate the relationship between asters and cleavage furrows. The normal number of asters is two, which specifies a single cleavage furrow dividing the cell into two progeny cells. Cells lacking asters do not develop a furrow and yield one large cell with two nuclei. Cells with three or four asters develop extra cleavage furrows and yield three or four progeny cells, respectively. Extra asters occur when more than one sperm fertilizes an egg, a highly unusual and abnormal situation.

Plant cells must construct a new cell wall that separates the two progeny cells. A structure consisting of microtubules, called a **phragmoplast,** forms between the progeny cells and entraps vesicles containing structural materials (cellulose fibers, other polysaccharides, and proteins) for the cell wall. The first sign of cell wall construction is a line separating the forming cells, called a **cell plate.** It is visible in the cell in telophase in figure 9.7. The layer of cellulose fibers embedded in surrounding material makes a strong and rigid wall that gives plant cells their rectangular shapes.

Usually, cytokinesis completes shortly after karyokinesis (fig. 9.8). In some situations, cytokinesis does not follow

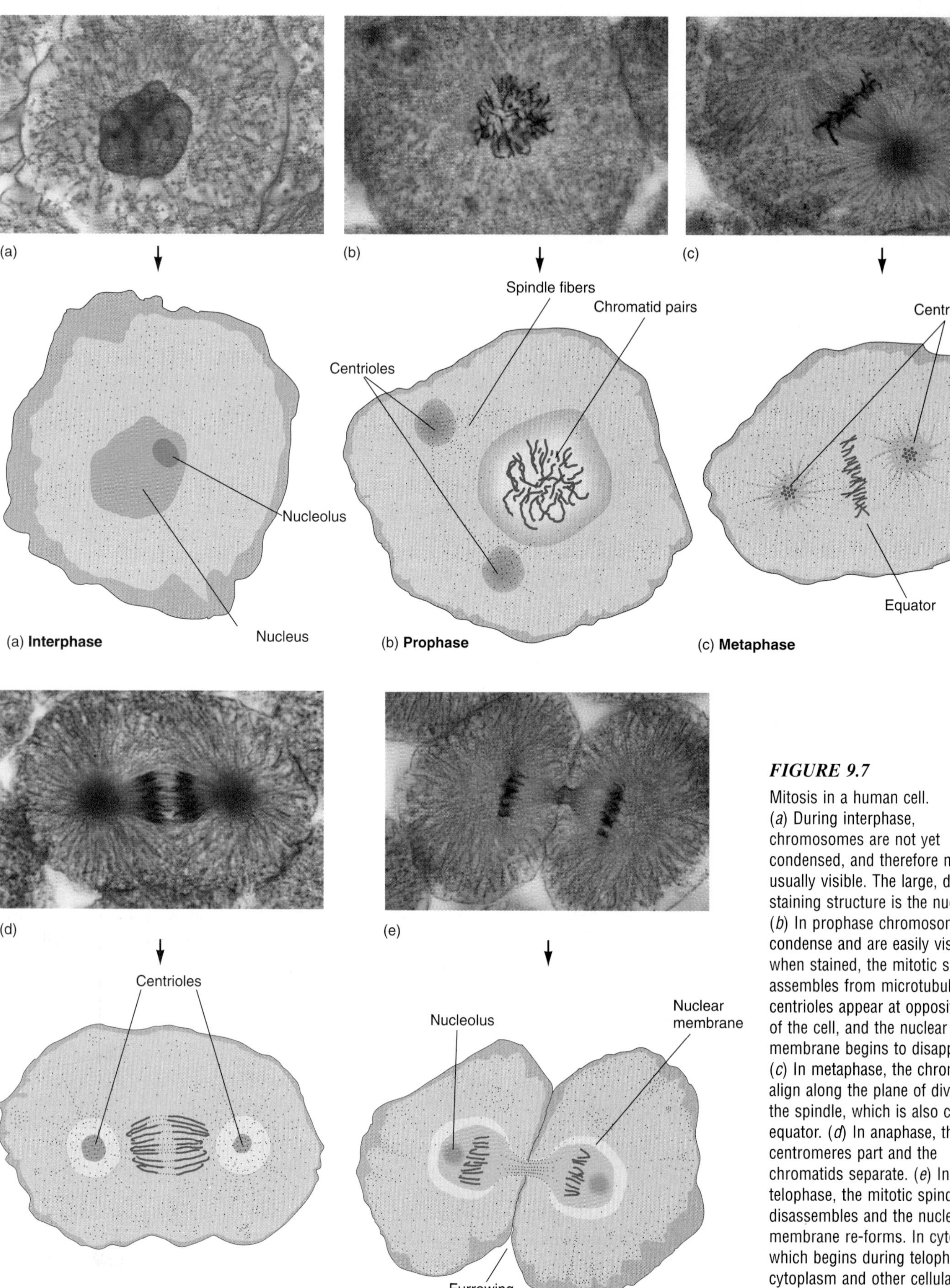

FIGURE 9.7

Mitosis in a human cell. (*a*) During interphase, chromosomes are not yet condensed, and therefore not usually visible. The large, dark-staining structure is the nucleus. (*b*) In prophase chromosomes condense and are easily visible when stained, the mitotic spindle assembles from microtubules, centrioles appear at opposite poles of the cell, and the nuclear membrane begins to disappear. (*c*) In metaphase, the chromosomes align along the plane of division of the spindle, which is also called the equator. (*d*) In anaphase, the centromeres part and the chromatids separate. (*e*) In telophase, the mitotic spindle disassembles and the nuclear membrane re-forms. In cytokinesis, which begins during telophase, the cytoplasm and other cellular structures pinch off into two progeny cells.

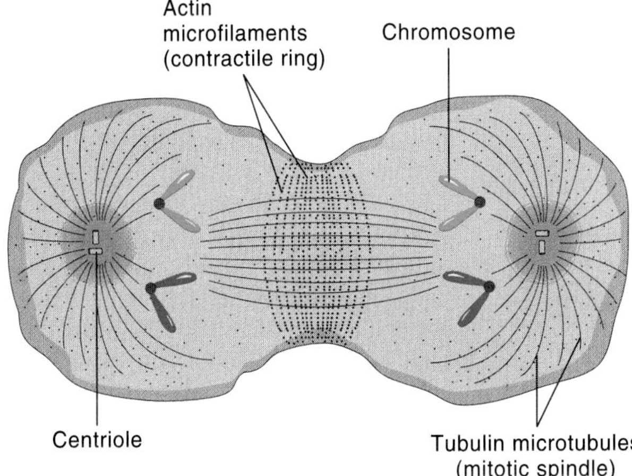

FIGURE 9.8

Karyokinesis and cytokinesis. Division of somatic cells entails duplication and separation of the chromosomes (karyokinesis) and of the cytoplasm and organelles (cytokinesis). The two events are coordinated.

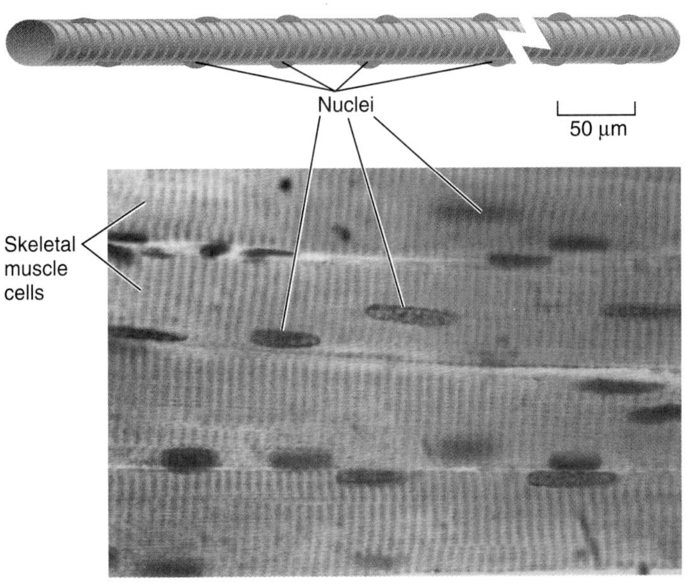

FIGURE 9.9

A skeletal muscle cell has many nuclei sharing cytoplasm. Such a cell may be very long. (× 2,000)

karyokinesis, resulting in a mass of multinucleated cells called a **syncytium.** Human skeletal (voluntary) muscle is a syncytium (fig. 9.9), with many nuclei scattered in the cytoplasm of very long cells. In some plants, endosperm tissue, which nourishes the developing embryo in a seed, consists of cells with many nuclei sharing a common cytoplasm. Some algae, molds, and fungi undergo mitosis without immediate cytokinesis, which produces multinucleated cells extending a meter or more in length.

Mastering Concepts

1. How do cell division and death sculpt the growth of an animal?
2. What events occur during the three stages of interphase?
3. How does the mitotic spindle form?
4. Distinguish between karyokinesis and cytokinesis.
5. How does mitosis maintain a species' characteristic chromosome number?
6. What happens during each stage of mitosis?

Control of Cell Division

What causes a somatic cell to divide or to cease dividing? Mechanisms that control the cell cycle and cell death operate at the molecular, cellular, and tissue levels. Understanding how and when cells respond to signals to divide, remain specialized but not divide, or die may reveal how abnormal states such as cancer arise.

A Cellular Clock—Telomeres

Laboratory-grown mammalian cells obey an internal "clock" that allows them to divide a maximum number of times. This is called the Hayflick limit, after its 1960s discoverer, Leonard Hayflick. A fibroblast (connective tissue cell) taken from a human fetus, for example, divides from 35 to 63 times, with an average of about 50 times. However, a fibroblast from an adult divides only 14 to 29 times. The younger the individual providing the cell, the greater the number of times it can divide. It is as if cells "know" how many times they have divided and how many more divisions they can undergo.

In the early 1990s, cell biologists discovered that the long-sought cellular clock resides in chromosome tips, called **telomeres.** The chromosomes of humans and nearly all vertebrates studied so far end in hundreds to thousands of repeats of a specific six-nucleotide DNA sequence. At each cell division, the telomeres in a cell collectively lose 50 to 200 DNA nucleotides from their ends. This means that chromosomes gradually shorten each time a cell divides. After about 50 divisions, a key amount of lost telo-mere DNA signals cell division to cease. The cell may remain functional but not divide further, or it may die.

More interesting than the many cells whose telomeres tick down are the few cell types whose telomeres do *not* shrink. For example, in humans certain blood cells and small intestinal lining cells divide more than 50 times, presumably because they must constantly renew these active tissues. The cells that give rise to human sperm cells also "ignore" the cellular clock, perhaps because they must continually produce huge numbers of sperm. Cancer cells maintain long telomeres. This may be how they continue to divide far beyond the capacity of the cells they derive from.

The mechanism of telomere shortening is somewhat indirect—cells with shrinking telomeres do *not* manufacture an enzyme called **telomerase.** When cells do make telomerase, their telomeres stay long. Telomerase can continually add DNA to chromosome tips because it includes a small piece of RNA that binds to chromosome ends, serving as a model, or template, for additional DNA nucleotides to add on (fig. 9.10). Telomerase levels

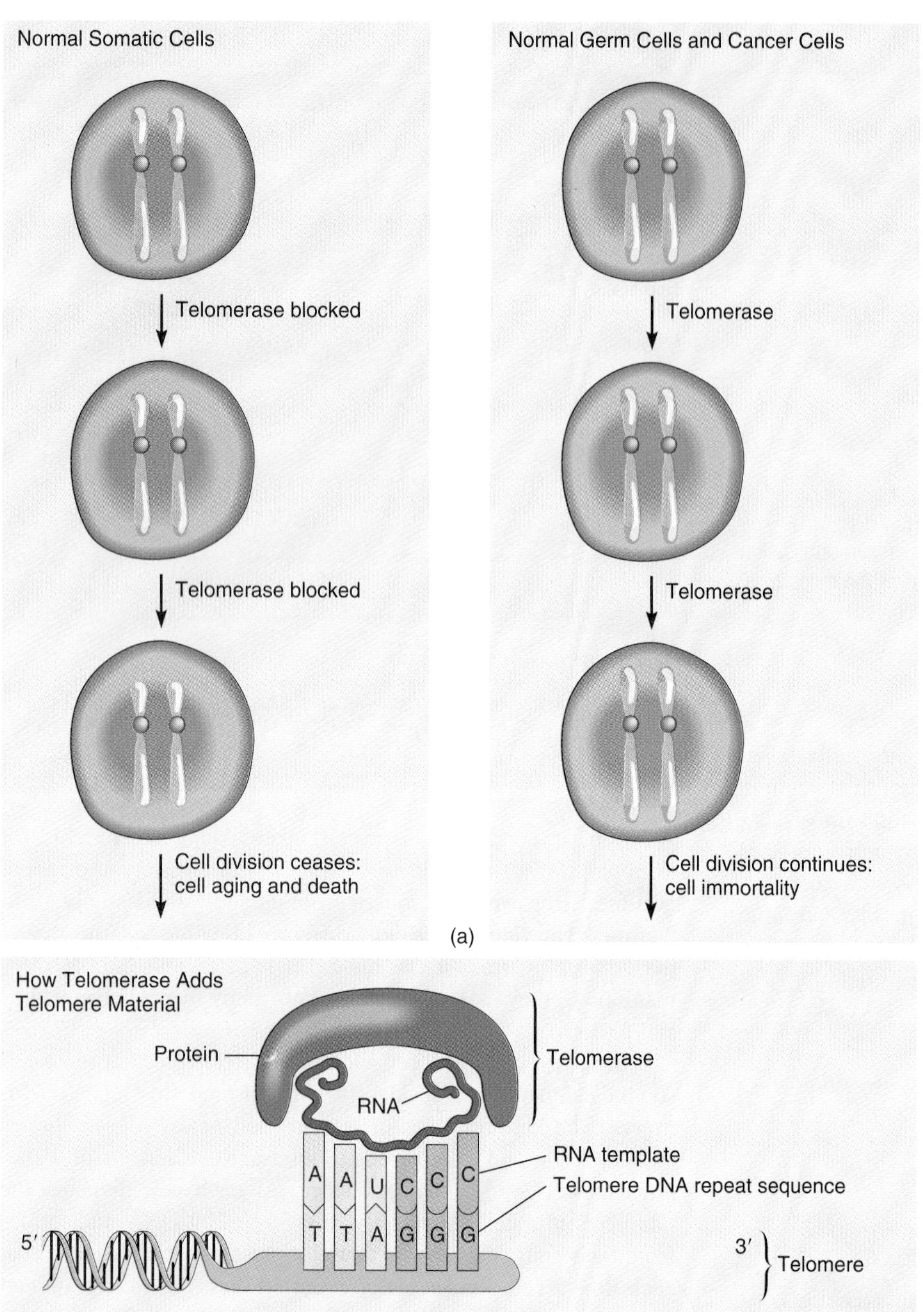

FIGURE 9.10

Telomeres. (*a*) In normal somatic cells in multicellular organisms, telomeres shorten with each cell division because the cells do not produce the enzyme telomerase. When the telomeres shrink to a certain point, the cell no longer divides. Germ cells and cancer cells produce telomerase and continually extend, constantly resetting the cell division clock. (*b*) Telomerase contains RNA, which includes a portion (the nucleotide sequence AAUCCC) that specifies the six-nucleotide DNA sequence TTAGGG, which repeats to form the telomere.

are high in sperm cells, blood cells, and digestive system lining cells. In cancer cells, the level of telomerase rises as the disease progresses. Much research focuses on finding ways to inactivate telomerase in cancer cells.

Cell Division Signals

Signals from outside the cell and from within affect the cell cycle.

Crowding

Crowding can slow or halt cell division. Normal cells growing in culture stop dividing when they form a one-cell-thick layer (a monolayer) lining their container. If the layer tears, remaining cells bordering the tear grow and divide to fill in the gap but stop dividing once it is filled (fig. 9.11). The term **contact inhibition** refers to the inhibiting effect of crowding on cell division.

Extracellular Signals—Hormones and Growth Factors

In an animal, a hormone is manufactured in a gland and travels in the bloodstream to another part of the body, where it exerts a specific effect. Cell division is one such consequence of a hormonal signal, as we saw in figure 2.5 for estrogen. At a certain time in the monthly hormonal cycle in the human female, estrogen levels

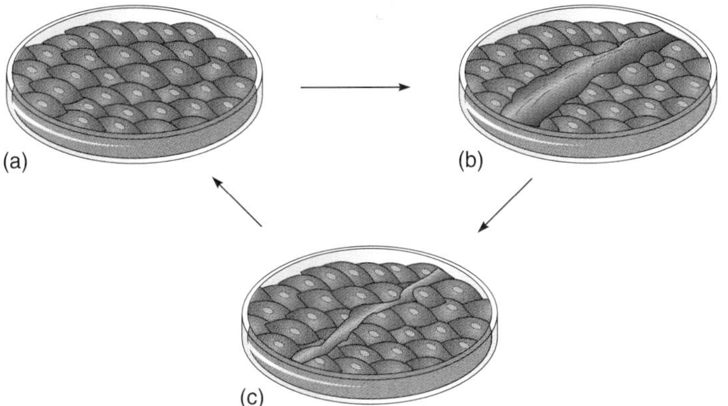

FIGURE 9.11

Contact inhibition. (*a*) Normal cells in culture divide until they line their container in a one-cell-thick sheet (a monolayer). If the monolayer tears (*b*), the cells at the wound site grow and divide, (*c*) filling the gap. However, the cells do not pile up on each other because of contact inhibition.

FIGURE 9.12

Hormones influence plant growth. The California poppy plant on the *right* was treated with the hormone gibberellin.

peak, stimulating the cells lining the uterus to divide, and build tissue in which a fertilized egg can implant. If an egg is not fertilized, another hormonal shift triggers cell death that breaks the lining down, resulting in menstruation. Hormones also control the cell proliferation required to rapidly convert a fatty breast into an active milk-producing gland.

Plant hormones coordinate growth and development, often by stimulating cell division in certain parts of the plant, such as roots, shoots, seeds, fruits, and young leaves. Figure 9.12 shows two plants, the larger of which has been treated with gibberellin, a hormone required for mitosis.

Growth factors are proteins that stimulate local cell division. At a wound site, growth factors increase the rate of cell division, which replaces damaged tissue. For example, epidermal growth factor (EGF) stimulates epithelium (lining tissue) to divide, which fills in new skin underneath a scab. Salivary glands produce EGF, which aids healing when an animal licks its wounds. Swallowed EGF also helps mend digestive system ulcers.

Another growth factor, fibroblast growth factor (FGF), stimulates formation of new blood vessels, which are important in wound healing because the increased blood supply brings oxygen and nutrients to nourish new tissue. FGF stimulates division of cells that line or form blood vessels and of fibroblasts, which secrete collagen, a protein that is part of blood vessels.

Growth factors are useful as drugs. EGF increases cell division rate to heal a torn cornea (the outside layer of the eyeball), speeds healing of skin ulcers, and helps the body accept skin grafts. FGF is used to fill in fibroblasts to heal surgical incisions and to build new blood vessel networks to repair damaged heart tissue.

Intracellular Signals—Kinases and Cyclins

Biochemicals inside cells also stimulate cell division. Pairs of proteins called **kinases** and **cyclins** activate genes whose protein products carry out cell division. Cyclins provide the checkpoint that operates in G_1 to determine whether or not a cell divides.

A kinase is a type of enzyme that activates other proteins by adding phosphates to them. The kinase controlling the cell cycle is present in all eukaryotic cells at all times. In contrast, levels of cyclin fluctuate, as its name implies. At one point in the cell cycle, cyclin levels plummet as if something is rapidly degrading it.

The relationship between kinase and cyclin controls the cell cycle (fig. 9.13). First, kinase molecules bind to cyclin molecules, which have accumulated during the previous interphase. An enzyme activates this kinase-cyclin complex and makes mitosis inevitable and stimulates production of enzymes that break down cyclin. As cyclin levels fall, levels of cyclin-degrading enzymes also drop, and cyclin accumulates again. When enough accumulates that cyclin again combines with the always-present kinase, cell division begins anew. Hence, a negative feedback loop controls the ebb and flow of cyclin.

Mastering Concepts

1. Do most cells divide continually, or reach a certain number of cell divisions and then cease dividing?
2. How does telomere length serve as a cell division clock?
3. How does crowding affect cell division?
4. What biochemicals external to cells affect cell division rate?
5. What biochemicals within cells affect cell division rate?

Apoptosis

The word "apoptosis" (pronounced ape-o-toe-sis) comes from the Greek for a tree shedding leaves no longer needed. Similarly, apoptosis in an animal's body refers to cell death that is part of normal development. One of the most striking illustrations of the precision of apoptosis is in the tiny roundworm *Caenorhabditis elegans.* Exactly 131 of its 1,090 larval cells die to produce the adult as a normal part of the animal's developmental program. We revisit this fascinating worm in chapter 11.

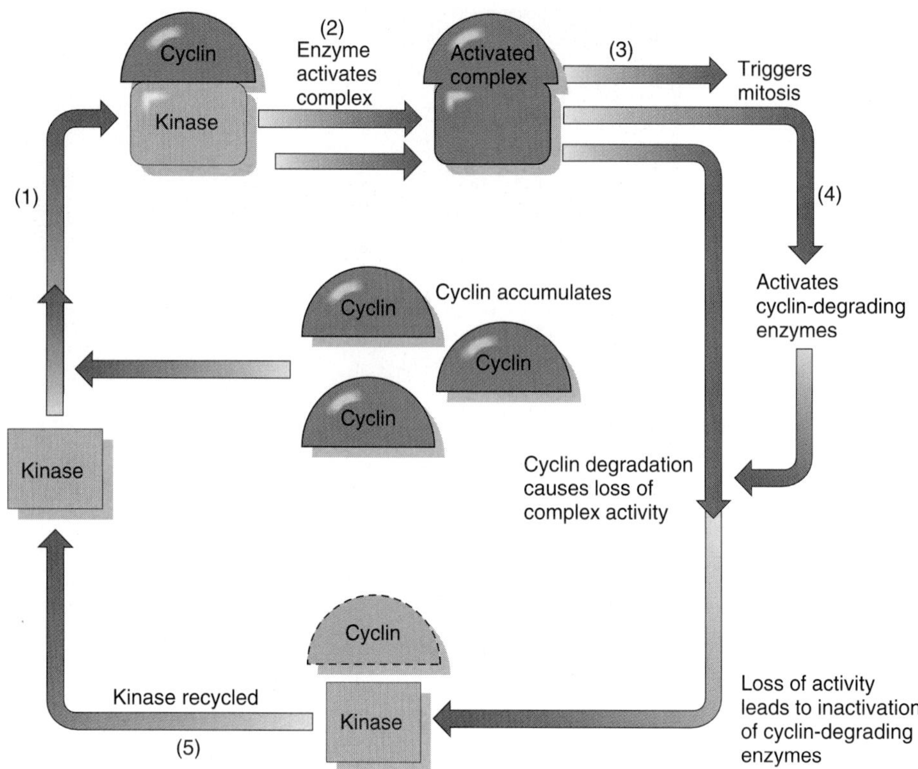

FIGURE 9.13

Cyclins and kinases pair to regulate the cell cycle. Kinase molecules are always present in the cell, and they join cyclin molecules that accumulate during interphase (*1*). Another enzyme activates the cyclin-kinase complexes (*2*). The activated complexes trigger mitosis (*3*) and also stimulate production of enzymes that degrade cyclin (*4*). Once the lack of cyclin limits the number of cyclin-kinase pairs (*5*), the cycle begins anew as more cyclin accumulates during the next interphase.

In the human body, apoptosis destroys certain cells as tissues and organs form. In the brain, for example, specific connections of nerve cells are vital for proper functioning. To establish these links from a large mass of neurons, certain cells release nerve growth factors (NGFs), which attract other neurons. Once a "receiving" neuron sends an extension towards a neuron beckoning it, the growth signal ceases, and no other neurons approach it. Apoptosis eliminates neurons not responding to NGF. This intentional death of nearly half of all brain cells sculpts a highly organized organ from what would otherwise be a meaningless tangle of neurons.

Apoptosis is also essential as the human body builds its immune system. Recall from chapter 5 that the immune system is a vast collection of cells and biochemicals that can distinguish between cell surfaces of the individual's body ("self") and all other cell surfaces. The specificity of immunity begins in the fetus, when the thymus gland produces many immature T cells. These cells recognize an enormous number of different cell surfaces, only a small percentage of which are self. A little-understood mechanism protects those T cells that can recognize self—apoptosis kills all others. This narrowing down of T cells is necessary because in an effective immune response, mature T cells must recognize self and take action when they detect nonself molecules.

The steps of apoptosis are distinctive (fig. 9.14 and 9.15). The cell rounds up and its membrane undulates, forming bulges called blebs. The nuclear membrane breaks down, chromatin condenses, and enzymes cut the chromosomes into many equal-sized pieces. Then the cell fragments. However, remaining pieces of membrane cover toxic cell contents, which prevents them from inducing inflammation. Finally, nearby cells consume the remains of the self-destructing cell. In contrast is **necrosis,** a form of cell death that is not a normal part of development but a response to injury. A necrotic cell swells and bursts, causing great inflammation.

Mastering Concepts

1. How is apoptosis a normal part of development in a general sense?
2. How does apoptosis contribute to development of the human brain and immune system?
3. What happens as apoptosis kills a cell?
4. How does necrosis differ from apoptosis?

Cell Division in a Tissue

Tissues must maintain their specialized natures as well as retain the capacity to generate new cells as the organism grows or repairs tissues damaged by injury or illness.

Stem Cells

To be able to produce new cells while remaining specialized, many tissues contain a few cells, called **stem cells,** that divide often. When a stem cell divides to yield two progeny cells, one remains a stem cell, able to divide again, while the other no longer divides and specializes to perform certain functions. Figure 9.16 shows stem cells in the basal (bottommost) layer of the epidermal skin

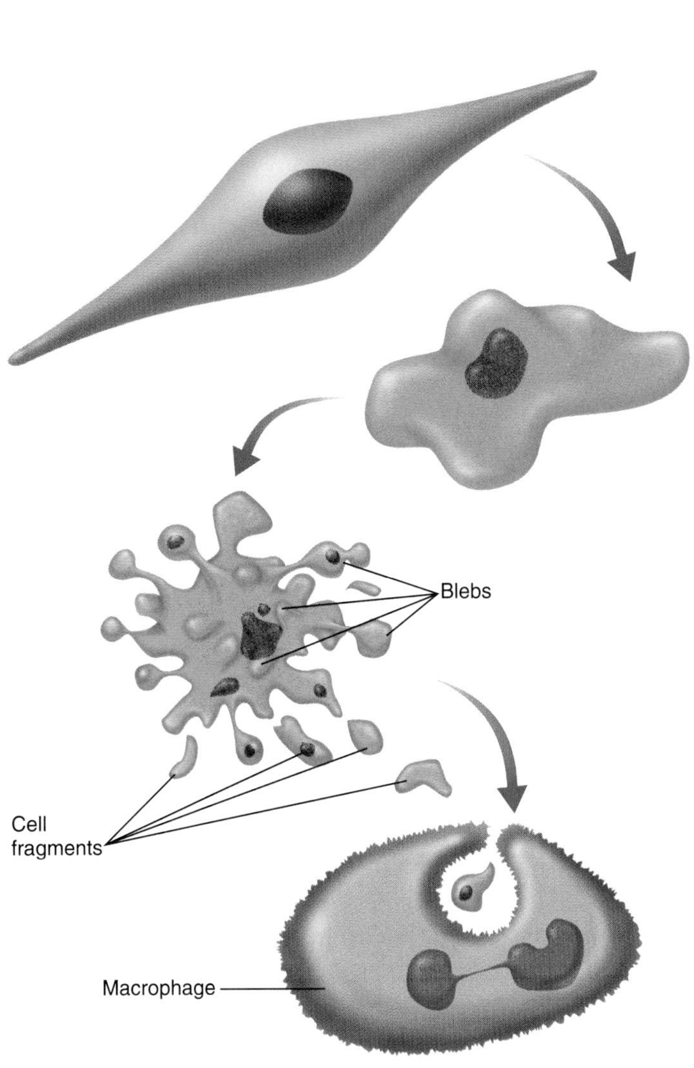

FIGURE 9.14

Death of a cell. A cell undergoing apoptosis loses its characteristic shape, forms blebs, and finally falls apart. Macrophages digest the remains.

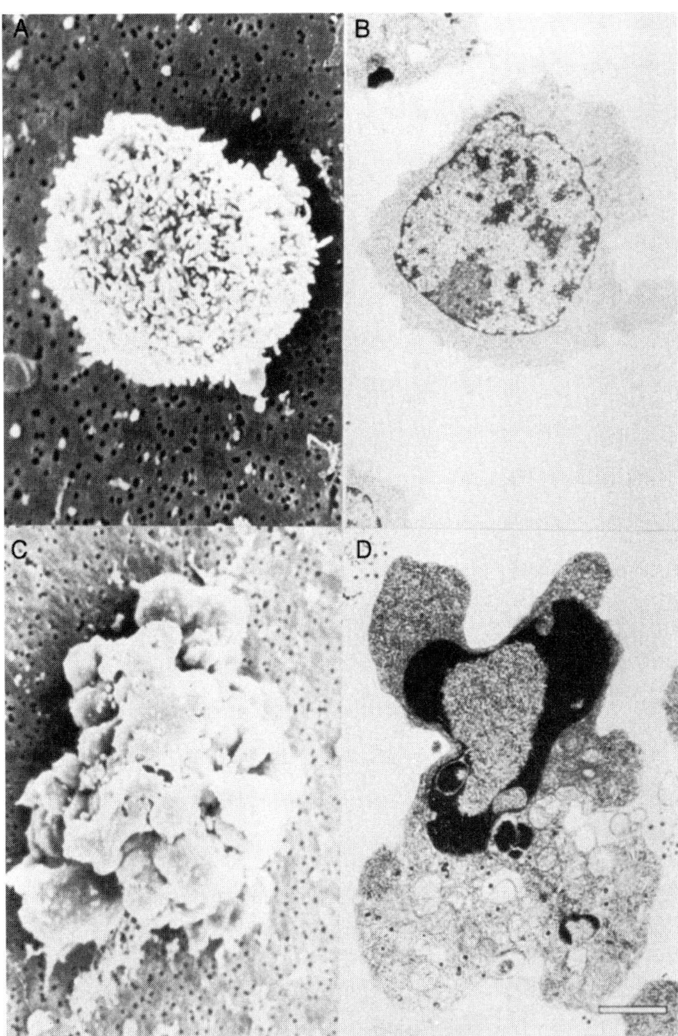

FIGURE 9.15

The dance of death. Apoptosis takes an hour or less. The T cell on top is alive and well, with a rounded form and extensive microvilli (projections) (*a*) and dispersed chromatin (*b*). When given a synthetic hormone that induces apoptosis, the cell forms blebs (*c*) and its chromatin condenses (*d*). (× 10,000)

layer in a human. These basal cells readily divide, but the cells above them, which divided from the basal stem cells, do not. The cells in the upper skin layers die via apoptosis—they flake off after a brisk toweling.

This organization of stem cells pushing more specialized cells upward also occurs in the folds of the small intestinal lining. Rapidly dividing cells, such as those in the skin and small intestine, are the cells that certain drugs and forms of radiation most easily damage. This is why radiation exposure from bombs and cancer therapy, as well as the drugs used in cancer chemotherapy, often cause skin and digestive problems.

Cell Populations

In tissues, cells form groups called **cell populations** based on the percentage of cells in particular stages of the cell cycle. In a **renewal cell population,** the cells are actively dividing. Renewal cell populations in animals maintain linings that are continually shed and rebuilt from dividing cells, such as the cell layers that line the inside of the digestive tract. In the human body, renewal cell populations replace many trillions of cells each day.

In an **expanding cell population,** up to 3% of the cells are dividing. The remaining cells of the expanding population are not actively dividing, but they can divide to repair an injury. The fast-growing tissues of young organisms, as well as adult kidney, liver, pancreas, and bone marrow tissues, consist of expanding cell populations.

Cells that are highly specialized and no longer divide make up **static cell populations.** Nerve and muscle cells form static cell populations and remain in G_1. These cells enlarge rather than divide. A single nerve cell may grow to a meter in length, but within

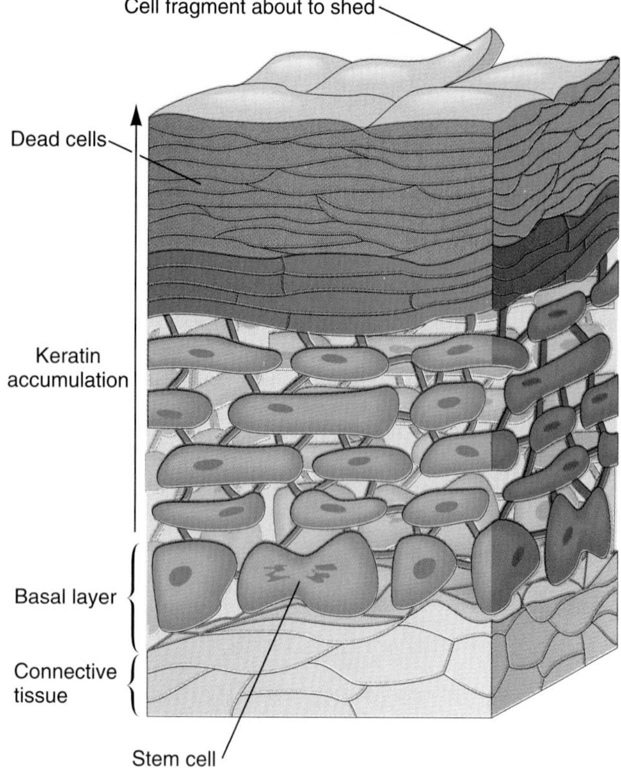

FIGURE 9.16

In some tissues, only cells in certain positions divide. The outer layer of human skin, the epidermis, has actively dividing stem cells in its basal (deepest) layer that push most of their progeny cells upward, yet they maintain a certain number of stem cells. The cells pushed upward accumulate so much of the protein keratin that eventually the nuclei and organelles are squeezed aside and degenerate. The cell remnant is flattened and shed from the skin's outer surface.

an organism it normally does not divide. Cells in static cell populations are not dividing, nor are they dying—they are in a specialized-cell holding pattern.

Mastering Concepts

1. How can a tissue be specialized yet retain the ability of its cells to divide?
2. How are cell populations defined to reflect the percentage of cells that are actively dividing?

Cancer—When the Cell Cycle Goes Awry

One out of three of us will develop cancer. In this group of disorders, certain body cells lose normal control over the rate and number of divisions they undergo, or they may ignore death (apoptosis) signals. Instead, the cell continues to divide, more often or more times than the surrounding cells, and grows into a mass called a **malignant tumor** or moves in the blood. Cancer begins with a single cell that breaks through its death and division controls.

A malignant tumor differs from a **benign tumor,** which is usually rounded, distinct from surrounding tissues, and does not travel to other locations in the body. If a benign tumor remains small, shrinks, or is surgically removed, it poses no health threat. If allowed to remain, it may grow and crowd healthy tissue, eventually causing symptoms and death if it grows large enough. A malignant tumor, in contrast, grows irregularly and sends tentacles in all directions. The word *cancer,* from the Greek for "crab," comes from the resemblance between malignant tumors and crabs (fig. 9.17).

Cancer cells probably arise from time to time in everyone because cell division occurs so frequently that an occasional cell is bound to escape from the mechanisms that normally control the process. In multicellular animals, certain cells, sometimes organized into immune systems, destroy cancer cells. Plants can develop abnormal growths called crown galls that are similar to cancer.

Characteristics of Cancer Cells

Cancer cells differ from normal cells in many ways. Overall, these distinctions give a cancer cell physical characteristics and behaviors like those of embryonic cells.

Division and Growth

Cancer cells can divide uncontrollably and eternally, given sufficient nutrients and space. The cervical cancer cells of a woman named Henrietta Lacks, who died in 1951, vividly illustrate these characteristics. Shortly before she died in 1951, she donated some of her cancer cells to a laboratory at Johns Hopkins University. Lack's cells grew so well in their laboratory homes, dividing so often, that they quickly became a favorite of cell biologists wishing to use cells in culture that would divide beyond the normal 50-division limit. Still used today, "HeLa" (for Henrietta Lacks) cells grow so well that if just a few of them contaminate a culture of different cells, within days they completely take over!

Cancer cells divide more times than the cells from which they derive. They are not necessarily the fastest-dividing cells. Even the fastest-dividing cancer cells, which complete a cell cycle every 18 to 24 hours, do not divide as often as some normal human embryo cells (table 9.3). Therefore, a cancer cell divides faster than the normal cell type it arises from, or it divides continuously at the normal rate. In a sense, a cancer cell is an immature cell that doesn't "know" when to stop dividing.

Some cancers grow at a very fast rate. The smallest detectable fast-growing tumor is half a centimeter in diameter and can contain a billion cells, dividing at a rate that produces a million or so new cells an hour. If 99.9% of a tumor's cells are destroyed, a million would still be left to proliferate. Other cancers are very slow to develop and may not produce symptoms for years. For example, lung cancer may take three to four decades to develop (fig. 9.18). However, any tumor's growth rate is slower at first because fewer cells are dividing. Once a tumor forms, it may grow faster than surrounding tissue simply because a larger proportion of its

(a)

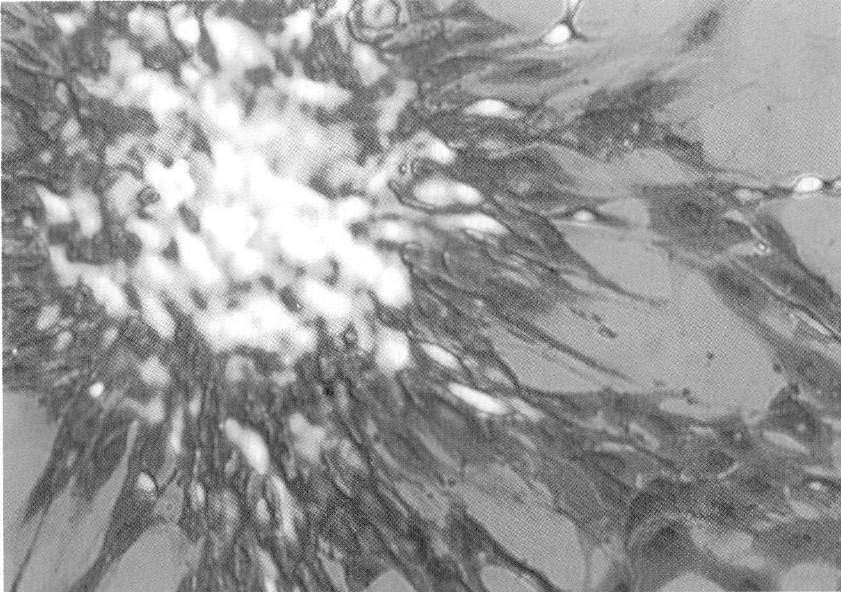

(b)

FIGURE 9.17

"It appears at length with turgid veins shooting out from it, so as to resemble the figure of a crab; or as others say, becomes like a crab; where it has once got, it is scarcely possible to drive it away." So did Hippocrates describe a cancer in the fifth century A.D. (*a*) This poster from the 1950s warns about cancer, using a crab as an image—but cancer is often curable. (*b*) This micrograph shows the crablike shape of a cancerous tumor.

Table 9.3 Time to Complete One Cell Cycle in Some Human Cells

Cell Type	Hours
Normal Cells	
Bone marrow precursor cells	18
Lining cells of large intestine	39
Lining cells of rectum	48
Fertilized ovum	36–60
Cancer Cells	
Stomach	72
Acute myeloblastic leukemia	80–84
Chronic myeloid leukemia	120
Lung (bronchus carcinoma)	196–260

cells are actively dividing. By the time the tumor is the size of a pea—when it is usually detectable—it consists of billions of dividing cells.

Heritability and Transplantability

When a cell becomes cancerous, it passes its loss of cell cycle control to its descendants. Therefore, cancer is said to be heritable. A cancer cell is also transplantable. A cancer cell injected into a healthy animal can spread the disease as the errant cell divides and forms more cancerous cells.

Form and Function

A cancer cell looks different from a normal cell. It is rounder because it is less adhesive than usual and because the cell membrane is unusually fluid. For example, a skin cancer cell is rounder and softer than the flattened, scaly, healthy skin cells above it in the epidermis. Cancer cells respond to crowding differently than do normal cells. They pile up because they lack contact inhibition and produce a tumor (fig. 9.19). Cancer cells often undergo genetic changes (fig. 9.20) and "dedifferentiate," losing the specializations of their parent cells.

Cancerous cells have surface structures that enable them to squeeze into, or invade, any available space. They anchor themselves to tissue boundaries and secrete chemicals that cut paths through healthy tissue. Eventually, cancer cells reach the bloodstream and move to other parts of the body, where they establish new tumors. By this time, the cells may have changed genetically, so that a treatment that shrank the original tumor may have no effect on this subsequent growth. The cells secrete biochemicals in their new locations that stimulate formation of new blood vessels that nourish the rapidly accumulating cells, a process called angiogenesis. The cancer has spread, or **metastasized** (from the Greek for "beyond standing still"). Table 9.4 reviews the characteristics of cancer cells.

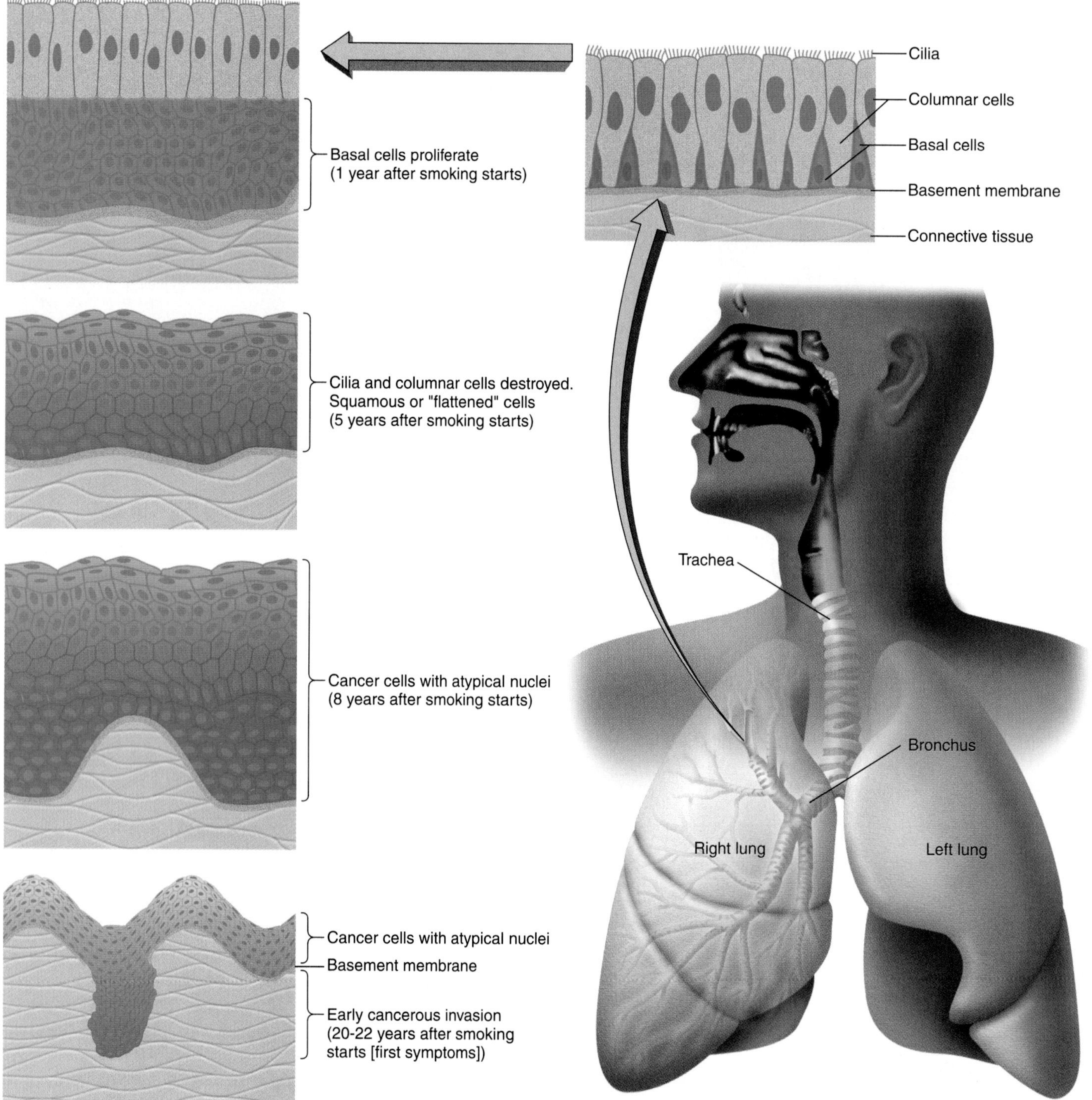

FIGURE 9.18

Some cancers take many years to spread. Lung cancer due to smoking begins with irritation of respiratory tubes. Ciliated cells die (but can be restored if smoking ceases), basal cells proliferate, and then, if the irritation continues, cancerous changes appear.

Mastering Concepts

1. How are the cell cycles of cancer cells abnormal?
2. Distinguish between a benign and a malignant tumor.
3. What are the characteristics of cancer cells?

The Causes of Cancer

Cancer arises from mistimed or misplaced cell division or absence of normal apoptosis (cell death). If cells in an adult liver divide at the rate or to the extent that embryo liver cells divide, the resulting overgrowth may lead to liver cancer. Because proteins such as

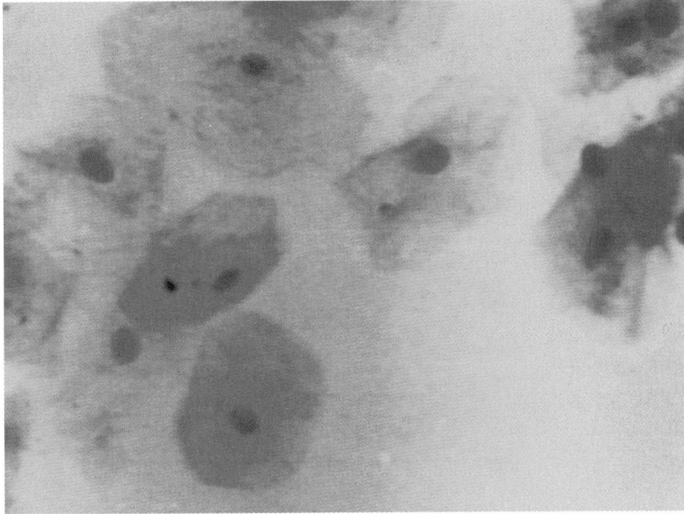

(a)

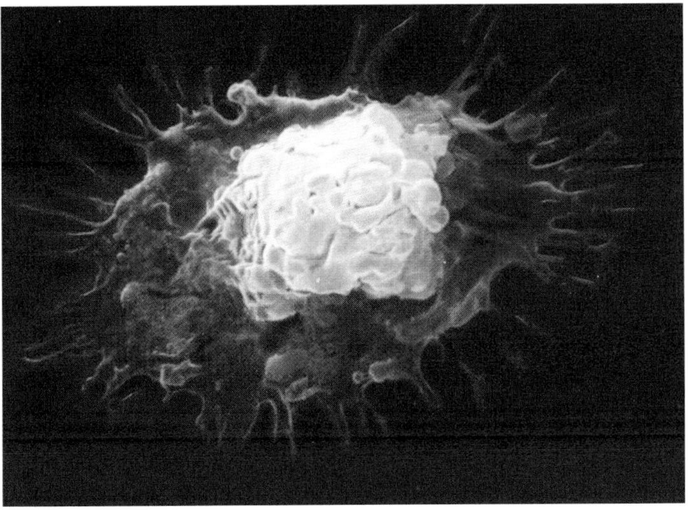

(b)

FIGURE 9.19

Cancer cells look different. (*a*) These cells are not cancerous. They are flattened and grow in neat sheets. (× 400) (*b*) These deranged cells are cancerous. Note their rounded appearance and haphazard growth atop each other. (× 3,000)

Table 9.4 Characteristics of Cancer Cells
Loss of cell cycle control
Heritability
Transplantability
Genetic mutability
Dedifferentiation
Loss of contact inhibition
Ability to induce local blood vessel formation (angiogenesis)
Invasiveness
Ability to spread (metastasis)

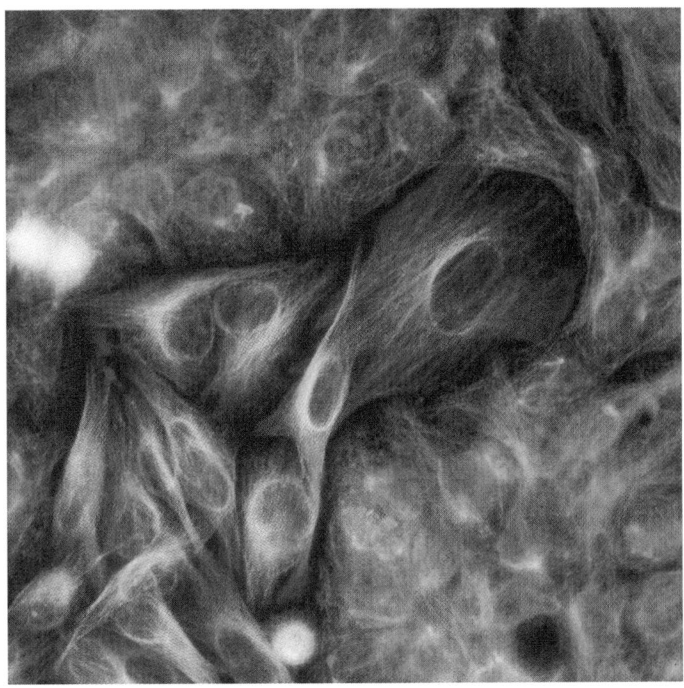

FIGURE 9.20

The orange cells are a melanoma (skin cancer) that is invading normal skin. Cancer cells, when stained for the presence of gene variants characteristic of cancer cells only, look very different from surrounding healthy tissue.

growth factors, kinases, and cyclins control the cell cycle, and because genes hold the information for cells to produce these proteins, genes play a role in causing cancer.

Genes That Cause Cancer

Two major classes of genes contribute to causing cancer. **Oncogenes** are versions of genes that normally trigger cell division, but they are overexpressed and cause the cell cycle acceleration that is cancer. For example, one human oncogene normally active at the site of a wound stimulates production of growth factors, which prompt cell division that heals the damage. When the oncogene is active at a site other than a wound, it still hikes growth factor production and therefore stimulates cell division, but, because there is no damaged tissue to replace, the new cells form a tumor.

In contrast to oncogenes are **tumor suppressor genes,** whose protein products normally prevent a cell from dividing or promote normal cell death. Inactivation or removal of a tumor suppressor gene can cause cancer. For example, a childhood kidney cancer called Wilms tumor is caused by the absence of a gene that normally halts mitosis in the rapidly developing kidney tubules in the fetus. Cell division does not cease on schedule, and the child's kidney retains pockets of cells that divide as frequently as if they were still in the fetus. In the child, these cells form tumors.

FIGURE 9.21

One of the earliest associations of an environmental insult and cancer causation was skin cancer of the scrotum in chimney sweeps.

We know of about 70 different oncogenes and a dozen tumor suppressor genes. Often a cancer results from a series of cancer-promoting genetic changes. For example, some forms of colon cancer occur because of a sequence of genetic abnormalities that include oncogene activation and tumor suppressor inactivation as well as possible environmental influences, such as diet and exercise habits.

Environmental Influences

Oncogenes may be activated or tumor suppressor genes inactivated in response to environmental triggers, such as sun exposure or cigarette smoking. Environmental influences that raise cancer risk include exposure to certain chemicals, radiation, and viruses as well as nutrient deficiencies.

Chemical carcinogens (cancer-causing chemicals) were recognized as long ago as 1761, when nasal cancer was linked to use of snuff, an inhaled form of tobacco. In 1775, British physician Sir Percival Potts suggested that the high rate of skin cancer in the scrotum among chimney sweeps was due to their exposure to a chemical in soot (fig. 9.21). The common cancer became known as "the soot wart". Other associations between cancers and environmental triggers followed. In 1795, physicians noted that lip cancer was more common among pipe smokers, and in 1879, lung cancer was linked to working in silver, cobalt, and uranium mines. It wasn't until 1950 that researchers established the link between lung cancer and cigarette smoking.

Table 9.5 lists several carcinogens, and table 9.6 lists some dietary situations that increase risk of developing specific types of cancer. Finally, Bioethics Connections 9.1 examines an ethical issue that arises in cancer treatment—how much a physician should tell a patient.

Table 9.5 Suspected Carcinogens

	Agent	Cancer Site
Drugs and food additives	Diethylstilbestrol (DES)	Female reproductive organs
	Immune suppressants	Blood
	Nitrites and nitrates in processed meats	Stomach
	Saccharin	Bladder
	Tamoxifen	Uterine lining
	Tobacco	Mouth, throat, lung
Environment	Radon	Lung
	Soot and fly ash	Lung
	Ultraviolet radiation	Skin
	X rays	Skin, thyroid, tonsils, blood, thymus
Industrial	Asbestos	Digestive system, lung
	Chromium, gold, silver, steel	Lung
	Nickel	Nasal cavity, sinuses, lung
	Solvents and dyes	Bladder, blood, urinary tract
	Tars and waxes	Skin
Agricultural	Herbicides (phenoxy type)	Blood
	Insecticides	Immature red blood cells

9.1 bioethics connections

How Much Should a Physician Tell a Cancer Patient?

The outcome of cancer depends upon its location. Thyroid and squamous cell skin cancers are nearly always curable, yet lung cancer is usually not curable. For breast, prostate, and colon cancers, the cure rate lies somewhere in between but is generally high for tumors detected early. Oncologists (physicians who care for patients with cancer) administer treatments, but they also try to dispense hope. Sometimes this can conflict with the desires of the patient. Consider two cases of the same type of cancer that doctors handled very differently.

Actor Michael Landon, known for his roles in the television programs *Bonanza, Little House on the Prairie,* and *Highway to Heaven,* first experienced chronic, severe indigestion early in 1990. To his shock, doctors diagnosed pancreatic cancer in April, and he announced his condition shortly afterwards on a late-night talk show. Landon was told of the rapid and nearly always fatal progression of his type of cancer. He chose to receive traditional and some alternative therapies and had time to take care of personal matters before he died that July.

Miklos Arato, a middle-aged electrical contractor, underwent kidney surgery in July 1980. During the surgery, the physician found a tumor in the man's pancreas and removed it. Afterwards, the surgeon told Mr. Arato that he thought he had removed all of the tumor and referred him to an oncologist. The oncologist told Mr. Arato that pancreatic tumors usually return, recurrences are not treatable, and that he could receive experimental radiation and chemotherapy, although it wasn't known whether these treatments would work. Mr. Arato chose to have the treatments. Neither the surgeon nor the oncologist informed the Arato family that only 5% of people with pancreatic cancer, treated or not, survive 5 years—an important piece of information that would become the basis of a lawsuit.

Mr. Arato's tumor grew back, and he died a year and 4 days after the kidney operation that had originally revealed the pancreatic cancer. The family was shocked and devastated. Mr. Arato, not expecting to die so soon, had not taken care of his business, which went bankrupt. His wife argued that had he known the slim chance of survival even with treatment, he would not have undergone the painful therapy.

Did the physicians act ethically in not telling the Arato family of the limited life expectancy? The physicians argued in their defense that in more than 70 office visits, Mr. Arato had never asked to know his prognosis or life expectancy, although on a questionnaire completed at his initial visit to the oncologist, he had answered "yes" to the question "If you are seriously ill now or in the future, do you want to be told the truth about it?" It wasn't clear, in retrospect, just how much of the truth he wished to know. The doctors also argued that they felt it was in the patient's best interest, in terms of maintaining hope, to not paint such a gloomy picture.

The California Supreme Court ruled in favor of the physicians, the Aratos appealed, a court of appeals reversed the decision, the physicians appealed, and the California Supreme Court upheld the appeals court's decision. That is, the court ruled that the physicians should have disclosed the poor chance of survival to Mr. Arato.

Table 9.6 Cancer and Diet Connections

Dietary Factor	Site of Increased Cancer Risk
Alcohol excess	Mouth, throat, liver
Fat excess	Colon, rectum
Fiber deficiency	Colon, rectum
Folic acid deficiency	Blood, cervix
Obesity	Colon, breast, prostate, gallbladder, ovary, uterus
Riboflavin deficiency	Esophagus
Selenium deficiency	Colon, rectum, breast
Vitamin A deficiency	Lung, bladder, larynx, cervix
Vitamin C deficiency	Stomach, cervix

Mastering Concepts

1. How do oncogenes cause cancer?
2. How do tumor suppressors cause cancer?
3. What environmental factors raise cancer risk?

SUMMARY

1. Cell division and cell death **(apoptosis)** shape organs, select certain cell types to persist in development, and protect tissues by ridding the body of cells that could become cancerous.

2. The **cell cycle** is a sequence of events that describes whether a cell is dividing its genetic material (**mitosis** or **karyokinesis**); dividing its cytoplasm, organelles, and macromolecules **(cytokinesis)**; or preparing to divide **(interphase).** The cell cycle has checkpoints to ensure the appropriate sequence of events.

The Anatomy of a Sunburn

continued from page 168.

Familiarity with the cell cycle is necessary to understand why peeling following a sunburn is protective—and only half the story.

Somehow, the p53 protein "knows" how damaging a sunburn is. If the damage is mild, p53 temporarily stops the cell cycle during G_1. During this pause, the cell can repair DNA damaged by exposure to ultraviolet radiation, possibly correcting genetic changes that could lead the cell on the path to cancer. It is a little like getting off a bicycle to repair a slipped gear, before the bike can no longer be ridden.

If the p53 protein deems a sun-exposed cell irreparably harmed, it avoids cancer by another route—jettisoning the cell by apoptosis. It is a little like junking a ruined bicycle. The burnt skin cell balloons, its chromatin condenses, and the cell membrane bulges out in blebs. The cell dies and peels away.

Pausing the cell cycle to repair DNA is a cellular-level protection from ultraviolet damage. Killing off damaged cells is a tissue-level strategy to maintain health. Clearly, control and coordination of cell division and death are critical to the functioning of a multicellular organism.

3. Interphase includes two gap periods, $\mathbf{G_1}$ and $\mathbf{G_2}$, when the cell makes proteins, carbohydrates, and lipids, and a synthesis period (**S**), when it replicates genetic material. G_1 includes a special checkpoint that sets a cell's fate—to divide, stay specialized but not divide, or die.

4. Microtubules synthesized during S phase assemble to build the **mitotic spindle.** It consists of microtubules that aggregate to form a **centriole,** enclosed in a **centrosome. Astral rays** extend from the centriole. Chromosomes attach to the mitotic spindle with microtubule assemblies called **kinetochores.**

5. Each replicated chromosome consists of two complete sets of genetic information, called **chromatids,** attached at a section of DNA called a **centromere** that also replicates.

6. Mitosis consists of four stages. In **prophase,** the chromosomes condense and become visible when stained, the nuclear membrane disassembles, and the mitotic spindle forms. In **metaphase,** spindle fibers align replicated chromosomes down the cell's equator. In **anaphase,** the chromatids of each replicated chromosome separate, sending a complete set of genetic instructions to each end of the cell. In **telophase,** the spindle breaks down and nuclear membranes form.

7. In cytokinesis in an animal cell, a **cleavage furrow** forms at the site of astral rays, marking formation of a band of microfilaments that will draw the two progeny cells apart. In plant cells, a **phragmoplast** provides space for a new cell wall to be laid down, separating the progeny cells. Cytokinesis usually begins during anaphase or telophase. Lack of cytokinesis results in a huge cell with many nuclei.

8. Human cells usually divide up to 50 times. Shrinking **telomeres** track the number of divisions a cell has undergone, and when telomeres reach a certain length, division ceases. Cancer cells and certain rapidly dividing somatic cells retain long telomeres and divide continually.

9. Crowding prevents normal cells from dividing. Extracellular signals (hormones and growth factors) and intracellular signals (**cyclins** and **kinases**) control cell division rate and number.

10. Apoptosis is a form of programmed cell death. It selects certain cells to die during prenatal development, particularly in the thymus and brain. An apoptotic cell rounds up, the cell membrane forms blebs, the nuclear membrane breaks down, chromatin condenses, and DNA is cut into many equal-sized pieces.

11. Stem cells actively divide, replenishing tissues. Different **cell populations** include specific proportions of cells in different stages of the cell cycle.

12. Cancer can result from excess cell division or deficient apoptosis. A cancer cell divides more often or more times than surrounding cells, has an altered surface, lacks specialization, and divides to yield other cancer cells. A malignant tumor infiltrates nearby tissues and **metastasizes** if it reaches the bloodstream.

13. Cancer can result from an overexpressed **oncogene** or an inactivated **tumor suppressor gene** and may be sensitive to environmental triggers.

TO REVIEW . . .

1. Identify the stage of mitosis the following illustrations depict:

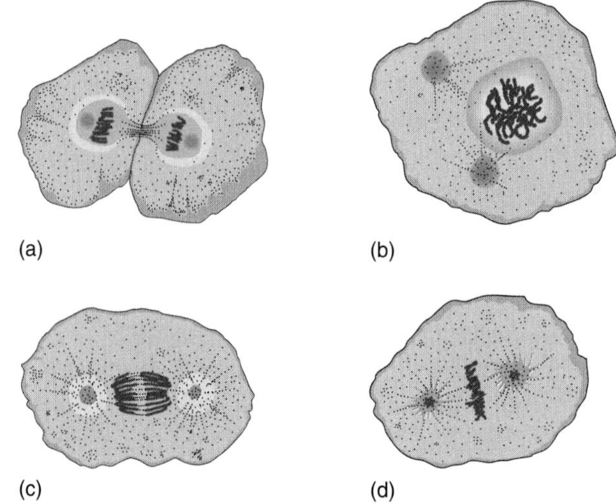

2. Distinguish between the following:
   - a. karyokinesis and cytokinesis
   - b. a replicated chromosome and an unreplicated chromosome
   - c. a centromere and a centrosome
   - d. an oncogene and a tumor suppressor gene
   - e. a cyclin and a kinase
   - f. interphase and mitosis

3. What are the roles of the following proteins in controlling cell division?
   - a. growth factors
   - b. tubulin
   - c. telomerase
   - d. cyclin
   - e. kinase

4. What might be the consequence of blocked cell division? Blocked apoptosis?

5. Describe the events that take place during mitosis.

6. Why is G_1 a crucial time in the life of a cell?

7. Why isn't interphase a time of cellular rest, as biologists once thought?

8. Give an example of how overly frequent or extensive cell division can harm health. Give an example of how too infrequent or too little cell division can harm health.

9. What is the consequence of karyokinesis without cytokinesis?

10. List four ways that cancer cells differ from normal cells.

11. How do telomeres in cancer cells keep their telomeres long?

12. What is contact inhibition?

13. How do biochemicals from inside and outside the cell control the cell cycle?

14. How can gene activation and gene inactivation each cause cancer?

TO THINK ABOUT . . .

1. A researcher classifies mouse tissue cells according to their stage in the cell cycle. She finds that 3% of the cells are dividing. Of the cells in interphase, 50% are in the G_1 phase, 40% in the S phase, and 10% in the G_2 phase. Based on this information, the investigator concludes that, for the cells in this tissue, the G_1 phase is of the longest duration, followed by the S phase. The cells spend the least amount of time in G_2. What is the basis for this interpretation?

2. A cell from a newborn human divides 19 times in culture and is then frozen for 10 years. Upon thawing, how many times is the cell likely to divide?

3. When the United States dropped atomic bombs on Japan in World War II, many people were not immediately killed but suffered slow, agonizing deaths from radiation sickness. The sudden, massive doses of radiation affected the cells in their bodies that proliferate at the highest rates. What types of cells and tissues were most affected?

4. Cytochalasin B is a drug that blocks cytokinesis by disrupting the microfilaments in the contractile ring. What effect would this drug have on cell division?

5. How might the observation that more advanced cancer cells have higher telomerase activity be developed into a test that could help physicians treat cancer patients?

6. How might the ability to grow cancer cells in laboratory culture help in cancer treatment?

7. A researcher removes a tumor from a mouse and breaks it into cells. He injects each cell into a different mouse. Although all of the mice in the experiment are genetically identical and were raised in the same environment, the animals develop cancers that spread at different rates. Some mice die quickly, some linger, and others recover. What do these results indicate about the cells that made up the original tumor?

8. A young boy had a tumor in his stomach. Because the doctors said the tumor was benign, his parents, thinking it was not dangerous, refused to allow them to surgically remove it. The boy died. Why?

9. Why would cancer developing in a stem cell in the basal layer of the skin's epidermis be potentially more harmful than cancer arising in a specialized cell closer to the skin's surface?

TO LEARN MORE . . .

References and Resources

Barinaga, Marcia. January 24, 1997. Designing therapies that target tumor blood vessels. *Science,* vol. 275. Robbing a tumor of its blood supply can kill it.

Glover, David M., Cayetano Gonzalez, and Jordan W. Raff. June 1993. The centrosome. *Scientific American.* A complex interplay between microtubules and other cytoskeletal elements helps cells divide.

Greider, Carol, and Elizabeth Blackburn. February 1996. Telomeres, telomerase and cancer. *Scientific American.* The tale of how an obscure pond dweller taught us about telomeres' control of cell division.

Lewis, Ricki. April 1, 1996. OncorMed Inc. Seeks to link bench to the patient's bedside. *Genetic Engineering News.* A test can predict inherited breast cancer—but who wants it?

Lewis, Ricki. February 19, 1996. Telomere findings may yield tips for treating cancer, geriatric disorders. *The Scientist.* Discoveries in telomere biology are rapidly finding clinical applications.

Lewis, Ricki. February 15, 1996. Telomerase level may aid in decisions on cancer therapy. *Genetic Engineering News.* The length of the telomeres of cancer cell chromosomes predicts severity of illness.

Lewis, Ricki. September 1995. Correcting sun damage the biotech way. *Photonics Spectra.* New gene-containing sunblocks boost DNA repair to lessen sun-induced skin damage.

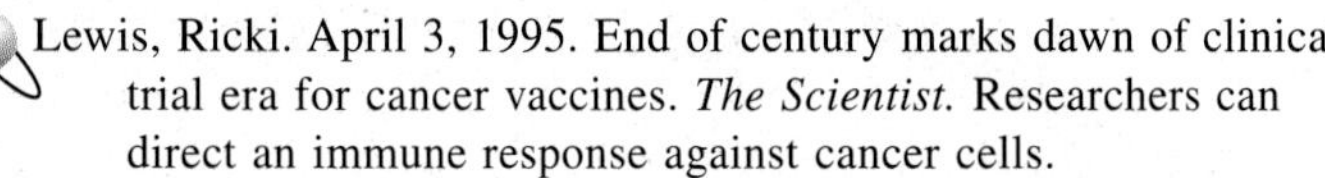
Lewis, Ricki. April 3, 1995. End of century marks dawn of clinical trial era for cancer vaccines. *The Scientist.* Researchers can direct an immune response against cancer cells.

Lewis, Ricki. February 6, 1995. Apoptosis activity: Cell death establishes itself as a lively research field. *The Scientist.* For many years much of life science research focused on cell division. Now cell death is in the limelight.

Lewis, Ricki. April 1994. Interleukin-2: New therapy for kidney cancer. *FDA Consumer.* Borrowing a biochemical from the immune system offers another chance for some people who have cancer.

Nicklas, R. Bruce. January 31, 1997. How cells get the right chromosomes. *Science,* vol. 275. Physical forces control chromosome delivery to progeny cells.

Pelech, Steven. July/August 1990. When cells divide. *The Sciences.* The history of our knowledge of the mitotic process and current thoughts on its pacemaker.

Strauss, Michael, Jiri Lukas, and Jiri Bartek. December 1995. Unrestricted cell cycling and cancer. *Nature Medicine,* vol. 1. A checkpoint in G_1 determines whether a cell will divide, remain in G_1, or die.

Theriot, Julie A. and Lisa L. Satterwhite. January 30, 1997. New wrinkles in cytokinesis. *Nature,* vol. 385. A new look at cytokinesis.

The December 6, 1996 issue of *Science* magazine has several articles on the cell cycle.

denotes articles by author Ricki Lewis that may be accessed via *Life*'s home page /http://www.mhhe.com/sciencemath/biology/life/.

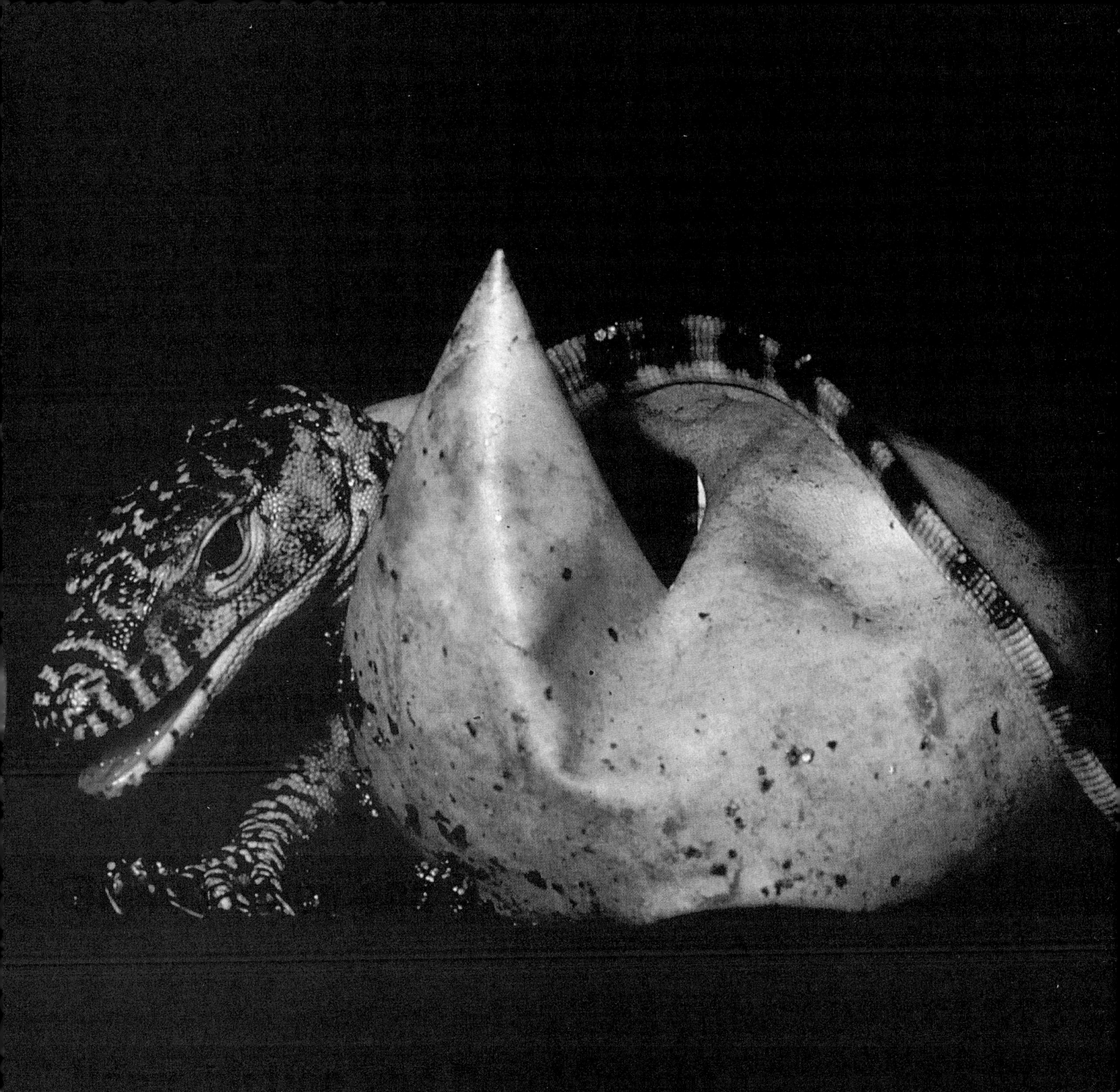

A Komodo dragon begins life.

Unit 3

Reproduction and Development

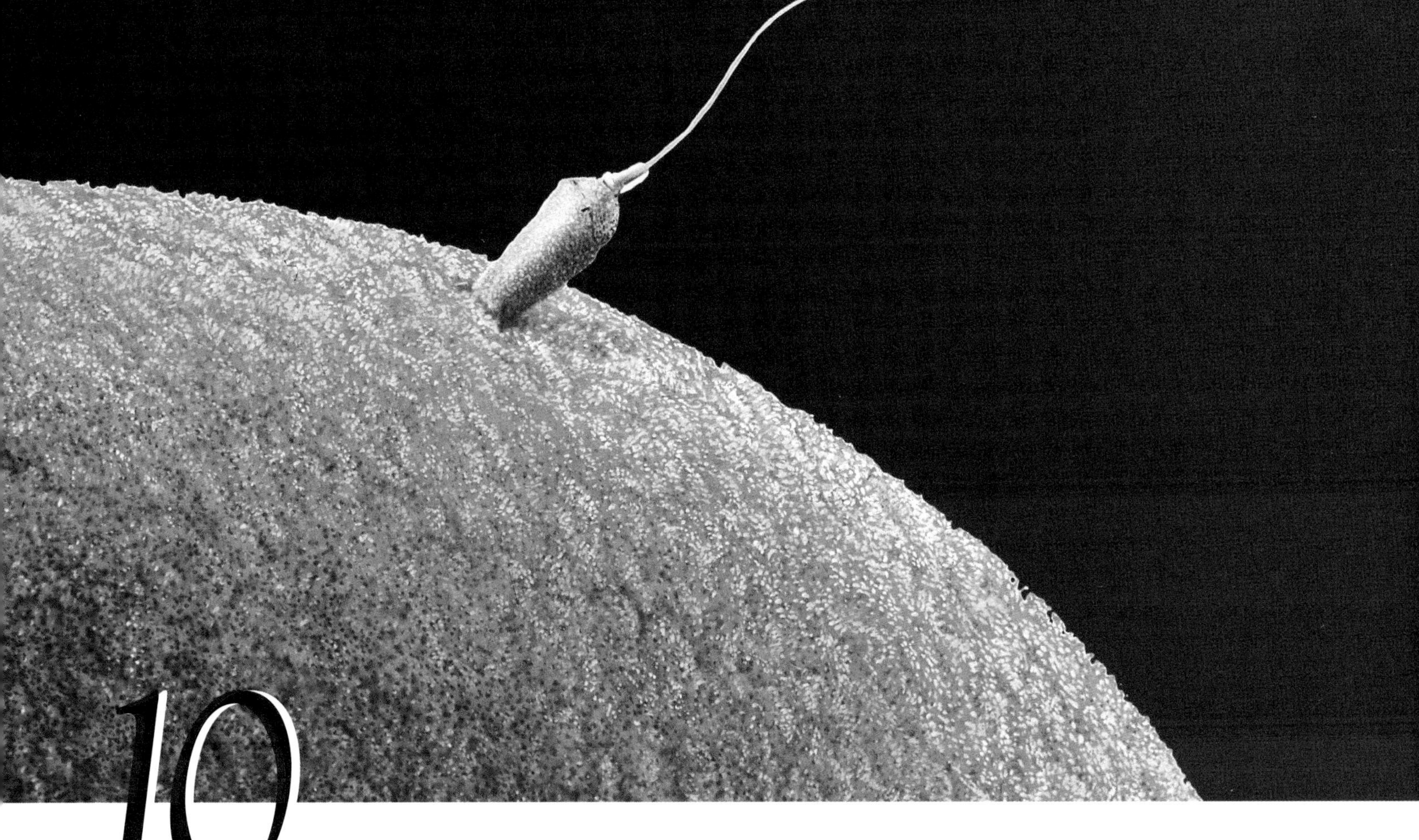

10 The Making of Gametes

Introducing the gametes: sperm and oocyte.

chapter outline

***Why Reproduce?* 192**

***Reproductive Diversity* 192**
Conjugation—Sex Without Reproduction
Cell Fusion—Sex with Reproduction

***Gamete Formation in the Human Body* 196**
The Human Male Reproductive System
The Human Female Reproductive System

***Meiosis* 198**

***Gamete Maturation* 201**
Sperm Development
Ovum Development

An Unusual Child

Medical journals refer to him only as "F. D.," and he is a most unusual child. Shortly after his birth in 1991, his parents noticed that his head was lopsided, with the left side skewed and drooping. Photographs of the boy from one side look completely normal, yet from the other side, he has a jutting chin, sunken eye, and oddly clenched jaw. Not obvious are a cleft palate and other abnormal throat structures. F. D. is relatively healthy, although he is slightly developmentally delayed and has aggressive outbursts.

Because chromosome abnormalities are often associated with unusual facial structures, doctors took a blood sample to examine the child's chromosomes. To their surprise, they found two X chromosomes in the blood cells, indicating a female. But F. D. looked like a boy. What had happened?

Suspecting that F. D. might be an "XX male"—a person with two X chromosomes, one of which bears a gene normally on the Y chromosome that determines maleness—researchers looked at chromosomes in other cell types. If F. D. was an XX male, all of his cells would be XX. But surprisingly, this was not the case. Cells shed in the urine and skin cells were XY, the normal chromosomes of a male. F. D.'s body apparently consists of two distinct cell populations, one XX and one XY.

It would take further studies and an understanding of the nuances of gamete (sex cell) formation to reveal that F. D. was a partial parthenogenote—that is, part of him formed from an egg cell that a sperm had never fertilized.

to be continued . . .

Why Reproduce?

Organisms must reproduce—generate other individuals like themselves—for a species to survive. The amoeba in figure 10.1*a* demonstrates a straightforward way to reproduce. It replicates its genetic material and then splits in two, apportioning the cellular contents of one individual into two. This ancient form of reproduction, called **binary fission,** is still common among single-celled organisms.

In an unchanging environment, the mass production of identical individuals, as in binary fission, makes sense. However, environmental conditions are rarely constant in the real world. If all organisms in a species were well suited to a hot, dry climate, the entire species might perish in a frost. A population of individuals of the same species with a diversity of characteristics is better able to survive in a changeable environment. Binary fission cannot create or maintain this diversity; **sexual reproduction** can. The persistence of sexual reproduction over time and its existence in many diverse species attest to its success.

Sexual reproduction provides genetic variability by producing novel combinations of traits in new individuals. The young organisms in figure 10.1*b* differ from each other because they were conceived by a sexual route. In contrast, the amoebae in figure 10.1*a,* the products of **asexual reproduction,** or reproduction without sex, are identical. Another form of reproduction is **parthenogenesis,** in which an offspring derives solely from one parent, typically the female. Biology in Action 10.1 examines parthenogenetic reproduction in an insect, the aphid. Parthenogenesis is not common.

Sexual reproduction has two essential qualities: it introduces new combinations of genes from different individuals (the parents), and it increases the number of individuals by producing offspring. The cells from the parents that combine to form the first cell of the offspring are called **gametes.** Often, gametes contain only one set of chromosomes and are said to be **haploid,** abbreviated **1*n*.** When two haploid gametes merge, they reconstitute the double, or **diploid (2*n*),** number of chromosome sets in the cells of the progeny.

Reproductive Diversity

Organisms reproduce in many ways. Some species exhibit both sexual and asexual phases. Consider the yeast *Saccharomyces cerevisiae,* which is a single-celled eukaryote that reproduces both asexually and sexually. A single diploid cell of the yeast can replicate its genetic material and "bud," yielding genetically identical diploid progeny cells. This is asexual reproduction, because one parent gives rise to one offspring. Alternatively, yeast can form a structure called an ascus, which gives rise to specialized haploid cells. Two of these haploid cells can fuse, restoring the diploid chromosome number. This is sexual reproduction, because two parents produce an offspring.

In plants, a complete individual life cycle includes both a diploid and a haploid stage. These stages are called generations, and because they alternate within a life cycle, sexually reproducing plants are said to undergo **alternation of generations** (fig. 10.2).

The diploid generation, or **sporophyte,** produces haploid spores through **meiosis,** a form of cell division that halves the number of chromosome sets. Haploid spores divide mitotically to produce a multicellular haploid individual called the **gametophyte.** Eventually, the gametophyte produces haploid gametes—eggs and sperm—which fuse to form a fertilized ovum, or **zygote.** The zygote grows into a sporophyte, and the cycle begins anew. Some plants can also reproduce asexually when a "cutting" from a larger plant grows into a new organism.

Conjugation—Sex Without Reproduction

Learning how diverse organisms reproduce and exchange genetic material today can provide clues to how sexual reproduction may have evolved. The earliest sexual process appeared about 3.5 billion years ago, according to fossil evidence, in a form of

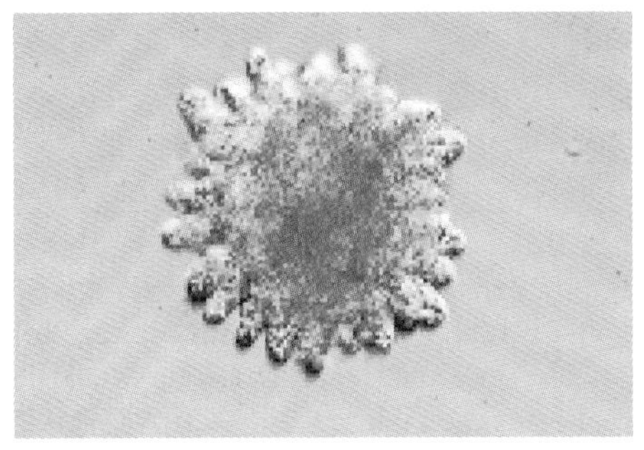

0 minutes

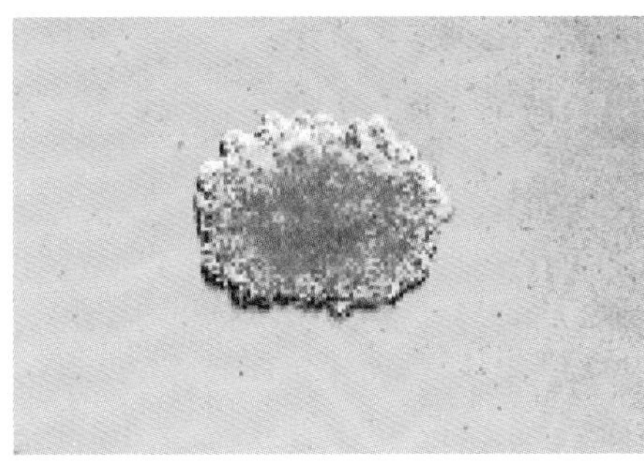

+ 6 minutes

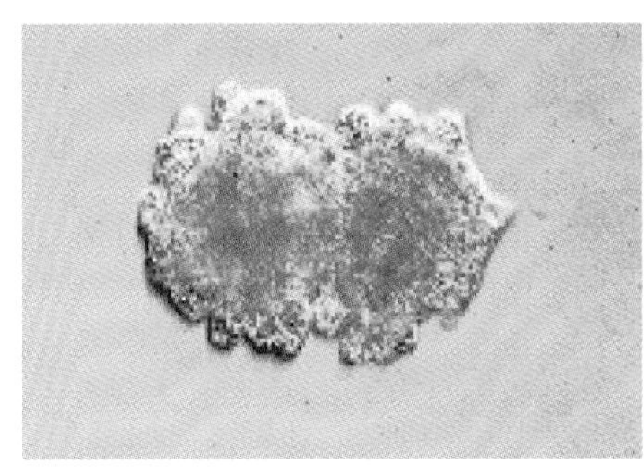

+ 8 minutes

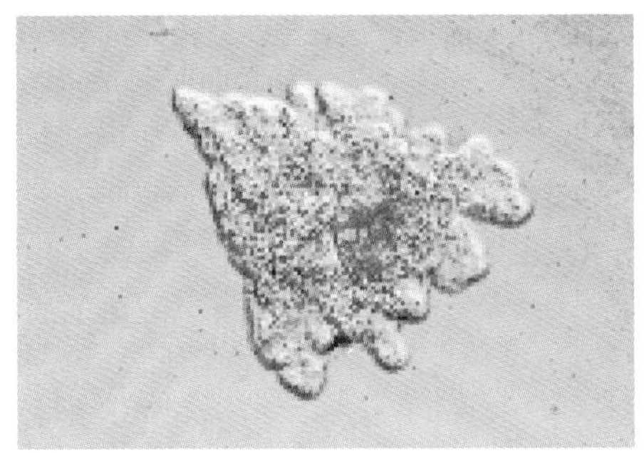

+ 13 minutes

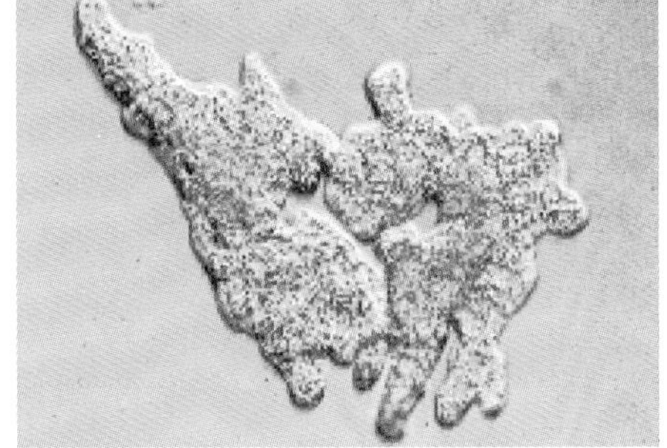

+ 18 minutes

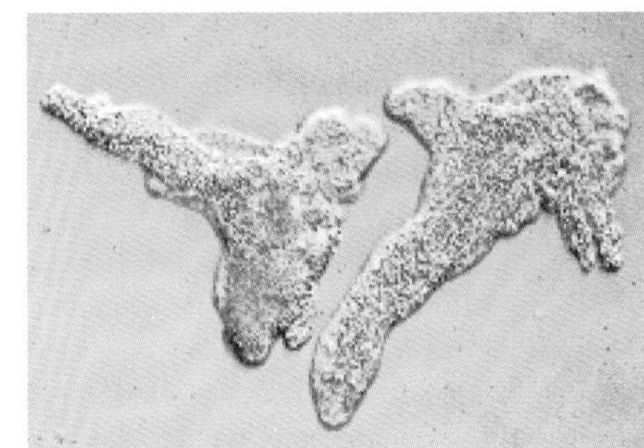

+ 21 minutes

(a)

(b)

FIGURE 10.1

Asexual and sexual reproduction. (*a*) The single-celled *Amoeba proteus* follows an asexual reproductive strategy by splitting in two (binary fission). (*b*) These siblings are obviously not exactly alike. The reason—each receives different combinations of the parents' genes, courtesy of sexual reproduction.

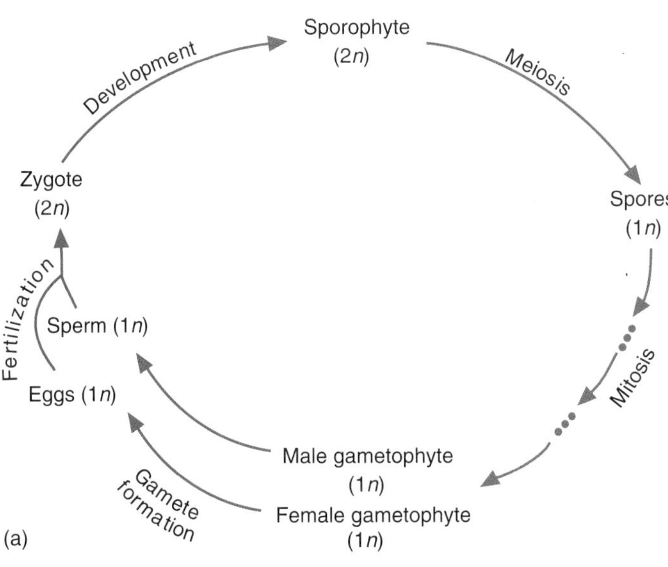

(b)

FIGURE 10.2

Reproduction in a moss. (*a*) Alternation of generations in a moss. (*b*) The gametophyte is the most noticeable part of a moss.

10.1 biology in action

The Complicated Sex Life of the Aphid

A sumac gall aphid reproduces through any of several strategies. These small insects inhabit diverse types of vegetation, including trees, mosses, ornamental flowers, and garden vegetables. Their unusual reproductive cycle begins in June, when small females called fundatrices hatch on tree bark and move to the leaves. There they burrow into the leaves, nestling into sacs called galls (fig. 10.A). As each fundatrix grows and matures within a gall, embryos begin to develop in her abdomen, without the aid of a male aphid (parthenogenetically).

By August, the gall that at first housed a single fundatrix has swollen to accommodate a thriving matriarchy of the original female and her daughters, all pale and wingless, and even some granddaughters (fig. 10.B). By autumn, the granddaughters emerge from the gall, mature, winged, and dark. They take flight, landing on mosses to live out their last few days. Sacs of embryos develop in the granddaughters, again without male input. As death nears, the granddaughters deposit groups of young on the moss.

Individuals of this fourth generation reproduce by parthenogenesis, grow, and live with their young protected beneath the waxy coat of the moss. Come springtime of the next year, or sometimes the year after that, the young aphids fly up from the mosses. But the aphid's curious multigenerational reproductive cycle isn't over yet.

The winged aphids alight on sumac trees, where they release offspring—this time of two types, male and female. The males and females mate. This rare sexual generation introduces genetic variability; individuals from different colonies, and therefore descended from different fundatrices, mate and pass their genes to the next generation in new combinations. The sexual females deposit fertilized eggs on sumac bark. Sometime later, the eggs hatch as a new generation of fundatrices, and the complex sequence of reproductive events begins anew.

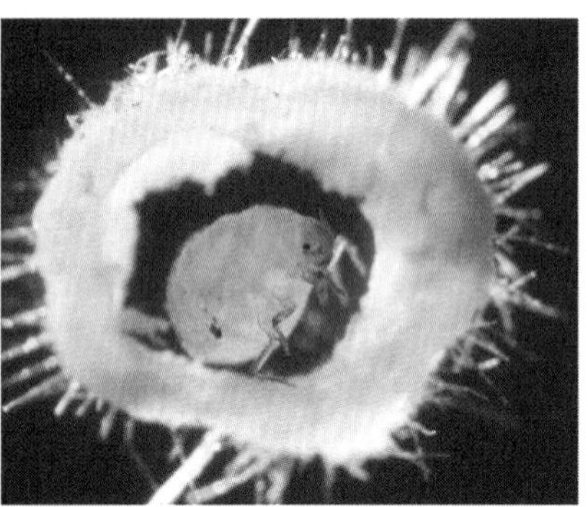

FIGURE 10.A
This cross section of a protective gall on a sumac tree shows a fundatrix, a female aphid who will reproduce clones of herself asexually, by parthenogenesis.

FIGURE 10.B
The fundatrix's daughters look like her—pale and wingless. Her granddaughters, destined to migrate, are dark and winged. Curiously, this colony is a clone—all of the individuals are genetically identical. However, the third generation expresses a different set of genes than the previous two generations, because these individuals look different.

bacterial gene transfer called **conjugation** that is still prevalent today. In conjugation, one bacterial cell uses an outgrowth called a **sex pilus** to transfer genetic material to another bacterial cell (fig. 10.3). Bacterial conjugation is sex without reproduction because it alters the genetic makeup of an already existing organism rather than producing a new individual.

Paramecium is a unicellular eukaryote that clearly separates reproduction and sex. Paramecia reproduce asexually by binary fission, yet two organisms can conjugate by aligning their oral cavities and forming a bridge of cytoplasm between them. Nuclei transfer across this bridge, resulting in new gene combinations but no new offspring.

Cell Fusion—Sex with Reproduction

Conjugation is sexual in the sense that it exchanges genetic material, but it is not reproduction, because the number of individuals does not increase. Unicellular green algae of the genus *Chlamydomonas* exhibit a simple form of true sexual reproduction that may be similar to the earliest sexual reproduction, which may have begun about 1.5 billion years ago.

In the more common nonreproductive phase, *Chlamydomonas* cells are haploid (fig. 10.4*a*). The cells are of two mating types, designated plus and minus. Under certain environmental conditions, a plus and minus cell join and form a single diploid cell.

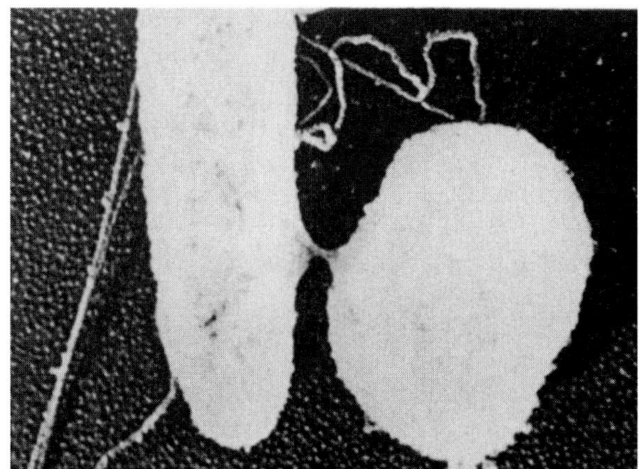

FIGURE 10.3

The bacterium *Escherichia coli* usually reproduces asexually by binary fission. *E. coli* can also transfer genetic material to another bacterium. A projection from the cell membrane from one cell, called a sex pilus, extends to the other cell. DNA transfer between bacterial cells, called conjugation, is similar to sexual reproduction in that it produces a new combination of genes. It is not reproduction, however, because a new individual does not form.

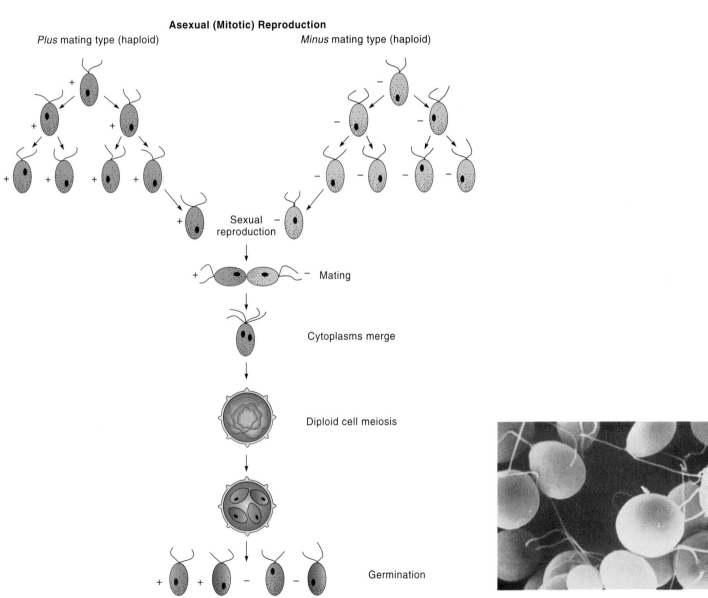

Asexual (Mitotic) Reproduction

Plus mating type (haploid)

Minus mating type (haploid)

Sexual reproduction

Mating

Cytoplasms merge

Diploid cell meiosis

Germination

(a) Two *plus* and two *minus* mating types (haploid)

(b)

FIGURE 10.4

Reproduction in *Chlamydomonas*. (*a*) The unicellular alga *Chlamydomonas* has two mating types, each haploid, that can reproduce asexually when they are separated or sexually when together. In some circumstances, cells of different mating types merge and form a diploid cell that undergoes meiosis, mixing up combinations of traits. Meiosis yields four haploid cells, two of each mating type. (*b*) *Chlamydomonas* cells.

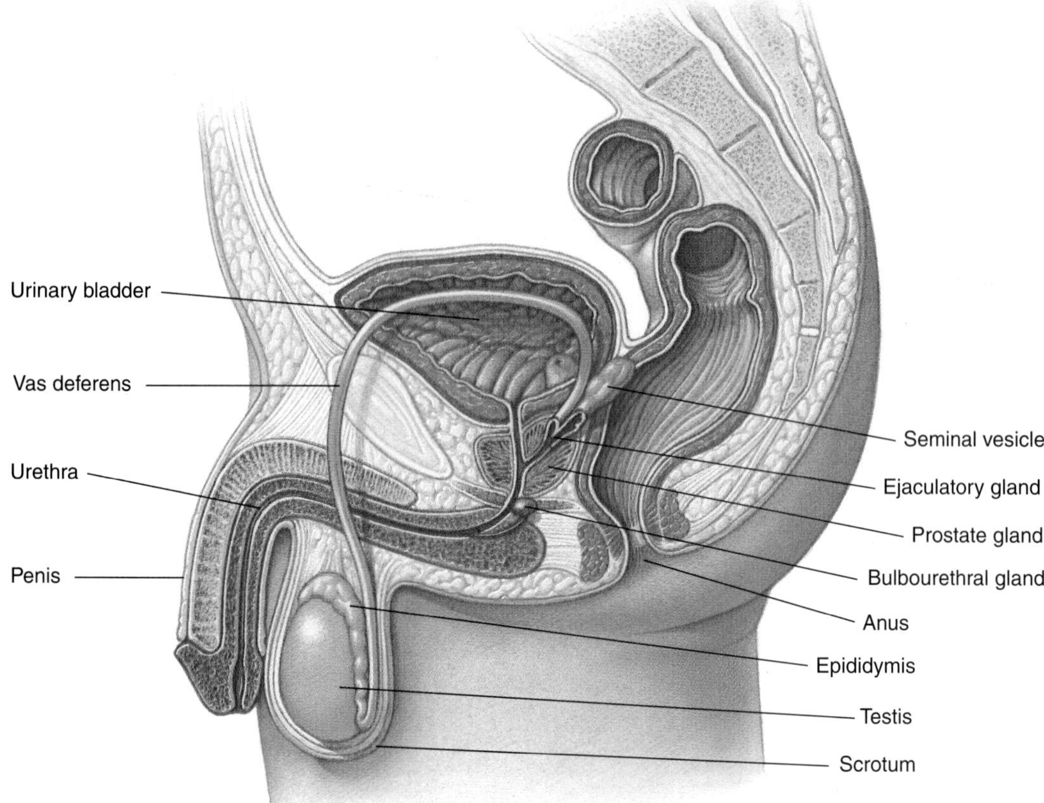

FIGURE 10.5

The human male reproductive system. Sperm cells are manufactured within the seminiferous tubules, which are tightly wound within the testes, which descend into the scrotum. Sperm mature and are stored in the epididymis and exit through the vas deferens. The paired vasa deferentia join in the urethra, through which seminal fluids exit the body. Secretions are added to the sperm cells from the prostate gland, the seminal vesicles, and the bulbourethral gland.

This cell then undergoes meiosis, which is a form of cell division that halves the number of replicated chromosomes, distributes them into two cells, and then duplicates those two cells to yield four haploid cells—two plus and two minus. Then the plus and minus cells join, thus meeting the two requirements of sexual reproduction—the number of individuals increases, and the genetic contributions of two parents are mixed and reapportioned in the offspring.

When reproduction depends upon the meeting and fusion of two (or more) types of cells, it is critical that the different types recognize each other. The two types of *Chlamydomonas* cells look identical to us, but the patterns of molecules on their surfaces differ, and they probably differ in other ways that are obvious to them (fig. 10.4*b*). However it happens, sexual reproduction in *Chlamydomonas* begins when molecules on the tail-like flagella of each type attract and draw the plus and minus cells into physical contact. Certain regions of the two cell membranes touch, like soap bubbles coalescing. The plus cell then extends a fingerlike appendage toward the minus cell and establishes a bridge of cytoplasm that further weds the two cells.

In more complex organisms, the two types of mating cells may differ more in appearance. Still, despite the differences—such as the enormity of eggs compared to sperm—each cell plays the same role as an emissary carrying genetic material to the next generation. Meiosis halves the chromosome number. An additional series of steps, called **maturation,** strips much of the cytoplasm and organelles from developing male gametes and, conversely, concentrates these supplies in female gametes. The cells that undergo meiosis and give rise to gametes (**sperm** and **oocytes**) are called **germ cells.** The entire process of producing sperm and oocytes, consisting of meiosis and maturation, is called **gametogenesis.** The remainder of the chapter considers gametogenesis in humans.

Mastering Concepts

1. Why must organisms reproduce?
2. How do sexual and asexual reproduction differ?
3. How do haploid and diploid cells differ?
4. Why is conjugation sex without reproduction?
5. Why is cell fusion sex with reproduction?
6. What are the two stages of gametogenesis?

Gamete Formation in the Human Body

The reproductive systems of the human male and female are similarly organized. Each system has paired structures, called **gonads,** in which the sperm and oocytes are manufactured; a network of tubes to transport these cells; and hormones and glandular secretions that control the entire process.

The Human Male Reproductive System

Sperm cells are manufactured within a 410-foot-long (125-meter-long) network of tubes called **seminiferous tubules,** which are packed into paired, oval organs called **testes** (sometimes called testicles) (fig. 10.5). The testes are the male gonads. They

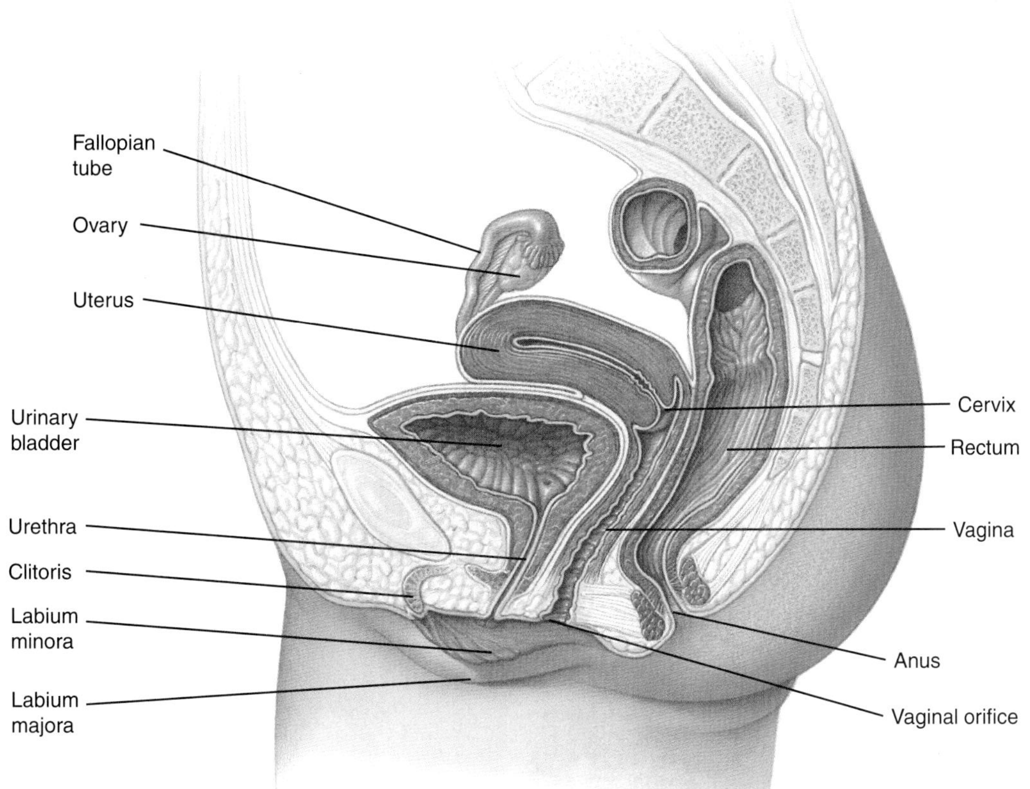

FIGURE 10.6

The human female reproductive system. Immature egg cells are packed into the paired ovaries. Once a month, one oocyte is released from an ovary and is drawn into the fingerlike projections of a nearby fallopian tube by ciliary movement. If the oocyte is fertilized by a sperm cell in the fallopian tube, it continues into the uterus, where it is nurtured for nine months as it develops into a new individual. If the ovum is not fertilized, it is expelled along with the built-up uterine lining through the cervix and then through the vagina. The external genitalia consist of inner and outer labia and the clitoris.

lie outside the abdomen within a sac called the **scrotum.** Their location outside of the abdominal cavity allows the testes to maintain a lower temperature than the rest of the body, which is necessary for sperm to develop properly. Leading from each testis is a tightly coiled tube, the **epididymis,** in which sperm cells mature and are stored. Each epididymis continues into another tube, the **vas deferens** (plural: vasa deferentia). Each vas deferens bends behind the bladder and joins the **urethra,** the tube that also carries urine out through the **penis.** Along the sperm's path, three glands produce secretions. The **prostate gland,** which produces a thin, milky, alkaline fluid that activates the sperm to swim, wraps around the vasa deferentia. Opening into each vas deferens is a duct from the **seminal vesicles.** These glands secrete the sugar fructose, an energy source, plus hormonelike prostaglandins, which may stimulate contractions in the female reproductive tract that help propel sperm. The **bulbourethral glands,** each about the size of a pea, attach to the urethra where it passes through the body wall. These glands secrete alkaline mucus, which coats the urethra before sperm are released. All of these secretions combine to form the **seminal fluid,** which the sperm cells travel in.

During sexual arousal, the penis becomes erect so that it can penetrate the vagina and deposit sperm in the female reproductive tract. At the peak of sexual stimulation, a pleasurable sensation called **orgasm** occurs, accompanied by rhythmic muscular contractions that eject the sperm from the vasa deferentia through the urethra and out the penis. The discharge of sperm from the penis is called **ejaculation.** One human ejaculation typically delivers about 100 million sperm cells.

The Human Female Reproductive System

The female sex cells develop within paired gonads in the abdomen called the **ovaries** (fig. 10.6). Within each ovary of a newborn female are about a million oocytes. Nourishing **follicle cells** surround each individual oocyte. Like a testis containing sperm cells in various stages of development, an ovary houses oocytes in different stages of development. Approximately once a month, beginning at puberty, one ovary releases the most mature oocyte. Beating cilia sweep the mature oocyte into the fingerlike projections of one of two **fallopian tubes.** The tube carries the oocyte into a muscular saclike organ, the **uterus,** or womb.

Once released from the ovary, an oocyte can live for about 72 hours, but a sperm can penetrate it only during the first 24 hours of this period—possibly even less. If the oocyte encounters a sperm cell in a fallopian tube and the cells combine and their nuclei fuse, the oocyte completes its development and becomes a fertilized ovum, or zygote. It then travels into the uterus and implants in the thick, blood-rich uterine lining that has built up over the preceding few weeks. The uterine lining is called the **endometrium.** Over the next 9 months, the fertilized ovum develops into a new human being. If the oocyte is not fertilized, both endometrium and oocyte are expelled as the **menstrual flow.**

The lower end of the uterus narrows to form the **cervix,** which opens into the tubelike **vagina,** which exits the body. Two pairs of fleshy folds protect the vaginal opening on the outside: the **labia majora** (major lips) and the thinner, underlying flaps of tissue they protect, called the **labia minora** (minor lips). At the upper juncture of both pairs of labia is a 1-inch-long (2-centimeter-long)

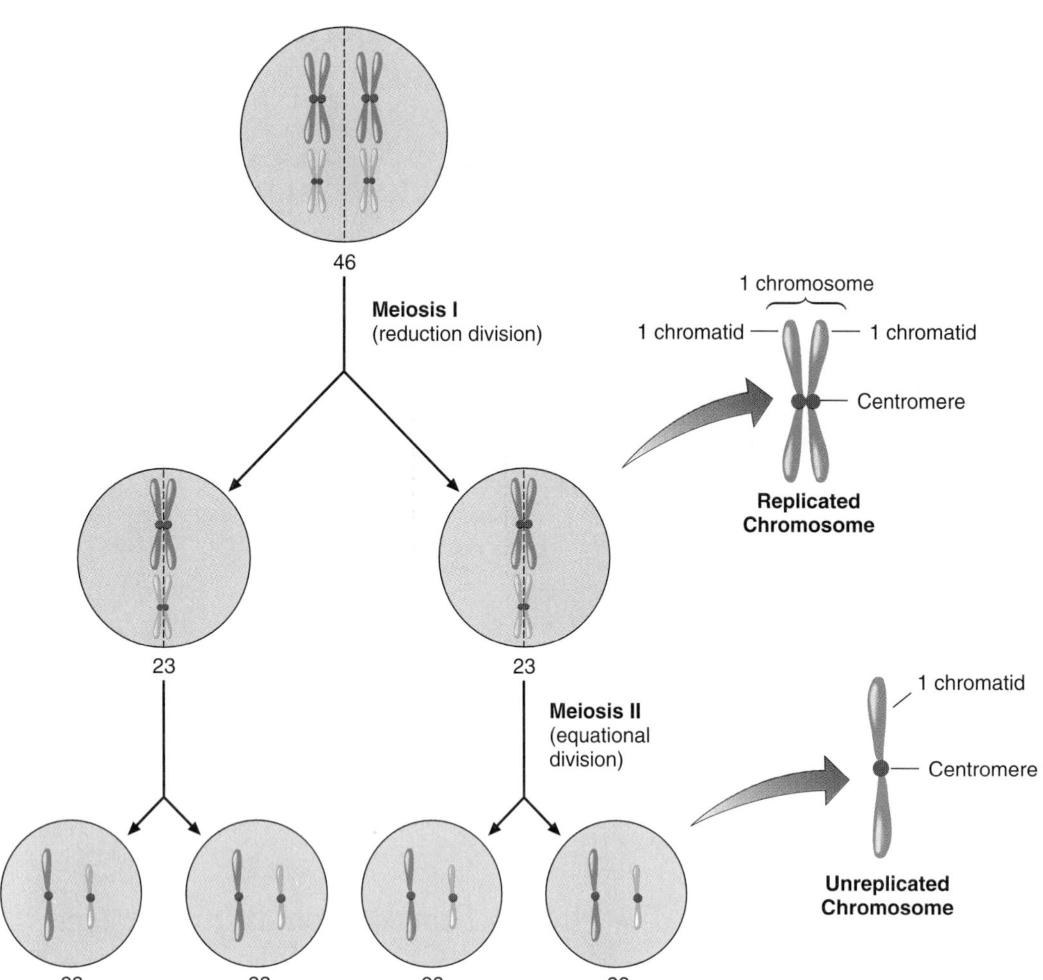

FIGURE 10.7

Meiosis is a special form of cell division in which certain cells are set aside to give rise to haploid gametes. In humans, the first meiotic division reduces the number of chromosomes to 23, all in the replicated form (*inset*). In the second meiotic division, each cell from the first division essentially undergoes mitosis. The result of the two divisions of meiosis is four haploid cells.

structure called the **clitoris,** which is anatomically analogous to the penis. Rubbing the clitoris stimulates females to experience orgasm. Hormones control the cycle of oocyte maturation in the ovaries and prepare the uterus to nurture a zygote.

Mastering Concepts

1. In what organ are sperm produced?
2. What are the other structures and associated glands of the male reproductive system?
3. In what organ are oocytes produced?
4. What are the other structures of the female reproductive system?

Meiosis

Meiosis entails two divisions of the genetic material. The first division is called **reduction division** (or meiosis I) because it reduces the number of chromosomes—in humans, from 46 to 23. The second division, called the **equational division** (or meiosis II), produces four cells from the two formed in the first division. The cells undergoing meiosis are diploid, abbreviated $2n$ for two sets of chromosomes. The products of meiosis, haploid gametes, are designated $1n$. Diploidy is reestablished when opposite gametes combine at fertilization.

Meiosis is preceded by an interphase period when DNA replicates (table 10.1). The cell in which meiosis begins has **homologous pairs** of chromosomes, or homologs for short. Homologs look alike and carry the genes for the same traits in the same sequence. A diploid human cell has 23 pairs of homologs. Of these, 22 pairs do not determine sex and are called **autosomes;** the other chromosomes (the X and the Y) determine sex and are known as **sex chromosomes.**

One homolog in a pair comes from the person's mother, and the other comes from the father. When meiosis begins, the DNA of each homolog replicates and forms two chromatids joined by a replicated centromere (fig. 10.7). The chromosomes are not yet condensed enough to be visible.

Prophase I (so called because it is the prophase of meiosis I) follows interphase. Early in prophase I, replicated chromosomes condense and become visible (fig. 10.8). Toward the middle of prophase I, the homologs line up next to one another, gene by gene, a phenomenon called **synapsis.** A conglomeration of RNA and protein joins the paired chromosomes.

Toward the end of prophase I, the synapsed chromosomes separate but remain attached at a few points along their lengths

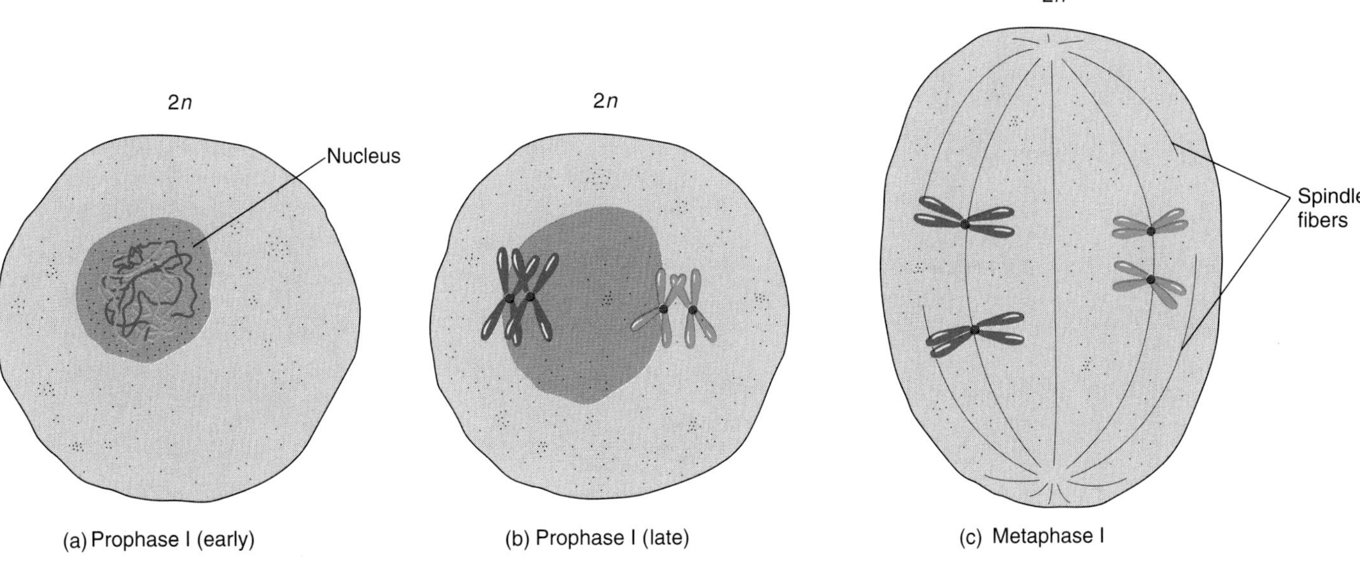

(a) Prophase I (early)

(b) Prophase I (late)

(c) Metaphase I

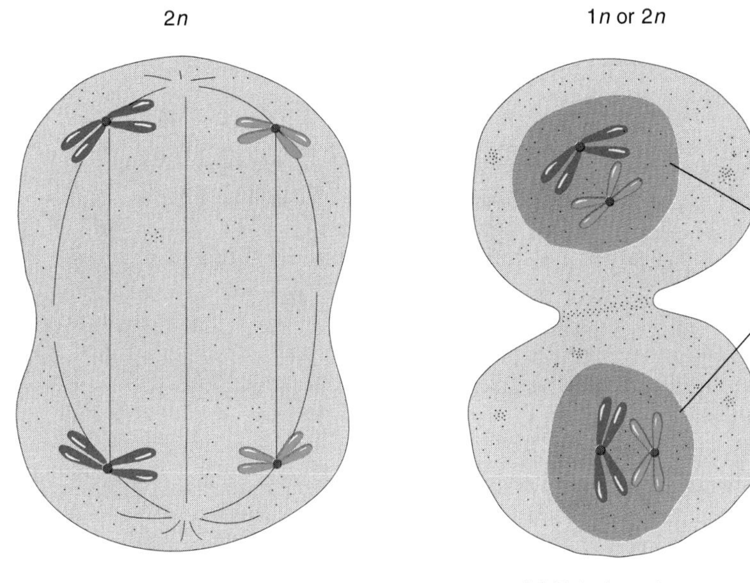

(d) Anaphase I

(e) Telophase I

FIGURE 10.8

Meiosis I. (*a*) In early prophase I, replicated chromosomes condense and become visible as a tangle within the nucleus. (*b*) By late prophase I, the pairs are aligned and the homologs cross over. (*c*) In metaphase I, spindle fibers align the homologs, and (*d*) in anaphase I, the homologs move to opposite poles. (*e*) In telophase I, the genetic material is partitioned into two progeny nuclei, each containing only one homolog from each pair. In most species, cytokinesis occurs between the meiotic divisions, forming two cells after telophase I.

Table 10.1 Comparison of Mitosis and Meiosis

Mitosis	Meiosis
One division	Two divisions
Two progeny cells per cycle	Four progeny cells per cycle
Progeny cells genetically identical	Progeny cells genetically different
Chromosome number in progeny cells same as in parent cell (2*n*)	Chromosome number in progeny cells half that in parent cell (1*n*)
Occurs in somatic cells	Occurs in germ cells
Occurs throughout life cycle	In humans, completed only after sexual maturity
Used for growth, repair, and asexual reproduction	Used for sexual reproduction, producing new gene combinations

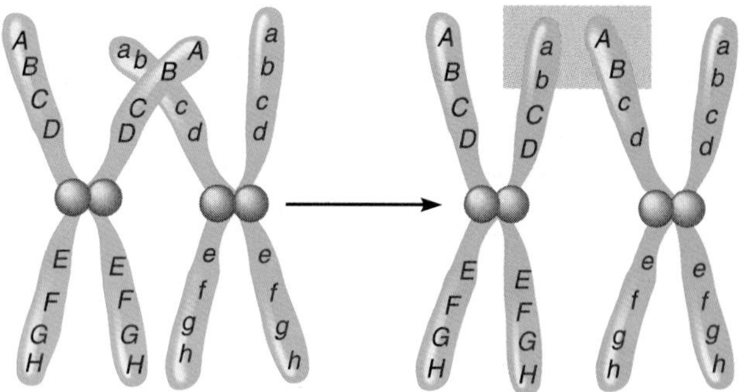

FIGURE 10.9

Crossing-over recombines genes. The capital and lowercase forms of the same letter represent different forms (alleles) of the same gene.

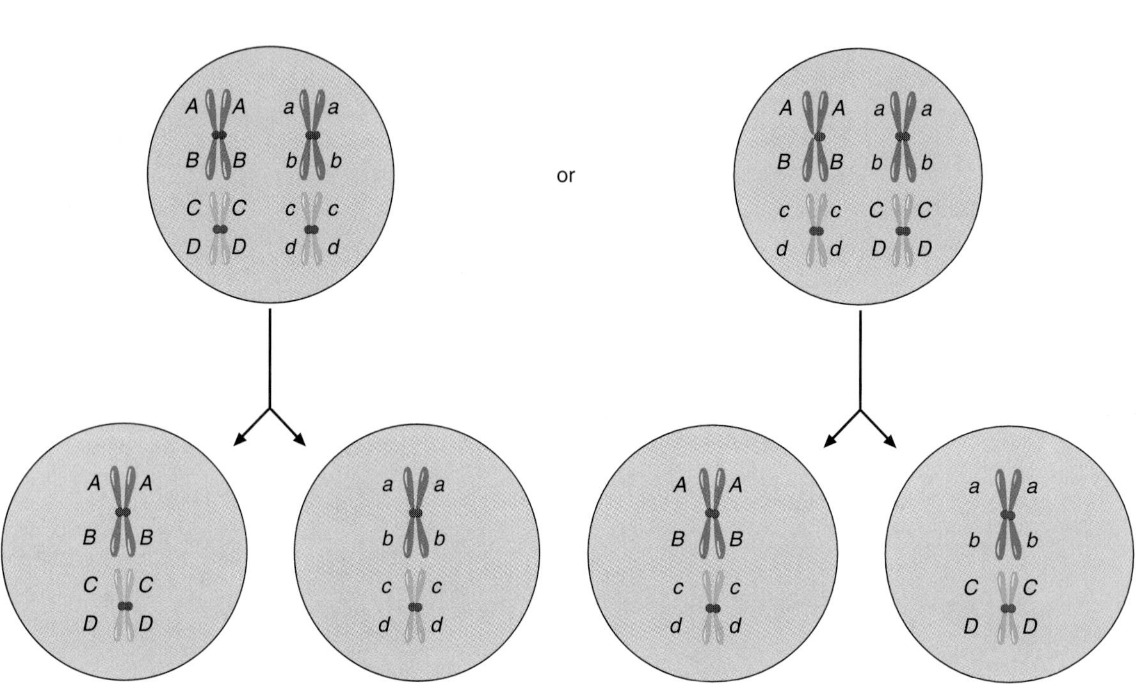

FIGURE 10.10

Independent assortment. The pattern in which homologs align during metaphase I determines the combination of chromosomes in the progeny cells. This illustration follows two chromosome pairs with different alleles of the same gene indicated by capital and lowercase forms of the same letter. Two pairs of chromosomes can align in two different ways to produce four different possibilities in the progeny cells. The potential variability generated by meiosis skyrockets when one considers all 23 chromosome pairs and the effects of crossing-over.

(fig. 10.9). Here, the homologs exchange chromosomal material in a process called **crossing-over.** Because each homolog comes from a different parent, crossing-over results in chromosomes that have some genes from a person's mother and some from the father. New gene combinations arise when the parents carry different forms of the same gene, which are called **alleles.** The mixing up of trait combinations resulting from crossing-over is one reason why siblings (except for identical twins) are never exactly alike genetically.

Consider a simplified example of how crossing-over mixes trait combinations. Suppose that one homolog carries genes for hair color, eye color, and finger length. One of the chromosomes in the homolog pair—perhaps the one that came from the person's father—carries alleles for blond hair, blue eyes, and short fingers. The homolog from the mother carries alleles for black hair, brown eyes, and long fingers. After crossing-over, one of the chromosomes might bear alleles for blond hair, brown eyes, and long fingers, and the other bear alleles for black hair, blue eyes, and short fingers.

Meiosis continues in **metaphase I,** when the homologs align down the center of the cell (see fig. 10.8*c*). A spindle forms, and each chromosome attaches to a spindle fiber stretching to the opposite pole. The chromosomes' alignment during metaphase I is important in generating genetic diversity. Within each homolog pair, the maternally derived and paternally derived members attach to each pole at random. The greater the number of chromosomes, the greater the genetic diversity; different combinations of maternal and paternal homologs move to each pole. It is a little like a double line of schoolchildren. Imagine the many different ways that 23 pairs of students could form a double line, while maintaining specific pairs.

For two pairs of homologs, four (2^2) different metaphase configurations are possible (fig. 10.10). For three pairs of homologs, eight (2^3) configurations can occur. Our 23 chromosome pairs can thus line up in 8,388,608 (2^{23}) different ways! (And a class of 23 students could line up in this many ways!) The random arrangement of homologs in a cell in metaphase I is called **independent assortment.** It accounts for a basic law of inheritance, which is discussed in the next unit.

Homologs separate in **anaphase I** (see fig. 10.8*d*), and they complete their movement to opposite poles in **telophase I** (see fig. 10.8*e*). In most species, the cell divides in two after telophase I. During a second interphase, the chromosomes unfold into very thin threads. (Some species skip a second interphase.) Proteins are manufactured, but the genetic material does not replicate a second time. The single DNA replication, followed by the double division of meiosis, is what halves the chromosome number.

Prophase II marks the start of the second meiotic division. The chromosomes again condense and become visible (fig. 10.11*a*). In **metaphase II** (fig. 10.11*b*), the replicated chromosomes align down the center of the cell. In **anaphase II** (fig. 10.11*c*), the centromeres part and the chromatids move to opposite poles. In **telophase II** (fig. 10.11*d*), nuclear envelopes form around the four nuclei containing the separated sets of chromosomes.

Cytokinesis separates the four nuclei into individual cells (fig. 10.11*e*). In most species, cytokinesis occurs between the first and second meiotic divisions, forming first two cells and then four following telophase II. In some species, cytokinesis occurs only following telophase II; the four nuclei enclosed in one large cell are separated into four smaller haploid cells.

Table 10.2 summarizes the stages of meiosis.

Meiosis generates astounding genetic variety. Any one of a person's 8,388,608 (2^{23}) possible chromosome combinations can combine with any one of the 8,388,608 combinations of his or her partner, raising potential variability to more than 70 trillion ($8{,}388{,}608^2$) genetically different individuals! Crossing-over contributes even more genetic variability to gametes.

Mastering Concepts

1. How do the numbers of rounds of DNA replication and cell division halve the chromosome number in meiosis?
2. What are homologs?
3. What event in prophase I contributes to genetic variability?
4. What event in metaphase I contributes to genetic variability?
5. What happens during the stages of meiosis I and meiosis II?

Gamete Maturation

Sperm and oocyte each have a haploid set of chromosomes, but they look quite distinct because each receives different amounts of other cellular components. A sperm is motile and lightweight; an oocyte is huge by comparison and packed with nutrients and organelles. We look first at gamete maturation in the male.

Table 10.2 The Stages of Meiosis

Stage	Events
Interphase	DNA replicates
Meiosis I (reduction division)	(Halves chromosome number)
Prophase I	Replicated chromosomes condense
	Homologs synapse and cross over
Metaphase I	Homologs align down equator
	Spindle forms and attaches each chromosome to one pole, with members of homologous pairs attached to opposite poles
Anaphase I	Homologs move apart
Telophase I	Homologs arrive at opposite poles
Interphase	DNA replicates
Meiosis II (equational division)	(Doubles number of progeny cells)
Prophase II	Chromosomes condense
Metaphase II	Chromosomes align down equator
Anaphase II	Centromeres part, separated chromatids move to opposite poles
Telophase II	Nuclear envelopes re-form as each of the two cells from meiosis I forms two cells

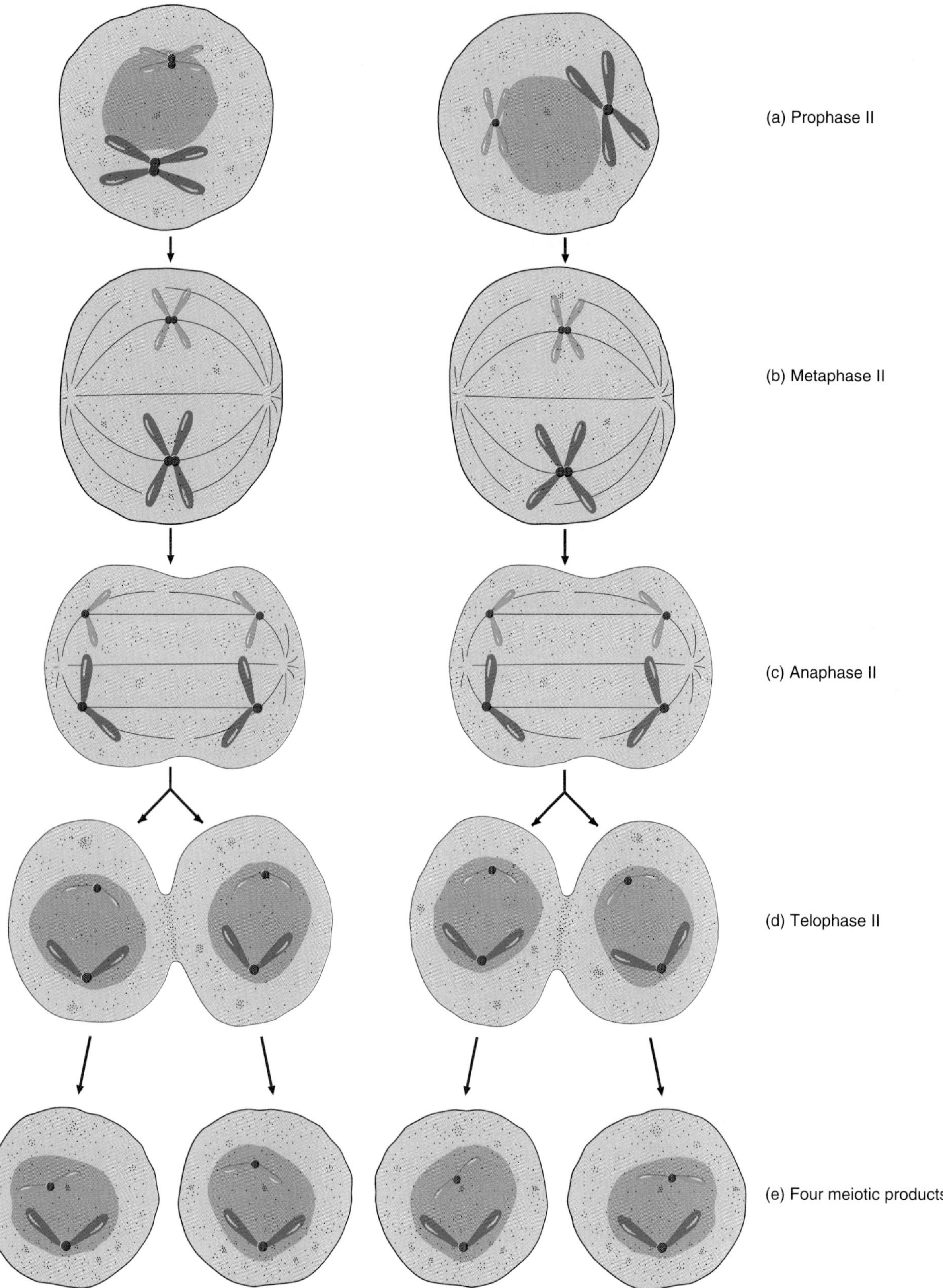

FIGURE 10.11

Meiosis II. The second meiotic division is very similar to mitosis. (*a*) In prophase II, the chromosomes are visible. (*b*) In metaphase II, the mitotic spindle aligns the chromosomes. (*c*). In anaphase II, the centromeres part, and each chromatid pair divides into two chromosomes, which are pulled toward opposite poles. (*d*) In telophase II in some species, each of the two separated sets of chromosomes is enclosed in a nuclear membrane, and (*e*) in cytokinesis, they are partitioned into individual haploid cells. The net yield from the entire process of meiosis: four haploid progeny cells.

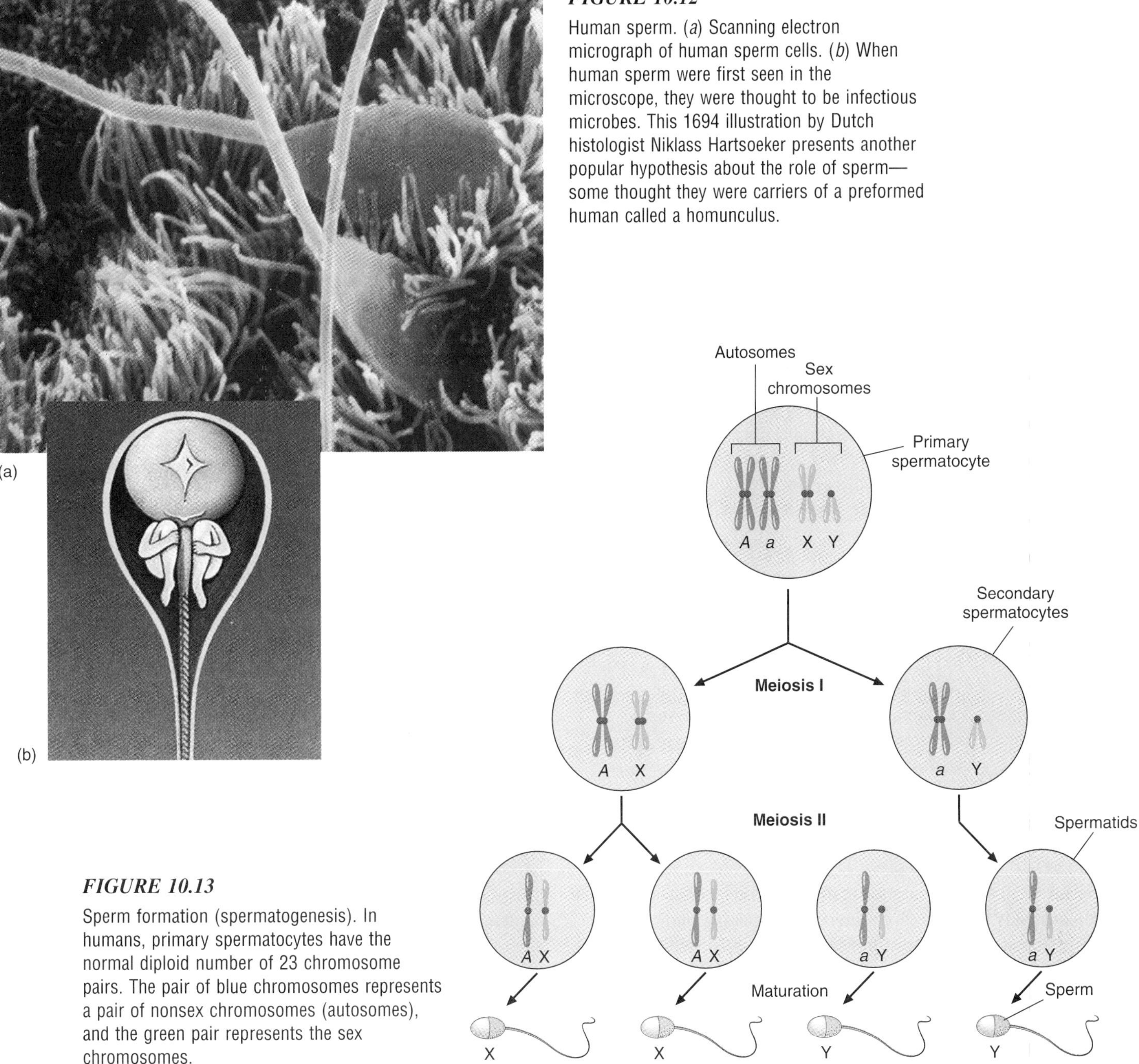

FIGURE 10.12

Human sperm. (*a*) Scanning electron micrograph of human sperm cells. (*b*) When human sperm were first seen in the microscope, they were thought to be infectious microbes. This 1694 illustration by Dutch histologist Niklass Hartsoeker presents another popular hypothesis about the role of sperm—some thought they were carriers of a preformed human called a homunculus.

FIGURE 10.13

Sperm formation (spermatogenesis). In humans, primary spermatocytes have the normal diploid number of 23 chromosome pairs. The pair of blue chromosomes represents a pair of nonsex chromosomes (autosomes), and the green pair represents the sex chromosomes.

Sperm Development

Sperm cells have long intrigued biologists, perhaps in part because the cells are so intriguing to watch under a microscope (fig. 10.12*a*). Sperm cells look a little like microorganisms; indeed, when Antonie van Leeuwenhoek first viewed sperm in 1678, he thought they were parasites in the semen. By 1685, he modified his view and suggested that each sperm is a seed containing a preformed being (fig. 10.12*b*) and requiring a period of nurturing in the female to develop into a new life.

In the 1770s, Italian biologist Lazzaro Spallanzani took an experimental approach to studying sperm by placing toad semen onto filter paper. He observed that the material that passed through the tiny holes in the filter paper—seminal fluid minus the sperm—could not fertilize eggs. Even though he had shown indirectly that sperm are required for fertilization, Spallanzani, like Leeuwenhoek before him, concluded that the sperm cells were contaminants. Several researchers in the nineteenth century finally showed that sperm were not microbial invaders but human cells that play a key role in reproduction.

Formation and specialization of sperm cells is called **spermatogenesis** (fig. 10.13). A diploid cell that divides mitotically, yielding diploid progeny cells destined to become sperm cells, is a **spermatogonium.** Several spermatogonia attach to each other by bridges of cytoplasm, and they undergo meiosis simultaneously. First, the spermatogonia accumulate cytoplasm and replicate their DNA. They are now called **primary spermatocytes.** Next, during reduction division (meiosis I), each primary spermatocyte divides, forming two equal-sized haploid cells called **secondary spermatocytes.**

In meiosis II, each secondary spermatocyte divides, yielding two equal-sized **spermatids.** Each spermatid then develops specialized structures, including the characteristic sperm tail, or flagellum. The base of the tail has many mitochondria and ATP molecules, forming an energy system that enables the sperm to swim inside the female reproductive tract. (A sperm, which is a mere 0.0023 inch long, must travel about 7 inches within the female reproductive system to reach an oocyte!) After the spermatids form, some of the cytoplasm connecting the cells falls away, leaving mature, tadpole-shaped **spermatozoa,** or sperm.

A mature sperm has three functional parts: a haploid nucleus, a locomotion system, and enzymes to penetrate an oocyte. Each sperm cell consists of a tail, body or midpiece, and head region. A small bump on the front end, the **acrosome,** contains enzymes that help the sperm cell penetrate the oocyte's outer membrane. Within the large sperm head, DNA wraps around proteins. A male manufactures trillions of sperm in his lifetime, yet only a very few will approach an oocyte.

In the male reproductive system, meiosis begins in the seminiferous tubules, where spermatogonia reside at the side of the tubule farthest from the lumen (the central cavity). When a spermatogonium divides, one progeny cell moves towards the lumen and accumulates cytoplasm, becoming a primary spermatocyte. The other progeny cell remains in the tubule wall. There, it acts as a stem cell—it remains unspecialized but continually gives rise to cells that specialize.

The developing sperm cells move towards the lumen of the seminiferous tubule, and by the time they reach it, they are spermatids (fig. 10.14). This sequential and synchronized sperm development is often compared to students at a high school. Just as the student body consists of freshmen, sophomores, juniors, and seniors, all present at the same time but usually in different classes, a seminiferous tubule houses spermatogonia (freshmen), primary spermatocytes (sophomores), secondary spermatocytes (juniors), and spermatids (seniors).

The epididymis stores spermatids. Here they complete their differentiation (specialization) into spermatozoa. The entire process, from spermatogonium to spermatozoan, takes 74 days in the human.

The epididymis contracts during ejaculation, sending sperm cells into the vasa deferentia. The sperm, along with the secretions from the accessory glands, form semen, which exits the body through the penis.

Meiosis in the male has some built-in protections against birth defects. Spermatogonia exposed to toxins may be so damaged that they never differentiate into mature sperm. Toxins can render sperm cells unable to swim. Toxic drugs carried in the seminal fluid but not actually damaging sperm can endanger an embryo or fetus if they enter a pregnant woman's reproductive tract. They can harm the uterus or enter a woman's circulation and reach the placenta, the organ connecting the woman to the fetus. Cocaine can affect a fetus by another route—it attaches to thousands of binding sites on the sperm, without harming the cells or impeding their movements. Therefore, sperm can ferry cocaine to a fertilized ovum.

Ovum Development

Meiosis in the female is called **oogenesis** (egg making). It begins, like spermatogenesis, with a diploid cell, an **oogonium.** Unlike spermatogonia, oogonia are not attached to each other, but each is surrounded by a layer of follicle cells. Each oogonium grows, accumulates cytoplasm, and replicates its DNA to become a **primary oocyte.** The ensuing meiotic division in oogenesis, unlike that in spermatogenesis, produces cells of different sizes.

In meiosis I, the primary oocyte divides into a small cell with very little cytoplasm, called a **polar body,** and a much larger cell called a **secondary oocyte** (fig. 10.15). Each cell is haploid but with replicated chromosomes. In meiosis II, the tiny polar body may divide to yield two polar bodies of equal size or may decompose. The secondary oocyte, however, divides unequally in meiosis II to produce a small second polar body and the mature egg cell or ovum, which contains a large amount of cytoplasm. Therefore, a cell undergoing meiosis in a female typically yields four cells, only one of which can become an ovum.

The secondary oocyte, in receiving the lion's share of the cytoplasm of four cells, is packed with biochemicals and organelles that the zygote will use until its genes begin to function. These biochemicals include proteins, RNA, and molecules that influence cell specialization in the early embryo. Some amphibian oocytes amass a trillion ribosomes during meiosis, providing a zygote with a jump-start on protein synthesis.

A woman's body absorbs polar bodies, which normally play no further role in development. However, sperm can fertilize polar bodies, and a mass of tissue that does not resemble an embryo grows, until the woman's body rejects it. A fertilized polar body, sometimes called a "blighted ovum," accounts for about 1 in 100 spontaneous abortions.

A test to analyze specific genes in polar bodies is being developed to determine whether or not an oocyte contains a disease-causing gene. This test is used to identify "healthy" oocytes that have been removed from a woman's body and will be fertilized in laboratory glassware. Such in vitro fertilization is discussed in chapter 12. If a woman knows that she is a carrier of a particular inherited condition, and a polar body has the disease-causing form of the gene, then it can be inferred that the oocyte does not have the gene, because the homologs that carry the genes separate as the polar body and oocyte form (fig. 10.16).

The timetable for oogenesis differs greatly from that of spermatogenesis. Three months after conception, the ovaries of a female fetus contain 2 million or more primary oocytes. From then on, the number of primary oocytes declines. A million are present by birth, and about 400,000 remain by the time of puberty. At birth, the oocytes arrest in prophase I. After puberty, one or a few oocytes complete meiosis I each month. These oocytes stop meiosis again, this time at metaphase II. Specific hormonal cues each month prompt an ovary to ovulate, or release a secondary oocyte.

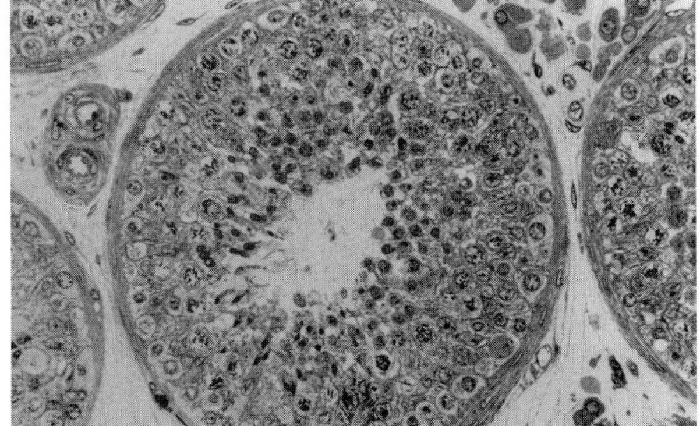

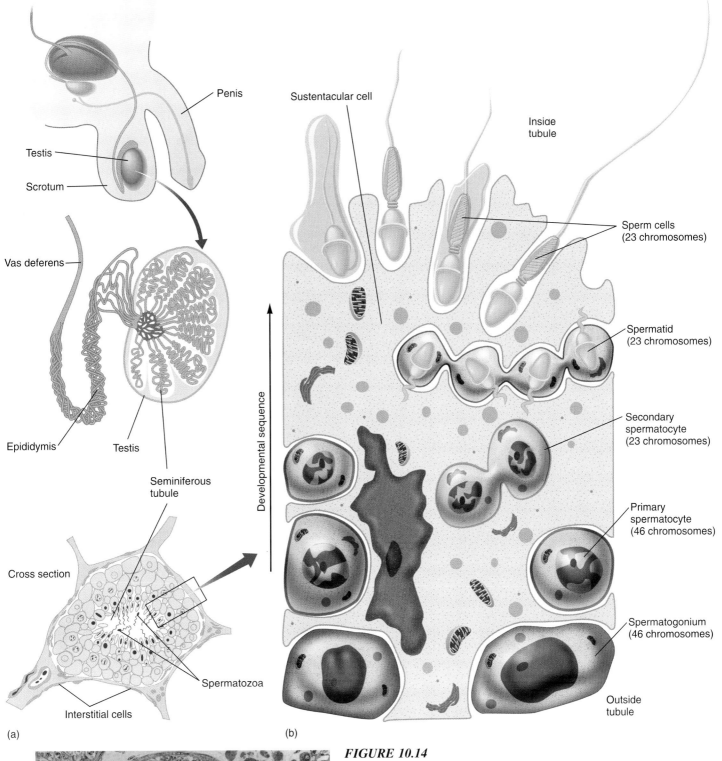

FIGURE 10.14

Developing sperm cells in the testis (*a*) can be seen in a cross section of a seminiferous tubule (*b*), with more mature forms nearer the inside of the tubule (the lumen). (*c*) A cross section of a human seminiferous tubule reveals sperm in several stages.

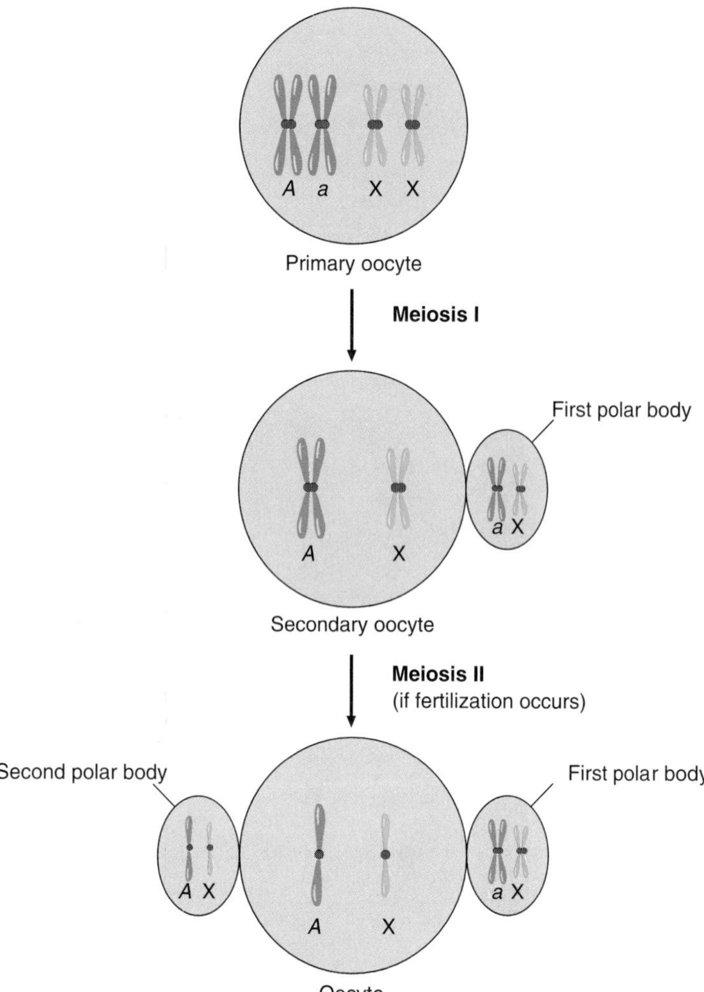

FIGURE 10.15

Ovum formation (oogenesis). In humans, primary oocytes have the normal diploid number of 23 chromosome pairs. The blue pair of chromosomes represents a pair of autosomes, and the green pair represents the sex chromosomes.

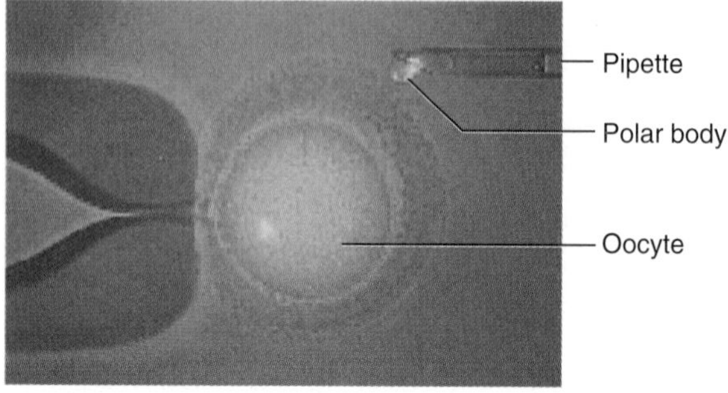

FIGURE 10.16

The fact that an oocyte shares a woman's divided genetic material with a much smaller companion, the polar body, is the basis for a new technique to select oocytes that are free of certain disease-causing genes.

If a sperm penetrates the oocyte membrane, meiosis in the oocyte completes, and the two nuclei approach slowly and combine to form the fertilized ovum. Therefore, female meiosis does not even finish unless fertilization occurs. If the secondary oocyte is not fertilized, it degenerates and leaves the body in the menstrual flow.

The stage of meiosis that occurs when a sperm fertilizes an oocyte varies with species. In many worms, fertilization occurs just as the primary oocyte forms, whereas in foxes and dogs, it occurs in an older primary oocyte. In shellfish, octopuses, and squids, fertilization takes place during metaphase I, and in the sea urchin, it follows the completion of meiosis.

A human female will only ovulate about 400 oocytes between puberty and menopause, and only a very few of these will likely contact a sperm cell. Only one in three fertilized ova will continue to grow, divide, and specialize to eventually form a new human life.

Mastering Concepts

1. What are the stages of sperm development and maturation?
2. What are the parts of a mature sperm cell?
3. Where does sperm development occur?
4. What are the stages of development and maturation of an oocyte?
5. How does the timetable for meiosis differ in males and females?

SUMMARY

1. Reproduction is essential to species' survival. **Asexual reproduction,** such as **binary fission,** can be successful in an unchanging environment. **Sexual reproduction** mixes traits and therefore provides species' protection in a changing environment. Sexual reproduction also increases the number of organisms. It occurs when **haploid gametes** fuse, restoring the **diploid** state.

2. Some species reproduce both asexually and sexually, including yeast and plants. Plant sexual reproduction involves an **alternation of generations** and both haploid and diploid phases.

3. Conjugation, a form of gene transfer in some microorganisms, is sexual because one individual transfers genetic material to another; however, it is not reproduction because no additional individual forms. *Chlamydomonas* undergoes cell fusion, which may have been a forerunner of sexual reproduction.

4. In humans, gametes are produced in the male and female reproductive systems. These systems include paired **gonads,** where **sperm** and **oocytes** are manufactured, networks of tubes, and glands. **Gametogenesis** includes **meiosis** and **maturation** and produces sperm and oocytes.

5. Developing sperm originate in **seminiferous tubules** within the paired **testes,** passing through different stages of maturation as they approach the lumen of the tubule. Sperm mature in the **epididymis** and **vasa deferentia,** and they exit the body through the **urethra** during sexual intercourse. The **prostate gland, seminal vesicles,** and **bulbourethral glands** add secretions to sperm.

An Unusual Child

continued from page 192.

Parthenogenesis is Greek for "virgin birth" and refers to the unusual situation of an oocyte activating and dividing to yield a viable offspring—without the input of sperm. It occurs in turkeys, lizards, snakes, flatworms, roundworms, and various small pond-dwellers. The first report of parthenogenesis is attributed to Swiss naturalist Charles Bonnet, who noticed that aphid eggs develop into aphids without benefit of sperm. Parthenogenesis is seen in some bee, wasp, and ant societies, where drones (males) develop from unfertilized eggs.

Parthenogenesis is advantageous when an organism is well adapted to its environment, because it leads to a sudden population explosion. Parthenogenesis explains how a pond forming overnight from a puddle becomes rapidly overrun with little swimming animals. In a changing environment, though, parthenogenesis becomes a liability. The organisms are genetically alike, and all perish if conditions become harsh. Because environments change, parthenogenesis is relatively rare.

In mice, parthenogenotes (individuals derived from parthenogenesis) survive only midway through gestation and then shrivel up and cease developing. Humans fare even worse. If a human oocyte replicates its genetic material but then cytokinesis fails, the result is a diploid cell. But rather than develop into an embryo, the diploid oocyte forms a benign growth of disorganized tissue called a teratoma. Could a parthenogenote derive from solely male tissue? Theoretically, yes. If an oocyte lacks a nucleus and is fertilized by a sperm whose nucleus subsequently doubles itself, a mass of placenta-like tissue develops instead of an embryo.

The bottom line: It takes a set of chromosomes from each sex to make an offspring. But could a person be a partial parthenogenote, with only some cells that derive from only one gamete?

This is apparently the case for F. D. The clue to the child's odd beginnings was that the chromosomes of his XX cells, and one set of chromosomes in the XY cells, are identical. This means that the two cell lines must have diverged before the chromosomes mixed up their parental contributions by crossing-over.

Researchers hypothesized several scenarios to explain F. D.'s beginnings. One explanation is that an activated secondary oocyte divided, yielding one cell that replicated its chromosomes but did not undergo cytokinesis (the XX cell line) and a second cell that was fertilized (the XY cell line). Remarkably, despite such a profound developmental detour, F. D. is expected to be able to father normal children, because he has normal XY cells that presumably gave rise to his reproductive system.

6. Oocytes originate in the **ovaries.** Each month after puberty, one ovary releases an oocyte into a **fallopian tube,** which leads to the uterus.

7. Meiosis halves the number of chromosomes in somatic cells to produce haploid gametes. A species' chromosome number stays constant because a gamete's DNA replicates once, but the cells divide twice.

8. Meiosis provides genetic variability by partitioning different combinations of genes into gametes through **independent assortment. Crossing-over** increases the variability.

9. **Spermatogenesis** begins with **spermatogonia,** which accumulate cytoplasm and replicate their DNA to become **primary spermatocytes.** After **reduction division** (meiosis I), the cells are haploid **secondary spermatocytes.** In an **equational division** (meiosis II) the secondary spermatocytes divide to each yield two **spermatids,** which then differentiate along the male reproductive tract.

10. In **oogenesis, oogonia** replicate their DNA to become **primary oocytes.** In meiosis I, the primary oocyte divides, apportioning cytoplasm to one large **secondary oocyte** and a much smaller **polar body.** In meiosis II, the secondary oocyte divides to yield the large ovum and another small polar body. The development of a sperm cell takes 74 days. Meiosis in the female completes at fertilization.

TO REVIEW . . .

1. What is the evidence that sexual reproduction has been successful?

2. Describe how a plant life cycle may include a halpoid and a diploid phase.

3. Why is conjugation sexual but not reproductive?

4. Define the following terms:
   - **a.** crossing-over
   - **b.** gamete
   - **c.** haploid
   - **d.** homolog
   - **e.** synapsis
   - **f.** diploid

5. How are the human male and female reproductive tracts similar? How are the structures of the testis and ovary similar?

6. How many sets of chromosomes are present in each of the following cell types?
   - **a.** an oogonium
   - **b.** a primary spermatocyte
   - **c.** a spermatid
   - **d.** a secondary oocyte
   - **e.** a polar body derived from a primary oocyte

7. Do the following examples illustrate asexual or sexual reproduction?
   - **a.** A fundatrix aphid produces dozens of daughters and granddaughters without a mate.
   - **b.** An amoeba divides, producing two from one.
   - **c.** A piece of a wild cucumber plant is replanted in a garden, where a new plant grows.
   - **d.** A brown guinea pig with a white tuft of hair on his head mates with a white guinea pig with a brown splotch on her back. Two of the offspring resemble the father. The third does not look like either parent but has several brown splotches in a white background.

8. Why must human gametes be haploid?

9. How does a human oocyte rid itself of excess genetic material? How does a human sperm shed excess cytoplasm?

10. How are the timetables different for male and female gametogenesis?

TO THINK ABOUT . . .

1. A dog has 39 pairs of chromosomes. Considering only independent assortment (the random lining up of maternally and paternally derived chromosomes), how many genetically different puppies are possible from the mating of two dogs? Is this number an underestimate or an overestimate? Why?

2. Men who have the genetic disorder cystic fibrosis often lack vasa deferentia. Why are they unable to father children?

3. Why is it extremely unlikely that a human child will be genetically identical to a parent?

4. Many male veterans of the Vietnam war claim that their children born years later have birth defects caused by a contaminant in the herbicide Agent Orange used as a defoliant in the conflict. What types of cells would the chemical have to affect in these men to cause birth defects years later? Explain your answer.

5. Why is a fertilized polar body unable to support the development of an embryo, while an ooocyte, which is genetically identical to it, can?

6. How do the structures of the male and female human gametes aid them in performing their functions?

7. The sperm of some infertile men can be injected into an oocyte, and conception sometimes occurs. Then the fertilized ovum is implanted into a woman, where development continues. Researchers in France and Spain took a similar approach on several men who produce spermatids that cannot mature into spermatozoa. It worked, and several healthy babies have been born. Why would this procedure work with spermatids but not with primary spermatocytes?

TO LEARN MORE . . .

References and Resources

Anderson, Alun. July 17, 1992. The evolution of sexes. *Science,* vol. 257. Did sexual reproduction evolve to temper competition between organelles?

Bell, Graham. February 1992. Dividing they stand. *Natural History. Paramecia* reproduce without sex and have sex without reproduction.

Gould, Stephen Jay. 1980. Dr. Down's syndrome. In *The Panda's Thumb.* New York: W. W. Norton. An entertaining essay on the importance of meiosis.

Lewis, Ricki. July 1996. Special delivery system for sperm. *Photonics Spectra.* Lasers are used to aim and deliver sperm into oocytes.

Lewis, Ricki. April 1996. Using technology to teach difficult genetic concepts. *The American Biology Teacher.* Polar bodies hold clues to health.

Moran, Nancy A. April 1992. Quantum leapers. *Natural History.* Sumac gall aphids shift from a parthenogenetic to a sexual lifestyle.

Roach, Mary. January 1997. Sperm futures. *Discover.* Sperm in the news.

Strain, Lisa, et al. October 1995. A human parthenogenetic chimaera. *Nature Genetics,* vol. 11. When a boy's blood cells tested as female, researchers knew that the child was most unusual.

denotes articles by author Ricki Lewis that may be accessed via *Life*'s home page /http://www.mhhe.com/sciencemath/biology/life/.

A polar bear parent and offspring.

11 Animal Development

chapter outline

A Descriptive Science Becomes Experimental 210
Preformation Versus Epigenesis
Early Experiments with Embryos
Differentiation and Selective Gene Activation

Two Beginnings: Early Development in a Worm and a Fly 213
A Worm—*Caenorhabditis elegans*
A Fly—*Drosophila melanogaster*

The Stages of Human Prenatal Development 216
The Preembryonic Stage
The Embryonic Stage
The Fetal Period
Labor and Birth

Postnatal Growth and Development 230

Aging 232
What Triggers Aging?

The Human Life Span 234
Causes of Death
Is Longevity Inherited?

Happy Birthday!

Reporters gathered in Arles, France, on March 1, 1995, to celebrate a very special birthday. Jeanne Calment, a woman with the distinction of being listed in *The Guinness Book of Records* as "oldest living human," was turning 120 years old. Although she was able to ride her bicycle daily up until her 100th birthday, after that Jeanne began to feel some physical effects of her many years. Her hearing diminished, and she lost her vision to cataracts. She hadn't been able to walk since breaking her hip in her 115th year. But this "oldest living human" still had a sharp wit. When reporters asked her what type of future she anticipated, she answered, "A very short one."

What factors contributed to Jeanne Calment's longevity?

to be continued . . .

A Descriptive Science Becomes Experimental

When kindergarten children gather daily around an incubator to watch a clutch of chicken eggs hatch in early spring, they observe, with wonder, precisely what Greek philosopher Aristotle did in the fourth century B.C. The children look at photographs to follow chick development (fig. 11.1); Aristotle broke open an egg each day and drew what he saw. The rapid transformation of a thin strip of tissue to a baby chick mesmerizes biologists of all ages.

Preformation Versus Epigenesis

Until the mid-eighteenth century, most biologists supported **preformation,** the idea that a fertilized egg (or even a gamete) contains a tiny, preformed organism, or "germ," that grows, like a homunculus (see fig. 10.12*b*). In 1759, German physiologist Kaspar Friedrich Wolff published a revolutionary pamphlet, which stated that a fertilized ovum contains no preformed germ but unspecialized tissue that gradually specializes. Other scientists doubted Wolff's hypothesis of **epigenesis** for the next 50 years.

Through the nineteenth century, biologists continued to observe and sketch embryos. A German scientist, Karl Ernst von Baer, studied dog, chick, frog, and salamander embryos; he was a founder of embryology. Baer watched, fascinated, as a fertilized egg divided into two cells, then four, then eight, gradually forming a solid ball that then hollowed out. The hollow balls would indent, then form two layers, then build up what Baer thought were four layers (later, scientists corrected this to three). Baer's observation—that specialized cell layers arise from nonlayered, unspecialized cells—confirmed Wolff's theory of epigenesis.

When Baer compared embryos of several vertebrate species, he noted that they were incredibly similar in early stages but then gradually developed species-distinct characteristics (fig. 11.2). For example, an early appearing bump looked similar on embryos from four species, but one developed as a paw, one an arm, another a wing, and another a flipper. Charles Darwin interpreted Baer's observations as evidence that different species descended from a common ancestor. Baer disagreed vehemently with Darwin on this matter for the rest of his days.

Later, German naturalist Ernst Haeckel extended Darwin's interpretation. Haeckel suggested that vertebrate embryos proceed through a series of stages that represent adult forms of the species that preceded them in evolutionary time. Today, most biologists reject the biogenetic law, as Haeckel's view is called. Structures in early embryos probably resemble each other simply because they have not yet specialized, and early in development, different species go through similar stages as cells organize into tissues and organs.

Early Experiments with Embryos

During the twentieth century, research went in a new direction. Embryologists began intervening in development, disrupting embryonic structures to see what happened. A student of Haeckel's, Wilhelm Roux, was one of the first to alter embryos. He shook them, suspended them upside down, destroyed certain cells with heat or cold, and observed fairly consistent results—disturbed embryos cease developing.

In the 1920s, German zoologist Hans Spemann experimented on newt and frog embryos. He teased apart the cells of a very early embryo and watched them develop, separately, into distinct but identical individuals. Each cell of the original embryo retained the potential to give rise to a complete individual, a capacity called **pluripotency.** In the 1980s, scientists repeated Spemann's experiment with a twist: they separated the cells of three different mice embryos and reassembled the cells to create one new embryo—which then developed into a mouse with six parents (fig. 11.3)! The mouse experiment only works at early stages of development; by the time a mouse embryo contains 1,000 cells, the cells have shut off the genes that would allow them to generate a complete organism.

Spemann also operated on embryos. In one famous experiment, he removed part of a region called the dorsal lip from a newt embryo. Then he transplanted the dorsal lip to the corresponding region of an embryo at the same stage but of a different newt species (fig. 11.4). A new embryo grew at the site where the dorsal lip tissue was grafted onto the second embryo! The dorsal lip, named Spemann's organizer, is therefore a vital trigger in embryonic development.

(a)

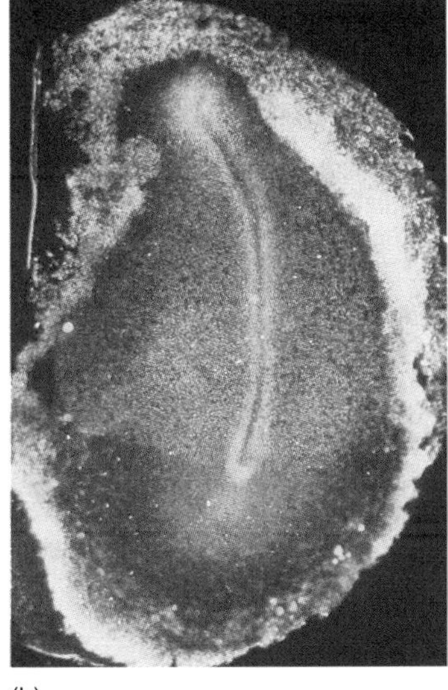

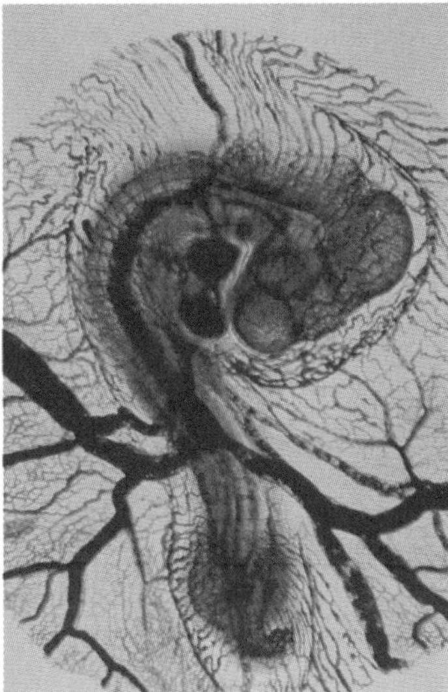

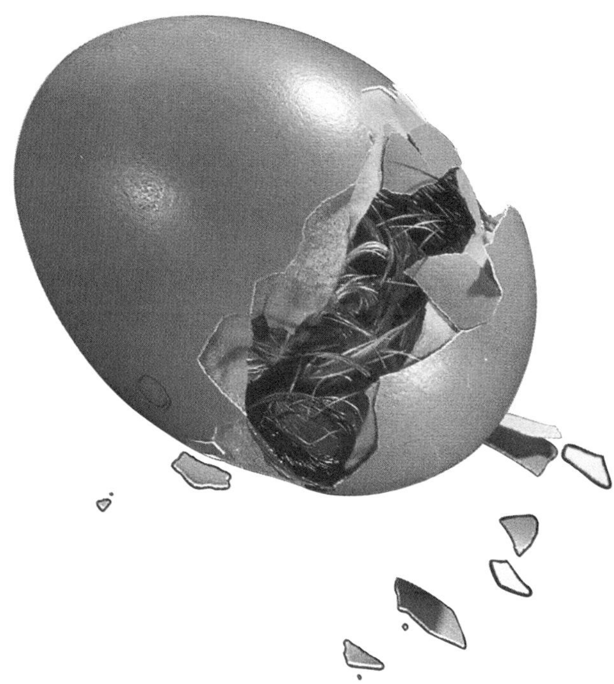

(b)

FIGURE 11.1

Watching embryos. (*a*) Using photographs, these children can follow the day-by-day development of chicks—echoing the observations of Aristotle and other biologists. (*b*) A developing chick first appears to be a streak of tissue. Layers and then distinct structures rapidly form. By the 21st day, a chick is ready to peck its way out of the shell.

Transplant experiments on many species helped researchers decipher the fates and functions of embryonic structures. These experimenters sought to determine when a cell commits to follow a specific **developmental pathway,** or series of events that culminates in formation of a particular tissue such as nerve or muscle. If transplanted cells develop in their new location as they would have if not moved, then their developmental pathway was set before the time of transplantation. If, on the other hand, a transplanted cell develops as other cells do in its new location, then the cell was still pluripotent when it was transplanted.

With the discovery of the structure of genetic material in 1953, embryologists—now known by the broader term developmental biologists—could use the ability to detect changes in gene activity as development proceeds. All cells contain all of the organism's genes, but as cells become specialized, or **differentiated,** they express only certain genes, whose products endow the cell with its distinctive characteristics. This differential gene expression underlies the transformation of a relatively formless clump of cells into a complex grouping of specialized tissues.

Differentiation and Selective Gene Activation

Elegant experiments conducted on the African clawed frog, *Xenopus laevis,* in the early 1960s demonstrated that differentiated cells contain a complete genetic package but only use some of the information in it. British developmental biologist J. B. Gurdon designed a way to "turn on" the genes that a differentiated cell normally "turns off."

Gurdon removed the nucleus from a differentiated cell lining the small intestine of a tadpole and injected it into an egg of the same species whose nucleus had been destroyed. Some of the altered eggs developed into tadpoles, and a few developed as far as normal adult frogs (fig. 11.5). The fact that a nucleus from a differentiated cell could support normal development from a single cell proved that such a nucleus contains complete genetic instructions for the species. Genes must therefore be inactivated—not lost, as some investigators once thought—as a multicellular organism develops. In 1997, Scottish researchers conducted a version of Gurdon's experiment with sheep. They successfully grew adult sheep by transferring nuclei from breast cells to oocytes that had had their nuclei removed.

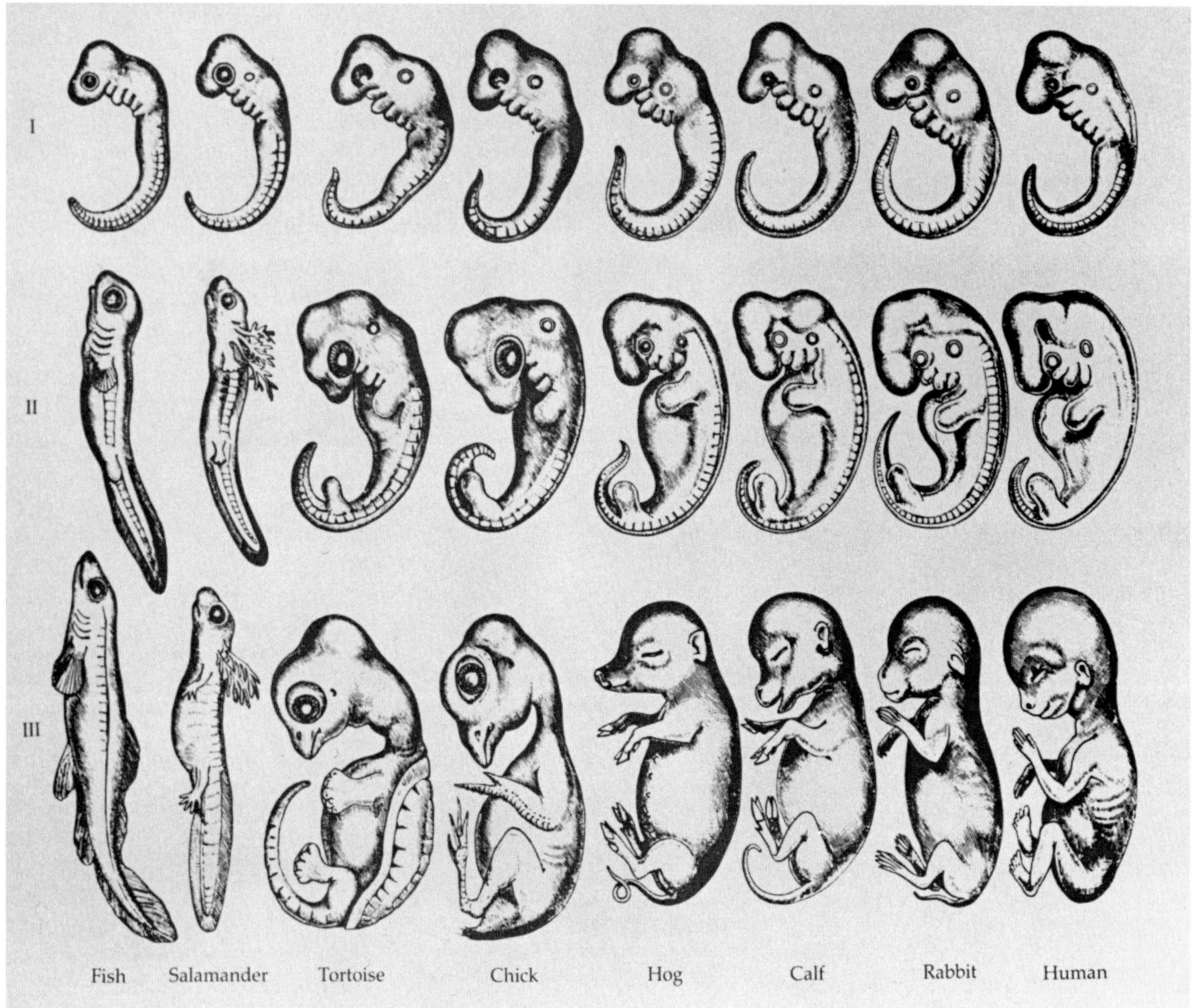

FIGURE 11.2

Embryonic resemblances. In 1828, embryologist Karl Ernst von Baer neglected to label two embryos in an experiment. The embryos were so similar in appearance that without the labels, he couldn't tell whether they were embryos of mammals, reptiles, or birds. Notice how similar the early embryos in the top row are—each has gill slits and a tail reminiscent of fish. Later in development, as depicted in the bottom row, the embryos become more distinctive. These sketches were drawn in 1901 and therefore have some inaccuracies.

Developmental biologists increasingly use molecular techniques to decipher the distributions of proteins that guide pluripotent cells to form tissues. This process requires growth (cell division) as well as differentiation. Consider muscle formation in mammals. A "master control gene" turns on dozens of genes, causing the cell to produce muscle-specific proteins. When the master control gene is placed in chick connective tissue cells growing in culture, the cells elongate and then pulsate—they become muscle!

The study of animal development is perhaps most exciting in our own species. But studying human development before birth is problematical for a variety of legal, ethical, and practical reasons. To avoid these problems, and because many developmental processes are very similar in different species, researchers often work with model organisms to learn more about human development. We now examine two widely used model organisms and then look at human development.

Mastering Concepts

1. How does epigenesis differ from preformation?
2. What is the significance of the similarity in appearance of early embryos of different species?
3. What are some of the ways that developmental biologists study embryos?

FIGURE 11.3

The tricolored mouse has six parents—a pair for each of the three original embryos combined to form the one individual. The tricolored mouse was bred to a white male mouse, and she gave birth to the three mice shown above her in the photograph. The fact that the tricolored mouse gave birth to solid-colored mice of the original three colors indicates that some of her gametes descended from each of the original three donor embryos.

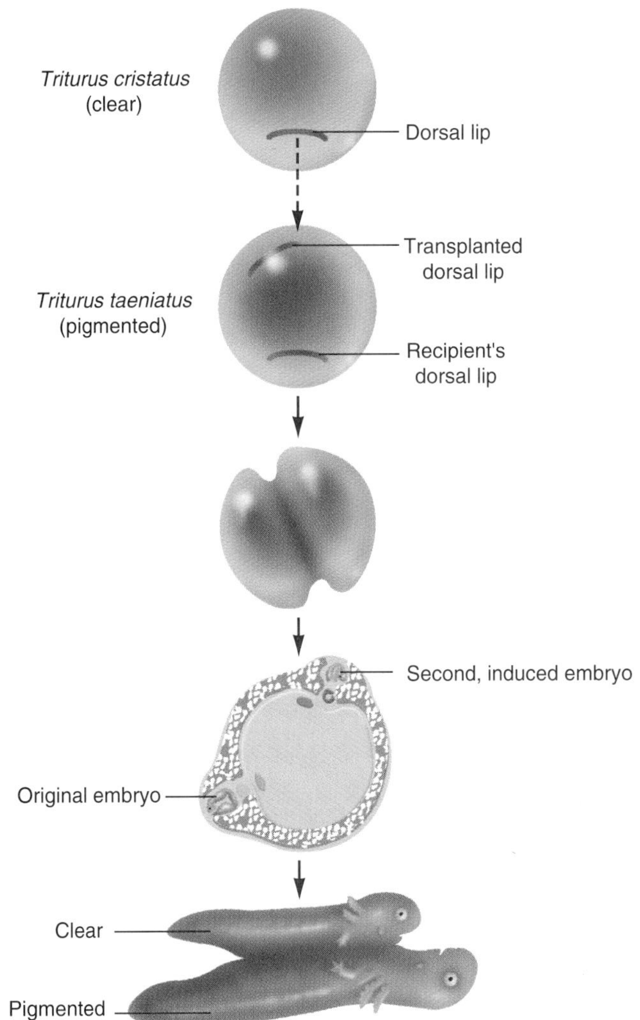

FIGURE 11.4

When Spemann transplanted the dorsal lip of an early embryo of one newt species to an embryo of a different species, the recipient developed into two embryos. This result indicates that the dorsal lip is very important in early development. The newt species are of different colors, which enabled Spemann to distinguish donor from recipient tissue.

Two Beginnings: Early Development in a Worm and a Fly

A millimeter-long worm, a fruit fly, and a human may not appear to have much in common. However, each animal meets the same challenges in development. It begins as a single cell, it undergoes rapid mitotic cell division, its cells commit to follow particular pathways, and, gradually, distinct structures begin to appear, specialize, interact, and organize.

A Worm—*Caenorhabditis elegans*

The nematode worm *Caenorhabditis elegans,* a 1-millimeter-long soil resident, is in many ways an ideal model organism. An adult worm consists of a mere 959 somatic cells, yet it has differentiated tissues just as complex animals do (fig. 11.6). The worms are easy to maintain in the laboratory and can even be frozen for later use. *C. elegans* grows from fertilized egg to adult in just 3 days and is completely transparent. With these worms, researchers can watch animal development unfold, counting and tracking cells as they divide, migrate, and interact.

That is exactly what Einhard Schierenberg of West Germany and John Sulston of England did from 1975 to 1983. They followed the fates of each cell under a light microscope and published a "fate map" showing all cell lineages—which cells give rise to which cells. By slicing the worms into very thin sections and observing them under an electron microscope, the researchers described every aspect of the adult animal's anatomy. For example, they identified 302 nerve cells and 7,600 connections the cells make with each other.

Development begins with fertilization. For the first few cell divisions, one cell in the worm looks much like another. By the fourth mitosis for some cells, and later for others, certain cells commit to follow certain developmental pathways—that is, to become part of the outer covering (cuticle), intestine, muscle, nerve, pharynx, vulva, egg, or gonad. These specialized structures do not actually appear until a certain number of cell divisions later. Sometime during the initial hours of rapid cell division, controls are set to transform the transparent bag of identical-appearing cells into a moving, coordinated multicellular organism.

A half day after the first cell division, the organism is a larva of 558 cells. Further rounds of cell division and the death of precisely

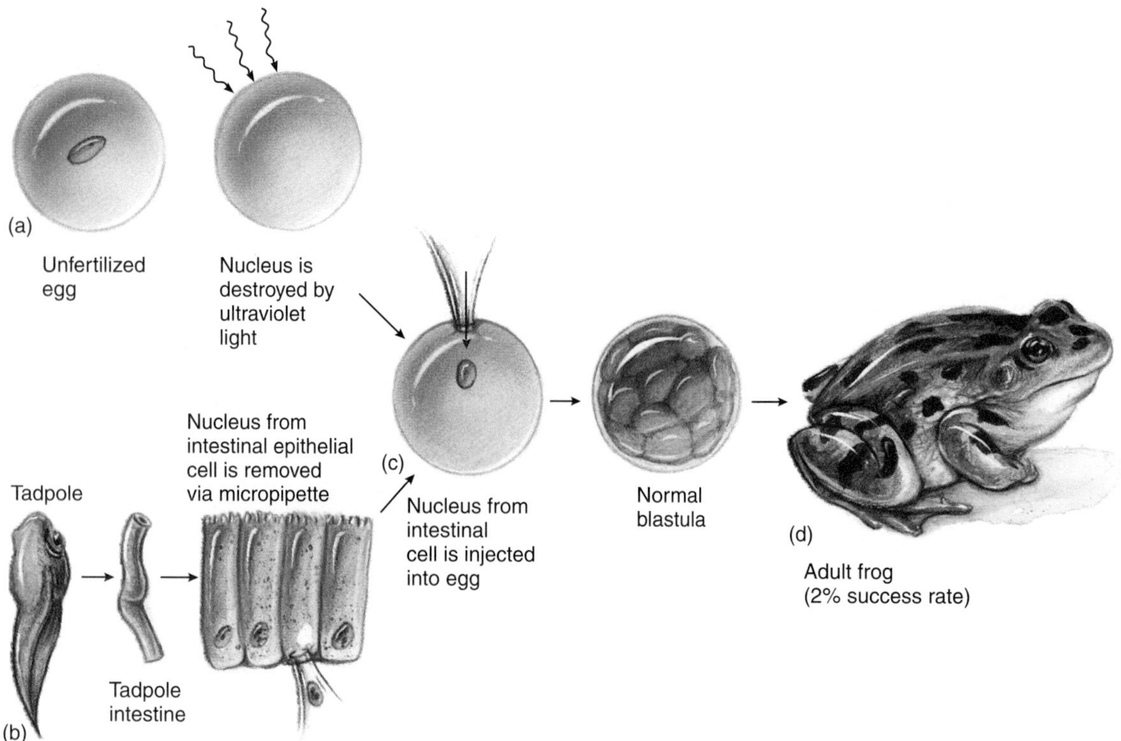

FIGURE 11.5

Nuclear transplantation shows that the nucleus of a differentiated cell can support complete development. (*a*) In the first step of the procedure, a frog egg nucleus is inactivated with radiation. (*b*) A nucleus is removed from a tadpole's differentiated cell. (*c*) Next, the nucleus from the differentiated cell is transferred to the enucleated egg. (*d*) The egg, controlled by its transplanted nucleus, develops into a tadpole and then a frog. Similar experiments have recently been conducted on sheep.

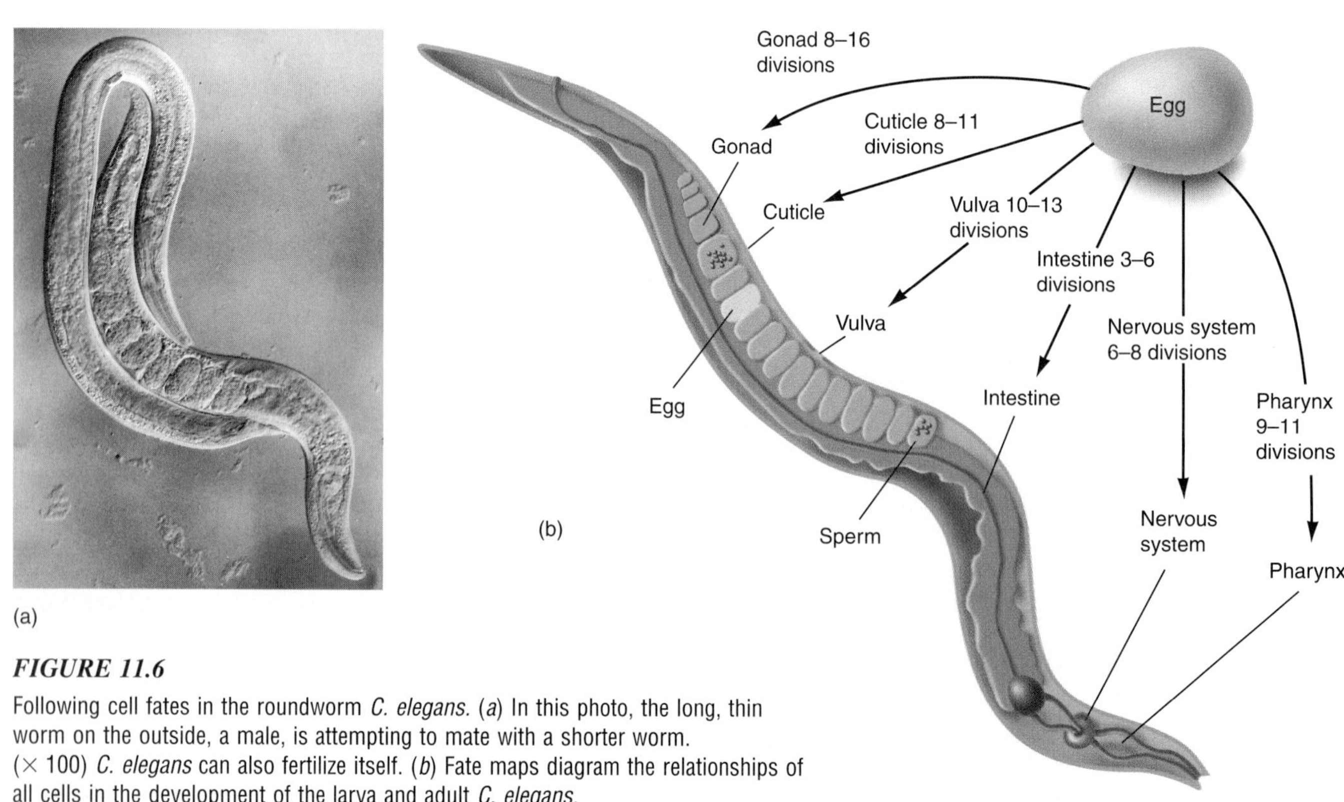

FIGURE 11.6

Following cell fates in the roundworm *C. elegans.* (*a*) In this photo, the long, thin worm on the outside, a male, is attempting to mate with a shorter worm. (× 100) *C. elegans* can also fertilize itself. (*b*) Fate maps diagram the relationships of all cells in the development of the larva and adult *C. elegans.*

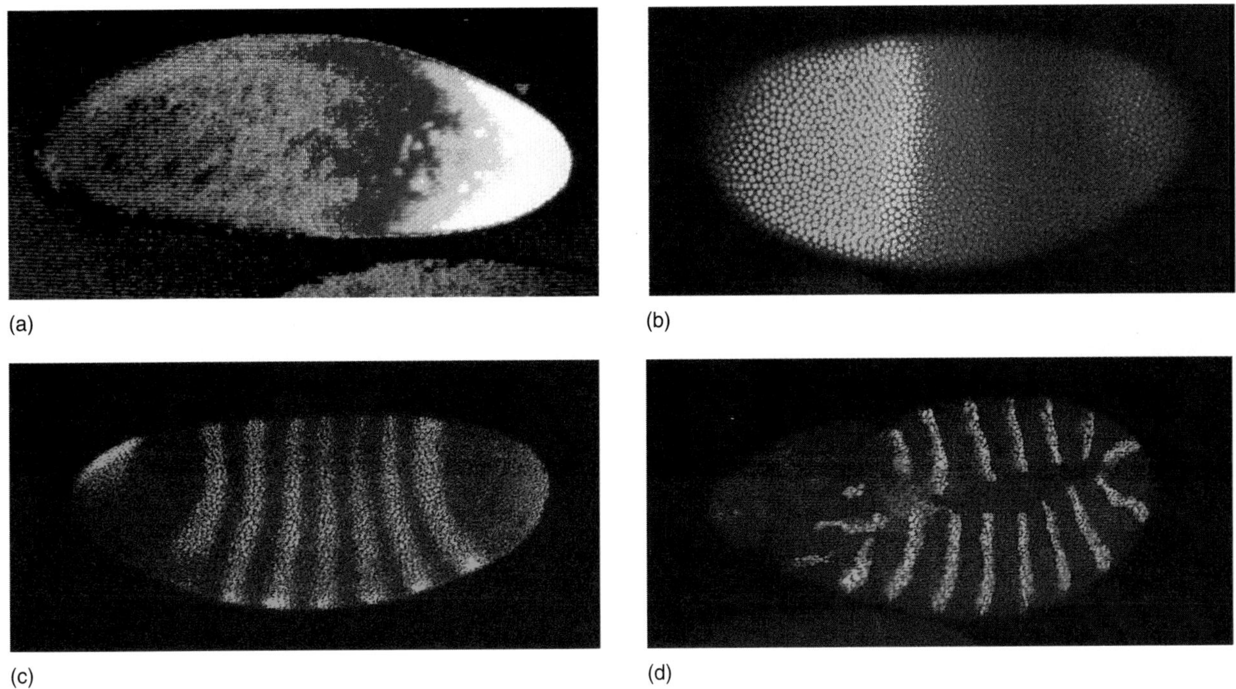

FIGURE 11.7

Biochemical gradients control early development in fruit flies. Very early in the development of a fruit fly, a gradient of bicoid protein, from the oocyte, distinguishes the embryo's head region from its rear. In the first panel of this computer-enhanced image (*a*), different colors represent different concentrations of bicoid. (*b*) Bicoid protein directs the synthesis of two other proteins, one shown in red and one in green. (*c*) A half hour later, another gene directs production of a protein that divides the embryo into seven stripes. (*d*) Yet another gene oversees dividing each existing section of the embryo in two. Overall, genes produce protein gradients that set up distinct biochemical environments in different parts of the embryo, and these biochemicals further influence differentiation.

113 cells sculpts the 959-celled adult worm. In addition, 2,000 cells are set aside as germ cells, which will give rise to sex cells.

Thanks to the efforts of Schierenberg, Sulston, and other "worm people," as *C. elegans* researchers call themselves, we now have a complete picture of an animal's development, providing clues to how more complex animals form. Imagine tracking development in the many-trillion-celled human!

A Fly—*Drosophila melanogaster*

For an embryo to form, cells must become "aware" of their position—of whether they are located at the front (anterior) or rear (posterior), back (dorsal) or belly (ventral), or left or right side of the forming body. Cells obtain this positional information from protein gradients. Recall that a gradient is a substance or activity that varies in amount or strength over a defined area. In a developing organism, cells at one end contain high levels of a particular protein, and cells at the other contain very little, with the cells in between having steadily diminishing levels.

The role of genes and the protein gradients they produce is well studied in fruit flies (*Drosophila melanogaster*). For example, a gene that produces a protein called bicoid instructs the embryo to develop a head end and a tail end. When the gene is abnormal (mutated), the embryo develops two rear ends! (Bicoid means "two tails.") Normally, a greater concentration of bicoid protein in the anterior end of the fly embryo causes it to differentiate the tissues characteristic of the head. Similarly, at the posterior of the animal, higher concentrations of a protein called nanos signal the abdominal tissues to differentiate. Proteins such as bicoid and nanos that are present in gradients that influence development are called **morphogens** (morphology means "form").

Gradients of several different morphogens orchestrate normal fruit fly development (fig. 11.7). The varying concentrations of different morphogens distinguish parts of the embryo, like using increasing intensities of several colors in a painting. The morphogen gradients are signals that stimulate cells to produce yet other proteins, which ultimately regulate the formation of the specific components of a differentiated structure. A cell destined to be part of the adult fly's antenna, for example, has different morphogen concentrations than a cell whose descendants will become part of the eye. Sometimes, when morphogen genes mutate, developmental signals go off in the wrong part of the animal—with striking results. The fly in figure 11.8*b* is an example. Called, appropriately, *Antennapedia,* it has legs in place of its antennae. Genes that, when mutated, lead to organisms with structures in the wrong places are called **homeotic.**

Although best studied in fruit flies, homeotic genes are widespread in life. Cows, earthworms, humans, frogs, lampreys, corn, mice, flour beetles, locusts, bacteria, yeast, chickens, roundworms, and mosquitoes are only some of the organisms that have homeotic genes. Homeotic mutations cause diseases in humans,

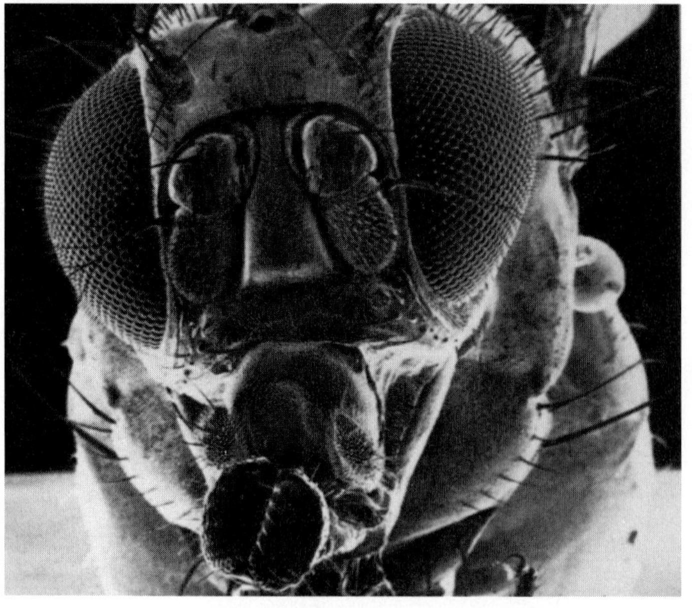

(a)

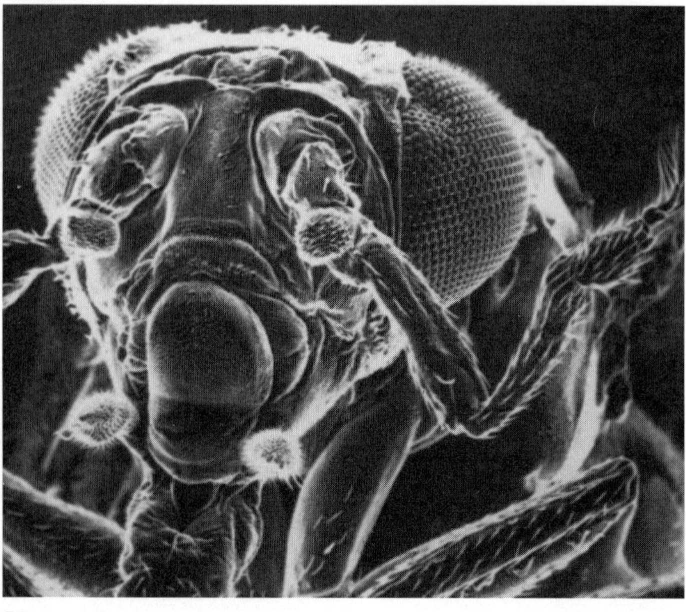

(b)

FIGURE 11.8

A homeotic mutation sends the wrong morphogen signals to the antenna of a fruit fly, directing the tissue to differentiate as leg. (*a*) A normal fly; (*b*) A homeotic fly.

Table 11.1 Stages of Animal Development

Stage	Events	Timing in Humans
Preembryo		
Gametogenesis	Manufacture of sperm and oocytes	Ongoing during much of adulthood
Fertilization	Sperm and oocyte meet	Within 24 hours after ovulation
Cleavage	Rapid cell division	Days 1–3
Embryo		
Gastrulation	Germ layers form	Week 2
Organogenesis	Cells specialize, forming organs	Weeks 3–8
Fetus		
Further development	Growth and tissue specialization	Week 9 throughout life

including a blood cancer, cleft palate, extra fingers and toes, lack of irises, and DiGeorge syndrome, a group of head and neck anomalies. The fact that homeotic genes are found in diverse species indicates that these controls of early development are both ancient and very important.

Mastering Concepts

1. How is development in diverse animals similar?
2. Why is *Caenorhabditis elegans* a good model system in which to study animal development?
3. How do protein gradients influence development?

The Stages of Human Prenatal Development

Developmental biologists recognize three stages of prenatal development, or gestation. The first 2 weeks, which occur before the three layers that define an embryo are distinct, have a variety of names, but we will call this period the **preembryonic stage.** It includes fertilization, rapid mitotic cell division of the zygote as it travels through the woman's fallopian tube towards the uterus; implantation into the uterine wall; and initial folding into layers.

The **embryonic stage** lasts from the end of the second week until the end of the eighth week. Cells of the three layers continue to divide, differentiate, and interact, forming tissues and organs. Structures that support the embryo—the **placenta, umbilical cord,** and **extraembryonic membranes**—also develop during this period.

The third stage of prenatal development is the **fetal period,** lasting from the beginning of the ninth week through the full 38 weeks of development. Organs begin to function and coordinate to form organ systems. Growth is very rapid. Prenatal development ends with **labor** and **parturition,** which is the birth of a baby. Table 11.1 lists the major stages in the prenatal development of humans and some other animals.

The Preembryonic Stage

It takes 2 weeks for the zygote to consist of enough cells to form the three folds that will develop further into the embryo.

Fertilization

The first step in prenatal development is the initial contact between sperm and secondary oocyte. Recall that the oocyte arrests in metaphase II. A thin, clear layer of proteins and carbohydrates, the **zona pellucida,** encases the oocyte, and a layer of cells called the **corona radiata** surrounds the zona pellucida. The sperm must penetrate these layers to fertilize the ovum (fig. 11.9).

Ejaculation deposits about 100 million sperm cells in the woman's body. A sperm cell can survive here for up to 6 days, but it can only fertilize the oocyte in the 12 to 24 hours after ovulation. A process called **capacitation** activates the sperm inside the woman, altering sperm cell surfaces in a way that enables them to enter an oocyte. In some species, the oocyte secretes a chemical that attracts sperm, although such an attractant has not been found in humans. The female's muscle contractions, the movement of the sperm tails, the surrounding mucus, and the waving cilia on the cells of the female reproductive tract all propel the sperm toward the oocyte. Despite this help, only 200 or so sperm near the oocyte, and only one will penetrate it (fig. 11.10). Even if a sperm touches an oocyte, fertilization may not occur.

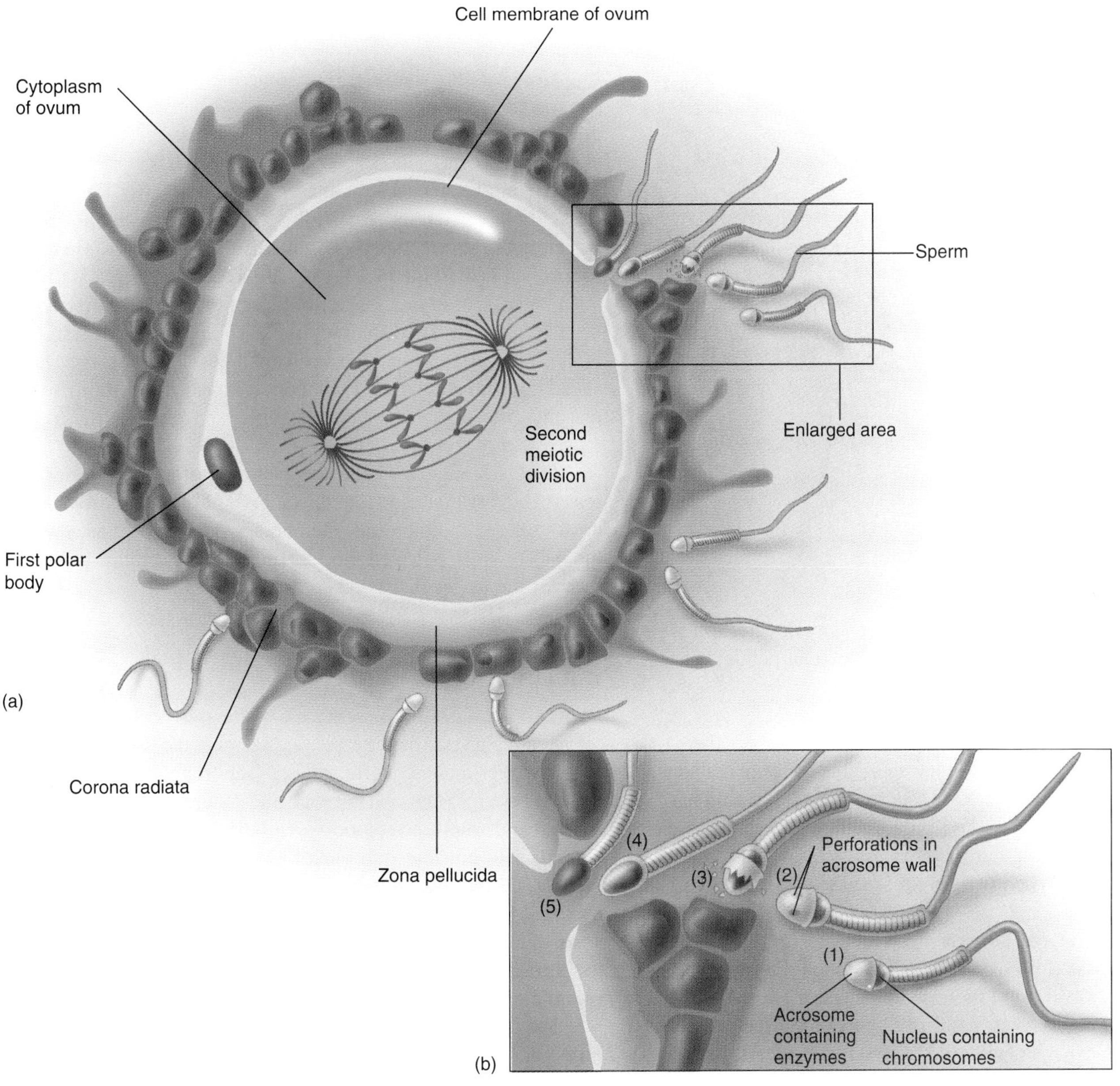

FIGURE 11.9

Fertilization. (*a*) At the moment of fertilization, the sperm's acrosome bursts, spilling forth enzymes that help the sperm's nucleus enter the oocyte (*b*).

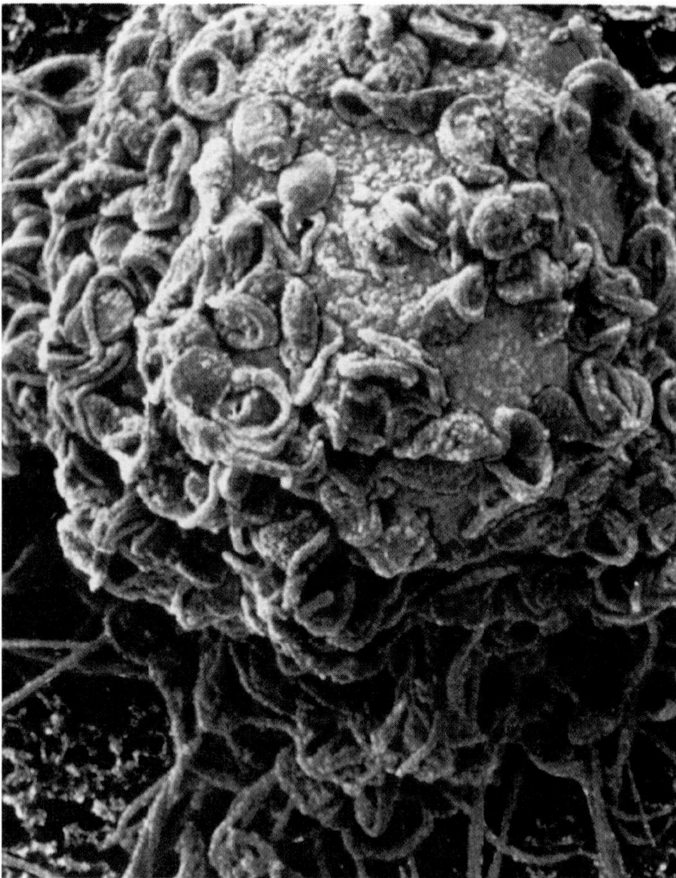

FIGURE 11.10

This electron micrograph shows many sperm near the oocyte's surface. Only one will penetrate.

When a particular sperm contacts a secondary oocyte, its acrosome bursts, spilling enzymes that digest the zona pellucida and corona radiata. Fertilization (conception) begins when the outer membranes of the sperm and secondary oocyte meet. A wave of electricity spreads physical and chemical changes across the entire oocyte surface, which keeps other sperm out. If more than one sperm were to enter a single oocyte, the resulting cell would have too much genetic material to develop normally. Very rarely, such fetuses survive until birth, but they have defects in many organs and die within days. When two sperm fertilize two oocytes, fraternal twins result.

As the sperm enters the secondary oocyte, the female cell completes meiosis and becomes a fertilized ovum. Fertilization is not complete, however, until the genetic packages, or **pronuclei,** of the sperm and ovum meet. Within 12 hours of the sperm's penetration, the nuclear membrane of the ovum disappears, and the two new sets of chromosomes mingle, forming the first cell of the zygote (fig. 11.11). This cell has 23 pairs of chromosomes—one chromosome of each pair from each parent. The cell is still within a fallopian tube.

Cleavage

About 3 hours after fertilization, the zygote divides for the first time, beginning a period of rapid mitotic cell division called **cleavage.** The progeny cells of cleavage divisions are called **blastomeres.** Sixty hours after fertilization, a second division occurs, and the zygote now consists of four cells. Cleavage continues until a solid ball of 16 or more cells forms. This ball is called a **morula,** Latin for "mulberry," which it resembles (figs. 11.12 and 11.13).

Three days after fertilization, the morula is usually still within the fallopian tube, but it is moving towards the uterus. It is about the same size as the fertilized ovum, because the initial cleavage divisions produce progeny cells that are about half the size of the parent cell. Soon the cellular size levels off, and the zygote begins to enlarge as cells accumulate. During cleavage, organelles and molecules from the secondary oocyte's cytoplasm still control most cellular activities, but some of the zygote's genes become active. The sperm contributes its nucleus, the centrosomes that organize cell division, and proteins that are necessary for cleavage to occur.

The morula reaches the uterus 3 to 6 days after fertilization. It then hollows out, its center filling with fluid that seeps in from the uterus. The preembryo is now also a **blastocyst,** Greek for "germ bag" (see Fig. 11.12 *e,* and *f*). The blastomeres are either in an outer layer or on the interior face of the blastocyst. The outer layer of cells is called the **trophoblast** (Greek for "nourishment of germ"). Certain trophoblast cells will develop into a membrane called the **chorion,** which eventually forms the fetal portion of the placenta, the organ that brings oxygen and nutrients to and removes wastes from the fetus.

The blastomeres inside the blastocyst form the **inner cell mass.** They will develop into the embryo plus its supportive structures, called extraembryonic membranes. The fluid-filled center of the ball of cells is called the **blastocyst cavity.** Table 11.2 summarizes the stages discussed so far.

Implantation

Between the fifth and seventh days after fertilization, the blastocyst attaches to the uterine lining, and the inner cell mass within settles against it (fig. 11.14). The trophoblast secretes digestive enzymes that eat through the outer layer of the uterine lining, and ruptured blood vessels surround the blastocyst, bathing it in nutrient-rich blood. This nestling of the blastocyst into the uterine lining is called **implantation,** and it completes by day 14. The trophoblast layer directly beneath the inner cell mass thickens and sends out fingerlike projections into the uterine lining at the site of implantation. These projections develop into the chorion.

The trophoblast cells now secrete a hormone, **human chorionic gonadotropin** (hCG), that causes other hormonal changes and prevents menstruation. In this way, the blastocyst helps to ensure its own survival, for if menstruation occurs, it would leave the woman's body along with the tissue in the uterus. The trophoblast cells produce hCG for about 10 weeks.

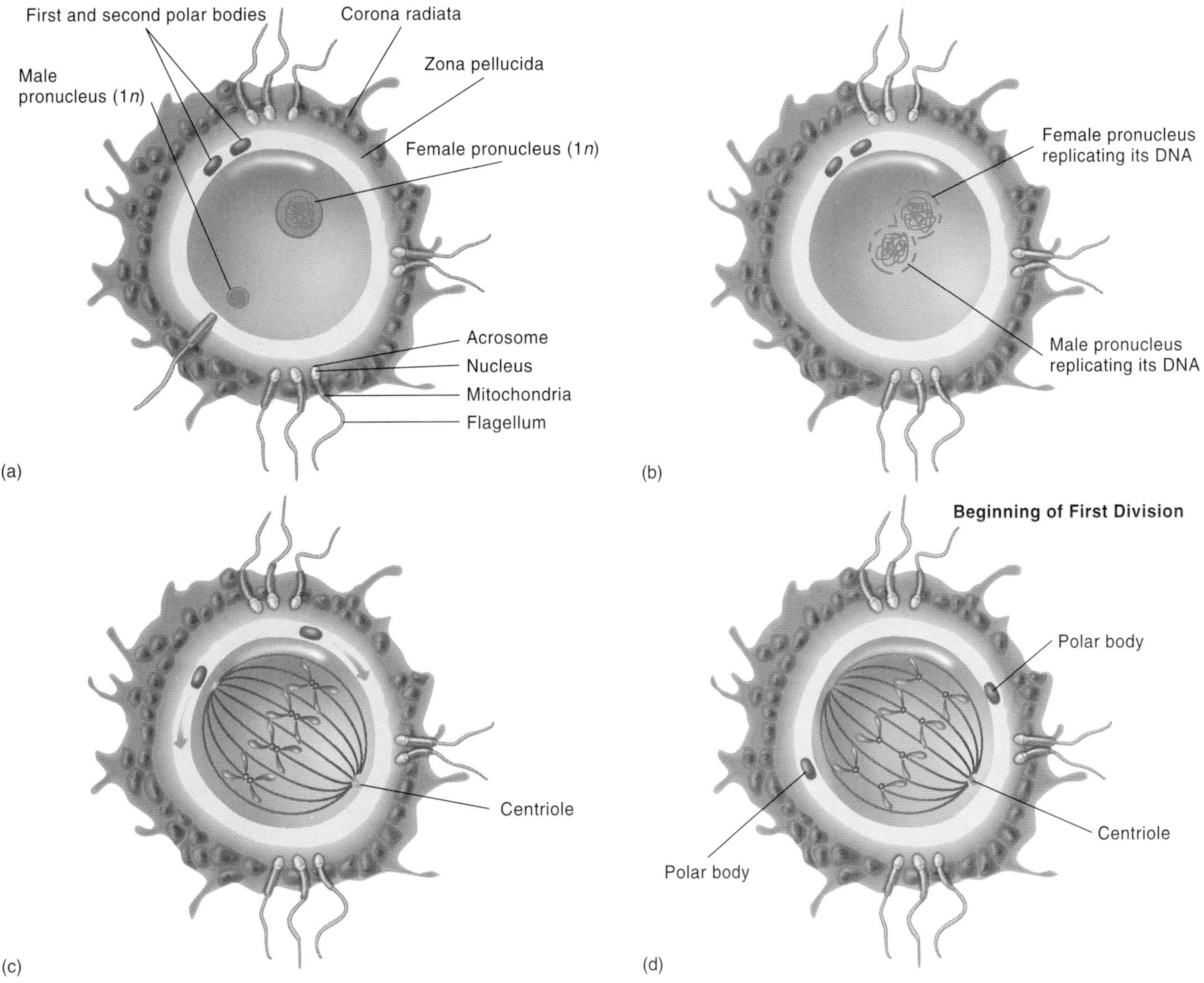

FIGURE 11.11

The zygote forms. (*a*) Once the sperm cell enters the oocyte, (*b*) its pronucleus approaches the oocyte's pronucleus and (*c*) finally brings the two sets of chromosomes together. (*d*) The first cell of the zygote begins to divide.

Table 11.2 Stages and Events of Early Human Prenatal Development

Stage	Time Period	Principal Events
Fertilized ovum	12–24 hours following ovulation	Oocyte fertilized; zygote has 23 pairs of chromosomes and is genetically distinct
Cleavage	30 hours to third day	Mitosis increases cell number
Morula	Third to fourth day	Solid ball of cells
Blastocyst	Fifth day through second week	Hollowed ball forms trophoblast (outside) and inner cell mass, which implants and flattens to form embryonic disc
Gastrula	End of second week	Primary germ layers form

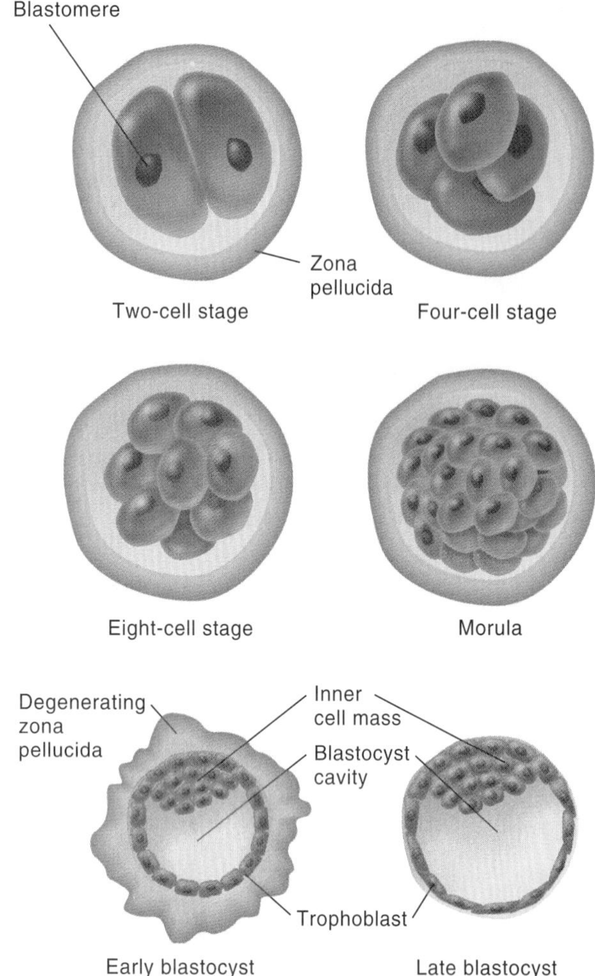

FIGURE 11.12

Cleavage divisions lead to formation of a mulberry-like structure, the morula, which then hollows out to form the blastocyst. The outer cell layer is the trophoblast, and the clump of cells on one side of the interior is the inner cell mass.

Gastrulation

During the second week of prenatal development, the blastocyst completes implantation, and the inner cell mass changes. A space called the **amniotic cavity** forms within a sac called the **amnion,** which lies between the inner cell mass and the portion of trophoblast that has nestled into the uterine lining. The inner cell mass then flattens and is called the **embryonic disc.**

The embryonic disc at first consists of two layers. The outer layer, nearest the amniotic cavity, is called the **ectoderm** (Greek for "outside skin"). The inner layer, closer to the blastocyst cavity, is the **endoderm** (Greek for "inside skin"). Shortly after, a third and middle layer called the **mesoderm** (Greek for "middle skin") forms. This three-layered structure is the primordial embryo, or **gastrula.** The process of forming the primordial embryo is called gastrulation, and the layers are called **germ layers** (fig. 11.15).

Gastrulation is important in prenatal development because a cell's location in a particular layer determines its fate (fig. 11.16). Ectoderm cells develop into the nervous system, sense organs, the outer skin layer (epidermis), hair, nails, and skin glands. Meso-

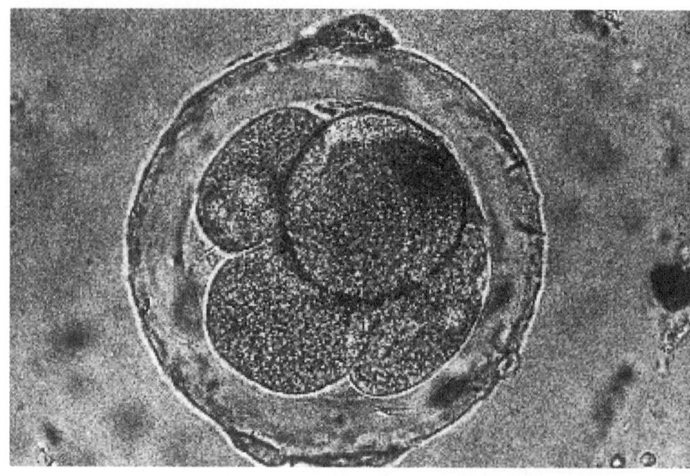

(a)

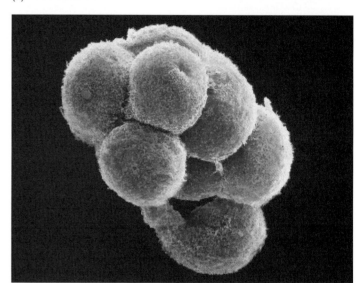

(b)

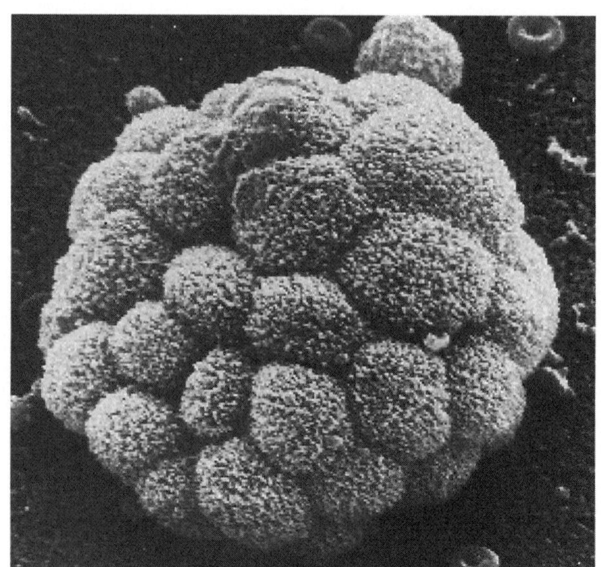

(c)

FIGURE 11.13

Electron micrographs of a human at the (*a*) 4-celled stage, (*b*) 16-celled stage, and (*c*) morula stage.

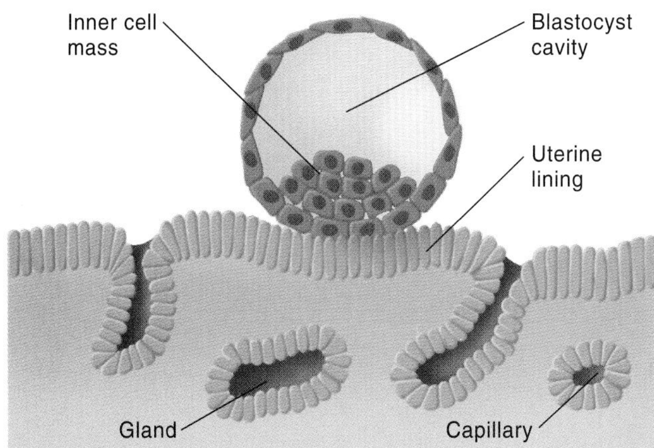

FIGURE 11.14

Implantation. By day 8 the blastocyst begins to implant in the uterine lining. By day 14, the process completes.

FIGURE 11.15

After 2 weeks, the primary germ layers form, and the embryonic stage of prenatal development begins.

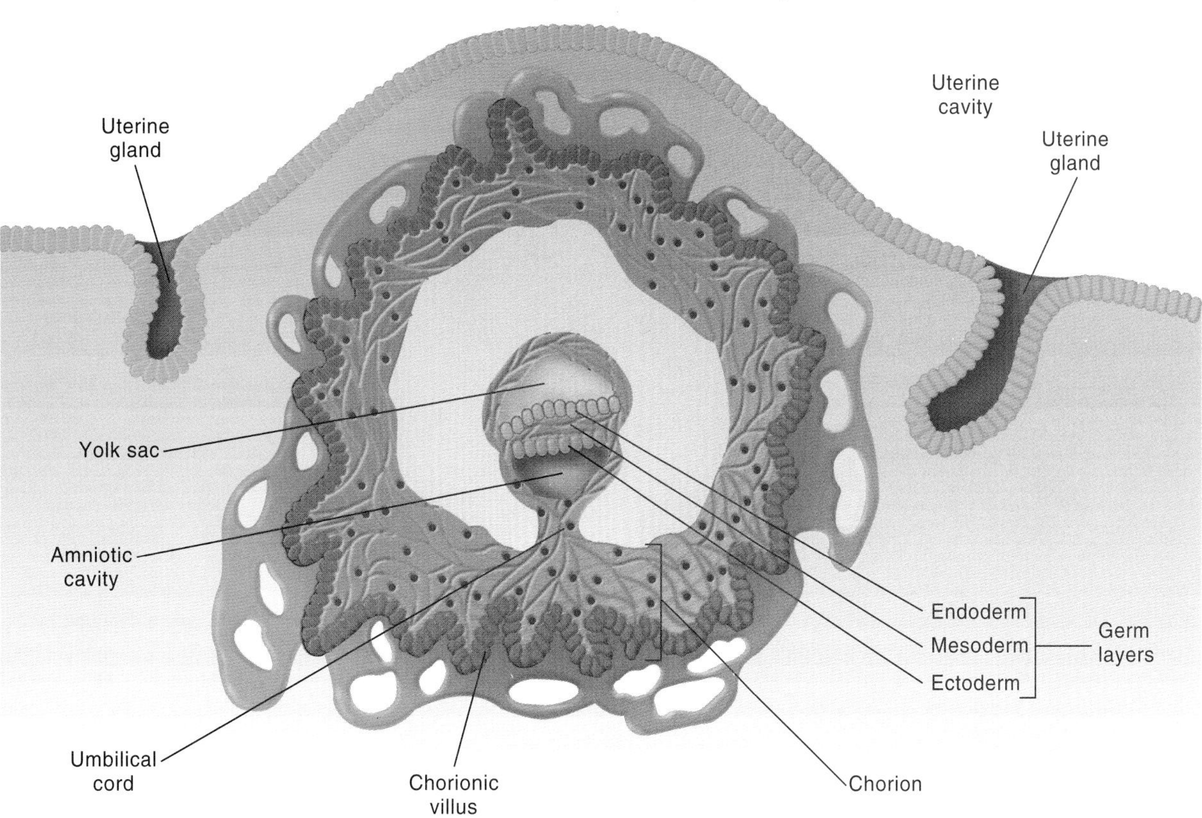

derm cells develop into bone, muscle, blood, the inner skin layer (dermis), reproductive organs, and connective tissue. Endoderm cells eventually form the organs and the linings of the digestive, respiratory, and urinary systems. Gastrulation marks the start of morphogenesis, the series of events that forms distinct structures.

The preembryonic stage ends after the second week of prenatal development. Although the woman has not yet missed her menstrual period, she might notice effects of her shifting hormones, such as swollen and tender breasts and fatigue. By now her urine contains enough hCG for an at-home pregnancy test to detect. Highly sensitive blood tests can detect hCG as early as 5 to 7 days after conception.

Bioethics Connections 11.1 discusses what the media and politicians commonly refer to as "embryo research" but is actually research on preembryos.

Mastering Concepts

1. List the events of the preembryonic stage of prenatal human development.
2. How are more than one sperm prevented from fertilizing an oocyte?
3. Why is the early cleavage preembryo the same size as the fertilized ovum?
4. How does the preembryo implant in the uterine lining?
5. At what point do cells of the preembryo form two distinct groups?
6. How is menstruation prevented when a woman becomes pregnant?

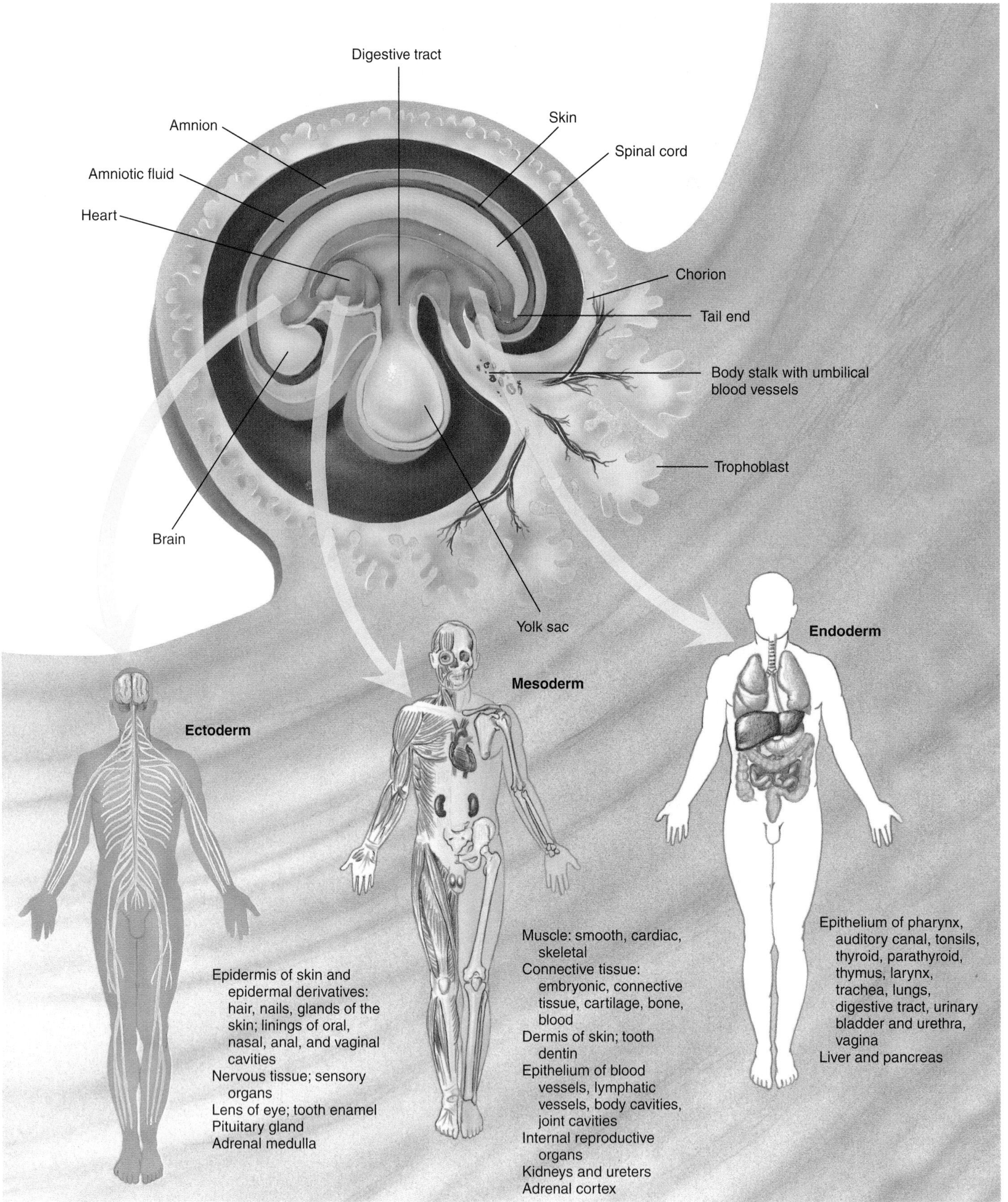
Digestive tract
Amnion
Skin
Spinal cord
Amniotic fluid
Heart
Chorion
Tail end
Body stalk with umbilical blood vessels
Trophoblast
Brain
Yolk sac
Endoderm
Mesoderm
Ectoderm
Epidermis of skin and epidermal derivatives: hair, nails, glands of the skin; linings of oral, nasal, anal, and vaginal cavities
Nervous tissue; sensory organs
Lens of eye; tooth enamel
Pituitary gland
Adrenal medulla
Muscle: smooth, cardiac, skeletal
Connective tissue: embryonic, connective tissue, cartilage, bone, blood
Dermis of skin; tooth dentin
Epithelium of blood vessels, lymphatic vessels, body cavities, joint cavities
Internal reproductive organs
Kidneys and ureters
Adrenal cortex
Epithelium of pharynx, auditory canal, tonsils, thyroid, parathyroid, thymus, larynx, trachea, lungs, digestive tract, urinary bladder and urethra, vagina
Liver and pancreas

FIGURE 11.16

These organ systems arise from each of the primary germ layers.

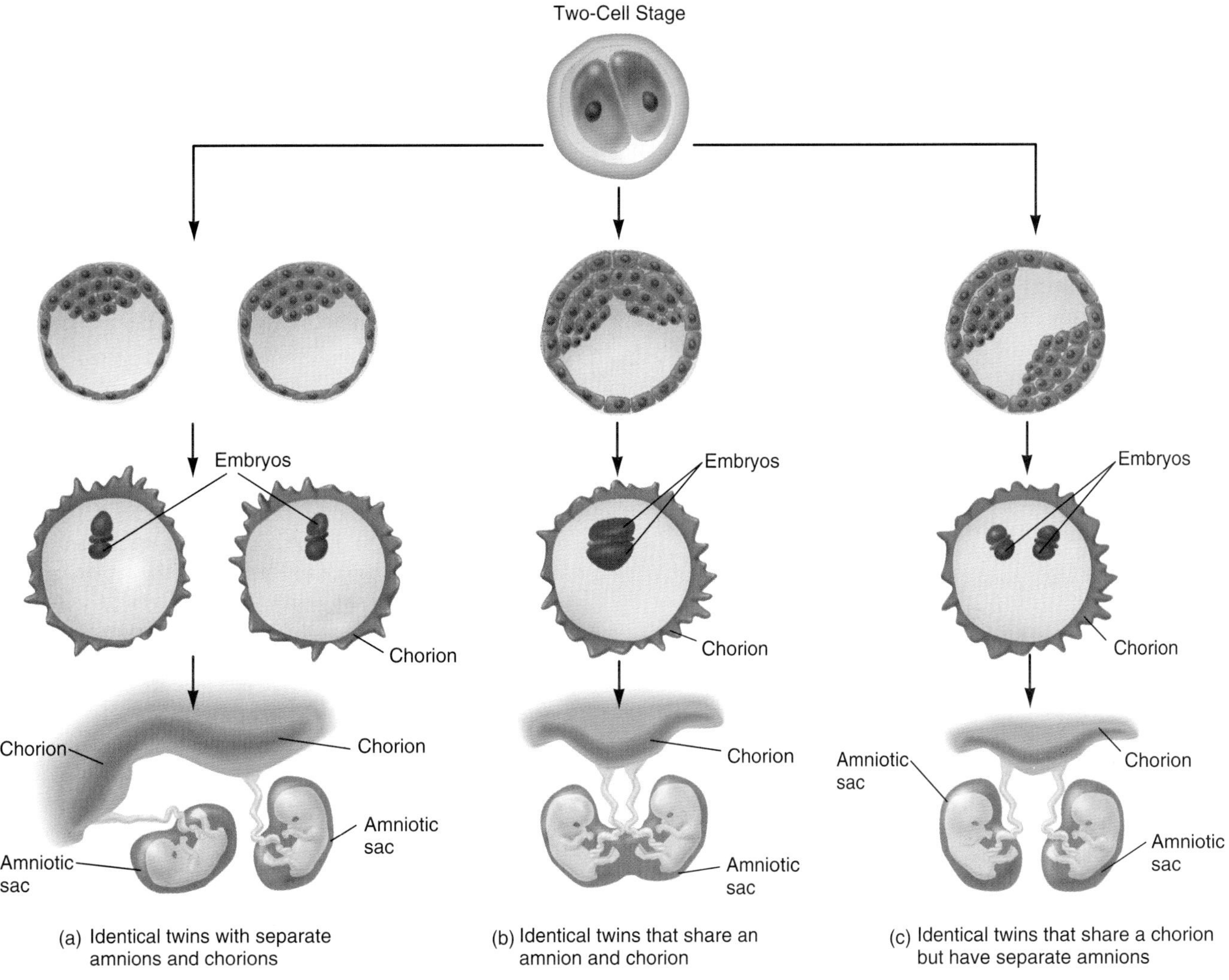

FIGURE 11.17

Facts about twins. Identical twins originate at three points in development. (*a*) In about one-third of identical twins, separation of cells into two groups occurs before the trophoblast forms on day 5. These twins have separate chorions and amnions. (*b*) About 1% of identical twins share a single amnion and chorion, because the tissue splits into two groups after these structures have already formed. (*c*) In about two-thirds of identical twins, the split occurs after day 5 but before day 9. These twins share a chorion but have separate amnions. Fraternal twins result from two sperm fertilizing two secondary oocytes. The twins develop their own amniotic sacs, yolk sacs, allantois, placentae, and umbilical cords. Fraternal twins can have two different fathers, if the mother ovulated twice and had intercourse with two men within a short period of time. Fraternal twins run in families. In about 55% of all twin conceptions, one twin dies before birth. Many times the parents do not know the surviving baby is a twin!

The Embryonic Stage

During the embryonic stage, organs begin to develop and structures form that will nurture and protect the developing organism. The embryo begins to organize around a central axis, and towards the end of this period, the embryo develops rapidly in a precisely regulated sequence of events. Supportive structures forming now are the placenta, the umbilical cord, and extraembryonic membranes called the **yolk sac** and **allantois.** The chorion and amnion are extraembryonic membranes that began to form during the preembryonic period. Three types of identical twins are distinguished by whether they share or have separate amnions and chorions (fig. 11.17).

Supportive Structures

By the end of the second week after fertilization, as the embryo folds into three layers, the fingerlike projections from the chorion, called **chorionic villi,** extend further into the uterine lining, establishing the beginnings of the placenta. One side of the placenta—the chorion tissue—comes from the embryo, and the other side consists of blood pools from the pregnant woman's circulation (fig. 11.18). The two blood systems are separate but lie side by side; the chorionic villi lie between the pools of maternal blood. This proximity enables nutrients and oxygen to diffuse from the woman's circulatory system across the chorionic villi cells to the embryo and for wastes to leave the embryo's circulation and enter the woman's, which eventually excretes them.

bioethics 11.1 connections

Preembryos as Research Subjects

The needle piercing a secondary oocyte in figure 11.A will deliver a sperm cell into the oocyte in an effort to help the people who donated these cells become parents. This procedure, discussed in chapter 12, would not have been possible without experiments that followed the fate of fertilized ova, for a few cell divisions, in laboratory glassware.

Biology or Politics?

Basic research on human prenatal development provides new infertility treatments and contraceptives. However, many people object to using prenatal humans in the experiments that make reproductive technologies possible. As a result of the controversy over the ethics of this type of experimentation, the U.S. government, at the time of this writing, does not fund such research.

In 1993, when President Clinton lifted a ban on government-funded human embryo research begun in 1980, the National Institutes of Health (NIH) established a Human Embryo Research Panel, consisting of scientists, ethicists, and attorneys. The panel outlined the goals and restrictions of the work:

- It distinguished "research embryos," intentionally created for research, from "spare embryos," which fertility clinics would otherwise discard because they are in excess or are abnormal.
- It defined when it is ethical to use research embryos. Conditions include when the work would be of outstanding value or could not occur any other way. The panel also advised using the fewest possible embryos for the shortest possible time.
- It determined that oocyte donors should be women undergoing abdominal surgery for other reasons. They must provide informed consent and not be paid for donating oocytes.
- It recommended that embryo research be restricted to the first 14 days of development, when implantation completes.

The NIH recommendations were soon tabled. Following the conservative outcome of the fall 1994 elections, President Clinton reinstituted the ban on federal funding but restricted it to embryos conceived solely for research purposes. Representative Robert K. Dornan (R-Calif.) led the congressional effort to halt the research, writing in a letter that more than two dozen members of the new congress signed: "The issue of human embryo research is extremely troubling to us. Given the new realities of Congress, many of the incoming freshmen will presumably feel the same way. So let it be known that we plan to use every legislative means available to prevent federal funds from being spent on grotesque research of this nature."

Figure 11.B shows the stage of human prenatal development that is meant by "human embryo." It is not yet, technically, an embryo.

Opinions vary widely on whether research on a prenatal human is ethical. The NIH panel concluded that moral consideration of the issue is appropriate, but that an undifferentiated collection of cells

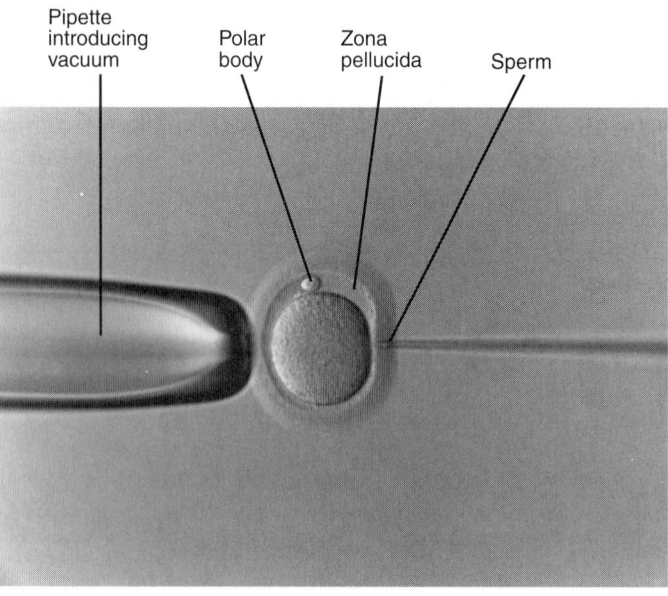

FIGURE 11.A
Injecting sperm into oocytes has helped some couples become parents. The technique was perfected with experiments that followed the fertilized ova's first few mitotic divisions.

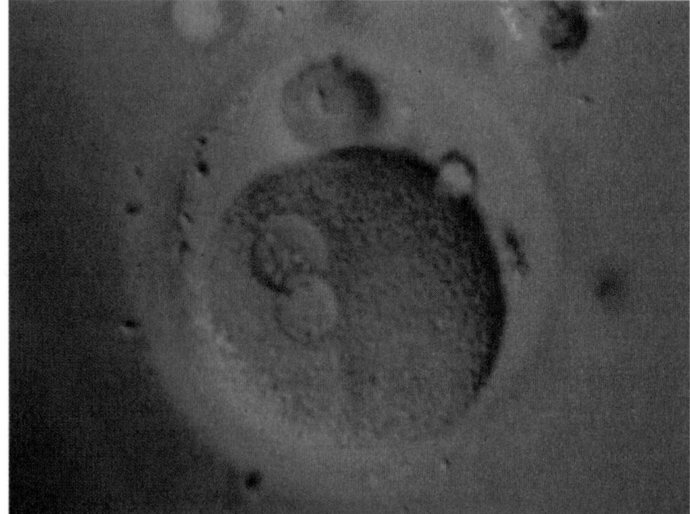

FIGURE 11.B
The subjects of "human embryo research" are often not technically embryos (which by definition have distinctive tissue layers), but preembryos, which consist of a few, unspecialized cells. This "research embryo"—created specifically for research purposes—is a single cell.

does not have the same moral status as a child or an adult. At the other extreme, one noted scientist compared the research to Nazi war crimes. Many scientists agree that education on the nature of prenatal stages of development might enable all people to develop more informed opinions. Reports of one particular experiment illustrate how easily the media can sensationalize and misinterpret developmental biology.

To Clone or Not to Clone

At a scientific meeting in 1993, George Washington University researchers Jerry Hall and Robert Stillman reported that they had performed "blastomere separation" on several human preembryos slated for discard at a fertility clinic because each had been fertilized by more than one sperm. The experiment was simple, a re-creation of what happens naturally when identical twins form—the researchers separated a two-celled preembryo and allowed each half to divide a few times (fig. 11.C). This had been done years earlier on frogs, mice, sheep, lambs, and calves, so biologists were not particularly impressed.

But the media went wild. Soon, headlines trumpeted the "cloning," with references to *Jurassic Park*'s cloned dinosaurs and *Brave New World*'s cloned babies. But cloning is the growing of a new individual from a differentiated somatic cell, as comedian Woody Allen did in the 1970s film *Sleeper,* in which a futuristic society clones a deceased leader from cells of his nose. Woody Allen, playing hero Miles Monroe, performed cloning. Hall and Stillman did not.

A Feat in Sheep

In, 1997, the media again painted images of mass-producing genetically identical humans when Scottish researchers cloned a sheep from the nucleus of a specialized (udder) cell. They used a technique similar to that depicted in figure 11.5. Politicians quite seriously called for "no clone zones," despite the great difficulty of cloning sheep and, possibly, humans. Lost in initial discussions was the fact that the ability to take an adult cell back to an earlier stage of development could perhaps be used to regenerate tissue, which could have medical applications such as treating spinal cord injuries.

On a more philosophical note, the sheep feat prompted many people to contemplate just how important DNA is in determining who we are. But that question has already been answered—just ask identical twins if they are truly identical in every way.

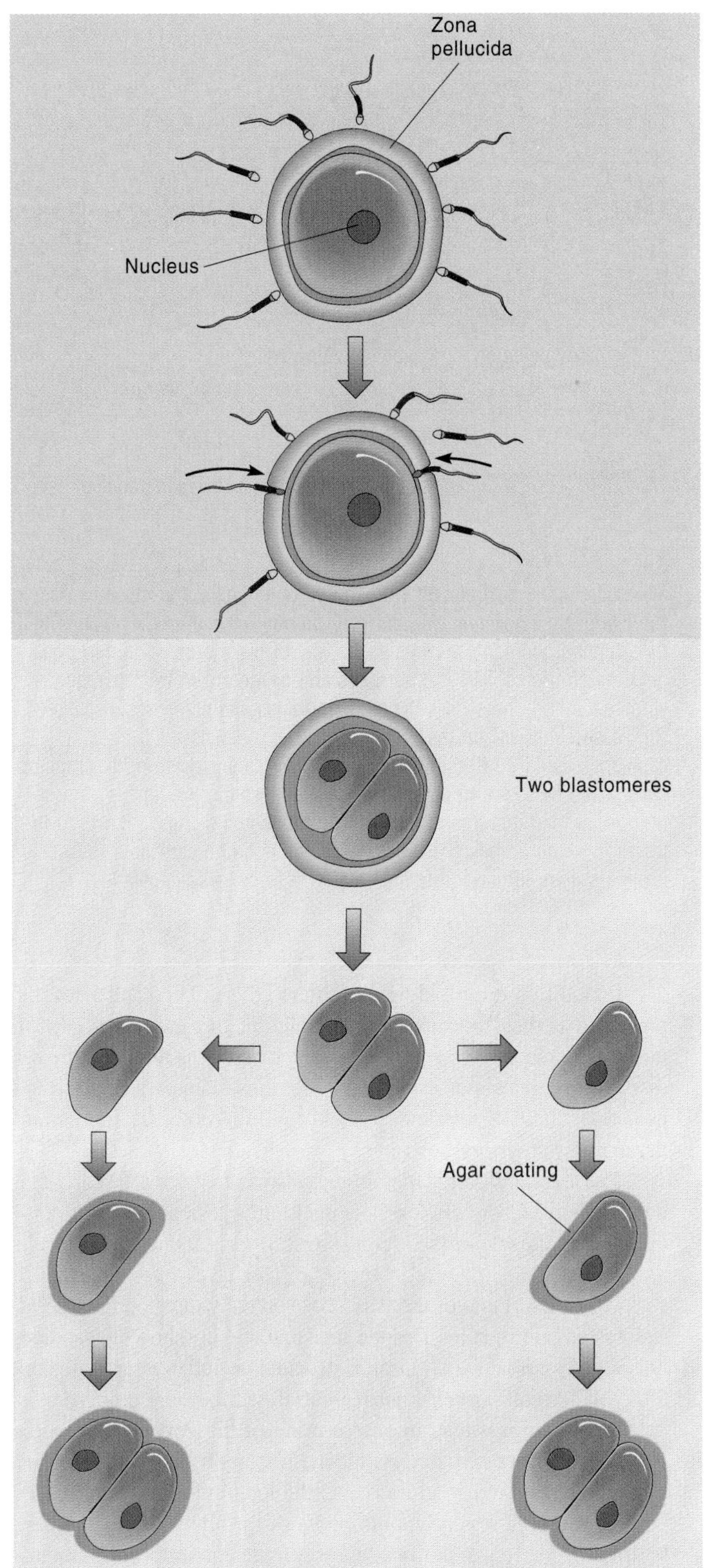

FIGURE 11.C
To clone or not to clone. The researchers who applied an enzyme to a two-celled human preembryo to detach the cells, and then allowed the cells to divide a few more times, called the procedure blastomere separation—an apt description. The media reported it as "cloning."

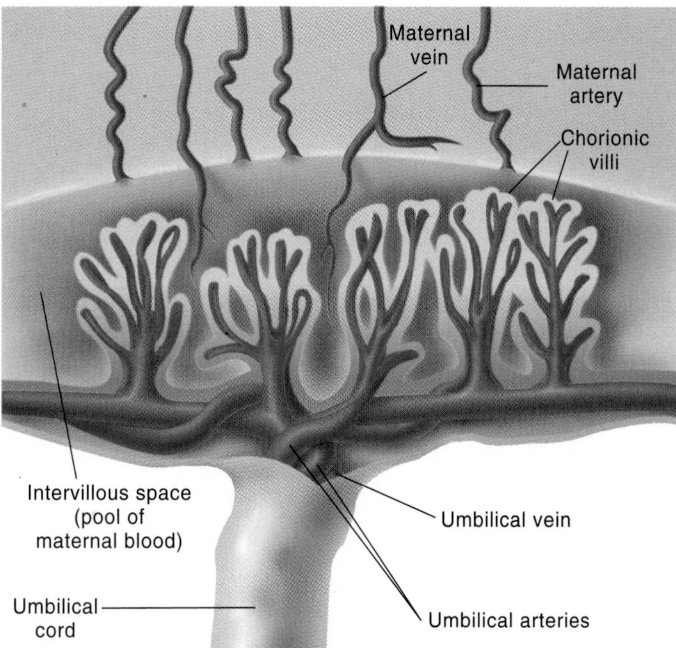

FIGURE 11.18

The circulations of pregnant woman and fetus come into close contact at the site of the chorionic villi, but they do not actually mix. Branches of the pregnant woman's arteries in the wall of her uterus open into pools near the chorionic villi. The nutrients and oxygen from the pregnant woman's blood cross the villi cell membranes and enter fetal capillaries (tiny blood vessels) on the other side of the villi cells. The fetal capillaries lead into the umbilical vein, which is enclosed in the umbilical cord. From here, the fresh blood circulates through the fetus's body. Blood that the fetus has depleted of nutrients and oxygen returns to the placenta in the umbilical arteries, which branch into capillaries. Waste products then diffuse to the maternal side from these capillaries.

In addition to providing a lifeline to the embryo and fetus, the placenta secretes hormones that maintain the pregnancy. At 10 weeks, the placenta is completely developed. It is a reddish brown disc that weighs about 2 pounds (900 grams). Chapter 12 discusses how toxins and viruses can pass through the placenta, sometimes damaging the embryo or fetus.

The extraembryonic membranes also form during the early embryonic period. The yolk sac begins to appear beneath the embryonic disc at the end of the second week. It manufactures blood cells until about the sixth week, when the liver takes over, and then it starts to shrink. Parts of the yolk sac eventually develop into the intestines and germ cells. Despite its name, the human yolk sac does not actually contain yolk. Similar structures in other animals do contain yolk, which provides nutrients to the developing embryo.

By the third week, an outpouching of the yolk sac forms the allantois, another extraembryonic membrane. It, too, manufactures blood cells, and it also gives rise to the fetal umbilical arteries and vein. During the second month, most of the allantois degenerates, but part of it persists in the adult as a ligament supporting the urinary bladder.

As the yolk sac and allantois develop, the amniotic cavity swells with fluid. This "bag of waters" cushions the embryo, maintains a constant temperature and pressure, and protects the embryo if the woman falls. The amniotic fluid comes from the woman's blood. It also contains fetal urine and cells from the amniotic sac, the placenta, and the fetus.

A human embryo is considered an "amniote egg," which means that the embryo is enclosed in a watery sac. The advantage of an amniote egg is that it has built-in water and food supplies, which free the female from dependency on a body of water in which to lay eggs. Amniote eggs are the result of an evolutionary giant step that may have occurred about 300 million years ago. Such structures appeared first in reptiles; today, reptiles, birds, and mammals all have embryos immersed in fluid. More primitive vertebrates such as amphibians must return to water to lay their eggs.

Toward the end of the embryonic period, as the yolk sac shrinks and the amniotic sac swells, the umbilical cord forms. The cord is 2 feet (0.6 meter) long and 1 inch (2.5 centimeters) in diameter. It houses two umbilical arteries, which transport oxygen-depleted blood from the embryo to the placenta, and one umbilical vein, which brings oxygen-rich blood to the embryo. The umbilical cord attaches to the center of the placenta. It twists because the umbilical vein is slightly longer than the umbilical arteries.

The structures that support and nourish an embryo are sometimes used in other ways. Fetal cells collected from amniotic fluid, or sampled chorionic villi cells, provide information on the genetic health of a fetus and are therefore the basis of many prenatal medical tests. Umbilical cord blood is a rich source of stem cells used in gene therapy and for bone marrow transplants needed later in life. Many hair products include extracts from human placentas.

The Embryo

As the days and weeks proceed, different rates of cell division in different parts of the embryo fold tissues into intricate patterns. In a process called **embryonic induction,** the specialization of one group of cells causes adjacent groups of cells to specialize. Gradually, these changes mold the three primary germ layers into organs and organ systems. **Organogenesis** is the term that describes the transformation of the structurally simple, three-layered embryo into a body with distinct organs. Developing organs are particularly sensitive to damage by environmental factors such as chemicals and viruses.

During the third week of prenatal development, a band called the **primitive streak** appears along the back of the embryonic disc. It gradually elongates to form an axis, which is an anatomical reference point that other structures organize around as they develop. The primitive streak eventually gives rise to connective tissue precursor cells and the **notochord,** a structure that forms the basic framework of the skeleton. The notochord induces overlying ectoderm to differentiate into a hollow **neural tube,** which develops into the brain and spinal cord (central nervous system). Formation of the neural tube, or **neurulation,** is a key event in early development because it marks the beginning of organ formation. Soon after neurulation ensues, a reddish bulge containing the heart appears. It begins to beat around day 18. Then the central nervous system starts to form. Figure 11.19 shows some early embryos undergoing these changes. Biology in Action 11.1 discusses organs that are on the wrong side of the body.

11.1 biology in action

A Heart in the Wrong Place

In the original *Star Trek* television series, Dr. McCoy would often complain when examining Mr. Spock that Vulcan organs weren't where they were supposed to be, based on human anatomy. The good doctor would have had a hard time examining humans with a condition called heterotaxy, in which certain normally asymmetrically located organs develop in the wrong place.

In a normal human body, certain organs lie either on the right or the left (fig. 11.D) of the body's midline. The heart, stomach, and spleen are on the left, and the liver is on the right. The right lung has three sections, or lobes; the left lung has two. Other organs twist and turn in either a right or left direction. All these organs originate in the center of the embryo, and then the embryo turns, and the organs migrate to their final locations.

Which way an organ moves depends upon decisions set very early in development, possibly even in the preembryo. Experiments show that damaging one end of a fertilized *Xenopus laevis* preembryo reverses the positions of the adult frog's organs. Similarly, altering the positions of cells in an eight-celled *Caenorhabditis elegans* roundworm preembryo, placing those on top in the middle, causes organs to develop in reverse orientation. Mice preembryos show differences in the genes that are expressed on the right and left sides as early as 48 hours after fertilization.

In humans, misplaced body parts are part of Kartagener syndrome, in which the heart, spleen, or stomach may be on the right, both lungs may have the same number of lobes, the small intestine may twist the wrong way, or the liver may span the center of the body (fig. 11.E). Many people with this syndrome die in childhood from heart abnormalities. Kartagener syndrome was first described in 1936 by a Swiss internist caring for a family with several members who had a strange collection of symptoms—chronic cough, sinus pain, a poor sense of smell, male infertility, and misplaced organs—usually a heart on the right. Many years later, researchers identified another anomaly, which would hold a clue to how the heart winds up on the wrong side of the chest.

All affected individuals lack dynein. Dynein is a protein that enables microtubules to slide past one another and generate motion. Without dynein, cilia cannot wave. In the upper respiratory tract, immobile cilia allow debris and mucus to accumulate, which explains the cough, clogged sinuses, and poor sense of smell. Lack of dynein also paralyzes sperm tails, accounting for male infertility. But how could dynein deficiency explain a heart that develops on the right instead of the left?

Dynein also helps establish the mitotic spindle, which determines the location of the cleavage furrow in dividing cells of the zygote. The cleavage furrow, in turn, determines where in three-dimensional space a particular progeny cell lies. Therefore, the dynein defect may, early on, set cells on a developmental pathway that diverts migration of the heart from the embryo's midline to the left.

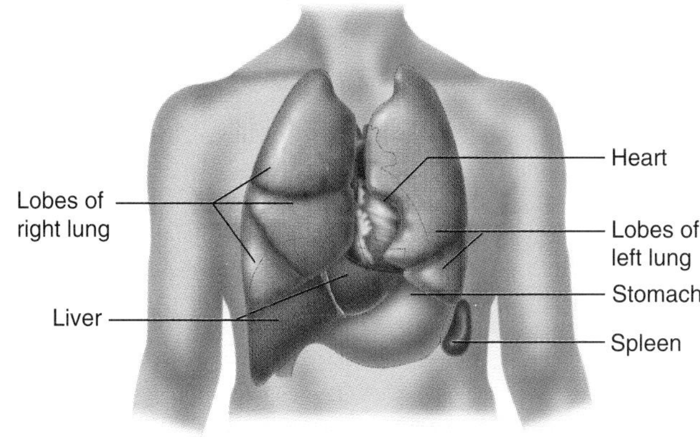

FIGURE 11.D
Normal position of the heart, lungs, liver, spleen, and stomach.

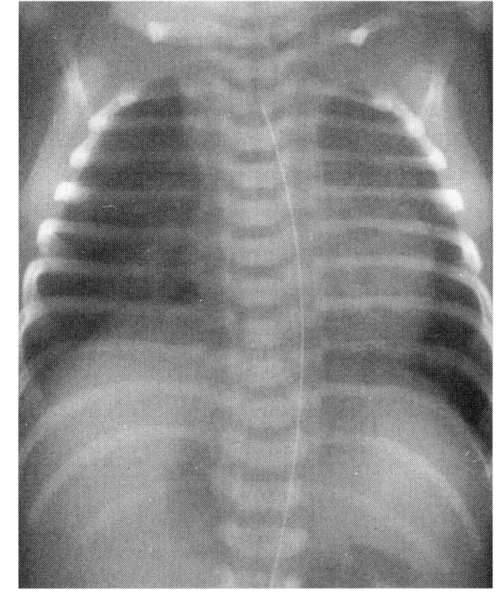

(a)

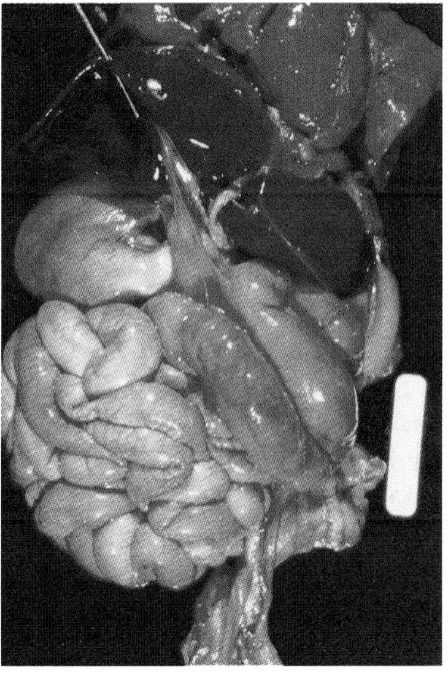

(b)

FIGURE 11.E
An X ray (*a*) and photograph (*b*) showing a child with abnormally located organs. The stomach is on the right, the liver too centrally placed, and the intestine turned the wrong way. The spleen is absent.

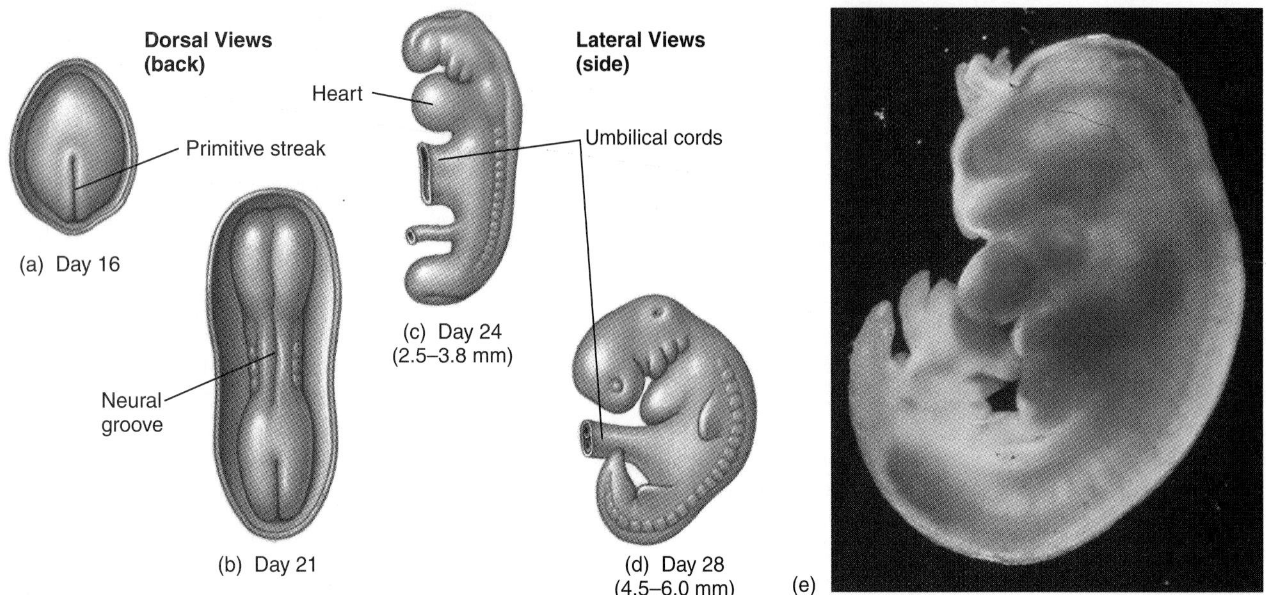

FIGURE 11.19

Early embryos. It takes about a month for the embryo to look like a "typical" embryo. At first all that can be distinguished is the primitive streak (*a*), but soon the central nervous system begins to form (*b*). By the 24th day, the heart becomes prominent as a bulge (*c*), and by the 28th day, the organism is beginning to look human (*d* and *e*).

The fourth week of the embryonic period is a time of rapid growth and differentiation. Blood cells begin to form and to fill primitive blood vessels. Immature lungs and kidneys appear. If the neural tube does not close normally at about day 28, a deformity called a neural tube defect results, which leaves open an area of the spine from which nervous tissue protrudes, causing paralysis. Small buds appear that will develop into arms and legs. The 4-week embryo has a distinct head and jaw and early evidence of eyes, ears, and nose. The rudiments of a digestive system appear as a long, hollow tube that will develop into the intestines. A woman carrying this embryo, which is now only 1/4 inch (0.6 centimeter) long, may suspect that she is pregnant because her menstrual period is about 2 weeks late.

By the fifth week, the embryo's head appears disproportionately large. Limbs extending from the body end in platelike structures. Tiny ridges run down the plates, and by week 6 the ridges deepen as certain cells die, molding fingers and toes. The eyes open, but they do not yet have eyelids or irises. Cells in the brain are rapidly differentiating. The embryo is now about 1/2 inch (1.3 centimeters) from head to buttocks.

During the seventh and eighth weeks, a cartilage skeleton appears. The placenta is now almost fully formed and functional, secreting hormones that maintain the blood-rich uterine lining. The embryo is about the size and weight of a paper clip. The eyes now seal shut and will stay that way until the seventh month. The nostrils are also closed. A neck appears as the head begins to make up proportionately less of the embryo, and the abdomen flattens somewhat (fig. 11.20).

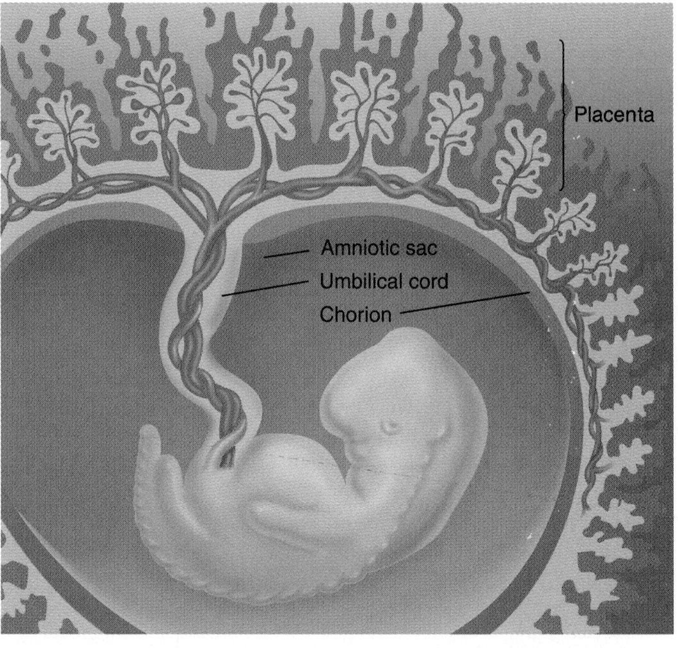

FIGURE 11.20

At 7 weeks of development, the chorionic villi and the tissue between them form the placenta. Blood from the embryo flows to and from the placenta in the umbilical vein and arteries.

Mastering Concepts

1. Which weeks of gestation make up the embryonic period?
2. What supportive structures and extraembryonic membranes develop during this period?
3. What is the relationship between the primitive streak, the notochord, and the neural tube?
4. What structures appear during weeks 4 through 8?

The Fetal Period

The body proportions of the fetus appear more like those of a newborn as time goes on. The ears lie low, and the eyes are widely spaced. Bone begins to form and will eventually replace most of the softer cartilage. Soon, as the nerves and muscles coordinate, the fetus will move its arms and legs.

Sex is determined at fertilization, when a sperm bearing an X or Y chromosome meets an oocyte, which always has an X chromosome. A gene on the Y chromosome, called SRY (for "sex-determining region of the Y") determines maleness. However, all early embryos have rudiments of both sexes, including unspecialized gonads and sets of tubes. At week 7, if the SRY gene is activated, male hormones stimulate differentiation of male reproductive organs and glands from the male precursor structures. If there is no SRY gene—that is, if the fetus is female—female reproductive structures develop. Physical differences between the sexes begin to appear after week 7.

By the 12th week, the fetus is obviously male or female. It now sucks its thumb, kicks, and makes fists and faces, and baby teeth begin to form in the gums. The fetus breathes the amniotic fluid and releases wastes into it. The first trimester (3 months) of pregnancy ends.

During the second trimester, the body proportions of the fetus become even more like those of a newborn. By the fourth month, it has hair, eyebrows, lashes, nipples, and nails. Some fetuses even scratch themselves before birth. Bone continues to replace the cartilage skeleton. The fetus's muscle movements become stronger, and the woman may begin to feel slight flutterings in her abdomen. By the end of the fourth month, the fetus is about 8 inches (20 centimeters) long and weighs about 6 ounces (170 grams).

During the fifth month, the fetus becomes covered with an oily substance called vernix caseosa that protects the skin. White, downy hair called lanugo holds the vernix in place. By 18 weeks, the vocal cords form, but the fetus makes no sounds because it does not breathe air. By the end of the fifth month, the fetus curls into the classic head-to-knees position. It weighs about 1 pound (454 grams) and is 9 inches (23 centimeters) long (fig. 11.21).

During the sixth month, the skin appears wrinkled because there isn't much fat beneath it. The skin turns pink as blood-filled capillaries extend into it. By the end of the second trimester, the woman feels distinct kicks and jabs and may even detect a fetal hiccup. The fetus is now about 12 inches (30.5 centimeters) long.

In the final trimester, fetal brain cells rapidly connect into networks, and organs differentiate further and grow. A layer of fat develops beneath the skin. The digestive and respiratory systems mature last, which is why infants born prematurely often have difficulty digesting milk and breathing. About 266 days after a single sperm burrowed into an oocyte, a baby is ready to be born.

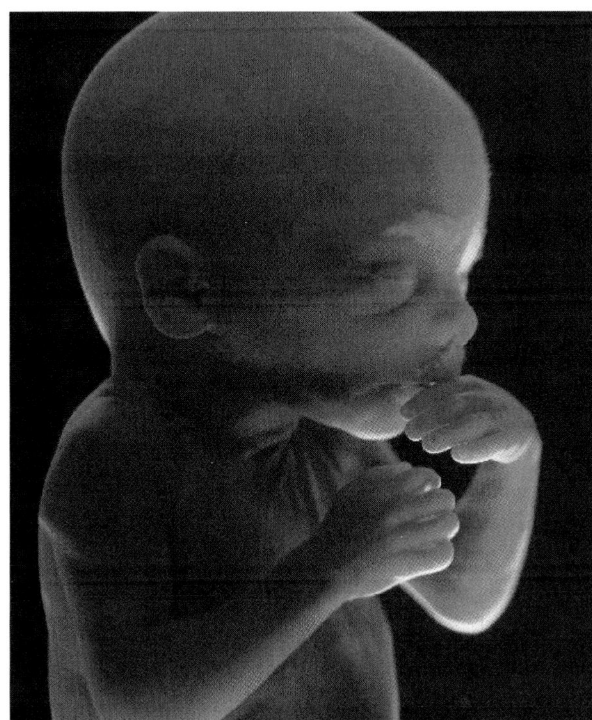

FIGURE 11.21
A fetus at 20 weeks.

Mastering Concepts

1. What happens to the skeleton during the fetal period?
2. When do sexual differences appear, and what triggers them?
3. What happens during the second trimester?
4. What happens during the third trimester?

Labor and Birth

"Labor" refers, appropriately, to the strenuous work a pregnant woman performs in the hours before giving birth. Labor may begin with an abrupt leaking of amniotic fluid as the fetus presses down and ruptures the sac ("water breaking"). This exposes the fetus to the environment; if birth doesn't occur within 24 hours, the baby may be born with an infection. Labor may also begin with a discharge of blood and mucus from the vagina, or a woman may feel mild contractions in her lower abdomen about every 20 minutes.

As labor proceeds, the hormone-prompted uterine contractions gradually increase in frequency and intensity. During the first stage of labor, the baby presses against the cervix with each contraction. The cervix dilates (opens) a little more each time. At the start of labor, the cervix is a thick, closed band of tissue. By the end of the first stage of labor, the cervix stretches open to about 10 centimeters. The cervix sometimes takes several days to open, with mild labor beginning well before a woman realizes it has started.

The second stage of labor is the delivery of the baby (fig. 11.22). It begins when the cervix completely dilates. The woman feels a tremendous urge to push. Within the next 2 hours, the baby descends through the cervix and vagina and is born. In the third and last stage of labor, uterine contractions expel the placenta and extraembryonic membranes from the woman's body.

A baby may be delivered surgically, through a procedure called a Caesarian section, for a variety of reasons. The baby may be too large to fit through the woman's pelvis; it may be

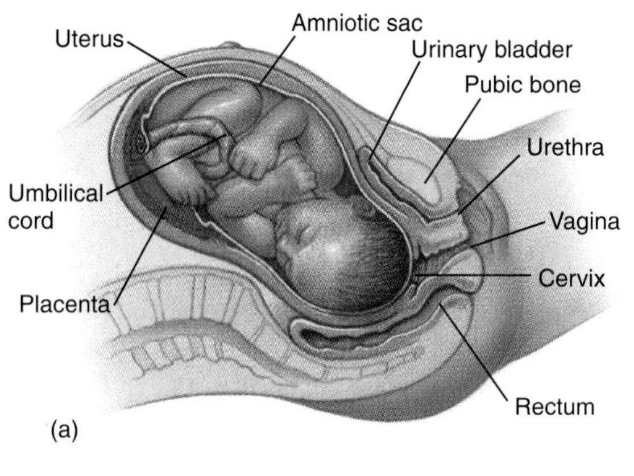

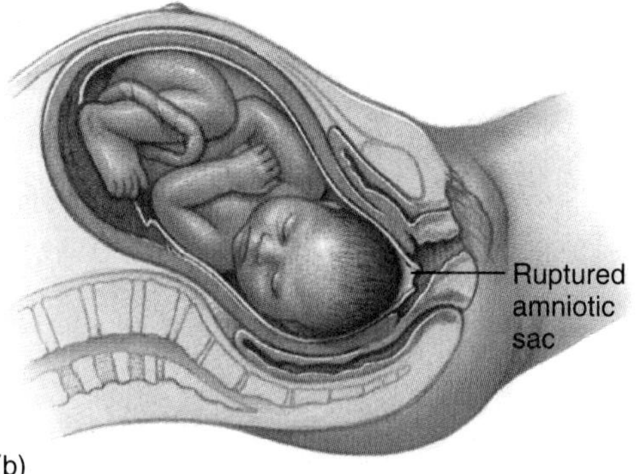

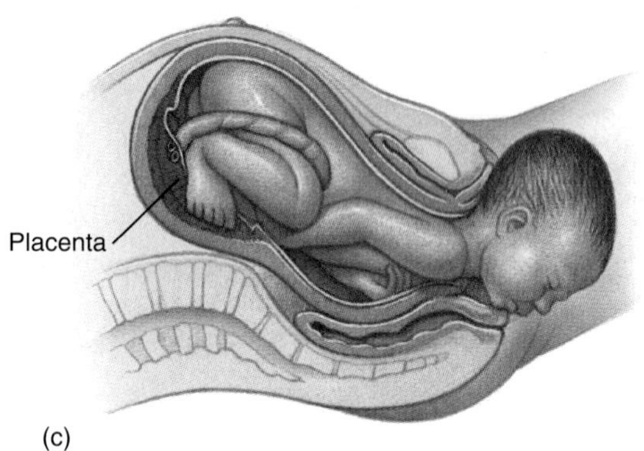

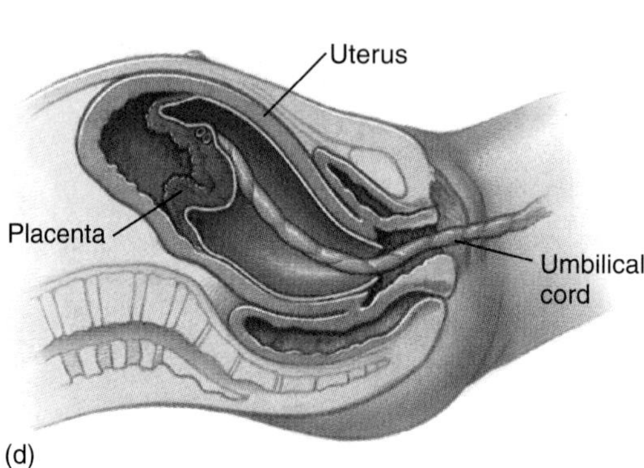

FIGURE 11.22

About 2 weeks before birth, the fetus "drops" in the woman's pelvis and the cervix may begin to dilate (*a*). At the onset of labor, the amniotic sac may break (*b*). The baby (*c*) and then the placenta (*d*) and extraembryonic membranes are pushed out of the birth canal.

positioned feet or buttocks down rather than head down (breech), so that the head may become caught if the baby is delivered vaginally; or the fetus may lie side-to-side. If the umbilical cord wraps around the fetus's neck as it moves into the birth canal, a Caesarian section can save the baby's life. About one in five births in the United States is surgical, and many people claim that some of these are unnecessary.

The birth of a live, healthy baby is against the odds, considering human development from the beginning. Of every 100 secondary oocytes exposed to sperm, 84 are fertilized. Of these, 69 implant in the uterus, 42 survive 1 week or longer, 37 survive 6 weeks or longer, and only 31 are born alive. Of those that do not survive, about half have chromosomal abnormalities too severe to maintain life.

Table 11.3 compares human gestation time to that of other mammals, and figure 11.23 describes how a female red kangaroo experiences motherhood—she nurtures an embryo, a newborn, and a toddler at the same time!

Table 11.3 Gestation Times in Placental Mammals

Species	Time (days)
Mice	21
Cats	60
Dogs	60
Bats	120–150
Humans	266
Cattle	280
Baleen whales	365
Elephants	660

Mastering Concepts

1. What are some signs that labor is imminent?
2. What events make up the three stages of labor?
3. Why are Caesarian sections performed?

Postnatal Growth and Development

Growth and development continue after birth, as the body rapidly enlarges while it retains the specialization of its tissues and organs. Some structures grow and their cells are replenished continually, such as hair, nails, skin, and the lining of the small intestine. Some structures and functions begin at or after birth, peak, and then decline, following specific timetables, yet changing in a coordinated manner so that the body as a whole operates efficiently. A human life can be considered in the stages depicted in fig. 11.24.

Birth triggers dramatic changes, as the newborn must suddenly breathe, eat, excrete, and regulate temperature on its own. The blood vessels that linked fetus to pregnant woman close off, and the baby's circulatory system now handles vital gas exchange. The human newborn is helpless, compared to newborns of some other mammals. Llamas, antelopes, and guinea pigs, for example, are born fully furred, and within minutes of birth, they stand and walk about. It is many months before a human infant stands and walks!

(a)

(b)

FIGURE 11.23

A female kangaroo typically mothers three offspring at a time, each in a different stage of development. (*a*) After developing for 33 days inside the female's body, the joey falls out of the female's cloaca, and in just 3 minutes pulls itself to the pouch and latches onto a teat. The female becomes sexually receptive, and soon conceives, but the joey's suckling sends hormonal signals that arrest development of the zygote when it consists of about 100 cells. (*b*) After 190 days, the joey leaves the pouch for increasingly frequent outings, returning periodically to nurse, until day 234, when it leaves. After 33 days, baby number two drops out of the cloaca and finds its teat, and baby number 3 is conceived.

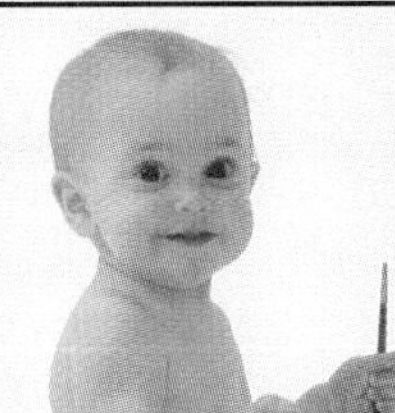

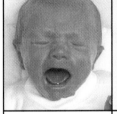

Newborn	Infant	Child	Adolescent	Adult
birth – 4 weeks	4 weeks – 2 years	2 – 12 years	13 – 18 years	18 years – death
Survival reflexes Start of respiration, digestion, kidney function, temper-ature control, Circulatory system changes	Adult hemoglobin Senses sharpen Muscles coordinate Rapid growth Teeth erupt Communication skills	Skeleton matures Immune system is completely active Brain cell division slows and stops Teeth erupt Bladder and bowel control	Sex hormones produced Secondary sexual characteristics appear Sex organs mature Growth spurt Emotional maturity Intellectual development	Organ systems function Repair of damaged tissue Muscle strength, senses, hair growth peak Aging-related changes begin in mid-thirties

FIGURE 11.24

Stages of human postnatal development.

Infancy (4 weeks to 2 years) is a time of incredibly rapid growth, as body weight triples. In the first 6 months, adult hemoglobin (the protein that carries oxygen in the blood) gradually replaces the fetal variety, which has a greater affinity for oxygen. The digestive system matures gradually, as evidenced by the increasing variety of foods that a baby can digest. Movements coordinate, immunity begins to build, and the baby learns to communicate.

During childhood (2 to 12 years), the bones complete hardening, the immune system becomes fully active, and primary teeth erupt, and some may even be replaced by secondary teeth. Bladder and bowel controls are mastered early in childhood. Adolescence, which may begin as early as age 10 or 11 in a girl and may start as late as 16 in either sex, is another period of rapid and profound change. As sex hormones begin to be produced, the sex organs mature, and secondary sexual characteristics (such as a deepening voice in a male and breast development in a female) appear. Growth is rapid, and a teen may eat voraciously. Social, emotional, and intellectual development occur rapidly too.

By early adulthood, all organ systems are fully developed and functioning. Unit 8 explores each of these systems in depth.

Mastering Concepts

1. What changes occur as the newborn adapts to conditions outside the uterus?
2. What events take place during infancy and childhood?
3. What bodily changes characterize adolescence?

Aging

Aging is part of living. Cell death even occurs before birth, as certain cells in the embryo undergo apoptosis. Figure 11.25 outlines some more familiar signs of aging.

By the age of 2 years, a child has as many brain cells as he or she will ever have, although connections between brain cells continue to form throughout life. The brain grows at the same rate over the first 2 years of life as it does in the last 6 months in the uterus. At age 10, a person's hearing is the best it will ever be. Shortly before adolescence, a small gland in the chest, the thymus, reaches its greatest size, about that of a walnut. It then slowly shrinks until it is almost microscopic by age 70. Because the thymus manufactures immune system cells, its declining activity may increase an older person's susceptibility to infection.

Muscle strength peaks in both sexes in the twenties. Hair is at its fullest in the thirties, and each hair is as thick as it will ever be. By the end of the third decade of life, the first obvious signs of aging may appear as the facial skin becomes less elastic and creates small wrinkles around the mouth and eyes. Height is already starting to decrease, but not yet at a detectable level.

Age 30 is a developmental turning point. Hearing often becomes less acute. Heart muscle begins to thicken. Ligaments between the small bones in the back become less elastic, setting the stage for the slumping posture characteristic of the later years. Some researchers estimate that beginning roughly at age 30, the human body becomes functionally less efficient by about 0.8% every year.

During their forties, many people weigh 10 to 20 pounds (4.5 to 9 kilograms) more than they did at age 20, thanks to slowing metabolism and sometimes to decreased activity levels. They may be 1/8 inch (0.3 centimeter) shorter. Hair may gray or thin, and the person may become farsighted or nearsighted. Immune system cells lose efficiency, increasing susceptibility to infection and cancer.

The early fifties bring further declines. Nail growth slows, taste buds die, and the skin continues to lose elasticity. For most people, the ability to see close objects becomes impaired. Decreased activity of the pancreas may lead to diabetes. By the decade's end, muscle mass and weight begin to decrease. Women stop menstruating (menopause), which may cause such conditions as vaginal dryness and hot flashes. A male produces less semen, though he is still fertile. His voice may become higher as his vocal cords degenerate. A man has half the strength in his arm muscles and half the lung function he did at age 25.

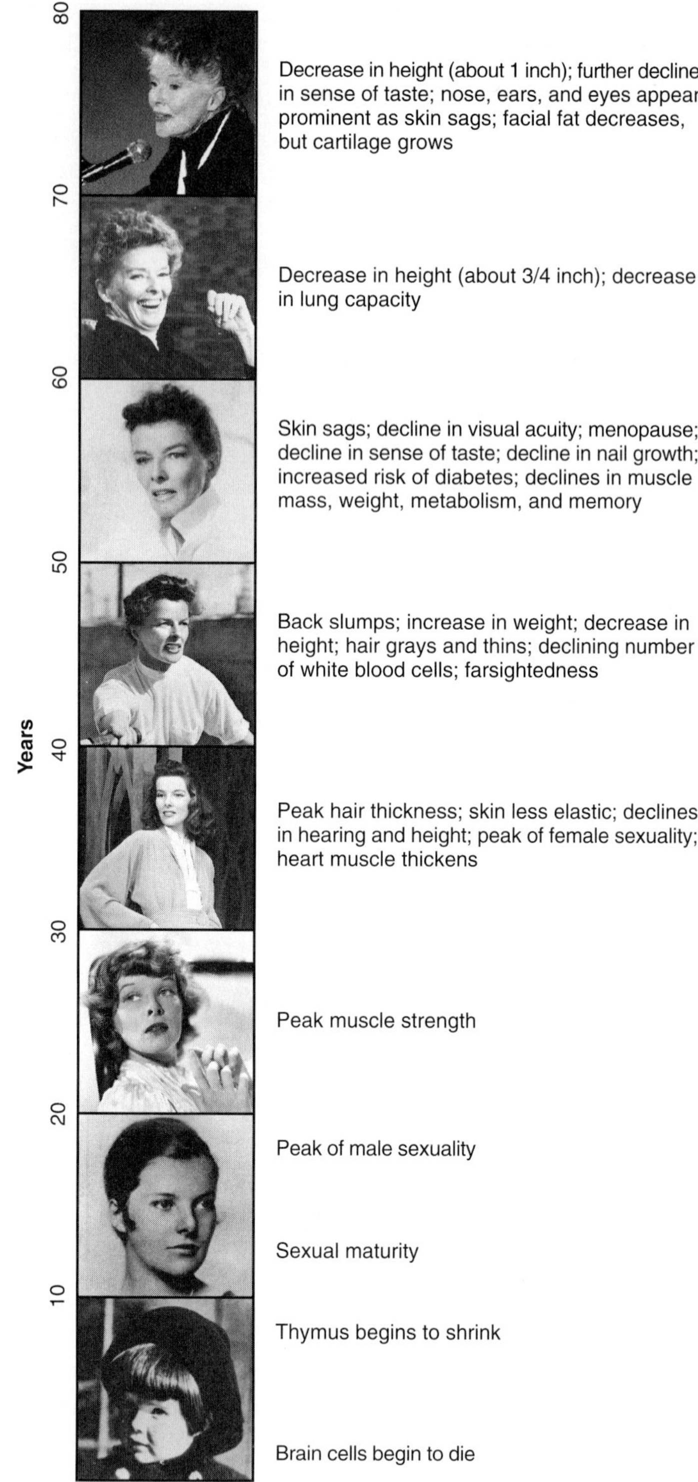
Years
80
70
60
50
40
30
20
10
Decrease in height (about 1 inch); further decline in sense of taste; nose, ears, and eyes appear prominent as skin sags; facial fat decreases, but cartilage grows
Decrease in height (about 3/4 inch); decrease in lung capacity
Skin sags; decline in visual acuity; menopause; decline in sense of taste; decline in nail growth; increased risk of diabetes; declines in muscle mass, weight, metabolism, and memory
Back slumps; increase in weight; decrease in height; hair grays and thins; declining number of white blood cells; farsightedness
Peak hair thickness; skin less elastic; declines in hearing and height; peak of female sexuality; heart muscle thickens
Peak muscle strength
Peak of male sexuality
Sexual maturity
Thymus begins to shrink
Brain cells begin to die

FIGURE 11.25

Although many biological changes ensue as we grow older, photographs of actress Katharine Hepburn at various stages of life indicate that we can age with grace and beauty.

The 60-year-old may experience minor memory losses. He or she may be about 3/4 inch (2 centimeters) shorter. A few million of the person's trillion or so brain cells have died over his or her lifetime, but for the most part, intellect remains sharp. By age 70, height decreases a full inch (2.5 centimeters). Sagging skin and loss of connective tissue, combined with continued cartilage growth, make the nose, ears, and eyes more prominent. For some people, life ceases when they are in their seventies.

What Triggers Aging?

The aging process is difficult to analyze because of the intricate interactions of the body's organ systems. One structure's breakdown ultimately affects the way others function. Aging has both passive and active components—existing structures break down, and new biochemicals and structures form.

Aging as a Passive Process

Aging is passive in the sense that structures break down and functions slow. At the molecular level, passive aging is seen in the degeneration of the connective tissue's elastin and collagen proteins. As these proteins fall apart, skin loses its elasticity and begins to sag, and muscle tissue slackens.

During a long lifetime, DNA sequence errors accumulate. Mistakes in a DNA sequence can occur when DNA replicates in dividing cells. Usually "repair" enzymes correct this damage immediately. But over many years, exposure to chemicals, viruses, and radiation disrupts DNA repair, so that the burden of fixing errors becomes too great. The cell may die as a result of faulty genetic instructions that go unrepaired.

Another sign of passive aging at the biochemical level is lipid breakdown. As aging cell and organelle membranes leak due to lipid degeneration, a fatty, brown pigment called lipofuscin accumulates. Lipofuscin does not cause aging, but it is a characteristic of old cells. Mitochondria also begin to break down in older cells, decreasing the supply of chemical energy available to power the cell's functions.

Highly reactive metabolic by-products called **free radicals** may stimulate the cellular degradation associated with aging. A free-radical molecule has an unpaired electron in its outermost valence shell, which attracts electrons from other molecules and destabilizes them. This sets into motion a chain reaction of chemical instability that could kill the cell. Exposure to toxins or radiation can also generate free radicals. Enzymes that catalyze reactions that inactivate free radicals before they damage cells diminish in number and activity in the later years. One such enzyme, superoxide dismutase, is promoted as an antiaging remedy at some health food stores. Even though this enzyme is a natural free-radical fighter, no evidence exists that it, or other natural antioxidants such as vitamins C and E and beta carotene, stalls aging on a whole-body level.

Aging as an Active Process

Aging can entail the active appearance of new substances or functions, as well as the passive breakdown of structure and function. One active aging substance may be lipofuscin granules that build up in aging muscle and nerve cells. Accumulation of lipofuscin illustrates both passive and active aging. Lipofuscin actively builds up with age, but it does so because of the passive breakdown of lipids. Another example of active aging is autoimmunity, in which the immune system turns against the body and attacks its cells as if they were infectious organisms. Rheumatoid arthritis and some forms of diabetes, both with onset in adulthood, are autoimmune disorders. Apoptosis is also an active aging process.

Accelerated Aging Disorders

Biologists decipher early developmental stages using model organisms such as roundworms and fruit flies. One way to study aging in humans is by examining inherited diseases that accelerate the aging timetable.

The most severe aging disorders are the **progerias,** a group of disorders (fig. 11.26). In Hutchinson-Gilford syndrome, a child appears normal at birth but by the first birthday is still very small. Within just a few years, the child ages dramatically, developing wrinkles, baldness, and the prominent facial features of advanced age. The body ages on the inside as well, as arteries clog with fatty deposits. The child usually dies of a heart attack or stroke by age 12, although some patients live into their twenties. Medical literature reports only a few dozen cases of this syndrome. In an "adult" form of progeria called Werner syndrome, which appears before the 20th birthday, death from old age usually occurs when the individual is in his or her forties.

Not surprisingly, the cells of progeria patients show aging-related changes. Recall that normal cells in culture divide only 50 times before dying. Cells from progeria patients die in culture after only 10 to 30 divisions, as if they were programmed to die prematurely. Certain structures seen in normal cultured cells as they near the 50-division limit (glycogen particles, lipofuscin granules, many lysosomes and vacuoles, and few ribosomes) appear in cells of progeria patients early on. Understanding the mechanisms that cause these diseased cells to race through the aging process may help us better understand the biological aspects of normal aging.

FIGURE 11.26

The Luciano brothers inherited progeria and appear far older than their years.

Mastering Concepts

1. How is aging a lifelong process?
2. At what age do signs of aging typically begin to appear?
3. What aging-related changes are apparent in each decade of life?
4. How is aging both passive and active?
5. What are some passive aspects of aging?
6. What are some active aspects of aging?

The Human Life Span

Wrote Jonathan Swift, of *Gulliver's Travels* fame, "Every man desires to live long; but no man would be old." In the age-old quest to prolong life while remaining healthy, people have sampled everything from turtle soup to owl meat to human blood. A Russian-French microbiologist, Ilya Mechnikov, believed that a human could attain a life span of 150 years with the help of a steady diet of milk cultured with bacteria. He thought that the bacteria would take up residence in the large intestine and somehow increase longevity. He died at 71. More recently, researchers have attempted to use hormone treatments to restore younger functions to older people. A 62-year-old woman has given birth, and men in their sixties have been treated with growth hormone to rebuild sagging tissues.

The human **life span**—theoretically, the longest length of time a human can live—is approximately 120 years, although Jeanne Calment, described in the beginning of the chapter, may surpass that. Of course, most people succumb to disease or injury long before they reach such an age. **Life expectancy** is a more realistic projection of how long an individual will live, based on epidemiological information and a person's current age. In the United States, life expectancy from birth is 79 years for males and 83 years for females.

In many nations, women outlive men by a few years. Developmental biologists do not know why this is so, because both males and females suffer from the same life-threatening illnesses. But the longer lives of females may reflect biology, because it is seen in a variety of species, including beetles, flies, mice, fishes, spiders, rats, birds, and butterflies. Differences that enable females to outlive males may begin very early in development. Experiments on human preembryos show that male cleavage embryos divide faster than female cleavage embryos. Perhaps mitotic clocks "tick" faster in males, reaching life's endpoint on average a few years sooner.

Causes of Death

Life expectancy draws closer to life span as technology conquers diseases. Both ends of the life span are lengthening. In many nations infant mortality has dropped significantly in this century. In the United States, only 1 in 2,000 infants dies. People are living longer too. In 1951, when the U.S. population was 44 million, 271 individuals celebrated their 100th or greater birthdays. In 1991, when the population was 51 million, the number of individuals 100 years and older was 4,500!

Medical advances have contributed greatly to improved life expectancy. Antibiotics have tamed some once-lethal infections, drugs enable many people with cancer to survive, and such advances as beta blocker drugs and coronary bypass surgery have extended the lives of people with heart disease. However, a look at the five leading causes of death in the world at three points in recent history—1900, 1986, and 1993—sounds a warning against complacency (table 11.4). The top-two killers in 1900, pneumonia and tuberculosis, are absent from the 1986 list. However, they appear again as numbers 4 and 5 in the 1993 list, in drug-resistant forms! The appearance of new or renewed infectious diseases, such as AIDS and measles, also indicates that we cannot conquer all killers. Chapter 18 explores these issues in depth.

Table 11.4

Five Leading Causes of Death Worldwide: 1900, 1986, and 1993

1900	1986	1993
1. Pneumonia and influenza	1. Heart disease	1. Heart and blood vessel disease
2. Tuberculosis	2. Cancer	2. Diarrheal diseases
3. Diarrheal diseases	3. Stroke	3. Cancer
4. Heart disease	4. Chronic respiratory disease	4. Pneumonia
5. Stroke	5. Injuries	5. Tuberculosis

Source: Data from Centers for Disease Control and Prevention, Atlanta, GA.

Is Longevity Inherited?

Aging reflects genetic activity plus a lifetime of environmental influences. Families with many very aged members usually have a fortuitous collection of genes, plus shared environmental influences, such as sound nutrition, excellent health care, devoted relatives, and a myriad of other nurturing advantages.

Identifying and isolating the inborn and environmental influences on life span is difficult. One approach compares the health of adopted individuals to that of their biological and adoptive parents. In one study, Danish adopted individuals with one biological parent who died of natural causes before age 50 were more than twice as likely to also die before age 50 as were adopted people whose biological parents lived beyond this age. This suggests an inherited component to longevity. Interestingly, adopted people whose biological parents died early due to infection were more than five times as likely to also die early of infection, perhaps because of inherited immune system deficiencies. The adoptive parents' age at death showed no correlation with the longevity of their adopted children, suggesting that heredity may play a more important role than environment in an individual's life expectancy.

Perhaps because many human populations are aging, researchers and physicians are paying more attention to the biological aspects of growing older. The field of gerontology

Happy Birthday!~

continued from page 210.

People over the age of 85 are the largest growing population group in the United States and many other nations. Medical researchers are finding that many of these "oldest old" people are extraordinarily healthy. The reason—the most common disorders strike before this age, essentially "weeding out" much of the population. Heart disease and cancers tend to occur between the fifties and the eighties, and people who have developed Alzheimer's disease or other neurological disorders usually succumb by their eighties.

The lucky "oldest old," researchers suspect, probably inherit combinations of genes that maintain their health as they age and enable them to fight off free radicals, repair DNA adequately, prevent cholesterol buildup in arteries, or resist certain cancers. But the oldest old may have more in common than lucky genes. Studies have shown that they handle stress well (such as living longer than your loved ones); live a moderate lifestyle; exercise regularly; and are well educated. Jeanne Calment, the oldest living human, explains her longevity. "I took pleasure when I could. I acted clearly and morally and without regret. I'm very lucky."

examines the changes of aging at the molecular, cellular, organismal, and population levels. As we learn more about the biology of aging, we can improve not only the length but the quality of life.

Mastering Concepts

1. What is the maximum human life span?
2. What are some common causes of death?
3. How has medical science extended the human life span?
4. What factors influence longevity?

SUMMARY

1. In the mid-eighteenth century, the theory of **preformation** gave way to the theory of **epigenesis** as biologists observed embryos specialize over time. Karl Ernst von Baer noted that early vertebrate embryos look alike but gradually become more distinctive. Ernst Haeckel proposed that embryonic stages repeat the adult structures of more primitive species. Modern scientists believe that vertebrate embryos simply share similar early stages.

2. Cells of the early embryo are **pluripotent.** Gradually, **morphogen** gradients establish biochemically distinct regions of the embryo, directing them to differentiate down specific **developmental pathways.**

3. A cell commits to a fate at a certain point in development. Before this time, a transplanted cell develops according to its new surroundings; after this point, it retains the specialization other cells exhibit in its original location. Differential gene expression underlies cell specialization.

4. Developmental biology has evolved from observing to disrupting development, often with a molecular focus. Model organisms are critical in studying development.

5. Human prenatal development begins at fertilization. A single, **capacitated** sperm cell at a secondary oocyte burrows through the **zona pellucida** and **corona radiata.** The two **pronuclei** join. **Cleavage** ensues, and a 16-celled **morula** forms. Between days 3 and 6, the morula arrives at the uterus and hollows to form a **blastocyst,** made up of individual **blastomeres.** The **trophoblast** layer and **inner cell mass** form. **Implantation** occurs between days 6 and 14. Trophoblast cells secrete **hCG,** which prevents menstruation.

6. During the second week, the **amniotic cavity** forms as the inner cell mass flattens, forming the **embryonic disc.** The **primitive streak** appears. **Ectoderm** and **endoderm** form, and then **mesoderm** appears, establishing the **germ layers** of the **gastrula.** Cells in a particular germ layer develop into parts of specific organ systems.

7. During the third week the **chorion** starts to develop into the **placenta,** and the **yolk sac, allantois,** and **umbilical cord** form as the amniotic sac swells with fluid. Organogenesis occurs throughout this embryonic period. Gradually structures appear, including the **notochord, neural tube,** arm and leg buds, heart, facial structures, skin specializations, and skeleton.

8. The **fetal period** begins in the third month. Organs specialize and body proportions come to more closely resemble those of a newborn. The fetus becomes active. During the fifth month, vernix caseosa and lanugo cover the fetus. In the last trimester, the brain develops rapidly and fat fills out the skin. The digestive and respiratory systems mature. At about 38 weeks, strong uterine contractions dilate the cervix and push the baby out.

9. Postnatal stages include the newborn, infant, child, adolescent, and adult. Drastic changes occur at birth, as the newborn takes on functions that the pregnant woman provided. Infancy and childhood are periods of rapid growth and maturation of organ systems. Hormonal changes dominate adolescence. Organ systems function during adulthood and begin to show signs of aging.

10. Aging is a part of life, and it begins even before birth. Various structures and functions reach their peaks at different ages.

11. In passive aging, structures break down, and DNA repair becomes less efficient. **Free radicals** build up, threatening cell stability, the thymus shrinks, and blood vessels narrow with cholesterol deposits. In active aging, lipofuscin accumulates in cells and cells die.

12. The theoretical maximum human **life span** is 120 years. The most common causes of death shift as technologies conquer various diseases.

TO REVIEW . . .

1. Why would the scientific method support the hypothesis of epigenesis better than the theory of preformation?

2. How can very different specialized cells have identical genes?

3. How do morphogen gradients cause cells to differentiate?

4. Describe an experiment in a model organism that illustrates how cells lose pluripotency as they develop.

5. Why did biologists choose to develop a complete fate map for *C. elegans,* rather than for an organism more like humans?

6. What role do homeotic genes play in early development?

7. Arrange these prenatal humans in order from youngest to oldest: morula, gastrula, zygote, fetus, blastocyst, and embryo.

8. What events must take place for fertilization to occur?

9. How do the developmental fates of a trophoblast cell and an inner cell mass cell differ? How do fates differ for cells in the ectoderm, endoderm, and mesoderm?

10. Toxins usually cause more severe medical problems if exposure occurs during the first 8 weeks of pregnancy rather than during the later weeks. Why?

11. What is the difference between life span and life expectancy?

12. What types of studies distinguish inherited from environmental influences on longevity?

13. How do less efficient enzymes that normally catalyze reactions carrying out DNA repair or free-radical scavenging contribute to aging?

TO THINK ABOUT . . .

1. Is the presence of a complete set of genetic instructions in a fertilized ovum consistent with the hypothesis of preformation or epigenesis?

2. When an isolated sperm is mechanically injected into an oocyte outside the woman's body, and the treated egg is then placed in the woman's body, no embryo develops. Why doesn't development ensue?

3. How would you produce
- **a.** a double newt embryo?
- **b.** a frog that forms from a somatic cell?
- **c.** a fruit fly embryo with two rear ends?

4. Cite two examples of work done in model organisms that also applies to a medical condition in humans.

5. Infertility patients in the past have been advised to abstain from intercourse until an ovulation test shows the woman has ovulated. What evidence indicates that this advice is incorrect?

6. A woman has been pregnant for 41 weeks. Her doctor performs a Caesarian section to deliver the child because biochemical tests indicate that the cells of the placenta are dying. Why is it vital to the child's survival to perform the Caesarian section?

7. What technology would be necessary to enable a fetus born in the fourth month to survive in a laboratory setting?

8. Do you think that the symptoms of menopause should be treated as an illness? Cite a reason for your answer.

9. Rats raised in an "enriched environment" (provided with toys and lots of attention from their keepers) live 33% longer than other rats, and their brains show cellular patterns similar to those of much younger rats. Other animal studies correlate low body temperature and low caloric intake with longevity. How might these findings be applied to the study of human aging?

10. When does aging begin? What structures exhibit aging long before a person appears to be elderly?

11. What factors do you think contribute to longevity?

12. Why might lists of leading causes of death differ in different parts of the world, or even in different communities within a nation?

13. What problems would a scientist likely encounter in studying the inheritance of longevity?

14. In the year 2050, 1 in every 20 persons in the United States will be over the age of 85. What provisions would you like to see made for older people of the future (especially if you will be one of them)? What can you do now to increase the probability that your final years will be healthy and enjoyable?

TO LEARN MORE . . .

References and Resources

Clarke, Cyril A., and Ursula Mittwoch. Spring 1994. Puzzles in longevity. *Perspectives in Biology and Medicine,* vol. 37. Developmental biologists are looking at many stages of development to better understand why women usually outlive men.

Dickman, Steven. August 1996. The left-hearted gene. *Discover.* Genes control body symmetry.

Gould, Stephen Jay. 1977. *Ever since Darwin.* New York: W. W. Norton. The essay "On Heroes and Fools in Science" argues that preformation once made sense.

Hall, Judith G. January 1997. Give the embryo a chance. *Nature Medicine.* Vitamin supplements prevent some neural tube defects.

Hayflick, Leonard. 1994. *How and why we age.* New York: Ballantine. Biological aspects of aging, written by a master.

Hodgkin, Jonathan, et al. October 20, 1995. The nematode *Caenorhabditis elegans* and its genome. *Science,* vol. 270. A detailed look at a very important worm.

Lewis, Ricki. April 1996. A human parthenogenote. *Photonics Spectra.* Half of the body of this unusual child developed without input from a sperm.

Lewis, Ricki. March 1996. From sugar and spice to PET and MRI. *Photonics Spectra.* Subtle brain differences distinguish the sexes.

Lewis, Ricki. April 1995. Lasers can be lifesavers for twins. *Photonics Spectra.* Laser treatment can correct a syndrome where one twin takes nutrients from the other before birth.

Stewart, Colin L. February 27, 1997. Send in the clones. *Nature,* vol. 385. Special sheep show that differentiated cells retain all their genes.

Wallis, Claudia. March 6, 1995. How to live to be 120. *Time.* Genetic luck and a healthy lifestyle can contribute to longevity, says a 120-year-old.

denotes articles by author Ricki Lewis that may be accessed via *Life's* home page /http://www.mhhe.com/sciencemath/biology/life/.

Babies are wonderful. Medical science offers many ways to help people become parents.

12 Human Reproductive and Developmental Concerns

chapter outline

Two Louises: Research and Reproductive Technology 238

Infertility 239
Infertility in the Male
Infertility in the Female
Infertility Tests

Assisted Reproductive Technologies 241
Donated Sperm—Artificial Insemination
A Donated Uterus—Surrogate Motherhood
In Vitro Fertilization
Gamete Intrafallopian Transfer
Oocyte Banking and Donation
Preimplantation Genetic Diagnosis

Problems in Pregnancy 245
Spontaneous Abortion
Born Too Soon—Prematurity
Birth Defects
Neonatology

Birth Control 250
Available Methods of Birth Control
Experimental Methods of Birth Control
Terminating a Pregnancy

Sexually Transmitted Diseases 253
Herpes Simplex
Chlamydia
Acquired Immune Deficiency Syndrome (AIDS)

Homosexuality 255
Psychological Aspects of Homosexuality
Biological Studies—Of Genes and Brains

Joined for Life

Patty Hensel's pregnancy in 1990 was uneventful. An ultrasound scan revealed an apparently normal fetus, although at one medical exam, Mike Hensel thought he heard two heartbeats, but he dismissed it as an error. Mike's ears weren't deceiving him though—he *had* heard two distinct heartbeats.

A Caesarean section was necessary because the baby was positioned bottom-first. To everyone's amazement, the baby had two heads and two necks, yet it appeared to share the rest of the body, with two legs and two arms in the correct places, and a third arm between the heads. The ultrasound had probably imaged the twins from an angle that superimposed one head on the other. Patty, dopey from medication, recalls hearing the word "Siamese" and thinking she had given birth to cats. She had delivered conjoined, or Siamese, twins.

The baby was actually two individuals, named Abigail and Brittany. Each twin had her own neck, head, heart, stomach, and gallbladder. Remarkably, each also had her own nervous system. The twins shared a large liver, a single bloodstream, and all organs below the navel, including the reproductive tract. They had three lungs and three kidneys.

Abby and Britty were strong and healthy. Doctors suggested surgery to separate the twins. Aware that only one child would likely survive surgery, Mike and Patty Hensel chose to let their daughters be. In England 2 years later, the parents of conjoined twins Eilish and Katie Holton faced the same agonizing choice, and they took the other option.

to be continued . . .

Two Louises: Research and Reproductive Technology

They are both named Louise, and each, in her special way, has contributed much to our knowledge of prenatal development (fig. 12.1). Louise Joy Brown was born in England in 1978, a normal little girl in every way except the way she was conceived—in a piece of laboratory glassware. "Adorable Louise" was born in 1985, a normal donkey in every way except the place where she developed prenatally—in the uterus of a horse.

Louise Brown was the first human conceived outside the body and transferred as a preembryo to her mother's uterus, where she continued development. Scientists had previously perfected the techniques that led to her birth in nonhuman animals, and since Louise's historic beginning, thousands of so-called test tube babies have been born.

Adorable Louise was conceived in a female donkey and then transferred to a horse's uterus 8 days later. Normally, the immune system of a female horse would abort the "foreign" donkey embryo, but researchers prevented rejection by exposing the horse to white blood cells from Louise's natural donkey parents before the embryo transfer. It worked. The horse's immune system, "tricked" into recognizing donkey cells as its own, accepted the donkey preembryo. A year later, the horse gave birth to Adorable Louise. Some women whose immune systems attack embryos as if they were foreign substances can now benefit from a similar treatment.

(a)

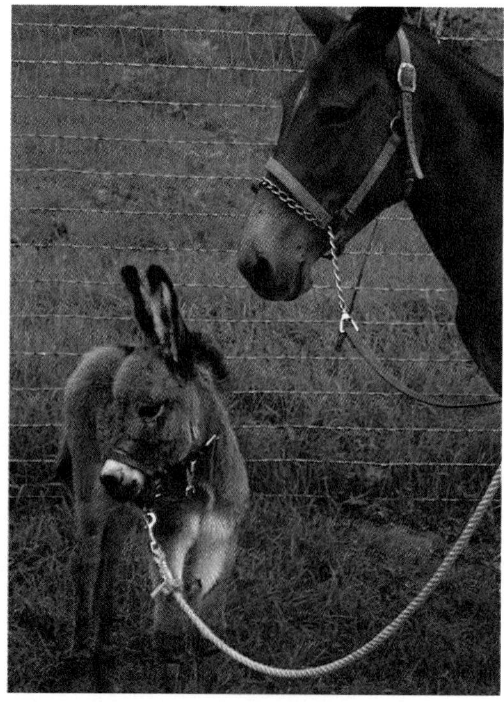

(b)

FIGURE 12.1

Two Louises. (*a*) Louise Joy Brown (bottom left) was conceived when her parents' gametes joined in a laboratory dish. (*b*) Adorable Louise looks like a normal donkey, but a horse gave birth to her following embryo transfer. The techniques used to gestate Adorable Louise in a horse led to treatments that can prevent some miscarriages in humans.

When doctors inject the women with their partners' white blood cells before pregnancy, their immune systems accept embryos.

The two Louises illustrate the goals of reproductive biology—to better understand animal development and to provide new medical technologies. The human reproductive system is unique among organ systems because it does not begin to function until several years after birth. It is also unique because an individual can survive without it. On the population level, however, the reproductive system is essential to the survival of a species. A healthy reproductive system can make life very fulfilling, both in providing the pleasures of sexuality and in producing new life. As medical scientists learn more about the complex biological interactions that form a newborn from a fertilized ovum, they discover more ways to enable or hinder conception, to prevent or treat birth defects, and to heal the fetus or newborn. This chapter addresses these issues, as well as reproductive health.

Infertility

For one out of every six couples in the United States, trying to conceive a child is a period of growing anxiety and frustration as these couples realize that they may be infertile. **Infertility** is the inability of the sperm nucleus to merge with the oocyte nucleus. Couples are considered infertile if they are unable to conceive after 1 year of frequent intercourse without using contraceptives. Physicians can identify a physical cause for infertility in 90% of all cases and can treat or correct many problems.

Among cases with an identifiable physical cause, 30% entail physical difficulties in the male and 50% entail problems in the female. In 20% of cases, both partners contribute to infertility. A common combination, for example, is a woman with an irregular menstrual cycle and a man with a low sperm count.

Infertility in the Male

Infertility in a male is easier to detect but sometimes harder to treat than in a female. Some men produce many fewer than the 100 million sperm in a normal ejaculate. If sperm are not plentiful, the probability of fertilization declines. Low sperm count can have several causes.

If a hormonal imbalance is lowering sperm count, administering the appropriate hormones may boost sperm output. Sometimes a man's immune system produces antibodies that cover the sperm and prevent them from binding to oocytes. A varicose vein in the scrotum can also cause infertility. This enlarged vein overheats the developing sperm so that they cannot mature. Surgeons can remove a scrotal varicose vein and improve fertility. Epidemiologists have noted a worldwide decrease in sperm counts, as discussed in chapter 2, and attribute it to estrogen-like pollutants. The lowered sperm counts have not yet caused noticeable declines in fertility.

For many men with low sperm counts, fatherhood is just a matter of time. If an ejaculate contains at least 60 million sperm cells, fertilization is likely to occur eventually. To speed conception, a man with a low sperm count can donate several semen samples over a period of weeks at a fertility clinic. The clinic keeps the samples in cold storage and then pools them. Some of the seminal fluid is withdrawn, leaving a sperm cell "concentrate," which is placed in the woman's reproductive tract. This procedure isn't very romantic, but it is highly effective at achieving pregnancy.

Sperm quality is important. Sperm cells that are unable to move—a common problem—cannot reach an oocyte. Hormone therapy can sometimes restore sperm motility, or spermatids can be injected into oocytes.

Infertility in the Female

Infertility in a woman can stem from abnormalities in any part of the reproductive system (fig 12.2). Many women who are infertile have irregular menstrual cycles, which makes it difficult to pinpoint when they are most likely to conceive. In an average 28-day menstrual cycle, ovulation usually occurs around the 14th day after menstruation begins, and this is when a woman is most fertile.

The average sexually active woman under 30 years old and not using birth control is likely to become pregnant within 3 or 4 months. Using an ovulation predictor test (which detects a hormone surge just before ovulation) or detecting a slight rise in body temperature can help her determine when ovulation will occur.

Many conditions can cause the hormonal imbalance that usually underlies irregular ovulation—a tumor in the ovary or in the brain's pituitary gland, which hormonally controls the reproductive system; an underactive thyroid gland; or using steroid-based drugs such as cortisone. Sometimes a woman produces too much prolactin, the hormone that normally promotes milk production and suppresses ovulation in new mothers. If prolactin is abundant in a nonpregnant woman, she will not ovulate and therefore cannot conceive.

Fertility drugs can stimulate ovulation, but they may also cause women to "superovulate" and produce more than one egg each month. This can result in multiple births. If a woman's ovaries are completely inactive or absent (due to a birth defect or surgery), she can become pregnant only if she uses a donor oocyte, discussed later in the chapter. Ovary transplants have only succeeded when donor and recipient are identical twins.

Blocked fallopian tubes are a common cause of infertility. Blockage can prevent sperm from reaching the oocyte or keep the fertilized ovum from descending into the uterus, resulting in an **ectopic** ("out of place") **pregnancy** in the tube. A birth defect or infection can block a tube. X rays or an ultrasound exam may reveal the blockage. A physician views the fallopian tubes through a telescope-like instrument, a laparoscope, that is inserted through a small incision made near the woman's navel. Surgery or even a blast of air can sometimes clear blocked tubes.

Excess tissue growing in the uterine lining can cause female infertility. Benign **fibroid** tumors, or **endometriosis,** a condition in which tissue builds up in and sometimes outside of the uterus, can make the uterine lining inhospitable to an embryo. In endometriosis, the excess tissue bleeds during menstruation, causing painful cramps. Endometriosis can make conception difficult, but curiously, once a woman with endometriosis has been pregnant, often the condition disappears.

Secretions in the vagina and cervix may harm sperm. Unusually thick or sticky mucus can entrap sperm, and acidic or

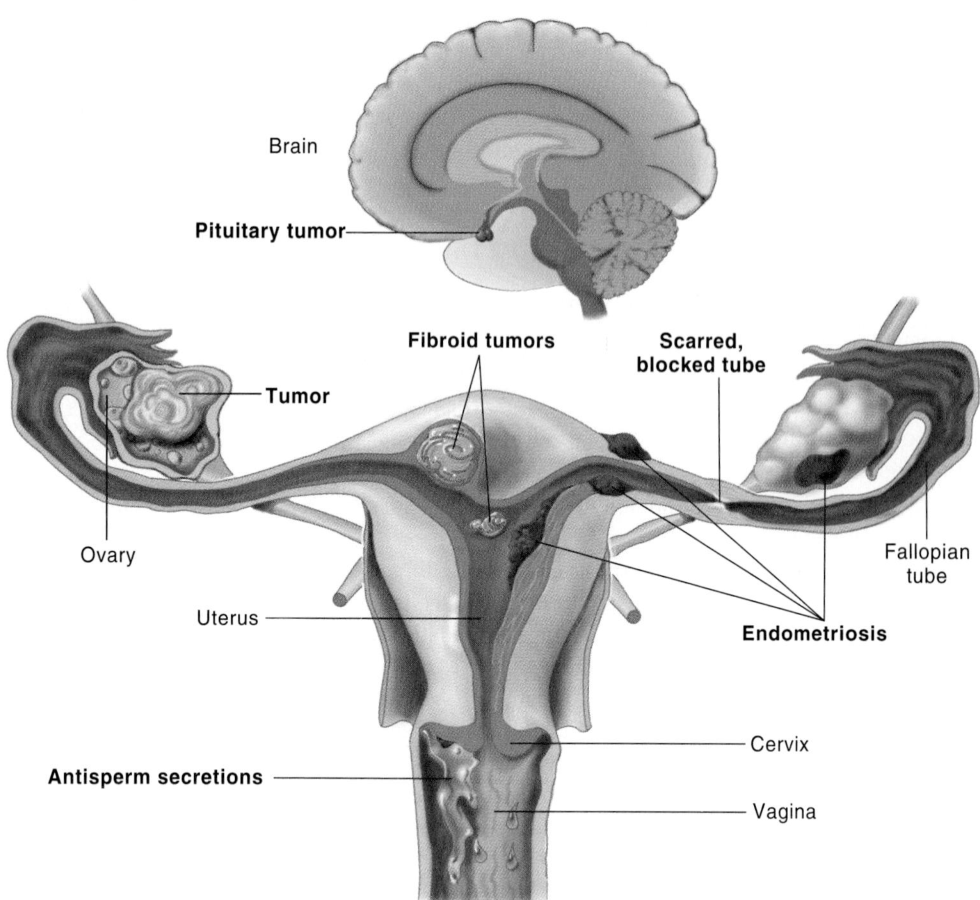

FIGURE 12.2
Infertility in a woman can result from medical problems affecting different parts of her reproductive tract.

alkaline vaginal secretions can weaken or kill them. Infertility specialists treat some mucous secretion problems with low doses of estrogen, or the woman can douche daily with an acidic solution, such as acetic acid (vinegar), or an alkaline solution, such as bicarbonate (baking soda), to alter the pH of the vagina and make it more receptive to sperm.

Infertility is more common among older women. They are more likely to produce oocytes with abnormal numbers of chromosomes, reflecting meiotic errors, because the oocytes have been around longer and have had greater exposure to harmful chemicals, viruses, and radiation. Usually too many or too few chromosomes devastates development, causing pregnancy loss so early that the woman never knows she was pregnant.

Table 12.1 Semen Analysis

Characteristic	Normal Value
Volume	1.5–5.0 milliliters/ejaculate
Sperm density	20 million cells/milliliter
Percent motile	> 40%
Motile sperm density	> 8 million/milliliter
Average velocity	> 20 micrometers/second
Motility	> 8 micrometers/second
Percent abnormal form	> 40%
White blood cells	> 5 million/milliliter

Infertility Tests

A number of medical tests seek to identify the cause or causes of a couple's infertility. Most specialists check the man first, because it is easier, less costly, and less painful to obtain sperm than oocytes. Sperm are checked for number (sperm count), motility, and morphology (shape). An ejaculate may contain abnormal forms numbering up to 40% and still be considered normal; more than 40% can interfere with fertility. Table 12.1 lists the normal test values.

If sperm tests do not reveal male infertility, the next step is for the woman to visit a gynecologist, who may perform several different tests. First, an ultrasound exam can quickly show if all of the woman's reproductive organs are present and in the correct place. Hormones are checked to see if the correct levels are present at the time of ovulation. Hormone supplements can restore normal levels. A postcoital test examines mucus from near the cervix to see if it is hostile to sperm. Several tests determine whether the fallopian tubes are blocked, and a sampling of uterine tissue reveals whether the lining can support an implanting zygote. Even with all of these tests, sometimes physicians are never able to pinpoint a particular couple's problem.

Mastering Concepts

1. What is infertility?
2. What causes infertility in a male?
3. What causes infertility in a female?
4. What are some tests and treatments for infertility?

Assisted Reproductive Technologies

Assisted reproductive technologies help couples conceive. The procedures usually involve a laboratory technique and sometimes participation of a third individual. Most of these techniques cost thousands of dollars and may require several attempts. Most assisted reproductive technologies were perfected in nonhuman animals (table 12.2).

Donated Sperm—Artificial Insemination

The oldest assisted reproductive technology is **artificial insemination,** in which a doctor places donated sperm in a woman's reproductive tract. A woman might seek artificial insemination if her partner is infertile or carries a gene for an inherited illness or if she desires to be a single parent. More than 250,000 babies have been born worldwide as a result of this procedure. The first human artificial inseminations by donor were done in the 1890s. For many years, physicians donated sperm, and this became a way for male medical students to earn a few extra dollars. By 1953, sperm could be frozen and stored for later use. Today, sperm banks freeze and store donated sperm and then provide it to physicians who perform artificial insemination.

Today, a woman or couple choosing artificial insemination can select sperm from a catalog that lists the personal characteristics

Table 12.2 Landmarks in Reproductive Technology

	In Nonhuman Animals	In Humans
1782	Artificial insemination in dogs	
1890s	Birth from transplanted rabbit embryo	Artificial insemination by donor
1949	Freezing and thawing animal sperm	
1951	First calf born after preembryo transplant	
1952	Live calf born after insemination with frozen sperm	
1953		First reported pregnancy after insemination with frozen sperm
1959	Live rabbit offspring from in vitro fertilization	
1972	Live offspring from frozen mouse preembryos	
1976		First reported commercial surrogate motherhood arrangement in United States
1978	Transplantation of ovaries between cows	Baby born after in vitro fertilization in United Kingdom
1980		Baby born after in vitro fertilization in Australia
1981	Calf born after in vitro fertilization	Baby born after in vitro fertilization in United States
1982	Sexing of embryos in rabbits	
	Cattle preembryos split to produce genetically identical twins	
1983		Preembryo transfer
1984		Baby born in Australia from frozen and thawed preembryo
1985		Baby born after gamete intrafallopian transfer (GIFT)
		First reported gestational surrogacy arrangement in United States
1986		Baby born in the United States from frozen and thawed preembryo
1988		First injections of sperm into oocytes
1989		First preimplantation genetic diagnosis
1992		First pregnancies from sperm injected into oocytes
1994		62-year-old woman gives birth
1997	Sheep develop from differentiated cell nuclei	

Source: Data from Office of Technology Assessment, *Infertility: Medical and Social Choices,* U.S. Congress, Government Printing Office, Washington, D.C., May 1988.

of the donors, including blood type; hair, skin and eye color; build; and even educational level and interests. Of course, not all of these traits are inherited. Although artificial insemination has helped many people to become parents, it, and other assisted reproductive technologies, have led to occasional disasters (table 12.3).

In the future, an alternative to artificial insemination may be a transplant of sperm-producing tissue in a male. Researchers have restored fertility to male mice with transplants of spermatogonia from fertile mice.

A Donated Uterus—Surrogate Motherhood

A man's role in reproductive technologies is simpler than a woman's. A man can be a genetic parent, providing half of his genetic self in sperm, but a woman can be both a genetic parent (providing the oocyte) and a gestational parent (providing the uterus).

If a man produces healthy sperm but his partner's uterus is absent or cannot maintain a pregnancy, a **surrogate mother** may be able to help them become parents. A surrogate mother is artificially inseminated with the man's donated sperm and carries the pregnancy, but the couple raises the child. In this situation, the surrogate is both the genetic and the gestational mother.

Another type of surrogate mother lends only her uterus; she receives a fertilized egg from a woman who has healthy ovaries but lacks a functional uterus. For example, Arlette Schweitzer lent her uterus to house her grandchildren-to-be. Her daughter, Christa, had been born without a uterus but had healthy ovaries. Christa's oocytes were fertilized in laboratory glassware with her husband's sperm, and the resulting preembryos were implanted into Arlette's uterus. The result—twins Chad and Chelsea.

Attorneys usually arrange surrogate mother relationships. The surrogate mother signs a statement signifying her intent to give up the baby, and she is paid for her 9-month job. More than 1,000 babies have been born by this method in the United States since 1988. The dilemma with surrogate motherhood is that a woman may not be able to predict her responses to pregnancy and childbirth as she signs paperwork in a lawyer's office. When a surrogate mother changes her mind, the results are wrenching for all, as Bioethics Connections 12.1 illustrates.

Table 12.3 Assisted Reproductive Disasters

1. A physician in California used his own sperm to artificially inseminate 15 patients and told them that he had used sperm from anonymous donors.
2. A plane crash killed the wealthy parents of two preembryos stored at −320°F (−195°C) in a hospital in Melbourne, Australia. Adult children of the couple were asked to share their estate with two eight-celled siblings-to-be.
3. Several couples in Chicago planning to marry discovered that they were half-siblings. Their mothers had been artificially inseminated with sperm from the same donor.
4. Two Rhode Island couples are suing a fertility clinic for misplacing several preembryos.
5. Several couples in California are suing a fertility clinic for implanting their oocytes or preembryos in other women without consent from the donors. One woman is requesting partial custody of the resulting children if her oocytes were taken and full custody if her preembryos were used, even though the children are of school age and she has never met them.
6. A man sued his ex-wife for possession of their frozen embryos as part of the divorce settlement.

In Vitro Fertilization

In **in vitro fertilization** (IVF), which means "fertilization in glass," sperm meets oocyte outside the woman's body. The fertilized ovum divides two or three times and is then introduced into the oocyte donor's (or another woman's) uterus. If all goes well, the preembryo will implant into the uterine lining, establishing a pregnancy.

A woman might undergo IVF if her ovaries and uterus work but her fallopian tubes are blocked. To begin, she takes a superovulation drug that hastens maturity of several oocytes. Using a laparoscope to view the ovaries and fallopian tubes, a physician removes a few of the largest oocytes and transfers them to a laboratory dish. Chemicals similar to those in the female reproductive tract, and sperm, are added to the oocytes.

If a sperm cannot penetrate the oocyte, it may be sucked up into a tiny syringe and injected using a tiny needle into the female cell (fig. 12.3). This variant of IVF, called **intracytoplasmic sperm injection,** is very successful, resulting in a 68% fertilization rate. It can help men with very low sperm counts, high numbers of abnormal sperm, or injuries or illnesses that prevent them from ejaculating. Minor surgery removes testicular tissue, from

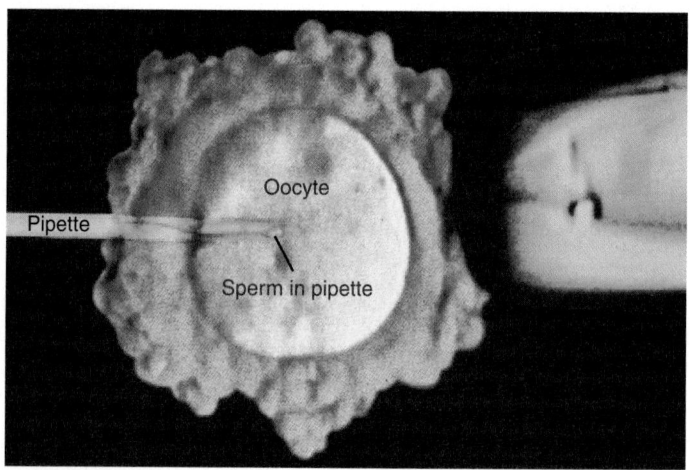

FIGURE 12.3

Intracytoplasmic sperm injection (ICSI) enables some infertile men and men with spinal cord injuries and other illnesses to become fathers. A single sperm cell is injected into the cytoplasm of an oocyte.

12.1 bioethics connections

Surrogates and the Definition of Motherhood

Surrogate mother relationships, when they go awry, can be devastating for all concerned. Consider what happened to Anna Johnson, Mark and Crispina Calvert, and the child whom the Calverts conceived and Johnson carried.

Crispina Calvert's uterus had been removed due to a medical condition, but her ovaries were functioning. The Calverts contracted Anna Johnson, a young, single nurse, to be a gestational surrogate and carry their fertilized ovum to term for $10,000. But as the birth neared, Johnson had misgivings. She went to court and asked to be declared the natural mother and to have visitation rights with the child.

The question in *Johnson v. Calvert and Calvert* boiled down to the very essence of motherhood—was Crispina Calvert baby Christopher's mother, or was Anna Johnson? Christopher wouldn't have been conceived without Calvert's oocyte, but he wouldn't have developed from a fertilized ovum to a baby without the nurturing uterus Johnson provided.

The California court ruled in favor of the genetic ties. Judge Richard N. Parslow Jr. told the courtroom:

> *Anna Johnson is the gestational carrier of the child, a host in a sense . . . she and the child are genetic hereditary strangers . . . Anna's relationship to the child is analogous to that of a foster parent providing care, protection, and nurture during the period of time that the natural mother, Crispina Calvert, was unable to care for the child.*

The judge based his ruling on several factors: scientific evidence indicating the importance of genes in establishing one's characteristics; the validity of the contract; and the best interests of the child. A court of appeals unanimously affirmed the ruling.

Not everyone agreed with the court's decision. Many a woman who has felt fetal movements and experienced childbirth would argue that this role qualifies her for motherhood more than donating a cell. The American College of Obstetricians and Gynecologists states that gestation, not genetics, defines motherhood. The American Academy of Pediatrics defines a surrogate mother as "a woman who carries a pregnancy for another woman," avoiding the genetic-gestation distinction. With such divergent views, we will certainly hear more about surrogate motherhood.

which viable sperm are isolated and injected into oocytes. A day or so later, a physician transfers some of the preembryos—balls of 8 or 16 cells—to the woman's uterus. If hCG, the pregnancy hormone, appears in her blood a few days later, and if its level rises, she is pregnant. The birth rate following IVF is about 17%, compared with 31% for natural conceptions.

To increase the chance that IVF will lead to a birth, several preembryos are usually transferred. This can result in multiple births, as Michele and Ray L'Esperance learned when their attempt to have one baby resulted in five (fig. 12.4)! Today, as IVF is increasingly successful, physicians implant fewer preembryos. If too many preembryos implant, some doctors remove some of them so that two or three have room to develop. Extra preembryos may be donated to an infertile couple or frozen for later use.

Gamete Intrafallopian Transfer

As the world marveled at the miracle of Louise Joy Brown and other test tube babies, disillusioned couples were learning that IVF is costly, is time-consuming, and rarely works. The reason for the frequent failure may be the artificial fertilization environment. A procedure called **GIFT,** which stands for **gamete intrafallopian transfer,** circumvents this problem by moving fertilization to the woman's body rather than in glassware.

In GIFT, a woman takes a superovulation drug for a week and then has several of her largest oocytes removed. A man donates a

FIGURE 12.4

Michele and Ray L'Esperance had IVF because Michele's diseased fallopian tubes had been removed. Physicians implanted five fertilized ova into her uterus, which developed into (left to right) Erica, Alexandria, Veronica, Danielle, and Raymond.

sperm sample, and a physician separates the most active cells. The collected oocytes and sperm are deposited together in the woman's fallopian tube, at a site past any obstruction so that implantation can occur. GIFT is 26% successful.

In **zygote intrafallopian transfer** (ZIFT), a physician places an in vitro fertilized ovum in a woman's fallopian tube. This is unlike IVF because the site of introduction is the fallopian tube and unlike GIFT because fertilization occurs in the laboratory. Allowing the fertilized ovum to make its own way to the uterus seems to increase the chance that it will implant. ZIFT is 23% successful.

Oocyte Banking and Donation

Can oocytes be frozen and stored, as sperm are? If so, a woman wishing to have a baby later in life, when fertility declines, could set aside oocytes while she is young. The ability to store healthy oocytes would also benefit women undergoing a medical treatment that could damage oocytes, such as cancer chemotherapy, and also women who work with gamete-damaging toxins, who are carriers for certain genetic diseases, or who have entered menopause prematurely.

Unfortunately, freezing oocytes isn't as easy as freezing sperm. Oocytes are frozen in liquid nitrogen at −86°F to −104°F (−30°C to −40°C) when they are at metaphase of meiosis II. At this time, the chromosomes are aligned along the spindle, which is sensitive to temperature extremes. If the spindle comes apart as the cell freezes, the oocyte may lose a chromosome, which can devastate development. Frozen oocytes may retain a polar body and form a diploid gamete.

Alternative approaches attempt to overcome the limitations of freezing oocytes. Researchers are developing ways to culture oocyte-packed, 1-millimeter-square sections of ovaries in the laboratory and offer fresh gametes to women who need them. Culturing oocytes in the lab requires a high level of oxygen and a complex combination of biochemicals.

Healthy young women may also donate healthy oocytes to other women. This is how 62-year-old Rosanna Della Corte conceived her son Riccardo (fig. 12.5). Donating oocytes isn't easy—the donor must receive several daily injections of a superovulation drug, have her blood checked each morning for 3 weeks, and then undergo laparoscopic surgery to collect the oocytes. The prospective father's sperm and the donor's oocytes are then placed in the recipient's uterus.

Preimplantation Genetic Diagnosis

Medical specialists can scrutinize technology-aided pregnancies using the same tests they use to check chromosomes in natural pregnancies. However, some assisted reproductive technologies also make possible earlier detection of genetic and chromosomal abnormalities, even before the preembryo implants in the uterus. Intervention at this very early stage is called **preimplantation genetic diagnosis.** It is still experimental.

In this technique, a doctor removes a cell from an eight-celled preembryo and examines its genetic material (fig. 12.6). If the DNA indicates that this particular preembryo has not inherited the condition in question, then the remaining seven-celled preembryo is implanted in the woman, where it completes development. Loss of one cell apparently does not harm the 7-celled preembryo.

One of the first beneficiaries of preimplantation genetic diagnosis was Chloe O'Brien. Chloe's brother has cystic fibrosis. Chloe was checked, as an eight-celled preembryo, to see if she had also inherited the disease. She hadn't; she was born, a healthy little girl, in March 1992. Since then the technique has been used to detect several inherited disorders. At this writing, however, the U.S. government has banned use of federal funds for conducting preimplantation genetic diagnosis.

Mastering Concepts

1. What is artificial insemination?
2. How can a surrogate mother assist an infertile couple?
3. How do GIFT and ZIFT differ from IVF and from each other?
4. Why is it difficult to freeze and thaw oocytes?
5. How does preimplantation genetic diagnosis screen preembryos?

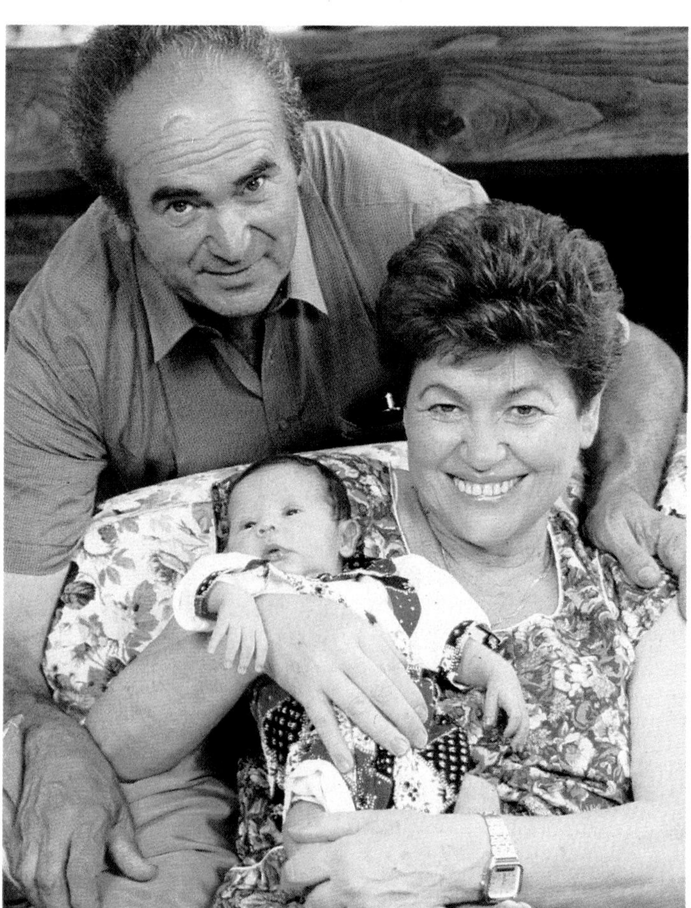

FIGURE 12.5

After hormone treatments, Rosanna Della Corte, age 62, became pregnant, with a donated oocyte her husband Mauro's sperm fertilized in a laboratory dish. Her pregnancy proved that the uterine lining of an older woman can support a fetus to term if given appropriate hormones.

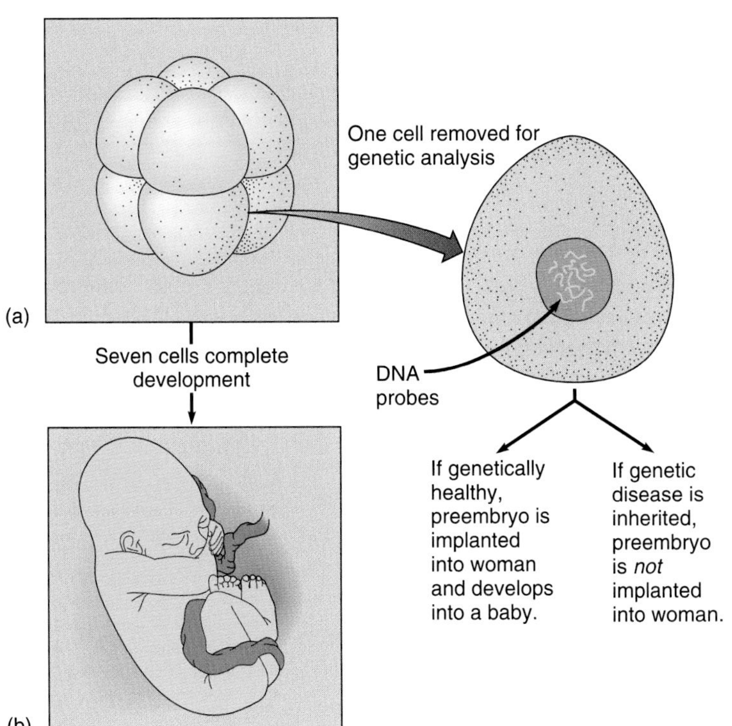
(a)
One cell removed for genetic analysis
Seven cells complete development
DNA probes
If genetically healthy, preembryo is implanted into woman and develops into a baby.
If genetic disease is inherited, preembryo is *not* implanted into woman.
(b)

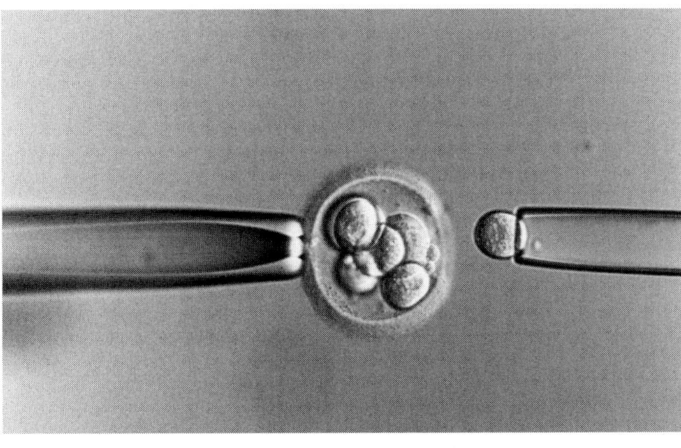

(c)

FIGURE 12.6

Preimplantation genetic diagnosis probes disease-causing genes in an eight-celled preembryo. A single cell is separated from the ball of cells and tested to see if it contains a disease-causing gene combination (*a*). If it doesn't, the remaining seven cells are implanted into the oocyte donor to complete development (*b*). The preembryo in (*c*) is held still by suction applied on the left. On the right, a pipette draws up a single blastomere. Fertilization took place 45 hours previously, in vitro.

Problems in Pregnancy

Prenatal development is so complex, with so many steps and intricately coordinated activities, that the birth of a healthy baby seems a miracle. Sometimes, development goes awry.

Spontaneous Abortion

In a **spontaneous abortion,** a pregnancy ends naturally before the embryo or fetus is developed enough to survive outside the woman's body. From 50 to 75% of all fertilized ova spontaneously abort, but most do so so early that the woman only notices a late menstrual period, with heavy bleeding and cramps. From 15 to 20% of recognized pregnancies end in spontaneous abortion. A spontaneous abortion that occurs past the first trimester is called a miscarriage.

Why don't all embryos and fetuses complete development? From 30 to 50% of all embryos cannot survive for more than a few weeks because they have an abnormal number of chromosomes. Sometimes the preembryo is normal, but the trophoblast cells that surround it do not produce enough hCG to suppress menstruation, thus preventing implantation. An embryo or fetus may literally fall out of a woman's body if the uterus is severely malformed or the cervix too loose. A physician can close an "incompetent" cervix with sutures and then remove them when labor begins.

Fibroid (uterine) tumors can cause spontaneous abortion if they crowd the developing embryo or fetus, stunting its growth and possibly blocking the nutrient supply. If the preembryo implants in a fallopian tube rather than in the uterus (an ectopic pregnancy), spontaneous abortion may occur. If it doesn't, an immediate therapeutic abortion is necessary to save the woman's life. Ectopic pregnancies, which occur in 2% of all pregnancies, are on the rise because of an increase in infections that damage the fallopian tubes and, ironically, because of the ability to treat them. Before doctors could diagnose pelvic infections and treat them promptly with antibiotics, the tubes would completely close with scar tissue, so pregnancy could not occur. Today, antibiotic treatment halts infection so that scar tissue only partially blocks the tubes. The scar tissue entraps fertilized ova.

A couple experiencing spontaneous abortion grieves for their expected child. Only 0.5 to 1% of women experience recurrent (three or more) consecutive pregnancy losses. From 80 to 90% of women who have an early spontaneous abortion deliver infants in their next pregnancies.

Born Too Soon—Prematurity

Babies born more than 4 weeks early are **premature** (fig. 12.7). These infants typically weigh less than 5 pounds (2.3 kilograms). The longer they remain in the uterus and the more they weigh, the greater their chances of survival. The smallest "preemies" are born at about 24 weeks of gestation and weigh less than a pound. Most infants born more than 12 weeks prematurely live only a few days. Premature births account for 85% of newborn deaths.

Each year about 10,000, or 10%, of babies born in the United States are premature. Twins and triplets are more likely to be born prematurely because of their cramped prenatal environment. Teens and malnourished women also are more likely to deliver prematurely. However, even a well-nourished, healthy woman can do so.

The most common cause of premature labor and birth is an undetected infection in the woman's reproductive tract. A type of infection called bacterial vaginosis produces no symptoms but changes the types of bacteria inhabiting the reproductive tract.

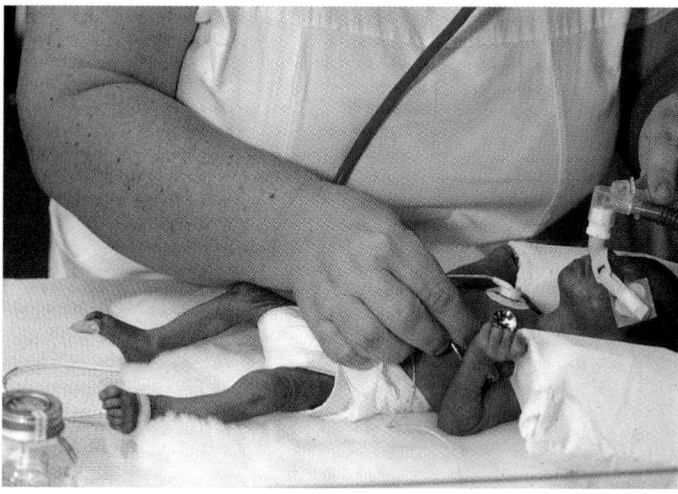

FIGURE 12.7

Babies born too soon must have special care at first, but by a year of age, most catch up to babies born on time.

Somehow, this triggers premature labor. Medical researchers are currently developing ways to screen all pregnant women for this condition, which is treatable with antibiotics.

Immature breathing and sucking reflexes are the most serious problems that premature infants face. The baby born too soon is often not yet able to breathe properly, because normally the placenta would still provide oxygen and remove carbon dioxide. **Respiratory distress syndrome** develops when the baby's lungs are not sufficiently covered with a soapy substance called surfactant. The small air sacs in the lungs cannot inflate without it. Straining to draw dry air into deflated lungs damages the delicate lung tissue. Fortunately, doctors can drip synthetic surfactant into a baby's windpipe to assist breathing and then place the infant on a respirator. Within minutes, a rosy blush spreads through the baby's skin, indicating a functioning respiratory system.

Often preemies cannot suck and must be fed either intravenously or through a stomach tube inserted through the nose. They are also more vulnerable to birth injuries and more prone to infection. Preemies must be kept warm because their small body size, low body fat, and immature temperature-control mechanisms make them lose heat rapidly.

A sign that premature labor may begin is when a protein called fibronectin appears in the woman's vaginal fluid. Fibronectin holds the amniotic sac to the uterine wall and does not normally appear in vaginal fluid. A woman with the telltale fetal fibronectin can be given drugs to prevent or halt premature labor.

Sometimes a fetus is carried full term but still weighs less than 5 pounds (2.3 kilograms) at birth. A "small for gestational age" infant usually has few problems beyond rapid heat loss due to low birth weight. Unlike the infant born too soon, this tiny tot has mature breathing and sucking reflexes. Women who are malnourished or who smoke during pregnancy are at higher risk of having full-term babies of low birth weight. Abnormal development of the placenta may also account for small for gestational age infants. Low birth weight, whether due to prematurity or not, increases the risk of illness in childhood and stunts growth.

Birth Defects

Genetic abnormalities or toxic exposures can disrupt prenatal development and cause birth defects. However, about 97% of newborns are apparently normal.

The Critical Period

The specific nature of a birth defect depends upon which structures are developing when the exposure occurs. The time when genetic abnormalities, toxic substances, or viruses can alter a specific structure is its **critical period** (fig. 12.8). Some body parts, such as fingers and toes, are sensitive for short periods of time. In contrast, the brain is vulnerable to damage throughout prenatal development, as well as during the first 2 years of life. Because of the brain's extensive critical period, many birth defect syndromes include mental retardation. The continuing sensitivity of the brain after birth explains why toddlers who eat lead-based paint suffer impaired learning.

About two-thirds of birth defects stem from a disruption during the embryonic period. Disruptions during the fetal period cause more subtle defects, such as learning disabilities. For example, damage during the first trimester might cause mental retardation, but in the seventh month of pregnancy, it might interfere with learning to read.

Some birth defects can be attributed to an abnormal gene that acts at a specific point in prenatal development. For example, in a rare inherited condition, phocomelia, an abnormal gene acts during the third to fifth weeks of embryonic development, when the limbs develop. The infant has flipperlike structures where arms and legs should be. Geneticists can predict the chance that a genetically caused birth defect will recur.

Toxins that pregnant women ingest cause some birth defects. These environmentally caused problems cannot pass to future generations and will not recur unless the exposure does. Chemicals or other agents that cause birth defects are called **teratogens** (Greek for "monster causing"). For example, a drug called thalidomide also acts during the third to fifth week of embryonic development and causes severe limb shortening, as the inherited phocomelia does (fig. 12.9). Table 12.4 lists some commonly used teratogens.

Alcohol

Alcohol is a teratogen. Aristotle noticed problems in children of alcoholic mothers more than 23 centuries ago. In the United States today, **fetal alcohol syndrome** (FAS) is the third most common cause of mental retardation in newborns, and 1 to 3 in every 1,000 infants has the syndrome—more than 40,000 born each year.

A pregnant woman who has just one or two drinks a day, or perhaps a large amount at a single crucial time in prenatal development, risks FAS. Because each woman metabolizes alcohol slightly differently, physicians advise that pregnant women avoid alcohol when pregnant or when trying to become pregnant.

A child with fetal alcohol syndrome has a small head, misshapen eyes, and a flat face and nose and grows slowly before and

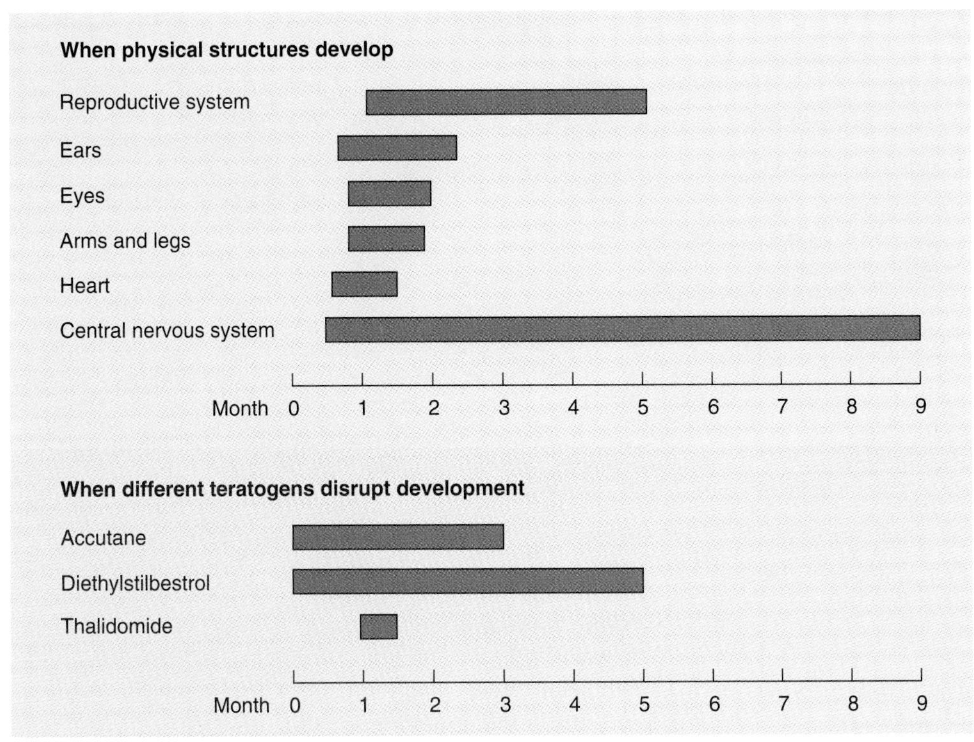

FIGURE 12.8

The nature of a birth defect resulting from drug exposure depends upon which structures are developing at the time of exposure. The time of sensitivity during development for a particular structure is called the critical period. Accutane is an acne medication. Diethylstilbestrol (DES) was used in the 1950s to prevent miscarriage (ineffectively), and thalidomide is a mild tranquilizer used in the 1960s in pregnant women in England and Germany.

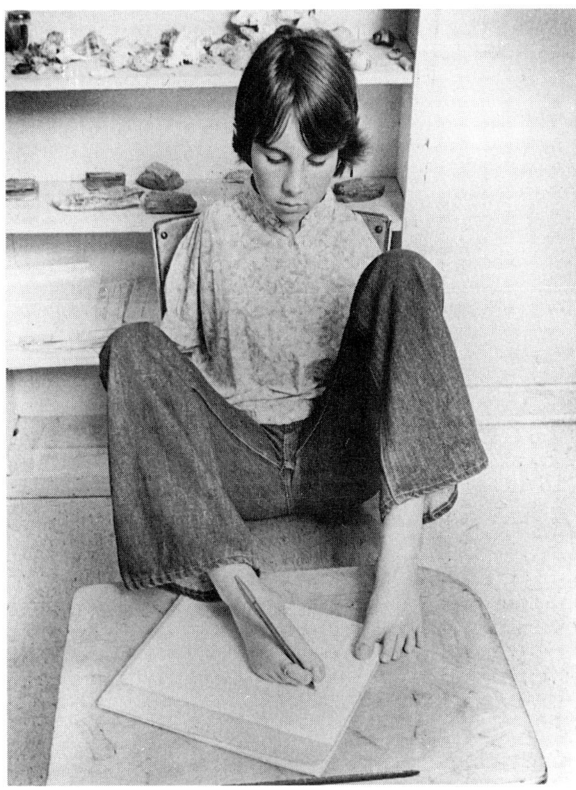

FIGURE 12.9

This child is 1 of about 10,000 children born between 1957 and 1961 in Europe to women who had taken the tranquilizer thalidomide early in pregnancy. The drug is a teratogen on developing limbs, stunting growth so that only stumps appear. However, thalidomide is still a very useful drug for certain conditions, such as leprosy and AIDS.

after birth. The child has impaired intellect, ranging from minor learning disabilities to mental retardation. Effects of FAS continue beyond childhood. Teens and young adults with the syndrome are short and have small heads. More than 80% of them retain facial characteristics of a young child with FAS, including abnormal lips, misaligned or malformed teeth, and a wide space between the upper lip and the nose. These facial traits make people with FAS look similar but not abnormal (fig. 12.10). The long-term mental effects of prenatal alcohol exposure are more severe than the physical vestiges. Many adults with FAS function at early grade school level. They often lack social and communication skills and cannot understand the consequences of actions, take initiative, or interpret social cues.

Cocaine

Cocaine is very dangerous to the unborn. It can induce a stroke in a fetus and cause spontaneous abortion. Cocaine-exposed infants who survive are distracted and unable to concentrate on their surroundings. Other health and behavioral problems arise as these children grow. A problem in evaluating the effects of cocaine is that affected children are often exposed to other substances and situations that could also account for their symptoms.

Cigarettes

Chemicals in cigarette smoke stress a fetus. Carbon monoxide crosses the placenta and plugs sites on fetal hemoglobin molecules that would normally bind oxygen, robbing rapidly growing fetal tissues of oxygen. Other chemicals in cigarette smoke prevent nutrients from reaching the fetus. The placentas of women who

Table 12.4 Teratogenic Drugs

Drug	Medical Use	Risk to Fetus
Alkylating agents	Cancer chemotherapy	Growth retardation
Aminopterin, methotrexate	Cancer chemotherapy	Skeletal and brain malformations
Coumadin derivatives	Seizure disorders	Tiny nose Hearing loss Bone defects Blindness
Danazol	Endometriosis	Masculinization of female structures
Diphenylhydantoin (Dilantin)	Seizures	Cleft lip, palate Heart defects Small head
Diethylstilbestrol (DES)	Repeat miscarriage	Vaginal cancer, vaginal adenosis Small penis
Isotretinoin (Accutane)	Severe acne	Cleft palate Heart defects Abnormal thymus Eye defects Brain malformation
Lithium	Bipolar affective disorder	Heart and blood vessel defects
Penicillamine	Rheumatoid arthritis Metabolic disorders	Connective tissue abnormalities
Progesterones in birth control pills	Contraception	Heart and blood vessel defects Masculinization of female structures
Tetracycline	Antibiotic	Stained teeth

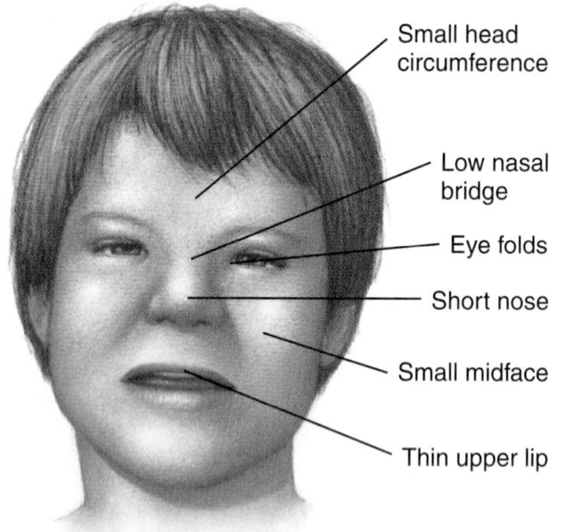

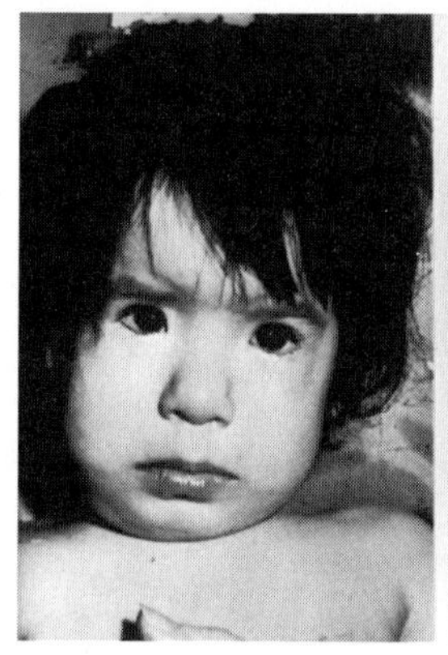

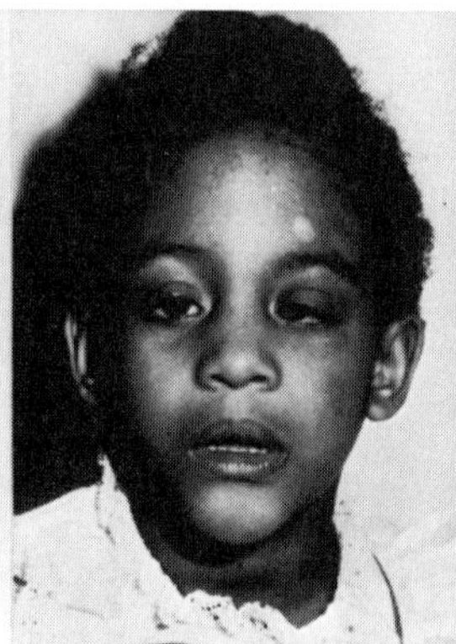

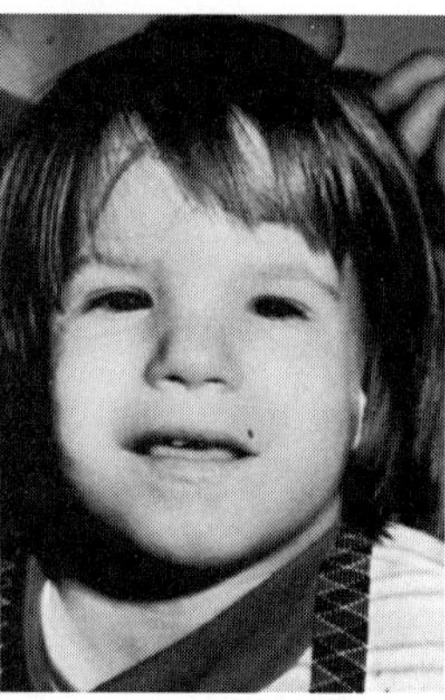

FIGURE 12.10

Fetal alcohol syndrome. A flat face and specific facial features are characteristic of children with FAS.

smoke lack important growth factors, thus lowering birth weight. Cigarette smoking during pregnancy increases risk of spontaneous abortion, stillbirth, and prematurity.

Nutrients

Certain nutrients ingested in large amounts, particularly vitamins, act as drugs in the human body. The acne medicine isotretinoin (Accutane), derived from vitamin A, causes spontaneous abortions and defects of the heart, nervous system, and face. The tragic effects of this drug were first noted in the early 1980s, exactly 9 months after dermatologists began prescribing it to young women. Today it is never prescribed to women who could be pregnant.

Another nutrient that can harm a fetus when the pregnant woman takes it in excess is vitamin C. The fetus becomes accustomed to the large doses and, after birth, when the supply drops, the baby develops symptoms of vitamin C deficiency (scurvy). The baby bruises easily and is prone to infection.

Malnutrition during pregnancy threatens a fetus because a pregnant woman requires extra calories. Obstetrical records of pregnant women before, during, and after World War II link inadequate nutrition in early pregnancy to an increase in the incidence of spontaneous abortion. The aborted fetuses had very little brain tissue. Poor nutrition later in pregnancy damages the placenta, causing low birth weight, short stature, tooth decay, delayed sexual development, learning disabilities, and possibly mental retardation.

Infections

Certain viral infections can pass to a fetus during pregnancy or to an infant during birth. Men can transmit infections to an embryo or fetus during sexual intercourse.

HIV, the virus that causes AIDS, infects 15 to 30% of infants born to infected (HIV positive) women. This risk can be cut sharply if an infected woman takes certain drugs used to treat AIDS while pregnant. Fetuses infected with HIV are at high risk for low birth weight, prematurity, and stillbirth.

The virus that causes rubella (German measles) is a powerful teratogen. Australian physicians first noted its effects in 1941, and a rubella epidemic in the United States in the early 1960s caused 20,000 birth defects and 30,000 stillbirths. Exposure in the first trimester leads to cataracts, deafness, and heart defects, and later exposure causes learning disabilities, speech and hearing problems, and juvenile-onset diabetes. Widespread vaccination has slashed the incidence of congenital rubella syndrome, and today it occurs only where people are not vaccinated. In 1991, for example, 14 of every 1,000 newborns among the Amish population in rural Pennsylvania had rubella exposure; the incidence in the general U.S. population is 0.006 per 1,000 births!

A herpes simplex viral infection can harm a fetus and newborn, whose immune systems are not yet fully functional. Forty percent of babies exposed to active vaginal lesions become infected, and half of these infants die from it. Of the surviving infants, 25% sustain severe nervous system damage, and another 25% have widespread skin sores. A Caesarean section can prevent an infected woman from transmitting the virus during the birth process.

Pregnant women are routinely checked for hepatitis B infection, which in adults causes liver inflammation, great fatigue, and other symptoms. Each year in the United States, 22,000 infants are infected with this virus during birth. These babies are healthy but are at high risk for developing serious liver problems as adults. By identifying infected women, a vaccine can be given to their newborns, which can help prevent future complications.

Occupational Hazards

People may encounter teratogens in the workplace. Women who work with textile dyes, lead, certain photographic chemicals, semiconductor materials, mercury, and cadmium face increased risk of spontaneous abortion and birth defects in their children. We do not know much about the role men play in transmitting environmentally caused birth defects. Men do have one built-in protection—a severely damaged sperm cell will not be able to move. However, men whose jobs expose them to sustained heat, such as smelter workers, glass manufacturers, and bakers, may produce sperm that can move, fertilize an oocyte, and possibly cause a spontaneous abortion or a birth defect.

Neonatology

Thirty years ago, if a woman went into labor after only 24 weeks of pregnancy, the result was almost always a spontaneous abortion or an infant who died within a few hours after birth. Today, nearly half of such pregnancies produce extremely small live infants. The medical technology of **neonatology** (study of the newborn) enables many of these infants to survive, although some are born with severe medical problems. Often new parents must decide whether or not to subject a severely ill newborn to corrective surgery and drug treatments.

Sometimes physicians can treat a fetus by administering drugs to the pregnant woman or by altering her diet. For example, a fetus that cannot produce adequate amounts of a specific vitamin can sometimes overcome the deficiency if the pregnant woman takes large doses of the vitamin. Physicians can treat some prenatal conditions directly. A tube inserted into the uterus can drain the dangerously swollen bladder of a fetus with a blocked urinary tract and relieve the problem until surgery can be performed shortly after birth. A similar procedure removes excess fluid from the brain of a hydrocephalic (fluid trapped in the brain) fetus. Drugs can reach the fetus, bypassing the mother, through a tube placed in the umbilical cord. Figure 12.11 shows fetal surgery.

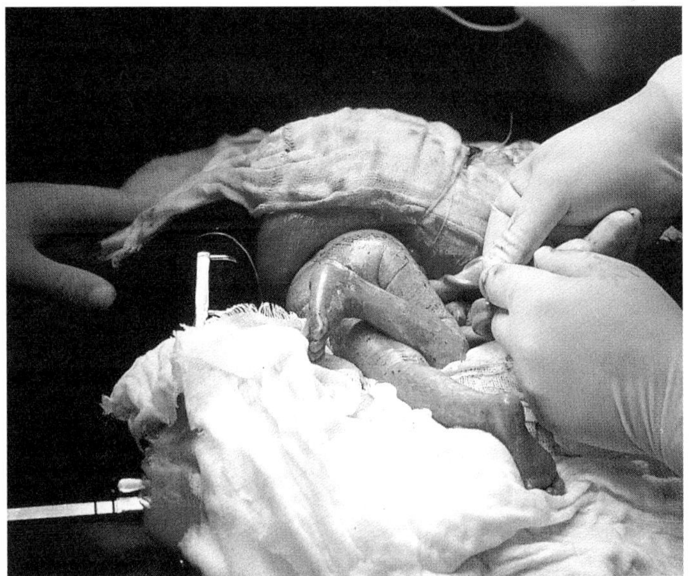

FIGURE 12.11

Baby Alexandra underwent surgery to repair a diaphragmatic hernia that was causing her intestines to protrude into her chest cavity, when she was a 7-month-old fetus.

Mastering Concepts

1. What are some causes of spontaneous abortion?
2. What are some problems that premature infants face?
3. What is the most common cause of premature birth?
4. What is a critical period?
5. Why do many birth defects affect the brain?
6. Name three teratogens.
7. What types of problems can physicians treat in a fetus?

Birth Control

Contraception is the use of devices or practices that work "against conception." Birth control methods either block the meeting of sperm and oocyte, or they make the female system's environment hostile to sperm or to a preembryo's implantation (table 12.5).

Available Methods of Birth Control

In 1960, the first contraceptive pill became available. The "combination pill" contains synthetic versions of the hormones estrogen and progesterone, which hinder conception in three ways: they suppress ovulation, alter the uterine lining so that a preembryo cannot implant, and thicken the mucus in the cervix so that sperm cells cannot penetrate.

Two newer contraceptive methods for women are a synthetic progesterone (Depo-Provera) and an implant of an estrogen-progesterone combination (Norplant). Depo-Provera is injected every 3 months. Norplant consists of six matchstick-shaped implants placed beneath the skin on the upper arm. It is effective for 5 years.

Contraceptives that provide a physical and/or chemical barrier between sperm and oocyte provide some protection against sexually transmitted diseases (STDs). A male **condom** is a sheath worn over the erect penis. A female condom is a baglike device inserted in the vagina. Either type blocks sperm from ascending the female reproductive tract. Spermicidal (sperm-killing) jellies, foams, creams, and suppositories are inserted into the vagina, and they can be used in conjunction with devices that block the cervix, such as the **diaphragm** or **cervical cap.** Most products use the spermicide nonoxynol-9.

"Natural" birth control or "fertility awareness" methods restrict sexual activity when conception is likely. In the **rhythm method,** the woman charts her menstrual cycle to estimate when she is most fertile and avoids intercourse on those days. However, because sperm can fertilize an oocyte days after intercourse, fertility awareness often fails.

Sterilization provides permanent birth control. A **vasectomy** cuts or burns the vasa deferentia so that sperm cannot leave the testicles. Developing sperm degenerate in the man's body. A **tubal ligation** cuts or ties a woman's fallopian tubes. Vasectomies and tubal ligations can sometimes be reversed. Figure 12.12 shows the sites of action of some birth control methods.

Experimental Methods of Birth Control

Progress in developing new contraceptives has been slow because many approaches that make sense in theory are too uncomfortable in practice. Consider using testosterone to "fool" a

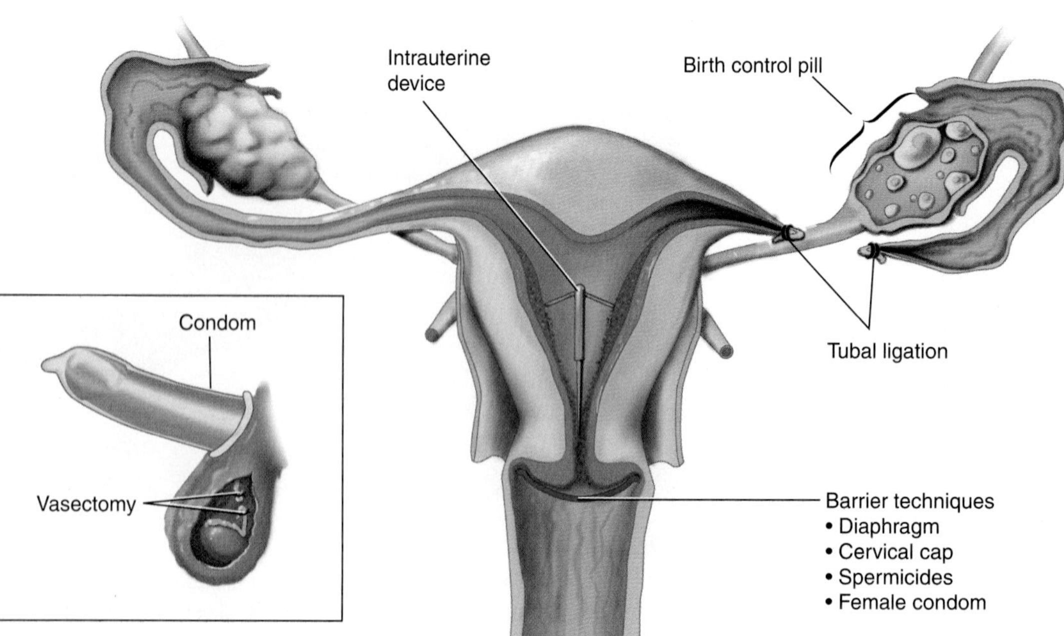

FIGURE 12.12
Sites of birth control action. Birth control pills prevent ovulation. A tubal ligation prevents oocytes from reaching the uterus. An intrauterine device embeds in the uterine wall, preventing a zygote from implanting. The diaphragm, female condom, and cervical cap block the cervix so that sperm cannot reach the uterus. Spermicidal jellies, creams, and suppositories kill sperm. A condom and vasectomy are male contraceptive strategies.

Table 12.5 Birth Control Methods

Method	Mechanism	Advantages	Disadvantages	Success
Barrier and Spermicidal				
Condom and spermicide	Worn over penis (male) or inserted into vagina (female), keeps sperm out of vagina, and kills sperm that escape	Protection against sexually transmitted diseases	Disrupts spontaneity, reduces sensation	95–98%
Diaphragm and spermicide	Kills sperm and blocks cervix	Inexpensive	Disrupts spontaneity, must be fitted	83–97%
Cervical cap and spermicide	Kills sperm and blocks cervix	Inexpensive, can be kept in for 24 hours	May slip out of place, must be fitted	80–95%
Spermicidal foam or jelly	Kills sperm and blocks cervix	Inexpensive	Messy	78–95%
Spermicidal suppository	Kills sperm and blocks cervix	Easy to use and carry	Irritates 25% of users	85–97%
Hormonal				
Combination birth control pill	Prevents ovulation and implantation, thickens cervical mucus	Does not interrupt spontaneity, lowers cancer risk, lightens menstrual flow	Raises risk of heart disease in some women, weight gain and breast tenderness	90–100%
Minipill	Blocks implantation, deactivates sperm, thickens cervical mucus	Fewer side effects	Weight gain	91–100%
Depo-Provera	Prevents ovulation, alters uterine lining	Easy to use, lasts 3 months	Menstrual changes, weight gain, injection	99%
Norplant	Prevents ovulation, thickens cervical mucus	Easy to use, lasts 5 years	Menstrual changes, doctor must implant	99.8%
Behavioral				
Rhythm method	No intercourse during fertile times	No cost	Difficult to do, hard to predict timing	79–87%
Withdrawal	Removal of penis from vagina before ejaculation	No cost	Difficult to do	75–91%
Surgical				
Vasectomy	Sperm cells never reach penis	Permanent, does not interrupt spontaneity	Requires minor surgery, difficult to reverse	99.85%
Tubal ligation	Oocytes never reach uterus	Permanent, does not interrupt spontaneity	Requires surgery, infection, difficult to reverse	99.6%
Other				
Intrauterine device	Prevents implantation	Does not interrupt spontaneity	Severe menstrual cramps, infection	95–99%

man's brain into halting synthesis of two other hormones that normally increase sperm production. Side effects include weight gain and acne, and the hormone must be injected weekly.

Several experimental contraceptives target the molecular events of fertilization (fig. 12.13). One approach is based on a sperm-binding carbohydrate on the zona pellucida. A synthetic version of that carbohydrate, applied to the female reproductive tracts of mice, blocks the sperm binding site and prevents fertilization. Another strategy targets the acrosomal reaction, in which the sperm tip releases enzymes that penetrate the zona. New drugs block channels in the sperm cell membrane that normally admit calcium ions necessary for the

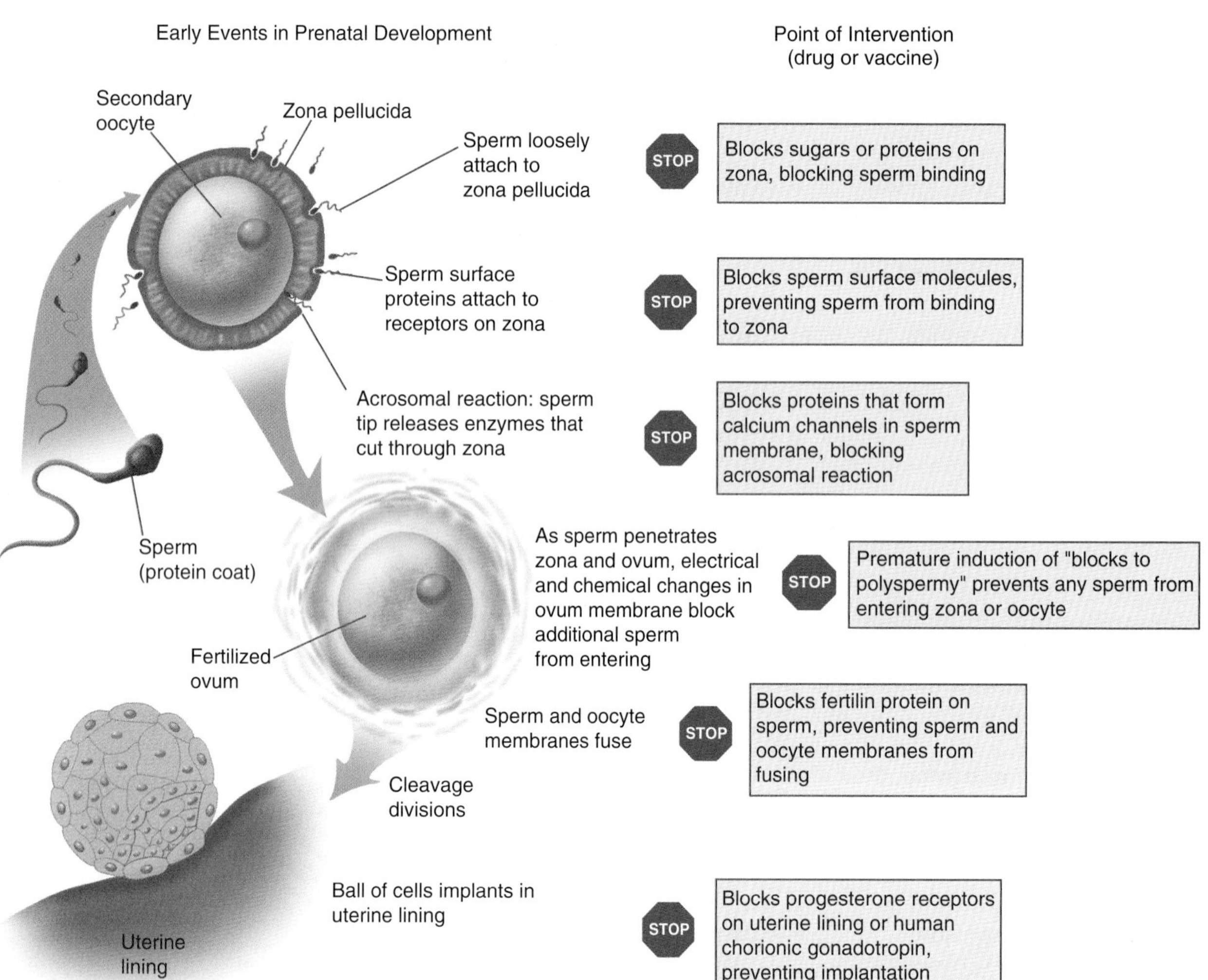
Early Events in Prenatal Development
Point of Intervention
(drug or vaccine)
Secondary
oocyte
Zona pellucida
Sperm loosely
attach to
zona pellucida
STOP
Blocks sugars or proteins on
zona, blocking sperm binding
Sperm surface
proteins attach to
receptors on zona
STOP
Blocks sperm surface molecules,
preventing sperm from binding
to zona
Acrosomal reaction: sperm
tip releases enzymes that
cut through zona
STOP
Blocks proteins that form
calcium channels in sperm
membrane, blocking
acrosomal reaction
Sperm
(protein coat)
As sperm penetrates
zona and ovum, electrical
and chemical changes in
ovum membrane block
additional sperm
from entering
STOP
Premature induction of "blocks to
polyspermy" prevents any sperm from
entering zona or oocyte
Fertilized
ovum
Sperm and oocyte
membranes fuse
STOP
Blocks fertilin protein on
sperm, preventing sperm and
oocyte membranes from
fusing
Cleavage
divisions
Ball of cells implants in
uterine lining
STOP
Blocks progesterone receptors
on uterine lining or human
chorionic gonadotropin,
preventing implantation
Uterine
lining

FIGURE 12.13

The molecular interactions of fertilization, implantation, and early prenatal development suggest new approaches to contraception.

reaction. Yet another approach blocks production of fertilin, a sperm membrane protein that brings the cell membranes of sperm and oocyte together.

An alternative to using drugs to block specific events of fertilization is to manipulate the immune system to produce antibody proteins that bind to, and thereby inactivate, key participants in the process. One such contraceptive causes a man's body to manufacture antibodies that inactivate a hormone the pituitary gland in the brain releases that normally stimulates testosterone production and sperm development. However, this approach also diminishes a man's sex drive. Another antibody approach in the male targets fertilin. Antibody-based contraceptives being developed for women target fertilin, sperm-binding proteins on the zona, and hCG, the "pregnancy hormone" that suppresses menstruation.

Terminating a Pregnancy

An unwanted pregnancy can be terminated by preventing the preembryo's implantation in the uterus or, later on, by removing an embryo or fetus from the uterus. First-trimester abortions use a sucking or scraping device to remove the embryo or fetus, while later abortions use more intense scraping, injection of a salt solution into the amniotic sac, or prostaglandin suppositories, which trigger labor within 24 hours. A drug developed in France, RU 486, induces abortion during the first 7 weeks of pregnancy. Because it is taken orally, it is safer than a surgical abortion. RU 486 blocks progesterone, which is necessary for implantation and early development. Combinations of certain drugs used for other purposes can also cause abortion.

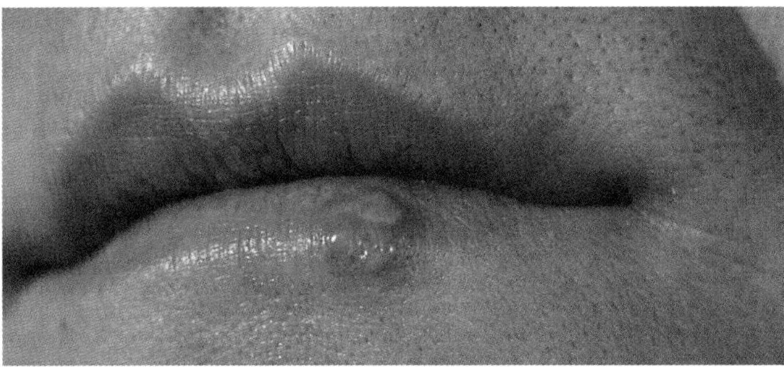

FIGURE 12.14
A mouth sore caused by a herpes infection.

Abortion differs from contraception because it intervenes after implantation. In the United States, abortion has been a controversial subject since it became legal in the early 1970s. In the former Soviet Union, however, two-thirds of all women have had at least one abortion because of the scarcity of contraceptives. In many nations, people choose abortion when a fetus has a devastating or untreatable medical condition or when pregnancy threatens the health or life of the woman.

Mastering Concepts

1. What are the two basic ways that contraceptives prevent pregnancy?
2. What are some currently available contraceptives, and how do they work?
3. How do experimental birth control methods work at the molecular level?

Sexually Transmitted Diseases

The 20 recognized STDs are often called "silent infections" because in the early stages they may not produce symptoms, especially in women. By the time symptoms appear, it is often too late to prevent complications or spread of the infection to sexual partners. Because many STDs have similar symptoms, and some of the symptoms may also arise from diseases or allergies that are unrelated to sexual activity, physicians suggest a medical exam if any of these symptoms appears:

1. burning sensation during urination;
2. pain in the lower abdomen;
3. fever or swollen glands in the neck;
4. abnormal discharge from the vagina or penis;
5. pain, itch, or inflammation in the genital or anal area;
6. pain during intercourse;
7. sores, blisters, bumps, or a rash anywhere on the body, particularly the mouth or genitals;
8. itchy, runny eyes.

Following is a closer look at three STDs.

Herpes Simplex

Infection by either of two types of herpes simplex virus, HSV-1 and HSV-2, initially causes tingling or itching at the infection site in the skin. Within 2 weeks, the area erupts into cold sores on or near the mouth or a rash of painful, clear genital blisters (fig. 12.14). The eruptions last about 3 weeks; during this time, the person can pass the virus to others by skin contact. Symptoms may be so mild or difficult to detect that the virus is passed unknowingly.

After the outbreak disappears, the virus enters nerve cell branches in the skin and moves further into the nervous system, where the immune system cannot detect it. The virus may cause a new eruption a week, a month, or a year later, or it may never appear again. Stress seems to precipitate recurrences. The drug acyclovir can shorten outbreaks.

Chlamydia

Five to 10% of us may harbor infections of the bacterium *Chlamydia trachomatis.* Most affected individuals are female, because the bacteria readily infect tissues covering the cervix. Symptoms of chlamydia infection include discharge from the genitals and painful urination. Often there are no symptoms.

Chlamydia and other bacterial infections can ascend in the female reproductive tract, causing pelvic inflammatory disease (PID). Symptoms progress from intermittent cramps to fever, chills, weakness, and more severe cramps. Intravenous antibiotics can stop the infection, but PID often scars the uterus and fallopian tubes, causing infertility and increasing risk of ectopic pregnancy. Two-thirds of infants whose mothers had chlamydia while pregnant are born with infections in their eyes or lungs and are at higher risk of developing lung problems later in life.

Acquired Immune Deficiency Syndrome (AIDS)

Unfortunately, one needn't read a biology textbook to learn about AIDS—this devastating viral infection has dominated headlines for more than a decade. Since the first reports of the disease in 1981, AIDS has killed more than a million people and HIV has infected many millions more. However, in 1995, the first glimmers of hope for treatments began to appear. We discuss AIDS briefly here in the context of its transmission and revisit it in later chapters on evolution, viruses, and the immune system.

HIV Infection

HIV is a retrovirus, which means that its genetic material is RNA. Inside a human cell, an enzyme called reverse transcriptase copies HIV's RNA into DNA, which inserts into the DNA of the cell. The cell then manufactures viral proteins and nucleic acids and becomes a minifactory of HIV production. Although the immune

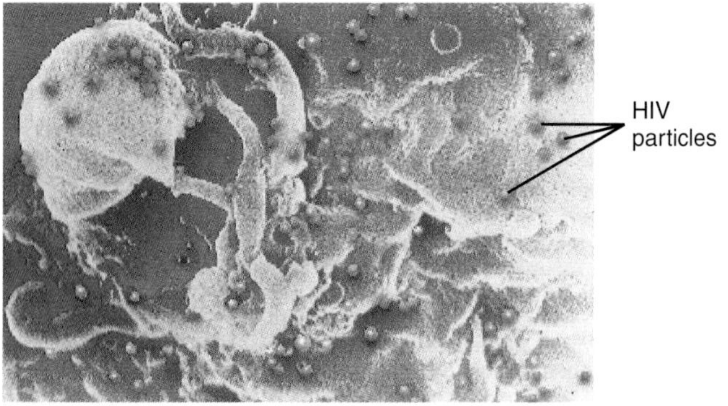

FIGURE 12.15

HIV particles (red) emerge from an infected white blood cell. (× 3,000)

Table 12.6 Some AIDS-Related Symptoms and Illnesses

Some AIDS-Related Symptoms and Illnesses
Persistent lymphadenopathy (swollen lymph glands)
Constant low-grade fever
Nausea and vomiting
Fatigue
Night sweats
Headaches
Wasting syndrome (persistent diarrhea, severe weight loss, weakness, fever)
Dementia (poor muscle control, confusion, apathy, inability to concentrate, sudden and strong emotions, memory loss, insomnia, headache, disorientation)
Cancers (Kaposi's sarcoma, cervical cancer, lymphoma, others)
Opportunistic infections (pneumonia, blindness, brain infection, diarrhea, spinal meningitis, tuberculosis, fungal infections, many others)

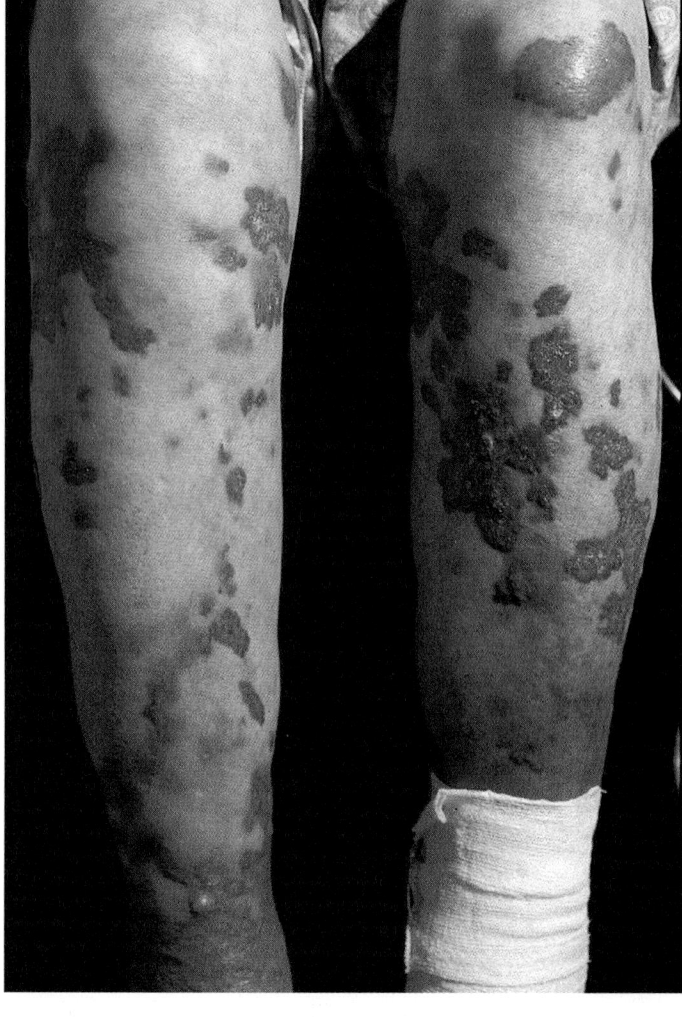

FIGURE 12.16

Kaposi's sarcoma, a common cancer among people with AIDS, was once seen only among elderly Jewish and Italian men. It began appearing in younger men when HIV infection became prevalent. Kaposi's sarcoma is caused by a virus other than HIV, but it is the immunity breakdown that enables this second virus to infect.

system initially mass-produces white blood cells to fight the virus, HIV wins as infected cells burst, releasing millions of new viruses at an ever escalating rate (fig. 12.15).

HIV initially enters a type of white blood cell called a T cell, which controls many aspects of immunity. Therefore, the signs and symptoms of AIDS reflect immunity breakdown, and different people may have different combinations of AIDS-related illnesses (table 12.6; fig. 12.16). The infection eventually spreads to other cell types.

HIV infection has three stages—initial symptoms, a latency period, and AIDS. The initial, acute phase may include weakness, recurrent fever, night sweats, swollen glands in the neck, and weight loss. This stage can vary in duration and severity. Often it lasts just a few days, and the person thinks it is just a bout with the flu. Then comes a latency period, typically lasting 5 to 10 years, when the person feels well, but this may be deceptive. In the lymph nodes and then the bloodstream, the battle intensifies as the immune system struggles to keep up with the growing HIV population. The third stage, AIDS, is marked by onset of opportunistic infections, so called because they appear as a result of a compromised immune system. The amount of virus in the bloodstream indicates the stage of infection.

Modes of Transmission

AIDS is transmitted in several ways, all requiring contact with a body fluid containing abundant HIV. Blood and semen are most likely to transmit HIV. Although the virus has been detected in sweat, tears, and saliva, levels are so low that transmission is highly unlikely. Whether or not a person becomes infected depends upon

Table 12.7 HIV Transmission
How HIV Is Transmitted
Sexual contact, particularly anal intercourse
Contaminated needles (intravenous drug use, accidental needle stick in medical setting)
During birth from infected mother
Receiving infected blood or other tissue (precautions usually prevent this)
How HIV Is NOT Transmitted
Casual contact (social kissing, hugging, handshakes)
Objects (toilet seats, deodorant sticks, doorknobs)
Mosquitoes
Sneezing and coughing
Saliva (sharing food)
Swimming in the same water

the amount of infected fluid contacted, the site of exposure in the body, and the health of the individual. Table 12.7 lists ways that HIV infection can and cannot be spread. The discussion of HIV and AIDS in chapter 39 looks at signs that the pandemic (worldwide epidemic) may finally be waning.

Better understanding of the biology of HIV infection, plus new drug weapons and clues from survivors, are providing something that has long been sorely lacking in the global fight against AIDS—hope.

Mastering Concepts

1. What are the general signs and symptoms of STDs?
2. What are some STDs and their symptoms and causes?
3. What are the three stages of HIV infection?

Homosexuality

No one really knows how or why we have feelings of being male or female or of being attracted to one sex or the other. A person who is homosexual is aroused by stimuli from the same sex but does not necessarily act on these feelings. Researchers believe that homosexuality is about 50% genetically controlled and 50% environmentally controlled, and that it is more common than many people think: it is found in all cultures examined and has been observed for thousands of years. Studies suggest that about 2% of people are homosexual.

Attitudes concerning whether homosexuality is abnormal or normal vary greatly. People in the Phillipines accept homosexuality as "nature's way." Conversely, in the United States, 20 states still have laws punishing homosexual practices. The French Napoleonic code, written in 1810 and followed in many nations today, states that if adults enter into a behavior voluntarily, it is not criminal. In 1980, the American Psychiatric Association removed "homosexuality" as an entry in its standard reference book defining behavioral disorders.

Psychological Aspects of Homosexuality

Psychologists have repeatedly attempted to identify psychiatric disorders that are more common among homosexuals—and failed. Studies that examine populations of deviant individuals, such as sex offenders, do not find that homosexuals are more heavily represented. Other studies find that homosexual people make just as fine parents as heterosexual individuals.

More useful, perhaps, are studies that examine homosexual feelings. Homosexual men generally report having felt "different" as children, later identifying homosexual feelings even though as children they did not know the term or concept of homosexuality. This suggests an inborn influence. If a child becomes aware of others' negative views of homosexuals, then feelings of shame, low self-esteem, and even self-hatred may arise. These feelings are not a symptom of homosexuality but a response to people's prejudices.

Biological Studies—Of Genes and Brains

Evidence for inheritance of homosexuality is indirect. Consider twin studies, often used to tease apart genetic from environmental influences on traits. Studies find that both members of twin pairs are more likely to be homosexual if the twins are identical (and therefore have the same genes) than if they are fraternal (and are no more genetically alike than any two siblings). These studies suggest an inherited component to homosexuality. However, identical twins may behave alike because they are often treated alike. Other studies show that homosexuality occurs repeatedly in some families, but again we cannot clearly distinguish between the influences of heredity and the environment.

Molecular approaches are also used to study sexual preference. In one study, 40 pairs of homosexual brothers were evaluated for five specific DNA sequences on their X chromosomes. These DNA sequences are highly variable in the population. Researchers interpreted the fact that 33 of the sibling pairs matched for the five "marker" sequences to mean that the X chromosome might bear genes predisposing to homosexuality. Heterosexual brothers of the homosexual pairs did not share the DNA sequences. However, the genetic markers may indicate traits the brothers share other than their homosexuality.

Do specific brain regions control homosexual feelings? Experiments focus on the hypothalamus, the seat of drives such as hunger, thirst, and sexual feelings. Two specific areas of the hypothalamus have been studied in humans. One, the suprachiasmatic nucleus (SCN), controls daily biochemical fluctuations. The SCN is about twice as large, and contains twice as many cells, in homosexual males as in heterosexual males. Another region, called the INAH 3, is half the size in homosexual males as in heterosexual males and is about the same size as in heterosexual women. Researchers studying these brain regions

Joined for Life~

continued from page 238.

Conjoined twins occur in 1 in 50,000 births, and about 40% are stillborn. Abby and Britty Hensel are rare among the rare, being joined in a manner seen only four times before. They are the result of incomplete twinning, which probably occurred during the first 2 weeks of gestation. Because the girls have duplicated tissue derived from ectoderm, mesoderm, and endoderm, the partial twinning event must have occurred before the three germ layers were established.

The term "Siamese twins" comes from Chang and Eng, who were born in Thailand, then called Siam, in 1811. They were joined by a ligament from the navel to the breastbone, which surgeons could easily correct today. Chang and Eng lived for 63 years, and each married.

Eilish and Katie Holton, born in Ireland in 1989, were similar to Abby and Britty but had four arms and a large, abnormal heart. In 1992, a team of 15 doctors separated them. Katie died 4 days after the surgery of heart failure. Eilish, now a healthy child who gets around quite well on an artificial leg, still looks over her shoulder for her missing twin.

In 1993, U.S. physicians attempted to separate Amy and Angela Lakeberg, infants who shared a liver and heart. Doctors determined that Angela had the better chance to survive, and so during surgery, they gave the heart to her. Amy died instantly. Angela lived until the following June, never able to breathe without the aid of a respirator.

Because Abby and Britty have separate hearts and nervous systems, they might fare better in surgery than the Holton or Lakeberg twins. But the girls are happy as they are, for now. They were particularly distressed to meet Eilish in 1995 to do a television program and learn the fate of Katie. Abby and Britty have very distinctive personalities and attend school, swim, ride bikes, and play, like any other kids.

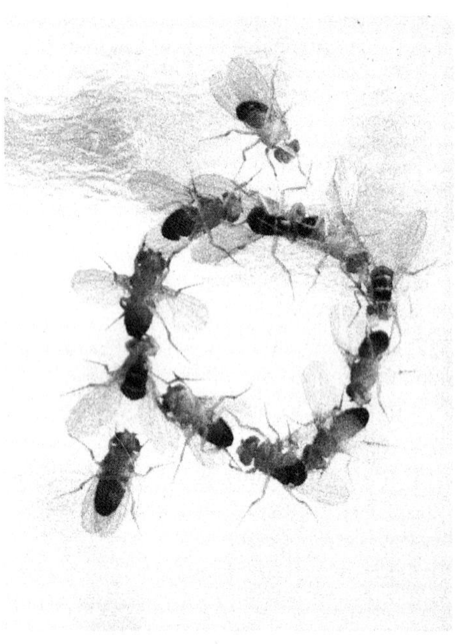

FIGURE 12.17

The ability to genetically alter male fruit flies in a way that causes them to display mating behavior towards each other adds to evidence that homosexuality is at least partially an inherited trait.

stress that the differences detected between homosexuals and heterosexuals are correlations, not proof of cause and effect.

Another brain area, the temporal lobe, may cause homosexuality. In a rare condition called Kluver-Bucy syndrome, damage to part of this area suddenly and dramatically changes a person's sexual orientation from heterosexual to homosexual. A man and a woman experienced this switch following viral brain infections, and other cases occurred in a man who suffered brain damage from a car accident and a man who had a portion of the temporal lobe cut to treat seizures. Although this evidence is anecdotal, it suggests another area for researchers to explore.

On another front, mutations in fruit flies induce homosexual-like behavior by altering levels of a neurotransmitter (serotonin) in the brain, which may explain mating behavior between males (fig. 12.17). Other animals exhibiting homosexual behavior have disruptions in serotonin levels too. Such animal studies provide further evidence that homosexuality is biological and widespread, rather than a social abberation.

Mastering Concepts

1. How common is homosexuality?
2. What are some views on the normalcy of homosexuality?
3. What is the evidence that homosexuality is inherited?

SUMMARY

1. Infertility is the inability to conceive a child after a year of trying. Causes in a male include low sperm count, a malfunctioning immune system, a varicose vein in the scrotum, heat exposure, structural sperm defects, and abnormal hormone levels. Infertility in a female can be due to absent or irregular ovulation, blocked fallopian tubes, an inhospitable uterus, or antisperm secretions.

2. Assisted reproductive technologies include **artificial insemination, surrogate mothers, IVF, GIFT, ZIFT,** and embryo transfer. **Preimplantation genetic diagnosis** screens IVF preembryos for disease-causing genes.

3. Spontaneous abortion is common and usually occurs because an embryo or fetus has an extra or missing chromosome or is due to a hormonal or structural abnormality in the woman's reproductive tract. Babies born **prematurely** often experience several health problems. Maternal infection may cause premature delivery. Environmental exposures **(teratogens)** or abnormal genes can cause birth defects. The nature of the defect reflects which structures were in **critical periods** when the insult occurred. **Neonatology** deals with the medical problems of newborns and includes fetal treatments.

4. Contraceptives block the meeting of sperm and oocyte or make the uterus inhospitable to implantation of a zygote. Contraceptives include hormone-containing pills, injections, or implants; barrier contraceptives; spermicides; and intrauterine devices. **Vasectomy** and **tubal ligation** are permanent methods. Experimental contraceptives target molecules involved in fertilization or manipulate the immune system.

5. The 20 STDs have similar symptoms but are easily spread because in their early stages they may not produce symptoms.

6. HIV causes AIDS by infecting immune system cells. HIV infects cells, replicates itself, and multiplies faster than the immune system can produce cells to fight it. As the immune system fails, opportunistic infections and cancer begin. The three stages are acute symptoms, a latent period, and then AIDS. The infection is transmitted by contact with infected body fluids.

7. Homosexuality is likely to have biological causes.

TO REVIEW . . .

1. Understanding the causes of infertility can be valuable in developing new birth control methods. Cite a type of contraceptive based on each of the following causes of infertility:

a. failure to ovulate due to a hormonal imbalance
b. a large fibroid tumor in the uterine lining
c. endometriosis tissue blocking fallopian tubes
d. low sperm count

2. A couple having trouble conceiving a child learns that the problem is twofold. The male has a sperm count of 30 million sperm per ejaculate, and the woman ovulates four times a year. Suggest a way to help the couple to conceive.

3. Big Tom is a bull with valuable genetic traits. His sperm are used to conceive 1,000 calves in many different cows. Mist, a dairy cow with exceptional milk output, is given a superovulation drug, and many oocytes are removed from her ovaries. Once fertilized, the oocytes are implanted into other cows. With their help, Mist becomes the genetic mother of 100 calves—far more than she could give birth to naturally. Which human technologies do the experiences of Big Tom and Mist mirror?

4. How can "fertility awareness" methods help a couple to conceive a child? How can they be used as contraceptive measures?

5. Ultrasound detects twin embryos in a woman who is 6 weeks pregnant. Seven months later, only one baby is born. What happened?

6. A few years ago, a doll manufacturer sold a cute, pudgy "preemie" model. Is this an accurate representation of an infant born too soon? Why or why not?

7. In Aldous Huxley's book *Brave New World,* oocytes are fertilized in vitro and develop assembly-line style. To render some of the embryos less intelligent, lab workers give them alcohol. What medical condition does this scenario invoke?

8. List three ways that HIV is transmitted and three ways that it is not.

9. Describe three lines of evidence suggesting a biological basis to homosexuality.

TO THINK ABOUT . . .

1. State who the genetic parents are and who the gestational mother is in each of the following scenarios:

a. A 26-year-old woman has her uterus removed because of cancer. However, her ovaries are intact and her oocytes are healthy. She takes drugs to superovulate and has oocytes removed, and they are fertilized in vitro with her husband's sperm. Two fertilized ova are implanted into the uterus of the woman's best friend.
b. Max and Tina had a child by IVF in 1986. At that time, they had three extra preembryos frozen. Two are thawed in 1994 and implanted into the uterus of Tina's sister, Karen. Karen's uterus is healthy, but she has ovarian cysts that prevent her from ovulating.
c. Forty-year-old Christensen von Wormer wanted children but not a mate. He donated sperm and hired an Indiana woman to undergo artificial insemination. On September 5, 1990, Wormer held his newly born daughter, Kelsey, for the first time.
d. Two men who live together want to raise a child. They go to a fertility clinic, have their sperm collected and mixed, and use it to artificially inseminate a female friend, who turns the baby over to them.

2. Lawyers and government regulatory agencies sometimes intervene in biological matters. What position do you think the law should take in the following procedures or situations?

a. Treatment of newborns with incurable, life-threatening conditions.
b. Use of "extra" fertilized ova or preembryos growing in vitro.
c. Malpractice liability if an embryo or fetus conceived by an assisted reproductive technology is spontaneously aborted or is born with a birth defect.
d. Custody of a child if a surrogate mother decides, after giving birth, to keep the child rather than surrendering it to its genetic parents.
e. Whether to inform a minor's parents that he or she has sought birth control.
f. Whether to approve a new contraceptive.
g. Which items of personal information a sperm donor should provide.

3. Methotrexate is a drug traditionally used to treat cancer, but it can also stop growth of an ectopic pregnancy and can cause abortion. How can one drug have such different applications?

4. Do you think an HIV test that a person can perform at home is a good idea? Why or why not?

5. Contraception literally means "against conception." According to this definition, is an intrauterine device a contraceptive? Why or why not?

6. What kinds of studies and information would be necessary to determine whether exposure to a potential teratogen during a war can cause birth defects years later? How would such an analysis differ if it was a man or a woman who was exposed?

TO LEARN MORE . . .

References and Resources

Annas, George J. February 6, 1992. Using genes to define motherhood—the California solution. *The New England Journal of Medicine,* vol. 326, no. 6. Not everyone agrees that genes alone define parenthood.

Edwards, Robert, and Patrick Steptoe. 1980. *A matter of life.* New York: William Morrow. The story of Louise Joy Brown—the first "test tube baby."

Gorman, Christine. February 12, 1996. Battling the AIDS virus. *Time.* An excellent review of the stages of HIV infection.

Hack, Maureen, and Irwin R. Merkatz. December 28, 1995. Preterm delivery and low birth weight—A dire legacy. *The New England Journal of Medicine,* vol. 333. Simple antibiotic treatment during pregnancy may prevent prematurity.

Howards, S. S. February 2, 1995. Current concepts: Treatment of male infertility. *The New England Journal of Medicine,* vol. 332. A comprehensive review of treatments for male infertility.

Lemonick, Michael. D. March 18, 1996. What's wrong with our sperm? *Time.* Are pollutants dropping human sperm counts?

Lewis, Ricki. March 1, 1995. Interferon must prove itself in market for genital warts. *Genetic Engineering News,* vol. 15. An immune system biochemical targets this widespread STD.

Lewis, Ricki. March 1993. Choosing a perfect child. *The World and I.* Preimplantation genetic diagnosis screens preembryos for inherited disease.

Wallace, Claudia. March 25, 1996. The most intimate bond. *Time.* Life as a conjoined twin is most difficult.

denotes articles by author Ricki Lewis that may be accessed via *Life's* home page /http://www.mhhe.com/sciencemath/biology/life/.

Unit 4

Genetics

People have noted family resemblances for thousands of years. Today, genetics is a part of such diverse fields as agriculture, forensics, and medical science.

13 Transmission Genetics

The laws of heredity apply to all diploid organisms. A mutation in a single gene causes achondroplastic dwarfism. Alexandre Schafnitz, 4 years old here, appears at a press conference to announce the discovery of the gene that causes his short stature.

chapter outline

The History of Genetics 262

Mendel's First Law (Segregation) Traces One Gene 262

How Biologists Describe Single Genes

Mendel's Laws and Meiosis

Mendelian Traits in Humans 267

Modes of Inheritance

Pedigrees

Mendel's Second Law (Independent Assortment) Traces Two Genes 269

When Gene Expression Appears to Alter Mendelian Ratios 272

Lethal Allele Combinations

Multiple Alleles

Different Dominance Relationships

Epistasis—Gene Masking at *General Hospital*

Penetrance and Expressivity

Environmental Influences

Pleiotropy

Phenocopies—When It's Not Really in the Genes

Complex Traits 276

Measuring Multifactorial Traits

Studying Multifactorial Traits

It Can't Be!

Erin and David Fitzgerald were thrilled at the birth of their daughter, Megan. Their three boys were robust and healthy, and Megan received a clean bill of health at birth. But in her first months, Megan seemed to have unusually frequent respiratory infections. Possibly because of her frequent illnesses, she gained weight slowly. When she developed bronchitis, her pediatrician began to suspect an inherited illness—cystic fibrosis (CF).

Before mentioning her suspicions to the parents, the doctor did an informal test—she kissed Megan. The telltale salt on the baby's skin was a distressing sign. The doctor ordered a "sweat test," which would check for extra salt in Megan's sweat, an indication of CF. She also ordered a genetic test, which would confirm a diagnosis. Then she took a detailed family history of the couple, asking questions about respiratory illnesses, digestive problems, and male infertility among siblings, parents, grandparents, other children, aunts, uncles, and cousins. The doctor told Erin and David some facts about CF—that it is variable and that many of the symptoms are treatable—and she also described the inheritance pattern of skipping generations.

The new parents, however, remained optimistic, because none of their many relatives had ever had the devastating disease. That couldn't possibly be the diagnosis. But it was.

to be continued . . .

The History of Genetics

Inherited similarities can be startling, such as the physical likeness between John Lennon and son Julian and in their distinctive singing voices (fig. 13.1). Interest in heredity is probably as old as humankind itself, as people throughout time have wondered at their similarities—and fought over their differences (fig. 13.2).

Awareness of heredity appears as early as the Talmud, an ancient book of Jewish law that tells of a woman whose three sisters had sons who died after circumcision. The woman's rabbi advised her not to circumcise her son, which reveals recognition of transmission of a blood-clotting disorder through a female carrier. Agriculture is built on genetics. Six thousand years ago, farmers in Mexico chose seed from the hardiest wild grass plants to sow the next season's crop. In this way, over many plant generations, they bred domesticated corn.

Nineteenth-century biologists thought that body parts controlled trait transmission and gave the units of inheritance, today called **genes,** such colorful names as pangens, idioblasts, bioblasts, plastidules, nuclein, plasomes, stirps, gemmules, or just characters. An investigator who used the term *elementen* made the most lasting impression on what would become the science of genetics. His name was Gregor Mendel.

As a child in what is now Czechoslovakia near the Polish border, Mendel learned farming from his family and tended fruit trees for the lord of a manor. He overcame extreme poverty to obtain a university education in science, mathematics, and teaching, and he eventually entered a monastery. There he learned more about plant breeding from the abbot, who had built a greenhouse for scientific research.

Mendel used peas to study heredity and bred them from 1857 to 1865. He deduced that consistent ratios of traits in the offspring indicated that the plants transmitted distinct units, or "elementen." Although Mendel did not understand how organisms physically transmit hereditary information, his two laws of inheritance eventually proved to apply to any diploid species, including humans.

(a)

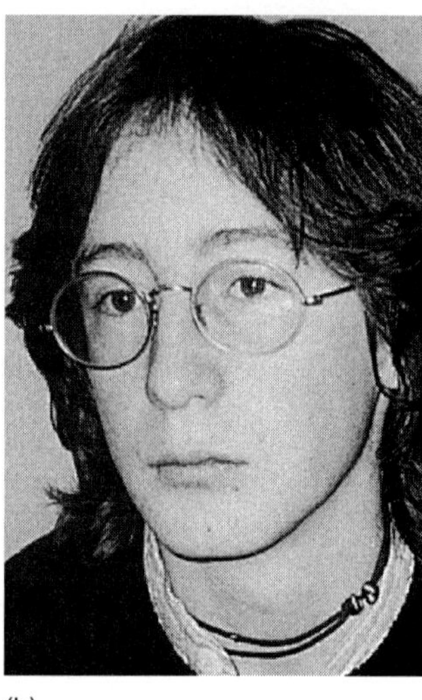

(b)

FIGURE 13.1

Heredity is apparent in appearances and talents, as illustrated by John (*a*) and Julian (*b*) Lennon.

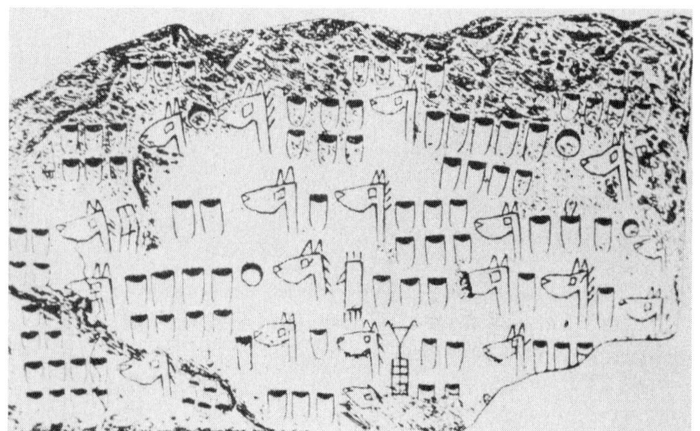

FIGURE 13.2

Ancient cultures recognized inheritance patterns. A horse breeder in Asia 4,000 years ago etched a record of his animal's physical characteristics in stone.

Mendel's First Law (Segregation) Traces One Gene

Mendel was especially interested in seven easily distinguishable characteristics (fig. 13.3). He noted that short plants bred with other

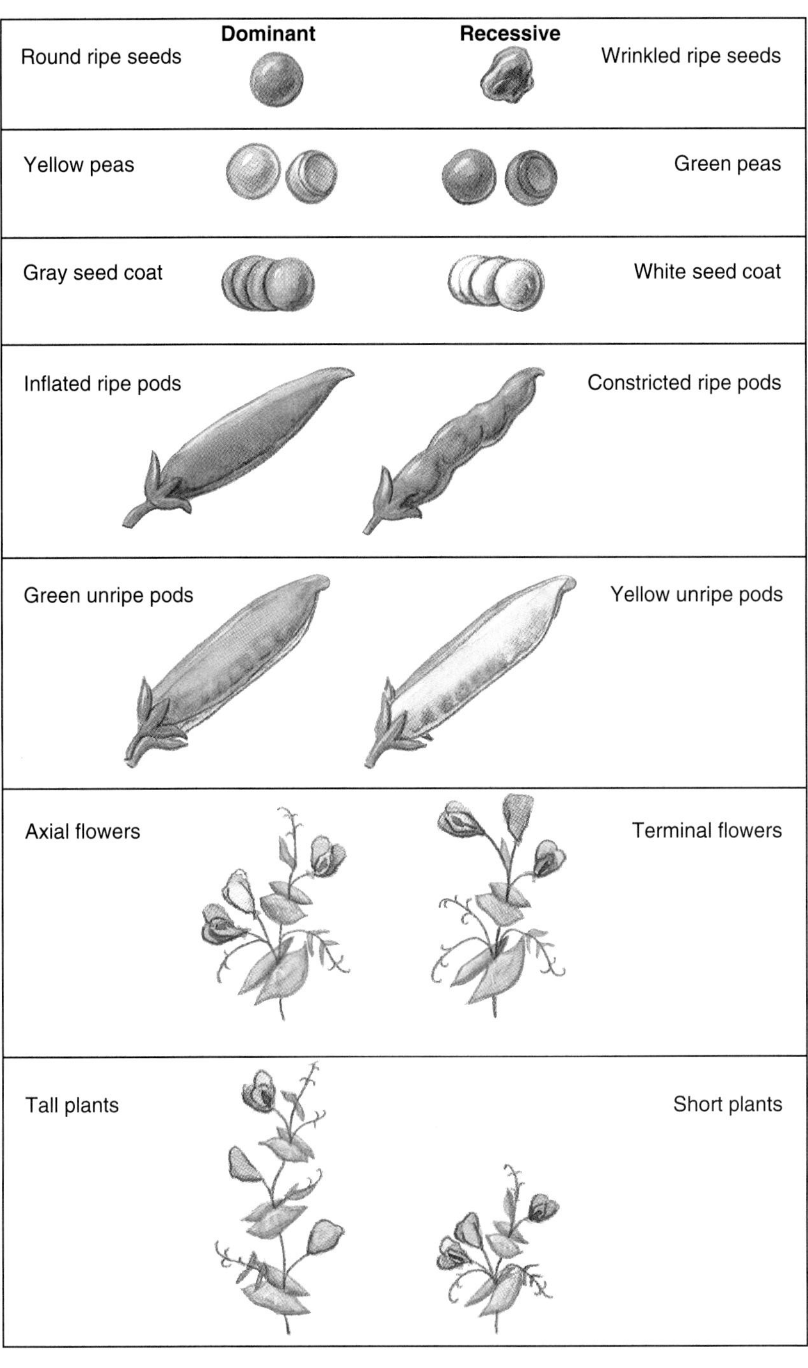

FIGURE 13.3

Gregor Mendel studied the transmission of seven traits in the pea plant. Each trait had two expressions, easily distinguished from each other. Seeds were either round or wrinkled; peas were either yellow or green; seed coats were either gray or white; ripe pods were either inflated or constricted; unripe pods were either green or yellow; flowers were either axial (arising sideways from the stems) or terminal (emanating only from the top of the plant); and plants were either tall (6 to 7 feet) or short (3/4 to 1 1/2 feet).

short plants were "true-breeding," always yielding seeds that gave rise only to more short plants. Tall plants were not always true-breeding. Some tall plants, when crossed with a short plant or another tall plant, produced only tall plants in the next generation. But when certain other tall plants were crossed with each other, about one-quarter of the plants in the next generation were short, suggesting that tallness could mask the potential to transmit shortness. Of the remaining plants, one-third (one-quarter of the total number of offspring) proved through further crosses with short plants to be "true-breeding tall." The other two-thirds produced some short plants in the next generation and were therefore "non-true-breeding"(fig. 13.4).

Mendel suggested that gametes distribute elementen, because these cells physically link generations. The elementen would separate from each other as gametes form. When gametes join at fertilization, the elementen would group into new combinations. Mendel reasoned that each elementen was packaged in a separate gamete, and if opposite-sex gametes combine at random, then he could mathematically explain the different ratios of traits produced from his pea plant crossings. Mendel's idea that elementen separate in the gametes would later be called the **law of segregation.** Mendel eventually turned his energies to monastery administration when other scientists did not recognize the importance of his work.

It wasn't until 1900 that botanists realized that Mendel's conclusions explained data from breeding experiments in many plant species. Also at this time, researchers noted that chromosomes behave much like Mendel's elementen. Both paired elementen

Short plants

(a) Short plants

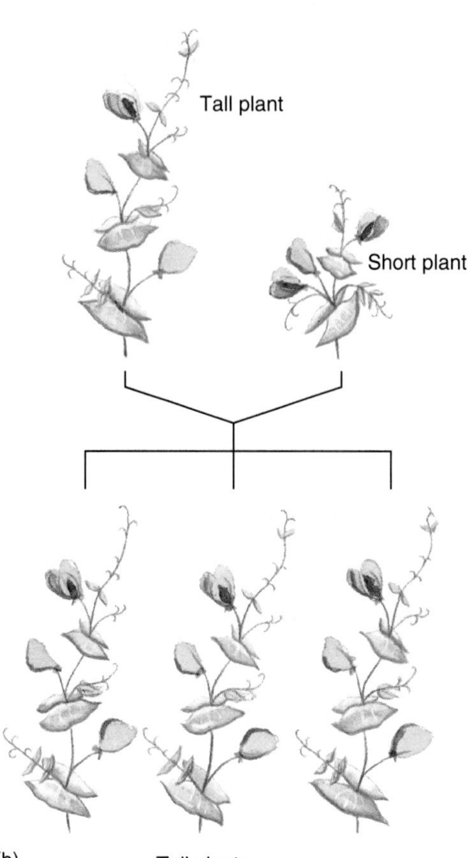

FIGURE 13.4

Mendel crossed short and tall pea plants. (*a*) When Mendel crossed short pea plants with short pea plants, all of the progeny were short. (*b*) Certain tall plants crossed with short plants produced all tall plants. (*c*) Other tall plants crossed with short plants produced some tall plants and some short plants. (*d*) When the tall plants that did not breed true were crossed with each other, one-quarter of the plants in the next generation were short, and three-quarters were tall. Of these tall plants, one-third were shown to be true-breeding, and the other two-thirds were not true-breeding.

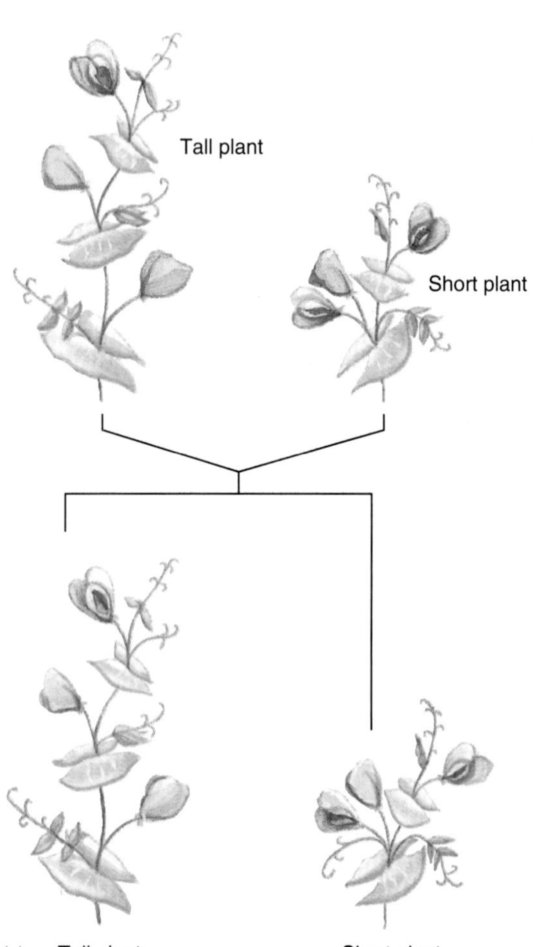

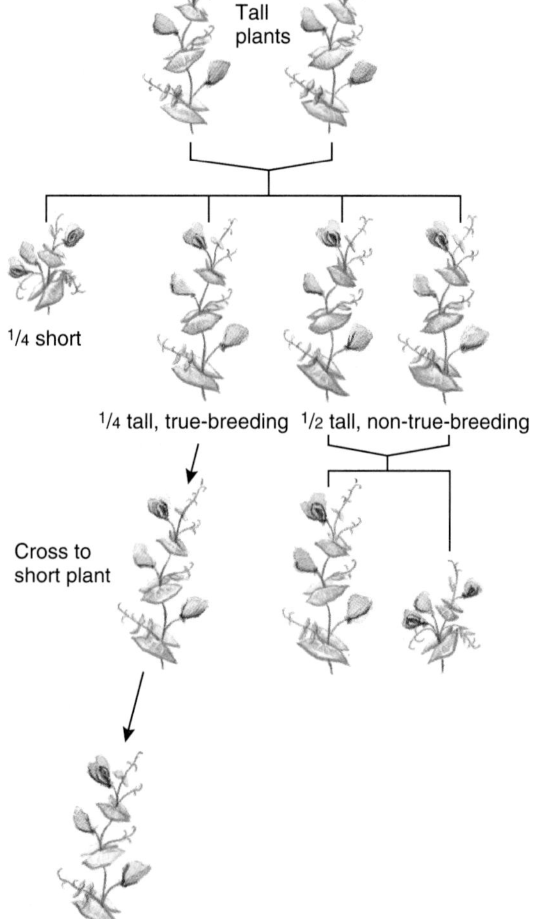

and paired chromosomes separate at each generation and pass—one from each parent—to offspring. Both elementen and chromosomes are inherited in random combinations. Chromosomes provided a physical mechanism for Mendel's hypotheses. In 1909, Mendel's elementen were renamed genes (Greek for "give birth to"), but for the next several years, the units of inheritance remained a mystery. By the 1940s, scientists began investigating the gene's chemical basis. We pick up the historical trail at this point in chapter 15.

How Biologists Describe Single Genes

Mendel observed two different expressions of a trait—for example, short and tall. Traits are expressed in different ways because a gene can exist in alternate forms, or **alleles.** An individual with two identical alleles for a gene is **homozygous** for that gene. An individual with two different alleles is **heterozygous.**

Mendel noted that for some genes, one variant could mask the expression of another. The allele that masks the other is completely **dominant,** and the masked allele is **recessive.** When Mendel crossed a true-breeding tall plant with a short plant, the tall allele was completely dominant to the short allele. The plants of the next generation were all tall.

When a gene has two alleles, it is common to symbolize the dominant allele with a capital letter and the recessive with the corresponding small letter. If both alleles are recessive, the individual is **homozygous recessive.** Two small letters, such as *tt* for short plants, symbolize this. An individual with two dominant alleles is **homozygous dominant.** Two capital letters, such as *TT* for tall pea plants, represent homozygous dominance. Another possible allele combination is one dominant and one recessive allele—*Tt* for non-true-breeding tall pea plants.

An organism's appearance does not always reveal its alleles. Both a *TT* and a *Tt* pea plant are tall, but the first is a homozygote and the second a heterozygote. The **genotype** describes the organism's alleles, while the **phenotype** describes the outward expression of an allele combination. A pea plant with a tall phenotype may have genotype *TT* or *Tt.* A **wild-type phenotype** is the most common expression of a particular gene in a population. A **mutant** phenotype is a variant of a gene's expression that arises when the gene undergoes a change, or **mutation.**

When analyzing genetic crosses, the first generation is the **parental generation,** or $\mathbf{P_1}$; the second generation is the **first filial generation,** or $\mathbf{F_1}$**;** the next generation is the **second filial generation,** or $\mathbf{F_2}$, and so on. If you considered your grandparents the P_1 generation, your parents would be the F_1 generation, and you and your siblings are the F_2 generation. Table 13.1 summarizes these and other commonly used genetic terms.

Mendel's Laws and Meiosis

Mendel's observations on inheritance of single genes reflect the events of meiosis. When a gamete is produced, the two copies

Table 13.1 A Glossary of Genetic Terms

Term	Definition
Allele	An alternate form of a gene
Dihybrid	An individual who is heterozygous for two particular genes
Dominant	An allele that masks the expression of another allele
F_1	The first filial generation; offspring
F_2	The second filial generation; offspring of offspring
Genotype	The allele combination in an individual
Heterozygous	Possessing different alleles of a gene
Homozygous	Possessing identical alleles of a gene
Independent assortment	Mendel's second law; a gene on one chromosome does not influence the inheritance of a gene on a different (nonhomologous) chromosome because meiosis packages chromosomes randomly into gametes
Monohybrid	An individual heterozygous for a particular gene
Mutant	A phenotype or allele resulting from a change (mutation) in a gene
Mutation	A change in a gene
P_1	The parental generation
Phenotype	The observable expression of an allele combination
Recessive	An allele whose expression is masked by another allele
Segregation	Mendel's first law; alleles of a gene separate into equal numbers of gametes
Sex-linked	A gene located on the X chromosome or a trait that results from the activity of a gene on the X chromosome
Wild type	The most common phenotype or allele for a gene in a population

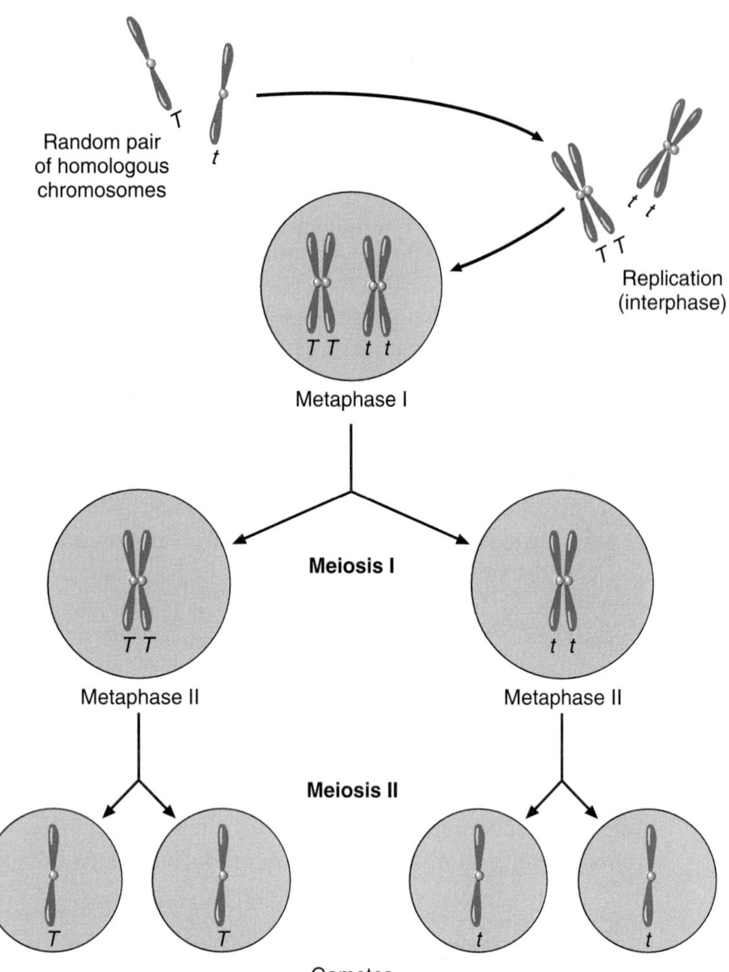

FIGURE 13.5

Mendel's first law—gene segregation. During meiosis, homologous pairs of chromosomes (and the genes that compose them) separate from one another and are packaged into separate gametes. At fertilization, gametes combine at random to form the individuals of a new generation.

		Female gametes	
		T	*t*
Male gametes	*T*	*TT*	*Tt*
	t	*Tt*	*tt*

FIGURE 13.6

A Punnett square, a diagram of gametes and how they can combine in a cross between two particular individuals, is helpful in following the transmission of traits. The different types of female gametes are listed along the top of the square; male gametes are listed on the left-hand side. Each compartment within the square contains the genotypes that result when gametes that correspond to that compartment join. The Punnett square here describes Mendel's monohybrid cross of two tall pea plants. In the progeny, tall plants outnumber short plants 3:1. Can you determine the genotypic ratio?

of a particular gene separate, as the homologs that carry them do. In a plant of genotype *Tt,* for example, gametes carrying either *T* or *t* form in equal numbers during anaphase I. When gametes meet to start the next generation, they combine at random. That is, a *t*-bearing oocyte is neither more nor less attractive to a *t*-bearing sperm than a *T*-bearing sperm. These two factors—equal allele distribution and random combinations of gametes—underlie Mendel's law (fig. 13.5).

Mendel crossed short plants (*tt*) with true-breeding tall plants (*TT*). The resulting seeds grew into F_1 plants of the same phenotype: tall (genotype *Tt*). Next, he crossed the F_1 plants with each other in a **monohybrid cross**, so called because one trait is followed by crossing two heterozygous, or hybrid, individuals. The three possible outcomes of such a cross are *TT, tt,* and *Tt.* A *TT* individual results when a *T* sperm fertilizes a *T* oocyte; a *tt* plant results when a *t* oocyte meets a *t* sperm; and a *Tt* individual results when either a *t* sperm fertilizes a *T* oocyte, or a *T* sperm fertilizes a *t* oocyte.

Because two of the four possible gamete combinations produce a heterozygote, and each of the others produces a homozygote, the **genotypic ratio** expected of a monohybrid cross is 1 *TT*: 2 *Tt*: 1 *tt*. The corresponding **phenotypic ratio** is three tall plants to one short plant, a 3:1 ratio. A diagram called a **Punnett square** helps to derive these ratios (fig. 13.6). Experiments yield numbers of offspring that approximate these ratios. When Mendel crossed *Tt* and other *Tt* pea plants, he observed 787 tall plants and 277 short plants—close to the predicted 3 tall to 1 short phenotypic ratio.

Mendel distinguished the two genotypes resulting in tall progeny—*TT* from *Tt*—with additional crosses. He bred tall plants of unknown genotype with short (*tt*) plants. If a tall plant crossed with a *tt* plant produced both tall and short progeny, Mendel knew it was genotype *Tt*; if it produced only tall plants, he knew it must be *TT*. Crossing an individual of unknown genotype with a homozygous recessive individual is called a **test cross.** The homozygous recessive is the only genotype that can be identified by the phenotype—that is, a short plant is only *tt*. The homozygous recessive is a "known" that can reveal the unknown genotype of another individual when the two are crossed. Figure 13.7 shows the results of monohybrid crosses in two other familiar species.

(a)

(b)

FIGURE 13.7

Monohybrid crosses. (*a*) Kernels on an ear of corn represent progeny of a single cross. When a plant with a dominant allele for purple kernels and a recessive allele for yellow kernels is self-crossed, the resulting ear has approximately three purple kernels for every yellow kernel. (*b*) Albinism can result from a monohybrid cross in a variety of organisms. A heterozygote, or "carrier," for albinism has one allele that directs the synthesis of an enzyme needed to manufacture the skin pigment melanin and one allele that fails to make the enzyme. Each child of two carriers has a one in four chance of inheriting the two deficient alleles and being unable to manufacture melanin.

Mastering Concepts

1. How did Mendel's experiments reveal how single genes are transmitted?
2. How do Mendel's observations reflect events of meiosis?
3. What is the law of segregation?
4. Distinguish between a heterozygote and a homozygote; phenotype and genotype; dominant and recessive; wild type and mutant.
5. What are the genotypic and phenotypic ratios expected of a monohybrid cross?

Table 13.2 Unusual Mendelian Traits

A compendium called *Mendelian Inheritance in Man* lists a wide variety of traits thought to be inherited, many by single genes. Here are a few odder ones.

Uncombable hair	Extra, missing, protruding, fused, shovel-shaped, or snowcapped teeth
Misshapen toes or teeth	Grooved tongue
Pigmented tongue tip	Duckbill-shaped lips
Inability to smell freesia flowers, musk, or skunk	Flared ears
Lack of teeth, brows, nasal bones, thumbnails	Egg-shaped pupils
White forelock	Three rows of eyelashes
Tone deafness	Spotted nails
Double toenails	Whorl in the eyebrow
Magenta urine after eating beets	Hairs that are triangular in cross section or have multiple hues
Hairy nosetips, knuckles, palms, soles	Sneezing fits in bright sunlight

Mendelian Traits in Humans

Mendel's laws apply to all diploid species, and characteristics determined by single genes are termed Mendelian traits. Figure 13.8 shows how Mendel's first law applies to a family in which the parents are heterozygotes (carriers) for a mutant allele causing cystic fibrosis, such as the family in the opening vignette. Each child faces a one in four chance of being homozygous recessive and inheriting the illness.

Mendelian disorders, even the most common ones, are rare, affecting only one person in many thousands. About 2,500 Mendelian disorders are known, and some 2,500 other conditions are suspected to be Mendelian, based on their recurrence patterns in large families. Table 13.2 lists some other interesting Mendelian traits in humans.

Modes of Inheritance

Mendelian traits are described by their transmission pattern in families, called the **mode of inheritance.** An **autosomal recessive** mode of inheritance means that the trait can affect both sexes, because its gene is carried on an autosome (a nonsex chromosome). Such a trait can skip generations, because heterozygotes (carriers) show no symptoms. Individuals whose children have an autosomal recessive condition must both be carriers. In contrast, an **autosomal dominant** trait can affect both sexes because it is autosomal, but an affected individual must have an affected parent, unless the trait arose by mutation. If a generation arises in which no individuals inherit a disease-causing autosomal dominant gene, transmission stops.

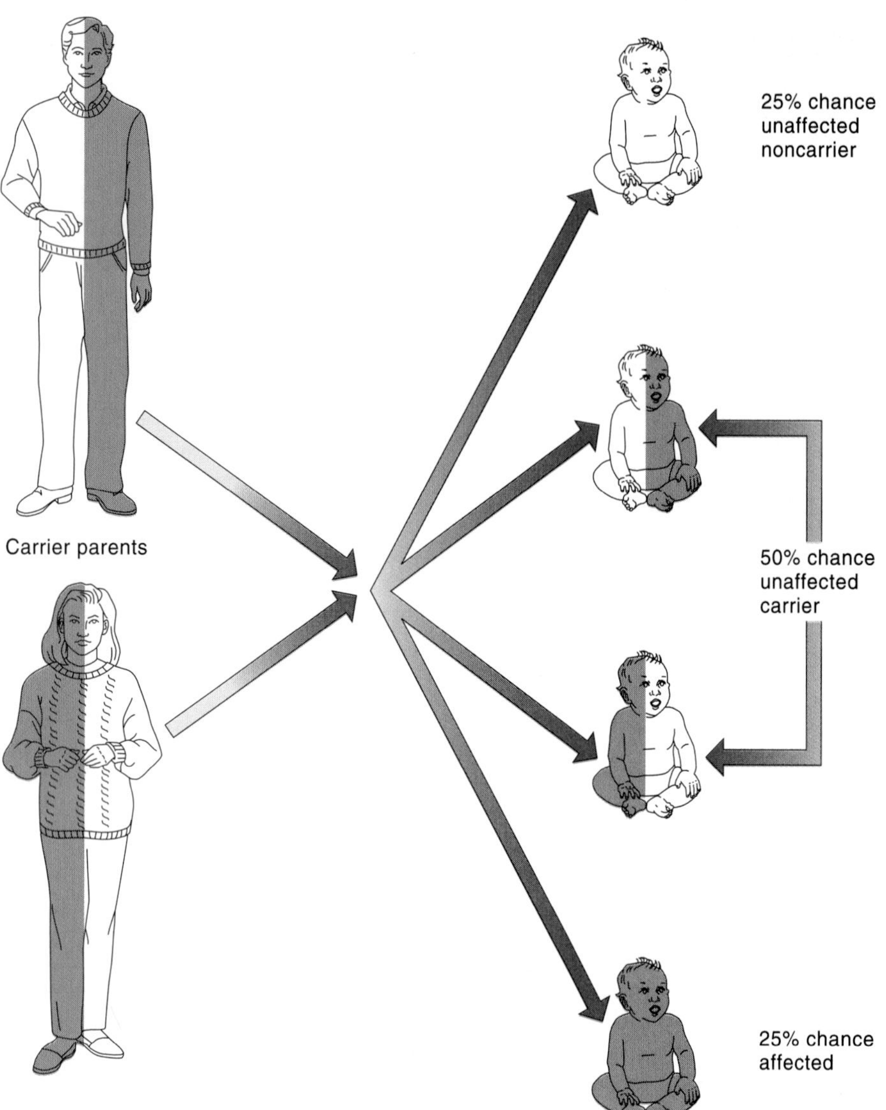

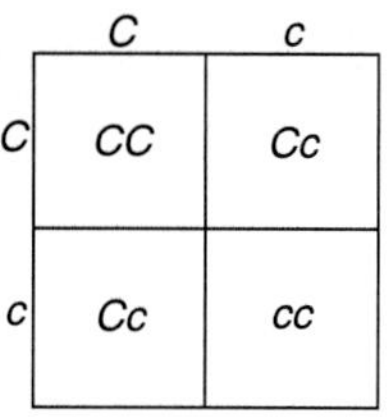

FIGURE 13.8

A 1:2:1 genotypic ratio results from a monohybrid cross, whether in peas or people. When parents are carriers for the same autosomal recessive trait or disorder, such as cystic fibrosis, each child faces a 25% risk of inheriting the condition (a sperm carrying a CF allele fertilizing an oocyte carrying a CF allele); a 50% chance of being a carrier like the parents (a CF-carrying sperm with a wild-type oocyte, and vice versa); and a 25% chance of inheriting two wild-type CF alleles.

The next chapter discusses two modes of inheritance that apply to genes on the X chromosome. Table 13.3 describes some autosomal recessive and autosomal dominant disorders.

Pedigrees

Charts called **pedigrees** depict family relationships and phenotypes and can reveal the mode of inheritance of a particular trait. A pedigree consists of lines connecting shapes. Vertical lines represent generations; horizontal lines connect parents; siblings are connected to their parents by vertical lines and joined by a horizontal line. Squares indicate males, circles females, and diamonds individuals of unknown sex. Colored shapes indicate individuals who express a particular trait, and half-filled shapes are known carriers. Figure 13.9 shows commonly used pedigree symbols.

The different modes of inheritance have characteristic pedigree patterns. An autosomal recessive condition can skip generations; an autosomal dominant condition cannot skip generations. Both modes of inheritance affect both sexes. Figure 13.10 depicts transmission of an autosomal recessive trait, albinism, to two of three children of two carriers. Figure 13.11 depicts transmission of an autosomal dominant condition. Notice that it does not skip generations.

A pedigree can be inconclusive—that is, either autosomal dominant or recessive—if the phenotype mildly affects health, so that autosomal recessive individuals can have children. Figure 13.12 shows how the same pedigree can be interpreted as autosomal recessive or autosomal dominant for alopecia, a form of hair loss. Two different genes can cause alopecia, one autosomal recessive and one autosomal dominant.

Pedigrees can be difficult to construct and interpret for several reasons. People sometimes hesitate to supply information because symptoms embarrass them. Adoption, children born out of wedlock, serial marriages and the resulting blended families, and assisted reproductive technologies complicate tracing family ties. Many people cannot trace their families back far enough to reveal modes of inheritance.

Table 13.3 Some Mendelian Disorders in Humans

Disorder	Symptoms
Autosomal Recessive	
Ataxia telangiectasis	Facial rash, poor muscular coordination, involuntary eye movements, high risk for cancer, sinus and lung infections
Cystic fibrosis	Lung infections and congestion, poor fat digestion, male infertility, poor weight gain, salty sweat
Familial hypertrophic cardiomyopathy	Overgrowth of heart muscle causes sudden death in young adults
Gaucher disease	Swollen liver and spleen, anemia, internal bleeding, poor balance
Hemochromatosis	Body retains iron; high risk of infection, liver damage, excess skin pigmentation, heart and pancreas damage
Maple syrup urine disease	Lethargy, vomiting, irritability, mental retardation, coma, and death in infancy
Phenylketonuria	Mental retardation, fair skin
Porphyria	Red urine, fever, abdominal pain, headache, coma, death
Sickle cell disease	Joint pain, spleen damage, high risk of infection
Tay-Sachs disease	Nervous system degeneration
Autosomal Dominant	
Achondroplasia	Dwarfism with short limbs, normal size head and trunk
Familial hypercholesterolemia	Very high serum cholesterol, heart disease
Huntington disease	Progressive uncontrollable movements and personality changes, beginning in middle age
Lactose intolerance	Inability to digest lactose causes cramps after eating this sugar
Marfan syndrome	Long limbs, sunken chest, lens dislocation, spindly fingers, weakened aorta
Myotonic dystrophy	Progressive muscle wasting
Neurofibromatosis (I)	Brown skin marks, benign tumors beneath skin
Polycystic kidney disease	Cysts in kidneys, bloody urine, high blood pressure, abdominal pain
Polydactyly	Extra fingers and/or toes

Mastering Concepts

1. What types of information do pedigrees include and convey?
2. How are Punnett squares helpful in following transmission of single genes?

Mendel's Second Law (Independent Assortment) Traces Two Genes

The law of segregation follows the inheritance pattern of two alleles of a single gene. In a second set of experiments with pea plants, Mendel examined the inheritance of two different traits, each of which has two different alleles: he looked at seed shape, which may be either round or wrinkled (determined by the R gene), and seed color, which may be either yellow or green (determined by the Y gene). When Mendel crossed plants with round, yellow seeds with plants with wrinkled, green seeds, all the progeny had round, yellow seeds. Therefore, he concluded that round is completely dominant to wrinkled, and yellow is completely dominant to green.

Next, Mendel took F_1 plants (genotype *RrYy*) and crossed them with each other. This is a **dihybrid cross,** because the individuals are heterozygous for two genes. Mendel found four types of seeds in the F_2 generation: round, yellow (315 plants); round, green (108 plants); wrinkled, yellow (101 plants); and wrinkled, green (32 plants). This is an approximate ratio of 9:3:3:1 (fig. 13.13).

Mendel took each plant from the F_2 generation and crossed it with wrinkled, green (*rryy*) plants. These test crosses established whether each F_2 plant was true-breeding for both genes (*RRYY* or *rryy*), true-breeding for one but heterozygous for the other (*RRYy, RrYY, rrYy,* or *Rryy*), or heterozygous for both genes (*RrYy*). Based upon the results of the dihybrid cross, Mendel proposed what is now known as the law of **independent assortment.** It states that a gene for one trait does not influence the transmission of a gene for another trait. This second law is true only for genes on different chromosomes, such as pea shape and color.

Symbols

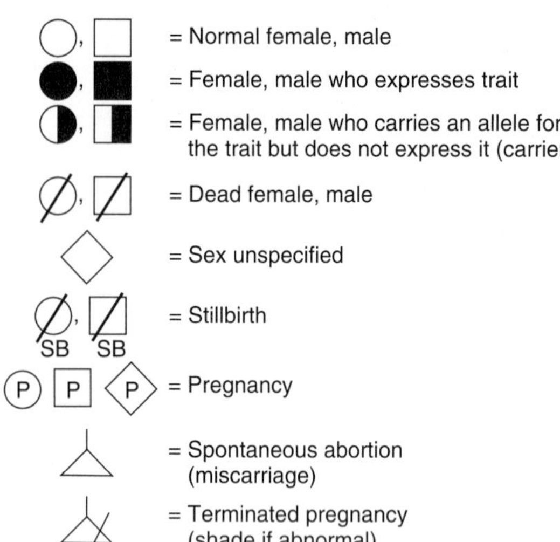

Lines

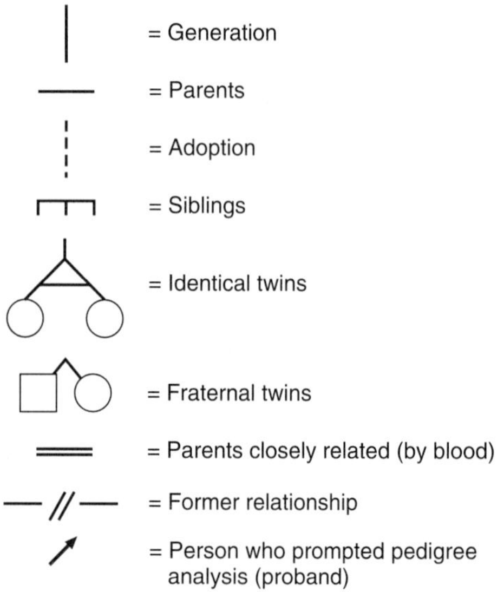

Numbers

Roman numerals = generations

Arabic numerals = individuals

FIGURE 13.9

Symbols used in pedigree construction are connected to form a pedigree chart, which displays the inheritance patterns of particular traits.

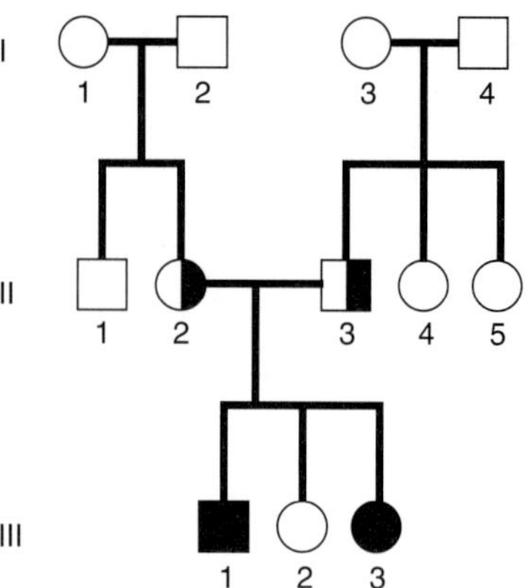

FIGURE 13.10

A pedigree for an autosomal recessive trait. Albinism does not show up in this family until the third generation. Individuals II-2 and II-3 must each be carriers.

In words:
"One couple has a son and daughter with normal pigmentation. Another couple has one son and two daughters with normal pigmentation. The daughter from the first couple has three children with the son of the second couple. Their son and one daughter have albinism; their other daughter has normal pigmentation."

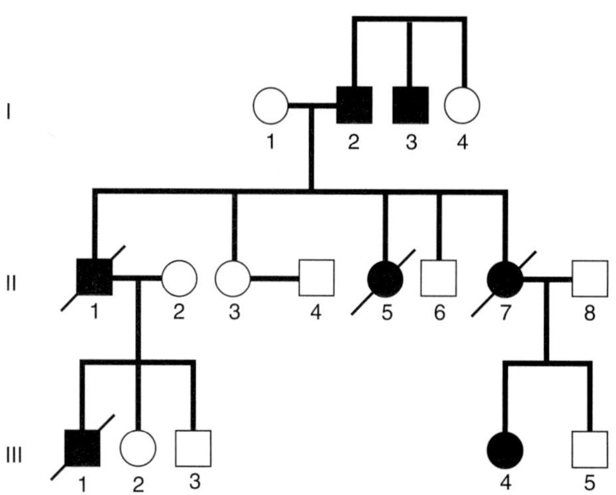

"When individual III-1 died suddenly of heart failure while playing basketball at age 19, his family and doctors were perplexed; he had seemed healthy. But the family history and microscopic examination of his heart tissue revealed familial hypertrophic cardiomyopathy, an inherited overgrowth of heart muscle. The boy's father had died young of heart failure too, as had two of the father's sisters. The 19-year-old's two cousins, whose mother had died of the condition, had tests for the mutant gene. One of them, a girl, inherited the gene. Restricting her exercise to swimming may extend her life. The 19-year-old's paternal grandfather and a paternal great-uncle also had the disorder."

FIGURE 13.11

A pedigree for an autosomal dominant trait. Familial hypertrophic cardiomyopathy appears in every generation in this family.

Mendel had again inferred a principle of inheritance based on meiosis. Independent assortment occurs because chromosomes from each parent combine in a random fashion (fig. 13.14). In Mendel's dihybrid cross, each parent produces equal numbers of gametes of four different types: *RY, Ry, rY,* and *ry.* (Note that each of these combinations has one gene for each trait.) A Punnett square for this cross (fig. 13.15) predicts that the four types of seeds—round, yellow (*RRYY, RrYY, RRYy,* and *RrYy*); round, green (*RRyy, Rryy*); wrinkled, yellow (*rrYY, rrYy*); and wrinkled, green (*rryy*)—will occur in the ratio 9:3:3:1, just as Mendel found.

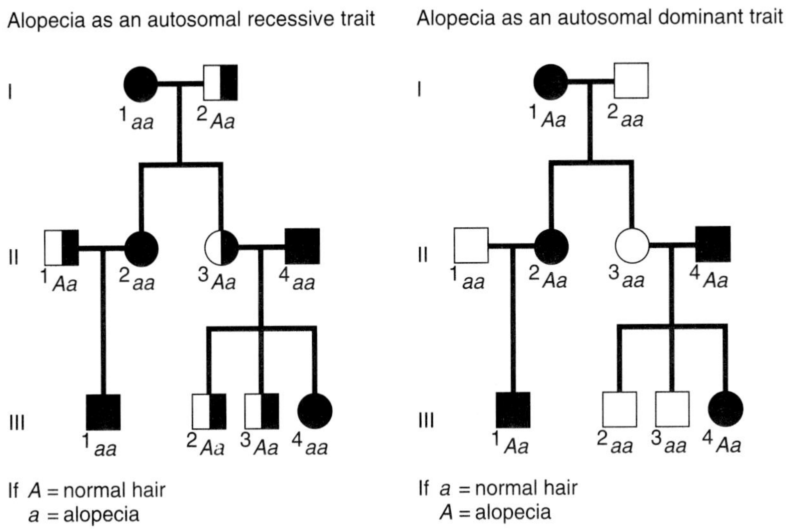

Alopecia as an autosomal recessive trait

I: 1 *aa*, 2 *Aa*
II: 1 *Aa*, 2 *aa*, 3 *Aa*, 4 *aa*
III: 1 *aa*, 2 *Aa*, 3 *Aa*, 4 *aa*

If *A* = normal hair
a = alopecia

Alopecia as an autosomal dominant trait

I: 1 *Aa*, 2 *aa*
II: 1 *aa*, 2 *Aa*, 3 *aa*, 4 *Aa*
III: 1 *Aa*, 2 *aa*, 3 *aa*, 4 *Aa*

If *a* = normal hair
A = alopecia

FIGURE 13.12

An inherited trait that does not impair health enough to prevent reproduction can result in an inconclusive pedigree. This is the case for alopecia, in which some hair on the head falls out, grows back, and falls out again. Alopecia can be inherited as an autosomal recessive or as an autosomal dominant trait. In the family depicted in this pedigree, the mode of inheritance cannot be discerned—either autosomal recessive or dominant may fit the data.

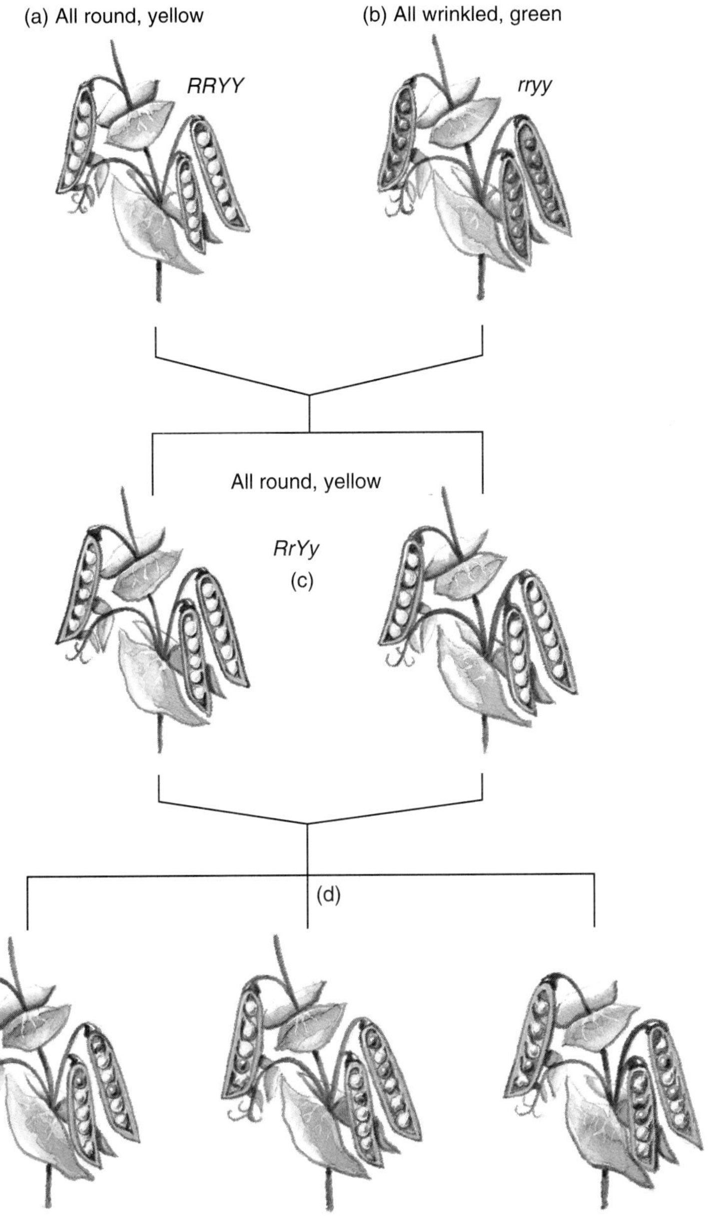

FIGURE 13.13

Mendel's crosses involving two genes. To study the inheritance pattern of two genes, Mendel crossed (*a*) a true-breeding plant with round, yellow seeds (*RRYY*) with (*b*) a plant with wrinkled, green seeds (*rryy*). (*c*) The peas of the F_1 generation were all round and yellow (*RrYy*). (*d*) When the F_1 dihybrid pea plants were crossed with each other, the F_2 generation exhibited the phenotypes round, yellow (*RrYr, RRYy, RrYY,* and *RRYY*); round, green (*Rryy* and *RRyy*); wrinkled, yellow (*rrYY* and *rrYy*); and wrinkled, green (*rryy*) in a 9:3:3:1 ratio.

Punnett squares become cumbersome when analyzing more than two genes—a Punnett square for three genes has 64 boxes; for four genes, 256 boxes. An easier way to predict genotypes and phenotypes is to use the mathematical laws of probability on which Punnett squares are based. Probability predicts the likelihood of an event. The **product rule** states that the chance that two independent events will both occur—such as two alleles both being inherited—equals the product of the individual chances that each event will occur.

The product rule can predict the chance of obtaining a wrinkled, green (*rryy*) plant from dihybrid (*RrYy*) parents. Consider the dihybrid individual one gene at a time. A Punnett square for *Rr* crossed with *Rr* shows that the probability that two *Rr* plants will produce *rr* progeny is 25%, or 1/4. Similarly, the chance of two *Yy* plants producing a *yy* individual is 1/4. According to the product rule, the chance of dihybrid parents (*RrYy*) producing homozygous recessive (*rryy*) offspring is 1/4 multiplied by 1/4, or 1/16. Now consult the 16-box Punnett square for Mendel's dihybrid cross. Only 1 of the 16 boxes is *rryy*.

Mastering Concepts

1. What is the law of independent assortment?
2. How is Mendel's second law based on meiosis?
3. What are the expected genotypic and phenotypic ratios of a dihybrid cross?
4. How can the product rule be used to predict the results of crosses involving more than one gene?

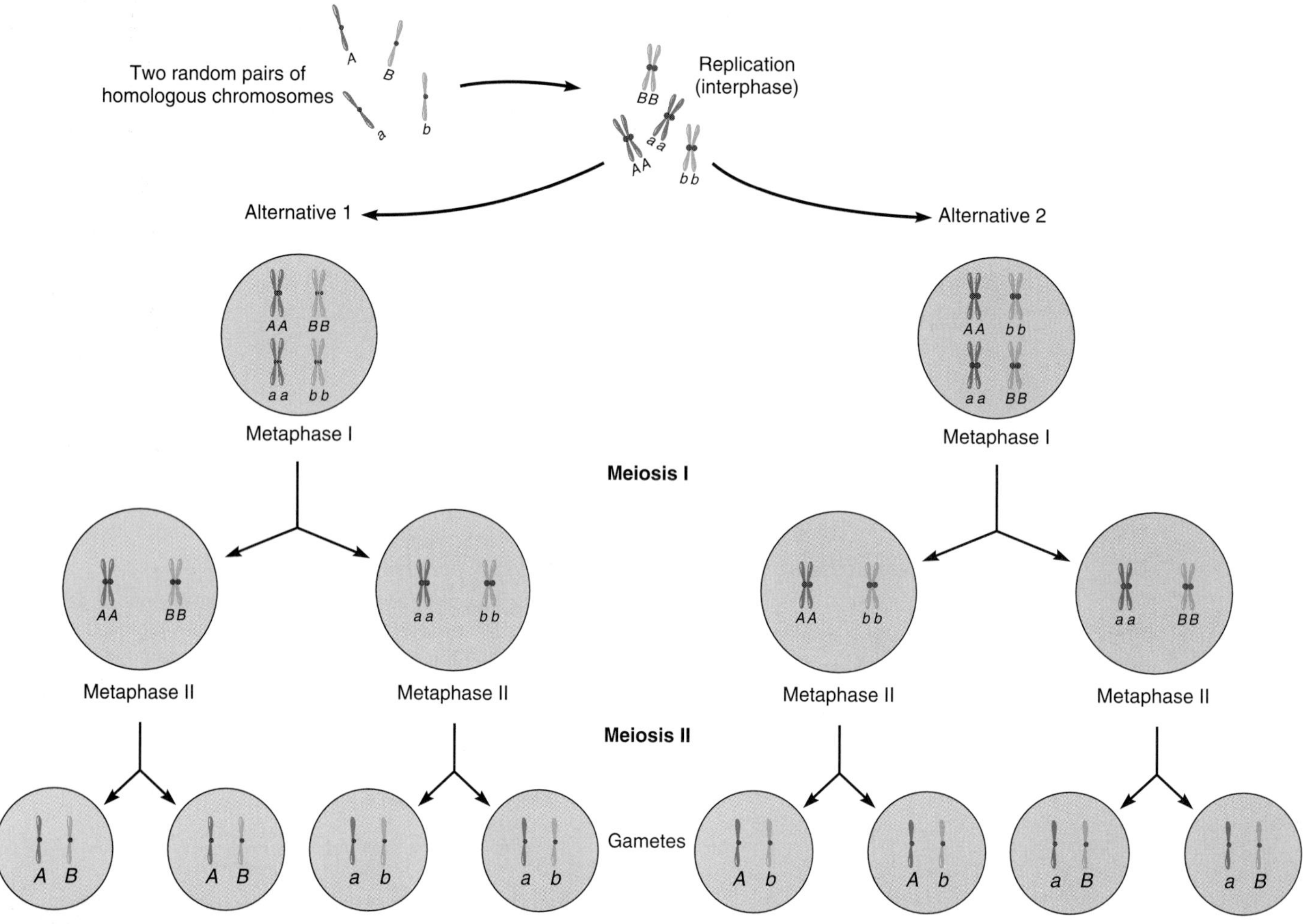

FIGURE 13.14

The independent assortment of genes carried on different chromosomes results from the random alignment of chromosome pairs during metaphase of meiosis I. An individual of genotype *AaBb,* for example, manufactures four types of gametes, containing the dominant alleles of both genes (*AB*), the recessive alleles of both genes (*ab*), and a dominant allele of one with a recessive allele of the other (*Ab* and *aB*). The allele combination depends upon which chromosomes are packaged together into the same gamete—and this happens at random.

When Gene Expression Appears to Alter Mendelian Ratios

Mendel's crosses yielded easily distinguishable offspring. A pea is either yellow or green, round or wrinkled; a plant is either tall or short. At times, however, offspring traits do not occur in the proportions Punnett squares or probabilities predict. It may appear that Mendel's laws do not apply—but they do. The underlying genotypic ratios are there, but the nature of the phenotype or other genes or the environment alter how traits appear.

Lethal Allele Combinations

Genes begin to function soon after fertilization. Some allele combinations cause problems so early in prenatal development that the individual is never born and therefore is never counted as a phenotypic class of offspring. An allele that causes such early death is called a **lethal allele.**

Lethal alleles are seen—or not seen—in plants of certain genotypes that die at fertilization, during seed development, or as seedlings. In humans, lethal alleles can cause spontaneous abortion (miscarriage). When a man and woman each carry a recessive lethal allele for the same gene, each pregnancy has a 25% chance of spontaneously aborting—a figure that represents the homozygous recessive class.

Multiple Alleles

A diploid cell has a "place," or locus, for two alleles of a given gene, one on each of a pair of homologous chromosomes. The alleles can be identical, producing a homozygous genotype, or different, producing a heterozygote. Even though an individual diploid cell has only two alleles for a particular gene, a gene can exist in more than two allelic forms. This is because a change, or mutation, in any of the hundreds or thousands of DNA bases of a gene constitutes a different allele. The change is detectable if the protein product the new allele encodes changes in a way that affects the phenotype.

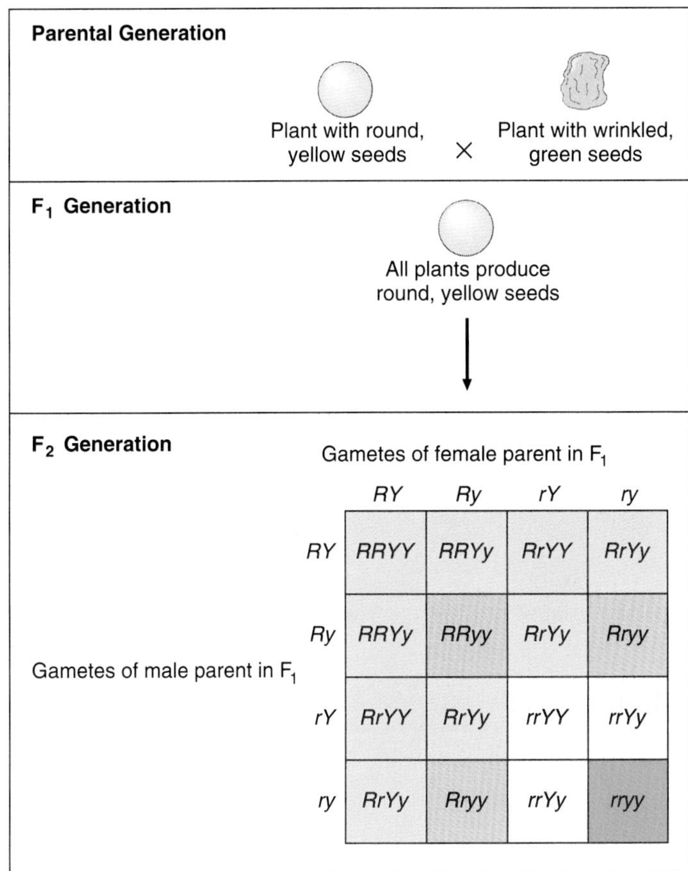

FIGURE 13.15

A Punnett square can be used to represent the random combinations of gametes produced by dihybrid individuals.

Multiple alleles can complicate interpretation of Mendel's laws because in different combinations they cause different variations of the phenotype. For example, in rabbits, four alleles determine coat color (table 13.4). Allele c is recessive; c^h is dominant to c; c^{ch} is dominant to c^h and c; and C is dominant to all alleles. In humans, the cystic fibrosis gene can mutate in hundreds of different ways and produce phenotypes ranging from frequent bronchitis and pneumonia to life-threatening, near-constant infections.

The more alleles a gene has, the more phenotypes and genotypes are associated with it. For a gene with two alleles, such as round (*R*) versus wrinkled (*r*) peas, three genotypes are possible—*RR*, *Rr*, and *rr*. A gene with three alleles has six genotypes. As rabbit coat color illustrates, a gene with four alleles has five associated genotypes.

Table 13.4 Phenotypes and Genotypes for Rabbit Coat Color

Phenotype	Possible Genotypes
Gray	CC, Cc^{ch}, Cc^h, Cc
Chinchilla	$c^{ch}c^{ch}$
Light gray	$c^{ch}c^h$, $c^{ch}c$
Himalayan	c^hc^h, c^hc
Albino	cc

Different Dominance Relationships

Some genes show **incomplete dominance,** with the heterozygous phenotype intermediate between the two homozygotes. The piebald trait in domestic cats confers varying numbers of white spots and shows incomplete dominance. Cats of genotype *SS* have many spots, *ss* cats have no spots, and *Ss* cats have an intermediate number of spots. Another example of incomplete dominance is the snapdragon plant (fig. 13.16). A red-flowered plant of genotype *RR* crossed with a white-flowered *rr* plant gives rise to a pink-flowered *Rr*. An intermediate amount of pigment in the heterozygote confers the pink color.

Different alleles that are both expressed in a heterozygote are **codominant.** Two alleles of the I gene, which determines human ABO blood type, are codominant. People of blood type A have a molecule called antigen A on the surfaces of their red blood cells. People of blood type B have red blood cells with antigen B. A

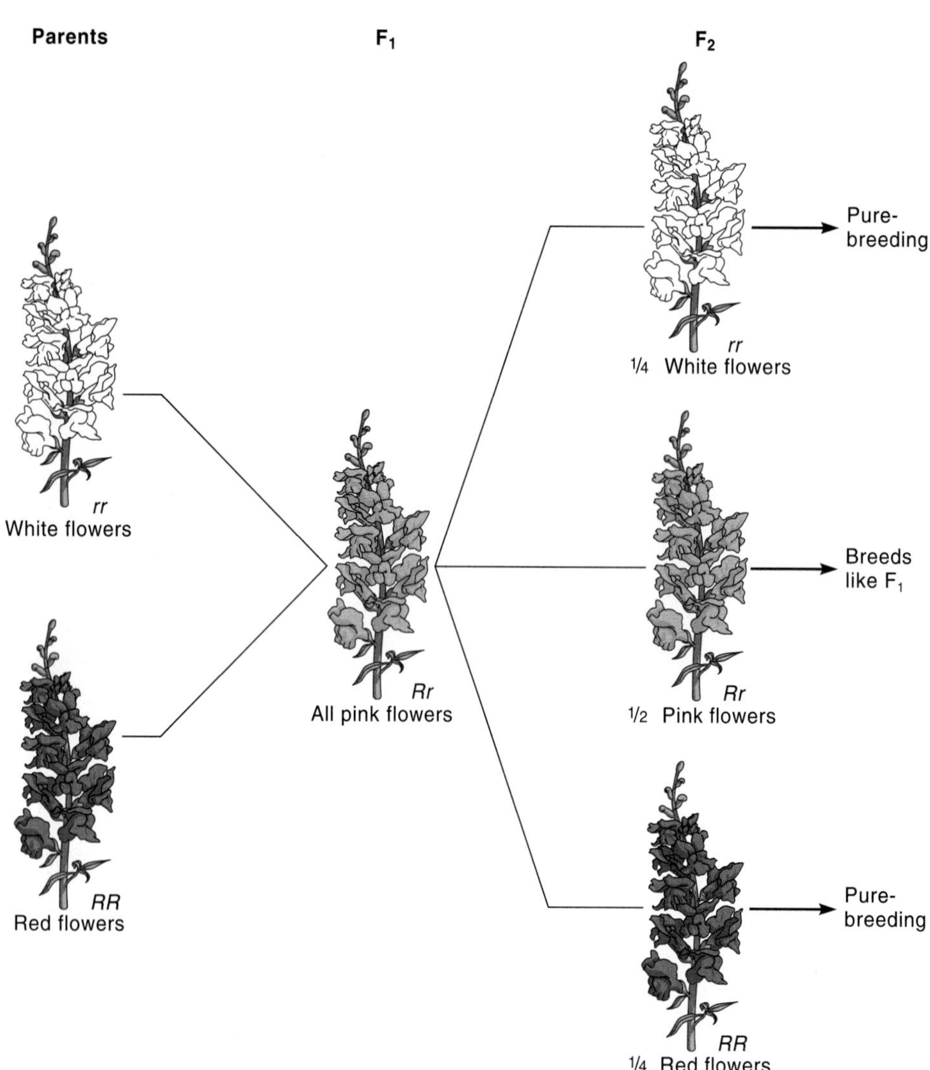

FIGURE 13.16

Incomplete dominance in snapdragon flowers. A cross between a homozygous dominant plant with red flowers (*RR*) and a homozygous recessive plant with white flowers (*rr*) produces a heterozygous plant with pink flowers (*Rr*). When *Rr* pollen fertilizes *Rr* egg cells, one-quarter of the progeny are red-flowered (*RR*), one-half are pink-flowered (*Rr*), and one-quarter are white-flowered (*rr*). The phenotypic ratio of this monohybrid cross is 1:2:1 (instead of the 3:1 seen in cases of complete dominance) because the heterozygous class has a phenotype different from that of the homozygous dominant class.

Table 13.5 Codominance in Human ABO Blood Types

Phenotype (Blood Type)	Genotype
A	I^AI^A or I^Ai
B	I^BI^B or I^Bi
AB	I^AI^B
O	ii

person with type AB has red blood cells with both the A and B antigens, and the red cells of a person with type O blood carry neither antigen.

The I gene has three alleles that encode the enzymes that place the A and B antigens on red blood cell surfaces. They are I^A, I^B, and i. People with type A blood may be either genotype I^AI^A or I^Ai; type B corresponds to I^BI^B or I^Bi; type AB to I^AI^B; and type O to ii. Even though the I^A and I^B alleles are codominant, they segregate between generations (fig. 13.17 and table 13.5).

Epistasis—Gene Masking at General Hospital

One gene masking the effect of another, a phenomenon called **epistasis,** may appear to disrupt operation of Mendel's laws. This is different from a dominant allele masking a recessive allele of the same gene. Epistasis was part of a plot in the television soap opera *General Hospital* (fig. 13.18). It concerned the Bombay phenotype, a result of interacting I and H genes. The relationship of these two genes affects the expression of the ABO blood type.

The *H* allele produces an enzyme that inserts a carbohydrate molecule onto a particular glycoprotein on the red blood cell surface. The A and B antigens then attach to this carbohydrate. The recessive *h* allele produces an inactive form of the enzyme, which cannot insert the carbohydrate. As long as at least one *H* allele is present, the ABO genotype dictates the person's blood type. However, in a person with genotype *hh,* the A and B antigens cannot adhere to the red blood cells. The person tests as blood type O but may actually have any ABO genotype.

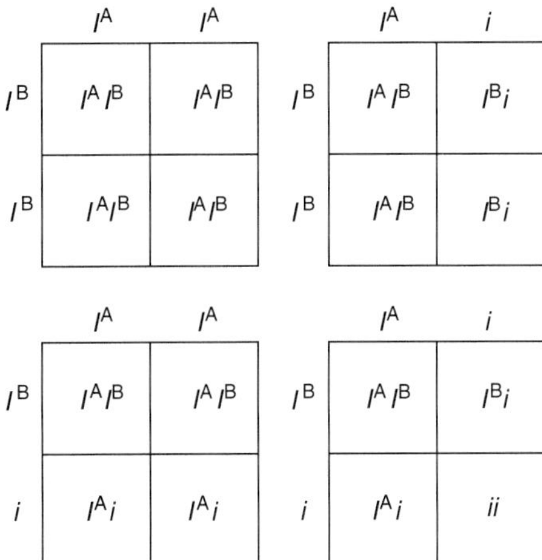

FIGURE 13.17

Even though the I^A and I^B alleles of the I gene are codominant, they still follow Mendel's law of segregation. These Punnett squares follow the genotypes that could result by crossing a person with type A blood with a person with type B blood. Is it possible for parents with type A and type B blood to have a child who is type O?

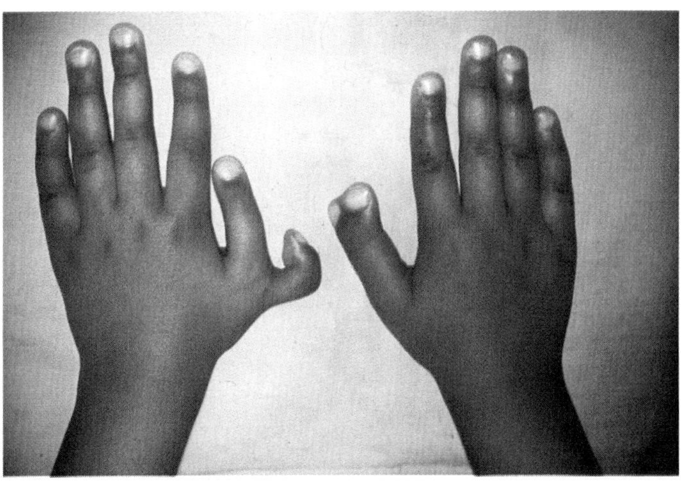

FIGURE 13.19

The dominant allele that causes polydactyly is incompletely penetrant and variably expressive. Some people who inherit the allele have the normal number of fingers and toes. Those in whom the allele is penetrant express the phenotype to differing degrees.

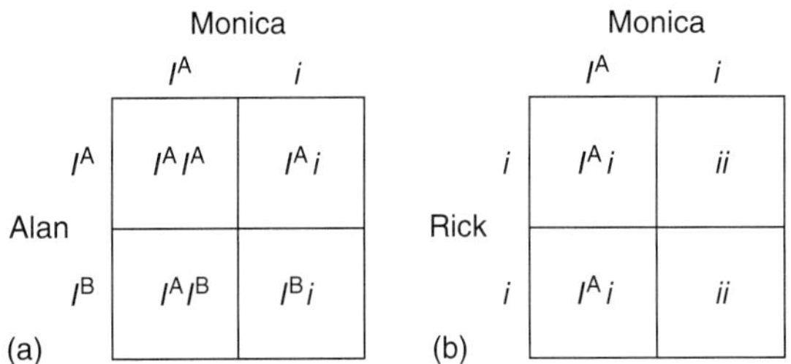

FIGURE 13.18

Gene masking at *General Hospital.* Monica has just had a baby. But who is the father—her husband, Alan, or friend and fellow doctor, Rick? Monica's blood type is A, Alan's is AB, and Rick's is O. The child's blood type is O. At first glance, it looks as if Rick is the father. (*a*) Whether Monica's genotype is I^Ai or I^AI^A, she cannot have a type O child with Alan, whose genotype is I^AI^B. (*b*) If Monica's genotype is I^Ai, she could have a type O child with Rick, whose genotype is *ii.* Unless . . . nosy nurse-in-training Amy has just learned about the Bombay phenotype and suggests that the baby have a blood test to see if he manufactures the H protein. She also advises the adults to look into their family histories to see if any other relatives' blood types are not predicted by their parents' blood types. Sure enough, the baby is of genotype *hh,* and Alan's family has had past incidences in which the blood type of a child did not match what was predicted by the parents' blood types. On further testing, both Monica and Alan are found to be *Hh,* but Rick is *HH.* Alan Jr. is *hh.* He has a type O phenotype, but his ABO genotype can be either I^AI^A, I^Ai, I^Bi, or I^AI^B. What he cannot be is Rick's son.

Penetrance and Expressivity

The same allele combination can produce different degrees of the phenotype in different individuals, even siblings, because of outside influences such as nutrition, toxic exposures, illnesses, and other genes. Most disease-causing allele combinations are completely penetrant, which means that all individuals who inherit the allele combination have some symptoms. A genotype is **incompletely penetrant** if some individuals do not express the phenotype. Polydactyly, having extra fingers or toes, is an incompletely penetrant trait seen in many types of mammals. Cats that inherit the dominant polydactyly allele have more than four toes on at least one paw, yet others who are known to have the allele (because they have an affected parent and offspring) have the normal number of toes. The penetrance of a gene is described numerically. If 80 of 100 cats that inherit the dominant polydactyly allele have extra digits, the allele is 80% penetrant. Figure 13.19 shows a person with polydactyly.

A phenotype is **variably expressive** if the symptoms vary in intensity in different individuals. One cat with polydactyly might have an extra digit on two paws; another might have two extra digits on all four paws; a third cat might have just one extra toe. Penetrance refers to the all-or-none expression of a genotype; expressivity refers to the severity of a phenotype.

Environmental Influences

Some gene expressions are exquisitely sensitive to the environment. Temperature influences gene expression in some familiar animals—Siamese cats and Himalayan rabbits have dark ears, noses, feet, and tails because these parts are colder than the animals' abdomens. Heat destroys pigment molecules that provide coat color.

Temperature-sensitive alleles also lead to striking phenotypes in the fruit fly (fig. 13.20). In flies mutant for a gene called proboscipedia, mouthparts develop as antennae when the flies are raised at low temperatures and as legs when the flies are raised at high temperatures. When raised at room temperature, the fly's mouthparts are mixtures of leg and antennal tissue. A phenotype that is expressed only under certain environmental conditions is termed conditional.

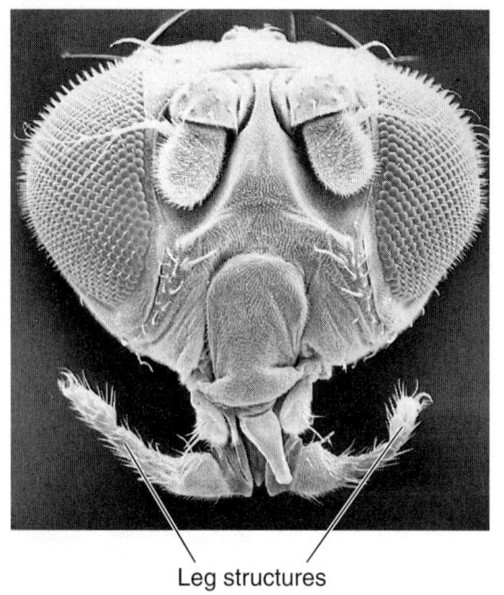

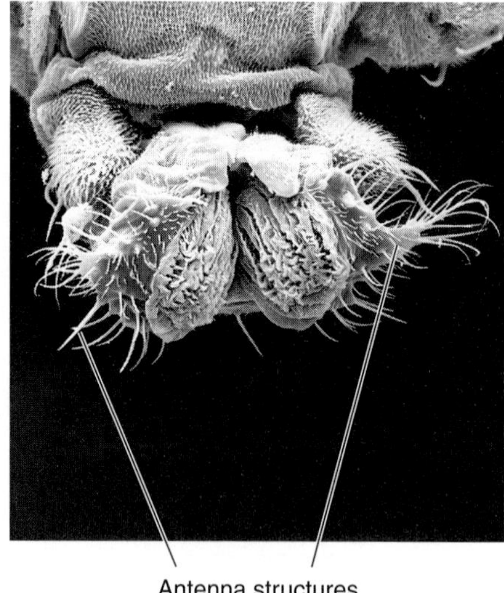

FIGURE 13.20

A temperature-sensitive homeotic gene in the fruit fly *Drosophila melanogaster* transforms mouthparts into leg structures at high temperature and into antennal structures at low temperature.

Pleiotropy

A Mendelian disorder with many symptoms is **pleiotropic.** Such conditions can be difficult to trace through families, because individuals with different subsets of symptoms may appear to have different disorders. In fact, some historians blame the American Revolution on a pleiotropic blood disorder, porphyria, that afflicted England's King George III.

Starting at age 50, King George III experienced recurrent bouts of abdominal pain and constipation, dark red urine, limb weakness, fever, hoarseness, racing pulse, insomnia, headache, visual problems, restlessness, delirium, convulsions, and finally stupor. Then he would recover. Because the king ripped off his wig and clothing and ran about at the peak of a fever, he was eventually deemed unfit to rule and was dethroned. History books call him the "mad king."

In George's time, doctors were permitted to do very little to the royal body and based their diagnoses on the king's comments. Twentieth-century researchers discovered that George's red urine resulted from porphyria. Lack of an enzyme results in a breakdown product of the red blood cell pigment hemoglobin being excreted in the urine, rather than being metabolized in cells. When urination cannot keep up with hemoglobin breakdown, the other symptoms begin. Examination of medical charts of George's relatives—easy to obtain for a royal family—showed that several of them had porphyria symptoms. But because individuals exhibited different combinations of the many symptoms, in the past no one suspected a single genetic cause behind the problems.

Pleiotropy occurs in genetic diseases that affect a single protein found in different parts of the body. For example, in Marfan syndrome, symptoms occur where an elastic connective tissue is absent or abnormal (see table 13.3).

Phenocopies—When It's Not Really in the Genes

A **phenocopy** is an environmentally caused trait that appears to be inherited because it affects several members of a family. The limb birth defect that the drug thalidomide causes, discussed in chapter 12, is a phenocopy of the inherited illness phocomelia. An infection can appear to be inherited when it affects several family members. However, unlike an inherited disorder, an infection does not recur with a predictable frequency. Its recurrence depends upon exposure to the infective agent.

Table 13.6 summarizes phenomena that can alter Mendelian phenotypic ratios.

Mastering Concepts

1. How do lethal alleles and epistasis decrease the number of phenotypic classes, and multiple alleles, incomplete dominance and codominance increase the number of phenotypic classes?
2. How do penetrance and expressivity differ?
3. How can the environment affect a phenotype?
4. How does pleiotropy differ from incomplete penetrance and variable expressivity?
5. How does a phenocopy differ from an inherited trait?

Complex Traits

A woman who is a prolific writer has a daughter who becomes a successful novelist. An overweight man and woman have obese children. A man whose father was an alcoholic is himself an alcoholic. Are these characteristics—writing talent, obesity, and alcoholism—inherited or imitated? It can be difficult to tell. For single-gene disorders, we can predict the probability that a certain family member will inherit the condition. Some traits and diseases, though, "run in families," appearing in a few relatives with no apparent pattern, because they are caused by more than one gene, the environment, or both.

Characteristics that do not follow Mendel's laws but have an inherited component are termed **complex traits.** These may be **polygenic** (determined by more than one gene) or **multifactor-**

Table 13.6 Factors That Alter Mendelian Phenotypic Ratios

Phenomenon	Effect on Phenotype	Example
Lethal alleles	A phenotypic class dies very early in development	Spontaneous abortion
Multiple alleles	Produces many variants of a phenotype	Rabbit coat color
Incomplete dominance	A heterozygote's phenotype is intermediate between those of two homozygotes	Snapdragon flower color
Codominance	A heterozygote's phenotype is distinct from and not intermediate between those of the two homozygotes	ABO blood types
Epistasis	One gene masks another's phenotype	Bombay phenotype
Penetrance	Some individuals inheriting a particular genotype do not have the associated phenotype	Polydactyly
Expressivity	A genotype is associated with a phenotype of varying intensity	Polydactyly
Conditional mutations	An environmental condition affects a gene's expression	Temperature-sensitive coat colors
Pleiotropy	The phenotype includes many symptoms, with different subsets in different individuals	Porphyria
Phenocopy	An environmentally caused condition whose symptoms and recurrence in a family make it appear to be inherited	Infection

Table 13.7 Polygenic Model of Skin Color Inheritance

Multilocus Genotypes	Number of Pigment Genes
AABBCC	6
AaBBCC, AABbCC, AABBCc	5
aaBBCC, AAbbCC, AABBcc, AaBbCC, AaBBCc, AABbCc	4
AaBbCc, aaBbCC, AAbbCc, AabbCC, AABbcc, aaBBCc, AaBBcc	3
AaBbcc, AabbCc, AAbbcc, aaBBcc, aabbCC	2
Aabbcc, aaBbcc, aabbCc	1
aabbcc	0

Studies that classify skin color by measuring the amount of light reflected from the skin surface suggest that three or four or more different genes, probably with several alleles each, interact to produce pigment in the skin. The greater the number of pigment-specifying genes, the darker the skin. The inheritance of skin color is most likely even more complicated than this three-gene model indicates.

ial (determined by one or more genes plus the environment). The term "environment" encompasses many influences, including position in the uterus, experiences, and exposure to infectious agents.

For polygenic (quantitative) traits, several genes each contribute to the overall phenotype in equal, small degrees. Their combined actions produce a continuum, or continuously varying expression, of the trait (table 13.7). The genes follow Mendel's laws individually (unless they are on the same chromosome), but they don't produce typical ratios because they contribute equally to the phenotype, and they are neither dominant nor recessive with respect to each other. Height and skin color are familiar examples of polygenic traits in humans (fig. 13.21). In plants, quantitative traits include flower color, petal length, blade length, and stomata density on the leaves.

Measuring Multifactorial Traits

Multifactorial traits are determined by one or more genes and the environment. For example, cardiovascular health (or disease) reflects the actions of genes controlling blood pressure, clotting ability of the blood, and lipid levels in the blood, as well as environmental factors such as diet, exercise habits, and stress. Body weight and intelligence are other multifactorial traits.

Information from population studies and from family relationships help describe the inherited component of multifactorial traits.

Empiric Risk

Empiric risk is a prediction of recurrence based on a multifactorial trait's incidence in a specific population. In general, empiric

FIGURE 13.21

Height is a continuously varying trait. When 175 soldiers were asked to line up according to their heights, they formed a characteristic bell-shaped, continuous distribution.

risk increases with the severity of the disorder, the number of affected family members, and how closely related the person is to affected individuals.

Empiric risk helps, for example, to predict the likelihood that a neural tube defect (NTD) will recur. Recall that a NTD is an opening or lesion in the brain or spinal cord that occurs on about the 28th day of prenatal development. In the United States, the overall population risk of carrying a fetus with a NTD is about 1 in 1,000. However, if a sibling has a NTD, the risk of recurrence increases to 3%, and if two siblings are affected, the risk to a third child is even greater. NTDs have a variety of causes, including mutation and folic acid deficiency. The environmental risk of NTDs may be greatly reduced by supplementing the diet with folic acid during early pregnancy.

Heritability—The Genetic Contribution to a Trait

A large biology class has students of many heights. Certainly heredity influences height, but nutrient intake, an environmental factor, also affects height by providing the raw material to build muscle and bone.

Heritability estimates the proportion of phenotypic variation in a group that can be attributed to genes. It is calculated as double the difference of the measurable variation between two groups of individuals. Approximately 80% of height variance is due to heredity, so this trait has a heritability of 0.80. Heritability values, however, are not always reliable. This is because researchers may define different phenotypes, particularly behavioral ones, differently. Also, specific genetic or environmental components may contribute to causing a complex trait to different degrees. Table 13.8 lists heritabilities for some traits.

Table 13.8 Heritabilities for Some Human Traits

Trait	Heritability
General intelligence	0.52
Extraversion	0.51
Verbal reasoning ability	0.50
Neuroticism	0.46
Vocational interests	0.42
Spatial reasoning ability	0.40
Scholastic achievement	0.38
Good memory	0.22

Heritability describes a particular group of individuals and is not a property of a gene. This means it can vary in different environments. For example, the influence nutrition has on egg production in chickens is greater in a starving population than in a well-fed one.

Coefficient of Relationship

The **coefficient of relationship** describes the percentage of genes two related individuals share (table 13.9). The closer the relationship between two individuals, the more genes they have in common, and the greater the probability that they will share a trait. For example, the risk that cleft lip will recur is 40% for the identical twin of an affected individual, but only 4% for a sibling, and less than 1% for a niece, nephew, or first cousin.

Studying Multifactorial Traits

Multifactorial inheritance analysis does not easily lend itself to the scientific method, which ideally examines one variable. Two types

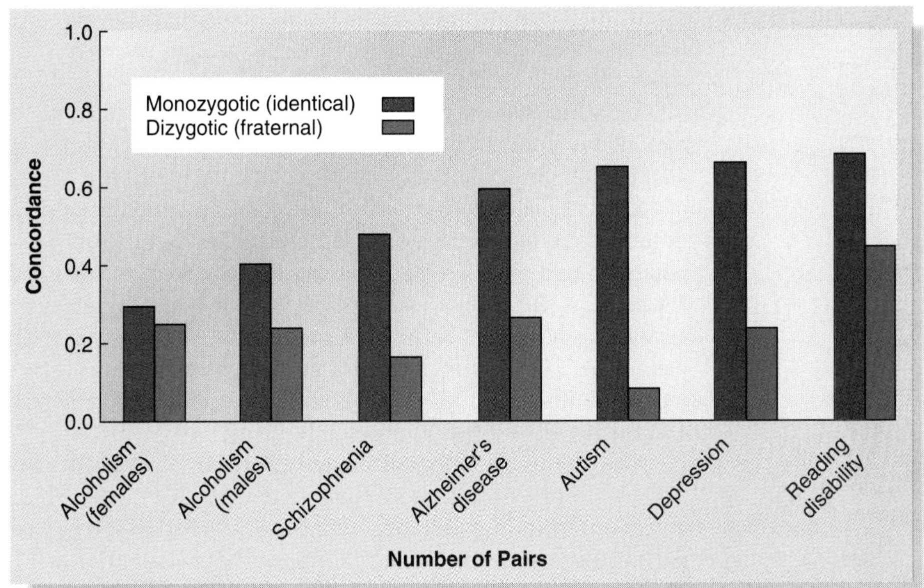

FIGURE 13.22

A trait that is more often present in both members of monozygotic twin pairs than it is in both members of dizygotic twin pairs has a significant inherited component.

Source: Robert Plomin, et al., "The genetic basis of complex human behaviors," *Science*, 17 June 1994, vol. 264, pp. 1733–1739. Copyright 1994 American Association for the Advancement of Science.

Table 13.9 Coefficient of Relationship and Shared Genes

Relationship	Degree of Relationship	Percent Shared Genes
Sibling to sibling	1°	50% (1/2)
Parent to child	1°	50% (1/2)
Uncle/aunt to niece/nephew	2°	25% (1/4)
First cousin to first cousin	3°	12 1/2% (1/8)

of people, however, help geneticists tease apart the genetic and environmental components of complex traits—adopted individuals and twins.

Adopted Individuals

A person adopted by nonrelatives shares environmental influences, but not genes, with his or her adoptive family. Conversely, adopted individuals share genes, but not the exact environment, with their biological parents. Therefore, biologists assume that similarities between adopted people and their adoptive parents reflect environmental influences, whereas similarities between adopted people and their biological parents mostly reflect genetic influences. Information on both sets of parents can reveal to what degree heredity and the environment contribute to a trait.

Many adoption studies use the Danish Adoption Register, a list of children adopted in Denmark from 1924 to 1947. One study examined causes of death among biological and adoptive parents and offspring. If a biological parent died of infection before age 50, the adopted child was five times more likely to die of infection at a young age than a similar person in the general population. This may be because inherited variants in immune system genes increase susceptibility to certain infections. In support of this hypothesis, the risk that adopted individuals would die young from infection did not correlate with adoptive parents' death from infection before age 50. The study also revealed that environment affects longevity. If adoptive parents died before age 50 of cardiovascular disease, their adopted children were three times as likely to die of heart and blood vessel disease as a person in the general population. Can you think of a factor that might explain this correlation?

Twins

Monozygotic (MZ), or identical, twins are always of the same sex and have identical genes, because they develop from one fertilized ovum. **Dizygotic** (DZ), or fraternal, twins are no more similar genetically than any other two siblings, although they share the same prenatal environment because they develop at the same time from two fertilized ova.

If a trait tends to occur more frequently in both members of identical twin pairs than in both members of fraternal twin pairs, it is at least partly controlled by heredity. Geneticists calculate the **concordance** of a trait, or the degree to which it is inherited, by calculating the percentage of twin pairs in which both members express the trait (fig. 13.22). Twins can be used to calculate heritability, which equals approximately double the difference between MZ and DZ concordance values for a trait.

Diseases caused by single genes, whether dominant or recessive, are always 100% concordant in MZ twins—that is, if one twin has it, so does the other. However, among DZ twins, concordance is 50% for a dominant trait and 25% for a recessive trait, the same Mendelian values that apply to any two siblings. For a trait determined by several genes, concordance values for MZ twins are significantly greater than for DZ twins. Finally, a trait molded mostly by the environment exhibits similar concordance values for both types of twins.

Separated at birth, the Mallifert twins meet accidentally.

FIGURE 13.23

Drawing by Chas. Addams; © 1981 The New Yorker Magazine, Inc.

Using twins to study genetic influence on complex traits dates to 1924, when German dermatologist Hermann Siemens compared school transcripts of identical versus fraternal twins. Noticing that grades and teachers' comments were much more alike for identical twins than for fraternal twins, he concluded that genes contribute to intelligence. Siemens also suggested that a better test would be to study identical twins who were separated at birth and then raised in very different environments.

Twins separated at birth provide natural experiments for distinguishing nature from nurture. Much of what they have in common can be attributed to genetics, especially if their environments have been very different (fig. 13.23). The twins' differences reflect differences in their upbringing.

Since 1979, more than 100 sets of twins and triplets separated at birth have visited the laboratories of Thomas Bouchard at the University of Minnesota. Each pair of twins undergoes a 6-day battery of tests that measure physical and behavioral traits, including 24 blood types, handedness, direction of hair growth, fingerprint pattern, height, weight, functioning of all organ systems, intelligence, allergies, and dental patterns. Researchers videotape the twins' facial expressions and body movements in different circumstances and probe their fears, vocational interests, and superstitions.

Identical twins separated at birth and reunited later can be remarkably similar, even when they grow up in very different adoptive families. Idiosyncrasies are particularly striking. A pair of identical male twins raised in different countries practicing different religions were astounded, when they were reunited as adults, to find that they both laugh when someone sneezes and flush the toilet before using it. Twins who met for the first time in their thirties responded identically to questions; each paused for 30 seconds, rotated a gold necklace she was wearing three times, and then answered the question. Coincidence, or genetics?

The "twins reared apart" approach is not a perfectly controlled way to separate nature from nurture. Identical twins share an environment in the uterus and possibly in early infancy that may affect later development. Siblings, whether adoptive or biological, do not always share identical home environments. Differences in age, sex, general health, school and peer experiences, temperament, and personality affect each individual's perception of such environmental influences as parental affection and discipline.

Adoption studies, likewise, are not perfect experiments. Adoption agencies often search for adoptive families who have socioeconomic or religious backgrounds similar to those of the biological parents. Thus, even when different families adopt separated twins, their environments may not be as different as they might be for two unrelated adopted people. However, at the present time, twins reared apart and adopted individuals provide intriguing insights into the body movements, psychological quirks, interests, behaviors, and other personality traits that seem to be rooted in our genes (Bioethics Connections 13.1).

Mastering Concepts

1. What are complex traits?
2. How do polygenic traits differ from multifactorial traits?
3. What is a continuously varying trait?
4. What three factors increase empiric risk of having a complex trait?
5. What does heritability estimate?
6. What does the coefficient of relationship indicate?
7. How do adopted individuals and twins help geneticists distinguish between genetic and environmental influences on traits?

13.1 bioethics connections

Blaming Genes

It is fashionable to blame genes for our shortcomings. A popular magazine's cover shouts "Infidelity: It May Be in Our Genes," advertising an article that actually has little to do with genetics. When researchers identified a gene that plays a role in fat metabolism, some people binged on chocolate and forsook their daily exercise, because, after all, if obesity is in their genes, there's nothing they can do to stop it. The first personality trait to be linked to a gene is "novelty seeking" (fig. 13.A). Researchers found that people who rate themselves as craving new experiences have a certain allele of a gene that controls how cells take up the neurotransmitter dopamine more frequently than chance would predict. The researchers caution, however, that such an inherited tendency could manifest itself in varied ways—such as a fondness for bungee-jumping or as violent actions.

Geneticists have only recently refocused attention on the biological roots of behavior. Early in the century, behavioral genetics was part of eugenics—the idea of improving a population's collection of genes, or gene pool. The horrific experiments and exterminations done in the name of eugenics in Nazi Germany, however, turned many geneticists away from studying the causes of behaviors. For many years, social scientists dominated behavioral research, and as a result, many behavioral disorders were attributed to mostly environmental influences until the 1960s. Then, armed with a clearer idea of what a gene is and what it can do, biologists reentered the debate. Today, researchers are identifying specific genes that predispose or cause a clearly defined behavior. Yet environment still plays a role.

Consider a Dutch family in which many members are mildly mentally retarded or display abnormal behavior, such as arson, rape, and exhibitionism. A mutation in affected family members appeared to explain the problems. Alteration of a single DNA base in the gene encoding an enzyme (monoamine oxidase A, or MAOA) rendered the enzyme nonfunctional. The enzyme normally metabolizes certain neurotransmitters (nerve cell messenger molecules). Perhaps, researchers suggest, the inherited enzyme deficiency causes slight mental impairment, and this interferes with the person's ability to cope with certain frustrating situations. The result—violence. Hence, the argument returns once again to genes interacting with the environment.

FIGURE 13.A

These air surfers were dropped from a helicopter over a mountain. Does a gene variant make them seek thrills?

It Can't Be!~

continued from page 262.

Both Erin and David Fitzgerald's family histories were indeed free of cystic fibrosis, at least in its most severe and familiar guise. In the two-thirds of cases that are severe, frequent and serious lung infections often prove deadly by young adulthood, and sometimes much earlier. Patients must receive daily physiotherapy to clear sticky mucus from the lungs, eat digestive enzymes to supplement their mucus-plugged pancreases, and take many medications daily. Life isn't easy, and Erin and David were certain that none of their relatives had gone through this.

A disease-causing autosomal recessive allele can go unnoticed in a family for many generations. Only when two mutant alleles appear in the same individual does the disease result. Erin could trace her family back to her great-grandparents, and no one had been affected, although several relatives had been or were carriers. Fortunately, none of them had had children with another carrier, so the disease never appeared. David discovered that an uncle, who was childless because he was missing the vasa deferentia and had suffered frequent bronchitis and pneumonia all his life, had a mild case of cystic fibrosis.

David's mother, the uncle's sister, was a carrier, and so was David, as genetic tests indicated. Erin, too, was a carrier but of the allele conferring a more severe phenotype. Two of their three sons were carriers; the youngest inherited Erin's mutant allele and the middle son inherited David's allele. Their eldest son was wild type. Megan, however, was unlucky in the genetic stakes and had inherited the disease-causing allele from each parent.

The family learned all about cystic fibrosis, and today they help Megan each day to do her exercises and encourage her through her infections. Megan is enrolled in a study of a gene therapy method to replace the protein missing in the cells lining her respiratory tract and pancreatic ducts. Because of greatly increased knowledge about cystic fibrosis and the gene that causes it, and several new drugs and treatment methods, Megan has a much brighter future than people with cystic fibrosis had in the past.

SUMMARY

1. Using pea plant crosses, Mendel followed transmission of one or two traits at a time. The **genes** for the traits he studied were carried on different chromosomes, and each trait had two forms, or **alleles**.

2. Mendel's first law, **segregation,** states that inherited "elementen" separate in meiosis. Each individual receives one copy of each elementen from each parent. Single genes cause Mendelian traits.

3. An allele whose expression masks another is **dominant;** an allele whose expression is masked by a dominant allele is **recessive.**

4. A **heterozygote** has two different alleles of a gene. A **homozygous recessive** individual has two recessive alleles. A **homozygous dominant** individual has two dominant alleles.

5. The combination of alleles constitutes a **genotype,** and the expression of a particular genotype is its **phenotype.** A **wild-type** allele is the most common one in a population. A change in a gene is a **mutation** and may result in a **mutant** allele and possibly phenotype.

6. The **parental generation** is designated $\mathbf{P_1}$, the next generation is the **first filial generation,** or $\mathbf{F_1}$, and the next is the **second filial generation,** or $\mathbf{F_2}$.

7. **Punnett squares,** based on the principles of probability, can be used to predict the outcomes of genetic crosses.

8. A **monohybrid cross** is between two heterozygotes and yields a genotypic ratio of 1:2:1 and a phenotypic ratio of 3:1. A **test cross** breeds an individual of unknown genotype to a homozygous recessive individual.

9. Geneticists use **pedigrees** to trace traits in families. Squares denote males, and circles, females. Horizontal lines represent either parents or siblings, and vertical lines depict generations. Pedigrees can reveal **mode of inheritance.** An **autosomal recessive** trait can appear in either sex, is passed from parents who are carriers or are affected, and can skip generations. An **autosomal dominant** trait affects both sexes and is inherited from one affected parent.

10. Mendel's second law, **independent assortment,** follows transmission of two or more characters on different chromosomes. Because maternally and paternally derived chromosomes (and the genes they carry) sort randomly in meiosis, different gametes have different combinations of genes.

11. A **dihybrid cross** breeds two heterozygotes, yielding a 9:3:3:1 phenotypic ratio.

12. The **product rule** can be used as an alternative to Punnett squares.

13. In some crosses, the ratio of progeny phenotypes does not seem to follow Mendel's laws. Individuals who have **lethal allele** combinations cease developing before they can be detected among progeny. A gene can have multiple alleles because its DNA sequence may be altered in many ways. Different types of dominance relationships influence phenotypic ratios among offspring. Heterozygotes of **incompletely dominant** alleles have phenotypes intermediate between those of the two homozygotes. **Codominant** alleles are both expressed. In **epistasis,** one gene masks the effect of another.

14. In an **incompletely penetrant** genotype, phenotype is not expressed in all individuals that inherit it. Phenotypes that vary in intensity are **variably expressive.** The environment can influence expression of conditional mutations. **Pleiotropic** genes have several expressions. A **phenocopy** appears to be inherited because it occurs repeatedly, but is environmentally caused.

15. **Complex traits** do not follow Mendel's laws but have an inherited component. A **polygenic trait** is determined by more than one gene and varies continuously in its expression. The environment and genes cause **multifactorial traits.**

16. **Empiric risk** measures the likelihood of a multifactorial trait recurring based on its prevalence. The risk rises with genetic closeness to an affected individual, severity of the phenotype, and number of affected relatives.

17. **Heritability** estimates the proportion of variation in a multifactorial trait due to genetics. It equals twice the difference of the measurable variability for that trait between two groups of individuals. Heritability is not a gene characteristic; it varies among populations.

18. The **coefficient of relationship** describes the proportion of genes different relatives share.

19. Characteristics adopted people share with their biological parents are mostly inherited, whereas similarities between adopted people and their adopted parents reflect environmental influences.

20. **Concordance** measures the expression of a trait in MZ or DZ twins. It equals the percentage of twin pairs in which both members express the trait. A high concordance value indicates that genes predominantly cause a trait.

TO REVIEW . . .

1. State how Mendel's two laws derive from meiotic events.

2. Distinguish between genotype and phenotype; homozygote and heterozygote; mutant and wild type; recessive and dominant.

3. How might pedigrees for an autosomal recessive and an autosomal dominant trait differ?

4. Draw a partial pedigree for the family described in the opening vignette, indicating known carriers and possible carriers of CF.

5. Mel and Barbara are healthy, but they know from a blood test that they are each carriers of Tay-Sachs disease. If their first three children are healthy, what is the probability that their fourth child will inherit the disease?

6. Calculate the genotypic and phenotypic ratios for the following crosses in peas for height, determined by the T gene:

- **a.** homozygous dominant crossed with homozygous recessive
- **b.** homozygous dominant crossed with heterozygous
- **c.** homozygous recessive crossed with heterozygous

7. A man and a woman each have dark eyes, dark hair, and freckles. The genes for these traits assort independently. The woman is heterozygous for each of these traits, but the man is homozygous. The dominance relationships of the alleles are as follows:

B = dark eyes; b = blue eyes
H = dark hair; h = blond hair
F = freckles; f = no freckles

a. What is the probability that their child will share the parents' phenotype?
b. What is the probability that the child will share the same genotype as the mother? as the father?

Use probability or a Punnett square to obtain your answers. Which method is easier?

8. This unusual pedigree depicts the Ptolemy dynasty, which began in 323 B.C. and ended with Cleopatra in 30 B.C. What is unusual about the pedigree?

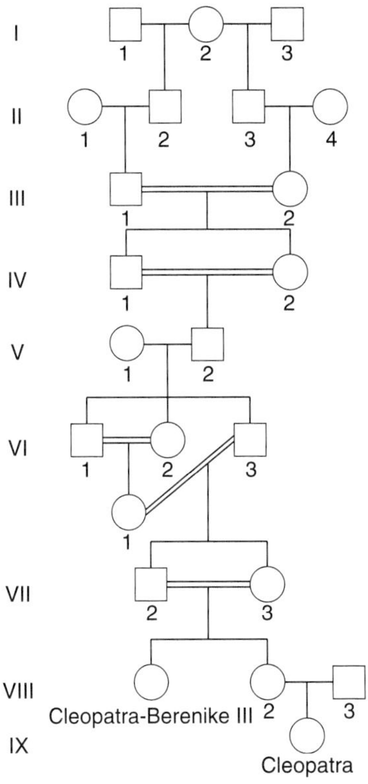

Source: Catriona Galbraith, Cherwell Scientific Publishing, Inc., Palo Alto, CA.

9. Which genetic principle do each of the following examples illustrate?

a. Many different genes affect hearing in humans. Being homozygous recessive for certain of these genes results in deafness, even if the other hearing genes are wild type.
b. A woman develops dark patches on her face, and her physician suggests that she has alkaptonuria, an inherited illness; however, a dermatologist discovers that the woman's skin problem is due to using a facial cream that contains a chemical known to cause dark patches in people with dark skin.

10. In cats with the Manx trait, the M allele causes a short or absent tail, whereas the m allele confers a normal, long tail. Cats of genotype MM die as embryos. If two Manx cats mate, what is the probability that each living kitten has a long tail?

11. Explain how each of the following appear to disrupt Mendelian ratios:

a. pleiotropy
b. variable expressivity
c. incomplete penetrance
d. lethal alleles

12. A man with type AB blood has children with a woman who has type O blood. What are the chances that a child they conceive will have type A blood? type B? AB? O?

13. The autosomal dominant disorder distal symphalangism causes the fingers and toes to be short with tiny nails. How can this disorder be both incompletely penetrant and variably expressive?

14. Domesticated hens with white feathers and large, single combs mate with roosters that have dark feathers and small combs. The offspring all resemble their mothers for these two traits.

a. Which alleles of each trait are dominant, and which are recessive?
b. If the F_1 are crossed with each other, what percentage of the F_2 would be expected to have dark feathers and large, single combs?

15. Two genes carried on different chromosomes impart pigment to potato skin. The D gene provides red color, and the P gene provides blue. If the dominant allele of the D gene is present, however, the potatoes are red, no matter which P gene allele is present. Which phenomenon that can alter Mendelian ratios does this illustrate?

16. Define complex, multifactorial, and polygenic traits.

17. A male cat has short hair, a stubby tail, and extra toes. A female cat has long hair, a long tail, and extra toes. The genes and alleles are as follows:

L = short hair; l = long hair
M = stubby tail (Manx); m = long tail
Pd = extra toes; pd = normal number of toes

The two cats have kittens. One has long hair, a long tail, and no extra toes. Another has short hair, a stubby tail, and extra toes. The third kitten has short hair, a long tail, and no extra toes. What is the genotype of the father?

18. A Mexican hairless dog has a dominant allele conferring the hairless phenotype. When two Mexican hairless dogs mate, one-third of the puppies are hairy, and two-thirds are hairless. Explain how these ratios could occur.

19. In cats, several genes control coat color. If a dominant allele is present for either the white gene (W) or the orange gene (O), the other pigment genes are not expressed. What phenomenon does this illustrate?

TO THINK ABOUT . . .

1. Why would Mendel's observations have differed if the genes he studied were part of the same chromosome?

2. If you wanted to confirm Mendel's law of independent assortment by recording the phenotypes and genotypes of the offspring of a dihybrid cross, what sorts of traits would you choose to examine, and in what organism?

3. A white woman with fair skin, blond hair, and blue eyes and a black man with dark brown skin, hair, and eyes have fraternal twins. One twin has blond hair, brown eyes, and light skin, and the other has dark hair, brown eyes, and dark skin. What Mendelian principle does this real-life case illustrate?

4. What are three explanations for a trait that shows a great deal of variability in different individuals?

5. How can parents carry recessive lethal genes and not know it? Why is there no physical evidence for the existence of dominant lethal mutations?

6. A man and woman want to have children, but each is a known carrier of Tay-Sachs disease. To avoid the possibility that they will conceive a child who will inherit this untreatable, fatal disorder, the woman is artificially inseminated with sperm from a man who does not carry the Tay-Sachs gene. Why is this procedure (discussed in chapter 12) a solution to the couple's problem?

7. Many plants have more than two sets of chromosomes. How would having four (rather than two) copies of a chromosome more effectively mask expression of a recessive allele?

8. In an attempt to breed winter barley that is resistant to barley mild mosaic virus, agricultural researchers cross a susceptible domesticated strain with a resistant wild strain. The F_1 plants are all susceptible, but when the F_1 are crossed with each other, some of the F_2 individuals are resistant. Is the viral resistance gene recessive or dominant? How do you know?

TO LEARN MORE . . .

References and Resources

Bouchard, T.J., Jr. June 17, 1994. Genes, environment, and personality. *Science,* vol. 264. Identical twins separated at birth offer clues to hereditary and environmental influences on behavior.

Derr, Mark. March 1996. The marathon mutt. *Natural History.* Alaskan huskies combine traits from many canine breeds.

Heim, Werner G. February 1991. What is a recessive allele? *The American Biology Teacher.* We now understand the pea plant traits Mendel studied at the molecular level.

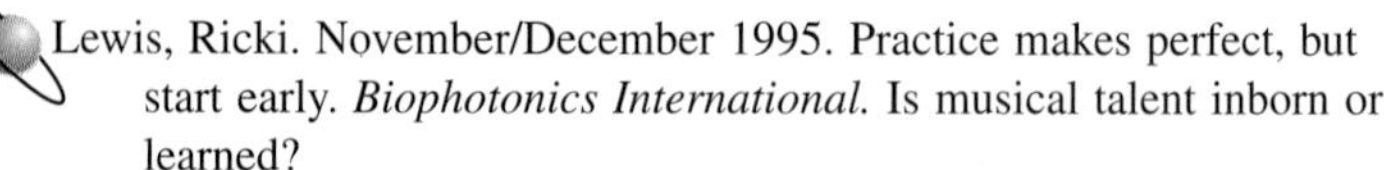
Lewis, Ricki. November/December 1995. Practice makes perfect, but start early. *Biophotonics International.* Is musical talent inborn or learned?

Lewis, Ricki. December 1994. The evolution of a classic genetic tool. *BioScience.* Computers have greatly increased the information content of pedigrees.

McKusick, Victor A. July 25, 1991. The defect in Marfan syndrome. *Nature,* vol. 352. Marfan syndrome is pleiotropic because of the widespread distribution of fibrillin in the body.

Plomin, R., M. J. Owen, and P. McGuffin. June 17, 1994. The genetic basis of complex human behaviors. *Science,* vol. 264. Separating nature from nurture can be very challenging.

Toufexis, Anastasia. January 15, 1996. What makes them do it? *Time.* Can genes mold daredevils?

denotes articles by author Ricki Lewis that may be accessed via *Life's* home page /http://www.mhhe.com/sciencemath/biology/life/.

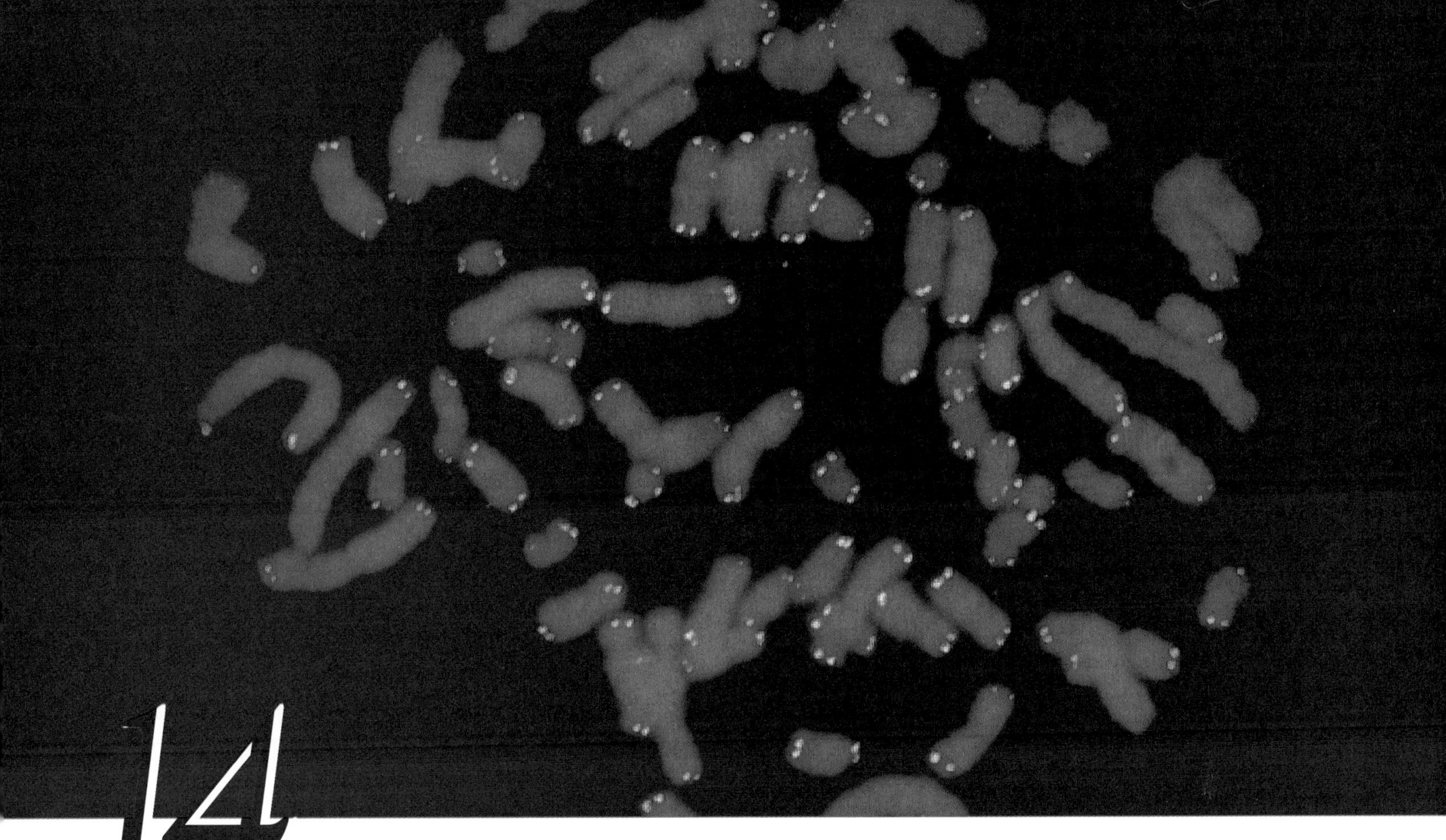

14 Chromosomes

Chromosome tips hold clues to aging and cancer.

chapter outline

Studying Chromosomes 286

Linked Genes Do Not Follow Mendel's Laws 286
Peas and Flies Reveal Linkage
Linkage Maps

Sex-Linkage and Sex Chromosomes 290
Sex Determination in Humans
Sex-Linked Recessive Inheritance
Sex-Linked Dominant Inheritance
Y Linkage

X Inactivation—Equalizing the Sexes 293

Chromosome Structure 296
Size and Shape
Stains and Probes
Evolution of the Karyotype

Abnormal Chromosome Numbers 297
Extra Chromosome Sets—Polyploidy
Extra and Missing Chromosomes—Aneuploidy

Smaller-Scale Chromosome Abnormalities 301
Deletions and Duplications
Chromosome Rearrangements

Chromosomes and Health 302

The Philadelphia Story

On August 13, 1958, two men entered hospitals in Philadelphia with the same symptom—weeks of fatigue. G. H., a 33-year-old African American man, and E. K., a 41-year-old Caucasian man, each had very high white blood cell counts and were diagnosed with a type of cancer called chronic myeloid leukemia (CML). Their blood had far too many immature white blood cells, which were crowding the healthy cells. Too few red blood cells, which carry oxygen to tissues, explained the men's fatigue.

Before G. H. and E. K. died, they donated blood samples to two young researchers at the University of Pennsylvania, Peter Nowell and David Hungerford. The two researchers noticed that some of the white blood cells were dividing—a requirement to view chromosomes then—and looked more closely. They knew that some cancer cells have unusual chromosomes. Would they find a specific problem in the cancerous white blood cells of the two men?

Nowell and Hungerford cultured the blood cells, arrested them in mitotic metaphase, stained them, and photographed them. But in 1960, the art and science of observing chromosomes was in its infancy. Researchers couldn't tell similar-sized chromosomes apart, and there were quite a few of these amidst the jumble of 46 chromosomes in a human somatic cell. But Nowell and Hungerford indeed found something unusual among the chromosomes of the two men with leukemia, a discovery that has affected many people with cancer ever since.

to be continued . . .

Studying Chromosomes

Impending parenthood makes many people acutely aware of chromosomes, because these rod-shaped bodies in cell nuclei are windows into the health of a fetus. If all goes well, a prenatal test will reveal 23 pairs of chromosomes, with no genetic material missing, extra, or in an unusual place. Although a normal chromosome count does not guarantee fetal health, it rules out many types of abnormalities. Figure 14.1 shows three ways that physicians sample and examine fetal chromosomes.

Studying human chromosomes is also used to diagnose certain types of cancers, detect a population's exposure to a particular toxin, and serve as a basis of comparison among species when investigating evolutionary relationships. Each species has a characteristic number of chromosomes, and the number has nothing to do with complexity of the organism. A mosquito has 6; a grasshopper, rice plant, and pine tree 24; a dog 78; and a carp 104.

The rise of cell biology early in the twentieth century joined Mendelian genetics to create the field of **cytogenetics,** which matches chromosomal aberrations to particular phenotypes (symptoms or traits). The primary tool of the cytogeneticist is a size-order chart of the chromosomes called a **karyotype.** Today's karyotypes are packed with information, sometimes showing locations of individual genes.

The evolution of cytogenetics in recent years can be compared to drawing maps of the United States with increasing detail, from just outlines of states, to showing major cities, and gradually to filling in smaller cities and towns and even streets. At the turn of the last century, however, when chromosomes were first visualized against the historical backdrop of rediscovering Mendel's laws, they were quite mysterious. This chapter first considers how the packaging of genes on chromosomes affects trait transmission and then examines chromosome structure, both normal and abnormal.

Linked Genes Do Not Follow Mendel's Laws

In 1902, German biologist Theodor Boveri and American graduate student Walter Sutton independently connected chromosomes to inherited traits, thus establishing the field of cytogenetics. Boveri and Sutton realized that chromosome movements during cell division explain Mendelian trait transmission.

Soon biologists realized that the number of observable inherited traits in an organism far exceeds the number of chromosomes. Fruit flies, for example, have four pairs of chromosomes but dozens of different bristle patterns, body colors, eye colors, wing shapes, and other characteristics. How might a few chromosomes control so many traits? The answer: Genes located on the same chromosome are **linked,** or inherited together, and do not assort independently, just as the cars of a train arrive at the same destination at the same time, whereas automobiles headed for the same place do not arrive exactly together. Genes on different chromosomes specified the seven traits Mendel followed in his pea plants. Had the same chromosome carried these genes near each other, Mendel would have generated markedly different results in his dihybrid crosses.

Peas and Flies Reveal Linkage

The different inheritance pattern of linked genes was first noticed in the early 1900s, when William Bateson and R. C. Punnett observed offspring ratios in pea plants different from what Mendel's laws predicted. They looked at different traits. Bateson and Punnett crossed true-breeding plants with purple flowers and long pollen grains (genotype *PPLL*) with true-breeding plants with red flowers and round pollen grains (genotype *ppll*). Then they crossed the F_1 plants, of genotype *PpLl,* with each other. Surprisingly, the F_2 generation did not show the expected 9:3:3:1 ratio for an independently assorting dihybrid cross (table 14.1).

Two types of F_2 peas—those with the same phenotypes as the parents, *P_L_* and *ppll*—were more abundant than predicted, while the other two progeny classes—*ppL_* and *P_ll*—were less common. Bateson and Punnett hypothesized that the prevalent parental allele combinations reflected genes transmitted on the same chromosome, which therefore did not separate during meiosis. Another meiotic event, crossing-over, explained the other two offspring classes. Recall that crossing-over is an exchange between homologs that mixes up maternal and paternal gene combinations in the gametes (fig. 14.2).

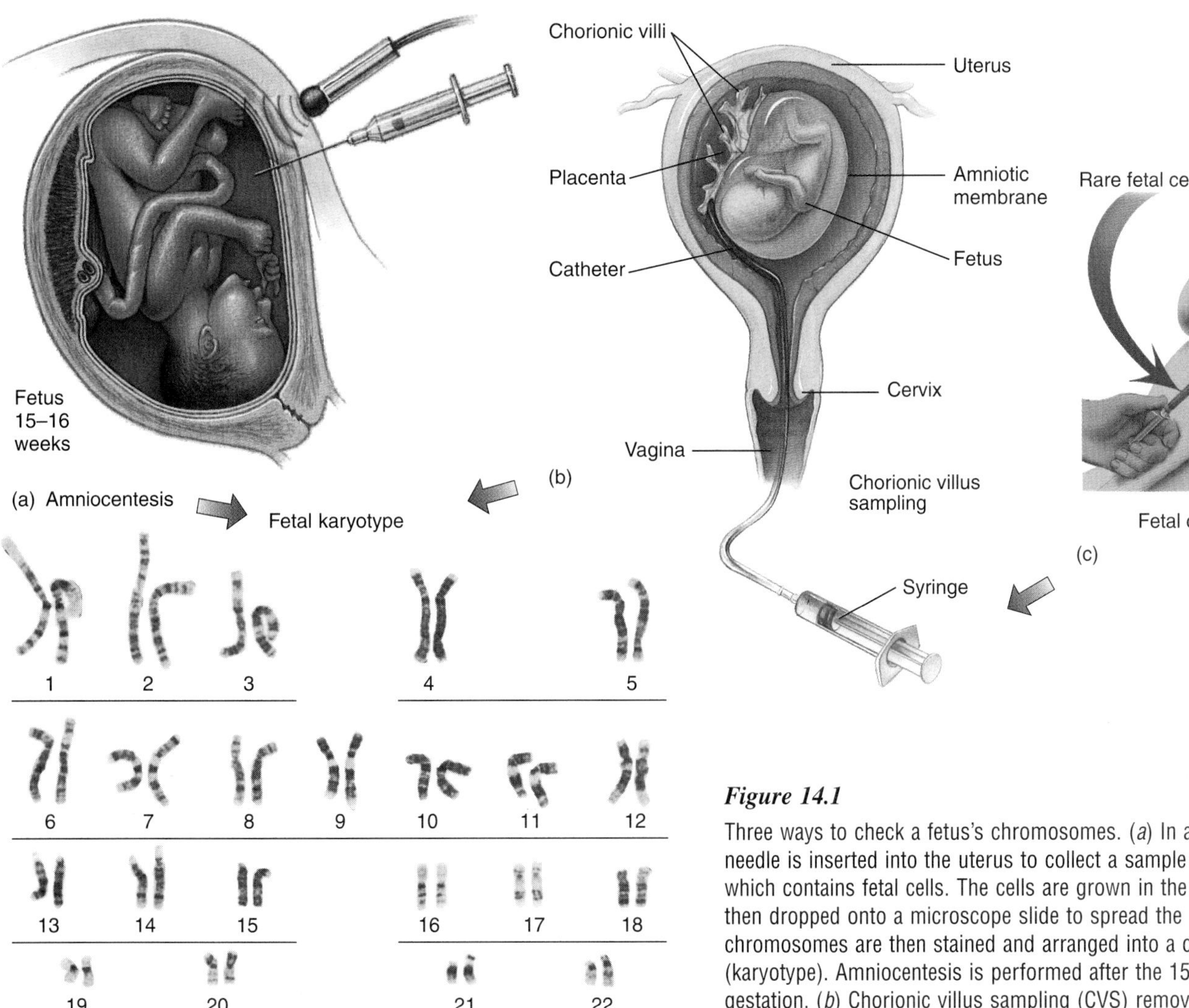

Figure 14.1

Three ways to check a fetus's chromosomes. (*a*) In amniocentesis, a needle is inserted into the uterus to collect a sample of amniotic fluid, which contains fetal cells. The cells are grown in the laboratory, and then dropped onto a microscope slide to spread the chromosomes. The chromosomes are then stained and arranged into a chromosome chart (karyotype). Amniocentesis is performed after the 15th week of gestation. (*b*) Chorionic villus sampling (CVS) removes cells of the chorionic villi, whose chromosomes match those of the fetus because they all descend from the fertilized ovum. CVS is usually performed earlier than amniocentesis. (*c*) Fetal cell sorting separates fetal cells in the woman's circulation. A genetic counselor interprets results of these tests—a fetal karyotype (*d*)—for patients.

Table 14.1 Observed and Expected Phenotypes for a Dihybrid Pea Plant Cross

Phenotype	Number of Observed Plants	Number of Expected Plants (9:3:3:1)
P = purple flower		
p = red flower		
L = long pollen grain		
l = round pollen grain		
Purple, long (*P_L_*)	284	215
Purple, round (*P_ll*)	21	71
Red, long (*ppL_*)	21	71
Red, round (*ppll*)	55	24
	381	381

Recombinant progeny result from the mixing of maternal and paternal alleles into new combinations in the gamete. **Parental** gametes retain the gene combinations from the parents. "Parental" and "recombinant" are relative terms, however, depending on the parents' allele combinations. Had the parents in Bateson and Punnett's crosses been of genotypes *ppL_* and *P_ll* (phenotypes red flowers, long pollen grains and purple flowers, round pollen grains), then *P_L_* and *ppll* would be the recombinant rather than the parental classes.

Two other terms describe the arrangement of linked genes in heterozygotes. Consider a pea plant with genotype *PpLl*. These alleles are on the chromosomes in two positions. If the two dominant alleles travel on one chromosome and the two recessive alleles on the other, the genes are said to be in coupling. In the opposite configuration, when one dominant allele is near one recessive allele on a chromosome, the two genes are in repulsion (fig. 14.3).

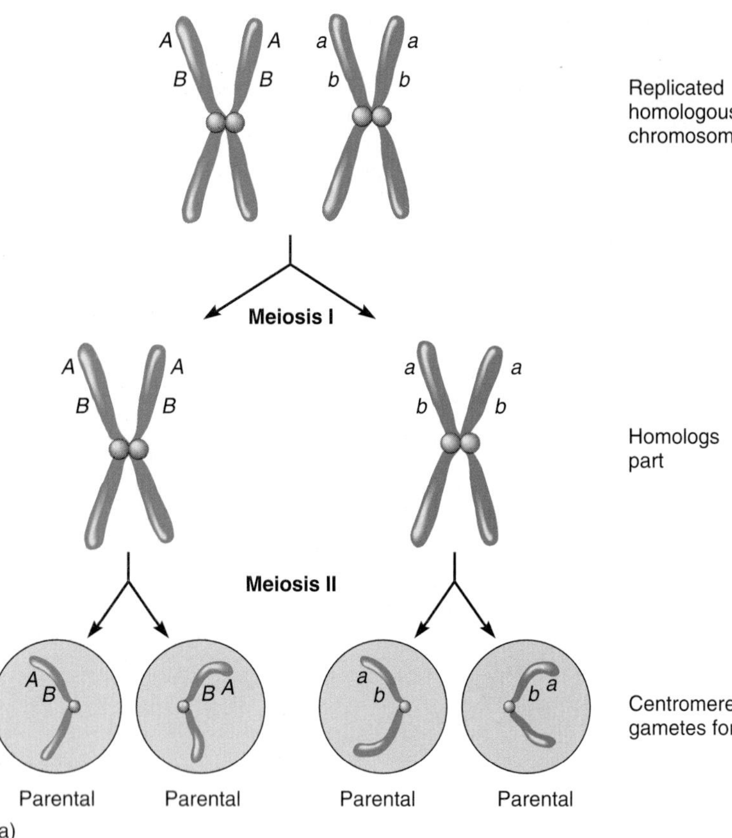

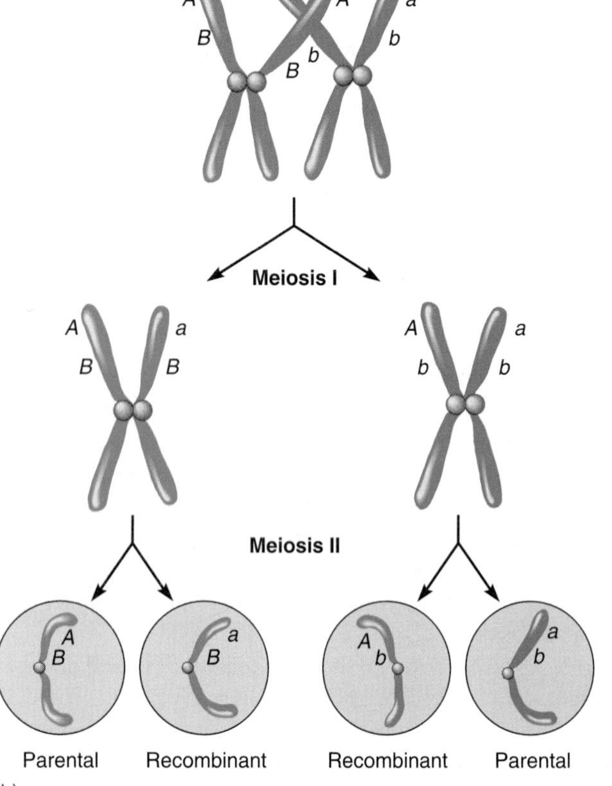

Figure 14.2

Crossing over. (*a*) Genes linked closely on the same chromosome are usually inherited together when that chromosome is packaged into a gamete. (*b*) Linkage between two genes can be interrupted if the chromosome they are located on crosses over with its homolog at a point between the two genes. This packages recombinant arrangements of the genes into gametes. (The centromere is exaggerated here; it is actually a constriction.)

Knowing the allele configuration is important. Alleles in coupling tend to be transmitted together because they reside on the same homolog. However, alleles on different homologs separate at meiosis and may not be passed on together. This is valuable information when tracing traits, because alleles in coupling tend to stick together, whereas alleles in repulsion separate with each generation. We return to medical implications of this distinction shortly.

As Bateson and Punnett were studying linkage in peas, Thomas Hunt Morgan at Columbia University was breeding the fruit fly *Drosophila melanogaster* to determine whether pairs of two traits were linked. As data accumulated, the trait pairs fell into four groups. Within each group, crossed dihybrids did not produce the proportions of offspring Mendel's second law predicts. This indicated four sets of linked genes—precisely the number of chromosome pairs in the fly. The traits fell into four groups because the genes specifying traits that are transmitted together are on the same chromosome.

Morgan wondered why the size of the recombinant classes varied among the crosses. Might the differences reflect different physical relationships between the genes on the chromosome? Alfred Sturtevant, Morgan's undergraduate assistant, explored this idea. In 1911, he developed a theory and technique that would profoundly affect the fledgling field of genetics in his day and the medical genetics of today. Sturtevant proposed that the farther apart two genes are on a chromosome, the more likely a crossover is to occur between them—simply because there is more space between the two genes. This idea became the basis for gene mapping.

Linkage Maps

Geneticists use the correlation between crossover frequency and the distance between genes to construct **linkage maps,** diagrams of gene order and spacing on chromosomes. By determining the proportion of recombinant offspring, investigators can infer the frequency with which two genes cross over and how far apart the genes are on the chromosome. Genes that occupy opposite ends of the same chromosome cross over often, generating a large recombinant class. In contrast, a crossover would rarely separate genes lying very close on the chromosome, and they would not generate a large recombinant class (fig. 14.4). It is a little like the ease of walking between two people standing at opposite ends of a room compared to the ease of walking between two people who are slow dancing.

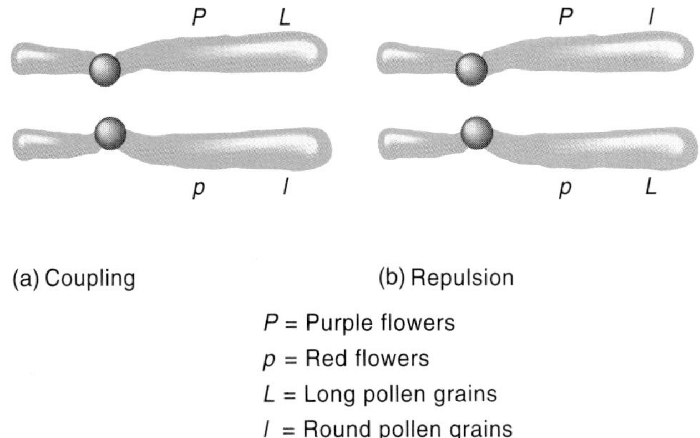

Figure 14.3

Parental type progeny can be distinguished from recombinant progeny only if one knows the allele configuration of the two genes—they are either in coupling *(a)* or in repulsion *(b).*

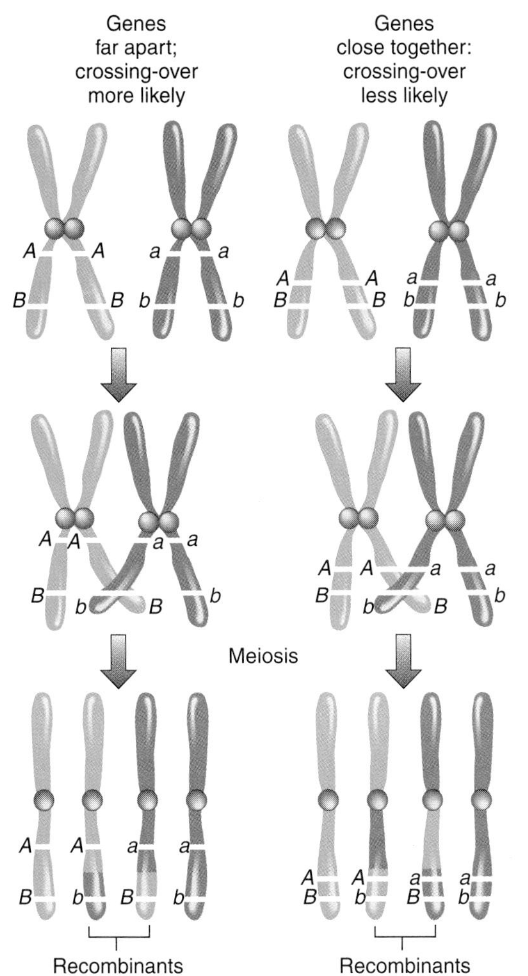

Figure 14.4

Crossing-over is more likely to occur between the widely spaced linked genes on the left than between the more closely spaced linked genes on the right, because there is more room for an exchange to occur.

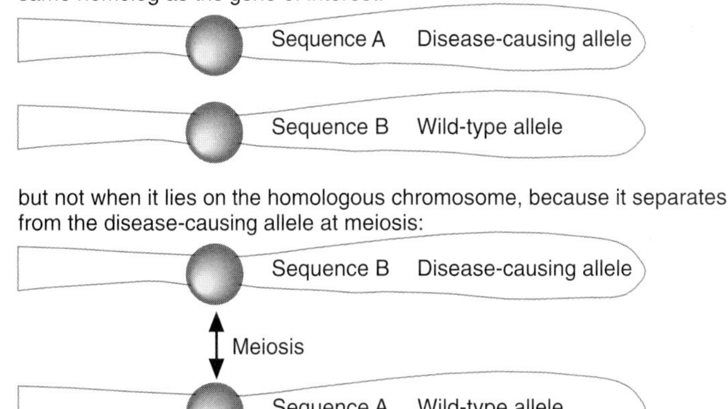

Figure 14.5

Genetic markers are based on meiosis. A genetic marker is a detectable DNA sequence that is closely linked to a known disease-causing gene and is therefore usually inherited with it in some families. Genetic markers make possible "diagnosis" of a genetic disorder before symptoms appear.

In 1913, Sturtevant published the first genetic linkage map, depicting the order of five genes on the X chromosome of the fruit fly. Researchers rapidly mapped genes on all four fly chromosomes. In humans, researchers mapped the X chromosome first. It is simpler to map than the autosomes because human males have only a single X chromosome. This means that X chromosome alleles in the human male are never masked and are therefore easier to study.

By 1950, genetic researchers began to contemplate the daunting task of mapping genes on the 22 human autosomes. The gene mapper needs a clue to begin matching a certain gene to its chromosome. To locate genes on autosomes, helpful clues come from individuals who have particular traits and particular chromosome abnormalities. That is, if people with a certain set of symptoms have an abnormality affecting the same part of a certain chromosome, then the gene causing the disorder is probably located on that chromosome.

Finding disease-chromosome links is rare. More often, researchers used the sorts of experiments Sturtevant conducted on his flies—calculating the percentage of recombination (crossovers) between two genes. Because individual humans do not produce hundreds of offspring, as fruit flies do, cytogeneticists had to devise other ways to detect linkage and recombination in humans. One way is to observe the same traits in many families and pool the data to obtain enough information to establish gene linkages. That is, if two specific traits nearly always occur together in many families, then the genes are probably linked and are usually inherited together, unless crossing-over separates them.

Linkage of specific pairs of genes is the basis of **genetic marker** technology, which can predict that a person will develop symptoms of an inherited illness. A genetic marker is a detectable DNA sequence that all family members who have a particular inherited illness share but that older healthy relatives do not have. The DNA sequence marks the presence of the disease-causing gene because the two are closely linked (fig. 14.5)—similar to the

14.1 biology in action

Sex Determination—Male or Female?

People thought about the origins of gender long before they knew about genes and chromosomes. Some ancient cultures thought the phase of the moon, the direction of the wind, or what the couple wore, ate, or said just prior to conception determined a child's sex. The ancient Greeks thought sex was determined by which testicle the sperm came from. Some European royal families perpetuated this idea—men tied off or removed their left testicles to guarantee conception of a son! Mendel was the first to suggest that sex is an inherited trait.

Sex determination mechanisms are diverse in the living world. Grasshoppers, for example, have only one type of sex chromosome, designated X. If two sex chromosomes are present (XX), the insect is a female; if only one is present (XO), it is a male.

In the fruit fly *Drosophila melanogaster,* the X-to-autosome ratio is important. Individuals with a pair of X chromosomes and a pair of each autosome are normal females; those with one X chromosome and a pair of each autosome are normal males. Although fertile male fruit flies are XY, it is not the Y chromosome that determines maleness but the ratio of the one X chromosome to the paired autosomes. An XO fruit fly is a sterile male (in humans, an XO is female), and an XXY fruit fly is female (in humans, an XXY is male).

Plants exhibit a variety of sex determination mechanisms. These include X-to-autosome ratios, active Y chromosomes, and autosomal genes determining sex.

Sex determination is more complex in some other species. Certain beetles have 12 X chromosomes, 6 Y chromosomes, and 18 autosomes. In other species, the number of total sets of chromosomes determines sex. In bees, a fertilized ovum becomes a diploid female; an unfertilized ovum develops into a haploid male. The queen bee actually determines sex: the oocytes that she fertilizes become females, and those she doesn't become males.

Some whiptail lizards (fig. 14.A) dispense with sex altogether. All individuals are females and lay eggs that hatch without male input. The lizards, however, engage in a behavior that resembles mating, perhaps a vestige from a sexual ancestor. The vignette to chapter 10 and Biology in Action 10.1 discuss other cases of parthenogenesis. Cichlid fish seem to have three sexes—a few bright-colored, sexually aggressive males, females, and inhibited males who do not behave sexually unless an aggressive male dies.

Figure 14.A
Whiptail lizards are parthenogenetic, but they engage in matinglike behavior.

concept of guilt by association. Most "gene hunts" begin with finding a marker, which tells researchers where on a particular chromosome to look for a linked gene of interest. Chapter 16 concludes with a look at the first genetic marker discovered, for Huntington disease.

Each fall, the journal *Science* publishes the latest human chromosome maps. Many of these maps originated as linkage maps. Researchers are supplementing linkage maps with analyses of DNA sequences at an ever-accelerating pace.

Mastering Concepts

1. What is the relationship between genes and chromosomes?
2. How are linked genes inherited differently from nonlinked genes?
3. How can linkage be used to map genes on chromosomes?

Sex-Linkage and Sex Chromosomes

In many species, males and females differ in the types of sex chromosomes. These chromosomes may determine sex because they include genes that direct development of reproductive structures of one sex, while suppressing development of structures of the other sex.

Sex Determination in Humans

In mammals, the sexes have equal numbers of autosomes, but males have one X and one Y chromosome, and females have two X chromosomes. The sex with two different sex chromosomes is the **heterogametic sex,** and the other, with two of the same sex chromosomes, is the **homogametic sex.** In all mammals, the male is the heterogametic sex. In birds, moths, and butterflies, the female is the heterogametic sex. Biology in Action 14.1 explores sex determination mechanisms in different species.

The human X chromosome carries more than 1,000 genes. The Y chromosome is much smaller than the X and carries far

Autosomal genes can influence sex determination. In fruit flies, a "transformer" mutation gives an XX individual a male phenotype. Different alleles can also determine sex. In the bread mold *Neurospora crassa,* the two sexes, called mating types, are specified by different alleles. Two individuals mate only if they have different alleles for this gene. In one species of wasp, a gene with nine alleles controls sex determination. All heterozygotes are female, and all homozygotes male.

In some species, the environment determines sex. For some turtles, the temperature of the land the eggs are laid on determines sex; eggs laid in the sun hatch females, and those laid in the shade produce males (fig. 14.B). In the sea-dwelling worm *Bonellia viridis,* larvae mature into females if no adult females are present. If an adult female is present, she secretes hormones that cause the larvae to develop as males. Some marine bivalves and fish can even change sex in adulthood (fig. 14.C). Compared to some of these species, human sex determination seems simple!

Figure 14.B
For some turtles, the environment determines sex.

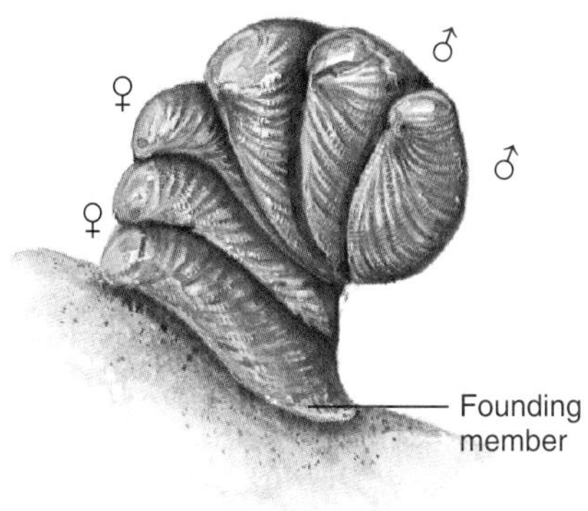

Figure 14.C
In the slipper limpet, *Crepidula fornicata,* position determines sex. These organisms aggregate in groups anchored to rocks and shells on a muddy sea bottom. The founding member is a small male. As he grows, he becomes a she, and a new small male enters at the top. When this second member grows large enough to become female, a third recruit enters. Thus, one's size and position in the pile determines sex.

fewer genes. A quarter of the genes on the Y chromosome correspond to genes on the X chromosome. This is called the pseudoautosomal region, and crossing-over can occur between these corresponding genes on the X and Y chromosomes. As researchers fill in specific genes on the Y chromosome, we will learn about traits that are distinctly male and about how sex evolved.

Researchers identified the part of the Y chromosome that determines maleness by studying interesting people—men who have two X chromosomes and women who have one X and one Y chromosome, the reverse of normal. A close look at these individuals' sex chromosomes revealed that each of the XX males actually had a small piece of a Y chromosome, and each of the XY females lacked a small part of the Y chromosome. The part of the Y chromosome present in the XX males was the same part that was missing in the XY females. It was only a tiny part of the Y chromosome, about half a percent of its total structure. Somewhere within this stretch of 300,000 DNA base pairs, investigators in England discovered the male sex-determining gene SRY (fig. 14.6).

The SRY gene produces a protein that switches on other genes that direct the embryo to develop male structures. Rudimentary testes begin to secrete testosterone, and cascades of other gene activities follow, shrinking breast tissue and promoting swellings that develop into male sex organs. The SRY protein also activates a gene that encodes a protein that destroys rudimentary female structures. "Femaleness" is not a default option if the SRY gene is not present but one of two possible gene-encoded choices in development.

Sex-Linked Recessive Inheritance

Any gene on the X chromosome of a male mammal is expressed in his phenotype, because he lacks a second allele for that gene, which would mask its expression. An allele on an X chromosome in a female may or may not be expressed, depending upon whether it is dominant or recessive and on the nature of the allele on the second X chromosome. The male is **hemizygous** for **sex-linked** traits because he has half the number of genes the female has. He either has the trait or does not—he cannot be a carrier.

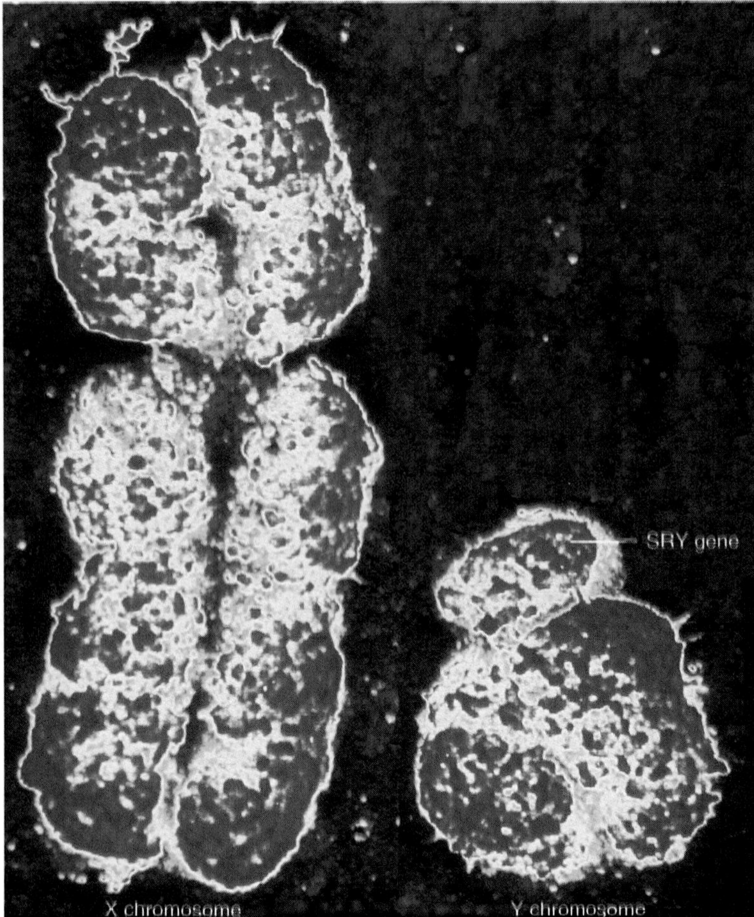

FIGURE 14.6

The human X and Y chromosomes. The SRY gene, at the top of the short arm of the Y chromosome, sets into motion the cascade of gene activity that directs male development.

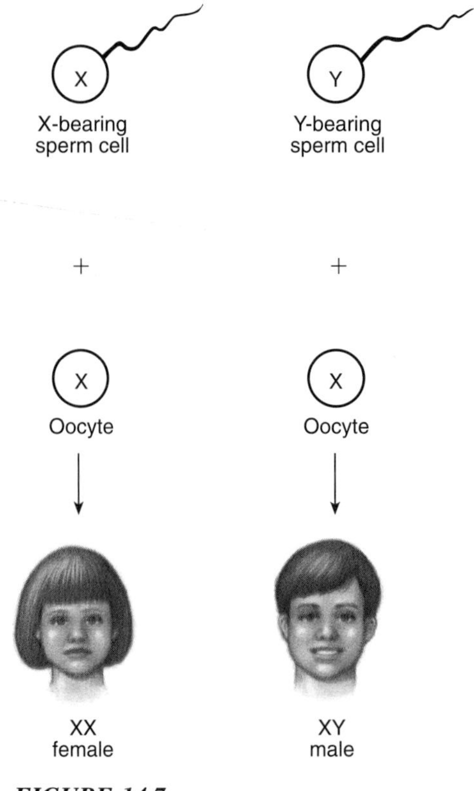

FIGURE 14.7

Sex determination in humans. An oocyte typically contains a single X chromosome. A sperm cell contains either an X chromosome or a Y chromosome. If a Y-bearing sperm cell with an SRY gene fertilizes an oocyte, the zygote is a male (XY). If an X-bearing sperm cell fertilizes an oocyte, then the zygote is a female (XX).

A male always inherits his Y chromosome from his father and his X chromosome from his mother (fig. 14.7). A female inherits one X chromosome from each parent. If a mother is heterozygous for a particular sex-linked gene, her son has a 50% chance of inheriting either allele from her. Sex-linked genes are therefore passed from mother to son. Because a male does not receive an X chromosome from his father, a man cannot pass a sex-linked trait to his son. Punnett squares and pedigrees are used to depict transmission of sex-linked traits, as figure 14.8 shows for the inheritance of hemophilia A in a famous family.

Hemophilia A is a sex-linked recessive disorder in which absence or deficiency of a protein clotting factor greatly slows blood clotting. Cuts are slow to stop bleeding, bruises happen easily, and internal bleeding may occur without the person's awareness. Receiving the missing clotting factor can control the illness.

The risk that a carrier mother will pass hemophilia A to her son is 50%, because he can inherit either her normal allele or the mutant one. The risk that a daughter will inherit the hemophilia allele and be a carrier like her mother is also 50%.

Consider a woman whose brother has hemophilia A but whose parents are healthy. The woman's chance of having inherited the hemophilia allele on the X chromosome she received from her mother and being a carrier is 50%, or one-half. The chance that the woman will conceive a son with hemophilia is one-half multiplied by one-half. This is because the chance that she is a carrier is one-half, and if she is a carrier, the chance that she will transmit the X chromosome bearing the hemophilia allele to a son is also one-half.

Females usually don't exhibit sex-linked traits because the wild-type allele on their second X chromosome masks the mutant allele. A female has the trait only if she inherits a recessive mutant allele from each parent. Unless she has a biochemical test, available for only a few disorders, a woman usually doesn't know she carries a sex-linked recessive trait unless she has an affected son. A sex-linked recessive trait appears more often in females if it isn't serious enough to prevent a man from surviving to father children. Color blindness is one such common sex-linked trait. Table 14.2 lists some sex-linked disorders.

Sex-Linked Dominant Inheritance

How is a dominant, disease-causing allele on the X chromosome expressed in each sex? A female who inherits a dominant allele has the associated trait or illness, but a male who inherits the allele is usually more severely affected. Consider incontinentia pig-

Table 14.2 Some Disease-Related Genes on the Human X Chromosome*

Condition	Description
Eye	
Green color blindness	Abnormal green cone pigments in retina
Ocular albinism	Eye lacks pigment
Red color blindness	Abnormal red cone pigments in retina
Retinitis pigmentosa	Constriction of visual field, night blindness, clumps of pigment in eye
Inborn Errors of Metabolism	
Chronic granulomatous disease	Skin and lung infections, enlarged liver and spleen
Diabetes insipidus	Copious urination
Gout	Inflamed, painful joints
G6PD deficiency and favism	Hemolytic anemia after eating fava beans
Hemophilia A	Absent clotting factor VIII
Hemophilia B	Absent clotting factor IX
Nerves and Muscles	
Hydrocephalus	Excess fluid in brain
Lesch-Nyhan syndrome	Mental retardation, self-mutilation, urinary stones, spastic cerebral palsy
Menkes disease	Kinky hair, abnormal copper transport, brain atrophy
Muscular dystrophy (Becker, Duchenne)	Progressive muscle weakness
Other	
Amelogenesis imperfecta	Abnormal tooth enamel
Hypohidrotic ectodermal dysplasia	Absence of teeth, hair, and sweat glands
Ichthyosis	Scaly skin on scalp, ears, neck, abdomen, limbs
Incontinentia pigmenti	Swirl skin color, hair loss, seizures, abnormal teeth

*Some of these conditions can also be transmitted through genes on the autosomes.

menti. Females have swirls of skin pigment that resemble swirls of paint or marble cake; males are so severely affected that they are never born.

Y Linkage

In Y-linked inheritance, a trait passes from father to son on the Y chromosome. This mode of inheritance is rare because the Y chromosome carries so few genes. A "hairy ear" trait seen in some men in Sri Lanka, India, Israel, and aboriginal Australia is thought to be Y-linked because it never appears in females, but it does appear in fathers and sons.

Mastering Concepts

1. How do the sex chromosomes determine sex in humans?
2. What are some forms of sex determination in different species?
3. How are genes on the X chromosome inherited?

X Inactivation—Equalizing the Sexes

Female mammals have two alleles for every gene on the X chromosome to a male's one. A mechanism called **X inactivation** helps balance this inequality by inactivating one X chromosome in each female cell during early prenatal development.

Whether the turned-off X chromosome comes from the mother or father is random. As a result, a female expresses the paternal X chromosome genes in some cells and the maternal genes in others. For unknown reasons, marsupials, the pouched mammals that include kangaroos, always shut off the X chromosome inherited from the father.

A specific part of the X chromosome, the **X-inactivation center**, shuts off the chromosome one gene at a time, leaving only a few genes active. Once an X chromosome is inactivated in one cell, all the cells that form when that cell divides have the same inactivated X chromosome (fig. 14.9). Because the inactivation occurs early in development, females have patches of tissue that differ in their expression of sex-linked genes. A female is thus a genetic mosaic for any heterozygous genes on the X chromosome because some cells express one allele, and other cells express the other allele.

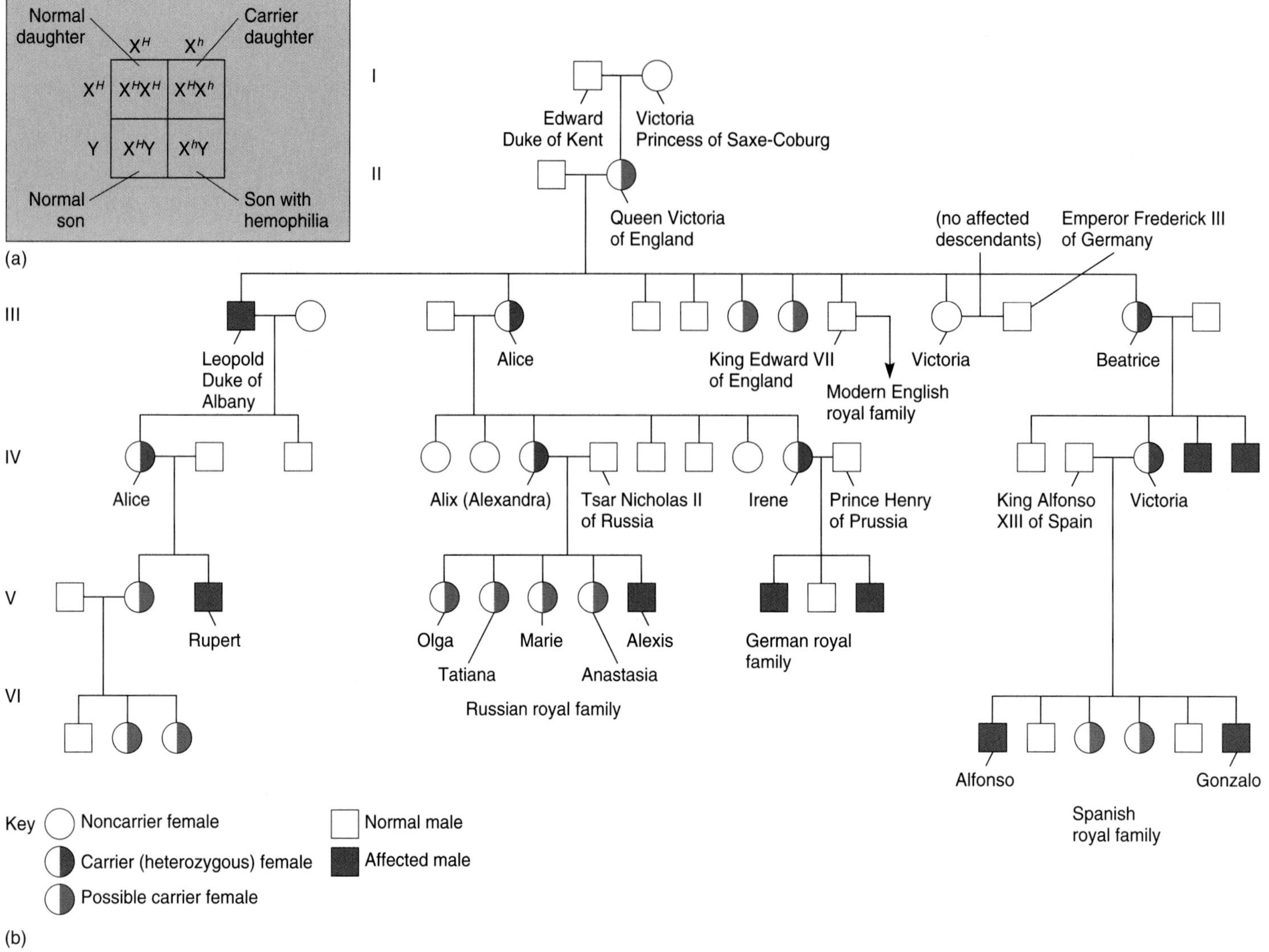

Figure 14.8

Hemophilia A. (a) This sex-linked recessive disease usually passes from a heterozygous woman (designated X^HX^h, where X^h is the hemophilia-causing allele) to heterozygous daughters or hemizygous sons. (*b*) The disorder has appeared in the royal families of England, Germany, Spain, and the former Soviet Union. The mutant allele apparently arose in Queen Victoria, who was either a carrier or produced an oocyte that mutated to carry the allele. She passed the allele to Alice and Beatrice, who were carriers, and to Leopold, who had a case so mild that he fathered children. In the fourth generation, Alexandra was a carrier and married Nicholas II, Tsar of Russia. Alexandra's sister Irene married Prince Henry of Prussia, passing the allele to the German royal family, and Beatrice's descendants passed it to the Spanish royal family. This figure depicts only part of the enormous pedigree. The modern royal family in England does not carry hemophilia.

For most traits, the cells with active wild-type alleles produce enough of the specified protein for the woman to be healthy. Rarely, a female who is heterozygous for a sex-linked gene can have symptoms if, by chance, the wild-type allele is inactivated in many of her cells that the condition affects. A carrier of hemophilia might experience slow blood clotting, and a carrier of muscular dystrophy, muscle weakness. Figure 14.10 explains how X inactivation of either of two alleles of a coat color gene causes the distinctive appearance of calico cats. The earlier X inactivation occurs, the larger the patches.

X inactivation is directly observable in cells, because the turned-off X chromosome absorbs a stain much more readily than an active X chromosome does (fig. 14.11). In the nucleus of a female cell in interphase, the dark-staining, inactivated X chromosome is called a **Barr body.** Murray Barr, a Canadian researcher, first noticed the bodies in 1949 in the nerve cells of female cats. A normal male cell has no Barr bodies because the one X chromosome remains active.

Mastering Concepts

1. How does X inactivation equalize the contributions of sex-linked genes between the sexes?
2. How is a woman a mosaic for sex-linked gene expression?
3. How could a carrier of a sex-linked condition have symptoms?
4. What is the observable evidence of X inactivation in a cell?

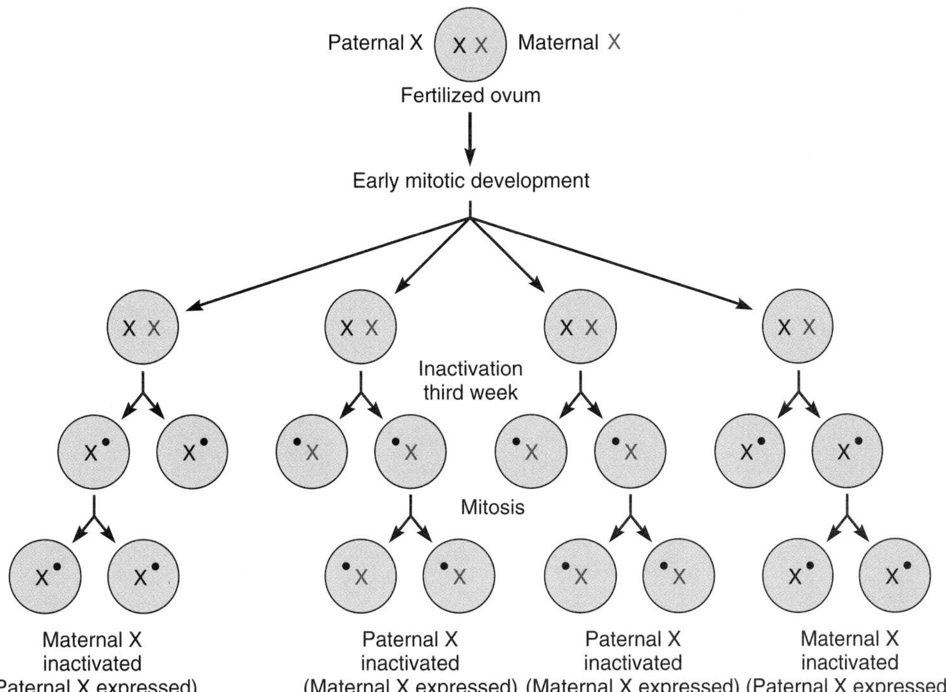

Figure 14.9

At about the third week of embryonic development in the human female, one X chromosome in each diploid cell is inactivated, and all progeny cells of these cells have the same X chromosome turned off. The inactivated X may come from the mother or father, resulting in a female who is mosaic at the cellular level for X chromosome gene expression.

(a)

(b)

Figure 14.10

A striking exhibition of X inactivation occurs in the calico cat. Each orange patch is made up of cells descended from a cell in which the X chromosome carrying the coat color allele for black was inactivated; each black patch is made of cells descended from a cell in which the X chromosome carrying the orange allele was turned off. (*a*) X inactivation happened early in development, resulting in large patches. (*b*) X inactivation occurred later in development, producing smaller patches.

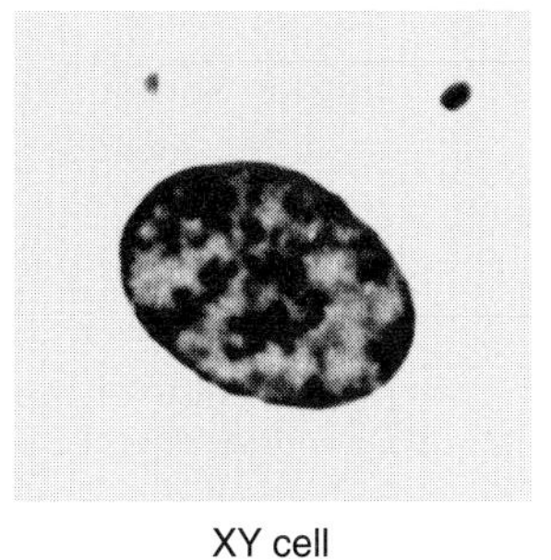

(a) XY cell No Barr bodies

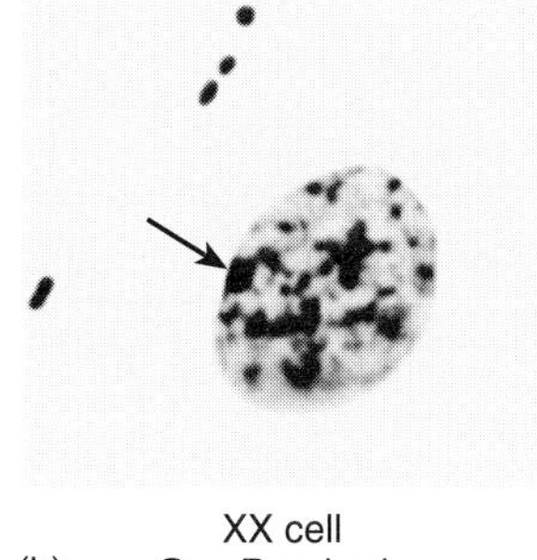

(b) XX cell One Barr body

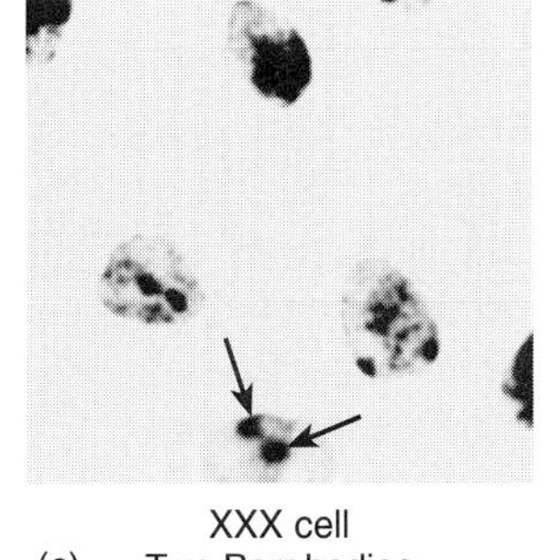

(c) XXX cell Two Barr bodies

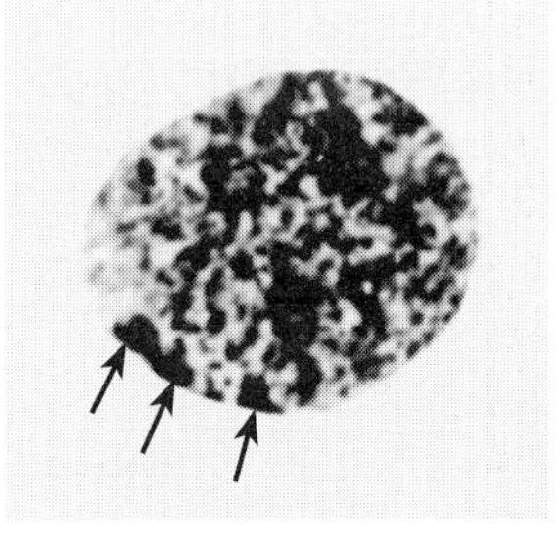

(d) XXXX cell Three Barr bodies

Figure 14.11

A cell of a normal male (*a*) has no Barr bodies, and a normal female cell has one Barr body (*b*). A triplo-X individual (*c*) has two Barr bodies. As the number of extra X chromosomes increases, mental ability declines. The cell in (*d*) has three Barr bodies—the person who donated it is XXXX.

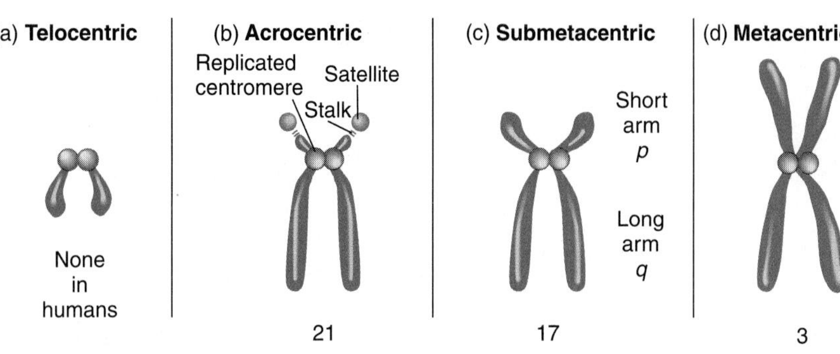

Figure 14.12

Centromere position is used to distinguish chromosomes. (*a*) A telocentric chromosome has the centromere very close to one end. (*b*) An acrocentric chromosome has a centromere near an end. (*c*) A submetacentric chromosome's centromere creates a long arm (*q*) and a short arm (*p*). (*d*) A metacentric chromosome's centromere creates equal-sized arms.

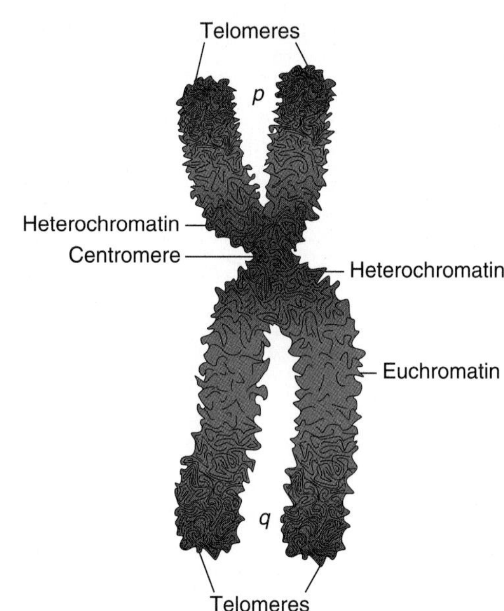

Figure 14.13

Anatomy of a chromosome. Dark-staining chromosomal material (heterochromatin) was once thought to be nonfunctional. Even though heterochromatin does not encode as many proteins as the lighter-staining euchromatin, it is important; it stabilizes the chromosome.

Chromosome Structure

A chromosome is more than a very long molecule of DNA. Several biochemicals emanate from the DNA, including RNA and various proteins that serve as DNA scaffolds, help replicate DNA, and help transcribe RNA. Chromosomes also include proteins specific to certain cell types. The DNA in chromosomes is highly coiled, particularly during mitosis (see fig. 15.10). Chromosomes are distinguished by size, shape, and the pattern in which they take up certain stains and form colored bands.

Size and Shape

Each chromosome has a characteristically located constriction, the **centromere,** which may help orient the chromosome during cell division. Centromere DNA replicates with the rest of a chromosome's DNA.

Centromere position distinguishes chromosomes. A chromosome is **metacentric** if the centromere divides it into two arms of approximately equal length, **submetacentric** if the centromere establishes one long arm and one short arm, **acrocentric** if the centromere pinches off only a small amount of material, and **telocentric** if the centromere appears very close to a chromosome tip (fig. 14.12). (Humans do not have any telocentric chromosomes.) The long arm of a submetacentric or acrocentric chromosome is termed *q,* and the short arm *p.* Some chromosomes also have blob-like ends called **satellites** that extend from a stalklike bridge from the rest of the chromosome.

Stains and Probes

The first stains used to impart color to chromosomes, in the late nineteenth century, were alkaline compounds that bound to the acidic DNA. Although these stains added color, the chromosomes looked alike, particularly those of similar size and shape. More useful are stains that create patterns of bands, which differ among the chromosomes. Most stains do this by binding preferentially to tightly wound DNA with many repetitive sequences, called **heterochromatin.** In contrast is **euchromatin,** which harbors more unique sequences and is looser. Heterochromatin comprises the telomeres (chromosome tips) and centromeres and may help maintain chromosomes' structural integrity. Euchromatin encodes proteins and may be more loosely wound so that its information is accessible (fig. 14.13).

Researchers are increasingly distinguishing chromosomes with a technique called FISH, which stands for fluorescence in situ hybridization. In FISH, a fluorescent dye is attached to a piece of DNA, called a **DNA probe,** that binds to its corresponding site on a particular chromosome. A flash of fluorescence after exposure to a laser indicates the location of the probed gene. Simultaneously applying probes specific to each chromosome yields a beautiful, glowing karyotype. FISH highlights telomeres in the chapter opener photograph.

Evolution of the Karyotype

Early karyotypes were crude and difficult to construct. The method of culturing cells, bursting them, and displaying stained chromosomes on a microscope slide was perfected in the 1950s, but chromosomes were still identified by belonging to a general size group, rather than having a distinctive number.

The 1970s brought methods to strip proteins from chromosomes, which, when combined with staining, produced more intricate banding patterns. Researchers then used scissors and tape to cut chromosomes from photographs to construct karyotypes. In the 1980s, automated karyotyping arrived. A computer attached to a microscope recognizes each chromosome's banding pattern, assembles a chart, and prints it, all in a few minutes. Figure 14.14 chronicles the increasing sophistication of karyotypes.

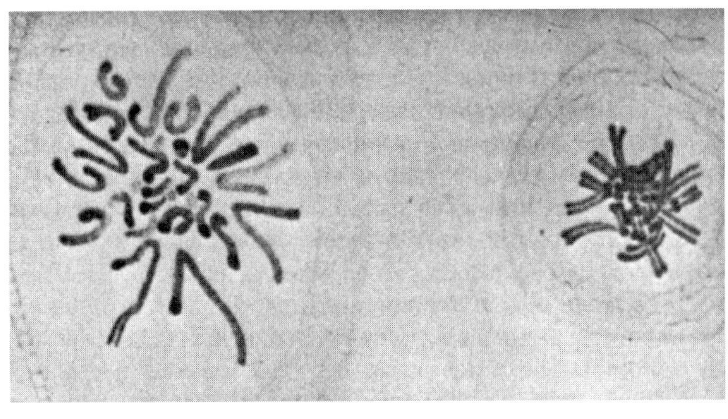

(a)

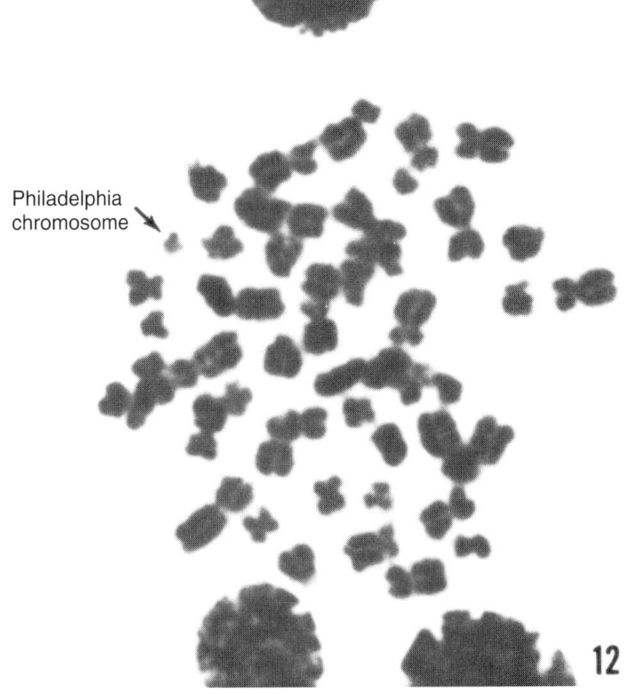

(b)

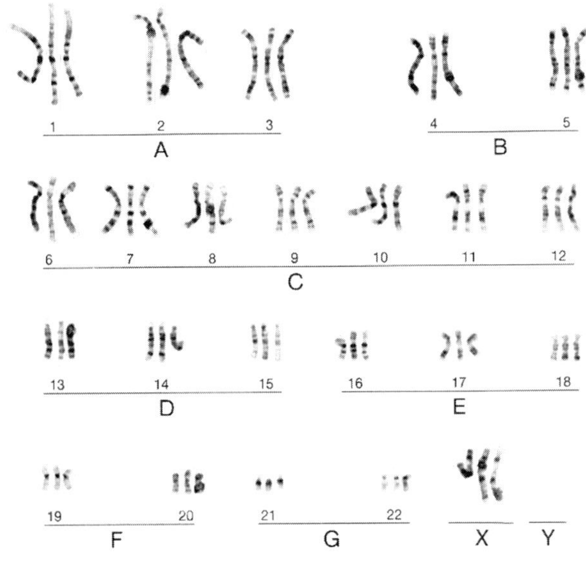

(c)

(d)

Figure 14.14

Evolution of the karyotype. (*a*) The earliest drawings of chromosomes date from 1882, by German biologist Walter Flemming. (*b*) This 1960 chromosome preparation reveals the Philadelphia chromosome, which causes leukemia, but reveals little else. (*c*) Stains create banding patterns in this karyotype. How is this karyotype abnormal? (*d*) This karyotype was constructed using FISH. Specific DNA probes tagged with distinctive combinations of fluorescent dyes enable a computer to identify the chromosomes and align them by size.

Mastering Concepts

1. What are the chemical components of a chromosome?
2. Cite several ways that chromosomes are distinguished from each other.
3. What are structural and functional differences between heterochromatin and euchromatin?
4. Distinguish between banding chromosomes with stains and using DNA probes to tell them apart.
5. How has karyotype construction changed over the years?

Abnormal Chromosome Numbers

An altered chromosome number drastically affects the phenotype because a chromosome consists of so many genes. Chromosomes can be abnormal in a variety of ways. Table 14.3, at the end of the chapter, reviews chromosome abnormalities.

Extra Chromosome Sets—Polyploidy

An error in meiosis can produce a gamete with an extra complete set of chromosomes, a condition called **polyploidy** (many sets). For example, if a sperm with the normal one copy of each chromosome fertilizes an oocyte with two copies of each chromosome,

Table 14.3 Chromosome Abnormalities

Type of Abnormality	Definition
Polyploidy	Extra chromosome sets
Aneuploidy	An extra or missing chromosome
Deletion	Part of a chromosome missing
Duplication	Part of a chromosome present twice
Inversion	Gene sequence reversed
Translocation	Two nonhomologous chromosomes join long arms or exchange parts

the resulting zygote will have three copies of each chromosome, a type of polyploidy called triploidy (see fig. 14.14*c*).

Most human polyploids cease dividing as embryos or fetuses, but occasionally one survives for a few days after birth, with defects in nearly all organs. However, about 30% of flowering plant species tolerate polyploidy well, and many crops are polyploids. The wheat in pasta is tetraploid—it has four sets of seven chromosomes—and the wheat species in bread is a hexaploid—it has six sets of seven chromosomes.

Extra and Missing Chromosomes—Aneuploidy

A **euploid** ("true set") cell has the normal chromosome number for the species. An **aneuploid** ("not true set"), by contrast, has a missing or extra chromosome. Symptoms or altered characteristics resulting from aneuploidy depend upon which chromosome is affected. In humans, autosomal aneuploidy often results in mental retardation, because so many genes contribute to brain function. Sex chromosome aneuploidy in humans is less severe.

Extra genetic material is apparently less dangerous than missing material. This is why most children born with the wrong number of chromosomes have an extra one (called a **trisomy**) rather than a missing one (a **monosomy**). Trisomies and monosomies are named according to the affected chromosome, and the syndromes they cause were originally named for the investigator who first described them.

Aneuploidy occurs due to a meiotic error called **nondisjunction** (fig. 14.15). Recall that in normal meiosis, pairs of homologous chromosomes separate so that each of the resulting sperm or oocyte contains only one member of each pair. In nondisjunction, a chromosome pair fails to separate, either at the first or second meiotic division. The result is a sperm or oocyte with either two copies of a particular chromosome or none at all, rather than the normal one copy. In humans, when such a gamete fuses with its opposite at fertilization, the resulting zygote has either 45 or 47 chromosomes instead of 46, the normal number for a human. Most aneuploids cease developing before birth and account for about 50% of all spontaneous abortions. In plants, aneuploidy may produce a seed (an embryo and its food supply) that does not germinate.

Aneuploidy and polyploidy can also arise during mitosis and produce groups of somatic cells with a chromosomal aberration. Only a few cells with an abnormal chromosome number may not impair health. A normally functioning human liver, for example, may have patches of polyploid cells. If mitotic aneuploidy occurs in a very early embryo, many cells descend from the original defective one, causing more serious problems. Some people who have mild Down syndrome, for example, are really chromosomal mosaics—that is, some of their cells carry the extra chromosome 21, and some are normal.

Following is a look at some syndromes in humans resulting from aneuploidy.

Autosomal Aneuploids

The most common autosomal aneuploid is **trisomy 21,** a form of Down syndrome. An affected individual has distinctive facial features, including a flat face, slanted eyes, straight, sparse hair, a protruding tongue, and thick lips. The individual has an abnormal pattern of hand creases, loose joints, and poor reflex and muscle tone. Children with trisomy 21 reach developmental milestones (such as sitting, standing, and walking) slowly, and toilet training may take several years. Their intelligence varies greatly; some have profound mental impairment, while others can attend college (fig. 14.16).

Prenatal tests can detect trisomy 21, but they cannot predict the severity of the syndrome. Nearly 50% of affected children die before their first birthdays, often of heart or kidney defects or a suppressed immune system that can make a bout of influenza deadly. Digestive system blockages are common and must be corrected surgically shortly after birth. Risk of developing leukemia is high, and people who live past age 40 develop Alzheimer disease.

The likelihood of giving birth to a child with trisomy 21 increases dramatically as a woman ages. For women under 30, the chances of conceiving a child with the syndrome are 1 in 3,000. For a woman of 48, the probability jumps to 1 in 9. The increased likelihood of nondisjunction in older oocytes may account for this age association.

After trisomy 21, trisomies 13 and 18 are the most common autosomal aneuploids. Affected fetuses usually cease developing before birth. An infant with trisomy 13 has an underdeveloped face, extra and fused fingers and toes, heart defects, small adrenal glands, and a cleft lip or palate. An infant with trisomy 18 suffers many of the problems seen in trisomy 13, plus a few distinctive features. These include a peculiar positioning of the fingers and flaps of extra abdominal skin called a "prune belly."

Sex Chromosome Aneuploids

People with the same chromosomal abnormality often share similar physical characteristics, as in trisomy 21. At a 1938 medical conference, a physician, Henry Turner, reported on seven young women, aged 15 to 23, who were short and sexually undeveloped, had folds of skin on the backs of their necks, and had malformed elbows. Other physicians soon began spotting patients with this disorder, named Turner syndrome. Doctors assumed that the odd collection of traits reflected a hormonal insufficiency. They were right, but there was more to the story—a chromosomal imbalance. In 1954, researchers discovered that the cells of Turner syndrome

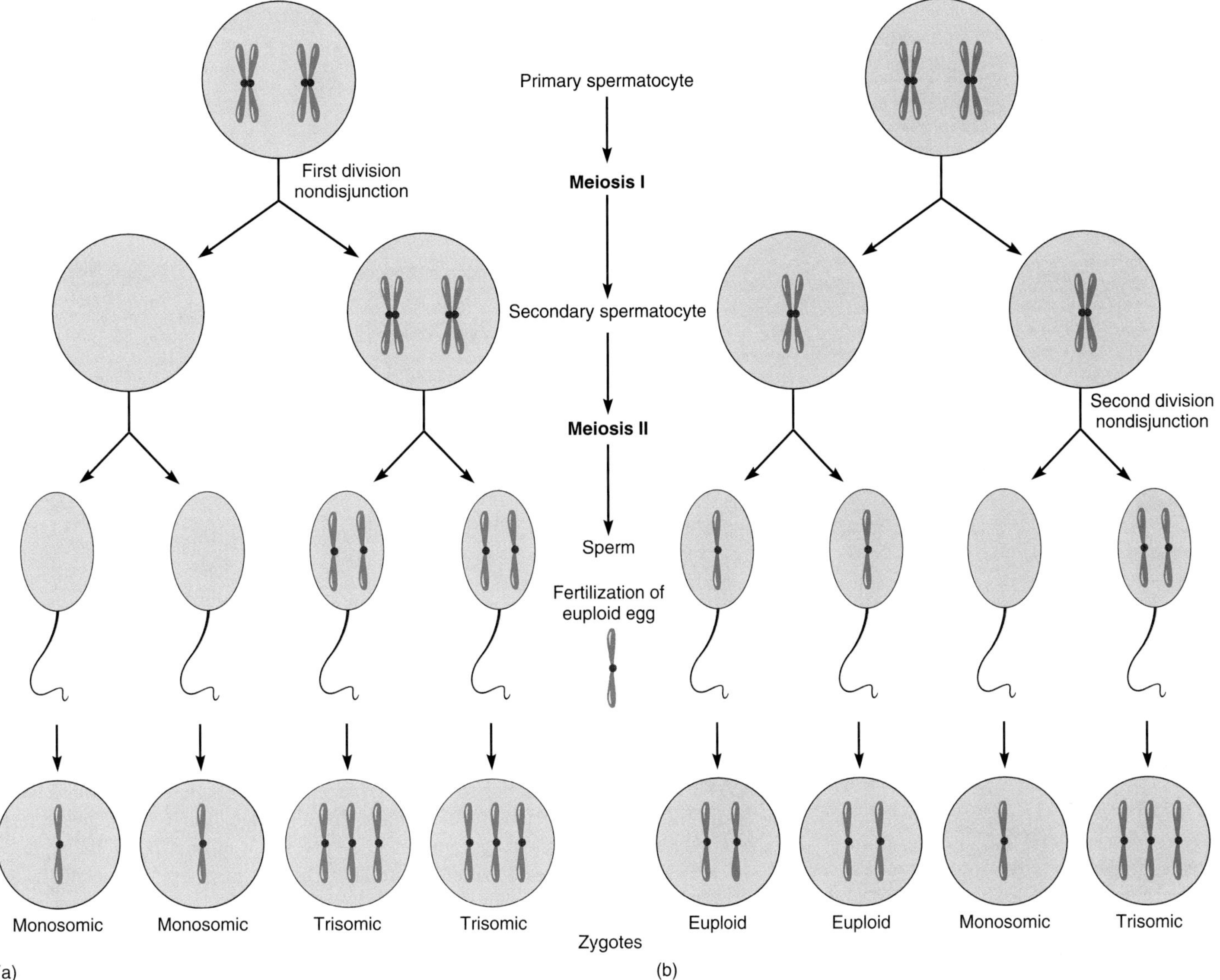

Figure 14.15

Extra and missing chromosomes—aneuploidy. Unequal division of chromosome pairs can occur at the first or second meiotic division. (*a*) A single pair of chromosomes unevenly partitioned into the two cells arising from the first division of meiosis in a male. The result: two sperm cells that have two copies of the chromosome, and two sperm cells that have no copies of that chromosome. When a sperm cell with two copies of the chromosome fertilizes a normal oocyte, the zygote is trisomic for that chromosome; when a sperm cell lacking the chromosome fertilizes a normal oocyte, the zygote is monosomic for that chromosome. Symptoms depend upon which chromosome is involved. (*b*) This nondisjunction occurs at the second meiotic division. Because the two products of the first division are unaffected, two of the mature sperm are normal and two are aneuploid. Oocytes can undergo nondisjunction as well, leading to zygotes with extra or missing chromosomes when normal sperm cells fertilize them.

patients lack Barr bodies and therefore have only one X chromosome. This deficit causes the symptoms, particularly the failure to mature sexually.

More than 99% of XO fetuses die before birth, and the other 1% account for the 1 in 2,000 newborn girls with Turner syndrome. Many affected young women do not know they have a chromosomal abnormality until they lag in sexual development. They are usually of normal intelligence and, if treated with hormone supplements, lead fairly normal lives, but they are infertile. Turner syndrome is the only aneuploid condition that is no more prevalent among offspring of older women.

About 1 in every 1,000 to 2,000 females is born with an extra X chromosome in each cell, a condition called triplo-X. The only symptoms are tallness, menstrual irregularities, and a normal-range IQ that is nevertheless slightly lower than that of other family members. However, a woman with triplo-X may produce some oocytes bearing two X chromosomes, which increases her risk of giving birth to triplo-X daughters or XXY sons. Can you see why?

Males with an extra X chromosome (XXY) have Klinefelter syndrome. They are sexually underdeveloped, with rudimentary testes and prostate glands and no pubic or facial hair. They also have very long arms and legs and large hands and feet, and they may

Figure 14.16

Wendy Weisz has trisomy 21. She enjoys studying art at Cuyahoga Community College.

develop breast tissue. Individuals with Klinefelter syndrome may be slow to learn, but they are usually not mentally retarded unless they have more than two X chromosomes, which is rare. Klinefelter syndrome occurs in 1 out of every 500 to 2,000 male births.

One male in 1,000 has an extra Y chromosome, a condition that once made headlines. In 1965, researcher Patricia Jacobs

Normal sequence of genes:
(a) *a b c d e f* *g h i j k l m n*

Deleted sequence of genes:
(b) *a b c* *g h i j k l m n*

Duplicated sequence of genes:
(c) *a b c d e f d e f* *g h i j k l m n*

Inverted sequence of genes:
(d) *a b c f e d* *g h i j k l m n*

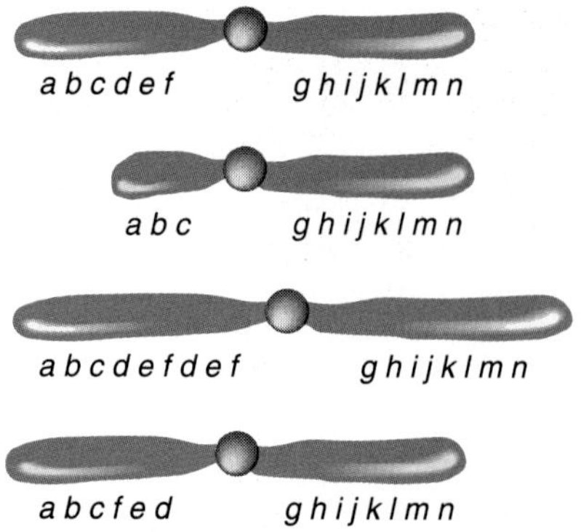

Figure 14.17

If a hypothetical normal gene sequence appears as shown in (*a*), then (*b*) represents a deletion, (*c*) a duplication, and (*d*) an inversion.

published results of a survey among inmates at a high-security mental facility in Scotland. Of 197 men, 12 had chromosome abnormalities, 7 of them an extra Y! Jacobs hypothesized that the extra Y chromosome might cause these men to be violent and aggressive. When other facilities reported similar incidences of XYY men, *Newsweek* magazine ran a cover story on "congenital criminals."

In the early 1970s, hospital nurseries in several nations began screening for XYY boys. Social workers and psychologists offered "anticipatory guidance" to parents in raising their toddling criminals. However, a correlation between two events does not necessarily signify cause and effect. By 1974, geneticists and others halted the program.

Today, we know that 96% of XYY males are apparently normal, the only consistent symptoms being great height, acne, and speech and reading problems. An explanation for the high prevalence of XYY among prison populations may be more psychological than biological. Teachers, employers, parents, and others may expect more of these physically large boys and men than of their peers, and a small percentage of them may cope with this stress by becoming aggressive.

Medical researchers have never reported a sex chromosome constitution of one Y and no X. When a zygote lacks any X chromosome, so much genetic material is missing that it probably cannot sustain more than a few cell divisions.

Mastering Concepts

1. How do polyploidy and aneuploidy differ?
2. What is the meiotic basis of polyploidy and aneuploidy?
3. How do monosomies and trisomies differ?
4. How do syndromes caused by autosomal and sex chromosome aneuploidy differ in severity?

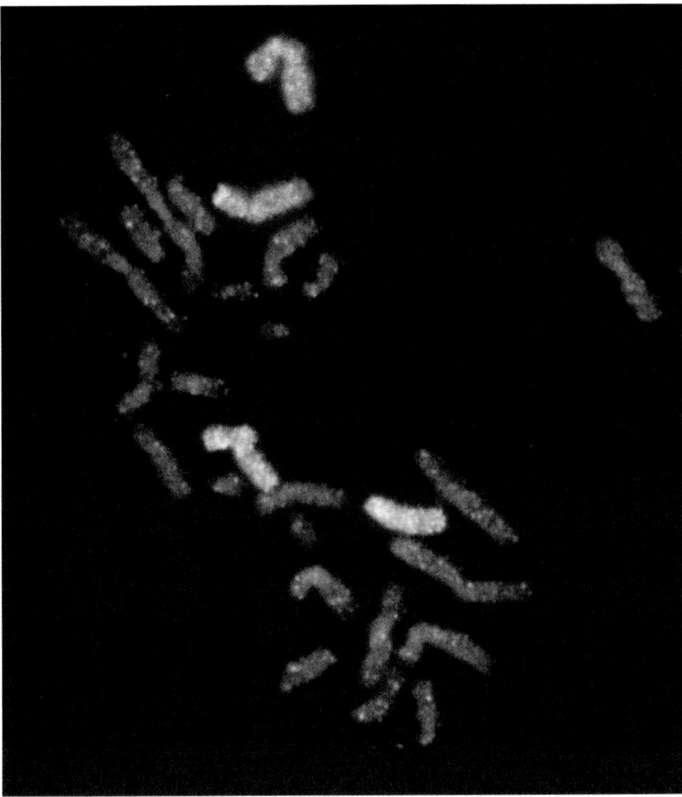

Figure 14.18

FISH (fluorescence in situ hybridization) reveals translocated chromosomes—the ones flashing pink and blue.

Smaller-Scale Chromosome Abnormalities

Because chromosomes consist of so many genes, even small changes—material missing, extra, inverted, or moved—can drastically affect the phenotype. Figure 14.17 illustrates these types of chromosomal changes.

Deletions and Duplications

Sometimes only part of a chromosome is missing (a **deletion**) or extra (a **duplication**). Large deletions and duplications appear as missing or extra bands on stained chromosomes. An example of a disorder caused by a deletion is cri du chat syndrome (French for "cat's cry"), which is associated with deletion of the short arm of chromosome 5. Affected children have an odd cry similar to the mewing of a cat and pinched facial features, and they are mentally retarded and developmentally delayed. The chromosome region responsible for the catlike cry is distinct from the region causing the mental and developmental symptoms, indicating that a deletion causes a syndrome by removing more than one gene.

Duplications, like deletions, are more likely to cause symptoms if they affect large amounts of genetic material. For example, duplications within chromosome 15 are common, but they do not produce a phenotype (seizures and mental retardation) unless they repeat several genes. Small duplications are usually not dangerous, but some "microdeletions" cause severe syndromes.

Chromosome Rearrangements

Deletions and duplications arise from chromosome rearrangements, which force chromosomes to pair in meiosis in ways that delete or duplicate genes.

Inversions

In an **inversion,** part of the chromosome flips and reinserts, rearranging the gene sequence. Studies in humans show that 5 to 10% of inversions harm health, probably because they disrupt vital genes.

An adult who is heterozygous for an inversion in one chromosome can be healthy, because all the genes are present, but have reproductive problems. This is because in meiosis (gamete formation), the inverted chromosome and its noninverted homolog twist around each other in a way that generates chromosomes with missing or extra genes, if crossing-over occurs. Spontaneous abortions or birth defects may be the result.

Translocations

In a **translocation,** different (nonhomologous) chromosomes exchange parts or combine. Exposure to certain viruses (such as the virus that causes the mumps), drugs (anticancer drugs), and radiation (medical X rays or ultraviolet radiation) can cause translocations, but we often do not know how they arise.

There are two types of translocations. In a **Robertsonian translocation,** the short arms of two nonhomologs break, leaving sticky ends that then join the two long arms. This forms a new, large chromosome. The person with the large, translocated chromosome, who is called a translocation carrier, may not have symptoms if a crucial gene has not been deleted or damaged. Even so, he or she may produce sperm or oocytes with too many or too few genes, which can lead to spontaneous abortion or birth defects. One in twenty cases of Down syndrome arises because a person has a Robertsonian translocation between chromosome 21 and another, usually chromosome 14, producing some gametes with extra chromosome 21 material.

In a **reciprocal translocation,** two nonhomologs exchange parts (fig. 14.18). If this action does not break any genes, then a person who has both translocated chromosomes is healthy and is also a translocation carrier. He or she has the normal amount of genetic material, but it is rearranged. A translocation carrier produces some gametes that have duplications or deletions of genes in the translocated chromosomes. This occurs if the sperm or oocyte receives one reciprocally translocated chromosome but not the other, causing a genetic imbalance. The resulting phenotype depends upon the particular genes that the chromosomal rearrangement disrupts.

A chromosomal aberration in a class by itself causes fragile X syndrome, a form of mental retardation affecting 1 in 1,500 males and 1 in 2,500 females. When X chromosomes from affected people are cultured in media low in folic acid, an X chromosome tip dangles, making it prone to breakage (fig. 14.19*a*). A person with fragile X syndrome characteristically has protruding ears and a long jaw (fig. 14.19*b*). Fragile X syndrome is caused by a gene that expands with each generation and is discussed further in chapter 16.

14.1 bioethics connections

Using Human Genome Information

Deciphering the human genome—the complete set of genetic instructions in a human somatic cell—is often compared to deciphering one thousand 1,000-page telephone directories. You wouldn't look up an item in a phone book by reading the entire document from the beginning but rather would search for key words located near the entry of interest. Likewise, the Human Genome Project has proceeded in a series of steps to make the huge amount of information more accessible.

Current gene mapping technologies evolved from linkage maps and from following cytogenetic abnormalities to localize genes. Today's gene searches still often begin with these classical tools to localize genes and then use more molecular-level techniques to obtain small pieces of DNA containing genes of interest (fig. 14.D). Researchers look at the accumulating information in several ways, including the anatomical depiction in figure 14.E.

The Human Genome Project will beat its original deadline of 2005, and it has already yielded many new diagnostic tests based on gene discoveries. The success of the project also raises ethical questions, mostly centering on access to a person's genetic information. Because of the potential misuse of genetic information, 5% of the project's annual budget is earmarked to explore ethical, legal, and social implications of the results.

Perhaps the most compelling practical questions to arise from the project concern the "healthy ill"—people who are healthy but have genes that are likely to one day make them sick. Will they experience discrimination in the workplace or in obtaining health insurance? Concern about stigmatization following revelation of one's genotype has already frightened people away from taking certain genetic tests. For example, many women requesting tests to detect breast cancer genes pay for it themselves and request that

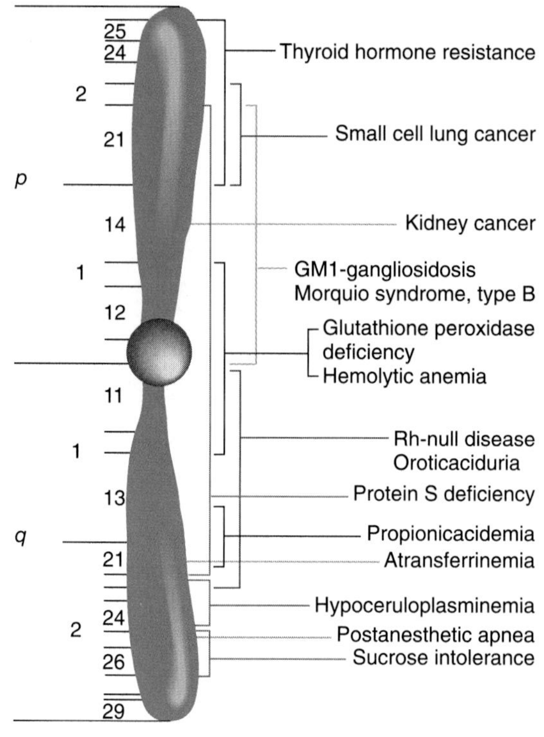

Figure 14.D
As the Human Genome Project proceeds, maps of individual chromosomes are becoming crowded with new information on gene locations. This illustration depicts only a few health-related genes on chromosome 3.

Mastering Concepts

1. How can deletions, duplications, inversions, and translocations cause symptoms?
2. Why do inversions and translocations result in reproductive problems?
3. Distinguish between a Robertsonian translocation and a reciprocal translocation.
4. What is fragile X syndrome?

Chromosomes and Health

Looking at chromosomes can be valuable at several levels. A chromosome check can distinguish between the two types of Down syndrome and tell a family if the condition is likely to recur (the rare translocation form) or not (trisomy 21). Recognizing an abnormal chromosome can save lives. In one New England family, several adults died from a very rare form of kidney cancer. Karyotypes showed a translocation between chromosomes 3 and 8 in every relative who had cancer. When two healthy young family members proved to have the translocation, physicians examined their kidneys. They both exhibited very early stages of the cancer! Subsequent surgeries prolonged their lives.

As the Human Genome Project nears completion, chromosome charts are rapidly filling in with information about specific genes. At the same time, many people are questioning how the information can be wisely used. Bioethics Connections 14.1 considers some of the implications of using genetic information.

Figure 14.E
One way to look at new information about genes is to consider the body parts they affect. The numbers in this illustration indicate the number of genes known to affect the particular body parts.

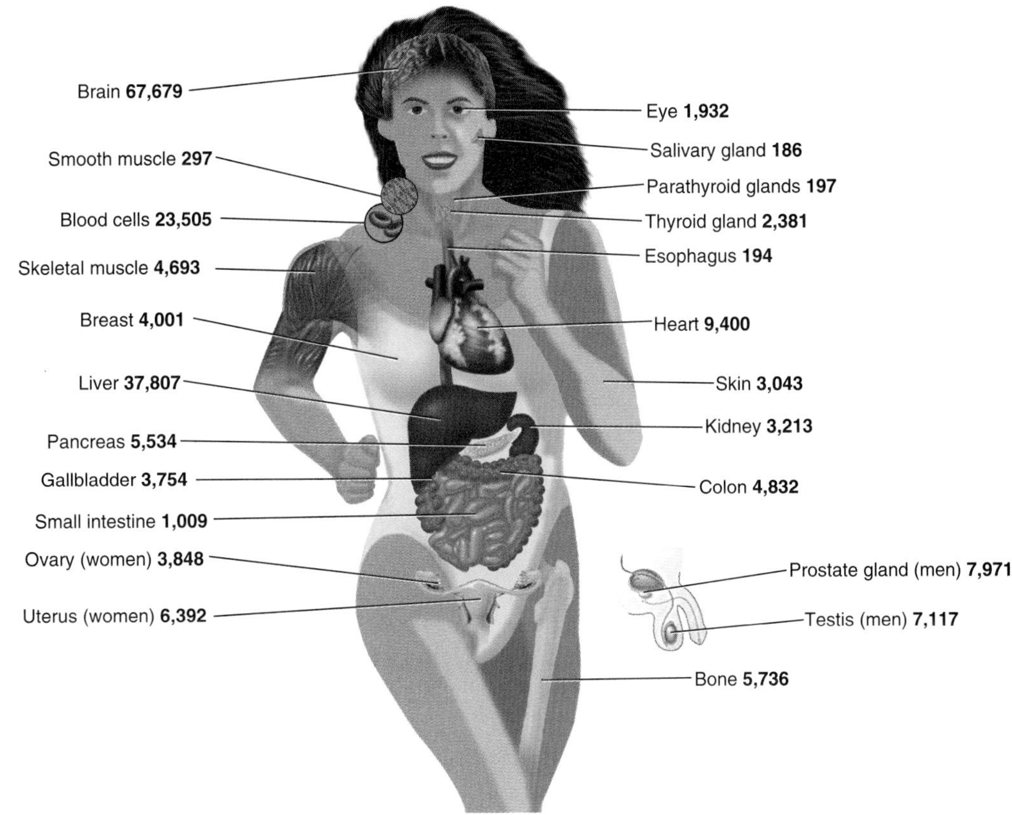

their insurers not see the test results. Yet, from another viewpoint, should an airline pilot or neurosurgeon who one day will inherit a disorder that makes the hands shake be compelled to leave that job early?

The problem of genetic discrimination may eventually solve itself. As our genes are sequenced, all of us will be found to harbor a disease-causing or disease-susceptibility allele—or two or even three. The insurance industry will have to alter the way it assigns risks to embrace knowledge of peoples' genotypes. It will be interesting to witness the impact of genome information on our lives in the coming century.

SUMMARY

1. Each species has a characteristic number of chromosomes. **Cytogenetics** links abnormal chromosomes to phenotypes. Chromosome charts, or **karyotypes,** provide information on genetic disease, exposure to environmental toxins, and evolutionary relationships between species.

2. Genes on the same chromosome are **linked;** rather than demonstrating independent assortment, they produce a large number of **parental** genotypes and a small number of **recombinant** genotypes.

3. Genotype predictions and **linkage maps** are derived from knowing allele configurations (coupling or repulsion) and crossover frequencies, which are directly proportional to distances between genes. The first linkage maps depicted genes on the X chromosome, which are easier to follow because they are present in only one copy in the **heterogametic sex.**

4. In humans, the male is **hemizygous** for genes on the Y chromosome. The female, with two X chromosomes, is **homogametic.** The SRY gene on the Y chromosome controls other genes that stimulate development of male structures and suppress development of female structures.

5. A **sex-linked** trait passes from mother to son because the male inherits his X chromosome from his mother and his Y chromosome from his father. A sex-linked allele may be dominant or recessive. Y-linked traits are very rare.

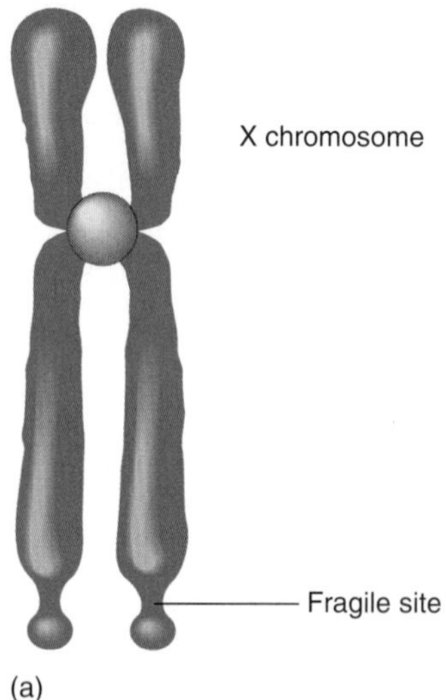

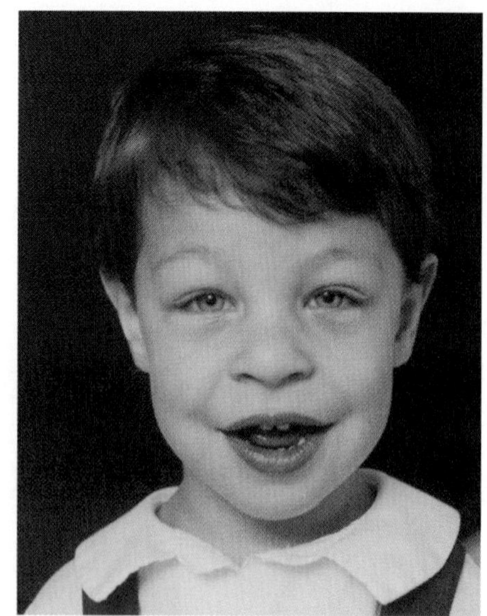

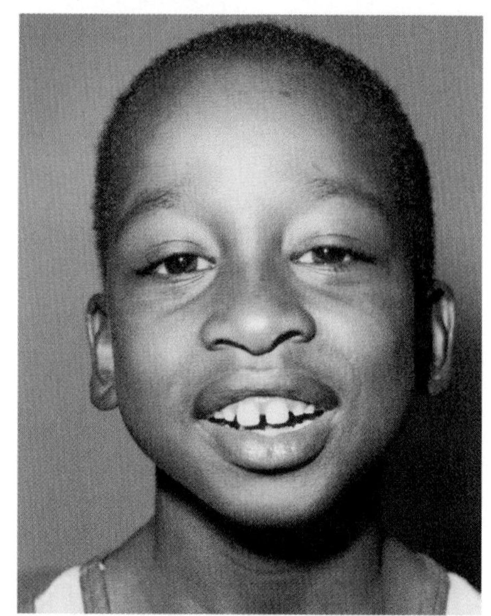

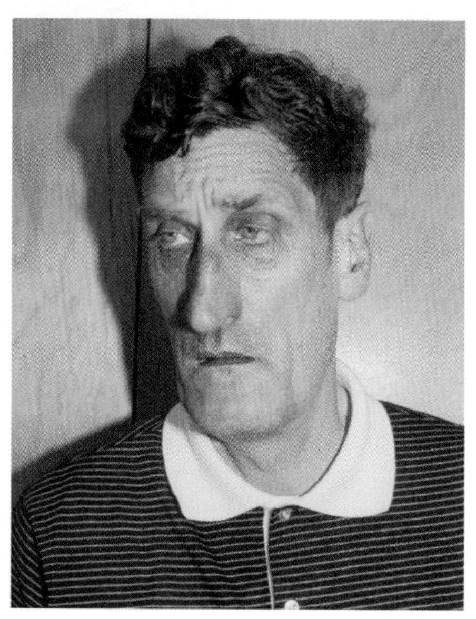

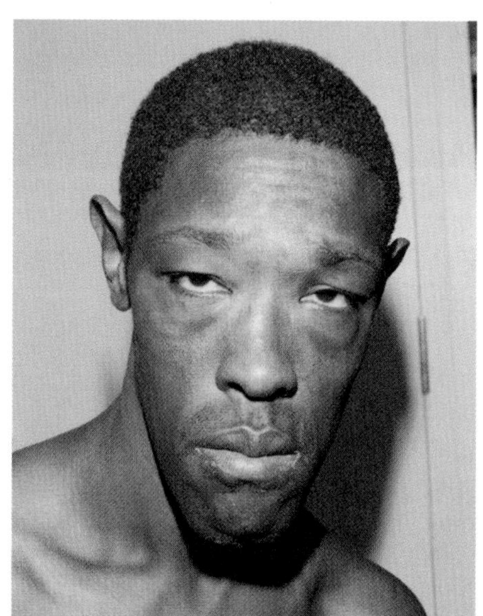

(b)

Figure 14.19

Fragile X-syndrome. (*a*) A constriction appears in the X chromosome of individuals who have symptoms of fragile X syndrome when their cells are grown in a medium that lacks thymidine and folic acid. (*b*) The characteristic facial structure and features of fragile X syndrome sufferers become more pronounced with age.

The Philadelphia Story~

continued from page 286.

Nowell and Hungerford studied many cells from G. H. and E. K. and found that each had an unusual "minute chromosome," which they thought at first to be a tiny Y chromosome. It wasn't until 1972 that researchers discovered that this "Philadelphia chromosome" was actually a combination of two chromosomes (fig. 14.14*b*). Researchers discovered in 1983 that it is the breaking and reattaching of these two chromosomes that causes cancer.

Back in the 1960s, though, the idea that an unusual chromosome caused leukemia was novel and not at first accepted. But the chromosome was found again and again in people with this type of leukemia, and the link grew.

Formation of the Philadelphia chromosome juxtaposes two genes that acquire a new function, turning the cell cancerous. The majority of patients who have the Philadelphia chromosome generally live from 3 1/2 to 4 years before entering a "blast crisis," when the immature white blood cells rapidly overtake other blood cell types. Patients survive in crisis only a few months. Those few patients with CML and normal chromosomes live only a year from the time of diagnosis.

For many types of cancer, identifying specific chromosomal abnormalities provides valuable clues not only to prognosis, as in the case of CML, but to what types of treatments are most likely to work. Identifying chromosome changes may also ultimately reveal the cause of the illness—powerful information for finding a cure.

6. X inactivation shuts off one X chromosome in the cells of female mammals, equalizing the number of active sex-linked genes in each sex. Early in development, each female cell inactivates one X chromosome. A female is mosaic for sex-linked heterozygous genes on the X chromosome because the X chromosomes are inactivated at random with respect to their origin.

7. A chromosome consists of DNA, RNA, and proteins. Chromosomes are distinguishable by size, **centromere** position, **satellites,** staining patterns, and **DNA probes.** Dark-staining DNA, called **heterochromatin,** maintains the chromosome's structure. It surrounds the centromere, comprises the telomeres, and is interspersed with light-staining, protein-encoding **euchromatin.**

8. Chromosomal abnormalities may affect number of chromosomes, missing **(deletion)** or extra **(duplication)** genetic material, and rearrangements. **Polyploid** cells have extra chromosome sets, and **aneuploids** have extra or missing individual chromosomes. A **trisomy** (one extra chromosome) is less harmful than a **monosomy** (one absent chromosome). Sex chromosome aneuploidy is less severe than autosomal aneuploidy. **Nondisjunction,** an uneven division of chromosomes in meiosis, causes aneuploidy.

9. Chromosomal rearrangements disrupt meiotic pairing, which can delete or duplicate genes. An inversion flips gene order, affecting the phenotype if it disrupts a vital gene. A **Robertsonian translocation** fuses the long arms of two nonhomologs. In a **reciprocal translocation** two nonhomologs exchange parts. Translocation carriers have reproductive problems because they produce some unbalanced gametes as a result of the unequal division of chromosome regions during meiosis. Fragile X syndrome causes mental retardation and is associated with a fragile site on the X chromosome.

TO REVIEW . . .

1. How are linked genes inherited differently than genes that are located on different chromosomes?

2. How are sex-linked genes inherited differently in male and female humans?

3. How is sex determination different in flies than it is in humans?

4. A normal-sighted woman with a normal-sighted mother and a color blind father marries a color blind man. What are the chances that their son will be color blind? their daughter?

5. Why is X inactivation necessary?

6. Explain how a woman who is a carrier of a sex-linked disorder can experience symptoms of the disorder.

7. Human chromosome 3 is metacentric, and chromosome 22 is acrocentric. What are two ways that these two chromosomes differ?

8. Why is FISH a more precise way of identifying chromosomes than stains?

9. How is this karyotype abnormal?

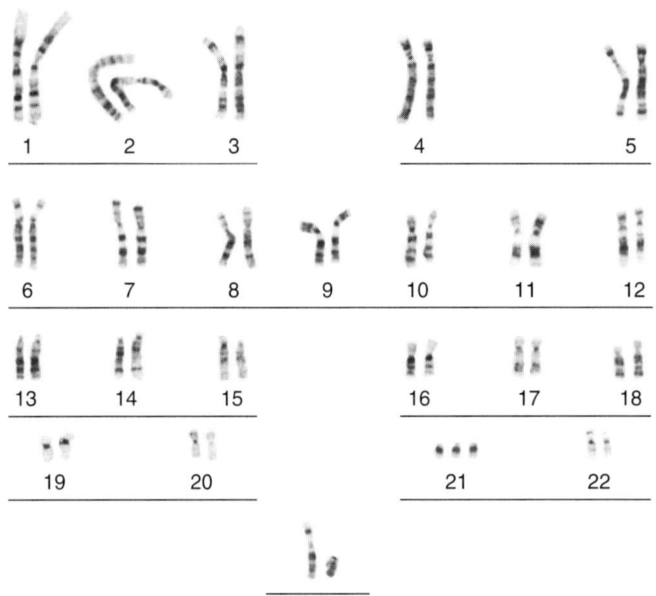

10. Distinguish between the following:

- **a.** heterochromatin and euchromatin
- **b.** a centromere and a telomere
- **c.** an aneuploid and a polyploid
- **d.** a monosomy and a trisomy
- **e.** a Robertsonian translocation and a reciprocal translocation

11. Describe an individual with each of the following chromosome constitutions. Mention the person's sex and possible phenotype.

- **a.** 47, XXX
- **b.** 45, XO
- **c.** 47, XX, trisomy 21
- **d.** 47, XXY

TO THINK ABOUT . . .

1. Why doesn't the inheritance pattern of linked genes disprove Mendel's laws?

2. In a new technology called sperm typing, researchers determine recombination frequencies between gene pairs in sperm cells, using DNA probes to detect specific alleles.

- **a.** What information would researchers need to know to determine whether a certain allele configuration in sperm is parental or recombinant?
- **b.** What are two advantages of working with sperm to determine recombination frequencies?

3. The tribble is a cute, furry creature ideal for embryonic experimentation. Cells from tribble embryos genetically destined to become brightly colored adults can be transplanted into tribble embryos that would otherwise develop into albino (pure white) adults. Investigators can follow the numbers and positions of cells descended from the original transplants by observing the pattern of colored patches in the adult tribble.

A gene in the tribble confers coat color. The brown allele *B* is dominant to the orange allele *b*. A single cell from a *Bb* female embryo at the 8-celled stage is transplanted into an albino embryo. The resulting adult tribble has both brown and orange patches on an albino background. But when a single cell from a *Bb* embryo at the 64-celled stage is transplanted into an albino embryo, only orange patches, or only brown patches, appear in the adult.

a. State the genetic principle demonstrated.

b. What is the difference between the two experiments?

4. A fetus dies in the uterus. Several of its cells are examined for chromosomal content. Approximately 75% of the cells are diploid, and 25% are tetraploid. What has happened, and when in development did it probably occur?

5. Why are there no male calico cats?

6. Which chromosomal anomaly might you expect to find more frequently among the members of the National Basketball Association than in the general population? Cite a reason for your answer.

7. For an exercise in a college genetics laboratory course, a healthy student takes a drop of her blood, separates out the white cells, stains her chromosomes, and constructs her own chromosome chart. She finds only one chromosome 3 and one chromosome 21, plus two unusual chromosomes that do not seem to have matching partners.

a. What type of chromosomal abnormality does she have?

b. Why doesn't she have any symptoms?

c. Would you expect any of her relatives to have any particular medical problems? If so, which ones?

8. A fetus has an inverted chromosome. What information might reveal whether the expected child will have health problems stemming from the inversion?

9. Patricia Jacobs concluded in her 1965 *Nature* article on XYY syndrome, "It is not yet clear whether the increased frequency of XYY males found in this institution is related to aggressive behavior or to their mental deficiency or to a combination of these factors." Why is this wording unscientific? What harm could it (or did it) do?

10. Stella Walsh won a gold medal in the 100-meter dash in the 1932 Olympics. In 1980, she was killed in a robbery. An autopsy revealed ambiguous genitalia and her karyotype was XY. Another supposedly female athlete, Ewa Klobukowska, who set a 1965 world record for the 100-meter dash, was also found to be XY. What was probably unusual about their Y chromosomes?

TO LEARN MORE . . .

References and Resources

Davies, Kevin. June 1994. The evolution of genetic maps. *Nature Genetics,* vol. 7. What began as an intellectual exercise is now part of medical genetics.

Gersh, M., et al. June 1995. Evidence for a distinct region causing a cat-like cry in patients with 5p deletions. *Am J. Hum. Genet.,* vol. 56. On chromosome 5, deletion of specific genes causes specific symptoms.

Gillis, Anna Maria. March 1994. Turning off the X chromosome. *BioScience.* Researchers are working out the molecular details of X inactivation.

Gould, Laura. 1996. *Cats are not peas: a calico history of genetics.* Copernicus Press, New York. Why are calico cats always female?

Gorman, Christine. January 20, 1993. Sizing up the sexes. *Time.* One's sex depends upon the species and its sex determination mechanism.

Lapham, E. V., et al. October 25, 1996. Genetic discrimination: Perspectives of consumers. *Science,* vol. 274. How should we use genetic information?

Lewis, Ricki. June 1996. Chromosome charting takes a giant step. *Photonics Spectra.* FISH yields spectacular karyotypes! But conventional banding remains very useful.

Lewis, Ricki. May 15, 1996. FISH technology ready to leap from lab to the clinic. *Genetic Engineering News.* Fluorescent dyes tagged to DNA pieces paint beautiful portraits of chromosomes.

Lewis, Ricki. March/April 1996. Chromosome tips in the spotlight. *Biophotonics International.* Telomeres control aging.

Lewis, Ricki. September/October 1995. Zeroing in on chromosomes. *Biophotonics International.* DNA probes help gene mapping.

Lewis, Ricki. January 1991. Genetic imprecision. *BioScience.* Checking fetal chromosomes provides a great deal of information—but not all abnormalities detected signify a health problem.

Masood, Ehsan. February 1, 1996. Gene tests: Who benefits from risk? *Nature,* vol. 379. How will insurers handle genetic information?

Suarez, Brian K., and Carol L. Hampe. March 1994. Linkage and association. *The American Journal of Human Genetics,* vol. 54. A short history of linkage analysis.

denotes articles by author Ricki Lewis that may be accessed via *Life's* home page /http://www.mhhe.com/sciencemath/biology/life/.

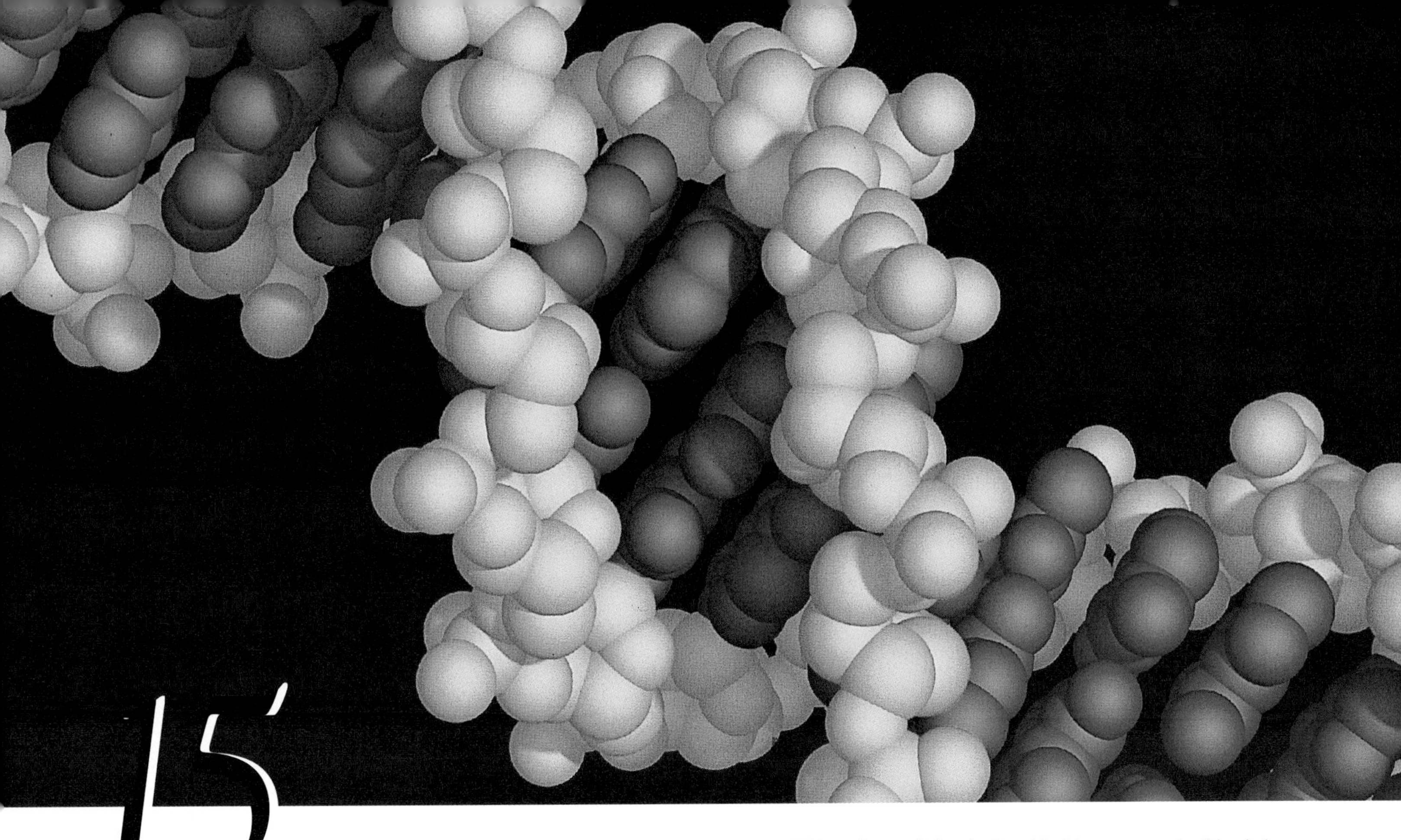

When DNA replicates, it "unzips" and builds a new nucleotide chain opposite each parental strand.

15 DNA Structure and Replication

chapter outline

A Multifunctional Molecule 308

Identifying and Describing the Genetic Material 308

DNA Transmits Traits

The Hereditary Material—DNA, Not Protein

Deciphering the Structure of DNA

Gene and Protein—An Important Partnership 312

DNA Structure Encodes Information

DNA Is Highly Coiled

DNA Replication—Maintaining Genetic Information 313

Semiconservative Replication

Steps and Participants in DNA Replication

DNA Repair 319

DNA Solves a Royal Mystery

One night in July 1918, Tsar Nicholas II of Russia and his family met gruesome deaths at the hands of Bolsheviks in a mountain town in Siberia called Ekaterinburg. Captors led the tsar, Tsarina Alexandra, their four daughters and one son, the family physician, and three servants to a cellar and shot and bayoneted them. The executioners then stripped the bodies and loaded them onto a truck. They planned to hurl them down a mine shaft, but the truck broke down. Instead, the killers placed the bodies in a shallow grave and poured sulfuric acid over them so they could not be identified.

In July 1991, two amateur historians found the grave and alerted the government that they might have unearthed the long-sought bodies of the Romanov family. Forensic researchers soon determined that the 1,000 pieces of bone at the scene came from nine individuals. The sizes of the skeletons indicated that three were children, and the porcelain, platinum, and gold in some of the teeth suggested that they were royalty. The bullets, bayonets, and acid had so destroyed the remains that some conventional forensic tests were not feasible. But one very valuable type of evidence survived—DNA.

Detecting Y chromosomes distinguished male from female skeletons. Mitochondrial DNA sequences revealed the mother and children, because mitochondria pass from mothers to children. But the researchers had to connect the skeletons with living royals. The tsarina's DNA matched that of Prince Philip.

The next step was to prove that a male skeleton with metallic teeth was the tsar. This proved more challenging, because of a surprise in his DNA.

to be continued . . .

A Multifunctional Molecule

Today, DNA is the subject matter of movies and headlines. Fictional dinosaurs are reconstructed from DNA preserved in an ancient mosquito's gut. Real-life trials hinge on DNA evidence, cloned sheep raise questions about the role of DNA in determining who we are, and we consider taking new genetic tests that will reveal our medical futures.

Far more important than DNA's role in society today is its role in life itself. Of all the characteristics that distinguish the living from the nonliving, the one most important to the continuance of life is the ability to reproduce. At the cellular level, reproduction duplicates a cell—be it a simple single-celled organism on the earth billions of years ago or a cell lining a person's intestine today. At the molecular level, reproduction depends upon a biochemical that has dual abilities: to direct the specific activities of that cell and to manufacture an exact replica of itself so that the instructions are perpetuated.

DNA is the multifunctional molecule that replicates as well as orchestrates cellular activities by controlling protein synthesis (fig. 15.1). But the recognition of DNA's vital role in life was a long time in coming.

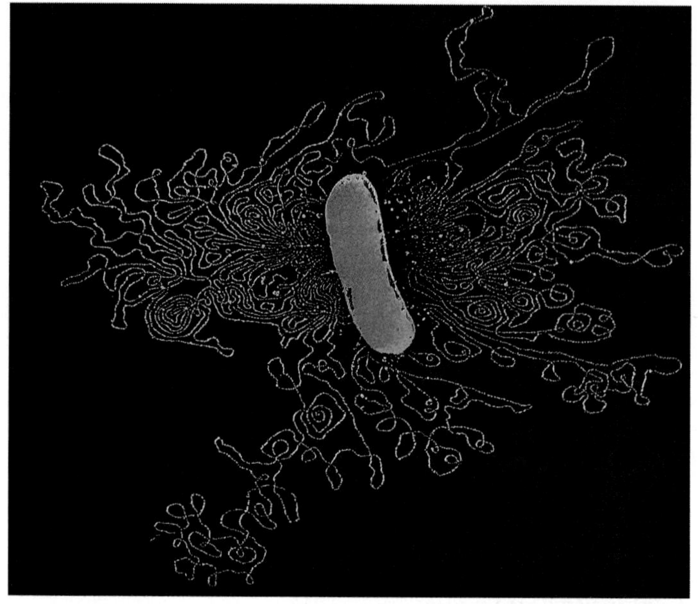

FIGURE 15.1

DNA bursts forth from this bacterial cell.

Identifying and Describing the Genetic Material

Swiss physician and biochemist Friedrich Miescher was the first investigator to chemically analyze the contents of a cell's nucleus. In 1869, he isolated the nuclei of white blood cells obtained from pus in soiled bandages. In the nuclei, he discovered an unusual acidic substance containing nitrogen and phosphorus. Miescher and others went on to find it in cells from a variety of sources. Because the material resided in cell nuclei, Miescher called it nuclein in his 1871 paper; subsequently it was called a nucleic acid.

Miescher's discovery, like those of his contemporary Gregor Mendel, was not appreciated for some years. Although investigators were researching inheritance, their work focused for several decades on the association between hereditary diseases and proteins.

In 1909, English physician Archibald Garrod was the first to link inheritance and protein. Garrod noted that people with certain inherited "inborn errors of metabolism" lacked certain enzymes. Other researchers added supporting evidence: they linked abnormal or missing enzymes to unusual eye color in fruit flies and nutritional deficiencies in bread mold variants. But how do enzyme deficiencies produce traits? Experiments in bacteria would answer the question and return, eventually, to Miescher's nuclein.

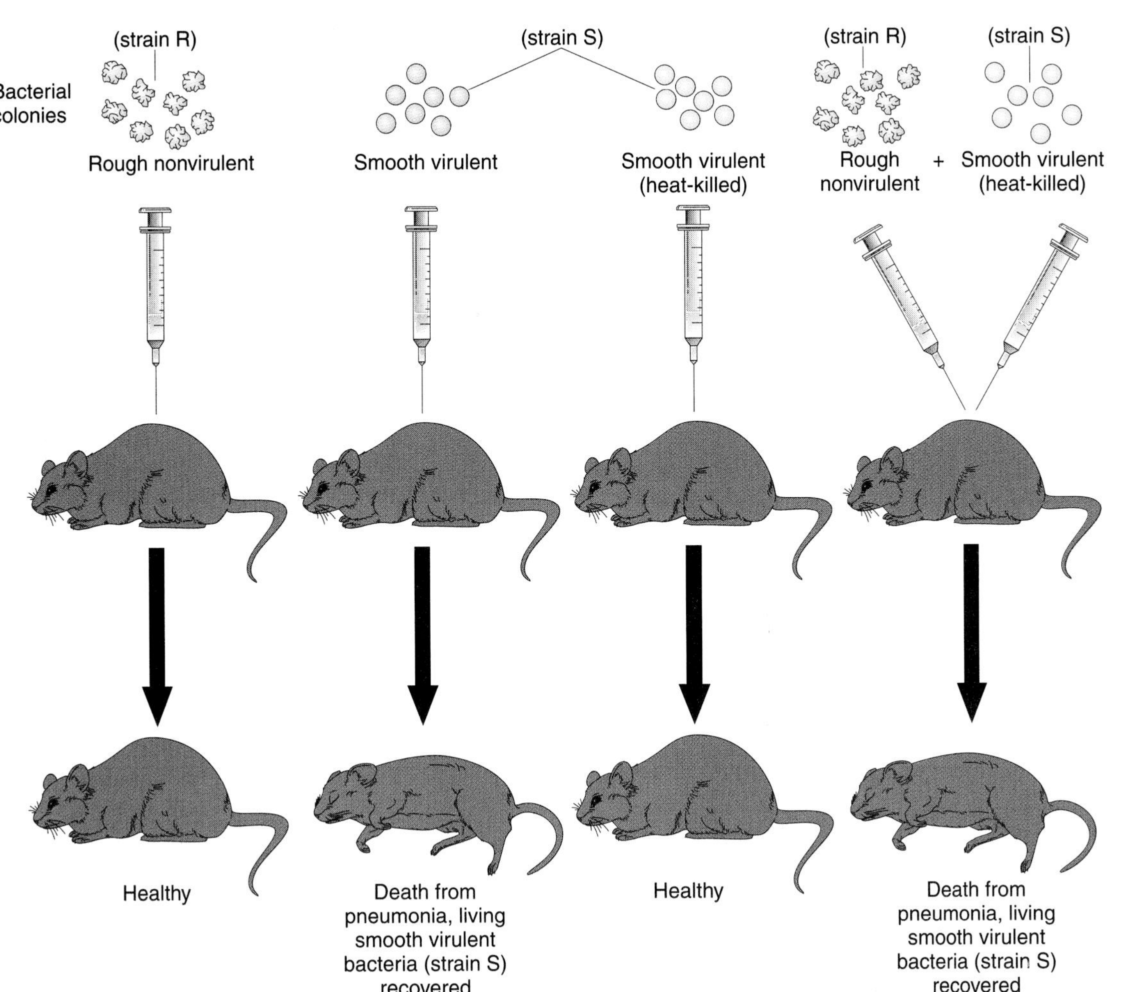

FIGURE 15.2

Griffith's experiments showed that a biochemical in a killer strain of bacteria can make a nonkilling strain deadly.

Source: Redrawn from Linda R. Maxson and Charles H. Daugherty, Human Genetics, 3d ed. Copyright © 1992 The McGraw-Hill Companies, Inc.

DNA Transmits Traits

In 1928, English microbiologist Frederick Griffith contributed the first step in identifying DNA as the genetic material. Griffith studied pneumonia in mice caused by a bacterium, *Diplococcus pneumoniae.* He identified two types of bacteria. Type S bacteria are smooth, because they are encased in a polysaccharide capsule. When injected into mice, type S bacteria cause pneumonia. Type R bacteria are rough in texture, and when injected into mice do not cause pneumonia. Therefore, the smooth polysaccharide coat seemed to be necessary for infection.

When Griffith heated type S bacteria ("heat-killing" them) and injected them into mice, they no longer caused pneumonia. However, when he injected mice with a mixture of type R bacteria plus heat-killed type S bacteria—neither able to cause pneumonia alone—the mice died of pneumonia (fig. 15.2). Their bodies contained live type S bacteria encased in polysaccharide. What was happening?

In the 1930s, U.S. physicians Oswald Avery, Colin MacLeod, and Maclyn McCarty offered an explanation. They hypothesized that something in the heat-killed type S bacteria entered and "transformed" the normally harmless type R strain into a killer. Was this "transforming principle" a protein? Injecting mice with a protein-destroying enzyme (a protease) failed to inhibit the type R strain from transforming into a killing strain. Therefore, a protein was not responsible for transmitting the killing trait. Injecting mice with a DNA-destroying enzyme, however, disrupted transformation. Could DNA transmit the killing trait?

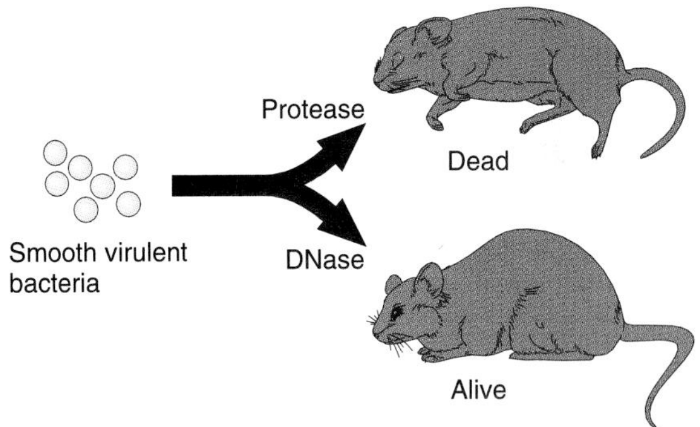

FIGURE 15.3

Avery, MacLeod, and McCarty identified Griffith's transforming principle as DNA. By adding either a protease or an enzyme that disrupts DNA to Griffith's experiment, they demonstrated that DNA transmits the ability to kill—and that protein does not.

Avery, MacLeod, and McCarty confirmed that DNA transformed the bacteria by isolating DNA from heat-killed type S bacteria and injecting it along with type R bacteria into mice (fig. 15.3). The mice died, and their bodies contained active type S

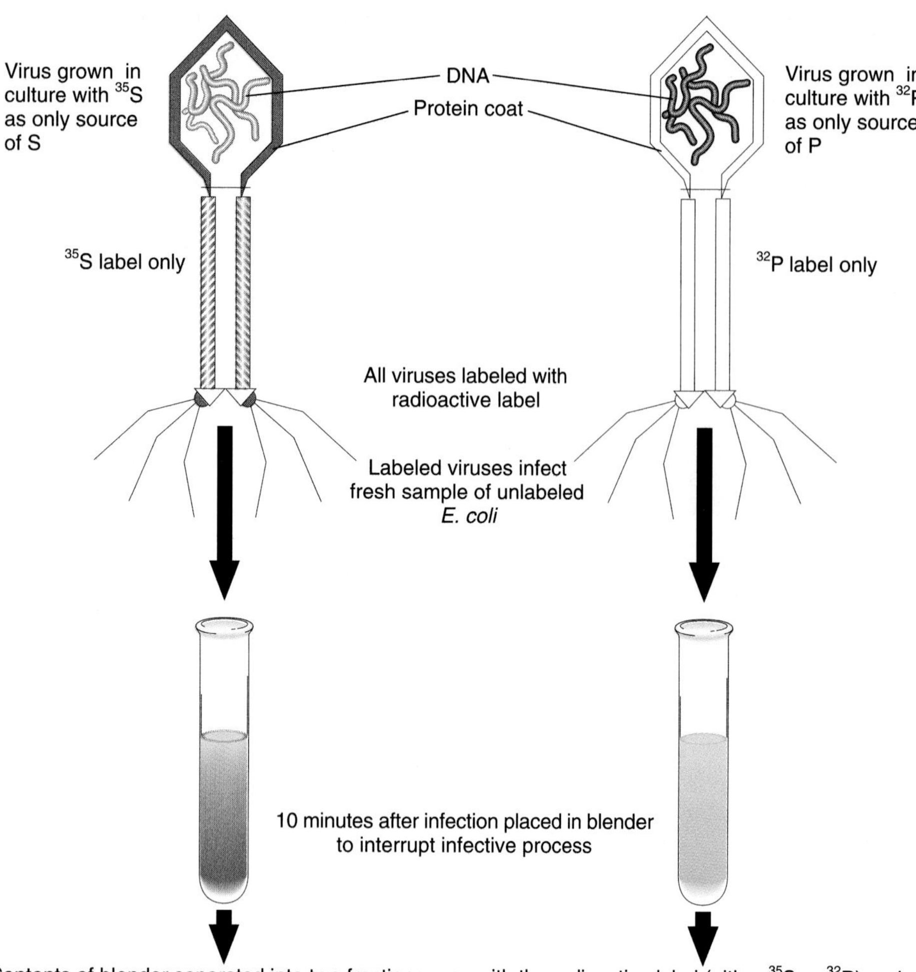
Virus grown in culture with ^{35}S as only source of S
DNA
Protein coat
Virus grown in culture with ^{32}P as only source of P
^{35}S label only
^{32}P label only
All viruses labeled with radioactive label
Labeled viruses infect fresh sample of unlabeled E. coli
10 minutes after infection placed in blender to interrupt infective process
Contents of blender separated into two fractions: one with the radioactive label (either ^{35}S or ^{32}P) and one fraction with no radioactivity. Each fraction put in petri dish with fresh media and scored for growth of E. coli and liberation of new virus.

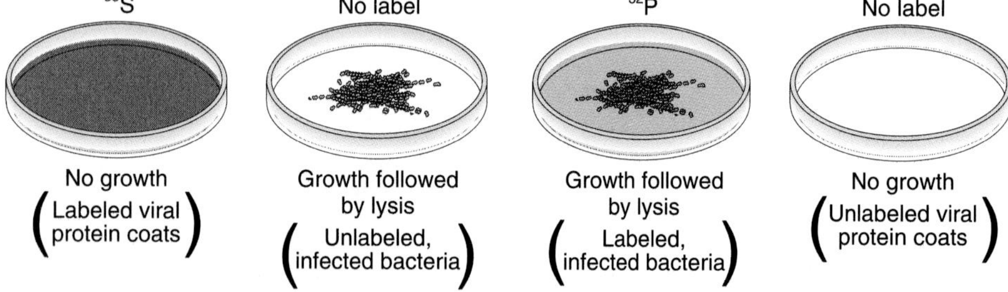
^{35}S
No label
^{32}P
No label
No growth (Labeled viral protein coats)
Growth followed by lysis (Unlabeled, infected bacteria)
Growth followed by lysis (Labeled, infected bacteria)
No growth (Unlabeled viral protein coats)

FIGURE 15.4

By labeling either sulfur (found in protein but not in DNA) or phosphorus (found in DNA but not in protein) in viruses used to infect *E. coli,* Hershey and Chase proved that DNA is the hereditary material—and protein is not.

Source: Redrawn from Linda R. Maxson and Charles H. Daugherty, Human Genetics, 3d ed. Copyright © 1992 The McGraw-Hill Companies, Inc.

bacteria. The conclusion: Type S DNA altered the type R bacteria, enabling them to manufacture the smooth coat necessary to cause infection.

The Hereditary Material—DNA, Not Protein

Perhaps because biologists knew much more about proteins than nucleic acids, accepting DNA as the biochemical of heredity required further proof. In 1950, U.S. microbiologists Alfred Hershey and Martha Chase confirmed that DNA—not protein—is the genetic material (fig. 15.4). Hershey and Chase infected *E. coli* bacteria with a virus. They knew that some part of the virus would enter the bacteria and direct them to produce more virus. Hershey and Chase asked which part of the virus controls its replication—the nucleic acid or the protein coat. Hershey and Chase already knew that virus grown with radioactive sulfur became radioactive, and the radioactivity was emitted from the protein coats. When they repeated the experiment using radioactive phosphorus, radioactivity came from the viral DNA instead. This showed that sulfur is in protein but not in nucleic acid and that phosphorus is in nucleic acid but not in protein. (Recall that Miescher had identified phosphorus in "nuclein" nearly a century earlier.)

Next, the researchers "labeled" two batches of virus, one with radioactive sulfur (which marked protein) and the other with radioactive phosphorus (which marked DNA). They used each type of labeled virus to infect a separate batch of bacteria and allowed several minutes for the virus particles to bind to the bacteria and

FIGURE 15.5
James Watson (*left*) and Francis Crick. Watson was the first director of the Human Genome Project, an ongoing worldwide effort to systematically determine the sequences of all human genes.

Table 15.1 The Road to the Double Helix

Investigator	Contribution
Friedrich Miescher	Isolated nuclein in white blood cell nuclei
Frederick Griffith	Killing ability in bacteria can be transferred between strains
Oswald Avery, Colin MacLeod, and Maclyn McCarthy	DNA transmits killing ability in bacteria
Alfred Hershey and Martha Chase	The part of a virus that infects and replicates is its nucleic acid and not its protein
Phoebus Levene, Erwin Chargaff, Maurice Wilkins, and Rosalind Franklin	DNA components, proportions, and positions
James Watson and Francis Crick	DNA's three-dimensional structure

inject their genetic material into them. Then they agitated each mixture in a blender, poured the mixtures into test tubes, and centrifuged them (spun them at high speed). This settled the heavier infected bacteria at the bottom of each test tube.

Hershey and Chase examined the contents of the bacteria that had settled to the bottom of each tube. In the test tube containing sulfur-labeled virus, the virus-infected bacteria were not radioactive. But in the other tube, where the virus contained radioactive phosphorus, the infected bacteria were radioactive. The "blender experiment" therefore showed that the part of the virus that could enter the bacteria and direct them to mass-produce more virus was the part with the phosphorus label—the DNA. The genetic material, therefore, was DNA and not protein.

Deciphering the Structure of DNA

Early in the twentieth century, Russian American biochemist Phoebus Levene continued Miescher's chemical analysis of nucleic acids. In 1909, Levene identified the five-carbon sugar **ribose** in some nucleic acids; in 1929, he discovered a similar sugar, **deoxyribose.** Levene's discovery revealed a major chemical distinction between the two types of nucleic acid, RNA (ribonucleic acid) and DNA (deoxyribonucleic acid).

Levene then determined that the three parts of a nucleic acid—carbohydrates, nitrogen-containing groups, and phosphorus-containing components—occur in equal proportions. He deduced that a nucleotide building block of a nucleic acid must include one of each component. Levene also found that while nucleotides always contain the same carbohydrate and phosphate portions, they may contain any one of four different nitrogen-containing bases. For several years thereafter, scientists erroneously thought that the nitrogenous bases occur in equal amounts. If this were the case, DNA bases could not encode much information, just as a sentence could not say much if it was restricted to equal numbers of certain letters.

In the early 1950s, two lines of evidence converged to reveal DNA's structure. Austrian American biochemist Erwin Chargaff showed that DNA contains equal amounts of the bases adenine (A) and thymine (T) and equal amounts of the bases guanine (G) and cytosine (C). Next, two English researchers, physicist Maurice Wilkins and chemist Rosalind Franklin, bombarded DNA with X rays using a technique called X-ray diffraction. The X-ray diffraction pattern revealed a regularly repeating structure of nucleotides.

In 1953, U.S. biochemist James Watson and English physicist Francis Crick combined these clues to guide them in building a replica of the DNA molecule using ball-and-stick models (fig. 15.5). Their model included equal amounts of G and C and of A and T, and it had the sleek symmetry the X-ray diffraction pattern revealed. Watson and Crick's model, which was based upon many others' experimental evidence, was the now familiar double helix (table 15.1).

Mastering Concepts

1. When was DNA first described?
2. What evidence linked enzyme deficiencies to inherited traits?
3. How did researchers demonstrate that DNA is the genetic material and protein is not?
4. What are the components of DNA?
5. What evidence enabled Watson and Crick to decipher the structure of DNA?

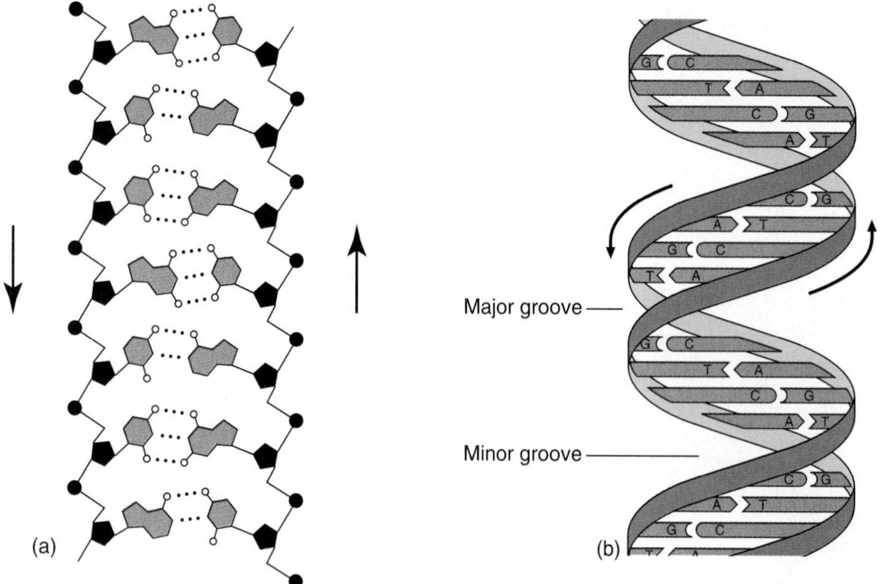

FIGURE 15.6

Different ways to represent the DNA double helix. (*a*) The helix is unwound to show the base pairs in color and the sugar-phosphate backbone in black. (*b*) The sugar-phosphate backbone is emphasized, along with the fact that the two strands run in opposite directions. (*c*) This representation shows the relationships of all of the atoms. (*d*) A schematic representation of an unwound section of the double helix outlines the relationship of the sugar-phosphate rails to the base pairs. The informational content of DNA lies in the sequence of bases. The sugar-phosphate rails are identical in all DNA molecules, and run in opposite directions in a double helix.

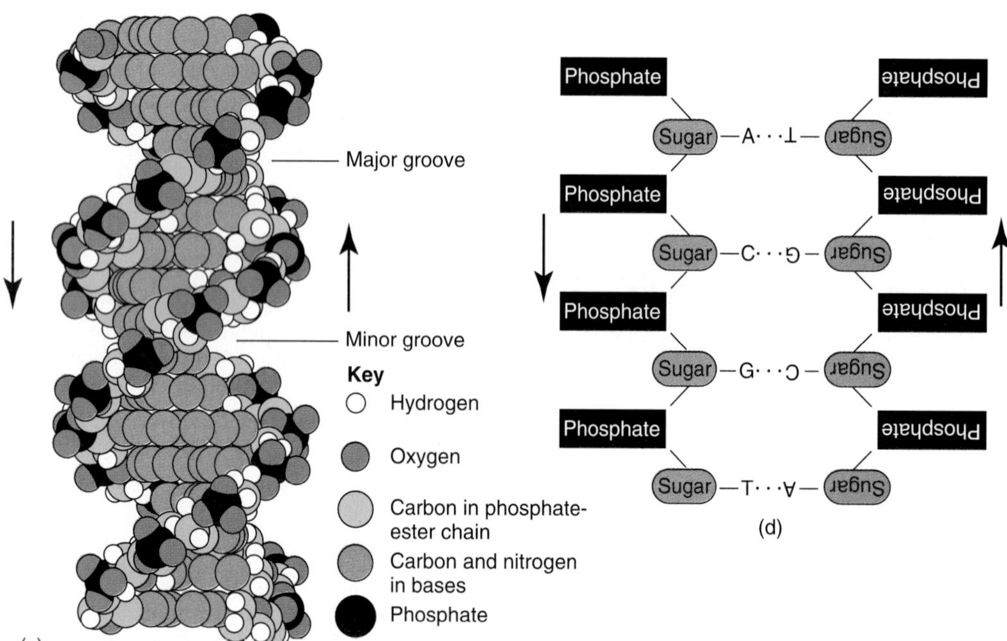

Gene and Protein—An Important Partnership

Genes and proteins encode information because they consist of sequences—genes of four types of DNA bases and proteins of 20 types of amino acids. Genes and proteins are large molecules, with genes consisting of thousands of base pairs and proteins of hundreds of amino acids.

Proteins provide inherited traits. Pigment proteins confer pea color; protein hormones control plant height. In animals, the amino acids that are assembled into proteins ultimately come from the diet. Enzymes are proteins that catalyze the chemical reactions of metabolism. Proteins such as collagen and elastin provide structural support in connective tissues. Tubulin protein builds the cytoskeleton, myoglobin and hemoglobin transport oxygen, and antibodies protect against infection. Malfunctioning or inactive proteins, which reflect genetic defects, can devastate health.

DNA Structure Encodes Information

If the DNA double helix is pictured as a twisted ladder, the rungs are A–T and G–C base pairs joined by hydrogen bonds, and the rails that make up the DNA's sugar-phosphate "backbone" are alternating units of deoxyribose and phosphate (PO_4) joined with covalent bonds. A nucleotide consists of one deoxyribose, one phosphate, and one base. Figure 15.6 depicts the DNA molecule.

DNA base sequences encode a gene's information. (Biology in Action 15.1 discusses DNA fingerprinting, which is based on

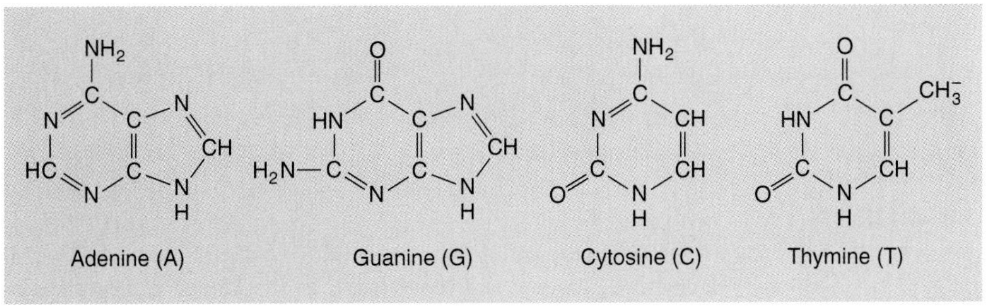

Adenine (A) Guanine (G) Cytosine (C) Thymine (T)

FIGURE 15.7

DNA bases. Adenine and guanine are purines, each composed of a six-member organic ring plus a five-member ring. Cytosine and thymine are pyrimidines, each built of a single six-member ring.

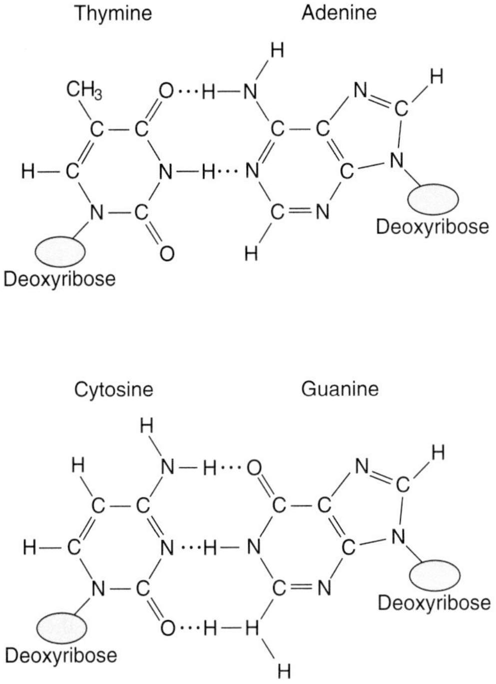

Thymine Adenine

Deoxyribose Deoxyribose

Cytosine Guanine

Deoxyribose Deoxyribose

FIGURE 15.8

DNA base pairs. The key to the constant width of the DNA double helix is the pairing of purines with pyrimidines. Specifically, adenine pairs with thymine with two hydrogen bonds, and cytosine pairs with guanine with three hydrogen bonds.

detecting DNA sequence differences between individuals.) Adenine and guanine are **purines,** which have a double organic ring structure. Cytosine and thymine are **pyrimidines,** which have a single organic ring structure (fig. 15.7). The pairing of A with T and G with C maintains the helix's width because each pair includes a single-ringed structure and a double-ringed structure. These specific purine-pyrimidine couples are termed **complementary base pairs** (fig. 15.8). Complementary base pairing is the basis of gene function, as the next chapter discusses.

The two chains of the DNA double helix run opposite to each other, somewhat like artist M. C. Escher's depiction of drawing hands (fig. 15.9*a*). This head-to-tail arrangement, called **antiparallelism,** is apparent when the deoxyribose carbons are numbered consecutively from right to left, according to chemical convention (fig. 15.9*b*). Where one chain ends in the 3′ carbon, the opposite chain ends in the 5′ carbon (fig. 15.9*c*). (The 3′ and 5′ designate the positions of the carbon atoms in the DNA molecule.)

DNA Is Highly Coiled

If the DNA bases of all 46 human chromosomes were typed as A, C, T, and G, the 3 billion letters would fill 4,000 books of 500 pages each! How can a cell only one-millionth of an inch across contain so much material? The explanation is that DNA is wrapped around proteins, much as a very long length of thread is wound around a wooden spool.

To form a chromosome, DNA 146 nucleotides long wraps twice around a structure of eight proteins, called histones, to form a **nucleosome,** which is 10 nanometers (nm, billionths of a meter) in diameter. A continuous thread of DNA connects nucleosomes like beads on a string, which in turn fold into structures that are 30 nanometers in diameter. Like thread, DNA must unwind to function (fig. 15.10). As sections of DNA unwind, the DNA in the more widely separated nucleosomes may be expressed. Whether the DNA rolls off the histone to be transcribed into RNA or remains tightly rolled, it maintains its structural integrity and sequence. Together, the DNA and histones constitute chromatin, the material that makes up chromosomes. Chromatin is the relatively unwound configuration of DNA during mitotic interphase.

Mastering Concepts

1. What is the relationship between DNA and proteins?
2. What is the structure of DNA?
3. How do large DNA molecules fit into cells?

DNA Replication—Maintaining Genetic Information

Every time a cell divides, its DNA must replicate, so that each progeny cell receives the same set of genetic instructions. DNA's structure makes its replication possible.

Semiconservative Replication

The structure of DNA suggests how it replicates. In their classic 1953 paper describing the genetic material, Watson and Crick wrote that "the specific pairing we have postulated immediately suggests a possible copying mechanism for the genetic material."

biology in action 15.1

DNA Fingerprinting: An Application of Understanding DNA Structure

DNA fingerprinting compares DNA sequences between individuals to reveal or disprove genetic relationships between them. A faster way to compare DNA sequences than actually determining the base order uses restriction enzymes, which cut DNA at specific sequences. If the DNA of two individuals differs at a cutting site for a restriction enzyme, then applying that enzyme to the DNA cuts the molecules into different-sized pieces. The pieces are then displayed in size order on a sheet of nylon and differences stand out (fig. 15.A).

The power of DNA fingerprinting in humans stems from the fact that there are far more ways for the 3 billion bases of the genome to vary than there are people. The process involves determining how likely it is that two DNA fingerprints match because the sources are the same rather than because two individuals resemble each other by chance. To make this distinction, the investigator consults population databases on allele frequencies to calculate the likelihood of a particular combination occurring in a certain population. If the allele combination being examined is fairly common, then a DNA match could happen by coincidence. This is why DNA fingerprinting examines DNA sequences that tend to vary greatly among individuals.

To generate the statistics that make DNA fingerprinting results meaningful, allele frequencies are multiplied, an application of the product rule. For example, if rare DNA sequence variants at five sites in the genome are exactly the same for sperm cells collected from a rape victim and blood from a suspect, then chances are very high that the same man provided both samples.

In the Courtroom

In 1986, DNA fingerprinting was unheard of outside of scientific circles. So rapist Tommie Lee Andrews thought he was being very meticulous in planning his crimes. He picked his victims months before he attacked, and watched them so that he knew exactly when they would be home alone. On a balmy Sunday night in May 1986, Andrews lay in wait for Nancy Hodge, a young computer operator at Disney World, at her home. Surprising her when she was in the bathroom, he covered her face, then raped and brutalized her repeatedly.

Andrews was very careful not to leave fingerprints, threads, hairs, or any other evidence. But he had not counted on DNA fingerprinting. Thanks to a clear-thinking crime victim and scientifically informed lawyers, Andrews was soon at the center of a trial not only of himself, but also of the technology that would eventually help convict him. When Andrews was arrested for another assault, DNA fingerprinting was done on his white blood cells. The pattern matched the DNA from the sperm sample taken from Hodge after she was raped. The match left little doubt that Tommie Lee Andrews was guilty.

DNA fingerprinting has grown into a valuable forensic tool. It saves time, money, and reputations. In the United States and England, a third of all rape suspects are released because DNA analysis vindicates them. The technique can also help to right wrong convictions. So far more than two dozen wrongly convicted people have been released from prison thanks to DNA evidence that revealed that they could not possibly have committed the crimes for which they stood trial.

In the Garden

Vitis vinifera—better known as the grape—was domesticated more than 6,000 years ago. Traditionally, grape tasters distinguish 14,000 to 24,000 cultivated strains, or cultivars, based on growth characteristics; vine branching pattern; leaf, fruit, and flower shape; and pollen grain diameter. Researchers are currently cataloging grape cultivars by DNA fingerprint pattern. The information will clarify the evolutionary relationships between different types of grapes, enable researchers to follow new combinations of characteristics, and provide a way to objectively identify cultivars to settle patent disputes.

DNA fingerprinting has revealed a close evolutionary relationship between sugar beets (*Beta vulgaris*), three cultivars (fodder beets, leaf beets, and garden beets), and a wild subspecies (*B. vulgaris maritima*). Not surprisingly, the native sugar beets and the cultivars are very similar genetically, whereas the wild strain is much more variable. Researchers are using this information to identify valuable characteristics in the wild strain, such as cold tolerance and disease resistance, that could be bred or genetically engineered into sugar beets.

FIGURE 15.A
DNA fingerprinting. A blood sample (*1*) is collected from the suspect. White blood cells containing DNA are extracted and burst open (*2*), releasing the DNA strands (*3*). The strands are snipped into fragments (*4*) using restriction enzymes. A technique called electrophoresis aligns the DNA pieces by size on a sheet of gel (*5*). Next, the resulting DNA fragments are transferred to a nylon sheet (*6*) and exposed to radioactively tagged DNA probes (*7*) that bind to the DNA areas used to establish identity. When the nylon sheet is placed against a piece of X-ray film (*8*) and processed, black bands appear where the probes stuck (*9*). This pattern of bands constitutes a DNA fingerprint (*10*).

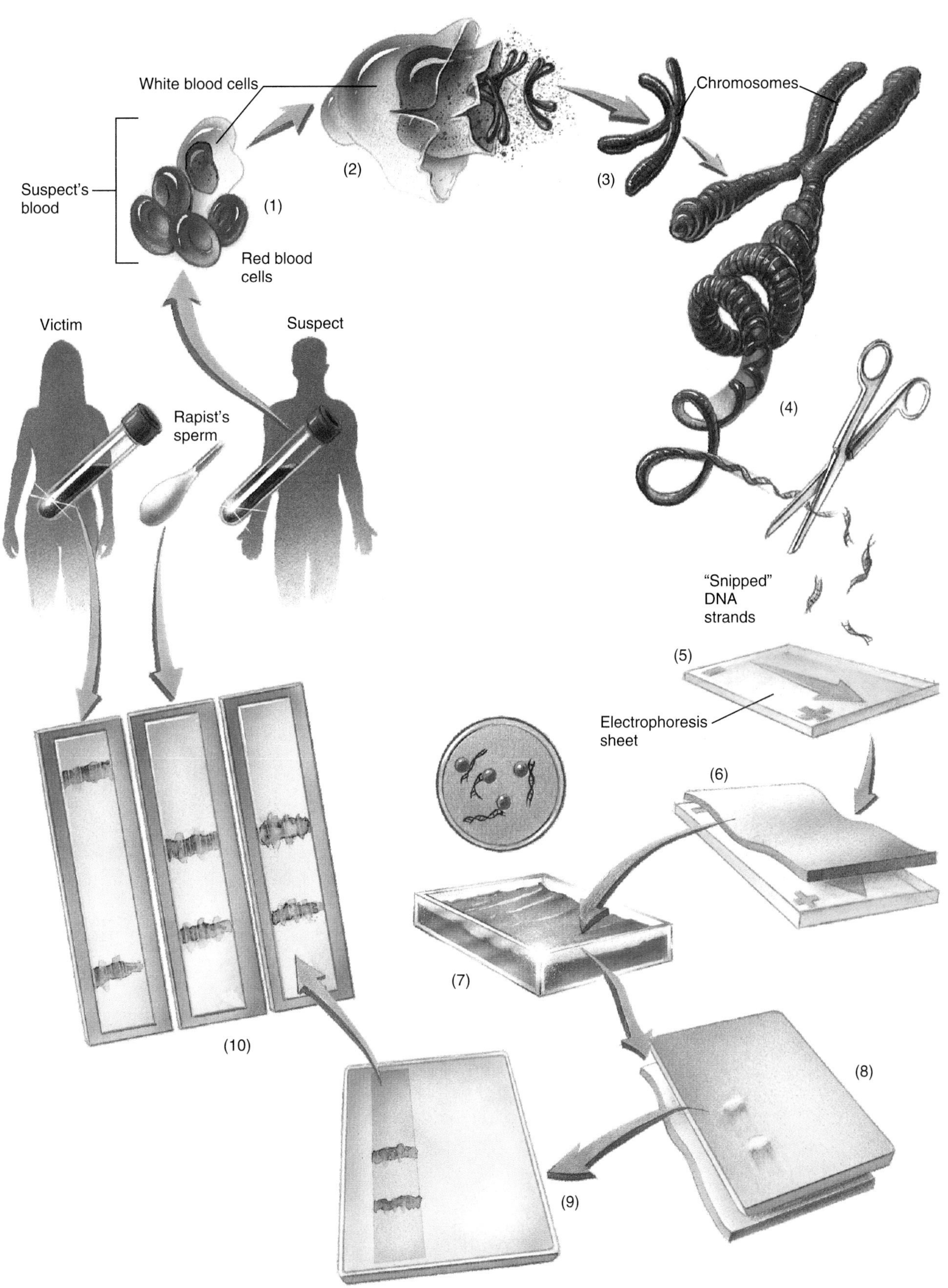
White blood cells
Suspect's blood
(1)
Red blood cells
(2)
(3)
Chromosomes
(4)
Victim
Suspect
Rapist's sperm
"Snipped" DNA strands
(5)
Electrophoresis sheet
(6)
(7)
(8)
(9)
(10)

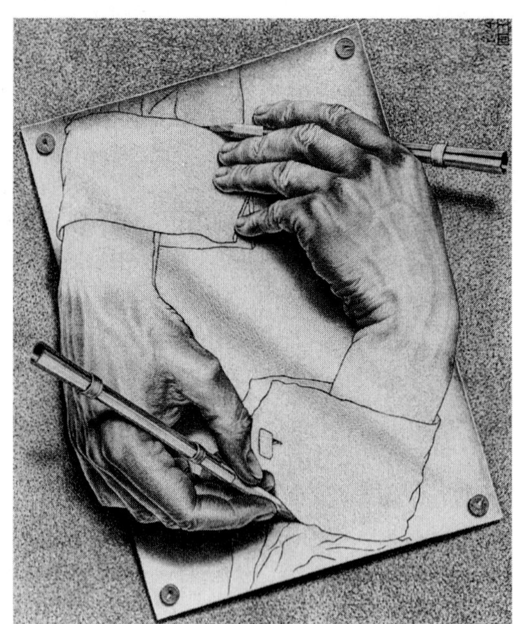

(a)

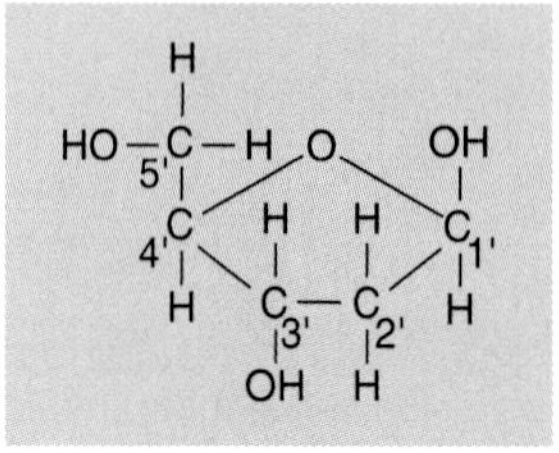

FIGURE 15.9

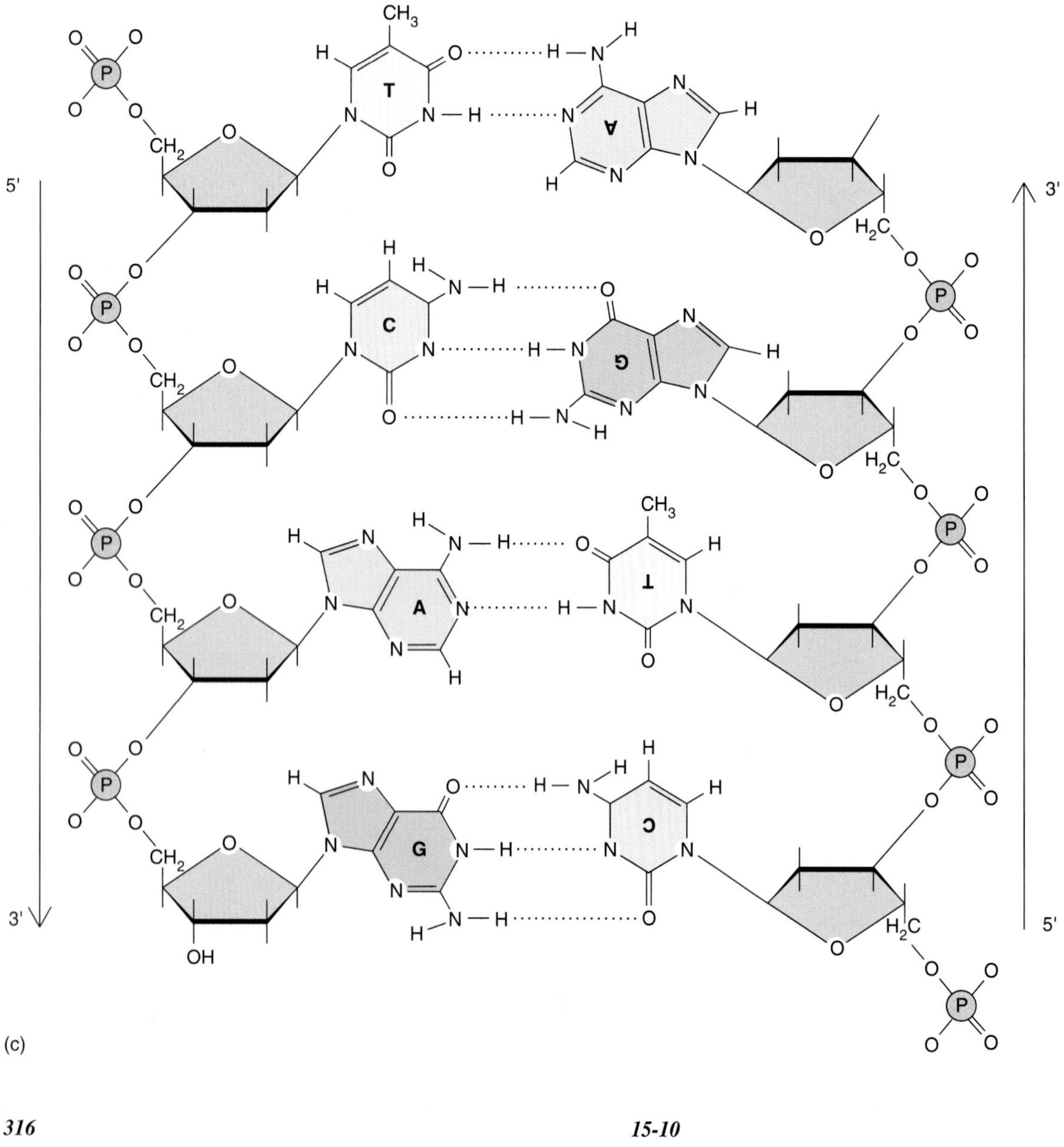

DNA strands are antiparallel. (*a*) The spatial relationship of these two hands resembles that of the two DNA chains that make up the DNA double helix. (*b*) Chemists assign numbers to differently positioned carbons in organic molecules. (*c*) The two strands of the DNA double helix run opposite one another in orientation. This arrangement is called antiparallelism. (a) © 1994 M.C. Escher/Cordon Art-Baarn-Holland. All rights reserved.

(b)

(c)

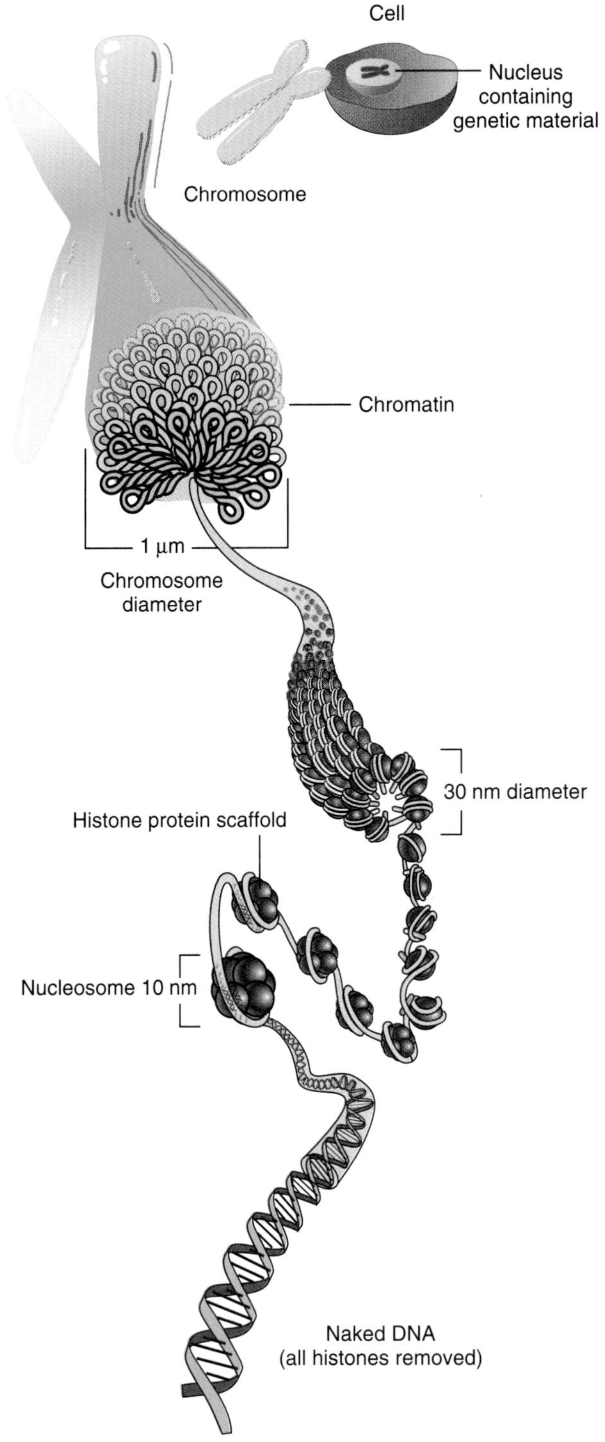

FIGURE 15.10

A chromosome consists of a very tightly wound molecule of DNA, plus associated proteins.

Watson and Crick envisioned a double helix disentangling, with each half serving as a template, or mold, for the assembly of a new half. Every cell contains unattached nucleotides from the diet and metabolism. The bases, Watson and Crick proposed, would hydrogen bond to complementary bases exposed on the single, unwound parental DNA strands. Covalent bonds then cement the sugar-phosphate backbone in the new DNA strands. This mechanism is called **semiconservative replication** because half ("semi") of each double helix is conserved from a preexisting double helix (see fig. 3.31). Experiments in several types of organisms that labeled newly replicated DNA demonstrated the semiconservative mechanism of DNA replication.

Steps and Participants in DNA Replication

To envision the task of replicating a cell's DNA, imagine sewing a hem on a pair of jeans using a doubled thread the length of a football field! The doubled thread would rapidly become hopelessly tangled. Maintaining the length of the thread would require continuously untwisting the tangles or periodically cutting the thread, untwisting it locally, and reattaching the cut ends. Conceptually speaking, this is what happens when a long length of DNA replicates. A contingent of enzymes carries out DNA replication, as figure 15.11 whimsically shows. Enzymes called **helicases** unwind and hold apart replicating DNA so that other enzymes can guide the assembly of new DNA strands.

A human chromosome replicates at hundreds of points along its length, and then the individual sections rejoin. While replicating, DNA resembles a fork in a road. Fittingly, the open portions of a replicating double helix are called **replication forks.** A human chromosome has many replication forks.

DNA replication begins when a helicase breaks the hydrogen bonds that connect a base pair. This first step occurs at an **initiation site.** Another type of enzyme, **RNA polymerase,** then attracts complementary RNA nucleotides to build a short piece of RNA, called an **RNA primer,** at the start of each DNA segment to be replicated. Next, the RNA primer attracts **DNA polymerase,** an enzyme that draws in DNA nucleotides to complement the exposed bases on the parental strand. The new DNA strand grows as hydrogen bonds form between the complementary bases.

DNA polymerase also "proofreads" as it goes, excising mismatched bases and inserting correct ones. At the same time, another enzyme removes the RNA primer and replaces it with the correct DNA bases. Enzymes called **ligases** knit together the sugar-phosphate backbone. Figure 15.12 depicts the replication process, and Biology in Action 15.2 discusses a widely used technology that uses DNA polymerase.

DNA polymerase works directionally, adding new nucleotides to the exposed 3′ end of the deoxyribose in the growing strand. Replication proceeds in a 5′ to 3′ direction. In order to follow this 5′ to 3′ "rule," DNA replication proceeds continuously (in one piece) on one strand but discontinuously (in short 5′ to 3′ pieces) on the other strand (fig. 15.13).

Mastering Concepts

1. Why must DNA replicate?
2. What is semiconservative replication?
3. What are the steps of DNA replication?

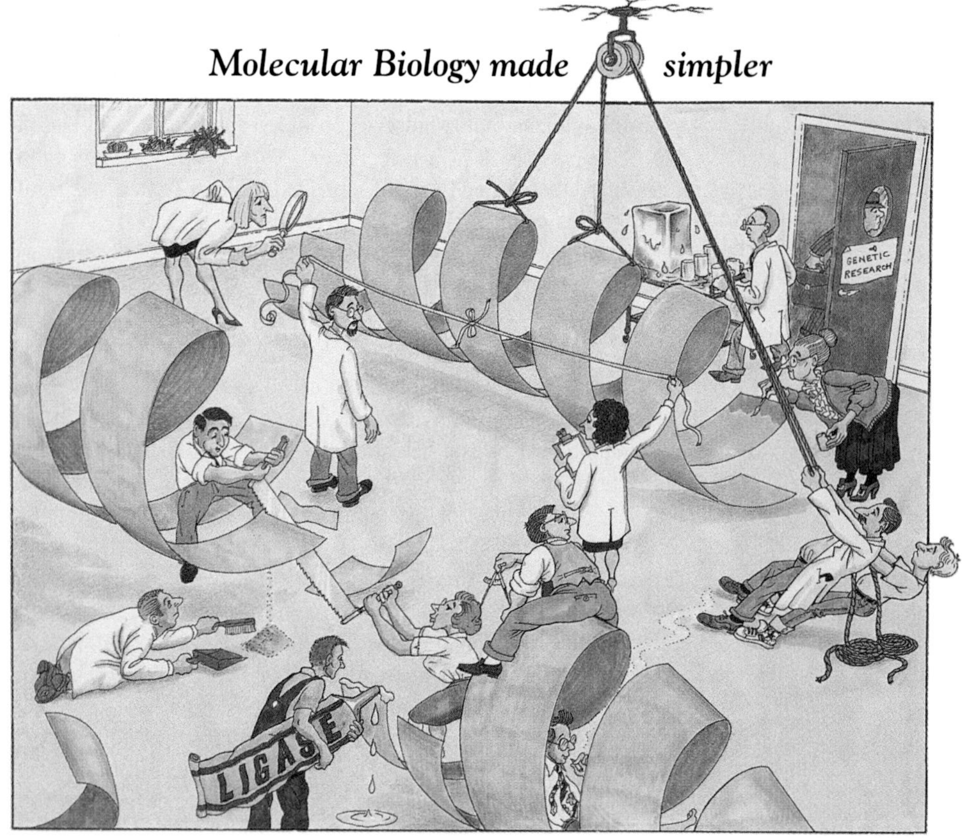

FIGURE 15.11

An army of enzymes replicates DNA. This advertisement for a company that sells DNA-cutting enzymes depicts the number of participants (enzymes) that help to replicate and repair DNA. What is wrong with this cartoon?

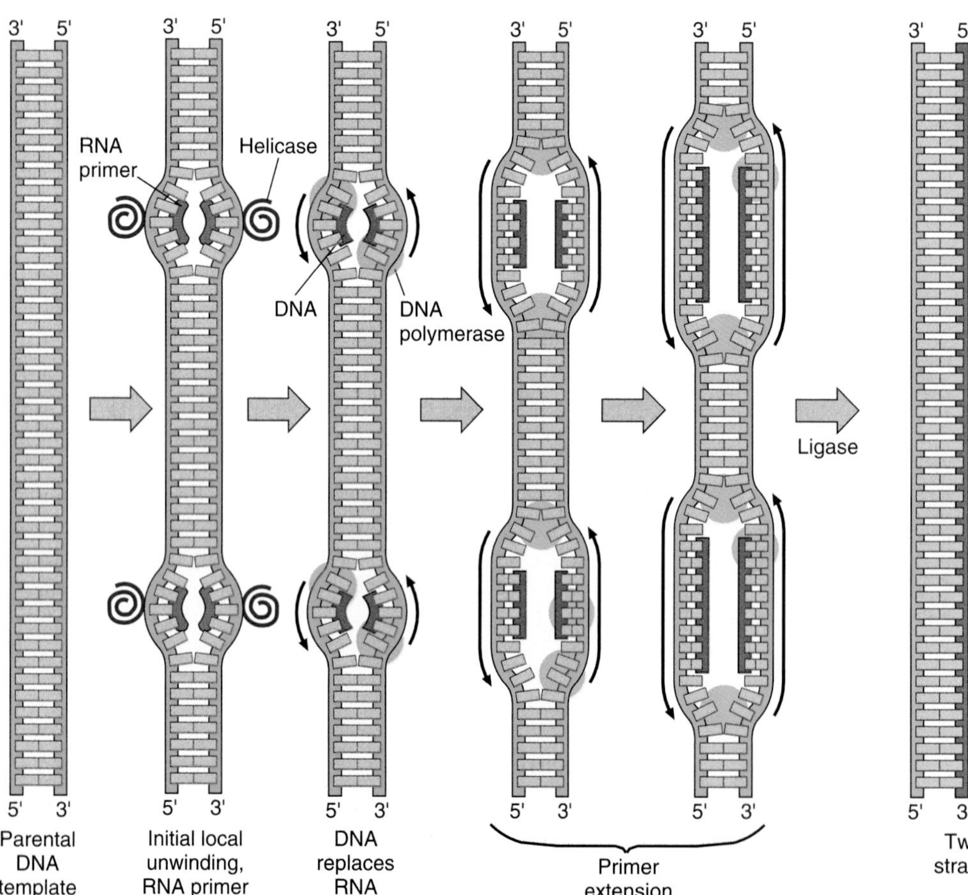

FIGURE 15.12

DNA replication takes several steps. Parental strands are blue and newly replicated DNA is red. Replication begins as the helix unwinds locally at many sites of origin, and RNA primers (purple) are synthesized. DNA polymerase extends the primers, DNA replaces the RNA primer, and the small, replicated portions of the chromosome join as ligase cements the sugar-phosphate backbone.

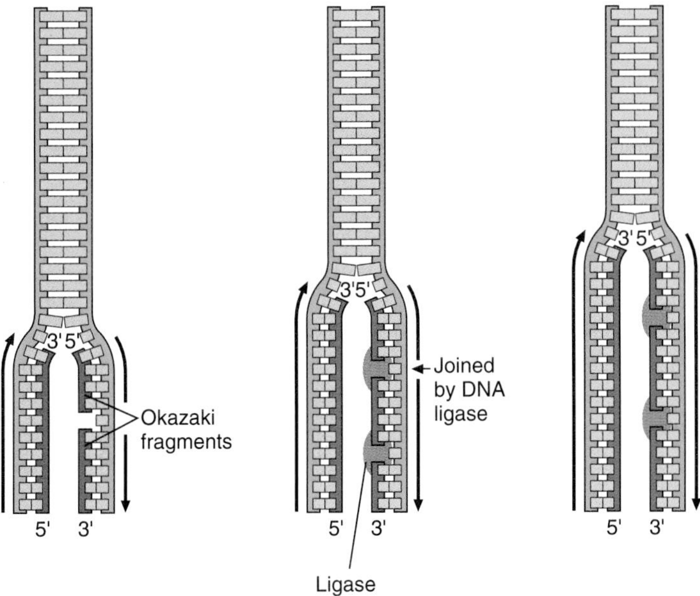

FIGURE 15.13

To maintain the 5′ to 3′ directionality of DNA replication, at least one strand must be replicated in short pieces, from the inner part of the fork outwards. Ligase joins the pieces.

FIGURE 15.14

DNA is repaired by photoreactivation, in which a pyrimidine dimer is split, or by excision repair, in which the pyrimidine dimer and a few surrounding bases are removed and replaced.

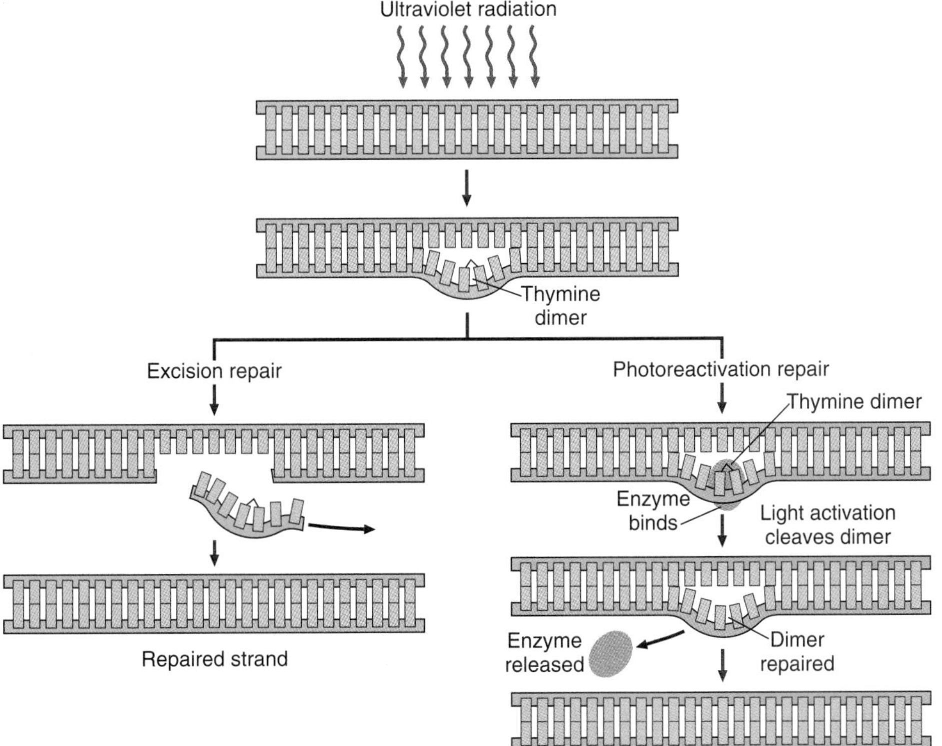

DNA Repair

Any manufacturing facility tests a product in several ways to see whether it has been assembled correctly. Production mistakes are rectified before the item goes on the market—most of the time. The same is true of a cell's DNA production.

DNA replication is incredibly accurate—only about 1 in 100,000 bases is incorporated incorrectly. DNA repair enzymes minimize replication errors. DNA damage can result from ultraviolet radiation and from errors in replication.

Ultraviolet radiation damages DNA by causing an extra covalent bond to form between adjacent pyrimidines on the same strand, particularly thymines. The linked thymines are called thymine dimers. This extra bond kinks the double helix, similar to two adjacent steps on an escalator collapsing. The extra bond is enough of an upset to disrupt replication and allow insertion of a noncomplementary base. In a type of repair called **photoreactivation,** enzymes called photolyases absorb energy from the blue part of visible light and use it to break the extra bond of the pyrimidine dimer (fig. 15.14).

Another type of DNA self-mending, **excision repair,** cuts the bond between the deoxyribose and the base and removes the pyrimidine dimer and surrounding bases. Then, a type of DNA polymerase fills in the correct nucleotides, using the exposed template as a guide.

In a third type of DNA repair called **mismatch repair,** enzymes proofread newly replicated DNA for small loops that indicate where the two strands are not precisely aligned, as they would be if complementary base pairing occurs at every point. Such mismatching tends to occur in chromosome regions where very short DNA sequences repeat. These repeated sequences, called microsatellites, are scattered throughout the genome. Their lengths can vary from person to person, but within an individual they are all the same length. Mismatch repair maintains microsatellite length in an individual.

Because the several forms of DNA repair use dozens of different enzymes, and therefore genes, several inherited disorders reflect faulty DNA repair (table 15.2). Such conditions usually produce chromosome breaks and a high susceptibility to cancer following exposure to ionizing radiation or chemicals that affect cell division. Life is very difficult for a person with such a repair disorder. One little girl suffering from xeroderma pigmentosum who hardly ever ventures outdoors drew a picture of a bright day, with the heading "the sun is a monster" (fig. 15.15). For her, it is.

Disorders of DNA repair reveal how vital precise DNA replication is to cell

15.2 biology in action

PCR: An Application of Understanding DNA Replication

Every time a cell divides, it replicates all its DNA. A technology called the polymerase chain reaction (PCR) enables researchers to use the cell's DNA copying machinery to rapidly replicate millions of copies of DNA sequences they particularly want to study. For example, PCR is now the best method to detect HIV infection. It amplifies viral nucleic acid to detectable levels, a much more direct approach than the previously used method of monitoring the immune system's response to infection.

Inspiration on a Starry Night

The polymerase chain reaction (PCR) was born in Kary Mullis's mind on a moonlit night in northern California in 1983. As he drove up and down the hills, Mullis, a molecular biologist, was thinking about the incredible precision and power of DNA replication. Suddenly, a way to tap into that power popped into his mind. He excitedly explained his idea to his girlfriend and then went home to think it through further. "It was difficult for me to sleep with deoxyribonuclear bombs exploding in my brain," he wrote much later.

The idea behind PCR was so stunningly straightforward that Mullis had trouble convincing his superiors at Cetus Corporation that he was really onto something. He spent the next year using the technique to amplify a well-studied gene so he could prove that his brainstorm was not just a flight of fancy. One by one, other researchers glimpsed Mullis's vision. After he convinced his colleagues at Cetus, Mullis published his landmark 1985 paper and filed patent applications, launching the era of gene amplification. Today, PCR, and other gene amplification techniques, are widely used in life science laboratories. Table 15.A lists just a few of the many and diverse applications of PCR.

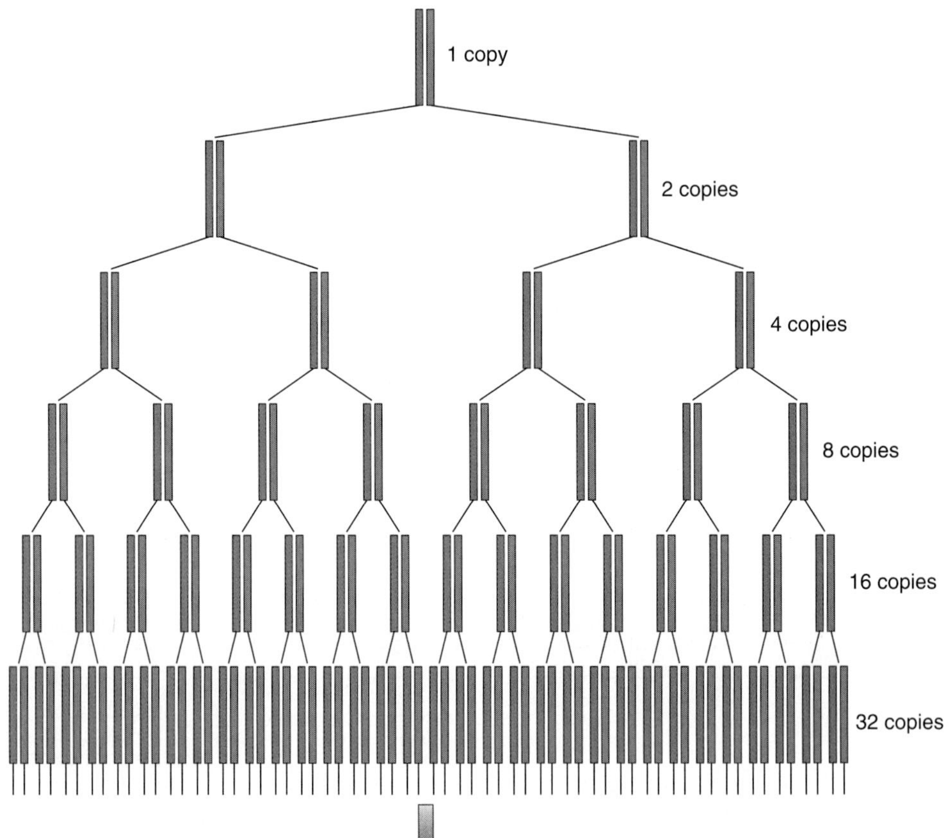

FIGURE 15.B
In the polymerase chain reaction, specific primers are used to bracket a DNA sequence of interest, along with a thermostable DNA polymerase and plenty of free nucleotides. The reaction rapidly builds up copies of the sequence.

Source: Kary Mullis, "Polymerase Chain Reaction," Scientific American, April 1990, p. 58.

Surprisingly Simple

PCR rapidly replicates a selected sequence of DNA in a test tube (fig. 15.B). It requires

--a known target DNA sequence to be amplified;

--two types of primers, which are short, single-stranded pieces of DNA with sequences complementary to opposite ends of the target sequence;

--a supply of the four types of DNA nucleotides;

--Taq1—a DNA polymerase from *Thermus aquaticus,* a microbe that inhabits hot springs. This enzyme is able to function in a hot environment and does not fall apart when DNA is heated, as other enzymes do.

To begin PCR, heat is applied to separate the two strands of the target DNA. Next, the temperature is lowered and the two short DNA primers are added. The primers base pair to the separated target strands. Then Taq1 DNA polymerase and bases are added. The polymerase adds bases to the primers and builds a sequence complementary to the target sequence. The newly synthesized strands are then templates for the next round of replication, which ensues by raising the temperature. All of this happens in an automated device called a thermal cycler that controls the key temperature changes.

DNA pieces accumulate geometrically. The number of amplified pieces of DNA equals 2^n, where n equals the number of temperature cycles. After just 20 cycles, 1 million copies of the original sequence accumulate in the test tube. A problem with PCR is its extreme sensitivity. Should a contaminating piece of DNA enter a PCR experiment, it can be amplified instead of the sequence of interest. PCR can also be used to amplifiy RNA, by using the enzyme reverse transcription to copy an RNA sample into DNA. A cell's RNAs reflect gene expression, and therefore it's specialization.

Table 15.A PCR Applications

PCR Has Been Used to Amplify These Materials:
DNA from poached moose meat in hamburger.
DNA from larvae in a rotting human corpse, to extrapolate the time of murder by determining the time an insect of a particular species deposited its larvae.
DNA in artificial knee joints, indicating infection.
DNA in a preserved quagga (a relative of the zebra) that became extinct 100 years ago.
DNA in sperm cells obtained from the body of a rape victim. Specific sequences were compared to DNA sequences from a suspect in the crime.
Genes from microorganisms that cannot be grown or maintained in culture for study.
Mitochondrial DNA from various modern human populations. Comparisons of mitochondrial DNA sequences indicate that *Homo sapiens* originated in Africa, supporting fossil evidence.
Genes from several organisms that are very similar in sequence. Comparing the extent of similarity reveals evolutionary relationships between species.
DNA from the brain of a 7,000-year-old human mummy. The DNA indicated that Native Americans were not the only people to dwell in North America long ago.
Genetic material from saliva, hair, skin, and excrement of organisms that we cannot catch to study. The prevalence of a rare DNA sequence among all of the bird droppings from a certain species in an area can be extrapolated to estimate the population size.
DNA in the digestive tracts of carnivores. This can reveal food web interactions.
DNA in deteriorated roadkills and carcasses washed ashore. This helps identify locally threatened species.
DNA in products illegally made from endangered species, such as powdered rhinoceros horn, which is sold as an aphrodisiac.
DNA sequences that are unique to the bacteria that cause Lyme disease. When these sequences are found in animals, they provide clues to how the disease is transmitted.
DNA from genetically altered microbes that are released in field tests, to trace their dispersion.
DNA from a cell of an eight-celled human preembryo, to diagnose cystic fibrosis.

survival and proliferation. Once these processes are completed, the cell begins to use the informational content of DNA. This is the subject of the next chapter.

Mastering Concepts

1. Why is DNA repair necessary?
2. How does ultraviolet radiation damage DNA?
3. What are different types of DNA repair?
4. What are the symptoms of disorders that result from defective DNA repair?

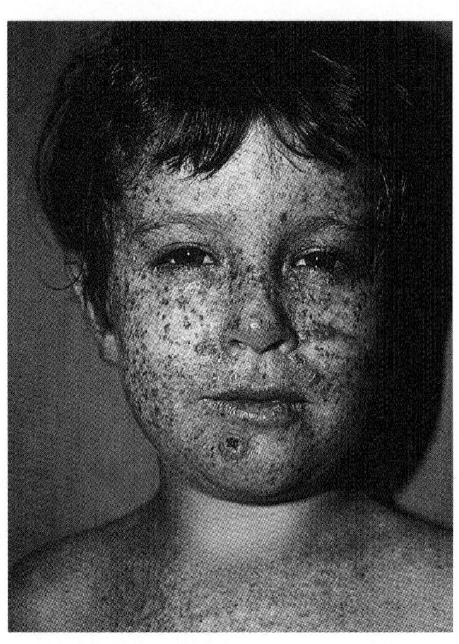

FIGURE 15.15

"The sun is a monster" wrote a 3-year old girl with xeroderma pigmentosum. Just a stray ray of sunlight painfully blisters the skin of such a child. These children grow up indoors, venturing out only at night, or they are covered in clothing and sunblock during the day. The marks on the child's face pictured here are a result of sun exposure. The large lesion on his chin is a skin cancer.

Table 15.2 DNA Replication and Repair Disorders

Disorder	Frequency	Defect
Ataxia telangiectasia	1/40,000	Deficiency in kinase that controls the cell cycle
Bloom syndrome (two types)	100 cases since 1950	DNA ligase is inactive or heat sensitive, slowing replication
Fanconi anemia (several types)	As high as 1/22,000 in some populations	Deficient excision repair
Hereditary nonpolyposis colon cancer	1/200	Deficient mismatch repair
Xeroderma pigmentosum (nine types)	1/250,000	Deficient excision repair

DNA *Solves a Royal Mystery*~

continued from page 308.

The problem in proving that the male remains with fancy dental work were once Tsar Nicholas II centered around nucleotide position 16169 of a mitochondrial gene that is highly variable in sequence among individuals. About 70% of bone cells examined from the remains had cytosine (C) at this position, and the remainder had thymine (T). Skeptics at first suspected contamination or a laboratory error, but when the odd result was repeated, researchers realized that this historical case had revealed a genetic phenomenon not seen before in human DNA. The bone cells apparently harbored two populations of mitochondria, one type with C at this position, the other with T.

The DNA of a living blood relative of the tsar, Countess Xenia Cheremeteff-Sfiri, had only T at nucleotide site 16169. Xenia is the great-granddaughter of Tsar Nicholas II's sister. However, DNA of Xenia and the murdered man matched at every other site. DNA of another living relative, the Duke of Fife, the great-grandson of Nicholas's maternal aunt, matched Xenia at the famed 16169 site. A closer relative, Nicholas's nephew Tikhon Kulikovsky, refused to lend his DNA, citing anger at the British for not assisting the tsar's family during the Bolshevik revolution.

But the story wasn't over. It would take an event in yet another July, in 1994, to clarify matters.

Attention turned to Nicholas's brother, Grand Duke of Russia Georgij Romanov. Georgij had died at age 28 in 1899 of tuberculosis. He was exhumed in July 1994, and researchers sequenced the troublesome mitochondrial gene in bone cells from his leg. They found a match! Georgij's mitochondrial DNA had the same double-base site as the man murdered in Siberia. He was, therefore, Tsar Nicholas II. The researchers calculated the probability that the remains are truly those of the tsar, rather than resembling Georgij by chance, as 130 million to 1. The murdered Russian royal family can finally rest in peace, thanks to DNA analysis.

SUMMARY

1. DNA must encode the information necessary for a cell's survival and specialization and be able to replicate.

2. Many experiments described DNA and showed it to be the genetic material. Miescher identified DNA in white blood cell nuclei. Garrod connected heredity to symptoms resulting from enzyme abnormalities. Griffith identified a substance that transmits pneumonia to bacteria; Avery, MacLeod, and McCarty showed that the transforming principle is DNA; Hershey and Chase confirmed that the genetic material is DNA and not protein. Levene described the proportions of nucleotide components. Chargaff discovered that A and T, and G and C, occur in equal proportions. Watson and Crick combined these clues to propose the double helix conformation of DNA.

3. The rungs of the DNA double helix consist of hydrogen-bonded complementary base pairs (A with T, and C with G). The rails are chains of alternating **deoxyriboses** and phosphates, which run **antiparallel** to each other. DNA is highly coiled. DNA replication is **semiconservative.**

4. DNA unwinds locally at several **initiation sites. Replication forks** form as hydrogen bonds break between an initial base pair. **RNA polymerase** builds a short **RNA primer,** which is eventually replaced with DNA. Next, **DNA polymerase** fills in DNA bases, and **ligase** seals the sugar-phosphate backbone. Replication proceeds in a 5′ to 3′ direction, necessitating that the process be discontinuous in short stretches on at least one strand.

5. DNA can repair itself in a variety of ways. **Photoreactivation** splits pyrimidine dimers. **Excision repair** cuts out the damaged area and replaces it with correct bases. **Mismatch repair** scans newly replicated DNA for mispairing and corrects the error.

TO REVIEW . . .

1. State the functions of the following enzymes that participate in DNA replication or repair:

a. RNA polymerase
b. DNA polymerase
c. ligase
d. helicase
e. photolyase

2. DNA specifies and regulates the cell's synthesis of protein. If a cell contains all the genetic material it must have to carry out protein synthesis, why must the DNA also be replicated?

3. Match the experiment or observation on the left to the concept it illustrates on the right:

a. Soiled bandages contain an acidic substance that includes nitrogen and phosphorus	(1) A "transforming principle" transmits the ability of bacteria to cause pneumonia in mice
b. "Blender experiments" showed that the part of a virus that infects bacteria contains phosphorus, but not sulfur	(2) Complementary base pairing is part of DNA structure and maintains a double helix of constant width
c. DNA contains equal amounts of G and C and of A and T	(3) Discovery of nuclein
d. Type R bacteria transmit a substance that restores the ability of heat-killed type S bacteria to cause pneumonia in mice	(4) DNA is the hereditary material

4. What part of the DNA molecule encodes information?

5. Write the complementary sequence of a new strand of DNA replicated from each of the following parental base sequences:

a. T C G A G A A T C T C G A T T
b. C C G T A T A G C C G G T A C
c. A T C G G A T C G C T A C T G

6. List the steps of DNA replication.

7. Choose an experiment mentioned in the chapter and analyze how it follows the scientific method.

8. Why is a disorder of DNA repair inherited?

9. How is DNA polymerase involved in both DNA replication and DNA repair?

TO THINK ABOUT . . .

1. To diagnose a rare form of encephalitis (brain inflammation), a researcher needs a million copies of a viral gene. She decides to use the polymerase chain reaction on a sample of cerebrospinal fluid, the fluid that bathes the person's infected brain. If one cycle of PCR takes 2 minutes, how long will it take the researcher to obtain her million-fold amplification?

2. Give an example from the chapter of when different types of experiments were used to address the same hypothesis. Why might this be necessary?

3. The experiments that revealed DNA structure and function used a variety of organisms—from peas, barley, fruit flies, and *E. coli* to yeast, molds, mice, and humans. How can such diverse organisms demonstrate the same genetic principles?

4. A person with deficient or abnormal ligase or excision repair may have an increased cancer risk and chromosomes that cannot heal breaks. The person is, nevertheless, alive. How long would an individual lacking DNA polymerase be likely to survive?

5. Until recently, HIV infection was diagnosed by detecting antibodies in a person's blood or documenting a decline in the number of the type of white blood cell that HIV initially infects. Why is PCR detection more accurate?

6. Cancerous cells from a person with hereditary nonpolyposis colon cancer have different length microsatellite sequences. What is the nature of the defect in DNA repair underlying the cancer?

TO LEARN MORE . . .

References and Resources

Debenham, Paul G. April 11, 1996. Heteroplasmy and the Tsar. *Nature,* vol. 380. A DNA quirk solved a royal mystery.

Jaroff, Leon. March 15, 1993. Happy birthday, double helix. *Time.* Watson and Crick shared a rare visit on the 40th anniversary of their monumental deciphering of DNA structure.

Lewis, Ricki. August 1996. UV damage to DNA revisited. *Photonics Spectra.* In some situations, cells ignore signals to repair UV-induced DNA damage. The result can be cancer.

Lewis, Ricki. May 1, 1996. PCR technology makes additional inroads in the clinic and in the field. *Genetic Engineering News.* PCR, the technique that borrows the DNA replication mechanism, continues to find new applications.

Lewis, Ricki. April 1996. Using technology to teach difficult genetic concepts. *The American Biology Teacher.* PCR, preimplantation diagnosis, and antisense technology are based on genetic concepts.

Lewis, Ricki. September 1995. Correcting sun damage the biotech way. *Photonics Spectra.* New gene-containing sunblocks boost DNA repair to lessen sun-induced skin damage.

Mullis, Kary B. April 1990. The unusual origin of the polymerase chain reaction. *Scientific American.* How PCR arose from a brainstorm.

Richards, Robert I., and Grant R. Sutherland. February 1994. Simple repeat DNA is not replicated simply. *Nature Genetics,* vol. 6. Short repeated DNA sequences can disrupt replication.

Sancar, Azis. April 5, 1996. No "end of history" for photolyases. *Science,* vol. 272. Blue light enables a repair enzyme to correct ultraviolet damage.

Watson, James D. 1968. *The double helix.* New York: New American Library. An exciting, personal account of the discovery of DNA structure.

Watson, James D., and F. H. C. Crick. April 25, 1953. Molecular structure of nucleic acids: A structure for deoxyribose nucleic acid. *Nature,* vol. 171, no. 4356. The original paper describing the structure of DNA.

denotes articles by author Ricki Lewis that may be accessed via *Life's* home page /http://www.mhhe.com/sciencemath/biology/life/.

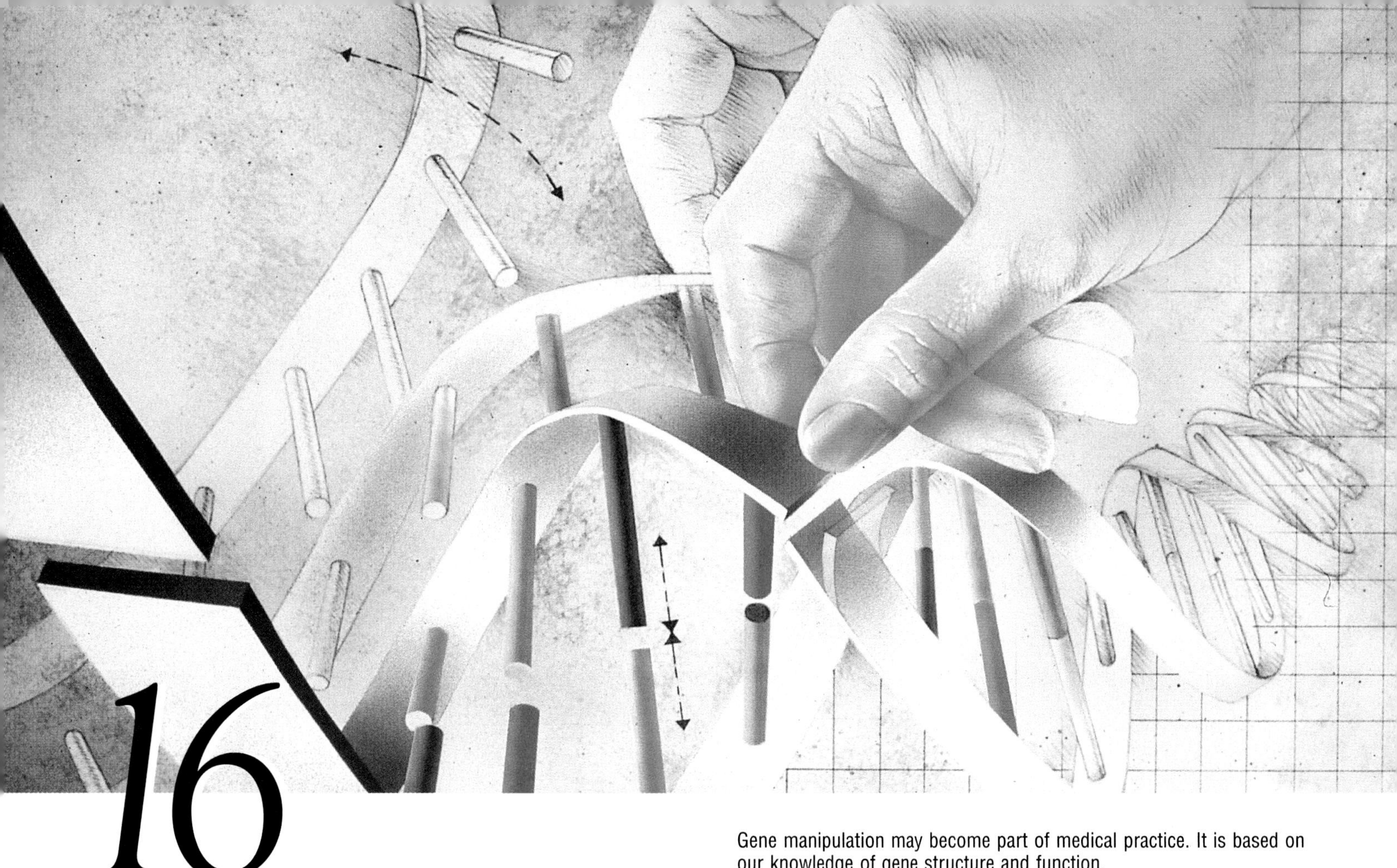

Gene manipulation may become part of medical practice. It is based on our knowledge of gene structure and function.

16 Gene Function

chapter outline

Transcription—The Genetic Crossroads 326
The Lac Operon Reveals Transcriptional Complexity
Transcription Factors
Steps of Transcription

Types of RNA 329
Messenger RNA (mRNA) Carries the Genetic Information
Ribosomal and Transfer RNAs Help Translate the Message
RNA Processing

Translation Expresses Genetic Information 331
The Genetic Code—From Genetic Message to Protein Product
Building a Protein
Protein Folding
Computers Coax Meaning from Genes

Mutation—Genetic Misinformation 338
Understanding Sickle Cell Disease
Causes of Mutation
Types of Mutations
Natural Protection Against Mutation

Genes to the Rescue

On September 14, 1990, at 12:52 P.M., 4-year-old Ashanti DeSilva sat up in bed at the National Institutes of Health in Bethesda, Maryland, and began receiving her own white blood cells through a needle in her arm. Earlier, doctors had removed the cells and added to them normal copies of a gene specifying the enzyme adenosine deaminase (ADA), which her cells lack because of an autosomal recessive condition. Soon after, 8-year-old Cynthia Cutshall received the same treatment. Both girls had been taking frequent enzyme supplements to prevent the complete lack of immunity that results from the enzyme deficiency.

In the years following, the girls stayed relatively healthy. They did require repeat treatments of their gene therapy as the altered white blood cells died out, as blood cells do.

Yet even as Ashanti and Cynthia were being treated, researchers were onto the next step—a longer-term correction of the inherited problem.

to be continued . . .

Transcription—The Genetic Crossroads

DNA replication preserves genetic information. **Translation** is the process that uses the information in a DNA sequence to construct a polypeptide's amino acid sequence. Between replication and translation lies **transcription,** a genetic crossroads that determines which genes are expressed in which tissues, at which stages of development. Transcription is the synthesis of a molecule of RNA that is complementary in sequence to one strand of the DNA double helix, called the **coding strand.** This RNA is essentially a mobile copy of a gene's information. Transcription and translation are sometimes summed up as "gene expression"—they carry out a gene's function. Figure 16.1 depicts the flow of genetic information from DNA to RNA to protein.

RNA differs from DNA in three ways: it contains the sugar ribose instead of deoxyribose, it contains the nitrogenous base uracil in place of thymine, and it is single- rather than double-stranded (fig. 16.2 and table 16.1).

The complexity of transcription lies in how a cell "knows" which genes to express and when to express them. What directs an immature red blood cell to transcribe the genes controlling hemoglobin synthesis rather than muscle-specific proteins? RNA is far more than a go-between that connects gene to protein—it participates in all aspects of gene expression.

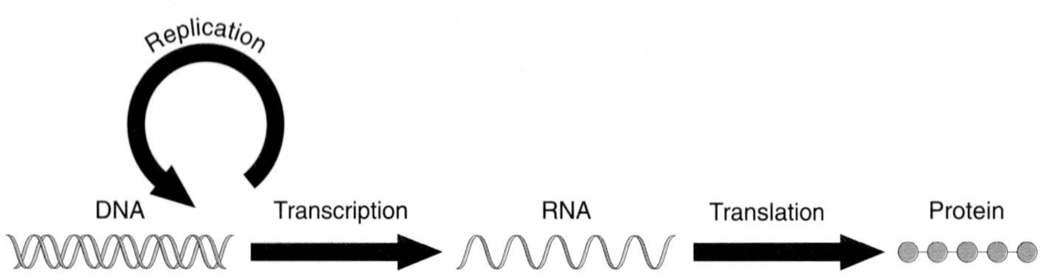

Figure 16.1

All organisms and most viruses use the directional flow of genetic information from DNA to RNA to protein. RNA also assists in DNA replication, and we are still discovering how RNA helps control gene expression. (A class of viruses called retroviruses, which includes HIV, has RNA as the genetic material and "reverse transcribes" it into DNA, which is opposite the usual direction of genetic information flow.)

The Lac Operon Reveals Transcriptional Complexity

Early evidence of the complexity of gene expression came from French biologists François Jacob and Jacques Monod in 1961. They described the remarkable ability of *E. coli* to produce the exact enzymes they require to metabolize the sugar lactose only

Table 16.1 Distinctions Between RNA and DNA

RNA	DNA
Single-stranded	Double-stranded
Has uracil as a base	Has thymine as a base
Ribose as the sugar	Deoxyribose as the sugar
Uses protein-encoding information	Maintains protein-encoding information

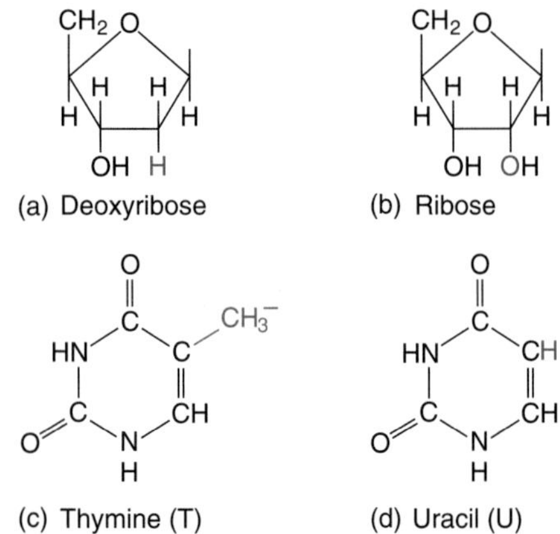

Figure 16.2

DNA and RNA differ structurally from each other in three ways. First, DNA nucleotides contain the sugar deoxyribose (*a*), whereas RNA contains ribose (*b*). Second, DNA nucleotides include the pyrimidine thymine (*c*), whereas RNA has uracil (*d*). And third, DNA is double-stranded, while RNA is generally single-stranded. The two types of nucleic acids also have different functions.

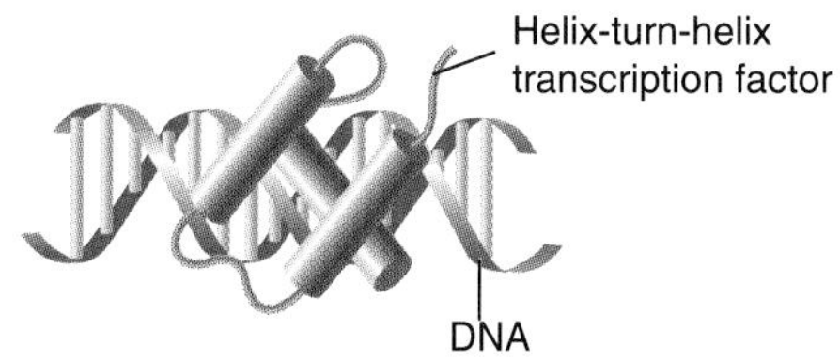

Figure 16.3

Transcription factors are proteins that bind to DNA and regulate gene expression by turning on or blocking transcription of particular genes. Motifs are regions named for the characteristic shapes of the DNA binding regions.

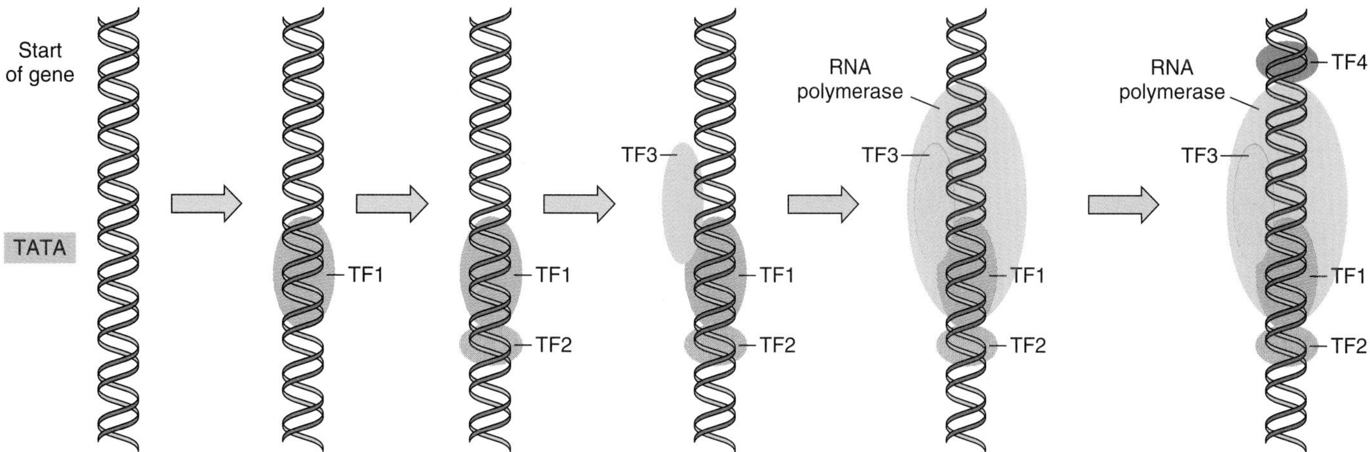

Figure 16.4

Transcription factors come together to start transcription of a gene. To begin transcription, four or more protein transcription factors bind near the start of the gene, guided by a "TATA" box. The bound transcription factors then enable an RNA polymerase to bind. The entire region where transcription factors and RNA polymerase bind is the promoter.

when lactose is present in the cell's surroundings. What "tells" a simple bacterial cell to transcribe the genes whose products metabolize lactose at precisely the right time?

The lactose itself is the trigger. The sugar attaches to a protein that normally binds to and suppresses the DNA sequence that contains the signal that tells the cell to begin transcribing the three enzymes required to break down the sugar. The lactose removes the blocking protein, and the cell synthesizes the enzymes. Lactose, in a sense, causes its own dismantling.

Jacob and Monod named the genes (and their controls) that produce the enzymes required for lactose metabolism an **operon.** Soon, geneticists discovered operons for the metabolism of other nutrients, and in other bacteria. Some operons, like the lac operon, negatively control transcription by removing a block. Others act positively, producing factors that turn on transcription.

Transcription Factors

In bacteria, operons function like switches, turning gene transcription on or off. In multicellular eukaryotes like ourselves, genetic control is more complex because different cell types express different subsets of genes. To manage such complexity, groups of proteins called **transcription factors** come together, forming an apparatus that binds DNA and initiates transcription at specific sites on the chromosome. The transcription factors, activated by signals from outside the cell, set the stage for transcription to begin by forming a pocket for **RNA polymerase** (RNAP)—the enzyme that actually builds an RNA chain.

Several types of transcription factors are required to transcribe a eukaryotic gene. Because transcription factors are proteins, they too are gene-encoded. Genes for transcription factors may be located near the genes they control or as far away as 40,000 bases. DNA may form loops so that the transcription factor genes encoding proteins that act together come near each other. Proteins called nuclear matrix proteins organize the cell's nucleus so that genes actively being transcribed, and transcription factors in use, are partitioned off together, much as books on a specialized topic might be grouped together in a library for easier access.

Hundreds of transcription factors are known. Many of them have regions in common called **motifs** that fold into similar three-dimensional shapes, or conformations. These motifs generally enable the transcription factor to bind DNA. They have very colorful names, such as "helix-turn-helix," "zinc fingers," and "leucine zippers," that reflect their distinctive shapes. Figure 16.3 shows how one transcription factor associates intimately with DNA, thereby controlling gene expression.

Steps of Transcription

How do transcription factors and RNA polymerase "know" where to bind to DNA to begin transcribing a specific gene? Transcription factors and RNA polymerase are attracted to a control sequence, called a **promoter,** near the start of the gene. The first transcription factor to bind, called a TATA binding protein, is attracted to a DNA sequence called a TATA box, which consists of the base sequence TATA surrounded by long stretches of G and C. Once the first transcription factor binds, it attracts others, and finally RNA polymerase joins the complex (fig. 16.4). Other sequences in or near the promoter regulate the frequency of transcription. The transcription factor complex guides RNA polymerase so that it binds just before the gene sequence starts.

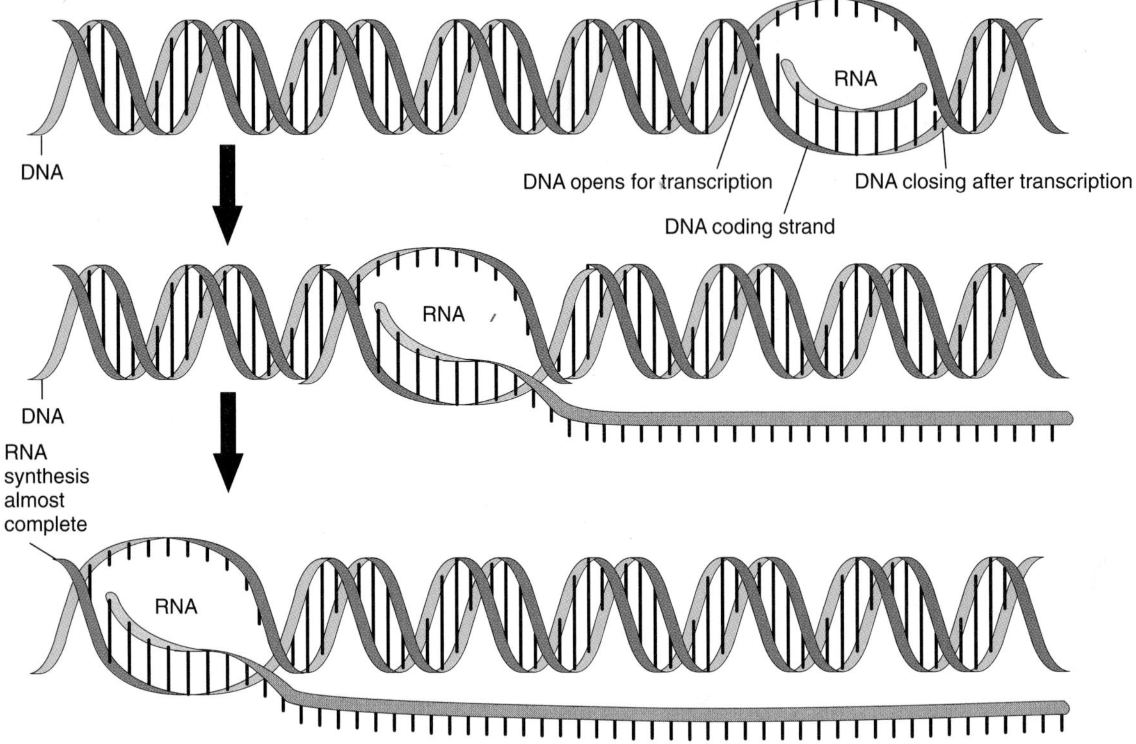

Figure 16.5

Transcription of RNA from DNA. The DNA double helix opens and closes, allowing RNA polymerase access to the DNA coding strand, from which RNA is transcribed.

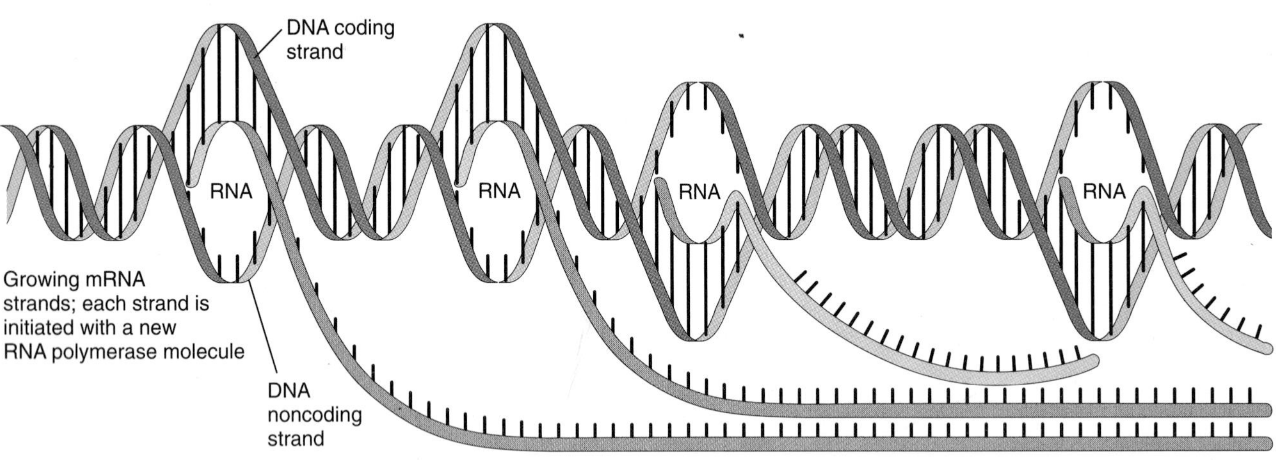

Figure 16.6

The same DNA coding sequence can be transcribed several times simultaneously.

Complementary base pairing underlies transcription, just as it does DNA replication. First, enzymes unwind the DNA double helix, and RNA nucleotides bond with exposed complementary bases on the coding DNA strand (fig. 16.5). RNA polymerase knits together the RNA nucleotides in the sequence the DNA specifies, moving along the DNA strand in a 3′ to 5′ direction. A terminator sequence in the DNA indicates where the gene's RNA-encoding region ends.

For a particular gene, RNA is transcribed from only one strand of the DNA double helix, the coding (or sense) strand. The other DNA strand that isn't transcribed is called the noncoding, or antisense, strand. RNA transcribed from the DNA coding strand is called **sense RNA.** Several RNAs may be transcribed from the same DNA sense strand simultaneously (fig. 16.6).

To determine the sequence of sense RNA bases transcribed from a gene, write the RNA bases that are complementary to the coding DNA strand, using uracil opposite adenine. For example, if a DNA coding strand has the sequence

C C T A G C T A C

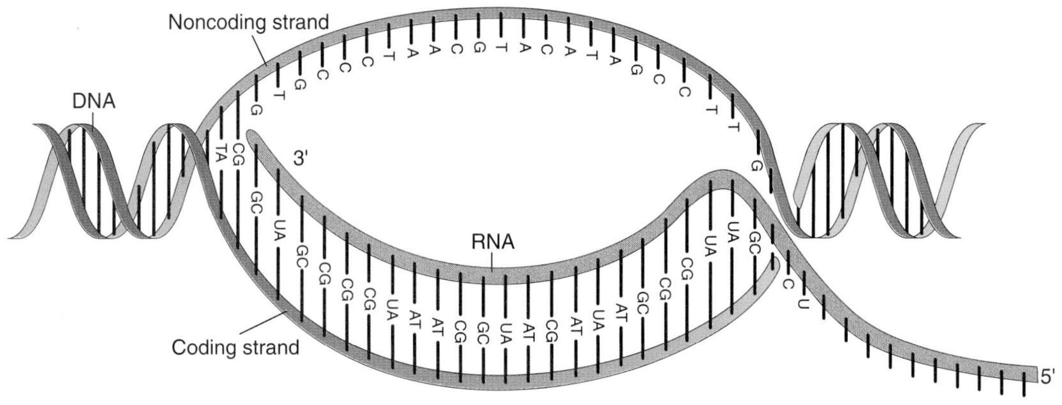

Figure 16.7

A single strand of mRNA is transcribed against the coding strand of a DNA double helix. Note that the RNA sequence is complementary to that of the DNA coding strand and therefore is the same sequence as the DNA noncoding (antisense) strand, with uracil (U) in place of thymine (T).

then it is transcribed into sense RNA with the sequence

G G A U C G A U G

Note the relationship between the sense RNA and the noncoding side of the DNA double helix—the sequences are the same, with T in the DNA wherever U is in the RNA. The noncoding DNA sequence is:

G G A T C G A T G

Figure 16.7 depicts the relationships between the two DNA strands and RNA.

Mastering Concepts

1. How do DNA replication, RNA transcription, and translation maintain and use genetic information?
2. How does lactose metabolism in *E. coli* illustrate control of gene expression?
3. How do transcription factors control gene expression?
4. What are the steps of transcription?

Types of RNA

As RNA is synthesized along DNA, it curls into three-dimensional shapes, or conformations, determined by complementary base pairing within the same RNA molecule. These three-dimensional shapes determine how RNA functions, as we shall soon see. Several types of RNA exist (table 16.2).

Messenger RNA (mRNA) Carries the Genetic Information

Messenger RNA (mRNA) carries the information specifying a particular protein product. Each three mRNA bases in a row forms a genetic code word, or **codon,** that specifies a particular amino acid. Because genes vary in length, so do mRNA molecules. Most mRNAs are 500 to 1,000 bases long.

Ribosomal and Transfer RNAs Help Translate the Message

Two other types of RNA help use the information carried in an mRNA.

Ribosomal RNA (rRNA) molecules range from 100 to nearly 3,000 nucleotides long. This type of RNA associates with certain proteins to form a ribosome. Recall from chapter 4 that a ribosome is a structural support for protein synthesis. A ribosome has two subunits that are separate in the cytoplasm but join at the site of protein synthesis (fig. 16.8). Certain rRNAs catalyze formation of bonds between amino acids. Such an RNA with enzymatic function is called a **ribozyme.**

Transfer RNA (tRNA) molecules are "connectors," binding mRNA codons at one end and specific amino acids at the other. A tRNA molecule is only 75 to 80 nucleotides long. Some of its bases form weak (hydrogen) bonds with each other, which fold the tRNA into a characteristic shape (fig. 16.9). One loop of the tRNA has three bases in a row that form the **anticodon,** which is complementary to

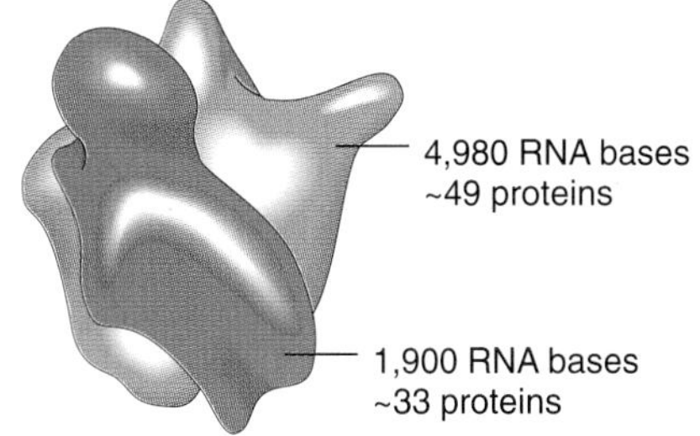

Figure 16.8

A ribosome from a eukaryote, shown here, has two subunits, containing 82 proteins and four rRNA molecules altogether. A prokaryotic ribosome is smaller.

Table 16.2 Major Types of RNA

Molecule	Size (number of nucleotides)	Function
mRNA	500–1,000	Codons encode amino acid sequence
rRNA	100–3,000	Associates with proteins to form ribosomes, which structurally support and catalyze protein synthesis
tRNA	75–80	Binds mRNA codon on one end, amino acid on the other, linking a gene's message to the amino acid sequence it encodes

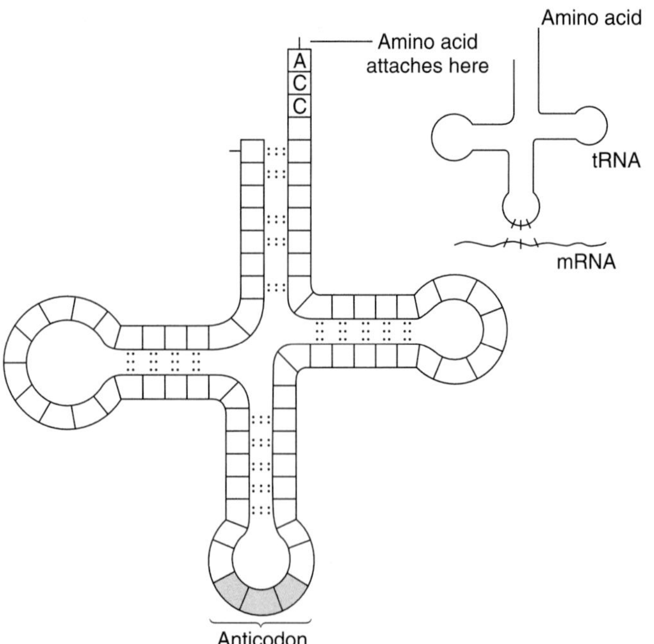

Figure 16.9

Transfer RNA. Certain nucleotide bases within a tRNA molecule hydrogen bond with each other to give the molecule a "cloverleaf" conformation that can be represented in two dimensions. The dotted lines indicate hydrogen bonds, and the filled-in bases at the bottom form the anticodon. Each tRNA terminates with the sequence CCA, and a particular amino acid covalently bonds with the RNA at this end.

an mRNA codon. The end of the tRNA opposite the anticodon forms a strong (covalent) bond to a specific amino acid. A tRNA with a particular anticodon always carries the same amino acid. For example, a tRNA with the anticodon sequence AAG always picks up the amino acid phenylalanine. Special enzymes attach amino acids to the tRNAs that bear the appropriate anticodons.

RNA Processing

In prokaryotes, RNA is translated into protein as soon as it is transcribed from DNA because a nucleus does not physically separate the two processes. In eukaryotic cells, mRNA must first exit the nucleus to enter the cytoplasm, where protein synthesis occurs. RNA is altered before it participates in protein synthesis in these more complex cells.

As mRNA is transcribed, a short sequence of modified nucleotides, called a cap, is added to the 5′ end of the molecule. At the 3′ end, 100 to 200 adenines bind, forming a "poly A tail." The cap and poly A tail "tell" the cell which mRNAs should exit the nucleus.

Not all of an mRNA is translated into an amino acid sequence. Parts of mRNAs called **introns** are transcribed but later removed. The ends of the remaining molecule are spliced together before the mRNA is translated. The parts of mRNA that are translated are called **exons** (fig. 16.10). Introns are excised by small ribozymes that associate in groups with proteins to cut introns out and knit exons together to form the mature mRNA that exits the nucleus.

Introns range in size from 65 to 100,000 bases. While the average exon is 100 to 300 bases long, the average intron is about 1,000 bases long. Many genes are riddled with introns—the human collagen gene, for example, contains 50 of them. The num-

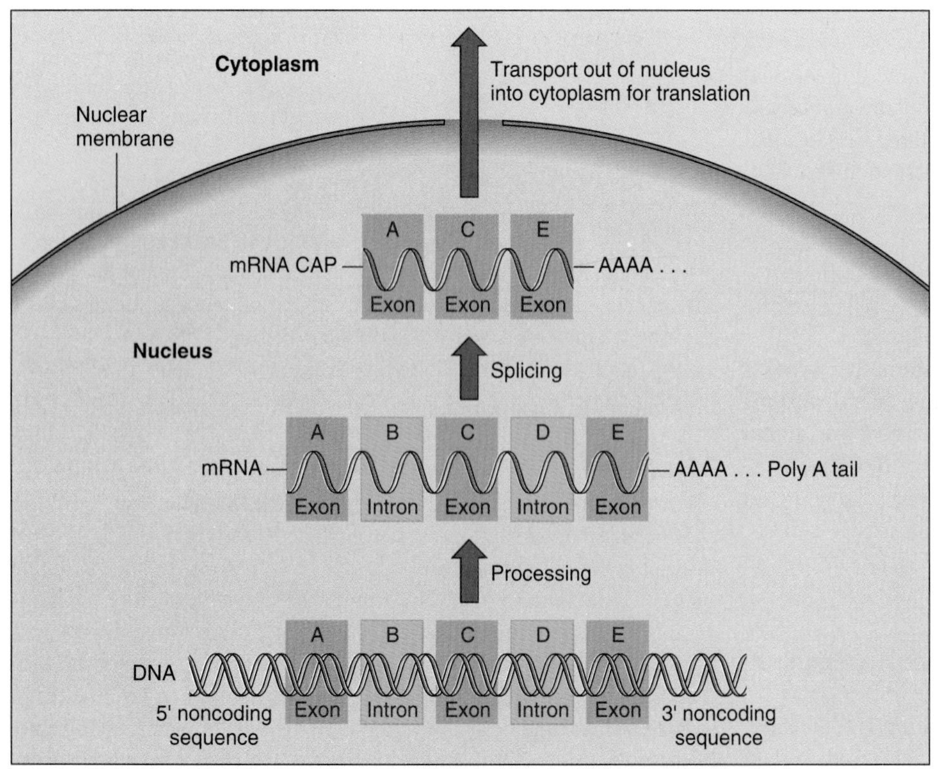

Figure 16.10

Messenger RNA processing—the maturing of the message. Several steps carve the mature mRNA. First, noncoding sequences surrounding bases in the DNA are not transcribed. A cap head and poly A tail are added, and introns are spliced out.

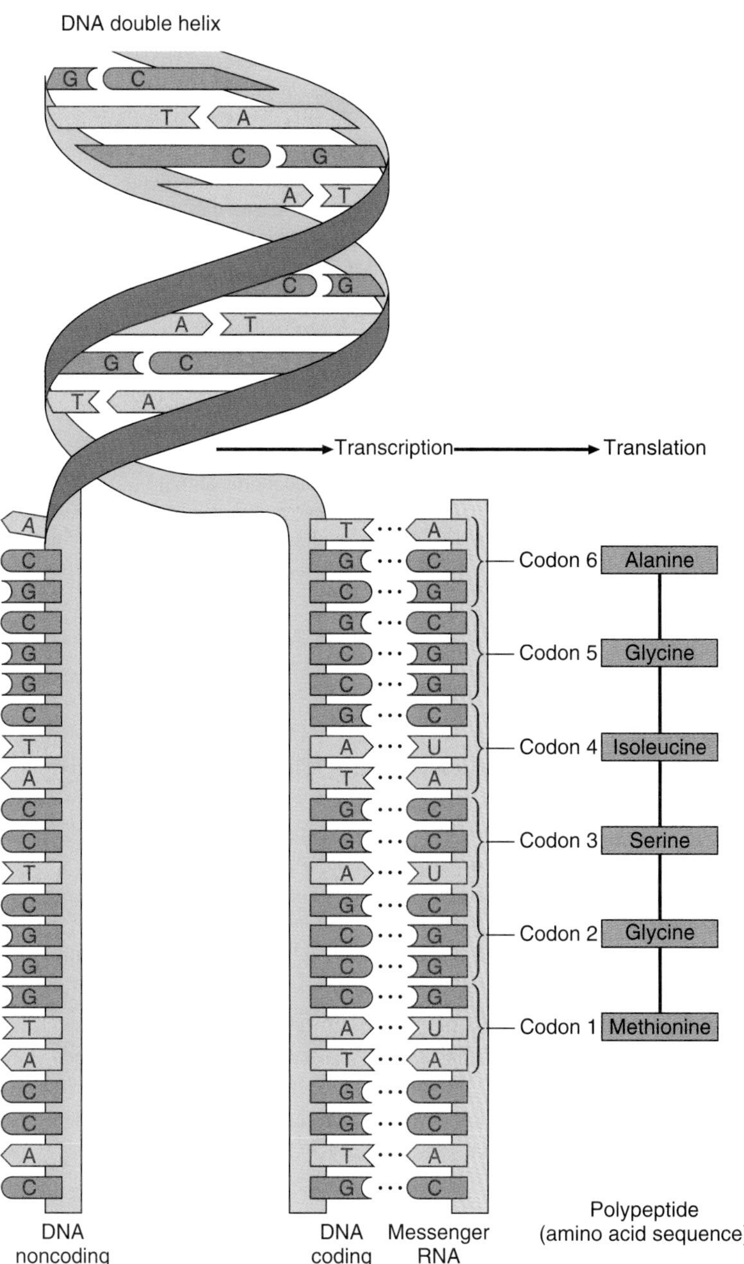

Figure 16.11

From DNA to RNA to protein. Messenger RNA is transcribed from a locally unwound portion of DNA. In translation, transfer RNA matches up mRNA codons with amino acids.

ber, size, and arrangement of introns vary from gene to gene. Once regarded as an oddity because their existence was not expected, introns are now known to make up large parts of the genomes of complex organisms. They are not common among prokaryotes.

Mastering Concepts

1. Why is RNA conformation important?
2. How are amino acids brought to mRNA codons?
3. What are the components and function of ribosomes?
4. What are the steps of mRNA maturation?

Size of a Genetic Code Word (Codon) To encode a protein alphabet of 20 amino acids:	**Logic**
mRNA genetic code of one letter U C A G	4 combinations: **not sufficient**
mRNA genetic code of two letters UU CC AA GG UC CU UA AU UG GU CA AC CG GC AG GA	16 combinations: **not sufficient**
mRNA genetic code of three letters UUU UUC UUA UUG UCU UCC UCA UCG UAU UAC UAA UAG UGU UGC UGA UGG CUU CUC CUA CUG CCU CCC CCA CCG CAU CAC CAA CAG CGU CGC CGA CGG AUU AUC AUA AUG ACU ACC ACA ACG AAU AAC AAA AAG AGU AGC AGA AGG GUU GUC GUA GUG GCU GCC GCA GCG GAU GAC GAA GAG GGU GGC GGA GGG	64 combinations: **sufficient**

Figure 16.12

A "thought" experiment in logic reveals the triplet nature of the genetic code.

Translation Expresses Genetic Information

Transcription copies the information encoded in a DNA base sequence into the complementary language of mRNA. The next step is translating this "message" into the specified sequence of amino acids. Particular mRNA codons (three bases in a row) correspond to particular amino acids (fig. 16.11). This correspondence between the chemical languages of mRNA and protein is called the **genetic code.** In the 1960s, many researchers collaborated to crack the code and determine which mRNA codons correspond to which amino acids. They used a combination of logic and experiments and began by posing certain questions.

The Genetic Code—From Genetic Message to Protein Product

Question 1—How many RNA bases specify one amino acid? Because the number of different protein building blocks (20) exceeds the number of different mRNA building blocks (4), each codon must contain more than one mRNA base. A genetic code in which codons consist of only one mRNA base could specify only four different amino acids, one corresponding to each of the four bases: A, C, G, and U (fig. 16.12). A code consisting of two-base codons could specify only 16 different amino acids. A code with a minimum of three-base codons is necessary to specify the 20 different amino acids that make up biological proteins.

Francis Crick and his coworkers conducted experiments that confirmed the triplet nature of the genetic code. They added one, two, or three bases within a gene with a known sequence and protein product. Altering the sequence by one or two bases greatly

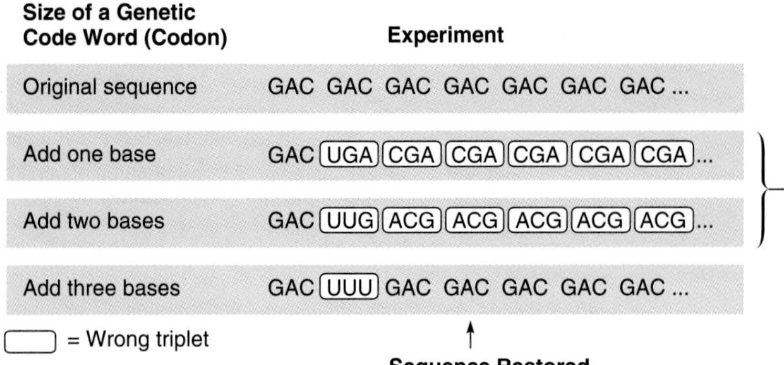

Figure 16.13

Francis Crick and coworkers confirmed the "thought experiment" in figure 16.12 showing that the genetic code is triplet by adding and subtracting one, two, and three bases to the sequence of a well-studied gene. Adding or subtracting one or two bases disrupted the reading frame, but adding or subtracting three bases restored it. Therefore, the code is triplet.

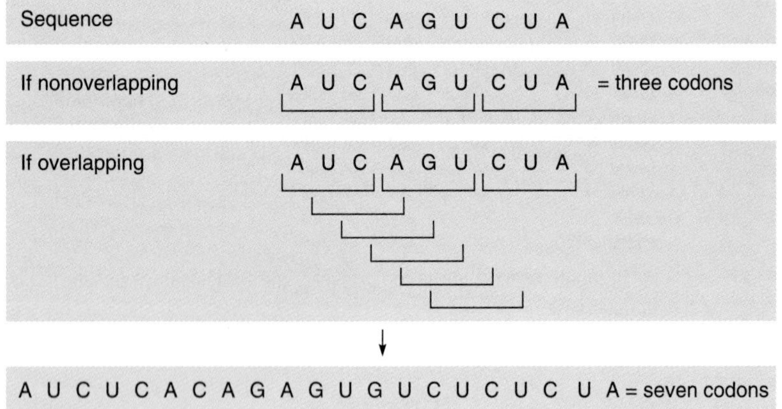

Figure 16.14

An overlapping genetic code may seem economical, but it is restrictive, dictating that certain amino acids must follow others in a protein's sequence. This does not happen; therefore, the genetic code is nonoverlapping.

disrupted the coded order of specified amino acids, known as the **reading frame.** This produced a different amino acid sequence. However, adding or deleting three bases in a row added or deleted only one amino acid in the protein product (fig. 16.13).

Question 2—Does the genetic code overlap? Consider the hypothetical mRNA sequence AUCAGUCUA. If the genetic code is triplet and does not overlap (that is, each three bases in a row forms a codon, but any one base is part of only one codon), then this sequence contains three codons: AUC, AGU, and CUA. If the code overlaps, the sequence contains seven codons: AUC, UCA, CAG, AGU, GUC, UCU, and CUA.

An overlapping code would pack maximal information into a limited number of bases (fig. 16.14). An overlapping code, however, constrains protein structure because certain amino acids would always follow certain others. For example, the amino acid the first codon specifies, AUC, would always be followed by an amino acid whose codon begins with UC. Experiments that determine the sequences of proteins show that no specific amino acid always follows another. The code does not overlap.

Question 3—Can mRNA codons signal anything other than amino acids? Chemical analysis eventually showed that the genetic code contains directions for starting and stopping translation. The codon AUG signals "start," and the codons UGA, UAA, and UAG each signify "stop." Another form of "punctuation" is a short sequence of bases at the start of each mRNA, called the **leader sequence,** which enables the mRNA to hydrogen bond with rRNA in a ribosome.

Question 4—Do all species use the same genetic code? The fact that all species use the same mRNA codons to specify the same amino acids is part of the abundant evidence that all life on earth evolved from a common ancestor. The genetic code is universal, with the exception of a few genes in the mitochondria of certain single-celled organisms. The ability of a cell from one species to translate mRNA from another species makes possible recombinant DNA technology, discussed in Biology in Action 16.1.

Question 5—Which codons specify which amino acids? In 1961, researchers at the National Institutes of Health began deciphering which codons specify which amino acids. An elegant series of experiments gave them the answer. First they synthesized mRNA molecules in the laboratory. Then they added them to test tubes containing all the chemicals and structures needed for translation, which they had extracted from *E. coli* cells.

The first synthetic mRNA tested had the sequence UUU UUU. . .. In the test tube, this translated into a polypeptide consisting entirely of one amino acid type: phenylalanine. Thus, the first entry in the genetic code dictionary was revealed: the codon UUU specifies the amino acid phenylalanine. The number of phenylalanines always equaled one-third of the number of mRNA bases, confirming that the genetic code is triplet and does not overlap. The next three experiments revealed that AAA codes for the amino acid lysine, GGG for glycine, and CCC for proline.

The next step was to synthesize chains of alternating bases. mRNA of the sequence AUAUAU . . . introduced codons AUA and UAU. When translated, the mRNA yielded an amino acid sequence of alternating isoleucines and tyrosines. But was AUA isoleucine and UAU tyrosine, or vice versa? Another experiment was needed.

An mRNA of sequence UUUAUAUUUAUA . . . encoded alternating phenylalanine and isoleucine. Because the first experiment showed that UUU codes for phenylalanine, the experimenters knew that AUA must code for isoleucine. If AUA codes for isoleucine, they reasoned, looking back at the previous experiment, then UAU must code for tyrosine. Table 16.3 summarizes some of these experiments.

By the end of the 1960s, researchers had deciphered the entire genetic code (table 16.4). Many research groups contributed

Table 16.3 Deciphering RNA Codons and the Amino Acids They Specify

Synthetic RNA	Encoded Amino Acid Chain	Puzzle Piece
UUUUUUUUUUUUUUUUUU	phe-phe-phe-phe-phe-phe	UUU = phe
AAAAAAAAAAAAAAAAAA	lys-lys-lys-lys-lys-lys	AAA = lys
GGGGGGGGGGGGGGGGGG	gly-gly-gly-gly-gly-gly	GGG = gly
CCCCCCCCCCCCCCCCCC	pro-pro-pro-pro-pro-pro	CCC = pro
AUAUAUAUAUAUAUAUAU	ilu-tyr-ilu-tyr-ily-tyr	AUA = ilu or tyr UAU = ilu or tyr
UUUAUAUUUAUAUUUAUA	phe-ilu-phe-ilu-phe-ilu	AUA = ilu UAU = tyr

Table 16.4 The Genetic Code

First Letter	Second Letter: U	Second Letter: C	Second Letter: A	Second Letter: G	Third Letter
U	UUU, UUC: phenylalanine (phe) UUA, UUG: leucine (leu)	UCU, UCC, UCA, UCG: serine (ser)	UAU, UAC: tyrosine (tyr) UAA: "stop" UAG: "stop"	UGU, UGC: cysteine (cys) UGA: "stop" UGG: tryptophan (try)	U C A G
C	CUU, CUC, CUA, CUG: leucine (leu)	CCU, CCC, CCA, CCG: proline (pro)	CAU, CAC: histidine (his) CAA, CAG: glutamine (gln)	CGU, CGC, CGA, CGG: arginine (arg)	U C A G
A	AUU, AUC, AUA: isoleucine (ilu) AUG: methionine (met) and "start"	ACU, ACC, ACA, ACG: threonine (thr)	AAU, AAC: asparagine (asn) AAA, AAG: lysine (lys)	AGU, AGC: serine (ser) AGA, AGG: arginine (arg)	U C A G
G	GUU, GUC, GUA, GUG: valine (val)	GCU, GCC, GCA, GCG: alanine (ala)	GAU, GAC: aspartic acid (asp) GAA, GAG: glutamic acid (glu)	GGU, GGC, GGA, GGG: glycine (gly)	U C A G

to this monumental task. Crick organized an "amino acid club" and inducted a new member whenever someone added a new piece to the puzzle of the genetic code.

Building a Protein

Protein synthesis requires mRNA, tRNAs carrying amino acids, ribosomes, energy-storing molecules such as adenosine triphosphate (ATP), and various protein factors. These pieces come together at the beginning of translation in a stage called initiation. First, the mRNA leader sequence hydrogen bonds with a short sequence of rRNA in a small ribosomal subunit. The first mRNA codon to specify an amino acid is always AUG, which attracts an initiator tRNA that carries the amino acid methionine (fig. 16.15). This methionine signifies the start of a polypeptide. The small ribosomal subunit, the mRNA bonded to it, and the initiator tRNA with its attached methionine form the **initiation complex.**

In the next stage of translation, elongation, a large ribosomal subunit attaches to the initiation complex. The codon adjacent to the initiation codon, which is GGA in figure 16.15, then bonds to its complementary anticodon, which is part of a free tRNA that carries the amino acid glycine. The two amino acids (methionine and glycine in the example), which are still attached to their tRNAs, align. A peptide bond forms between them, and the first tRNA is released. It will pick up another amino acid and be used again. The ribosome and its attached mRNA now bind to a single tRNA, with two amino acids extending from it. This begins the polypeptide.

Next, the ribosome moves down the mRNA by one codon. A third tRNA enters, carrying its amino acid (cysteine in fig. 16.15). This third amino acid aligns with the other two and forms a peptide bond to the second amino acid in the growing chain. The tRNA attached to the second amino acid is released and recycled. The polypeptide continues to build, one amino acid at a time, as new tRNAs deliver their cargo. The tRNAs carry anticodons that correspond to the mRNA codons.

16.1 biology in action

From Genetics to Biotechnology

Through the 1960s and 1970s, researchers elucidated DNA structure and function. Since then, that knowledge has been put to practical use and biotechnologies have been created with applications in such diverse fields as health and veterinary care, agriculture, and forensics. Here is a look at biotechnologies that spawned from basic research on gene structure and function (fig. 16.A).

Antisense Technology

Is it wasteful to transcribe only one side of the DNA double helix? It may seem so, but the unexpressed, "antisense" side may control gene expression. As the details of transcription and translation were discovered, researchers realized that mRNA or DNA complementary in base sequence to sense DNA or sense RNA could block gene expression by physically binding to the sequence on its way to being used. Use of complementary nucleic acid sequences to block expression of other sequences is called antisense technology.

Antisense technology first targeted plants. "FlavrSavr" tomatoes owed their longer shelf lives to an antisense sequence stamped on the gene encoding a ripening enzyme. With ripening slowed, the manipulated tomatoes remain firm and red in supermarket bins for days longer than their unaltered counterparts. Antisense sequences directed against pigment genes in orchids created new floral variants. In human health care, antisense-based drugs are being developed to treat disorders in which genes are overexpressed, such as certain cancers and autoimmune disorders, or to squelch expression of genes from infectious viruses. Antisense technology might also be used to silence a mutant gene, and then another technology used to replace it.

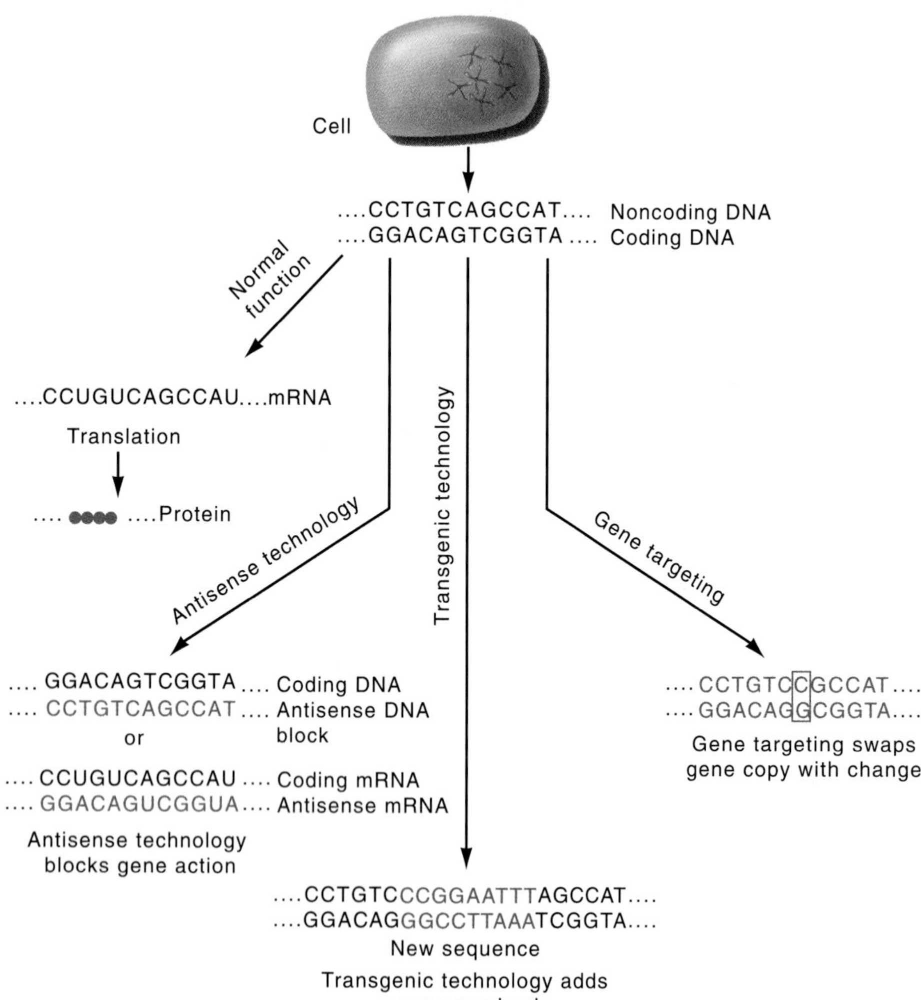

Figure 16.A
Biotechnologies. Antisense technology uses either antisense DNA or RNA complementary to sense RNA to selectively block gene action. Transgenic technology introduces a genetic change into a gamete or fertilized ovum, so that the change is perpetuated in all cells of the resulting individual. Gene targeting swaps one gene for another. It is more precise than transgenic technology.

Recombinant DNA

The fact that all species use the same genetic code means that one type of organism can express a gene from another. Recombinant DNA technology accomplishes this for single-celled organisms or cells growing in culture. To create such a cell, a researcher cuts a gene of interest as well as the host's DNA with restriction enzymes, which cut at specific DNA sequences. Using these enzymes gives the DNA "sticky ends," which may result in a host cell picking up some of the "foreign" DNA in its own DNA and then expressing it.

Recombinant DNA technology is used to create drugs. The first was human insulin, to treat diabetes. Inserting the human gene for insulin into bacteria gave pure and plentiful supplies of the drug, which became increasingly important as the AIDS epidemic made other sources of human proteins unsafe. More than two dozen drugs are now made this way. Recombinant DNA technology also gives common bacteria new talents. For example, *E. coli* can produce indigo dye, used to color blue jeans, thanks to a gene from another type of bacterium. This saves the endangered plant previously used to supply indigo dye.

Transgenic Technology

The equivalent of recombinant DNA technology in a multicellular organism is transgenic technology. Genes of interest are introduced into a gamete or fertilized ovum, and the organism that develops from it carries the inserted gene. Researchers have clever ways of adding a gene—zapping the gamete or fertilized ovum with electricity to open temporary holes, shooting it in, sending it in in a fatty bubble (a liposome), injecting it, or hitching it to an infecting virus. The human gene therapies currently being tested do not produce transgenic humans but instead correct abnormalities in specific cells or tissues (fig. 16.B).

Transgenic technology focuses on species used in agriculture, in research, or as new drug sources. Transgenic goats manufacture human tPA, a drug used to limit damage in heart attacks and strokes. A sheep named Nancy makes an enzyme used to treat hereditary emphysema in humans, and a sheep named Ethel manufactures a human clotting factor. Crops are transgenically altered to resist pests and overcome harsh conditions. Transgenic fish given growth hormone genes from larger species reach market size incredibly fast. A transgenic tobacco plant manufactures human hemoglobin, the four-subunit protein that carries oxygen in the blood.

Gene Targeting Creates "Knockout" Organisms

Transgenic technology isn't very precise, because the gene of interest enters the host cell's genome at random sites. More controllable is gene targeting, which is based on crossing-over. Just as homologs cross over during meiosis, in gene targeting, an introduced gene exchanges places with its counterpart on a host's chromosome. Like transgenic technology, gene targeting is conducted on gametes or fertilized ova, and an organism with the genetic change develops.

When gene targeting is used to replace a gene with a nonfunctional version, researchers can observe the effects of inactivating, or "knocking out," a gene. Knockout mice provide models of human genetic disease, and experimental therapies can be tried on them before human trials. The technology has had surprising results when researchers inactivate a gene thought to be vital, and the resulting mice are robust and healthy! This usually means that the gene is not the only one performing its function.

Figure 16.B
In gene therapy, functional genes are added to tissues directly affected in an illness or to nearby cells that secrete essential biochemicals. Disabled viruses or lipid-based compounds are used to deliver therapeutic genes, either outside the body for later implantation or directly in the body.

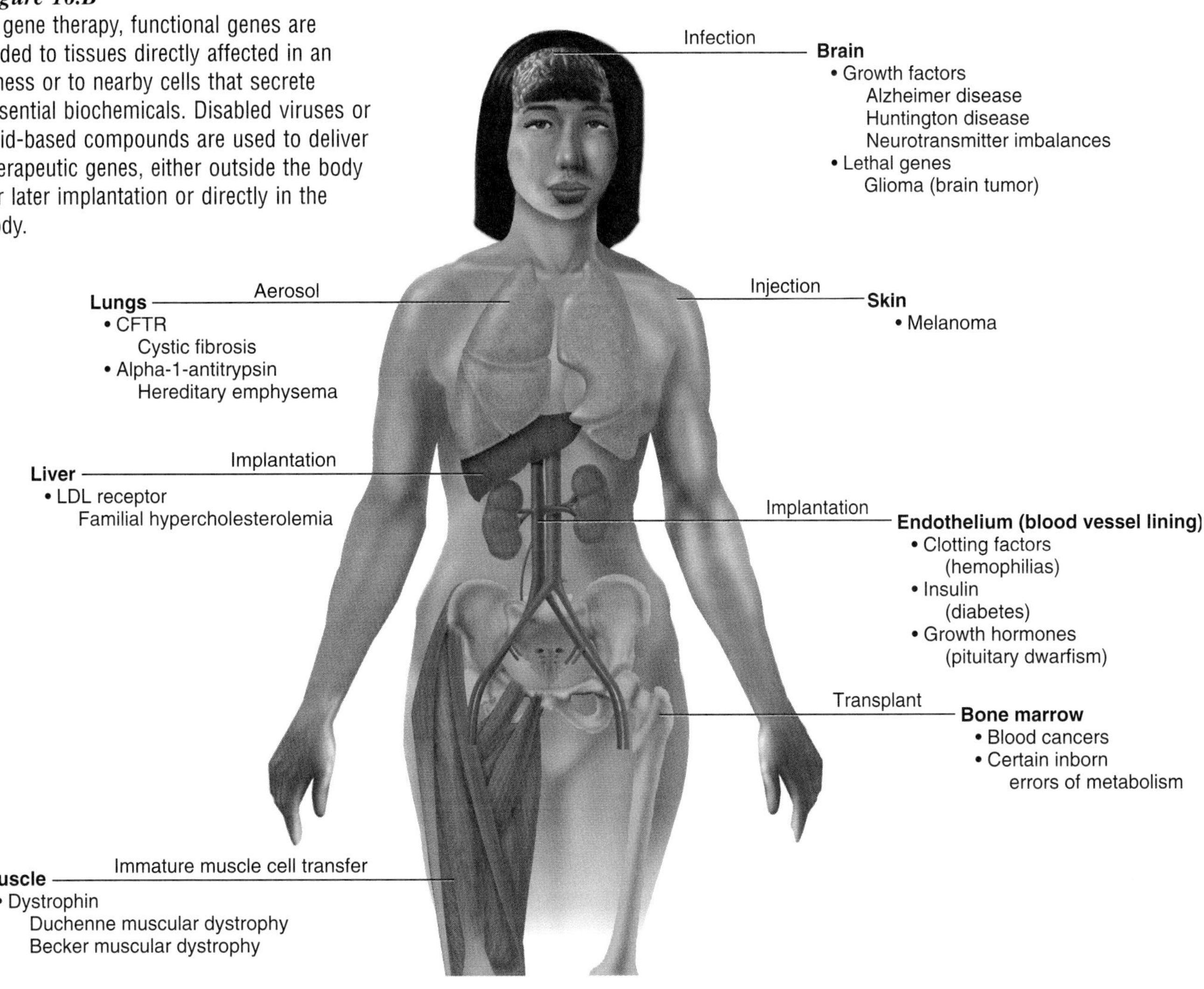

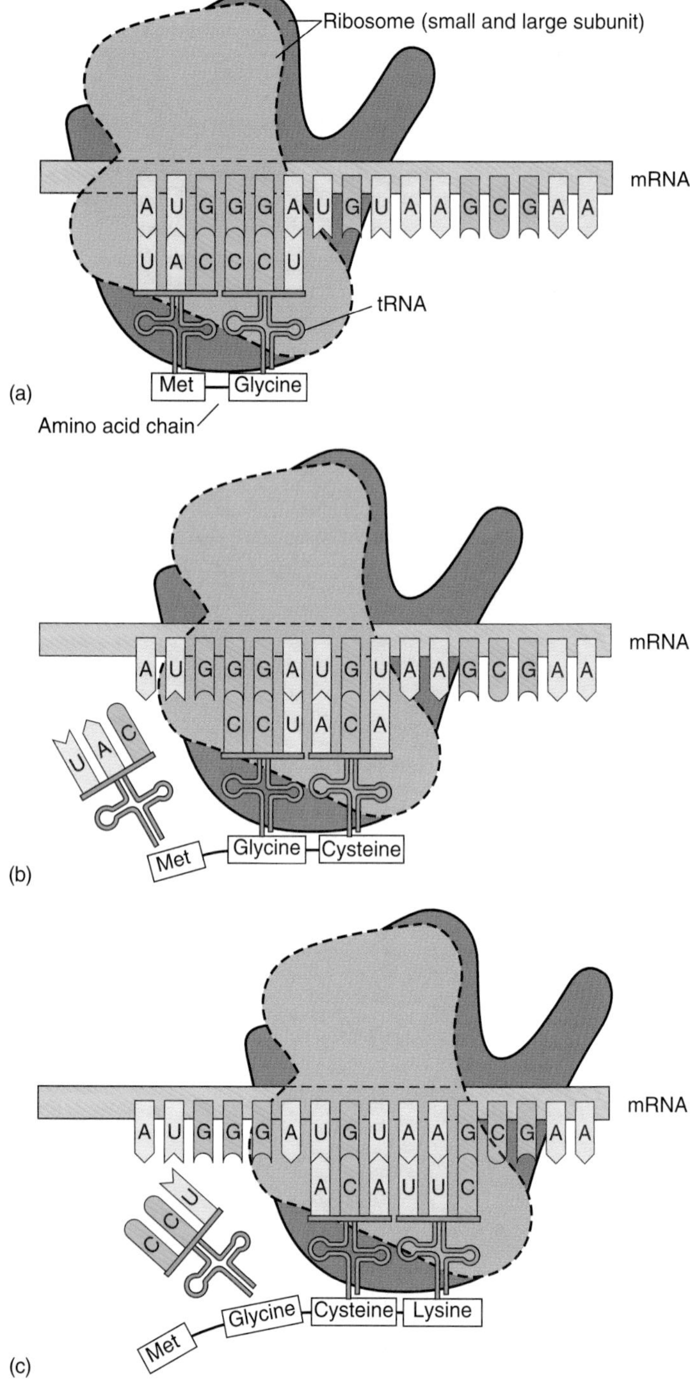

Figure 16.15

Translating a polypeptide. Translation begins when an mRNA binds to an rRNA that is part of a small ribosomal subunit. The anticodon of a tRNA bearing methionine (met) hydrogen bonds to the initiation codon (AUG) on the mRNA, forming the initiation complex. Next, a large ribosomal subunit binds to the complex, and a tRNA bearing a second amino acid (glycine, in this example) forms hydrogen bonds between its anticodon and the second mRNA's codon (*a*). The met brought in by the first tRNA forms a dipeptide bond with the amino acid brought in by the second tRNA, and the first tRNA detaches. The ribosome moves down the mRNA by one codon, and a third tRNA arrives—in this example, carrying the amino acid cysteine (*b*). A fourth amino acid is linked to the growing polypeptide chain (*c*), and the process continues until a termination codon is reached.

Elongation halts at an mRNA "stop" codon (UGA, UAG, or UAA) because no tRNA molecules correspond to these codons. The last tRNA is released from the ribosome, the ribosomal subunits separate from each other and are recycled, and the new polypeptide is released.

Protein synthesis is economical. A cell can produce large amounts of a particular protein from just one or two copies of a gene. A plasma cell in the human immune system, for example, manufactures 2,000 identical antibody molecules per second. To mass-produce on this scale, RNA, ribosomes, enzymes, and other proteins must be continually recycled. Many mRNAs can be transcribed from a single gene. Several ribosomes can simultaneously translate an mRNA, each at a different point along the message, and produce polypeptides of different lengths that branch from them (fig. 16.16).

Protein Folding

For many years, biochemists thought that protein folding was straightforward; the amino acid sequence dictated specific attractions and repulsions between parts of a protein, tangling it into its final form as it emerged from the ribosome (see fig. 2.11*a*). But these attractions and repulsions are not sufficient to fold the polypeptide into the highly specific form essential to its function. A protein apparently needs help to fold correctly.

An amino acid chain may start to fold as it emerges from the ribosome (fig. 16.17). Localized regions of shape form, and possibly break apart and form again, as translation proceeds. Experiments that isolate proteins as they are synthesized show that other proteins oversee the process. These accessory proteins include enzymes that foster chemical bonds and **chaperone proteins,** which stabilize partially folded regions that are important to the molecule's final form.

Just as repair enzymes check newly replicated DNA for errors, proteins scrutinize a folding protein to detect and dismantle incorrectly folded regions. Errors in protein folding can cause illness. Some mutations that cause cystic fibrosis, for example, prevent the CFTR protein from assuming its final form and anchoring in the cell membrane, where it normally controls the flow of chloride ions. Alzheimer disease is associated with a protein called amyloid that forms an abnormal, gummy mass instead of remaining as distinct molecules, because of improper folding.

In addition to folding, certain proteins must be altered further before they become functional. Sometimes enzymes must shorten a polypeptide chain for it to become active. Insulin, which is 51 amino acids long, for example, is initially translated as the polypeptide proinsulin, which is 80 amino acids long. Some polypeptides must join others to form larger protein molecules. The blood protein hemoglobin, for example, consists of four polypeptide chains.

Computers Coax Meaning from Genes

The linguistic nature of molecular genetics makes it ideal for computer analysis. The view of DNA sequences as a language emerged in the 1960s, as experiments revealed the linear relationship between nucleic acid sequences in genes and amino acid sequences in proteins. Yet the "rules" by which DNA sequences specify protein shapes are not well understood.

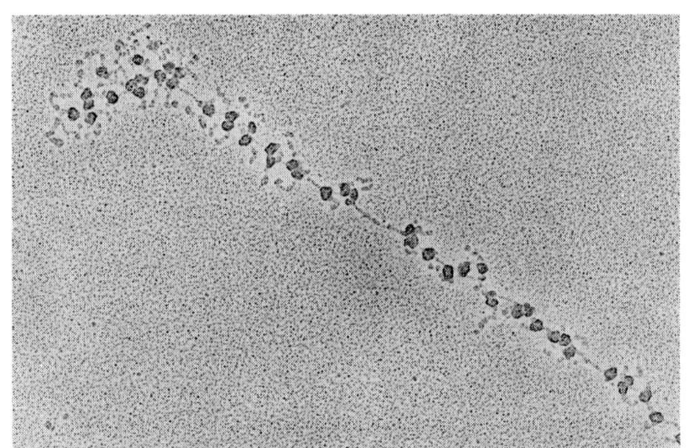

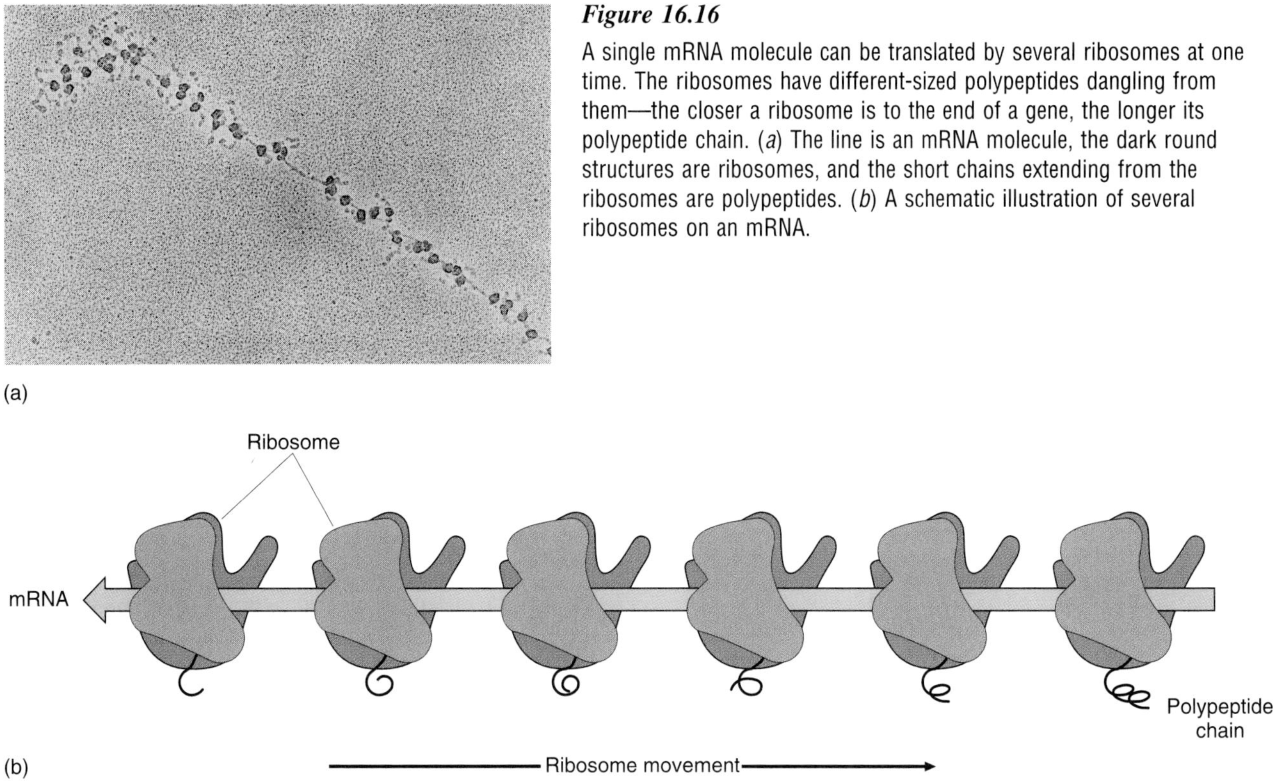

(a)

(b)

Figure 16.16

A single mRNA molecule can be translated by several ribosomes at one time. The ribosomes have different-sized polypeptides dangling from them—the closer a ribosome is to the end of a gene, the longer its polypeptide chain. (*a*) The line is an mRNA molecule, the dark round structures are ribosomes, and the short chains extending from the ribosomes are polypeptides. (*b*) A schematic illustration of several ribosomes on an mRNA.

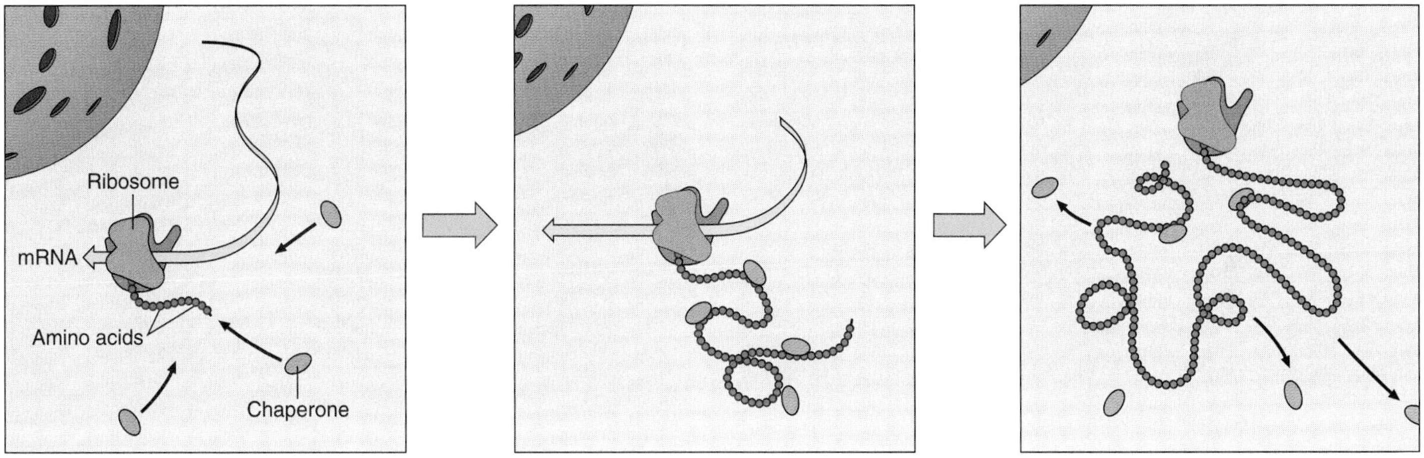

Figure 16.17

Protein folding. As this amino acid chain extends from the ribosome it is synthesized on, chaperone proteins bind the portions that will reside on the molecule's interior, leaving the other regions free to interact. When the final form is assumed, the chaperone proteins leave.

Researchers today send newly discovered gene and protein sequences to a clearinghouse, such as GenBank in the United States. A researcher can compare a newly discovered sequence to those on file to identify regions in common that might suggest the molecule's function. For example, proteins embedded in cell membranes are often anchored by seven distinctive loops formed by certain amino acids hydrogen bonding with each other. A new protein with these seven loops is probably a membrane protein (fig. 16.18). Similarly, a certain type of helix in a protein suggests a role in signal transduction. This approach of looking for clues in existing sequences is called "database homology searching."

Mastering Concepts

1. What is the genetic code?
2. How did logic and experiments help determine that the genetic code is triplet and nonoverlapping?
3. How did researchers determine which codons specify which amino acids?
4. What are the steps of protein synthesis?
5. Why is protein folding important?

Table 16.5 Effects of Mutations

Effect	Example
Prevents a protein from forming	Lack of dystrophin causes muscle cells to collapse in Duchenne muscular dystrophy
Lowers amount of a protein	Blood clots very slowly due to too little clotting factor in hemophilia A
Alters a protein	Skin blisters because amino acid substitution alters protein filaments that hold skin layers together, in epidermolysis bullosa
Adds a function to a protein	Addition of bases to the Huntington disease gene adds a stretch of amino acids to the protein product that gives it a new function that somehow leads to brain degeneration

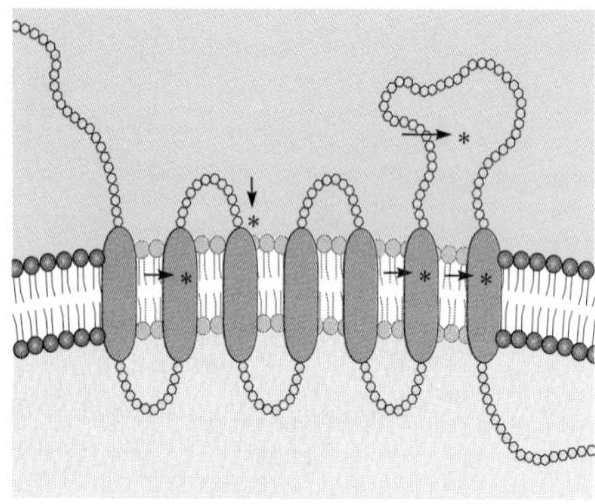

Figure 16.18

Researchers use clues from known proteins to decipher the structure and identify the possible function of a newly discovered protein. This seven-domain motif is characteristic of a protein that spans a cell membrane. Cystic fibrosis and Alzheimer disease are associated with such proteins.

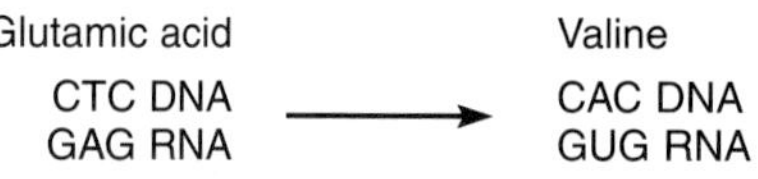

(a)

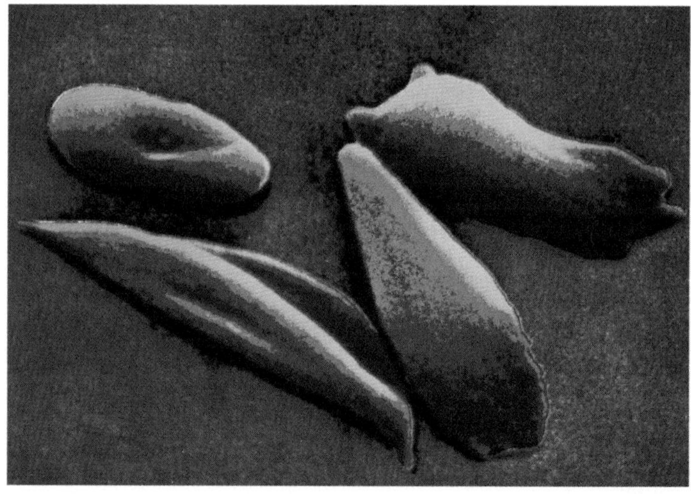
(b)

Figure 16.19

Normal and sickled red blood cells. (*a*) A normal red blood cell is a concave disc containing about 200 million molecules of the protein hemoglobin. (*b*) A mutation in the beta globin gene results in abnormal hemoglobin that crystallizes when oxygen tension is low and bends the red blood cells into sickle shapes. These abnormally shaped cells obstruct circulation, causing pain and loss of function in various organs. A single DNA nucleotide substitution on both homologs causes sickle cell disease.

Mutation—Genetic Misinformation

A gene can change, or mutate, in many ways. This is hardly surprising, given DNA's informational content. Think of how many ways the words on this page can be altered!

A mutation is a physical change in the genetic material. The change can be a single DNA base substituted for another, the addition or deletion of one or more DNA nucleotides, or even a transfer of nucleotides between chromosomes. A mutation can occur in the part of a gene that encodes amino acids, in a sequence that controls transcription, in an intron, or at a site critical to intron removal and exon splicing. A mutation may or may not cause a mutant (different) phenotype. This depends upon how the alteration affects the gene's product or activity. Table 16.5 lists some general ways that mutations can cause inherited illness.

Understanding Sickle Cell Disease

The first genetic illness researchers traced to a specific mutation was sickle cell disease. In the 1940s, scientists showed that sickle-shaped red blood cells (fig. 16.19) accompanied this inherited form of anemia (weakness and fatigue caused by a deficient number of red blood cells). In 1949, a team led by Linus Pauling discovered

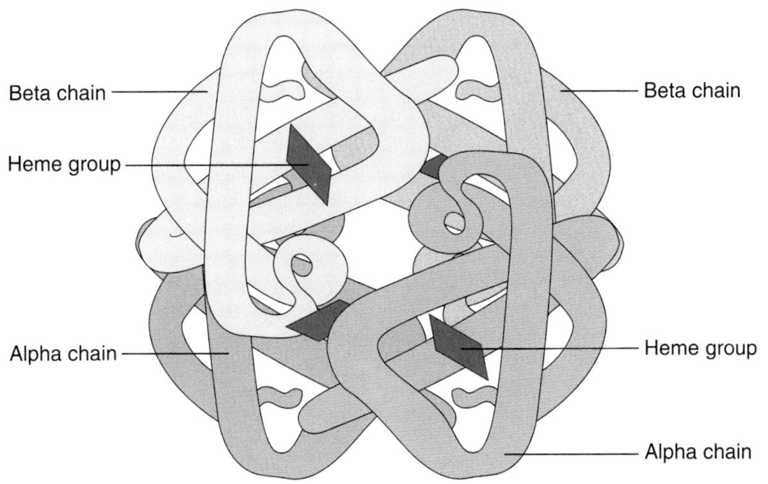

Figure 16.20

Defects in the beta globin genes. Adult hemoglobin is built of two beta chains and two alpha chains, each of which binds an iron atom. More than 300 different mutations of the human beta globin gene are known. Mutations can disrupt binding of the globin chains to each other or to the iron groups; change a stop codon to one specifying an amino acid, elongating the beta globin chain; or change an amino-acid-specifying codon into a stop codon, stunting a beta chain. Many beta chain mutations are silent because they change a codon into one specifying the same amino acid or a similar one or they occur in a part of the chain that is not essential to function.

that hemoglobin (the oxygen-carrying molecule in red blood cells) functioned differently depending on whether it came from healthy people or people with sickle cell disease. When Pauling placed the hemoglobin in a solution in an electrically charged field (a technique called electrophoresis), each type of hemoglobin moved to a different position. Hemoglobin from sickle cell carriers showed both movement patterns.

The researchers suspected that a physical difference accounted for the difference in electrophoretic mobility between normal and sickle hemoglobin. But how could they identify the protein portion of hemoglobin that sickle cell disease alters? Hemoglobin is a very large molecule. It consists of four globular-shaped polypeptide chains, each surrounding an iron atom (fig. 16.20).

Protein chemist V. M. Ingram tackled the problem by cutting normal and sickle hemoglobin with a protein-digesting enzyme and then separating the pieces, staining them, and displaying them on paper. The patterns of fragments differed. This meant, Ingram deduced, that the two molecules must differ in amino acid sequence. One piece of the molecule occupied a different position in the two types of hemoglobin. Ingram concentrated on this peptide (just 8 amino acids long, rather than the 146 of a full beta globin sequence) to find the site of the mutation. It was in the sixth amino acid in the gene sequence. The amino acid valine appeared in the sickle cell hemoglobin, where glutamic acid appeared in the normal chain. At the DNA level, a single base caused the mutation—a CAC replaced the normal CTC. This tiny change causes devastating effects in the human body.

Causes of Mutation

A mutation can form spontaneously or be induced by chemical or radiation exposure. An agent that causes mutation is called a **mutagen.**

Spontaneous Mutation

Two healthy people of normal height have a child with short stature. The parents are surprised to learn that an autosomal dominant mutation caused their son's achondroplasia (dwarfism). Each of their son's children will face a 50% chance of inheriting the condition. How could this happen when no one else in the family is affected? The boy's achondroplasia arose from a new, or *de novo,* mutation in either his mother's or father's gamete. Such a **spontaneous mutation** usually originates as a DNA replication error.

Sometimes spontaneous mutation stems from chemical changes in DNA's nitrogenous bases. Free nitrogenous bases exist in two slightly different forms called tautomers. For a short time, each base takes on an unstable tautomeric form. If such an unstable base inserts by chance into newly forming DNA, an error will be perpetuated. Figure 16.21 shows how a spontaneous mutation affects DNA replication.

The spontaneous mutation rate varies in different organisms and for different genes. Mutations occur more frequently in bacteria and viruses than in more complex organisms because their DNA replicates more often. In more complex organisms, the larger the gene, the more likely it is to spontaneously mutate. The dominant genes that cause neurofibromatosis type I and Duchenne muscular dystrophy are huge, and they mutate at the highest rates known for human disorders, arising in 40 to 100 of every million gametes. Each human gene has about a 1 in 100,000 chance of mutating. Each of us probably carries at least a few spontaneously mutated genes, but most are recessive, so they do not affect the phenotype.

A gene's sequence also influences mutation rate. Genes with repeated base sequences, such as GCGCGC . . . , are more likely to mutate because bases from a single strand may pair with each other as the double helix unwinds to replicate. It is as if the molecules that guide and carry out replication become "confused" by short, repeated sequences, as an editor scanning a manuscript might miss the spelling errors in the words "happpiness" and "bananana" (fig. 16.22).

Genes containing repeated sequences may mutate by misaligning during replication. The alpha globin genes that make up part of the hemoglobin molecule are especially prone to mutate because they are repeated in their entirety next to each other on each human chromosome 16. If mispairing occurs in meiosis, chromosomes may result that have more or fewer than the normal two copies of the gene (fig. 16.23).

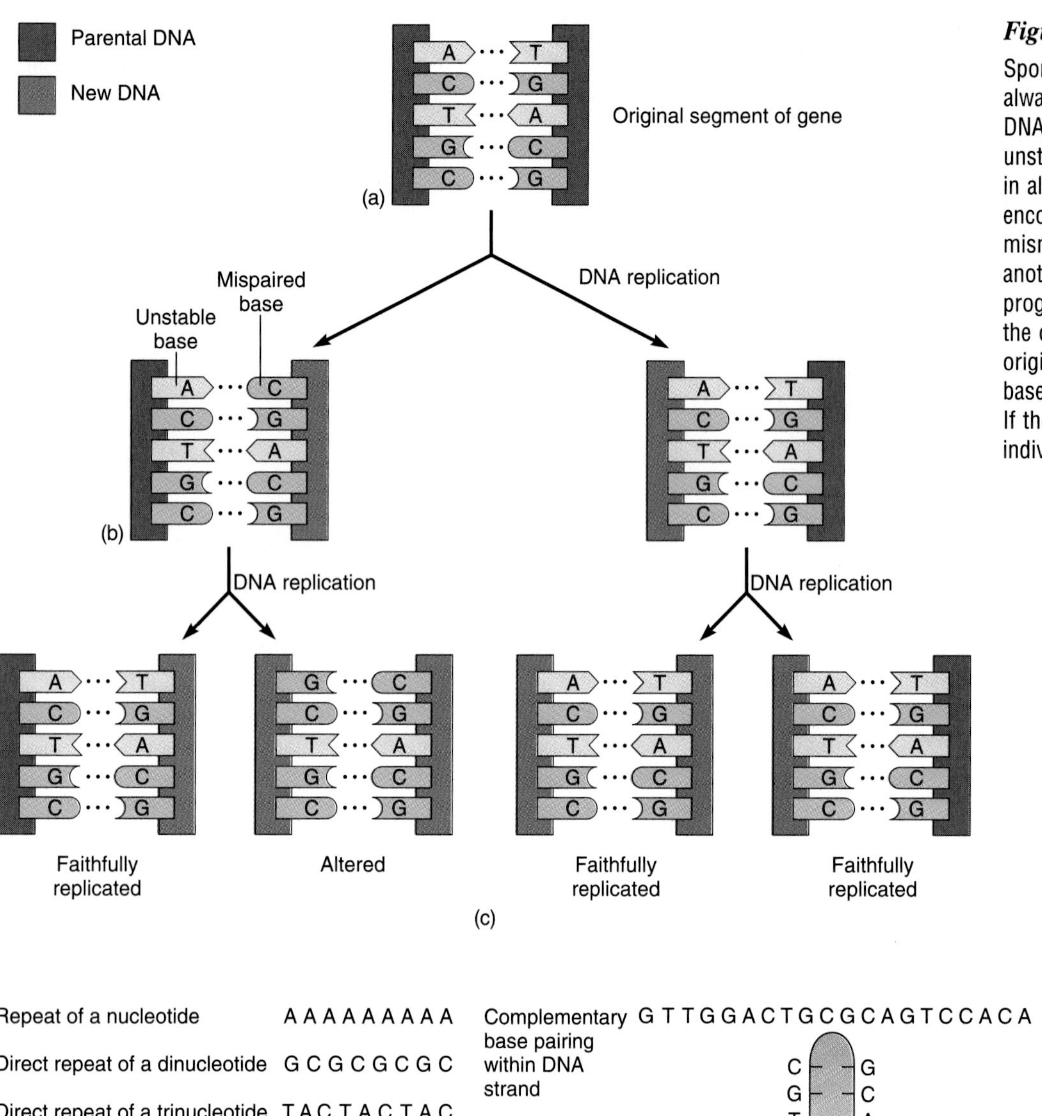

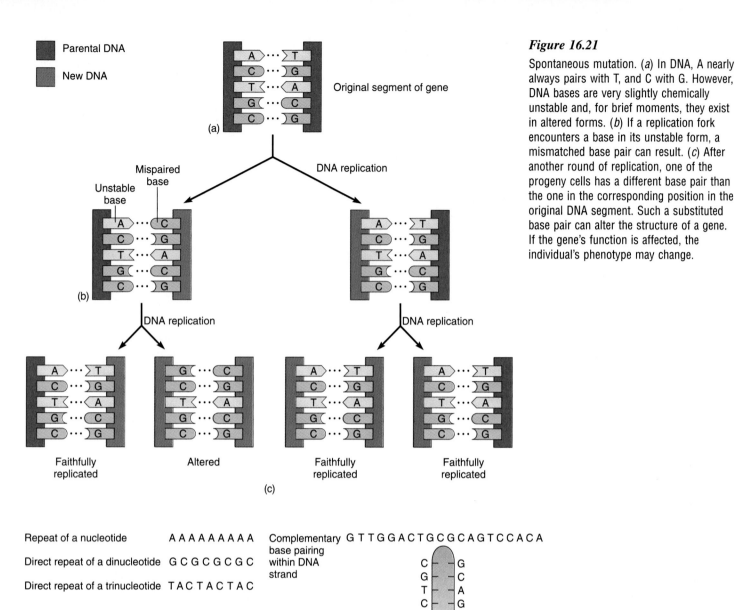

Figure 16.21

Spontaneous mutation. (*a*) In DNA, A nearly always pairs with T, and C with G. However, DNA bases are very slightly chemically unstable and, for brief moments, they exist in altered forms. (*b*) If a replication fork encounters a base in its unstable form, a mismatched base pair can result. (*c*) After another round of replication, one of the progeny cells has a different base pair than the one in the corresponding position in the original DNA segment. Such a substituted base pair can alter the structure of a gene. If the gene's function is affected, the individual's phenotype may change.

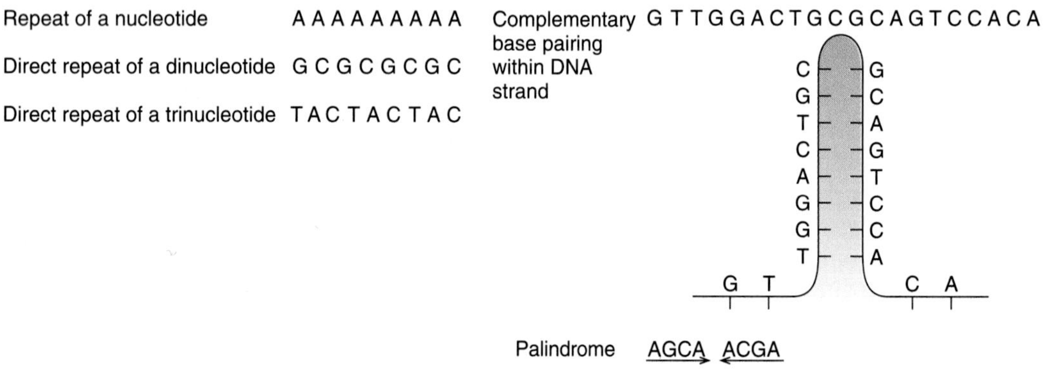

Repeat of a nucleotide A A A A A A A A A

Direct repeat of a dinucleotide G C G C G C G C

Direct repeat of a trinucleotide T A C T A C T A C

Complementary base pairing within DNA strand G T T G G A C T G C G C A G T C C A C A

Palindrome AGCA ACGA

Figure 16.22

DNA symmetry may increase the likelihood of mutation. These examples show repetitive DNA sequences that may "confuse" replication enzymes and cause errors.

Induced Mutation

The spontaneous mutation rate for most genes is far too low to provide enough genetic variants for experiments, so researchers use mutagens to help the process along. Chemicals called alkylating agents, for example, remove a DNA base; since any one of the four types of bases may replace it, and three of them will be a mismatch, an alkylating agent is very likely to produce a mutation. Other chemical mutagens act in different ways, such as adding or deleting several nucleotides in a row. X rays and other forms of radiation break DNA, causing both large- and small-scale mutations in DNA.

Some commonly encountered substances may cause mutations in cells growing in culture, such as sodium nitrite in smoked

Two copies of alpha globin gene

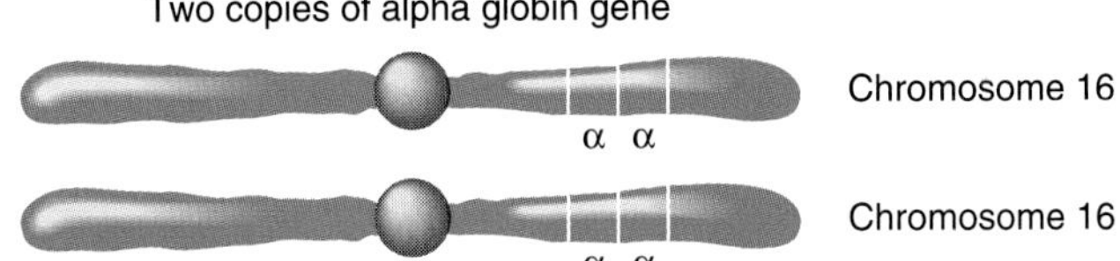

Normal chromosome arrangement of alpha globin genes

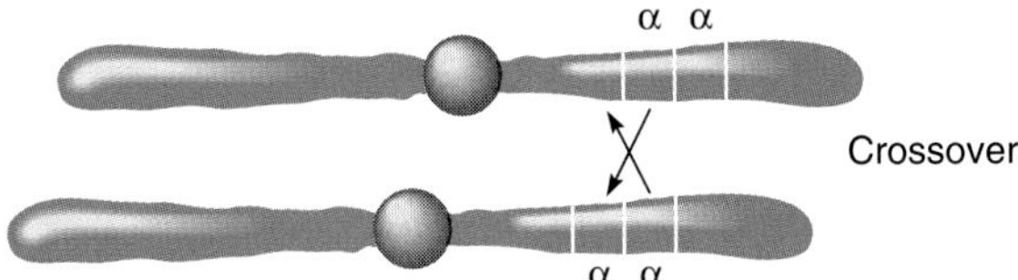

Misalignment of alpha globin genes during meiosis I

Chromosome resulting from crossover with three alpha globin genes

Chromosome resulting from crossover with one alpha globin gene

Figure 16.23

The repeated nature of the alpha globin genes makes them prone to mutation by mispairing during meiosis.

Table 16.6 Commonly Encountered Mutagens

Mutagen	Source
Aflatoxin B	Fungi growing on peanuts and other foods
2-amino 5-nitrophenol	Hair dye components
2,4-diaminoanisole	
2,5-diaminoanisole	
2,4-diaminotoluene	
p-phenylenediamine	
Caffeine	Cola, tea, coffee, pain relievers
Furylfuramide	Food additive
Nitrosamines	Pesticides, herbicides, cigarette smoke
Proflavine	Veterinary antiseptic
Sodium nitrite	Smoked meats
Tris (2,3-dibromopropyl phosphate)	Flame retardant in fabrics

meats (table 16.6). Natural mutagens include radiation from rocks, cosmic rays, and the ultraviolet light in sunlight. Nuclear accidents, bomb tests, and mustard gas also cause mutations (Biology in Action 16.2).

Types of Mutations

Mutations can affect somatic cells or sex cells, and they may alter DNA to different extents.

Somatic Versus Germinal Mutations

A **germinal mutation** occurs during the DNA replication period preceding meiosis. The change appears in the resulting gamete and all the cells that descend from it following fertilization. A **somatic mutation** can occur during the DNA replication period just before mitosis. In this case, the genetic change is perpetuated in the progeny cells of the original mutated cell. Because a somatic mutation occurs in somatic cells rather than gametes, it affects a subset of a multicellular organism's cells.

Point Mutations

A **point mutation** is a change in a single DNA base. The two types of point mutations are distinguished by their consequences. A **missense mutation** changes a codon that normally specifies a particular amino acid into one that encodes a different amino acid. If the substituted amino acid alters the protein's conformation sufficiently at a site critical to its function, the phenotype changes. A missense mutation can greatly affect a gene's product if it alters a site controlling intron removal. The encoded protein would have additional amino acids, even though the causative mutation only affects a single DNA base.

A **nonsense mutation** is a single base change that changes a codon specifying an amino acid into a "stop" codon. This shortens the protein product, which can profoundly influence the phenotype. If a normal stop codon mutates into a codon that specifies an amino acid, the resulting protein would contain extra amino acids.

Alterations in the Number of DNA Bases

In genes, the number three is very important, because triplets of DNA bases specify amino acids. Adding or deleting bases by any number other than a multiple of three devastates a gene's function. It disrupts the reading frame, and therefore it also disrupts the sequence of amino acids. Such a change is called a **frameshift mutation.** Sometimes, even adding or deleting multiples of three can alter a phenotype.

16.2 biology in action

Chernobyl: Radiation Raises Mutation Rates

Between 1:23 and 1:24 A.M. on April 26, 1986, Reactor 4 at the Chernobyl Nuclear Power Station in Ukraine exploded, sending a great plume of radioactive isotopes into the air that spread for thousands of miles. The reactor had been undergoing a test, its safety systems temporarily disabled, when it became overloaded and rapidly raged out of control (fig. 16.C). Thirty-one people died instantly, and several hundred others died in the following weeks of acute radiation sickness. Estimates of total human lives lost in the decade following the disaster, from a variety of related effects, average about 32,000.

Rates of thyroid cancer among children living in nearby Belarus have risen 10-fold. This effect was not surprising, because the thyroid glands of young people soak up iodine, which in a radioactive form bathed the area in the first days after the explosion. What was unexpected was the great effect of the disaster on the mutation rate.

The first studies assessing genetic effects focused on voles (a type of rodent) and humans. In voles living right next to the reactor, researchers examined the amino acid sequence of cytochrome b, a protein encoded by a gene in the mitochondria. The sequence of cytochrome b is usually the same among individuals, but so far it has differed in every Chernobyl vole studied! This is not the case for voles living elsewhere. The mutation rate calculated for this mitochondrial gene is two base changes for every 10,000 bases. This is 100 times the normal rate for mitochondrial genes, which tend to mutate faster than genes in the nucleus.

Figure 16.C
Protective gear shields investigators in the destroyed nuclear reactor at Chernobyl. Ten years after the disaster, mutation rate increases were already evident in nearby rodent and human populations.

To study humans, researchers compared lengths of minisatellite sequences in children born in 1994 and their parents, who lived in the Mogilev district of Belarus at the time of the accident and have remained there (fig. 16.D). Minisatellites are highly repeated DNA sequences that are the same length within an individual. A minisatellite size in a child that does not match the size of either parent indicates that a mutation occurred in that parent's gamete. This was twice as likely to occur in exposed families than in control families living elsewhere. Like mitochondrial genes, minisatellite sequences mutate faster than genes in the nucleus and therefore enable researchers to chart genetic change.

The increased mutation rates seen in voles and humans near Chernobyl are put into perspective by comparison to studies of survivors of the World War II atomic bomb blasts at Hiroshima and Nagasaki and their descendants. Various studies on the Japanese, including minisatellite analyses very similar to the Chernobyl study, do not show an increased mutation rate.

The Chernobyl nuclear reactor meltdown, perhaps the worst environmental disaster ever, is serving as an accidental experiment, revealing the effects on genes of radioactive isotopes in nuclear waste.

Figure 16.D
People living in Belarus are the most seriously affected by the radiation exposure.

Source: Yuri Dubrova, et al., "Human Minisatellite Mutation Rate After the Chernobyl Accident" Nature, vol. 380, no. 657b, p. 683, April 25, 1990.

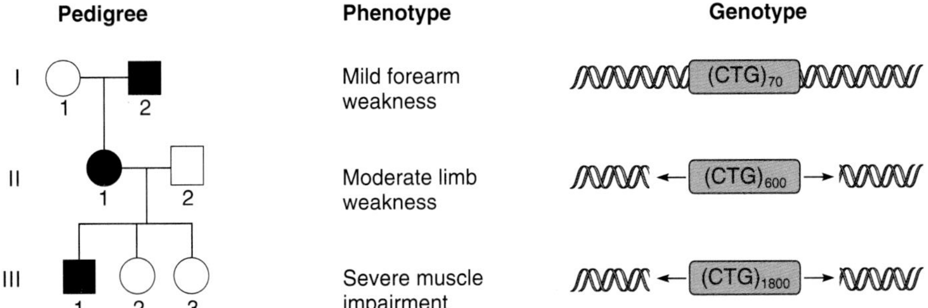

Figure 16.24

Certain disorders worsen with each generation because a repeated part of the causative gene expands.

Missing genetic material—sometimes even a single base—is a mutation that can greatly alter gene function. A DNA deletion can be so large that detectable sections of chromosomes are missing, or it can be so small that only a few genes or parts of a gene are lacking. The mutation that causes severe cystic fibrosis deletes only a single codon, and the resulting protein lacks just one amino acid.

Just as deleting a DNA base can upset a gene's reading frame, adding a base can, too. A recently discovered type of mutation is an expanding gene, in which the number of repeats of a three-nucleotide sequence increases over several generations. Consider myotonic dystrophy, an autosomal dominant form of muscular dystrophy that worsens from one generation to the next. A grandfather might experience only mild weakness in his forearms. His daughter might have moderate arm and leg weakness. Her children, if they inherited the genes, might have severe muscle impairments.

For many years, doctors attributed the worsening symptoms to people worrying about their family histories and then reporting increasingly severe symptoms. However, once researchers could sequence genes, they found startling evidence that the disease was indeed worsening with each generation because the gene was expanding (fig. 16.24)! The gene, on chromosome 19, has an area rich in the trinucleotide CTG. This area repeats over and over, almost like a molecular stutter. A person who does not have myotonic dystrophy has from 2 to 28 repeats, whereas a person with the disorder has from 28 to 50 copies of the sequence.

Expanding genes underlie several well-known inherited disorders, including Huntington disease and fragile X syndrome. Most of these disorders affect brain function, but the precise connection between genotype and phenotype is not yet well understood. Researchers hypothesize that these mutations add a function, because individuals who have a deletion for the gene do not have symptoms, but people with the extended gene do.

Jumping Genes

Many types of organisms have some genes that move from one chromosomal location to another. About 10% of the human genome consists of these "jumping genes," or **transposable elements.** These pieces of DNA can block gene expression by inserting into a gene and preventing its transcription. For example, a boy with hemophilia A was found to have a transposable element in his factor VIII gene. The element was found in his mother's cells but on her chromosome 22! Apparently as the oocyte that would eventually be fertilized formed, the transposable element jumped into the factor VIII gene on the X chromosome, disabled it, and eventually caused the hemophilia in the boy.

Viruses are in a sense transposable elements because they insert their genetic material into the host cell's chromosomes. Transposable elements in bacteria carry genes confering resistance to certain antibiotics, a problem that is discussed further in chapter 18.

Table 16.7 summarizes the different types of mutations using an English sentence analogy.

Table 16.7 Types of Mutations

A sentence of three-letter words can serve as an analogy to demonstrate the effects of mutations on gene sequence:

Wild type	THE ONE BIG FLY HAD ONE RED EYE
Missense	THQ ONE BIG FLY HAD ONE RED EYE
Nonsense	THE ONE BIG
Frameshift	THE ONE QBI GFL YHA DON ERE DEY
Deletion	THE ONE BIG HAD ONE RED EYE
Duplication	THE ONE BIG FLY FLY HAD ONE RED EYE
Insertion	THE ONE BIG WET FLY HAD ONE RED EYE
Expanding mutation	P_1 THE ONE BIG FLY HAD ONE RED EYE
	F_1 THE ONE BIG FLY FLY FLY HAD ONE RED EYE
	F_2 THE ONE BIG FLY FLY FLY FLY FLY FLY HAD ONE RED EYE

Figure 16.25

A mutation is a variant. (*a*) This boy has a mutation in a collagen gene that confers unusual stretchability to his skin. (*b*) This girl also has an unusual inherited variant, red hair. Mutations and mutants do not always mean illness.

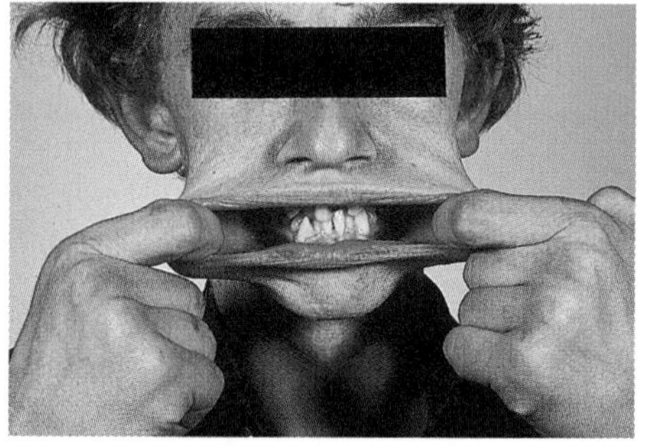

(a)

(b)

Natural Protection Against Mutation

The natural repair systems discussed in chapter 15 are only some of the built-in protections against mutation.

The Genetic Code

The genetic code seems to have too much information—61 types of codons specify 20 types of amino acids. Different codons that encode the same amino acid are termed **degenerate codons.** The redundancy of the genetic code helps protect against mutation because many DNA alterations of the third codon position specify the same amino acid and are therefore "silent." For example, both CAA and CAG specify glutamine. A change from one to the other does not alter the designated amino acid, so it would not alter a protein containing that amino acid and would not affect the phenotype.

Mutations in the second codon position often cause one amino acid to replace another with a similar conformation, which may not disrupt the protein's form drastically. For example, if a GCC codon mutates to GGC, glycine replaces alanine; both are very small amino acids.

Position in the Protein

A mutation's effect on the phenotype depends upon how it alters the conformation of the gene product. A mutation replacing an amino acid with a very dissimilar one may not affect the phenotype if the change occurs in a part of the protein that is not critical to its function. Certain mutations in the beta globin gene, for example, do not cause anemia or otherwise affect how a person feels, but they may slightly alter how hemoglobin migrates in an electric field.

Some proteins are more vulnerable to disruption by mutation than others. Collagen, a major constituent of connective tissue, is unusually symmetrical. The slightest change in its DNA sequence can greatly disrupt its overall structure and lead to a disorder of connective tissue. The boy in figure 16.25*a* has such a connective tissue disorder.

Although mutations can cause disease, they also provide much of the variation that makes life interesting, as the red-haired girl in figure 16.25*b* demonstrates. The next unit explores how genetic variation is essential for the perpetuation and evolution of life.

Mastering Concepts

1. What is a mutation?
2. What causes a spontaneous mutation?
3. How do different types of mutations alter DNA?
4. What are some natural protections against mutation?

SUMMARY

1. A gene's information must be **transcribed** into RNA before it is **translated** into a sequence of amino acids. RNA is a single-stranded nucleic acid similar to DNA, but it contains uracil and ribose rather than thymine and deoxyribose.

2. In bacteria, **operons** coordinate expression of genes whose encoded proteins participate in the same metabolic pathway. In multicellular organisms, **transcription factors** regulate which genes are transcribed and when in a particular cell type. These factors have certain common regions called **motifs.**

3. Transcription begins when transcription factors help **RNA polymerase** (RNAP) bind to a promoter on the DNA **coding strand.** RNAP then adds complementary RNA nucleotides to the **RNA sense strand. Ribozymes** cut **introns** out of RNA and splice together the remaining **exons.**

4. Several types of RNA participate in translation. **Messenger RNA** (mRNA) carries a protein-encoding gene's information. **Ribosomal RNA** (rRNA) associates with certain proteins to form ribosomes, which support and help catalyze protein synthesis. **Transfer RNA** (tRNA) has an **anticodon** sequence complementary to a particular mRNA **codon** on one end and a particular amino acid at the other end.

5. Each three consecutive mRNA bases is a codon that

Genes to the Rescue~

continued from page 326.

Ashanti and Cynthia's treatment of their inherited enzyme deficiency was successful, but short-lived, because the gene therapy worked on mature white blood cells. When the cells died, so did the therapy. Wouldn't the effect last longer if the therapy altered immature blood cells, the stem cells that continue to give rise to other cells? Stem cells reside in the bone marrow, where they account for only one in a million or so cells, and the type of stem cell that produces T cells is even rarer, accounting for one in several billion marrow cells. How could researchers harvest enough stem cells to alter?

The answer came from a different source of stem cells—umbilical cord blood. If fetuses with the inherited ADA deficiency could be identified and the parents agreed, then the appropriate stem cells could be separated from their cord blood at birth, given ADA genes, and reinfused.

Crystal and Leonard Gobea had already lost a 5-month-old baby to ADA deficiency when amniocentesis revealed that a second fetus was affected. They and two other couples in the same situation were asked to participate in the experiment. In early spring of 1993, Andrew Gobea and two other newborns received their bolstered blood cells on the fourth day after birth. They also received ADA to prevent symptoms while the gene therapy took hold. The plan was to monitor the babies frequently to see if T cells carrying normal ADA genes would appear in the blood.

The experiment was a success, although the altered T cells are appearing slowly. After a few months, in each child, about 1 in 10,000 T cells had the genetic alteration. But after a year, 1 in 100 did! By 18 months of age, with the genetically altered T cell population still rising, researchers halved the ADA dose. The babies remained healthy. By the summer of 1995, the three toddlers each had about 3 in 100 T cells carrying the ADA gene. Gradually, the healthier, bolstered cells are replacing the enzyme-deficient ones.

ADA deficiency is quite rare. It received researchers' first efforts at gene therapy because they knew the gene and its specified protein, and they had a way to introduce the correct gene into the body. But the success stories of Ashanti, Cynthia, Andrew, and others pioneering gene therapy will have repercussions on many of us. We began the twentieth century regarding the unit of inheritance as a mysterious "black box." We will begin the twenty-first century using genes as medicines to prevent and treat many kinds of illnesses.

specifies a particular amino acid. The correspondence between codons and amino acids constitutes the **genetic code.** Of the 64 different codons, 61 specify amino acids and three signal the end of polypeptide synthesis. More than one codon may encode a single amino acid type. The genetic code is nonoverlapping, triplet, and identical in nearly all species. Scientists deciphered the code with synthetic RNA molecules by exposing them to the contents of *E. coli* cells and noting which amino acids formed peptide chains. Other experiments confirmed the triplet nature of the code.

6. Translation requires tRNA, ribosomes, energy storage molecules, enzymes, and protein factors. An **initiation complex** forms when mRNA, a small ribosomal subunit, and a tRNA carrying methionine join. A large ribosomal subunit joins the small one. Next, a second tRNA binds by its anticodon to the next mRNA codon, and its amino acid bonds with the methionine the first tRNA brought in. tRNAs continue to add amino acids, building a polypeptide. The ribosome moves down the mRNA as the chain grows.

7. When the ribosome reaches a "stop" codon, it separates into its subunits and is released, and the new polypeptide breaks free. **Chaperone proteins** help fold the polypeptide, and it may be shortened or combined with others.

8. A mutation is a change in DNA (genotype) that adds, deletes, alters, or moves nucleotides. A phenotype that a mutation alters is mutant. A gene can mutate spontaneously, particularly if it contains regions of repetitive DNA sequences. **Mutagens** are chemicals or radiation that increase the mutation rate.

9. A **germinal mutation** originates in meiosis and affects all cells of an individual. A **somatic mutation** originates in mitosis and affects a subset of cells. A **point mutation** alters a single DNA base. This mutation may be **missense** (substituting one amino acid for another) or **nonsense** (substituting a "stop" codon for an amino-acid-coding codon). Altering the number of bases in a gene may disrupt the **reading frame,** altering the amino acid sequence of the gene's product. Expanding genes cause some inherited illnesses. **Transposable elements** move, possibly disrupting a gene's function.

10. Some mutations are silent. A mutation in the third position of a degenerate codon can substitute the same amino acid. A mutation in the second codon position can replace an amino acid with a similarly shaped one. A mutation in a nonvital part of a protein may not affect function.

TO REVIEW . . .

1. Define and distinguish between transcription and translation.

2. List the differences between RNA and DNA.

3. Where do DNA replication, transcription, and translation occur in a cell?

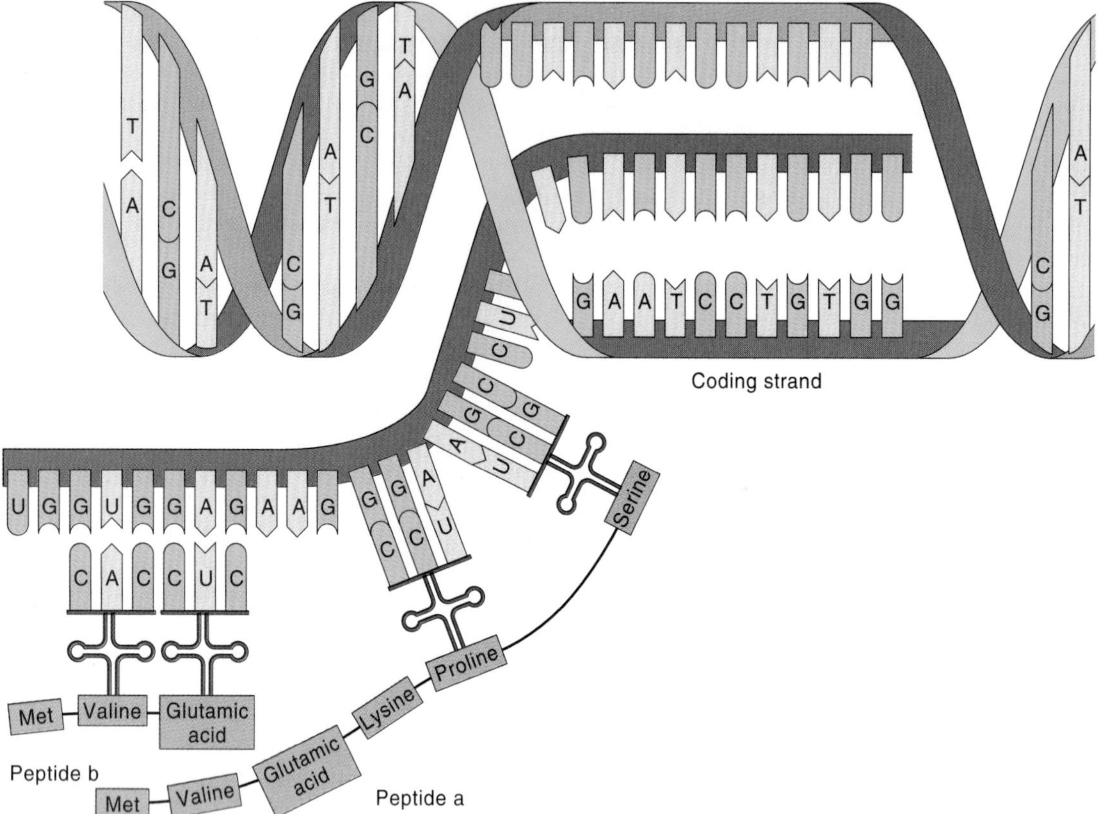

4. List the three major types of RNA and their functions.

5. How are transcription and translation economical?

6. List the sequences of the RNA sense strands transcribed from the following DNA sequences:

- **a.** TTACACTTGCTTGAGAGTT
- **b.** ACTTGGGCTATGCTCATTA
- **c.** GGAATACGTCTAGCTAGCA

7. Given the following partial mRNA sequences, reconstruct the corresponding DNA coding sequences:

- **a.** GCUAUCUGUCAUAAAAGAGGA
- **b.** GUGGCGUAUUCUUUUCCGGGUAGG
- **c.** AGGAAAACCCCUCUUAUUAUAGAU

8. Refer to the figure above to answer the following questions:

- **a.** Label the mRNA and the tRNA molecules, and draw in the ribosomes.
- **b.** What are the next three amino acids to be added to peptide *b*?
- **c.** Fill in the correct codons in the mRNA opposite the coding strand.
- **d.** What is the sequence of the DNA antisense strand (as much as can be determined from the figure)?
- **e.** Is the end of the peptide encoded by this gene indicated in the figure? How can you tell?
- **f.** What might happen to peptide *b* after it is terminated and released from the ribosome?

9. List three mRNA sequences that could encode this amino acid sequence:

histidine-alanine-arginine-serine-leucine-valine-cysteine

10. Write a DNA sequence that would encode this amino acid sequence:

valine-tryptophan-lysine-proline-phenylalanine-threonine

11. When Crick and his coworkers examined synthetic RNA of sequence ACACACACACACACA . . . , they found that it encoded the amino acid sequence thr-his-thr-his-thr . . . . What would a next step be in determining the codon assignments for ACA and CAC?

12. To answer the following questions, refer to this DNA sequence, which is transcribed:
GCAAAACCGCGATTATCATGCTTC.

- **a.** What is the sequence of the noncoding DNA strand?
- **b.** What is the mRNA sequence?
- **c.** What amino acid sequence does the transcribed DNA sequence specify?
- **d.** What would the amino acid sequence be if the genetic code were completely overlapping?
- **e.** If the 15th DNA base mutates to thymine, what happens to the amino acid chain?

13. How is complementary base pairing responsible for

- **a.** the structure of the DNA double helix?
- **b.** DNA replication?
- **c.** transcription of RNA from DNA?
- **d.** mRNA's attachment to a ribosome?
- **e.** codon/anticodon pairing?
- **f.** tRNA conformation?

14. What changes does RNA undergo after transcription and a protein undergo after translation?

15. How can a defect in protein folding cause disease?

16. A protein-encoding region of a gene has the following DNA sequence:

G T A G C G T C A C A A A C A A A T C A G C T C

Determine how each of the following mutations alters the amino acid sequence:

- **a.** substitution of a T for the C in the 10th DNA base
- **b.** substitution of a G for the C in the 19th DNA base
- **c.** insertion of a T between the 4th and 5th DNA bases
- **d.** insertion of a GTA between the 12th and 13th DNA bases
- **e.** deletion of the first DNA base

17. State the sequence of an antisense RNA that might be used to fight a virus that has a partial DNA coding sequence of TTCGCTAAAGACTGT.

TO THINK ABOUT . . .

1. Many articles on genetics in the popular press state, "We are now beginning to crack the genetic code," or "Everyone has his or her own unique genetic code." What is inaccurate about these statements?

2. How can a mutation alter the sequence of DNA bases in a gene but not produce a noticeable change in the gene's polypeptide product? How can a mutation alter the amino acid sequence of a polypeptide yet not alter the phenotype?

3. Why do different cell types have different rates of transcription and translation?

4. Why would two-nucleotide codons not be able to encode the number of amino acids in biological proteins?

5. A missense mutation causes a form of Alzheimer disease by replacing a histidine with an arginine. Determine how a codon is altered in this mutation.

6. Cite two ways that proteins assist in their own synthesis. How does RNA assist in its own synthesis?

7. Many antibiotic drugs work by interfering with protein synthesis in bacteria that cause infection. Explain how each of the following antibiotic mechanisms disrupts genetic function in bacteria:

- **a.** tRNAs misread mRNA codons, binding with the incorrect codon and bringing in the wrong amino acid
- **b.** methionine is released from the initiation complex before translation can begin
- **c.** tRNA cannot bind to the ribosome
- **d.** ribosomes cannot move
- **e.** a tRNA picks up the wrong amino acid

8. Which biotechnology might be able to accomplish the following goals? More than one answer may be possible.

- **a.** Develop a mouse that has a gene causing breast cancer in humans in each of its cells to serve as a model for that disease.
- **b.** Shut off HIV genes integrated into the chromosomes of people with HIV infection.
- **c.** Create bacteria that produce human growth hormone, used to treat extremely short stature.

9. How could a mutation involving one DNA base be more devastating than a mutation involving three contiguous DNA bases?

10. What are three ways that mutation can add amino acids to a protein?

TO LEARN MORE . . .

References and Resources

Askari, Frederick, and W. Michael McDonnell. February 1, 1996. Antisense-oligonucleotide therapy. *The New England Journal of Medicine,* vol. 334. Antisense RNAs may make powerful drugs.

Benson, Dennis A., et al. January 1996. GenBank. *Nucleic Acids Research,* vol. 24. Determining protein structures is as much an informational science as it is a life science.

Blalock, J. Edwin. September 1995. Genetic origins of protein shape and interaction rules. *Nature Medicine,* vol. 1. The genetic code holds clues to protein folding, but they are neither simple nor obvious.

Boguski, Mark S. September 7, 1995. Hunting for genes in computer data bases. *The New England Journal of Medicine,* vol. 333. Clues from known genes and proteins help researchers describe new discoveries.

Gilbert, Walter. February 9, 1978. Why genes in pieces? *Nature,* vol. 271. A classic and insightful look at the enigma of introns.

Hartl, Daniel L. March 12, 1996. The most unkindest cut of all. *Nature Genetics,* vol. 12. Two neurological disorders in humans are caused by transposable elements—from insects!

Haseltine, William A. March 1997. Discovering genes for new medicines. *Scientific American.* The human genome project promises to spawn many new drugs.

Ingram, V. M. 1957. Gene mutations in human hemoglobin: The chemical difference between normal and sickle cell hemoglobin. *Nature,* vol. 180, p.1326. The classic paper explaining the molecular basis for sickle cell disease.

Lewis, Ricki. October 1996. Spider silk. *BioScience.* Spider silk has a number of uses—with a little help from genetic manipulation.

Lewis, Ricki. July 22, 1996. Software helps researchers in sorting through the human genome. *The Scientist.* Because molecular genetics is informational, computer analysis is a valuable tool.

Lewis, Ricki. May/June 1996. A royal mystery solved. *Biophotonics International.* How DNA fingerprinting, and an unexpected result, led to identification of the remains of Tsar Nicholas II.

Lewis, Ricki. April 1, 1996. OncorMed seeks to link bench to the patient's bedside. *Genetic Engineering News.* A test can predict inherited breast cancer—but who wants it?

Lewis, Ricki. April 1996. Using technology to teach difficult genetic concepts. *American Biology Teacher.* PCR and antisense technology are based on genetic concepts.

Lewis, Ricki. February 1996. On cracked codes, cell walls, and human fungi. *The American Biology Teacher.* Newspapers, magazines, and films often misuse the term "genetic code."

Lewis, Ricki. June 1988. Witness for the prosecution. *Discover.* The story of the first DNA fingerprinting case.

Marden, Michael C. et al. March 6, 1997. Haemoglobin from transgenic tobacco. *Nature,* vol. 385. A blood protein from tobacco.

Matthews, Kathleen. March 1, 1996. The whole lactose repressor. *Science,* vol. 271. Computer modeling reveals the lac operon.

Rosenberg, R. N. October 17, 1996. DNA triplet repeats and neurologic disease. *The New England Journal of Medicine,* vol. 335. Some genes grow, causing disease.

Rosenthal, Nadia. July 7, 1994. DNA and the genetic code. *The New England Journal of Medicine,* vol. 331. A very concise overview of gene function.

Snow, Allison and Pedro Moran Palma. February 1997. Commercialization of transgenic plants: potential ecological risks. *BioScience,* vol. 47. Plants genetically altered to provide characteristics important to humans may affect other plant species.

Surridge, Christopher. March 28, 1996. The core curriculum. *Nature,* vol. 380. How transcription factors regulate gene expression.

Taubes, Gary. March 15, 1996. Misfolding the way to disease. *Science,* vol. 271. Alzheimer disease may be caused by a misfolded protein.

Velander, William H., Henryk Lubon, and William N. Drohan. January 1997. Transgenic livestock as drug factories. *Scientific American.* Genetic alterations enable domesticated animals to produce human proteins that have therapeutic value.

denotes articles by author Ricki Lewis that may be accessed via *Life's* home page /http://www.mhhe.com/sciencemath/biology/life/.

This 25-million-year-old piece of amber entombed several termites.

Unit 5

Evolution

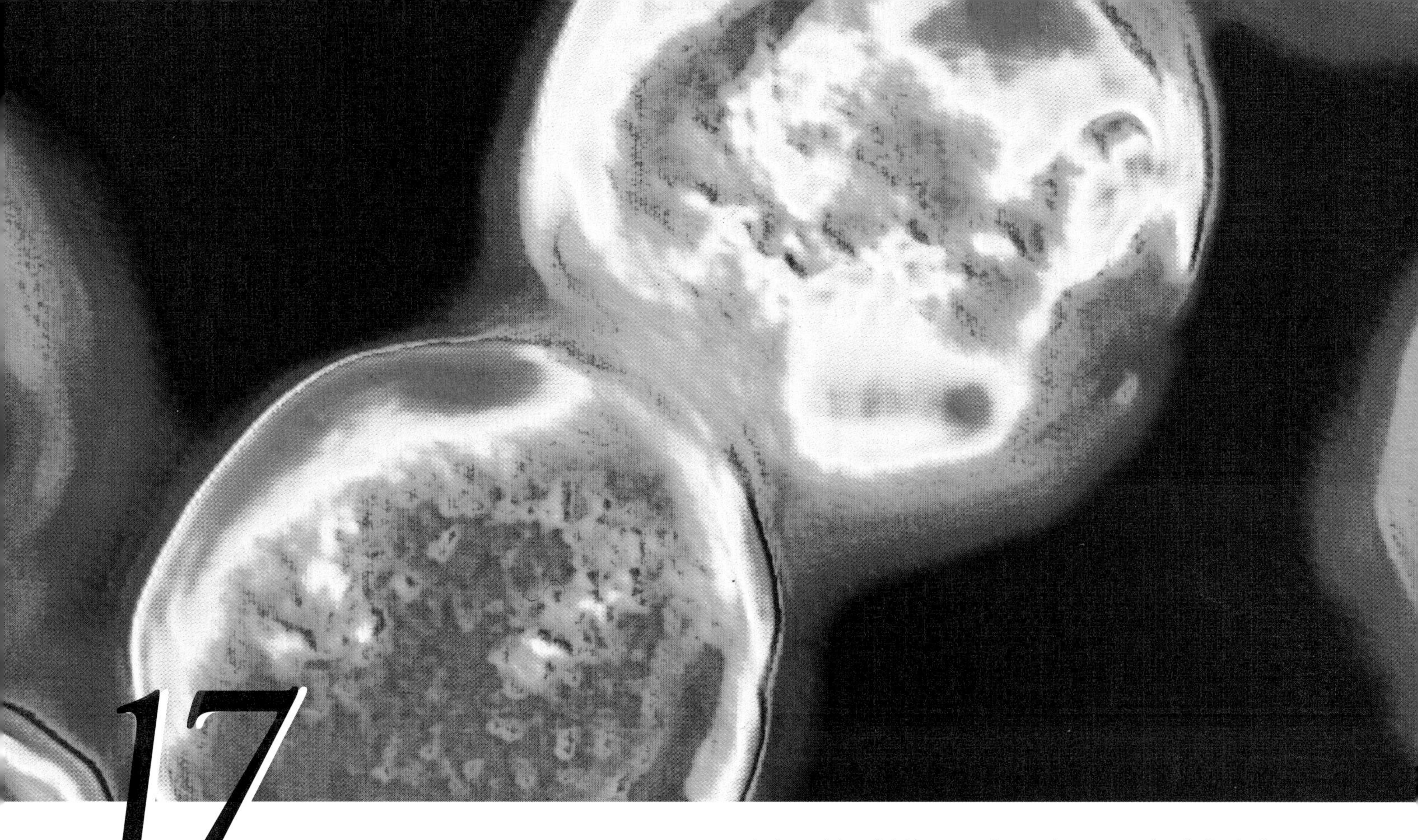

17 The Origin and Diversification of Life

Archaeoglobus fulgidus, a modern archaean, may be similar to the earliest cells.

chapter outline

***Setting the Scene* 352**

***Where Did Life Come From?* 352**
Spontaneous Generation
Life from Space
Chemical Evolution

***The RNA World* 355**

***The Origin of Metabolism* 357**

***The History of Life on Earth* 359**
Precambrian Time
The Paleozoic Era
The Mesozoic Era
The Cenozoic Era
The Road to Humanity

Spontaneous Generation

Theories explain observations and experimental results, to make sense of nature, dispel fear of the unknown, or both. Theories often reflect beliefs characteristic of a certain time or place. This was especially true of experiments that addressed the question, How does life begin?

The Scene: The eighteenth century. Invention of the microscope has revealed microscopic life. The idea that life arises from the nonliving—spontaneous generation—is well accepted.

The Players: English naturalist and priest John T. Needham and, later, Lazzaro Spallanzani, an Italian biologist and priest.

The Question: How does microscopic life appear in a container?

Round 1: Needham cooks meat to produce broth and then pours the heated broth into a bottle and corks it. Days later, bacteria appear in the bottle. **Conclusion:** Life arose from the nonliving broth.

Round 2: Spallanzani claims that Needham did not boil the broth, hence the bacteria. Spallanzani sets up flasks with broth with their necks sealed and boils the contents. These flasks do not contain bacteria. A second set of flasks, with boiled broth but with corks, have bacteria. **Conclusion:** Life does not appear spontaneously because sealing keeps it out.

Round 3: Needham and others are unconvinced. He and French naturalist Georges Buffon propose that a "vegetative force" forms life and boiling weakens it. **Conclusion:** Spallanzani's experiments actually support the theory of spontaneous generation because they weaken the force.

to be continued . . .

Setting the Scene

About 4.5 billion years ago, solid matter began to condense out of a vast expanse of dust and gas, forming the earth and other planets. For the next billion years, comets, meteors, and asteroids pummeled the red-hot ball that was to become earth, boiling off the seas and vaporizing rock to carve the features of the fledgling world. Just 700 million years after the earth formed, organisms left traces of cell-like forms in the rocks of Greenland and Australia (fig. 17.1.)

How might something as complex and organized as life form in the midst of geologic chaos? How could chemicals aggregate and interact so that they could use energy from their surroundings to reproduce? Where might life have originated? Did it happen only once? Life probably didn't start in the "warm little pond" peacefully brewing life's biochemical precursors, as Charles Darwin envisioned. Nor does it seem likely that life, at least as we know it today, could have begun in the intense heat of a volcano, on land frequently demolished by seismic upheaval or cosmic debris, or in an atmosphere that let in cell-damaging ultraviolet radiation.

Reconstructing life's start is like reading all the chapters of a novel except the first. A reader can get an idea of the events and setting of the opening chapter from clues throughout the novel. Similarly, scattered clues from life through the ages reflect events that may have led to the origin of life. This evidence includes experiments that simulate chemical reactions that may have occurred on the early earth; fossilized microorganisms that might have been among earth's first residents; exploring how clays or minerals might have molded biochemical building blocks into polymers (chains) important in life; and identifying ancient gene and protein sequences retained in the cells of modern organisms.

Mastering Concepts

1. What events led to the formation of the earth?
2. What types of experiments and evidence do biologists use to study the origin of life?

FIGURE 17.1

Between 2.5 billion years ago and 600 million years ago, bacteria flourished in the shallow seas. Cyanobacteria aggregated in huge colonies, forming great submerged mats that sediments infiltrated, hardening to preserve replicas of the ancient organisms. These fossils, called stromatolites, are found today in only a few places, such as Shark Bay in Western Australia, where these came from.

Where Did Life Come From?

Fossils indicate that life evolved, or changed, through great stretches of time and diversified. Yet organisms today consist of the same chemicals and use the same biochemical information systems as did their oldest ancestors. The chemical similarities of modern organisms, biologists hypothesize, indicate that all life descended from a first type of organism that partitioned a subset of chemicals from the environment. But even before people knew about chemistry, they had ideas about how life began.

FIGURE 17.2

According to the theory of spontaneous generation, this larva arose out of nothingness.

FIGURE 17.3

Some meteorites contain water-soluble phosphorus compounds and amino acids. Might meteorites have brought phosphate precursors to the prebiotic earth, a prerequisite for forming nucleic acids?

Spontaneous Generation

In the seventeenth and eighteenth centuries, a popular theory of life's origins was **spontaneous generation,** the appearance of the living from the nonliving. "Evidence" abounded: Beetles and wasps seemed to sprout from cow dung. Egyptian mice arose from the mud of the Nile. Even such noted scientists as Newton, Harvey, and Descartes accepted this theory that life could arise from the nonliving (fig. 17.2).

A few skeptics questioned spontaneous generation. In the mid-seventeenth century, Italian physician Francesco Redi conducted the first of a series of experiments that would eventually disprove spontaneous generation. First, Redi put meat in eight flasks—four that were open to the air and flies, and four that were closed. Maggots appeared in the open flasks a few days later but not in the closed ones. The maggots matured into flies. Next, Redi placed meat in four open flasks and covered them with gauze to exclude flies. No maggots appeared. Redi's conclusion: flies—not some mysterious life-generating substance in the meat—were responsible for the maggots. Despite Redi's experiment, the theory of spontaneous generation persisted. Finally, demonstrations that excluded bacteria from meat broth, described in the chapter opener, disproved the idea.

Life from Space

Life drifting to the earth from space has inspired science fiction, scientific hypotheses, and experiments. The basis of this idea is that asteroids, meteorites, comets, and interstellar dust contain chemicals like those in organisms. For example, the well-studied Murchison meteorite contains phosphorus compounds that are soluble in water (fig. 17.3). On the early earth, most phosphorus was in water-insoluble compounds unavailable for use in nucleic acids or the phosphates that form the backbones of biological membranes. Meteorites and comets may have provided usable phosphorus compounds to the prebiotic earth.

How might biochemicals come here from space? According to one scenario, simple organic molecules in interstellar dust clouds formed complex organic compounds in comets. Exploding stars released heat that melted ice in the comets, providing water that brewed the chemicals into a living "soup." Pieces of comets crashing to earth eons ago may have seeded the planet with the molecules that became life. Alternatively, organic molecules in extraterrestrial objects may simply be the result of chemical reactions not associated with life. The finding of such molecules in a meteorite from Mars, plus evidence of water on Mars, has researchers (and the public) wondering about life on the red planet.

Could life have come here already formed? Swedish physical chemist Svante Arrhenius suggested in his 1908 book *Worlds in the Making* that life arrived from spores that are abundant in the cosmos. Two major objections arose to this idea, called panspermia. First, spores would not survive exposure to radiation in space. Second, panspermia does not answer the question of how life originated—it just moves life's birthplace to an extraterrestrial site. Neither spontaneous generation nor what we know of life from space satisfactorily explains how life began.

Chemical Evolution

Most origin-of-life researchers today take a chemical approach, targeting that time long ago when complex collections of chemicals became the first cells.

Primordial Soup Brews Life's Building Blocks

How can we envision the beginnings of life, short of inventing a time machine? One approach is to reconstruct the chemicals and conditions on the early earth and see if simple components can react to yield the large and complex informational molecules of life. In 1953, a graduate student named Stanley Miller, now a professor at the University of California at San Diego, conducted the first origin-of-life simulation. He showed how simple chemicals plus energy can produce more complex chemicals—including some biochemicals.

The roots of Miller's experiment go back to 1929, when English chemist John B. S. Haldane hypothesized that when life began, the earth's atmosphere differed markedly from today's. One of Haldane's contemporaries, Russian chemist Alexander I. Oparin, proposed that methane (CH_4), ammonia (NH_3), water (H_2O), and hydrogen gas (H_2) were the raw materials for the formation of amino acids and nucleotides. (That idea was based on geological information of the time. More recent evidence points to an early atmosphere rich in carbon dioxide.)

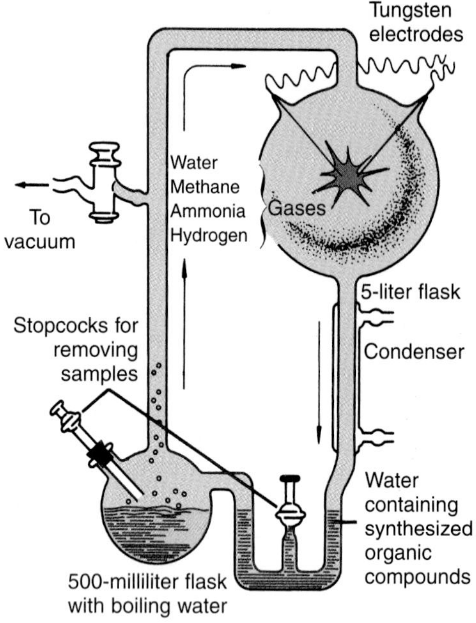

FIGURE 17.4

When Stanley Miller passed an electrical spark through heated gases, he generated amino acids—a more complex type of chemical that may have played a role in the origin of life.

Oparin realized that to synthesize large molecules from small ones requires energy. On the early earth, possible energy sources included lightning, hot springs, volcanoes, earthquakes, deep-sea hydrothermal vents, and unfiltered solar radiation (today, the ozone layer filters the sun's rays). Heat energy would have kept water in liquid form, sped the rates of some chemical reactions, and eventually made possible dehydration synthesis (evaporation or condensation) to form polymers.

Miller's mentor, chemist Harold C. Urey at the University of Chicago, suggested that Miller try to re-create conditions on the prebiotic earth to see what chemicals would form. Miller set up a simple but clever apparatus. In a large glass container, he mixed the gases Oparin had suggested and then exposed them to electric sparks to simulate lightning. Next, he condensed the gases in a narrow tube and passed them over an electric heater, a laboratory version of a volcano (fig. 17.4). This prebiotic soup brewed for a week—and produced four amino acids! The experiment showed that under certain conditions, chemical building blocks could react and link to build more complex molecules that are important in life.

Many researchers repeated Miller's experiment—perhaps too successfully, because several variations on the theme also produced biochemicals. In another scenario, ammonia and methane formed clouds of hydrogen cyanide (HCN) in the ancient atmosphere and polymerized when exposed to ultraviolet radiation from

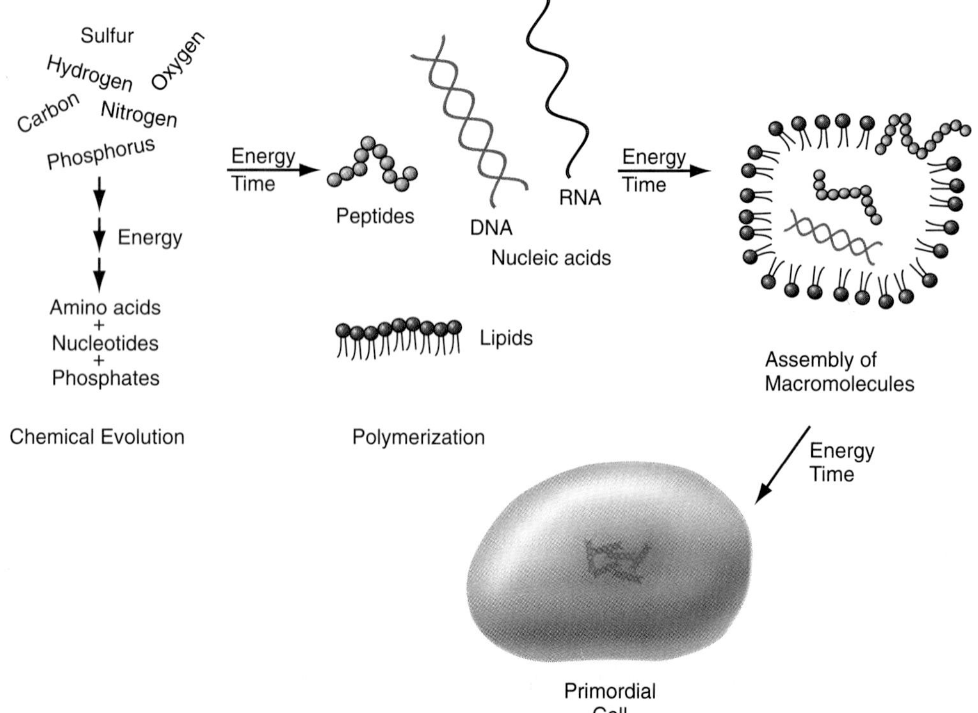

FIGURE 17.5

Likely stages in the origin of life. Figure 17.7 presents a more detailed view of this scenario.

the sun. When hydrogen cyanide chains fell into the seas, they reacted with water to produce amino acids. Simulations of these events in laboratory glassware yield a yellow-brown, sticky substance containing six different amino acids. In yet other experiments, prebiotic soup including phosphorus compounds produces nucleotides, the building blocks of nucleic acids and the source of biological energy, ATP.

"Crepes" Promote Polymerization

If the first stage of chemical evolution is described as a primordial soup of building blocks, then the second stage, when those building blocks polymerized into peptides and nucleic acids, is aptly described as French pancakes, or crepes (fig. 17.5). These pancakes are made by pouring liquid dough onto a hot surface, where it dehydrates to form the solid pancake. Similarly, organic contents of primordial soups may have dried on hot mineral or clay surfaces that served as molds. The heat caused dehydration synthesis, which condensed polymers from free amino acids or nucleotides.

The reason for the "crepe" dish following the "soup" first course is chemical. In an aqueous environment, hydrolysis is more likely to occur than its opposite reaction, dehydration synthesis (see fig. 3.24). The result is that strings of amino acids and nucleotides break apart faster than they form, countering the build-up of polymers necessary for life to begin. Certain clays and minerals not only may have provided molds to build polymers and physically supported growing molecules at one end, but may even have supplied energy (fig. 17.6). Formation of iron pyrite (fool's gold), for example, releases electrons that could have powered polymerization. Minerals and clays fringed with the growing polymers would have furnished many of the molecules that would eventually build cells.

Mastering Concepts

1. How did the theory of spontaneous generation account for the origin of life, and how was it disproved?
2. What is the evidence that life might have originated extraterrestrially?
3. What type of information do early earth simulations provide about the origin of life?
4. How might the chemical building blocks of life have polymerized?

The RNA World

The first cells may have emerged from a biochemical world where RNA was the informational molecule. Unlike DNA, which only carries genetic information, RNA carries that information and uses it to manufacture proteins and control gene expression. It can also act as an enzyme. Researchers call the time when RNA may have been pivotal in the origin of life the "RNA world."

Another biochemical that may have been very important in life's beginnings is the enzyme **reverse transcriptase.** This enzyme catalyzes construction of a DNA sequence from an RNA sequence. Today, reverse transcriptase enables retroviruses, which use RNA as their genetic material, to convert their RNA to DNA in a host cell.

Figure 17.7 outlines a hypothetical sequence of events, involving RNA and reverse transcriptase, that could have led to the origin of life. The process may have begun with short RNAs encoding sequences of amino acids. Eventually, an RNA encoded reverse transcriptase, which then copied RNA sequences into DNA

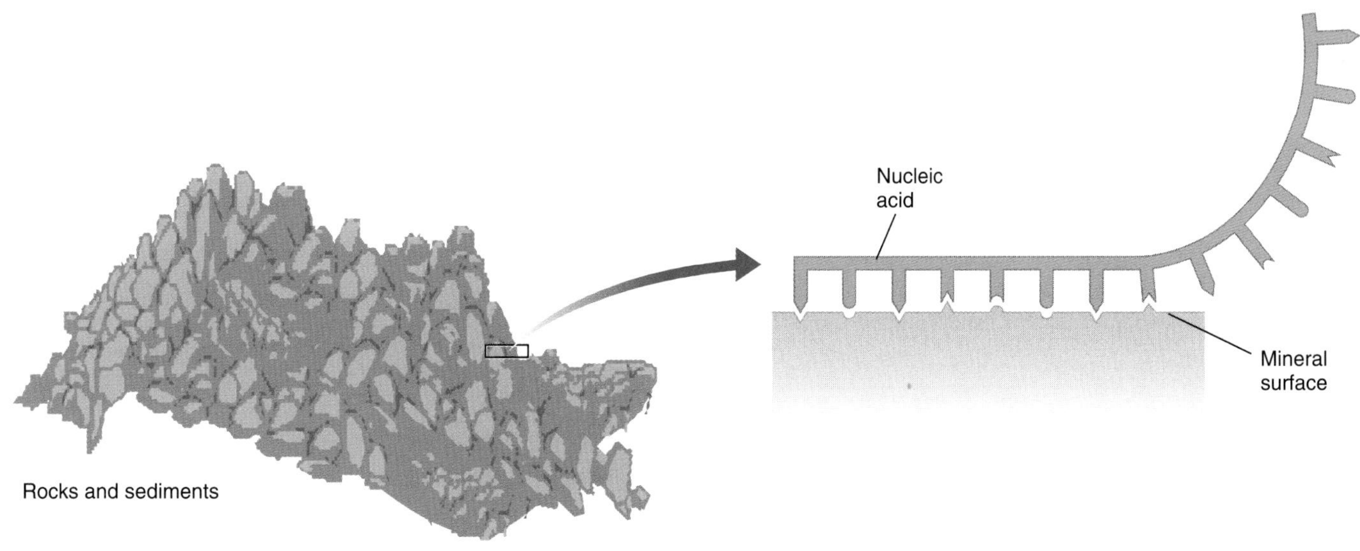

FIGURE 17.6
The surfaces of certain minerals and clays may have provided templates (molds), energy, and structural supports for nucleotides to polymerize (join) to form nucleic acids.

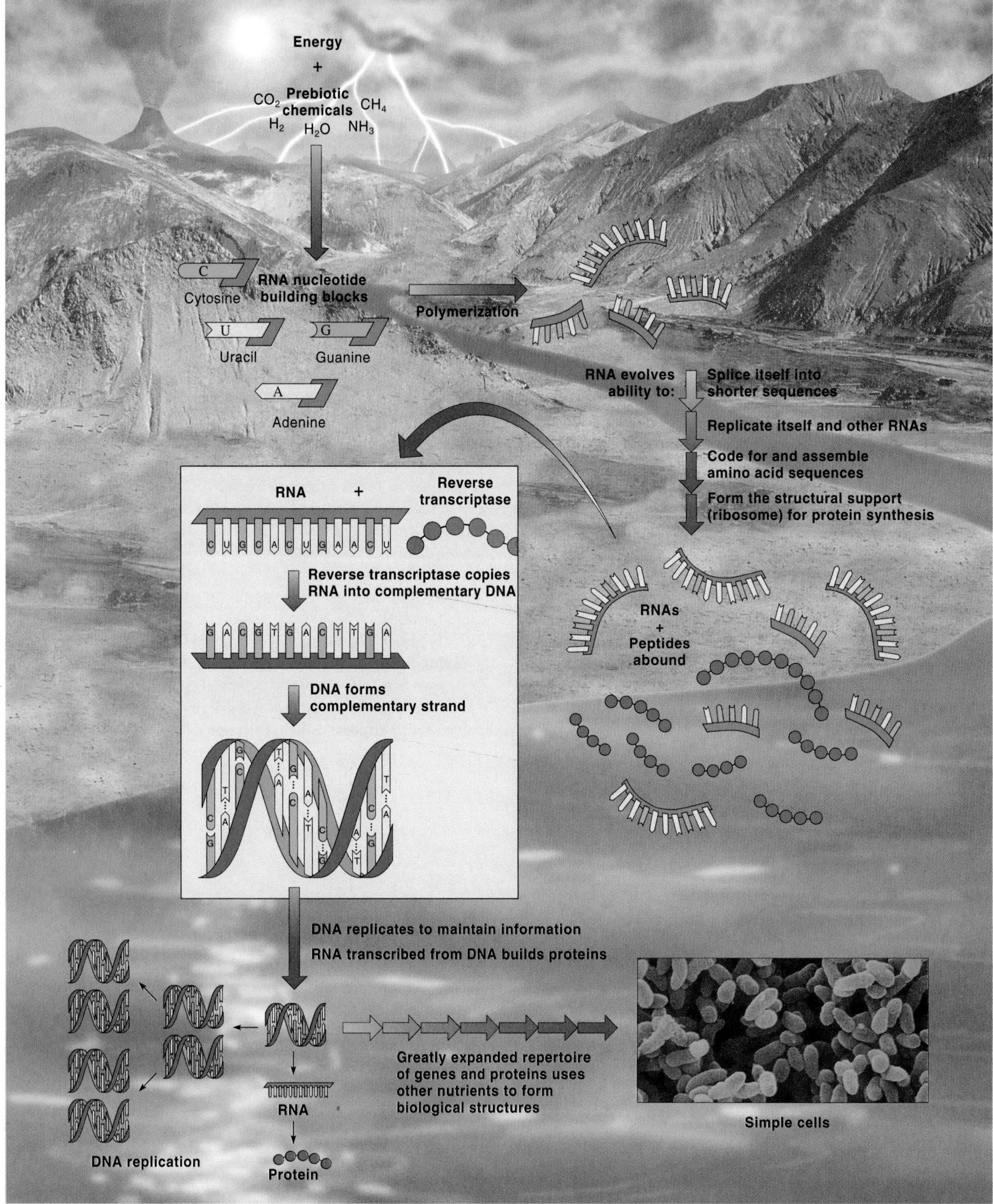

FIGURE 17.7

A scenario for life's beginnings—the RNA world. Chemical evolution in the presence of energy led to formation of nitrogen-containing nucleotide bases in an RNA-like molecule. The phosphate that tied the bases together might have been liberated from rocks when erupting volcanoes released energy. RNA evolved capabilities to replicate and encode proteins. A protein, perhaps similar to reverse transcriptase, copied the information in RNA into DNA. The DNA then polymerized a complementary strand, fashioning the first DNA double helices. As nucleic acids, proteins, carbohydrates, and lipids formed, the structures that would become parts of cells took shape. At some point, life arose from these structures.

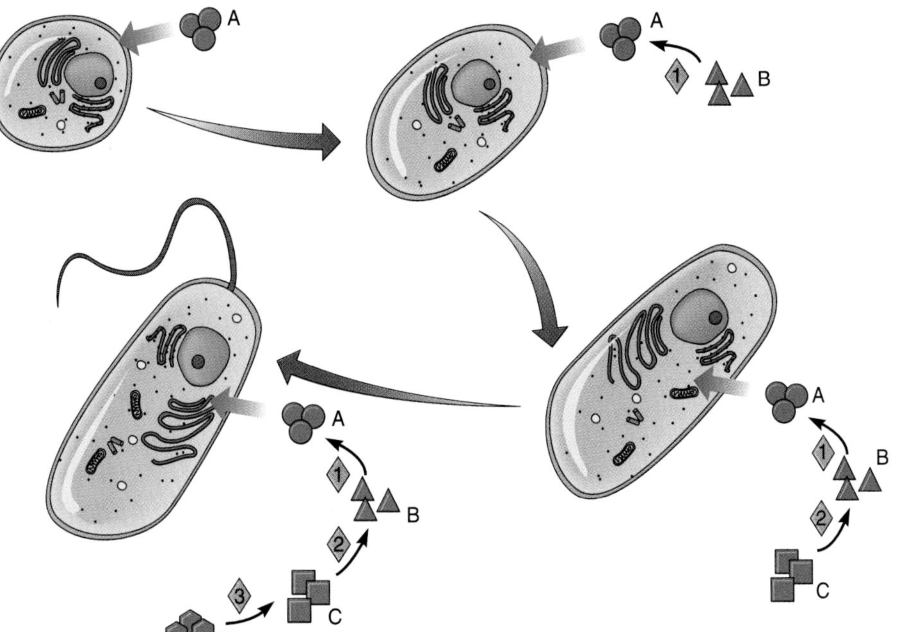

FIGURE 17.8

Evolution of metabolic pathways. The evolution of new enzymes probably expanded the number of different nutrients certain progenotes could use, thus giving them a survival advantage over others competing for limited food. The units marked *1, 2,* and *3* are enzymes; *A* through *D* are nutrients.

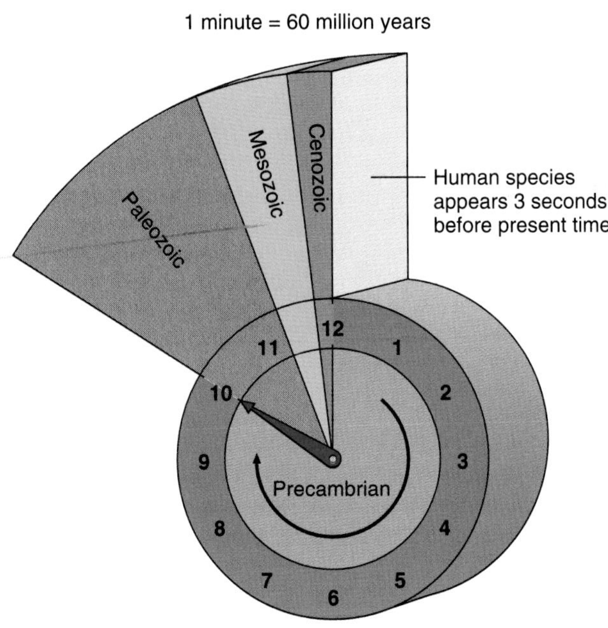

FIGURE 17.9

Precambrian time accounts for five-sixths of earth's history. The time when we think life on earth has been abundant—the Paleozoic, Mesozoic, and Cenozoic eras—accounts for only one-sixth of the planet's history.

sequences. Once information could be retained as DNA, yet translated into proteins by RNA, the chemical blueprints for life were in place.

Early groupings of nucleic acids and proteins would have eventually been set off and protected from the surroundings—which the cell membrane does today. Under certain temperature and pH conditions, and with the necessary precursor chemicals, phospholipids could have assembled into membranelike structures. Such membrane precursors are indeed found in sediments. Experiments using lipids show that pieces of membrane can grow on structural supports and break free when they reach a certain size and form a bubble. Might an ancient membrane bubble have enclosed neighboring nucleic acids and proteins to form a cell-like structure (see fig. 17.5)?

These ancient collections of RNA, DNA, protein, and lipid, precursors of cells but not yet nearly as complex, are termed **progenotes.** The next step in the origin of life was the evolution of metabolism.

Mastering Concepts

1. Why might RNA have been pivotal in the origin of life?
2. What role might reverse transcriptase have played in life's origin?
3. How might cell membranes have originated?
4. Describe the chemical interactions of nucleic acids, proteins, and lipids that might have preceded formation of the first cells.

The Origin of Metabolism

Metabolism is the ability of organisms to acquire and use energy to maintain the organization necessary for life. The capacity of nucleic acids to mutate may have enabled progenotes to become increasingly self-sufficient, giving rise eventually to the reaction pathways of metabolism. Here is how it might have happened.

Imagine a progenote that fed on a molecule, nutrient A, that was abundant in its environment. As the progenote reproduced, its food source dwindled. The ancient seas probably held more than one genetic variety of progenote. One type might have had an enzyme that could convert another nutrient, B, into the original nutrient A. The second progenote type would have a nutritional advantage, because it could extract energy from food sources B and A.

Soon the first type of progenote, totally dependent on nutrient A, would die out as its food vanished. For a while, the variant

biology in action 17.1

Handedness in Biomolecules

Before there was life, there were collections of chemicals. But a living entity is more than a subset of chemicals. One way that the chemistry of cells differs from that of nonliving matter is in the symmetry of certain organic molecules.

Nineteen of the 20 amino acids of life, and the nucleic acid sugars ribose and deoxyribose, are asymmetric. They can exist in two forms that are mirror images of each other but are not superimposable—like a right and left hand (fig. 17.A). This situation, called chirality, occurs when a central carbon bonds to four different chemical groups (fig. 17.B). When chiral molecules are synthesized in the laboratory, half of any given batch is one configuration, half the other. But in organisms, the 19 amino acid types are all left-handed (the exception, glycine, does not have four different groups around the central carbon); ribose and deoxyribose are always right-handed. How, and why, is this so?

The handedness of the proteins and nucleic acids of life has been the subject of murder mysteries and science fiction. Biologists and chemists have puzzled over it for 150 years.

French physicist Jean Baptiste Biot discovered chirality in 1815. He passed beams of light through certain organic chemicals in solution and found that some solutions of a particular compound moved the light in a counterclockwise direction, some in a clockwise direction, and some not at all. He suggested that a quality of the molecules affected the way they bend polarized light. French

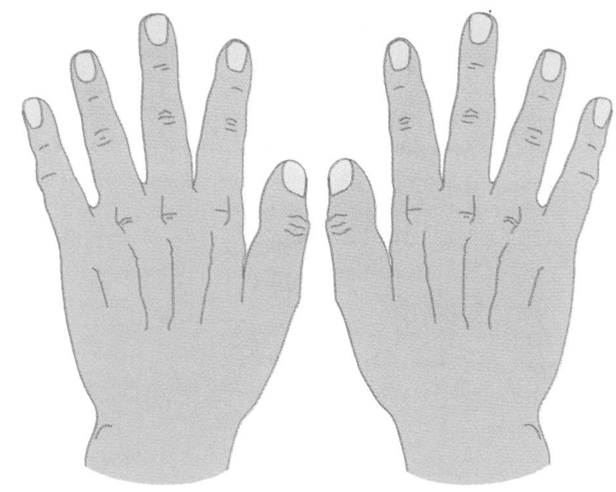

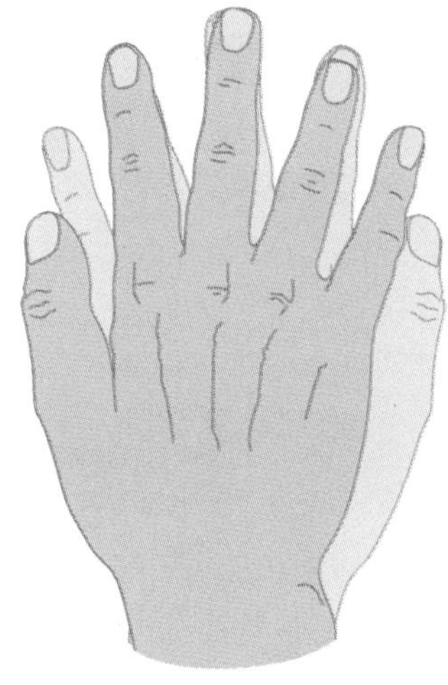

FIGURE 17.A
A left and right hand cannot be exactly superimposed because of differences in symmetry. Neither can certain forms of the same molecule. Such molecules are said to be chiral.

progenote that could convert nutrient B to A would flourish. But, in time, nutrient B would also become scarce. Perhaps a new type of progenote would then arise, with an additional enzyme that converted a nutrient C to B (and then, using the first enzyme, B to A). In time, this progenote would flourish; then another, and yet another. Over time, an enzyme-catalyzed sequence of connected chemical reactions—a biochemical pathway linking D to C to B to A—would arise (fig. 17.8). More pathways would form. As intermediates of one pathway spawned others, metabolism evolved.

Many different associations of proteins, nucleic acids, and lipids must have formed spontaneously over many millions of years, and somehow a certain combination, or combinations, persisted and diversified. It probably wasn't a random process, as Biology in Action 17.1 discusses. But with so many chance experiments of nature, it seems almost inevitable that, eventually, a cell-like structure, with the ability to replicate and change, would emerge. Biological evolution would begin.

Mastering Concepts

1. Why was the ability to mutate necessary for the evolution of metabolism?
2. How might metabolic pathways have evolved?
3. What characteristic would a cell require to persist?

chemist Louis Pasteur extended Biot's observation. He synthesized crystals of various organic compounds and then examined them under a microscope. He discovered two types of crystals that differed subtly in their handedness. When he separated a 50:50 mixture of the two types of crystals by handedness, he found that the right-handed crystals rotated light in a clockwise direction, and the left-handed crystals rotated light counterclockwise. Solutions with equal numbers of types cancel each other out, and the path of light doesn't rotate.

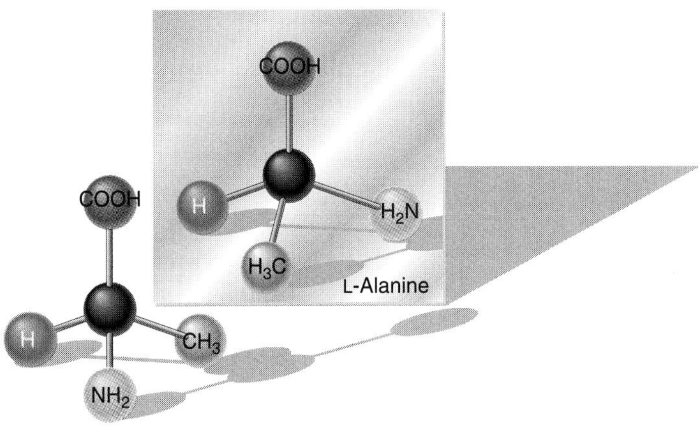

FIGURE 17.B
Nineteen of the 20 types of amino acids in organisms are chiral, but only the left-handed versions occur in organisms. Similarly, ribose and deoxyribose, the sugars in RNA and DNA, are only right-handed in organisms, although both configurations are chemically possible. What does this mean? Scientists do not know.

Ten years after the crystal-picking experiment, Pasteur discovered that chiral organic molecules in organisms are only one form. Today, life scientists still argue about what this asymmetry means, and how and when it happened. Three hypotheses are:

1. Meteorites or comets carrying just one form of chiral molecules seeded the earth with the precursors of life. The observation that the Murchison meteorite has predominantly the left-handed form of the amino acid alanine is evidence for this explanation.
2. Chirality was part of the origin of life. As various combinations of primordial self-replicating molecules formed, one type, including either a left-handed or right-handed version of a key constituent, had an advantage—a better ability to catalyze or replicate, perhaps—and therefore outcompeted other forms and persisted.
3. Chirality occurred after the origin of life. Evidence for this explanation is that the forms of ribose and deoxyribose in nucleic acids in organisms break down when alone. Chirality may have stabilized early nucleic acids.

Perhaps the answer to why the biomolecules of life are chiral will come from space explorations. If amino acids are left-handed and nucleic acid sugars right-handed because of a chance choice long ago, then life elsewhere might be the opposite. But if the handedness on earth is a prerequisite for life, then it should be this way beyond the earth—if life there is as we know it.

The History of Life on Earth

Organisms flourished and diversified. A textbook can hardly do justice to the vast parade of life that followed the long-ago progenotes, but following are a few glimpses into times past.

The backdrop for life history is geological history. Geological activity began before life did and continues today. The **geological time scale** is divided into major **eras** of biological and geological activity that lasted long stretches of time. Often the boundaries mark periods of known devastation or mass extinctions. The eras are subdivided into **periods,** and within periods, **epochs.** The eras begin with the Archean (4,600 to 2,500 million years ago) and then the Proterozoic (until 550 million years ago). (These two eras are also considered "Precambrian time" because the time afterwards, the Cambrian period, saw an explosion of life.) Next comes the Paleozoic era (until 240 million years ago), the Mesozoic (until 65 million years ago), and finally the ongoing Cenozoic (figs. 17.9 and 17.10).

Precambrian Time

By about 3 billion years ago, the geological tumult had calmed somewhat, and unicellular organisms inhabited nearly every part of the planet. Some of these ancient microorganisms contained pigments that enabled them to extract energy from sunlight via photosynthesis. This activity gradually added oxygen to the atmosphere, which forever altered the pattern of life on earth.

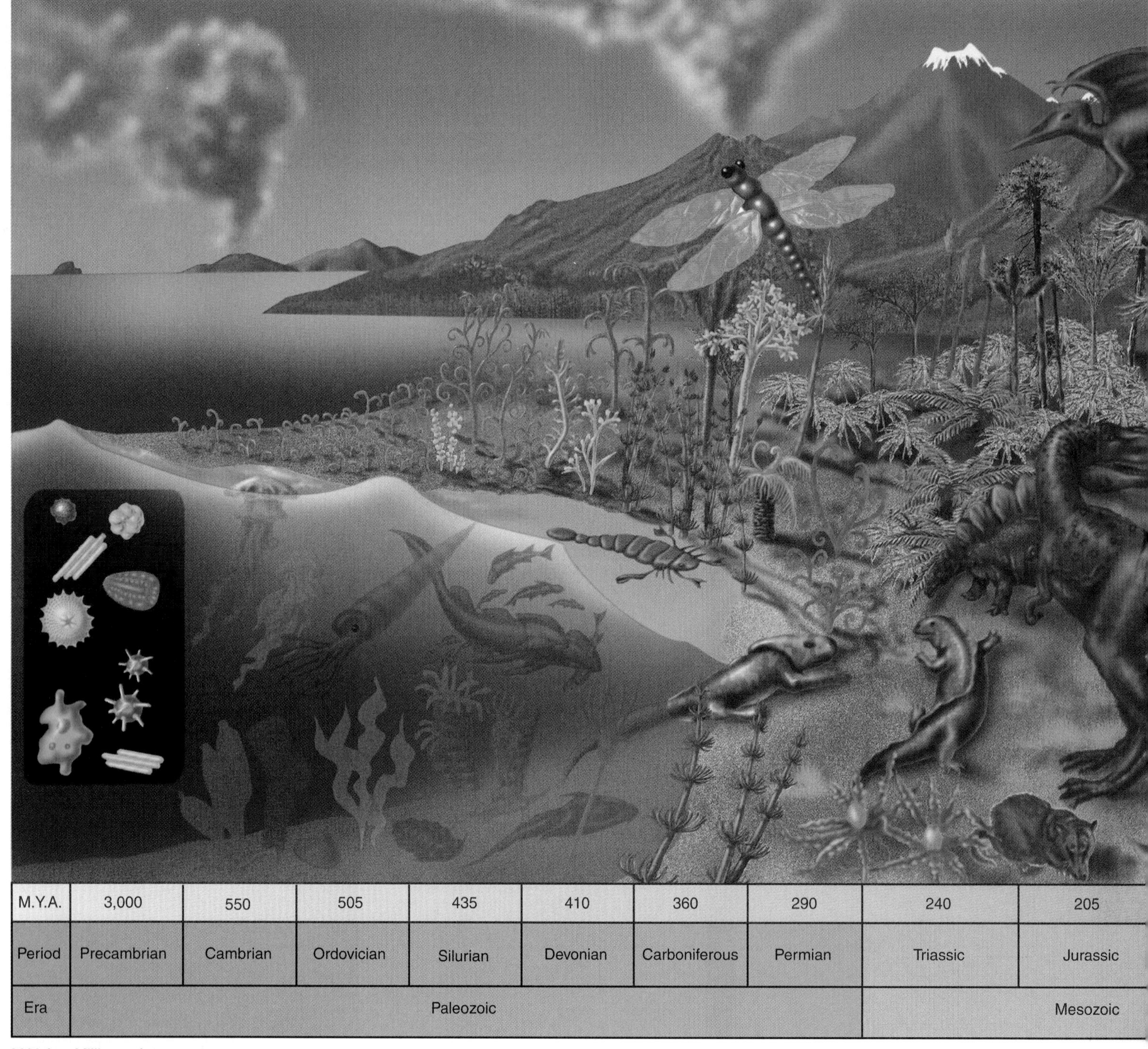

M.Y.A.
3,000
550
505
435
410
360
290
240
205
Period
Precambrian
Cambrian
Ordovician
Silurian
Devonian
Carboniferous
Permian
Triassic
Jurassic
Era
Paleozoic
Mesozoic

M.Y.A. = Millions of years ago

Fossil evidence before 600 to 700 million years ago is sparse because intense heat and pressure destroyed most remains. Researchers have found organic molecules in 3.85-billion-year-old minerals from Greenland that have not been altered by heat or pressure. The types and ratios of carbon isotopes in these deposits suggest that the organic molecules may once have been part of a progenote. Early fossils of organisms are from the Fig Tree sediments in South Africa from 3.5 billion years ago. The sizes and intricate folds resemble modern prokaryotes. Fossils from Rhodesia from 2.8 billion years ago contain breakdown products of chlorophyll, suggesting that microbes then photosynthesized.

The earliest fossils of eukaryotic cells found so far date from 1.4 to 1.9 billion years ago and come from China (fig. 17.11*a*). These organisms may have been single-celled algae. Australian fossils consisting of an organic residue 1.69 billion years old are chemically similar to eukaryotic membrane components and may come from a very early unicellular eukaryote. We know very little of how archaea, prokaryotes, and eukaryotes diversified from the first cells. Chapter 21 discusses

138	65	54	38	25	5	1.65	0.01
etaceous	Paleocene	Eocene	Oligocene	Miocene	Pliocene	Pleistocene	Recent
	Tertiary epoch					Quaternary epoch	
	Cenozoic						

FIGURE 17.10

A glimpse of the sequence of appearance of different life-forms on earth, not drawn to scale. We know little about the earliest types of organisms, but they were single-celled. Life has increased in complexity, yet simpler organisms, such as bacteria, still flourish. Note how recently humans appear.

the relationships of these major groups of organisms in greater detail.

About 1.2 billion years ago, multicellular life appeared. Exactly how life proceeded from the single-celled to the many-celled is a mystery. Perhaps many unicellular organisms united, or a large single-celled organism divided into many units (see fig. 26.2).

The earliest fossils of multicellular life are from a red alga that lived about 950 million to 1.25 billion years ago in Canada (fig. 17.11*b*). Abundant fossil evidence of multicellular algae comes from Siberia, dating from 1 billion years ago (fig. 17.11*c*). This particular time period saw an explosion of soft-bodied (invertebrate) eukaryotic life that left fossils in the rocks of southern Australia and the Canadian arctic.

The Paleozoic Era

Fossils from the period beginning about 550 million years ago are abundant, earning the name "Cambrian explosion." Ancestors of all modern animal groups debuted then, plus many now-extinct

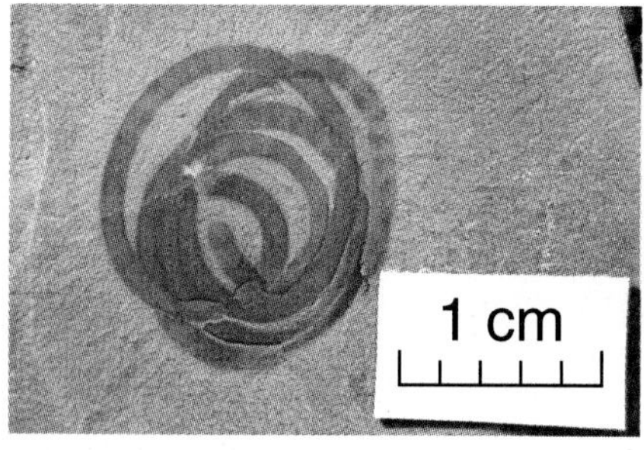

(a)

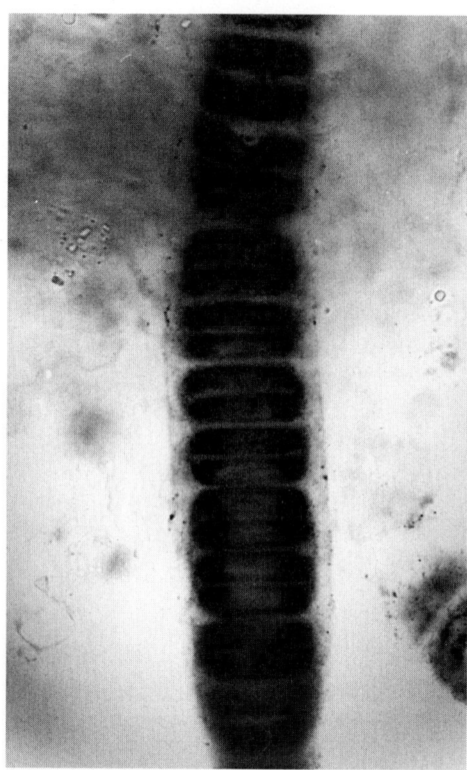

(b)

(c)

FIGURE 17.11

Early eukaryotes. (*a*) Researchers believe this spiral of organic matter found in 1.4-billion-year-old rock in China represents *Grypania spiralis,* a very early eukaryote. (*b*) This early red alga, found in 950-million-year-old to 1.25-billion-year-old rock in Canada, is the earliest evidence of multicellularity. (*c*) By the time these multicellular algae were preserved, about 900 million to 1 billion years ago, multicellular life was more widespread.

species. The seas held algae, sponges, jellyfish, and worms, as well as the earliest known organisms with hard parts, such as shellfish, insectlike trilobites, nautiloids, scorpion-like eurypterids, and brachiopods, which resembled clams (fig. 17.12).

The Ordovician period followed the Cambrian period 505 to 435 million years ago. The seas continued to support enormous communities of algae and invertebrates. Organisms called graptolites were common, named for their fossils, which resemble pencil markings. The first vertebrates to leave fossils, the jawless fishes, date from this time.

At the end of the Ordovician period and during the Silurian period, from 435 to 410 million years ago, the first organisms appeared on land—plants called psilophytes. They had bare stems from which spores scattered and underground branches, but no leaves or roots. The first animals to leave fossils on land resembled scorpions.

The Devonian period of 410 to 360 million years ago was the "Age of Fishes." The seas continued to support most life, including the now prevalent invertebrates, plus fishes with skeletons of cartilage or bone. Corals and animals called crinoids that resembled flowers were abundant. The land was still relatively barren, inhabited by scorpions, millipedes, and other invertebrates.

The fresh waters of the Devonian were home to the lobe-finned fishes (fig. 17.13*a*). These animals had fleshy, powerful fins and could obtain oxygen through gills as well as through primitive lunglike structures. (Chapter 36 discusses how gills work.) The lobe-finned fishes first appeared about 410 million years ago and persist today as lungfish.

By 360 million years ago, the first four-limbed animals, or tetrapods, appeared. The earliest tetrapods had a mix of adaptations to living in water and on land, and they provide a study in transition. The 3-foot-long (.9-meter-long) *Acanthostega* had a fin on its tail like a fish and used its powerful tail to move underwater (fig. 17.13*b*). Yet it also had hips, legs, and toes. Its ribcage wasn't strong enough to support internal organs without the buoyancy of water, but preserved footprints indicate that the animal could venture briefly onto land. Its feet were especially interesting—flat to move easily through the water but with eight toes! A contemporary of *Acanthostega,* called *Ichthyostega,* was more like an amphibian than a fish (fig. 17.13*c*). Its legs were more powerful and its ribcage stronger, yet it had a skull shape and finned tail reminiscent of fish ancestors.

By 360 million years ago, the beginning of the Carboniferous "Age of Amphibians," the descendants of the lobe-finned fishes were spending more time on land. Their legs were too weak to support full-time walking, so they had to return for frequent dips to wet their skins and to lay eggs, so that the embryos within could get sufficient moisture and nutrients.

Between 300 and 350 million years ago, as amphibians flourished, some types arose that coated their eggs with a hard shell. These animals branched from the other amphibians to eventually give rise to reptiles, birds, and mammals. The first animals capable of living totally on land, primitive reptiles, appeared about 300 million years ago. Earth then was one large, green landmass within a vast

(a)

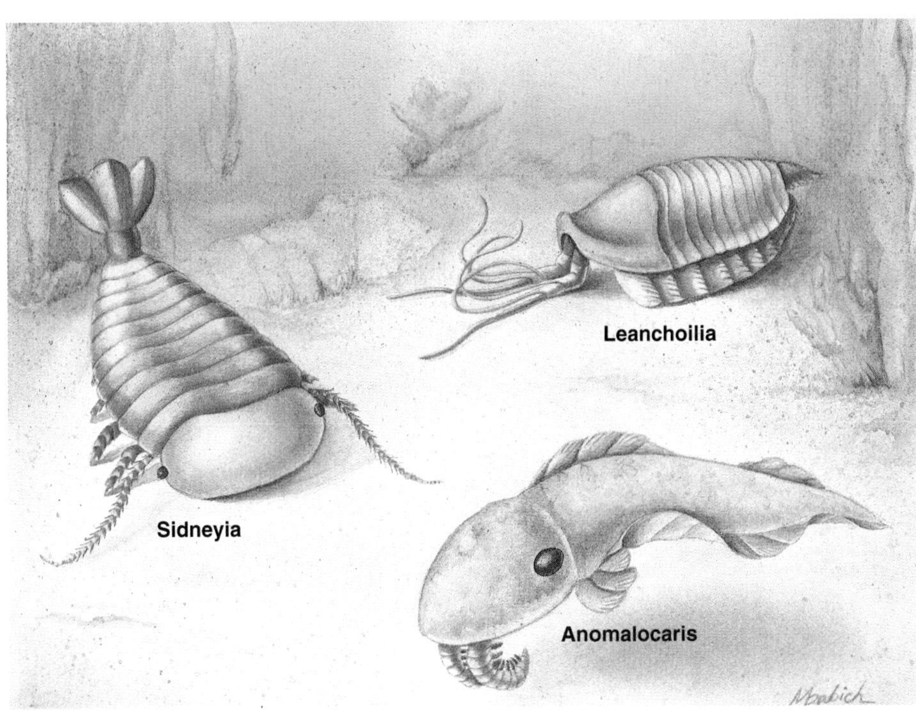

(b)

FIGURE 17.12

The seas of the Cambrian were home to an explosion of life-forms. (*a*) This trilobite was a distant forerunner of insects. Trilobites ranged in size from that of a pea to an automobile. They left many fossils because of their abundance, wide geographical distribution, and hard parts. (*b*) The Burgess shale in British Columbia contains fossils of many soft-bodied invertebrates not seen anywhere else in the world, apparently fossilized when a mudslide buried them.

ocean, the waters and swamps brimming with life. The air, warm year-round, was alive with the sounds of dragonflies, grasshoppers, and crickets, some of them giant versions of their familiar modern descendants. Invertebrates flourished in the sediments.

The swamps of the Carboniferous had fernlike plants and majestic conifers towering to 130 feet (40 meters). By the end of the period, many of the plants had died, buried beneath the swamps to form, over the coming millennia, coal beds (fig. 17.14). Because of these bountiful reminders of the ancient first forests, this time is also called the Coal Age. Today, a split piece of coal will sometimes reveal an impression left by an ancient fern.

The Paleozoic era ended dramatically with the Permian period (290 to 240 million years ago), when half the families of vertebrates and more than 90% of species in the shallow seas became extinct. Reptiles were becoming more prevalent, as amphibian populations shrank.

The reptile introduced a new biological structure, the **amniote egg,** in which an embryo could develop completely on dry land.

(a) Lobe-finned fishes

(b) *Acanthostega*

(c) *Ichthyostega*

FIGURE 17.13

From an aquatic to a terrestrial existence. (*a*) Lobe-finned fishes lived during the Devonian period and had an adaptation that would permit their descendants to venture onto the land—lungs. (*b*) *Acanthostega* lived as the Devonian period ended. It stayed mostly in the water but had adaptations that permitted it to spend short periods of time on land—including legs with eight toes! (*c*) A contemporary, *Ichthyostega,* could spend longer periods of time on land because its ribcage was stronger. It retained the skull shape and finned tail of its fish ancestors, however.

(a)

(b)

FIGURE 17.14

The Carboniferous, or Coal Age. (*a*) Forests were so lush that they were preserved in massive coal beds. The fernlike plants in the foreground are very ancient seed-bearing gymnosperms. The thin tree on the right is an extinct ancestor of the modern horsetail, and the thicker trunks on the left, also extinct, gave rise to modern club mosses and ground pines. (*b*) In Indiana and Illinois, a piece of coal will occasionally contain the imprint of a fern from the Coal Age.

An amniote egg contains nutrient-rich yolk and protective extraembryonic membranes. The amnion encloses the embryo in a sac of fluid that is chemically very similar to seawater, which nurtured so many ancient species. Amniote eggs persist today in reptiles, birds, and mammals.

The Permian period foreshadowed the dawn of the dinosaur age. Cotylosaurs were early Permian reptiles that gave rise to the dinosaurs, as well as to modern reptiles, birds, and mammals. They coexisted with their immediate descendants, the pelycosaurs, or sailed lizards.

The Mesozoic Era

During the Triassic period from 240 to 205 million years ago, small animals called thecodonts flourished, occupying a place in time between the Permian cotylosaurs and the great dinosaurs to come. Thecodonts shared the forest of cycads, ginkgos, and conifers with other animals called therapsids. These were reptiles, but their posture and teeth were mammal-like (fig. 17.15). By the end of the Mesozoic era, the therapsids would evolve into small, hair-covered animals, most of whom lived on the forest floor—the first mammals.

At the close of the Triassic period, thecodonts and therapsids were becoming rarer as much larger animals began to infiltrate a wide range of habitats. These new, well-adapted animals were the dinosaurs, and they would dominate for the next 120 million years. The dinosaurs had characteristics of the reptiles that preceded them and of the birds that would follow.

By the Jurassic period of 205 to 138 million years ago, the dinosaurs were everywhere. Ichthyosaurs swam in the seas, archaeopteryx took to the air, and apatosaurs, stegosaurs, and allosaurs roamed the land in the second half of the period. The first flowering plants (angiosperms) appeared, but forests still consisted largely of tall ferns and conifers, club mosses, and horsetails.

FIGURE 17.15

The dawn of the Mesozoic era saw many small animals, called thecodonts, that had characteristics of reptiles and mammals. Mammals did not suddenly appear with the demise of the dinosaurs; they just became more prevalent.

The Cretaceous period (138 to 65 million years ago) was a time of great biological change. Flowering plants spread in spectacular diversity, and the number of dinosaur species declined. Although this was the beginning of the end for the dinosaurs, many species soared to quite healthy numbers. Duck-billed maiasaurs, for example, traveled in herds of thousands in what is now Montana (fig. 17.16). Huge herds of apatosaurs migrated on the plains of Alberta to the Arctic, northern Europe, and Asia, which were then joined. By the end of the period, triceratops were so widespread that some paleontologists call them the "cockroaches of the Cretaceous."

The reign of the dinosaurs ended about 65 million years ago. We do not know what caused their demise, but it is clear that their disappearance opened up habitats for many other species, including the primates that eventually gave rise to our own.

The Cenozoic Era

At the start of the Cenozoic era 65 million years ago, diverse hoofed mammals grazed the grassy Americas. Many were **marsupials** (pouched mammals) or egg-laying **monotremes,** ancestors of the platypus. Their young were helpless, born tiny, hairless, and blind. They crawled along the mother's fur to sweat glands modified to form tiny nipples, which secreted milk.

Gradually a new type of mammal appeared whose young were better protected and therefore more likely to survive, reproduce,

FIGURE 17.16

Duck-billed dinosaurs called maiasaurs left fossil evidence of their lifestyle on a hill in Choteau, Montana, named Egg Mountain. A sudden volcanic eruption 80 million years ago preserved in time their community, including nine inhabited nests. "Dinosaur upchuck" near the nests suggested that parents fed offspring by chewing food for them, much as birds do today.

(a) *Dryopithecus*

(b) *Australopithecus*

(c) *Homo erectus*

FIGURE 17.17

Human forerunners. (*a*) The "oak ape" *Dryopithecus,* who lived from 22 to 32 million years ago, was more dextrous than its predecessors. (*b*) Several species of *Australopithecus* lived from 2 to 4 million years ago. These hominids walked upright on the plains. (*c*) *Homo erectus* made tools out of bone and stone, used fire, and dwelled communally in caves from 1.6 million years ago to possibly as recently as 35,000 years ago.

and persist. These newcomers were the **placental mammals.** The young remain within the female's body longer, where the placenta nurtures them. Mammary glands are well developed.

Placental mammals eventually replaced most of the marsupials in North America. Then they invaded South America, thanks to changing geography. Until 2 to 3 million years ago, South America was an island. During the Tertiary period, several types of large marsupials thrived there because the placental mammals, which had originated in North America, could not reach the enormous southern island.

Then the Bering land bridge rose and connected Asia to North America. Many mammals, including our ancestors, probably journeyed from Asia to Alaska and southward through North America. The isthmus of Panama formed and provided a route for northern placental mammals to invade the southern marsupial communities. The first animals to arrive, rodents and ground sloths, crossed when the land bridge was only a string of islands. Arrival of the placental mammals and their use of natural resources drove nearly all the South American marsupials to extinction. Many of the species found there today, including peccaries, llamas, alpacas, deer, tapirs, and jaguars, are descendants of immigrants from the north. A few species, including the opossum, armadillo, and porcupine, traveled in the opposite direction.

The Road to Humanity

Our species, *Homo sapiens* ("the wise human"), probably first appeared during the Pleistocene epoch, about 200,000 years ago. The earliest of our order, Primates, were rodentlike insect eaters that lived about 60 million years ago. The ability of primates to grasp and to perceive depth provided the flexibility and coordination necessary to dominate the treetops. They diversified into many new species.

About 30 to 40 million years ago, a monkeylike animal about the size of a cat, *Aegyptopithecus,* lived in the lush tropical forests of Africa. Although the animal probably spent most of its time in the trees, fossils of limb bones indicate that it could run on the ground too. Fossils of different individuals found together indicate that these were social animals. *Aegyptopithecus* had fangs that it might have used for defense. This animal and a few monkeylike contemporaries are possible ancestors of gibbons, apes, and humans.

The Hominoids

From 22 to 32 million years ago lived the first **hominoids,** primates thought to be ancestral to apes and humans only. Most hominoid evidence comes from Africa, but an animal living in Europe at that time was named *Dryopithecus.* Its name means "oak ape," because its fossils were found with oak leaves (fig. 17.17*a*). The way the bones fit together suggests that this ape lived in the trees but could walk farther than *Aegyptopithecus.*

Several species of apes lived 11 to 16 million years ago in Europe, Asia, and the Middle East. These apes were about the size of a human 7-year-old and had small brains and pointy snouts. Because of competition in the treetops and forests shrinking due to climate shifts, these apes ventured onto the grasslands. One species of Miocene ape survived to give rise to humans and African apes.

The Hominids

Hominoid and **hominid** (ancestral to humans only) fossils from 4 to 19 million years ago are scarce. This was the time that the stooped, large-brained ape became the upright, smaller-brained ape-human. A rare 10-million-year-old fossilized face from northern

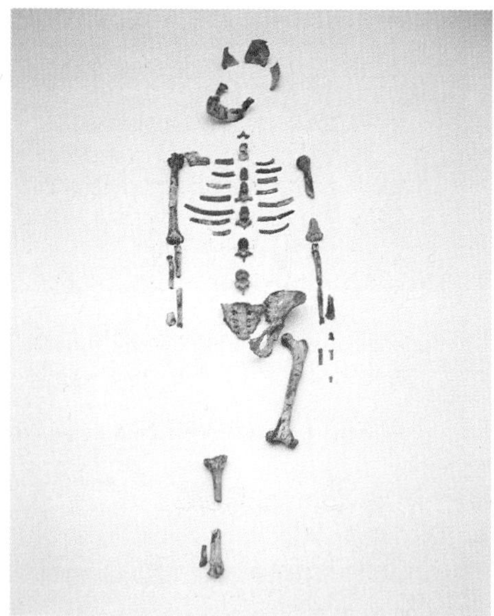

FIGURE 17.18

About 3.6 million years ago, a small-brained human ancestor walked upright in the grasses along a lake in the Afar region of Ethiopia. She skimmed the shores for crabs, turtles, and crocodile eggs to eat. She died at the age of 20, with severe arthritis in the backbone. Nearly 40% of her skeleton was discovered in 1974. Donald Johanson of the Cleveland Museum of Natural History and Timothy White of the University of California at Berkeley named her "Lucy"; they were listening to the Beatles song "Lucy in the Sky with Diamonds" when they found her. Unfortunately, Lucy lacked most of her skull. But in 1994, Johanson and William H. Kimbel discovered a male skull about a mile from where Lucy was found. He lived about 200,000 years later than Lucy but is believed to be the same species of primate—*Australopithecus afarensis.*

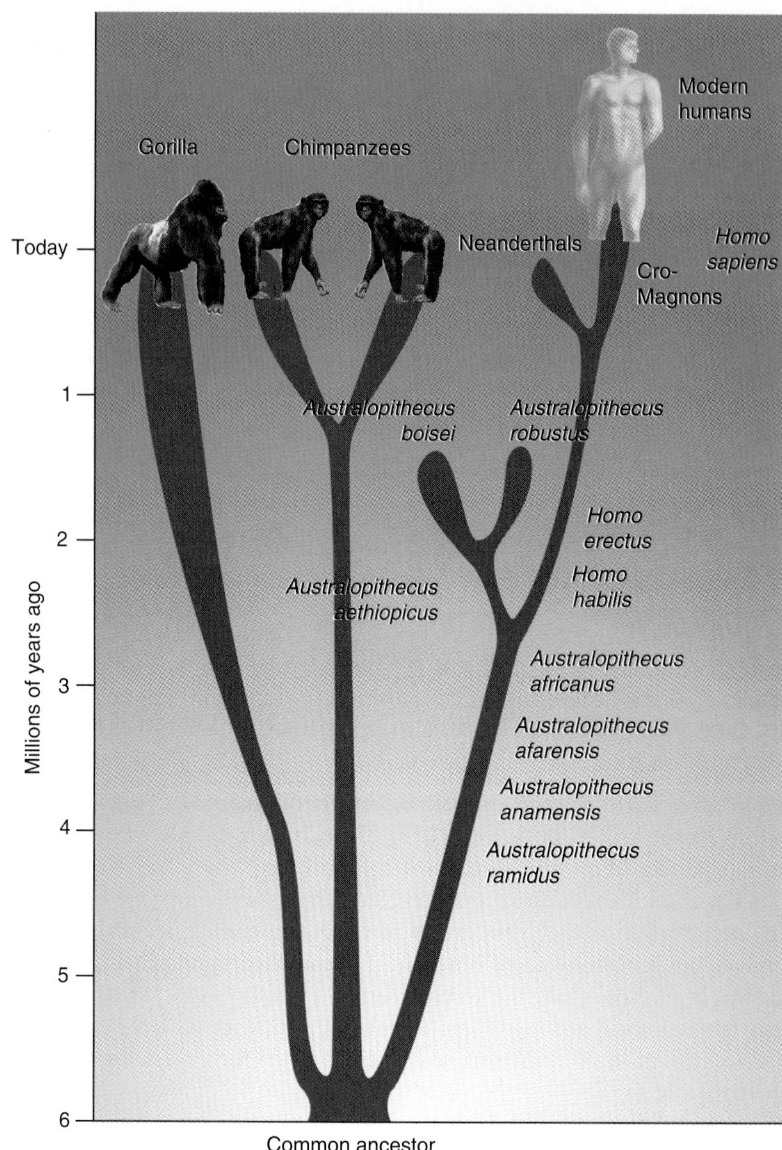

FIGURE 17.19

An evolutionary tree diagram depicting the relationships in time of the primates leading to our own species, *Homo sapiens.*

Greece has small canine teeth and thick tooth enamel, which are more humanlike. This animal could have been an immediate forerunner of hominids.

Four million years ago, **bipedalism**—the ability to walk upright—opened vast new habitats on the plains to hominids. Several species of *Australopithecus* lived at this time (fig. 14.17*b*). These animals had flat skull bases, a feature seen in all modern primates except humans. They stood from 4 to 5 feet (1.2 to 1.5 meters) tall and had brains about the size of a gorilla's and humanlike teeth. The angle of preserved pelvic bones, plus the finding of fossils near those of grazing animals, suggests that this ape-human had left the forest. Figure 17.18 describes a famous Australopithecine named after a Beatles song.

Two million years ago, Australopithecines coexisted with *Homo habilis*—a more humanlike primate who lived communally in caves and cared intensely for young. "Habilis" means handy, and this primate is the first to evidence extensive tool use. *H. habilis* coexisted with and eventually was replaced by *Homo erectus.* Fossilized teeth and jaws suggest that *H. erectus* ate meat. They were the first primates to have an angled skull base that permitted a greater range of sounds, making speech possible. *H. erectus* fossils have been found in China, tropical Africa, and southeast Asia, indicating that they may have been the first hominids to leave Africa. The distribution of fossils suggests that *H. erectus* lived in families of male-female pairs (most primates have harems). They left fossil evidence of cooperation, social organization, and tool use, including the use of fire (fig. 17.17*c*).

H. erectus lived from the end of the Australopithecine reign some 1.6 million years ago to possibly as recently as 35,000 years

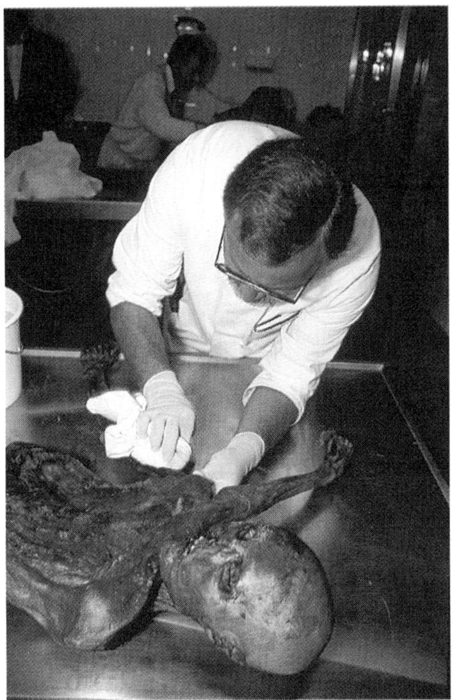

(a)

(b)

FIGURE 17.20

The "ice man" Otzi. (*a*) The oldest known intact human, at 5,300 years of age, lay frozen in the Alps between Austria and Italy, where hikers found him in 1991. (*b*) He wore well-made clothing, had intricate arrows indicating familiarity with ballistics and engineering, and carried mushrooms with antibiotic properties. He had tattoos, indentations in his ears that suggest he wore earrings, evidence of a haircut, and a hat. This depiction was derived from the evidence on and near the preserved body.

ago and probably coexisted with the first *Homo sapiens.* Several pockets of ancient peoples may have been dispersed throughout the world at that time.

The Neanderthals, contemporaries of *H. erectus* and members of *Homo sapiens,* appeared in Europe about 150,000 years ago. By 70,000 years ago, they had spread to western Asia. The Neanderthals had slightly larger brains than us, gaps between certain teeth, very muscular jaws, and large, barrel-shaped chests. They take their name from Neander Valley, Germany, where their fossils were discovered in 1856. Other skeletons buried with flowers suggest that the Neanderthals recognized important events.

From 30,000 to 40,000 years ago, the Neanderthals coexisted with the lighter-weight, finer-boned, and less hairy Cro-Magnons. The Cro-Magnons also had high foreheads and well-developed frontal brain regions. The first Cro-Magnon fossils found were the skeletons of five adults and a baby arranged in what appeared to be a communal grave in a French cave. Nearby were seashells pierced in a way that suggested they may have been used as jewelry. Intricate art decorated the cave walls.

Anthropologists once thought that the Cro-Magnons replaced the Neanderthals and arrived soon after their mysterious disappearance. Fossils indicate, however, that these two types of humans coexisted, but did not interbreed, in the same region of the Mideast that links Africa and Eurasia. What kept them from mating with each other? Perhaps it was incompatible genitalia or a geographical separation. Maybe they migrated, occupying the same areas at different times. Another theory is behavioral—perhaps the great facial differences between the large-boned Neanderthals and more delicate Cro-Magnons made it impossible for them to exchange mating cues.

We still don't know how it happened, but by 30,000 years ago, the Neanderthals were no longer, and the Cro-Magnons presumably continued on the path to humanity. Figure 17.19 summarizes the relationships in time of the hominids that we know about.

Evolution continued, taking on a cultural aspect as social and communication skills improved. Cave art from about 14,000 years ago indicates that our ancestors had developed fine hand coordination and could use symbols. By 10,000 years ago, people had migrated from the Middle East across Europe. They brought agriculture and may have replaced hunter-gatherers.

We have evidence of what humans were like 5,300 years ago from Otzi, an ancient man whom hikers in the Alps found preserved in 1991 (fig. 17.20). Otzi was apparently tending sheep high on a mountaintop when he perished. He was dressed for the weather. Berries found with him place the season as late summer or early fall. He probably feel into a ditch, where he froze to death and was soon buried in snow, sealing his body quickly in a

17.1 bioethics connections

Does Creationism Belong in a Biology Classroom?

When studying the origin of life, students often ask their professors to explain how the ideas of biblical creation and biological evolution relate to one another. One answer is that both ways of thinking may be valid, but one is science, and the other is not.

A standard dictionary defines science as "the observation, identification, description, experimental investigation, and theoretical explanation of natural phenomena." Science is a thought process that has a starting point and proceeds logically, based on accumulating experimental and observational evidence, to a conclusion or idea. Additional information may modify the conclusion at any time. What is *not* science is a belief or acceptance that something is so, without convincing evidence or an explanation.

Both religion and natural science seek to explain the wonders of nature, and sometimes they clash over how life originated. Religion evokes a creator, a force or being that placed life on earth. For many people, this settles the issue and explains the mysterious. Science attempts to explain the origin of life more methodically, by studying fossils, seeking clues in biological information molecules, and simulating chemical evolution under early earth conditions. Neither religion nor science answers all the questions. Neither approach is right or wrong, good or evil—what they are is different.

Many biologists feel that creationism—the biblical version of the origin of life—has no place in the science classroom simply because it is not science. To counter this argument, and to justify teaching the biblical view of life's origin with evolution, some people use terms such as "creation science" and "intelligent design theory" to present the biblical view as science. Yet the Pope acknowledges natural selection as a reality. Because creationism relies on belief and faith, and not rigorous adherence to the scientific method, it is not science. Teaching creationism may be appropriate in philosophy, history, comparative religion, and perhaps social science classes.

Table 17.A lists some key events in the continuing controversy over the place in education of the biblical view of life's origin.

Table 17.A A Short History of the Evolution/Creationism Debate in the United States

1900–1920	Botanist and evangelical Christian Asa Gray writes about how evolution and religious views can coexist. Biology textbooks cover evolutionary theory.
1920s	Increased interest in fundamentalism leads some states to rule as illegal teaching of life's origin as other than the biblical view. This trend culminates in the trial of biology teacher John T. Scopes in 1925 in Dayton, Tennessee. Scopes stands trial for using a textbook that teaches evolutionary theory. Prosecutor William Jennings Bryant objects to teaching the idea that humans are animals. Scopes loses.
1925–1950s	In the aftermath of the Scopes trial, evolutionary ideas are scarce to nonexistent in biology textbooks.
1957	With the successful launch of the Soviet satellite *Sputnik,* U.S. educators realize that students aren't learning much science.
1968	In *Epperson v. Arkansas,* the Supreme Court rules that teaching biblical creation in biology class violates the First Amendment's "establishment clause" because doing so supports a particular religion.
1971	Standards are set for evaluating when state laws or school board rulings overstep the 1968 Supreme Court ruling.
1971–present	Various groups try to introduce creationism into life science curricula, calling it "creation science" or other scientific-sounding names and not overtly discussing a supreme being. Biologists maintain that creationism, by any name, is not science.

natural tomb. Otzi now resides in a simulated glacier (a freezer) at the University of Innsbruck. DNA analysis of his tissues indicates that he belongs to the same gene pool as modern people living in the area—not surprising, since 5,300 years is but a flicker of evolutionary time.

This chapter has given brief and sweeping glimpses of life through the ages. Bioethics Connections 17.1 discusses how the biblical view of creation fits—or doesn't fit—with the study of the origin and diversification of life. The following chapters in this unit explain the forces and mechanisms behind the changes in life over time—evolution.

Mastering Concepts

1. What is the geologic time scale?
2. What do fossils reveal about life in Precambrian time?
3. What types of organisms lived during the Paleozoic, Mesozoic, and Cenozoic eras?
4. What types of animals led to the evolution of humans?
5. What adaptations enabled life to change in drastic ways?

Spontaneous Generation~

continued from page 352.

A concession from John Needham sent Lazzaro Spallanzani back to the laboratory bench. Needham admitted that his "vegetative force," which was strong enough to create life, could probably survive a very short period of boiling.

Round Four: Spallanzani set up more flasks with broth but boiled the contents for different time periods. After boiling for just a few minutes, the samples were pure. **Conclusion:** Spontaneous generation, even from a vegetative force, does not account for the appearance of bacteria in flasks of broth—previously present life causes it.

Finally, Spallanzani's explanation was accepted, the theory of spontaneous generation fell into disfavor, and Needham retired. Further experiments confirmed that spontaneous generation does not occur. German physiologist Theodor Schwann used hot air instead of sealing flasks, and French chemist Louis Pasteur used thin, S-shaped flask necks to keep bacteria out. Taking another approach, Spallanzani watched bacteria for hours, in tiny droplets of pond water, until he observed them divide.

In the twentieth century, Haldane and Oparin and then Miller and Urey took Needham and Spallanzani's question to another level—that of the origin of life. It will be interesting to see what future thinkers bring to the compelling question of how life began.

SUMMARY

1. The solar system formed about 4.5 billion years ago, and life left evidence on earth by 700 million years after that. Conditions when life originated differed from today.

2. Life does not **spontaneously generate.** Comets and meteorites contain chemicals also in organisms.

3. Origin of life simulations combine simple chemicals that include the elements in organisms in the presence of energy. More complex biochemicals form.

4. Amino acids may have polymerized into peptides and nucleotides into nucleic acids on hot clay or mineral surfaces that served as templates and structural supports and supplied energy.

5. The "RNA world" refers to the idea that RNA was the initial informational molecule of life. **Reverse transcriptase** could have copied RNA's information into more permanent DNA. RNA is the only biochemical that can carry genetic information and function as an enzyme, two properties important in the origin of life.

6. Phospholipids growing on supports and forming bubbles around nearby chemicals may have formed biological membranes.

7. Metabolic pathways may have originated when cell precursors, called **progenotes,** mutated in ways that enabled them to use additional nutrients.

8. The **geological time scale** is divided into **eras, periods,** and **epochs.**

9. Fossils from more than 3.5 billion years ago are unknown, because conditions then were harsh. The earliest fossils are of prokaryotes; the oldest eukaryotic fossils, of algae, date from 1.9 billion years ago. Evidence of multicellularity dates to 1.2 billion years ago.

10. The Cambrian explosion introduced many species. Life began in the seas. Amphibian-like animals ventured onto land about 360 million years ago. By 300 million years ago, reptiles had appeared and then diverged, eventually evolving also into birds and mammals. Invertebrates, ferns, and forests flourished. The Paleozoic era ended with mass extinctions.

11. Dinosaurs prevailed throughout the Mesozoic era, when forests were largely cycads, ginkgos, and conifers. In the middle of the era, flowering plants became prevalent. When the dinosaurs died out 65 million years ago, resources opened up for mammals.

12. Replacement of **marsupials** and **monotremes** with **placental mammals** in South America during the Cenozoic era illustrates the great effects of geographic changes on life.

13. *Aegyptopithecus* and other primates preceded the **hominoids,** which were ancestral to apes and humans. **Hominids** were ancestral to humans only. Four million years ago several species of *Australopithecus* existed and were gradually replaced with *Homo habilis* and then *Homo erectus.* The Neanderthals and Cro-Magnons were *Homo sapiens,* as are modern humans.

TO REVIEW . . .

1. About how soon after the earth formed did life first appear, according to fossil evidence?

2. How was the scientific method used to disprove spontaneous generation?

3. What chemical elements had to have been in primordial soup to generate nucleic acids and proteins?

4. What event must have taken place between the formation of biochemical building blocks (amino acids and nucleotides) and formation of progenotes?

5. What role did hydrolysis and dehydration synthesis play in chemical evolution and the origin of life?

6. What roles might clays and minerals have played in the origin of life?

7. Cite two reasons why RNA is a more likely candidate for the most important biochemical in the origin of life than DNA or protein.

8. List three examples of geographical changes that affected the history or distribution of life on earth.

9. List in the order of their appearance:
    a. lobe-finned fishes, crinoids, tetrapods
    b. Neanderthals, *Australopithecus afarensis, Dryopithecus, Homo habilis, Homo erectus*
    c. cycads, algae, flowering plants

10. What is the difference between a hominoid and a hominid?

TO THINK ABOUT . . .

1. How did the following adaptations affect the history of life on earth?
- **a.** multicellularity
- **b.** the amniote egg
- **c.** placental mammals
- **d.** bipedalism

2. The study of the origin of life requires speculation, because the oldest fossils have probably been destroyed. Moreover, we cannot know if biochemical experiments to re-create conditions on the early earth are accurate. Do you think that because of these uncertainties we should abandon research on the origin of life? Cite reasons for your answer.

3. Louis Pasteur is often credited with disproving the theory of spontaneous generation. Stanley Miller often receives credit for "brewing" primordial soup. What previous discoveries and hypotheses influenced Pasteur and Miller?

4. Why isn't the presence of amino acids in comets, meteorites, and primordial soup experiments proof of how life originated?

5. Cite three types of organisms that combine characteristics of two types of organisms and therefore may represent one species evolving into another.

6. What is the evidence that the amniote egg has been a biological success?

TO LEARN MORE . . .

References and Resources

Angel, J. Roger P., and Neville J. Woolf. April 1996. Searching for life on other planets. *Scientific American.* We may learn more about the origin of life on earth by discovering life beyond our planet.

Bada, Jeffrey L. February 14, 1997. Extraterrestrial handedness? *Science,* vol. 275.

Darnell, J. E., and W. F. Doolittle. March 1986. Speculations on the early course of evolution. *Proceedings of the National Academy of Sciences, USA,* vol. 83. A classic paper outlining the steps that could have bridged chemical and biological evolution.

Hayes, John M. November 7, 1996. The earliest memories of life on earth. *Nature,* vol. 384. Chemical traces of life go back nearly 4 billion years.

Kay, Richard F., Callum Ross, and Blythe A. Williams. February 7, 1997. Anthropoid origins. *Science,* vol. 275. A look at recent fossil evidence of our distant forebears.

Kerr, Ricard A. August 16, 1996. Ancient life on Mars? *Science,* vol. 273. Are organic residues on a Martian meteorite signs of past life?

Kiedrowski, Gunter von. May 2, 1996. Primordial soup or crepes? *Nature,* vol. 381. Clays and minerals may have polymerized the building blocks of life.

Lewis, Ricki. March 31, 1997. Scientists debate RNA's role at beginning of life on Earth. *The Scientist.* The first molecule on the road to life might have been RNA—or something like it.

Wood, Bernard. January 23, 1997. The oldest whodunit in the world. *Nature,* vol. 385. Stone tool manufacture dates back 2.5 million years.

Zimmer, Carl. June 1995. Coming on the land. *Discover.* Rare fossils reveal the transition from life in the water to life on land.

denotes articles by author Ricki Lewis that may be accessed via *Life's* home page /http://www.mhhe.com/sciencemath/biology/life/.

18 Darwinian Evolution

The land iguana is well adapted to conditions in its home, the Galápagos Islands.

chapter outline

A Tale of Three Islands 374
Mauritius
The Hawaiian Islands
The Galápagos Islands

Macroevolution and Microevolution 375

The Influence of Geology—Clues to Evolution in Rock Layers 376

The Voyage of the HMS* Beagle *379
Darwin's Observations on the Voyage
After the Voyage

Natural Selection—The Mechanism Behind Darwinian Evolution 382
Sexual Selection
The Direction of Natural Selection Can Change

Evolution and Epidemiology 384
Emerging Infectious Diseases
An Evolving Infectious Illness—Ebola
The Rise of Antibiotic Resistance

Natural Selection of Snake Venom

A pit viper is a type of snake that has two depressions in its skull, one beneath each eye. These pits lead to nerves that enable the reptile to sense a slight rise in temperature—such as might occur when a potential meal is near. The pit viper's long, hollow fangs extend only when it needs to immobilize prey. The snake holds the world record for eating the largest meal in relation to its weight—it is the equivalent of a 100-pound person consuming a 160-pound brownie!

The Malayan pit viper, *Calloselasma rhodostoma,* eats other reptiles, amphibians, birds, and mammals and bites humans in southeast Asia. Enzymes in its potent venom begin to digest a victim's body even before the snake finishes swallowing. The observation that specific symptoms of this snake's bite vary among geographic regions led researchers to hypothesize that the venom itself varies chemically in different areas.

To investigate whether snake venom varies geographically, researchers at the University of Wales collected venom from 67 wild *C. rhodostoma* pit vipers in 36 locations in Vietnam, Thailand, Malaysia, and Java and determined the types of proteins that made up the venom. They found 53 venom variants; duplicates always came from snakes collected in the same area.

The researchers next wanted to know the basis of the difference in geographically distinct venom. They had three hypotheses:

Hypothesis 1: If the snake venoms differ solely based on geography, then the farther apart two snakes live, the more dissimilar their venoms should be.

Hypothesis 2: If the snake venoms reflect different genetic subgroups of the snakes, then the closer the DNA sequence of a highly variable gene is between two snakes, the more alike their venom should be.

Hypothesis 3: If the composition of snake venom is an adaptation to the types of prey that the snake encounters, then snakes that eat similar diets should produce similar venoms.

To disprove hypothesis 1, the researchers compared snake venom composition to the snakes' homes. They found no correlation and concluded that venom variability does not reflect geography. To disprove hypothesis 2, they compared venom composition to the DNA sequence of a well-studied gene. Again, they found no correlation. Venom types do not reflect genetic subtypes.

Disproving hypothesis 3 was more challenging. The researchers had to squeeze bowel movements from snakes and probe them for remnants of meals, and examine stomach contents.

to be continued . . .

A Tale of Three Islands

Evolution means change. **Biological evolution** is the process by which the genetic compositions of populations of organisms change over time. In sexually reproducing organisms, genetic changes accumulate that eventually prevent members of one group, or population, from successfully breeding with members of another population and creating different species. Species are groups of organisms that do not normally interbreed. The genetic changes that underlie evolution also occur in single-celled organisms that reproduce asexually, and in viruses.

Islands, because they are isolated, provide a microcosm for viewing evolution at an accelerated rate. They also powerfully illustrate **biogeography,** the different distributions of organisms in different parts of the planet. The formation and disappearance of habitats that are part of biogeography have been crucially important in the evolution of life on earth.

If we know when an island formed, and have evidence indicating which species were the first to colonize it, and we know which species live there today, we can try to reconstruct events that took place in between. Over time, certain individuals were better able to produce offspring because they had inherited trait variants that were helpful, or **adaptive,** in that particular environment. The isolated island populations accumulated so many inherited traits adaptive to their specific environments that they eventually could no longer successfully mate with members of the original population. In this way, new species arose. The collections of traits distinct to certain island populations persisted if the organisms could not go to other land masses. Evidence of evolution in island populations is abundant. Three examples follow.

Mauritius

About 8 million years ago, the island of Mauritius rose in volcanic fury from the depths of the Indian Ocean. By the sixteenth century, the island teemed with dense, tall forests, colorful birds, scurrying insects, and basking reptiles, all well adapted to the island and to one another.

One inhabitant of Mauritius was the flightless, 3-foot- (1-meter-) tall, 30-pound (14-kilogram) dodo bird, described by one scientist as "a magnificently overweight pigeon" (fig. 18.1). The dodo ate the hard fruits of the abundant *Calvaria* tree. The tree also depended on the dodo; the bird's digestive tract destroyed the hard outer layer of the fruits, enabling the seeds to disperse and germinate. This process of two types of organisms becoming dependent upon each other is termed **coevolution.**

Life flourished on Mauritius until the 1500s, when Spanish and Portuguese sailors arrived. The men ate dodo meat, and their pet monkeys and pigs ate dodo eggs. Nearly all the *Calvaria* trees died as their means of dispersing and germinating seeds vanished with the dodo. Rats and mice swam ashore from ships and attacked native insects and reptiles. The sailors' Indian myna birds inhabited nests of the native echo parakeet, while Brazilian purple guava and oriental privet plants crowded the seedlings of native trees undisturbed for centuries. Soon, the forest that had been home mostly to birds, insects, and reptiles became a mammalian

FIGURE 18.1

Humans caused extinction of the dodo. Most of what we know about these birds comes from remains of 400 individuals and this 1848 description from fossils: "These birds were of large size and grotesque proportions, the wings too short and feeble for flight, the plumage loose and decomposed, and the general aspect suggestive of gigantic immaturity. . . . So rapid and complete was their extinction that the vague descriptions given of them by early navigators were long regarded as fabulous or exaggerated, and these birds, almost contemporary of our great-grandfathers, became associated in the minds of many persons with the Griffin and Phoenix of mythological antiquity."

haven. By the mid-1600s, only 11 of the original 33 species of native birds remained. The dodo was exterminated by 1681, although a few birds were taken to zoos. The animal became the first recorded extinction due to direct human intervention.

The Hawaiian Islands

About 3 million years after Mauritius erupted from the Indian Ocean, the first of a chain of other volcanic islands arose in the central Pacific, 2,000 miles (3,218 kilometers) from the closest continents. After the new land cooled, life arrived, seeds floating in on mats of vegetation or drifting in on the wind or on migrating birds. Insects and a lone type of mammal, a bat, flew in. From this original gathering of plants and animals a diverse collection of new varieties evolved over the millennia. The original 15 species of land birds diversified into more than 80 species; the 20 founding land mollusks are ancestral to more than 1,000 species living there today; one or two colonizing fruit fly varieties led to hundreds of modern species.

Islands continued to erupt from the ocean floor, and the main island of Hawaii appeared a mere half million years ago. Its five volcanoes frequently disrupt Hawaii's habitats with fresh lava flows. The island is home to hundreds of species seen nowhere else on the planet.

The Hawaiian Islands, like Mauritius, have seen many species vanish. Some species cannot compete with immigrants for food and shelter. Humans who arrived in the Hawaiian Islands by A.D. 400 transformed the lush lowland forests into fields, destroying habitats of many resident species. Settlers ate the large, flightless birds and killed the vibrantly colored ones to adorn their clothing with feathers. Europeans who arrived a dozen centuries later further altered the land, hastening the extinctions of even more species.

The Galápagos Islands

About 40 million years ago, a group of small volcanoes arose from the sea just off the coast of Ecuador. Over the years, the volcanic islands cooled, and organisms living on the South American mainland arrived by flight or on floating vegetation. Gradually, the sea covered these first islands as newer islands formed from volcanic eruptions farther away from the mainland. Perhaps some of the islands coexisted for a time, allowing some organisms to move from one to another.

Today's Galápagos archipelago is a group of volcanic islands 600 miles (965 kilometers) west of Ecuador. The particular species that chanced to inhabit these islands were a subset of all those on the mainland. They survived only if they, and their descendants, adapted to the particular environmental challenges the island habitat posed. The modern Galápagos Islands are home to an unusual assortment of organisms. The name *Galápagos* comes from the Spanish for "tortoise," one of the islands most noticeable residents (fig. 18.2).

Macroevolution and Microevolution

Evolution is a continuing process that explains the history of life on earth, as well as the diversity of life today. Biological evolution includes large-scale events, such as **speciation,** the appearance of new species. Such large changes constitute **macroevolution.** Evolution also includes changes in individual allele (gene variant) frequencies within a population, termed **microevolution.** Macroevolutionary events tend to span very long periods of time; microevolutionary changes can happen rapidly. Chapter 20 discusses the mechanisms that drive macroevolution and microevolution.

The rapid adaptation of wormlike invertebrates (animals without backbones) to environmental pollution illustrates microevolution in action. Individuals whose genetic makeups render them vulnerable to pollution die out or do not reproduce, eventually disappearing from the population. This happened in Foundry Cove, part of the Hudson River about 50 miles (80 kilometers) north of New York City. The cove has a richly toxic history. During the Revolutionary War, a forge at the site manufactured metal chains that were placed in the river to stop British ships. During the Civil War, bullets were produced near the cove. Then, in the 1950s, manufacturing

FIGURE 18.2

A Galápagos tortoise. When he visited the Galápagos Islands, Charles Darwin found that natives could tell which island a particular tortoise came from by the length of its neck and other obvious adaptations. This tortoise subspecies has a long neck, which enables individuals to reach the vegetation they eat.

facilities at Foundry Cove made batteries. With such a record of heavy metal manufacture, it's not surprising that the sediments at the bottom of the river contained up to 25% cadmium.

Curiously, the polluted river area swarmed with invertebrate life, as did neighboring coves. But when animals from the nearby regions with clearer sediments were moved to the polluted areas, they died. Had the Foundry Cove animals inherited the ability to survive heavy metal poisoning? To find out, biologists took animals from the areas with polluted sediments and bred them in the laboratory for several generations. They then returned the descendants to the polluted cove, where they survived—indicating that they had inherited toxin resistance.

Next, researchers hypothesized that the population had evolved in a way that enabled the animals to withstand the pollution. This may have taken as little as 30 years. To test their hypothesis, researchers tried to re-create the cove's history in the laboratory (fig. 18.3). They sampled invertebrates from a clean cove and exposed them to cadmium-rich mud in the laboratory. Then they bred the survivors with each other, building a population where most individuals could tolerate cadmium. When placed in polluted waters, they survived. Evolution had occurred, and was occurring, due to the more likely survival and reproduction of animals resistant to heavy metal poisoning.

Evolution occurred among the Foundry Cove invertebrates because the toxin-resisting trait was inherited. Back in the nineteenth century, biologists realized that evolution entails inherited (rather than acquired) traits, even before Gregor Mendel described the patterns of inheritance (fig. 18.4).

The idea of evolution as gradual change came from many observations (Biology in Action 18.1). Charles Darwin crystallized these concepts into a coherent theory in the late nineteenth century. Darwin's theory was radical for its time, but compelling, based on many observations in nature. Darwin's ideas trace their roots to geology, the study of rocks.

Mastering Concepts

1. What is biological evolution?
2. How do island populations reveal the results of evolution?
3. Distinguish between macroevolution and microevolution.
4. What is the role of adaptation in evolution?

The Influence of Geology—Clues to Evolution in Rock Layers

Geology and the evolution of life are intimately related. In the latter half of the nineteenth century, people puzzled over the layers (strata) in the earth that ditches, cliffs, and mountains revealed (fig. 18.5). One explanation that would become important in studying biological evolution was the **principle of superposition,** which states that lower rock layers are older than those above them. Several interesting theories attempted to explain how rock layers came to be.

Neptunism, named after the Roman god of the sea, held that a single great flood organized the earth's surface, with the waters receding to reveal the mountains, valleys, plains, and rock strata present today. A variation of Neptunism, **mosaic catastrophism,** proposed that a series of great floods molded the earth's features, killing some life-forms and allowing others to survive. Mosaic catastrophism explained why preserved remains of sea-dwelling organisms were sometimes found in different rock strata. Neptunism's single flood could not account for this observation.

In contrast was the theory of **uniformitarianism,** the continual remolding of the earth's surface. One supporter of this idea was a physician and farmer, James Hutton, who walked in the Scottish countryside, pondering nature. Similar rocks in different areas set him thinking about the history of the earth. Why did pebbles the tide brought in resemble rocks in the hills overlooking the ocean? Why did the rock layers in mountains contain evidence of organisms that clearly had lived in the sea? Hutton noted weathering of exposed rock on the land and gradual deposition of

Evolution in Action

Observation: A heavily polluted portion of a river teems with invertebrate life, but similar animals from clean waters transferred to the polluted waters die.

Hypothesis: The invertebrates living in the polluted water are adapted in a way that enables them to survive.

Experiment: Re-create the evolutionary process resulting in selection of toxin-resistant individuals.

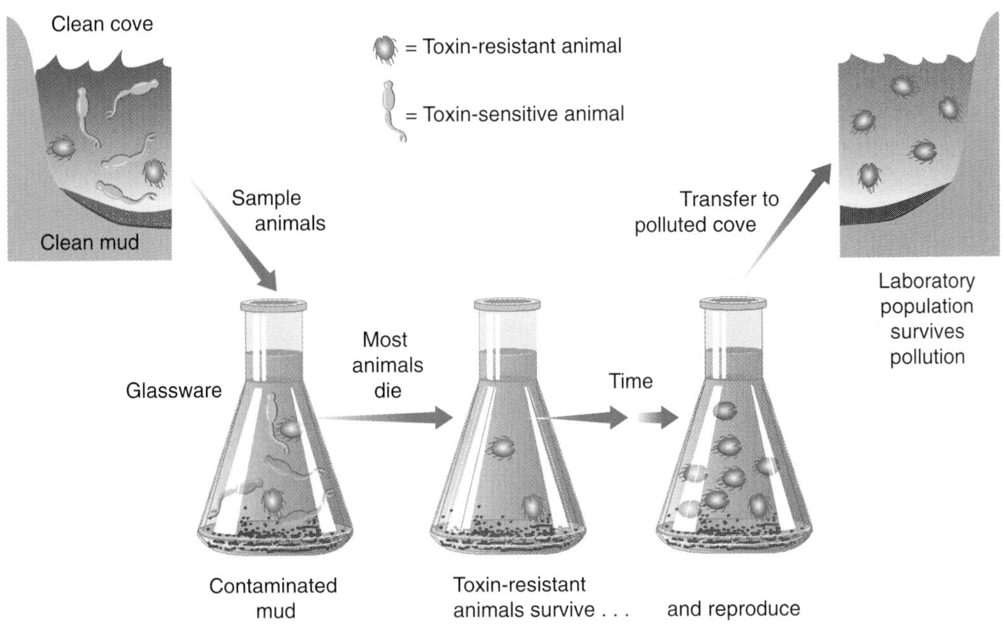

Result: Naturally occurring variants that can survive in contaminated mud reproduce, founding a population that can survive in the polluted cove.

Conclusion: Evolution has happened.

FIGURE 18.3

Evolution can occur rapidly enough for us to observe it, as the animal life in Foundry Cove illustrates.

FIGURE 18.4

Inherited variants are the raw material of evolution. The monster beech is a variant of the European beech (*Fagus sylvatica*) seen in France, Germany, Northern Ireland, Sweden, and Norway. But are its twisted roots, trunk, and branches inherited? Past theories attributed the tree's appearance to being put here by extraterrestrials or a response to a mineral imbalance in soil.

Experiments revealed that the tree's unusual appearance is indeed inherited. A piece of a monster beech grafted onto a normal European beech produces gnarly new growth. Seeds also pass on the phenotype. Some biologists hypothesize that the odd variant is rare because the trees have been selected against when people cut them down because it is too difficult to obtain lumber from them.

18.1 biology in action

Evolutionary Thought Through History

People have long pondered how the great diversity of life came to be and how types of organisms—species—are related to one another. Many people advanced hypotheses to explain the evolution of life.

One of the first thinkers to place human beginnings into a scheme with the origin of other organisms was fourth-century B.C. Greek physician Empedocles. He proposed that the appearance of animals was the most recent stage in the evolution of the universe. First, parts of animal bodies appeared, and then the parts joined randomly to form animals. The first combinations were poorly adapted monsters that did not survive or reproduce and hence did not perpetuate. Eventually, argued Empedocles, the animals we see today formed and survived.

Like Empedocles, Aristotle, who lived a century later, thought animals appeared after nonliving matter evolved. He believed inorganic matter evolved into organic matter, which led to soft living matter, which gave rise to perfect life-forms. But Aristotle also thought each species arose independently and did not change. He envisioned a "great chain of being," a detailed ordering of many life-forms arranged into a single line of descent. The chain of being was uneven—it did not represent all types of organisms in the same degree of detail, and it omitted entire groups of organisms. Human races were considered distinct types of organisms, and their "order," from less advanced to more advanced, often reflected the ethnic background of the orderer. Like many plans before it, the chain of being traced the evolution of the supposedly imperfect to the perfect. This differs markedly from the modern view of evolutionary change, which reflects adaptations to particular environments, rather than a progression towards a subjective, universal perfection.

In the eighteenth century, Swiss naturalist Charles Bonnet coined the term "evolution" in reference to the appearance of life-forms over time. He envisioned earth's natural history as a series of major catastrophes. After each event, the organisms of the past period became fossils, and new and more complex types of organisms appeared. The most recent event, according to Bonnet, catapulted apes to human status.

French taxonomist Jean Baptiste Lamarck was the first to propose a theory to explain the sometimes arbitrary orderings of organisms that others had suggested. In the late eighteenth and early nineteenth centuries, he originated some of the ideas Darwin later adopted, such as the evolution of species from preexisting ones and the ability of animals to adapt to a changing environment. He is most remembered, however, for his theory that organisms inherit acquired characteristics, which implied that evolutionary change results as individuals acquire new body parts or functions in response to need. According to this view, characteristics an individual gains during its lifetime pass to its offspring. An oft-quoted example is the giraffe's long neck. According to Lamarck's theory, the animal "needed" a long neck to reach its treetop food, so the neck grew with each generation. Once acquired, the trait was passed on to the next generation. Lamarck is also noted for his law of use and disuse. A used organ would be maintained from generation to generation; an unused organ would gradually disappear. Although Lamarck was incorrect in some ways, he was among the first not only to state that evolution occurred but also to suggest how it might have happened.

Darwin did not know how natural variations arose. Evolutionary theory had to wait until the early twentieth century, when scientists learned about genetics, to supply the missing information.

sediments in bodies of water. He proposed that the earth's surface features had not formed in one event, or even in several, but were continually building up and breaking down.

According to Hutton, the seas receded to reveal new, uplifted sediments, and water and wind eroded ancient mountains to send sediments to the seas. Uniformitarianism inferred that the earth had existed for a very long time because the forces of geological change—erosion and deposition of sediments—work very slowly, yet vast geological change is clearly evident in many places. In 1947, Hutton proposed a theory of evolution startlingly like the one that Charles Darwin would describe in the next century.

A contemporary of Hutton's, William Smith, also noted remnants of organisms in rock layers. Smith was a surveyor who dug canals in the English countryside in the late 1700s. He noticed that each stratum had a characteristic set of fossils. Even if the layer meandered, or was sharply disrupted for a distance, when it straightened out or reappeared it always contained fossils of the same types of organisms. Smith used his ability to track strata by their fossils in his surveying, but others made inferences about how the fossils got there. Because lower strata were older than strata closer to the surface, the positions of certain fossils within rock layers roughly indicated the times when certain organisms existed. Fossils could thus be placed into a relative time frame.

Geologist Charles Lyell added detail to the theory of uniformitarianism. He suggested that sandstone was once sand, shale rock was once mud, and islands were once active volcanoes. He described these ideas in his three-volume *Principles of Geology,* a set of books that would profoundly influence Charles Darwin years later.

Mastering Concepts

1. How does the principle of superposition provide a framework for evaluating evidence of past life?
2. Distinguish between the theories of Neptunism, mosaic catastrophism, and uniformitarianism.
3. Why is uniformitarianism consistent with the distribution of fossils in rock strata?

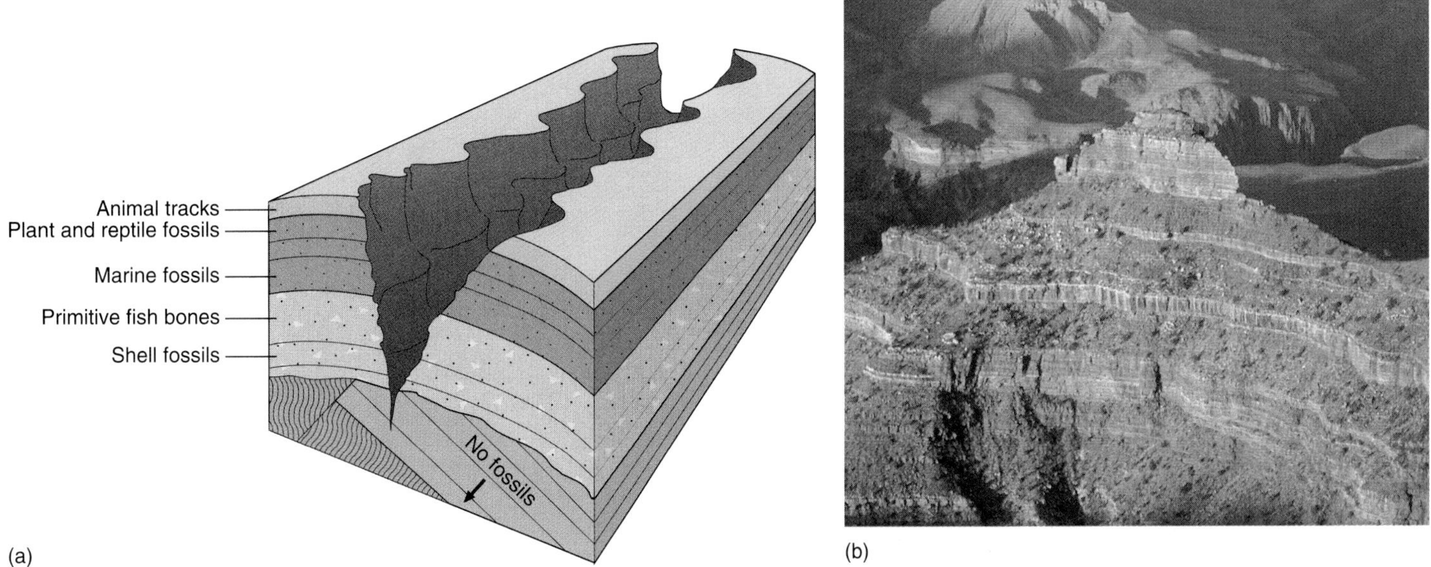

FIGURE 18.5

Rock layers reveal earth history and sometimes life history. (*a*) Layers of sedimentary rock form from sand, mud, and gravel deposited in ancient seas. The rock layers on the bottom are older than those on top. Rock strata sometimes contain fossil evidence of organisms that lived when the layer was formed and provide clues about when the organism lived. (*b*) Sediment layers are visible along the Grand Canyon. Hiking there is like taking a journey through time. Although the rim now rises over 6,500 feet (2,000 meters) above sea level, it has repeatedly been submerged and uplifted.

The Voyage of the HMS *Beagle*

Charles Darwin was born in 1809, the son of a physician and grandson of noted physician and poet Erasmus Darwin. Young Charles was a poor student. He preferred to wander the countryside examining rock outcroppings, collecting shells, and observing birds and insects. Under family pressure, he began to study medicine but abandoned it because he could not stand to watch children undergoing surgery without anesthesia. Next he studied for the clergy, but he grew bored, leading his father to comment, "You're good for nothing but shooting guns, and rat catching, and you'll be a disgrace to yourself and all of your family."

While Darwin reluctantly explored the vocations his family preferred, he also followed his own interests. He joined several geological field trips and met geology professors. Eventually Darwin was offered a position as captain's companion and naturalist aboard the HMS *Beagle.* Before the ship set sail for its 5-year voyage 2 days after Christmas in 1831 (fig. 18.6), the botany professor who had arranged Darwin's position gave the young man the first volume of Lyell's *Principles of Geology.* Darwin picked up the second two volumes in South America.

Darwin read Lyell's volumes while battling continual seasickness. By the time he finished reading, Darwin was an avid proponent of uniformitarianism. Lyell's ideas meshed with his observations prior to boarding the HMS *Beagle.* The combination of Darwin's keen observational skills, Lyell's ideas, and a voyage to some of the most unusual and undisturbed places on the planet set the stage for the young man to collect bountiful evidence on which to construct the theory of evolution by natural selection.

Darwin's Observations on the Voyage

Darwin recorded his observations as the ship journeyed down the coast of South America. He noted forces that uplifted new land, such as earthquakes and volcanoes, and the constant erosion that wore it down. He marveled at the intermingling of sea and land in the earth's layers, at forest plant fossils interspersed with sea sediments, and at shell fossils in a mountain cave. The layers of earth revealed gradual changes in life. Darwin suggested that fossils, like sediments, appeared in a chronological sequence, with those most like present forms in the uppermost layers. Like paleontologists today, Darwin tried to reconstruct the past from contemporary observations and wondered how an organism had arrived where he found it (biogeography).

Darwin was particularly aware of similarities and differences between organisms, on a global scale and in localized environments. If there was a one-time period of special creation, as the Bible held, then why was one sort of animal or plant created to live on a mountaintop or frigid plain and another to reside in a warmer habitat?

Even more puzzling to Darwin was the resemblance of organisms living in similar habitats but in different parts of the world. Such species have undergone **convergent evolution,** adapting in similar ways to similar environmental conditions. The giant carnivorous dinosaurs described at the start of chapter 19 illustrate convergent evolution—they lived on different continents but their fossilized bones are remarkably alike. Large, flightless birds that inhabit Africa, Australia, and South America are contemporary evidence of convergent evolution (fig. 18.7). The ostrich lives on the plains and deserts of Africa and stands 8 feet tall (2.4 meters) and weighs

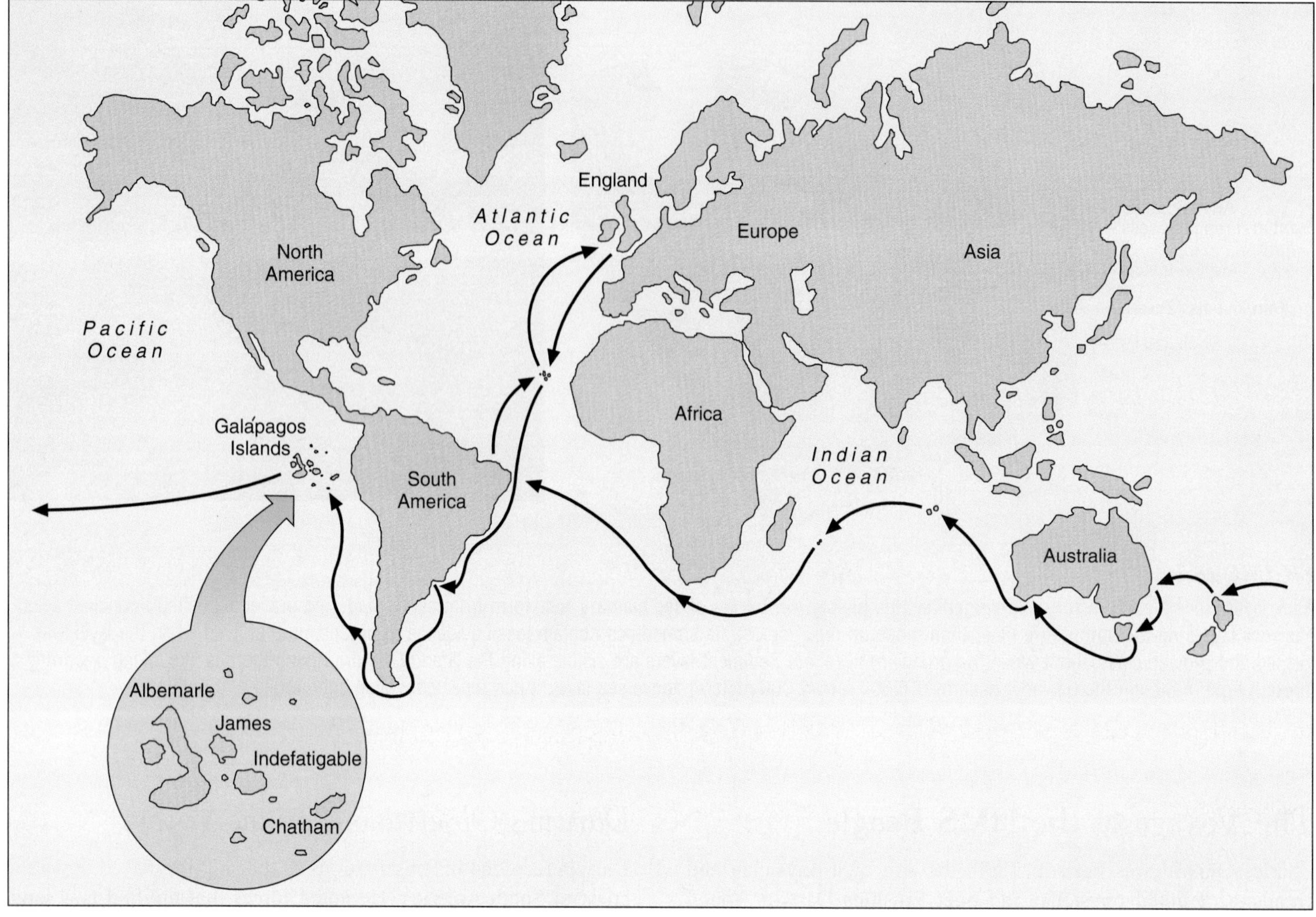

FIGURE 18.6

The 5-year journey of the HMS *Beagle.* During his voyage, Charles Darwin recorded his observations of geological formations and the distribution of organisms. His voluminous notes would serve as the basis for his theory of evolution by natural selection, synthesized years later.

345 pounds (156 kilograms). The emu is Australian and stands 5 1/2 feet (1.7 meters) tall and weighs 100 pounds (45 kilograms), and the rhea lives on the grasslands of South America, reaching 5 feet (1.5 meters) and weighing 50 pounds (23 kilograms).

Darwin kept detailed notes on organisms he saw and on geological formations. He wrote after the voyage that wherever he found a barrier—a desert, river, or mountain—different types of organisms populated the areas on each side. Islands lacked the types of organisms that could not travel to them. Why would a Creator have put only a few types of organisms on islands, Darwin wondered?

In the fourth year of the voyage, the HMS *Beagle* spent a month in the Galápagos Islands. Although Darwin spent half of this time shipbound due to illness and visited only 1 of the 11 islands, the notes and samples he brought back would form the seed of his theory of evolution.

After the Voyage

Darwin returned to England in 1836 and published the first account of the voyage, *A Naturalist's Voyage on the Beagle,* three years later. During that time he spoke with other naturalists and geologists whose observations and interpretations helped mold his thoughts on the evolutionary process.

In March 1837, Darwin spoke with an ornithologist (an expert on birds) who was very excited by the finches the *Beagle* brought back from the Galápagos Islands. The ornithologist could tell from beak structures that some of the birds ate small seeds, some large seeds, others leaves or fruits, and some insects. In all, Darwin had brought back or described 14 distinct types of finch, each different from the finches on the mainland. A special creation of slightly different finches on each island did not make sense; Darwin thought it more likely that the different varieties of finch on the Galápagos had descended from a single ancestral type of bird that dwelled on the mainland. Some birds flew to the islands and, finding a relatively unoccupied new habitat, flourished. Gradually, the finch population branched in several directions, with different groups eating insects, fruits and leaves, and seeds of different sizes. Darwin noted similar changes in the length of the Galápagos tortoise's neck. He called this gradual change from an ancestral type "descent with modification."

In September 1838, Darwin read a work that enabled him to understand the diversity of finches on the Galápagos Islands. Economist and theologian Thomas Malthus's "Essay on the Principle of

FIGURE 18.7

Convergent evolution. (*a*) The ostrich, emu, and rhea look alike but evolved in very different parts of the world. (*b*) The North American wolf and Tasmanian wolf look almost identical and live in similar environments, yet their methods of nurturing their unborn young are vastly different. The North American wolf is a placental mammal. The Tasmanian wolf, a marsupial, nurtures young in the mother's pouch. (*c*) Each of these three animals has a back fin, flippers, a tail with two lobes, and a streamlined body, all adaptations to aquatic habitats. Yet the animal on the left is an ancient reptile (an ichthyosaur), the middle animal is a modern fish, and the animal on the right is a dolphin, a mammal.

Population," written 40 years earlier, stated that food availability, disease, and war limit the size of a human population. Wouldn't populations of other organisms face similar limitations? If so, then individuals who could not obtain essential resources would die. Individuals better able to obtain resources were more likely to survive and to reproduce. This would explain the observation that more individuals are produced in a generation than survive. Over time, environmental challenges would "select" out the more poorly equipped variants, and gradually the population would change. Darwin called this "weeding out" of less adaptive variants natural selection. He got the idea of natural selection from thinking about **artificial selection,** the breeding strategies used to create types of pigeons with specific inherited traits. Artificial selection is responsible for many of the breeds of domestic dogs and cats (see Biology in Action 20.1).

Natural selection elegantly explained the diversity of finches on the Galápagos. Originally, some finches flew from the mainland to one island. When that first island population grew too large for all individuals to obtain enough small seeds, those who could eat nothing else starved. But birds who could eat other things, perhaps because of an inherited quirk in beak structure, ate other foods. Since the new food was plentiful, these once-unusual birds gradually came to make up more of the population. Because each of the islands had slightly different habitats, different varieties of finches predominated on each one. The divergence of several new types from a single ancestral type is called **adaptive radiation.**

An alternate explanation of the diversity of finches on several islands is that several varieties evolved separately on the islands. This alternate explanation is possible because researchers did not observe the actual events but instead must attempt to reconstruct them from modern evidence. Biologists are continuing to observe evolutionary changes in finch populations. Figure 18.8 depicts what may be an example of speciation-in-progress today as two finch species interbreed.

Species 1

Hybrid

Species 2

FIGURE 18.8

Since the early 1970s, Peter and Rosemary Grant, of Princeton University, have continued Darwin's observations of changing finch populations. Because of natural selection, the average beak size in a population can change appreciably, altering dietary possibilities, in as short a time as a year. The bird in the center is a fertile hybrid of the birds depicted on either side of it. These hybrids seem to be better adapted to their environment on the island of Daphne Major than are the two parental species. If the environment does not change, the hybrids may soon be the dominant finch species there. Evolution is happening.

Darwin unveiled his theory of evolution by natural selection in a 35-page sketch published in 1842 and 2 years after that as a 230-page analysis. In 1856, with his treatise expanded but not yet published, Darwin was disturbed to receive a 4,000-word manuscript from British naturalist Alfred Russell Wallace. Its title was rather long-winded: *On the Tendency of Varieties to Depart Indefinitely from the Original Type.* It was as if Wallace had peered into Darwin's mind. Actually, Wallace, like Darwin, had observed the principles of evolution demonstrated among the diverse life-forms of South America.

Darwin and Wallace presented their ideas at a scientific meeting later that year. In 1859, Darwin finally published 490 pages of the even longer-titled *On the Origin of Species by Means of Natural Selection, or Preservation of Favoured Races in the Struggle for Life.* It would form the underpinning of life science.

Mastering Concepts

1. What did Darwin observe about the distribution of organisms and geology that led him to ask the questions that would form his ideas on biological evolution?
2. What is convergent evolution?
3. How does the idea of descent with modification explain differences among the Galápagos finches?

Natural Selection—The Mechanism Behind Darwinian Evolution

Darwin observed that organisms within species vary. More specifically, certain variants are more prevalent than others in different environments. Darwin envisioned this natural variation as the raw material of evolution. Depending upon environmental conditions and competition for resources, certain variants would be less fit than others and therefore less likely to have offspring healthy enough to produce offspring of their own. The traits that contribute to reproductive success need not directly affect reproduction. Any trait that ensures an individual's survival can make reproduction more likely.

Sexual Selection

Sexual selection is a type of natural selection that directly affects traits that increase an individual's chance of reproducing. Sexually selected traits include elaborate feathers in male birds, horns in species ranging from beetles to giant elk, and the courtship songs various insects and birds sing to attract mates (fig. 18.9). So important is successful reproduction that an adaptation that gives a male a mating advantage can kill him, as many a male spider has discovered.

Animals seem to be able to recognize inherited traits associated with reproductive success. A female long-horned wood-boring beetle, for example, chooses a mate whose appearance or behavior enable him to obtain a large territory, a succulent saguaro fruit, which can feed young. Similarly, female barn swallows select mates whose tail feathers are symmetric in length. Preferring symmetric tails is not as superficial as it may seem. Experiments reveal that male barn swallows with even tails resist parasitic infections better than birds with more ragged rears. Somehow, females know this.

The Direction of Natural Selection Can Change

Natural selection does not lead to perfection but reflects transient adaptation to a prevailing environmental condition. Selective forces act upon preexisting characteristics; they do not create new, perfectly adapted variants. The furry fox that survives a frigid winter to bear pups in the spring is not "perfect" and would certainly be less adapted to an unusually mild winter or sweltering summer. Similarly, bighorn sheep whose dwindling numbers

(a)

(b)

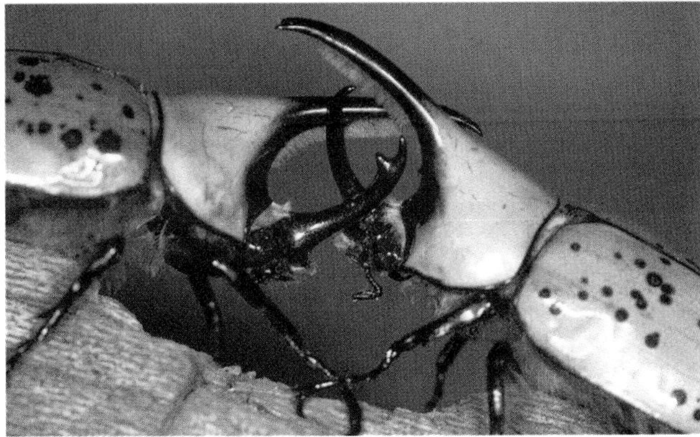
(c)

(d)

FIGURE 18.9

Sexual selection. The individual who is most successful in attracting a mate and reproducing is the most fit, according to Darwin, even if a male dies in the process. (*a*) The male bowerbird builds intricate towers of sticks and grass to attract a mate. (*b*) The male bird of paradise displays bright plumes and capes in his quest for sexual success. (*c*) Diverse organisms use horns or antlers to battle rivals for access to the hornless sex, such as these Hercules beetles. (*d*) A female long-horned, wood-boring beetle chooses her mate based on the size of his territory, a succulent saguaro fruit. A better territory means more food for offspring.

Table 18.1 Darwin's Main Ideas

Facts Based on Observations of Nature:

1. Organisms are varied, and some variations are inherited. Within a species, no two individuals (except identical siblings) are exactly alike.
2. More individuals are born than survive to reproduce.
3. Individuals compete with one another for the resources that enable them to survive.

Inferences from Observations:

4. Within populations, the characteristics of some individuals make them more able to survive and reproduce in the face of certain environmental conditions.
5. As a result of the environment's selection against nonadaptive traits, or "survival of the fittest," only individuals with adaptive traits live long enough to transmit traits beneficial in that environment. Over time, natural selection can change the characteristics of populations, even molding new species.

grace the crags of the northern Rockies were far more plentiful at the height of the ice age, when they flourished in the cooler climate.

A phenotype that seems to be "perfect" in one place and time may be a liability in another place and time. Natural selection can take the same population in different directions under different environmental circumstances. Consider the finches on the tiny Galápagos Island of Daphne Major. In a very dry season in the early 1980s, birds with large beaks were more likely to survive because most of the seeds that they eat are large and dry. In 1983, 8 months of extremely heavy rainfall allowed many small seeds to accumulate. Over the next 2 years, finches with small beaks, who could easily eat the tiny seeds, came to predominate.

A population may undergo sexual selection under one circumstance and natural selection for survival under another. This happens in some guppy populations. When predatory fish are present, the guppies are mostly drab, an adaptation that enables them to blend in with the murky stream bottom and escape predation. Natural selection (being eaten) weeds out the colorful guppies. When the stream lacks predators, however, vibrantly colored fish gradually come to make up more of the population. Here, the color reflects sexual selection, because female guppies prefer colorful mates.

Table 18.1 presents Darwin's main ideas.

Mastering Concepts

1. What is natural selection?
2. What is sexual selection?
3. Why does natural selection not produce perfection?
4. How can natural selection favor different traits within the same population at different times?

Evolution and Epidemiology

So far this chapter's view of evolution has been historical in flavor. Biological evolution, however, is continual and ongoing and is evidenced today in several familiar ways. We look now at two areas of evolution in action—the changeable nature of infectious diseases and the growing resistance of bacteria to antibiotic drugs.

Emerging Infectious Diseases

Epidemiology, the study of infectious disease origin and transmission in populations, includes microevolution. Anyone who has ever had more than one cold (upper respiratory infection) is reacting to evolutionary change. The coats of cold-causing viruses change faster than our immune systems can adapt to fight them. Rapid genetic changes that alter the pattern of proteins on the surfaces of influenza viruses render one season's flu vaccine ineffective the next year.

Infectious diseases change in severity and type over easily observed spans of time because the causative organisms or viruses have very short generation times. A human generation takes 20 to 30 years; a bacterial generation takes 20 to 40 minutes. Given the success of antibiotic drugs in shortening the course of infections and vaccines in preventing them, we grew complacent about our power to defeat these diseases. We banished one scourge, smallpox, completely. But the resurgence of some diseases, including measles, dengue fever, cholera, diphtheria, and tuberculosis, and the appearance of apparently new ones, especially AIDS, has made us all too aware that the organisms that cause infectious diseases in humans are ever-evolving (Biology in Action 18.2).

In the past two decades, toxic shock syndrome, Legionnaires' disease, Lyme disease, and AIDS have arisen (fig. 18.10). The bacteria or viruses causing these "new" illnesses may not be new at all; rather, genetic change led to emergence of the associated

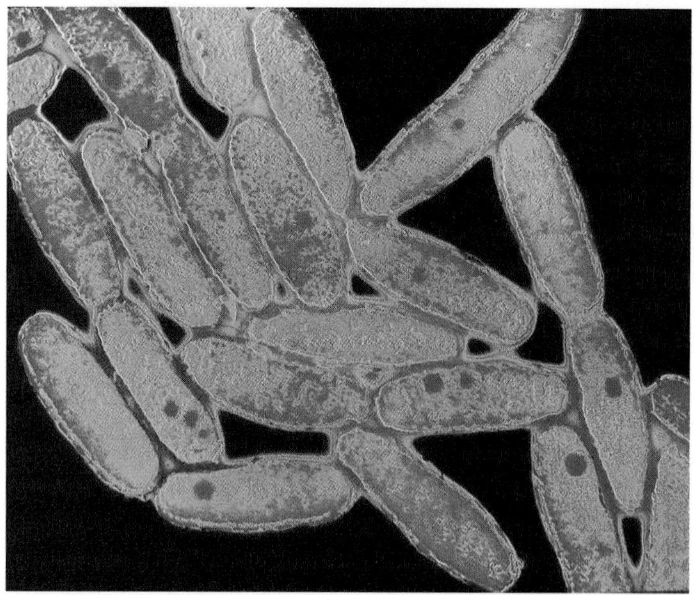

(a)

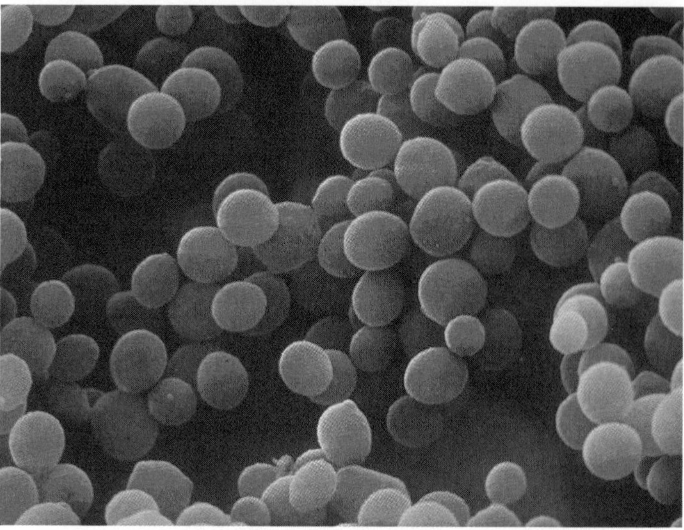

(b)

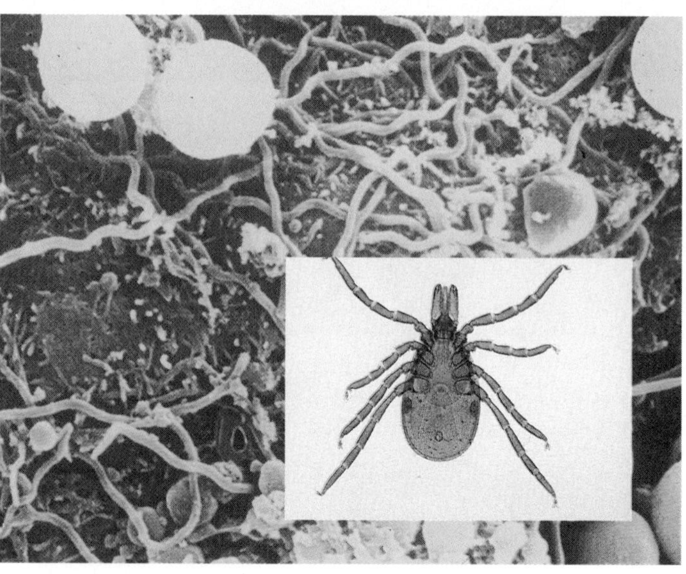

(c)

FIGURE 18.10

A trio of "new" bacterial diseases. (*a*) Legionnaires' disease is a form of pneumonia caused by the bacterium *Legionella pneumophila,* which was implicated in the 1970s after it struck American Legion conventioneers in Philadelphia by traveling through air-conditioning ducts. (*b*) Toxic shock syndrome first appeared in the early 1980s mostly in women who used a certain brand of tampon. A material in the tampons entrapped the common bacterium *Staphylococcus aureus,* provoking it to produce a toxin that causes vomiting, diarrhea, plummeting blood pressure, skin peeling, and death. (*c*) Lyme disease produces a rash, aches, and sometimes arthritis and nervous system complications. A spiral-shaped bacterium, *Borrelia burgdorferi,* causes the illness and is transmitted to humans by the bite of a deer tick.

18.2 biology in action

Crowd Diseases

When Europeans first explored the New World, they inadvertently brought a lethal weapon—bacteria and viruses to which their immune systems had adapted. The immune systems of Native Americans had never encountered these pathogens before and were unprepared.

Smallpox, a viral infection (figure 18.A), decimated the Aztec population in Mexico from 20 million in 1519, when conquistador Hernán Cortés arrived from Spain, to 10 million by 1521, when Cortés attacked again. By 1618, the Aztec nation was just 1.6 million. The Incas in Peru and northern populations were also dying of smallpox. When explorers visited what is now the southeast United States, they found abandoned towns where natives had died from smallpox, measles, pertussis, typhus, and influenza.

The diseases that so easily killed Native Americans are known as "crowd" diseases, because they arise with the spread of agriculture and urbanization and affect many people. Crowd diseases swept Europe and Asia as expanding trade routes spread bacteria and viruses along with silk and spices. More recently, air travel has spread crowd diseases. Penicillin-resistant gonorrhea came by jet from southeast Asia to the United States during the Vietnam war, and more recently, a jet transported cholera from Peru to Los Angeles.

Crowd diseases tend to pass from conquerors who live in large, intercommunicating societies to smaller, more isolated and more susceptible populations, and not vice versa. When Columbus arrived in the New World, the large populations of Europe and Asia had existed far longer than American settlements. In Europe and Asia, infectious diseases had time to become established and for human populations to adapt to them. In contrast, an unfamiliar infectious disease can quickly wipe out an isolated tribe, leaving no one behind to give the illness to invaders.

Crowd diseases are often zoonoses, which are diseases that pass to humans from other types of animals. Typically, a zoonose begins when a person picks up a microbe from close contact with an animal, such as a bird, cow, dog, or sheep. The microbe mutates in a way that enables it to infect humans, possibly causing symptoms. Another reason for the Old World to New World transmission of infectious diseases was that Native Americans had fewer domesticated animals to begin the chain of infection than Europeans did. A modern zoonose may be AIDS, if HIV originated in a monkey.

Fortunately, most crowd diseases vanish quickly, in any of several ways. Medical researchers may develop vaccines or treatments to stop the transmission. People may alter their behaviors to avoid contracting the infection, or the disease may be so devastating that infected individuals die before they can pass the disease on. Sometimes, we don't know why a disease vanishes or becomes less severe.

We may be able to treat and control newly evolving infectious diseases one at a time, with new drugs and vaccines. But the process that continually spawns new genetic variants in microbe populations—resulting in evolution—means that the battle against infectious disease will continue.

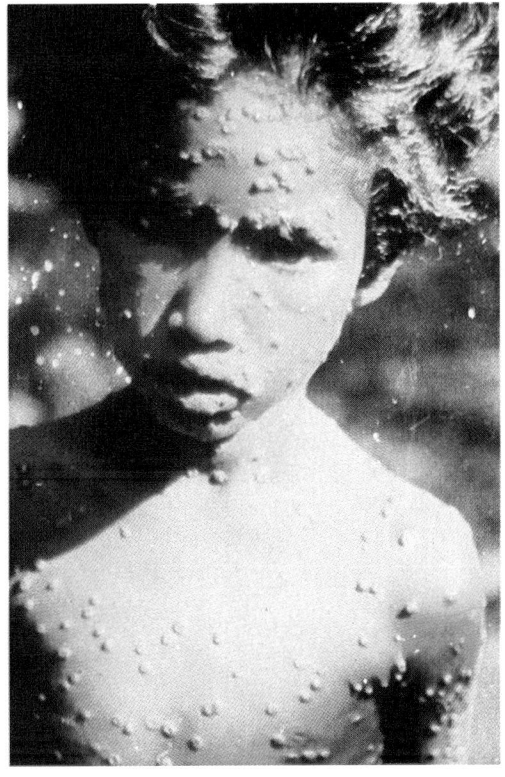

FIGURE 18.A
Thanks to vaccination, smallpox vanished in October 1977.

illness. A mutation may enable an infection to spread by a different route, affect a new species, or cause different symptoms. For example, hantavirus caused an outbreak of a fatal respiratory syndrome in the southwest United States in 1994, yet in Korea in the past, and today in the former Yugoslavia, a variant of the virus causes a hemorrhagic fever causing bleeding and kidney failure. The infection is changing and becoming two types.

HIV infection is changing too. On a population level, more people are living longer with the infection. On an individual level, natural selection acts on the "swarm" of viral particles that infect a body. As the virus replicates, its genetic material mutates. Viral genetic diversity is low as infection begins, as rapidly replicating viruses predominate. Then during the 2- to 15-year latency period, the person's immune system selects out certain viral variants. Gradually, hardier viral variants accumulate and symptoms appear. In HIV infection, evolution usually benefits the virus.

Another factor that can favor once-rare disease-causing variants is a change in the environment, such as deforestation or

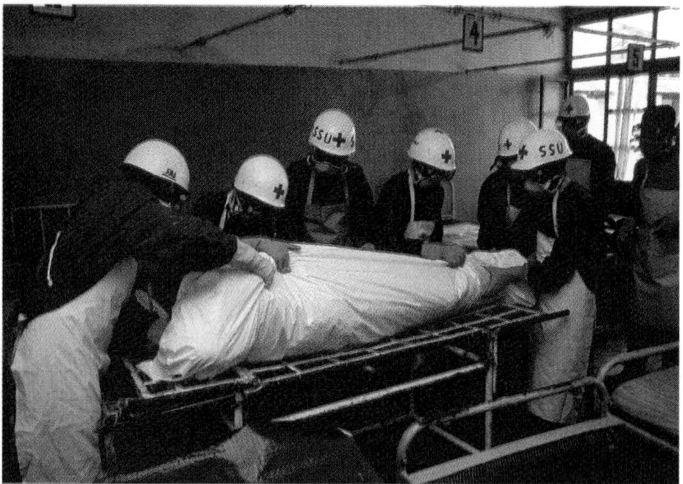

FIGURE 18.11

Because viral hemorrhagic fever from Ebola-Zaire spreads by contact with infected body fluids, health care workers must take extreme precautions when disposing of the dead.

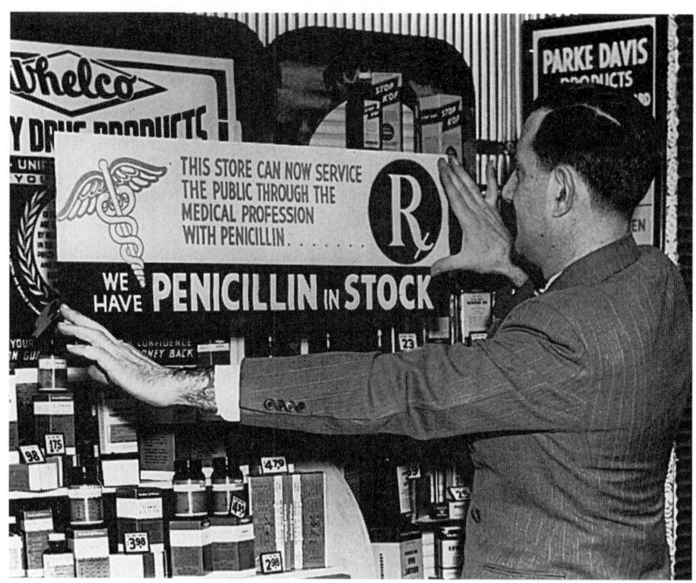

FIGURE 18.12

Penicillin, a fungus-produced antibiotic compound, revolutionized health care during World War II by enabling people to survive otherwise fatal bacterial infections. Penicillin destroys the cell walls of susceptible bacteria. Today, many antibiotics are losing their effectiveness because bacteria have become drug resistant. The reason for antibiotic resistance lies in the basic concepts of evolution.

drought. Consider Lyme disease, a bacterial infection that causes a flulike illness and arthritis (joint inflammation). When the first recognized cases of the disease appeared in Lyme, Connecticut, in the early 1970s, physicians thought it was a new form of apparently infectious juvenile arthritis. But the disease was actually commonplace in colonial times, when communities were more heavily forested. The bacterium that causes the illness, *Borrelia burgdorferi,* lives in ticks that feed on deer that inhabit forests. As colonists cleared land to build settlements, the habitat of the deer, ticks, and bacteria shrank, and the arthritis became rare. Then in the northeast people began abandoning farms. Forests regrew, deer returned, and ticks once again transferred *Borrelia burgdorferi* to people enjoying the outdoors. Lyme disease returned.

A mutation that enables a microorganism to produce a new toxin, or a change in the environment that favors a microorganism that produces a toxin, can cause infectious disease. Recently, people have died after eating hamburger or drinking juice that contained a strain of *E. coli,* a normally harmless bacterium, that produced a toxin that causes bloody diarrhea and kidney failure. In the early 1980s, toxic shock syndrome struck women using a new type of tampon. A chemical in the product enabled a strain of *Staphylococcus aureus* that produces a toxin to flourish. The disease-causing bacteria were present, but rare, until an environmental circumstance opened a niche for them, and their numbers drastically increased. That's evolution.

In a more general sense, the types of symptoms that define infectious diseases are actually adaptations that confer a survival advantage to the infectious agents. The symptoms that an infected human body produces help the microbes spread. Sneezing, bleeding, and diarrhea help spread viruses and bacteria. Ebola virus (see fig. 4.16) provides a vivid illustration of how an infectious agent "uses" a human body to reproduce.

An Evolving Infectious Illness—Ebola

Ebola virus was discovered as the cause of a hemorrhagic (bleeding) fever in humans in 1976 in Zaire, with a 90% fatality rate. Outbreaks occurred in Africa every few years, and then the illness would mysteriously vanish. A person picks up the infection by contacting body fluids of an infected individual. Symptoms progress rapidly over a 2-week or shorter period from headache to rash, vomiting blood, and a massive breakdown of internal tissues until the victim "bleeds out," expelling tissues and secretions from all orifices.

Given this frightening picture, when in 1989 a few monkeys imported from the Philippines and housed in an animal facility in Reston, Virginia, began dying of an airborne infection identified as Ebola, public health authorities feared the worst—a deadly virus transmitted from monkeys to humans *in the air*. But this illness seemed to be different. The monkeys died of extreme lethargy and failure to eat or drink, not the dramatic "bleeding out" of a human Ebola victim. The few humans infected with what came to be known as "Ebola-Reston" did not get very sick and fully recovered. The virus had changed from its African form or perhaps was a different variant from the Philippines. But the deadly African form was still very much in existence.

In the spring of 1995, Ebola hemorrhagic fever swept the city of Kikwit, Zaire. The virus infected 315 people before it could be contained, with a 79% fatality rate—lower than earlier outbreaks probably because of swift medical attention. As in 1976, the infection spread by body fluid contact (fig. 18.11).

In 1996, Ebola-Reston emerged again, this time in laboratory monkeys imported from the Philippines being kept at a facility in Texas. Like the Reston incident, the virus was airborne and produced different symptoms. Genetic analysis showed that the virus in the Texas monkeys matched that from the Virginia monkeys in 98.9% of its nucleotide sequence.

The two types of Ebola virus—from Africa causing hemorrhagic fever in humans and from the Philippines causing lethargy in monkeys—illustrate evolution in action in its most stripped-down form, a changing virus.

The Rise of Antibiotic Resistance

The use of antibiotic drugs revolutionized medical care in the twentieth century and enabled people to survive many once-deadly bacterial infections (fig. 18.12). As we near the twenty-first century, however, some infectious diseases are becoming resistant to the antibiotic drugs that once vanquished them. The reason for the rise in antibiotic resistance is evolution.

Any natural population of organisms includes unusual variants. If an antibiotic dismantles a component of a bacterial cell wall, for example, a resistant mutant might have a slightly different molecule in its cell wall that the antibiotic cannot alter. When a person takes an antibiotic, the drug kills all the susceptible bacteria and leaves behind those that can resist it. The resistant bacteria multiply, increasing their numbers a million-fold in a day, and the person returns to the doctor, seeking an alternate cure. The antibiotic does not create the bacterial resistance but creates a situation where an already existing variant can flourish. If this sounds a lot like natural selection, it should. It is.

Antibiotic-resistant bacteria appeared just 4 years after the drugs entered medical practice in the late 1940s, and researchers responded with new drugs. But the microbes kept pace, as evolution tends to do. Today, 40% of hospital *Staphylococcus* infections resist all antibiotics but one, and already laboratory strains resist that one drug. We may be able to slow antibiotic resistance by using these powerful drugs only when it is appropriate.

New and returning infectious diseases, and the growing resistance of bacteria to antibiotic drugs, offer compelling proof that evolution, driven by natural selection and accumulating genetic change, is an ongoing process. The next chapter examines the forces of evolutionary change.

Mastering Concepts

1. How can a genetic change alter the characteristics of an infectious disease?
2. How can an environmental change alter an infectious disease?
3. What factors have contributed to the rise in resistance of bacteria to antibiotic drugs?

SUMMARY

1. **Biological evolution** is change in the genetic characteristics of a population, which can lead to new species. Evolution has occurred in the past and is ongoing.

2. **Biogeography** considers the physical distribution of organisms. Islands provide evidence of evolution because their geographical isolation separates individuals with particular combinations of traits from ancestral organisms. Interdependence between species illustrates **coevolution.**

3. Evolution consists of large-scale, species-level changes of **macroevolution** and gene-by-gene changes of **microevolution.**

4. Geology laid the groundwork for evolutionary thought. The **principle of superposition** states that lower rock strata are older, suggesting a time frame for fossils within them. Some people explained the distribution of rock strata with **Neptunism** (a single flood) or **mosaic catastrophism** (a series of floods), but **uniformitarianism** (continual remolding of the earth's surface) became widely accepted.

5. Charles Darwin observed the distribution of organisms in many diverse habitats as naturalist aboard the HMS *Beagle.* Afterwards, he synthesized his theory of the origin of species by means of natural selection.

6. The theory of natural selection states that individuals least adapted to their environments are less likely to survive to produce fertile offspring. Better adapted individuals with

Natural Selection of Snake Venom~

continued from page 374.

After searching through the bowel movements and stomach contents of 67 Malayan pit vipers, the researchers identified 216 types of prey. Then they determined the snakes' dietary preferences with pie diagrams indicating the proportion of their meals that were reptilian, amphibian, or warm-blooded organism (bird or mammal). The snakes had diverse palates. One snake ate only reptiles, another 80% birds and mammals with 20% reptiles, yet another feasting on 50% reptiles, with the remainder split between amphibians and birds and mammals. This time, considering diet and venom composition, a correlation emerged. The more alike the diets of two snakes, the more alike their venom.

A logical conclusion: Natural selection has molded the characteristics of this type of snake's venom. Over time, snakes whose venom was the most effective would eat more and be more likely to survive to reproduce. The venom recipe, consisting of gene-encoded enzyme variants, would come to predominate in the population. Evidence supporting this conclusion came from an entirely different type of experiment. Researchers raised Malayan pit vipers in captivity and fed them laboratory fare quite different from their natural diets. The snakes' venom remained the same as that from their parents, confirming that venom composition is inherited. Evolution has molded the snake's venom.

different variations of inherited traits reproduce more successfully and flourish. Eventually, enough changes accumulate that members of the population can no longer mate with members of the ancestral population, and a new species arises. Natural selection does not produce perfection, but adaptation.

7. **Sexual selection** is a form of natural selection in which certain inherited traits make an individual more likely to mate and produce viable offspring.

8. Emerging and returning infectious illnesses and increasing resistance of bacteria to antibiotic drugs illustrate evolution in action.

TO REVIEW . . .

1. What is biological evolution?

2. How can populations of organisms on islands illustrate evolution?

3. What is the difference between microevolutionary and macroevolutionary change?

4. How did James Hutton, William Smith, Charles Lyell, and Thomas Malthus influence Charles Darwin's thinking?

5. How can natural selection explain evolutionary change?

6. Why must natural selection work on inherited, rather than on acquired, traits in order to drive evolution?

7. What sorts of traits lead to "fitness" in the Darwinian sense?

8. The term *highly evolved* is often interpreted to mean "perfect." Why is this definition inaccurate if applied to Darwinian evolution?

9. How do the appearances of new infectious diseases or resurgences of old ones illustrate continuing evolution?

10. How does the rise of antibiotic resistance among disease-causing bacteria illustrate evolution in action?

TO THINK ABOUT . . .

1. Many islands have populations of large, flightless birds found nowhere else in the world. How might these birds have gotten to the islands?

2. Which theory best explains Darwinian evolution—Neptunism, mosaic catastrophism, or uniformitarianism? Give a reason for your answer.

3. An early and major objection to Darwin's concept of natural selection as the mechanism behind evolution was that it refutes the pleasant idea that we humans are more advanced and special than other species. How does Darwinian evolution do this?

4. You have a pet cocker spaniel and believe that if you snip off the end of his tail, and then breed him, his puppies will be born with snipped-off tails. Is this idea consistent with Darwinian evolution? Why or why not?

5. In the former Soviet Union, a toxic strain of the bacterial infection diphtheria has reemerged. Diphtheria causes a white coating on the throat, fever, cough, and damage to the skin, heart, kidneys, and nerves. The World Health Organization cites several reasons for the resurgence of this infection: failure of parts of the population to be vaccinated, crowded living conditions, and forced migrations of people. How might these factors have influenced the reappearance of this illness?

6. The strain of hantavirus that causes a lethal hemorrhagic fever is present all the time (endemic) in Korea, but outbreaks of disease in humans occur only when soldiers from other nations occupy the area. Why might this be so?

7. How is HIV infection similar to the rise of antibiotic resistance among bacteria, in terms of the role of natural selection?

8. If snake venom is adapted to the snake's prey, what would happen to the snake population if the types of prey were to suddenly change?

TO LEARN MORE . . .

References and Resources

Caldwell, Mark. January 1996. Ebola tamed for now. *Discover.* Disease outbreaks caused by the Ebola virus come and go.

Darwin, Charles. 1958 (1859). *On the origin of species by means of natural selection.* New York: Mentor Books. Darwin's writing is very detailed but offers fascinating examples and interpretations of the natural world.

Diamond, Jared. October 1992. The arrow of disease. *Discover.* The evolution of infectious disease parallels history in intriguing ways.

Greenwood, Jeremy J. D. April 22, 1993. Theory fits the bill in the Galápagos Islands. *Nature,* vol. 362. Rapid adaptation of bird bills to seed size illustrates natural selection in action.

Lewis, Ricki. September 1995. The rise of antibiotic-resistant infections. *FDA Consumer.* Once treatable bacterial infections are becoming resistant to the antibiotic drugs that once easily vanquished them. The reason—evolution.

Markotic, A., et al. March 1996. Hantaviruses are a likely threat to NATO forces in Bosnia and Herzegovina and Croatia. *Nature Medicine,* vol. 2. Hantavirus causes a hemorrhagic fever in Korea and eastern Europe but a respiratory illness in the United States. It is an evolving virus.

Moore, Randy. February 1997. The persuasive Mr. Darwin. *BioScience.* Few people have read Darwin's original work.

Patlack, Margie. April 1996. Book reopened on infectious diseases. *FDA Consumer.* Many infectious diseases are reemerging.

Preston, Richard. 1994. *The Hot Zone.* New York: Anchor Books, Doubleday. The true story of an outbreak of Ebola virus infection among monkeys in an animal facility in Reston, Virginia.

Quammen, David. 1996. *The song of the dodo.* New York: Scribner Books. Birds and islands tell powerful evolutionary tales.

denotes articles by author Ricki Lewis that may be accessed via *Life's* home page /http://www.mhhe.com/sciencemath/biology/life/.

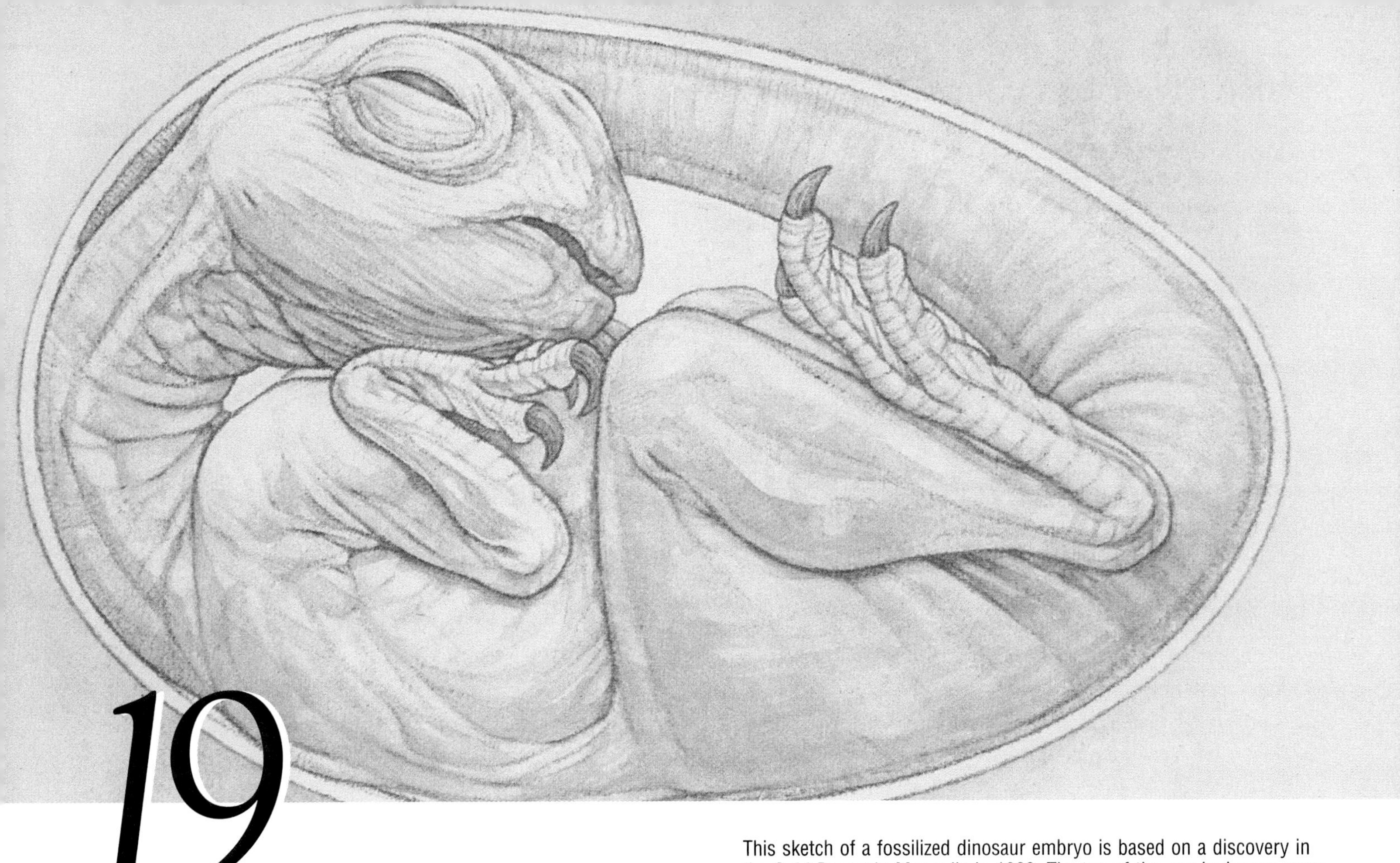

19 Evidence for Evolution

This sketch of a fossilized dinosaur embryo is based on a discovery in the Gobi Desert in Mongolia in 1993. The top of the egg had worn away, revealing the tiny skeleton within. The dinosaur was a theropod, or meat-eater, related to *Tyrannosaurus rex* and velociraptors. Unfolded, the embryo would stretch to 8 inches (20 centimeters). Dinosaur eggs and embryos are rare. Seventy years earlier, the first dinosaur eggs were discovered, also in the Gobi Desert.

chapter outline

The Stories of Life 390

Fossils 390
How Fossils Form
Determining the Age of a Fossil

Comparing Structures in Modern Species 393
Comparative Anatomy
Vestigial Organs
Comparative Embryology

Molecular Evolution 395
Comparing Chromosomes
Comparing Protein Sequences
Comparing DNA Sequences
Molecular Trees
Molecular Clocks
Reconciling Fossil, Biogeographical, and Molecular Evidence

New Dinosaurs

Today the Kem Kem region of Morocco is a stretch of parched red sandstone, a part of the Sahara Desert with the Atlas Mountains in the distance. Daytime temperatures typically reach 120°F (49°C). But 93 million years ago, this same land was a huge floodplain, with rivers leading to towering coniferous forests. It was ideal for dinosaurs to roam.

In the summer of 1995, on the 41st day of a grueling hunt for dinosaur fossils, a U.S. writer, Gabrielle Lyon, spotted bones jutting from a cliff, about 500 feet up. Paul Sereno, the head of the international exploration team, soon determined that the bones were from an unknown dinosaur. They named it *Deltadromeus agilis,* meaning "agile delta runner," in honor of the animal's long, thin bones.

The team then found a huge skull and several 5-inch-long teeth of *Carcharodontosaurus saharicus,* meaning "shark-toothed reptile from the Sahara." The giant dinosaur had been identified in 1927 from a few teeth from western Algeria. In September, a research group in Argentina made another spectacular dinosaur find, remnants of an animal called *Giganotosaurus carolinii,* meaning "giant reptile."

Carcharodontosaurus saharicus and *Giganotosaurus carolinii* shared more than their enormity. Specific bones are similarly shaped, such as the broad bones around the eyes and the squarish front lower jaw. But *Carcharodontosaurus* lived in Africa, and *Giganotosaurus* in South America. Why were they so alike?

to be continued . . .

The Stories of Life

Imagine being asked to tell a story from the following facts:

1. A pumpkin turns into a coach.
2. A prince is giving a ball.
3. A poor but beautiful young woman with a nice voice and small feet has two mean stepsisters who are ugly and have large feet.

Chances are that unless you're familiar with the fairy tale *Cinderella,* you wouldn't come up with that exact story. In fact, 10 people given the same set of facts might construct 10 very different tales. But if more information was available, the stories might become more and more alike. So it is with the sparse evidence that we have of the evolution of life on earth—pieces of a puzzle in time, some out of sequence, many missing (fig. 19.1).

Traditionally, the science of **paleontology** examines the fossil record to glimpse past life. Researchers also consider similarities between ancient and modern species and compare anatomy, physiology, behavior, and the sequences of informational molecules. For example, the close evolutionary relationship between humans and chimpanzees is revealed in the similar ways that their bones and muscles connect and function, by similar chromosome organization, and by many nearly identical gene and protein sequences. Researchers use these types of clues to construct **phylogenies,** which are diagrams that depict the lineages, or evolutionary relationships, among species.

The evolution of life on earth has not been a single line leading to an ill-defined "perfection" but a hierarchical diversion of many, many branches, figuratively like a tree. Humans and chimpanzees represent only a tiny branch of the tree of all life-forms that have existed on earth—many types, of course, that we do not even know about. An area of evolutionary biology called **cladistics** attempts to decipher how ancestral species diversified and branched into different types of organisms as time progressed. Researchers consult different types of evidence to construct diagrams called cladograms that group organisms by their similarities that reflect probable descent from a shared ancestor (fig. 19.2). A clade is a group of organisms that, because of their shared characteristics, are more like each other than other organisms.

Mastering Concepts

1. What do phylogenies depict?
2. What does cladistics examine?

Fossils

One of the greatest challenges of studying past life is to reconcile the molecular, fossil, geological, and geographical information the earth holds. We begin our look at evidence for evolution with the more traditional approach: fossils.

How Fossils Form

A fossil is any evidence of an organism. Some paleontologists consider an object to be a fossil only if it is older than 10,000 years. Fossils need not be remains of actual organisms. Dinosaur footprints and worm borings reveal how the animals that made them traveled. Pigments in fossilized dinosaur dung offer clues to what the great reptiles ate. Microscopes can aid paleontologists in identifying small fossils, such as pollen, plant fragments, and the coatings of cyanobacteria that lined ancient seafloors and formed outcroppings of rocks called stromatolites (see fig. 17.1).

Many fossils are hard parts of organisms in which minerals have slowly replaced the original tissue. An inch-long horn coral dies on an ancient sea bottom in what is now Indiana and is covered by sand and mud. Gradually, the sediment hardens into rock. Meanwhile, the hard part of the horn coral dissolves, leaving an impression of its shell. Millions of years later, a person walking along land that was once that sea bottom sees the impression. Perhaps the mold was filled in with more mud, which also hardened into rock. The explorer may then find a cast of the horn coral, a rocky replica of the ancient animal (fig. 19.3*a*). Similarly, minerals can replace the living tissues of trees, producing petrified wood that reveals cellular structures when sliced thin and viewed under a microscope (fig. 19.3*b*). Evidence of our recent relatives often consists mostly of teeth, which are even harder than bone and are the likeliest anatomical parts to be preserved (fig. 19.3*c*). Bones are fossilized when minerals replace cells and intercellular materials.

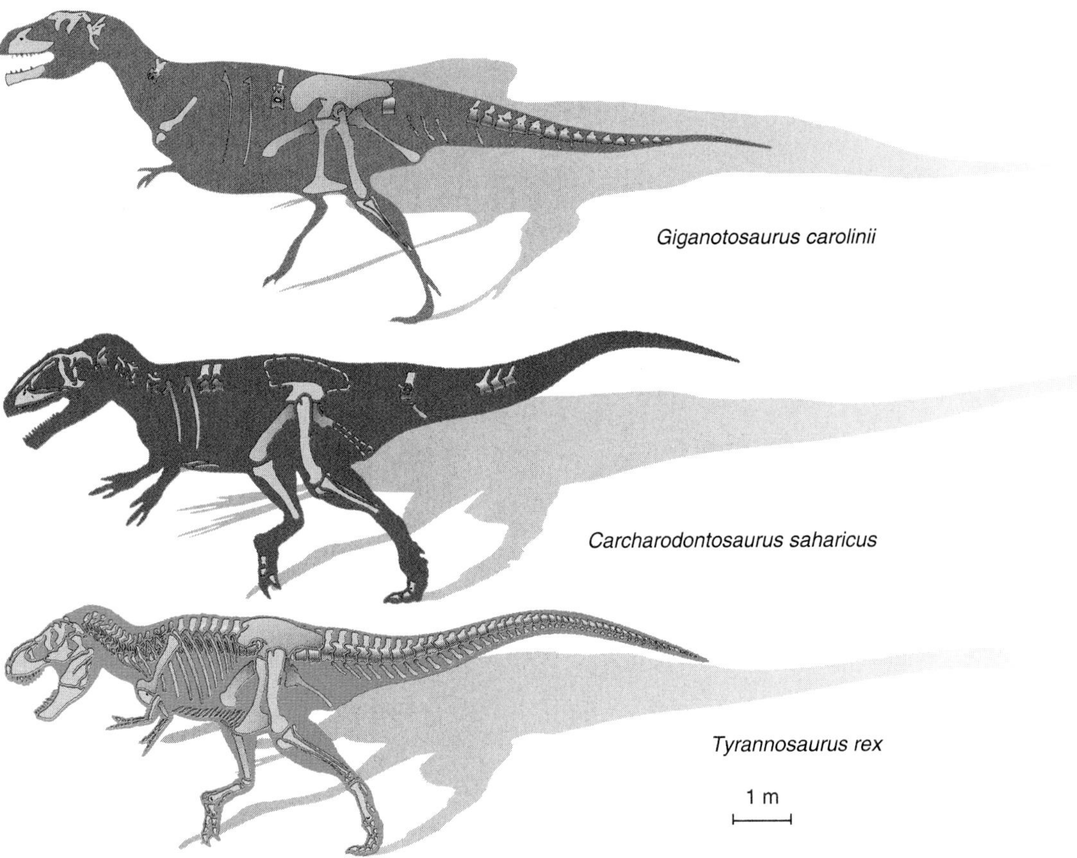

FIGURE 19.1

Paleontologists reconstruct the forms of extinct species from fossils. Such knowledge is often incomplete. Researchers had enough preserved bones from *Giganotosaurus carolinii* and *Carcharodontosaurus saharicus* to realize that they were ancestors of the more familiar *Tyrannosaurus rex.*

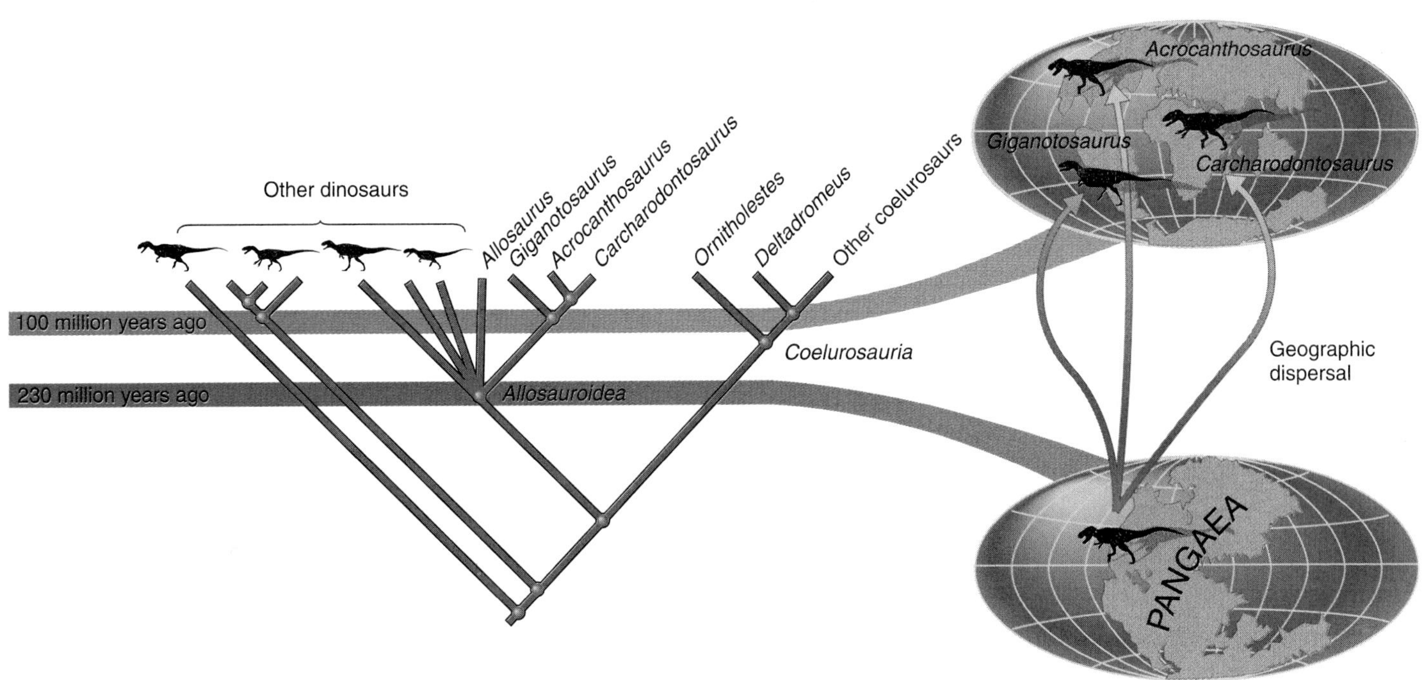

FIGURE 19.2

This cladogram depicts possible evolutionary relationships between dinosaurs discussed in the opening vignette. *Giganotosaurus, Acrocanthosaurus,* and *Carcharodontosaurus* are giant meat-eaters that lived about 90 million years ago in, respectively, South America, North America, and Africa. Their ancestor *Allosaurus* left more geographically widespread fossils because it lived in the early Jurassic period, when the earth had one large landmass called Pangaea. *Tyrannosaurus rex* is not on this cladogram because it lived more recently (70 million years ago) than this tiny part of the great tree of life shows.

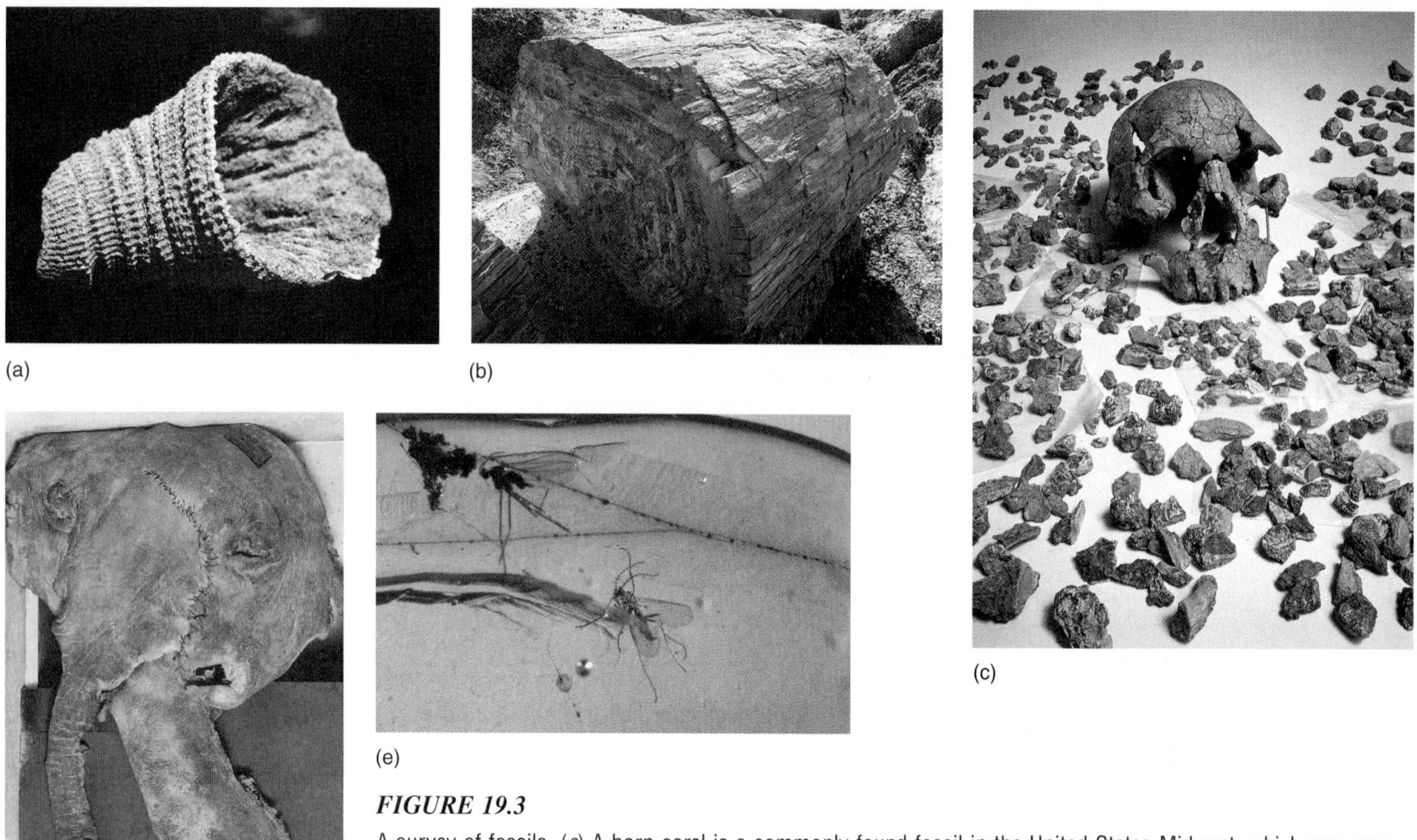

FIGURE 19.3

A survey of fossils. (*a*) A horn coral is a commonly found fossil in the United States Midwest, which was once under water. (*b*) This petrified tree trunk is in Yellowstone National Park, but 50 million years ago it was part of a lush forest. Burial under volcanic debris enabled minerals to replace the tree's tissues. (*c*) Reconstructing a hominid skull is difficult—this *Homo habilis* skull was in 150 pieces! (*d*) "Dima," a frozen mammoth found in Siberia, is preserved in a refrigerator at the American Museum of Natural History. (*e*) Amber captured in time these two gall midges, caught in a spider's web 25 million years ago in what is now the Dominican Republic.

Unusual circumstances may preserve biological material. Gold miners found a perfectly preserved baby mammoth in the ice of the Arctic circle in Siberia. "Dima" was about 9 months old when she perished 40,000 years ago (fig. 19.3*d*). She stood 3 feet (1 meter) tall and weighed about 140 pounds (64 kilograms). The Los Angeles La Brea tar pits preserved vertebrates. Elsewhere, sticky tree resin entrapped insects and frogs and hardened into translucent amber tombs (fig. 19.3*e*). These finds can reveal behavior and ecology by catching insects in the acts of mating, entrapping prey, or laying eggs. One spectacular piece of amber about the size of a half dollar contains 62 insects, many of them whole. When amber hardens, the entrapped material dehydrates, which preserves its DNA. Researchers can extract DNA from ancient organisms preserved in amber and determine its sequence.

The most striking fossils form when sudden catastrophes, such as muds and floods, rapidly bury organisms in an oxygen-poor environment. Without oxygen, tissue damage is minimal, and scavengers cannot reach the dead, enabling even soft-bodied organisms to leave exquisitely detailed anatomical portraits. A block-long section of the Canadian Rockies called the Burgess shale houses an incredibly varied collection of 530-million-year-old invertebrates, preserved, the evidence suggests, in a sudden mudslide (see fig. 17.12*b*). Similarly, a sudden flood preserved a family of rhinoceros-like styracosaurs living along the Milk River in Montana about 75 million years ago.

Fossils offer only fleeting glimpses into life's history. The formidable forces of nature—erosion, tides, volcanoes, continental shifts, earthquakes, storms, and even meteor impacts—destroy much evidence of life. The fossil record is also incomplete because some species lived for such short periods of time, or dwelled in such restricted areas, that their fossils, if they left any, have never been discovered.

Determining the Age of a Fossil

Scientists can estimate when an organism that left fossils lived in two ways. **Relative dating** is based on the principle of superposition, with lower rock strata presumed to be older. The farther down a fossil is, the longer ago the organism it represents lived—a little like a memo at the bottom of a pile of papers on a desk being older than one nearer the top. Relative dating provides an age relative to other fossils in strata above and below. Researchers obtain more precise dates by using the constant and measurable rate of decay of certain radioactive minerals.

Radiometric dating uses natural radioactivity as a "clock." Isotopes of certain elements are naturally unstable, which causes them to emit radiation. As the isotopes release radiation (or "radioactively decay"), they change into different isotopes of the same element, or they become a different element. (Recall from chapter 3 that an isotope is a form of an element distinguished by a different number of neutrons.) Each radioactive isotope decays to its other form at a characteristic and unalterable rate called its **half-life.** The half-life is the time it takes for half of the isotopes in a sample of the original element to decay into the second form. If we know the half-life of a radioactive isotope and the amounts of its "before" and "after" forms in a rock or tissue sample, we can deduce when the sample formed.

Two radioactive isotopes often used to assign dates to fossils are potassium 40 and carbon 14. Potassium 40 radioactively decays to argon 40 with a half-life of 1.3 billion years, so it is valuable in dating very old rocks containing traces of both isotopes (fig. 19.4). Chemical analyses can detect argon 40 in amounts small enough to correspond to fossils that are about 300,000 years old or older. Many sedimentary (layered) rocks, however, the source of most fossils, contain argon 40 that was already part of much older rock when the fossil formed. A potassium 40 date for these rocks would not correspond to the age of the fossils within them.

Fossils up to 40,000 years old are radiometrically dated by measuring the proportion of carbon 12 to the rarer carbon 14. Carbon 14 is a radioactive isotope that forms naturally in the atmosphere when nonradioactive nitrogen 14 is bombarded with cosmic rays (radiation from space). Organisms accumulate carbon 14 as they assimilate carbon during photosynthesis or by eating organic matter. When an organism dies, however, its intake of carbon 14 stops. From then on, carbon 14 decays to the more stable carbon 12 with a half-life of 5,710 years. Radioactive carbon dating was used to determine the age of fossils of vultures that lived in the Grand Canyon. The birds' remains have about one-fourth the carbon 14 to carbon 12 ratio of a living organism. Therefore, about two half-lives, or about 11,420 years, passed since the animals died—it took 5,710 years for half of the carbon 14 to decay, and another 5,710 years for half of what was left to decay to carbon 12.

One limitation of potassium-argon and carbon 14 dating is that they leave a gap, resulting from the different half-lives of the radioactive isotopes. Carbon 14 dates extend up to 40,000 years ago, but potassium-argon dates do not begin reliably until 300,000 years ago. Several new techniques cover the missing years, which include the time of human origin.

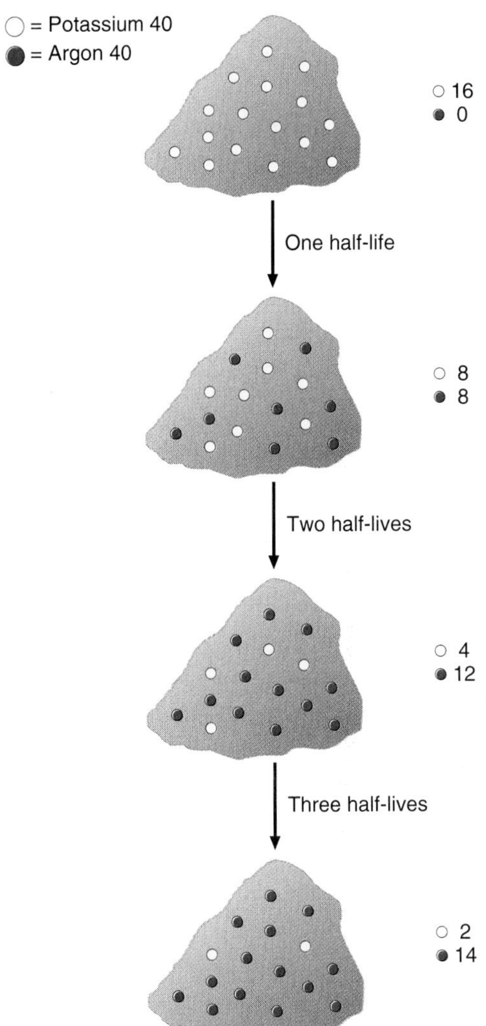

FIGURE 19.4

Radiometric dating. Scientists can determine the approximate age of some fossils by determining the proportion of potassium 40 to argon 40 in the rock. After a half-life of 1.3 billion years, half of the potassium 40 in a sample has decayed to argon 40. After two half-lives, a quarter of the potassium 40 remains, and after three half-lives, one-eighth of the original isotope is left.

Mastering Concepts

1. What are some types of fossils?
2. Why is the fossil record incomplete?
3. Why is radiometric dating more exact than relative dating?

Comparing Structures in Modern Species

Some clues to the biological past come from the biological present. Such techniques compare anatomy and biochemistry among modern species.

Comparative Anatomy

A compelling explanation for anatomical similarities among different modern species is that the organisms descended from a common ancestor. For example, all vertebrate skeletons consist of the same parts and support the body (fig. 19.5). Presumably, an ancestor to modern vertebrates originated this skeletal organization, which then gradually became modified as more recent organisms (amphibians, reptiles, birds, fishes, and mammals) adapted to their environments.

Structures that are similar in different types of organisms because they are inherited from a common ancestor, such as the vertebrate skeleton, are **homologous.** The structures may or may not have similar functions, but they share a common origin. The middle ear bones of mammals, for example, originated as bones that support the jaws of primitive fishes, and they still exist as such in simpler vertebrates.

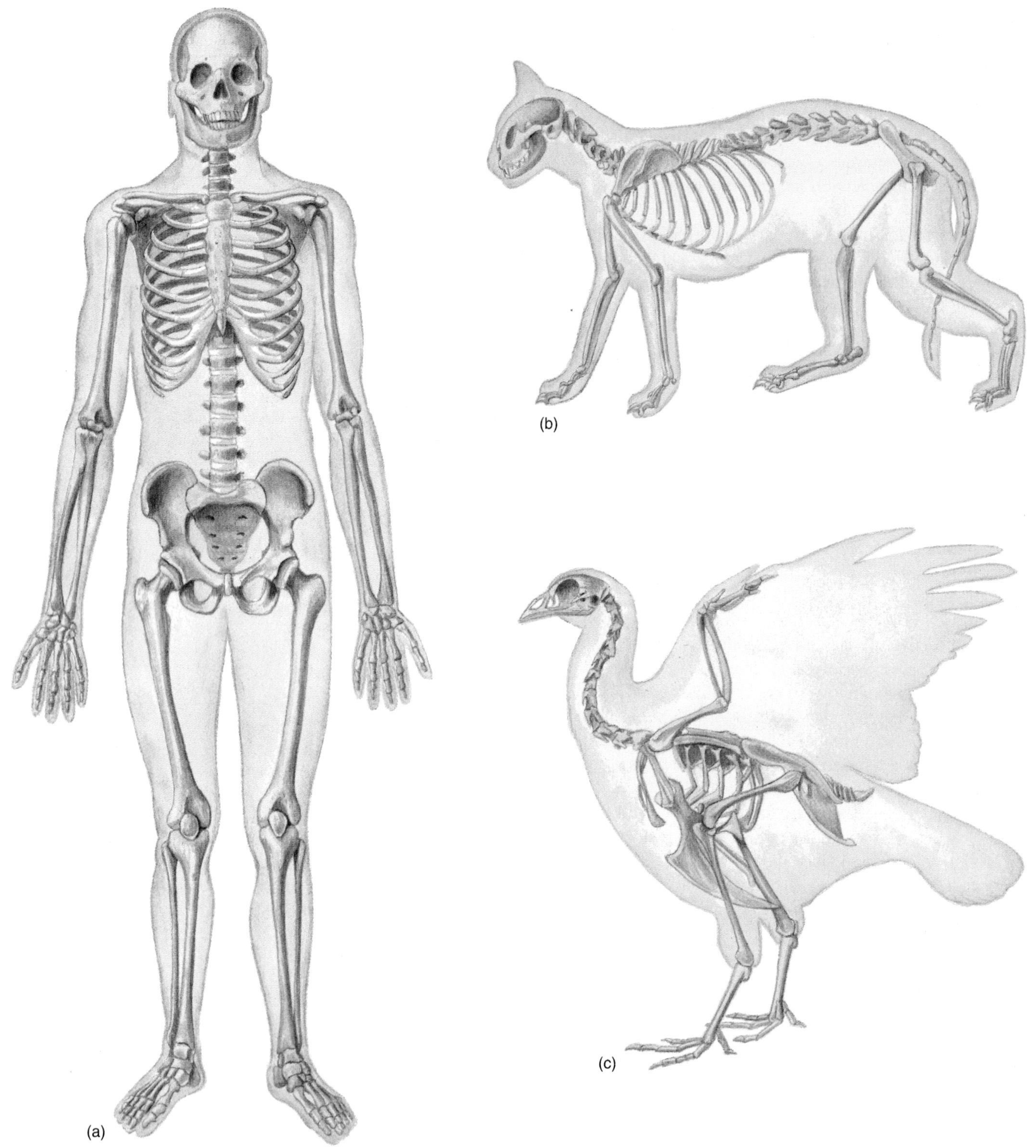

FIGURE 19.5

Vertebrate skeletal organization reflects common ancestry. Although a human walks erect (*a*), a cat walks on all fours (*b*), and a bird flies (*c*), all have a similar skeletal organization of skull, shoulder, spine, hips, and limbs.

Structures that are similar among species but that evolved independently, perhaps in response to similar environmental challenges, are **analogous.** Birds and insects, for example, have wings, but the bird's wing is a modification of vertebrate limb bones, whereas the insect's wing is an outgrowth of the cuticle that covers its body.

Determining whether particular body parts are homologous or analogous can be difficult. In general, though, analogous structures tend to resemble one another only superficially, whereas the similarities in body parts between two species related by common descent—homologies—tend to be complex and numerous.

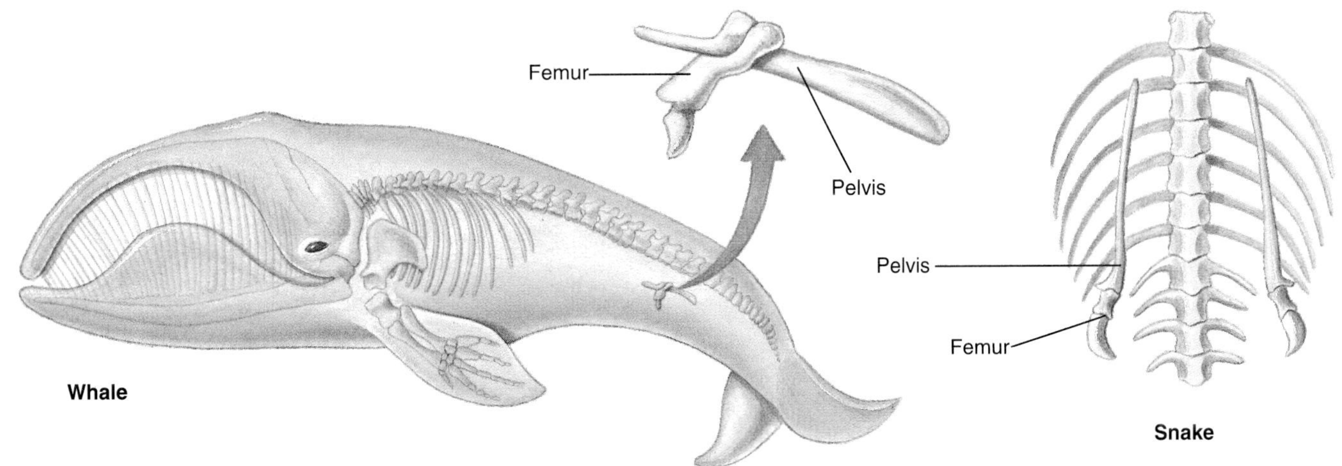

FIGURE 19.6

Vestigial organs. The whale and the snake have tiny femurs (leg bones) apparently of no use to them. These femurs are vestigial organs.

Vestigial Organs

Evolution is not a perfect process. As environmental changes select against certain structures, others are retained, sometimes persisting even if they are not used. A structure that seems to have no function in one species, yet is homologous to a functional organ in another species, is termed **vestigial.** Darwin compared vestigial organs to silent letters in a word—they are not pronounced, but they offer clues to the word's origin.

In whales and some snakes, leg bones are vestigial, retained from vertebrate ancestors who used their legs (fig. 19.6). A horse has two tiny leg bones on each foot that do not end in toes, as do the feet of deer and antelope. These bones are useful to the horse for strength and muscle attachment, but it is interesting to note that the horse's ancestors used their feet more like modern deer and pigs do. These animals all have three toes compared to the modern horse's one toe on each foot. The two tiny leg bones in the horse are vestiges of a different type of gait from times past.

Comparative Embryology

Embryos of a human, monkey, dog, cat, rabbit, and mouse look very much alike. Darwin suggested that the striking similarities among vertebrate embryos reflect adaptations to their similar environments—floating in a watery bubble, either in a uterus or in an egg (see fig. 11.2). These embryos may at first express the same sets of ancestral genes; but as embryos develop into fetuses, they express other genes that specify characteristics of their particular species. Ancestral embryonic genes are retained because natural selection cannot act on them in the protected prenatal environment.

Common ancestral early genes might explain such prenatal features as the gill slits and rudimentary tail seen in a 7-week-old human embryo. German naturalist Ernst Haeckel, a contemporary of Darwin, erroneously thought that an embryo actually repeats the evolution of its ancestors in stages of prenatal development.

Mastering Concepts

1. How do homologous and analogous structures differ?
2. What is a vestigial structure?
3. Why do early embryos of different vertebrate species resemble each other?

Molecular Evolution

Clues to the past also come from within the cell, where the molecules of life change over time, a process termed **molecular evolution.**

The techniques used to study molecular evolution compare gene and protein building block sequences in different species. Proteins and the genes that encode them are composed of many bits of information. It is highly unlikely that two unrelated species would evolve precisely the same gene (nucleotide) or protein (amino acid) sequences by chance. It is more likely that the similarities were inherited from a common ancestor and that differences developed after the species diverged from the ancestral type.

An underlying assumption of molecular evolution studies is that the greater the molecular similarities between two modern species, the closer their evolutionary relationship. Chromosome banding patterns and gene or protein sequences that are identical or very similar in different species are said to be **highly conserved.**

Comparing Chromosomes

Researchers compare the number of chromosomes, chromosome band patterns, and gene orders on chromosomes to measure species relatedness. For example, human chromosome banding patterns most closely match those of chimpanzees, then gorillas, and then orangutans. If one human chromosome were broken in half, we would have 48 chromosomes instead of 46, and the 48 would very closely resemble those of the three species of apes. Some primate chromosomes differ in organization rather than

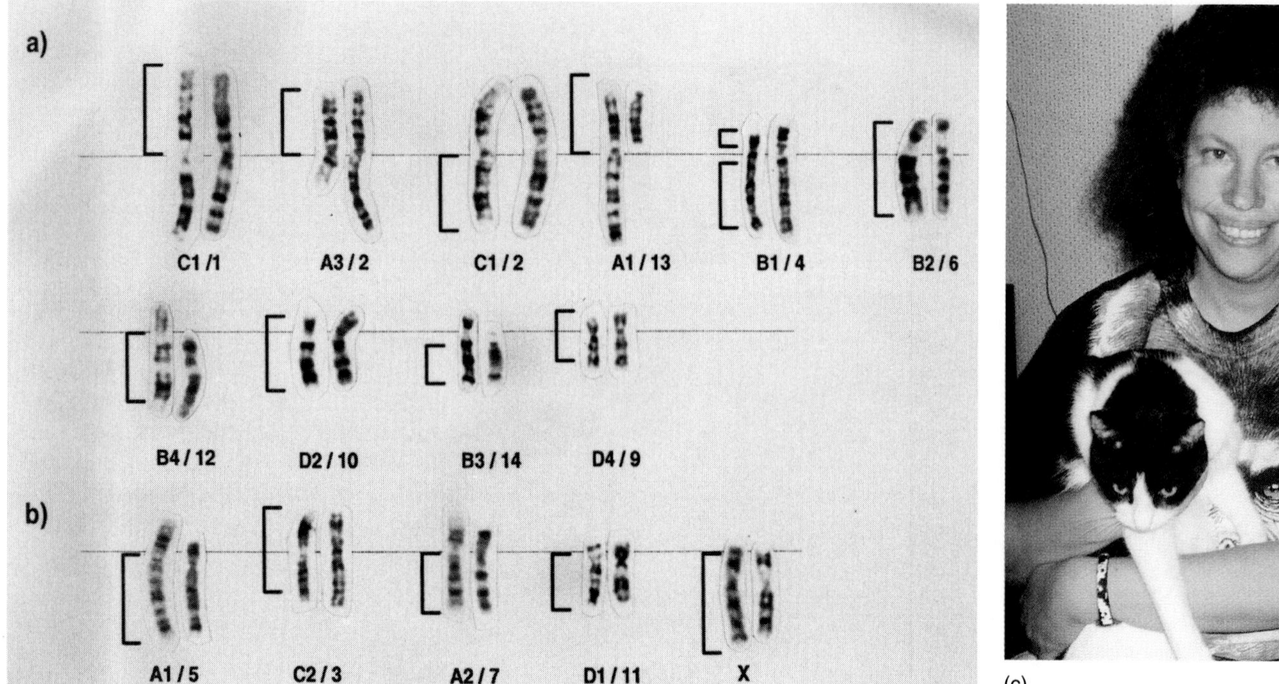

FIGURE 19.7

Conserved regions of human and cat chromosomes. In each pair, the chromosome on the left is from a cat, and the chromosome on the right is from a human. The brackets indicate apparently corresponding areas. The chromosomes in (*a*) have similar banding patterns generated from traditional stains, which do not target specific genes but chromosome regions with generally similar DNA base content. The chromosomes in (*b*) do not look alike, but DNA probes to specific genes indicate that the pairs do indeed share many specific DNA sequences. (*c*) The author and one of her cats, Nirvana.

content, with gene sequences inverted or moved to other chromosomes. For example, the banding pattern of chromosome 1 in humans, chimps, gorillas, and orangutans matches the pattern in two small chromosomes in the African green monkey. Because fossil evidence indicates that monkeys are more ancient than apes, the two small chromosomes may have originated in green monkeys and then fused as hominoids evolved.

Chromosome patterns can also be compared between species that are not as closely related as are humans and other primates (table 19.1). All mammals, for example, have identically banded X chromosomes. Figure 19.7 shows similarities between the karyotypes of a human and a cat.

Chromosome band similarities can be striking, but they do not measure species relatedness as precisely as do a gene's nucleotide or a protein's amino acid sequence. Chromosome bands resulting from traditional stains may superficially resemble each other in different species, without corresponding on a gene-by-gene basis. Incorporating DNA probes into chromosome banding enables researchers to compare individual genes (see fig. 14.14). The goal is to identify regions of **synteny,** or identical sequences of genes along parts of chromosomes. For example, 11 genes are closely linked on human chromosome 21, on mouse chromosome 16, and on chromosome U10 in cows. Several genes on human chromosome 3, however, are located in mice and cows on chromosomes corresponding to human chromosome 21. A hypothesis based on this observation is that a mammal ancestral to all three species carried all of these genes on one chromosome. In humans, some of the genes dispersed to an extra chromosome. Chromosome similarities are thus puzzle pieces of the past.

Table 19.1 Percent of Common Chromosome Bands Between Humans and Other Species

Species	Percent
Chimpanzees	99+%
Gorillas	99+%
Orangutans	99+%
African green monkeys	95%
Domestic cats	35%
Mice	7%

Comparing Protein Sequences

The fact that all species use the same genetic code to synthesize proteins argues for a common ancestry to all life on earth. Additional evidence for evolution is that many different types of organisms use the same proteins, with only slight variations in amino acid sequence. Comparisons of amino acid sequences in different species often support fossil and anatomical evidence.

Organism	Number of Amino Acid Differences from Humans
Chimpanzee	0
Rhesus monkey	1
Rabbit	9
Cow	10
Pigeon	12
Bullfrog	20
Fruit fly	24
Wheat germ	37
Yeast	42

(a)

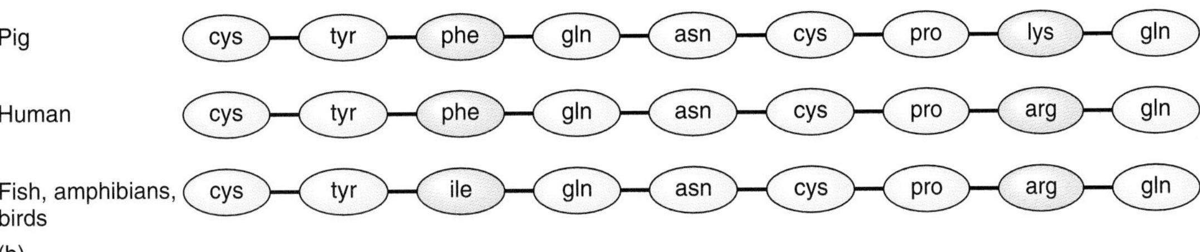

(b)

FIGURE 19.8

Amino acid sequence similarities are a measure of evolutionary relatedness. (*a*) Similarities in amino acid sequence for the respiratory protein cytochrome c in humans and other species parallel degree of relatedness between them. (*b*) Antidiuretic hormone is a peptide hormone nine amino acids long that signals the kidneys to conserve water. Its sequence differs only slightly between major groups of vertebrates.

An ancient and well-studied protein is cytochrome c, which is part of the electron transport chain in cellular respiration. Twenty of its 104 amino acids occupy identical positions in the cytochrome c of all eukaryotes. The more closely related two species are, the more alike is their cytochrome c amino acid sequence (fig. 19.8). It is identical in humans and chimpanzees.

Comparing DNA Sequences

Similarities in two species' DNA sequences can be assessed for a stretch of nucleotides, for a single gene, or for an entire genome. Because many genes have counterparts in different species, researchers can seek clues to human genes in the genes of other types of organisms. For example, when a new human gene is discovered and its sequence determined, the first step in learning its function is to compare its DNA sequence to a database of gene sequences from many species. When the computer identifies known genes in other species, researchers can

- estimate the gene's importance. A gene with counterparts in many distantly related species—such as in flies, worms, and mice—is ancient and important;
- get an idea of what type of protein the gene encodes;
- discover an animal "model" of a human genetic disease, useful in testing new medical treatments.

Sometimes similar genes in different species are evident in striking phenotypes. People with Waardenburg syndrome, for example, have a characteristic white forelock of hair, wide-spaced, light-colored eyes, and hearing impairment. The causative gene is very similar in sequence to one in cats that have white coats and blue eyes and are deaf (fig. 19.9). Horses, mice, and minks also have this combination of traits. The phenotype may result from abnormal movements of pigment cells in the embryo's outermost layer.

A technique called **DNA hybridization** uses complementary base pairing to estimate how similar are the genomes of two species. DNA double helices from two species are unwound and mixed. The rate at which hybrid DNA double helices—that is, DNA molecules containing one helix from each species—form is a direct measure of how similar they are in sequence. The faster the DNA from two species forms hybrids, the more of the sequence they share, and the more closely related they are presumed to be (fig. 19.10). DNA hybridization shows, for example, that human DNA differs in 1.8% of its base pairs from chimpanzee DNA, in 2.3% from gorilla DNA, and in 3.7% from orangutan DNA.

Molecular Trees

Sequence information can help biologists construct **molecular tree diagrams,** which are phylogenies (depictions of evolutionary relationships) based on molecular evidence. Molecular

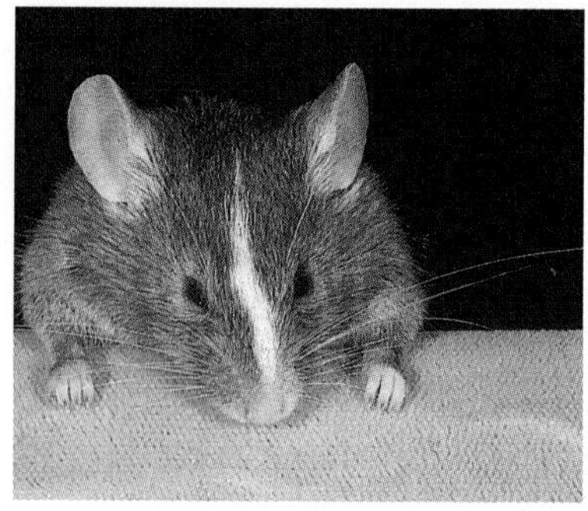

FIGURE 19.9

A gene in mice, cats, humans, and other species causes light eye color, hearing or other neurological impairment, and a fair forelock of hair. Geneticists are trying to determine whether the genes that cause these similar phenotypes are homologous.

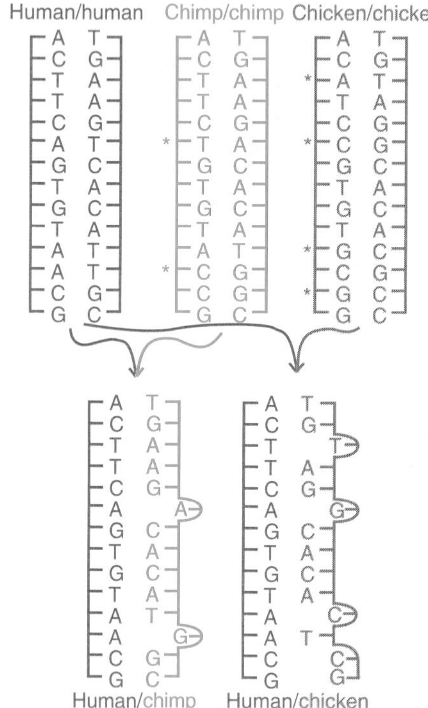

FIGURE 19.10

The rate of DNA hybridization reflects the degree of evolutionary relatedness. This highly schematic diagram shows why DNA from a human hybridizes more rapidly with chimpanzee DNA than it does with chicken DNA.

information can help fill in gaps in our knowledge. This is the case for bats, a diverse group with a very poor fossil record.

The two major types of bats, megabats and microbats, have very similar wing structures, yet the megabats also have nervous systems and male genitalia very similar to those of primates. One megabat, called a "flying fox," looks very monkeylike. Are the two types of bats close relatives, or are the megabats really primates? A comparison of the DNA sequences encoding a blood protein, epsilon globin, among 17 mammalian species produced a tree diagram that indicates that the bats are closely related to each other. They are probably not flying primates (fig. 19.11).

Researchers can study DNA from rare preserved remains of extinct organisms, such as a 40,000-year-old frozen mammoth, an extinct relative of the horse called a quagga (fig. 19.12), and leaves and insects preserved in amber. So far, the oldest "fossil" gene comes from a weevil preserved in amber more than 120 million years ago.

Molecular Clocks

A clock measures the passage of time by moving its hands through a certain portion of a circle in a specific and constant interval of time—a second or a minute. Similarly, a polymeric molecule (one with many building blocks) can be used as a **molecular clock** if its components are replaced at a known and constant rate. Molecular clock studies, however, are not quite as straightforward as glancing at a wristwatch.

If biologists know the mutation rate for the bases in a particular gene, plus the number of differences in the DNA sequences for that gene in two species, they can estimate the time when the organisms diverged from a common ancestor. For example, many human and chimpanzee genes differ in about 4 to 6% of their bases, and substitutions occur at a rate of 1% per 1 million years. Therefore, about 4 to 6 million years have passed since the two species diverged, based on analysis of these genes.

A complication of molecular clock studies is that DNA base substitutions (microevolution) occur in different genes at different rates. This is because some DNA sequences, such as highly repeated ones, are more prone to replication errors than others. Once mutations have occurred, natural selection acts on them to different extents, depending upon the phenotype and the environment. Also, a particular gene may exert a drastic effect on evolution, such as the genes described in Biology in Action 20.2, yet others may exert little effect. Therefore, to construct molecular tree diagrams incorporating molecular clocks, researchers often consider DNA sequences that are not subject to natural selection, such as genes encoding ribosomal RNAs.

Time scales based on fossil evidence and molecular clocks can be superimposed on molecular tree diagrams constructed from sequence data. When building evolutionary trees, however, a single set of data can be arranged into many different tree configurations. A computer can depict the data on 17 mammalian species, for exam-

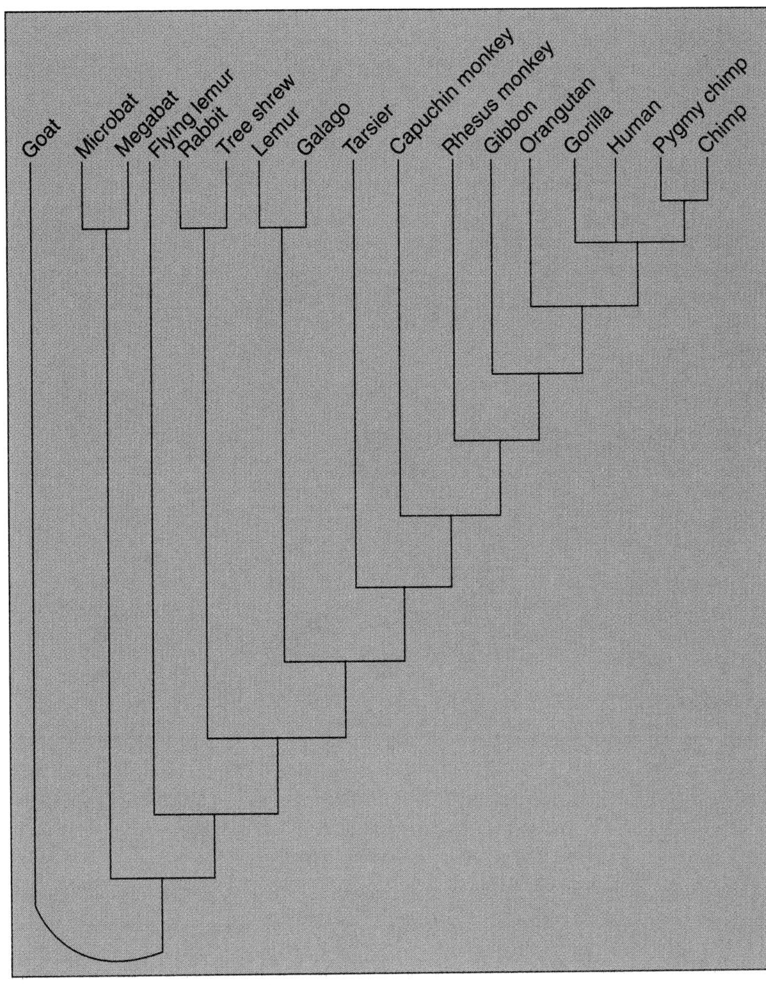

FIGURE 19.11

What is a bat? When the sequences of the epsilon globin gene are compared among 17 mammalian species, and the simplest tree diagram is constructed, bats do not appear to be "flying primates," as the similarities between bat and primate nervous systems suggest. Rather, they represent a separate lineage. In the diagram, species are placed closest to their closest relatives. For example, a rabbit is most closely related to the tree shrew and then to the galago and lemur, which are primates.

Source: Data from Wendy Baily, et al., "Rejection of the 'Flying Primate' Hypothesis by Phylogenetic Evidence of E-globin," in Science, vol. 256, April 3, 1992.

FIGURE 19.12

Probing the molecules of extinct organisms. The last quagga, a relative of the horse and zebra, died in captivity in Amsterdam in 1883. Geneticists are now deciphering DNA extracted from this preserved quagga to see how closely related the extinct animal was to its still-living relatives.

ple, in 10,395 different ways! The order in which data are entered into a tree-building computer program influences the tree's shape, which in turn affects the interpretation of species relationships. With every new bit of sequence information, the tree possibilities change.

Parsimony analysis is a statistical method that helps identify a molecular tree most likely to represent what really happened. A computer connects all evolutionary tree sequence data using the fewest possible number of mutations to account for observed DNA base sequence differences. The most parsimonious (simplest) tree is the most likely because it requires the fewest chance mutations.

A special type of molecular clock measures mutations in mitochondrial DNA. Recall from chapters 4 and 7 that mitochondria are organelles that house the biochemical reactions of cellular energy acquisition. Mitochondria contain about 17,000 DNA bases, which encode RNA molecules and a few enzymes. Mitochondrial DNA (mtDNA) is valuable in tracking recent evolutionary time because its molecular clock ticks 5 to 10 times faster than the nuclear DNA clock—that is, mtDNA mutates 5 to 10 times faster than DNA in the nucleus. This is because mitochondrial DNA cannot repair errors in DNA replication.

Just as with other measures of evolutionary relatedness, the more similar the mtDNA sequence between two individuals, the more recently they are presumed to share a common ancestor. Mitochondrial DNA, however, presents an interesting problem because it is inherited from mothers only, in oocytes. The part of the sperm cell that enters the oocyte at fertilization usually does not contribute mitochondria, thus preventing males from contributing mitochondria to offspring.

Mitochondrial DNA comparisons are used to trace human origins. For example, many studies show that people from Africa have the most numerous and diverse mitochondrial mutations. This suggests that Africans have existed longer than other modern peoples, because it takes time for mutations to accumulate. Additional evidence in support of the African origin of all modern humans comes from molecular trees constructed by parsimony analysis. In many of these trees, the groups of people whose DNA sequences form the bases are from Africa. This idea that early humans originated in Africa and then migrated, eventually replacing humans elsewhere, is called the single origin hypothesis. An alternative explanation, the multiregional hypothesis, suggests that separate populations of early humans lived throughout the world.

Possibly, some combination of the single origin and multiregional models is closest to the truth. Perhaps African peoples migrated to Asia and Europe, where they met and mixed with other early humans, forming the gene pools from which we ultimately arose.

19.1 biology in action

What Exactly Is a Velvet Worm?

From its appearance, the velvet worm (*Onychophora*) seems to be part segmented worm, part insect. Is it a "missing link" of sorts between the more primitive annelid (segmented) worms and the arthropods, a huge and more complex group including many invertebrates with segmented legs?

A velvet worm looks like a worm that locomotes on short legs. Like the arthropods, the worm molts its soft exterior and has a heart, antennae, and brain. The animal's locomotion and simple head structure, however, are more like those of the annelid worms, and the velvet worm lacks the chitinous outer covering of arthropods such as insects, spiders, scorpions, and lobsters (fig. 19.A). What is it?

Fossil evidence suggests that velvet worms diverged from a common ancestor of the annelids and arthropods about 500 million years ago. Because neither the velvet worm's anatomy, development, nor fossil remains clarifies its relationship to the annelids and arthropods, researchers looked to molecular evidence to try to trace the animal's evolutionary relationships.

Comparing the nucleotide sequences of ribosomal RNAs from velvet worms with the corresponding sequences in many modern annelids and arthropods reveals that velvet worms are more like arthropods. This additional information, as is often the case in scientific research, raises as many questions as it provides answers. If the velvet worm is an arthropod, then it is a primitive one because of its structural simplicity. But is it truly more primitive than other arthropods, or did it become less complex over time? We still have much to learn about the mysterious velvet worm.

FIGURE 19.A
This velvet worm from Australia resembles a slug with legs. Molecular evidence is helping evolutionary biologists and taxonomists determine its relationship to two major groups of animals—the annelid worms and the more complex and diverse arthropods.

Reconciling Fossil, Biogeographical, and Molecular Evidence

Just as fossil evidence has limitations, so does molecular evidence of evolution. Molecular data paint gene-by-gene comparisons, and they cannot tell us directly about extinct life, unless we can obtain preserved DNA and use it to produce a phenotype in a living organism. Molecular data also cannot explain the evolution of many obvious physical differences between organisms whose genes are very similar. Humans and chimpanzees, for example, share many similar if not identical gene sequences, but you would hardly mistake one for the other.

What molecular evidence *can* tell us is how species are related to each other. Fossil and biogeographical evidence introduce the element of time—that is, when species most likely diverged from common ancestors, against the backdrop of other events happening on earth.

It is exciting when molecular evidence supports other kinds of evidence for evolution. This is so for the hundreds of species of cichlid fishes that have spread during the past several hundreds of thousands of years in the African great lakes. As new lakes formed, they isolated populations of the fishes, and the animals eventually accumulated sufficient genetic changes to constitute new species. The approximate times when DNA sequences diverged, according to mitochondrial DNA sequence comparisons, coincide with geographic evidence of when earthquakes occurred. The earthquakes may have formed the lakes that separated the ancient gene pools of cichlid fishes.

Sometimes different types of evolutionary evidence do not agree. For example, most fossils of simple multicelled animals called metazoa date to 500 to 600 million years ago, with a very few dating back a billion years. DNA and protein sequence data on modern descendants of the first metazoa, however, clearly indicate that these early animals diverged into several species at least a billion years ago. A possible explanation for the disagreement between molecular and fossil evidence is that metazoa may have been abundant a billion years ago, as the sequence data suggests, but left few fossils during that time. If they were soft-bodied and subjected to long periods of geologic upheaval, chances are fossils would not have formed or persisted. Biology in Action 19.1 discusses another disagreement between molecular, fossil, and modern anatomical differences that led to an interesting hypothesis.

Mastering Concepts

1. What types of biochemicals hold clues to the evolutionary past?
2. How do DNA hybridization experiments estimate species relatedness?
3. How is a time scale applied to molecular evidence?
4. Why are mitochondrial DNA sequences valuable in molecular evolution studies?
5. Why do different types of evidence for evolution sometimes not agree?

New Dinosaurs~

continued from page 390.

Our reconstructions of past life change as fossil evidence accumulates. This is the case for the African dinosaur *Carcharodontosaurus* and the South American *Giganotosaurus.*

When Rodolfo Coria and Leonardo Salgado discovered *Giganotosaurus,* they did not know of the near-simultaneous dinosaur finds in Morocco. Coria and Salgado conclude in their paper introducing *Giganotosaurus:* "That the enormous size of *Tyrannosaurus* and *Giganotosaurus* evolved independently suggests that gigantism may be linked to common environmental conditions of their ecosystems."

But the discovery of the *Carcharodontosaurus* skeleton made it a little less likely that three giants had such similar adaptations yet lived so far apart. An alternate hypothesis arose—perhaps the immediate ancestor of these semi-upright carnivores, called theropods, lived when the continents were joined. By the time *Carcharodontosaurus* lived in Africa and *Giganotosaurus* lived in South America, and a little later *Tyrannosaurus* in North America, the continents were separate. By then, Africa had a unique collection of dinosaur types.

Geographical evidence supports this hypothesis of dinosaur dispersal. The purported ancestor would have lived at the end of the Jurassic period, when the earth had two large landmasses, the northern Laurasia and the southern Gondwana. In the early Cretaceous period, about 140 million years ago, the large predatory dinosaurs spread all over the globe. By the late Cretaceous, theropods had settled on specific continents—rather suddenly, according to the fossil record. Researchers interpret this to indicate that at this time, the last land bridges connecting northern and southern landmasses sunk, and the oceans separated the southern continents.

SUMMARY

1. Many types of evidence enable researchers to construct **phylogenies,** which depict possible evolutionary relationships among species. **Paleontology** is the study of fossils. Other evidence comes from comparing anatomical and biochemical characteristics of modern species. **Cladistics** is the study of species diversification from shared ancestors.

2. A fossil may form when mineral replaces tissue, developing gradually or after a sudden catastrophe, or may be indirect evidence such as footprints. Ice or tar can preserve actual remains. Remains in amber dehydrate, which preserves DNA. The fossil record is incomplete. Few individuals are preserved and few of these escape erosion or are discovered.

3. Scientists estimate fossil age in relative and absolute terms. The rock layer a fossil is in provides a **relative date.** The ratio of a stable radioactive isotope to its breakdown product provides an estimate of an absolute year date. This is **radiometric dating.** Potassium 40 can date fossils older than 300,000 years, and carbon 14 dating is used on fossils younger than 40,000 years; new techniques fill the gap.

4. **Homologous** structures are inherited from a shared ancestor but may differ in function. **Analogous** structures are similar in function but are not inherited from a common ancestor. They are similar responses to the same environmental challenge. **Vestigial** organs and similar embryonic structures in different species reflect genes retained from ancestors.

5. **Molecular evolution** considers similarities and differences between sequences of chromosome bands, a protein's amino acids, or a gene's DNA bases. Presumably, these sequences contain so many bits of information that it is unlikely similarities evolved by chance.

6. Similar chromosome band patterns may or may not reflect similarities at the gene level. DNA probes compare chromosomes of different species and identify **syntenic,** or corresponding, sections. Many genes and proteins are **highly conserved** among species.

7. A **molecular tree diagram** shows evolutionary relationships between species, based on a gene's mutation rate. A **molecular clock** estimates the time when two species diverged from a common ancestor. One set of data can yield many molecular trees. **Parsimony analysis** identifies the trees requiring the fewest mutations. Molecular clocks based on mitochondrial DNA are used to date recent events because mtDNA mutates 5 to 10 times faster than nuclear DNA.

TO REVIEW . . .

1. Why are a preserved dinosaur bone, a ripple in a rock made by an ancient worm's movements, limestone composed of the shells of crustaceans, and insects preserved in amber all fossils?

2. Cite three factors that explain why the fossil record is incomplete.

3. What does a cladogram depict?

4. What assumptions underlie the following:
- **a.** relative dating based on a fossil's position in rock strata
- **b.** absolute dating based on radiometric data
- **c.** the mitochondrial clock

5. An elephant uses its trunk to bring food to its mouth; a human uses his or her hand. What information would help to distinguish whether an elephant's trunk and a human's hand are analogous or homologous structures?

6. Cite a limitation of comparing chromosome banding patterns to define species relationships.

7. How does the type of data that molecular sequence data provide differ from the type of information that relative and absolute dating provide?

8. Why does a human genome researcher compare a newly discovered human gene's sequence to gene sequences from other organisms?

9. What types of information are required to construct a molecular tree diagram? What are the limitations of these diagrams?

TO THINK ABOUT . . .

1. A geneticist aboard a federation starship must determine the evolutionary relationships among Humans, Klingons, Romulans, and Betazoids. Each organism walks on two legs, lives in complex societies, uses tools and technologies, looks similar, and reproduces in the same manner. Each can interbreed with any of the others. The geneticist finds the following data:

Klingons and Romulans each have 44 chromosomes. Humans have 46 chromosomes. Human chromosomes 15 and 17 each resemble part of the same large chromosome in Klingons and Romulans.

Humans and Klingons have 97% of their chromosome bands in common. Humans and Romulans have 98% of their chromosome bands in common, and Humans and Betazoids show 100% correspondence. Humans and Betazoids differ only in an extra segment on chromosome 11, which appears to be a duplication.

The cytochrome c amino acid sequence is identical between Humans and Betazoids. It differs in Humans and Romulans by one amino acid and in Humans and Klingons by two amino acids.

The gene for collagen contains 50 introns in Humans, 50 introns in Betazoids, 62 introns in Romulans, and 74 introns in Klingons.

Mitochondrial DNA analysis reveals many more differences among Klingons and Romulans than among Humans or Betazoids.

a. Hypothesize how chromosome differences among these organisms may have arisen.

b. Which are our closest relatives among the Klingons, Romulans, and Betazoids? What evidence supports your answer?

c. Are Klingons, Romulans, Humans, and Betazoids distinct species? What information supports your answer?

d. Which of these molecular tree diagrams is consistent with the data?

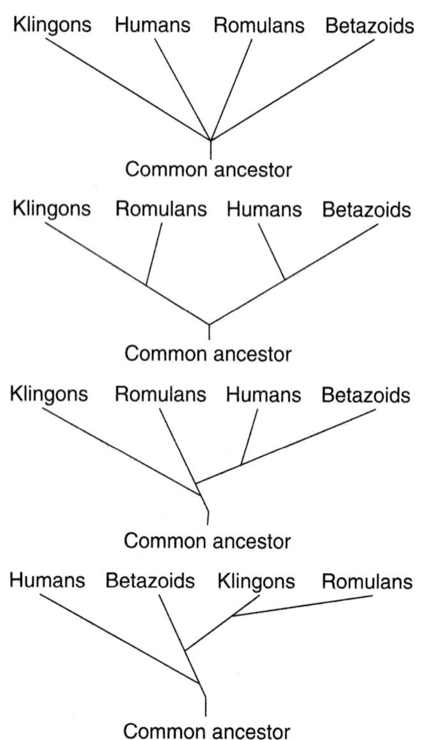

2. How is species diversification among cichlid fishes in the African great lakes similar to that among the Galápagos Island finches, discussed in chapter 18?

3. This cladogram shows the evolutionary relationships among a tree species with leaves preserved in amber, *Hymenaea protera,* three living relatives, and six other types of plants:

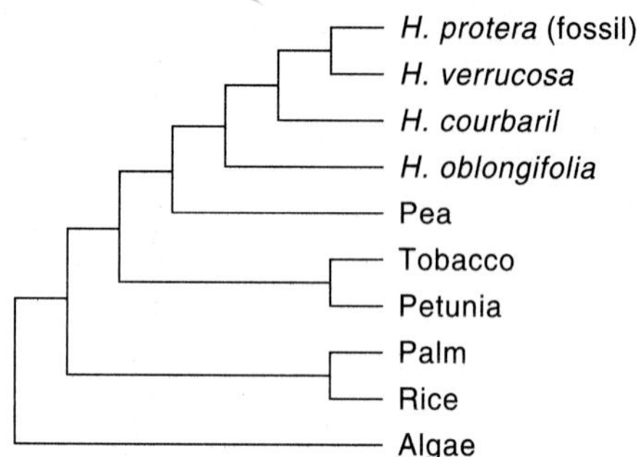

a. Which organism is *H. protera*'s closest relative?

b. Is *H. protera* more closely related to tobacco or to palm?

c. Which organism depicted is ancestral to all the others?

4. Some genes are more alike between human and chimp than other genes are from person to person. Does this mean that chimps are humans or that humans with different alleles are different species? What other explanation fits the facts?

5. Inheritance can be traced through females by tracking mitochondrial DNA sequences. Suggest a way that male inheritance can be followed.

6. Why is the DNA sequence of one gene a less accurate indicator of the evolutionary relationship between two species than a DNA hybridization experiment comparing large portions of the two genomes?

7. What present environmental conditions might preserve organisms to form tomorrow's fossils?

TO LEARN MORE . . .

References and Resources*

Burrows, Wes, and Oliver A. Ryder. January 9, 1997. *Nature,* vol. 385. Can Y chromosomes be studied to trace evolutionary relationships?

Cooper, Henry S. F. June 1996. Origins: The backbone of evolution. *Natural History.* A visit to the American Museum of Natural History is like a walk through time.

Currie, Philip J. May 17, 1996. Out of Africa: Meat-eating dinosaurs that challenge *Tyrannosaurus rex. Science,* vol. 272. Did large carnivorous dinosaurs from different continents share ancestors when the earth had one land mass?

Doolittle, Russell F., et. al. January 26, 1996. Determining divergence times of the major kingdoms of living organisms with a protein clock. *Science,* vol. 271. Sequence data from 57 enzymes trace the origins of the major groups of modern organisms.

Fortey, Richard A., and Richard H. Thomas. January 21, 1993. The case of the velvet worm. *Nature,* vol. 361. Molecular evidence reconciles fossil and anatomical data.

Hillis, David M. April 11, 1997. Biology recapitulates phylogeny. *Science,* vol. 276. The concept of phylogeny was first proposed in 1866, and is currently experiencing a resurgence in interest.

Lewin, Roger. 1997. *Patterns in evolution:the new molecular view.* New York: Scientific American Library. Informational polymers hold clues to evolution.

**Life's* website features frequently updated references.

Natural selection is largely responsible for the superb camouflage of the mourning gecko.

20 The Forces of Evolutionary Change

chapter outline

Bullwinkle's Story—Sticking to One's Own Species   404

Evolution After Darwin—The Genetics of Populations   404

When Gene Frequencies Stay Constant—Hardy-Weinberg Equilibrium   405

When Gene Frequencies Change—Evolution   406
Nonrandom Mating
Migration
Genetic Drift
Mutation
Natural Selection

How Species Arise   413
Premating Reproductive Isolation
Postmating Reproductive Isolation

The Pace of Evolution—Slow or Fast?   415

How Species Become Extinct   416
What Causes Mass Extinctions?
Mass Extinctions Through Geologic Time

The Taming of Tuberculosis

The spread of tuberculosis (TB) in the Plains Indians of the Qu'Appelle Valley Reservation in Saskatchewan, Canada, and its resurgence today in a multidrug-resistant form, illustrates evolution at work.

When TB first appeared on the reservation in the mid-1880s, it struck swiftly and lethally, infecting many organs. Ten percent of the population died. By 1921, TB tended to affect only the lungs, and the death rate had dropped to 7%. By 1950, the disease had changed further, mortality now down to 0.2%.

Outbreaks of TB ran similar courses in other human populations. In the 1700s, TB raged through the cities of Europe. Immigrants brought it to the United States in the early 1800s, where it likewise swept the crowded cities. The disease was so widespread that many thought it was hereditary, until German bacteriologist Robert Koch identified the causative bacterium in 1882. TB killed Emily Brontë, Anton Chekhov, John Keats, Edgar Allan Poe, Henry David Thoreau, and many other citizens of the time.

TB incidence and virulence fell dramatically not only on the Plains Indian reservation but in the cities of the industrialized world during the first half of the twentieth century—before antibiotic drugs were developed. What tamed tuberculosis?

to be continued . . .

Bullwinkle's Story—Sticking to One's Own Species

Bullwinkle J. Moose first ambled onto farms in upstate New York in 1980. A refugee from the dense moose populations in nearby New England and Canada, Bullwinkle was one of only 20 moose in New York state, most of them males. Because Bullwinkle couldn't find a cow of his own species, he became attracted to a similar species—domestic cows (fig 20.1). The moose's vision was poor, and from a distance the cows resembled moose cows. Their low-pitched moos sounded like the moans that female moose utter to lure mates. The scent of the cows was enough like moose cow scent that Bullwinkle dug his hooves into the ground and urinated around a small area, marking a territory as a prelude to mating. He then approached a cow.

Once Bullwinkle mounted her, the courtship ended. A moose is a moose and a cow is a cow—members of two different species. The moose's genitalia did not fit the cow's, so they could not mate. Bullwinkle continued to frequent upstate New York farms in search of cows until 1987, when he finally found a mate of his own species near Tupper Lake.

FIGURE 20.1

Members of two species cannot successfully mate with one another. In some areas of New England and upstate New York, male moose will occasionally approach a cow with romance in mind.

Not all species are as different in appearance as the moose and the cow. For example, two species of fruit fly, *Drosophila persimilis* and *Drosophila pseudoobscura,* look identical, to human observers, down to the tiniest bristle. Yet the flies will not mate with each other, for reasons not obvious to us.

How does a group of organisms become able to mate successfully only among themselves? In other words, how do species form? Charles Darwin theorized that species evolve as particular environments select against different traits. Eventually, accumulated adaptations change a group of organisms enough that they cannot mate with others outside the group. On a small scale, then, evolution is the changing frequencies of alleles (gene variants) among groups of individuals. On a large scale, evolution is the formation of species.

Evolution After Darwin—The Genetics of Populations

The "raw material" of evolution is inherited variation. Genes influence evolution at all levels above that of the individual, including the smallest, the **population.** Population is a general term for any group of organisms sharing a geographic area and able to interbreed. Chapter 41 examines populations from an ecological perspective.

All the genes in a population constitute its **gene pool.** The movement of alleles between populations is called **gene flow.** The proportion of different alleles for each gene determines the characteristics of that population. A Swedish population, for example, might include a large proportion of hair color alleles conferring blondness; a population of black Africans would have very few, if any, such alleles but would have many alleles conferring darker hair. Shifting allele frequencies in populations are the small steps of change that collectively drive evolution.

Microevolution occurs when the frequency of an allele in a population changes. A gene's frequency may be altered when

- mutation introduces new alleles into a population;
- individuals migrate between populations;
- individuals remain in closed groups, mating among themselves within a larger population;
- natural selection acts differently on variant phenotypes;
- genes are eliminated because individuals with certain genotypes do not produce fertile offspring.

Because these situations are common, evolution is not only possible but unavoidable. This chapter focuses on these common phenomena that make evolution very much a part of life—past, present, and future.

When Gene Frequencies Stay Constant—Hardy-Weinberg Equilibrium

Mathematics can describe the theoretical state when gene frequencies do not change from one generation to the next. This state is called **Hardy-Weinberg equilibrium.** It serves as a basis of comparison for more realistic situations in populations. Hardy-Weinberg equilibrium is only possible if a population is large, its members mate at random, and there is no migration, mutation, or natural selection—conditions that in actuality do not occur together.

In 1908 mathematician H. H. Hardy and physician W. Weinberg independently proposed that the expression $p + q = 1$ could be used to represent the frequency of all of the alleles for a particular gene in a population of diploid organisms, if only two alleles exist for that gene (fig. 20.2). This expression means that p and q are all the alleles in the population. If 70% of the members of a ferret population have an allele conferring dark fur, and the gene has only two alleles, then the allele frequency for the dark allele p is 0.7, and that for the alternate allele q is 0.3.

Once we know allele frequencies, we can calculate genotype frequencies. This is because the proportion of alleles equals the proportions of gametes, as a consequence of Mendel's first law. If p represents the frequency of the D allele for dark fur, then the proportion of the population that is homozygous D (genotype DD) equals p multiplied by p, or p^2, which equals 0.7 times 0.7, or 0.49. Likewise, the proportion of population members that are homozygous recessive (pp) equals 0.3 times 0.3, or 0.09. We can calculate the proportion of the population that is heterozygous by subtracting the total proportion of homozygotes ($0.49 + 0.09 = 0.58$) from 1, which gives 0.42. Put another way, 42% of the ferrets in the population have dark fur and are heterozygous for the fur color gene.

In algebraic terms, $2pq$ represents the heterozygous class. This is because a heterozygote has one p allele and one q allele, and this combination can arise in two ways—p from the mother and q from the father, and vice versa. (This is a population-level demonstration of Mendel's first law—recall a Punnett square for a monohybrid cross.)

The Hardy-Weinberg expression $p^2 + 2pq + q^2 = 1$ represents the proportions of genotypes that make up the population. If conditions of Hardy-Weinberg equilibrium are met, allele and genotype frequencies will not change as generations continue. We can see this by deriving allele frequencies back from the genotype frequencies, assuming that mating is at random, that there is no migration or mutation, and that natural selection doesn't weed out any phenotypes.

The population of 49% DD, 9% dd, and 42% Dd individuals manufactures gametes. The proportion of "D" alleles equals that of the homozygous dominant (DD) individuals (0.49) plus one-half of the gametes from the heterozygotes (Dd), which equals one-half of 0.42, or 0.21. Therefore, the proportion of D alleles is $p = 0.49 + 0.21 = 0.70$. If p equals 0.7, then q equals 1 minus p, or 0.3. We are back at the beginning. Hardy-Weinberg equilibrium persists, and evolution is not occurring. These calculations also reveal how recessive alleles remain in a population, even if homozygous recessive individuals are at a disadvantage. Heterozygotes maintain recessive alleles.

Table 20.1 organizes these calculations. It depicts the random mating that underlies Hardy-Weinberg equilibrium. The "male" and "female" columns show every possible combination of gametes, and the other columns show how genotype frequencies of the next generation derive from the random matings.

Algebraic Expression	What It Means
$p + q = 1$ (allele frequencies)	All dominant alleles plus all recessive alleles add up to all of the alleles for a particular gene in a population.
$p^2 + 2pq + q^2 = 1$ (genotype frequencies)	For a particular gene, all homozygous dominant individuals (p^2) plus all heterozygotes ($2pq$) plus all homozygous recessives (q^2) add up to all of the individuals in the population.

FIGURE 20.2

The Hardy-Weinberg equation: using algebra to follow gene frequencies.

Table 20.1 Hardy-Weinberg Equilibrium—When Allele Frequencies Stay Constant

Genotype Frequencies				F$_1$ Genotypes Produced		
Male		Female	Proportion of Crosses	*DD*	*Dd*	*dd*
0.49 *DD*	×	0.49 *DD*	0.2401	0.2401		
0.49 *DD*	×	0.42 *Dd*	0.2058	0.1029	0.1029	
0.49 *DD*	×	0.09 *dd*	0.0441		0.0441	
0.42 *Dd*	×	0.49 *DD*	0.2058	0.1029	0.1029	
0.42 *Dd*	×	0.42 *Dd*	0.1764	0.0441	0.0882	0.0441
0.42 *Dd*	×	0.09 *dd*	0.0378		0.0189	0.0189
0.09 *dd*	×	0.49 *DD*	0.0441		0.0441	
0.09 *dd*	×	0.42 *Dd*	0.0378		0.0189	0.0189
0.09 *dd*	×	0.09 *dd*	0.0081			0.0081
				0.49	0.42	0.09
				DD	*Dd*	*dd* = 1

Mastering Concepts

1. What is a population?
2. How do allele frequencies underlie evolution?
3. Why doesn't evolution occur if Hardy-Weinberg equilibrium exists?
4. Why doesn't Hardy-Weinberg equilibrium occur in real populations?

When Gene Frequencies Change—Evolution

The conditions necessary for Hardy-Weinberg equilibrium (unchanging gene frequencies) are rarely, if ever, met in natural populations, particularly human populations. We give great consideration to selecting mates—it is hardly a random process. Migration disrupts allele frequencies, as we move from place to place and mix, yet at the same time we create population pockets of ethnicity by living in the same areas and having children with people like ourselves. We counter natural selection by medically correcting phenotypes that would otherwise prevent some of us from having children. A cougar with poor vision cannot hunt and thus starves; a human with poor vision simply wears glasses, and life goes on.

Let's look now at how frequencies in real populations change. We will focus especially on human populations, for we know the most about ourselves.

Nonrandom Mating

The nonrandom mating that is a requirement for Hardy-Weinberg equilibrium refers to genes considered one at a time—something we hardly think about when choosing a mate. Humans don't generally mate randomly; we choose mates based on appearance, ethnic background, and intelligence. Economics, culture, and even political events affect mating patterns. Sometimes, certain individuals contribute disproportionately to the next generation. Agriculture is an extreme example of nonrandom mating. An animal or plant with valuable characteristics may be extensively bred. Semen from one prize bull, for example, may artificially inseminate thousands of cows. Breeds of popular domesticated animals also illustrate nonrandom mating (Biology in Action 20.1).

Occasionally, a man in a human population fathers an unusually large number of children. In the Cape population of South Africa, for example, a Chinese immigrant known as Arnold had a very rare dominant genetic disease that causes the teeth to fall out before age 20. Arnold had seven wives and was very fertile. Of his 356 living descendants, 70 have the dental disorder. The frequency of this allele in the Cape population is exceptionally high, compared to elsewhere in the world, thanks to Arnold's disproportionate contribution to the gene pool.

The high frequency of people with albinism among Arizona's Hopi Indians also reflects nonrandom mating. This autosomal recessive condition, marked by lack of skin and hair pigmentation, affects about 1 in 12,500 people in the general U.S. population but about 1 in 200 Hopi Indians. The reason is cultural. Hopi men with albinism often stay back and help the women, rather than risk severe sunburn working in the fields with the other men. They contribute disproportionately to the next generation because they have greater contact with the women.

Historical events often influence nonrandom mating patterns. When one group of people is subservient to another, genes tend to flow from ruling class to underclass because ruling class males mate with underclass females. Historical records and studies of DNA sequences on the Y chromosome, which enable researchers to track male gene transmission, support this observation.

Despite our personal mating choices and the effects of cultural and historical trends, many traits mix randomly in each human generation. This may be because we are unaware of these

biology in action 20.1

Dogs and Cats—Products of Artificial Selection

The pampered poodle and graceful greyhound may win in the show ring, but they are poor specimens in terms of genetics and evolution. Human notions of attractiveness can lead to bizarre breeds that may never have evolved naturally. Behind carefully bred traits lurk small gene pools and extensive inbreeding—all of which may harm the health of highly prized and highly priced show animals. Purebred dogs suffer from more than 300 types of inherited disorders.

The sad eyes of the basset hound make this dog a favorite in advertisements, but these runny eyes can be quite painful. Short legs make the dog prone to arthritis, the long abdomen encourages back injuries, and the characteristic floppy ears often hide ear infections. The eyeballs of the Pekingese protrude so much that a mild bump can pop them out of their sockets. The tiny jaws and massive teeth of pugdogs and bulldogs cause dental and breathing problems, as well as sinusitis, bad colds, and their notorious "dog breath." Folds of skin on their abdomens easily become infected. Larger breeds, such as the Saint Bernard, have bone problems and short life spans. A Newfoundland or a Great Dane may suddenly die at a young age, its heart overworked from years of supporting a large body (table 20A).

We artificially select natural oddities in cats too. One of every 10 New England cats has six or seven toes on each paw, thanks to a multitoed ancestor in colonial Boston (fig. 20a, *left*). Elsewhere, these cats are quite rare. The sizes of the blotched tabby populations in New England, Canada, Australia, and New Zealand correlate with the time that has passed since cat-loving Britons colonized each region. The Vikings brought the orange tabby to the islands off the coast of Scotland, rural Iceland, and the Isle of Man, where these feline favorites flourish today.

A more modern breed appealing to cat fanciers is the American curl cat, whose origin is traced to a stray female who wandered into the home of a cat-loving family in Lakewood, California, in 1981. This cat passed her unusual, curled-up ears to kittens in several litters (fig. 20.A, *right*). A dominant gene that makes extra cartilage grow along the outer ear causes the trait. Cat breeders attempting to fashion this natural peculiarity into an official show animal are hoping that the gene does not have other, less lovable effects. Cats with floppy ears, for example, are known to have large feet, stubbed tails, and lazy natures.

All these examples make one genetic truth clear: You may be able to breed desired characteristics into a dog or cat, but you can't always breed other traits out.

Table 20.A Purebred Plights

Breed	Health Problems
Cocker spaniel	Nervousness
	Ear infections
	Hernias
	Kidney problems
Collie	Blindness
	Bald spots
	Seizures
German shepherd	Hip dysplasia
Golden retriever	Lymphatic cancer
	Muscular dystrophy
	Skin allergies
	Hip dysplasia
	Absence of one testicle
Great Dane	Heart failure
	Bone cancer
Labrador retriever	Dwarfism
	Blindness
Shar-pei	Skin disorders

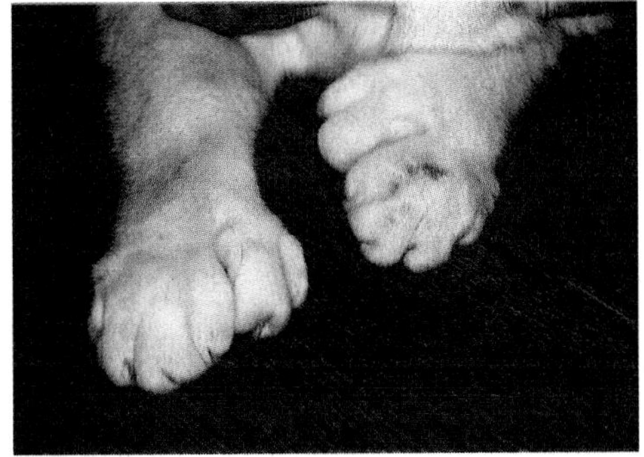

FIGURE 20.A
Cat varieties. Left: Multitoed cats are common in New England but rare elsewhere. Right: American curl cat.

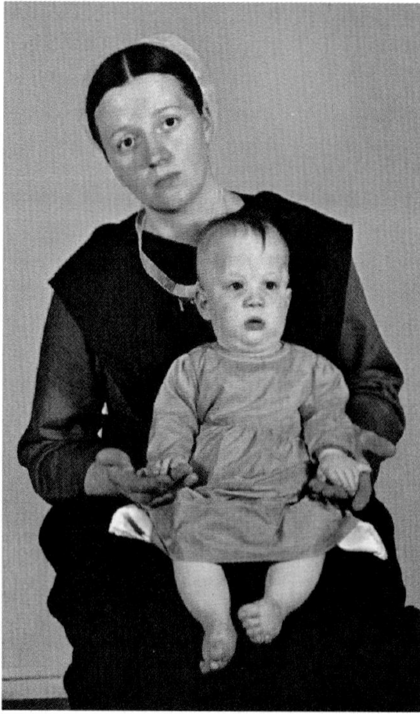

FIGURE 20.3

This Amish child from Lancaster County, Pennsylvania, has inherited Ellis-van Creveld syndrome. He has short-limbed dwarfism, extra fingers, heart disease, and fused wrist bones, and he had teeth at birth. Ellis-van Creveld is autosomal recessive and occurs in 7% of the people of this Amish community—a high figure, because they marry among themselves.

characteristics, and so cannot choose or reject them, or it may be because we do not consider them important in mate selection. We hardly choose life partners, for example, on the basis of blood type! However, sometimes the opposite occurs, and people with particular alleles mate more often than they would by chance. This happens when families with a particular inherited illness meet each other through activities sponsored by patient organizations, and marriages occur between carriers for the same condition.

Migration

Large cities, with their pockets of ethnicity, defy Hardy-Weinberg equilibrium by their very existence. Waves of immigration built the population of New York City, for example. The original Dutch settlers of the 1600s lacked many of the gene variants present in today's metropolis; English, Irish, Slavic, African, Hispanic, Italian, Asian, and many other types of immigrants introduced them.

We can trace migration-caused microevolution by correlating allele frequencies in present-day populations to events in history. For example, the frequency of ABO blood types in certain parts of the world reflects Arab rule. The ABO distribution is very similar in northern Africa, the Near East, and southern Spain, precisely the regions the Arabs ruled until 1492. Even over half a millennium later, these alleles have not greatly dispersed.

Directional gene flow also occurred when nomadic peoples with a hunter-gatherer lifestyle encountered a more stable group of people. For example, in the eighteenth century, nomadic European Caucasians called trekboers migrated to the Cape area of South Africa. The men had children with the native women of the Nama tribe and settled in the area. The resulting mixed society remained fairly isolated and led to the present-day "Richtersveld coloureds" of the Cape region.

Because geographical barriers greatly influence migration patterns, allele frequencies sometimes differ between relatively close geographical regions, such as on either side of a mountain range. In Europe, geographical barriers are important factors in creating **clines,** which are allele frequencies that progressively change across a region.

Genetic Drift

Gene frequencies can change when a small group of individuals separates from a larger population, a phenomenon called **genetic drift.** By chance, the small group may not accurately represent the whole. The students who failed a biology class, for example, do not accurately represent the academic ability of the entire group—neither do all the A students.

A common type of genetic drift in human populations is the **founder effect,** which occurs when small groups of people leave their homes to found new settlements. The new colony may have different genotype frequencies than the original population and may amplify traits that were rare in that larger group. For example, certain groups of North American Native Americans do not have type B blood, though type B appears in the Asian populations the Indians descended from. The founding band of Asian settlers who crossed the Bering Strait to North America thousands of years ago may not have included a person with type B or AB blood. If it did, that individual may not have had children survive to pass on the trait. In either case, the founder effect is responsible for the absence of type B blood in their descendants.

People do not have to leave home for genetic drift to occur. It can happen when members of a small community choose to mate only among themselves, which keeps genetic variants within their group. The Amish people are an extreme example of this type of genetic drift. They have a higher incidence of certain traits than do other populations because people with these traits or who carry the genes for them marry within the group (fig. 20.3). Similarly, some cities, such as Pittsburgh, Pennsylvania, are divided into neighborhoods where residents are genetically more like one another than like others in the city.

Genetic drift is striking in the Dunker community of Germantown, Pennsylvania. The Dunkers left Germany between 1719 and 1729 to settle in the New World. Today, the frequencies of some genotypes are different among the Dunkers than among either their non-Dunker neighbors and/or the people living in their ancestral German village (fig. 20.4). For example, the frequency of type A blood in the United States is 40% and in Germany 45%. Yet among the Dunkers, the frequency of type A blood is 60%. The original settlers included a disproportionate number of individuals with type A or AB blood.

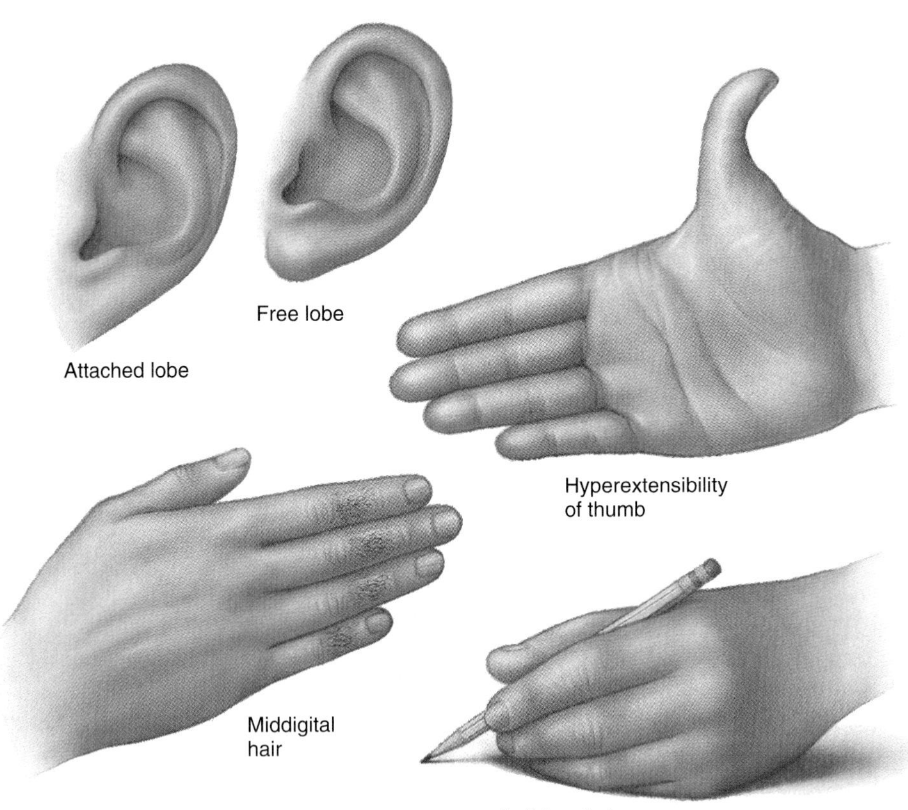

FIGURE 20.4

Four traits that occur in unique frequencies among the Dunkers are attached earlobes, the ability to bend the thumb backwards, hair on the middle of the fingers, and left-handedness.

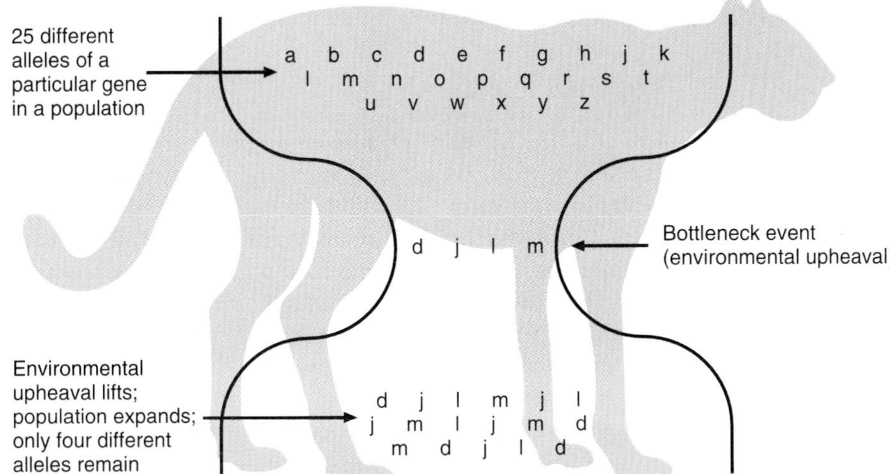

FIGURE 20.5

A population bottleneck occurs when the size of a genetically diverse population drastically falls and a few individuals mate to restore it. The rebuilt population loses some genetic diversity because different alleles are lost in the bottleneck event. The two dwindling cheetah populations in South and East Africa vividly illustrate a population bottleneck. The cheetahs are more genetically alike than mice bred specifically for genetic uniformity. Cheetahs are difficult to breed in zoos because sperm quality is poor and many newborns die—both due to this lack of genetic diversity in the surviving cheetah populations of the world.

Population Bottlenecks

Genetic drift also may result from a **population bottleneck.** This occurs when many members of a population die, and a few remaining individuals mate to restore their numbers. The new population has a much more restricted gene pool than the larger ancestral population.

Population bottlenecks sometimes occur when people (or other animals) colonize islands. An extreme example of this type of genetic drift occurred among the Pingelapese people of the eastern Caroline islands. Between 4 and 10% of these people are born with "Pingelapese blindness," an autosomal recessive combination of color blindness, nearsightedness, and cataracts (clouding of the lens). The prevalence of this otherwise very rare condition traces to a typhoon that decimated the population in 1780. Only nine males survived, and they founded the present-day population. This severe population bottleneck combined with geographical isolation to produce the high frequency of the blindness gene.

The world's cheetahs are currently undergoing a population bottleneck (fig. 20.5). Until 10,000 years go, these cats were prevalent in many areas. Today, just two isolated populations live in South and East Africa, numbering only a few thousand animals between them. Protein sequences from 55 cheetahs are uniform, indicating a population bottleneck. The South African cheetahs are so genetically alike that even unrelated animals can accept skin grafts from each other. Researchers attribute the genetic uniformity

(a)

(b)

FIGURE 20.6

Industrial melanism (pigment adaptations) in the peppered moth. (*a*) At one time, most peppered moths in England were light-colored. Because pale lichens grew on the bark of the trees where the moths rested during daylight hours, light-colored insects blended into the background, whereas darker-colored moths were highly visible to birds that ate them. (*b*) The smoke and soot of the Industrial Revolution killed the lichens and darkened the tree bark in and around cities. In such areas, the numbers of dark-colored moths increased dramatically and the light ones all but disappeared. In the unpolluted countryside, however, where lichens remained on the tree bark, light-colored moths continued to flourish and the dark ones remained rare.

of cheetahs to two bottlenecks—one that occurred at the end of the most recent ice age, when habitats changed drastically, and another when humans slaughtered massive numbers of cheetahs during the nineteenth century.

Mutation

All populations have some alleles that are harmful if homozygous, collectively called the **genetic load.** Harmful, or deleterious, alleles persist in a population for two reasons: they are introduced in new mutations, and they are maintained in heterozygotes, who "carry" the alleles yet can still successfully reproduce. For example, the allele that, when homozygous, causes Tay-Sachs disease is eliminated from a population when affected young children die. Some of their heterozygous siblings, however, keep the allele in the population. They do not develop the illness, yet they may pass the causative allele to their children.

Each person probably has four to eight deleterious recessive alleles. It may take a very long time, however, for mutant alleles to be detectable, because two individuals must carry the same mutant allele to produce a homozygous child, and each of their children has only a 25% chance of inheriting the recessive allele from both parents. Many recessive mutations occurred from radiation released from atomic bombs dropped over Japan in 1945. Geneticists estimate that it will take 30 generations for those mutations to appear in homozygous form. The same is true for radiation-exposed human populations near the Chernobyl nuclear plant that melted down in 1986 (see Biology in Action 16.2).

When mating occurs among blood relatives, the chance of conceiving an individual who is homozygous recessive for a deleterious allele increases dramatically because the parents have a high probability of carrying the same mutant allele, inherited from a shared ancestor. This is why some states forbid first cousins to wed. Because they have one-eighth of their genes in common, each offspring of first cousins faces a 1 in 64 ($1/8 \times 1/8$) chance of inheriting a matched pair of harmful recessive alleles.

Natural Selection

Gene frequencies may change in response to environmental change. For example, populations of approximately 100 insect species have undergone color changes enabling them to blend into polluted backgrounds. This adaptive response, called **industrial melanism,** is a vivid example of natural selection at work (fig. 20.6). The classic example of industrial melanism took place centuries ago in England; it is occurring today in Detroit.

Before the Industrial Revolution in the United Kingdom, dark peppered moths were rare because they were easy prey for birds, who spotted them against the background of tree bark whitened by the growth of lichens. As industrialization spread, air pollution killed the lichens, darkening the tree bark. The dark moths could then hide more easily than the white moths and were thus better able to survive and reproduce. Changes in moth coloration began in the 1700s. By 1956, 95% of the moth population was dark—the reverse of the moths' coloration before the environmental change.

The predominant moth color varies with the degree of industrialization in a particular geographic region. Dark moth populations are most common in Liverpool, a sooty factory town, and are rarely seen in pristine, rural northern Wales. Today, only small pockets of northern England have populations of mostly dark moths—a trend attributed to 20 years of active pollution control.

Types of Natural Selection

Industrial melanism illustrates **directional selection,** in which a changing environment selects against one genotype, allowing another to gradually become more prevalent. Natural selection can have other effects. In **disruptive selection,** two extreme expressions of a trait are the most fit. A population of marine snails, for example, lives among tan rocks encrusted with white barnacles. The snails are either white and camouflaged while near the barnacles or tan and hidden while on the bare rock. The snails that are not white or tan, or that lie against the opposite-colored background, are more often seen and eaten by predatory shorebirds.

In **stabilizing selection,** extreme phenotypes are less adaptive, and an intermediate phenotype has greater survival and reproductive success. Human birth weight illustrates this tendency to stabilize. Newborns who are under 5 pounds (2.27 kilograms) or over 10 pounds (4.54 kilograms) are less likely to survive than babies weighing between 5 and 10 pounds.

Balanced Polymorphism

A form of stabilizing selection called **balanced polymorphism** maintains a genetic disease in a population even though the illness clearly diminishes the fitness of affected individuals. (Polymorphism means genetic variant.) The disease persists because carriers have some health advantage over individuals who have two copies of the wild-type allele. Balanced polymorphism links certain inherited and infectious diseases.

Sickle Cell Disease and Malaria Sickle cell disease is an autosomal recessive disorder that causes anemia, joint pain, a swollen spleen, and frequent, severe infections. Individuals usually do not feel well enough, or live long enough, to reproduce. Sickle cell disease carriers (heterozygotes) are resistant to malaria, which is an infection by any of four species of the protozoan genus *Plasmodium* that causes an agonizing cycle of severe chills and fever. Mosquito bites transmit the malaria parasite to humans.

In 1949, British geneticist Anthony Allison found that the frequency of sickle cell carriers in tropical Africa was unusually high in regions where malaria raged. Blood tests of children hospitalized with malaria revealed that nearly all of them were homozygous for the wild-type sickle cell allele. The few sickle cell carriers among them had the mildest cases of malaria. Was malaria somehow selecting against the wild-type sickle cell allele? The fact that sickle cell disease is far less common in the United States, where malaria is very rare, supports the idea that sickle cell heterozygosity protects against malaria.

Further evidence supports the hypothesis that being a sickle cell carrier in a malaria-ridden environment is an advantage. The rise of sickle cell disease parallels cultivation of crops that provide breeding grounds for the malaria-carrying *Anopheles gambiae* mosquito. About 1000 B.C., sailors from southeast Asia traveled in canoes to East Africa, bringing new crops of bananas, yams, taros, and coconuts. When the jungle was cleared to grow these crops, mosquitoes came, offering a habitat for the malaria parasite in the early part of its life cycle.

When an infected mosquito bites a human, the malaria parasite enters the person's red blood cells; these cells eventually burst, releasing the parasite throughout the body. Something about the red blood cells of sickle cell carriers is inhospitable to the parasite. When malaria first invaded East Africa, sickle cell carriers, who remained healthier, had more children and passed the protective allele to approximately half of them. Gradually, over 35 generations, the frequency of the sickle cell allele in East Africa rose from 0.1% to a spectacular 45%. However, whenever two carriers produced a child who suffered from sickle cell disease—a homozygote—the child paid the price for this genetic protection.

A cycle set in. Settlements with large numbers of sickle cell carriers escaped debilitating malaria. Their residents were therefore strong enough to clear even more land to grow food—and support the disease-bearing mosquitoes. Even today, sickle cell disease is more prevalent in agricultural societies than among people who hunt and gather their food.

Cystic Fibrosis and Cholera The cellular defect that underlies cystic fibrosis (CF) protects against diarrheal diseases such as cholera. Epidemics of cholera have devastated many human populations, causing rampant death in just days. In the summer of 1831, an epidemic killed 10% of the population of St. Louis. In 1994, thousands of people fleeing Rwanda died within days of cholera.

Certain *Vibrio cholerae* bacteria harbor a virus that produces a toxin that causes diarrhea, which rapidly dehydrates the body. A victim can lose his or her body weight in water in a week. Dehydration leads to shock, kidney failure, and heart failure. The toxin opens chloride channels in cells lining the small intestine. As salt (NaCl) leaves the cells, water follows to dilute the salt. Water rushing out of intestinal cells leaves the body as diarrhea, because the large intestine cannot reabsorb the fluid fast enough. The chapter 22 opener vignette discusses how *V. cholerae* become infectious.

In 1989, when researchers identified the CF gene and described its protein product as a chloride channel regulator in certain secretory cells, a possible explanation for the prevalence of CF became apparent. Cholera opens chloride channels, allowing chloride and water to leave cells. The abnormal CF protein does just the opposite; it closes chloride channels, trapping salt and water in cells and drying out mucus and other secretions. A person with CF cannot contract cholera, because the toxin cannot open the chloride channels in the small intestine cells.

Carriers of CF enjoy the mixed blessing of balanced polymorphism. They do not have enough abnormal chloride channels to suffer the labored breathing and clogged pancreas CF causes, but they do have a sufficient defect to prevent cholera toxin from opening chloride channels. During the devastating cholera epidemics that have peppered history, individuals who carried mutant CF alleles had a selective advantage, and they disproportionately transmitted those alleles to future generations. CF endures.

Table 20.2 lists several examples of balanced polymorphism, and figure 20.7 summarizes the forces of evolutionary change.

Mastering Concepts

1. How do nonrandom mating, migration, genetic drift, mutation, and natural selection alter gene frequencies?
2. What is the founder effect?
3. What is a population bottleneck?
4. What two factors maintain deleterious recessive alleles in populations?
5. What are different types of natural selection?

Table 20.2 Balanced Polymorphism

Inherited Disease	Infectious Disease	Possible Mechanism
Cystic fibrosis	Diarrheal disease (cholera)	Carriers have too few chloride channels in intestinal cells, blocking toxin
G6PD deficiency	Malaria	Red blood cells inhospitable to malaria parasite
Phenylketonuria (PKU)	Spontaneous abortion	Excess amino acid (phenylalanine) in carriers inactivates ochratoxin A, a fungal toxin that causes miscarriage
Sickle cell disease	Malaria	Red blood cells inhospitable to malaria parasite
Tay-Sachs disease	Tuberculosis	Unknown

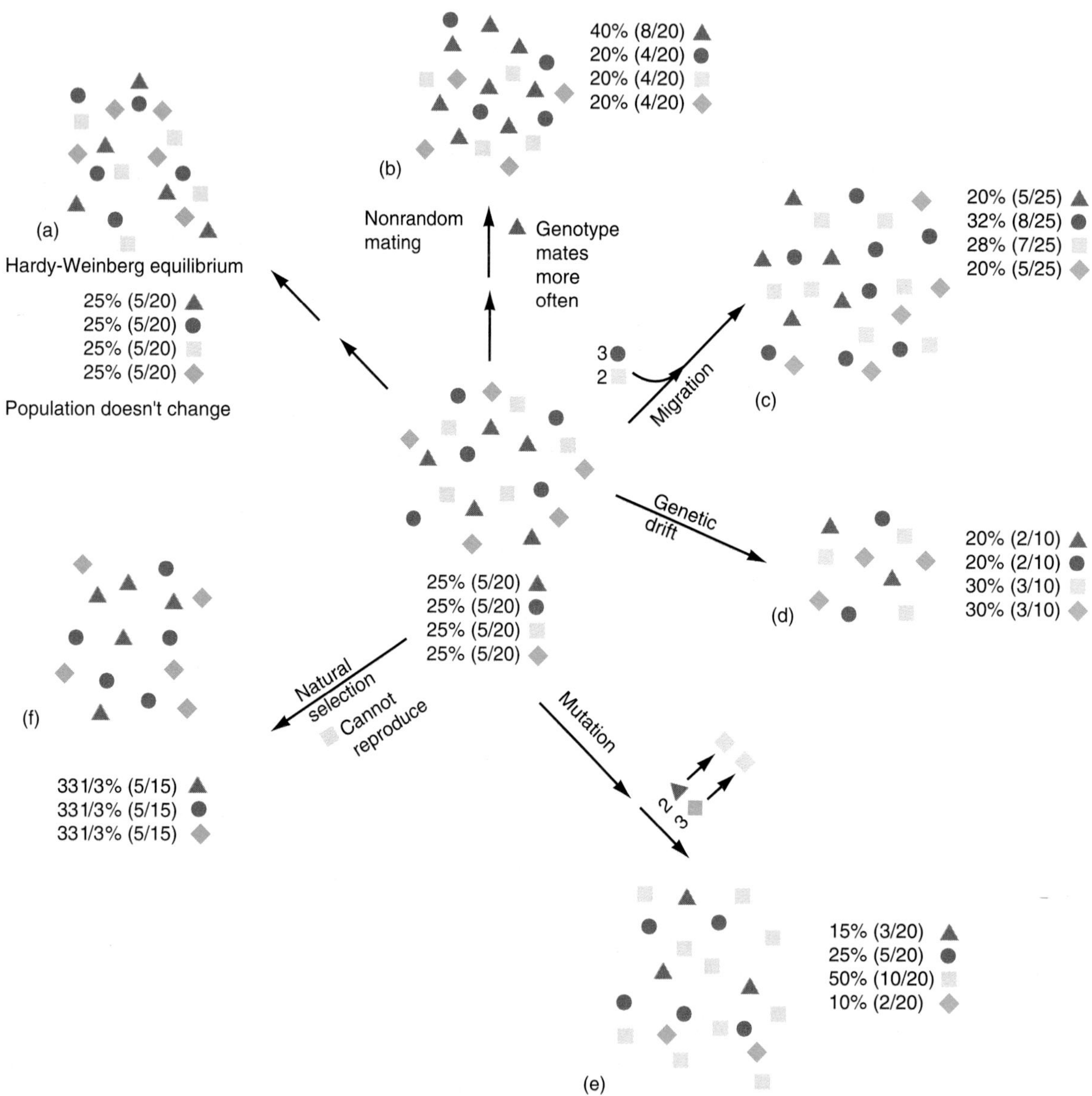

FIGURE 20.7

Several factors alter gene frequencies and thereby contribute to evolution. The different-colored shapes represent individuals with distinctive genotypes. (*a*) In Hardy-Weinberg equilibrium, gene frequencies stay constant. (*b*) Nonrandom mating increases some gene frequencies and decreases others because individuals with certain genotypes mate more often than others. (*c*) Migration removes alleles from or adds alleles to populations. (*d*) Genetic drift samples a portion of a population, altering allele frequencies. (*e*) Mutation changes some alleles into others. (*f*) Natural selection operates when environmental conditions prevent individuals of certain genotypes from reproducing successfully.

How Species Arise

The history of life on earth includes times when many species arose. Bursts of speciation may reflect response to a drastically altered environment, which may have opened new habitats as other species declined, or a profound change in life, such as the acquisition of multicellularity, the ability to use oxygen in the atmosphere, or sexual reproduction (see fig. 17.10). Fossil evidence provides glimpses of some periods of great species diversification. Recall from previous chapters the increasing prevalence of single-celled eukaryotes from 1 to 1.4 billion years ago, the "Cambrian explosion" beginning 550 million years ago that populated the seas, appearance of the first amphibians 360 million years ago, and appearance of the first reptiles 290 million years ago.

Natural selection, genetic drift, migration, nonrandom mating, mutation, and biogeography influence speciation in two stages. First, a population becomes geographically divided, so that members of the two newly formed groups cannot interact. Then, over time, the populations come to have different allele frequencies as the forces of natural selection, nonrandom mating, migration, and mutation act independently in each group.

The bluish gray pupfish, which inhabits a warm spring at the base of a mountain near Death Valley, Nevada, illustrates how geographic isolation contributes to speciation. The summer pupfish population numbers about 600 individuals, and the winter residents number only 200. What makes these minnow-sized fish of evolutionary note is that their home has been isolated from other bodies of water for only 50,000 years. In that time, the genetic makeup of the fish has shifted so greatly that they can no longer mate successfully with pupfish from populations beyond the spring. An entire new species has evolved in what is a mere flicker of geologic time compared to the total time that life has existed on earth.

For a new species to form, geographic isolation must be followed by reproductive isolation—that is, members of the two separated populations must not be able to reproduce successfully with each other if the geographic barrier is lifted. The failure to produce viable offspring can occur before or after mating.

Premating Reproductive Isolation

Premating reproductive isolation mechanisms—obstacles that prevent two individuals from mating—are not always as obvious as the mismatched genitals of the male moose and the female cow. In **ecological isolation** members of two populations prefer to mate in different habitats. In **temporal isolation,** they have different mating seasons. In **behavioral isolation,** the organisms perform different courtship cues.

Two species of toad descended from a single ancestral type illustrate premating reproductive isolation. *Bufo americanus* breeds in the early spring in small, shallow puddles or nearby dry creeks, whereas *Bufo fowleri* breeds in the late spring in large pools and streams. Each type of toad also has a unique mating call. The two are thus ecologically, temporally, and behaviorally isolated.

FIGURE 20.8

Hybrids between two species are usually infertile. The liger, the offspring of a male lion and a female tiger, is not fertile. Only four such animals exist in the United States. The first liger we know of was bred by accident, in 1950, at a circus that undoubtedly kept tigers and lions together. In the wild, these great cats will not normally mate.

Postmating Reproductive Isolation

Sometimes organisms mate but do not produce fertile offspring. The incompatibility can occur as early as fertilization if the genetic material of the two species is packaged into chromosomes differently. A dog's gamete, with 39 chromosomes, could not form a healthy zygote with a cat's gamete, which has 19 chromosomes. A further barrier to dog-cat mating is that the enzymes released from a dog's sperm cannot penetrate a cat's oocyte. The behavior of these two species probably also contributes to their inability to mate.

Rarely, hybrid offspring result when individuals of different species mate, but these offspring tend to be infertile and so have no evolutionary impact. A mule, for example, is an infertile hybrid between a horse and a donkey. Sometimes two animals that do not mate in the wild do so in captivity. Tigers and lions sometimes mate in zoos, but their hybrid offspring are not fertile (fig. 20.8). The same is true for horses and zebras.

Chromosome Incompatibility

Chromosome incompatibility can reproductively isolate individuals of the same species if an event creates subgroups with different organizations of genetic material. This happened in the plant *Clarkia rubicunda,* common along the coast of central California.

A severe drought in the Golden Gate Bridge region in San Francisco nearly decimated the local population of *C. rubicunda.* The only survivors had several chromosomal abnormalities. These plants cross-fertilized among themselves and established a new population in which the chromosomal aberrations were the norm. When the drought ended, *C. rubicunda* plants encroached from

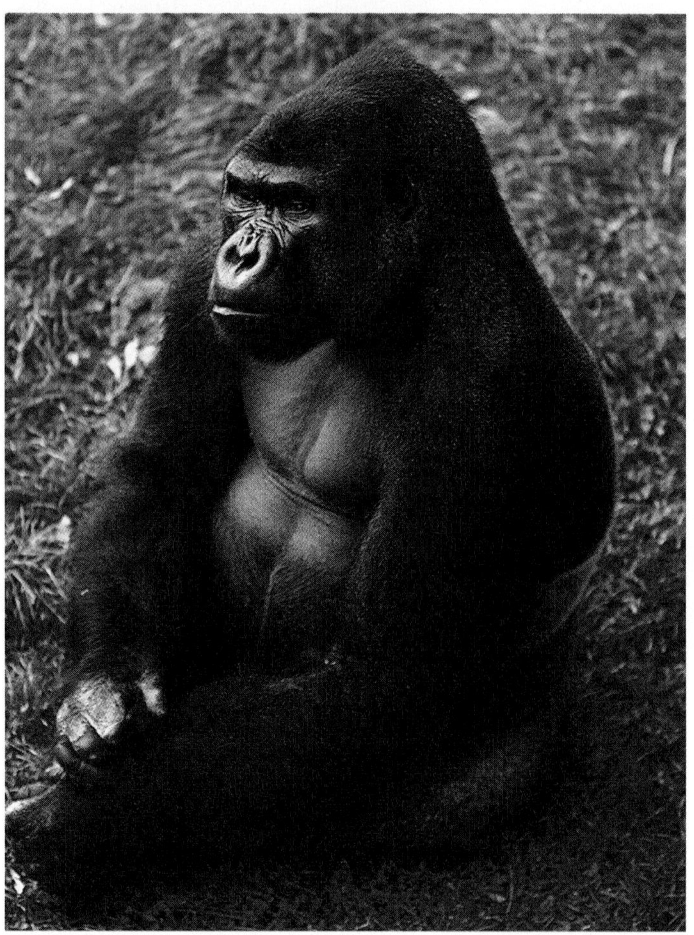

FIGURE 20.9

Speciation in action. The mountain gorilla and western lowland gorilla look alike, but on a molecular level, they are distinct enough to suggest that they might be separate species. Anthropologists are now observing the animals' behavior to see if they have diverged into two reproductively isolated groups. Analysis of DNA in chimpanzee hairs suggests that they too may be diverging to yield a new species.

the surrounding regions, but they could not reproduce with the Golden Gate group. The gametes of the two groups were too dissimilar to unite, although both types of plants descended from the same ancestors. A new species, *C. franciscana,* had arisen.

Polyploidy

Instant reproductive isolation results from **polyploidy,** when the number of chromosome sets increases. Polyploidy can occur when meiosis fails, producing, for example, diploid sex cells in a diploid individual. If diploid sex cells in a plant self-fertilize, a tetraploid individual (with four sets of chromosomes) results.

When extra chromosome sets derive from the same species, the organism is called an **autopolyploid.** Autopolyploids of the rose have 14, 21, 28, 35, 42, or 56 chromosomes—presumably all descended from an ancestral species having 7 chromosomes. **Allopolyploids** form when gametes from two different species fuse. For example, an "Old World" species of cotton has 26 large chromosomes, whereas a species found in Central and South America has 26 small chromosomes. The type of cotton commonly cultivated for cloth is an allopolyploid of the Old World and American types. It has 52 chromosomes—26 large and 26 small.

Polyploidy may occur in the wild in response to unusually low temperatures. Agriculturalists often induce polyploidy intentionally because such plants generally have larger leaves, flowers, and fruits. The drug colchicine, an extract of the autumn crocus plant, induces polyploidy by dismantling the mitotic spindle, which normally aligns and then separates the chromosomes in dividing cells. Many new crop varieties are induced polyploids, including alfalfa, apples, bananas, barley, potatoes, and peanuts.

The genetic differences between a polyploid and the plant it arises from are so great that geographic isolation is not even necessary for speciation. Nearly half of all flowering plant species are natural polyploids, which indicates the importance of this form of reproductive isolation in plant evolution.

Polyploidy rarely appears in animals because the disruption in sex chromosome constitution usually causes infertility. One exception is the grey tree frog *Hyla versicolor,* a tetraploid probably derived from the identical-looking diploid *Hyla chrysoscelis.* Once in a great while, a triploid human infant is born, but massive birth defects usually end its life within days. Certain cells in the healthy human body, however, are normally polyploid. Some liver cells, for example, are octoploid, with eight sets of chromosomes.

Speciation is ongoing—it isn't only a phenomenon of the distant past. Armed with evidence of evolution at the molecular level, researchers are applying this idea to identify species in the process of diverging into new species (fig. 20.9). Biology in Action 20.2 describes speciation occurring in a greenhouse.

Mastering Concepts

1. What two events are necessary for speciation to occur?
2. What are some premating and postmating reproductive isolation mechanisms?
3. How do changes in chromosome distribution underlie evolution?

biology in 20.2 action

Sunflower Speciation in the Laboratory

Plant biologists can watch speciation occur among sunflowers in a greenhouse experiment. The diploid sunflower *Helianthus anomalus,* which lives among sand dunes in Utah and Arizona, arose about 100,000 years ago—a fairly recent arrival. Examination of its chromosomes suggests that *H. anomalus* is a hybrid of two other, much older, species, *H. annuus* and *H. petiolaris* (fig. 20.B). Researchers knew that hybrids form from the two older species in the wild, but the new plants are usually infertile because of chromosome rearrangements that make fertilization and cleavage impossible. A new species could arise, theoretically and naturally, if several hybrid individuals had the same chromosome rearrangements and would therefore be compatible with each other.

To entice hybrid formation and speciation to occur, researchers bred the two ancestral plants—*H. annuus* and *H. petriolaris*—in a greenhouse. They followed three particular hybrid lines, breeding them in various ways with each other and with the parental species. After five generations, the investigators examined the genetic makeup of the hybrids to determine which genes the new hybrids retained. The result: Certain genes, obviously essential, remained from the ancestral species in each hybrid.

The researchers extrapolated from their greenhouse experiment what might have happened in the deserts of the U.S. Southwest: a group of individual plants arose with the same chromosome rearrangements—so they were now the norm and not aberrant—and genes necessary for survival and reproduction remained. This set the stage for speciation.

FIGURE 20.B
Two species of sunflower (top) were crossed to create a third (bottom)—in a greenhouse.

The Pace of Evolution—Slow or Fast?

Charles Darwin thought that evolution occurred as one life-form transformed into another through a series of intermediate stages. The pace was gradual, although not necessarily constant. Many steps of this gradual evolution, or **gradualism,** did not leave fossil evidence, so we do not know of many intermediate or transitional forms. Reasons for the sparse fossil record include poor preservation of biological material, natural forces that destroyed fossils, and the fact that we haven't discovered all there is to discover.

Another explanation for the absence of some transitional forms is that they did not exist. Perhaps some biological changes occurred quickly, due to a single genetic step that had a profound effect on the phenotype.

In 1944, paleontologist George Gaylord Simpson suggested that some of the leaps in the fossil record may actually be true gaps, representing the sudden appearance of a species followed by periods of little change. Simpson estimated that perhaps 10% of all speciation falls into this "quantum evolution" category. In 1944, strict Darwinian gradualists did not accept Simpson's concept of evolution by "leaps and starts."

In 1972, two young paleontologists, Stephen Jay Gould of Harvard University and Niles Eldredge of the American Museum of Natural History, again raised the idea of swift speciation interspersed with long periods of stability (stasis). They came to this conclusion from investigation of the fossil record—Gould for land snails in Bermuda, and Eldredge for trilobites in New York. Gould and Eldredge coined the term **punctuated equilibrium** to reflect the fast evolutionary changes that interrupted long periods of stasis. The fossil record, they claimed, lacks some transitional forms because they never existed.

The fossil record reveals that both punctuated equilibrium and gradualism occur. Consider the fossil records for bryozoa, which are sea-dwelling animals that resemble lacy plants, and cyanobacteria. Bryozoa appear unchanged for millions of years and then "suddenly" (over 100,000 to 200,000 years) split to yield new species that coexist with the original species (fig. 20.10*a*). Yet cyanobacteria appear much the same today as they were a billion years ago (fig. 20.10*b*). They may owe their apparent stasis to being extraordinarily well adapted to environmental changes. Support for this hypothesis is that cyanobacteria have superb DNA repair systems, which might have enabled them to survive environmental disasters that released DNA-damaging radiation.

Bryozoa

Fossil species

Cyanobacteria

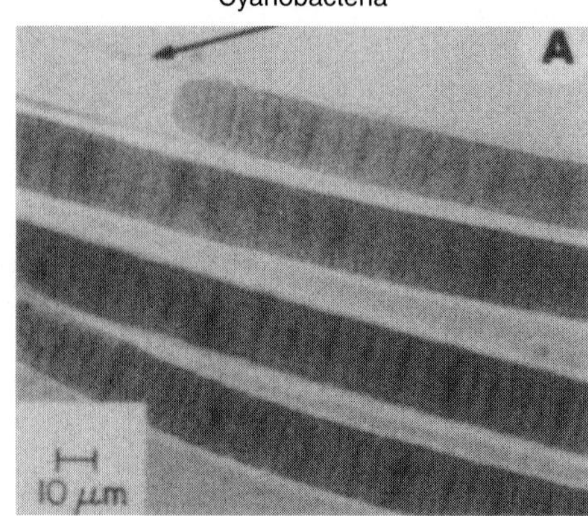

Fossil species

(a) Modern species

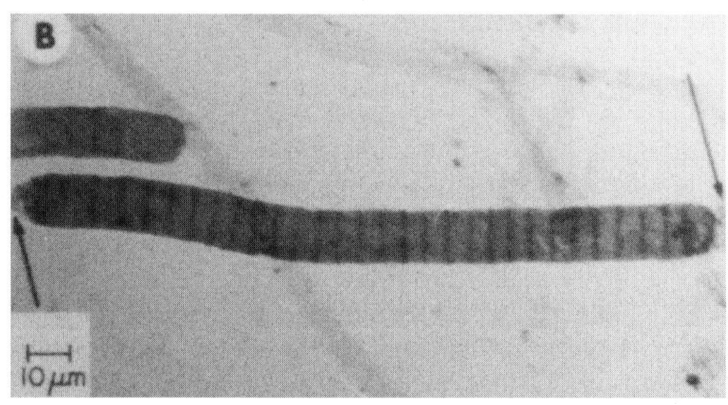

(b) Modern species

FIGURE 20.10

The pace of evolution—stasis and speciation. (*a*) According to the fossil record, bryozoa changed little over millions of years and then suddenly split, forming a new species. As a result, modern species often differ from fossil species. (*b*) In contrast, many species of modern cyanobacteria are remarkably like their ancestors; compare modern *Lyngbya* (*bottom*) to the 95-million-year-old *Palaeolyngbya* (*top*).

Evidence of very rapid evolution supports the concept of punctuated equilibrium. Bacterial resistance to antibiotic drugs, the reemergence of infectious diseases, and industrial melanism and other adaptations to pollution all illustrate evolution occurring fast enough to observe in a human lifetime.

A short generation time is one factor that enables fast genetic changes to occur. Another mechanism of rapid evolution is a single genetic change that greatly alters the anatomy or physiology of an organism. A gene that changes the timing of early developmental events, for example, may cause obvious changes in the adult. An inherited delay in pigmentation in the embryo could greatly change an adult's external color pattern, which in turn could be important if survival depends upon protective coloration. Similarly, a single genetic "switch" altering the timing of cell division could have produced the prolonged brain growth characteristic of our own species.

An unusual laboratory experiment demonstrated that a single genetic change can alter the phenotype in a way that may be of evolutionary significance. By blocking expression of one gene, researchers forced a chicken's foot to develop to resemble a duck's foot (fig. 20.11). A chicken's foot typically is scale-covered and has four distinct digits. In contrast, a duck's foot is webbed and covered with feathers. These differences enable a duck to swim and fly long distances, which a chicken cannot do.

Mastering Concepts

1. How can gradualism and punctuated equilibrium both occur?
2. How can a single genetic change cause rapid evolution?

How Species Become Extinct

Not all species emerge in healthy numbers from population bottlenecks or flourish after geographic isolation—some become extinct. Extinction reflects the inability of organisms to adapt to a particular environmental challenge. It usually occurs over an expanse of time that may seem long to us—a million or so years—but is a brief portion of the total time that life has been present on earth.

Table 20.3 Mass Extinctions

Time	Species Affected	Suggested Cause
3 b.y.a.* (Precambrian)	Anaerobic bacteria	Oxygen in atmosphere
545 m.y.a.† (end of Cambrian)	Trilobites, other marine invertebrates	Meteor impact
440 m.y.a. (Ordovician)	Marine invertebrates	Gondwana formed
370 m.y.a. (end of Devonian)	Most fish and invertebrates	Meteor impact, Gondwana moved, asteroid shower
250 m.y.a. (Permian)	90% of marine species	Pangaea formed
200 m.y.a. (end of Triassic)	75% of marine invertebrates	Meteor impact
140 m.y.a. (end of Jurassic)	Marine species	Not known
90 m.y.a. (mid-Cretaceous)	Dinosaurs	Rise of flowering plants
60 m.y.a. (Cretaceous/Tertiary boundary)	Dinosaurs, marine species	One or more meteor impacts
11,000 years ago	Large mammals	Drought, infectious disease

*Billion years ago.

†Million years ago.

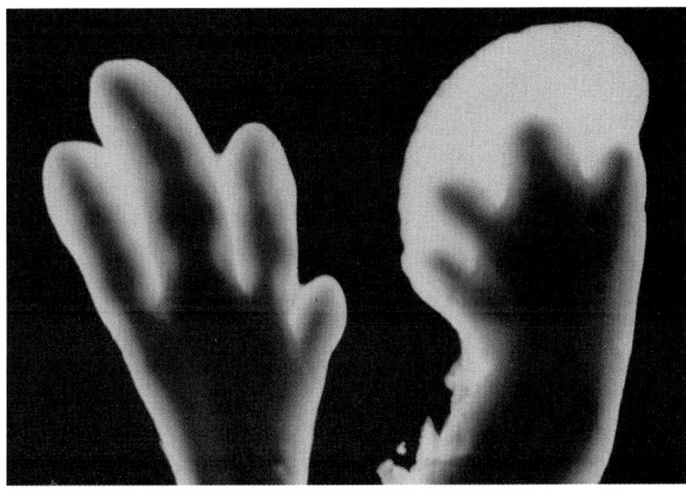

FIGURE 20.11

From chicken legs to duck legs with a single genetic switch. The leg on the left is a normal, untreated chicken leg. The leg on the right, from the same animal, has blocked cell surface receptors that prevent certain cells from receiving the genetic message to die, which would normally carve digits out of webbing. The blocked leg also developed feathers instead of scales. The chicken-to-duck experiment shows that a single genetic change can exert a profound effect on phenotype—one great enough to have influenced evolution.

What Causes Mass Extinctions?

At least a dozen periods of mass extinction mark earth's history (table 20.3). During these times, many species disappeared over relatively short expanses of time. Paleontologists study clues in the earth's sediments to the catastrophic events that contributed to mass extinctions. Two general hypotheses have emerged in recent years to explain these events, although several processes have probably contributed to mass extinctions.

The impact theory suggests that a meteorite or comet crashing to earth sent dust, soot, and other debris into the sky, blocking the sun and setting into motion a deadly chain reaction. Without sunlight, plants, unable to photosynthesize, would die. The animals that eat plants, and the animals that eat those animals, would then die. An extraterrestrial object landing in the ocean would also be devastating because it would mix water layers. Oxygen-poor deeper waters would rise in the turbulence, and upper-dwelling organisms adapted to oxygen in their usual surroundings would die of oxygen starvation.

Evidence for the impact theory includes centimeter-thin layers of earth that are rich in iridium, an element rare on earth but common in meteors. Quartz crystals in iridium deposits are cracked at angles that suggest an explosion. The impact theory of mass extinction may explain why layers of rock unusually devoid of fossils lie near an iridium layer. Scientists have recently found another type of chemical known to form in meteors near iridium deposits—large molecules of carbon called fullerenes.

Alternatively, the restlessness of the planet's rocks may explain some mass extinctions. The geological theory of plate tectonics views the earth's surface as several rigid layers, called *tectonic plates,* that can move, like layers of ice on a lake. These plates continually drift away from oceanic ridges, where new molten rock bubbles forth. Older regions of tectonic plates sink into the earth's interior at huge trenches.

According to the plate tectonics theory, continents that drifted, coalesced, or broke apart thrust profound environmental changes upon life. Suddenly organisms that had thrived in certain habitats were competing with unfamiliar species for limited resources. Weather conditions changed; ice ages and droughts killed many. Shifting continents altered shorelines, diminishing shallow sea areas packed with life. The shrinking habitats of large, meat-eating dinosaurs that accompanied continent formation, described in the chapter 19 opener, vividly illustrate the effects of changing landmasses on biogeography.

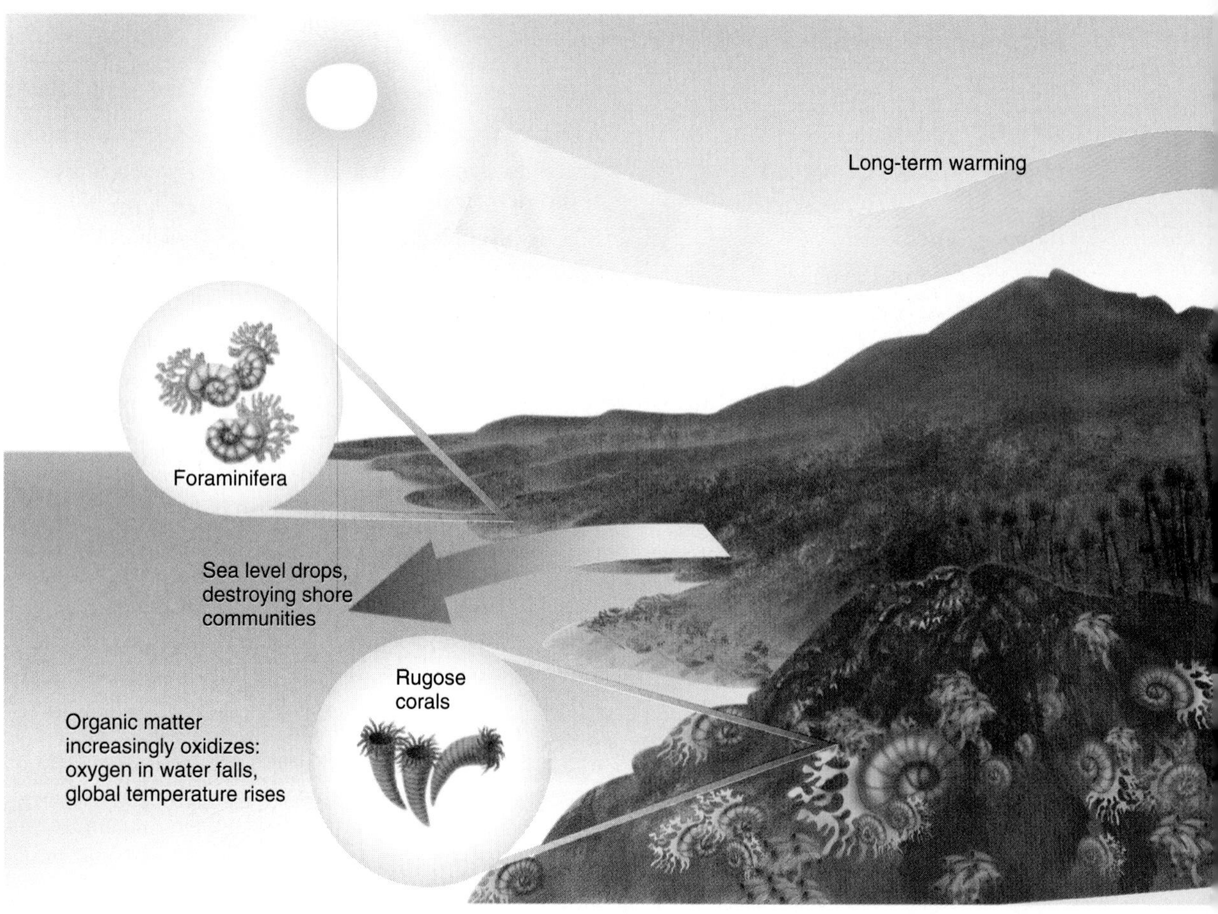

FIGURE 20.12

The Permian extinctions. A series of environmental upheavals probably caused the Permian mass extinctions of 250 million years ago. These changes may have included a drop in sea level, oxygen depletion in the seas, global warming, volcanic eruptions, and rise in sea level.

Mass Extinctions Through Geologic Time

Clues from fossils and biogeography help paleontologists reconstruct scenarios of past mass extinctions. Extinctions probably date to the dawn of life, as progenotes gave way to more efficient successors. After each mass extinction, the survivors repopulated the earth. It is interesting to ponder our own fate had different subsets of species vanished or survived!

An early mass extinction reflected the increase in atmospheric oxygen as photosynthesis arose. The oxygen released was toxic to some anaerobic species, and many died out as the more efficient energy users flourished.

About 440 million years ago, a huge continent, Gondwana, formed and covered the South Pole, causing an ice age and severely disrupting life in the seas. The glaciers drew water from the oceans and destroyed many habitats. Then, 370 million years ago, geological upheaval occurred again. Sulfur-containing minerals in disturbed rock layers in the Canadian Rockies indicate that ancient ocean layers mixed at about this time. Iridium in these layers suggests a possible meteor impact, and rock formations suggest glaciation. Many fish species and nearly 75% of all marine invertebrates perished.

The end of the Permian period, about 250 million years ago, saw what paleontologists call "the mother of mass extinctions." Over about 8 million years, 90% of marine species vanished in several waves of death. Most vulnerable were animals that clung to the seafloor, such as corals, sea lilies, brachiopods, and bryozoa. Abundant fossils of fungal spores suggest that dying plants had freed habitats that fungi then occupied. On the land, 30% of insect orders, and two-thirds of the families of amphibians and reptiles disappeared.

Paleontologists believe the Permian extinctions were the result of a sequence of events (fig. 20.12). The first might have been a drop in sea level, which dried out coastline communities of the one landmass of the time, Pangaea. Next carbon dioxide accumulated in the atmosphere from oxidation of organic molecules, which raised global temperature and depleted oxygen dissolved in water. These events could have devastated sea life. Volcanic eruptions 255 million years ago, lasting a few million years, further altered global climate. Finally, sea level rose, drowning coastline communities.

Forty million years after the Permian extinctions, the environment changed again, and 75% of marine invertebrate species vanished. A crater in Quebec, Canada, roughly half the size of Connecticut, may be evidence of a meteor impact that had devastating repercussions at this time. Alternatively, shifting tectonic plates could have caused these extinctions.

Another meteor impact may be responsible for the mass extinctions that peaked about 65 million years ago, when 75% of all

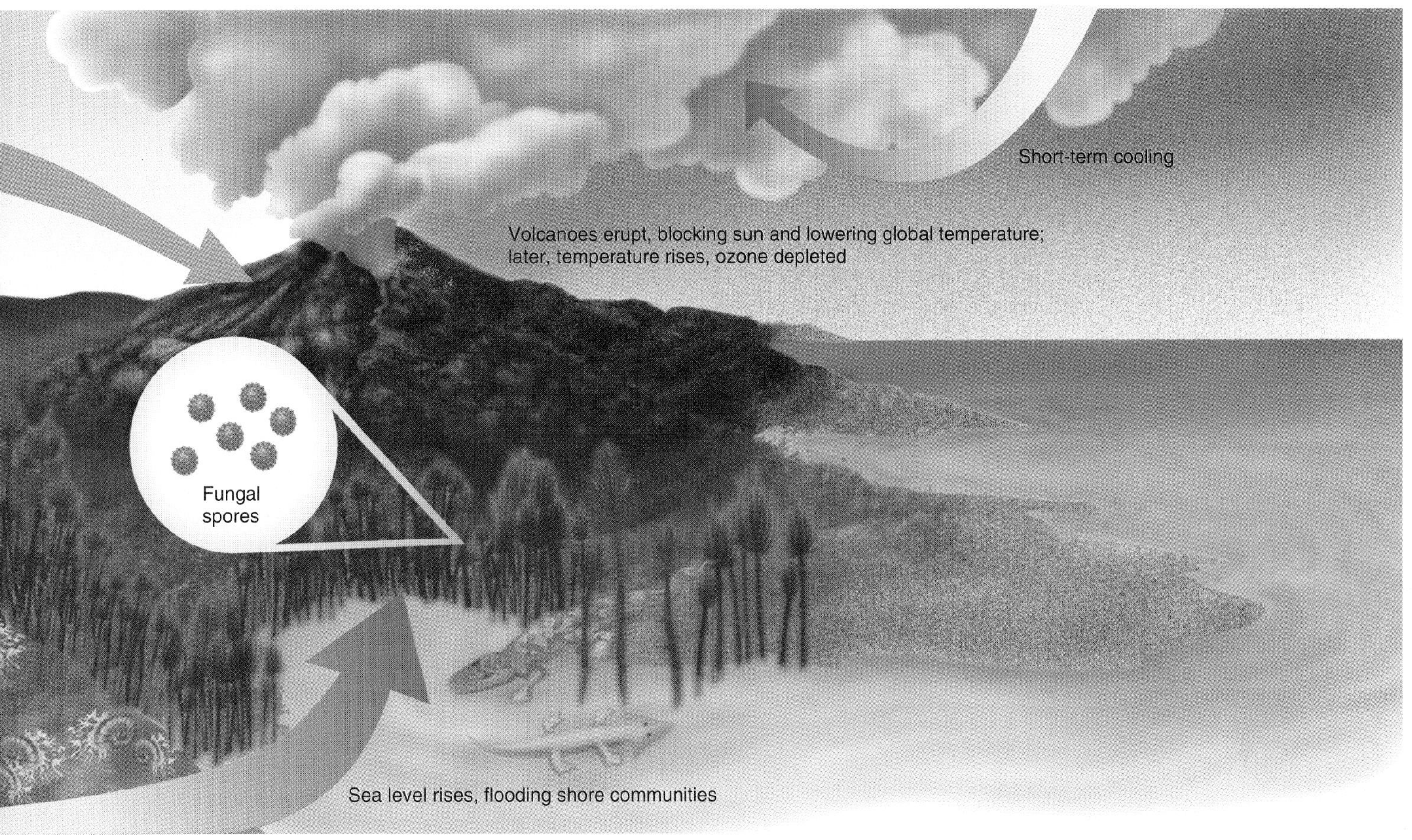
Short-term cooling
Volcanoes erupt, blocking sun and lowering global temperature; later, temperature rises, ozone depleted
Fungal spores
Sea level rises, flooding shore communities

plant and animal species, including dinosaurs, disappeared. In the seas, plankton, which provides microscopic food for many larger marine dwellers, died as well. These extinctions did not all happen at once. Ocean chemistry changes span 2 million years around this time, suggesting a series of impacts rather than a single cataclysmic event.

Extinctions in more modern times have affected species of large animals. Between 10,500 and 11,000 years ago, many such species vanished from the North American plains, including saber-toothed tigers, mastodonts, huge sloths, and giant birds. The Americas lost 135 species that we know of in that short time, nearly all of them animals weighing more than 100 pounds (45 kilograms). Similar vanishings occurred, on a global scale, between 1,000 and 40,000 years ago.

Paleontologists proposed two hypotheses to explain these extinctions. Global warming as the most recent ice age ended was one, but opponents pointed out that large animal species in Europe survived the warming, and that similar extinctions occurred in New Zealand after the ice age and in Australia before it.

The second hypothesis was that humans destroyed habitats and hunted large animals to extinction, based on the observation that spread of humans coincided with the decline in large animal species. Hunting species to extinction was unlikely, because modern hunter-gatherer societies do not kill huge numbers of one type of animal, yet spare others. Collections of mammoth bones thought to be evidence of mass hunting were found to be remains of mammoths that died naturally.

The recent emergence of new and resurging infectious diseases suggests a third scenario for why certain animal species vanished 11,000 years ago. Humans may have transmitted infections that killed the animals, especially when they traveled extensively, bringing diseases to populations that had no immunity.

Like all aspects of evolution, extinctions continue today. Recently extinct species include the 10-foot (3-meter) moa bird, the zebralike quagga, the passenger pigeon, and the dodo bird (see fig. 18.1). Within the next 25 years, a million more species may vanish.

Mastering Concepts

1. How does the impact theory explain mass extinctions?
2. How might plate tectonics account for extinctions?
3. What are some specific events associated with mass extinctions?

The Taming of Tuberculosis~

continued from page 404.

Natural selection, operating on both the bacterial and human populations, explains the decline in virulence of tuberculosis. Some people inherited resistance and passed it to offspring. The most virulent bacteria killed their hosts so quickly that there was no time to pass the microbes to others. As the deadliest bacteria were selected out of the population, and as people who inherited resistance mutations contributed disproportionately to each successive generation, TB's effects on human health grew more benign. The disease evolved from an acute, systemic infection to an increasingly rare chronic lung infection. This was true until the late 1980s.

Over the past decade, a series of unrelated events has enabled TB to resurge in a form resistant to many of the dozen drugs used to treat it. The return of TB is due partly to complacency. Some pharmaceutical companies, believing the disease is conquered because we can treat it, halted programs to find new drug treatments. Some existing drugs worked so fast that patients stopped taking them too soon and unknowingly continued to spread live bacteria for up to 18 months. In the 1950s, patients were isolated in rest homes for a year or longer, which contained the infection.

In the 1980s, the *Mycobacterium tuberculosis* microbes that cause TB survived in patients who stopped taking antibiotics and mutated. Antibiotic-resistant forms arose. Because people with poor health care living in crowded, unsanitary conditions are especially susceptible to infection, TB became alarmingly prevalent and deadly among prison inmates and the homeless in several large cities. Today, one in seven new cases of TB resists treatment.

Another reservoir of new TB infection developed in HIV-infected people. TB develops so quickly in individuals with suppressed immunity that it can kill before physicians have determined which drugs to use—and some physicians have died of the infection while trying to treat patients. An especially frightening aspect of the newly evolved TB is that it is easily spread when an infected person coughs or sneezes. The resurgence of this killer should remind us never to underestimate the fact that evolution operates continually—and unpredictably—in all organisms.

SUMMARY

1. A **population** is a group of organisms that share a geographic area and can mate only among themselves. A **gene pool** includes all the genes in a population. Allele movement between populations is **gene flow.** Inherited characteristics of the individuals in a population reflect allele frequencies.

2. In **Hardy-Weinberg equilibrium,** evolution is not occurring because allele frequencies do not change from generation to generation. In this idealized state, we can calculate the proportion of genotypes and phenotypes in a population by inserting known allele frequencies into an algebraic equation: $p^2 + 2pq + q^2$. The equation also can reveal allele frequency changes when we know the proportion of genotypes in a population.

3. When gene frequencies change, evolution occurs. Nonrandom mating causes certain alleles to predominate. In **genetic drift,** small populations split from larger ancestral populations and establish a new gene pool that includes only a subset of the original alleles. The **founder effect** and **population bottlenecks** are forms of genetic drift. Mutation alters allele frequencies by changing one allele into another. Harmful recessive alleles are selected against in homozygotes, but heterozygotes maintain them and mutation reintroduces them into the gene pool. The deleterious alleles in a population constitute its **genetic load.**

4. Natural selection is a major driving force behind evolution. In **directional selection,** a trait shifts in one direction. In **disruptive selection,** extreme expressions survive at the expense of intermediate forms. In **stabilizing selection,** an intermediate phenotype has an advantage. **Balanced polymorphism** maintains deleterious recessive alleles because heterozygotes are protected against another medical condition.

5. Speciation bursts reflect a drastic environmental change, or a great change in biological strategy. Speciation usually results from changing allele frequencies and geographic isolation of two populations. Eventually, this results in two groups whose members can no longer reproduce successfully with each other.

6. Premating reproductive isolation prevents two individuals from mating due to ecological, temporal, or behavioral differences. Postmating reproductive isolation results when members of the species can mate, but they have incompatible chromosomes. If hybrid offspring of two species survive, they are usually infertile. **Polyploidy** causes rapid speciation in plants by introducing immediate reproductive isolation.

7. Evolutionary change may be gradual (**gradualism**) or include periods of rapid change and stasis (**punctuated equilibrium**).

8. Extinction results when a species is unable to adapt to a changing environment. The earth has had several periods of mass extinction, which may be related to meteor impacts and geological upheavals such as continental shifts.

TO REVIEW . . .

1. Define the following:
  a. population
  b. gene pool
  c. gene flow

2. Explain how the algebraic expression $p^2 + 2pq + q^2$ is used to represent genotypes.

3. How are the founder effect and population bottlenecks genetic drift?

4. Explain an experiment discussed in this chapter or elsewhere in this unit that demonstrates that evolution is occurring.

5. What are two reasons that deleterious recessive alleles persist in populations, even though they prevent individuals from reproducing when present in two copies?

6. How did natural selection alter the colors of moths in populations in the United Kingdom and the virulence of tuberculosis in the United States?

7. Distinguish between directional, disruptive, and stabilizing selection.

8. What is genetic load?

9. What two events are necessary for speciation to occur?

10. Distinguish between premating and postmating mechanisms of reproductive isolation.

11. What two general types of changes can lead to speciation bursts?

12. What is the difference between gradualism and punctuated equilibrium?

13. Why might a speciation burst follow a mass extinction?

14. Give an example of a change in a single gene that can affect evolution.

15. In gene therapy, a functioning gene replaces a defective gene. In what part of an organism would this have to be performed to influence evolution? Why?

TO THINK ABOUT . . .

1. How do the following situations or practices disrupt Hardy-Weinberg equilibrium?

- **a.** People who find out that they are heterozygotes for the same illness decide to not have children together.
- **b.** Several dozen young adults in a large midwestern city discover they are half-siblings. Each was conceived by artificial insemination, with sperm from the same donor.
- **c.** Members of a very close-knit Amish community are forbidden to marry outside the community.
- **d.** A new viral illness kills only people who have a certain blood type.

2. Use the information in this chapter to explain the following:

- **a.** Certain populations of Native Americans lack individuals who have type B blood.
- **b.** Many people among the Cape population in South Africa lose their teeth before age 20.
- **c.** Dwarfism combined with polydactyly and heart defects is more common among the Pennsylvania Amish than in other nearby populations.
- **d.** Cheetah populations are declining.
- **e.** Mongrel dogs are often healthier and live longer than purebreds, which can have characteristic health problems.
- **f.** The Pingelapese people of the Pacific Islands have a very high incidence of a particular type of blindness.

3. A thin-shelled crab can more readily move to escape a predator than can a thick-shelled crab, but it is more vulnerable to predators that drill through the shell. As a result of these opposing forces, shell thickness for many types of crabs has remained within a narrow range, over a long time. What type of natural selection does crab shell thickness illustrate?

4. The fraggles are a population of mythical, mouselike creatures that live in underground tunnels and chambers beneath a large vegetable garden that supplies their food. Of the 100 fraggles in this population, 84 have green fur and 16 have gray fur. A dominant allele F confers green fur and a recessive allele f confers gray fur. Assuming Hardy-Weinberg equilibrium is operating, answer the following questions:

- **a.** What is the frequency of the gray allele f?
- **b.** What is the frequency of the green allele F?
- **c.** How many fraggles are heterozygotes (Ff)?
- **d.** How many fraggles are homozygous recessive (ff)?
- **e.** How many fraggles are homozygous dominant (FF)?

5. One spring, a dust storm blankets the usually green garden of the fraggles in gray. Under these conditions, the green fraggles become very visible to the Gorgs, who tend the gardens and try to kill the fraggles to protect their crops. The gray fraggles, however, blend into the dusty background and find that they can easily steal radishes from the garden. How might this event affect microevolution in this population of fraggles?

6. Which factors contributing to evolution discussed in this chapter do the following science fiction film plots illustrate?

- **a.** In *When Worlds Collide,* the earth is about to be destroyed. One hundred people, chosen for their intelligence and fertility, leave to colonize a new planet.
- **b.** In *The Time Machine,* set in the distant future on earth, one group of people is forced to live on the planet's surface and another group is forced to live in caves. After many years, they look and behave differently. The Morlocks, who live belowground, have dark skin, dark hair, and are very aggressive, whereas the Eloi, who live aboveground, are blond, fair-skinned, and meek.
- **c.** In *Children of the Damned,* genetically identical beings from another planet impregnate all the women in a small town.
- **d.** In *The War of the Worlds,* Martians cannot survive on earth because they are vulnerable to infection by terrestrial microbes.

7. When the American Kennel Club formed more than a century ago, its stated goal was "to do everything to advance the study, breeding, exhibiting, running and maintenance of purity of thoroughbred dogs." How is the idea of controlled breeding to emphasize certain traits genetically unwise?

8. Biotechnology companies that develop new drugs are using an approach called "directed molecular evolution." From a large batch of molecules, researchers repeatedly select those that perform an activity of interest. They then use those molecules to derive a new batch of molecules. How does this procedure differ from evolution by natural selection?

TO LEARN MORE . . .

References and Resources*

Carroll, Robert L. May 2, 1996. Revealing the patterns of macroevolution. *Nature,* vol. 381. Macroevolutionary change can result from a series of microevolutionary changes.

Coyne, Jerry. May 3, 1996. Speciation in action. *Science,* vol. 272. Researchers re-create sunflower evolution in a greenhouse.

Erwin, Douglas E. July 1996. The mother of mass extinctions. *Scientific American.* A series of environmental disasters 250 million years ago drove most of earth's resident species into extinction.

Kettlewell, B. 1973. *The evolution of melanism.* Oxford: Clarendon. Protective coloration in moths vividly displays natural selection at work.

Stebbins, G. Ledyard, and Francisco J. Ayala. August 28, 1981. Is a new evolutionary synthesis necessary? *Science,* vol. 213. A classic paper placing rapid evolution in the framework of traditional Darwinian theory.

Yoon, Carol Kaesuk. November 12, 1996. Parallel plots in classic of evolution. *The New York Times.* The tale of peppered moth populations changing color to adapt to pollution occurred in Detroit as well as in England.

Zimmer, Carl. July 1995. Carriers of extinction. *Discover.* Extinction of many species of large mammals 11,000 years ago may have been due to infectious diseases spread by humans.

Zou, Hongyan, and Lee Niswander. May 3, 1996. BMP signaling in interdigital apoptosis and scale formation. *Science,* vol. 272. A single genetic switch can make the difference between a chicken's digits and scales and a duck's webbing and feathers.

**Life's* website features frequently updated references.

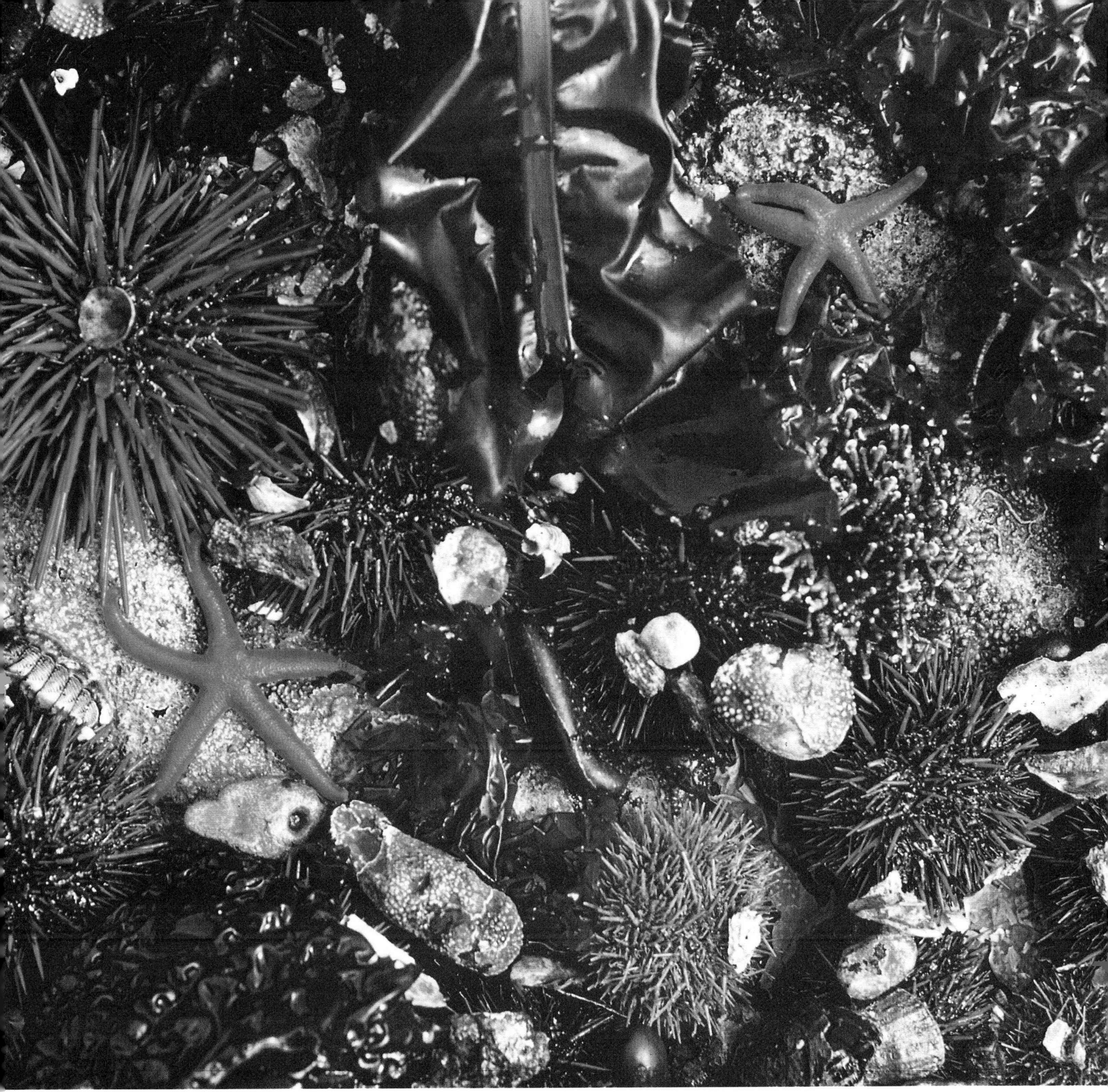

Many species inhabit this tide pool in the Pacific Northwest.

Unit 6

The Diversity of Life

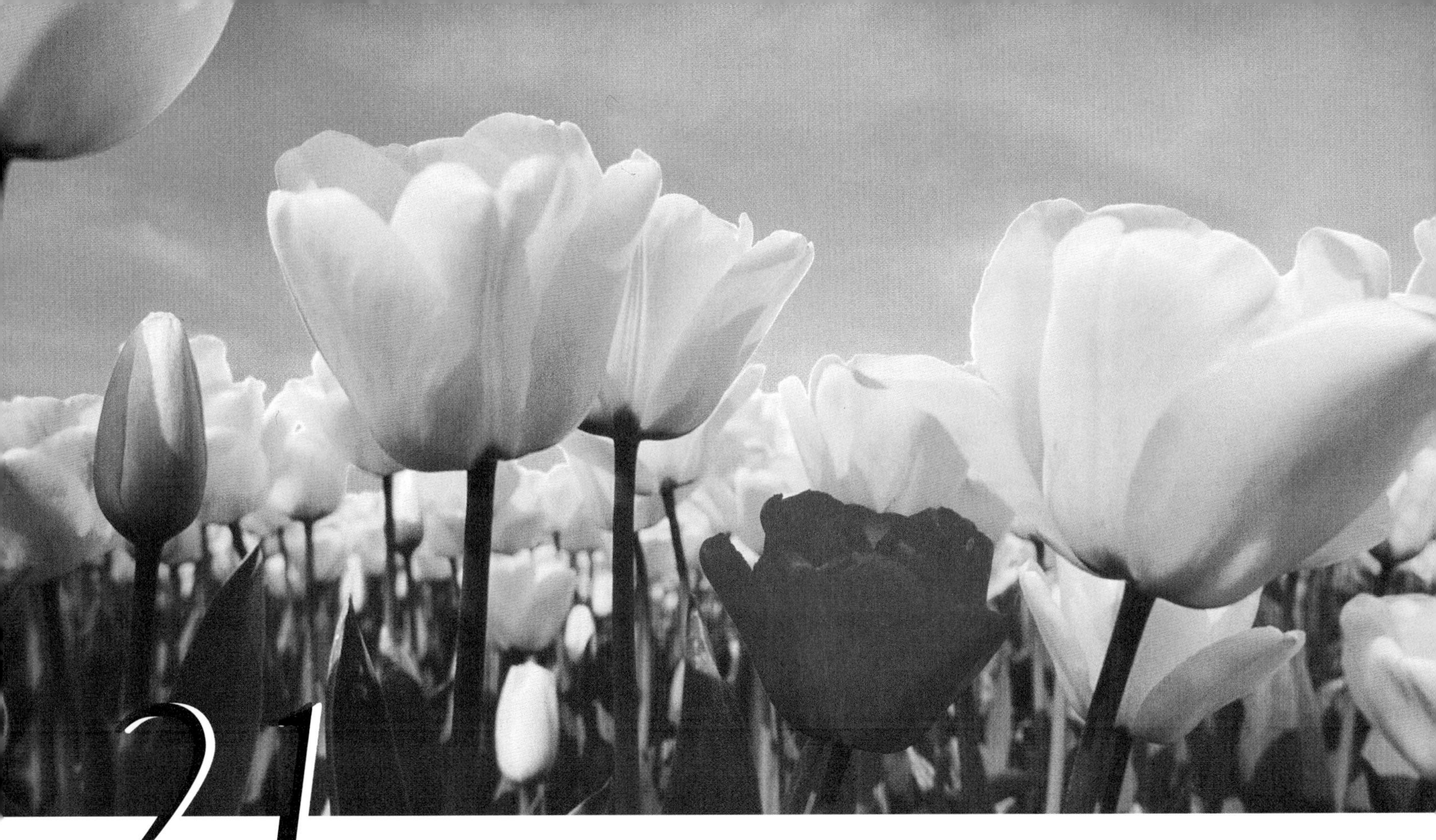

These tulips belong to Kingdom Plantae.

21 Classifying Life

chapter outline

On the Meaning of Names 426

Taxonomy Then and Now 427

Characteristics to Consider 428

The Evolution of Taxonomy 429
The History of the Kingdom Concept
The Introduction of Domains
More Restrictive Taxonomic Levels
Challenges of Taxonomy

How Many Species Live on Earth? 435
Estimating Biodiversity

Taxonomy in Turmoil

Most people think that a guinea pig is a rodent. But some life scientists maintain that the guinea pig, *Cavia porcellus,* belongs to a different group that includes 17 other types of South American small mammals.

The problem in placing guinea pigs within the vast tree of life arose when researchers began looking beyond obvious physical traits to consider DNA and protein sequences as characteristics useful in defining species. A study published in the journal *Nature* in 1991, entitled "Is the Guinea Pig a Rodent?", opened the debate. It compared 15 gene and protein sequences among various mammals and concluded that members of order Rodentia apparently descend from two types of ancestors. By definition, members of an order descend from a single ancestor. Some researchers questioned the 1991 study because it traced only 15 molecules.

In 1996, a report appeared in *Nature* from a different research group, with the title "The Guinea Pig Is Not a Rodent." These investigators sequenced all 16,000 bases of the mitochondrial DNA for 16 types of mammals, including guinea pigs, mice, rats, cows, humans, opossums, gray seals, horses, whales, and hedgehogs. Evolutionary tree diagrams clearly placed guinea pigs on a separate branch from rats and mice.

Yet other researchers disagreed. Although the study compared more DNA sequences, it lacked breadth in another way.

to be continued . . .

On the Meaning of Names

Most names have meanings. A surname may reflect its owner's occupation, such as Taylor, Fishman, or Cook. An unusual moniker, such as Chastity, America, or Dweezil, might reflect a parental whim. Names may also mirror marital history—witness soap opera queen Erica Kane Martin Brent Cudahy Montgomery Chandler Montgomery Marick Marick.

Biological names have meaning. A binomial name—genus and species, such as *Homo sapiens*—abbreviates a much longer name that summarizes an organism's past. A human's full taxonomic name is Animalia Chordata Mammalia Primates Hominidae *Homo sapiens,* which corresponds to the increasingly restrictive classification groups of kingdom, phylum or division, class, order, family, and the familiar genus and species. The name not only identifies a distinct type, as names generally do, but tells how a type of organism is related to others.

Consider the gray squirrel. It is closely related to humans, and its full taxonomic name reflects this. A squirrel is Animalia Chordata Mammalia Rodentia Sciuridae *Sciurus carolinensis.* This means that a squirrel is a member of the animal kingdom; it belongs to a large group of animals called chordates that share key embryonic characteristics; and it is a mammal, with its trademark fur and milk secretion. The categories of Rodentia Sciuridae *Sciurus* and *carolinensis* refer to more specific characteristics that further describe and distinguish this organism—features that humans do not share with it.

Table 21.1 Taxonomic Classifications of Four Animals

Category	Human	Gorilla	Squirrel	Katydid
Kingdom	Animalia	Animalia	Animalia	Animalia
Phylum	Chordata	Chordata	Chordata	Arthropoda
Class	Mammalia	Mammalia	Mammalia	Insecta
Order	Primates	Primates	Rodentia	Orthoptera
Family	Hominidae	Pongidae	Sciuridae	Tettigoniidae
Genus	*Homo*	*Gorilla*	*Sciurus*	*Scudderia*
Species	*sapiens*	*gorilla*	*carolinensis*	*furcata*

FIGURE 21.1

The Tamandua anteater of Venezuela has many distinguishing characteristics. The organization of its organ systems, its highly developed brain, separate sexes, mammary glands, and internal fertilization define it as a mammal. Its lack of teeth place it among order Edentata—the toothless mammals. The animal also has short, coarse hair and a long, thin tail that it uses to manipulate things, and it eats termites and ants in the treetops.

Biological names also reveal how we think types of organisms are related to one another in time. The more taxonomic levels two organisms have in common, the more recently it is hypothesized they diverged from a shared ancestor. Thus, a taxonomic name is shorthand for an organism's past. The full taxonomic names listed in table 21.1 indicate that of these organisms, humans are most closely related to gorillas, then squirrels, then katydids. We are far less closely related to bacteria, amoebae, mushrooms, and trees, which belong to different kingdoms. Ultimately, all organisms share a common ancestor—some ancient, single-celled type of organism that persisted, changed, and diversified.

Taxonomy Then and Now

Biologists have classified organisms for centuries. Until recently, designations depended on easily observed similarities and differences between types of organisms (fig. 21.1). Many classifications today include gene and protein sequence data, as the chapter opener on guinea pig classification illustrates.

Our ability to classify organisms depends, to an extent, on technology. Before the invention of microscopes, people thought the earth was inhabited only by what they could see—plants and animals. Microscopes revealed the vast world of microorganisms. Today, computers help taxonomists get information much faster than was possible in the past, enabling them to more quickly, and more accurately, classify newly discovered organisms.

For centuries, biologists kept meticulous records of biodiversity, from the long lists of plants, called "floras," published in the seventeenth century to today's field guides to birds, wildflowers, and insects. Computerized databases enable a biologist to obtain information on a species' habitat and location, its complete taxonomic name, and its distinguishing characteristics and references to experts and literature, with a few keystrokes (fig. 21.2). In the past, a query about the location and habits of a rare bird might have required months of searching libraries, letter writing, telephoning, and travel.

Botanical gardens, zoos, and natural history museums have been entering information on their collections into computerized databases since the 1970s, but the task is daunting, because of the size of collections. The Smithsonian National Museum of Natural History, for example, has so far listed only 21 million of its 120 million species specimens.

Biologists classify organisms to sort out and order a vast amount of information and also to provide a basis for hypotheses about evolutionary relationships among species, which is a broader field called **systematics.** The field of biodiversity is an outgrowth of systematics and considers the different types of organisms occupying an area.

Classifying organisms has practical uses too. For example, researchers combine taxonomic information indicating species relationships with ecological data to identify endangered species and those that might become endangered. Decisions on

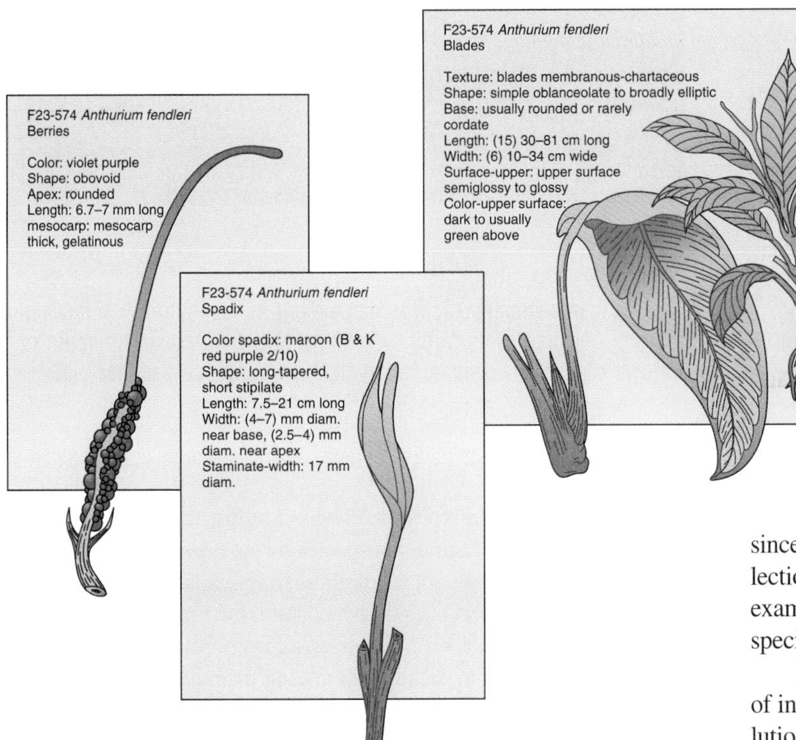

FIGURE 21.2

Computers help taxonomists classify organisms. To access information about a plant species, all a botanist need do is press a few computer keys. Field exploration and studying museum specimens, however, are still vital to taxonomic research.

21.1 biology in action

Taxonomic Finds

In the Wild

The Peruvian wildlife preserve called Tambopata is only 2 square miles (approximately 5 square kilometers) in area, yet it is home to perhaps the most diverse assortment of organisms anywhere on earth. Its seven types of forests contain a huge number of species of plants, birds, and insects. Yet Tambopata reveals only a slice of the diversity of life in the tropical rain forest, a part of the world where half of the known living species of plants and animals now reside. Three million species populate the treetops alone.

The Peruvian forest canopy teems with insects. Entomologists, life scientists specializing in insects, are trying to describe and classify the millions of previously unseen species. Areas 13 yards (approximately 12 meters) square are roped off and draped with cone-shaped nets. Then the researchers release an insecticide fog into the area, and insects fall into the nets and are funneled into collecting bottles. The fog kills only insects, and insects from surrounding regions replace them in about 10 days. The "rain" of insects from a single fogging usually numbers about 10,000 specimens. The collectors transport the specimens to laboratories and museums, where they are meticulously probed and information on their distinguishing characteristics filed in computer databases.

In the Market

For a select few of the half million fish displayed in New York City's Fulton Fish Market each day, the final destination isn't a restaurant but a display case at the American Museum of Natural History (fig. 21.A). The fish come to the market from the Gulf Coast, the West Coast, South America, Africa, the North Atlantic, the Caribbean, the Mediterranean, and the Pacific. Among the familiar and perhaps not-so-familiar animals, a sharp-eyed ichthyologist (fish expert) can spot a rare or even unknown species.

Language among fish market taxonomists is symptomatic of problems in terminology in the discipline in general—too many names for the same organism. What an ichthyologist recognizes as a red snapper (*Lutjanus campechanus*), for example, is a red porgy to the fish wholesaler, a pink snapper to a restaurant owner, and a white snapper to the person who finally eats it.

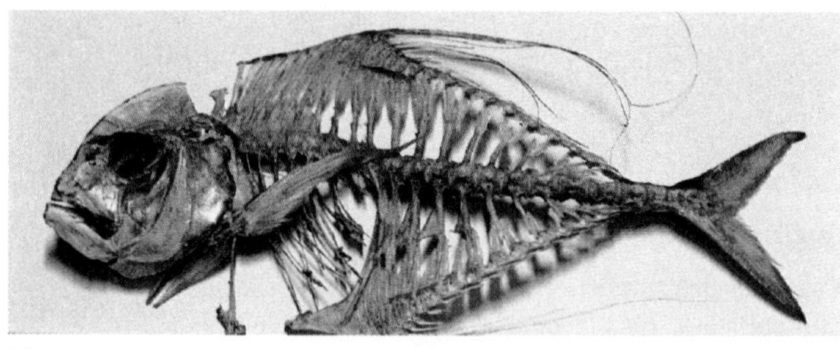

(1)

(2)

FIGURE 21.A

This African pompany (*1*) was discovered at the Fulton Fish Market (*2*) and because of its rarity, wound up at the American Museum of Natural History on display, rather than on someone's dinner plate.

where to build a housing development, or which land to leave wild, are based on such predictions of vulnerable species.

Despite the value of computers and molecular analyses, discoveries of life's diversity still rely very much on work in the outdoors. Organisms never before cataloged are still found in unexpected places (Biology in Action 21.1).

Characteristics to Consider

The task of classifying, in a meaningful way, the 500 million existing species, as well as the many millions now extinct, is daunting. Through history, taxonomists have considered many types of characteristics to distinguish organisms, from skeletal organization and eye position to toes per hoof or number of seed leaves. An ornithologist might classify a bird by nest-building behavior, song, feather color and pattern, and types of flowers it pollinates. An algae expert (phycologist) might distinguish species by the types of pigments they produce, and a microbiologist might distinguish microorganisms by types of cellular appendages. Erect posture and brain size are among the characteristics that set humans apart from other primates. Table 21.2 lists traits used to distinguish among primates.

Many biologists have traditionally separated the three groups of familiar multicellular organisms—plants, fungi, and animals—by the way they obtain nutrients. An evolutionary biologist attempting to estimate when such broad groups of organisms shared a common ancestor might consider morphological (form) and behavioral characteristics, fossil evidence, and molecular sequence data for several genes and proteins, sometimes fleshed out with ecological and geological information. The resulting taxonomies are information-packed tools that try to explain the past, present, and future of life on earth.

Table 21.2 Some Characteristics Used to Classify Primates

Characteristic	Human	Orangutan	Chimpanzee	Gorilla	Gibbon
Short snout	Yes	Yes	Yes	Yes	Yes
Trapezoid-shaped nostrils	Yes	Yes	Yes	Yes	No
Thick molar enamel	Yes	Yes	No	No	No
Long hair	Yes	Yes	No	No	No
Wide-set nipples	Yes	Yes	No	No	No
Mating throughout menstrual cycle	Yes	Yes	No	No	No
Knuckle-walking	No	No	Yes	Yes	No

Mastering Concepts

1. What is taxonomy?
2. How can taxonomy be subjective?
3. What is systematics?
4. What types of traits do taxonomists consider in classifying organisms?

The Evolution of Taxonomy

Noting similarities and differences among organisms has been a popular pursuit since Aristotle assembled his "great chain of being" in the third century B.C. Because life scientists are always discovering never-before-seen organisms—even major new types of organisms—and because we cannot be certain that we know all of life's diversity, taxonomy is a very active field.

The History of the Kingdom Concept

From the time of Aristotle through the mid-nineteenth century, biologists recognized two major groups of organisms—plants (kingdom **Plantae**) and animals (kingdom **Animalia**) (fig. 21.3*a*). Organisms that are neither plant nor animal, such as fungi, were classified into whichever kingdom they most closely resembled. Since then, several paradigms have suggested as few as one, and as many as thirteen, kingdoms of life.

Swedish botanist Carolus Linnaeus originated the idea of classifying organisms according to a hierarchy of increasingly restrictive levels, or taxa. Such a system would ensure that researchers in different parts of the world could communicate. Linnaeus recognized the kingdoms Plantae and Animalia, but as biologists studied more species, they noted that all did not fit the definition of plant or animal, particularly microscopic life. To account for the growing number of exceptions, biologists proposed at least four taxonomic schemes between 1860 and 1866, each including a third kingdom for microscopic life-forms.

German naturalist Ernst Haeckel proposed the predominant three-kingdom taxonomy in 1866 (fig. 21.3*b*). His third kingdom, **Protista,** included all unicellular life and lumped the prokaryotes and single-celled eukaryotes together, although prokaryotic and eukaryotic cells are vastly different. Haeckel's three-kingdom scheme also did not distinguish the fungi as a separate kingdom and instead placed them among the plants or the protista, depending upon whether they were composed of one cell or many. Ironically, molecular data place the fungi closer to animals than to either plants or protista.

The latter half of the nineteenth century was a pivotal time in taxonomy and in biology in general. Discoveries of many fossils revealed a parade of life that preceded modern organisms. Charles Darwin's theory of natural selection provided a framework for explaining how life changed over long periods of time. Gregor Mendel's lucid descriptions of inheritance, although not appreciated until the next century, helped to explain how natural selection might occur. As scientists realized that biodiversity had changed over the ages, the biological science of taxonomy, at first a mere cataloging of contemporary life, acquired an evolutionary flavor. A species name became far more than a name—it implied a history.

Perhaps with these developments in mind, ecologist Robert Whittaker proposed a four-kingdom system in 1959 (fig. 21.3*c*). This scheme recognized plants, fungi, animals, and protista, still lumping together prokaryotes and single-celled eukaryotes. An alternate view

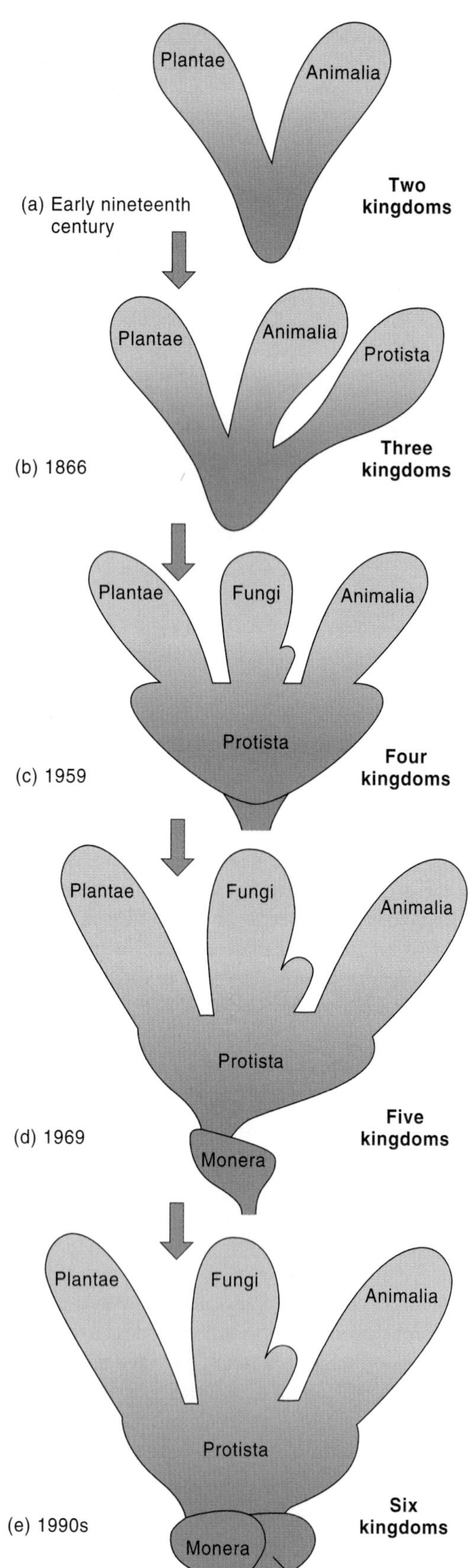

FIGURE 21.3

Evolution of the biological kingdom concept. For centuries, people thought organisms were either plants or animals. The two-kingdom scheme in (*a*) classes fungi, bacteria, and protista with plants. Between 1860 and 1866, several biologists suggested three-kingdom schemes (*b*). Haeckel's popular conception separated protista from plants and animals but lumped prokaryotes and unicellular eukaryotes together. In 1959, Whittaker outlined a four-kingdom system that placed fungi in their own kingdom, distinguishing plants, animals, and fungi by their distinct modes of obtaining nutrition (*c*). By 1969, he added a kingdom, distinguishing between the prokaryotic monera and eukaryotic protista (*d*). The discovery of archaea prompted proposals of six-kingdom schemes, even though the evolutionary relationship between Monera and Archaea is not yet completely known (*e*).

at the time, proposed by American biologist Herbert F. Copeland, recognized plants and animals and separated the **Monera** (prokaryotic unicellular life) from the **Protoctista** (nucleated cells that are not plant or animal). In Copeland's plan, fungi belonged to Protoctista, along with protozoa and red and brown algae.

In 1969, Whittaker devised a five-kingdom plan that separated prokaryotes and unicellular eukaryotes into Monera and Protista (fig. 21.3*d*). Whittaker's five kingdoms—Monera, Protista, Plantae, Fungi, and Animalia—combined his four-kingdom scheme with Copeland's distinction between prokaryotes and unicellular eukaryotes. More recently, some taxonomists suggested a sixth kingdom to include the archaea, organisms that do not fit existing kingdom categories (fig. 21.3*e*). Discovery of these organisms stimulated a major rethinking of taxonomic designations and systems.

The Introduction of Domains

In the late 1970s, biologists began to describe a type of microorganism not seen before. The archaea, at first called archaebacteria, are similar to prokaryotes in cellular structure because they lack nuclei. The structures and sequences of certain molecules in these microbes are as different from those in bacteria as they are from those in eukaryotes, however, and in some cases they actually more closely resemble molecules in eukaryotes. In other words, the archaea do not fit into any kingdom in the five-kingdom scheme. They are not prokaryotes or eukaryotes, but a third form of life.

Where should taxonomists place the archaea? Some biologists include them with Monera, although they are quite different from bacteria. Placing them in a sixth kingdom is not a satisfying solution either, because their differences are greater than those that distinguish members of other kingdoms from each other. In 1990, researchers Carl Woese, Otto Kandler, and Mark Wheelis took a bold step—if the life-forms couldn't fit the prevailing taxonomy, then alter the taxonomy! So they grouped the kingdoms into three broader taxa, the **domains** of **Bacteria, Archaea,** and **Eukarya.** Biologists distinguish organisms within the domains by cell complexity, sequences of ribosomal RNA molecules, and the types of lipids that make up cell membranes. Researchers often add new molecular criteria.

Before the discovery of archaea, a two-domain system—prokaryotes and eukaryotes—might have sufficed. A three-domain system, however, accounts for the differences that set the archaea

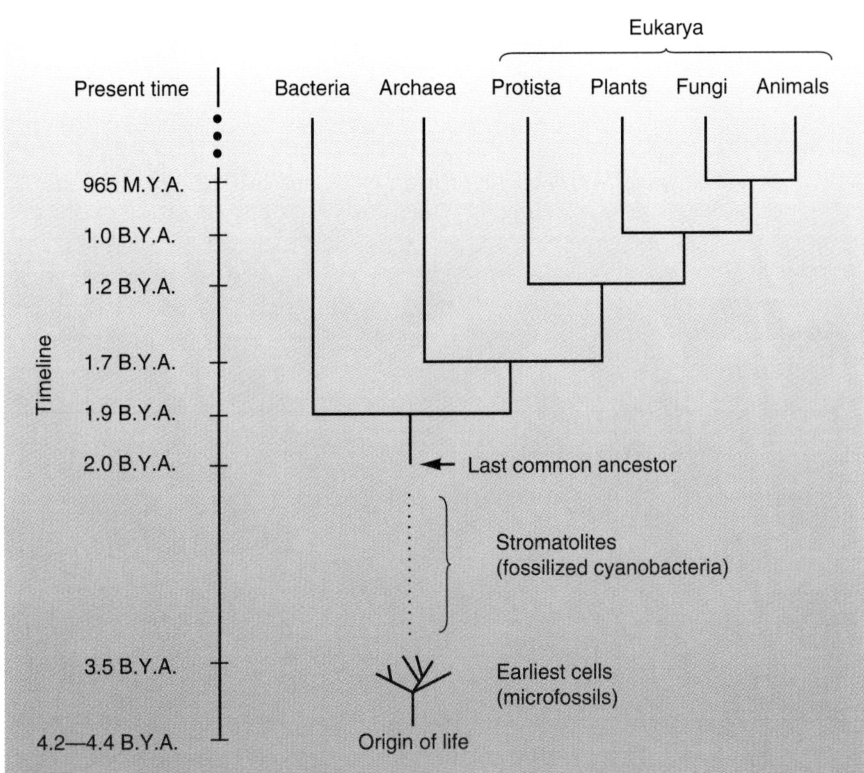

FIGURE 21.4

A tree of life showing possible evolutionary relationships among the major groups of organisms. Life scientists recognize three domains (Bacteria, Archaea, and Eukarya) and either five or six kingdoms. The six-kingdom scheme counts Archaea as a kingdom. The time scale is based on fossil evidence, and on evidence from protein amino acid sequence differences among modern species. The precise evolutionary relationship between Bacteria, Archaea, and Eukarya is still unclear. Based on molecular evidence, the Archaea and Eukarya are more closely related to each other than either is to the Bacteria.

Table 21.3 Human (*Homo sapiens*) Taxonomy

Category	Name	Description
Kingdom	Animalia	Complex cells; multicellular; nervous tissue
Phylum	Chordata	Body consisting of head, trunk, and tail; highly developed organ systems; three tissue layers in embryo; internal cavity; notochord
Class	Mammalia	Hair; mammary glands; internal fertilization; large skull; homeotherm or endotherm (warm-blooded); extraembryonic membranes
Order	Primates	Complex brain; flexible toes and fingers; excellent vision
Family	Hominidae	Upright posture; small face; large brain; V-shaped jaw
Genus	*Homo*	Large brain; relatively short arms; lightweight jaws; small teeth; large thumbs
Species	*Homo sapiens*	Only living species of genus *Homo*

apart from prokaryotes and eukaryotes. Studies of highly conserved ribosomal RNA molecules and proteins among members of the three domains suggest that the bacteria first diverged from a common ancestor of archaea and eukarya nearly 1.9 billion years ago. Divergence between archaea and eukarya then occurred about 1.2 billion years ago (fig. 21.4). The most ancient bacteria for which we have microfossil evidence are approximately 3.5 billion years old.

Many biologists today recognize the three domains but also adhere to the long-accepted five-kingdom scheme. This presents a contradiction, however—the archaea, considered distinct enough to warrant a separate domain, have no place in the five-kingdom system, which places them with the bacteria, which they may only superficially resemble. The domain system is still evolving—don't be surprised if terms change. For example, the term *archaea* implies that these organisms appeared before all others, and the term *prokaryote* implies that these organisms preceded eukaryotes. We do not yet know enough about the origins of these three basic types of organisms to determine which came first. It will be interesting to see how taxonomy shifts to accommodate new information on the archaea.

The three-domain system considers mostly subcellular and molecular differences between species—differences we cannot easily see. Much work needs to be done, however, to reclassify already known organisms in the three-domain system. The following chapters use the five-kingdom system, which includes more traditional macroscopic criteria and integrates molecular data whenever possible.

Table 21.4 Corn (*Zea mays*) Taxonomy

Category	Name	Description
Kingdom	Plantae	Land-dwelling, multicellular, eukaryotic organisms with cellulose cell walls; photosynthesize using chlorophyll a and b
Division	Anthophyta	Vascular plants with seeds and flowers; ovules enclosed in an ovary and mature seeds in fruits
Class	Monocotyledoneae	One seed-leaf
Order	Commelinales	Fibrous leaves
Family	Poaceae	Grasses
Genus	*Zea*	Separate male and female flowers
Species	*Zea mays*	Corn

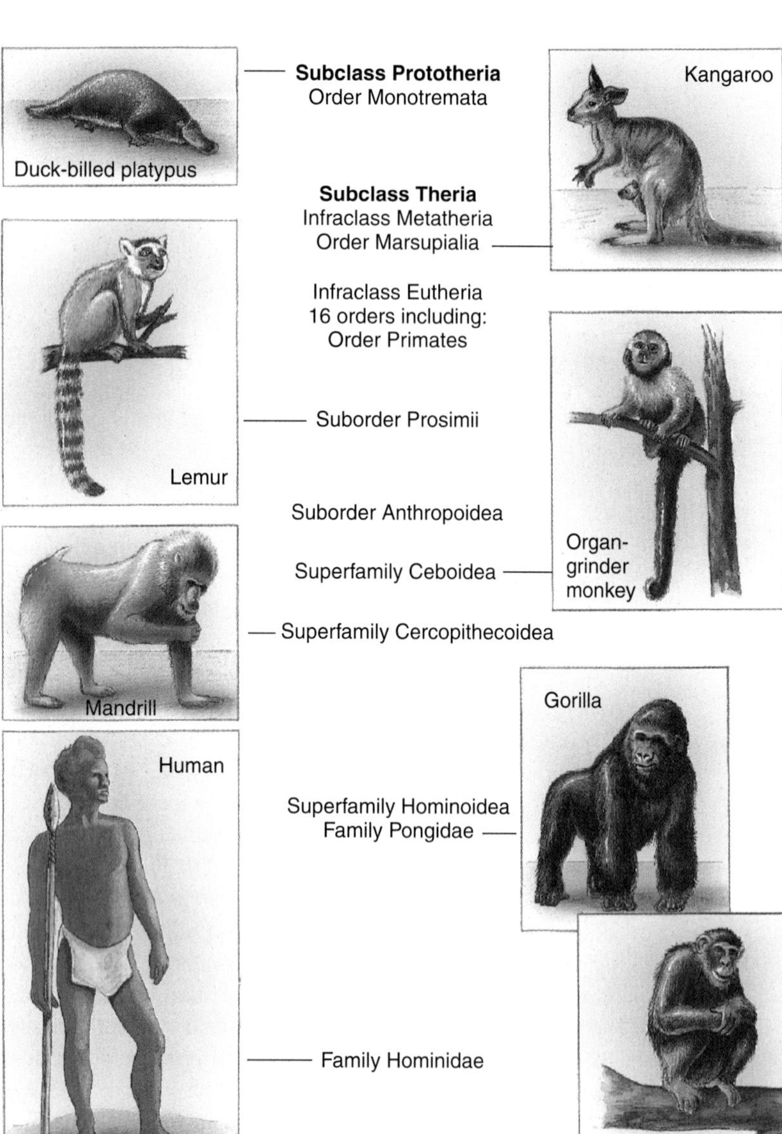

FIGURE 21.5
Taxonomic classification of humans.

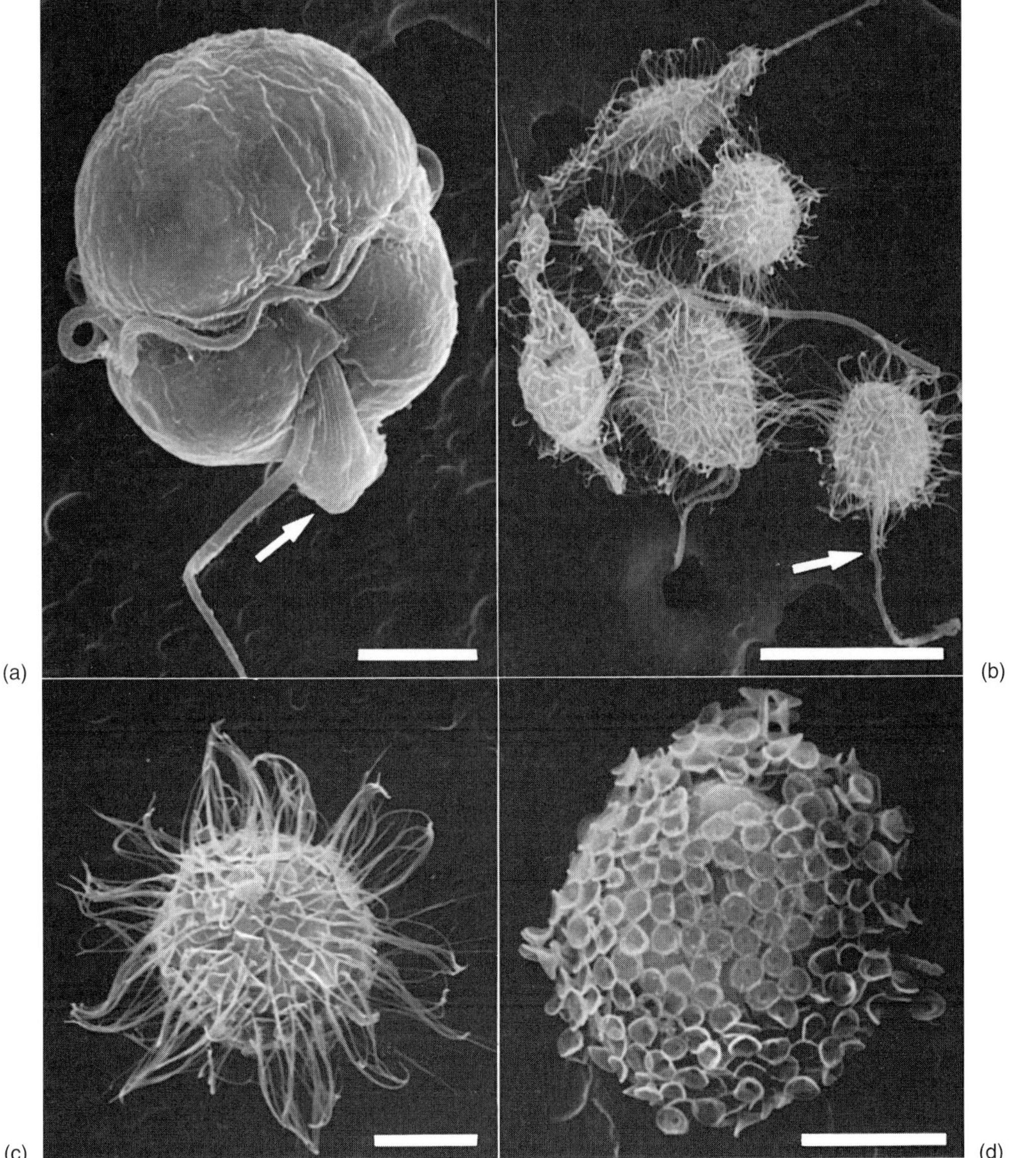

FIGURE 21.6

Describing and classifying a newly found species is especially difficult when it undergoes several stages in its life cycle. This dinoflagellate alga, *Pfiesteria piscimorte*, kills and eats fish in mid-Atlantic estuaries (areas where seawater meets fresh water). The algae in the photographs were observed in an aquarium in the presence of dying fish. At first, each alga swells and extends a structure that sucks up dead fish (*a;* size bar 3 μm). Two hours later (*b*), when all the fish are dead, the algae form cysts (size bar 10 μm). The arrow points to an individual with its feeding tube extended. A day later (*c*), all of the algae settle into cysts with bristlelike extensions (size bar 3 μm). A month later, the cysts remain, minus the bristles (*d;* size bar 5 μm). The algae kill fish by secreting a neurotoxin (nerve poison).

More Restrictive Taxonomic Levels

Chapter 1 briefly outlined the taxa that make up an organism's full name. After domains and kingdoms, plants are classified into divisions, and the other types of organisms are classified into phyla. Then follow classes, orders, families, genera, and species. The more closely related two species are, the more taxa they share, as table 21.1 illustrates for humans, gorillas, squirrels, and katydids. Tables 21.3 and 21.4 list the full taxonomic names of two very different organisms, humans and corn, and figure 21.5 classifies humans. Note that some taxonomic levels are further subdivided, such as into subclass and superfamily.

Challenges of Taxonomy

Because taxonomy is subjective, taxonomists often disagree over classification schemes and the placement of particular organisms. So-called "splitters" perceive many differences among related kinds of organisms and thus assign them to many taxonomic groups; "lumpers" see broad commonalities and assign the same organisms to fewer groups. A lumper might classify all finch species in a single group, whereas a splitter might discern many subgroups among these birds.

A good illustration of the limitations of our powers of observation in classifying life is when an organism's appearance differs

biology in action 21.2

Symbiosis

Life exists in and on other life. These symbiotic relationships can sometimes complicate a biologist's ability to catalog life. Determining what each member of a symbiotic pair gains from the relationship—if anything—can reveal fascinating facts about the individual organisms.

Mutualism

In the type of symbiosis called **mutualism,** both partners benefit; in some cases, one might not be able to survive without the other. On sea bottoms, hermit crabs often carry sea anemones on the snail shells they inhabit (fig. 21.B). The anemones protect the crabs by stinging hungry potential predators, such as octopuses. Meanwhile, the crabs transport the otherwise immobile sea anemones, taking them along as they scavenge for food.

FIGURE 21.B
This hermit crab lives within an abandoned whelk shell. Several sea anemones cling to the shell. The anemones protect the crab while obtaining transportation to new food sources.

FIGURE 21.C
Painted in the 1500s, Pieter Bruegel's *Triumph of Death* portrayed the unimaginable horror of bubonic plague, which from 1347 to 1351 killed one in three Europeans in certain regions.

Biologists distinguish termite species by the types of microorganisms that live in the insects' intestines. These symbiont types can number into the hundreds. Were it not for these resident protozoa and bacteria, the insects could not digest the cellulose in their wood food. The microbes gain a habitat with plentiful food.

Commensalism

In **commensalism,** one member of the pair benefits without affecting the other. The human body, for example, hosts many microorganisms that do not threaten our health.

In precolonial times, Native Americans used a commensal relationship between a turtle and a small animal called a suckerfish to catch the turtle. They would tie the suckerfish onto a string and hurl

markedly during its life cycle or between the sexes. A caterpillar and the butterfly it metamorphoses into, for example, appear to be two different types of organisms. The alga *Pfiesteria piscimorte,* which lives on the mid-Atlantic coast, has at least 15 different-appearing life stages! Figure 21.6 shows a few of its guises. At different times, the organism photosynthesizes like a plant and swims and attacks fish like an animal. Yet it is a protist. Biology in Action 21.2 discusses how species' interactions can impede taxonomic efforts—when one species lives on or in another, a biological relationship called **symbiosis.**

Another problem in biological classification is how to weigh the importance of different characteristics in determining species relatedness. Different criteria can sometimes lead to different conclusions. For example, consider the question of which of the apes is most closely related to humans. That is, which ape shares the most recent common ancestor with us? If we consider only gene and protein sequences, our closest relative is unquestionably the chimpanzee. If we consider the bone hardening pattern as the sole criterion for classification, we are equally related to chimpanzees, gorillas, orangutans, and gibbons. By yet another measure—analyzing sexual structures and behaviors—we are most closely related to the orangutans.

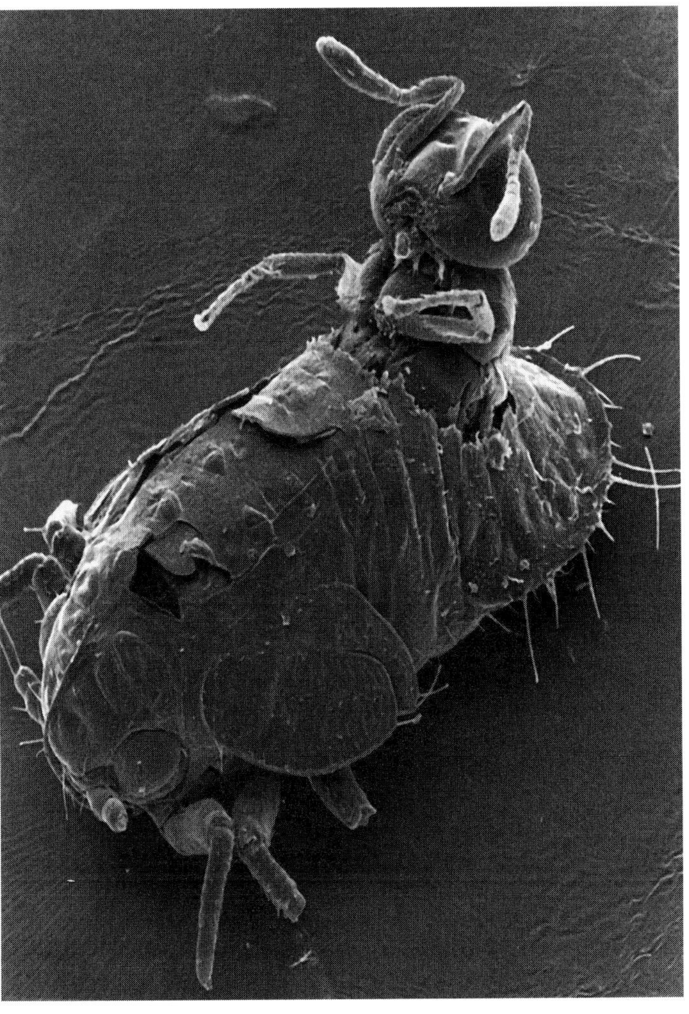

FIGURE 21.D

The encyrtid wasp emerging from this corpse of a fruit tree pest called psylla is a parasitoid that feeds on its host, eventually killing it from within.

it into a stream. When the suckerfish stuck to a turtle to eat scraps from the turtle's meal, the people pulled the turtle in and then ate it.

Parasitism

In **parasitism,** one organism—the parasite—derives benefit from and harms the other—the host. Usually the host does not die.

In humans, parasitic organisms cause such varied ills as athlete's foot, acne, tuberculosis, strep throat, malaria, and sleeping sickness. Many societies have suffered greatly from diseases caused by parasites. The most devastating epidemics were of the "black death," or bubonic plague (fig. 21.C). It killed a third of the population of Europe (40 million people) from 1338 to 1348 and swept the European continent in five waves before the end of the fourteenth century. Another outbreak of bubonic plague in London in 1665 killed 70,000 people, and at the end of the nineteenth century, the disease killed more than half a million people a year in India.

Toxins produced by the bacterium *Yersinia pestis* cause plague symptoms. People contract the bacteria from the bites of fleas that are carried on rats. The three types of bubonic plague have primary symptoms of swollen lymph nodes (called buboes), infected lungs, or infection in the blood. Untreated, bubonic plague kills in days.

Sometimes a microorganism that is usually commensalistic becomes parasitic when the host's immune system is damaged and cannot regulate the size of the microorganism population. The result is an opportunistic infection. These infections are often one of the early signs of AIDS.

A **parasitoid** benefits from the host but kills it. Figure 21.D vividly shows a parasitoid, an encyrtid wasp, emerging from its host. The wasp injects its egg into an immature form of psylla, an insect that causes great damage in fruit orchards by draining the sap from fruit tree leaves. The wasp larva hatches from the egg and eats the psylla host surrounding it. Just before the gorged larva secretes a cocoon about itself, the psylla dies, leaving a protective shell for the parasitoid. A few weeks later, the adult wasp chews its way out of its former host.

Mastering Concepts

1. Why were some taxonomic schemes used in the past inadequate?
2. What is the justification for adding the domain taxon to supercede kingdoms?
3. How do taxa reflect the closeness of evolutionary relationships?
4. What are some of the problems encountered in classifying organisms?

How Many Species Live on Earth?

Charles Darwin wrote in his classic *On the Origin of Species* that in nature there is "no limit to the amount of profitable diversification of structure and therefore no limit to the numbers of species which might be produced." Despite this vast proclamation, many biologists are attempting to quantify life's diversity.

Biologists estimate that as many as 30 million species inhabit the earth. Perhaps because of our perspective as one of those organisms, our lists tend to include many birds and mammals (revealing our bias toward "feathers and fur") but relatively few

Taxonomy in Turmoil~

continued from page 426.

Taxonomists raised several objections to using mitochondrial DNA sequence comparisons to remove guinea pigs from order Rodentia. Many researchers point to abundant evidence from fossils, modern anatomy and physiology, and details of embryonic development that place the guinea pigs with other rodents. Traditional classification of the animals considers the shapes of their skulls, muscles in their jaws that enable them to chew with their molars while they gnaw with their incisors, enamel on the front of the incisors so these teeth can grow fast and stay sharp, and distinctive fetal membranes.

The new placement of guinea pigs may also be flawed on a statistical basis—it reflects too small a sampling. About 2,000 modern species belong to order Rodentia. They are grouped into three suborders, with guinea pigs in a group with porcupines. The study using molecular evidence to separate guinea pigs from Rodentia looked at only 3 species among those 2,000—a little like judging the academic abilities of a high school graduating class of 2,000 students from the grade point averages of just 3.

The taxonomic status of guinea pigs is probably not of earth-shattering importance to most people. The animals will continue to be delicacies in South America, pets in North America, and valuable laboratory subjects throughout the world. But the debate over the types of evidence to use in classifying organisms is widespread.

Shortly before the latest chapter in guinea pig origins appeared in *Nature,* another paper, also citing molecular data, claimed that rabbits and rodents, long considered to be closely related orders within the "superorder" Glires, are not nearly as close as we had thought. Other studies use molecular data to "split" taxonomic designations. For example, some ornithologists now suggest that molecular evidence supports the existence of 20,000 recognized bird species, twice the number defined by such features as feather pattern and beak size.

The importance of molecular sequence data depends upon how we define species. A long-accepted definition is that of a "biological" species, a group of interbreeding individuals that are reproductively isolated from others. In contrast is a "phylogenetic" species, defined as the smallest group of organisms descended from a common ancestor. Visible characteristics are used as criteria for a biological species; gene and protein sequences provide information to classify organisms phylogenetically.

Perhaps researchers both in the field and in the biochemistry lab can find a way to embrace both types of evidence—the macroscopic and microscopic—and develop a frame of reference, an overall perspective, so that trait distinctions have more meaning.

Table 21.5 Known and Estimated Numbers of Microbial Species

Group	Known Species	Estimated Total Species	Percentage of Known Species
Viruses	5,000	130,000	4%
Bacteria	4,760	40,000	12%
Fungi	69,000	1,500,000	5%
Algae	40,000	60,000	67%
Protozoa	30,800	100,000	31%

bacteria, fungi, and animals lacking backbones (invertebrates). Human subjectivity in choosing organisms for study may be one reason it took so long for biologists to recognize the archaea. Microbiologists tended to use only a few types of bacteria in experiments, sometimes erroneously assuming that these organisms represented all prokaryotes. Imagine how a bacterium, if it could, might categorize life!

Estimating Biodiversity

It would be impossible to systematically canvas the entire earth and record every type of organism. To bring the problem of estimating biodiversity to a workable level, field biologists select a site and collect specimens from it. The collectors determine the percentage of specimens of unknown species, extrapolate, and then estimate the total number of such species in the world.

In a small area of tropical rain forest in Sulawesi, Indonesia, for example, researchers collected 1,690 species of "true bugs," insects called hemipterans. Of these, 37% were known species. Next, the researchers extrapolated to consider all 900,000 known species of these insects. If these 900,000 represent 37% of all true bugs, then there must actually be about 2 to 3 million true bug species (900,000/0.37) in the world. Similar calculations led to the species numbers listed in table 21.5.

Extrapolating estimates of species number is very imprecise. The calculations assume that the percent of known species in one sample represents the proportion among larger populations. A second source of error is the subjectivity of taxonomy itself, for the reasons discussed earlier in the chapter. Some differently named species may actually be the same organism at different stages of its life cycle. Conversely, different species may be considered to be the same if we cannot detect their distinguishing characteristics.

The spectrum of species that biologists attempt to catalog today differs markedly from what they might have encountered 10 million years ago, or 2 billion years ago, and also probably differs from what the spectrum will include millions of years in the future. The next five chapters examine the classification and diversity of life on earth today.

Mastering Concepts

1. Why are our estimates of species numbers subjective and rough estimates?
2. How do taxonomists extrapolate species' numbers?

SUMMARY

1. Taxonomy is the classification of life-forms into groups, or taxa. Taxonomic names reflect what we know about an organism's evolutionary relationships. The more closely related two species, the more taxa they share.

2. Taxonomic designations are based on similarities between species, including visible characteristics, behavior, microscopic features, and sequences of genes and proteins. Computers assist taxonomists by rapidly supplying information. Field work and museums are still important parts of taxonomy.

3. Taxonomy has changed as technology enabled biologists to discern more intricate details of cellular and molecular structure, anatomy and physiology, and behavior. The initial two-kingdom system that classified all organisms as plants or animals expanded to a three-kingdom plan recognizing fungi and then to various four-kingdom schemes, some acknowledging the great difference between prokaryotes and eukaryotes. The five-kingdom system, classifying life-forms as plants, animals, fungi, protista, and monera, takes both of these changes into account. Archaea are usually considered to be a part of Monera or a sixth kingdom.

4. A system proposing three **domains** divides life into **Archaea, Bacteria,** and **Eukarya.** The domain system contradicts the five-kingdom system because it includes a separate grouping for the archaea.

5. Taxa more restrictive than kingdom include division or phylum, class, order, family, genus, and species.

6. Taxonomy is subjective because biologists perceive different degrees of difference between types of organisms and may consider different classification criteria with different weights. Classification is confounded when an organism looks very different at different stages of its life cycle, when the sexes look very different, or if it takes part in a **symbiotic** relationship.

7. Taxonomists' attempts to number all species on earth are biased by observing only certain species. Biologists estimate the number of species by extrapolating from the percentage of new species in small samples.

TO REVIEW . . .

1. Taxonomy of scientists: What types of organisms do the following types of life scientists study?
- **a.** ichthyologist
- **b.** phycologist
- **c.** ornithologist
- **d.** entomologist
- **e.** microbiologist
- **f.** botanist

2. What types of information do taxonomic designations reveal?

3. Cite two ways technology has influenced taxonomy.

4. Cite the limitations of taxonomic schemes based on two, three, and four kingdoms. Why is the five-kingdom system probably insufficient to account for all life that we know of on earth?

5. How did Whittaker and Copeland's four-kingdom taxonomic schemes differ?

6. What is the reason for proposing three domains to supercede kingdoms?

7. Indicate whether each of the following biological relationships illustrates mutualism, commensalism, or parasitism:
- **a.** Bluejays locate mounds of earth that are formic ant nests. These insects squirt formic acid from glands in their abdomens to protect themselves. A bluejay grabs a formic ant and tucks it under its wing. The ant releases formic acid, which kills feather mites that disturb the bird.
- **b.** Fleas live on dogs, where they bite the skin and consume blood.
- **c.** The cattle egret is a large bird that follows herds of zebra, gazelles, and antelopes. As the mammals move, grasshoppers pop up, and the birds eat them. (In North America, similar birds follow tractors for the same reason). Birds sense predators before the mammals do, and screech and flap their wings noisily.
- **d.** The oxpecker is a bird that lives in Asia and Africa. With its claws, the bird latches onto a warthog or buffalo, where it digs out ticks and drinks the host's blood. If a predator nears, the bird yells. If the warthog or buffalo ignores the bird, the bird hits the host's head with its beak.

TO THINK ABOUT . . .

1. Taxonomy once dealt only with the past and the present. How might it address the future?

2. How do the five-kingdom system of biological classification and the three-domain system conflict?

3. If molecular data conflicted with physical characteristics used to classify two species, which type of evidence would you support? Why?

4. How is the tree of life depicted in figure 21.4 based on both molecular and fossil evidence?

5. What limits biologists in estimating the number of species of a particular type of organism?

6. Imagine you are an organism from another planet landing on earth. What criteria might you use to classify the abundant life-forms you encounter if you land in a tropical rain forest? in the arctic?

7. The poo-uli bird is classified as a Hawaiian honeycreeper on the basis of its feather colors and the resemblance of its skeleton to fossils of honeycreeper ancestors. The poo-uli's song differs from those of honeycreepers, however, and its tongue is round and smooth, while the honeycreepers have fringed, narrow tongues. Furthermore, the poo-uli eats snails and insects, whereas the honeycreeper uses its tongue to eat nectar, and the poo-uli has a very distinctive odor. Because of these differences, classification of the poo-uli as a honeycreeper is now in question. What further evidence might a biologist consider in more accurately classifying the poo-uli?

TO LEARN MORE . . .

References and Resources*

Altaba, Cristian. April 11, 1996. Counting species names. *Nature,* vol. 380. Estimating the number of species is a very inexact science.

**Life's* website features frequently updated references.

Angier, Natalie. June 13, 1996. Guinea pigs not rodents? Scientific panel says DNA is the key. *The New York Times.* Molecular evidence alters the taxonomy of guinea pigs, say some scientists.

Blackmore, Stephen. October 4, 1996. Knowing the earth's biodiversity: Challenges for the infrastructure of systematic biology. *Science,* vol. 274. We know little of the planet's biodiversity.

Cohn, Jeffrey P. September 1995. Connecting by computer to collections. *BioScience.* Computers bring the living world to taxonomists.

May, Robert M. October 1992. How many species inhabit the earth? *Scientific American.* Taxonomists estimate biodiversity by canvassing a small area and then extrapolating to global proportions.

Mooers, Arne O., and Rosemary J. Redfield. February 15, 1996. Digging up the roots of life. *Nature,* vol. 379. When did bacteria, archaea, and eukarya last share an ancestor?

Morell, Virginia. August 23, 1996. Life's last domain. *Science,* vol. 273. Researchers have sequenced the entire genome of an archaeon, and it holds many secrets.

Natural History. November 1996. This issue has several articles on discovering species in the oceans.

Pace, Norman R. September 1996. New perspective on the natural microbial world: molecular microbial ecology. *American Society of Microbiology News,* vol. 62. A phylogeny of the three domains of life based on ribosomal RNA sequences.

Stutz, Bruce. August 1986. Fish market taxonomy. *Natural History.* Taxonomists are not confined to jungle floors or museum halls. Fascinating fish species can be discovered in the bustle of the Fulton Fish Market.

Woese, Carl R., Otto Kandler, and Mark L. Wheelis. June 1990. Towards a natural system of organisms: Proposal for the domains Archaea, Bacteria, and Eucarya. *Scientific American.* The three-domain system accounts for archaea, which have no logical place in the five-kingdom scheme.

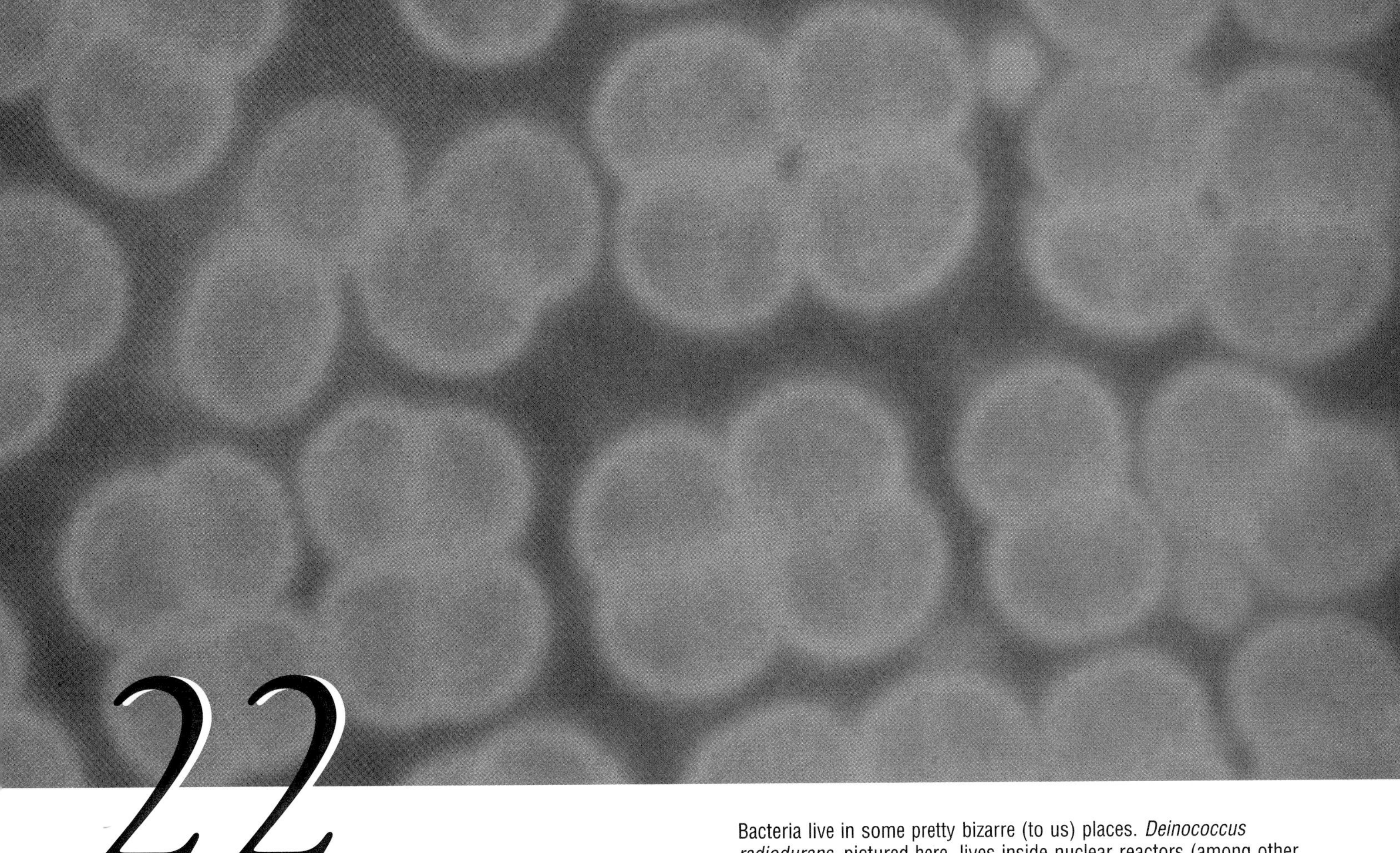

22 Monera

Bacteria live in some pretty bizarre (to us) places. *Deinococcus radiodurans,* pictured here, lives inside nuclear reactors (among other places). Its unusual cell wall somehow enables it to survive in the presence of high levels of ionizing radiation, which would harm or kill most other cells. (The average cell diameter is 2.5 micrometers.)

chapter outline

The Monera—A Biological Success 440
Bacteria and Archaea
Microbial Ecology

Studying Moneran Diversity 442
Laboratory Culture
Molecular Methods

Moneran Characteristics 443
Classifying Monera
Distinguishing Bacteria from Archaea
Energy Acquisition

A Closer Look at the Bacteria 446
Distinguishing Characteristics
Reproduction
Bacteria as Human Pathogens

The Archaea—Life in the Extremes 450
Classification

Monera—Crucial to Life on Earth 452
Industrial Microbiology

Sick Bacteria?

Cholera can kill in mere days. Cramps, vomiting, fever, and watery diarrhea that can deplete a human body of 2 to 3 gallons (10 to 15 liters) of fluid rapidly progress to circulatory shock and collapse. Because cholera is a food- or water-borne bacterial infectious disease, it isn't surprising that it has devastated towns and cities. Many cholera epidemics have swept Asia, the Middle East, Europe, and Africa, and today the disease is epidemic in areas of Central and South America and has appeared in Texas and Louisiana.

The bacterium *Vibrio cholerae* causes cholera. The infection spreads when fecal matter from infected persons contaminates a water supply, or when people eat shellfish that came from contaminated water. The bacteria use hairlike structures on their surfaces, called pili, to adhere to the mucosal lining of the person's small intestine, where they secrete a toxin (poison). When the toxin enters intestinal cells, it causes water and chloride ions to leak out and prevents sodium ions from entering. The result—massive, life-threatening diarrhea.

V. cholerae causes cholera because the bacterium itself is infected.

to be continued . . .

The Monera—A Biological Success

Anyone who has eaten yogurt, cheese, or bread, smelled spoiled milk, or suffered from a strep throat has encountered Monera, more commonly called bacteria (see fig. 4.8*b*). Bacteria familiar to us represent a tiny fraction of the members of this ancient and diverse kingdom.

Bacteria are everywhere. We have already mentioned several of the extreme habitats where bacteria live—in sulfur or iron-rich hot springs where temperatures exceed 212°F (100°C) (see Biology in Action 1.1); in highly acidic environments that would burn the skin off of most animals (see "Chemistry Explains Biology," chapter 3); in rocks (chapter 3 opener), and inside other organisms (see Biology in Action 21.2). They are abundant in more familiar surroundings too.

Bacteria and Archaea

Kingdom Monera includes the **bacteria** and, for now, the **archaea** (fig. 22.1). The bacteria include cyanobacteria, which have characteristic bluish green pigments, in addition to chlorophyll, that enable them to photosynthesize (see figs. 4.8*c* and 8.7*c* and *d*). The archaea comprise the separate domain Archaea, and in the past have been placed in the five-kingdom classification scheme with bacteria because both types of cells lack nuclei. As we learn more about archaea, however, their taxonomic status may change. For now, we consider them as part of Monera.

People have been studying bacteria since Antonie van Leeuwenhoek scraped them from his teeth and viewed them under his microscope in 1673. Today we know the structures of many types of bacterial cells in great detail, yet researchers are still determining the basic chemical structures of many components of archaeal cells. A major reason for the difference in our knowledge of these two types of microorganisms is that bacteria are widespread in nature, living in such disparate places as ocean bottoms, arctic tundra, and polar bear hairs. In contrast, many of the archaea identified so far live in extreme environments where we are less likely to discover them.

Medical researchers discover new types of bacteria too. For example, it is only in recent years that physicians have recognized that gastric ulcers result from bacterial infection. For many years

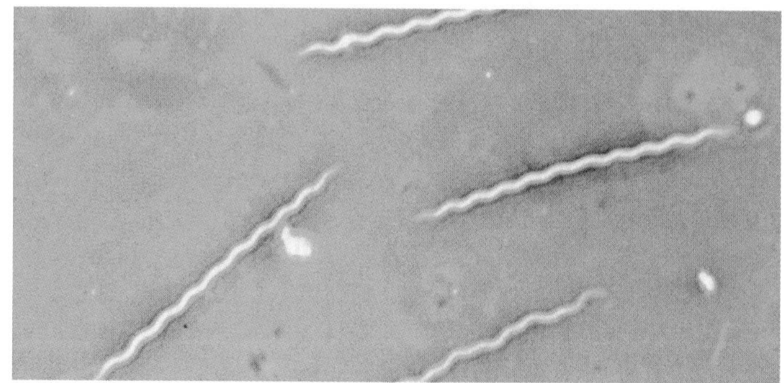

(a)

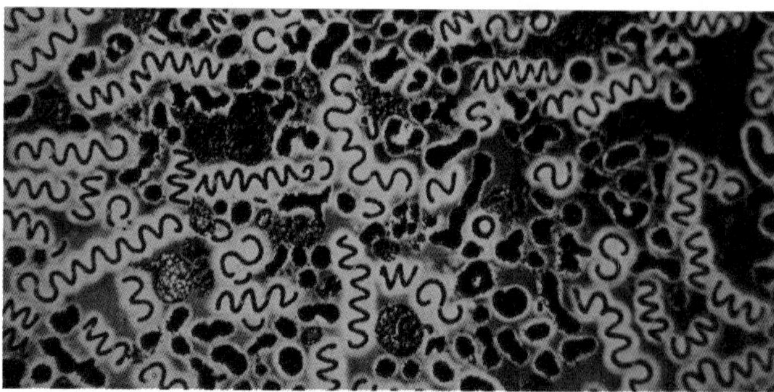

(b)

(c)

FIGURE 22.1

Types of monera. (*a*) *Treponema pallidum* is a bacterium that causes the sexually transmitted disease syphilis. (×1,000) (*b*) The cyanobacterium *Microcystis aeruginoda.* (×1,000) (*c*) Archaea *Halobacterium salinarium.* (The bar equals 1 micrometer.)

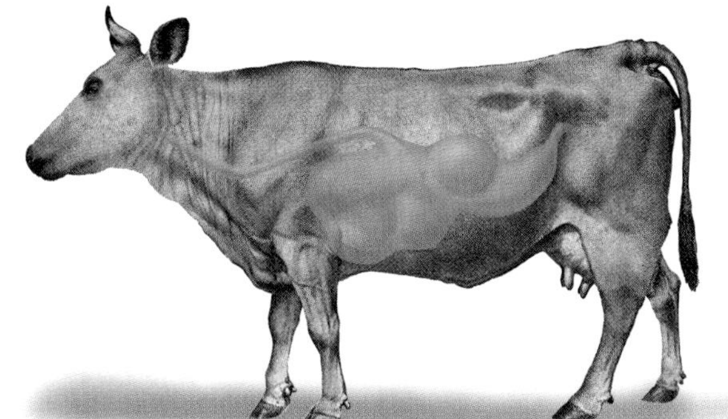

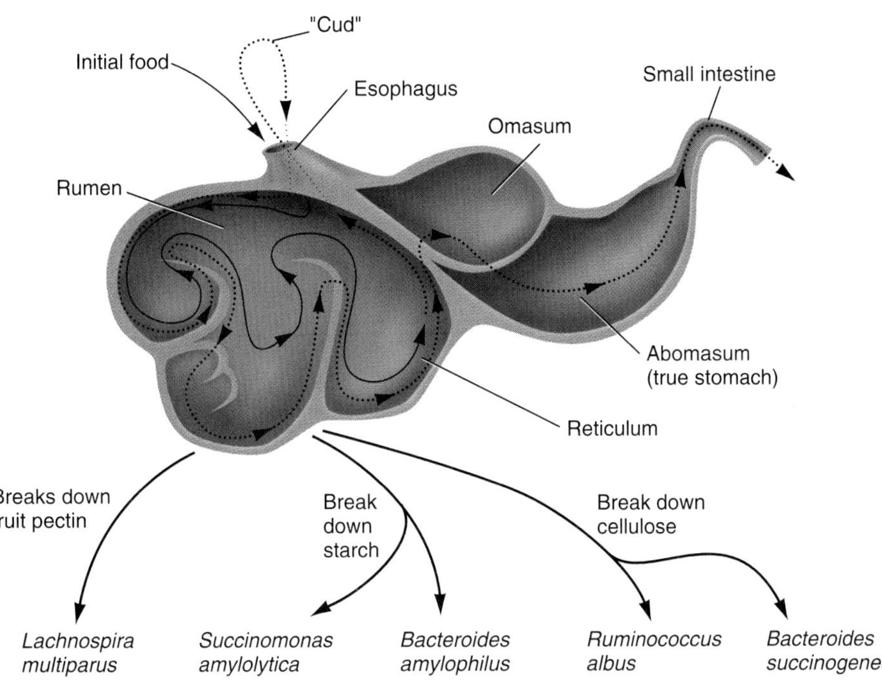

FIGURE 22.2

Populations of bacteria in a cow's rumen change to reflect its diet, thanks to natural selection.

most doctors thought that bacteria could not live in the very low pH of the human stomach—but *Helicobacter pylori* thrives there, protected in the mucus-rich lining.

Microbial Ecology

Monera and other unicellular organisms are essential components of the global ecosystem. Without microscopic life, macroscopic life would not exist, because microorganisms capture energy from the nonliving environment and form the bases of food webs. Photosynthesis harnesses much of this energy and bacteria that can photosynthesize support vast living communities in many habitats.

Yet even in areas far from sunlight, monera support life by harnessing chemical energy from the nonliving environment and using it to synthesize compounds that are nutrients for other organisms. This is the case in a community of 48 species living in a cave in southern Romania. All species depend upon mats of bacteria in parts of the cave that are underwater. The bacteria incorporate carbon from the limestone cave walls and use hydrogen sulfide as an energy source.

The first organisms on earth were most likely monera, and life on earth today still depends on them to maintain delicate ecological balances. Metabolic activities of microorganisms control global levels of oxygen, carbon dioxide, methane, and other chemicals. Bacteria, along with other single-celled organisms and multicellular fungi, serve as decomposers, which break down the biochemicals in dead and decaying organisms and recycle their nitrogen, sulfur, phosphorus, and carbon back to the environment. Bacteria even utilize remains of ancient life in deeply-buried sedimentary rock.

In nature, monera typically live alongside many other types of organisms, and many live in symbiotic relationships. A cow's digestive tract, for example, houses a variety of bacteria that decompose the animal's food (fig. 22.2). The section of the four-part stomach called the rumen houses a trillion bacteria per milliliter of food! Here, *Bacteroides succinogenes* and *Ruminococcus albus* break down plant cellulose to simple carbohydrates (sugars), which assists the cow in digesting grass. Two other types of bacteria, *Bacteroides amylophilus* and *Succinomonas amylolytica,* degrade complex carbohydrates (starches), which plays a more prominent role when the cow eats grains. Yet another bacterium,

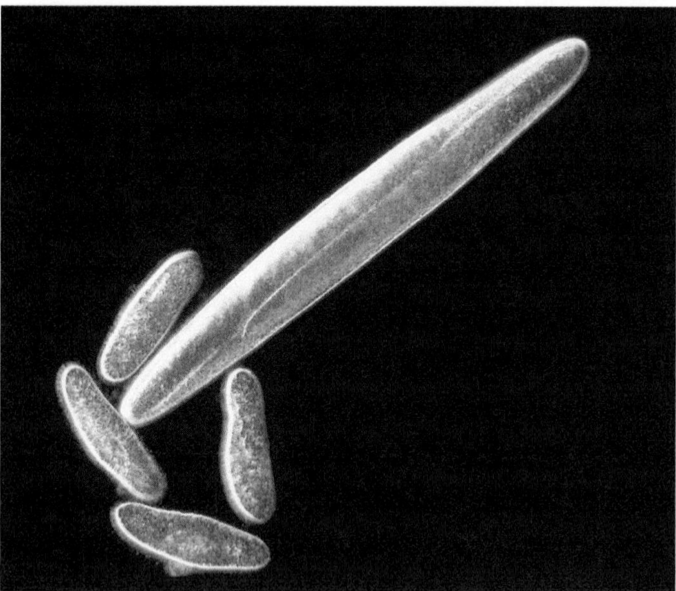

FIGURE 22.3

The "giant" bacterium *Epulopiscium fishelsoni* is a million times larger than many other bacteria, and it is even larger than the single-celled eukaryote *Paramecium.* (×200)

Lachnospira multiparus, produces enzymes that break down fruit pectin, which help a cow with a sweet tooth. Because of natural selection, the sizes of specific bacterial populations in a cow's rumen come to correspond to its diet.

Monera use a wide range of organic and inorganic materials to obtain energy and reducing power, a versatility that enables them to live in habitats no other organisms could withstand. Consider *Deinococcus radiodurans,* a species of bacteria that lives inside nuclear reactors (see chapter opening photograph). These bacteria can resist the harmful effects of ionizing radiation produced from the radioactive materials used to fuel nuclear power plants. Ionizing radiation in high doses is lethal to most organisms because it damages proteins, lipids, and DNA. Researchers aren't certain how this bacterium can survive intense radiation, but its unusually complex cell wall structure may be protective.

Mastering Concepts

1. What two types of microorganisms make up the kingdom Monera?
2. Why are these two groups placed together, and why might this grouping change?
3. Why is microscopic life vital for all life on earth?
4. What are some of the places that monera live?

Studying Moneran Diversity

Tremendous diversity exists within the moneran world. For each of the 16,000 or so identified bacterial species, microbiologists estimate that we have yet to identify at least 10 others. The proportion of archaea awaiting discovery is probably much higher.

Figure 22.3 illustrates that even large bacteria can sometimes escape our notice. The giant bacterium *Epulopiscium fishelsoni* was discovered in 1993 living in the intestine of a large fish in the Red Sea. It is nearly a million times larger than the human intestinal bacterial inhabitant *Escherichia coli* (see fig. 4.8*a*) and is larger even than the familiar eukaryote *Paramecium.*

A group of bacteria called mycoplasmas also illustrates how moneran diversity probably extends far beyond what we know. Mycoplasma live within eukaryotic cells. Members of one very large genus, *Spiroplasma,* associate with many species of insects. Since we are constantly identifying new insect species, we must also be unaware of many mycoplasma species that live within unidentified insects. Aquatic habitats are another vast, untapped resource of moneran life.

Laboratory Culture

Microbiologists culture microorganisms in the laboratory to study biochemical and physiological properties. Monera grow in petri dishes containing agar (a gelatin-like seaweed extract) and nutrients. A technique called **enrichment culturing** uses certain nutrient combinations to encourage growth of specific organisms of interest.

Culturing microorganisms is challenging because it can be difficult to precisely mimic a natural habitat—especially unusual or out-of-the-way ones, such as the inside of an insect cell, a fish's intestine, or a nuclear reactor. For this reason, many monera die in culture. The giant bacterium *E. fishelsoni,* for example, lives only 20 minutes outside its fish host.

Despite this limitation, microbiologists have assembled a huge collection of microorganisms that particularly interest them, which noted microbiologist Carl Woese describes as "a zoo of laboratory freaks that perform the physiological feats required of them." How these laboratory-nurtured microbes fit into the real world of microbial ecology, where nutrients may be limited and a diversity of species live, is not clear.

Molecular Methods

Molecular evidence is especially valuable when characteristics such as shape, color, and mode of reproduction cannot be observed because certain monera do not survive in laboratory culture. For example, molecular evidence helped researchers realize that *E. fishelsoni* was a bacterium and not a eukaryote, as they had originally thought based on its enormity. The organism's ribosomal RNA gene sequences were clearly like those of bacteria. Researchers were able to overcome the difficulty of growing *E. fishelsoni* in the laboratory by sampling its DNA and using the polymerase chain reaction to mass-produce the DNA, rather than attempting to culture enough of the organism to extract large amounts of its genetic material.

Comparing gene sequences among microorganisms has practical as well as taxonomic value. Computer analysis reveals that many of the genes that encode ribosomal RNA contain **signature sequences,** which are short stretches of nucleotides unique to certain taxonomic groups. This information is useful in classifying newly discovered microorganisms. In the diagnosis of infections,

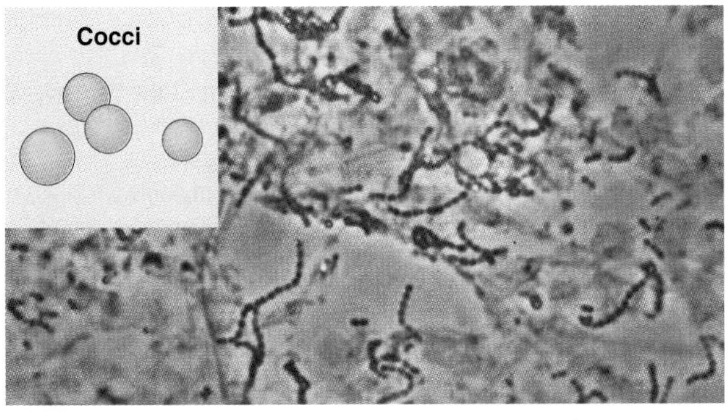

(a)

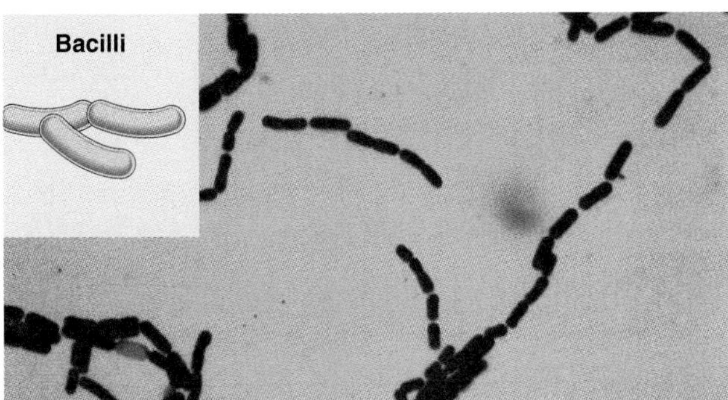

(b)

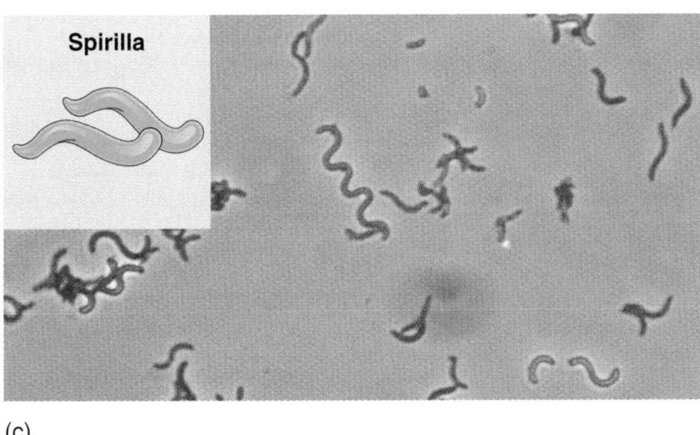

(c)

FIGURE 22.4

Bacteria are classified by shape. *Enterococcus faecalis* (*a*) are round (cocci). *Bacillus megaterium* (*b*) are rods (bacilli) that aggregate into chains. *Rhodospirillum rubrum* (*c*) are spiral-shaped (spirilla).

DNA probes consisting of signature sequences enable physicians to rapidly determine the species that causes a patient's symptoms. Ecologists use similar DNA probes to analyze the diversity of microscopic life in a particular habitat.

Mastering Concepts

1. Why do we have a limited view of microbial diversity?
2. What is enrichment culturing?
3. How can molecular methods help us to study the monera?

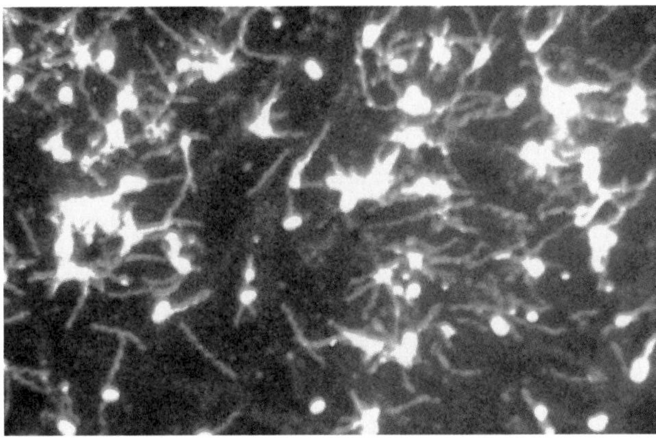

(a)

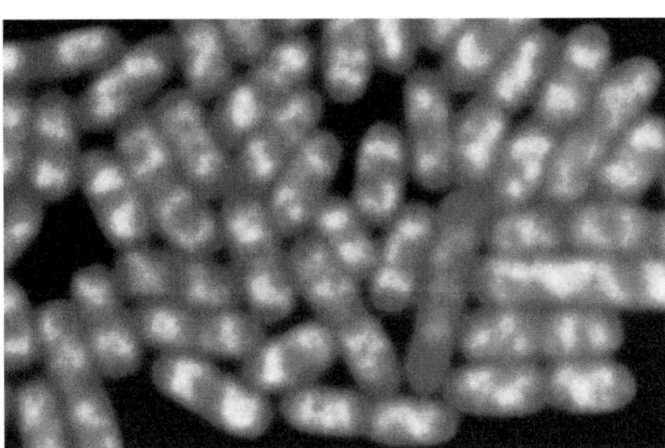

(b)

FIGURE 22.5

Fluorescent stains highlight *Salmonella enteritidis* (*a*), which causes food poisoning (gastroenteritis), and *Salmonella typhi* (*b*), which causes typhoid fever.

Moneran Characteristics

Microbiologists classify monera by form, staining characteristics, and biochemistry.

Classifying Monera

Monera take a variety of forms. Three of the most common are round (**cocci**), rod-shaped (**bacilli**), spiral (**spirilla**) (fig. 22.4). Microbes exhibit many variations on this theme, however. Some bacteria are also classified by whether the cells aggregate into strips or sheets.

To visualize monera under a microscope, microbiologists apply a variety of chemical compounds that provide contrast. In simple staining, a single compound creates contrast. Methylene blue, for example, is a positively charged dye that binds to negatively charged proteins on a bacterial cell surface. Another simple staining technique uses India ink. Because the ink cannot enter cells, bacteria on a microscope slide appear as clear structures against an inky background. Very striking are fluorescent stains used to highlight microorganisms (fig. 22.5).

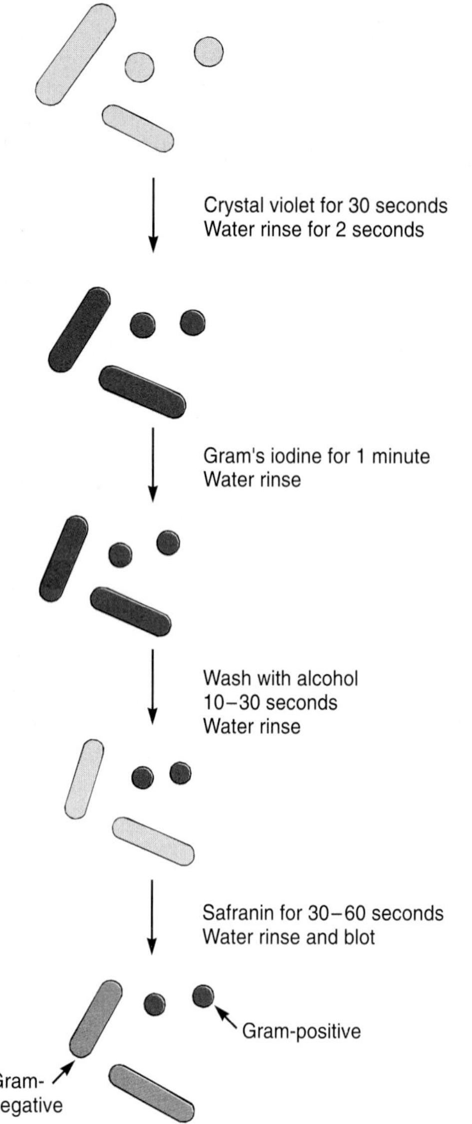

FIGURE 22.6

The gram staining procedure is the most common way to divide bacteria into two general classes. Archaea stain either positive or negative too but on a different physical basis than do the bacteria because their cell walls are different.

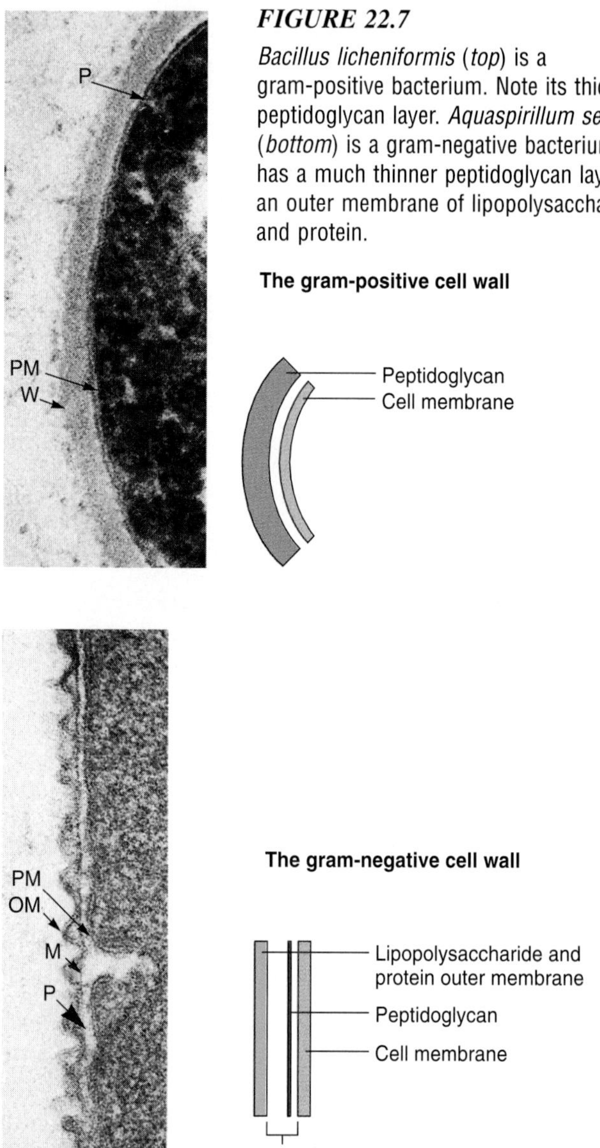

FIGURE 22.7

Bacillus licheniformis (*top*) is a gram-positive bacterium. Note its thick peptidoglycan layer. *Aquaspirillum serpens* (*bottom*) is a gram-negative bacterium. It has a much thinner peptidoglycan layer, plus an outer membrane of lipopolysaccharide and protein.

Differential staining divides bacteria into groups based on differing abilities to take up stains, which reflect distinctions in cell wall structure. The most common differential staining technique, devised in 1884, distinguishes gram-negative from gram-positive cells. Many bacteria and archaea stain either gram-negative or gram-positive.

To determine whether a cell is gram-negative or gram-positive, cells are first stained with a purplish blue dye called crystal violet. They are then exposed to iodine, which enhances uptake of the violet dye (fig 22.6). Next, cells are washed with alcohol, which removes the crystal violet. Finally, cells are exposed to a second dye, safranin, which imparts a reddish pink color. Gram-positive cells retain the crystal violet and appear purplish blue. Gram-negative cells lose the violet hue, become colorless, and then take up safranin and appear pink.

Differential gram staining reflects cell wall structure (fig. 22.7). Most bacterial cell walls are composed partially of a polysaccharide called peptidoglycan that consists of linked monomers of two types of nitrogen-containing monosaccharides. The other components of the cell wall vary among different species. In gram-positive bacteria, the cell wall is primarily a thick layer of peptidoglycan. Cell walls of gram-negative bacteria, in contrast, have a thinner inner layer of peptidoglycan and an outer covering of lipid, polysaccharide, and proteins.

Distinguishing Bacteria from Archaea

The functions of the cell wall in bacteria and archaea are probably similar—providing shape and protection and blocking entry of water by osmosis, which would burst the cell. The chemical

Table 22.1 Carbon, Energy, and Hydrogen/ Electron Sources

Carbon Sources	
Autotrophs	CO_2 sole or major carbon source for biosynthesis
Heterotrophs	Reduced, preexisting organic molecules from other organisms
Energy Sources	
Phototrophs	Light
Chemotrophs	Oxidation of organic or inorganic compounds
Hydrogen/Electron Sources	
Lithotrophs	Reduced inorganic molecules
Organotrophs	Organic molecules

Source: Lansing M. Prescott, et al., *Microbiology,* 3d ed., copyright 1996 The McGraw-Hill Companies, Inc.

components of the cell walls and their organization, however, differ greatly between bacteria and archaea. Most bacterial cell walls contain peptidoglycan, whereas most archaeal cell walls consist primarily of glycoprotein. Furthermore, the lipids that form cell membranes in bacteria and archaea differ in the types of covalent bonds that link them, and in the organization of lipid bilayers.

Another difference between bacteria and archaea is in the structure of RNA polymerase, the enzyme that all cells use to transcribe DNA into messenger RNA. Bacteria use a single type of RNA polymerase consisting of four polypeptide subunits. The archaeal equivalent of this enzyme occurs in several forms that are far more complex than the bacterial RNA polymerase.

Energy Acquisition

Unicellular organisms devote most of their energy to obtaining nutrients that enable them to reproduce. Biochemical pathways extract energy from the environment or nutrients, store it in ATP high-energy phosphate bonds, and use the energy to carry on life's activities, including metabolism and the duplication of cell contents and DNA necessary for reproduction.

Monera are grouped by how they acquire energy (table 22.1). The most fundamental distinction reflects carbon source. Recall from chapter 8 that autotrophs extract energy from carbon dioxide (CO_2) and heterotrophs use more reduced and complex organic molecules, typically from other organisms. A rare type of bacterium, called a **mixotroph,** combines characteristics of autotrophs and heterotrophs and uses inorganic molecules for energy and organic molecules for carbon.

Heterotrophs are further distinguished by their source of organic compounds. **Saprobes** obtain nutrients from dead plants and animals. Symbionts live within living organisms and acquire nutrients from them (Biology in Action 22.1).

An autotroph or heterotroph may also be a **phototroph** or **chemotroph,** which refers to the energy source. Phototrophs derive energy from the sun, and chemotrophs get energy by oxidizing chemicals.

Finally, monera are classified by the source of hydrogen or electrons, which are important in energy acquisition. **Lithotrophs** obtain electrons from reduced inorganic compounds, such as the "rock-eating" bacteria described in the chapter 3 opener. **Organotrophs** obtain hydrogen or electrons from organic compounds.

Combining these terms describes how specific microorganisms function and fit into the environment. Cyanobacteria, for example, are photolithotrophic autotrophs—which means that they use sunlight (photo), water (litho), and CO_2 (auto) for energy. Chemoorganotrophic heterotrophs, often abbreviated just to "heterotrophs," use molecules from other organisms to obtain all their requirements—energy, hydrogen, electrons, and carbon—often from the same molecule. This group includes many of the bacteria that cause disease.

Monera are also classified by their oxygen requirements. **Obligate aerobes** require oxygen and harness it in carrier molecules forming an electron transport chain on the inner face of the cell membrane. **Facultative anaerobes** can either use oxygen or not, and they obtain energy from fermentation. For **obligate anaerobes,** oxygen is toxic, and they live in habitats that lack it.

Mastering Concepts

1. What are some shapes of bacterial cells?
2. How does the gram stain procedure distinguish among bacteria?
3. What are some biochemical differences between bacteria and archaea?
4. Distinguish among autotrophs, heterotrophs, and mixotrophs; between phototrophs and chemotrophs; between lithotrophs and organotrophs.
5. Distinguish among obligate aerobes, facultative anaerobes, and obligate anaerobes.

22.1 biology in action

Life in a Sponge

Nearly 3.1 miles (5,000 meters) beneath the sea, near the lip of a mud volcano in the Barbados Trench, live hundreds of sponges, each about the size of a large dog. The sponges, of the family *Cladorhizidae,* were known to live only as solitary individuals, but the clustering of the sponges of the Barbados Trench suggests that they form a biological community.

Unlike other sponges, *Cladorhizidae* cannot suck up nutrients through internal pore systems. These sponges are carnivores of sorts, with branches of sharp spicules that stab passing crustaceans (shellfish), which the sponge then consumes. But spiking passing crabs isn't sufficient to support the large sponges. They get extra calories from at least two types of monera that live both between and within their cells (fig. 22.A).

Researchers recently discovered two types of bacteria, one coccoid, with stacked membranes in its cytoplasm, the other smaller and rod- or oval-shaped, living in the sponges. Both types of monera live between sponge cells, where they apparently coexist without doing harm or being harmed, as well as within the sponge cells. Researchers do not yet know what benefit the bacteria derive from their hosts. Monera inside the host's cells, however, appear to be degrading, suggesting that the sponge uses the bacteria for food—sometimes. Both types of bacteria also appear in flat cells on the surfaces of sponge embryos, but here, they are not degraded. This finding suggests that sponges transmit their bacterial symbionts to progeny, perhaps as a nutritional boost.

The sponge-dwelling bacteria have enzymes that enable them to metabolize methane, the gas that the nearby volcano spews. With bacteria that can tap the energy of the volcano, and sponges that tap the energy within the monera, an undersea community thrives.

(1) Sponge

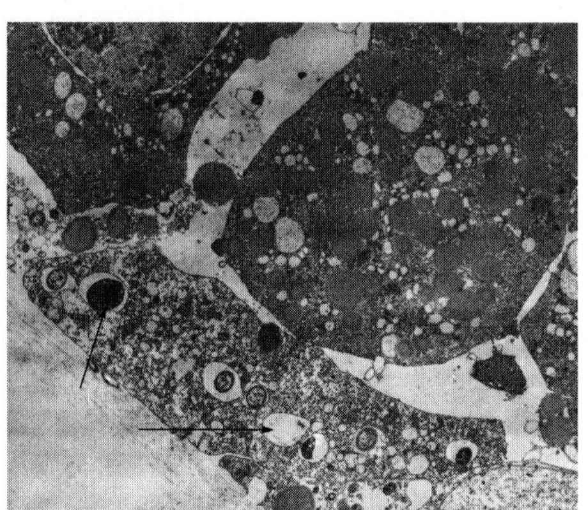

(2) Bacteria

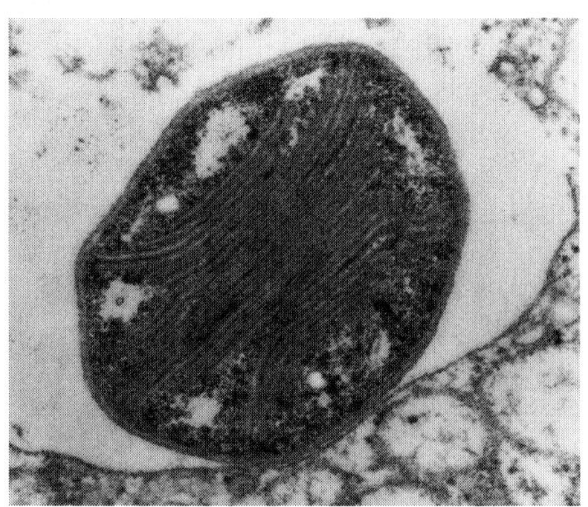

(3) Bacterium

FIGURE 22.A
Newly identified bacteria that use methane from an undersea volcano are symbionts of a carnivorous sponge (*1*). Colonies of the sponges were found by a submersible at the rim of the mud volcano in the Barbados Trench. The sponge derives nourishment from at least two types of bacteria (*2*). A close-up of one type of bacterium is shown in (*3*).

A Closer Look at the Bacteria

A bacterial cell is prokaryotic (see figs. 4.7 and 22.8). Its DNA is a continuous molecule free of proteins, located in a region of the cell called the nucleoid. A typical bacterium is about 2 micrometers long—about 5,000 of them could fit across a pencil eraser.

Distinguishing Characteristics

Bacteria have several distinguishing characteristics. Not all bacteria have these structures, and some cells have more than one type.

Pili are short projections on bacterial cells, resembling hairs, that enable the cells to attach to objects (fig. 22.9*a*). The bacterium that causes cholera attaches to a human's intestinal wall using pili. Another bacterial structure is a **flagellum,** which is an extension that rotates, moving the cell (fig. 22.9*b*). A flagellum is thin, only about 10 to 30 nanometers in diameter. It is very different chemically from a flagellum on a human cell, such as that on a sperm cell.

A bacterium's cell wall may have a sticky layer called a **glycocalyx,** which is composed of proteins and/or polysaccharides. A loose glycocalyx is called a slime layer, and a firm glycocalyx is a capsule. The glycocalyx enables the cell to adhere to various surfaces (fig. 22.9*c*).

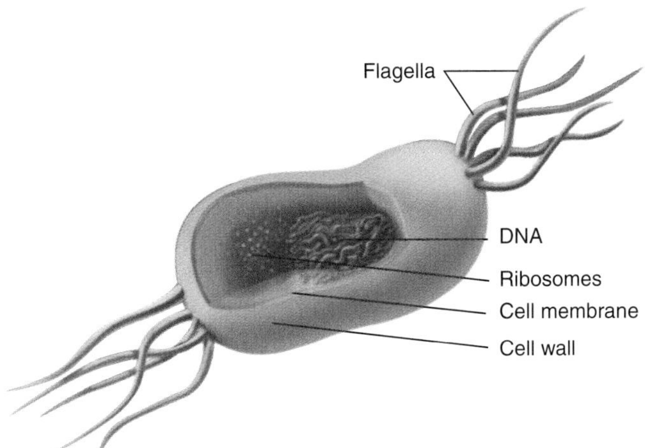

FIGURE 22.8

A bacterial cell is prokaryotic. Its nucleic acid is in a diffuse area called the nucleoid. Free ribosomes are in the cytoplasm. Some bacteria have flagella, which enable them to rotate and move.

FIGURE 22.9

Bacterial characteristics. (*a*) The *E. coli* on the right is covered with pili. It is sending DNA to the cell on the left, by way of a structure called a sex pilus, in a process called conjugation. (*b*) The extensions emanating from *Legionella pneumophila* are flagella. (*c*) A glycocalyx is a sticky layer on the cell wall that enables bacteria to adhere to various surfaces. Here, bacteria stick to each other and to a person's intestinal lining. (×17,500) (*d*) A bacterium (*Clostridium pectinovovum*) emerging from an endospore (the bar equals 0.5 micrometers). (*e*) These yellow rings are plasmids isolated from *E. coli.* Plasmids are circles of DNA, and they can transfer genes among bacteria.

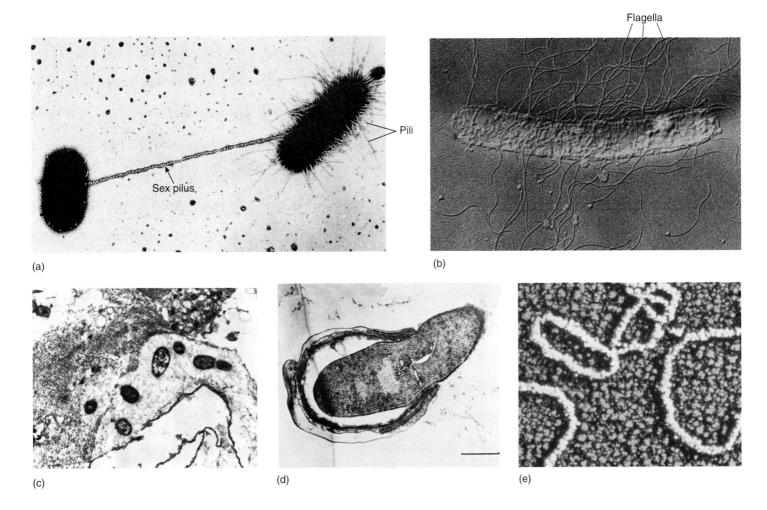

Pili

Sex pilus

(a)

Flagella

(b)

(c)

(d)

(e)

Bacteria can form structures called **endospores** that enable them to survive harsh conditions. An endospore is a walled structure that forms around the nucleus and a small amount of cytoplasm. The normal cellular form returns when environmental conditions improve (fig. 22.9*d*). Endospores can withstand boiling and drying. Because many pathogenic (disease-causing) bacteria form endospores, food manufacturing processes typically include a 10- to 15-minute pressurized steam treatment to destroy them.

Plasmids are circles of DNA that can pass among bacterial cells, carrying genes and thereby providing new functions (fig. 22.9*e*). Recombinant DNA technology, discussed in Biology in

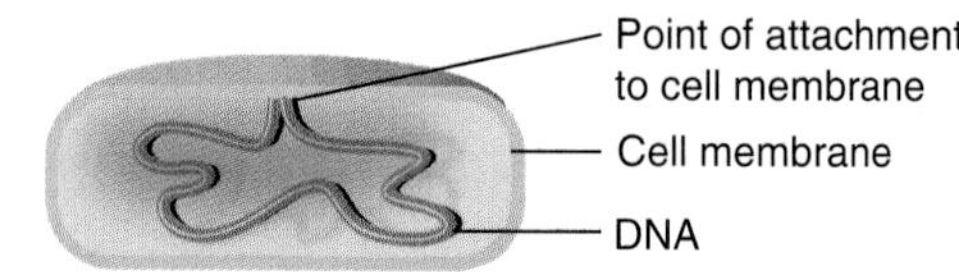
Point of attachment
to cell membrane
Cell membrane
DNA
DNA attaches to cell membrane.

DNA replicates.

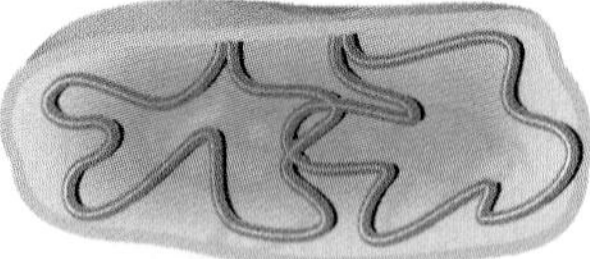
Cell membrane grows, pulling DNA molecules apart.

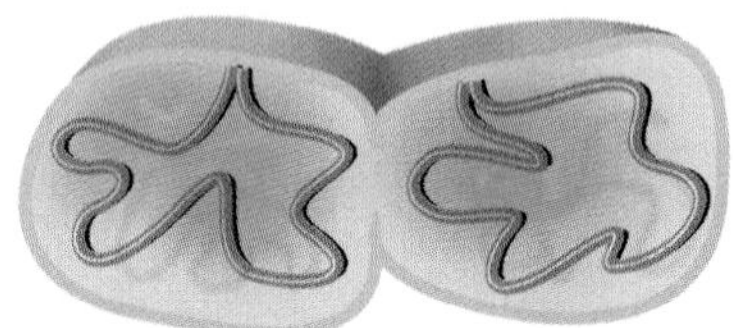
Cell membrane indents.

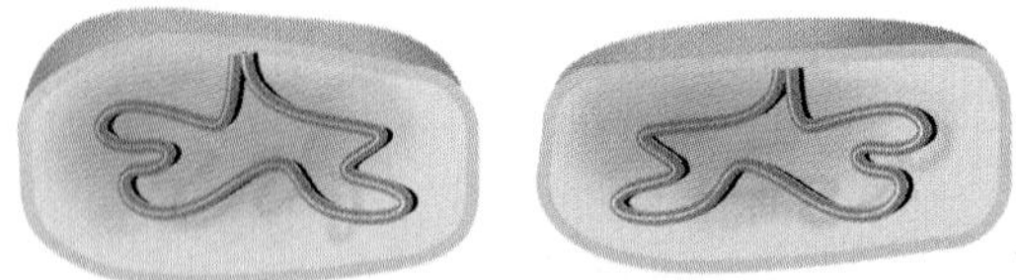
Cell divides; two progeny cells form.

FIGURE 22.10

In binary fission, the cell membrane grows and then indents as the DNA replicates, separating one cell into two.

Action 16.1, enables bacterial cells to produce proteins of use to humans by transferring human genes to them on plasmids.

Reproduction

Bacteria reproduce by **binary fission,** a process that copies DNA and distributes it and other cellular contents to progeny cells. Binary fission resembles mitosis, but it is not the same, because it lacks spindle fibers and protein-coated chromosomes.

In binary fission, DNA first attaches to a point on the inner face of the cell membrane (fig. 22.10). The DNA then begins to replicate, while the cell membrane grows between the parental and new DNAs and separates them. Then the cell membrane dips inward, pinching off two progeny cells from the original one. Formation of cell walls completes the process.

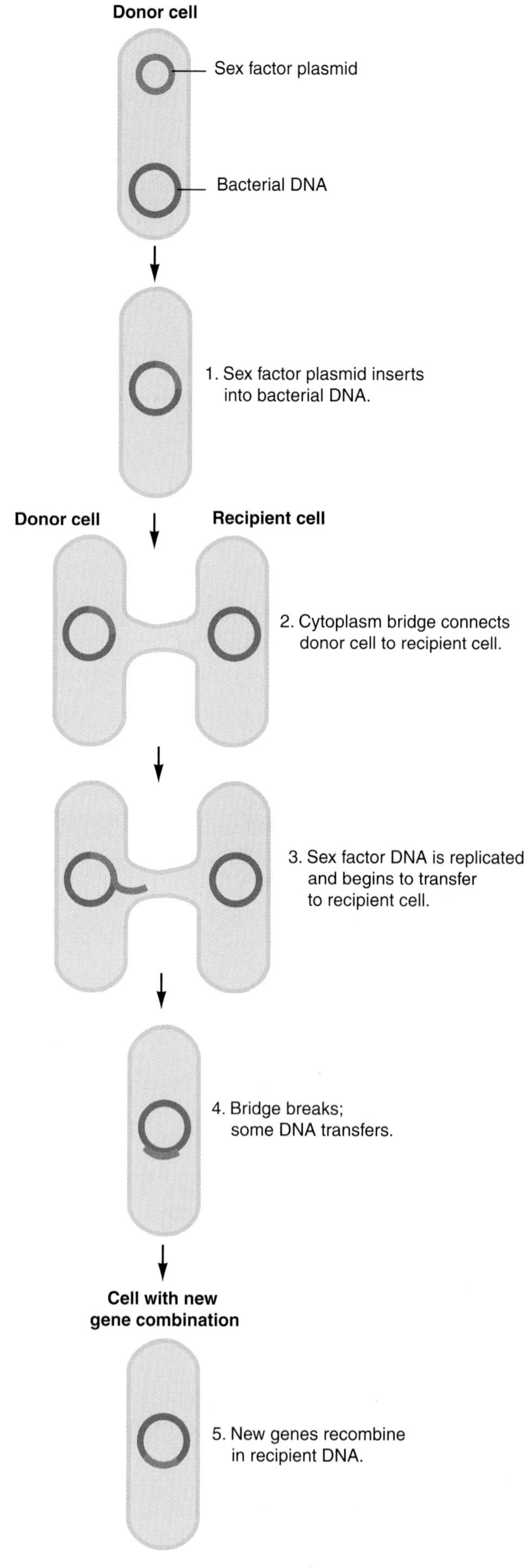
Donor cell
Sex factor plasmid
Bacterial DNA
1. Sex factor plasmid inserts into bacterial DNA.
Donor cell
Recipient cell
2. Cytoplasm bridge connects donor cell to recipient cell.
3. Sex factor DNA is replicated and begins to transfer to recipient cell.
4. Bridge breaks; some DNA transfers.
Cell with new gene combination
5. New genes recombine in recipient DNA.

FIGURE 22.11

In bacterial conjugation, a plasmid ferries a copy of some of one cell's DNA into another, where it replaces the corresponding sequences. Conjugation mixes gene combinations.

Table 22.2 Some Bacterial Infection Outbreaks

Bacterium	Symptoms	Example of Outbreak
Borrelia hermsii	"Relapsing fever" of cycling chills, fever, headache, sweats, muscle ache	Campers in Grand Canyon North Rim infected by spirochetes from ticks in rodent nests
Brucella species	Brucellosis—fever, loss of appetite, chills, sweats, headache, muscle and joint pain, fatigue, weight loss	Employees of pork processing plant in North Carolina picked up bacteria in pig-slaughtering area
Chlamydia psittaci	"Parrot fever"—fever; malaise, sore throat, pneumonia	Pet parakeets and cockatiels from a Mississippi bird dealer
Clostridium botulinum	Nausea, weakness, diarrhea, headache, dizziness, blurred vision, hoarseness, paralysis	Thirty-six people ate sandwiches made with garlic butter that had not been refrigerated; soil bacteria were in the garlic
Clostridium perfringens	Diarrhea, abdominal cramps, vomiting	Contaminated, unrefrigerated corned beef in St. Patrick's Day meals from a Cleveland deli
Clostridium tetani	Tetanus—facial muscle spasm and pain	A Vermont woman who had not been immunized in many years stubbed her toe in her garden
Salmonella enteritidis	Diarrhea, cramps, fever, nausea, vomiting	Six people ate improperly refrigerated egg-based foods at a Los Angeles diner
Vibrio cholerae	Severe diarrhea, back pain	A Los Angeles native traveling in India ate contaminated shrimp and drank unbottled water
Vibrio vulnificus	Blood poisoning in people with liver disease or impaired immunity	Vulnerable people ate contaminated raw oysters in Florida

Bacteria do not reproduce sexually, but a process called **conjugation** enables them to recombine genes. In conjugation, a bacterial cell with a plasmid that carries a gene called a sex factor can pass this gene to a cell lacking the sex factor (see figs. 22.9*a* and 22.11), along a bridge of cytoplasm called a **sex pilus** that forms between the two cells. Recombination occurs when the cell with the sex factor copies some of its DNA, which moves to the other cell as the sex factor is transferred. The cytoplasm bridge usually collapses before all of the sending cell's DNA transfers. The new genes arriving in the recipient cell trade places with their counterparts, which enzymes destroy. As a result, the receiving cell is left with some of its own genes and some from the other cell. Such new combinations of genes are important in driving evolutionary change.

Bacteria as Human Pathogens

Most bacteria are not pathogenic, but many familiar diseases of humans are bacterial infections. Diagnosing a bacterial infection on the basis of symptoms is difficult, because many illnesses appear alike. If the set of symptoms is distinct enough, a doctor may begin antibiotic therapy before a laboratory confirms the diagnosis. Often, however, it is necessary to determine the specific type of bacterium causing a patient's illness, so that effective treatment can begin promptly. This is done in several ways:

—culturing the microorganisms from a body fluid or tissue sample;

—detecting a toxin specific to a pathogen;

—detecting antibodies that the host's immune system produces in response to the presence of a specific pathogen;

—using DNA probes to detect signature sequences of specific bacteria.

Because bacterial infections pass from person to person, sometimes the circumstances of individuals just before symptoms appear provide clues to the identity of the causative bacterium. Table 22.2 lists some examples of bacterial infection outbreaks.

Route of transmission is one criterion used to classify bacterial infections. An infection may be airborne, arthropod (insect, tick, or spider) borne, transmitted by direct contact, or passed in water or food.

Legionella pneumophila causes legionellosis, an airborne bacterial infection (see figs. 18.10 and 22.9*b*). This infection first came to public attention, as Legionnaires' disease, when a group of American Legion conventioneers staying at a Philadelphia hotel in 1976 suddenly developed cough, fever, headache, and pneumonia. Several of them died, as epidemiologists frantically traced the outbreak to an aerobic, gram-negative, rod-shaped bacterium that spread through the hotel's air conditioning system. The identification of this bacterium explained several earlier outbreaks of respiratory illness whose source had never been identified.

Pertussis (whooping cough) and diphtheria are airborne bacterial infections that had become rare but are now on the rise. Pertussis has resurfaced because growing numbers of people refuse to vaccinate their children in fear of the very slight risk

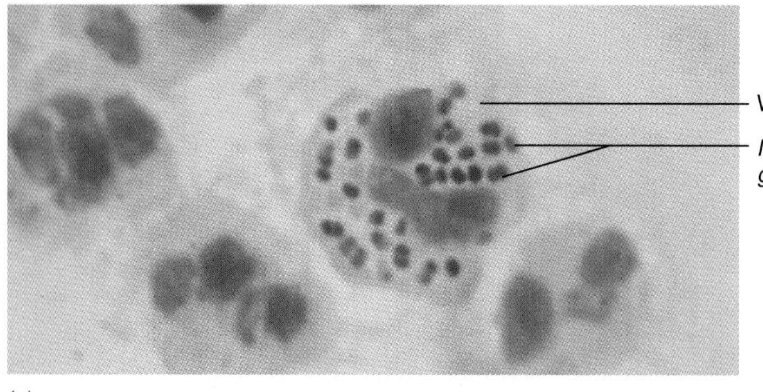

(a)

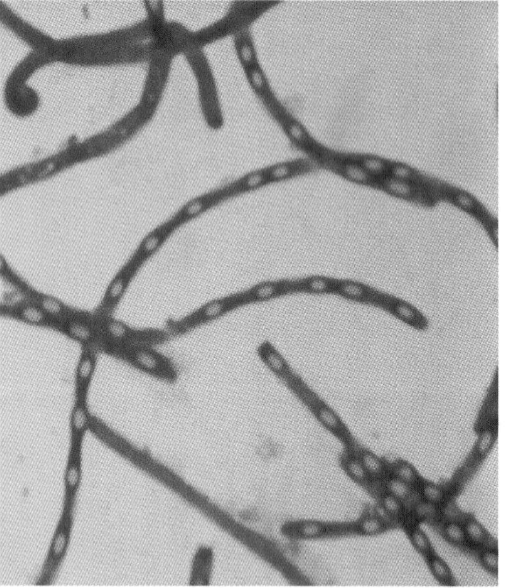
(b)

FIGURE 22.12

Direct-contact bacterial infections. (*a*) *Neisseria gonorrhoeae* causes the sexually transmitted disease gonorrhea. (× 500) (*b*) *Bacillus anthracis* spores survive in soil or in cattle, goat, or sheep excrement. Bacteria enter the human body through skin sores and then produce a toxin that causes headache, fever, and nausea. If the bacteria are not stopped from entering the circulation, anthrax can kill.

of a complication. Diphtheria is on the upswing in the former Soviet Union due to lapses in vaccination and crowded living conditions. The infection produces a fever, cough, and thick nasal discharge.

Arthropods such as spiders and ticks can transmit bacterial infections. A spirochete that lives in deer ticks causes Lyme disease (see fig. 18.10) in hikers, producing malaise, flulike symptoms, and sometimes arthritis. The bacterium that causes bubonic plague, *Yersinia pestis,* has usually passed from rats to fleas to humans (see fig. 21.C). Today, the 25 or so cases of bubonic plague that occur each year in the southwest United States are usually contracted through fleas from dogs and cats or through eating wild game. Ironically, the bacteria proliferate in the host's phagocytic cells, the very cells that usually destroy bacterial pathogens.

Some bacterial infections enter the host directly, usually through the skin. In the sexually transmitted disease gonorrhea, bacteria attach by pili to hairlike microvilli that fringe cells lining the human urethra. White blood cells engulf the bacteria, which persist inside the cells, where they are shielded from immune system biochemicals (fig. 22.12*a*). Meanwhile, the area in the urethra where the bacteria attach becomes inflamed, and fibrous tissue forms, and blocks urine flow. Painful urination is often the first sign of gonorrhea.

Anthrax is another direct-contact bacterial infection. *Bacillus anthracis* produces spores that survive in soil and in cattle, goat, and sheep excrement (fig. 22.12*b*). The bacteria enter the human body through skin sores and then produce a toxin that causes headache, fever, and nausea. The infection can be deadly if it enters the bloodstream before treatment begins.

Many familiar bacterial infections fall into the food- and waterborne category. Several types of bacteria cause food poisoning, producing symptoms of abdominal cramps and diarrhea. Each year in the United States, 65 million people develop food poisoning, and 9,000 die.

Salmonella enteritidas causes a common type of food poisoning called salmonellosis (see fig. 22.5*a*). The infection often spreads from eggs, raw chicken, or undercooked beef. Salmonellosis also spreads when people clean the cages of pet reptiles, which can pass the bacteria in their feces. Children have contracted salmonellosis from turtles, snakes, and lizards, as well as from a nonreptile pet, the African pygmy hedgehog. Cholera and *E. coli*-associated diarrheal illness are food- or waterborne bacterial infections caused by toxins. Table 22.3 summarizes common routes of bacterial infection.

Mastering Concepts

1. What are some distinguishing characteristics of bacteria?
2. How do endospores protect bacteria?
3. How do bacteria reproduce, and how do they recombine genes?
4. How are bacterial infections diagnosed and classified?
5. What are some routes of transmission of bacterial infection?

The Archaea—Life in the Extremes

The archaea appear to be a diverse group in several ways. Their cells may be gram-negative or gram-positive. They come in a va-

Table 22.3 Routes of Transmission of Bacterial Infection

Route	Infection	Bacterium
Airborne	Legionellosis	*Legionella pneumophila*
	Pertussis	*Bordetella pertussis*
	Diphtheria	*Corynebacterium diphtheriae*
Arthropod	Lyme disease	*Borrelia burgdorferi*
	Bubonic plague	*Yersinia pestis*
Direct contact	Gonorrhea	*Neisseria gonorrhoeae*
	Anthrax	*Bacillus anthracis*
Food or waterborne	Salmonellosis	*Salmonella enteritidas*
	Cholera	*Vibrio cholerae*
	Diarrhea and kidney failure	*Escherichia coli*

riety of shapes, including rods, spheres, spirals, lobes, and plates, as well as hard-to-define and changeable forms. Most archaea are between 0.1 and 15 micrometers in diameter, although some have very long extensions. They have a variety of reproductive strategies, including binary fission, fragmenting to yield new individuals, and budding off progeny cells from parental cells. To acquire energy, archaea may be obligate aerobes, obligate anaerobes, or facultative anaerobes.

Many archaea are well adapted to extreme surroundings, such as very high or low pH, temperature, mineral content, or salinity (salt content). Often these conditions occur together, such as sulfur- or iron-rich acidic hot springs, where temperatures exceed 212°F (100°C) and pH approaches 1, or in salty habitats where the pH is 11. The highest known habitat temperature is 235°F (113°C), but researchers suspect the upper limit may approach 302°F (150°C). The proteins in archaea that live in these extreme places are especially stable under conditions that would quickly denature other proteins.

Archaea are also found in some less forbidding places. Some occupy the digestive tracts of animals, and nearly a third of the moneran life in the surface waters of the Antarctic are archaea.

Classification

Researchers are still discussing how to classify archaea. One scheme groups them according to habitat and metabolism. This view recognizes **methanogens** (which produce methane as a by-product of metabolism), **halophiles** (which live in extremely salty environments), **hyperthermophiles** (which use sulfur and tolerate extreme heat), and **thermophiles** (organisms of genus *Thermoplasma,* which live under high heat and acid conditions). This approach, however, allows for some overlap. Many methanogens, for example, are also thermophiles or hyperthermophiles.

Another way to classify archaea is according to protein and nucleic acid sequence similarities. On this basis, some researchers recognize two kingdoms: the Crenarchaeota, which are all thermophiles and hyperthermophiles, and the Euryarchaeota, which include all others. Obviously, taxonomy of the archaea is an evolving science!

The Methanogens

The methanogens are a group of strictly anaerobic organisms whose metabolic activities produce methane (CH_4). Their habitats are widespread. *Methanosarcina* and *Methanotrix* live in lake and swamp sediments, where they degrade dead vegetation. Some methanogens, such as *Methanobrevibacter ruminantium,* live in the rumens of herbivores and digest cellulose. Others live deep in the oceans, where they use organic matter from dead and decaying aquatic organisms. Still other methanogens live near undersea volcanic vents at very high temperatures and pressures, where they synthesize organic molecules from CO_2 and H_2 (hydrogen gas). These organic compounds sustain other organisms that inhabit the vents. Methanogens also live in raw sewage.

Halophiles

Halophiles, such as *Halobacterium* and its relatives, live in environments where the salt concentration is high enough to pickle most other organisms (see fig. 22.1*c*). Salty habitats include tidal pools that have dried over time, as well as the Dead Sea and the Great Salt Lake, where the salinity is nearly 10 times greater than that of the oceans. (The term "halophile" refers to "halogen," the group of chemicals including chlorine, which is a component of table salt.)

Before refrigeration was widely available, people used salt as a preservative because it killed microorganisms. We still use

FIGURE 22.13

Certain archaea live in hypersaline environments in marine salterns in the San Francisco Bay area.

salt-preserved meats such as sausage and salt pork. Halophilic archaea, however, thrive in salty surroundings. But even if archaea could survive in the high-salt environment of a hot dog, the conditions in the human digestive tract, such as very low pH in the stomach, would destroy them.

A good place to observe *Halobacteria* is in the San Francisco Bay area. Here, huge containers called marine salterns are filled with drying seawater; they produce sodium chloride and other salts with commercial value. Over time, as water evaporates, salt is left behind. When the salt concentration reaches a certain level, the salterns become reddish purple, because of pigments in the massive growth of halophiles (fig. 22.13).

Hyperthermophiles

Thousands of meters below the earth's surface, in deep oil reservoirs where temperatures and pressures are extremely high, huge numbers of hyperthermophilic archaea flourish. An expedition of scientists studying oil quality and recovery rates from oil reserves in the North Sea and the Alaskan North Slope discovered these abundant hyperthermophiles, along with bacteria that are also extremely heat tolerant.

Hyperthermophiles reduce sulfate (SO_4^{2-}) in their oily environments to hydrogen sulfide (H_2S), which lowers the oil quality and recovery rate and eats away the iron and steel alloys that hold offshore oil platforms together. Engineers have known about the problem of H_2S on oil platforms for a long time but have only recently recognized the role hyperthermophiles play in producing it.

Hyperthermophiles live in other areas that are rich in sulfur or sulfide deposits, such as the hot springs of Yellowstone National Park in Wyoming (see Biology in Action 1.1) and deep-sea thermal vents. Such environments are usually very acidic because the archaea produce sulfuric acid when they metabolize sulfur. Most hyperthermophiles live under anaerobic conditions.

Deep-sea hydrothermal vents can support abundant life, including not only microorganisms but invertebrates. The hyperthermophilic archaea here support the more complex species. Consider the tube worm, a 6.5-foot-long (2-meter-long) animal living near hydrothermal vents (fig. 22.14). Archaea live symbiotically with the worms. The animals consume dead archaea and the chemicals they release when they oxidize sulfide. Giant clams and mussels use archaea too.

Thermophiles

Archaea in genus *Thermoplasma* lack cell walls and resemble mycoplasma, a type of bacteria. *Thermoplasma* species thrive under the hot, acidic conditions of coal refuse piles, which spontaneously produce heat. Only one *Thermoplasma* species has been isolated outside of the hot coal environment.

Mastering Concepts

1. What are some ways that archaea are diverse?
2. What are two ways of classifying archaea?
3. What are some of the extreme conditions in which archaea live?

Monera—Crucial to Life on Earth

The monera helped shape earth's natural history. The huge iron deposits in ancient sea sediments, for example, represent the metabolic work of photosynthetic purple nonsulfur bacteria that oxidized iron as part of their carbon assimilation pathways—an ability they passed to their modern counterparts. Similarly, massive microbial "blooms," or bursts of bacterial growth, deposited iron in ocean floors some 2 to 3 billion years ago.

Moneran metabolism affects many species. **Nitrogen fixation** is a microbial process that is very important in agriculture. In this process, nitrogen-fixing bacteria of genus *Rhizobium* in soil or water reduce nitrogen gas in the atmosphere (N_2) to ammonia (NH_3). The significance of this action is that other organisms cannot use N_2, but they can incorporate the nitrogen in NH_3 into organic molecules that are starting materials for biosynthetic pathways. Some of these bacteria that live symbiotically in plant roots share the nitrogen that they "fix" (make biologically accessible) with their plant hosts. For example, when *Rhizobium* infects roots of a leguminous plant, the plant forms nodules that can convert atmospheric nitrogen into organic nitrogen (fig. 22.15). Nitrogen fixation in plants increases the levels of organic nitrogen in the soil, which offers the host plant a selective growth advantage in fertilizer-poor soils. Spontaneous events such as lightning can also fix nitrogen but are very rare. Only bacteria can regularly convert nitrogen into biologically useful forms.

Industrial Microbiology

Biotechnology and pharmaceutical laboratories commonly use bacteria to mass-produce proteins from genes inserted into their

(a)

(b)

FIGURE 22.14

A hydrothermal vent called a black smoker (*a*) and archaea support a thriving community of larger organisms, such as the tube worms in (*b*).

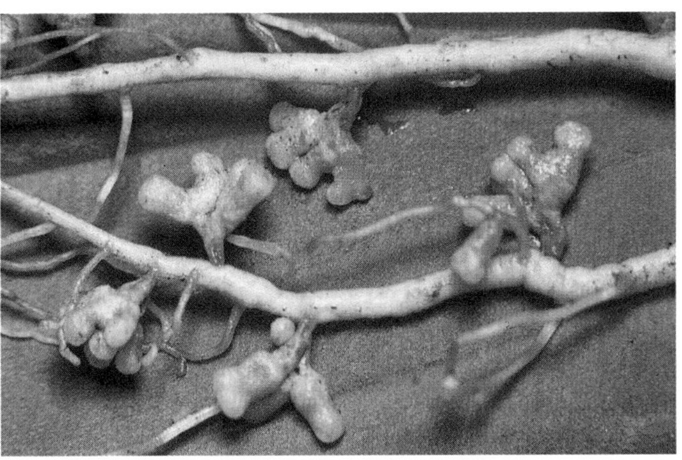

FIGURE 22.15

Rhizobium infects these sweet clover roots, producing root nodules where nitrogen fixation occurs.

FIGURE 22.16

Bacteria of genus *Lactococcus* are used to manufacture cheddar cheese from fermenting milk. The name of the cheese comes from Cheddar, a village in England where the cheese originated.

genomes using recombinant DNA technology (see chapter 16). Many important drugs, such as human insulin, growth factors, and interferon, are produced with this technology. In the molecular biology laboratory, *E. coli* provides a vehicle through which researchers can study genes of other organisms. Much of what we know of how genes work comes from studies done on *E. coli.*

Humans have used bacteria in such applications as brewing, baking, and cheesemaking for centuries (fig. 22.16). More recently, scientists have explored archaea as a source of highly stable biochemicals. For example, the polymerase chain reaction (PCR) commonly used to amplify genes of interest was inefficient until its inventor began using a DNA polymerase from *Thermus aquaticus* (see Biology in Action 15.2). Because PCR depends upon constantly shifting temperatures of the replicating

DNA, this heat-tolerant archaean enzyme works throughout the temperature shifts. Previously, researchers had to constantly add new polymerase as the heat destroyed it.

Bioremediation, another area of industrial microbiology, uses microorganisms that naturally metabolize what to us is toxic to clean the environment. For example, when the *Exxon Valdez* ran aground and dumped oil into the Prince William Sound in Alaska, biologists recruited oil-eating bacteria living on the shoreline to clean the oil (see fig. 44.1). Scientists fed native bacteria inorganic nitrogen and phosphorus, which stimulated their growth without further damaging the environment. Within 2 weeks, the bacteria had consumed much of the oil, degrading it into recyclable products—CO_2, water, and biomass. Microbial bioremediation played a major role in cleaning up this devastating oil spill.

Researchers have put bacteria to work in many other cleanup projects. Through natural selection, populations of microorganisms living at toxic waste sites develop biochemical abilities to use toxins, sometimes degrading them to more benign chemicals. For example, anaerobic bacteria in deep sediments detoxify chlorinated compounds such as polychlorinated biphenyls (PCBs) that industrial plants dumped into rivers under lax environmental regulations. The bacteria strip the chlorines from the molecules. Although PCB dumping is no longer permitted in most states, the chlorinated solvents remain in riverbeds and soils. Using bacteria to detoxify PCBs and other chlorinated compounds can save millions of dollars in cleanup costs and is safer than dredging the contaminated sediments and trucking them away.

Archaea may help remove highly toxic halogens (chlorine and bromine) from the environment. By using archaea that produce enzymes that are both heat-stable and can break down halogens, industrial chemists hope to be able to manufacture certain compounds without generating these toxic by-products.

Certain anaerobic bacteria produce biochemicals that can chelate (bind to and surround) heavy metals, which helps to remove them from water. Heavy metals such as cadmium, lead, copper, zinc, silver, and mercury from old mining and milling operations contaminate many areas in the western United States. Because the bacteria can grow on wastes, bioremediation is less costly, more effective, and safer than traditional chemical methods of heavy metal removal.

Mastering Concepts

1. How did bacteria contribute to formation of iron deposits in the earth?
2. What is the role of bacteria in nitrogen fixation?
3. What are some industrial uses of monera?

SUMMARY

1. Monera are abundant, widespread, and occupy diverse, sometimes extreme, habitats. They include the **bacteria** and **archaea.** Monera support other species by forming the bases of food webs, recycling chemicals to the environment, and controlling global levels of essential gases.

2. Enrichment culturing enables us to study certain species but does not accurately replicate natural conditions.

3. Signature sequences are DNA sequences unique to specific taxonomic groups.

4. Characteristics used to classify monera include shape, color, biochemical characteristics and sequences, and staining properties. Gram staining is based on cell wall structure.

5. Archaea have different cell wall components and organization and different RNA polymerases than bacteria.

Sick Bacteria?~

continued from page 440.

Sometimes infection by *V. cholerae* has no effect on a person or produces only mild symptoms. At other times, the infection is swiftly and painfully lethal. Why the difference? *V. cholerae* that kill do so because they harbor a viral infection themselves.

Bacterial cells, like other cells, can become infected with viruses. Viruses that infect bacteria are called bacteriophages, or phages for short. Researchers recently discovered that *V. cholerae* that are pathogenic harbor a phage bearing a gene that enables them to manufacture a toxin—and this toxin causes the symptoms of cholera in humans. The phage also passes along genes to the bacteria that enable phage to spread to other bacteria.

Cholera vaccines have never worked well, and the discovery that it isn't the bacterium itself, but a genetic stowaway, that causes the illness explains why. In fact, researchers working on developing a cholera vaccine discovered the phage because their work was not progressing well.

A vaccine is a disabled form of a pathogen, or even just a piece of it that is sufficient to stimulate a human immune response. The observation that various bits of *V. cholerae* did not protect against the deadly diarrhea suggested that the bacterium was acquiring its virulence elsewhere. Using an electron microscope, vaccine researchers identified a newly seen phage. It grabs the bacterium's pili and inserts itself into the cell, along with its deadly gene.

The discovery of the pili route to infection explained previous research showing that the same bacterial gene controls both cholera toxin production and pili structure. Experiments further revealed that when this gene is mutated, the cell fails to grow pili—but also never manufactures cholera toxin. Without pili, the virus can't enter, and the bacterium never receives instructions to produce the cholera toxin.

Identifying the agent directly responsible for cholera, and learning how it enters bacterial cells, is expected to finally set vaccine developers on the right track. They can now focus on the true cause of cholera—not a bacterium, but a virus.

6. Autotrophs get energy from CO_2, and heterotrophs from reduced organic molecules from other organisms. **Mixotrophs** use inorganic molecules for energy and organic molecules for carbon. **Saprobes** are decomposers, whereas symbionts live in or on other organisms. **Phototrophs** capture light energy, and **chemotrophs** harness chemical energy. **Lithotrophs** get hydrogen or electrons from reduced inorganic molecules, and **organotrophs** do so from organic molecules. **Obligate aerobes** use oxygen, **obligate anaerobes** are poisoned by oxygen, and **facultative anaerobes** can use oxygen or not.

7. Bacteria may have the following characteristics: **pili** (short projections that provide adhesion), **flagella** (extensions that provide motility); and/or a **glycocalyx** (sticky outer layer). An **endospore** protects a bacterial cell in stressful conditions.

8. In **binary fission,** monera replicate DNA and duplicate other structures and distribute their contents into two cells. The cell membrane grows between the two copies of DNA and pinches off.

9. In **conjugation,** a **sex pilus** passes the sex factor as well as some copied genes from a donor cell to a recipient cell, where the incoming genes swap places with their counterparts. Conjugation mixes gene combinations.

10. A bacterial infection can be diagnosed by culturing the microorganism, detecting a toxin or antibody response, or using a DNA probe bearing a signature sequence.

11. Bacterial infections may be transmitted in the air, by arthropods, or through direct contact or passed in water or food.

12. Archaea are diverse and many live in extreme conditions. They are classified by habitat and metabolism and by molecular sequence similarities. **Methanogens** are anaerobic and produce methane. **Halophiles** live in high-salt conditions. **Hyperthermophiles** live in hot, sulfur-rich environments. **Thermophiles** include members of genus *Thermoplasma.*

13. Monera affect the earth's mineral deposits, and their metabolism affects many species. Some bacteria fix atmospheric nitrogen into a biologically useful form. Industrial microbiology uses microbial metabolism to detoxify substances and to produce useful biochemicals.

TO REVIEW . . .

1. Cite three ways that monera are essential to life on earth.

2. What is an advantage and a limitation of enrichment culturing?

3. State four characteristics used to classify monera.

4. What is a signature sequence?

5. What is the physical basis for the distinction between gram-negative and gram-positive cells?

6. Distinguish between the following:
- **a.** a saprobe and a symbiont
- **b.** an autotroph and a heterotroph
- **c.** a phototroph and a chemotroph
- **d.** a lithotroph and an organotroph
- **e.** an obligate aerobe, an obligate anaerobe, and a facultative anaerobe

7. How is the ability to form an endospore adaptive?

8. How do bacterial and archaeal cell walls differ?

9. Compare the ways a person contracts legionellosis, bubonic plague, Lyme disease, and cholera.

10. *Capnocytophaga canimorsus* passes to humans in a dog's bite and can cause fatal blood poisoning in some people. The microorganism is gram-negative and requires CO_2 for growth. It has peptidoglycan in its cell wall. Is it a bacterium or an archaean? How do you know this?

11. The bacteria that cause Lyme disease and syphilis are spirochetes. What does this mean?

TO THINK ABOUT . . .

1. A young child develops a very high fever and an extremely painful sore throat. Knowing that the child could have an infection with a strain of *Streptococcus* that could be deadly, the physician seeks a very specific diagnosis. What three approaches might the doctor (or a laboratory) use to tell whether this infection is viral or bacterial and, if the latter, identify the bacterium?

2. How can the polymerase chain reaction be useful in studying monera that do not survive in laboratory culture?

3. Cite two places that bacteria and archaea coexist.

4. How could a methanogen also be a thermophile?

5. What evidence has prompted microbiologists to hypothesize that archaea probably are abundant in the earth's crust?

6. Do you think the microorganisms living in the sponges described in Biology in Action 22.1 are bacteria or archaea? What characteristics would you examine to distinguish between these possibilities?

7. Suggest an alteration that might render *Vibrio cholerae* unable to cause cholera.

8. In a crowded airport, 37 people suddenly become ill and rush to bathrooms with abdominal cramps and diarrhea. What questions would you ask to try to determine the cause of the outbreak?

9. In previous chapters, we discussed classifying organisms by evidence of shared common ancestry. Why might classifying archaea by habitat not be consistent with this evolutionary approach?

TO LEARN MORE . . .

References and Resources

Bult, Carol J. et al August 23, 1996. Complete genome sequence of the methanogenic archaeon, *Methanococcus jannaschii. Science*, vol 273.

de Kruif, Paul. 1966. *Microbe Hunters.* New York: Harcourt Brace Jovanovich. This engrossing historical account of major discoveries in microbiology highlights the effects of monera on human health.

Koprowski, Hilary and Michael B.A. Oldstone, eds. 1996. *Microbe Hunters—Then and Now.* New York: Medi-Ed Press. Similar to Paul de Kruif's classic, this book traces discoveries about infectious diseases, written by the discoverers.

Lewis, Ricki. June 1994. Getting Lyme disease to take a hike. *FDA Consumer.* Humans, deer, mice, ticks, and bacteria interact in the transmission of Lyme disease.

L'Haridon, S., et al. September 21, 1995. Hot subterranean biosphere in a continental oil reservoir. *Nature,* vol. 377. Archaea are more abundant and widespread than we first thought.

Losick, Richard and Dale Kaiser. February 1997. Why and how bacteria communicate. *Scientific American.* Bacterial simplicity is deceiving.

Sarbu, Serban, et al. June 28, 1996. A chemoautotrophically based cave ecosystem. *Science,* vol. 272. In a submerged portion of a cave, bacteria derive energy from rock walls.

Vacelet, Jean, et al. September 28, 1995. A methanotrophic carnivorous sponge. *Nature,* vol. 377. New bacteria are identified in an unusual sponge colony.

Williams, Nigel. June 28, 1996. Phage transfer: A new player turns up in cholera infection. *Science,* vol. 272. Cholera results from an infection—of bacteria.

denotes articles by author Ricki Lewis that may be accessed via *Life's* home page /http://www.mhhe.com/sciencemath/biology/life/.

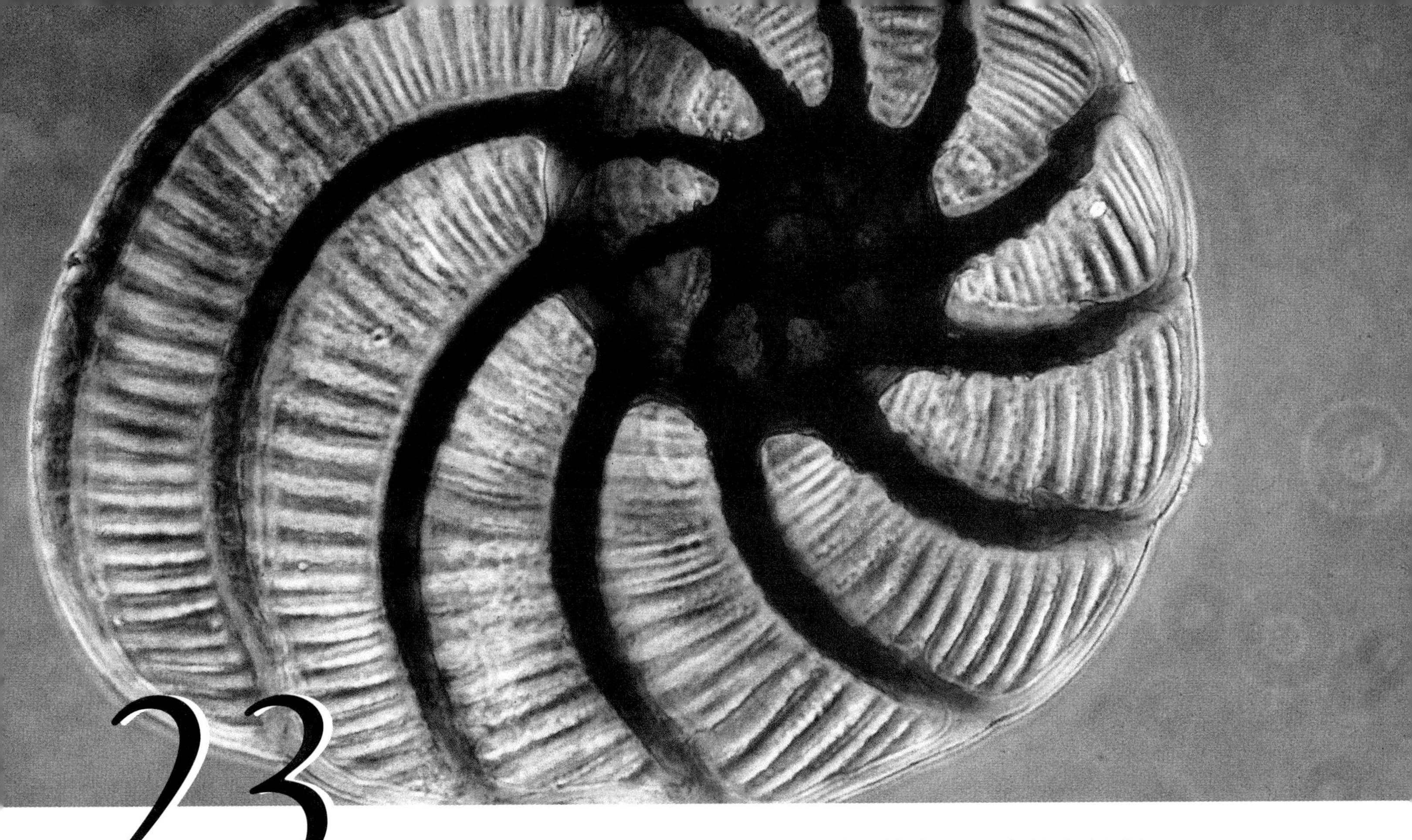

Foraminifera have highly symmetrical tests (shells).

23 Protista

chapter outline

Protista—The Simplest Eukaryotes 458

Protozoa 458
Protozoan Classification
Types of Protozoa
Protozoa and Human Disease

Algae 463
Algal Classification
Types of Algae

Chytrids, Water Molds, and Slime Molds 469
Chytrids
Water Molds
Slime Molds

Living Sands

On ocean bottoms in the tropics and semitropics lies what appears to be colorful sand. On closer scrutiny, this sand consists of tiny, intricate starry shapes, resembling snowflakes, visible with the naked eye. These "living sands" are really intimate, symbiotic combinations of two members of kingdom Protista—organisms called foraminifera housing several types of algae, including green algae, diatoms, and dinoflagellates.

Foraminifera provide the starlike shapes to living sands, which are their calcium-containing shells. These organisms were abundant in the past. Many compressed layers of them form the White Cliffs of Dover in England and the material used to build the Egyptian pyramids. Today, limestone homes are foraminiferan graveyards. Limestone is white because the foraminifera no longer have their live, colorful algal guests. But in the oceans, where the chambers of a foraminiferan's shell house algae, the symbiont's pigments impart hues of green, yellow, red, and brown, depending upon the species present.

What do protistan partners gain from the living sand relationship?

to be continued . . .

Protista—The Simplest Eukaryotes

Kingdom Protista includes the simplest eukaryotic organisms: **protozoa, algae, chytrids, water molds, slime molds,** and a few other groups of organisms (fig. 23.1 and table 23.1). Most members of the 27 protistan phyla are unicellular, and in this regard they resemble bacteria. Protistan cell structures and functions, however, are far more complex than those of bacteria, which are prokaryotes. Some species of protista form many-celled colonies, and some green, brown, and red algae are truly multicellular.

Protista were probably ancestral to the other three kingdoms of eukaryotes—the fungi, plants, and animals. Protistan structure, biochemistry, and early developmental stages, however, are distinct from those of other eukaryotes.

Protozoa

Protozoa are usually unicellular, but they are nonetheless quite complex, even more so than the cells of some multicellular organisms. A sample of soil or drop of pond water is certain to contain many representatives of the seven phyla of protozoa. These organisms first appeared about 1.2 billion years ago, and we know of about 64,000 species today. They are important in a global sense, because many protozoan species form **plankton,** from the Greek for "wanderer." Plankton are usually microscopic organisms that float or drift in large numbers in salt or fresh water. Protozoa, along with some simple animals, make up **zooplankton** ("zoo" means animal). Algae are **phytoplankton** ("phyto" means plant). Plankton are important ecologically because they occupy positions near the bases of food webs, thereby supporting many other types of organisms that eat them.

Most protozoa live freely in fresh or salt water or in very moist terrestrial habitats. About 10,000 protozoan species live symbiotically with other organisms. Some protozoa are parasites, causing illness in host organisms, yet some are beneficial symbionts. In a termite's gut, for example, protozoa of genus *Trichonympha* secrete enzymes that digest wood, the insect's food. Protozoa can even eat each other (fig. 23.2).

Protozoa come in a variety of forms. Some secrete calcium-containing shells, called **tests,** that resemble elaborate spaceships or stars. The foraminifera discussed in the chapter opening have magnificent tests, as do radiolaria, which live suspended in water.

Table 23.1 Kingdom Protista

Type	Characteristics
Protozoa	A diverse group of unicellular eukaryotes that usually obtain nourishment from other organisms.
Algae	Photosynthetic eukaryotes that generally lack roots, stems, leaves, conducting vessels, and complex sex organs. May be unicellular or multicellular.
Chytrids	Unicellular organisms with a single flagellum and chitin cell walls, which may be parasitic.
Water molds	A diverse group with cellulose cell walls. Water molds may aggregate into forms that have many nuclei in a shared cytoplasm.
Slime molds	Molds that form acellular or cellular mobile masses that enable cells to collectively migrate to locate food.

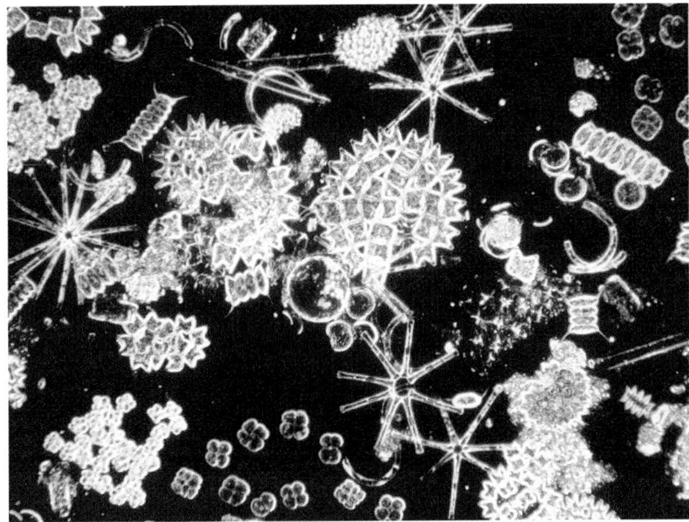

FIGURE 23.1
A drop of pond water contains varied forms of protistan life. (×40)

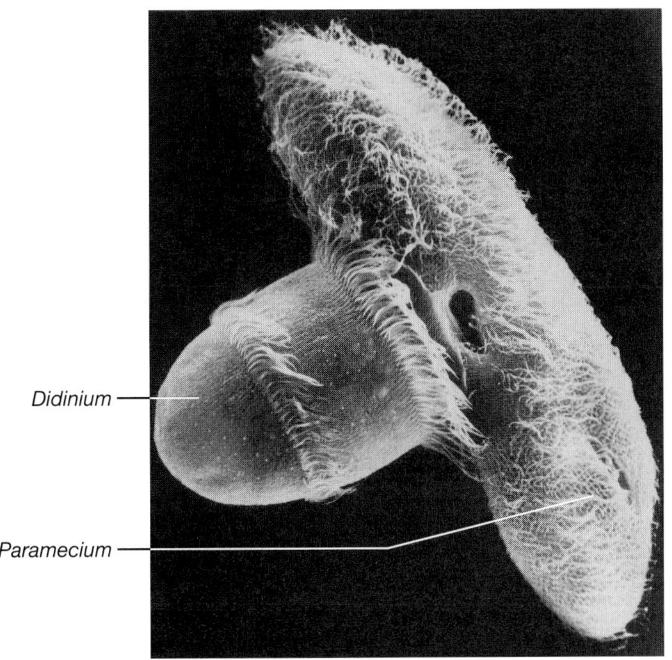

FIGURE 23.2
Didinium consumes another protozoan, *Paramecium.*

Protozoan Classification

Protozoa have traditionally been classified by such characteristics as cell shape, mode of reproduction, and presence of certain biochemicals, such as starches and pigments. As in other taxonomic groups, addition of DNA, RNA, and protein sequence information has altered some classifications. For example, *Pneumocystis carinii* causes pneumonia in people with AIDS. The organism was considered a protozoan, but ribosomal RNA sequence analysis places it closer to the fungi. Researchers are still discussing its classification.

For an inhabitant of the human intestine called *Blastocystis hominis,* molecular evidence helped confirm its classification as a protozoan. *B. hominis* has confused taxonomists for more than a century because it assumes different forms in the human intestine. Sometimes it resembles an amoeba (a free-moving cell), at other times it has flagella, yet it can also resemble yeast, which is a fungus. Its ribosomal RNA gene sequences, however, are like those of protozoa, which is evidence that this organism belongs to kingdom Protista.

Biologists consider several characteristics in classifying protozoa: mode of locomotion, nutrition, reproduction, type and number of nuclei, and whether they are symbiotic or free-living.

A protozoan cell may contain a single nucleus or two nuclei. If it has two, they may be identical, or one may be large (a **macronucleus**) and one small (a **micronucleus**). Protozoa have three types of vacuoles with distinct functions. A **contractile vacuole** pumps out water, a necessary function for protozoa living in a hypotonic environment (see fig. 5.8). A **phagocytic vacuole** forms when the cell engulfs food particles. A **secretory vacuole** contains enzymes with a variety of functions. Structures that provide motility include waving flagella, beating cilia, **pseudopods** (false feet) of cytoplasm that drag the cell, and amoeboid motion (see figs. 34.14 and 23.3).

Protozoa obtain nutrients in several ways. These organisms may be autotrophs (photosynthetic) or heterotrophs. A heterotrophic protozoan may be **holozoic,** taking in food through phagocytic vacuoles, or **saprozoic,** obtaining nutrients across the cell membrane by other mechanisms. These include diffusion, facilitated diffusion, active transport, or a process called **pinocytosis** ("cell drinking") in which small amounts of fluid surrounding a cell are pinched inward and brought into the cell.

Many protozoa can enter a dormant stage by forming a **cyst** that has a cell wall and lowered metabolism. Cysts enable a protozoan to survive a temporarily harsh environment, they provide a setting for asexual reproduction, and they are a means of transport from one host to another if the protozoan is a symbiont. When environmental conditions are more favorable—a drought ends, food becomes available, or the organism is in an appropriate host—the protozoan bursts from its protective cyst.

All protozoa reproduce asexually, and some reproduce by sexual routes too. Asexual reproduction can occur in several ways (fig. 23.4). In binary fission, an individual splits to yield two genetically identical offspring. In **budding,** a parent cell divides to form a second, smaller progeny cell, which then grows. Some protozoa reproduce asexually by **multiple fission,** in which several divisions of nuclei precede a division of the cytoplasm, yielding several progeny at once.

Sexual reproduction exchanges the genetic material of two individuals, which may produce progeny with novel combinations of their parents' genes. Many protozoa sexually reproduce by **conjugation.** The ciliated protozoan *Paramecium caudatum,* for example, conjugates by a complex series of steps that entail divisions of both a macronucleus and a micronucleus. Two paramecia of opposite mating types produce eight progeny.

Pseudopodium or false foot

Nucleus

Direction of movement

(a) *Amoeba*

(b) *Paramecium*

Waving cilia

Gullet

Eyespot

Flagellum

Nucleus

Chloroplasts

(c) *Euglena*

FIGURE 23.3

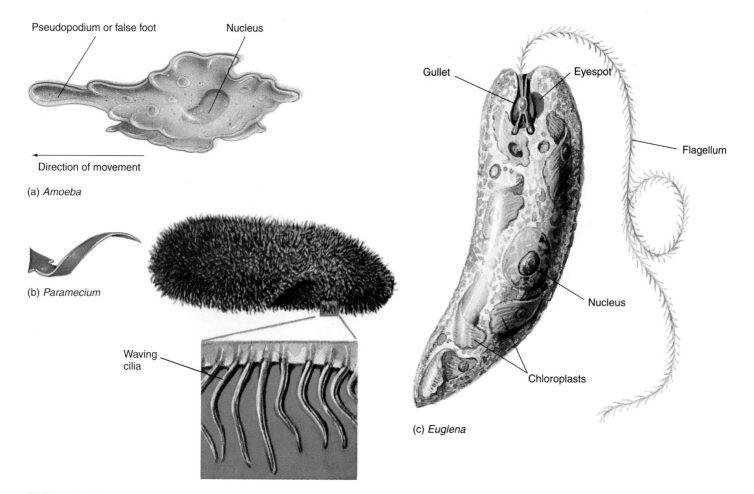

Protistan locomotion. The amoeba (*a*) moves by molding its body into footlike pseudopods, the paramecium (*b*) by waving cilia, and the euglena (*c*) by waving a flagellum.

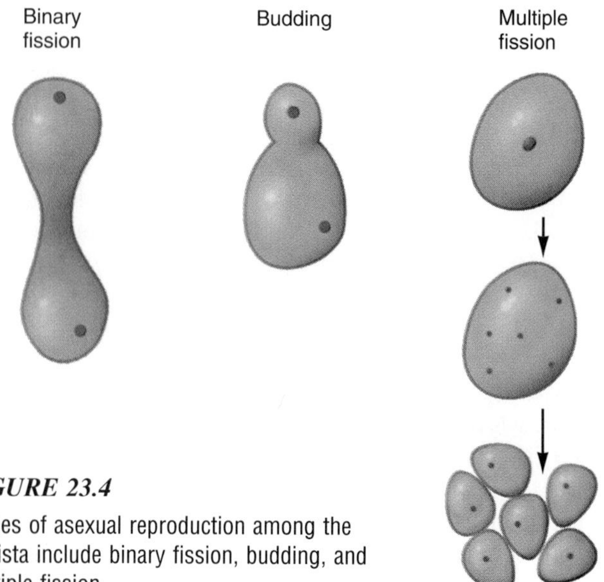

FIGURE 23.4

Modes of asexual reproduction among the protista include binary fission, budding, and multiple fission.

Types of Protozoa

The seven protozoan phyla are *Sarcomastigophora, Labyrinthomorpha, Ascetospora, Apicomplexa, Myxozoa, Microspora,* and *Ciliophora.*

Sarcomastigophora

This phylum is divided into two subphyla. The *Mastigophora* have flagella, and the *Sarcodina* have pseudopods. Members of both phyla have a single type of nucleus.

The *Mastigophora* are of two types. Phytoflagellates have chloroplasts and therefore obtain nutrients from photosynthesis. Zooflagellates lack chlorophyll and are holozoic, saprozoic, or symbiotic. Zooflagellates cause many human infectious diseases.

Sarcodina includes a great diversity of amoebae. They include the foraminifera and radiolaria, as well as amoebae that parasitize cockroaches and termites, and those that cause amoebic dysentery in humans.

Labyrinthomorpha

This small phylum includes species with elongated cells. Most types live in the ocean, where they are either saprozoic or parasitic on algae. Protozoa of genus *Labyrinthula* made headlines when they killed much of the eelgrass on the Atlantic coast, threatening populations of ducks that eat the grass.

Table 23.2 Human Protozoan Infections

Organism	Illness	Symptoms
Cryptosporidium parvum	Cryptosporidiosis	Diarrhea; waterborne
Giardia lamblia	Giardiasis	Diarrhea
Trichomonas vaginalis	Trichomoniasis	In female, infection of vagina and urethra, producing itching and discharge; in male, infection of prostate, seminal vesicles, and urethra, with no symptoms
Trypanosoma brucei rhodesiense and T. brucei gambiense	African sleeping sickness	Degeneration of lymph nodes and small blood vessels in brain and heart; transmitted by tsetse flies
Entamoeba histolytica	Amoebic dysentery	Severe diarrhea, abscesses in liver, lungs, and brain
Toxoplasma	Toxoplasmosis	Fatigue or no symptoms in adults; very harmful to fetuses
Plasmodium	Malaria	Cycle of fever and chills
Acanthamoeba	Keratitis	Cornea inflammation, often from contaminated soft contact lenses
Leishmania	Leishmaniasis	Lesions of mouth, nose, throat, and skin; transmitted by sand flies

Ascetospora

These parasitic protista have characteristically shaped spores. Their hosts are usually mollusks.

Apicomplexa

Also called sporozoa, these organisms have an elongated anterior end called an apical complex that consists of microtubules, fibrils, vacuoles, and other organelles. Adult sporozoa are not motile—they lack locomotive structures. Their life cycle includes multiple fission, sexual reproduction involving haploid and diploid stages, and passage through mammalian and insect hosts. Sporozoa cause malaria and toxoplasmosis in humans.

Myxozoa

The myxozoa parasitize fish. They cause "whirling disease" in trout and salmon by infecting the fishes' nervous systems and hearing organs and upsetting their sense of balance, causing them to swim in a whirling fashion. These protista have a characteristic spore form.

Microspora

These small organisms live within the cells of other organisms, particularly insects. The microsporan *Nosema bombicus* causes a disease called pebrine in silkworms; an outbreak can greatly damage the silk industry. The infection spreads as silkworms ingest contaminated feces from other silkworms on the leaves they eat. *Nosema apis* is a microsporan that causes dysentery in honeybees. Some members of this phylum are used to control insect pests.

Ciliophora

The ciliates are unicellular, heterotrophic protista with rows of beating cilia that enable them to move with great precision. These organisms also use cilia to eat by waving food particles towards the buccal cavity, or mouth. *Paramecium* is a familiar ciliate that reproduces by binary fission and conjugation.

Protozoa and Human Disease

Human pathogens appear in all groups of protozoa, but zooflagellates and sporozoa cause the most severe diseases (table 23.2). All of the sporozoa are obligate parasites that become nonmotile as adults and remain inside their hosts, absorbing nutrients from host cells. Because they are nonmotile, sporozoan parasites require an intermediary called a vector, usually a blood-sucking insect, to carry them in their proliferative forms to new hosts. Biology in Action 23.1 explores malaria, a sporozoan infection.

Hikers in the Adirondack Mountains of upstate New York are often warned to boil water before they drink it, because many streams and ponds are habitats for *Giardia lamblia,* which causes stomach and intestinal illness. When a person drinks water containing *Giardia* cysts, the cysts burst in the individual's duodenum (upper small intestine) as forms called **trophozoites.** They then attach by disc-shaped suckers to the intestinal lining, where they reproduce rapidly, hampering the intestine's ability to absorb nutrients (fig. 23.5). The protozoa also produce a toxin that causes stomach pain, diarrhea, intestinal gas, weight loss, and lack of appetite.

Most *Giardia* trophozoites cling tenaciously to intestinal cells, but some become encysted and exit the body ın feces. Antonie van Leeuwenhoek saw these cysts in his lenses in a sample of his own feces. About 200 million people worldwide carry *Giardia* without developing symptoms. From 5 to 15% of diaper-clad children in U.S. day-care facilities are carriers, and as much as 10% of the general population may be infected. Only about 30,000 people in the United States, however, become ill from *Giardia* each year.

The AIDS epidemic has sharpened our awareness of protozoan infections, because immunity breakdown opens the human body to protozoa that normally infect other animals, reside in the human body without causing illness, or usually cause mild, treatable symptoms. For example, cryptosporidiosis, a waterborne infection by *Cryptosporidium parvum,* was very rare between 1976 and 1982 and then began to rise in 1982, as HIV infection made people more susceptible to it.

Toxoplasma gondii is a protozoan that causes toxoplasmosis, with symptoms so mild (fatigue) that many otherwise healthy people are infected and never know it. The infection came to public

23.1 biology in action

Malaria

Malaria is a serious, even deadly protozoan infection that has left its mark on human history. Today, in many tropical parts of the world, it continues to threaten human health. Worldwide, 150 million people a year suffer from malaria. In tropical Africa, more than a million children die each year of the infection.

Symptoms and Course of Infection

Malaria starts with chills and violent trembling and then progresses to high fever and delirium. The person sweats profusely, is completely exhausted, and has a dangerously enlarged spleen. The disease strikes in a relentless cycle, with symptoms returning every 2 to 4 days. Within weeks, the sufferer either dies of circulatory system collapse or overcomes the parasite. A person whose immune system fights off the infection may feel well for months and then become ill again. The waxing and waning symptoms reflect the parasite's life cycle. The frequent crises cause anemia as the person's red blood cell supplies plummet and strain the liver and spleen, both of which process broken-down red cells.

A cycle of malaria begins when an infected female mosquito of any of 60 *Anopheles* species bites a human. The insect's saliva contains an anticlotting agent, as well as sporozoites, which are small, haploid cells of *Plasmodium falciparum, vivax, malariae, ovale,* and others. The sporozoites enter the human host's liver cells, where they divide rapidly and form structures called merozoites. Some merozoites reinfect liver cells, and others infect blood cells. For the next 2 to 3 days, the merozoites enlarge and divide, finally bursting from the red blood cells and infecting other cells. The rush of merozoites, burst red blood cells, and toxins causes severe chills and fever.

Meanwhile, a few *Plasmodia* specialize into gametes (sexual forms). When mosquitoes drink the gametes from an infected person's blood, the gametes unite in the insect's stomach. After several additional steps, sporozoites form, which move to the mosquito's salivary glands, ready to enter a new host when the insect seeks its next blood meal.

Plasmodium falciparum causes a particularly deadly form of malaria that affects the brain. Only two drugs can treat certain strains of *P. falciparum.* Infectious disease experts fear that if the parasite evolves resistances to these drugs, severe malaria may become more prevalent in some parts of the world.

The type of vegetation in an area determines whether the mosquito and its parasite will thrive. Agriculture often creates conditions that encourage malaria by replacing dense forests with damp rice fields, a haven for moisture-loving mosquitoes. One reason that malaria is relatively uncommon in the United States is that in many places warm weather doesn't last long enough for the mosquitoes to survive and perpetuate the infective cycle.

History

For centuries, people have recognized the link between swamps and recurring malaria. The name of the disease comes from seventeenth-century Italy, where people living near smelly swamps outside of Rome developed mal'aria—"bad air." As long ago as the fourth century B.C., however, Greeks noted that people living near swamps had bouts of fever and enlarged spleens. It took centuries to identify the protozoan that causes malaria.

In 1880, French army surgeon Charles Louis Alphonse Laveran saw the parasites in a sick person's blood (fig. 23.A). By the end of the century, researchers had discovered the role the *Anopheles* mosquito plays in transmitting the disease. In 1902, British army surgeon Sir Ronald Ross, using birds as models, deciphered the entire complex cycle of malaria.

Eradication

People were able to treat malaria long before they understood the infective cycle. In the sixteenth century, Peruvian natives gave Jesuit missionaries on their way to Europe their secret malaria remedy—the bark of the cinchona tree. It was not until 1834, though, that French chemist Pierre Joseph Pelletier extracted the active ingredient from cinchona bark: quinine, which is still used in some forms, in addition to many other drugs developed to keep pace with *Plasmodium*'s evolving resistance to various drugs. In 1955, the World Health Organization announced an eradication campaign against malaria; by 1976, it admitted failure. Malaria had actually spread through developing nations as people cleared land for farming.

Although we have moderately effective ways to prevent and treat malaria, stemming the illness is very challenging, for biological as well as sociological reasons. Not only does *Plasmodium* continue to develop drug resistance, but nations troubled by poverty and civil unrest struggle to distribute drugs to prevent or treat malaria. Still, certain measures can lower the risk of contracting malaria:

- a drug called mefloquine can prevent malaria when administered weekly for 2 years and can kill parasites within 48 hours;
- bednets soaked in insect repellent keep mosquitoes away;
- screens on windows and doors help keep out mosquitoes;
- wearing long pants and long sleeves, especially during the evening, helps prevent mosquito bites.

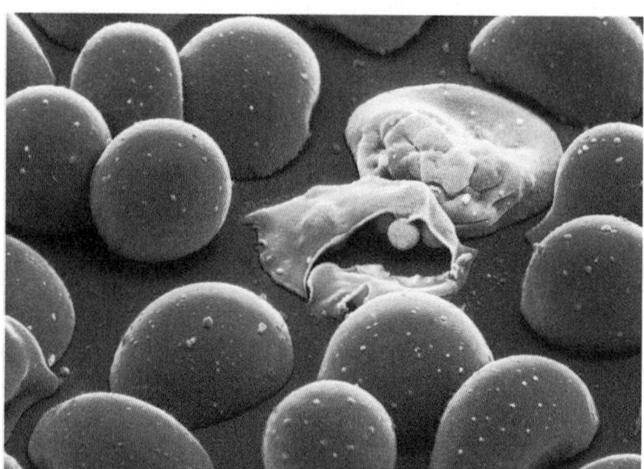

FIGURE 23.A

Plasmodia that cause malaria burst from these red blood cells in a laboratory culture dish.

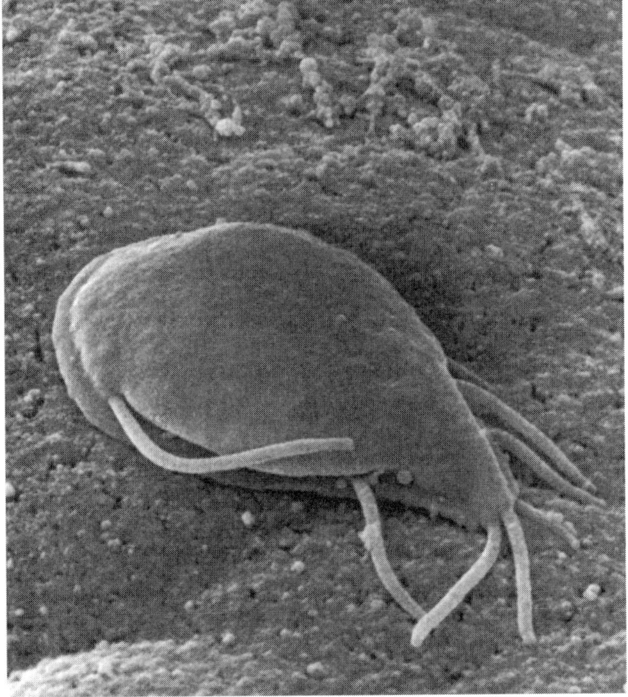

(a)

(b)

FIGURE 23.5

Giardia lamblia are protozoa that cause painful giardiasis. The organism attaches to the lining of a person's small intestine (*a*) and leaves an impression when it detaches (*b*).

attention when physicians traced fetal defects to pregnant women who contracted toxoplasmosis from cat feces in litterboxes. Symptoms in the fetus include an enlarged liver and spleen, a rash, eye inflammation causing blindness, and brain damage. The women had no symptoms when pregnant—only an immune system test revealed the infection. The maternal infection must occur early in pregnancy to affect the fetus, and only a small percentage of exposed fetuses actually develop symptoms.

People also may contract toxoplasmosis from eating raw or undercooked meat, undergoing an organ transplant, or receiving a blood transfusion. An infected person takes in cysts through the nose or mouth.

In people with AIDS, toxoplasmosis is not a vague, mild disorder, but a deadly disease. The protozoa quickly spread to the brain, causing inflammation (encephalitis), and also travel to the lungs, heart, liver, and eyes.

Mastering Concepts

1. How do protista differ from monera and from other eukaryotes?
2. What are some habitats of protozoa?
3. What are the types of nuclei and vacuoles in protozoa?
4. How do protozoa obtain nutrients?
5. What are some types of protozoa?

Algae

The algae are a very diverse group of organisms that have in common use of pigments to capture solar energy. They range in size from single cells a mere 100 micrometers long to enormous multicellular kelps more than 82 yards (75 meters) long. Like plants, algae are eukaryotes that have a variety of pigments that enable them to photosynthesize. Unlike plants, algae lack roots, stems, leaves, and complex sex organs. In addition, the types of pigments and carbohydrates in algal cells differ from those in plant cells, and some algae have cell walls not composed of cellulose, the major component of plant cell walls.

Ancient algae were probably ancestors of modern plants, but the algae are so varied that they might not all descend from a recent shared ancestor. The algae include cyanobacteria in kingdom Monera and seven divisions within kingdom Protista. Some botanists, however, consider the brown and red algae to be plants.

Algal Classification

One way of distinguishing types of algae is by habitat. Most algae are aquatic, living in fresh water, seawater, and brackish waters. Other algae live in a great variety of habitats, including moist rocks or wood, soil, tree bark, in the air, in crevices of shellfishes' exoskeletons, and even lending a greenish hue to mammals by taking up residence in their fur (fig. 23.6). Some algae are endosymbionts of plants, mollusks, protozoa, corals, or worms. Brown and red algae, in the form of kelps, form dense "forests" in oceans, supporting vast and diverse living communities. An alga may combine with a fungus to form a unique composite organism called a lichen, discussed in the next chapter.

Aquatic algae are phytoplankton. Algae that live on the bottoms of bodies of water are **benthic,** and those that live where the water meets the air are **neustonic.**

FIGURE 23.6
Algae tinge this polar bear's fur green.

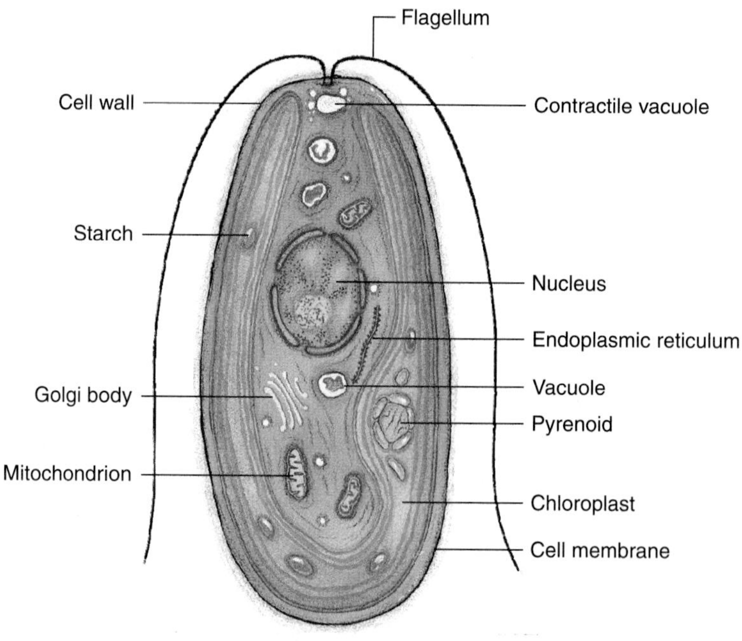

FIGURE 23.7
An algal cell. Note the typical eukaryotic organelles, coupled with the specialized algal structures, such as a pyrenoid and contractile vacuole.

Other criteria for classifying algae include:

Cell structure. Algal cells contain typical eukaryotic organelles (fig. 23.7) but may differ from each other by the thickness and composition of the cell wall, if one is present.

Form of nutrient storage. Algae may store photosynthate as any of several types of carbohydrate, including common starches, as well as molecules unique to certain types of algae.

Photosynthetic pigments. Algae contain chlorophylls, carotenoids, and other pigments that expand the range of light wavelengths from which they can capture energy.

Number of flagella. Algae differ in how and where flagella arise from their cells.

Vegetative body form. The nonsexual parts of algae, called **thalli** (singular: thallus), may take several forms (fig. 23.8). Thalli may resemble long filaments, tubules, sheets, or blades. They may be unicellular or colonial, consisting of several aggregated individual cells. Unicellular algae may be able to move (motile) or not (nonmotile).

Reproduction and life cycle. Algae can reproduce asexually and sexually. Asexual reproduction takes any of three forms: binary fission; fragmentation, in which the thallus breaks into pieces that develop into separate individuals; and spores that form from vegetative cells and develop into new individuals. In sexual reproduction, eggs develop in vegetative cells called **oogonia,** and sperm develop in male structures called **antheridia.** Egg and sperm fuse, forming a zygote.

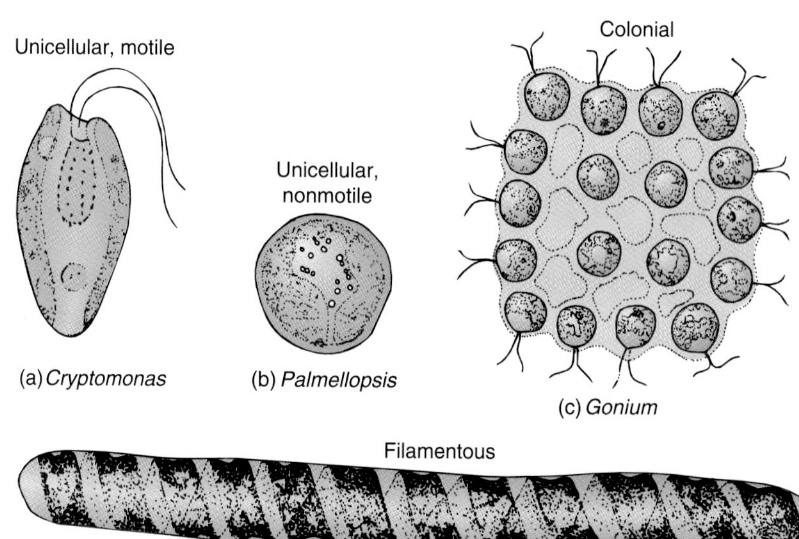

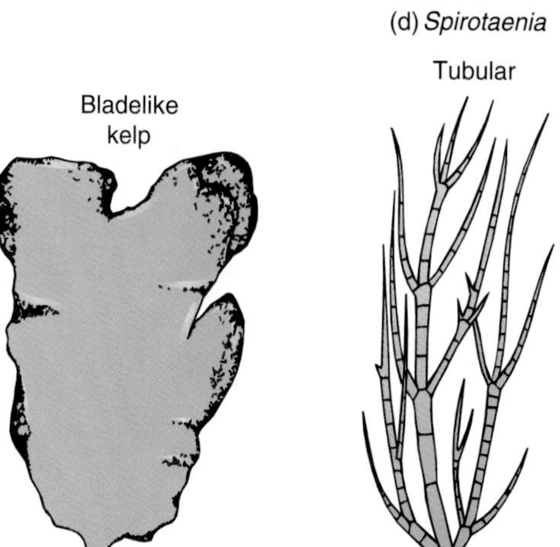

FIGURE 23.8
Algal body forms.

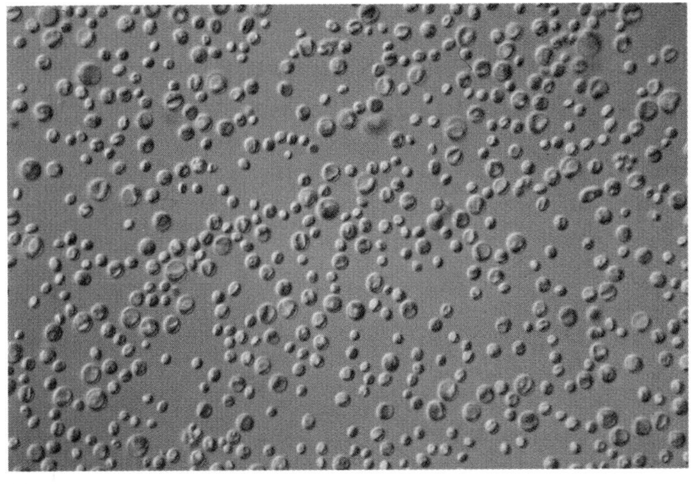

(a)

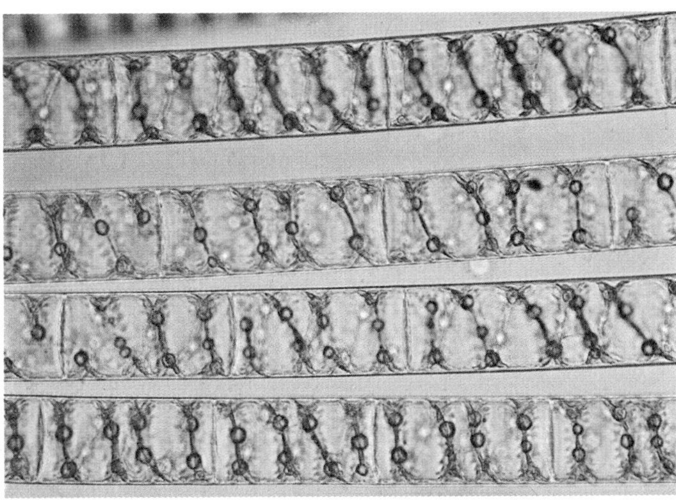

(b)

(c)

FIGURE 23.9

Green algae take a variety of forms. (*a*) *Chlorella* is unicellular and nonmotile. (×160) (*b*) This view shows four filaments of *Spirogyra* (×100), so called for its spiral-shaped chloroplasts. (*c*) *Acetabularia* has a plantlike form.

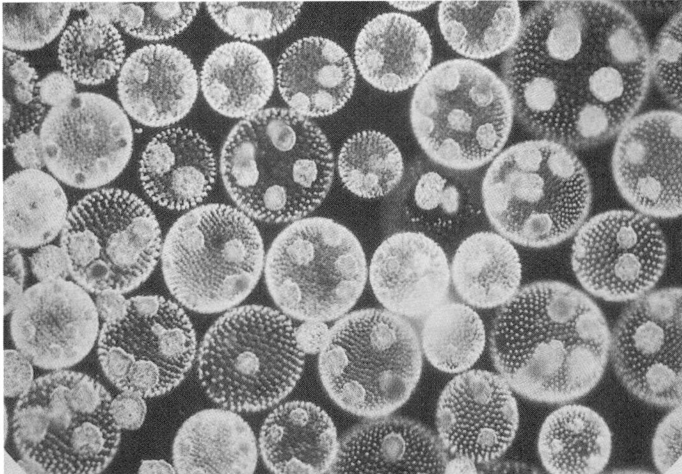

FIGURE 23.10

Volvox is a colonial green alga. (×50)

Types of Algae

Algae are grouped into seven divisions.

Chlorophyta (green algae)

The 7,500 species of **green algae** exhibit the most diverse body forms and habitats (fig. 23.9). Many algal cells have two flagella, a single haploid nucleus, and one large chloroplast, which contains a starch storage region called a **pyrenoid.** A green alga has chlorophyll a and b and carotenoid pigments. It reproduces asexually and sexually.

One well-studied green alga is *Chlamydomonas.* Figure 10.4 shows the organism and its mechanisms of asexual and sexual reproduction. *Chlamydomonas* survives drought by pulling in its flagella and entering a dormant state. When water returns, it regrows flagella and reproduces. *Chlamydomonas* tinges puddles green following a heavy rain. It may have descended from an ancient alga that diverged through evolutionary time and gave rise to nonmotile, single-celled green algae, as well as to an intriguing genus called *Volvox.*

Unlike *Chlamydomonas, Volvox* is colonial, consisting of 500 to 60,000 single cells that aggregate to form a hollow ball just barely visible to the human eye (fig. 23.10). The cells wave their

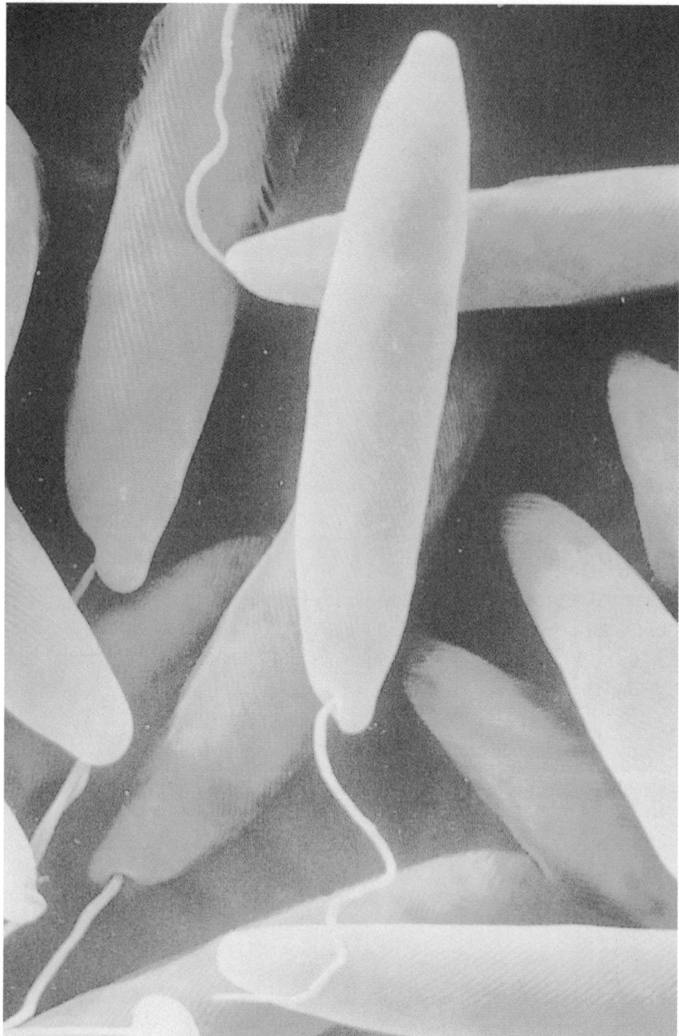

FIGURE 23.11

Euglena (×2,800) have flexible periplast layers beneath the cell membrane.

flagella in a coordinated fashion, which enables the sphere to move. *Volvox* also demonstrate cellular specialization. Reproductive cells cluster at one end of the sphere, almost as if it were a true multicellular organism. Because of this organization, *Volvox* may represent a bridge between the single-celled and many-celled way of life.

Euglenophyta (euglenoids)

The 800 species of **euglenoids** (see figs. 23.3*c* and 23.11) are elongated algal cells that have no cell walls. They have protein strips that form a structure called a **periplast** on the inside face of the cell membrane, which provides motility while maintaining cell shape. Euglenoids aggregate into "blooms" on pond bottoms and also may live in other aquatic habitats and in soil. They have chlorophyll a and b and carotenoids and several chloroplasts per cell.

FIGURE 23.12

The branched filaments and leaflike structures make this stonewort, of genus *Chara,* resemble a plant. The brown structures are oogonia and antheridia.

Euglenoids have a unique storage carbohydrate called **paramylon.** Large contractile vacuoles enable euglenoids to pump out water that enters from their hypotonic aquatic environments. Most euglenoids have two flagella, one of which emerges from the cell and beats. These protista reproduce through a mitosis-like cell division.

Charophyta (stoneworts or brittleworts)

These algae derive their popular names, **stoneworts** or **brittleworts,** from their limestonelike coverings. *Charophyta* construct these characteristic coverings by precipitating magnesium carbonate and calcium carbonate from water. Charophytes live on the bottoms of shallow ponds. These organisms superficially resemble plants, with branched filaments resembling stems and broad structures that look like leaves (fig. 23.12). Their size, up to a meter in length, is also reminiscent of plants. *Charophyta*'s tissues, however, are less complex than those of plants.

Pyrrhophyta (dinoflagellates)

Dinoflagellates are named for their characteristic two flagella of uneven length. These organisms may have distinctive outer plates composed of silica-covered cellulose, and they range in color from yellowish green to brown, depending upon the relative amounts of chlorophyll a and c, carotenoids, and xanthophyll pigments they contain. The pigments in species of *Noctiluca, Pyrodinium,* and *Gonyaulax* lend an eerie glow to the ocean at night.

Dinoflagellates are important ecologically because they are planktonic (fig. 23.13). Populations of *Gymnodinium* and *Gonyaulax,* however, sometimes "bloom," or overgrow, and appear as red tides because of their pigments (see fig. 8.7*a* and *b*). Red tides occur along the coast of central California and on Florida's Gulf coast. Under these conditions, the dinoflagellates release a powerful neurotoxin that inhibits sodium ion transport in the nerve cells that stimulate

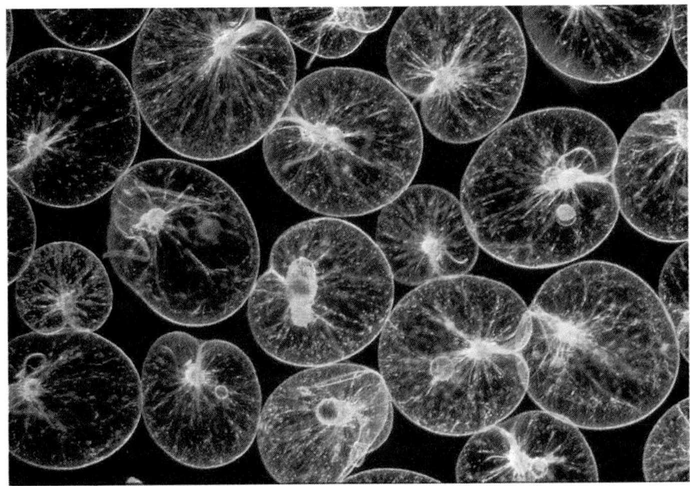

FIGURE 23.13

These dinoflagellates (*Noctilluca scintillans*) comprise phytoplankton. (×12)

respiratory muscles in vertebrates. In humans, the toxin causes the numb mouth, lips, face, and limbs of paralytic shellfish poisoning. The toxin does not harm the shellfish.

A red tide is one type of algal bloom, which is usually caused by warming or phosphate pollution. The precise cause of red tides, however, is not known. More familiar algal blooms occur in murky, stagnant ponds. When overgrown algae begin to die, they can no longer supply sufficient oxygen to support the community, and many organisms die.

Chrysophyta (golden-brown and yellow-green algae; diatoms)

These algae, which live mostly in fresh water, may impart a foul taste to drinking water. The **golden-brown algae** owe their distinctive colors to a pigment called **fucoxanthin. Diatoms** have two-part silica walls called **frustules** that give them a variety of intricate shapes (fig. 23.14). Diatomaceous earth contains frustules, which add abrasiveness. Such deposits are used as swimming pool filters to, ironically, remove other algae. The abrasiveness of frustules also makes diatoms useful ingredients in detergents, polishes, oils, paint remover, toothpaste, and soundproofing and insulation materials. Diatoms may also impart the distinct quality of paint used in reflective road signs.

Diatoms are an abundant form of phytoplankton: a liter of seawater may contain a million diatoms. These organisms undergo a form of asexual reproduction that decreases cell size at each generation. When cells reach a certain minimum size, a sexual form of reproduction occurs.

Phaeophyta (brown algae)

These multicellular seaweeds, called **brown algae,** live along shorelines in cool climates all over the world and also in freshwater habitats. The *Phaeophyta* are all multicellular and are the most complex algae. They owe their brown to olive-green colors to fucoxanthin pigment, and they have a storage carbohydrate called **laminarin.** The huge kelps that give the Sargasso Sea, near the Caribbean, its name belong to this group. The cuisines of many countries include brown algae.

Brown algae have a distinctive body form, consisting of a holdfast organ, a stemlike region called a **stipe,** a balloonlike area called a bladder, and emanating from the bladder, blades (fig. 23.15*a* and *b*). Their motile cells have a very distinctive pair of flagella on one side. The longer flagellum has projections and points to the anterior end of the cell; the shorter flagellum points toward the posterior end (fig. 23.15*c*). The position and coordinated movement of the flagella move the cell. All brown algae have some cells with these flagella at some point in their life cycles.

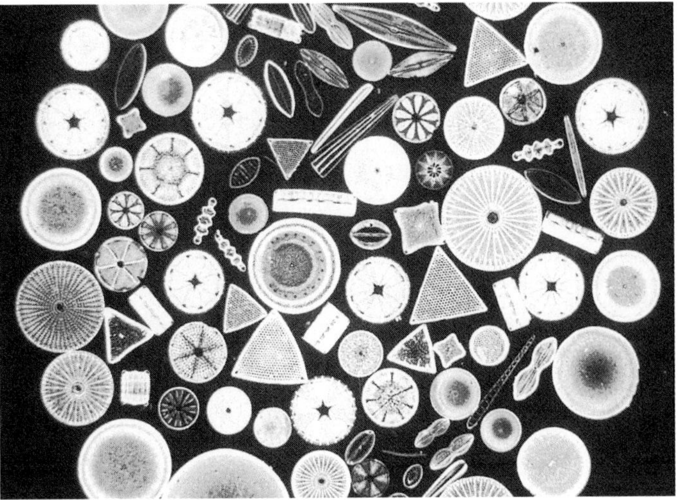

(a)

(b)

FIGURE 23.14

Diatoms. (*a*) Diatoms exhibit a dazzling variety of forms. (×900) (*b*) Diatomaceous earth has many uses because of its abrasiveness, provided by frustules of huge numbers of diatoms. This deposit is in Lompoc, California.

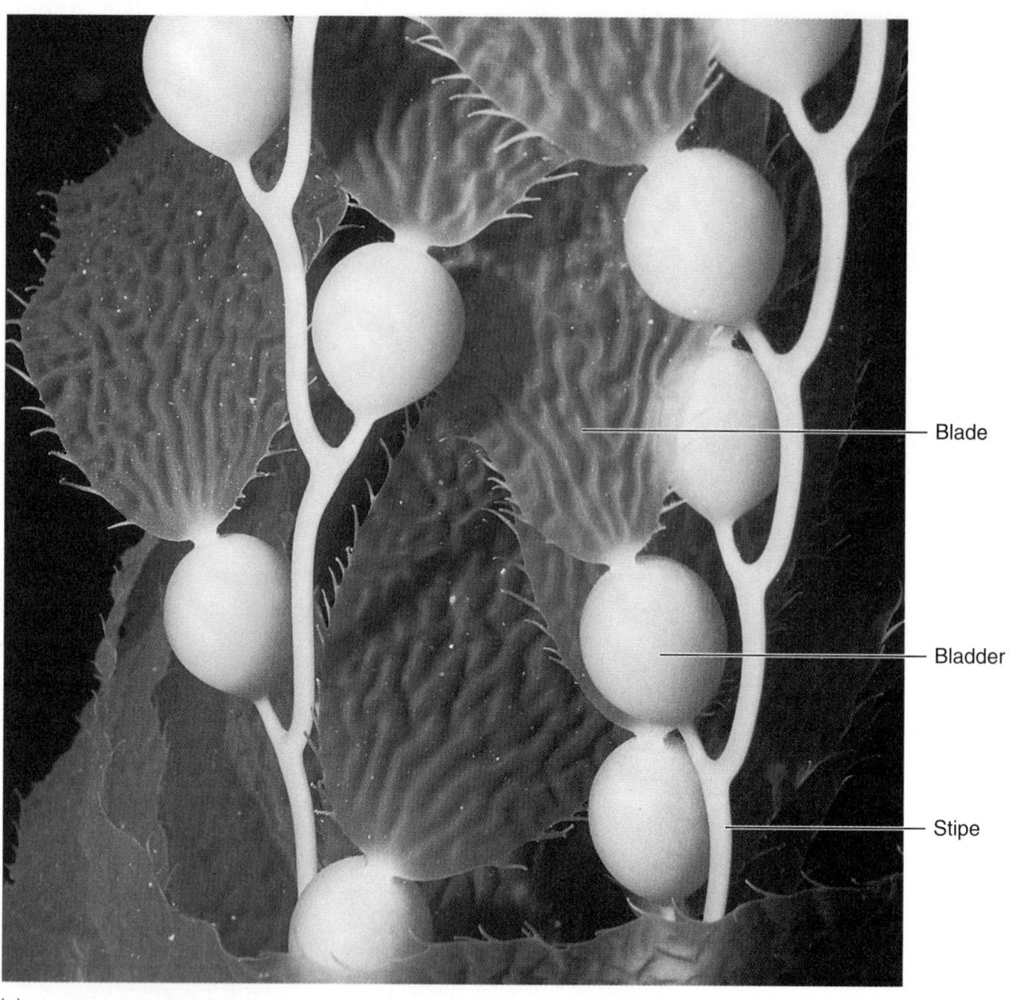

(a)

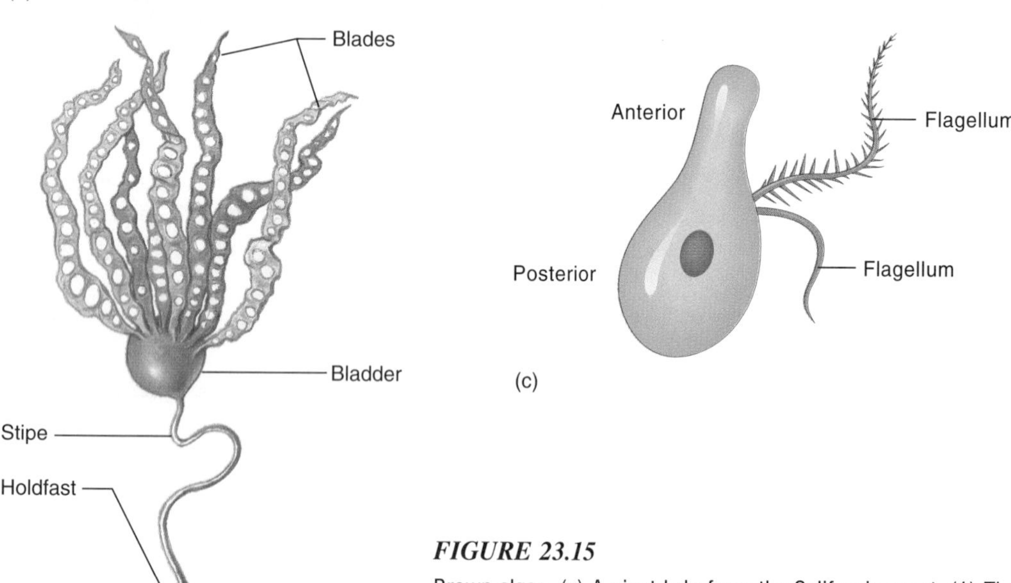

(b)

(c)

FIGURE 23.15

Brown algae. (*a*) A giant kelp from the California coast. (*b*) The holdfast organ anchors the alga to surfaces when tides and currents are strong. The bladder is a gas-filled float that enables the upper parts of the organism to rise above the holdfast. (*c*) Motile brown algal cells have a characteristically located pair of flagella.

FIGURE 23.16

Note the filament-like branches of the red alga *Bossiella.*

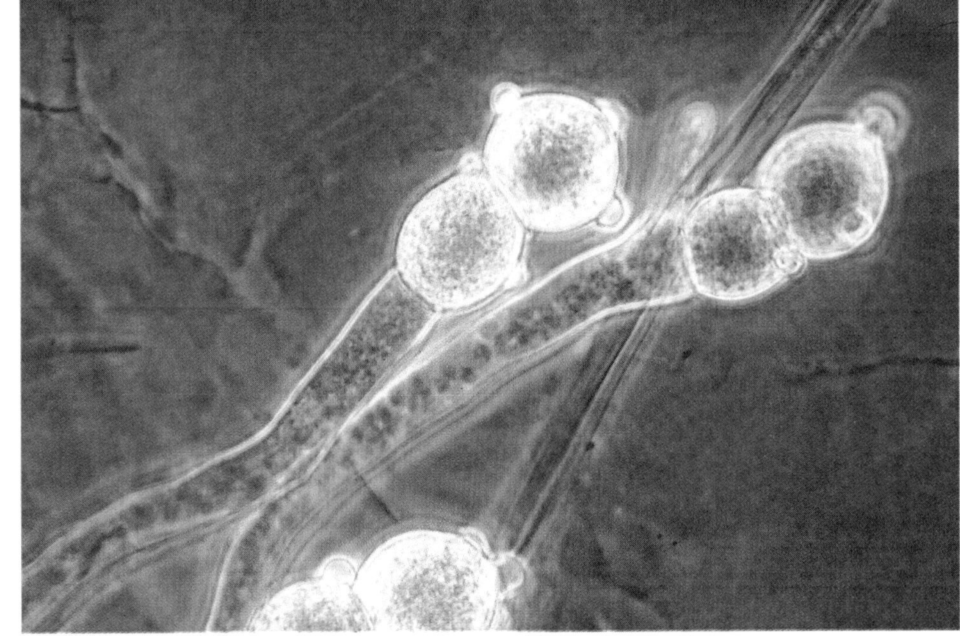

FIGURE 23.17

Allomyces, a chytrid, absorbs nutrients from its surroundings.

Rhodophyta (red algae)

Many seaweeds belong to this group of 3,900 species of **red algae.** The red seaweeds are typically multicellular with branches up to a meter long (fig. 23.16). Their form of carbohydrate is called **floridean starch.** Red algae have pigments called **phycobilins,** which include a red form (phycoerythrin) and a blue form (phycocyanin). These pigments allow a seaweed far below the ocean's surface to absorb wavelengths of light that chlorophyll in plants cannot capture. The algal pigments pass solar energy to chlorophyll molecules.

Deep in the ocean, *Rhodophyta* appear red, but when they wash ashore or near the surface, light destroys the red pigment, and the seaweeds look blue, brown, or dark green. Their slippery feel comes from a microfibril scaffolding within a gluey matrix clinging to the inner surface of the cell walls. One biochemical in the matrix is agar, a gelatinous substance used as a culture medium for microorganisms, as an inert substance in medications, to gel canned meats, and to thicken ice cream and yogurt. Another useful red algal product is the polysaccharide carrageenan, which is used to emulsify fats in chocolate bars and to stabilize paints, cosmetics, and creamy foods.

Mastering Concepts

1. What are algae?
2. What are some characteristics used to distinguish among algae?
3. What are the major types of algae and their characteristics?

Chytrids, Water Molds, and Slime Molds

Chytrids, water molds, and slime molds are classified more by what they are not than by what they are. They do not photosynthesize, so they are not algae or plants, but neither are they animals or fungi. Their modes of reproduction, life cycles, and cellular organization are most like those of protista.

Chytrids

The 750 species of chytrids (phylum *Chytridiomycota*) are varied. All have a single flagellum and cell walls composed of chitin, a carbohydrate that is also in the cell walls of fungi and in the exoskeletons of insects (fig. 23.17). Some chytrids are parasites of water molds, algae, or plants.

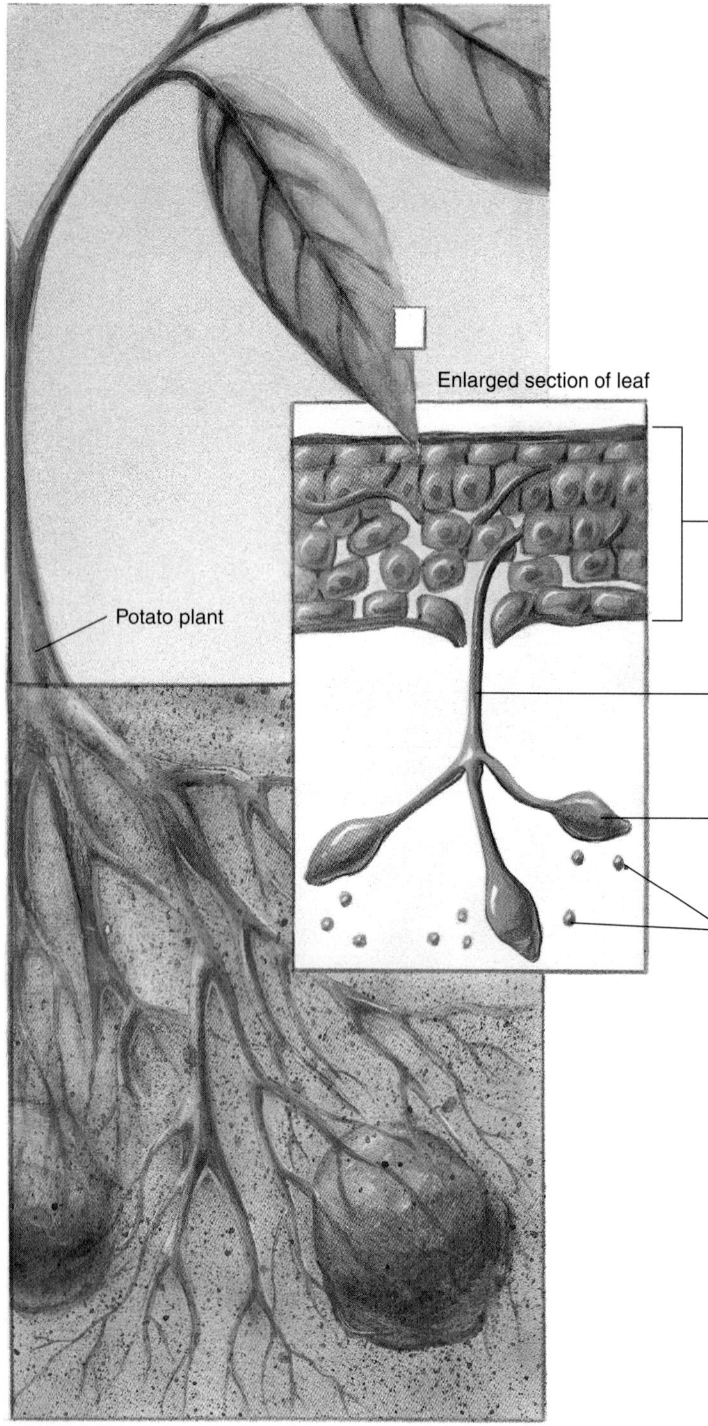

FIGURE 23.18

The water mold *Phytophthora infestans* entangles itself in potatoes. It caused the Irish potato famine of 1845–47 and has recently reappeared in several Northeast states in the United States.

Water Molds

Water molds (phylum *Oomycota*) live in fresh water or moist soil. Some of the 475 species of water molds are unicellular with two flagella, and others form long strands with finely branched filaments containing many nuclei, resembling fungi. Water molds have cellulose cell walls, like plants.

The best-known water molds are those that have thwarted agriculture. Perhaps the most notorious is *Phytophthora infestans,* which means "plant destroyer." It causes an infection called late potato blight and also affects tomatoes. *P. infestans* caused the 1845–47 potato famine in Ireland, when more than a million people starved (fig. 23.18). Although the famine lasted for years, the mold wiped out most of the potatoes in a week. The Irish potato famine followed several rainy seasons, which fostered growth of the mold.

Ruined potato crops have recently resulted from *P. infestans* outbreaks in New Jersey, New York, Pennsylvania, and Delaware. The water mold transforms entire fields of potatoes into rotting wastelands in just a week. New strains of the water mold are resistant to pesticides formerly used to control outbreaks. Many of the molds are in spores now because of dry conditions, but a few years of heavy rains could result in a potato blight in the United States.

Other water molds include *Saprolegnia* and *Achlya,* which resemble cotton and grow on small dead animals and algae; *Plasmopara viticola,* which causes grape downy mildew and can destroy vineyards; and *Peronospora hyoscyami,* which causes blue mold on tobacco. Water molds are sometimes classified as fungus, which is why some accounts of the Irish potato famine attribute it to a fungus.

Slime Molds

The major difference between the two types of slime molds is reflected in their names—acellular (phylum *Myxomycota*) and cellular (phylum *Acrasiomycota*) slime molds. Both types of organisms exist as single cells (amoebae) and also as multicellular masses that respond as if they were single organisms.

In the acellular slime molds, the multinucleated mass, called a **plasmodium,** migrates in soil or along the forest floor, over leaves, debris, and rotting logs, consuming bacteria as it moves (fig. 23.19).

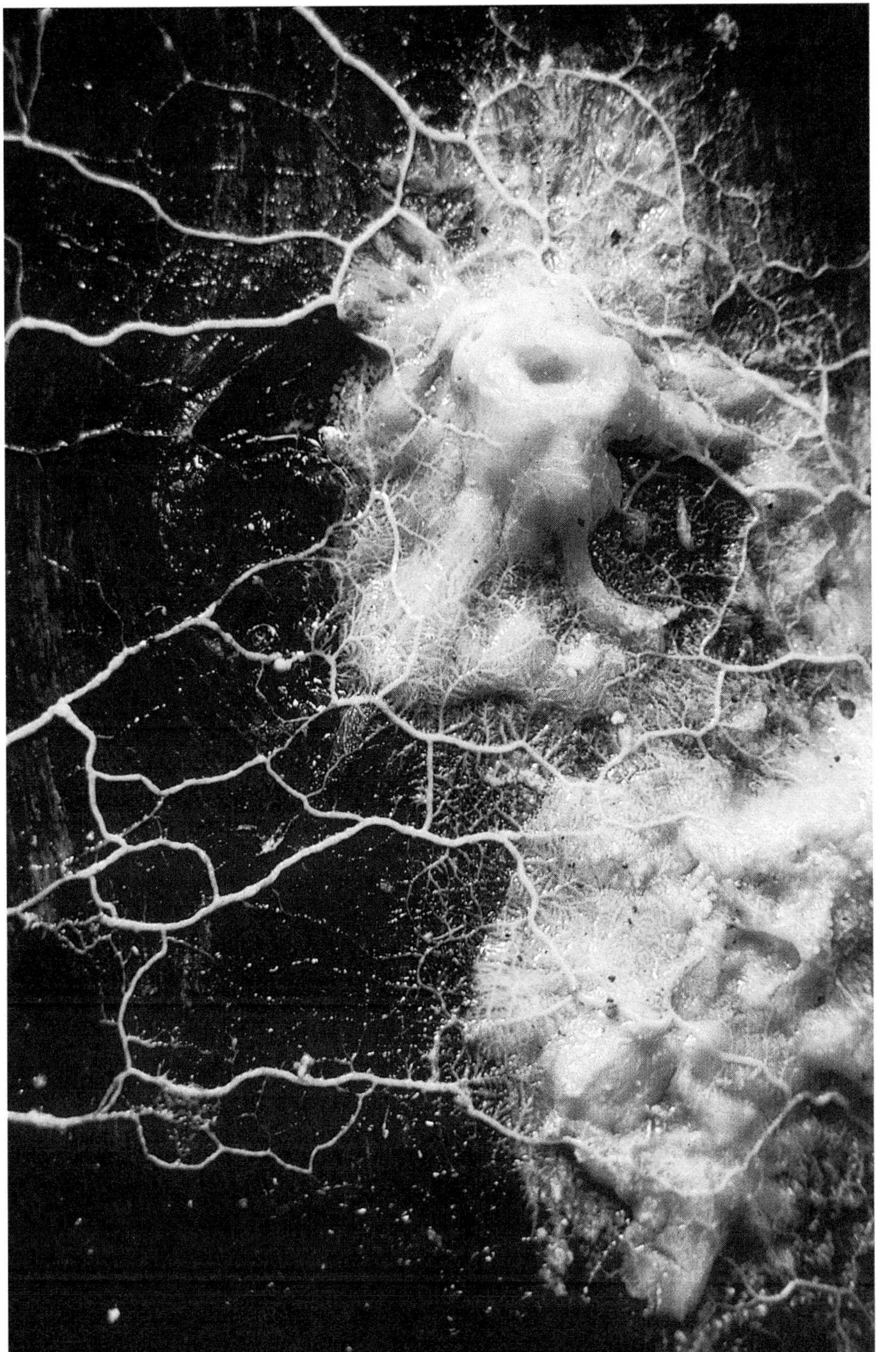

FIGURE 23.19
Streaming masses of the acellular slime mold *Physarum* can move. (×175)

The plasmodium (not to be confused with the organism causing malaria) contains many nuclei, but it is not divided into separate cells. Cytoplasm streams around the nuclei and produces an effect that resembles cells lacking outer boundaries, a little like an office with many desks but no partitions between individual workplaces.

An acellular slime mold on the move may grow to the size of a grapefruit. As the plasmodium enlarges, nuclei divide. In contrast, the individual cells of a cellular slime mold retain their separate identities when they collect into a **pseudoplasmodium,** or multicellular mass.

The slime molds switch body form in response to environmental change. When the habitat is moist and their bacterial and yeast foods are plentiful, acellular slime molds swarm as huge plasmodia. In times of drought or famine, they halt and form stalks topped with reproductive structures that ultimately release single cells (amoebae), starting the cycle anew.

In the cellular slime mold, when food grows scarce, amoebae secrete chemical attractants, which stimulate the amoebae to aggregate, forming a sluglike pseudoplasmodium (fig. 23.20). Cellular differentiation distinguishes the front and back cells of the slug. The structure moves towards light. Upon locating food, the slug stops, forming a stalk topped by a fruiting body that contains reproductive spores. The cells of the stalk perish, allowing the spores to survive. If environmental conditions are favorable, the spores form haploid amoebae. The cycle renews.

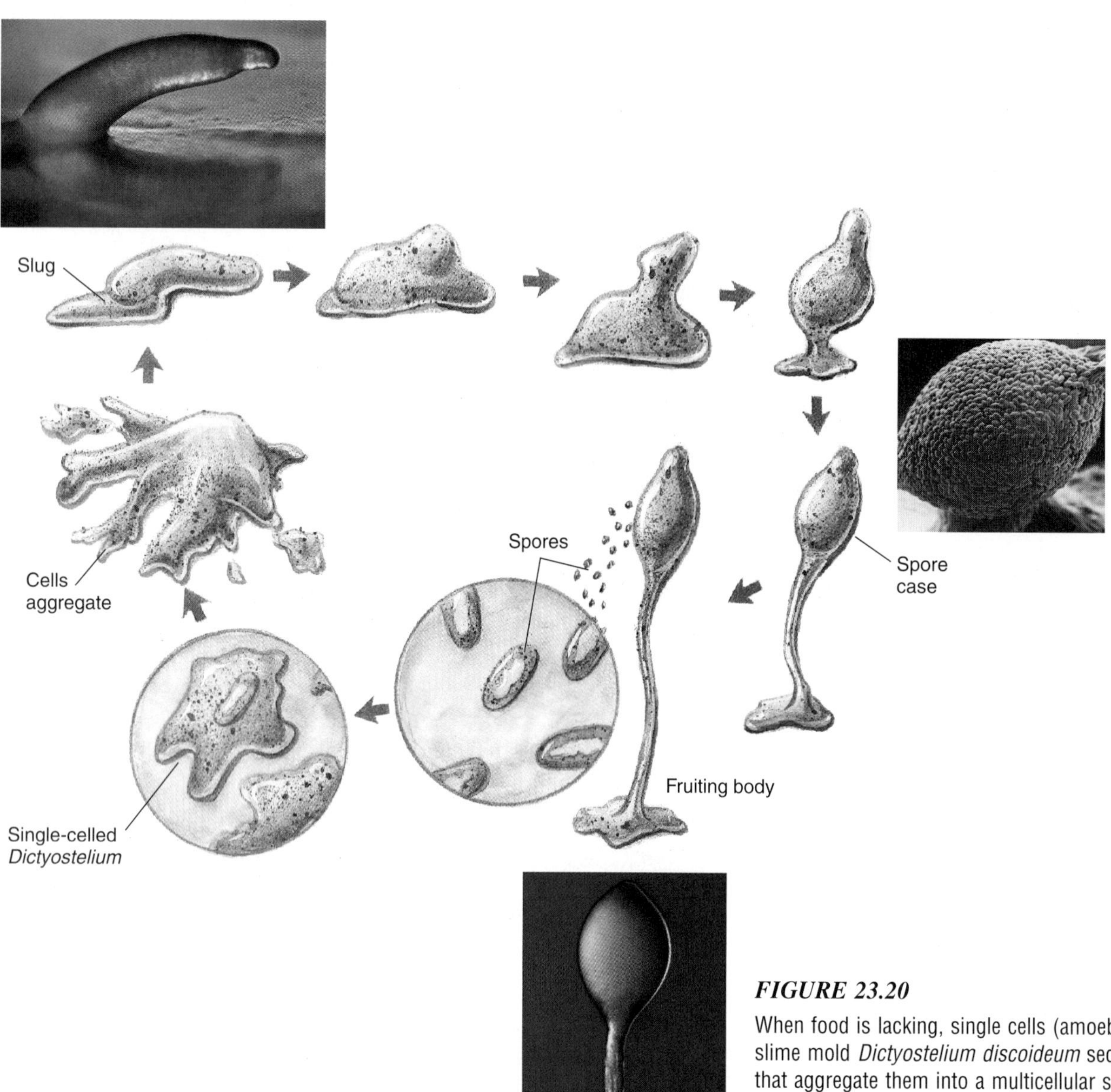

FIGURE 23.20

When food is lacking, single cells (amoebae) of the cellular slime mold *Dictyostelium discoideum* secrete biochemicals that aggregate them into a multicellular slug. Attracted to light, the slug travels until it locates food (bacteria) and moisture. Then the slug halts, forming a stalk with a fruiting body on top containing reproductive spores. The stalk cells perish, leaving spores that release single cells. If food grows scarce, the switch to multicellularity occurs again.

The cellular slime mold *Dictyostelium discoideum* has long been a favorite of biologists because its unique life cycle offers the opportunity to study how cells specialize. On a more philosophical note, the cellular slime mold inspires us to think about such nonbiological concepts as cooperation, communication, altruism, and what it means to be an individual as well as part of a group.

Mastering Concepts

1. What are the characteristics of chytrids?
2. Describe water molds, and give an example of how they can affect agriculture.
3. What do acellular and cellular slime molds have in common, and how do they differ?

SUMMARY

1. Kingdom Protista includes many simple eukaryotes. Most are unicellular, but some may be colonial or multicellular. The kingdom includes **protozoa, algae, chytrids, water molds,** and **slime molds.** Many protista form **plankton,** which consists of microscopic organisms that form the bases of food webs.

2. Protozoa are unicellular protista that move by using cilia, **pseudopods,** flagella, or amoeboid motion. They have one or two nuclei; if they have two, one may be a **macronucleus** and one a **micronucleus. Holozoic** protozoa take in food by phagocytosis, and **saprozoic** protozoa do so by diffusion, facilitated diffusion, active transport, or **pinocytosis. Contractile vacuoles** pump out water, and **secretory granules**

Living Sands~

continued from page 458.

The relationship between the components of living sand are not well understood. Clearly the ability of algae to photosynthesize is an advantage for the foraminifera, providing a live-in food source. Laboratory evidence supports this hypothesis—adding an herbicide to a living sand in culture kills both partners.

What do the algae gain from the foraminifera? Ecologists hypothesize that the algae, ensconced in their protected chambers of the hosts' shells, obtain fresh seawater, dissolved gases, and nutrients, particularly those containing phosphorus and nitrogen, which are scarce in the environment.

An alga entering a foraminiferan vacuole follows one of two fates. If the host's vacuole merges with a lysosome, the captured alga becomes dinner. But if the vacuole becomes a lining called a symbiosome, the alga becomes a well-cared-for tenant.

One of the biggest mysteries about living sands is how and why particular foraminera species associate with certain algal species. The host does not randomly consume algae—rather, the same 20 or so algal species reside in the nooks and crannies of the shell. Biologists do not know how the foraminiferan chooses its algal cargo, but experiments show that it is picky. If the foraminiferan *Amphistegina lobifera* grows in the laboratory with herbicide to kill its algal guests, and is then exposed to various algal species, the host eats most of the algae. The species of algae that repopulate the foraminiferan are precisely those that lived there before! Even more puzzling is that the algal species that live in the foraminifera are often quite rare, or absent, in the surroundings. Perhaps the hosts inherit them.

On a larger scale, living sands reflect environmental change. In the distant past, living sands have grown enormously in times of global warming, when seas fill. A single-celled foraminiferan can grow to 4 inches (10 centimeters)!

Today pollution harms these delicate living partnerships. Living sands are noticeably absent in polluted areas where fossilized remains indicate they were once abundant. Regions neighboring such polluted sites lack the brilliant hues of living sands, but they are sometimes dotted with white and brown patches, legacies of past symbioses. The white areas are foraminifera lacking their photosynthesizing algae, and the brown spots are concentrations of free algae. Ecologists hypothesize that pollution, which can overwhelm an area of ocean with nutrients not usually present, upsets the delicate nurturing relationship between the foraminiferan and its algal guest.

contain enzymes. A **cyst** form enables certain protozoa to survive harsh environmental conditions. All protozoa reproduce asexually, by binary fission, **budding,** or **multiple fission,** and may reproduce sexually. Some protozoa form **zooplankton.**

3. The protozoa include seven phyla. Phylum *Sarcomastigophora* includes the flagellated *Mastigophora* and the *Sarcodina,* whose amoebae have pseudopods. *Labyrinthomorpha* are elongated and are either saprozoic or parasitic on algae. *Ascetospora* live in mollusks, *Myxozoa* in fish, and *Microspora* in insects. *Apicomplexa* have a unique apical complex and mode of reproduction, are nonmotile, and have a complex life cycle. *Ciliophora* locomote by waving rows of cilia and may reproduce by conjugation. Protozoa cause many diseases.

4. Algae include cyanobacteria and seven protistan divisions. Protistan algae may be unicellular or multicellular, and they contain a variety of photosynthetic pigments and storage carbohydrates. Algae resemble plants because they photosynthesize, but they lack roots, stems, and leaves. They are probably ancestral to plants. Some algae form **phytoplankton.**

5. Algae are classified by habitat, cell structure, number of flagella, thallus form, reproductive mechanism, and life cycle characteristics.

6. Green algae (*Chlorophyta*) are very diverse. Many have two flagella, a single haploid nucleus, chlorophyll a and b, a large chloroplast, and a starch-storing **pyrenoid.**

7. Euglenoids (*Euglenophyta*) are elongated, unicellular, and lack cell walls, but they have a supportive flexible layer called a **periplast** and a storage carbohydrate, **paramylon. Stoneworts** (*Charophyta*) have a limestonelike covering and resemble plants. **Dinoflagellates** (*Pyrrhophyta*) have two flagella of differing lengths, outer plates, and many pigments. **Golden-brown algae, yellow-green algae,** and **diatoms** (*Chrysophyta*) live in fresh water and have distinctive coverings and pigments. **Brown algae** (*Phaeophyta*) are complex, multicellular seaweeds. Their body form includes a holdfast, stemlike **stipe,** bladder, and blades. Motile cells have a set of flagella. **Red algae** (*Rhodophyta*) are multicellular seaweeds with **floridean starch** and **phycobilin** pigments.

8. Chytrids have a flagellum and chitin cell walls. **Water molds** live in soil or fresh water and may have one cell or many nuclei in a shared cytoplasm. Their cell walls consist of cellulose. **Slime molds** are single-celled amoebae when food is plentiful, but they aggregate to form large, migrating masses when conditions are harsh. This enables them to move to find food. Slime molds may be acellular or cellular.

TO REVIEW . . .

1. How do protista differ from members of the other kingdoms?

2. Name a protist that has each of the following:
- **a.** periplast
- **b.** pyrenoid
- **c.** apical complex
- **d.** thallus
- **e.** macronucleus and a micronucleus
- **f.** frustule
- **g.** plasmodium
- **h.** test

3. Name three types of algal pigments and describe the colors they impart.

4. How do algae differ from plants?

5. Distinguish between a sporozoite, a merozoite, and a trophozoite.

6. How do phytoplankton differ from zooplankton?

7. Describe three forms of asexual reproduction in algae.

8. List three forms of protozoan motility.

9. Distinguish between a holozoic and a saprozoic method of obtaining nutrients.

10. How does binary fission differ from multiple fission?

11. Which organism causes the following:
 a. malaria
 b. whirling disease
 c. paralytic shellfish poisoning
 d. toxoplasmosis
 e. grape downy mildew
 f. late potato blight

12. Name the type of alga that has each of the following characteristics:
 a. used as an abrasive
 b. forms red tides
 c. extremely diverse body forms
 d. a periplast
 e. a limestonelike covering
 f. a holdfast and stipe
 g. phycobilin pigments

13. What do water molds have in common with plants? with fungi?

TO THINK ABOUT . . .

1. Protista are small and many cause disease. Why aren't they considered bacteria, which also have these characteristics?

2. What strategies enable cellular slime molds and *Giardia lamblia* to survive a temporarily harsh environment?

3. How can plankton include members of different kingdoms?

4. HIV infection makes people extremely susceptible to developing protozoan infections. How might this endanger people who are not infected with HIV?

5. In November 1992, 402 refugees from the Kampuchia-Vietnamese border settled in North Carolina. Nearly half of them were infected with *Plasmodium,* the organism that causes malaria. The illness did not spread beyond the refugee community. What might have limited the spread of malaria?

6. What type of information has changed the classification of *Pneumocystis carinii* from a protozoan to a fungus?

7. What do *Volvox* and the cellular slime molds have in common?

8. What changes would have to have occurred for algae to evolve into plants?

9. Would a foraminiferan be likely to contact an alga that is benthic or neustonic?

10. Why are plankton important in ecosystems? What organisms mentioned in the chapter are planktonic?

11. Why is it adaptive for an alga to have pigments in addition to chlorophyll?

12. If you were to take a sample of scummy material from the surface of a pond and view it under a microscope, what might you look for to tell if you are viewing an alga, protozoan, plant, or bacterial colony?

13. How is the lifestyle of slime molds adaptive? How is it similar to the function of cysts in protozoa?

TO LEARN MORE . . .

References and Resources*

Aronowitz, Paul. August 1996. Fever without a cause. *Discover.* Amoebae can cause illness in humans.

Austad, Steven. February 1994. Reflections on slime. *Natural History.* The cellular slime mold gives new meaning to our concepts of individuality and cooperation.

Bonner, John Tyler. 1994. *Life cycles: Reflections of an evolutionary biologist.* Princeton, N.J.: Princeton University Press. Bonner has spent a lifetime probing the life cycle of the cellular slime mold.

Mack, Walter N., and Elizabeth A. Leistikow. August 1996. Sands of the world. *Scientific American.* The composition of sands reflects geology and biology.

MacLean, Marlie, Joe Anderson, and Catherine Davies. January 1997. Making malaria research bite.. *Nature Medicine.* Malaria is a major cause of death in many parts of the world.

Nagel, Ronald L. July 1991. Malaria's genetic billiards game. *Natural History.* Mosquitoes shuttle the protozoa that cause malaria.

Prescott, Lansing M., John P. Harley, and Donald A. Klein. 1996. *Microbiology,* 3d ed. Dubuque, Iowa: Wm. C. Brown Publishers. Chapters 26, 27, 28, and 39 discuss protista.

Silberman, Jeffrey D. April 4, 1996. Human parasite finds taxonomic home. *Nature,* vol. 380. Molecular evidence places *Blastocystis hominis* among the protista.

**Life*'s website features frequently updated references.

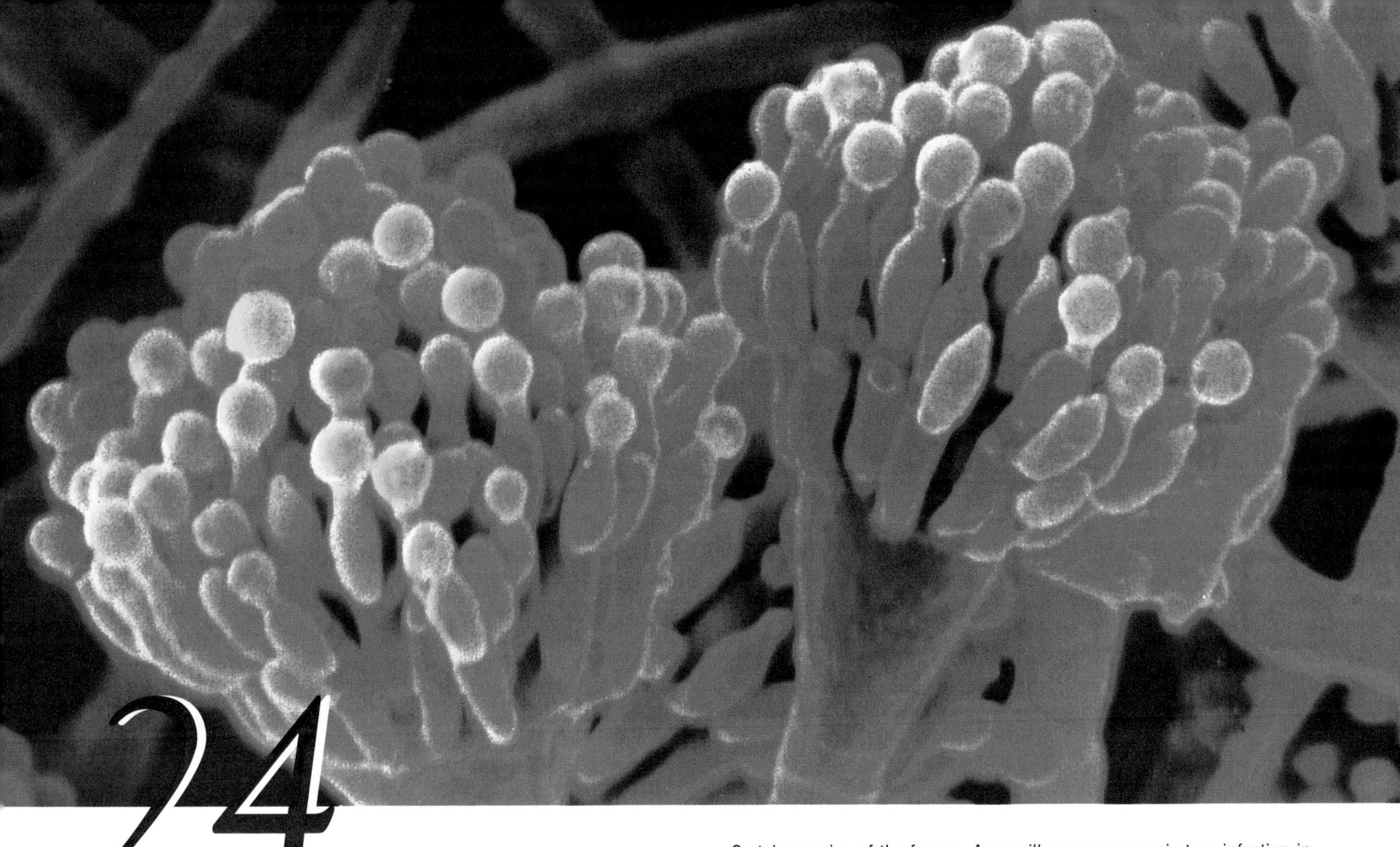

24 Fungi

Certain species of the fungus *Aspergillus* cause a respiratory infection in humans, and some produce a cancer-causing compound called aflatoxin. Yet other *Aspergillus* species are used to ferment soy sauce. (×1,200)

chapter outline

A Killer Fungus 476

Characteristics of Fungi 476
Differences and Similarities from Plants and Animals
The Fungal Body
Reproduction

Fungal Diversity 480

Fungal Divisions 480
Zygomycetes
Ascomycetes
Basidiomycetes
Deuteromycetes
Lichens

The Impact of Fungi on the Living World 484
Evolution
Ecology
Diseases Fungi Cause

A Rain of Fungal Spores

Valley fever is a systemic (bodywide) fungal infection caused by *Coccidioides immitis,* a fungus that lives in parts of Arizona, Nevada, California, New Mexico, Texas, and Utah. The fungus's spores, so small that they penetrate protective masks, are normally quiescent, lying undisturbed in dust and rarely entering human lungs. The spores burst into the air when rain follows a drought. People can then inhale the spores and become ill. The infection doesn't pass from person to person but from the disturbed environment to human lungs.

While usually extremely rare, Valley fever began increasing in prevalence in 1991, particularly in California's Kern County, in the San Joaquin Valley. Two simultaneous events triggered the outbreak—the end of a six-year drought and increased construction. Beating rain and excavation exposed and stirred up dust and released many millions of fungal spores. The 428 annual cases of Valley fever reported in California surged to more than 4,000 in the early 1990s.

In November 1993, Kern County Health Department officials hosted a national meeting of public health experts to discuss how to control and prevent Valley fever. Little did they know how soon their skills would be tested—to a degree they couldn't have imagined.

to be continued . . .

A Killer Fungus

A nematode worm moves through decomposing wood on the forest floor using its strong sense of smell, for it is normally sightless. A very sweet smell beckons, and the worm moves toward it. The worm locates the source of the enticing odor, but it is a trap, a biochemical emanating from the fungus *Arthrobotrys anchonia.* When the worm touches the sweet-smelling threads of the fungus, it sticks. Instantaneously, ringlike extensions from the fungus encircle the worm and constrict, swiftly crushing it (fig. 24.1). Then, the fungal threads extend into the worm's body, releasing digestive enzymes and eating it from within.

We don't usually think of fungi as killers. More familiar fungi are mushrooms, colorful growths on food abandoned in the refrigerator, and mildew clinging to bathroom walls. Fungi cause illnesses in all kinds of organisms. They are plant pathogens, they harm domestic animals, and they may be partially responsible for declining amphibian populations. In humans, fungi cause athlete's foot, ringworm, yeast infections, and life-threatening infections in people with impaired immunity.

Fungi also help us. They are used to manufacture a range of products, including cheese, bread, alcoholic beverages, dyes, plastics, photographic developers, soaps, toothpaste, medicinal drugs, vitamins, soy sauce, pesticides, flavorings, fuel, and food coloring.

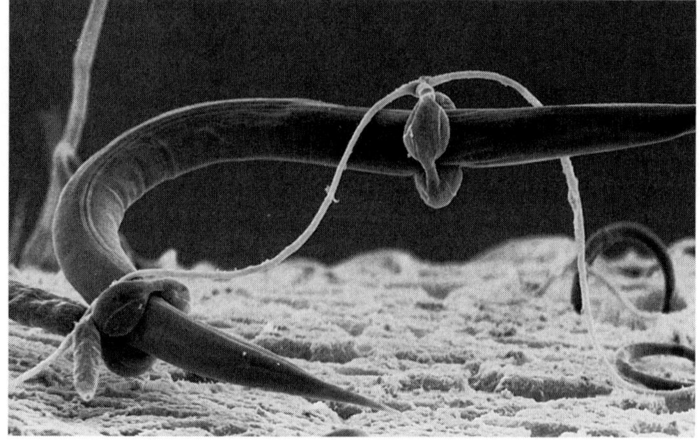

FIGURE 24.1

A carnivorous fungus. The fungus *Arthrobotrys* (×750) lures a nematode worm with a chemical attractant and then weaves itself around the worm. Next, the fungal rings constrict, crushing the worm, and tentacle-like hyphae then extend into the worm's body, digesting it from within.

Characteristics of Fungi

We give fungi picturesque nicknames based on appearance. A walk in the woods in New England may reveal the long, black spores of dead man's fingers, the thimblelike structures of bird's nest fungi, onion-shaped earthstars that emit what looks like smoke when disturbed, witches'-broom engulfing a birch tree, or the flattened, blue, turkey tail bracket fungi that cling tenaciously to trees (fig. 24.2). The study of fungi is called **mycology;** fungi experts are mycologists, and diseases fungi cause are mycoses.

Differences and Similarities from Plants and Animals

Fungi are eukaryotes with filamentous shapes. For many years they were classified with plants, and many people still think that they are plants. In "tree of life" depictions circa 1925, fungi occupy a lower branch amid the plants. This classification was based on rather superficial criteria: fungi have cell walls and appear to be rooted to the ground, like plants.

Biologists gave fungi their own kingdom as their differences from plants become better known. Fungal cell walls, for example, include the carbohydrate chitin, which plant cell walls lack. A major difference from plants is that fungi do not photosynthesize.

A mushroom may not seem very much like a human, but fungi indeed share some characteristics with animals. Their primary storage carbohydrate is glycogen, as it is for animals. Fungi lack the organized nervous tissue of animals, but fungi sense their environment and can react to it in very visible ways, as *Arthrobotrys anchonia*'s attack on the nematode worm vividly illustrates. Unlike either plants or animals, fungal cells retain a nuclear envelope when they divide.

Many fungi are **saprophytes,** with a characteristic absorptive mode of nutrition. They release digestive enzymes into the immediate environment that break down macromolecules into di-

(a)

(b)

FIGURE 24.2

Fungi have vivid common names, such as bird's nest fungi (*a*) and turkey tail bracket fungi (*b*).

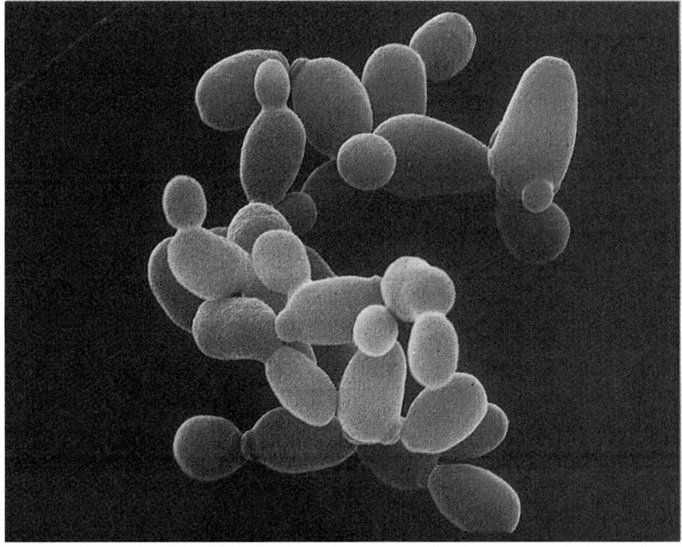

(a)

(b)

FIGURE 24.3

Fungi are unicellular and multicellular. (*a*) The yeast *Saccharomyces cerevisiae* makes bread dough rise and is a model organism for genetic research. Note that these yeast cells are budding, a form of asexual reproduction. (×21,000) (*b*) The puffball mushroom of genus *Lycoperdon* is obviously multicellular.

gestible nutrients. Then they absorb the nutrients through their cell walls. Even carnivorous fungi obtain nutrients by absorption, because their digestive enzymes act outside their bodies. We will return to the fungal saprophytic mode of nutrition later in the chapter. Many fungi are also symbionts of a variety of types of organisms.

The Fungal Body

Most fungi are multicellular, but many have a unicellular, or **yeast,** form (fig. 24.3). **Dimorphic** (two body type) species have both multicellular and unicellular phases of the life cycle. For example, *Blastomyces dermatitidis* is a yeast when it infects the human body, but it is multicellular when it grows in the laboratory or in soil. A few species of fungus, such as the familiar bread yeast *Saccharomyces cerevisiae,* are solely unicellular, although their single cells may aggregate into long strands that superficially resemble the threads of multicellular fungi.

A fungal thread, or **hypha** (plural: hyphae), is the basic structural unit of a multicellular fungus. A hypha grows at its tip and may be very long. The diameter of a hypha ranges from 0.5 to 100 micrometers. Thousands of microscopic hyphae may align, branch, or fuse, forming visible strands. An individual fungus consists of an assemblage of hyphae called a **mycelium.** A mushroom, for example, is a mycelium of many tightly packed hyphae. The visible part of a fungus, extending above the ground, is called a **thallus** (plural: thalli). It is a vegetative (nonsexual) structure.

Hyphae are of two types (fig. 24.4). In a **coenocytic** hypha, protoplasm streams continuously along the length of the hypha

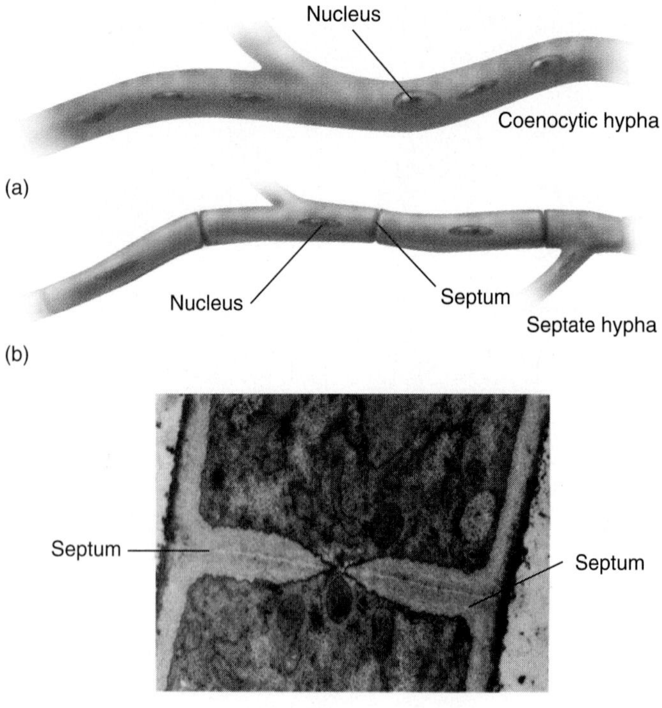

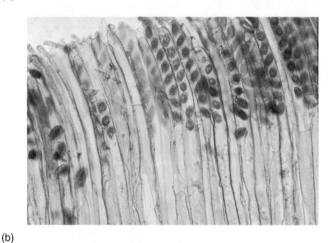

FIGURE 24.4

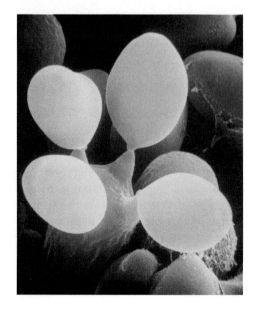

Hyphae may consist of free-flowing cytoplasm among many nuclei (coenocytic) (*a*) or be divided into cells by septa (septate) (*b*). The electron micrograph in (*c*) is a cross section showing a septum and a narrow channel that permits cytoplasm to flow between adjacent cells.

and encircles many nuclei. In a **septate** hypha, dividers called septae partition the hypha into distinct cells, which may have one, two, or several nuclei. The septae have pores that allow protoplasm to pass from cell to cell.

Reproduction

Biologists primarily distinguish fungi by reproductive structures and processes (fig. 24.5). The three major divisions have both asexual and sexual modes of reproduction. They are the **zygomycetes, ascomycetes,** and **basidiomycetes,** named for their distinguishing types of reproductive structures. A fourth group, the **deuteromycetes,** do not have a recognizable sexual phase (table 24.1).

The designation deuteromycetes is somewhat artificial and reflects the limitations of taxonomy rather than a distinctive set of characteristics defining a type of organism. Fungi are classified as deuteromycetes if we cannot observe their sexual reproduction or sexual structures. Oftentimes a fungus initially classified as a deuteromycete is subsequently grouped elsewhere, usually the ascomycetes, when researchers learn more about it.

A fifth type of fungus, a lichen, is a symbiotic organism. It is part fungus and part alga or cyanobacterium.

Fungi reproduce asexually by a variety of methods. One cell may replicate its genetic material and duplicate organelles and then split, yielding two progeny cells. Alternatively, a yeast cell may

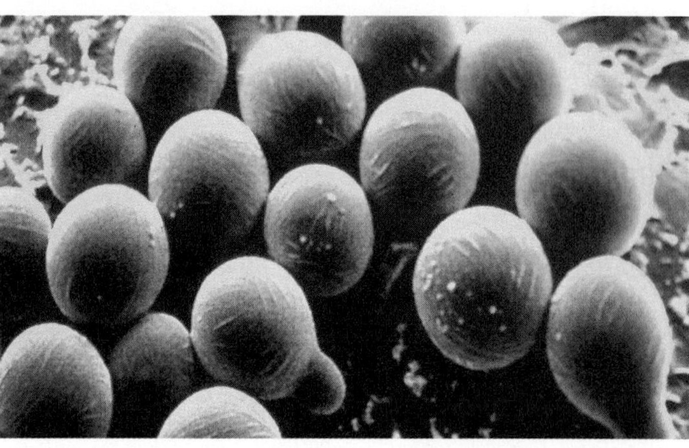

(a)

(b)

(c)

FIGURE 24.5

Fungi are classified by reproductive structures. The zygomycetes (*a*) have sexual resting spores, as in this mold that grows on bread, *Rhizopus stolonifer.* (*b*) The ascomycetes carry sexual spores in saclike asci, such as this cup fungus *Peziza.* (× 400) (*c*) The basidiomycetes have spore-bearing structures called basidia, as seen in this *Psilocybe mexicana* fungus.

Table 24.1 Kingdom Fungi

Major divisions are distinguished by reproductive mechanisms.

Division	Reproductive Mode
Zygomycetes	Thick, black zygospores
Ascomycetes (sac fungi)	Saclike asci house ascospores
Basidiomycetes (club fungi)	Club-shaped hyphae tips contain spore-bearing basidia
Deuteromycetes (imperfect fungi)	Sexual reproduction unknown

"bud," giving rise to a smaller cell (see fig. 24.3*a*). Many fungi produce vegetative (asexual) spores, which may bud or fragment from hyphae or form in a sac.

When fungi reproduce sexually, haploid cells of compatible mating types (a fungal version of sexes) join. These sex cells may come from the same mycelium or be positioned where two mycelia make contact. A sexually reproducing fungus grows distinctive mating structures in one phase of the life cycle; these structures join in another phase and produce a different form.

The zygote that forms from sexual reproduction may become encapsulated to form a spore. A spore can withstand environmental extremes and, when dispersed, gives rise to a new individual.

Flower Mimicry Aids Reproduction

Some fungi use insects to transfer sex cells from one individual to another, much as sexually reproducing plants use insects to pollinate flowers. A professor at the University of Manitoba, A. H. Reginald Butler, discovered the role of insects in fungal sexual reproduction inadvertently in 1927, when he noticed a lone fly buzzing about the greenhouse where he was studying stem rust fungus growing on sunflower plants. The fly flew from plant to plant, alighting briefly exactly where fungus grew. Butler hypothesized that the fly assisted the fungi in combining sex cells—and proved this experimentally by reenacting the situation. He eventually discovered that the fungi secrete a nectarlike substance that attracts flies.

Evolution has elaborated on the theme of insects delivering fungal sex cells. The anther-smut fungus fills flowers with its spores. When an insect touches the flower, the spores burst forth and some stick to the insect, which unwittingly delivers them to fungi on other plants. A fungus called *Monilinia* grows on blueberry plants, crumpling leaves and causing them to emit an odor that attracts insects. Under ultraviolet light, which insects can detect, these crumpled leaves resemble flowers. The insects go to these pseudoflowers as readily as to real ones.

The rust fungus *Puccinia monoica* is a master of flower mimicry (fig. 24.6). In late summer, wind carries its spores to plants of the mustard family. Over the winter, the fungi grow into the plant's meristem (growing) tissue. This drastically alters the plant's appearance. It may grow twice as tall as normal or produce twice as many leaves. Then, bright yellow fungus grows on clusters of leaves on stems, secreting a sticky, sugary substance. The resulting structure looks amazingly like a flower—but like a buttercup, not a mustard flower. When researchers placed buttercups, mustard plants, insects, and rust fungi together in a field, the insects preferred the fungal "flowers," revealing how adaptive is a fungus's flower mimicry.

Mastering Concepts

1. What characteristics distinguish fungi from plants and animals?
2. What is a dimorphic fungus?
3. What are the parts of a fungus?
4. What mechanisms and structures do fungi use to reproduce?

Fungal Diversity

Fungi live nearly everywhere—in soil, in and on plants and insects, in water, in dung, and in the intestines of diverse animals. Cryptoendolithic ("hidden within rock") fungi live inside rocks in Antarctica. Most fungi live on land and are aerobes. Fungi living in intestines, where there is little oxygen, are anaerobes.

Fungi range greatly in size. They may be microscopic or extend many meters beneath the leaf litter of a forest floor (fig. 24.7). Hyphae will extend in all directions if nothing blocks their growth. Fungal genetic diversity is great. A tiny patch of bark on an elm tree may harbor many different genotypes of *Ophiostoma novo-ulmi,* the fungus that causes Dutch elm disease. Yet forests in Michigan and Montana are home to huge fungal associations of just one genotype. A single individual of the tree root pathogen *Armillaria luteobubalina* in northern Michigan, for example, covers 37 acres (15 hectares), weighs more than 22,000 pounds (10,000 kilograms), and is estimated to be 1,500 years old—easily rivaling the giant sequoia and the blue whale as among the largest organisms! This giant fungus's hyphae align into great bundles that extend across the forest.

Despite the fungi's widespread distribution, we know very little of their diversity. Taxonomists have described nearly 100,000 species of fungi—600 zygomycetes, 35,000 ascomycetes, 30,000 basidiomycetes, and 30,000 deuteromycetes. This total, however, may represent only half of the number of fungal species. Taxonomists have not yet accounted for the many types of fungi that live in and on insects, which are themselves incredibly diverse. Fungi have been studied most extensively in the British Isles, but even there, mycologists have yet to catalog all the fungal residents of one site.

FIGURE 24.6

Even biologists are sometimes fooled by these "pseudoflowers." They are not flowers at all but the rust fungus *Puccinia monoica,* growing here on a mustard plant. The fungus secretes a nectarlike substance that attracts insects that pollinate flowers. The insects assist sexual reproduction in the fungus by distributing haploid fungal sex cells among other fungi.

Mastering Concepts

1. Where do fungi live?
2. How do fungi range in size?
3. About how many species of fungi are known?

Fungal Divisions

Zygomycetes

The 600 species of zygomycetes (division Zygomycota) undergo both asexual and sexual reproduction. In asexual reproduction,

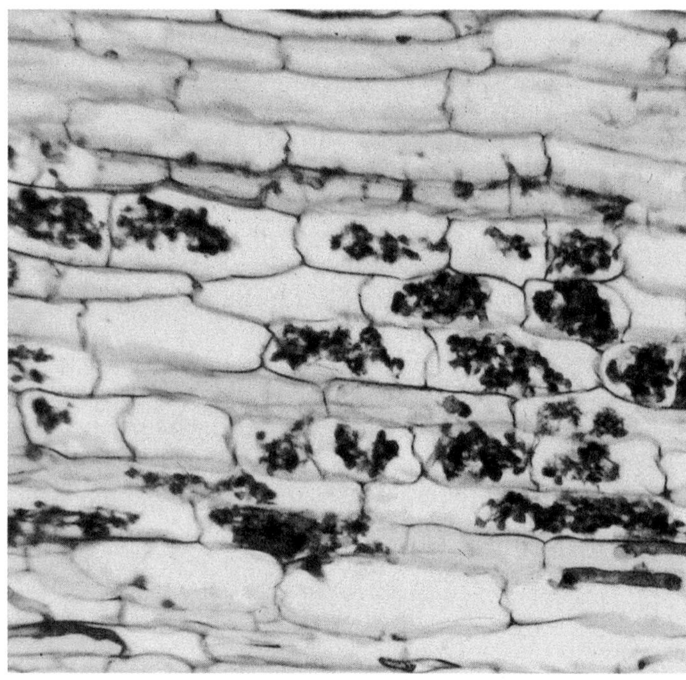

FIGURE 24.7

Some fungi are so small that they inhabit the cells of other organisms. The dark structures are the fungus *Gigaspora margarita,* growing in cotton cells.

FIGURE 24.8
Entomophthora is a zygomycete that grows on insects.

spores that are genetically identical to the parents arise from swellings at the tips of hyphae. Sexual reproduction involves two mating types. Zygomycetes' sexual structures, called **zygospores,** consist of thick black coats surrounding zygotes. Asexual reproduction prevails when food is available; food scarcity triggers sexual reproduction. This phenomenon reflects the adaptive value of sexual reproduction in mixing inherited traits, perhaps generating variants that can withstand an environmental threat—such as lack of a particular food.

Most zygomycetes feed on decaying plant and animal matter, but some parasitize plants and insects. Zygomycetes are particularly valuable to plants when they grow on roots, a point we will return to soon. Most zygomycetes are coenocytic. The black mold *Rhizopus stolonifer,* which grows on bread, fruits, and vegetables, is a familiar zygomycete (see fig. 24.5*a*). Figure 24.8 shows a zygomycete growing on a housefly.

Ascomycetes

The "sac fungi" of division Ascomycota include many familiar species among their 30,000 members. Ascomycetes reproduce both asexually and sexually. Asexual spores pinch off of the ends of exposed hyphae. Their saclike sexual reproductive structures are called **asci.** An individual ascus contains structures called **ascospores,** often in rows of eight haploid nuclei. Thousands of asci align and, when touched, burst with a smoky emission of spores.

Many of the red, brown, and bluish green growths on spoiled foods are ascomycetes. Other ascomycetes cause Dutch elm disease, chestnut blight, powdery plant mildew, strawberry stem rot, apple scab rust, and brown rot of apricots, plums, and peaches. In humans and some other animals, ascomycetes cause athlete's foot and the mouth infection thrush. Other ascomycetes include the bread mold *Neurospora crassa,* used in genetics experiments, and the delicacies truffles and morels (fig. 24.9). The yeast *Saccharomyces cerevisiae* is used to ferment carbohydrates to produce alcohol and carbon dioxide, which underlies the brewing and baking industries. Other ascomycetes are used to manufacture soy sauce and tofu (bean curd). Bioethics Connections 24.1 considers a hallucinogenic drug, LSD, that is derived from an ascomycete.

(a)

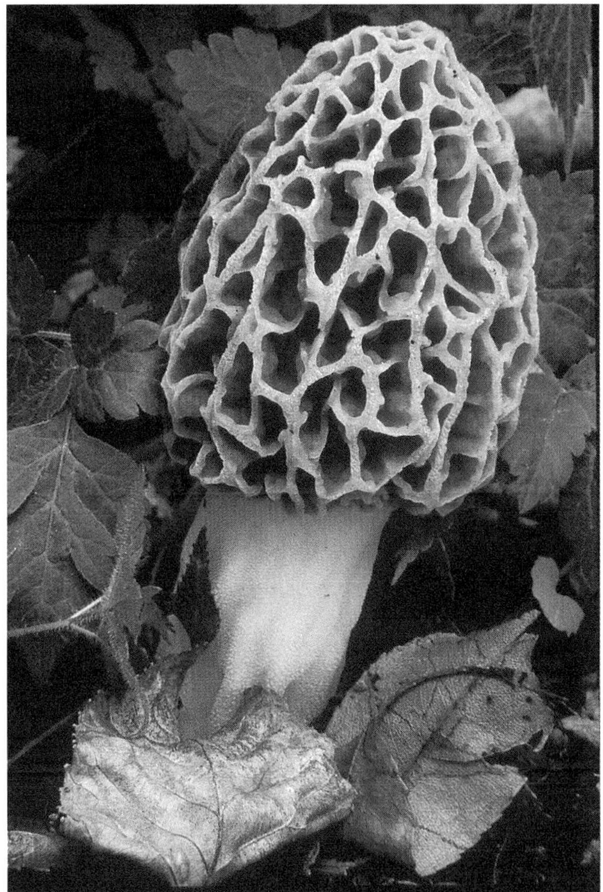

(b)

FIGURE 24.9
Familiar ascomycetes include (*a*) truffles (*Tuber melanosporum*) and (*b*) morels (*Morchella esculenta*).

bioethics 24.1 connections

LSD

Lysergic acid diethylamide (LSD) is a drug that became widely known in the 1960s, when it was used recreationally because of its hallucinogenic effects. LSD is a derivative of lysergic acid, a biochemical found in the ascomycete *Claviceps purpurea.* LSD was first synthesized in 1938, and its mind-altering effects first described in 1943, but the reputation of this fungus to affect human perceptions goes much farther back in time.

Claviceps purpurea parasitizes rye and other grasses, causing a disease of plants called ergot (fig. 24.A). People who eat bread made from infected grain can develop an illness called ergotism, which causes convulsions, gangrene, and psychotic delusions. Ergotism, also known as St. Anthony's fire because of a burning sensation in the hands, killed tens of thousands of people in Europe during the Middle Ages. Some of the women in Salem, Massachusetts, who were burned at the stake as witches in the late 1690s may have been suffering from ergotism after eating infected grain; people thought their uncontrollable movements meant they were in the throes of demonic possession. More recently, in 1951, an outbreak of ergotism in a French town made 30 infected people race about insisting that snakes and devils were chasing them—5 people died.

C. purpurea and its extracts affect the human body in several ways—dilating the pupils and producing sleeplessness, tremors, dry mouth, elevated temperature, profuse sweating, and lack of appetite. Even short-term users of LSD may experience flashbacks, in which the sensations and perceptions of a "trip" recur without actually ingesting more of the chemical. Long-term use of LSD can lead to psychoses, including depression and schizophrenia.

Association of LSD with medieval deaths, Salem witches, and hippies on bad trips has hampered investigation of the compound as a medicinal drug. For example, researchers must get approval from the Drug Enforcement Administration to study LSD. The Food and Drug Administration requires researchers to document that LSD used in experiments comes from a manufacturing facility that meets certain standards—but most pharmaceutical companies do not produce the drug. LSD for research comes largely from university laboratories and from European pharmaceutical companies. Finally, studies using LSD often require participants who have used the drug recreationally. Since LSD was declared a controlled substance in 1970, users are not likely to confess to an illegal activity.

A controlled substance is a drug that has a high potential for abuse, presents an unacceptable safety risk, and has no approved medical use. LSD, however, does have medical uses. From 1946 through 1970, researchers studied its effects on the brain and planned to examine its efficacy in treating various mental illnesses. This research stopped by the mid-1970s, as the drug's reputation as a psychedelic mushroomed.

If LSD can overcome its social stigma, it may find therapeutic uses. Extracts of *C. purpurea* have been used for four centuries in various cultures to induce labor, possibly because it increases muscle contractions. Because LSD dilates veins, it can lower blood pressure and possibly treat migraine headaches. As researchers once again recognize the potential of LSD as a bona fide therapeutic drug, they are continuing studies begun in the 1960s. Current research is investigating LSD's ability to treat addiction to opium, alcohol, heroin, and sedatives and its potential as a painkiller for cancer patients.

FIGURE 24.A
Claviceps purpurea infects this grain.

Basidiomycetes

The 25,000 species of division Basidiomycota are called club fungi because their hyphae tips form spore-bearing swellings, called **basidia.** The basidiomycetes have characteristic links between hyphal cells, called **clamp connections** (fig. 24.10).

Familiar basidiomycetes include mushrooms, toadstools, puffballs, stinkhorns, shelf fungi, rusts, bird's nest fungi, and smuts (fig. 24.11). The fungi may grow quite large—a puffball may be more than a meter in diameter and release trillions of spores. Some mushrooms are edible, some are deadly, and others are hallucinogenic. The Aztecs regarded the mind-altering

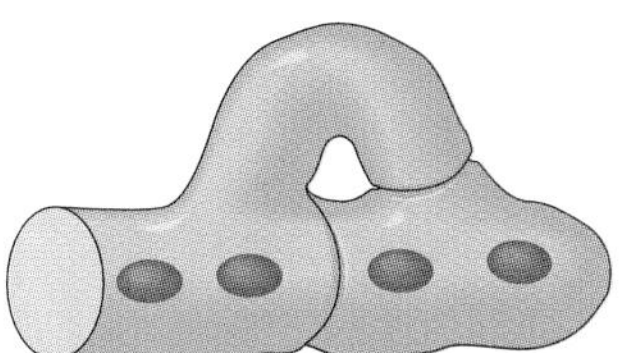

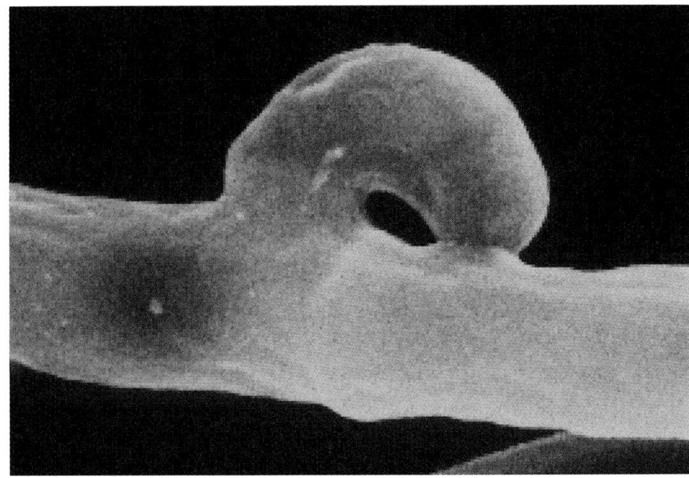

FIGURE 24.10
Clamp connections link basidiomycete cells.

(a) (b)

FIGURE 24.11
Basidiomycetes include (*a*) puffballs (genus *Lycoperdon*) and (*b*) stinkhorns (*Phallus impudicus*).

mushroom *Psilocybe mexicana* as sacred. Animals other than humans can become intoxicated too from eating hallucinogenic mushrooms.

Smuts and rusts are plant pathogens that can destroy cereal crops. Another basidiomycete, *Cryptococcus neoformans,* infects human lungs and the central nervous system and is especially devastating in people with AIDS.

Deuteromycetes

The 17,000 species of deuteromycetes are grouped together based upon lack of information or evidence of sexual reproduction. Most deuteromycetes are terrestrial, and most are saprophytes or plant parasites. *Arthrobotrys anchonia,* the fungus that attacks nematode worms, described at the beginning of the chapter, is a deuteromycete.

Researchers disagree on the placement of many deuteromycetes. Some investigators, for example, consider *Penicillium* and *Aspergillus* to be deuteromycetes, and some consider them to be ascomycetes (fig. 24.12). However they are classified, these two genera of fungus have great effects on humans.

Penicillium includes the fungi that produce penicillin antibiotics, as well as two species, *Penicillium roquefortii* and *Penicillium camembertii,* used to ripen the cheeses named after them (Roquefort and Camembert). Different species of *Aspergillus* cause a respiratory infection in humans and are used to ferment soy sauce. *Aspergillus flavus* and *Aspergillus parasiticus* produce a biochemical on certain plants, called aflatoxin, that can cause cancer in humans.

(a)

(b)

FIGURE 24.12
The deuteromycetes include *Penicillium* and *Aspergillus.* (*a*) *Penicillium notatum* produces penicillin. (*b*) *Aspergillus fumigatus* causes a respiratory infection in humans.

Lichens

A symbiotic relationship between certain fungi and algae or cyanobacteria forms a unique type of organism, a lichen. Many lichens look quite different from their constituents, suggesting that the composite organism carries out biochemical activities that neither partner does alone. For example, some green algae produce certain types of alcohols only when they are part of a lichen.

The algal or cyanobacterial part of a lichen contributes the ability to photosynthesize. The fungus contributes the ability to manufacture certain biochemicals, many of whose functions aren't known.

Experiments suggest that a lichen benefits the fungus at the expense of the alga. If the algal component of a lichen is grown separately, it flourishes compared to its growth rate in the lichen. Conversely, the fungal component cultured alone forms slow-growing, small colonies. Perhaps the fungus is simply "borrowing" the photosynthetic capabilities of an alga (fig. 24.13).

Lichens grow on trees and rocks, in minute crevices between mineral grains in rocks, on the backs of beetles, in the driest deserts, and in the wettest tropical rain forests. Lichens can survive extreme environmental conditions because they can dry out. Once wet, they explosively grow. They spring up as if by magic in the damp soil of a fresh road cut. Lichens are called plant pioneers because they break rock down into biologically useful soil, which can then support plant growth.

Given lichens' environmental versatility, it is interesting to note one kind of habitat that cannot support them—polluted areas. Lichens absorb toxins, such as sulfur dioxide from rainfall, but cannot excrete them. Toxin buildup hampers photosynthesis, and the lichen dies. Disappearance of native lichens is a sign that encroaching pollution is disturbing the environment. This is why tree bark in the countryside is rich in lichens, but the trees in urban parks and along city streets are noticeably devoid of them.

Lichens have left their mark on human history. The "manna lichen" *Lecanora esculenta* blows in the wind in the Near East, sometimes accumulating in drifts so large people can use it to make bread. People in eastern Turkey call this lichen "wonder grain," and Kurds call it "bread from heaven." Might this lichen also have been the "manna from heaven" that sustained the starving Israelites in the desert during biblical times?

In North America, the Dakota Indians used *Usnea* lichen to dye porcupine quills, and the Wylackie tribe used it to tan leather. The Achomawi Indians soaked arrowheads in bright green "wolf moss" lichen for a year, rendering the arrowheads quite lethal. Native peoples of western Canada used the hair-like, dark brown *Bryoria* lichen to make a bitter-tasting black paste that sustained them in times of famine. The Nez Perce Indians ate the same lichen to treat stomachaches, and the Okanagan applied it to the navels of newborns to guard against infection. The Thompson Indians twisted strands of lichen together to fashion clothing.

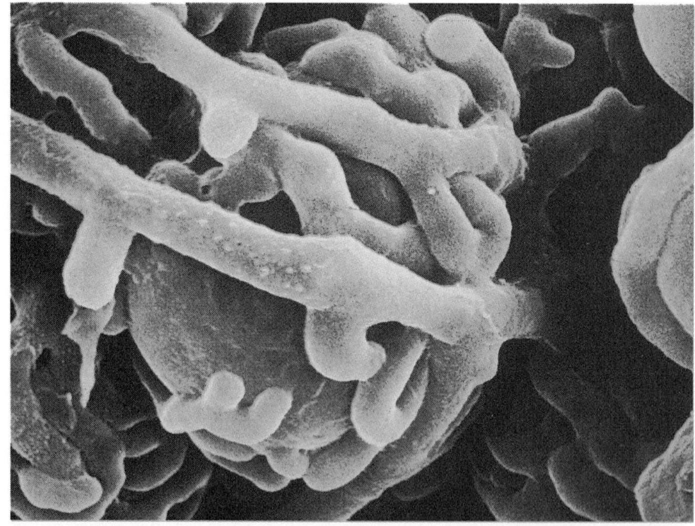

(a)

(b)

FIGURE 24.13

In this British soldier lichen, *Cladonia cristatella,* fungal hyphae surround algal cells (*a*). On a larger scale, the lichen looks like a single organism (*b*).

Mastering Concepts

1. What are the distinguishing characteristics of zygomycetes, ascomycetes, and basidiomycetes?
2. What is the basis for grouping fungi as deuteromycetes?
3. What are the components of a lichen?

The Impact of Fungi on the Living World

Fungi have been key players in evolution and are vital components of ecosystems today. They affect humans and other organisms in many ways.

Evolution

The fungal fossil record is sparse, largely because of the lack of hard parts. Fossils are usually of spores and hyphae, but a rare, preserved mushroom was discovered in 1990 in a piece of amber from the Dominican Republic. The mushroom is about 35 to 40 million years old. Because it resembles the modern inky cap mushrooms of genus *Coprinus,* the extinct species was named *Coprinus dominicana.* Yet fungi are far older than this lone mushroom fossil suggests.

Some of the earliest fossil evidence of fungi are filamentous hyphae that appear in Cambrian period rocks from 500 to 580 million years ago. These fungi lived in habitats similar to those of contemporary lichens consisting of an ascomycete and a cyanobacterium. Other fossil fungi from the Cambrian period show evidence of wind-transported asci. Because of their role in recycling nutrients and forming soil, early fungi were probably pivotal in the evolution of land plants. Fungi growing within vines and stems reveal that the intimate association between fungi and plants that persists today existed as early as 400 million years ago.

We know little of the beginnings of the fungal kingdom. In the past, mycologists had hypothesized that zygomycetes were more primitive than ascomycetes or basidiomycetes because their reproduction is simpler. More recently, molecular evidence has supported this hypothesis. Molecular studies have also revealed that fungi diverged from animals, their most recent common ancestor, about a billion years ago.

Ecology

The ways that fungi obtain nutrients impact ecosystems. Fungi obtain their nutrients either from or in conjunction with other organisms in their surroundings.

FIGURE 24.14

Ectomycorrhizae form sheaths around root tips. These *Ceanothus* (buckbrush) roots have been washed to reveal the fungal associations.

Saprophytes

Saprophytic fungi are vital decomposers in ecosystems, releasing carbon, oxygen, nitrogen, and phosphorus from dead plants and animals in soil. Much of the 85 billion tons of carbon recycled to the environment each year passes through fungi. Saprophytes obtain sugars by breaking down complex carbohydrates in rotting wood. They obtain amino acids by consuming proteins in microbes, other fungi, and small animals, such as the unlucky worm in figure 24.1.

Mycorrhizae

Eighty percent of vascular (vessel-containing) plants have fungi associated with their roots. The root-fungus structures, called **mycorrhizae,** constitute a mutualistic relationship. The fungi help the roots absorb water and minerals, while obtaining sugars from the plant. An acre of trees may contain 2,500 pounds (1,125 kilograms) of mycorrhizae.

Mycorrhizae are of two types. **Ectomycorrhizae** (fig. 24.14) wrap around the plants' roots but do not greatly penetrate the plant body. Basidiomycetes, ascomycetes, and zygomycetes form ectomycorrhizae, mostly in temperate trees and shrubs, such as beech, oak, birch, and conifers. The other type, **endomycorrhizae,** extend within the outer cells of roots in grasses, tomatoes, oranges, apples, beans, corn, wheat, and other crop plants. Basidiomycetes form most endomycorrhizae.

Diseases Fungi Cause

Infectious diseases fungi cause in plants and animals can be difficult to treat compared to bacterial infections because fungi, plants, and animals are all eukaryotes. Thus, a drug that harms a fungus, perhaps by interfering with nucleic acid or protein synthesis, may also disrupt that function in the infected plant or animal. Much current mycological research focuses on identifying unique fungal structures that drugs can target. These include cell wall components and a sterol unique to fungal cell membranes.

Table 24.2 lists fungal pathogens, and figure 24.15 illustrates fungal infections in a plant and in an animal.

Table 24.2 Fungal Plant Pathogens

Fungus	Plant Infection
Cochliobolus heterostrophus	Corn leaf blight
Cryphonectria parasitica	Chestnut blight
Erysiphe graminis	Barley powdery mildew
Magnaporthe grisea	Rice blast
Mycosphaerella graminicola	Wheat leaf blotch
Ophiostoma novo-ulmi	Dutch elm disease
Puccinia graminis	Wheat rust
Rhynchosporium secalis	Barley blotch
Sclerotinia sclerotiorum	Canola stem rot

(a)

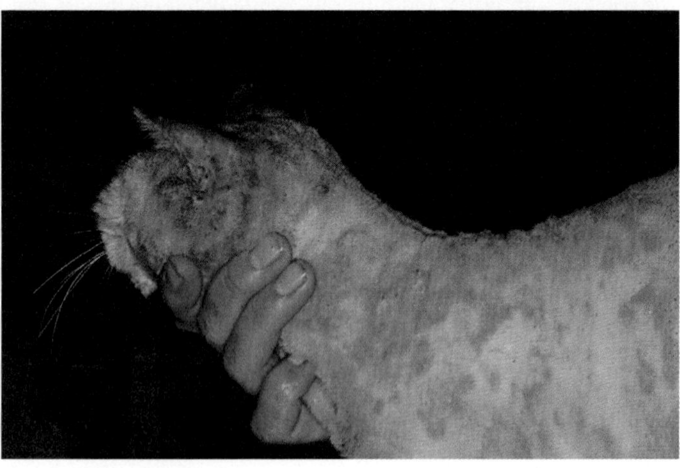

(b)

FIGURE 24.15

Fungal diseases. (*a*) Apple scab infection begins in the spring and becomes visible on fruits and leaves in midsummer. The lesions on this apple produce spores capable of spreading the disease further. (*b*) This kitten has a ringworm infection caused by *Microsporum canis.*

Human Mycoses

Human fungal infections (mycoses) are on the rise for several reasons, many related to advances in medical technology.

Many fungi are normal residents of the human body, but in small numbers. Drugs used to treat bacterial infections may kill the bacteria that usually keep fungal populations low. Without the bacteria to constrain their reproduction, the fungal population grows, eventually producing fungal infection. People receiving certain cancer chemotherapies, immunosuppressant drugs, or extended antibiotic therapy are especially susceptible to this problem. Sixteen percent of individuals undergoing bone marrow transplants develop fungal infections because of medications that dampen their immune responses to prevent their bodies from rejecting the transplant. Invasive procedures, such as catheter and prosthesis insertions, may introduce fungi. People with AIDS are particularly prone to develop severe fungal infections.

Table 24.3 Human Fungal Infections

Type	Area Affected	Example
Superficial	Hair and outer skin layer	Black piedra of hair
Cutaneous	Deeper skin layers	Ringworm
Subcutaneous	Beneath the skin	*Sporotrichosis*
Systemic	Lungs, bloodstream, possibly elsewhere	Valley fever
Opportunistic	Widespread—anywhere in body	*Candida albicans*

Physicians classify the 50 or so recognized human mycoses by the extent to which they affect the body (table 24.3 and fig. 24.16). Least serious is a superficial infection that affects the outermost skin layer or hair. Doctors simply remove the affected area. Superficial fungal infections are common in the tropics but are rare in North America.

A cutaneous infection affects deeper skin layers. The ringworms are cutaneous fungal infections. Domesticated animals usually transmit them to humans, and these infections are more likely to spread in areas of overcrowding and poor hygiene. Treatment is usually a topical ointment.

In a subcutaneous infection, fungi live beneath the skin. A person contracts such an infection through a puncture wound, because the fungi must have a way to penetrate the skin. We can easily pick up a subcutaneous fungal infection walking barefoot on a farm—if skin on the bottom of the foot is scratched, soil fungi can enter the wound. A subcutaneous infection often takes years to become noticeable, finally appearing as an ulcerating nodule at the contact site. Then fungi spread in the lymphatic system and form nodules elsewhere on the body. A combination of antifungal drugs and surgical excision can treat these infections.

A systemic infection infiltrates the respiratory and circulatory systems and travels to other organs if untreated. Such an infection can spread to the urinary tract, esophagus, kidneys, bones, liver, spleen, heart valves, or central nervous system. A common systemic fungal infection is histoplasmosis. *Histoplasma capsulatum,* which causes the infection, grows within the macrophages (immune system cells) that normally fight infection. These macrophages then harm rather than help by spreading their fungal stowaways throughout the body.

Bird droppings transmit histoplasmosis, which is widespread in Mississippi, Kentucky, Tennessee, and Ohio. It produces symptoms in humans and bats but not in the birds that transmit it, because their high body temperatures prevent symptoms from occurring. Most people experience fever, cough, and joint pain that clears up on its own. Exploring caves or working with bird droppings increases the risk of contracting this mild infection.

Sometimes a respiratory illness that occurs suddenly and cannot be traced to an irritant or bacterial infection is due to inhaling fungal spores. This happened to eight teens at a party in Wisconsin, who chewed and sniffed puffball mushrooms. Puffballs grow tall in the fall, and can be eaten then, but in the spring they dry

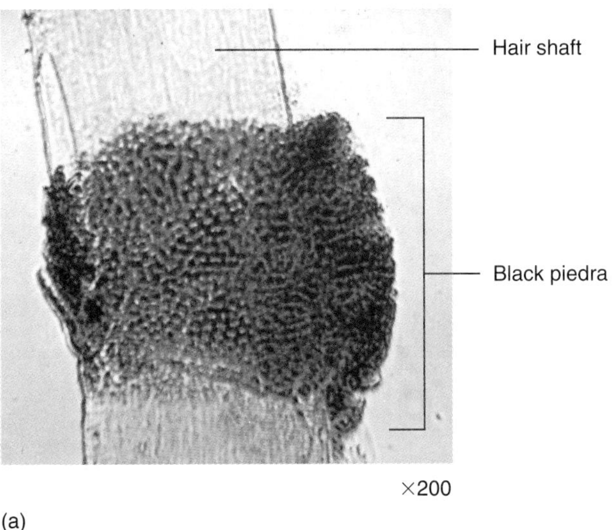

(a)

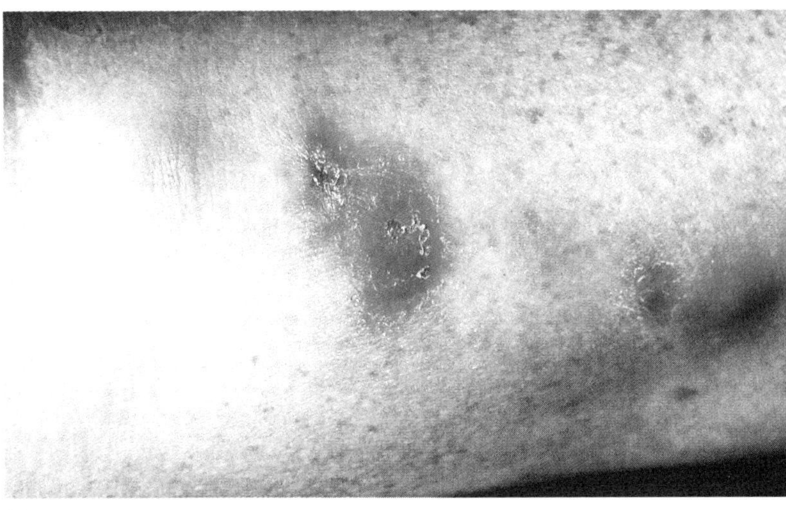

(b)

(c)

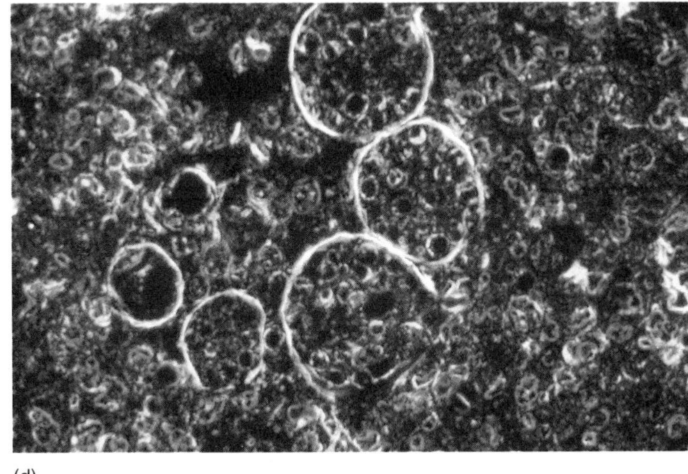

(d)

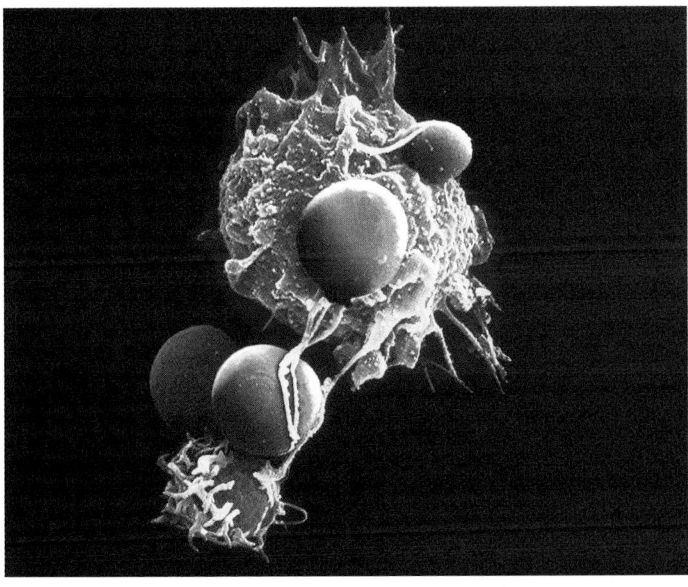

(e)

FIGURE 24.16

Human mycoses. In the superficial fungal infection black piedra (*a*), hard black nodules of *Piedraia hortae* form on hair shafts on the scalp. (×200) Sporotrichosis is a subcutaneous mycosis affecting the arm (*b*), caused by *Sporothrix schenckii.* This agricultural worker (*c*) in the San Joaquin Valley is at risk of contracting Valley fever, a systemic infection, from *Coccidioides immitis* spores (*d*) in the dust. (*e*) Human macrophages attack the yeast *Candida albicans.*

out and release spores when shaken. The teens, apparently thinking the puffballs were of species *Lycoperdon marginatum,* known to be hallucinogenic, mistakenly inhaled spores of *Lycoperdon perlatum.* Instead of the sought-after high, they experienced nausea and vomiting and then a cough, high fever, shortness of breath, fatigue, and muscle aches. They were ill from 3 to 7 days, and five of them were hospitalized, two intubated to assist their breathing. Samples of their lung tissue revealed signs of inflammation and structures resembling yeast—these were spores of *L. perlatum.* A few weeks of steroid drugs to dampen the inflammation, and antifungal drugs, cured the teens.

In the fifth type of human mycosis, opportunistic infection, the fungal pathogen is an opportunist, taking advantage of a crippled

biology in 24.1 action

Yeast—"A Stripped-Down Human Cell"

The yeast *Saccharomyces cerevisiae* is best known as an essential ingredient of pizza dough and baked goods (see fig. 24.3*a*). Since it was originally domesticated in Mesopotamia 6,000 years ago, most human cultures have used this single-celled ascomycete to manufacture bread, beer, and wine. Human genome researchers, however, have another interest in yeast—many of the organism's genes have counterparts in humans. Matching yeast genes to human genes provides shortcuts for researchers attempting to discover the functions of human genes.

A yeast's genome is 1/200 the size of a human's, consisting of 12.5 million nucleotide pairs, on 16 chromosomes, and forming approximately 6,000 genes. It took 147 researchers, working in 35 laboratories in 17 nations, to determine the DNA sequence of just one yeast chromosome, which consists of 315,356 base pairs.

The yeast genome project, yielding a preliminary map of all of the genes, was completed in 1996 and took slightly more than a decade to construct. More than a third of yeast's genes have human counterparts, or homologs. For example, yeast produces the contractile proteins actin and myosin. Instead of contracting muscles, as these proteins do in humans and other animals, in yeast actin and myosin move organelles within the cell.

Yeast also has protein kinases that begin cell division. The observation that a human protein kinase placed into a yeast cell can trigger cell division demonstrates the similarity of these proteins in the two organisms. Human disease-causing genes, such as those behind cystic fibrosis and neurofibromatosis, also have yeast homologs, although we do not know what these genes do in the far simpler fungus. Yeast also has a gene for nitrogen fixation found in certain bacteria.

Why would yeast have genes that also function in an organism as complex as a human? Biologists hypothesize that the yeast genome is a basic set of instructions to build a complex, eukaryotic cell.

immune system. Opportunistic infections occur in people with AIDS, certain types of cancer, alcoholism, malnutrition, or diabetes. People who recently have undergone surgery, been injured, or taken antibiotics are also at increased risk for developing an opportunistic infection. Drugs that treat these infections often have severe side effects. A drug commonly used in AIDS patients, amphotericin, is so toxic that patients call it "amphoterrible."

The two most common opportunistic fungal infections are aspergillosis and candidiasis. In aspergillosis, a person inhales *Aspergillus fumigatus* or a related species. Mycelia grow into "fungus balls" in the lungs, which produce asthma and bronchitis.

Candida albicans, which causes candidiasis, normally lives in the mouth, vagina, and gastrointestinal and respiratory tracts, where resident bacteria keep the populations low enough that we are generally unaware of the fungus. When something suppresses immunity, or alters bacterial populations, the *Candida* population grows and causes symptoms.

A familiar manifestation of candidiasis is thrush, which produces white patches in the mouth and on the tongue. The patches consist of *Candida* yeast, white blood cells, and cellular debris. Some newborns develop thrush because they pick up *Candida* in the birth canal from the mother but have not yet acquired the mouth bacteria that check fungal growth. *Candida* infection can also appear as white subcutaneous lesions on the fingers, toes, and nails, often occurring when hands or feet have been submerged for long periods of time. Yet another form of candidiasis occurs in moist folds of the body, such as in a baby whose diaper is not changed often enough. Candidiasis may cause itchy, burning vesicles to form in the flap of tissue at the end of an uncircumcised penis. A female may develop candidiasis in the folds of the external genitalia and in the vagina, producing the curdlike, yellow-white discharge of a "yeast infection."

Biologists have had a difficult time classifying fungi. Ironically, one of the best studied and understood organisms today is from this kingdom. Biology in Action 24.1 takes a closer look at the yeast *Saccharomyces cerevisiae.*

Mastering Concepts

1. Why is fossil evidence of fungi scarce?
2. How does the fungal saprophytic lifestyle impact other organisms?
3. What are mycorrhizae?
4. What are different types of fungal infections?

SUMMARY

1. Fungi are eukaryotes with chitinous cell walls. They do not photosynthesize, do not have nervous tissue, use glycogen as a carbohydrate source, and are symbionts or **saprophytes.**

2. Fungi may be unicellular **yeasts,** multicellular with **hyphae** strands organized into **mycelia,** or **dimorphic,** undergoing both phases during the life cycle. The visible portion of a fungus is a **thallus.** Hyphae may be **coenocytic** (continuous) or **septate** (consisting of distinct cells).

3. Fungal reproduction may be asexual, as in budding or when a parental cell divides in two, or sexual, so that haploid sex cells combine, either in one mycelium or between two individuals. Insects may assist fungi in distributing sex cells.

4. Fungi occupy many habitats, range greatly in size, and are genetically diverse. Molecular data place them most closely related in an evolutionary sense to animals.

A Rain of Fungal Spores~

continued from page 476.

At 4:31 A.M. Monday morning, January 17, 1994, an earthquake measuring 6.6 on the Richter scale shook central and southern California. As mountains rose, freeways fractured, gas lines and oil mains broke, and buildings collapsed, another upheaval that would make hundreds of thousands of people ill was taking place—the release of untold trillions of fungal spores, as dust was forcefully expelled into the air. The more than 1,000 aftershocks following the earthquake created repeated showers of fungal spores.

Valley fever, technically known as coccidioidomycosis, produces a range of symptoms, in varying severities. Only 40% of affected individuals develop symptoms, perhaps because they have been previously exposed to the spores and do not know it. For the others, the illness may be as mild as an upper respiratory infection that lasts a few weeks and then clears up, conferring immunity to further infection, or serious enough to cause a prolonged high fever, bad headache, bumpy rash, swollen feet, fatigue, joint pain, severe pneumonia, or meningitis (inflammation of the membranes around the brain).

Valley fever takes 1 to 4 weeks to incubate. Because of this, many people visiting California at the time of the earthquake and in the weeks after became ill after returning home. Their physicians, unfamiliar with Valley fever, often were unable to diagnose the condition quickly. Fortunately, antifungal drugs treated most cases of the illness.

5. Type of reproductive structure distinguishes the zygomycetes, ascomycetes, and basidiomycetes. **Zygomycetes** have **zygospores** and feed on decaying plant and animal matter. **Ascomycetes** have **asci** containing eight haploid nuclei each. Fungi growing on spoiled foods are ascomycetes. The **basidiomycetes,** including mushrooms, have clublike hyphal tips. The **deuteromycetes** cannot be classified by reproductive structures. Lichens consist of a fungus associated with an alga or cyanobacterium. A lichen has properties its constituent organisms lack.

6. Fungi are decomposers, forming soil and recycling carbon and other elements. Fungal fossils are rare. This kingdom split from other eukaryotes about a billion years ago.

7. Saprophytic fungi secrete digestive enzymes that break down macromolecules in the environment, which the fungus then absorbs. **Mycorrhizae** are associations between fungi and the roots of vascular plants. **Ectomycorrhizae** surround roots; **endomycorrhizae** infiltrate deeper into the plant. Both help plant roots absorb water and minerals, while obtaining sugars from the plant.

8. Mycoses, or fungal infections, affect many types of organisms. Human mycoses, classified by their effect on the body, may be superficial, cutaneous, subcutaneous, systemic, or opportunistic.

TO REVIEW . . .

1. What does a fungus have in common, biochemically, with an animal?

2. Name three distinguishing characteristics of fungi, including one that is unique to them.

3. Many people think that fungi are a type of plant. Cite two reasons why this is incorrect.

4. Distinguish between a hypha, mycelium, and thallus.

5. What is a yeast?

6. How is the fungal mode of nutrition different from the mode of nutrition in humans? How is it similar?

7. Distinguish between coenocytic and septate hyphae.

8. How do reproductive structures differ among zygomycetes, ascomycetes, and basidiomycetes?

9. How have fungi influenced plant evolution?

10. Where do fungi obtain protein and carbohydrates?

11. Name the fungus that causes each of the following:
- **a.** ergotism
- **b.** Valley fever
- **c.** thrush
- **d.** histoplasmosis
- **e.** Dutch elm disease

12. What is the basis for classification of human mycoses?

TO THINK ABOUT . . .

1. To study the reproduction of *Coccidioides immitis,* the fungus that causes Valley fever, researchers examined 14 DNA sequences in fungal cells taken from 25 people with the illness. Each patient had a unique combination of the 25 DNA regions studied. Does this indicate that the fungus reproduces asexually or sexually? Cite a reason for your answer.

2. What do the fungi that cause athlete's foot, chestnut blight, Dutch elm disease, and thrush have in common?

3. Why might classification of a fungus as a deuteromycete reflect human ideas rather than a true biological phenomenon?

4. Name a fungus structure that might serve as an effective drug target by minimizing the chance that the drug would harm a plant or animal fighting a fungal infection.

5. Design an experiment to reveal what a fungus contributes to a lichen.

6. Researchers are split between classifying *Pneumocystis carinii,* the organism that causes pneumonia in people with AIDS, as a protist or as a fungus. Cite three characteristics that taxonomists might consider to make this distinction.

7. Penicillin, derived from a fungus, kills certain bacteria by destroying their cell walls. Why doesn't penicillin harm fungal cell walls?

8. Fifteen percent of people with AIDS develop *Cryptococcus neoformans* infection, which is usually transmitted in bird droppings. The infection starts in the respiratory tract and spreads to the brain, causing meningitis. Into which two categories of human mycoses can this infection be classified?

9. How can the use of antibiotics to fight a bacterial infection increase the risk of contracting a fungal infection?

10. What occupations might put a person at risk of contracting each of the following?
- **a.** Valley fever
- **b.** histoplasmosis
- **c.** ringworm
- **d.** candidiasis of the fingers

11. Which fungal illness mentioned in the chapter can be sexually transmitted?

12. Why are fungal infections more difficult to treat than bacterial infections?

13. How can a fungus that is normally present in the human body sometimes cause an illness?

14. How might knowing the sequence of the yeast genome benefit human health care?

TO LEARN MORE . . .

References and Resources

Barron, George. March 1992. Jekyll-Hyde mushrooms. *Natural History.* Fungi can be aggressive carnivores.

Bussey, Howard. March/April 1996. Chain of being. *The Sciences.* The yeast genome has been mapped—and it's full of surprises.

Georgopapadakou, Nafsika H., and Thomas J. Walsh. April 15, 1994. Human mycoses: Drugs and targets for emerging pathogens. *Science,* vol. 264. Finding drugs that stifle fungal infections without causing severe side effects is quite a challenge.

Gould, Stephen Jay. July 1992. A humungous fungus among us. *Natural History.* Even mycologists were amazed at the size of these midwestern fungi.

Johnson, Marguerite. February 1, 1993. Valley fever. *Time.* Weather, construction, and a massive earthquake shook fungal spores loose from the California dust, spreading illness.

Kurtzweil, Paula. September 1995. Medical possibilities for psychedelic drugs. *FDA Consumer.* Researchers are rediscovering the medicinal value of LSD, but studies are still difficult to carry out.

Lewis, Ricki. June 1994. A new place for fungi. *BioScience.* A review of fungal taxonomy—and some startling new information.

McDonald, Bruce A., and Joseph M. McDermott. May 1993. Population genetics of plant pathogenic fungi. *BioScience.* From ancient Rome to the present, farmers have feared crop destruction by fungi.

Roy, B. A. March 4, 1993. Floral mimicry by a plant pathogen. *Nature,* vol. 362. Fungi that resemble flowers lure insects into distributing their sex cells.

Smith, Garriet W., et al. October 10, 1996. Caribbean sea-fan mortalities. *Nature,* vol. 383. Deaths of sea-fans were traced to a newly identified species of *Aspergillus*—a fungus thought to be strictly a land-dweller.

Watling, Roy. January 23, 1997. The business of fructification. *Nature,* vol. 385. Fungi as food.

denotes articles by author Ricki Lewis that may be accessed via *Life*'s home page /http://www.mhhe.com/sciencemath/biology/life/.

25 Plantae

Nearly every terrestrial habitat on earth has many representatives of the plant kingdom, which numbers more than 275,000 recognized species. Shown here is a narrow beech fern.

chapter outline

The Value of Diversity 492

Basic Plant Characteristics 492

Bryophytes 492
Bryophyte Diversity
Bryophyte Ecology
Uses of Bryophytes

Seedless Vascular Plants 496
Seedless Vascular Plant Characteristics
Seedless Vascular Plant Diversity
Uses of Seedless Vascular Plants

Seed-Producing Vascular Plants 498
Gymnosperms—Naked Seed Plants
Angiosperms—Flowering Plants

Nature's Botanical Medicine Cabinet 504

Preserving Plant Diversity 504

Paper

Our uses of paper seem endless. We use it to package food, as currency, to communicate, to decorate walls, and even to mop up various bodily secretions. This versatile product consists of cellulose fibers from plant cell walls, mixed with water and pressed through a fine mesh. The fibers entangle, adjacent cellulose molecules bond, the water evaporates, and paper forms. Manufacturing paper today demands a constant supply of wood pulp from millions of trees. Over the centuries, paper has been made from cotton, hemp, jute, bamboo, sugarcane, wheat, rice straw, aspen, beech, birch, fir gum, oak, pine, hemlock, and spruce.

The word "paper" comes from "papyrus," which refers to reeds that the ancient Egyptians used to make paper. A Chinese man named Cai Lun, who served in the Emperor's Court in A.D. 105, invented modern paper from the inner bark of the mulberry tree.

Arabs who captured Chinese learned the art of papermaking, and soon Baghdad had a thriving paper industry. The Crusades and Moor conquest of north Africa and Spain brought papermaking to Europe.

The first papermill in the United States was built near Philadelphia in 1690. Paper for newspapers came from linen or cotton rags. When demand exceeded supply, people found inventive rag sources—such as the wrappings on Egyptian mummies. When mummies became hard to obtain, investors began looking elsewhere for sources of paper.

to be continued . . .

The Value of Diversity

The importance of plant diversity is apparent to anyone who has ever nurtured a vegetable garden. Plant only potatoes in a garden, and a beetle infestation spells ruin. Plant only corn, and crows consume the entire crop. Plant only broccoli, and an early frost triggers flower formation in place of edible parts. But plant a mix of potatoes, corn, and broccoli and only a simultaneous onslaught of beetles, crows, and frost could ruin the entire garden.

Maintaining diversity is a key part of successful agriculture. In Java, farmers plant vegetable gardens of more than 600 crop species, ensuring a food supply. Large-scale farms also benefit from growing more than one plant species (fig. 25.1).

Yet the plants we cultivate are only a tiny percentage of the vastly diverse kingdom Plantae. This chapter briefly introduces the major types of plants and their distinguishing features.

FIGURE 25.1

Planting several varieties of a crop helps increase the odds of a successful harvest. This rice paddy is planted with several types of rice. Should disease strike, at least some rice plants may survive.

Basic Plant Characteristics

Plants are multicellular eukaryotes with cellulose-rich cell walls, chloroplasts containing chlorophyll a and b and carotenoids, and starch as their primary food reserve. They have a characteristic life cycle called alternation of generations, in which an individual goes through a diploid stage, called a sporophyte, and a haploid stage, termed a gametophyte. Each stage is called a generation, and they alternate. Figure 10.2 summarizes alternation of generations, and specific plant life cycles are depicted in chapter 28.

Plants live in almost all habitats, including soil, air, and arctic ice. Wherever they live, plants usually influence—even govern—all other life in the habitat because of their role as autotrophs, harnessing solar energy in a usable form. True plants are nearly absent in the oceans, however, and those that live in fresh water seem to be highly specialized.

At the most fundamental level, plants can be classified by whether or not they have vessels to transport water and nutrients and whether or not they have seeds. The major types of plants include the **bryophytes,** which lack vessels, and the vascular plants, which have vessels. Vascular plants may be seedless or seed-producing vascular plants. The seed-producing plants may have uncovered or "naked" seeds (the **gymnosperms**) or flowering plants, with covered seeds (the **angiosperms**). Table 25.1 lists the major types of plants, and figure 25.2 depicts probable evolutionary relationships among the major types of plants.

Mastering Concepts

1. Why does agriculture benefit from growing a variety of plant species?
2. What are the general characteristics of plants?
3. What are the major types of plants?

Bryophytes

Plants probably evolved from green algae. The earliest plants were likely simple organisms with no organized vascular systems. They may have resembled modern bryophytes.

Bryophytes are small, compact, green plants that do not have vascular tissue. They also lack supporting tissue, but the carbo-

Table 25.1 The Plant Kingdom

Type	Characteristics
Bryophytes	Small and compact, lacking vascular tissue. Gametophyte (haploid) stage is dominant. Includes mosses, liverworts, and hornworts.
Seedless vascular plants	Have vascular tissue but lack seeds. Sporophyte (diploid) stage is dominant. Includes whisk ferns, club mosses, horsetails, and ferns.
Seed-producing vascular plants	Have vascular tissue and produce seeds. Sporophyte (diploid) stage is dominant. Includes gymnosperms (naked seed plants) and angiosperms (flowering plants).

hydrate **lignin** hardens them. Bryophytes grow close to the ground, where they absorb water and nutrients. They also obtain nutrients from dust, rainwater, and chemicals dissolved in water at the soil's surface. Hairlike extensions called **rhizoids** along a bryophyte's lower surface anchor the plant and absorb water and minerals. Although bryophytes do not have true leaves and stems, many have structures that are functionally equivalent to these structures.

Most reproduction in bryophytes is asexual; it occurs by fragmentation, when a plant breaks apart and each piece grows into a new plant. The Japanese use tiny fragments of *Polytrichum* moss to create their lavish moss gardens. In nature, moss fragmentation is so extensive that a cubic meter of arctic snow can contain more than 500 bryophyte fragments, each of which can form a separate plant.

Bryophytes also reproduce sexually. Gametes produced during the haploid generation of the life cycle (the gametophyte stage) form by mitosis in multicellular structures called **antheridia** (male) and **archegonia** (female). Each flask-shaped archegonium produces one egg, and each saclike antheridium produces many sperm (fig. 28.3). A protective, sterile sheath of cells surrounds each gamete-producing structure. Because sperm swim through water to eggs, bryophytes require free water for sexual reproduction. Sperm released from an antheridium do not swim randomly but are attracted to a biochemical that a nearby archegonium produces. The sperm fertilizes an egg and forms a diploid zygote, thus beginning the diploid (sporophyte) generation of the life cycle. In bryophytes, the haploid (gametophyte) generation is dominant.

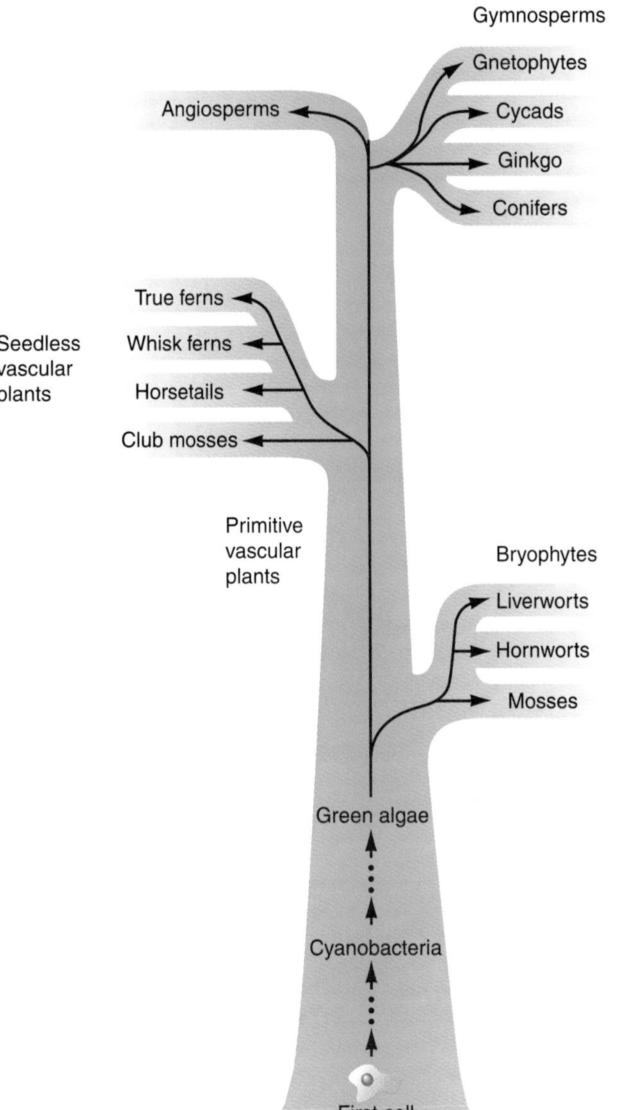

FIGURE 25.2

A schematic evolutionary tree diagram depicting the major plant groups.

Bryophyte Diversity

Bryophytes are classified in three divisions: **mosses, liverworts,** and **hornworts.**

Mosses

The 12,000 species of mosses are the largest and most familiar group of bryophytes. They live in diverse habitats, including waterfalls, lava beds, at the mouths of caves, and on rocks (fig. 25.3). Mosses often grow alongside flowering plants.

A moss gametophyte is radially symmetrical (emanating from a central point) and often has protective scales along the lower surfaces. The gametophytes may resemble leaves and are in fact referred to as leaves, but unlike the complex leaves of vascular plants, they are usually only one cell thick.

Interestingly, not all organisms called mosses are, in fact, mosses. For example, Irish moss is a red alga, reindeer moss is a lichen, and Spanish moss is a flowering plant related to pineapple.

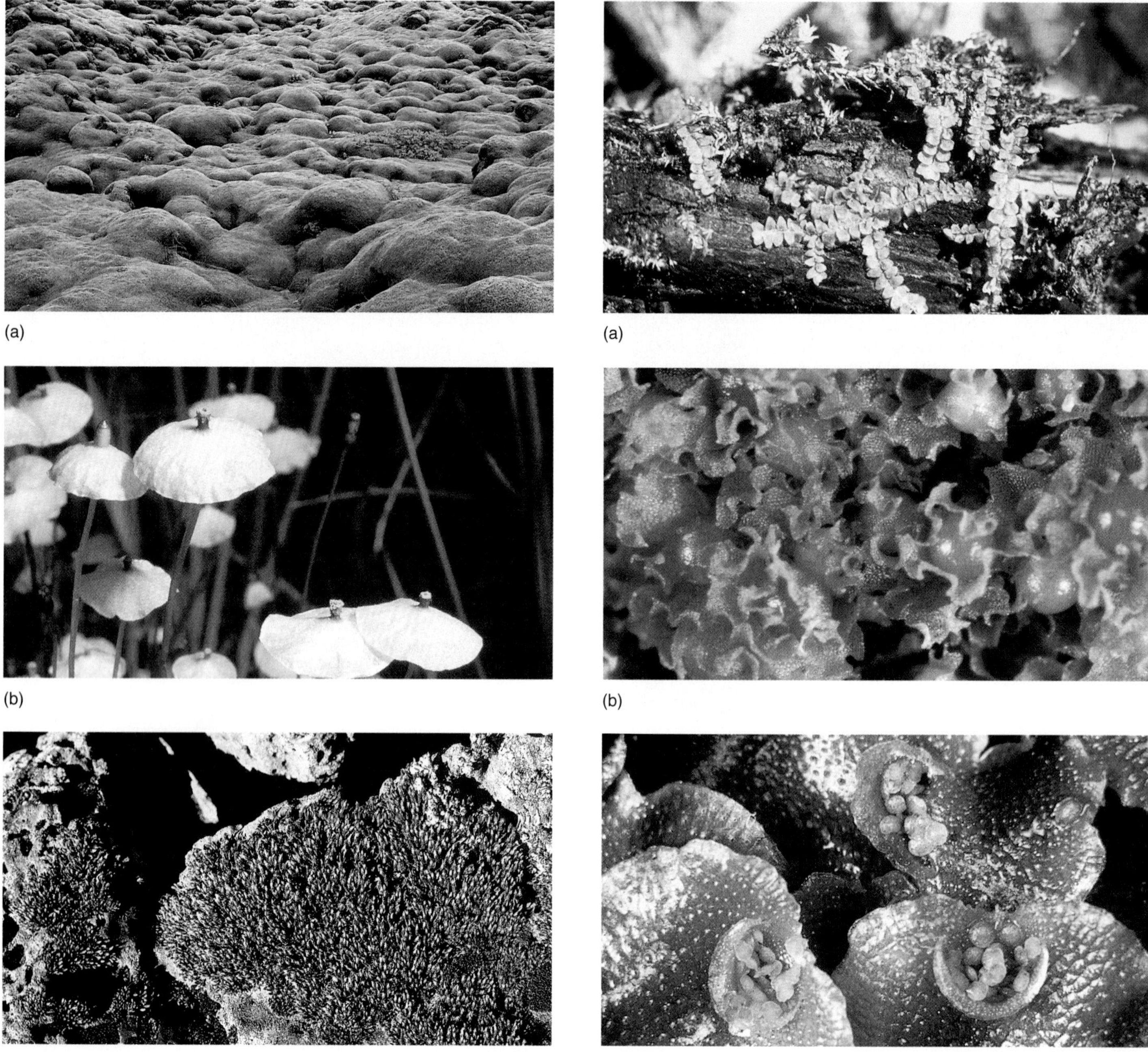

FIGURE 25.3

Moss diversity. (*a*) *Rhacomitrium* moss colonizes a lava bed in Iceland. (*b*) *Splachnum luteum* moss looks like a flowering plant. (*c*) Moss in genus *Grimmia* flourishes on bare rocks in bright sunlight.

FIGURE 25.4

Liverworts. (*a*) The gametophyte of *Calypogeia muelleriana* resembles small leaves. (*b*) Note the ribbonlike shape of *Fossombronia cristula.* (*c*) Gemmae cups on a liverwort, *Lunularia,* are involved in sexual reproduction. Raindrops splash gemmae from the cups. In their new habitat, each gemma can form a new plant.

Liverworts

Liverworts take their name from the resemblance of some species to a human liver, which has lobes. "Wort" means herb. In the fifteenth and sixteenth centuries, people thought that a plant that resembled a part of the human body could treat a disorder there, and so liverworts were thought to heal liver ailments.

The 8,500 species of liverworts vary in size and shape. A leaflike liverwort may be less than half a millimeter in diameter, or a broader liverwort may extend more than 8 inches (20 centimeters) in diameter. Liverwort gametophytes are of two general shapes—leafy, and thallose, which are flat and resemble ribbons (fig. 25.4*a* and *b*).

Liverworts reproduce asexually by producing pieces of tissue called **gemmae.** These form in small "splash cups" on the upper surface of the gametophyte. Falling raindrops detach the gemmae, which can then grow into new plants (fig. 25.4*c*).

biology in action 25.1

Buried in the Bog

In the 1950s, several countries in northern Europe began mining peat (a precursor to coal) from bogs. Peat not only provided valuable fuel but also yielded some 2,000-year-old secrets as grotesque as they were valuable. Horrified workers uncovered several hundred human bodies long buried in the bogs. Tanned by the bog's acids, much as we use acids to tan leather, many of the bodies were so well preserved that the workers thought they had only recently been deposited in the swamp. Workers also noticed that many of the bodies had been mutilated—several had slit throats, severed vertebrae, or nooses around their necks.

Further study revealed that the remains were actually 2,000 to 3,000 years old. The acidity and anaerobic environment of the bogs preserved the bodies by inhibiting growth of bacteria and fungi, which are decomposers. Archaeologists named specific preserved bodies according to the geographical location of the embalming bog. The most intensively studied bog person is Lindow man, who lived about the time of Aristotle (fig. 25.A). His preserved stomach contents revealed his last meal, of linseed gruel and barley. Thanks to the preservative effects of the bog, archaeologists also have learned about musical instruments and household items common in the 2,000-year-old civilizations and determined that bogs were the sites of human religious sacrifice.

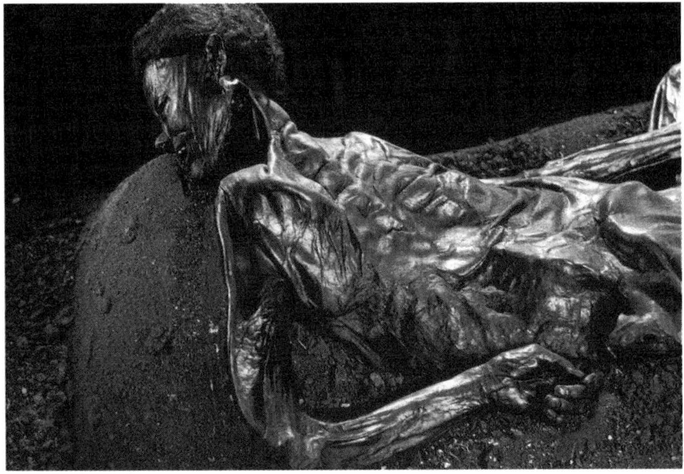

FIGURE 25.A
Lindow man.

FIGURE 25.5
The hornwort *Anthoceros*.

Hornworts

Hornworts are the smallest group of bryophytes, with only about 100 species. A hornwort's sporophyte is shaped like a tapered horn, hence the name (fig. 25.5). Unlike the archegonia of other bryophytes, hornwort archegonia are not discrete organs but are embedded in the plant. Also unlike other bryophytes, hornwort cells have only one chloroplast.

Bryophyte Ecology

Bryophytes grow in diverse habitats, including some too extreme for other types of plants—moist soil, rooftops, the faces of cliffs, tombstones, volcanically heated soil (up to 131°F, or 55°C), and in Antarctica, where summer temperatures seldom exceed 14°F (−10°C). Bryophytes are often the first plants to invade an area after fire, and they can withstand many years of dehydration. They also contribute to and grow around bogs (swamps) (Biology in Action 25.1). Many bryophytes are notoriously sensitive to pollution. As a result, bryophytes tend to be rare in cities and industrialized areas.

Uses of Bryophytes

Mosses are used to stuff furniture, condition soil, and absorb oil spills. Florists use peat moss as a damp cushion around plants designated for shipment, and aboriginal people use *Sphagnum* moss for diapers and as a disinfectant. Peat moss may also be an important source of renewable energy. It is abundant in the United States: the more than 60 billion tons of peat (a precursor to coal) are equivalent to about 240 million barrels of oil. Countries of the former Soviet Union annually harvest more than 200 million tons of peat and use it to fuel more than 80 power plants.

Bryophytes are usually not edible, but people in the 1500s soaked liverworts in wine and ate them. Today, the earthy aroma of Scotch whiskey is partly due to the peat moss that surrounds the storage containers.

(a) (b) (c) (d)

FIGURE 25.6

Seedless vascular plants. The whisk fern *Psilotum nudum* (*a*), club moss *Lycopodium selage* (*b*), horsetail *Equisetum fluviatile* (*c*), and the narrow beech fern (*d*).

Mastering Concepts

1. In what way are bryophytes similar to the earliest plants?
2. Describe the types of bryophytes.
3. What are some habitats of bryophytes?

Seedless Vascular Plants

The earliest plants were aquatic, as were the algae from which plants descended. Life on land required several adaptations that enabled plants to survive and reproduce without a constant water supply. Plants were able to conquer dry land because of

- a dominant sporophyte (diploid) stage that is usually long-lived and large and protects the gametophyte until conditions are favorable for the start of a new individual;
- conducting (vascular) tissues that enable plants to transport water and nutrients rapidly and efficiently throughout the larger plant body; **xylem** tissue carries dissolved ions and water, and **phloem** tissue carries the products of photosynthesis;
- lignin, which strengthens and supports cells, enables large plants to grow tall and upright. In long-ago swamps, the strength that lignin provided increased exposure of leaves to the sun;
- a waterproof cuticle and stomata that control gas exchange and minimize water loss.

Although the earliest vascular plants are extinct, their successful strategies for adapting to land persist in their descendants. The more than 250,000 species of vascular plants are divided into two groups: those that do not produce seeds and those that do.

Seedless Vascular Plant Characteristics

Seedless vascular plants share several features with bryophytes, including the same types of pigments, the same basic life cycle of alternation of generations, and stored starch as a primary food reserve. Seedless vascular plants, by definition, also contain vascular tissue. The evolution of this tissue enabled them to invade and dominate the drier land habitats more effectively than bryophytes could.

(a)

(b)

FIGURE 25.7

The scale-leaf spike moss *Selaginella lepidophylla* forms a dried-up ball (*a*) in times of drought, losing up to 97% of its moisture. Yet it can expand and turn green once rain comes (*b*).

Seedless vascular plants such as ferns have a life cycle similar to that of bryophytes. The dominant sporophyte (diploid phase) produces spores by meiosis. Each spore germinates and grows into a gametophyte that produces gametes by mitosis. Archegonia and antheridia produce the female and male gametes, which fuse to form diploid zygotes. Figure 28.4 depicts alternation of generations in a fern.

Unlike bryophytes, seedless vascular plants have a well-developed cuticle that minimizes water loss, as well as stomata that regulate gas exchange.

Seedless Vascular Plant Diversity

The seedless vascular plants include the **whisk ferns, club mosses, horsetails,** and **true ferns.** Most varieties of these plants live in tropical areas. Their popular names often reflect their appearances.

FIGURE 25.8

Quillworts (genus *Isoetes*) take their popular name from their quill-like leaves.

Whisk Ferns

Whisk ferns are the simplest vascular plants, primarily because they lack roots and because most species have no obvious leaves. Instead of roots, whisk ferns have **rhizomes,** which are fleshy, underground stems that run horizontally.

The name "whisk fern" comes from the highly branched, broomlike stems of *Psilotum,* which resemble whisk brooms (fig. 25.6*a*). *Psilotum,* the largest of the two genera in this group, is widespread in subtropical regions of the southern United States and Asia. These plants are easily cultivated and grown in greenhouses worldwide. The other genus of whisk ferns, *Tmesipteris,* is rarely cultivated and grows only on islands in the South Pacific.

Club Mosses

Club mosses, also called lycopods, consist of more than 1,100 species that live in various habitats worldwide. Most of the species are included in two genera, the club mosses (*Lycopodium*) and the spike mosses (*Selaginella*), which get their names from their club- or spike-shaped reproductive structures. Club mosses have leaves, stems, and roots (fig. 25.6*b*). In some species, the roots grow from rhizomes extending from a central point, and they may be quite extensive. One such ring structure of a *Lycopodium* species was found to have begun growing in 1839!

An interesting *Selaginella* species is the resurrection plant (fig. 25.7). During drought, this plant rolls up in a tight, dried-up ball. When rain comes, its branches expand and photosynthesize. The club mosses also include the quillworts, a group of aquatic plants with narrow, quill-like leaves (fig. 25.8). Club mosses in the Carboniferous period were enormous, towering 130 feet (40 meters) and resembling trees.

Horsetails

Horsetails include only one living genus, *Equisetum,* with 15 species. Some of these plants have branched stems that look

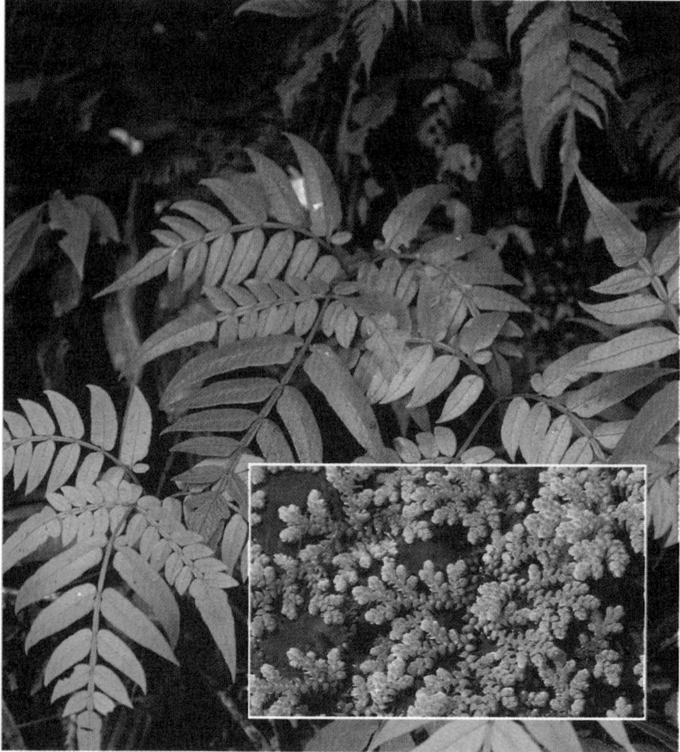

FIGURE 25.9

Ferns range in size from huge (*Marattia*) to tiny (*Azolla,* inset).

somewhat like horses' tails (see fig. 25.6*c*). These plants grow worldwide, usually along streams or at the borders of forests. Their rhizomes are highly branched. Horsetails have aerial stems that are green and photosynthetic and can poison livestock that eat them. The plants also have brown, nonphotosynthetic stems with reproductive structures.

Horsetails are also called scouring rushes because their epidermal (surface) tissue contains abrasive glass particles. Native Americans used horsetails to polish bows and arrows, and early colonists and pioneers used them to scrub pots and pans.

True Ferns

Ferns are by far the largest group of seedless vascular plants, with 12,000 living species. Ferns are primarily tropical plants, but some grow in temperate regions, and others grow in deserts. North or south of the tropics, the number of fern species decreases because of decreasing moisture. In Guam, for example, about 12% of the species of vascular plants are ferns, but in California, only about 2% are ferns.

The leaves, or **fronds,** of ferns are their most obvious feature (see fig. 25.6*d*). Some genera of ferns have leaves that are the largest and most complex in the plant kingdom. For example, one species of tree fern has leaves up to 30 feet (9 meters) long and 16 feet (5 meters) wide, which is nearly the size of a two-car garage! At the other extreme, the aquatic fern *Azolla* has relatively tiny leaves (fig. 25.9).

Each year, new fronds of true ferns grow from rhizomes. The new leaves are at first lopsided, because their upper surfaces grow more slowly than their lower surfaces. This growth pattern coils the new leaves into structures called **fiddleheads** (fig. 25.10*a*). As each growing season begins, new fiddleheads arise near the growing tip of the rhizome.

Most ferns have dark spots on their undersides called **sori** (singular: sorus) (fig. 25.10*b*). Sori are collections of sporangia (structures that produce spores) that are often enclosed in a protective covering. Sori can produce as many as 50 million spores per plant each season. The drying sporangium catapults the spores away from the plant.

Uses of Seedless Vascular Plants

Many seedless vascular plants, especially ferns, are grown in greenhouses or as houseplants. Many more of these plants were once useful in ways that are either no longer in vogue or are only important in economically deprived parts of the world.

Before the invention of flashbulbs, for example, photographers used flash powder that consisted almost entirely of dried *Lycopodium* spores. Chinese people who cannot afford petroleum-based fertilizers grow *Azolla* in rice paddies. This aquatic fern hosts a bacterium that fixes nitrogen from the air, thereby acting as a fertilizer to replenish the nitrates the rice plants remove from the soil.

Mastering Concepts

1. How do the two types of vascular plants differ from each other?
2. What do seedless vascular plants have in common with bryophytes?
3. What are some types of seedless vascular plants?

Seed-Producing Vascular Plants

The ancestors of modern seedless vascular plants dominated the earth's vegetation for more than 250 million years, eventually giving way to the seed plants. Two major groups of vascular plants produce seeds: gymnosperms (naked seed plants) and angiosperms (flowering plants). One of the most significant events in the evolution of vascular plants was the origin of the seed, because it eased plants' survival on dry land.

A seed-bearing plant has two spore types. One type develops into a female gametophyte, giving rise to eggs. The other type is a pollen grain, which develops into a male gametophyte. Once fertilized, the female gametophyte develops into an embryo, which is packaged with nutritive materials into a tough outer coat to form a seed.

Fossil evidence of the first seeds dates to 360 million years ago, and to the first pollen, 150 million years ago. Seed plants arose as the cool, dry Permian period followed the wet, swampy Devonian and Carboniferous periods. Many seeded plant species arose and then became extinct as the environment continued to change. A group called the seed ferns dominated for about 70 million years and then disappeared as towering trees flourished, particularly **cycads, conifers,** and **cycadeoids.** So dominant were these trees by the Jurassic that botanists call this period the Age of Cycads, in contrast to the more familiar Age of Dinosaurs. Today, only a few types of cycads and conifers remain.

(a)

(b)

FIGURE 25.10

Fern features.
(*a*) Fiddleheads are rolled-up leaf buds, such as these fiddleheads of *Blechnum.* (*b*) Sori are accumulations of sporangia on the undersides of fronds.

Gymnosperms—Naked Seed Plants

The term "gymnosperm" derives from the Greek words *gymnos,* meaning "naked," and *sperma,* meaning "seed." The seeds of these plants are termed "naked" because they are not enclosed in fruits. That is, gymnosperms are fruitless seed plants.

Gymnosperms produce secondary growth that usually forms woody trees or shrubs, although a few species are more vinelike. The xylem tissue of most gymnosperms is not organized into vessels, similar to seedless vascular plants.

Gymnosperms, as well as angiosperms, differ from the seedless plants because they do not require water for sperm to swim in to reach the egg. Only the cycads and the maidenhair tree, both gymnosperms, produce swimming sperm. A watery medium, however, is not necessary for fertilization—animals, the wind, or moving water can carry pollen to the female gametophyte. Gymnosperms produce huge amounts of pollen, which is an adaptation to increase the likelihood of fertilization when pollen dispersal is inefficient. A male pine cone annually releases about 2 million pollen grains!

Living gymnosperms are remarkably diverse in reproductive structures and leaf types. Leaves range from simple, flat blades to needles, large fernlike leaves, and tiny, reduced leaves.

Gymnosperm Diversity

The 720 species of gymnosperms are grouped into four divisions: the maidenhair tree (***Ginkgo***), cycads, conifers, and the ***Gnetophyta.***

The maidenhair tree, *Ginkgo biloba,* the only living representative of this division, is a popular cultivated tree. It no longer grows wild in nature. The ginkgo's distinctive, fan-shaped leaves have remained virtually unchanged for 80 million years (fig. 25.11*a*). All living ginkgo trees descend from plants grown in the temple gardens of China and Japan.

A ginkgo tree is either male or female. The seeds are very fleshy, resembling berries or small plums, but they produce a foul odor and may irritate a person's skin. In some parts of Asia, however, pickled ginkgo seeds are a delicacy.

Cycads comprise about 100 species that are distributed primarily in the tropical and subtropical regions of the world. Cycads have palmlike leaves that bear no resemblance to the leaves of other living gymnosperms, but they do produce large cones (fig. 25.11*b*). Although cycads are planted as ornamentals, only two species are native to the United States. Both are in the genus *Zamia,* and neither lives in the wild outside of Florida. Many cycad species are near extinction.

Conifers have needlelike or scalelike leaves and produce cones (fig. 25.11*c*). The pine needles occur in groups, called **fascicles.** Most species have fascicles with two to five needles, but some types have only one and some have as many as eight. At any given time, a few fascicles are shed, so that a conifer is not truly "evergreen" but turns over its needle supply about every 5 years. Bristlecone pine needles are exceptionally long-lived—a needle may last 25 to 30 years! Conifers have a second type of leaf. These are narrow and do not occur in fascicles.

The pines (*Pinus*) are the largest group of conifers and are the most abundant trees in the northern hemisphere. Included in this genus is the bristlecone pine (*Pinus longaeva*) of the western United States, some of which are among the oldest known plants. One tree, named Methuselah, is about 4,725 years old. Other conifers include fir, larch, spruce, and the coastal redwood. The Pacific yew is a source of taxol, a biochemical used to treat certain cancers. Figure 28.5 depicts the pine life cycle.

(a)

(b)

(c)

(d)

FIGURE 25.11

Gymnosperm diversity. (*a*) *Ginkgo biloba* is also called the maidenhair tree. Plants have either male or female flowers. Male trees are preferred for cultivation, because the fruits of female trees emit an odor of rot. (*b*) *Zamia* is a cycad that produces large cones. (*c*) The bristlecone pine (*Pinus longaeva*) is one of the most ancient plants known. (*d*) *Welwitschia* lives only in the Namib and Mossamedes Deserts of southwest Africa. The woody stem may extend 5 feet (1.5 meters) in diameter.

Gnetophytes include some of the most distinctive (if not bizarre) of all seed plants: *Ephedra, Gnetum,* and *Welwitschia. Ephedra,* known as Mormon tea, is used as a medicinal tea. Its stimulatory effects on the human nervous system led it to be included in a dangerous drug called herbal ecstasy. *Gnetum* lives in tropical forests, either as climbing vines or trees.

Welwitschia is a slow-growing plant that lives in African deserts, where it gets most of its water from fog that rolls in from the ocean at night (fig. 25.11d). Mature plants have a pair of large, strap-shaped leaves that persist throughout the life of the plant.

Uses of Gymnosperms

The gymnosperms are second only to the angiosperms in their widespread daily effect on human activities. Conifers produce about 75% of the world's timber and virtually all the pulp we use to make paper. The chief source of pulpwood for newsprint and other paper in North America is the white spruce (*Picea glauca*). A single midweek issue of a large metropolitan newspaper may use an entire year's growth of 124 acres (50 hectares) of these trees; that amount may double for weekend editions. Spruce wood is also used to make musical instruments such as violins.

Important wood products besides lumber and pulp include resin, the sticky, aromatic substance that oozes from wounded conifers. In the plant, resin is in special ducts and stores waste. Resin is a combination of a liquid solvent called turpentine and a waxy substance called rosin. These compounds often are referred to as naval stores, a term that originated when the British Royal Navy used large amounts of resin for caulking and sealing their sailing ships and waterproofing wood, rope, and canvas. Egyptians sealed their mummy wrappings with pine resin, and the Greeks lined their clay wine vessels with pine resin to prevent leakage.

The pine needle–like taste of gin comes from juniper berries. Pine flavoring is still added to Greek wines, giving retsina its distinctive flavor. The unappreciative liken the wine's taste to turpentine.

Turpentine is the premier paint and varnish solvent and also is used to make deodorants, shaving lotions, drugs, and limonene—the lemon flavoring in lemonade, lemon pudding, and lemon meringue pie. Ballerinas dip their shoes in rosin to improve their footing on the stage, and violinists drag their bows across blocks of rosin to increase the bow's friction against the strings. Baseball pitchers use rosin to help them hold onto the ball, and batters apply pine tar to bat handles to firm their grips.

Angiosperms—Flowering Plants

Angiosperms are plants that produce flowers (see fig. 28.7). Unlike gymnosperms, which produce naked seeds, angiosperms produce seeds encased in fruits. Figure 28.8 illustrates sperm and egg production in a flowering plant.

Angiosperms are the dominant plant group on the planet, in both abundance and diversity, and include more than 250,000 species. These plants live in almost all surface habitats, excluding most marine habitats. Except for coniferous forests and moss-lichen tundras, angiosperms dominate major terrestrial zones of vegetation.

Angiosperms include some of the largest and some of the smallest plants, and certainly some of the most familiar (fig. 25.12). Lilies, oaks, lawn grasses, cacti, broccoli, and magnolias

(a)

(b)

(c)

(d)

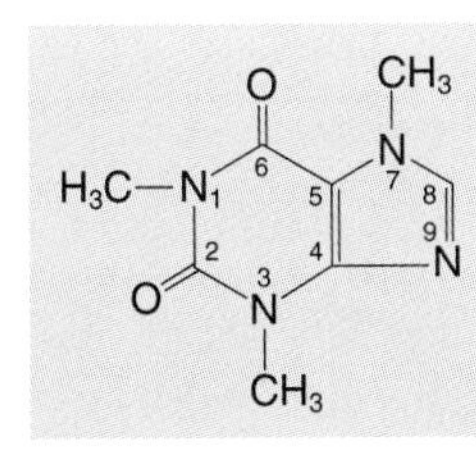

(e)

FIGURE 25.12

Familiar angiosperms—bananas, coconuts, coffee, and tea. (*a*) People have cultivated bananas in tropical Africa for 2,000 years. (*b*) The 50 to 100 coconuts that grow on a palm tree each year provide proteins, oils, and carbohydrates, while the trunks, leaves, and coconut shells make excellent building materials. (*c*) Coffee originally was cultivated in the mountains of Ethiopia; Arab traders spread it elsewhere. Today, *Coffee arabica* is grown for export in 50 nations. (*d*) Tea comes from *Camellia sinensis* and originated in the mountains of subtropical Asia. (*e*) Both coffee and tea contain caffeine, which stimulates the human central nervous system.

25.2 biology in action

How Does a Plant Become a Fossil?

Archaeologists trying to piece together evidence of past civilizations search for telltale shards of pottery and other artifacts. Paleobotanists seek similar direct evidence in fossils. Plant fossils may outwardly resemble the organisms they represent, or they may be fossil fuels, such as coal, oil, or natural gas.

Different types of fossils form, depending upon where the organisms lived and how fast sediments buried them. A compression fossil forms when heavy sediments bury pieces of plants that fell into water long ago and flatten the plant parts, squeezing out water and leaving only a thin film of tissue (fig. 25.B(*1*)). Cellular structures rarely survived the compression process, but well-preserved cuticles occasionally have appeared in compression fossil deposits. Intact cells, and even DNA, have also been found in compressions.

Other types of fossils usually lack plant tissue. An impression, for example, is the imprint an organism leaves behind when its organic remains are completely destroyed. Only the contour of the plant persists. Pieces of coal can house beautiful imprints of ferns (fig. 25.B(*2*)). In fact, a coal deposit is actually an ancient swamp—a habitat where ferns once flourished.

In a third type of fossilization, hardened sediment surrounds tissues, which then decay and leave behind a space or mold where the plant once lay. When the mold fills with other sediments, a cast forms.

In a poorly understood route to fossilization, minerals sometimes replace cell contents. This transforms the organic material into rock. Such fossils, called petrifications, make spectacular displays (fig. 25.B(*3*)). Arizona's Petrified Forest is famous among botanists and tourists alike.

(1)

(2)

(3)

FIGURE 25.B
Plant fossils. (*1*) A compression fossil forms when surrounding sediment and rock layers squeeze moisture from plant remains, leaving a film of tissue that was once a leaf. (*2*) An impression fossil forms when organic remains dissolve, leaving a distinctly shaped space where the leaf once was. (*3*) In petrifaction, mineral replaces tissue. This is a petrified tree.

are all flowering plants. Such plants surround us and affect many aspects of our daily lives. Consequently, this group of plants attracts the most attention from scientists and the public alike.

Angiosperm Origins

More than a century ago, Charles Darwin referred to the sudden and abundant appearance of angiosperms in the fossil record as "an abominable mystery." Despite much study, we still do not know where angiosperms originated or how they evolved so quickly into such a diverse group compared to other seed plants. Molecular tree diagrams suggest that angiosperms first became distinct from other seed plants and then rapidly diversified.

The first angiosperm fossils are no older than about 135 million years; these fossils share features that occur in several modern families of flowering plants. The angiosperms, however, probably arose much earlier—perhaps as early as 200 million years ago. Fossils of that age are unknown, probably because organisms died in dry uplands that were not conducive to preservation (Biology in Action 25.2). The rapid evolution of angiosperms into the most diverse group of plants on earth occurred much later, in the lowlands. Scientists hypothesize that this swift diversification may have been due largely to a combination of adaptive features that appeared and persist in the angiosperms:

- dispersal of seeds in fruit;
- short life cycles;
- rapid growth;
- use of insects and other animals as pollinators.

Angiosperms and Dinosaurs

Dinosaurs and cycads were contemporaries, and the great reptiles continued to flourish during the rise of the angiosperms. Of these three groups of organisms—dinosaurs, cycads, and angiosperms—only the angiosperms remain as a highly diversified and dominant life-form today.

Dinosaurs were extinct by the early Tertiary period, a time when the number of different cycads and other gymnosperms also had begun to diminish. How might gymnosperms, angiosperms, and dinosaurs have interacted and perhaps influenced each other's evolutionary histories? More specifically, how did angiosperms gain prominence at the apparent expense of gymnosperms? One hypothesis is that the dietary habits of dinosaurs may have played a role in the rise of the angiosperms.

The largest herbivores ever to walk the earth were the enormous sauropods, such as *Apatosaurus* and *Brachiosaurus.* These animals probably roamed in great herds, browsing on the tops of conifers and other gymnosperms. This nibbling would not have killed large trees or harmed the smaller seedlings, which the giants could not reach. Thus, gymnosperms could flourish alongside the vegetation-munching reptilian giants.

The coexistence of large, herbivorous dinosaurs and tall gymnosperms began to crumble at the beginning of the Cretaceous period, when dinosaur body size mysteriously began to shrink. By the late Cretaceous, smaller browsers had begun to replace the large herbivores. These new dinosaurs had relatively large, muscular heads with flat teeth suited for chewing plant tissues.

A long-term decrease in dinosaur body size and increase in their diversity coincided with the spread of angiosperms and decline of gymnosperms. The low browsers probably devastated the gymnosperms by eating small seedlings before the plants could reach maturity and produce seeds. Because the first angiosperms were smaller and herbaceous, they probably grew and reproduced more rapidly than the woody gymnosperms and therefore were more likely to produce seeds before they were eaten. Furthermore, small dinosaurs eating gymnosperms opened up new habitats for angiosperm invasion and evolution (fig. 25.13). When dinosaurs became extinct in the Tertiary period, angiosperms exploded in abundance.

An alternate explanation for the rise of angiosperms is that the appearance of insects made pollination more efficient. It may be that both of these explanations are correct—that dinosaurs and insects both influenced the rapid evolution of angiosperms in the

FIGURE 25.13

Tall gymnosperms flourished when dinosaurs were giants, because the reptiles didn't eat the lower portions of the plants. As smaller dinosaurs appeared and began eating younger gymnosperms, the plants began to die out, opening habitats for angiosperms.

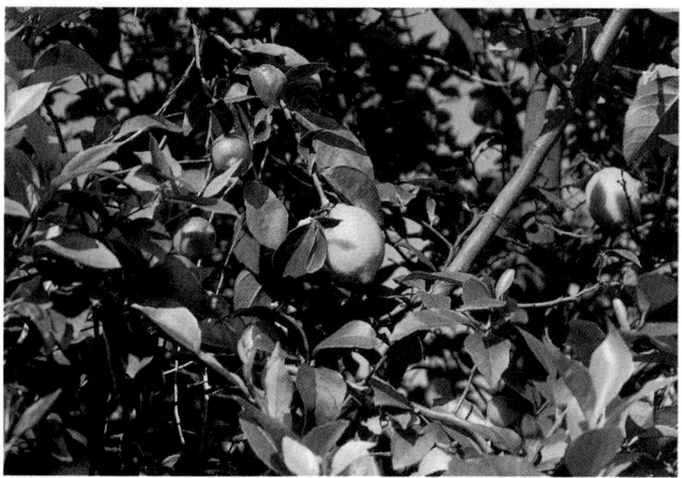

FIGURE 25.14

The tart fruit of the lemon has had various uses in human health care. The juice promotes urination and perspiration and has been used to stimulate these body functions to treat colds and malaria. The astringent (skin-pore closing) nature of lemon juice makes it soothe sunburn; it is also an effective gargle for a sore throat and, it has been claimed, a cure for hiccups. Lemon juice kills bacteria on teeth and gums, and chewing the rind can reduce tartar buildup. The high vitamin C content of the fruit prevents scurvy and the lemon plant leaf makes a sleep-promoting tea.

FIGURE 25.15

The U.S. Department of Agriculture stores seeds of rare lettuces at a gene bank in California.

Cretaceous period. Whatever the precise triggers, the spurt of angiosperm diversity in the Cretaceous set the stage for their spectacular global diversification.

Mastering Concepts

1. How did the evolution of the seed stimulate plant diversification?
2. How do gymnosperms differ from angiosperms?
3. What are some types of gymnosperms?
4. What are some types of angiosperms?

Nature's Botanical Medicine Cabinet

The film *Medicine Man* tells of a jungle plant containing a biochemical that cures cancer—a plant the protagonist tragically loses. Many plants contain chemicals that have healing properties (fig. 25.14). For example, a chemical in bark of the Indian neem tree keeps desert locusts away, but the people of Serengeti National Park in East Africa chew the tree's twigs to prevent tooth decay. The idea of obtaining powerful drugs from "natural" plant products is ancient as well as contemporary and is pursued in many cultures. Chemists seek to duplicate, or even improve upon, the effects of botanical biochemicals in synthetic formulations.

Herbal medicine practices may have begun in prehistoric times. Clay tablets carved 4,000 years ago in Sumari list several plant-based medicines, as do records from ancient Egypt and China. Roman philosopher Pliny the Elder wrote in the first century A.D., "If remedies were sought in the kitchen garden, none of the arts would become cheaper than the art of medicine."

Today, 25% of prescription drugs in the United States are derived from plants. Alkaloids, for example, come from the periwinkle plant and other species and have helped revolutionize treatment of some blood cancers. Alkaloid narcotics derived from the opium poppy, including morphine, are potent painkillers. The many medicines we humans "borrow" from the plant kingdom provide a powerful argument against continuing to destroy the world's tropical rain forests. In these forests, plant life is lush, abundant, and diverse—we know only a fraction of the resident species. Who knows what valuable species we may unknowingly destroy?

Preserving Plant Diversity

Just as plants affect human existence, we affect plants. Agriculture selects certain strains of certain species and propogates them, treating others as weeds and destroying them. In times of war and famine, hungry people eat seeds and thereby destroy future crops.

Storing seeds from many species can help to preserve plant diversity (fig. 25.15). Today the International Plant Genetics Resources Institute in Rome tracks seed banks in 129 countries, which include more than 10,000 species representing more than 2,000 genera. The entries include written descriptions of a plant's characteristics as well as seeds stored in laminated foil pouches.

A look at seed banks around the world provides a view of plant diversity, even when restricted to those species that interest humans. A bank in Indonesia has 240 mango species and 1,700 sugarcane species. A bank in Taiwan has 40,000 types of cabbage, onion, garlic, shallot, tomato, soybean, pepper and mung bean; and in India, the entries are mostly sorghum, pigeon pea, peanut, chickpea, and pearl millet. Sweden's bank has more than 24,000 Nordic plant species, and Mexico City's plant bank has 13,000 types of corn!

Paper~

continued from page 492.

Wood became the new source of paper. The Germans had been using wood to make paper since 1840, and in 1854, the English began to use wood too. Several chemical steps were added to the papermaking process to improve the transformation of wood pulp into paper. On August 23, 1873, the *New York Times* published its first all-wood issue.

Today, people in the United States use almost 200,000 tons of paper each day—enough to cover 1,350 square miles (an area the size of Long Island). In time, the increasing manufacture of these products will outstrip the wood pulp supply.

How can we help conserve wood resources? Many publishers now print magazines and books on recycled paper, and napkins, towels, and paperboard come from recycled paper. To recycle paper, old newspapers, magazines, and junk mail are placed into huge tanks called pulpers, to which solvents and detergents are added to remove inks. The fibers are then reassembled into new paper.

One large American publishing company took a clever approach to saving paper—it trimmed 2.5 centimeters from the width of toilet paper rolls in its building. The employees still used the same number of rolls each month. Trimming all rolls of toilet paper in the United States would save a million trees a year. Other inventive ideas could undoubtedly save millions more.

Even seed banks aren't immune to human damage. In Somalia starving people ate the contents of a seed bank. In Ethiopia, a combination of drought and political unrest led people to eat the stores in a regional seed bank, using up all of the nation's zera zera, a type of sorghum used in research. In the terrible Soviet winter of 1941–42, townspeople invaded a national institute for plant preservation that stored seeds. The curators managed to save the seeds, but nine of them starved to death protecting the remnants of rice, potatoes, beans, and corn.

Today, many seed banks are suffering from lack of funding. Let's hope it doesn't take famine to make the public and legislators realize the value of plant diversity.

Mastering Concepts

1. What are some plant-derived drugs?
2. How does agriculture affect plant diversity?
3. Describe a seed bank.

SUMMARY

1. Plants are multicellular eukaryotes that have cellulose cell walls, chloroplasts, and photosynthetic pigments and that use starch as a nutrient reserve. They have alternation of a **sporophyte** (diploid) phase and a **gametophyte** (haploid) phase in each generation. Plants are widely distributed and are essential to the survival of many other species.

2. Plants are classified by presence or absence of vessels and seeds. **Bryophytes** lack vessels. Vascular plants have vessels and may be seedless or produce seeds. **Gymnosperms** have naked seeds, and **angiosperms** have covered seeds.

3. Bryophytes are small green plants lacking vascular tissue, supportive tissue, and true leaves and stems. They resemble the earliest plants, which arose from algae. **Lignin** hardens bryophytes and **rhizoids** anchor them to the ground, where they absorb water and nutrients. Bryophytes reproduce both asexually and sexually. The gametophyte is dominant, and sexual reproduction requires water for sperm to travel through. Sperm from an **antheridium** travel to eggs in an **archegonium.** The three divisions of bryophytes are **mosses, liverworts,** and **hornworts.**

4. Seedless vascular plants share the same pigments, reproductive cycles, and starch storage mechanisms as bryophytes, but they differ in having conducting tissue and a dominant sporophyte generation. **Xylem** tissue carries dissolved nutrients and water, and **phloem** carries sugars. Seedless vascular plants are adapted to life on land and include **whisk ferns, club mosses, horsetails,** and **true ferns.**

5. The seed-producing vascular plants are the gymnosperms and angiosperms. Neither requires water for sperm to meet eggs. Gymnosperm seeds are not enclosed in fruits, and many lack vessels in the xylem. They have diverse leaves and reproductive structures. Gymnosperms include **ginkgos, cycads, conifers,** and **gnetophytes.** Angiosperms appear rather suddenly in the fossil record and rapidly diversified.

TO REVIEW . . .

1. In what types of plants do the following structures appear?
- **a.** gemmae
- **b.** sori
- **c.** rhizoids
- **d.** rhizomes
- **e.** fiddleheads
- **f.** flowers
- **g.** fascicles
- **h.** fronds

2. What does "alternation of generations" mean?

3. What are the two basic criteria used to distinguish among major groups of plants?

4. Describe the three groups of bryophytes.

5. Cite four characteristics that enabled plants to live on dry land.

6. What is the role of lignin in bryophytes and in vascular plants?

7. How are seedless vascular plants similar to bryophytes, and how do they differ from them?

8. Describe the four types of seedless vascular plants.

9. How did forests in the Jurassic differ from forests today?

10. Describe the four divisions of gymnosperms.

11. How do angiosperms differ from gymnosperms?

12. Why are angiosperms considered to be a successful species? What adaptations and factors may have contributed to angiosperm success?

TO THINK ABOUT . . .

1. From which type of plant might you obtain the following products?
- **a.** turpentine
- **b.** rosin
- **c.** a stuffing material for furniture
- **d.** fertilizer
- **e.** paper
- **f.** the drug taxol
- **g.** the drug morphine

2. Why are bryophytes much smaller than most vascular plants?

3. Mosses and liverworts have been used extensively to monitor radioactive fallout from the Chernobyl reactor accident that occurred in the former Soviet Union in April 1986. What features of these organisms make them ideal for such study?

4. Although bryophytes require free water for sexual reproduction, several bryophytes grow in deserts. How do you think these plants reproduce in their mostly dry environment?

5. What evolutionary developments enabled seedless vascular plants to invade new habitats?

6. What advantage did the development of seeds offer in early plant evolution?

7. Some scientists claim that a geological catastrophe, such as an asteroid impact, killed the dinosaurs. If this is true, how might angiosperms have survived such a catastrophe?

8. What are some ways that you, personally, might help preserve wood resources? How might nations help preserve diversity of native plant species?

9. Michael Crichton's novel and its film adaptation, *Jurassic Park,* featured large herbivorous dinosaurs and huge ferns. What organisms might have been featured if the film had been Cretaceous Park? What if it had been Tertiary Park?

10. Should nations experiencing famine allow people to eat the contents of seed banks? Why or why not?

TO LEARN MORE . . .

References and Resources*

Doyle, Jeffrey J. June 1993. DNA, phylogeny, and the flowering of plant systematics. *BioScience.* A look at angiosperm origins.

Heywood, V. H. 1993. *Flowering plants of the world.* Oxford, England: Oxford University Press. A colorful book about angiosperms of the world, including their various adaptations.

Holland, Bart K. June 30, 1994. Prospecting for drugs in ancient texts. *Nature,* vol. 369. People have used plants to heal for centuries.

Jones, D. 1993. *Cycads of the world.* Washington, D.C.: Smithsonian. A beautifully illustrated book about cycads.

Moore, Randolph, et al. 1996. *Botany.* Dubuque, Iowa: William C. Brown Publishers. An excellent basic botany textbook.

Nee, Sean, and Paul H. Harvey. June 19, 1994. Getting to the roots of flowering plant diversity. *Science,* vol. 264. Evolutionary tree diagrams indicate that angiosperms have diversified more than once.

Powledge, Fred. April 1995. The food supply's safety net. *BioScience.* Seed banks are essential for preserving plant diversity.

Raven, Peter. 1995. *The biology of plants.* New York: Worth Publishers. Another excellent botany text.

**Life*'s website features frequently updated references.

26 Animalia

Animals aren't always what we expect. These giant tube worms (*Lamellibrachia*) live in the Gulf of Mexico. Researchers only recently discovered their embryonic and larval forms.

chapter outline

***What Is an Animal?* 508**

***Blueprint of an Animal* 508**

Levels of Organization
Development
Body Cavity
Symmetry

***Patterns of Animal Development* 512**

***Animal Phyla* 514**

Phylum Porifera: Sponges
Phylum Cnidaria: Hydras, Jellyfish, Sea Anemones, Corals, and Others
Phylum Platyhelminthes: Flatworms, Including Planarians, Flukes, and Tapeworms
Phylum Nematoda: Roundworms
Phylum Mollusca: Clams, Snails, Octopuses, Squid, and Others
Phylum Annelida: Segmented Worms, Including Oligochaetes, Polychaetes, and Leeches
Phylum Arthropoda: Insects, Crustaceans, Arachnids, Myriapods, and Others
Phylum Echinodermata: Sea Stars, Brittle Stars, Sea Urchins, Sand Dollars, Sea Lilies, Sea Cucumbers, and Others
Phylum Chordata: Lancelets, Tunicates, and Vertebrates

Animal Discoveries—From Tube Worms to Lobster Lips

Imagine collecting sex cells from an organism found on another planet, mixing them in a glass dish, and watching what develops. This is what researchers did when they approached vestimentiferan tube worms (pictured in the chapter opening photograph) in the Gulf of Mexico.

Using a deep-sea submersible, the researchers collected two types of worms 1,970 feet (600 meters) beneath the surface, dissected them immediately, gathered gametes from the reproductive tubes, and, back in the lab, conducted in vitro fertilization. Soon, never-before-seen embryos and larvae developed of these species that biologists have learned little about since their discovery in 1965. The gutless worms, a meter or two long, live near deep-sea thermal vents or cool drafts, obtaining energy from hydrogen sulfide in the immediate environment that symbiotic bacteria harness.

By nurturing tube worms in the laboratory, researchers gathered enough clues about their embryonic development to identify the animals' probable closest relatives. This isn't true for an animal called *Symbion pandora,* recently discovered in a most unusual place—on lobster lips.

to be continued . . .

What Is an Animal?

A few years ago, a South Carolina company that supplies animals to researchers caught the attention of an animal rights activist group. When a protester entered the premises and learned that the company bred, raised, and shipped leeches for biomedical research and surgical reattachment of severed fingers and toes, he quickly apologized and left. His organization had thought the company was using animals, not leeches.

A leech is, of course, an animal but not a familiar fuzzy one (fig. 26.1). Many of us utter phrases such as "humans and animals" or distinguish "animal research" from experiments conducted on humans, implying that we occupy a taxonomic status above mere animals. In fact, it is the close biological relationship between *Homo sapiens* and other animals that makes research on nonhuman animals valuable (Bioethics Connections 26.1).

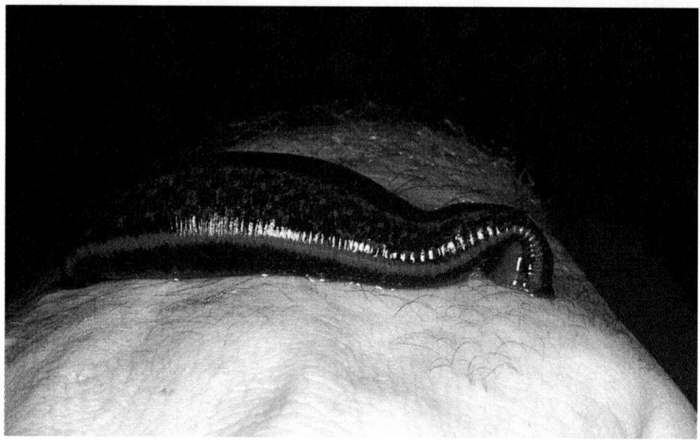

FIGURE 26.1

Animal rights activists changed their minds about protecting the animals at a South Carolina company when they discovered that the animals the facility bred were leeches. As the activists discovered, not all members of kingdom Animalia are cute and fuzzy.

Which characteristics unite within one kingdom such diverse organisms as humans, leeches, squid, birds, insects, and a seemingly endless variety of worms (table 26.1)? Despite this diversity, "animal" has a succinct definition: A multicellular organism with eukaryotic cells that lack cell walls. An animal is a heterotroph, which obtains energy from nutrients, in contrast to an autotroph, which manufactures nutrients with energy obtained from the sun or earth.

Scientists have discovered and named well over a million animal species, yet possibly tens of millions more remain to be cataloged. About 45,000 known animal species are **vertebrates** (animals with backbones). Because most vertebrates are relatively large, it is unusual to discover a new species. **Invertebrates** (animals without backbones) make up most of the kingdom. Phylum Arthropoda, for example, includes more than 750,000 identified species of insects, plus many shellfish, spiders, and other organisms.

The earliest animal fossils are about 700 million years old, dating from the Precambrian period. Animals became abundant about 550 to 600 million years ago, during the Cambrian period. Animals probably descended from unicellular protista, but precisely which species directly preceded the animals is a matter of debate. An animal ancestor may have been a colonial organism whose cells differentiated, eventually forming a multicellular body (fig. 26.2*a*). Or, a ciliated protozoan with many nuclei, such as a paramecium, may have given rise to a multicellular animal by partitioning parts of itself, dividing labor, and specializing functions (fig. 26.2*b*). A third possibility is that different lineages of animals arose from more than one kind of protist ancestor.

However animals originated, they have since split into many groups. Figure 26.3 is a phylogenetic tree of animal phyla that indicates how modern species may have diverged from common ancestors.

Mastering Concepts

1. What are the defining characteristics of an animal?
2. Distinguish between a vertebrate and an invertebrate.
3. What are two ways that animals might have originated from protista?

Blueprint of an Animal

In classifying an organism as an animal, taxonomists consider several aspects of body form. These defining characteristics include level of organization or complexity, type of development, presence of a body cavity, and symmetry.

26.1 bioethics connections

Animal Research: From *Science* to *People*

AUGUST 9, 1996: *People* Magazine—*People* reports the case of Deborah Zimmerman, a troubled waitress who, on March 16, drank herself silly in a bar and then gave birth that evening. In the hospital, she tore fetal monitors from her abdomen, shrieking, "If you don't keep me here, I'm just going to go home and keep drinking and drink myself to death. And I'm going to kill this thing because I don't want it anyways."

But baby Meagan lived, although she was underweight, had a blood alcohol level double that of legal intoxication, and showed the facial signs and behavioral responses of fetal alcohol syndrome (see chapter 12). Zimmerman became the first woman in the United States to be charged with a crime—attempted murder of her fetus—because of behavior while pregnant.

Zimmerman's case was extreme—she drank steadily during her pregnancy, fully aware of the danger to her fetus. Many fear that if found guilty, her case could set a dangerous precedent for women who drink alcohol before they know they are pregnant or drink lightly or moderately while pregnant, believing limited exposure to be harmless. A crucial piece of information to evaluate the potential harm of a woman's drinking while pregnant is the relationship between the amount of alcohol and the eventual effect on the child.

AUGUST 9, 1996: *Science* Magazine—Ironically, the same day that *People* magazine reported on Deborah and Meagan Zimmerman, the journal *Science* published a study conducted on rats that finally clarified the issue of whether limited prenatal alcohol exposure can be dangerous. The answer: Yes.

Researchers at Harvard Medical School isolated cells from rat brains that correspond to cells in humans who have an inherited disorder resembling fetal alcohol syndrome. When alcohol at the level of legal (human) intoxication was applied to the rat cells, they lost their characteristic stickiness, which hampered their ability to connect with each other and therefore to function. This was the first demonstration that alcohol can directly alter brain cell function.

Extrapolation to humans suggests that even low exposure to alcohol can affect brain cells. This finding supports clinical observations that fetal alcohol syndrome is only the most extreme expression of what physicians are beginning to call "alcohol-related neurodevelopmental disorders."

The simultaneous stories in *People* and *Science* illustrate the practical and immediate value of research on animals. Virtually every section of that issue of *Science* mentions discoveries tied to animal experimentation (table 26.A). See table 2.1 for a more traditional look at the value of animal experimentation over the past century.

Table 26.A Animals in Research as Reported in an Issue of *Science* Magazine

Section of Magazine	Report
Letters	Links a study immunizing pregnant rats and tracking resulting immunity in offspring to vaccine strategies in developing nations.
Research News	"Forging a Path to Cell Death" mentions cell death research on roundworms, rats, and chickens. This research impacts on human cancer and degenerative disorders.
Policy Forum	"The Public Health Risk of Animal Organ and Tissue Transplantation into Humans" discusses direct use of nonhuman animal tissue.
Perspectives	"Is the New Variant of Creutzfeldt-Jakob Disease from Mad Cows?" Cows are sacrificed to study how, and if, this disorder jumps from raw cow brains to humans.
Reports	Four articles use mice to investigate —regulation of heart function; —immune deficiency; —hypertension; —nerve cell function.

Source: *Science,* August 9, 1996.

Levels of Organization

Some animals, such as sponges, have poorly defined tissues and no real organs. Others have well-defined tissues but few organs, such as jellyfish, and still others have partial organ systems, such as planarians (flatworms). Some invertebrates, such as segmented worms and mollusks, have complex organ systems similar to those of humans. Complexity, however, does not imply that an organism (including humans) is "advanced" compared to animals with fewer organs or organ systems. The measure of biological success is simply how well adapted an organism is to its environment.

Development

In animals with well-defined tissues, adult body tissues may develop from two germ layers in the embryo (a **diploblastic**

Table 26.1 Some Animal Phyla

Phylum	Characteristics	Examples	Known Species
Porifera	Simple multicellular body; filter water and nutrients through individual cells	Sponges	9,000 +
Cnidaria	Hollow two-layered body; interior jellylike; radial symmetry	Jellyfish, hydras, sea anemones	9,000 +
Platyhelminthes	Three-layered body; organs; bilateral symmetry	Flatworms	20,000
Nemertena	Three-layered body; one-way digestive tract; long muscular "nose" to catch prey	Ribbonworms	900
Nematoda	Three-layered body; one-way digestive tract; unsegmented; thick cuticle; nervous system	Roundworms	12,000
Annelida	Three-layered body; one-way digestive tract; body cavity; segmented; circulatory, excretory, and nervous systems; sensory cells	Segmented worms	15,000
Mollusca	Soft body; hard shell; three body regions; muscular foot, circulatory, excretory, and nervous systems	Clams, oysters, snails, squid, octopuses	100,000
Arthropoda	Three body regions; segmented; outer jointed skeleton that is molted; blood in body cavities; complex nervous system	Insects, crabs, lobsters, spiders	1,000,000
Echinodermata	Five-part body plan; radial symmetry; spiny outer covering	Sea urchins, starfish, sand dollars, sea lilies	20,000
Chordata	Notochord (rod down back); hollow nerve cord above notochord; gill slits	Fishes, amphibians, reptiles, birds, mammals	50,000

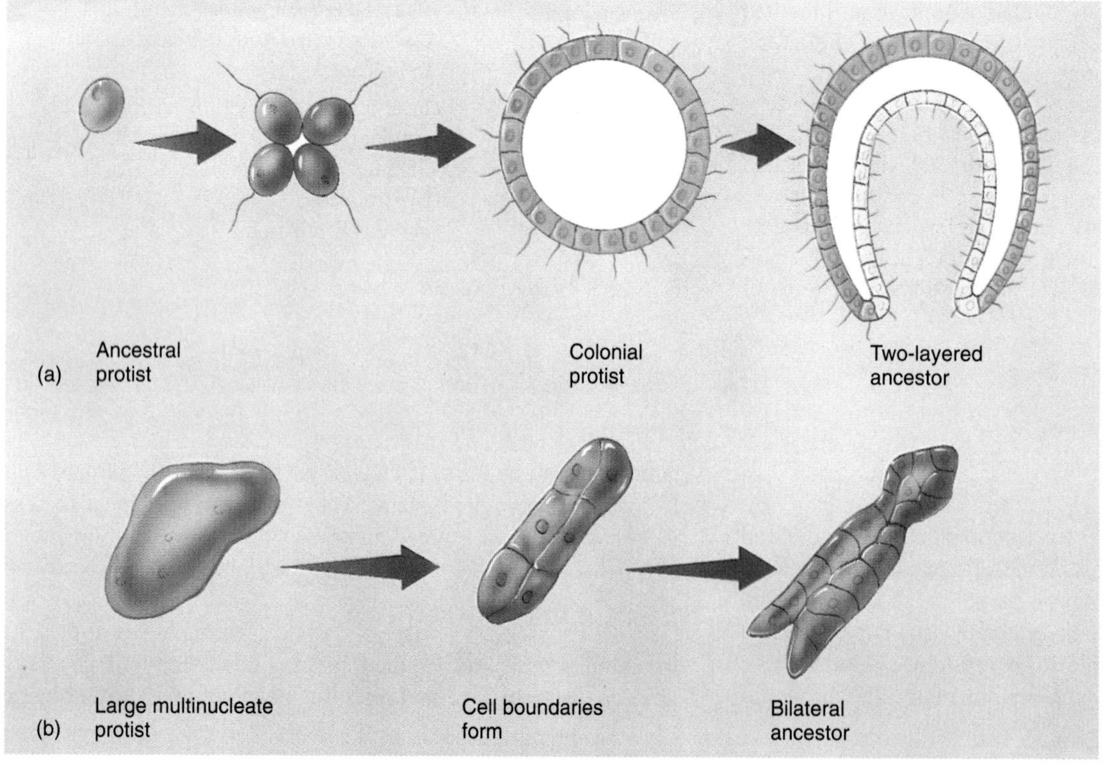

FIGURE 26.2

Multicellular animals may have arisen when aggregates of single-celled protista eventually folded to form layers (*a*), or when cell boundaries formed in a large protistan cell with several nuclei (*b*). Sponges today may echo how cells long ago assembled into multicellular organisms. If sponges from different species are broken into individual cells, and the cells mixed, they assemble with their own kind into whole sponges. Molecules on sponge cell surfaces enable cells to recognize each other.

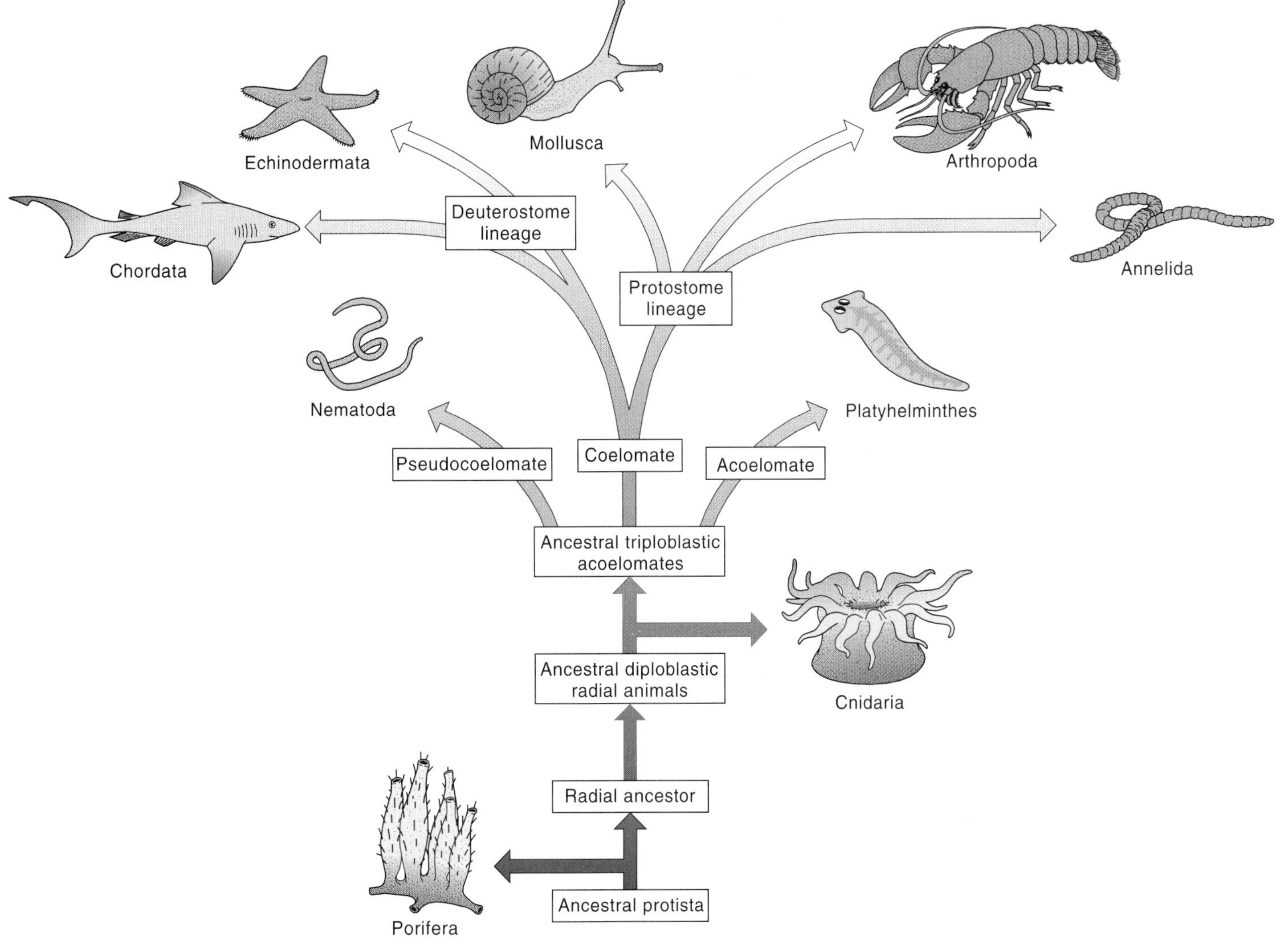

FIGURE 26.3

A phylogenetic tree of the animal kingdom based on hypothesis *b* in Figure 26.2.

embryo) or from three layers (**triploblastic**). The immature form of a jellyfish, for example, consists of two layers—an ectoderm and an endoderm—that sandwich a layer of noncellular material. The jellyfish embryo is therefore diploblastic. A human embryo, in contrast, has three germ layers—ectoderm and endoderm, sandwiching the cellular mesoderm (see figs. 11.15 and 11.16). A human embryo is thus triploblastic.

Body Cavity

Taxonomists distinguish animals by the presence or absence of a body cavity, and whether such a cavity is lined. A true body cavity is called a **coelom.** It is fluid-filled and surrounds the digestive tract (gut). A coelom has several functions, including buffering shocks to internal organs, aiding peristalsis (waves of muscle contraction that move food along a digestive tract), and enhancing body flexibility.

The simple bodies of some animals, such as hydras and planaria (flatworms), do not have true body cavities, although they may be hollow. Such organisms lacking a coelom are called **acoelomate** (fig. 26.4). Some animals, such as roundworms, are **pseudocoelomate,** which means that they retain a body cavity from an early embryonic stage, but it is not lined with mesoderm. A true coelom, such as that seen in vertebrates, is a body cavity formed within mesoderm.

Symmetry

Symmetry refers to the organization of body parts around an axis (fig. 26.5). Body organization is adaptive. The long and flat shape of a planarian worm, for example, enables it to pull itself along.

Some species, such as certain sponges, are bloblike, lacking symmetry. Others exhibit a high degree of symmetry. In a **radially symmetrical** animal, any plane passing from the **oral end** (the mouth) to the **aboral end** (opposite the mouth) divides the body into mirror images. For example, a wheel cut in two has radial symmetry. Wherever it is cut, the halves are mirror images of one another. Hydras are animals that have radial symmetry.

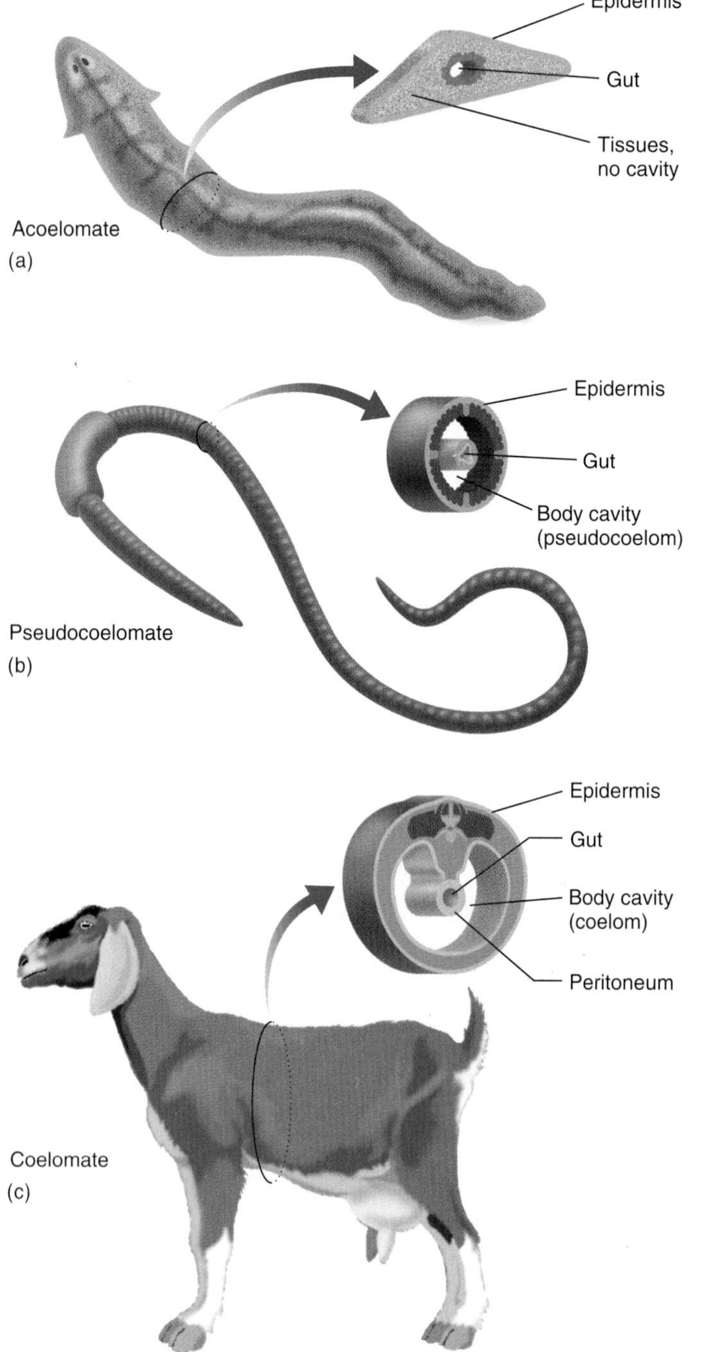

FIGURE 26.4

Presence or absence of a body cavity distinguishes animals. Certain worms are acoelomates (*a*), which lack body cavities. Some other worms are pseudocoelomates (*b*), with body cavities that are not lined with mesoderm. Vertebrates and many other types of animals are coelomates (*c*), with lined body cavities.

In a **bilaterally symmetrical** animal, such as a crayfish, only one cut or plane divides the animal into mirror images. Bilaterally symmetrical animals are **cephalized,** which means that they have head ends. Bilaterally symmetrical body forms also have back sides (**dorsal**) and belly or undersides (**ventral**). Table 26.2 defines these terms, which are used later in the chapter to describe structures in animal bodies.

Table 26.2 Anatomical Terms Related to Body Symmetry in Animals

Radial Animals	
Oral	The mouth end
Aboral	The end opposite the mouth
Bilateral Animals	
Anterior	The front (head) end
Posterior	The rear (tail) end
Dorsal	The top (back) side
Ventral	The bottom (belly) side
Lateral	The side(s)

Mastering Concepts

1. What is the range of tissue complexity among animals?
2. Distinguish between diploblastic and triploblastic animals.
3. What is a coelom?
4. Distinguish between types of symmetry of animal bodies.

Patterns of Animal Development

Development from fertilized egg to adult may occur in either of two ways. Animals that undergo **direct development** hatch from eggs or are born resembling adults of the species. A newborn elephant, for example, looks just like an adult elephant (fig. 26.6*a*). Animals that directly develop have no larval stage. A larva is an immature stage that does not resemble the adult of the species.

Larvae undergo a developmental process called **metamorphosis,** in which they change greatly (fig. 26.6*b*). Familiar larvae are tadpoles and caterpillars, which become frogs and butterflies, respectively. Many aquatic animals metamorphose, some developing through several larval stages before they resemble the adult.

Animals are also classified according to differences in early embryonic development, as the chapter opening text illustrates. Some embryonic characteristics important in taxonomy include:

- the amount of nutritive yolk in the egg;
- the pattern of early cleavage divisions;
- cell movements during development;
- the number of primary germ layers in the embryo;
- whether and how a body cavity forms.

Biologists use such differences in embryonic development to define major lineages (lines of descent) in the animal kingdom. The two major lineages among the coelomates are called **protostomes** and **deuterostomes.** Protostomes include **mollusks**

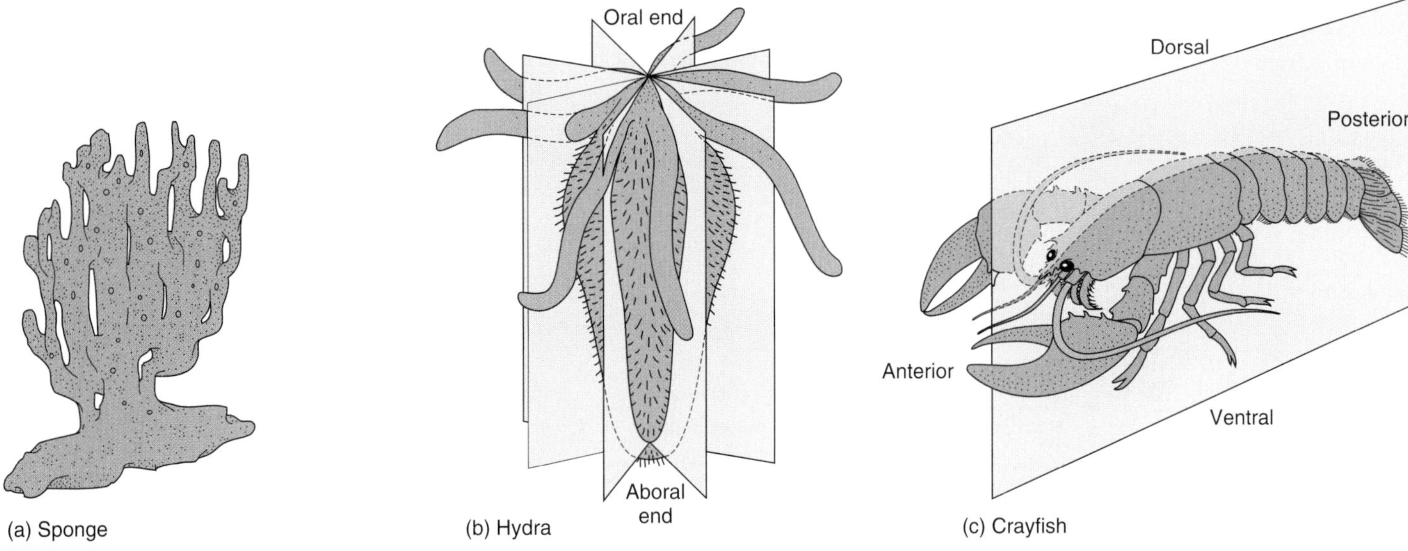

FIGURE 26.5

Animal body forms. (*a*) Some sponges have an asymmetrical body form. (*b*) The radially symmetrical body form of hydra. Any plane passing through the oral-aboral axis divides the animal into mirror images. (*c*) The bilaterally symmetrical body form of a crayfish. Only one plane divides the animal into mirror images.

(a)

(b)

FIGURE 26.6

Type of development distinguishes animals. (*a*) Orangutans exhibit direct development—a juvenile resembles an adult. (*b*) The larva-to-butterfly transformation is a familiar example of metamorphosis.

(including snails, clams, slugs, and octopuses), **annelids** (segmented worms), and **arthropods** (including insects). Deuterostomes include **echinoderms** (spiny-skinned marine organisms) and **chordates,** the group to which vertebrates belong.

Protostomes and deuterostomes show major distinctions very early in development. The first difference is the pattern of cells at the eight-celled stage. In protostomes, the top layer of four cells is offset from the bottom layer, giving the embryo a spiral appearance (**spiral cleavage**). In contrast, the two four-celled layers in the eight-celled deuterostome are directly aligned, demonstrating **radial cleavage** (fig. 26.7*a*).

Later in development, at the gastrula stage, a second major difference between protostomes and deuterostomes emerges. In protostomes, an indentation called the **blastopore** develops into a mouth; the word protostome means "first mouth" in Greek. In deuterostomes, the blastopore is a different shape and develops into an anus (fig. 26.7*b*). Deuterostome means "second mouth."

A third major difference between protostomes and deuterostomes is the developmental potential of the first cells resulting from early cleavage divisions. In protostomes, if an early cell is removed, development ceases, leaving an incomplete mass

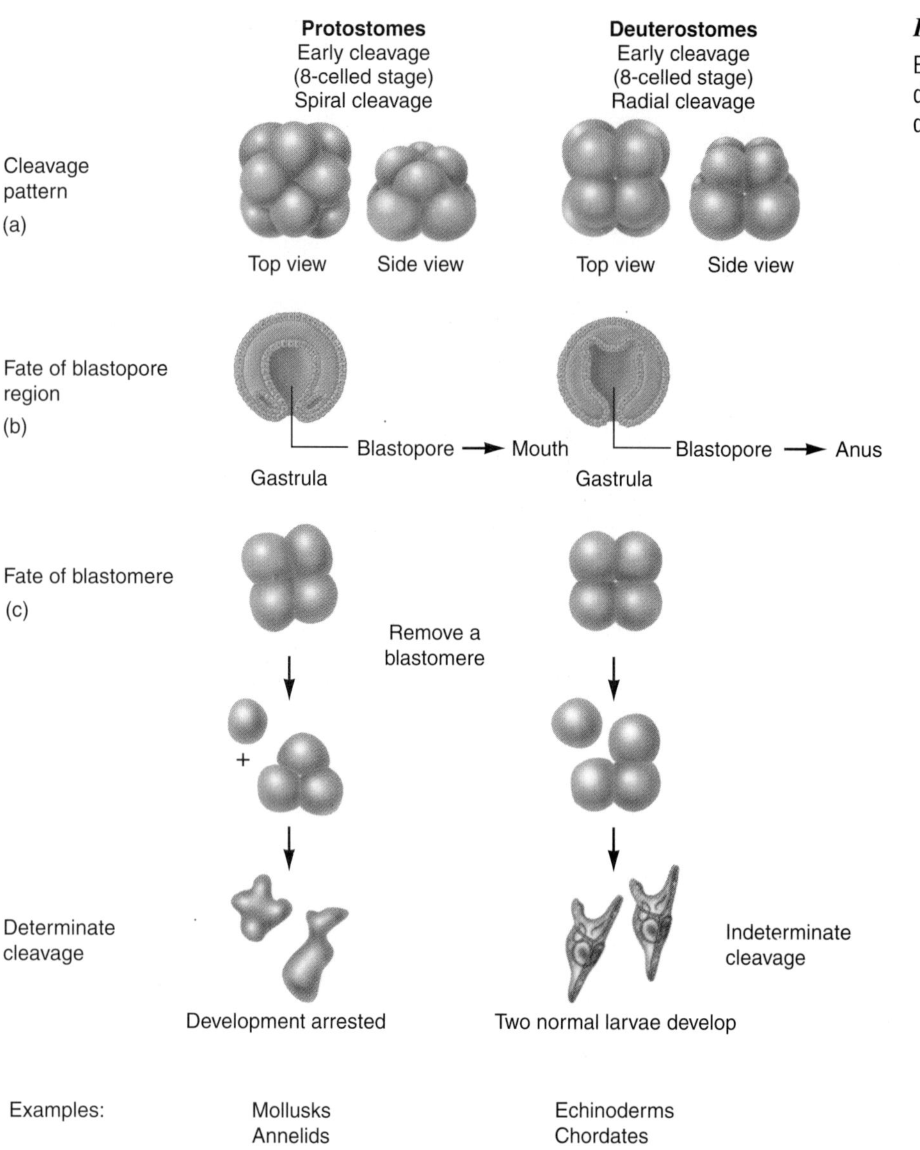

FIGURE 26.7

Early developmental events, as well as other differences, distinguish protostomes from deuterostomes.

(fig. 26.7*c*). This phenomenon, called **determinate cleavage,** occurs because the fate of each cell of the embryo is predetermined. In deuterostomes, if one cell is removed from the two- or four-celled stage, both the remaining three cells and the separated fourth cell form complete, genetically identical embryos. This is called **indeterminate cleavage,** because the fate of each cell is not predetermined at this early stage—that is, a cell separated this early in a deuterostome preembryo retains the potential to yield a complete embryo.

Human indeterminate cleavage makes possible the technology of preimplantation genetic diagnosis, depicted in figure 12.6. Researchers remove a single cell from an eight-celled human preembryo developing in a laboratory dish and test it for a disease-causing gene. If the abnormal gene is not detected, the remaining seven-celled preembryo is implanted into a woman, where it completes development into an individual free of the inherited disease.

Mastering Concepts

1. Distinguish between direct development and metamorphosis.
2. In what three ways do protostomes and deuterostomes differ?

Animal Phyla

To fully describe all of the animal phyla would require a book—and a long one. Here we consider animal phyla that include more than 5,000 species, from simplest to most complex. The common name for members of a phylum often derives from the scientific name. For example, members of phylum Nematoda are commonly called nematodes; those of phylum Chordata, chordates.

(a)

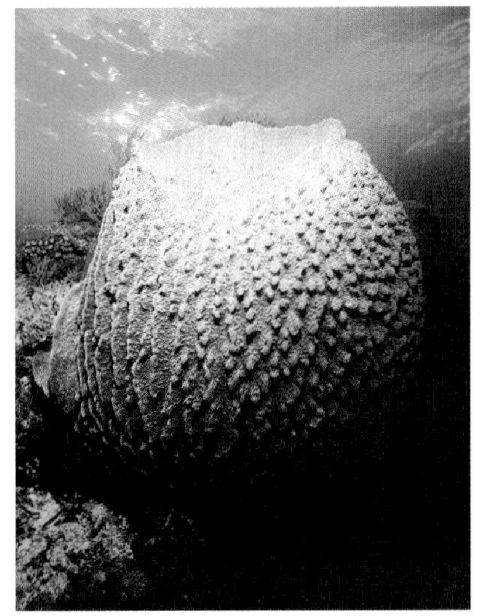

(b)

FIGURE 26.8

Sponge types. (*a*) An asymmetrical bath sponge. (*b*) A radially symmetrical sponge.

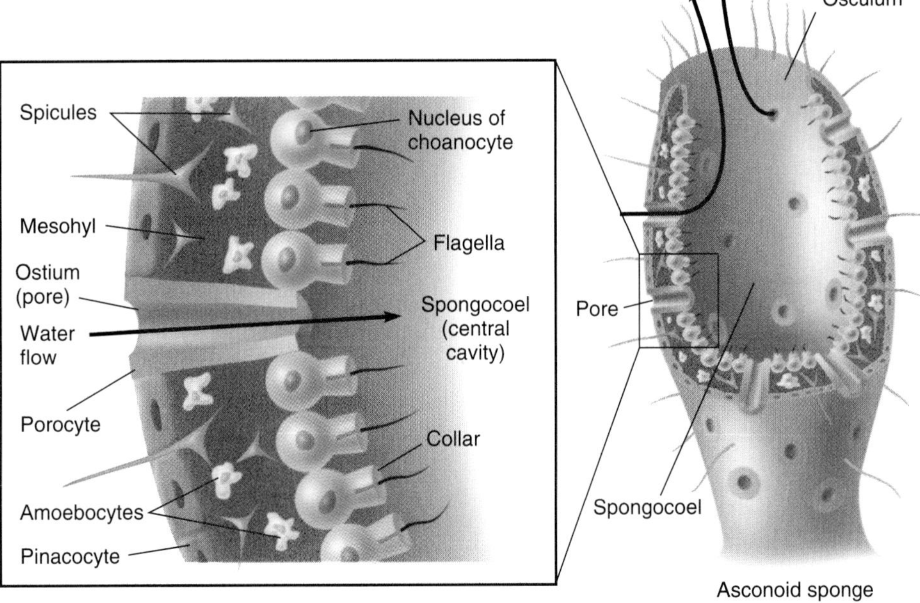

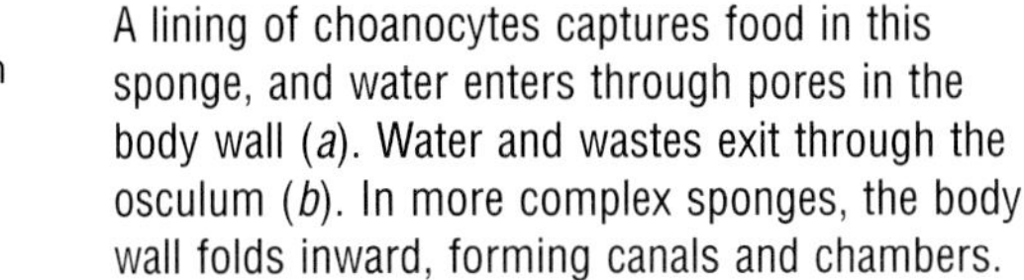

FIGURE 26.9

A lining of choanocytes captures food in this sponge, and water enters through pores in the body wall (*a*). Water and wastes exit through the osculum (*b*). In more complex sponges, the body wall folds inward, forming canals and chambers.

Phylum Porifera: Sponges

Sponges belong to phylum Porifera, which means "pore-bearers"—an apt description of these simple animals. Sponges are sessile (nonmotile) and either asymmetrical or radially symmetrical (fig. 26.8). They are all aquatic, living mostly in the ocean, and are **suspension feeders,** capturing food particles suspended in the water.

The sponge body form is quite different from that of other animals. The body wall consists of cells, called **pinacocytes,** embedded in a gelatinous matrix. The **mesohyl** region of a sponge contains **amoebocytes,** which are multifunctional, mobile cells. Amoebocytes can digest, store and transport nutrients, divide, or secrete skeletal components.

The simplest sponges are hollow and feed by drawing water through many body surface pores, called **ostia,** into the central cavity, or **spongocoel** (fig. 26.9). To accomplish this, flagellated cells called **choanocytes,** or "collar cells," produce a current that draws food and water into the spongocoel. Food, usually particles of organic matter and small unicellular organisms, moves to the cell body and then is ingested in a food vacuole, where it is partially broken down. The food particles then pass to amoebocytes, which complete digestion. Water and wastes exit the sponge through a large hole at the top, the **osculum.**

In more complex sponges, the body wall folds into finger-shaped incurrent and radial canals. Choanocytes are only located in radial canals. In other sponges, body wall folding obliterates the spongocoel. In these sponges, choanocytes occupy only small cavities, called choanocyte chambers, and several merging canals lead to several oscula.

The varied body forms of sponges reflect increasing complexity in the path water takes through the animal. Body wall folding increases the surface area-to-volume ratio. This increased ratio allows the sponge to move water, capture food, and excrete wastes more efficiently, which preserves energy and enables the sponge to grow.

Sponge skeletons are unique; they consist of individual slivers of glassy or limy material called **spicules** or an organic mesh called **spongin**—or both or neither (see fig. 34.3*b*). Amoebocytes secrete spicules, which come in many shapes, resembling toothpicks, umbrellas, dumbbells, and sunbursts. Biologists classify sponges by skeleton type. Bath sponges have no spicules and only spongin, and therefore are soft enough to use to bathe.

Sponges can reproduce either sexually with gametes or asexually by budding or regenerating fragments. Freshwater and some marine sponges asexually produce "resting bodies" called **gemmules** when their habitat dries up or freezes. Gemmules contain amoebocytes and lie within spicules. When the environment becomes habitable, amoebocytes leave gemmules through a small opening and grow into new sponges.

Sponges are abundant in diverse marine environments. They fend off predators with their spicules and with toxic chemicals. Sponges frequently house symbionts that avoid predators in their spiky host. Fishes, for example, hide in spongocoels, and worms and shrimps occupy sponge canals. The decorator crab attaches sponges to its shell to camouflage itself from predators! See Biology in Action 22.1 for a description of a carnivorous sponge.

Mastering Concepts

1. How are sponges distinct from other types of animals?
2. What are the major cell types of a sponge?
3. How is a sponge's form an adaptation to acquiring nutrient-containing water?
4. How do gemmules protect sponges in harsh environments?

Phylum Cnidaria: Hydras, Jellyfish, Sea Anemones, Corals, and Others

Many ocean swimmers fear stings from cnidaria such as the sea nettle, sea wasp, fire coral, and Portuguese man-of-war. To sting, a cnidarian discharges microscopic capsules called **nematocysts** embedded in cells called **cnidocytes.** These animals use different kinds of nematocysts for defense, locomotion, attachment, and capturing prey. Figure 26.10 shows the location and function of nematocysts in the body wall of a hydra.

Cnidaria are all aquatic and mostly marine. Many, such as corals, are sessile, but others (for example, jellyfish and Portuguese men-of-war) are free-swimming. All cnidaria are predators, and many are prey for other predators. Small cnidaria eat zooplankton, and larger ones eat animals as large as fishes.

Cnidaria have radial symmetry around an oral-aboral axis, extending from the mouth to a point opposite the mouth. They are

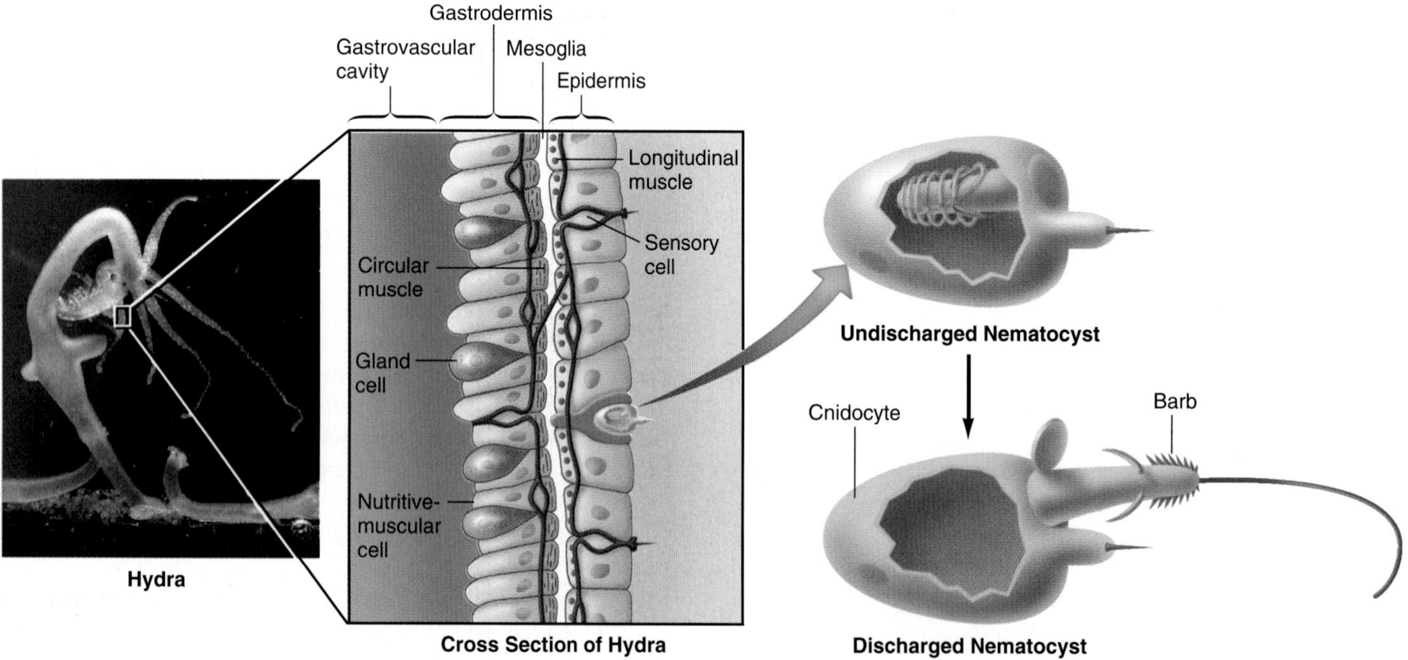

FIGURE 26.10

Cnidaria such as this hydra characteristically have stinging cells, or cnidocytes, in their body walls. These cells contain wound-up stingers called nematocysts.

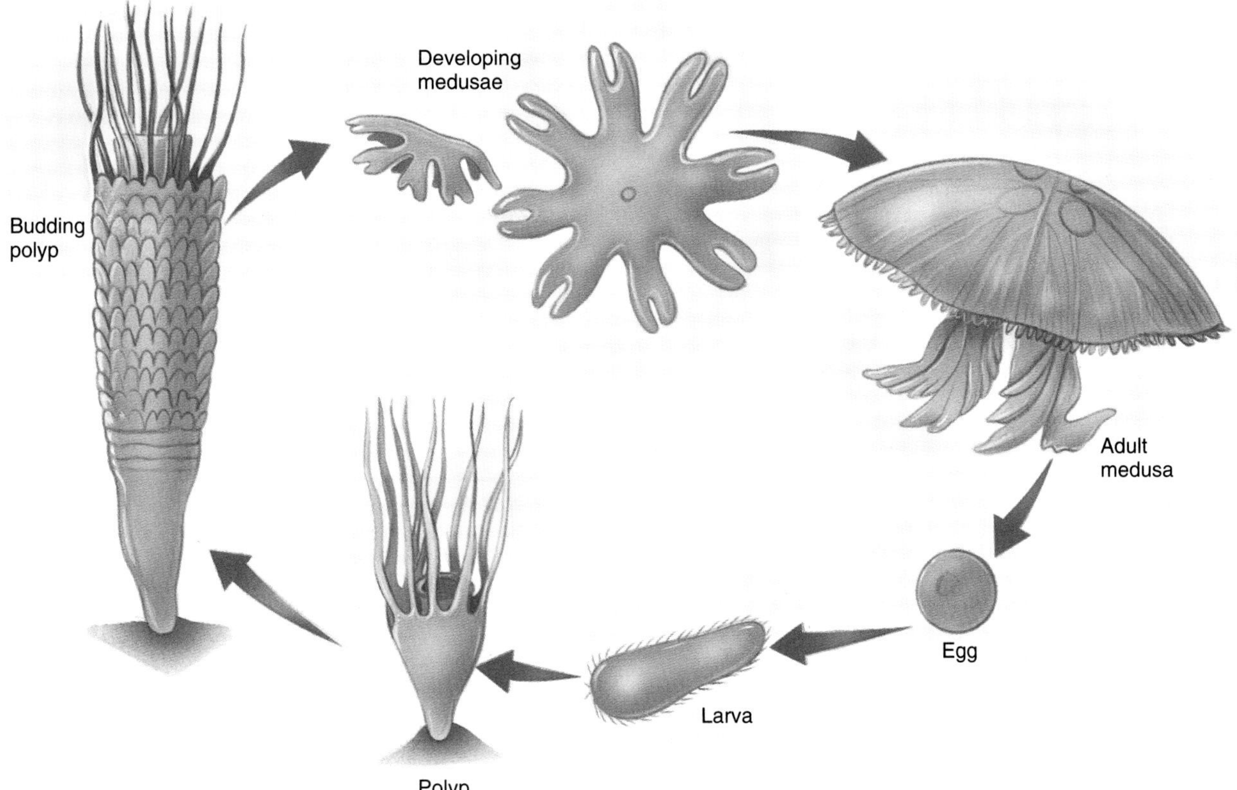

FIGURE 26.11

In the cnidarian *Aurelia,* the medusa is the sexually reproducing form. Eggs develop into larvae and then polyps. The polyps bud, yielding medusae, and the cycle continues.

diploblastic, consisting of **epidermis** and **gastrodermis** tissue layers sandwiching a jellylike, noncellular substance called **mesoglia.** The thick mesoglia of jellyfish gives them their "jelly." Extracellular (outside of cells) digestion occurs in the **gastrovascular cavity.** The animal ejects wastes through the mouth (see fig. 37.6*b*).

The two types of cnidarian body forms are a sessile **polyp** and a free-swimming **medusa.** Hydras and sea anemones are polyps. Adult jellyfish are medusae. Cnidarian species that undergo alternation of generations have both polyp and medusa stages (fig. 26.11). The medusa is the mature, sexually reproducing stage, with male and female individuals. Taxonomists classify cnidaria partly on the basis of their dominant stage in the life cycle.

Cnidaria may aggregate into colonies. The Portuguese man-of-war (*Physalia*) is actually a free-swimming colony of individual medusae and polyps. The floating portion, or **pneumatophore,** is a gas-filled structure that keeps the colony at the ocean's surface. Corals are colonies of polyps that reproduce asexually and sexually.

Some cnidaria have a hydrostatic skeleton, a fluid-filled compartment that provides support. Other cnidaria secrete a limy exoskeleton or have an endoskeleton of protein or the carbohydrate chitin. As successive generations of colonial coral polyps die and new individuals grow, their limy exoskeletons form magnificent coral reefs. Red algae symbionts add spectacular colors.

Mastering Concepts

1. How do cnidaria sting prey?
2. Where do cnidaria live?
3. Distinguish between polyps and medusae.

Phylum Platyhelminthes: Flatworms, Including Planarians, Flukes, and Tapeworms

Flatworms are flat worms. They are triploblastic, lack coeloms, and are bilaterally symmetrical.

Variations within this phylum reflect adaptations to either a free-living or parasitic lifestyle. Parasitic flatworms have greater reproductive capacity, as evidenced by their greater number or size of reproductive organs compared to those of free-living species. Many flatworms parasitize humans and domestic animals and are usually contracted by eating undercooked meat.

Flukes are flatworms that parasitize vertebrates. They feed on their hosts' blood and tissues through an anterior mouth surrounded by an adhesive oral sucker that attaches to the host. A muscular pharynx pulls food into two large, unbranched, saclike **intestinal ceca** for digestion and absorption.

Tapeworms parasitize vertebrate intestines. They lack mouths and absorb food through the **tegument,** a protective body covering also found in flukes. This covering has microscopic, finger-shaped projections that increase the surface area for absorption. What looks like the head of a tapeworm is a holdfast organ, the **scolex,** that helps the worm attach to the host's intestine (fig. 26.12). Since tapeworms absorb digested food from their hosts, they lack digestive systems.

Free-living flatworms usually are predators or scavengers in aquatic ecosystems. A planarian, for example, has a structure called an eversible pharynx that flicks out of the mouth to catch prey or eat dead animals. This worm has a highly branched gut for extracellular digestion and ejects wastes through its mouth. The digestive system is incomplete because there is no anus.

Some flatworms have concentrations of nerve cell bodies in their heads, which form a rudimentary brain, and nerve cords. Others have a simple nerve net. Chemosensory (taste and smell) cells in some freshwater planaria cluster on earlike structures called auricles. Eyespots detect light but do not form images.

Protonephridia are structures in flatworms that maintain internal water balance and form a simple excretory system. Because of their shape and beating cilia, protonephridial cells are often called flame cells (see figure 38.12). Most excretion occurs by simple diffusion through the body wall. Flatworms have no circulatory or respiratory systems—diffusion accomplishes these functions (see fig. 35.1*c*).

Many flatworms reproduce both sexually and asexually. Free-living species may simply pinch in half and regenerate the missing parts asexually. Free-living flatworms are hermaphroditic (they carry the organs of two sexes in one individual), but they cross-fertilize (they mate with each other). Young that resemble adults hatch from eggs, except in some marine species that have ciliated larvae.

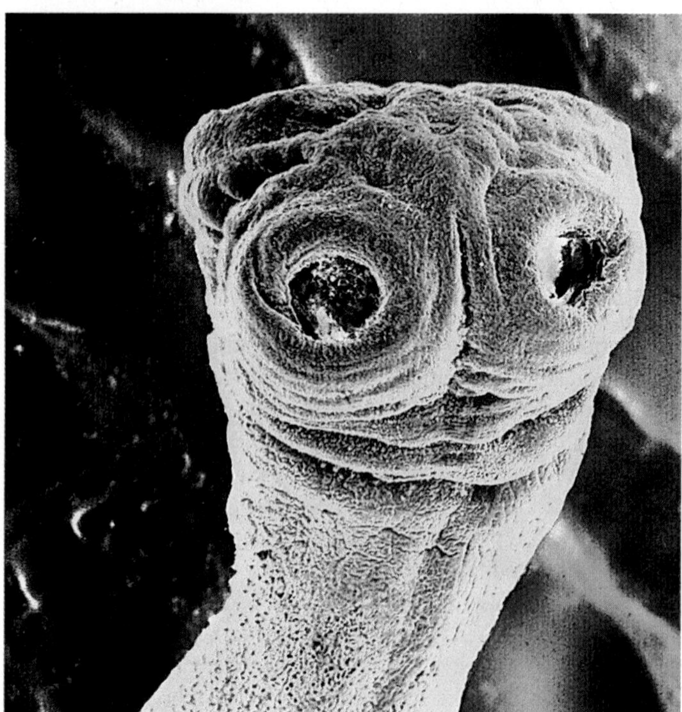

FIGURE 26.12

The dog tapeworm *Taenia pisiformis* has hooks and suckers that enable it to latch onto its host's intestine.

Most flukes are hermaphroditic and cross-fertilize inside the host. Eggs pass out of the host with urine or feces and must land in water to hatch. The intermediate host—the animal that houses the larval form of the parasite—is usually a snail. Another organism may host the fluke before it moves to its final host to develop into an adult.

Tapeworms are reproductive machines. They contain many structures, called **proglottids,** each of which houses a complete set of male and female reproductive structures. The animals can self-fertilize or cross-fertilize. After eggs mature and are fertilized, individual proglottids break off and leave the host along with its feces. When another appropriate host swallows the proglottids, eggs hatch and start a new generation.

Mastering Concepts

1. What are some adaptations of flatworms for a parasitic existence?
2. What types of specialized cells and tissues do flatworms have?
3. How do flatworms reproduce?

Phylum Nematoda: Roundworms

Roundworms are cylindrical worms without segments. Most are barely visible to the naked eye and are free-living in soil or in sediments at the bottoms of aquatic ecosystems. Some roundworms parasitize plants or animals and are acquired in a variety of ways (table 26.3 and fig. 26.13). Many developmental biologists use the nematode *Caenorhabditis elegans* as a model organism to trace cell lineages (see fig. 11.6*a*).

Roundworms have only longitudinal (lengthwise) muscles, which limits their movement to thrashing—they cannot crawl. They have complete digestive systems, and are thus more com-

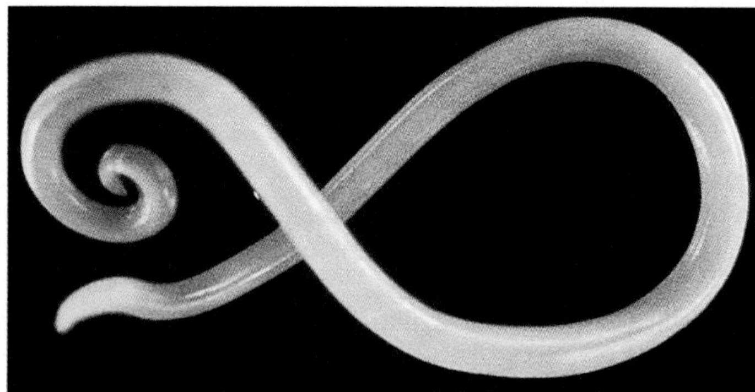

FIGURE 26.13

Millions of humans harbor the giant intestinal roundworm *Ascaris lumbricoides.* The parasite is acquired through contact with human feces containing eggs. When a person consumes infected fecal matter, the eggs hatch in the intestine. The larvae burrow through the intestinal wall and enter the circulation, which carries them to the lungs. Here, the larvae molt twice, move up the trachea, and are swallowed. Once the worms return to the intestine, they become sexually mature, mate, and lay eggs, which exit in the feces. This is a mature male giant intestinal roundworm.

Table 26.3 Nematode Parasites of Humans and Domestic Animals

Worm (Common Name)	Host	Mode of Infection	Distribution
Enterobius vermicularis (pinworm)	Humans only	Inhaling eggs; fingers to anus	Worldwide;* most common worm parasite in United States
Necator americanus (New World hookworm)	Humans only	Larvae burrow into feet from contaminated soil; can cause anemia	New World;† southeastern United States
Ascaris lumbricoides (giant human intestinal roundworm)	Humans only	Eating infective stage eggs in contaminated food	Worldwide;* Appalachia and southeastern United States
Trichinella spiralis (trichina worm)	Humans, pigs, dogs, cats, rats	Eating encysted larvae in undercooked meat; can be fatal	Worldwide;* throughout United States
Wuchereria bancrofti (filarial worm)	Humans	Mosquito bite; can cause elephantiasis (severe swelling)	Tropics
Dirofilaria immitis (heartworm)	Dogs	Mosquito bite; can be fatal	Worldwide

* Distribution is scattered throughout the world; appears to varying degrees in different places.

† Distribution is scattered throughout the New World; appears to varying degrees in different places.

plex than flatworms, but they still have poorly developed heads and lack circulatory or respiratory organs. A roundworm's hydrostatic skeleton consists of a fluid-filled **pseudocoel.** The **hypodermis,** a surrounding layer of tissue, secretes a protective cuticle.

Nematodes have separate sexes. Females of parasitic species produce large numbers of eggs that are extremely resistant to harsh environmental conditions such as dryness and chemical exposure. Most nematodes undergo direct development, and most parasitic forms live in only one host. Insect bites can transmit certain roundworm infections.

Mastering Concepts

1. Describe a roundworm.
2. How is a roundworm more complex than a flatworm?
3. How do roundworms reproduce?

Phylum Mollusca: Clams, Snails, Octopuses, Squid, and Others

Phylum Mollusca is the second largest phylum in the animal kingdom, with many species living in the oceans, in fresh water, and on land. The giant squid is the largest invertebrate known, ranging up to 65 feet (20 meters) in length, including its massive arms.

Mollusks are among the most complex invertebrates and possess all the organ systems vertebrates have. The mollusk circulatory system is usually open; blood circulates throughout the body cavity instead of within vessels. Aquatic species have gills, and terrestrial species have one lung derived from a space called the mantle cavity. Mollusks are protostomes, and most are bilaterally symmetrical.

Mollusks have several body structures in common (fig. 26.14), although these animals are quite diverse in appearance—compare a giant clam, a snail, and an octopus (fig. 26.15)! One of these body parts is the **mantle,** a dorsal fold of tissue that secretes a shell. Many mollusks have two-part shells and are therefore called **bivalves.** Mollusks locomote using a ventrally located organ called a **foot.** The **visceral mass** contains the digestive and reproductive systems. Many mollusks have a chitinous, tonguelike structure called a **radula** that they use to scrape algae or plant matter into their mouths.

Classification of mollusks often is based on the shapes of the shell and foot. Most bivalves, for example, have a hatchet-shaped foot for digging into the sediment. **Gastropods,** including snails with spiral shells and slugs with small internal shells, have a broad, flat foot for crawling. Snails and slugs secrete a mucous trail they glide on, which prevents the epidermis from peeling away as the animals move. A snail's body twists as it develops and becomes asymmetrical. Gastropods are very diverse mollusks that live in marine, freshwater, and terrestrial environments. **Cephalopods** such as the octopus and squid have reduced or absent shells.

The molluscan nervous system varies from simple and ladderlike to very complex and cephalized. An octopus's nervous system is so sophisticated that neurobiologists studying how learning occurs use the animal in laboratory experiments. The octopus also has very highly developed eyes.

Most mollusks have separate sexes. Bivalves shed eggs or sperm into the water, where external fertilization occurs. Gastropods such as snails, slugs, limpets, whelks, and conches have copulatory organs and fertilize eggs internally. A male cephalopod reaches into his mantle cavity with an armlike appendage to draw out a package of sperm and then delivers the sperm to the female's mantle cavity, where it fertilizes eggs.

Many marine mollusks have a ciliated larval stage called a **trochophore,** which may develop into a larva with a tiny mantle, shell, foot, and swimming organ. The trochophore larva settles to

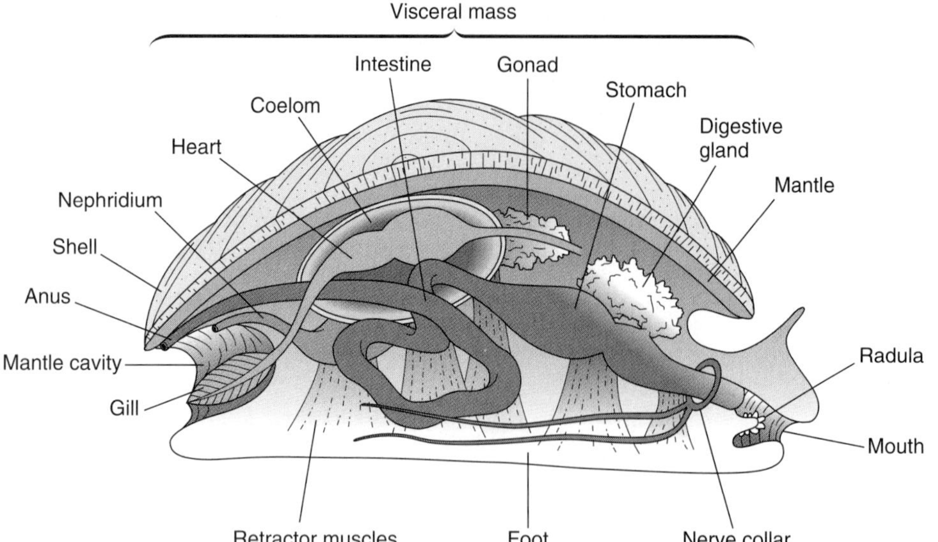

FIGURE 26.14

A generalized molluscan body form.

(a)

(b)

(c)

FIGURE 26.15

Molluscan diversity. (*a*) The Pacific giant clam belongs to class Bivalvia. (*b*) The land snail is a member of class Gastropoda. Note the eyes on its tentacles. (*c*) The octopus is also a mollusk.

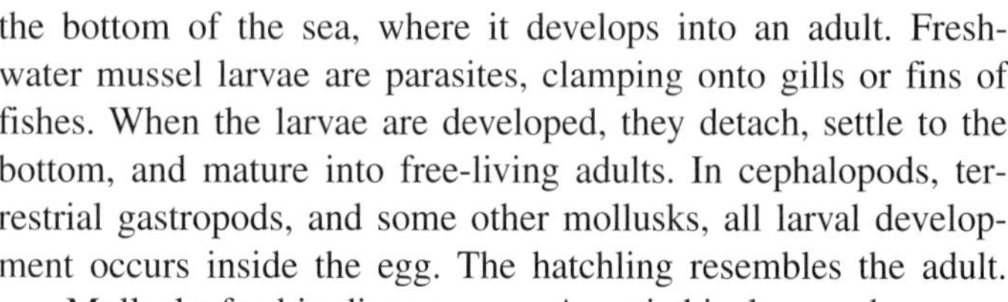

the bottom of the sea, where it develops into an adult. Freshwater mussel larvae are parasites, clamping onto gills or fins of fishes. When the larvae are developed, they detach, settle to the bottom, and mature into free-living adults. In cephalopods, terrestrial gastropods, and some other mollusks, all larval development occurs inside the egg. The hatchling resembles the adult.

Mollusks feed in diverse ways. Aquatic bivalves such as oysters, clams, scallops, and mussels are suspension feeders, eating organic particles and small organisms they strain out of the water. Mollusks concentrate pollutants or toxins produced by dinoflagellates and may become poisonous to human seafood eaters. Slugs consume vegetation so voraciously that they can rapidly destroy a garden. Some marine snails drill holes in bivalve shells to eat them. Cephalopods are active predators. Their keen eyes, closed circulatory systems, and ability to move by "jet propulsion," squirting water out of modified feet, allow them to detect and catch fast-moving prey such as fishes. The octopus and chambered nautilus scavenge dead fishes.

Mastering Concepts

1. What structures do mollusks share?
2. Why are mollusks complex invertebrates?
3. How do mollusks locomote, reproduce, and feed?

Phylum Annelida: Segmented Worms, Including Oligochaetes, Polychaetes, and Leeches

The animal rights activist who thought a leech was not an animal was wrong—leeches belong to phylum Annelida, the segmented worms. The annelids' main identifying characteristic is their body form of repeated segments, called **metamerism.**

Phylum Annelida includes three easily recognized classes. **Oligochaetes,** such as earthworms, have a few usually small bristles called **setae** on their sides (fig. 26.16*a*). A part of the outer covering called the **clitellum** secretes mucus when the worms copulate and a cocoon that protects the animal early in development. See figure 34.2 for an illustration of earthworm locomotion.

Marine segmented worms, the **polychaetes** (fig. 26.16*b*), have pairs of fleshy appendages called **parapodia** that they use for locomotion. Parapodia are located on the sides of segments and have many long setae embedded in them. Setae anchor annelids and keep them from slipping backwards when they move. The third class of annelids, the **hirudinea,** include the leeches. They lack parapodia and setae and have suckers and demarcations called **annuli** between their segments.

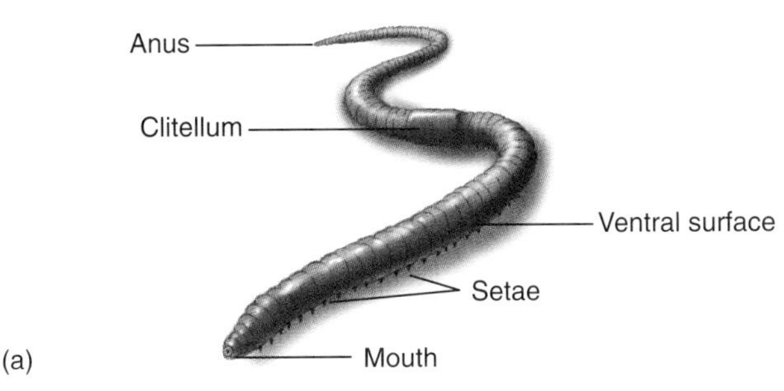

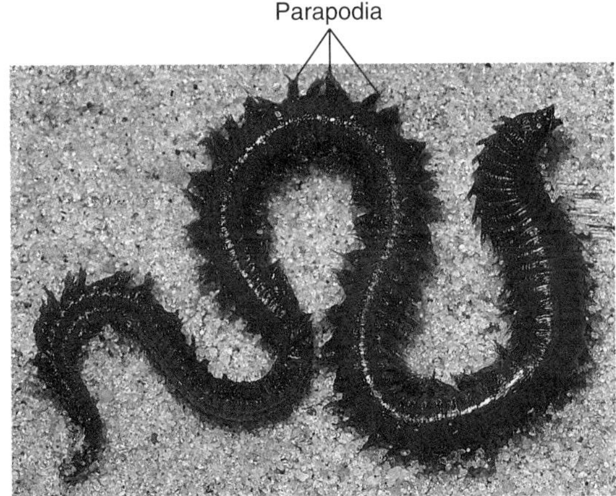

FIGURE 26.16

Annelid worms. (*a*) Oligochaetes, such as this earthworm (*Lumbricus terrestris*), show the metamerism (repeated body parts) characteristic of phylum Annelida. (*b*) *Nereis virens,* the sandworm, is a predaceous polychaete that has parapodia with setae and whose sensory organs lie at the anterior end. Biology in Action 35.2 focuses on and illustrates the third type of annelid, a leech.

Annelids feed in several ways. Earthworms and several other annelids are **deposit feeders**, which take in large quantities of soil or aquatic sediment and strain out organic material for food. Many polychaetes are suspension feeders, using their crowns of ciliated structures called **radioles** to filter small organisms and organic particles from seawater. Some polychaetes are predators with formidable jaws. Many leeches suck blood from vertebrates, but more eat small organisms such as arthropods or other annelids. Because leeches release a blood-thinning chemical when they bite, they are used to reattach severed human digits and ears, as Biology in Action 35.2 describes.

Marine worms (polychaetes) have various external structures that carry out gas exchange. Earthworms (oligochaetes) and leeches (hirudinea) respire more simply, by diffusion through the body wall. An annelid's circulatory system is closed. The coelom serves as a hydrostatic skeleton; muscles work against it as the worm crawls, burrows, or swims. The nervous system consists of a brain connected around the digestive tract to a ventral nerve cord, with lateral nerves running through each segment. See figure 38.13 for an illustration of the excretory structures in an earthworm's segments.

Leeches and oligochaetes reproduce differently than polychaetes. Leeches and oligochaetes are hermaphroditic and also cross-fertilize. Two individuals copulate, each discharging sperm that its partner temporarily stores. Eggs and sperm then are shed into a cocoon that each worm's clitellum secretes. In earthworms, the clitellum is visible at all times, but in leeches it is only seen during breeding season. Juvenile leeches that resemble adults develop within the cocoon and hatch from the fertilized eggs.

In contrast, polychaetes have separate sexes; eggs and sperm are usually shed into the ocean, where fertilization occurs. The *Palolo* worm has a particularly interesting method of reproduction, using specialized posterior body segments. On one night with a full moon in October or November, these segments detach, swarm to the surface of the ocean, and break open, releasing many gametes. The anterior part of the body survives and can regenerate the posterior portion the following year.

Annelids, like mollusks, have trochophore larvae. This is a distinguishing characteristic of protostomes.

Mastering Concepts

1. Describe the annelid body form.
2. What are the distinguishing features of the three phyla of annelids?
3. How do annelids feed, respire, exchange gases, respond to the environment, and reproduce?

Phylum Arthropoda: Insects, Crustaceans, Arachnids, Myriapods, and Others

If biological success is judged in terms of diversity, perseverance, and sheer numbers, then the arthropods certainly qualify as the most successful type of organism. Biologists hypothesize that more than 75% of all animal species are arthropods, the phylum that includes insects, crustaceans, and arachnids.

Arthropods are protostomes with bilateral symmetry. They have segmented bodies, skeletons composed of chitin, and jointed appendages (arthropoda means "jointed foot"). These animals molt their exoskeletons to grow, and many metamorphose.

Arthropods have open circulatory systems. Their blood, called **hemolymph,** circulates freely through the body cavity, or **hemocoel.** The heart is dorsal and has holes that blood enters. In many terrestrial arthropods, the circulatory system plays only a limited role in gas exchange. The arthropod respiratory system consists of body wall openings called **spiracles** that open into a series of branching tubes—the tracheae and smaller tracheoles. The smallest tracheoles serve individual cells. The tracheal system very efficiently transports oxygen and carbon dioxide to and from tissues.

An arthropod's nervous system is similar to an annelid's, with a dorsal brain and a ventral nerve cord. Many arthropods have large compound (faceted) eyes. See figure 31.11 for an illustration of the increasing complexity of the nervous systems of cnidaria, flatworms, annelids, and arthropods.

Different types of arthropods vary in number of major body regions and appendages, whether mouthparts chew or pierce, and the number of body appendage branches. An appendage with two branches is **biramous;** with one lobe, **uniramous.** Many arthropods have three major body regions: head, thorax (chest), and abdomen (fig. 26.17).

Because of its great diversity, phylum Arthropoda is divided into four subphyla. The evolution of subphyla and classes within the phylum has caused controversy among biologists. Older references say that modern arthropods descended from animals called **trilobites** that dominated the oceans of the early Paleozoic era but were extinct before the Cenozoic era began. Trilobites had biramous appendages and a three-lobed body (fig. 26.18*a*). Today, however, biologists believe that trilobites were a dead end on the arthropod family tree, with no direct descendants.

Subphylum Chelicerata includes marine horseshoe crabs (fig. 26.18*b*) and the mainly terrestrial arachnids, such as mites, ticks, spiders, and scorpions. Their mouthparts pierce, they have four pairs of legs, and they lack antennae. Arachnids and horseshoe crabs have two major body segments: a cephalothorax and an abdomen.

Subphylum Crustacea includes a wide variety of mostly aquatic arthropods, including lobsters (fig. 26.18c), crayfish, crabs, and shrimp, and many smaller forms such as waterfleas (*Daphnia*) and copepods. Barnacles are unusual crustaceans in that they are sessile. Isopods, commonly known as pillbugs, are terrestrial crustaceans. Crustaceans have gills, biramous appendages, mandibles (jaws), two pairs of antennae, and two or three major body segments.

Subphylum Uniramia includes insects (fig. 26.18*d*) and **myriapods** (centipedes and millipedes). Myriapods are terrestrial, as are most insects, but some insects live in fresh water. The ocean is about the only place devoid of insects. As their name suggests, uniramians have uniramous appendages, one pair of antennae, and mandibles. Insects have a head, thorax, and abdomen, and myriapods have a head and trunk. Most insects also have six legs and usually two pairs of wings. Millipedes ("thousand leggers") do not really have a thousand legs, but they differ from centipedes because they have two pairs of legs per body segment rather than one pair. Insects range in size from wingless soil species less than 1 millimeter long to fist-sized beetles, foot-long walkingsticks, and others with foot-wide wingspans. Fossil dragonflies were even larger—one had a wingspan of about 30 inches (about 76 centimeters)!

Table 26.4 lists some arthropods that threaten human health.

Mastering Concepts

1. What are the distinguishing characteristics of arthropods?
2. What types of organisms are arthropods?
3. Describe an arthropod's organ systems.

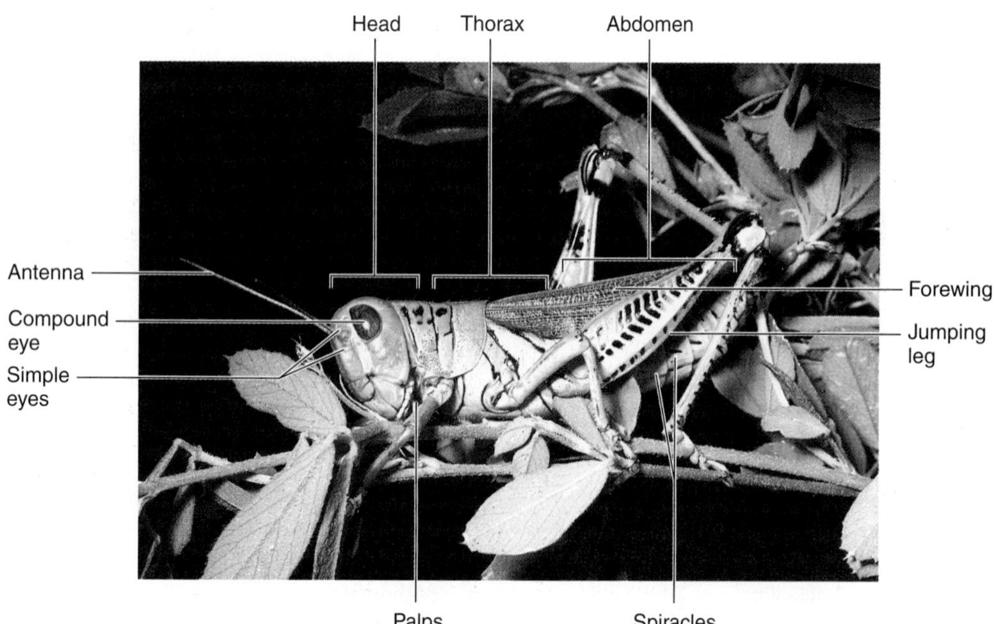

FIGURE 26.17

The grasshopper displays typical insect characteristics.

Table 26.4 Arthropods of Medical Importance in the United States

Arthropod	Effect on Human Health
Spiders	
Latrodectus species (black widows)	Venomous bite
Loxosceles reclusa (brown recluse or violin spider)	Venomous bite
Mites	
Trombiculid mites (chiggers)	Dermatitis
Itch mite	Dermatitis
Ticks	
Ixodes dammini (deer tick)	Bite transmits Lyme disease
Dermacentor species (dog tick, wood tick)	Bite transmits Rocky Mountain spotted fever
Scorpions	
	Venomous sting (not dangerous in U.S. species)
Centipedes	
	Venomous bite (not dangerous in U.S. species)
Insects	
Mosquitoes	Female bite transmits disease (encephalitis, filarial worms)
Horseflies, deerflies	Female has painful bite
Houseflies and relatives	Many transmit bacteria, viruses, worms to food or water
Fleas	Dermatitis; bite transmits plague, tapeworms
Bees, wasps, ants	Venomous stings (single stings not dangerous unless person is allergic)

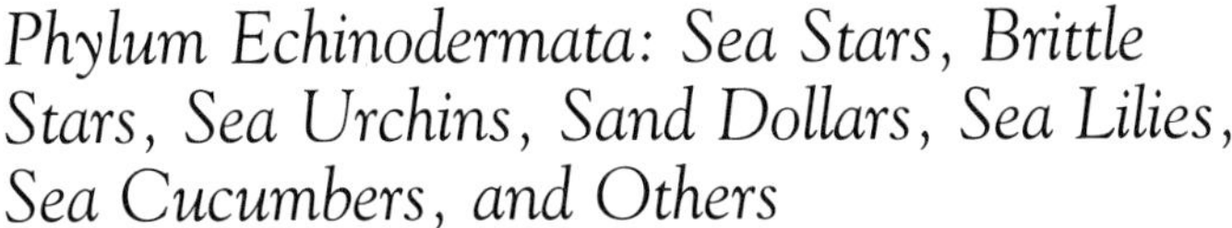

Phylum Echinodermata: Sea Stars, Brittle Stars, Sea Urchins, Sand Dollars, Sea Lilies, Sea Cucumbers, and Others

Echinoderms derive their name from their characteristic spiny (echino) coverings (derm) (fig. 26.19*a*). Biologists classify these animals by variations in endoskeleton and general body form. Some echinoderms (such as the sea cucumber) have small, separate calcarious (calcium-containing) structures called **ossicles** embedded in their leathery skin; others (for example, the sea star) have a more elaborate system of calcarious spines and endoskeletal plates. Sea urchins and sand dollars exhibit yet another endoskeletal variant—their plates are completely fused into a shell called a **test.**

(a)

(b)

(c)

(d)

FIGURE 26.18

Four arthropod subphyla. (*a*) Trilobites were marine arthropods named for their three body lobes—two lateral and one central. (*b*) Horseshoe crabs are marine scavengers placed in subphylum Chelicerata. (*c*) Subphylum Crustacea includes familiar shellfish such as this lobster (*Homarus americanus*). (*d*) The jumping black specks you might see on your cat or dog during flea season exhibit the arthropod's characteristic jointed legs.

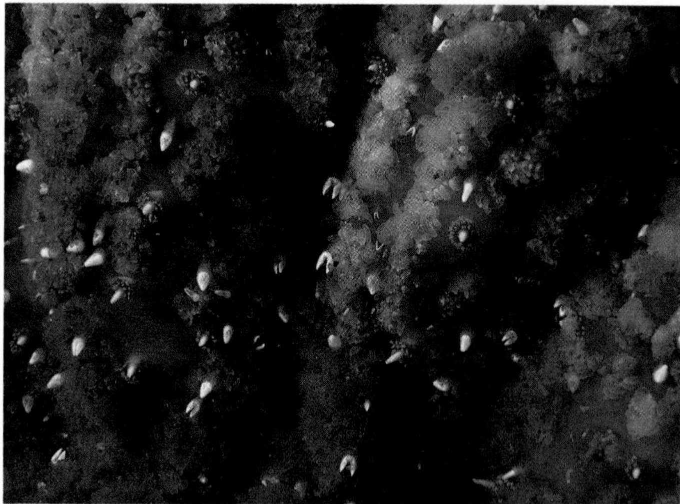

(a)

(b)

FIGURE 26.19

Echinoderms. (*a*) The aboral body surface of a sea star shows spines with pedicellariae (pincers) at their tips. (*b*) A sea urchin vividly displays the "spiny skin" that gives the echinoderms their name.

Echinoderms have a variety of body forms and locomote in different ways. Many species (such as sea stars and brittle stars) have five arms in a form called **pentaradial symmetry.** Feather stars and sea lilies have branched arms adapted for suspension feeding, while sea urchins and sand dollars have no arms and "walk" on stiltlike spines. Echinoderms are among the most colorful of sea inhabitants. Most are bottom dwellers that crawl on coral reefs or other hard surfaces. Sand dollars and some sea cucumbers burrow into soft sediments. Some brittle stars can swim using their appendages.

A unique echinoderm feature is **pedicellariae,** tiny pincers around the base of the spines on sea stars and sea urchins (fig. 26.19*b*). Pedicellariae clean the skin surface of debris and larvae. Another unique characteristic is the echinoderm's **water vascular system;** seawater enters a series of enclosed canals ending in suction-cup-like structures called **tube feet.** When muscles contract, tube feet extend, forcing water in. When the muscles relax and water exits, the sucker at the bottom of each tube foot clamps to the seafloor. The wavelike pumping of water in and out of the tube feet allows the animal to locomote in slow, gliding movements.

Tube feet have other functions. *Asterias,* a sea star, uses its tube feet to open bivalve shelves so it can eat the soft flesh within. The sea star attaches its tube feet to prey and steadily pulls until the muscles of the bivalve tire and the shell opens. The sea star then everts its stomach through its mouth into the bivalve, secretes digestive enzymes, and absorbs the liquified food.

A sea cucumber's tentacles are modified tube feet used to catch food suspended in the water. The animals then insert their food-covered tentacles into their mouths. A brittle star's tube feet lack suckers and are used mainly for feeding, whereas the sea urchin's suckerless tube feet are mainly respiratory.

Due to their relatively sedentary lifestyle, echinoderms generally lack heads and have few sensory or nervous specializations. They have no excretory systems and reduced circulatory systems. Sea stars, for example, respire by means of tiny skin gills. In sea cucumbers, a network of passageways called a respiratory tree projects internally from the cloaca. This tree exchanges gases and eliminates nitrogenous wastes.

Echinoderms are closely related to members of phylum Chordata, which includes humans. Although chordates may seem to have little in common with sea stars and sand dollars, the early development of all these organisms is quite similar, suggesting possible descent from a common ancestor. Specifically, cleavage division patterns and early embryonic stages are very similar among members of both phyla. In addition, echinoderm larvae are bilaterally symmetrical, like those of chordates.

Mastering Concepts

1. What features classify animals as echinoderms?
2. What are some characteristics unique to echinoderms?
3. What are some functions of tube feet?
4. How are echinoderms similar to chordates?

Phylum Chordata: Lancelets, Tunicates, and Vertebrates

The chordates share several characteristics, although some may occur only in a prenatal stage in a particular species. These animals derive their name from their distinguishing **notochord,** a semirigid rod that runs the length of the body (see fig. 31.12). In vertebrates, the notochord tends to be prominent only early

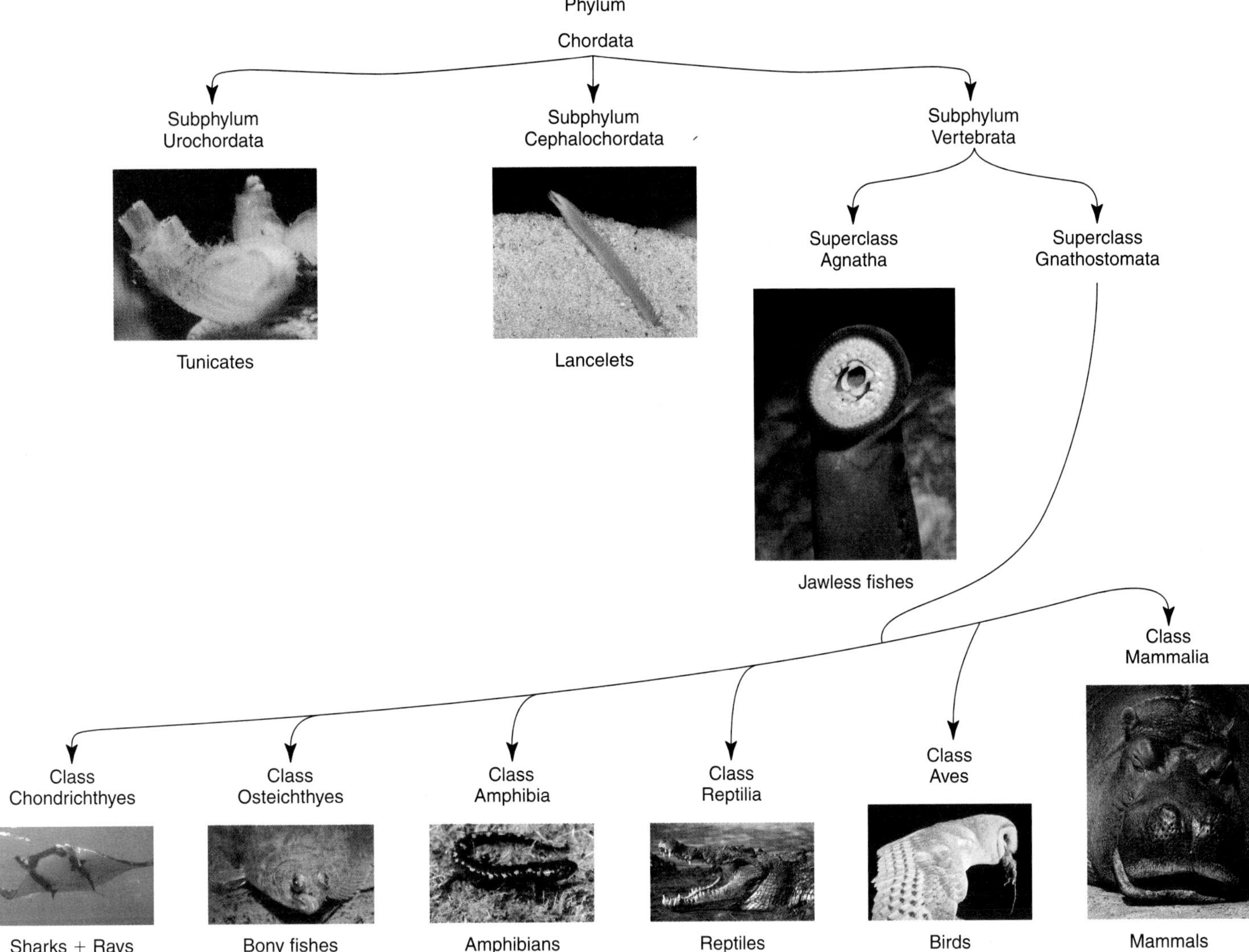

FIGURE 26.20
Chordate diversity.

in development. Then it stops growing, leaving remnants in the structures between vertebrae (central intervertebral discs). In other chordates, the notochord provides support and points of attachment for muscles.

A second chordate characteristic is the position of the nerve cord. It lies dorsal to the digestive tract, in contrast to its ventral position in other phyla.

Chordates have tails, as well as structures called **pharyngeal gill slits** that can function as feeding or respiratory organs at some stages of development. From their ancestors, chordates retain a three-layered embryo, bilateral symmetry, a coelom, and a prominent head.

The chordates are subdivided into three groups. Two of them, the **tunicates** (sea squirts, subphylum Urochordata) and the **lancelets** (subphylum Cephalochordata), are small groups of primitive invertebrates. Members of the third subphylum, Vertebrata, have bone or cartilage around the spinal cord. Subphylum Vertebrata is further divided into two superclasses. The jawless fishes constitute superclass Agnatha. The jawed fishes, along with all four-limbed animals (tetrapods) make up a second superclass called Gnathostomata. Gnathostomata's six classes are familiar—the sharks, skates, and rays (Chondrichthyes); the bony fishes (Osteichthyes); amphibians (Amphibia); reptiles (Reptilia); birds (Aves); and mammals (Mammalia). Figure 26.20 outlines the groups that make up the phylum Chordata.

Subphylum Cephalochordata—The Lancelets

The lancelets resemble small, eyeless fishes. These animals bury their tails in coastal sand and ingest particles suspended in the water using ciliated structures around their mouths

(a)

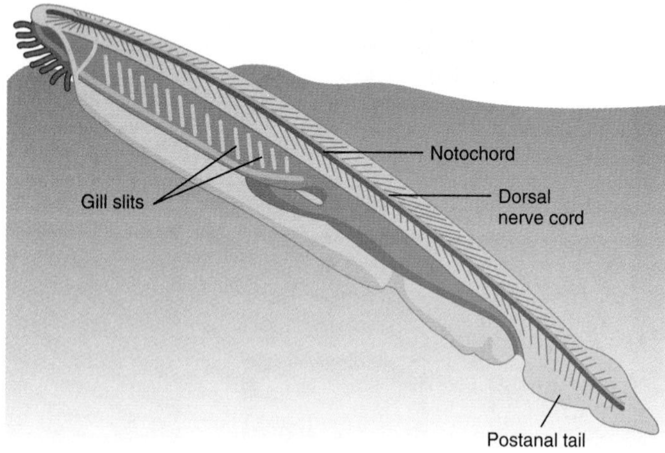

(b)

FIGURE 26.21

Amphioxus, the lancelet, of subphylum Cephalochordata. (*a*) This amphioxus is in suspension-feeding posture, with its head sticking up out of the substrate and its tail buried. (*b*) The internal anatomy of *Amphioxus* reveals the four chordate characteristics.

(fig. 26.21*a*). Cilia on the gills move water from the mouth into the pharynx. A ciliated ventral groove catches food particles and passes them to the intestine, where the food continues to the **hepatic cecum** for digestion and absorption. Water passes out through gill slits and leaves the body through the **atriopore.**

Lancelets display all four chordate characteristics (fig. 26.21*b*) at all stages of development. They have separate sexes, external fertilization, and a ciliated larval stage.

Subphylum Urochordata—The Tunicates

Tunicates (fig. 26.22) are named for their nonliving covering, or tunic, which contains cellulose. Like lancelets, tunicates are marine. Their tadpole larvae show all chordate characteristics, but most adults retain only gill slits. Most adult tunicates are sessile and are commonly called sea squirts because they shoot water out of their **excurrent siphons** when disturbed. Tunicates are suspension feeders. They are hermaphroditic and cross-fertilize externally. Sessile tunicates sometimes grow to great numbers on submerged parts of boats.

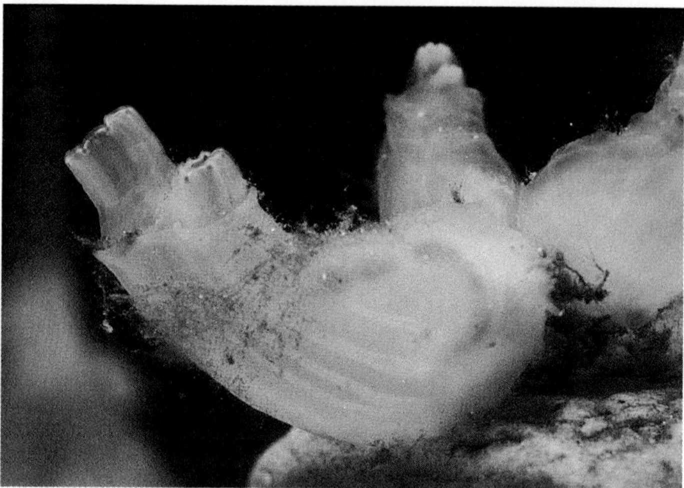

FIGURE 26.22

The adult tunicate (subphylum Urochordata) is sessile and shows all chordate-defining characteristics except gill slits. The free-swimming tadpole does have gill slits.

Researchers recently discovered a mutant tunicate that lacks chordate characteristics. The fact that a single genetic change can erase this group of traits suggests that they evolved rapidly.

Subphylum Vertebrata—The Vertebrates

Chordates may have descended from a sessile, filter-feeding, echinoderm-like ancestor in which gills eventually replaced the tentacles. The first chordates may have resembled free-swimming tunicate larvae. The vertebral column would have appeared after the lancelets branched off, and jaws may have evolved from gill arches after jawless fishes diverged.

Amphibians were probably the first tetrapods (see fig. 17.13). Amphibians are restricted to moist habitats because of their wet, easily dehydrated skin and because their eggs are fertilized in the water, externally, as is true of most fishes. The larval stages have gills and are aquatic. After metamorphosis, most adults have lungs and can leave the water.

The amniote egg evolved in reptiles. Recall from chapters 11 and 17 that an amniote egg contains yolk and extraembryonic membranes, which nourish and protect the embryo and enable the egg to survive on dry land. Dry, scaly skin and tough, leathery eggs are other reptilian adaptations to life on land. Their fertilization is internal.

Birds and mammals evolved from reptiles—birds from the same ancestors that gave rise to dinosaurs. Birds and mammals are endotherms (generating body heat internally), and dinosaurs may have been, too. Birds are well adapted to flight, with wings, feathers, lightweight hollow bones, and a sternum with a keel where flight muscles attach. Bird and mammal fertilization is internal. Birds lay amniote eggs with calcareous shells.

A Closer Look at Mammals

Mammals began to evolve about 150 million years ago from a reptilian ancestor and exploded in diversity between 7 and 70

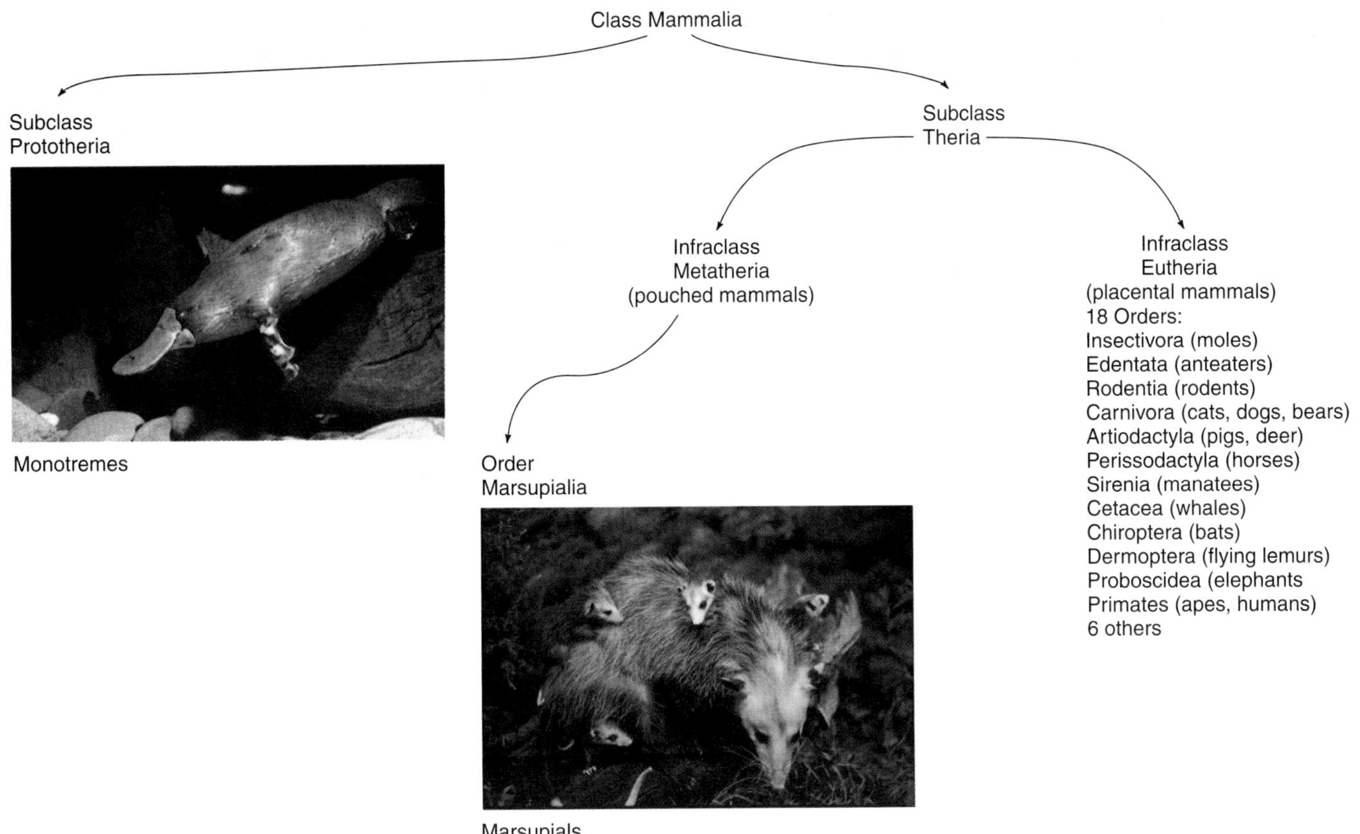

FIGURE 26.23

Mammalian diversity.

million years ago (see fig. 17.15). Today, mammals remain a diverse group, due largely to their adaptations to varied environments and ways of obtaining food. Figure 26.23 outlines mammalian diversity.

The name "mammal" comes from one identifying characteristic of the group—milk-secreting mammary glands. These glands are well developed in the female but rudimentary in the male. Mammals have hairy bodies, large braincases, internal fertilization, and a constant body temperature.

Class Mammalia is divided into subclasses and infraclasses based upon the ways mothers carry and bear young. Subclass Prototheria includes the **monotremes,** or egg-laying mammals, such as the duck-billed platypus and the spiny anteater. The large subclass Theria includes the **marsupials** and the **placental mammals.** Marsupials (infraclass Metatheria) are pouched mammals, such as the kangaroo and the opossum, in which the embryo leaves the female's body when it is very immature and climbs along her fur to her pouch to complete early development (see fig. 11.23). Placental mammals (infraclass Eutheria) nourish their unborn offspring within the female's body through a placenta.

The 18 orders of placental mammals include many familiar animals, such as dogs, cats, whales, and hippos. Members of the orders are distinguished by dental patterns; specializations of limbs, toes, claws, and hooves; and complexity of nervous systems.

Order Primates includes the lemurs, monkeys, apes, and humans (see fig. 21.5). Primates have larger brains than other mammals, five digits (fingers and toes) on each limb, and an opposable thumb that can fold against the other fingers, enabling the animal to grasp.

Within the order Primates, members of suborder Prosimii, small squirrel-like primates such as lemurs, tarsiers, and lorises, are tree dwellers. Other modern primates constitute suborder Anthropoidea, which means "resembling man." Superfamilies within this suborder are distinguished by nose structure, thumb position, tail use, and other characteristics. New World monkeys (superfamily Ceboidea) include the spider and howler monkeys, and Old World monkeys (superfamily Cercopithecoidea) include the mandrill and rhesus monkey.

The third superfamily of anthropoid primates, Hominoidea, consists of two families: Pongidae, which includes gibbons, orangutans, chimpanzees, and gorillas; and Hominidae, which includes only humans (fossil and human), *Homo sapiens.* Humans are distinguished from other primates by more erect posture, reduced eyebrow ridges, lighter jaws, smaller front teeth, increased thumb use, shorter arms, complex speech, and (we think) greater intelligence.

The animal we call *Homo sapiens*—the human animal—is, like other living species, the result of billions of years of evolution. We reflect a heritage of accumulating characteristics that have shaped life on earth from the simple, single-celled organisms that

Animal Discoveries—From Tube Worms to Lobster Lips~

continued from page 508.

Discovering and describing never-before-seen organisms is nothing new to biologists. Finding *Symbion pandora* was unusual not only because of its habitat on the mouthparts of Norwegian lobsters, but because its characteristics do not fit into any known animal phylum. Until biologists learn more about it, *Symbion pandora* has been assigned to a new phylum, Cycliophora. The name in Greek means "small wheel" and refers to the circular, ciliated ring that is the animal's mouth.

S. pandora is tiny, less than a millimeter long. During the feeding stage of its life cycle, it has a saclike body that adheres by suction to a lobster. The gut continues from the ciliated mouth ring, opening as an anus near the mouth. In addition to large female and smaller male feeding forms, a larval stage moves but doesn't feed. Larvae result from asexual reproduction, in which a series of buds forms, each containing organs that degenerate and are then replaced. Sexual reproduction occurs when the animals molt.

What exactly is *Symbion pandora?* It is acoelomate (lacking a cavity between the gut and the body wall), and the structure of the ciliated mouth places it among the protostomes. The unusual internal budding that gives rise to larvae is reminiscent of two types of animals, the entoprocta (an obscure group) and the ectoprocta (including bryozoa). With so little anatomical and physiological evidence to consider, gene and protein sequence data should be especially helpful in enabling researchers to place these animals among the 35 known phyla—or to confirm that it is truly unique enough to warrant its own phylum.

swarmed in the ancient seas to the diverse and highly complex organisms that live on earth today. But taxonomy and the study of biodiversity are sciences that do not end with the present day. What types of animals, and other organisms, will occupy earth in the future? Will we be among them? What role (if any) will *Homo sapiens* play in determining the spectrum of life on earth 10,000 or a million or a billion years hence? Will earth even exist then? Our knowledge of life's diversity on earth yesterday and today will enable us to speculate intelligently upon—and perhaps even to influence—the living earth of tomorrow.

Mastering Concepts

1. What are the four features common to chordates?
2. What are the adaptive values of amniote eggs and placentas?
3. What are the types of mammals?
4. What are characteristics of primates?

SUMMARY

1. Animals are eukaryotic, multicellular heterotrophs whose cells do not have cell walls. More than 1 million animal species are known, most of them **invertebrates.** The oldest animal fossils date from about 700 million years ago. Animals may have descended from colonial and/or multinucleate protista.

2. Animals differ in development and body form. Distinguishing characteristics include number of primary germ layers (**diploblastic** or **triploblastic**), presence and type of body cavity (**coelomate, acoelomate,** and **pseudocoelomate**), and body symmetry (**radial symmetry,** with **oral** and **aboral** ends, or bilateral symmetry, with **cephalization**).

3. Animals undergo **direct development** or **metamorphosis.** Several major lineages of coelomates differ in development patterns. **Protostomes** exhibit **spiral cleavage,** a **blastopore** developing into a mouth, and **determinate cleavage. Deuterostomes** have **radial cleavage,** a blastopore developing into an anus, and **indeterminate cleavage.**

4. Sponges (phylum Porifera) are sessile and draw in food through pores and water canals. Their **amoebocytes** have a variety of functions. Sponge skeletons may consist of **spongin** or **spicules.** They may bud asexually using **gemmules,** or reproduce sexually.

5. Cnidaria (phylum Cnidaria) capture prey with stinging **nematocysts.** Cnidaria include jellyfish, corals, and Portuguese men-of-war. They have radial symmetry, are diploblastic, digest extracellularly, and occur in **polyp** or **medusa** forms.

6. Most flatworms (phylum Platyhelminthes) are parasitic and have great reproductive capacity. They are triploblastic, lack coeloms, and have bilateral symmetry. **Protonephridia** maintain internal water balance. Flatworms reproduce asexually and sexually, and some are hermaphrodites.

7. Roundworms (phylum Nematoda) include parasites and free-living species in soil or aquatic sediments. They are cylindrical and segmentless and have complete digestive systems and separate sexes.

8. Phylum Mollusca, the second largest phylum, includes diverse animals that have a shell, **foot,** and **visceral mass** and complex organ systems. They are protostomes with bilateral symmetry and have **trochophore** larvae. **Gastropods** and **cephalopods** are mollusks. Sexes are separate.

9. Phylum Annelida, the segmented worms, includes **oligochaetes,** characterized by **setae** bristles and covered with a **clitellum; polychaetes,** with **parapodia** as appendages; and the **hirudinea,** which are leeches with **annuli** between their segments. Annelids feed in diverse ways and have organ systems. Leeches and oligochaetes are hermaphroditic; polychaetes have separate sexes.

10. Phylum Arthropoda is the largest phylum. Arthropods are protostomes with bilateral symmetry, and they are segmented with jointed appendages and a chitinous exoskeleton. They have open circulatory systems, **spiracles** to respire, and a nervous system. Four subphyla include trilobites, horseshoe crabs and arachnids, crustaceans, and insects and **myriapods.**

11. Phylum Echinodermata includes spiny-skinned marine animals with **pentaradial symmetry** that move using **water vascular systems** with **tube feet.** Early development is similar to that of **chordates.**

12. All chordates (phylum Chordata) share four characteristics at some stage of their life cycle: a dorsal tubular nerve cord, a **notochord, pharyngeal gill slits,** and a postanal tail. Subphylum Cephalochordata includes the **lancelets** and subphylum Urochordata the **tunicates,** both primitive invertebrates. Subphylum Vertebrata includes superclass Agnatha (jawless fishes) and superclass Gnathostomata (jawed fishes and tetrapods).

13. Class Mammalia includes **monotremes, marsupials,** and **placental mammals,** distinguished by how young are carried. Humans belong to order Primates.

TO REVIEW . . .

1. Cite the taxonomic criteria used to distinguish each of the following:
- **a.** animals from other organisms
- **b.** protostomes from deuterostomes
- **c.** mammals from other vertebrates
- **d.** humans from other primates

2. Distinguish between the following:
- **a.** radial, bilateral, and pentaradial symmetry
- **b.** direct development and metamorphosis
- **c.** spiral and radial cleavage
- **d.** determinate and indeterminate cleavage
- **e.** biramous and uniramous appendages
- **f.** diploblastic and triploblastic embryos
- **g.** deposit and suspension feeding
- **h.** coelomate, acoelomate, and pseudocoelomate body forms
- **i.** vertebrates and invertebrates

3. Identify an animal that has the following cell types:
- **a.** cnidocytes
- **b.** flame cells
- **c.** choanocytes

4. Identify an animal that has the following tissues:
- **a.** epidermis and gastrodermis
- **b.** pseudocoel and hypodermis
- **c.** hemolymph

5. What characteristics are used to distinguish among species within the following phyla?
- **a.** cnidaria
- **b.** arthropods
- **c.** platyhelminthes
- **d.** mollusks
- **e.** annelids

6. On an episode of the television series *The X-Files,* two FBI agents chase a bizarre creature that has the form of a human but with a scolex atop its head. From what animal phylum did the writer get inspiration for including a scolex?

7. List three adaptations of platyhelminthes to a parasitic way of life.

8. Distinguish between cephalization and metamerism.

9. How are cnidaria and echinoderms adapted to a sessile lifestyle?

10. What characteristics do mollusks, chordates, and echinoderms have in common?

11. Name the type of animal that has each of the following structures, and describe each structure's function:
- **a.** proglottids
- **b.** gemmules
- **c.** pedicellariae
- **d.** protonephridia
- **e.** pneumatophores
- **f.** trochophore larvae
- **g.** ossicles
- **h.** excurrent siphon
- **i.** annuli
- **j.** atriopore
- **k.** mantle
- **l.** setae
- **m.** spiracles
- **n.** intestinal ceca
- **o.** tegument

12. What do monotremes, marsupials, and placental mammals have in common, and how do they differ?

TO THINK ABOUT . . .

1. A letter in *Time* magazine responding to a story about life on Mars states, in part, "What we do know is that billions of creatures, both human and animal, exist here on earth." What is incorrect about this oft-heard statement?

2. A researcher separates the cells of a four-celled frog preembryo and regenerates four complete frogs from them. How does this information reveal whether a frog is a protostome or a deuterostome?

3. Why wouldn't preimplantation genetic diagnosis work if cleavage in humans was determinate?

4. Do you think that humans are "more advanced" than other animals? What criteria might you use to propose the hypothesis that humans are superior to other animals?

5. To what phylum are the chordates most closely related? What evidence supports this?

6. Platyhelminthes, nematodes, and annelids are all worms. How might you tell them apart?

7. A dinner in a seafood restaurant includes clams, squid, and swordfish. What phyla are represented?

TO LEARN MORE . . .

References and Resources*

Braun, Stephen. August 9, 1996. New experiments underscore warnings on maternal drinking. *Science,* vol. 273. Experiments on rats were crucial in showing that alcohol damages nerve cells.

**Life*'s website features frequently updated references.

Conniff, Richard. April 1993. On the lowly worm we earthlings pin our loftiest dreams. *Smithsonian.* Worms assist us in gardening, fishing, composting, and a host of other activities.

Hickman, Cleveland P., Jr., Larry S. Roberts, and Allan Larson. 1996. *Integrated principles of zoology,* 10th ed. St. Louis: C. V. Mosby. A detailed examination of all animal phyla.

Morris, Simon Conway. December 14, 1995. A new phylum from the lobster's lips. *Nature,* vol. 378. Researchers identified an organism living on lobster mouthparts that has such an unusual combination of traits that they placed it in a new phylum.

Natural History. April 1994. This issue contains more than a dozen articles devoted to the mammals.

O'Reilly, James C. March 20, 1997. Lifelines: imitate the action of the earthworm. *Nature,* vol. 386. Caecilians are amphibians that move just like earthworms.

Pennisi, Elizabeth. November 15, 1996. Tracing backbone evolution through a tunicate's lost trail. *Science,* vol. 274. A mutant tunicate called "manx," which lacks chordate features, suggests that the chordate form may have evolved rapidly.

Popescu, Octavian and Gradimir N. Misevic. March 20, 1997. Sticky sponges. *Nature,* vol. 386. Each sponge species has a unique pattern of cell surface molecules, which enables them to recognize each other.

Young, Craig M. June 6, 1996. Embryology of vestimentiferan tube worms from deep-sea methane/sulphide seeps. *Nature,* vol. 381. Tube worm embryos are revealed.

A prickly pear cactus.

Unit 7

Plant Life

27 Plant Form and Function

Secondary growth produces the enormous girth of this yellow-barked acacia tree.

chapter outline

A Garden of Plant Cells and Tissues   534

Primary Growth   534

Meristems
Ground Tissue
Dermal Tissue
Vascular Tissues

Parts of the Plant Body   540

Stems
Leaves
Roots

Secondary Growth   550

Vascular Cambium
Cork Cambium

Tree Rings Reveal the Past

Studying tree ring patterns—a technique called dendrochronology—is helping to fill gaps in our knowledge of ancient civilizations. Rings arise in trees that grow in temperate climates, where wood cells that form in the spring are larger than those that form in summer, creating a distinctive pattern in the wood grain.

Tree rings also provide information on climate. The larger the ring, the more plentiful the rainfall that year. A fire leaves behind a charred "burn scar." Two rings fall very close together when voracious caterpillars or locusts eat the season's early leaves. With fewer leaves, the tree cannot obtain sufficient nutrition from photosynthesis, and tree rings compress. Tree ring structures and patterns also provide clues about light availability, altitude, temperature, and length of the growing season in times past.

Researchers at Cornell University have recently combined dendrochronology with other types of evidence to "anchor" a nearly 1,500-year sequence in the history of a little-understood society—that of ancient Turkey—into an absolute time frame. They used wood and charcoal samples from the Midas burial mound that were once juniper trees. The evidence may actually rewrite some ancient history.

to be continued . . .

A Garden of Plant Cells and Tissues

A summer garden offers a spectacular display of plant life. Towering, flowering green bean plants curl about each other; the tiniest tomatoes peek out where vibrant yellow flowers bloomed a few days earlier. Melon and cucumber vines snake along the ground, and cabbages begin to form tight knots of leaves. Feathery green tufts in one corner herald the growing carrots beneath the surface, as robust, leafy plants indicate potatoes forming below. In the evergreens that ring the garden, birds gather, eager to attack succulent ears of corn wrapped in their tight, leafy jackets.

Although these plants look different from one another, each consists of similar cells, tissues, and organs. These components, like those of other types of multicellular organisms, are evolutionary adaptations that provide such basic requirements as protection, structural support, water and nutrient transport and storage, and waste removal. This chapter explores how plant cells and tissues interact to perform the activities of life, with a focus on the angiosperms (flowering plants), the most familiar, diverse, and abundant plants.

Primary Growth

Plants consist of four basic primary tissues, each composed of distinctive cell types (table 27.1 and fig. 27.1). **Meristematic tissue** adds new cells and enables a plant to grow and specialize. **Dermal tissue** covers and protects the plant, thus controlling gas exchange between the plant and the environment. **Ground tissue** makes up the bulk of the plant and stores nutrients or photosynthesizes. **Vascular tissues,** including **xylem** and **phloem,** conduct water and dissolved nutrients, respectively.

The primary body of a flowering plant is an axis consisting of a **root** (usually belowground) and a **shoot** (aboveground). The extensive shoot and root systems of many plants increase the surface area available for the chemical reactions of life. Shoots and roots support one another. Shoots, through photosynthesis, provide carbohydrates to the roots below. Roots gather water and minerals and transport them to the shoots.

Table 27.1 Tissue Types in Angiosperms

Tissue	Function
Meristem	Cell division and growth
Ground	Bulk of interiors of roots, stems, and leaves
Dermal	Protects plant; controls gas exchange
Vascular	Conducts water and dissolved nutrients

Meristematic tissue is located primarily in shoot and root tips. Dermal, ground, and vascular tissues form three types of organs: stems, leaves, and roots. Leaves, which extend from stems, are photosynthetic factories.

Meristems

Unlike an animal, which grows explosively very early in its development, a plant has parts—roots, shoots, and buds—that enable it to grow rapidly during what may be a very long lifetime. Meristems may be thought of as embryonic tissue that persists throughout the life of the plant.

Meristems are localized regions that undergo cell division and are the ultimate source of all the cells of a plant. Some cell types descended from meristems can themselves divide. The cells farthest from meristems are the most mature and differentiated, and they may be many times larger than the meristematic cells that gave rise to them. Meristems function throughout a plant's life; because of them, plants never stop growing.

Apical meristems occur near the tips of roots and shoots in all plants. Cells in these regions are small and unspecialized (fig. 27.2). When these meristematic cells divide, the plant lengthens in what is called **primary growth.** Apical meristems produce three primary meristems, which form the other three primary tissues—ground, dermal, and vascular tissues.

Lateral meristems grow outward, thickening the plant, in a process called **secondary growth,** discussed at the end of the

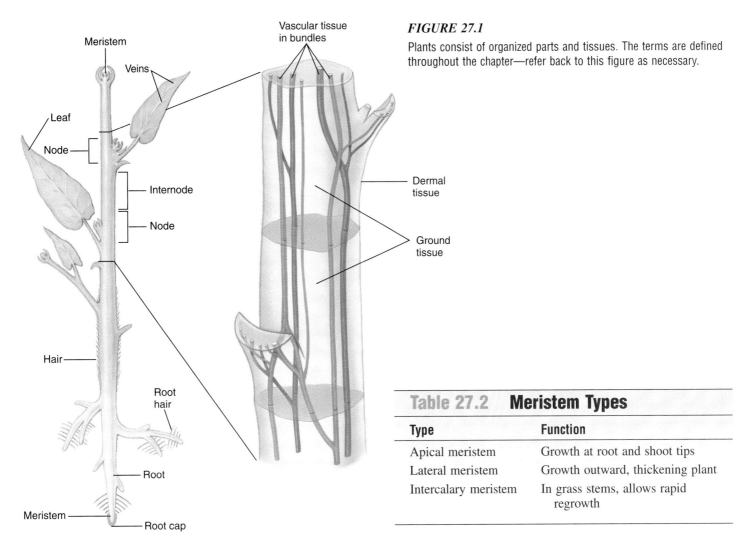

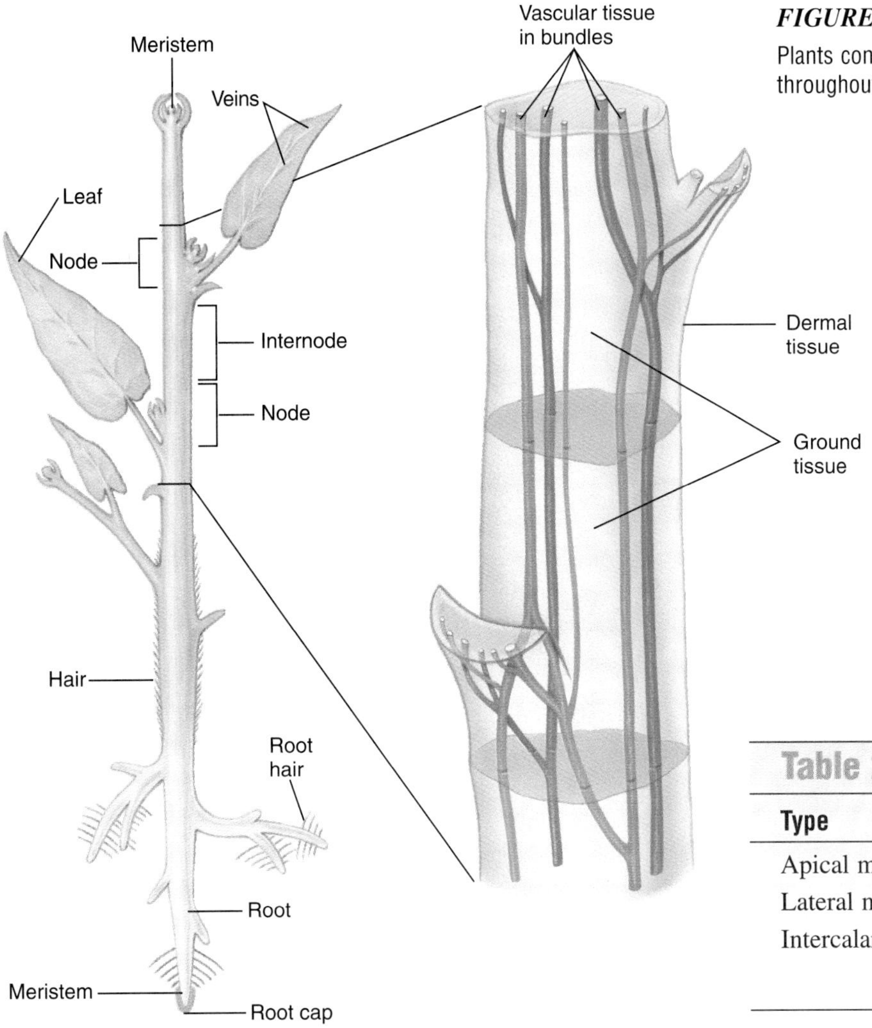

FIGURE 27.1

Plants consist of organized parts and tissues. The terms are defined throughout the chapter—refer back to this figure as necessary.

Table 27.2 Meristem Types

Type	Function
Apical meristem	Growth at root and shoot tips
Lateral meristem	Growth outward, thickening plant
Intercalary meristem	In grass stems, allows rapid regrowth

chapter. Wood forms from secondary growth. Unlike primary growth, secondary growth does not occur in all flowering plants.

Intercalary meristem is an unusual type of dividing tissue found in grass blades. If the tip of a stem or leaf is torn off—by a lawn mower, for example—intercalary meristems re-form the structure of the plant from that point up. This is why grass blades grow back so quickly after a lawn is mowed. Intercalary meristems enable grasses to survive the constant munching of herbivores.

Table 27.2 summarizes meristem types.

Ground Tissue

Ground tissue makes up most of the primary body of a flowering plant, filling much of the interior of roots, stems, and leaves. These cells have many functions, including storage, support, and basic metabolism. Ground tissue consists of three cell types: parenchyma, collenchyma, and sclerenchyma. Table 27.3 summarizes these and other plant cell types.

Parenchyma cells are the most abundant plant cells. They keep the plant erect when it is filled with water. Parenchyma cells are relatively unspecialized and in an evolutionary sense likely gave rise to more specialized cell types. Mature parenchyma cells are alive and can divide, which enables the tissue to specialize in

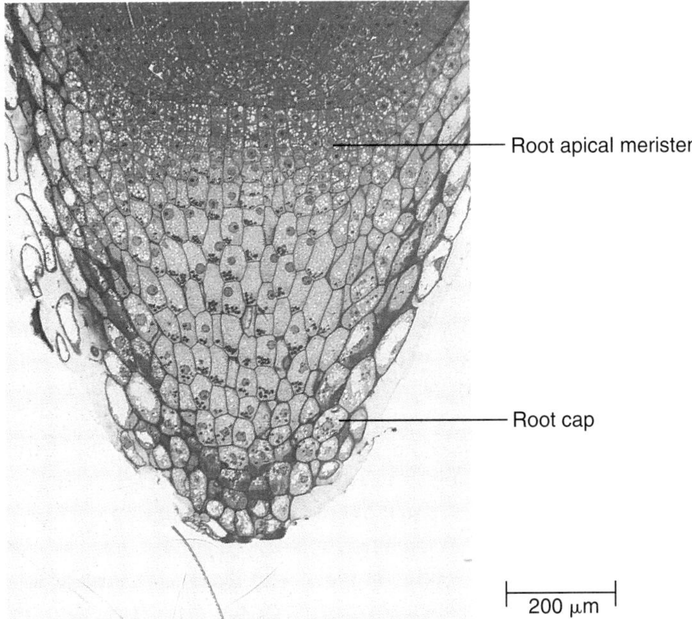

FIGURE 27.2

Cells of the apical meristem divide, and the plant elongates. This is a root tip.

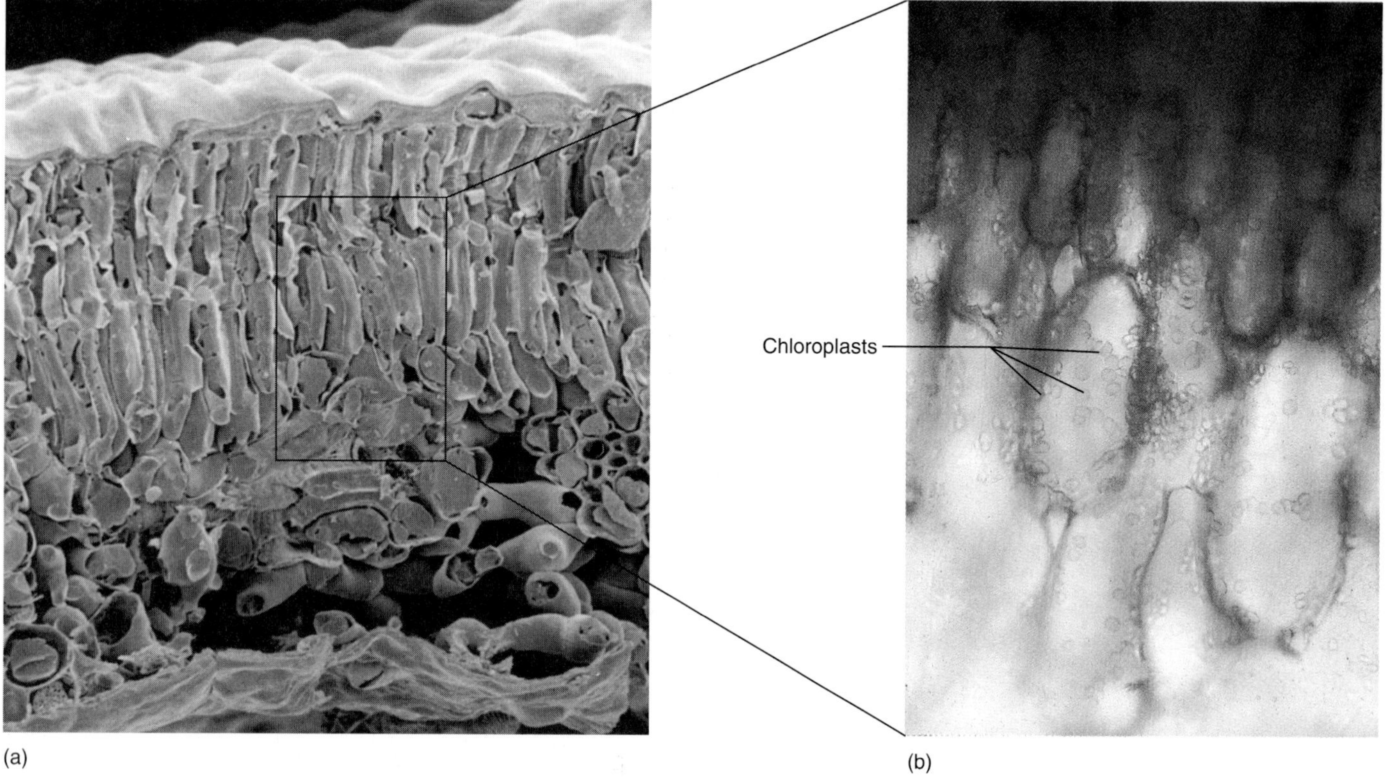

FIGURE 27.3
A cross section of an apple leaf (*a*; ×950) reveals chlorenchyma cells, distinguished by their chloroplasts (*b*).

response to a changing situation. For example, parenchyma cells can help a plant survive injury or adapt to a new environmental condition. These cells can also return to an unspecialized state.

Parenchyma cells store the biochemicals that give plants their familiar edible parts, such as the carbohydrates in a potato or an ear of corn. These cells also store fragrant oils, salts, pigments, and organic acids. Parenchyma cells of oranges and lemons, for example, store citric acid, which gives them their tart taste. These cells also conduct vital functions, such as photosynthesis, cellular respiration, and protein synthesis. **Chlorenchyma** are parenchyma that photosynthesize. Their chloroplasts impart the green color to leaves (fig. 27.3).

Collenchyma cells are elongated living cells that differentiate from parenchyma and support the growing regions of shoots. Collenchyma cells have unevenly thickened primary (outer) cell walls that can stretch and elongate the cells. As a result, collenchyma supports without interfering with growth of young stems or expanding leaves.

Sclerenchyma cells are long with thick, nonstretchable secondary cell walls (a trilayered structure inside the outer, or primary, cell wall). These cells support parts of plants that are no longer growing and are usually dead at maturity. Two types of sclerenchyma form from parenchyma: sclereids and fibers.

Sclereids have many shapes and occur singly or in small groups. Small groups of sclereids create a pear's gritty texture (fig. 27.4*a*). Sclereids may form hard layers, such as the hulls of peanuts. **Fibers** are elongated cells that usually occur in strands that vary from a few to a few hundred millimeters long. Many textiles are sclerenchyma fibers. Humans now cultivate more than 40 families of plants for fibers and have fashioned cords from fibers since 8000 B.C. The fibers of *Agave sisalana,* commonly known as sisal, or the century plant, are used to make brooms, brushes, and twines (see fig. 29.14). Linen comes from the fibers of *Linum usitatissimum,* or flax. Figure 27.4*b* shows another source of fiber, Manila hemp.

Dermal Tissue

Dermal tissue covers the plant. The **epidermis,** usually only one cell thick, covers the primary plant body. Epidermal cells are flat, transparent, and tightly packed. Special features of the epidermis provide a variety of functions.

The **cuticle** is a covering over all but the roots of a plant that protects the plant and conserves water (fig. 27.5). The cuticle consists primarily of **cutin,** a fatty material epidermal cells produce. This covering retains water and prevents desiccation. As a result, plants can maintain a watery internal environment—a prerequisite to survival on dry land. The cuticle and underlying epidermal layer also are a first line of defense against predators and infectious agents. In many plants, a smooth white layer of wax covers the cuticle; when it is thick, it is visible on leaves and fruits. The layer on the undersides of wax palm leaves may be more than 5 millimeters thick. It is harvested and used to manufacture polishes and lipstick.

Since an impermeable cuticle covers the tightly packed epidermal cells, how do plants exchange water and gases with the atmosphere? Stomata are pores that solve this problem (see figs. 8.11 and 27.6). **Guard cells** control the opening and closing of the pores, which regulate gas and water exchange. Stomata control the amount of carbon dioxide that diffuses into a leaf for pho-

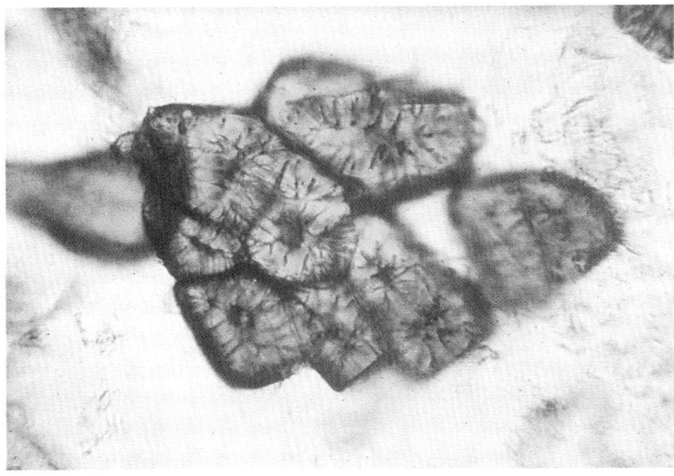

(a)

(b)

FIGURE 27.4

Sclereids and fibers. (*a*) The gritty texture of pear (*Pyrus communis*) flesh is due to sclereids. (*b*) Fibers from Manila hemp (*Musa textilis*) have a variety of uses.

FIGURE 27.5

A waterproof cuticle protects a plant's epidermis.

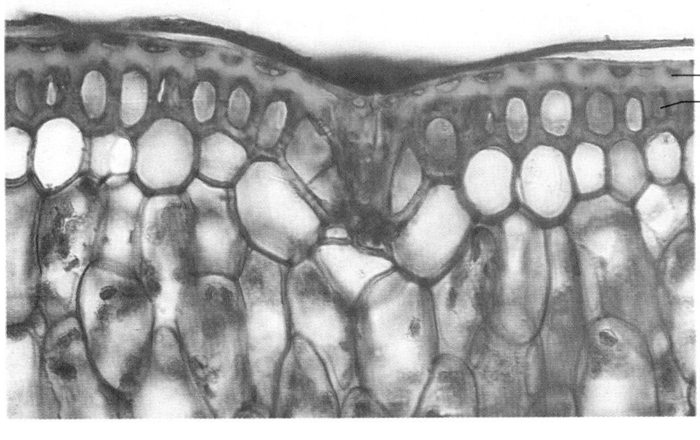

tosynthesis and the amount of water that evaporates from leaves, a process called **transpiration.** Stomata may be very numerous. The underside of a black oak leaf, for example, has 100,000 or so stomata per square centimeter! Because stomata help plants conserve water, they are an essential adaptation for life on land.

Trichomes are outgrowths of epidermal cells in almost all plants. These structures deter predators in interesting ways. Hook-shaped trichomes may impale marauding animals. In other plants, predators may inadvertently break off the tips of trichomes, releasing a sticky substance that traps the invading animal (fig. 27.7). Stinging nettle trichomes have spherical tips that break off and penetrate a predator's body, injecting their poisonous contents into the wounds. Trichomes of carnivorous plants such as the Venus's flytrap secrete enzymes to digest trapped animals. These trichomes then absorb the digested prey.

Many trichomes are economically important. Cotton fibers, for example, are trichomes from the epidermis of cotton seeds. Menthol comes from peppermint trichomes, and hashish, a powerful narcotic, is purified resin from *Cannabis* trichomes.

Vascular Tissues

Vascular tissues are specialized conducting tissues that form the veins in leaves. The two types of plant vascular tissues, xylem and phloem, each form during both primary and secondary growth.

Vascular tissue evolved as an adaptation to increasing plant size and to existence on land, an event that occurred about 400 million years ago. In the unicellular algae that were the probable ancestors of plants, diffusion was sufficient to transport materials in and out of the cell. Dissolved nutrients could easily enter from the aquatic environment, and wastes leave. Multicellular life, particularly on dry land, requires a more elaborate transport system

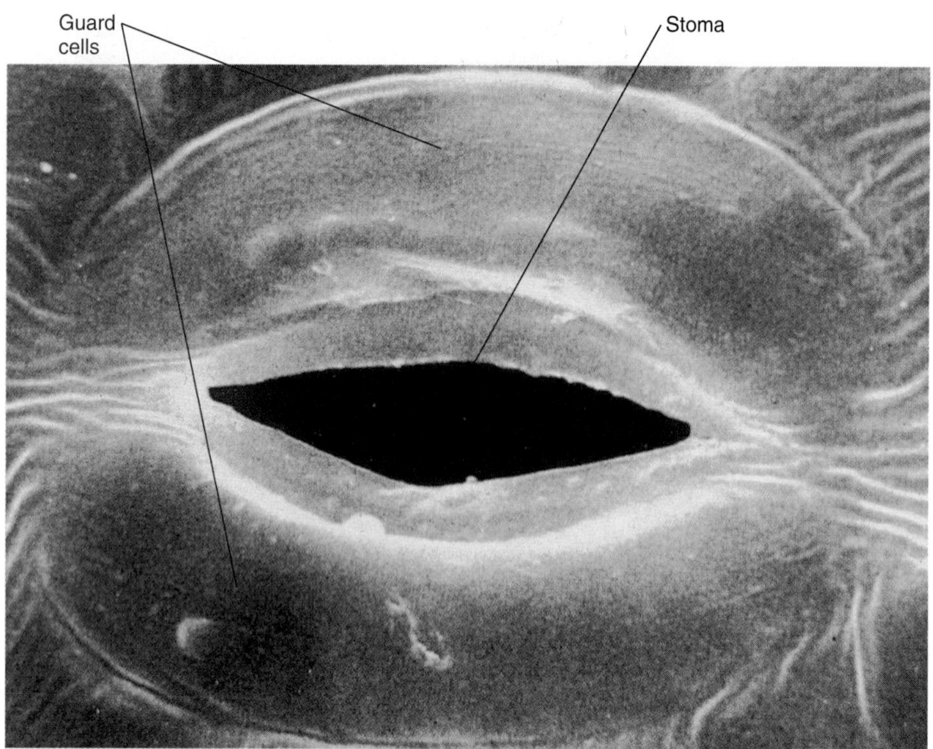

FIGURE 27.6

A stoma on a *Mascagnia macroptera* leaf hair. Guard cells hug the stomatal pore.

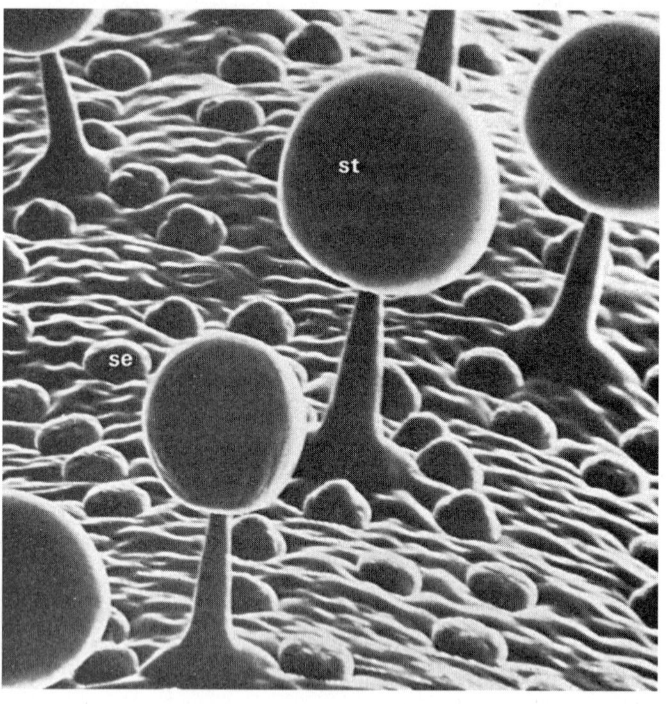

FIGURE 27.7

Stalked (*st*) and sessile (*se*) trichomes on the leaf of a butterwort (*Pinguicula grandiflora*) secrete biochemicals that trap and digest animals.

plus ways to replace the tremendous amounts of water lost through transpiration. A plant's epidermis and stomata are adaptations to conserve water; vascular tissue is an adaptation to transport water and dissolved nutrients. Water is vital to the plant for several reasons: to synthesize biochemicals; as a solvent for nutrients; and to create turgor pressure, which physically supports the plant body.

Leaves have many spaces that provide a tremendous surface area from which water evaporates. A corn plant, for example, loses approximately 110 gallons (500 liters) of water to transpiration in its 4-month growing season. Veins carrying vascular tissues permeate leaves and distribute water to these surfaces.

Xylem

Xylem, from the Greek for wood, forms a continuous system that transports water and dissolved minerals from the roots to all parts of the plant, in an upward direction only. This water replaces water lost in transpiration through stomata.

The two kinds of conducting cells in xylem are called **tracheids** and **vessel elements** (fig. 27.8). Both are elongated, dead at maturity, and have thick walls, characteristics that are adaptive to their transport function. The fact that these cells are dead means that organelles do not block water flow. The thick cell walls prevent tracheids and vessel elements from collapsing as water rises through the plant under negative pressure (suction).

Tracheids, the least specialized conducting cells, are long and narrow, overlapping at their tapered ends. Water moves from tracheid to tracheid through thin areas in cell walls called pits. The water movement is slow because of the slender shapes of the tracheids. Tracheids are probably ancient cells, because they are the only means of water conduction in most nonflowering plants, which are more ancient than flowering plants.

Vessel elements are more specialized cells that evolved from tracheids. Unlike tracheids, vessel elements are short, wide, barrel-shaped cells. Vessel elements stack end to end, and their

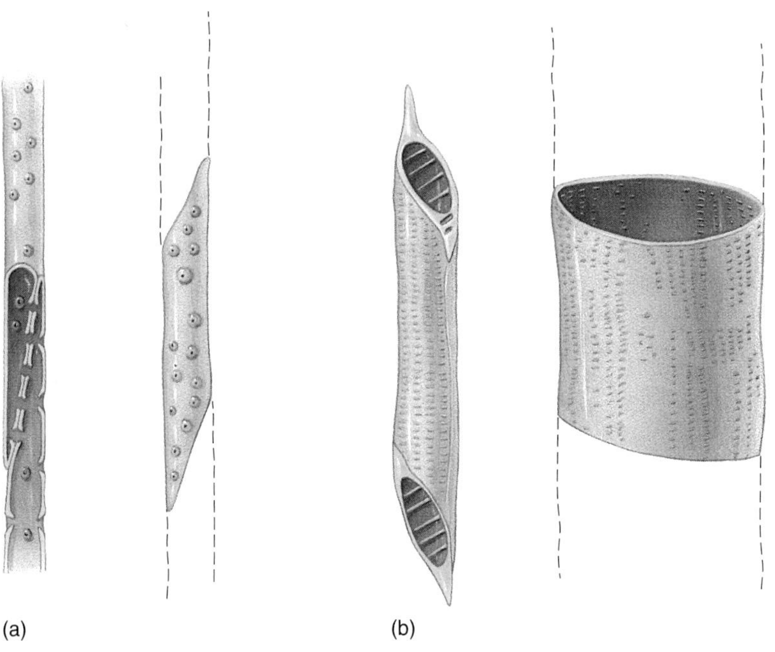

FIGURE 27.8

Xylem: (*a*) tracheids and (*b*) vessel elements. Xylem transports water and dissolved minerals from roots to shoots. Tracheids are long, narrow, and less specialized; vessel elements are short, wide, and more specialized.

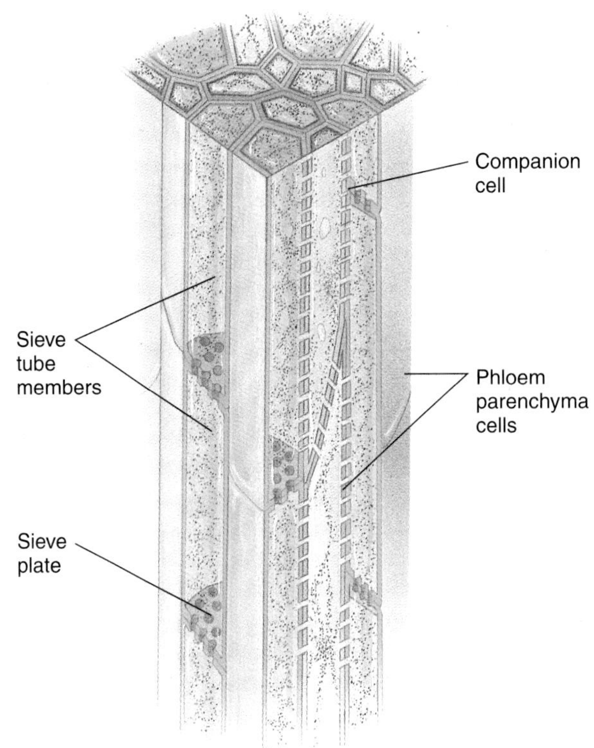

FIGURE 27.9

A longitudinal view of phloem in a tobacco stem. Phloem transports water and dissolved sugars in all directions in plants.

Table 27.3 Major Plant Cell Types

Cell	Tissue Type	Function
Parenchyma	Ground	Storage, division, metabolism
Collenchyma	Ground	Support
Sclerenchyma	Ground	Support
Tracheids	Vascular (xylem)	Water transport
Vessel elements	Vascular (xylem)	Water transport
Sieve cells	Vascular (phloem)	Water and nutrient transport
Sieve tube members	Vascular (phloem)	Water and nutrient transport

end walls usually dissolve, forming hollow tubes, or vessels, that may extend from a centimeter to 3 meters long. Vessel elements are like cellulose pipes, and water in them moves much faster than in the narrower tracheids. The greater width of vessel elements and the fact that water can pass directly from one cell to another accounts for its more efficient water conduction. Biology in Action 27.1 discusses how water rises to treetops.

Phloem

Phloem transports primarily dissolved carbohydrates throughout a plant. In some plants phloem is so sugary that it forms a thick sap. Phloem also transports hormones, alkaloids, viruses, and inorganic ions. Unlike xylem, which transports water upward under negative pressure, like a drinking straw, phloem transports substances under positive pressure, which is like water flowing through a hose when the spigot is turned on. Thus, water and dissolved sugars can move through phloem in all directions. Also, the conducting cells of phloem, unlike those of xylem, are alive at maturity. Their cell walls have thin areas perforated by many sieve pores, and solutes move through these pores from cell to cell.

Phloem has two kinds of conducting cells: sieve cells and sieve tube members. **Sieve cells** are the more primitive (that is, less specialized) conducting cells. They are elongated with tapered, overlapping ends. Sieve pores permeate all sieve cell walls. **Sieve tube members** are more complex, shorter, and wider than sieve cells, and they lie end to end, forming long sieve tubes. The pores of individual cells cluster on contacting end walls and form an area called a **sieve plate** (fig. 27.9). This organization allows faster, more efficient nutrient transport. Mature sieve tube members lack nuclei but contain living protoplasm—a unique feature among plant cells. Near sieve tube members are **companion cells,** which help transfer carbohydrates into and out of the seive tube members.

Table 27. 3 reviews major cell types in flowering plants.

biology in 27.1 action

How Water Reaches Treetops

Some of the giant redwoods towering in the forests of northern California reach more than 328 feet (100 meters) tall (see fig. 3.10*a*). How can water rise so far from roots? Understanding how this happens requires a return to the concept of potential energy (see chapter 6).

In plants, water moves passively—that is, without energy input. The water exerts a pressure, called the water potential. Theoretically, water pressure is the potential energy of water in a particular system (the plant), compared to pure water at atmospheric pressure and at the same temperature. Differences in water potential in different parts of a plant determine the direction of water flow. Water always flows from regions of high water potential to low water potential.

A likely explanation of how water moves in xylem was proposed more than a century ago and is called the transpiration-cohesion hypothesis (fig. 27.A). The following events occur:

- The sun causes transpiration in the plant, which dries out leaf parenchyma.
- Water loss through cell walls lowers the cells' water potential. Water enters from neighboring cells that have a higher water potential because they are farther away from the air spaces.
- Cells even farther from the sites of transpiration have greater water potentials, so water moves from cell to cell, toward the air spaces in the transpiring leaves.
- Cells bordering tracheids replace their water from xylem. Loss of water from xylem creates negative pressure (suction) and lifts the water column up the plant.
- The negative pressure decreases the water potential all the way down to the root tips, even in the tallest trees. This tension can lower the water potential in the root xylem so much that water flows into the plant passively from the soil, to be pulled up the xylem to leaves and replace water lost in transpiration.

Transpiration can lift a column of water to a treetop because of the water potential gradient (difference) that favors the movement of water from soil to plant to atmosphere. Water in xylem under constant negative pressure forms a continuous hydraulic system that rises in columns through the plant body. These columns can withstand the tensions necessary to reach the treetops because of water's tensile strength, cohesiveness, and adhesiveness (see chapter 3). These qualities overcome gravity, and water rises in the plant.

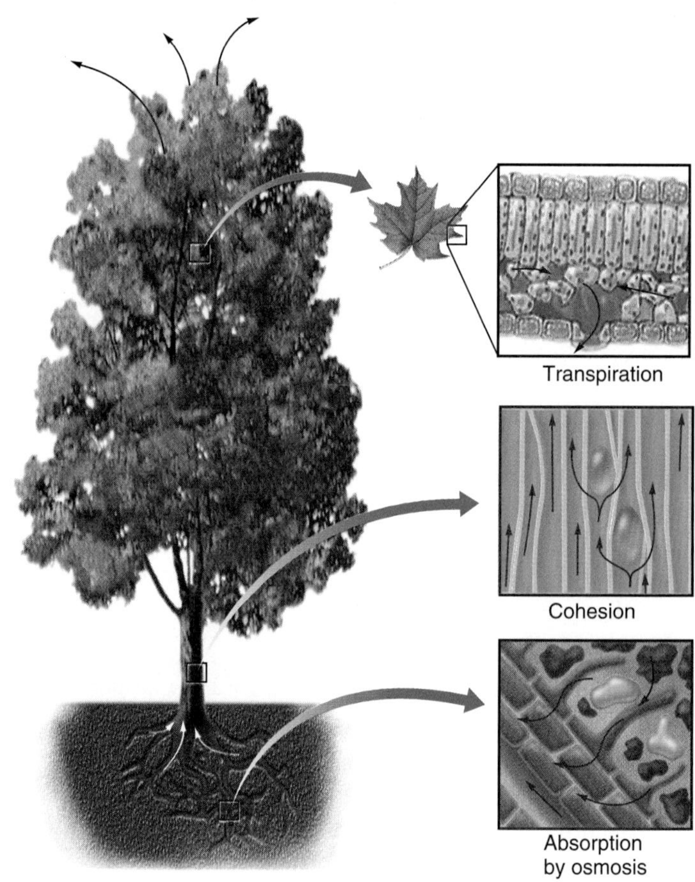

FIGURE 27.A
According to the transpiration-cohesion hypothesis, water ascends in plants from the roots because of differences in water potential generated from the distance from transpiring leaf cells.

Mastering Concepts

1. What is the function of meristematic tissue?
2. What cell types make up ground tissue?
3. How does dermal tissue protect a plant but enable gas exchange to occur?
4. What are the functions of the two types of vascular tissues in plants?
5. What types of cells make up vascular tissues?

Parts of the Plant Body

The stems, leaves, and roots of a plant are all composed of meristematic, dermal, ground, and vascular tissues. Yet plants come in a wide variety of forms—from a prickly cactus to a lush vine to the majestic sunflower. Natural selection sculpted different organizations of plant tissues as certain forms became better adapted to particular environments. We look now at the basic parts of a flowering plant.

Stems

The central axis of a shoot system is the stem, which consists of **nodes** (areas of leaf attachment) and **internodes** (portions of stem between the nodes) (see fig. 27.1). The region of stem just above the point where the leaf attaches is the leaf **axil.** Axillary buds are undeveloped shoots that form in leaf axils. Although axillary buds can elongate to form a branch or flower, many remain small and dormant.

Normally, stem elongation occurs in the internodal regions; however, some plants have stems called **rosettes** that do not elongate. Rosettes have short internodes and overlapping leaves. A banana tree is a rosette—much of its "trunk" region is made of large, tightly packed leaves.

The epidermis surrounding a stem is a transparent, unicellular layer. It contains stomata, but fewer than are in a leaf's epidermis. The epidermis of a stem also may have protective trichomes.

Vascular tissues in stems are organized into groups called **vascular bundles** that branch into leaves at the nodes. Phloem forms on the outer portions of a bundle, whereas xylem forms on the inside. Often, thick-walled sclerenchyma fibers associate with vascular bundles and strengthen the vascular tissue.

Vascular bundles are organized differently in different types of plants. Consider the two classes of flowering plants: monocotyledons (**monocots** for short), which have one first, or "seed," leaf; and dicotyledons **(dicots)**, which have two seed leaves. Monocots such as corn have vascular bundles scattered throughout their ground tissue, whereas dicots such as sunflowers have a single ring of vascular bundles (fig. 27.10).

The ground tissue that fills the area between the epidermis and vascular tissue in a stem is called the **cortex.** This area is mostly parenchyma but may include a few supportive collenchyma strands. Some cortical cells are photosynthetic and store starch. The centrally located ground tissue in dicots is called **pith.**

Stems have several functions. They support leaves, produce and store nutrients, and transport nutrients and water between roots and leaves. Stems may provide food for other species—asparagus, for example, is an edible stem. Many plant stems are modified for special functions such as reproduction, climbing, protection, and storage (fig. 27.11). Some specialized stems are as follows:

- **Stolons,** or runners, are stems that grow along the soil surface. New plants form from their nodes. Strawberry plants develop stolons after they flower, and several plants can arise from the original one.
- **Thorns** often are stems modified for protection, such as on grapefruit plants. The "thorns" on a rosebush are actually prickles, which are extensions of the epidermis.
- **Succulent stems** of plants such as cacti are fleshy and store large amounts of water.
- **Tendrils** support plants by coiling around objects, sometimes attaching by their adhesive tips. Tendrils enable a plant to maximize sun exposure. The stem tendrils of green bean plants readily entwine around anything they can grab onto. Grapevines also have many tendrils.
- **Tubers** are swollen regions of underground stems that store nutrients. Irish potatoes are tubers produced on burrowing stolons.

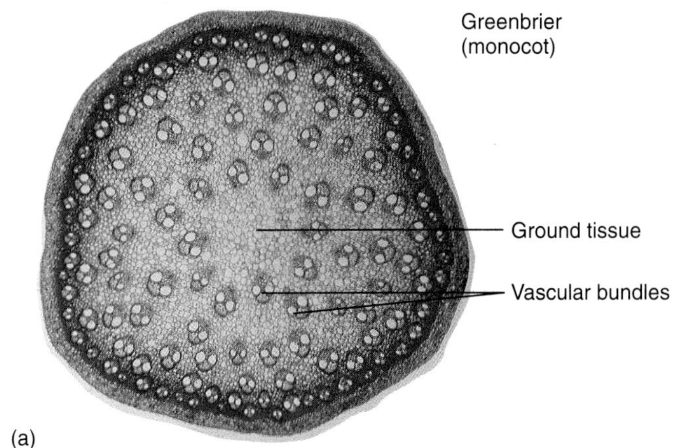

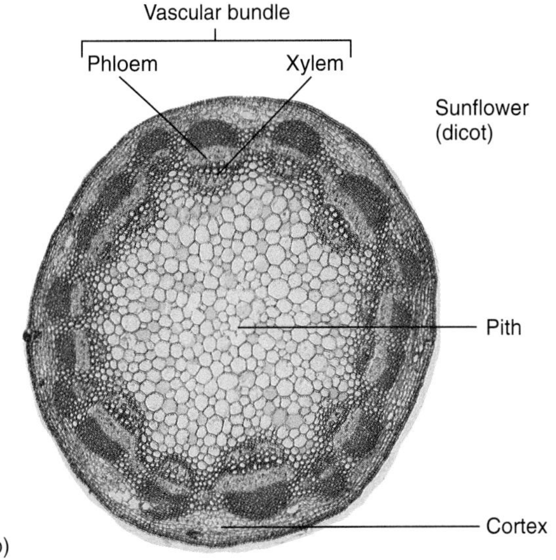

FIGURE 27.10

Cross sections of a (*a*) monocot (greenbrier) and a (*b*) dicot (sunflower). (×10) Notice the ring of vascular bundles in the dicot and the scattered vascular bundles in the monocot.

Leaves

In addition to the stem, a shoot system has leaves. Like stems, leaves consist of epidermal, vascular, and ground tissues. Leaves are the primary photosynthetic organs of most plants. They provide an enormous surface area for the plant to capture solar energy. For example, a maple tree with a 1-meter-wide trunk has approximately 100,000 leaves, with a total surface area that would cover the area of six basketball courts.

Leaves are extremely diverse in form. The leaves of tropical palms may be 65 feet (20 meters) long, whereas the leaves of *Wolffia,* an aquatic plant, are no larger than a pinhead. A mature American elm may have several million leaves, whereas the desert plant *Welwitschia mirabilis* produces only two leaves during its entire lifetime (see fig. 25.11*d*). Leaves may be needlelike, feathery, waxy, or smooth. No two species have identical leaves.

Botanists classify leaves according to their basic forms as well as their patterns on stems (fig. 27.12). Most leaves consist of a flattened **blade** and a supporting, stalklike **petiole.** The large vein

(a)

(b)

(c)

(d)

(e)

FIGURE 27.11

Modified stems. (*a*) Stolons of the beach strawberry (*Fragavia chilensis*) run parallel to the ground. (*b*) Thorns are protective stem outgrowths and protect the plant (Honey locust). (*c*) The succulent leaves of a stone plant (*Lithops*) store water. (*d*) Tendrils coil around objects, supporting and anchoring plants. (*e*) The potato is a tuber. Sprouts grow from its eyes and form new plants.

down the center is called the **midrib.** Leaves may be simple or compound. **Simple** leaves have flat, undivided blades. Elm, maple, and zinnia have simple leaves. **Compound** leaves are divided into leaflets and are further distinguished by leaflet position. **Pinnate** compound leaflets are paired along a central line and include ash, rose, and walnut. **Palmate** compound leaflets all attach to one point at the top of the petiole, like fingers on a hand, and include lupine, four-leaf clover, and shamrock.

Leaves grow in different patterns on stems to maximize sun exposure (fig. 27.13). Plants with one leaf per node have **alternate,** or spiral, organization. Oak and birch have this pattern. Plants with two leaves per node have **opposite** patterns and include maple and ash. Plants with three or more leaves per node have **whorled** patterns. These include oleander and horsetails (see fig. 25.6*c*). With so many variable features, leaves come in a great variety of shapes.

Leaf epidermis, the "skin" covering the leaf, consists of tightly packed transparent cells. The epidermis is usually nonphotosynthetic and contains many stomata—more than 11 million in a cab-

(a)

(b)

(c)

FIGURE 27.12

Leaf forms. (*a*) This simple leaf of *Thottea* attaches to the stem by its petiole. (*b*) This pinnately compound leaf of a mimosa plant has many leaflets. (*c*) The alfalfa plant has palmately compound leaves, with three leaflets.

bage leaf, for example. Water loss is minimized in many species because stomata are most abundant on the protected undersides of leaves.

Vascular tissues in leaves occur in strands, the veins. Xylem forms on the upper side of a vein, and phloem forms on the lower side. Sclerenchyma fibers and parenchyma cells support leaf veins. Leaf veins may be of two types: **netted,** with minor veins branching off from larger, prominent midveins, or **parallel,** with several major parallel veins connected by smaller minor veins (fig. 27.14). Most dicots have netted veins, and many monocots have parallel veins. **Vein endings** are the blind ends of minor veins, where water and solutes move in and out.

Leaf ground tissue, which is called **mesophyll,** is made up of parenchyma cells. Some of these cells are chlorenchyma and therefore photosynthesize and produce sugars. Horizontally oriented leaves have two types of chlorenchyma. The long, columnar cells along the upper side of a leaf, the **palisade mesophyll cells,** are specialized for light absorption. Below the palisade layer are **spongy mesophyll cells,** irregularly shaped chlorenchyma cells

(a)

(b)

FIGURE 27.13

Leaf patterns on stems. (*a*) The leaves of *Costus* are spiral. (*b*) *Coleus* has two leaves per node, which is an opposite leaf pattern.

(a)

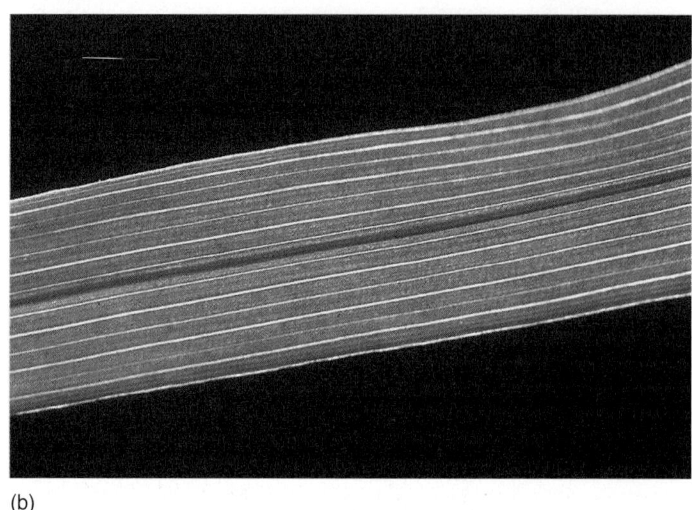

(b)

FIGURE 27.14

Leaves in dicots and monocots. (*a*) Leaves of dicots, such as this pumpkin, have a netted venation. (*b*) Leaves of monocots, such as this lily, have parallel venation.

separated by large air spaces (fig. 27.15). These cells are specialized for gas exchange and can also photosynthesize. In contrast to horizontally oriented leaves, vertically oriented leaves have uniformly shaped chlorenchyma cells and are simply called mesophyll cells.

In addition to photosynthesizing, leaves provide support, protection, and nutrient procurement and storage, with the following specializations:

- **Tendrils** are modified leaves that wrap around nearby objects to support climbing plants. Pea plants growing in a garden will "grab" a fence with leaf tendrils. (Both leaves and stems can be modified into tendrils.)
- **Spines** of plants such as cacti are leaves modified to protect the plant from predators.
- **Bracts** are floral leaves that protect developing flowers. They are colorful in some plants, such as poinsettia.
- **Storage leaves** are fleshy and store food. Onion bulbs are the bases of such leaves.
- **Insect-trapping leaves** are found in about 200 types of carnivorous plants and are adapted for attracting, capturing, and digesting prey. Some leaves have sticky "flypaper" surfaces, whereas others form water-filled chambers that drown insects. The trigger hairs of a Venus's flytrap respond to a visiting insect's movements by stimulating the two halves of the leaves to snap together. The insect is trapped and then destroyed as the leaves secrete digestive enzymes (fig. 27.16).
- **Cotyledons** are embryonic leaves that store carbohydrates, which supply energy for germination.

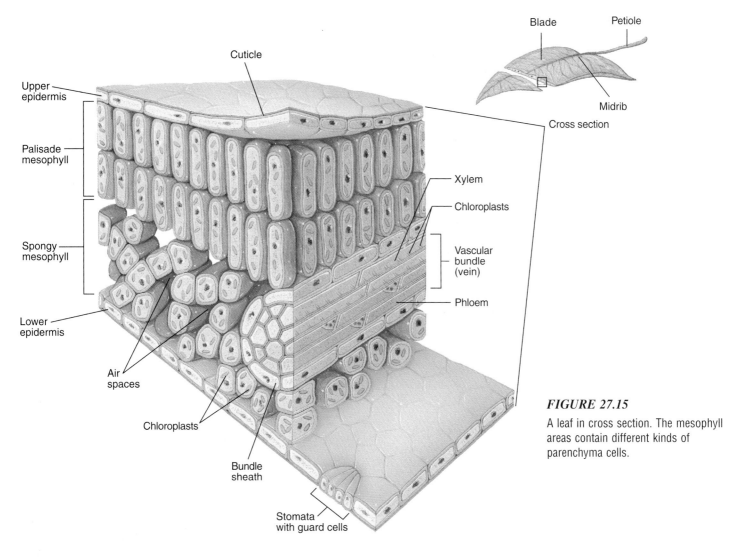

Blade
Petiole
Midrib
Cross section
Cuticle
Upper epidermis
Palisade mesophyll
Spongy mesophyll
Lower epidermis
Air spaces
Chloroplasts
Bundle sheath
Stomata with guard cells
Xylem
Chloroplasts
Vascular bundle (vein)
Phloem

FIGURE 27.15

A leaf in cross section. The mesophyll areas contain different kinds of parenchyma cells.

(a)

(b)

FIGURE 27.16

Carnivorous plants have leaves modified to trap insects. (*a*) The leaves of the Venus's flytrap snap shut to catch prey. This plant lives only in North and South Carolina. (*b*) Leaves of the pitcher plant *Sarracenia* trap and kill insects, which the plant digests.

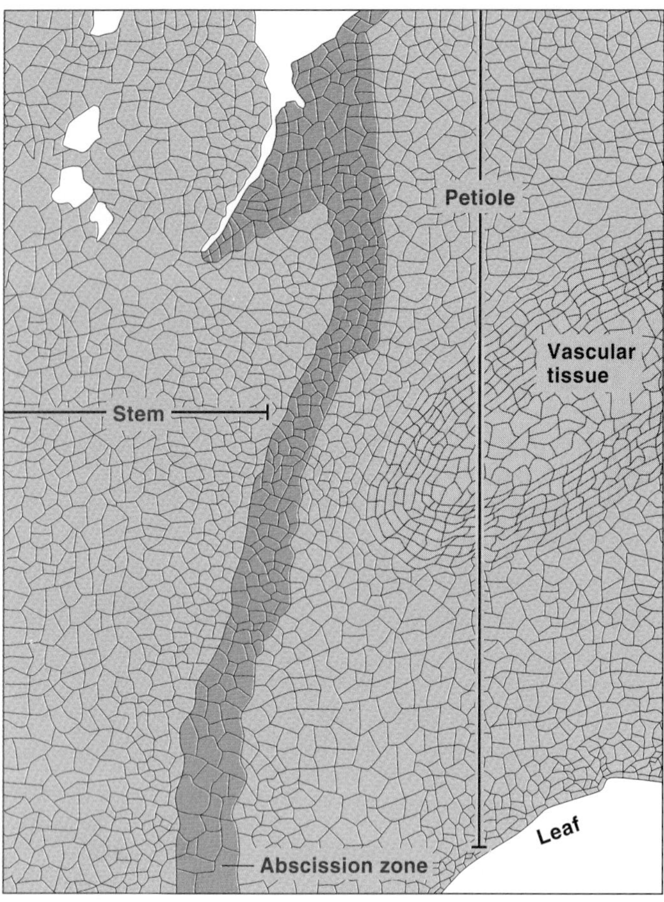

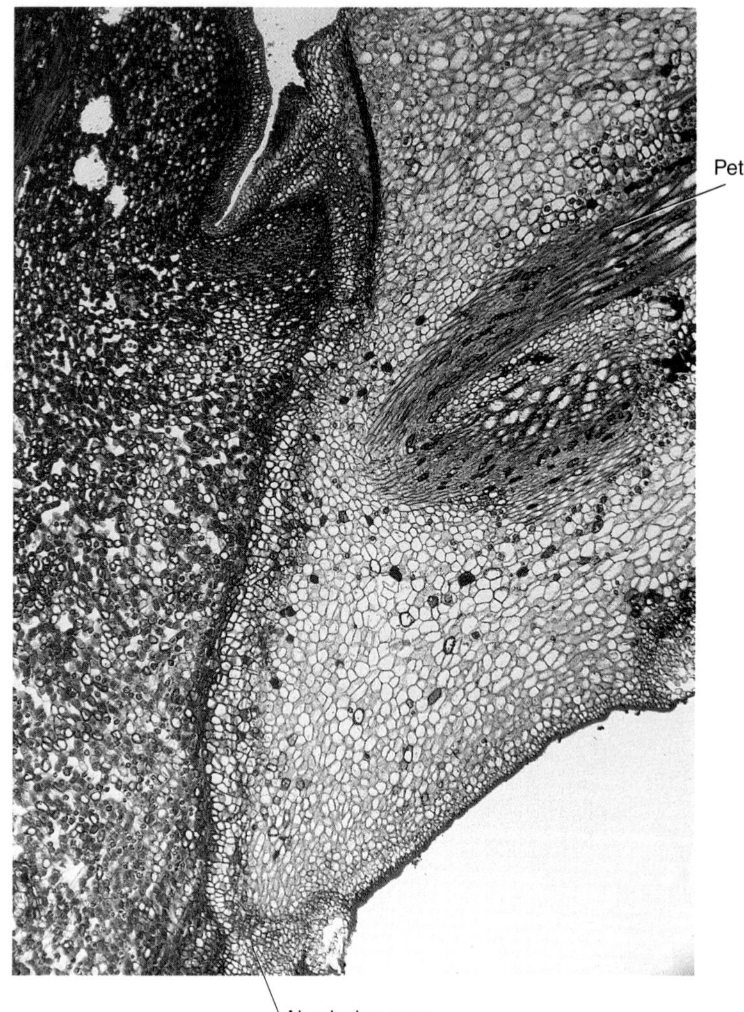

FIGURE 27.17

The abscission zone is a region of separation that forms near the petiole base. This zone minimizes risk of infection and nutrient loss when the leaf is shed.

Anyone who has ever raked a lawn knows well that leaves have a limited life span. **Leaf abscission** is the normal process by which a plant sheds its leaves. **Deciduous trees** shed leaves at the end of a growing season. Evergreens retain leaves for several years but shed a few leaves each year.

Leaves shed from an **abscission zone,** a region at the base of the petiole (fig. 27.17). In response to environmental cues such as shortening days or cooler temperatures, a separation layer forms in the abscission zone, isolating the dying leaf from the stem. Eventually, wind, rain or some other disturbance, such as a scurrying squirrel, breaks the dead leaf from the stem. Leaf abscission is important because it removes injured or dying leaves from the plant.

Roots

Plants are immobile, but they are biologically active, especially underground, where roots grow. Roots are so indispensable to plant growth and photosynthesis that annual root production often consumes more than half of a plant's energy and may account for a substantial portion of its body. Roots anchor plants and absorb, transport, and store water and nutrients. They absorb oxygen from between soil particles; roots pushing through firmly packed soil may die from lack of oxygen.

The first root to emerge from a seed is the **radicle.** The length of life of the radicle differs in two types of root systems. In a **taproot system,** the radicle enlarges to form a major root that persists throughout the life of the plant (fig. 27.18*a*). Taproots grow fast and deep, maximizing support and enabling a plant to use materials deep in the soil. Engineers once found a mesquite root 174 feet (53 meters) below the earth's surface! Most dicots develop taproot systems.

A **fibrous root system,** conversely, has a short-lived radicle that becomes replaced with **adventitious roots,** which form on stems or leaves (fig. 27.18*b*). Because they are relatively shallow, fibrous root systems rapidly absorb materials near the soil surface and prevent soil erosion. Most monocots have fibrous root systems. The adventitious roots form an extensive system of similarly sized roots.

Rapidly growing roots lose as many as 10,000 cells per day as they push through soil. The **root cap,** a thimble-shaped structure, covers the tips of roots and protects them as apical meristem replaces cells. The root apical meristem continually pushes cells forward. Eventually, these cells reach the outside of the root, where they are sloughed off as the root grows. The root cap also produces **mucigel,** a slimy substance that protects root tips from desiccation. Mucigel lubricates root tips as they extend between soil particles and helps absorb water and nutrients from the soil.

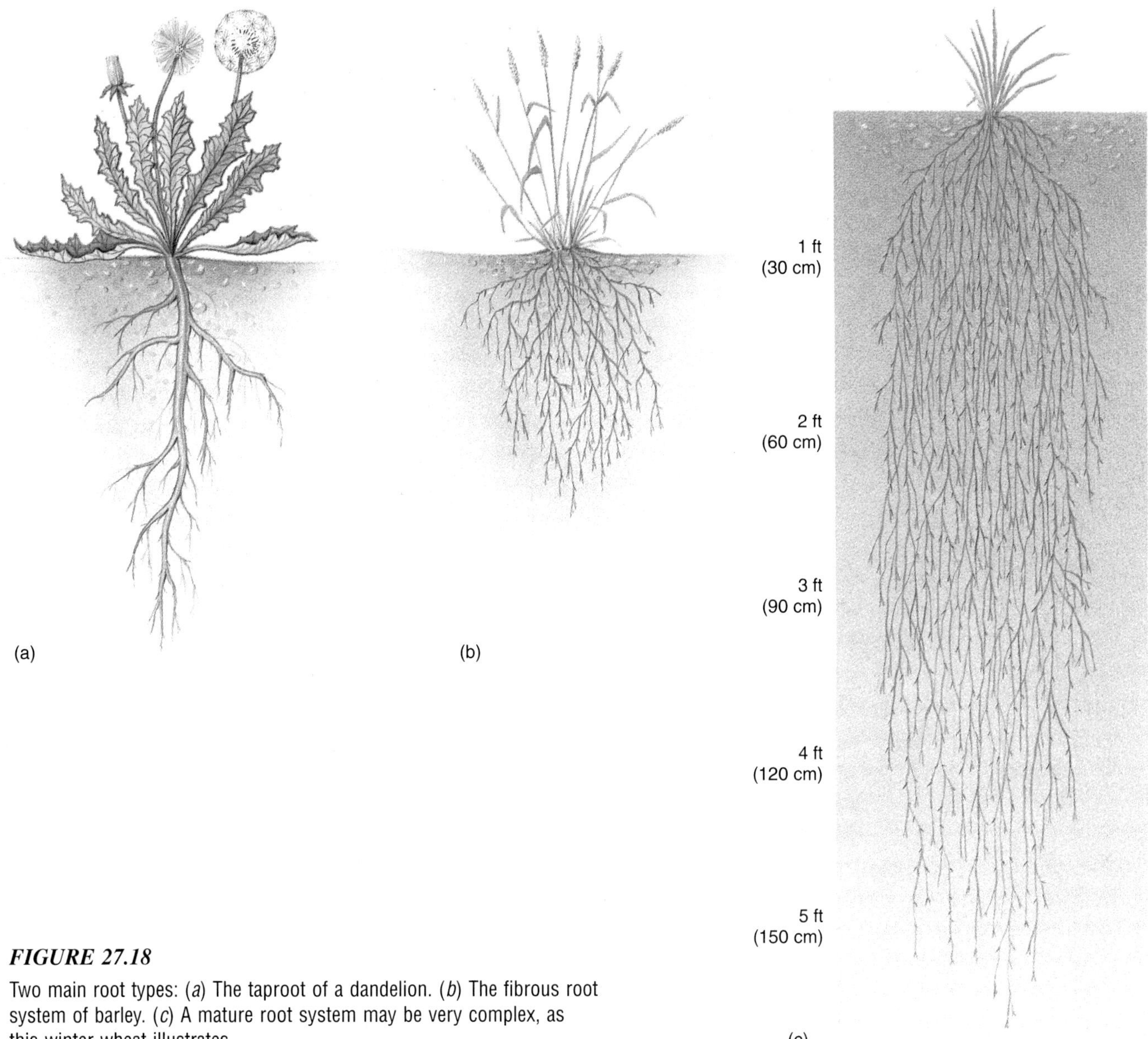

FIGURE 27.18

Two main root types: (*a*) The taproot of a dandelion. (*b*) The fibrous root system of barley. (*c*) A mature root system may be very complex, as this winter wheat illustrates.

Just behind the root cap is a cluster of seemingly inactive cells called the **quiescent center.** This region is actually part of the root apical meristem, and it functions as a reservoir, replacing damaged cells in the adjacent meristem. The rest of the root, the portion above the tip, anchors the plant in the ground. The region immediately behind the root cap is called the **subapical region.** It is loosely divided into three zones (fig. 27.19). The **zone of cell division** is meristematic. Cells in this region, which surrounds the quiescent center, divide as rapidly as every 12 to 36 hours. The **zone of cell elongation** lies behind the zone of cell division. Cells in this region elongate by as much as 150-fold as their vacuoles fill with water and push the root rapidly through the soil.

Cells mature and differentiate in the **zone of cell differentiation.** This is also called the root hair zone, because many tiny root hairs protrude from epidermal cells (fig. 27.20*a*). A single plant may grow several billion root hairs, which greatly increase the surface area available for absorbing water. Root hairs give this portion of the root a downy, fuzzy appearance (fig. 27.20*b*). Transplanting a garden plant destroys many root hairs. It is only after they begin to grow back that the plant regains its vigor.

The zone of cell differentiation is also where primary tissues such as the epidermis and cortex develop. The epidermis surrounds the entire root except the root cap. Root epidermal cells have a very thin cuticle or none at all and are thus well adapted for absorbing water and minerals. These cells usually do not have stomata.

The root epidermis surrounds the entire cortex, which consists of three layers: hypodermis, storage parenchyma cells, and endodermis. The **hypodermis** is the outermost, protective layer of the cortex. It consists mostly of loosely spaced storage parenchyma cells. These cells form a vast collecting system that absorbs water and minerals moving through the epidermis and stores them for future growth.

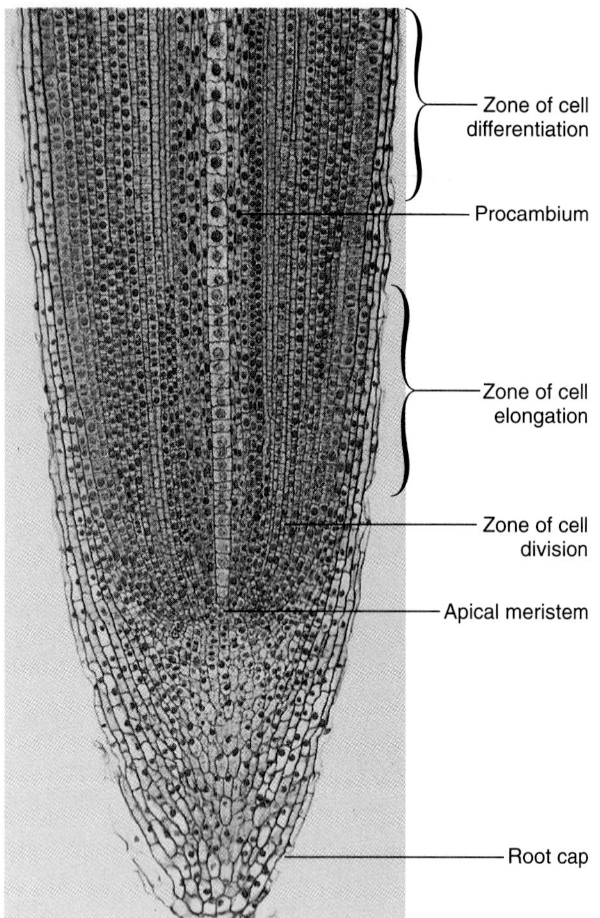

FIGURE 27.19

The subapical region of a dicot root has three distinct regions: the zone of cell division, zone of cell elongation, and zone of cell differentiation.

The **endodermis** is the innermost ring of the cortex (fig. 27.21). It includes a single layer of tightly packed cells, whose walls contain deposits of lignin and a waxy, waterproof material called **suberin.** The waxy deposits form a rubberband-like structure called a **Casparian strip.** The endodermal cells resemble bricks in a wall, and the Casparian strip is like the mortar surrounding each brick (fig. 27.22). This organization enables the endodermis to regulate the movement of nutrients and water into and out of the central vascular tissue of the root.

Inside the cortex, the **pericycle** is the ring of parenchyma cells that produces branch roots that burst through the cortex and epidermis and reach into the soil. The root's vascular tissues are interior to the pericycle, where bundles of xylem and phloem alternate. Roots absorb water and minerals from the soil and transport them to shoots via the xylem. In return, roots receive organic nutrients from the shoots via the phloem. Cortex cells store these organic nutrients.

Like stems and leaves, roots are modified for special functions (fig. 27.23):

- Storage is a familiar root specialization. Beet and carrot roots store starch, sweet potato roots store sugar, and desert plant roots store water.

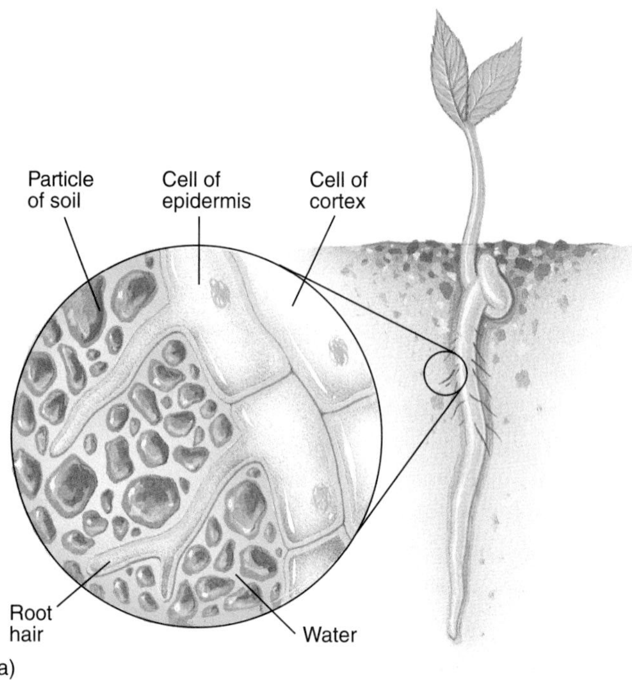

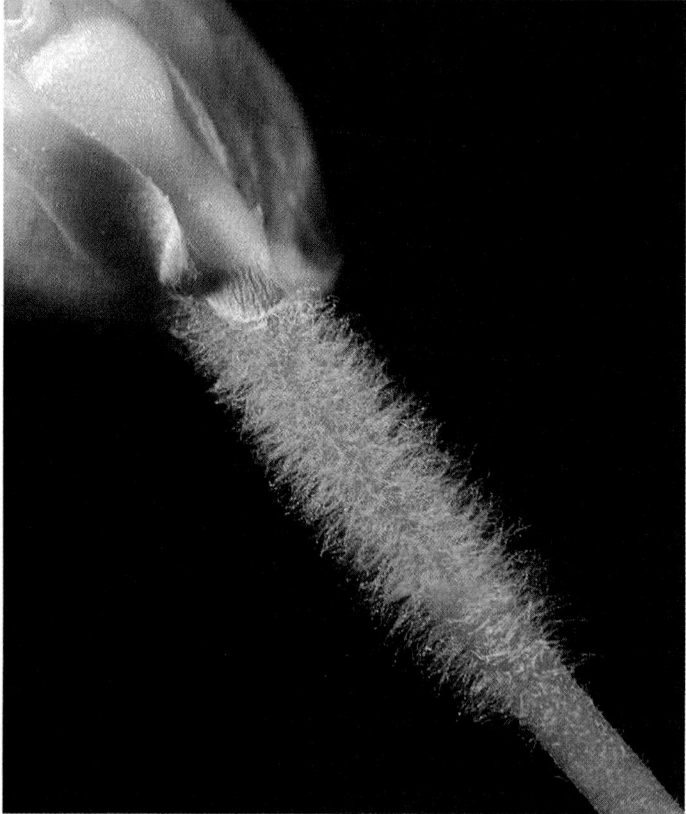

FIGURE 27.20

Root hairs. (*a*) These epidermal cell outgrowths extend through the soil, greatly increasing the absorptive surface area of the root. (*b*) This corn seedling has many root hairs, which increase the surface area for absorption of water carrying dissolved minerals.

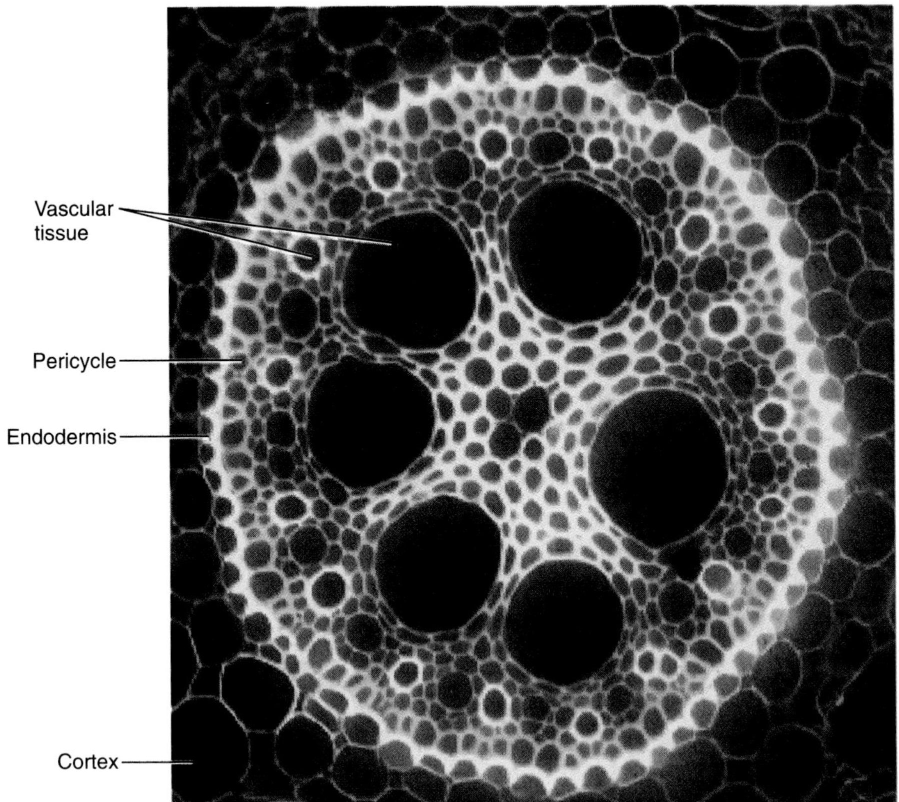

FIGURE 27.21

Cross section of a monocot (corn) root, showing the vascular tissue, pericycle, endodermis, and cortex.

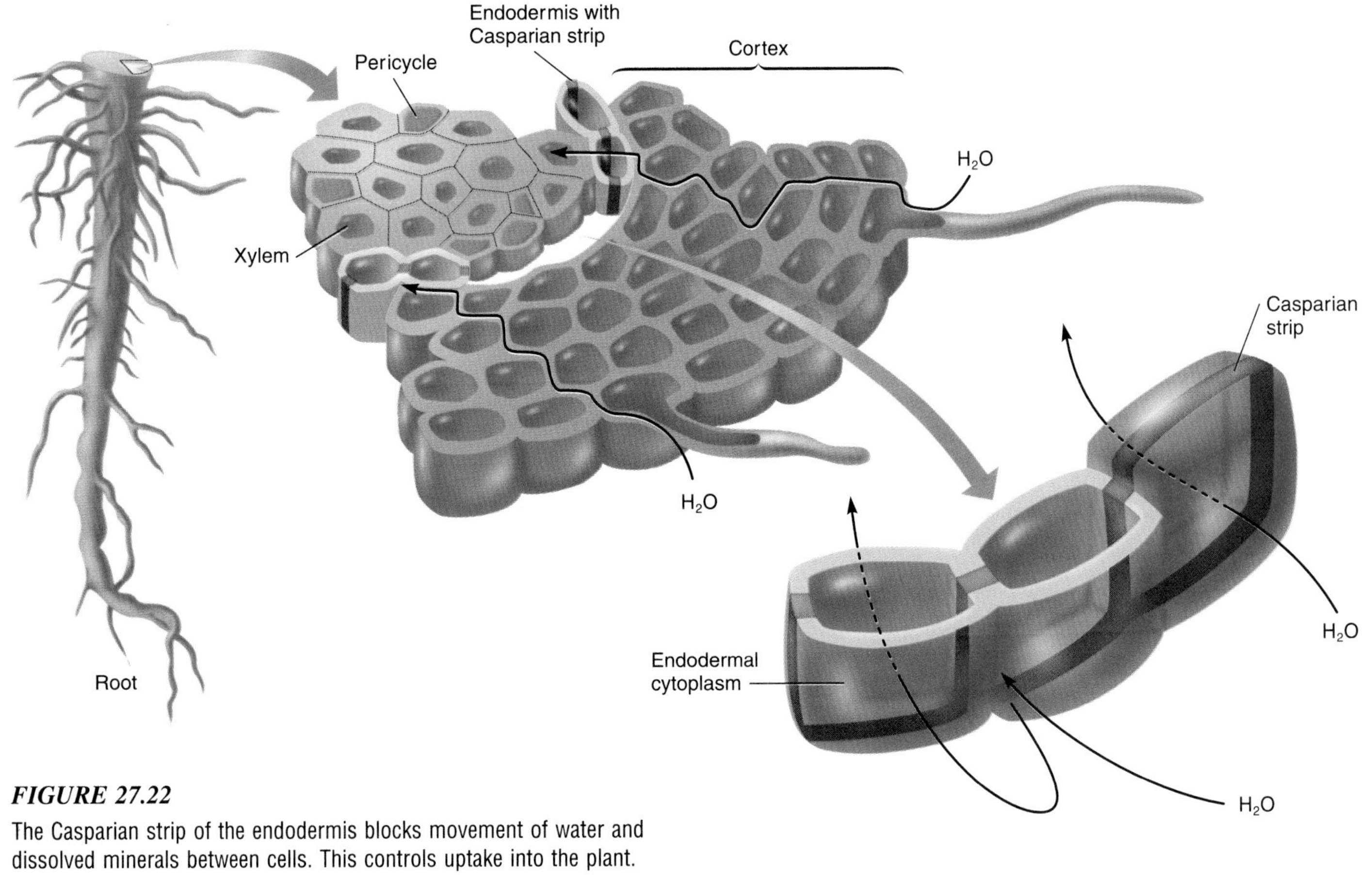

FIGURE 27.22

The Casparian strip of the endodermis blocks movement of water and dissolved minerals between cells. This controls uptake into the plant.

(a)

(b)

(c)

(d)

FIGURE 27.23

Root modifications. (*a*) Yams are fleshy roots that store carbohydrate. (*b*) The banyan tree has aerial roots growing out of its branches. (*c*) This tropical fig tree has buttress roots so enormous they resemble a trunk. (*d*) Prop roots on corn arise from the stem and support the plant.

- **Pneumatophores** are specialized roots that form on plants growing in oxygen-poor environments, such as swamps. Black mangrove trees have pneumatophores. These roots form underground and grow up into the air, allowing oxygen to diffuse into the plant body.
- **Aerial roots** are adventitious roots that form and grow in the air. Mistletoe and orchids have aerial roots.
- Thick, enormous **buttress roots** at the base of a tree provide support, as do **prop roots** that arise from the stem, as seen in corn.

A plant's roots may interact with other organisms. Recall from chapter 24 that many roots form mycorrhizae with beneficial fungi. The fungi absorb nutrients from soil, while the plants provide nutrients to the fungi. Roots of legume plants, such as peas, are often infected with bacteria of genus *Rhizobium.* The roots form nodules in response to the infection. The bacteria function as built-in fertilizer, providing the plant with nitrogen "fixed" into compounds it can use.

Mastering Concepts

1. What are the parts and tissues of a stem?
2. What are the functions of stems?
3. What are the structures and functions of leaves?
4. How do the two types of root systems differ?
5. What are the regions and structures of a root?
6. What are some special modifications of stems, leaves, and roots?

Secondary Growth

Plants compete to intercept light. Growing taller than neighboring plants is one way to maximize light absorption. Continued elongation poses a problem, however, because primary tissues cannot adequately support tall plants. Lateral meristems, which increase the girth of stems and roots by secondary growth, address this problem. These meristems are called the **vascular cambium** and **cork cambium,** and they produce the secondary plant body.

Secondary growth can be impressive. A 2,000-year-old tule tree in Oaxaca, Mexico, is 148 feet (45 meters) in circumference and only 131 feet (40 meters) tall. A 328-foot-tall (100-meter) giant sequoia in northern California is more than 23 feet (7 meters) in diameter. To support all of this extra tissue, a plant's transport system must also become more complex and powerful.

Vascular Cambium

The vascular cambium produces most of the secondary plant body (the plant's diameter). It is a thin cylinder of meristematic tissue in roots and stems. Generally, vascular cambium forms only in plants that exhibit secondary growth—primarily dicots and gymnosperms. Meristematic cells produce secondary xylem on the inner side of the vascular cambium and secondary phloem on the outer side. Overall, the vascular cambium produces much more secondary xylem than secondary phloem. Secondary xylem is more commonly known as wood (fig. 27.24).

The vascular cambium produces wood during the spring and summer. During the moist days of spring, wood is made of large cells and is specialized for conduction. During the drier days of summer, the vascular cambium produces summer wood that has small cells and is specialized for support. These seasonal differences in wood cell sizes generate visible demarcations called **growth rings** (fig. 27.25). Secondary xylem indicates the passage of time because spring wood cells are large and appear light-colored, and summer wood cells are smaller and darker-colored. The contrast between the summer wood of one year and the spring wood of the next creates the characteristic annual tree ring. The

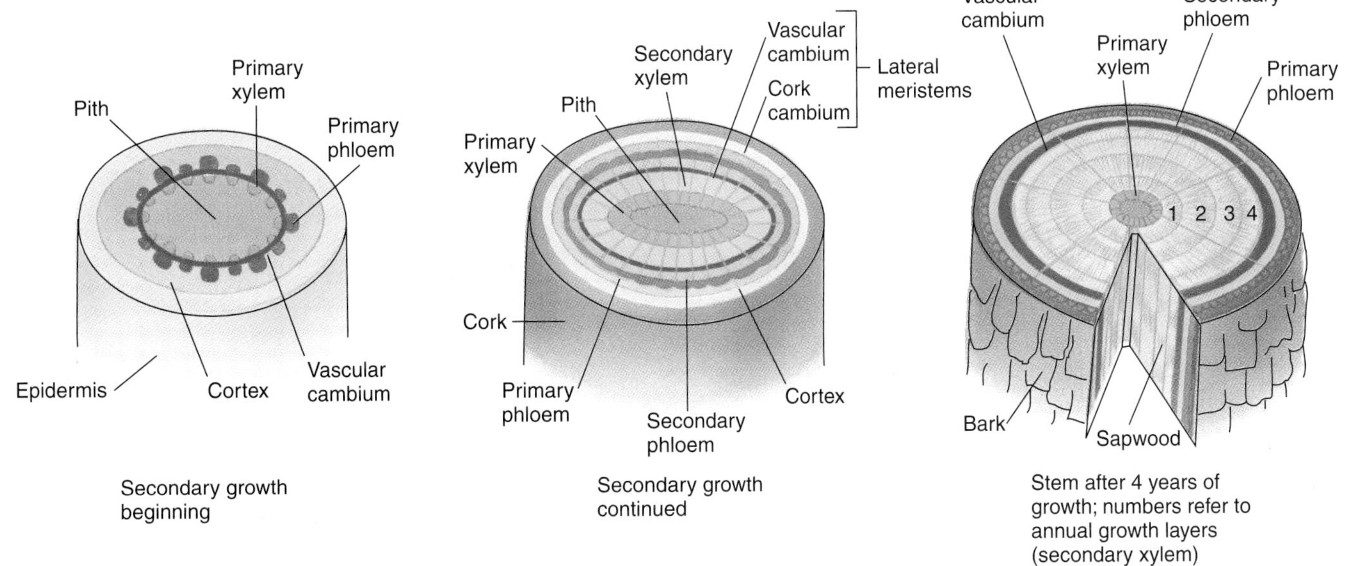

FIGURE 27.24

The secondary growth of a woody stem involves the activities of two lateral meristems—vascular cambium and cork cambium. The apical meristem produces primary xylem and phloem; vascular cambium produces secondary xylem and phloem.

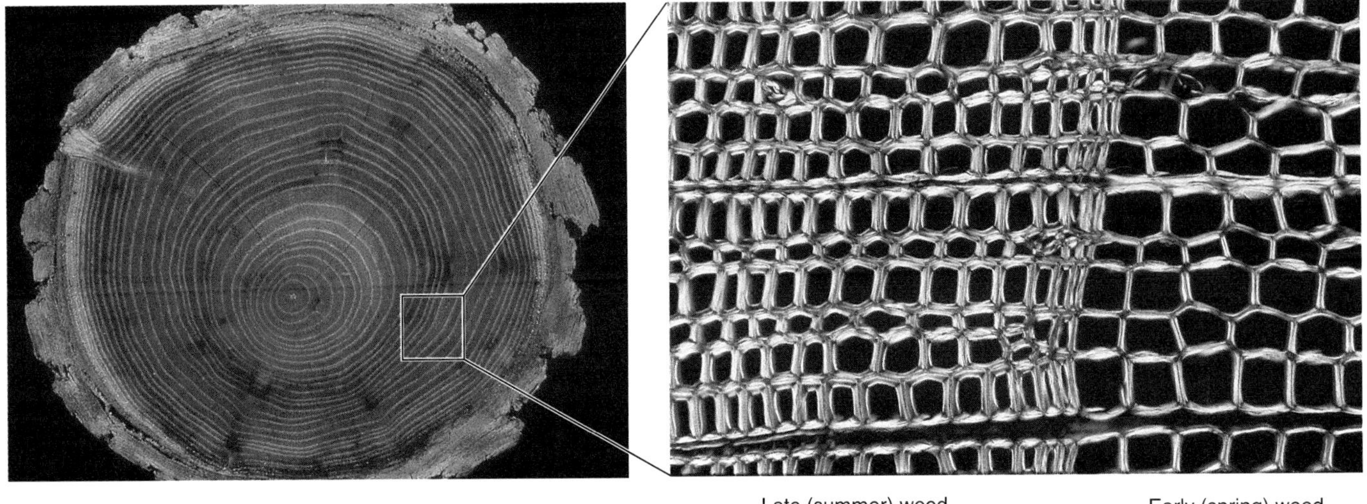

FIGURE 27.25

Tree rings. The differences in moisture when wood forms in summer and spring create different-sized cells, which are visible as tree rings.

most recently formed ring is next to the microscopic vascular cambium. Tropical species, which have secondary growth all year long, do not have tree rings. Figure 27.26 shows how dendrochronologists align tree ring data to correlate to historical events, as described in the chapter opener.

Woods differ in hardness. **Hardwoods** come from dicots such as oak, maple, and ash. **Softwoods** are the woods of gymnosperms such as pine, spruce, and fir. Hardwoods contain tracheids, vessels, and supportive fibers, whereas softwoods are 90% tracheids. As a result, hardwoods are usually stronger and denser than softwoods. In fact, hardwoods are often too hard to penetrate with nails so are seldom used for construction.

Different regions of wood are specialized for different functions. **Heartwood,** the wood in the center of a tree, collects a plant's wastes. It forms a solid, rodlike center that supports the tree. **Sapwood,** located nearest the vascular cambium, transports water and dissolved nutrients.

Bark includes all of the tissues outside the vascular cambium. Secondary phloem forms the inner layer of bark, but only the innermost secondary phloem transports fluid and nutrients. The **periderm** is the outer protective covering of bark on mature stems and roots. It forms after secondary growth breaks through the epidermis. The periderm consists of the cork cambium, cork, and phelloderm.

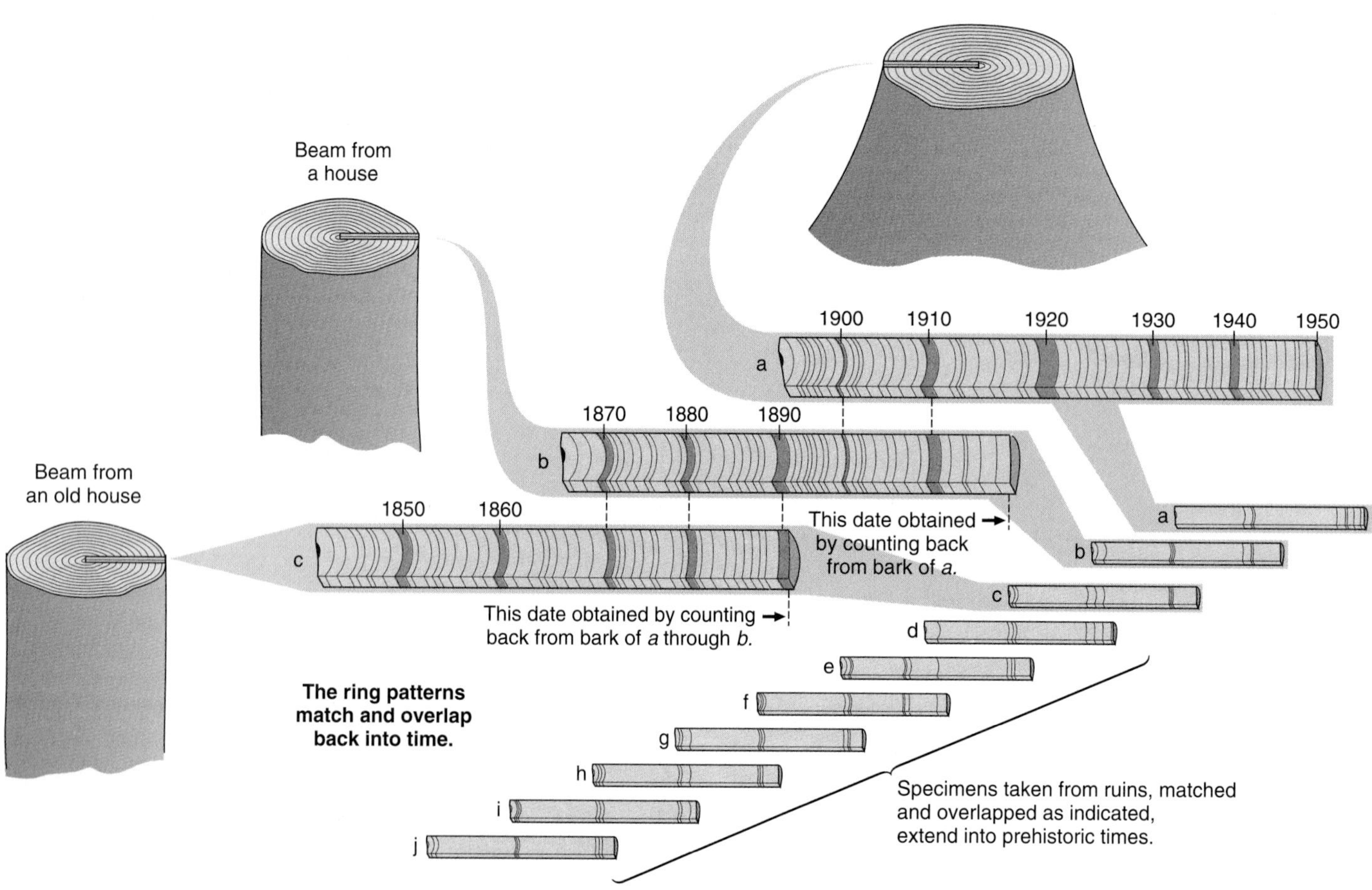

FIGURE 27.26

Researchers infer past climate patterns by studying tree ring patterns. To establish a chronology, dendrochronologists overlap patterns. In the study depicted here, a tree cut down in 1950 establishes a starting date. This tree's rings are compared to those of a tree cut down and used as a wooden beam in a newer house and then to those of a tree used as a beam in an older house. This tree ring data begins shortly after 1840 and extends until 1950.

Cork Cambium

The cork cambium is the lateral meristem that produces the periderm. The cork cambium produces cork cells to the outside; it produces phelloderm to the inside. **Cork cells** are waxy, densely packed cells covering the surfaces of mature stems and roots. They are dead at maturity and form waterproof, insulating layers that protect plants. Areas of loosely packed cells penetrate cork layers and enable gas exchange to occur. **Phelloderm** consists of living parenchyma cells, which may be photosynthetic and store nutrients. A familiar example of cork is the "skin" of a potato. The cork used to stopper wine bottles comes from a cork oak tree that grows in the Mediterranean. Every 10 years, harvesters remove much of the cork cambium and cork, which grows back.

Ninety percent of a typical tree is secondary xylem. U.S. forests produce 18.6 million cubic meters of secondary growth per day. We can thank the secondary growth of plants for such diverse products as rubber, chewing gum, turpentine, cardboard, rayon, synthetic cattle food, and ice cream fillers.

Mastering Concepts

1. How does secondary growth affect the plant body?
2. What is the function of vascular cambium?
3. How are different regions of wood adaptive?
4. What are the components of bark?

SUMMARY

1. The tissues of a flowering plant are **meristems, ground tissue, dermal tissue,** and **vascular tissue,** which includes **phloem** and **xylem.** The plant body consists of a **shoot** and a **root.**

2. Meristems are localized collections of cells that divide throughout the life of the plant. **Apical meristems** located at the plant's tips provide **primary growth,** and **lateral meristems** add girth, or **secondary growth.**

3. Most of the plant body is ground tissue. It includes **parenchyma** cells, which can divide and store substances.

Tree Rings Reveal the Past~

continued from page 534.

When researchers can align overlaps in the patterns among several trees, tree ring data can span many years. As long as one of the trees has a ring next to the vascular cambium, so that the dendrochronologist can determine the most recent year, such comparisons can yield a very accurate "master chronology." For example, the oldest known living tree is a bristlecone pine (*Pinus longaeva*) growing in the White Mountains of California. It is 4,906 years old. Combining its tree ring information with information from other trees in the area has provided rainfall data going back 8,200 years!

The juniper remains from the Midas Mound in ancient Turkey, however, had no link to the present, and so the Cornell researchers had a "floating" dendrochronology, a sequence of events with no anchoring end. But the researchers used two clever approaches to assign absolute dates.

First, in a technique called wiggle-matching, the researchers correlated irregularities in the tree ring data to irregularities in radioactive carbon dating and established an approximate date. Then they aligned an unusual wideness in the tree rings to similar anomalies in patterns from trees in the United States and Europe, which had been linked to a volcanic eruption on the island of Thera, near Crete, in 1628 B.C. With the new tree ring data, anthropologists now have a time frame in which to place objects found with the wooden remains in ancient Turkey. Tree rings may have much to tell us about early history.

Chlorenchyma is a type of parenchyma that photosynthesizes. **Collenchyma** supports growing shoots and **sclerenchyma** supports nonliving plant parts. Sclerenchyma includes **sclereids,** which form hard coverings, and **fibers,** which form strands.

4. Dermal tissue includes the **epidermis,** a single cell layer covering the plant, and the **cuticle,** which the epidermis secretes. Gas and water exchange occur through epidermal pores (stomata) and much water is lost through **transpiration.** **Trichomes** are epidermal outgrowths that deter predators.

5. Vascular tissue is an adaptation to multicellularity and life on land. Xylem transports water and dissolved minerals from roots upwards. Xylem cells are elongated with thick walls and are dead. They include the long, narrow, less-specialized **tracheids** and the more specialized, barrel-shaped **vessel elements.**

6. Phloem transports dissolved carbohydrates throughout a plant and includes **sieve cells** and the more specialized **sieve tube members.** Pores of sieve tube member cells cluster at **sieve plates,** allowing more efficient nutrient transport. **Companion cells** help transfer carbohydrates.

7. A stem is the central axis of the shoot and consists of **nodes,** where leaves attach, and **internodes** between leaves, where the stem elongates. **Vascular bundles** in the stem contain xylem and phloem, which are scattered in **monocots** and form a ring in **dicots.** Between a stem's epidermis and vascular tissue lies the **cortex,** made of ground tissue. **Pith** is storage ground tissue in the center of a stem. Stem modifications include **stolons, thorns, succulent stems, tendrils,** and **tubers.**

8. Simple leaves have undivided blades, and **compound** leaves form leaflets, which may be **pinnate** (with a central axis) or **palmate** (extend from a common point). Leaf organization may be **alternate, opposite,** or **whorled.** Leaves are the sites of photosynthesis.

9. Leaf epidermis is tightly packed, transparent, and nonphotosynthetic. Veins may be in either **netted** or **parallel** formation. Leaf ground tissue includes **palisade** and **spongy mesophyll cells.** Leaf modifications include **tendrils, spines, bracts, storage leaves, insect-trapping leaves,** and **cotyledons.** Leaves are shed from an **abscission zone** in response to environmental cues.

10. Roots absorb water and dissolved minerals. A plant's first root is the **radicle. Taproot systems** have a large, persistent major root, whereas **fibrous root systems** are shallow, branched, and shorter-lived. A **root cap** protects the tip as an apical meristem replaces cells lost during rapid extension. The **subapical region** behind the root cap is divided into three areas: the **zones of cell division, cell elongation,** and **cell differentiation.** This last zone includes root hairs, epidermis, and the cortex. The cortex consists of **hypodermis,** storage parenchyma, and **endodermis.** Some roots are specialized for storage or adapted to low-oxygen environments.

11. Secondary tissues increase a tall plant's girth and enable plants to reach sunlight. Two lateral meristems, **vascular cambium** and **cork cambium,** each produce secondary xylem and phloem to provide support. The vascular cambium forms a tree's wood, while cork cambium produces part of its **bark.**

TO REVIEW . . .

1. Which tissue provides a plant's unceasing growth?

2. Which plant tissue is the most abundant?

3. What are the functions of the following substances?
a. cutin
b. mucigel
c. suberin

4. How are the two types of xylem similar to the two types of phloem?

5. List two ways that a plant's anatomy is adaptive to existence on land.

6. Cite two ways that leaves maximize surface area.

7. Is the trunk of a banana plant a stem or leaves? How might you demonstrate which type of structure it is?

8. What are some of the ways in which we classify leaves?

9. Cite a stem specialization and a leaf specialization that provide protection.

10. Corn is a monocot and cucumber a dicot. How do these plants differ in the following?

- **a.** stem structure
- **b.** leaf venation
- **c.** root organization

TO THINK ABOUT . . .

1. If an overanxious tomato picker tears off some tomato plant stems and leaves, the plant regrows these parts in a few weeks. Which tissue type is responsible for this regrowth?

2. What plant structures are adaptations to conserve and transport water?

3. Why can phloem transport materials in all directions, while xylem can transport materials upward only?

4. How does a tree "know" when to shed its leaves?

5. How can roots grow in compact soils, such as clays?

6. What would happen (or not happen) if a plant's quiescent centers were destroyed?

7. Many drugs used to treat cancer are derived from plant compounds. Ironically, we are destroying many of the habitats where such plants live. Suggest ways that researchers might catalog and preserve plant compounds that could be useful drugs.

TO LEARN MORE . . .

References and Resources*

Jacoby, Gordon C., et al. August 9, 1996. Mongolian tree rings and 20th-century warming. *Science,* vol. 273. Tree rings provide evidence of periods of global warming.

Moore, Randy, and W. Dennis Clark. 1994. Botany. Dubuque, Iowa: William C. Brown Publishers. A detailed yet easy-to-read look at the basic parts of the plant body.

Niklas, Karl J. February 1996. How to build a tree. *Natural History.* Trees give architects tips on how to construct tall buildings.

Renfrew, Colin. June 27, 1996. Kings, tree rings and the Old World. *Nature,* vol. 381. Tree ring dating supplements historical information.

Roach, Mary. June 1996. Bamboo solution. *Discover.* Bamboo has many uses, but it is not strong enough to support a building greater than one story.

Steudle, Ernst. December 14, 1995. Trees under tension. *Science,* vol. 378. Proof of a century-old hypothesis—that water's properties enable it to be drawn to a tree's top.

**Life*'s website features frequently updated references.

28 Plant Life Cycles

The opium poppy is the source of powerful painkillers called morphine alkaloids, which people have used for at least 3,500 years. Morphine comes from the green ovary capsule, which protects the growing seeds. The capsule becomes visible 80 days after planting, when the petals fall off. A whitish fluid within the capsule contains the active alkaloid. Opium processors collect this juice and dry it in the air to form a black semisolid, which is then used to manufacture narcotic drugs.

chapter outline

Plant Reproduction and the Wandering Wasp 556

Asexual Reproduction 556

Alternation of Generations 557

The Pine Life Cycle 560

Flowering Plant Life Cycle 563

Flower Structure
Gamete Formation
Pollination
Fertilization
Seed Development
Seed Dormancy
Fruit Formation
Fruit and Seed Dispersal
Seed Germination
Plant Development
A Comparison of Gymnosperm and Angiosperm Life Cycles

Imperiled Pollinators

All life on earth depends, ultimately, on plants, because of photosynthesis. Yet animals make much of plant life possible by transferring pollen (male sex cells). Bees are particularly proficient pollinators and may store pollen from dozens of plant species.

European honeybees (*Apis mellifera*) pollinate many flowering plant species in the United States. Europeans introduced these bees in Jamestown, Virginia, in 1621. Because honeybees can use nectar and pollen from a variety of plants in a variety of habitats, they rapidly displaced native bee species. They reached a population pinnacle of nearly 6 million colonies in 1947.

Several adaptations enabled honeybees to flourish. With their superb senses they spot rare flowers peeking out of otherwise barren landscapes more readily than can other pollinators. Because of their eclectic tasks, honeybees flock to many angiosperm species, even in disrupted areas, while other bees look for specific plants, satisfied with mustard or mistletoe, daisies or lilies, and many others. But the honeybees' wide-ranging tastes make them less-than-efficient pollinators, because they often deposit pollen from one species on plants of another species. Still, the large numbers of honeybees maintain many flowering plant populations.

Clearly, honeybees are important for reproduction in both crops and wild populations of flowering plants. But bee populations worldwide are plummeting.

to be continued . . .

Plant Reproduction and the Wandering Wasp

Newly emerged from his cocoon, the male wasp seeks a mate, although the females will not emerge for another week. The male wasp flies among orchid plants of genus *Ophrys*. Suddenly, he becomes intensely excited: the unmistakable fragrance of a sexually receptive female wasp wafts from the orchid flowers! The sexually stimulated wasp approaches the blooms more closely. To his eyes, which can discern ultraviolet light invisible to humans, one petal of each flower looks like a female wasp with eyes, antennae, and wings.

The male alights on the petal, which not only looks and smells like a female but whose fuzzy surface feels like a potential mate. He begins the motions of copulation, and in the process, ends up with small packets of pollen on his head. The agitated male soon moves to another flower. He tries again to mate with a petal, and in doing so, deposits some of the first flower's pollen (fig. 28.1).

The behavior of the male wasp seeking a mate and finding a flower may seem to waste biological energy, but it is an amazingly efficient mechanism for delivering pollen—the male sex cells of flowering plants—from one plant to another. Sexual reproduction in orchids and other flowering plants is intimately linked to certain animal behaviors. Flowers are complex biological structures that facilitate sexual reproduction, and thereby the recombination of genetic material, in certain plants. Plants can also reproduce asexually.

FIGURE 28.1

A male wasp is attracted to this orchid because, to him, it resembles a female wasp. The wasp facilitates plant reproduction by transferring pollen from flower to flower.

Asexual Reproduction

Many plants reproduce asexually and form new individuals by mitotic cell division. Asexual reproduction does not involve meiosis, gametes, and fertilization, but it is still a complex biological process that we do not fully understand. In asexual reproduction, a parent organism produces progeny that are genetically identical to it and to each other—they are **clones.**

Asexual reproduction is advantageous when environmental conditions are stable and plants are well adapted to their unchanging surroundings. Why risk losing a favorable combination of genes when asexual reproduction will produce clones well suited to the environment? Perhaps the most stable and predictable environment is a laboratory, where researchers can control conditions. Biology in Action 28.1 describes several biotechnologies that are based on asexual reproduction in plants.

Plants often reproduce vegetatively (asexually) by forming new plants from portions of their roots, stems, or leaves (fig. 28.2). For example, adventitious buds form on the roots of cherry, pear, apple, and black locust plants, and when they sprout, aerial shoots grow upward. These shoots are called "suckers" because they "suck" nutrients from the parent plant. If separated from the par-

(a)

(b)

(c)

FIGURE 28.2

Vegetative propagation takes many forms. New shoots can grow from underground, as in bamboo (*a*), or new plantlets can form on leaves, as in a fern (*b*). In the laboratory, greenhouse, or kitchen window, cuttings can grow roots and even flower in water or in a growth medium, as this pussy willow has grown roots in water (*c*).

ent plant, suckers become new individuals. Sometimes, however, the new plants remain attached to the parent and produce huge individuals. For example, in the Wasatch Mountains of Utah, an aspen tree named *Pando* (Latin for "I spread") covers more than 100 acres of land! That plant, produced by asexual reproduction, consists of 47,000 tree trunks (each with an ordinary tree's usual complement of leaves and branches) and weighs more than 13 million pounds, making it the world's most massive organism.

A few plants use leaves to reproduce asexually. When the leaves of a walking fern or maternity plant lie on a moist surface, roots and shoots develop at their edges. These plantlets become new individuals when the parent's leaves shed them.

Mastering Concepts

1. What is the function of flowers?
2. When are sexual and asexual reproduction each adaptive?
3. How does a plant reproduce asexually?

Alternation of Generations

The life cycle of a plant differs from that of an animal. In sexually reproducing organisms, special haploid cells from two individuals combine to form a diploid offspring. In sexually reproducing plants, however, a complete individual life cycle is an alternation between a diploid and a haploid stage.

The diploid generation, or sporophyte, produces haploid spores through meiosis. Haploid spores divide mitotically to produce a multicellular haploid individual called the gametophyte. Eventually, the gametophyte produces haploid gametes—eggs and sperm—which fuse to form a zygote. The zygote grows and develops, and its cells divide, becoming a sporophyte, and the cycle begins anew.

In most vascular plants, gametophytes produce eggs and sperm. Because the female egg-producing gametophytes are often larger than the male sperm-producing gametophytes, the female structures are called **megagametophytes,** and the male structures are called **microgametophytes.** Male and female gametophytes arise from two types of spores called **megaspores** and **microspores.** These spores form in two different types of structures, **megasporangia** and **microsporangia.**

The gametophyte stage dominates some plant life cycles, and the sporophyte stage dominates others (figs. 28.3 and 28.4). The gametophyte stage is the more obvious phase in bryophytes such as mosses, whereas vascular plants have a reduced gametophyte phase and a dominant sporophyte phase. Familiar vascular plants such as ferns, trees, and grasses are sporophytes. We now take a closer look at sexual reproduction in the gymnosperms (naked seed plants) and angiosperms (flowering plants).

Mastering Concepts

1. What is alternation of generations?
2. How do female gametophytes usually differ from male gametophytes?
3. Which generation is dominant in bryophytes, and which generation is dominant in vascular plants?

biology in 28.1 action

New Routes to Plant Reproduction

Traditional plant breeding introduces new varieties through sexual reproduction: pollen carrying genes for certain traits fertilizes egg cells carrying a different set of traits. Rather than beginning with sperm and egg, which have a half set of genetic instructions each, most plant biotechnologies start with somatic cells from nonsexual parts of the plant, including leaves, stems, or embryos. These cells contain a complete set of genetic instructions. Plants regenerated from somatic cells do not have the unpredictable mixture of characteristics of sexually derived plants because the somatic cells are clones.

Protoplast fusion and cell culture are plant biotechnologies that use somatic cells.

Protoplast Fusion

Protoplast fusion combines cells from different species and regenerates a mature plant hybrid. Cells may fuse spontaneously or after treatment with digestive enzymes, polyethylene glycol (an antifreeze), electricity, or a laser (fig. 28.A). A single gram (about 1/28 of an ounce) of plant tissue can yield up to 4 million protoplasts. A plant regenerated from protoplasts of different species is called a somatic hybrid.

Not all protoplast fusions yield mature plants, and not all that do are agriculturally useful. A protoplast fusion of radish and cabbage yields a plant that grows radish leaves and cabbage roots! Protoplast fusion is more successful when the parent cells come from closely related species. Consider fusion of a protoplast from a potato plant that is normally killed by the herbicide triazine with a protoplast from wild black nightshade, a relative that resists the herbicide. The resulting hybrid grows well in triazine-treated soil.

Cell Culture

When tiny pieces of plant tissue called explants are nurtured in a dish with nutrients and hormones, the cells lose their special characteristics after a few days and form a white lump called a callus. The callus grows, its cells dividing, for a few weeks. Then certain callus cells develop into either a tiny plantlet with shoots and roots or a tiny embryo. Such an embryo is called a somatic embryo because it derives from somatic, rather than sexual, tissue. Researchers are not sure how or why the calli of some species give rise to somatic embryos, others form plantlets, and still others never develop beyond a lump of tissue (fig. 28.B). Callus growth is unique

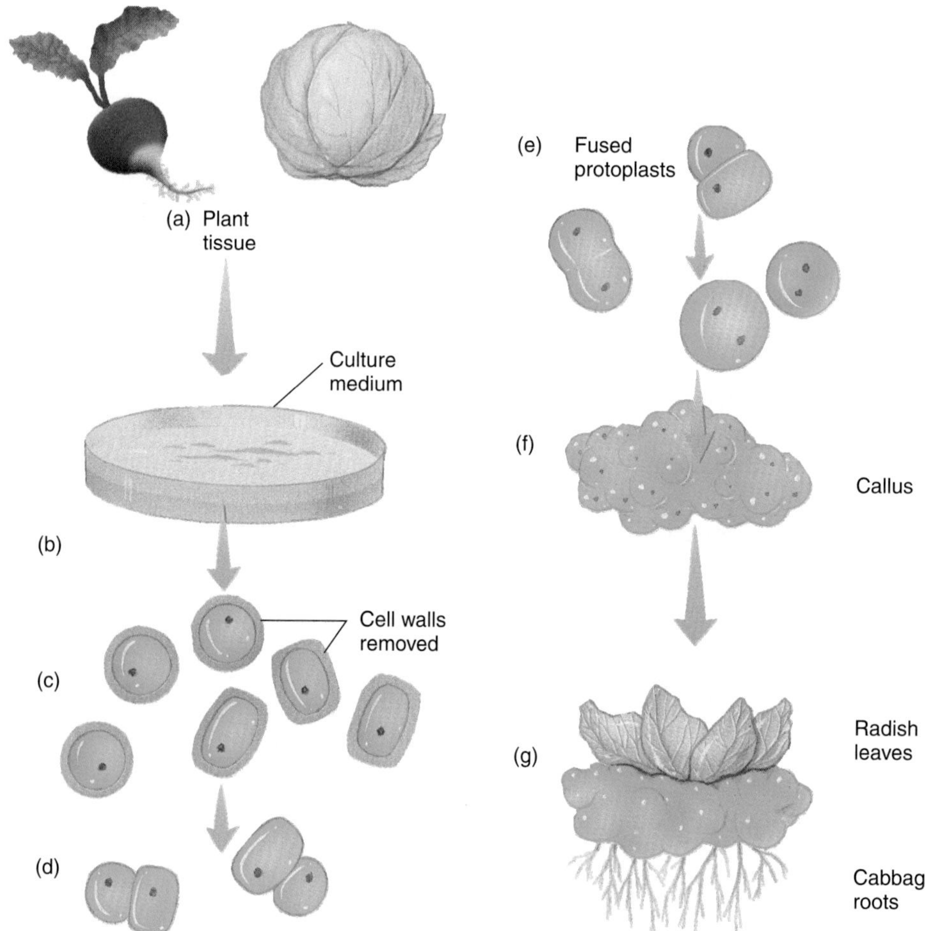

FIGURE 28.A
Protoplast fusion—making one cell from two. Protoplast fusion can yield plants with the traits of two species. Tissues from different plants (*a*) are placed in tissue culture (*b*), and the cell walls are dissolved (*c*). The resulting protoplasts are mixed and stimulated to fuse (*d*). Among the fusion products may be hybrid protoplasts (*e*), which are selected and grown into an unspecialized lump of tissue called a callus (*f*). Sometimes, plants regenerated from the callus exhibit new combinations of traits (*g*).

to plants. The human equivalent would be a cultured skin cell multiplying into a blob of unspecialized tissue and then sprouting tiny humans or human embryos!

Embryos or plantlets grown from a single callus are usually clones, but sometimes they differ from each other because certain callus cells mutate. Researchers can to some extent control whether somatic embryos or plantlets are identical or variant by altering the nutrients and hormones in the culture medium. Either clones or variants may have uses.

Identical cultured somatic embryos are the basis of artificial seed technology. A natural seed is a plant embryo and its food supply packaged in a protective shell. An artificial seed is a somatic embryo suspended in a transparent gel containing nutrients and hormones. A biodegradable outer polymer coat provides protection and shape. Researchers can package somatic embryos with pesticides, fertilizer, nitrogen-fixing bacteria, and even microscopic parasite-destroying worms! So far, however, artificial seeds are too costly to be practical.

The occasional variant plantlets that grow from a callus are called somaclonal variants because they derive from a single somatic cell and were therefore originally clones. Normally, new plant varieties that arise from spontaneous mutation are literally one in a million. Somaclonal variants occur much more frequently. In one experiment, researchers chopped up a tomato leaf and cultured it into a callus and regenerated 230 plantlets. Thirteen of the plantlets were new variants—one had light orange fruits; two lacked joints between stems and tomatoes, which makes harvesting easier; and two others had a high solids content, which is important in food processing. Other intriguing somaclonal variants include stringless celery, crunchier carrots, and popcorn with built-in buttery taste.

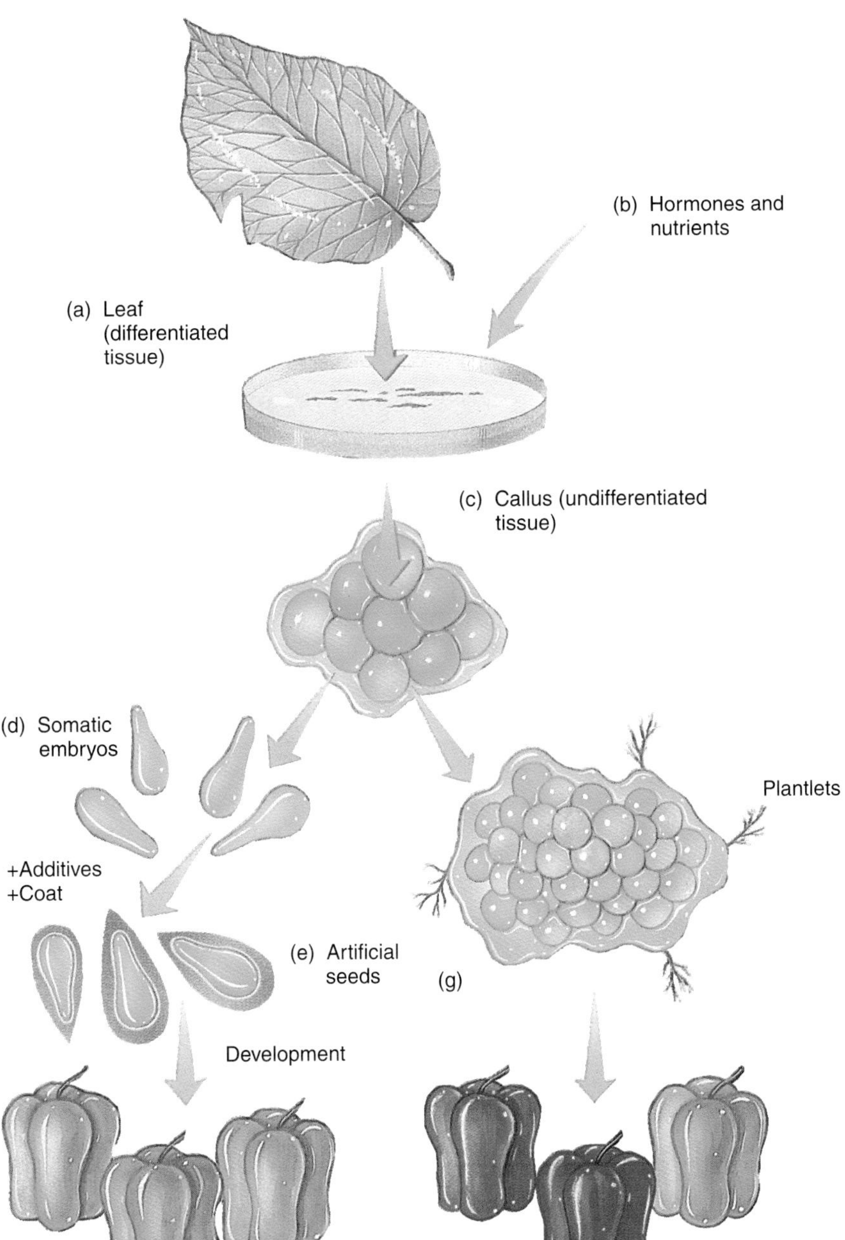

FIGURE 28.B
The fate of a callus—embryos or plantlets. New plants grown from cell culture are valuable either for their uniformity or their variation. A leaf explant (*a*) is cultured with hormones and nutrients in a dish (*b*), giving rise to a callus (*c*). Under certain conditions, the callus produces somatic embryos (*d*), which when encapsulated become artificial seeds (*e*). When planted, artificial seeds grow into genetically identical plants, producing a uniform crop (*f*). Under other conditions, callus cells undergo somatic mutation, producing somaclonal variant plantlets (*g*), which may yield new crop varieties (*h*).

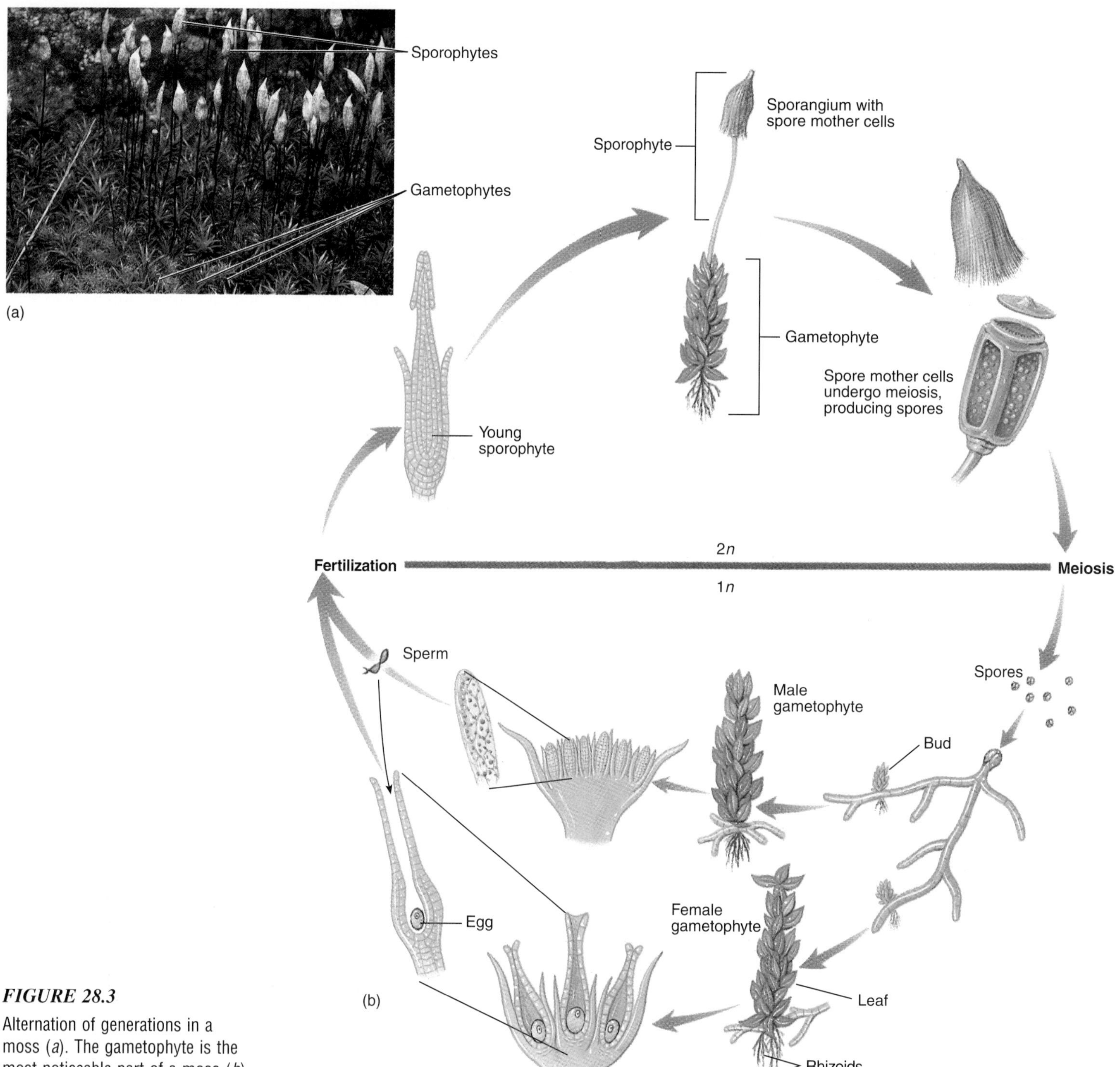

FIGURE 28.3

Alternation of generations in a moss (*a*). The gametophyte is the most noticeable part of a moss (*b*).

The Pine Life Cycle

Gymnosperms are vascular plants with unprotected seeds. Recall from chapter 25 that conifers are the best-known group of gymnosperms and include pines, spruces, firs, and redwoods.

Cones are the reproductive structures of pines. Large female cones bear two ovules (megasporangia) on the upper surface of each **scale.** Through meiosis, each ovule produces four haploid megaspores, three that degenerate and one that continues to develop into a female gametophyte. Over many months, the female gametophyte undergoes mitosis. Finally, two to six structures called **archegonia** form, each of which houses an egg ready to be fertilized (fig. 28.5).

Small male cones bear pairs of microsporangia on thin, delicate scales. These microsporangia produce microspores through meiosis, and each microspore eventually becomes a four-celled **pollen grain** (microgametophyte). The microsporangia burst, releasing millions of winged pollen grains. Tapping a male cone in the spring releases a halo of pollen (fig. 28.6). Most of these pollen grains, however, never reach a female cone.

Pollination occurs when airborne pollen grains drift between the scales of female cones and adhere to drops of a sticky secretion. Female cones are above male cones on most species, which makes it unlikely that pollen from the same tree will drift upwards to land on female cones on the same tree. This is an adap-

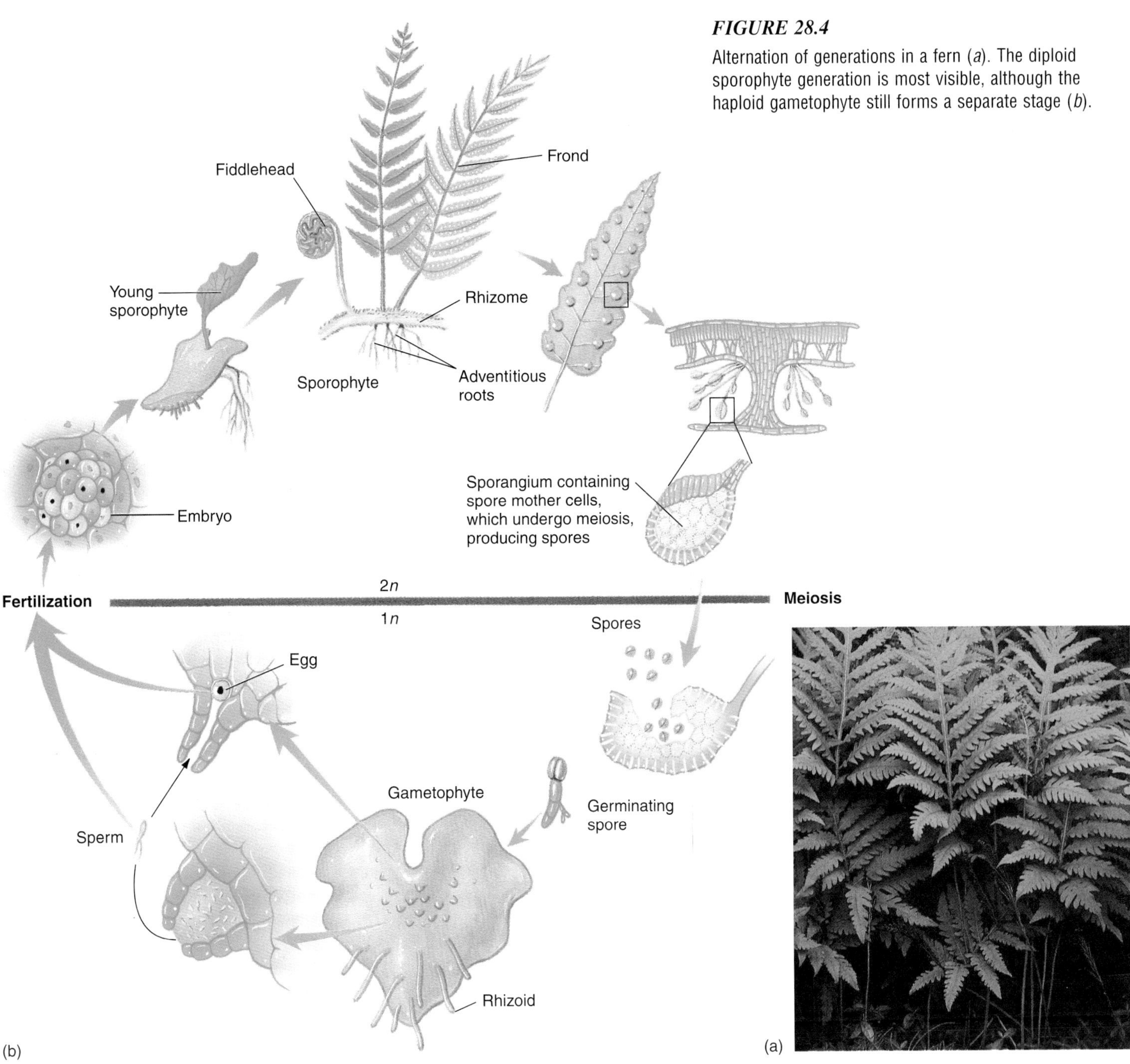

FIGURE 28.4

Alternation of generations in a fern (*a*). The diploid sporophyte generation is most visible, although the haploid gametophyte still forms a separate stage (*b*).

tation that encourages outcrossing (mating between different individuals), which fosters new combinations of genes in the next generation.

After pollination occurs, the cone scales grow together, and a structure called a **pollen tube** begins growing through the ovule toward the egg. But the pollen is not yet mature—before the pollen tube reaches the egg, the pollen grain must divide twice more to become a mature, six-celled microgametophyte. Two of six cells become active sperm cells, and one of them fertilizes the egg cell. The whole process is so slow that fertilization occurs about 15 months after pollination.

Within the ovule, the haploid tissue of the megagametophyte nourishes the developing embryo. Following a period of metabolic activity, the embryo becomes dormant and the ovule develops a tough, protective seed coat. It may remain in this state for another year. Eventually, the ovule is shed as a **seed.** If conditions are favorable, the seed germinates, giving rise to a new tree.

Mastering Concepts

1. Describe the steps in the development of the female and male cells in the pine life cycle.
2. What happens during and after pollination in gymnosperms?
3. Why is it adaptive for male cones to be below female cones on a tree?

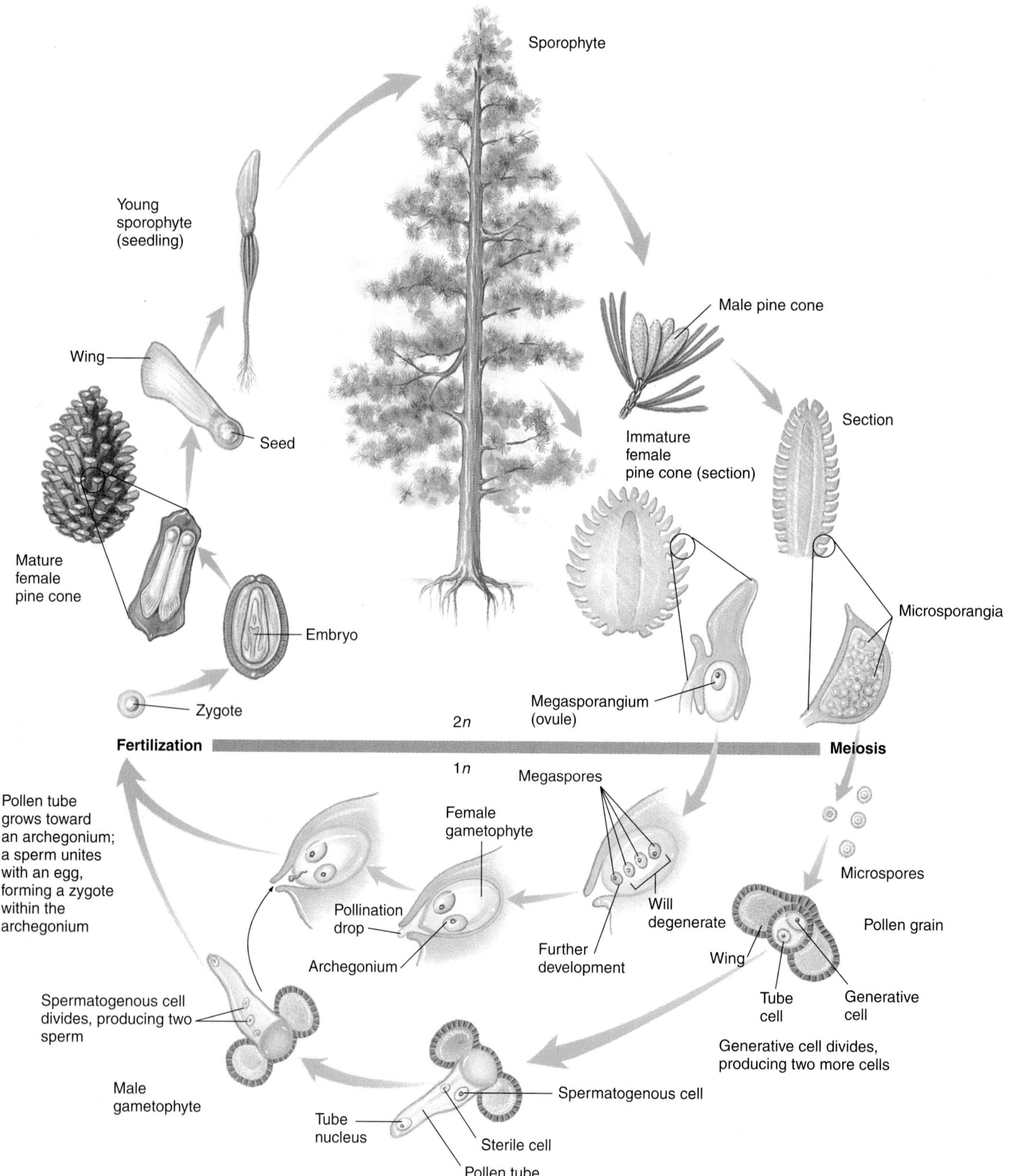

FIGURE 28.5

Pine life cycle. Notice the alternation of generations. Pines have no flowers, but they do produce male and female cones, which in turn produce pollen and eggs. The sporophyte stage is dominant (a huge tree), whereas the gametophyte stage consists of only a few cells.

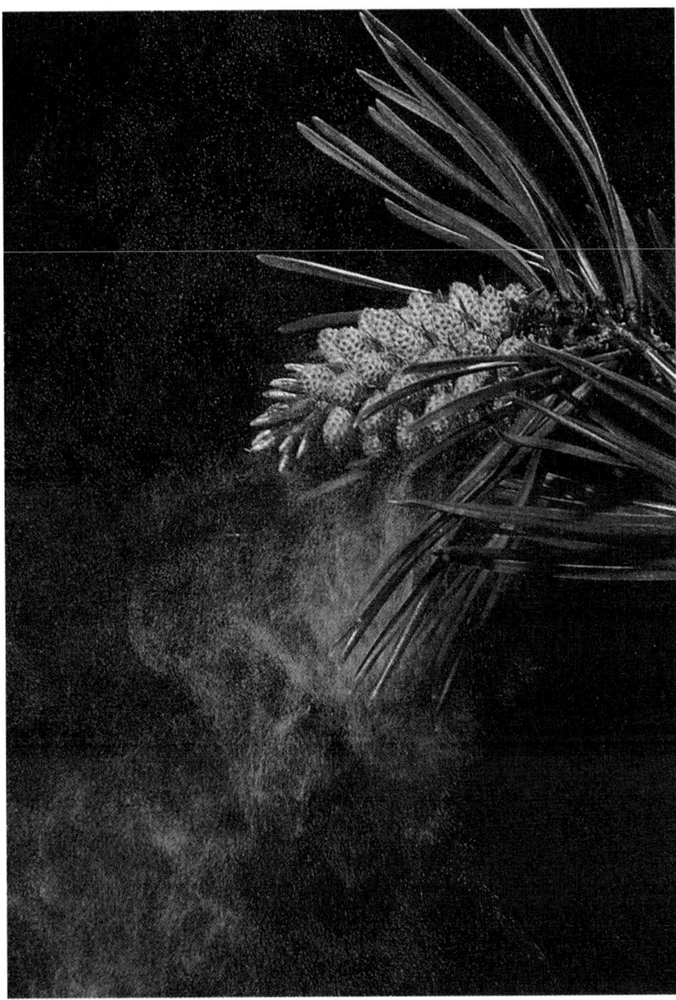

FIGURE 28.6

Pollen blows from male cones.

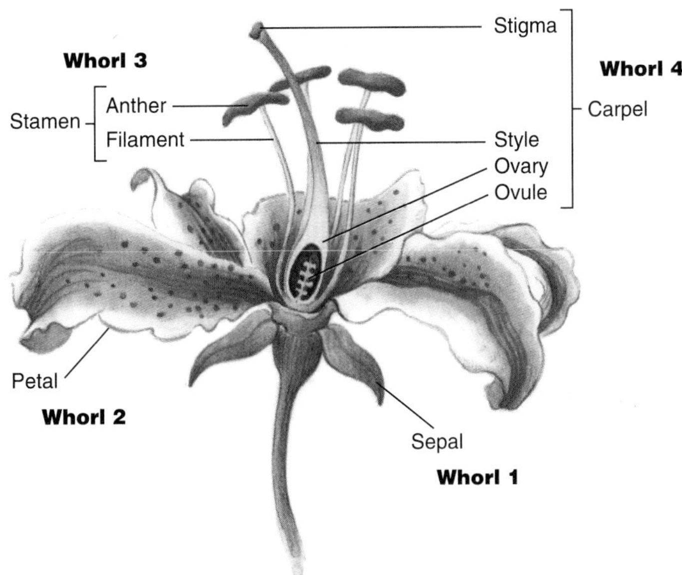

FIGURE 28.7

A flower is made up of four levels, or whorls. The outermost whorl (whorl 1) is sepals, the next whorl (whorl 2) is petals, the next whorl (whorl 3) is stamens, and the innermost whorl (whorl 4) is the carpel.

Flowering Plant Life Cycle

Angiosperms are the dominant group of vascular plants. Flowers are the reproductive organs of angiosperms.

Flower Structure

Complete flowers have four types of structures, all of which are modified leaves. Each type of structure forms a whorl, or circle, around the end of the flower stalk (fig. 28.7). The **calyx** is the outermost whorl of a flower. It consists of green, leaflike **sepals,** which enclose and protect the inner floral parts. Inside the calyx is the **corolla,** which is a whorl of **petals.** Large, colorful petals attract pollinators to the flower. The calyx and the corolla do not play a direct role in sexual reproduction and are therefore considered accessory parts of the flower.

The two innermost whorls of a flower, the **androecium** and the **gynoecium,** are essential for sexual reproduction. The whorl within the corolla, the androecium, consists of male reproductive structures called **stamens,** which are stalklike filaments that bear pollen-producing oval bodies called **anthers** at their tips. The whorl at the center of a flower, the gynoecium, consists of the female reproductive structures, or **pistil.** The gynoecium forms from one or more **carpels,** which are leaflike structures enclosing the **ovules.** The bases of carpels and their enclosed ovules make up the **ovary.** The upper part of the carpels are stalklike **styles** that bear **stigmas** at their tips. Stigmas receive pollen.

Biology in Action 28.2 describes how mutant genes misdirect development so that flowers have their parts in the wrong places.

Gamete Formation

Both male and female reproductive organs produce haploid gametes. Figure 28.8 illustrates the development and meeting of both gamete types.

Microspores, the male gametes, form in the anthers at the tips of the stamens. Each anther contains four microsporangia called **pollen sacs.** Pollen sacs contain **microspore mother cells**, which each divide meiotically to produce four haploid microspores. Each microspore then divides mitotically and produces a two-celled structure containing a haploid **generative nucleus** and a haploid **tube nucleus**. A thick, resistant wall forms around each two-celled structure, forming male microgametophytes more familiarly known as pollen grains. The generative cell divides to form two sperm cells, either before or after the pollen sac sheds the pollen grains. A mature pollen sac bursting open releases millions of pollen grains.

Megaspores, the female gametes, form in the ovary of the flower. Each ovule within the ovary is a megasporangium containing a **megaspore mother cell**, which divides meiotically to produce four haploid cells (megaspores) in most species. Three of these cells quickly disintegrate, leaving one large haploid megaspore. The

28.2 biology in action

Unraveling the Genetic Controls of Flower Formation

Flowers appear rather suddenly in the fossil record, from 120 to 150 million years ago. Such rapid evolutionary change can happen when genes that control development mutate—as we know well from the striking homeotic mutants of the fruit fly *Drosophila melanogaster* (see fig. 11.8), which have mixed-up body parts. Might similar mutations affect flowers?

Developmentally, flower formation is straightforward. Meristematic tissue in a stem tip differentiates into structures of one of a flower's four whorls—sepal, petal, stamen, or carpel. If single genes control decisions on such specialization, then mutations in those genes should produce flowers with parts in the wrong places but in predictable patterns. This is indeed the case.

Many plant geneticists work with the mustard weed *Arabidopsis thaliana.* In the early 1980s, California Institute of Technology researcher Elliot Meyerowitz alerted colleagues that he was looking for abnormal *Arabidopsis* flowers. He and his coworkers also mutagenized seeds to screen for mutant flowers. Soon, they had a collection of unusual flowers that fell into three categories. Instead of the normal sequence, from outside inwards, of sepals-petals-stamens-carpels (fig. 28C), the mutant flowers were either carpals-stamens-stamens-carpels, sepals-sepals-carpels-carpels, or sepals-petals-petals (fig. 28.D).

Geneticists like to figure out how genes interact to account for certain phenotypes. Meyerowitz came up with an elegant hypothesis to explain how a few single gene mutations could produce each of the three mutant phenotypes he observed. Here are his "rules," followed by explanations of how the mutant classes arise:

Rule 1: Three groups of genes control flower formation, A, B, and C. Each class of genes is expressed in certain parts of the flower:

- In sepals, "A" genes are expressed.
- In petals, "A" and "B" genes are expressed.
- In stamens, "B" and "C" genes are expressed.
- In carpels, "C" genes are expressed.

Put another way, "A" genes are expressed in the outer two whorls (sepals and petals), "B" genes are expressed in the middle two whorls (petals and stamens), and "C" genes are expressed in the inner two whorls (stamens and carpels).

Rule 2: When an "A" gene is inactive, a "C" gene replaces it.

Rule 3: When a "C" gene is inactive, an "A" gene replaces it.

After geneticists devise rules, they create mutants to see if the model accurately predicts the phenotype that arises. So Meyerowitz inactivated each class of gene to see what effect this had on the phenotype. The results supported his hypothesis.

Class A Mutations

(Remember that if an "A" gene is inactivated, a "C" gene takes over.)

- The first whorl (normally sepals) sees "C" instead of "A" and develops as carpels.
- The second whorl (normally petals) sees "C" and "B" instead of "A" and "B" and develops as stamens.
- The third whorl (normally stamens) still sees "B" and "C" and develops normally as stamens.
- The fourth whorl (normally carpels) still sees "C" and develops normally as carpels.

Carpels
Stamens

FIGURE 28.C
Two views of a wild-type (normal) flower of *Arabidopsis thaliana.*

The resulting flower whorl pattern is carpels-stamens-stamens-carpels. The flower lacks sepals and petals.

Class B Mutations

- The first whorl (normally sepals) still sees "A" and develops normally as sepals.
- The second whorl (normally petals) only sees "A" and therefore also forms sepals.
- The third whorl (normally stamens) sees only "C" instead of "B" and "C" and therefore develops as carpels.
- The fourth whorl (normally carpels) still sees only "C" and develops normally as carpels.

The resulting flower whorl pattern is sepals-sepals-carpels-carpels. The flower lacks petals and stamens.

Class C Mutations

(Remember that if a "C" gene is inactivated, an "A" gene takes over.)

- The first whorl (normally sepals) still sees "A" and develops normally as sepals.
- The second whorl (normally petals) still sees "A" and "B" and therefore forms petals.
- The third whorl (normally stamens) instead of "B" and "C" sees "B" and "A" and develops petals.
- The fourth whorl (normally carpels), for reasons unknown, does not develop.

Deletion of "C" also apparently lifts control on whorl formation. This mutant flower develops many whorls, in the pattern sepals-petals-petals.

Meyerowitz then made double mutants, inactivating two gene classes at a time. Again, his predictions of flower structure were accurate. Other researchers soon found that the same broken genetic rules underlie mutant flowers in other species. The picture is even more complex than explained here, because yet other genes control the A, B, and C genes.

Not surprisingly, the A, B and C genes encode transcription factors, which are proteins that activate other genes. Just as the homeotic genes that control animal bodies share a DNA sequence in common (the homeobox), so do the A, B, and C genes that control flower formation. Although the sequences in plants and animals are not the same, these two types of organisms do share a fundamental aspect of their development—control by a very few genes.

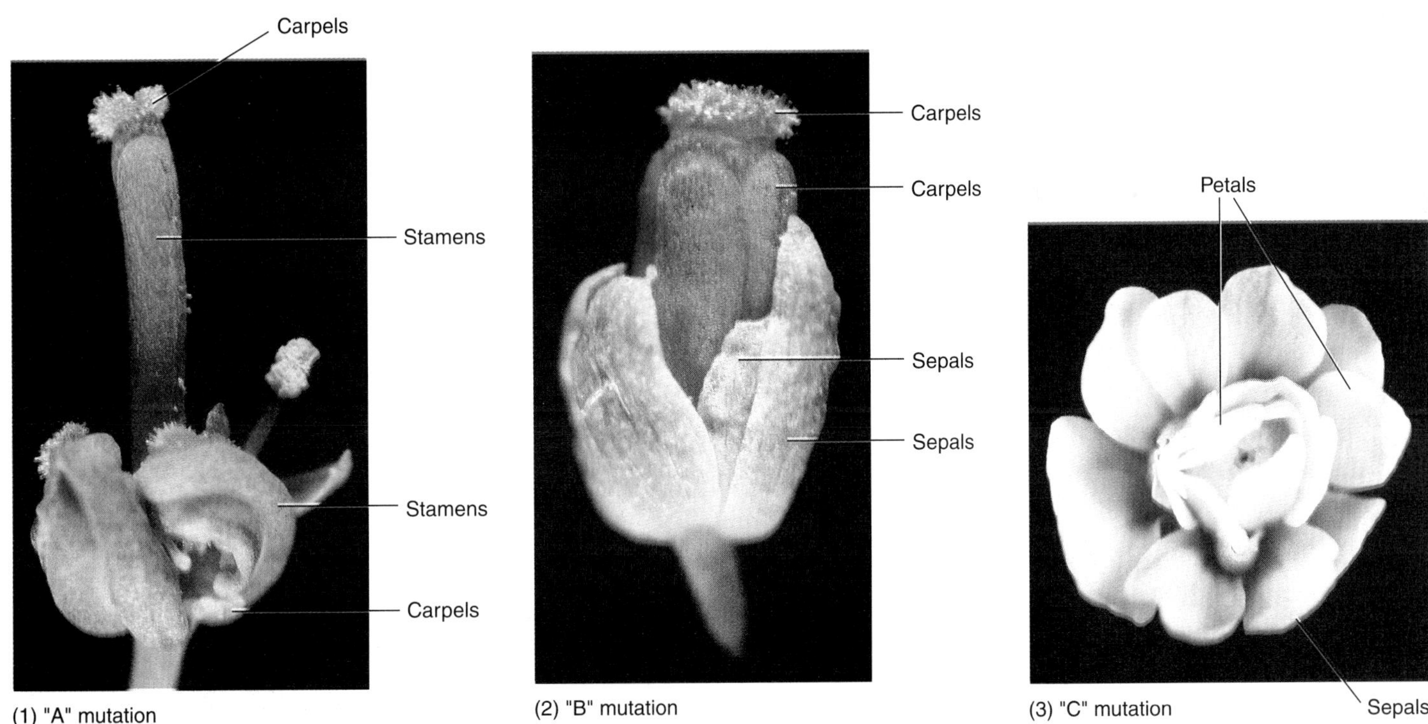

FIGURE 28.D
Mutant flowers. (*1*) A mutation in an "A" gene yields a flower that consists of carpels-stamens-stamens-carpels. (*2*) A mutation in a "B" gene yields a flower that consists of sepals-sepals-carpels-carpels. (*3*) A mutation in a "C" gene yields a flower that consists of several repeats of sepals-petals-petals.

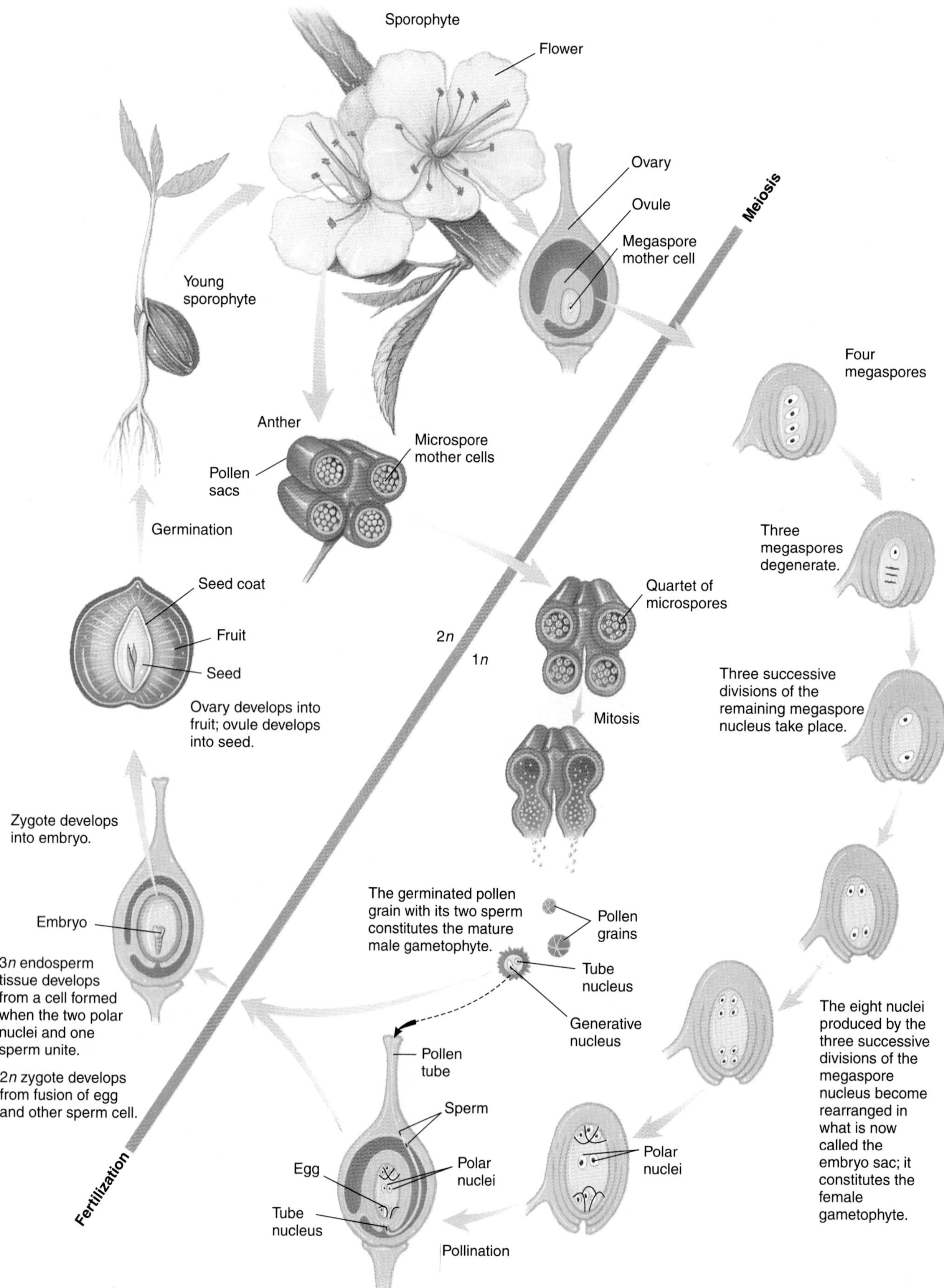

Sporophyte
Flower
Ovary
Ovule
Megaspore mother cell
Meiosis
Young sporophyte
Four megaspores
Anther
Microspore mother cells
Pollen sacs
Germination
Three megaspores degenerate.
Seed coat
Quartet of microspores
Fruit
2n
1n
Seed
Three successive divisions of the remaining megaspore nucleus take place.
Ovary develops into fruit; ovule develops into seed.
Mitosis
Zygote develops into embryo.
The germinated pollen grain with its two sperm constitutes the mature male gametophyte.
Pollen grains
Embryo
Tube nucleus
3n endosperm tissue develops from a cell formed when the two polar nuclei and one sperm unite.
Generative nucleus
The eight nuclei produced by the three successive divisions of the megaspore nucleus become rearranged in what is now called the embryo sac; it constitutes the female gametophyte.
Pollen tube
2n zygote develops from fusion of egg and other sperm cell.
Sperm
Polar nuclei
Egg
Polar nuclei
Fertilization
Tube nucleus
Pollination

FIGURE 28.8

Sperm and egg production and their place in the life cycle of a flowering plant.

(a)

(b)

(c)

FIGURE 28.9

Animal pollinators include insects, birds, bats, and many others that brush past flowering plants. Bumblebees (*a*) and hummingbirds (*b*) are efficient pollinators. Certain moths, such as the striped morning sphinx moth (*Hyles lineata*) (*c*), strikingly resemble hummingbirds, a case of convergent evolution reflecting similar adaptations to efficient pollination. The moth has a long tongue that it uses to gather nectar from flowers, whereas the hummingbird uses its beak.

megaspore undergoes three mitotic divisions, forming a female megagametophyte with eight nuclei but only seven cells—because one large cell has two nuclei called **polar nuclei.** In a mature megagametophyte, also called an **embryo sac**, one of the cells with a single nucleus is the egg.

Pollination

Pollination is the transfer of pollen from an anther to a receptive stigma. Some angiosperms are self-pollinating, meaning that pollen grains are transferred from the anther to the stigma of the same flower. Other angiosperms are outcrossing species. In these plants, pollen grains from one flower are carried to the stigma of another flower.

Self-fertilization reduces genetic variability and is more common among plants in tropical regions, where environmental conditions are relatively uniform. Outcrossing is crucial in a changing environment because it produces a genetically variable population that can adapt to changing conditions. Some angiosperms promote outcrossing by producing physically isolated anthers and stigmas. Other angiosperms lack the androecium or gynoecium, which makes self-fertilization impossible. The magnolia, one of the most primitive flowering plants, ensures cross-pollination by releasing pollen only after its eggs have been fertilized by pollen from other individuals.

Animals often transfer pollen from plant to plant. Common pollinators are insects, birds, and bats (fig. 28.9); rarer pollinators include geckos, gliding possums, lemurs, flying foxes, and even an occasional human hiker who picks up pollen by brushing against flowers and transports it to other plants of the same species. Many pollinators benefit from their association with plants—they obtain food in the form of pollen or nectar, they may seek shelter among petals, or they use the flower for a mating ground. Nectar is a sweet-tasting substance that has no known function in plants other than attracting pollinators.

Flower color, shape, and odor attract particular animals to particular flowers, such as when orchids draw male wasps by resembling female wasps. Some flowers produce heat and release their aromatic molecules as fragrances. Different types of insects have characteristic floral preferences. Bees are attracted to blue or yellow sweet-smelling blooms, whereas beetles respond to dull-colored flowers with spicier scents. Bees are the most common pollinators, accounting for 40,000 of the 130,000 to 200,000 pollinating species known worldwide, although, as the chapter opener points out, their numbers are diminishing. Bees initially locate food by its fragrance and then by its color. Because bees can detect ultraviolet light, bee-pollinated flowers often have markings that are visible only under this wavelength of light (fig. 28.10).

Birds are attracted to red flowers, a color that insects cannot distinguish. The structure of the flower almost ensures that a visiting bird will transport pollen. A hummingbird flitting about a California fuchsia plant, for example, sticks its long, thin beak into the flower. The bird catches its bill in a network of fine threads, and pollen grains rain down. With the pollen on its bill, the hummingbird ferries the cells to the next plant it visits.

Moths pollinate white or yellow heavily scented flowers, which are easy to locate at night, when these insects are most

FIGURE 28.10

Insects can "see" ultraviolet light, which makes these black-eyed Susans appear purple. To us, they appear yellow.

active. The flowers may have flat surfaces that their pollinators use as landing strips. In the tropics, where more flowers are open at night, bats are important pollinators.

Not all floral scents are pleasing to the human nose. In South Africa, the "carrion flowers" of stapelia plants smell remarkably like rotting flesh, a repulsive scent to humans but a highly attractive scent to flies. As if the stench were not sufficient to beckon the insects, the stapelia's leaves are wrinkled and reddish brown, resembling decayed meat.

Wind-pollinated angiosperms such as oaks, cottonwoods, ragweed, and grasses shed large quantities of pollen that breezes disperse. The wind, however, does not carry pollen very far. This is why wind-pollinated plants must grow closely together and manufacture abundant pollen. Wind pollination is also far less precise than animal pollination. The wind drops pollen where and when it slackens, whereas an animal delivers pollen directly to a particular type of plant. The flowers of wind-pollinated angiosperms are small, greenish, and odorless. Rather than invest energy in developing flowers, it is more efficient for these plants to invest energy in pollen production.

Fertilization

After a pollen grain lands on a stigma (pollination), it produces a growing pollen tube (see fig. 28.8). The pollen grain's two sperm cells enter the pollen tube as it grows through the tissues of the stigma and the style towards the ovary. When the pollen tube reaches an ovule, it discharges its two sperm cells into the embryo sac. One sperm nucleus fuses with the egg nucleus and forms a diploid zygote. After a series of cell divisions, the zygote becomes the embryo.

The second sperm cell nucleus fuses with the two polar nuclei and forms a triploid nucleus. Recall that a triploid cell has three complete sets of chromosomes. The triploid nucleus divides to form a tissue called **endosperm,** which is stored food for the developing embryo. Familiar endosperms are coconut milk and the fleshy part of a kernel of corn. Notice that the egg and the polar nuclei both are fertilized, a phenomenon termed double fertilization. The bottom and left-hand side of figure 28.8 depicts fertilization and subsequent development of the seed and fruit.

Seed Development

A seed is a dormant sporophyte within a tough protective coat. Immediately after fertilization, the ovule contains an embryo sac with a diploid zygote and a triploid endosperm nucleus, both of which are encased in layers of maternal tissue. Initially, the endosperm nucleus divides more rapidly than the zygote and forms a large mass of endosperm. The developing embryo forms cotyledons, or seed leaves. Angiosperms that have one cotyledon are monocots, and those with two cotyledons are dicots (figs. 28.11 and 28.12).

Further development in monocot and dicot seeds differs. In monocots, the cotyledon does not absorb the endosperm but transfers it to the nonleaf part of the embryo during germination. In many dicots, the cotyledons become thick and fleshy as they absorb the endosperm.

Apical meristems form early in embryonic development. Recall that these are regions of cellular division that will provide primary growth throughout the plant's entire life. The shoot apical meristem forms at the tip of the **epicotyl,** the stemlike region above the cotyledons. The root apical meristem differentiates near the tip of the embryonic root, or radicle. When one or more embryonic leaves form on the epicotyl, the epicotyl plus its young leaves is called a **plumule.** The stemlike region below the cotyledons is the **hypocotyl.**

In monocots, a sheathlike structure called the **coleoptile** covers the plumule. Also, the ovary wall remains attached to a monocot seed and forms a fruit. A fruit is a ripened ovary, which usually houscs a seed.

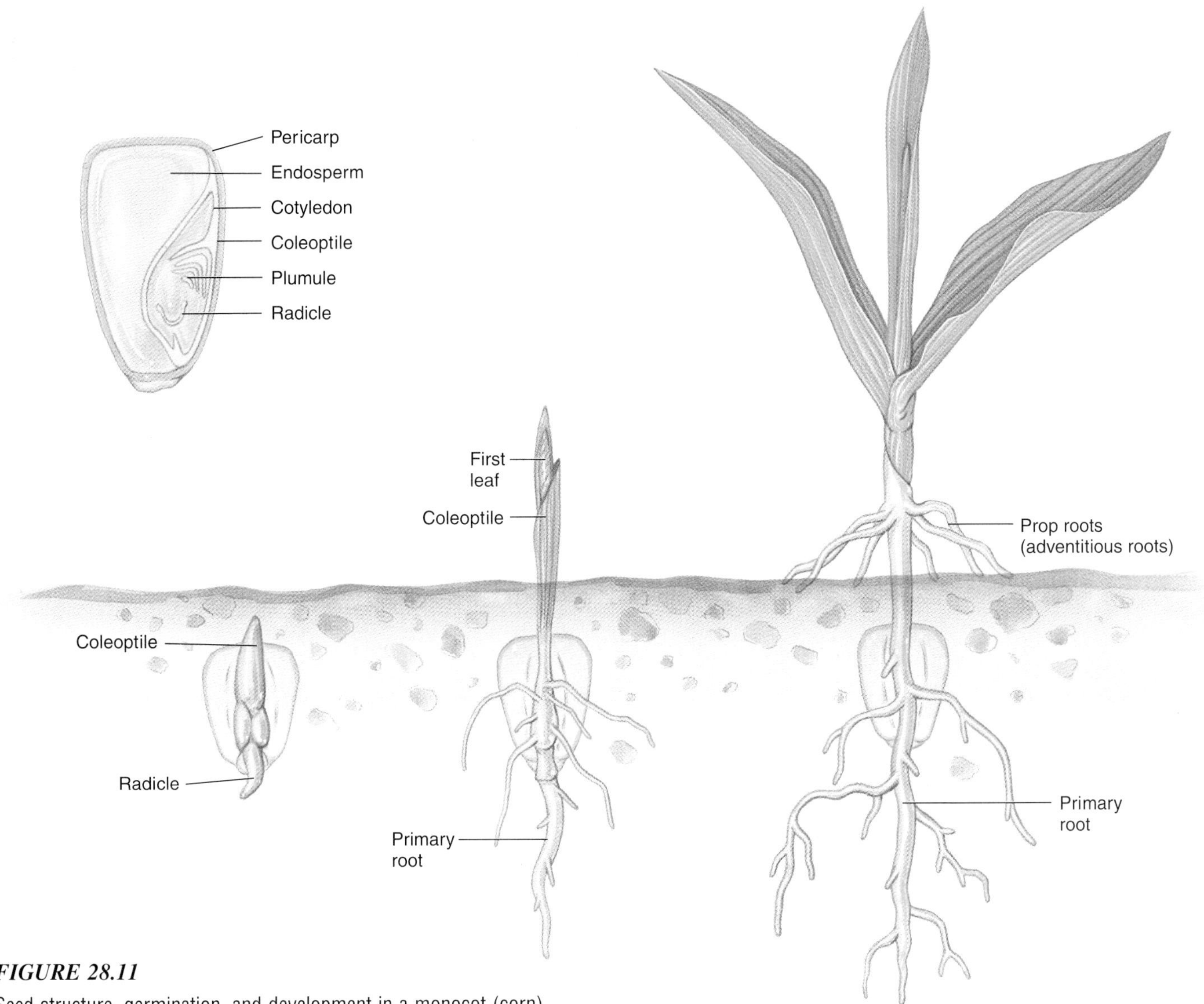

FIGURE 28.11
Seed structure, germination, and development in a monocot (corn).

Seed Dormancy

At a certain point in embryonic development, cell division and growth stop, and the embryo becomes dormant. A tough outer layer, the **seed coat,** protects the dormant plant embryo and its food supply. Together, the plant embryo, stored food, and seed coat make up the seed.

Why should a plant embryo simply stop growing? Seed dormancy is a crucial adaptation that enables seeds to postpone development when the environment is unfavorable, such as during a drought or frost. Some seeds can delay development for centuries! The oldest seed to germinate was a 1,288-year-old lotus seed from China, which germinated in 1996 in just 4 days. Favorable conditions trigger growth to resume when young plants are more likely to survive. Dormant seeds are also more likely to disperse into new environments.

Fruit Formation

A flower begins to change as seeds form. When a pollen tube begins growing, the stigma produces large amounts of **ethylene,** a simple organic molecule that is a plant hormone. (Recall that a hormone is a biochemical produced in one part of an organism that exerts a physiological effect on some other part of the organism.) Ethylene triggers senescence (aging) of the flower. Floral parts that are no longer vital—usually all parts except the ovary—wither and fall to the ground. Sometimes the ovary swells and develops into a fruit.

Fruits come in several varieties. A **drupe** is a fleshy fruit with a hard pit surrounding a single seed, such as a peach (fig. 28.13). A **berry** is fleshy and contains many seeds. Tomatoes, peppers, and eggplants are berries, although we mistakenly think of them as vegetables. Ironically, blackberries, raspberries, and strawberries are not true berries. They are aggregate fruits, which consist of several clustered small fruits. An apple is a **pome,** a fleshy fruit that develops mostly from tissues surrounding the ovary.

Plant hormones continue to influence fruit development. Seeds within fruits synthesize the hormone **auxin,** which stimulates fruit growth. Ethylene hastens fruit ripening in many species, including tomatoes, apples, and pears. Fruit falls from the plant when auxin levels drop and ethylene levels rise. The next chapter discusses plant hormones in more detail.

FIGURE 28.12

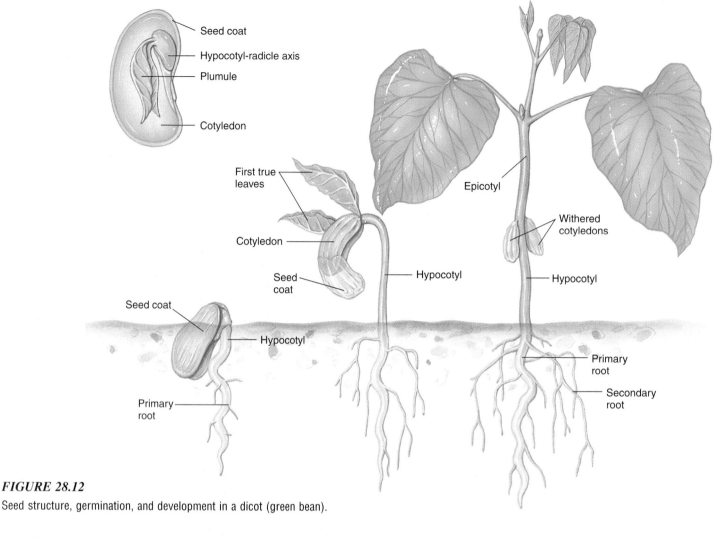

Seed structure, germination, and development in a dicot (green bean).

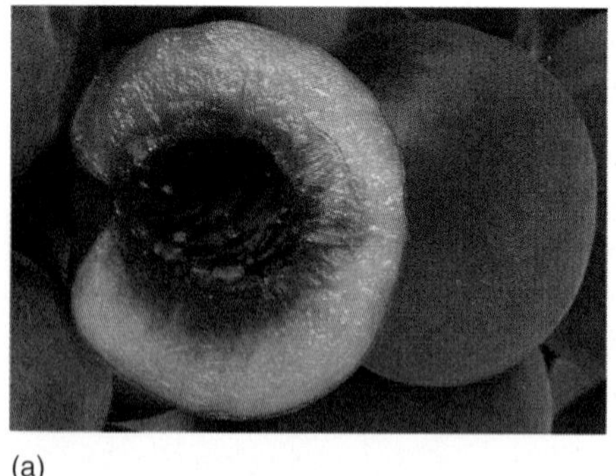

(a)

(b)

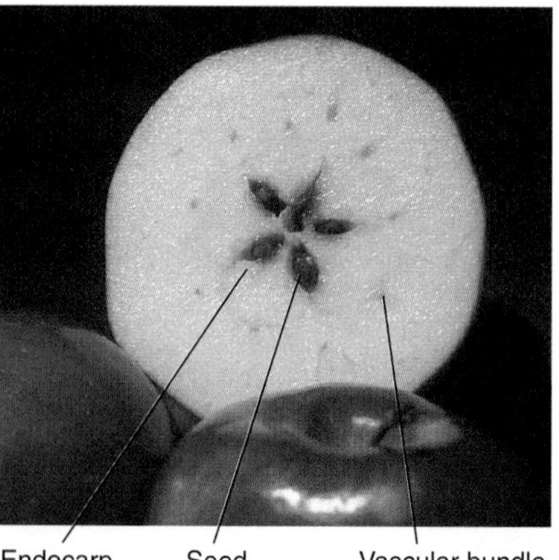

(c)

FIGURE 28.13

Fruits are ripened ovaries that usually enclose seeds. Fruits include drupes (*a*), berries (*b*), and pomes (*c*).

Fruit and Seed Dispersal

Fruits protect seeds from desiccation and destruction. This is why an unripe fruit is hard. Fruits also facilitate seed dispersal. Shiny berries attract birds and other animals, which carry the ingested seeds to new locations, where they are released in the animals' feces. Some seeds germinate only after passing through the intestines of birds or mammals.

Mammals also spread seeds from place to place when fruits bearing hooked spines attach to their fur. The fruit of the burdock

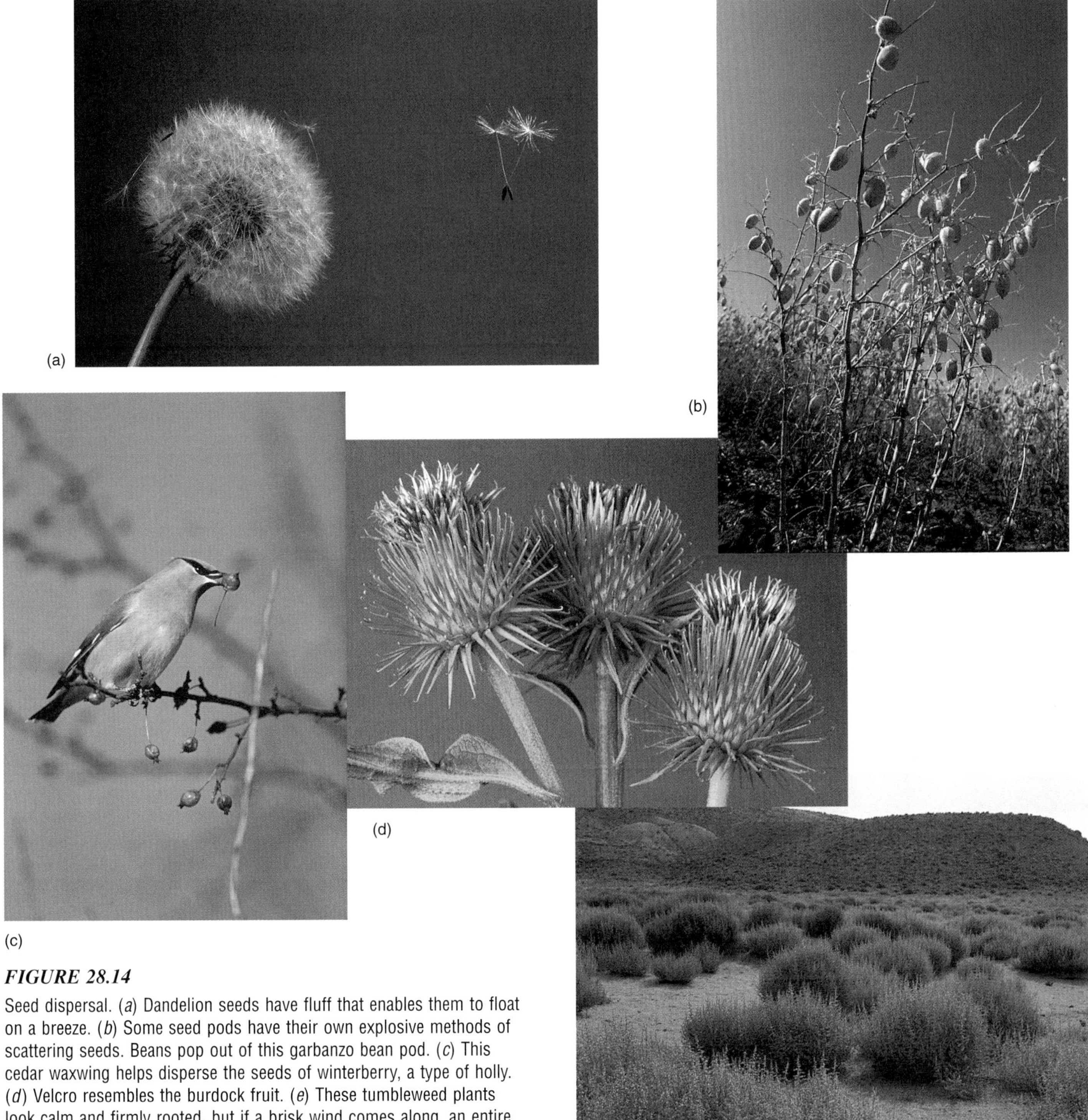

FIGURE 28.14

Seed dispersal. (*a*) Dandelion seeds have fluff that enables them to float on a breeze. (*b*) Some seed pods have their own explosive methods of scattering seeds. Beans pop out of this garbanzo bean pod. (*c*) This cedar waxwing helps disperse the seeds of winterberry, a type of holly. (*d*) Velcro resembles the burdock fruit. (*e*) These tumbleweed plants look calm and firmly rooted, but if a brisk wind comes along, an entire plant can blow away, shedding seeds.

plant, for example, has barbed hooks that cling to a passing deer or a hiker's jeans. The inspiration for Velcro, a fuzzy fabric that sticks to other fabrics, came from the strong attachment of burdock fruits (fig. 28.14).

Wind-dispersed fruits such as those of dandelions and maples have wings or other structures that enable them to ride on air currents and land far from the original plant. Coconuts are water-dispersed fruits that travel long distances before colonizing distant islands.

Seed Germination

Germination is the resumption of growth and development. Germination usually requires water, oxygen, and a source of energy. First, the seed imbibes (absorbs) water. In some seeds, imbibition causes the embryo to release hormones that stimulate the endosperm's breakdown. The starch in the endosperm breaks down to sugars that the embryo can use for energy. Imbibition also swells a seed, eventually rupturing the seed coat

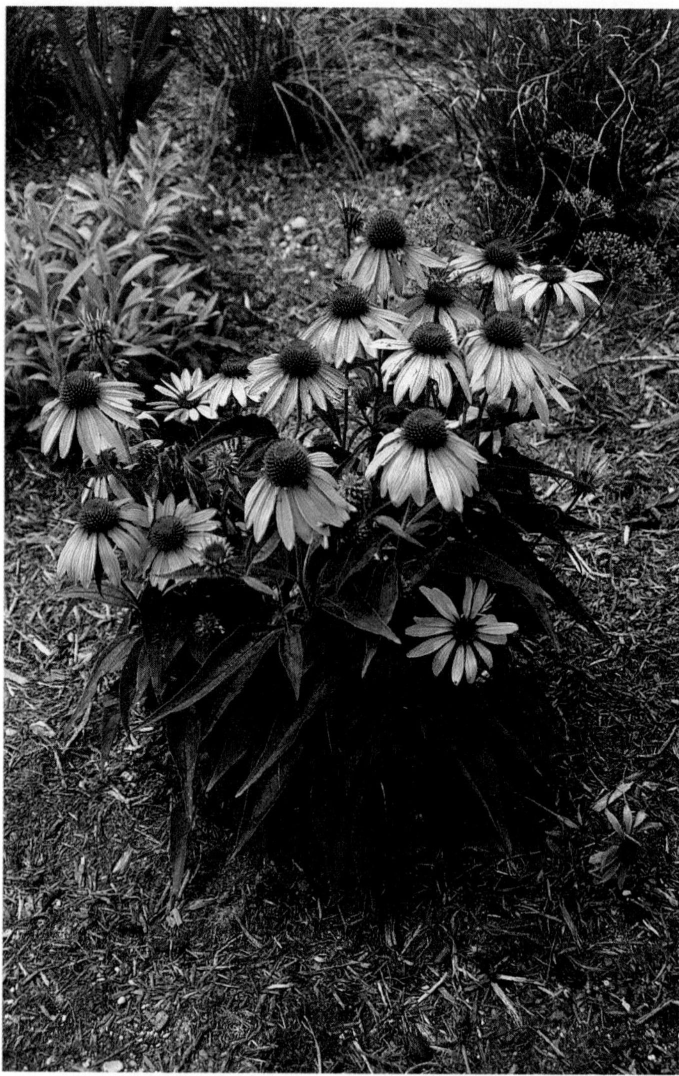

FIGURE 28.15

Juice from flowers of *Echinacea purpurea* is the top-selling herbal medicine in the United States. Two polysaccharides among the many organic compounds in the flowers stimulate immune system cells growing in culture. Clinical studies have shown echinacea extract can speed recovery from bronchitis, colds, and influenza and increase resistance to recurrent vaginal yeast infections. Extracts are used instead of purified compounds because researchers do not yet know whether the polysaccharides alone stimulate immunity or if a combination of biochemicals in the plant do, some of which occur in such low concentrations that we do not yet know their identities.

and exposing the plant embryo to oxygen. At this point, an embryo may resume growth; however, seeds of some plants normally germinate only after additional stimulation, such as exposure to light of a certain intensity or several days at a particular temperature.

Plant Development

After the growing embryo bursts out of the seed coat, further growth and development depend upon the root and shoot apical meristems. The hypocotyl, with its attached radicle, emerges first from the seed. In response to gravity, the radicle grows downward and anchors the plant in the soil. Root systems develop rapidly, providing the continual water supply that plants require to grow.

The shoot emerges from the soil in different ways, depending on the species. In most dicots, the elongating hypocotyl forms an arch that breaks through the soil and straightens in response to light, pulling the cotyledons and epicotyl up out of the soil. In most monocots, the coleoptile-protected epicotyl begins growing upward shortly after root growth starts. The single cotyledon remains underground.

A plant produces its first chloroplasts and begins photosynthesizing by the time its shoot emerges from the soil. When they have exhausted their embryonic food reserves, plants are already producing their own food photosynthetically.

A Comparison of Gymnosperm and Angiosperm Life Cycles

The gymnosperm life cycle differs from the angiosperm life cycle in several ways. Gymnosperms

- have cones instead of flowers as reproductive structures;
- have ovules that are exposed on female reproductive structures rather than embedded in their tissues;
- do not produce fruits;
- undergo single fertilization rather than double fertilization;
- nourish embryos with haploid tissue of the female gametophyte, rather than with a triploid endosperm.

In addition to enabling plants to reproduce sexually, flowers add much to our lives because of their great beauty. They are also important sources of medicines, as the chapter opener photograph indicates for the opiate drugs. A flower with healing properties that we do not yet fully understand is *Echinacea purpurea* (fig. 28.15), which may stimulate the human immune system to fight certain infections. It will be interesting to see what other valuable substances we can obtain from flowers and other plant parts in the future.

Mastering Concepts

1. What are the parts of a flower?
2. How do the female and male gametes of an angiosperm form?
3. What are some mechanisms of pollination?
4. Describe the events of angiosperm fertilization.
5. How does seed development differ in monocots and dicots?
6. What is a fruit?
7. What happens during germination?
8. How is the gymnosperm life cycle different from the angiosperm life cycle?

Imperiled Pollinators~

continued from page 556.

Several factors have contributed to the decline of honeybee populations in North America. Increased use of pesticides to grow crops began killing bees shortly after World War II and had depleted more than a million colonies by 1972. A study of *Cereus* cactus in an area along the Mexico and Arizona border revealed the effects of herbicide use to clear land to build new homes. Where herbicide was used, 27% of the plants monitored were pollinated and 5% developed fruit. Without herbicide, 60 to 100% of the cacti were pollinated and 75 to 100% developed fruit.

Herbicide use is only one threat to bee pollinators. Tracheal mites first appeared in Florida bee colonies in 1984 and rapidly spread throughout the United States, killing up to half of the bees in individual hives. Another type of mite, the *Varroa* mite, began invading hives in Wisconsin in 1987 and likewise spread, eventually destroying 60% of honeybee colonies in the United States. Caterpillars with an appetite for beeswax plagued many hives. Finally, Africanized bees invaded at the end of the 1980s, bringing deadly mites to many honeybee hives.

Bee experts—called apiculturists—predict that the falling bee populations will soon lead to diminished crop yields. European apiculturists have more experience than North American bee authorities, and they recommend that we can avert a crisis by preserving, or restoring, areas of native vegetation. An experiment on Staten Island in New York City suggests that the Europeans are correct. Staten Island is home to a huge landfill. Ecologists have begun to plant small sections of it with once-native trees and shrubs. In a short time, honeybees as well as other types of bees have come back and are pollinating the new plants. The closer the new areas are to undisturbed woodlands, the more bees gather.

SUMMARY

1. In asexual reproduction, a parent plant gives rise to **clones,** which can develop from roots, stems, or leaves. Asexual reproduction is advantageous in a stable environment where plants are well adapted to their surroundings.

2. Sexually reproducing plants undergo alternation of generations. The diploid sporophyte generation produces spores through meiosis. The spores divide mitotically to produce the haploid gametophyte, which in turn produces haploid gametes. In different species, either the sporophyte or gametophyte generation dominates. Female gametophytes are **megagametophytes,** and male structures are **microgametophytes.** They arise from **megaspores** and **microspores,** respectively, which form in **megasporangia** and **microsporangia.**

3. In pines, large female **cones** bear two ovules on each **scale.** Each ovule yields three haploid cells that degenerate and one haploid megaspore that develops into a female gametophyte. Microsporangia grow on the small scales of male cones. Through meiosis, the cones produce microspores, which become **pollen grains.** Pollen lands on sticky female cones, and a **pollen tube** grows towards the egg. The pollen divides twice, and two of the resulting cells are sperm. One sperm fertilizes the egg. The megagametophyte nourishes the embryo, which becomes dormant within a tough seed coat. Location of female cones above male cones encourages outcrossing.

4. Flowers are reproductive structures. The **calyx,** made of **sepals,** and the **corolla,** made of **petals,** are accessory parts. Inside the corolla, the **androecium** consists of **stamens** and their pollen-containing **anthers.** At the center of the flower the **gynoecium** encloses the ovary. The **stigma** extends from the ovary and captures pollen.

5. Male and female structures produce gametes. In the anther, four pollen sacs contain **microspore mother cells.** Each divides meiotically to yield four haploid microspores, which each divide mitotically to yield a haploid **generative nucleus** and a haploid **tube nucleus;** these two cells and their covering are a pollen grain. Sperm cells arise from the generative nucleus. In the ovary, **megaspore mother cells** divide meiotically to yield four haploid cells, one of which persists as a haploid megaspore that divides mitotically three times. The resulting megagametophyte, or **embryo sac,** contains eight cells. One is the egg.

6. Animals or wind transfers pollen from the anthers of one plant to its own or another plant's stigma. Flower structures and odors are adapted to encourage animal or wind **pollination.** Once on a stigma, a pollen grain grows a pollen tube, and its two sperm move through the tube towards the ovary. In the embryo sac, one sperm fertilizes the egg to form the zygote, and the second sperm fertilizes the polar nuclei to form the **endosperm,** which nourishes the embryo. Cotyledons develop and transfer nutrients in monocots and absorb the endosperm in dicots. Apical meristems promote the growth of shoot and root in the embryo.

7. After fertilization, nonessential floral parts fall off, and hormones may influence the ovary to develop into a fruit. A **drupe** is a fleshy fruit with a hard pit; a **berry** is fleshy with many seeds; blackberries are aggregate fruits; and a **pome** is a fleshy fruit that develops from tissues surrounding the ovary. A seed germinates in response to environmental cues. Germination requires oxygen, energy, and water. When the embryo bursts from the **seed coat,** the plant's primary growth ensues.

TO REVIEW . . .

1. Why doesn't sexual reproduction generate clones?

2. Give an example of asexual reproduction in a plant.

3. List and define the structures that carry out sexual reproduction in a pine.

4. What function do flowers perform? Which floral structures participate directly in this function?

5. State whether the following floral parts are male, female, or accessory structures:
- **a.** sepals
- **b.** carpels
- **c.** stigmas

d. calyx
e. styles
f. petals
g. stamens
h. corolla
i. pistil
j. anthers

6. In what two ways is animal pollination more effective than wind pollination?

7. List three adaptations in plants that encourage outcrossing.

8. What type of pollinator might be attracted to a plant with each of the following?
   a. red sweet-smelling flowers
   b. dull-colored spicy-smelling flowers
   c. yellow heavily scented flowers

9. Name tissues or cells in an angiosperm that are haploid, diploid, and triploid.

10. Gardeners know to plant many more seeds than the number of plants they desire because not all of the seeds will germinate. Why might seeds fail to germinate?

11. How are the life cycles of a corn plant and a spruce tree similar, and how are they different?

12. How does early development differ in a monocot and a dicot?

13. Give an example of a drupe, a berry, and a pome.

14. Why are the biotechnologies of protoplast fusion and somaclonal variation unpredictable?

TO THINK ABOUT . . .

1. Chefs consider a plant food a fruit or a vegetable according to how it is prepared and eaten and how sweet it tastes. How does this differ from the biological definition of a fruit?

2. Why is it more adaptive for a plant species if an insect species that pollinates it does not also pollinate other types of plants?

3. How could you tell if a portion of a plant is part of the sporophyte generation or the gametophyte generation?

4. Humans and plants each have diploid as well as haploid cells. How do the life cycles of humans and flowering plants differ?

5. How are petals, which are considered accessory structures, nevertheless necessary for reproduction?

6. Why might a flower smell like putrefied meat?

7. What would a flower look like that had its class "A" and "B" genes inactivated? That is, only genes of class "C" function.

8. Outline the steps of the scientific method that Elliot Meyerowitz used to test his hypothesis that three genes control flower development.

TO LEARN MORE . . .

References and Resources*

Bradley, Desmond, January 3, 1997. Inflorescence commitment and architecture in *Arabidopsis. Science,* vol. 275. Mutations alter flower structure.

Buchmann, Stephen L., and Gary Paul Nabhan. July/August 1996. The pollination crisis. *The Sciences.* A decline in bee populations threatens many plant populations.

Kearns, Carol Ann and David Inouye. May 1997. Pollinators, flowering plants, and conservation biology. *Bioscience,* vol. 47. Plant-pollinator relationships are complex.

Rawls, Rebecca. September 23, 1996. Europe's strong herbal brew. *Chemical & Engineering News,* vol. 74. Fruits and flowers provide many "natural" medicines.

Rousch, Wade. September 6, 1996. Probing flowers' genetic past. *Science,* vol. 273. A single genetic switch may have made the difference between a nonflowering plant and a flowering plant.

Science, vol. 268. May 5, 1995. This issue has a special section devoted to plant biotechnology.

**Life*'s website features frequently updated references.

Sunflowers facing the sun dramatically illustrate the responsiveness of plants to environmental stimuli.

29 Plant Responses to Stimuli

chapter outline

The Sundew Plant's Response to a Stimulus 576

Plant Hormones 576

Auxins
Gibberellins
Cytokinins
Ethylene
Abscisic Acid
Hormonal Interactions

Tropisms—Growth Movements 578

Phototropism—Response to Unidirectional Light
Gravitropism—Response to Gravity
Thigmotropism—Response to Touch

Nastic Movements—Nongrowth Movements 581

Thigmonasty
Photonasty
Thigmomorphogenesis

Seasonal Responses of Plants to the Environment 582

Flowering—Response to Photoperiod
Do Plants Measure Day or Night?
Phytochrome—A Pigment Controlling Photoperiodism
Other Responses Influenced by Photoperiod and Phytochrome
Senescence
Dormancy

Circadian Rhythms 587

Plant Clocks

Chronobiology is the study of genetically controlled internal mechanisms that keep biochemical activities and physiological functions on a regular schedule. Many aspects of chronobiology are familiar—jet lag, and disturbances due to altered biological clocks in people who change work shifts.

Biological clocks seem to be common. Hamsters with mutations in "clock" genes snooze when their normal cage mates run on the treadmill during the night and run when normal hamsters rest. In many animals, hormone levels, blood pressure, and immune system functions wax and wane throughout the day. Fruit flies, bread molds, and even cyanobacteria have such day-based, or circadian, rhythms too.

Do plants have biological clocks? Botanists suspected so but knew it would be harder to demonstrate in plants than in animals. Photography reveals daily cycles in such activities as leaf movements in bean plants and sunflowers facing towards the sun. But researcher Steven Kay and graduate student Andrew Millar, at the University of Virginia, sought a more dramatic demonstration of biological clocks in plants. For help, they turned to an unusual source—fireflies.

to be continued . . .

The Sundew Plant's Response to a Stimulus

To a human observer, the sundew plant is a magnificent member of a swamp community. To an insect, however, the plant's beckoning club-shaped leaves, bearing nectar-covered tiny tentacles, are treacherous. Once the insect alights on the plant and begins to enjoy its sweet, sticky meal, the surrounding hairlike tentacles begin to move. Gradually they fold inward, entrapping the helpless visitor and forcing it down toward the leaf's center. Here, powerful digestive enzymes dismantle the insect's body and release its component nutrients. After 18 hours, the leaves open. All that remains of the previous day's six-legged guest is a few bits of indigestible matter (fig. 29.1).

The sundew's insect capture dramatically illustrates a plant's response to its immediate environment. Other plant responses may not be as obvious, but they are nonetheless based on a complex interplay of biochemicals and little-understood biological activities.

Unlike animals, most plants cannot move away from unfavorable environmental conditions. Plants adapt to the environment by growing or becoming dormant, processes influenced by both internal and external factors. A plant's DNA contains the blueprints for organizing specialized cells into tissues and organs, and these genetic instructions can include several options. External factors, such as light intensity or temperature, may determine which option a plant expresses. The result can be a finely tuned growth response that enables the plant to adapt to its surroundings and survive.

FIGURE 29.1

This sundew plant grows in the swamps of upstate New York and the Pine Barrens of New Jersey. Its leaves are insect traps.

Plant Hormones

Hormones regulate many aspects of plant growth. These biochemicals are synthesized in small quantities in one part of the organism and transported to another, where they stimulate or inhibit growth. Hormones may have other effects, such as maintaining and repairing tissue.

The five major classes of plant hormones are auxins, gibberellins, cytokinins, ethylene, and abscisic acid (table 29.1). Their influence depends upon other hormones present and the sensitivity of the tissue to the hormones. As a result, a single hormone can elicit numerous responses.

Table 29.1 Major Classes of Plant Hormones

Class	Principal Actions
Auxins	Elongate cells in seedlings, shoot tips, embryos, leaves
Gibberellins	Elongate and divide cells in seeds, roots, shoots, young leaves
Cytokinins	Stimulate cytokinesis in seeds, roots, young leaves, fruits
Ethylene	Hastens fruit ripening
Abscisic acid	Inhibits growth; closes stomata

FIGURE 29.2
Effects of auxin. The pea plants on the left were not treated with auxin; those on the right were.

FIGURE 29.3
Effects of gibberellin. Note the size difference between these two California poppy plants. Can you tell which one was treated with gibberellin?

Auxins

Auxin was the first plant hormone scientists described. In the late 1870s, decades before researchers determined the chemical structures of plant hormones, Charles Darwin and his son Francis learned that a plant-produced "influence" caused plants to grow towards light (fig. 29.2). The Darwins were describing auxin, a plant hormone that stimulates cell elongation in grass seedlings and herbs. Auxin stimulates cells to elongate by altering the plasticity of cell walls, so that the walls stretch by taking in more water. Auxin acts rapidly, spurring noticeable growth in a grass seedling in minutes.

Three auxins occur naturally. The most active, indoleacetic acid (IAA), is produced in shoot tips, embryos, young leaves, flowers, fruits, and pollen. Synthetic compounds having auxinlike effects, such as 2,4-D (2,4-dichlorophenoxyacetic acid), are important commercially. 2,4-D is used extensively as an herbicide.

Gibberellins

In the late 1920s, Japanese botanists studying "foolish seedling disease" in rice discovered **gibberellin,** another plant hormone that causes shoot elongation. Plants suffering from foolish seedling disease grow rapidly because they are infected by a fungus that produces gibberellin. The infected plants become so spindly that they fall over and die (fig. 29.3).

Gibberellin, abbreviated GA, has several functions. It stimulates shoot elongation in regions of elongation in trees, shrubs, and a few grasses. It is also found in immature seeds, apices of roots and shoots, and the young leaves of flowering plants. The hormone induces both cell division and elongation and also stimulates seed germination. GA-induced elongation occurs after about a 1-hour delay, which is much slower than auxin-induced growth. We know of more than 80 naturally occurring gibberellins.

Cytokinins

As early as 1913, scientists knew that a biochemical stimulates plant cells to divide. It was not until 1964 that researchers discovered the first naturally occurring **cytokinin,** in corn kernels. Since then, researchers have isolated other cytokinins and have synthesized several artificial forms.

Cytokinins earned their name because they stimulate cytokinesis (the division of the cell after the genetic material has replicated and separated) by pushing cells into mitosis. Cytokinins cannot work alone—auxin must be present before mitosis begins. The effects of cytokinins are similar to some of the effects of auxin and gibberellin—they promote cell division and growth and participate in development, differentiation, and senescence.

In flowering plants, most cytokinins act in roots and developing organs such as seeds, fruits, and young leaves. Synthetic cytokinin-like compounds have a variety of uses, including stunting the growth of wheat stalks to prevent them from blowing over, shortening shrubs to a manageable level, and extending the shelf life of lettuce.

Ethylene

The effects of ethylene were noticed long before its discovery. In 1910, scientists in Japan observed that bananas ripened prematurely when stored with oranges. Apparently, the oranges produced something that induced rapid ripening. By 1934, scientists realized ethylene gas was hastening ripening.

Ethylene ripens fruit in many species. Ripening is a complex process that includes pigment synthesis, fruit softening, and breakdown of starches to sugars. A tomato picked from the garden when it is hard and green with streaks of pale orange will turn, aided by ethylene, into a soft, red, succulent tomato (fig. 29.4).

Ethylene also helps ensure that a plant will survive injury or infection. When a damaged plant produces the hormone, it hastens aging of the affected part so that it can be shed before the problem spreads to other regions of the plant.

All parts of flowering plants synthesize ethylene, particularly roots, the shoot apical meristem, nodes, and ripening fruits. The dark spots on a ripening banana peel are concentrated pockets of melanin, produced under the influence of ethylene. Because ethylene is a gas, its effects can be contagious. The expression "one bad apple spoils the whole batch" refers to the way ethylene released from one apple can hasten the ripening, and eventual spoiling, of others nearby.

Although ethylene's chemical structure is simple, a plant takes several biochemical steps to synthesize it. Biotechnologists can

FIGURE 29.4

As ethylene accumulates, these tomatoes ripen.

block the gene that specifies the enzyme necessary for the very last step in ethylene production. Blocking ethylene production extends a crop plant's shelf life.

Abscisic Acid

Can plants manufacture substances that inhibit growth? By the 1940s, botanists thought so. Twenty years later, researchers isolated such a compound.

Abscisic acid, or ABA, inhibits the growth-stimulating effects of many other hormones. It inhibits seed germination, countering the effects of gibberellins. ABA also closes stomata, which helps plants conserve water during drought. This hormone also causes leaf, branch, and fruit abscission (shedding). ABA is used commercially to inhibit the growth of nursery plants so that shipping is less likely to damage them.

Hormonal Interactions

Plant hormones seldom function alone. Several plant hormones, for example, influence abscission, the shedding of leaves or fruit. Senescence (aging and death of a plant or plant part) precedes abscission. Recall that during senescence, an abscission zone forms at the base of the organ that will be shed (see fig. 27.17). Eventually, the leaf or fruit separates from the plant at this abscission zone.

Auxin in actively growing leaves and fruits and cytokinin and gibberellin in roots normally retard senescence. A drop in auxin production in response to environmental stimuli, such as an injury or shorter days in autumn, normally triggers senescence. During senescence, cells in the abscission zone begin producing ethylene, which swells the cells and signals them to produce enzymes that digest cell walls. As a result, the cells separate and the leaf or fruit drops off the plant.

If lowered auxin levels lead to abscission, then the application of synthetic auxins should prevent it. Indeed, synthetic auxins are often used to prevent preharvest fruit drop in orchards, to hold berries on holly, and to coordinate fruit abscission at harvest time.

The coordinated production of several hormones allows a plant to survive the changing seasons as well as extreme weather conditions. This gives the plant time to develop its reproductive organs and structures, which ensures the perpetuation of the species.

The search to identify new plant hormones continues, such as those that control flowering and responses to infection. Biology in Action 29.2, at the end of the chapter, discusses how techniques of molecular biology are enabling researchers to study plant communication biochemicals that are not as abundant as the well-known auxins, gibberellins, cytokinins, and ethylene.

Mastering Concepts

1. What are the five major classes of plant hormones?
2. What are the functions of the five types of plant hormones?
3. Give an example of how plant hormones interact.

Tropisms—Growth Movements

Despite their obvious immobility, plants respond to many environmental signals. The glorious heads of sunflowers turn towards the sun, and roots grow down in response to gravity. Some plant responses are short term. For example, a Venus's flytrap closes in less than a second, and a plant stem usually curves towards light in only a few hours. Other behaviors, such as flowering, are long-term responses to changing seasons.

The term **tropism** refers to plant growth toward or away from environmental stimuli, such as light or gravity. Hormones cause many tropisms. Tropisms are named for the stimuli that elicit the responses. **Phototropism,** for example, is a growth response to unidirectional light, and **gravitropism** is a growth response to gravity. Table 29.2 lists some tropisms.

Tropisms result from differential growth, in which one side of the responding organ grows faster than the other, curving the organ. When an organ curves towards the stimulus, as a stem grows toward light, it is called a positive tropism. When an organ curves away from a stimulus, it is a negative tropism. Roots are negatively phototropic—they grow away from light.

Phototropism—Response to Unidirectional Light

In phototropism, cells on the shaded side of a stem elongate more than cells on the lighted side of the stem. Auxin from the apex controls the rapid elongation of cells along the shaded side of coleoptiles (protective sheaths around young grass shoots).

In the 1950s, Winslow Briggs and his colleagues were the first to discover precisely how auxin controls phototropism. First, they determined that the amount of auxin coleoptiles grown in the light produce is the same as the amount coleoptiles grown in the dark produce—that is, they established that light does not destroy auxin. They then discovered that they could collect more auxin from the shaded side of coleoptiles than from the lighted side, which suggested that light causes auxin to migrate to the shaded side of the stem. More recent experiments support this finding. Auxin labeled with radioactive carbon (^{14}C) and exposed to unidirectional light moves to the shaded side of coleoptiles. Cells in the shade then elongate more than cells in the light, curving the coleoptile toward the light (fig. 29.5).

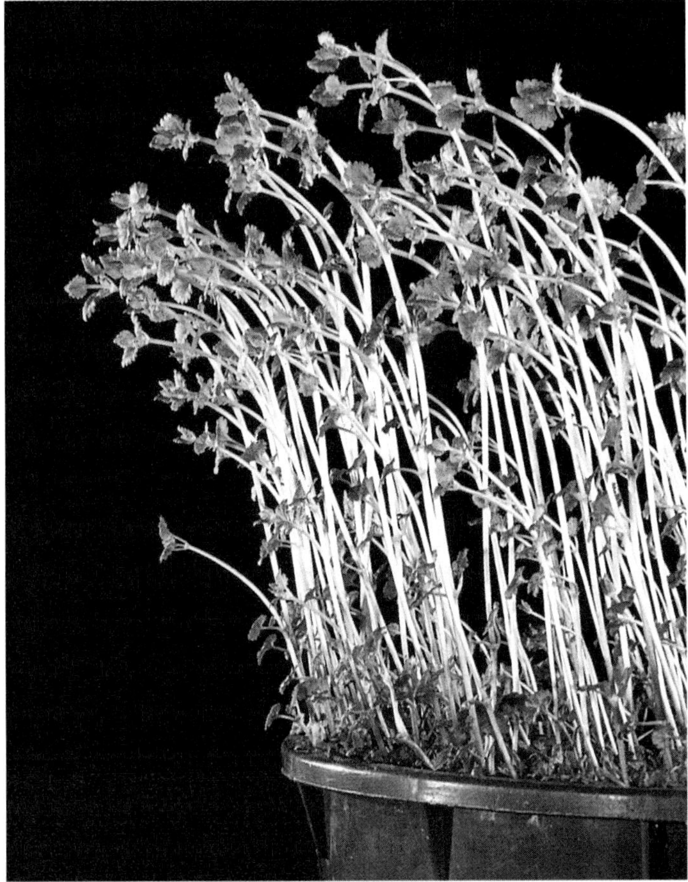

FIGURE 29.5
Positive phototropism. Seedlings bend towards the light.

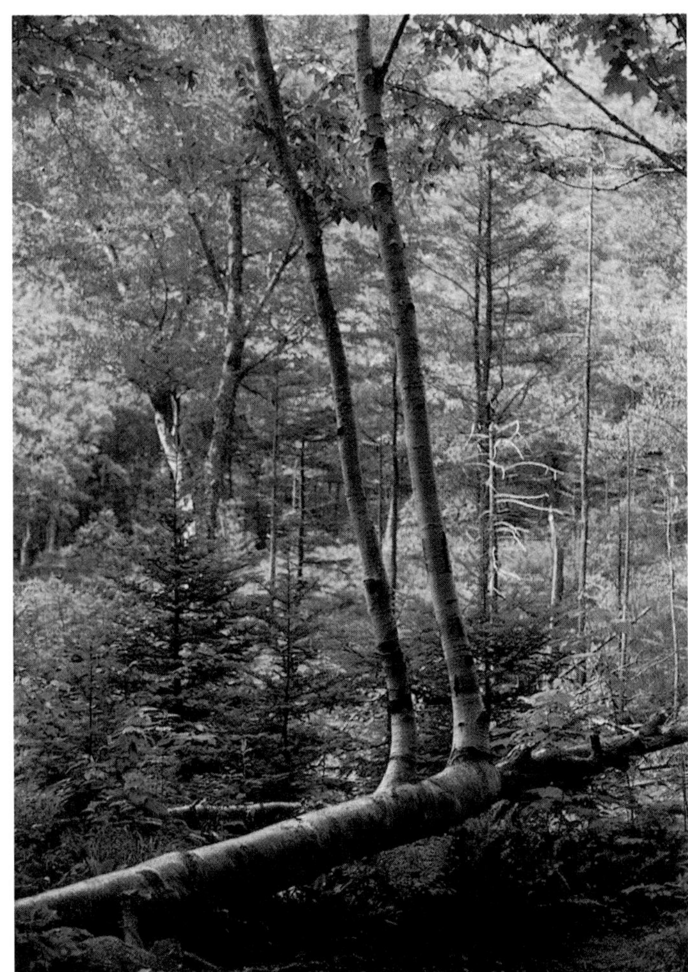

FIGURE 29.6
Formerly horizontal branches of this fallen tree now grow upward due to gravitropism.

How does a plant detect light coming from one direction? Only blue light with a wavelength less than 500 nanometers effectively induces phototropism. The yellow pigment flavin is probably the photoreceptor molecule for phototropism. Flavin alters auxin transport to the shaded side of the stem or coleoptile.

Gravitropism—Response to Gravity

Charles Darwin and his son Francis also studied gravitropism, a plant's growth response to gravity. They wondered how roots would grow if their tips were removed. The Darwins discovered that decapped roots grew, but not downwards, in response to gravity. Therefore, the root cap is necessary for gravitropism. It is interesting to examine root growth in the absence of gravity, as when plants grow aboard orbiting spacecraft in minimal gravity (Biology in Action 29.1).

A plant's shoot is negatively gravitropic, because it grows upwards, and its roots are positively gravitropic, extending downward along the gravitational force. Curiously, auxin accumulation seems to provoke both of these opposite responses. In a horizontal stem or shoot, auxin accumulates on the lower side and stimulates differential growth. As a result, the structure bends upward (fig. 29.6). If a root is held horizontally, however, auxin accumulates in the lower regions and causes downward growth. How can this be? One hypothesis is that root tissue is more sensitive to auxin than shoot tissue, and the amount of auxin that accumulates is so great that it inhibits cell elongation. When the cells in the upper portion of a horizontal root grow faster, the root curves downward.

Thigmotropism—Response to Touch

The coiling tendrils of twining plants such as morning glory and bindweed exhibit **thigmotropism,** a response to touch. When they hang free, tendrils often grow in a spiral fashion, which increases their chances of contacting an object they can cling to (fig. 29.7). A plant detects contact with an object with specialized epidermal cells, which induce differential growth of the tendril. In only 5 to 10 minutes the tendril completely encircles the object. Thigmotropism is often long-lasting. Stroking the tendril of a garden pea plant for only a few minutes induces a curling response that lasts for several days. Auxin and ethylene control thigmotropism.

Interestingly, tendrils seem to "remember" touch. Tendrils touched in the dark do not respond until they are illuminated. They apparently store sensory information in the dark but do not respond until light is present.

29.1 biology in action

Plants in Space

Plants will play a key part in space colonization, providing food and oxygen. But how will organisms that have evolved under constant gravity function in its near absence beyond the earth?

Researchers are studying the effects of microgravity on a variety of species by sending them on space shuttle voyages. These experiments can more realistically assess plant growth and development in space than simulations conducted on earth, which use a rotating device called a clinostat to diminish gravitational force. So far, it appears that lack of gravity greatly affects plants, in everything from subcellular structural organization to the functioning of the organism as a whole. These responses will present interesting challenges to future space farmers.

Subcellular Responses to Microgravity

Plant cells grown in space have fewer starch grains and more abundant lipid-containing bodies than their earthly counterparts, indicating a change in energy balance. Organelle organization is also grossly altered; endoplasmic reticula occur in randomly spaced bunches and mitochondria swell. Nuclei enlarge, and chromosomes break. Chloroplasts have enlarged thylakoid membranes and small grana.

Interesting effects occur in amyloplasts, which are starch-containing granules in certain root cells. On earth, amyloplasts aggregate at the bottoms of these cells, but in space, they occur throughout the cytoplasm. For many years, botanists hypothesized that sinking starch granules in these cells are gravity receptors. A more recent view, though, holds that cells sense gravity by detecting the difference in pressure on the cell membrane at the top versus the bottom of the cell. This difference results from protoplasm sinking in response to gravity, so there is less of a difference as gravitational force falls. The actual "sensing" may be carried out by a cell membrane protein called an integrin, and the starch granules may amplify the difference in pressure by sinking in the presence of gravity. Whatever the precise mechanism of gravity sensation, a root tip cannot elongate downward normally in microgravity (fig. 29.A).

Cell Division

Microgravity halts mitosis, usually at telophase, which produces cells with more than one nucleus. Oat seedlings germinated in space have only one-tenth as many dividing cells as seedlings germinated on earth. Microgravity also disrupts the spindle apparatus that pulls chromosome sets apart during mitosis.

Cell walls formed in space are considerably thinner than their terrestrial counterparts, with less cellulose and lignin. Microgravity also inhibits regeneration and alters cell distribution. A decapped root will regenerate in 2 to 3 days on earth but not at all in space. Lettuce roots have a shortened elongating zone when grown in space.

Growth and Development

Germination is less likely to occur in space than on earth because of chromosome damage, but it does happen. Early growth seems to depend upon the species—bean, oat, and pine seedlings grow more slowly than on earth, and lettuce, garden cress, and cucumbers grow faster. Many species, including wheat and peas, cease growing and die before they flower. In 1982, however, the mustard weed *Arabidopsis* successfully completed a life cycle in space—indicating that human space colonies containing plant companions may indeed be possible.

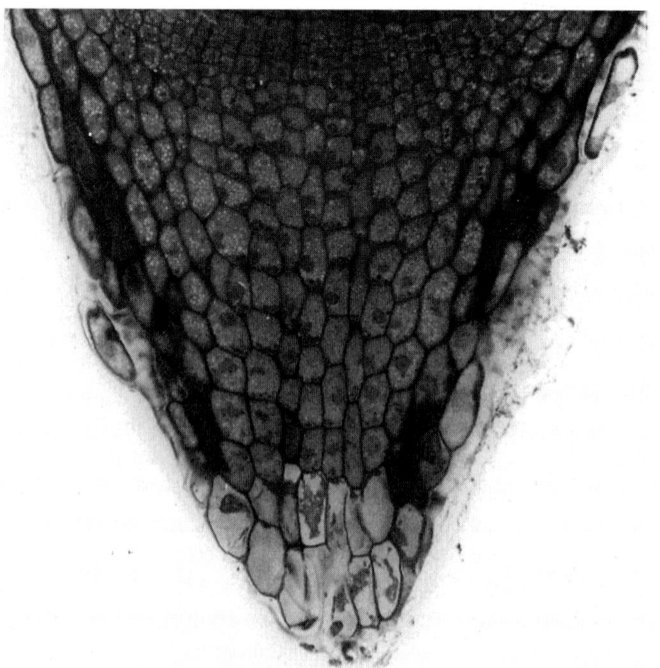

(1)

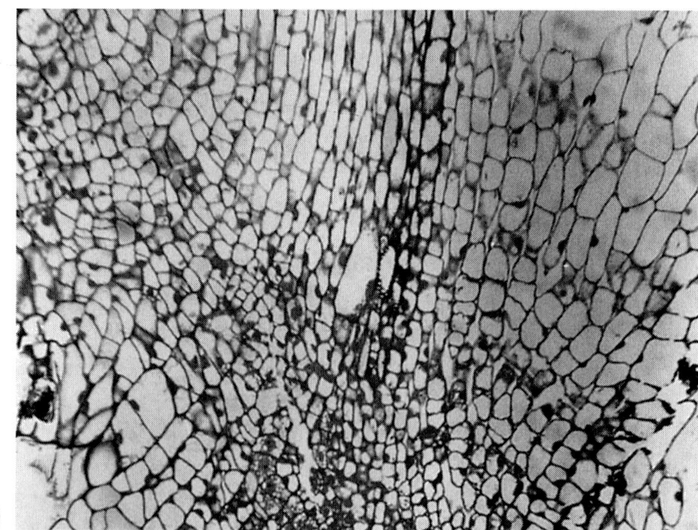

(2)

FIGURE 29.A
Plant growth in space. (*1*) On earth, a root whose cap has been removed regenerates an organized, functional root cap. (*2*) In the microgravity of space, however, regrowth of a decapped root is disorganized. Clearly, gravitational cues help direct normal regeneration.

FIGURE 29.7
Thigmotropism. A tendril's twining is a rapid response to touch. This tendril of a passion vine wraps around a blackberry stem.

Mastering Concepts
1. What is the physical basis of a tropism?
2. How does auxin cause phototropism?
3. How is gravitropism adaptive?

Nastic Movements—Nongrowth Movements

Plant movements that are not oriented with respect to a stimulus are called **nastic movements.** They include some of the most fascinating responses seen in the plant kingdom (table 29.2).

Thigmonasty

A nastic movement resulting from physical contact or mechanical disturbance is **thigmonasty.** Thigmonastic movements depend upon a plant's ability to rapidly transmit a stimulus from touch-sensitive cells to responding cells elsewhere. Consider leaf movement in the "sensitive plant," mimosa. When the leaves of the mimosa are touched, the leaflets fold and the petiole droops. Touch causes the motor cell membrane at the base of the leaflet to become more permeable to K^+ and other ions. As ions move out of the motor cells, the osmotic potential in the surrounding area decreases, causing water to move out of the motor cells by osmosis. The water loss shrinks the motor cells and causes thigmonastic movement. Reversal of the process causes the leaves to unfold in approximately 15 to 30 minutes (fig. 29.8). This electrical stimulation and response is considerably faster than a hormone-induced action.

Table 29.2 Plant Responses to Stimuli

Tropism: Growth towards or away from an environmental stimulus.
Nastic movement: Movement not oriented toward or away from an environmental stimulus.

Stimulus	Tropism	Nastic Movement
Light/dark	Phototropism	Photonasty
Gravity	Gravitropism	None
Touch	Thigmotropism	Thigmonasty
Temperature	Thermotropism	Thermonasty
Chemical	Chemotropism	Chemonasty
Water	Hydrotropism	Hydronasty

How are thigmonastic movements adaptive? Such movements may defend sensitive plants. For example, in some plants, closing leaflets expose sharp prickles and stimulate certain motor cells to secrete noxious substances called tannins. These responses discourage hungry animals from eating the plant. The Venus's flytrap is famous for its dramatic thigmonastic response (see fig. 27.16*a*). Unlike the sensitive plant, in which thigmonastic movements result from reversible changes in turgor pressure, the Venus's flytrap's movements result from increased cell size, which begins when the cell walls acidify to pH 4.5 and below. The leafy traps each consist of two lobes, and each lobe has three sensitive "trigger" hairs overlying motor cells. When a meandering animal touches two of these hairs, it signals the plant's motor cells, which then initiate H^+ transport to epidermal cell walls along the trap's outer surface. The resulting acidification of these cell walls expands the outer epidermal cells along the central portion of the leaf. Since epidermal cells along the inner surface of the leaf do not change volume, the flytrap shuts.

The Venus's flytrap takes 1 to 2 seconds to close and requires a large expenditure of ATP. Motor cells use almost one-third of their ATP to pump the H^+ that acidifies the cell walls and closes the trap. The trapping mechanism is quite sophisticated; it will not close unless two of its trigger hairs are touched in succession or one hair is touched twice within about 15 seconds. By limiting responses to two stimuli instead of one, the plant can distinguish potential prey from other objects, such as falling leaves.

The closed leaves press the captive animal against digestive enzyme-secreting glands on the inner surface of the trap. Then the plant digests the prey to use the nutrients that make up the insect's body. Digestion may take 2 days or longer. Empty traps usually open after 8 to 12 hours as epidermal cells along the midrib of the inner surface of the leaf expand. The sundew in figure 29.1 is another thigmonastic carnivorous plant.

(a)

(c)

(b)

FIGURE 29.8

Mimosa pudica is called the "sensitive plant" because of its thigmonastic movements. Leaves are erect in undisturbed plants (*a*), but touching a leaf (*b*) causes the leaflets to fold and the petiole to droop (*c*).

Photonasty

Photonasty, the nastic response to daily rhythms of light and dark, is also known as "sleep movement." The prayer plant *maranta* is an ornamental houseplant that exhibits such a response. Prayer plant leaves orient horizontally during the day, which maximizes their interception of sunlight. At night, the leaves fold vertically into a configuration resembling a pair of hands in prayer (fig. 29.9).

The prayer plant's leaves move in response to light and dark as the turgor pressure changes in motor cells at the base of each leaf. In the dark, K^+ moves out of cells along the upper side and into cells along the lower side of a leaf base. This moves water, via osmosis, into cells along the lower side of the leaf base, swelling them, as cells along the upper side lose water and shrink. As the cellular volume changes, the leaf stands vertically. At sunrise, the process reverses and the leaf again lies horizontally. Changes in leaf position can conserve water and heat.

Sorrel and legumes such as beans have similar sleep movements that occur at the same time each day. Carl Linnaeus, a famous Swedish botanist, made clever use of these regular movements. He filled wedge-shaped portions of a circular garden with plants that had sleep movements at different times. By checking to see which plants in his so-called *horologium florae* (flower clock) were "asleep," Linnaeus could tell the time of day.

Thigmomorphogenesis

Plants are extremely sensitive to mechanical disturbances such as rain, hail, wind, animals, and falling objects. In response, they typically inhibit cellular elongation, remaining short and stocky, and produce large amounts of thick-walled supportive tissue (collenchyma and sclerenchyma fibers). Some growers brush their crops to keep them short and bushy. Ethylene controls this slower response to mechanical disturbances, called **thigmomorphogenesis.**

Mastering Concepts

1. What is a nastic movement?
2. Why is a nastic movement fast?
3. Describe two types of nastic movements.
4. What is thigmomorphogenesis?

Seasonal Responses of Plants to the Environment

Seasonal changes affect plant responses in many ways. Autumn brings cooler nights and shorter days, which produce beautifully colored leaves, dormant buds, and decreased growth. In the spring, buds resume growth and rapidly transform a barren forest into a dynamic, photosynthetic community.

Flowering—Response to Photoperiod

Flowering reflects seasonal changes. Many plants flower only during certain times of the year. Clover and iris flower during the long days of summer, whereas poinsettias and asters bloom in the short days of early spring or fall.

FIGURE 29.9

The prayer plant, *Maranta,* exhibits photonasty. When the sun goes down, the prayer plant's leaves fold inward.

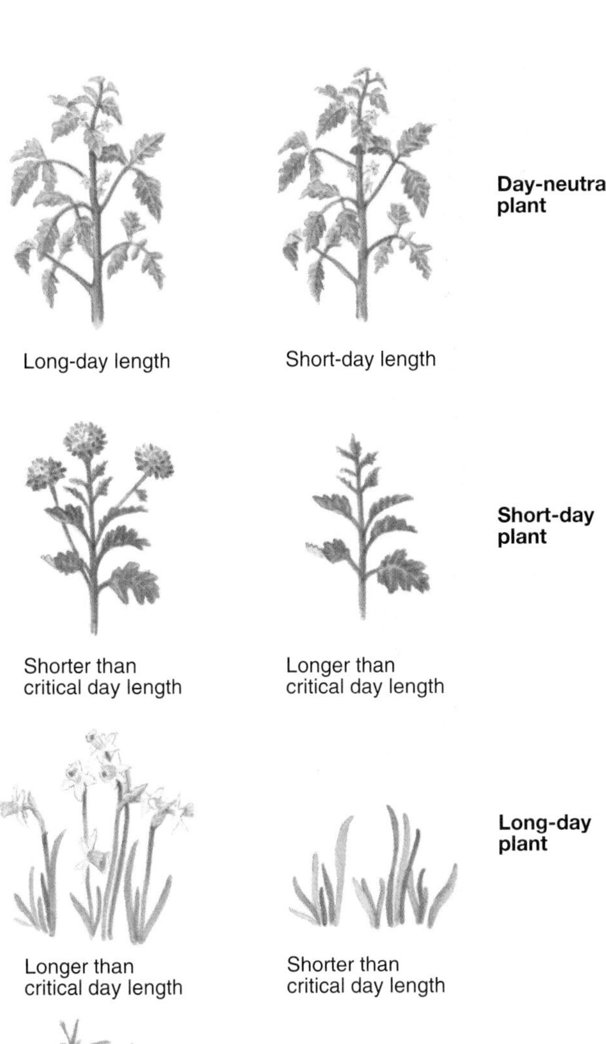

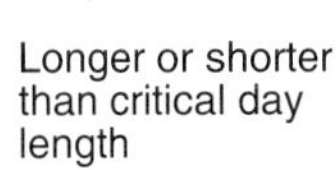

FIGURE 29.10

Flowering responses to day length. For a long time botanists thought that day length controls flowering. Flowering in day-neutral plants does not rely on response to day length. Short-day plants respond to a light period shorter than a critical length. Long-day plants flower in response to light periods longer than a critical period. Intermediate-day plants require an intermediate day length to trigger flowering. In most plants, flowering responds to the length of the dark period.

Studies of how seasonal changes influence flowering began in the early 1900s. W. W. Garner and H. A. Allard at a U.S. Department of Agriculture research center in Maryland were studying tobacco, which flowers during late summer in Maryland. One group of tobacco mutants did not flower but continued to grow vegetatively into autumn. These mutants became large, leading Garner and Allard to name them Maryland Mammoth. Since these oversized mutants had the potential for increasing tobacco crop yields, Garner and Allard moved their Mammoth plants into the greenhouse to protect them from winter cold and continued to observe their growth. To their surprise, the mutants finally flowered in December!

Could the plants somehow measure day length? To test this hypothesis, Garner and Allard set up several experimental plots of the tobacco, each planted approximately a week apart. All of the plants flowered at the same time, despite the fact that the staggered planting times resulted in plants of different ages and sizes. Garner and Allard suggested that the plants were exhibiting **photoperiodism,** the ability to measure seasonal changes by day length.

Photoperiodism attunes plant responses to a changing environment. Plants measure and respond to day length rather than other climatic factors, such as rainfall or temperature, because weather is unpredictable from year to year. Day length is consistent due to the position of the earth as it moves around the sun.

Plants are classified into four groups, depending upon their responses to photoperiod (duration of daylight) (fig. 29.10). **Day-neutral plants** do not rely on photoperiod to stimulate flowering. These include roses, snapdragons, cotton, carnations, dandelions, sunflowers, tomatoes, cucumbers, and many weeds. **Short-day**

plants require light periods shorter than some critical length. These plants usually flower in late summer or fall. For example, ragweed plants flower only when exposed to 14 hours or less of light per day. Asters, strawberries, poinsettias, potatoes, soybeans, and goldenrods are short-day plants.

Long-day plants flower when light periods are longer than a critical length, usually 9 to 16 hours. These plants usually bloom in the spring or early summer and include lettuce, spinach, beets, clover, corn, and iris. **Intermediate-day plants** flower only when exposed to days of intermediate length, growing vegetatively at other times. They include sugarcane and purple nutsedge.

The season when a plant flowers depends upon whether the photoperiod is longer or shorter than some species-specific critical length. For example, ragweed and spinach both flower when exposed to exactly 14 hours of daylight, yet ragweed is a short-day plant while spinach is a long-day plant. If the photoperiod is shorter, ragweed will flower and spinach won't; if it is longer, only the spinach flowers. Thus, spinach flowers in the long days of summer, while ragweed blooms in the shorter days of fall.

Flowering response to photoperiod greatly influences the geographical distribution of plants. For example, many short-day plants do not grow in the tropics, where daylight is always too long to induce flowering. The measuring system in many plants is remarkably sensitive. Henbane, a long-day plant, flowers when exposed to light periods of 10.3 hours but not when the light period is 10.0 hours. Leaves sense photoperiod. Remove a plant's leaves, and it does not respond to changes in photoperiod.

Some plants will not bloom unless they are exposed to the correct photoperiod. These plants exhibit **obligate photoperiodism.** For plants such as soybeans, the requirement for an inductive photoperiod is absolute—these short-day plants will not flower unless the nights are long. Conversely, other plants, including marijuana and Christmas cactus, will eventually flower even without an inductive photoperiod; for them, an inductive photoperiod merely hastens flowering. This response is called **facultative photoperiodism.** In some plants, other factors influence the photoperiodic requirement for flowering. For example, poinsettias are short-day plants at high temperatures and long-day plants at lower temperatures.

Do Plants Measure Day or Night?

Plant physiologists Karl Hamner and James Bonner continued Garner's and Allard's work by studying the photoperiodism of the cocklebur, a short-day plant requiring 15 or fewer hours of light to flower. Hamner and Bonner used controlled-environment growth chambers to manipulate photoperiods. They were startled to discover that plants responded to the length of the dark period rather than light period. The cocklebur plants flowered only when the dark period exceeded 9 hours.

Hamner and Bonner also discovered that flowering did not occur if a 1-minute flash of light interrupted the dark period, even if darkness exceeded the required 9 hours. Similar experiments in which darkness interrupted the light period had no effect on flowering. Furthermore, a long-day plant flowering on a photoperiod of 16 hours light to 8 hours dark will also flower on a photoperiod of 8 hours light to 16 hours dark if a 1-minute exposure to light interrupts the dark period. Other experiments with long- and short-day plants confirmed that flowering requires a specific period of uninterrupted dark, rather than uninterrupted light (fig. 29.11). Thus, short-day plants are really long-night plants, because they flower only if their uninterrupted dark period exceeds a critical length. Similarly, long-day plants are really short-night plants.

Phytochrome—A Pigment Controlling Photoperiodism

Because photoperiodism is a response to light, botanists suspected that it might be carried out by a pigment molecule whose structure changes when it absorbs light of a particular wavelength (fig. 29.12). The existence of such a pigment was suggested by the observation that red light inhibits flowering when it is used to interrupt the dark period. This inhibition can be reversed if the red-light interruption is immediately followed by far-red light, a form of red light corresponding to a wavelength at the edge of the visible portion of the electromagnetic spectrum (see fig. 8.2). From

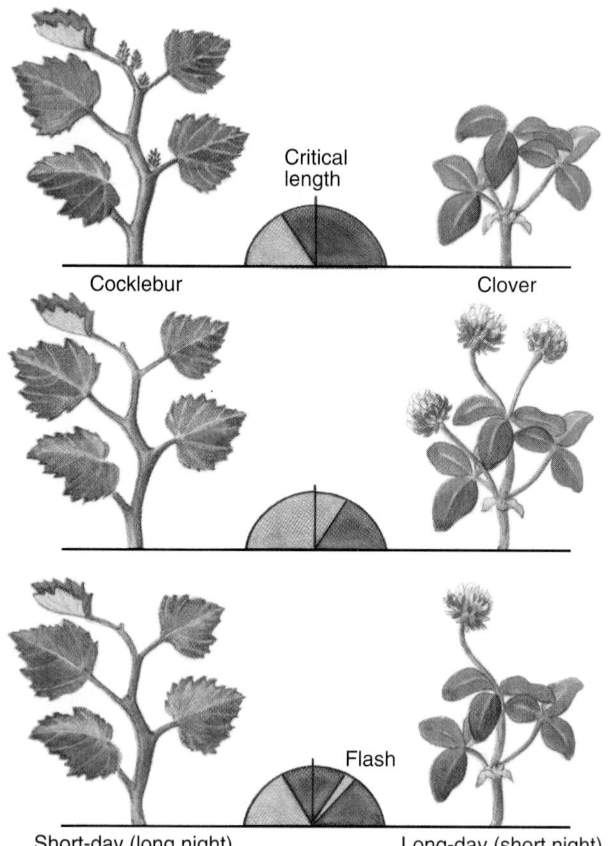

FIGURE 29.11

Length of day and night influences flowering in long-day plants such as clover and short-day plants such as cocklebur. Short-day plants require an uninterrupted dark period longer than a critical period, whereas long-day plants require a dark period shorter than a critical length. Interrupting the dark period of a short-day plant inhibits flowering.

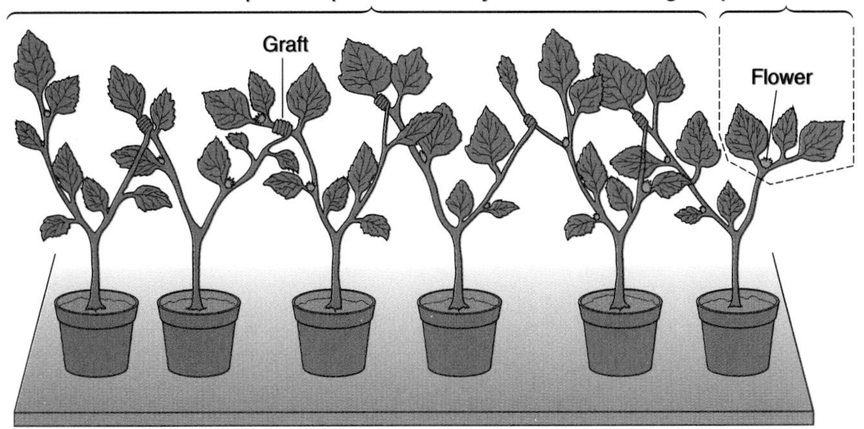

FIGURE 29.12

An experiment shows that induction of flowering requires a biochemical message. Although a branch of only one plant had the right amount of light and darkness, all of the plants flowered. This is because they were grafted together and could exchange sap.

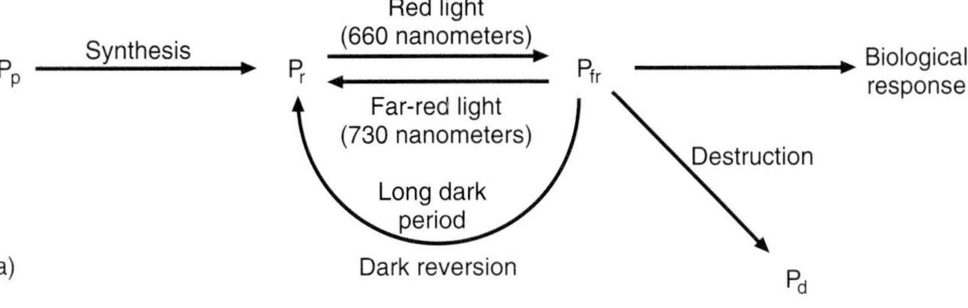

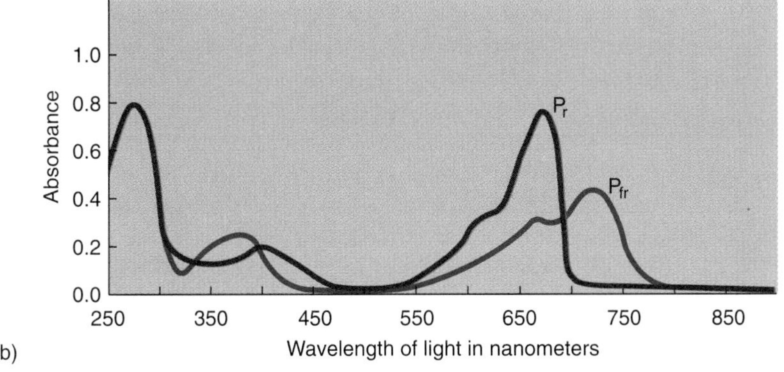

FIGURE 29.13

Two forms of phytochrome: P_r and P_{fr}. (*a*) P_{fr} stimulates biological responses, whereas P_r does not. (*b*) The two different forms of phytochrome absorb different wavelengths of light.

these observations, botanists concluded that the pigment exists in two forms, one that absorbs red light and the other that absorbs far-red light.

In 1959, researchers isolated and identified a pale blue pigment and called it **phytochrome.** As they hypothesized, phytochrome exists in two interconvertible forms, P_r and P_{fr}. The inactive form of phytochrome is synthesized as P_r. When P_r absorbs red light, it is converted to P_{fr}, the form of phytochrome that is active in flowering. P_{fr} promotes flowering in long-day plants and inhibits flowering in short-day plants. P_{fr} is converted back to P_r when it absorbs far-red light (fig. 29.13).

How does the P_r to P_{fr} ratio provide information on photoperiod? Sunlight has proportionately more red light than far-red light. Therefore, during the day, P_r is converted to P_{fr}. This abundance of P_{fr} could signal the plant that it is absorbing light. When the plant is placed in darkness, P_{fr} slowly is converted to P_r. This dark conversion of phytochrome to P_r was originally believed to initiate a set of reactions that enabled the plant to measure the length of darkness relative to the length of light. Indeed, this reasoning would explain why an uninterrupted period of dark (rather than light) controls photoperiodism. The dark reversion of P_{fr} to P_r requires only 3 to 4 hours, however, so this does not fully explain the light-dark sensing mechanism that controls flowering.

Some internal clocklike mechanism probably also influences flowering. A plant can respond to a photoperiodic initiation of flowering only if it is reproductively mature. Reaching reproductive maturity can take from a few days (as in the Japanese morning glory) or weeks (as in annuals) to several years. Some species of the century plant require decades of development before they flower (fig. 29.14), and the giant bamboo of Asia

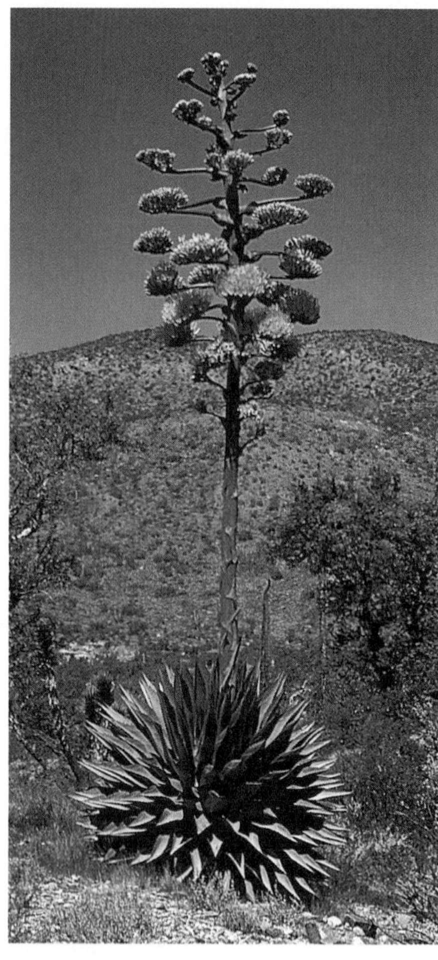

FIGURE 29.14

The century plant. These plants flower only once in 12 years and then die.

FIGURE 29.15

Etiolated seedlings (*left*) have had insufficient light for normal growth.

flowers only about every 33 or 66 years! Attaining reproductive maturity before flowering ensures that the plant has stored enough food to completely form and maintain its flowers. Moisture, soil conditions, nearby plants, and temperature may also influence flowering.

Other Responses Influenced by Photoperiod and Phytochrome

Phytochrome also affects seed germination. In seeds of lettuce and many weeds, red light stimulates germination, and far-red light inhibits it. Seeds alternately exposed to red and far-red light are affected only by the last exposure. Thus, germination occurs after exposure to red/far-red/red/far-red/red light, just as it does after a single exposure to red light. Treatments with far-red and red/far-red/red/far-red/red/far-red light result in no germination. Therefore, the phytochrome system can "inform" a seed whether sunlight is available for photosynthesis and thus promote germination under favorable conditions or inhibit germination until good conditions prevail. If seeds are buried too deeply in the soil, P_{fr} is absent (due to lack of sunlight), and germination does not occur.

Phytochrome also controls early seedling growth. Seedlings grown in the dark are **etiolated**—they have abnormally elongated stems, small roots and leaves, a pale color, and a spindly appearance (fig. 29.15). Bean sprouts used in Chinese cooking are etiolated. Etiolated plants rapidly elongate towards light before exhausting their food reserves. Normal growth replaces etiolated growth once the plants are exposed to the light. Red light controls transformation from etiolated to normal growth, and etiolated plants are very sensitive to red light. Normal growth ensues after exposure to only 1 minute of red light. If red light is followed immediately by exposure to far-red light, however, P_{fr} is converted to P_r, and etiolated growth continues.

Phytochrome may also help direct shoot phototropism, in addition to the influence of auxin. Light coming from only one direction would presumably create a gradient of P_r and P_{fr} across the stem. P_{fr} would be most abundant on the illuminated side of the stem, and P_r would be most abundant on the shaded side of the stem. This phytochrome gradient could bend a shoot as P_r promotes stem elongation and P_{fr} inhibits it.

Yet another function of phytochrome is to provide information when a plant is in the shade of taller plants. Chlorophyll in a plant canopy absorbs much of the red light of sunlight. By the time the light reaches underlying plants, such as those on a forest floor, less red light is available, and therefore less P_r is converted to P_{fr}. Since P_r promotes stem elongation (as in etiolated plants), the information that a plant is in shade that the phytochrome system provides can help a plant reach sunlight more rapidly.

Horticulturists use photoperiodism to produce flowers as the market demands. For example, chrysanthemums (a short-day plant) can be made available year-round by using shades to cre-

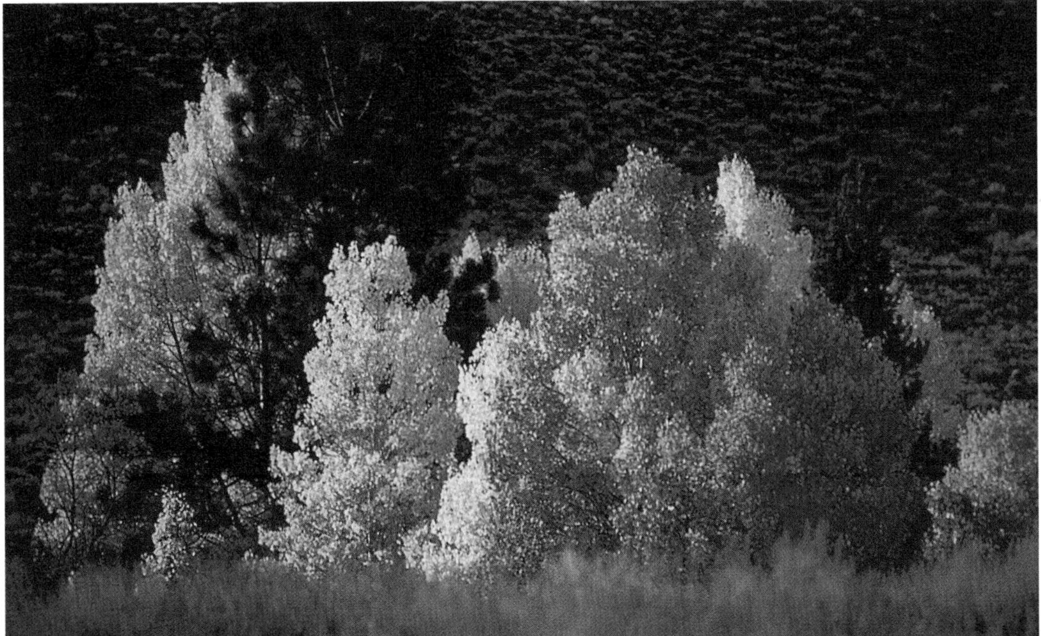

FIGURE 29.16

Fall leaves. Senescent leaves lose their chlorophyll, and carotenoid and anthocyanin pigments become visible, producing brilliant fall colors.

ate an inductive, short-day photoperiod. Similarly, poinsettias can be induced to flower near Christmas if they are kept in darkness for at least 14 hours a day for at least a month ahead of time.

Senescence

Senescence, or aging, is also a seasonal response of plants. Aging occurs at different rates in different species. The flowers of plants such as wood sorrel and heron's bill shrivel and die only a few hours after blooming. Slower senescence is seen in the colorful changes in leaves in autumn.

Whatever its duration, senescence is not merely a gradual cessation of growth but an energy-requiring process brought about by new metabolic activities. Leaf senescence begins during the shortening days of summer, as nutrients are mobilized. By the time a leaf is shed, most of its nutrients have long since been transported to the roots for storage. Fallen leaves are little more than cell walls, wastes, and remnants of nutrient-depleted protoplasm.

Destruction of chlorophyll in leaves is part of senescence. In autumn, the yellow, orange, and red **carotenoid pigments,** previously masked by chlorophyll, become visible. Senescing cells also produce pigments called **anthocyanins.** Loss of chlorophyll, visibility of carotenoids, and production of anthocyanins combine to create the spectacular colors of autumn leaves (fig. 29.16).

Dormancy

Before the onset of harsh environmental conditions such as cold or drought, plants often become **dormant** and enter a state of decreased metabolism. Like leaf senescence, dormancy entails structural and chemical changes. Cells synthesize sugars and amino acids, which function as antifreeze, preventing or minimizing cold damage. Growth inhibitors accumulate in buds, transforming them into winter buds covered by thick, protective scales. These changes in preparation for winter are called acclimation.

Growth resumes in the spring as a response to changes in photoperiod and/or temperature. Lengthening spring days awaken birch and red oak trees from dormancy. Fruit trees such as apple and cherry have a cold requirement to resume growth. This is why a warm period in December, before temperatures have really plummeted, will not stimulate apple and cherry trees to bud, but a similar warm-up in late February will induce growth. Apple and cherry trees transplanted to warm climates become late bloomers in the spring. The exact mechanism by which photoperiod or cold breaks dormancy is unknown, although hormonal changes are probably involved.

In some plants, factors other than photoperiod or temperature trigger dormancy. In many desert plants, for example, rainfall alone releases the plant from dormancy. In contrast, potatoes require a dry period before renewing growth.

Mastering Concepts

1. Define the four types of plants in relation to photoperiod.
2. Do plants respond to duration of light or dark?
3. Explain how phytochrome controls photoperiodism and seed germination.
4. What events occur during senescence and dormancy?

Circadian Rhythms

Some rhythmic responses in plants are not seasonal. Consider the common four-o'clock, which opens its flowers only in late afternoon, or the yellow flowers of evening primrose, which open only at nightfall. Similarly, the photonastic movements of prayer plants occur at the same time every day. Other daily, or **circadian, rhythms** in plants include cell division, stomatal opening, protein synthesis, nectar secretion, and growth regulator synthesis. Many eukaryotic organisms, including humans, have circadian rhythms.

29.2 biology in action

Plant Communication from Within

Plant communication goes beyond the well-studied hormones. For example, researchers recently discovered that virally-infected tobacco plants emit oil of wintergreen, which acts as an airborne signal that stimulates nearby plants to manufacture infection-fighting peptides. Oil of wintergreen is a familiar plant product. The tools of molecular biology are revealing previously unrecognized biochemical mechanisms that are also important internal signals in plant growth and development. Here are a few.

Steroids

We usually associate steroid hormones with their muscle-building effects in humans. Plants have steroids too. Botanists recognize a plant steroid called brassinolide but until recently didn't know its function.

Understanding the role of brassinolide began when geneticists working with the mustard weed *Arabidopsis thaliana* identified a gene similar in nucleotide sequence to a gene encoding an enzyme in mammals essential for producing male steroid sex hormones. Plants with this gene mutated cannot use cues from light to regulate growth and development. As a result, plants are stunted and too green (fig. 29.B). When researchers supplied the mutant plants with brassinolide, however, the plants developed normally. This result suggests that brassinolide stands in for the biochemical whose synthesis the inactivated gene's enzyme product normally catalyzes. Therefore, the steroid controls plant growth and development.

FIGURE 29.B
Steroids. The stunted *Arabidopsis thaliana* plant (*right*) cannot manufacture a steroid hormone that normally enables it to grow and develop in response to light. The dwarfed plant is dark green because of excess chlorophyll production, and it cannot sexually reproduce.

Peptides

Peptides have long been known to transmit messages within animal bodies. We now know that plants use these short amino acid chains too.

Researchers isolated the first plant communication peptide in 1991—a single microgram of a molecule called systemin from 30,000 tomato plants! Systemin enables plants to fight insect pathogens. The second plant communication peptide discovered, called ENOD40, enables certain root cells to ignore signals from auxin that would normally prevent them from growing. Under ENOD40's influence,

How do plants measure a day? Does a plant detect passage of a day internally, or does it use external, environmental factors? Evidence suggests at least some internal control. Several experiments demonstrate that circadian rhythms in many species do not coincide with a 24-hour day but may be a few hours longer or shorter. In addition, circadian rhythms often continue under laboratory conditions of constant light or dark.

Environmental factors, such as a change in photoperiod, however, can affect a plant's circadian rhythms. This environmentally controlled resynchronization of the biological clock is called **entrainment.** However, entrainment to a new environment is limited. If the new photoperiod differs too much from a plant's biological clock, the plant reverts to its internal rhythms. Also, a plant maintained in a modified photoperiod over a long period of time reverts to its natural rhythms when placed in constant light.

Biological clocks allow plants to synchronize their activities. Flowers of a particular species can open when pollinators are most likely to visit. For some plants, this timing is quite precise. Bats pollinate the flowers of genus *Cereus;* the blossoms must therefore open at night when bats are active. Furthermore, different sexes must flower within a few days of each other, because *Cereus* flowers persist for only a week. Figure 29.17 illustrates the ex-

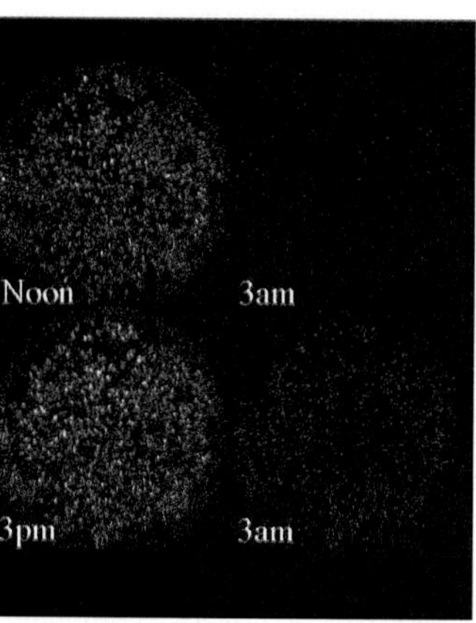

FIGURE 29.17
Demonstration of a circadian rhythm in plants. Researchers linked a photosynthesis gene expressed only during the day to a gene encoding luciferase, which causes firefly bioluminescence. A glowing seedling indicates the photosynthesis gene being expressed. In normal *Arabidopsis* plants (*top*), seedlings (growing in a petri dish) do not glow at 3 A.M. Mutant seedlings, however, do glow in the middle of the night. They can be seen as the few bright points in the petri dish on the lower right.

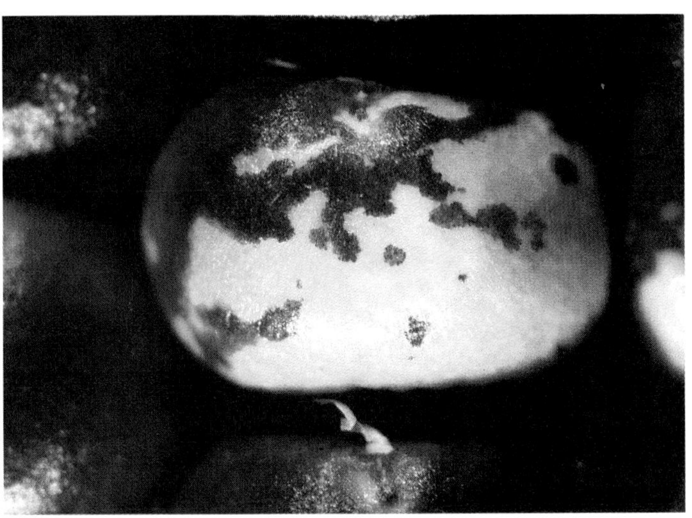

FIGURE 29.C

Peptides. The aleurone layer of this corn kernel does not completely cover the embryo because it did not receive sufficient peptide messages to finish migrating during early development.

these cells divide, forming root nodules in the presence of nitrogen-fixing bacteria. ENOD40 is a mere 13 amino acids long!

Corn plants with crinkled leaves and abnormal kernels led researchers to a gene encoding a receptor for a kinase. Kinases are enzymes that add phosphate groups to other molecules, and they are vital in signal transduction, the biochemical pathways that cells use to receive messages. The crinkled kernel (fig. 29.C) results from a failure to communicate—cells forming the aleurone layer on the embryo cover only half of the embryo, because of a block in the signal to migrate during development.

Genes for receptor kinases have interesting functions in several plant species. They control stem thickness in *Arabidopsis* and resistance to bacterial infection in rice. In broccoli and cauliflower, a kinase receptor prevents self-pollination. Pollen sends a signal to pistils on the same plant, and in response, the pistils somehow block pollen grains from sticking to them.

Methylation

Turning off certain genes at certain times is crucial to development. In humans and mice, a process called methylation places methyl (CH_3) groups on certain genes, which inactivates them. *Arabidopsis* mutants that cannot methylate sufficiently are quite deranged, more so in proportion to the degree of undermethylation. Mutant plants grow up to 27 extra leaves, take 47 days to flower compared with the normal 26 days, and develop five times the normal number of flower stalks!

Some researchers think that methylation functions as an internal clock in plants, because its absence in *Arabidopsis* mutants sends development out of sync. Meristems do not receive messages to cease division, and plant parts continue to form when they shouldn't, resulting in plants that, for example, have many extra petals. In addition, in the same region of a mutant plant, some cells differentiate as leaves, and others as flowers, as if the flower cells are developmentally ahead of the leaf cells.

periment described in the chapter opener, which shows a biological clock that controls the light reactions of photosynthesis.

This chapter has focused on plants' responses to environmental stimuli. To survive, communication is also essential within a plant's body. Biology in Action 29.2 details how genetic researchers are discovering communication molecules and mechanisms in plants that are surprisingly similar to those in animals—demonstrating, once again, that we do not know all there is to know about life.

SUMMARY

1. Plant hormones affect growth and development and may interact in complex ways. Auxins stimulate cell elongation in shoot tips, embryos, young leaves, flowers, fruits, and pollen. **Gibberellins** stimulate cell division and elongation but act more slowly than auxins. **Cytokinins** stimulate mitosis in actively developing plant parts. Ethylene speeds ripening. **Abscisic acid** inhibits the growth-inducing effects of other hormones.

2. A **tropism** is a growth response toward or away from an environmental stimulus, usually caused when different parts of an organ or structure grow at different rates. In **phototropism,** light sends auxin to the shaded portion of the plant, stimulating growth towards the light. Shoot growth is a negative **gravitropism,** and root growth is a positive gravitropism. Auxin accumulation can cause these opposite responses because the responding tissues differ in their sensitivities to it. **Thigmotropism** is a response to touch.

3. Nastic movements are not oriented toward a stimulus. **Thigmonasty** is response to contact. Nastic response to light and dark is **photonasty,** caused by osmotic changes that differentially alter cell volume. **Thigmomorphogenesis** is a growth-inhibiting response to mechanical disturbance.

4. Plants sense seasonal and other environmental changes. **Photoperiodism** is the ability of a plant to measure length of day and night. Flowering can depend upon photoperiodism. **Short-day plants** flower only when the duration of light is less than a critical length, whereas **long-day plants** require a light period longer than a critical length. **Intermediate-day plants** require days of intermediate length to flower. **Day-neutral plants** do not use light or dark cues to flower. The type of plant

Plant Clocks~

continued from page 576.

Kay and Millar knew that daily light/dark cycles regulate the light-dependent reactions of photosynthesis. They also knew of a gene, which encodes chlorophyll-a/b-binding protein (CAB), in *Arabidopsis thaliana* that is active during the day, but not at night, and presumably functions in the light-dependent reactions of photosynthesis. But how could the researchers see the gene in action? That's where the fireflies come in.

As everyone knows, fireflies glow. Biology in Action 6.1 explains how, in the firefly, a biochemical called luciferin reacts with ATP to yield another biochemical that, when acted upon by luciferase in the presence of oxygen, produces a flash of light. Kay and Millar linked the CAB gene to the firefly luciferase gene. They then exposed *Arabidopsis* seeds to mutagens, grew the seedlings, sprayed them with luciferin, and watched and photographed what happened. Collecting the data was easy—looking for petri dishes that glowed when most of the others were "asleep."

The experiment worked brilliantly, quite literally. Most of the 4,500 seedlings glowed during the day but not at night, as expected. Twenty-six seedlings glowed in the middle of the night, however, indicating that they could no longer receive, or respond to, environmental cues not to express the CAB gene. The mutant plants activated the gene in cycles ranging from 21 to 28 hours—but not the normal 24.

determines the season when it flowers. Plants requiring a precise photoperiod display **obligate photoperiodism,** whereas plants whose flowering is hastened by a certain photoperiod display **facultative photoperiodism.** Plants may respond to length of darkness rather than length of daylight.

5. A plant pigment, **phytochrome,** controls response to light. The inactive form, P_r, absorbs red light to become P_{fr}, the active form. P_{fr} promotes flowering of long-day plants and inhibits flowering of short-day plants. It is reconverted to P_r by absorbing far-red light. The ratio between these two forms provides information about daylight because sunlight has more red than far-red light. Phytochrome also controls early seedling growth, provides information about shading, and helps direct shoot phototropism and seed germination.

6. Senescence is an active and passive cessation of growth. Growth becomes dormant during cold or dry times and resumes when environmental conditions are more favorable. Internal biological clocks control daily responses, or **circadian rhythms.** Environmental changes can alter, or **entrain,** these clocks.

TO REVIEW . . .

1. Why have some plant hormones been known for decades, but other biochemicals that plants use for communication were discovered only recently?

2. How do the actions of gibberellin and auxin differ?

3. A tendril's coiling pattern can be due to thigmotropism or thigmonastic coiling. What is the difference between the two responses?

4. Describe how ion movement influences leaf movement in three plant species.

5. Give examples of how a tropism, nastic response, and flowering response to photoperiod are adaptive.

6. What experimental evidence supports the idea that a short-day plant could be more accurately described as a long-night plant?

7. How do auxin and phytochrome affect shoot phototropism?

TO THINK ABOUT . . .

1. Describe the steps of the scientific method that researchers used to demonstrate a circadian rhythm controlling photosynthesis in *Arabidopsis thaliana.*

2. How can farmers use different plant hormones, or their synthetic equivalents, in their crops?

3. You want to make a fruit salad for a barbecue tomorrow, but the bananas are not ripe yet. Which plant biochemical could you use to hasten ripening?

4. Why doesn't rain cause the leaves of a Venus's flytrap to close?

5. Although spinach and ragweed each require 14 hours of sunlight to flower, spinach flowers in the summer and ragweed (as hay fever sufferers can attest) flowers in the fall. Explain the seasonal difference in flowering.

TO LEARN MORE . . .

References and Resources*

Laferriere, Joseph E. September 1993. Competitive gravitropic and phototropic stimuli in *Coleus*. *The American Biology Teacher,* vol. 55. How to demonstrate response to gravity and light in a common houseplant.

Lewis, Ricki. December 11, 1995. Chronobiology researchers say their field's time has come. *The Scientist.* Diverse species exhibit circadian rhythms.

Marx, Jean. September 6, 1996. Plants, like animals, may make use of peptide signals. *Science,* vol. 273. Inner communication is important to plant survival.

Millar, Andrew J., et al. February 24, 1995. Circadian clock mutants in *Arabidopsis* identified by luciferase imaging. *Science,* vol. 267. Borrowing a firefly gene helped researchers watch the rhythms of photosynthesis.

Pennisi, Elizabeth, August 2, 1996. Chemical shackles for genes? *Science,* vol. 373. In normal development, certain genes must shut down at certain times.

Russell, David W. April 19, 1996. Green light for steroid hormones. *Science,* vol. 272. Plants use steroid hormones.

Seymour, Roger S. March 1997. Plants that warm themselves. *Scientific American.* Some plants can alter their metabolism to produce heat.

Shulaev, Vladimir et al. February 20, 1997. Airborne signalling by methyl salicylate in plant pathogen resistance. *Nature,* vol. 385. Infected plants that give off wintergreen scent may be signalling other plants of the threat.

**Life*'s website features frequently updated references.

Hippopotamus affection.

Unit 8

Animal Life

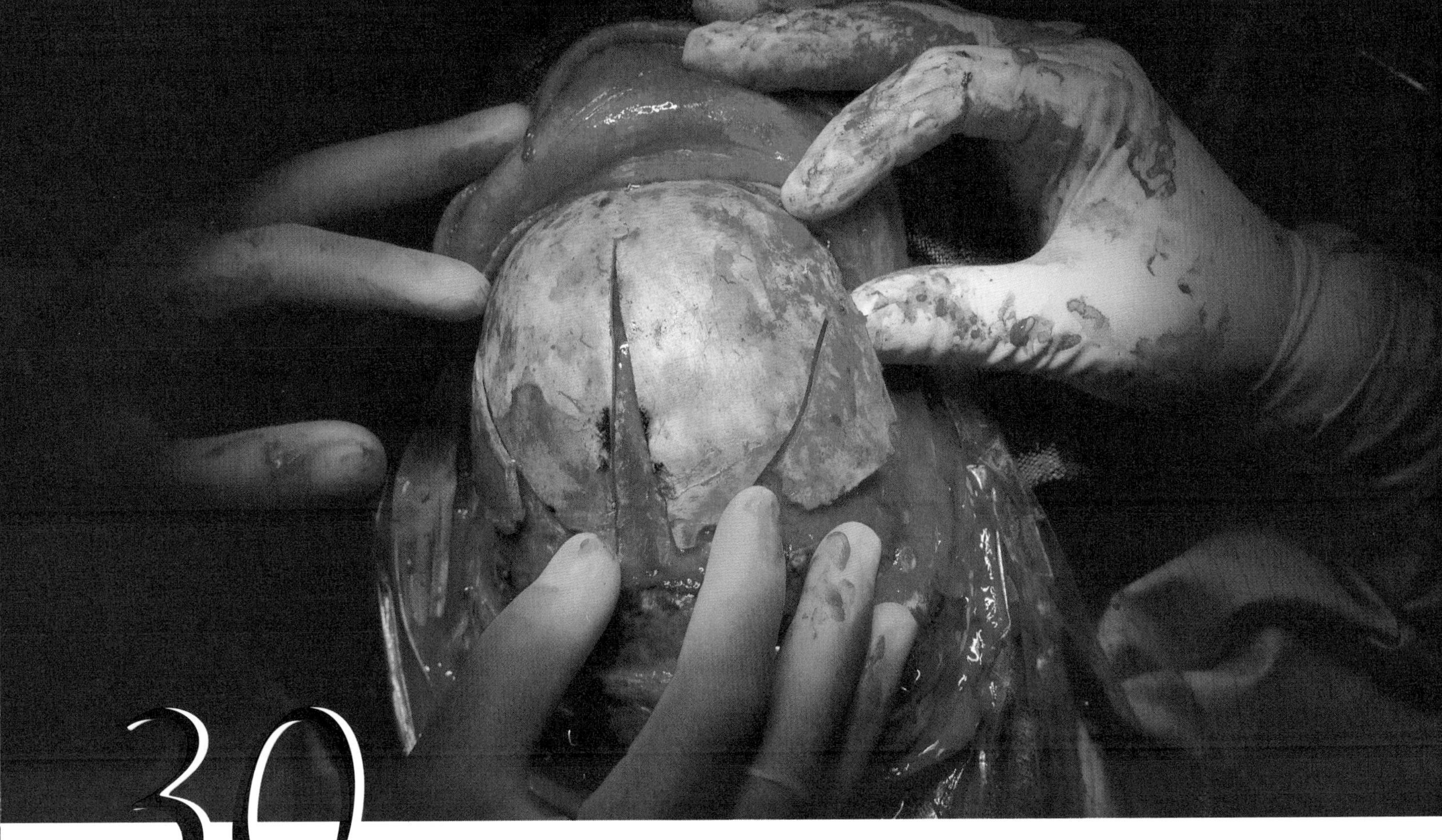

Surgical procedures reveal layers of tissues. This surgery is being performed on a 2-year-old's skull and brain.

30 Animal Tissues

chapter outline

Specialized Cells 594
Epithelium
Connective Tissues
Nervous Tissue
Muscle Tissue

A Sample Organ System—The Integument 598
A Closer Look at Human Skin
Specializations of the Integument

Replacing Damaged Tissues 603
Transplanted Tissues and Organs
Tissue Engineering

One in a Million

Organ systems interact. When one fails, others malfunction too. Amy Porter learned this quite painfully, when a standard antibiotic drug taken to treat a minor infection had a drastic, delayed side effect—Amy's entire epidermis fell off!

In early October 1995, the 28-year-old Amy noticed a minor skin rash. Within hours, the rash turned to blisters that soon covered her entire body. Amy also had a raging fever. Her husband took her to an emergency room, where a dermatologist diagnosed a severe form of Stevens-Johnson syndrome. In short, Amy's skin was reacting as if she had suffered devastating burns everywhere. Amy's prognosis was bleak: Only one person in 2 million develops Stevens-Johnson syndrome; only 10% survive. Because of her severe case, Amy's odds were even worse.

to be continued . . .

Specialized Cells

Animal cells are variations on a central theme of a sac of interacting organelles. Cells have diverse structures and functions that reflect which genes they express. Specialized, or differentiated, cells that function together constitute tissues. In complex animals, tissues build organs, and organs connect and function coordinately, forming organ systems (fig. 30.1).

The chapters in this unit are also, in a sense, variations on a theme. Each chapter examines a particular human organ system, after considering how some other animals meet the challenges that the organ system addresses. This initial chapter introduces the basic tissue types of animals, focusing on the human.

A human body consists of about 200 different types of specialized cells that form four basic tissue types:

- **Epithelial tissue** covers and lines organs.
- **Connective tissue** provides support and other functions.
- **Nervous tissue** forms rapid communication networks between cells.
- **Muscular tissue** contracts, powering the movements of life.

Figure 30.2 summarizes tissue types.

Epithelium

An animal's body has many surfaces, from the most obvious, the skin or **integument,** to those on the outsides and insides of organs. Epithelium lines structures with closely aggregated cell sheets and tubules. Cell shapes and number of cell layers distinguish different types of epithelia.

A layer of epithelium one cell thick is called **simple epithelium.** Layers that are two or more cells thick form **stratified epithelium.** A single layer of cells whose nuclei are at different levels gives the illusion of stratification (layering) and is therefore termed **pseudostratified epithelium.**

Flat epithelial cells are termed **squamous.** In the top layer of human skin, squamous epithelium accumulates a hard protein called **keratin.** Cells in the uppermost layer become so thin that they easily flake off, such as after a brisk rubbing with a towel. Squamous cells lining the mouth, esophagus, and vagina, however, do not produce much keratin and are therefore softer. Epithelial cells can also be cube-shaped (**cuboidal**) or tall (**columnar**). Table 30.1 illustrates types of epithelium.

Connective Tissues

Much of the human body consists of connective tissues. These tissues fill spaces, attach epithelium to other tissues, protect and cushion organs, and provide mechanical support. All connective tissues consist of cells embedded in a nonliving substance called a **matrix.**

A **fibroblast** is a common type of connective tissue cell. Fibroblasts manufacture two types of protein fibers that become part of the matrix—**collagen,** a flexible white protein that resists stretching, and **elastin,** a yellowish protein that stretches readily, like a rubber band. The matrix also includes a thin gel of proteoglycans, which are complex carbohydrates linked to proteins.

The major types of connective tissues in the human body are **loose connective tissue, dense connective tissue, adipose tissue, blood, cartilage,** and **bone.** Matrix composition, types of fibers, cell specializations, and the proportion of cells to matrix distinguish each type of connective tissue. Table 30.2 summarizes and illustrates types of connective tissue.

Loose connective tissue is the "glue" of the body. It consists of widely spaced fibroblasts and a few **adipocytes** (fat cells) within a meshwork of collagen and elastin fibers. In contrast, dense connective tissue consists of more tightly packed tracts of collagen. Dense connective tissue forms ligaments, which bind bones to each other, and tendons, which connect muscles to bones, and much of the middle layer of skin.

Adipocytes form from fibroblasts that accumulate enormous amounts of lipid (fat), which pushes the nuclei to the cell periphery. Adipose tissue insulates, cushions joints, protects organs, and stores energy.

Blood is a complex mixture of different cell types suspended in a liquid matrix called **plasma.** Blood cells circulate through the body, whereas other connective tissue cells do not migrate. **Red blood cells** transport oxygen and constitute the bulk of the blood cells. **White blood cells** are less numerous than red blood cells and are of several varieties. White blood cells protect against infection and help clear the body of cells that have worn out or be-

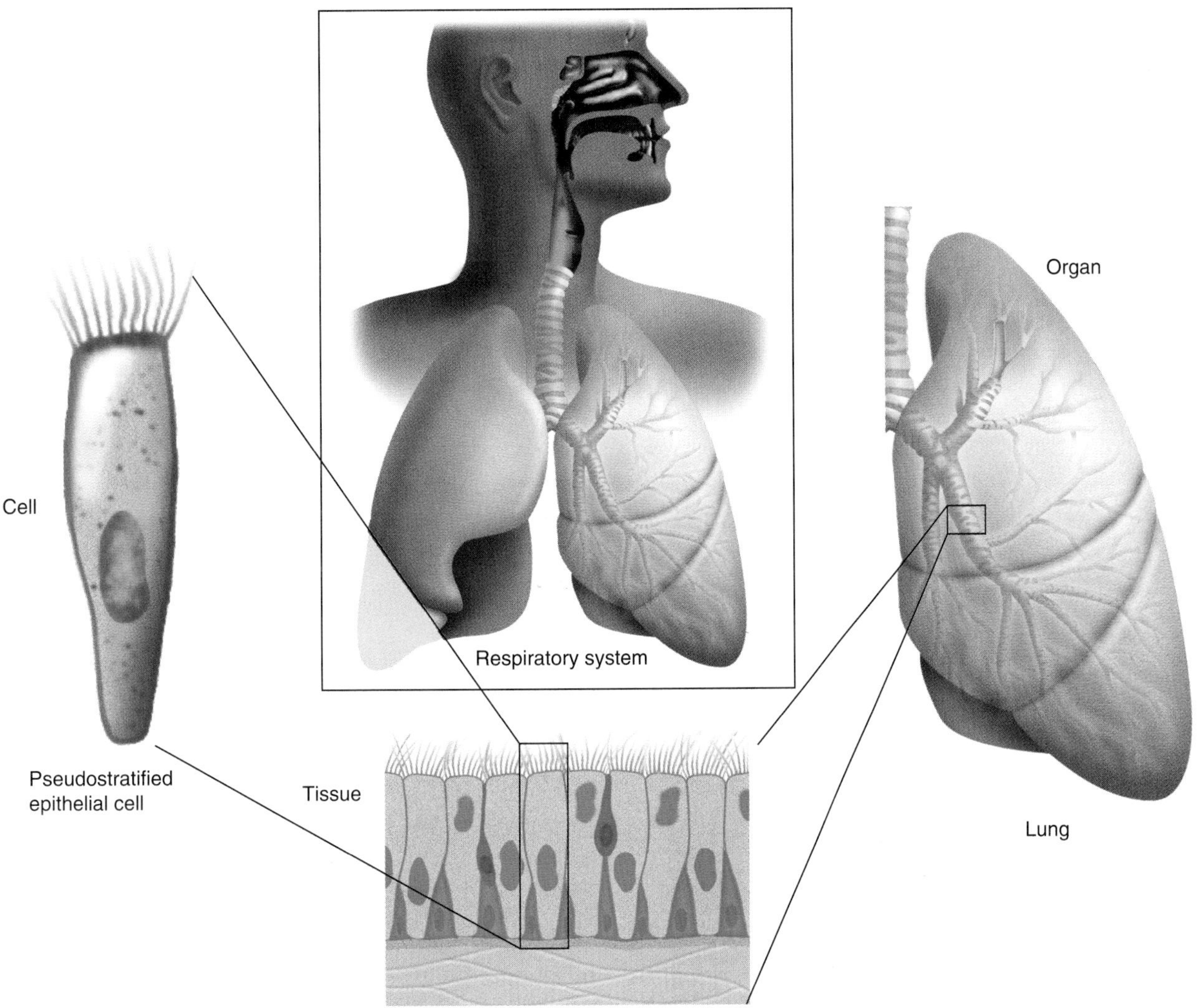

FIGURE 30.1
Differentiated cells interact to form tissues. Different tissues make up organs. Several interacting organs form organ systems.

come abnormal. Blood also contains cell fragments called **platelets,** which release chemicals that promote blood clotting. In a healthy individual, the different types of blood cells are present in specific proportions. Alterations in blood composition can signify illness. Chapters 35 ("The Circulatory System") and 39 ("The Immune System") discuss blood cells further.

Cartilage cushions organs and forms a structural framework that keeps tubular organs from collapsing, such as in the ear, nose, and respiratory passages. In joints, cartilage can sustain weight while allowing bones to move against one another. Cartilage consists of a single cell type, the **chondrocyte.** These cells lodge within oblong spaces called **lacunae** that are embedded in a collagen matrix. Cartilage grows as chondrocytes secrete collagen. Some cartilage also contains elastin. Chondrocytes have large nuclei and extensive endoplasmic reticula, a sign of their high protein secretion.

Bone also consists of cells within a matrix but with an important addition—mineral salts. The mineral hydroxyapatite, which contains calcium and phosphate, constitutes most of the mineral part, or phase, of bone. A labyrinth of tunnels forms as the hard mineral is deposited. The organic phase of bone consists almost entirely of collagen. Cells called **osteocytes** occupy lacunae (spaces) in the bone. Long narrow passageways, called **canaliculi,** connect lacunae. Osteocytes extend through canaliculi and touch each other.

Bone is an unusual tissue in that it is continually under reconstruction. Several cell types (other than osteocytes) remodel bone. **Osteoblasts** secrete bone matrix. **Osteoclasts** degrade bone matrix. They are large cells with many nuclei. **Osteoprogenitor cells** line bone passageways, and they differentiate into osteoblasts if growth or injury requires additional bone to form. Chapter 34 considers cartilage and bone further.

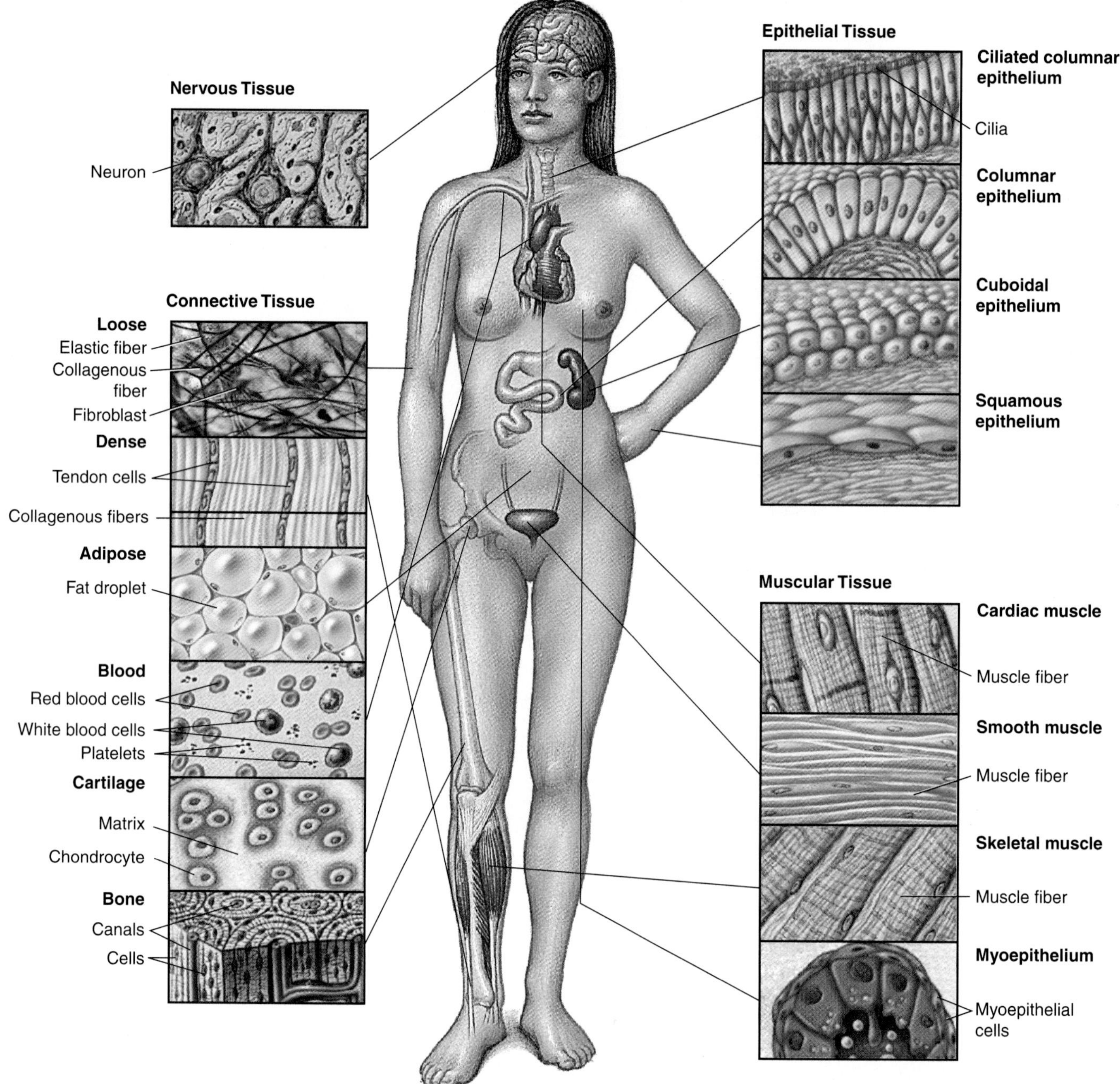

FIGURE 30.2

Human cell types. Organs in the human body consist of four basic tissue types. Epithelial tissue lines organs and other structures. Connective tissues provide form and support and include loose and dense connective tissue, adipose tissue, blood, cartilage, and bone. Nervous tissue provides cell-to-cell communication, and muscle cells contract, providing locomotion and finer movements.

Nervous Tissue

Nerve cells, called **neurons,** and associated cells called **neuroglia** make up nervous tissue. This tissue forms a communication network within an animal's body that enables it to receive information from the outside environment and respond to it.

Neurons can be quite long, with intricate branches. A typical neuron consists of an enlarged portion called the **cell body** (which contains the nucleus), a thick branch called the **axon,** and several thinner branches called **dendrites.** Most dendrites receive information from other neurons in the form of chemical

Table 30.1	Types of Epithelium
Type	**Examples**
Simple squamous	Lining of the heart
Simple cuboidal	Lining of the kidney tubules
Simple columnar	Lining of the oviduct
Stratified squamous	Lining of the vagina
Pseudostratified	Lining of the trachea

neurotransmitters. Environmental stimuli such as light, heat, or pressure may activate neurons specialized as **sensory receptors,** located in the skin and in sense organs. Arrival of a neurotransmitter or sensory stimulation alters the permeability of the dendrite's cell membrane and triggers an electrochemical impulse that then passes to another neuron or to a muscle or gland cell (fig. 30.3).

Neurons impinge upon one another to form intricate nerve networks. A single neuron in the human brain, for example, might receive thousands of incoming messages from other neurons at any one time.

Several types of neuroglia, as well as connective tissue, are found around and between neurons. One abundant type of neuroglia, **Schwann cells,** have very fatty cell membranes that wrap around axons and form insulating sheaths of the lipid myelin, which speeds nerve impulse conduction. Other neuroglia form scaffoldings that support highly branched neurons, and some supply nutrients and growth factors to neurons or remove ions and neurotransmitters that accumulate in synapses. Chapters 31 and 32 examine nervous tissue.

Muscle Tissue

Muscle tissue provides voluntary and involuntary movements. Muscle cells contract when two types of filaments, composed of the proteins **actin** and **myosin,** slide past one another, thus

Table 30.2 Types of Connective Tissue

Tissue Type	Cells	Matrix Composition	Cells to Matrix	Site
Loose connective tissue	Fibroblasts, adipose and white blood cells	Loose elastin and collagen networks	High	Under skin
Dense connective tissue	Fibroblasts	Dense elastin and collagen networks	Low	Ligaments and tendons
Adipose tissue	Adipocytes	Cells with abundant lipid	High	Beneath skin, between muscles, around heart and joints
Blood	Red and white blood cells, platelets	Plasma	High	Throughout the body
Cartilage	Chondrocytes	Collagen	Low	Ears, joints, bone ends, respiratory passages, embryonic skeleton
Bone	Osteoclasts, osteoblasts, osteocytes, osteoprogenitor cells	Collagen, minerals	Low	Skeleton

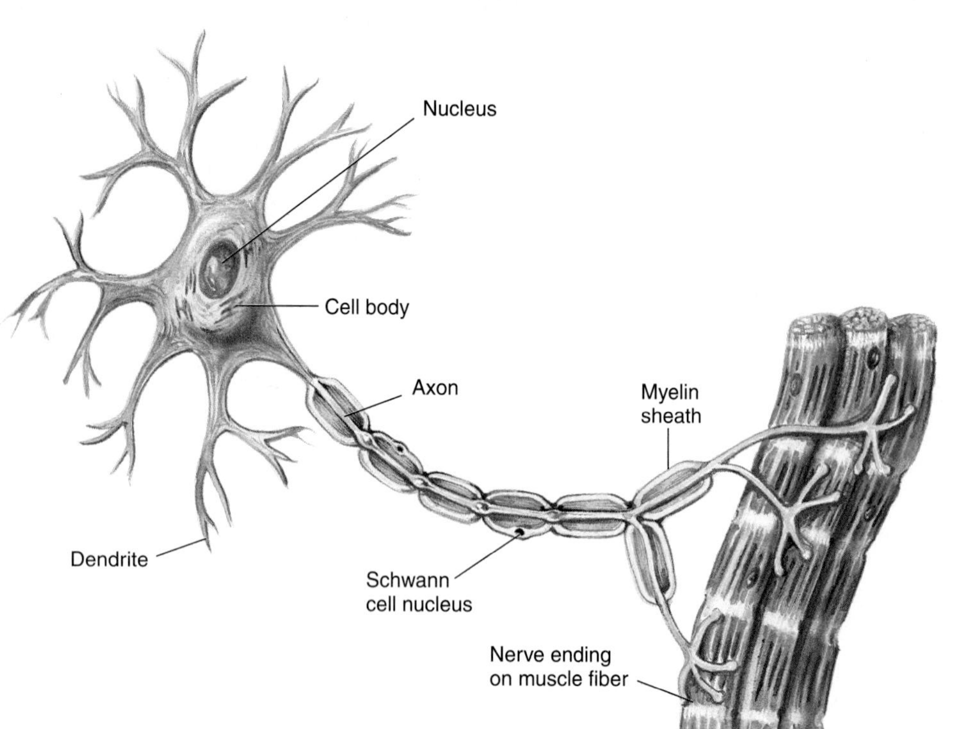

FIGURE 30.3

Some highly branched neurons impinge upon muscle cells, transmitting the message to contract.

shortening their total length. Other, less abundant proteins are necessary for muscle contraction, too. Muscle cells have many mitochondria, which provide the energy for contraction.

Contractile cells are of four types: **skeletal, cardiac,** and **smooth muscle cells** (see fig. 34.13) and **myoepithelial cells,** which have characteristics of both epithelium and muscle.

A fiber of skeletal muscle consists of one long cell with many nuclei. It appears striped, or striated, when viewed under a microscope because the actin and myosin proteins align. Skeletal muscle provides voluntary movements. Cardiac muscle, found only in the heart, is striated, but cells have single nuclei. Disclike structures join cardiac muscle cells to each other. Smooth muscle is not striated, and its involuntary contractions are slower than those of other contractile cells. Smooth muscle cells pulsate along the digestive tract, helping to move food along.

Myoepithelial cells, which are not striated, are in epithelium such as skin. These cells both form linings and contract, expelling secretory products, such as milk, saliva, and sweat from glands.

Mastering Concepts

1. What are the four types of tissues in humans?
2. Describe the types of epithelium.
3. In what way are the different types of connective tissue alike?

A Sample Organ System—The Integument

The chapters of this unit examine the organ systems defined in table 30.3 and depicted in figure 30.4. To illustrate how cells and tissues make up organs, we look now at the integumentary system—the coverings of animal bodies.

The integumentary system has several functions. As the boundary between an animal and its surroundings, the integument obviously protects. In many complex animals, immune system cells in the integument add to the barrier function by recognizing and destroying infectious microorganisms and initiating an inflammatory response to injury. The immune system is the subject of chapter 39.

A second function of skin is sensory, the topic of chapter 32. Sensory receptors in the skin enable us to distinguish hot from cold and a pinch from a caress. Skin also prevents water loss and helps to regulate body temperature. Glands that are part of the skin secrete sweat, oils, and scents.

A Closer Look at Human Skin

The structure of human skin illustrates how cells aggregate and form tissues. Human skin has several layers (fig. 30.5). The outermost layer, the **epidermis,** is predominantly stratified squamous epithelium. Scattered cells called **melanocytes** produce melanin pigment, which imparts color to skin (fig. 30.6). Epidermal cells produce and accumulate keratin, which hard-

Table 30.3 Human Organ Systems

System	Functions
Nervous system	Detects, interprets, and responds to stimuli from outside and within the body. Coordinates all organ functions.
Endocrine system	Produces hormones and functions coordinately with the nervous system to control many body functions, including reproduction, response to stress, and metabolism.
Skeletal system	Provides framework for muscles to attach, making movement possible. Houses bone marrow. Protects soft organs. Stores minerals.
Muscular system	Enables body to locomote and provides finer movements, heartbeat, digestion, lung function.
Circulatory system	Vessels carry blood throughout the body, nourishing cells, delivering oxygen, and removing wastes.
Respiratory system	Delivers oxygen to blood and removes carbon dioxide from cells. Controls blood pH.
Digestive system	Breaks down nutrients into chemical components that are small enough to enter the circulation. Eliminates undigested food.
Urinary system	Excretes nitrogenous wastes and maintains volume and composition of body fluids.
Immune system	Protects body from infection, injury, and cancer.
Reproductive system	Enables an individual to manufacture gametes and the female to carry and give birth to offspring.
Integumentary system	Protects the body, controls temperature, and conserves water.

ens them into a protective scaliness. Cells in the deepest epidermal layer can divide, and they continually push progeny cells upwards. As a result, the outermost skin layer constantly renews itself.

The middle layer of skin, the **dermis,** is mostly dense connective tissue, plus various epithelial derivatives. Myoepithelial cells in the dermis move hairs and propel secretions from glands. The dermis also contains blood vessels and nervous tissue. Neurons stimulate glands to secrete and convey information from sensory receptors to the brain for interpretation. Blood vessels carry nutrients and oxygen to the dermis and constrict to conserve heat and dilate to dissipate heat.

A thin layer called a **basement membrane** separates the epidermis from the dermis. Here, a type of protein called **laminin** binds the layers, much as a staple seals two sheets of cardboard together. Beneath the dermis lies a **subcutaneous layer** consisting of loose connective tissue and adipose tissue. The subcutaneous layer is not technically part of the skin. Altogether, the skin is an extensive organ system, comprising about 15% by weight of an adult's body.

The integrity of the epidermis and dermis is crucial to the skin's ability to protect and retain fluids. This is why severe burns threaten life. A group of inherited disorders, called **epidermolysis bullosa** (EB), further demonstrates the importance of adhesion between the skin layers (fig. 30.7).

EB causes the skin to blister from even slight touching. The different variants of the disease reflect abnormalities in different skin proteins that lie in different layers. EB caused by abnormal keratin is the mildest form and affects the epidermis. It causes blistering but does not permanently scar the skin. EB caused by abnormal collagen affects the dermis and scars the skin. The most severe form, junctional EB, affects laminin and disrupts the basement membrane. The skin literally comes apart. Affected individuals usually die within the first weeks of life as they lose fluids through the torn skin and become infected. These children have skinless patches where the layers rubbed off before birth.

Specializations of the Integument

Many animals have integuments consisting of an epidermis and a dermis. Animals also share specialized integumental structures that are similar in composition, but they appear as different as a hair, a feather, and a great variety of claws, nails, horns, beaks, and scales. These structures can have widely different functions. Hair insulates, horns defend, feathers provide locomotion and insulate, and claws enable an animal to scamper up a tree to escape a predator.

Epidermal Specializations

A hair grows from a group of cells in the epidermis called a **hair follicle,** which is anchored in the dermis and nourished by blood vessels there. Epidermal cells at the base of a hair follicle divide, pushing progeny cells up. These cells stiffen with keratin and die (fig. 30.8). Hairs are really just dead epidermal cells.

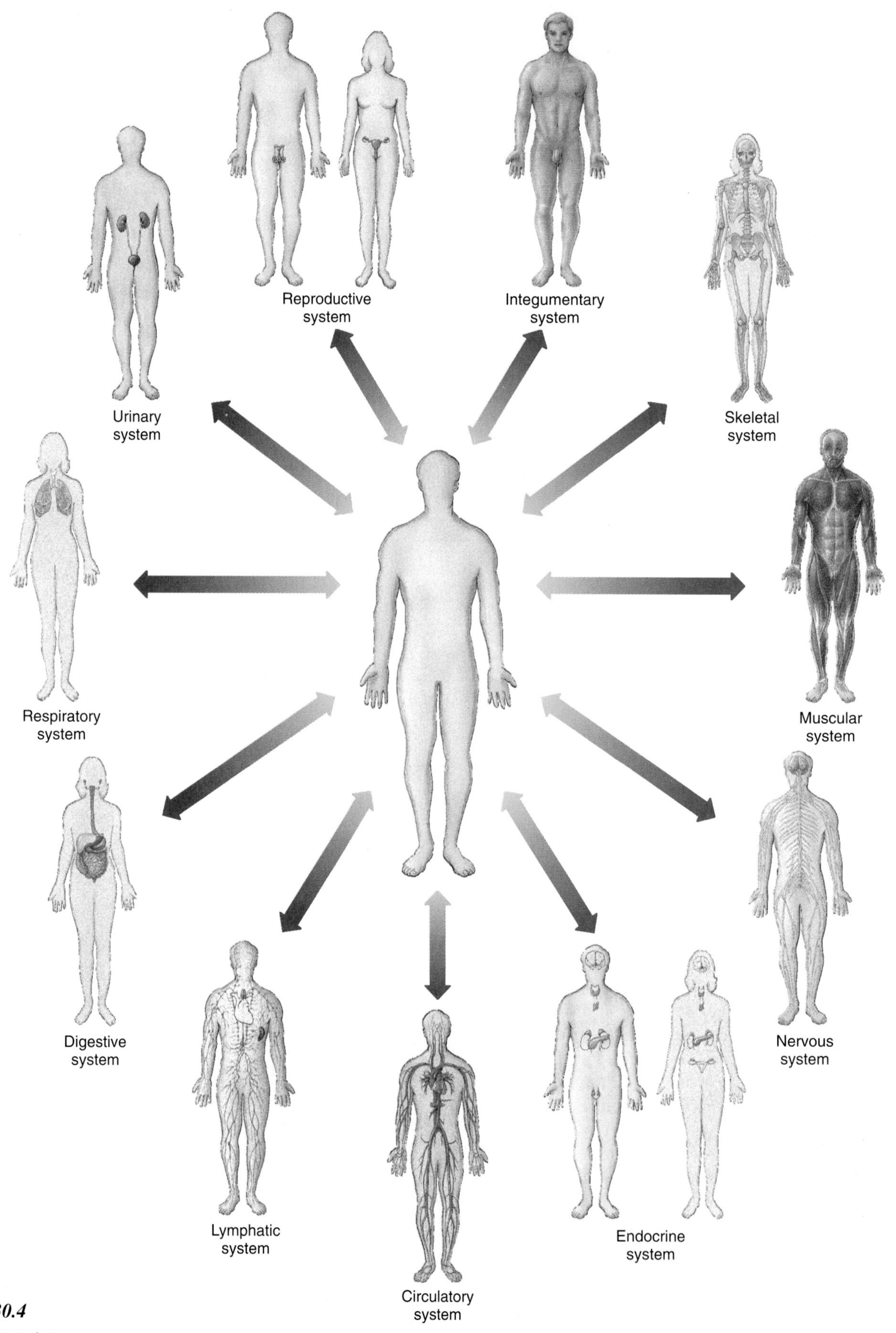

FIGURE 30.4
Human organ systems.

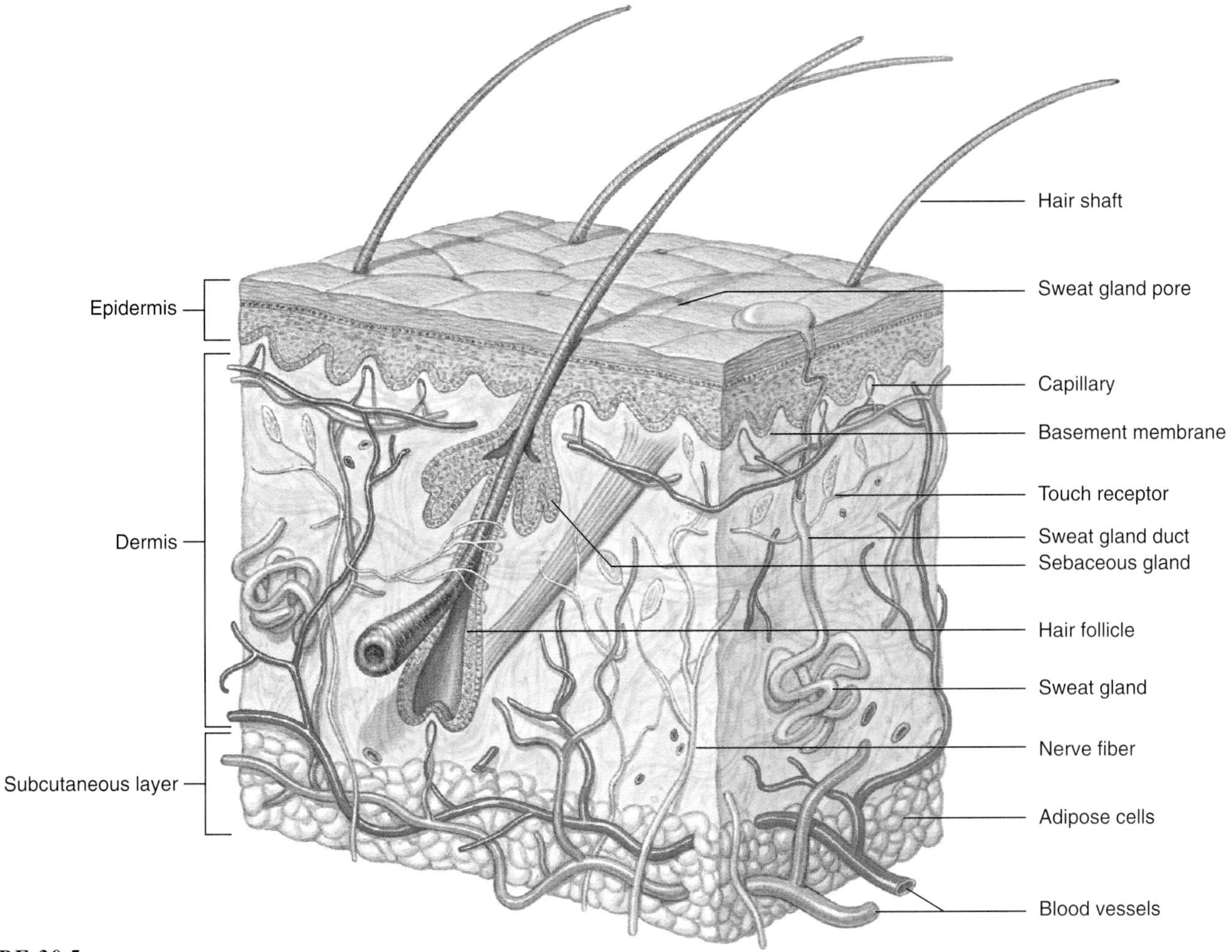

Hair shaft
Sweat gland pore
Epidermis
Capillary
Basement membrane
Touch receptor
Dermis
Sweat gland duct
Sebaceous gland
Hair follicle
Sweat gland
Nerve fiber
Subcutaneous layer
Adipose cells
Blood vessels

FIGURE 30.5

Human skin consists of the epidermis and the dermis, with the subcutaneous layer beneath it. A basement membrane separates the epidermis and the dermis. Associated structures include hair follicles, sweat glands, and sebaceous (oil) glands.

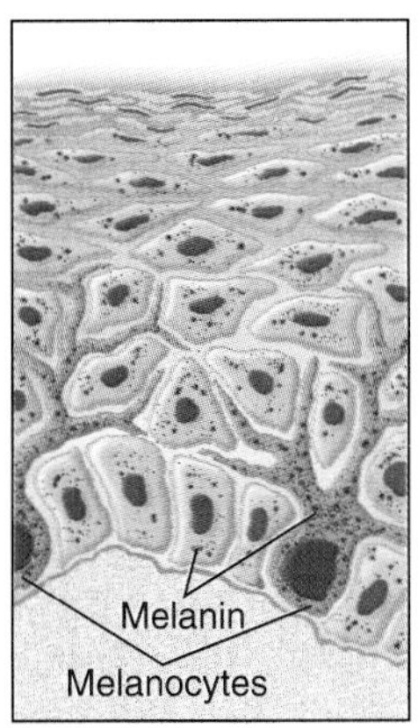

Light skin

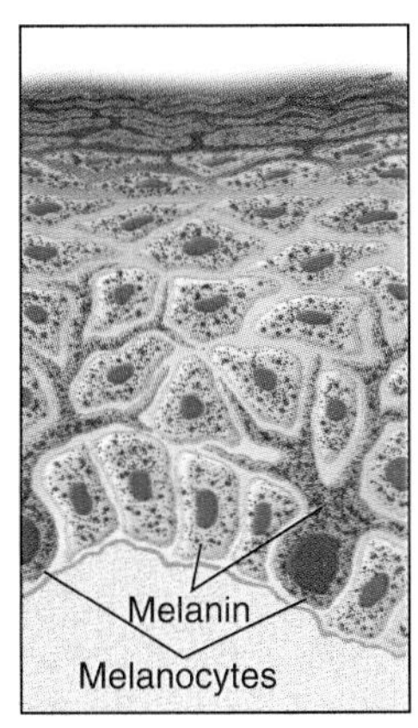

Dark skin

FIGURE 30.6

Melanocytes release the pigment melanin, which spreads through the epidermis and imparts color to the skin.

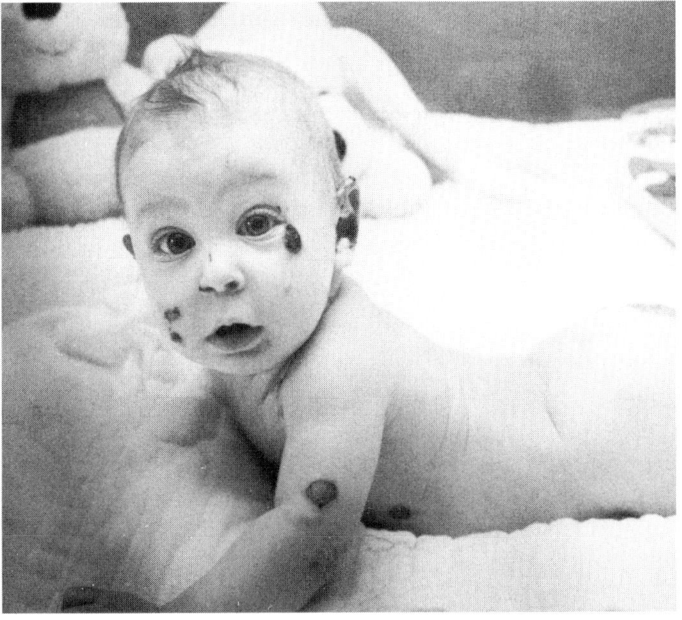

FIGURE 30.7

This child has epidermolysis bullosa. The skin blisters very easily because collagen in the dermis is abnormal and cannot form a tight network.

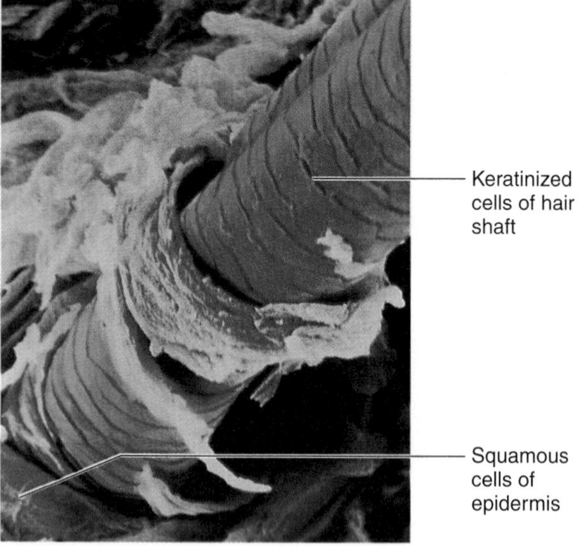

FIGURE 30.8

A scanning electron micrograph of a human hair emerging from the epidermis. ($\times$ 340) Note the scaliness of the cells that make up the hair.

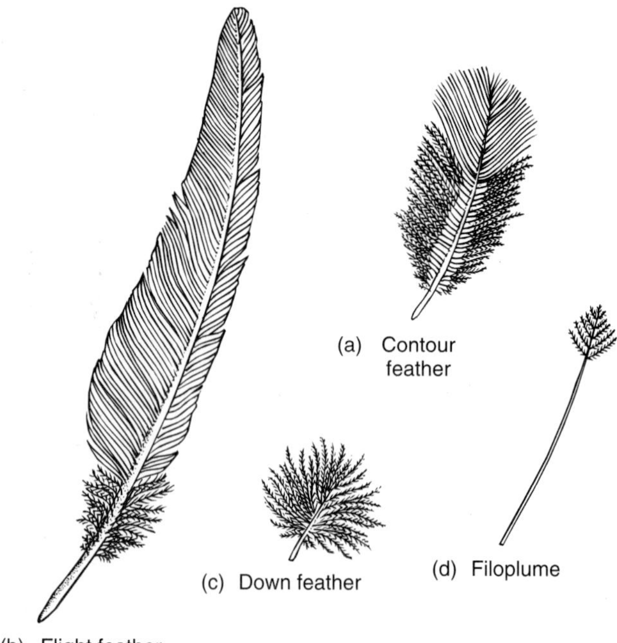

FIGURE 30.9

Feathers derive from a bird's epidermis. (*a*) Contour feathers provide the animal's shape. (*b*) Flight feathers make locomotion possible. (*c*) Down feathers insulate, and (*d*) filoplumes are displayed in various behaviors.

A feather is also an outgrowth of the epidermis. Like a hair, a feather is keratinized, but it has no blood or nerve supply. Feathers develop from **feather follicles,** which are extensions of the epidermis dipping into the dermis, much like hair follicles.

Feathers have different roles, depending upon their size, structure, and location (fig. 30.9). Contour feathers mold a bird's overall shape, which is important to its ability to fly. Flight feathers provide direct surfaces that make flying possible. In contrast, down feathers are the small, soft underlying feathers that insulate the bird. Filoplumes are colorful feathers used in mating displays.

Nails, claws, and hooves are keratinized, stratified squamous epithelium that are part of the epidermis. They differ in their position on the animal's digits (fig. 30.10). Nails are plates of very tightly packed keratinized cells on the fingers and toes of primates. Claws are curved and project beyond the digit in some amphibians and in most birds, reptiles, and mammals. Hooves are large keratinized plates on the ends of digits in deer, elk, and moose.

Dermal Specializations

Dermis is composed largely of sheets of collagen fibers. How tightly these fibers pack against each other, and their orientation, determines the integument's flexibility. Consider a shark.

A shark's skin is extremely tight, which enables the animal to swim rapidly through water without wrinkling, which would cause some turbulence. In shark's skin, collagen bundles lie at 45-degree angles to each other, much like in woven cloth (fig. 30.11). This pattern is adaptive in both sharks and garments. When the collagen or material fibers are pulled from above or the side, they do not deform much, if at all. If pulled from an oblique angle, however, the patch of collagen or material readily moves. This provides flexibility to the animal—and to a person moving about in a garment. Collagen bundles in human skin are not nearly as highly ordered, and as a result our skin sags and wrinkles easily, especially as dermal collagen stores fall as we age.

The dermis of birds differs from that of sharks and humans. It contains many blood vessels, sensory nerve endings, and smooth muscles. These tissues heat the dermis in the breast area of a bird, which is adaptive for brooding eggs.

Perhaps the most distinctive dermal derivative is **baleen,** which hangs in plates from the roof of certain whales' mouths. Baleen screens microscopic organisms (krill) from the water, which the whale consumes (see fig. 37.3). Baleen plates originate in the dermis and extend down into the epidermis, where they are covered with keratin.

Glands are part of the dermis, and their types, numbers, and distributions distinguish animal integuments. **Sweat glands** are epidermal invaginations into the dermis. Humans have two types of sweat glands. One type is present before puberty, produces a thin secretion, and is not associated with a hair follicle. This type of sweat gland cools the body when its secretion evaporates. The second type of sweat gland in humans is near a hair follicle, produces a thick, pungent form of sweat, and develops at or after puberty.

Not all mammals have sweat glands, and they appear in different places in different species. Humans and chimpanzees have widespread sweat glands. Elephants lack them. Rabbits have sweat glands on their lips, deer at the base of the tail, and duck-billed platypuses on the snout.

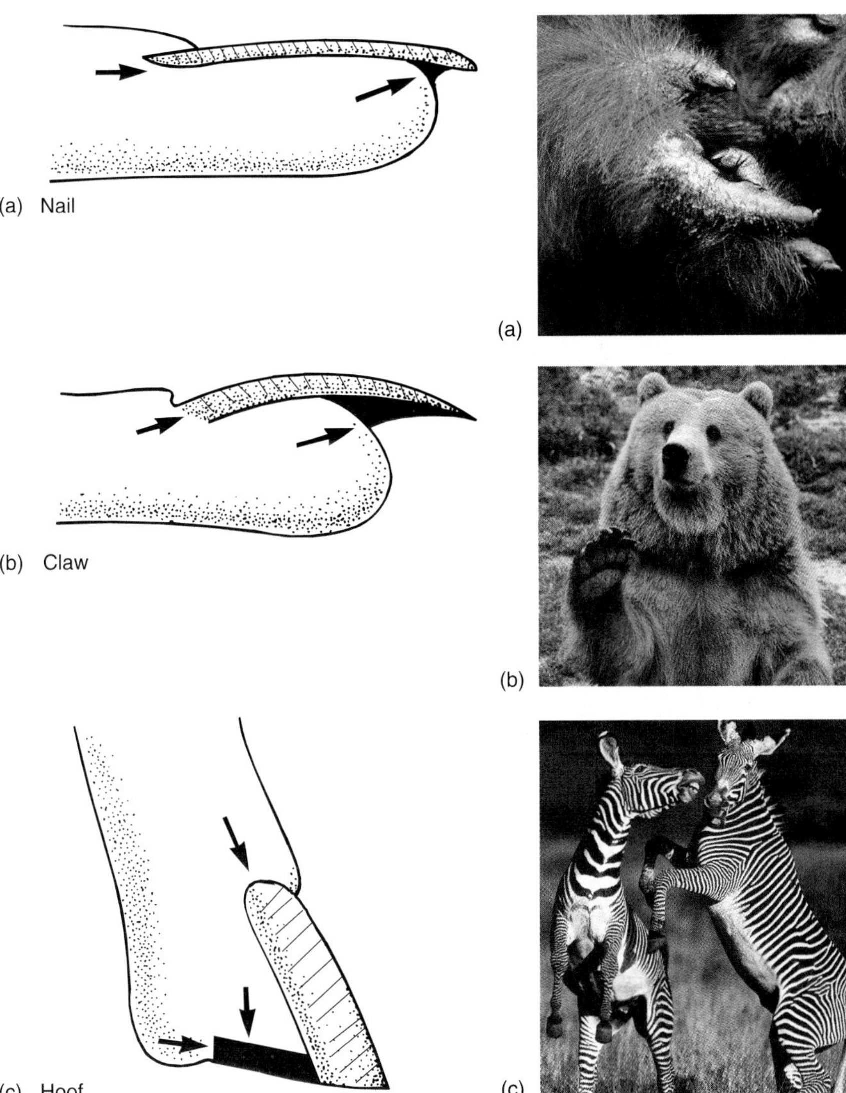

FIGURE 30.10
Nails, claws, and hooves are epidermal structures, each differing in its position in relation to the animal's digit.

Scent glands derive from sweat glands, and they too occur in different parts of the body in different species. Rodents, cats, and dogs have scent glands near the anus; camels on the back; rabbits on the chin; alligators on the lower jaw; and cattle on the feet and legs. Scents are important in communication, discussed in chapter 40. Many animals, including humans, have **sebaceous glands,** which secrete an oily substance called sebum that softens the skin and hair. **Mammary glands,** which produce milk in mammals, also derive from sweat glands.

Skin and Bone Combinations

Horns and antlers form when bone pushes the integument upwards, creating a skin or fur-covered protrusion (fig. 30.12). Horns are hollow; antlers are solid. In horns, the integument forms a tough coating on the bone. Sheep, cows, antelope, goats, wildebeests, and bison have horns. These outgrowths enlarge and persist for the lifetime of the animal. In some larger species, both sexes have horns, but they are typically larger in the males, who use them for displays of strength. In smaller species, females may lack horns.

Antlers are covered with an integument that is rich in blood vessels, creating a velvety texture. The skin falls away, leaving bare bone. Male deer, elk, and moose grow antlers, which are shed yearly.

Mastering Concepts

1. What functions does an integumentary system provide?
2. What are the layers of the human integumentary system?
3. What are some epidermal and dermal specializations?

Replacing Damaged Tissues

Damaged or diseased tissues can repair and/or regrow to differing extents. Medical technology can help this process by replacing damaged tissues to restore organ function and health. **Transplantation** uses organs from other individuals, and **tissue engineering** combines cells with synthetic materials to fashion implants.

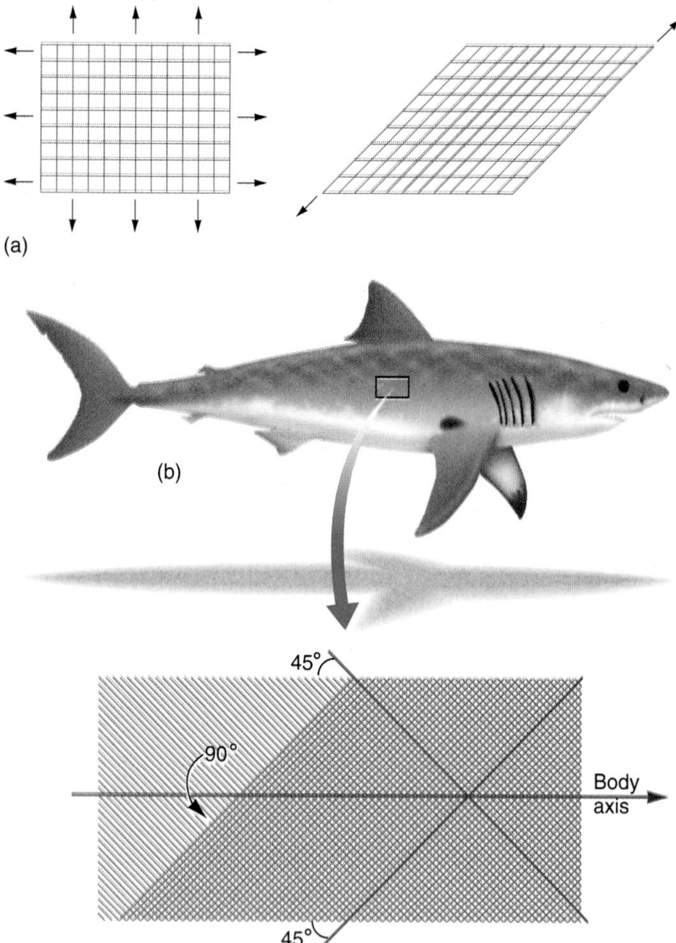

FIGURE 30.11

The collagen fibers in a shark's skin are very tightly arrayed, in a pattern similar to that of woven cloth (*a*). This organization results in a skin that is not easily deformed but remains flexible (*b*).

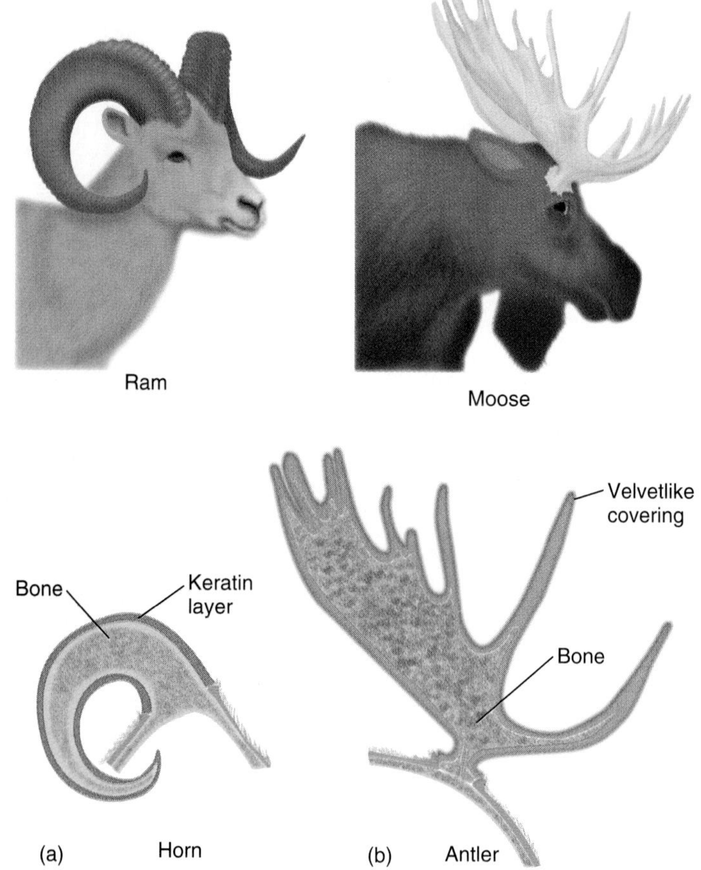

FIGURE 30.12

Horns and antlers form from skin and underlying bone. (*a*) In a horn, a hollow keratinized skin layer covers the bone. (*b*) In an antler, a velvety, blood-rich integument covers the bone for part of the year.

Transplanted Tissues and Organs

Transplanted tissues and organs include corneas, pancreases, kidneys, skin, livers, lungs, bone marrow, and hearts. The greatest challenge in an organ transplant—after obtaining the organ—is to prevent the recipient's immune system from rejecting a transplant, described at the beginning of chapter 5. Physicians minimize rejection by matching the cell surface pattern of donor to recipient and giving the recipient drugs to suppress immunity.

Transplant Types

The relationship of donor to recipient distinguishes types of transplants (table 30.4). In an **autograft,** tissue from one part of a person's body is transferred to another part of the body. Technically, therefore, it is not truly a transplant. A skin graft from the thigh to replace burned skin on the chest is an autograft. The immune system does not reject the graft, because it is from "self." An **isograft** is tissue from an identical twin. Because such twins are genetically identical, the recipient does not reject the transplant.

Table 30.4 Transplant Types

Type	Donor
Autograft	Self
Isograft	Identical twin
Allograft	Same species
Xenograft	Different species

An **allograft** comes from an individual who is not genetically identical to the recipient but is the same species. A kidney transplant from a relative or suitable donor is an allograft. In a **xenograft** tissue from one species is transplanted into another. Heart valves from pigs transplanted to humans are xenografts.

A transplant may be an entire organ, such as a kidney or a heart, or a small amount of tissue. For example, since 1990, more than 150 people with diabetes have received transplants of pancreas cells that produce insulin. The cells come from cadavers and are infused into a vein in the recipient's liver, where they secrete

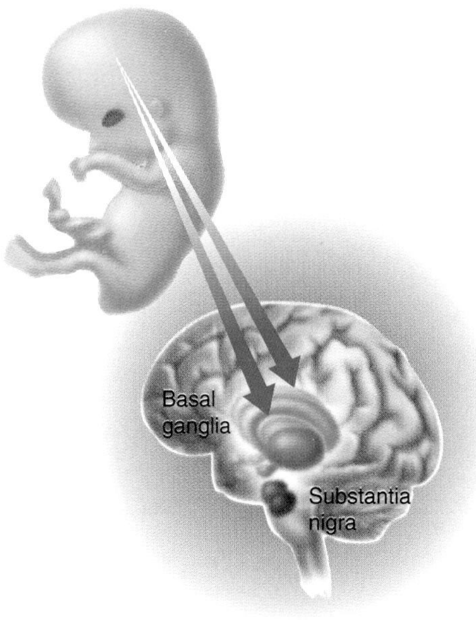

FIGURE 30.13

Tissue from the brains of human fetuses is used experimentally as transplants to alleviate the shaking movements of Parkinson disease in adults. In this disorder, a part of the brain called the substantia nigra does not make sufficient dopamine, a neurotransmitter. Normal movement requires that another brain region, the basal ganglia, obtain dopamine. Fetal brain tissue grafted to the basal ganglia supplies dopamine without alerting the recipient's immune system, leading to some improvement in symptoms.

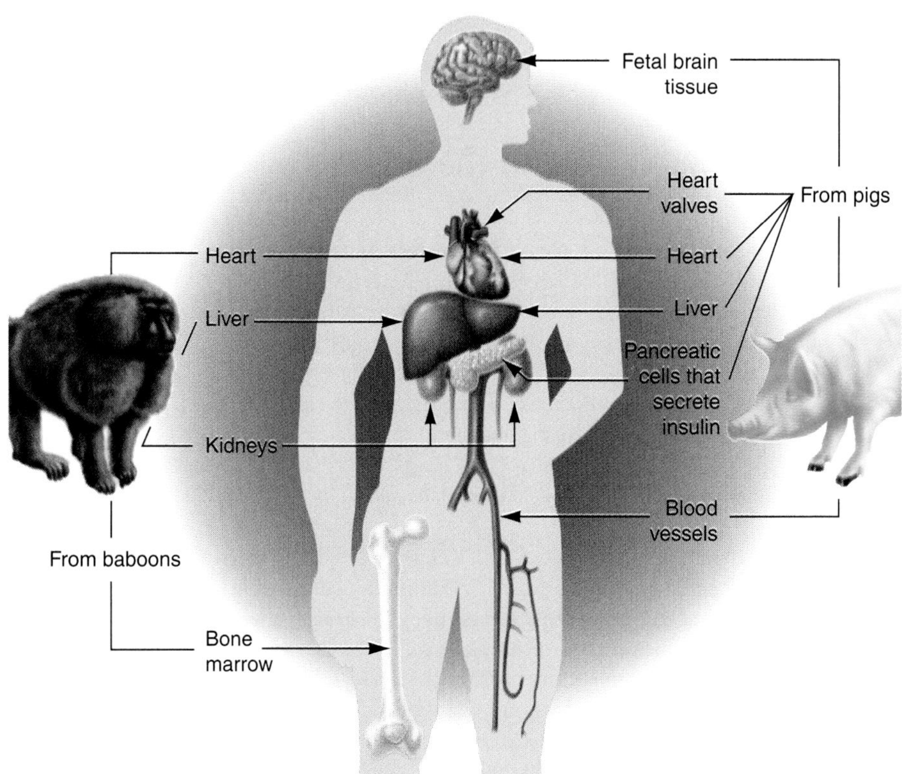

FIGURE 30.14

Baboons and pigs can provide tissues and organs for transplant.

insulin. Some transplants are parts of donated organs. One liver, for example, can be transplanted into several recipients, whose bodies regenerate full organs.

Controversial Transplants

Because of severe organ shortages, researchers are seeking alternate sources of donor tissue. Some sources of transplantable tissue are highly controversial—human fetal tissue, organs from newborns with fatal birth defects, and organs from nonhuman animals. Bioethics Connections 30.1, at the end of the chapter, explores another controversial use of human tissues.

From a biological viewpoint, fetal tissue is more suitable transplant material than adult tissue. Fetal cells have not yet formed all the cell surface molecules that stimulate tissue rejection. Plus, fetal cells can divide, whereas some of their adult counterparts cannot. For example, fetal neurons might be able to repopulate depleted brain tissue in people with Parkinson disease or to repair damaged spinal cords (fig. 30.13).

Because healthy fetal tissue is required (spontaneously aborted tissue is often abnormal), some people fear that women will intentionally become pregnant to sell healthy fetal tissue.

Another source of tissue for transplant is from babies with anencephaly. In this condition, much of the brain is missing, including the areas that control reasoning, awareness, and pain perception. The brain stem, which controls heartbeat and respiration, functions, so that the baby may live for a few days with breathing assistance but dies when life support ceases.

The controversy over this source of transplant tissue arises from the observation that if the baby dies naturally, lack of oxygen will damage its organs and render them useless for transplant. But discontinuing life support to harvest vital organs for transplant is killing the anencephalic newborn. In several cases parents of such children have pleaded to end life support and donate the organs, but physicians and courts have denied their requests. Only one transplant from an anencephalic infant has taken place. It saved an infant born in 1987 with a fatal heart defect.

Interest in using nonhuman animal parts for transplants has resurged since researchers have developed ways to remove rejection-provoking molecules from cell surfaces and even to genetically engineer pig cell surfaces to display human molecules. The pig is a good candidate for heart transplants because its heart is about the same size as that of a human, and its cardiovascular system is quite similar to ours (fig. 30.14). Pancreas cells from pigs could be used to treat diabetes, but it would take ten pigs to treat one human. Another experimental approach is to transplant pancreas cells from a type of freshwater fish called tilapia. This fish, which is farmed in the tropics, is genetically altered to produce human insulin. The baboon is another possible source of donor tissue.

Many people object to intentionally creating and nurturing animals to be sacrificed for their parts. Another concern is that use of tissue from other species will expose humans to viruses that do not normally infect a human, possibly causing new diseases that could then spread.

30.1 bioethics connections

From Death Row to the Internet

Health care professionals must memorize many details of human anatomy and physiology. Traditionally students have relied on atlases of drawings and dissecting cadavers. In 1994, the National Library of Medicine introduced a new way to study human anatomy and physiology, from every conceivable angle. In the Visible Human Project, a newly deceased man was presented on the Internet as 1,800 one-millimeter thick sections taken using magnetic resonance imaging (MRI) and computerized tomography (CT) scans (fig. 30.A). The cause of the anonymous Visible Man's death was officially listed as "drug overdose."

In 1996, word got out of the Visible Man's identity. He was Joseph Paul Jernigan, executed in 1993 at age 39 for killing a man in a burglary attempt. Jernigan had voluntarily left his body to science, a wish on file with the Anatomical Board of the State of Texas. He was executed by lethal injection—the "drug overdose."

After the execution, Jernigan's body was flown to the University of Colorado Medical Sciences Center, where, within 8 hours of death, the imaging began. Researchers saw it as a rare opportunity to obtain a healthy body soon after death. Medical ethicists concerned with the rights of death row inmates, however, saw the situation differently.

At the heart of the issue is the concept of informed consent. When Jernigan willed his body to science, did he imagine that every nuance of his anatomy would appear on computer screens throughout the world? More likely he envisioned a fate on a dissection table in a medical school, ethicists argue. But Jernigan's lawyer maintains that his client was not coerced into his donation, and that he willed his body to science in an honest attempt to make amends for his crime. Relatives of the man he murdered also applaud Jernigan's final act.

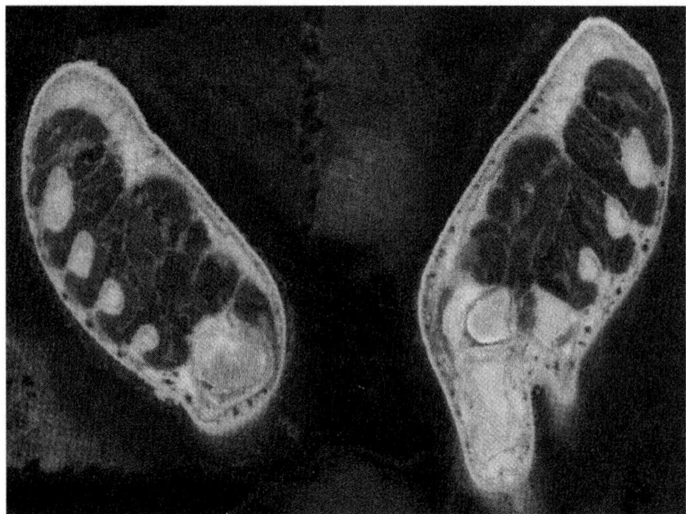

FIGURE 30.A
Images of the feet of executed man Joseph Paul Jernigan. He donated his body to science, but probably didn't anticipate appearing on the Internet.

Postscript: The Visible Man now has a partner on the Internet—images of a 59-year-old woman who died of a heart attack.

Tissue Engineering

An engineered tissue combines living tissue and a synthetic chemical. It begins with a **cell implant**—a few cells "seeded" in or on a synthetic scaffolding that the body accepts. The synthetic component may be a polymer membrane with holes of a particular size. The holes are large enough to allow nutrients and possibly drugs into the tissue, but not so large that immune system biochemicals can enter and cause rejection. The cells in the implant divide and secrete biochemicals as they normally would, replacing damaged tissue. For example, encapsulated pancreas cells form implants to treat diabetes.

Cell implants may be genetically altered to overproduce their natural secreted products or to synthesize entirely different ones with therapeutic benefit. For example, a cell implant might manufacture growth factors that enhance its own growth. Cell implants may also be delivered to novel locations. In the brain of a person with Parkinson disease, a cell implant could supply the missing neurotransmitter dopamine, circumventing the blood-brain barrier that normally keeps dopamine and other large molecules out of the brain.

Other engineered tissues are being developed. See figure 35.4*b* for a description of a "living blood vessel replacement" that combines collagen with cells. Burns may be treated with a bioengineered skin consisting of a patient's epidermal cells placed in sheets over dermal cells grown in culture, both layers supported from beneath with a nylon mesh framework. Such semisynthetic skin is already used in place of live animals to test cosmetic ingredients.

As we learn more about how cells interact to form tissues, we will develop other medical technologies to replace or supplement damaged or diseased tissues.

Mastering Concepts

1. What are the types of transplants?
2. Why does the immune system reject only some transplanted tissue?
3. What are some controversial sources of donor tissue?

One in a Million~

continued from page 594.

Amy's bizarre skin peeling was excruciating and terrifying. At the University of Texas at Galveston's skin bank, physicians covered her entire body with donor skin and anchored it in place with a thousand staples. Over the next several weeks, Amy endured painful daily baths. She received drugs to dull the pain and to impair her memory of the treatment. Amy beat the odds—gradually the skin grafts wore off as new, healthy skin from beneath replaced them. Today, Amy is fine.

The underlying cause of Stevens-Johnson syndrome is an abnormal response of the immune system to a drug, often an antibiotic. Yet the effects were felt in different organ systems—the integumentary (skin) system, the nervous system, and the organs that control body temperature and fluid levels.

SUMMARY

1. Specialized cells express different genes. These cells aggregate and function together to form tissues. The four basic tissue types are **epithelial tissue, connective tissues, nervous tissue,** and **muscle tissue.**

2. Epithelium is lining tissue. It may be **simple** (one layer), **stratified** (more than one layer), or **pseudostratified** (one layer appearing as more than one). Epithelial cells may be **squamous** (flat), **cuboidal** (cube-shaped), or **columnar** (tall). **Keratin** hardens squamous epithelium.

3. Connective tissues consist of cells within a **matrix. Fibroblasts** secrete **collagen** and **elastin. Loose connective tissue** has loose collagen strands, whereas **dense connective tissue** has tightly packed collagen fibrils. **Adipose tissue** consists of **adipocytes** hugely swelled with lipid. The matrix of blood is **plasma,** which carries **red blood cells, white blood cells,** and **platelets.** Cartilage consists of **chondrocytes** in **lacunae,** in a collagen matrix. In bone, **osteocytes** communicate through **canaliculi. Osteoclasts** and **osteoblasts** tear down and rebuild bone tissue. **Osteoprogenitor cells** can become osteoblasts. Bone matrix consists of collagen and minerals.

4. Neurons and **neuroglia** constitute nervous tissue. A neuron has a **cell body,** an **axon,** and **dendrites** and functions in communication. Neuroglia support neurons. Muscle tissue provides movement when filaments of **actin** and **myosin** slide past one another. Contractile cells include **skeletal, cardiac,** and **smooth muscle cells** and **myoepithelial cells.**

5. The **integument** in many animals consists of an **epidermis** over a **dermis,** plus specialized structures, such as hairs, feathers, claws, and glands. In humans a **basement membrane** joins the epidermis to the dermis, and a **subcutaneous layer** underlies the dermis.

6. Transplants, cell implants, and **tissue engineering** replace damaged tissue. An **autograft** transfers tissue within an individual. An **isograft** occurs between identical twins. An **allograft** uses tissue from the same species. A **xenograft** uses tissue from a different species. Controversial tissue sources include fetuses, newborns on the verge of death, and nonhuman animals.

TO REVIEW . . .

1. Identify and describe the four basic tissue types.

2. What are the two criteria used to distinguish types of epithelium?

3. Draw pictures of the following:
- **a.** simple squamous epithelium
- **b.** stratified squamous epithelium
- **c.** pseudostratified epithelium

4. Match the following cell types and structures to the tissues they are part of:

Cell Types	*Tissue*
a. platelets	(1) cartilage
b. chondrocytes	(2) blood
c. neurons	(3) loose connective tissue
d. osteocytes	(4) muscle
e. fibroblasts	(5) adipose tissue
f. neuroglia	(6) bone
g. adipocytes	(7) nervous tissue
h. actin and myosin filaments	(8) dense connective tissue

5. State the tissues that include the following proteins:
- **a.** collagen and elastin
- **b.** actin and myosin
- **c.** keratin

6. Describe where collagen, keratin, and laminin are located within human skin.

7. How are hair follicles and feather follicles similar, and how are they different?

8. Indicate whether each of the following procedures is an autograft, isograft, allograft, or xenograft:
- **a.** A nerve from a rat's leg patches its damaged spinal cord.
- **b.** A parent donates a lobe of her lung to her child, who has cystic fibrosis.
- **c.** Cow cartilage replaces a man's damaged knee cartilage.
- **d.** A young woman receives a kidney from a man who was killed in a car accident and is not a relative.
- **e.** A woman donates an ovary to her identical twin.

TO THINK ABOUT . . .

1. What do all connective tissues have in common, and how do they differ?

2. Dense connective tissue, cartilage, and bone all contain collagen. How, then, do these tissues differ?

3. Do you think that using fetal tissue for treating diseases in adults is linked to the issue of abortion? If so, what safeguards, if any, might help minimize such conflicts?

4. Do you think that organs of anencephalic infants should be harvested for donation by turning off life support? What further information might you need to make such a decision?

5. Collagen injections can temporarily fill out facial skin, diminishing wrinkles. Into which skin layer should collagen be injected? Why isn't the treatment permanent?

TO LEARN MORE . . .

References and Resources*

Aguilera-Hellweg, Max. July 1996. The hidden world of surgery. *Scientific American.* A surgeon cuts through layers of tissues.

Gorman, Christine. January 15, 1996. Are animal organs safe for transplant? *Time.* Xenotransplantation may work.

Lacy, Paul E. July 1995. Treating diabetes with transplanted cells. *Scientific American.* Can packets of shielded pancreatic islet cells treat diabetes?

Lewis, Ricki. July 24, 1995. Tissue engineering now coming into its own as a scientific field. *The Scientist.* Synthetic materials and cells are combined to mimic tissues.

Mackenzie, Debora. November 16, 1996. Doctors farm fish for insulin. *New Scientist.* Fish cell implants may help people with diabetes.

Platt, Jeffrey L. January 1997. Xenotransplanting hepatocytes: The triumph of a cup half full. *Nature Medicine.* The human body might accept transplants of liver cells from other species because these cells are highly adaptable to where they are transplanted.

Schwab, Martin E. September 1996. Bridging the gap in spinal cord regeneration. *Nature Medicine,* vol. 2. Peripheral nervous tissue may patch spinal cord injuries.

Stoye, Jonathan P. March 13, 1997. Proviruses pose potential problems. *Nature,* vol. 386. Pig tissue transplants may introduce viruses into humans.

Wadman, Meredith. August 22, 1996. Ethics worries over execution twist to Internet's "visible man." *Nature,* vol. 382. Was it ethical to put images of an executed man on the Internet, without his complete knowledge?

*Life's website features frequently updated references.

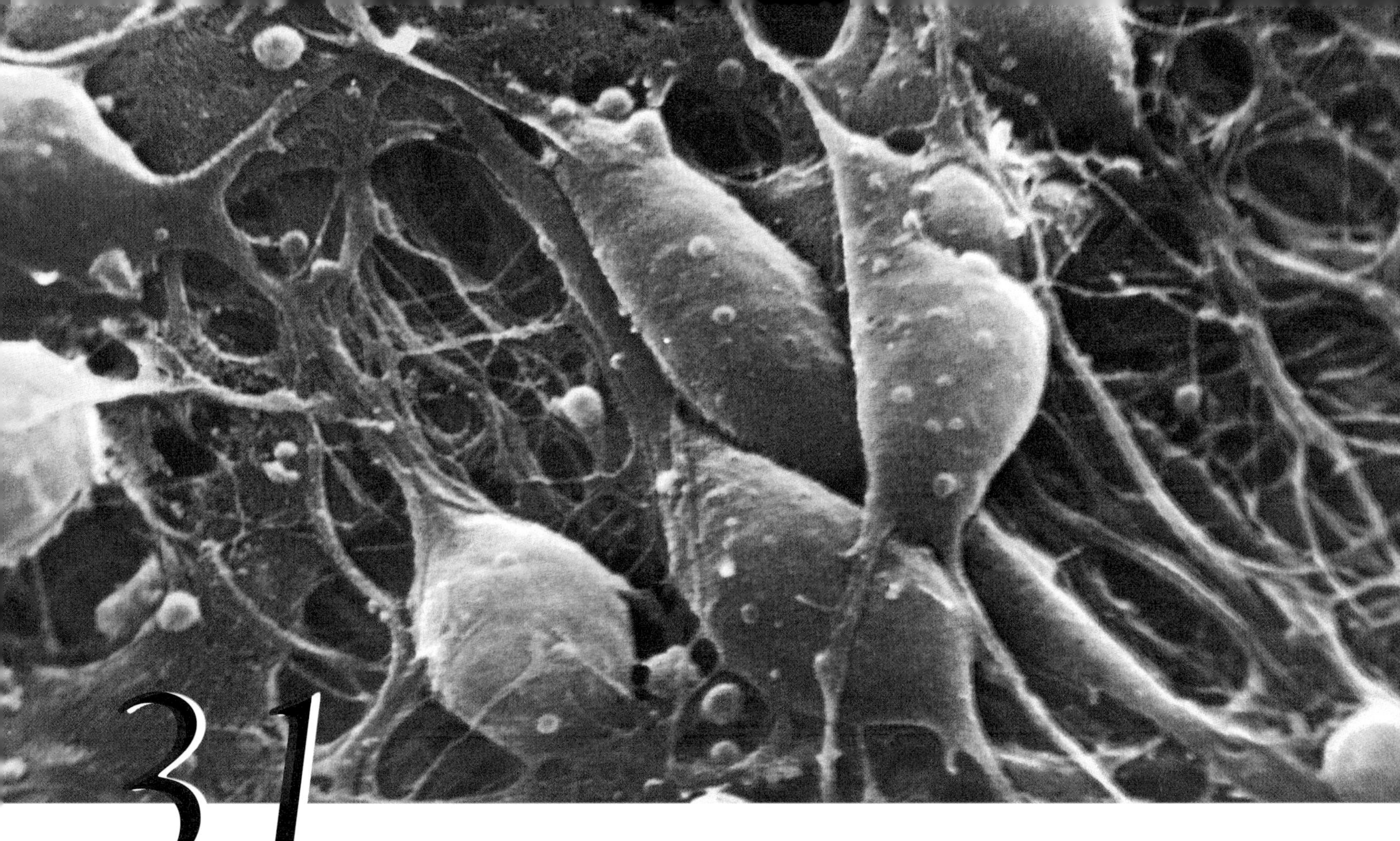

Neurons in a human's cerebral cortex. (×162)

31 The Nervous System

chapter outline

The Nervous System—Always in Control 610
Evolutionary Trends in Nervous Systems

Neurons—Functional Units of the Nervous System 611

A Neuron's Message 611
The Resting Potential
The Action Potential
The Myelin Sheath and Saltatory Conduction

Synaptic Transmission 616
Disposal of Neurotransmitters
Synaptic Integration—A Neuron's Response
Types of Neurotransmitters

A Comparative View of Nervous Systems 619
Invertebrate Nervous Systems
Vertebrate Nervous Systems

The Human Central Nervous System 622
The Spinal Cord
The Brain

The Cerebrum 629
The Cerebral Cortex
Specializations of the Cerebral Hemispheres

Memory 630

The Peripheral Nervous System 630
The Somatic Nervous System
The Autonomic Nervous System

The Unhealthy Nervous System 631
Molecular Causes
Environmental Causes

Protection of the Nervous System 635

Head Games

Phineas Gage worked in Vermont evening out terrain for railroad tracks. To blast away rock, he would drill a hole, fill it with gunpowder, cover that with sand, insert a fuse, and then press down with an iron rod called a tamping iron. The explosion would penetrate the rock.

On September 13, 1848, Gage pounded on the tamping iron before his coworker put down sand. The gunpowder exploded outward, slamming the 1-inch-thick, 40-inch-long iron rod through Gage's skull and out the other side of his head. Gage stood up a few moments later, fully conscious and apparently unharmed.

The freak accident *had* harmed Gage, but in ways so subtle that they were not at first evident. His friends reported that "Gage was no longer Gage," as the young man underwent a profound personality transformation. Once a trusted, honest, and dedicated worker, he became irresponsible, shirking work, cursing, and pursuing what his doctor termed "animal propensities."

As long ago as 1868, researchers hypothesized that the tamping iron had ripped out a part of Gage's brain controlling personality. In 1994, computer analysis more precisely pinpointed the damage to the famous brain of Phineas Gage, which is in a museum at Harvard University. The tamping iron pierced two small areas in the front of the brain that control rational decision making and processing of emotion.

In 1994, after consuming a few too many beers, a young man named Roger and a friend decided to play Russian roulette with Roger's father's pistol. Roger held the barrel of the gun to his temple and squeezed the trigger. It fired, and left Roger with an injury not unlike that of Phineas Gage—but with results that were at the same time similar yet opposite.

to be continued . . .

The Nervous System—Always in Control

The lynx is one of the few inhabitants of the snowy midwinter northern Montana plain. The cat is acutely attuned to its frigid environment, able to hear the quiet swish of a snowshoe hare on the icy cover before it sees its well-camouflaged prey (fig. 31.1). The lynx watches. The hare doesn't realize it is being observed until the lynx bursts from behind a tree. The cat's snowshoelike feet enable it to quickly overtake and capture its fleeing prey.

The interaction between the lynx and the hare is orchestrated by each animal's nervous system. The cat's adaptations as a predator—inch-long claws, knifelike fangs, compact powerful body, and snowshoe feet—require a nervous system that enables it to spot and attack prey. Similarly, the hare's nervous system triggers its swift flight—although too late in this case.

FIGURE 31.1

Nervous systems in action. Networks of nerve cells control this lynx's attack—and the hare's attempt to escape.

Overall, the function of the nervous system, and of all animal organ systems, is to maintain a range of internal constancy, or homeostasis, by detecting and reacting to changes in the environment. The nervous system maintains homeostasis by regulating virtually all other organ systems. The lynx's nervous system controls its respiratory system, which brings oxygen to the circulatory system, which delivers the oxygen plus nutrients to all cells. The nervous system also controls the excretory system, as cell wastes move through the circulatory system to the kidneys. The nervous system enables muscles to act on bones so that the lynx can move; keeps the digestive system providing nutrients to cells for energy conversion and work; causes the endocrine system to secrete adrenaline, which sharpens the animal's awareness and coordinates the many systems for rapid action; and keeps the immune system on alert against infection.

The lynx's pursuit of the hare demonstrates three major roles of the nervous system—sensory input, sensory integration, and motor response. Sensory information arrives as sound and visual images. The lynx compares this information to past hunting experiences and decides how to act. The motor systems coordinate muscles to move the lynx into position to catch the hare. These events are initiated at the conscious level of the brain. Meanwhile, the heart beats, the lungs inhale and exhale, and the cat maintains its balance through sensory input, decision making, and motor output at the unconscious level.

Recall from chapter 30 that the nervous system includes neurons and neuroglia. Neurons make possible many sensations, actions, emotions, and experiences. Networks of interacting nerve cells control mood, appetite, blood pressure, coordination, and perception of pain and pleasure. The unique ability of neurons to communicate rapidly enables animals to be aware of and react to the environment and to screen out unimportant stimuli, to learn, and to remember. Yet despite their diverse functions, all neurons communicate in a similar manner.

Evolutionary Trends in Nervous Systems

Multicellularity posed a profound new challenge to organisms—how would cells communicate with each other? In a single-celled organism, the cell membrane serves as a selective gateway to the outside environment. In a multicellular organism, cells maintain this gatekeeper function; they react to messages that must pass between cells to coordinate the cells' functions. These messages are electrical and chemical changes that transmit information by stimulating or inhibiting cell surface receptor molecules.

Most animals conduct electrochemical messages over cell surfaces in a three-part response that helps them react to changes in their environments. First, sensory cells or specialized sensory tissues detect environmental cues, or stimuli. The sensory tissues then cause electrochemical changes in the conducting cells that make up nervous tissues. Finally, nervous tissues stimulate **effector** structures (muscles or glands) that respond to the change. These response systems help animals maintain internal constancy in the face of changing external conditions, an ability necessary for survival. Homeostasis is thus not a static condition but a dynamic, ever-changing response to a changing world.

All nervous systems consist of organizations of interacting cells. Comparing nervous systems among diverse modern species reveals that nervous systems probably increased in complexity as new species evolved.

Different organizations of nervous systems are adaptive in particular environments. Radially symmetrical animals with wheel-like bodies, such as sea anemones, must respond to stimuli that can come from any direction. If these animals had brains or head ends, this would concentrate their responsiveness in one area, leaving them open to danger from other directions. Instead, radially symmetrical animals usually have simple, diffuse nerve nets that make all parts of their bodies equally receptive to stimuli.

Other animals have more intricate nerve cell organizations in certain areas of the body. Bilaterally symmetrical, or "two-sided," animals such as vertebrates are more active and occupy diverse environments. Their more intricate nervous systems enable them to respond to these environments. The nervous tissue is centralized or concentrated in certain areas of the body, where large numbers of nerve cells maintain highly intricate interconnections. This increases the number and complexity of possible responses. Animals became cephalized, with accumulation of nervous tissue into a brain and the development of sensory structures to form a head. The most adaptive location for nervous tissue to concentrate is in the end of the animal that takes in the most sensory information from the environment.

Mastering Concepts

1. How does a nervous system maintain homeostasis?
2. How is the organization of a species' nervous system adaptive?
3. What are the three components of nervous system action?
4. How do nervous systems transmit information?

Neurons—Functional Units of the Nervous System

All neurons have the same basic parts, but they vary considerably in shape and size. A neuron consists of a rounded central portion with many emanating long, fine extensions, a form well adapted to receiving, integrating, and conducting messages over long distances (fig. 31.2). The central portion of the neuron, the **cell body,** does most of the neuron's metabolic work. It contains the usual organelles: a nucleus, extensive endoplasmic reticulum, mitochondria to supply energy, and ribosomes to manufacture proteins necessary to convey messages.

A neuron's extensions are of two types. The shorter, branched, and more numerous extensions are **dendrites.** They receive information from other neurons and transmit it toward the cell body. The many branching dendrites can receive input from many other neurons. The second type of extension from the cell body is the **axon,** which conducts the message away from the cell body and towards another cell. Because a nerve's message may have to reach to a cell quite far away, an axon is usually longer than a dendrite—sometimes surprisingly so. An axon that permits a person to wiggle a big toe, for example, extends from the base of the spinal cord to the toe. An axon is usually thicker than a dendrite, and a neuron usually has only one axon. Axons are sometimes called nerve fibers.

To picture the relative sizes of a motor neuron's parts, imagine that the cell body is the size of a tennis ball. The axon might then be 1 mile (1.6 kilometers) long and half an inch (1.27 centimeters) thick. The dendrites would fill an average-size living room.

One way to classify neurons is into three groups according to their general functions (fig. 31.3). A neuron that brings information toward the central nervous system (the brain and spinal cord) is called a **sensory** (or afferent) **neuron.** It has remote dendrites that carry its message from a body part, such as from the skin, toward the cell body, located just outside the spinal cord. A sensory neuron's axon is relatively short, because it delivers the message to another neuron whose dendrites are located nearby within the spinal cord.

A **motor** (or efferent) **neuron** conducts its message outward, from the **central nervous system (CNS)** (brain and spinal cord) toward muscle or gland cells. It has a long axon to reach the effector (the muscle or gland) and short dendrites. When a motor neuron stimulates a muscle cell, it contracts, and when a neuron stimulates a gland, it secretes. A third type of neuron, an **interneuron,** connects one neuron to another within the CNS to integrate information from many sources and coordinate responses. Large, complex networks of interneurons receive information from sensory neurons, process and store this information, and generate the messages that the motor neurons carry to effector organs.

Mastering Concepts

1. What are the parts of a neuron?
2. What are the three types of neurons, and how do they differ from each other?

A Neuron's Message

The message that a neuron conducts is called a nerve impulse. This is an electrochemical change that occurs when ions move across the cell membrane. A measurement called an **action**

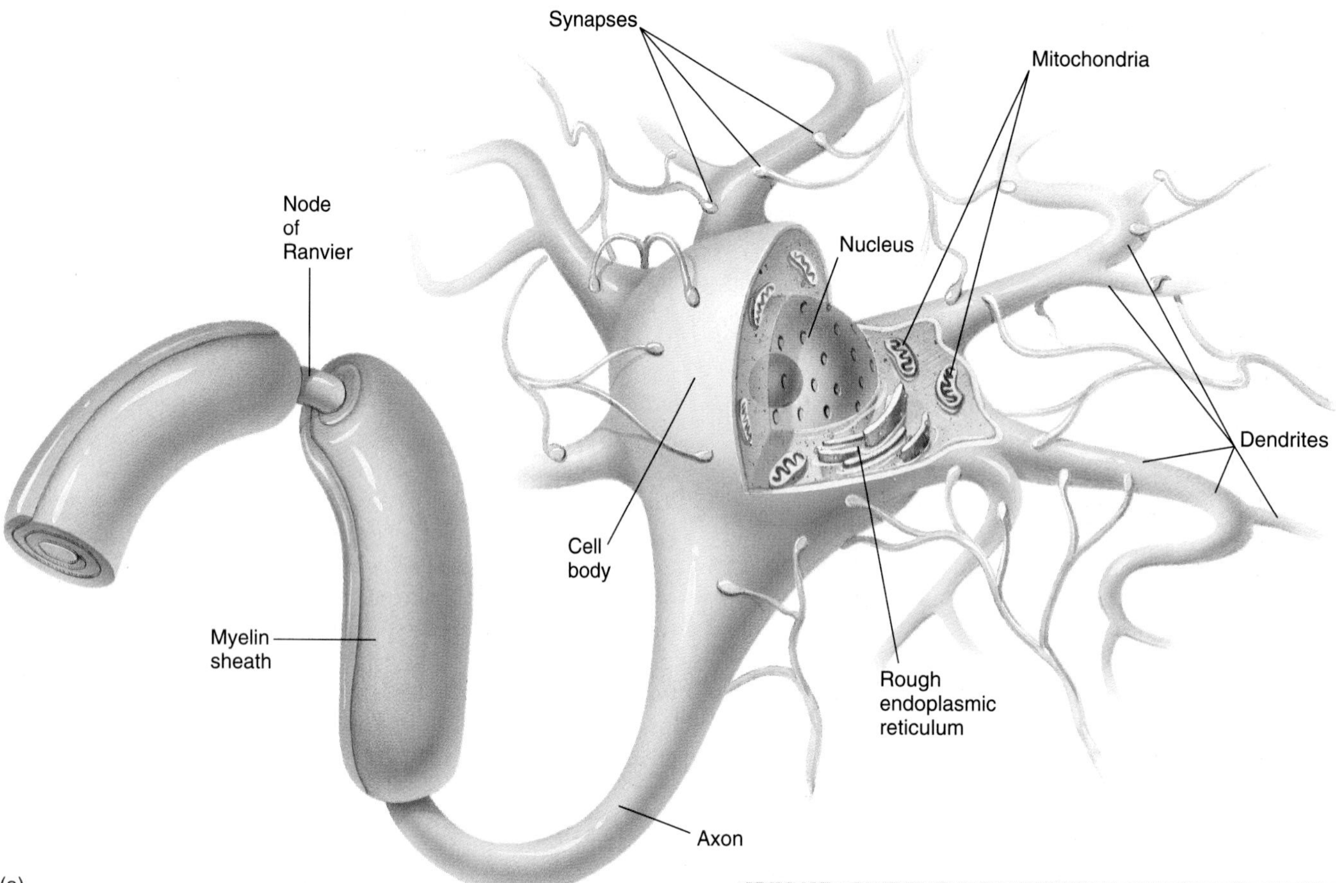

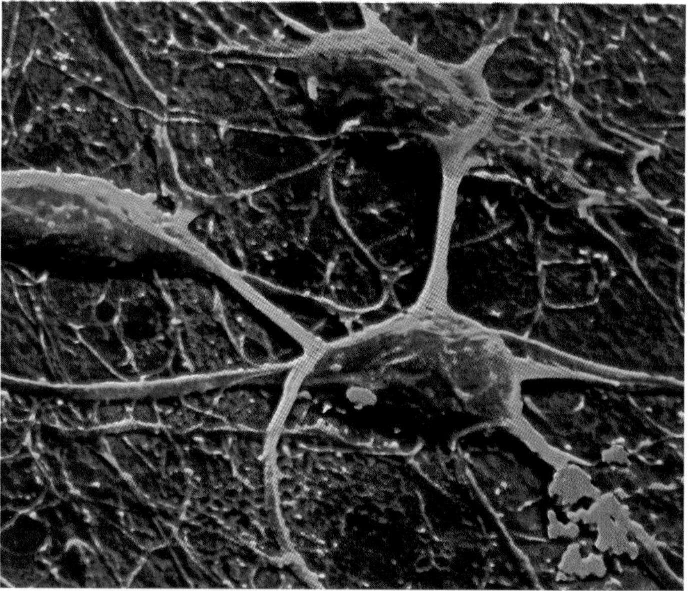

FIGURE 31.2

Parts of a neuron. (*a*) A neuron consists of a rounded cell body, "receiving" branches called dendrites, and a "sending" branch called an axon. The space between an axon terminal of one neuron and a dendrite of an adjacent neuron is a synapse. Many axons are encased in fatty myelin sheaths. Unmyelinated regions between adjacent myelin sheath cells are called nodes of Ranvier. (*b*) These neurons from the cortex of a human brain are magnified 500 times. Note their entangled axons and dendrites.

potential describes the change that is the nerve impulse. A nerve impulse is the spread of electrochemical change (action potentials) along an axon. To understand how and why ions move in an action potential, it helps to be familiar with the **resting potential,** the state a neuron is in when it is not conducting an impulse.

The Resting Potential

The membrane of a resting neuron is **polarized,** that is, the inside carries a slightly negative electrical charge relative to the outside. This separation of charge creates an electrical "potential" that measures −65 millivolts. (A volt measures the difference in electrical charge between two points. The minus sign indicates that the inside of the cell is negative.) The charge differences across the membrane result from the unequal distribution of ions (fig. 31.4).

How is this unequal distribution of ions established and maintained? First, the cell membrane is **selectively permeable;** it admits some molecules but not others. Ions move through the membrane through small pores called channels. Some channels are always open, but others open or close like gates, depending on proteins that change shape. A few of these gates are voltage regulated—whether a gate opens or closes depends upon the electrical charge of the membrane. Other gates open and close in response to certain chemicals. Some membrane channels are specific for sodium ions (Na^+), and others are specific for potassium ions (K^+).

Another property of the membrane that establishes and maintains ion distribution is the **sodium-potassium pump.** Recall from

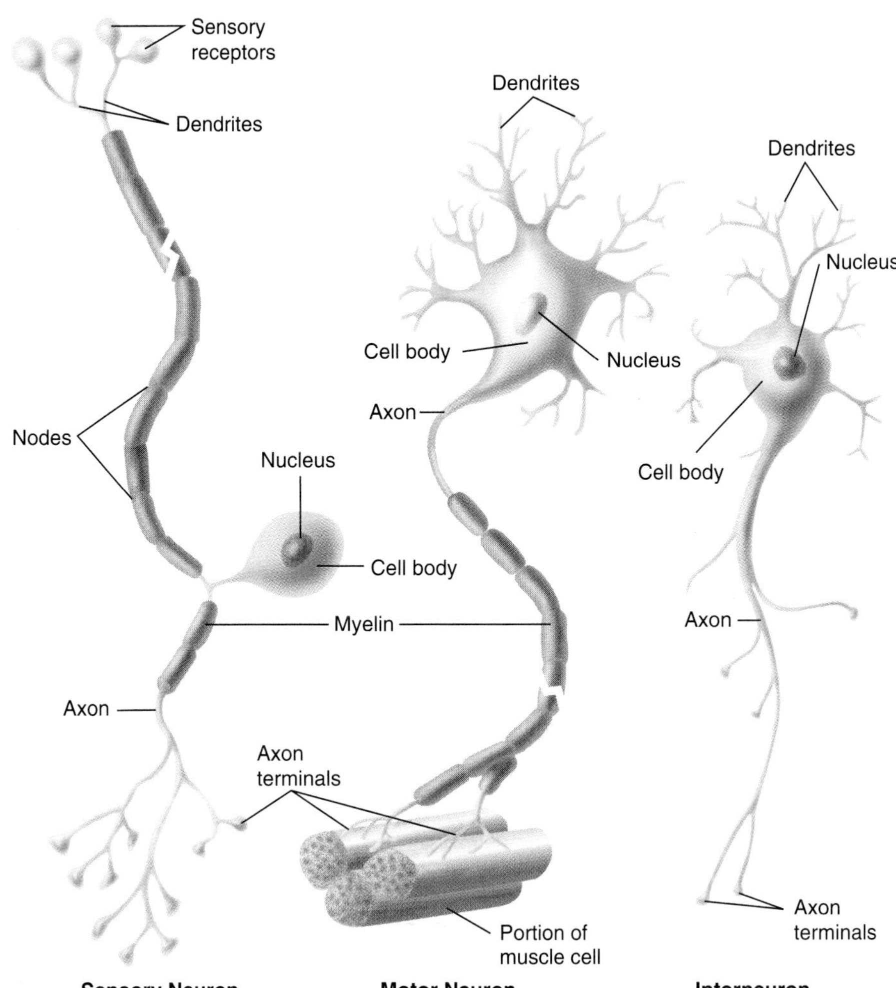

FIGURE 31.3

Categories of neurons based on their functions. Sensory neurons transmit information from sensory receptors in contact with the environment to the central nervous system. They have remote dendrites and short axons. Motor neurons send information from the central nervous system to muscles or glands; they have long axons and short dendrites. Interneurons connect other neurons.

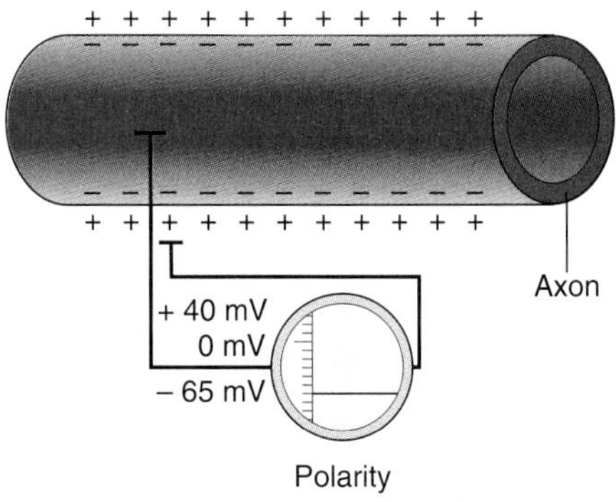

FIGURE 31.4

The resting potential. An oscilloscope measures the difference in electrical potential between two electrodes. When one electrode is placed inside an axon at rest and one is placed outside, the electrical potential inside the cell is −65 millivolts (mV) relative to the outside. This potential difference is due to the separation of positive (+) and negative (−) charges along the membrane.

chapter 5 (see fig. 5.12) that this pump is a mechanism that uses cellular energy (ATP) to transport Na^+ out of and K^+ into the cell. The sodium-potassium pump uses active transport to move Na^+ and K^+ against their concentration gradients. Because of this pump, the concentration of K^+ is 30 times greater inside the cell than outside, and the concentration of Na^+ is 10 times greater outside than inside.

Ions distribute themselves in response to two forces. First, ions follow an electrical gradient. Like charges (negative and negative; positive and positive) tend to repel one another. Unlike charges (negative and positive) attract. Second, ions follow a concentration gradient and passively diffuse from an area in which they are highly concentrated toward an area of lower concentration. Therefore, a particular ion enters or exits the cell at a rate determined by its permeability and concentration gradient.

In summary, three mechanisms establish and maintain the resting potential. First, the sodium-potassium pump concentrates K^+ inside the cell and Na^+ outside. The pump ejects three Na^+ for every two K^+ it pumps in. Second, large, negatively charged proteins and other negative ions are trapped inside the cell because the cell membrane is not permeable to them. Third, the membrane in the resting state is 40 times more permeable to K^+ than to Na^+.

Because of the concentration gradient and high permeability, K^+ is able to diffuse out of the cell. As K^+ moves through the membrane to the outside of the cell, it carries a positive charge, leaving behind large, negatively charged molecules. A charge or potential is therefore established across the membrane; positive on the outside and negative on the inside. The magnitude of the charge is determined by the balance of opposing forces acting on K^+. The concentration gradient drives K^+ outward, and the negative charge inside the cell holds K^+ in. When these two opposing forces are equal, no net movement of K^+ occurs, and the cell is in equilibrium. At this time, the cell is uniformly charged over its entire surface, establishing the resting potential.

The importance of the sodium-potassium pump in maintaining the resting potential becomes evident when a metabolic poison such as cyanide disables the pump. K^+ slowly diffuses out and Na^+ in, destroying the concentration gradients. Nerve transmission is then impossible because a charge no longer exists across the membrane. Death occurs in minutes.

It is curious that a neuron uses more energy while resting than it does conducting an impulse. Presumably, expending energy to

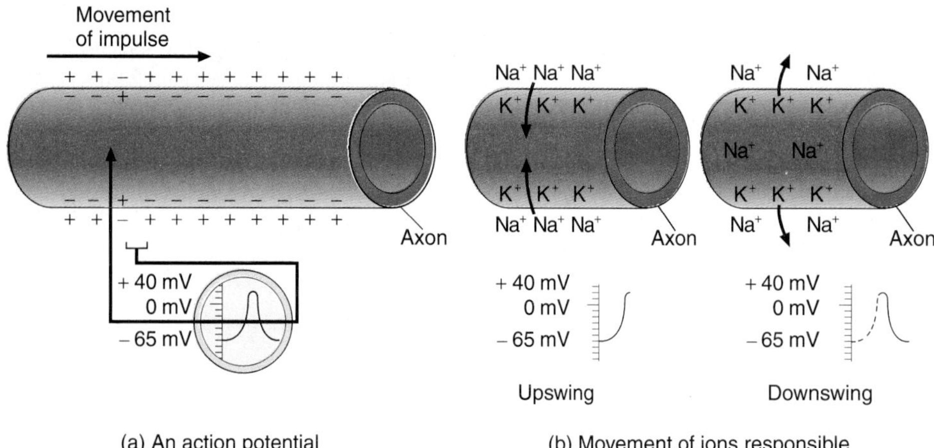

FIGURE 31.5

An action potential is a wave of localized depolarization that travels down the axon. (*a*) The action potential causes a change in electrical potential as it moves along the axon. (*b*) The movements of Na^+ and K^+ cause the electrical changes. These two positive ions move in opposite directions according to their concentration gradients. Gated channels regulate their permeabilities.

maintain the resting potential allows the neuron to respond more quickly than it could if it had to generate a potential difference across the membrane each time it received a stimulus. This is analogous to holding back the string on a bow to be continuously ready to shoot an arrow.

The Action Potential

During an action potential, Na^+ and K^+ quickly redistribute across a small patch of the cell membrane, creating an electrochemical change that moves like a wave along the nerve fiber. An action potential begins when a stimulus (a change in pH, a touch, or a signal from another neuron) changes the permeability of the membrane so that some Na^+ begins to leak into the cell (fig. 31.5). As Na^+ enters the neuron, the interior becomes less negative. The membrane is depolarized, or loses its charge, because of the influx of Na^+. When enough Na^+ enters to depolarize the membrane to a certain point (the "threshold"), the sodium gates sensitive to charge in that area of the membrane open, increasing permeability to Na^+. Driven by both the electrical gradient and the concentration gradient, enough Na^+ enters the cell to positively charge the interior. Na^+ influx continues until the positive charge peaks.

At this peak of the action potential, membrane permeability changes again. Permeability to Na^+ halts as sodium gates close, but permeability to K^+ suddenly increases as potassium gates open. Now Na^+ cannot enter in large numbers; however, exit of K^+ begins, driven by both electrical and concentration gradients. K^+ flows outward because it is more concentrated inside than outside and because the inside of the membrane is now positively charged due to the influx of Na^+.

The loss of positively charged K^+ restores the negative charge to the interior of the cell, repolarizing the cell membrane. The electrical potential fleetingly drops below the resting value, because the K^+ gates stay open slightly longer than the Na^+ gates. Quickly, the ion concentrations return to normal as the sodium-potassium pump "resets" the membrane potential to the resting state.

While the Na^+ and then the K^+ gates are open, a second action potential cannot begin. Still, an action potential takes only 1 to 5 milliseconds, and it spreads rapidly to adjacent areas of membrane. This capacity to rapidly transmit nerve impulses makes the nervous system an effective communication network.

The characteristic changes in membrane permeability that constitute the action potential travel along the neuron, usually from dendrite to cell body and down an axon. The action potential spreads because some of the Na^+ rushing into the cell at a particular point moves to the neighboring part of the neuron and causes it to reach threshold. This triggers an influx of Na^+ there, carrying the action potential forward. An action potential is an all-or-none phenomenon—it either happens or it doesn't.

Neurons discern the intensity of a stimulus from the frequency of action potentials. Whereas a light touch to nerve endings in the skin might produce 10 impulses in a given time period, a hard hit might generate 100 impulses, intensifying the sensation. Although all action potentials are the same, neurons also discern the type of stimulation. We can distinguish light from sound because the neurons that light stimulates transmit impulses to a different place in the brain than sound-generated impulses reach.

The Myelin Sheath and Saltatory Conduction

Not all nerve fibers conduct action potentials at the same speed. Speed of conduction depends on certain characteristics of the fiber. The greater the diameter of the fiber, the faster it conducts an action potential; however, thin vertebrate nerve fibers can conduct action potentials very rapidly when coated with a fatty material called a **myelin sheath.**

Outside the brain and spinal cord, **Schwann cells,** which contain enormous amounts of lipid, form myelin sheaths. A Schwann cell wraps around an axon many times, forming a whitish coating (fig. 31.6). Each of many Schwann cells wraps around a small segment of the axon. Between each Schwann cell is a short region of exposed axon called a **node of Ranvier.** Some neurons in the brain and spinal cord are wrapped in myelin that cells called oligodendrocytes produce.

When an action potential travels along a myelinated axon, it "jumps" from node to node in a type of transmission called **saltatory conduction** (fig. 31.7). The impulse appears to leap from node to node because the myelin insulation prevents ion flow across the membrane, but a small electrical current spreads instantly between nodes. Because an action potential moves faster

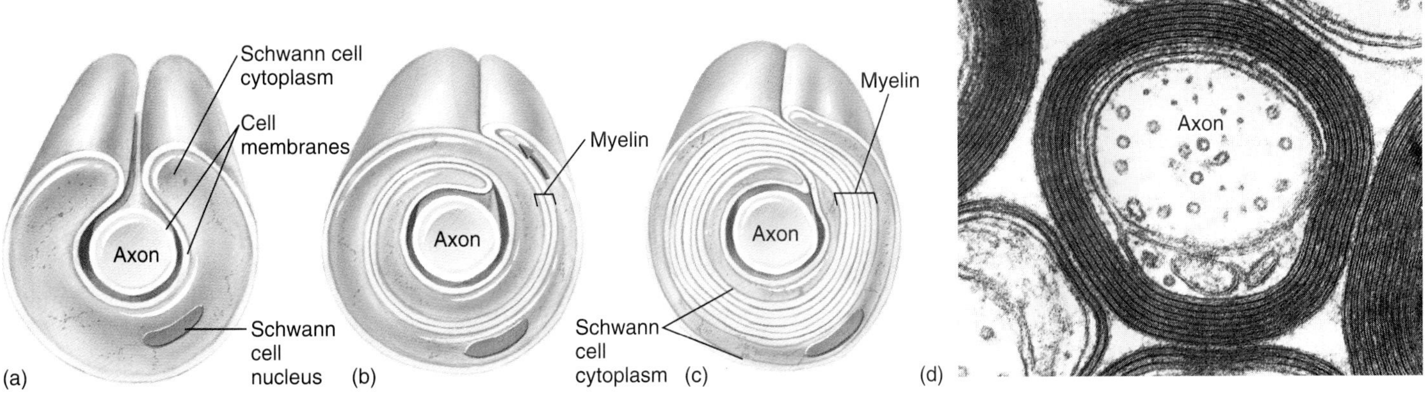

FIGURE 31.6

Sheaths of fatty myelin encase many axons of vertebrate nerve cells. This covering forms when a Schwann cell (*a*) winds around an axon in a spiral (*b*), so that several layers of its lipid-rich cell membrane surround the axon. The Schwann cell's cytoplasm and nucleus are often squeezed into the periphery of the sheath (*c*). (*d*) A myelinated axon in cross section.

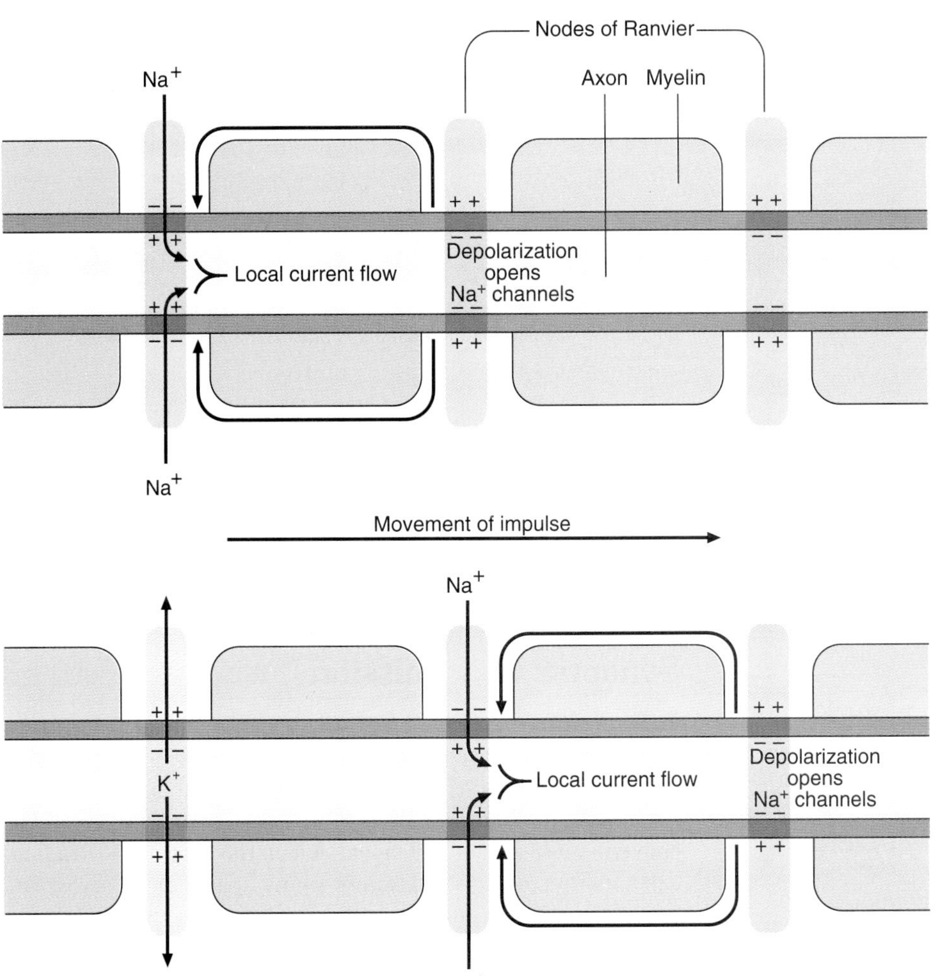

FIGURE 31.7

Saltatory conduction. In myelinated axons, the inward diffusion of Na^+ can only occur at the nodes of Ranvier. Thus, action potentials appear to "jump" from one node to the next.

when it jumps from node to node, saltatory conduction speeds nerve impulse transmission. Myelinated axons may conduct action potentials 100 times faster than unmyelinated axons, at speeds of up to 394 feet (120 meters) per second (an astounding 270 miles [435 kilometers] an hour)! This means that a sensory message travels from the toe to the spinal cord in less than 1/100 of a second.

Myelinated fibers are found in pathways that transmit impulses over long distances. They make up the white matter of the nervous system. Cell bodies and interneurons that lack myelin usually specialize in interpreting multiple messages. These unmyelinated fibers, which make up the gray matter of the nervous system, form much of the nerve tissue in the brain and spinal cord.

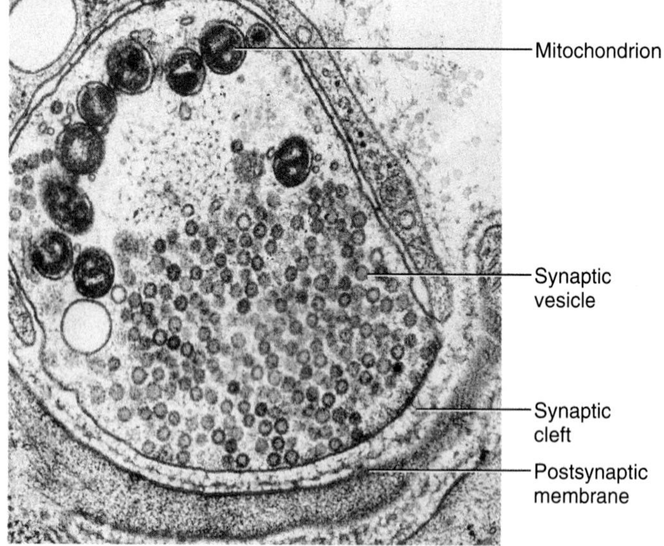

FIGURE 31.8

The synapse—where neurons meet. (*a*) A neuron may form synapses with hundreds or even thousands of other cells. (*b*) A neurotransmitter released from the presynaptic cell in response to the action potential traverses the synaptic cleft and binds to receptors on the postsynaptic cell's membrane. If sufficient excitatory neurotransmitters bind to the postsynaptic cell, it will "fire" an action potential. (*c*) Synaptic vesicles filled with neurotransmitter.

Mastering Concepts

1. What three mechanisms establish and maintain the resting potential?
2. Why does a neuron maintain a resting potential?
3. How do changing cell membrane ion permeabilities generate and transmit a nerve impulse?
4. How do action potentials indicate stimulus intensity?
5. How do fatty coverings on neurons speed nerve impulse transmission?

Synaptic Transmission

To form a communication network, a neuron must convey an action potential to another neuron or to a muscle or gland cell. Neurons do not touch each other, so the action potential cannot travel directly from cell to cell. Instead, the action potential is converted into a chemical signal, which travels from a "sending" cell to a "receiving" cell across a tiny space. Once across this space, this **neurotransmitter** chemical alters the permeability of the receiving cell's membrane, either provoking or preventing an action potential.

The space between neurons is called a **synapse.** The end of an axon has tiny branches that enlarge at the tips to form synaptic knobs, and these knobs contain many synaptic vesicles, small sacs that hold neurotransmitter molecules. An action potential passes down the axon of the **presynaptic cell,** the cell sending the message. When the action potential reaches the membrane near the space, or **synaptic cleft,** the permeability of the membrane

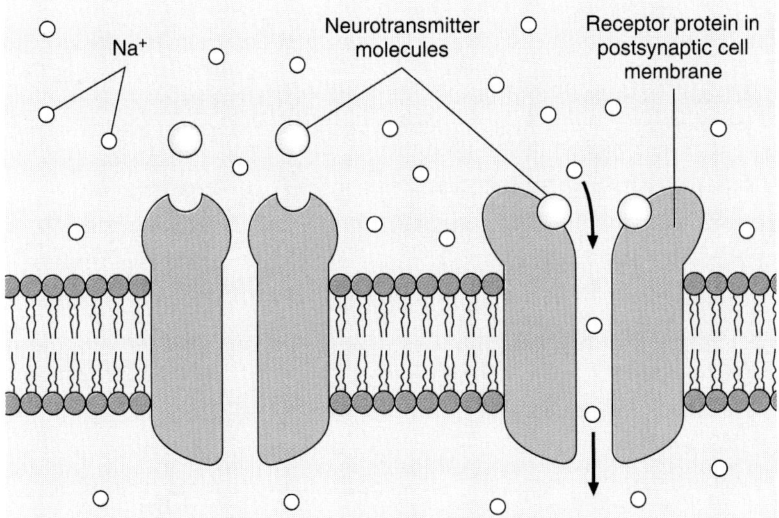

FIGURE 31.9

Receptors on the postsynaptic membrane. The presynaptic cell synthesizes a neurotransmitter and packages it into vesicles. When an action potential fires, the cell releases the neurotransmitter, which crosses the synaptic cleft and binds to receptor proteins on the postsynaptic membrane. This opens channels in the membrane and permits sodium ions to enter. If enough Na^+ enters the postsynaptic cell, an action potential begins. The neuron's "message" moves on.

changes and calcium ions enter the cell. Calcium ions cause the loaded vesicles to move toward the synaptic membrane, fuse with it, and dump their neurotransmitter contents into the synaptic cleft by exocytosis (fig. 31.8). Figures 5.14 and 5.C show other views of neurotransmitter transport.

Neurotransmitter molecules diffuse across the cleft and attach to protein receptors on the membrane of the receiving neuron (the **postsynaptic cell**) (fig. 31.9). A particular neurotransmitter fits only into a specific receptor type, as a key fits only a certain lock. When the neurotransmitter contacts the receptor, the shape of the receptor changes. This binding opens channels in the postsynaptic membrane, admitting specific ions to flow through and changing the probability that an action potential will occur.

Disposal of Neurotransmitters

If a neurotransmitter stayed in the synapse, its effect on the receiving cell would be continuous, perhaps causing it to fire unceasingly and bombard the nervous system with stimuli. A neurotransmitter, however, is either destroyed by an enzyme or taken back into the presynaptic axon soon after its release. This prevents chemical chaos. Nerve gases and certain insecticides work by blocking reuptake of the neurotransmitter acetylcholine. The resulting excess acetylcholine activity stimulates skeletal muscle to contract continuously, and the person convulses and dies. The twitching legs of a cockroach sprayed with insecticide also demonstrate the effects of blocking acetylcholine breakdown.

Synaptic Integration—A Neuron's Response

The nervous system has two types of synapses, excitatory and inhibitory. The combination of excitatory and inhibitory synapses provides finer control over a neuron's activities.

Excitatory synapses depolarize the postsynaptic membrane, and **inhibitory synapses** increase the polarization (hyperpolarize it). A neurotransmitter that acts at an excitatory synapse increases the probability that an action potential will be generated in the second neuron by slightly depolarizing it. For example, when acetylcholine binds to the receptors at an excitatory synapse, channels open that admit Na^+ into the postsynaptic cell. In a millisecond, half a million sodium ions flow in. If enough Na^+ enters to reach a threshold level of depolarization, it triggers an action potential in the postsynaptic cell.

On the other hand, a neurotransmitter may inhibit an action potential in the postsynaptic cell by making the cell's interior more negative than the usual resting potential. In this case, extra Na^+ must enter before the membrane depolarizes enough to generate an action potential.

A single neuron in the nervous system may receive input from tens of thousands of other neurons, some excitatory and others inhibitory. Nearly half of a neuron's receiving surface adjoins synapses. Whether that neuron transmits an action potential depends on the sum of the excitatory and inhibitory impulses it receives. If it receives more excitatory impulses, the postsynaptic cell is stimulated; if inhibitory messages predominate, it is not. A neuron's evaluation of impinging nerve messages, which determines whether an action potential is "fired," is termed **neural** or **synaptic integration.**

Synapses markedly increase the informational content of the nervous system. The human brain has a trillion neurons, each of which can be viewed as carrying bits of information. But if a synapse is also considered a unit of information, then the informational capacity of the brain increases a thousandfold, because a typical brain neuron has synaptic connections to a thousand other neurons, each sending or receiving messages hundreds of times per second.

Table 31.1 Mechanisms of Drug Action

Drug	Neurotransmitter Affected	Mechanism of Action	Effect
Tryptophan	Serotonin	Stimulates neurotransmitter synthesis	Sleepiness
Reserpine	Noradrenaline	Packages neurotransmitter into vesicles	Limb tremors
Curare	Acetylcholine	Decreases neurotransmitter in synaptic cleft	Muscle paralysis
Valium	GABA	Enhances receptor binding	Decreases anxiety
Nicotine	Dopamine	Stimulates release of neurotransmitter	Increases alertness
Cocaine	Dopamine	Binds dopamine transporters	Euphoria
Tricyclic antidepressants	Noradrenaline	Blocks reuptake	Mood elevation
Monoamine oxidase inhibitors	Noradrenaline	Blocks enzymatic degradation of neurotransmitter in presynaptic cell	Mood elevation
Ritalin	Dopamine	Binds dopamine transporters	Treats attention deficit disorder

Table 31.2 Disorders Associated with Neurotransmitter Imbalances

Condition	Symptoms	Imbalance of Neurotransmitter in Brain
Alzheimer disease	Memory loss, depression, disorientation, dementia, hallucinations, death	Deficient acetylcholine
Depression	Debilitating, inexplicable sadness	Deficient noradrenaline and/or serotonin
Epilepsy	Seizures, loss of consciousness	Excess GABA leads to excess noradrenaline and dopamine
Huntington disease	Personality changes, loss of coordination, uncontrollable movement, death	Deficient GABA
Hypersomnia	Excessive sleeping	Excess serotonin
Insomnia	Inability to sleep	Deficient serotonin
Myasthenia gravis	Progressive muscular weakness	Deficient acetylcholine at neuromuscular junctions
Parkinson disease	Tremors of hands, slowed movements, muscle rigidity	Deficient dopamine
Schizophrenia	Inappropriate emotional responses, hallucinations	Deficient GABA leads to excess dopamine

Types of Neurotransmitters

The **peripheral nervous system,** or **PNS** (the part outside the brain and spinal cord), uses three neurotransmitters: acetylcholine, noradrenaline, and adrenaline. The CNS uses at least 40 neurotransmitters, including dopamine, serotonin, the inhibitory transmitter GABA (gamma amino butyric acid), and internal opiates called endorphins. A single neuron may produce only one neurotransmitter type, or more than one, releasing the same combination at each synapse. Different neurotransmitters seem to be associated with particular responses. For example, noradrenaline released in the PNS increases heart rate, whereas adrenaline in the CNS is inhibitory.

Table 31.1 lists drugs that alter specific neurotransmitter levels, and table 31.2 describes illnesses that result from neurotransmitter imbalances.

Mastering Concepts

1. How do neurotransmitters transmit action potentials from a neuron to another neuron, a muscle or a gland cell?
2. What happens to a neurotransmitter after it is released?
3. What is neural integration?

FIGURE 31.10

This dark purple jellyfish first appeared on beaches at Baja, California, in July 1989. It is 20 feet (6 meters) in length. Jellyfish move through the water propelled by waves of muscular contraction, which their relatively simple nervous systems stimulate. The diffuse organization of their nervous systems enables them to respond to stimuli coming from any direction.

A Comparative View of Nervous Systems

Nervous systems are highly organized collections of neurons and supporting cells that enable an animal to detect and react to environmental stimuli. A nervous system might be a loose network of relatively few cells or a structure as complex as the human brain.

Invertebrate Nervous Systems

Some simple animals, such as sponges, have nervous responses without a defined nervous system. Somewhat more complex are the simplest nervous systems, which are diffuse networks of neurons, called nerve nets. In body walls of hydra, jellyfish, and sea anemones, nerve nets synapse with muscle cells near the body surface, enabling each animal to swim and to move its tentacles. A stimulus at any point on the body spreads over the entire body surface. Cnidaria can detect touch, certain chemicals, and light and maintain balance.

This diffuse nervous system organization and the ability of each neuron to conduct impulses in both directions allow cnidaria to react to stimuli, such as the diver in figure 31.10, that approach from any direction. If an animal is on the bottom of a pond where danger and food may approach from any direction, a nerve net is adaptive. Figure 31.11*a* illustrates the nerve net of a hydra.

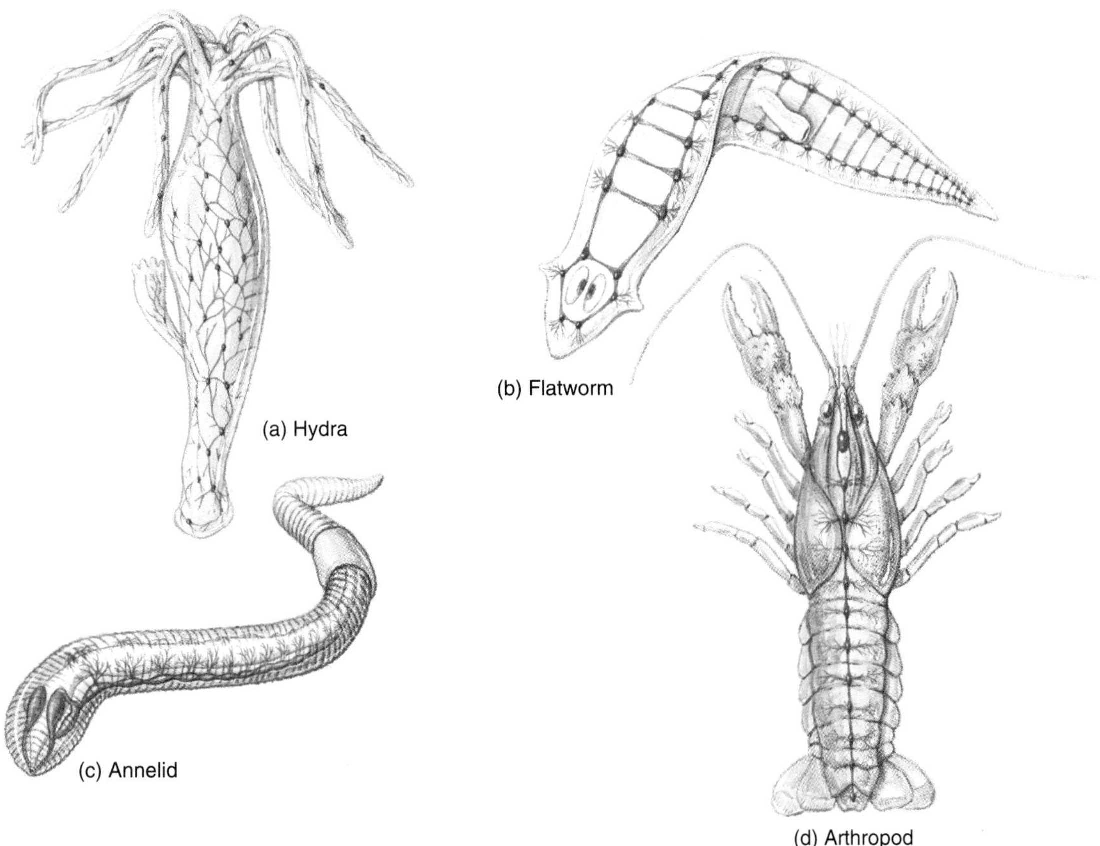

FIGURE 31.11

Nervous system diversity in invertebrates including (*a*) hydra, (*b*) flatworms, (*c*) annelids, and (*d*) arthropods.

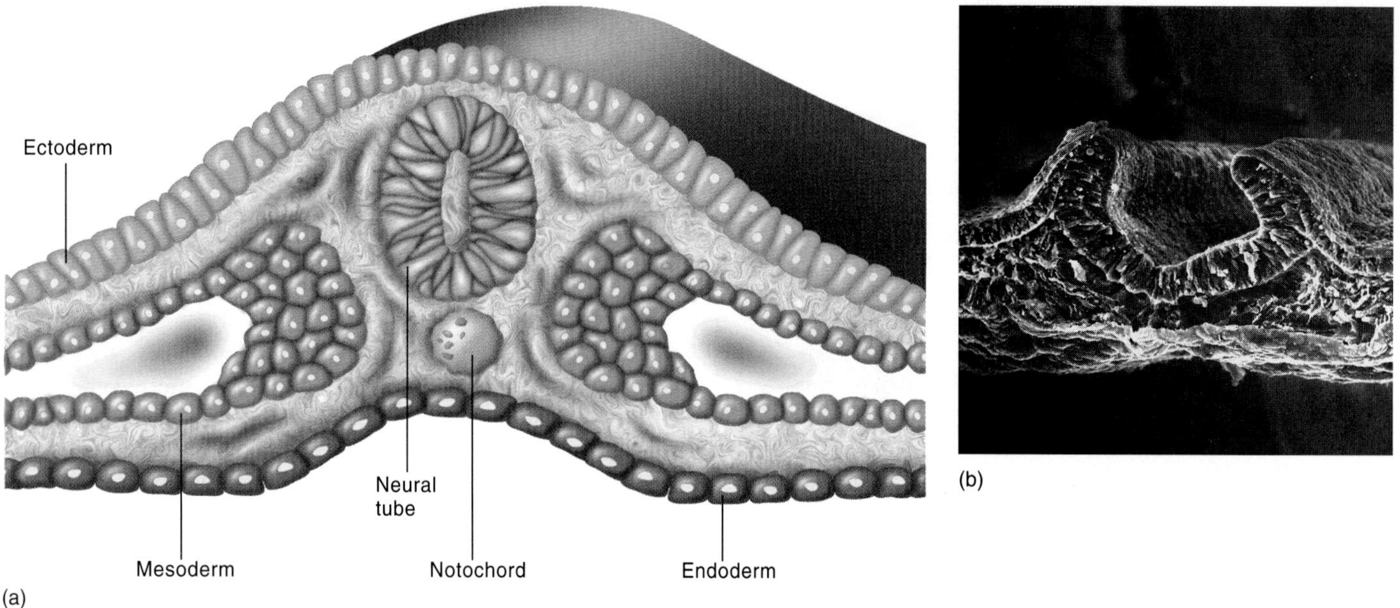

FIGURE 31.12

The notochord and neural tube. (*a*) The notochord is a flexible rod derived from mesoderm shown here in cross section. It induces the adjacent ectoderm to form the neural tube. (*b*) At a signal from the notochord, ectoderm folds into the neural tube, which will gradually form the brain and spinal cord. This chick embryo neural tube is well into the process of forming.

Flatworms have brains that consist of two clusters of nerve cell bodies and two nerve cords that extend down the body and connect to each other to form a "ladder" type of nervous system (fig. 31.11*b*). Motor structures and the neurons that control them are paired, which allows the worm to move in a coordinated forward motion. Paired symmetrical sense organs at the head end allow the flatworm to detect a stimulus and crawl toward it. Bilaterally paired receptors enable the animal to determine what direction stimulation is coming from by comparing the stimulus intensity at each receptor. The receptor with the strongest response is closest to the stimulus.

Segmented worms have a more elaborate ladder organization and a larger brain (fig. 31.11*c*). Two nerve cords run down the body on the ventral side beneath the digestive tract. These cords are solid and fused to each other. Peripheral nerves branch from the central fused ladder, and neuron clusters along the nerve cords coordinate functions in particular regions of the body through interneurons.

The nervous systems of arthropods (including insects, spiders, lobsters, and scorpions) also consist of a brain and a ventral, fused nerve cord (fig. 31.11*d*). In addition, arthropod nervous systems have highly developed sensory organs for touch, smell, hearing, chemical reception, and balance. These animals have complex behaviors and even rudimentary learning.

Vertebrate Nervous Systems

Less variation is seen among vertebrate nervous systems than among invertebrate systems. Chordates in particular have complex brains and well-developed senses and behaviors. The vertebrate nervous system is a modification of the symmetry in the simpler worms and insects. A tubular spinal cord extends lengthwise and expands at the top to form the brain. The cord conducts information to and from the brain and centralizes rapid involuntary responses.

Embryonic Development

A vertebrate embryo has simple beginnings of a nervous system. Recall that the early embryo has three layers—ectoderm and endoderm, and the mesoderm between them. As these layers form, a cord of mesoderm, the notochord, develops just underneath the back of the ectoderm. All vertebrate embryos have notochords, but these structures shrink greatly in adults of most species.

The notochord touching nearby ectoderm triggers **neurulation,** the formation of the nervous system. In amphibians, reptiles, birds, and mammals, the notochord causes a sheet of ectoderm running down the back to fold into a hollow tube called the **neural tube.** (The nervous system of a fish forms differently—a solid tube hollows out to form the neural tube.) As the embryo develops, the neural tube contorts to form the brain and spinal cord. Figure 31.12 shows the relationship between the notochord and the neural tube.

The neural tube begins to close at the front, forming head structures as the back end is still folding into the three germ layers. The head end swells and constricts as cell populations within it aggregate. Ectodermal cells specialize into the neurons and glia that will compose the three major parts of the brain—the **hindbrain, midbrain,** and **forebrain.** For a time, the vertebrate nervous system is merely an open tube with a swollen top.

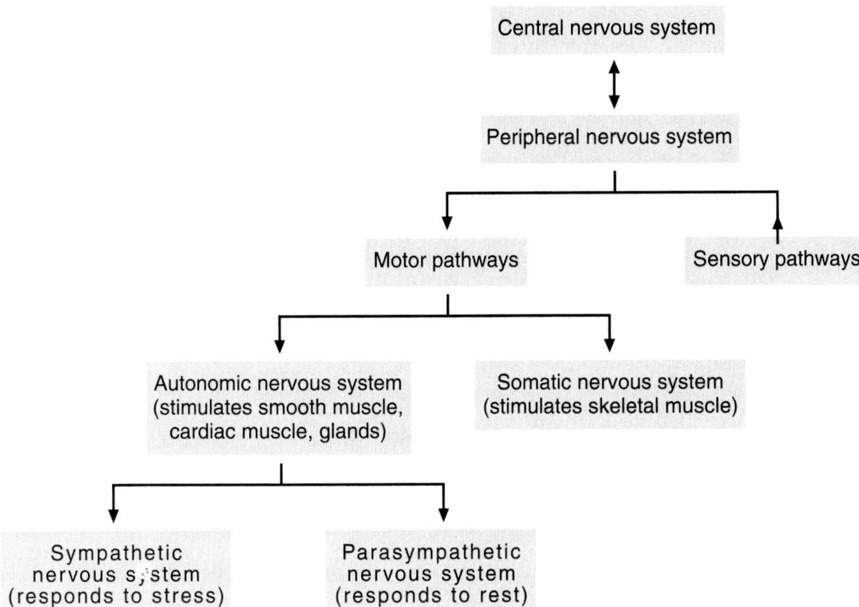

FIGURE 31.13

The peripheral nervous system consists of all nervous tissue outside the central nervous system. Most generally, it can be divided into motor and sensory pathways. The motor pathways are further subdivided into the somatic nervous system, which innervates skeletal (voluntary) muscle, and the autonomic nervous system, which stimulates (involuntary) effectors. The autonomic nervous system is further subdivided into the sympathetic nervous system, which predominates in threatening situations, and the parasympathetic nervous system, which predominates in restful circumstances.

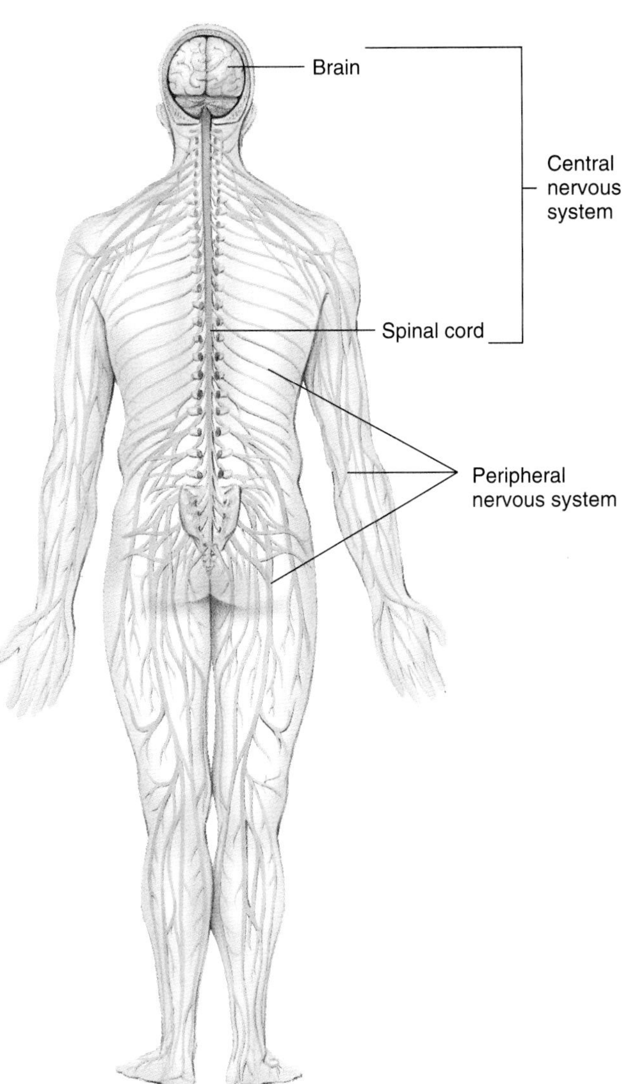

FIGURE 31.14

Major divisions of the human nervous system. The central nervous system (CNS) consists of the brain and the spinal cord, which integrate and interpret incoming information. All other nervous tissue is part of the peripheral nervous system, which sends impulses from sensory receptors to the CNS and delivers impulses from the CNS to effectors.

Closure of the openings of the vertebrate nervous system is a key developmental event. In human embryos, closure occurs on the 27th day after conception—usually. In 1 in 10 embryos, closure is incomplete. In a few cases (1 in 1,000) the incomplete closure causes a neural tube defect, and nerve tissue bulges from the tube at the site of the opening. If the opening is along the spinal cord, the baby is born with spina bifida ("open spine"). Children with spina bifida have paralysis, the severity depending on the size and location of the opening. A neural tube defect at the anterior end, called anencephaly, is more devastating. Higher brain structures do not develop, but the fetus usually continues to develop because structures in the hindbrain that control vital functions are present. Affected individuals usually die shortly before or after birth.

After neural tube closure, spaces called ventricles form deep within the anterior swelling that will become the brain. From the ventricles, neurons grow outward to eventually form the outer brain layers. Scaffolding of glia guide developing neurons to their permanent sites of the brain.

Organization of the Vertebrate Nervous System

As the brain and spinal cord develop, cells from the area where the neural tube contacts ectoderm develop into the peripheral

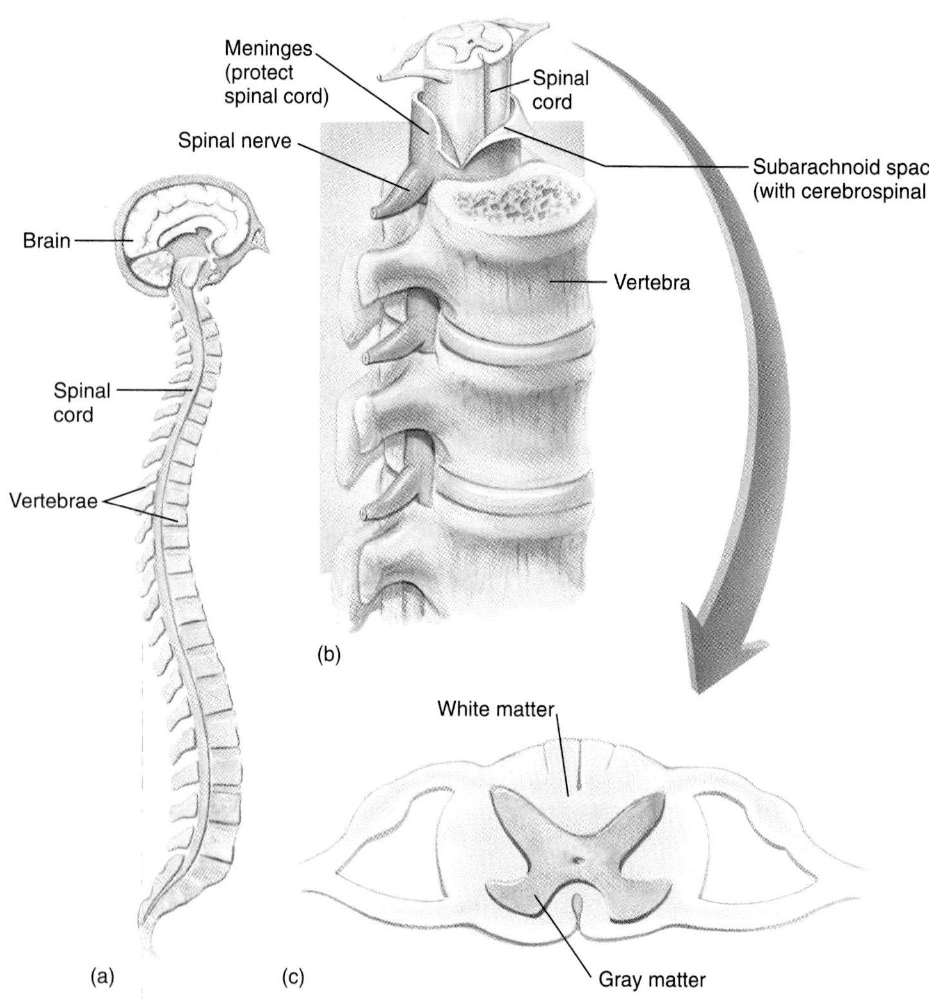

FIGURE 31.15

The spinal cord. (*a*) The CNS consists of the brain, which is encased in the bony skull, and the spinal cord, which is protected by the bony vertebrae. (*b*) Three membranes collectively called the meninges surround the spinal cord. (*c*) Like the brain, the spinal cord contains both gray and white matter. The gray central region, which in cross section resembles the letter *H*, contains short interneurons, unmyelinated fibers, motor neuron cell bodies, and glial cells. The white matter surrounding the *H* consists of myelinated nerve tracts.

nervous system (PNS) (figs. 31.13 and 31.14). The PNS consists of **cranial nerves** that exit the CNS from the brain and **spinal nerves** that extend from the spinal cord as well as collections of cell bodies. The PNS consists of the sensory (or afferent) pathways that carry information to the CNS and the motor (or efferent) pathways that transmit action potentials from the CNS to muscle or gland cells. The motor pathways in turn consist of the **somatic** (voluntary) **nervous system,** which leads to skeletal muscles, and the **autonomic** (involuntary) **nervous system,** which goes to smooth muscle, cardiac muscle, and glands. Finally, the autonomic nervous system consists of the **sympathetic nervous system,** which mobilizes the body to respond to threatening environmental stimuli, and the **parasympathetic nervous system,** which controls more mundane functions such as digestion, respiration, and heart rate at rest. The autonomic nervous system is well developed in reptiles, birds, and mammals but less so in fishes and amphibians.

The most complex nervous systems are in primates. The remainder of the chapter focuses on the structures and functions of the human nervous system.

Mastering Concepts

1. Contrast the complexity of invertebrate and vertebrate nervous systems.
2. What events occur in a vertebrate embryo as the nervous system develops?
3. What are the subdivisions of the nervous system?

The Human Central Nervous System

The human CNS is like that of many other vertebrates.

The Spinal Cord

The spinal cord is a tube of neural tissue encased in the bony armor of the vertebral column (backbone). The vertebral column is composed of individual vertebrae and is partially derived

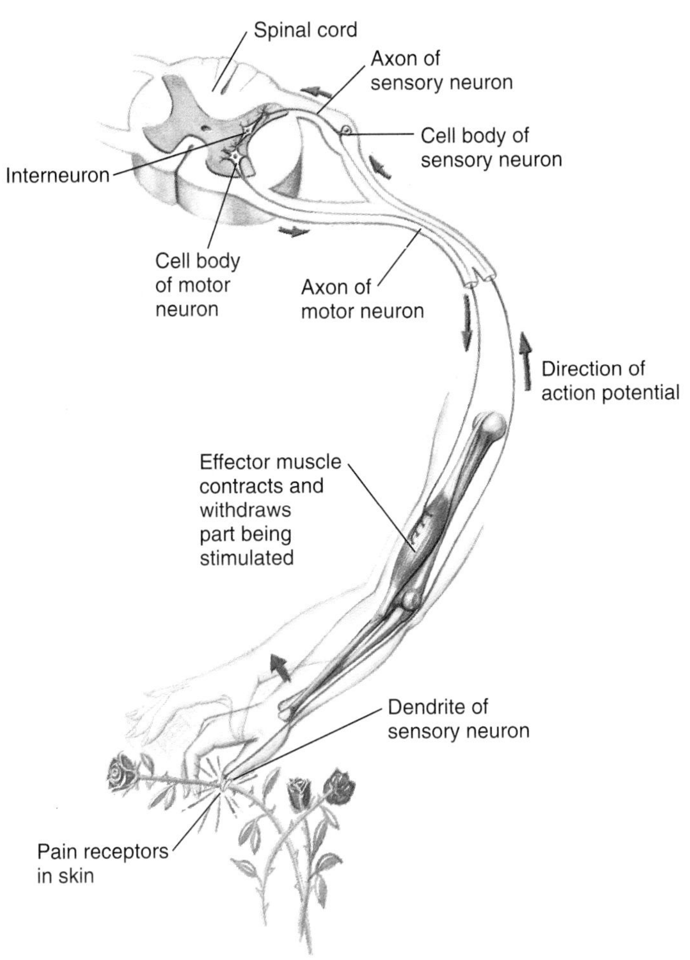

FIGURE 31.16

Neuron types within a reflex arc. A reflex arc links a sensory receptor to an effector. A sensory receptor—perhaps a sensory neuron dendrite that ends in the skin—senses an environmental stimulus and relays an action potential to its cell body in the spinal cord. This neuron sends the information to an interneuron, which relays the action potential to a motor neuron; this neuron, in turn, stimulates a muscle cell to contract. Motor neurons may activate other neurons or gland or muscle cells.

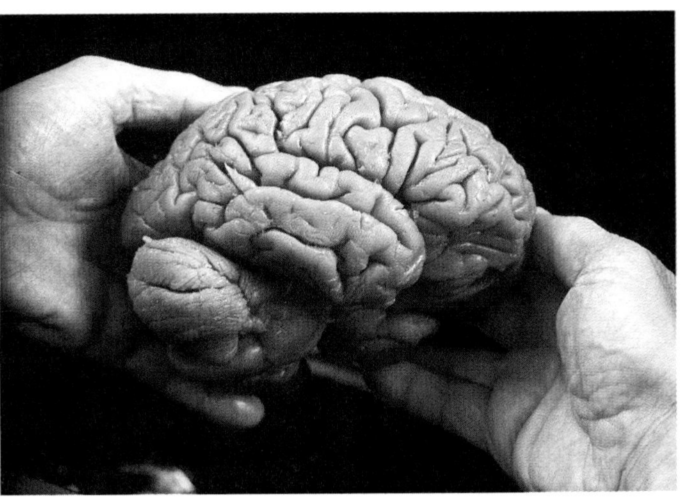

FIGURE 31.17

The human brain.

from the notochord. The vertebral column protects the delicate nervous tissue and provides points of attachment for muscles.

The spinal cord extends about 17 inches (43 centimeters) from the base of the brain to an inch or so below the last rib (fig. 31.15). It carries impulses to and from the brain and is the site where spinal reflex neurons interact. A **reflex** is a rapid, involuntary response to a stimulus that may come from within or outside the body.

The spinal cord communicates with and receives information from the rest of the body through 31 pairs of spinal nerves, numbered according to their position along the cord. Sensory information travels to the rear of the spinal cord, and instructions for activity of skeletal muscles pass outward from the front of the cord. The sensory and motor nerve fibers form a single cable, which passes through the opening between the vertebrae.

The spinal cord conducts information to and from the brain via myelinated fibers that form white matter at the periphery of the cord. Specific types of information travel in particular tracts (axons and dendrites) of the white matter. Ascending tracts carry sensory information to the brain, and descending tracts carry motor information from the brain to muscles and glands. The gray matter interior of the spinal cord consists of motor neuron cell bodies, unmyelinated fibers, interneurons, and glial cells.

The spinal cord handles reflexes without interacting with the brain (although impulses must be relayed to the brain for awareness to occur). Two or more neurons may carry out reflex reactions. A simple neural connection of sensory neuron to interneuron to motor neuron allows the body to respond more rapidly than it could if the information had to travel to the brain. Reflexes are often protective.

A **reflex arc** is a neural pathway that links a sensory receptor and an effector such as a muscle. The three neurons that cause you to pull your hand away from a painful stimulus such as a thorn make up a reflex arc (fig. 31.16). The arc begins when a sensory receptor, perhaps a sensory neuron whose dendrite extends all the way to your fingertip, detects the sharp thorn. An action potential is generated along its axon and is transmitted to the dorsal (back) side of the spinal cord. Within the spinal cord, the sensory neuron's axon synapses with an interneuron; within the gray matter, the interneuron synapses with a motor neuron. The motor neuron's axon exits the spinal cord on the ventral side, and its action potential stimulates a skeletal muscle cell to contract. When a sufficient number of muscle fibers contract, you pull your hand away from the thorn. The original sensory neuron synapses with other interneurons, too. Some of these send action potentials to the brain, perhaps prompting you to yell in pain. Biology in Action 31.1 discusses the spinal cord injury of a well-known individual, Christopher Reeve.

31.1 biology in action

Spinal Cord Injuries

Christopher Reeve's life changed forever in a split second on May 27, 1995. Reeve, best known for his portrayal of Superman in four films, was one of 300 equestrians competing on that bright Saturday in Culpeper County, Virginia. He and his horse, Buck, were poised to clear the third of 15 hurdles in a 2-mile event. The horse's front legs went over the hurdle, and Reeve's back arched as he propelled himself forward, in sync with the horse, as they had done many times in practice. But this time was different.

Buck stopped, his back legs never clearing the fence. Reeve hurled forward, pitching over the horse's head, striking his head on the fence. He landed on the grass—unconscious, not moving or breathing.

Reeve had broken the first and second cervical vertebrae, between the neck and the brain stem. Someone performed mouth-to-mouth resuscitation until paramedics inserted a breathing tube and then stabilized him on a board. At a nearby hospital, Reeve received methylprednisolone, a drug that diminishes the extremely damaging swelling that occurs as the immune system responds to the injury. If given within 8 hours of the accident, this drug can save a fifth of the damaged neurons. Reeve was then flown to a larger medical center, where he was sedated and fluid suctioned from his lungs.

The first few days following a spinal cord injury are devastating (fig. 31.A). At first, the vertebrae are compressed and may break, which sets off action potentials in neurons, many of which soon die. The massive neuron death releases calcium ions, which attract tissue-degrading enzymes. Then white blood cells arrive and produce inflammation that can destroy healthy as well as damaged neurons. Axons tear, myelin coatings are stripped off, and vital connections between nerves and muscles are cut. The tissue cannot regenerate.

A week after the accident, Reeve could move a few muscles in his chest and back, which indicated that his spine had not been severed. Three weeks after the accident, he could sit and had some sensation in his upper body. On June 5, he underwent surgery to implant U-shaped wires in his neck to limit further damage. A month later he went to a rehabilitation facility to begin therapy.

Progress has been slow. It wasn't until January 1996 that Reeve could breathe unaided for longer than a few minutes, which he compared to climbing a very steep hill. A year after the injury, he could breathe unaided for close to an hour, which indicated that some healing had occurred. By 1997, Reeve was providing voiceovers, directing and acting in roles requiring a wheelchair-bound character.

Each year, 11,000 people in the United States sustain spinal cord injuries, most in automobile accidents, sports, or violent actions. More than 200,000 people live with these injuries, 55% with partial paralysis due to damage in the lower back, and 45% with damage to the upper spine and near-total paralysis.

Ironically, just as Reeve campaigned tirelessly to increase funding for spinal cord injury research, investigators at the Karolinska Institute in Sweden showed for the first time in experimental animals that a spinal cord injury can partially heal. The researchers severed the spinal cords of 23 rats, so that the animals could not move their legs. Then, they took nervous tissue from the rats' chests and placed it between the severed regions of the spinal cord. After 3 months, the rats began to move their legs and crawl. A year after the surgery they could move about their cages. Although there are differences in injuries to human spinal cords and intentional damage to rats, and despite the fact that the rats did not fully recover, their limited movement is good news for the thousands of people living with spinal cord injuries.

FIGURE 31.A (opposite page)
A spinal cord injury is just the start of a process that continues to damage delicate neural tissue.

The Brain

The human brain weighs a mere 3 pounds (1.36 kilograms) and looks and feels like grayish pudding (fig. 31.17). Magnetic resonance imaging and positron emission tomography are techniques used to study brain structure and function, respectively (fig. 31.18).

The brain requires a large and constant energy supply to oversee organ systems and to provide the qualities of "mind"—learning, reasoning, and memory. At any given time, brain activity consumes 20% of the body's oxygen and 15% of its blood glucose. Permanent brain damage occurs after just 5 minutes of oxygen deprivation.

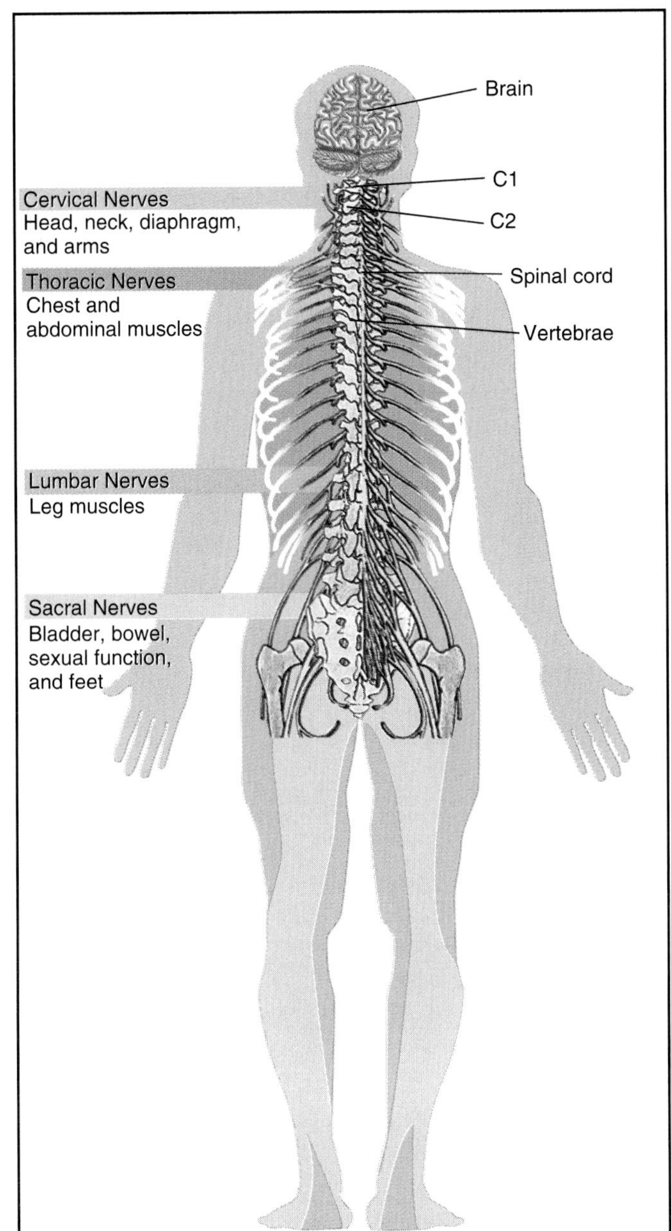

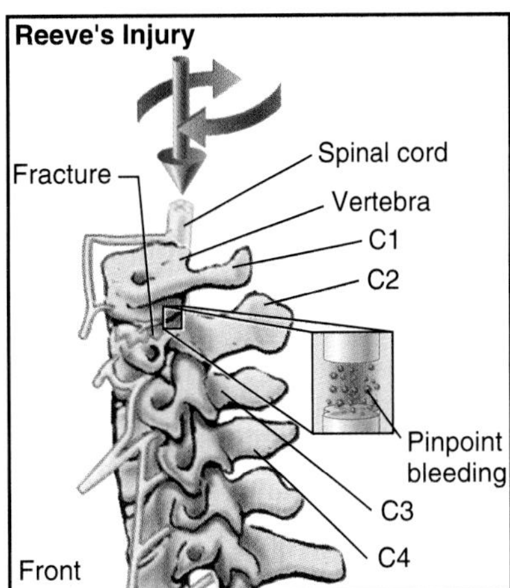

C1 and C2 vertebrae compressed, twisted, C2 fractured. Pressure builds, bleeding begins.

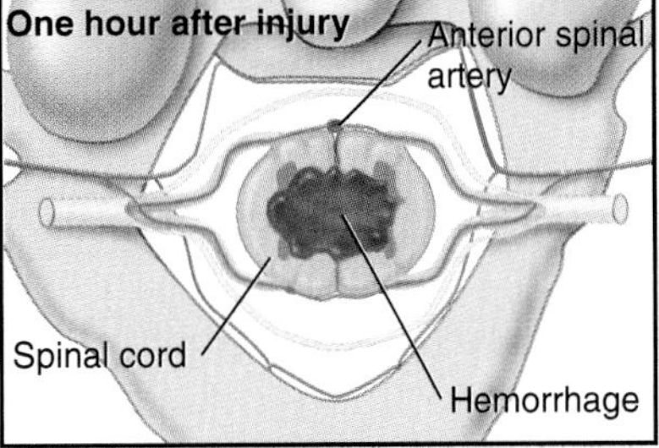

Tissue-destroying enzymes attracted to calcium ions that dying neurons release, triggering oxidative damage. Inner portion of spinal cord degenerates.

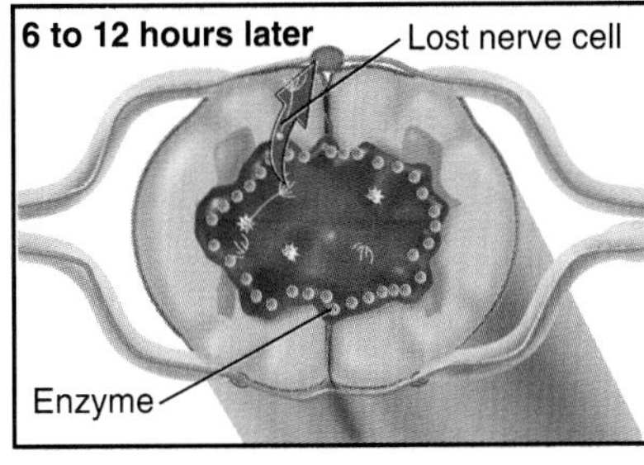

Immune system causes great inflammation, damaging injured area further.

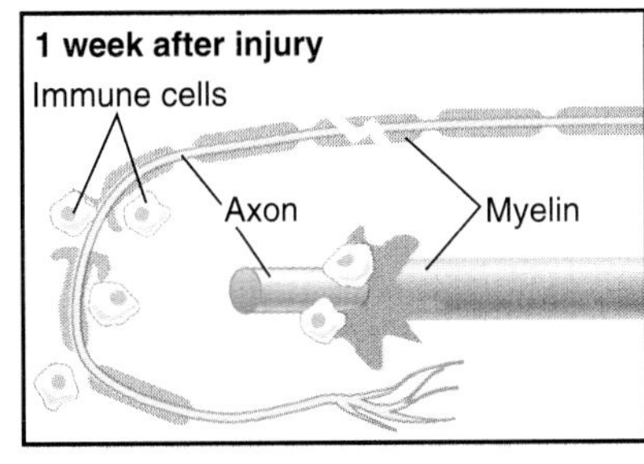

Axons tear and degenerate, losing their myelin sheaths. Nerve transmission to body parts beneath neck ceases further.

Brain anatomy can be considered in terms of the three subdivisions that first appear in the early embryo—the hindbrain, the midbrain, and the forebrain (fig. 31.19). Within each region lie major structures and a few smaller (but nonetheless important) structures. The three largest parts are the **brain stem,** the **cerebellum,** and the **cerebrum.** The brain stem, which supports the hindbrain, and the cerebellum, which lies in the hindbrain, are considered more primitive brain structures because they appear in species less complex than humans. The cerebrum, located in the forebrain, is most highly developed in the primates.

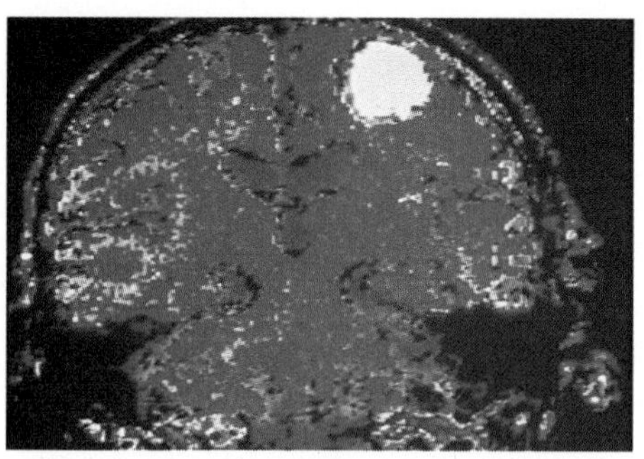

(a)

FIGURE 31.18

Brain scans. (*a*) Magnetic resonance imaging (MRI) uses radio waves in a magnetic field to depict brain structures in great detail. The tumor in this brain is clearly seen as a white area. (*b*) A positron emission tomography (PET) scan highlights brain function by following a radioactively labeled compound as it is metabolized in the brain. A PET scan can be superimposed on an MRI to show where a function occurs. These particular PET scans demonstrate learning. Each column shows different parts of the same brain. To obtain the "naive" scan on the left, experimenters gave the person a list of nouns and asked the individual to spontaneously visualize the object each word represents. The middle scans were obtained after the person practiced the task and could picture the nouns with less brain activity. In the third scan, the subject was given the same task but a new list of nouns. The learning centers in the brain show increased activity. PET and MRI scans are used not only in learning research but to diagnose brain disorders, such as tumors and epilepsy.

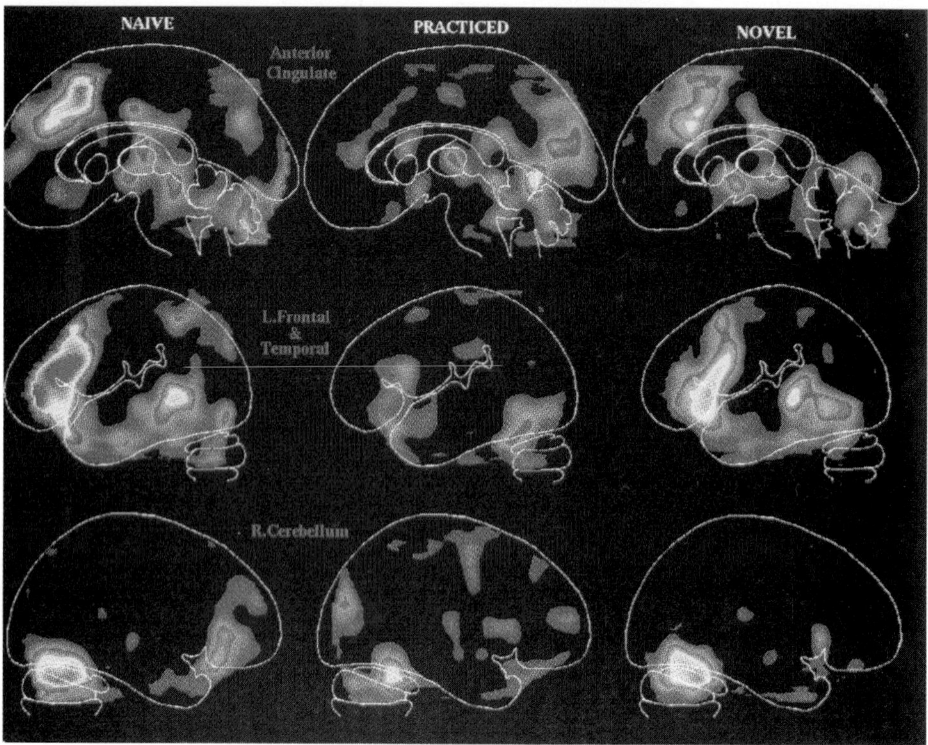

(b)

Corpus callosum
Forebrain: Cerebrum, Thalamus, Hypothalamus
Midbrain: Tectum
Hindbrain: Pons, Medulla oblongata, Cerebellum
Spinal cord

FIGURE 31.19

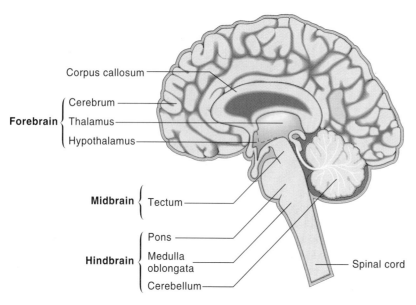

The three major areas of the vertebrate brain—the hindbrain, midbrain, and forebrain.

The Hindbrain

The hindbrain is located toward the back of the skull. It contains the brain stem, which includes the **medulla oblongata** (often just "medulla"), the cerebellum, and the **pons.** Anatomically, the brain stem is a continuation of the spinal cord within the skull. It is aptly named for its appearance: it supports the brain as a stem supports a head of cauliflower. But it does much more. The brain stem controls vital functions.

The section of brain stem closest to the spinal cord is the medulla oblongata, which regulates essential physiological processes including breathing, blood pressure, heart rate, and contractility. The medulla adjusts these activities to suit the body's varying requirements, increasing heart and breathing rates and blood pressure during vigorous activity and slowing them during restful periods. In addition, the medulla contains reflex centers for vomiting, coughing, sneezing, urinating, defecating, swallowing, and hiccoughing.

The medulla is also a thoroughfare. Most messages entering or leaving the brain must pass through the medulla oblongata. As axons and dendrites carrying sensory or motor messages traverse this region, most of them cross from one side of the body to the other. In this way, the left side of the brain receives impressions from and directs the motor activity of the right side of the body, and vice versa.

The cerebellum ("little brain") is an outgrowth of and is connected to the brain stem. The cerebellum is gray on the outside, white with gray patches on the inside, and divided into two hemispheres. As the second largest structure in the human brain, it accounts for almost one-eighth of the brain's mass.

The neurons of the cerebellum refine motor messages and coordinate muscular movements subconsciously. The cerebellum receives sensory input from the cerebral cortex (the outer portion of the cerebrum) and the PNS. It then compares the action the cerebrum intended with the actual movement and makes corrections so that the two agree. For example, the cerebellum acts when we try to bring the tips of our two index fingers together without looking. The cerebellum quickly recognizes and corrects the initial miss.

Many of our conscious activities have subconscious components that the cerebellum governs. If you had to consciously plan every movement in catching a ball, for instance, you would probably not be quick enough to prevent it from striking your head. Training and practice result in an automatic program of movements, many of which the cerebellum controls at a subconscious level.

The area above the medulla is the pons, which means "bridge." This is a suitable name for this oval mass because white matter tracts in the pons form a two-way conduction system that connects higher brain centers with the spinal cord and connects the pons to the cerebellum. In addition, gray matter in the pons controls some aspects of respiration.

The medulla and pons have changed very little through vertebrate evolution. The cerebellum, however, is greatly enlarged in more complex vertebrate species.

The Midbrain

A narrow region above the pons, the midbrain, is also part of the brain stem. The midbrain receives sensory information from touch, sound, visual, and other receptors and passes it to the forebrain. Neurons in the midbrain interpret sensations as perceptions. Much of this integrating activity occurs in a thickened area of gray matter at the roof of the midbrain called the **tectum.** In amphibians and fishes, the tectum is the largest part of the brain. This may be an adaptation related to their heavy reliance on their senses for survival.

The Forebrain

The part of the brain that has changed the most through vertebrate evolution is the forebrain. The forebrain includes two major regions: the **telencephalon,** which includes the cerebrum, and the **diencephalon,** which lies in front of the midbrain and includes the **thalamus, hypothalamus,** pineal gland, and pituitary gland. (Chapter 33 discusses the latter two structures.) Forebrain structures are important in complex behaviors such as learning, memory, language, motivation, and emotion.

The hypothalamus weighs less than 4 grams and occupies less than 1% of the brain volume, but it regulates a surprising number of vital functions, including body temperature, heartbeat, water balance, and blood pressure. Groups of nerve cell bodies in the hypothalamus control hunger, thirst, sexual arousal, and feelings of pain, pleasure, anger, and fear. The hypothalamus also regulates hormone secretion from the pituitary gland at the base of the brain. Thus, the hypothalamus links the nervous and endocrine systems, the body's two communication systems.

The thalamus is a gray, tight package of nerve cell bodies and their associated glial cells located beneath the cerebrum. It acts as a relay station for sensory input, processing incoming information and sending it to the appropriate part of the cerebrum.

The **reticular activating system** (RAS) is not really localized to hindbrain, midbrain, or forebrain but is a diffuse network of cell bodies and nerve tracts that extends through the brain stem and into the thalamus. The name *reticular* ("little net") alludes to the RAS's role in screening sensory information so that only certain impulses reach the cerebrum. If the RAS did not do so, the senses would be overwhelmed—you would be acutely aware of every cough, sneeze, and rustle of paper from those around you, every scent, every movement, in the classroom; you would even be aware of the touch of your clothing against your skin.

As its name implies, the reticular activating system is also important in overall activation and arousal. When certain neurons within the RAS are active, you are awake. When other neurons inhibit them, you sleep.

The thalamus is a gateway between the RAS and regions of the cerebral cortex. On a cellular level, sleep is a state of synchrony, when many neurons in all of these regions are inhibited from firing action potentials at the same time.

The electrical activity of the brain—the brain waves detectable in an electroencephalogram (EEG) tracing—differs during wakefulness and sleep (fig. 31.20). We experience two types of sleep—REM sleep, named for the rapid eye movements that occur under closed lids, and nonREM sleep. During REM sleep, the nervous system is quite active, although movement is suppressed. The autonomic nervous system is alert, and heart rate and respiration are elevated. Brain temperature rises as its blood flow and oxygen consumption increase. Action potentials zip about the brain more rapidly than during nonREM sleep or even during wakefulness.

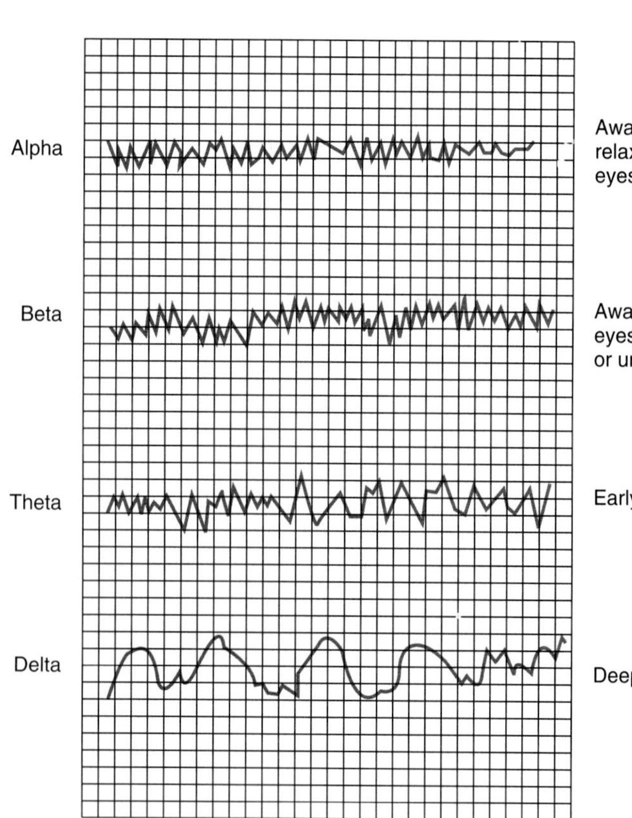

FIGURE 31.20

Characteristic brain-wave patterns represent different states of consciousness.

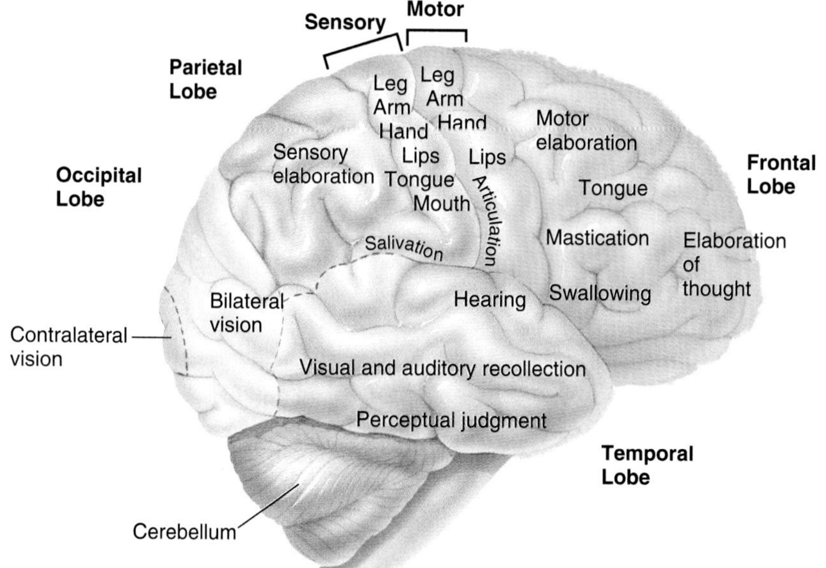

FIGURE 31.21

Cerebral specialization. The cerebrum contains four main sections visible from the surface—the frontal, temporal, occipital, and parietal lobes. Experiments have shown that different parts of the cerebrum carry out different specific functions. Each hemisphere of the cerebral cortex contains a band of tissue, the primary sensory area, that receives sensory input from the muscles, joints, bones, and skin. Each also has a band of tissue called the primary motor area that controls voluntary muscles. Association areas are located in the front of the frontal lobe and in parts of the occipital, parietal, and temporal lobes.

REM sleep is a time of vivid dreaming. NonREM sleep proceeds through four continuous stages, from light sleep to deep sleep, and the corresponding EEG wave tracings become less frequent but higher in amplitude (size).

Sleep follows a predictable pattern—70 to 90 minutes of non-REM followed by 5 to 15 minutes of REM, repeated many times a night. If a person is deprived of REM sleep and therefore doesn't dream, the next night's pattern compensates to deliver more REM sleep.

Mastering Concepts

1. Describe the functions of the neurons that form a reflex arc.
2. What are the three largest parts of the brain, and what functions do they oversee?
3. What are the major structures in the hindbrain, midbrain, and forebrain?
4. How do REM and nonREM sleep differ?

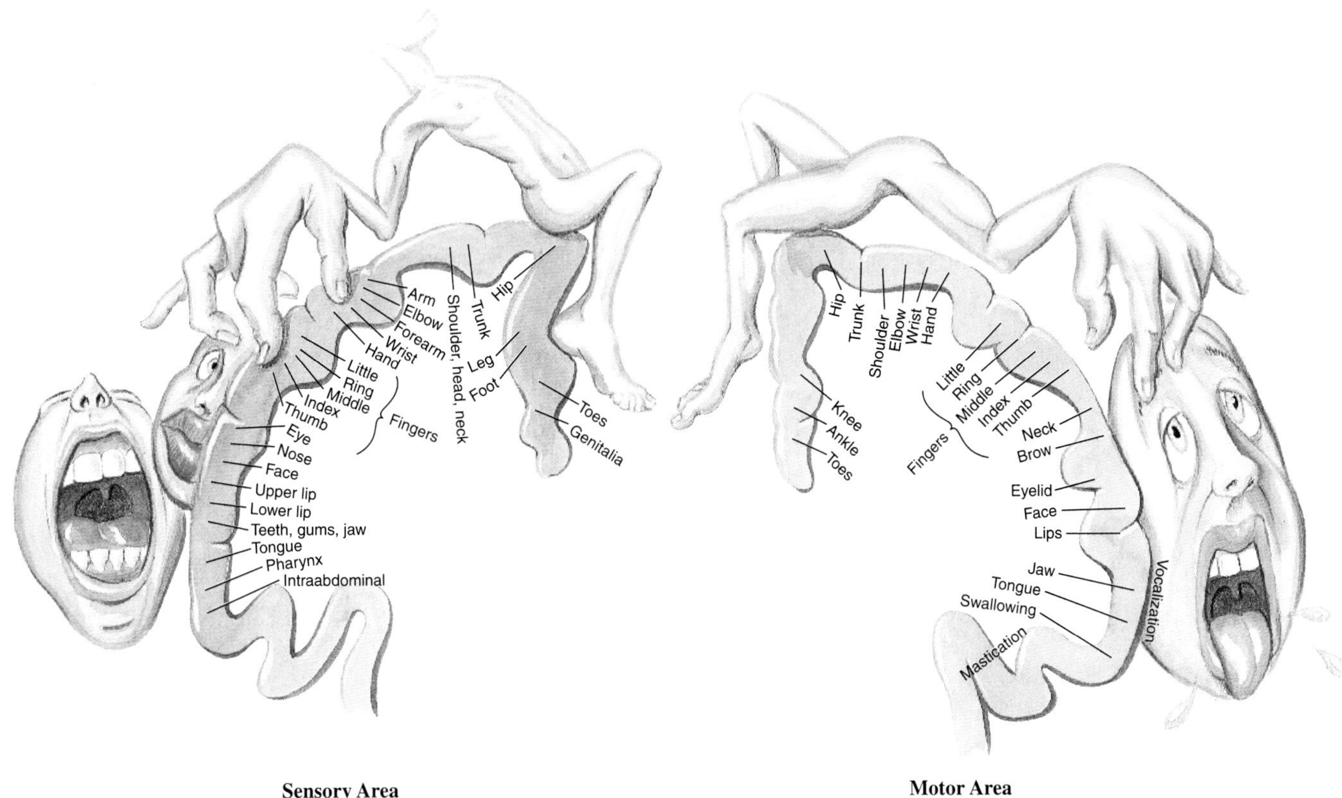

FIGURE 31.22

Different body parts are innervated to different degrees. The amount of surface area devoted to a particular body part depends not upon the size of the part but upon how precisely the nervous system must control it. A distorted human figure superimposed on this diagram of the primary sensory and motor areas indicates the areas that are finely innervated, such as the face and hands.

The Cerebrum

The human brain has a large, highly developed cerebrum that controls the qualities of "mind," including intelligence, learning, perception, and emotion. The two hemispheres of the cerebrum make up 80% of total human brain volume.

The outer layer of the cerebrum, the cortex, consists of gray matter that integrates incoming information. In mammals, the cerebral cortex increases in size and complexity as species become more complex. Through evolutionary time, the cortex expanded so rapidly that it folded back on itself to fit into the skull, forming characteristic convolutions.

The Cerebral Cortex

The cerebral cortex contains sensory, motor, and association areas (fig. 31.21). Sensory areas receive and interpret messages from sense organs about temperature, body movement, pain, touch, taste, smell, sight, and sound. Motor areas send impulses to skeletal muscles. Association areas do not appear to be either sensory or motor, but they are the seats of learning and creativity.

A band of cerebral cortex extending from ear to ear across the top of the head, called the primary motor cortex, controls voluntary muscles. Just behind it is the primary sensory cortex, which receives sensory input from the skin. Nearly every part of the body is represented in both the sensory cortex and motor cortex. The surface area devoted to a particular body part is proportional to the degree of sensitivity and motor activity in that area. For example, the hands, tongue, and face are extensively represented in both the sensory and motor regions of the cortex (fig. 31.22).

Specializations of the Cerebral Hemispheres

Each cerebral hemisphere has some specific functions, yet the hemispheres work together. In most people, parts of the left hemisphere are associated with speech, linguistic skills, mathematical ability, and reasoning, while the right hemisphere specializes in spatial, intuitive, musical, and artistic abilities. Under normal conditions, both cerebral hemispheres gather and process information simultaneously.

Mastering Concepts

1. How does the cerebral cortex differ among mammalian species?
2. Distinguish between the functions of sensory and motor areas of the cerebral cortex.

Memory

Why is it that you cannot remember facts your biology professor mentioned an hour ago, but you can easily recite lyrics to a song you haven't heard in years? Memory is one of the most astonishing capabilities of the nervous system, particularly in humans. It is closely associated with learning. Whereas learning is the acquisition of new knowledge, memory is the retention of learning with the ability to access it at a later time. Although we usually think of learning and memory as human abilities, we all know of pets who recognize people. Even an animal as simple as a sea snail can remember.

Researchers have recognized two types of memory, short-term and long-term, for many years, but they now think that the two types differ in characteristics other than duration. Research today focuses on how neurons in different parts of the brain encode memories and how short-term memories are converted to long-term memories.

Short-term memories are thought to be electrical in nature. Neurons may connect in a circuit so that the last in the series restimulates the first. As long as the pattern of stimulation continues, you remember the thought. When the reverberation ceases, so does the memory—unless it enters long-term memory.

Long-term memory probably changes neuron structure or function in a way that enhances synaptic transmission. A long-term memory may be a certain synaptic connection pattern. Such synaptic patterns fulfill two requirements of long-term memory. First, there are enough synapses in a human's brain to encode an almost limitless number of memories. Each of the 10 billion neurons in the cortex can make tens of thousands of synaptic connections to dendrites of other neurons, forming 60 trillion synapses. Second, a certain pattern of synapses can persist for years.

According to a theory called long-term synaptic potentiation, frequent stimulation of the neurons in an area of the cerebral cortex called the **hippocampus** strengthens their synaptic connections. This strengthening could occur because the action potentials triggered in postsynaptic cells in response to the repeated stimuli undergo greater depolarization. For example, certain neurons are consistently stimulated when a 2-year-old first learns what an orange is. The child's senses tell the brain that the object is round, orange, and rough in texture and that it smells and tastes sweet. With repetition, the stimulation of the same sets of neurons communicating different aspects of "orange" become, somehow, associated into a single memory. Soon, just the smell of an orange or a drawing of one summons up a mental image—the memory—of the real thing.

Mastering Concepts

1. Distinguish between short-term and long-term memory.
2. How might neurons encode memories?
3. What part of the cerebral cortex is the seat of memory formation?

The Peripheral Nervous System

Recall that the PNS is divided into sensory pathways and motor pathways. The motor pathways, in turn, are divided into the somatic (voluntary) nervous system and the autonomic (involuntary) nervous system.

The Somatic Nervous System

The nerves of the somatic nervous system send impulses to the muscles, sense organs, and glands associated with the body's surface. They also conduct action potentials to the CNS from muscles and glands.

Somatic nerves are classified by their site of origin. Twelve pairs of cranial nerves arise from the brain or brain stem. Thirty-one pairs of spinal nerves exit the spinal cord and emerge between the vertebrae. Eleven of the cranial nerve pairs innervate portions of the head or neck, while the exception, the vagus nerve, leads to internal organs. Each pair of spinal nerves innervates a section of the body near its point of departure from the spinal cord. Loss of feeling in a particular patch of skin can indicate damage to the spinal nerves that lead to that area.

The Autonomic Nervous System

The autonomic nervous system enables internal organs to function properly without our conscious awareness by receiving sensory information and transmitting impulses to smooth muscle, cardiac muscle, and glands. The autonomic nervous system is subdivided into the sympathetic and parasympathetic nervous systems. When an individual is under stress the sympathetic nervous system dominates; during more relaxed times, the parasympathetic nervous system is in charge. Although both the sympathetic and parasympathetic nervous systems often innervate the same organ, they usually have opposite effects on it (fig. 31.23). For example, the parasympathetic nervous system stimulates salivation, intestinal activity, and pupil constriction. In contrast, the sympathetic nervous system inhibits salivation and intestinal activity and dilates the pupils.

The sympathetic nervous system prepares the body to face emergencies—it accelerates heart rate and breathing rate; shunts blood to where it is required most, such as the heart, brain, and the skeletal muscles necessary for "fight or flight" (as opposed to "rest and repose"); and dilates airways, easing gas exchange. Figure 31.24 depicts the actions of the sympathetic nervous system.

The parasympathetic nervous system is activated when the body returns to rest. After an emergency, heart rate and respiration slow and digestion resumes. Different organs recover at different rates, because the organization of the parasympathetic nerves permits independent organ control.

The activities of the autonomic nervous system are involuntary, but we can control them to an extent, if we are aware of them. Biofeedback is a practice that uses a mechanical device to detect and amplify an autonomically controlled body function, such as blood pressure or heart rate. Once made aware of an unhealthy measurement, a person can consciously try to bring the body function within a normal range.

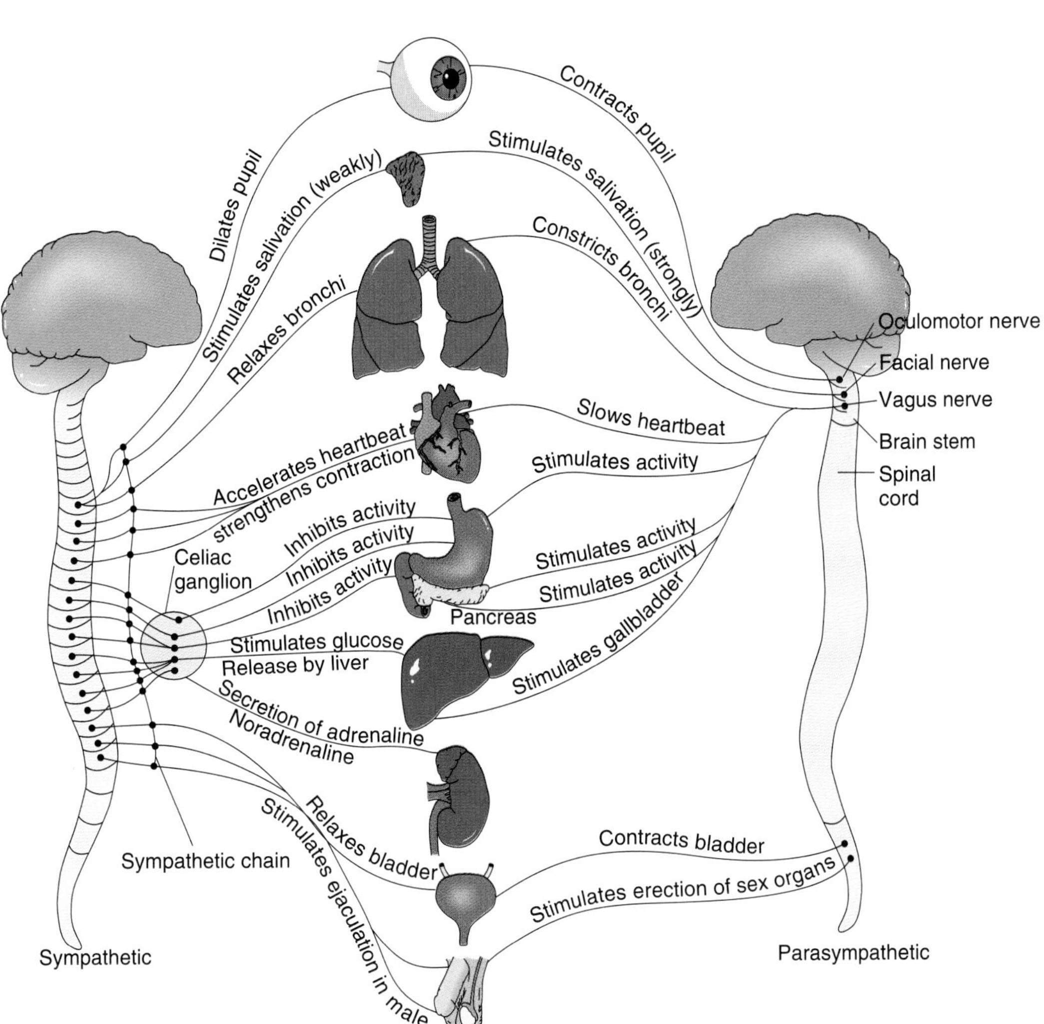

FIGURE 31.23

The autonomic nervous system. Neurons from the parasympathetic and sympathetic divisions of the autonomic nervous system innervate the internal organs. Note that in the parasympathetic division, the nerves emerging from the central nervous system are long and those near the organ are short. The reverse is true in the sympathetic nervous system. Each system has an opposite effect on the same organs.

Mastering Concepts

1. What are the pathways of the peripheral nervous system?
2. Which division of the autonomic nervous system dominates during stress, and which division dominates when the situation returns to normal?
3. What is biofeedback?

The Unhealthy Nervous System

Disorders of the nervous system may affect such hard-to-define qualities as thinking and feeling but nonetheless have biochemical underpinnings (table 31.3). Often, treatments for neurological disorders are based on understanding the physical causes of the symptoms.

Molecular Causes

Some of the earliest links discovered between neurological disorders and biochemistry were abnormal levels of specific neurotransmitters, as table 31.2 illustrates. More recently, genetic research has revealed other biochemical mechanisms of nervous system malfunction. These include impaired signal transduction (how cells receive incoming messages), defects in cell adhesion (how cells interact), and damage from the toxic by-products of oxygen metabolism called free radicals. Several degenerative neurological conditions are caused by mutant genes in which a specific three-base sequence repeats many times more than normal (see fig. 16.24). The "triplet repeat" may somehow destabilize a nearby gene that controls a nervous system function.

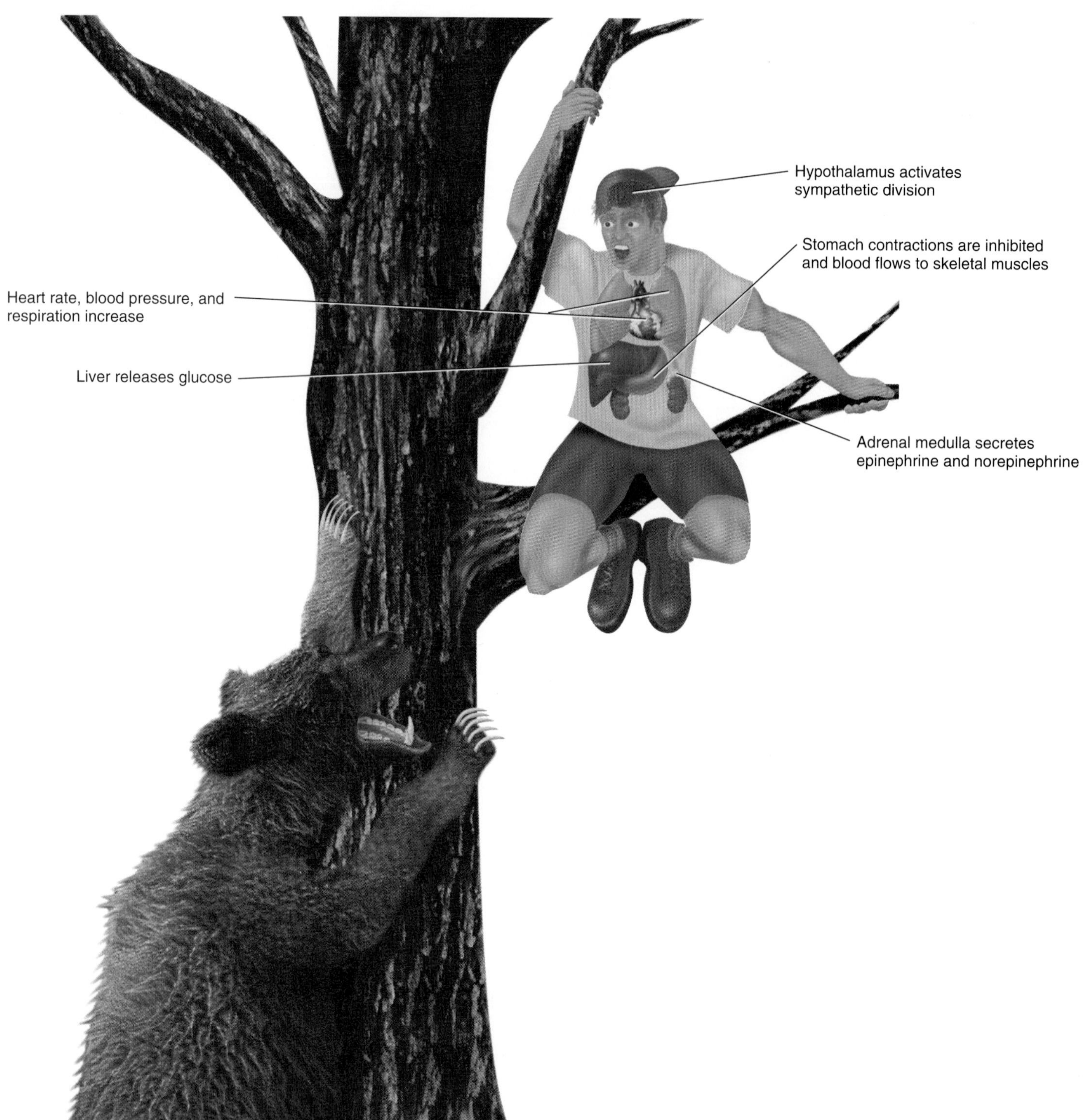
Hypothalamus activates sympathetic division
Stomach contractions are inhibited and blood flows to skeletal muscles
Heart rate, blood pressure, and respiration increase
Liver releases glucose
Adrenal medulla secretes epinephrine and norepinephrine

FIGURE 31.24

Danger! This individual's sympathetic nervous system is activated as he senses that the bear may attack. Heart rate and respiration speed up, blood flows to the muscles, and digestion slows.

Table 31.3 Molecular Mechanisms of Nervous System Disease

Mechanism	Representative Disorder	Symptoms/Causes
Adhesion protein abnormality	Kallmann syndrome	Embryonic neural cells cannot migrate to parts of brain controlling smell perception and gonad maturation. Symptoms are lack of smell and immature gonads.
Enzyme deficiency	Tay-Sachs disease	Nervous system degenerates in early childhood due to excess myelin on nerves. Deficient enzyme cannot break down myelin.
Neurotransmitter imbalance	Myasthenia gravis	Muscles progressively weaken because of deficient acetylcholine at neuromuscular junctions caused by autoimmune attack.
Oxygen free radical damage	Amyotrophic lateral sclerosis (Lou Gehrig disease)	Abnormal superoxide dismutase cannot stop oxygen free radical damage, which causes adult-onset progressive muscle weakness and paralysis due to degeneration of lower motor neurons in spinal cord and upper neurons in cerebral cortex.
Protein excess	Familial Alzheimer disease	Amyloid protein deposits build up in cerebrum, impairing function.
Signal transduction block	Lissencephaly ("smooth brain")	Cerebral cortex is smooth because lack of signal transduction protein prevents certain neurons from migrating to area during prenatal development.
Unstable triplet repeats	Huntington disease	Extra copies of a DNA base triplet occur in a gene on chromosome 4. Symptoms are personality changes and uncontrollable, dancelike movements.

Table 31.4 Common Nervous System Dysfunctions

Dsyfunction	Causes and/or Symptoms
Amnesia	Disorientation and memory loss
Cerebral palsy	Damage to motor cortex, basal ganglia, and/or cerebellum before, during, or shortly after birth; paralysis and absence of muscular control
Cerebrovascular accident	A blood clot in an artery of the brain (stroke), hemorrhage in the brain, hardening or plaque in cerebral arteries, or aneurysm (ballooning out of artery) in brain; variety of motor and other effects
Coma	Unconsciousness from which a person cannot be aroused
Concussion	Brief unconsciousness followed by confusion; caused by sudden brain movement due to violent blow to head
Dyslexia	Language skill disorder; person cannot easily recognize sequences of letters as meaningful words
Epilepsy	Seizures caused by electrical disturbances in the brain
Fainting	Brief loss of consciousness caused by blow to head, rapid reduction in blood pressure, or sudden shock
Headache	Pain sometimes caused by dilation of blood vessels in meninges (membranes surrounding brain)
Meningitis	Viral or bacterial infection of meninges around brain or spinal cord; deafness, neurological damage
Polio	Viral infection of nerve cell bodies in anterior of spinal cord; symptoms progress from fever, headache, and pain to loss of reflexes and muscle paralysis and atrophy

31.2 biology in action

Addiction!

Drug abuse and addiction are ancient as well as contemporary problems. A 3,500-year-old Egyptian document decries that society's reliance on opium. In the 1600s, a smokable form of opium enslaved many Chinese, and the Japanese and Europeans discovered the addictive nature of nicotine. During the American Civil War, morphine was a widely used painkiller; cocaine was introduced a short time later to relieve veterans addicted to morphine. Today, we continue to abuse drugs intended for medical use. LSD was originally used in psychotherapy but was abused in the 1960s as a hallucinogen (see Biology in Action 24.1). PCP was an anesthetic before being used in the 1980s. Why do people become addicted to certain drugs? Answers lie in the complex interactions of neurons, drugs, and individual behaviors.

The Role of Receptors

Eating hot fudge sundaes is highly enjoyable, but we usually don't feel driven to consume them repeatedly. Why do certain drugs compel a person to repeatedly use them in steadily increasing amounts—the definition of addiction? The biology of neurotransmission helps to explain how we, and other animals (fig. 31.B), become addicted to certain drugs.

Understanding how neurotransmitters fit receptors can explain the actions of certain drugs. When a drug alters the activity of a neurotransmitter on a postsynaptic neuron, it either halts or enhances synaptic transmission. A drug that binds to a receptor, blocking a neurotransmitter from binding there, is called an antagonist. A drug that activates the receptor, triggering an action potential, or that helps a neurotransmitter to bind, is called an agonist. The effect of a drug depends upon whether it is an antagonist or an agonist and on the particular behaviors the affected neurotransmitter normally regulates.

Neural pathways that use the neurotransmitter noradrenaline control arousal, dreaming, and mood. Amphetamine enhances noradrenaline activity, thereby heightening alertness and mood. Amphetamine's structure is so similar to that of noradrenaline that it binds to noradrenaline receptors and triggers the same changes in the postsynaptic membrane.

Cocaine has a complex mechanism of action, both blocking reuptake of noradrenaline and binding to molecules that transport dopamine to postsynaptic cells. Cocaine's rapid and short-lived "high" reflects its rapid stay in the brain—its uptake takes just 4 to 6 minutes, and within 20 minutes the drug loses half its activity.

GABA is an inhibitory neurotransmitter used in a third of the brain's synapses. The drug valium causes relaxation and inhibits seizures and anxiety by helping GABA bind to receptors on postsynaptic neurons. Valium is therefore a GABA agonist.

Opiates in the Human Body

Opiate drugs, such as morphine, heroin, codeine, and opium, are potent painkillers derived from the poppy plant. These drugs alter pain perception, making pain easier to tolerate, and they also elevate mood. When taken repeatedly by healthy individuals, opiate drugs are addictive. When taken to relieve intense pain, opiates are usually not addictive.

The human body produces its own opiates, called **endorphins** (for "endogenous morphine"). These peptides are a type of neurotransmitter. Like the poppy-derived opiates they structurally resemble, endorphins influence mood and perception of pain. Opiates and endorphins bind the same receptors in the human brain.

Humans produce several types of endorphins, which are released in response to stress or pain. Endorphins are released during a "runner's high," as a mother and child experience childbirth, and during acupuncture.

Endorphins help explain why some people addicted to an opiate drug such as heroin experience withdrawal pain when they stop taking the drug. Initially, the body interprets the frequent binding of heroin to its endorphin receptors as excess endorphins. To bring the binding down, the body slows its own production of endorphins. Then, when the person stops taking heroin, the body is caught short of opiates (heroin as well as endorphins). The result is pain.

Drug addiction is a powerful force, its causes rooted in both biology and psychology and its effects a sociological and political problem. Fighting an addiction is far more difficult than "just saying no" to temptation, as one might to a hot fudge sundae.

FIGURE 31.B
A rat bred to be prone to addiction helps researchers identify compounds likely to be highly addictive in humans. The rat self-administers drugs it likes to use.

Environmental Causes

Some neurological disorders may have environmental causes. Consider Parkinson disease, which usually first appears after age 60 and causes slowed movements, muscle rigidity, tremors, and loss of fine motor coordination. The symptoms arise from lack of the neurotransmitter dopamine.

Chemical exposures may cause Parkinson disease. Some people exposed to a chemical called MPTP, either in a laboratory setting or illicitly, develop severe Parkinson-like symptoms in early adulthood. Because MPTP is chemically similar to certain potent herbicides, researchers hypothesized that years of exposure to herbicides might cause Parkinson disease in susceptible individuals. Epidemiological studies support this hypothesis by showing an increase in incidence of Parkinson disease in areas that have used MPTP-related herbicides for many years. The illness may also be inherited, or caused by repeated blows to the head.

Bioethics Connections 5.1 and Biology in Action 31.2 explore drug addiction, a common problem affecting the nervous system.

Mastering Concepts

1. Describe some molecular bases of neurological disease.
2. What are two possible environmental causes of nervous system disorders?

Protection of the Nervous System

Because the nervous system interacts so closely with other organ systems, injuries or illnesses can be devastating (table 31.4). Many adaptations protect the nervous system or enable the body to survive damage to it.

Several structures protect the CNS. First, bones of the skull and vertebral column shield the delicate nervous tissue from bumps and blows. Trilayered membranes called **meninges** jacket the CNS for further protection. The **blood-brain barrier,** formed by specialized brain capillaries (see fig. 1.3) helps protect the brain from chemical insult. The tilelike cells that form these capillaries fit so tightly together that only certain chemicals can cross into the **cerebrospinal fluid** that bathes and cushions the brain and spinal cord. This fluid, made by cells that line ventricles (spaces) in the brain, further insulates the CNS from injury.

Despite these protections, disease or injury can harm the nervous system. Many nervous system problems are difficult to treat because mature neurons usually cannot divide to repair damaged tissue. In addition, the blood-brain barrier blocks many drugs used to combat CNS infections. How, then, can the nervous system compensate for damage?

Neurons of the PNS can regenerate. Schwann cells at the tip of a damaged peripheral nerve cell extend and provide a pathway for new growth. A crushed peripheral nerve can grow back, reforming its synaptic connections.

In the CNS, healing is minimal. Damaged unmyelinated neurons can sprout new axons, but these do not extend as far as the old ones. Crushed or damaged neurons in a spinal tract are replaced with a scar consisting of glial cells, connective tissue, and random axons, rather than with a new set of regrown, connected axons.

Still, surviving neurons can help maintain overall functions when some neurons die. Undamaged neighboring neurons extend new terminals to the cells that previously synapsed with the destroyed cell, much as office workers assume the duties of a departed coworker.

The brain, like other organs, loses efficiency over time. In the aging brain, some synaptic connections deteriorate, some postsynaptic receptors become less sensitive to neurotransmitters, and levels of some neurotransmitters fall. By age 60, the human brain decreases in weight by about 3 ounces (85 grams). Although an older person's memory may be less reliable and learning more difficult, he or she can continue to be mentally alert and enjoy the sensations, feelings, and perceptions the nervous system makes possible for many years.

Mastering Concepts

1. List three ways that the nervous system is protected from damage.
2. To what extent can the nervous system regenerate?
3. How does the brain change with age?

Head Games~

continued from page 610.

Roger had been a "troublemaker," frequently cutting school and getting failing grades when he did attend. He was uncommunicative or nasty to his father, who was raising him alone.

The gunblast sent powerful mechanical waves through the soft matter at the front of 16-year-old Roger's brain, but because the bullet was small and not very fast, Roger survived. Surgery to remove the bullet required incisions in the white matter of both parts of the prefrontal lobes. The surgeon removed pressure on the brain from inflammation, cleared away metal fragments and pulverized bone, and washed the area in antibiotics.

Recovery was long and slow. At first Roger could not follow simple directions or speak, but gradually he began to understand what people said to him. He cooperated with the therapists in the rehabilitation center where he lived for several months. Eight months after the injury, Roger returned to school. It quickly became apparent that he was a changed person.

Roger attended school regularly, did his work, and earned decent grades. He was friendly and cooperative. Roger was no longer Roger.

The true stories of Phineas Gage and Roger illustrate how tragic accidents provide information that enables researchers to link certain brain regions to nervous system function. Yet these events also highlight how little we understand of how masses of mushy brain tissue encode the many components of a human personality.

SUMMARY

1. Nervous tissue and systems maintain homeostasis by enabling animals to respond to environmental changes. Nervous tissue includes neurons and glial. Neurons transmit electrochemical changes that stimulate or inhibit cell surface receptors. The complexity of a species' nervous tissue or system reflects adaptations to the environment.

2. A neuron has a **cell body; dendrites,** which receive impulses and transmit them toward the cell body; and an **axon,** which conducts impulses away from the cell body. A **sensory neuron** carries information towards the brain and spinal cord (**central nervous system,** or CNS). A **motor neuron** carries information from the CNS and stimulates an **effector** (a muscle or gland). An **interneuron** conducts information between two neurons and coordinates responses.

3. A neuron's message, or nerve impulse, is measured as an **action potential** that occurs as an electrical change. In a neuron at rest, K^+ concentration is 30 times greater inside the cell than outside, and the Na^+ concentration is 10 times greater outside than inside. The resulting diffusional gradients combined with negatively charged proteins within the cell give the interior a negative charge. When an action potential begins, membrane channels open and allow Na^+ in, creating a positive charge inside. The positively charged Na^+ depolarizes the membrane. At the peak of depolarization, Na^+ channels close. Repolarization occurs as K^+ leaves the cell, restoring the negatively charged **resting potential.** The action potential spreads along the nerve fiber.

4. Myelination increases the speed of nerve impulse transmission. A **myelin sheath** forms around some nerve fibers as fatty **Schwann cells** wrap around the fiber. The gaps between these insulating cells are **nodes of Ranvier.** In a myelinated fiber, the nerve impulse "jumps" from one node to the next, increasing speed of conduction. Transmission of an action potential along a myelinated fiber is called **saltatory conduction.**

5. When an action potential reaches the end of an axon, it causes vesicles in the terminal portion of the **presynaptic cell** to move toward the membrane and release **neurotransmitters** into the **synaptic cleft.** These chemicals diffuse across the cleft and bind receptors on the **postsynaptic cell.** If the neurotransmitter is **excitatory,** it slightly depolarizes the postsynaptic membrane, making an action potential more probable. An **inhibitory,** neurotransmitter hyperpolarizes the membrane, making an action potential less likely. This **neural integration,** the summing of excitatory and inhibitory messages, finely controls neuron activity. After it functions, a neurotransmitter is enzymatically destroyed or reabsorbed into the presynaptic cell.

6. The simplest invertebrate nervous systems are nerve nets, which detect stimuli from any direction. Flatworms have simple brains, paired sense organs, and two nerve cord ladders that allow localized motor control. Annelids have a more elaborate brain and ladder organization, with acute senses. Arthropods have even more complex sensory organs and behaviors.

7. In the vertebrate nervous system, the CNS begins developing when the embryonic notochord stimulates ectoderm to develop into the **neural tube.** The neural tube closes at both ends, and then the anterior end swells, forming a brain. Neurons extend from the brain ventricles outward to form the outer brain layers. Other ectoderm forms the **peripheral nervous system** (PNS), which includes all neural tissue outside of the CNS.

8. The human spinal cord is a tube of neural tissue encased in the vertebral canal. The white matter on the periphery of the cord conducts impulses to and from the brain. The spinal cord is a reflex center. A **reflex** is a quick, automatic, protective response that travels through a **reflex arc.** A reflex arc usually consists of a sensory receptor, a sensory neuron, a spinal interneuron, a motor neuron, and an effector, such as a muscle or gland.

9. The brain has three regions. The hindbrain includes the **medulla oblongata,** which controls many vital functions; the **cerebellum,** which coordinates unconscious movements; and the **pons,** which bridges the medulla and higher brain regions and connects the cerebellum to the cerebrum. The midbrain processes visual and auditory sensory information; the **tectum** integrates this information. The forebrain consists of the **telencephalon,** which includes the **cerebrum** and the

diencephalon, which contains the **thalamus,** a relay station between lower and higher brain regions; the **hypothalamus,** which regulates many vital physiological processes and regulates levels of some pituitary hormones; and the pineal and pituitary glands. The **reticular activating system** filters sensory input and is important for arousal.

10. The cerebrum has an inner layer of white matter and an outer layer of convoluted gray matter. It has two hemispheres, which receive sensory input from and direct motor responses to the opposite side of the body. The left hemisphere specializes in language and analytical reasoning, and the right hemisphere regulates spatial, intuitive, and creative abilities.

11. Short-term memory may depend on temporal electrical activity in neuronal circuits. Short-term memories may consolidate into long-term memories, which depend on permanent chemical or structural changes in neurons.

12. The PNS consists of the **somatic** (voluntary) **nervous system** and the **autonomic** (involuntary) **nervous system.** The somatic nervous system includes cranial and spinal nerves that transmit sensations from sensory receptors or stimulation to voluntary muscles. The autonomic nervous system receives sensory information and conveys impulses to smooth muscle, cardiac muscle, and glands. Within the autonomic nervous system, the **sympathetic nervous system** controls physical responses to stressful events, while the **parasympathetic nervous system** dominates during rest.

13. Nervous system disorders may have molecular and/or environmental causes.

14. Bones of the skull and vertebrae, **cerebrospinal fluid,** the **blood-brain barrier,** and **meninges** protect the CNS. PNS neurons can regenerate, but those in the CNS cannot. Undamaged neurons can sometimes take over some functions of damaged neurons.

TO REVIEW . . .

1. Sketch a neuron and label the cell body, dendrites, and axon. What functions does each perform?

2. How do the functions of sensory, motor, and interneurons differ?

3. Describe the distribution of charges in the membrane of a resting neuron.

4. What ionic events cause the wave of depolarization and repolarization constituting an action potential?

5. How does myelin alter conduction of an action potential along a nerve fiber?

6. Describe the differences between transmission of an action potential and synaptic transmission.

7. Sketch a synapse. Include the following structures: presynaptic membrane, vesicles, receptors, postsynaptic membrane.

8. List 10 skills you use in everyday life that depend upon some part of your nervous system. If you can, indicate what part of the nervous system controls each skill.

9. What role does the cerebellum play in athletics?

10. A person who has narcolepsy (uncontrollably falling asleep) probably has a defect in what part of the brain?

11. In what part of the brain do the qualities of "mind" lie?

12. How does the structure of the cerebral cortex provide different degrees of sensitivity in different body parts?

13. You are walking home from an evening class when you hear footsteps and heavy breathing and then see a tall shadow approaching from behind you. What is your nervous system likely to do?

TO THINK ABOUT . . .

1. If a student gets only 3 hours of sleep for several days, how might her sleeping pattern differ from normal on her next full night of sleep?

2. All human brains are about the same size, contain the same major structures, and function in similar ways. How, then, does each of us develop a distinct personality?

3. Smoking marijuana kills some brain neurons, which cannot regenerate. Why don't individuals usually suffer noticeable brain damage from smoking marijuana?

4. Now that you know a little about the physical basis of memory, can you think of some ways to make learning biology a little easier?

5. Fossilized skulls of humans and our immediate ancestors reveal that the brain has increased in capacity over 2 million years from 440 cubic centimeters to the present-day range of 1,350 to 1,400 cubic centimeters. How might you test the hypothesis that brain size increases as intellectual skills grow?

6. In an episode of Star Trek, crew members of a starship are deprived of REM sleep for many consecutive nights. What symptoms would they likely exhibit?

7. Nicotine is a highly addictive drug, as is cocaine. One is legal, the other is not. What, if anything, do you think could be done about these designations to help young people from becoming addicted to either drug?

TO LEARN MORE . . .

References and Resources

Barde, Yves-Alain. January 30, 1997. Help from within for damaged axons. *Nature,* vol. 385. Neurons can regenerate to a small extent.

Damasio, Hanna, et al. May 20, 1994. The return of Phineas Gage: Clues about the brain from the skull of a famous patient. *Science,* vol. 264. In 1848, young Phineas Gage suffered a brain injury in a freak accident.

Frank, Eric. January 17, 1997. Synapse elimination: For nerves it's all or nothing. *Science,* vol. 275 Muscles of young vertebrates are more highly innverted than muscles of older individuals.

Hart, Stephen. March 1997. Cone snail toxins take off. *BioScience,* New drugs may come from a snail's neurotoxin.

Hill, C. Stratton. December 20, 1995. When will adequate pain treatment be the norm? *The Journal of the American Medical Association,* vol. 274. Physicians should relieve patients of chronic or cancer-related pain.

Koch, Christof. January 16, 1997. Computation and the single neuron. *Nature,* vol. 385. Neural networks mimic the nervous system.

Lemonick, Michael D. September 2, 1996. No wonder you can't resist. *Time.* Can you be addicted to chocolate?

Lewis, Ricki. June 1997. Imaging depression. *Photonics Spectra.* People with depression have detectable brain abnormalities.

Lewis, Ricki. July/August 1996. MRI brings back pallidotomy. *Biophotonics International.* Surgery can sometimes treat Parkinson disease.

Lewis, Ricki. June 1996. Evening out the ups and downs of manic-depressive illness. *FDA Consumer.* Two drugs help people with bipolar disorder.

Lewis, Ricki. May 13, 1996. Neurogastroenterologists combine old and new research approaches. *The Scientist.* The digestive system has its own nervous system.

Lewis, Ricki. May 1996. Imaging cocaine in action. *Photonics Spectra.* Viewing cocaine's action at synapses explains its effects.

Lewis, Ricki. March 1996. From sugar and spice to PET and MRI. *Photonics Spectra.* Subtle brain differences distinguish the sexes.

Lewis, Ricki. October 1995. Fractured vision reveals brain functions. *Photonics Spectra.* Bizarre visual disturbances are traced to the brain.

Rosenblatt, Roger. August 26, 1996. New hopes, new dreams. *Time.* A riveting account of Christopher Reeve's spinal cord injury.

Youdim, Moussa B. H., and Peter Riederer. January 1997. Understanding Parkinson's disease. *Scientific American.* Parkinson disease has several causes—faulty genes, injury, and environmental toxins.

Young, Wise. July 26, 1996. Spinal cord regeneration. *Science,* vol. 273. Experiments in rats reveal that spinal cord injuries may be treatable.

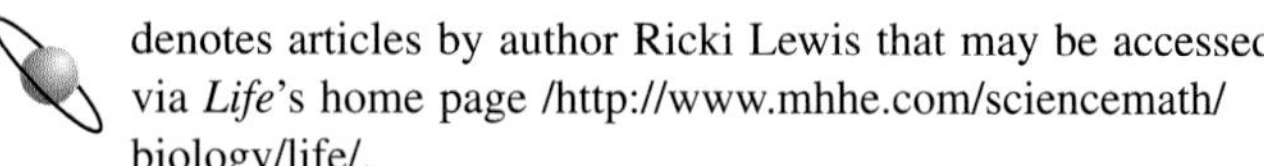

denotes articles by author Ricki Lewis that may be accessed via *Life*'s home page /http://www.mhhe.com/sciencemath/biology/life/.

32 The Senses

The platypus uses its ears and eyes when its head is above water but uses touch and electroreception to find food on the rocky river bottom.

chapter outline

The Senses Paint Pictures of the World 640

Sensory Systems and Adaptation 640

General Principles of Sensory Reception 641
Sensory Receptors—Detecting, Transducing, and Amplifying Stimuli
Sensory Adaptation

Sensory System Diversity 644
Invertebrate Senses
Vertebrate Sensory Systems

Human Sensory Systems 645
Chemoreception
Photoreception
Mechanoreception

Platypus Electroreception

Like most animals, a platypus uses its senses to find food; however, it has one sense that we humans lack—electroreception.

A platypus is a monotreme, a primitive type of mammal. It is furry and warm-blooded, but unlike most mammals, it lays eggs. A platypus feeds often, navigating along the river bottoms of its native Australia, nudging rocks aside with its bill to find crayfish, worms, insect larvae, and sometimes a frog or a fish to eat. The animal consumes about half its body weight each day. How can it possibly find so much food? The platypus uses the electroreceptors in its bill to locate prey.

While looking for food, the platypus closes its ears and eyes, but its bill uses thousands of sensory nerve endings. For many years, zoologists thought the highly innervated bill located food by touch. But in 1927, platypus expert Harry Burrell hypothesized that the animal must use something in addition to sense of touch to locate and identify food. One possibility was that the platypus could detect the minute electrical fields organisms emit in water.

to be continued . . .

The Senses Paint Pictures of the World

Every animal inhabits a world of its own, a world whose character depends on the animal's sense organs. The sensory abilities of some animals are vastly different from (and sometimes amazingly more acute than) human senses. If a person and a dog were to walk through the New England woods in the fall, the person's sensations would be primarily visual, yet the dog more likely would detect odors that paint a picture human eyes miss—fox urine left as a calling card near an oak and the scent in a shrub revealing a deer's bed.

If this is an evening stroll, another animal, the little brown bat, experiences yet another view of the scene. The bat's view of the world is not just an eye view but an ear view. Bats emit high-frequency pulses of sound and analyze the resulting echoes to picture the world around them. We cannot experience the bat's world because our ears cannot hear such high-frequency sounds. Although a bat emits sound pulses that are about as intense as a passing train, to our ears these are silent screams.

Why do animals have such different sensory worlds? Each species senses the range of stimuli most critical to its survival. Cells in a frog's eyes respond selectively to small moving objects which help the animal to locate food. Male and female butterflies that look alike to us appear quite different to one another (fig. 32.1). The ears of female cricket frogs respond to the frequencies in the croaks of males of the species but cannot hear other frequencies. Biology in Action 32.1 describes senses in snakes, butterflies, and hippos that are unlike those in humans.

(a) (b)

FIGURE 32.1

To us, the male and female "dogface" butterflies *Colias cesonia* look alike—both are yellow and black (*a*). The insects, however, can see in the ultraviolet range of light; to them, the female appears dark blue, and the male has reflectant patches on his wings (*b*). This color difference is important in courtship behavior; however, this photograph is still a human visual perception of a butterfly under UV illumination. A butterfly probably sees much less precise outlines.

Sensory Systems and Adaptation

Animals live in virtually all habitats on earth and are active at all times of the day and night. In all of these habitats and conditions, animals must maintain homeostasis. This requires that they respond to changes in their external and internal environments. **Sensory receptors** are specialized neurons that detect and pass stimulus information to sensory neurons.

Generally, a stimulus alters the shape of a protein embedded in a sensory receptor's cell membrane in a way that changes the membrane's permeability. If the stimulus is strong enough, resulting changes in ion movement across the membrane trigger action potentials. Figure 32.2 shows types of sensory receptors, and figure 32.3 and table 32.1 depict the information flow from environmental stimulus to the brain that underlies sensation and perception (interpreting sensation).

Natural selection has molded sensory receptors in every group of animals in ways that reflect their environments, the time of day when they are active, the foods they eat, their reproductive mechanisms, and other characteristics. The receptors a bat uses to find its food are very different from, but no more or less specialized than, those an eagle uses. But while some receptors are highly complex and gathered into sense organs, other receptors are simpler and more widely distributed, such as touch receptors.

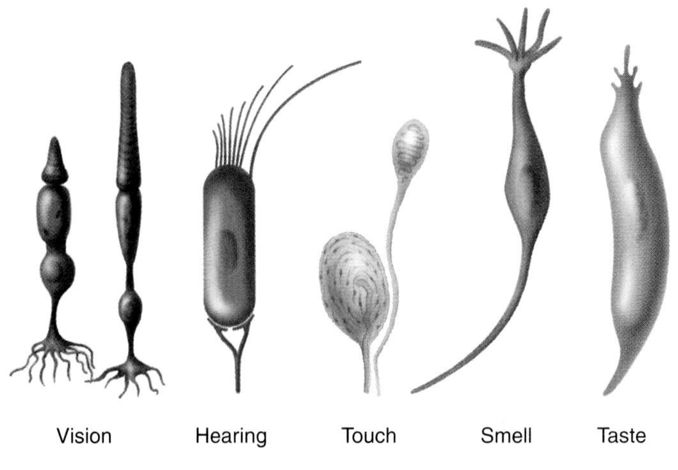

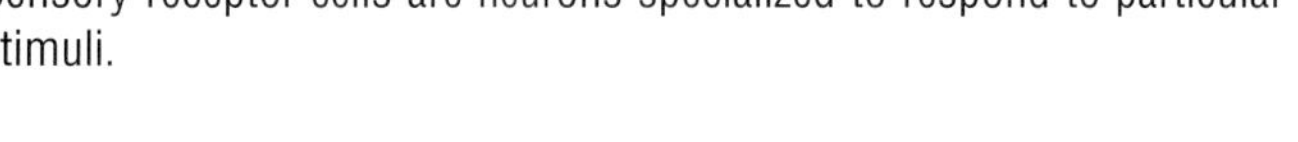

FIGURE 32.2

Sensory receptor cells are neurons specialized to respond to particular stimuli.

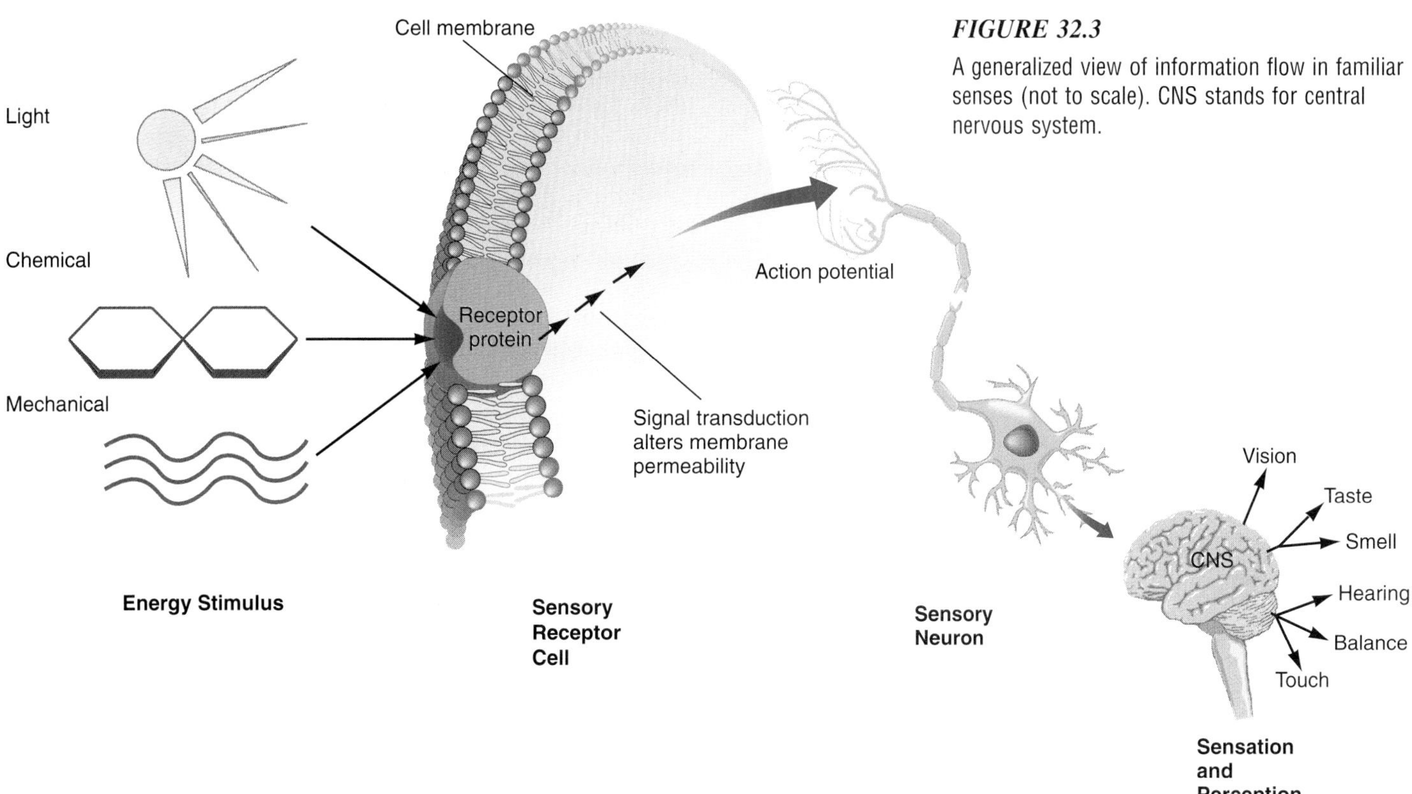

FIGURE 32.3

A generalized view of information flow in familiar senses (not to scale). CNS stands for central nervous system.

The senses are vital to survival; many lifesaving adaptations "fool" the senses of potential predators. A green beetle against a leafy background evades a bird's usually sharp eyes. A mockingbird confuses its keen-eared potential predators with multiple songs. A viceroy butterfly would taste good to a bird, but most won't touch it—it closely resembles the foul-tasting monarch butterfly.

General Principles of Sensory Reception

The brain receives sensory input, integrates the information, and interprets it by consulting memories to form a perception—the individual's particular view of environmental phenomena. Sensory receptors, then, are portals through which nervous systems experience the world.

Sensory Receptors—Detecting, Transducing, and Amplifying Stimuli

If all sensory receptors send information as action potentials, and the action potentials on the sensory neurons are all alike, how can an animal distinguish sight from sound? The answer is that sensory receptors are selective and they use different neural pathways. They absorb a particular type of energy—light

32.1 biology in action

Snake Tongues, Butterfly Rears, and Hippo Jaws

Not all species see with their eyes or hear with their ears.

The Snake's Forked Tongue

The forked tongue of the serpent is a metaphor for deceit, but to the animal, it is a powerful sense organ, able to trace scent trails (fig. 32.A).

FIGURE 32.A
A snake's forked tongue enables it to detect chemical gradients in the environment and thereby follow a scent trail towards or away from another animal.

Human observers over the centuries have hypothesized that the snake's tongue is forked to pluck dirt from its nose, to catch flies, or to experience two tastes at one time.

Because the tongue is two-pronged, it can touch two spots simultaneously. If a scent diminishes or strengthens over a distance, a forked tongue can detect this by sensing at two points. The forks of the tongue bring odor molecules into the mouth. The molecules then pass through openings in the palate and contact chemosensors in the snout called vomeronasal organs. Receptor cells in the organs pass action potentials to a center in the brain, which compares the dual sensory input and interprets whether the trail is weakening or strengthening. From these cues, the serpent can determine whether it is approaching a potential mate or a meal.

A Rear-End View

Arthropods have light-sensitive cells (photoreceptors) on various body parts. Butterflies have two pairs of neurons that function as photoreceptors on their genitals (fig. 32.B). Japanese researchers hypothesized that the photoreceptors enable the Japanese yellow swallowtail butterflies, *Papilio xuthus,* to align their bodies during mating. In an experiment, the researchers removed the photoreceptors or painted them black—and mating efficiency plummeted. The fault was the males'. When parts align properly, there is no space for light to enter, and the photoreceptors are not stimulated. If the parts are misaligned, light activates the photoreceptors, and the males know to move until he fits she.

Hippo Hearing

A hippo lives at an interface between air and water, typically remaining submerged for up to 5 minutes at a time before lifting the nostrils to take a breath (fig. 32.C). The animal makes different sounds in its two different environments—bellows, rumbles, groans, and

Table 32.1 Information Flow from the Environment Through the Nervous System

Information Flow	Smell	Taste	Sight	Hearing
Sense receptors	Olfactory cells in nose	Taste bud receptor cells	Rods and cones in retina	Hair cells in cochlea
↓	↓	↓	↓	↓
Stimulation of nerve	Olfactory nerve	Sensory fiber in taste bud	Optic nerve	Auditory nerve
↓	↓	↓	↓	↓
Impulse transmission to CNS	Cortex	Cortex	Midbrain and visual cortex	Midbrain and cortex
↓	↓	↓	↓	↓
Sensation (memory, experience)	A pleasant smell	A sweet taste	A small, round, red object	A crunching sound
↓	↓	↓	↓	↓
Perception	The smell of an apple	The taste of an apple	The sight of an apple	The sound of biting into an apple

FIGURE 32.B
In some butterfly species, activated photoreceptors on the genitals signal a male that the position for mating is not quite correct. Only when male and female parts align in a way that blocks light from stimulating genital photoreceptors can mating occur.

FIGURE 32.C
Hippos hear above the water with their ears but detect sounds underwater with their jaws.

grunts above, and softer moans, squeals, clicks, and clacks below the water's surface When a hippo bellows from above, other hippos react from below. How do they hear underwater?

To answer this question, researcher Bill Barklow thought of dolphins, who have a form of sensory reception called "jaw hearing." In the porpoise ear, the middle and inner segments dangle from ligaments, rather than being fused to other bony structures, which prevents extra bone from conducting sound. Each middle ear also contacts a line of fatty tissue that runs along the inside of the jaw, ending at a thin, oval indentation. The porpoise hears underwater as sound impinges on the jaw and is transmitted through the fat conduit to the ears. The fat conducts sound because it is close to the density of water. Underwater sounds arrive at the ears separately, and the animal orients until the sounds arrive simultaneously. The brain then uses these cues to track objects.

Barklow went to the Smithsonian Institute to examine a huge hippo head that Teddy Roosevelt had donated. Sure enough, the middle and inner ears were suspended from ligaments. But the middle ear had an additional attachment—to a part of the skull that touches the lower jaw. Barklow hypothesizes that hippos hear underwater in a way similar to dolphins—through their jaws.

energy, for example—and transduce (convert) it into the electrochemical energy of an action potential. The action potentials a specific type of receptor generates travel to the brain over a specific sensory pathway. Different qualities of sensation, such as light, sound, or pressure, are the brain's interpretation of the input from different pathways. Receptors also amplify input, so that a single photon or a single odor molecule produces an impulse that has inherently more energy than the original stimulus. Thus, sensory receptors detect, transduce, and amplify sensory stimuli.

In addition to transmitting information on the type of stimulus, sensory receptors send messages about the strength of the stimulus. One way stimulus intensity is signaled to the brain is in nerve impulse frequency; the more intense the stimulus, the more action potentials are generated in a given time period. Sensory receptors respond to an environmental change with a **receptor potential,** which is a local change in electrical potential that precedes generation of an action potential. Like an action potential, redistribution of ions triggers a receptor potential. Unlike an action potential, however, a receptor potential is not "all-or-none." The magnitude of a receptor potential varies with the strength of the stimulus. For example, louder sound may cause greater depolarization, resulting in a larger receptor potential. The change in the receptor potential, in turn, influences the likelihood that an action potential will be generated. In some receptors, the strength and duration of the receptor potential determine the rate at which action potentials are generated and how long they last. The brain interprets this as an increase in stimulus intensity.

A second way information about stimulus strength reaches the brain is through variations in the number of sensory neurons carrying the message. As the stimulus strengthens, more receptors are stimulated, and these activate more sensory neurons.

Sensory Adaptation

Many sensory receptors detect changes in input—if a sensation remains constant, it becomes less noticeable. This mechanism, called **sensory adaptation,** keeps the nervous system from becoming overly sensitive. The strong smell of a fish market, for example, may be overpowering to anyone who has just entered, but to the people working there, it is hardly noticeable. Of all the senses, smell adapts most quickly. Similarly, a tub filled with hot water may seem too hot at first, but it soon becomes quite tolerable. Without sensory adaptation, we would be distinctly aware of the touch of clothing and every sight and sound, almost to the point of pain. Concentrating on a single stimulus, such as a person speaking, would be difficult; however, some sensory receptors are very slow to adapt—and this is also adaptive. For example, being very aware of continuous pain helps ensure action to remove its source.

Humans have multiple receptors, providing a rich tapestry of sensations. A person's skin contains 4 million pain receptors, 500,000 pressure receptors, 150,000 receptors specific for cold, and 16,000 specific for heat. Specialized receptors in the joints, tendons, ligaments, and muscles provide information that, when combined with information from sense organs such as the eyes and ears, gives us a feeling of where certain body parts are relative to the rest of the body. Although the five most familiar senses are vision, hearing, taste, smell, and touch, our nervous systems blend these into a far larger number of sensations.

Sense organs and receptors fall into a few major classes. They include:

chemoreception (smell and taste);

photoreception (sight);

mechanoreception (hearing, equilibrium and balance, and touch).

The remainder of the chapter explores these groups of sensory receptors first in invertebrates and then in vertebrates, including humans.

Mastering Concepts

1. Why do different animal species sometimes have different types of senses?
2. How is sensation translated into action potentials?
3. How does a receptor potential differ from an action potential?

Sensory System Diversity

Sensory systems are evolutionary adaptations; they reflect an animal's surroundings and behavior. Human senses are no more "advanced" than the light-sensing cells that enable an earthworm to distinguish between bright light (which it avoids) and dim light (which it is attracted to).

Invertebrate Senses

Chemoreception

Chemoreceptors are sensory receptors that detect chemical stimuli and perceive them as smell and/or taste. Chemical senses of invertebrates enable them to detect food. The sensory cells on a flatworm's head detect a potential meal of algae clinging to rocks, dead animals, or other live invertebrates. Crayfish have chemoreceptors on their mouthparts, appendages, and antennae that allow them to taste food and detect mating pheromones. Insects use chemical senses to establish social organization and select food, mates, and places to lay eggs.

Photoreception

Animals detect light with photoreceptors, which include a pigment molecule associated with a membrane. When the pigment absorbs light, its structure changes, which alters the charge across the associated membrane. These membrane changes may generate action potentials.

Vision is interesting from an evolutionary standpoint because the pairing of photoreceptor cells and pigment molecules appears to be quite ancient, yet the visual abilities of different species are adapted to particular ways of life. Nearly all multicellular species with photoreceptor cells use the pigment molecule **rhodopsin,** and split it to provoke a nerve impulse in response to light. Even some unicellular organisms use rhodopsin, such as the alga *Chlamydomonas.* A photosynthetic bacterium, *Halobacterium halobium,* uses a rhodopsin variant in place of chlorophyll for photosynthesis.

Different species use rhodopsin in very different structures with varying degrees of acuity. A turbellarian flatworm's photoreceptor cells are gathered into two eyespots, or **ocelli.** These simple structures enable the flatworm to detect shadows, which is sufficient light sensation for the animal to orient itself in its environment. Similarly, a starfish has cells at the tip of each arm that detect differences in light intensity, which is adaptive in its underwater habitat. Arthropods have compound eyes consisting of closely packed photosensitive units called **ommatidia.** Each ommatidium contains a lens that transmits light to its own or nearby photoreceptor cells, generating a tiny view of the world. The animal's nervous system then integrates the many images from each ommatidium. Animals with many ommatidia may see as clear an image as we do.

Some arthropods have combinations of ocelli and compound eyes, or they have different photoreceptors at different stages of development. A crayfish's compound eyes, located on two movable eyestalks, may contain from 25 to 14,000 ommatidia; but earlier in life, the wormlike crayfish larva has ocelli that enable it to detect light at the water's surface, where its single-celled food lives. Insects have compound eyes of up to 28,000 ommatidia each (fig. 32.4*a*) and may also have up to three ocelli, each consisting of 500 to 1,000 receptor cells grouped under a lens. Ocelli may complement visual information from the compound eyes. While compound eyes detect movement, enabling insects to capture moving prey, ocelli detect shadows, which may help insects to escape approaching predators.

(a)

(b)

FIGURE 32.4

The invertebrate eye. (*a*) The compound eyes of the southern aeshna dragonfly contain more ommatidia than any other insect—about 28,000 each. (*b*) The visual system of the octopus is remarkably like our own.

The cephalopods (octopus, squid, and cuttlefish) have visual systems much like our own (fig. 32.4*b*). Their eyes form images, distinguish shapes, and discriminate colors.

Mechanoreception

Mechanoreceptors enable an organism to respond to physical stimuli (touch), to sound waves (hearing), and to maintain balance and equilibrium. The sense of hearing operates when pressure hit and move receptors. In insects, hairlike structures called **setae** at the base of the antennae vibrate in response to sound. The movement stimulates sensory cells that transmit the information to the brain.

Some insects also have drumlike **tympanal organs,** thin membranes stretched over large air sacs that form resonating chambers for sound. Sensory cells beneath the membrane detect sound pressure waves and relay information to the brain. Crickets and katydids have tympanal organs on their legs, while grasshoppers and some moths have them on their abdomens. These organs are usually on both sides of an insect's body, which enables it to sense the direction as well as the origin of sound.

Setae and tympanal organs are highly sensitive; insects can detect much higher pitches than the human ear can hear. The ability of certain moths to escape predatory bats vividly displays the sensitivity of insect hearing. Moths can detect the high-pitched cries of bats and dive down when the sound approaches. This confuses the bats' echolocation, the ability to locate airborne insects by bouncing ultrasound waves off them. When moths fly down near the ground, the bat's echolocation bounces off of the moths and the ground at about the same time. Thus, the bat can no longer pick the moths out.

Mechanoreception also includes a sense of balance and equilibrium. In many animals, structures called **statocysts** provide this sense. A statocyst is a fluid-filled cavity containing sensory hairs and bits of mineral called **statoliths.** When the animal's body tilts, the statoliths touch the sensory hairs, stimulating the nervous system. In response, the animal can orient itself with respect to gravity. Such diverse invertebrates as jellyfish, snails, squid, crayfish, and worms use statocysts for balance. Invertebrate statocysts are similar to human structures that carry out the same functions.

Vertebrate Sensory Systems

Vertebrate sensory systems—like those of invertebrates—reflect species-specific adaptations. The sensory systems of vertebrates may differ in location. A catfish tastes with its whiskers, and an owl can pinpoint the height a sound emanates from because one ear is higher than the other and one points up, while the other points down. Because it would be virtually impossible to describe all the variations in vertebrate sensory systems, we will focus on a familiar representative: our own species.

Mastering Concepts

1. What are the functions of chemoreceptors in invertebrates?
2. Name two types of photoreceptors in invertebrates.
3. What senses do mechanoreceptors provide?

Human Sensory Systems

We are accustomed to our five senses and the enormous variety of sensations they provide. Biology in Action 32.2 looks at a condition that mixes sensations and perceptions.

32.2 biology in action

Mixed-up Senses—Synesthesia

"The song was full of glittering orange diamonds."
"The paint smelled blue."
"The sunset was salty."
"The pickle tasted like a rectangle."

To 1 in 500,000 people with a condition called synesthesia, sensation and perception mix, so that the brain perceives a stimulus to one sense as coming from another. Most commonly, visions that have characteristic smells or sounds are linked to specific colors. These associations are involuntary and persist over a lifetime.

We do not know what causes synesthesia, although it does seem to be inherited. People have sheepishly reported the condition to psychologists and physicians for at least 200 years. Synesthesia has been attributed to an immature nervous system that cannot sort out sensory stimuli; altered brain circuitry that routes stimuli to the wrong part of the cerebral cortex; and simply an exaggerated use of metaphor, in which such descriptions as "sharp flavor" are taken too literally.

PET (positron emission tomography) scanning reveals a physical basis to synesthesia. A scan of the brain of a woman who tasted in geometric shapes discovered that blood flow in her left temporal lobe plummeted 18% during a tasting experience. The left hemisphere is the site of the language center. In addition, people with temporal lobe epilepsy are often synesthetic. Based on PET studies, researchers hypothesize that the condition reflects a breakdown in the translation of a perception into language.

Chemoreception

Chemoreceptors in humans, like those of other organisms, detect smell and taste. These sensations may be closely tied to texture, temperature, pain, and even appearance. Taste is intimately tied to smell, as anyone whose sense of taste has been dulled by a stuffy nose can attest. For the millions of people who cannot smell normally, eating is a tasteless experience. This is because about 75 to 80% of flavor derives from smell (olfaction).

Smell

Smell is the ability of specialized **olfactory receptor cells** to detect certain molecules, called odorant molecules, that impart smells. These "smell cells" are located in an inch-square patch of tissue high in the nasal cavity called the **olfactory epithelium** (fig. 32.5). To trigger the sense of smell, an odorant molecule must bind receptor proteins on cilia in this patch of tissue. Odorant molecules, usually in gaseous form, pass over the olfactory epithelium during breathing.

To stimulate the olfactory receptor cells, an odorant molecule must also dissolve in the mucous layer covering the cilia. In the nose, dissolved odorant molecules diffuse through the mucus until they reach the receptors.

Part of each olfactory receptor cell passes through the skull and synapses with neurons in the brain's olfactory bulb. This structure relays the message to the cerebral cortex, which interprets the message as an odor and identifies it. Genetic researchers are currently identifying the proteins that participate in olfaction. These molecules include hundreds of odorant receptor proteins that bind specific odorant molecules, as well as the signaling proteins that "tell" a receptor cell that an odorant molecule has bound to it. In the mid-1980s, researchers identified a general odorant-binding molecule that attracts odorant molecules and gathers enough of them to stimulate olfactory receptor cells. This concentration of odorant molecules may explain why a scent must reach a certain strength before we can detect it.

Researchers hypothesize that different receptor cells have different combinations of odorant receptor proteins, thus generating specificity in the pattern of activated neurons that stimulate the brain. Olfactory receptor cells with the same receptor proteins are not necessarily near each other, but their axons meet in the olfactory bulb. The brain links a particular pattern of neurons that have the same odorant receptor proteins activated and associates the pattern with a particular smell.

For example, if there are 10 types of odor receptors, banana smell might stimulate receptors 2, 4, and 7; garlic, receptors 1, 5, and 9. Humans have about 12 million receptor cells, each with 10 to 20 cilia that increase the surface area for receiving smells. We can detect a single molecule of the chemical that gives a green pepper its odor in 3 trillion molecules of air! Yet other species have even more acute senses of smell. A bloodhound has 4 billion receptor cells, and its olfactory epithelium, if it were unfolded, would cover 59 square inches, compared with our 1 to 1 1/2 square inches! This is why we use bloodhounds to trace the scent of a missing person—the dog can smell the skin cells that people shed and distinguish between individuals. The animal kingdom has other expert sniffers. A polar bear can smell a dead sea lion 12 miles away; a male emperor moth can detect one molecule of a female's scent 3 miles away; and a shark can smell blood in the water 500 yards away.

Sensory information from olfactory receptors also travels to the limbic system, the brain center for memory and emotion. This is why we may become nostalgic over a scent from the past. A whiff of the perfume Grandma used to wear may bring back a flood of memories. Olfactory input to the limbic system also explains

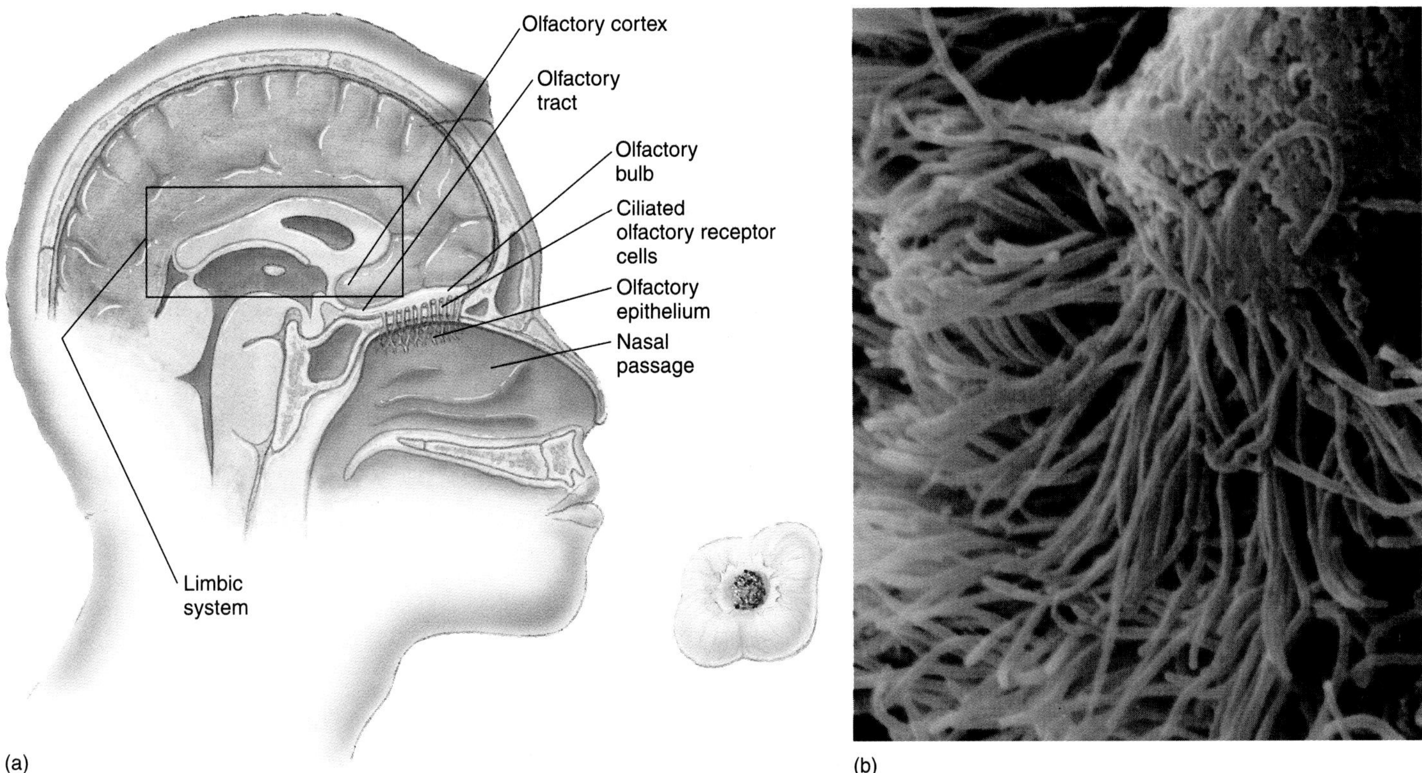

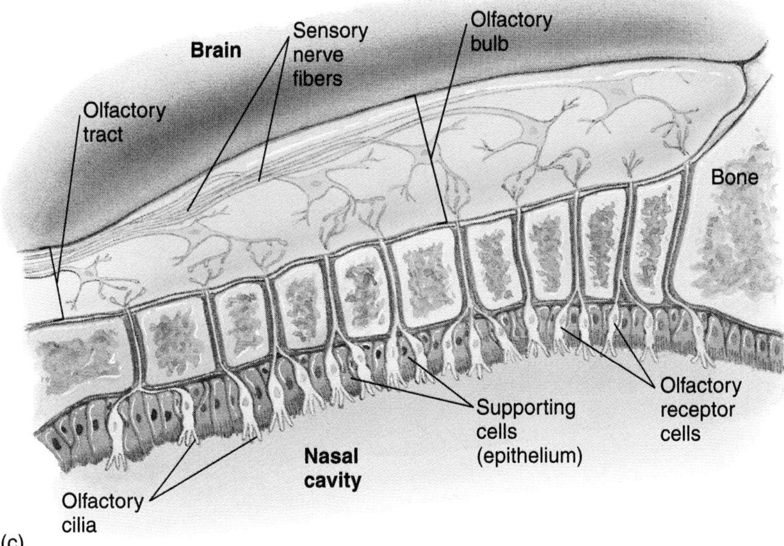

FIGURE 32.5

How smells signal the brain. (*a*) The powerful sense of smell derives from an inch-square section of sensitive ciliated olfactory receptor cells in each nostril. (*b*) This close-up of the olfactory epithelium shows the cilia that fringe receptor cells. (*c*) An olfactory receptor cell physically binds an odorant molecule (such as garlic) at its ciliated end and transmits an action potential to fibers of the olfactory nerve. This nerve passes the information to the brain, which interprets the sensation as a particular smell. Some action potentials reach the limbic system, the brain's seat of memory and emotion. This is one reason why a particular food or smell may evoke powerful feelings.

why odors can alter mood so easily. For example, the scent of new-mown hay or rain on a summer's morning makes us feel good.

Taste

Taste receptors are in the parts of the body that contact chemical stimuli, including food. Lobsters and insects taste with receptors on their legs. An octopus samples food with taste receptors on its tentacles. In humans, taste is centered primarily in the mouth in clusters of cells called **taste buds** that resemble the sections of an orange. Humans have about 10,000 taste buds; cattle have up to 25,000.

Although taste buds are lightly scattered around all the places that food and drink are likely to touch, including the cheeks, throat, and roof of the mouth, they are concentrated on the tongue in the grooves around certain of the bumps, or papillae, on the tongue's upper surface. Each human taste bud contains 60 to 100 receptor cells, which generate action potentials when food molecules bind to them. These receptor cells live for only about 3 days and are then replaced by new ones.

For many years, people identified four primary taste sensations—sweet, sour, salty, and bitter. More recently, scientists have proposed that many more taste sensations exist, with our sense of taste depending upon the pattern of activity across all taste neurons. Most taste buds sense all four of the "primary" tastes, but each responds most strongly to one or two (fig. 32.6). Although individual "tongue maps" vary considerably, we can make some

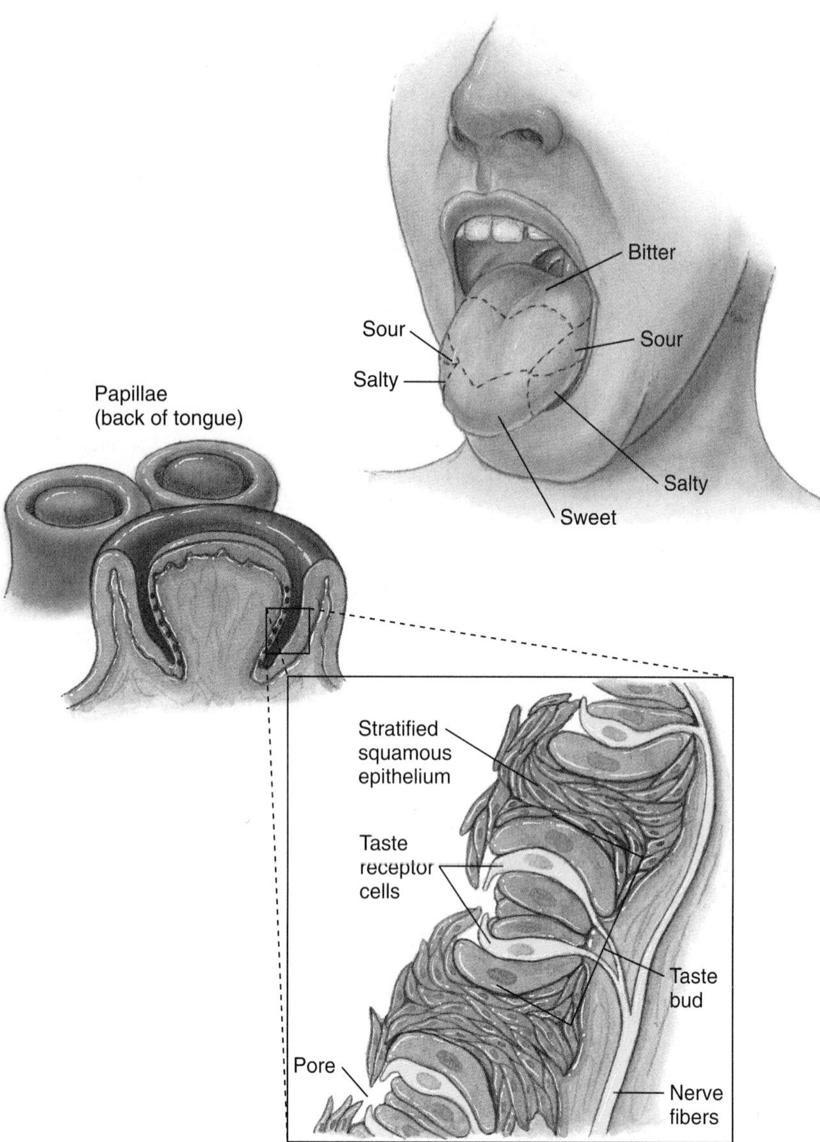

FIGURE 32.6

Taste receptors scattered over the tongue detect particular tastes.

generalizations. Receptors that respond most strongly to sugars concentrate on the tip of the tongue. Sour receptors are most common along the sides of the tongue. Receptors along the rim of the tongue respond most strongly to salts, and receptors at the back of the tongue are most sensitive to bitter substances.

On the basis of information the taste receptors transmit to our brains, we decide whether to swallow substances in our mouths. We generally perceive sweet tastes as pleasant and quickly consume a sweet food. Bitter tastes, on the other hand, we often refuse. Rejecting a bitter substance can be lifesaving; the poisonous alkaloids in wild herbs, for example, often taste bitter. Figure 32.7 shows how taste and smell sensations travel to the brain.

Photoreception

The Human Visual System

Our eyes, like those of all vertebrates, work as a camera does. Both a camera and the human eye have a **lens** system that focuses incoming light into an image on a light-detecting surface. In a camera, this surface is the film; in the eye, it is a sheet of photoreceptors called the **retina.** Other structures support and nourish the photoreceptive area.

The human eyeball is a fluid-filled sphere made up of three distinct layers. The white of the eye is the outermost layer, or **sclera,** which protects the inner structures (fig. 32.8). Towards the front of the eye, the sclera is modified into the **cornea,** a transparent curved window that admits light. This curve bends incoming light rays, which helps to focus them on the photoreceptor cells.

The middle layer of the eyeball, the **choroid coat,** is rich in blood vessels that nourish the eye. The choroid coat contains a dark pigment that absorbs light and prevents it from reflecting off the retina, much as black paint inside a camera absorbs light. Like the sclera, the choroid coat is specialized into different structures at the front of the eyeball. Near the front of the eye, the choroid thickens into a highly folded structure called the **ciliary body,** which houses the **ciliary muscle.** This muscle alters the shape of the lens to adjust the focus of the image. In front of the ciliary body, the choroid becomes the thin, opaque **iris,** which provides eye color. Like the diaphragm in a camera, the iris regulates the amount of light entering the eye. An opening in the iris, the **pupil,** admits light. In bright light, the pupil constricts, shielding the retina from excess stimulation. In dim light, the pupil dilates, letting more light strike the retina. Behind the iris, the lens focuses the light into an image at the back of the eyeball onto the retina.

The third and innermost layer of the eyeball, the retina, is itself built of several cell layers (fig. 32.9). The middle layer contains the photoreceptor cells, called **rods** and **cones** (fig. 32.10). The human eye contains about 125 million rod cells and 7 million cone cells. Rod cells, which are concentrated around the edges of the retina, provide black-and-white vision in dim light and enable us to see at night. Cone cells, which are concentrated toward the center of the retina, detect color. An indentation directly opposite the lens, the **fovea centralis,** contains only cones. The number of cones in this area is a measure of the acuity of an animal's sight. Humans have about 150,000 cones per square millimeter, whereas some predatory birds have a million. Birds of prey owe their incredible eyesight to a second fovea. Other animals have other adaptations. For example, a cat's eyes glow in the dark because an extra layer behind the retina reflects back the light that enters the retina. This increases visual acuity, especially at night.

The distribution of rods and cones helps explain why human vision works as it does. To see detail in bright light, it is best to look directly at the object because this focuses light on the central retina area that is dense with cones. At night, however, we can see an object more clearly from the corner of the eye. From this perspective, the light bouncing off the object stimulates the rod-rich region of the eye. Only rods are sensitive enough to respond to dim light.

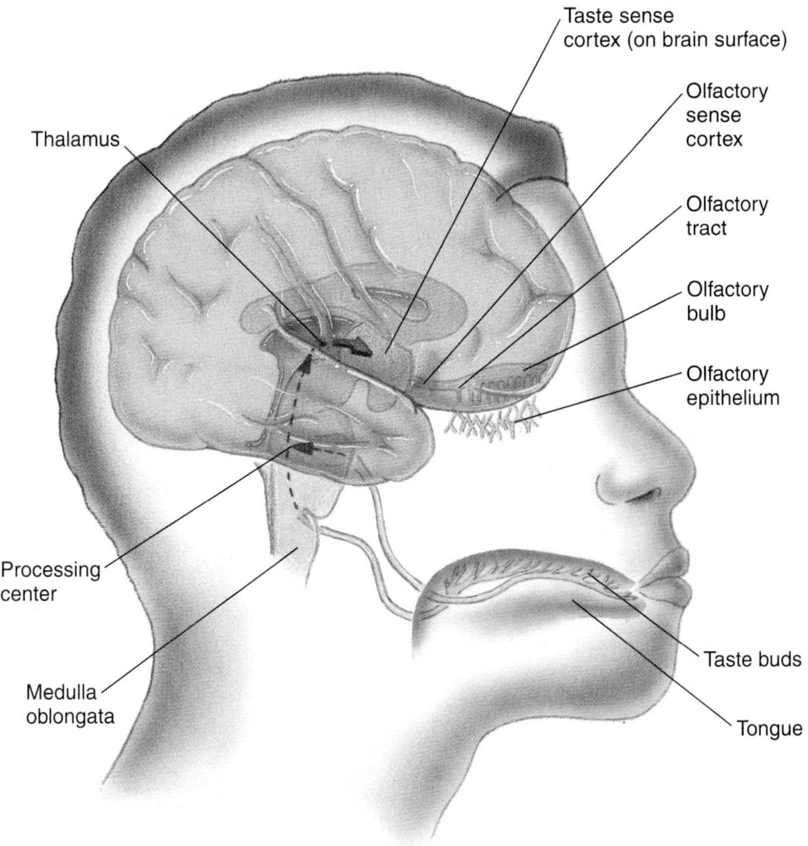

FIGURE 32.7

Neural connections for taste and smell. The sense of taste begins when receptors in the tongue (taste buds) detect chemical stimuli. They generate action potentials that travel through the medulla oblongata, processing center, and thalamus to the taste areas of the brain's cortex. Action potentials from olfactory receptors have a more direct route to the cortex and do not go through the thalamus. These cortex regions interpret the messages as smell and taste.

Rods and cones synapse with bipolar neurons, which form another layer of the retina. The bipolar neurons in turn synapse upon cells of the next retinal layer, the ganglionic cell layer. Finally, the fibers of the ganglion cells form the **optic nerve,** which leads to the visual cortex in the brain.

The eyeball is filled with fluid. Behind the lens, making up most of the volume of the eyeball, is the jellylike mass of **vitreous humor.** Between the cornea and the lens is the watery **aqueous humor.** Both humors help bend light rays to focus them on the retina. The aqueous humor cleanses and nourishes the cornea and lens and presses against the sclera to maintain the shape of the eyeball. A fresh supply of aqueous humor is secreted and absorbed every 4 hours. Some aqueous humor diffuses to the back of the eye and bathes the rods and cones.

Focusing the Light

Vision begins when light rays pass through the cornea and lens and are focused on the retina. Because most objects that we see are considerably larger than our retinas, the light rays must bend to form an image on the fovea centralis. The cornea, the lens, and the humors of the eye bend the light rays.

The lens changes shape to focus the light reflecting from distant or close objects onto the retina. This adjustment to suit the distance between the eye and the viewed object is called **accommodation.** To focus on a very close object, the ciliary muscle contracts, rounding the lens so that it can bend incoming light rays at sharper angles. Close work often causes eyestrain because the ciliary muscle must remain contracted to view nearby objects. When you view a faraway object, the ciliary muscle relaxes and the lens flattens. That is why gazing into the distance can relieve eyestrain.

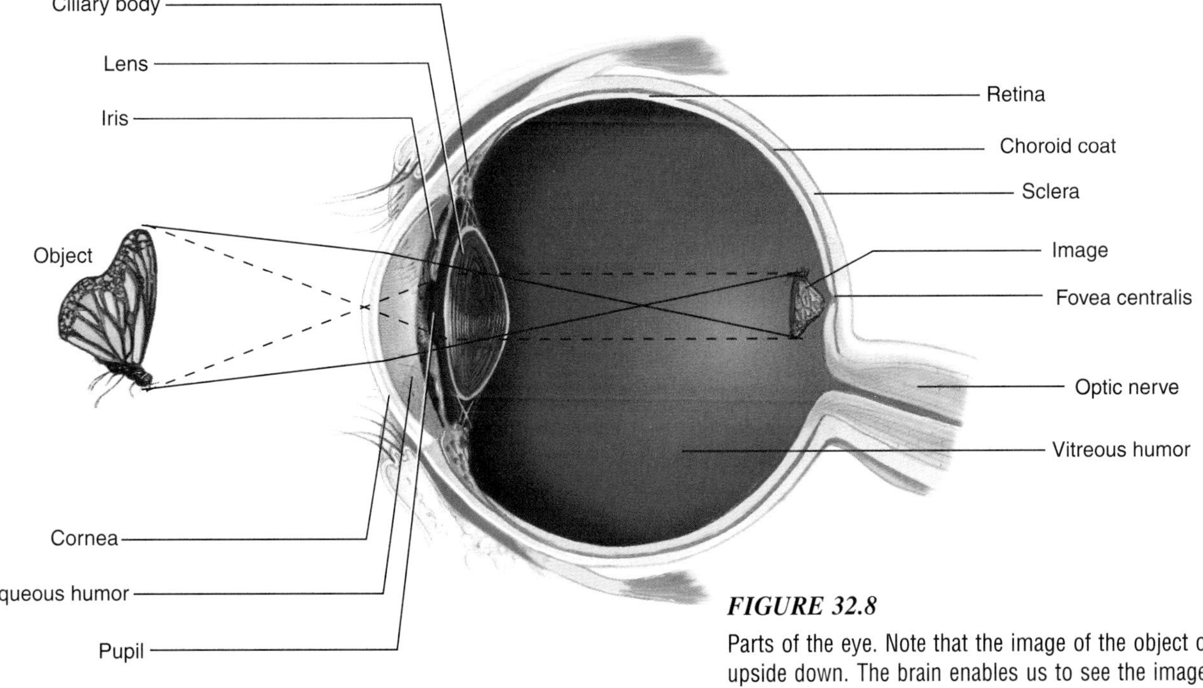

FIGURE 32.8

Parts of the eye. Note that the image of the object on the retina is upside down. The brain enables us to see the image right side up.

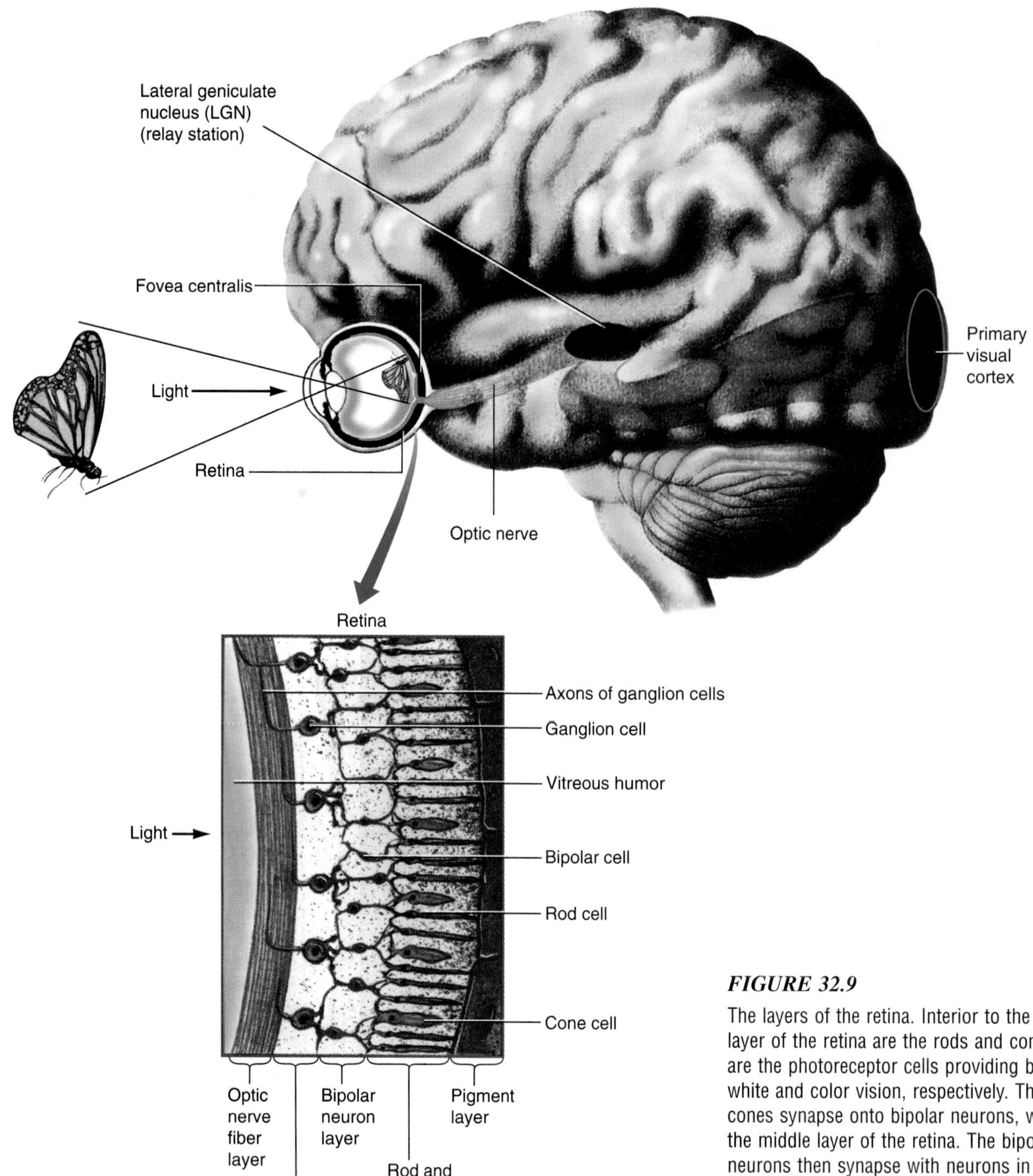

FIGURE 32.9

The layers of the retina. Interior to the pigment layer of the retina are the rods and cones, which are the photoreceptor cells providing black-and-white and color vision, respectively. The rods and cones synapse onto bipolar neurons, which form the middle layer of the retina. The bipolar neurons then synapse with neurons in the ganglionic cell layer. The axons of the ganglion cells form the optic nerve.

Sometimes the shape of the eyeball makes it difficult for the lens to focus light rays on the retina. Figure 32.11 discusses nearsightedness, farsightedness, and astigmatism.

The image projected onto the retina is actually upside down. The brain processes this information so that we perceive a right-side-up world.

Converting Light Energy to Neural Messages

Pigment molecules within the rods and cones change to convert visual information to receptor potentials. In a rod cell, about 100 million molecules of rhodopsin are stored in a highly folded membrane at one end of the cell. Rhodopsin contains **retinal,** a derivative of vitamin A, and it is the retinal that actually absorbs light. Light causes retinal to elongate, which distorts a protein called opsin, to which retinal is attached. The activated rhodopsin triggers a cascade of signal transduction (see chapter 5), which changes ion permeability of the cell membrane, which triggers an action potential. Figure 32.12 illustrates rhodopsin. In the dark, it takes several seconds for rhodopsin to assume its inactive form; this is why it takes our eyes several seconds to make out shapes in a very dark room.

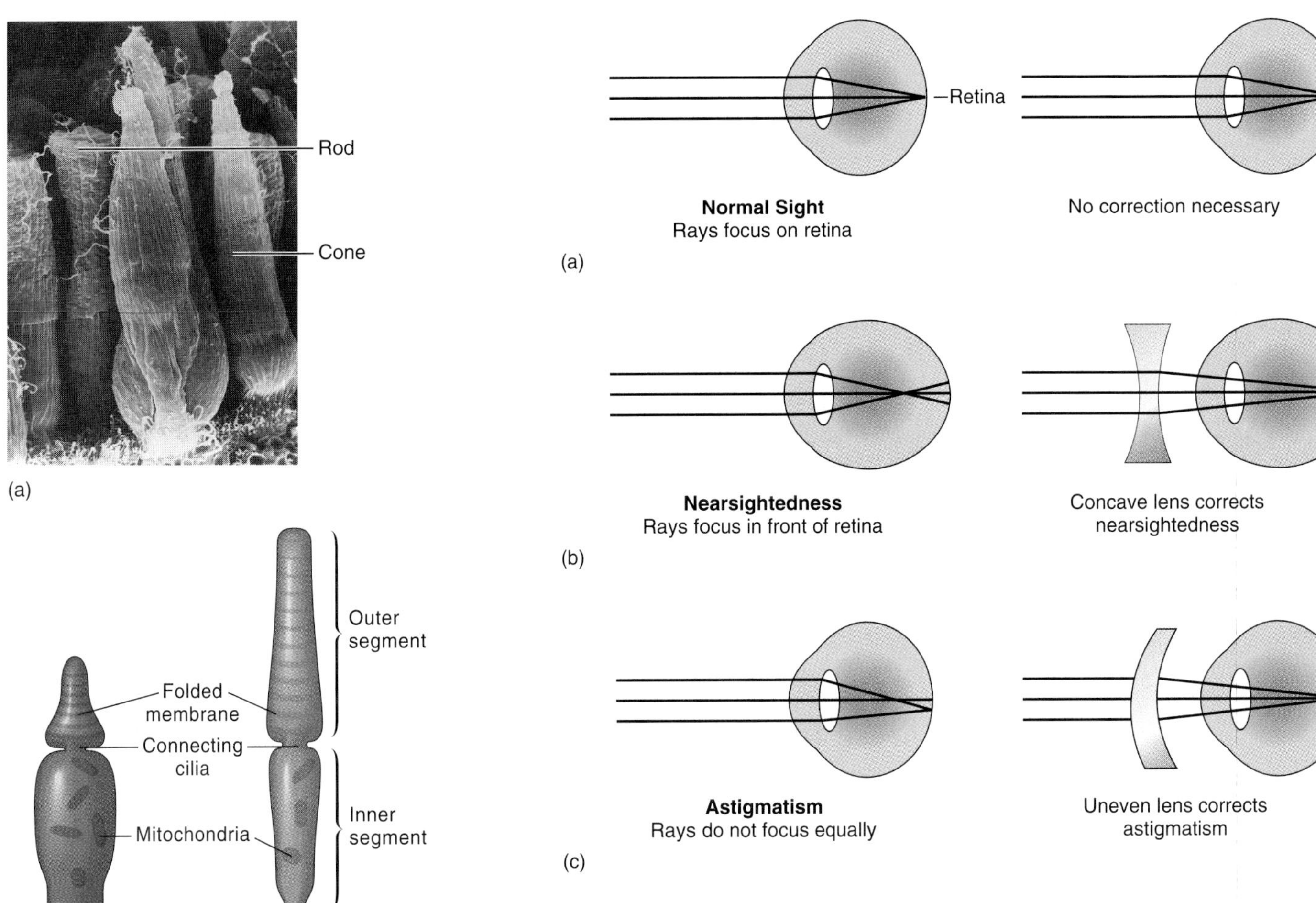

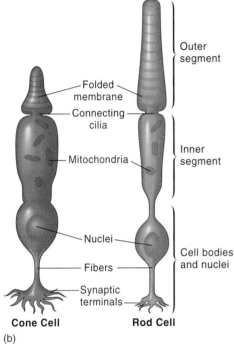

FIGURE 32.10

Rods and cones. (*a*) The outer segment of rod and cone cells consists of highly folded membranes that contain rhodopsin. (*b*) The inner segment of the cells contains organelles, and the cell body portion houses the nuclei.

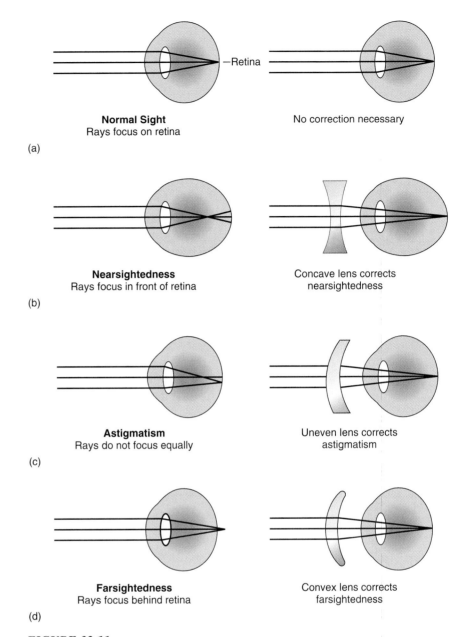

FIGURE 32.11

The shape of the eyeball affects the ability of the cornea and lens to focus light rays into an image on the retina. (*a*) When the eyeball is shaped normally, no correction is required to focus light rays on the retina. (*b*) If the eyeball is unusually elongated, incoming light rays converge before they reach the retina, impairing ability to see distant objects (nearsightedness). A concave eyeglass lens can alter the point of focus to the retina. (*c*) Sometimes the lens or cornea is shaped so unusually that incoming light rays do not focus evenly, a condition called astigmatism. An irregular shaped eyeglass lens can compensate for the eyeball's abnormal shape and focus light rays on the retina. (*d*) If the eyeball is too short, light rays focus beyond the retina, and the person has difficulty seeing close objects (farsightedness). A convex lens can correct this problem.

Color vision is possible because the cone cells contain different pigments, and each absorbs light of different wavelengths (colors). The molecules that provide color vision also consist of opsins bound to retinal. One pigment absorbs the wavelengths around red, another green, and the third blue. Although an individual cone cell contains only one type of pigment, the wavelengths each type of cone absorbs overlap.

When a cone absorbs light, its pigment molecule changes chemically and a receptor potential is generated. The brain interprets the ratio of activity among the cone types as light of a particular color. If one or more cone types is missing, an individual cannot distinguish among all colors and is color blind (table 32.2).

Because our eyes are located close to one another on the front of our heads, the visual fields of each eye overlap. You can see an

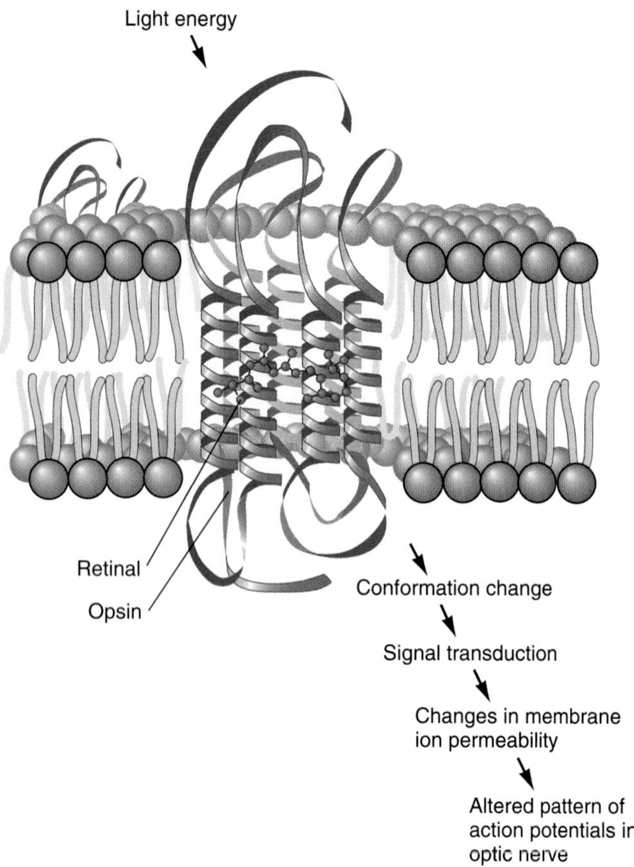

FIGURE 32.12

In rod cells, retinal portions of rhodopsin absorb light, altering the conformation in a way that triggers an action potential.

object directly in front of you with either eye closed, although each eye sees it from a slightly different angle. When both eyes function, the visual cortex integrates and interprets the information from each eye to produce three-dimensional perceptions. Depth perception enabled our primate ancestors to see branches in three dimensions.

The visual cortex fuses images formed at discrete moments of time into a fluid perception of motion. In 1980 researchers in Germany reported on a woman who could not perceive motion after a stroke damaged part of her visual cortex. She saw movement as a series of separate, static images. Her deficit had profound effects on her life. She could not pour a drink, because she could not tell when the cup would overflow. She could not cross a street because she could not detect cars moving towards her. Many animals can only detect prey if it moves. If she were a member of a different species, the inability to detect motion could be deadly.

Mechanoreception

Mechanoreceptors convert mechanical energy to action potentials. This usually occurs when the receptor is bent or compressed in some way. Specialized mechanoreceptors provide our senses of hearing, balance, and touch.

Hearing

The clatter of a train, the sounds of a symphony, a child's wail—what do they have in common? All sounds, regardless of the source, originate when something vibrates. The vibrating object creates repeating pressure waves in the surrounding medium, which, for humans, is usually air. The size and energy of these pressure waves determine the intensity or loudness of the sound, while the number of waves (cycles) per second determines the sound's frequency or pitch. The more cycles per second, the higher the pitch.

Table 32.2 Visual Problems

Problem	Cause	Treatment
Astigmatism	Asymmetrical cornea or lens; light rays do not focus evenly to form sharp images	Uneven corrective lens focuses image on retina
Cataracts	Lens is opaque due to birth defect, injury that perforates cornea, complication of diabetes or glaucoma, or aging	Surgical removal of cataracts
Farsightedness (cannot see close objects clearly)	Light focuses behind retina because eyeball is too compact	Corrective convex lens focuses light on retina
Glaucoma	Aqueous humor cannot drain near the iris; fluid buildup presses capillaries that nourish nerve fibers in retina, damaging optic nerve, causing blindness	Drugs to inhibit synthesis of aqueous humor; drugs that thin iris, improving fluid drainage Surgical removal of piece of iris
Nearsightedness (cannot see distant objects clearly)	Light focuses in front of retina because eyeball is too elongated	Corrective concave lens focuses light on retina
Red-green color blindness	Red or green cones absent	None

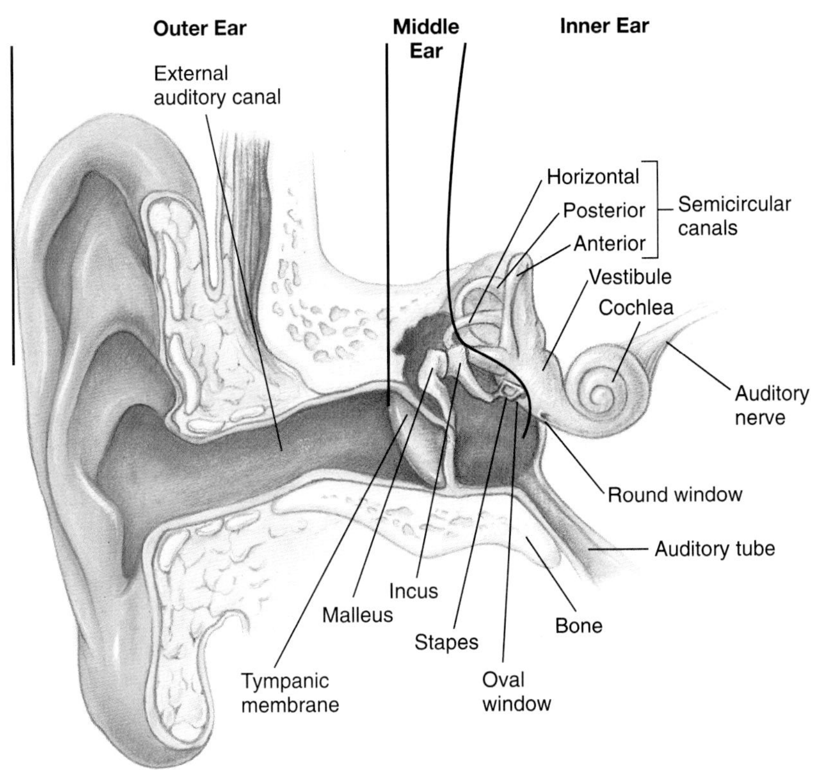

FIGURE 32.13

Anatomy of the ear. Sound enters the outer ear and, upon reaching the middle ear, impinges upon the tympanic membrane, which vibrates three bones (malleus, incus, and stapes). The vibrating stapes hits the oval window, and hair cells in the cochlea convert the vibrations into action potentials, which follow the auditory nerve to the brain. Hair cells in the semicircular canals and in the vestibule sense balance. The auditory tube connects the middle ear to the throat, equalizing air pressure.

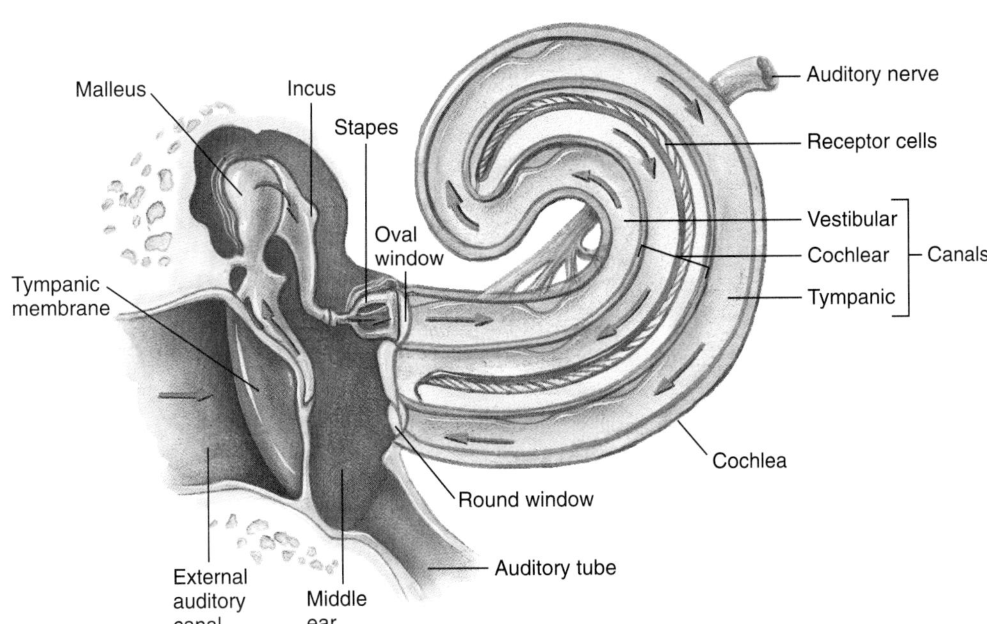

FIGURE 32.14

The cochlea is a spiral-shaped structure consisting of three fluid-filled canals. When the stapes moves against the oval window, the fluid in the vestibular canal vibrates. This vibration eventually converts the sound wave to an action potential.

The fleshy outer part of the human ear traps sound waves and funnels them down a wax- and hair-lined ear canal to the **tympanic membrane,** or eardrum (fig. 32.13). The sound pressure waves cause the eardrum to vibrate, which moves three small bones, the **malleus, incus,** and **stapes,** located in the middle ear. These bones transmit the incoming sound and amplify it 20 times. From the middle ear, the vibrations of the stapes are transmitted through the **oval window,** a membrane that opens into the inner ear.

The inner ear is a fluid-filled chamber that houses both balance and hearing structures. Two parts of the inner ear, the **semicircular canals** and the **vestibule,** control balance and are discussed in the next section. The remaining portion of the inner ear is the snail-shaped **cochlea.** The spirals of the cochlea are three fluid-filled canals called the vestibular, cochlear, and tympanic canals. The vestibular and tympanic canals actually form a continuous U-shaped tube (fig. 32.14). Between them lies the cochlear canal. When the last bone of the middle ear, the stapes, moves, it pushes on the oval window, transferring the vibration to the fluid in the vestibular canal. The pressure of each vibration is dissipated by the movement of the round window at the other end of the U. The vibration of the fluid also moves the **basilar membrane,** which forms the lower wall of the cochlear space, and initiates the change of mechanical energy to receptor potentials.

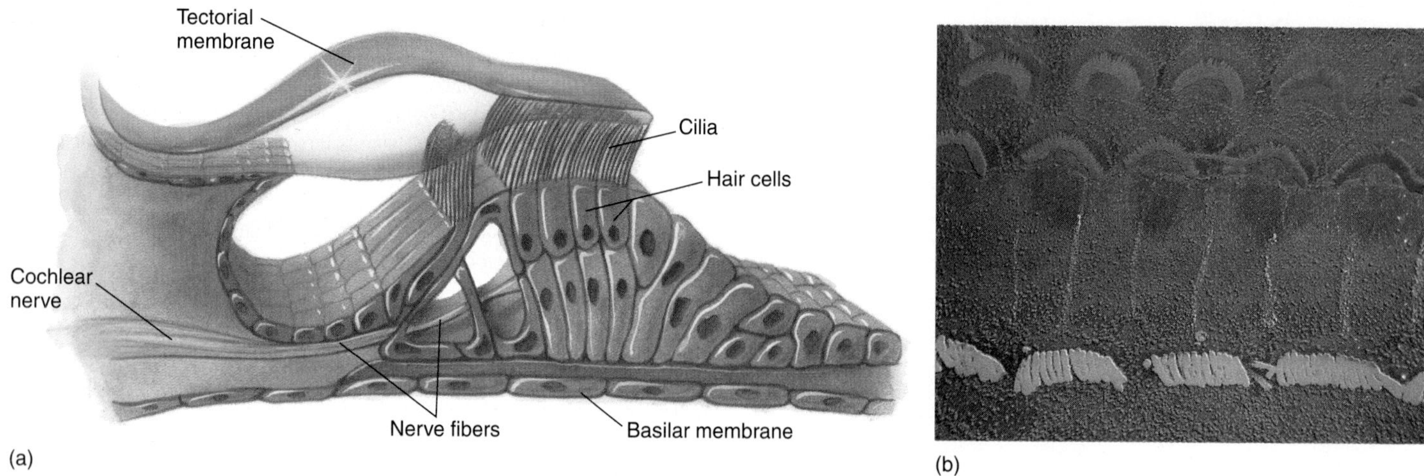

FIGURE 32.15

From vibration to nerve impulse. (*a*) Hair cells lie between the tectorial and basilar membranes. Vibrations along particular portions of the basilar membrane correspond to different sound frequencies. The vibration exerts pressure on the hair cells, which in turn push against the tectorial membrane, bending the hairs. This triggers an action potential in the auditory nerve. (*b*) Groups of cilia at the tops of hair cells.

Table 32.3 Hearing Impairment

Problem	Cause	Result
Conductive deafness	Impaired transmission of sound through middle ear due to infection, earwax buildup, damaged eardrum	Inability to clearly hear all pitches of sound
Sensory (neural) deafness	Inability to generate action potentials in the cochlea, blocked communication between cochlea and brain, or brain's inability to interpret sensory messages	Inability to hear some pitches
Tinnitus	Damaged hair cells in inner ear due to ear infection, brain tumor, earwax buildup, or loud noise	Ringing in the ears

How is sound translated into the universal language of the nervous system? Within the cochlea, specialized hair cells (the mechanoreceptors) lie between the basilar membrane below and another sheet of tissue, the **tectorial membrane,** above (fig. 32.15). The basilar membrane is narrow and rigid at the base of the cochlea and widens and becomes more flexible nearer the tip. Because of this variation in width and flexibility, different areas of the basilar membrane vibrate more intensely when exposed to different frequencies. The high-pitched tinkle of a bell stimulates the narrow region of the basilar membrane at the base of the cochlea, while the low-pitched tones of a tugboat whistle stimulate the wide end.

When a region of the basilar membrane vibrates, the hair cells are displaced relative to the tectorial membrane, and this initiates action potentials in fibers of the **auditory nerve.** This nerve carries the impulses to the brain, which interprets the input from different regions as sounds of different pitches (fig. 32.16). The louder the sound, the greater the vibration of the basilar membrane and, consequently, the more hair cells stimulated. The brain interprets the resulting increase in the rate and number of neurons firing as an increase in amplitude, or loudness.

Hearing loss may be temporary or permanent. It can result from conductive deafness (blocked transmission of sound waves through the middle ear) or from sensory deafness (damage to the nervous system) (table 32.3). Many cases of hearing loss result from exposure to loud noise. Sound intensity is measured in decibels, with each 10-decibel increase representing a tenfold increase in sound intensity. Damage to the inner ear's hair cells begins to occur at about 80 decibels, which is as loud as heavy traffic. The degree of damage depends both upon decibel level and duration of exposure to the sound.

Many rock stars of the 1960s are going deaf in the 1990s! The damage begins as the hair cells develop blisterlike bulges that eventually pop. The tissue beneath the hair cells swells and softens until the hair cells and sometimes the neurons leaving the cochlea become blanketed with scar tissue and degenerate.

Equilibrium and Balance

The semicircular canals and the vestibule of the inner ear regulate the sense of balance (fig. 32.17). The semicircular canals tell us when the head is rotating and help us maintain the position of the head in response to sudden movement. The enlarged bases of the semicircular canals, the **ampullae,** are lined with small, ciliated hair cells. Because the semicircular canals are perpendicular to each other, the fluid that fills a particular semicircular canal may or may not swish back and forth in response to a movement, depending on its direction. When the fluid in a canal moves, it bends the cilia on the hair cells in that ampulla, which in turn stimulates those cells to generate action potentials in a nearby cranial nerve. The brain interprets these impulses as body movements.

Information from the vestibule (fig. 32.17) senses the position of the head with respect to gravity. In addition, the vestibule senses changes in velocity when traveling in a straight line. When riding in a car, for example, we can sense acceleration and deceleration. The vestibule functions in a similar way to the semicircular canals. It contains two pouches, the **utricle** and the **saccule,** which are filled with jellylike fluid and lined with ciliated hair cells. Granules of calcium carbonate, called **otoliths,** float on the fluid. The granules in the utricle move in response to acceleration or to tilting of the head or body, bending the cilia on the hair cells (fig. 32.18). When the cilia bend in one direction, the rate of sensory impulses to the brain increases. A shift in the opposite direction inhibits the sensory neuron. The brain interprets this information as a change in velocity or a change in body position.

Motion sickness results from contradictory signals. The inner ear signals the brain that the person is not accelerating. At the same

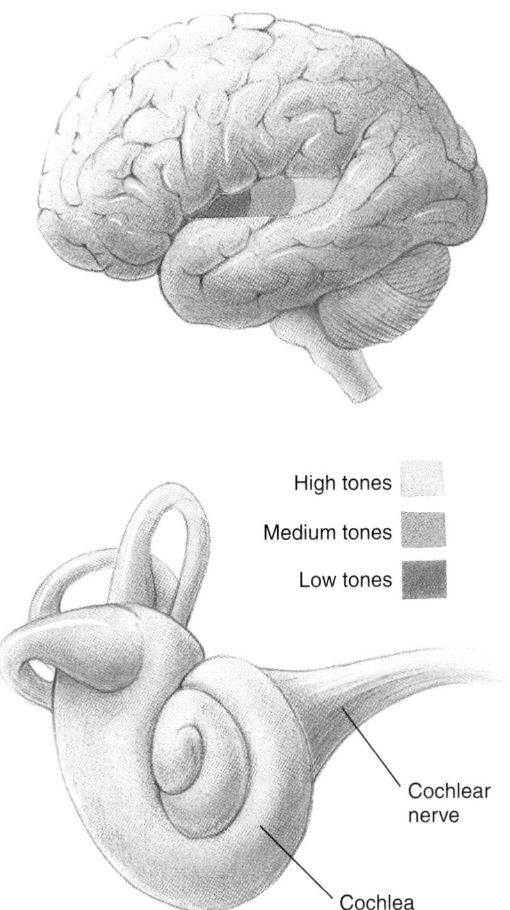

FIGURE 32.16

Correspondence between cochlea and cortex. Sounds of different frequencies (pitches) excite different sensory neurons in the cochlea. These neurons, in turn, send their input to different regions of the auditory cortex.

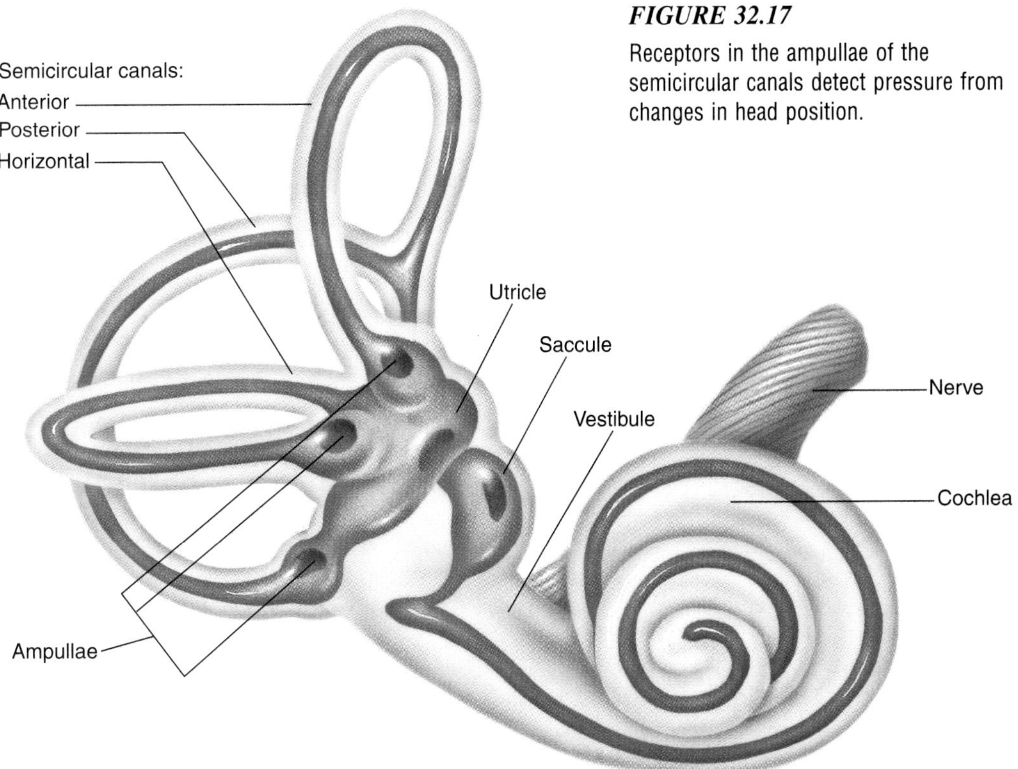

FIGURE 32.17

Receptors in the ampullae of the semicircular canals detect pressure from changes in head position.

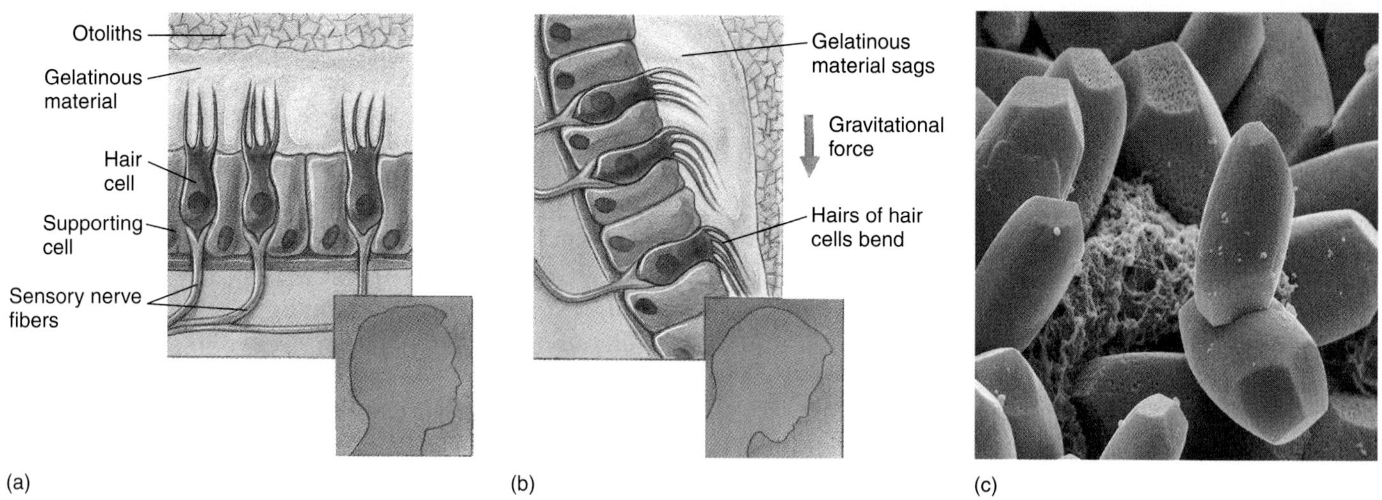

FIGURE 32.18

Sense of balance. Gravitational forces on the head move the otoliths, providing a sense of balance. (*a*) When the head is upright, the otoliths balance atop the cilia of hair cells. (*b*) When the head bends, the otoliths move, bending the cilia. This provokes a nerve impulse. (*c*) A scanning electron microscope reveals the crystalline nature of otoliths.

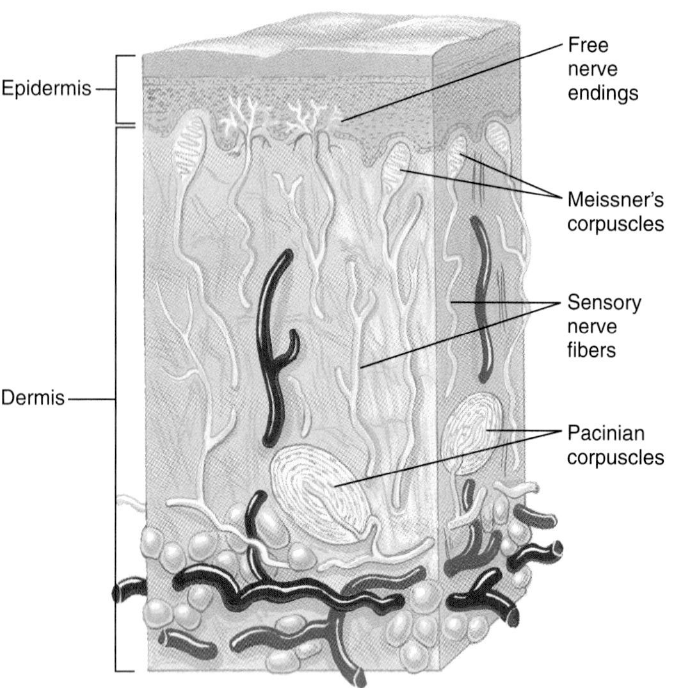

FIGURE 32.19

Touch receptors include free nerve endings, which respond to touch, pressure, and pain; Meissner's corpuscles, which respond to light touch; and Pacinian corpuscles, which detect firm pressure.

time, the eyes detect passing scenery and signal the brain that the person is moving. The result of these mixed signals is nausea.

Touch

Different species have different structures to detect touch—a cat uses its whiskers, an insect its antennae, and a spider uses it legs to feel vibrations transmitted along the strands of its web when something is caught. Human sensitivity to touch comes from several types of receptors in the skin—**Pacinian** and **Meissner's corpuscles** and free nerve endings (fig. 32.19). A corpuscle is encapsulated and resembles an onion—it consists of tissue surrounding a single nerve fiber. A touch pushes the flexible sides of the corpuscle inwards, generating an action potential in the nerve fiber.

Firm pressure, such as a bear hug, stimulates Pacinian corpuscles. On the other hand, a light touch stimulates Meissner's corpuscles, particularly a moving touch, as in a gentle caress. The free ends of sensory neurons in epithelial surfaces, including the skin, are sensitive to touch, pressure, and pain. In signaling pain, they may respond to pressure, temperature, or chemicals associated with tissue damage.

Senses are not only essential to our survival, but they help make life enjoyable. Fortunately, because our senses both complement and overlap, we can still live full, functional lives if one of our senses is impaired; we can compensate with one sense for another. A person who cannot see a garden can smell the flowers and touch the plants to gain a detailed perception of it. A person who cannot hear can read lips or use sign language to converse. Thanks to the blend of our senses, the world is an exciting, multidimensional place.

Mastering Concepts

1. Which cells and molecules impart our sense of smell?
2. How does a taste bud function?
3. What are the parts of the eye?
4. How does the eye transduce incoming light into an image?
5. What are the parts of the ear, and how do they transmit sound?
6. How do structures in the ear provide equilibrium and balance?
7. Which structures provide the sense of touch?

Platypus Electroreception~

continued from page 640.

Scientists knew that fish and amphibians use electroreceptors, but they had never seen evidence of this sense in reptiles, birds, or mammals.

To test the hypothesis that platypuses detect and respond to electrical fields in water, Australian researchers placed hungry platypuses in pools with live and dead batteries. The animals investigated only the live batteries. Next, the researchers recorded platypuses' brain waves while stimulating their bills electrically. As the hypothesis would predict, the brain waves responded. Finally, the zoologists took a close look at the platypus's bill. Deep within the pits that form a type of mucous gland, they found tiny, thread-like ends of nerve fibers. These were the hypothesized electroreceptors.

Further observations revealed how the platypus finds sufficient food. While under water, the animal continually swishes its head to and fro. The tiniest flick of the tiniest shrimp's tail can stimulate the platypus's electroreceptors to trigger 20 to 50 action potentials a second. This sense, unusual in a mammal, is adaptive in the platypus's environment—and a striking example of sensory system diversity.

SUMMARY

1. **Sensory receptors** detect, transduce, and amplify stimuli, allowing an animal to perceive its environment. Senses are adaptive. Sensory receptors may form sense organs.

2. A sensory receptor selectively responds to a single form of energy and converts it to **receptor potentials,** which change membrane potential in proportion to stimulus strength. If a stimulus remains constant, a sensory receptor ceases to respond **(sensory adaptation).**

3. Sensory receptors include **chemoreceptors, photoreceptors,** and **mechanoreceptors.**

4. Invertebrates have chemoreceptors to detect food and perform other functions. Invertebrate photoreceptors contain pigments associated with membranes. Light stimulation alters the pigment and changes the charge across the membrane, which may generate an action potential. Visual systems range from simple **ocelli** to compound eyes to complex lens systems. Invertebrate hearing depends on **setae** and **tympanal organs,** and balance and equilibrium depend on **statolith** crystals within **statocysts.**

5. Vertebrate sensory systems are similar in structure and function in different species, but they may be located in different parts of the body. Some species have senses humans do not have.

6. Humans perceive smell when odorant molecules bind receptors in the **olfactory epithelium** of the nasal passages. The brain perceives a smell by evaluating the pattern of **olfactory receptor cells** that bind odorant molecules. Humans perceive taste when chemicals stimulate receptors within **taste buds.** Humans perceive vision with a complex **lens** system in the eye.

7. The human eye contains three layers. The outer layer, the **sclera,** protects. It forms the transparent **cornea** in the front of the eyeball. The next layer, the **choroid coat,** is pigmented, located toward the rear of the eye, and absorbs light. In the front of the eye, the choroid coat forms the **ciliary body,** which controls the shape of the lens that focuses light on the photoreceptors, and the opaque **iris.** The **pupil** constricts or dilates to adjust the amount of light entering the eye.

8. The innermost eye layer is the multilayered **retina.** Beneath a pigment layer lie photoreceptors: **rods** for black-and-white vision in dim light and **cones** for color vision in brighter light. These cells synapse with **bipolar cells** that form the middle retinal layer. The bipolar cells, in turn, synapse with ganglion cells whose fibers leave the retina as the **optic nerve.** This nerve carries the neural messages to the brain for interpretation. Light activates **rhodopsin** in a rod cell in a way that alters ion permeability of the cell membrane, generating action potentials. Three types of cones each contain a pigment that maximally absorbs light of a particular wavelength. The brain interprets the ratio of the activities of the three cone types as a color.

9. Mechanoreceptors bend in response to sound, body movement, or touch. In human hearing, sound enters the auditory canal, vibrating in the **tympanic membrane.** These vibrations are transmitted through the middle ear and amplified by three bones, the **malleus, incus,** and **stapes.** The movements of these bones change the pressure in fluid within the **cochlea,** which in turn vibrates the **basilar membrane.** As the basilar membrane moves, it pushes hair cells against the **tectorial membrane,** which signals the brain to perceive the pitch of the sound through the location of the moving hair cells. The brain determines the sound's loudness from the frequency of action potentials and the number of stimulated hair cells.

10. The **semicircular canals** and the **vestibule** in the inner ear sense body position and movement. Fluid movement within these areas stimulates sensory hair cells, and the brain interprets this information, providing a sense of equilibrium.

11. **Pacinian corpuscles, Meissner's corpuscles,** and free nerve endings are mechanoreceptors that detect touch. Pacinian corpuscles respond to intense pressure, Meissner's corpuscles to gentle pressure, and free nerve endings to touch, pressure, and pain.

TO REVIEW . . .

1. What is a sense organ?

2. What are the three general types of senses?

3. How can a pigment associated with a protein transduce an environmental stimulus into a sensation?

4. Distinguish between sensation and perception.

5. Distinguish between ocelli and ommatidia.

6. How does the nervous system detect different smells?

7. In what way are the senses of smell and taste similar?

8. List the structures of the human eye and their functions.

9. In humans, otolith movement provides equilibrium and balance. What is the corresponding structure in an invertebrate?

10. Why doesn't burning your tongue on a piping hot slice of pizza permanently damage your sense of taste?

11. How does vision in humans differ from that of an insect?

12. How do we see colors other than red, blue, or green?

13. In what ways do the three major structures of the inner ear (cochlea, semicircular canals, and vestibule) function similarly?

14. Sensation results from an anatomical pattern of many sensory cells that transmit impulses to a nerve leading to a specific part of the brain. Describe the organization of neurons in three human senses that fit this pattern of funneling incoming information from many sensory neurons to a nerve fiber leading to the brain.

TO THINK ABOUT . . .

1. Loss of the sense of smell is an early symptom of Parkinson disease and Alzheimer disease. How might this information be used to manage these conditions? What additional information is necessary to make medical use of this association?

2. People can inherit an inability to smell certain substances, such as freesia flowers, jasmine, skunk scent, and hydrogen cyanide. How could the hypothesis that odors are encoded in combinations of stimulated receptors explain these conditions?

3. Why is dietary vitamin A good for eyesight?

4. Cite two examples of how people who lack one sense can compensate by relying more heavily on another.

5. People who are hearing impaired due to cochlea damage do not suffer from motion sickness. Why not?

6. We have relatively few sensory systems. How, then, do we experience such a diversity of sensory perceptions?

7. Humans love sucrose (table sugar), but armadillos, hedgehogs, lions, and seagulls do not respond to it. Opossums love lactose (milk sugar) but rats avoid it, and chickens hate the sugar xylose, while cattle love it. In what way might these diverse tastes in the animal kingdom be adaptive?

8. Genetic researchers discovered odorant receptor proteins by searching for proteins in nasal epithelium that resemble rhodopsin. What does the similarity of odorant receptor proteins to the visual molecule rhodopsin suggest about the origin of senses?

TO LEARN MORE . . .

References and Resources

Arikawa, K., et al. July 11, 1996. Light on butterfly mating. *Nature,* vol. 382. Photoreceptors help butterflies mate.

Gilbert, Avery N., and Charles J. Wysocki. October 1987. The smell survey. *National Geographic.* In 1986, 1.5 million Americans filled out a scratch-and-sniff survey on what they could and could not smell.

Gold, Geoffrey H., and Edward N. Pugh Jr. February 20, 1997. The nose leads to the eye. *Nature,* vol. 385. The molecular details of how our sense of smell adapts to ignore a persistent odor.

Gregory, Ed. May 1991. Tuned-in, turned-on platypus. *Natural History.* The platypus can detect electrical fields given off by its food.

Lewis, Ricki. November/December 1995. Practice makes perfect, but start early. *Biophotonics International.* Is musical talent inborn?

Lewis, Ricki. November 1995. Correcting nearsightedness without glasses. *Photonics Spectra.* Surgery may replace eyeglasses.

Lewis, Ricki. October 1995. Fractured vision reveals brain functions. *Photonics Spectra.* Visual disturbances are traced to the brain.

Lewis, Ricki. November 1991. *FDA Consumer.* When smell and taste go awry. It is hard to imagine life without the chemical senses.

Sachs, Jessica Snyder. March 1996. The fake smell of death. *Discover.* Dogs have acute senses of smell.

Schwartz, David M. March 1996. Snatching scientific secrets from the hippo's gaping jaws. *Smithsonian.* Hippos hear underwater with their jaws.

Schwenk, Kurt. March 18, 1994. Why snakes have forked tongues. *Science,* vol. 263. A snake's forked tongue enables it to sense chemical gradients.

denotes articles by author Ricki Lewis that may be accessed via *Life*'s home page /http://www.mhhe.com/sciencemath/biology/life/.

The endocrine system orchestrates the "fight or flight" response to a threat to life.

33 The Endocrine System

chapter outline

The Human Metamorphosis—Puberty 660

Hormones—Chemical Messengers and Regulators 660

The Endocrine and Nervous Systems Maintain Homeostasis 661

Pheromones 662

How Hormones Exert Their Effects 662
Water-Soluble Hormones—Activating Second Messengers
Fat-Soluble Hormones—Directing DNA Action
Controlling Hormone Levels

Diversity of Hormones and Glands 664
Hormones in Invertebrates
Hormones and Glands in Vertebrates

Pituitary Hormones 667
Anterior Pituitary Hormones
Posterior Pituitary Hormones
Intermediate Region Hormones

Endocrine Control of Metabolism 669
The Thyroid Gland
The Parathyroid Glands
The Adrenal Glands
The Pancreas

Endocrine Control of Reproduction 674
The Menstrual Cycle
The Testes

The Pineal Gland 675

Prostaglandins 676

Neuroendocrine-Immune Interactions 677

History of an Illness—Diabetes Mellitus

The sweet-smelling urine that is the hallmark of diabetes mellitus was noted as far back as an Egyptian papyrus from 1500 B.C. In A.D. 96 in Greece, Aretaeus of Cappadocia described the condition as a "melting down of limbs and flesh into urine."

Type I (juvenile-onset) diabetes mellitus results from failure of the pancreas to secrete the hormone insulin. Normally, insulin enables cells to take up glucose from the bloodstream. Without the hormone, glucose enters the urine. Diabetes causes weakness and weight loss as the body breaks down tissues to obtain energy. Eventually toxins enter the bloodstream, and coma and death follow. For many years "treatment" was a starvation diet and exercise—which hastened death. Then a series of elegant experiments changed everything.

In October 1920, a young surgeon named Frederick Banting was lecturing medical students at the University of Toronto. While describing an experiment performed in 1889 that demonstrated that removing a dog's pancreas caused it to dramatically weaken, lose weight, and die, Banting wondered whether a pancreatic extract might restore such a dog's health—and also wondered if such an approach could help people with the same symptoms. Banting asked his department chair for help and received a tiny lab, an assistant named Charles Best, and 10 dogs.

to be continued . . .

The Human Metamorphosis—Puberty

Puberty is a time that most adults would like to forget. The changes that swept through our bodies simultaneously seemed too sudden and too slow. We can thank hormones for our transfiguration. At puberty, in the brain, the hypothalamus began sending chemical messages to the pea-sized pituitary gland. In turn, the pituitary's hormonal messages began to stimulate the ovaries or testes to produce sex hormones.

Soon our reproductive organs began to enlarge and mature. Hair appeared in new places. New muscle growth broadened male shoulders; thickening vocal cords and an enlarging larynx deepened their voices. Females broadened in the hips and developed mature breasts. Other changes probably escaped notice—alterations in bone growth, calcium metabolism, and red blood cell formation, to name a few. New feelings and attitudes also ripened. As if in protest of all this change, at a time when appearance seems so critical, acne erupted.

Puberty is probably the time when we are most aware of hormones, but they affect us throughout life. The pervasive effects of the sex hormones are typical of hormones in general.

Hormones—Chemical Messengers and Regulators

Our sex hormones and their hard-to-ignore effects are just one component of the **endocrine system.** Overall, this system—hormones and the glands that secrete them—helps maintain homeostasis by integrating and coordinating many diverse physiological functions. Preparing the body for reproduction, for example, involves far more than initiating sperm or egg development. Many developmental changes, some activating genes that have lain dormant for a dozen years, must ready the body for reproductive maturity. In many animals, growth, maturation, and reproduction must coincide with the time of year when climatic conditions are favorable and food is available. Hormones help coordinate the timing of these events, which is crucial to their success.

The endocrine system also coordinates with the nervous system to perform another vital function—communication within the body. Although the endocrine system communicates much more slowly than the nervous system, its effects are generally more prolonged. These two systems work together to produce a variety of responses, from sexual and reproductive behavior, to control of growth and development, to adjusting the delicate chemical balance of body fluids. Their functions overlap, and they control each other.

The endocrine system communicates through hormones. A hormone is a biochemical that certain cells release into the bloodstream, which carries it to another part of the body where it alters other cells' activities. For example, a baby sucking on its mother's nipple stimulates neurons that cause the mother's pituitary gland to release the hormone oxytocin. The oxytocin enters the mother's circulation and travels to myoepithelial cells surrounding milk-filled cells in the breast. There, the hormone stimulates the myoepithelial cells to contract, which squirts milk out of small holes in the nipple. Although a hormone is a single substance, it can exert diverse effects when it controls a complex process.

Traditionally, hormones are considered to be products of endocrine glands and are secreted into the bloodstream. In recent years, however, the term "hormone" has also referred to **pheromones** and **prostaglandins,** which exert their effects locally.

An endocrine gland consists of groups of cells organized into cords or plates, connected with a large blood supply to carry away the hormones the gland produces. Figure 33.1 illustrates the location of endocrine glands in the human body, and figure 33.2 shows a typical cellular organization within a gland. Whereas endocrine glands secrete their hormones directly into the circulatory system, **exocrine glands,** such as sweat and salivary glands, secrete their products into ducts that release the substances onto the body's surface (fig. 33.3).

Dozens of hormones circulate to all cells of the human body, but only certain cells respond to a particular hormone. This specificity exists because cells have membrane-receptor molecules that only bind certain chemicals. A cell that binds a particular hormone is called that hormone's **target cell.** A target cell's receptor physically fits the molecular shape of the hormone, and the hormone-receptor complex initiates the cell's response.

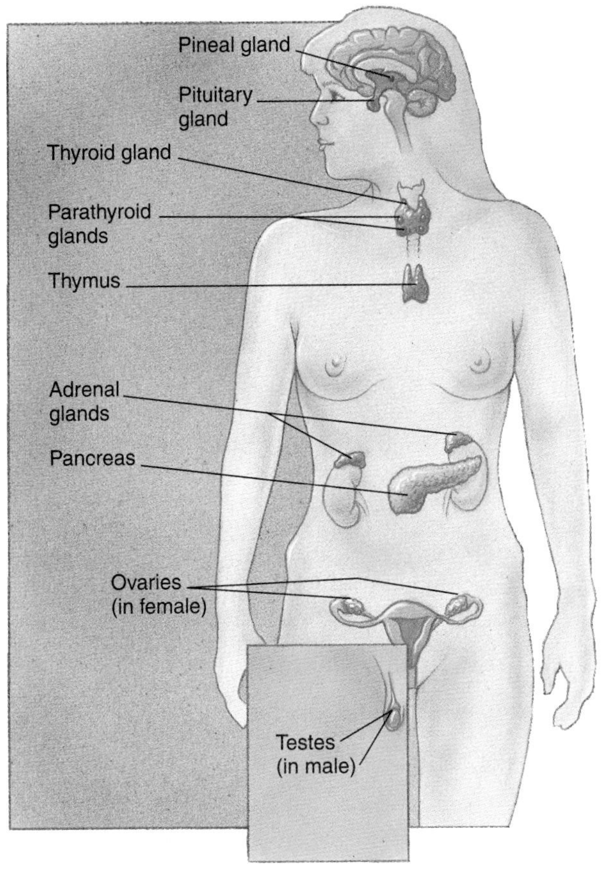

FIGURE 33.1

Some human endocrine glands. The endocrine system includes several glands that contain specialized cells that secrete hormones, as well as cells scattered among the other organ systems that also secrete hormones. (The parathyroids are not actually visible, as depicted here—they are behind the thyroid.)

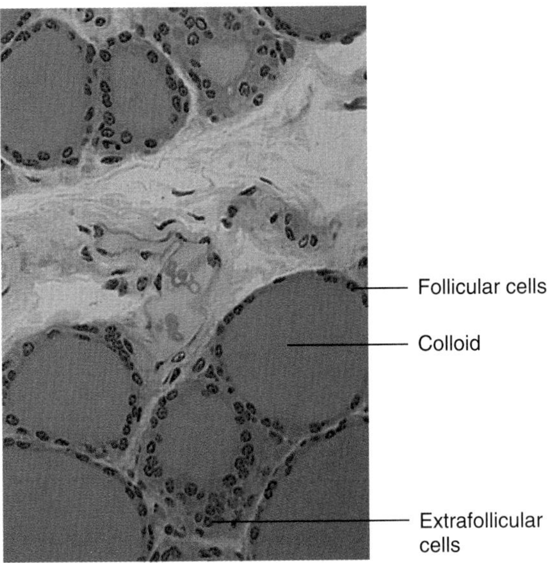

FIGURE 33.2

The organization of the thyroid is typical of endocrine gland structure. The thyroid consists of thousands of tiny spheres. Each sphere, or follicle, consists of epithelial cells, which secrete the hormone thyroxine into the central space for storage or send it directly into the bloodstream.

Mastering Concepts

1. What is the overall function of the endocrine system?
2. What is a hormone?
3. How do endocrine responses differ from nervous system responses?
4. How do exocrine glands differ from endocrine glands?
5. How does a target cell recognize a particular hormone?

The Endocrine and Nervous Systems Maintain Homeostasis

Homeostasis, or maintaining a relatively constant range of internal conditions no matter what the external environment, is a prerequisite for life. The nervous and endocrine systems work in tandem to maintain homeostasis. Neural regulation tends to be local—each neuron affects an adjacent neuron, muscle, or gland cell. Hormonal regulation tends to occur at a distance.

The fact that chemical regulation is present in diverse modern organisms indicates that it is an ancient adaptation. For example, the messenger molecule **cyclic adenosine monophosphate,** or **cAMP,** detects a hormone's arrival at a cell in diverse species. In slime molds, cAMP initiates aggregation of single cells into a multicellular organism. In the simple aquatic organism *Hydra,* cAMP conveys signals to regenerate damaged tissue. In insects, it controls metamorphosis from larva to adult. In our own bodies, cAMP relays signals that control cell division rate.

One way in which the nervous and endocrine systems are linked, both physiologically and in an evolutionary sense, is by a process called neurosecretion. Neurons called **neurosecretory cells,** rather than releasing neurotransmitters locally, release hormones that travel to exert their effect.

Not all hormones originate in cells within a gland, however, and not all hormones travel very far to evoke a response. Individual cells scattered throughout a tissue or organ may produce localized hormones. For example, the mucus-rich stomach and intestine lining contains cells that secrete "gut hormones" that regulate movement of the digestive tract and secretion rates of other digestive biochemicals. Similarly, scattered cells in the mammalian heart secrete **atrial natriuretic hormone,** which regulates blood pressure. The kidney and placenta also contain hormone-secreting cells, and white blood cells produce and release hormones.

Hormone systems likely arose as scattered hormone-secreting cells in the simpler animals. As animal life grew more complex, such cells aggregated to form glands. Today endocrine systems include networks of glands but also scattered hormone-secreting cells. As technology reveals more widely dispersed hormone-producing cells, the list of endocrine system components will grow.

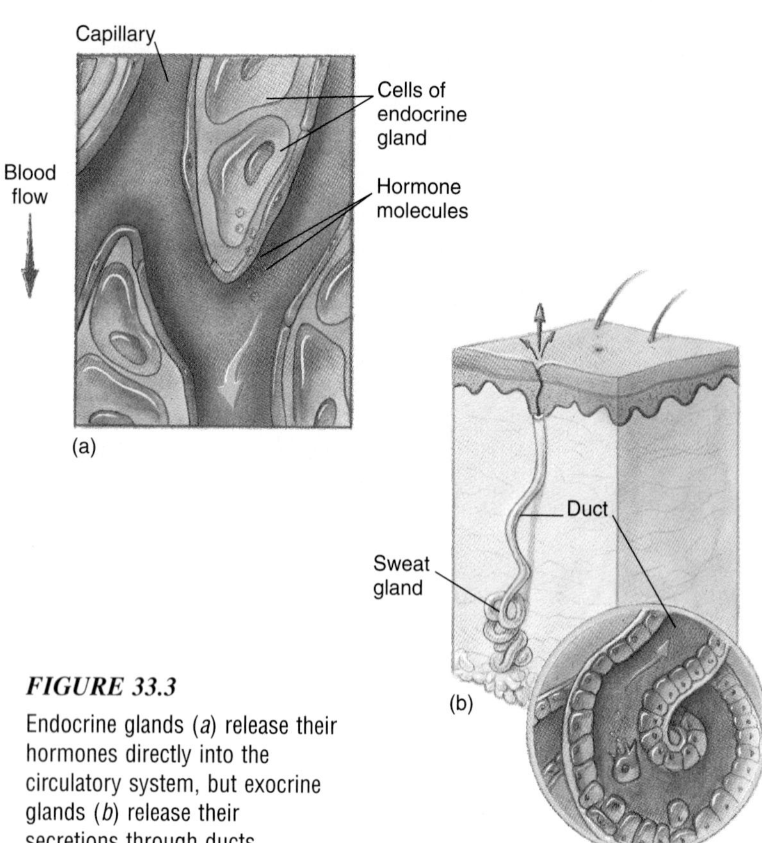

FIGURE 33.3

Endocrine glands (*a*) release their hormones directly into the circulatory system, but exocrine glands (*b*) release their secretions through ducts.

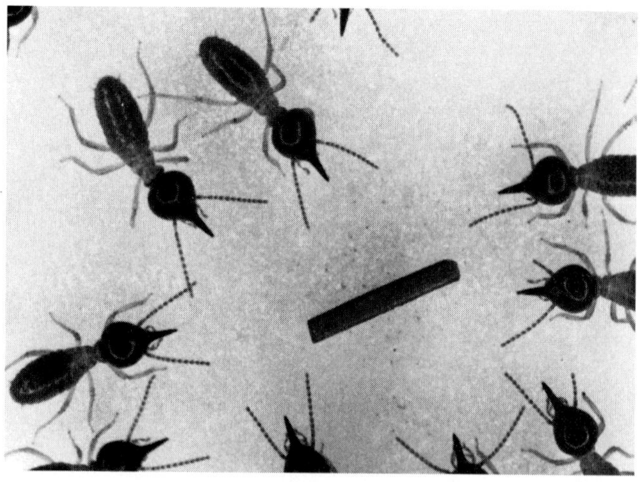

FIGURE 33.4

Pheromones. The small metal bar in the center has been sprayed with a secretion from a termite soldier. The pheromone signals "alarm" to the other termites, who rush to the scene.

Pheromones

Pheromones are substances that an individual secretes that stimulate a physiological or behavioral response in another individual of the same species. Pheromones are similar to hormones because they affect reproduction and behavior, but they differ because they transmit information to other individuals, rather than within an organism. Unlike hormones, which may have similar actions on several species, pheromones are highly species-specific. A female cockroach secreting a pheromone attracts a male of the same species only. Pheromones reach their "targets" when the target individuals breathe, ingest, or absorb them.

Researchers discovered pheromones in 1959 and first studied them in insects. Since then, researchers have discovered pheromones in birds, fish, and mammals. Biochemicals that attract opposite mating types in yeast and in some ciliates are pheromones. Human pheromones may exist, although they have yet to be identified. Figure 33.4 illustrates the attraction of a termite pheromone.

Mastering Concepts

1. How do neural and hormonal regulation differ?
2. Where do hormones come from, in addition to glands?
3. What is the evidence that hormonal systems are ancient?
4. How do pheromones differ from hormones?

How Hormones Exert Their Effects

One way hormones can be classified is as either water soluble or lipid soluble. Water-soluble hormones include **peptide hormones,** which are short chains of amino acids. Lipid-soluble hormones include **steroid hormones.** These two general types of hormones have different mechanisms of action.

Water-Soluble Hormones—Activating Second Messengers

Peptide hormones are water soluble, so the watery plasma of the bloodstream easily carries them to virtually all of an animal's cells. Because peptide hormones are lipid insoluble, however, they cannot cross cell membranes. Thus, a peptide hormone generally does not enter the target cell itself (one exception may be thyroid hormone), but instead it binds to one of the hundreds or even thousands of receptors on the target cell that it fits as a key fits a lock. When a peptide hormone binds its receptor, it initiates the target cell's response. For example, specific ion channels in the membrane might open. This would alter the concentration of ions within the cell, which might change the rate of certain cellular activities.

Usually when a hormone and a membrane receptor bind to form a complex, it activates an intracellular biochemical called a second messenger that triggers the cellular response. (Chapter 5 introduced second messengers.) This intracellular molecule is called a second messenger because the hormone first activates it and it then activates the enzymes that produce the effects associated with the hormone.

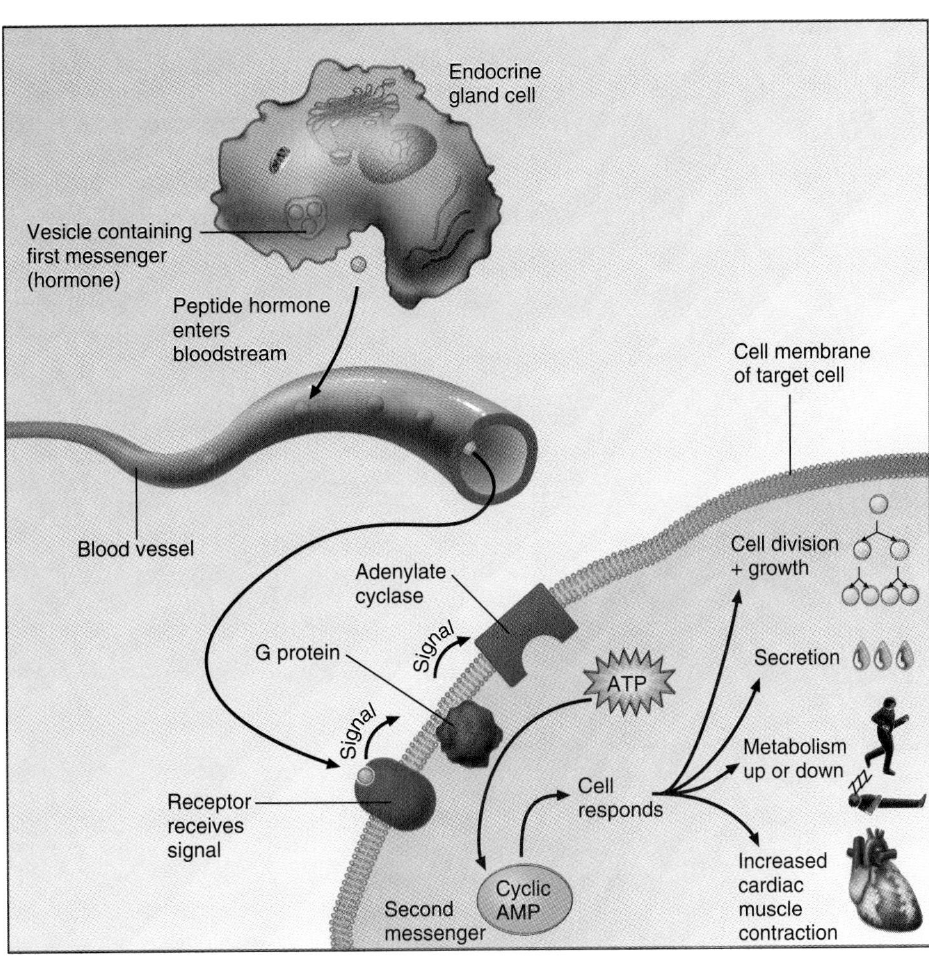

FIGURE 33.5

Peptide hormone action. An endocrine cell secretes a peptide hormone, which enters a capillary and travels in the bloodstream to the tissue fluid. The hormone binds to a cell that has surface receptors for it and triggers the enzymatic conversion of ATP to cyclic AMP (cAMP) on the inner face of the target cell's membrane. This reaction stimulates other chemical reactions, which produces the physiological effect associated with the particular hormone. In this case, the hormone is the first messenger; cAMP is the second messenger. The endocrine cell, capillary, and target cell are not drawn to scale.

Cyclic AMP is a common second messenger (see fig. 5.28). When a hormone binds to a receptor, it activates an enzyme called adenylate cyclase on the inner face of the membrane (fig. 33.5). Adenylate cyclase then catalyzes conversion of ATP into cAMP. (Recall that ATP is the energy currency of the cell.) In turn, cAMP activates specific enzymes within the cell. These enzymes catalyze the reactions that produce the response to the hormone, such as cell division, secretion, contraction, or metabolic changes. Different hormones can stimulate cAMP formation. The specific biochemical events following cAMP formation trigger the effects characteristic of different peptide hormones.

Parathyroid hormone is one type of peptide hormone. Cells of the **parathyroid glands** in the neck synthesize and secrete the hormone into the bloodstream. The hormone eventually binds to receptors on specific kidney cells, where it converts ATP to cAMP. The cAMP sets into motion a series of biochemical reactions that cause the kidney tubules to reabsorb calcium from the forming urine.

Fat-Soluble Hormones—Directing DNA Action

Steroid hormones, such as estrogen and **testosterone,** are fat soluble, and they can easily pass through the membranes of target cells. Once inside a cell, the steroid hormone binds to a receptor in the cytoplasm or nucleus. In the nucleus, the hormone-receptor complex stimulates particular genes to direct the manufacture of particular proteins (fig. 33.6). The steroid hormone testosterone, for example, activates muscle cell genes that direct muscle protein synthesis. The body synthesizes steroid hormones from cholesterol—which is one reason why some cholesterol is essential for good health.

Controlling Hormone Levels

To maintain homeostasis, the body must strictly regulate the levels of specific hormones in the bloodstream at a given time. Feedback interactions between the hormone (or its effects) and the gland often regulate levels of hormones in the blood. In a negative feedback loop, the accumulation of a certain biochemical switches off hormone synthesis.

A negative feedback loop controls blood glucose levels. After a meal rich in carbohydrates, the digestive system breaks down the complex carbohydrates into the simple carbohydrate glucose, which enters the circulation in the walls of the small intestine. The resulting rise in blood sugar level stimulates beta cells in the pancreas to secrete the hormone **insulin**. Insulin stimulates target cells to admit glucose, which the cell then stores or breaks down to release energy. As glucose leaves the bloodstream to enter cells, the pancreas slows insulin secretion. Rising insulin levels thus eventually turn off insulin production as the hormone completes its "job." Negative feedback also controls levels of the hormone

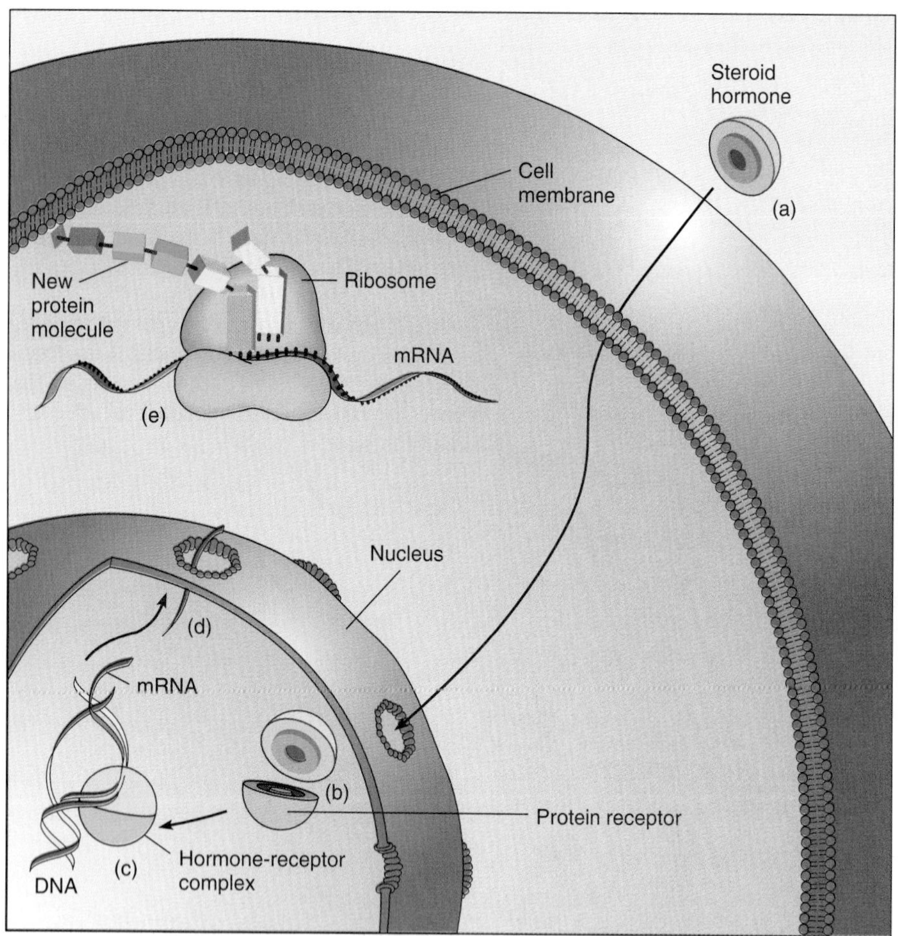

FIGURE 33.6

A steroid hormone enters a target cell and binds a receptor, usually in the nucleus. The hormone-receptor complex activates transcription of certain genes, which leads to synthesis of proteins that provide the characteristic responses to the particular hormone.

glucagon but with effects opposite those of insulin (fig. 33.7). Alpha cells in the pancreas secrete glucagon in response to lowering levels of glucose in the bloodstream.

In a **positive feedback loop,** an accumulating biochemical increases its own production. For example, at the onset of labor, the uterus contracts, releasing the hormone **oxytocin** and prostaglandins. These substances intensify the uterine contractions, and the contractions further stimulate oxytocin and prostaglandin production. When the baby is born, other hormonal changes stop the cycle.

Mastering Concepts

1. How do water-soluble hormones function?
2. How do steroid hormones exert their effects?
3. Distinguish between the ways that negative and positive feedback loops control hormone secretion.

Diversity of Hormones and Glands

Hormones control and coordinate timing and expression of complex biological functions that may entail many biochemical reactions. All animals have some hormonal control. Neurosecretory cells accomplish this in simpler invertebrates; similar cells organized into glands control hormone synthesis in more complex invertebrates and in vertebrates.

Hormones in Invertebrates

In animals without backbones, hormones control a variety of functions, including regeneration in jellyfishes and flatworms; gonad maturation in flatworms, mollusks, and arthropods; and regulation of blood sugar level in segmented worms. A familiar hormone-orchestrated event of invertebrate life is a larval insect's metamorphosis into its adult form (see fig. 26.6*b*). The physiological and biochemical changes that sculpt a butterfly from a caterpillar, a fly from a larva, or a beetle from a grub are among the most profound in the animal kingdom.

From Larva to Adult

An insect larva is a streamlined eating machine—a rapidly expanding bag with jaws at one end. The larva is equipped to eat but lacks such structures as sex organs and wings. It grows at a spectacular rate, increasing its size several thousandfold in a few days. This creates a problem—how can the young insect contain its increasingly bulging body? The solution is to molt—to periodically shed its cuticle (outer covering) to allow for rapid growth.

After several larval molts, the outer covering splits to reveal a silky cocoon produced by silk glands. Larval cells die, as other cells that were set aside in tiny disclike packets in early larval life obtain energy from the degenerating larval cells. The reawakened cells divide and project outwards, like a painting springing into

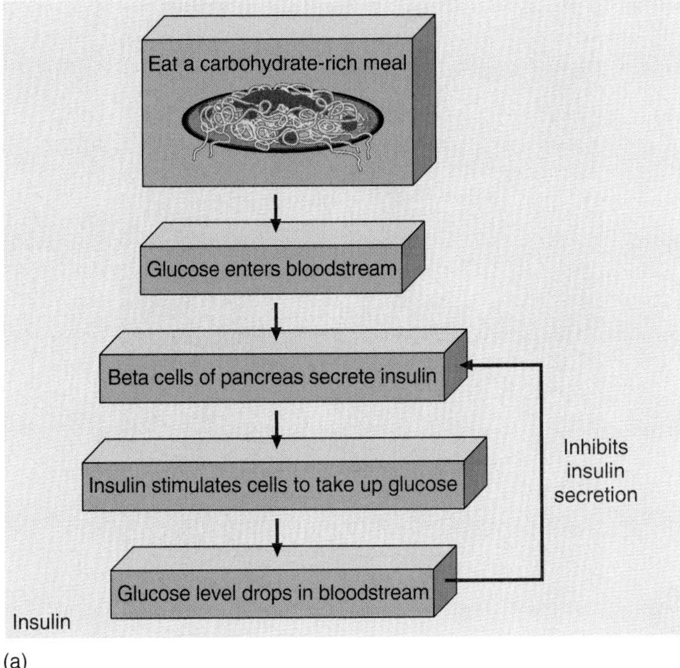

(a)

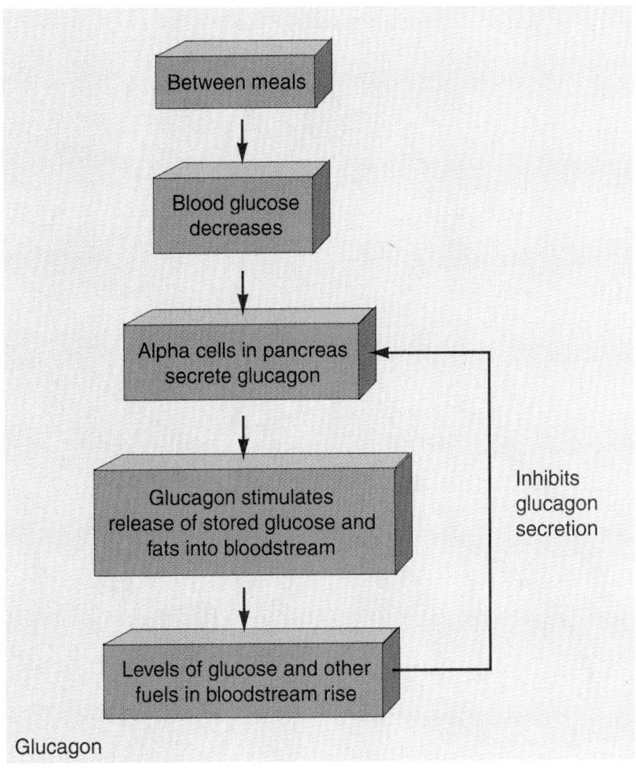

(b)

FIGURE 33.7

Insulin (*a*) and glucagon (*b*) regulate levels of glucose in the bloodstream.

three-dimensional life. Within a few days, the cells form small adult body parts—legs, antennae, wings. Eventually the cocoon splits, and an adult emerges.

The adult insect is wet and compacted as it emerges from the cocoon, its wings plastered against its abdomen. It takes a few hours to dry out and for fluid to flow into veins in the wings and expand them. Soon the insect is free to embark on an existence very different from its life as a larva. Rather than living to eat, the adult insect lives to mate.

Insect metamorphosis is a complex, highly coordinated, and intricately timed process. Changing levels of two hormones—**juvenile hormone** and **molting hormone**—control the process. In the early larva, juvenile hormone predominates, suppressing production of molting hormone (fig. 33.8). As the larva grows, the juvenile hormone level drops and the molting hormone level rises. In a midsize larva, the two hormone levels are nearly equal. Then, just before the cocoon forms, the neurosecretory cells producing juvenile hormone shut off, and molting hormone predominates. It will oversee the metamorphosis to follow.

Hormones and Glands in Vertebrates

It is difficult to generalize about hormones and glands across vertebrate species, because the same hormone may have different functions in different species. This may reflect differences in lifestyles. In humans, for example, the thyroid gland hormone **thyroxine** sets the pace of metabolism, but in frogs, thyroxine controls metamorphosis. Add thyroxine to water, and a tadpole becomes a frog too soon. Block a tadpole's thyroxine secretion, and it remains a tadpole. In birds, which have metabolic requirements different from those of humans and frogs, thyroxine controls heat production and oxygen consumption. Other diversities in hormone function are harder to explain. For example, **antidiuretic hormone (ADH)** causes the human kidney to reabsorb water into the blood. In the prairie vole, a rodent, the same hormone is responsible for a male's monogamy (dedication to a single mate) and caring for its young!

Sexual Seasons

Reproductive behavior illustrates how hormonally controlled processes can vary among vertebrate species. The human female is unusual among animals in that she is sexually receptive at all times. Because of her monthly cycle, she is referred to as a **cyclic ovulator.** Her sexual receptivity is not limited to the time when she ovulates.

By contrast, most female mammals are **seasonal ovulators.** They go through periods of estrus (heat) when they are both sexually receptive and ovulating. The correlation of sexual behavior and peak fertility may be an adaptation to maximize the chance of conceiving. During estrus, the seasonal ovulator experiences a surge in estrogen production. Higher estrogen levels make her intensively seek sex and display her interest in potential mates by her scent, appearance, posture, and behavior. When she is not in estrus, she is unreceptive to mating behavior of any sort. In some species, females are actually physically incapable of copulation when they are not in estrus. The African bush baby, for example, grows a covering of skin over her vagina in the sexual off-season. The male adapts to the female pattern. In some species, males manufacture sperm only when females are in estrus.

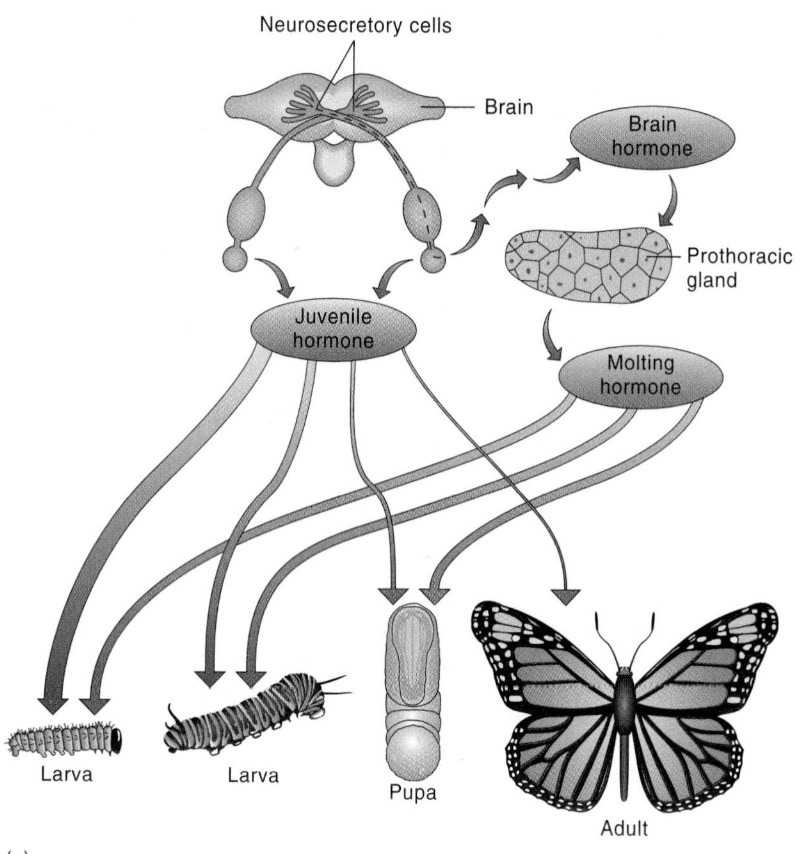

FIGURE 33.8

Metamorphosis. (*a*) Two hormones control insect metamorphosis. Juvenile hormone predominates in the larva. Molting hormone levels rise toward the end of larval life, finally triggering pupation (cocoon formation) and then metamorphosis into the adult. (*b*) A monarch butterfly emerges from its cocoon.

Even greater reproductive economy is seen among **induced ovulators.** In sheep and goats, for example, the presence of a male induces ovulation, and in rabbits and domestic cats, copulation induces it. The female mink ovulates when her mate bites her on the back of the neck.

In mice, a male can disrupt a female's hormones even after she becomes pregnant. If a female is less than 4 days pregnant when she meets a "new" male (a male other than the father of the embryos), his presence causes her to abort her embryos and to become sexually receptive to him. To trigger this "pregnancy block," the male emits a pheromone that causes the female's hypothalamus to stimulate her **pituitary gland** to release **follicle-stimulating hormone (FSH)** and **luteinizing hormone (LH).** This induces estrus and ovulation, drastically lowering blood levels of the hormone prolactin. When prolactin levels are inadequate, embryos cannot remain implanted in the uterus.

Neuroendocrine Control—The Hypothalamus and the Pituitary

All vertebrate endocrine systems are regulated in large part by the hypothalamus, the neuroendocrine control center in the brain, and the pituitary, the nearby pea-sized gland. The hypothalamus directly or indirectly controls most other vertebrate endocrine glands as well.

The hypothalamus links the nervous and endocrine systems by controlling pituitary secretions. The pituitary gland attaches to a stalk extending from the hypothalamus and is actually two glands, the anterior (front) and posterior (back) pituitary. Each half of the pituitary releases a different set of hormones and is regulated by the hypothalamus in a different manner. Between the anterior and posterior lobes of the pituitary is an intermediate region known to produce at least one hormone.

The posterior hypothalamus develops from the central nervous system and maintains a close relationship with the brain, with nerve fibers touching all parts of the brain. In the embryo, the posterior pituitary develops from the hypothalamus. Later in development, neurosecretory cells in the hypothalamus synthesize ADH and oxytocin, the hormones that the posterior pituitary releases. ADH and oxytocin move through the stalk along the axons of the cells that produced them to the posterior lobe of the pituitary, where they are stored within the cell endings. When neural activity in the brain stimulates the neurosecretory cells, they release their hormones.

The hypothalamus also controls the anterior lobe of the pituitary, but in a different manner. Other neurosecretory cells in the hypothalamus secrete hormones that either stimulate or inhibit the production of anterior pituitary hormones. These hypothalamic hormones reach the anterior pituitary through a specialized system of vessels. Hormones that one gland produces to influence another gland's hormone production are called **tropic hormones.** When the hypothalamus produces such hormones, they are called **releasing hormones** or **inhibiting hormones** (fig. 33.9).

Vertebrate endocrine systems may vary. Here we explore the human system as an example.

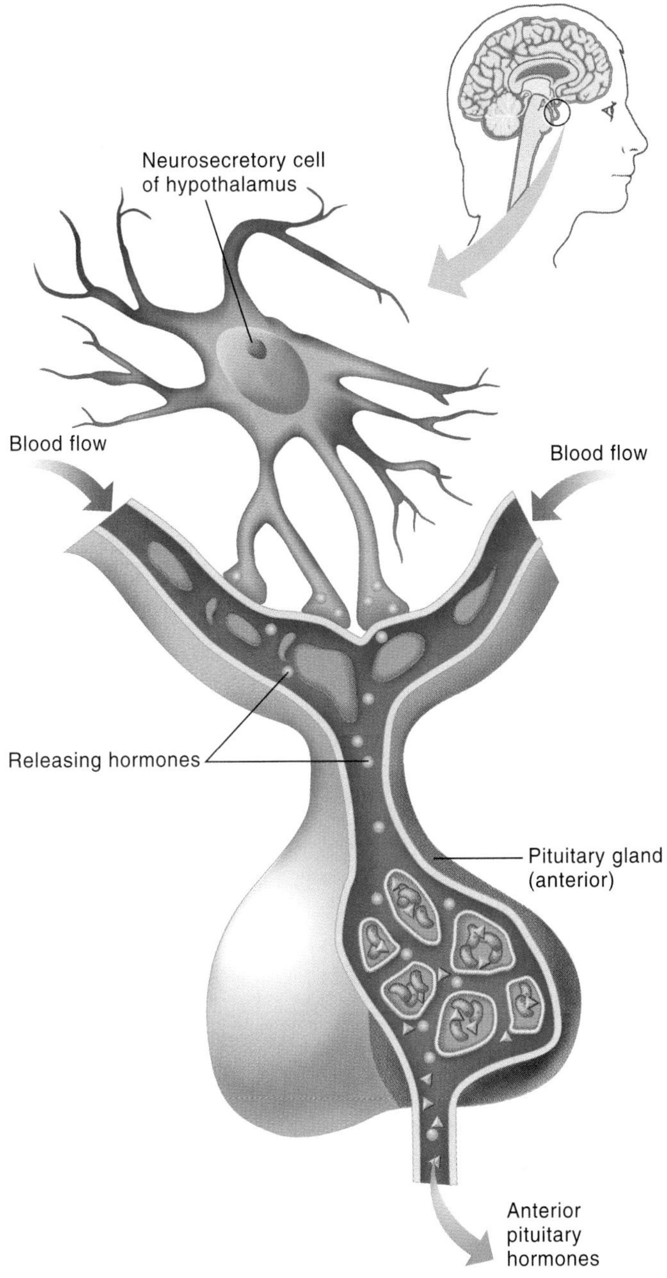

FIGURE 33.9

Neurosecretory cells in the hypothalamus send releasing hormones to the anterior pituitary gland, where they affect its activity.

Mastering Concepts

1. Explain how changing levels of two hormones control insect metamorphosis.
2. In what way is a female vertebrate's ovulation pattern adaptive?
3. How do the hypothalamus and the pituitary gland interact?

Pituitary Hormones

The pituitary gland is really two glands in one, both structurally and functionally.

Anterior Pituitary Hormones

In humans, the anterior pituitary gland produces hormones that directly affect body parts and hormones that affect other endocrine glands (fig. 33.10). **Growth hormone (GH)** promotes growth and development of all tissues by increasing protein synthesis and cell division rates. GH stimulates cells to take up more amino acids, to mobilize fats, and to release glucose from the liver to supply energy. GH also increases the cell division rate in cartilage and bone, thereby promoting height.

Levels of GH peak in the preteen years and, together with rising levels of sex hormones, cause adolescent growth spurts. Abnormal levels of GH have noticeable effects. A severe deficiency of the hormone during childhood leads to **pituitary dwarfism,** which produces extremely short stature. This condition can be treated with human growth hormone manufactured using recombinant DNA technology. Excess GH causes overgrowth of many tissues. In a child, a pituitary tumor produces a **pituitary giant.** In an adult, excess GH does not affect overall height because the growth regions in the long bones are no longer active, but it causes **acromegaly.** This is a thickening of the bones that can respond to GH, most noticeably in the hands and face (fig. 33.11).

Several anterior pituitary hormones affect reproduction. **Prolactin** stimulates milk production in a woman's breasts after she gives birth. In males and in women who are not breast-feeding, an inhibitory hormone produced in the hypothalamus suppresses prolactin synthesis in the anterior pituitary. The inhibition is overcome in nursing mothers by nerve impulses generated when the infant sucks on the nipples.

The remaining anterior pituitary gland hormones influence hormone secretion by other endocrine glands (fig. 33.12). The **gonadotropic hormones** affect the ovaries and testes. In the human female, follicle-stimulating hormone (FSH) initiates development of ovarian follicles, oocyte maturation, and release of estrogen from the follicles. In the male, FSH promotes the development of the testes and manufacture of sperm cells. In the female, luteinizing hormone (LH) stimulates an ovary to release an oocyte each month. In the male, LH is called interstitial-cell-stimulating hormone (ICSH), and it stimulates the testes to produce the hormone testosterone, which is necessary for sperm development.

Thyroid-stimulating hormone (TSH) causes the thyroid gland in the neck to release two hormones that control metabolism. **Adrenocorticotropic hormone (ACTH)** stimulates release of the **glucocorticoid hormones** from the outer portion (cortex) of the **adrenal glands,** which are located on top of the kidneys. The glucocorticoids increase blood glucose level during stress.

Posterior Pituitary Hormones

Two hormones are synthesized in the hypothalamus and stored in and released from the posterior pituitary. One, antidiuretic hormone, contracts smooth muscle cells lining blood vessels, which

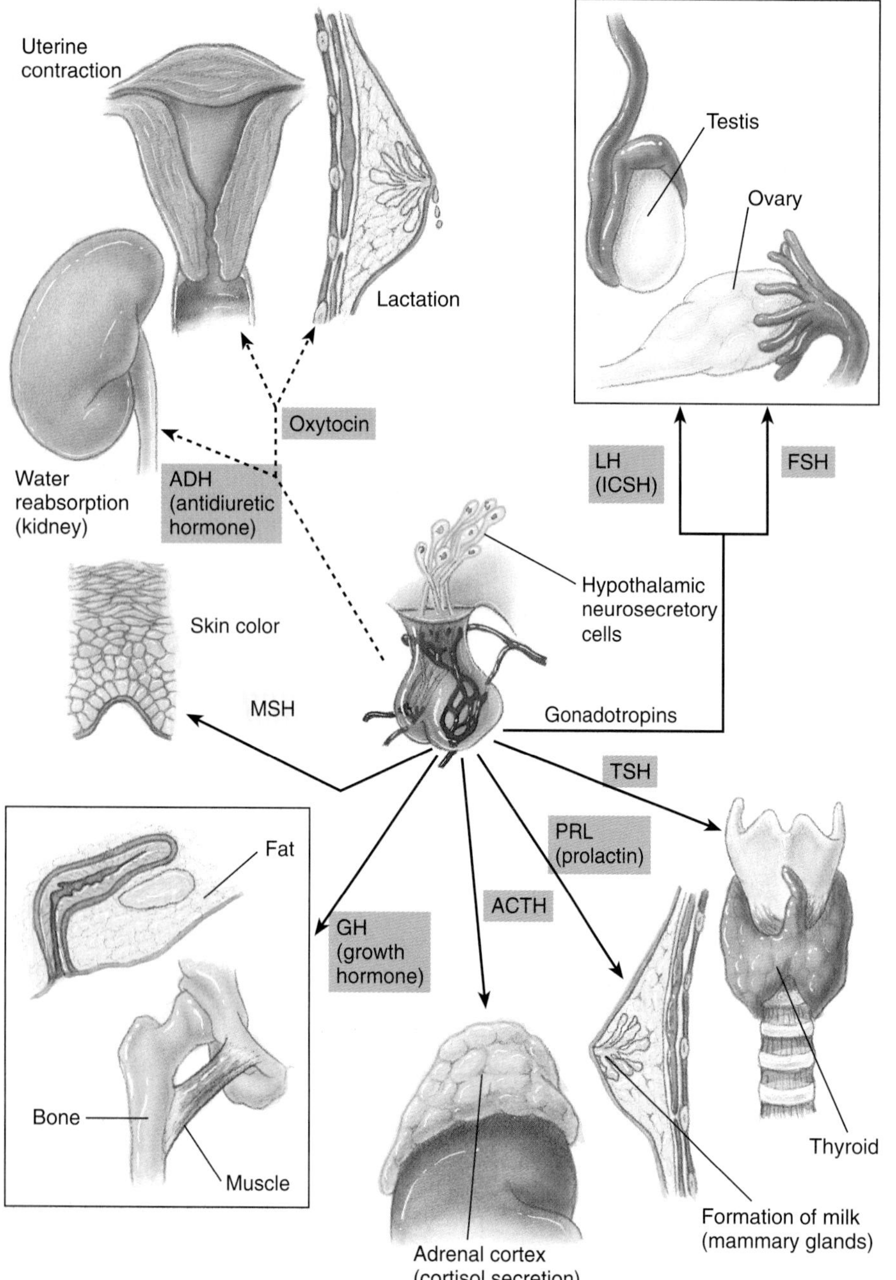

FIGURE 33.10

The hormones of the pituitary gland have many functions. The anterior lobe of the pituitary gland secretes six hormones—thyroid-stimulating hormone (TSH), prolactin (PRL), follicle-stimulating hormone (FSH), luteinizing hormone (LH) (called interstitial-cell-stimulating hormone [ICSH] in males), adrenocorticotropic hormone (ACTH), and growth hormone (GH). The posterior lobe of the pituitary gland secretes oxytocin and antidiuretic hormone (ADH). The intermediate lobe produces melanocyte-stimulating hormone (MSH).

FIGURE 33.11

Growth hormone overproduction that begins in adulthood causes acromegaly. Adult limb bones no longer respond to growth hormone; however, the bones of the hands and face still respond. When GH is overproduced, these bones enlarge considerably, as the changes in the facial features of this woman show as she ages from (*a*) 9 years to (*b*) 16 years to (*c*) 33 years to (*d*) 52 years.

(a)

(b)

(c)

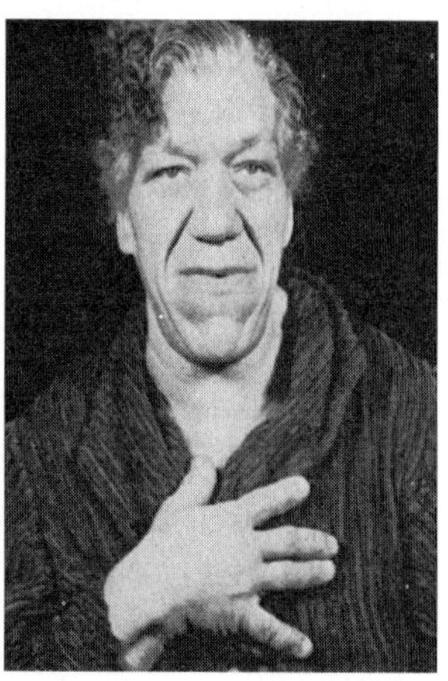

(d)

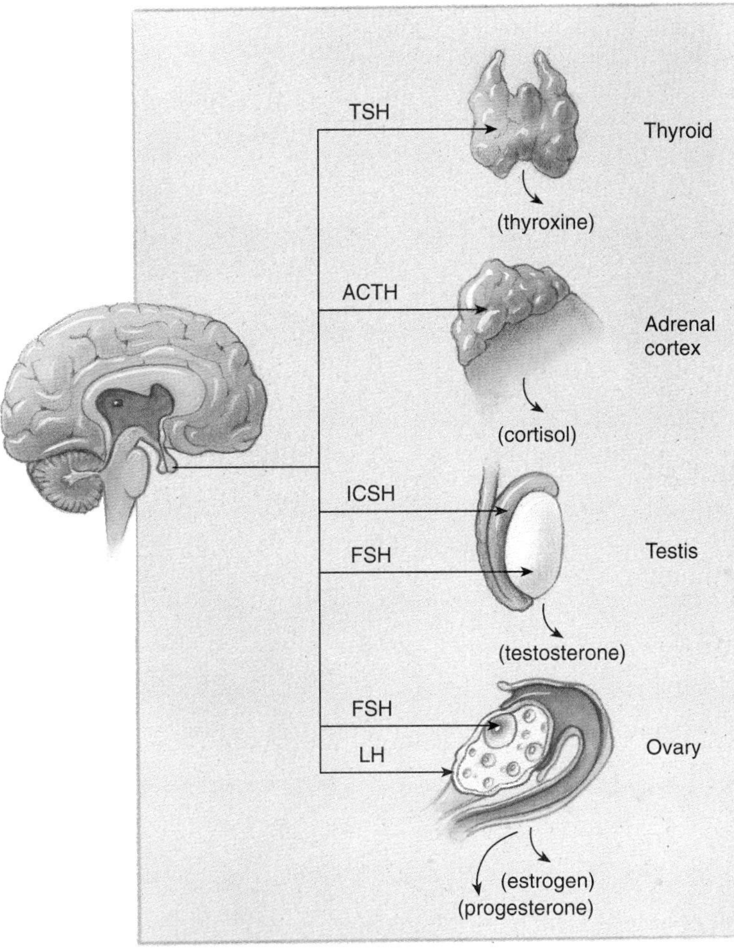

FIGURE 33.12

Several anterior pituitary hormones regulate hormone production by other endocrine glands.

helps to control blood pressure. It is called antidiuretic ("diuretic" means increasing urine flow) hormone because it makes the kidneys' collecting ducts more permeable to water so that the body can reabsorb more water.

Control of ADH secretion helps maintain the chemical balance of body fluids. Specialized cells in the hypothalamus called osmoreceptors monitor the amount of water in the blood. If the blood has too little water, the osmoreceptor cells stimulate the posterior pituitary to release more ADH. As ADH increases water reabsorption within the kidneys and dilutes the blood, the osmoreceptor cells signal the ADH-manufacturing cells in the hypothalamus to decrease their activity. See figure 5.10 for an illustration of ADH function.

Oxytocin is the other hypothalamus-produced hormone stored in and released from the posterior pituitary. This hormone contracts cells in the breasts, causing them to release milk when a baby nurses, and contracts cells in the uterus, which pushes a baby out during labor. Synthetic oxytocin can induce labor or accelerate contractions in a woman who is exhausted from a long effort at giving birth.

Intermediate Region Hormones

Melanocyte-stimulating hormone (MSH) is produced in the region of the pituitary between the anterior and posterior lobes. MSH causes the color changes that many vertebrates undergo. When the hormone binds to receptors on melanocytes in the skin, pigment granules disperse, darkening the skin.

In fishes, amphibians, reptiles, and some mammals, MSH is an adaptation that provides camouflage. In humans, the intermediate lobe is poorly developed, and significant levels of MSH are normally found only in children and pregnant women. Some researchers hypothesize that excess sunlight may trigger MSH secretion, which protects the skin against sun damage.

Mastering Concepts

1. Which hormones does the anterior pituitary produce, and what are their functions?
2. What are the functions of ADH and oxytocin?
3. What is the function of melanocyte-stimulating hormone?

Endocrine Control of Metabolism

The thyroid gland, parathyroid glands, adrenal glands, and the pancreas secrete hormones that control metabolism.

The Thyroid Gland

The thyroid is a two-lobed gland located at the front of the larynx and trachea in the neck (fig. 33.13*a*). Two thyroid hormones, thyroxine and **triiodothyronine,** increase the rate of cellular metabolism. Other cells in the thyroid produce a third hormone, **calcitonin,** which lowers blood calcium levels under certain conditions. Dietary iodine is required to manufacture thyroxine and triiodothyronine. The gland actively absorbs and concentrates iodine, making its iodine levels 25 times higher than levels in the blood.

Thyroid hormones bind to many different cell types, where they speed metabolism by increasing supplies of enzymes required for cellular respiration. The specific nature of this activity depends upon the cell type. Under thyroid stimulation, the lungs exchange gases faster, the small intestine absorbs nutrients more readily, and fat levels in cells and in blood plasma diminish.

An underactive thyroid gland (hypothyroidism) slows metabolic rate. Because the body burns fewer calories, weight increases despite a poor appetite. Heartbeat slows, and blood pressure and body temperature fall. Hypothyroidism beginning at birth produces **cretinism,** leaving the child physically and mentally retarded. If cretinism is detected at birth and treated with thyroxine before 3 months of age, however, the child develops normal intelligence. Hypothyroidism beginning in adulthood is termed **myxedema** and causes lethargy, a puffy face, and dry, sparse hair. It, too, is treatable with thyroxine.

Hypothyroidism due to lack of iodine in the diet causes the thyroid gland to swell and form a lump in the neck called a **goiter** (fig. 33.13*b*). Ancient Egyptian doctors treated this type of goiter with seaweed, not realizing that iodine in the seaweed

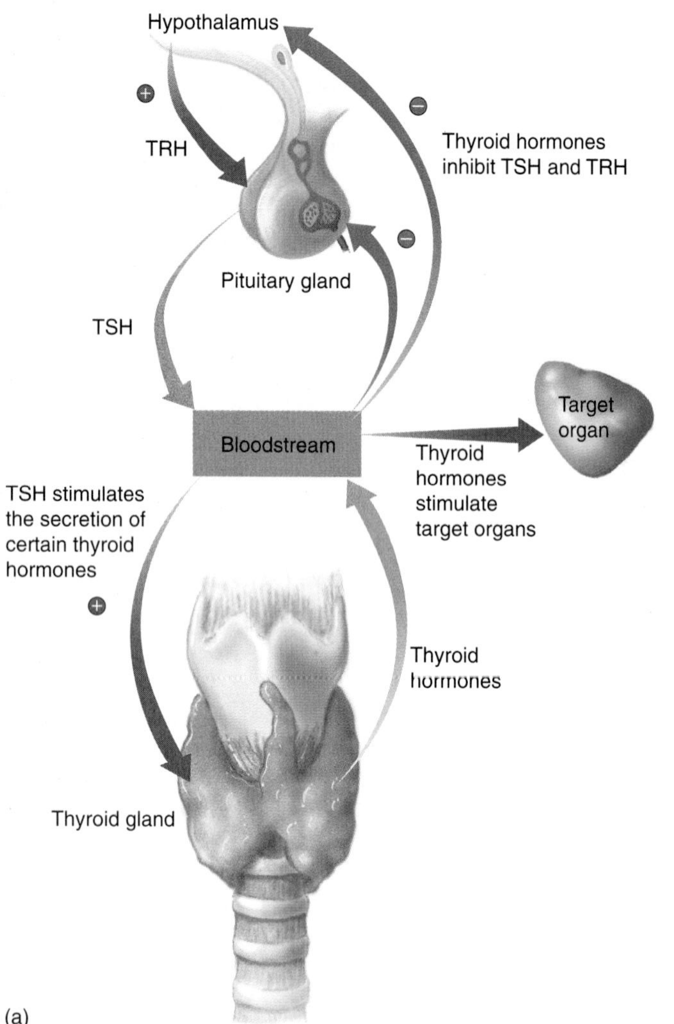

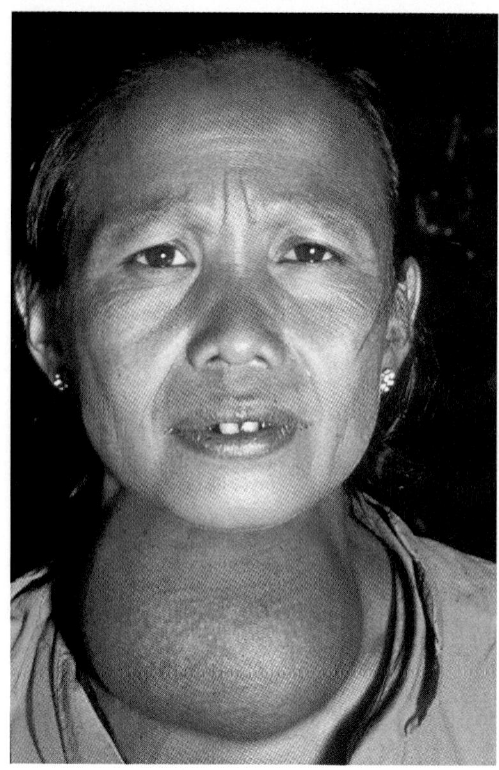

FIGURE 33.13

The thyroid gland. (*a*) The thyroid surrounds the front of the voice box (larynx) and the windpipe (trachea). The gland produces hormones that increase the metabolic rate of most cells. A negative feedback loop inhibits thyroid-stimulating hormone from the anterior pituitary and thereby regulates thyroid hormone production. (*b*) A woman with a goiter—thyroid swelling due to lack of iodine in the diet.

reversed the condition by enabling the thyroid to produce more hormone. Today iodine-deficient goiter is rare in nations where iodine is added to table salt.

An overactive thyroid, called **hyperthyroidism,** can also cause a goiter. Other symptoms reflect accelerated metabolism, including a very short attention span, irritability and hyperactivity, and elevated heart rate, blood pressure, and temperature. Appetite is great, but the high metabolic rate keeps weight off. The chapter 4 opener discusses thyroid cancer.

The Parathyroid Glands

Embedded in the back of the thyroid gland are four small groups of cells called the parathyroid glands. These glands secrete **parathyroid hormone (PTH),** which maintains calcium levels in blood and tissue fluid by releasing calcium from bones and by enhancing calcium absorption through the digestive tract and kidneys (fig. 33.14). Parathyroid hormone action opposes that of calcitonin and also inhibits the kidneys from reabsorbing phosphate.

Calcium is vital to muscle contraction, nerve impulse conduction, blood clotting, bone formation, and the activities of many enzymes. Underactivity of the parathyroids can swiftly be fatal. Excess parathyroid hormone causes calcium to leave bones faster than it accumulates, resulting in easily broken bones. This condition, called **osteoporosis,** is most common in women who have reached menopause (cessation of menstrual periods). The estrogen decrease that accompanies menopause makes bone-forming cells more sensitive to parathyroid hormone, which depletes bone mass.

The Adrenal Glands

The paired adrenal glands ("ad" means near, "renal" means kidney) sit on top of the kidneys (fig. 33.15). The inner portion of each gland is the **adrenal medulla,** and the outer portion is the **adrenal cortex.** These sections differ in structure, function, and embryonic origin and can be considered separate but physically attached glands.

The adrenal medulla hormones, **epinephrine** (also known as adrenaline) and **norepinephrine** (also called noradrenaline), help the body respond to and survive emergencies by increasing heart and breathing rates and blood flow. These hormones, called **catecholamines,** are under the control of the sympathetic nervous system.

The adrenal cortex secretes three types of steroid hormones: **mineralocorticoids, glucocorticoids,** and **sex hormones.** The

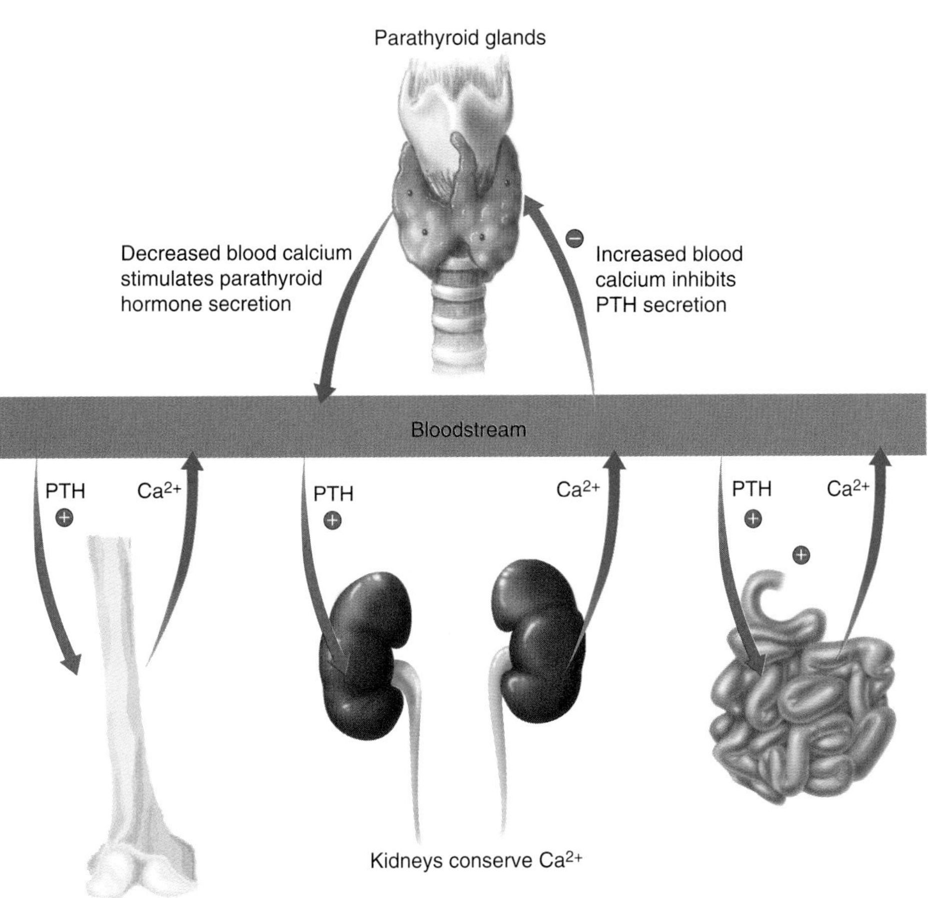

FIGURE 33.14

Parathyroid hormone stimulates bone to release calcium (Ca^{2+}) and the kidneys to conserve calcium. PTH indirectly stimulates the intestine to absorb calcium. The resulting increase in blood calcium concentration inhibits secretion of PTH.

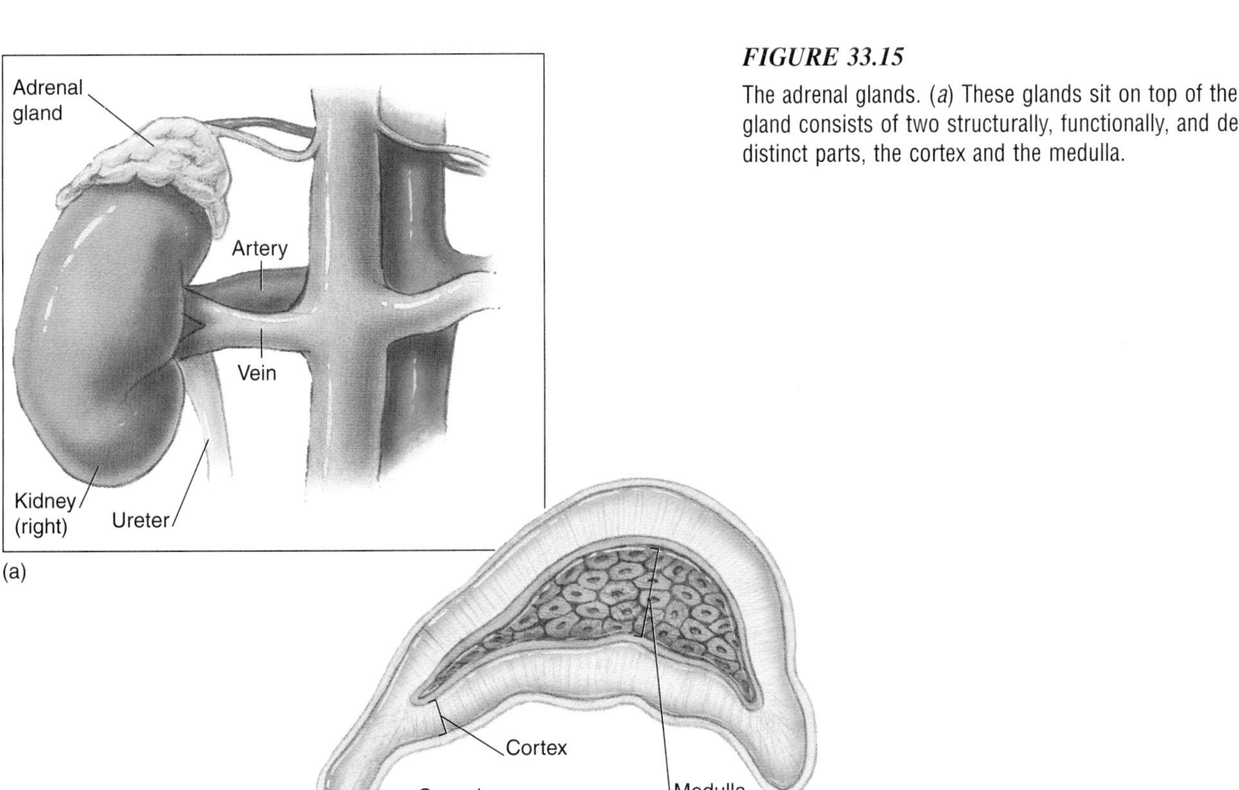

FIGURE 33.15

The adrenal glands. (*a*) These glands sit on top of the kidneys. (*b*) Each gland consists of two structurally, functionally, and developmentally distinct parts, the cortex and the medulla.

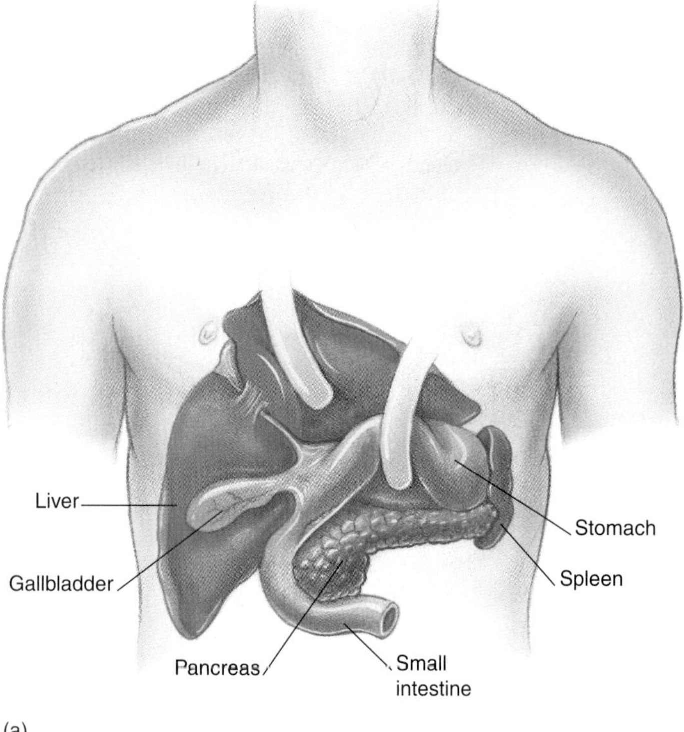

(a)

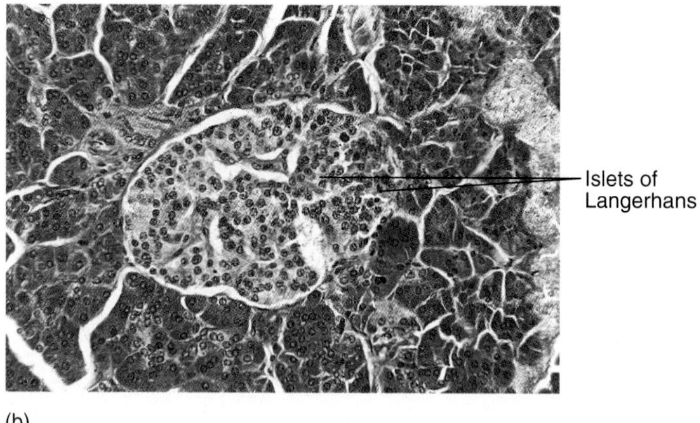

(b)

FIGURE 33.16

The pancreas. (*a*) This structure lies above the small intestine and beneath the stomach. It consists of many lobes and produces digestive enzymes and hormones. (*b*) Light micrograph of islets of Langerhans within the pancreas.

mineralocorticoids and glucocorticoids function together in times of prolonged stress to mobilize energy reserves while stabilizing blood volume and composition.

The mineralocorticoids maintain blood volume and electrolyte balance by stimulating the kidney to return sodium ions and water to the blood while excreting potassium ions. The major mineralocorticoid, **aldosterone,** maintains the level of sodium ions in the blood by increasing the amount the kidneys reabsorb. Since water reabsorption follows sodium ion reabsorption, aldosterone also affects blood volume. This action is particularly important in compensating for fluid loss from severe bleeding.

The glucocorticoids, the most important of which is **cortisol,** are essential in the body's response to prolonged stress. Whereas the adrenal medulla braces the body against an immediate danger, the glucocorticoids take over for longer periods, for instance during an infection. Glucocorticoids affect carbohydrate, protein, and lipid metabolism. They break down proteins into amino acids and then stimulate the liver to synthesize glucose from parts of these freed amino acids. The newly synthesized glucose supplies energy for healing after the immediate supply of glucose is depleted. Glucocorticoids also indirectly constrict blood vessels, which slows blood loss and prevents tissue inflammation, important responses to injury.

A deficiency in adrenal cortex hormones causes **Addison's disease.** Lack of mineralocorticoids causes an ion imbalance that decreases water retention so that the person has low blood pressure and is dehydrated. Symptoms include weight loss, mental fatigue, weakness, and impaired resistance to stress. Lack of glucocorticoids in Addison's disease darkens the skin. Replacing the deficient hormones can treat Addison's disease. Former president John F. Kennedy had the condition.

Excess glucocorticoids causes **Cushing's syndrome,** which redistributes body fat. A person with the syndrome may have thin legs but excess fat in the face, behind the shoulder blades, and in the abdomen. The person bruises easily and heals slowly.

The adrenal cortex secretes small amounts of both male and female sex hormones. The cortex produces more male sex hormone (testosterone) than female sex hormones (estrogen and progesterone). Adrenal hormones are particularly important in postmenopausal women.

The Pancreas

The **pancreas** is a large gland located between the spleen and the small intestine (fig. 33.16*a*). It produces digestive enzymes and hormones. Dispersed among the cells that produce digestive enzymes are clusters of cells called **islets of Langerhans** (fig. 33.16*b*), which are the endocrine portion of the pancreas. The hormones they secrete are polypeptides that regulate the body's use of nutrients.

The islets produce three hormones, each secreted by a different type of cell: **insulin** (Latin for "island"), **glucagon,** and **somatostatin.** Figure 33.7 illustrates how the actions of insulin and glucagon oppose one another in regulating blood glucose levels. Somatostatin controls the rate at which the blood absorbs nutrients.

Diabetes Mellitus

Insulin deficiency, or an inability of the body to use insulin, causes **diabetes mellitus,** which is the subject of this chapter's opener. Because the body cannot use or store glucose properly in diabetes mellitus, sugar builds up in the blood and appears in the urine.

Untreated diabetes leads to a condition similar to starvation (fig. 33.17). Of the nearly 11 million people in the United States who suffer from diabetes mellitus, 15% have type I, or insulin-dependent, diabetes, which usually begins in childhood. Symptoms include thirst, blurred vision, weakness, fatigue, irritability, nausea, and weight loss. Untreated, a severe insulin deficiency can result in a lethal diabetic coma. Insulin-dependent diabetes may be an autoimmune disease, in which the body attacks insulin-producing beta cells as if they are foreign.

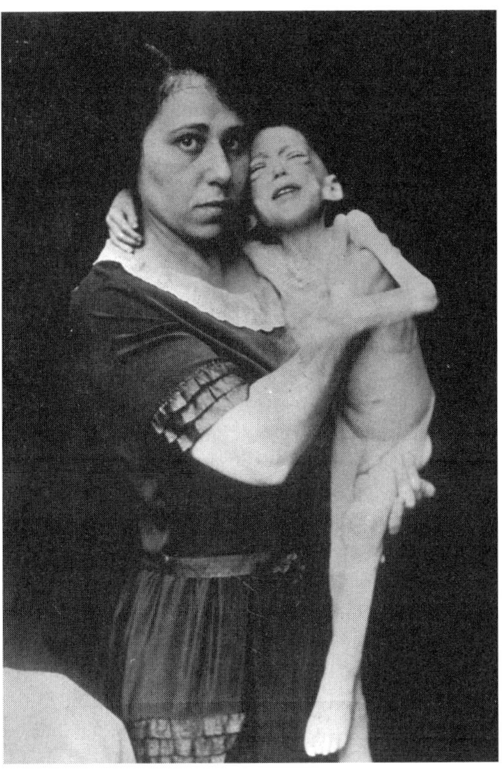

(a)

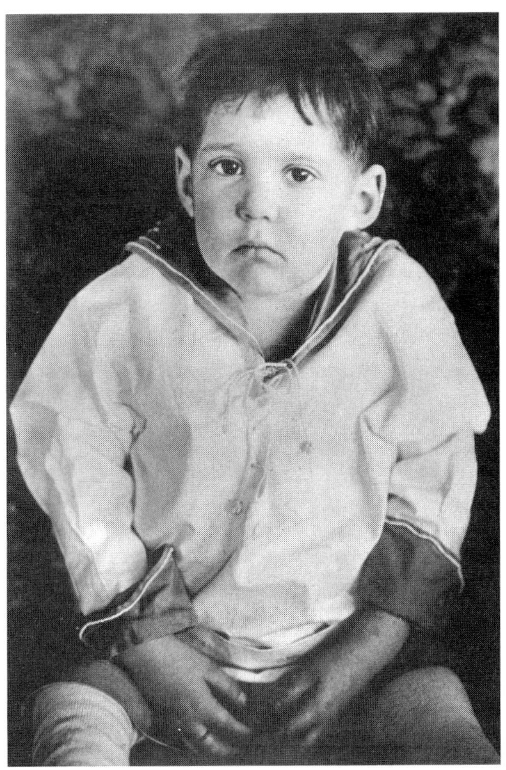

(b)

FIGURE 33.17

Insulin treatment. (*a*) A mother holds her son, who was 3 years old and weighed 15 pounds when this photo was taken in December 1922. He suffered from diabetes mellitus. (*b*) The same boy 2 months later; after insulin treatment, his weight increased to 29 pounds.

FIGURE 33.18

Diabetes mellitus is the result of abnormal insulin function. It can arise from a defect at any of the following points in the synthesis or functioning of insulin. (*a*) Pancreatic islet cells must synthesize and store insulin. (*b*) The islet cells must secrete the insulin as blood glucose levels rise. (*c*) The insulin must reach receptor cells. (*d*) Receptors in the target cells must bind the insulin, increasing membrane permeability so glucose can enter (*e*). Glucose must be metabolized for energy or stored as glycogen (*f*). A defect in any of these steps causes diabetes.

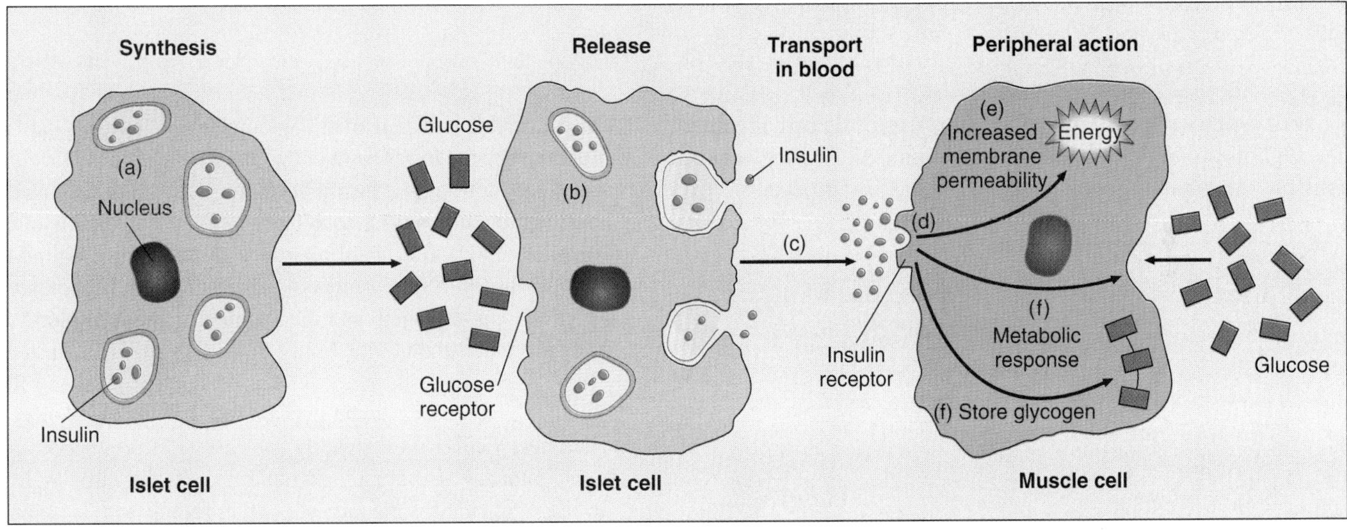

The more common form of the disorder, insulin-independent diabetes, usually begins in adulthood. In this form, the body produces insulin but is unable to use it, perhaps due to defective insulin receptors on cell surfaces or an abnormality in a protein that regulates insulin activity. Symptoms of insulin-independent diabetes mellitus include fatigue, itchy skin, blurred vision, slow healing, and poor circulation. The condition affects twice as many women as men and more blacks than whites, and it is more often seen in overweight individuals. Diet and exercise can help about 90% of these people, but the remainder must take drugs that enable the body to more effectively use insulin.

Figure 33.18 outlines the steps in glucose use at the cellular level that diabetes mellitus may alter. The complexity of the process suggests that diabetes symptoms may develop in several ways.

Hypoglycemia

Hypoglycemia is a deficient level of glucose in the blood, indicating excess insulin production. A person with this condition feels weak, sweaty, anxious, and shaky. Hypoglycemia is most likely to occur when a person with diabetes injects too much insulin or when a tumor grows in beta cells. A healthy person might experience transient hypoglycemia following very strenuous exercise.

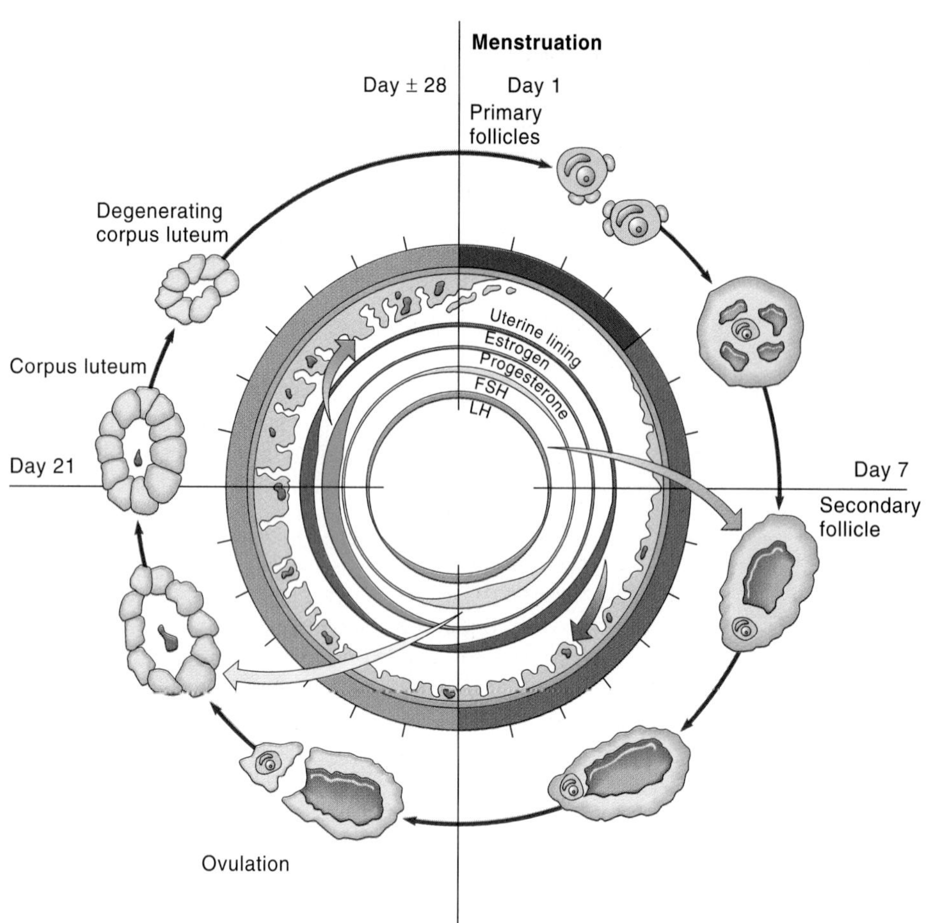

FIGURE 33.19

Three types of biological changes make up the menstrual cycle. The outer part of the illustration traces growth of the ovarian follicle. The center portion depicts the changes in thickness of the uterine lining. The inner four circles show changing levels of the hormones estrogen, progesterone, follicle-stimulating hormone (FSH), and luteinizing hormone (LH) throughout the cycle. On the first day of the cycle, the oocyte is within its follicle, the uterine lining is actively being shed, and levels of all four hormones are low. The low hormone levels signal the hypothalamus to release GnRH, which increases the anterior pituitary's output of FSH and LH. By the midpoint of the cycle, around day 14, LH level surges, which prompts the oocyte to burst out of its follicle. Meanwhile, the follicle has been producing estrogen to prepare the uterine lining for possible pregnancy. After ovulation, if conception does not occur, the ruptured follicle produces estrogen and progesterone. The uterine lining continues to build as the corpus luteum (the follicle without its oocyte) first enlarges and then degenerates. Once it degenerates, estrogen and progesterone production slow, and by the end of the cycle, the decreasing hormone levels cause the padding of blood and tissue in the uterine lining to deteriorate and exit the body as the menstrual flow.

Following a diet of frequent, small meals low in carbohydrates and high in protein can help relieve symptoms of hypoglycemia. This prevents insulin surges that lower blood sugar level.

Mastering Concepts

1. What are the three thyroid hormones, and what do they control?
2. What is the function of parathyroid hormone?
3. How do the functions of hormones secreted by the adrenal cortex and the adrenal medulla differ?
4. What are the functions of insulin, glucagon, and somatostatin?

Endocrine Control of Reproduction

The Menstrual Cycle

In a woman's menstrual cycle, the hypothalamus stimulates the pituitary to release hormones that stimulate the ovaries to produce estrogen and progesterone. Estrogen increases cell division rates in the vagina, uterus, and breasts and triggers fat buildup beneath a woman's skin. Progesterone controls secretion patterns associated with reproductive function.

Figure 33.19 tracks changes in the ovarian follicle, the uterine lining, and the levels of four hormones during the menstrual cycle. On the first day of menstrual bleeding, blood levels of estrogen and progesterone are low. The hypothalamus detects these low levels and sends gonadotropin-releasing hormone (GnRH) to the anterior pituitary, stimulating it to release large amounts of follicle-stimulating hormone (FSH) and small amounts of luteinizing hormone (LH). The increase in FSH prompts the largest follicle in the ovary to begin producing estrogen, and the ensuing increase in blood estrogen level stimulates the uterine lining to thicken in preparation for pregnancy. The rising estrogen level also lowers the level of FSH in the blood, which stimulates the hypothalamus to produce more GnRH. The rising GnRH level steadily increases LH production in the anterior pituitary. Eventually, a rapid increase occurs in LH blood levels, which causes the largest ovarian follicle to release an oocyte. This is **ovulation,** when a woman is most fertile. Ovulation occurs 14 days before the end of a monthly cycle.

After ovulation, in response to the LH surge, the follicle enlarges to form a gland, the **corpus luteum,** which produces estrogen and progesterone. The increasing levels of these hormones inhibit LH and FSH production and stimulate the buildup of the uterine lining.

Nine days after ovulation, if a fertilized ovum has implanted in the uterine lining, the developing preembryo secretes a new hormone, **human chorionic gonadotropin (hCG),** into the woman's blood. (hCG detection is the basis for pregnancy tests.) If conception does not occur, the progesterone level in the blood continues to rise, which inhibits GnRH production. As a result, estrogen and progesterone levels in the blood fall, which breaks down the thick uterine lining. This tissue passes out of the vagina as the menstrual flow. Estrogen and progesterone levels in the blood are low, and the cycle begins anew. The menstrual cycle usually begins at about age 12 and ceases around age 45.

The Testes

In the human male, hormone levels are constant. GnRH from the hypothalamus stimulates the anterior pituitary to release FSH and interstitial-cell-stimulating hormone (ICSH). FSH and ICSH travel in the bloodstream to the testes. There, FSH stimulates the early stages of sperm formation, and ICSH completes sperm production and stimulates the testes to synthesize the male steroid hormone testosterone. Testosterone stimulates development of male characteristics during embryonic existence and promotes later development of male secondary sexual characteristics, including facial hair, deepening of the voice, and increased muscle growth. Some athletes take synthetic versions of testosterone to increase muscular strength—a dangerous practice (Biology in Action 33.1).

The Prostate Gland

Along with stimulating the development of male characteristics, testosterone controls growth and activity of the **prostate gland,** a walnut-sized, two-lobed structure beneath the urinary bladder (see fig. 10.5). The prostate wraps around the first inch of the urethra, the tube that carries urine from the bladder through the penis to the outside. The prostate produces a thin, milky fluid with an alkaline pH. Myoepithelial cells within the gland eject the fluid into the urethra, where it contributes to the seminal fluid; the alkalinity neutralizes acid in the sperm's metabolic wastes. The prostate secretion also helps the sperm to swim. Because the prostate does not secrete into the bloodstream, it is technically not an endocrine gland, but an endocrine secretion—testosterone—controls it.

The location of the prostate gland often causes health problems. When the prostate enlarges, it presses on the urethra—a condition that occurs in half of all men over age 50. Symptoms of an enlarged prostate include pressure on the bladder and frequent urination. Because the enlarged prostate prevents the bladder from emptying completely, the retained urine can cause infection and inflammation, bladder stones, or kidney disease.

Medical researchers do not know what enlarges the prostate. Risk factors include a fatty diet, having undergone a vasectomy, and possibly occupational exposure to batteries or the metal cadmium. The enlargement may be benign or cancerous. Because prostate cancer is highly treatable if detected early, it is imperative that men have regular prostate examinations. Diagnostic tests include a rectal exam as well as a blood test to measure the amount of prostate-specific antigen (PSA), a cell surface protein normally on prostate cells. Elevated blood PSA levels may indicate an enlarged prostate, which may be caused by a benign or cancerous growth. An ultrasound scan may provide further information. Table 33.1 summarizes diagnostic tests and treatments for an enlarged prostate. The particular combination of approaches depends upon the individual man.

Mastering Concepts

1. Describe the hormonal fluctuations of the menstrual cycle.
2. What are the functions of testosterone?
3. What are the functions of the fluid that the prostate gland produces?

The Pineal Gland

The **pineal gland** is a small oval structure in the brain near the hypothalamus. It produces **melatonin,** a hormone that helps regulate reproduction in certain mammals by inhibiting the anterior pituitary hormones that regulate gonadal activity.

Melatonin production depends upon the pattern of light and dark an individual is exposed to. Light inhibits melatonin synthesis; darkness stimulates it. Because melatonin synthesis depends upon the absence of light, changing levels of the hormone may signal an organism to prepare for changing seasons—perhaps by growing a heavier coat or initiating mating behavior.

Experiments show that melatonin influences other endocrine glands. In rodents, melatonin injections lower the levels of thyroid hormone, melanocyte-stimulating hormone, and sex hormones.

The role of the human pineal gland is not well understood, possibly because we alter natural light-dark cycles with artificial lighting. Mood swings may be linked to abnormal melatonin secretion patterns, particularly a form of depression called seasonal affective disorder (SAD). Exposing such individuals to additional

Table 33.1 Medical Approaches to Treating an Enlarged Prostate Gland

Diagnostic Tests	Treatments
Rectal exam	Surgical removal of prostate
Blood test for prostate-specific antigen	Radiation
Ultrasound	Drugs to block testosterone's growth-stimulating effects on prostate (finasteride)
	Microwave energy delivered through a probe inserted into the urethra or rectum
	Balloon inserted into urethra inflated with liquid
	Tumor frozen with liquid nitrogen delivered by probe through skin
	Device (stent) inserted between lobes of prostate to relieve pressure on urethra

33.1 biology in action

Steroids and Athletes—An Unhealthy Combination

Canadian track star Ben Johnson flew past his competitors in the 100-meter run at the 1988 summer Olympics in Seoul (fig. 33.A). But 72 hours later, the gold medal awarded for his record-breaking time of 9.79 seconds was rescinded after tests detected traces of the drug stanozolol in his urine.

Stanozolol is one of several synthetic versions of the steroid hormone testosterone. Like testosterone, these drugs promote signs of masculinity (androgenic effect) and increase synthesis of muscle proteins (anabolic effect). Steroids have been used in the past to treat a few medical conditions—anemia and breast cancer among them—and athletes use them to build muscle tissue.

Steroid users may improve performance and appearance in the short term, but in the long run they may suffer. Steroids hasten adulthood, stunting height and causing early hair loss, and may cause males to develop breast tissue and females to develop a deepened voice, hairiness, and a male physique. Steroids can damage the kidney, liver, and heart and cause atherosclerosis. In males, the body mistakes synthetic steroids for the natural hormone and lowers its own production of testosterone, causing infertility later.

Steroid use began in Nazi Germany, where Hitler used the drugs to fashion his "super race." Ironically, steroids were used shortly thereafter to build up the emaciated bodies of concentration camp survivors. In the 1950s, Soviet athletes began using steroids in the Olympics, and a decade later, U.S. athletes did so too. In 1976, the International Olympic Committee banned the use of steroids and mandated urine tests to detect them. Finally, the Anabolic Steroids Act of 1990 made possession with intent to distribute the drugs a federal offense.

Ben Johnson was caught in his tracks by a urine test, refined by 1988 to detect parts-per-billion traces of synthetic steroids even weeks after ingestion. Although Johnson at first claimed the stanozolol in his urine was the result of a spiked drink of an approved antiinflammatory drug used for his ankle, a test of his natural testosterone level showed he had only 15% of the normal amount of the hormone—a sure sign that this athlete had been taking steroids for a long time.

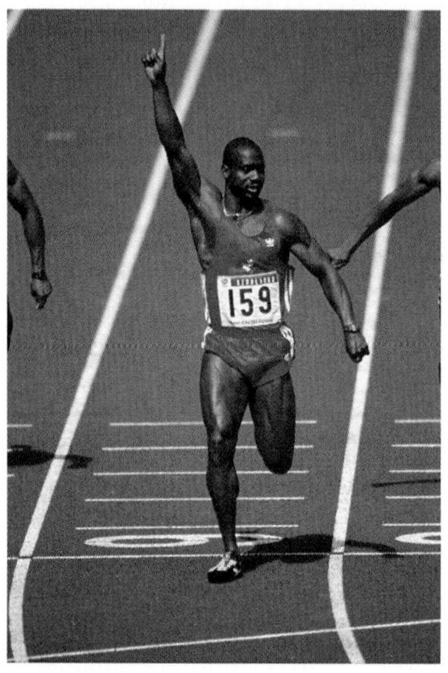

FIGURE 33.A

Canadian track star Ben Johnson ran away with the gold medal in the 100-meter race at the 1988 summer Olympics—then had to return the award when traces of a steroid drug showed up in his urine.

hours of daylight can elevate their moods. Even though we know very little about the role of melatonin in the human body, this hormone has been "hyped" as a cure for many conditions. Bioethics Connections 33.1 discusses the controversy surrounding self-medication with melatonin.

Mastering Concepts

1. Which gland secretes melatonin?
2. How do darkness and light affect melatonin secretion?
3. What medical conditions might melatonin treat?

Prostaglandins

Prostaglandins are little-understood biochemicals that exert profound effects on various human tissues and organs, often by altering hormone levels. Prostaglandins are lipids that appear locally and transiently when cells are disturbed. When a cell membrane is disrupted by an injury, binding of a hormone, or an immune system attack, the damaged membrane releases certain fatty acids into the cytoplasm. These fatty acids provide the substrate for enzymes that catalyze prostaglandin formation.

Prostaglandins do not fit the classical definition of a hormone for several reasons. Prostaglandins function at their synthesis site, rather than traveling in the bloodstream to a target tissue, as hormones do. In addition, different types of stimuli initiate prostaglandin secretion and hormone release. Also, prostaglandins are in every mammalian tissue type thus far examined, whereas hormones are produced in significant amounts only by specialized endocrine system cells. Nonetheless, prostaglandins, like hormones, are chemical regulators.

Prostaglandins have a variety of functions: they affect smooth muscle contraction, secretion, blood flow, reproduction, blood clotting, respiration, transmission of nerve impulses, fat metabolism, the immune response, and inflammation. Consider the varied effects of prostaglandins on smooth muscle. Two types stimulate the smooth muscle in blood vessel walls to adjust blood flow

33.1 bioethics connections

Misrepresenting Melatonin

If something sounds too good to be true, it usually is. So it may be with melatonin, a hormone that the pineal gland of vertebrates produces. Although widely trumpeted in the press and available on the shelves of health food stores, melatonin has never been tested for safety or efficacy to treat anything in humans. It is sold as a food supplement, and not a drug, and therefore does not require Food and Drug Administration approval. Yet people have claimed melatonin can cure Alzheimer disease, diabetes, epilepsy, schizophrenia, autism, and many other conditions.

Researchers discovered a pineal hormone in the early 1950s, but investigations stalled until work on circadian rhythms sparked interest in melatonin in the 1970s. The pineal synthesizes melatonin during periods of darkness, but not in the light. As a result, vertebrates use melatonin levels to distinguish day from night and to tell the season by determining relative lengths of day and night. Melatonin levels cue vertebrates to molt, mate, migrate, and hibernate. We know little about the role of melatonin in the human body. We do know that melatonin levels decrease with age, and from this observation arose the hypothesis that taking extra melatonin might prevent diseases associated with aging. One study did find that melatonin helps older people with insomnia to sleep, but less sleep may be a normal part of aging.

A book published in 1995 catapulted melatonin into the headlines, calling the hormone an "amazing natural pill that can reset the body's aging clock and restore youth." The book claimed that melatonin "rejuvenates sexual organs" (although in some rodent studies melatonin shrunk gonads) and "is an extremely useful tool in helping to prevent heart disease" (although clinical trials in human subjects have never been conducted).

The book, many endocrinology researchers claim, is an excellent example of misunderstanding the scientific method. It repeatedly presents hypotheses as established fact, extrapolates results on nonhuman animals to humans, ignores data that contradicts cure-all claims, and presents anecdotes as valid scientific evidence.

Melatonin is a hormone, and hormones are powerful biochemicals, as the opener to chapter 2 reveals for synthetic estrogens. Because something is sold in a "health" store, or touted in headlines and on talk shows, does not mean that it is safe or effective. Consumers can protect themselves from taking an unproven "food supplement" by asking for, and evaluating, clinical trial data that proves safety and efficacy. There are no miracle cures.

to regions of the body. Other prostaglandins affect smooth muscle in the walls of airways within the lungs, altering the ease of oxygen delivery. Still other prostaglandins contract uterine muscles. Depending on the circumstance, these uterine contractions may assist sperm in their journey to the oocyte, propel a fetus out of the uterus, or cause menstrual cramps.

Learning how different prostaglandins exert specific effects has explained the mechanisms of certain drugs and led to some interesting new medical applications. For example, some prostaglandins promote inflammation, the immune system's attempt to fight infection at the site of a wound by sending in fluid and white blood cells. Aspirin relieves the pain of inflammation by inactivating an enzyme required to synthesize several prostaglandins. Aspirin also blocks synthesis of prostaglandins that stimulate pain receptors in the nervous system.

Prostaglandins' functions in blood clotting and controlling blood vessel diameter may also play a part in heart attacks. Daily doses of aspirin may reduce heart attack risk by altering prostaglandin activity. Prostaglandins that dilate blood vessels may treat hypertension (high blood pressure), which results from constricted blood vessels.

Mastering Concepts

1. How are prostaglandins similar to hormones, yet different from them?
2. What are some functions of prostaglandins?
3. How does prostaglandin function explain aspirin's effects?

Neuroendocrine-Immune Interactions

The nervous and endocrine systems also interact with the immune system, enabling an animal to survive infection or injury. Inflammation illustrates the connection between the endocrine and immune systems. The immune system rushes white blood cells and biochemicals to infected or injured areas, causing redness and swelling. Although inflammation fights infection, it can damage the body if it is excessive. To prevent this, the endocrine system tempers inflammation by increasing pituitary and adrenal gland secretions, perhaps through nervous system action.

We are beginning to discover that all organ systems exert such checks and balances on each other, ensuring that the animal functions in a coordinated, integrated manner. Tables 33.2 and 33.3 summarize the actions of the endocrine system.

SUMMARY

1. The **endocrine system** includes several glands and scattered cells and the hormones they secrete into the bloodstream. A hormone exerts a specific physiological effect on **target cells,** which have receptors for it. **Exocrine glands** deliver substances into ducts, which open onto body surfaces.

2. The nervous and endocrine systems interact to maintain homeostasis. The nervous system acts faster and more locally than the endocrine system. **Neurosecretory cells,** which are neurons that secrete hormones, are a physical and evolutionary link between the two systems.

Table 33.2 Functions of the Endocrine Glands

Gland	General Functions
Pituitary	Hormones of the anterior lobe increase growth and development of all tissues (growth hormone); concentrate pigment granules in melanocytes (melanocyte-stimulating hormone); stimulate secretions from the thyroid (thyroid-stimulating hormone), the adrenal cortex (adrenocorticotropic hormone), and the gonads (follicle-stimulating hormone, luteinizing hormone, and interstitial-cell-stimulating hormone); and initiate milk production (prolactin) Hormones manufactured in the hypothalamus but released from the posterior pituitary raise blood pressure (antidiuretic hormone) and contract uterine muscles and milk-secreting cells (oxytocin)
Thyroid	Regulates metabolic rate and body temperature (thyroxine and triiodothyronine); lowers blood calcium level (calcitonin)
Parathyroids	Raises blood calcium level (parathyroid hormone)
Adrenal cortex	Mobilizes energy reserves while maintaining blood volume and composition (mineralocorticoids and glucocorticoids)
Adrenal medulla	Prepares body to cope with emergency situations (adrenaline and noradrenaline)
Pineal gland	Regulates effects of light-dark cycle on other glands (melatonin)
Pancreas	Regulates body's use of nutrients (insulin, glucagon, somatostatin)
Ovary	Stimulates maturation of oocyte and ovulation (follicle-stimulating hormone, luteinizing hormone, estrogen, progesterone)
Testis	Stimulates sperm maturation, secondary sexual characteristics (follicle-stimulating hormone, interstitial-cell-stimulating hormone, testosterone)

3. Diverse species have hormonelike regulation, and many use the signaling molecule **cAMP.**

4. The gastrointestinal tract secretes several hormones that regulate digestion. In mammals, the atrium of the heart secretes **atrial natriuretic hormone,** which regulates blood pressure, blood volume, and the excretion of sodium ions, potassium ions, and water. **Pheromones** are hormonelike biochemical signals between individuals.

5. Peptide hormones are water soluble and bind to the surface receptors of target cells, stimulating conversion of ATP to cAMP on the inner face of the cell membrane. The cAMP then triggers a specific metabolic effect. The lipid-soluble **steroid hormones** cross target cell membranes, bind receptors in the cytoplasm or nucleus, and activate genes to direct synthesis of proteins that provide the cell's response.

6. In a negative feedback loop, excess of a hormone or the product of a hormone-induced response suppresses further synthesis or release of that hormone until levels return to normal. In a **positive feedback loop,** the hormone causes an event that increases its production. Ions or nutrient levels near the endocrine cells, input from the nervous system, and other hormones control feedback loops.

7. Simpler invertebrates may have only neurosecretory cells. More complex invertebrates have interacting hormones. Changing levels of **juvenile** and **molting hormone** control insect metamorphosis. The same hormones may function differently in different vertebrate species. Reproductive cycles are hormonally controlled and are diverse.

8. The endocrine system functions in a hierarchy. The **hypothalamus** produces **releasing hormones,** which travel in

History of an Illness—Diabetes Mellitus

continued from page 660.

In May 1921, Banting and Best's experiments began. They removed the pancreas from one dog and tied off the duct from the second dog's pancreas that releases digestive enzymes. Within a week the first dog lay dying; the second couldn't digest very well. Banting and Best removed the dried-up pancreas from the second dog, mashed it up, and extracted fluid from it. They then injected the fluid into the dog dying from diabetes. An hour later, the diabetic dog was walking about wagging its tail!

But the treated dog died the next day. Further experiments showed that dogs require daily insulin to prevent diabetes. Banting soon thought of a better source of insulin—calf fetuses removed from cows at slaughter and discarded. Fetal pancreases produce insulin but not digestive enzymes, which makes the insulin more effective.

The next step was to give insulin to people. The first volunteer was a physician-friend of Dr. Banting who was dying of diabetes—the treatment saved his life. Then a 14-year-old boy was saved. Banting conducted a large-scale trial on residents of an institution for injured and ill soldiers, again achieving astounding success. Soon, people with diabetes were flocking to Toronto. By mid-1922, insulin treatment became widely available.

Biochemists produced purer insulin from cows and pigs, but some people were allergic to the nonhuman hormones. In 1982, geneticists stitched the human insulin gene into *E. coli,* and soon pure and plentiful human insulin became available thanks to recombinant DNA technology.

Today, people with diabetes have choices—human insulin, implanted insulin pumps, and implants of insulin-producing cell clusters. Despite these advances, we still cannot exactly duplicate the function of the pancreas in secreting its hormone product.

Table 33.3 Summary of Major Hormone Functions in the Human Endocrine System

Hormone	Site of Synthesis	Target	Effects
Releasing hormones	Hypothalamus	Anterior pituitary	Stimulate release of hormones from anterior pituitary
Growth hormone (GH)	Anterior pituitary	All cells	Increases cell division, protein synthesis
Thyroid-stimulating hormone (TSH)	Anterior pituitary	Thyroid gland	Stimulates secretion of thyroxine and triiodothyronine
Adrenocorticotropic hormone (ACTH)	Anterior pituitary	Adrenal cortex	Stimulates secretion of glucocorticoids and mineralocorticoids
Prolactin	Anterior pituitary	Cells in mammary glands	Stimulates milk secretion
Melanocyte-stimulating hormone (MSH)	Anterior pituitary	Skin	Increases skin pigmentation
Follicle-stimulating hormone	Anterior pituitary	Ovaries	Stimulates follicle development, oocyte maturation, release of estrogen
Luteinizing or interstitial-cell-stimulating hormone (LH or ICSH)	Anterior pituitary	Ovaries, testes	Promotes ovulation; stimulates late sperm cell development, testosterone synthesis
Antidiuretic hormone (ADH)	Hypothalamus (released from posterior pituitary)	Kidneys, smooth muscle cells of blood vessels	Helps maintain composition of body fluids
Oxytocin	Hypothalamus (released from posterior pituitary)	Uterus, cells of mammary glands	Stimulates muscle contraction
Thyroxine, triiodothyronine	Thyroid gland	All cells	Increase metabolic rate
Calcitonin	Thyroid gland	Bone	Increases rate of calcium deposition in bone
Parathyroid hormone	Parathyroid glands	Bone, digestive organs	Releases calcium from bone, increases calcium absorption in digestive tract and kidneys
Glucocorticoids	Adrenal cortex	All cells	Increase glucose levels in blood and brain
Mineralocorticoids	Adrenal cortex	Kidneys	Maintain blood volume and electrolyte balance
Catecholamines	Adrenal medulla	Blood vessels	Raise blood pressure, constrict blood vessels, slow digestion
Melatonin	Pineal gland	Other endocrine glands	Regulates effects of light-dark cycle on other glands
Insulin	Pancreas	All cells	Increases cellular glucose uptake and conversion of glucose to glycogen
Glucagon	Pancreas	All cells	Stimulates breakdown of glycogen to glucose
Somatostatin	Pancreas	Small intestine	Regulates nutrient absorption rate
Progesterone	Ovaries	Uterine lining	Controls monthly secretion patterns
Estrogen	Ovaries	Uterine lining	Increases rate of mitosis
Human chorionic gonadotropin (hCG)	Fertilized ovum	Uterine lining	Interrupts menstrual cycle and maintains pregnancy
Testosterone	Testes	Skin, muscles	Maintains secondary sexual characteristics

neurosecretory cells to the anterior lobe of the **pituitary gland.** There they stimulate release of anterior pituitary hormones: **growth hormone,** which stimulates cell division, protein synthesis, and growth in all cells; **thyroid-stimulating hormone,** which prompts the **thyroid gland** to release **thyroxine** and **triiodothyronine,** which regulate metabolism; **adrenocorticotropic hormone,** which stimulates the **adrenal cortex** to release hormones that enable the body to cope with a serious threat; **prolactin,** which stimulates milk production; and the sex hormones, which control sex cell development. The hypothalamus also manufactures two hormones that are stored in and released from the posterior lobe of the pituitary gland—**antidiuretic hormone,** which regulates body fluid composition, and **oxytocin,** which contracts the uterus and milk ducts. In many vertebrate species, the region between the anterior and posterior lobes of the pituitary secretes **melanocyte-stimulating hormone,** which colors skin.

9. The **parathyroid glands** secrete **parathyroid hormone,** which increases blood calcium level by releasing calcium from bone and increasing its absorption in the gastrointestinal tract and kidneys. Another hormone that affects calcium metabolism is **calcitonin,** which the thyroid gland secretes. Calcitonin lowers the level of calcium in the blood by increasing the rate at which it enters bone.

10. The **adrenal cortex** secretes **mineralocorticoids** and **glucocorticoids,** which mobilize energy reserves during times of stress and maintain blood volume and blood composition. The **adrenal medulla** secretes **epinephrine** and **norepinephrine,** which ready the body to cope with an emergency.

11. The endocrine portion of the **pancreas** secretes **insulin,** which stimulates cells to take up glucose; **glucagon,** which breaks down glycogen to glucose; and **somatostatin,** which regulates insulin and glucagon production. Lack of insulin or the inability to use it causes **diabetes mellitus.**

12. The female gonads, the ovaries, secrete estrogen and progesterone, hormones that stimulate development of female sexual characteristics and together with GnRH, FSH, and LH control the menstrual cycle. The male gonads, the testes, secrete **testosterone,** which stimulates the development of secondary sexual characteristics and controls **prostate gland** function.

13. The **pineal gland** may regulate the responses of other glands to light-dark cycles through its hormone, **melatonin.**

14. Prostaglandins are lipids that form enzymatically when disturbed cell membranes release fatty acids. Prostaglandins function at the site where they are released, and the different types exert different effects on the body.

TO REVIEW . . .

1. Why doesn't a hormone affect all body cells?

2. How do the effects of peptide hormones and steroid hormones differ?

3. How do hormones regulate their own levels in the body? Describe the regulation of a specific hormone.

4. What effects might a tumor on the anterior lobe of the pituitary gland have? the posterior lobe of the pituitary gland? the hypothalamus?

5. List two hormones that exert opposite effects.

6. Why would someone with an insufficient supply of thyroid hormones feel tired all the time?

7. Describe how insulin and glucagon interact to regulate blood glucose levels.

8. Identify the gland that secretes each of the following hormones:

- **a.** melatonin
- **b.** oxytocin
- **c.** calcitonin
- **d.** aldosterone
- **e.** follicle-stimulating hormone
- **f.** growth hormone
- **g.** epinephrine

9. Which glands are each of the following disorders associated with?

- **a.** cretinism
- **b.** acromegaly
- **c.** osteoporosis
- **d.** diabetes mellitus
- **e.** Addison's disease

10. A queen honeybee secretes a substance from a gland in her jaw that inhibits the development of ovaries in worker bees. Is this substance most likely a hormone, prostaglandin, or pheromone? Cite a reason for your answer.

11. Since the government deemed possession of anabolic steroids a federal offense in 1990, some athletes have sought to ingest other biochemicals to build muscle. Some have used growth hormone. What might the side effects of this action be?

TO THINK ABOUT . . .

1. Imagine you are a researcher who has just found a small glandlike structure in a human. How would you determine whether it is an endocrine gland? If it is, how might you identify the gland's role in the endocrine system?

2. Is it dangerous for a person to attribute nervousness to hypoglycemia without a physician's confirmation? Why or why not?

3. Do you think GH should be given to healthy, but very short, children so that they can be more successful athletes?

4. Why might an insulin infusion pump inserted to treat diabetes be more effective than one or two daily injections of insulin?

5. Many endocrine glands contain cells that produce cytokines, which are immune system biochemicals. Similarly, some white blood cells, which are components of the immune system, can manufacture hormones. What is the significance of this overlap in function?

6. Should symptoms associated with the menstrual cycle or menopause be considered signs of illness?

7. Do you think that melatonin should be available without a prescription and without FDA approval, or should it be regulated as a drug? Cite a reason for your answer.

TO LEARN MORE . . .

References and Resources*

Brzezinski, Amnon. January 16, 1997. Melatonin in humans. *The New England Journal of Medicine,* vol. 336. Melatonin affects the sleep-wake cycle in humans.

Eisenbarth, George S., and Mark Stegall. September 19, 1996. Islet and pancreatic transplantation—autoimmunity and alloimmunity. *The New England Journal of Medicine,* vol. 335. Getting the body to accept pancreatic islet implants is difficult.

Finnell, Rebecca B. December 1995. A collection of articles on human milk. *Natural History.* Breast feeding has biological benefits.

Gura, Trisha. February 7, 1997. Obesity sheds its secrets. *Science,* vol. 275. Several hormones interact to control body weight.

Oelkers, Wolfgang. October 17, 1996. Adrenal insufficiency. *The New England Journal of Medicine,* vol. 335. Too little of an adrenal hormone has multiple effects.

Reichlin, Seymour. October 21, 1993. Neuroendocrine-immune interactions. *The New England Journal of Medicine,* vol. 329. The nervous, endocrine, and immune systems are intricately linked, which helps maintain homeostasis.

**Life*'s website features frequently updated references.

34 The Musculoskeletal System

Astronaut Shannon Lucid was able to walk immediately after returning to earth after 188 days in space—which only 30% of astronauts can do. The 53-year-old Lucid minimized the decrease in muscle mass and depletion of bone mineral that commonly occurs in microgravity conditions by logging 400 hours on a treadmill and stationary bicycle.

chapter outline

The Musculoskeletal System and Movement 682

Responding to the Environment 682
Musculoskeletal Plasticity

Skeletal System Diversity 683
Hydrostatic Skeletons
Exoskeletons
Endoskeletons

Vertebrate Musculoskeletal Systems—A Closer Look 684
Functions
Structure

Cells and Tissues of the Vertebrate Skeleton 687
Cartilage
Bone

Biological Movement 689
Muscle Cell Types
Amoeboid Movement
Invertebrate Muscles
Vertebrate Skeletal Muscle

Muscle and Bone Working Together 700
Joints—Where Bone Meets Bone
Lever Systems
Muscle Tone
Effects of Exercise on Muscle

Polio—Then and Now

During the summers and falls of the 1940s and early 1950s, a viral infection, acute paralytic poliomyelitis, suddenly struck many children, sometimes progressing from fever to paralysis in just days. Polio could kill, leave lasting disability, or vanish without any effects. When the fever first struck, the outcome was unknown.

The polio virus can survive for days in food and water, spreading easily. The virus lives in the lining of the throat or the small intestine, where it reproduces and travels to the lymph nodes and tonsils. It may enter the blood. Polio only becomes more than a very mild illness if the virus enters the central nervous system and concentrates in spinal cord cells that control muscular contraction. As parts of the spinal cord deteriorate, muscle function disappears.

Vaccines have eliminated polio from most developed nations, and the hope is to eradicate it globally by the year 2000. But before 1956, polio struck a reported 25,000 to 50,000 children and young people a year in the United States.

Polio survivors recall that treatment was as agonizing as the illness. Because of the infectious nature of polio, children were quarantined. Patients had their limbs splinted or were wrapped in body casts, and orthopedic surgery and braces attempted to restore muscle function. An early type of respirator called an iron lung enabled patients to breathe when their respiratory muscles could not work.

People who survived severe cases of polio learned how to live with their persisting disabilities by using different muscle groups to compensate for permanently weakened ones. The memories of their childhood illness may have faded, but the fear of relapse never left. Then, decades later, some survivors began to experience new muscular symptoms. Was polio back?

to be continued . . .

The Musculoskeletal System and Movement

The dancers line up nervously, each perfectly still, waiting for the choreographer's signal. At the cue, they suddenly come alive, jumping and turning in unison, their graceful movements precisely timed to the music (fig. 34.1*a*). Their ability to move depends upon connected bones and muscles, just as the leap of a gray tree frog does (fig. 34.1*b*).

Muscle attached to bone is one type of musculoskeletal system. An earthworm's skeleton, for example, isn't bone but water-filled compartments. Together with perpendicular muscles in its body wall, this skeleton helps the worm cling tenaciously to its burrow.

Responding to the Environment

An animal's ability to react to stimuli is important for homeostasis. Sensing a stimulus and responding through a nerve impulse or hormone, however, may be futile unless the organism changes its activity in some way. The animal may use effectors to move away from or toward the stimulus, move food along its digestive tract, or perform any other function. Effectors are cells, tissues, or organs that respond to nerve or hormone stimulation with an action. Usually, effectors are muscles.

The first organisms were probably able to move by a mechanism similar to the one that powers our muscles today—protein filaments that slide past each other, shortening and thereby contracting the cells containing them. In multicellular animals, contractile tissues are organized into tissues or organs called muscles.

Some muscles operate independently of skeletal structures. A vertebrate's digestive tract muscles, for example, mix and propel food without skeletal support. When large animals move, however, muscles pull against rigid bones to form lever systems, which increase the strength and efficiency of movement. This close association of skeletal and muscular systems was adaptive as animals ventured onto land and had to move against gravity without the buoyancy of water. The skeletal and muscular systems have evolved as closely allied, interacting organ systems.

Musculoskeletal Plasticity

Muscles and skeletal structures help maintain homeostasis because they are dynamic—that is, they can change. Muscle change is evident at the individual as well as at the species level.

A person beginning an intense weight-lifting program will gradually note that muscles enlarge as they respond to increased use. Bones change too and become denser with use. In the opposite situation, an athlete restricted to bed for an extended period of time may notice muscular deterioration since certain muscles are no longer used. Similarly, astronauts lose bone density if they are in a weightless environment for a long time, because their musculoskeletal systems don't have to work as hard in the absence of gravitational force.

The musculoskeletal system's contribution to an individual's homeostasis also varies over time. A vertebrate's skeleton is under perpetual biological renovation, as bones break down and rebuild. This is how a bone retains its shape yet enlarges as an individual grows. Similarly, arthropods such as insects and certain shellfish periodically shed the skeletons they wear on the outside and grow new ones, which accommodate their enlarging bodies.

At the species level, the skeletal system's capacity to change is striking, especially among vertebrates. The skeletons of the hummingbird and the hippo, the cheetah and the sloth, and the salmon and the whale are markedly different, yet they are composed of the same types of cells and have similar spatial patterns. The skeleton, in turn, influences the distribution of the musculature. The characteristics of each species' musculoskeletal system reflect its common ancestry and changes reflecting natural selection.

(a)

(b)

FIGURE 34.1

Muscles and bones make movement possible.(*a*) The graceful movements of these dancers are possible because muscles contract against bones in a lever system. (*b*) The leap of this gray tree frog also depends on the musculoskeletal system.

Mastering Concepts

1. Why is it important for an organism to respond to environmental stimuli?
2. What are effector structures?
3. How do bones and muscles change in response to changing conditions?

Skeletal System Diversity

A skeleton is a supporting structure or framework. It gives an animal's body shape, protects internal organs, and provides a firm surface for muscles to pull against, permitting movement. Several types of skeletons perform these functions for different species.

Hydrostatic Skeletons

The simplest type of skeleton is a **hydrostatic skeleton** ("hydro" means water), which consists of liquid within a layer of flexible tissue. Hydrostatic skeletons are found in jellyfishes, hydra, squid, sea anemones, slugs, and certain worms, including the familiar earthworm. The tension of the constrained fluid supports and helps to determine shape, as it does in a water-filled balloon.

To see how a hydrostatic skeleton permits movement, consider earthworm locomotion. The earthworm has two layers of muscle in the body wall (fig. 34.2). One layer encircles the body and the other runs lengthwise. When the circular muscles contract against the fluid in the body cavity, the worm lengthens, just as a water-filled balloon would if squeezed at one end. The worm then sticks its bristles (setae) into the ground as an anchor. When the longitudinal muscles contract, the fluid shifts again and the animal shortens, pulling forward.

Exoskeletons

A more complex skeleton is a **braced framework,** which consists of solid structural components strong enough to resist pulling forces. Muscles can attach to the surfaces of the framework, facilitating movement.

An **exoskeleton** ("exo" means outside) is a braced framework that protects an organism from the outside, much like a suit of armor. Many groups of invertebrates have exoskeletons, including arthropods (such as lobsters and insects) and mollusks (such as snails). Figure 34.3 shows the jointed exoskeleton of a flea, which consists of the complex carbohydrate chitin, and the exoskeleton of a sponge, composed of spiky projections called spicules. A store-bought natural sponge is actually the animal's exoskeleton.

Clams and snails use readily available minerals to produce hard, calcium-containing shells. These rigid shells and spikelike structures thicken as the organism grows, remaining with it for life. Fossil evidence indicates that exoskeletons became prevalent rather suddenly some 570 million years ago.

The jointed suits of armor of a cockroach and a lobster, like the flea's, are composed of chitin. These exoskeletons, however, have a shortcoming in common with armor—the animal may outgrow them. Animals must periodically shed, or molt, the exoskeleton and grow a slightly larger new one. Until the new exoskeleton has formed and hardened, the animal is in greater danger of being eaten by predators. Seafood lovers may be familiar with the results of molting: A soft-shelled crab has just begun to harden its exoskeleton, and a hard-shelled crab has completed the process.

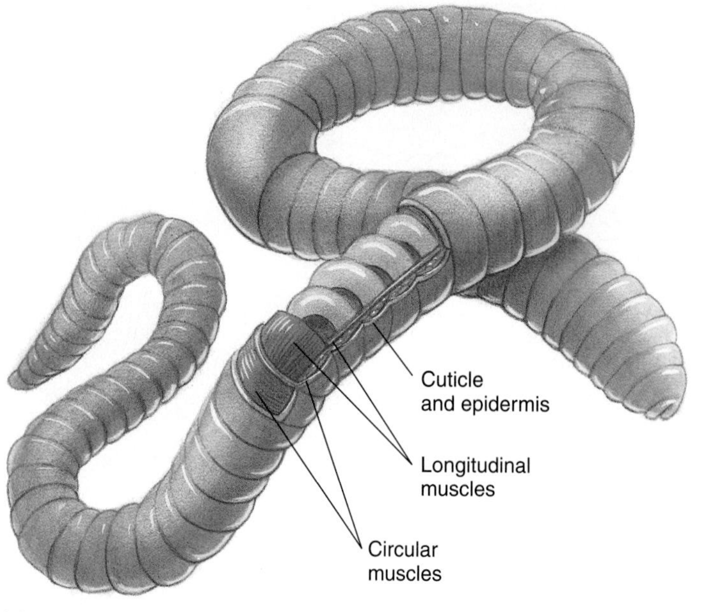

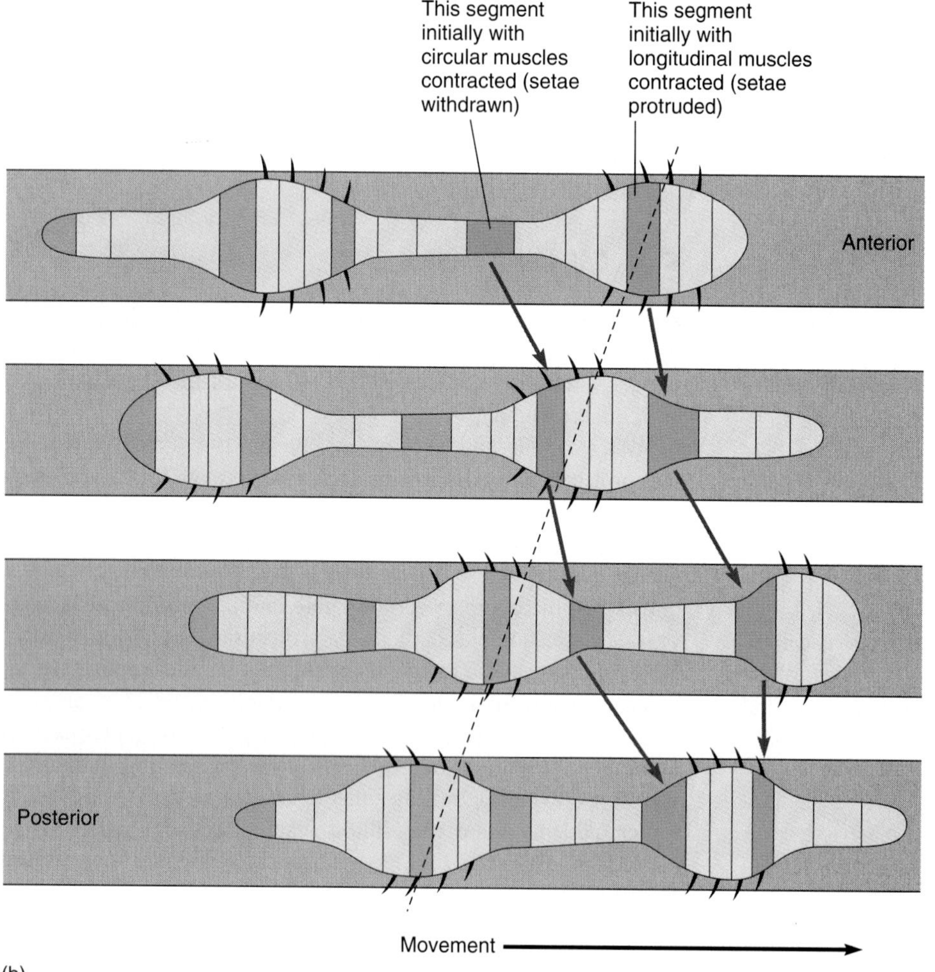

FIGURE 34.2

Earthworm locomotion. (*a*) The earthworm has a hydrostatic skeleton. (*b*) Coordinated contraction of the longitudinal and circular muscles moves the worm.

Endoskeletons

An **endoskeleton** ("endo" means inner) is an internal braced framework and is found in vertebrates and echinoderms, such as starfish. An endoskeleton offers several advantages over an exo-skeleton—it does not restrain growth and, because it consumes less of an organism's total body mass, it increases mobility.

Some fishes, including sharks, have endoskeletons made of cartilage. The endoskeletons of most vertebrates, including humans, are composed primarily of bone.

Mastering Concepts

1. Describe and name the simplest type of skeleton.
2. How does a braced framework skeleton differ from a hydrostatic skeleton?
3. What are two ways that endoskeletons and exoskeletons differ?

Vertebrate Musculoskeletal Systems—A Closer Look

Functions

The vertebrate endoskeleton has several functions that promote homeostasis:

- *Support*—The skeleton is a framework that supports an animal's body against gravity. To a large extent it provides the body's shape.
- *Lever system for movement*—The vertebrate skeleton is a system of muscle-operated levers. Typically, the two ends of a skeletal (voluntary) muscle attach to different bones that connect in a structure called a **joint.** When the muscle contracts, one bone is pulled towards the other.
- *Protection of internal structures*—Some bones form cages protecting soft internal organs. The backbone surrounds and shields the spinal cord, the skull protects the brain, and ribs protect the heart and lungs.
- *Production of blood cells*—Many bones, such as the long bones of the human arm and leg, contain and protect red marrow, a tissue where red blood cells, white blood cells, and platelets are produced.
- *Mineral storage*—The skeleton stores calcium and phosphorus, minerals required for a number of important activities. The minerals in bone are continually withdrawn for use elsewhere in the body but are later replaced.

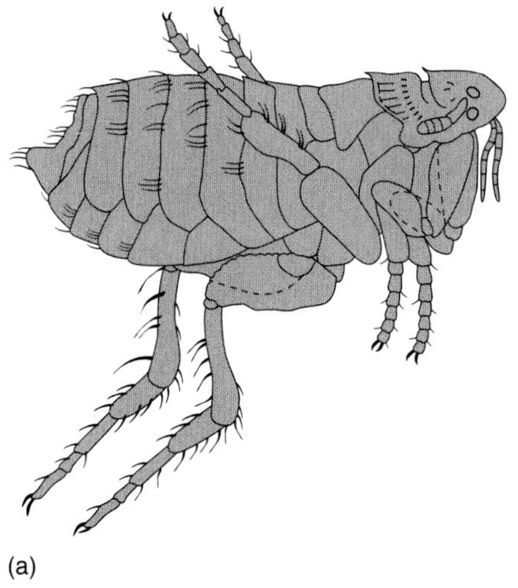

(a)

(b)

FIGURE 34.3

Exoskeletons. (*a*) A flea has a jointed exoskeleton. (*b*) A sponge's exoskeleton may consist of the protein collagen, silica, or calcium carbonate.

About 99% of the calcium in the human body is stored in the bones. Some of the remaining 1% is carried in blood plasma. Calcium is vital for the activity of some enzymes, muscle contraction, blood clotting, cell adhesion, and cell membrane permeability. To maintain the blood concentrations of calcium necessary to these body functions, the body constantly shuttles calcium between blood and bone.

Structure

The bones of the vertebrate skeleton are grouped into two parts. The **axial skeleton,** so named because it is located in the longitudinal central axis of the body, consists of the skull, vertebral column, ribs, and sternum (breastbone). The **appendicular skeleton** consists of the limbs and bones that support them; the clavicles (collarbones) and scapulae (shoulder blades) form the **pectoral girdle,** which supports the forelimbs, and the **pelvic girdle** supports the hind limbs. Both parts of the skeleton effectively absorb the tremendous shocks generated in locomotion. Figure 34.4 depicts the human axial and appendicular skeletons. The number of bones differs among vertebrate species, but the organization is similar.

The Human Skeleton

The human axial skeleton shields soft body parts. The skull protects the brain and many of the sense organs. It consists of hard, dense bones that fit like puzzle pieces at immovable boundaries called sutures (fig. 34.5). Sutures also help dissipate the shock of a blow to the head. Inside the skull, nooks and crannies fit snugly around the brain and eyes. All the head bones are attached with immovable joints except for the lower jaw and the middle ear. These bones must move to enable chewing, speech, and hearing. Spaces within the bones of the head, called sinuses, help lighten the skull and create a resonating voice chamber. The foramen magnum, a hole at the base of the skull, allows the nervous tissue of the brain to join the spinal cord.

The **vertebral column** protects the spinal cord down to the lower back (fig. 34.6). The position of the vertebral column and its flexible, multipart construction make locomotion possible. A human vertebral column consists of 33 vertebrae. The 7 **cervical vertebrae** in the neck are the smallest and allow the widest range of motion so that we can turn our heads in many directions. In the upper back are 12 heavier **thoracic vertebrae.** They are specialized with indentations on each side that help anchor the ribs in place. The 5 **lumbar vertebrae** in the small of the back are the largest of the spinal bones, presumably because they support most of the body's weight. The 5 fused pelvic vertebrae make up the **sacrum,** which forms a wedge between the two hipbones. The final 4 vertebrae fuse to form the **coccyx,** or tailbone, which has no known function in humans.

The **rib cage** protects the heart and lungs. Humans have 10 pairs of ribs attached to the sternum, or breastbone, and two additional, unattached pairs that "float." All the ribs also attach to the vertebral column. The flexibility of the cartilage between the ribs and other bones allows muscles to elevate the ribs, a movement important in breathing.

The long and short bones of the appendicular skeleton function as lever systems, which, when combined with muscles, power movement. The greater the number of bones connected by movable joints, the finer and more variable the movements an animal can make.

The lower limb bones attach to the pelvic girdle. This girdle is made up of the two hipbones, each of which is actually three

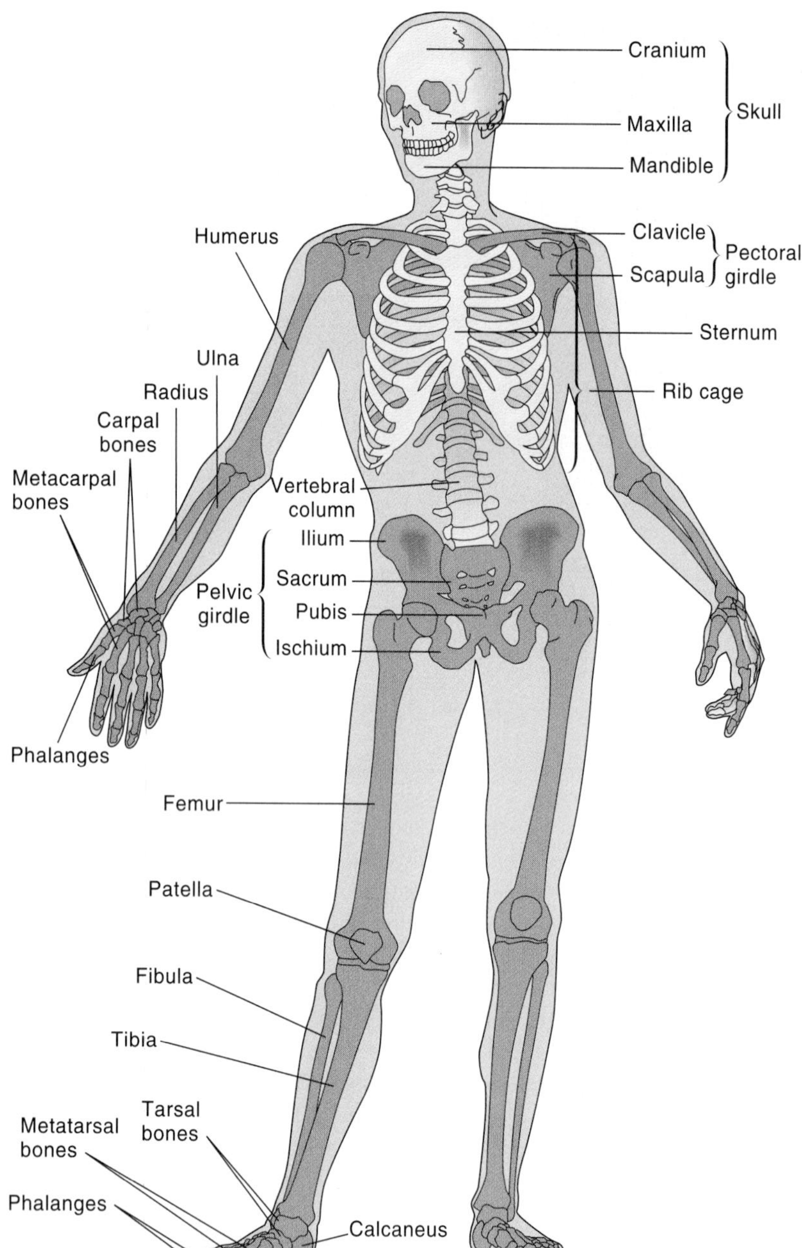

FIGURE 34.4

The human skeleton is divided into the axial skeleton (indicated in the lighter color) and the appendicular skeleton (indicated in the darker areas). The axial skeleton in humans includes the skull, vertebral column, and rib cage. The bones of the limbs and those that support them make up the appendicular skeleton.

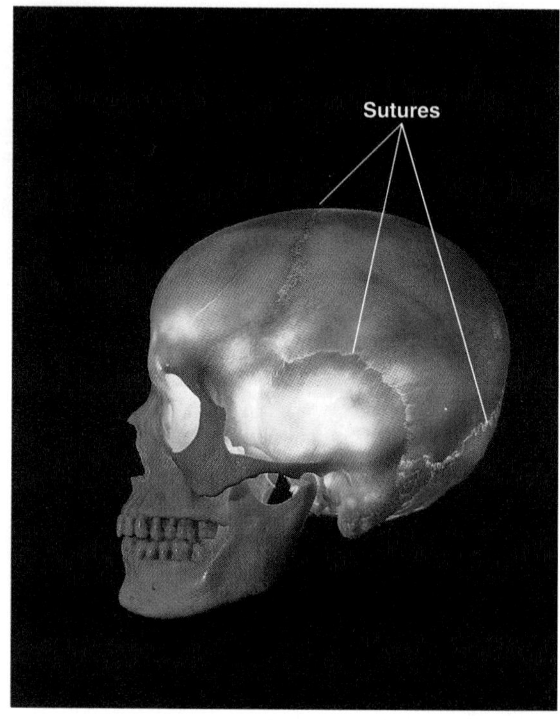

FIGURE 34.5

The skull forms a protective case for the brain and eyes. The braincase is composed of several bones held together by immovable joints called sutures, some of which are shown here.

separate but fused bones (see fig. 34.4). The hipbones join the sacrum in the rear and meet each other in front, creating a bowl-like pelvic cavity. (The term "pelvis" is Latin for "basin.") The bony pelvis protects the lower digestive organs, the bladder, and some of the reproductive structures (especially in the female). The front of the female pelvis is broader than the male's, and it is larger and has a wider bottom opening. These differences in the female pelvis are adaptations for childbirth. In the later months of pregnancy, hormones loosen the ligaments holding the pelvic bones together, allowing them greater flexibility when the newborn passes through.

Variations on the Vertebrate Skeletal Theme

The general properties and organization of the vertebrate musculoskeletal system can vary to accommodate animals that live in vastly different environments—for example, the owl and the seal (fig. 34.7).

Birds descended from dinosaurs that lived about 200 million years ago, and today they share many skeletal characteristics with reptiles. Natural selection has modified the owl's forelimb into a wing, providing the support yet light weight necessary for flight. In addition, the digits of the ancestral hand gradually fused, forming a shape better able to support flight. The owl's forelimb bones—the radius, ulna, and humerus—are thin but sturdy. Its bones, joints, and muscles connect in a way that promotes flapping movements.

The fur seal's humerus is as long as the owl's, but it is at least twice its width and denser. The seal's bone weighs more than five times that of the owl. The short, heavy forelimb of the fur seal is an adaptation to its habitat, which includes both sea and land. The seal's heavy limb bones enable the animal to forage on the ocean bottom for food without expending enormous muscular energy to fight the buoyancy of water. The bone's density provides the

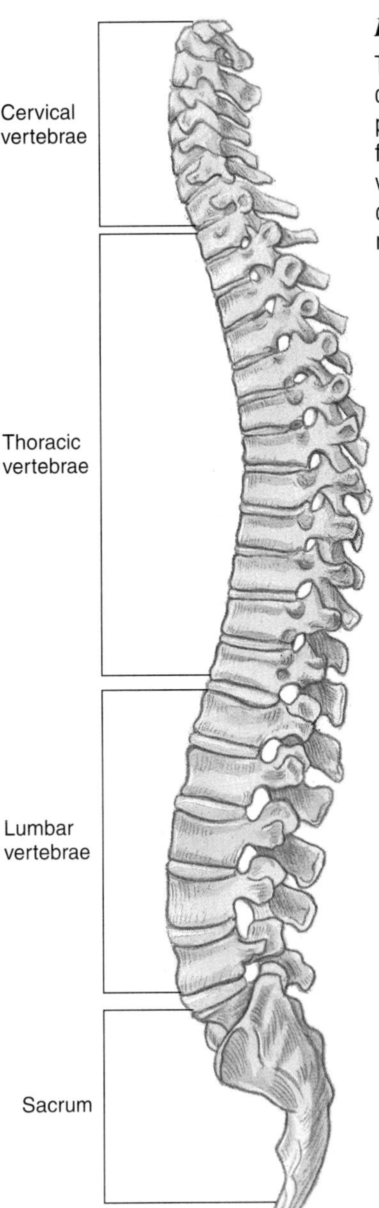

FIGURE 34.6

The human vertebral column consists of 33 vertebrae that provide a flexible protective case for the spinal cord. The vertebrae are specialized for different functions in various regions of the spinal column.

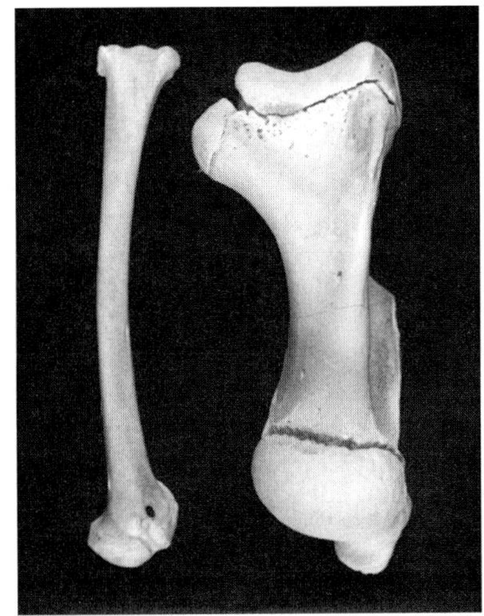

FIGURE 34.7

The length of the humerus in the barred owl (*left*) and the northern fur seal (*right*) is the same, but the two bones differ markedly in width and density.

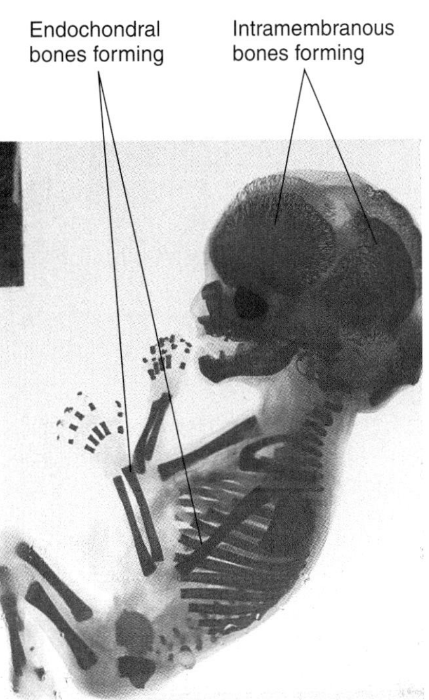

FIGURE 34.8

The cartilage model of a fetal skeleton. During fetal development, the model will re-form in bone.

strength required to move against water resistance, a force considerably greater than the air resistance birds encounter. At the same time, the joints of the seal's musculoskeletal system provide enough flexibility to permit the animal to drag itself ashore.

Mastering Concepts

1. What are some functions of the vertebrate endoskeleton?
2. What are the two major parts of the vertebrate skeleton?
3. What are the major groups of bones that make up the human skeleton?
4. How are bones in different species adapted to different ways of life?

Cells and Tissues of the Vertebrate Skeleton

The skeletal system consists primarily of cartilage and bone. These two types of connective tissue are specialized to serve as an internal framework.

Cartilage

Cartilage has many functions in the skeleton. During embryonic development, the skeleton first forms in cartilage, providing a mold for the bone that will later replace it (fig. 34.8 and Biology in Action 34.1). Cartilage is widely dispersed in the adult body. The flexibility of elastic cartilage helps the rib cage expand when the lungs inflate. Another type of cartilage, in the pads between the

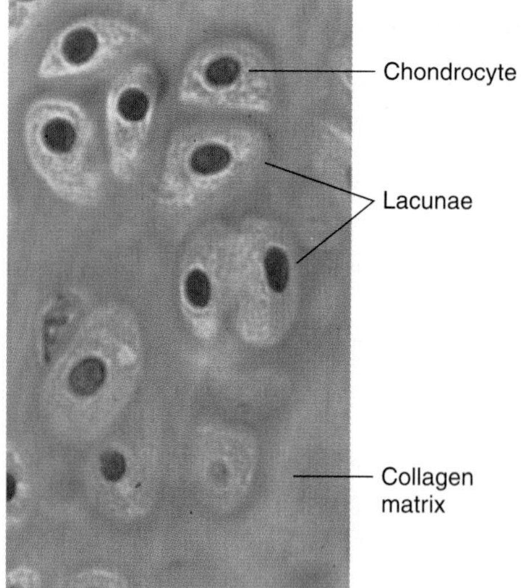

FIGURE 34.9

Cartilage consists of chondrocytes (protein-producing cells) in lacunae (spaces) within a matrix of collagen and elastin.

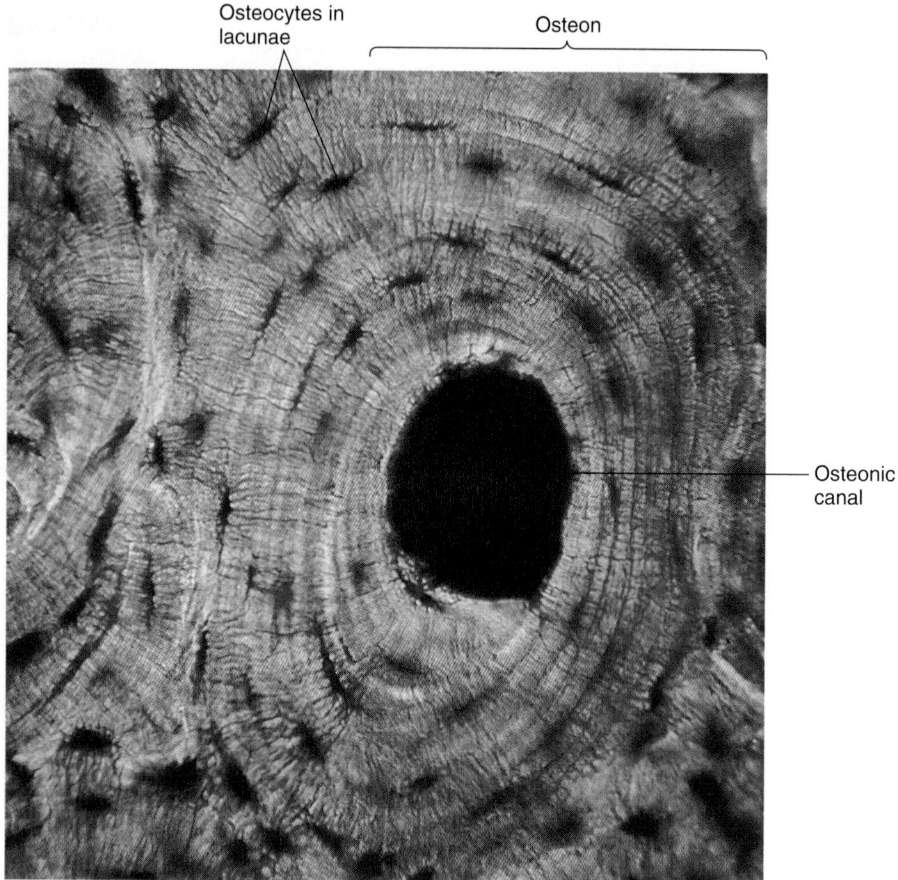

FIGURE 34.10

Bone consists of osteons, which are rings of osteocytes within lacunae in a matrix of collagen and mineral. Osteons surround an osteonic canal.

bones that make up the backbone, can withstand compression. In the appendages, smooth cartilage coverings allow the surfaces of adjoining bones to slide smoothly past one another.

Recall from chapter 30 that cartilage consists of chondrocytes housed in lacunae within a matrix (fig. 34.9). Chondrocytes secrete proteins that give the tissue its firmness and flexibility. Strong networks of collagen fibers enable the tissue to resist breakage and stretching even when bearing great weight. Elastin provides flexibility. The protein network in cartilage also entraps a great deal of water, which provides support and firmness. The high water content of the matrix makes cartilage an excellent shock absorber, which is its primary function between many bones. The water motion within cartilage that accompanies body movement also cleanses the bloodless tissue and bathes it with dissolved nutrients.

Bone

Bone packages maximal strength into a lightweight form. One cubic inch of bone can bear loads up to 19,000 pounds (approximately 8,618 kilograms). The strength of bone rivals that of light steel, but a steel skeleton would weigh four or five times as much.

Bone derives strength from collagen and hardness from minerals. A 20-pound (9-kilogram) weight could dangle from a collagen strand a mere millimeter thick without causing it to snap. Collagen's strength is an asset to bone, but it is flexible and elastic. The hardness and rigidity of bone comes from the minerals, primarily calcium and phosphate, which precipitate out of body fluids coating the collagen fibers. If the minerals were removed from bone, it could be tied in a knot like a garden hose; however, removal of the collagen would make the bone brittle and easy to shatter.

Embedded in the bone matrix are living cells responsible for growth and repair. Nutrients and wastes cannot diffuse through the matrix to and from the cells; instead, blood reaches the cells through small canals that penetrate the matrix. The predominant bone cells, osteocytes, lie in spaces (lacunae) in concentric rings called **osteons** (fig. 34.10). Osteons surround central portals that contain the blood supply, the **osteonic canals.** Communicating canals link adjacent osteonic canals, and narrow passageways, the canaliculi, connect lacunae. Osteocyte extensions reach through the canaliculi, maintaining contact between widely separated osteocytes. Materials pass from the blood, and from osteocyte to osteocyte, through a "bucket brigade" of up to 15 cells. Still other canals connect the entire labyrinth to the outer surface of the bone and to the marrow cavity within. A layer of connective tissue called the periosteum surrounds the entire complex structure of bone (fig. 34.11).

Bones are lightweight as well as strong because they are porous rather than solid. Most of the irregularly shaped flat bones and almost all of the bulbous tips of long bones are made of **spongy bone.** This type of bone is hard but has many large spaces between a web of bony struts that align with the lines of stress, increasing the bone's strength. The spaces within spongy bone are filled with **red marrow,** a nursery for blood cells and platelets that releases more than a million cells a second. Spongy bone is covered with a layer of more solid **compact bone.** The weight of long bones, such as limb bones, is further reduced by the **marrow cavity,** a space in the shaft that contains yellow marrow consisting primarily of fat cells (fig. 34.12). Because a hollow tube is more resistant to certain types of stress than a solid rod of equal mass, this cavity allows for maximum strength with minimum weight.

Mastering Concepts

1. What properties of cartilage make it an excellent supportive tissue in the skeletal system?
2. What characteristics of bone enable it to form an internal support system?
3. Describe the organization of bone tissue.

Biological Movement

Contractility is an ancient characteristic of cells. Diverse animal cells contract, and they use the same types of motor molecules—actin and myosin—and the same mechanism—sliding protein filaments.

An individual muscle cell can be in either of two states—contracted or relaxed. The highly coordinated movements of animals result from specific combinations of contracted and relaxed muscle cells.

Muscle Cell Types

Recall from chapter 30 that three types of cells form muscle tissues (fig. 34.13). Smooth muscle cells are long and tapered at each end. Each cell has a single nucleus. Because animals cannot consciously control smooth muscle cells, these cells form involuntary muscles. Smooth muscle is found in many parts of the vertebrate body. In the walls of certain blood vessels, smooth muscle contracts, regulating blood pressure and directing blood flow. Smooth muscle in the intestinal wall contracts, pushing food through the digestive tract. Smooth muscle forms sheetlike linings in several hollow organs and either pushes material through or maintains the diameter of a tube.

Cardiac muscle cells are unique to the heart, where they contract, propelling blood throughout the body. Cardiac muscle cells are striated, their control is involuntary, and they have one nucleus

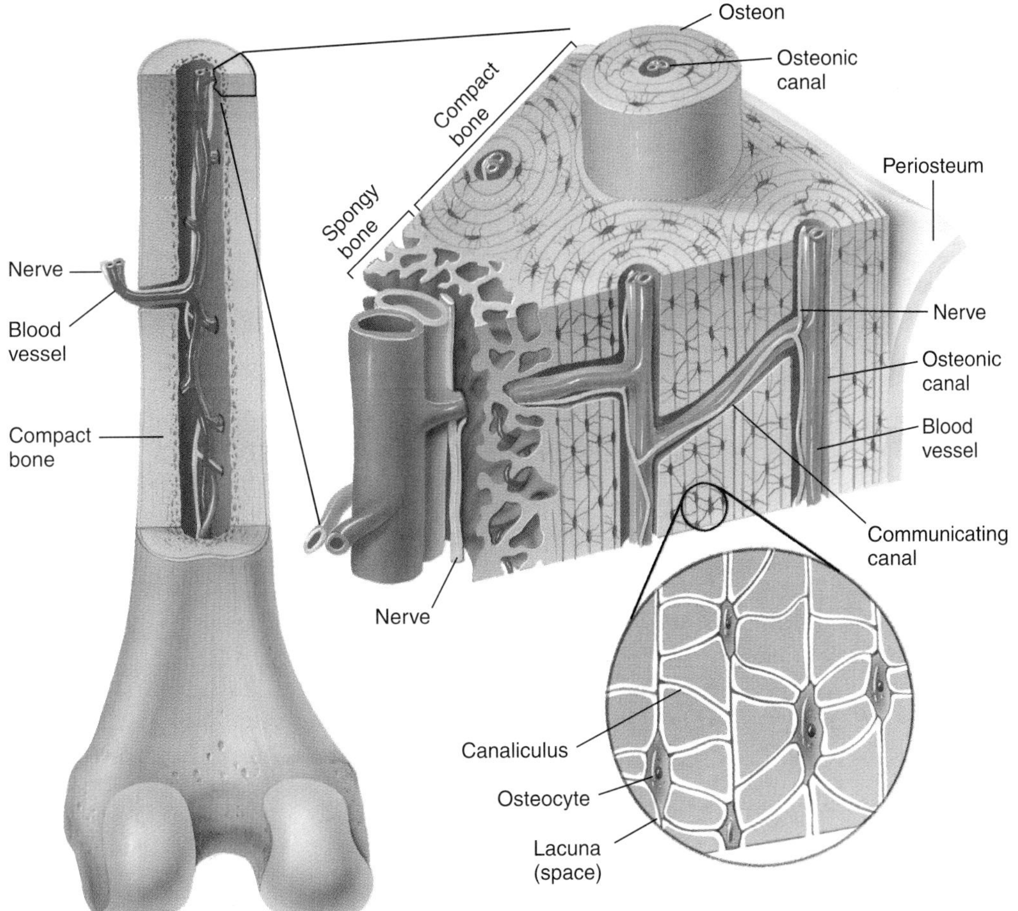

FIGURE 34.11

Bone cells form concentric circles around a central passageway, the osteonic canal, that contains the blood supply.

34.1 biology in action

The Human Skeleton over a Lifetime

Before Birth

Cartilage and bone interact throughout development, sculpting the ever-changing skeleton. Most bones originate in the embryo as cartilage models that bone tissue gradually replaces (fig. 34.A). Cartilage models are shaped like the bones they will become. A layer of connective tissue surrounds the models.

In a 4-week embryo, cells just beneath this connective tissue secrete a "collar" of compact bone around the central shaft. Meanwhile, cartilage cells and their lacunae within the shaft enlarge, squeezing the cartilage matrix into thin spicules. The compressed cartilage matrix hardens with calcium salts, which kills the cartilage cells. The cartilage matrix degenerates, and capillaries replace it. Bone cells enter and secrete matrix, establishing the first internal region of new bone.

During Childhood

A baby's skull bones still contain dense connective tissue (rather than bone) at the sutures. The regions between the bones pulsate from underlying blood vessels. The skull bones of the newborn, not yet fused, can slide together as the infant squeezes through the birth canal. The soft spots in the head usually fill in with bone tissue by 18 months of age.

Within a few months of birth, bone growth becomes centered near the ends of the long bones in thin disks of cartilage called the epiphyseal plates. These cartilage disks enable bones to elongate as the child grows. The cartilage cells in the plate divide, pushing progeny cells toward the shaft, where they calcify. The dividing cartilage cells keep the disk relatively constant in thickness, but the calcified progeny cells become part of the shaft, elongating it. Growth continues until the late teens, when bone tissue begins to replace the cartilage plates. By the early twenties, bone growth is complete. Only a line remains marking the position of a cartilage disk.

Fracture Repair

Cartilage helps heal a bone break. The immediate reaction to a fracture is bleeding, followed rapidly by blood clot formation at the site of the break. Then dense connective tissue replaces the blood clot. Next, cartilage cells enter the connective tissue and form a fibrous "callus" that fills the gap the injury left. New spongy bone replaces the cartilage, closing the gap. Exercising the injured bone stimulates bone cells in the healing area to secrete collagen, which compacts the newly formed bone.

Osteoporosis

Decreasing bone mass due to loss of calcium salts is a normal part of aging, but it can lead to osteoporosis (holes in the bones), in which bones are likely to break (fig. 34.B). Ten percent of women have severe enough osteoporosis by age 50 to suffer bone fractures. Nearly 25% of all women develop osteoporosis by age 60, and by age 90, one-third of all women and one-sixth of all men have it. Osteoporosis causes shrinking stature, chronic back pain, and frequent fractures.

Females are more likely to suffer from osteoporosis than males for several reasons. Women's longer life span contributes to their overrepresentation among those with osteoporosis. Furthermore, the bone mass of the average woman is about 30% less than a man's, so her bones are more easily depleted of calcium stores. A decline in the level of estrogen also makes a woman more prone to osteoporosis. For this reason, doctors advise all women to take 1,000 to 1,500 milligrams of calcium daily and to exercise regularly.

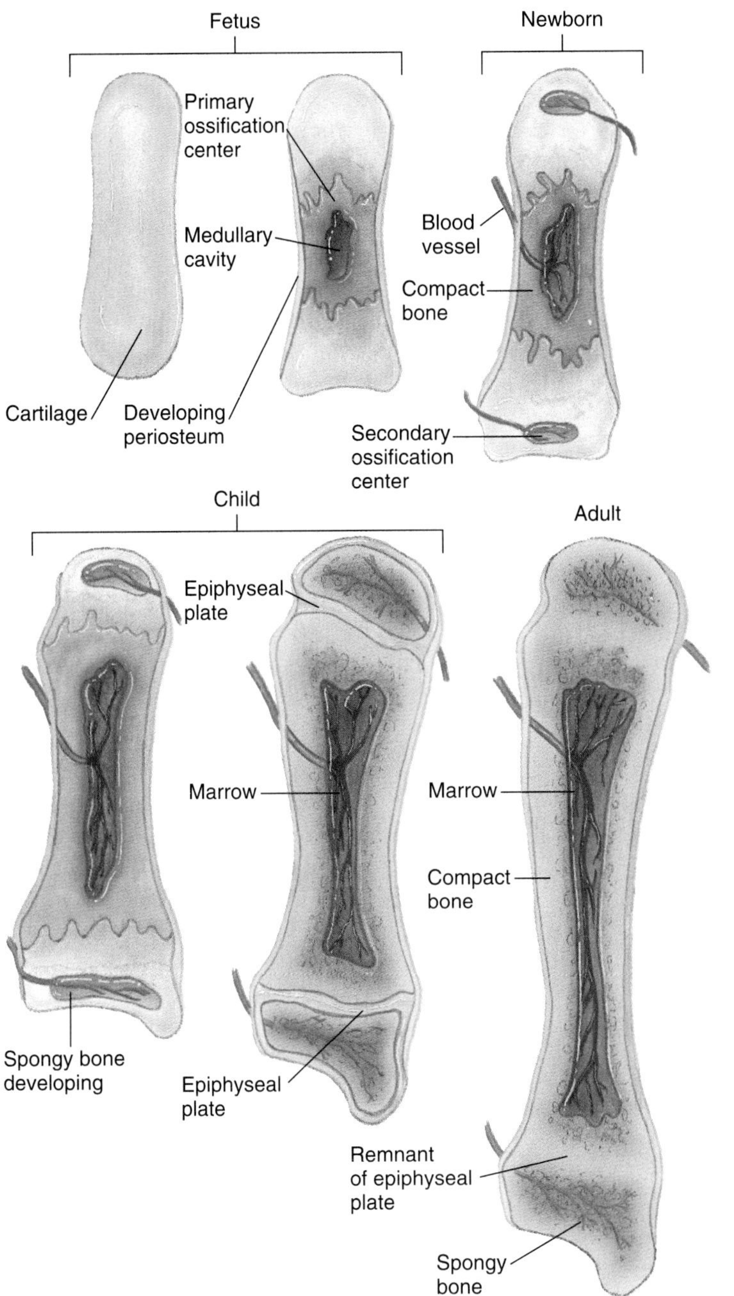

FIGURE 34.A
Bone growth. During fetal development, bone gradually replaces the cartilage skeleton. Bone forms first around the central shaft as cartilage cells die within a calcifying matrix. Later, bones elongate from the epiphyseal plates, which are cartilage disks located between the shaft and the knob at the end of the bone.

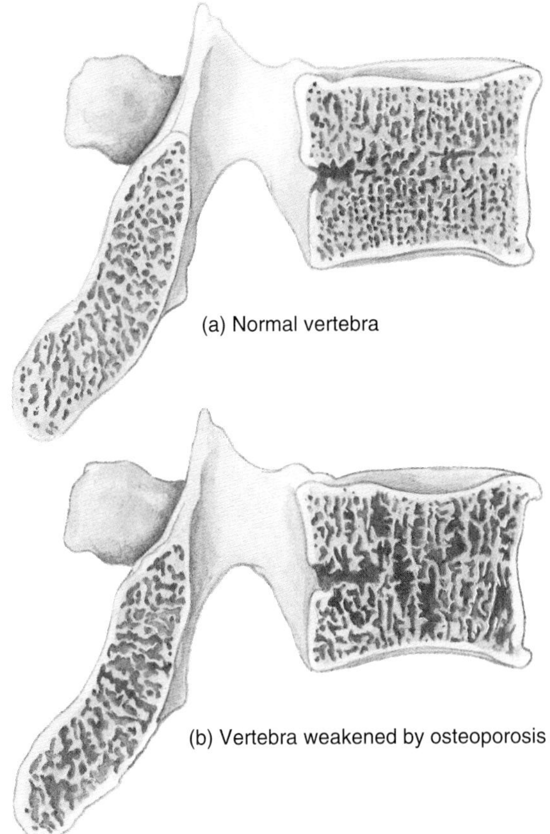

FIGURE 34.B
In osteoporosis, calcium loss weakens bones.
(*a*) Normal vertebra and (*b*) vertebra weakened by osteoporosis.

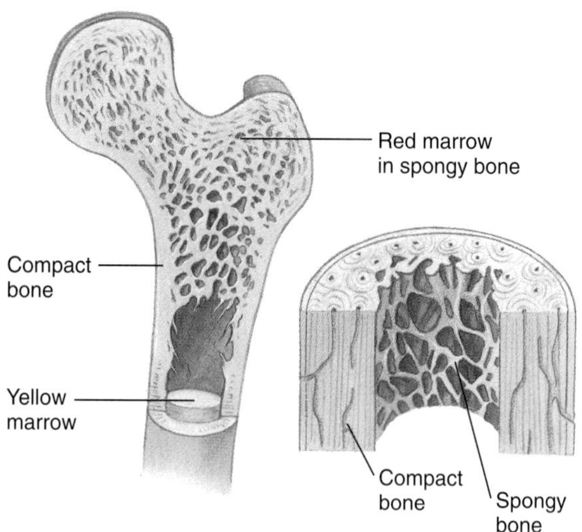

FIGURE 34.12

The structure of a long bone. The shaft of a bone is hollow and contains yellow marrow. Surrounding this is a layer of spongy bone. The outer coat consists of compact bone. The knobby ends of the bone are spongy bone coated with compact bone.

per cell. Cardiac muscle cells branch, forming a netlike structure, and strong but permeable membrane foldings called **intercalated disks** join them.

Skeletal muscle cells, so named because they attach to bones, are under voluntary control. Skeletal muscle cells are very long and appear striated under the microscope because of the pattern of contractile filaments within them. A single skeletal muscle cell has many nuclei, which are pushed to the outer surface. We briefly consider muscles in nonvertebrates and then take a closer look at skeletal muscles in the human body.

Amoeboid Movement

A simple illustration of biological movement is the amoeboid motion of the protozoan *Amoeba proteus.* Figure 34.14*b* shows an amoeba wrapping around a small, ciliated cell that it will engulf and consume.

The amoeba moves by extending **pseudopods** ("fake feet") and dragging the rest of the cell to catch up. This movement depends on the cytoplasm's ability to assume two forms—an outer layer of thick, gel-like material called **ectoplasm** and an inner, more fluid, cytoplasm called **endoplasm** (fig. 34.14*c*). A pseudopod extends from any point of the cell when endoplasm flows forward and thickens into ectoplasm. At the same time, what used to be ectoplasm loosens into endoplasm. This endoplasm then pushes forward.

Actin and myosin provide an amoeba's movement. If amoebae are exposed to chemicals that destroy either of these proteins, they can no longer move. If researchers place extracts of an amoeba's cytoplasm in a test tube with ATP and calcium ions, both of which are necessary for vertebrate muscle contraction, the cytoplasm contracts in the test tube!

Invertebrate Muscles

The movements of some invertebrates are so fluid and complex that they are the envy of engineers trying to design machines to explore rugged terrain. A crab, for example, must coordinate the movements of 10 legs to scurry sideways to escape danger. Arthropods are able to move well because of sense organs within their exoskeletons that function as strain gauges, detecting stress on the animals' outsides. These sense organs are linked to sensory neurons beneath the cuticles (fig. 34.15). When a falling rock slightly dents a crab's leg, for example, the sensory neurons beneath the dent fire action potentials to the central nervous system, which then signals certain muscles to contract in a pattern that enables the crab to quickly leave the danger scene.

Invertebrates use combinations of smooth and striated muscles. A scallop, for example, uses striated muscle to snap its shell shut, propelling it forward. But its retractor muscle, which can close the shell for days, is smooth muscle.

Insect flight illustrates frenetic muscular activity—some flies beat their wings a thousand times a second! Striated muscle produces fast muscle contractions in the insect wing. The elastic tissue surrounding the flight muscles helps sustain a rapid contraction rate. When a group of muscles called depressors lower the wing by contracting, they stretch another group, the elevator muscles. These muscles then contract, raising the wing. When the elevator muscles contract, they in turn stretch the depressor muscles, stimulating them to contract. The result of this oscillation: A single nerve impulse can trigger 20 to 30 contractions of the wing muscles. A fly whizzes by.

Vertebrate Skeletal Muscle

At the microscopic level, muscle structure and function is similar in different vertebrate species. On a macroscopic level, however, differences become apparent. As was true of the skeletal system, these differences reflect natural selection pressures.

The muscles of fish and amphibians, for example, lie parallel to the vertebrae, which is adaptive for an aquatic habitat. In mammals and some reptiles, skeletal muscles are specialized into groups that are adaptive for moving on land. The difference in gait among different land-dwelling animal species also illustrates the musculoskeletal system's adaptation. Camels, rabbits, monkeys, and dogs all walk differently. The mountain goat illustrates an interesting adaptation in the way it turns around on a very narrow ledge. It pulls its back legs over its head in a reverse backbend that is possible because of its short legs, low center of gravity, small haunches, and muscular front end.

Sometimes fossils reveal how musculoskeletal structures changed over time, as with the modern horse, *Equus.* Over many millions of years, the limb bones and muscles changed in a way

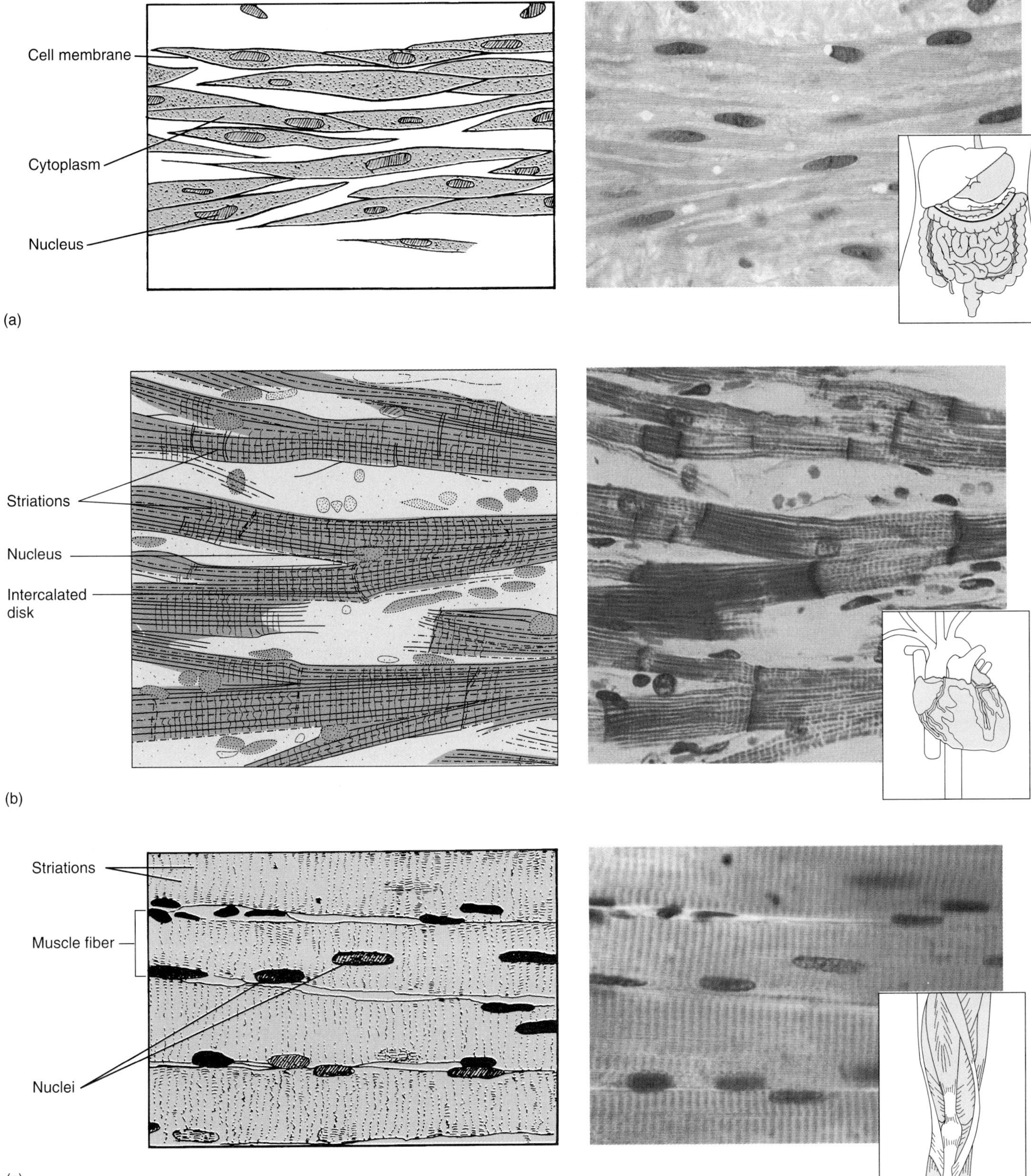

FIGURE 34.13

Three types of muscle cells and tissues. (*a*) Smooth muscle is composed of spindle-shaped cells with one nucleus that often form layers. (×250) Smooth muscle contraction is involuntary. (*b*) Cardiac muscle is unique to the heart. Its striated cells join at connections called intercalated disks. The cells form a branching network, and their contraction is involuntary. (×400) (*c*) Skeletal muscle is under voluntary control, and the cells have many nuclei. (×250) The cells are striated due to the orderly organization of contractile proteins.

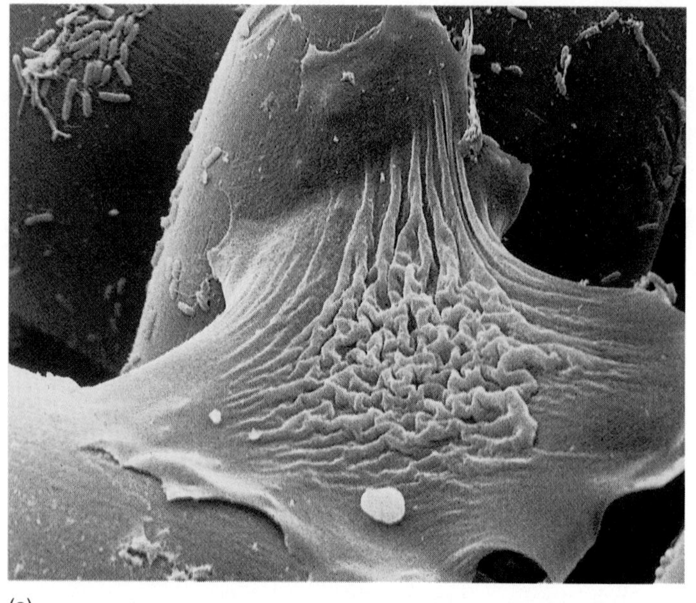

(a)

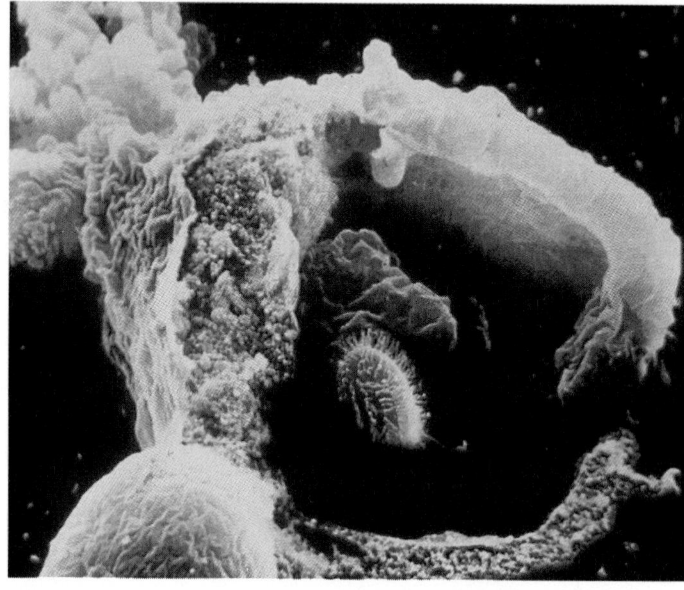

(b)

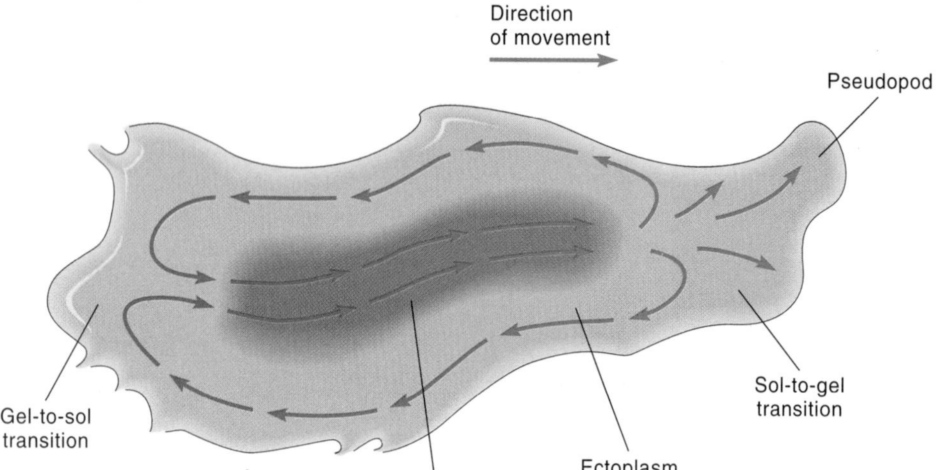

(c)

FIGURE 34.14

The amoeba moves by extending pseudopods (*a*) and can engulf smaller cells such as this protozoan (*b*). (*c*) Amoeboid movement depends on the cytoplasm's conversion from a gel consistency to a more fluid state and back again. The process is probably powered by sliding actin and myosin filaments.

that enabled the animal to run swiftly and gracefully. The ancient horse gradually lost its side toes, increased the ratio of length to width in its long limb bones, and acquired an altered foot stance that places weight at the tip of the foot (fig. 34.16). These adaptations increased speed and stride length. As limb muscles became capable of rapid flexion and extension, bones lost much of their ability to rotate around the joints in the limbs. This adaptation protects the animal from twisting a leg.

Skeletal Muscle Organization

The powerful arm muscles that propel a tennis player's racquet can be viewed at several levels (fig. 34.17). The most familiar level is the whole muscle. Each muscle lies along the length of a bone. A heavy band of fibrous connective tissue called a **tendon** attaches each end of the muscle to a different bone. At the microscopic level, the long muscle cells are called **muscle fibers.** These are organized into many larger bundles, or **muscle fasciculi.** A muscle consists of clusters of fasciculi encased in a connective tissue sheath, as is each individual fasciculus and muscle fiber too.

Blood vessels and nerves run between the bundles of muscle fibers. The rich blood supply provides muscle cells with nutrients and oxygen to power contraction and removes cellular wastes. The nerves trigger and control muscle contraction.

Muscle fibers—that is, individual muscle cells—can be more than 1 foot (.3 meter) long. Most of the volume of a muscle fiber consists of hundreds of thousands of cylindrical subunits called **myofibrils** (fig. 34.18). The rest of the cell contains the usual organelles. The cell membrane and endoplasmic reticulum (ER) are extensive and have special names—**sarcolemma** for the cell membrane and **sarcoplasmic reticulum** for the ER. Both membrane networks fold against one another at many points. The parts of the outer membrane that jut into the inner membrane are called the **transverse,** or **T, tubules.**

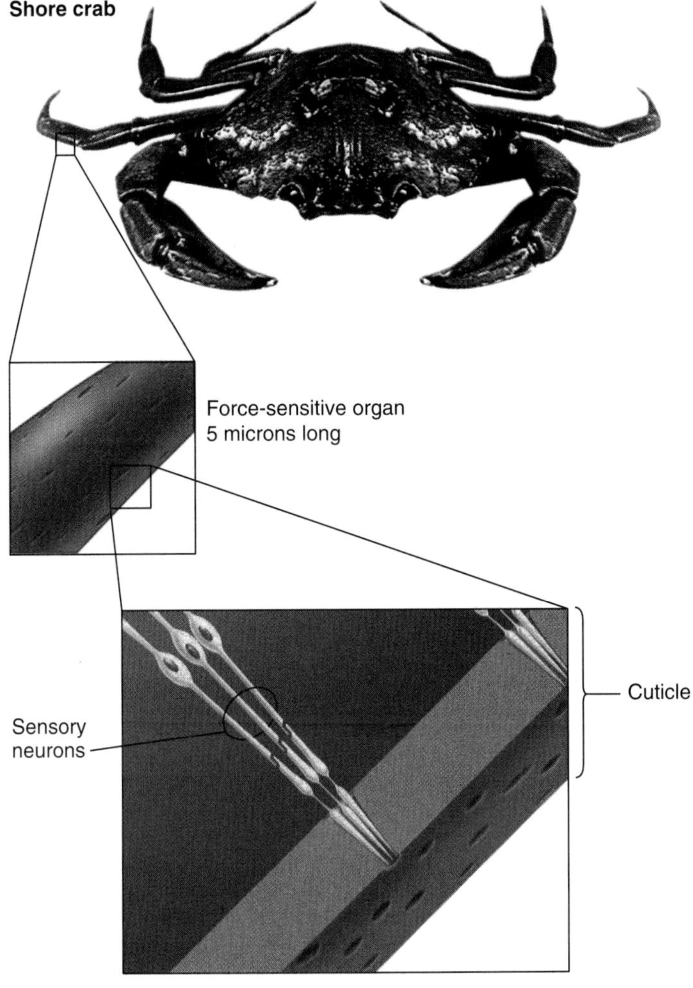

FIGURE 34.15

A touch sensor in a crab's exoskeleton.

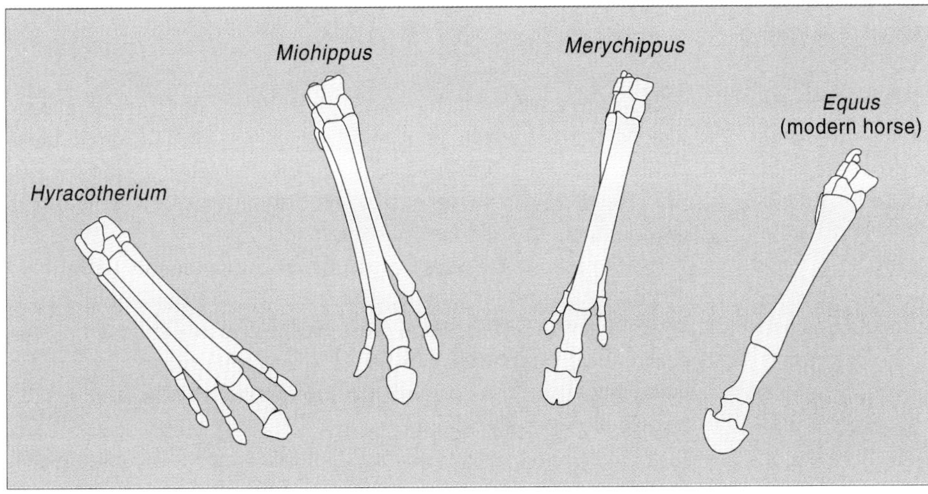

FIGURE 34.16

A distant ancestor of the modern horse, *Equus, Hyracotherium* ("dawn horse"), lived 55 million years ago and was about the size of a small dog. Note how different its front foot was from a modern horse's front hoof. *Miohippus* and *Merychippus* were two related species that lived in time between *Hyracotherium* and *Equus.*

Myofibrils contain even finer "strings," the **myofilaments,** which consist of the contractile proteins. Myofilaments are of two types—thin and thick. Each thin myofilament is composed primarily of actin and also contains two proteins important in controlling muscle contraction, **troponin** and **tropomyosin.**

The thick myofilaments are composed of myosin. Each myosin molecule is shaped like a golf club. The "shafts" of myosin molecules adhere to one another, forming a bundle that may consist of several hundred molecules. The "heads" protrude from both ends of the bundle in a spiral pattern. Each thick myosin myofilament is surrounded by six thin actin myofilaments (fig. 34.19). This organization allows connections between myosin and actin to form, break apart, and form again, which is essential for muscle contraction.

Muscle contains other proteins too. A muscle protein need not be very abundant to be vital. For example, Duchenne muscular dystrophy results from lack of or abnormal **dystrophin,** protein which makes up only 0.2% of the total muscle protein. Dystrophin normally supports the inside face of muscle cell membranes, enabling them to withstand the force of contraction. When a child lacks dystrophin, muscles gradually fail.

How Skeletal Muscle Contracts

The striations of skeletal muscle derive from the organization of the thick and thin myofilaments. The stripes occur in a pattern of repeated units called **sarcomeres** (muscle units). Figure 34.20 shows a microscopic view of a sarcomere and a diagrammatic representation of how the myofilaments interact to produce the various shadings. Different areas within a sarcomere have specific names.

Muscle contracts when the thin myofilaments slide between the thick ones. This motion shortens the sarcomere without shortening its protein components. You can illustrate this using your hands as a model: Hold your hands sideways in front of you so that you are looking at your palms and your fingertips are touching (fig. 34.21*a*). Now slide the fingers of one hand between those of the other. The distance between your palms shortens even though the length of your fingers remains the same (fig. 34.21*b*). The movement of protein myofilaments to shorten skeletal muscle cells is called the **sliding filament model** of muscle contraction.

To understand how actin and myosin are propelled past one another, it is necessary to understand the action of the myosin heads. The club-shaped heads attach to the shafts with hingelike connections that allow the head to rock back and forth. This swiveling motion is similar to the path of an oar as a stroke propels a boat through the water. The movement pulls on actin and causes it to slide past myosin in the same way the oar's motion moves a boat. Myosin heads can bind both actin and ATP. Splitting ATP provides the energy that powers muscle contraction. Muscle contraction also requires calcium ions (Ca^{2+}).

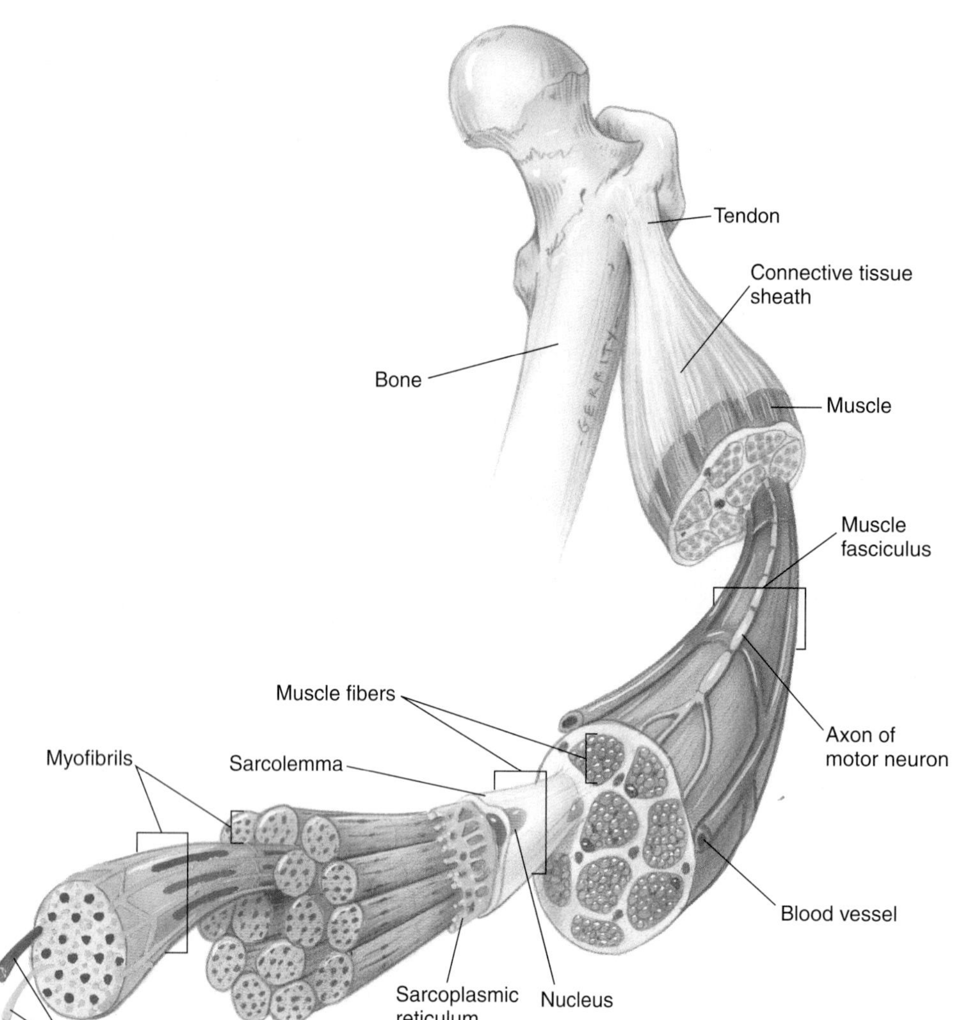

FIGURE 34.17

Levels of skeletal organization. A skeletal muscle is composed of bundles of muscle fibers. Each muscle fiber, or muscle cell, contains many myofilaments.

For contraction to occur, actin and myosin must touch. Because a myosin head can swing out to contact an actin molecule it is also called a cross bridge. In a resting muscle, troponin holds tropomyosin in a groove between the two entwined actin strands, which blocks myosin cross bridges from binding to actin.

The directive to contract begins when a motor nerve cell that extends from the brain or spinal cord to a muscle cell releases the neurotransmitter acetylcholine at a neuromuscular junction, where a neuron and muscle cell meet (fig. 34.22). When the neurotransmitter binds to the outer membrane of the muscle cell, it induces an electrical wave that races to storage sacs within the sarcoplasmic reticulum. The sacs release calcium ions, which flood the cytoplasm. The calcium binds to the troponin molecules attached to the actin myofilaments. The troponin then moves, taking the attached tropomyosin molecules to the interior of the thin myofilament, where they no longer block actin from contacting the myosin of the surrounding thick filaments. When tropomyosin moves, myosin cross bridges attach to the exposed actin.

Long before the nerve impulse arrived, ATP slipped into its site on the myosin head (fig. 34.23), where it was enzymatically split to yield ADP and inorganic phosphate, releasing the energy that activates myosin cross bridges. When the activated cross bridges bind to actin, this energy is released, causing the myosin head to forcefully swing back to its original position. This forces the filaments to slide past one another.

After this so-called power stroke, ADP and the inorganic phosphate pop off the myosin head, leaving it free to bind a new ATP. When a new ATP binds, the cross bridge releases from actin even before the ATP splits. Then, as ATP releases its energy, the myosin head flips back to its original position, ready to bind a new actin further along on the chain.

This sliding repeats about a hundred times a second on each of the hundreds of myosin molecules of a thick filament. Although the movement of filaments due to the interactions of single actin and myosin filaments would be no more astounding than the forward progress of a boat with a single person rowing, a skeletal muscle contracts quickly and forcefully due to the efforts of many thousands of "rowers."

Skeletal muscle must also relax on cue. Myofilament sliding stops when calcium ions are no longer available. Shortly after mo-

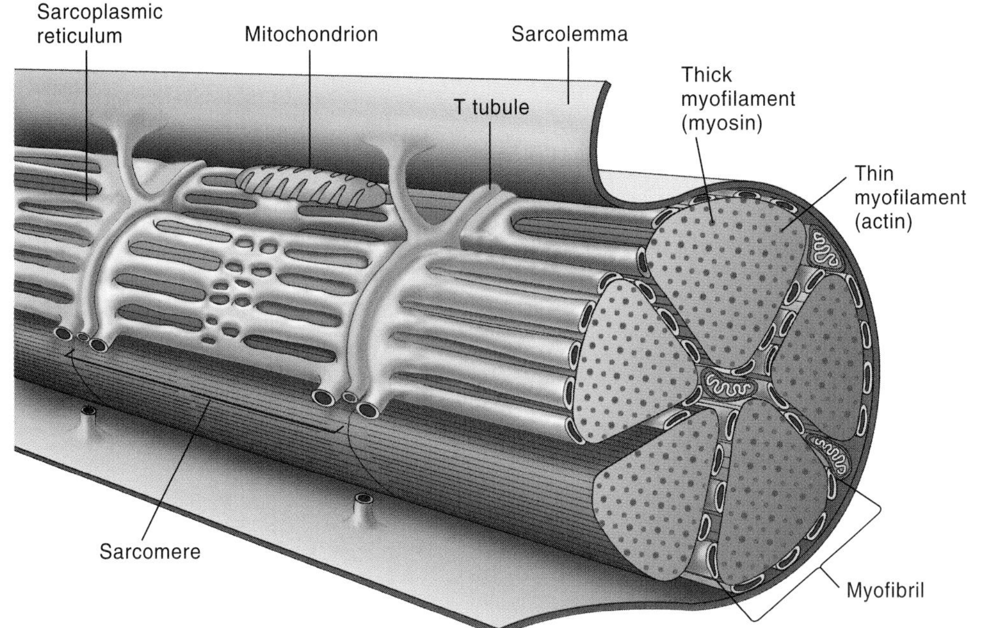

FIGURE 34.18

Anatomy of a muscle fiber. The sarcoplasmic reticulum is a specialized endoplasmic reticulum in a muscle cell. It is highly folded to store the calcium ions that regulate muscle contraction. Transverse tubules (T tubules) are membranes that fold inward between contractile units of the muscle cell. They conduct the nerve impulse's electrical charge to the interior of the cell.

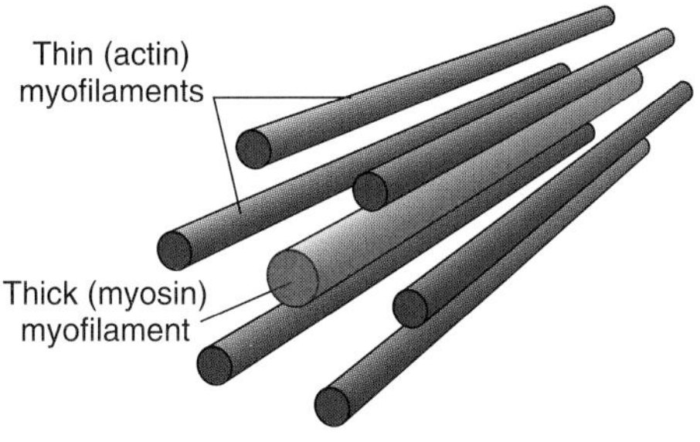

FIGURE 34.19

A schematic view of muscle microscopic structure—six actin myofilaments surround each myosin myofilament.

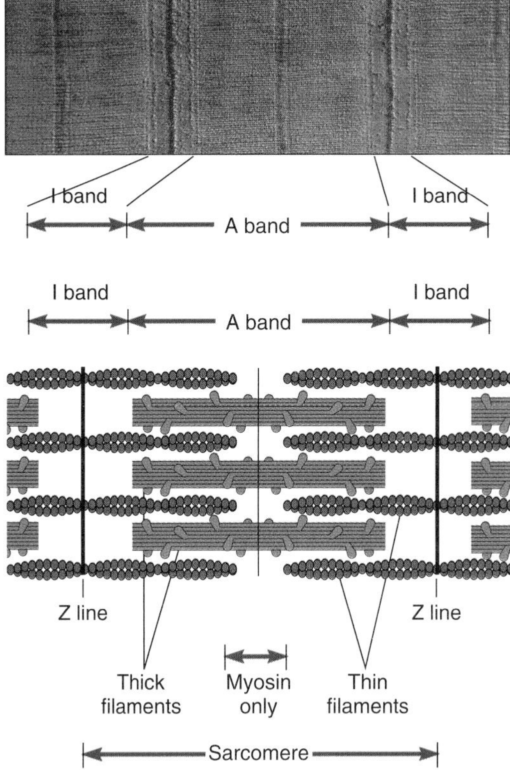

FIGURE 34.20

The boundaries of a sarcomere form by membranes called Z lines, which the actin myofilaments attach to. The area on either side of a Z line, occupied only by actin myofilaments, is the I band. It appears light in color (here green) when viewed under a light microscope. The entire area containing myosin is the A band. The outer borders of the A band look dark under a light microscope (here purple) because the thick and thin myofilaments overlap in this region. The midregion of the A band is occupied by myosin alone and is light in color. The light and dark bands the myofilaments form give skeletal muscle its striated appearance. (Note: four of the actin filaments surrounding each myosin filament are not visible in the plane of this drawing.)

tor nerve stimulation causes the release of calcium ions, they are actively transported back to the sarcoplasmic reticulum. Tropomyosin returns to its original position, blocking actin from interacting with myosin heads, and the sarcomere relaxes.

Energy for Muscle Contraction

Skeletal muscle contraction requires huge amounts of ATP, both to power the return of Ca^{2+} to the sarcoplasmic reticulum and to break the connection between actin and myosin, allowing a new cross bridge to form. After death, muscles run out of ATP. The cross bridges cannot release from actin, and the muscles remain in a stiff position. This stiffening is called rigor mortis.

The body has several ways to generate ATP. When muscle activity begins, ATP can be replenished as rapidly as it is depleted when a molecule called **creatine phosphate,** which is stored in muscle fibers, donates a high-energy phosphate to ADP to regenerate ATP. Obtaining ATP from creatine phosphate requires only one enzymatic step, which is why this source of ATP is rapid, but

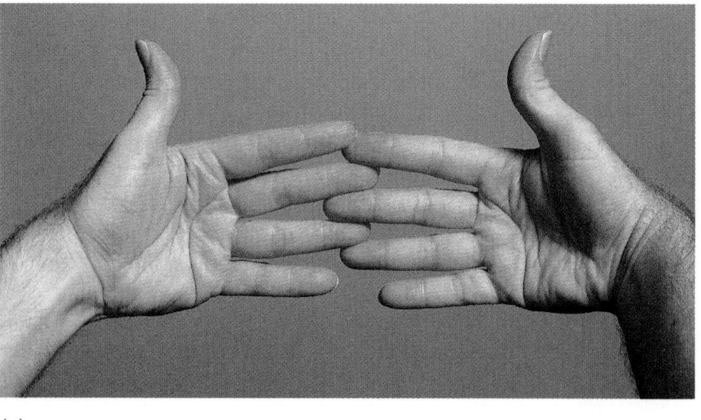

(a)

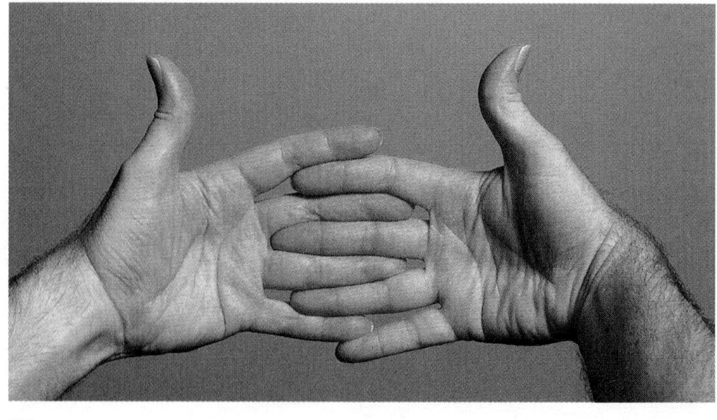

(b)

FIGURE 34.21

Hands can illustrate the sliding filament model of muscle contraction. As fingers slide between one another (*a*), the distance between the palms shortens (*b*). This is analogous to the movements of thick and thin myofilaments during muscle contraction.

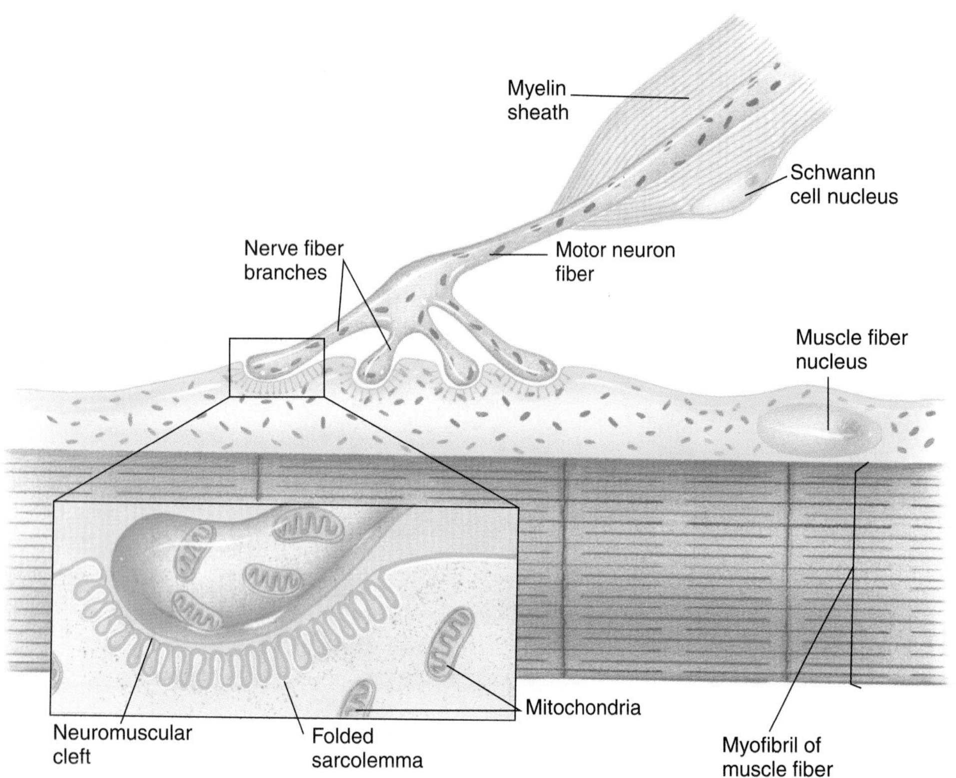

FIGURE 34.22

A neuromuscular junction. A motor axon branches, forming specialized junctions with muscle fibers. The axon releases a neurotransmitter that triggers electrical changes in the muscle cell membrane, which releases calcium ions inside the muscle cell, leading to contraction.

it is also depleted first. After the supply of creatine phosphate falls, the body must generate ATP from other sources. As long as enough oxygen can reach the muscle tissue, the mitochondria form ATP by aerobic respiration (see chapter 6). The fuel sources are fatty acids and carbohydrates. The muscle stores glycogen, a glucose polymer that can break down to provide a ready supply of glucose for metabolism.

When aerobic respiration and creatine phosphate can no longer provide enough ATP, the muscle cell obtains ATP from anaerobic respiration, a less efficient metabolic route that rapidly leads to muscle fatigue.

Macroscopic Structure and Function of Skeletal Muscle

The components of an individual muscle fiber work well as a unit. If nerve stimulation releases calcium within the fiber, all its sarcomeres contract and the entire fiber shortens. Therefore, the muscle fiber responds in an all-or-none fashion.

A muscle's contraction strength depends on how many individual fibers contract within the muscle. Thus, the muscle as a whole does not respond in an all-or-none manner, because normally not all of the fibers contract at one time.

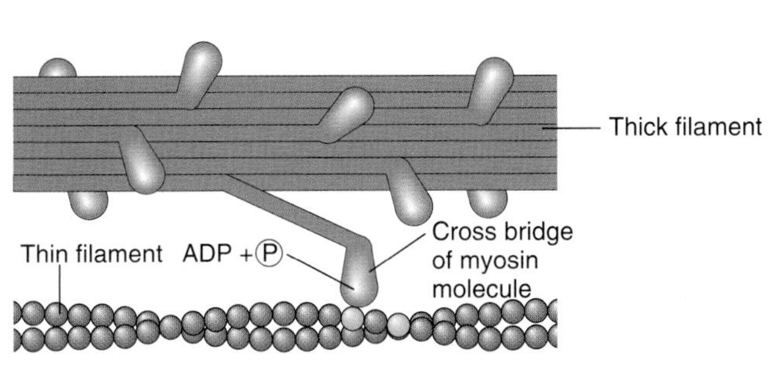

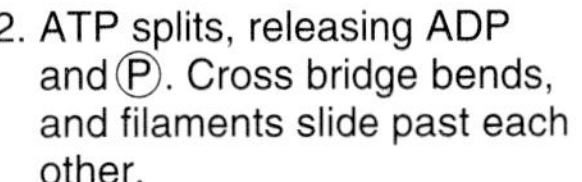

Cross bridge of myosin molecule
1. Cross bridge of thick filament attaches to a monomer of actin on thin filament.

ADP + Ⓟ
2. ATP splits, releasing ADP and Ⓟ. Cross bridge bends, and filaments slide past each other.

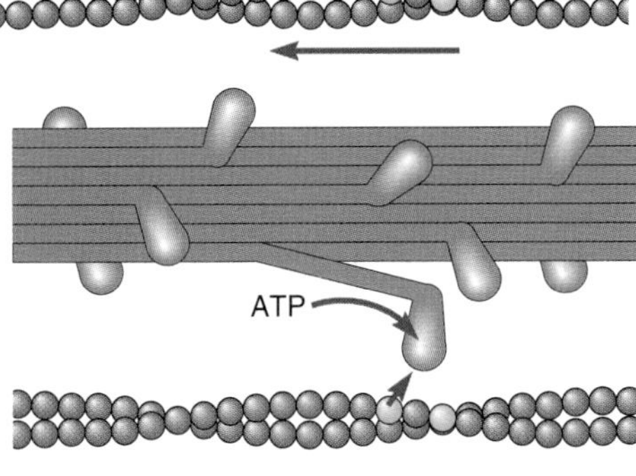

ATP
3. ATP attaches to cross bridge, causing it to release from actin monomer.

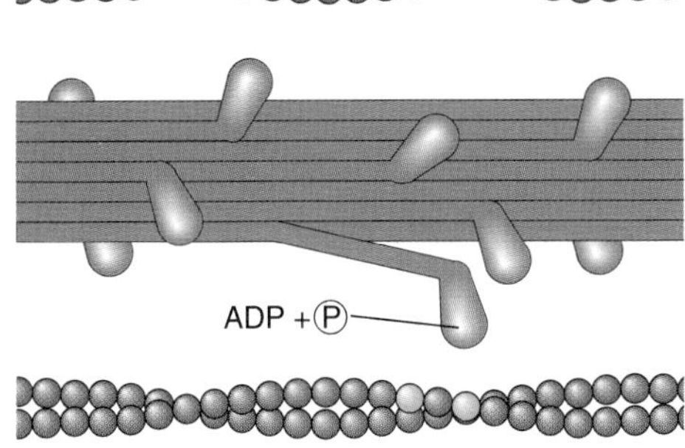

ADP + Ⓟ
4. ATP splits, and cross bridge is "reset" to original position.

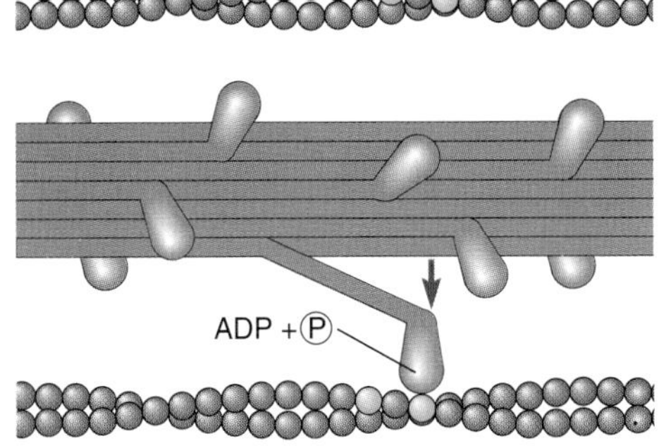

ADP + Ⓟ
5. Cross bridge binds new actin monomer, farther down the thin filament, beginning a new cycle.

FIGURE 34.23

The sequence of molecular interactions of the sliding filament model of muscle contraction.

A nerve cell and all the muscle fibers it contacts form a **motor unit** (fig. 34.24). The neuron may exert either a localized or a large effect on a muscle depending on how many muscle fibers its nerve ends touch. A motor nerve cell reaching the eye, for example, may only control a few muscle fibers, producing fine, small-scale movements. In contrast, a single nerve cell in the arm may have hundreds of endings that each contact a different muscle cell. As a result, when that neuron fires, many muscle cells in the arm respond, contributing significantly to the strength needed to lift a heavy weight. More active motor units provide greater strength.

The stimulation rate influences the muscle response pattern. A single stimulation causes the muscle to contract quickly and then relax. This response is called a **twitch.** It is followed by a refractory period of 2/100 of a second when the muscle fiber cannot respond to further stimulation. If the muscle is stimulated again, after the refractory period but before it is fully relaxed, the second contraction adds to the remains of the first one, increasing the contraction strength. If a muscle receives repeated strong stimulation without time to relax, the muscle contraction strength builds until a smooth and continuous contraction called a **tetanus**

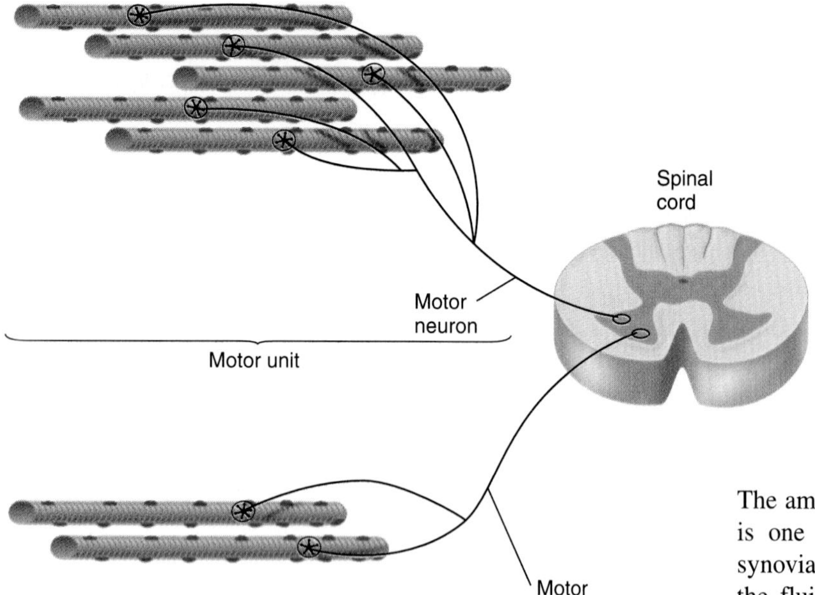

FIGURE 34.24

A motor unit consists of a motor neuron axon and the muscle fibers it innervates. If one nerve cell stimulates many muscle cells, they all contract at once in a strong movement. If a nerve cell stimulates only a few muscle cells, it generates finer movements.

occurs. Tetanus ceases when the muscle uses up its chemical energy reserves and is no longer able to contract, even when further stimulation is applied.

Mastering Concepts

1. What are the three muscle cell types?
2. How do amoebae and invertebrates move?
3. Describe the levels of organization of a vertebrate muscle.
4. Describe how protein interactions move muscles.
5. What is the source of energy that powers muscle contraction?

Muscle and Bone Working Together

The vertebrate musculoskeletal system is a highly organized framework of bones and muscles and the tissues and structures that connect and support them. All the components work together to produce movement.

Joints—Where Bone Meets Bone

Most of the vertebrate skeleton consists of freely movable bones separated by cartilage but joined by a capsule of fibrous connective tissue called a **synovial joint** (fig. 34.25). This kind of joint is filled with fluid, which, when combined with the slipperiness of the cartilage on the bone ends, allows bones to move against each other in a nearly friction-free environment. Cartilage on cartilage in a healthy synovial joint generates only 20% of the friction of ice on ice.

Tough bands of fibrous connective tissue called **ligaments** form the joint capsule. Lining the interior of the cavity is the **synovial membrane,** which secretes the lubricating synovial fluid. The amount of synovial fluid in joints decreases with age, which is one reason why older people have stiffer joints. Within a synovial joint, small membrane-lined packets called **bursae** store the fluid and help to reduce friction between bones and other nearby structures such as skin and muscle. Calcium deposits in the bursae can cause an inflammation known as **bursitis.** Tennis elbow is a familiar example of bursitis.

Freely movable joints are a prime target for disease and injury. The most common type of joint problem is **arthritis,** or joint inflammation. The most serious form, rheumatoid arthritis, is an inflammation of the synovial membrane, usually of the small joints of the hands and feet (fig. 34.26). Rheumatoid arthritis may be caused by a faulty immune system that attacks the synovial membranes. In the more common osteoarthritis, joint cartilage wears away. As the bone is exposed, small bumps of new bone begin to form. Osteoarthritis usually manifests itself as stiffness and soreness in certain joints after age 40.

Lever Systems

Muscles, joints, and bones interacting to generate movement form biological lever systems (fig. 34.27). A lever is a structure that can pivot around a point, called the fulcrum, when force is applied. In the body, bones function as levers, joints are fulcrums, and skeletal muscles supply the force. Tendons pass force generated in muscles to bones.

A tendon attaches a skeletal muscle to one end of each of two bones. The muscle's **origin** is the end on the bone that does not move, and the muscle's **insertion** is the end on the movable bone.

Each movable bone is flanked by two muscles or muscle groups that move it in opposite directions and are therefore termed **antagonistic muscles.** The muscles of the upper arm, for example, form an antagonistic pair. When you contract the biceps (the muscle that bulges when you "make a muscle"), your arm bends at the elbow. When you contract the triceps, the muscle at the back of the upper arm, your arm straightens.

Typically, one member of an antagonistic muscle pair is stronger than the other. For example, the muscles that close the jaw are more powerful than those that open it. (If you ever have to wrestle an alligator, remember that it is easier to hold its mouth closed than to keep it open.) In the arm, the biceps, which is used for lifting, is generally stronger than the triceps. Likewise, the

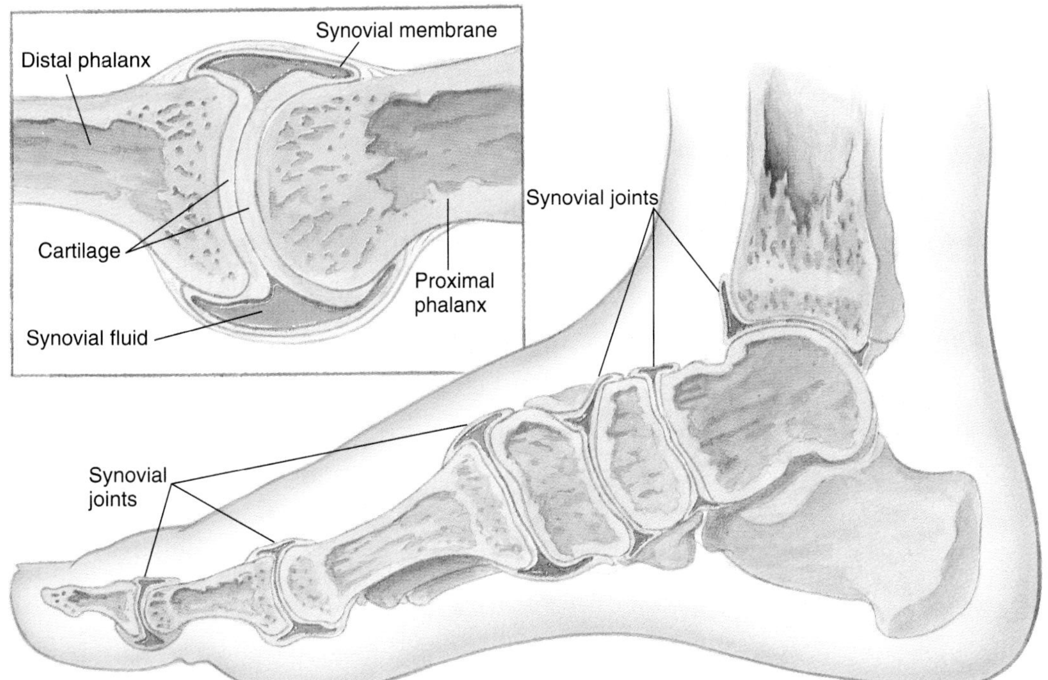

FIGURE 34.25

The synovial joint—where bone meets bone. A synovial joint is a fluid-filled capsule of fibrous connective tissue ligaments. This illustration shows the synovial joints in the foot and ankle. The enlargement shows phalanges (bones) of the big toe.

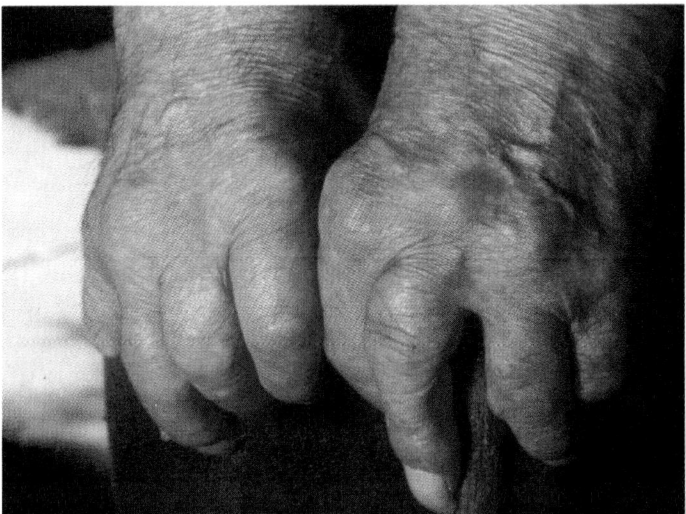

(a)

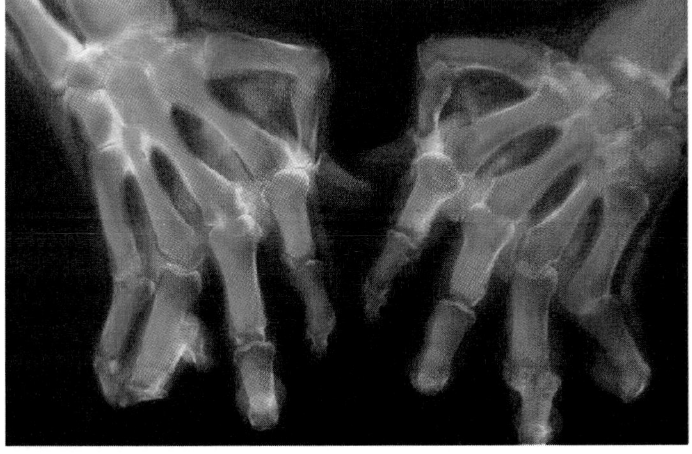

(b)

FIGURE 34.26

Arthritis inflames joints. (*a*) The characteristic gnarled hands of a person who has rheumatoid arthritis. (*b*) Note the swollen joints, visible in this x-ray.

calf muscle, which must lift the weight of the body with every step, is more powerful than the muscles near the shin, which lift only the foot.

Muscle Tone

Just functioning in the presence of gravity requires muscle contraction. At any given time, some fibers in skeletal muscle are contracted, giving muscle a property called **muscle tone.** Receptors called **muscle spindles** monitor the degree of tone in a muscle. Sensory information from the spindles travels to the brain, which adjusts muscle tone (fig. 34.28). When we faint, we temporarily lose the constant interplay between muscle spindles and the nervous system, and we lose muscle tone and collapse.

Effects of Exercise on Muscle

Regular exercise strengthens the muscular system and enables it to use energy more efficiently.

Consider someone who runs 6 miles six times a week. When an individual runs, muscles consume more than 90% of the total energy generated in the body. During the few months after the runner begins training, gradually increasing to 6 miles, leg muscles noticeably enlarge. This exercise-induced increase in muscle mass, or **hypertrophy,** is attributable to an increase in the size of individual skeletal muscle cells rather than to an increase in their number.

Getting into good physical shape makes obvious changes at the microscopic level, too (Biology in Action 34.2). The enzymes

FIGURE 34.27

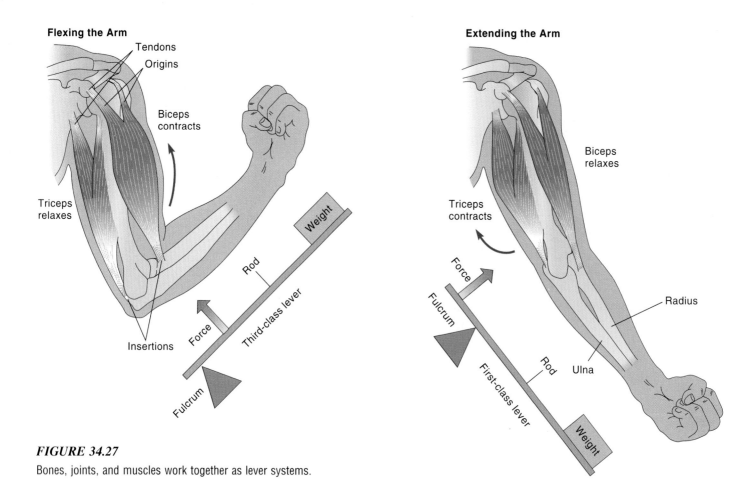

Bones, joints, and muscles work together as lever systems.

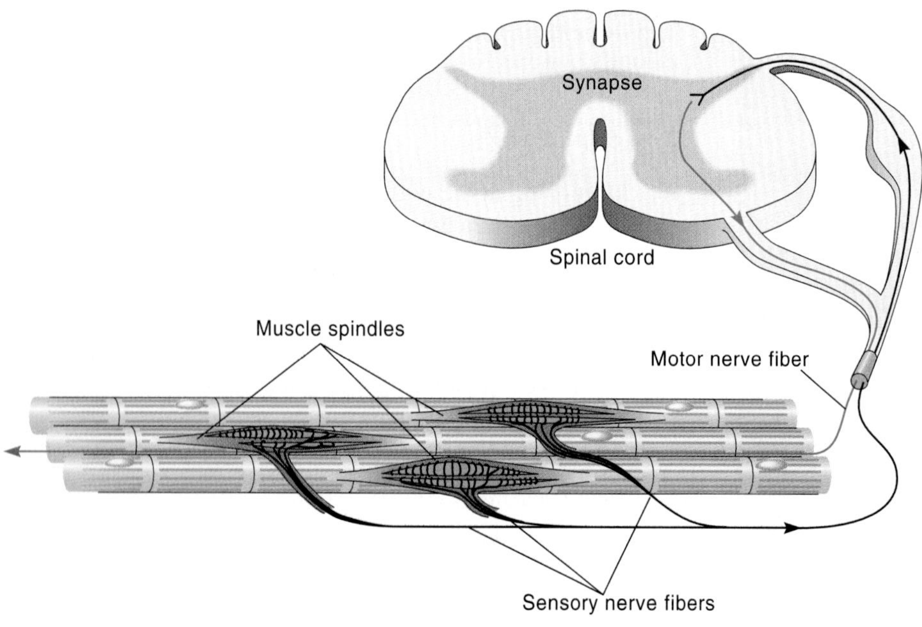

FIGURE 34.28

Muscle tone results when some of the muscle fibers within a muscle contract. Sensory receptors called muscle spindles detect the stretch of a muscle and send the information to the central nervous system to adjust muscle tone. This illustration shows the reflex pathway responsible for muscle tone.

within a trained runner's muscle fibers are both more active and more numerous, and the mitochondria are larger and more abundant than the corresponding structures in a sedentary person. As a result, the runner's muscles can withstand far more exertion than the untrained person's before they respire anaerobically. The athlete's muscles also receive more blood flow and store more glycogen than those of an untrained person.

Exercise-induced hypertrophy is even more pronounced in a weight lifter, because the muscles are greatly stressed by the resistance of the weights. The opposite condition, muscle degeneration resulting from lack of use or immobilization in a cast, is called **atrophy.**

The changes in muscles due to regular exercise disappear quickly if activity stops. After just 2 days of inactivity, mitochon-

34.2 biology in action

Inborn Athletic Ability and Muscle Fiber Types

What is a "natural athlete"? Are some individuals born with certain body features that give them a competitive edge? One way this might occur is when a person inherits specific types of muscle fibers.

Most skeletal muscles contain fibers of three different twitch types, distinguished by how quickly they contract and tire (fig. 34.C). **Slow-twitch** (fatigue-resistant) **fibers** contract slowly because the myosin heads split ATP slowly. They resist fatigue, however, because they are well supplied with oxygen. This oxygen is bound to a large number of molecules of oxygen-carrying myoglobin or is delivered by the extensive blood supply serving the muscle fibers. Myoglobin is similar to hemoglobin, the oxygen carrier in the blood supply, and like hemoglobin is reddish. Therefore, these slow-twitch, slow-fatiguing muscles are called red or dark fibers. They are abundant in muscles specialized for endurance. For example, the breast meat (flight muscles) of ducks and geese, which require muscle endurance for long, migratory flights, is reddish "dark meat."

The second type of muscle fiber, **fast-twitch** (fatigable) **fibers,** splits ATP quickly but does not have much oxygen. These fibers are white because they lack myoglobin and a rich blood supply. They are specialized for short bouts of rapid contraction. The very white breast muscle of the domesticated chicken, for example, can power barnyard flapping for a short time but cannot support sustained long-distance flight.

Intermediate fibers have the fast ATP-splitting characteristic of fast-twitch fibers but enjoy a better blood supply. The speed and fatigue properties are intermediate between the two main types.

The proportion of fast-twitch to slow-twitch fibers within particular muscles affects athletic performance. Most people have about equal numbers of fast- and slow-twitch muscle fibers. Those who have a higher proportion of slow-twitch fibers excel at endurance sports, such as long-distance biking, running, and swimming. Consider Alberto Salazar, who holds many first-place finishes in the marathon but does not do nearly as well in the 400-meter dash. Salazar's leg muscles consist of about 93% slow-twitch fibers. By contrast, athletes who have a higher proportion of fast-twitch fibers tend to perform best at short, fast events such as sprinting, weight lifting, and shot putting. In some European nations, ratio of fast- to slow-twitch fibers is used as a predictor of athletic success in certain events. Athletes give small samples of muscle tissue to test for twitch fiber proportions.

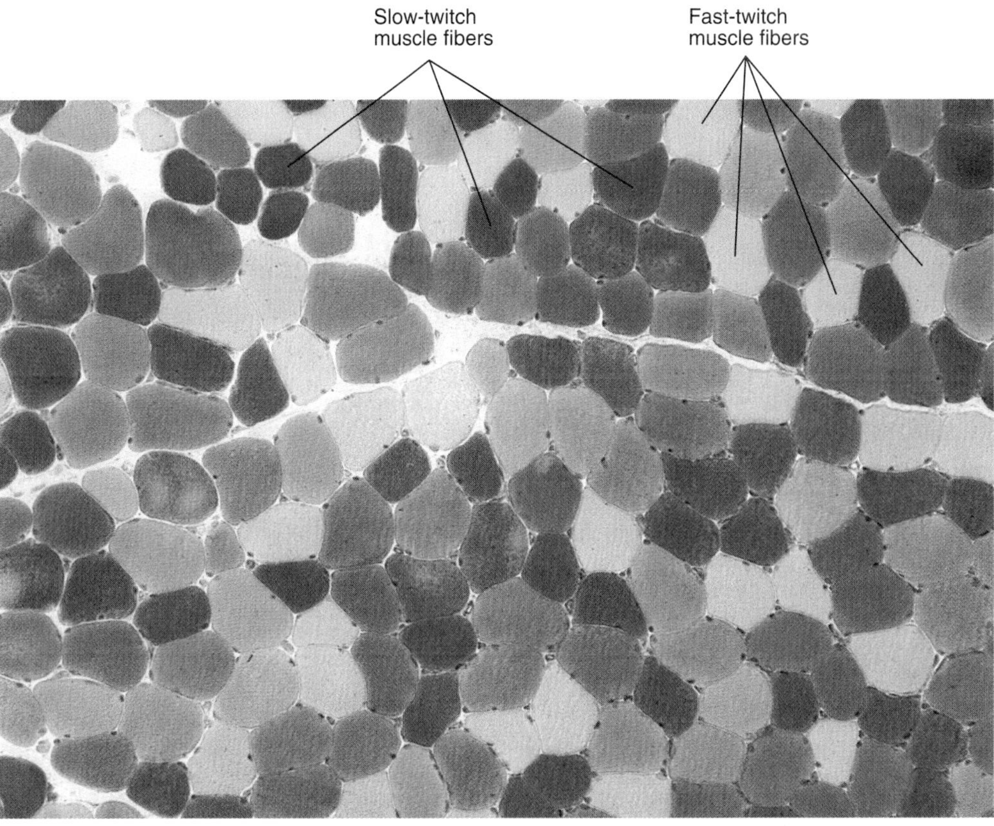

FIGURE 34.C
Types of skeletal muscle fibers. The darker fibers are the slow-twitch fibers, which are specialized for endurance. The dark color comes from myoglobin, a molecule that binds oxygen. The lighter fibers are fast-twitch fibers. They tire more quickly than the dark fibers.

Polio—Then and Now~

continued from page 682.

In 1982, a physician and public health specialist named Marilyn Fletcher went to the National Institutes of Health, begging that the biomedical community investigate an apparently new illness in people who had survived polio. Dr. Fletcher, herself affected, described the great fatigue, new muscle weakness, muscle and joint pain, difficulty sleeping and breathing, and headache that triggered memories of polio. Most frustrating, she reported, was the fact that physicians simply did not take the complaints seriously or misdiagnosed them. Many doctors today are unfamiliar with polio, as well as with its aftermath, called postpolio syndrome.

Dr. Fletcher's persistence finally directed attention to the condition, stimulating scientific meetings, research grants, and recognition and financial aid to sufferers. Government agencies cooperated to identify people claiming disability from polio by sending out questionnaires about symptoms. Gradually, a description of this new illness emerged. Although we still do not know why some survivors develop postpolio syndrome and others do not, the growing recognition should help people who lost control of part of their musculoskeletal systems in the 1950s not repeat their nightmare.

drial enzyme activity drops in skeletal muscle cells. After a week without exercise, aerobic respiration efficiency falls by 50%. The number of small blood vessels surrounding muscle fibers declines, lowering the body's ability to deliver oxygen to the muscle. Lactic acid metabolism becomes less efficient, and glycogen reserves fall. After 2 or 3 months of inactivity, the benefits regular exercise provides have all but disappeared.

Along with the muscular and nervous systems, the skeletal system endows us with the capacity to carry out an enormous variety of bodily motions. To ensure that this masterpiece of natural engineering continues to serve its functions throughout the life span, we must use it regularly and provide it with the nutrients necessary to sustain the continual microscopic ebb and flow of its components.

Mastering Concepts

1. What structures form a joint?
2. How do muscles, joints, and bones form levers?
3. What is muscle tone?
4. How does exercise affect muscle?

SUMMARY

1. The musculoskeletal system helps to maintain homeostasis by enabling an animal to move in response to environmental stimuli.

2. An animal's skeleton supports its body, protects soft tissues, and enables the animal to move. A **hydrostatic skeleton** consists of tissue containing constrained fluid. A **braced framework,** which has solid components, can be on the organism's exterior as an **exoskeleton** or within the body as an **endoskeleton.** Exoskeletons include shells that thicken and grow throughout life and jointed suits of armor that are periodically shed and replaced. An endoskeleton grows along with the organism.

3. The vertebrate skeleton supports, protects, and attaches to muscles and stores minerals. The **axial skeleton** consists of the **skull, vertebral column,** breastbone, and ribs. The **appendicular skeleton** includes the limbs and the limb girdles (**pectoral** and **pelvic**) that support them. Bones in different vertebrate species can differ in width and density.

4. Cartilage and bone make up skeletons. Cartilage entraps a great deal of water, which makes it an excellent shock absorber. Bone has a rigid matrix and derives its strength from collagen and its hardness from minerals (calcium and phosphate). In compact bone, osteocytes form osteons. The central channel for blood supply is an **osteonic canal.** Canaliculi connect lacunaes that house bone cells and the osteonic canal. Bone cells extend through canaliculi and pass materials from cell to cell. Bone continually degenerates and builds up.

5. Smooth muscle cells are spindle-shaped and involuntary, and they line organs, push food through the digestive tract, and regulate blood flow and pressure. **Cardiac muscle** cells, in the heart, are striated and involuntary and are joined by **intercalated disks** into a branching pattern. Skeletal muscle cells are striated, voluntary, and move bones.

6. Sliding protein filaments in muscles provide movement in diverse organisms. **Tendons** attach an intact whole skeletal muscle to bones. Within a muscle are bundles of **muscle fibers.** Each muscle fiber is a long cylindrical cell composed primarily of thick and thin **myofilaments.** The thick myofilaments are bundles of myosin molecules, each with a head (cross bridge) attached to a shaft. The heads project outward from each end of a myosin bundle. The thin myofilaments are composed of actin plus **troponin** and **tropomyosin.**

7. A muscle fiber cell is a chain of many contractile units called **sarcomeres.** Myofilaments within a sarcomere give the tissue its striated appearance. According to the **sliding filament model,** muscle contraction occurs when the thick and thin myofilaments are pulled past one another so that they overlap more.

8. When a motor neuron stimulates a muscle fiber, the neurotransmitter acetylcholine is released at a neuromuscular junction. Electrical charges enter the muscle cell causing the **sarcoplasmic reticulum** to release calcium ions, which bind to the troponin molecules on actin myofilaments. As a result, troponin moves and no longer prevents actin from binding myosin. Once myosin cross bridges touch actin, ATP attached to the myosin head splits and the head moves, causing the actin myofilament to slide past the myosin myofilament. ADP and inorganic phosphate are released. A new ATP binds to the myosin head and the cross bridge to actin breaks, the myosin head returns to its original position, and a new cross bridge forms further along the actin myofilament.

9. After the nerve impulse, calcium ions are actively pumped back into the sarcoplasmic reticulum and tropomyosin moves to prevent actin and myosin interactions. This leads to muscle relaxation.

The energy for muscle contraction comes first from stored ATP, which is quickly exhausted. For a short time new ATP can form from **creatine phosphate** stored in muscle cells, and then aerobic respiration generates ATP. If ATP use exceeds the ability of muscle cells to produce ATP aerobically, the cells switch to anaerobic respiration. An important source of energy for muscle contraction is glycogen.

10. When stimulated, a muscle cell responds in an all-or-none fashion, although not all of the many muscle cells in a whole muscle contract. When a muscle cell is stimulated once, it contracts and relaxes, a response called a **twitch.** If the stimulation rate increases, the muscle cell does not completely relax between pulses, and the response strengthens. At a high rate of stimulation, muscle cells reach a sustained state of maximal contraction called **tetanus.**

11. **Joints** attach bones. Some joints, such as those holding the skull bones in place, are immovable. Freely moving **synovial joints** consist of cartilage and connective tissue **ligaments** that contain lubricating synovial fluid.

12. Most voluntary muscles attach to bones, forming lever systems. When a muscle contracts, a bone moves. The muscle end attached to the stationary bone is its **origin.** The end attached to the movable bone is called the **insertion.** Muscles form **antagonistic pairs,** enabling bones to move in two directions.

13. Muscles are always partially contracted, generating **muscle tone. Muscle spindles** control muscle tone by sending impulses to the brain or spinal cord when a muscle is stretched.

14. A muscle exercised regularly increases in size (**hypertrophy**) because each muscle cell thickens. An unused muscle shrinks (**atrophies**). Regular exercise causes microscopic changes in muscle cells that enable them to use energy more efficiently.

TO REVIEW . . .

1. Distinguish between a hydrostatic skeleton, an exoskeleton, and an endoskeleton. Give an example of an animal with each.

2. What advantages and disadvantages does a jointed exoskeleton have over a shell? What advantage does an endoskeleton have over a jointed exoskeleton?

3. List five functions of the human skeleton.

4. What role does cartilage play in the skeletal system?

5. What are the two major components of bone matrix? How do they work together to give bone its characteristics?

6. How is a bone cell nourished?

7. What are the structural differences between spongy bone and compact bone and between red marrow and yellow marrow?

8. How do bones elongate throughout childhood?

9. List four muscle proteins and their functions.

10. How is an amoeba's movement similar to human locomotion?

11. Define the following terms:

a. sarcolemma
b. sarcomere
c. sarcoplasmic reticulum

12. How does the muscular system interact with each of the following?

a. the nervous system
b. the skeletal system
c. the circulatory system

13. How does a skeletal muscle cell get ATP?

14. How do antagonistic muscle pairs move bones? Give an example of such a pair.

TO THINK ABOUT . . .

1. How can the musculoskeletal systems of diverse vertebrate species consist of the same molecules, cells, and tissues, and have similar organization, yet also be adapted to a particular species' way of life?

2. Osteogenesis imperfecta is an inherited disorder that affects the skeleton. Children with the condition lack a form of collagen that is normally part of bone. Their bones break very easily, even when they make normal movements, and may even break before birth. How can the molecular defect causing osteogenesis imperfecta explain the symptoms?

3. Archaeologists have uncovered skeletal remains of our humanlike ancestors in Ethiopia. Examination of some of the bones suggests that the remains represent four types of humans. Specifically, microscopic examination shows that the bone masses of some skeletons are about 30% less than those of the other type, and the skeletons with the lower bone mass have broader front pelvic bones than those of the heavier bone mass group. Within the two groups defined by bone mass and pelvic bone shape, smaller skeletons have bones with evidence of epiphyseal plates, but larger bones have only a thin line where the epiphyseal plates should be. What might the four types of individuals in this find be?

4. Cartilage cells produce a chemical that suppresses the growth of blood vessels. A tumor requires a large blood supply to grow and spread. How might doctors use the cartilage chemical to fight cancer? Why would a shark be a better source of cartilage for this substance than a rat?

5. If a bedridden hospital patient is not moved for a long time, the patient's cartilage begins to degenerate. What is the molecular basis for this?

6. What roles does fluid play in hydroskeletons, cartilage, and synovial joints?

7. Spondylitis is a painful condition in which ligaments in the backbone harden. How could this impair the functioning of the backbone? What might the symptoms be?

8. What functions might hollow bones have served in large dinosaurs?

9. A man exercises intensively, building up his muscles. He believes he will pass this hypertrophy on to his future children. Why is he incorrect?

10. The aerobics instructor chants, "Just concentrate, and feel your muscles expand and contract." Is her statement an accurate description of muscle action? Why or why not?

11. A father wants to test his healthy daughter's muscles to determine the percentage of slow-twitch and fast-twitch muscle fibers to decide if she should try out for soccer or cross-country (long-distance) running. Do you think this is a valid reason to test muscle tissue? Why or why not?

TO LEARN MORE . . .

References and Resources*

Allen, William H. June 1995. Animals and their models do their locomotions. *BioScience,* vol. 45. Researchers study locomotion in crabs, cockroaches, and centipedes running on tiny treadmills.

Karpati, George. January 1997. Utrophin muscles in on the action. *Nature Medicine.* Researchers are discovering new muscle proteins.

Kearney, Jay T. June 1996. Training the Olympic athlete. *Scientific American.* Athletic success depends on an optimally functioning musculoskeletal system.

Lewis, Ricki. July/August 1991. Arthritis: Modern treatments for that old pain in the joints. *FDA Consumer.* New hope for the many people who suffer from arthritis.

Mooseker, Mark. 1993. A multitude of myosins. *Current Biology,* vol. 3, no. 4. Yeast and humans use myosins for movement.

Pennisi, Elizabeth. February 21, 1997. A new view of how leg muscles operate on the run. *Science,* vol. 275. How do turkeys trot?

Zill, Sasha N., and Ernst-August Seyfarth. July 1996. Exoskeletal sensors for walking. *Scientific American.* Specialized sense organs enable arthropods to coordinate complex movements.

Zuckerman, J. D. June 6, 1996. Hip fracture. *The New England Journal of Medicine,* vol. 334. Hip fractures disable many people.

**Life*'s website features frequently updated references.

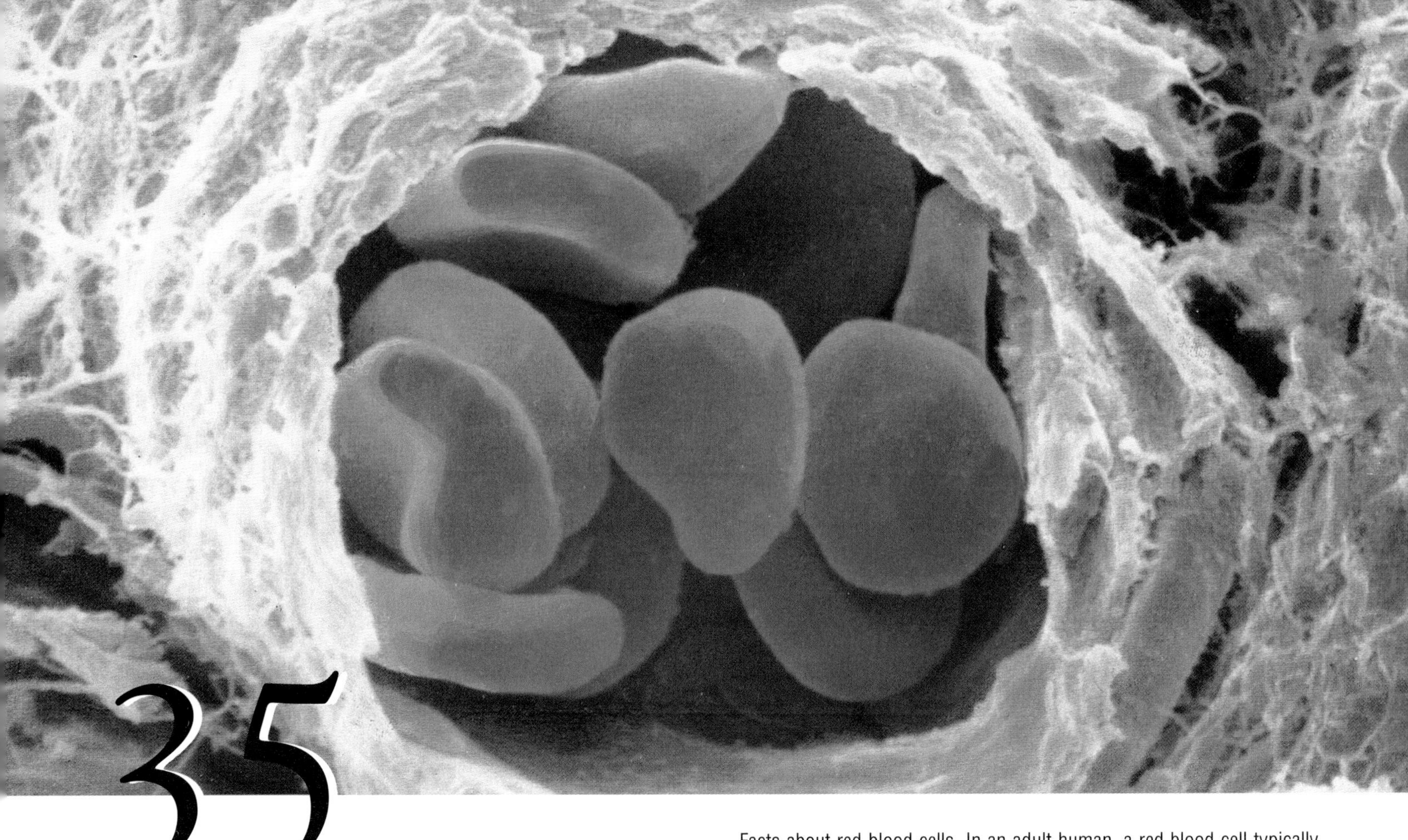

35 The Circulatory System

Facts about red blood cells. In an adult human, a red blood cell typically lives 120 days and travels through the bloodstream some 50,000 times. The 25 trillion red blood cells, if spread out, would cover 4,784 square yards (4,000 square meters). Mature red blood cells in mammals are biconcave discs that lack nuclei. In other vertebrates, red blood cells are oval and have nuclei. An exception is the red blood cell of a camel—it is oval, but lacks a nucleus.

chapter outline

***Circulatory Systems* 708**

Types of Circulatory Systems
Respiratory Pigments
Vertebrate Circulatory Systems

***The Human Circulatory System* 710**

Blood Composition
The Human Heart
Blood Vessels
Blood Vessels and Circulation
Blood Pressure
Control of Circulatory Functions

***Exercise and the Circulatory System* 727**

***The Lymphatic System* 727**

Treating Sickle Cell Disease

Having sickle cell disease is like suffering from several disorders. Sluggish blood cuts off circulation, causing the equivalent of "ministrokes" anywhere in the body. Blocked circulation in a joint causes pain that one boy described as "having your hand or foot smashed in a slamming car door." In the brain, blockage can cause paralysis; in the legs, painful ulcers; in the eyes, blindness. As the spleen works to get rid of abnormally shaped red blood cells, anemia (too few red blood cells) develops, and susceptibility to infection increases. Over the years, the intermittently blocked circulation steadily damages the kidneys, lungs, liver, and bones.

Sickle cell disease results from just one substituted amino acid in the beta globin chains of the oxygen-carrying protein hemoglobin. The abnormal beta chains link in a way that deforms the red blood cell containing them into a sickle shape. The misshapen cells lodge in narrow blood vessels when the surroundings are low in oxygen. Once oxygen is replenished at the lungs, hemoglobin binds oxygen, and the red blood cells pop back into their normal doughnut shapes.

The only cure for sickle cell disease is a bone marrow transplant, but it is dangerous. Daily antibiotics begun in childhood can prevent infections, and painkillers and fluid replacement can ease painful "crises." In 1995 came a breakthrough. An already available drug was found to halve the frequency of sickle cell crises. The drug, hydroxyurea, works in an unusual way that is based upon the function of the circulatory system—to replenish the blood with oxygen.

to be continued . . .

Circulatory Systems

The first European settlements in the New World were not much more than makeshift shelters and, after some months, log cabins. As towns grew into cities, the settlers had to find ways to distribute supplies arriving by sea, to trade crops, and to keep the colony functioning. In short, people needed effective transportation systems.

Animals are like towns and cities—they require transportation systems to bring in supplies (nutrients and oxygen for energy) and remove garbage (metabolic wastes) without disturbing their internal environments. Just as small towns can accomplish these tasks with simpler systems than can larger cities, animals with simpler bodies can sustain life with less complex systems. Diffusion across the body surface suffices in some animals; in more complex animals, circulatory systems transport energy supplies and wastes to and from individual cells, and wastes can diffuse across the body surface to sustain life. In a protozoan (fig. 35.1*a*), molecules enter and leave by diffusing across the cell membrane, and move inside the cell by cytoplasmic streaming. A hydra's central cavity has lining cells that exchange gases with and release wastes into incoming water (fig. 35.1*b*). In flatworms, a central cavity branches so that all cells lie close to a branch or to the body surface. This enables the animal to use diffusion to meet metabolic demands (fig. 35.1*c*). In hydra and flatworms, fluid moves through the central cavity as muscles within the body wall contract.

A starfish has a fluid-filled body cavity, a coelom, that contacts internal structures (fig. 35.1*d*). Openings on the animal's surface link its coelom directly to the outside and allow nutrients in seawater to enter the body. The coelomic fluid washes through the body as cilia lining the coelom beat. This simple mechanism is sufficient for the starfish's sedentary lifestyle on the ocean floor.

In animals more active than flatworms, cells require more raw materials and oxygen and generate more waste. More active animals, however, are usually larger, and the increase in size decreases the animal's surface-to-volume ratio. This makes diffusion inefficient for distributing raw materials throughout the organism. Circulatory systems in large, active animals deliver vital materials to cells and remove wastes efficiently enough to permit the animal to take in sufficient energy to move, yet maintain its internal environment.

Types of Circulatory Systems

A circulatory system transports fluid in one direction, powered by a pump that forces the fluid through vessels that reach throughout the body. Circulatory systems evolved in conjunction with respiratory systems. That is, animals that have organs (lungs or gills) that compartmentalize gas exchange also have systems that transport oxygen throughout the body.

In an **open circulatory system,** the fluid is not contained in vessels. Most mollusks and arthropods have open circulatory systems. These systems consist of a heart and blood vessels that lead to spaces where the fluid, hemolymph, directly bathes cells before returning to the heart (fig. 35.2). Invertebrates with open circulatory systems can be fairly active because their respiratory systems branch in a way that allows the outside environment to come close to internal tissues. Insects, for example, are among the most active animals, and their respiratory systems operate independently of their open circulatory systems.

In a **closed circulatory system,** blood remains within vessels. Large vessels called **arteries** conduct blood away from the heart and branch into smaller vessels, called **arterioles,** which then diverge into a network of very tiny, thin vessels called capillaries. Materials diffuse between cells and blood across the walls of capillaries. Blood then collects into slightly larger vessels, called **venules,** which unite to form still larger vessels, the **veins,** which carry blood back to the heart. Annelids are the simplest animals with a closed circulatory system.

Squids, which are mollusks, have an interesting variation on a closed circulatory system. They have two extra hearts in their extensive network of gills, which boost blood circulation through the dense vessels in that region. The extra pumping action and very large gills help the animal to move very quickly.

Respiratory Pigments

In smaller, less active invertebrates, the fluid that transports nutrients, dissolved gases, and wastes is clear and watery. It contains salts, proteins, and phagocytes (protective cells that engulf smaller

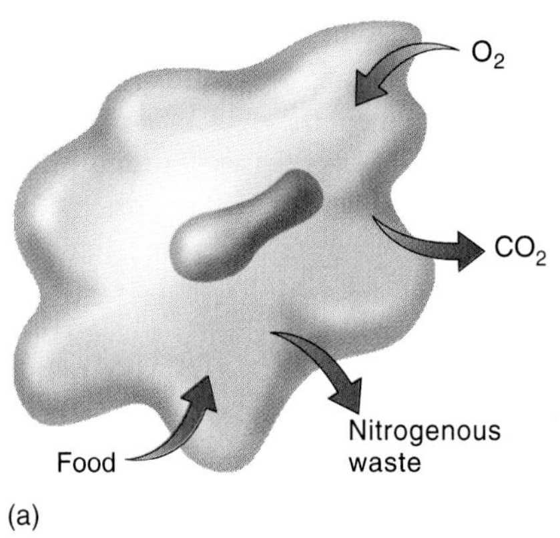

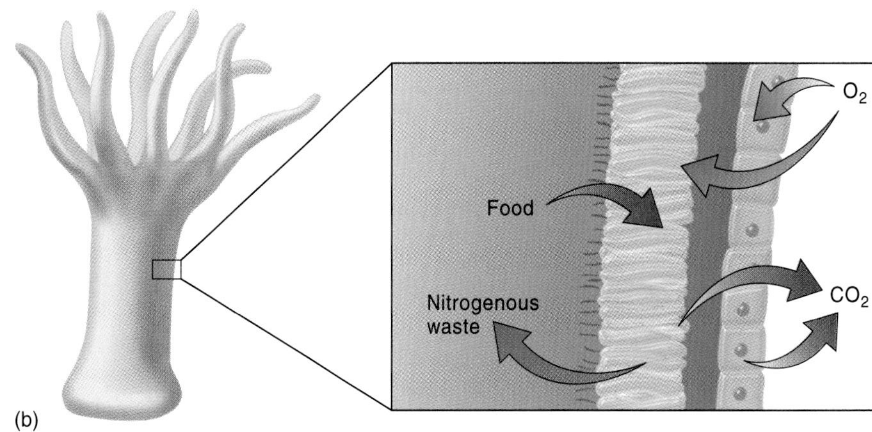

FIGURE 35.1

Exchange of material by diffusion across the body surface. The cells of small organisms such as protozoa (*a*), cnidaria such as the hydra (*b*), flatworms (*c*), and starfishes (*d*) maintain close enough contact with the environment that the animals do not require a circulatory system.

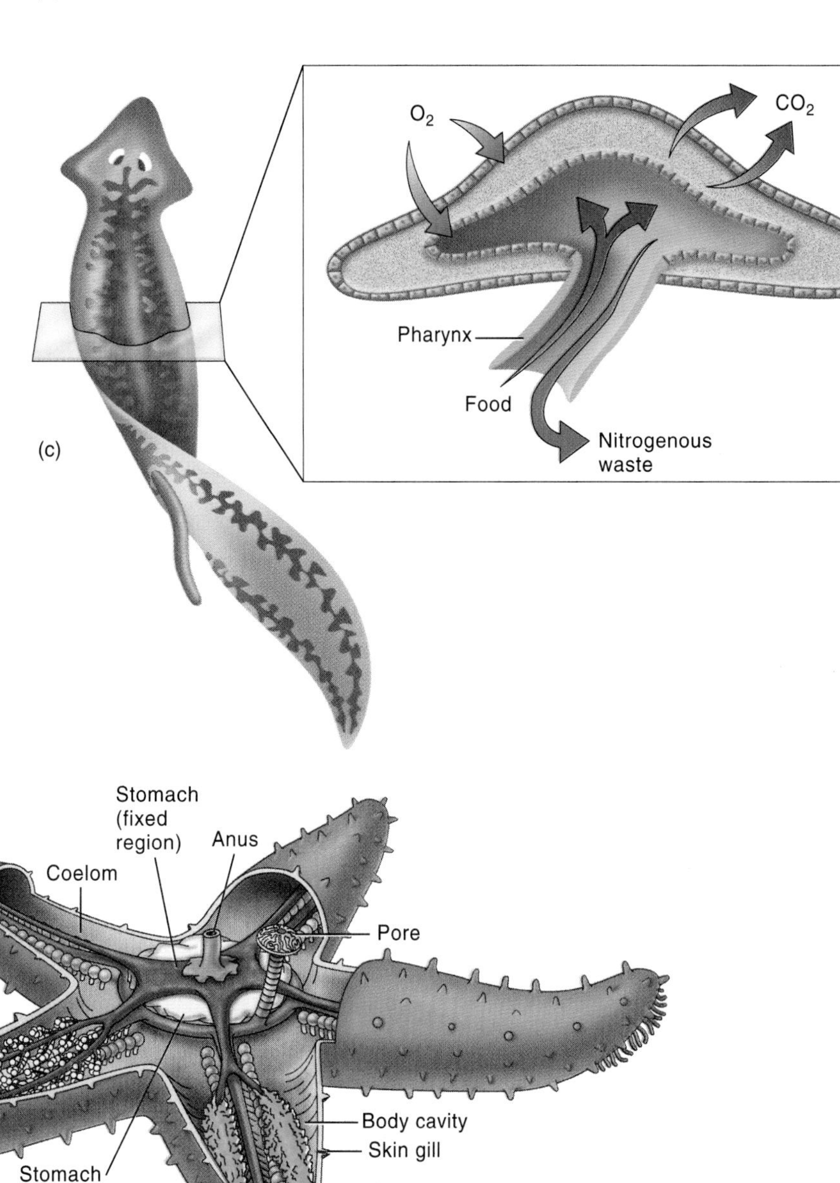

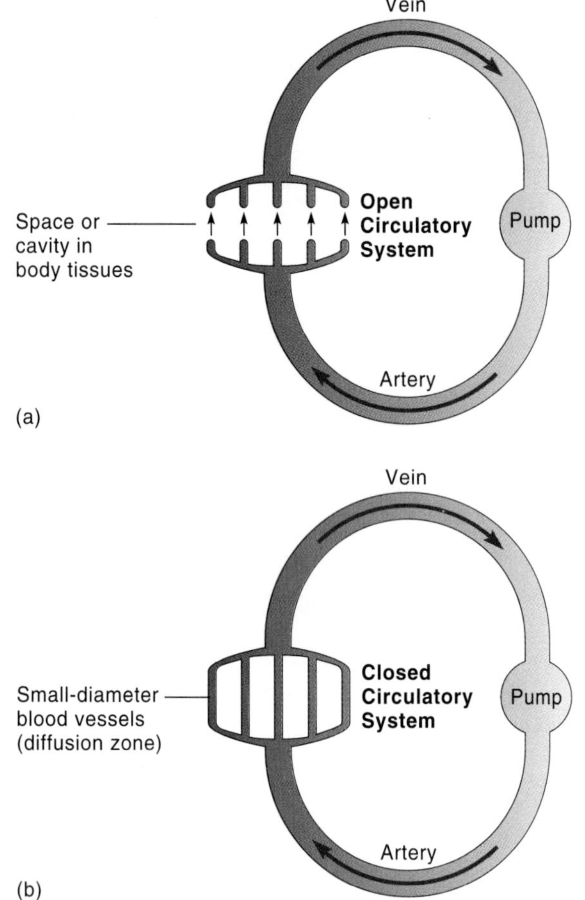

FIGURE 35.2

General scheme of open and closed circulatory systems. (*a*) In an open circulatory system, blood bathes cells directly. (*b*) In a closed circulatory system, blood flows within vessels.

cells and particles). In larger and more active animals, the circulatory system fluid contains pigment molecules that carry additional oxygen. Respiratory pigments must be able to bind oxygen where it is plentiful, yet readily release it in oxygen-depleted tissues, where it is required. **Hemoglobin** is the respiratory pigment in mammals, birds, reptiles, amphibians, fishes, annelids, and some mollusks. Because of hemoglobin, a milliliter of a mammal's blood carries 25 times as much oxygen as a milliliter of seawater.

Another respiratory pigment, **hemocyanin,** carries three times as much oxygen as seawater. Some mollusks, such as the horseshoe crab, have hemocyanin. Hemoglobin contains iron, which gives blood its rich red color, and hemocyanin contains copper, which gives the horseshoe crab's blood a metallic blue color. In some animals, such as the annelids, hemoglobin is free in the blood; in vertebrates, cells contain hemoglobin. A single human red blood cell holds about 265 million hemoglobin molecules, comprising a third of the cell's mass.

Vertebrate Circulatory Systems

In vertebrates the circulatory system has a highly specialized heart, blood containing red and white cells in a closed system, and hemoglobin in cells. Circulatory systems become increasingly complex among the fishes, amphibians, reptiles, birds, and mammals, probably because of adaptations to the different physical demands as animals moved onto land. Again, the increasing complexity of the circulatory system parallels the evolution of respiratory structures. Aquatic species exchange gases through gills, which are outfoldings of the integument (skin). Other vertebrates exchange gases through lungs, which are highly folded internal structures.

A fish's blood flows through the heart only once in each circuit around the body (fig. 35.3*a*). The heart pumps the blood through the gills and then to the rest of the body before it returns to the heart. Blood flow slows through the gills, where it is under very low pressure. A fish's heart has two chambers—one **atrium** (where blood enters) and one **ventricle** (from which blood exits).

In amphibians and reptiles, blood makes two trips through the heart in each circuit around the body. The three-chambered amphibian heart has a left and a right atrium and one ventricle. A reptile's heart also has two atria and one ventricle, but the ventricle is partially divided. The division helps to separate blood that has picked up oxygen (oxygenated blood) from blood depleted of oxygen (deoxygenated blood). An amphibian does not require a divided ventricle because it obtains much of its oxygen through the mouth surface and skin.

Birds and mammals have four-chambered hearts with two atria and two ventricles that circulate blood separately through the lungs and the rest of the body. Such double circulation is adaptive because the ventricles can divide the work, ensuring that blood circulates through the entire body, including the extremities. Also, separating oxygenated from deoxygenated blood maximizes the amount of oxygen reaching tissues.

Mastering Concepts

1. By what process do single-celled organisms obtain nutrients and rid themselves of wastes?
2. How does the organization of cells in a simple multicellular animal enable it to obtain nutrients and dispose of wastes by diffusion?
3. What are the general components of a circulatory system?
4. How do open and closed circulatory systems differ from each other?
5. What is the function of a respiratory pigment?
6. How are the circulatory systems of fish, amphibians and reptiles, and birds and mammals increasingly complex?

The Human Circulatory System

The human circulatory system consists of a central pump, the heart, and a continuous network of tubes, the blood vessels. This system is also called the cardiovascular system ("cardio" refers to the heart, "vascular" to the vessels). The circulatory system transports blood, a complex fluid that nourishes cells, removes cellular waste, and helps to protect against infection. The circulatory system is so extensive that no cell is more than a few cell layers away from one of the system's smallest branches. A defect or malfunction in the heart or a major vessel can be fatal (Biology in Action 35.1). For this reason, as figure 35.4 points out, researchers are developing circulatory system replacement parts.

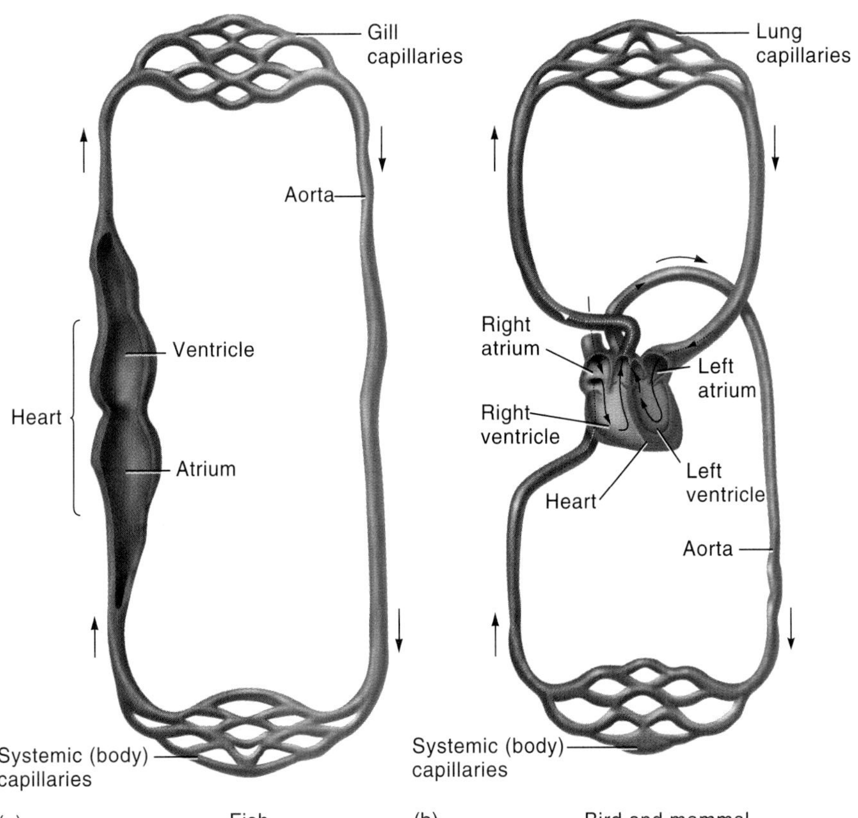

FIGURE 35.3

Heart structural diversity. (*a*) A fish's two-chambered heart allows blood to flow in a single circuit around the body. (*b*) A bird or mammal has a four-chambered heart, supporting two trips through the heart in each circuit. This more powerful pumping system enables birds and mammals to be more active.

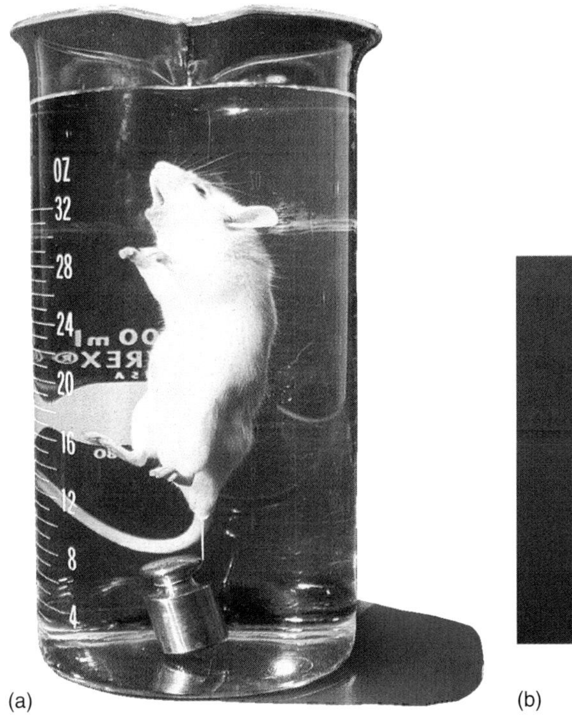

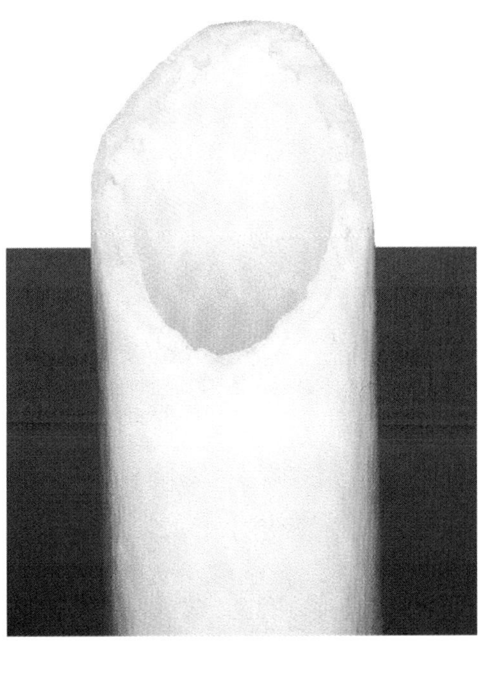

FIGURE 35.4

Cardiovascular "spare parts." (*a*) This mouse is "breathing" a fluorocarbon liquid saturated with oxygen that researchers tested as a blood substitute. (*b*) This blood vessel replacement consists of an inner layer of endothelium, a middle layer of smooth muscle cells, and an outer layer of connective tissue. Cell surface molecules likely to cause an immune reaction are removed, and an internal, synthetic mesh is woven into the structure to give it strength. The blood vessel equivalent can be stitched next to natural blood vessels so smoothly that blood clots do not form as they do in synthetic grafts.

Blood Composition

Blood is a complex mixture that has a number of functions in the body (table 35.1). Blood follows the general connective tissue scheme of cells within a matrix. The matrix of blood is a liquid called **plasma** that contains many suspended or dissolved biochemicals. Blood cells or cell fragments are called **formed elements.**

Plasma

Blood plasma, which comprises more than half of the blood's volume, is 90 to 92% water (fig. 35.5). It is about 1% dissolved molecules, including salts, nutrients, vitamins, hormones, metabolic wastes, and gases. Although the concentrations of the dissolved molecules are low, they are critical. For example, the blood is

35.1 biology in action

The Unhealthy Circulatory System

Coronary Heart Disease

Diseases of the heart and blood vessels account for 53% of all deaths in the United States each year. Nearly 87% of these deaths are due to coronary heart disease, which affects the coronary arteries that nourish the heart. Coronary heart disease often begins with atherosclerosis, when fatty plaques accumulate inside the walls of the coronary arteries and reduce blood flow to the heart muscle. In some people, the fatty plaques build up because their cells cannot remove enough cholesterol from the bloodstream. In others, a lifetime of eating fatty food overwhelms the cell surface receptors that transport cholesterol into cells, and the excess backs up into the blood. Blood vessels also become blocked from overgrowth of smooth muscle cells, blood clot formation, and malfunction of inner lining cells.

Although high levels of cholesterol in the blood increase risk of atherosclerosis, total blood cholesterol level does not tell the whole story. Some types of cholesterol are not as harmful as others. Because cholesterol, a fat, is insoluble in water, a carrier protein must transport it. One carrier, the low-density lipoprotein (LDL), transports cholesterol to artery walls and deposits it. High-density lipoproteins (HDLs) transport cholesterol to the liver for elimination. Exercising and avoiding saturated fats in the diet can raise HDL levels. Guarding against obesity and high blood pressure and eliminating cigarette smoking also help maintain heart health. Atherosclerosis can eventually cause heart attack, angina pectoris, arrhythmia, and congestive heart failure.

Heart Attack

Blocked blood flow in a coronary artery kills the muscle in the region. This is a heart attack, and it is often caused by a blood clot that forms in vessels narrowed by coronary heart disease (fig. 35.A*1*). The dying portion of the heart develops altered electrical activity, which may lead to fibrillation, a dangerous sudden irregularity in the heartbeat. The person may initially feel pressure, fullness, or a squeezing sensation in the chest that spreads to the shoulder, neck, and arms and lasts for at least 2 minutes. He or she may also feel sweaty, short of breath, dizzy, or nauseated.

A heart attack can come on suddenly. The outcome depends to some extent on the availability of help. About two-thirds of heart attack patients die within minutes or hours of the attack. Once help arrives, the person is given a drug to slow the heartbeat and oxygen to keep as much of the heart muscle functioning as possible. In the ambulance, an electrocardiogram locates and assesses the damage while monitoring heartbeat. If the heart continues to beat erratically, an electric shock to the heart may restore a normal beat.

At the hospital in a coronary care unit, doctors monitor heart rate and blood pressure. Dye is injected into the coronary arteries to see which are blocked, blood chemistry is evaluated to see if clots form too quickly, and enzyme measurements tell the extent of heart muscle damage. Because damage continues over the next 72 hours, the patient begins drug therapy to minimize it and to ward off a second heart attack. A "clot-busting" drug, such as tissue plasminogen activator (tPA) or streptokinase, can save a life and eliminate further treatment if the patient receives it within 4 hours.

If a person survives the few weeks following the initial heart attack, and perhaps takes drugs to prevent a second attack, coronary circulation can be restored as other arteries deliver blood to the damaged part of the heart. This natural healing is called collateral circulation (fig. 35.A*2*). If the arteries fail to take over or if atherosclerosis is too severe, a coronary bypass operation (fig. 35.A*3*) may give blood an alternate route to the heart. Otherwise, plaque buildup may continue.

Angina Pectoris

Decreased oxygen flow in the heart due to a partially blocked coronary blood vessel can cause a gripping, viselike pain called **angina pectoris.** It is often difficult to tell whether such pain is a warning of an impending heart attack, an isolated event, or indigestion. The circumstances of the attack provide clues to its origin. *Angina at rest* occurs during sleep and is due to a calcium imbalance or a blood clot rather than atherosclerosis. *Angina of exertion* follows sudden activity in a sedentary person and also is not usually a result of atherosclerosis. *Unstable angina pectoris* is the dangerous variety, producing chest pain either when the person is at rest but awake or is engaged in minimal exertion, such as housecleaning or brisk walking.

Arrhythmia

An arrhythmia is an abnormal heartbeat. Some arrhythmias may be transient flutters or racing that lasts only a few seconds. These arrhythmias usually are centered in the atria. They may reflect malfunctioning pacemaker cells, which slow ventricular pumping so that blood flow is too slow to support life. An electronic pacemaker can supplement faulty pacemaker cells. In **ventricular fibrillation,** pacemaker cells lose control, causing the heart muscle to suddenly twitch wildly. The heart pumps ineffectively, blood pressure plummets and death occurs. Fibrillation causes 77% of in-hospital deaths following heart attack.

Congestive Heart Failure

In **congestive heart failure,** the heart is too weak to maintain circulation. Fluid accumulates in tissues, swelling them. The 5 million U.S. citizens who have congestive heart failure also experience fatigue, weakness, and shortness of breath, and 20,000 of them die each year. Atherosclerosis or defective heart valves can cause congestive heart failure. Treatment includes drugs—digitalis to strengthen the heart muscle and diuretics to rid the body of excess fluid.

Hypertension

Consistently elevated blood pressure, or hypertension, affects 15 to 20% of adults in industrialized nations. Blood pressure of 140/90 is considered high; 200/100 is dangerously high. Hypertension strains the heart, raising the risk of heart attack or failure. The high pressure may cause a stroke and deposit lipids in arteries, which elevates blood pressure further. Because hypertension often has no symptoms, it is important for medical exams to include blood pressure measurement (fig. 35.21).

About 10% of the time, a malfunctioning organ or gland, such as the kidneys, pituitary, thyroid, or adrenal cortex, causes hypertension. Narrowed or hardened arteries account for many cases. Rarely, a faulty gene causes too many nerves to surround small blood vessels, constricting them too much, which elevates blood pressure. A fatty diet, obesity, and smoking increase the risk of developing hypertension.

Stroke

In a stroke, a blood clot or bleeding interrupts circulation in the brain, producing numbness, weakness, loss of balance, or paralysis. Stroke is the third leading cause of death in the United States and the number-one cause of disability. Treating a stroke caused by a blood clot within 3 hours with the "clot-busting" drug tPA improves full recovery rate by 50%. Other drugs limit the damaging effects of the response of cells surrounding the site of the stroke.

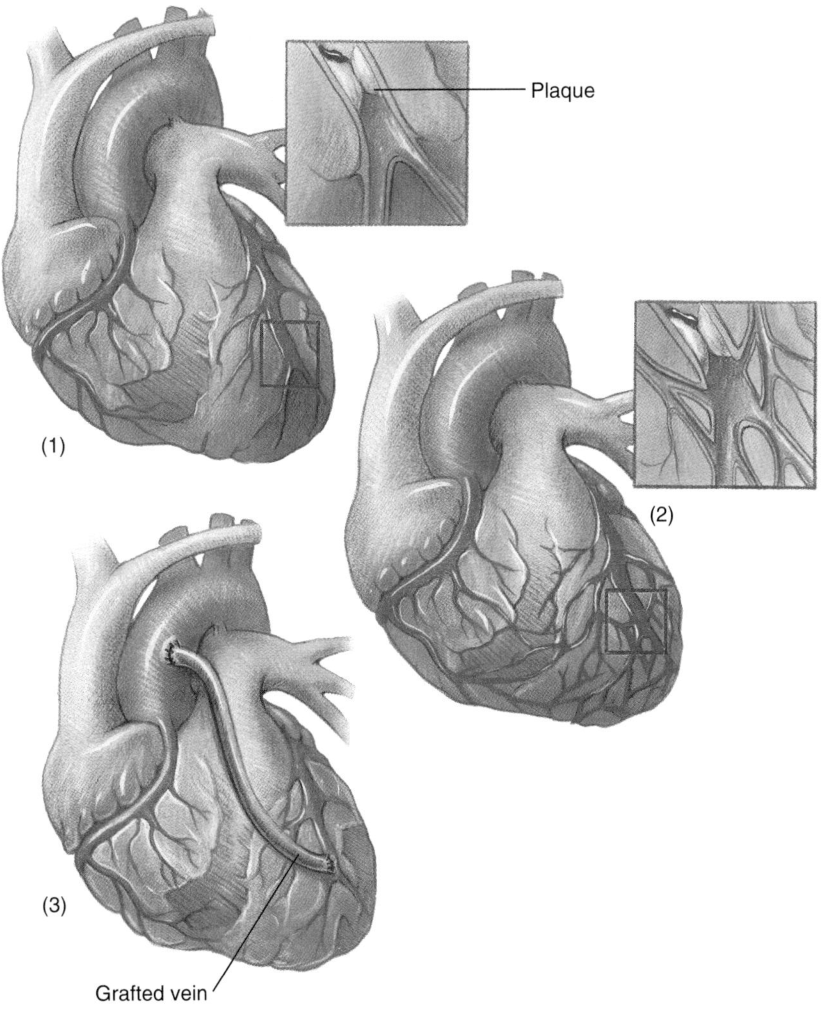

FIGURE 35.A
Bypassing a blocked coronary artery. (*1*) A heart attack has occurred—an atherosclerotic plaque has blocked a coronary artery, and the tissue that artery normally serves, robbed of nutrients and oxygen, dies. (*2*) If the person recovers, other arteries may grow and attach to nonblocked arteries, establishing a collateral circulation to the damaged area. (*3*) A coronary bypass operation uses a vein removed from the person's leg or chest to fashion an alternative pathway for blood past the obstruction.

FIGURE 35.5

Blood composition. Human blood is a complex mixture of formed elements (platelets, red blood cells, and white blood cells) suspended in a liquid plasma that is 90 to 92% water, 7% proteins, and 1% salts, wastes, nutrients, hormones, and dissolved gases. Alterations in blood composition may signify illness.

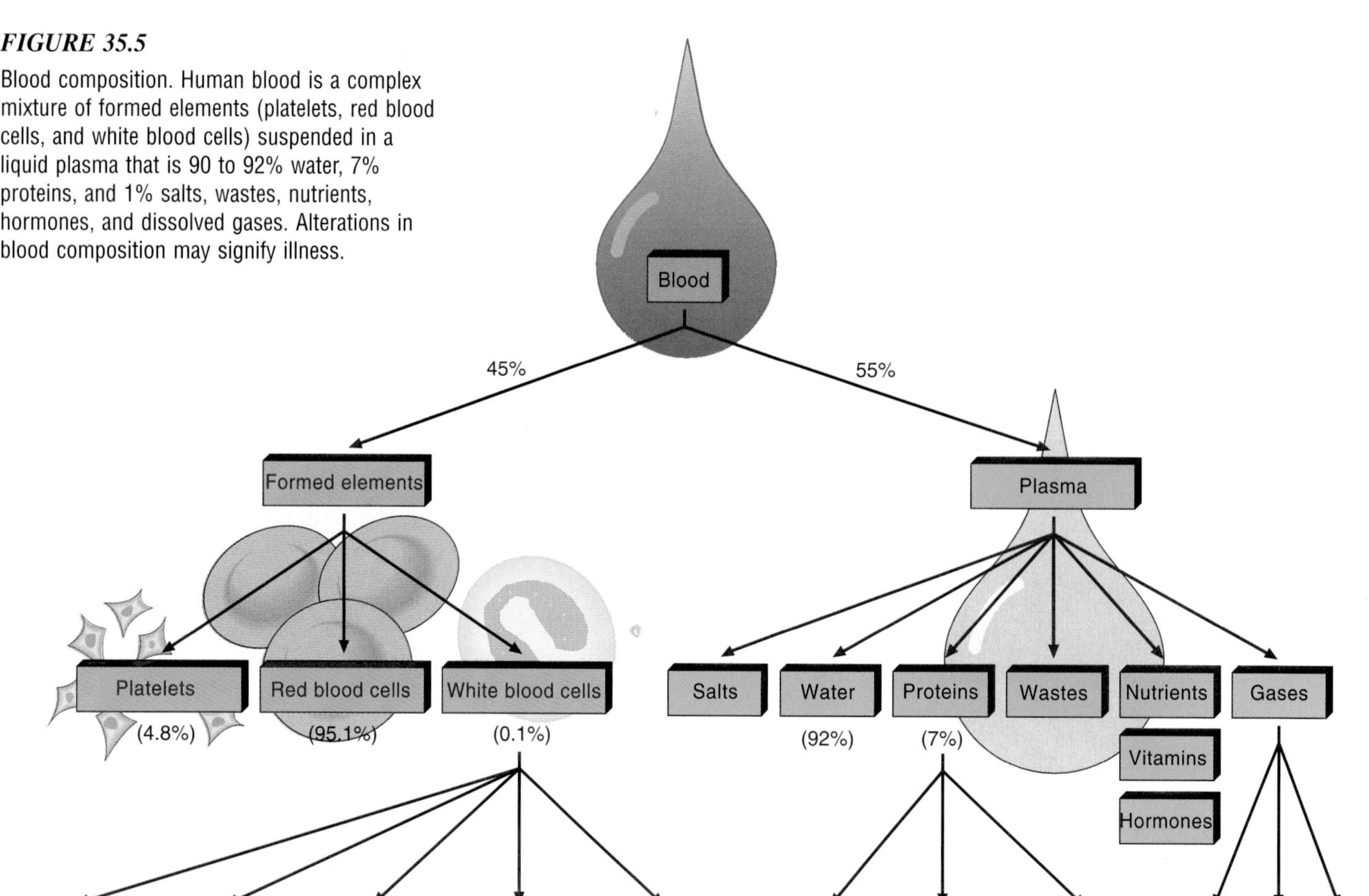

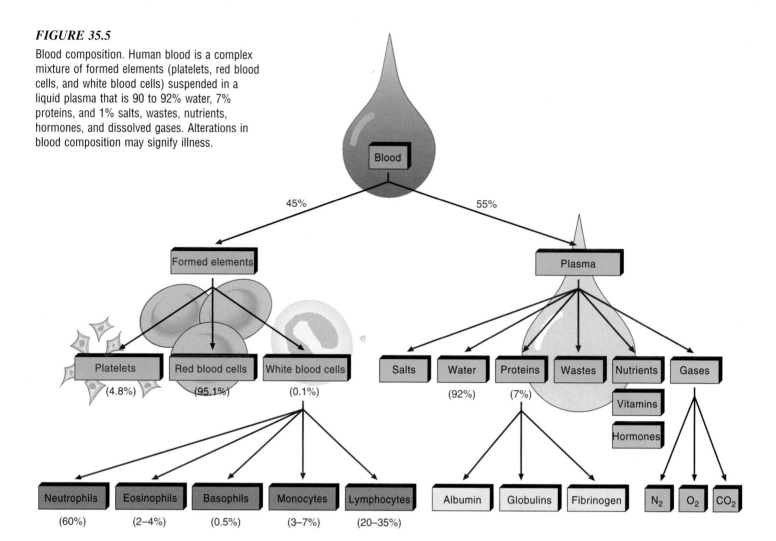

Table 35.1 Blood Functions
1. Carries oxygen, nutrients, and hormones to tissues.
2. Carries carbon dioxide (CO_2) as bicarbonate ion (HCO_3^-) to the lungs for exhalation.
3. Transports urea (a waste product of protein metabolism) from the liver, which produces it, to the kidneys, which send it to the bladder to be excreted in urine.
4. Influences the composition of the tissue fluid that surrounds cells. Tissue fluid forms from blood plasma.
5. Maintains homeostasis by helping to regulate pH at 7.4, regulating cells' water content, generating a pressure gradient that keeps plasma in capillaries, and absorbing heat and dissipating it at the body's surface, keeping body temperature constant.
6. Protects from injury by clotting to stem leaks, transporting white blood cells to destroy foreign particles, and sending white blood cells in the inflammatory response, which dilutes toxins at an injury site.

usually about 0.1% glucose; if glucose levels fall to 0.06%, convulsions begin. Plasma also contains about 7 to 8% dissolved proteins of more than 70 different types. Some, such as the albumins, maintain osmotic pressure. The globulins provide immune responses, and some transport lipids or metals, including zinc, iron, and copper. The third major plasma protein, **fibrinogen,** plays a key role in blood clotting.

Formed Elements

The three varieties of formed elements are **red blood cells** (erythrocytes), **white blood cells** (leukocytes), and **platelets** (fig. 35.6). A cubic millimeter of blood normally contains about 5 million red blood cells, 7,000 white blood cells, and 250,000 platelets.

Red Blood Cells Red blood cells are by far the most numerous of the formed elements. Their precursors form within red bone marrow at the rate of 2 to 3 million per second. As they develop, red blood cells lose their nuclei, ribosomes, and mitochondria. Therefore, they are unable to reproduce or carry out most cellular metabolism. (They do, however, have many glycolytic enzymes that provide ATP for energy to perform their functions.)

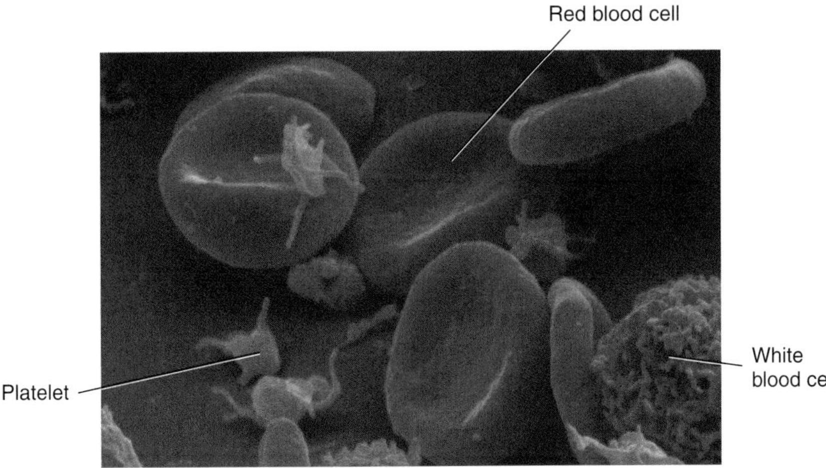

FIGURE 35.6
Blood consists of formed elements (red blood cells, white blood cells, and platelets) within a watery matrix, plasma.

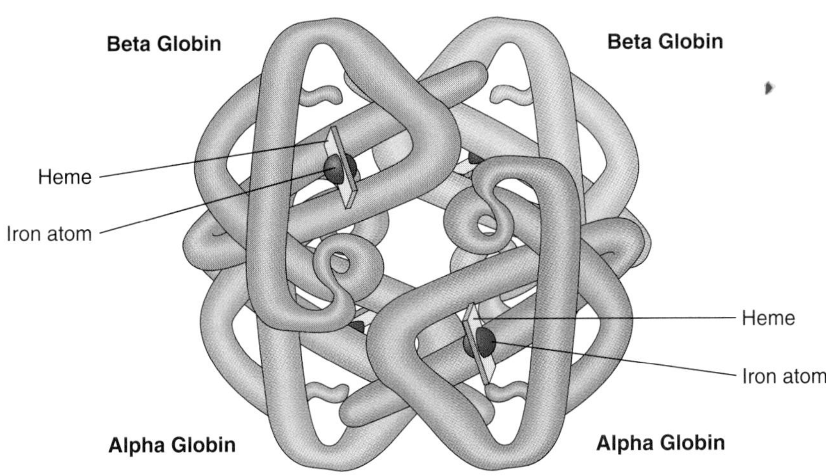

FIGURE 35.7
A hemoglobin molecule consists of two alpha globins and two beta globins. Each globin chain surrounds a heme group, which contains an iron atom that can irreversibly bind oxygen (only two hemes and two irons are labeled).

Mature red blood cells are biconcave discs packed with hemoglobin. During its short life span, a red blood cell pounds against artery walls and squeezes through tiny capillaries. Finally, the cell is destroyed in the liver or spleen, and most of its components recycled for future use.

The red blood cell's shape and content are adapted to transport oxygen. The thin, easily bent cell can squeeze through narrow passageways, and its biconcave shape increases its surface area for gas exchange. The combined surface area of all the red blood cells in the human body is roughly 2,000 times as great as the body's exterior surface.

A hemoglobin molecule is composed of four polypeptide chains called globins. Each globin is bound to an iron-containing complex called a **heme group**—hence the name hemoglobin (fig. 35.7). Hemoglobin exists in two functional forms. **Oxyhemoglobin** is bright red and compacted, with four globin chains wrapped closely around four iron-containing heme groups that bind oxygen picked up in the lungs. Then the globin chains of oxyhemoglobin unfold, releasing oxygen to the cells. The molecule becomes the deep red **deoxyhemoglobin.** Conversion of deoxyhemoglobin back to oxyhemoglobin takes only 30 millionths of a second. If all of the oxygen-binding sites of the body's hemoglobin molecules are blocked, death follows swiftly. Carbon monoxide (CO) poisoning occurs because CO binds more strongly to heme groups than oxygen does, although at a different site.

Adequate oxygenation of the body's tissues depends on a sufficient number of red blood cells carrying enough hemoglobin. The proportion of red blood cells in the blood adapts to the environment. At high altitudes, where the oxygen content of air is low, red blood cell production increases to maintain an adequate supply of oxygen to the tissues. A similar adaptation occurs as a result of regular aerobic exercise, when the muscles' oxygen use increases. Conversely, a sedentary individual may have fewer than the average number of red blood cells.

The proportion of red blood cells can change because red blood cell maturation rate and hemoglobin synthesis respond to oxygen availability. When certain cells in the kidney do not receive enough oxygen, they produce a substance that combines with a plasma protein to form the hormone **erythropoietin.** This hormone stimulates red blood cell production in red bone marrow.

Anemia is a reduction in red blood cells or in the amount of hemoglobin. These conditions decrease the amount of oxygen delivered to cells. Because oxygen is necessary to maximize the

Table 35.2 Types of Anemia

Type	Cause	Defect
Aplastic anemia	Toxic chemicals, radiation	Damaged bone marrow
Hemolytic anemia	Toxic chemicals	Red blood cells destroyed
Iron deficiency anemia	Lack of dietary iron	Hemoglobin deficiency
Pernicious anemia	Inability to absorb vitamin B_{12}	Excess of immature cells
Sickle cell disease	Defective gene	Red blood cells abnormally shaped
Thalassemia	Defective gene	Hemoglobin deficiency; red blood cells short-lived

Table 35.3 White Blood Cell Alterations

White Blood Cell Population Change	Illness
Elevated lymphocytes	Tuberculosis
	Whooping cough
Elevated eosinophils	Tapeworm infection
	Hookworm infection
Elevated monocytes	Typhoid fever
	Malaria
	Mononucleosis
Too few helper T cells	AIDS

energy extracted from nutrients to make energy-laden ATP, the symptoms of anemia are fatigue and lack of tolerance to cold. Anemia can result from red blood cells that are too small or too few, that contain too little hemoglobin, or that are manufactured too slowly or die too quickly. Table 35.2 describes several types of anemia. The most common cause of anemia is iron deficiency, treated by eating more iron-rich foods or taking iron supplements.

White Blood Cells The five types of white blood cells—neutrophils, eosinophils, basophils, lymphocytes (B cells and T cells), and monocytes—protect against toxins, infectious organisms, and to some extent, cancer. White blood cells are larger than red blood cells and retain their nuclei. They originate in the bone marrow and typically live about 1 year, spending only 3 or 4 days in the blood. White blood cells leave the circulation and wander

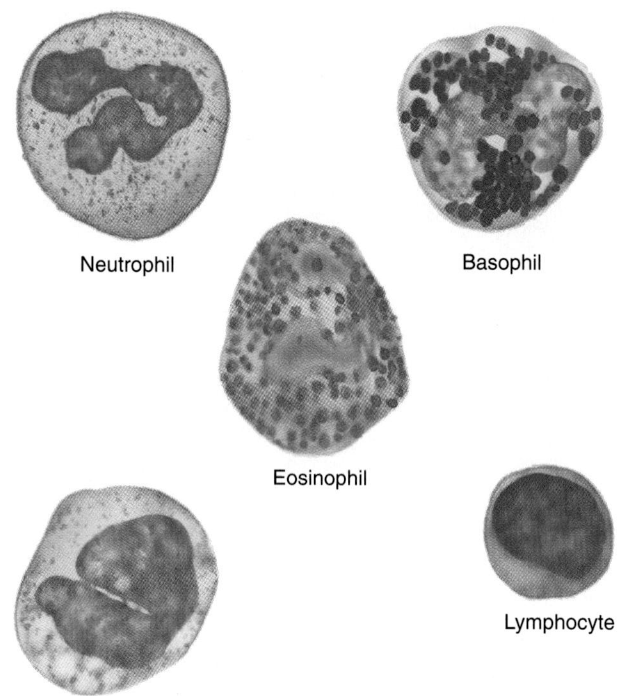

FIGURE 35.8

Types of white blood cells. Neutrophils surround, engulf, and destroy infectious microorganisms. Eosinophils attack parasites that the neutrophils do not affect. Basophils release chemicals that take part in the inflammatory response. Monocytes attack organisms resistant to neutrophils. Lymphocytes produce antibodies and cytokines.

through tissues. The types of white blood cells are distinguished by size, life span, location in the body, nucleus shape, number of granules in the cytoplasm, and staining properties (fig. 35.8).

In general, white blood cells defend the body against infection or injury. They cause the warmth and swelling of inflammation, surround microbes and destroy them, and produce proteins called antibodies that help attract other white blood cells to destroy foreign microbes or particles. The specific functions and interactions of white blood cells and their products are discussed further in chapter 39.

The proportions of white blood cell types are a window on health, because elevated or diminished numbers of specific types may provide a clue to the type of infection or disease present (table 35.3). For example, the leukemias are cancers of the white blood cells in which abnormal white blood cells greatly predominate because they divide more often than the other blood cells. The abnormal cells form at the expense of red blood cells, so when the patient's "white cell count" rises, the "red cell count" falls. Thus, leukemia also causes anemia.

Having too few white blood cells is also devastating. Acute exposure to radiation or toxic chemicals can severely damage bone marrow, killing many white blood cells. Unless the cells are immediately replaced, death occurs in a day or two from rampant lung and digestive tract infection.

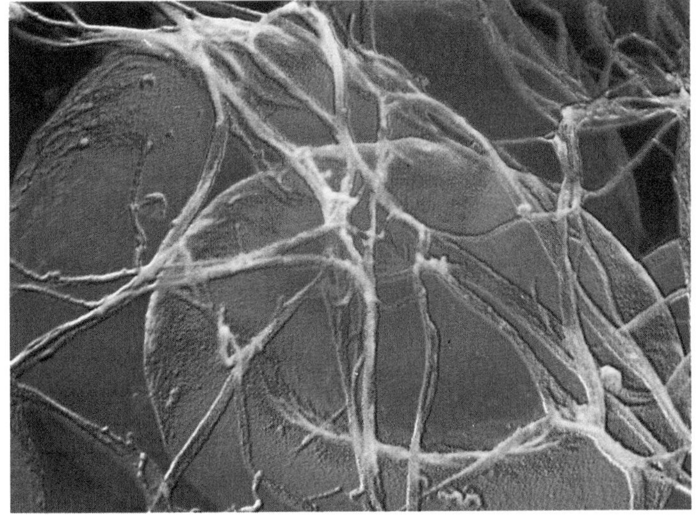

(a)

(b)

FIGURE 35.9

Cells of the blood vessel wall initiate a chemical cascade that results in synthesis of fibrin, which threads around and between red blood cells (*a*), eventually knitting a clot (*b*).

Platelets In mammals, platelets are small, colorless cell fragments that live about 1 week and initiate blood clotting. A platelet originates as part of a huge bone marrow cell called a **megakaryocyte.** This giant cell has rows of vesicles that divide the cytoplasm into distinct regions, like a sheet of stamps. The vesicles enlarge and coalesce, "shedding" fragments that become platelets. Vertebrates other than mammals have small cells with nuclei, called thrombocytes, that carry out clotting.

In a healthy circulatory system, platelets travel freely within the vessels. But when a wound nicks a blood vessel, or a blood vessel's normally smooth inner lining becomes obstructed, platelets catch on the bumps, shattering and releasing biochemicals from secretory granules. These biochemicals combine with plasma proteins to start a complex series of reactions that results in blood clotting.

Clotting begins as soon as an injury occurs, at the same time as the injured vessel constricts, slowing the flow of blood. Within seconds, platelets collect at the site, change shape, and connect to each other. Meanwhile, cells from the blood vessel wall release a protein, **thromboplastin,** which activates a series of plasma proteins that eventually convert the blood protein **prothrombin** into **thrombin.** Thrombin, in turn, initiates a series of chemical reactions that ultimately converts the plasma protein fibrinogen into **fibrin** (fig. 35.9). The threadlike fibrin molecules form a meshwork that entraps red blood cells and more platelets. The platelets trapped in the clot release more thromboplastin, perpetuating the clotting cycle. Platelets in clots also polymerize actin and myosin into a contractile apparatus (somewhat like skeletal muscle) that shrinks the clot as new tissue forms to fill in the wound.

Clotting that occurs too slowly or too rapidly seriously affects health. Inherited bleeding disorders called hemophilias are caused by absent or abnormal clotting factors, the proteins essential for blood clotting. Deficiencies of vitamins C or K can also dangerously slow clotting because these vitamins are required to synthesize clotting factors.

Blood that clots too readily is extremely dangerous. In atherosclerosis, clots may form as platelets snag on rough spots in blood vessel linings. A clot that blocks a blood vessel is a **thrombus.** It can cut off circulation, ultimately causing death. A clot that travels in the bloodstream to another location is an **embolus.** An embolus that obstructs a blood vessel in a lung, called a pulmonary embolism, can kill. Biology in Action 35.2 describes a rediscovered approach to controlling clotting during recovery from microsurgery—using leeches.

The Origin of Blood Cells

Blood cells originate in bone marrow from pluripotent stem cells. "Pluripotent" means that these cells have the potential to differentiate into any of a number of specialized cell types. As these stem cells divide, the progeny cells respond to **colony-stimulating factors.** These are secreted growth factors that turn on some genes and turn off others, sculpting the distinctive characteristics of the different blood cell types (fig. 35.10).

Because pluripotent stem cells can produce blood cells, they are a very valuable medical resource. The challenge lies in obtaining them, because they make up less than 0.1% of the cells in bone marrow. Because of the rarity of stem cells, a bone marrow transplant must deliver some 40 billion cells to give a patient with a life-threatening blood disorder enough stem cells to rebuild the circulatory system.

Because bone marrow transplants are very dangerous to the recipient, researchers are developing new sources of stem cells. Giving a colony-stimulating factor as a drug, for example, sends stem cells into a person's bloodstream, where they replenish

35.2 biology in action

The Return of the Medicinal Leech

It took surgeon Joseph Upton 10 hours to sew the 5-year-old's ear back on after a dog had bitten it off. The operation appeared to be a success, but after 4 days, blood flow in the ear was blocked. The arteries the surgeon had repaired were fine, but the smaller veins were becoming congested. Dr. Upton decided to try an experimental technique—he applied 24 leeches to the wound area.

The leeches latched on for up to an hour each, drinking the boy's blood. Leech saliva contains a potent anticlotting agent called hirudin in honor of its source—the medicinal leech, *Hirudo medicinalis.* Unlike conventional, short-acting anticlotting drugs such as heparin, hirudin works for up to 24 hours after the leech drops off. Hirudin blocks thrombin in veins, giving a wound time to heal before clotting begins.

Leeches have long been part of medical practice (fig. 35.B); they were used in ancient Egypt 2,500 years ago. During the nineteenth century, French physicians used more than a billion leeches a year to drain "bad humours" from the body to cure nearly every ill. Leech use fell in the latter half of the nineteenth century. In 1960, plastic surgeons in the former Yugoslavia rediscovered the medicinal uses of the leech. French microsurgeons followed in the early 1980s. In 1985, Dr. Upton brought leeches into U.S. headlines by saving the boy's ear at Children's Hospital in Boston (fig. 35.C). The return of the leech to medical practice is largely due to the efforts of zoologist Roy Sawyer, founder of a company that sells the animals.

A leech's bite doesn't hurt, patients say. But for those unwilling to have a 3-inch long, slimy, green-gray invertebrate picnicking on a wound, a drug version of hirudin is available, thanks to biotechnology.

FIGURE 35.B
For centuries, bloodletting with leeches was believed to cure many ills. This woman in seventeenth-century Belgium applies a medicinal leech to her arm.

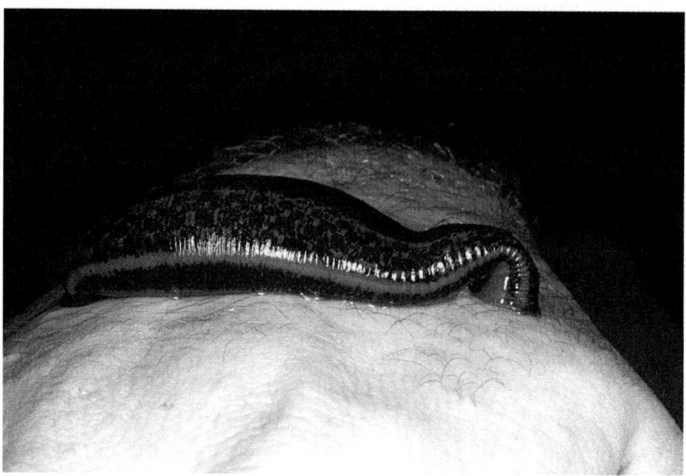

FIGURE 35.C
Microsurgeons sometimes use leeches to help maintain blood flow through human veins after reattaching severed ears or digits. An anticoagulant in the leech's saliva keeps the blood thin enough to flow.

supplies of deficient blood cell types. Umbilical cord blood is an especially rich source of stem cells. In the future, all newborns may have umbilical cord stem cells stored, to provide perfectly compatible transplant material to treat blood disorders later in life.

The Human Heart

Each day, the human heart sends more than 7,000 liters of blood through the body, and it contracts more than 2.5 billion times in a lifetime. This fist-sized muscular pump is specialized to ensure that blood flows in one direction on its journey.

Structure of the Pump

Contraction of the cardiac muscle that makes up most of the heart walls provides the force that propels blood. Cardiac muscle cells are striated because they are packed with highly organized contractile proteins. The cells branch, forming an almost netlike pattern. As a result, cardiac muscle contracts in two dimensions, "wringing" the blood out of the heart (fig. 35.11). The heart is enclosed in a tough connective tissue sac, the **pericardium** ("around the heart"), that protects the heart but allows it to move even during vigorous beating.

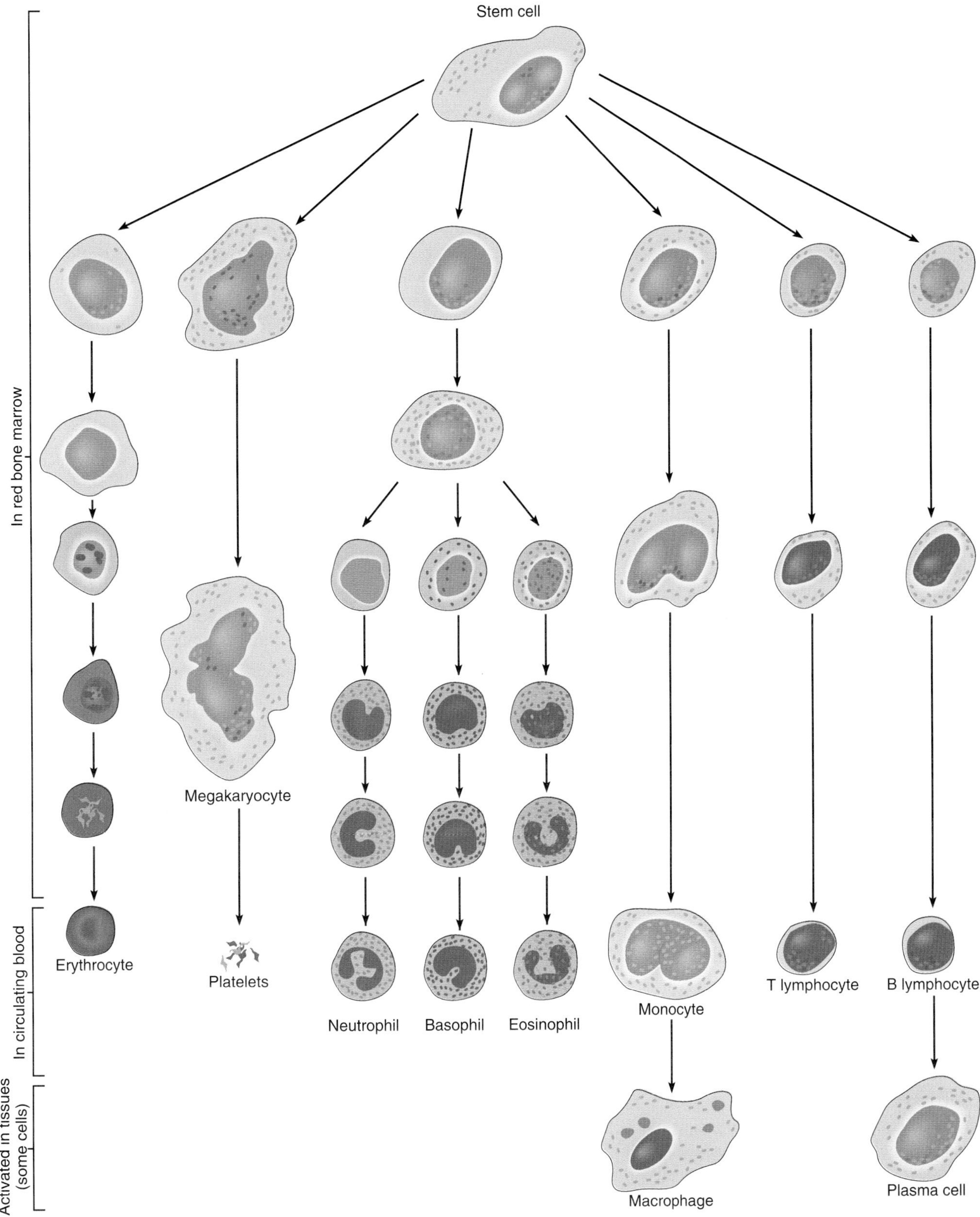

FIGURE 35.10

A single type of stem cell in bone marrow gives rise to all mature blood cell types, to platelets, and to scavenger cells (macrophages) and cells involved in allergic responses (mast cells).

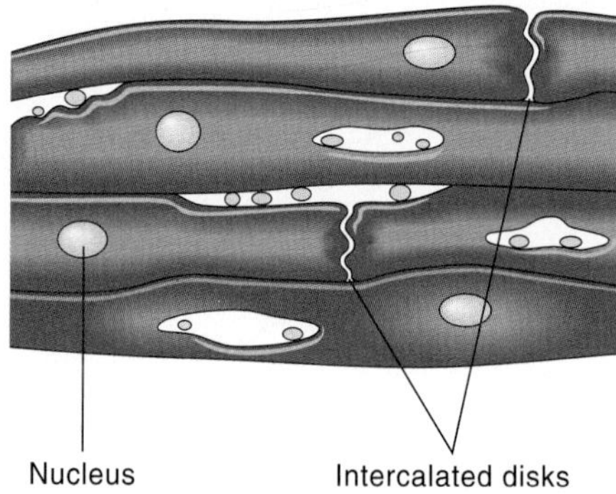

FIGURE 35.11

Cardiac muscle. The structure of cardiac muscle is well suited to its function. The branching network shortens the muscle in two dimensions, creating a "pumping" action. The intercalated disks are connections that allow electrical activity to spread quickly from cell to cell.

The human heart's four chambers pump blood to and from the lungs to pick up oxygen and then distribute it to the rest of the body (fig. 35.12). Most of the blood entering the heart moves through the upper chambers (atria) into the relaxed lower chambers (ventricles). The atria are primer pumps that finish filling the much larger ventricles. The thick, muscular walls of the ventricles generate enough force to push blood to even the most remote body parts.

Heart Valves

Two pairs of **heart valves** keep blood flowing in a single direction. The **atrioventricular valves (AV valves)** are located between each atrium and ventricle. These thin flaps of tissue move in response to pressure changes that accompany the ventricle's contraction. The AV valve on the right side of the heart is the **tricuspid valve;** the AV valve on the left side is the **bicuspid** (or **mitral**) **valve.** When the ventricle contracts, pressure increases, pushing the flaps up and preventing blood from moving back into the atrium. The valves cannot be pushed into the atria because strings of connective tissue anchor them to muscles extending from the wall of the ventricle.

The **semilunar valves** consist of three pocketlike flaps forming a ring in the arteries just outside each ventricle. Blood leaving the ventricle moves past these valves easily. Blood moving backward toward the ventricle, however, fills these pockets and causes the valves to bulge, like a parachute filling with air (fig. 35.13). This blocks the blood from flowing back.

Two sets of heart valves closing generate the "lub-dup" sound of the heartbeat. The "lub" corresponds to the closing AV valves, and the "dup" to the semilunar valves closing. A heart murmur, which is a variation on the "lub-dup" sound, often reflects abnormally functioning valves. It could mean a dangerous backflow of blood, forcing the heart to work harder, or it could be completely harmless.

Each year in the United States, about 75,000 people receive replacement heart valves made of synthetic materials such as metal or ceramic or of tissue from a cow or pig. About 10% of the population has a condition called mitral valve prolapse, in which the mitral valve stretches and balloons back into the atrium when the ventricle contracts. This can cause palpitations, a heart murmur, or occasional chest pain but often has no symptoms. Mitral valve prolapse is not dangerous unless blood flows backward through the valve.

Blood Flow in the Heart

The heart is sometimes considered to be two pumps that beat in unison because it pumps blood through two closed circuits of blood vessels—the right side sends blood to the lungs and the left side powers the vast systemic circulation to the rest of the body. Both sides of the heart function in the same way, but the blood from each side has a different destination.

Dark red, oxygen-poor blood from the systemic circulation enters the heart at the right atrium and is delivered by the two largest veins in the body, the **superior vena cava** and the **inferior vena cava.** A smaller vein that enters the right atrium, the coronary sinus, returns blood that has been circulating within the heart.

From the right atrium blood passes through an AV valve into the right ventricle. Next, the blood flows out of the right ventricle, past the **pulmonary semilunar vaive,** and through the **pulmonary arteries** to the lungs, where it picks up oxygen. The oxygen-rich blood, now bright red in color, returns through four **pulmonary veins** to the heart, where it enters the left atrium. The oxygenated blood then flows from the left atrium through the bicuspid valve and into the left ventricle. The massive force of contraction of the left ventricle, the most powerful heart chamber, sends blood through the **aortic semilunar valve** and into the **aorta,** the largest artery in the body. From there, the blood circulates throughout the body before returning to the heart. Two branches off of the aorta, the coronary arteries, provide blood to the heart muscle.

Heartbeat

The actions of the two sides of the heart are highly synchronized. Each beat of the heart, called a **cardiac cycle,** consists of a sequence of contraction **(systole)** and relaxation **(diastole).** The heartbeat begins on the 23rd day following conception, and continues until death.

The heartbeat originates in the cardiac muscle cells themselves. These cells can contract on their own, without nervous stimulation. When grown outside the body, isolated heart cells beat at different rates. As two cardiac muscle cells make contact, however, they begin to beat synchronously. Intercalated disks, the tight junctions between the cardiac muscle cells, help spread electrical impulses from cell to cell, allowing depolarization and repolarization to spread throughout the heart. Many heart cells contracting in unison generate enough force to powerfully eject blood.

Specialized cardiac muscle cells initiate each beat and coordinate contractions so that excitation reaches each region of the heart and contraction is forceful. The beat begins in cells of a region in the **sinoatrial (SA) node** in the wall of the right atrium slightly below the opening of the superior vena cava. Because the SA node sets the tempo of the beat, it is called the **pacemaker.**

Left atrium
Right atrium
Left ventricle
Right ventricle
(a)

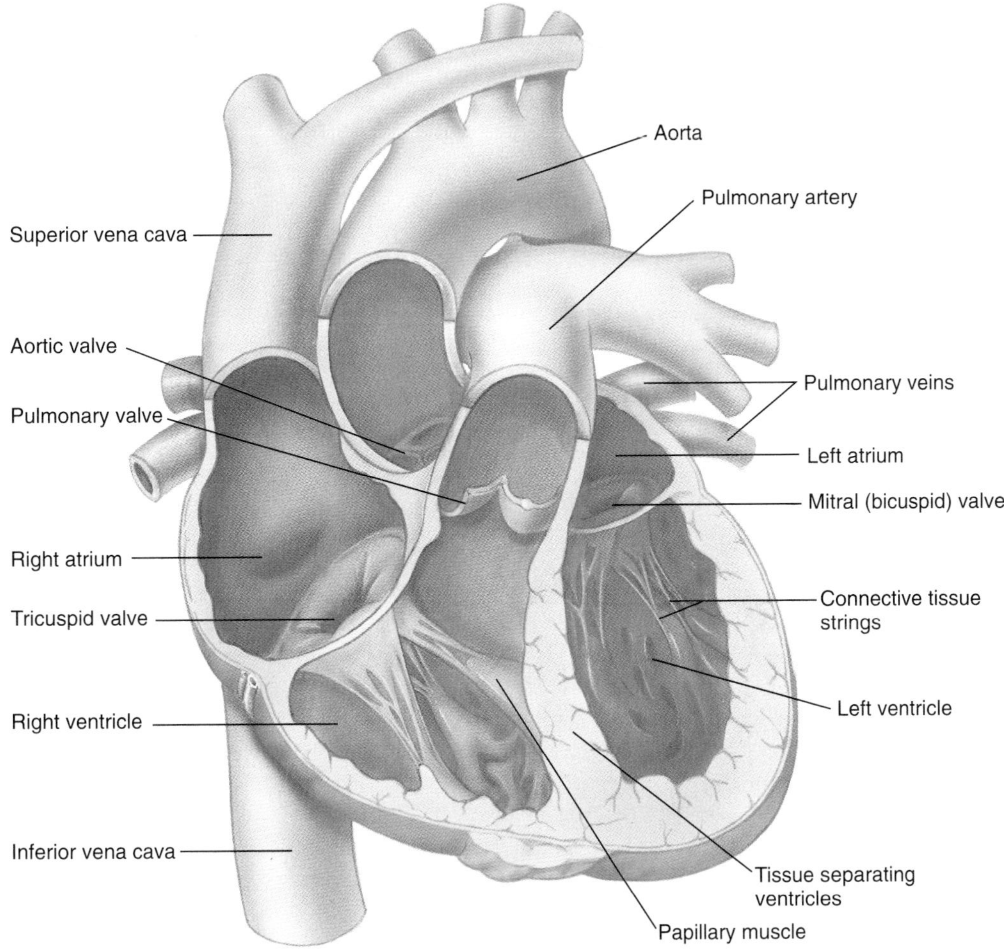

FIGURE 35.12

A human heart. (*a*) Notice the locations of the four chambers. (*b*) This illustration depicts the four chambers, the valves, and the major blood vessels of the human heart.

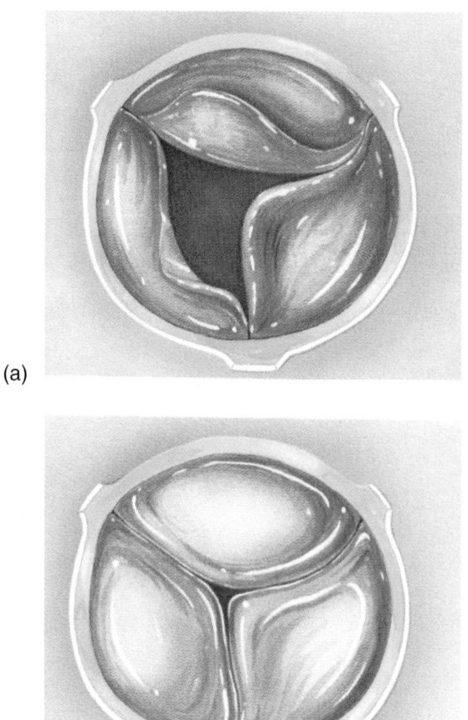

FIGURE 35.13

Semilunar heart valves. (*a*) Semilunar valve open. (*b*) Valve nearly closed.

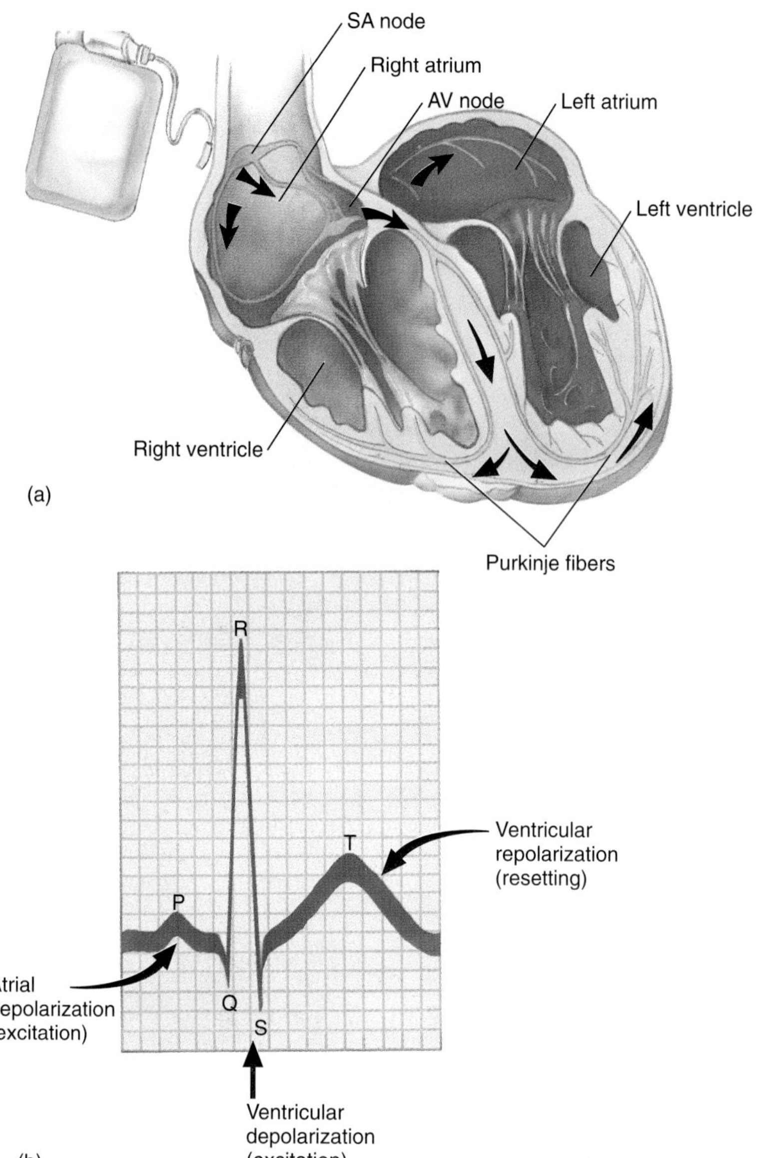

FIGURE 35.14

Electrical activity in the heart. (*a*) Specialized muscle cells form a conduction network through the heart. Electrical changes start in the SA node and travel through the atrial wall to the AV node and then to the ventricle walls through the Purkinje fibers. (*b*) An electrocardiogram (ECG) is a printout of electrical changes on the body's surface in response to electrical changes as the heart contracts. An ECG has a characteristic pattern in a healthy heart. The P wave indicates atrial excitation. The QRS complex charts the spread of depolarization through the ventricles (a small electrical change that resets the atria also occurs here). The T wave results from ventricular repolarization.

Once kindled, the impulses triggering contraction race across the atrial wall to another region of specialized muscle cells, the **atrioventricular (AV) node.** The AV node branches into specialized **Purkinje fibers,** which are cardiac muscle fibers that conduct electrical stimulation six times faster than other parts of heart muscle. The AV node delays spread of the impulse briefly, which allows the ventricles to fill before they contract. It takes only 8/10 of a second from the time the SA node produces its electrical stimulation until the ventricles contract.

We can observe the electrical changes that accompany the contraction of the heart using a device called an electrocardiograph (fig. 35.14). An electrocardiogram (ECG) can detect an irregular heartbeat, called an arrhythmia. A normal heartbeat is regular and predictable. Highly irregular patterns may result in a dangerous arrhythmic condition called ventricular fibrillation (fig. 35.15).

Blood Vessels

The structures of blood vessels are well adapted to their functions.

Arteries

The outermost layer of an artery consists of connective tissue, the middle layer is elastic tissue and smooth muscle, and the inner layer is a smooth, one-cell-thick lining of **endothelium** (fig. 35.16). When heart muscle contracts, a wave of blood pounds against arterial walls, and the arteries dilate, accommodating this surge. Then, while the heart relaxes, the arteries elastically recoil, helping to push the blood along.

The largest arteries have such thick walls that they contain smaller blood vessels to nourish the vessel. As arteries branch farther from the heart, they become thinner and the outermost layer may taper away. In arterioles, the smooth muscle layer is predominant, which enables these small vessels to constrict or dilate, which helps to regulate blood pressure.

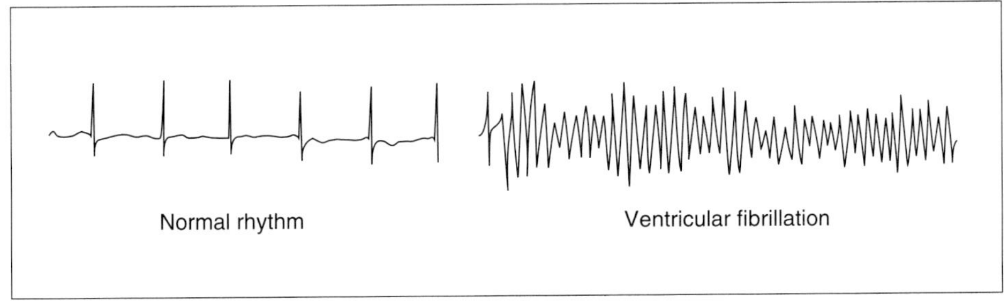

(a)

(b)

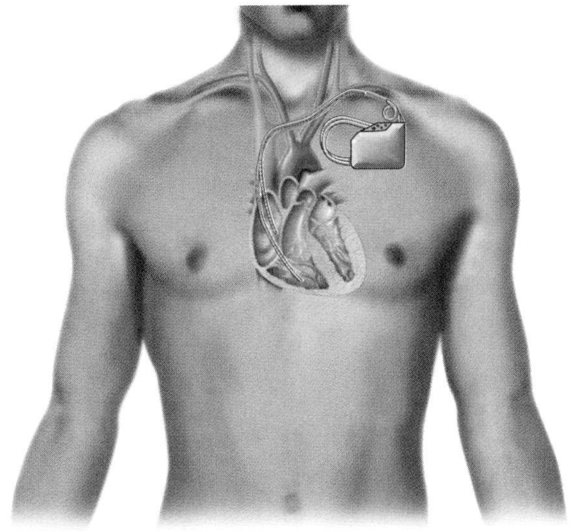
(c)

FIGURE 35.15

An electrocardiogram (ECG) showing (*a*) a normal pattern and (*b*) ventricular fibrillation, the abnormal electrical pattern associated with impending cardiac arrest. (*c*) Implantable defibrillators, such as the one shown here, save lives. Ventricular fibrillation can cause death in minutes. For patients known to have arrhythmias, an implanted defibrillator responds to abnormal heart rhythm by shocking it back to normal.

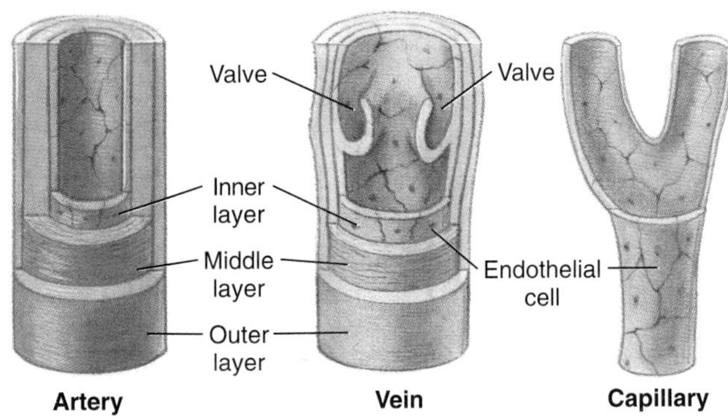

FIGURE 35.16

Long sections of blood vessels. The walls of arteries and veins have three layers: the outermost layer is connective tissue; the middle layer is elastic and muscular; the inner layer is smooth endothelium. Walls of arteries are much thicker than those of veins. The middle layer is greatly reduced in veins. The capillary wall consists only of endothelium. Many veins have valves, which keep blood flowing in one direction.

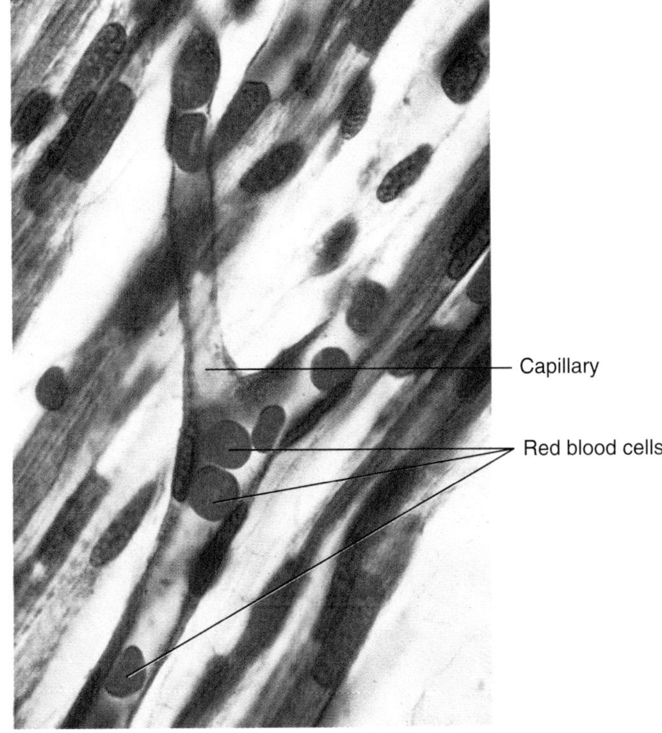

FIGURE 35.17

Red blood cells in capillaries. The pumping action of the heart and rebound against artery walls propel red blood cells through extensive networks of microscopic capillaries.

Capillaries

Capillaries consist only of endothelium, the single-celled layer that lines the interior of arteries and veins (see fig. 1.3). This simple structure enables capillaries to be the sites of gas exchange. Capillaries are very numerous, providing extensive surface area (fig. 35.17). All the capillaries in a human body laid end to end would extend 25,000 to 60,000 miles.

Capillaries form networks called capillary beds that connect an arteriole and a venule (fig. 35.18). Not all capillary beds in the body are open at one time—there simply is not enough blood to fill them all. The smooth muscle in an arteriole can contract to reduce blood flow to the capillary bed it feeds or relax to open the area.

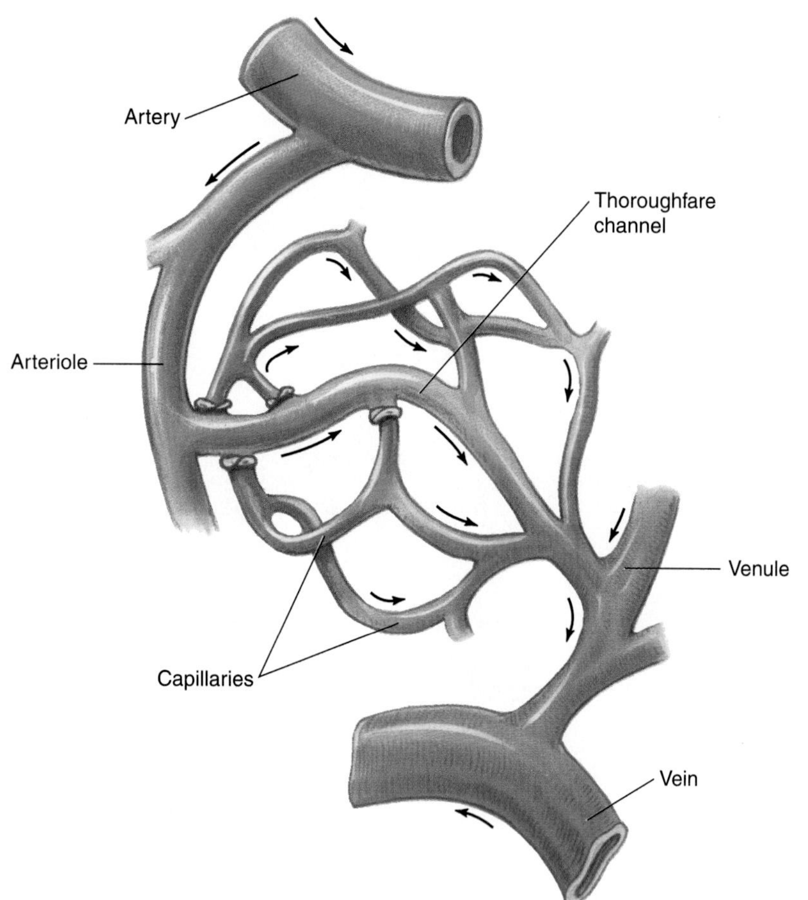

FIGURE 35.18

A capillary bed is a network of tiny vessels that lies between an arteriole and a venule. Blood does not flow through all a person's capillaries at one time; it flows through one capillary called a thoroughfare channel.

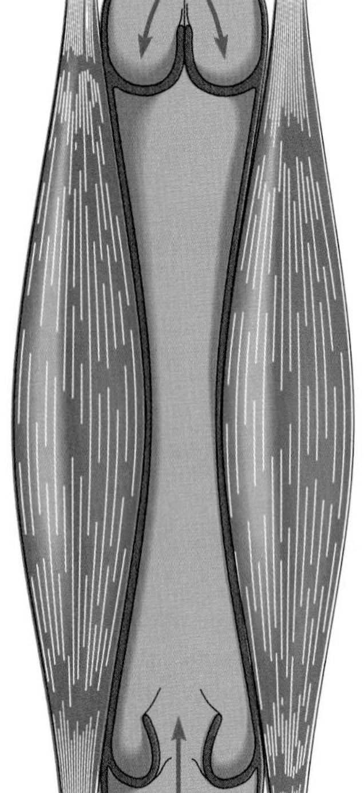

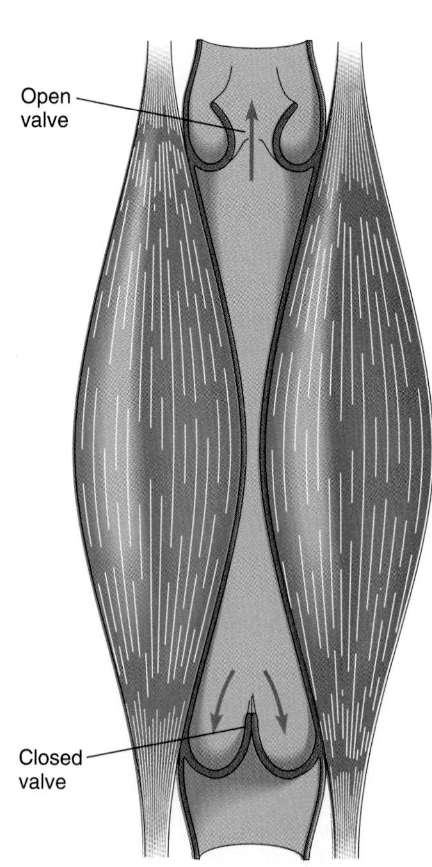

FIGURE 35.19

Valves in veins help return blood to the heart. When skeletal muscles relax, valves prevent blood from flowing backward. When skeletal muscles contract, they thicken and squeeze the veins, propelling blood through the open valves. Gravity helps blood circulate in the legs. After prolonged weightlessness on space flights, astronauts' circulations become disturbed. Blood pools in the middle region of the body, where sensors interpret higher-than-normal blood volume, and increase urination. When the astronaut returns to earth, blood volume has fallen by 10 to 20 percent. If blood vessels in the legs cannot constrict enough to accommodate the sudden drop in blood volume, the astronaut wobbles and falls down. Special spacesuits that squeeze the legs can prevent this "orthostatic hypotension."

An individual's circulatory requirements vary with activity and environmental conditions. The ability of the arterioles to dilate and constrict allows the body to adapt to exercise, when blood supply to muscles increases at the expense of organs not in immediate use, such as those in the digestive tract. Blood must redistribute for other reasons, too. A cold-water drowning victim, for example, may survive until help comes if blood is shunted from the extremities to the heart and brain.

Veins

Veins are thinner and less elastic than arteries, because these characteristics are not critical in a vessel that conducts blood with so little pressure. The middle layer of a vein is thus much reduced or even absent (see fig. 35.16). Veins not filled with blood collapse.

Because the blood pressure in veins is low, some medium and large veins—particularly in the lower legs, where blood must move against gravity to return to the heart—have flaps called **venous valves** that keep blood flow in one direction. As skeletal muscles contract, they squeeze veins and propel blood through the open valves in the only direction it can move: toward the heart (fig. 35.19).

Veins are also blood reservoirs. The sympathetic nervous system can stimulate smooth muscle in vein walls to contract, which constricts the vessels and displaces the extra blood they contain. Because valves prevent backflow, the displaced blood moves toward the heart, filling the atria. This mechanism can supply the heart with extra blood to pump.

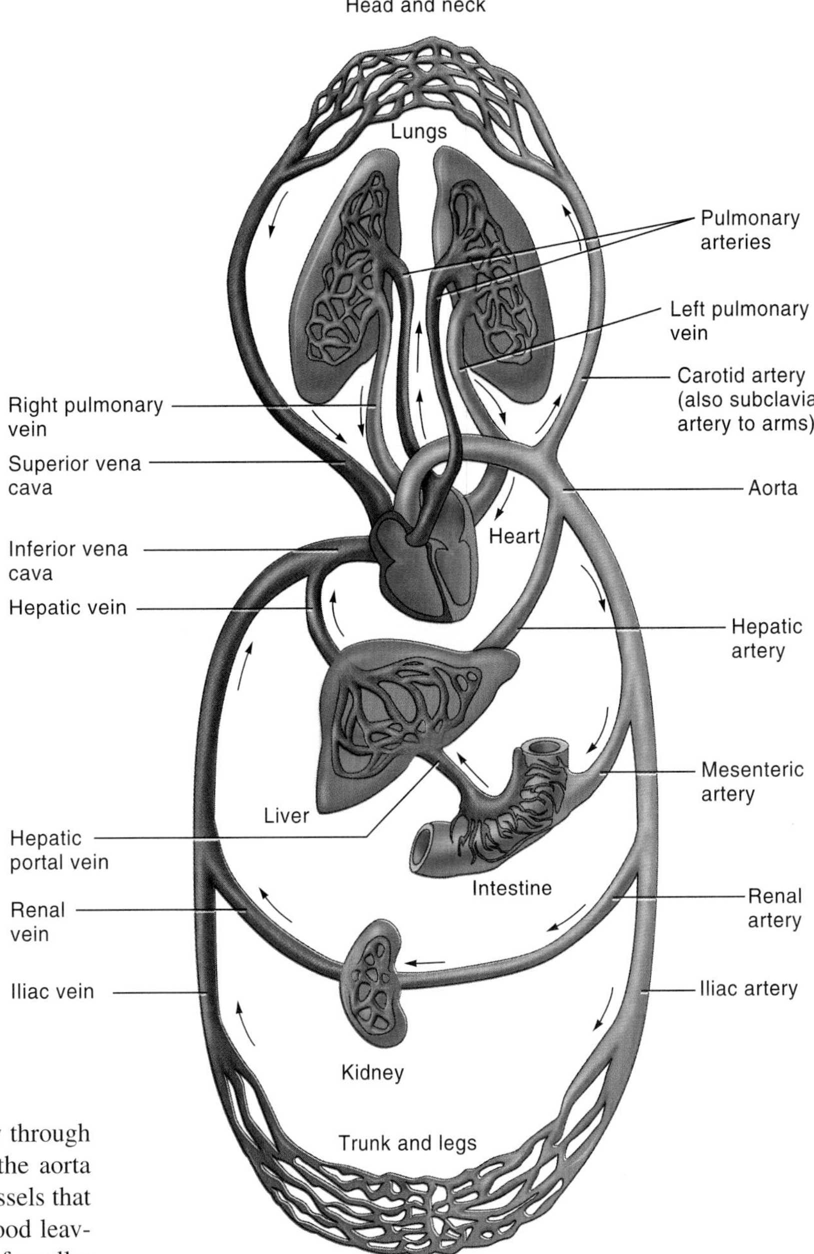

FIGURE 35.20

Pulmonary and systemic circulation—the journey of blood. Red depicts arteries and blue depicts veins, except in the pulmonary circuit, where arteries carry deoxygenated blood and veins carry oxygenated blood. In the latter, the colors are reversed.

Blood Vessels and Circulation

Blood journeys through the body in a particular pathway through specific blood vessels. Blood leaves the heart through the aorta (fig. 35.20). Among the first branches of the aorta are vessels that nourish the heart itself, which carry about 4% of the blood leaving the heart. These arteries lead to an extensive network of smaller blood vessels that nourish the active heart muscle cells.

Other branches of the aorta diverge to supply the rest of the body. At rest, about a quarter of the blood travels to the kidneys, and another quarter delivers oxygen and nutrients to muscles. About 15% of the blood goes to the abdominal organs, 10% to the liver, 8% to the brain, and the remainder to other parts of the body.

In the liver, circulating blood detours from the usual route of capillary to venule to vein. The **hepatic portal system** ("hepatic" means liver) is a special division of the circulatory system that helps quickly harness the chemical energy in digested food. Capillaries from the stomach, small intestine, pancreas, and spleen lead into four veins. These veins converge into the **hepatic portal vein,** which leads to the liver. In the liver, the hepatic portal vein diverges into venules and then capillaries, and these capillaries reconverge into the hepatic vein. The hepatic vein empties into the inferior vena cava and returns to the heart. The unusual position of the hepatic portal vein between two beds of capillaries allows dissolved nutrients that enter the bloodstream at the digestive organs to reach the liver rapidly, where they are metabolized to supply energy. The liver also removes toxins and microorganisms from blood from the digestive system.

Blood Pressure

The force that drives blood through vessels can be felt as a "pulse" pressing outward on the blood vessel walls. This **blood pressure** results from the heart's pumping action and the degree of blood vessel narrowing (vasoconstriction) in the peripheral circulation.

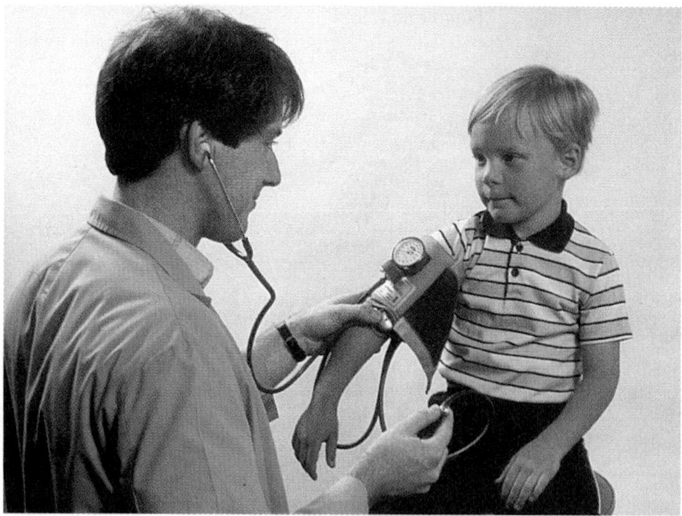

(a)

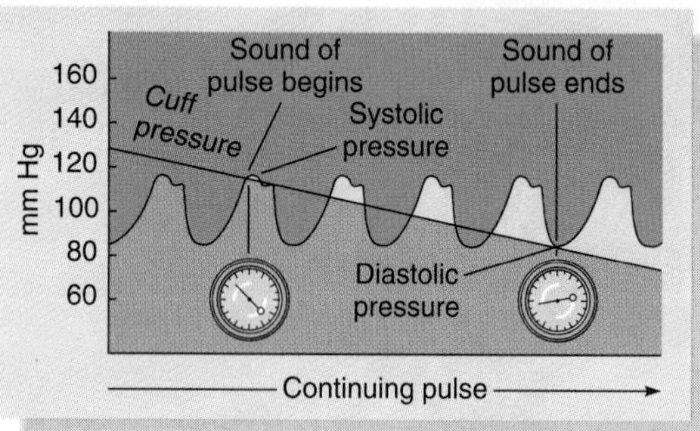

(b)

FIGURE 35.21

Blood pressure. (*a*) A sphygmomanometer, which is an inflatable cuff attached to a pressure gauge, measures blood pressure. The cuff is wrapped around the upper arm and inflated until no pulse is felt in the wrist, which signifies that circulation to the lower arm has been temporarily cut off. A stethoscope is placed on the arm just below the cuff, so the sound of returning blood flow can be heard when the cuff is slowly deflated. When a thumping is heard through the stethoscope, the listener notes the pressure on the gauge. This sound is the blood rushing through the arteries past the deflating cuff as the pressure peaks due to ventricular contraction. (*b*) The value on the gauge when the sound begins is the systolic blood pressure. The sound fades until it disappears, and the pressure reading at this point is the diastolic blood pressure.

The peak of the blood pressure pulse occurs when the ventricles contract and is called the systolic pressure (or systole). A low point called the diastolic pressure (or diastole) occurs when the ventricles relax. A device called a sphygmomanometer measures how far up a tube the blood pressure can push the heavy liquid mercury, sometimes using a simple pressure gauge (fig. 35.21). A normal blood pressure for a young adult is 120 millimeters of mercury for the systolic pressure and 80 millimeters for the diastolic pressure, expressed as "120 over 80" (written 120/80). "Normal" blood pressure varies with age, sex, and race.

The value 120/80 reflects the arterial blood pressure, in arteries near the heart. Pressure drops with distance from the heart, so it is highest in the arteries, lower in the capillaries, and lower still in the veins. Much like water flowing down a mountain stream, blood flows in a circuit from where the arteries leave the heart to where the veins return to the heart. Blood pressure is so low in veins that often skeletal muscle contraction and venous valves must help return blood to the heart.

Blood does not flow at the same speed throughout the circulatory system. If a mountain stream widens and branches, the water flows less rapidly. In much the same way, branching of blood vessels further from the heart increases their total cross-sectional area, slowing the speed of the flowing blood. By the time the blood reaches the capillaries, the cross-sectional area is 600 to 800 times that of the aorta. Thus, blood moves the slowest through the capillaries, facilitating nutrient and waste exchange. Past the capillaries, veins merge and the cross-sectional area decreases, helping to speed blood flow back to the heart.

Arterial blood pressure rises when blood vessels narrow (vasoconstriction) and lowers when they open (vasodilation). The number of arterioles that are constricted or dilated depends on several factors, including blood volume, temperature, blood chemistry, and activity level of the tissues surrounding the arterioles. Emotions such as anger and fear can also alter blood pressure. **Hypotension,** which is blood pressure significantly lower than normal, may cause fainting. **Hypertension,** which is blood pressure significantly higher than normal, does not produce symptoms directly, but it may severely damage the circulatory system and other organs.

Control of Circulatory Functions

The dynamic nature of the circulatory system requires constant regulation. The vasomotor center in the medulla of the brain regulates heart rate and the diameters of arterioles, via the autonomic nervous system. An increased rate of impulses from the sympathetic nervous system constricts arterioles and increases heart rate. These actions may occur in a person with low blood pressure, blood loss, or blood that carries excess carbon dioxide. When blood pressure is elevated, brain centers decrease sympathetic nervous system input, causing vasodilation and decreased heart rate, while impulses from the parasympathetic nervous system slow the heart. Pressure receptors within the walls of certain major arteries continually monitor blood pressure, detecting changes and sending information to the medulla. This negative feedback leads to responses that maintain circulatory homeostasis.

Mastering Concepts

1. What are the components of blood and their proportions and functions?
2. Where do blood cells originate?
3. How does the organization of cells in the heart provide contractility?
4. Describe the path of blood through the heart's chambers and valves, to the lungs and back through the systemic circulation.
5. How does heartbeat originate and spread?
6. How are the structures of blood vessel types adaptive?
7. What is blood pressure?

Exercise and the Circulatory System

The normal heart pumps out as much blood as enters it from the veins. If more blood enters the heart, the strength of the next ventricular contraction increases to handle the increased load. Exercise enhances this ability of the heart, so that the amount of blood that the heart pumps each minute, called the **cardiac output,** is maintained at a lower heart rate. Thus, the heart of a person dedicated to exercise performs more efficiently than a sedentary person's heart, pumping the same amount of blood at 50 beats per minute that the sedentary person's heart pumps at 75 beats per minute. (The lowest heart rate ever recorded in an athlete was 25 beats per minute!) The number of red blood cells increases in response to regular exercise, and these cells are packed with more hemoglobin to deliver more oxygen to tissues.

Exercise provides several other cardiovascular benefits. It can lower blood pressure significantly. It elevates the level of high-density lipoproteins (HDLs), the cholesterol carriers that pick up cholesterol and transport it to the liver for elimination. Exercise also spurs the development of collateral circulation, or extra blood vessels, which may help prevent a heart attack by providing alternative pathways for blood flow to the heart muscle.

To achieve the benefits of exercise, the heart rate must be elevated to 70 to 85% of its "theoretical maximum" for at least half an hour three times a week. You can calculate your theoretical maximum by subtracting your age from 220. If you are 18 years old, your theoretical maximum is 202 beats per minute. Seventy to 85% of this value is 141 to 172 beats per minute. Tennis, skating, skiing, handball, vigorous dancing, hockey, basketball, biking, and brisk walking can all elevate heart rate to this level.

It is wise to consult a physician before starting an exercise program. People over the age of 30 are advised to have a stress test, which is an electrocardiogram taken while exercising. (The standard electrocardiogram is taken at rest.) An arrhythmia that appears only during exercise may indicate heart disease that has not yet produced symptoms.

Mastering Concepts

1. How does regular exercise affect cardiac output and the oxygen-carrying capacity of blood?
2. How does regular exercise affect blood pressure and cholesterol profile?
3. How much exercise is necessary to benefit the circulatory system?

The Lymphatic System

The **lymphatic system** is another transport system (fig. 35.22). **Lymph,** the fluid that fills these capillaries and vessels, is made up of tissue fluid, formed from blood plasma minus certain proteins too large to leave blood capillaries. Vessels of the lymphatic system begin as dead-end **lymph capillaries** that are highly permeable to tissue fluid that seeps out of blood capillaries. The lymph capillaries gradually converge to form larger **lymph vessels** that eventually empty into veins.

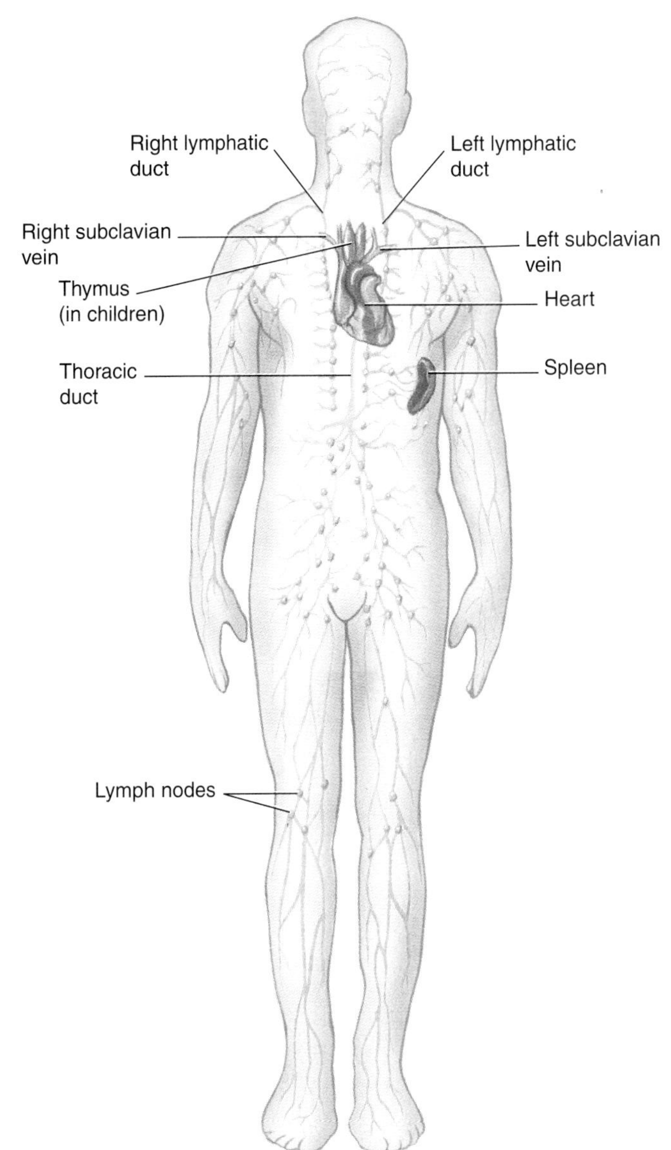

FIGURE 35.22

The lymphatic system is a network of vessels that collect excess fluid and proteins that leak from the blood capillaries and return them to the blood. At places along the lymph vessels, lymph nodes filter lymph. Lymph vessels have valves, and muscle contraction moves lymph.

Contractions of surrounding skeletal muscles and valves in lymph vessels help move the sluggish lymph fluid. The lymphatic system normally returns less than 1 ounce (30 milliliters) of fluid per minute to the veins, in contrast to the 4 to 5 quarts (5 or 6 liters) pumped through the circulatory system in the same amount of time. Tissue fluid enters lymphatic vessels and flows toward the chest region, where it returns to the blood. Special lymph capillaries in the small intestine absorb dietary fats.

Breakdown in lymphatic functioning causes the body to swell with the excess tissue fluid, a condition called **edema.** As the slow leakage of protein and fluid from the blood circulatory system continues, the swelling may become severe. In a disease called elephantiasis, parasitic worms reside in the lymph vessels, blocking the flow of lymph and causing grotesque swelling of tissues.

The lymphatic system has other important functions. Structures called **lymph nodes** contain great numbers of white blood cells that protect against infection. The kidney-shaped lymph nodes are located primarily where the limbs meet the trunk and in the neck. Lymph nodes may be microscopic or much larger masses, such as those that can be felt as "swollen glands" along the sides of the neck during respiratory infections.

A lymph node consists of fibrous tissue with many pockets, each filled with millions of lymphocytes that efficiently filter cellular debris and bacteria from the lymph flow. The fluid that enters the nodes is often packed with infectious particles, 99.9% of which are destroyed in the node. Although lymphocytes in the lymph nodes can detect and destroy cancer cells, if a cancer originates in or spreads to a lymph node and is not destroyed, it may spread to the rest of the body in the lymph fluid.

The lymph nodes continually add lymphocytes to the lymph, which transports them to the blood. Lymphocytes are manufactured by other parts of the lymphatic system too—the **spleen,** the **thymus,** and possibly the **tonsils.** B lymphocytes (B cells) secrete antibody proteins. T lymphocytes (T cells) secrete biochemicals called cytokines, discussed further in chapter 39.

The largest component of the lymphatic system is the spleen, which is to the left of and behind the stomach. The thymus is a lymphatic organ in the upper chest near the neck that is prominent in children but begins to degenerate in early childhood. The thymus "educates" certain lymphocytes in the fetus to distinguish body cells from foreign cells. Tonsils are collections of lymphatic tissue high in the throat. Tonsils do not filter lymph, but they may help protect against infection in other ways.

The blood circulatory system and the lymphatic system work together. The blood carries a continuous supply of nutrients and oxygen to cells and removes wastes from them, and the lymphatic system cleanses fluid that has leaked from blood capillaries, removing bacteria, debris, and cancer cells and returning the filtered fluid to the blood. The blood circulatory system and the lymphatic system nurture even cells far removed from the body's surface, ensuring that the many trillions of cells of the human body can carry out the functions of life.

Mastering Concepts

1. Where does lymph come from?
2. What are the components of the lymphatic system?
3. What do lymph nodes do?
4. How do the blood and lymph circulatory systems work together?

SUMMARY

1. A circulatory system consists of fluid, a network of vessels, and a pump. The fluid delivers nutrients and oxygen to tissues and removes metabolic wastes.

2. In simple animals, open body cavities give interior structures direct contact with fluids from the environment so that nutrients and oxygen can diffuse into cells and wastes can diffuse out. In more complex animals, a heart pumps the fluid to the body cells. In an **open circulatory system,** the fluid bathes tissues directly in open spaces before returning to the heart. In a **closed circulatory system,** the heart pumps the fluid (blood or hemolymph) through a system of vessels to cells and back.

3. Pigments such as **hemoglobin** and **hemocyanin** increase the oxygen-carrying capacity of blood or hemolymph.

4. Vertebrates have closed circulatory systems that increase in complexity in fishes, amphibians, reptiles, birds, and mammals. A double heart and double circulation that separates oxygenated from deoxygenated blood enables land animals to be active.

5. Human blood is a mixture of cells and cell fragments (collectively called **formed elements**), proteins, and molecules that are dissolved or suspended in **plasma.** The formed elements include **red blood cells, white blood cells,** and **platelets.** Red blood cells carry oxygen. White blood cells protect against

Treating Sickle Cell Disease~

continued from page 708.

Hemoglobin is an especially interesting molecule because it exists in a different form in the fetus. Where the two beta chains will be after birth are two slightly different delta globin chains, which bind oxygen differently. The resulting fetal hemoglobin is an adaptation to the fact that fetuses do not use their lungs to breathe. The genes encoding delta globin shut off a few weeks before birth, as the beta globin genes turn on. Normally, by 6 months of age, a baby no longer manufactures fetal hemoglobin.

The first clue that turning on dormant fetal hemoglobin genes might treat sickle cell disease (and some other blood disorders) came from people who have "hereditary persistence of fetal hemoglobin." Their delta globin genes never shut off—but the people do not have symptoms! Fortunately, researchers already knew of a few drugs that could reawaken the fetal genes. One was hydroxyurea, used to treat certain cancers.

From 1984 through 1992, experiments in monkeys, healthy humans, and a small number of people with sickle cell disease showed that hydroxyurea raises fetal hemoglobin levels. Then a trial began on 299 severely affected patients to see if the drug could improve their symptoms. Half of the group received the drug, and half received a placebo. By 1995, researchers halted the study, because the experimental group had half the number of painful crises, and needed far fewer blood transfusions, than did those receiving placebo.

Hydroxyurea works by delaying sickling, which enables cells carrying sickled hemoglobin to reach the lungs faster, where fresh oxygen maintains the red blood cells' normal shapes. Formation of sickle hemoglobin slows because the normal delta globin chains compete with the abnormal beta globin chains for binding to alpha chains.

By understanding circulation at the molecular, cellular, and organ system levels, an effective treatment for sickle cell disease is in the making.

infection. Platelets break and collect near a wound, releasing chemicals that trigger blood clotting. Blood cells originate in bone marrow.

6. The heart is the muscular pump that drives the human circulatory system. The heart has two **atria** that receive blood and two **ventricles** that propel blood in the body. **Heart valves** separate the right atrium and right ventricle, the left atrium and left ventricle, and the site where the pulmonary and systemic blood pathways leave the heart. Heart valves ensure one-way blood flow.

7. The **pacemaker** or **sinoatrial (SA) node,** a collection of specialized cardiac muscle cells in the right atrium, sets the heart rate. From there, heartbeat spreads to the **atrioventricular (AV) node** and then along **Purkinje fibers** through the ventricles.

8. The circulatory system leads to and from the lungs and to and from the rest of the body. Blood leaves the heart through the **aorta** and travels in increasingly narrower **arteries** and **arterioles** to the **capillaries,** where nutrient and waste exchange occur. Blood flows from the capillaries to **venules** and then to **veins,** and it reenters the heart through the venae cavae. Arteries have thicker, more elastic walls than veins.

9. The pumping of the heart and constriction of blood vessels produces **blood pressure.** Systolic pressure is the pressure exerted on blood vessel walls when the ventricles contract. The low point, diastolic pressure, occurs when the ventricles relax.

10. The circulatory system is controlled to help maintain homeostasis. When the volume of blood entering the heart changes, ventricular contraction changes in response to adjust **cardiac output.** The autonomic nervous system, under the influence of the brain's vasomotor center, speeds or slows heart rate and dilates or constricts blood vessels in a way that adjusts blood pressure.

11. The **lymphatic system** is a network of vessels that collect fluid from the body's tissues, purify it, and return it to the blood. **Lymph nodes** filter lymph, and the **spleen, thymus,** and possibly the **tonsils** manufacture lymphocytes. The blood and lymphatic systems continually supply tissues with nutrients and oxygen and remove or destroy wastes.

TO REVIEW . . .

1. Why do large, active animals require circulatory systems? Give three examples of circulatory systems that are adapted to the activity level of the organism.

2. How do open and closed circulatory systems differ?

3. What role do pigments play in the respiratory and circulatory systems of vertebrates?

4. Maintaining the proper proportions of formed elements in the blood is essential for health. What can happen when the blood contains too few or too many red blood cells, white blood cells, or platelets?

5. How is a red blood cell adapted for its function?

6. It usually takes 13 seconds for prothrombin to be converted to thrombin and thrombin to trigger the formation of fibrin. When a person's diet lacks vitamin K, however, these chemical reactions take 30 seconds or longer. What circulatory system function does this delay impair? What symptoms might a dietary deficiency of vitamin K cause?

7. Name three ways that the circulatory system helps to maintain homeostasis.

8. What causes the heartbeat? Why is the heart sometimes referred to as two hearts that beat in unison?

9. What types of changes in blood vessels would raise blood pressure?

10. Name three illnesses doctors could treat if medical technology could isolate, mass produce, and reimplant a person's healthy bone marrow stem cells.

11. Where does lymph originate? What propels it through the lymphatic vessels?

TO THINK ABOUT . . .

1. Some members of a Scandinavian family have inherited a condition that causes them to overproduce erythropoietin (EPO). One young man from this family has won several Olympic medals for skiing. How might the extra EPO account for his athletic prowess? When some athletes learned of this skier's biological abnormality, they attempted to obtain EPO as a drug (normally used to help surgical patients rebuild blood supplies). Why would athletes want to take EPO? What might a side effect of EPO abuse be?

2. HIV colonizes lymph nodes before it appears in the bloodstream. Why is the presence of HIV in lymph nodes dangerous?

3. Chemotherapy used to treat cancer often severely damages bone marrow. What biochemical mentioned in the chapter might help patients receiving chemotherapy to recover their bone marrow function?

4. Athletes tend to be slim and strong, have low blood pressure, do not smoke, and alleviate stress through exercise. How might these characteristics complicate a study to assess the effects of exercise on the cardiovascular system?

5. A man feels "flutters" in his chest, so he has a physical exam. The doctor says, "Your diastolic pressure is elevated. You seem to have angina of exertion, but the stress test does not show an arrhythmia or a blocked coronary artery. Your leukocyte count and clotting time are normal, but your red count is a bit low." What is—and is not—wrong with the man's circulatory system?

6. Why is it important to use clot-busting drugs such as tPA within a few hours after a heart attack or stroke?

7. Cigarette smoke contains thousands of chemicals, including nicotine and carbon monoxide. Nicotine constricts blood vessels. Carbon monoxide prevents oxygen from binding to hemoglobin. How would these two components of smoke affect the circulatory system?

TO LEARN MORE . . .

References and Resources

Barinaga, Marcia. May 3, 1996. Finding new drugs to treat stroke. *Science,* vol. 272. Fast use of a clot-busting drug can limit damage from a stroke.

Blood. 1993. An excellent collection of wonderfully illustrated articles, available from the Howard Hughes Medical Institute, 4000 Jones Bridge Road, Chevy Chase, Maryland, 20815-6789.

Brown, Michael S., and Joseph L. Goldstein. May 3, 1996. Heart attacks: Gone with the century? *Science,* vol. 272. By lowering LDL, we can avoid heart attacks.

Dieryck, Wilfrid et al. March 6, 1997. Human haemoglobin from transgenic tobacco. *Nature,* vol. 386. Human hemoglobin from transgenic plants would be pure and plentiful.

Gillum, Richard F. November 21, 1996. The epidemiology of cardiovascular disease in black Americans. *The New England Journal of Medicine,* vol. 335. Why are African Americans at higher risk for cardiovascular disease?

Hayden, Michael R., and Michael Reidy. January 1995. Many roads lead to atheroma. *Nature Medicine,* vol. 1. Fatty diets are linked to atherosclerosis, but so is abnormality of the endothelium and its control of the chemical environment of the blood vessel lining.

Lange, Richard A., and L. David Hillis. October 24, 1996. Thrombolysis—The preferred treatment. *The New England Journal of Medicine,* vol. 335. Busting blood clots is a recommended treatment for a heart attack.

Lewis, Ricki. July 15, 1997. Cloaked rbcs and transgenic tobacco—blood substitutes? *Genetic Engineering News.* Experimental blood substitutes have unusual sources.

Lewis, Ricki. April 1995. Lasers can be lifesavers for twins. *Photonics Spectra.* Laser treatment can correct a syndrome where one twin's circulatory system takes nutrients from the other before birth.

Mayfield, Eleanor. May 1996. New hope for people with sickle cell anemia. *FDA Consumer.* Reactivating fetal hemoglobin can lessen the severity of sickle cell disease.

Wickelgren, Ingrid. May 3, 1996. New devices are helping transform coronary care. *Science,* vol. 272. Mechanical hearts, defibrillators, and stents save many lives.

Zamir, Mair. September/October 1996. Secrets of the heart. *The Sciences.* The heart thrives on exercise.

denotes articles by author Ricki Lewis that may be accessed via *Life*'s home page /http://www.mhhe.com/sciencemath/biology/life/.

The cilia that line the human trachea help remove particles. Smoking destroys these cilia.

36 The Respiratory System

chapter outline

Breathing and Respiration 732

Evolution and Diversity Among Gas Exchange Systems 732
Body Surface
Tracheal Systems
Gills
Lungs

The Human Respiratory System 735
The Nose
The Pharynx and the Larynx
The Trachea, Bronchi, and Bronchioles
The Alveoli
The Lungs

The Breathing Mechanism 742

Gas Transport 742
Oxygen
Carbon Dioxide

Control of Respiration 744
The Role of Blood CO_2 Level
Breathing at High Altitudes

The Unhealthy Respiratory System 746

A Case of Poisoning

When the 78-year-old man arrived in the emergency room, physicians at first thought he was suffering from heart failure. He reported a pain in the left side of his chest, shortness of breath, and fatigue. He was also confused and upset, with an irregular pulse, fast respirations, and abnormal heart rhythm. A narrowed or blocked coronary artery, preventing sufficient blood and therefore oxygen from reaching the heart, would explain the man's symptoms.

Physicians treated the man with oxygen, aspirin, and nitroglycerin, assuming that he was suffering from a heart condition. Then they noticed that his wife was lying on a stretcher, complaining of the same symptoms!

The woman reported that her symptoms had developed about 2 hours after her husband's had. Her chest pain began in her left arm and intensified as she served him tea. Although the woman's heart signs were normal, she, too, was treated as if she had suffered heart damage.

The couple was admitted to the hospital, even though their symptoms improved after 2 hours in the emergency room. They felt well when they returned to their apartment 2 days later, but 2 days after that, the couple was back in the hospital, reporting the same symptoms. This time, the doctors suspected that something in their home environment was causing the symptoms. The first guess, that they were misusing a medication, proved incorrect.

to be continued . . .

Breathing and Respiration

The first breath of life is the toughest. A baby often emerges from the birth canal with a bluish color that may be frightening to new parents. Soon, as the baby musters up 15 to 20 times the strength needed for subsequent breaths, the millions of tiny sacs in the lungs, each only partially inflated, fill with air for the very first time. As oxygen rapidly diffuses into the bloodstream and reaches the tissues, the infant turns a robust pink and lets out a yowl (fig. 36.1).

The lungs of an adult who has spent years breathing polluted air and smoking cigarettes are quite different from the healthy lungs of a newborn. The passageways to the lungs are dotted with bare patches where dense cilia, which move particles up and out of the respiratory tract, once waved. Deep within the lungs, the pattern of bare patches continues, and sections of air sacs are deflated or altogether gone. While the newborn's lung linings are pure pink, the smoker's are a sooty black. The tissues that capture life-giving oxygen to be distributed to the adult's tissues have been ravaged beyond repair. The adult is made aware of the damage with each hacking cough.

Breathing, technically termed external respiration, is actually one of three forms of respiration. External respiration is the process by which animals exchange oxygen and carbon dioxide (CO_2) between respiratory organs and the blood. Internal respiration exchanges these gases between the blood and cells. Cellular respiration, discussed in chapter 6, refers to oxygen-utilizing biochemical pathways that store chemical energy as ATP.

The gas exchange that breathing makes possible enables cells to harness the energy held in the chemical bonds of nutrient molecules. Without gas exchange, cells die.

Recall from chapter 6 that energy is slowly liberated from food molecules when electrons are stripped off and channeled through a series of electron carriers, each at a lower energy level than the previous one. The energy released at each step is used to form ATP, producing water. Oxygen is the final acceptor of the low-energy electrons at the end of the chain of acceptors. Without oxygen, electrons cannot pass along this series of acceptors; this halts breakdown of organic molecules much earlier and leaves a great deal of energy locked in the food molecule. Oxygen also combines with carbons cleaved from nutrient molecules in the Krebs cycle, producing carbon dioxide.

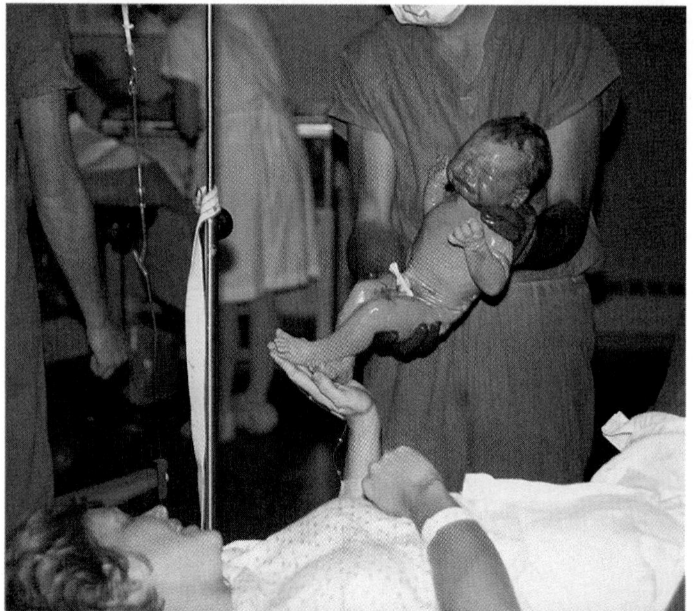

FIGURE 36.1
The first breath of life is the toughest.

Evolution and Diversity Among Gas Exchange Systems

The most ancient organisms did not respire as animals do today because the atmosphere before photosynthesis arose lacked sufficient oxygen. These earliest life-forms were probably anaerobic microorganisms that obtained energy through relatively inefficient glycolysis. The gaseous oxygen that accumulated with the rise of photosynthesis eventually became essential for life.

One-celled organisms can exchange gases using simple diffusion. Larger organisms, however, require respiratory systems

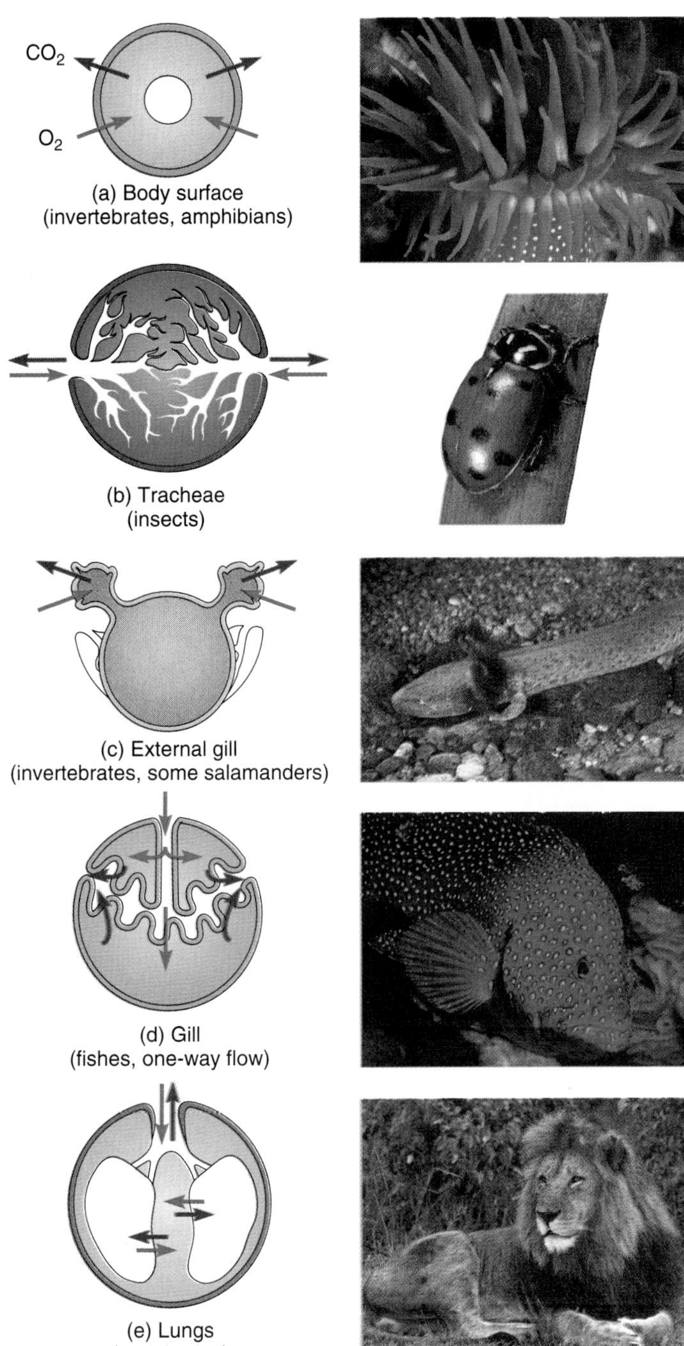

FIGURE 36.2

Diversity in respiratory surfaces. (*a*) The sea anemone has the simplest type of respiratory surface—diffusion across a thin layer. (*b*) Many land-dwelling insects exchange gases using an extensive network of tracheae. (*c*) A mud puppy uses an external gill.(*d*) A fish uses an internalized gill. (*e*) Terrestrial vertebrates use lungs.

with moist membranes that provide enough diffusional surface area to meet metabolic requirements. Because diffusion can only effectively distribute substances over about 0.5 millimeters, more active animals also require a circulatory system to transport gases between the respiratory membrane and cells.

Water-dwelling animals have respiratory structures that can extract oxygen from a viscous medium holding small quantities of oxygen. Water saturated with oxygen contains only about one-thirtieth of the oxygen present in an equal volume of air, and the warmer or saltier the water, the less oxygen is present. Furthermore, oxygen takes about 300,000 times longer to diffuse through water than through air, so it takes longer to replenish oxygen used for cellular respiration. This means that to obtain the same quantity of oxygen, an aquatic organism must move a much greater quantity of water over its respiratory surface than the volume of air a land dweller must move across its respiratory surface. In addition, because water is more dense and more viscous than air, an animal in an aquatic environment must use more energy to keep oxygen-rich water flowing over its respiratory system. As a result, an aquatic organism requires a larger respiratory surface than a terrestrial animal of similar size and with similar energy requirements. Land dwellers have their own challenge, however: keeping the respiratory membrane from drying out. The respiratory surface of terrestrial animals is thus inside the body, where it can be kept moist.

Respiratory surfaces include body surface, tracheal systems, gills, and lungs. Each of these specialized structures is an adaptation to a specific environment.

Body Surface

In the simplest gas exchange mechanism, all cells are in close enough contact with the environment that gases can diffuse across the cell membrane and reach all cells. A single-celled organism can respire by simple diffusion because its cell membrane provides enough surface area for sufficient oxygen to diffuse into the cell.

Multicellular organisms use different strategies for gas exchange. One way to keep cells in contact with the oxygen-rich environment is to have a flat body. For example, cnidaria such as hydra have only two cell layers, so all their cells contact the outside environment. Flatworms (phylum Platyhelminthes) are thicker, but they are still thin enough for gas exchange and distribution to occur by diffusion. The sea anemone (fig. 36.2*a*) respires by diffusion across thin body parts.

In larger animals, diffusion cannot effectively distribute gases. A circulatory system transports gases between body cells and the respiratory membrane. Some organisms, such as earthworms (phylum Annelida) and amphibians, still use the body surface for gas exchange but have a circulatory system to transport gases between cells and the body surface. The respiratory surface and the circulatory system in such animals are intimately and extensively linked; just beneath the skin surface, many tiny blood vessels allow gases to diffuse across the skin into or out of the circulatory system.

Tracheal Systems

In terrestrial arthropods, including insects, centipedes, millipedes, and some spiders, tough, waterproof exoskeletons dip inward to form an extensively branching system of tubules, called **tracheae,** that bring the outside environment close enough to every cell for gases to diffuse (fig. 36.2*b*). Valves guard openings to the tracheae and close them in very dry environments to keep cells from drying out. In small arthropods, the tracheae are short enough for

FIGURE 36.3

Gills. (*a*) Removing the protective operculum reveals the feathery gills of this fish. Each side of the head has four gill arches, and each arch consists of many filaments. A filament houses capillaries within lamellae. Note that the direction of water flow opposes that of blood flow. This is called countercurrent flow. (*b*) The countercurrent relationship of blood flow to water flow ensures that the blood receives the maximal amount of oxygen.

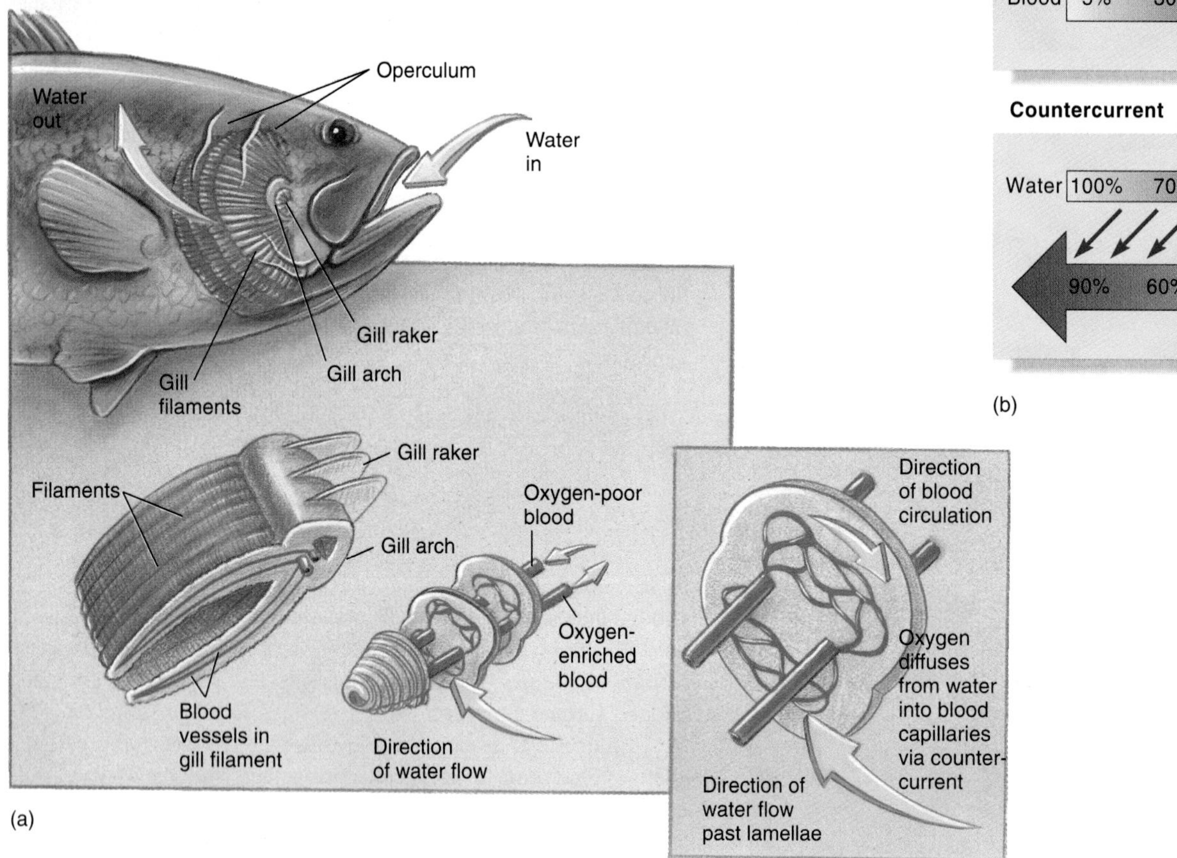

diffusion to effectively deliver oxygen to cells. Large, active arthropods, however, must move air in and out of the tracheal system to refresh the oxygen supply. They use their abdominal and flight muscles to do this. A foraging honeybee, for example, pumps segments of the abdomen back and forth to move air.

Gills

Large, active organisms require a gas exchange surface greater than the skin surface. In aquatic organisms, the vast respiratory surface usually takes the form of **gills.** Gills are extensions of the body wall that are so highly folded and branched that they may superficially resemble feathers. Within each gill is a dense network of capillaries. Oxygen diffuses from the water across delicate gill membranes and thin capillary walls into the blood, which delivers it to cells. Carbon dioxide diffuses in the opposite direction.

Amphibian larvae have external gills that close off as lungs begin working after metamorphosis. Salamanders retain external gills through adulthood (fig. 36.2*c*). The fish's complex gills create the extensive surface area necessary to extract maximal oxygen from water and deliver it to cells (figs. 36.2*d* and 36.3*a*). Gases are exchanged across a very thin respiratory membrane and the single layer of capillary cells and transported by the circulatory system. A structure called an **operculum** protects the gills while pumping water over their surfaces. Maximal gas exchange requires that a continuous supply of fresh water flows from the mouth over the gills. This is more energy efficient than moving water into a gill chamber, only to reverse the direction of flow and push it out again.

Blood flows through the capillaries in the opposite direction that the water flows over the gill membrane. This **countercurrent flow** maximizes the amount of oxygen extracted. Recall that substances diffuse from areas of higher concentration to areas of lower concentration. Because of the countercurrent flow, blood leaving the gills, which has a relatively high oxygen concentration, moves past water entering the gills, which has the highest available oxygen concentration. Because of this gradient, oxygen diffuses into the blood, bringing oxygen in the water and blood closer to equilibrium as the blood leaves the gills. The oxygen level in the water drops as it continues to flow over the gills. The blood vessels that water encounters have even less oxygen, however, so there is always a gradient for diffusion (fig. 36.3*b*).

FIGURE 36.4

Obtaining oxygen. (*a*) Elephant seals store extra oxygen in their blood. An elephant seal can spend 85% of its time diving, with only short recovery periods, because of its tremendous blood supply (two and one-half times that of humans, for its weight). The elephant seal's blood is packed with oxygen-carrying red blood cells. Beneath 130 feet, the lungs actually collapse. (*b*) This green sea turtle, swimming off the coast of Mexico, has enough oxygen to take long dives despite its small lungs because its muscle is packed with oxygen-carrying myoglobin.

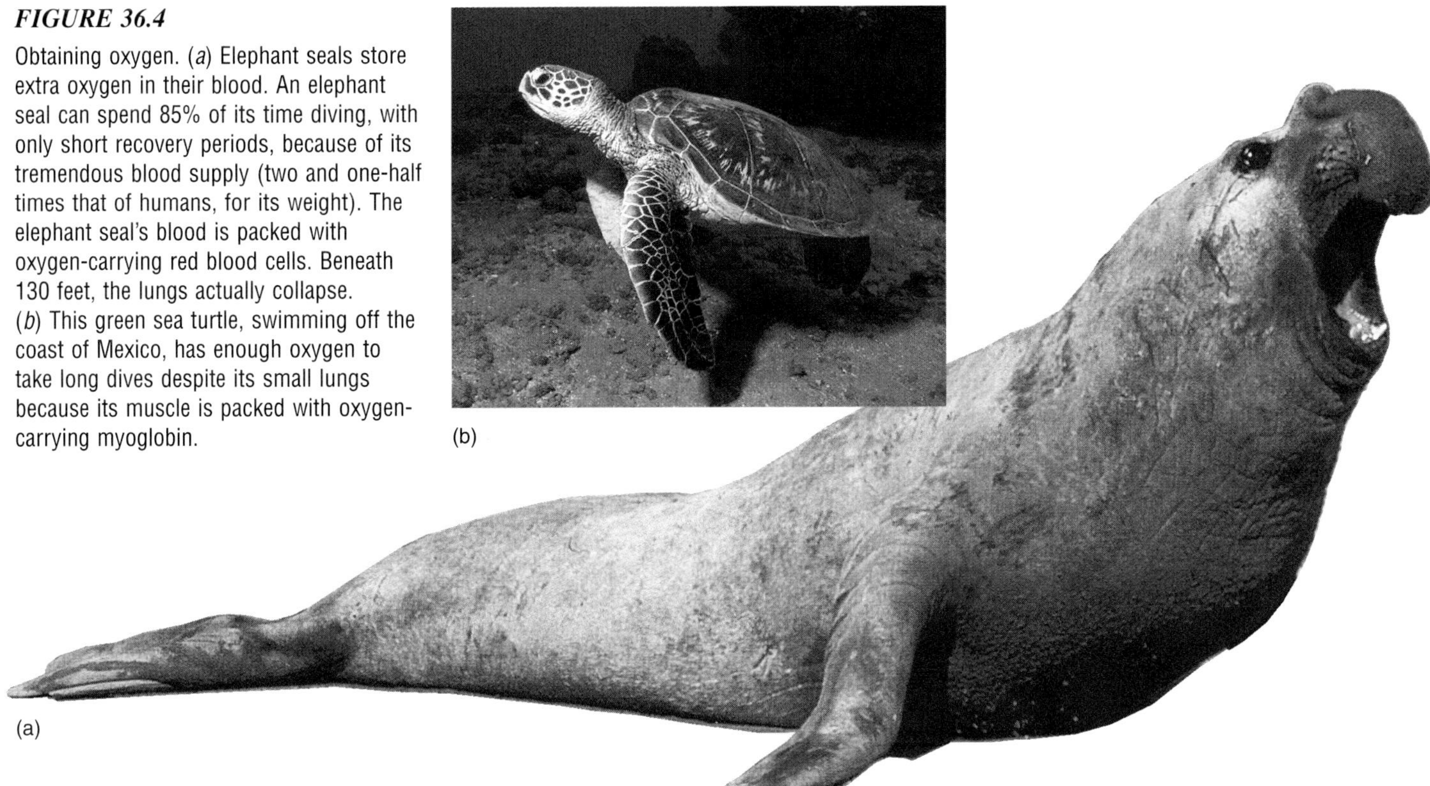

Lungs

A **lung** is an internal respiratory surface that usually closely associates with the circulatory system (fig. 36.2*e*). One advantage of lungs, particularly for terrestrial animals, is that the respiratory surface is kept moist, a necessity for gas exchange. A disadvantage is that the lung must move air in and out to renew the oxygen supply, rather than let air simply flow in one direction over the respiratory membrane as water does over fish gills.

Lung complexity varies. The simplest lungs are little more than air-filled pouches some present-day but primitive fishes use to supplement their gill systems. The bichir, for example, lives in shallow, oxygen-poor waters in Africa, where it occasionally swims to the surface to gulp air. Oxygen diffuses into the lush network of capillaries that pervade its pouchlike lung.

Amphibians and reptiles have some infoldings in their lungs, which increase surface area for gas exchange. This maximization of internal respiratory surface area is spectacular in the mammals, whose lungs are subdivided into sections and contain millions of microscopic, capillary-surrounded air sacs.

Other adaptations can increase oxygen delivery to cells. An elephant seal and a leatherback turtle can obtain oxygen to support frequent diving because of their abundant, oxygen-rich blood supplies (fig. 36.4). Many animals use hemoglobin to supply oxygen to tissues by way of the bloodstream, and myoglobin to supply oxygen directly to the muscle.

Birds have an exceptionally extensive respiratory system that provides the continuous oxygen supply required to power flight. A bird's lungs consist of many narrow tubules where gas exchange occurs. These tubules connect to air sacs that are located throughout the bird's body. Because of the added space that the air sacs provide, oxygen remains in the respiratory system even during exhalation, providing a continual oxygen supply.

Mastering Concepts

1. What are the three forms of respiration?
2. Why do animals require oxygen?
3. What is the role of a circulatory system in respiration?
4. Why do aquatic animals require a larger respiratory surface than terrestrial animals?
5. Describe how tracheal systems, gills, and lungs participate in gas exchange.

The Human Respiratory System

The human respiratory system is representative of mammalian systems. Respiration begins as air enters the nose.

The Nose

In the human body, air must be cleaned, warmed, and humidified before the lungs can extract oxygen from it. Millions of airborne bacteria and dust particles enter the nose each day. Winter temperatures in many parts of the world would freeze lung tissues, and exposure to dry air would kill lung cells.

Ciliated epithelium in the nose forms a mucous membrane that filters, moistens, and warms incoming air. The nose is subdivided

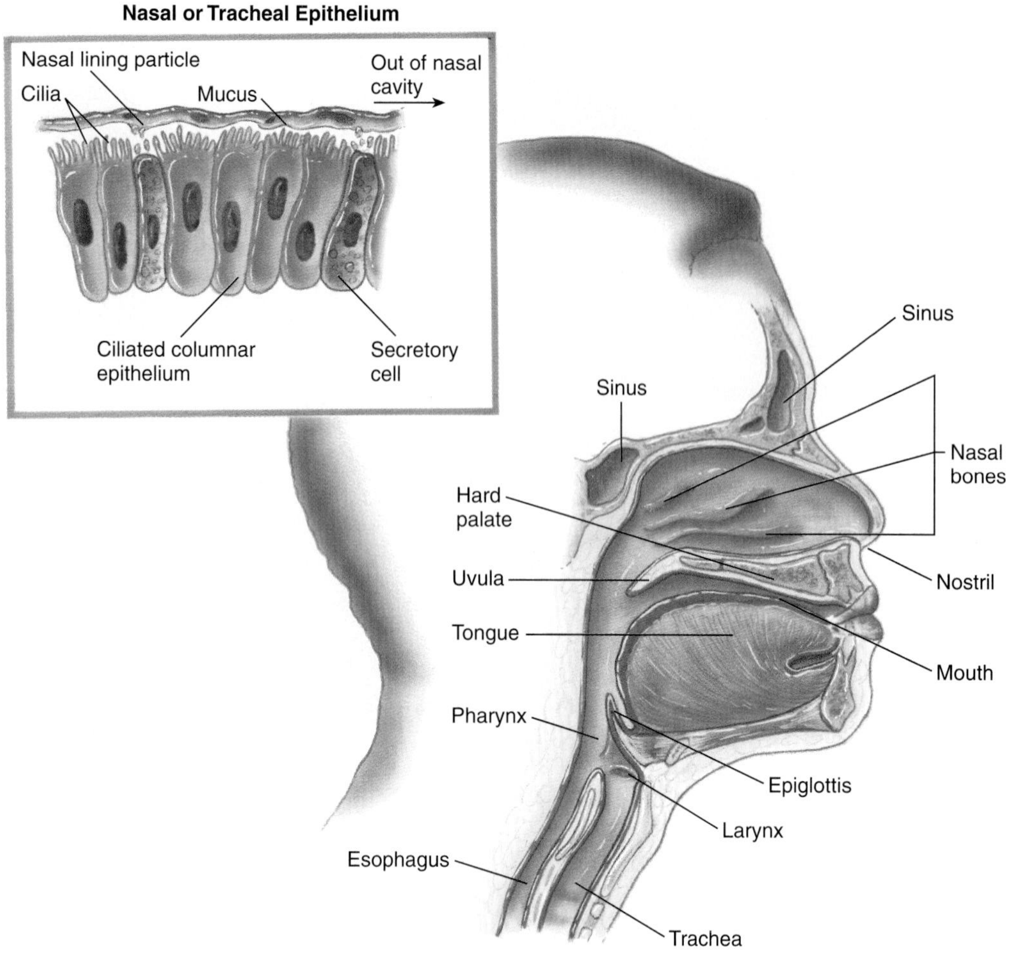

FIGURE 36.5

Human upper respiratory structures. The nose is divided in two, and the nasal bones further divide each half into chambers. The inset shows how mucus and cells lining the nose propel particles outward. Mucous membranes are shown in red.

into compartments, forming many surfaces for ciliated cells to cover. Three shelflike nasal bones partition each of the two nasal cavities into channels. The entire surface is also rich with blood vessels (fig. 36.5).

Hairs at the entrance of the nasal cavities first filter incoming air. Most bacteria and particles that pass the hairs catch in sticky mucus or, if inhaled further, are trapped by mucus lower in the respiratory tract and swept back out by waving cilia. A large inhaled particle may trigger a sensory cell in the nose that signals the brain to orchestrate a sneeze, which forcefully ejects the particle. Sneeze-borne particles can travel up to 100 miles (161 kilometers) per hour.

The nose also protects lung tissue by adjusting the temperature and humidity of incoming air. Blood within the many vessels permeating the mucous membrane warms the air, and the mucus moisturizes it. Cigarette smoke disrupts all of these vital functions of the nose by a variety of mechanisms, including destruction of cilia (see Biology in Action 36.2).

The Pharynx and the Larynx

The back of the nose leads into the **pharynx,** or throat, a 4.5-inch (12-centimeter) tube that conducts food and air. The **larynx,** or Adam's apple, is a boxlike structure below and in front of the pharynx that passes materials through and also produces the voice. A reflex action steers food and fluid toward the digestive system during swallowing. Inhaled air passes through the **glottis,** the opening to the larynx. Swallowing moves the larynx upward, flipping down a piece of cartilage called the **epiglottis** that covers the glottis like a trapdoor (fig. 36.6). This prevents food from entering the respiratory tract (Biology in Action 36.1).

Stretched over the glottis are two elastic bands of tissue, the **vocal cords,** that vibrate when air passes them. The vibrations produce sounds that can be molded into speech. Sound resonates in the pharynx, the nasal cavities, and the mouth and can be molded by moving the tongue and lips. The deepening voice of a male during puberty is caused by expansion of the Adam's apple and lengthening of the vocal cords.

The Trachea, Bronchi, and Bronchioles

The larynx sits atop the windpipe, or **trachea,** a 5-inch- (13-centimeter-) long tube about 1 inch (roughly 2.5 centimeters) in diameter (fig. 36.7). Horseshoe-shaped rings of cartilage hold the trachea open in spite of the negative pressure inhalation creates. You can feel these rings in the lower portion of your neck. The inside surface of the trachea is ciliated and secretes mucus. This helps to filter, warm, and moisten incoming air.

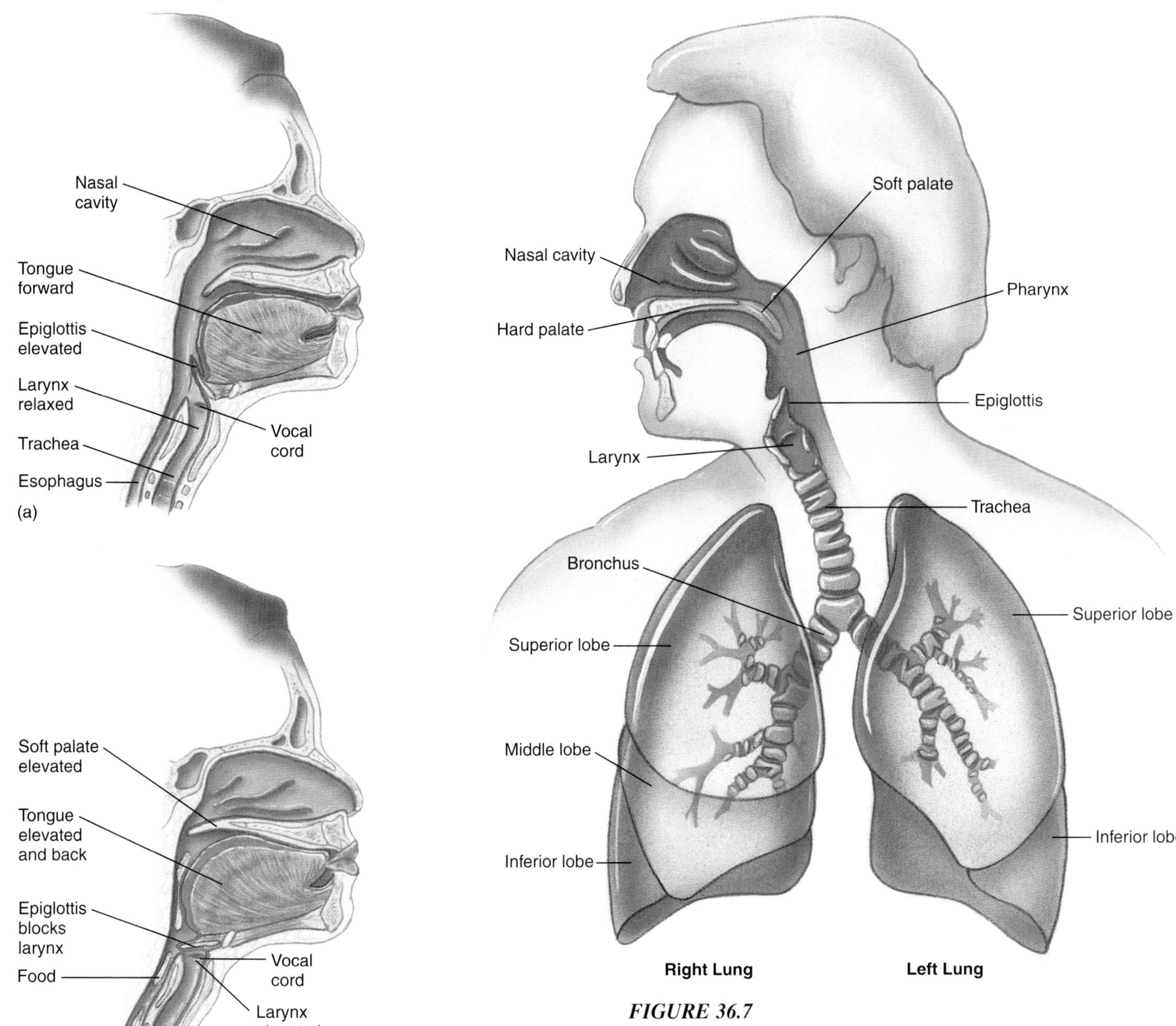

FIGURE 36.7

The trachea leads from the upper respiratory structures to the lungs.

FIGURE 36.6

Separating air from food. (*a*) When we breathe, air passes through the glottis, which is ringed by the epiglottis, down the trachea. (*b*) When we swallow, the epiglottis blocks entry to the trachea, routing food to the digestive system.

As the trachea approaches the lungs, it branches into two **bronchi.** Each bronchus continues the tracheal organization of C-shaped cartilage rings, an inner ciliated epithelial layer, and an outer layer of smooth muscle. The bronchi branch repeatedly, each branch decreasing in diameter until it consists of only a thin layer of smooth muscle and elastic tissue. At this microscopic level, the tubes are called **bronchioles** (little bronchi). The respiratory passageways within the lung are called the "bronchial tree" because their branching pattern resembles an upside-down tree. The 10 million cilia lining the bronchial tree beat hundreds of times a minute to remove inhaled particles.

The bronchioles have no cartilage, but their walls contain smooth muscle. The autonomic nervous system controls contraction of these muscles, adjusting airflow to suit metabolic demands. During a stressful or emergency situation, certain muscle cells may require more oxygen to produce the ATP that fuels the emergency response. Under these conditions, the sympathetic nervous system dilates the bronchioles, increasing their diameters.

Contraction of the bronchial wall muscle is usually adjusted to the body's condition. Sometimes, however, as in **asthma,** the bronchial muscles spasm, making airflow exceedingly difficult (fig. 36.8). Asthma is chronic inflammation of the bronchial passages and episodic wheezing. An allergy to pollen, dog or cat

36.1 biology in action

Food Inhalation and the Heimlich Maneuver

The reflex movements of the larynx normally prevent food from entering the air passageways in the respiratory system. Accidents occur, however, more frequently than many of us realize. Approximately 8,000 to 10,000 people choke to death in the United States each year. This is more likely to happen when people talk with the mouth full of food, hiccup, or are intoxicated. People who are choking may act as if they are having a heart attack. If air passageways are totally blocked, a choking person is unable to speak.

What should you do if you witness a choking incident? First, determine whether the air passages are completely blocked. Most choking incidents do not completely block the air passageways. Therefore, if the person remains calm; it may be possible to move enough air in and out to keep the individual alive until he or she can get to a hospital. If the person can breathe, get help. If the person is turning blue, the air passages are completely blocked and immediate action is necessary.

The best way to assist an upright, severely choking person is to use the Heimlich maneuver (fig. 36.A). Push inward and upward with your clenched fists under the rib cage from the center of the abdomen. This creates a sudden burst of pressure as air is pushed out of the lungs, hopefully with enough force to dislodge the blockage. If the person is lying on the ground, push inward and upward in the anterior part of the abdomen. If you are alone and you begin to choke, you can "Heimlich" yourself by throwing your upper abdominal region against the rounded corner of a table or other stationary object.

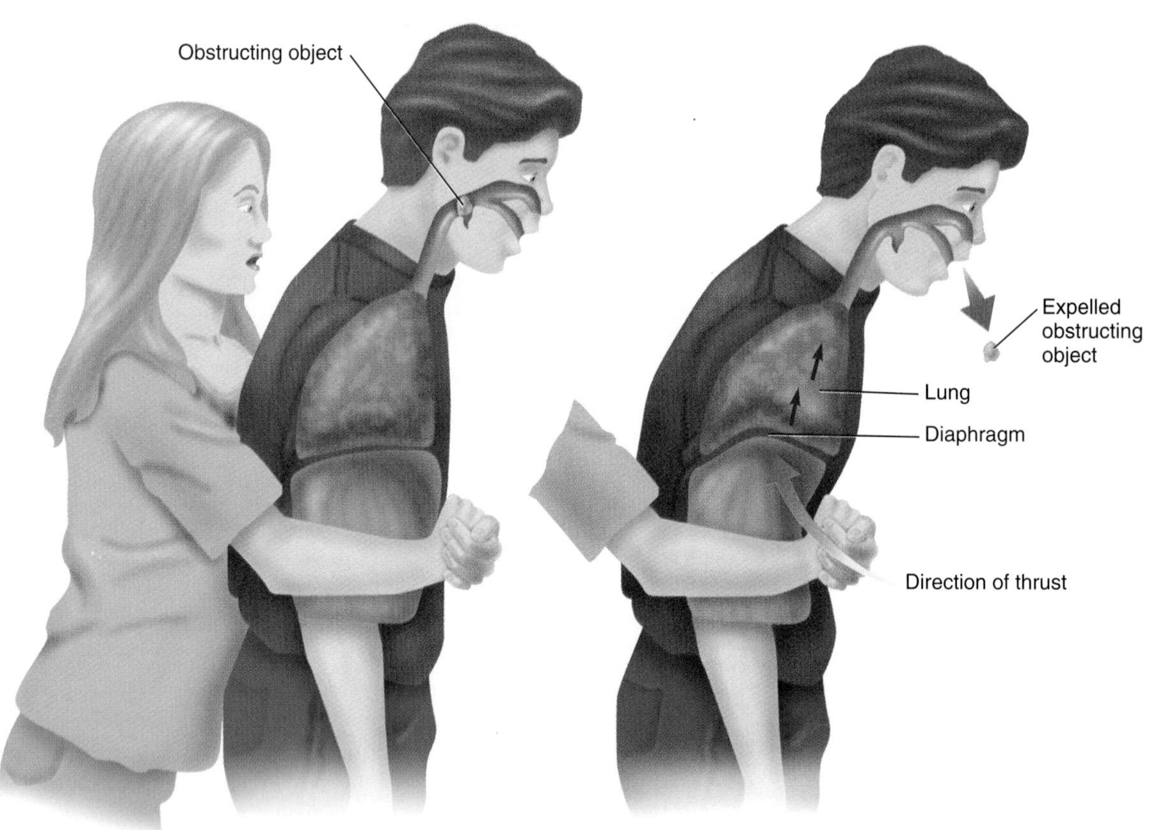

FIGURE 36.A
The Heimlich maneuver expels food stuck in the trachea.

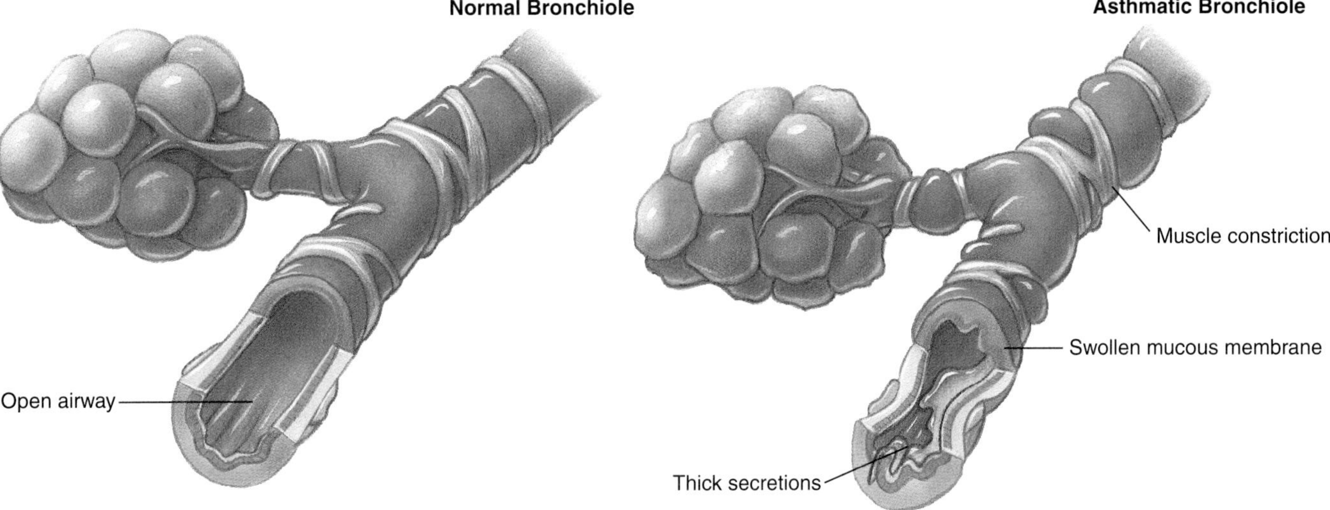

FIGURE 36.8

During an asthma attack, temporarily constricted bronchioles impair breathing. Powerful muscles surrounding the bronchial passages spasm, constricting airways. Mucous membranes lining the bronchioles swell, and mucus secretions thicken and increase. The combination of airway constriction, swollen membranes, and thick secretions can severely curtail the lungs' ability to move air in and out.

dander (skin particles), or tiny mites in house dust trigger most asthma attacks. Inhalant drugs that treat asthma usually control chronic bronchial and lung inflammation and relax the bronchial muscles. Some aerosol inhalers spray adrenaline onto the bronchial walls, which mimics the effect of the sympathetic nervous system and dilates the bronchioles. Asthma is on the rise.

The Alveoli

Each bronchiole narrows into several **alveolar ducts,** and each duct opens into a grapelike cluster of **alveoli** (fig. 36.9). Each alveolus is a tiny sac with a one-cell-thick wall. Most of the lung is composed of some 300 million alveoli, which makes the lung structure similar to that of foam rubber. The total surface area the alveoli in one set of human lungs creates would cover about half a tennis court!

A vast network of capillaries surrounds each cluster of alveoli. Gas diffuses through the thin walls of the alveoli and the neighboring capillaries. Oxygen enters and carbon dioxide leaves the blood within the alveoli.

The alveoli must expand to efficiently use their extensive surface area for gas exchange. A mixture of phospholipid molecules called **human lung surfactant** keeps the alveoli open by occupying the space between the watery film that lines the inner alveoli surfaces and the air within the alveoli. Without surfactant, the watery lining exerts surface tension, the consequence of cohesive forces between water molecules. This surface tension generates a contractile force that would collapse the alveoli, as the inside of a moist balloon sticks together and is difficult to expand with air. Surfactant counters this force, so the alveoli remain open and available for gas exchange (fig. 36.10).

Human lung surfactant is especially important at birth, when a baby's lungs must inflate to take the first breath. The infant must overcome surface tension, which makes the first breath 15 to 20 times more difficult than ventilating expanded alveoli. Once the first breath expands the alveoli, however, surfactant, which has been collecting for the preceding few weeks, keeps them open and eases subsequent breaths.

Many premature babies have not yet produced enough surfactant to prevent the alveoli from collapsing after each breath, which causes infantile respiratory distress syndrome. The newborn must fight as hard for every breath as a normal newborn does for the first one. Synthetic surfactant dripped into the respiratory tract can enable these tiny patients to breathe.

The Lungs

The bronchial tree and the alveoli are housed within the paired lungs. Each human lung weighs about 1 pound (454 grams). A newborn's lung tissues are characteristically pink, but they darken with age as we breathe in environmental pollutants, particularly cigarette smoke (Biology in Action 36.2). The left lung is divided into two lobes, and the right is divided into three. The heart lies between the lungs.

Mastering Concepts

1. How do structures in the nose clean and warm inhaled air?
2. What structures prevent food from entering the respiratory system?
3. Describe the structures and functions of the trachea, bronchi, bronchioles, and alveoli.
4. What is the function of surfactant in the lungs?

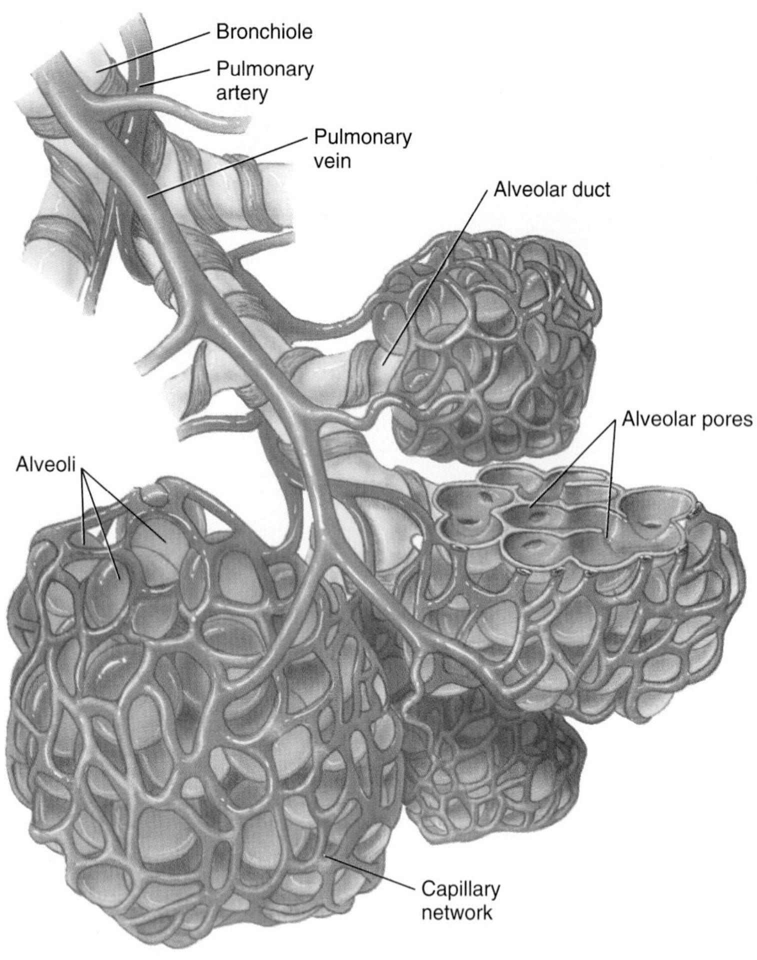

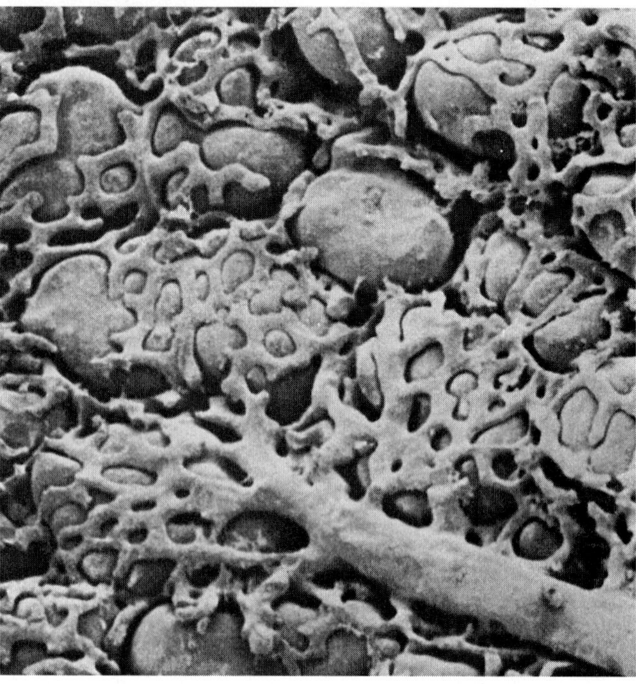

FIGURE 36.9

Alveoli. (*a*) Bronchioles narrow to form alveolar ducts, which lead to clusters of alveoli. A lush capillary supply surrounds each cluster. Alveoli greatly increase the surface area available for gas exchange. (*b*) This electron micrograph was made by filling alveoli and their capillaries with resin. When the tissue decomposed, the resin left behind a cast of the respiratory structures.

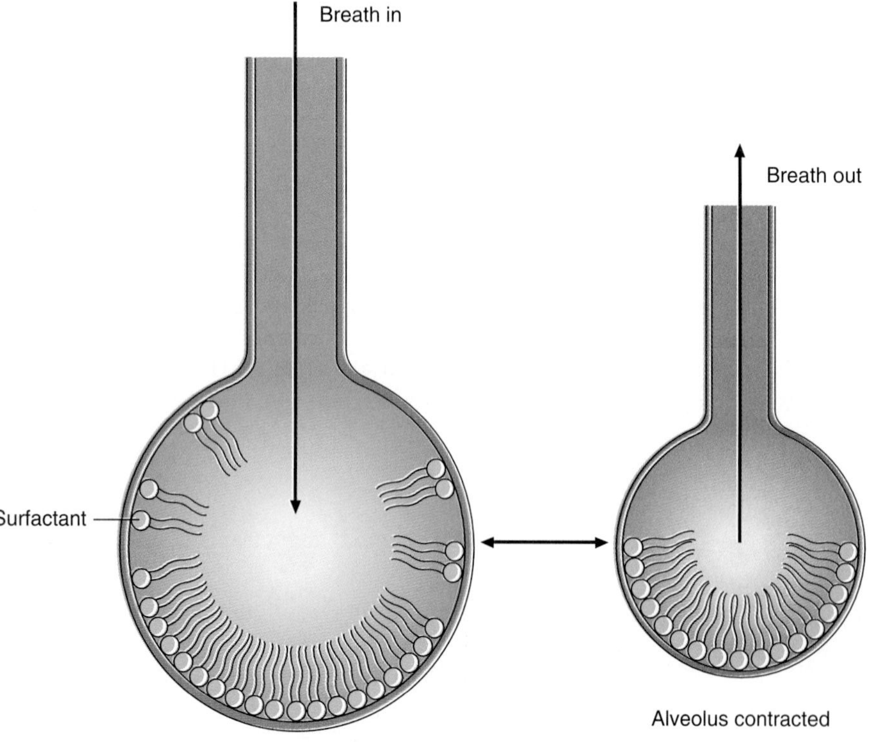

FIGURE 36.10

Lung surfactant allows alveoli to expand. Human lung surfactant is a mixture of phospholipids that insert between the watery lining and air on the inside of alveoli. This reduces the surface tension of the liquid lining, allowing the alveoli to expand.

36.2 biology in action

The Effects of Cigarette Smoking on the Respiratory System

Damage to the respiratory system from cigarette smoking is slow, progressive, and deadly. A healthy respiratory system is continuously cleansed. Mucus traps dirt and microorganisms, and cilia sweep these foreign particles out. Cigarette smoke severely impairs the respiratory system's ability to cleanse itself. With the very first drag, the beating of the cilia slows. With time, the cilia become paralyzed and, eventually, vanish. When the cilia can no longer remove mucus, the individual must cough to clear particles from the airways. Smoker's cough is usually worse in the morning because mucus accumulates during sleep.

To make matters worse, smoking produces excess mucus that clogs the air passageways. Disease organisms that are normally removed now have easier access to respiratory surfaces, and lung congestion favors their growth. This is why smokers are more susceptible to respiratory infections. In addition, a lethal chain reaction begins. Smoker's cough leads to chronic bronchitis; mucus production increases and the linings of the bronchioles thicken, making breathing difficult. As the bronchioles lose elasticity, they are no longer able to absorb the pressure changes that accompany coughing. As a result, a cough can increase the air pressure within the alveoli enough to rupture the delicate alveolar walls, producing emphysema. Burst alveoli cause a worsening cough, fatigue, wheezing, and impaired breathing. Emphysema is 15 times more common among heavy smokers than among nonsmokers.

As structural changes progress to emphysema, cellular changes may lead to lung cancer. First, cells in the outer border of the bronchial lining begin to divide more rapidly than usual, displacing ciliated cells. If smoking continues, these cells may eventually break through the basement membrane and begin dividing within the lung tissue, forming a tumor that could spread throughout the lung tissue and beyond (figs. 36.B and 36.C). Eighty-five percent of lung cancer cases correlate with cigarette smoking. Unfortunately, only 13% of lung cancer patients live as long as 5 years after the initial diagnosis.

It pays to quit smoking. Much of the structural damage to the respiratory system can heal. Cilia may reappear, and the thickening of alveolar walls can reverse, although ruptured alveoli are gone forever.

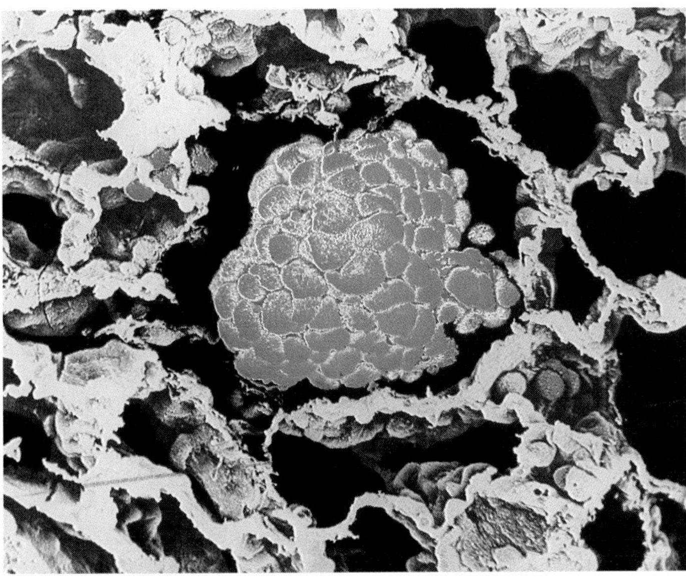

FIGURE 36.B
A tumor invades a microscopic air sac (alveolus) in the lung.

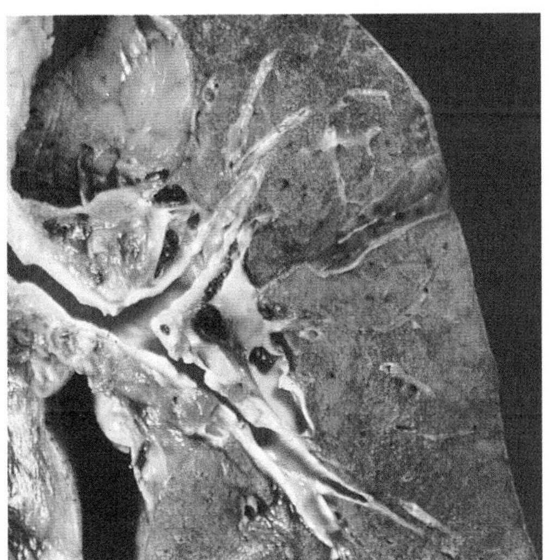

(1)

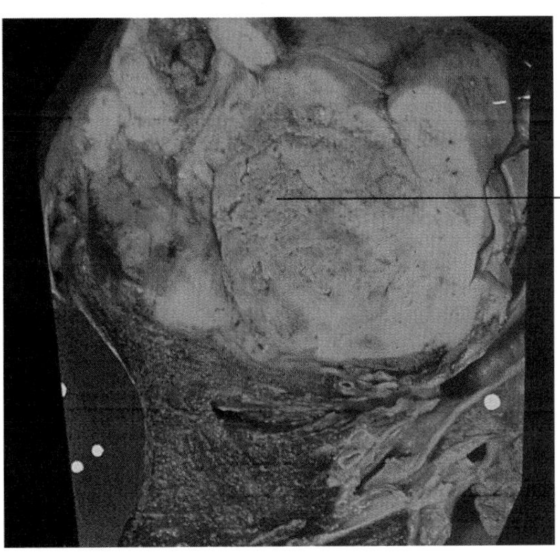

(2)

FIGURE 36.C
(*1*) Normal lung tissue. (*2*) In this advanced case of lung cancer, a tumor takes up nearly half of the lung space.

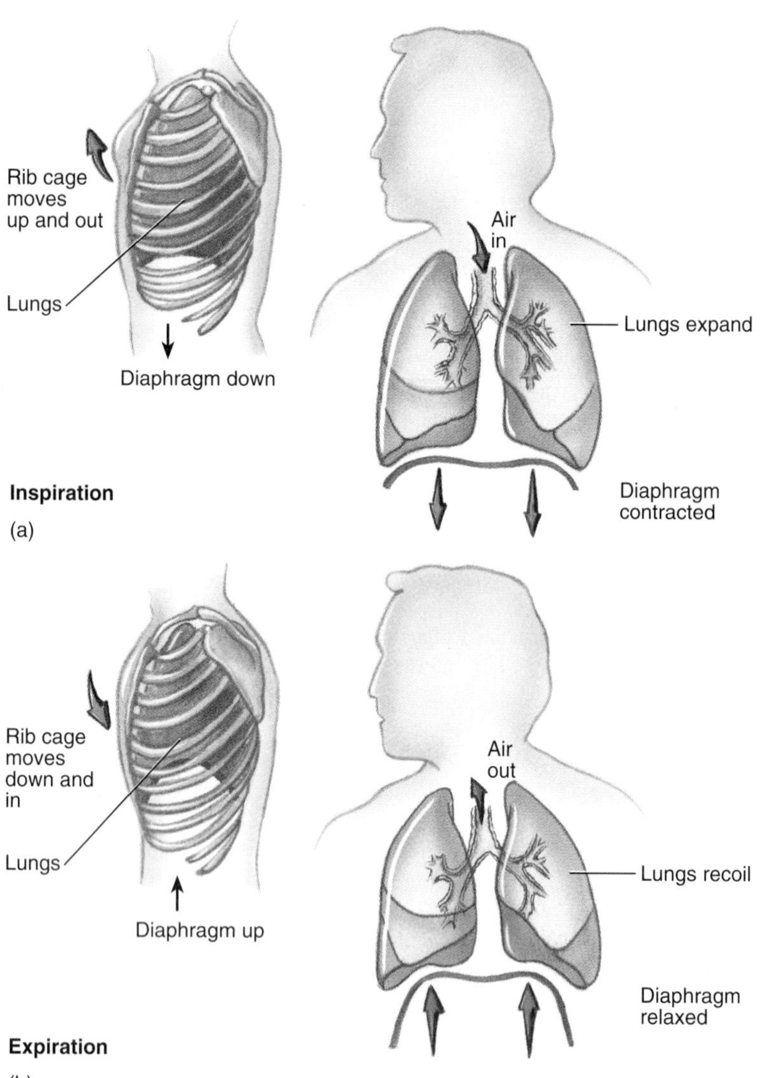

FIGURE 36.11

How we breathe. (*a*) When we inhale, the diaphragm flattens and the rib cage rises, expanding the thoracic cavity and lowering the air pressure in the pleural cavity. This expands the lungs, drawing in outside air. (*b*) When we exhale, the diaphragm relaxes and the rib cage lowers, reversing the process and pushing air out of the lungs.

The Breathing Mechanism

In breathing air moves between the atmosphere and the lungs in response to pressure differences. Air moves in when the air pressure in the lungs is lower than the pressure outside the body, and air moves out when the pressure in the lungs is greater than the atmospheric pressure.

The anatomy of the thoracic (chest) cavity explains the generation of the pressure changes responsible for pulmonary ventilation. The thoracic cavity has no direct connection to the outside of the body. A broad sheet of skeletal muscle, the **diaphragm,** separates it from the abdominal cavity. Membranes coat all the surfaces in the thoracic cavity, which the lungs essentially fill. Changing the size of the thoracic cavity creates pressure changes that move air in and out of the lungs.

When the inspiratory muscles of the rib cage and the diaphragm contract, the thoracic cavity enlarges and draws air into the lungs. This process is called inspiration, or inhalation. When the rib muscles contract, they pull the rib cage upward and forward. Elevation of the rib cage expands the thoracic cavity from front to back, and contraction of the diaphragm elongates it. The increase in size of the thoracic cavity lowers the air pressure within the space between the lungs and the outer wall of the thorax. This causes the lungs to expand, lowering pressure in the alveoli, and air rushes in (fig. 36.11).

When the muscles of the rib cage and the diaphragm relax, the thoracic cavity shrinks, causing expiration, or exhalation. Now the rib cage falls to its former position, the diaphragm rests up in the thoracic cavity again, and the elastic tissues of the lung recoil. As the lungs squeeze back to their resting volume, the pressure within them increases. When the pressure in the lungs exceeds atmospheric pressure, air moves out.

Not all of the respiratory tract actively participates in gas exchange. The part not used in gas exchange—the nose, pharynx, trachea, and bronchi—is called **dead space.**

One inspiration and one expiration constitute a **respiratory cycle.** The volume of air inhaled or exhaled during a single respiratory cycle is called the **tidal volume;** when breathing quietly, you move roughly 1 pint (about 0.5 liter) of air in and out with each breath. During exercise, tidal volume increases. If you were to take the deepest breath possible and exhale until you could not force any more air from your lungs, you could measure your **vital capacity**—the maximal amount of air you can force in and out of the lungs during one breath. Even after a maximal expiration, a volume of air called **residual air** remains in the lungs and inflates the alveoli.

Mastering Concepts

1. What force drives air in and out of the respiratory system?
2. Describe the events of inspiration and expiration.
3. Distinguish between dead space, tidal volume, and residual air.

Gas Transport

Oxygen

From the alveoli, blood carries oxygen throughout the body. Oxygen delivery is quite efficient, for about 99% of the oxygen that reaches cells binds to hemoglobin in red blood cells. The remaining 1% of the oxygen the cells receive is dissolved in plasma. Encapsulating hemoglobin in red blood cells is adaptive because the pigment can carry more oxygen than cells would receive from oxygen suspended in plasma. A red blood cell spending only a second or two in the alveolar capillaries becomes almost completely saturated with oxygen. About 98% of the hemoglobin molecules leaving the lungs are saturated with oxygen. This percentage falls as cells use the oxygen.

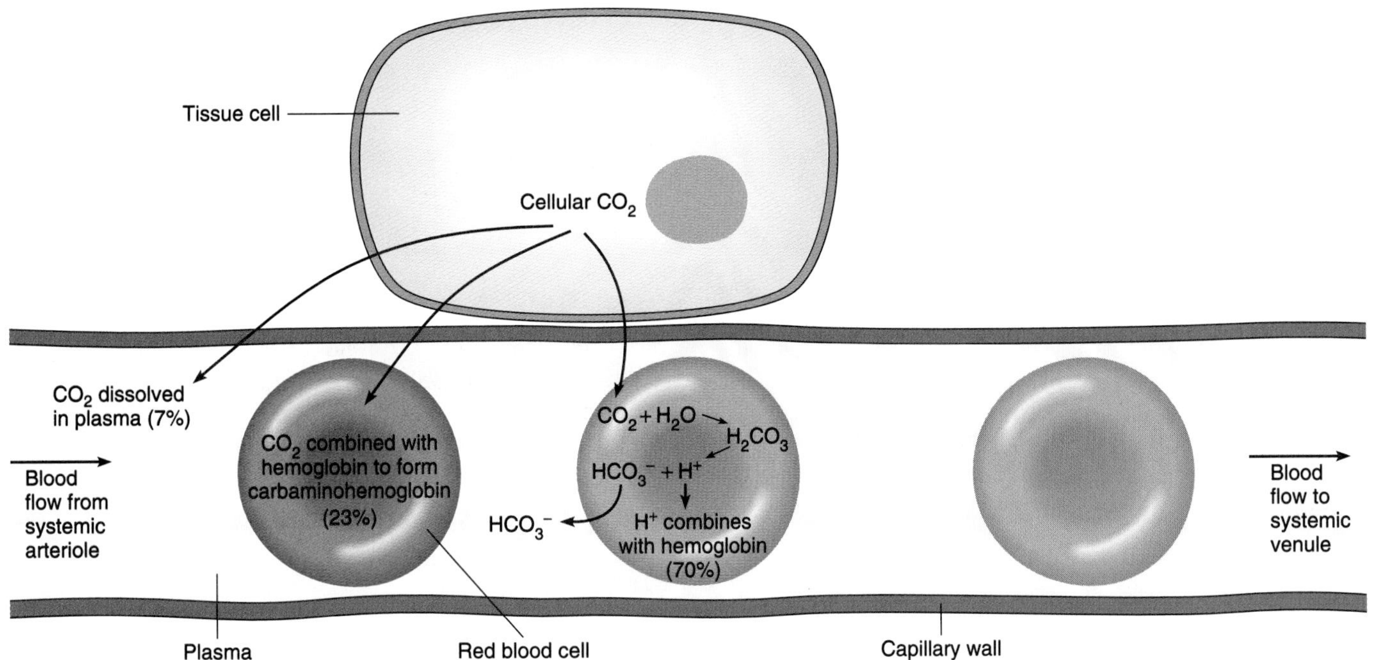

FIGURE 36.12

Carbon dioxide from cells moves in the blood in three ways: dissolved in plasma, combined with hemoglobin, or as bicarbonate ions (HCO_3^-).

Oxygen delivery is amazingly responsive to the requirements of the cells and helps to maintain homeostasis. The amount of oxygen delivered to the cells can more than triple without increasing blood flow rate. Exercise, for example, causes muscle cells to require more oxygen than usual to produce ATP. Under these metabolic conditions, hemoglobin releases more of its oxygen load to the cells. Hemoglobin does this in response to an exercise-induced rise in acidity in the blood. Strenuous exercise can prompt muscle cells to respire anaerobically and release lactic acid. In addition, a very active muscle cell produces more carbon dioxide (CO_2) as waste. The CO_2 forms carbonic acid when it dissolves in the water of the tissue fluid and blood, a point we will soon return to.

Active tissues also produce heat as a by-product of energy metabolism. Blood flowing through warmer parts of the body releases more of its oxygen. Once again, more oxygen is released to cells that use oxygen at a higher rate (table 36.1).

Table 36.1 Volume of Air Inhaled During Selected Activities

Activity	Liters Inhaled per Minute
Bed rest	6
Sitting	16
Walking	24
Running	50

Carbon Dioxide

Besides delivering oxygen, blood must also remove CO_2 that cells produce. CO_2 transport occurs in three ways (fig. 36.12). A small amount of carbon dioxide, about 7%, is transported dissolved in plasma. When CO_2 reaches the lungs, it simply diffuses into the alveoli and is exhaled. Hemoglobin molecules transport slightly less than a quarter (23%) of the CO_2. CO_2 combined with hemoglobin is called carbaminohemoglobin.

About 70% of CO_2 is transported as bicarbonate ion (HCO_3^-) dissolved in plasma. The bicarbonate ion forms when CO_2 that cells produce diffuses into the blood and into the red blood cells. The CO_2 reacts with water to form carbonic acid (H_2CO_3), which dissociates to hydrogen ions (H^+) and bicarbonate ions; within the red blood cell, an enzyme called carbonic anhydrase speeds this reaction. The H^+ then combines with hemoglobin, and the bicarbonate ions diffuse out of the red blood cell into the plasma. When the blood reaches the lungs, the whole process reverses (fig. 36.13). In the presence of carbonic anhydrase within the red blood cell, carbonic acid re-forms CO_2 and H_2O. The carbon dioxide then diffuses into the alveolar air and is exhaled.

In the tissues, CO_2 enters the bloodstream, where its concentration is lower. In the lungs, CO_2 moves from the bloodstream to the alveoli because the concentration in the bloodstream is higher.

Mastering Concepts

1. What parts of blood deliver oxygen to cells?
2. How does oxygen delivery adapt to energy requirements?
3. What are the three ways that the blood transports CO_2?

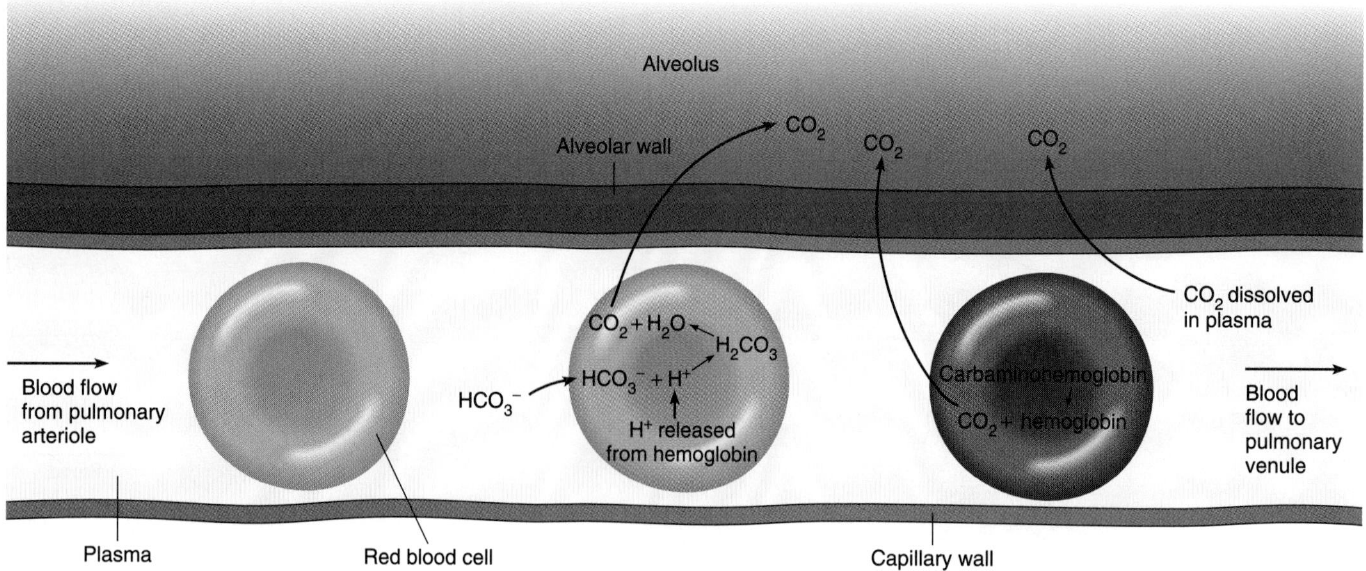

FIGURE 36.13
In the lungs, CO_2 diffuses from the blood to the alveoli.

Control of Respiration

Awake or asleep, without conscious effort, we breathe an average of 12 times a minute. A rhythmicity center in the brain's medulla controls this steady breathing. During quiet breathing, part of this region, called the inspiratory center, is in control. It periodically triggers the diaphragm to contract and thus initiates inspiration. Expiration is the passive result of the lungs elastic recoil that forces air out of the respiratory passageways.

During strenuous activity, the respiratory cycle changes so that exhalation becomes an active process. The inspiratory center sends impulses to another part of the medulla's rhythmicity center called the expiratory center. This contracts muscles that pull the rib cage down and the sternum inward and also contracts the abdominal muscles, causing the abdominal organs to push the diaphragm farther into the thorax than usual. Air is quickly pushed out of the lungs.

Other brain areas may modify the basic breathing pattern. We can voluntarily alter our breathing pattern through impulses originating in the cerebral cortex. For example, we can control breathing when we speak or play a musical instrument, and we can voluntarily pant like a dog, sigh, or hold our breath.

The Role of Blood CO_2 Level

The most important factor in regulating breathing rate is the blood CO_2 level, or, more precisely, the number of H^+ formed when CO_2 reacts with H_2O to form carbonic acid. Because under most circumstances CO_2 is a by-product of aerobic respiration, monitoring blood CO_2 levels is a way to determine how quickly cells use oxygen.

Chemoreceptors in the medulla, and to some extent those in the aorta and carotid arteries, monitor blood CO_2 levels (fig. 36.14). These receptors are near the surface of the medulla, where they are bathed in cerebrospinal fluid. Since cerebrospinal fluid lacks the blood's ability to buffer pH changes, dissolved CO_2 changes the pH of cerebrospinal fluid more than that of blood. When the blood CO_2 level increases, it raises the H^+ concentration of the cerebrospinal fluid. Chemoreceptors are then stimulated to send messages to the medulla's inspiratory center. As a result, breathing rate increases.

Although the CO_2 concentration in the air we breathe is low, small changes can trigger a tremendous increase in breathing rate. For example, an increase to 5% CO_2 in the air causes us to breath 10 times faster.

Oxygen level is actually less important than CO_2 level in regulating breathing. In fact, oxygen level does not regulate breathing rate unless it falls dangerously low. This does not normally happen because the great amount of O_2 bound to hemoglobin provides a large safety margin. Activation of the inspiratory center due to low oxygen levels is a last-ditch effort to increase breathing rate and restore normal oxygen levels. Sudden infant death syndrome (SIDS), in which a baby dies while asleep, may be caused by failure of chemoreceptors in major arteries and/or the brain's respiratory centers to respond to low oxygen levels in arterial blood (fig. 36.15).

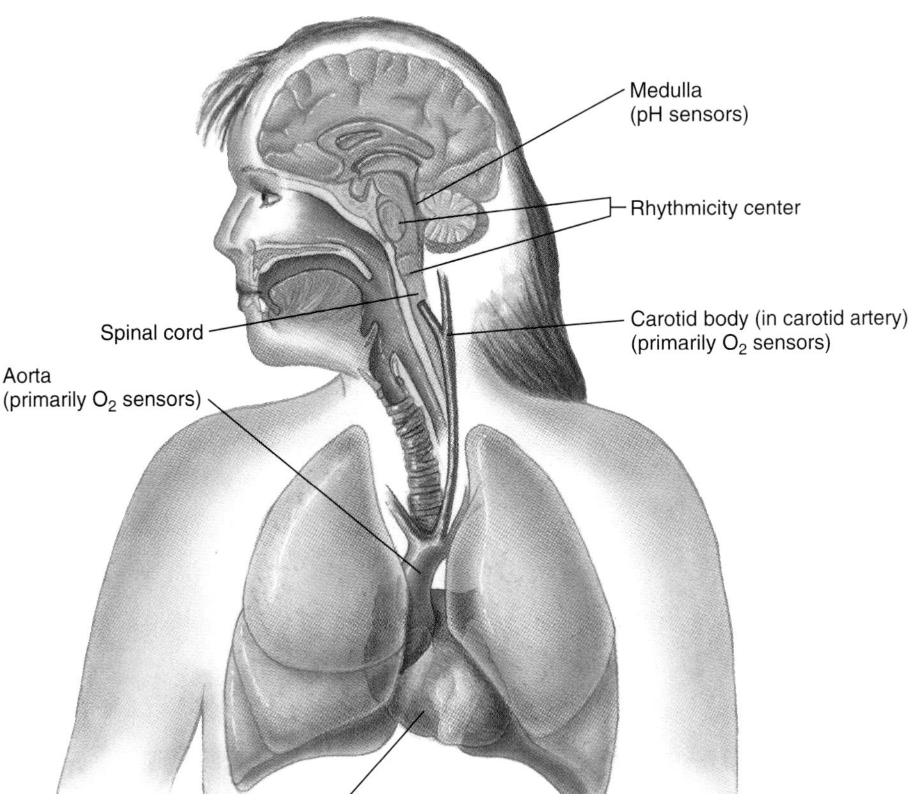

FIGURE 36.14

Breathing control. The medulla has receptors that monitor H^+ concentrations in cerebrospinal fluid. Other receptors in the aorta and carotid arteries sense oxygen levels in the bloodstream, but they also respond to CO_2.

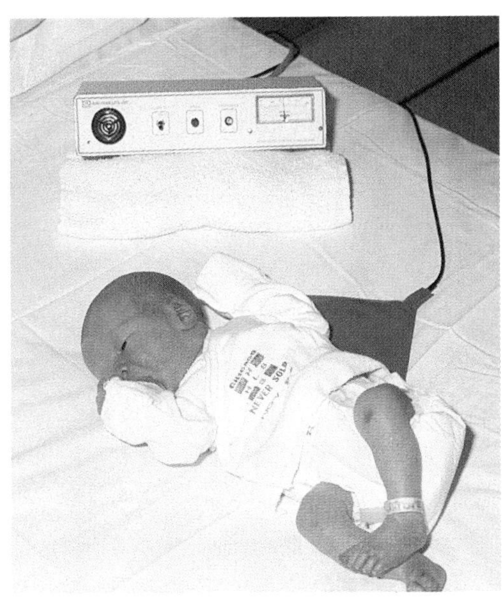

FIGURE 36.15

Sleep apnea. Each year, 10,000 babies in the United States die because they stop breathing in their sleep, a condition called sleep apnea that leads to sudden infant death syndrome (SIDS). Babies who exhibit sleep apnea just after birth are often sent home with monitors, which sound an alarm when the child stops breathing, alerting parents to resuscitate the infant. The position in which the baby sleeps seems to affect the risk of sleep apnea—sleeping on the back is safest. Parental smoking near the infant also raises the risk of SIDS. Sleep apnea in adults causes snoring.

Table 36.2 The Effects of High Altitude

Condition	Altitude	Symptoms
Acute mountain sickness	5,900 feet (1,800 m)	Headache, weakness, nausea, poor sleep, shortness of breath
High-altitude pulmonary edema	9,000 feet (2,700 m)	Severe shortness of breath, cough, gurgle in chest, stupor, weakness; person can drown in accumulated fluid in lungs
Cerebral edema	13,000 feet (4,000 m)	Brain swells, causing severe headache, vomiting, loss of coordination, hallucinations, coma, and death

Breathing at High Altitudes

The human respiratory system functions best at the 21% oxygen concentration near sea level. At high elevations, the percentage of oxygen in the air falls gradually, so that at an elevation of almost 10,000 feet (about 3,000 meters) above sea level, an individual inhales a third less oxygen with each breath. Each year, 100,000 mountain climbers and high-altitude exercisers experience varying degrees of altitude sickness. This occurs when the body's effort to get more oxygen—by increasing breathing and heart rate and stepping up production of red blood cells and hemoglobin—cannot keep pace with declining oxygen concentration. Altitude sickness can be cured by descending slowly at the first appearance of symptoms. Table 36.2 indicates the symptoms associated with increasing altitude.

Table 36.3 The Unhealthy Respiratory System

Condition	Cause	Symptoms
Bronchitis	Bacterial or viral infection or irritation destroys cilia	Persistent heavy cough
Cystic fibrosis	Secretions thicken	Severe lung congestion and infection, impaired digestion
Emphysema	Loss of elastic fibers around alveoli overinflates them	Cough, fatigue, wheezing, difficulty breathing
Lung cancer	Inhalation of asbestos, cigarette smoking	Fever, bloody cough, hoarseness, wheezing
Pleurisy	Inflammation of pleural linings due to infection	Chest pain, fever, dry cough
Pneumonia	Inflammation of bronchioles and alveoli due to bacterial or viral infection	Fatigue, fever, bloody cough, chest pain, difficulty breathing
Respiratory distress syndrome (infantile)	Deficiency of surfactant in lungs of newborn	Newborn cannot breathe
Streptococcal sore throat (strep throat)	Bacterial infection of throat	Painful swallowing, fever
Sudden infant death syndrome (crib death)	Unknown	Breathing of infant suddenly stops (apnea)
Pulmonary tuberculosis	Bacterial infection of lungs	Fever, emaciation, cough

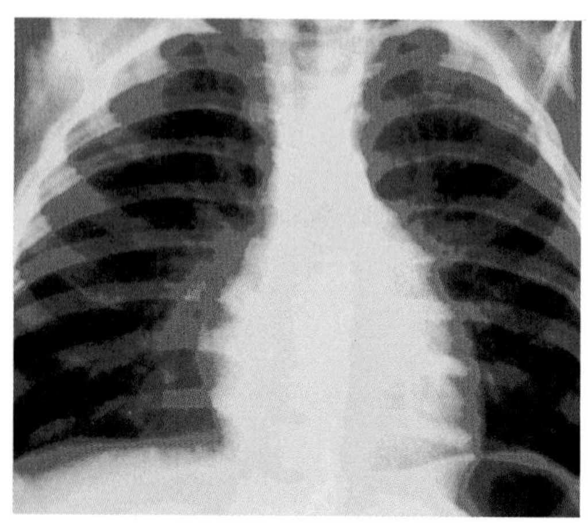

(a)

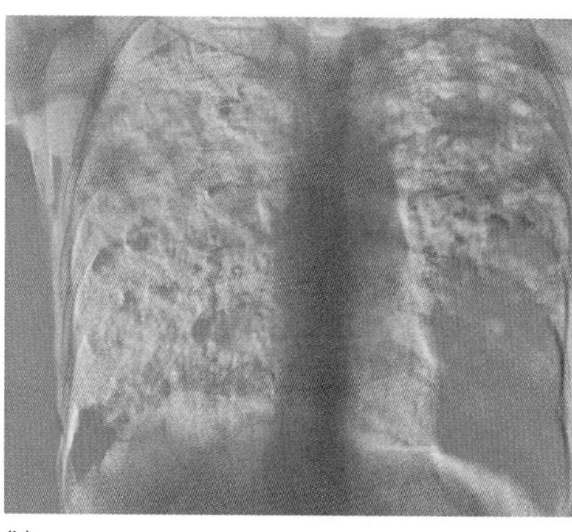

(b)

FIGURE 36.16

Tuberculosis. (*a*) Healthy lungs are dark and clear on X-ray film. (*b*) Lungs with tuberculosis have cloudy areas where the body attempts to wall off infected regions.

Mastering Concepts

1. What parts of the brain control breathing?
2. How does strenuous activity alter the respiratory cycle?
3. How is blood CO_2 level detected and responded to?

The Unhealthy Respiratory System

Some of the most common human illnesses affect the respiratory system (table 36.3). Nearly everyone has suffered through a cold (upper respiratory infection) or influenza, and many of us have had bronchitis, pneumonia, or pharyngitis (a sore throat). People who have frequent bouts of pneumonia or bronchitis may actually have a form of cystic fibrosis, the inherited illness that impairs breathing and digestion (see fig. 5.4).

A respiratory infection on the rise is a multidrug-resistant form of tuberculosis, which is a bacterial infection (fig. 36.16). Because tuberculosis is transmitted by airborne droplets propelled by sneezes or coughs, it spreads quickly in crowded places, such as prisons and shelters for the homeless. People with AIDS are especially prone to tuberculosis and can spread it easily, although AIDS is not easily spread.

Fortunately, we can actively maintain respiratory health. Regular exercise, avoiding cigarette smoke, eating properly, and good personal hygiene habits can help make every breath a pleasant, painless experience.

A Case of Poisoning~

continued from page 732.

Nurses asked the couple detailed questions about how and where they lived, and it soon emerged that they had spent the winter in a small, unventilated apartment that was heated by two kerosene space heaters. The source of their medical problem became clear—carbon monoxide (CO) poisoning.

A test for carboxyhemoglobin yielded levels of 13 to 14%, compared with the normal 2% of this form of hemoglobin in healthy nonsmokers. The couple's levels were probably nearer 20 to 25% when they were in the apartment. The table in Biology in Action 3.1 lists the symptoms associated with increasing CO levels. CO binds tightly to hemoglobin, preventing oxygen from binding, and thereby robs tissues of oxygen delivered in the blood.

Carbon monoxide poisoning results from incomplete combustion of carbon-containing fuel. Most often this occurs in wood stoves, kerosene heaters, appliances, gasoline engines, blocked chimneys, or malfunctioning fireplaces. A few reports cite faulty ice resurfacing machines in public ice skating rinks, construction workers using gas-powered equipment in closed warehouses and garages, and indoor tractor pull events as sources of carbon monoxide poisoning.

The man and woman were fortunate to have developed the symptoms suggestive of heart failure, because it sent them to seek help. They could have easily fallen unconscious, gone into a coma, and died. CO poisoning is known as "the great imitator," because its symptoms are easily attributed to other causes.

A medical journal published an account of the carbon monoxide poisoning of the couple, which led two physicians to respond in a letter that they had begun measuring carboxyhemoglobin levels among patients with chest pain. Of 104 patients, 7 had carboxyhemoglobin levels between 10 and 20%—enough to account for their symptoms.

SUMMARY

1. In external respiration, animals exchange oxygen and carbon dioxide between respiratory organs and blood. In internal respiration, these gases exchange between blood and cells. Cellular respiration uses oxygen to store nutrient energy in ATP.

2. CO_2 forms as a by-product of aerobic ATP production and must be eliminated from the body. Gas exchange brings oxygen to and removes CO_2 from cells.

3. In respiration, oxygen and CO_2 exchange by diffusion across a moist membrane. An animal's size, metabolic requirements, and habitat have affected the evolution of respiratory systems. Simple organisms exchange oxygen and CO_2 directly across the body surface. Larger or more active animals have circulatory systems to transport gases between cells and the environment. Terrestrial arthropods bring the environment into contact with almost every cell through a highly branched system of **tracheae.** Complex aquatic animals exchange gases across **gill** membranes, body surface infoldings. In bony fishes' gills, water flows over the gills in the direction opposite blood flow. Vertebrate **lungs** create an extensive internal surface for gas exchange.

4. In humans, the nose, purifies, warms, and moisturizes inhaled air. The air then flows through the **pharynx, larynx,** and **trachea,** which is held open by cartilage rings. The trachea divides to form **bronchi** that deliver air to the lungs. The bronchi branch extensively to form tinier air tubules, **bronchioles,** which end in clusters of thin-walled, saclike **alveoli.**

5. Many capillaries surround the alveoli. Oxygen diffuses into the blood from the alveolar air, while CO_2 diffuses from the blood into the alveolar air. **Human lung surfactant,** a chemical mixture that reduces surface tension, helps keep alveoli open.

6. Breathing brings oxygen-rich air into the lungs and removes air high in CO_2. When the **diaphragm** and rib cage muscles contract, the thoracic cavity expands. This reduces air pressure in the lungs, drawing air in. Expiration results when these muscles relax and the thoracic cavity shrinks, raising air pressure in the lungs and pushing out air.

7. Air in **dead space** in the nose, pharynx, trachea, and bronchi is not used in gas exchange. **Residual air** remains in the lungs after expiration. The volume of air moved in and out of the lungs during a **respiratory cycle** is the **tidal volume.** The amount of air that can be exhaled after a maximal inspiration is the **vital capacity.**

8. Almost all oxygen transported to cells is bound to hemoglobin in red blood cells. Increase in blood acidity, usually due to a rise in CO_2 level, or elevated temperature due to metabolism, increases the amount of oxygen reaching cells.

9. Some CO_2 in the blood is carried bound to hemoglobin or dissolved in plasma. Most CO_2 is transported as bicarbonate ion, generated from carbonic acid that forms when CO_2 reacts with water and ionizes. Carbonic anhydrase speeds this reaction. The process reverses in the lungs, releasing CO_2 for expiration.

10. During quiet breathing, the inspiratory center within the medulla's rhythmicity center spontaneously generates impulses that trigger inhalation. During heavy breathing, the inspiratory center remains active and the expiratory center becomes active, causing active expiration. Other brain regions can alter the basic breathing pattern.

11. Breathing rate adjusts to the body's demands chemically. When CO_2 levels rise, chemoreceptors in the medulla, aorta, and carotid arteries sense the rise in blood acidity and increase breathing rate. Oxygen levels do not change breathing rate unless levels are critically low.

TO REVIEW . . .

1. What are the two major functions of external respiration?

2. What is the connection between external respiration and aerobic, cellular respiration?

3. How does an animal's size, activity level, and environment influence the structure and function of its respiratory system?

4. Trace the path of an oxygen molecule from a person's nose to a red blood cell.

5. How is air cleaned before it reaches the lungs?

6. Describe inspiration. Which muscles are active in the process?

7. How does the body transport oxygen to cells?

8. What mechanisms direct more oxygen to metabolically active cells?

9. How does the respiratory system transport most CO_2? In what other ways is CO_2 transported?

10. How does the brain establish breathing rhythm?

11. What chemical change in the blood is most important in altering breathing rate?

TO THINK ABOUT . . .

1. It is below 0°F outside, but the dedicated runner bundles up and hits the roads anyway. "You're crazy," shouts a neighbor. "Your lungs will freeze." Why is the well-meaning neighbor wrong?

2. Why does breathing through the mouth instead of the nose dry the throat out?

3. What color might the insides of prehistoric human lungs have been? Why?

4. On December 13, 1799, George Washington spent the day walking on his estate in a freezing rain. The next day, he had trouble breathing and swallowing. Several doctors were called in. One suggested a tracheostomy, or cutting a hole in the president's throat so he could breathe. He was voted down. The other physicians suggested bleeding the patient, plastering his throat with bran and honey, and placing beetles on his legs to produce blisters. Within a few hours, Washington's voice became muffled, his breathing more labored, and he grew restless. For a short time he seemed euphoric; then he died. Washington had epiglottitis, a swelling of the epiglottis to 10 times its normal size. How does this diagnosis explain Washington's symptoms? Which suggested treatment might have worked?

5. Why can't you commit suicide by holding your breath?

6. Why do you think the times of endurance events at the 1968 Olympics, held in 7,218-foot-high (2,200-meter-) Mexico City, were rather slow?

7. Why might it be dangerous for a heavy smoker to use a cough suppressant?

8. The most successful treatment for lung cancer is surgery performed when a tumor is small and confined to one lung. The entire lung or the lobe containing the tumor may be removed. Many physicians, however, will not attempt such surgery on a patient who is a heavy smoker because of a second illness. What might this illness be?

TO LEARN MORE . . .

References and Resources*

Bartecchi, Carl E., et. al. May 1995. The global tobacco epidemic. *Scientific American.* Lung cancer rates will remain high until people stop smoking.

Eckert, Scott A. March 1992. Bound for deep water. *Natural History.* Leatherback turtles have small lungs, but they can survive deep dives because their muscles have high concentrations of oxygen-carrying myoglobin.

Flieger, Ken. November 1996. Controlling asthma. *FDA Consumer.* Asthma incidence is on the rise.

Lewis, Ricki. June 1993. Cystic fibrosis: Test, treatments improve survival. *FDA Consumer.* Several new treatments, including gene therapy, are easing breathing for people with cystic fibrosis.

Lewis, Ricki. June 1992 Defeating adult sleep apnea. *FDA Consumer.* In sleep apnea, a person ceases to breathe for 10 to 20 seconds hundreds of times a night.

Luce, John M. December 19, 1996. Reducing the use of mechanical ventilation. *The New England Journal of Medicine,* vol. 335. Mechanical ventilation can temporarily take over for the respiratory system.

Patlack, Margie. June 1995. Treating lung cancer. *FDA Consumer.* Successful treatment of lung cancer is rare.

Poets, C. F., and D. P. Southall. August 5, 1993. Prone sleeping position and sudden infant death. *The New England Journal of Medicine,* vol. 329. Sleeping position may alter the risk of sudden infant death syndrome.

**Life*'s website features frequently updated references.

The long beak of this Galápagos mockingbird enables it to find, extract, attack, vanquish, and finally consume a long centipede.

37 Digestion and Nutrition

chapter outline

Why Animals Eat ***750***
The Evolution of Eating

Digestive Diversity ***752***
Types of Digestive Systems
Specific Digestive System Adaptations

Human Nutrition ***755***
Nutrient Deficiencies

An Overview of the Human Digestive System ***757***

Food's Journey Through the Human Body ***760***
The Mouth and the Esophagus
The Stomach
The Small Intestine
The Large Intestine
The Pancreas, Liver, and Gallbladder

What We Can't—and Can—Control ***765***

A New View of Ulcers

Physicians and medical researchers are often hesitant to abandon a long-entrenched idea. So it was with the discovery that gastric ulcers, a common, painful digestive disorder, can result from bacterial infection.

Until recently, ulcers were thought to be the direct result of excess stomach acid secretion, dating from a German researcher's concise pronouncement in 1910, "No acid, no ulcer." The finding in the 1970s that drugs that block acid production relieve ulcers (albeit temporarily) seemed to confirm the hypothesis that excess acid causes ulcers. Acid buildup was attributed to stress and eating spicy foods. For decades, the standard treatment for ulcers in the stomach or small intestine was a bland diet, stress reduction, acid-blocking drugs, or surgery. Over the many years that ulcers typically persist, cost of this treatment mounts. Considering this background, the idea that an ulcer was actually a bacterial infection that could be easily, cheaply, and permanently cured seemed preposterous.

Helicobacter pylori—the ulcer bacterium—was identified in the laboratory of J. Robin Warren at Royal Perth Hospital in western Australia. Warren had found bacteria in stomach tissue samples from people suffering from gastritis, an inflammation of the stomach. He didn't know which way to phrase the hypothesis—were the bacteria attracted to inflamed tissue, or did they cause the inflammation?

Warren's assistant, medical resident Barry Marshall, helped choose between the hypotheses. Marshall knew he had a healthy stomach and had never had gastritis or an ulcer. If the bacteria caused the irritation, they would do so in him. So on a hot July day in 1984, Marshall concocted some "swamp water"—a brew of a billion or so of the microbes—and drank it.

to be continued . . .

Why Animals Eat

The early bird catches the worm, the saying goes, but sometimes it catches much more. The American robin (fig. 37.1) is a "seasonal frugivore," meaning that it is an animal whose tastes vary with the season. It eats mostly insects, spiders, and worms in the fall, winter, and spring but relies more on fruits and berries in the summer (table 37.1).

The robin's digestive system, like that of many birds, is highly adapted to its eclectic diet. Its digestive biochemicals can break down the protein and fat that make up most of its animal-based meals, yet its stomach is specialized to rapidly digest fruits. In the stomach, large seeds are separated from a fruit meal and regurgitated; the pulp is then digested quickly. The intestine absorbs the nutrients, and the fruit's peel and remaining seeds move on to become a bird dropping.

For the robin to use the proteins and fats in its worm meal, or the sugars in fruits, its body must dismantle these nutrient molecules into smaller molecules—proteins to amino acids, complex carbohydrates (starches) to simple carbohydrates (sugars), and fats to fatty acids and glycerol. Only such small molecules can enter the cells lining the digestive tract and then leave them to enter the circulation. Once inside their destination cells, nutrient molecules are broken down to release the energy in their chemical bonds that holds them together. Nonnutritive parts of food are eliminated from the body as feces.

Animals eat to replace energy they expend in the activities of life and to provide raw material for the structural components required for growth, repair, and maintenance of the body. All types of animals use the same types of nutrient molecules. Digestion is both mechanical—breaking food into small pieces—and chemical. Hydrolytic (water-splitting) enzymes add water molecules between the building blocks of large nutrient molecules, thereby splitting them apart (fig. 37.2).

Animals obtain food in diverse ways. Humans grow, hunt, and purchase food. A protozoan engulfs plankton, surrounding it with a bubble of membrane where enzymes complete digestion. Many

FIGURE 37.1

A robin's digestive system is adapted to eating fruits and berries, as well as insects, spiders, and worms.

Table 37.1 Types of Feeders

Term	Type of Food
Carnivore	Animals
Frugivore	Fruits and berries
Herbivore	Plants
Insectivore	Insects
Omnivore	Plants and animals

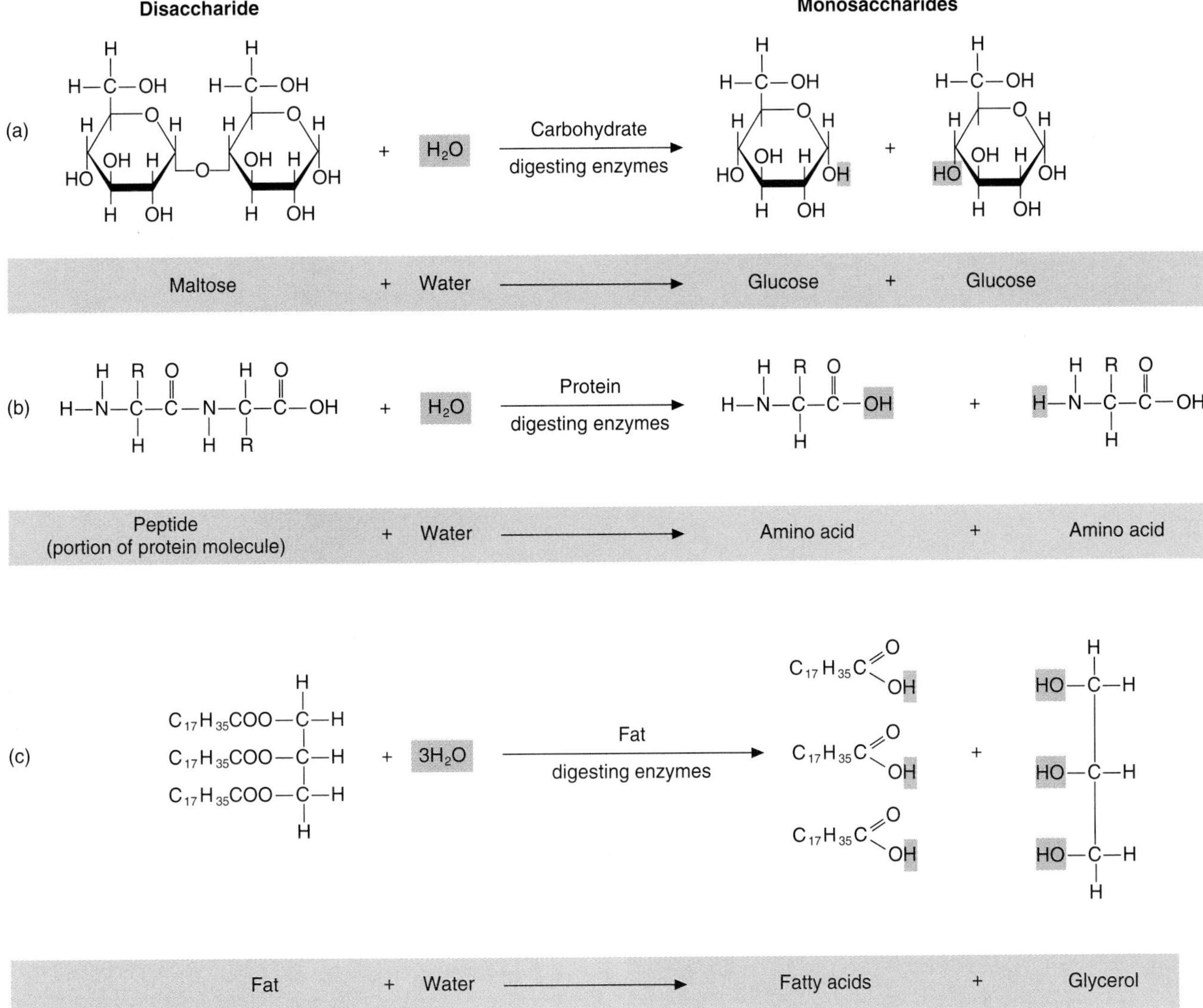

FIGURE 37.2

Chemical digestion. (*a*) Hydrolytic enzymes break complex carbohydrates down to maltose and then break maltose down to glucose. (*b*) Proteins are broken down to peptides, and then the peptides are broken down to their constituent amino acids. (*c*) Fats are chemically digested to fatty acids and glycerol.

animals use teeth or other chewing mouthparts to grab food, and some, such as snakes and lizards, swallow mammal meals whole. Some animals drink their food—a mosquito alights on human skin to take a "blood meal." Other organisms digest their food alive, as a spider consumes an insect, or a starfish dines on a clam.

Several types of aquatic animals are **filter feeders.** They use ciliated tissue surfaces to create waves that usher food particles into their mouths. Filter feeding is particularly adaptive in animals that do not move about much, because it brings the food to them, although the menu is limited to organisms that live in the vicinity. Bivalve mollusks such as clams and mussels, for example, sweep food towards their mouths using cilia on the feathery surfaces of their gills. Some motile animals also filter feed, including certain small crustaceans, fishes such as the herring, and the most dramatic example, the baleen whale. This behemoth opens its huge mouth and gulps in a few thousand fish and crustaceans as it moves forward. Smaller dietary fare, such as the plentiful, tiny shrimplike animals called krill, adhere to some 300 horny baleen plates that dangle from the roof of the whale's mouth like a shredded shower curtain (fig. 37.3). The whale uses its tongue to lick the food off the baleen and then swallows it.

The Evolution of Eating

A nutrient is a food substance used in the body to promote growth, maintenance, and repair. Obtaining nutrients from the environment, as opposed to synthesizing them, is termed **heterotrophy.**

The earliest heterotrophs probably engulfed organic matter and used it to supply energy and building blocks for proteins and nucleic acids. The evolution of photosynthesis and chemosynthesis, which enabled organisms to manufacture useful biochemicals from solar or geothermal energy, had two important implications for early life-forms. First, some organisms could now synthesize the complex organic molecules they required, freeing them from depending on an uncertain and possibly dwindling food supply. Second, the nutrient molecules obtained from photosynthesis and

FIGURE 37.3

A whale can swim through a school of krill and filter them through 60 tons of water in just a few seconds, thanks to the hanging curtain of epithelial outgrowths called baleen plates.

FIGURE 37.4

The giant panda consumes enormous quantities of bamboo.

chemosynthesis, called **primary production,** eventually built up a reservoir of food that would make heterotrophy a sustainable lifestyle over the long term.

Heterotrophy may have evolved because many organisms never acquired the ability to synthesize their own biochemicals. In terms of natural selection, heterotrophy is advantageous if it takes less energy to obtain a nutrient from the environment than to synthesize it. This may be especially true for a heterotroph that has a nearly constant supply of nutrients, such as an animal that has an abundant supply of a particular amino acid in its diet. Over time the species could lose the ability to synthesize this amino acid with no harmful consequence, and divert the energy required for its synthesis to other functions.

Over hundreds of millions of years, considering the great diversity of life, it is not surprising that many animals have become very specialized feeders, relying on one or a few kinds of food to supply nutrients. Consider the giant panda (fig. 37.4). It has a short digestive tract characteristic of its meat-eating ancestors, with less capacity to absorb nutrients. The panda lacks cellulase to break down bulky plant matter and also lacks an organ to store food. The poorly adapted panda must eat almost constantly. Even though it consumes 6% of its body weight in bamboo each day (compared with 2% for most plant-eating mammals), it just barely meets its basic nutritional requirements. Animals with more varied diets, such as the robin and human, can live in more widespread habitats.

Mastering Concepts

1. Why must digestion break down nutrient molecules?
2. Why do animals eat?
3. How can heterotrophy be adaptive?

Digestive Diversity

Animals' varied ways of obtaining and digesting food reflect specializations for their nutrient requirements, their habitats, the kinds of food available, and each species' evolutionary background. Even the animals in a particular taxonomic group may have diverse feeding habits and digestive adaptations.

Types of Digestive Systems

Because animals and nutrient molecules are composed of the same types of chemicals, digestive enzymes could just as easily attack an animal's body as its food. To prevent this, digestion occurs within specialized compartments. Even in protozoa, digestion is separated from other functions.

Protozoa absorb dissolved nutrients by active transport through the cell membrane or by endocytosis, in which the cell membrane engulfs a food particle, drawing it into the cell and surrounding it as a **food vacuole.** Some protozoa have special-

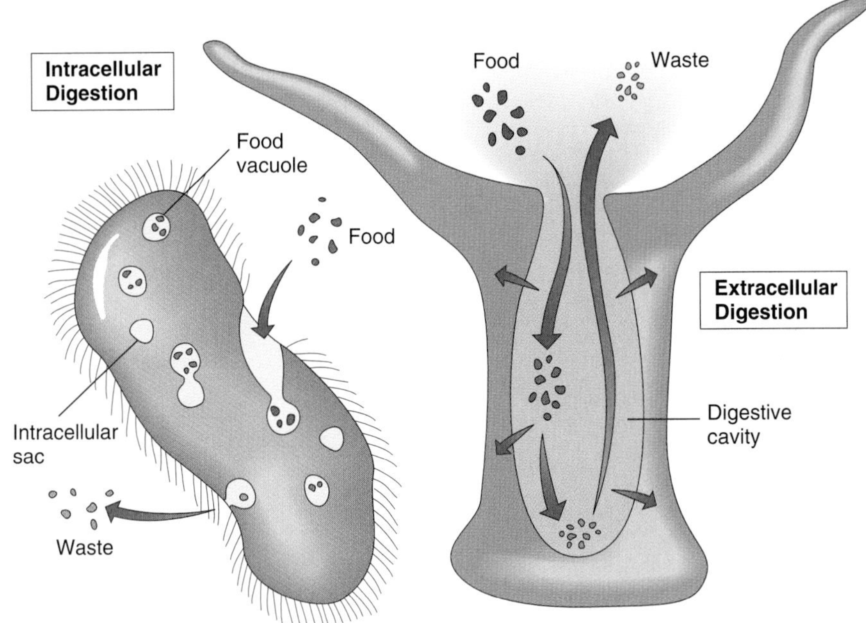

FIGURE 37.5

Intracellular and extracellular digestion. (*a*) Intracellular digestion is an ancient form of obtaining nutrients that single-celled organisms such as the paramecium use. Food is digested when the food vacuole that contains it fuses with an intracellular sac containing digestive enzymes. (*b*) In simple multicellular animals such as the hydra, food is broken down in a cavity, and the cells lining the cavity absorb the nutrients. This is an example of extracellular digestion.

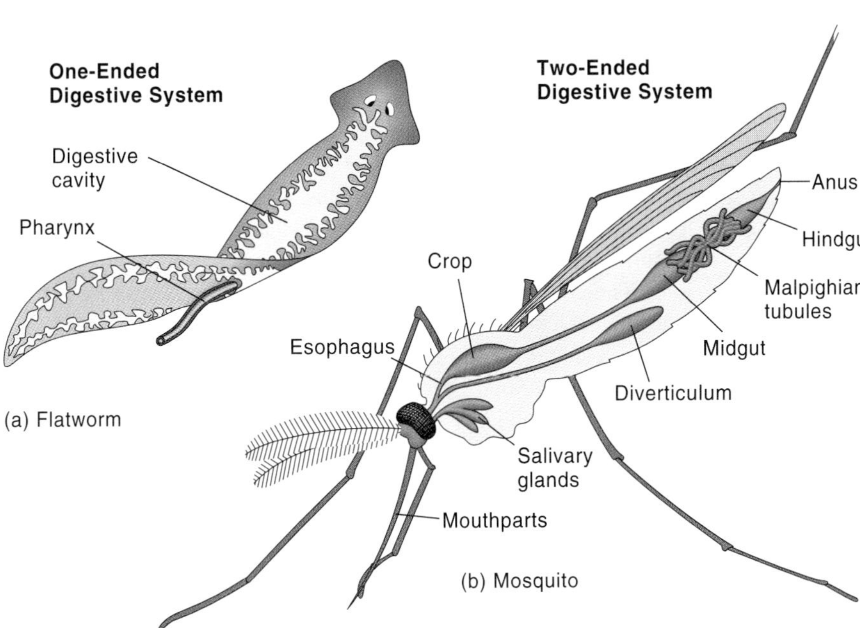

FIGURE 37.6

One-opening and two-opening digestive systems. In a digestive system with a single opening, such as in the flatworm (*a*), food and wastes mix. Cells lining the digestive cavity absorb partially digested nutrients. Further digestion occurs in these cells. (*b*) A mosquito's digestive system is open at the mouth and anus. Digestive systems with two openings are often specialized into compartments.

ized mouthlike openings that food passes through en route to food vacuoles. Once loaded, the food vacuole fuses with another sac containing digestive enzymes that break down large nutrient molecules. Digested nutrients exit the food vacuole and are used in other parts of the cell. The cell extrudes waste. Digestion within a cell's food vacuoles is **intracellular digestion** (fig. 37.5*a*).

Intracellular digestion can handle only very small food particles. In **extracellular digestion,** hydrolytic enzymes in a cavity connected with the outside world (the vertebrate stomach, for example) dismantle larger food particles outside the cells that will use them (fig. 37.5*b*). When nutrient molecules are small enough, they enter cells lining the cavity, where chemical breakdown may continue. Extracellular digestion eases waste removal and is more efficient and specialized than intracellular digestion. The cavity in which extracellular digestion occurs constitutes a **digestive system.**

Digestive systems may have one or two openings (fig. 37.6). Some animals, such as the hydra and the flatworm, have digestive systems with a single opening. Indigestible food exits the digestive system through the same opening it entered. Consequently, food must be digested and the residue expelled before the next meal can begin. This two-way traffic makes any specialized compartments for storing, digesting, or absorbing nutrients impossible. In these organisms, the digestive cavity doubles as a circulatory system, which distributes the products of digestion to the body cells for use.

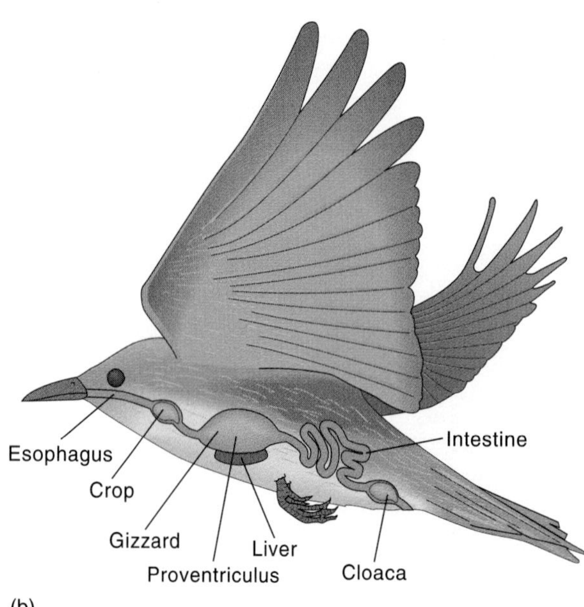

(a) (b)

FIGURE 37.7

Birds' beaks are adapted to handle different types of foods. (*a*) The flattened beak of an owl couldn't do this! (*b*) A bird's digestive tract.

The digestive systems of many familiar animals, such as segmented worms, mollusks, insects, starfishes, and vertebrates, have two openings, a separate entrance and exit in a tubelike structure that allows one-way traffic. Hydrolytic enzymes are released into this tube, and the products of digestion are absorbed and delivered to cells via a circulatory system. Indigestible food remains in the tube and leaves the body as part of the feces. A major advantage of a two-opening digestive system is that regions of the tube can become specialized for different functions: breaking food into smaller particles, storage, chemical digestion, and absorption.

Specific Digestive System Adaptations

The vertebrates exhibit great digestive diversity. Birds metabolize nutrients very rapidly, and they must eat continually and digest food quickly to fuel their lightweight bodies. Their mouths have adapted to their diets. The stork's bill easily scoops fish (fig. 37.7*a*); the strong, short beak of the finch cracks and removes hulls from seeds; the vulture's hooked beak rips carrion into manageable chunks. A bird's esophagus has an enlargement called a **crop,** which temporarily stores food. The stomach is divided into two sections. The first, the **proventriculus,** secretes gastric juice that chemically digests food. The second part of the stomach, the **gizzard,** is a muscular organ lined with ridges. It mechanically digests food with the aid of sand and small pebbles the bird swallows. The digestive system of a bird ends at the **cloaca,** an opening common to the digestive, reproductive, and excretory systems (fig. 37.7*b*).

The intestines of some types of birds show a remarkable adaptation—they enlarge when the animal eats huge amounts of food, such as just before a long migration. In the wren, rufous-sided towhee, rock ptarmigan, and spruce grouse, the intestine lengthens up to 22%. This adaptation enables birds to store more food than usual. Because it takes longer for food to travel through the lengthened intestine, more nutrients are absorbed.

Many mammals have digestive system specializations that enable them to break down the cellulose in the cell walls of plant foods. The cow has a four-sectioned stomach. Grass enters the first section, the rumen, where certain bacteria break it down into balls of cud (see fig. 22.2). The cow regurgitates the cud back up to its mouth, where it chews to mechanically break the food down further. The cow's teeth can mash the cud forward and backward, up and down, and sideways. After chewing the cud, the cow swallows again, but this time the food bypasses the rumen, continuing digestion in the other three sections of the stomach. Sheep, deer, buffalo, and elephants also have quadruple stomachs.

Elephants have particular difficulty digesting their meals of leaves and twigs. Because they have only tusks and molars, a lot of wood lands in their massive stomachs. It remains there for over 2 days, churning about amidst digestive juices and cellulose-degrading bacteria. Still, elephant dung contains many undigested twigs.

Rabbits have their own way of getting the most out of a meal. Like cows and elephants, rabbits eat leaves that spend time in their stomachs. Instead of regurgitating the partially digested plant matter, rabbits excrete it in moist pellets. Then they eat the moist pellets, which continue digestion in the stomach and intestines. This time, the rabbit excretes dry pellets, which it does not eat.

Mastering Concepts

1. Distinguish between intracellular and extracellular digestion.
2. What is the limitation of a one-way digestive system?
3. What are some diverse ways that animals obtain and digest nutrients?

Human Nutrition

Carbohydrates, proteins, and lipids are called **macronutrients** because humans require them in large amounts or **energy nutrients** because they supply energy. Humans require vitamins and minerals in very small amounts, and hence these nutrients are called **micronutrients.** Water is also a vital part of the diet.

Carbohydrates, which include the sugars and starches, are our major energy source. Cells extract this energy from the covalent bonds of glucose. Digestion breaks down complex carbohydrates and disaccharides into glucose, which is metabolized to yield ATP. Proteins and lipids, particularly fats, can also supply energy under certain conditions, but these nutrients have other major functions. Tables 37.2 through 37.5 provide details on specific vitamins and minerals. Recall that vitamins function as coenzymes and that minerals are essential to many biochemical pathways.

Essential nutrients must come from food because the body cannot synthesize them. Essential nutrients vary among species. Vitamin C is an essential nutrient in humans, guinea pigs, Indian fruit bats, and certain monkeys but not in most other animals. If a guinea pig eats only rabbit chow, which lacks vitamin C because rabbits produce it, the guinea pig will develop vitamin C deficiency. **Nonessential nutrients** come from foods but are also made in the body. For example, the adult human body synthesizes 12 of the 20 types of amino acids, which are therefore nonessential.

The amount of energy released from a nutrient is measured in **kilocalories (kcal),** often called simply calories. A food's caloric content is determined by burning it in a bomb calorimeter,

Table 37.2 Vitamins and Health

Vitamin	Function	Food Sources	Deficiency Symptoms
Water-Soluble Vitamins			
Thiamine (vitamin B_1)	Growth, fertility, digestion, nerve cell function, milk production	Pork, beans, peas, nuts, whole grains	Beriberi (neurological disorder), loss of appetite, swelling, poor growth, heart problems
Riboflavin (vitamin B_2)	Energy use	Liver, leafy vegetables, dairy products, whole grains	Hypersensitivity of eyes to light, lip sores, oily dermatitis
Pantothenic acid	Growth, cell maintenance	Liver, eggs, peas, potatoes, peanuts	Headache, fatigue, poor muscle control, nausea, cramps*
Niacin	Growth	Liver, meat, peas, beans, whole grains, fish	Dark rough skin, diarrhea, mouth sores, mental confusion (pellagra)
Pyridoxine (vitamin B_6)	Protein use	Red meat, liver, corn, potatoes, whole grains, green vegetables	Mouth sores, dizziness, nausea, weight loss, neurological disorders*
Folic acid	Manufacture of red blood cells, metabolism	Liver, navy beans, dark green vegetables	Anemia, cancer
Biotin	Metabolism	Meat, milk, eggs	Skin disorders, muscle pain, insomnia, depression*
Cyanocobalamin (vitamin B_{12})	Manufacture of red blood cells, growth, cell maintenance	Meat, organ meats, fish, shellfish, milk	Pernicious anemia
Ascorbic acid (vitamin C)	Growth, tissue repair, bone and cartilage formation	Citrus fruits, tomatoes, peppers, strawberries, cabbage	Weakness, gum bleeding, weight loss (scurvy)
Fat-Soluble Vitamins			
Retinol (vitamin A)	Night vision, new cell growth	Liver, dairy products, egg yolk, vegetables, fruit	Night blindness, rough dry skin
Cholecalciferol (vitamin D)	Bone formation	Fish-liver oil, milk, egg yolk	Skeletal deformation (rickets)
Tocopherol (vitamin E)	Prevents certain compounds from being oxidized	Vegetable oil, nuts, beans	Anemia in premature infants*
Vitamin K	Blood clotting	Liver, egg yolk, green vegetables	Bleeding, liver problems*

*Deficiencies of these vitamins are rare in humans, but they have been observed in experimental animals.

Table 37.3 Vitamin Fallacies and Facts

Fallacy	Fact
Dietary vitamins are superior to those in tablet form.	A vitamin molecule is chemically the same, whether it comes from a bean or a bottle.
The more vitamins, the better.	Too much of a water-soluble vitamin is excreted; too much of a fat-soluble vitamin can harm health.
A varied diet provides all required vitamins.	Many people require vitamin supplements, particularly pregnant and breast-feeding women.
Vitamins provide energy.	Vitamins do not directly supply energy; they aid in the release of energy from carbohydrates, fats, and proteins.

Table 37.4 Bulk Minerals in the Human Diet

Mineral	Food Sources	Functions in the Human Body
Calcium	Milk products, green leafy vegetables	Bone and tooth structure, blood clotting, hormone release, nerve transmission, muscle contraction
Chloride	Table salt, meat, fish, eggs, poultry, milk	Digestion in stomach
Magnesium	Green leafy vegetables, beans, fruits, peanuts, whole grains	Muscle contraction, nucleic acid synthesis, enzyme activity
Phosphorus	Meat, fish, eggs, poultry, whole grains	Bone and tooth structure
Potassium	Fruits, potatoes, meat, fish, eggs, poultry, milk	Body fluid balance, nerve transmission, muscle contraction, nucleic acid synthesis
Sodium	Table salt, meat, fish, eggs, poultry, milk	Body fluid balance, nerve transmission, muscle contraction
Sulfur	Meat, fish, eggs, poultry	Hair, skin, and nail structure, blood clotting, energy transfer, detoxification

Table 37.5 Trace Minerals Important to Human Health

Mineral	Food Sources	Functions in the Human Body
Chromium	Yeast, pork kidneys	Regulates glucose use
Cobalt	Meat, eggs, dairy products	Part of vitamin B_{12}
Copper	Organ meats, nuts, shellfish, beans	Part of many enzymes, storage and release of iron in red blood cells
Fluorine	Water (in some areas)	Maintains dental health
Iodine	Seafood, iodized salt	Part of thyroid hormone
Iron	Meat, liver, fish, shellfish, egg yolk, peas, beans, dried fruit, whole grains	Transport and use of oxygen (as part of hemoglobin and myoglobin), part of certain enzymes
Manganese	Bran, coffee, tea, nuts, peas, beans	Part of certain enzymes, bone and tendon structure
Selenium	Meat, milk, grains, onions	Part of certain enzymes, heart function
Zinc	Meat, fish, egg yolk, milk, nuts, some whole grains	Part of certain enzymes, nucleic acid synthesis

a chamber immersed in water. When burning food is placed in the chamber, the energy released raises the water temperature, and this energy is measured in kilocalories. One kilocalorie is the energy needed to raise 1 kilogram of water 1°C under controlled conditions. Bomb calorimetry studies have shown that 1 gram of carbohydrate yields 4 kilocalories, 1 gram of protein yields 4 kilocalories, and 1 gram of fat yields 9 kilocalories. (A teaspoon of dried food weighs approximately 5 grams.) Although approximate, these values help to explain why a fatty diet may cause weight gain; fats supply more energy than most people can use.

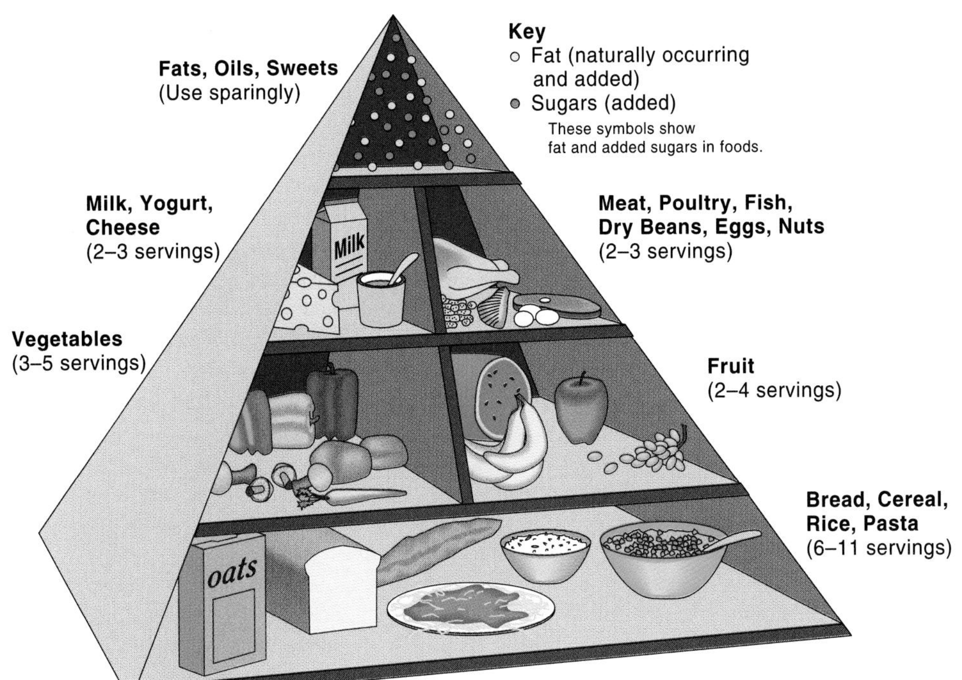

FIGURE 37.8

The U.S. Department of Agriculture's food pyramid of human nutritional requirements emphasizes whole grains, fruits, and vegetables. The four-food-group plan devised in the 1950s suggested that humans need nearly as many daily servings of meat as of grains, dairy products, and fruits and vegetables. In the 1940s, the department divided food into eight groups, including separate groups for butter and margarine and for eggs—foods now associated with the development of heart disease. In the 1920s, an entire food group was devoted to sweets! As our knowledge of nutrition grows, our concept of a healthy diet changes.

Sources: U.S. Department of Agriculture/U.S. Department of Health and Human Services.

Good nutrition is a matter of balance. When we take in more kilocalories than we expend, weight increases; those who consume fewer kilocalories than expended lose weight—and, if taken to an extreme, may even starve. Balancing vitamins and minerals is important to health, too. Eating a variety of foods helps meet nutritional requirements. The dietary suggestions pictured in a food pyramid (fig. 37.8) can help a person make healthful food choices.

Nutrient Deficiencies

Obtaining an adequate number of kilocalories each day from a variety of foods is sometimes difficult. A student may eat nothing but pizza during exam week; a busy working person may skip meals or eat on the run; many individuals may not be able to afford enough milk or meat. All these situations may lead to **primary nutrient deficiencies,** or deficiencies caused by diet. **Secondary nutrient deficiencies,** in contrast, result from an inborn metabolic condition that causes the body to malabsorb, overexcrete, or destroy a particular nutrient.

Micronutrient deficiencies develop more slowly than the obvious weight loss that accompanies macronutrient deficiency. Vitamin or mineral deficiencies cause subtle changes before health is noticeably affected. For example, the blood of a vegetarian who does not eat many iron-containing foods may have an abnormally low number of red blood cells or small red blood cells with little hemoglobin. The person feels fine, temporarily, as the body uses its iron stores. After several weeks, however, the signs of anemia appear—weakness, fatigue, frequent headaches, and a pale complexion.

Biology in Action 37.1 looks at a prevalent problem that takes many guises—starvation.

Mastering Concepts

1. Why are carbohydrates, proteins, and lipids called macronutrients and vitamins and minerals called micronutrients?
2. What is an essential nutrient?
3. What are the effects of macronutrient and micronutrient deficiencies?

An Overview of the Human Digestive System

The human digestive system consists of a continuous tube called the **gastrointestinal tract** (including the mouth, pharynx, **esophagus,** stomach, small intestine, and large intestine, or colon) and accessory structures (fig. 37.9). These structures aid food breakdown either mechanically (for example, the teeth and tongue) or chemically by secreting digestive enzymes (the salivary glands and pancreas). Other accessory structures, the **liver** and **gallbladder,** produce and store **bile,** which assists fat digestion. The **mesentery,** an epithelial (lining) sheet reinforced by connective tissue, supports the digestive organs and glands.

The digestive lining begins at the mouth and ends at the anus. The moist innermost layer secretes mucus that lubricates the tube so that food slips through easily, while also protecting underlying cells from rough materials in food and from digestive enzymes. In some regions of the digestive system, cells in the innermost lining also secrete digestive enzymes. Beneath the lining a layer of connective tissue contains blood vessels and nerves. The blood nourishes the cells of the digestive system and in some regions picks up and transports digested nutrients.

biology in 37.1 action

Starvation

A healthy human can survive for 50 to 70 days without food. In prehistoric times, this margin allowed survival during seasonal famines. In some areas of the world today, famine is not a seasonal event but a constant condition, and millions of people are starving to death. Hunger strikes, inhumane treatment of prisoners, and eating disorders can also cause starvation.

Whatever the cause, the starving human body begins to digest itself. After only one day without food, the body's reserves of sugar and starch are gone, and it starts to extract energy from fat and then from muscle protein. By the third day, hunger ceases as the body uses energy from fat reserves. Gradually, metabolism slows to conserve energy, blood pressure drops, the pulse slows, and chill sets in. Skin becomes dry and hair falls out as the proteins in these structures are broken down and release amino acids required for the vital functions of the brain, heart, and lungs. When the body dismantles the immune system's antibody proteins for their amino acids, protection against infection declines. Mouth sores and anemia develop, the heart beats irregularly, and bone begins to degenerate. After several weeks without food, coordination deteriorates. Near the end, the starving human is blind, deaf, and emaciated (fig. 37.A).

Marasmus and Kwashiorkor

Starvation can take different forms. When individuals lack all nutrients, they develop marasmus, a condition that causes a skeletal appearance. Children under the age of two with marasmus often die of measles or other infections; their immune systems are too weak to fight normally mild viral illnesses.

Some starving children do not look skeletal but instead have protruding bellies. These youngsters suffer from a form of protein starvation called kwashiorkor, which in the language of Ghana means "the evil spirit which infects the first child when the second child is born." Kwashiorkor typically appears in a child who has recently been weaned from the breast, usually because of the birth of a sibling. The switch from protein-rich breast milk to the protein-poor gruel that is the staple of many developing nations causes protein deficiency. The children's bellies swell with the fluid that builds up when protein is lacking, and their skin may develop lesions. Infections overwhelm the body as antibodies are broken down.

Anorexia Nervosa

Anorexia nervosa is self-imposed starvation. The condition is reported to affect 1 out of 250 adolescents, 95% of them female. The sufferer, typically a well-behaved adolescent from an affluent family, perceives herself as overweight and eats barely enough to survive, losing as much as 25% of her original body weight. She may further lose weight by vomiting, taking laxatives and diuretics, or engaging in intense exercise.

Anorexia eventually leads to low blood pressure, slowed or irregular heartbeat, constipation, and constant chills. A woman stops menstruating as body fat level plunges. Hair becomes brittle, the skin dries out, and soft, pale, fine body hair called lanugo, normally seen only on a developing fetus, grows to preserve body heat.

A person with anorexia who becomes emaciated may be hospitalized, so that intravenous feedings can prevent sudden death from heart failure due to a mineral imbalance. Psychotherapy and nutritional counseling may help identify and remedy the underlying cause of the abnormal eating behavior. Despite these efforts, 15 to 21% of people with anorexia die.

Bulimia

A person suffering from bulimia eats, often in huge amounts, and then gets rid of the thousands of extra kilocalories by vomiting, taking laxatives, or exercising frantically. A person with bulimia often binges in privacy, eating well beyond the point of pain.

Sometimes a dentist is the first to identify bulimia by observing teeth decayed from frequent vomiting. The backs of the hands may bear telltale scratches from efforts to induce vomiting, and the throat may be raw and the stomach lining ulcerated from the stomach acid forced forward by vomiting. The binge-and-purge cycle is very hard to break, even with psychotherapy and nutritional counseling.

Health services at many colleges and universities screen students for eating disorders. One in 10 college students suffers from an eating disorder.

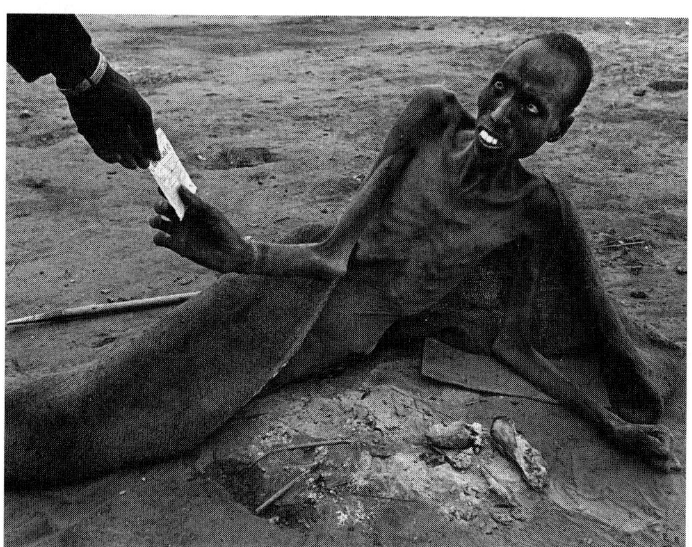

FIGURE 37.A
This starving man in the Sudan reaches for a packet of rehydrating salts provided by a relief agency—but after months with little or no food, it is far too late to save his life.

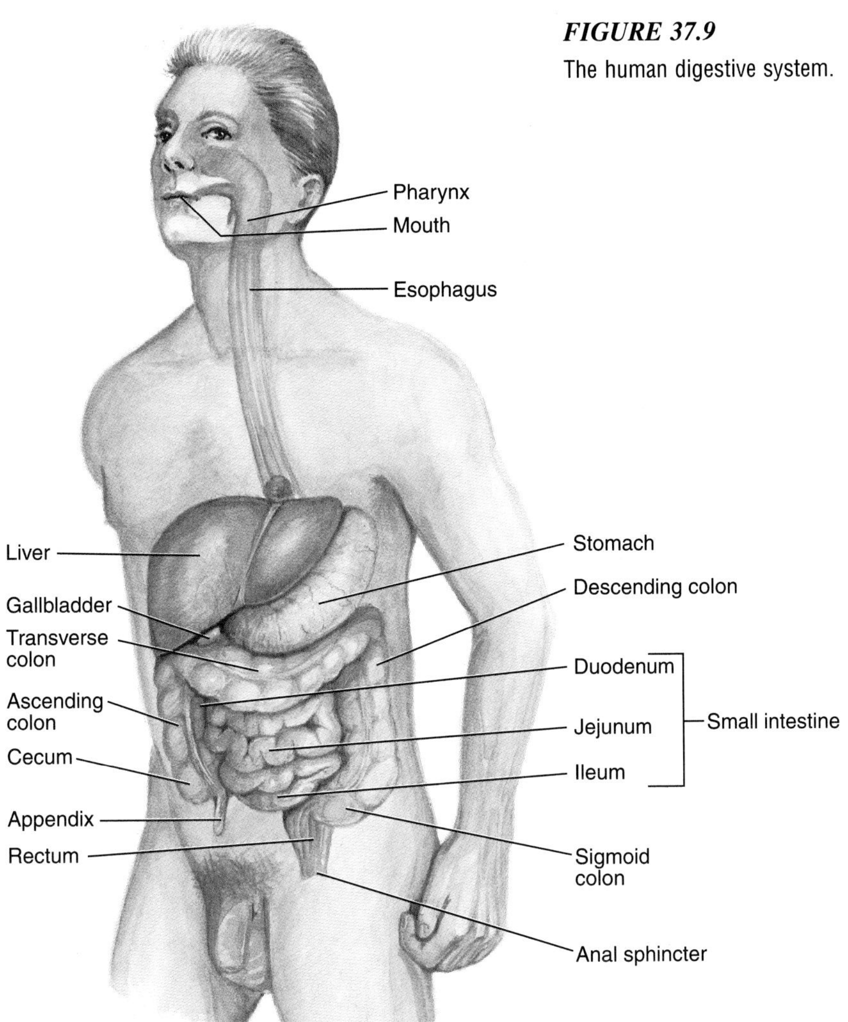

FIGURE 37.9
The human digestive system.

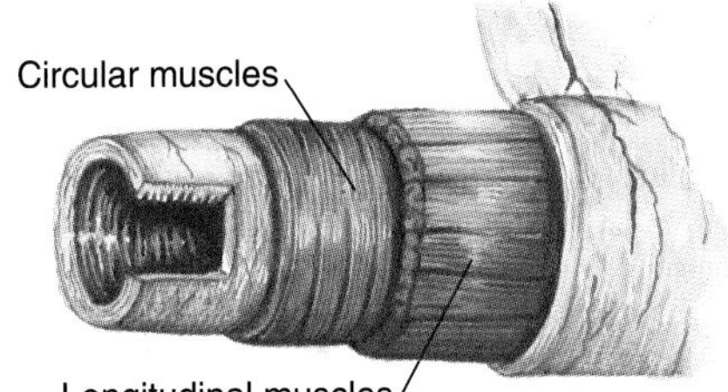

FIGURE 37.10
Muscles move food. The two muscle layers of the digestive tract coordinate contractions to move food in one direction. These contractions occur about three times per minute, even between meals. Note that the muscle layers are perpendicular to each other. (In the stomach a third layer lies obliquely.)

The next layers around the digestive tract are muscular. Except for the stomach, which has three layers, two layers produce the movements of the digestive tract. The muscles of the inner layer circle the tube, constricting it when they contract, and the muscles in the outer layer run lengthwise, shortening the tube when they contract (fig. 37.10). This contraction pattern churns food in the tract, mixing it with enzymes into a liquid. Waves of contraction called **peristalsis** propel food through the digestive tract. When food distends the walls of the tube, the circular muscles immediately behind it contract, squeezing the mass forward (fig. 37.11).

Mastering Concepts

1. What are the components of the human digestive system?
2. What are the linings of the human digestive system?
3. What is peristalsis?

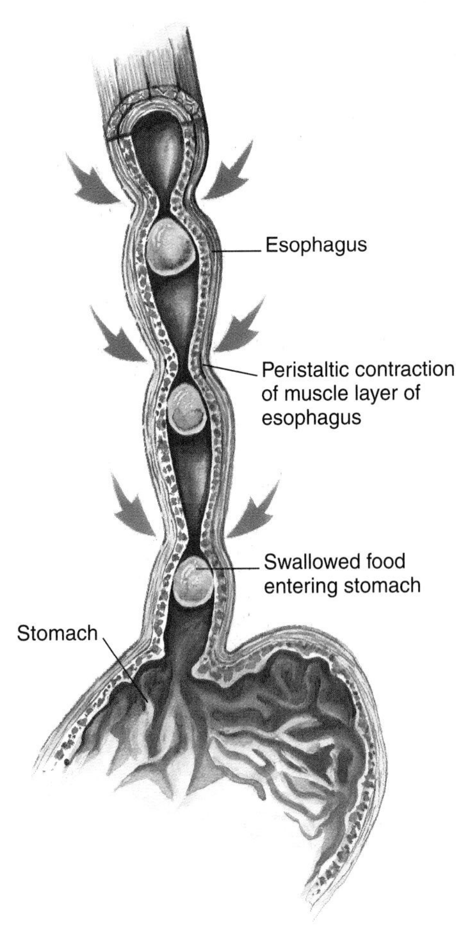

FIGURE 37.11
Peristalsis. The muscular contractions of peristalsis move food along the digestive tract.

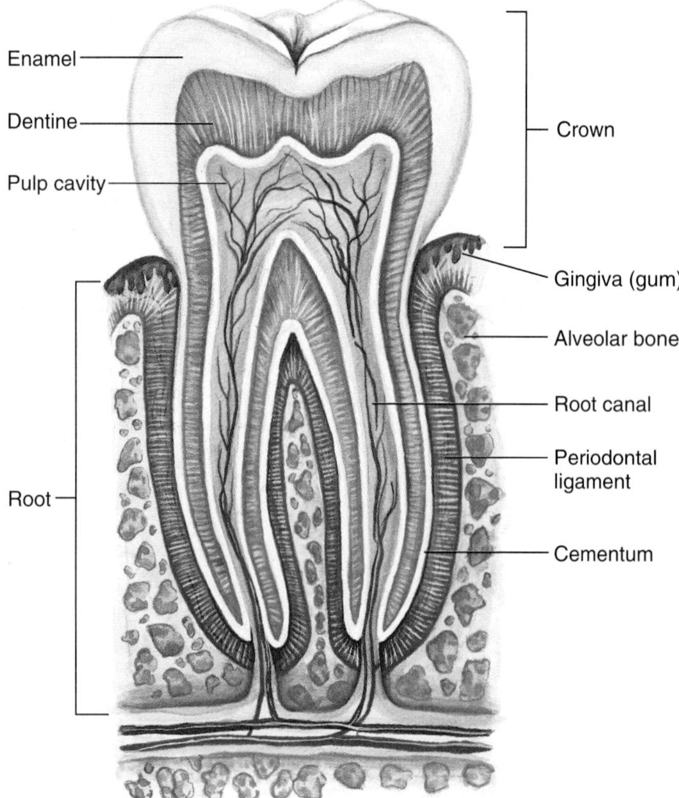

FIGURE 37.12
Anatomy of a tooth.

Food's Journey Through the Human Body

The Mouth and the Esophagus

In humans, the thought, smell, or taste of food triggers three pairs of salivary glands near the mouth to secrete saliva. Chemical digestion begins as **salivary amylase,** an enzyme in saliva, starts to break down starch into maltose. You can test the action of salivary amylase by holding a piece of bread in your mouth. Within a minute, it begins to taste sweet as the enzyme breaks down starch to sugar.

Mechanical digestion begins as the teeth chew the food. Water and mucus in saliva aid the teeth as they tear food into small pieces, thereby increasing the surface area available for chemical digestion. The thick **enamel** that covers a tooth is the hardest substance in the human body. Beneath the enamel is the bonelike **dentine,** and beneath that, the soft inner **pulp,** which contains connective tissue, blood vessels, and nerves (fig. 37.12). Two layers on the outside of the tooth, the **periodontal ligament** and the **cementum,** anchor the tooth to the gum and jawbone. The visible part of a tooth is the **crown,** and the part below the surface is the **root.** Calcium compounds harden teeth, as they do bones.

The tongue rolls chewed food into a lump called a **bolus** and pushes it to the back of the mouth for swallowing. The bolus passes first through the pharynx and then through the esophagus, a muscular tube leading to the stomach. During swallowing, the epiglottis covers the passageway to the lungs, routing food to the digestive tract.

Food does not merely slide down the esophagus due to gravity—contracting esophageal muscles push it along in a wave of peristalsis. This is why it is possible to swallow while standing on your head. Muscle contractions continue, propelling the food down the esophagus toward the stomach, where the next stage in digestion takes place.

The Stomach

From the esophagus, food progresses to the stomach, a J-shaped bag about 12 inches (30 centimeters) long and 6 inches (15 centimeters) wide. The stomach has three important functions: storage, some digestion, and pushing food into the small intestine (fig. 37.13).

The stomach is the size of a large sausage when empty, but it can expand to hold as much as 3 or 4 quarts (approximately 3 or 4 liters) of food. Folds in the stomach's mucosa, called **rugae,** can unfold like the pleats of an accordion to accommodate a large meal. Muscular rings called **sphincters** control entry to and exit from the stomach, pinching shut to contain the stomach's contents.

The stomach has four regions: the entry to (or "neck" of) the stomach is the **cardia,** the domelike top is the **fundus,** the midsection is the **body,** and the bottom is the **pylorus.** Fats usually remain in the stomach from 3 to 6 hours, proteins for up to 3 hours, and carbohydrates 1 to 2 hours.

Mechanical and chemical digestion occur in the stomach. Waves of peristalsis carry out mechanical digestion, pushing food against the stomach bottom, churning it backwards, and mixing it with **gastric juice** to produce a semifluid mass called **chyme.**

Gastric juice is responsible for chemical digestion in the stomach. About 40 million cells lining the stomach's interior secrete 2 to 3 quarts (approximately 2 to 3 liters) of gastric juice per day. Gastric juice consists of water, mucus, salts, hydrochloric acid, and enzymes. The hydrochloric acid creates a highly acidic environment, which activates the enzyme **pepsin** from its precursor, **pepsinogen.** Pepsin breaks down proteins to yield polypeptides, a first step in protein digestion. Cells called **chief cells** secrete pepsinogen and **parietal cells** secrete hydrochloric acid from indentations in the stomach lining called **gastric pits** (fig. 37.14).

Nerves and hormones regulate gastric juice secretion. The thought or taste of food initiates nerve impulses that stimulate the stomach to secrete between 6 and 7 ounces (about 200 milliliters) of gastric juice. This prepares the stomach to receive food. When food arrives in the stomach, cells called gastric cells release a hormone, **gastrin,** that stimulates secretion of another 20 ounces (about 600 milliliters) of gastric juice.

Why doesn't the stomach digest the protein of its own cells along with the protein in food? The organ has built-in triple protection. First, the stomach secretes little gastric juice until food is present for it to work on. Secondly, some stomach cells secrete mucus, which coats and protects the stomach lining from the corrosive gastric juice. Finally, the stomach produces pepsin in an inactive form, pepsinogen, and cannot digest protein until hydrochloric acid is present.

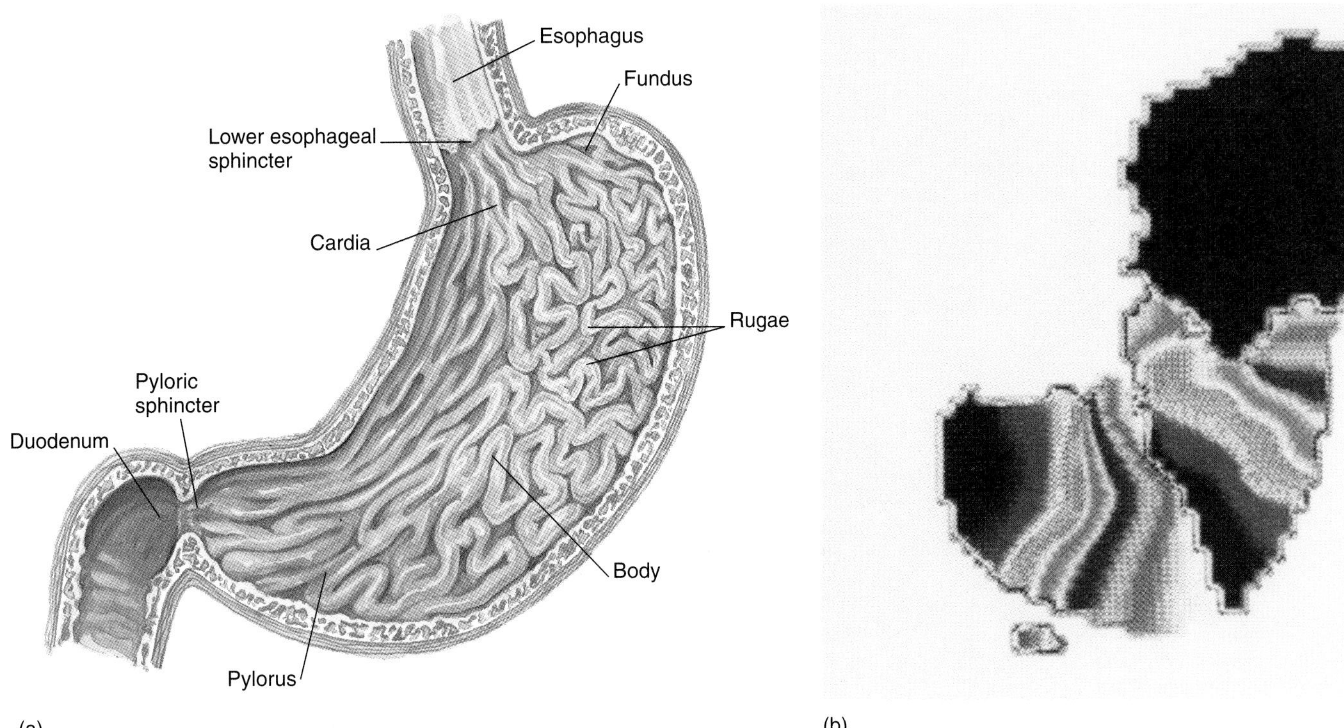

FIGURE 37.13

The stomach is a J-shaped bag that stores and mixes food until it is fluid enough to move on to the small intestine. (*a*) The stomach has four regions, two sphincters, and folds called rugae that increase its capacity. (*b*) The isotopic radiogastrogram is a new way to image stomach functioning and is helpful in diagnosing ailments such as chronic indigestion. A patient eats a scrambled egg sandwich containing a radioactive isotope, which allows a special camera to detect the food's journey through the digestive tract. A computer assembles hundreds of camera shots into an image, color-coded to trace the stomach's movements. Here, the fundus and cardia are black because the food has already passed this point. Color in the body and pylorus, however, indicates the wavelike motion of peristalsis—what the technique's inventor calls the "stomachbeat."

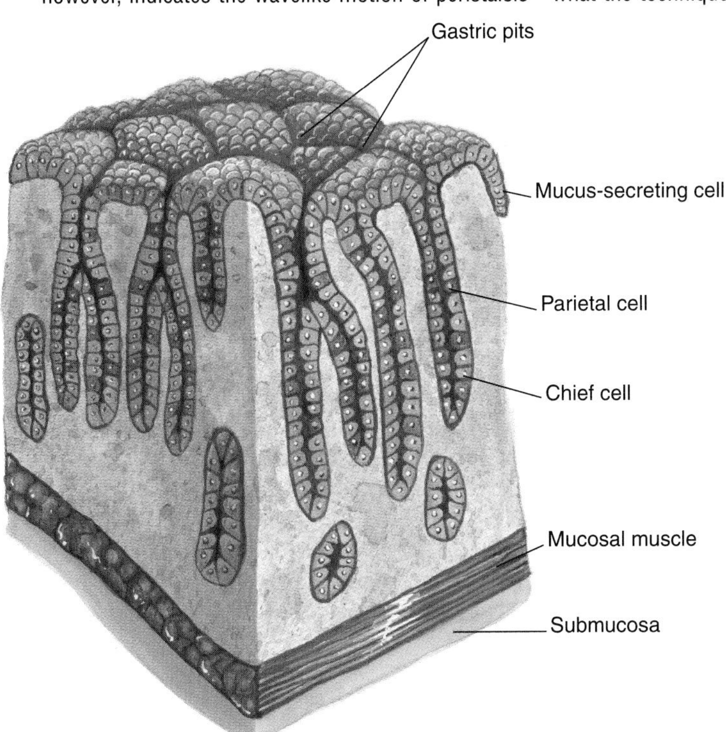

FIGURE 37.14

The lining of the stomach contains indentations called gastric pits, where chief cells secrete pepsinogen and parietal cells secrete hydrochloric acid. Still other lining cells secrete the abundant mucus that coats the stomach lining, preventing it from digesting itself.

The stomach absorbs very few nutrients; most food has not yet been digested sufficiently. However, the stomach can absorb some water and salts (electrolytes), a few drugs (such as aspirin), and, like the rest of the digestive tract, alcohol. This is why aspirin can irritate the stomach lining and why we feel alcohol's intoxicating effects quickly.

After the appropriate amount of time, depending upon the composition of a meal, stretching of the stomach wall or release of gastrin triggers neural messages that relax the pyloric sphincter at the stomach's exit. The stomach then squirts chyme through the sphincter into the small intestine. Nerve impulses from the first part of the small intestine then control the stomach's actions. When the upper part of the small intestine is full, receptor cells on the outside of the intestine are stimulated and the pyloric sphincter tightens. Once pressure on the small intestine lessens, the sphincter opens, the stomach contracts, and more chyme enters.

The Small Intestine

The small intestine, a 23-foot (approximately 7-meter) tubular organ, completes digestion and absorbs the resulting nutrients. The first 10 inches (25 centimeters) of the organ forms the **duodenum,** the next two-fifths the **jejunum,** and the remainder (almost three-fifths) the **ileum.** Localized muscle contractions carry out mechanical digestion in the small intestine in a process called **segmentation** (fig. 37.15). During this process, the chyme sloshes

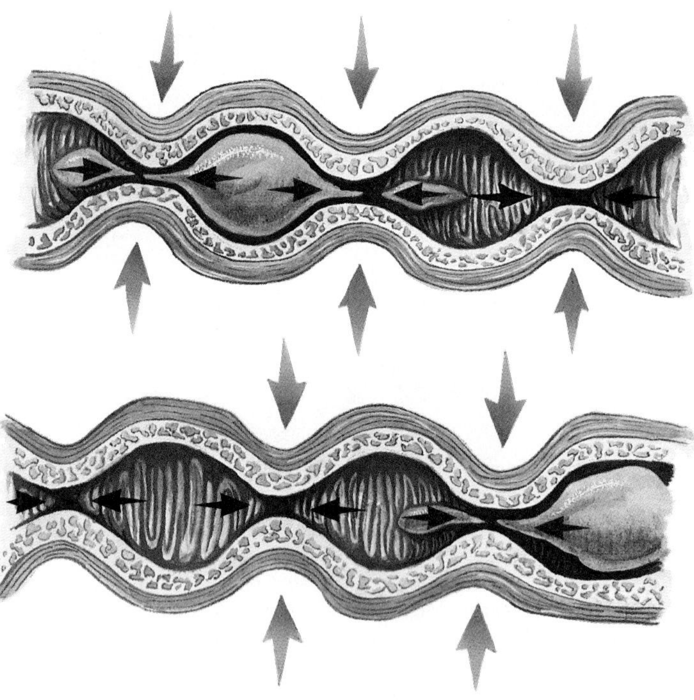

FIGURE 37.15

Segmentation. Coordinated muscular contraction of different parts of the small intestine mixes the chyme with mucus and digestive enzymes.

back and forth between segments of the small intestine that form when bands of circular muscle temporarily contract. Peristalsis moves the food lengthwise along the intestine.

Digestive secretions from the small intestine, the liver, and the pancreas contribute mucus, water, and enzymes. The intestinal enzymes bind to the cell membranes of the lining cells. These cells may protect underlying cells from being digested by the enzymes.

Chemical digestion in the small intestine acts on all three major types of nutrient molecules. The small intestine continues the protein breakdown the stomach began. **Trypsin** and **chymotrypsin** from the pancreas break down polypeptides into peptides, and **peptidases** secreted by small intestinal cells break peptides into tripeptides, dipeptides, and single amino acids small enough to enter the lining cells (fig. 37.16).

Fats present an interesting challenge to the digestive system. **Lipases,** the enzymes that chemically digest fats, are water soluble, but fats are not. Therefore, lipase can only act at the surface of a fat droplet, where it contacts water. Bile from the liver emulsifies fats, breaking them into many droplets, which expose greater surface area to lipase. Most fats in our diet are triglycerides, which are digested into fatty acids and glycerol.

Carbohydrate digestion also continues in the small intestine. The pancreas sends **pancreatic amylase** to the small intestine to digest starches that salivary amylase missed in the mouth. In addition, the small intestine produces **carbohydrases,** which chemically break down certain disaccharides into monosaccharides. Sucrase, for example, is a carbohydrase that breaks down the disaccharide sucrose into the monosaccharides glucose and fructose.

Deficiency of a particular carbohydrase can cause digestive distress when a person eats certain foods. Many adults have difficulty digesting milk products because of **lactose intolerance,** which results from the absence of the enzyme lactase in the small intestine. This enzyme breaks down the disaccharide lactose (milk sugar) into the monosaccharides glucose and galactose. Bacteria ferment undigested lactose in the large intestine, producing abdominal pain, gas, diarrhea, bloating, and cramps. A person with lactose intolerance can avoid these symptoms by consuming fermented dairy products such as yogurt, buttermilk, and cheese instead of fresh dairy products, because lactose has already been broken down. Taking lactase tablets can also prevent symptoms of lactose intolerance.

Hormones coordinate the small intestine's various digestive activities. When the stomach squirts acidic chyme into the small intestine, intestinal cells release the hormone **secretin.** This triggers the release of bicarbonate from the pancreas, which quickly neutralizes the acidity. Secretin also stimulates the liver to secrete bile. When chyme contains fats, intestinal cells secrete another hormone, **cholecystokinin (CCK),** which signals release of biochemicals that digest fats. CCK stimulates the liver to continue bile secretion and triggers the gallbladder to contract, sending stored bile to the small intestine. CCK also releases pancreatic enzymes, including lipase, into the small intestine.

Hormonal regulation of digestion protects intestinal cells because digestive biochemicals are produced only when they have food to work on. As an additional safeguard, other glands in the small intestine secrete mucus, which protects the intestinal wall from digestive juices and neutralizes stomach acid. Mucus offers limited protection, however. Many intestinal lining cells succumb to the caustic contents. Fortunately, these epithelial cells have a high division rate and replace the lining every 36 hours. Nearly one-quarter of the bulk of feces consists of dead epithelial cells from the small intestine.

At the end of digestion, carbohydrates have been digested to monosaccharides, proteins to amino acids, and fats to fatty acids and glycerol. Cells lining the small intestine absorb these products, which then enter the circulation.

Nutrient absorption must take place at a surface. The small intestine illustrates maximization of surface area. The organ is very long and highly folded, enabling it to fit into the abdomen. In addition, the innermost layer is corrugated with circular ridges almost half an inch high. The surface of every hill and valley of the lining looks velvety due to additional folds—about 6 million tiny projections called **villi** (fig. 37.17*a*). The tall epithelial cells on the surface of each villus bristle with projections of their own called **microvilli** (fig. 37.17*b*). Each villus cell and its 500 microvilli increase the surface area of the small intestine at least 600 times, creating an absorption area of 200 to 300 square meters—about the size of a tennis court (fig. 37.18)!

Within each villus a capillary network absorbs amino acids, monosaccharides, and water, as well as some vitamins and minerals. These digested nutrients are then distributed to cells. Fatty acids and glycerol are reassembled into triglycerides within the small intestinal lining cells, and they are coated with proteins to make them soluble before they enter the lymphatic system. The triglycerides then enter a lymph vessel called a **lacteal** that is surrounded by capillaries in the intestinal villi. Triglycerides enter the blood where lymphatic vessels join the circulatory system.

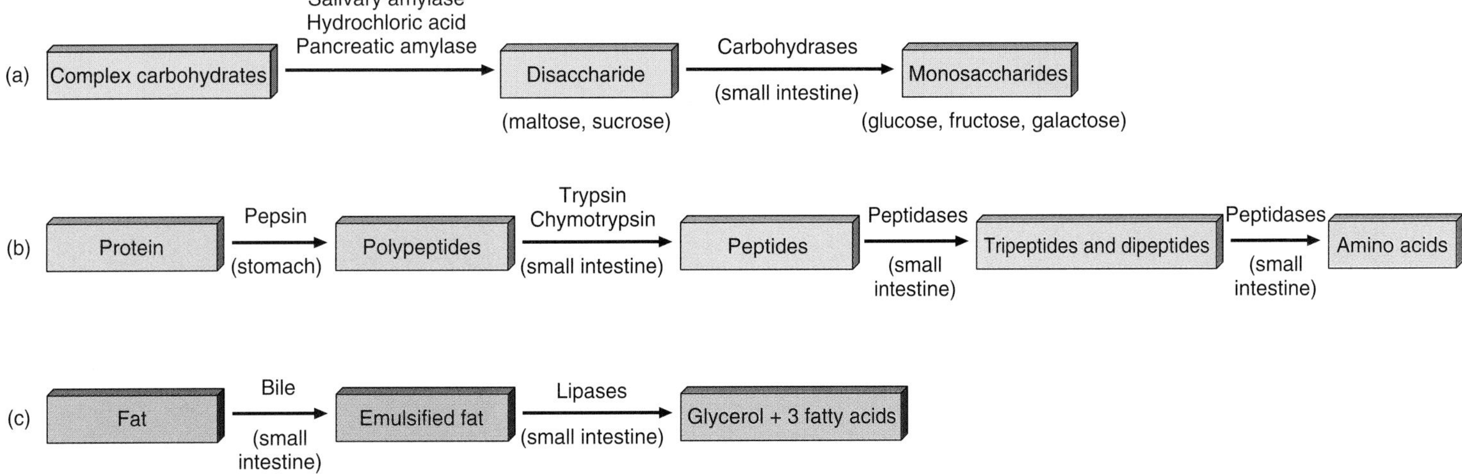

FIGURE 37.16

Digestion gradually breaks down large molecules. (*a*) Salivary amylase, hydrochloric acid, and pancreatic amylase act on starches, and carbohydrases digest disaccharides to monosaccharides to complete the process. (*b*) Protein digestion begins in the stomach, where pepsin breaks down proteins into polypeptides. In the small intestine, trypsin and chymotrypsin continue chemical digestion, breaking polypeptides down to peptides. Then peptidases break peptides into tripeptides and dipeptides, and eventually single amino acids, which are small enough to enter capillaries. (*c*) Bile emulsifies fats; then lipases break them down to glycerol and fatty acids.

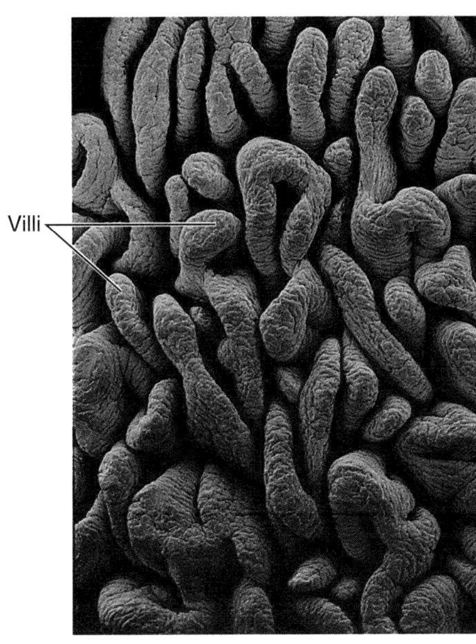

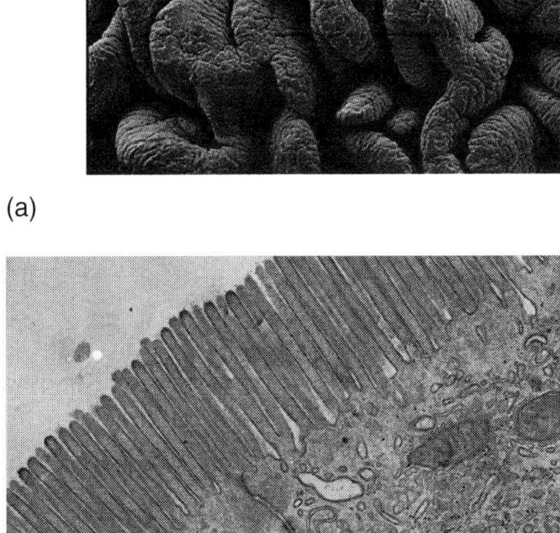

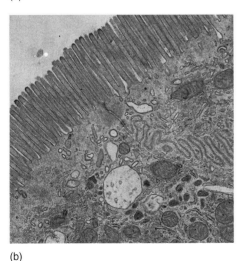

FIGURE 37.17

The small intestine lining maximizes surface area. (*a*) The lining of the small intestine is folded into many villi. (*b*) The cell membrane of a villus cell, in turn, folds into hundreds of microvilli.

Intestinal glands between villi secrete a watery fluid lacking digestive enzymes. This fluid helps transport digested nutrients through villus cells.

The Large Intestine

The material remaining in the small intestine after nutrients are absorbed enters the large intestine, or **colon.** The large intestine is shorter than the small intestine, but its 2.5-inch (6.5-centimeter) diameter is greater. The 5-foot (1.5-meter) tube surrounds the convoluted mass of the small intestine, roughly in the shape of a question mark. At the start of the large intestine is a pouch called the **cecum** (fig. 37.19).

Dangling from the cecum is the **appendix,** a thin wormlike tube. The appendix has an immune function in some vertebrates but its role in humans is not known. In our primate ancestors, the appendix may have helped digest fibrous plant matter. If bacteria or undigested food become trapped in the appendix, the area can become irritated and inflamed and infection can set in, producing severe abdominal pain. If this happens, the appendix must be promptly removed before it bursts and spills its contents into the abdominal cavity and spreads the infection.

The large intestine absorbs most of the water, electrolytes and minerals from chyme, leaving solid or semisolid feces. Billions of bacteria of about 500 different species are normal inhabitants of the healthy human large intestine. These "intestinal flora" decompose any nutrients that escaped absorption in the small intestine; they produce vitamins B_1, B_2, B_6, B_{12}, K, folic acid, and biotin; and they break down bile and foreign chemicals such as certain drugs.

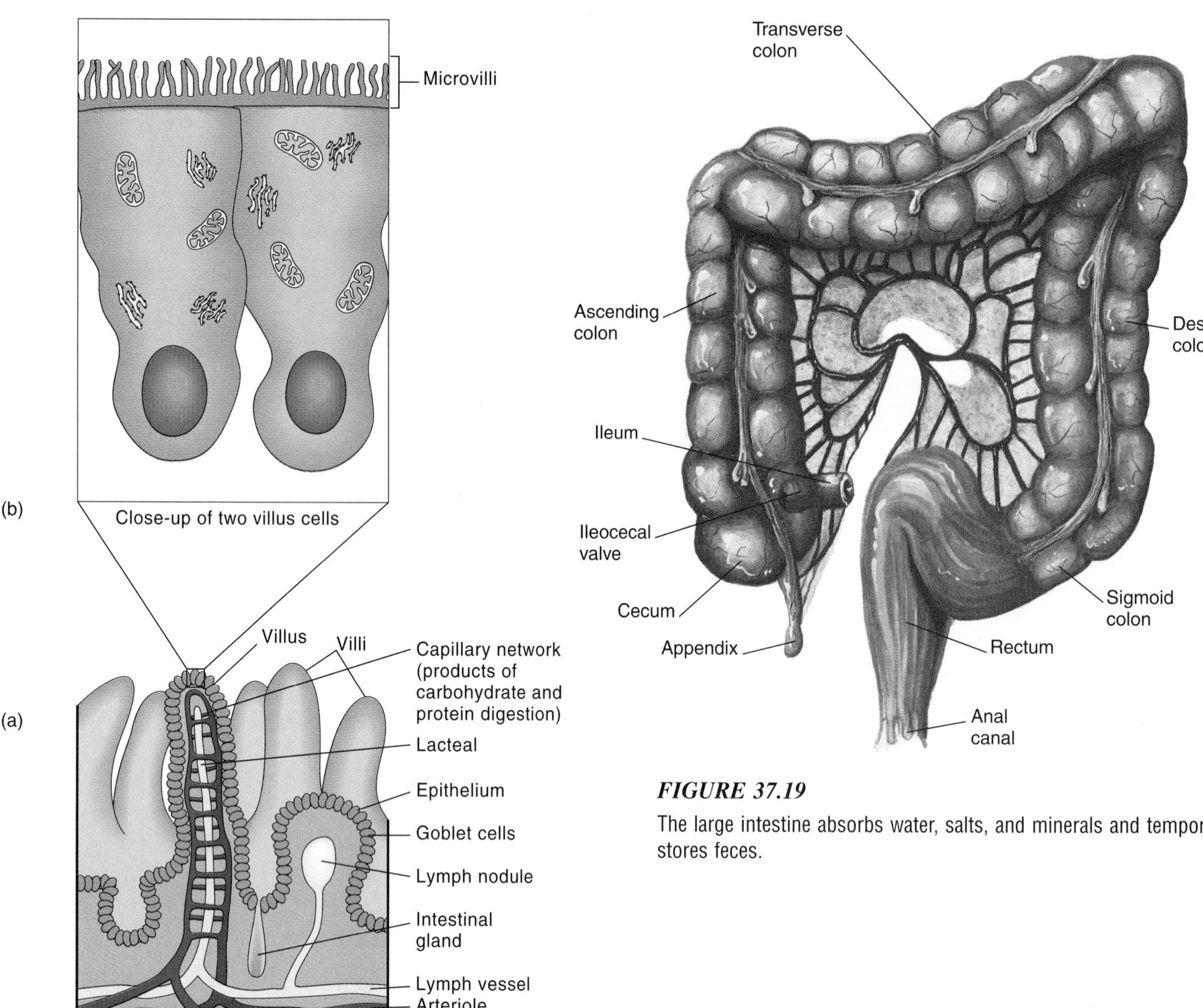

FIGURE 37.18

The small intestinal surface has ridges, and the inner lining is folded into villi (*a*). Each villus cell, in turn, has extensive surface area because its cell membrane is folded into microvilli (*b*). Capillaries absorb digested carbohydrates and proteins, and the lacteal absorbs digested fats.

FIGURE 37.19

The large intestine absorbs water, salts, and minerals and temporarily stores feces.

Intestinal flora produce foul-smelling compounds that cause the characteristic odors of intestinal gas and feces. These bacteria also help prevent infection by other microorganisms. Antibiotic drugs often kill the normal bacteria and allow other microorganisms, especially yeasts, to grow. This alteration in the intestinal flora causes the diarrhea that is sometimes a side effect of taking antibiotic drugs.

The remnants of digestion—cellulose, bacteria, bile, and intestinal cells—collect as feces in the **rectum,** a 6- to 8-inch-(15- to 20-centimeter) long region. Within the rectum is the 1-inch-(2.5-centimeter-) long anal canal. The opening to the anal canal, the **anus,** is usually closed by two sets of sphincters, an inner smooth-muscle sphincter under involuntary control and an outer skeletal-muscle sphincter, which is voluntary. When the rectum is full, receptor cells trigger a reflex that eliminates the feces.

Figure 37.20 summarizes the fate of food in the human digestive system. Other glands and organs assist digestion.

The Pancreas, Liver, and Gallbladder

The pancreas is a multifunctional structure associated with the digestive tract (fig. 37.21). It sends about a liter of fluid to the duodenum each day, including trypsin and chymotrypsin to digest polypeptides, pancreatic amylase to digest carbohydrates, pancreatic lipase to further break down emulsified fats, and **nucleases** to degrade DNA and RNA. Pancreatic "juice" also contains sodium bicarbonate to neutralize the acidity that hydrochloric acid pro-

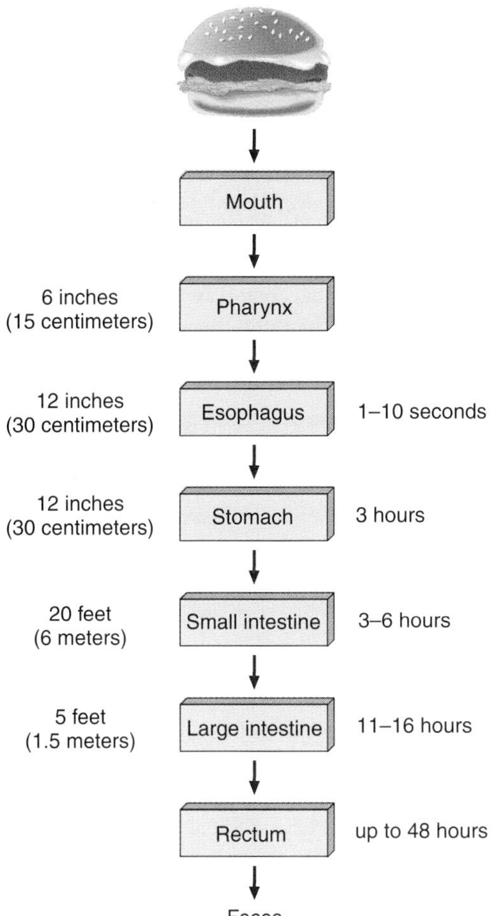

FIGURE 37.20

Food's journey through the human digestive tract. Food enters at the mouth and then moves through a system of tubes and hollow organs that mechanically and chemically degrade the food. Digested nutrients enter the blood or lymphatic systems at the small intestine. The remaining material enters the large intestine for further absorption or decomposition; it collects in the rectum and exits the body as feces.

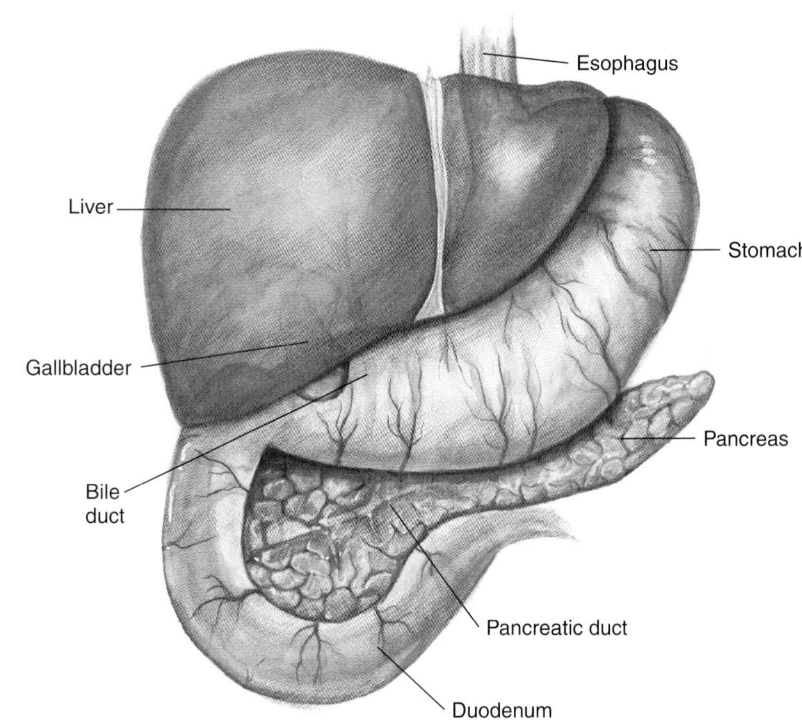

FIGURE 37.21

Associated structures of the digestive system include the liver, gallbladder, and pancreas.

duces in the stomach. Recall from chapter 33 that the pancreas also functions as an endocrine gland, regulating blood sugar level.

At a weight of about 3 pounds (1.4 kilograms), the liver is the largest solid organ in the body. It has more than 200 functions, including detoxifying harmful substances in the blood, storing glycogen and fat-soluble vitamins, and synthesizing blood proteins. The liver's contribution to digestion is the production of greenish yellow bile. Bile is stored in the gallbladder until fatty chyme in the small intestine triggers its release. The cholesterol in bile can crystallize, forming gallstones that partially or completely block the duct to the small intestine. Gallstones are very painful and may require the gallbladder to be removed, which can be done with laser surgery.

Bile consists of pigments derived from the breakdown products of hemoglobin and bile salts from the breakdown of cholesterol. This colorful substance is responsible for the brown color of feces and the pale yellow of blood plasma and urine. It also creates the abnormal yellow complexion of jaundice, a condition that deposits excess bile pigments in the skin.

Mastering Concepts

1. What structures in the mouth and throat participate in digestion?
2. Describe the mechanical and chemical digestion that occurs in the stomach.
3. How is the small intestine specialized to maximize surface area?
4. What are the products of digestion?
5. What occurs in the colon?
6. How do secretions of the pancreas, liver, and gallbladder aid digestion?

What We Can't—and Can—Control

Digestion is a biological process somewhat beyond our conscious control—we eat a meal, our stomachs churn, our small intestines secrete. We can, however, control nutrition—what we put into our digestive systems. Regulating food intake is not only important in maintaining a desired body weight but in keeping us healthy (Biology in Action 37.2). We are probably the only species with so many choices and so much conscious control over our food intake.

37.2 biology in action

Obesity

On her television talk show in 1988, Oprah Winfrey pulled in a red wagon heaped with an unsightly white glob—67 pounds of animal fat—to illustrate the weight that she had lost on a near-starvation diet. Alas, by the Daytime Emmy Awards in 1992, Oprah sat in the audience hoping she wouldn't win because she was embarrassed. She had gained back the lost weight plus considerably more, topping the scales at 237 pounds. Along the way, Oprah had tried many popular diets, only to see shed pounds return. The drastic diet plans didn't work because the body interprets such efforts as starvation—and when regular meals return, the weight piles on.

By 1996, Oprah seemed to have conquered her weight problem. The secret—a lifelong change in diet and exercise habits and, most important, a recognition of why she overeats.

Oprah Winfrey is far from alone. In the United States, 35% of women, 31% of men, and 25% of children and adolescents are obese. Technically, obesity is measured with the body mass index (BMI), which accounts for a person's height as well as weight. BMI equals weight in kilograms divided by height in meters squared. Obesity for males is a BMI above 27.8 and for females above 27.3. Weight gain occurs when a person takes in more kilocalories than the body uses for energy.

Tens of millions of people at any one time are in weight loss programs, and the U.S. population spends $33 billion a year on weight loss products and services—usually to no avail. Many people try to lose weight to improve their appearance, but obesity is a bona-fide medical condition. It raises the risk of developing certain chronic illnesses, including hypertension, diabetes, heart disease, and some cancers. The discrimination directed at overweight people can result in depression, low self-esteem, and eating disorders.

Despite recent advances in understanding and treating obesity—including discovery of biochemicals that control appetite, development of "fake fats," and availability of antiobesity drugs—obesity remains epidemic. The reason, according to the Institute of Medicine, is that overweight individuals do not make the lifestyle changes necessary to keep weight off. The Centers for Disease Control (CDC) found that of more than 25,000 people interviewed who claimed to be trying to lose weight, 37% engaged in no physical activity at all.

The Institute of Medicine identifies five approaches to weight loss:

1. diet
2. physical activity
3. behavior modification
4. drugs
5. gastric surgery

The roles of diet and physical activity are well known. Behavior modification identifies events and feelings that trigger eating so that a person is better able to avoid overeating. Weight loss drugs and gastric surgery ("stomach stapling") are extreme medical measures.

It may be easier to lose weight by the conventional diet and exercise route if you have a cook and personal trainer, as Oprah Winfrey does. But the rest of us can try to follow her example of limiting food choices and portion sizes, exercising at least 10 hours a week, and recognizing why we overeat.

A New View of Ulcers

continued from page 750.

Barry Marshall got sick from drinking the bacterial brew, but he was lucky—he suffered for a few days from gastritis, which then cleared up without treatment. A second volunteer, however, was ill for several months. Although a dose of bismuth subcitrate (Pepto Bismol, the pink stuff) gave him temporary relief, the pain returned. Only when he took two antibiotic drugs for a few weeks did the gastritis vanish completely.

Other researchers, using laboratory animals rather than themselves, soon confirmed the causative link between *H. pylori* and gastritis and then a link to ulcers, too. Still, it took until 1994—a full decade after Marshall's self-experiment—for the National Institutes of Health to publish a consensus statement advising doctors that a 2-week course of two antibiotic drugs plus an antacid become the standard treatment for gastritis and ulcers in patients with an infection. Yet a year later, a study found that about 90% of surveyed physicians still prescribed a bland diet, stress reduction, and acid-blocking drugs as a first-line treatment. These measures can't hurt, but for many patients they mean expense and recurrent pain.

By 1996, with a new subdiscipline in the biomedical sciences devoted to stomach bacteria and medical journals frequently reporting on *H. pylori's* activities in the human digestive tract, the idea that gastritis and ulcers are usually the consequence of bacterial infection had finally been accepted. Although the precise interactions between stress, diet, stomach acid production, and bacterial infection are still not fully understood, for many patients, our knowledge so far may mean that gastritis and ulcers are a pain of the past.

SUMMARY

1. A **digestive system** breaks food molecules into their components and absorbs them. Mechanical digestion breaks food into smaller pieces to increase the surface area exposed to chemical digestion; in chemical digestion, hydrolytic enzymes further break food down into component molecules.

2. Digestion breaks proteins down to amino acids, fats (triglycerides) to fatty acids and glycerol, and starches to monosaccharides. The circulatory system distributes the products to cells, where they are oxidized to provide energy for cellular activities, used to make new cellular materials, or stored.

3. Animals have a variety of mechanisms for obtaining food. **Heterotrophy** allowed organisms to obtain nutrients from the environment.

4. Simple organisms have **intracellular digestion;** more complex organisms have **extracellular digestion** in a special cavity outside the cells. Digestive cavities can have one opening or two. In a system with two openings, food enters through the mouth and is digested and absorbed; undigested material leaves through the second opening, the anus.

5. Nutrients promote growth, maintenance, and repair of body tissues. **Macronutrients,** or **energy nutrients,** include carbohydrates, proteins, and fats. **Micronutrients** include vitamins and minerals. Water is also vital. **Kilocalories** measure the energy food provides. One gram of carbohydrate or protein yields 4 kilocalories, and 1 gram of fat yields 9. An inadequate diet causes a **primary nutrient deficiency.** A **secondary nutrient deficiency** reflects a metabolic defect.

6. In humans, digestion begins in the mouth, where teeth break food into smaller pieces. Salivary glands produce saliva, which moistens food and begins starch digestion. Swallowed food moves through the **esophagus** to the stomach. Waves of contraction called **peristalsis** move food along the digestive tract.

7. The stomach stores food, churns it until it liquefies, and mixes it with **gastric juice.** Hydrochloric acid in the gastric juice activates **pepsinogen,** forming the protein-splitting enzyme **pepsin.**

8. In the small intestine, chemical digestion continues. Enzymes produced by the small intestine digest proteins and carbohydrates. The small intestine absorbs digested nutrients through its tremendous surface area, provided by circular folds, **villi,** and **microvilli.** In each villus, amino acids and monosaccharides enter capillaries, and digested fats enter **lacteals.**

9. Material remaining after absorption in the small intestine passes to the **colon,** which absorbs water, minerals, and salts, leaving feces. Many bacteria digest remaining nutrients and produce useful vitamins that are then absorbed. Feces exit the body through the **anus.**

10. The **pancreas, liver,** and **gallbladder** aid digestion. The pancreas supplies **pancreatic amylase, trypsin, chymotrypsin, lipases,** and **nucleases.** The liver supplies **bile,** which emulsifies fat to prepare it for lipase action, and the gallbladder stores bile. The nervous and endocrine systems regulate digestive secretion.

TO REVIEW . . .

1. What is a nutrient?

2. How do the circulatory, muscular, nervous, and endocrine systems take part in digestion?

3. What are the digestive products of carbohydrates, proteins, and fats?

4. Identify the part of the digestive system that includes the following:
- **a.** ileum, jejunum, duodenum
- **b.** cecum, appendix, rectum, anus
- **c.** villi and microvilli
- **d.** rugae, body, pylorus, cardia, fundus

5. Name an organism that has each of the following:
- **a.** extracellular digestion
- **b.** intracellular digestion
- **c.** a digestive tract with one opening
- **d.** a digestive tract with two openings
- **e.** filter feeding

6. Describe how the digestive systems of two types of mammals are adapted to their diets.

7. How does mechanical digestion facilitate chemical digestion?

8. Why don't the stomach and small intestine digest themselves?

9. How is surface area maximized in the stomach and in the small intestine? Why is it necessary for a digestive system to have extensive surface area?

10. Is the presence of bacteria in the large intestine a sign of infection? Why or why not?

11. Zinc deficiency causes an enlarged liver and spleen, poor growth, delayed sexual development, poor wound healing, and loss of the ability to taste and smell. A teenager following a poor diet has zinc deficiency symptoms. Another person has a zinc deficiency due to an inborn inability to absorb this micronutrient. Which of these persons suffers from a primary nutrient deficiency and which from a secondary nutrient deficiency?

12. Select a food, and trace the digestion of its component nutrients in the various organs of the digestive system.

13. How do the liver, pancreas, and gallbladder contribute to digestion?

14. Name three hormones that are associated with digestion and their functions.

TO THINK ABOUT . . .

1. How can people consume vastly different diets yet obtain adequate nourishment?

2. Several years ago, health food stores marketed a protein derived from kidney beans as a "starch blocker." It worked by inhibiting amylase action. Supposedly, one could consume huge amounts of starch without digesting it. After a summer in which millions of people took the pills, studies found that the starch blocker did not work. Many people developed constipation, diarrhea, gas, nausea, and vomiting. Why do you think the starch blocker failed? What could have caused the side effects?

3. Why can't a person gain nutritional benefits by finely grinding meat and injecting it into the bloodstream?

4. Name an organelle in human cells that has a function similar to protozoan intracellular digestion.

5. Why can certain tripeptides cross intestinal lining cells, but others must be dismantled to dipeptides or even single amino acids to do so?

6. The protein in a hamburger is mostly myosin. Why doesn't the human body simply use the cow version of myosin to build muscle tissue? What happens instead?

7. So-called fat-free cakes and cookies contain diglycerides, which are technically not fats. In the small intestine, the diglycerides are repackaged into triglycerides. In what part of a villus would this probably occur?

TO LEARN MORE . . .

References and Resources

Diamant, Nicholas E. November 1996. Firing up the swallowing mechanism. *Nature Medicine.* Researchers are learning to restore swallowing to brain-damaged individuals.

Gould, Stephen Jay. May/June 1997. Pulling teeth. *The Sciences.* Teeth are a part of the skeletal system vital to the digestive system.

Karasov, William H. November 1993. In the belly of the bird. *Natural History.* Birds that gorge before migrating have a fascinating adaptation—their guts lengthen.

Kumanyika, Shiriki. September 5, 1996. Improving our diet—Still a long way to go. *The New England Journal of Medicine,* vol. 335. Not all of us follow dietary recommendations to lower fat and increase fiber intake.

Kurtzweil, Paula. May-June 1997. Today's special—nutrition information. *FDA Consumer.* Restaurants must provide nutrition information.

Larkin, Marilynn. January/February 1996. Losing weight safely. *FDA Consumer.* Weight loss requires a long-term commitment.

Lewis, Ricki. May 13, 1996. Neurogastroenterologists combine old and new research approaches. *The Scientist.* The digestive system has its own mini–nervous system.

Lewis, Ricki, February 1996. Stomach rumblings on camera. *Photonics Spectra.* Imaging the stomach in action can help diagnose digestive woes.

Lewis, Ricki. December 1994. The ulcer bug. *FDA Consumer.* Most peptic ulcers are really bacterial infections.

Thomas, Paul R., ed. 1995. *Weighing the options.* Washington, D.C.: National Academy Press. Tips on how to evaluate weight loss programs, from the Institute of Medicine.

denotes articles by author Ricki Lewis that may be accessed via *Life*'s home page /http://www.mhhe.com/sciencemath/biology/life/.

A Sally Lightfoot crab (*Grapsus grapsus*) shoots fluid from antennal glands near its eyes. The crab "relieves itself" in this manner to maintain fluid volume and ion balance but does not excrete nitrogenous wastes this way. Ammonia exits by diffusing across the animal's gills.

38 Regulation of Temperature and Body Fluids

chapter outline

The Camel—Playing It Cool 770

Temperature Regulation 771
Source of Body Heat—External or Internal?
Body Temperature—Constant or Variable?
Coping with Cold
Beating the Heat

Osmoregulation 776

Nitrogenous Waste Removal 776
Nephridia

The Human Urinary System—An Overview 780

The Kidney 780

The Nephron 781
The Glomerular Capsule
The Proximal Convoluted Tubule
The Nephron Loop
The Distal Convoluted Tubule
The Collecting Duct

Control of Kidney Function 784

The Unhealthy Urinary System 785
Bedwetting
Urinary Tract Infections
Kidney Stones
Kidney Failure

Hemolytic Uremic Syndrome

Case 1: On a trip in mid-July 1994, several children from a Virginia summer camp and their counselors cooked hamburgers over an open fire. Starting 5 days later, 18 campers and 2 counselors experienced abdominal cramps and bloody diarrhea. One child developed a complication called hemolytic uremic syndrome (HUS), with bloody urine, severe anemia, and lack of blood platelets. Fortunately, the child recovered. The diarrhea and HUS were caused by a toxin produced by the common intestinal bacterium *Escherichia coli.*

Case 2: Health officials in Adelaide, southern Australia, reported 23 cases of HUS in children in January 1995. Sixteen of the children required kidney dialysis to temporarily take over their kidney function, and one child died. Of the 23 children, 22 had diarrhea before blood appeared in their urine. All of the children had eaten uncooked sausage preserved through a fermentation process.

Case 3: In 1996, a medical journal reported an unusual case of a 6-year-old girl with HUS. She entered the hospital hours after developing severe abdominal pain and pain upon urination, but unlike other patients with the condition, she did not previously have diarrhea. Protein, blood cells, bacteria, and kidney lining cells in her urine indicated an infection. By her third day in the hospital, the red color of her urine, and her damaged blood—with torn red blood cells, excess white blood cells, and too few platelets—indicated severe HUS. An ultrasound scan revealed an abscess in her lower left kidney. After 11 days of dialysis and blood transfusions, she recovered.

The little girl's case differed by the route of infection—it started in her urinary system. The Virginia campers and Australian sausage-eaters had first developed gastrointestinal pain from a bacterial toxin in undercooked meat. The circulatory system then transported the toxin to the kidneys.

Although these three outbreaks differed in location, route of infection, and even the strain of causative bacteria, cellular and molecular events were similar.

to be continued . . .

The Camel—Playing It Cool

The camel is a walking definition of adaptation; everything from its hump to its behavior helps the animal conserve water and tolerate heat (fig. 38.1).

Standing up to 7 feet (2.1 meters) tall and weighing up to 1,000 pounds (454 kilograms), the camel stays cooler longer than a smaller desert dweller. Its characteristic hump, rich in fatty tissue, absorbs heat, preventing some of it from reaching the rest of the body; its long, thin legs radiate heat. The camel's bulk casts a large shadow that provides shade, especially when several camels squat together.

Internal adaptations also help the camel survive in the desert heat. Just beneath the brain, a pool of cool venous blood surrounds arteries that carry warmer blood to the brain in a circulatory specialization called a **carotid rete.** This mechanism is a lifesaving adaptation because the brain is especially sensitive to heat. A carotid rete is a type of **countercurrent exchange system,** in which blood in one vessel transfers heat to blood in a nearby vessel.

The camel's blood composition is also adapted to surviving heat. It is more watery than the blood of mammals living in cooler climates. The camel's red blood cells are oval, compared to other mammals' rounder cells that shrivel in great heat, which blocks blood flow to the surface necessary to dissipate body heat. Oval cells remain oval, allowing blood flow to continue even if the camel loses a third of its body water.

The camel's metabolism is also adapted to the heat, slowing in the summer rather than speeding up as in most mammals. Water loss (dehydration) suppresses activity of the camel's thyroid gland, lowering metabolic rate. Lowered metabolism decreases respiratory rate, so water that would leave through respiration is conserved. The urinary system conserves water, too, with urine just watery enough to excrete wastes. A camel's droppings are very dry, which also conserves water.

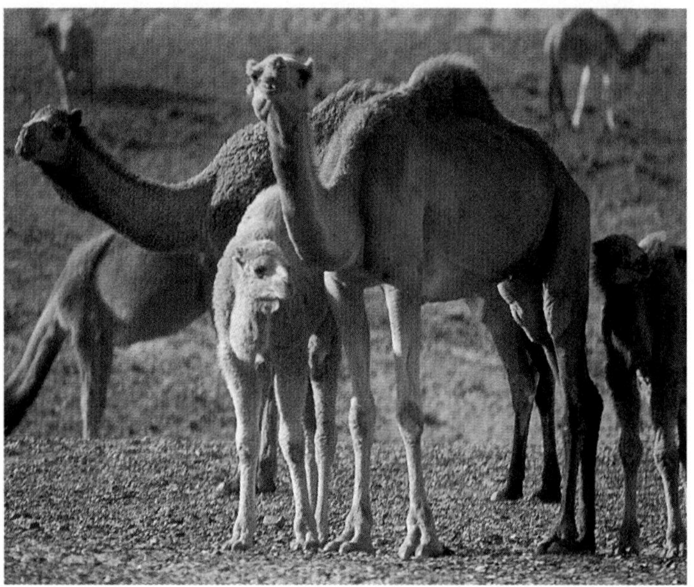

FIGURE 38.1

The camel has many adaptations that help it tolerate high temperatures. Its long, thin legs radiate heat, and its fatty hump absorbs heat that would otherwise harm internal organs. The young camel on the right exhibits another adaptation—it is taking advantage of the shadow thrown by the larger animal next to it.

Even the camel's behavior has a cooling effect. The animal sits much of the time, minimizing energy expenditure. Like many other animals, a camel urinates on its legs, enjoying the cooling effects of evaporating liquid.

The mechanisms for temperature regulation, water conservation, and waste excretion in diverse organisms are evolutionary adaptations for life in varied environments. The extremely dry, scorchingly hot home of a camel contrasts sharply with a frigid arctic habitat or with the perpetual humidity of a tropical rain forest. In hot, dry areas, an animal may have difficulty conserving enough water to meet excretory demands and also produce sufficient sweat to tolerate heat. In arctic freshwater habitats, an animal must continually pump excess water out of the body, conserve ions, and survive the cold. Most animals live in environments that are intermediate between these extremes, but each species has specific adaptations that enable it to maintain homeostasis in its environment.

Temperature Regulation

The ability of an animal to balance heat loss and gain within its particular environment is called **thermoregulation.**

It is important that an animal's body temperature remain within limits because temperature influences the rate of the chemical reactions that sustain life. An animal's enzymes, for example, work best at its customary body temperature. Should cellular temperatures vary from this optimum, the enzymes function less efficiently, and vital biochemical reactions occur more slowly. If enzymes in the same biochemical pathway respond differently to temperature changes, the coordination that enables the pathway to function breaks down.

Extreme temperatures alter biological molecules. For example, a drastic temperature increase alters a protein's three-dimensional shape and disrupts its function. An extreme temperature drop solidifies lipids. Since lipids are an integral component of biological membranes, such a change would impede many functions, especially those involving transport of molecules across membranes.

Source of Body Heat—External or Internal?

Although all animals exchange heat with their environments, they may be classified according to their major source of body heat—external or internal. **Ectotherms** obtain heat primarily from their external surroundings. A snake or a lizard warms itself by basking in the sun on a rock. By carefully choosing a sunny spot and positioning itself properly, an ectotherm can absorb enough solar heat to considerably increase its body temperature, even on a chilly morning. When it becomes too warm, the animal moves out of the sun's rays by finding shade or burrowing. At night or on overcast days, an ectotherm cannot warm itself, and its activity slows. Because its defense reactions are also slowed, the animal often seeks shelter from predators. An ectotherm thus regulates its body temperature by its behavior, moving to areas where it can gain or lose external heat. The marine iguana lizard is a master at moving about to maintain a relatively stable body temperature (Biology in Action 38.1).

An **endotherm** obtains most of its heat internally from its metabolism. The metabolic rate of an endotherm is generally five times that of an ectotherm of similar size and body temperature. Layers of insulation in the form of fat, feathers, or fur help retain body heat.

Ectothermic and endothermic ways of life each have advantages and disadvantages. The ectotherm depends upon a continuous ability to obtain or escape environmental heat. An injured iguana that could not squeeze into a crevice to escape the broiling noon sun would cook to death. In contrast, an endotherm's internally generated heat enables it to maintain body heat even in the middle of the night. This internal constancy comes at a cost, however. The endotherm must eat much more food and must use 80% of the energy from that food just to maintain its temperature.

Body Temperature—Constant or Variable?

Animals can also be classified according to whether their body temperature is constant (a **homeotherm**) or variable (a **heterotherm**). Most endotherms are also homeotherms; they maintain a constant body temperature by balancing heat generated in metabolism with heat lost to the environment. Most ectotherms are also heterotherms; changing external factors control their fluctuating body temperatures. There are exceptions to these trends. A reptile is a homeotherm and an ectotherm, because it maintains a constant body temperature by moving around to gain or lose heat. A hummingbird is a heterotherm and an endotherm because it generates occasional high body temperatures internally.

Adaptations that enable animals to regulate their temperatures fall into three general categories: anatomical and physiological; metabolic and hormonal; and behavioral. Clusters of neurons in the brain's hypothalamus control many of an animal's thermoregulatory responses by reacting to feedback from sensory cells elsewhere in the body. Thermoregulation depends on the interactions of the musculoskeletal, nervous, endocrine, respiratory, integumentary, and circulatory systems. Animals have adapted to a variety of environmental extremes.

Coping with Cold

Environmental water temperature determines the body temperatures of many fishes. Yet, these animals must maintain body temperatures warm enough to permit survival. To do this, many fishes have blood chemicals that function as antifreeze, lowering the point at which the blood freezes so that it stays liquid even when the water is very cold.

The great white shark and bluefin tuna have another type of adaptation to cold water, a countercurrent exchange system in which venous blood coming from warm interior muscles warms colder blood carried in arteries from near the body surface. Marine reptiles such as sea turtles and sea snakes shunt warmer blood to the centers of their bodies, preferentially protecting vital organs. Amphibians, which tend to lose what little metabolic heat they generate through their moist skin, conserve heat by restricting their habitats to warm, moist places.

Birds and mammals can function for long periods at low temperatures, thanks to their adaptations to the cold. Feathers retain warmth by creating pockets that trap air near the skin. In many mammals, countercurrent exchange systems transfer the heat in some blood vessels to cooler blood in nearby vessels, retaining heat in the body rather than allowing it to escape to the environment through the extremities. Such countercurrent exchange systems allow penguins to spend hours on ice or snow or in frigid water and enable arctic mammals such as wolves to hunt in extreme cold (fig. 38.2).

38.1 biology in action

A Day in the Life of a Marine Iguana

Iguana lizards spend much of their time regulating their body temperatures, which are about the same as, or slightly higher than, a human's body temperature. Marine iguanas on the Galápagos Islands begin their days basking in the rising sun, draped on boulders and hardened lava, sunning their backs and sides. After about an hour, they turn and aim their undersides at the sun.

By midmorning, the air temperature is rising rapidly. Iguanas cannot sweat; instead, they must escape the blazing sun. They lift their bodies by extending their short legs, which removes their bellies from the hot rocks and allows breezes to fan them. But by noon, these push-ups are insufficient to stay cool. The iguanas retreat to the shade of rock crevices (fig. 38.A).

By midday, the animals are hungry. The iguanas dive into the ocean, although it is too cold for them to stay in for more than a few minutes. They eat green algae on the ocean floor or seaweed by the shore, hanging off rocks to reach it. The water is so cold that the lizards' body temperatures rapidly drop. Arteries near their body surfaces constrict to help conserve heat.

Because the iguana's food comes from the ocean, it is very salty. To rid its body of excess dietary salt, the iguana has a multilobed structure called a salt gland in its head (fig. 38.A, *inset*). Cells lining the interior of the gland secrete a fluid that travels in branching tubules to empty through the nostrils. This fluid carries excess salts with it.

After feeding, the iguanas stretch out on the rocks again, warming sufficiently to digest their meal. They continue basking as the day ends, absorbing enough heat to sustain them through the cooler night temperatures until a new day begins.

FIGURE 38.A
Marine iguanas retreat to shade to escape the heat of the noonday sun.

Blubber helps whales and seals maintain a constant body temperature. In their flippers and tails, which lack blubber, countercurrent heat exchange blood vessels keep the heat in. Mammals also have several behavioral strategies for retaining heat, including migrating, hibernating, and simply huddling together.

Mammals generate much of their metabolic heat in a process called **thermogenesis.** In one type of thermogenesis, fatty acids in brown fat are oxidized to release heat energy. Brown fat is abundant in animals that hibernate, and it also helps many newborn mammals, including humans, to stay warm.

Shivering is another form of thermogenesis. Shivering contracts muscles, releasing heat as ATP splits and actin and myosin filaments slide. Hormones also provide body heat. Falling body temperatures stimulate the thyroid gland to produce more thyroxine, which increases cell permeability to sodium ions (Na^+). When cell membranes pump Na^+ out of the cells, ATP splits, releasing heat energy. This hormone-directed internal heating mechanism is called **nonshivering thermogenesis.**

Beating the Heat

Invertebrates and vertebrates share many adaptations for surviving in extreme heat and drought. These include evaporative cooling, circulatory adaptations, and adaptive body structures and behaviors.

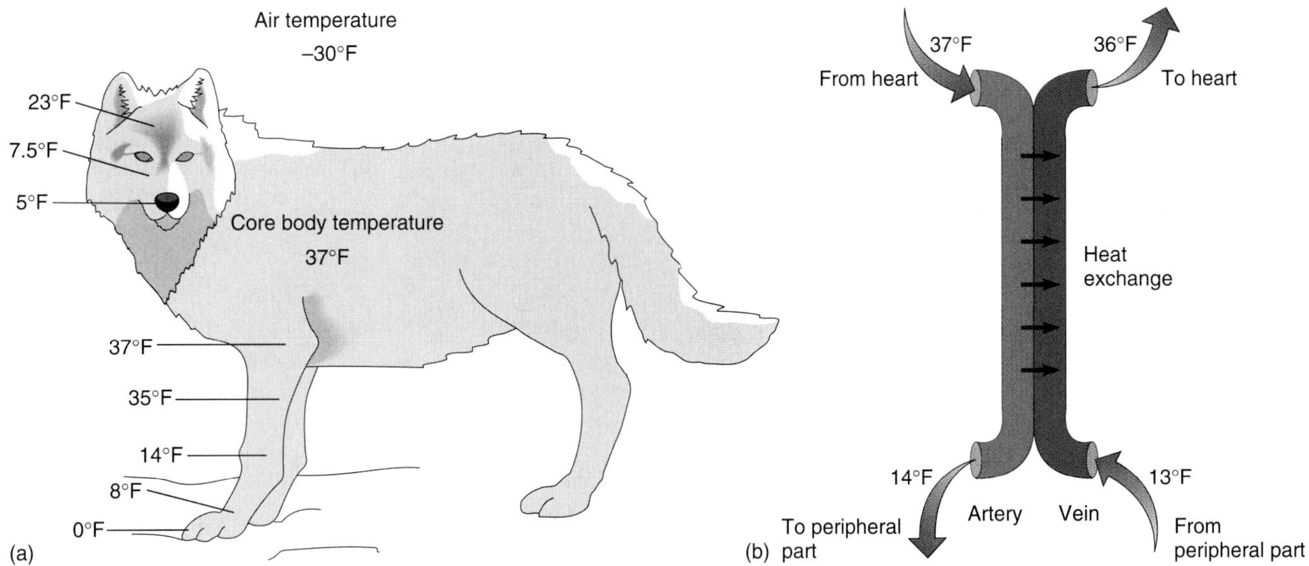

FIGURE 38.2

Anatomy enables an animal to tolerate heat. (*a*) A mammal's limbs, ears, snout, and tail chill more easily than interior body parts. (*b*) In a countercurrent exchange system, the close proximity of arteries carrying warm blood to veins carrying cooler blood conserves heat, rather than losing it to the environment.

Evaporative Cooling

One common strategy for lowering body temperature is **evaporative cooling,** in which air moves over a moist surface, evaporating the moisture and cooling that part of the body. A kangaroo rat salivates onto its throat, which cools its head until it can seek shelter from the blazing sun. When a honeybee's temperature reaches 113°F (45°C), a heat sensor in its brain stimulates it to regurgitate nectar stored in an organ called a honey-crop. The nectar spills onto the tongue, and the bee licks itself, cooling as its nectar bath evaporates. Australian sawfly larvae smear rectal fluid on their bodies, which evaporates and enables them to survive another hour in 115°F (46°C) heat.

Humans sweat. Some invertebrates sweat in a sense, too. Consider a type of cicada (an insect) that lives in the Sonoran Desert in Africa, where the temperature can reach 105°F (41°C) in the shade. Water from the cicada's blood moves through tubules to the body's surface, where it exits through ducts and evaporates (fig. 38.3). By "sweating," the cicada can lower its body temperature as much as 18°F to 27°F (10°C to 15°C) below that of surrounding air.

A cicada can lose 20 to 35% of its body weight in fluid when the outside temperature hits 115°F (46°C). Most insects perish at a 20% loss, and humans cannot survive a 10% loss of body fluid. The African cicada can survive such extreme dehydration because fluid loss is balanced by intake—the insect imbibes tremendous amounts of fluid in its succulent plant food. Interestingly, cicadas in the United States have fewer evaporative cooling pores, and cicadas living in the cool alpine forests of Australia have no pores at all.

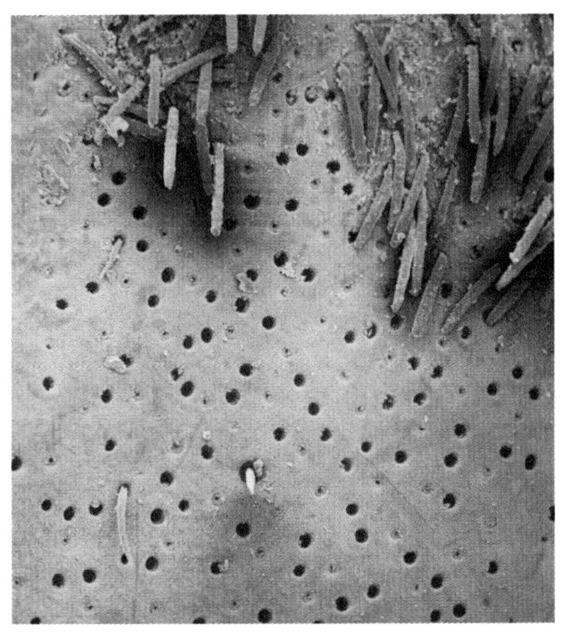

FIGURE 38.3

Cicadas living in the desert "sweat." These sweat pores in the cicada's cuticle are magnified 330 times.

Circulatory Adaptations to Heat Tolerance

In many mammals, evaporative cooling is linked to circulatory system specializations that route cooled blood past warmer blood. The carotid rete in many mammals consists of warm-blood-bearing arteries that run through a pool of cool venous blood at the base of the brain (fig. 38.4).

The ancient Greeks were the first to describe the carotid rete. They named it the *rete mirabile,* which means "wonderful net." Cats, dogs, seals, sea lions, antelopes, sheep, goats, and cows all have carotid retia. These animals also have long snouts and tend to pant. The snout interiors have a rich blood supply and many glands that secrete fluids that keep the area moist. Each time the

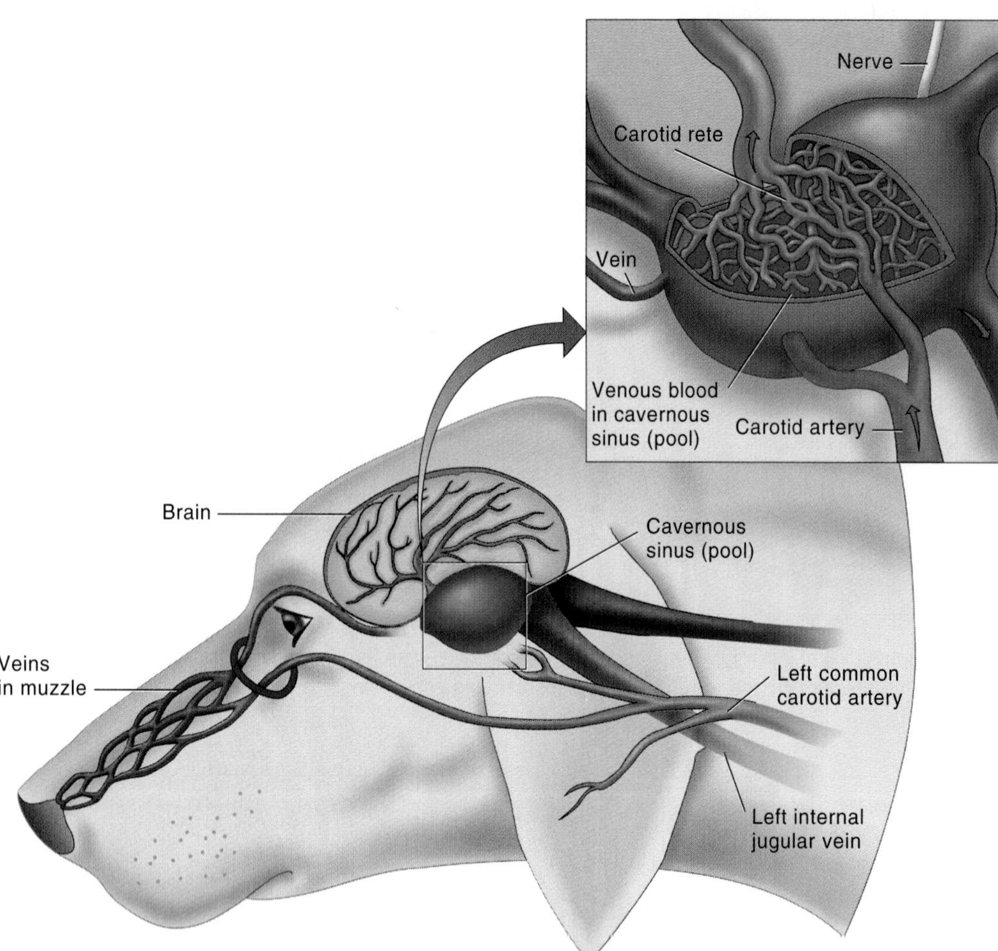

FIGURE 38.4

The carotid rete. The brain of a hoofed mammal or carnivore stays cool thanks to the carotid rete, a network of arteries that cool the blood they deliver to the brain by passing first through a pool of cool venous blood beneath the brain. The venous blood comes from veins in the animal's muzzle.

animal pants, water from the secretions evaporates, cooling blood in nearby vessels. Veins take the cooled blood to the pool beneath the brain where the carotid arteries bring warm blood. Cooling just the brain—the area that needs it most—requires less energy than cooling the entire body.

Humans lack a carotid rete but have a different circulatory adaptation for keeping a cool head. A cooling mechanism is necessary because a fever exceeding 107°F (42°C) can cause convulsions, hallucinations, brain damage, and death. To prevent this, tiny "emissary" veins in the face and scalp reroute blood cooled near the body's surface through the braincase to cool brain tissue. This adaptation, called a **cranial radiator** (fig. 38.5), causes the flushed face of a person vigorously exercising. Studies of fossilized skulls reveal that the cranial radiator first appeared in human ancestors about 2 million years ago. This adaptation may have helped them to walk erect and to develop larger brains.

Body Form and Behavior

Behavior and body form provide powerful adaptations to extreme heat and drought, in vertebrates and invertebrates. Consider the Namib ant, which lives in the desert of southwest Africa, where sand temperatures can reach 150°F (66°C). The temperature drops with elevation, however, even if the elevation is just the height of an ant. The Namib ant's long legs place its head high enough that the surrounding temperature is a mere 131°F (55°C). The ant has other behaviors that help it tolerate heat. By clinging to a blade of grass or a leaf for short intervals, or resting atop the dead body of a less lucky insect, an ant can cool off sufficiently to race across another stretch of blisteringly hot sand (fig. 38.6). This combination of body form and behavior enables the Namib ant to forage for food when individuals of less well-adapted species must seek shade.

The human's upright stance and relatively hairless body surface may be adaptations to heat. Like the ant that climbs high enough to meet cooler air, a human's head is the highest part of the body and is therefore in a slightly cooler environment. Our bipedal (walking on two limbs) posture reduces the solar radiation hitting the body to 60% less than the radiation we would absorb if we were quadrupedal (walking on all fours). Bipedalism is also adaptive because it requires less energy than walking on the knuckles or on all fours. In quadrupedal mammals, fur reflects some heat. Perhaps human bipedalism made fur as a heat radiator less important. Gradually, erect, less hairy individuals may have evolved.

Animals combine evaporative, circulatory, body, and behavioral strategies to cool their bodies and retain fluids. Turtles, for example, salivate, urinate on their back legs, and pant. Their heart rates change and their peripheral blood vessels dilate, dissipating heat at the body's surface. A turtle's shell absorbs heat, protecting vital inner organs, and its large bladder stores water. Turtles also burrow into the ground, or seek shade, to avoid heat. Figure 38.7 illustrates several adaptations to heat in the eland, an antelope that lives on the hot plains of central Africa.

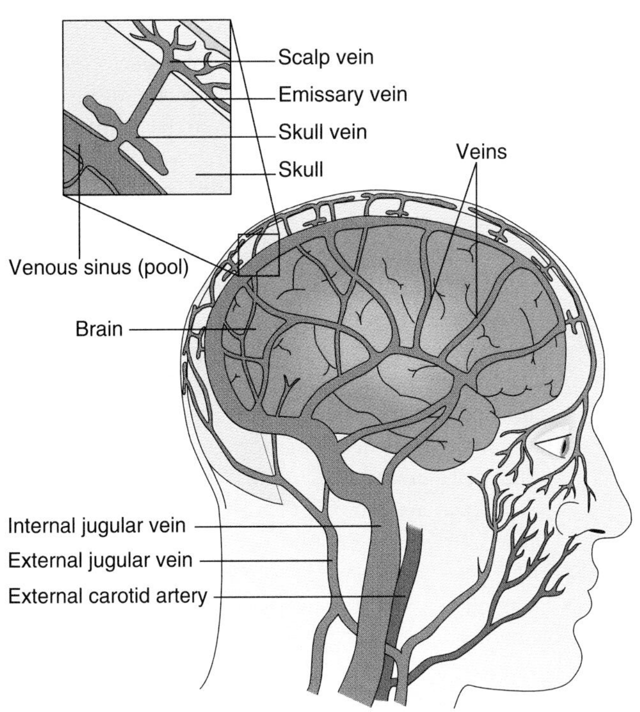

FIGURE 38.5

The cranial radiator. A human's brain cools by a circulatory adaptation called a cranial radiator. Under normal temperatures, blood from the scalp and face enters the external jugular vein to return directly to the heart. If body temperature is greatly elevated, blood from the scalp and face detours, entering the braincase through emissary veins to cool the brain and exiting through the internal jugular vein to return to the heart.

FIGURE 38.6

An ant's stance can be lifesaving. A Namib ant can scavenge for food on scorching sand by taking breaks perched on a blade of grass, a leaf, or on the body of a dead ant.

FIGURE 38.7

The eland dwells on the hot plains of central Africa, with several adaptations to maintain its body temperature in the heat with a minimum of water loss.

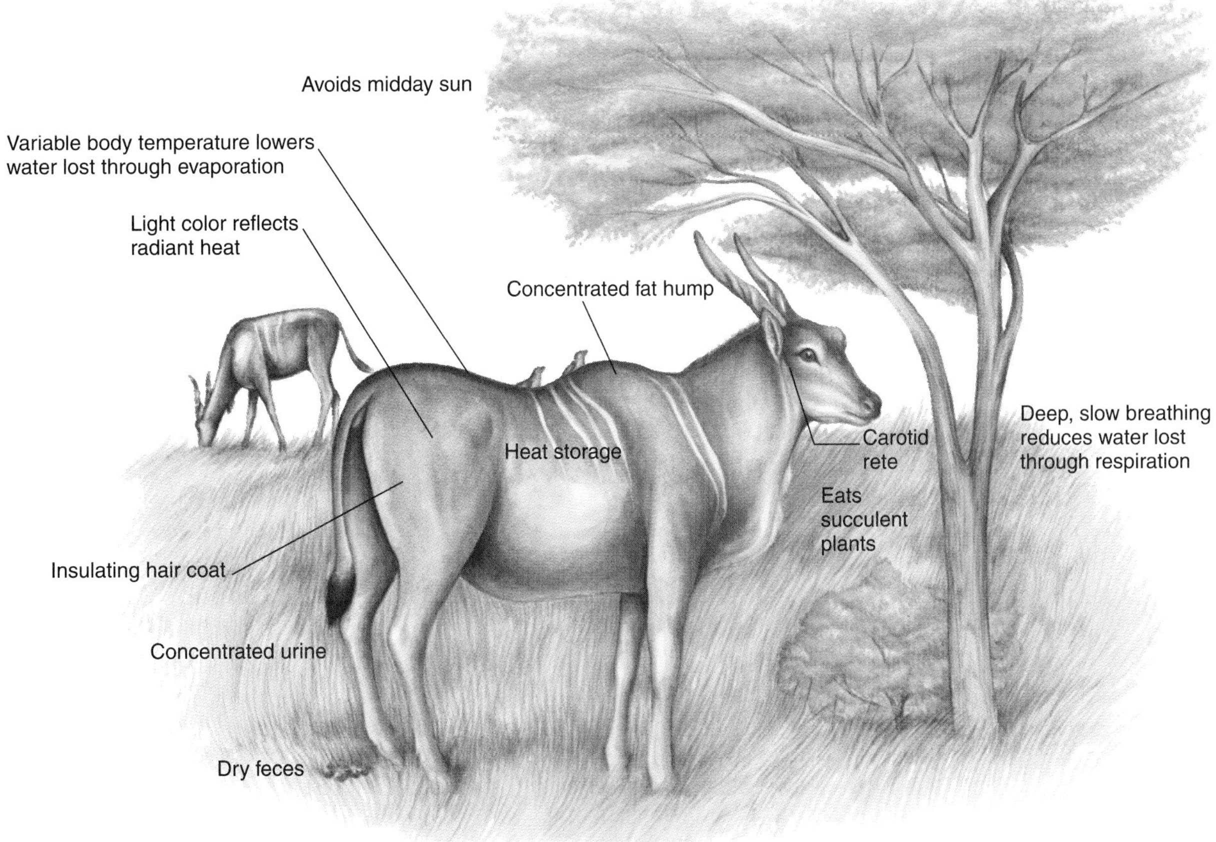

Table 38.1 Water Gain and Loss in a Human and a Kangaroo Rat

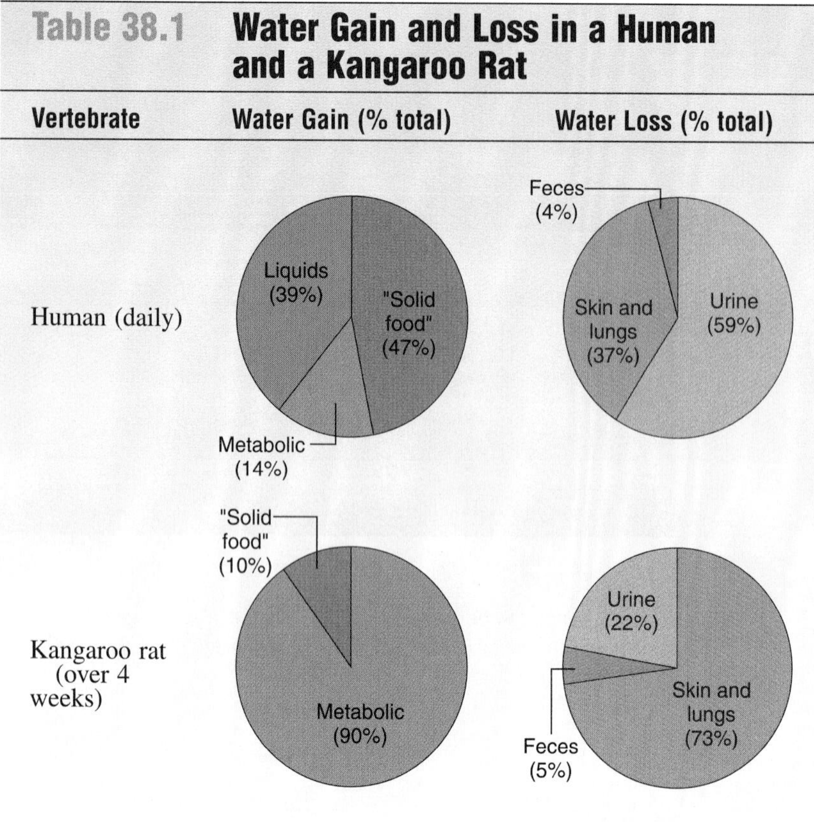

Source: Data from Stephen Miller and John Harley, *Zoology*, 2nd edition, 1994, Wm. C. Brown Communications, Inc., Dubuque, Iowa.

Mastering Concepts

1. Why is thermoregulation important?
2. Distinguish between ectotherms and endotherms and between homeotherms and heterotherms.
3. What are some adaptations to extreme cold and extreme heat?

Osmoregulation

Different environments present different challenges to organisms not only in regulating body temperature but in maintaining ion concentrations in body fluids. **Osmoregulation** is the control of water and ion balance in an organism.

Osmoregulation is highly attuned to the environment. Recall that most cell membranes are permeable to small ions and highly permeable to water molecules. A fish in seawater must handle the influx of salts (ions) into its body and the loss of water to the environment. (Salts diffuse in and water out by following their concentration gradients.) A fish in a lake faces the opposite problem—it must conserve salts, yet rid its body of excess water.

Animals osmoregulate to different extents. Oceanic invertebrates expend little energy maintaining body fluid composition because they are **osmoconformers—**the ion concentrations in their body fluids are similar to those in the surrounding water. Because their body surfaces are permeable to water and salts, their fluid compositions would change with the composition of their surroundings. But they can survive because the ion concentrations in the ocean are fairly constant. Some crustaceans and marine worms are osmoconformers.

Invertebrates living in coastal environments, where salinity changes with the tide as fresh water enters from rivers, are **osmoregulators.** They more actively keep their ion concentrations constant. Crabs living in estuaries (where saltwater and fresh water meet), for example, have glands in their heads that pump out excess water that enters the body when the surrounding water becomes less salty. In the opposite situation, when too many salts exit the crab's body through the gills and in urine, specialized salt-secreting cells in the gills remove needed ions from incoming seawater and transport them to the blood.

Fishes are osmoregulators, but their specific adaptations reflect the relative salinity of the environment (fig. 38.8). Fishes in the ocean replace the water that their bodies lose by drinking seawater. The salts and water from the seawater enter the bloodstream from the digestive tract. The fishes then excrete excess salt in several ways: through secretory cells in the gills, in feces, and in urine. In contrast, freshwater fishes must retain salts and rid themselves of water. Their kidneys excrete excess water in dilute urine and retain valuable ions to return them to the bloodstream. Specialized salt-absorbing cells in the gills help replace lost ions.

Animals in terrestrial environments must conserve water. They obtain water by eating and drinking and as a by-product of metabolism; they lose water through evaporation from lungs and body surfaces, in feces, and in urine. Different animals use different combinations of strategies to obtain and conserve water. A human takes in most of its water in food and drink. A kangaroo rat, in contrast, derives most of its water from its metabolism (table 38.1).

Mastering Concepts

1. What is osmoregulation?
2. Distinguish between osmoconformers and osmoregulators.
3. How does osmoregulation differ between fish living in oceans and in fresh water?

Nitrogenous Waste Removal

Animals must rid their bodies of nitrogen-containing (nitrogenous) wastes that result when proteins and nucleic acids break down. Protein destruction generates three types of nitrogenous wastes: **ammonia, uric acid,** and **urea** (fig. 38.9). As a protein degrades, amino groups ($-NH_2$) are released from amino acids. Each amino group then picks up a hydrogen ion and becomes ammonia, NH_3. Excreting ammonia directly is energetically efficient because it requires only one chemical reaction. Because ammonia is very toxic, however, it must be excreted in a dilute solution.

Some freshwater animals take in water from the environment to dilute ammonia. Land-dwelling animals, which must conserve

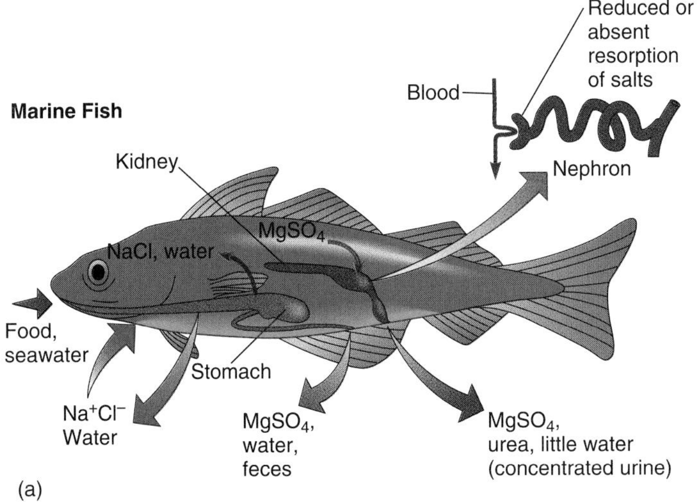

(a)

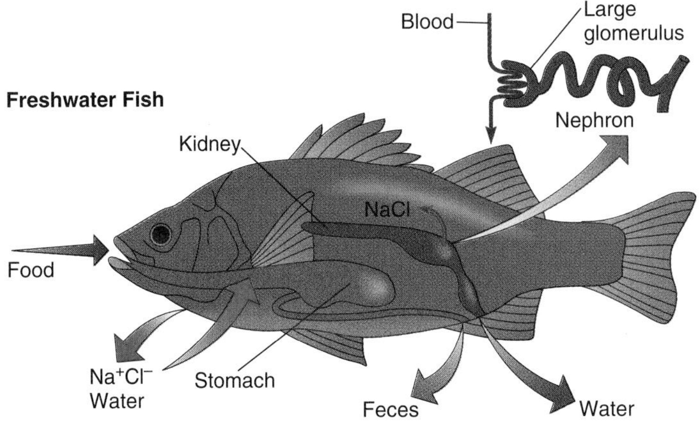

(b)

FIGURE 38.8

Fishes in the ocean and in freshwater lakes and streams face different challenges in regulating the ion concentrations in their body fluids. (*a*) The marine fish must rid its body of excess salts that diffuse in from the environment and that it drinks in seawater. Secretory cells in the gills actively pump out sodium ions (Na^+) and chloride ions (Cl^-). Magnesium sulfate ($MgSO_4$) that remains in the intestines is eliminated in feces and excreted from the kidneys in concentrated urine. (*b*) In contrast, the freshwater fish must conserve its body salts, which tend to diffuse out into the more dilute environment. Its kidneys shed excess water in dilute urine. Salt-absorbing cells in the gills actively absorb Na^+ and K^+. The kidneys have a more highly developed reabsorbing capacity, returning salts to the bloodstream. (Some of the terms in the figure are defined later in the chapter.)

water, use energy to convert ammonia to less toxic substances that can be stored and excreted in a relatively concentrated form. Adult amphibians and mammals convert ammonia to urea as their primary nitrogenous waste. Insects, reptiles, and birds convert most of their nitrogenous wastes to uric acid, which can be excreted in an almost solid form. In birds, uric acid mixes with undigested food in a storage organ called the cloaca to form a pasty substance called guano—the familiar "bird dropping" (fig. 38.10).

Humans produce very small amounts of uric acid. Unlike the uric acid of birds and reptiles, which comes from protein

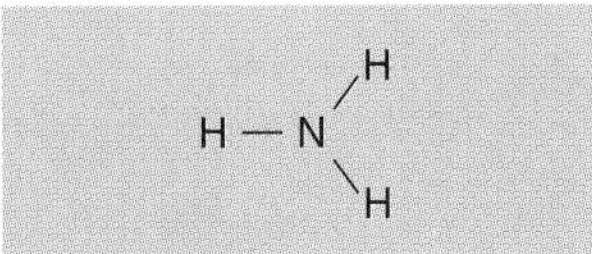

(a) Ammonia

$$H_2N-\overset{\overset{\large O}{\|}}{C}-NH_2$$

(b) Urea

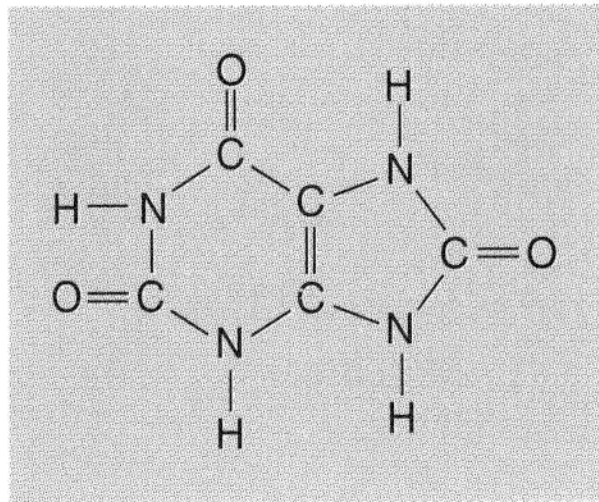

(c) Uric acid

FIGURE 38.9

Nitrogenous wastes. (*a*) Ammonia is a toxic nitrogenous waste. (*b*) Mammals and sharks excrete nitrogen-containing wastes primarily in the form of urea, a water-soluble compound that forms in the liver, travels in the bloodstream, and is excreted from the body in the urinary tract. (*c*) Birds, reptiles, and insects excrete most of their nitrogenous waste as uric acid.

FIGURE 38.10

The islands off the coast of South America are covered with guano, bird droppings rich in solid uric acid. Birds nest on the islands, building them up with their waste. Humans use guano deposits as fertilizer; guano from these islands was an economic commodity in the late 1800s.

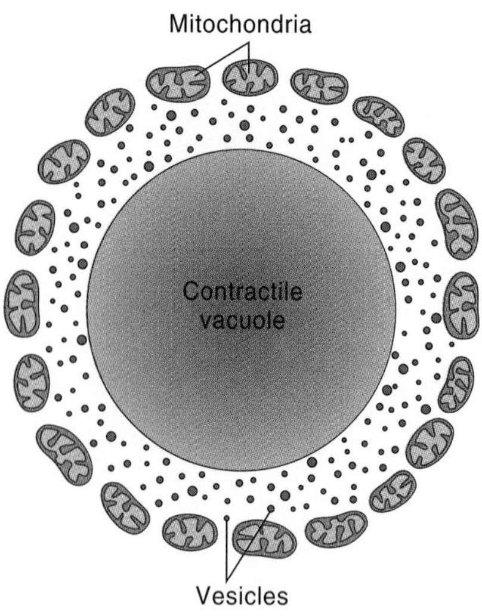

FIGURE 38.11

Freshwater protozoa and sponge cells have contractile vacuoles to get rid of excess water that enters to dilute their salty interiors. The vesicles surrounding the contractile vacuole fill with water and then empty into the vacuole. Mitochondria encircling the contractile vacuole probably provide the energy required for the vacuole to move to the cell membrane, contract, and expel water.

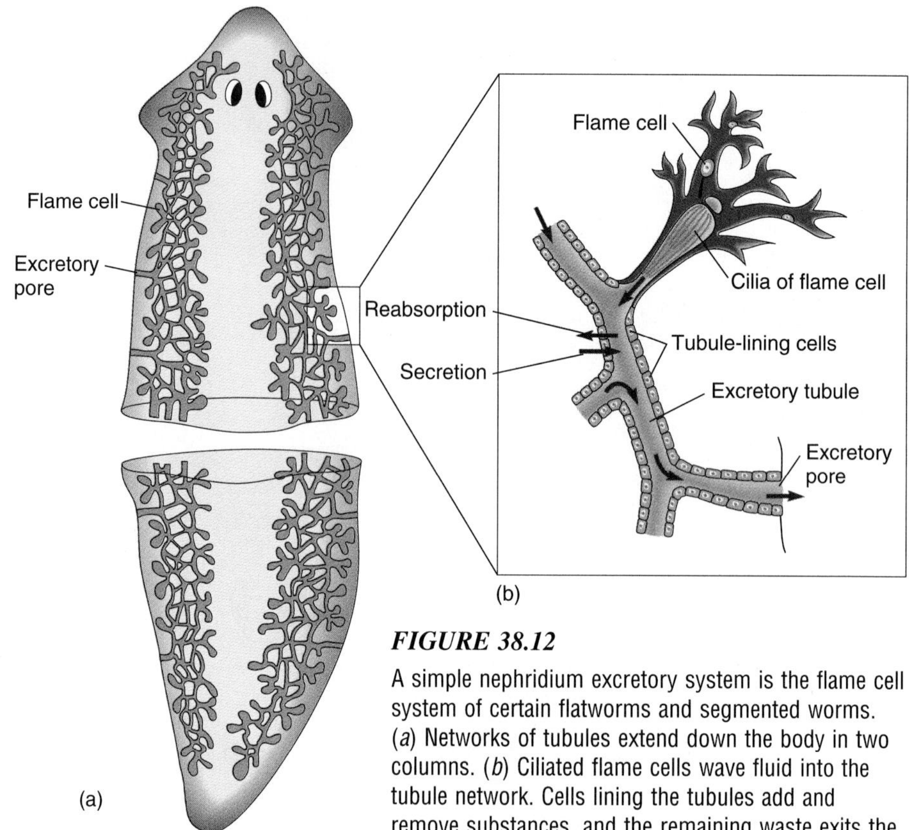

FIGURE 38.12

A simple nephridium excretory system is the flame cell system of certain flatworms and segmented worms. (*a*) Networks of tubules extend down the body in two columns. (*b*) Ciliated flame cells wave fluid into the tubule network. Cells lining the tubules add and remove substances, and the remaining waste exits the body through excretory pores.

breakdown, human uric acid comes from breakdown of DNA and RNA purine bases. In the human metabolic disorder called gout, excess uric acid crystallizes in the joints, causing pain.

Animals and certain protozoa have several types of structures and organs that remove nitrogenous wastes. Figure 5.8 illustrates how paramecia use contractile vacuoles to pump out excess water, and figure 38.11 shows a closer view of how vesicles fill the vacuole. Freshwater sponges also have contractile vacuoles. Toxins and wastes diffuse out of these cells across the cell membranes. Marine protozoa don't require contractile vacuoles because their solutes are at the same concentrations as those in the seawater, so they don't gain excess water.

Nephridia

Many invertebrates have networks of tubules called **nephridia** that form excretory systems. A simple type of nephridium excretory system is the planarian (flatworm) **flame cell system,** which consists of two branched tubules that extend the length of the body (fig. 38.12). Short side branches called flame cells (because their extensions resemble leaping flames) project into the body cavity. The cilia on the flame cells wave fluid into the tubules, generating a negative pressure that draws more fluid into the tubules. Cells lining the tubules add some substances **(secretion)** and remove others **(reabsorption).** The tubules feed into pores that open onto the body surface. Flame cell systems reach most body cells because planarian circulatory systems are not developed enough to effectively deliver wastes to the excretory system. Nitrogenous wastes diffuse directly out the surface of the body through the excretory pores.

Some mollusks, segmented worms (annelids), and marine worms have a more complex nephridium excretory system. The earthworm has two individual nephridia in most of its segments (fig. 38.13). A single earthworm nephridium consists of a ciliated opening called a **nephrostome,** which leads to a coiled tubule surrounded by capillaries. The close association between an excretory tubule and a blood supply enables the circulation to reabsorb valuable biochemicals as urine forms. Urine moves to the bladder for storage and then leaves the earthworm's body at an opening called a **nephridiopore,** located in the adjacent segment.

Insects' and spiders' excretory systems consist of **Malpighian tubules** (fig. 38.14). Ions, especially potassium (K^+), are actively secreted into the tubules. When K^+ exits, it draws water as well as nitrogenous wastes with it. The Malpighian tubules join the intestine, where rectal glands reabsorb most of the K^+ and water. The nitrogenous waste leaves the body as uric acid.

Mastering Concepts

1. Where do nitrogenous wastes originate in animal bodies?
2. How do animals detoxify ammonia?
3. Describe three types of structures that animals use to rid their bodies of nitrogenous wastes.

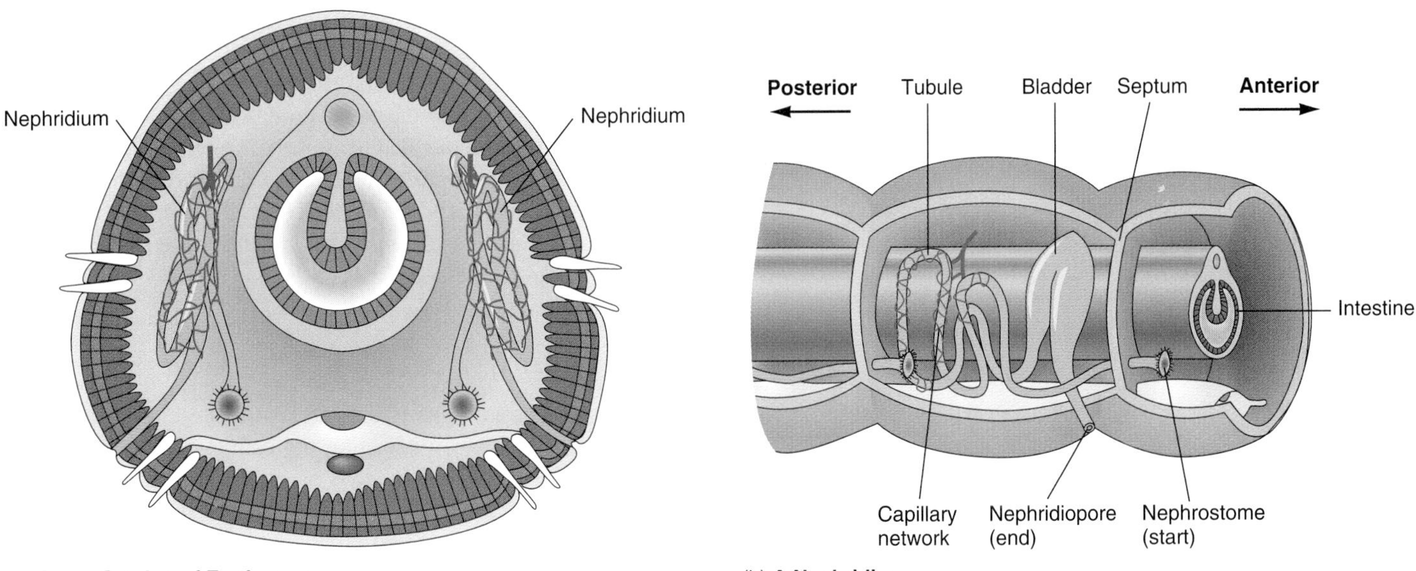

(a) **Cross Section of Earthworm** (b) **A Nephridium**

FIGURE 38.13

Earthworms excrete waste through paired nephridia in nearly all body segments. A nephridium begins at an opening, the nephrostome and then continues as a coiled tubule surrounded by capillaries. A bladder stores waste. The system opens to the outside at a nephridiopore on an adjacent segment.

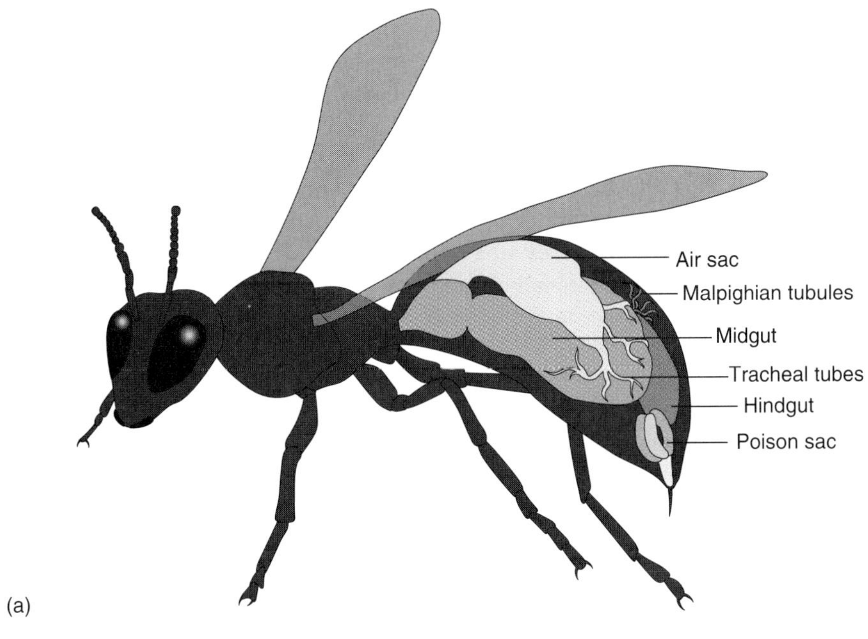

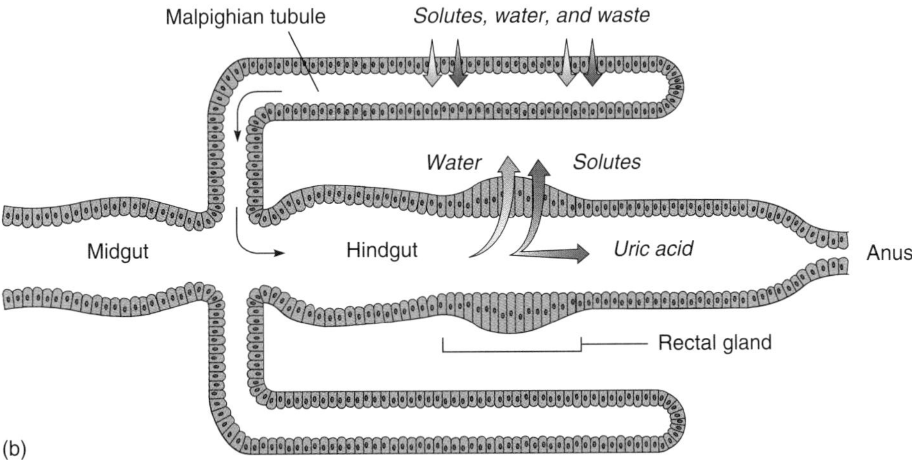

FIGURE 38.14

Excretion in insects. (*a*) Insect excretory organs, called Malpighian tubules, join the digestive tract. (*b*) Solutes, including sodium (Na^+) and potassium (K^+), are actively transported into the Malpighian tubules, and water, amino acids, sugars, and nitrogenous waste (uric acid) follow passively. Useful substances are reabsorbed in the hindgut and the bases of the Malpighian tubules. Uric acid leaves the body through the anus.

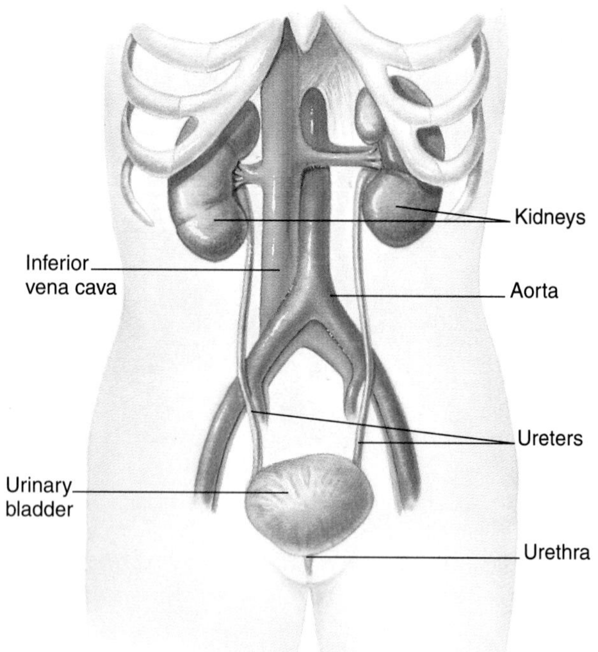

FIGURE 38.15

The human urinary system includes the kidneys, ureters, urinary bladder, and urethra.

The Human Urinary System—An Overview

The major excretory structures of vertebrates are **kidneys,** paired organs, located along the dorsal body wall. They conserve water and essential nutrients and remove nitrogenous wastes. While the organizations of tubules and ducts that remove wastes differ among vertebrate groups, all kidneys share functional units called **nephrons.** We consider the human urinary system to illustrate the vertebrate's approach to waste removal.

The human kidneys are the major organs responsible for excretion and osmoregulation (fig. 38.15), although the lungs excrete carbon dioxide. Each kidney, located against the wall of the lower back within the abdominal cavity, is about the size of an adult fist and weighs about half a pound. Urine forms within each kidney and drains into a muscular tube about 11 inches (28 centimeters) long called a **ureter.** Waves of muscle contraction squeeze urine along the ureters and squirt it into a saclike, muscular **urinary bladder.** A flap of tissue at the junction where each ureter meets the bladder keeps urine flowing in one direction—out. Urine drains from the bladder and exits the body through the **urethra.** In females, this tube is about 1 inch (2 to 3 centimeters) long and opens between the clitoris and vagina. In males, the urethra is about 8 inches (20 centimeters) long and extends the length of the penis.

Two rings of muscle (sphincters) guard the exit from the bladder. Both sphincters must relax before urine can leave the bladder. A spinal reflex involuntarily controls the innermost sphincter. Most people learn to consciously control relaxation of the outer sphincter at about 2 years of age.

The adult bladder can hold about 20 ounces (600 milliliters) of urine; however, as little as 10 ounces (300 milliliters) of urine accumulating stimulates stretch receptors in the bladder. The receptors send impulses to the spinal cord, which stimulates sensory neurons that contract the bladder muscles, generating a strong urge to urinate. We can suppress the urge to urinate for a short time by contracting the external sphincter. Eventually the cerebral cortex directs the sphincters to relax, and bladder muscle contractions force urine out of the body.

The Kidney

The kidneys form an incredibly efficient blood-cleansing mechanism by simultaneously adjusting the composition, pH, and volume of the blood. As the kidneys process blood, urine forms.

Each kidney is packed with 1.3 million microscopic tubular structures, the nephrons. An individual nephron consists of a tuft of filtering capillaries (the **glomerulus**), a continuous **renal tubule** ("renal" means "of the kidney") plus the **peritubular capillaries** that entwine around it. A stretched-out nephron would be about a half inch (approximately 12 millimeters) long.

The same regions of each nephron lie at similar positions within the kidney (fig. 38.16). The outermost part of the kidney, the **renal cortex,** contains the glomeruli and is grainy in appearance. The middle section, the **renal medulla,** resembles a collection of aligned strings. This region corresponds to the long sections of the renal tubules. The innermost portion of the kidney, the **renal pelvis,** collects the urine each nephron produces before the urine leaves the kidney to enter a ureter.

The kidney illustrates a familiar biological organization: extensive surface area (afforded by the renal tubules) packed into a relatively small volume (the kidneys). Specific portions of the nephrons carry out specific functions.

The body's entire blood supply courses through the kidney's blood vessels every 5 minutes. At that rate, the equivalent of 425 to 525 gallons (1,600 to 2,000 liters) of blood passes through the kidneys each day. Yet a person excretes only about 0.4 gallons (1.5 liters) of urine daily. Most of the blood components that the nephrons process are reabsorbed into the blood rather than excreted in urine.

The kidney selectively retains and recycles important dissolved chemicals and water through three processes. First, it filters wastes, nutrients, and water from the blood into the renal tubule, leaving large structures such as proteins and blood cells in the blood. Second, the kidney reabsorbs salts, water, and nutrients into the peritubular capillaries. Finally, the tubules transport toxic substances out of the peritubular capillaries for excretion. Overall, the nephrons extract wastes and recycle valuable nutrients, ions, and water to the bloodstream.

Mastering Concepts

1. Distinguish between a kidney and a nephron.
2. What are the components of the human urinary system?
3. What are the components of a nephron?
4. What are the regions of a kidney?

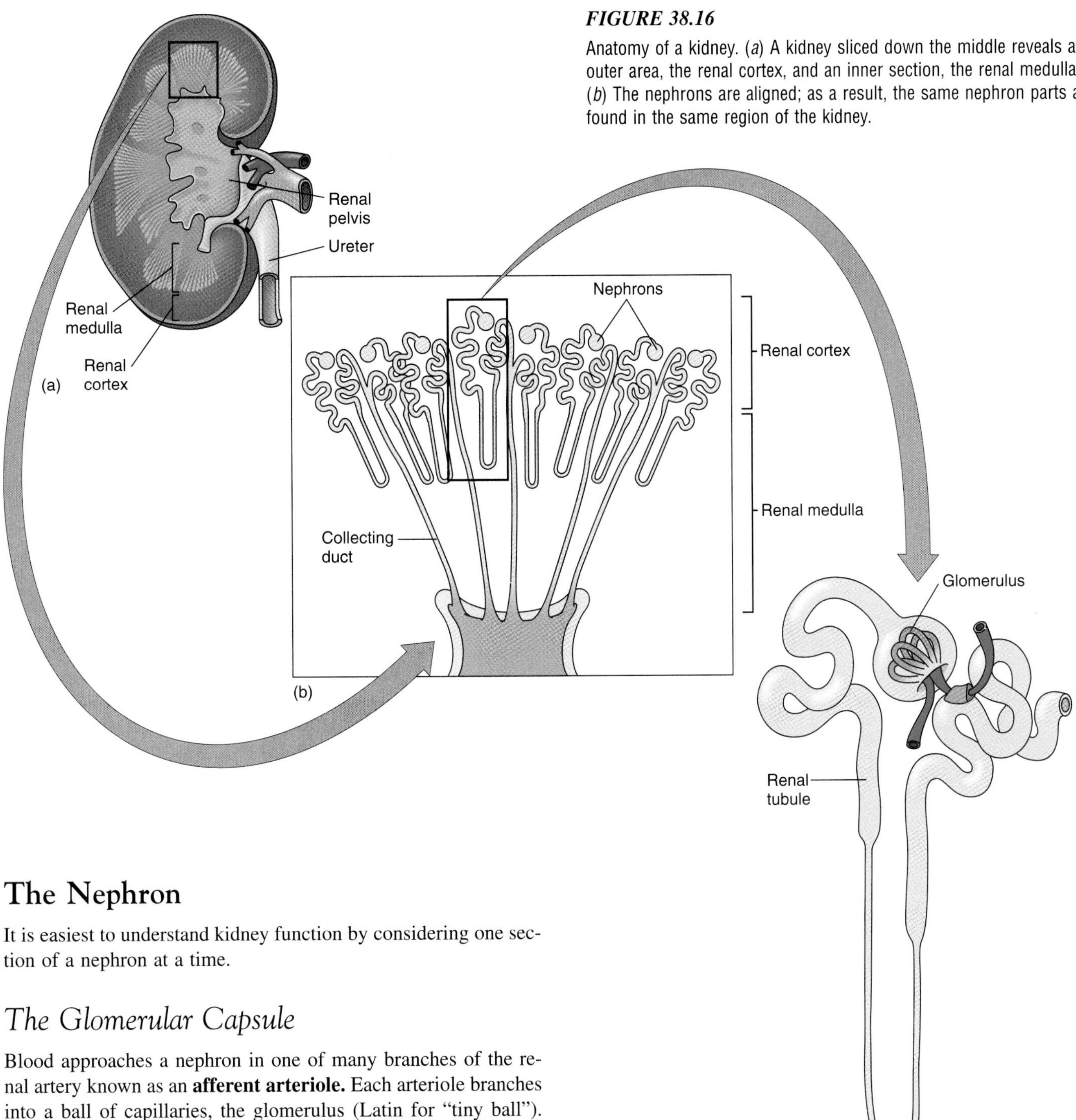

FIGURE 38.16

Anatomy of a kidney. (*a*) A kidney sliced down the middle reveals an outer area, the renal cortex, and an inner section, the renal medulla. (*b*) The nephrons are aligned; as a result, the same nephron parts are found in the same region of the kidney.

The Nephron

It is easiest to understand kidney function by considering one section of a nephron at a time.

The Glomerular Capsule

Blood approaches a nephron in one of many branches of the renal artery known as an **afferent arteriole.** Each arteriole branches into a ball of capillaries, the glomerulus (Latin for "tiny ball"). These capillaries then join again to form an **efferent arteriole.** The glomerulus is surrounded by a cup-shaped structure at the end of the renal tubule known as the **glomerular capsule** (fig. 38.17). Glomerular capsules, one at the beginning of each tubule, lie in the cortex of the kidney.

The glomerular capillaries filter blood, and anything that fits through the pores in the glomerulus and is not repelled by the charge on the membrane passes into the glomerular capsule. Like a catcher's mitt catching a ball, the glomerular capsule surrounds the glomerulus and captures all of this material.

Blood pressure provides the force that drives substances out of the glomerulus. Plus, the afferent arteriole has a larger diameter than the efferent arteriole. The resulting pressure forces fluid and small dissolved molecules across the capillary walls. Large molecules such as plasma proteins, blood cells, and platelets remain in the bloodstream and leave the glomerulus in the efferent arteriole, which ultimately leads into the capillary network surrounding the nephron. These capillaries then empty into a venule, which runs into the renal vein, which finally joins the inferior vena cava.

The material in the glomerular capsule is chemically similar to blood plasma. Because only some substances from the blood

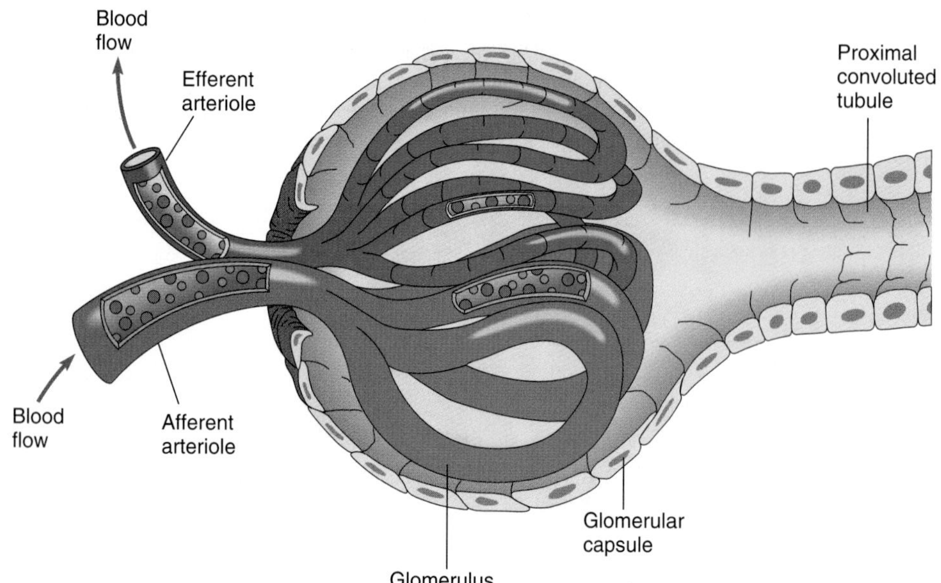

FIGURE 38.17

The glomerulus and glomerular capsule. Small molecules dissolved in blood plasma cross the capillary walls leaving some plasma, larger proteins, blood cells, and platelets in the bloodstream to exit through the efferent arteriole.

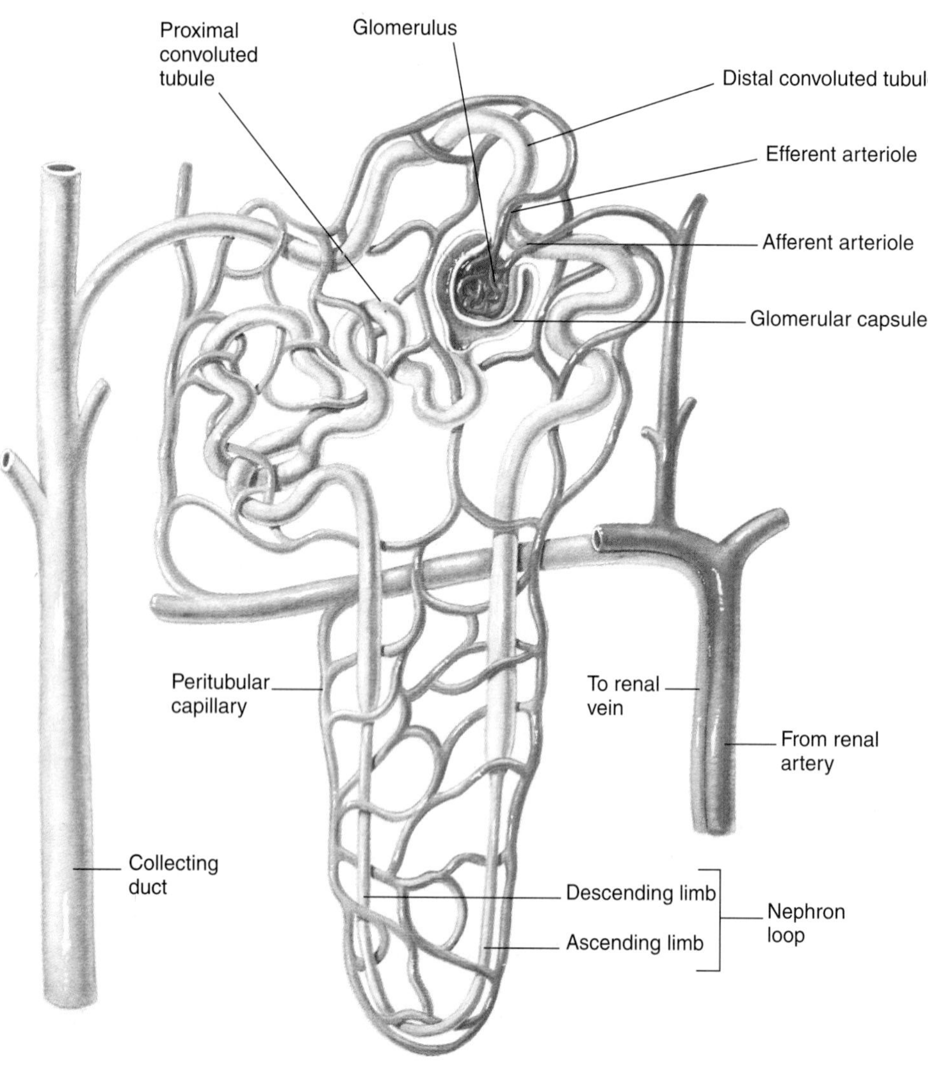

FIGURE 38.18

The nephron. The nephron is tan; the peritubular capillaries are shown in blue, purple, and red.

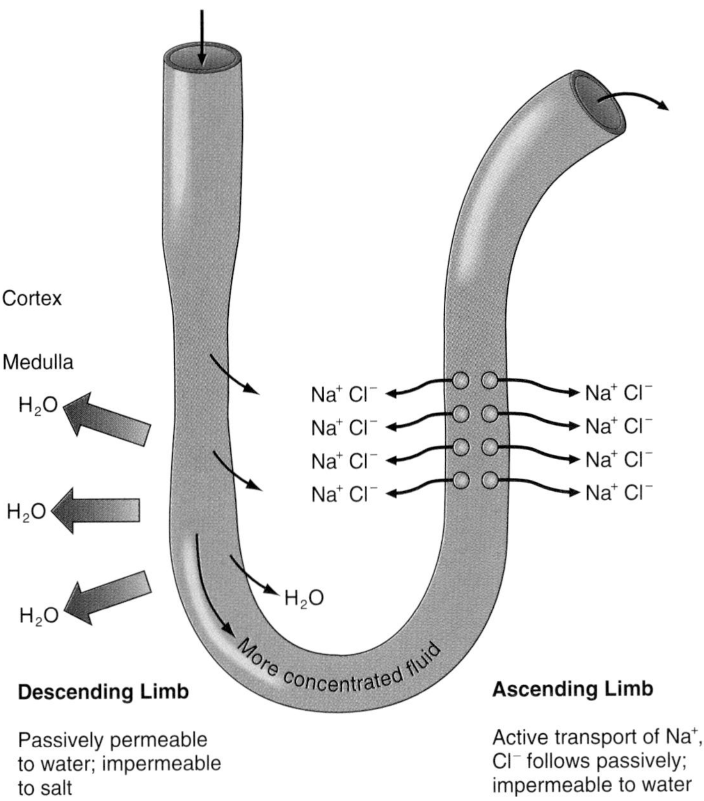

FIGURE 38.19

The nephron loop conserves water. The active transport of Na^+ from the ascending limb causes water to diffuse out of the loop and eventually enter the surrounding capillaries.

enter the nephron here, however, the material passing into the glomerular capsule is called the **glomerular filtrate.** This filtrate is the product of the first step in urine manufacture. The 425 to 525 gallons (1,600 to 2,000 liters) of blood per day that pass through the kidneys produce approximately 45 gallons (180 liters) of glomerular filtrate.

The remainder of the renal tubule consists of a winding passageway with four functional regions: the **proximal convoluted tubule,** the **nephron loop,** the **distal convoluted tubule,** and the **collecting duct** (fig. 38.18).

The Proximal Convoluted Tubule

The glomerular filtrate passes from the glomerular capsule into the proximal convoluted tubule. This is an important site for selective reabsorption, a process that returns useful components of the glomerular filtrate to the blood. The wall of the proximal convoluted tubule is highly folded, which increases its surface area, and its cells have many mitochondria to power the active transport of dissolved nutrient molecules back into the blood. All along the tubule, specialized cells actively reabsorb all the glucose and vitamins and about 75% of the amino acids. These cells also reabsorb important ions (electrolytes). Some ions are actively transported, and others tag along passively, attracted by the electrical charge on the transported ion. Water follows the electrolytes by osmosis and is also reabsorbed.

The Nephron Loop

After the proximal convoluted tubule, the renal tubule of a mammal forms the nephron loop, which conserves water and concentrates urine. The loop consists of a **descending limb** and then an **ascending limb.** The loop dips into the medulla region of the kidney. A special mechanism, described next, causes the bottom of the loop to sit in a salty brew, which forces water to diffuse out of the renal tubule and into the surrounding capillaries (fig. 38.19).

The cells that form the ascending limb are impermeable to water, and they actively transport sodium ions (Na^+) into the space in the renal medulla not occupied by nephrons (some chloride and potassium ions are transported as well, but they are omitted here for simplicity). Because of the Na^+ movement, the medulla builds up a higher concentration of Na^+ than the interior of the descending limb. The cells of the descending limb are impermeable to ions but permeable to water, so water passively diffuses into the medulla, following its concentration gradient. The descending limb thus conserves water by sending it into the renal medulla, where it can diffuse into the peritubular capillaries and back into the circulation.

As water leaves the descending limb, the Na^+ concentration inside the tubule rises until it reaches osmotic equilibrium with the outside. This concentrated tubular fluid moves along the loop until it reaches the ascending limb. The cycle continues as the ascending limb transports a now even greater amount of Na^+. Each time this circuit is completed, the concentration of Na^+ in the descending limb is increased, or multiplied. For this reason, the movement of Na^+ and water between the nephron loop limbs and the medullary space is called a **countercurrent multiplier system.** *Countercurrent* refers to the opposite directions the fluid flows in the two limbs (down and then up). A portion of the peritubular capillaries called the **vasa recta** closely parallels the loop and allows the descending limb to reach equilibrium with the medullary fluid.

The nephron loop is an adaptation of the mammalian nephron to conserve water. The loop's relative length differs among mammalian species. The longer the loop, the more water is reabsorbed into the bloodstream and the more concentrated the urine. The beaver and the Australian hopping mouse are at two extremes. The beaver lives surrounded by water and so does not have to conserve it. This animal has a short nephron loop and excretes a watery urine only up to twice as concentrated as its blood plasma. The Australian hopping mouse, in contrast, is a desert dweller with the longest mammalian nephron loop relative to its body size. Its urine is up to 22 times as concentrated as its blood plasma. In comparison, human urine is up to 4.2 times as concentrated as blood plasma.

The Distal Convoluted Tubule

Because of active transport in the ascending limb, once the glomerular filtrate passes through the nephron loop, it is considerably more dilute than it was at the tip of the loop. In the next region of the renal tubule, the distal convoluted tubule in the kidney's cortex, sodium is actively reabsorbed into the peritubular capillaries. The steroid hormone aldosterone stimulates Na^+ reabsorption in the distal tubule. The accumulation of Na^+ outside the distal convoluted tubule stimulates water to diffuse passively out of the tubule into the capillaries.

Most of the potassium ions (K^+) in the glomerular filtrate are actively reabsorbed in the proximal convoluted tubule. Some of the remaining K^+ is secreted in the distal convoluted tubule and collecting duct. K^+ is attracted to negatively charged regions of the tubule that form as Na^+ leaves. Aldosterone also stimulates K^+ secretion, almost as if it were swapping one ion for the other.

The distal convoluted tubule also helps to maintain the pH of blood between 6 and 8. Any variance from this range is deadly. Blood is more likely to become too acidic than too alkaline, largely because most metabolic wastes are acids. The distal convoluted tubules can raise a too-low blood pH by secreting H^+ into the urine. It can lower a too-high blood pH by inhibiting H^+ secretion in the tubules.

The distal convoluted tubule also secretes waste molecules such as creatinine (a by-product of muscle contraction) and drugs such as penicillin.

The Collecting Duct

The fourth portion of the renal tubule, the collecting duct, descends into the medulla, as does the nephron loop. Several renal tubules drain into a common collecting duct. Some of the urea diffuses out of the collecting duct as it passes through the medulla, where it contributes to the high solute concentration surrounding the nephron loops.

After reabsorption and secretion, the filtrate is urine. From the collecting duct, urine accumulates in the renal pelvis before peristalsis carries it through the ureter to the urinary bladder and it finally moves out of the body through the urethra. Table 38.2 summarizes the functions of each part of the nephron.

Mastering Concepts

1. How does urine formation begin?
2. What occurs in the proximal convoluted tubule?
3. What is the function of the countercurrent multiplier in the nephron loop?
4. What happens in the distal convoluted tubule?
5. How does urine exit the body?

Table 38.2 Summary of Functions of Each Part of the Renal Tubule

Part	Site in Kidney	Function
Glomerular capsule	Cortex	Filtration
Proximal convoluted tubule	Cortex	Selective reabsorption of water, amino acids, electrolytes
Nephron loop	Medulla	Water conservation; urine concentration
Distal convoluted tubule	Cortex	Secretion of large molecules
Collecting duct	Medulla	Further reabsorption of water

Control of Kidney Function

The maintenance of blood pH, volume, and composition within an optimum range is essential to the functioning of each cell. Since the kidney plays a key role in regulating these factors, its activities are continuously monitored and adjusted to meet the body's requirements.

The amount of water reabsorbed from the filtrate influences two important characteristics of blood: the concentrations of plasma solutes (osmotic pressure) and volume. Osmotic pressure affects many cellular activities, particularly the exchange of materials between cells and blood. Blood volume influences blood pressure and, therefore, affects cardiovascular health.

The solute concentration of the blood remains constant despite variations in the amount of water we consume in food or drink. When we drink too much, our kidneys allow more fluid to pass into the urine. If water is scarce, however, our kidneys conserve it by producing concentrated urine.

Osmoreceptor cells within the hypothalamus determine how much water to retain (see fig. 5.10). When blood plasma becomes too concentrated, increasing osmotic pressure, the osmoreceptor cells send impulses to the posterior pituitary gland in the brain, which then secretes antidiuretic hormone (ADH). ADH increases water permeability in the distal convoluted tubule and the collecting duct. Because of the concentration gradient created by the countercurrent multiplier, increased water permeability increases water reabsorption into the blood.

Conversely, if blood plasma is too dilute—if the osmotic pressure is too low—the same cells in the hypothalamus signal the posterior pituitary gland to stop producing ADH. As a result, the distal convoluted tubule and collecting ducts become less permeable to water. More water is excreted in the urine, and the osmotic pressure of the plasma and tissue fluids returns to normal. In a disease called diabetes insipidus, ADH activity is insufficient, and the person urinates from 1 to 2 gallons (4 to 8 liters) per day. Thirst is intense, yet water consumed rapidly leaves through the kidneys.

38.2 biology in action

Urinalysis—Clues to Health

Urine has long been a part of many folk remedies. It has been used as a mouthwash, a toothache treatment, and a cure for sore eyes.

Hippocrates (460–377 B.C.) was the first to observe that the condition of the urine can reflect health, noting that frothy urine denoted kidney disease. During the Middle Ages, health practitioners frequently consulted charts that matched urine colors to certain diseases. In the seventeenth century, British physicians diagnosed diabetes mellitus by having medical students taste sugar in a patient's urine!

Today, urine composition is still used as an indicator of health. The urine of a healthy individual contains water, urea, creatinine, uric acid, ammonia, amino acids, and several salts, in particular proportions. Urine should be pale yellow, with a pH of 4.8 to 8.0. These characteristics may fluctuate slightly due to diet. In some situations, urine is tested for traces of biochemicals from illegal drugs.

What can urine reveal about health? Blood in the urine indicates bleeding in the urinary tract. More than a trace of glucose may be a sign of diabetes mellitus. Carrier molecules actively transport glucose out of the filtrate in the kidney tubules and into the blood. If there are more glucose molecules than carrier molecules can handle, due to deficient insulin, the excess glucose remains in the filtrate and appears in the urine. Excess glucose in the blood may also indicate a high-carbohydrate diet or stress. Stress causes secretion of excess adrenaline, which stimulates the liver to break down more glycogen into glucose.

Albumin in the urine may be a sign of damaged nephrons, because this plasma protein is too large to normally fit through the pores at the entrance to the kidneys' microscopic tubules. Pus and a complete absence of glucose in the urine indicates infection somewhere in the urinary tract. The pus consists of white blood cells, which fight infection, and the infecting bacteria, which consume all glucose.

The ethyl alcohol in alcoholic beverages is a diuretic; this means it increases the volume of urine. Alcohol stimulates urine production by decreasing ADH secretion, thereby increasing the permeability of the tubules to water. Because it increases water loss to urine, an alcoholic beverage actually intensifies thirst. Dehydration resulting from drinking too much alcohol causes the discomfort of a hangover. Caffeine is also a diuretic, exerting its effects by decreasing Na^+ (and therefore water) reabsorption from the renal tubule. Similarly, many diuretic drugs (water pills) increase urine volume by decreasing reabsorption of Na^+ in the proximal tubules. Diuretics lower blood pressure and relieve edema (tissue swelling).

The adrenal hormone aldosterone enhances Na^+ reabsorption in the distal convoluted tubules as well as in the salivary glands, sweat glands, and large intestine. When the sodium level of the blood falls, or when blood pressure or volume declines, aldosterone synthesis increases and more Na^+ enters the bloodstream. Groups of specialized cells in the afferent arterioles also sense decreased blood volume and pressure. They respond by releasing another hormone, **renin,** which initiates a series of chemical reactions that eventually boost aldosterone levels.

Mastering Concepts

1. Why must kidney function be constantly monitored?
2. How does the solute concentration of the blood remain constant, despite fluctuations in water intake?
3. How do aldosterone and renin control blood volume and sodium level?

The Unhealthy Urinary System

Bedwetting

Nighttime bedwetting is an age-old problem, noted as long ago as 1500 B.C. Treatments have ranged from drinking the broth from boiled hens combs, to blocking the urethra at night, to punishment and ridicule based on the erroneous assumption that urinating while asleep is bad behavior. We now know that a common form of bedwetting is inherited. Nighttime bedwetting, technically called nocturnal enuresis, affects 10% of all 7-year-olds. In the inherited form, a person has never been able to remain dry at night. Fortunately, the problem usually clears up with time.

Urinary Tract Infections

Urinary tract infections (UTI) are a common health problem. UTIs most often affect women because the urethra's location near the vaginal and anal openings provides an open pathway for bacteria to enter the urinary system. Older men sometimes develop a UTI when the prostate gland swells and blocks urination. A UTI produces frequent, painful urination, fever, and lower abdominal pain. Typically, the urine is cloudy and has a foul odor. Analysis of a urine sample can reveal infection or other health conditions (Biology in Action 38.2). Antibiotic drugs prevent a UTI from spreading to the kidneys.

Kidney Stones

Kidney stones form when calcium salts or uric acid precipitate out of newly formed urine and accumulate in the kidney tubules and renal pelvis. Kidney stones can be caused by excessive milk consumption, infection, dehydration, hormonal problems, or inherited metabolic disorders.

Kidney stones cause fever, chills, frequent urination, and kidney pain. If the stones pass through the ureters, they can cause excruciating pain. Treatment for large stones has traditionally been surgery, but today the most common treatment is extracorporeal shock wave lithotripsy. In this less invasive procedure, shock waves are focused on the stone and passed through a water cushion held to the patient's body. The waves shatter the stones, and the much smaller fragments then pass in the urine.

Kidney Failure

Kidney failure results from damaged nephrons. Elevated blood pressure in the glomerulus causes kidney failure by straining the filtering capabilities of the glomerular capillaries. As this part of the nephron is overworked, it hardens and eventually cannot function. As individual nephrons progressively shut down, the remaining nephrons compensate for the loss, working harder. Gradually, these nephrons become overworked too. The cycle culminates with the failure of the entire kidney.

Malfunctioning kidneys cause rapid buildup of toxins in the blood, alter ion concentrations in the tissues, and retain water, causing potentially painful swelling. If both kidneys are failing, a kidney transplant can be lifesaving. Alternatively, about 20,000 people in the United States undergo a procedure called dialysis for a few hours twice a week. The patient's blood passes through a tube separated from a balanced salt solution by an artificial semipermeable membrane. Wastes and toxins diffuse into the salt solution, cleansing the blood.

Another form of dialysis uses the peritoneal membrane lining the body cavity, instead of an artificial membrane. The balanced salt solution is placed into the patient's abdominal cavity, and waste products and excess water diffuse from the blood into the fluid, which is suctioned out. A new treatment for kidney failure is atrial natriuretic peptide, a hormone that the heart secretes. It dilates the afferent arterioles and constricts the efferent arterioles, which increases glomerular filtration rate.

Mastering Concepts

1. What can cause nighttime bedwetting?
2. Why are UTIs more common in women?
3. How do kidney stones form?
4. What happens microscopically to cause kidney failure?

SUMMARY

1. Adaptations permit animals to regulate body temperatures, the amount of water in their bodies, and solute concentrations in body fluids, which enable them to populate diverse habitats.

2. **Ectotherms** regulate body temperature by seeking an environment with the appropriate temperature. **Endotherms** use internal metabolism to generate heat. **Homeotherms** maintain constant body temperature, and **heterotherms** have temperatures that fluctuate with their surroundings.

Hemolytic Uremic Syndrome~

continued from page 770.

Hemolytic uremic syndrome is a complication of exposure to Shiga toxin. The toxin is associated with several types of *E. coli,* because the genes encoding it are carried on a bacteriophage, which is a virus that passes easily among host bacterial cells. *E. coli* strain 0157:H7 was first associated with HUS in 1982. Since then, outbreaks have occurred in many nations, and the condition has been associated with more than 100 different bacterial strains.

Shiga toxin typically causes diarrhea, which is often bloody, but in only 6% of cases it produces bloody urine (uremia), a hallmark of kidney damage. HUS occurs almost exclusively in children. It is a leading cause of acute renal failure, killing 3 to 5% of affected children.

Understanding how the toxin destroys kidney lining cells illustrates how molecular and cellular events cause the illness. First, *E. coli* produces and releases Shiga toxin, which is a two-part protein. Toxin molecules bind to specific glycolipid receptors on cells forming glomeruli. The cell membranes form pits that engulf and draw the toxin into the cell. In the process, the two protein parts of the toxin molecules separate, with one type becoming activated. The activated toxin protein inactivates the subunits of ribosomes, grinding the cell's protein synthesis to a halt—which kills it.

With the glomeruli damaged, large molecules such as proteins that would normally be excluded from the filtrate, plus blood cells and even dead kidney cells, pass into urine. Platelets also aggregate at the damage site, initiating blood clotting and eventually healing the damage. Most of the time, new cells form and repair the site of the toxin's attack.

A clue to why children are more prone to develop HUS than adults came from rabbits. Rabbits also develop bloody diarrhea from Shiga toxin, but they do not suffer kidney damage because their kidney cells lack receptors for the toxin. Similarly, the kidney cells of human adults do not have toxin receptors, but those of children do.

3. Animals regulate their body temperatures metabolically, behaviorally, and with anatomical and physiological adaptations. Adaptations to cold include shunting blood to interior organs; **countercurrent exchange systems** of blood vessels; antifreeze-like chemicals in blood; heat-seeking behaviors; blubber, feathers, and oxidation of fatty acids in brown fat; shivering; and **nonshivering thermogenesis.** Heat-resisting adaptations include **evaporative cooling;** circulatory specializations such as the **carotid rete** and **cranial radiators;** body forms and behaviors.

4. Osmoregulation is the control of ion concentrations in body fluids. Specific osmoregulatory adaptations depend on habitat. **Osmoconformers** are aquatic organisms that have permeable body surfaces. Their body fluids have solute concentrations similar to those of the surrounding water. **Osmoregulators** include aquatic animals in coastal regions that conserve ions under some conditions and lose excess ions in others. Terrestrial animals have many adaptations to conserve water and regulate ion composition.

5. Animals must remove wastes. Most nitrogenous wastes come from protein breakdown and include **ammonia, urea,** and **uric acid.** Ammonia formation requires the least energy but releases the most water. Urea requires more energy but is less toxic and helps conserve water. Uric acid requires the most energy to produce but can be excreted in an almost solid form, saving water. Animals have diverse excretory systems to remove nitrogenous wastes, including contractile vacuoles, **nephridia, Malpighian tubules,** and **kidneys.**

6. The human urinary system excretes nitrogenous waste (mostly urea) and regulates water and electrolyte levels. The kidneys, each of which drains into a **ureter,** produce urine. Ureters drain urine into the **urinary bladder** for storage. Urine leaves the body through the **urethra.**

7. The functional unit of the kidney is the **nephron.** Blood is filtered from a tuft of capillaries (the **glomerulus**) into the **glomerular capsule.** As the resulting filtrate moves along the renal tubule, its composition is adjusted. Important materials are reabsorbed from the filtrate back to the blood and other substances are secreted into the urine. The adjustment begins in the **proximal convoluted tubule.** The filtrate moves into the **nephron loop** that dips into and then out of the center, or **medulla,** of the kidney. The nephron loop helps to concentrate urine. Water leaves the filtrate because an osmotic gradient forms in the fluid around the nephron loop by a **countercurrent multiplier** built by Na^+ and urea reabsorption. In the **distal convoluted tubule,** more reabsorption and secretion occur. The reabsorption of Na^+ helps control Na^+ balance. Secretion of H^+ helps regulate blood pH. The filtrate then moves to the **collecting duct** and finally to the central cavity of the kidney, the **renal pelvis,** before draining through the ureter.

8. Antidiuretic hormone (ADH) regulates the amount of water reabsorbed from the distal convoluted tubule. The posterior pituitary gland secretes ADH in response to signals from the hypothalamus, which senses the osmotic pressure of blood. ADH increases permeability of the distal convoluted tubule and the collecting duct so more water is reabsorbed and urine is concentrated. Aldosterone increases the kidneys' salt retention. The adrenal glands release aldosterone in response to either low sodium concentration in the plasma or low blood pressure. Some water reabsorption always follows Na^+ reabsorption.

9. Problems in the urinary system include nighttime bedwetting, urinary tract infections, kidney stones, and kidney failure. If the kidneys fail, a kidney transplant or dialysis becomes necessary.

TO REVIEW . . .

1. How does a homeothermic endotherm differ from a heterothermic ectotherm?

2. Cite three adaptations that help keep an animal's feet warm.

3. Explain how a carotid rete cools a mammal's brain.

4. How do humans and iguana lizards differ in body temperature regulation?

5. What are the three types of nitrogenous waste, and where does each come from?

6. Draw a nephron and label the parts. Indicate which regions of the renal tubule are specialized for each of the three processes that form urine.

7. How is urine concentrated?

8. What are some medical conditions that urinalysis might detect?

9. How do ADH and aldosterone control kidney function?

10. Why can't you live without at least one kidney unless you undergo dialysis?

TO THINK ABOUT . . .

1. Cicadas living in the tropics have many more sweat pores than cicadas in cooler climates. Suggest how these differences might have arisen.

2. Imagine you are adrift at sea. Why would you dehydrate more quickly if you drank seawater to quench your thirst instead of fresh water?

3. An infant suffers from severe and frequent vomiting and diarrhea for 24 hours. When her father takes her to the doctor, he is surprised when the doctor hospitalizes the child and orders intravenous fluids. Why does the doctor take such drastic action?

4. Urinary tract infections frequently accompany sexually transmitted diseases, particularly in women. Why?

5. Would an excess or deficiency of renin be likely to cause hypertension (high blood pressure)? Cite a reason for your answer.

6. Why is protein in the urine a sign of kidney damage? What structures in the kidney are probably affected?

7. How could very low blood pressure impair kidney function?

TO LEARN MORE . . .

References and Resources

Baker, Mary Ann. August 1993. A wonderful safety net for mammals. *Natural History.* The carotid rete keeps many a mammal's brain cool.

Cloudsley-Thompson, John. August 1993. When the going gets hot, the tortoise gets frothy. *Natural History.* Tortoises use evaporative cooling to maintain body temperature.

Falk, Dean. August 1993. A good brain is hard to cool. *Natural History.* Cranial radiators enable our brains to survive high temperatures.

Humes, H. David. March 20, 1997. Acute renal failure—the promise of new therapies. *The New England Journal of Medicine,* vol. 336. Cell implants may be useful in treating kidney failure.

Lewis, Ricki. May/June 1997. Kidney donation becomes easier. *Biophotonics International.* A new surgical procedure eases kidney donation.

Lewis, Ricki. April 15, 1997. Varied bacterial toxins continue to challenge the developers of diagnostic tests. *Genetic Engineering News,* vol. 17. Strains of *E. coli* produce a toxin that damages childrens' kidneys.

Lewis, Ricki. April 1994. Interleukin-2: New therapy for kidney cancer. *FDA Consumer.* Borrowing a biochemical from the immune system offers another chance for some people who have kidney cancer.

Marsh, Alan C. August 1993. Ants that are not too hot to trot. *Natural History.* Namib ants scuttle across desert sand by pausing briefly on blades of grass.

Ronald, Allan. August 15, 1996. Sex and urinary tract infections. *The New England Journal of Medicine,* vol. 335. Sexual activity in women increases the risk of contracting a UTI.

Rondeau, Eric, and Marie-Noelle Peraldi. August 29, 1996. *Escherichia coli* and the hemolytic-uremic syndrome. *The New England Journal of Medicine,* vol. 335. Toxins produced by certain strains of *E. coli* can damage childrens' kidneys.

Toolson, Eric C. August 1993. In the Sonoran desert, cicadas court, mate, and waste water. *Natural History.* Zoologists were surprised to discover that an insect sweats.

Williams, Rebecca D. July/August 1996. Preventing dehydration in children. *FDA Consumer.* Childhood diarrhea is a serious illness.

Yagil, Reuven. August 1993. From its blood to its hump, the camel adapts to the desert. *Natural History.* The camel is well adapted to life in dry heat.

denotes articles by author Ricki Lewis that may be accessed via *Life*'s home page /http://www.mhhe.com/sciencemath/biology/life/.

AIDS destroys the human immune system. The AIDS quilt is a tribute to people who have died of the disease. Each square represents a life lost.

39 The Immune System

chapter outline

The Immune System Protects 790

Nonspecific Defenses 790
Phagocytosis
Inflammation
Antimicrobial Substances
Fever

Specific Defenses 792
Immune System Characteristics
Steps in the Immune Response

Development of the Human Immune System 798

Immunity Breakdown 798
AIDS
Severe Combined Immune Deficiency
Chronic Fatigue Syndrome

Autoimmunity and Allergy 800
Autoimmunity—Attacking Self
Allergies—Attacking the Innocuous

Altering Immune Function 802
Vaccines—Augmenting Immunity
Transplants—Suppressing Immunity
Immunotherapy

Dying Lions

At first, the scene looked normal to the tourists flying in a hot air balloon over the Serengeti Plain of Africa. The plains were dotted with grazing and moving herds of wildebeests and antelope, with smaller groups of elephants and giraffes. Then the tourists saw what they had been looking for—a lion. But something was terribly wrong. The great cat staggered about as if intoxicated, jerking its limbs and struggling to keep its balance.

Alarmed, the balloonist radioed the chief government veterinarian at Tanzania National Parks. When the veterinarian arrived, the shaking lion tried to stand but collapsed into a heap. The veterinarian observed the lion for many hours. The animal shook with seizures, then lay exhausted, and then shook again. The lion died a day later.

That was February 1994. By March, 19 more sick lions were spotted. By May, a third of the 250 lions that researchers track in the southeast Serengeti were dead, and many of those surviving had telltale twitches in their faces, signs of nerve damage. By year's end, nearly 3,000 lions were dead of this rapidly spreading illness.

Then, a graduate student recognized what was happening.

to be continued . . .

The Immune System Protects

Picture the human body as a castle surrounded by a moat and patrolled inside by an army. The moat is the skin and the mucous membranes are a physical-chemical barrier against the outside environment (see fig. 30.5). Backing up the skin are secretions in and around natural openings (such as the nose and mouth) that also keep microorganisms out of the body. The internal army is an enormous contingent of cells and biochemicals that distinguish "self" antigens from foreign or "nonself" antigens (see chapter 5).

An antigen can be any molecule that elicits an immune response in another individual. An antigen is usually a carbohydrate or a protein, and it can be as small as a few amino acids linked on the surface of a microorganism.

All animals have molecules and/or cells that enable them to detect nonself antigens—this seems to be a prerequisite for successfully sharing the planet with other organisms. The first step in mounting a defense is recognizing the enemy. Single-celled eukaryotes may recognize patterns of molecules on the surfaces of their cells or on cell surfaces of others of their species. Similarly, animals can distinguish self from nonself. In simpler invertebrates, such as sponges, corals, anemones, and some flatworms, amebocytes move through the body, engulfing, or phagocytizing, foreign tissue. In animals with circulatory systems, similar cells patrol the tissues and the bloodstream. These phagocytes recognize surface molecules on foreign cells and then engulf them, ultimately destroying them in lysosomes (organelles containing digestive enzymes) (fig. 39.1). Vertebrates have complex immune systems that include **antibody** proteins that provide specific defenses and T cells and their products. Plants can also distinguish self from nonself.

Phagocytes carry out a generalized, nonspecific immune response; that is, a phagocyte isn't choosy about what it engulfs, as long as the pattern of surface molecules on the target does not signal "self." Other parts of vertebrate immune systems are highly specific. The remainder of the chapter focuses on the human immune system.

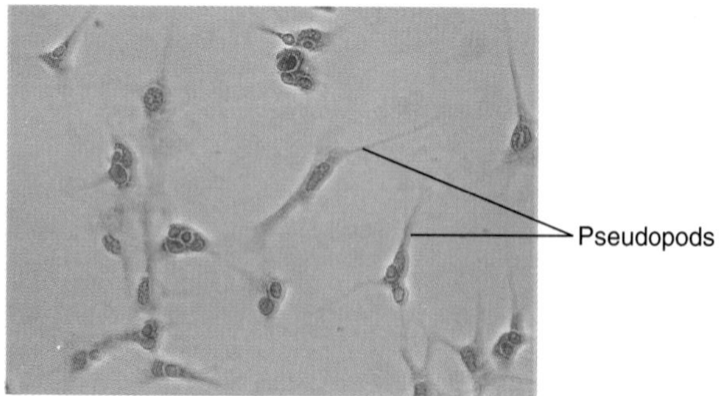

FIGURE 39.1

Multicellular animals have phagocytes, cells that can engulf and destroy foreign matter. These phagocytes are hemocytes (white blood cells) from the land snail *Helix pomatia.* The cells send out pseudopods onto a glass surface in the laboratory, as they would in the snail to reach for foreign material and destroy it.

Mastering Concepts

1. What is the general role of an immune system?
2. How do single-celled eukaryotes have immunity?
3. What are the functions of amebocytes and phagocytes?

Nonspecific Defenses

Immunity is both nonspecific (general) and highly specific. Figure 39.2 illustrates the components of the human immune system. Barriers nonspecifically prevent microbes from entering the body—that is, they block all microbes, without distinguishing whether they are harmful or not. Unpunctured skin is the most pervasive and obvious wall, but there are others. Mucus in the nose traps inhaled dust particles; tears wash chemical irritants from the eyes and contain lysozyme, a biochemical that kills bacteria by rupturing their cell walls; wax traps dust particles in the ears. Most microorganisms that pass these barriers and reach the stomach die in a vat of acidic secretions. Bacteria that enter the respiratory system are swept out of the airways by cilia and then swallowed.

Phagocytosis

Phagocytes are scavenger cells that engulf and digest foreign cells or substances (fig. 39.3). Some phagocytes are anchored in tissues; others, such as neutrophils and **macrophages,** move. Like

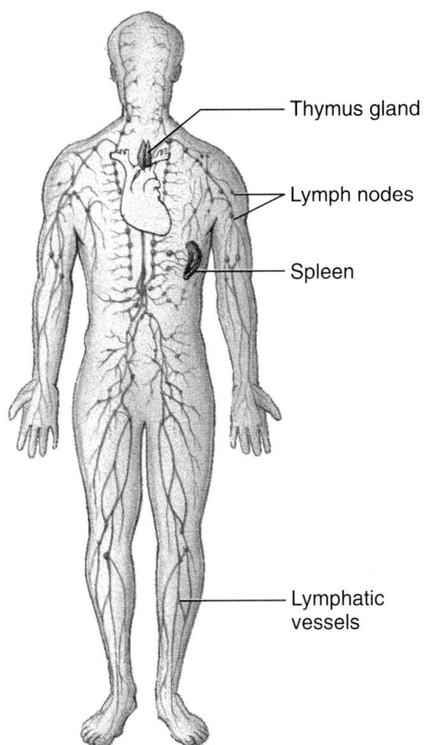

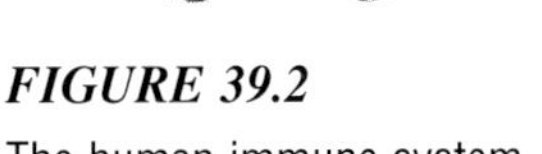

FIGURE 39.2
The human immune system.

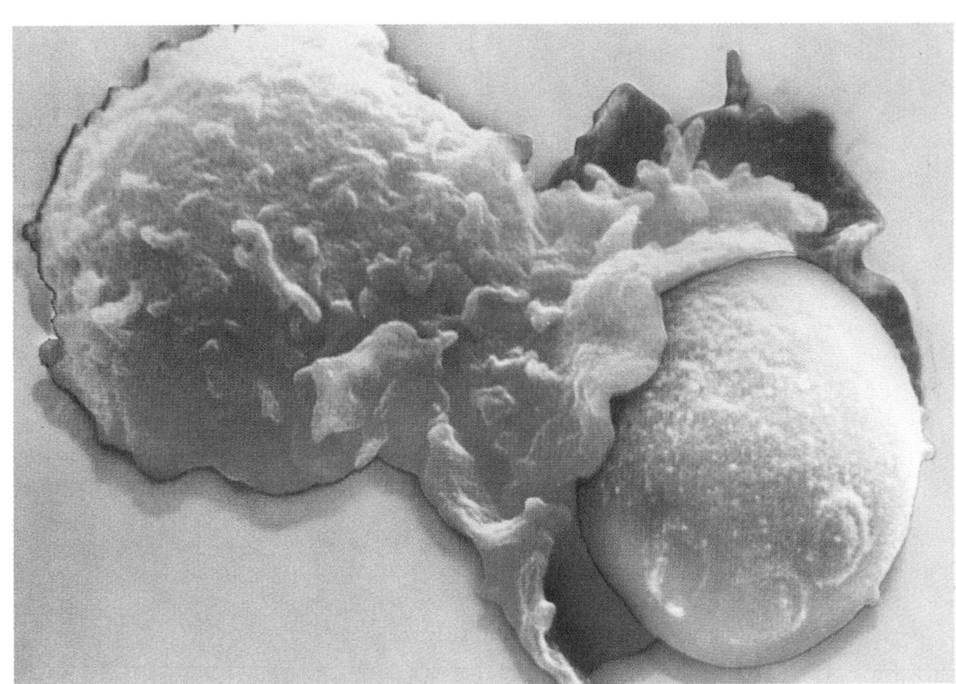

FIGURE 39.3
A human phagocyte devouring a yeast cell.

soldiers on the front lines, many of these cells die in battle. A neutrophil can engulf only about 20 bacteria before it dies, but a macrophage can engulf up to 100. Dead phagocytes are one component of pus, a sure sign of infection.

Inflammation

Inflammation is a nonspecific defense that creates an environment hostile to microorganisms at the site of injury (fig. 39.4). Consider a puncture wound, which triggers the inflammatory response. First, phagocytes infiltrate the injured skin, attacking entering bacteria. Plasma accumulates at the wound site, diluting toxins that bacteria secrete and bringing antimicrobial substances to the site. Increased blood flow warms the area, turning it swollen and red.

Antimicrobial Substances

Antimicrobial biochemicals are part of the nonspecific immune response. Certain animal cells infected with viruses produce polypeptides called **cytokines,** which carry out communication between immune system cells. One type of cytokine, an **interferon,** diffuses to healthy cells and stimulates them to produce biochemicals that block viral replication. When these cells become infected, the viruses are unable to reproduce, halting spread of the infection. Table 39.1 lists some types of cytokines.

The **complement system** is a group of proteins that assist, or complement, several of the body's other defense mechanisms. Some complement proteins trigger a chain reaction that punctures bacterial cell membranes, bursting the cells. Other complement proteins assist inflammation by causing **mast cells** to release a biochemical called **histamine.** Histamine dilates (widens) blood vessels, easing entry of white blood cells and fluid to the injured area. Still other complement proteins attract phagocytes to an injured area.

Table 39.1 Types of Cytokines

Cytokine	Function
Colony-stimulating factors	Stimulate bone marrow to produce lymphocytes
Interferons	Block viral replication, stimulate macrophages to engulf viruses, stimulate B cells to produce antibodies, attack cancer cells
Interleukins	Control lymphocyte differentiation and growth
Tumor necrosis factor	Stops tumor growth, releases growth factors, causes fever that accompanies bacterial infection, stimulates lymphocyte differentiation

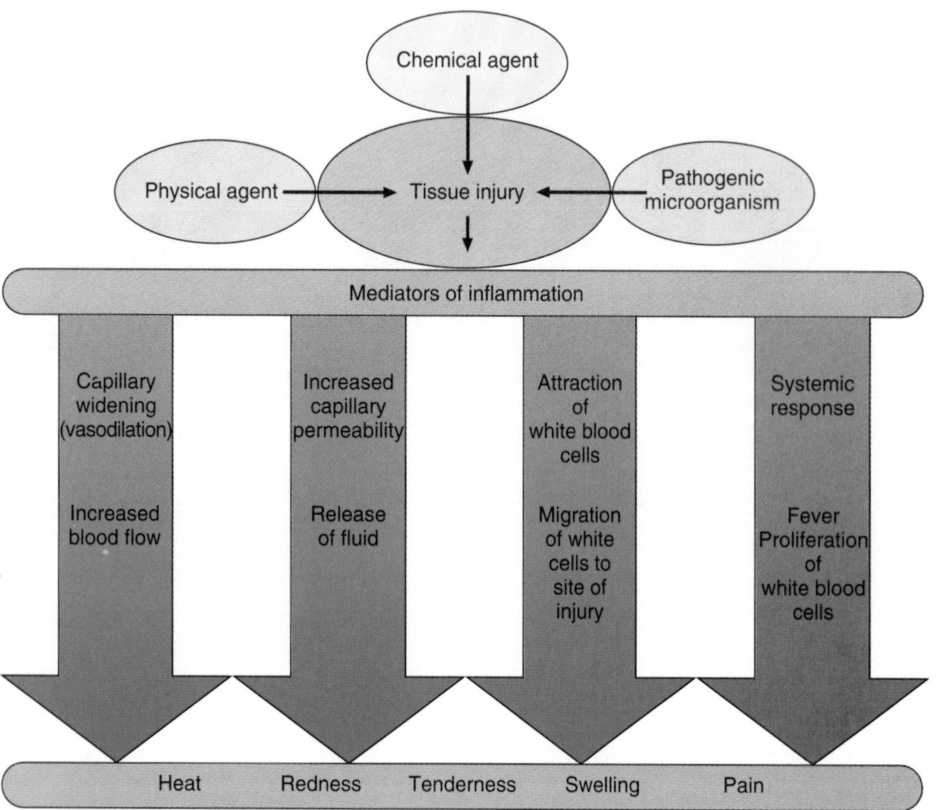

(a)

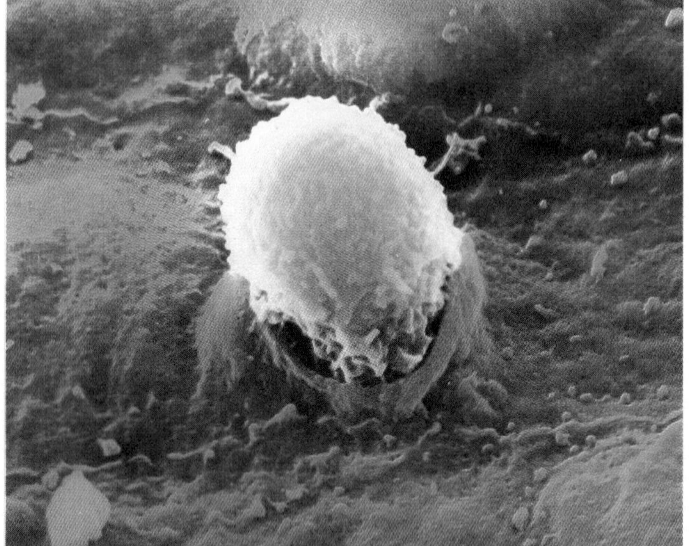
(b)

FIGURE 39.4

Inflammation. (*a*) An injury sets into motion the several steps of the inflammatory response. (*b*) This white blood cell is migrating through the endothelial wall of a capillary, perhaps on its way to the site of an injury.

Fever

A fever protects nonspecifically. Fever begins when a viral or bacterial infection stimulates white blood cells to proliferate. Some of these cells secrete **interleukin** proteins, which reset the thermoregulatory center in the hypothalamus to maintain a higher body temperature. The heat kills some infecting bacteria and viruses directly. Fever also counters microbial growth indirectly, because higher body temperature reduces the iron level in the blood. Because bacteria and fungi require more iron as the temperature rises, a fever-ridden body stops their growth. Phagocytes also attack microorganisms more vigorously when the temperature rises.

Mastering Concepts

1. What are some nonspecific barriers to infection in the human body?
2. What happens during an inflammatory response?
3. What are some antimicrobial biochemicals?
4. How is fever protective?

Specific Defenses

About 2 trillion lymphocytes and wandering macrophages provide immunity against specific nonself cells and biochemicals (see fig. 35.10). The two types of lymphocytes are made in the bone marrow and migrate to the thymus, lymph nodes, spleen, tonsils, and mucous membranes. Some lymphocytes also circulate in the blood and tissue fluid. B lymphocytes (B cells) secrete antibody proteins in response to foreign antigens. T lymphocytes (T cells) stimulate B cells to produce antibodies and release cytokines (see fig. 5.1). Figure 39.5 depicts B cells, T cells, and macrophages.

Immune System Characteristics

Four qualities of the immune system make it an effective defender. First, the immune response is swift—many infections halt even before symptoms arise. Second, immunity is specific; the immune

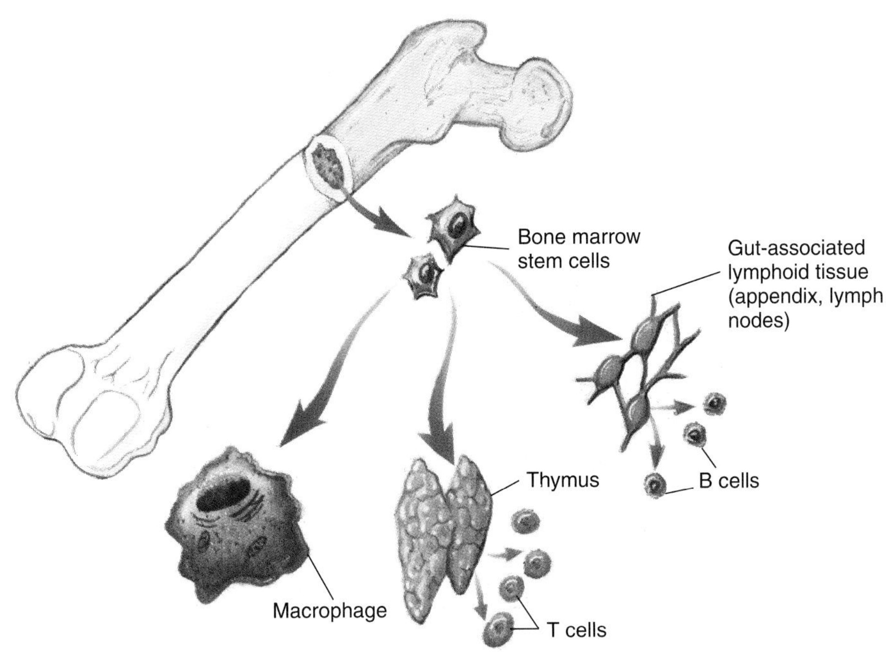

FIGURE 39.5

Macrophages, T cells, and B cells interact to generate immune responses. All three cell types originate in the bone marrow and migrate into the blood. Macrophages engulf bacteria and also stimulate helper T cells to proliferate and activate B cells. T cells mature in the thymus gland, in the small intestine, and in the skin, and some manufacture cytokines that are released in response to viral infections. T cells also attack cancer cells and transplanted tissues. B cells are released from lymphoid tissues such as the appendix and lymph nodes, and they secrete antibodies into the bloodstream in the humoral immune response.

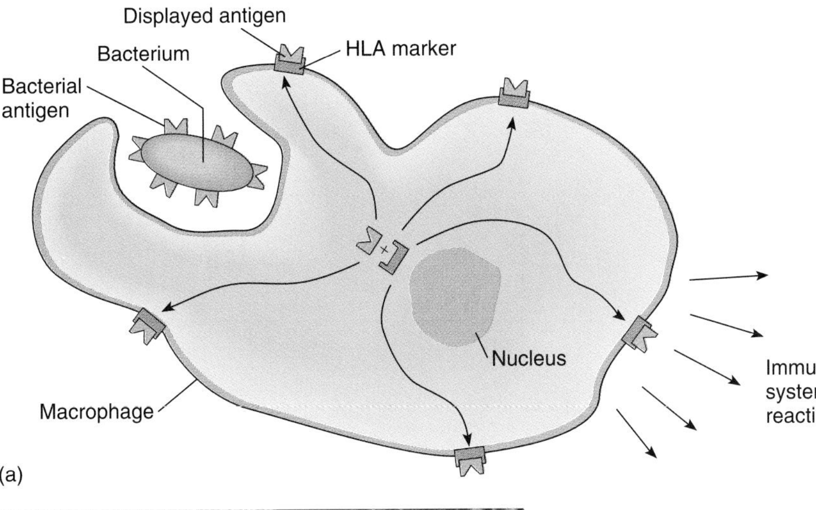

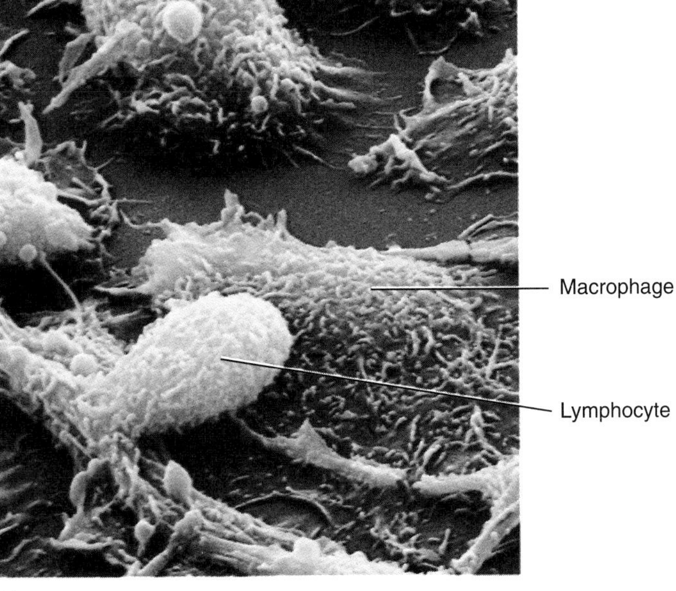

FIGURE 39.6

Macrophages and lymphocytes. A macrophage internalizes a bacterium and then dismantles it. (*a*) A piece of a bacterial antigen attaches to an HLA protein inside the macrophage and picks up a sugar, and the complex moves to the cell's surface. The combination of the foreign antigen bound to the self HLA glycoprotein alerts lymphocytes (*b*) and other immune system components.

system cells and biochemicals inactivate pathogenic bacteria, but not those that normally inhabit the body. Third, the immune system is diverse, fighting a tremendous variety of infectious agents and even cancer. Finally, the immune system "remembers" each foreign antigen and responds quickly should it return. As a result, many infectious illnesses occur only once.

Steps in the Immune Response

Step 1: Macrophages Alert Lymphocytes

One of the first cell types to respond to infection is the macrophage. These scavengers engulf foreign matter and cellular debris and activate lymphocytes.

Macrophages alert lymphocytes by displaying an antigen from the engulfed invader (fig. 39.6). The antigen is held to the macrophage surface by a self protein specified by genes of the human leukocyte antigen (HLA) complex (see chapter 5). Displaying

FIGURE 39.7

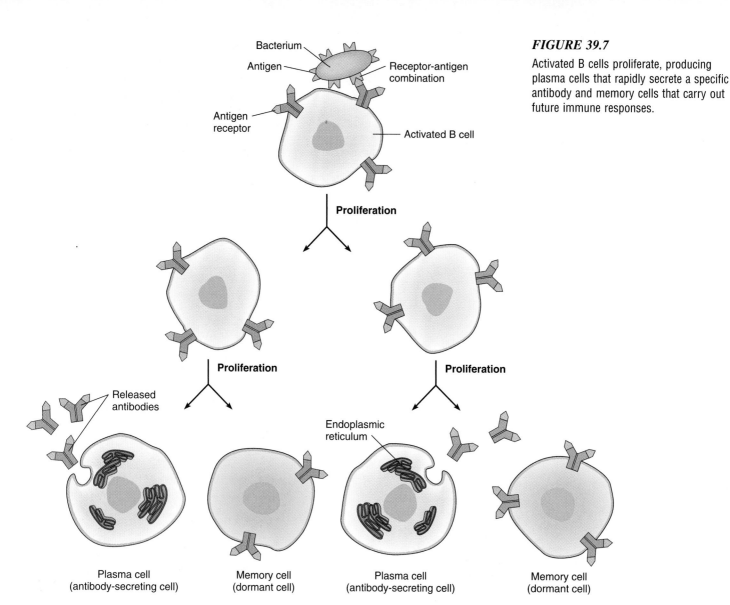

Activated B cells proliferate, producing plasma cells that rapidly secrete a specific antibody and memory cells that carry out future immune responses.

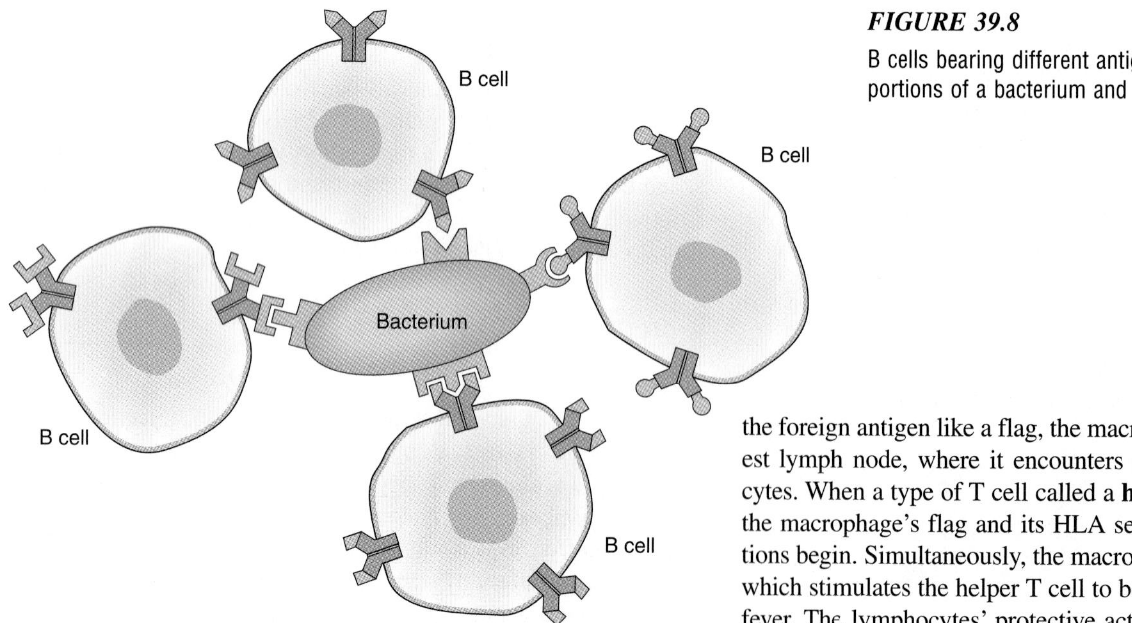

FIGURE 39.8

B cells bearing different antigen receptors bind to different portions of a bacterium and produce different antibodies.

the foreign antigen like a flag, the macrophage migrates to the nearest lymph node, where it encounters several varieties of lymphocytes. When a type of T cell called a **helper T cell** recognizes both the macrophage's flag and its HLA self label, other immune reactions begin. Simultaneously, the macrophage secretes interleukin-1, which stimulates the helper T cell to begin replicating and causes a fever. The lymphocytes' protective actions are intricate and highly coordinated.

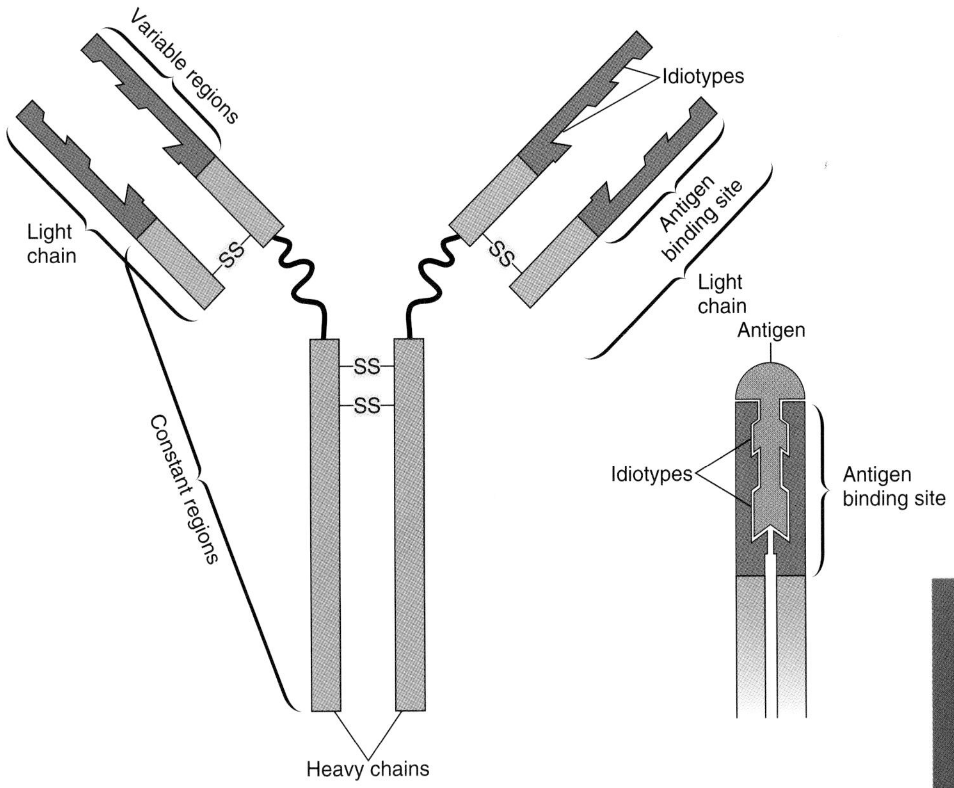

Variable regions
Idiotypes
Light chain
Antigen binding site
Light chain
Antigen
Constant regions
Idiotypes
Antigen binding site
Heavy chains

FIGURE 39.9

Antibody structure. The simplest antibody molecule consists of four polypeptide chains, two heavy and two light, joined by disulfide bonds. Part of each polypeptide chain has a constant sequence of amino acids, and the remainder of the sequence is variable. The tops of the Y-shaped molecules form antigen binding sites. The inset shows how an antigen fits the antigen binding site.

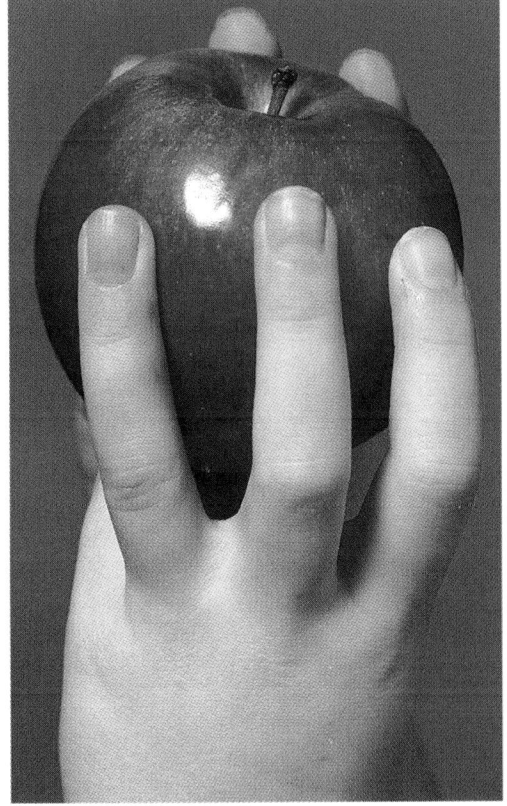

FIGURE 39.10

An antibody's antigen binding site hugs an antigen as a hand holds an object.

Step 2: The Humoral Immune Response—B Cells Produce Antibodies

B cells secrete antibodies into the bloodstream in the **humoral immune response** ("humor" means fluid). The response begins when a T cell reacts to a foreign antigen "displayed" on a macrophage by secreting cytokines and activating B cells. The stimulated B cells then divide, producing clones (groups of identical cells) that can identify the foreign antigen. These descendant B cells mature into either **plasma cells** or **memory cells** (fig. 39.7).

A plasma cell secretes up to 2,000 identical antibodies per second during its few days of existence. These antibodies surround, bind, and inactivate foreign antigens. Plasma cells derived from different B cells secrete different antibodies, and each antibody type corresponds to a specific portion of the foreign antigen (fig. 39.8). The different types of antibodies secreted in response to different portions of an antigen are collectively called a polyclonal antibody response.

Memory B cells respond to the antigen quickly and forcefully should it be encountered again in the future. The **primary immune response** is the immune system's reaction to its first meeting with a nonself antigen, and it takes a few days. Subsequent encounters stimulate a **secondary immune response,** which is so much faster that a person may not even be aware of it. Memory B cells function in the secondary immune response.

Antibodies (also called immunoglobulins) are large, multipolypeptide proteins. Different genes encode the different polypeptides that form an antibody. The simplest antibody molecule consists of four polypeptide chains connected by disulfide (sulfur-sulfur) bonds, forming a shape like the letter Y (fig. 39.9 and 39.10). The two larger polypeptides are called **heavy chains,** and the other two are called **light chains.** The lower portion of each chain is an amino acid sequence that is very similar in all antibody molecules, even in different species. These areas are called **constant regions.** The amino acid sequence of the upper portions of each polypeptide chain, termed **variable regions,** can differ a great deal between antibodies. Thus, an antibody subunit has two heavy chains and two light chains, and each chain is partly constant and partly variable in its amino acid sequence.

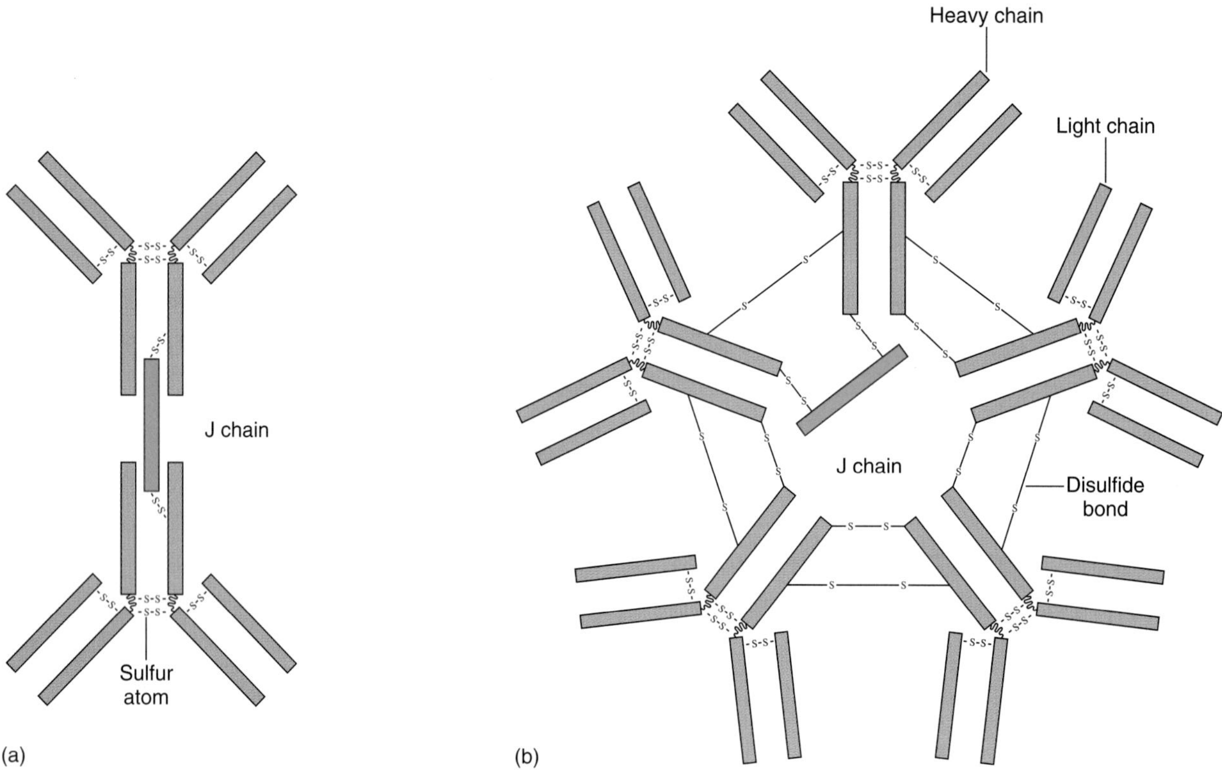

FIGURE 39.11

The five classes of antibodies are distinguished by specific conformations of the heavy chain. Antibodies of class IgA and IgM consist of multiple antibody molecules held together by short peptides called J (for "joining") chains. (*a*) IgA, found in several secretions, is composed of two such subunits. (*b*) IgM, secreted into the blood during a primary immune response, consists of five subunits. Blood antibodies A and B of the ABO typing system are IgMs. IgD, IgE, and IgG each consist of only one four-polypeptide molecule.

Table 39.2 Types of Antibodies

Type*	Location	Functions
IgA	Milk, saliva, and tears; respiratory and digestive secretions	Protects against microorganisms at points of entry into body
IgD	B cells in blood	Stimulates B cells to make other antibodies (little is known about IgD)
IgE	Mast cells in tissues and serum	Provides receptors for antigens that cause mast cells to secrete allergy mediators
IgG	White blood cells and macrophages in blood, serum	Binds bacteria and macrophages at same time, assisting macrophage in engulfing bacteria; activates complement; participates in secondary immune response; passes from pregnant woman to fetus to protect until fetus manufactures antibodies
IgM	B cells in blood, serum	Activates complement in primary immune response

*The letters A, D, E, G, and M refer to the specific conformations of heavy chains characteristic of each class of antibody.

Antibody molecules can aggregate into complexes (fig. 39.11), in which short peptides called J (for "joining") chains hold antibodies together. Five major classes of antibodies are distinguished by location in the body and function (table 39.2).

Antibodies can bind to specific antigens because of the three-dimensional shapes of their variable regions. The antibody contorts slightly to form a pocket around the antigen. The specialized ends of the variable regions of the antibody molecule are called **antigen binding sites,** and the particular parts that actually bind the antigen are called **idiotypes** (see fig. 39.9).

Antibody-antigen binding inactivates a microbe or neutralizes the toxins it produces. Antibodies can cause pathogens to clump, making them more visible to macrophages, which then destroy them. Antibodies also activate complement, which destroys microorganisms.

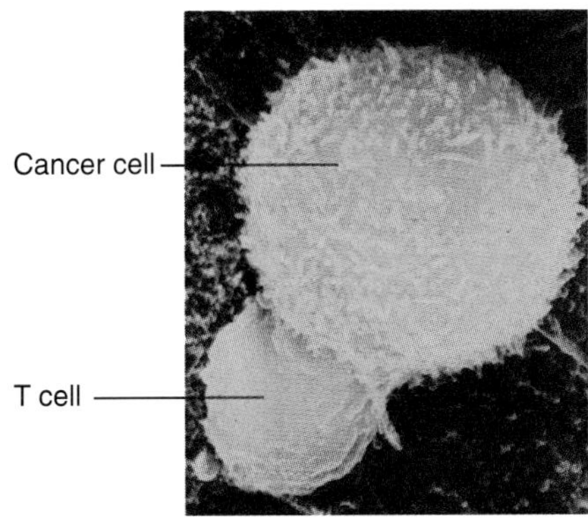

FIGURE 39.12

Cytotoxic T cell action. The smaller cell is a T cell, which homes in on the surface of the large cancer cell above it. The T cell will literally break the cancer cell apart, leaving nothing behind but scattered fibers.

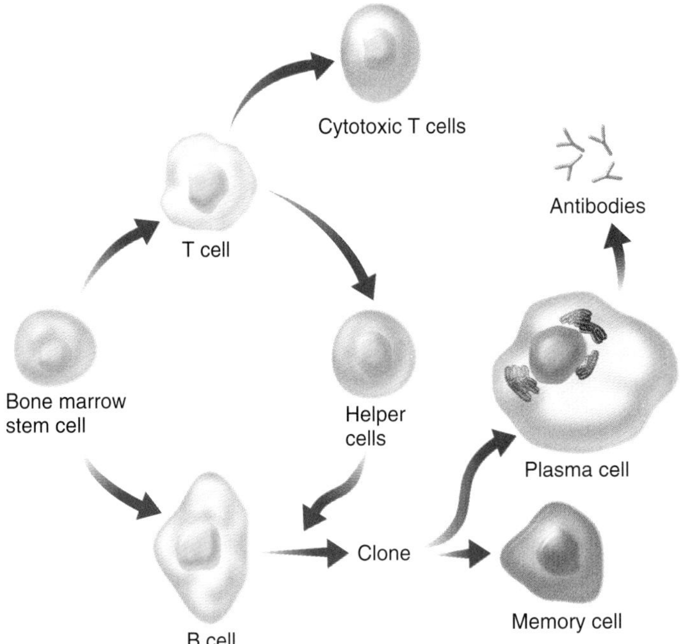

FIGURE 39.13

All lymphocytes descend from stem cells in the bone marrow. T cells, which spend time in the thymus, account for 65 to 85% of all lymphocytes. B cells do not pass through the thymus. Upon stimulation by helper T cells responding to a foreign antigen, B cells divide and differentiate into plasma cells, which secrete antibodies, and memory cells. Helper T cells stimulate B cells. Cytotoxic T cells attack nonself cells directly.

The human body can manufacture an apparently limitless number of different antibodies because different antibody gene products combine. As B cells develop, sections of their antibody genes move to other chromosomal locations, creating new genetic instructions for antibodies. Antibody diversity is analogous to using the limited number of words in a language to compose an infinite variety of stories. Antibody genes combine in many ways to generate an enormous diversity of antibodies.

Step 3: The Cellular Immune Response—T Cells Secrete, Stimulate, and Attack

At the same time that B cells are specializing into plasma cells and secreting antibodies, T cells provide the **cellular immune response.** It is termed "cellular" because the cells themselves move to where they are required, unlike B cells, which stay where they are and secrete antibodies into the bloodstream. In the thymus, T cells acquire the ability to recognize self cell surfaces and molecules. Several types of T cells have distinct functions.

Helper T cells are crucial to a multifaceted immune response. They stimulate B cells to produce antibodies, secrete cytokines, and activate another type of T cell called a **cytotoxic T cell** (sometimes called a killer T cell). Helper T cells initiate a specific immune response. An immune reaction begins when a helper T cell with a surface protein called a **cluster-of-differentiation (CD4) antigen** recognizes a macrophage with a nonself antigen bound to its surface (see fig. 39.6). The stimulated helper T cell, called a CD4 T cell, then causes B cells to mature and secrete a specific antibody. The CD4 T cell also releases cytokines, including interleukins, interferons, and **tumor necrosis factor (TNF),** which elicit biological responses that destroy the nonself antigen.

Cytotoxic T cells attack nonself cells by attaching to them and releasing biochemicals. To do so, they use **T cell receptors** on their surfaces that bind foreign antigens. When a cytotoxic T cell encounters a nonself cell—a cancer cell, for example—its T cell receptors contact the foreign cell. The T cell then releases a biochemical that cuts holes in the foreign cell's membrane. This disrupts the flow of biochemicals in and out of the foreign cell and kills it (fig. 39.12). Cytotoxic T cells are also attracted by their receptors to body cells to which certain viruses have attached. The T cells destroy the attacked cells before the viruses on them can enter, replicate, and spread the infection.

Turning off an immune response once an infection has been halted is as important as turning it on, because the powerful immune biochemicals can attack the body's healthy tissues. Certain cytokine combinations shut off the immune response. Figure 39.13 summarizes relationships between the lymphocytes, whose balance is crucial to health.

Mastering Concepts

1. What are the four basic characteristics of the immune response?
2. How does a macrophage display a foreign antigen?
3. What two cell types do stimulated B cells produce, and what are their functions?
4. Describe the structure and function of an antibody.
5. Why are helper T cells central to the immune response?
6. Why must an immune response cease?

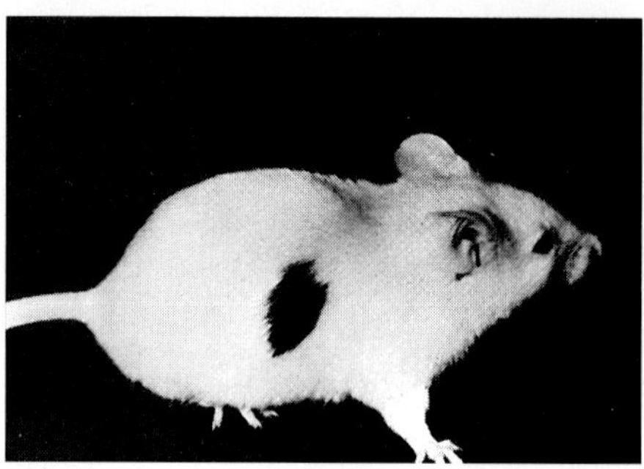

FIGURE 39.14

The immune system learns to distinguish self from nonself during fetal existence. A white mouse exposed to the blood of a brown mouse while in the uterus will accept a transplant of skin from the brown mouse after birth because its immune system recognizes tissue from the brown mouse as self.

Development of the Human Immune System

The fetus's fledgling immune system "catalogs" its own cells and learns to distinguish self from nonself. Before this happens, any antigen a fetus is exposed to will be cataloged as self. Encounters with that antigen later in life will not evoke immune rejection (fig. 39.14).

A newborn has temporary, or **passive, immunity,** from maternal antibodies received before birth. Human milk provides additional antibody protection. Shortly before and after giving birth, a woman's milk contains a yellow substance called **colostrum** rich in antibodies that protect the baby from certain digestive and respiratory infections. In a few days, mature milk, which contains antibodies against some intestinal parasites, replaces colostrum.

A newborn manufactures its first antibodies in response to bacteria and viruses transmitted by whoever is in closest physical contact. This is **active immunity.** The mother's body also makes antibodies to the bacteria and viruses on her baby's skin (faster than the baby can) and passes those antibodies to the baby in her milk.

Gradual exposure to foreign antigens seems to be part of normal immune system development. By 6 months of age, a baby can produce a great enough variety of antibodies to overcome most infections, although the immune system isn't mature until 18 months. Some infections, such as those caused by the herpes simplex viruses, cannot trigger an immune response until the second year of life.

The immune system begins to decline early in life. The thymus gland reaches its maximal size in early adolescence and then slowly degenerates. By age 70, the thymus is one-tenth the size it was at the age of 10, and the immune system is only 25% as powerful. The declining immune response explains why elderly persons often succumb to infections that are minor in younger individuals.

Mastering Concepts

1. How does human milk provide immune protection?
2. What stimulates a newborn to manufacture antibodies?
3. At what age does the immune response begin to weaken?

Immunity Breakdown

Immune system breakdown devastates health. Such immune deficiency can be acquired or inherited.

AIDS

Imagine that each day, 3 billion ants appear at your front door. The first day, you kill a billion of them. The next day, 3 billion more ants arrive, but even after you kill another billion, 4 billion remain. Each day you kill about a third of the ants, yet still they replenish their numbers. Soon you tire. You can only kill a fourth of them, then a fifth. Before long, you can no longer wage the battle, and ants overrun your house. As they come through the windows, under the doors, and down the chimney, they open up routes for other unwanted visitors, who damage the house in other ways.

This scenario is not unlike what happens when human immunodeficiency virus (HIV) infects a human body. The first stage is what one researcher calls "a titanic struggle between the virus and the immune system." For some time, ranging from months to a decade or more, the body can produce enough new T cells daily to remove 100 million to a billion viruses. When the T cell number begins to fall, immunity falters, and the infections and cancers of AIDS begin. The infections of AIDS are termed "opportunistic" because they do not threaten a person with a healthy immune system. Kaposi's sarcoma (fig. 39.15) and pneumonia caused by *Pneumocystis carinii* are two common AIDS-associated illnesses.

HIV

HIV is a retrovirus; its genetic material is RNA. The virus attaches to CD4 receptors on helper T cells and sends in its RNA. A viral enzyme, **reverse transcriptase,** then catalyzes construction of a DNA strand complementary to the viral RNA. The initial viral DNA strand replicates to form a DNA double helix, which enters the T cell's nucleus.

HIV replicates, filling the host cell with viral RNA and proteins (figs. 39.16 and 4.16*b*). Not only can the dying T cell no longer provide immunity, but it also bursts, unleashing many new HIV particles. Maturation of new T cells to fight the virus cannot keep pace with viral replication, and gradually the numbers of CD4 helper T cells drop.

So genetically diverse is HIV that, within days of the initial infection, variants arise that resist the drugs used to treat AIDS. All it takes is a change in a single RNA base for a drug-resistant variant to appear. HIV's changeable nature has important clinical implications. Drugs that block viral replication must be adminis-

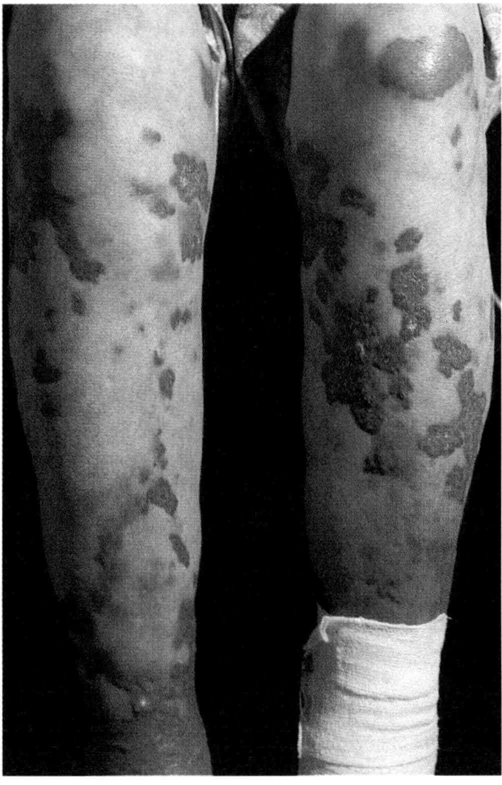

FIGURE 39.15

Prior to the appearance of AIDS, Kaposi's sarcoma was a rare cancer seen only in elderly Jewish and Italian men and in people with suppressed immune systems. In these groups, it produces purplish patches on the legs, but in AIDS patients, patches appear all over the body and sometimes internally, too. This patient displays characteristic lesions. Kaposi's sarcoma is itself caused by a virus, which HIV prevents the immune system from vanquishing.

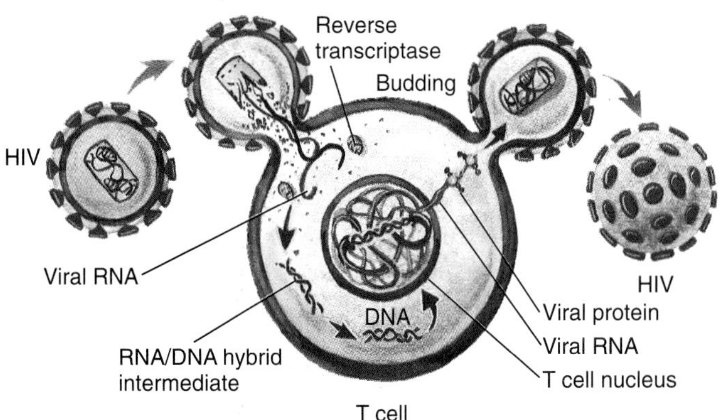

FIGURE 39.16

How HIV infects a helper T cell. HIV sends its RNA and reverse transcriptase into a helper T cell by docking at a CD4 receptor. Once inside, reverse transcriptase builds a complementary DNA chain; after replicating, the viral genetic information, now encoded in DNA, enters the nucleus. When infection by another type of virus stimulates the T cell to divide and manufacture cytokines, the HIV genes are activated too. Instead of assisting B cells and secreting cytokines, the helper T cell becomes an HIV factory.

tered as soon as possible. Combining drugs that act in different ways has the greatest chance of slowing the disease process. For example, two types of drugs, protease inhibitors and nucleoside analogs, intervene at different stages of HIV replication and, when teamed early in infection, may greatly slow its progress.

AIDS in the 1990s

HIV infection is spread by direct blood contact with infected body fluids, such as blood or semen. As HIV infection has spread to diverse segments of the population, public health definitions have expanded. In 1993, the definition of AIDS was amended to include pulmonary tuberculosis, recurrent pneumonia, and invasive cervical cancer, which led to inclusion of more female AIDS patients. In 1994, public health officials amended the AIDS definition again to account for differences between children under age 13 and older people.

Physicians can determine the stage of infection directly by the amount of virus in the bloodstream or indirectly by a CD4 cell count or presence of antibodies to HIV.

An End to AIDS?

Not everyone exposed to HIV becomes infected or progresses to AIDS. About 5% of people infected with HIV remain healthy 12 to 15 years after infection. Some prostitutes in Africa who have been exposed to HIV hundreds of times never become infected. Some infants born infected with HIV "lose" the virus. A detailed study of such long-term nonprogressors revealed a lucky combination of circumstances that may explain their persistent health—a strong immune response plus infection by weakened strains of HIV. They have abundant antibodies and helper T cells. In two of the patients, HIV was missing part of a particular gene, which apparently weakens it. In 1996, researchers discovered that about 1% of caucasians in the U.S. have a mutation that alters certain receptors on CD4 cells that bind HIV. These people cannot become infected because HIV cannot enter their cells.

Long-term nonprogressors, faulty HIV receptors, and successful drug combinations are all signs that the AIDS epidemic may finally be on the wane, while suggesting new ways to fight the infection. Incidence of AIDS is declining in many places. Biology in Action 39.1 compares the AIDS epidemic in humans to the courses of immunodeficiency illnesses in other species.

Severe Combined Immune Deficiency

AIDS is acquired. Immune deficiency can also be inherited. Each year, a few children are born defenseless against infection due to severe combined immune deficiency (SCID), in which neither T nor B cells function. David Vetter was one such youngster (fig. 39.17). Born in Texas in 1971, David had no thymus gland and spent the 12 years of his life in a vinyl bubble, awaiting a treatment that never

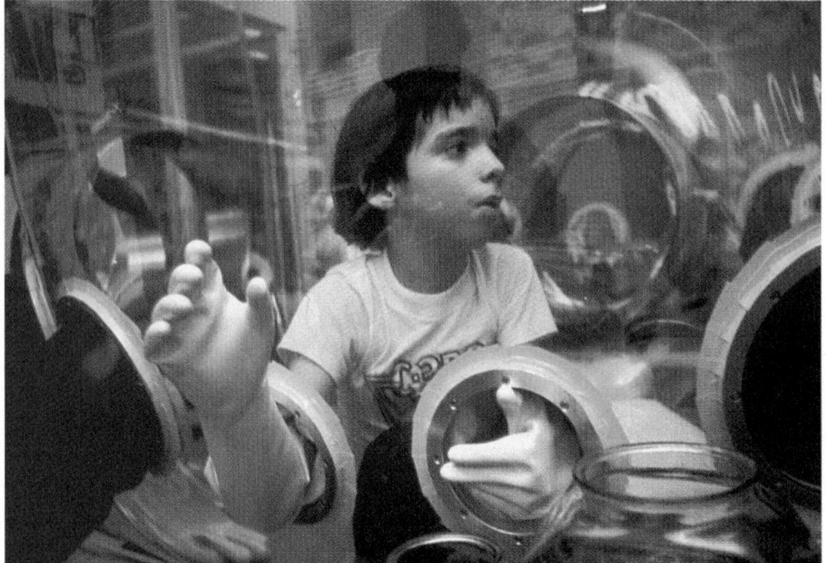

FIGURE 39.17

David Vetter, the "bubble boy," was born without a thymus gland. Because his T cells could not mature, he was virtually defenseless against infection.

came. As David reached adolescence, he wanted to leave his bubble. An experimental bone marrow transplant was unsuccessful—soon afterward, David began vomiting and developed diarrhea, signs of infection. David left the bubble, but he died within days of a massive infection.

Today, some children with SCID caused by an inherited enzyme deficiency have several treatment options that were unavailable to David. The opening text in chapter 16 tells the story of the first recipients of gene therapy to treat inherited immune deficiency.

Chronic Fatigue Syndrome

Some 2 to 5 million people in the United States suffer from a poorly understood immune system imbalance called chronic fatigue syndrome. The condition begins suddenly, producing fatigue so debilitating that getting out of bed is an effort. Chills, fever, sore throat, swollen glands, muscle and joint pain, and headaches are other symptoms.

The disabling aches and pains of chronic fatigue syndrome reflect an overactive immune system. People with chronic fatigue syndrome have up to 40 times the normal amount of interleukin-2 and excess cytotoxic T cells but too little interferon. It is as if the immune system mounts a defense and doesn't know when to shut it off. Chronic fatigue syndrome may be a viral infection.

Mastering Concepts

1. How does HIV replicate?
2. Why are some people unable to become infected with HIV?
3. How does severe combined immune deficiency differ from HIV infection?
4. What is chronic fatigue syndrome?

Autoimmunity and Allergy

Immune deficiency destroys health. An overactive or misdirected immune system is also dangerous.

Autoimmunity—Attacking Self

Sometimes the immune system turns against the host, manufacturing **autoantibodies** that attack the body's own cells and cause **autoimmunity.** The specific nature of an autoimmune disorder depends upon the cell types the immune system attacks. Table 39.3 lists some autoimmune disorders.

Why might the immune system turn on itself? Perhaps a virus, while replicating within a human cell, "borrows" proteins from the host cell's surface and incorporates them onto its own surface. When the immune system "learns" the surface of the virus to destroy it, it also learns to attack the human cells that normally bear that particular protein. Another possible explanation of autoimmunity is that T cells never learn to distinguish self from nonself in the thymus and at some point in life, attack certain self cells. A third possible route of autoimmunity is when a nonself antigen coincidentally resembles a self antigen, such as antibodies attacking heart valve cells that have antigens that resemble those on certain pathogenic bacteria.

Allergies—Attacking the Innocuous

In an **allergy,** the immune system is overly sensitive and attacks harmless substances. The triggering substances, called **allergens,** activate IgE antibodies, which stimulate **mast cells** to explosively release **allergy mediators,** which include histamine and heparin (fig. 39.18). Allergy mediators cause symptoms by increasing the permeability of capillaries and venules, contracting smooth muscle, and stimulating mucous glands. Because histamine causes many allergy symptoms, drugs with antihistamine action can relieve symptoms.

biology in action 39.1

The Origin of AIDS

Understanding and conquering HIV infection may be possible by tracing its beginnings in species other than humans.

Immunodeficiency viruses are known to infect cattle (bovine immunodeficiency virus, or BIV), cats (feline immunodeficiency virus, or FIV), and monkeys (simian immunodeficiency virus, or SIV). By sequencing the genetic material of BIV, FIV, SIV, and HIV, and arranging them into a molecular tree diagram that depicts possible evolutionary relationships, researchers can hypothesize a time sequence in which the four viral variants appeared.

Cattle to Felines to Monkeys to Humans?

The first immunodeficiency virus appears to have been BIV, originating in an ancestor of buffalo. Perhaps a lion ancestor contracted BIV while eating an infected buffalo. The virus persisted among lions and then other felines; it became FIV and for a time killed many cats. An infected wild feline attacking a monkey could have passed the virus to primates, and the virus might have, over long periods of time, changed sufficiently among monkeys to become a simian variant, SIV. Then, in more recent time, a monkey biting a human may have transmitted the virus to us. It changed further and became HIV.

The origin and spread of immunodeficiency viruses may have taken a short time, in evolutionary terms, because of the nature of viral change. Viruses can replicate frequently, and therefore they have many opportunities for faulty replication of genetic material to introduce mutations. Because viruses cannot repair mutations, the changes accumulate, possibly enabling the virus to jump from species to species.

Clues from Cats

While some researchers try to decipher the natural history of the immunodeficiency viruses, others are looking to feline populations for clues that we can use to overcome AIDS. FIV was discovered in 1987, when a woman took her pet cat to a laboratory at the University of California, Davis, where researchers were studying HIV. Her cat, she insisted, had AIDS—it was lethargic and had many infections. Researchers indeed found the pet cat to be infected with what looked like a feline version of HIV, and they called it the feline immunodeficiency virus.

By 1990, virologists had discovered antibodies to FIV in 25 wildcat species, including ocelots, panthers, leopards, cheetahs, and lions—in seemingly healthy animals. This meant that these species had encountered FIV, yet they did not develop AIDS-like illness. Serengeti lions fall desperately ill from canine distemper, as the chapter opener describes, but they remain healthy when infected with FIV. In contrast, domestic cats become immunodeficient from FIV in much the same pattern as people become ill with AIDS. Both people and domestic cats suffer an initial acute illness, years of apparent health, and then a series of opportunistic infections.

RNA sequencing comparisons of subtypes of FIV suggest that the virus first appeared about 3 to 6 million years ago in an ancestor of domestic cats. The animals' biting behavior may have spread the infection. Over millions of years, all of the susceptible individuals died, leaving resistant animals in today's populations. This is why lions, ocelots, and panthers do not die from FIV. Infection in domestic cats, however, is fairly recent, and so susceptible animals are still present and become ill (fig. 39.A). As susceptible individuals die from the illness and a greater proportion of the population is resistant, a peaceful coexistence of sorts—a symbiosis—prevails. Epidemiologists call this end to epidemics an "adaptive episode."

People infected with HIV today don't have time to wait for a natural "adaptive episode." But signs that this may be slowly beginning lie in the few HIV-infected individuals who do not become ill, in people who are repeatedly exposed to HIV but never become infected, and in children who rid their bodies of the virus. By identifying the precise molecular variants that protect wild cats and resistant people, researchers should finally be able to develop a drug treatment that mimics the natural protective condition.

On a more philosophical note, the existence of immunodeficiency viruses in species other than humans is proof, if proof was needed, that HIV is not a punishment directed at humans.

(1)

(2)

FIGURE 39.A
The fact that many wildcat species (*1*) become infected with FIV but do not develop symptoms but infected domestic cats become ill (*2*) may hold clues to conquering AIDS.

Table 39.3 Autoimmune Disorders

Disorder	Symptoms	Targets of Antibody Attack
Glomerulonephritis	Lower back pain	Kidney cell antigens that resemble *Streptococcus* antigens
Grave's disease	Restlessness, weight loss, irritability, increased heart rate and blood pressure	Thyroid gland antigens
Juvenile diabetes	Thirst, hunger, weakness, emaciation	Pancreatic beta cells
Hemolytic anemia	Fatigue and weakness	Red blood cells
Myasthenia gravis	Muscle weakness	Nerve message receptors on skeletal muscle cells
Pernicious anemia	Fatigue and weakness	Binding site for vitamin B on cells lining stomach
Rheumatic fever	Weakness, shortness of breath	Heart valve cell antigens that resemble *Streptococcus* antigens
Rheumatoid arthritis	Joint pain and deformity	Cells lining joints
Scleroderma	Thick, hard, pigmented skin patches	Connective tissue cells
Systemic lupus erythematosus	Red rash on face, prolonged fever, weakness, kidney damage	DNA, neurons, blood cells
Ulcerative colitis	Lower abdominal pain	Colon cells

The symptoms of a specific allergic response depend upon the site in the body where mast cells release mediators. Because many mast cells are in the skin, respiratory passages, and digestive tract, allergies tend to affect these organs, producing hives, runny nose and eyes, asthma, nausea, vomiting, and diarrhea.

Common allergens are foods, dust mites, pollen, and fur (fig. 39.19). Many people are allergic to more than one substance and must restrict their diets and activities to avoid allergens. Allergies run in families. The mind can also influence an allergic reaction. A person allergic to tulips, for example, may experience allergy symptoms when presented with an artificial tulip!

To identify the cause of an allergy, a physician injects extracts of suspected allergens beneath a patient's skin on the upper arm or back. If a red bump develops, the substance injected at that point is an allergen for that person.

A treatment called desensitization can help hay fever and asthma sufferers. A physician periodically injects small amounts of the allergen under the skin. The doses are not enough to stimulate IgE production, but they do stimulate IgG production. When the person next encounters the allergen in a natural setting—pollen during a camping trip, for example—IgG binds to the allergen before IgE can. Because only IgE can stimulate mast cells to release allergy mediators, the attack is prevented.

Not all allergic reactions are as benign (although uncomfortable) as hives and watery eyes. Some individuals may react to certain stimuli with a terrifying and potentially life-threatening reaction called anaphylactic shock, in which mast cells release mediators throughout the body. The person may at first feel an inexplicable apprehension and then suddenly the entire body itches and erupts in hives. He or she may vomit and have diarrhea. The face, tongue, and larynx begin to swell, and breathing becomes difficult. Unless the person receives an injection of adrenaline and sometimes a tracheotomy (an incision into the windpipe to restore breathing), he or she will lose consciousness and die within 5 minutes. Anaphylactic shock most often results from an allergy to penicillin or insect stings. Fortunately, thanks to prompt medical attention and allergen avoidance, fewer than 100 people a year die of anaphylactic shock.

Anaphylactic shock is a severe response to something as nonthreatening as an insect bite or penicillin. One theory of allergy origin, and particularly of anaphylactic shock, is that allergies evolved long ago when insect bites and common substances such as penicillin might have threatened human survival. Evidence for this theory is that IgE also protects against roundworm and flatworm infections, which may have been a greater threat in the past. Perhaps IgE is a holdover from times past, when the environment presented different challenges to the human immune system than it does today.

Mastering Concepts

1. What events might lead to autoimmunity?
2. How is an allergic reaction a misplaced immune response?
3. Which cells and biochemicals participate in an allergic reaction?

Altering Immune Function

Because the immune system controls the vulnerability of the human body to disease and injury, it isn't surprising that part of medical practice aims to alter immune function. Depending upon the goal, we can help or hinder the immune response.

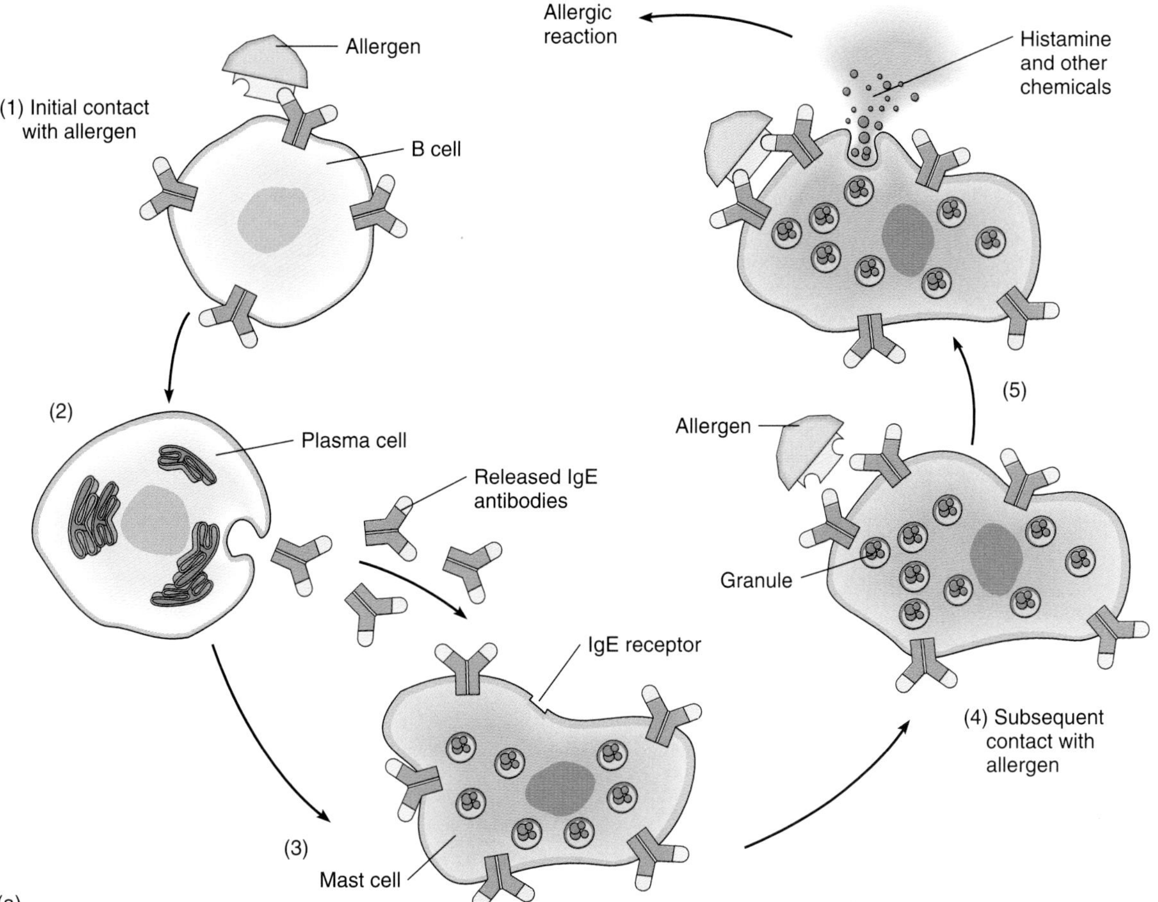

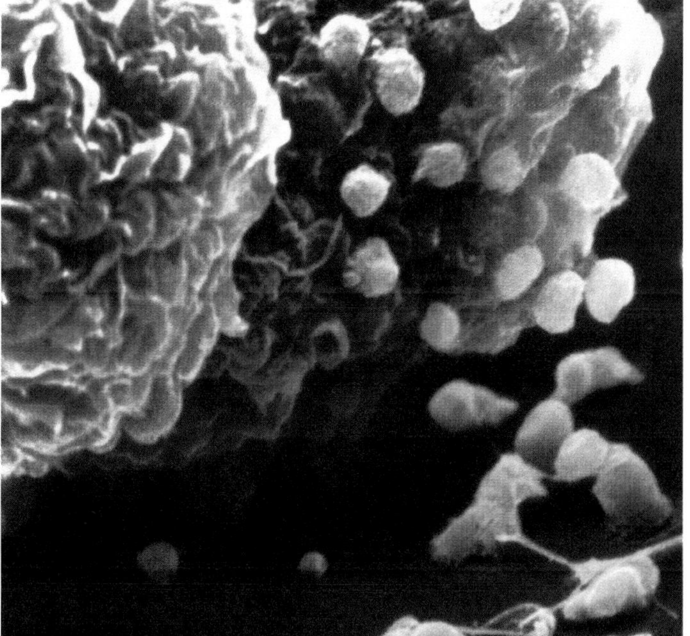

FIGURE 39.18

Allergy. (*a*) In an allergic reaction, (*1*) B cells are activated when they contact an allergen. (*2*) An activated B cell differentiates into an antibody-secreting plasma cell. (*3*) Antibodies attach to mast cells. (*4*) When allergens are encountered again, they combine with the antibodies on the mast cells. (*5*) The mast cells release allergy mediators, which cause the symptoms of the allergy attack. (*b*) A mast cell releases histamine granules. (× 3,000)

Vaccines—Augmenting Immunity

A **vaccine** is an inactive or partial form of a pathogen that the immune system responds to by producing memory B cells. If the true invader is met later, the immune system mounts a secondary immune response, as if it has been previously infected.

Some vaccines are entire viruses or bacteria, in inactivated forms. Such vaccines can cause illness if they mutate to a pathogenic form, but this is rare. A safer vaccine uses only the part of the viral or bacterial surface that elicits an immune response. A multiple vaccine may include genetic information from several pathogens. It encodes proteins from the pathogens, thereby protecting against several disorders. A new type of vaccine immunizes against cancer by exposing the immune system to antigens found on cancer cells but not on healthy cells. We've certainly come a long way in vaccine technology since the Middle Ages, when people tried to avoid smallpox by stuffing crusts from others' smallpox lesions into their nostrils! Biology in Action 39.2 chronicles the development of two successful vaccines.

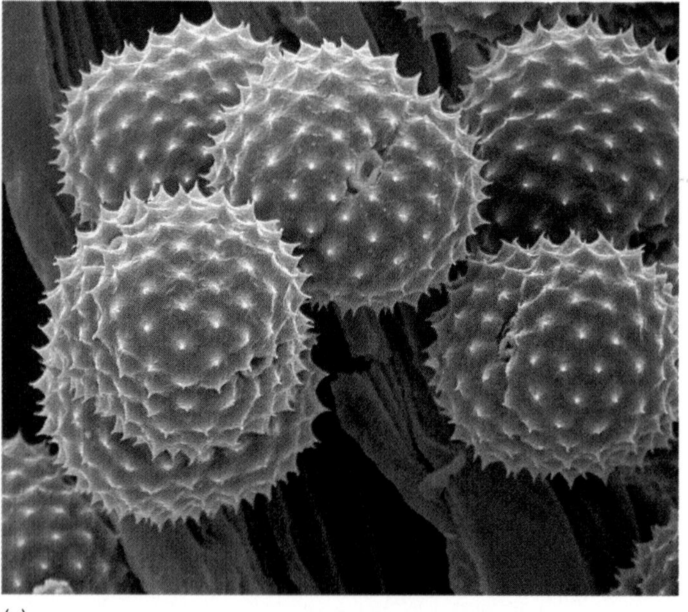
(a)

(b)

FIGURE 39.19

Causes of two common allergies: (*a*) ragweed pollen causes hay fever, and (*b*) dust mites cause year-round runny nose and eyes and asthma. (*a.* Magnification, × 325.)

Transplants—Suppressing Immunity

In 1905, doctors transplanted the cornea of an 11-year-old boy who lost an eye in an accident into a man whose cornea had been destroyed by a splash of a caustic chemical. This early transplant worked because the cornea is one of the few parts of the body free of self antigens. Early transplant attempts with other organs failed.

The first kidney transplants were performed in the 1940s, but they often failed because tissues were not adequately matched. Doctors attempted such drastic surgeries on young people suffering from kidney failure who had no other hope of survival, although biologists warned of probable tissue rejection. By the late 1950s, doctors began considering blood types to predict how successful a transplant would be.

In 1967, South African surgeon Christiaan Barnard performed the first heart transplant. The patient lived for 18 days, and a flurry of transplants followed. But by the mid-1970s, heart transplants still only extended life a few months. Immune system rejection, surgical complications, and failure to alleviate the underlying medical condition all dampened the initial excitement over transplants. Drugs that suppressed immunity sufficiently to prevent rejection also dampened defenses against infection and cancer. Some medical centers banned transplants.

By the 1980s, improved surgical techniques, biochemical techniques to better match donor to recipient and to strip antigens from donated tissue, and safer immunity-suppressing drugs revived interest in transplants. Most important was the drug cyclosporin, a fungal product originally discovered in a soil sample from Switzerland. Cyclosporin suppresses the T cells that reject transplanted tissue but not those that attack cancer cells or stimulate B cells to produce antibodies. Since then several other immunosuppressant drugs have been in use. Chapter 30 reviews types of transplants and controversial sources of tissue for transplant.

Immunotherapy

The immune system is remarkably effective at keeping potentially infectious bacteria, viruses, and tumor cells from taking over our bodies. Can we improve on nature? The idea of immunotherapy—amplifying or redirecting the immune response—was born late in the 1890s, when New York surgeon William Coley gave cancer patients killed bacteria. He had noticed that some cancer patients spontaneously recovered following a bacterial infection. Sometimes it worked—apparently the immune response against the bacteria also killed the cancer cells.

Today, many immunotherapies are in clinical trials. A few are already part of standard medical practice.

Boosting Humoral Immunity—Monoclonal Antibody Technology

When a single B cell recognizes a single foreign antigen, it manufactures a single, or monoclonal, type of antibody. A large amount of a single antibody type would make a powerful medicine because of its great specificity. It could be used to target a particular pathogenic microorganism or virus, or cancer.

In 1975, British researchers Cesar Milstein and Georges Köhler devised **monoclonal antibody** (MAb) technology, which amplifies the specificity of a single B cell. First, they injected a mouse with a sheep's red blood cells (fig. 39.20). They then isolated a single B cell from the mouse's spleen and fused it with a cancerous white blood cell from a mouse. The fused cell, called a **hybridoma,** had a valuable pair of talents. Like the B cell, it produced large

biology in 39.2 action

Two Vaccine Success Stories

The Smallpox Vaccine

In 1796, Edward Jenner, an English physician, developed the first successful vaccine. The vaccine protected against smallpox, an illness that once killed a third of the people who contracted it, leaving survivors horribly scarred. The luckiest people developed mild cases; they were left with relatively smooth skin and could not contract the disease a second time.

Jenner investigated the common observation that people who worked with cows did not develop smallpox. Milkmaids whose faces had the characteristic lesions of a mild disease called cowpox did not contract smallpox, while others in the same areas who did not work with animals did. We know today that the cowpox virus is so similar to the smallpox virus that an immune system exposed to one reacts against the other as well.

Jenner found a young milkmaid with active cowpox. He took fluid from a blister and injected it under the skin of a healthy boy. The boy contracted cowpox. Two months later, Jenner gave him fluid from a smallpox blister. The boy did not develop smallpox—he had become immune to it. It took Jenner 2 years to find another milkmaid with active cowpox and another young volunteer (see fig. 2.13). He repeated his experiment and published his results. Although Jenner was ridiculed, his vaccine, which he named *vaccinia* from the Latin for "cow," eventually freed the world from smallpox.

For the first few years, the smallpox vaccine was administered in an arm-to-arm fashion—a recently vaccinated person would rub the resulting skin lesion against an open scratch on the arm of another person. But this crude method spread other infectious diseases. In 1860, scientists began to manufacture the smallpox vaccine in calf lymph. In 1950, a freeze-dried variety was developed that was very easy to transport. This vaccine brought an end to smallpox in many developing nations. In 1959, the World Health Organization began its smallpox eradication campaign, and the disease was virtually gone from the United States and Europe by 1971, from Asia by 1975, and from Africa by 1977. Today, smallpox virus exists only in laboratories, and the vaccine is no longer necessary (see Bioethics Connections 4.1).

The Polio Vaccine

In the early 1950s, microbiologist Jonas Salk extended Jenner's idea to conquer another disease. He developed the first vaccine against the crippling disease poliomyelitis using a killed form of the virus. His first batch of polio vaccine was tried in 1952 on children who had already recovered from the disease. The vaccine raised the children's antibody levels, indicating that Salk was on the right track. The next step—a daring one—was to inject healthy children with the vaccine. Would they develop polio when exposed to the killed virus? They did not, which proved that the vaccine was safe. By 1954, youngsters all over the United States were lining up for the first polio shots.

A few years later, another microbiologist, Albert Sabin, developed a live oral polio vaccine. Sabin tried his vaccine first on laboratory animals, then on himself, and then on volunteer prison inmates. It, too, worked. The oral vaccine is safer because it goes to the small intestine, where the polio virus normally establishes itself before it travels in the bloodstream to nerves and causes paralysis. In contrast, the killed polio vaccine halts the virus in the bloodstream but still allows the virus to colonize the small intestine. Although a recipient of the killed vaccine cannot contract polio, he or she can harbor the virus in the small intestine, and spread it through bowel movements to other people.

The use of polio vaccines has drastically reduced the incidence of polio in many nations, but about 250,000 cases occur each year, mostly in the developing countries. The World Health Organization expects to eradicate polio everywhere by the year 2000. On a single day in 1996, for example, more than 120 million preschoolers in India were vaccinated. The chapter 34 opener vignette describes postpolio syndrome, which affects some polio survivors.

amounts of a single antibody type. Like the cancer cell, it divided continuously. A hybridoma is a specific antibody-making machine.

Today, MAbs are used in basic research, veterinary and human health care, agriculture, forestry, food technology, and forensics. Researchers have developed techniques to make them more like human antibodies—the original mouse versions cause allergic reactions in many people.

Diagnostic MAb kits that detect tiny amounts of a molecule are used to diagnose everything from strep throat to turf grass disease. One common use is a home pregnancy test. A woman places drops of her urine onto a paper strip impregnated with an MAb that binds to hCG, the hormone present only during pregnancy. A color change ensues if the MAb binds its target—the woman is pregnant.

MAbs can detect cancer earlier than other methods can. The MAb is attached to a fluorescent dye and injected into a patient or applied to a sample of tissue or body fluid. If the MAb binds its target—an antigen found mostly or only on cancer cells—the fluorescence is detected with a scanning technology or fluorescence microscopy. MAbs linked to radioactive isotopes or to drugs can be used in a similar fashion to ferry treatment to cancer cells.

Boosting Cellular Immunity—Harnessing Cytokines

As coordinators of immunity, cytokines are used to treat a variety of conditions, including cancer, multiple sclerosis, genital warts, AIDS, tuberculosis, and parasitic infections. It has been difficult, however, to develop cytokines into drugs for several reasons. They have side effects, remain active only for short periods, and must be delivered precisely where they are needed.

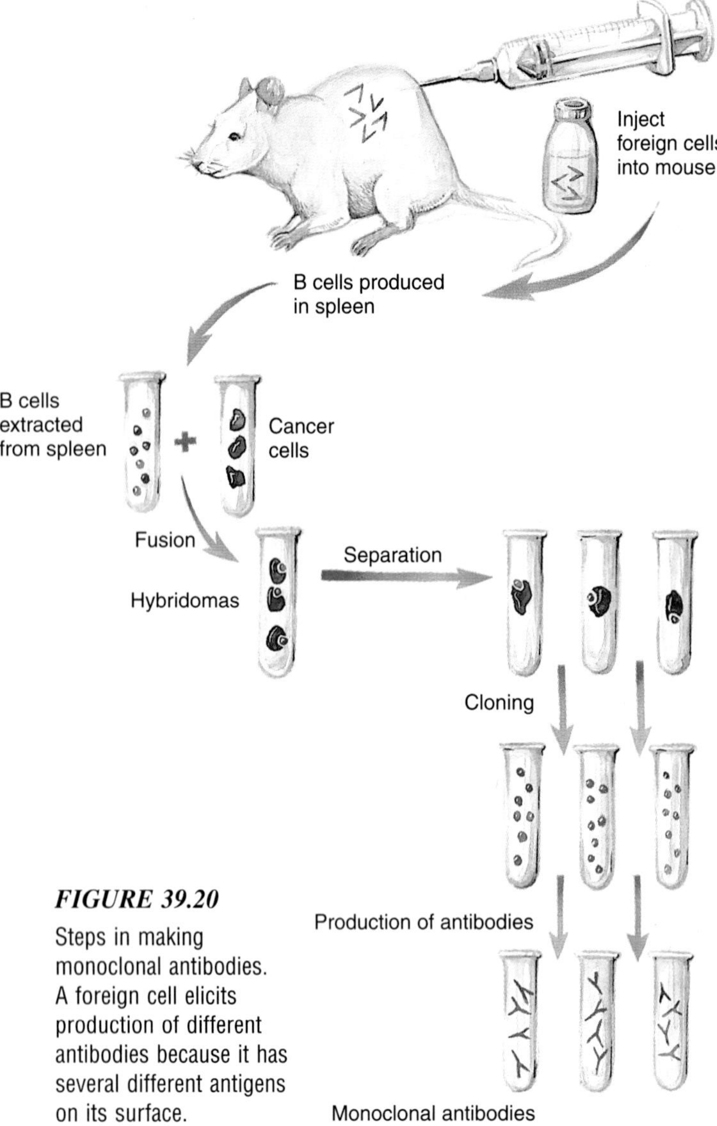

FIGURE 39.20

Steps in making monoclonal antibodies. A foreign cell elicits production of different antibodies because it has several different antigens on its surface.

Interferon (IF) was the first cytokine to be tested on a large scale. When researchers discovered it in the 1950s, IF was erroneously hailed as a cure-all wonder drug. Although it did not live up to early expectations, IF is used today to treat a dozen or so conditions, including a few types of cancer, genital warts, and multiple sclerosis.

Interleukin-2 (IL-2) is administered intravenously to treat kidney cancer recurrence. In another approach, IL-2 is applied to T cells removed from a patient's tumor; then the stimulated T cells, along with IL-2, are reinfused into the patient. Much ongoing research is focused on discovering uses for the many other interleukins. Colony-stimulating factors, which cause white blood cells to mature and differentiate, are used to boost white blood cell supplies in people with suppressed immune systems. People with AIDS or those undergoing cancer chemotherapy might benefit from colony-stimulating factors.

So logical and promising is immunotherapy that cancer treatment may soon include combinations of immune system cells and biochemicals, plus standard therapies. Immunotherapy can enable a patient to withstand higher doses of a conventional drug or destroy cancer cells that remain after standard treatment.

The human immune system has had to continually adapt to new challenges. In the past decade, we've seen several new viral and bacterial illnesses arise and old ones resurge, often in deadly, drug-resistant forms. Let us hope that when we conquer AIDS, what we learn about immune system functions will have repercussions in treating other conditions—and in understanding how this fascinating organ system works.

Mastering Concepts

1. What is a vaccine?
2. What developments have improved the success rate of organ transplants?
3. Explain how monoclonal antibodies and cytokines might be medically useful.

Dying Lions~

continued from page 790.

The sick lions of the Serengeti were suffering from canine distemper. The signs and symptoms that led the graduate student to this diagnosis were confirmed when pathologists examined tissues sampled from lions. Statistics showed a recent and highly virulent infection—90% of Serengeti lions had been infected in 1994, but between 1981 and 1993, none had been infected. How did the plague begin?

Canine distemper, as its name implies, infects dogs, and many thousands of dogs ring the Serengeti, as pets and to help farmers round up livestock. An RNA virus called a morbillivirus causes the disease and spreads easily when an infected animal sneezes. But because surviving a morbillivirus confers lifelong immunity, the virus can persist, and spread, only by finding large populations of susceptible animals. One way the "promiscuous" virus does this is by jumping from species to species—canine distemper also affects dolphins, seals, and ferrets.

How did the virus jump from domestic dogs, who live in human settlements, to lions, who live on the plains? Because lions and dogs don't normally cross paths, epidemiologists hypothesize that intermediaries pass the infection. Leopards, hyenas, and other carnivores attack dogs and drag the kill out into the park, where lions may join the feast and become infected.

Investigators link the canine distemper epidemic among Serengeti lions to a past epidemic of another type of virus called rinderpest. This virus causes lethal diarrhea and decimated many populations, of many species, from 1890 until the 1930s, when it disappeared in all of Africa except the Serengeti. It remained there because of the large number of animals to support it. In the early 1960s, a massive vaccination campaign eliminated rinderpest virus among cattle, and gradually it vanished from wild populations too. Without rinderpest, populations soared, and human settlements grew too, along with the increasing availability of food. People raised many dogs—which became the reservoir for the next scourge to come along, the morbillivirus that causes distemper.

Again, a vaccination plan has come to the rescue. Project Life Lion has led to the vaccination of thousands of domestic dogs near the Serengeti, with the goal of saving the lions. But what viral epidemic will be next?

SUMMARY

1. All animals have immune defenses that distinguish self from nonself and make survival possible in an environment with parasitic or infectious organisms. Immunity also fights cancer. Simple animals have amebocytes, and more complex animals have circulating phagocytes that engulf and destroy invaders. An immune response is fast, specific, diverse, and "remembers."

2. Nonspecific human defenses prevent infectious agents from entering the body. These include barriers such as skin, mucous membranes, secretions, and cilia. If microbes penetrate the barriers, phagocytes engulf and digest them. **Inflammation,** antimicrobial substances, and fever create an environment hostile to pathogens.

3. Macrophages quickly phagocytize infectious microbes. The macrophage displays a piece of the microbe's surface on its own surface, with an HLA glycoprotein, which alerts and activates **helper T cells,** which in turn activate **B cells.**

4. Activated B cells carry out the **humoral immune response.** They mature into **plasma cells** and secrete **antibodies.** A first reaction to a particular foreign antigen is a **primary immune response; memory cells** formed from some of the activated B cells provide a quick **secondary immune response** when an antigen is encountered again.

5. An antibody is a Y-shaped protein composed of two **heavy chains** and two **light chains.** Each chain has a region with a constant amino acid sequence and a region of variable sequence. The tips of the Y-shaped antibody form an **antigen binding site,** where a foreign antigen binds. Antibodies bind antigens and form complexes that attract other immune system components. The human body can produce a tremendous variety of antibody molecules because DNA segments shuffle during early B cell development.

6. T cells carry out the **cellular immune response.** They are educated in the thymus gland to recognize self. Helper T cells activate other T cells and B cells. A helper T cell's **CD4 antigen** binds to macrophages presenting antigens. **Cytotoxic T cells** release biochemicals that bore into bacteria, kill them, and also destroy body cells covered with viruses. T cells secrete **cytokines,** which control communication within the immune system.

7. The fetal immune system learns to recognize self but does not mature until about 18 months after birth. The fetus and newborn receive **passive immunity** via the placenta and breast milk. **Active immunity** builds during the first 2 years. The immune system begins to lose function in adolescence.

8. Immune system breakdown may be acquired or inherited. HIV overtakes helper T cells, preventing them from activating B cells. HIV replicates using **reverse transcriptase;** the T cells burst, releasing many viruses. Immunity is suppressed and opportunistic infections set in. HIV is affecting more diverse population groups, but long-term survivors may provide information that can lead to a cure. People who are resistant to HIV infection have abnormal HIV receptors on CD4 helper T cells. In severe combined immune deficiency (SCID), a person is born lacking T and B cells. Chronic fatigue syndrome reflects an overactive immune system.

9. In an **autoimmune** disorder, the body manufactures **autoantibodies** against its own cells. Autoimmunity may result from a virus that has borrowed a self antigen, from T cells that cannot distinguish self from nonself, or from bacterial or cancer cells that have antigens that resemble a self antigen.

10. In an **allergy,** immunity attacks a harmless substance, called an **allergen.** When a person contacts an allergen, IgE antibodies bind **mast cells,** which release **allergy mediators** that cause symptoms. Anaphylactic shock is a systemic, severe allergic reaction.

11. Several technologies alter immune system function. **Vaccines** are killed, weakened, or partial pathogens that stimulate antibody production without causing disease. Transplants can sometimes succeed if immune attack against nonself tissue is shut off. Individual activated B cells fused with cancer cells form **hybridomas,** which secrete **monoclonal antibodies.** Cytokines and monoclonal antibodies are useful in diagnosing and treating some diseases.

TO REVIEW . . .

1. Describe the immune defenses of three types of organisms.

2. Which parts of the immune system do the following conditions affect?
- **a.** AIDS
- **b.** hay fever
- **c.** anaphylactic shock
- **d.** myasthenia gravis

3. How do each of the following cells interact with other immune system cells?
- **a.** mast cells
- **b.** macrophages
- **c.** cytotoxic T cells
- **d.** helper T cells
- **e.** B cells

4. List the cells and biochemicals that provide the humoral and cellular immune responses.

5. What do a plasma cell and a memory cell descended from the same B cell have in common, and how do they differ?

6. What part do antibodies play in allergic reactions and in autoimmune disorders?

7. How might a drug advertised to be a "histamine blocker" relieve allergy symptoms?

8. A young man eats gooseberry pie and soon begins to feel uneasy and warm. Then itchy hives pop out on his skin. The itching intensifies, and suddenly he heads for the bathroom, where he vomits and has diarrhea. His throat swells, and he seeks help. What is happening to him?

9. How can a vaccine cause the illness it is intended to prevent?

10. Why is a polyclonal antibody response valuable in the body but a monoclonal antibody valuable as a diagnostic tool?

TO THINK ABOUT . . .

1. How might a fever be adaptive?

2. Vaccines must be tested on healthy individuals to see if they protect against disease. Often, researchers vaccinate themselves, or people who have already had the illness, and then look for a boost in antibody levels as a sign of success. How do you think AIDS vaccines should be evaluated?

3. Rasmussen's encephalitis is a rare and severe form of epilepsy that causes children to have 100 or more seizures a day. Affected children have antibodies that attack brain cell receptors that normally receive neurotransmitters. Is this condition an inherited immune deficiency, an acquired immune deficiency, an autoimmune disorder, or an allergy?

4. A therapeutic approach to combat AIDS is to block CD4 receptors on T cells. How would such an agent fight AIDS?

5. One out of every 310,000 children who receives the vaccine for pertussis (whopping cough) develops permanent brain damage. The risk of suffering such damage from pertussis is about 1 in 30,000. Some parents refuse to vaccinate their children because of the few reported adverse cases. What are the dangers, both to the individual and to the population, when parents refuse to allow their children to be vaccinated against pertussis?

6. Soon after a heart attack, damaged cardiac muscle tissue releases small amounts of myosin into the surrounding tissue fluid. Considering this information, devise a way to use monoclonal antibodies to assess the extent of heart attack damage.

7. Sharks seem to be very resistant to cancer and to bacterial and viral infections. A shark's skeleton consists of cartilage. Many health food stores sell shark cartilage extracts and claim that it boosts immunity. Why might this claim be unfounded?

TO LEARN MORE . . .

References and Resources

Beck, Gregory, and Gail S. Habicht. November 1996. Immunity and the invertebrates. *Scientific American.* A great diversity of animals have immunity.

Hill, C. Mark, and Dan R. Littman. August 22, 1996. Natural resistance to HIV? *Nature,* vol. 382. Some people cannot become infected with HIV because they have a mutation.

Kubic, Mike. January-February 1997. New ways to prevent and treat AIDS. *FDA Consumer.* People with AIDS now have options.

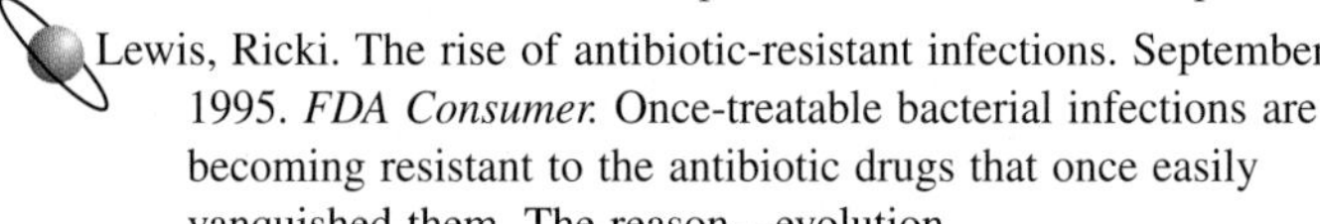

Lewis, Ricki. The rise of antibiotic-resistant infections. September 1995. *FDA Consumer.* Once-treatable bacterial infections are becoming resistant to the antibiotic drugs that once easily vanquished them. The reason—evolution.

Lewis, Ricki. April 3, 1995. End of century marks dawn of clinical trial era for cancer vaccines. *The Scientist.* Cancer vaccines bolster the immune system to combat cancer cells.

Lewis, Ricki. March 1, 1995. Interferon must prove itself in market for genital warts. *Genetic Engineering News.* An immune system biochemical targets this widespread STD.

Lewis, Ricki. June 1994. Getting Lyme disease to take a hike. *FDA Consumer.* Humans, deer, mice, ticks, and bacteria interact in the transmission of Lyme disease.

Licinio, Julio, and Ma-Li Wong. December 1996. Interleukin 1B and fever. *Nature Medicine.* An interleukin controls fever.

Mestel, Rosie. September 1996. Sharks' healing powers. *Natural History.* Sharks hold clues to the evolution of immunity.

Morell, Virginia. July 1995. The killer cat virus that doesn't kill cats. *Discover.* FIV may hold clues to conquering HIV.

Murphey-Corb, Michael. January 1997. Live-attenuated HIV vaccines. How safe is safe enough? *Nature Medicine.* An HIV vaccine in monkeys seems to work, but is deadly in high doses.

Packer, Craig. June 1996. Coping with a lion killer. *Natural History.* Canine distemper is killing lions in the Serengeti.

Stoye, Jonathan P. March 13, 1997. Xenotransplantation: Proviruses pose potential problems. *Nature,* vol. 386. Using nonhuman animals for replacement parts could introduce new viruses.

Wain-Hobson, Simon. May 8, 1997. Down or out in blood and lymph? *Nature,* vol. 387. HIV cannot take over all cells.

Wakelin, Derek. January 1997. Parasites and the immune system. *BioScience.* Parasites challenge the immune system.

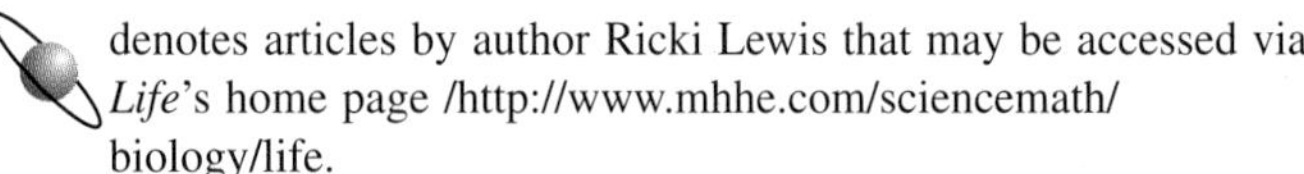

denotes articles by author Ricki Lewis that may be accessed via *Life*'s home page /http://www.mhhe.com/sciencemath/biology/life.

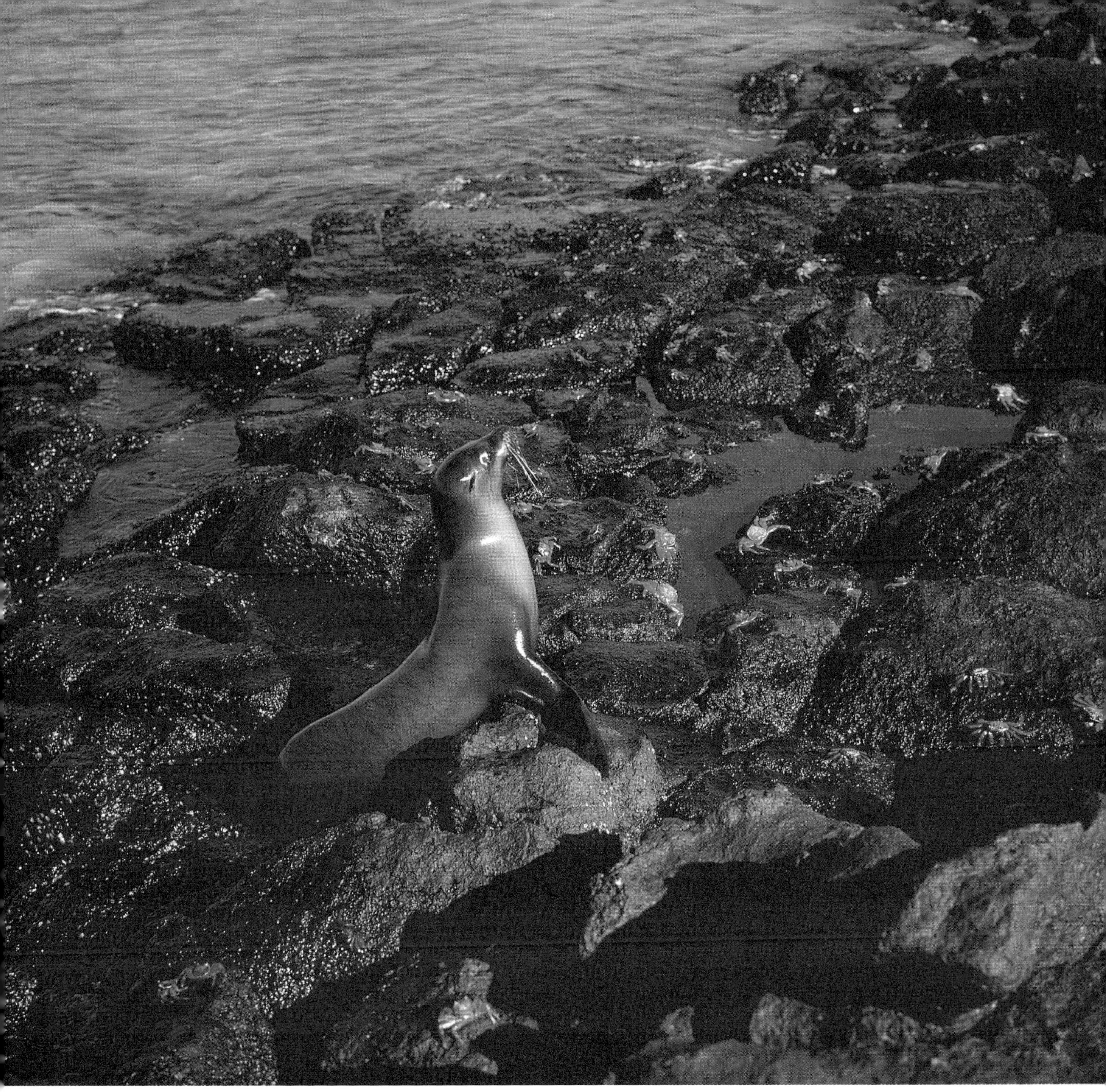

Unit 9

Behavior and Ecology

A sea lion and scarlet crabs are part of this ecosystem on an island in the Galápagos.

40 Behavior

Snapping shrimp (*Synalpheus regalis*) live in social groups in sponges in Caribbean coral reefs. Each colony has a single reproductive queen, shown here in the center. Some of the shrimp around her display their characteristic claws, which they use to protect the colony—sponge homes are in high demand.

chapter outline

Studying Behavior 812

Genes and Experience Shape Behavior 812

Innate Behavior 813
Fixed Action Patterns
Releasers
Reaction Chains

Learning 814
Habituation
Classical Conditioning
Operant Conditioning
Imprinting
Insight Learning
Latent Learning

Orientation and Navigation 817
The Compass Sense
Homing

Biological Societies 819

Advantages and Disadvantages of Group Living 821
More Favorable Surroundings
Better Defense
Enhanced Reproductive Success
Improved Foraging Efficiency
Improved Learning
Disadvantages of Group Living

Social Structure and the Situation 823

Communication and Group Cohesion 824
Chemical Messages
Auditory Messages
Tactile Messages
Visual Messages

Altruism 826

Aggression 827
Territoriality
Dominance Hierarchies
Cannibalism

Mating Behavior 829
Mating Systems
Two Views of Mating Behavior—Spiders and Sparrows
Primate Mating Behavior

Shrimp Society

Animal societies are well known among insects and primates, and a few years ago, researchers described the complex group living of naked mole rats. A newly recognized social species is the snapping shrimp, *Synalpheus regalis.*

Snapping shrimp are so named for their powerful fighting claws. These crustaceans occupy an unusual habitat—cavities within two types of sponges that live among coral reefs in the Caribbean. Shrimp colonies within sponges average 149 individuals but may include up to 300.

Each shrimp colony has only one female that reproduces, which she does very often. Young undergo direct development—there are no larvae. The members of the colony are therefore half or full siblings. The spongy home protects the shrimp and provides food, which floats into the colony from the sponge's internal canal system.

Collecting and dissecting sponges easily reveals two of the three defining characteristics of an animal society. First, generations overlap, as shown in the groups of young of different size and presumably different age. Second, shrimp colonies have a division of labor, as evidenced in the lone reproductive female. Demonstrating the third characteristic of an animal society—cooperative care of young—required more than simply dissecting sponges and observing their crustacean inhabitants. An experiment was necessary.

to be continued . . .

Studying Behavior

The ground squirrel appeared to sense danger as soon as she neared her burrow and saw her neighbor's frenzied tail flagging and darting about. She had been out foraging for food and had left behind three pups. The squirrel's behavior sent an unmistakable message—rattlesnake!

A snake commandeering a burrow to keep itself warm could trap and eat a newborn squirrel. But this squirrel's pups were standing their ground well. Although they were too young to have encountered a rattlesnake before, or to have learned from their mother how to defend themselves, the pups instinctively reacted to the threat. They flagged their tails, pushed out their snouts, jumped back and forth as if anticipating a snake's strike, and vigorously kicked dirt and sand at the reptile. The arrival of the mother sent the snake slithering into the temporarily abandoned burrow. Mother and pups immediately kicked sand and dirt into the burrow. They were safe—for a while (fig. 40.1).

The encounter between the squirrels and the snake vividly illustrates animal behavior. **Ethology,** the study of animal behavior, was officially recognized in 1973, when zoologists Karl von Frisch, Konrad Lorenz, and Niko Tinbergen won a Nobel Prize for their groundbreaking studies over four decades. Ethology examines adaptive behaviors that enable animals to survive and reproduce in their natural habitats. Also in 1973, evolutionary biologist John Maynard Smith compared behaviors that promote survival to strategies in a game in which one individual's behavior depends upon the actions of others. Survival strategies are subject to natural selection.

Long before 1973, biologists painstakingly observed animals in the wild (fig. 40.2). The mostly observational studies that formed the foundation of ethology are today complemented by experiments that examine the components of and driving forces behind particular behaviors. In addition, biologists now use DNA typing to identify and follow particular individuals whereas in the past, determining how individuals were related was inferred from observing animals' interactions. All of these approaches are used to reveal how and why animals behave in characteristic ways.

Genes and Experience Shape Behavior

Although squirrels instinctively can deter a snake, predators still corner and devour many squirrel newborns. Experience teaches a pup to refine its inborn skills to evade snakes and other predators. Heredity (nature) as well as the environment (nurture) determine all behaviors, with each influence contributing to particular responses to different degrees. Behavior is complex. Even the simplest actions reflect input from several genes, and heredity and environmental influences can be very difficult to identify and distinguish.

An animal's anatomical systems for detecting stimuli, integrating the input, and responding limit behavior. Genes determine how sophisticated and sensitive these anatomical systems are. On the other hand, learning can be important in mastering complex behaviors, such as reading, speaking a language, or following the rules of a game.

A behavior is usually primarily genetically determined, or **innate,** when it is critical that the animal perform the action correctly the first time. Escape behaviors, such as the squirrel's sand-kicking stance, tend to be under fairly precise genetic control. If an individual fails to respond appropriately in its first encounter with a predator, it is eaten and never gets a second chance to refine the response. Animals that respond correctly to predators live to pass their genes to future generations.

Genetic constraints on learning are many, as animal trainers well know. One group of trainers attempted to teach raccoons, roosters, and pigs to put coins into a piggy bank for a food reward. The animals were not always cooperative. Roosters scratched their feet on the floor, pigs rooted on the bare floor, and raccoons fondled their coins rather then depositing them in the bank. The animals were behaving instinctively to food deprivation—roosters scratch the ground to uncover grain, pigs root around for food, and raccoons wash their food in water, rubbing it with their fingers.

A bird's song reveals the influence genes exert on learning. Young male, white-crowned sparrows learn to sing by hearing the melodies that their fathers sing. If isolated from older males of their species, they sing abnormal songs as adults. Learning

FIGURE 40.1

Ground squirrels effectively defend against rattlesnakes and are rarely bitten. When they are, the squirrels' blood proteins detoxify the venom. Many of the squirrels' protective behaviors are innate. Pups who have never encountered rattlers move, kick sand, and flag their tails as if they were quite experienced at evading snakes.

FIGURE 40.2

Jane Goodall, a British zoologist, spent many years observing chimpanzees at very close range, discovering much of what we know about their behavior.

doesn't completely control singing, however, because the sparrow can only learn the song of its own species. If a young isolated bird hears recordings of a variety of birdsongs, including a normal white-crowned sparrow song, it learns to sing properly. The bird innately recognizes its species' song and then perfects the performance through experience.

Mastering Concepts

1. How are behaviors adaptive?
2. What are some ways that ethologists study animal behavior?
3. What are the roles of heredity and the environment in determining behaviors?
4. Why is it difficult to isolate the contributing factors to a behavior?

Innate Behavior

Natural selection molds innate behaviors because they are largely genetically determined. Genes that predispose an individual to a behavior that lowers the chances of survival to reproduce will ultimately diminish in the population.

Fixed Action Patterns

A **fixed action pattern (FAP)** is an innate, stereotyped behavior, such as a dog digging on the floor as if trying to bury a bone or a kitten pouncing on a rustling leaf as if it were a mouse. Members of a species perform a FAP nearly identically, with little environmental modification.

The egg-rolling response of a female greylag goose is a FAP (fig. 40.3). When a brooding female sees an egg outside her nest, she retrieves it in a characteristic manner: she stretches her bill just beyond the egg and then scoops it toward her, repeating this motion until the egg is back in the nest. She adjusts her movements as the egg wobbles over uneven terrain. This action is adaptive, because an unincubated egg will not hatch.

The goose will retrieve a beer bottle as readily as an egg. Apparently she has only a vague sense of what an egg is and automatically responds to all small, rounded objects outside the nest. If the egg is removed after she has begun rolling it back to the nest, she continues her retrieval motions until the imaginary egg is in the nest. Once a FAP begins, the action continues until completion, even without appropriate feedback.

FIGURE 40.3

Like all fixed action patterns, egg retrieval in the greylag goose is always the same. She extends her neck so that it is just beyond the egg and then scoops the egg toward her, rolling it back to the nest.

Releasers

The specific factor that triggers a FAP in a species is called a **releaser.** An animal's senses are bombarded constantly with many more stimuli than it could possibly respond to. The animal must select and respond to only the few key stimuli that are reliable cues. Even a single object or organism has many aspects to it, such as color, shape, odor, taste, sound, and movement.

To identify the specific releaser for a behavior, ethologists build models that isolate a single stimulus, such as color or scent. The researcher shows the model to an organism in an appropriate physiological state and observes whether it responds as it would to the natural stimulus. For example, a male stickleback fish ignores a model that looks exactly like a rival male except that it lacks a red underbelly; however, a male will attack a model of almost any shape if its bottom half is painted red. The color red, then, releases aggressive behavior in the male stickleback.

Releasers are important in human parent-infant interactions. When an infant gazes and smiles at her exhausted parents, the adults respond with affection. But an infant up to 2 months old will smile at anything with two dots resembling eyes! A model of a face lacking eyes will not elicit a smile.

Many releasers are auditory (sensed by hearing). A mosquito is attracted to his mate by the buzz of her wings. A model of the stimulus, a tuning fork that vibrates at the same frequency as a female's wings, also attracts him. The pulsing chirps of crickets, the clatter of katydids, and the droning of a bullfrog are all auditory signals that release mating behavior.

Releasers may be tactile (sensed by touch). During the mating ritual of the stickleback fish, the female enters the nest, which stimulates the male to thrust his snout against her rump in a series of quick, rhythmic trembling movements. These thrusts release her spawning behavior. Prodding the female with a glass rod mimics the releaser.

Pheromones are chemical releasers, and they include sex attractants of some insects, crustaceans, fishes, and salamanders. To show that a pheromone is a releaser of mating behavior in the Canadian red-sided garter snake, researchers extracted lipids from the skins of sexually mature female and male snakes. Lipids from the females were absorbed onto paper towels, and the towels placed near males. The snakes became stimulated, some even attempting to mate with the pheromone-soaked paper. When lipids that males normally secrete were added to the towel, the male snakes backed off. Scientists working with pheromones may become releasers themselves, because just a trace of a sex attractant can be powerful. One researcher was besieged by amorous male moths at a football game, even though he had showered.

A **supernormal releaser** is a model that exaggerates a releaser and elicits a stronger response than the natural object does. Birds such as the oystercatcher and the herring gull, for example, prefer to sit on large eggs. The female will try diligently to sit on an egg that is more immense than she could have possibly laid, even though her own egg is nearby.

Reaction Chains

A complex behavior called a **reaction chain** develops when a sequence of releasers joins several behaviors. A reaction chain may be a series of actions an individual takes, such as nest building, or an exchange of releasers between two individuals, such as courtship and mating behaviors. Consider the elaborate mating rituals of the three-spined stickleback fish (fig. 40.4). The male prepares the nest, attracts a female, and entices her into the nest. He then enters the nest and fertilizes the eggs she has laid. The sequence of behaviors is always performed in the same order because the completion of each step is the releaser for the next.

Mastering Concepts

1. What is a fixed action pattern?
2. What is the function of a releaser?
3. Give examples of some types of releasers.

Learning

Learning is a change in behavior as a result of experience. A young bee innately recognizes a flower pattern of light petals and dark center. An older bee, with experience, recognizes specific flowers. Ethologists recognize several types of learning, and table 40.1 gives familiar examples of learning in humans.

(a)

(b)

(c)

(d)

(e)

FIGURE 40.4

Steps of mating behavior. (*a*) Mating in the three-spined stickleback is mostly innate. In early spring, in shallow ponds and lakes, male sticklebacks prepare to mate. A male leaves the group, stakes out a territory, and builds a nest of algal weeds, coating the structure with a gooey material from his kidneys. He digs a hole through the mass and then his chin turns red and his back turns a bluish white. Spying a female, he zigzags toward her, and she swims toward him. (*b*) The male swims to the nest and pushes his snout into it. He flips on his side and presents the fins across his back to the female. (*c*) The female swims into the burrow in the nest, followed by her mate, who touches the base of her tail. In response, she lays 50 to 100 eggs. (*d*) The male enters the nest to fertilize the eggs and then moves water over them, bringing in oxygen. (*e*) Fertilization accomplished, the male leaves the nest.

Habituation

In the simplest form of learning, **habituation,** an animal learns not to respond to certain irrelevant stimuli. When a stimulus occurs many times without any consequence, the animal usually decreases its response, perhaps eliminating it completely. For example, young chicks learn not to run from innocuous, common stimuli, such as blowing leaves and other nonpredatory birds. Responses to predators are retained because they are rarer and more essential for survival.

Habituation also helps ensure that aggression is appropriate and adaptive. If seabirds that often nest within a few feet of one another did not habituate to the presence of neighbors, they would waste much time on aggressive encounters. Instead, they learn not to show aggression toward neighbors as long as they remain outside a certain territorial limit. They do retain aggression toward other birds that they encounter less frequently.

Classical Conditioning

In **classical conditioning,** an animal learns to respond in a customary way to a new stimulus. A new association forms between a stimulus and a response when the new or **conditioned stimulus** is repeatedly presented immediately before a familiar or **unconditioned stimulus** that normally triggers the response. After the stimuli have been paired in a series of trials, the new stimulus alone elicits the response.

The most familiar example of classical conditioning is a dog associating the sound of a bell with feeding. When fed many times following a bell ring, the dog salivates at the sound of a bell. If, however, the bell rings several times without ensuing food, the dog will no longer salivate in response to the bell. This loss of a conditioned response is called **extinction.**

Table 40.1 Learning in Humans

Type of Learning	Example
Conditioned learning	Craving popcorn in a movie theater
Habituation	Being able to concentrate on written work with music playing in the background but jumping when the phone rings
Imprinting	Puppies and kittens become tame if people handle them in the first few weeks of life
Insight learning	A child places a box on roller skates to build a vehicle
Latent learning	A medical student watches a physician give an injection and then does it herself
Operant conditioning	A student learns what to do to earn high grades in school

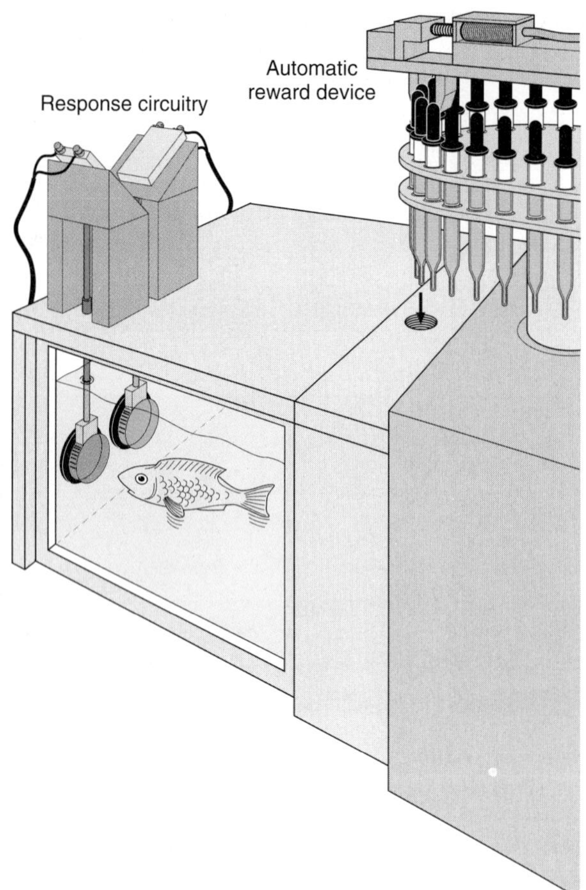

FIGURE 40.5

Operant conditioning of a fish. In this discrimination tank, a variation on a Skinner box, a fish is shown two colors. When it taps the one the experimenter desires, it receives a treat—a worm tidbit.

Operant Conditioning

Operant conditioning is a trial-and-error type of learning in which an animal voluntarily repeats behavior that offers a reward (positive reinforcement) or avoids a painful stimulus (negative reinforcement). Reinforcement increases the probability that the animal will repeat the behavior. Trial-and-error learning can refine natural behaviors, such as predatory skills. A grizzly bear, for example, might learn that splashing about in a stream does not yield a salmon dinner. Staying still and quiet in one place is much more effective.

Ethologist B. F. Skinner designed an apparatus to demonstrate operant conditioning. When a hungry rat is placed in a Skinner box, it explores its surroundings and eventually presses a lever by accident, which releases a food pellet. This food reward increases the probability that the rat will press the lever again. Figure 40.5 shows a variation on a Skinner box.

Animal trainers use operant conditioning. They first reinforce any behavior that vaguely resembles the desired one and then restrict the reward to better approximations. Pigeons learn to play table tennis in this way. Hungry pigeons are taught to use their beaks to hit a ball. When the ball falls in the trough on either side of the table, the trainer delivers a food pellet to the pigeon that won the point.

FIGURE 40.6

Many baby birds imprint on the first moving object they see. This is usually the mother but may be an ethologist, such as Konrad Lorenz.

Imprinting

Imprinting is a type of learning that occurs quickly, during a limited time (called the **critical period**) in an animal's life, and usually is performed without obvious reinforcement. Young chicks, goslings, or ducklings are imprinted to follow the first moving object they see, which in the wild is their mother. This "following" response is adaptive, because the mother typically would lead them to a safe place where they are likely to find food.

Ethologist Konrad Lorenz is famous for his studies of imprinting. Lorenz hatched eggs in an incubator and then had the baby birds waddle after him. Thereafter they showed the same responses to him as they would have to their mother (fig. 40.6). A recent film chronicled a flock of geese that imprinted on a 13-year-old girl in Canada and even followed her as she flew in an airplane south for the first migration.

Imprinting is important in the development of attachment between mother and newborns in several species, including goats, sheep, and the Alaska fur seal. In the first few critical minutes after a goat kid's birth, the mother learns to identify her offspring by their odor. She will accept and nurse any young that she smelled during the critical period and reject any youngsters she does not recognize.

Insight Learning

Insight learning, or reasoning, is the ability to apply prior learning to a new situation without trial-and-error activity. For example, in one experiment, chimpanzees were put into a room with boxes and a banana hanging from the ceiling. The chimpanzees reasoned that stacking the boxes would enable them to reach the banana—and that is exactly what they did.

Latent Learning

Latent learning occurs without obvious reward or punishment and is not apparent until after the learning experience. Even without reinforcement, a rat that has been allowed to run freely through

FIGURE 40.7

These long-eared owls are congregating before they start their annual migration across Lake Ontario.

a maze will master its twists and turns more quickly than a rat with no previous experience in the maze. In the wild, animals may learn the details of their surroundings during daily explorations. This information may not be of immediate use, but knowing where to quickly find hiding places may make the difference between life and death when a predator strikes.

Mastering Concepts

1. What is learning?
2. Describe the types of learning.

Orientation and Navigation

Each spring, North American skies fill with wings as birds migrate north from the tropics. Using the position of the sun or stars, subtle shifts in the earth's magnetic field, sights, sounds, and winds and the aromas they carry, these animals cover vast distances and arrive with great accuracy. Their lives depend on it.

Migration is a regularly repeated journey from one specific geographic region to another. Bird migrations are well studied (fig. 40.7). Arctic terns fly nearly from pole to pole, about 11,000 miles (17,700 kilometers). Some Swainson's hawks cover 7,000 miles (11,272 kilometers) from Alaska to Argentina. Birds are not the only migrants. Atlantic salmon that begin life in New England rivers migrate through the sea to Greenland to feed and then return to the exact river or stream where they were born to spawn. Some insects, crustaceans, and salamanders also orient (move in a specific direction) and navigate (follow a specific course).

The simplest type of navigation relies on recognizing landmarks in a random search. When bank swallows are moved from their nesting burrows, they find home by looking around until they recognize a familiar landmark. The further the birds are displaced, the less likely they are to return home. Biology in Action 40.1 discusses turtles that migrate 2,800 miles (4,505 kilometers) to lay eggs!

The Compass Sense

Most migrating species use environmental cues as a compass to orient them in one direction. Caged birds denied the opportunity to migrate still orient in the directions that free members of their species do at approximately the right flight times.

The sun can serve as a compass. It rises in the east and moves across the sky at an average rate of 15° an hour, setting in the west. An animal's biological clock tells it the time of day. When the animal combines this information with the position of the sun, it can orient in any direction.

The earth's magnetic field, which runs in a generally north-south direction, can also serve as a compass. A pigeon can find its way home if released up to 1,000 miles (1,609 kilometers) away; however, bar magnets placed on a pigeon's wings impair its ability to find its way home on cloudy days. On sunny days, when the sun can provide cues, the magnets have no effect. Sham magnets (nonmagnetic bars of equal size and weight) never disrupt orientation, indicating that the magnetism, and not the bar itself, disrupts normal navigational cues.

Homing

The most complex navigational skill is the ability to home—that is, to return to a given spot using no environmental cues after being displaced to an unfamiliar location. Homing requires both a compass sense for direction and a map sense telling the animal where it is relative to home. We know very little about the map sense. It may depend on regular variations in the strength of the earth's magnetic field, which is about twice as strong at the poles as it is at the equator. An animal very sensitive to the strength of a magnetic field would know how far north or south it is. Pigeons become disoriented when sunspots disrupt magnetic fields.

Mammals that miraculously find their way home suggest that familiar senses may help foster a map sense. A wolf that had spent her whole life in a pen in Barrow, Alaska, was taken 175 miles (282 kilometers) away and released. She found her way home. A deer taken 350 miles (563 kilometers) from its wildlife refuge on the Gulf of Mexico found his way back. Closer scrutiny revealed how the animals may have navigated. The wandering wolf had grown up next to an airport and was accustomed to the sounds of jets taking off and landing. Such loud sounds travel far on the bleak Alaskan landscape, and the animal may have traveled so that the jets became louder. (Another wolf taken on the same trip returned to the wrong airport!) The deer seeking his wildlife refuge may have homed in on scents wafting on gulf breezes.

40.1 biology in action

DNA Testing Solves the Mystery of Green Turtle Migration

Imagine traveling 1,400 miles (2,253 kilometers) to have babies and then turning around and going back. This is exactly what females in a population of green turtles do every few years (fig. 40.A). Each breeding season, a female turtle migrates from feeding grounds off the coast of Brazil to a tiny peak in the middle of the Atlantic Ocean called Ascension Island. She digs a trench on the beach, deposits about 100 eggs, swims near shore for 2 weeks, lays another clutch of eggs, and continues this pattern for 3 months. Then she swims 1,400 miles back to Brazil, not eating during the long journey (fig. 40.B).

Why do these particular turtles do this, when others of their species breed near where they feed? In the 1960s, turtle expert Archie Carr proposed an intriguing hypothesis.

Carr considered the location of Ascension Island. It lies directly above the mid-Atlantic ridge, which is a buckling in the ocean floor where hot rock pours forth from the earth's interior, forming islands that move outwards and are eventually submerged. Ascension Island, then, is constantly being renewed. But it was not always so far from Brazil, nor from Africa. About 80 million years ago, the two continents were connected, and then they began to drift apart. Carr hypothesized that the turtles' ancestors nested on an Ascension Island that was not very far from the Brazilian coast. Over the years, as the continents continued to separate, the turtles instinctively continued to migrate to the island, as they still do today.

Since 1965, researchers have tagged and followed 28,000 migrating turtles. The turtles always return to Ascension Island to lay eggs. If the behavior is innate, implying a genetic basis, then the Ascension turtles should be genetically quite distinct from other populations. In 1994, researchers tested this hypothesis. They collected eggs from the island and analyzed mitochondrial DNA (mtDNA) from the embryos. Recall that mitochondrial DNA traces the maternal lineage. Surprisingly, the studies indicated that the Ascension turtles are not very different from green turtles in other populations. Therefore, the drive to return to the tiny island is probably *not* an instinct handed down from when the ocean was narrower.

If the green turtles' homing isn't innate, could it be learned? Many researchers today think so. They hypothesize that baby turtles imprint on some feature of the island, and this is what drives them to return there to nest. This is a more adaptive behavior than being locked into an invariant innate drive, because if some force of nature should destroy the nesting site, the turtles could successfully lay eggs elsewhere.

Mysteries remain concerning these green turtles. Just what do baby turtles recognize on the island that makes such a strong and early impression? What cues do they use to find their way home? The Ascension turtle population has been endangered since seafaring Britains began capturing them to make soup in the last century. Let's hope that they survive and that we can one day understand their astonishing migration behavior.

FIGURE 40.A
Green turtles migrate far to lay their eggs.

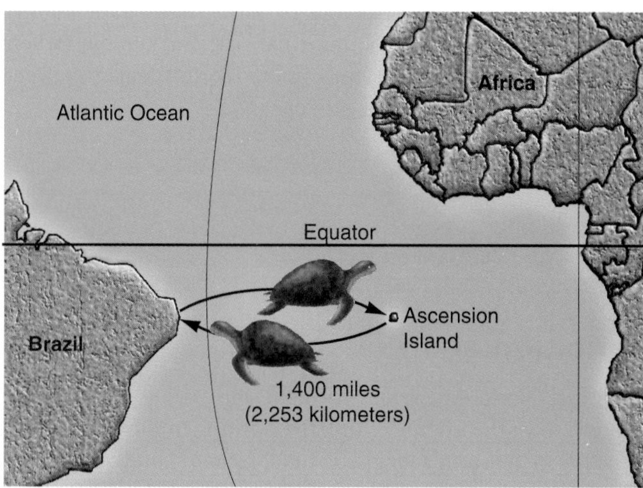

FIGURE 40.B
Female green turtles (*Chelonia mydas*) in a certain population migrate 1,400 miles (2,253 kilometers) to a tiny island in the Atlantic to lay their eggs.

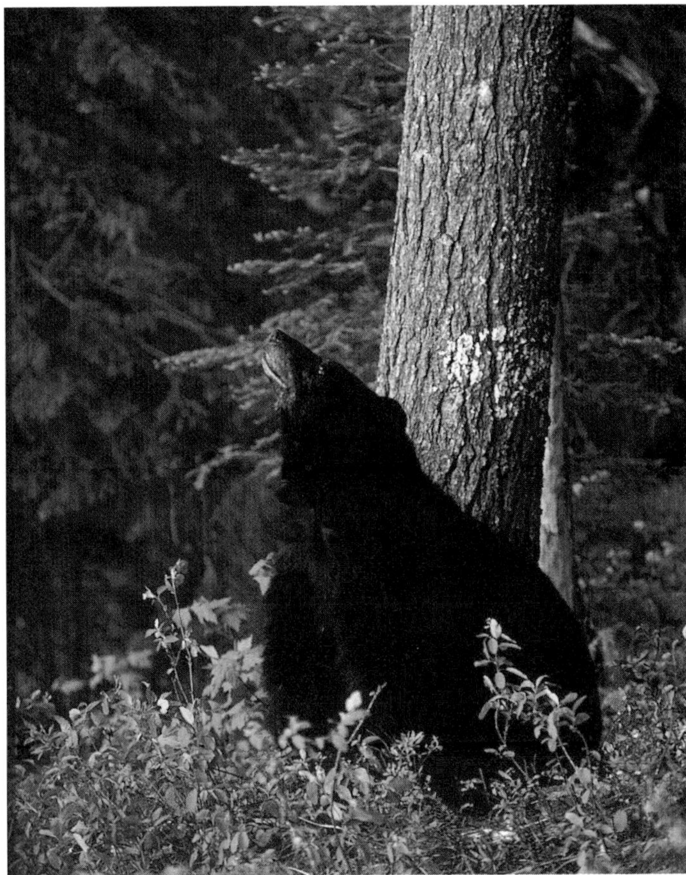

FIGURE 40.8

Sniffing the way home. Bears provide much information on animal navigation because many are removed from towns as "nuisances," tagged, and transported beyond the 40-mile (64-kilometer) radius of their home turf. Odors may help bears find familiar surroundings. This bear is sniffing the breeze.

Many tales of mammal navigation remain inexplicable. Big Mac, a 450-pound (204-kilogram) black bear tracked by a radio collar, found himself far from home, in search of berries to eat. With cold weather imminent, he stopped foraging and tirelessly tramped directly homeward, through forests, backyards, farms, and roads, without stars or sun for guidance, as snow began to fall. His pace slowed and he started to meander only when he was within 40 miles (64 kilometers) of home, the normal radius a bear travels in. Finally, he arrived—to a tiny crevice in a hillside, 126 miles (203 kilometers) from where his berry hunt had taken him (fig. 40.8).

Mastering Concepts

1. What types of environmental cues do animals use to navigate?
2. What is homing?
3. What skills does an animal require to home?

Biological Societies

On a damp, spring morning, all is still in the field. Suddenly, a swarm of red rises from one of the foot-tall mounds that dot the land. Then a second, smaller splash of red follows. The second cloud consists of female red fire ants, and as they pass through the first contingent of tiny males, they copulate, each female picking up a packet of 12 million sperm (fig. 40.9). The successful males immediately die. Worker ants on the ground may eat their deceased brethren.

FIGURE 40.9

A fire ant.

The females drop to the ground, too, shedding their wings. In groups of up to 20, these newly minted queens dig burrows and deposit their eggs within. The first 15 or so eggs from each queen will not develop—they provide food for the others. The next 15 to 20 eggs from each female hatch into larvae. They will be the workers of the next generation and feed their queen. By the time the larvae are thriving, all but one queen has died. She feeds the larvae a rich broth made of her own degraded flight muscles, which she regurgitates to them.

The fire ants that inhabit the southern United States superbly illustrate a biological society. Such a group is termed **eusocial** if it exhibits three characteristics:

1. cooperative care of the young;
2. overlapping generations;
3. division of labor.

Communication among members is also a requirement of eusocial groups, but species that do not have such groups also communicate.

Division of labor may make possible the evolutionary success of social insect colonies. Coordinated groups of individuals make up the colony, each group with a specific function. One group might locate new food sources, while another assesses the sugar content of prospective foods. Other groups care for young or dispose of dead members. Many individuals each performing only a few tasks minimizes errors.

Several factors influence a social insect's role. These include nutrition, the temperature of the nest, pheromones, and age. For many species, younger members stay in the interior of the hive or nest, where they care for the queen and eggs. As the insects age,

40.2 biology in action

Three Animal Societies

Naked Mole Rats

They live in vast, subterranean cities, digging tunnels and chambers in the soils of Kenya, Ethiopia, and Somalia. Roads above them collapse; crops that get in their way, such as yams, they eat. The naked mole rats of Africa live in societies remarkably like those of eusocial paper wasps and sweat bees, with a powerful queen ruling a colony of fewer than 300 individuals. Unlike most mammalian social groups, in which many females reproduce, the naked mole rat queen is the sole sexually active female.

Other roles in the naked mole rat community are well defined. The "janitors," the smallest and youngest males and females, patrol the colony's tunnels. They keep the walls smooth by rubbing them with their soft, hairless bodies. Janitors eat roots hanging from ceilings and confine excrement to widened dead ends.

The oldest and largest mole rats serve sentinel duty, their senses sharpened to detect intruders (fig. 40.C). A sentinel will stop and feel the air currents on its nearly sightless eyeballs. A change in current indicates an invader. The sentinel also senses the low-frequency sound of footsteps overhead and detects many odors. If these signs, for example, indicate that a snake has poked its head into a tunnel, the sentinels signal to the others to descend to wider tunnels.

Mole rats dig new tunnels in early morning and late afternoon. "Dirt carriers" form groups; the head animal is a "digger," and a "kicker" works at the back to get rid of the dirt. Food carriers use their long incisors to break off bits of roots and tubers but often sneak snacks before returning to the group.

The queen patrols the tunnels several times an hour. Every 3 months, she gives birth to a litter of up to 27 pups, nursing them for a month while three male workers bring her food. For the next 2 months, these workers care for the young, who then graduate to worker status. Apparently the queen's presence is crucial to the integrity of this social colony. In the laboratory, when the queen is removed, chaos ensues. Jobs vanish, and each individual fares for himself or herself.

FIGURE 40.C
Naked mole rats from different colonies attack each other viciously, each trying to drag the enemy into its own territory to sink its teeth into the other animal's flesh.

Inside a Beehive

The inside of a honeybee hive is an efficient living machine, with each individual taking a specific role. At the summit of the organizational ladder is the queen, a large, specialized female who lays about 1,000 eggs each day. Eggs fertilized by a male drone develop through larval and pupal stages into female workers. Unfertilized eggs develop into males. Workers secrete a substance called royal jelly onto a very few eggs. These eggs hatch to yield larvae that eat much more royal jelly than do other larvae, which places them on a developmental pathway toward eventual queendom.

The 20,000 to 80,000 workers vastly outnumber the hundred or so drones, and only one queen reigns per hive. If the queen leaves, workers chemically detect her absence and hasten development of the young potential queens (fig. 40.D). The first new queen to emerge from her pupa case kills the others and then takes a "nuptial flight"

they venture out, first helping to build the nest, then guarding the colony, and finally foraging, the most dangerous activity. The insect society is thus a series of **temporal castes**—groups whose roles change with time.

The proportion of an insect society's population that carries out a specific task changes to meet the colony's needs, even as individual members proceed through the temporal castes. In laboratory studies, if some members of a caste are removed, remaining members compensate by adding tasks to their own jobs. When removed insects are returned, everyone resumes their original tasks.

The best-studied eusocieties are those of ants, termites, and some species of bees and wasps. Fossil evidence of insects in groups suggests that insect societies existed 200 million years ago. Biology in Action 40.2 considers three animal societies.

Mastering Concepts

1. What are the defining characteristics of a eusocial group?
2. How does division of labor help an animal society?
3. What is a temporal caste?

FIGURE 40.D
Honeybee workers build a chamber called a queen cell for a potential queen to develop in.

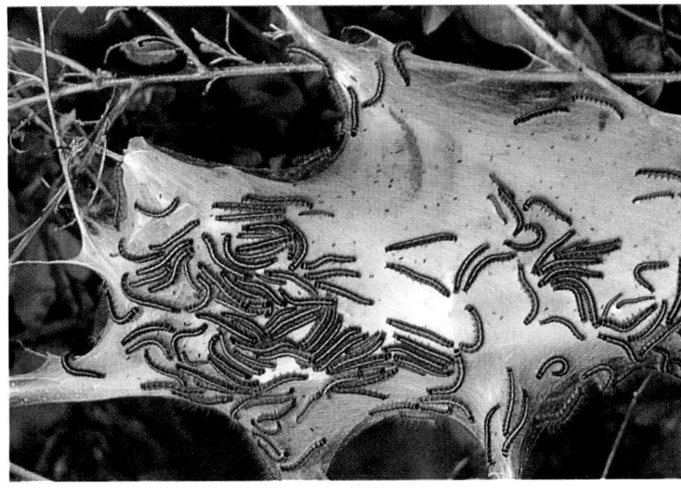

FIGURE 40.E
Eastern tent caterpillars cooperate to locate and advertise a food source.

to attract drones. She collects and stores their sperm and then later uses it to fertilize eggs. Males die when they inseminate the queen, or if they survive, they are forced out of the hive when food becomes scarce or autumn arrives.

A worker's existence is more complex, including several stages and specializations. Newly hatched workers feed the others. One-week-old females make and maintain the wax cells of the hive, where larvae and pupae develop. Some females are undertakers, ridding the hive of dead bees. Some older workers collect food.

Orchestrating the overall social order of a beehive is a pheromone that the queen secretes from glands in her jaw. This "queen mandibular pheromone" is actually a mixture of biochemicals. It has different functions in different contexts. The pheromone inhibits workers from developing ovaries and from rearing additional queens. Inside the hive, queen mandibular hormone inhibits males from mating, yet outside the hive, the pheromone has an opposite effect, stimulating males to mate.

A Tent Caterpillar Community

The silken treetop communities of eastern tent caterpillars are centers for exchanging information about food locations. Several times a day, contingents of 200 caterpillars leave to explore new food sources (fig. 40.E). On the way out, they leave an "exploratory trail" of silk strands permeated with a steroid they secrete from their abdomens. Should they find food, such as an apple tree, they return to the tent along the trail, dragging their hind ends through the original markings. The path is now a "recruitment trail" that the others at the tent know will lead to food.

Tent caterpillars also behave in a way that keeps the community warm. The animals line up next to each other to retain body heat.

Advantages and Disadvantages of Group Living

Groups of animals, from the dozen or so lions in a pride to the thousands of individuals in insect societies, acquire food, protection, and care for their young more easily than solitary animals. Animal societies have been so successful in securing both individual and species survival that they have evolved independently many times.

More Favorable Surroundings

By forming groups, animals may change their environments to their advantage. For example, animals group to conserve moisture or heat. During the winter pigs sleep in heaps. They move and squeal periodically as those on the perimeter seek the warmth of body heat inside the pile, and the central pig becomes too hot and moves to the outside (fig. 40.10).

A group also can physically alter its surroundings. Water fleas cannot survive in alkaline water, but when large numbers congregate, carbon dioxide from their respiration decreases the water's alkalinity to acceptable levels. Similarly, fruit flies fare poorly when they lay too few eggs in an area because there are not enough larvae to break up the fruit and make it soft enough to eat.

FIGURE 40.10
A pile of pigs conserves heat.

FIGURE 40.11
Ostriches protect their eggs. A major hen will sit patiently on eggs not her own. But she places her own eggs directly beneath her body, leaving the eggs of others more available to predators. One of this ostrich's eggs can be seen under her body, to the right part of the photograph.

Better Defense

There is safety in numbers. A predator can more easily fell a lone prey than pick one from a group where a few alert individuals warn the others. Vervet monkeys use different calls to warn of the approach of an eagle, a leopard, or a poisonous snake. These different signals are adaptive because the route of escape differs depending upon whether the predator strikes from sky or land.

Group defense can also be passive—that is, it can avoid confrontation. For example, some small fishes swim at specific distances and angles from each other, forming a school. When a predator faces a school, it often suffers a **confusion effect** and is unable to decide which fish to attack. In the presence of a faster-swimming enemy, the school may use a **fountain effect,** splitting in two and regrouping behind the baffled predator.

Ostriches use the **dilution effect** of passive grouping to protect their young. Their societies consist of bonded pairs of a cock and a major hen, as well as minor hens who mate with many males. Minor hens deposit their eggs in the nest of a bonded pair, who sit on them, with their own eggs, for 6 weeks. Why would a bonded pair take in other eggs? Possibly because if not all the eggs are theirs, predators will consume fewer of their eggs (fig. 40.11). The major hen moves the minor hen's eggs to the edges of the clutch so they are more visible than her own. This is called a dilution effect because the pair's own eggs no longer make up a large percentage of the clutch.

Sometimes the best defense is a good offense, with help from others. Many prey species engage in mobbing behavior, where adults bother a predator that is larger than any individual of the group. Redwing blackbirds make hit-and-run attacks against much larger owls, uttering shrill calls. Baboons and chimpanzees use a similar strategy against leopards. Screaming, they charge and retreat and maybe even throw sticks at the leopard.

Forming a circle with the most formidable parts of the body facing outwards is a common defense strategy. When young catfish are disturbed, they mass together with their large pectoral fins with poisonous spines projecting in all directions. Similarly, adult musk oxen form a ring with their heads pointed outward toward the attacker. The older animals face the threat, while young find shelter within the woolly wall.

Enhanced Reproductive Success

Social behavior may increase the probability that an individual will find a mate. An increase in the number of animals in an area may even trigger physiological changes necessary for successful reproduction. For example, a pig's ovaries may not develop normally unless she hears and smells a boar.

The sights, sounds, and scents of other courting individuals enhance and synchronize breeding. Spotted salamanders provide an extreme example of explicitly timed group mating. Each spring in many parts of North America, for two to seven consecutive nights, the salamanders perform courtship dances on pond bottoms, with males depositing bundles of sperm and coaxing females to take them. At other times, the yellow-dotted black amphibians do not seem to interact socially at all.

Reproduction that occurs simultaneously throughout a population can have other benefits. Predators may become glutted with food for a short time and eat fewer offspring. Synchronous breeding also may be adaptive if it coincides with abundance of seasonally available food.

Improved Foraging Efficiency

If food is plentiful in some places and scarce elsewhere, it helps enormously to have others around to locate good feeding sites. When one member of a flock of starlings finds a food source, for example, other birds rapidly change their searching strategies and head for the meal.

(a)

(b)

FIGURE 40.12

Paternal behavior. (*a*) The dominant male brown capuchin monkey allows an infant near his food—probably because he is the father. (*b*) White-fronted capuchin youngsters stick together. Adult males show little, if any, interest in infants and juveniles, possibly because males do not know which offspring are theirs.

Spiderwebs vividly illustrate the advantage of group living in securing a meal. The community spiders of genus *Stegodyphus* that live in South and East Africa build vast silken empires that house hundreds of spiders. At night during the wet early autumn, the spiders repair their netting, torn in the daytime by visitors, rain, and wind. Maintaining the webs is essential, for these structures trap the spiders' food of gnats, termites, ants, stinkbugs, beetles, and grasshoppers. By cooperating to erect this food trap, the spiders enjoy a rich and varied menu that an individual could not subdue on its own. An added benefit is that the web entraps some predators.

Improved Learning

Social behavior may enhance learning and the passage of tradition. Acquiring the navigational skills necessary for migration, for example, entails learning. Learning can be adaptive by enabling others to benefit from a behavior an individual discovers. Many animals acquire information by watching others. In Japan, a monkey washed sweet potatoes and wheat in the sea to clean and season the food. Others in her troop imitated her. Evolution of our own species has relied heavily on rapid transmission and assimilation of new adaptive behaviors.

Disadvantages of Group Living

Individuals also may suffer as a result of group living, although the group is not endangered. An ostrich egg carried off by a vulture or a young wildebeest felled by a predator as it straggles at the back of its herd certainly does not benefit from their species' social behavior.

Colonial living also makes it easier for infections to spread. Consider mite infestations in honeybee hives in the United States. The microscopic tracheal mites collect in the breathing tubes of the bees. Mated female mites crawl from the tubes and lie on the tips of the bees' hairs, where they are easily picked up by another bee. Beehives infested with tracheal mites may be morgues by winter's end.

Mastering Concepts

1. What is the evidence that group living is successful?
2. What are some advantages of group living?
3. How can group living benefit the group yet harm an individual?

Social Structure and the Situation

A species' social structure is influenced both by the animals' physiology and their surroundings. When closely related species occupy similar habitats, yet behave differently, the reason for the diversity in behavior may not be obvious to us.

Great differences in social behavior are seen, for example, between two related groups of monkeys that live among the treetops of Peru (fig. 40.12). Brown capuchin monkeys are the larger and stronger of the two species. They live in groups led by a dominant male. One female is also dominant and sticks close by the male. When in estrus, she aggressively pursues him until he mates with her. The other two or three females are submissive. Other males are submissive to the head male; females seek only the head male for mating.

When food is plentiful, such as in a fig tree, the brown capuchins live peacefully. When food is scarce, the dominant male locates tough, fibrous fruits and opens them. But he is very particular about sharing his dinner. When close to being full, he allows the dominant female and youngsters to feed. Next come the nondominant females. Only when the head male moves away from the food can the submissive males safely approach. If anyone threatens an infant, the dominant male becomes violent.

Life among the white-fronted capuchins is considerably more tranquil. Their slightly larger social groups, with up to 20 individuals, consist of three dominant males, plus females and offspring. Any adult may mate with any other. Although the adults show little aggression toward one another, dominant males ignore or are hostile to the juveniles, who are very submissive to them. The white-fronted capuchins feed farther up in the forest canopy than the brown capuchins, on trees abundant with soft fruits.

Food availability and mating practices are key to the differences in aggression between these two species of monkey. A spotty food supply fosters aggression among brown capuchins. The dominant male protects his offspring because he is their father and is ensuring enough food for their survival. In contrast, with abundant, easy-to-eat food, aggression is rare among the white-fronted capuchins. The males, however, usually do not bother with the juveniles. Perhaps this is because the male adults cannot know which offspring they have fathered.

Mastering Concepts

1. How might food availability affect the structure of a society or the behavior of its members toward each other?
2. How does mating behavior affect a society's structure?
3. Among capuchin monkeys, how does a male's knowledge of whether or not he fathered a particular juvenile influence his behavior toward it?

Communication and Group Cohesion

Communication is necessary for social behaviors. This may include chemical (taste or smell), auditory, tactile, or visual signals. The best way to send a message depends upon the environment. Beneath the sea, for example, light does not travel far, so visual communication is not useful. Sound, however, travels easily under water. The songs of whales can be heard hundreds of miles away.

Chemical Messages

Many animals use chemicals to communicate. Lobsters are practically blind, but their bodies are covered with millions of tiny hairs that function as chemoreceptors, providing a sense of smell 1,000 to 1 million times more sensitive than ours. Lobster mating depends upon chemical cues (fig. 40.13). When a female enters a crevice and faces its male occupant, the two stand face to face, flicking their antennules to sense whether the other is ready to mate. After several stereotyped motions, the female ejects a fluid containing heavily scented urine. The male senses the stream in his antennules, and, in response, he fans the odor outward. Then, as the female shrinks and sheds her shell, he flips her onto her back, and they mate.

Auditory Messages

Elephants are masters of auditory communication. They live in matriarchal herds consisting of sisters, cousins, and their offspring. Males roam alone or in other groups (see Biology in Action 2.1). Related families greet each other with a cacophony of rumbles, trumpets, and screams. When a female approaches a male to mate, she bellows her intent. Her relatives watch, trumpeting loudly.

Sperm whales live in groups of 10 to 12 relatives, including mothers, daughters, aunts, and young males. They converse in a language of clicks and pauses that resembles Morse code. Researchers describe each pattern of clicks and pauses as a coda. For example, a 6 + 1 + 1 coda corresponds to

click-click-click-click-click-click-pause-click-pause-click

A coda in the sperm whale's language is the equivalent of a letter in our alphabet.

Certain codas often follow others, suggesting that they are not random utterances but meaningful messages. Ethologists observe how sperm whales behave when they use certain codas to try to discern meaning. Sperm whales thousands of miles apart use the same codas. Linguistic differences between sperm whales in the Pacific and in the Caribbean relate to the frequencies of certain codas and not to the "message" of the codas themselves. The clicks and pauses of sperm whale communication are more like human language in their consistency and their association with specific behaviors than are the whistles, moans, grunts, and squeals of other whale species.

Tactile Messages

Physical contact can cement some social bonds. Touching, for example, can reduce tensions and signify greeting. Members of a wolf pack may surround the dominant male and lick his face and poke his mouth with their muzzles. This ceremony occurs when it is useful to reinforce social ties, such as upon awakening, after separations, and just before hunting.

Touch is common among nonhuman primates, who often sit with their arms about one another. These animals groom each other's fur, using their hands, teeth, and tongues. Grooming removes parasitic insects, prevents infection, and also promotes social acceptance.

Visual Messages

Ethologist Karl von Frisch described the importance of vision in the foraging behavior of honeybees. In 1910, he read a paper stating that bees are color-blind. Why, he wondered, are flowers so brightly colored if not to attract the bees that pollinate them? To demonstrate color vision in bees, von Frisch placed sugar water on a blue cardboard disc near a hive. The bees drank the water. Next, he placed blue and red discs near the hive; the bees went only to the blue disc. Similar trials with different colors and shades showed that bees not only see color (except red), but they can also detect ultraviolet and polarized light, which humans cannot see. Von Frisch deciphered the "dances" that bees use to communicate food source locations.

As a forager or scout bee dances, recruit bees cluster around her, sensing her movements with their antennae. If the food is close to the hive, the scout dances in tight little circles on the face of the comb. This "round dance" incites the others to search for food close

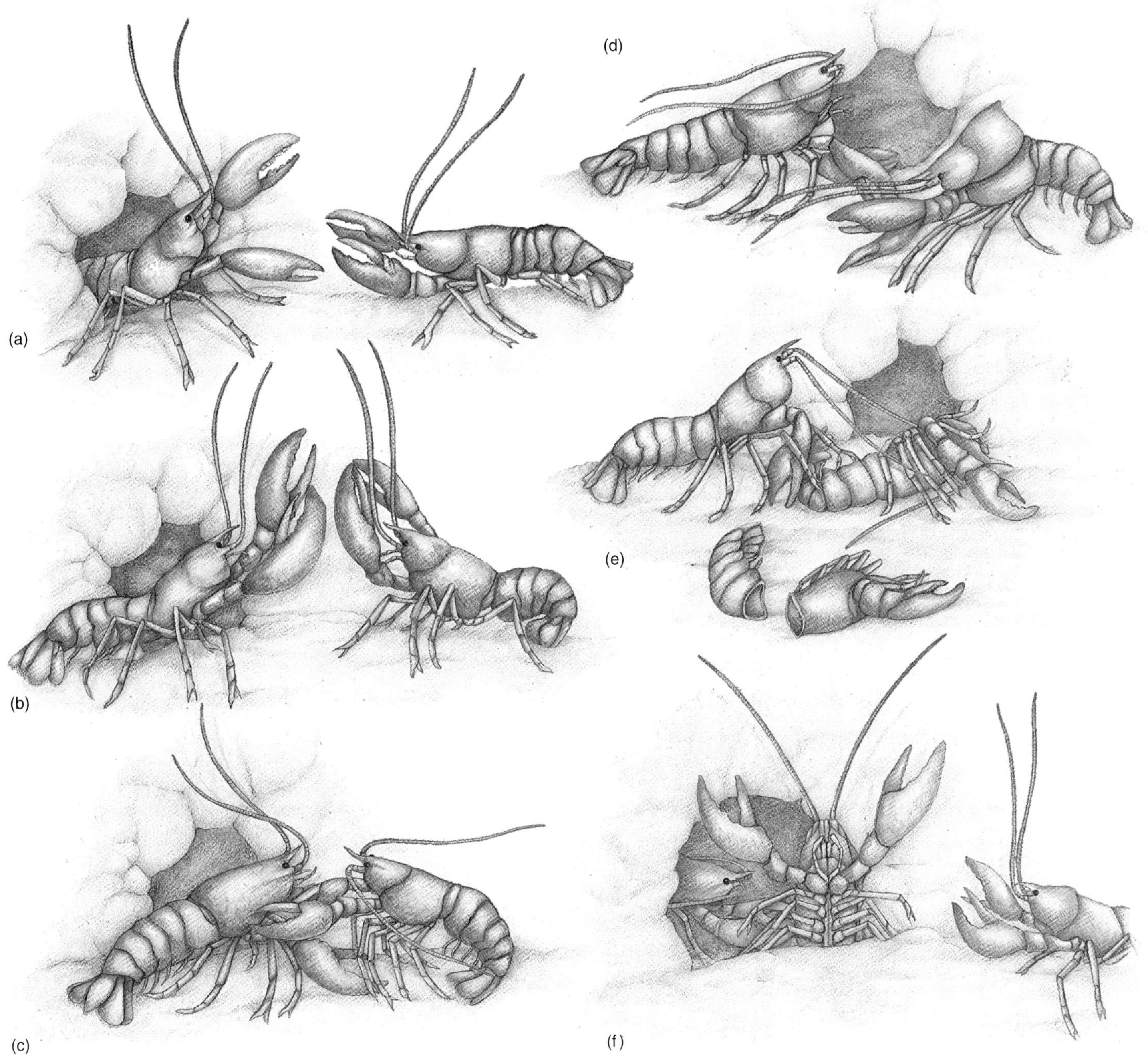

FIGURE 40.13

Chemical cues are important in lobster mating. (*a*) Courtship starts when a female approaches a male in his crevice. (*b*) She "knights" him by raising her claws, announcing her intent. (*c*) The female's body shrinks, and (*d*) she sheds her shell. (*e*) The male turns her gently on her back, and they mate. (*f*) For the next week, he shelters her as her new shell hardens. At certain points in the courtship, the female emits a fluid rich in urine, which excites her selected male—and possibly other males as well.

to the hive (fig. 40.14*a*). The longer the dance, the sweeter the food. A "waggle dance" signifies that food is farther from the hive (fig. 40.14*b*). The speed of the dance, the number of waggles on the straightaway, and the duration of the buzzing sound correlate with distance to the food. The orientation of the straight part of the run indicates the direction of the food source relative to the sun.

Researchers have deciphered the finer nuances of bee dances using models to entice them into performing. A robot bee accepted in the hive revealed that sound is also important in communication. The computerized bee dances, vibrates, and delivers sugar water from its front end, taking into account the angle of the sun. The bees respond.

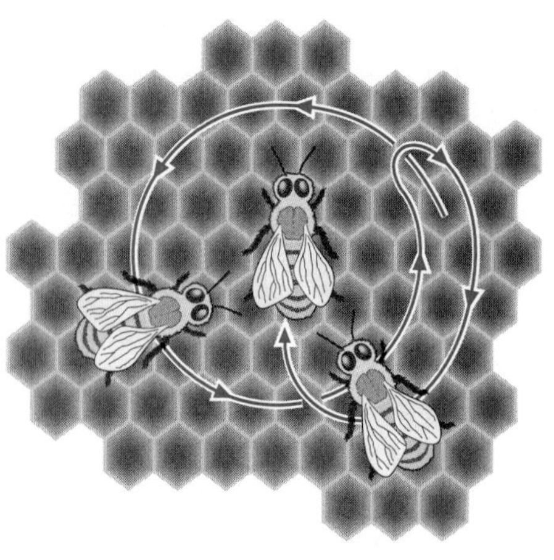

(a)

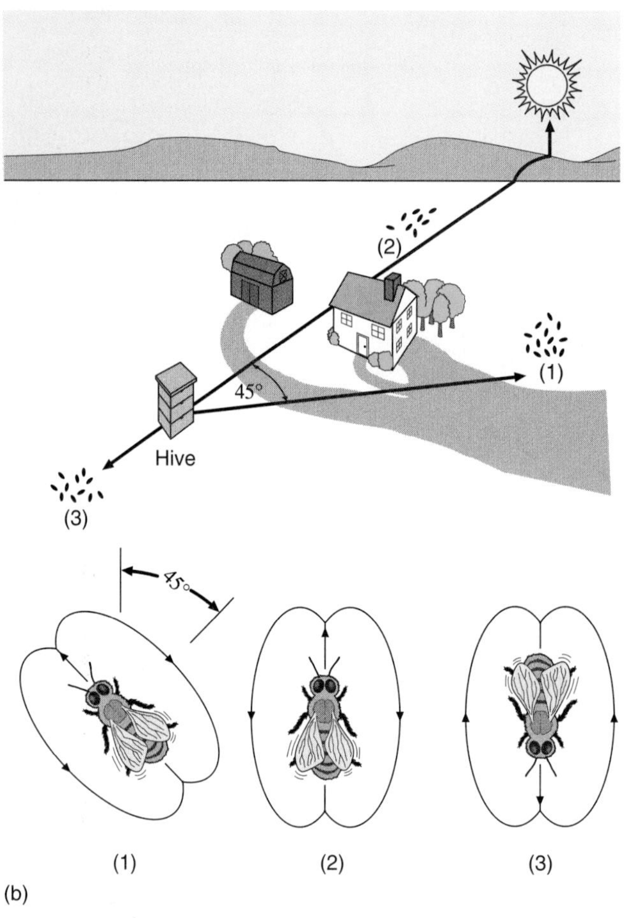

(b)

FIGURE 40.14

Bee dances. (*a*) A round dance informs recruits that food is close by but does not point them in the right direction. (*b*) A waggle dance tells recruits the distance and direction to the food source. The dancer loops once to the right, waggles her abdomen along a straight path, and then loops to the left, repeating these movements. Distance is indicated by the number of waggles and the duration of buzzing during the waggle run and the speed of the dance. The angle of the waggle run relative to gravity indicates direction. This angle corresponds to the angle that a worker must assume with the sun as she flies to the food source. If the angle formed by the sun, the hive, and the food source is 45° to the right of the sun, then the angle of the waggle run relative to gravity is 45° to the right of vertical (*1*). If the worker should fly directly toward the sun, the scout dances so that the waggle run is straight up (*2*). When the worker should fly directly away from the sun to reach the food, the waggle run points down (*3*).

Mastering Concepts

1. Which senses are important in communication among animals?
2. Describe how lobsters, elephants, whales, and bees communicate.

Altruism

Dictionaries define *altruism* as concern for the welfare of others. In the study of animal behavior, altruism has a more specific meaning—sacrificing one's ability to reproduce so that one's relatives can do so.

Biological altruism provides evidence for a theoretical idea called **inclusive fitness.** This hypothesis holds that an individual may sacrifice its own reproductive capacity in order to perpetuate copies of its own genes through a relative's offspring. For example, the white-fronted bee-eater lives in extended family groups of five to nine birds. In each group, "cooperative breeders" do not raise their own young but assist parents, who are usually their siblings or half-siblings, in rearing their young. Individuals may breed in one season and help in another.

Since close relatives inherit a greater proportion of their genes from a common ancestor, inclusive fitness theory predicts that the degree of altruism should be proportional to the degree of relatedness. Arizona tiger salamanders vividly demonstrate the ability to recognize close relatives in order to enhance their survival and reproductive success.

The tiger salamander has two larval (tadpole) forms. The more common type eats invertebrates. When the salamander population explodes, a few larvae grow very rapidly, developing large teeth and jaws. These larvae become fierce cannibals, gobbling the others (fig. 40.15). The more closely related a tadpole is to a cannibal, however, the less likely the cannibal is to eat it. Researchers demonstrated this by observing larvae in tanks and keeping track of the genetic relationships between individuals. In 80 tanks each housing 16 sibling larvae, 40% became cannibals and devoured the others. But in 40 tanks each containing 8 sibling larvae and 8 larvae from another brood (sometimes cousins), 85% became cannibals. This percentage also occurred in 40 tanks each containing 16 larvae from a mixture of eight broods. The presence of close relatives seemed to suppress the cannibalistic behavior that crowding would otherwise induce.

(a)

FIGURE 40.15

Cannibalism and kin. (*a*) The tiger salamander larva develops the jaws and tearing structures of a cannibal when its habitat becomes crowded. Cannibalistic tendencies are tempered, however, if nearby larvae are relatives. (*b*) Cannibalistic salamanders recognize kin. Arizona tiger salamander larvae are more likely to develop as cannibals if they are crowded into a tank with individuals who are not relatives or who are distant relatives. They are less likely to eat their own if their neighbors are siblings.

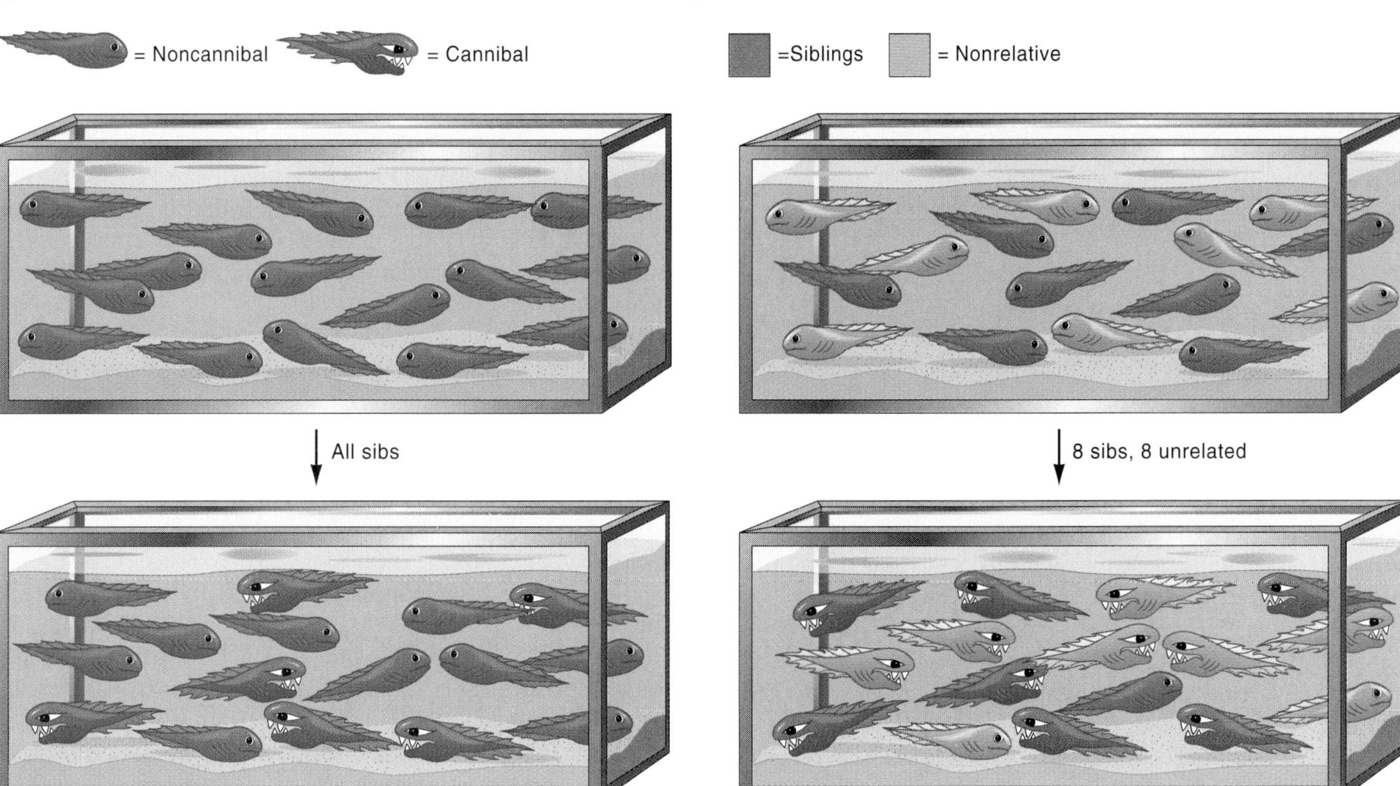

(b)

Mastering Concepts

1. What is the biological definition of altruism?
2. How is altruism adaptive on a population level but not on an individual level?

Aggression

Animals often display aggressive behavior when members of the same species compete for resources, such as mates, food, shelter, and nesting sites. Competition can be fierce because the stakes are high. Winners survive and leave more offspring than losers do.

Territoriality

One way that animals acquire resources with minimal aggression is to defend an area of their habitat, or territory, against invasion. Different types of territories form for different functions, including sexual selection, foraging, defense, and protection of young. An animal may defend the territory against other species, against

FIGURE 40.16
These two tigers are respecting a territory boundary.

all members of the species, or against only those of the same sex. A territory may be fixed, such as prairie dog burrows, or moving, such as baboon settlements. Individuals hold some territories; groups hold others.

Behavior that defends one's territory, termed **territoriality,** is common. Male birds often occupy territories and repel other males while enticing females to enter. Male insects sometimes take over a food resource that attracts females. A territory may serve groups of animals, who defend it against others of their species. Wolf packs, hamadryas baboons, and pack rats maintain group territories.

Some animals defend territories only during the breeding season, but others hold them year-round. Territory size also may vary tremendously. A flightless cormorant defends only the area it can reach while sitting on its nest. Yet African weaver ants defend territories up to 2,153 square yards (1,800 square meters). Figure 40.16 shows two tigers respecting a territorial boundary.

Territorial Markers

If territorial boundaries are to be respected, an animal must indicate ownership in some way. Although fighting may accompany territory establishment, boundaries are generally respected without contest once they are set. Animals usually avoid actual combat. An animal becomes less aggressive when it crosses the boundary to leave its territory. The point where an animal is as likely to attack an intruder as to flee from it marks the borderline of the territory.

An animal may advertize its territory with a threat posture, which often emphasizes natural weapons. The sight of a male baboon's fangs, for example, deters many potential opponents. Some animals vocally proclaim ownership of a territory. A toadfish's rumbling, a bull alligator's roar, and a wolf's howling at the moon announce possession of an area. Many animals mark their territories with strong scents. Badgers, martins, and mongeese have scent glands at the bases of their tails, and rabbits have chin and anal glands. Urine is a common territorial marker. The giant galago, a primate, urinates on its hands, rubs the urine on its feet, and walks around to mark its territory. A male hippopotamus defines its territory by "dung showering"—rotating its tail to fling solid waste along the perimeter of its turf.

Traversing Territories

When animals ignore territorial markers, threat behavior follows. A horse tosses its head and kicks. A sea anemone withdraws its tentacles and inflates a collarlike structure. If one animal is clearly superior, the other may leave without a challenge.

At any time during a battle, one combatant can end the fight with a display of submission, without provoking retaliation from the victor. Wolves expose their vulnerable throats as a sign of appeasement. In contrast to a threat posture, in which many organisms attempt to appear larger, appeasing animals try to seem smaller.

Diversion is another response to a threat. The "loser" behaves in a way unrelated to the situation to distract the aggressor. A dog might roll over, paws up, eliciting parental feelings rather than aggression from the other dog. In displacement behavior, another response to a threat, both parties cannot decide what to do, so they pursue another activity. Two dogs at a standoff might investigate a fire hydrant instead of attacking each other.

When competition does lead to combat, it is more like a tournament with a set of rules than a bloody fight to the death. Although roe deer could easily gore an opponent with their antlers, they do not use them in this way against members of their species. Two males face each other, antler to antler, and push in a ritualized fight. In some species, however, battling over a territory can sometimes lead to injury or death.

Dominance Hierarchies

A **dominance hierarchy** is another way to distribute resources with minimal aggression. This is a social ranking of each adult group member of the same sex. In a linear hierarchy, one animal dominates, a second animal dominates all but the first, a third dominates all but the first two, and so on. Researchers first studied this type of relationship in the pecking order of domestic hens. Each hen knows whom it may peck and whom it may be pecked by. Hierarchies in other organisms are often more complex. Female Amboseli baboons, for example, are ranked according to the age of their mothers and the age differences between them and their sisters.

A dominance hierarchy forms when members of the group first meet and may involve many threats and an occasional fight. When

FIGURE 40.17

Tadpoles can become victims of cannibalism when they are too developed to swim easily, yet not as agile as adult frogs. Young tadpoles dragged this metamorphosing one from vegetation that it was clinging to with its partially formed limbs.

pairs of Bewick's swans return from Siberia to their home estuary in western England, they flap their wings and make buglelike sounds to quickly establish the dominance hierarchy. Once the status of each individual is set, life within the group is generally peaceful.

Cannibalism

More than 1,000 species practice cannibalism. In Crystal Lake, Vancouver, traveling bands of 300 or so female cannibal stickleback fishes attack lone males guarding nests and devour the eggs. Alligators in shallow rivers in Florida's Everglades become too crowded during droughts as their habitats shrink. When elders eat younger alligators, the remaining population has more food and space.

Cannibalism can be adaptive—for the cannibals, of course, and for the population, but not for the victims. In some fishes and amphibians, eating one's own kind causes faster growth and development, which enables an individual to have more offspring. In the Costa Rican rain forest after a heavy rain, meadow tree frogs mate frenetically, the females depositing so many clutches of eggs in transient puddles that the habitat very rapidly becomes overcrowded. When the tadpoles hatch, they compete intensely for limited plant food. The crowding and dwindling resources trigger some tadpoles to become cannibals. By the time of metamorphosis, the cannibals are much larger than any remaining vegetarian tadpoles (fig. 40.17). When the puddles dry up, the larger frogs are more likely to survive because they are more adept at getting food and avoiding becoming food.

To see whether frogs grew faster from eating their own kind, or just from eating meat, researchers set up an experiment. They fed meadow frog tadpoles mashed tadpoles of either their own or a different species. Only the cannibals were bigger than normal at metamorphosis. A possible explanation is that the frogs eat an extra dose of the hormone thyroxine, which regulates metamorphosis, that is tailored to their species.

(a)

(b)

FIGURE 40.18

Mating behavior. (a) Bewick's swans pair for life, yet the male performs this courtship display each year. (*b*) Part of mating stickleback style is dorsal pricking, in which the male touches the female's underside with his spines. This is a more subdued courtship behavior than the zigzag dance depicted in figure 40.4. Dorsal pricking may be adaptive because it is less noticeable to cannibal females, who rely on male mating cues to find nests with eggs to eat.

Mastering Concepts

1. What events or situations trigger aggression?
2. How does territoriality limit aggression?
3. How do dominance hierarchies limit aggression?
4. How can cannibalism benefit a population?

Mating Behavior

With dominance relationships and territorial boundaries keeping them apart, how do animals come close enough to mate? Courtship rituals are stereotyped, elaborate, and conspicuous behaviors that overcome aggression long enough for mating to occur. A male swan rises up and flaps his wings as a prelude to mating, and a male stickleback rubs his mate in a certain way (fig. 40.18). Both are demonstrating courtship rituals.

Males behave to reduce aggression in the female. A male orb weaving spider may stand at the edge of a female's web and tug on the suspension cord before venturing in. A male wolf spider

(a)

(b)

(c)

FIGURE 40.19

Approach of a stranger alerts a male prairie vole (*a*). When the invader advances, the prairie vole shows a threat display (*b*). Finally, when the second animal ignores the threat, a fight ensues (*c*).

cautiously approaches a female and, keeping a safe distance, signals his intent to mate by waving specially adorned appendages.

Courtship displays are specific, preventing the costly error of mating with the wrong species. Male fiddler crabs, for example, wave their large claws to lure females into their burrows, but the exact pattern of movement is unique to different species. Many courtship behaviors hold attractions that we humans simply cannot appreciate. A female jackdaw bird, for example, stuffs regurgitated worms into the ear of her mate.

Mating Systems

Animals have several types of mating systems. In **monogamy,** a female and male associate exclusively with one another for a period of time—a few days, a season, or a lifetime. Pairs of golden-rumped elephant shrews are monogamous, but they spend most of their time apart, foraging alone on the forest floor and meeting to mate only every 6 to 8 weeks. In contrast, Bewick's swans also pair for life but are nearly inseparable. Unlike the definition of a monogamous relationship between humans, monogamy in the broader biological sense means that the pair spend time together to the exclusion of others, but they may also occasionally mate with others. Only 7% of mammalian species are known to be monogamous. These animals tend to intensely care for and protect their young, as the male prairie vole in figure 40.19 does.

When a parent devotes time, energy, and resources to ensuring that offspring survive, it is termed **parental investment.** Such behavior is highly adaptive because it increases the chances of an individual's offspring surviving. Parental investment may be greater when mating is monogamous because the male somehow knows that he is the only father.

Several types of mating systems involve multiple partners. In **polygamy,** a member of one sex associates with several members of the opposite sex. The most common polygamous behavior is **polygyny,** in which one male mates with several females. Competition among polygynous males is often intense. In **polyandry,** a female mates with several males. Still other species, including frogs, fruit flies, and antelopes, engage in communal mating displays, called **polygynandry.**

A species' mating system is related to the degree of difference in appearance between the sexes, which is called **sexual dimorphism.** Polygamous species are usually highly sexually dimorphic (very different), whereas the sexes in monogamous species usually look more alike.

Two Views of Mating Behavior—Spiders and Sparrows

Mating behavior is diverse among animal species. Consider such behavior in the Sierra dome spider compared to that of the hedge sparrow.

A female Sierra dome spider has several mates, which increases the chances that at least one of them will be fertile enough to pass on his fertility to her offspring. Natural selection favors behaviors that enable her to recognize and select such a mate.

Male spiders seek a female's web because it provides food, protection from predators, and a potential mate. When a female is about

FIGURE 40.20

A sex attractant lures two male Sierra dome spiders to the web of a sexually receptive female, at the upper right. Only one male will succeed in this initial contest, but the female may mate with others later.

to become sexually receptive, she secretes a pheromone. Males arrive and battle for entry to her web. The winner becomes her primary mate and fertilizes 70 to 80% of her eggs (fig. 40.20). When the female is finished with the first male, she plucks the web to signal that she wishes him to leave. Then she mates with other males.

Females select mates based on their body size and stamina during mating. Spider sex takes 2 to 7 hours and consists of hundreds of couplings of only a few seconds duration. (In between coupling, the female eats; the male is too exhausted to compete with her for food.) Because males that are large and vigorous mates also tend to be healthy, the female chooses the most fit partners—an example of sexual selection (see fig. 18.9).

Unlike the Sierra dome spider, the hedge sparrow displays all types of mating systems. Females maintain territories in thick hedges, but one territory may house a monogamous pair, another a female with two males, yet another a single male who flits between several females' lairs. Two or three males may mate in groups with two to four females. Parental investment in these birds varies with type of mating behavior. In monogamous pairs of hedge sparrows, the male cares for all of the chicks—and they are all his. In groups of two males and one female, the proportion of time each male spends feeding young is about the same as the proportion of time that he mates with the female.

The diversity of mating styles among hedge sparrows arose, ethologists hypothesize, because the sexes seek different strategies. For a female hedge sparrow, mating is most adaptive when male participation in chick-rearing is maximized, because more chicks will survive. Monogamy or polyandry, where she is the sole female, each accomplishes this. For a male, however, mating with several females (polygyny) maximizes the number of offspring he produces. Polyandry is the least efficient mating style for a male, because he may care for chicks that aren't his.

The sexes of hedge sparrows are pitted against each other in a dominance struggle. If the male wins, polygyny proceeds. If the female wins, polyandry rules—and she takes a second mate. If neither male nor female can get a second mate, monogamy is a default option. Finally, polygynandry—group sex—may represent a stalemate situation, when one female cannot chase away another, and one male can't get rid of another. (Of course, this is just a human view of the birds' mating styles.) Figure 40.21 shows another interesting sexual behavior of hedge sparrows—males can ensure that it is their sperm that fertilizes eggs.

Female releases droplet containing sperm from past matings

FIGURE 40.21

Before a male hedge sparrow mates, he pecks the female on her cloacal opening. In response, she releases a droplet containing sperm stored from previous matings. Only then, with his competition for fatherhood eliminated, does the male complete mating.

Primate Mating Behavior

Primate species have diverse sexual behaviors. They form male-female pairs, one-sex bands, mixed-sex troops, harems, homosexual pairs, loners, and social groups with frequently changing members (fig. 40.22). Devoted monogamous gibbon pairs sing together in the early morning to defend their territory. In bonobo societies,

(a)

(b)

FIGURE 40.22

Primate sexual behavior is diverse. (*a*) Mountain gorillas form heterosexual pairs with deep and lasting bonds. (*b*) Once thought to be a type of chimpanzee, the bonobo is actually a great ape. Unlike chimp and human social groups, bonobo females dominate and may form what appear to us to be lifelong friendships with other females, sometimes engaging in homosexual behavior. Bonobo society is very peaceful compared to other primate groups.

females are dominant and often form intense relationships with each other. Orangutans are polygynous, with a female mating only once every few years.

Primate sexual behavior varies even within species. An aggressive male mountain gorilla mates with and kills an older female who cannot keep up with the troop; yet other mountain gorillas form long-lasting bonds. Some chimpanzee males prefer to mate with an unfamiliar female. Other males take turns mating with a single female. Still other chimpanzees spend their lives in monogamy. Humans have also practiced every mating system observed in nonhuman primates.

It can be difficult and disturbing to scrutinize human behavior as we would the behavior of other members of the animal kingdom. The fact that behavior in many other species is partially genetically determined, however—from simple fixed action patterns to complex social interactions—suggests that the same is true for us. Such a supposition is frightening when applied to humanity because it has spawned much prejudice and persecution when one group of people presumes that another is genetically inferior. It may help to realize that human behavior, perhaps even more than other animal behavior, is molded by the environment as well as directed by genes. This leaves much room for modification.

Mastering Concepts

1. What is a courtship ritual?
2. Describe different mating systems in animal species.
3. How might the type of mating system affect parental investment?

Shrimp Society~

continued from page 812.

Do shrimp living in sponges cooperatively care for young? Such behavior might be revealed by identifying a way that adults could help juveniles.

Unlike some other animal societies, young shrimp obtain food without adult assistance—their meals just float in. Nor must adults build a nest—the host sponge provides a secluded home. Researcher J. Emmett Duffy, of the Virginia Institute of Marine Science, hypothesized that adults care for young by protecting their hold on the sponge homestead. He based this hypothesis on the observation that he could not find any sponges that did not house shrimp—the sponges are valuable real estate.

If competition exists for sponge homes, then introducing a threatening situation should provoke protective behavior among resident shrimp. Duffy set up two experimental situations in sponges reared in his laboratory. In the first experiment, a shrimp colony encountered shrimp of another species. In the second experiment, a shrimp colony encountered shrimp of the same species that had been removed from the colony beforehand. Results were striking. When confronted with unrelated shrimp that competed for sponge space, the resident snapping shrimp fought valiantly, chopping up the competition. In contrast, the resident shrimp did not attack their former colony mates. More telling, only the larger individuals defended the colony—which Duffy concludes is evidence of cooperative care of young. The shrimp are the first animal society discovered in the oceans.

SUMMARY

1. Ethology considers innate behavior in the wild. The field was founded on observational studies, but today it includes several experimental approaches.

2. Genes and the environment (learning) shape behavior, which is usually adaptive. **Innate** (genetic) influences dominate when it is essential for a behavior to occur correctly the first time. Learning is important in mastering certain complex behaviors. Anatomy and physiology limit possible behaviors for a species.

3. A **fixed action pattern** is an innate, stereotyped response that continues without feedback. A **releaser** is a stimulus that triggers a FAP, and it may be visual, auditory, chemical, or tactile. A **supernormal releaser** is more effective than the natural stimulus in triggering the response. **Reaction chains** can form from a sequence of FAPs. Individuals can exchange releasers.

4. Different types of learning overlap. In **habituation,** an animal ignores a highly repeated stimulus. In **classical conditioning,** an animal gives an old response to a new, **conditioned stimulus** instead of the older, **unconditioned stimulus.** In **operant conditioning,** a behavior increases in frequency because it is positively or negatively reinforced. **Imprinting** occurs during a **critical period** and doesn't require reinforcement. **Insight learning** applies prior learning to new situations without trial-and-error activity. **Latent learning** uses past observations to perform a new activity.

5. Orientation and navigation rely on responses to environmental cues. The simplest form of navigation recognizes landmarks. Birds use the sun or stars as a compass and possibly the earth's magnetic field. Homing depends on a compass sense to set direction and a map sense to direct the navigator to a particular place. The map sense may detect magnetic fluctuations or odors.

6. A **eusocial** group has cooperative care of young, overlapping generations, and division of labor. Insect societies have **temporal castes,** in which age determines role in the colony.

7. Advantages of group living include ability to alter the environment, defense, improved reproductive success, more effective foraging, and opportunity to learn. Group living is disadvantageous because of increased competition for resources and ease of spreading infection.

8. A species' social structure reflects physiology as well as the environment. Many behaviors are based on communication, which may be chemical, auditory, tactile, or visual.

9. Altruism toward a relative increases one's **inclusive fitness,** thereby ensuring that some of an individual's genes persist, if not the individual.

10. Animals are most likely to fight with members of their own species because they compete for resources. **Territoriality** and **dominance hierarchies** minimize aggression. Threats and appeasement, diversion, or displacement responses often prevent combat; if combat occurs, it is usually restrained. Overcrowding may trigger cannibalism, which improves the habitat for survivors.

11. Courtship rituals calm aggression so that physiologically ready individuals of the same species can mate. Mating systems include devoted pairs **(monogamy),** a male with multiple partners **(polygyny),** a female with several partners **(polyandry),** and group mating **(polygynandry). Polygamy** is a general term for mating involving multiple partners.

12. Parental investment is effort to ensure survival of one's offspring and increases in proportion to how frequently a particular male mates with the mother of his offspring. Polygamous species are highly **sexually dimorphic**—the sexes appear different. Primates have diverse mating systems.

TO REVIEW . . .

1. Distinguish between an innate and a learned behavior.

2. How can the following behaviors be adaptive (further the likelihood of reproductive success for some individuals)?
- **a.** territoriality
- **b.** cannibalism
- **c.** altruism
- **d.** sexual selection
- **e.** dominance hierarchies

3. Define the following terms:
- **a.** sexual dimorphism
- **b.** parental investment
- **c.** inclusive fitness

4. Distinguish between confusion, fountain, and dilution effects in deterring predators.

5. Identify the type of learning displayed in the following situations:
- **a.** A guinea pig chirps whenever he hears the refrigerator door open, because this sound always precedes his owner giving him lettuce.
- **b.** Newborn geese see a little girl when they are born. The girl raises them, and they follow her everywhere, as if she is their mother.
- **c.** A teen watches her older brothers ski for several seasons. Then, one year, she tries it and does well.
- **d.** A man prepares a meal with a parrot sitting on his shoulder. The bird chatters and moves about, but the man doesn't notice the activity.
- **e.** A child walks to school without a sweater or jacket in October and is very cold. Finally, he begins wearing a jacket.
- **f.** Children of a certain age can learn a foreign language with much less effort than an adult.
- **g.** In the film *Apollo 13,* a group of scientists are given a pile of materials and asked to fashion a device that will solve a problem aboard a spacecraft.

6. How does altruism differ in its literal and biological senses?

7. What are the advantages of group living?

8. How do the behaviors of the brown and white-fronted capuchin monkeys support the theory of inclusive fitness? How does the egg-sitting behavior of the major ostrich hen support or contradict the theory?

9. How is communication important in the following situations?
- **a.** maintaining a eusocial group
- **b.** minimizing aggression
- **c.** promoting courtship and mating behavior

TO THINK ABOUT . . .

1. This chapter describes many experiments. Choose one of the following questions, and explain how an experiment helped to interpret why or how something happened.

- **a.** How does an insect society adjust when many members of a caste die?
- **b.** Is migration of female green turtles mostly innate or mostly learned?
- **c.** Do tadpoles cannibalize any tadpoles or those they are not related to?
- **d.** What information do bee dances communicate?

2. An infant who is blind smiles at her mother. Why is this evidence that smiling is an innate behavior?

3. Give examples of humans engaged in threat, appeasement, diversion, and displacement behaviors.

4. The modern feral (wild) horse was domesticated 5,000 years ago but has returned to the wild behavior of its ancestors. Even a modern broken horse will revert to wild behavior if allowed to range freely, with stallions stampeding mares and mock fighting. What type of behavior might explain the primitive actions of the free-ranging modern horse?

5. People from the United States and people of the Minangkabau culture in western Sumatra, Indonesia, have the same facial expressions for disgust, anger, sadness, and fear and experience the same physiological changes with these feelings. Does this suggest that nature (genetics) or nurture (culture) plays a more dominant role in determining facial expressions?

6. A dog adopted between 4 and 8 weeks of age often develops closer ties to its owner than one adopted at an older age. What type of behavior accounts for this observation?

7. A taxonomist working in a South American tropical rain forest uncovers a new species of ant. What signs should she look for to determine whether to classify the species as eusocial?

8. In laboratories, naked mole rats are studied in elaborate Plexiglas colonies; bees establish hives in glass-encased honeycombs built into a window; spotted salamanders mate in huge bathtubs. Do you think that researchers using these setups can accurately assess animal behavior? Why or why not? Can you suggest alternative approaches for observing these organisms?

TO LEARN MORE . . .

References and Resources

Bowen, Brian W., and John C. Avise. December 1994. Tracking turtles through time. *Natural History.* Why do green turtles travel 1,400 miles to lay their eggs—and then go back?

Davies, Nicholas B. April 1995. Backyard battle of the sexes. *Natural History.* Sparrows have a variety of mating styles.

Duffy, J. Emmett. June 6, 1996. Eusociality in a coral-reef shrimp. *Nature,* vol. 381. Colonies of snapping shrimp live in sponges in the Caribbean.

Dugatkin, Lee Alan. June 1997. The evolution of cooperation. *Bioscience,* vol. 47. Cooperative behavior is complex.

Gordon, Deborah M. March 14, 1996. The organization of work in social insect colonies. *Nature,* vol. 380. An insect society is fluid. Individuals take on different tasks depending upon the situation.

Greenspan, Ralph J. April 1995. Understanding the genetic construction of behavior. *Scientific American.* Even simple innate behaviors are complex.

Lewis, Ricki. October 1996. The unraveling of spider silk. *BioScience.* Many behaviors in spiders center around their webs.

Lewis, Ricki. December 11, 1995. Chronobiology researchers say their field's time has come. *The Scientist.* Much of human behavior is linked to daily rhythms in physiology.

Lewis, Ricki. December 1995. Adjusting the circadian clock. *Photonics Spectra.* Altering the lights for nighttime shift workers can vastly improve their health.

Robinson, Gene E. March 29, 1996. Chemical communication in honeybees. *Science,* vol. 271. Queen mandibular pheromone maintains social order in a beehive.

Roy, Tui De. April 1997. Where giants roam. *Natural History.* Galapagos tortoises migrate daily.

Watson, Paul J. March 1995. Dancing in the dome. *Natural History.* A female Sierra dome spider's web offers many advantages to a male.

denotes articles by author Ricki Lewis that may be accessed via *Life*'s home page /http://www.mhhe.com/sciencemath/biology/life/.

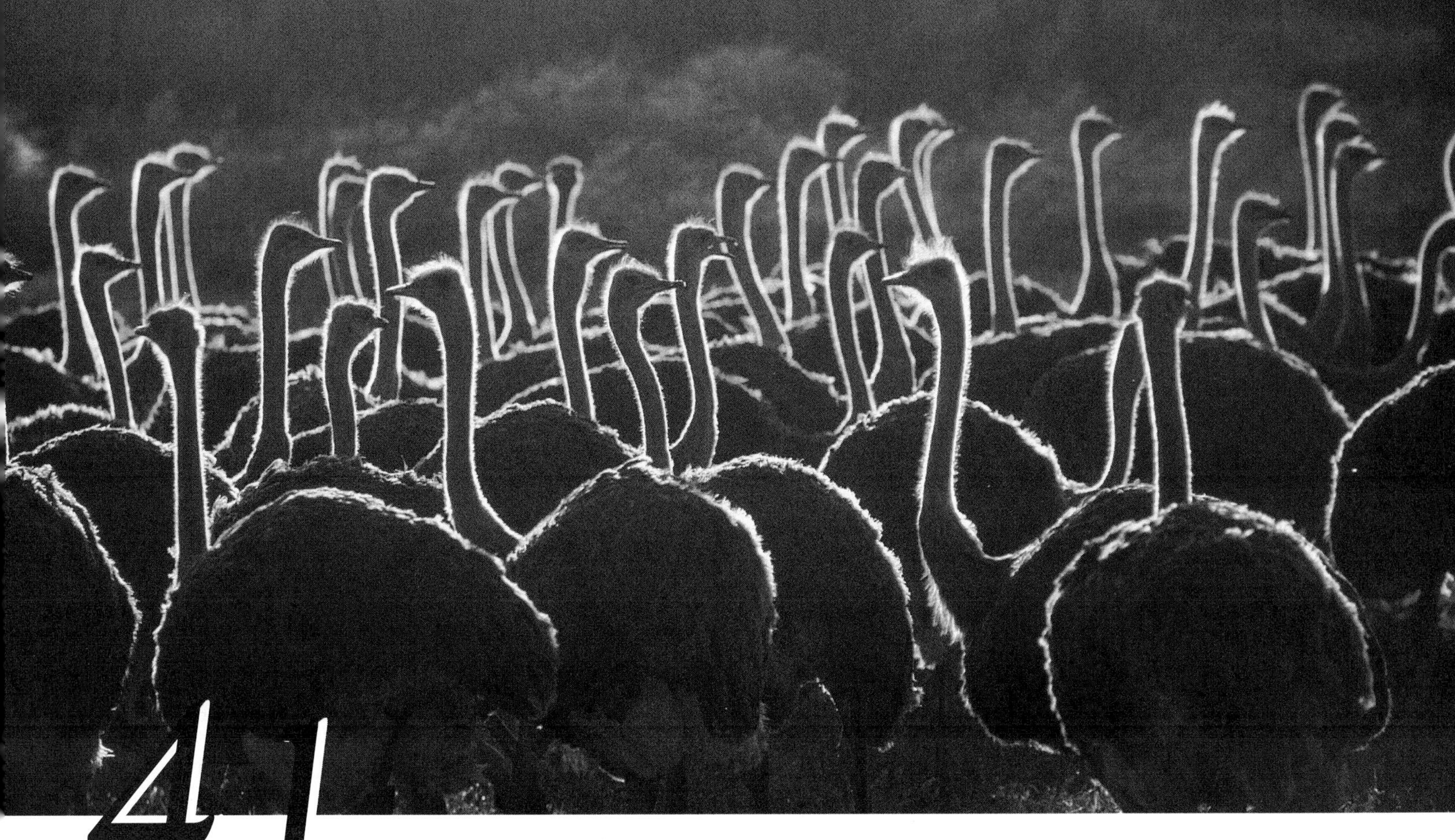

A population of ostriches, on an ostrich farm in South Africa.

41 Populations

chapter outline

Effects of Overpopulation 836

Ecology: An Organism's Place in the World 836

Habitat and Niche 836

Characteristics of Populations 838

Potential Population Growth 840
Intrinsic Rate of Increase and Exponential Growth
Survivorship, Fecundity, and Age Structure

Regulation of Population Size 842
Environmental Resistance and the Carrying Capacity
Density-Independent and Density-Dependent Factors
Logistic Growth Curve
"Boom and Bust" Cycles
K-selection Versus r-selection

Interspecific Interactions 845
Interspecific Competition
Predator-Prey and Parasite-Host Interactions

Human Population Growth 848

Where Have All the Frogs Gone?

Catching tadpoles in early springtime was once a yearly rite for many children. Today, formerly frog-laden ponds and lakes in many areas are nearly devoid of amphibian life. Alarmed that diminishing amphibian populations may reflect a change in the environment that could affect other species, life scientists the world over are actively counting groups of amphibians to assess the extent of the problem and to identify the cause or causes of population declines.

Scientists became aware of falling amphibian populations in the 1960s and 1970s. Only populations that lay their eggs in shallow water exposed to the sun were affected. Also, adult amphibians were becoming rarer, but eggs were abundant, and in some species, eggs were decaying in large numbers. Researchers disproved one possible explanation, water pollution, by sampling amphibian eggs in the wild and growing them in polluted pond water in the laboratory—the eggs hatched and developed normally. Plus, eggs were dying even in clean natural habitats. What was destroying amphibian eggs?

to be continued . . .

Effects of Overpopulation

A population is a group of organisms of the same species in a given geographic location. Because many different species typically inhabit an area and may compete for resources, populations interact. Too little food and a population plummets. Yet abundant food and the resulting exploding population size can ultimately be equally disastrous. We begin the chapter with a look at the dangers of overpopulation in a group of reindeer on St. Matthew Island in the Bering Sea, south of Alaska, and conclude the chapter with a look at overpopulation in our own species.

In 1944, 24 female and 5 male reindeer were introduced to the 128-square-mile (332-square-kilometer) St. Matthew Island, which was free of large predators and provided an optimal habitat for reindeer. By 1957, the population had grown rapidly to 1,350 apparently healthy animals. By the summer of 1963 the population had reached about 6,000 animals, but adults were smaller and the proportion of young animals fell. Food was becoming scarce. Mats of lichens, the main winter food of reindeer, were almost completely gone from areas where reindeer fed in winter. In the following winter, which was cold and snowy, all but 42 animals starved (fig. 41.1). Examination of carcasses showed that their usual winter foods—lichens, willows, grasses, and sedges—were absent. The density of reindeer just prior to the population crash was 47 animals per square mile—over three times the estimated capacity of arctic winter range to support stable populations of reindeer.

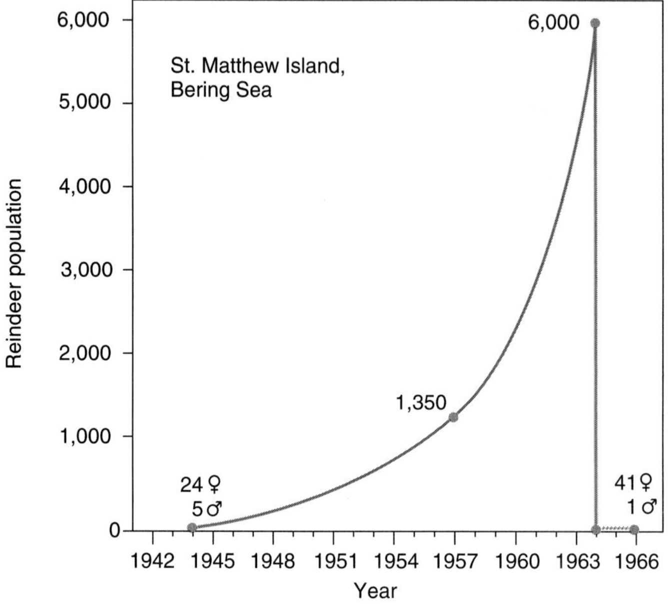

FIGURE 41.1

A population explosion and crash. Reindeer introduced to an optimal habitat on St. Matthew Island in the Bering Sea grew exponentially until the winter of 1963–64, when severe weather and a depleted food supply led to mass starvation.

Source: David R. Klein, *Journal of Wildlife Management,* Vol. 32, No. 2., April 1968, p. 352.

Ecology: An Organism's Place in the World

The interactions among individuals in populations are part of the life science of **ecology.** The word *ecology* comes from the Greek words for "home" (*oikos*) and "study" (*logos*). Ecology is the study of relationships between organisms and their environments—their "homes" in the broadest sense. The environment includes nonliving (abiotic) features and living (biotic) features. These abiotic and biotic factors determine the distribution and abundance of species and the kinds of species that coexist in a given situation.

This chapter examines interactions of individuals of a single species and then investigates interactions among individuals of different species. The next chapter looks at how these interactions translate into patterns of overall biotic composition and processes of energy exchange and nutrient cycling in different types of environments.

Habitat and Niche

Each species has a characteristic home and way of life. Its home, or **habitat,** includes the physical and biotic conditions of the places where individuals live. Habitat includes features of the en-

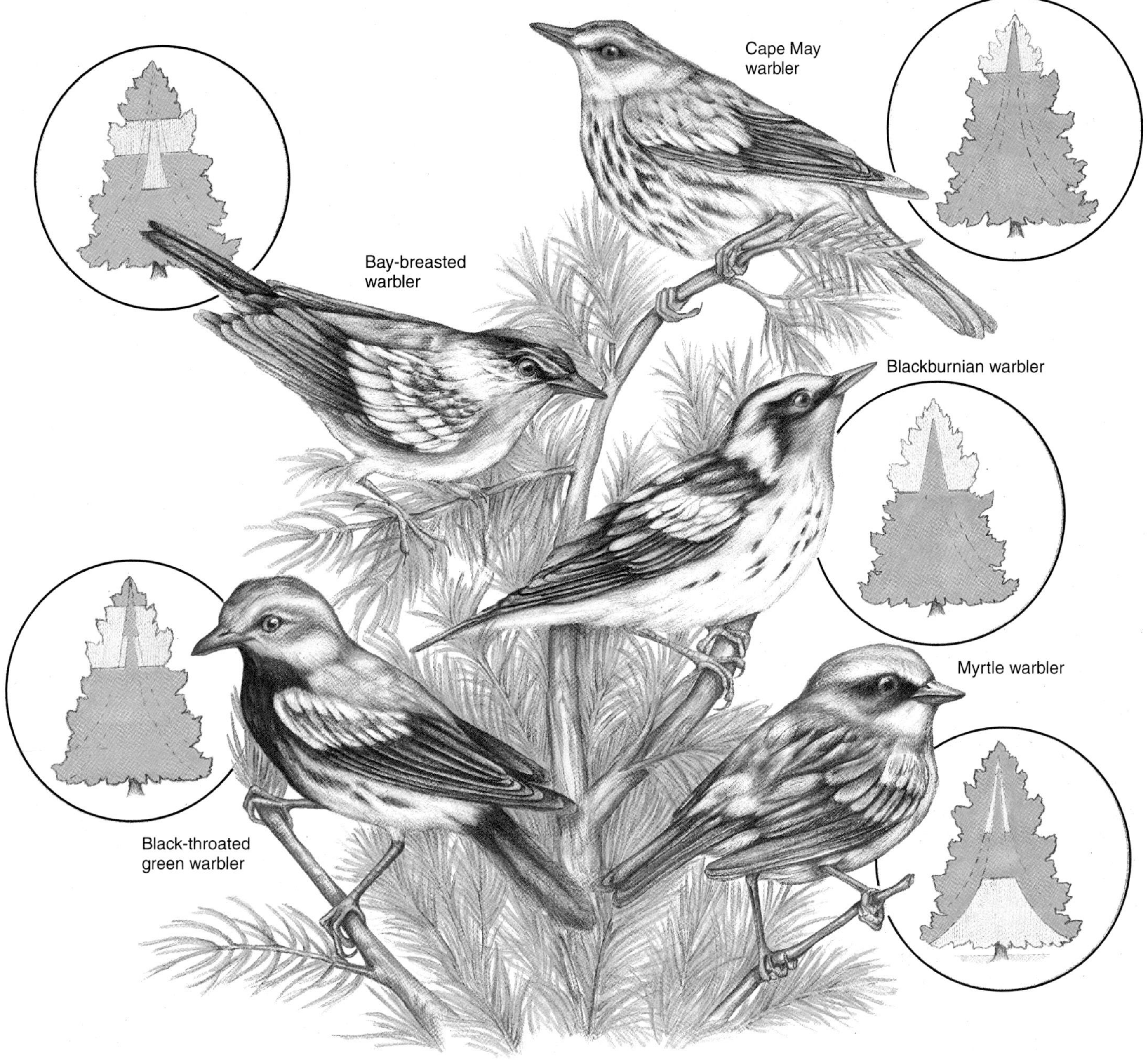

FIGURE 41.2

Populations coexist by reducing competition. These five species of North American warblers live together in conifer forests because each forages in a different portion of the tree and uses different foraging behavior. Yellow-shaded areas of the tree indicate primary feeding zones.

vironment that affect the organism's behavior or physiology, but the organism's use of these features does not reduce their availability to other organisms. For a terrestrial animal, habitat might include temperature, humidity, and structure of the vegetation. For an aquatic organism, habitat conditions might include salinity, current, and the structure created by submerged plants. The edge of a pond, a desert, and a rain forest are major types of habitats. An organism's habitat might even be another organism, on or in which it lives (See Biology in Action 21.2).

Many species live together in major habitats, yet species that are very similar often do not live in the same places. To understand the reasons for this, it helps to distinguish between the place where a species lives and the way it makes its living. A species' way of life is its **ecological niche.** The ecological niche is the way that a species exploits resources for survival, growth, and reproduction. Resources define an ecological niche. These are features of the environment that an organism uses in a way that makes them unavailable to other organisms. Therefore, organisms compete for resources, such as food, water, nesting sites, and space.

Different species can occupy different niches in the same habitat. For example, several species of seals share the seas around Antarctica, each using the environment differently. The crabeater seal eats krill, a small, shrimplike organism, and breeds along the edges of ice sheets. The elephant seal dives deep to eat squid and breeds on open beaches, whereas the fur seal prefers small, hidden beaches and rocky shores. The Weddell seal eats mostly fish, squid, and octopus. It winters beneath the ice, cutting breathing holes with its teeth. Safe beneath the ice, Weddell seal pups escape predation by yet another seal, the leopard seal, which eats fish, penguins, and other seals. Similarly, in New England, five species of warblers, all small insect eaters, coexist in the same forest stands, each type feeding in different ways and in different parts of the tree (fig. 41.2).

Different species can live together when they use different resources within the habitat. Two or more species that compete for resources may have to narrow their niches or be unable to coexist. A species that can use a wider variety of resources will be better able to coexist with a species that has similar requirements.

Two terms describe resource use in a niche. A species' **fundamental niche** includes all the resources a species could possibly use. If two or more species with similar but not identical fundamental niches live in the same area, competition between them may restrict each species to only some of the resources available. Or, the species may be restricted to using resources in different habitats. Therefore, due to competition, a species' **realized niche,** the one it actually fills, may be smaller than its fundamental niche.

Two species of barnacles, for example, live in the intertidal zone along Scotland's shoreline. Both species share the same fundamental niche, the entire intertidal zone. When present alone, either type of barnacle can live throughout the entire zone. When both are present, however, one species grows faster than the other in the lower, moister region of the intertidal zone and crowds out its competitor. The other species, however, better tolerates dehydration when exposed to the air while the tide is out and can more efficiently use resources of the upper region of the intertidal zone.

Mastering Concepts

1. What is the ultimate consequence of overpopulation?
2. What is ecology?
3. Distinguish between habitat and niche.
4. Distinguish between fundamental and realized niche.

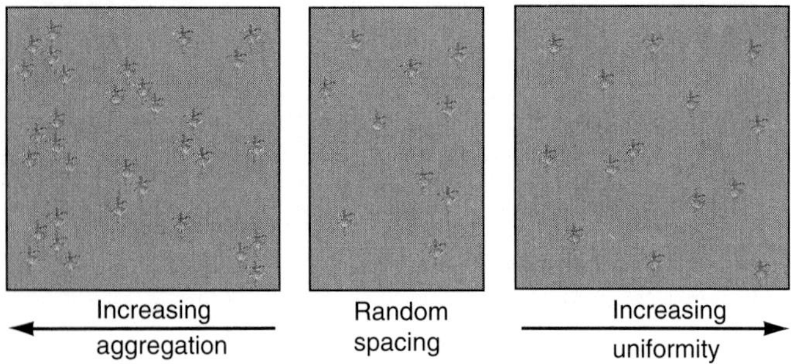

FIGURE 41.3

The dispersion of individuals in habitat space is a basic characteristic of populations. The pattern of dispersion of the plants in these fields may be random but more often shows some uniformity or aggregation. Note that uniformity and aggregation may vary from being only slightly different from random to being perfectly uniform or very tightly aggregated.

Characteristics of Populations

The term *population* may refer to all individuals of a species over the species' entire range or to some of these individuals that live in one part of the range. A population has many descriptive features that reflect its ecological relationships.

Population density is the number of individuals of a species per unit area or unit volume of habitat. **Population dispersion** is the pattern of individuals scattered through habitat space (fig. 41.3). Organisms may be dispersed in a random pattern if no strong ecological influences determine where individuals can live. A tendency toward uniform spacing may occur if individuals respond negatively to each other's presence. Most often, however, organisms show some degree of aggregation, coexisting where habitat conditions are most favorable, seeking out each other due to social attraction, or clumping in the vicinity of their parents. Other descriptive features of populations include **sex ratio** (proportion of males to females), **age structure** (distribution of ages), and, for many species, distribution of body sizes. Biology in Action 41.1 discusses some ways that biologists measure population density.

Populations are always changing. Individuals enter populations by reproduction and leave by **mortality** (death). They may also enter a population by immigration from other locations, and they may leave a population by emigration to other locations.

A species is rarely distributed continuously and in the same density over broad geographic areas within its range. More often it exists as scattered local populations that may exchange individuals by immigration and emigration. When this happens, the connected populations form a regional **metapopulation** that behaves in a very complex fashion.

Mastering Concepts

1. What is population density?
2. Distinguish between the types of dispersion in a habitat.
3. What forces add individuals to populations and remove them?

biology in 41.1 action

Conducting a Wildlife Census

In Washington, D.C.'s Lafayette Park, biologist Vagn Flyger and his helpers place nest boxes in trees and then collect them when grey squirrels wander in. They anesthetize the animals and record their vital statistics—weight, sex, age, and health status. The researchers also tattoo the squirrels so their whereabouts can be monitored and then release them. Once a certain number of squirrels have been tattooed (M), a "recapture" sample of a certain size (R) is taken and the fraction of tattooed animals (f) noted. The number of marked squirrels (M) divided by this fraction (f) estimates the total squirrel population size.

Wildlife biologists also use a variation of this "capture-mark-recapture" approach on deer. Researchers place bright orange collars on a small number of deer. Then, an observer spots a group of deer from a plane and takes an aerial photograph. From the easily seen percentage of collared animals in the photo, biologists calculate the total number of animals in the group.

Counting droppings is another way to take a wildlife census, as is done for deer and moose in the Adirondack Mountains of upstate New York. The numbers of excrement piles are correlated with the number of animals. A wildlife refuge on the Gulf Coast combines the capture-mark-count approach with dropping detection. Scientists capture otters and inject them with a radioactive, harmless chemical, which the animals eliminate in feces. The scientists then collect droppings and extrapolate population size from the proportion of radioactively labeled droppings.

Many wildlife surveys count migrating herds of mammals and flocks of waterfowl from the air (fig. 41.A). Counts are also done from the ground. Every year, the North American Breeding Bird Survey obtains counts along more than 2,000 routes. Perhaps most inventive is a device used in Montana. Rotting meat or fresh sardines are draped in a tree, along with an infrared heat detector attached to a camera. When a bear saunters over to collect the treat, the detector picks up the bear's body heat and snaps its picture (fig. 41.B). Tags are used to trace bear identities.

Describing wildlife populations is useful in evaluating the effects of environmental catastrophes, such as fires, volcanic eruptions, or oil spills, by providing information on population dynamics and habitats before the disaster. Hunting, trapping, and fishing regulations and schedules are based on these data, as well as decisions on where to build—or not to build—houses, dams, bridges, and pipelines. The status of wildlife populations also reflects the overall well-being of an ecosystem.

FIGURE 41.A
Each spring, U.S. and Canadian waterfowl spotters traverse 1.3 million square miles (3.4 million square kilometers) in the air, trying to count the number of animals in flocks, such as these snow geese.

FIGURE 41.B
A male grizzly bear takes his own picture while stealing food from a tree with a camera triggered by the bear's body heat. From tags or physical features visible in such photos, biologists can obtain information on the overall bear population in a region.

Table 41.1 Population Parameters

Parameter	Meaning
Growth rate	Increase in a population per unit time
Specific growth rate	Per-capita increase in a population per unit time
Intrinsic rate of natural increase	Specific population growth rate under ideal, unlimited conditions
Carrying capacity	The population that can be sustained indefinitely in a particular habitat

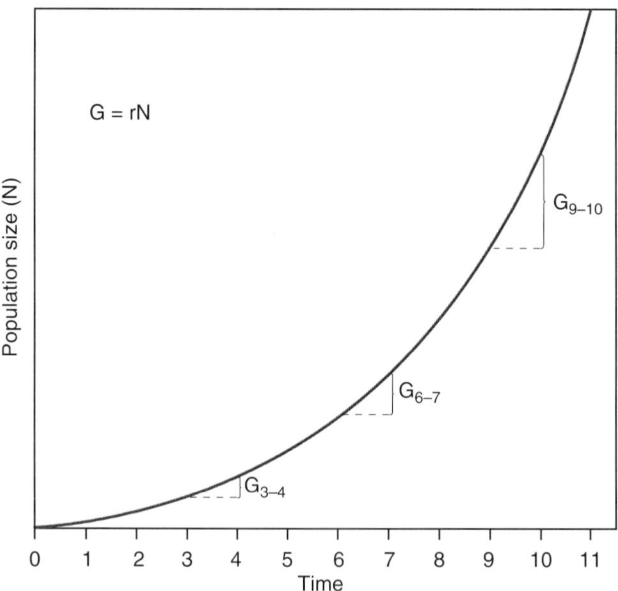

FIGURE 41.4

Exponential population growth. During each time interval, the growth, *G*, is the product of the intrinsic rate of natural increase, *r*, and the number of individuals in the population at the beginning of the interval, *N*. With each successive interval the actual growth increases.

Potential Population Growth

Reproduction and mortality are often the most important determinants of population growth. In a local population, however, organisms moving into or out of the area may have a substantial effect. For example, immigration caused 33% of human population growth in the United States during 1993.

We can describe reproduction, mortality, and growth in populations by changes in absolute numbers. The fact that a population increased by 500 individuals, however, does not predict how large the population will be next week or next year. Knowing how long it took to add those 500 individuals, and the size of the population before the 500 were added, are essential to put growth into perspective. Five hundred individuals added to a population of 1,000 have a much greater impact than 500 individuals added to a population of 100,000.

A population change is usually described as a rate, or the change in the number of individuals during a set interval of time. More precisely, a **specific rate of population growth** is the number of events (such as births, deaths, or growth) divided by the number of individuals in the population during a specific time period. For example, population growth of birds and mammals is usually expressed as the number of individuals added to the population per individual per year.

The larger a population, the faster it can potentially grow. If the additional individuals reproduce, a population may grow much the way compound interest accumulates in the bank—the more you have, the more you get. Even if the specific rate of population growth remains the same, an increase in population size means that the number of future members can increase even faster, because more individuals can contribute to the next generation. Table 41.1 lists ways to describe population growth.

Intrinsic Rate of Increase and Exponential Growth

A population in an environment with unlimited space, shelter, and resources would grow exponentially. That is, the population increases by a constant multiplier during each time interval. Actual growth might be slow at first, but the increase in numbers would accelerate continuously. Growth resulting from repeated doubling (1, 2, 4, 8, 16 . . .), for example, is exponential.

Plotting the number of individuals in a population over time as it grows exponentially yields a characteristic **J-shaped curve** like that in figure 41.4. A population unrestricted by the environment and with maximal survival and reproduction grows at its **intrinsic rate of increase.** Until the winter of 1963–64, the reindeer population on St. Matthew Island, discussed earlier in this chapter, grew in nearly this fashion.

Under ideal conditions, a population will show its maximum specific growth rate, termed the **intrinsic rate of natural increase** and symbolized by *r*. This rate is determined by how quickly individuals reproduce and the average (mean) number of offspring they produce during their lives under these ideal conditions. Factors that influence the intrinsic rate of natural increase include number of offspring produced at one time, the length of time an individual remains fertile, and how early in life reproduction begins.

The earlier reproduction begins the faster the population will grow. For example, a multigenerational human family in which women give birth in their teens will be considerably larger than a family in which the women delay motherhood until they are in their mid- to late thirties (fig. 41.5).

The exponential growth potential for even the slowest-breeding species is incredibly high and seldom attained. Although the gestation period for elephants is 22 months, if elephants met

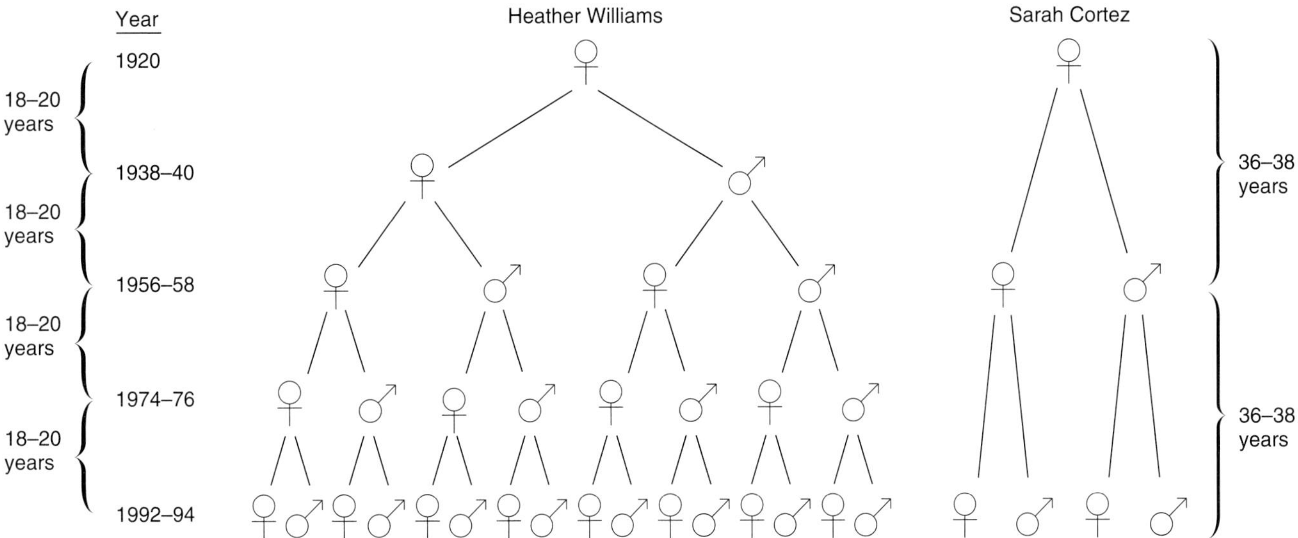

FIGURE 41.5

Early reproduction leads to faster population growth. In the hypothetical Williams family depicted on the left, women, including those marrying into the family, have two children each. The women become mothers when they are between the ages of 18 and 20. The women of the Cortez family, depicted on the right, do not have children until ages 36 and 38. By the time that Sarah, the matriarch of the Cortez family, is a grandmother of four, Heather, the head of the Williams family, is already a great-great-grandmother of sixteen!

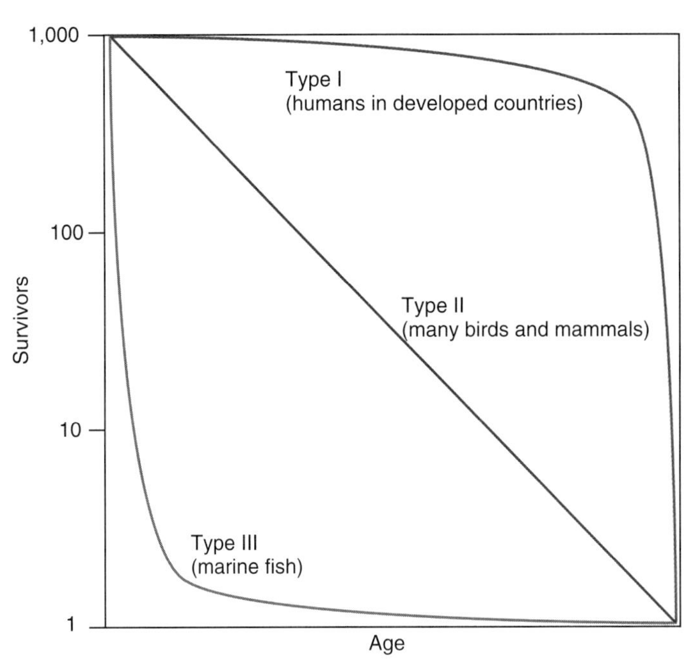

FIGURE 41.6

Survivorship curves show extremes among animal species in relation to age. This graph depicts the number of survivors out of 1,000 individuals beginning life as age increases. The scale of survivors is logarithmic, which means that straight-line portions of the curves reflect a constant survivorship rate.

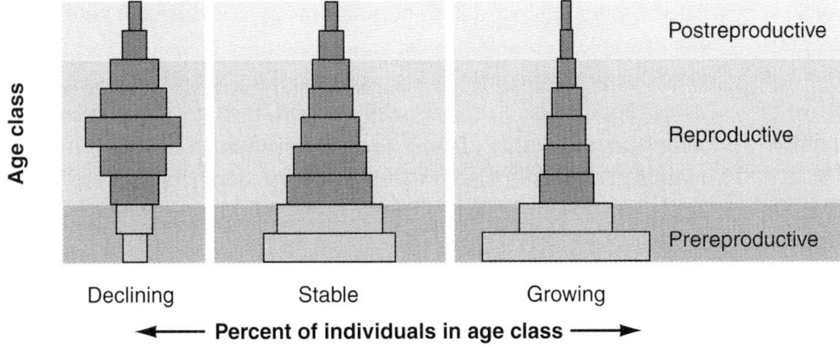

FIGURE 41.7

In this diagram of age structure, the width of each bar is proportional to the percent of individuals in that age class. In many populations, individuals younger than reproductive age (prereproductive), of reproductive age, or past reproductive age (postreproductive) can be distinguished. A stable population will show a gradual decline in the percent of individuals in classes of increasing age. If most of its members are in postreproductive age classes, the population is likely to decline. If unusually large percentages of individuals are in prereproductive age classes, the population will grow fast as these members reach reproductive age.

Source: Arthur S. Boughey, *Ecology of Populations,* 2d ed., © 1973, p. 617.

their reproductive potential, a single pair could leave 19 million descendants in 750 years! The biotic potential for rapidly breeding organisms is astounding. Under ideal conditions, a single *E. coli* could produce 5 trillion billion offspring in one day! Human biotic potential is also impressive. Imagine a couple who starts a family as soon as physiologically possible and continues having one baby as frequently as possible until menopause. If all their children reproduced in the same manner, within five generations (about a century), the couple would have more than half a million descendants.

Survivorship, Fecundity, and Age Structure

In a group of individuals that begin life at the same time, **survivorship** is the fraction that survive to various ages. By graphing survivorship against age for various species, three patterns emerge (fig. 41.6). Type I species, such as humans in developed countries, produce few offspring, but these survive to old age, when the mortality rate becomes high. Type II species, including many birds and mammals, produce more offspring and show a constant survivorship rate through life. Type III species, such as most ocean fish, produce many offspring, but most do not survive long.

Fecundity is the number of offspring an individual produces. Like survivorship, fecundity varies with age. A **fecundity schedule** is the average number of offspring individuals produce during each age interval.

Survivorship plus the fecundity schedule determine whether the population is growing, stable, or declining and defines its age structure (fig. 41.7). We can consider the members of a population in three age classes: prereproductive, reproductive, and postreproductive. A stable population has a certain ratio of individuals in these three classes, reflecting the pattern of survivorship and fecundity of the species.

If a particular population has a much greater fraction of prereproductive individuals but the same survivorship, the population grows as individuals enter their reproductive years. Even if a pair produces only two offspring to replace themselves, this mostly younger population would continue to grow as the proportion of individuals entering the reproductive years increases until the proportion of individuals in the prereproductive years declines. Two-thirds of China's 1 billion people, for example, are under the age of 24, meaning that the population will continue to grow even if the specific rate of population growth slows. By the year 2000, the Chinese will have 200 million more mouths to feed.

Assuming that fecundity and survivorship remain the same, a population consisting of mostly older individuals will decline as the members die, due to a higher death rate among the older individuals. In such a population, the proportion of females able to bear young is small. Therefore, even if each female bears the same number of offspring as in the past, the total number of young added to the population will be smaller.

Mastering Concepts

1. What is the specific rate of population growth?
2. What conditions support exponential population growth?
3. How do survivorship patterns, fecundity, and age structure influence population growth?

Regulation of Population Size

Fortunately, populations do not grow as rapidly as is theoretically possible. **Environmental resistance** are factors that check population growth by reducing reproduction and immigration or by increasing mortality and emigration.

Environmental Resistance and the Carrying Capacity

Environmental resistance includes shortage of food or other resources, predation, disease, and parasitism. It also includes natural or human-caused environmental disasters, pollution that impairs health and ability to reproduce, and any other factor that reduces population size.

The theoretical maximum number of individuals that an environment can support for an indefinite time period is its **carrying capacity.** In reality, the features of the environment that determine the carrying capacity for a species vary through time. A population at its carrying capacity has equal birth and death rates; therefore, it does not grow—it remains stable.

How populations respond as the carrying capacity approaches varies. At one extreme, population growth may slow and stabilize as the carrying capacity nears. At the other extreme, the population tends to grow exponentially and fluctuate wildly about the carrying capacity.

Density-Independent and Density-Dependent Factors

Environmental resistance results from factors that differ depending upon population density. **Density-independent factors** exert effects that are unrelated to population density. A severe cold snap, for example, might kill a certain percentage of fish in two populations, even if the populations differ in size. Natural disasters such as earthquakes and severe weather conditions are typical density-independent factors. Consequences of human-caused environmental disasters are also independent of population density. Oil spills, for example, kill whatever life becomes entrapped in the suffocating black mess.

A spectacular density-independent event occurred on May 18, 1980, when the north face of Mount St. Helens in Washington state began to bulge and rumble. Underground, water superheated by molten rocks was turning to steam. At 8:32 A.M., Mount St. Helens erupted. The 900°F (477°C) blast pulverized rocks and

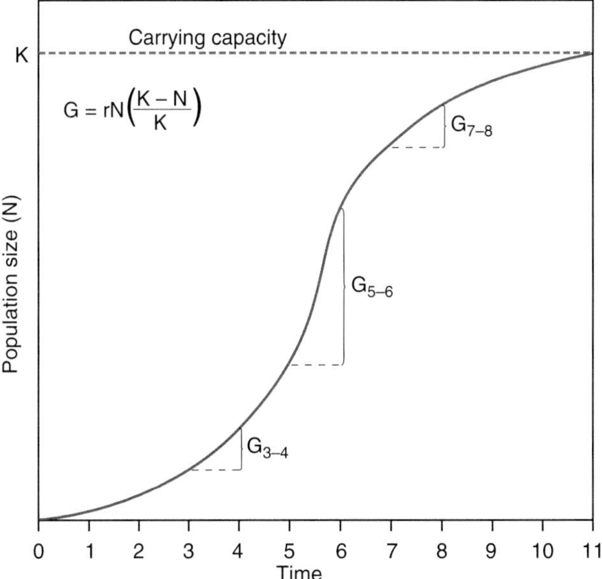

FIGURE 41.8

The logistic, or S-shaped, curve of population growth approaches the environmental carrying capacity, *K*. Note that the actual growth per time interval is small at the beginning of the curve, peaks at the midpoint, and declines as the curve approaches the carrying capacity.

trees, killing nearly everything within a 6-mile (9.7-kilometer) radius and triggering earthquakes and avalanches. Ten miles (16 kilometers) away, trees fell; 16 miles (25.7 kilometers) away, trees were scorched. Millions of dead insects rained from the ash-ridden sky. A 25-foot- (nearly 8-meter-) high tide of mud and molten rock flowed out of control, joined by melting snow and ice in the higher elevations, and buried everything in its path. As the top of the mountain caved in, jets of gas and steam rocketed 65,000 feet (19,812 kilometers) into the air, spraying a fine gray dust over nearly half the state. A forest of five-century-old fir trees disappeared instantly. Hundreds of thousands of tons of new plant growth were suddenly buried or torn away. The animal death toll included some 5,200 elk, 6,000 black-tailed deer, 200 black bears, 11,000 hares, 15 mountain lions, 300 bobcats, 1,400 coyotes, 27,000 grouse, 11 million fish, and uncountable numbers of insects and microorganisms.

A density-independent event that is disastrous for some species might benefit others. This happened in 1993, when a flooding Mississippi River sent deer, raccoons, rodents, skunks, and opossums scurrying to higher ground but provided rich new shallow-water habitats for legions of microscopic algae and plankton and the fishes and insects that feed on them. Wading birds ate fish trapped in the shallower floodwaters and muck. Floodwaters carried wild gourds and zebra mussels to new habitats. Duck populations soared as the ravaged river banks provided more nesting sites and places to hide from predators. Other birds weren't as lucky. The least tern, for example, already endangered, became even more so as the waters washed away its nests, eggs, and fledglings.

In contrast, effects of **density-dependent factors** increase as the size or density of a population increases. Density-dependent factors include crowding, competition, and predation. In response to crowding, for example, some animals cease mating, neglect young, and become aggressive. Physiological responses to crowding include increased rate of spontaneous abortion, delayed maturation, and hormonal changes. Such responses ultimately slow population growth.

Competition for space and food is a common density-dependent factor that affects population growth. In **scramble competition,** individuals compete directly for a limited resource. When many individuals share limited food, none of them may eat enough to be able to reproduce, and population growth slows. Insect larvae compete in scramble fashion. If several hundred fruit fly larvae hatch on a very small apple already riddled with larvae, they soon deplete the food supply. Larvae that do not eat enough starve to death or metamorphose into small adults too weak to reproduce. Population growth slows.

Alternatively, members of a population may compete for a resource indirectly. In **contest competition,** animals compete for social dominance or possession of a territory, factors that guarantee the winners an adequate supply of a limited resource. The losers get less of the resource or none at all. Population growth slows because losers of the competition breed less successfully than the winners.

Logistic Growth Curve

When density-dependent relationships are strong, and few density-independent disturbances occur, population growth slows as it approaches the carrying capacity in response to environmental resistance. This leveling off produces a characteristic **S-shaped curve,** or logistic growth curve. This curve follows a simple equation that gives the expected growth during a particular time interval (fig. 41.8):

$$\text{Growth} = rN\,(K - N)/K$$

In this equation, N is the number of individuals in the population at a given time, K is the carrying capacity, and r is the specific population growth rate under nonlimiting conditions. The right side of the equation states that growth is equal to rN multiplied by $(K - N)/K$. The term rN is the potential growth that would occur if resources were not limiting, based on the number, N, of individuals present and the growth potential, r, associated with each. $(K - N)/K$ applies a "degree of realization" to potential growth. This term simulates the increasing intensity of density-dependent factors as the population approaches carrying capacity. When N is very small relative to K, this expression yields a numerical value near 1.0; most of the potential growth is thus

biology in 41.2 action

Booming and Busting Sheep

In 1932, the Marquis of Bute brought 107 feral sheep to the island Hirta near Scotland. By 1942, their numbers had grown to 500, and by 1952, to 1,114 animals. A few summers later, when the population exceeded 1,400, the tiny island became littered with sheep corpses. More than a third of the animals had starved to death. Curiously, populations of sheep on other islands, and of other large animals on Hirta, did not decline.

The nature of the Soay sheep's reproductive cycle explained the boom and bust cycle. A combination of factors enables the population to grow much too quickly for its own good. Females can become pregnant before they are a year old, and 20% of them deliver twins. Infant mortality is low. Babies stop nursing in June, when food is abundant, which gives their mothers sufficient time to gain enough weight to conceive in the fall. The Soay sheep also lack predators and cannot leave the island (fig. 41.C).

Characteristics of the populations of other large mammals on Hirta allow them to maintain their sizes. Red deer, for example, feed their young for longer periods, which causes the mothers to miss a reproductive cycle. This spaces births, and the population does not grow as fast. Red deer also have few twin pregnancies and high infant mortality.

FIGURE 41.C
Soay sheep on the island of Hirta graze in rocky sheltered areas. As their population explodes, due to early and frequent pregnancies, the sheep eat all the vegetation. Every 4 years, many sheep starve, but the population recovers rapidly.

realized. When N is near K, the value of the expression drops to near zero; most of the potential growth is not realized. Thus, the logistic equation can be stated:

realized growth = potential growth × degree of realization

"Boom and Bust" Cycles

In populations with high reproductive rates, exponential growth may overshoot the carrying capacity. Then the population drastically drops or crashes. A population that repeatedly and regularly increases (booms) and decreases (busts) in size over a given period of time exhibits a **boom and bust cycle.** Food supply, predation, and changes in population age structure influence boom and bust cycles. These cycles can occur when density-dependent factors are delayed, which allows population size to reach a level that cannot continue.

Snowshoe hare populations in Canada and Alaska follow a boom and bust cycle, with their numbers peaking every 8 to 11 years. When a snowshoe hare population is at its densest, up to four hares may live on an acre of land. In the leanest years, one hare may be the only one of its species on 200 acres. The population densities of animals that eat hares fluctuate along with the changing hare population.

Until recently, boom and bust cycles were only known in small mammals such as lemmings, hares, and voles. Biology in Action 41.2 discusses boom and bust cycles in a population of feral (wild) sheep on a small island off the coast of Scotland.

Human population booms and busts often mirror historical events. The large proportion of middle-aged U.S. citizens today reflects a "baby boom" after soldiers returned from World War II. From an all-time low in population growth in the depression years of the early 1930s, fertility peaked in 1957 at 3.7 children per woman of childbearing age. Schools in the United States are currently severely overcrowded because of a "baby boomlet" of children of the baby boomers.

K-selection Versus r-selection

Natural selection has adapted the life histories of species to the regime of density-independent and density-dependent factors that most strongly influence them. At one extreme is a pattern of natural selection termed **K-selection.** Density-dependent factors regulate K-selected populations close to the carrying capacity. Individuals tend to be long-lived, to be late-maturing, and to produce a small number of offspring that receive extended parental care. Many large mammals and birds live in K-selected populations.

FIGURE 41.9

Several adaptations enable this butterfly, *Cithaerias menander,* to avoid predators. These include transparent portions of the wings, which make its detection difficult, and eyespots, which may frighten predators.

At the other extreme is **r-selection.** Density-independent factors strongly influence r-selected populations, which rarely near the carrying capacity. Individuals of most such species are short-lived, reproduce at an early age, and have many offspring that receive little care. Insects and other invertebrates live in strongly r-selected populations.

Many species are intermediate between K- and r-selection. Sea turtles, for example, are long-lived, late-maturing species that tend to produce many offspring but give them little parental care.

For decades, ecologists thought that logistic growth and close regulation of populations to a stable carrying capacity was the general rule in nature. Long-term studies of populations of many plants and animals, however, reveal that such patterns are the exception rather than the rule. Even populations of large animals that show K-selected life histories fluctuate greatly in response to changes in their environments that range from gradual and progressive to sudden and catastrophic.

Mastering Concepts

1. How does environmental resistance keep populations from reaching their carrying capacities?
2. Distinguish between density-independent and density-dependent effects on population size.
3. What conditions are necessary for S-shaped (logistic) growth to occur?
4. What is a boom and bust cycle?
5. Distinguish between K-selected and r-selected populations.

Interspecific Interactions

Individuals of a species interact not only with others of their own kind but with individuals of many other species. These interactions are as ecologically significant as those within populations of single species.

Interspecific Competition

Members of different species living in the same area may compete for the same resource. This interaction can be a serious problem in farming. In soybean fields, for example, yellow nutsedge plants control the size of soybean populations by crowding the young bean plants and depleting the soil of nutrients the soybean requires. Homeowners who try to maintain weed-free lawns are also familiar with interspecific competition.

According to the **principle of competitive exclusion,** two or more species that compete for the same limited resources cannot indefinitely occupy the same habitat. If conditions continually favor one species, it will eventually replace the other. For example, on Africa's Serengeti Plain, wild dogs and hyenas hunt zebras, wildebeests, and other antelopes. Hyenas are larger and more aggressive than wild dogs, and they often displace wild dogs from their kills. The hyena's superior hunting skills have contributed to the decline of the wild dog population.

Similarly, when a new species enters an area, competition may drive a native species to extinction. This happened when starlings were released in New York City's Central Park in 1891. The starlings robbed the native bluebirds of their nesting sites.

Predator-Prey and Parasite-Host Interactions

The interactions of predators and prey influence populations of both types of species. Effects on the prey population are obvious—their numbers diminish as they are eaten. Prey populations are often maintained by high reproduction rates that compensate for the loss to predation. Predation also influences prey populations by eliminating the weakest individuals, who are more easily captured.

Prey species use hiding, camouflage, and natural weapons to avoid being eaten. Many species of butterflies have evasive flight maneuvers to escape hungry birds. The butterfly in figure 41.9 has several adaptations that enable it to avoid predation. The butterfly remains close to the forest floor, where birds can't swoop down to eat it without crashing to the ground. The butterfly's transparent wings resemble the wings of another species that birds avoid because of its unpalatable taste.

A sea urchin is also adapted to avoid predation. It digs into the ocean bottom with its spines, but it may use them as a weapon should a predator approach (fig. 41.10*a*). Small white structures aligned between the spines inject poison into potential predators. Similarly, a sand dollar uses spines to dig into its sandy background to hide (fig. 41.10*b*).

From the predator's perspective, inability to find food may mean death. Some predators are specialists and some are generalists. The short-tailed weasel, for example, must consume small rodents and birds at least every 4 hours to fuel its very high metabolism and eats particular species. In contrast, the South American gray fox eats a varied diet of goose eggs, beetles, mice, scorpions, hares, birds, frogs, and lizards.

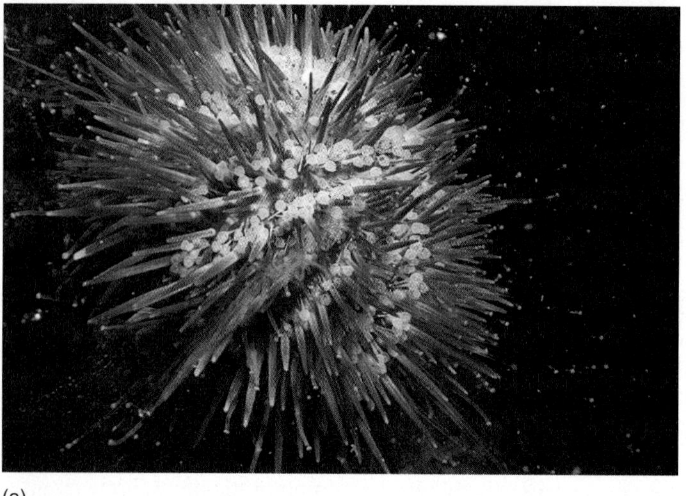
(a)

(b)

FIGURE 41.10

The variegated sea urchin (*a*) survives predation in the Atlantic Ocean with the help of its spines and pincerlike structures that inject poison into predators. The sand dollar (*b*) uses its fringe of moving spines to burrow into the sand, where predators cannot see it.

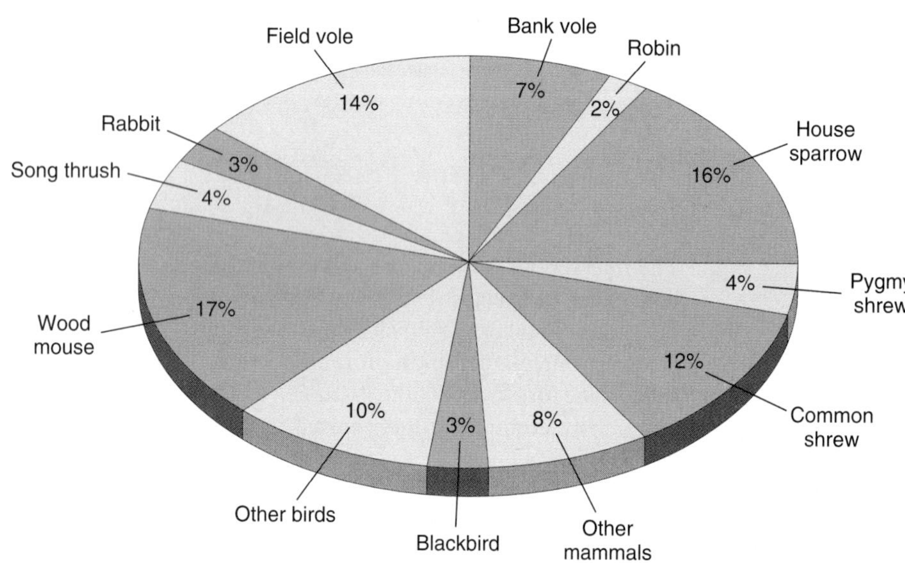

FIGURE 41.11

Thanks to the prey-collecting efforts of 78 cat-owning families in a small English village, researchers described the predation pattern of house cats.

Reproduction among predators is intimately tied to the density of their prey populations. Many animals that cannot find sufficient food become temporarily infertile or have small litters. When a female bobcat has trouble finding food, she has only one or two kittens, instead of the usual three, and those few may not survive their first 10 months, when the mother must care for and feed them. In southeast Idaho, declining bobcat litter size parallels declines in the populations of cottontail and jackrabbits, their prey.

The type of prey a species eats usually depends simply on what is available. For example, the diet of bobcats varies depending upon where they live. In Washington state, they primarily eat mountain beaver; in Minnesota, they prey on deer and snowshoe hare; in the central and southern states, they eat rabbits, muskrats, voles, and birds.

Some predators kill with great precision. A study of monkeys killed by eagles showed that the birds always pierce the monkeys' hearts in the same quickly lethal way. Cat lovers whose pets proudly deposit captured birds and chipmunks on the doorstep are witnessing the results of predation. A study of house cats in an English village found that the pets, although well fed at home, killed a significant number of local wildlife (fig. 41.11). To participate in the study, the cat owners saved every tidbit their pets brought home for 1 year. Other interesting ways that researchers investigate predator-prey interactions include probing gut contents from roadkills or examining animal droppings.

Many predators hunt in groups. The Harris's hawks perched in a tree in figure 41.12*a*, for example, are scanning the area for prey. When they spot a rabbit, the hawks take turns diving at it. Should the rabbit escape into a burrow, the birds again take turns

(a)

(b)

FIGURE 41.12

Predators sometimes cooperate. The hunt begins when the group of Harris's hawks assembles on a lookout (*a*). Smaller teams of two or three birds make short flights to nearby perches, scanning the area for prey. After the hunt (*b*), the hawks share the meal.

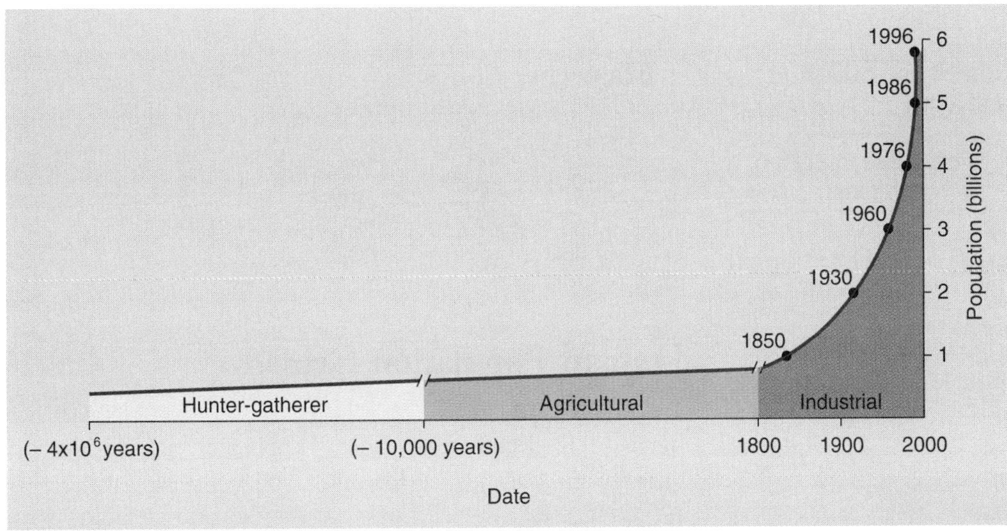

FIGURE 41.13

Human population growth is on the rise, as indicated by the J-shaped curve.

Source: Data from Alexander Leaf, *The New England Journal of Medicine,* Volume 1321, No. 23, Page 1581.

going in after it. When the rabbit runs out, the other hawks, waiting in a circle, kill it. Afterwards, the hawk hunting team shares the meal (fig. 41.12*b*).

Parasites interact with their hosts in a manner somewhat similar to predators with their prey, except that a parasite may not kill its host, or at least not kill it quickly. Recall that parasites are organisms that live on or in a host and derive nutrition from it. They may also lay eggs in a host. Larvae hatch and consume the host's tissues, eventually killing it. The North American brown-headed cowbird exhibits another form of parasitism. It lays eggs in the nests of another bird species.

Species that interact intensely influence the evolution of each other's life history characteristics. Some species of milkweeds, for example, emit chemicals that deter most insect herbivores. The monarch butterfly, however, has evolved in a way that enables it to tolerate these chemicals, and it eats only milkweeds. In Central America, the bull's-horn acacia plant and a species of ant have evolved a close mutual set of adaptations. The acacia has swollen thorns that the ants hollow out, occupy, and eat. The ants, in turn, viciously defend the acacia against other herbivores.

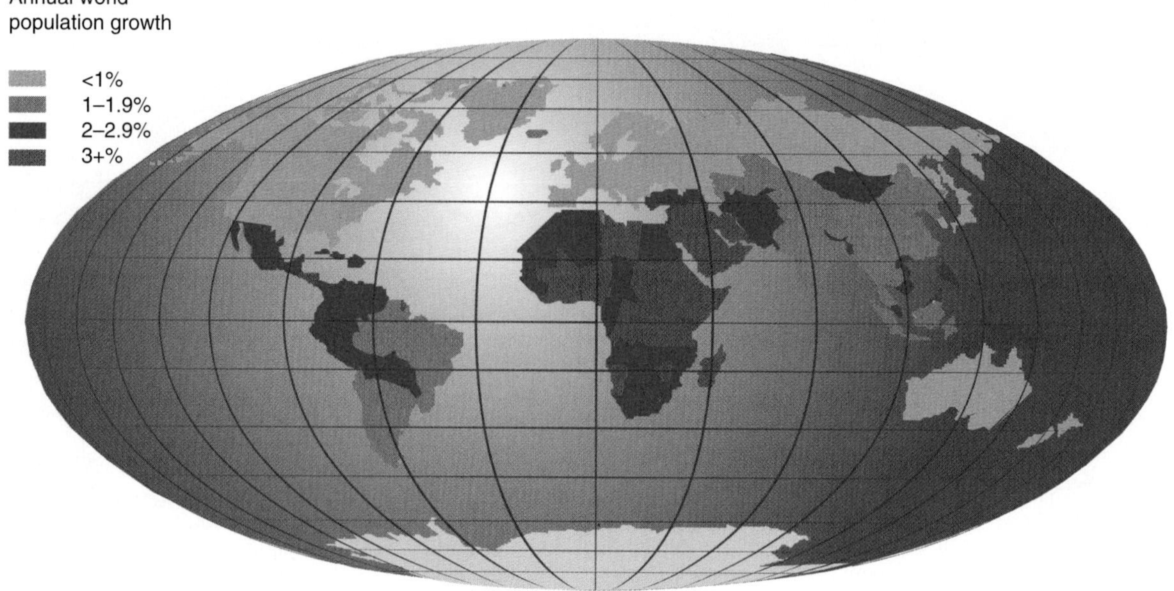

FIGURE 41.14

Human population growth rates are not evenly distributed among nations. The highest growth rates are in Africa, and the lowest are in Europe and North America.

Source: Diane Defrancesco, Population Reference Bureau.

Table 41.2 Projected Growth of Some Human Populations

Area	% Growth (1994)	Population in 2025 (millions)
Central Africa	153	190
East Africa	149	516
West Africa	148	510
Middle East	105	286
North Africa	90	280
Southern Africa	88	85
Central America	68	199
Southeast Asia	55	715
South America	48	452
Australia + New Zealand	40	29
Canada	38	38
United States	25	322
East Asia	24	1,762
Former Soviet Union	21	344
Eastern Europe	10	107
Western Europe	4	434

Source: *The Independent* (London), 5 September 1994

Mastering Concepts

1. What is the principle of competitive exclusion?
2. In what ways do predator and prey populations influence each other?
3. How is a parasitic relationship similar to a predator-prey relationship?

Human Population Growth

The human population is growing at an alarming rate, as figure 41.13 depicts. Each hour, 11,000 more people are born than die. As a result, 264,000 more people are alive today than yesterday.

The rate of population increase varies greatly in different parts of the world (fig. 41.14 and table 41.2). This unequal growth has led to sometimes opposite governmental attempts to control reproduction. In Thailand, for example, population growth has fallen over the past 15 years from 3.2% to 1.6% because of increasing availability of contraceptives. China has also controlled a runaway population growth, but with somewhat drastic measures, as discussed in Bioethics Connections 41.1. Yet the governments of France and the Canadian province of Quebec offer financial incentives and extended new parent leaves from work to encourage citizens to have babies to bolster the population. Table 41.3 shows another way to express differences in population growth—comparing the time it will take a population, growing at the present rate, to double.

Global human population growth has not been steady. It has accelerated in recent years, thanks in part to medical advances and

41.1 bioethics connections

Human Population Control in China—Did It Go Too Far?

In the year 2020, China will have a serious population problem—too few young women. A million or more young men reaching marriageable age will not be able to find wives. The coming scarcity of women is not only predictable, it was planned. Government policies that limited reproduction, and long-held ideas about the values of males and females, have led to the impending male-dominated society.

Statistics for the number of each sex in China are striking. The sex ratio at birth (SRB) is the number of live male births per 100 live female births in a particular time period, such as a year. In most human populations, without intervention to alter the numbers of boys or girls born, the SRB is about 105. In China in 1990 the SRB was 113.8, and it is rising. This overabundance of boys reflects peoples' desire to have sons and their willingness to avoid having daughters. Female fetuses identified using prenatal diagnostic techniques are often terminated, infanticide is practiced on newborn girls, and some female births simply are not reported.

China's selection of one sex has its roots in the one-child program, which began in 1979 (fig. 41.D). Births among government workers have been closely monitored, and "granny police" have watched homes to report unregistered births. First-born children receive benefits, such as free education and extra paid leave for their mothers. If a second child is born, the first child's benefits are revoked. Therefore, people take the above-mentioned measures to increase the chance that a first birth is male.

The coming shortage of females in China illustrates the complexity and interrelatedness of population dynamics, as well as the fact that a population intervention today can have a drastic impact tomorrow. Perhaps the scarcity of females will reverse the nation's priorities.

FIGURE 41.D
This class in Beijing teaches Chinese citizens about birth control.

Table 41.3 Population Doubling Times

Area	Time for Population to Double
Africa	24 years
Asia	35 years
Central America	35 years
North America	98 years
Europe	1,025 years

our ability to manipulate the environment. From the appearance of modern humankind some 40,000 years ago until 1850, our population grew relatively slowly, reaching 1 billion. Yet only 80 years later, by 1930, we doubled our number to 2 billion. In another 30 years, by 1960, 3 billion humans populated the earth. The human population hit 4 billion in 1976, 5 billion in 1986, and 5.8 billion in 1996. It is expected to reach 6 billion by the turn of the century.

Although the rate of global human population increase peaked in the 1960s at slightly more than 2% per year, and is now about 1.6%, the total number of people is so large that the increase in years to come will still be great. Demographers project that by the year 2050, the human population will fall to between 7.8 and 12.5 billion individuals. Even if all women, from now until then, have the "replacement rate" of 2.1 children each, the population in 2050 will still reach 7.7 billion.

When the study of population dynamics applies to humans, fields other than biology come into play—such as politics, history, economics, anthropology, and sociology. Shifting the focus of studying populations from a global view to a family view helped reveal factors that contribute to rapid population growth.

Following a decade of studies on the structures of families in many societies, attendees at the 1994 United Nations International Conference on Human Population Growth, held in Cairo, structured a new hypothesis to explain the drive behind burgeoning human populations. In many cultures, they found, families are large because people want many children to help obtain increasingly scarce resources, such as wood for fuel, clean drinking water, and food (fig. 41.15). Hence, population growth, poverty, and environmental decline (which limits resources) are intricately interrelated.

The key to limiting family size, according to the United Nations conference, is to find ways to give women in developing

Where Have All the Frogs Gone?~

continued from page 836.

After a decade of recording the precarious fates of amphibian populations, a possible explanation emerged. In the late 1980s, researchers documented thinning in the ozone layer, which would allow more ultraviolet radiation to reach the earth's surface, where it could damage DNA, impair immunity, cause cancer, and disrupt reproduction and development in many species.

Ecologist Andrew Blaustein and his students at Oregon State University conducted experiments to address the hypothesis that ultraviolet radiation was damaging amphibian eggs. Knowing that DNA repair enzymes enable animals to correct ultraviolet-induced genetic damage, they predicted that species with low levels of these enzymes would be more vulnerable to damage. They collected eggs from Cascades frogs, western toads, and northwestern salamanders, all with low levels of repair enzymes, and Pacific tree frogs, which have very high levels of the enzymes. The eggs were placed in screened containers in their natural habitats. The researchers used three types of containers—covered with clear plastic that admitted ultraviolet rays, with clear plastic that blocked the rays, or with no shield. After the eggs hatched, the researchers tallied the results.

As predicted, many of the eggs of the Cascades frogs, western toads, and northwestern salamanders failed to hatch when ultraviolet rays reached them, but nearly all their eggs hatched when plastic blocked the radiation. In contrast, nearly all of the Pacific tree frogs' eggs hatched in all three types of containers. Ultraviolet radiation presumably did not prevent their eggs from hatching because these frogs have very high levels of DNA repair enzymes.

These experiments elegantly demonstrate how one facet of the environment—ultraviolet radiation—can have drastic effects on populations. Yet other observations and experiments suggest that several factors contribute to decline of amphibian populations. These factors include habitat destruction, fungal and bacterial infections, our use of amphibians for food and drugs, and introduction of nonnative species into natural habitats. As in many matters ecological, causes are complex, multiple, and interacting.

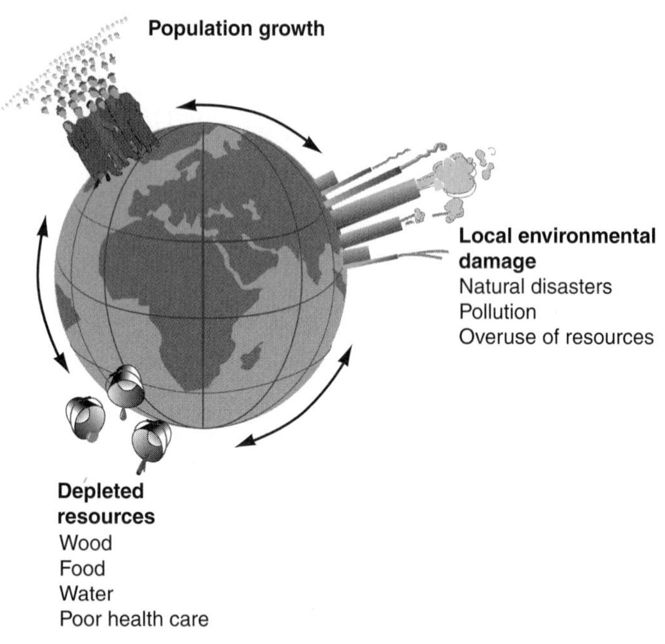

FIGURE 41.15

Human population growth may be part of a cycle. Resources are depleted by overuse, natural disasters such as floods and droughts, and human-caused factors such as pollution. To obtain as many resources as possible, families have more children to help. The increasing population depletes resources further, and the cycle continues.

nations options other than continual reproduction. This suggestion is based on many studies that correlate large families to women who are illiterate and/or who work for no pay. Studies also show that as the percent earnings that a mother contributes to the family income rises, number of children falls. Specific recommendations include educating women for trades, increasing availability of contraceptives, and improving health care.

Despite the influences of culture on human population dynamics, our population ups and downs are basically like those of other species. Changing numbers reflect birth and death rates, as well as diverse environmental influences. The next two chapters explore in greater detail the interactions of organisms and the living and nonliving environment—ecology.

Mastering Concepts

1. Which parts of the world have the highest and lowest rates of population growth?
2. What are some ways to measure or study population growth?
3. What factors affect the growth rates of human populations?

SUMMARY

1. A population is a group of organisms of the same species living in a geographic region. **Ecology** considers relationships between organisms and their living and nonliving environments. It includes the relations of individuals in populations and their interactions with individuals of other species.

2. Each species has characteristic conditions where it lives **(habitat)** and resources necessary for its life activities **(niche).** Slight differences in the niche allow different species to share surroundings. Because of competition, a species' **realized niche** is often smaller than its **fundamental niche.**

3. A population grows when more individuals are added through birth or immigration than are subtracted due to death or emigration. Population growth depends upon the initial size of the population, how many individuals are added and at what rate, the age at which individuals begin to reproduce, and the **age structure** of the population.

4. Unrestrained growth is exponential and produces a **J-shaped curve. Environmental resistance** counters unrestrained growth.

5. Environmental resistance includes **density-independent factors,** which kill a fraction of the population regardless of its size. **Density-dependent factors,** which have a greater effect on large populations can regulate population size. Competition for a limited resource is density-dependent when members of the same species compete directly for it **(scramble competition).** Everyone gets less of the resource and reproduction declines. When members of a species contest territories or social dominance **(contest competition),** the winner usually receives the resource.

6. Carrying capacity is the number of individuals an environment can indefinitely support. After a period of exponential growth, a population may overshoot the carrying capacity and crash, or because of density-dependent factors, slow and level off at the carrying capacity, producing an **S-shaped curve.** Populations that regularly increase and decrease in size have a **boom and bust cycle.**

7. In **K-selection,** density-dependent factors maintain a population close to its carrying capacity. In **r-selection,** density-independent factors predominate, rarely reaching carrying capacity.

8. Several interspecific interactions influence population size. These include competition, predation, parasitism, and symbiosis. Competition between species may drive out all but the best competitors. In a natural environment, both predator and prey usually survive because of coevolved adaptations.

9. Many human populations are growing rapidly. Human population growth has not been steady and occurs unevenly in different parts of the world. Global human population growth will level off towards the end of the next century.

TO REVIEW . . .

1. How do a habitat and niche differ? How does a fundamental niche differ from a realized niche? Describe your habitat and niche.

2. What resources are necessary for a population to grow?

3. How can different species share the same habitat?

4. State three descriptive features of population structure.

5. Define the following terms:
- **a.** specific rate of population growth
- **b.** carrying capacity
- **c.** intrinsic rate of natural increase
- **d.** environmental resistance

6. Distinguish between scramble competition and contest competition.

7. What are the differences between the conditions that result in a J-shaped compared to an S-shaped population growth curve?

8. Why do Soay sheep populations on the island of Hirta undergo boom and bust cycles but populations of red deer on the same island remain relatively stable?

9. Distinguish between K-selection and r-selection.

10. Why don't prey populations completely decimate prey populations?

11. How do survivorship, fecundity, and the age structure of a population control its growth?

12. Why will the global human population continue to increase for several more years even though the rate of increase peaked three decades ago?

TO THINK ABOUT . . .

1. In some nations, well-educated, well-to-do families tend to have fewer children, and poorer families tend to have more children. How might such a trend affect a population in the short and the long term?

2. Many children growing up in China today will have no brothers or sisters, and their children will have few aunts, uncles, or cousins. The alternative, statistics predict, is mass starvation in the next century. Do you think China's "one-child" policy is an effective way to control population growth? Why or why not?

3. Suggest an experiment to test the hypothesis that crowding makes humans more aggressive and violent.

4. Cite three environmental upheavals that may have had a density-independent impact on wildlife populations.

5. In Pakistan, 43% of the population is under age 15 and 4% is over 65. In the United Kingdom, 19% of the people are under 15 and 15% are over 65. Which population will increase faster in the future? Why?

6. Describe a method to estimate the number of elk in a particular area.

7. If human population growth were to continue at the rate that it was in 1990, there would be 694 billion people by the year 2150. Why is this unlikely to happen?

8. The suggestions made at the 1994 United Nations population conference largely targeted women. What can men do to help limit human population growth?

TO LEARN MORE . . .

References and Resources*

Allen, William H. December 1993. The great flood of 1993. *BioScience.* The 1993 flood in the midwestern United States displaced many deer, mice, and humans but enabled other populations to grow.

Blaustein, Andrew R., and David B. Wake. April 1995. The puzzle of declining amphibian populations. *Scientific American.* Several factors contribute to the scarcity of certain amphibians.

Clutton-Brock, Tim. March 1994. Counting sheep. *Natural History.* Sheep on an isolated island have boom and bust cycles because of their rapid reproduction.

Coghlan, Andy. October 5, 1996. Doomsday has been postponed. *New Scientist.* Global human population is expected to peak by the end of the twenty-first century.

Cohen, Joel E. April 1996. Ten myths of population. *Discover.* A concise review of the principles of population biology applied to humans.

Cohen, Joel E. 1995. *How many people can the earth support?* New York: W. W. Norton. The history of human population growth and efforts to predict it, with an analysis of the relationships that define the earth's carrying capacity.

Dasgupta, Partha S. February 1995. Population, poverty and the local environment. *Scientific American.* Growing human populations are part of a cycle.

Godfray, Charles, and Michael Hassell. January 17, 1997. Chaotic beetles. *Science,* vol. 275. Computer modeling is used to study a stressed beetle population.

Levin, Simon A., et al. January 17, 1997. Mathematical and computational challenges in population biology and ecosystems science. *Science,* vol. 275. Mathematics is used to study population ecology.

Line, Les. April 1, 1997. Winter devastates island's moose. *The New York Times.* Severe weather can trigger a population crash, as it did to moose in Isle National Park in Lake Superior.

Nichols, James D. February 1992. Capture-recapture models. *BioScience.* Ecologists use clever techniques to monitor wild populations.

Tuljapurkar, Shripad, Nan Li, and Marcus W. Feldman. February 10, 1995. High sex ratios in China's future. *Science,* vol. 267. A policy to control population growth will have repercussions for many years.

**Life*'s website features frequently updated references.

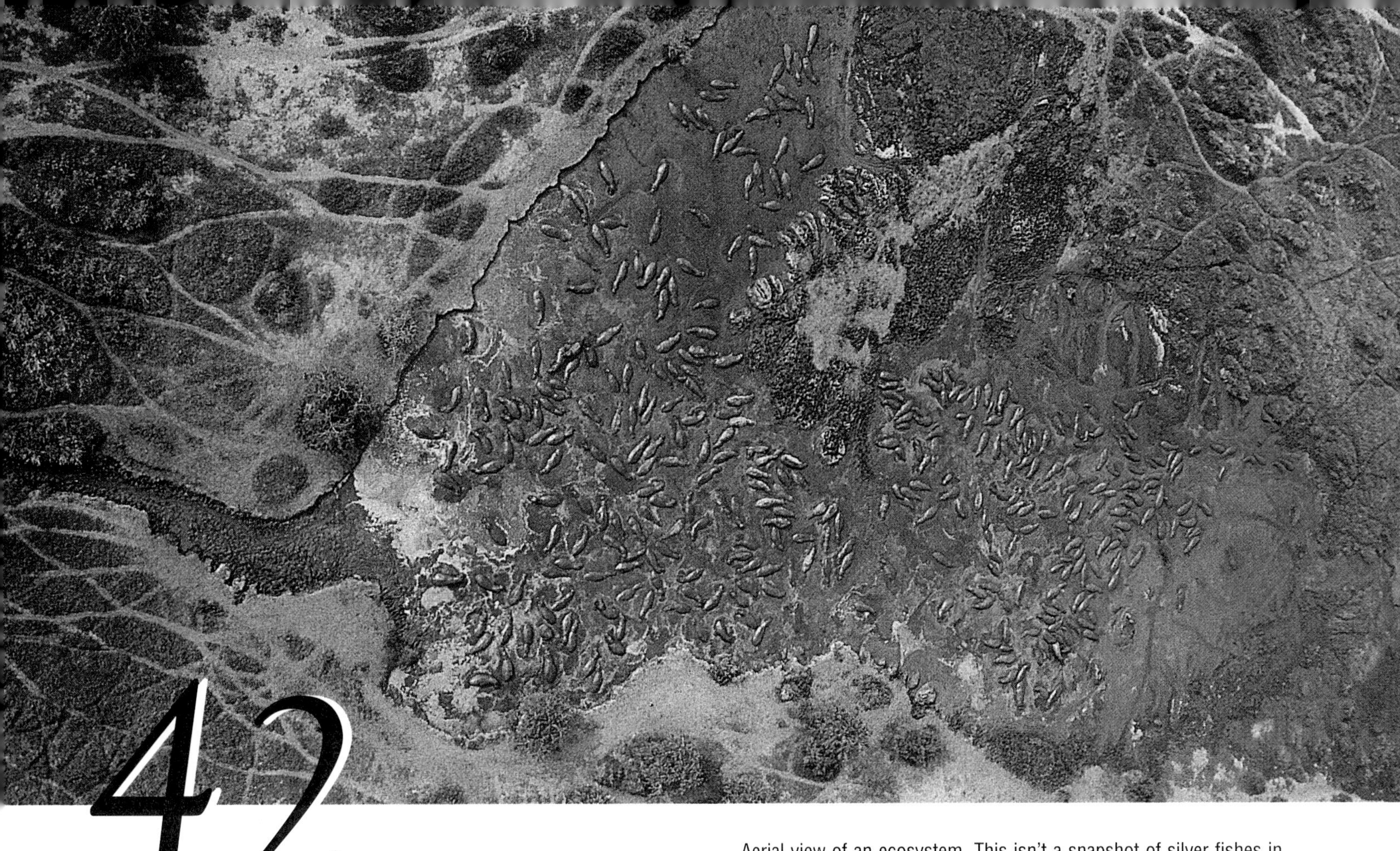

42 Communities and Ecosystems

Aerial view of an ecosystem. This isn't a snapshot of silver fishes in a tiny pond but a glimpse from the air of hippos lolling in a wallow. The lines emanating from the water are trails that the animals follow when they leave the water to forage for food at night.

chapter outline

From Krill to Whale—The Antarctic Ecosystem *854*

Communities and Their Characteristics *854*

Biotic Succession *856*
Primary and Secondary Succession
Mechanisms of Succession
Disturbance and Disequilibrium

Ecosystems and Their Characteristics *858*
Energy Flow Through an Ecosystem
Stable Isotope Tracing—Deciphering Food Webs
Ecological Pyramids

Biogeochemical Cycles *864*
The Water Cycle
The Carbon Cycle
The Nitrogen Cycle
The Phosphorus Cycle

Bioaccumulation and Biomagnification *866*
DDT
Mercury

Nonnative New Yorkers—Peregrine Falcons

The peregrine falcon (*Falco peregrinus*) is the swiftest and most powerful bird of prey. From ledges in cliffs and mountainsides, the falcon spots its prey, takes flight, and then suddenly swoops down at 200 miles (320 kilometers) per hour. Its potential meal is a moving target, another bird. In flight, the falcon grasps or hits the prey with its feet and then bites the victim's head or neck, quickly killing it. In the 1920s, when people used carrier pigeons for communication and for racing, mountain climbers often came upon a peregrine falcon site containing hundreds of pigeon feet, complete with the rings used to identify them—remains of many meals.

The female peregrine falcon is about 17 inches (43 centimeters) long at maturity, and the male about 2 inches (5 centimeters) shorter. The female lays eggs on ledges or in holes in cliffs and doesn't build a nest. She sits on the eggs for 33 days. Then the male feeds the fledglings for the first 5 weeks, after which they attempt to leave the site. Many perish, but those that survive typically live more than a decade.

Peregrine falcons have very specific habitats. Then why have they taken up residence in the bridges around New York City?

to be continued . . .

From Krill to Whale—The Antarctic Ecosystem

An adult blue whale is by far the largest animal ever to live on earth—25 times as heavy as an elephant and 98 feet (30 meters) long. The whale lives part-time in the southern Antarctic Ocean. Its existence is intimately tied to another organism, 2.4-inch-(6-centimeter-) long shrimplike krill.

The Antarctic krill, *Euphausia superba,* is a crustacean that provides the great blue whale's diet. Krill travel in great underwater clouds, circling Antarctica and concentrating in eddies where north-flowing cold surface water meets warmer south-flowing water. Krill lay eggs at the ocean's surface from January through March. The eggs sink and hatch about 820 yards (750 meters) down. The larvae rise as they develop and produce a constant stream of organic material as they periodically shed their outer coverings.

Blue whales cruise the Antarctic Ocean every summer. They glide through clouds of krill, their giant jaws partially opened, capturing vast quantities of the tiny animals. The whale pushes water out of its mouth through sheets of feathery, hornlike baleen that descend from the roof of its mouth like a shower curtain (see fig. 37.3). The krill are trapped by this filter and swallowed. If a concentrated cloud of krill is not available, the hungry whale stirs one up. By diving beneath a dispersed cloud and then swimming up in a spiral pattern, the whale initiates an underwater minivortex of krill and swims through its whirling meal.

Krill directly or indirectly support nearly all life in Antarctica (fig. 42.1). They provide food for 6 species of baleen whales, 20 species of squid, 120 species of fishes, 35 species of birds (mostly penguins), and 7 species of seals. Krill eat algae, protozoa, small crustaceans, and fish eggs and larvae. They feed around and below ice floes and within tiny channels in the ice that support rich microbial communities. Some krill have pigmented intestines, from recent feasts on "blooms" of brightly colored phytoplankton.

Communities and Their Characteristics

The plankton, bacteria, crustaceans, squid, fishes, birds, seals, and whales of Antarctica compose a **biotic community.** Such a biotic community includes all of the organisms, sometimes hundreds of species, in a given area. For example, Right Whale Bay in the South Georgia region of Antarctica is a community of fur seals, king penguins, macaroni penguins, dominican gulls, elephant seals, and all the marine organisms that these mammals eat.

Many species can coexist within a community by using different resources, as the warblers occupying the same tree in figure 41.2 demonstrate. Consider a downed Douglas fir (fig. 42.2), which is a bustling biotic community. Some resident species prefer dry, shaded branches, others the soggy, rotting roots, and still others, the deepest heartwood. Soon after the tree falls, certain insects invade the inner bark, and then other species enter the sapwood. Still later, different species attack the heartwood. Deep within this decomposing tree, bark beetles whittle a labyrinth-like "egg gallery" where they rear the next generation; the mother beetle guards the entrance. This chamber is also home to more than 100 other types of beetles. Tiny wasps drill in from the outside and lay eggs on the bark beetle larvae. When wasp larvae hatch, they eat the beetle larvae. Scavenging beetles eat any dead bark beetle larvae left over from the wasp's meal. Other insects eat the fungus that grows in indentations in the bark beetle's exoskeleton.

Biotic communities are distributed unevenly on the planet. Biodiversity is highest near the equator, possibly due to the favorable climate, high biological productivity, and infrequent catastrophies such as continental glaciation. Diversity declines as climate fluctuates more toward the poles. In general, similar types

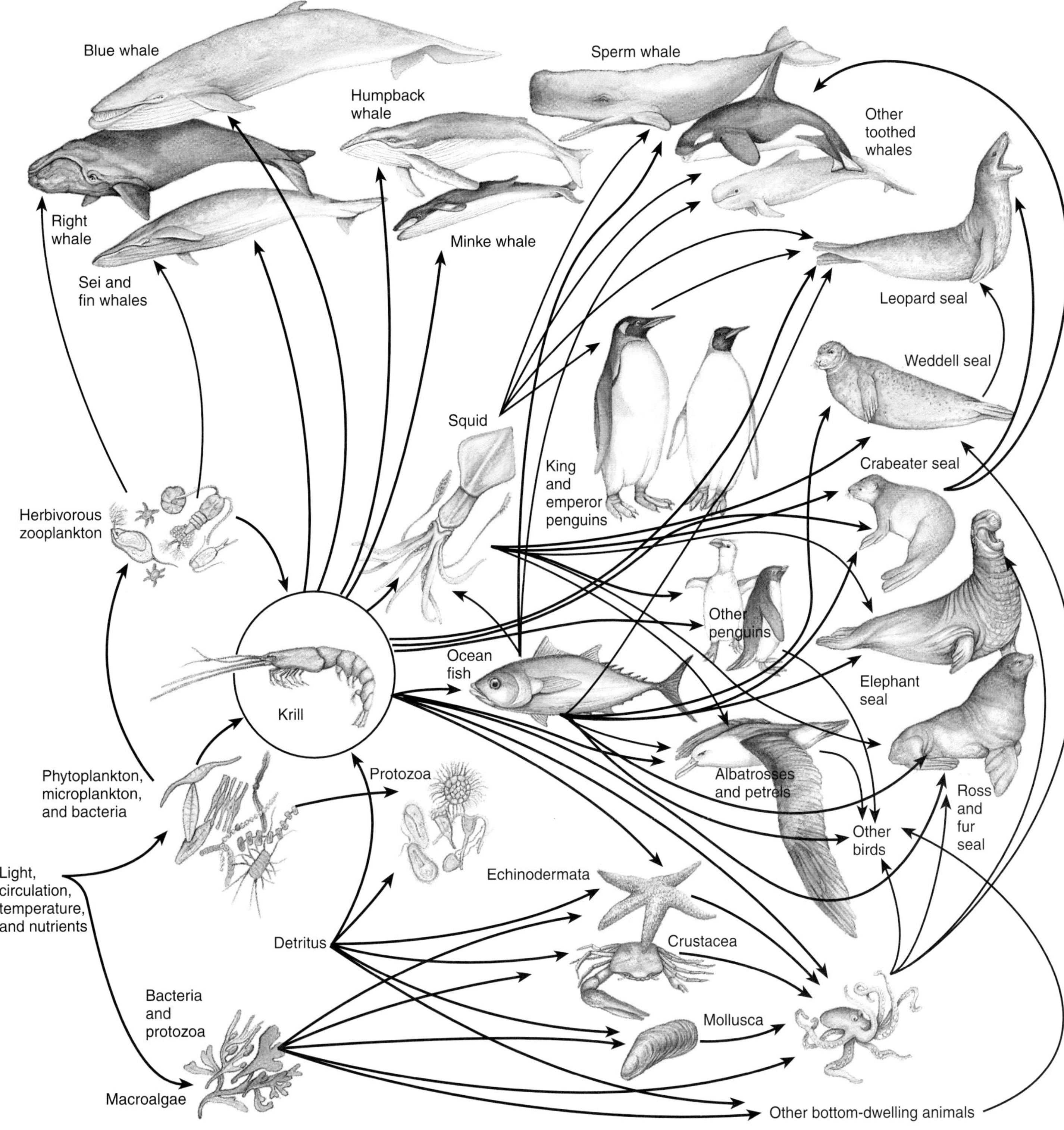

FIGURE 42.1

The Antarctic web of life. Nearly all species in Antarctica depend, directly or indirectly, upon great numbers of krill.

of biological communities appear at corresponding latitudes because they have similar climates. The steamy heat of South American and African rain forests is likely to be home to species with similar adaptations. Differences among communities at the same latitude in different parts of the world reflect regional climatic characteristics or geographical influences.

Mastering Concepts

1. What is a biotic community?
2. How can many species occupy a biotic community?
3. Why is biological diversity unevenly distributed on the planet?

Biotic Succession

Biotic communities change through time in response to many influences—climatic change, disturbance, and species invading from other areas. A community also changes as a result of its own interaction with its physical environment. This gradual and directional process of change in the community, in which species replace each other and the biotic structure may change considerably, is called **biotic succession.** Succession often leads toward a **climax community,** one that may remain fairly constant as long as climate does not change and major disturbances do not strike.

FIGURE 42.2

A fallen tree holds a thriving community of organisms. This newly fallen tree will be home to a changing community of hundreds of insect species.

Primary and Secondary Succession

Primary succession occurs in an area where no community previously existed. On a patch of bare rock in New England, for example, the first species to invade, the **pioneer species,** are hardy organisms such as lichens and mosses that can grow on smooth rock. The fungal component of a lichen (see fig. 24.13) holds onto the rock and obtains water, while the algal part photosynthesizes and provides nutrients. As lichens produce organic acids that erode the rock, sand accumulates in the crevices. Decomposing lichens add organic material to the sand, eventually forming a thin covering of soil. Then rooted plants, such as herbs and grasses, invade. Soil continues to form, and larger plants, such as shrubs, appear. Next come aspens and conifers, such as jack pines or black spruces. The soil finally becomes rich enough to support other deciduous trees. After many years, a climax community of an oak-hickory forest may develop (fig. 42.3).

Secondary succession occurs where a community is disturbed, such as by fire or agriculture (fig. 42.4). Secondary succession is often faster than primary succession because a well-developed soil already exists.

An abandoned farm illustrates secondary succession (fig. 42.5). This "old field" succession begins when the original deciduous forest is cut down for farmland. As long as crops are cultivated, natural succession stops. When the land is no longer farmed, fast-growing

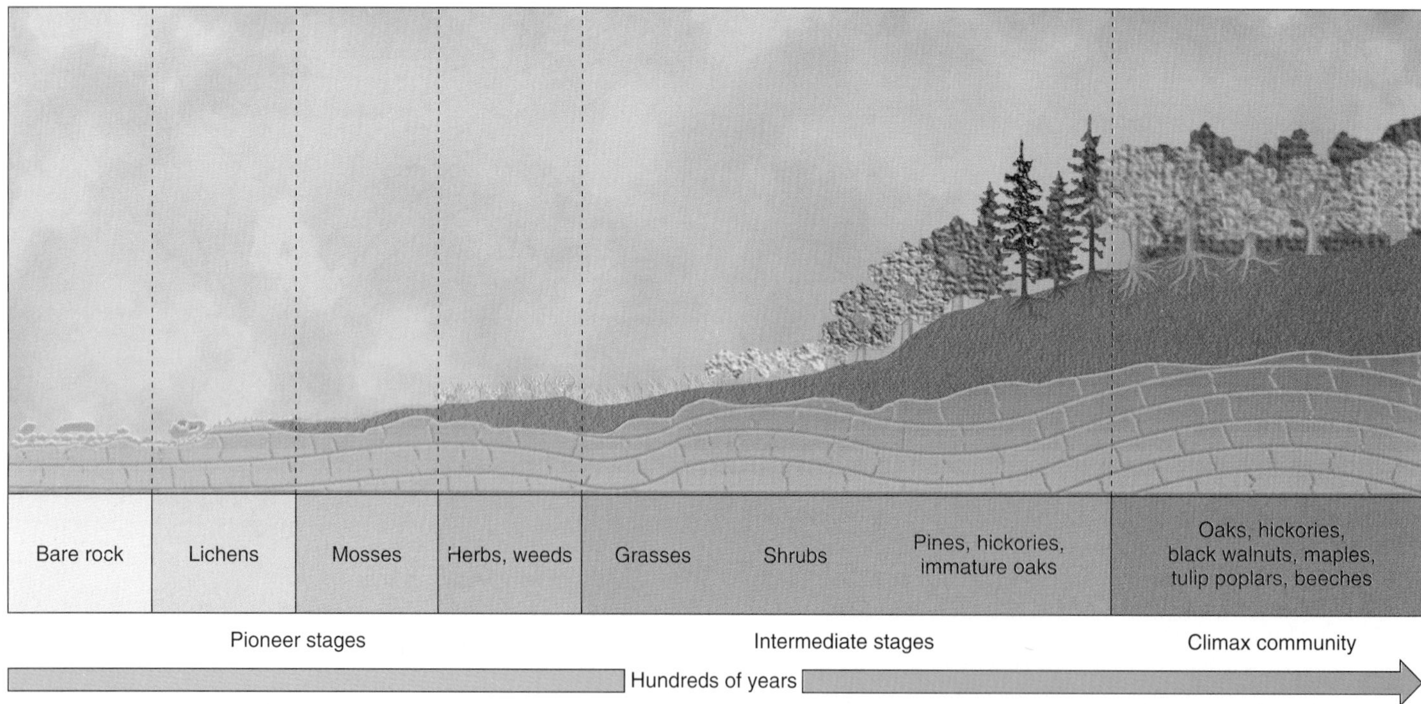

FIGURE 42.3

Primary succession. Lichens arrive and produce acids that begin to break down rock into a thin layer of soil. Next, mosses, fungi, small worms, insects, bacteria, and protozoa colonize the scanty soil. Tiny herbs and weeds grow, grasses and shrubs move in, and trees come many years later.

FIGURE 42.4

Secondary succession. Just a few weeks after a fire, new plants grow.

pioneer species, such as black mustard, wild carrot, and dandelion, move in, followed by slower-growing, taller goldenrod and perennial grasses. In a few years, pioneer trees such as pin cherries and aspens arrive and are eventually replaced by pine and oak. A century or so later, the climax community of beech and maple may again be well developed.

Secondary succession is dramatic in the Mount St. Helens area of Washington state, where a volcano erupted in 1980. Dead animals and animal droppings provided nutrients, which gophers mixed into the soil. Gopher mounds became the first areas to sprout new plant life. Small flowers and ferns grew in protected areas between fallen trees.

Just 1 month after the volcanic eruption, the moonlike landscape was dotted with purple and yellow aster blossoms, their buds saved from the eruption by snow cover; green bunches of thimbleberry and fireweed sprouts peeked out from their pumice coats; and spreading fungus carpeted the land in pink. Four months after the blast, insects were back. Bluebirds and woodpeckers built nests in dead trees, and

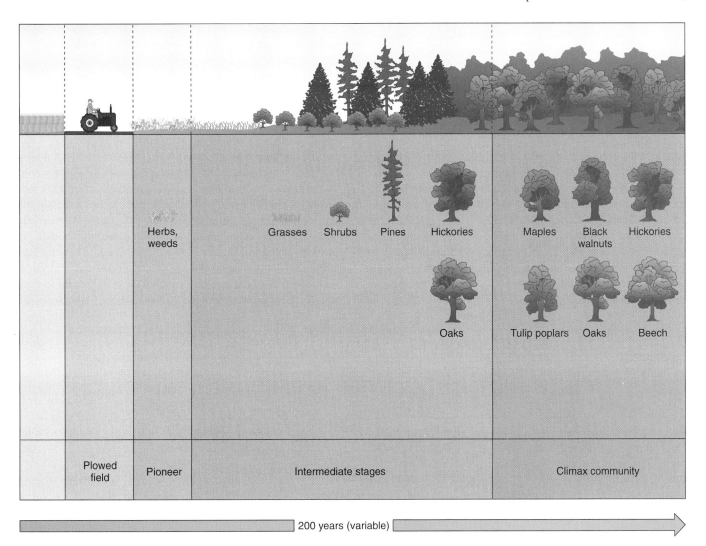

FIGURE 42.5

Secondary succession. Last summer it was a farmer's field, with row after row of corn, beans, and sunflowers, in the southeastern United States. This year it was plowed and then abandoned. But the land does not stay bare for long. Weeds and herbs appear first and then grasses and shrubs. Then small versions of the trees that ring the area will grow. Over many years, tree species will appear in a certain sequence, until the area stabilizes as a climax community.

geologists exploring the area were often attacked by hummingbirds mistaking their bright clothing for flowers. These pioneer species could invade the disturbed region quickly because they were adapted to life in an area where disturbances are common.

In the years following the eruption, animals from surrounding areas wandered in, some bringing seeds on their coats. Other seeds blew in and germinated. The new inhabitants returned nutrients to the soil in their excrement and decomposing carcasses, and rain mixed the soil. Over the next three decades, a new forest will begin to grow. Forty years from now, scattered trees should stand, each 40 to 50 feet (12 to 15 meters) tall. By 2080, a dense new forest should flourish—barring further eruptions.

Mechanisms of Succession

Succession occurs in four stages. First, pioneer species colonize a bare or disturbed site. Then, species with less efficient dispersal mechanisms arrive, and they begin to change the community. Some early colonists do not survive the challenges of the new environment, which further alters the community. Throughout the process, organisms alter the physical conditions, often in ways that enable other species to become established. Lastly, interactions between species change the composition of the biotic community. When pine trees invade a site, for example, they support species that grow or feed on pines yet shade out lower-growing plants that appeared earlier in succession.

Disturbance and Disequilibrium

For many years, ecologists thought that stable climax communities were common. They realize now, however, that few communities ever reach a perfect steady state. Climate shifts are almost always in progress, and such factors as fire, disease, and severe storms can influence successional patterns for centuries. In reality, true "climax" conditions rarely develop. What usually happens is that the rate of change slows late in the successional process but never ceases. In the Pacific Northwest, for example, old-growth forests are 500 to 1,000 years old, yet they are still changing in their structure and composition.

Mastering Concepts

1. What is biotic succession?
2. Distinguish between primary and secondary succession.
3. What processes and events contribute to succession?
4. How common are climax communities?

Ecosystems and Their Characteristics

The Antarctic marine biotic community and its physical environment form an **ecosystem.** An ecosystem includes all the living (biotic) organisms and nonliving (abiotic) things within a defined area. It may be as large as the whole earth (the **biosphere**), a small part of it, such as a stretch of creek, or even the community and its physical habitat beneath a single large boulder (fig. 42.6). Even a single leaf is an ecosystem. An ecosystem is terrestrial if it is on land and aquatic if it is in water (table 42.1). Biology in Action 42.1 visits some diverse ecosystems.

Ecosystems are open systems, which means that they exchange matter and energy with their surroundings (see chapter 6). Some ecosystems, such as a lake, are relatively self-contained—more materials cycle within than enter or leave. In some other ecosystems, such as a section of stream, materials rapidly flow in and out. All ecosystems rely on energy from some outside source. This energy ultimately derives from the sun (solar energy) or from the earth's interior (geothermal energy).

Ecosystems can and do change. The Antarctic krill population increases by 13-fold in the summer; some of the ice cover melts, which gives the organisms more space in which to breed. A moose's dietary preferences can gradually change a forest of mostly deciduous trees to mostly conifers (fig. 42.7). Biotic succession leads to change not only in the community, but in many characteristics of the physical environment.

Ecosystems interact with one another. Insects, mosses, and fungi on the surface of a downed tree interact with organisms in the adjacent soil ecosystem. A terrestrial ecosystem may contain an aquatic

Table 42.1 Examples of Ecosystems

			Biota	
Ecosystem	**Sample Location**	**Physical Environment**	**Plants**	**Animals**
Alpine tundra	Cascade Mountains, northwest Washington	Strong winds, stony soil, heavy winter snow	Grasses, sedges, broad-leafed herbs	Mountain goats, marmots, pikas
Boreal forest	Denali National Park, Alaska	Long, cold winter with short days. Short summer with long days	Spruce, aspen, birch, shrubby willows	Moose, snowshoe hares, lynx, wolves
Hot desert	U.S. Southwest	Hot summers, low and variable rains	Shrubs, cacti, annual flowering plants	Rodents, snakes, lizards, kit foxes, peccaries
Antarctic Ocean	Surrounding Antarctica	Cold, upwelling seawater rich in nutrients	Phytoplankton, ice algae	Krill, fish, squid, seals, penguins, whales
Lake	U.S. Northeast	Fresh water, low nutrient content, seasonal change in temperature	Phytoplankton, rooted aquatic flowering plants	Zooplankton, insects, lake trout, waterfowl

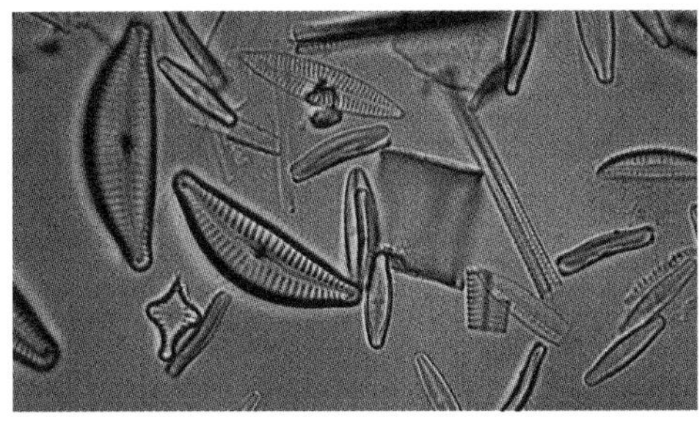

(a)

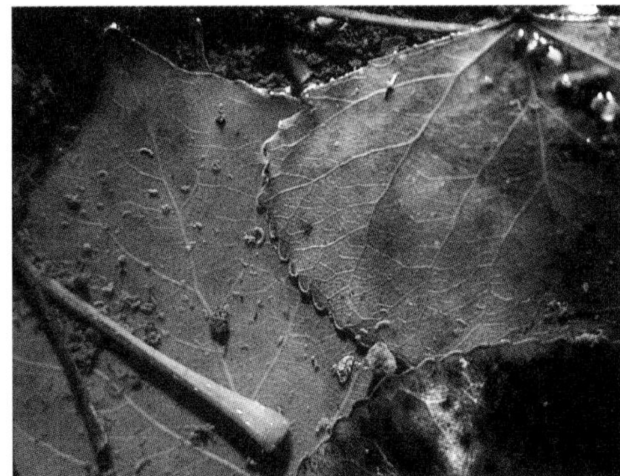

(b)

(c)

(d)

(e)

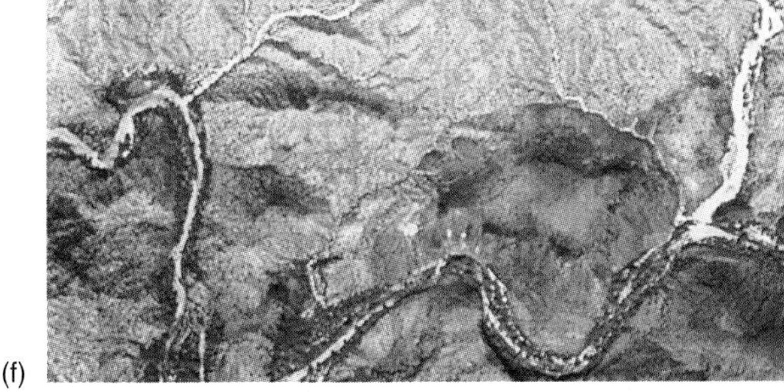

(f)

FIGURE 42.6

An ecosystem can be studied at several levels, as this sequence of photos of the Sycamore Creek drainage basin near Phoenix, Arizona, shows. Fourteen streams feed into this body of water. (*a*) The water abounds with microscopic diatoms. (*b*) Fallen leaves provide food for invertebrates. (*c*) Snails and insects feed on carpets of algae. (*d*) A flash flood is not necessarily a catastrophe—invertebrates survive under rocks and in crevices, fish swim into eddies and protected pools, and decimated insect populations recolonize the stream quickly because of their high reproductive rate. Aerial photographs (*e* and *f*) reveal the pattern of stream channels and the relation of the stream to vegetation of its floodplain and watershed. In these photos, dense vegetation appears bright red.

one, such as a pond or marsh, or an aquatic ecosystem, such as a bay, may include islands or sandbars. Ecosystems receive seeds in the air and dissolved nutrients from rivers and streams. Animals continually ferry nutrients and organisms between ecosystems. Even physically distant ecosystems interact through the air and waters. Smokestacks in the American Midwest, for example, pollute and alter the acidity of lakes in the Northeast.

Terrestrial ecosystems that have a predominant plant species, or aquatic ecosystems that have characteristic physical and chemical conditions, form a larger ecological unit, the **biome.** Biomes include deserts, grasslands, deciduous forests, freshwater lakes, and oceans.

Ecosystems are so interconnected that the entire planet can be viewed as one huge, interacting ecosystem, the biosphere. This is the part of the planet where life exists, reaching about 6.2 miles (10 kilometers) into the atmosphere and down about the same distance into the deepest ocean trenches.

Energy Flow Through an Ecosystem

Energy flows through an ecosystem in one direction, beginning with solar energy that photosynthetic organisms convert to chemical energy. This stored energy then passes through a **food chain,** a series of organisms that successively eat one another. The **trophic level,** or feeding level, of an organism is the number of food chain links between it and the ecosystem's energy source.

Primary producers form the first link in the food chain and the base of the trophic structure. Primary producers ultimately provide energy for all other organisms; they use inorganic materials and energy to produce all the organic material they require. Primary producers include plants, algae, and bacteria that derive energy from the earth.

The other levels in a food chain are both producers, which provide food energy for the next level, and consumers, which obtain

42.1 biology in action

A Sampling of Ecosystems

The Sea Surface Microlayer

The sea surface, perhaps the richest ecosystem on the earth, covers approximately 71% of the planet. Life is diverse and abundant in a microlayer only half an inch (1.27 centimeters) thick (fig. 42.A). Ecologists study the sea surface microlayer by rolling a cylinder atop a stretch of ocean. The roller gently lifts the biologically packed top layer and transfers it into glass containers.

Organisms form layers in the water. The top 0.002 inch (0.025 millimeters) contains amino acids, fatty acids, and sugars from plankton below that feed bacteria at the very top. Beneath the top layer are eggs and larvae of diverse aquatic species, and jellyfish poke up from below.

The sea surface microlayer is incredibly thin compared to the vast depth of the ocean, but it is of great ecological importance. It is a transitional zone between ocean and atmosphere and is the base of a huge food web, with birds from above and larger animals from below feeding on its abundant, concentrated life. The sea surface microlayer is also a reservoir for pollutants, including pesticide residue, sewage, by-products of combustion from planes and cars, and wastes from incinerators and power plants.

Lake Torrens—An Instant Ecosystem

Lake Torrens is not usually a lake at all but a 11,600-square-mile (18,676-square-kilometer) depression in the South Australian terrain. Every few centuries, heavy rains fill the natural basin, creating a lake. This happened in March 1989. The water depth ranged from 1 inch (2.54 centimeters) to 3 feet (0.915 meter)—enough for long-desiccated eggs of crustaceans, such as brine shrimp, to hatch and reproduce.

The rich living broth soon attracted thousands of birds to feast, lay eggs, and nurture young. A flock of 130,000 banded stilts nested on tiny islands within the lake, where the water kept out predators (except for other birds, who ate about 60% of the eggs) (fig. 42.B).

Niagara Escarpment

A sheer 98-foot (30-meter) cliff face doesn't seem a likely habitat, but at least two hardy species occupy the Niagara Escarpment in Ontario, Canada. The climate there is much more like the dry cold of the Arctic tundra than the surrounding wet and temperate Canadian forest.

Eastern white cedars of species *Thuja occidentalis* emerge from cracks in the cliff face (fig. 42.C). These trees are superbly adapted to their unusual environment. All of the trees are stunted, so that they cannot easily blow over. They grow very slowly, which may be a response to scarce nutrients and little space for root growth. A tree may grow only 0.002 inches (0.05 millimeter) in width a year.

Life exists not only between the rocks of the escarpment but within them. Colored layers just beneath the rock surface consist of algae, fungi, and lichens (fig. 42.D). This sparse, cryptoendolithic ("hidden inside rocks") ecosystem survives because the translucent rock lets sufficient sunlight through to allow photosynthesis to occur.

The Subnivean Space—Under the Snow

Small mammals and some flowering plants thrive in spaces beneath snow covers in the northeastern United States (fig. 42.E). Squirrels and voles peek out to look for food. The temperature usually remains a stable 30°F to 32°F (−1.0°C to 0°C). As the top layers of snow melt and refreeze throughout the winter, ice crystals shrink, which al-

FIGURE 42.A
A jellyfish is a conspicuous resident of the sea surface. This region also brims with abundant microscopic life.

FIGURE 42.B
Banded stilts lay their eggs on an island in the middle of the transient Lake Torrens.

lows more light through as spring approaches. In response to lengthening daylight, mammals awaken from hibernation to reproduce. At winter's end, some flowering plants blossom beneath the snow.

Fiords—Where Fresh Water Meets Seawater

In the southwest corner of New Zealand's national park, narrow glacier-cut valleys called fiords extend from the Tasman Sea 10 miles (16.1 kilometers) distant, ending between the Southern Alps. The fiords flood, but in an unusual way—seawater pours in, but near-daily rains send rivulets of fresh water down the steep mountain slopes. The result is a body of water that is fresh water for 10 to 12 feet (3.1 to 3.7 meters) on top but seawater below. Fresh water rises because it is lighter and because the fiords are so narrow that waves, which would mix the layers, cannot form.

The spectrum of life in the fiords ranges from a freshwater to a seawater ecosystem. In the surface layer, freshwater mussels and barnacles abound. The water here is stained yellow-brown from leaf litter. The color blocks much sunlight, so the sealike portion of the fiord lacks the light-dependent seaweeds and other algae common in the ocean. But dark-adapted sea life is plentiful and includes sea slugs, corals, and sponges. Brachiopods, crinoids ("sea lilies"), and other shelled animals cling to the cliff walls (fig. 42.F).

Predatory starfish hover at the interface between fresh and salty water and shift position as the interface moves. When a day without rain raises the boundary, freshwater animals are trapped in sudden salinity—giving the waiting starfish an easily captured meal. As soon as the rain returns, the starfish move down, for they cannot tolerate fresh water.

FIGURE 42.C
Eastern white cedar trees are adapted to life on the Niagara Escarpment.

FIGURE 42.D
A layer of green algae is located just beneath the surface of the dolomite rock making up the cliff.

FIGURE 42.E
In the northeastern United States, small mammals and dormant buds of flowering plants survive the winter beneath the snow.

FIGURE 42.F
Crinoids with feathery arms, together with many other species, live in a fiord.

energy from the preceding level. Herbivores (plant-eaters) are **primary consumers** and form the second trophic level. Carnivores (meat-eaters) are **secondary consumers,** animals that eat herbivores. Carnivores that eat other carnivores form a fourth trophic level; they are **tertiary consumers.** These animals expend a great deal of energy capturing their prey. **Scavengers,** such as vultures, eat remains of another's meal, and **decomposers,** such as certain fungi, bacteria, insects, and worms, break down dead organisms and feces.

Food chains interconnect, forming complex **food webs,** such as the Antarctic web in figure 42.1. Webs form when one species eats or is eaten by several other species and when one species functions at more than one trophic level. A person eating a tuna sandwich, for example, is both a primary consumer (a herbivore, eating bread) and a tertiary consumer (a carnivore eating another carnivore, the tuna).

FIGURE 42.7

Foraging moose can change an ecosystem. When moose preferentially eat foliage of aspen, balsam poplar, and birch, these trees are replaced by the less palatable spruce and firs. This was demonstrated by fencing off parts of the forest on Isle Royale in Michigan. Where moose were free to munch, conifers predominated. In areas without moose, aspen, balsam poplar, and birch thrived.

Stable Isotope Tracing—Deciphering Food Webs

Ecologists can describe food webs by simply watching which organisms eat which organisms. A method called **stable isotope tracing,** borrowed from geology, can reveal who eats whom from afar. The technique analyzes proportions of certain isotopes in tissue samples. (Recall that isotopes are atoms of the same element that have different numbers of neutrons.) Stable isotope tracing relies on the fact that primary producers each have a characteristic ratio of carbon-13 to carbon-12, or a "carbon signature" (fig. 42.8*a*).

By determining the $^{13}C/^{12}C$ ratio in a tissue sample, such as a nail paring from a polar bear, researchers can identify the species at the base of the food chain (fig. 42.8*b*). The tissue is thoroughly cleaned and then burned until all that remains is carbon dioxide gas. A device called a mass spectrometer then measures the relative amounts of ^{13}C and ^{12}C in the gas.

Stable isotope tracing has shown that the primary producers in many coastal ecosystems are not phytoplankton, as ecologists previously thought, but vast underwater forests of kelp. The technique is also used to map whale migrations by examining how the $^{13}C/^{12}C$ ratio varies along their baleen, which are constantly growing and hence accumulating carbon. Because the $^{13}C/^{12}C$ ratio of whales' plankton food varies with latitude, changes in the ratio along the baleen reveal where the whale has traveled and eaten.

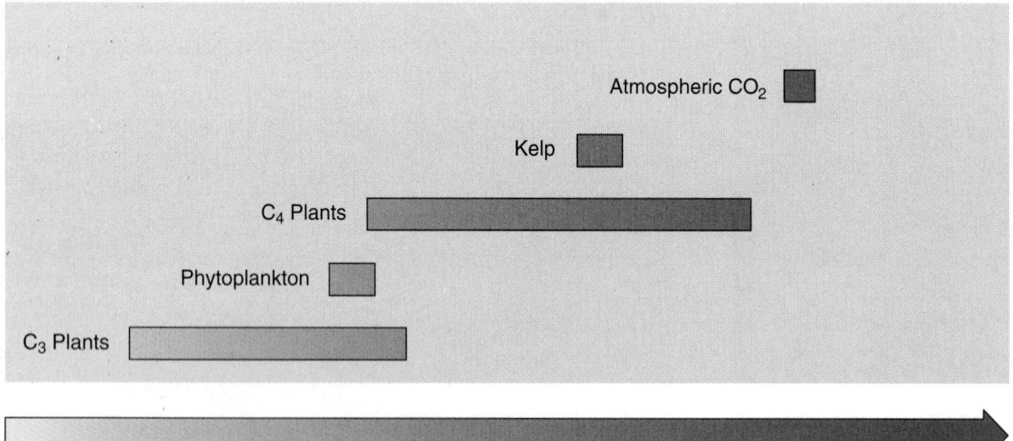

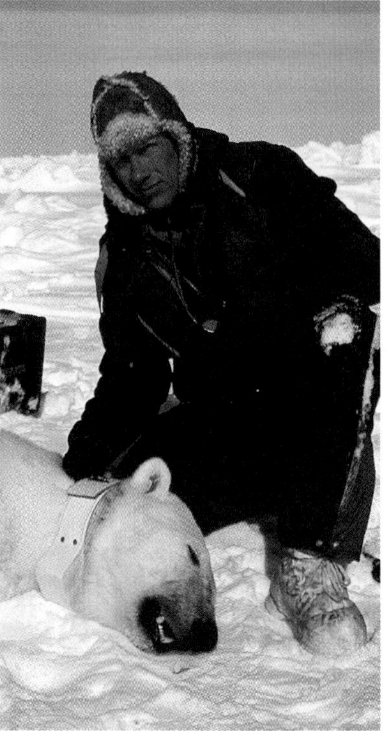

FIGURE 42.8

Deciphering food webs using chemistry. (*a*) Stable isotope tracing reveals the ratio of carbon-13 to carbon-12 in a burned tissue sample (the darker the shading in the bars, the greater the ratio of carbon-13 to carbon-12). Because each group of primary producers has a distinctive range of isotope ratios, researchers can identify the producer group that higher members of the food chain consumed by testing these consumers' tissues. (Recall from chapter 8 that C_3 and C_4 plants differ in some of their metabolic reactions.) (*b*) Ecologist Steven C. Amstrup sedates and radio-collars polar bears. Parings from the bears' claws reveal a $^{13}C/^{12}C$ ratio that identifies the primary producer in its food web.

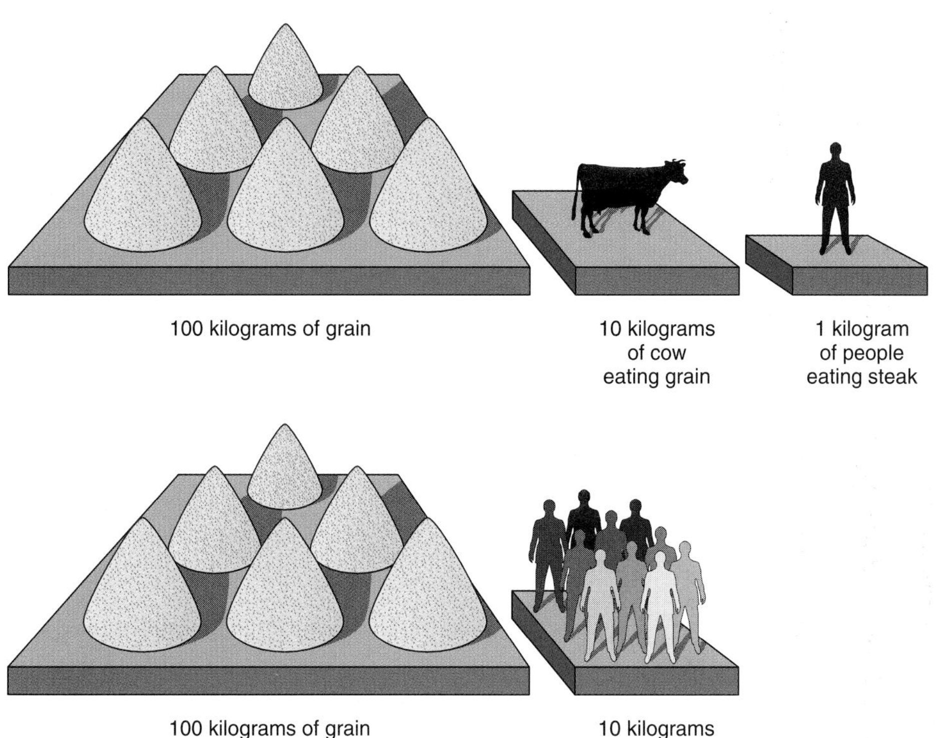

FIGURE 42.9

On a population level, the 10% rule of energy transfer along food chains means that more consumers, such as humans, can be supported as herbivores than as carnivores.

Ecological Pyramids

The total amount of solar or geothermal energy converted to chemical energy in a certain time period in a given region is called **gross primary production.** This includes the solar energy that photosynthetic organisms convert to chemical energy, plus the amount of organic matter organisms synthesize from inorganic geothermal compounds in water. **Net primary production** is energy left over for growth and reproduction after plant cellular respiration.

According to the "rule of tens," approximately 10% of the potential energy stored in the bonds of organic molecules at one trophic level fuels growth and reproduction of organisms at the next trophic level. If a food chain consists of four organisms—such as an owl eating a snake that has eaten a bird, which ate seeds—the owl can only use 1/1,000 (1/10 × 1/10 × 1/10) of the energy in the seeds for its growth and development. This inefficiency of energy transfer is why food chains rarely extend beyond four trophic levels. One way to maximize energy obtained from food is to lower the number of trophic levels. A person can do this by getting protein from beans instead of meat (fig. 42.9).

If each trophic level is represented by a volume directly proportional to the energy stored in new tissues per unit time, these volumes can be stacked to form a steep-sided **energy pyramid** (fig. 42.10). Consider a simplified example. In Cayuga Lake in New York, algae store 1,500 kilocalories of energy in a period of time. The algae feed small aquatic animals, which store about 150 kilocalories in their growth and development. When a smelt fish eats these small aquatic animals, it derives perhaps 15 kilocalories for growth and development. A human eating the smelt would convert only about 1.5 kilocalories of the original 1,500 kilocalories to new growth.

Other types of pyramids describe ecosystems. A **pyramid of numbers** shows the number of organisms at each trophic level. The shape of this type of pyramid depends on the size and number of the producers and consumers. The pyramid of numbers in a grassland community has a broad base because the producers—grasses—are small and numerous (fig. 42.11*a*). In a forest, in contrast, the pyramid of numbers stands on a very narrow base, because a single tree can feed many herbivores. Such an inverted pyramid can also represent a single large animal that supports a large number of parasites, such as the field mouse and its ticks depicted in figure 42.11*b*.

A **pyramid of biomass** takes into account size or weight or organisms. **Biomass** is the total dry weight of organisms in an area. Many pyramids of biomass are wide at the bottom and narrow at the top, because energy that can be converted to biomass leaves with each trophic level. In some aquatic ecosystems, however, the pyramid is inverted, because the biomass of the primary producers (phytoplankton) is smaller than that of the primary consumers or herbivores (zooplankton). This is because biomass is measured at one time. Phytoplankton reproduce quickly, but zooplankton eat them almost immediately. In contrast, zooplankton live longer, so more of them are present at one time. If we consider the biomass of all the phytoplankton that zooplankton consume during their life span, the pyramid would be upright.

Mastering Concepts

1. How do ecosystems range in size?
2. Identify the types of organisms that form trophic levels.
3. What does the rule of tens state about efficiency of energy transfer in food webs?
4. Distinguish between gross and net primary production.
5. List three types of pyramids that describe ecosystems.

FIGURE 42.10

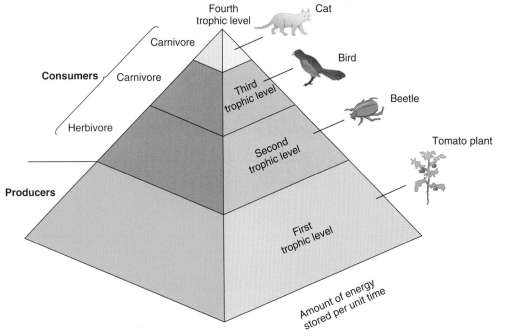

Energy flow through an ecosystem—an energy pyramid and a food chain. The cat eats the bird that eats the beetle that eats the tomato plant. Only 10% of the energy stored at one trophic level per unit time is transferred and stored in new growth and reproduction at the next level.

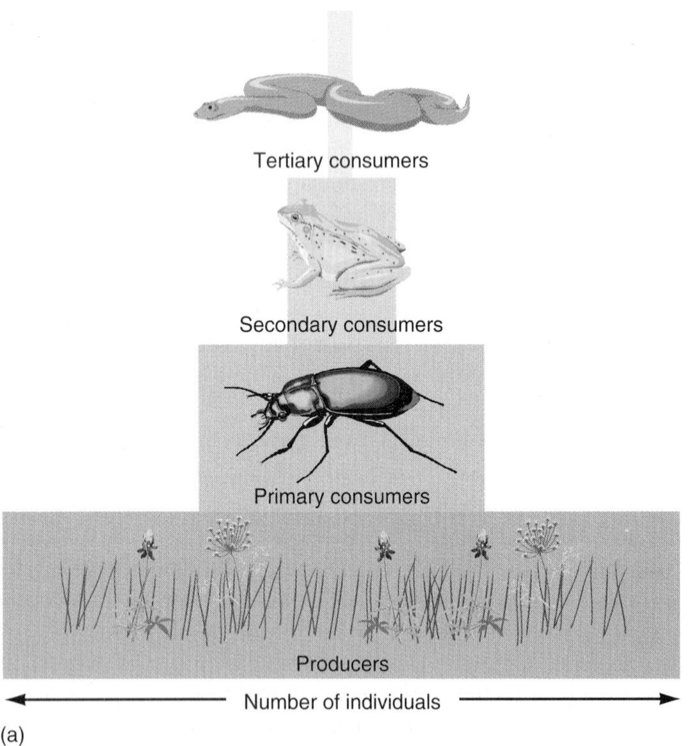

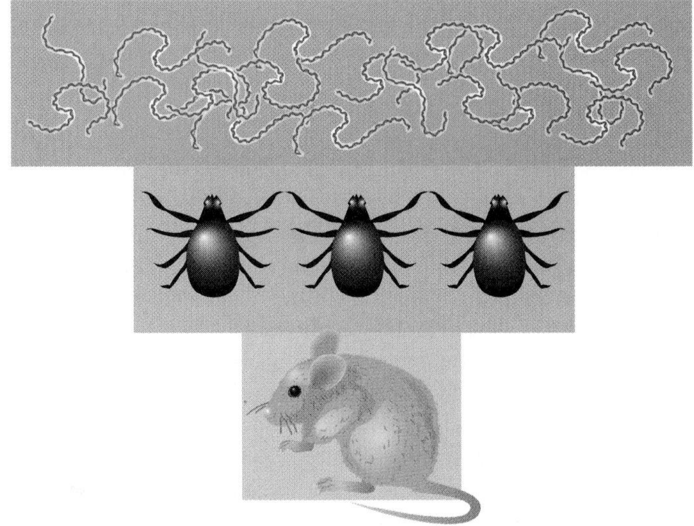

FIGURE 42.11

Pyramids of numbers. (*a*) A pyramid of numbers that represents a grassland has a broad base. (*b*) Transmission of Lyme disease illustrates an inverted pyramid of numbers. A field mouse may carry several deer ticks in its fur, which in turn are parasitized by many spirochaete bacteria (*Borrelia burgdorferi*). The bacteria cause Lyme disease in humans.

Biogeochemical Cycles

All life, through all time, must use the elements present when the earth formed. These elements continuously recycle through the interactions of organisms and their environments. If not for this constant recycling, these elements would have been depleted as they became bound in the bodies of organisms that lived eons ago. Because recycling chemicals essential to life involves both geological and biological processes, these pathways are called **biogeochemical cycles.**

Each chemical element in an ecosystem has a characteristic biogeochemical cycle, but all cycles have steps in common. Generally, the element is first taken from the environment and incorporated into the tissues of photosynthetic organisms, such as plants. If an animal eats the plant, the element may become part of animal tissue. If another animal eats this animal, the element may be incorporated into the second animal's body. All organisms die, and decomposers break their tissues down, which releases the elements back into the environment.

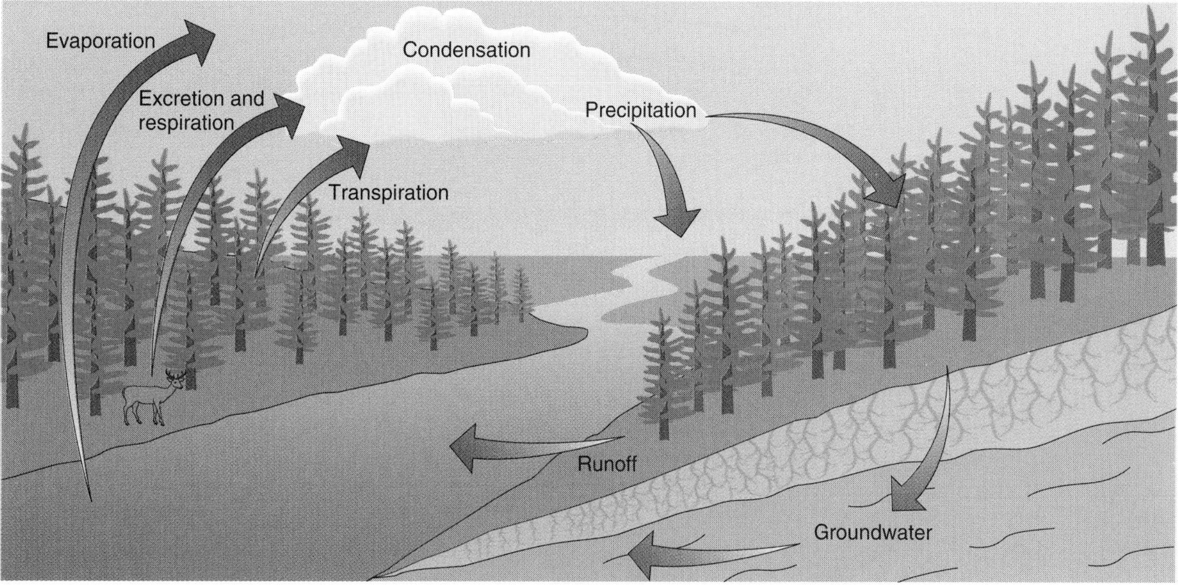

FIGURE 42.12

The water cycle. Water falls to earth as precipitation. Organisms use some water, and the remainder evaporates, runs off into streams, or enters the ground. Animals return water to the environment by respiring and excreting, and plants do so by transpiring.

The Water Cycle

Water covers much of the earth's surface, primarily as oceans but also as lakes, rivers, streams, ponds, swamps, snow, and ice. Water is also below the land surface as groundwater. Water's ability to exist in all three states of matter—solid (ice), liquid, and gas (vapor)—in the temperature range of the biosphere allows it to cycle efficiently from the earth's surface to the atmosphere and back again. (fig. 42.12)

The sun heats land and water, causing evaporation. Plants absorb water from soil, use it, and release some of it from their leaves in transpiration. Animals return water to the environment through evaporation and excretion. Water vapor may rise on warm air currents, cooling and forming clouds. If air currents carry this moisture higher or over cold water, more cooling occurs, and the vapor condenses into water droplets. Depending on the temperature and atmospheric pressure, the droplets fall as rain, snow, hail, fog, sleet, or freezing rain.

Most precipitation enters oceans or other bodies of water, but some also falls on land, where it either soaks into the ground and porous rock to restore soil moisture and groundwater or runs along the surface following the natural contours of the land. Rivulets join and form streams that unite into rivers. Most rivers eventually lead back to the ocean, where the sun's energy again heats the surface, evaporating the water and continuing the cycle. Water in the ground and porous rocks creates an aquifer and water table, which consists of water trapped between the surface and impervious rock. This underground water supplies wells and feeds springs that many species use. Spring water may evaporate or flow into streams, linking the groundwater system to the overall water cycle.

The Carbon Cycle

In the carbon cycle, photosynthetic organisms capture the sun's energy and use it with atmospheric carbon dioxide (CO_2) to synthesize organic compounds that are incorporated into plant tissues (fig. 42.13). Cellular respiration releases carbon to the atmosphere as CO_2. Dead organisms and excrement also return carbon to soil or water. Invertebrates feed on and fragment the dead, and bacterial and fungal decomposers complete the breakdown of these organic compounds to release simple carbon compounds into the soil, air, and water.

Certain geological deposits contain carbon from past life. Limestone, for example, consists mostly of skeletons and shells of ancient sea inhabitants. Fossil fuels, such as coal and petroleum, form from the remains of plants and animals. When these fuels burn, carbon returns to the atmosphere as CO_2.

The Nitrogen Cycle

Nitrogen is an essential component of proteins and nucleic acids, as well as other parts of living cells. Although the atmosphere is 79% nitrogen gas, most organisms cannot use this nitrogen to manufacture biochemicals. They depend on nitrogen-fixing bacteria that convert atmospheric nitrogen into ammonia (NH_3), which can be incorporated into plant tissue. Decomposers release some ammonia. **Nitrifying bacteria** convert ammonia from dead organisms to nitrites (NO_2) and eventually to nitrates (NO_3), which plants can use and then pass to animals. (Some nitrate is also produced when lightning fixes atmospheric nitrogen.) Finally, nitrogen returns to the atmosphere when **denitrifying bacteria** convert nitrites and nitrates to nitrogen gas (fig. 42.14).

The enzyme nitrogenase enables nitrogen-fixing bacteria to convert nitrogen to nitrogen-containing compounds. Because oxygen inactivates this enzyme, nitrogen-fixing microbes are typically anaerobic or shield nitrogenase from oxygen. For example, *Rhizobium* bacteria live in nodules on the roots of legumes such as beans, peas, and clover. Farmers rotate legumes with nonleguminous crops, such as corn, so the soil is continually enriched with biologically fixed nitrogen.

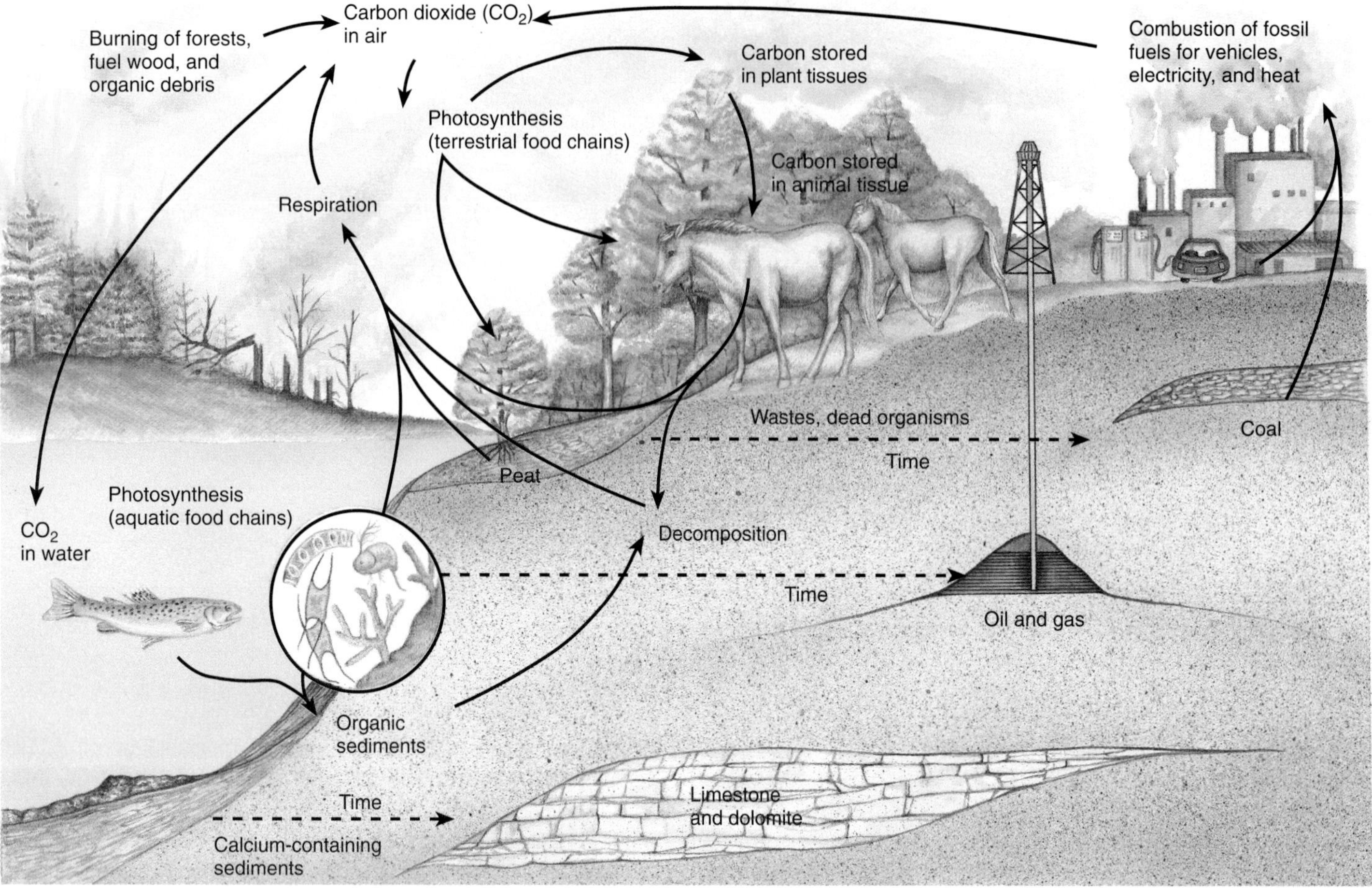

FIGURE 42.13

The carbon cycle. Carbon dioxide in the air and water enters ecosystems through photosynthesis and then passes along food chains. Respiration, decomposition, and combustion return carbon to the abiotic environment. Carbon is retained in geological formations for long periods of time.

The Phosphorus Cycle

Phosphorus is a vital component of genetic material, ATP, and membrane phospholipids. Lack of phosphorus limits growth. Most of the earth's phosphorus is stored in rocks. Erosion of these rocks releases phosphate (PO_4) that organisms can use to form tissues. Decomposers return phosphorus to soil and water (fig. 42.15).

Mastering Concepts

1. What are the basic steps of a biogeochemical cycle?
2. Describe the steps of the water, carbon, nitrogen, and phosphorus cycles.

Bioaccumulation and Biomagnification

Because cells admit some chemicals but not others, certain chemicals become more concentrated within cells than in the surrounding environment. This process, termed **bioaccumulation,** can concentrate particular elements or compounds to thousands or millions of times their concentration in the environment. Some substances, including certain synthetic compounds not normally found in ecosystems, become increasingly concentrated in organisms at successively higher trophic levels. This process, **biomagnification,** occurs because the chemical passes to the next consumer rather than being metabolized or excreted.

DDT

DDT is a pesticide that was once widely used because it damages the nervous systems of insects, such as disease-carrying mosquitoes. DDT soon proved to harm many species. The United States banned use of DDT in 1971, after much evidence showed it to cause cancer, disrupt wildlife, bioaccumulate, and become biomagnified (figs. 2.1 and 42.16).

Despite the ban, DDT spread in the atmosphere and has touched nearly all life on earth. DDT and its breakdown products have even been found in the fat of Antarctic penguins, who live where DDT was never used. Most of the DDT entering an animal remains in its fat and then passes through successive trophic levels. The pesticide concentrates with each trophic level and most severely affects organisms at the top of the energy pyramid. By the fourth trophic level, DDT concentrations may be 2,000 times greater than in organisms at the base of the food web (fig. 42.17).

FIGURE 42.14

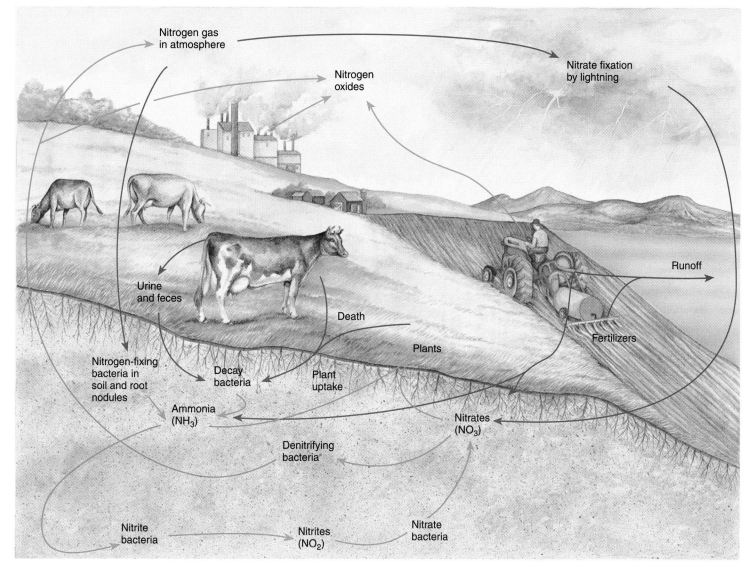

The nitrogen cycle. Plants and bacteria use nitrogen to synthesize amino acids and nucleic acids and pass these biochemicals along food chains. Nitrogen returns to the abiotic environment in urine and feces and by decomposition of dead matter. Specific groups of bacteria convert other nitrogen compounds.

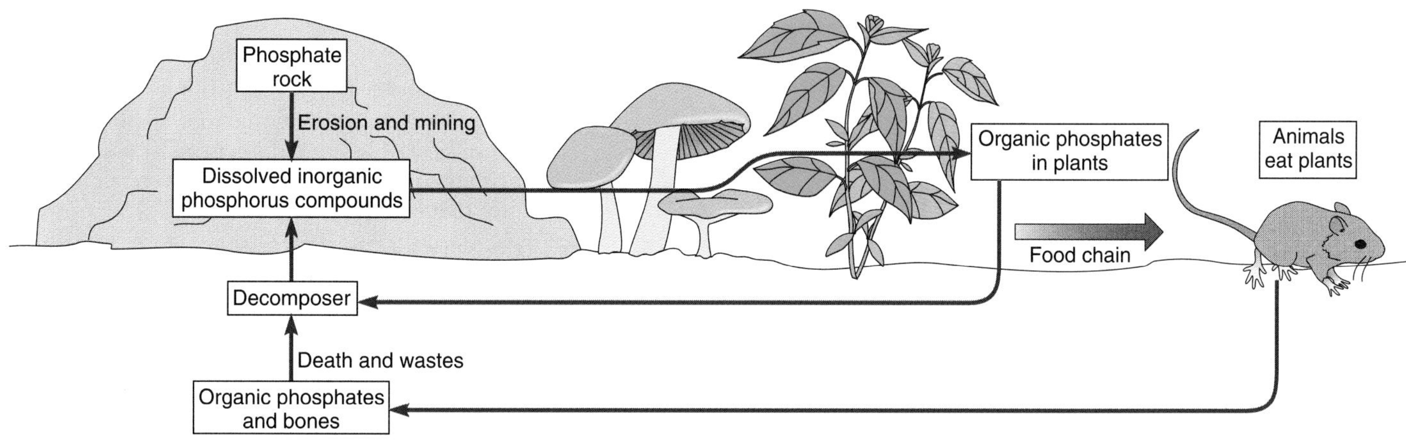

FIGURE 42.15

The phosphorus cycle. Phosphorus ultimately comes from rock and is then taken up by plants and passed up food chains. Decomposers return phosphorus to the abiotic environment.

Nonnative New Yorkers—Peregrine Falcons~

continued from page 854.

New York City's peregrine falcons have claimed their niche, circling and shrieking should a curious human venture near. Pairs of the birds live atop the Brooklyn, Verrazano-Narrows, Marine Parkway, and Throgg's Neck Bridges. A pair lives atop a Wall Street building, another on a midtown insurance agency building, and another on New York Hospital.

Peregrine falcons are endangered. They were once abundant nearly everywhere, but in the 1960s, widespread use of pesticides, particularly DDT, killed many of the birds. In the 1970s, wildlife biologists began raising peregrine falcons in captivity, with the goal of reintroducing them into the wild. New York state researchers hand-raised 171 birds and released them. They expected the birds to settle along the Palisades cliffs that parallel the lower Hudson River, or perhaps among the Adirondack Mountains upstate. But the birds had other ideas.

In 1983, a pair of falcons claimed territory on top of a 700-foot (213-meter) tower on the Verrazano-Narrows Bridge, which links Brooklyn to Staten Island. Other falcons lived between riveted steel beams and underneath the spans of that bridge and the Throgg's Neck Bridge. Researchers began to tag the birds and their fledglings, and have traced them as far as Wisconsin, although many remain in the Big Apple.

Why would falcons seek huge suspension bridges, hardly natural environments? To the birds, researchers hypothesize, bridges are similar to their more typical habitats. They provide nooks and crannies and many ledges. Like an isolated mountain, a bridge is relatively remote. Predators can't reach the falcons, yet food is abundant, particularly the plump pigeons so prevalent in New York City. The bridge falcons also eat 60 or so bird species that migrate past their lair.

The peregrine falcons of New York City illustrate a recurring theme in ecology—even in the most unusual or unexpected circumstances, life finds a way to continue.

FIGURE 42.16

The corpse of this about-to-hatch peregrine falcon, discovered amidst broken, empty eggs in Scotland in 1971, is testament to the concentration of DDT at the top of the food web, a position the animal occupies as a predatory bird. The pesticide caused birds' livers to abnormally metabolize the hormones required to secrete a firm, calcium-rich eggshell.

Mercury

Mercury is a naturally occurring element that bioaccumulates and biomagnifies. Plants growing near volcanoes or thermal springs, such as in parts of California, Hawaii, and Mexico, assimilate airborne mercury particles through leaves and mercury in the soil through roots. Phytoplankton bioaccumulate mercury from fresh water or ocean water. Biomagnification of mercury along food chains may create levels in fish that are hazardous to humans and top carnivores such as birds of prey and marine mammals. Pregnant women who eat mercury-tainted fish give birth to severely deformed infants.

This chapter has presented the basic concepts, terms, and processes of ecology. The final two chapters apply these ideas.

Mastering Concepts

1. What is bioaccumulation?
2. What is biomagnification?
3. How has DDT become widespread among organisms?

SUMMARY

1. Biotic communities are made up of coexisting species that are specialized in where, when, and how they live. Communities are most diverse near the equator.

2. As species interact with each other and their physical habitats, they change the community, a process called **biotic succession. Primary succession** occurs in a previously unoccupied habitat; **secondary succession** occurs after a disturbance. Succession leads toward a stable **climax community,** but complete stability is rare—most communities continue to change.

3. An **ecosystem** is the biotic community and its physical habitat. An ecosystem can range from a very small area to the entire **biosphere.** Ecosystems interact and change. Related ecosystems form **biomes,** which can be terrestrial or aquatic.

4. A **food chain** begins when **primary producers** harness energy from the sun or earth, forming the first **trophic level.** The total amount of energy converted to chemical energy in food is **gross primary production.** The energy remaining after metabolism is **net primary production.**

5. At the next trophic level, **primary consumers** (herbivores) eat the primary producers. A **secondary consumer** may eat the primary consumer, and a **tertiary consumer** may eat the secondary consumer. **Decomposers** break down nonliving organic material into recyclable nutrients. Food chains rarely extend beyond four trophic levels because only about 10% of the energy one trophic level takes in transfers to the next level. Ecological pyramids measure energy, numbers of organisms, or **biomass** in a food chain. **Stable isotope tracing** measures $^{13}C/^{12}C$ ratios to identify primary producers.

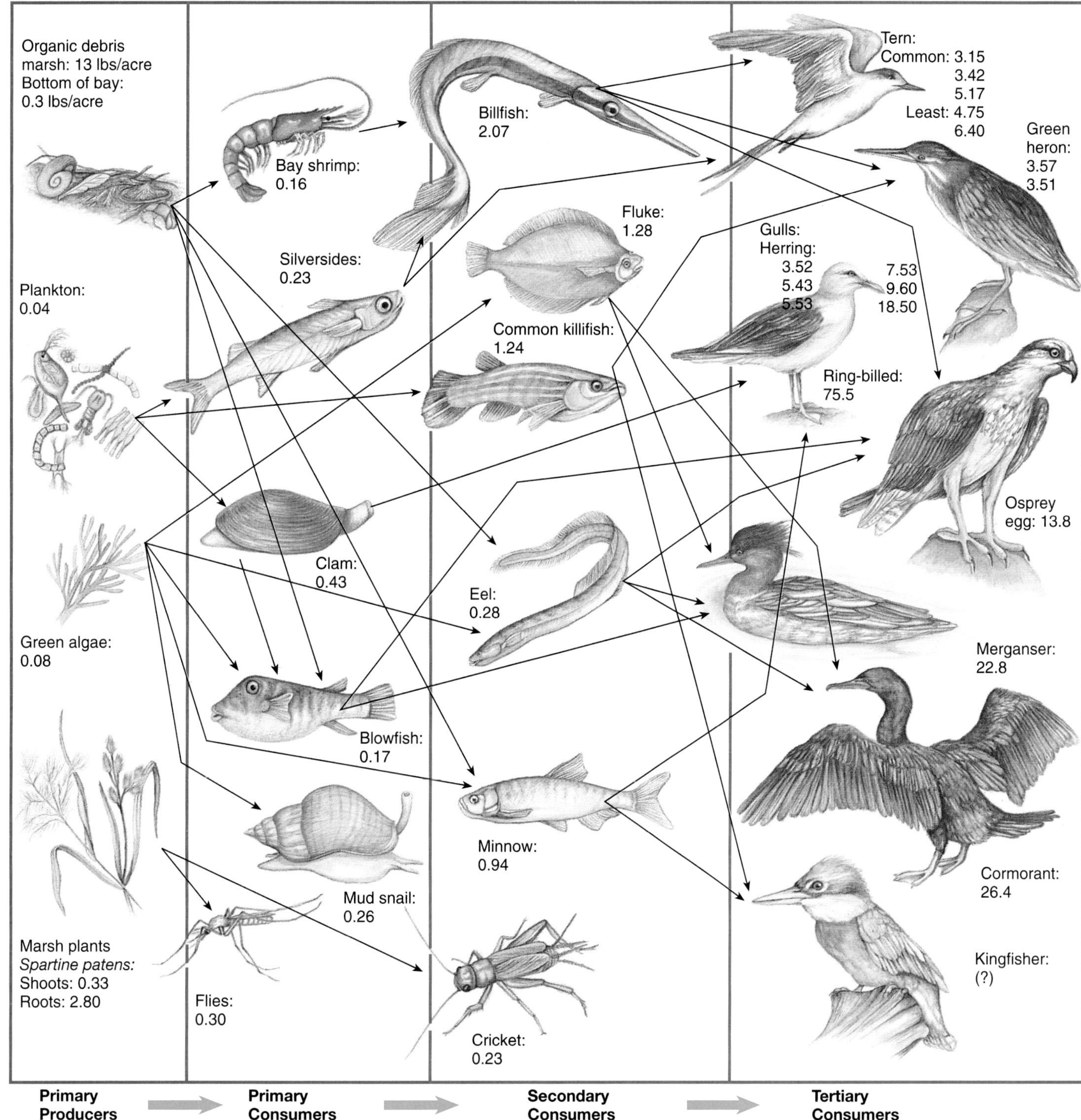

FIGURE 42.17

Biomagnification. The concentration of DDT pesticide in organisms' bodies increases up the food web. Values are levels of DDT per unit of tissue, measured in parts per million (ppm).

6. Biogeochemical cycles are geological and chemical processes that recycle chemicals essential to life. Water cycles from the atmosphere as precipitation, into the ground, and into organisms that release water in metabolism. Photosynthesis uses atmospheric carbon in CO_2 to manufacture carbohydrates. Cellular respiration and burning fossil fuels release CO_2. Decomposers release carbon from once-living tissues. Bacteria convert atmospheric nitrogen to ammonia and then to usable nitrates. Decomposers convert the nitrogen in dead organisms to ammonia. **Nitrifying**

bacteria convert the ammonia to nitrites, and other bacteria convert nitrites to nitrates. **Denitrifying bacteria** convert nitrates and nitrites to nitrogen gas. As rain falls over land, rocks release phosphorus as useable phosphates. **Decomposers** return phosphorus to the soil.

7. Bioaccumulation concentrates chemicals in cells compared to the surroundings. **Biomagnification** concentrates chemicals to a greater degree at successive trophic levels because the chemical passes to the next consumer rather than being metabolized or excreted.

TO REVIEW . . .

1. Why isn't life distributed evenly across the planet?

2. Distinguish between primary and secondary succession.

3. Describe the stages of succession.

4. What are the components of ecosystems?

5. What is a biome?

6. Why is a food web a more accurate description of predator-prey relationships in an ecosystem than a food chain?

7. At what trophic level does a vegetarian eat compared to a person who eats meat?

8. What are two ways that researchers decipher food webs?

9. Distinguish between gross and net primary production.

10. How can an organism be both a producer and a consumer?

11. What do ecological pyramids of energy, numbers, and biomass measure?

12. How do organisms return water, carbon, nitrogen, and phosphorus to the abiotic environment?

TO THINK ABOUT . . .

1. Some people have suggested that we "farm" the krill in Antarctica and use it to feed people who are starving elsewhere. What effects might krill farming have on the Antarctic ecosystem?

2. How is the inside of the human mouth an ecosystem?

3. Women who have breast cancer have 35% more DDT in their blood than women of the same age who do not have breast cancer. If DDT use was stopped in 1971, how can these women have it in their tissues today?

4. The buffalo-bur is a weed that grows in overgrazed areas of Colorado and poisons livestock. Striped Colorado potato beetle larvae eat the buffalo-bur. In 1859, European settlers introduced potatoes to Colorado. As potatoes flourished, so did the beetles. The insect's natural predators—toads, stinkbugs, birds, and snakes—could not control the beetle population. By 1874, the beetles had spread across the United States and to Europe. By 1930, potatoes all over the world were infested. Pesticides eventually controlled growth of potato beetle populations but killed other insects too. The populations that these insects controlled rose to pest levels. Using DDT also selected against sensitive strains of potato beetles, and resistant strains accumulated.

a. Draw a food web that includes all of the organisms mentioned.

b. What evolutionary process does the changing potato beetle population illustrate?

c. Suggest an intervention that might benefit agriculture other than use of a chemical pesticide.

5. After fires destroyed much of Yellowstone National Park in 1988, forest managers suggested humans could help the areas recover by feeding deer, bringing in plants, and planting trees. What are some of the advantages and disadvantages of intervening in recovery from a natural disaster?

6. How is natural selection apparent in succession and in vertical stratification?

TO LEARN MORE . . .

References and Resources*

Blaustein, Andrew R. October 1994. Amphibians in a bad light. *Natural History.* Many amphibian populations are on the decline.

DesMeules, Mark P. and Philip Nothnagle. May 1997. *Natural History.* Transient pools harbor ecosystems.

Grange, Ken, and Walter Goldberg. March 1993. Fiords down under. *Natural History.* In flooded valleys, freshwater and marine life occupy different zones of the same body of water.

Jordan, Thomas E., and Donald E. Weller. October 1996. Human contributions to terrestrial nitrogen flux. *BioScience.* How humans fit into the nitrogen cycle.

Mechaber, Wendy L. May 1996. Mapping leaf surface landscapes. *Proceedings of the National Academy of Sciences,* vol. 93. A single leaf houses a complete ecosystem.

Sinclair, A. R. E., and P. Arcese, eds. 1995. *Serengeti II: Dynamics, management, and conservation of an ecosystem.* Chicago: University of Chicago Press. A new synthesis of information on the world's most remarkable wildlife ecosystem.

Sprugel, D. G. 1991. Distance, equilibrium, and environmental variability: What is "natural" vegetation in a changing environment? *Biological Conservation,* vol. 58. Disturbances operate on different time scales to unbalance ecosystems.

**Life*'s website features frequently updated references.

Caribou cross a body of water that cuts through a ridge of sandy gravel in the Canadian arctic. The ridge is a formation called an esker.

43 Biomes

chapter outline

Terrestrial Biomes 872

Tropical Rain Forest
Temperate Deciduous Forest
Temperate Coniferous Forest
Taiga
Grasslands and Savannas
Tundra
Desert

Freshwater Biomes 879

Lakes and Ponds
Rivers and Streams

Marine Biomes 881

The Coast
The Ocean

El Niño Affects Ecosystems and Biomes 884

Infectious Disease Patterns in the Next Century

Infectious diseases are a consequence of our sharing the planet with other organisms—those that cause illness as well as the vectors that transmit infectious organisms. Some particularly debilitating infectious diseases occur in the tropics:

- River blindness (onchocerciasis) is a worm infection that black flies transmit. Symptoms include swollen lymph nodes, inflamed skin, worm-filled skin nodules, and immature worms in the eyes, causing blindness.
- Schistosomiasis causes abdominal cramps and bloody urine. Snails spread the fluke worms that cause this illness.
- Dengue, "breakbone fever," begins with severe headache and then pain in the back, joints, and behind the eyes. On the fourth day symptoms abate and then return with a red rash and skin peeling for two more days before recovery. A rare form, called dengue hemorrhagic fever, causes internal bleeding in children and may be fatal. Mosquitoes transmit arboviruses that cause dengue.
- African sleeping sickness (trypanosomiasis) results from protozoa passed by tsetse flies. Over several weeks, the parasites enter the bloodstream and then the central nervous system. Symptoms include high fever, rapid heartbeat, sleepiness, emaciation, and personality and gait changes.
- Biology in Action 23.1 discusses the most prevalent vector-borne tropical infectious illness, malaria.

River blindness, schistosomiasis, dengue, African sleeping sickness, and malaria are endemic to certain tropical regions, meaning that they are always there. But these illnesses are expected to invade new areas in the coming century.

to be continued . . .

Terrestrial Biomes

Biomes are distinctive major types of communities and ecosystems in large geographic areas. Biomes have characteristic species and climax communities. Terrestrial biomes are land-based, and aquatic biomes occur in fresh water, in the oceans, and in the regions where salt and fresh water meet.

Plants define terrestrial biomes not only because they are primary producers but because they form a vegetational matrix that largely determines which animals and microorganisms can also exist. The types of plants in a biome reflect long-term interactions of plants with regional climate and soils.

Climate varies in predictable ways with latitude and altitude (fig. 43.1). The world's major climatic regions include the tropics, the temperate zone, the subarctic, and the arctic (fig. 43.2). Terrestrial biomes within these regions also differ according to moisture availability. For example, tropical biomes include the very wet tropical rain forest, the drier seasonal forest, the still drier savanna, the semidesert, and the hot, dry desert.

Soils differ in composition, depending upon the materials from which they form. Some common sources of soil are igneous rocks, sedimentary rocks, glacial materials, and volcanic ash. Climate influences soil development in many ways. Heavy precipitation may leach soluble materials from surface layers and deposit them in deeper layers, or it may remove them entirely from the soil system. Temperature and moisture conditions may also determine the fate of organic matter from living organisms. Under a warm, moist climate, rapid decomposition may leave little humus (organic material) in the soil. In cold, damp climates, undecomposed peat may accumulate in the soil.

The following discussion considers a few representative terrestrial biomes.

Tropical Rain Forest

Tropical rain forests occur where the climate is almost constantly warm and moist. Rainfall is typically between 79 and 157 inches (200 and 400 centimeters) per year, and the soils are often poor in nutrients because water leaches them away. The tropical rain forest of the Amazon Basin in South America is incredibly vast; its meandering rivers and dense foliage cover an area 90% the size of the continental United States. Within the lush maze of intertwined branches and moss-covered vines live a staggering diversity of species. An area of tropical rain forest 4 miles square (10.4 kilometers square) is likely to house, among its 750 tree species, 60 species of amphibians, 100 of reptiles, 125 of mammals, and 400 of birds. One tree alone may support 400 insect species.

From the air, a tropical rain forest appears as a solid, endless canopy of green, consisting of treetops 50 to 200 feet (15 to 66 meters) above the forest floor. Plants beneath the canopy compete for sunlight and form layers of different types of organisms. This layering of different plant species in the forest canopy is called **vertical stratification.**

Plants are adapted in different ways to capture sunlight. Very tall trees poke through the canopy. Other trees have broad, flattened crowns that maximize their sun exposure in this equatorial region where the sun's rays are almost perpendicular to the earth's surface. Vines and epiphytes (small plants that grow on the branches, bark, or leaves of another plant) grow on tall trees. Epiphytes may have aerial roots that penetrate masses of rich soil on branches. The ground of the tropical rain forest looks rather bare beneath the overwhelming canopy, but it is a habitat for countless shade-adapted species.

The lush vegetation feeds a variety of herbivores. Insects, sloths, and tapirs devour leaves. Small deer and peccaries, mon-

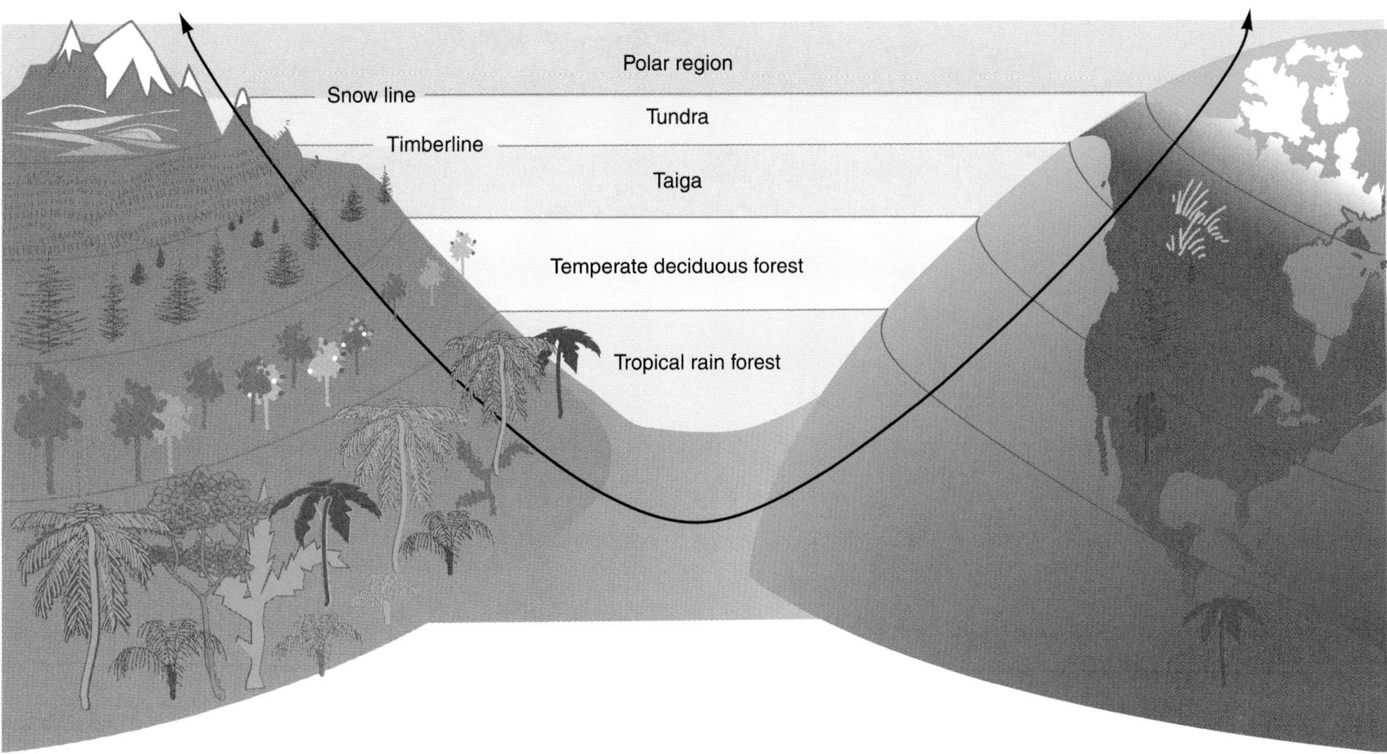

FIGURE 43.1

Latitude and altitude influence plant communities. With increasing elevation on a mountain, the dominant types of plants parallel the types that appear in different biomes as distance from the equator (latitude) increases.

keys, rodents, bats, and birds primarily eat fruit (fig. 43.3). Many carnivores, such as large cats, birds of prey, and snakes, consume the herbivores. Most animals are small or medium in size, but a few are enormous. The pirarucu fish of the Amazon River is more than 7 feet (2.1 meters) long, and the wingspans of many insects are several inches.

Nutrients cycle rapidly in the tropical rain forest. Frequent torrential rains wash away nutrients released in the soil from dead and decaying organisms. The heat and humidity speed decomposition. Termites recycle nutrients with the help of protozoa in their guts, which break down cellulose and release its atoms back to the environment. Roots of giant trees absorb nutrients so efficiently that most decaying plant material is saved. Essential in a tree's recycling program are fine roots that permeate the upper 3 inches (7.6 centimeters) of ground and attach to dead leaves. These roots absorb nutrients from falling dust and rain. Fungi on rootlets quickly absorb nutrients released from dead leaves. Animals recycle nutrients too. Certain leaf-cutting ants, for example, farm gardens of fungi that hasten decomposition.

Temperate Deciduous Forest

Hardwood trees that shed leaves in winter thrive where the growing season lasts at least 120 days, annual rainfall ranges from 28 to 59 inches (70 to 150 centimeters), and the soil is rich. Vertical stratification is evident here, too, although species diversity is less than in the tropical rain forest.

Usually one or two species of trees predominate in thc **temperate deciduous forest,** as in oak-hickory or beech-maple forests. The dominant trees, victors in the competition for light, are also best suited for the average rainfall. Trees in the deciduous forest often have ball-shaped tops that maximize light absorption for the angle of the sun's rays at these latitudes. Shrubs grow beneath the towering trees. Below them, herbaceous flowering plants grow mostly in early spring, when light penetrates the leafless tree canopy. Mosses and liverworts coat damp spots on the forest floor.

Decomposers in temperate forests—nematodes, earthworms, and fungi—break down leaf litter and create a rich soil. Herbivores include whitetail deer, ruffed grouse, and gray squirrels. Red fox and raccoon are common carnivores (fig. 43.4*a*). All residents of the deciduous forest must cope with seasonal changes. In winter, some insects enter an inactive state called diapause, and some mammals hibernate. Some birds migrate to warmer climates. Many animals survive winter by fattening up, growing thicker coats, and hiding food.

Temperate Coniferous Forest

The **temperate coniferous forest** has a dense understory of shrubs such as alder and hazelnut. Herbivores include the whitetail deer,

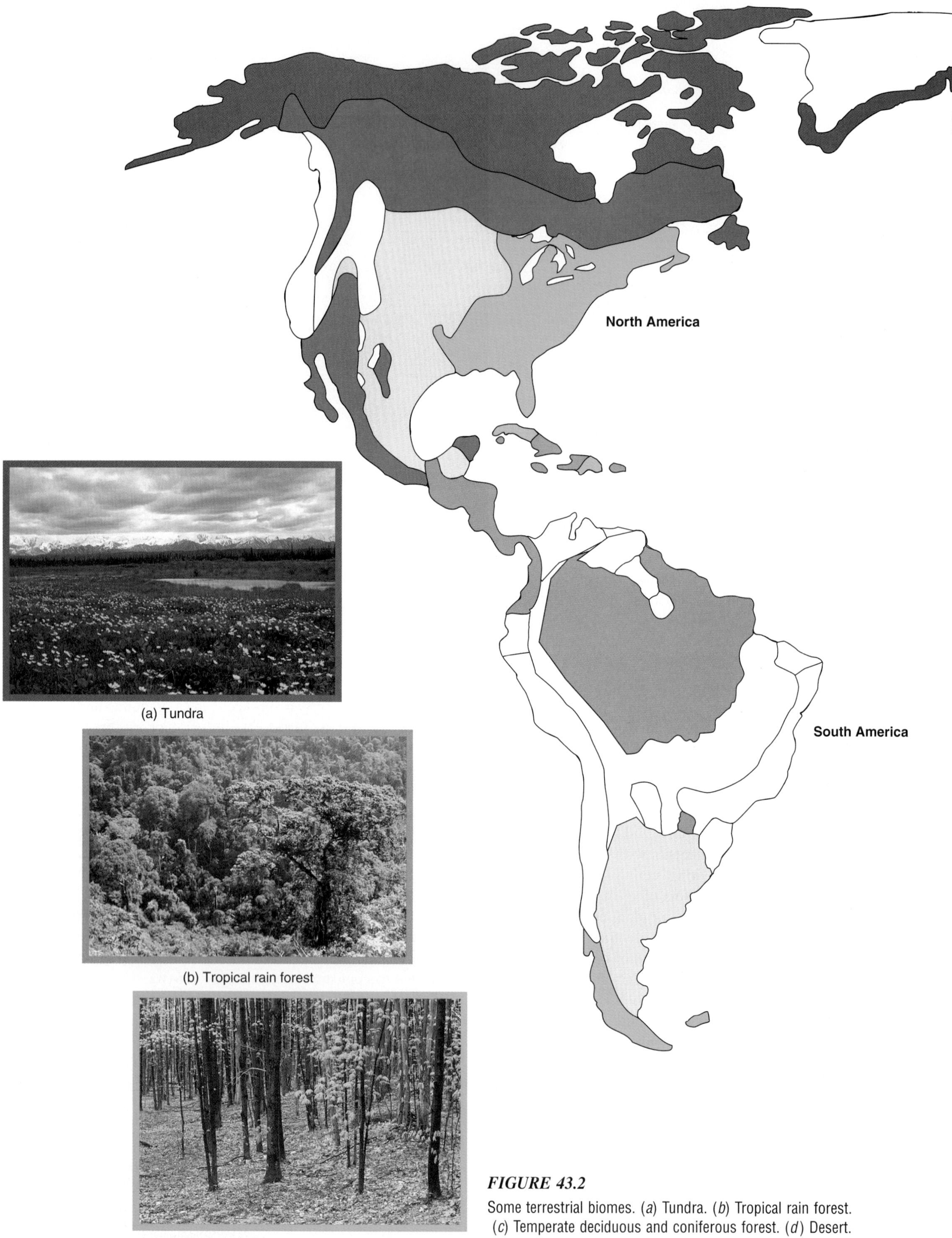

FIGURE 43.2

Some terrestrial biomes. (*a*) Tundra. (*b*) Tropical rain forest. (*c*) Temperate deciduous and coniferous forest. (*d*) Desert. (*e*) Temperate grassland. (*f*) Taiga.

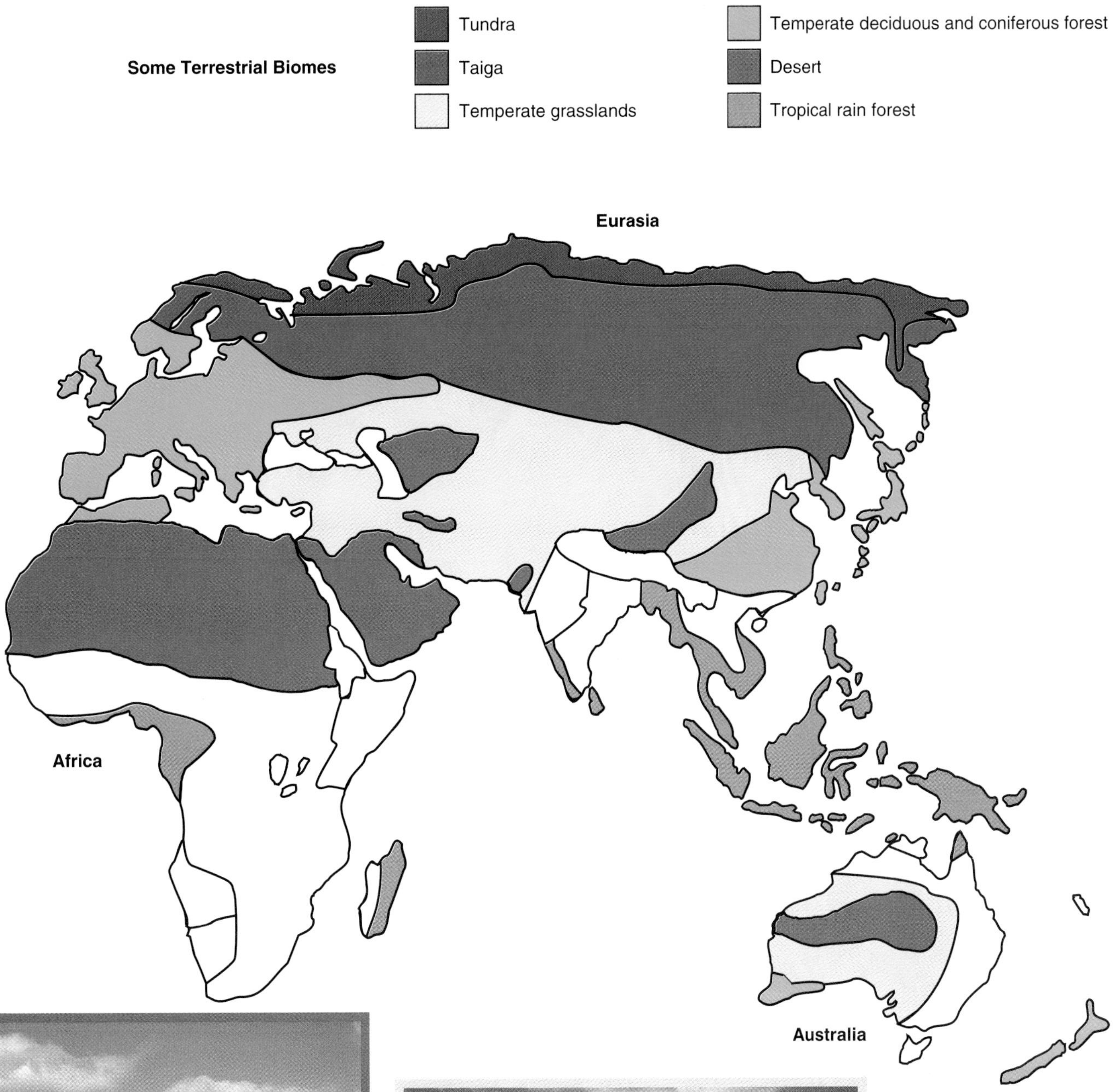

Eurasia

Africa

Australia

(d) Desert

(e) Temperate grasslands

(f) Taiga

FIGURE 43.3
Squirrel monkeys live in the tropical rain forest of Central and South America.

(a)

(b)

FIGURE 43.4
Residents of temperate forests. (*a*) The raccoon lives in a temperate deciduous forest. (*b*) Moose live in temperate coniferous forests.

red squirrel, spruce grouse, and moose (fig. 43.4*b*). The bobcat and black bear are typical carnivores.

Most coniferous trees are evergreen. They lose leaves a few at a time, so most conifers are green all year. Many coniferous forest soils are thin, acidic, and poor in nutrients. Spruces, pines, firs, and hemlock are the dominant trees. The shortage of nutrients favors evergreen plants over deciduous species, which require these resources to replace all their leaves in a short time.

Conifers are also adapted to areas that have recurring fires, and in fact fires are a natural part of a dynamic temperate coniferous forest. Fire recycles nutrients and selects against some species, opening new niches. The thick bark on certain pine trees resists flames, and some pine cones liberate their seeds only after exposure to the extreme temperature of a forest fire.

Parts of New Jersey's Pine Barrens near Atlantic City burn about every 30 years. The sandy soil, the pine needles and branches that litter the ground, and the "pitch" that seals pine cones shut and acts as a gasoline accelerant all create conditions just right for fire. In the last big fire in the Pine Barrens, in 1994, squirrels, deer, reptiles, and rabbits fled the affected area. The fire cleared the forest floor of three decades of accumulated plant debris, leaving space for future growth. The cones that the fire liberated soon yielded abundant buds against the backdrop of blackened trees. By the year 2,000, ecologists predict, the forest will look as if a fire had never happened.

Taiga

North of the temperate zone is the cold, snowy **taiga,** home to a few well-adapted species (fig. 43.5). Soils are cold, damp, and acidic. Decomposition is slow, allowing a dense litter layer to accumulate. Spruces dominate, with a few deciduous trees near lakes, streams, and rivers. Conifers are well adapted to an environment so cold that water is usually frozen throughout the winter. With a

FIGURE 43.5
Woodland caribou forage for food in the Canadian taiga.

FIGURE 43.6
This east African savanna is a grassland.

scant amount of sap, a conifer has little liquid to freeze. The needle shape, waxy coat, and minimal stomata set in deep pits are adaptations of conifer leaves to conserve water. The conical tree shape helps capture the oblique rays of light in northern latitudes and prevents damaging snow and ice buildup.

The taiga forest floor is often boggy, with scattered shrubs and mats of mosses and lichens. Fungi are common decomposers in the taiga. Typical herbivores include the woodland caribou, moose, and spruce grouse and carnivores include lynx and gray wolf. As is true nearly everywhere else on the planet, insect populations can at times grow very dense.

Grasslands and Savannas

Grasslands and **savannas** have 10 to 30 inches (25 to 75 centimeters) of rainfall annually, which is often not sufficient to support trees. These biomes usually have one or two severe dry seasons, when the vegetation becomes dry and flammable, so that fire is an important ecological factor.

The North American prairie is a temperate grassland. The height of the grasses reflects local moisture. Grasses reach 4 to 8 feet (1.2 to 2.4 meters) around the Mississippi Valley, where moisture from the Great Lakes and Gulf of Mexico contributes to an annual rainfall of about 39 inches (100 centimeters). Westward, toward the Rocky Mountains, annual rainfall decreases, and shorter grass species dominate the landscape. Grasses can do quite well with little water, however. Varied blade texture, surface, and shape are adaptations that enable grasses to conserve water. Root systems are extensive, with some species sending roots 6 feet (1.8 meters) underground to reach water. The mat of roots holds soil together and prevents it from blowing away during drought. Growth response to rain is rapid.

Unlike trees, perennial grasses easily survive damage caused by fire and herbivores. The perennial buds of grasses lie below the soil surface and are protected from flames. Because these plants grow from below, an animal's (or a lawnmower's) removal of the blade does not hinder growth. Even chunks of grass kicked up by a grazing animal can reroot. In contrast, the growing regions of trees—the tips of branches—are the first structures that fire or herbivores destroy.

Grasslands are rich in plant and animal species. In North America, bison, elk, and pronghorn antelope were originally the important grazing herbivores. Their predators were the gray wolf and coyote. Antelopes used speed to escape these predators. The size and power of elk and bison were sufficient protection against natural predators, but they were no match for humans with guns. Other herbivores include rodents, prairie chickens, and insects. Many rodents, such as prairie dogs, retreat into burrows to escape predators.

Tropical savannas are grasslands with scattered trees or shrubs and bands of woody vegetation along stream courses. In Africa, zebras, giraffes, wildebeest, gazelles, and many other kinds of antelopes graze the land, and lions, cheetahs, wild dogs, and hyenas prey on them (fig. 43.6). Vultures and other scavengers soon consume a leftover meal. In many savannas, termites are major herbivores, and the landscape is dotted with their huge nests. The particular vegetation pattern of the savanna depends on a complex interaction of soils, rainfall, animal feeding activity, and fire.

Tundra

A band of **tundra** runs across the northern parts of Asia, Europe, and North America. Winter is bitterly cold, with typical temperatures about −26°F (−3°C). The ground remains frozen, in a zone called **permafrost,** even during summer, when temperatures range from 40°F to 70°F (4.5°C to 21°C). Permafrost begins at 18 inches (46 centimeters) below the surface and extends 300 feet (91.5 meters) down. Because the permafrost blocks water infiltration, spring runoff from ice and snow drains rapidly into rivers or accumulates, forming bogs and small, stagnant ponds.

The shallow tundra soil supports dwarf shrubs, low-growing perennial plants such as sedges and broad-leafed herbs, and reindeer lichens. It is difficult for annual plants to germinate, grow, and flower in the short growing season, which may be less than 60 days, so most plants are perennials. Tundra plants tend to be low and flat, a shape that lets the wind blow over them and protects them under snow. The plants often clump together, which helps to conserve moisture and warmth, and they are often buried amid rocks, which block the wind. Some plants have protective hairs that insulate them and help break the wind. The dark green color of many plants allows them to absorb more light for photosynthesis.

Animal inhabitants of the tundra include caribou, musk oxen, reindeer, lemmings, snowy owls, foxes, and wolverines (fig. 43.7). Polar bears sometimes visit coastal areas of the tundra to den. In the summer, migratory birds stop in tundra areas to raise their young and feed on the insects that flourish in the tundra and its ponds.

Like plants, the animals of the tundra have adapted to the harsh climate. Both the hunters and the hunted benefit from camouflage. White winter colors make the arctic fox, ptarmigan, ermine, and arctic hare as inconspicuous against snow as their brown summer colors make them against the snow-free landscape. These animals often have short extremities, a form that helps to conserve heat. The snowshoe hare's big feet are natural snowshoes. The shallow soil, short growing season, and slow decomposition of the tundra make it a very fragile environment.

Relieving the bleak tundra landscape are regions called **eskers,** from the Irish for "ridge." Eskers are extensive ridges of gravel and sand that rivers running under glaciers deposited long ago. Because the soil particles of an esker are so loose, water can reach the surface during the summer, unlike the situation for permafrost. The chapter opening photo shows an esker in the Canadian arctic, a region called, appropriately, the Barren Lands. In the short summer, the Canadian esker supports herds of caribou and many foxes, bears, musk oxen, and wolves. Ground squirrels build extensive dens in the gravel. Bountiful blueberries stain blue the tongues of migrating birds who stop to feed. Come winter, the land is indeed barren. An English expedition exploring the area in 1821 was reduced to cannibalism to survive the brutal winter.

FIGURE 43.7
Polar bears and arctic foxes live in the tundra.

Desert

Deserts receive less than 8 inches (20 centimeters) of rainfall per year. The days can be searingly hot because few clouds block or filter the sun's strongest rays. Nights are cool, sometimes 86°F (30°C) below the daytime temperature, because heat radiates rapidly to the clear night sky.

Deserts vary in biodiversity. Although many desert habitats support only a few species, others, such as the upland desert of southern Arizona, are rich in species (fig. 43.8). In the driest and largest desert in the world, Africa's great Sahara, rainfall is less than 1 inch (2 centimeters) per year. Many areas are almost devoid of life.

Desert plants and animals show many striking adaptations to heat and drought. Annual plants grow quickly, squeezing their entire life cycles into wet periods between droughts. Most of their seeds germinate only after a soaking rain, which rinses growth inhibitors from their coats. The thick-skinned, fleshy leaves of succulent perennials help hold precious water. In rainy times, annual and perennial flowers bloom magnificently.

Cacti minimize water loss by not having leaves. After a rainfall, the stems of barrel and saguaro cacti expand as they take up and store water. Spines guard their succulent tissues, although some desert animals still eat them. The root system of a cactus is shallow but widespread—that of a large cactus extends up to 55 to 65 yards (50 to 60 meters). Some desert plants produce chemicals that inhibit the growth of other plants nearby, decreasing competition for water.

Desert animals also cope with water scarcity. Body coverings, such as a scorpion's exoskeleton or a reptile's leathery skin, minimize water loss. Some small mammals, such as the kangaroo rat, have very concentrated urine, which saves water. Few animals face the midday sun. Most burrow or seek shelter during the day and become active when the sun goes down and the risk of water loss lessens.

(a)

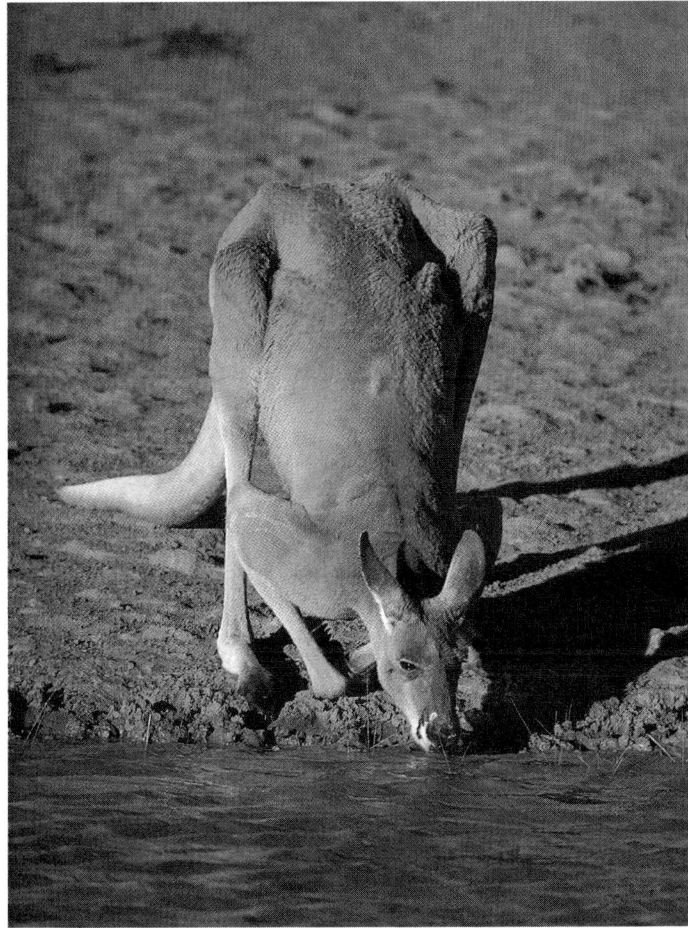

(b)

FIGURE 43.8

Desert residents. (*a*) Israel's Negev Desert appears to be lifeless except for lichens on the undersides of rocks. Trails of slime, however, reveal snails that only appear at night and sunrise, when they eat the lichens and even some rocks. The feeding snails continually expose new rocky habitats for the lichens. (*b*) A red kangaroo can go for 2 weeks without drinking. But when it finds a water hole, it drinks heartily. This desert is in the Australian outback.

Mastering Concepts

1. What are the major climatic regions of the world?
2. How do climate and soil composition determine characteristics of biomes?
3. Describe the rainfall patterns, climates, nutrient cycling, and inhabitants of terrestrial biomes.

Freshwater Biomes

The earth's waters house diverse life-forms adapted to the temperature, light, current, and nutrient availability of their surroundings. Aquatic biomes are distinguished by physical and chemical factors such as current pattern and degree of salinity. Two types of freshwater biomes are standing water, such as lakes, swamps, and ponds **(lentic systems),** and running water, such as rivers and streams **(lotic systems).**

Lakes and Ponds

Light penetrates the regions of a lake to differing degrees. These differences determine the types of plants that live in particular areas.

The **littoral zone** is the shallow region along the shore where light reaches the bottom sufficiently for photosynthesis. Some photosynthetic organisms in this region are free-floating; others are rooted to the bottom and have submerged, floating, or emergent leaves (fig. 43.9). The littoral zone is the richest area of a lake or pond. Producers include free-floating and attached cyanobacteria, green algae, and diatoms. Floating plants such as water lilies and emergent plants such as cattails, reeds, and rushes are also part of the flora. Animal life is diverse and includes damselfly and dragonfly nymphs, crayfish, rotifers, flatworms, hydra, snails, snakes, turtles, frogs, and the young of some species of deep-water fish.

The **limnetic zone** is the layer of open water where light penetrates. Phytoplankton, zooplankton, and fish inhabit this area. The **profundal zone** is the deep region beneath the limnetic zone where light does not penetrate. Organisms here, which rely on the rain of organic material from above, include mostly scavengers and decomposers such as insect larvae and bacteria.

Oxygen and mineral nutrients in a lake are distributed unevenly. Oxygen concentration is usually greater in the upper layers, where it comes from the atmosphere and photosynthesis. Decomposed organic matter that sinks to the bottom releases phosphates and nitrates into the lower layers of the lake and consumes oxygen. In a shallow lake, wind blowing across the surface mixes the water, redistributes nutrients, and restores oxygen to bottom waters.

Deeper lakes in temperate regions often develop layers with very different water temperatures and densities. This **thermal stratification** prevents free circulation of nutrients and oxygen in the lake. Degree of thermal stratification varies with the season.

In the summer, the sun heats the surface layer, but the lower layers remain cold. Between these two layers is a third region, the **thermocline,** where water temperature drops quickly. In the fall, the temperature in the surface layer drops as the air cools. Gradually, water temperature becomes the same throughout the lake.

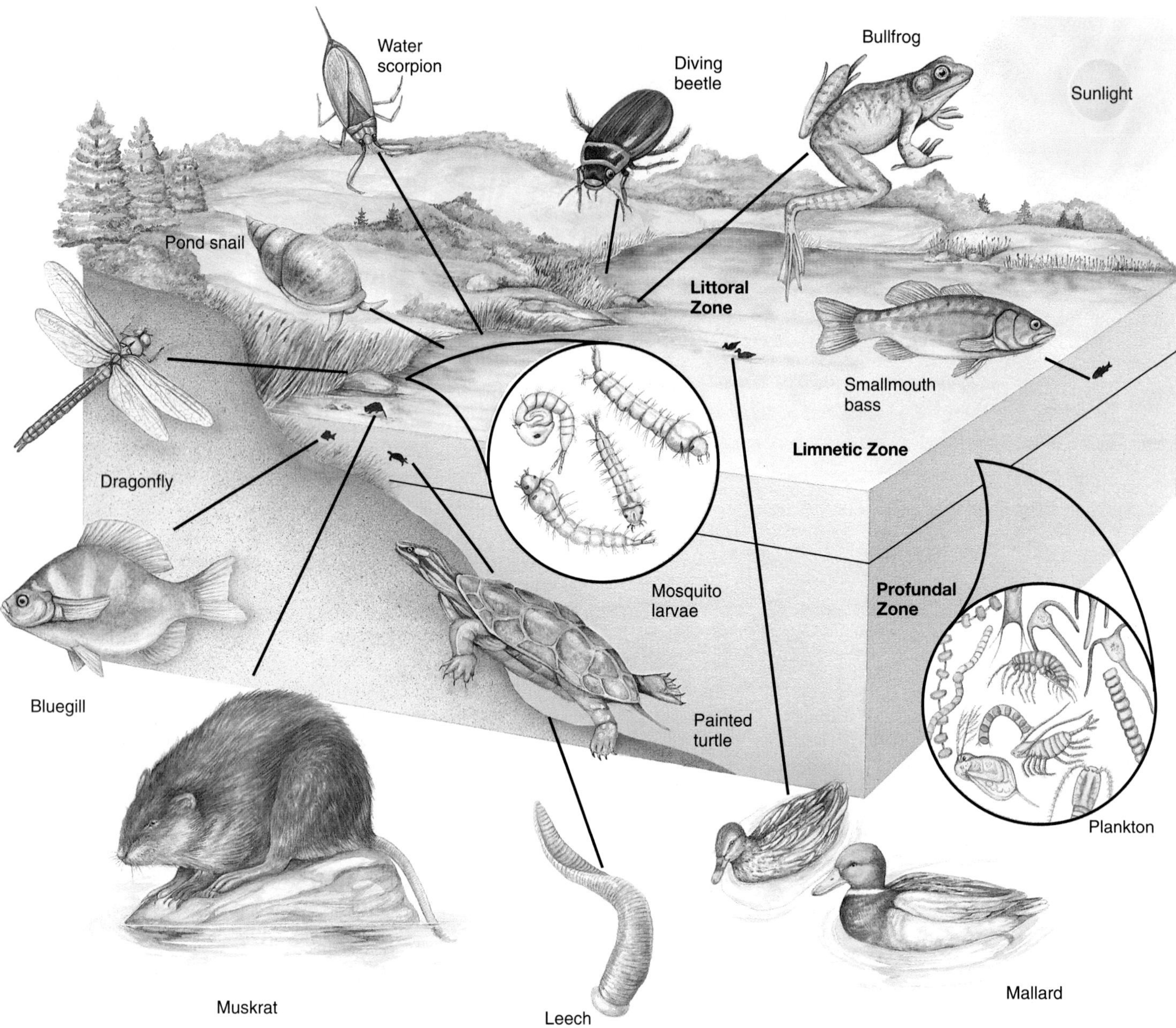

FIGURE 43.9
Zones and life in a lake.

Wind then mixes the upper and lower layers, creating a **fall turnover** that redistributes nutrients and oxygen throughout the lake. During winter, surface water cools. When water cools to 39°F (4°C), the temperature at which it is most dense, it sinks. Water colder than this floats above the 39°F layer and may freeze, giving the lake an ice cover. In the spring, when the surface layer warms to 39°F, a **spring turnover** occurs, again redistributing nutrients and oxygen. After the spring turnover, algae bloom in the warming, nutrient-rich surface water.

Lakes age. Younger lakes are often deep, steep-sided, and low in nutrient content. The deep zone of bottom water stores a large quantity of oxygen, which is rarely depleted. These lakes are said to be **oligotrophic,** which means they are low in fertility and productivity. They are clear and sparkling blue, because phytoplankton aren't abundant enough to cloud the water. Lake trout and other organisms that thrive in cold, oxygen-rich deep water are numerous.

As a lake ages, organic material from decaying organisms and sediment begins to fill it in, and nutrients accumulate. Shallow water mixes more easily and absorbs more light. These lakes are termed **eutrophic,** which means they are nutrient-rich and high in productivity. The rich algal growth makes the water green and murky. Decomposing matter in the deeper waters depletes oxygen during the summer. Fish and plankton communities change, and fish tolerant of low oxygen replace species such as lake trout. In time, the lake becomes a bog or marsh and, eventually, dry land. Discharge of nutrient-rich urban wastewater and runoff carrying phosphate-rich fertilizers from cultivated lands can speed conversion of oligotrophic lakes to eutrophic lakes. This transformation is termed **eutrophication.**

Rivers and Streams

Rivers and streams depend heavily on the land, for water and nutrients. Dead leaves and other organic material that fall into a river add to the nutrients that resident organisms recycle. Rivers also restore nutrients to land. Many rivers flood each year, swelling with meltwater and spring runoff and spreading nutrient-rich silt onto their floodplains. When a river approaches the ocean, its current diminishes, which deposits fine, rich soil that forms new delta lands.

Oxygen in flowing water is usually abundant because turbulence mixes air and water. Many organisms that live in streams require a high oxygen level. Organic material, such as sewage, dumped into a river or stream can increase decomposition and exhaust the oxygen supply that river residents require.

Different organisms live in different parts of a river, with each species adapted to local conditions. In a swift current, some organisms hold onto any available stationary surface, such as rocks or logs. Algae, diatoms, and mosses, and snails that graze on them, live here. Many larval and adult insects adhere to undersides of rocks with hooks or suckers. The streamlined bodies of organisms that live in flowing water enable them to glide through currents and flatten to squeeze under rocks. As a river or stream is fed by tributaries, it widens and current slows. Such slower-moving rivers and streams support more diverse life, including crayfish, snails, bass, and catfish. Worms burrow in the murky bottom, and plants line the banks.

Mastering Concepts

1. Describe the types of organisms that live in each zone of a lentic system.
2. What factors distribute oxygen and minerals in lakes?
3. What is eutrophication?
4. What adaptations enable organisms to survive in moving water?

Marine Biomes

The ocean, covering 70% of the earth's surface and running 7 miles (11.2 kilometers) deep in places, is the largest and most stable biome. Here, biome divisions are based on proximity to land.

The Coast

At the margin of the land, where the fresh water of a river meets the salty ocean, is an **estuary.** Life in an estuary must be adapted to a range of chemical and physical conditions. The water is brackish, which means that it is a mixture of fresh water and salt water; however, the salinity fluctuates. When the tide is out, the water may not be much saltier than water in the river. The returning tide, however, may make the water nearly as salty as the sea. As the tide ebbs and flows, nearshore areas of the estuary are alternately exposed to drying air and then flood.

FIGURE 43.10

Life in the intertidal zone. Where the sea laps the land, assemblages of organisms that can stay in place, such as seaweeds, crabs, mussels, sea urchins, sea anemones, snails, and starfishes, gather.

Organisms able to withstand these environmental extremes enjoy daily deliveries of nutrients, from the slowing river as well as from the tides. Photosynthesis occurs in shallow water. An estuary houses a very productive ecosystem, its rocks slippery with algae, its shores lush with salt marsh vegetation, and its water teeming with plankton. Almost half of an estuary's photosynthetic products go out with the tide and nourish coastal communities. In tropical regions, **mangrove swamps** bridge marine environments and forests (Biology in Action 43.1).

Estuaries are nurseries for many sea animals. More than half of commercially important fish and shellfish spend some part of their life cycle in an estuary. Migratory waterfowl feed and nest here as well. The Chesapeake Bay in the United States is one of the world's richest, but also most polluted, estuaries.

Along coastlines, in the littoral zone, lie the rocky or sandy areas of the **intertidal zone.** This region is alternately exposed and covered with water as the tide ebbs and flows. The organisms in a rocky intertidal zone often attach to rocks so that wave action doesn't carry them away (fig. 43.10). Holdfasts attach large marine algae (seaweeds) to rocks. Threads or suction fasten mussels to rocks. Sea anemones, sea urchins, snails, and starfish live in pools of water that form between rocks as the tide ebbs. The organisms of the sandy beach, such as mole crabs, burrow to escape the pounding waves that would wash them away. Sandy beaches have very little primary production.

Colorful and highly productive coral reefs border some tropical coastlines. Coral reefs are vast, underwater structures of calcium carbonate whose nooks and crannies provide habitats for hundreds of species. Residents include protista, bacteria, algae, invertebrates such as snails, sea stars, octopus, and sea urchins, and many varieties of vibrantly colored tropical fish (fig. 43.11*a*). Food is abundant. The sun penetrates the shallow water, allowing photosynthesis to occur, and the constant wave action brings in nutrients. Coral reefs are rich in biodiversity, but they are fragile.

43.1 biology in action

Mangrove Swamps

A mangrove swamp appears to be rather underpopulated, as biological communities go. Yet appearances are deceptive. This area where sea becomes land is home to a unique assemblage of species. A mangrove swamp provides a variety of minienvironments, from its treetops to deeply submerged roots.

The general term "mangrove" refers to plants that are adapted to survive in shallow, salty water, typically with aerial roots. About 40 species of trees are considered to be mangrove. Mangrove swamps mark the transitional zone between forest and ocean and are located in many areas of the tropics (fig. 43.A). Within them, salinity varies from the salty ocean, to the brackish estuary region, to the fresh water of the forest.

Mangrove swamps exhibit vertical stratification. Life is the least abundant in the treetops, where sun exposure is greatest and water availability the lowest. Snakes, lizards, birds, and many insects live here. A hollow mangrove branch may house a thriving community of scorpions, termites, spiders, mites, roaches, beetles, moths, and ants.

Aerial roots of mangroves provide the middle region of the swamp's vertical stratification. Here, roots are alternately exposed and submerged as the tide goes in and out. Barnacles, oysters, crabs, and red algae cling to the roots (fig. 43.B). Lower down lies the root region of the mangrove swamp, populated by sea anemones, sponges, crabs, oysters, algae, and bacteria. The algal slime that coats roots discourages hungry animals.

Root bottoms form the lowest region of the mangrove swamp ecosystem. Here live sea grasses, polychaete worms, crustaceans, jellyfish, the ever-present algae, and an occasional manatee. Ecologists estimate that up to 30% of the resident species here are unknown.

Unfortunately, many mangrove swamps are in prime vacation spots for *Homo sapiens*—which means habitat destruction. When people cut down mangrove trees, small shrubs that can tolerate salt grow in the area, and trees cannot grow back. The diverse mangrove ecosystem vanishes.

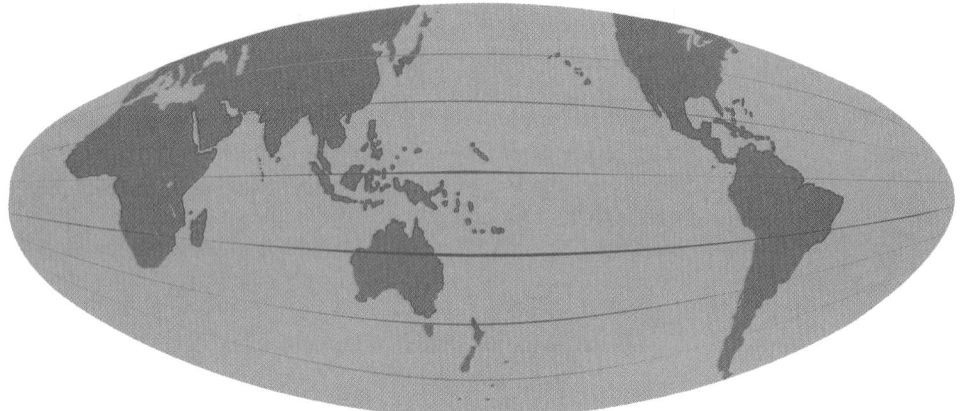

FIGURE 43.A
Mangrove swamps hug many tropical shores, bridging oceans and forests.

FIGURE 43.B
An oyster clings to the algae-coated exposed roots of a mangrove tree.

(a) (b)

FIGURE 43.11
Corals provide living habitats. (*a*) A coral reef and (*b*) tentacles of a *Pocillopora* coral polyp that house algae.

Coral animals build the reefs. The individual organisms, called polyps, house unicellular algae called zooxanthellae. The algae live symbiotically within cells lining coral polyps' digestive tracts, where they photosynthesize and can sometimes be seen. Figure 43.11*b* shows algae in the tentacles of a *Pocillopora* coral polyp. Disease, unusual heat, pollution, and storms can kill the algae. This causes the corals to "bleach," or develop white patches. Without their symbiotic algae, corals must survive on plankton that the stinging cells of their tentacles capture. Many die.

The Ocean

Beyond the continental shelf are the deep, open seas of the **oceanic zone** (fig. 43.12). While the depth of the coastal, or **neritic, zone** averages about 0.33 mile (0.5 kilometer), the oceanic zone may dip below 1.8 miles (3 kilometers). The bottom of the ocean, the **benthic zone,** is home to crabs, starfish, and many other invertebrates. The benthic zone covers all depths. The part of the benthic zone that light never reaches, below 6,500 feet (2,000 meters), is called the **abyssal zone.** The water above the ocean floor is the **pelagic zone.** Within the pelagic zone are the coastal ocean, where the water is less than 650 feet (200 meters) deep, and the open ocean, where the water is deeper.

In the sunlit waters of the pelagic zone where photosynthesis occurs, also called the **euphotic zone,** the most abundant producers are algae. Herbivores are zooplankton such as copepods. Fishes eat each other and zooplankton. The remains of prey sink to the bottom, where decomposers release their nutrients.

Very productive ocean environments arise where cooler, nutrient-rich bottom layers move upward in a process called **upwelling.** The resulting sudden influx of nutrients causes phytoplankton to "bloom," and with this widening of the food web base, many ocean populations grow. Upwelling generally occurs on the western side of continents, where wind blows off the land and out to sea, such as along the coasts of southern California, parts of Africa, and the Antarctic.

The importance of upwelling to sea communities is vividly displayed in **El Niño,** a periodic slack in the trade winds that prevents the upwelling that normally occurs along the coast of Peru. An ecological chain reaction occurs, with populations plummeting throughout food webs in response to widespread changes in wind patterns, rainfall, and temperature. We conclude the chapter with a closer look at El Niño.

Mastering Concepts

1. What is the basis for marine biome divisions?
2. Describe conditions in an estuary.
3. What adaptations enable organisms to survive in intertidal zones?
4. Describe the zones of an ocean.

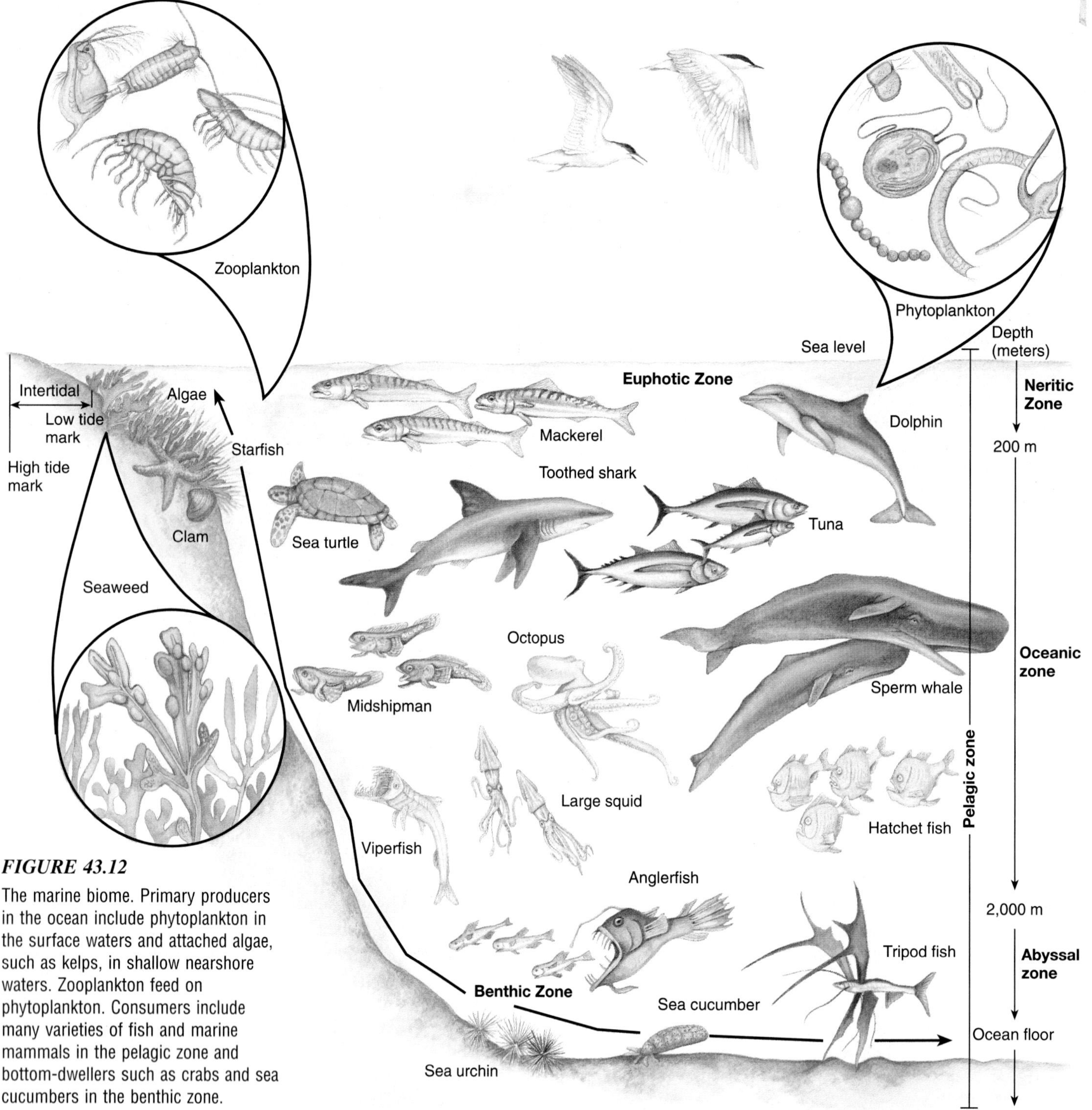

FIGURE 43.12

The marine biome. Primary producers in the ocean include phytoplankton in the surface waters and attached algae, such as kelps, in shallow nearshore waters. Zooplankton feed on phytoplankton. Consumers include many varieties of fish and marine mammals in the pelagic zone and bottom-dwellers such as crabs and sea cucumbers in the benthic zone.

El Niño Affects Ecosystems and Biomes

People who fish along the coasts of Peru and Ecuador have long recognized the effects of El Niño. Every few years, the waters warm in late December, and stay warm until late spring. Shrimp and scallop populations grow, yet many types of fish become so scarce that the fishing industries in Peru, Ecuador, and the Pacific Northwest temporarily collapse. The changes reverberate farther up food webs, as birds, seals, and sea lions starve.

The name *El Niño* means "child" in Spanish and refers to the Yuletide beginnings of the phenomenon (fig. 43.13). Nine El Niños have occurred in the past 40 years. The events occur every 2 to 7 years and typically last a year or two.

What triggers an El Niño? One hypothesis suggests that spurts of activity beneath the tectonic plates of the ocean floor send huge plumes of hot water to the surface. Supportive evidence is that

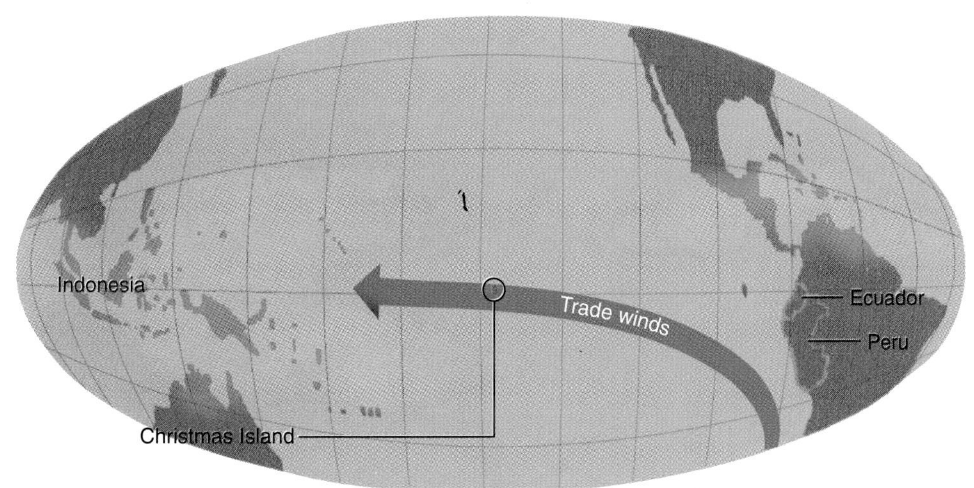

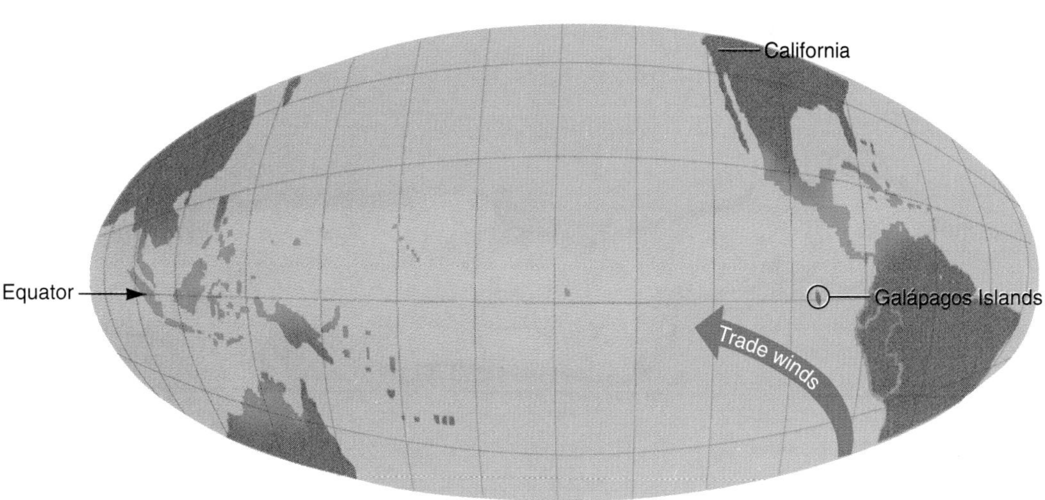

FIGURE 43.13

The temperature at the Pacific Ocean surface is much higher, over a wide area, during an El Niño, as the pink color indicates.

earthquakes and volcanic activity parallel El Niño events (fig. 43.14). Further research is necessary, however, to determine if this correlation represents a cause-and-effect relationship.

A severe El Niño occurred from autumn 1982 until summer 1983. The first sign was weakening of the easterly trade winds that blow across the Pacific Ocean from the Galápagos Islands to Indonesia. In some places, the wind shifted direction, causing powerful storms and raising sea surface temperature along a 5,000-mile (8,050-kilometer) belt in the Pacific around the equator. Fish that were adapted to tropical and subtropical waters followed the warmth towards the poles. Many shore populations starved as changing sea levels and temperatures drastically disrupted ecosystems.

Severe and misplaced weather patterns also affected the distribution of life. Monsoons in the central Pacific, typhoons in Hawaii, and forest fires raging through Indonesia and Australia decimated many local populations. In northern Peru and Ecuador, torrential rains, caused by the unusually warm Pacific waters, turned deserts into muddy grasslands. Here, insects flourished, providing food for burgeoning toad and bird populations. Fish populations suffered, however, as the animals migrated from the swollen ocean to pockets of salinity in the transient lakes that appeared in what was usually desert. When conditions began to reverse, fishes became trapped and died as the pools dried up.

In the United States, the 1982–83 El Niño caused flooding in southern California, blizzards in the Rockies, a warm winter in the Northeast, and rising chicken prices, as the fishes used in chicken feed migrated from their usual habitats in Peru and Ecuador south and escaped fishing nets. Yet at the same time, areas of Mexico, India, the Philippines, Australia, and southern Africa lost countless plants and animals to severe droughts.

The extreme, widespread, and interconnected climatic changes of El Niño greatly upset the upwelling that normally occurs in a 100-mile-wide band along the western coasts of North and South America. The changing winds, by moving surface water westward, altered the level of the thermocline, which prevented upwelling. Without the normal constant supply of nutrients from below, plankton populations at the ocean's surface died, triggering a cascade of starvation up the food web. In addition, lack of upwelling prevented sea surface cooling, which added to the unusual ocean warmth that triggered the abundant rains. For a time, El Niño was a self-sustaining system.

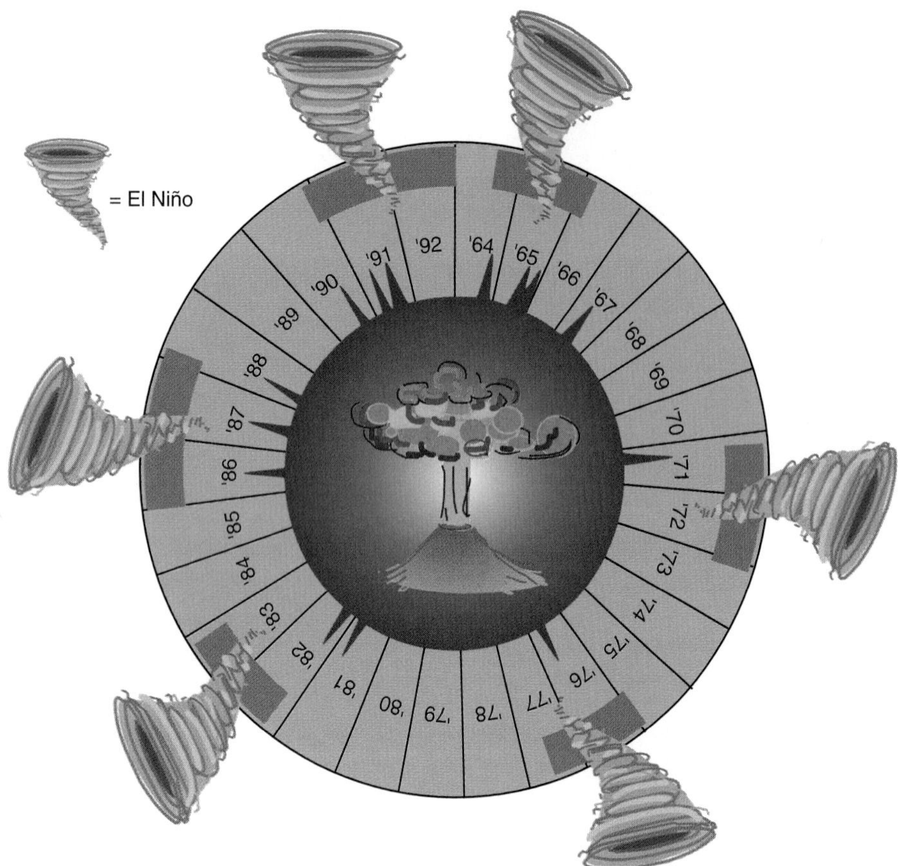

FIGURE 43.14

Coincidence or cause and effect? A controversial hypothesis suggests that seismic activity beneath the ocean's surface spews hot water upwards, beginning the chain of events of El Niño. A more accepted view is that El Niño is a response to normal cyclical fluctuations in air pressure.

El Niños vividly demonstrate the complex links between ecosystems and biomes and the chain reaction of changes in biotic communities in response to drastic, large-scale natural environmental fluctuations. We humans can also alter ecosystems, often with unpredictable consequences that stem from the interdependency of species. The next and final chapter explores some of the impacts that humans have had on planet earth.

Mastering Concepts

1. What physical changes begin an El Niño?
2. How do populations change in response to an El Niño?
3. What is the role of altered upwelling in the biological changes associated with an El Niño?

Infectious Disease Patterns in the Next Century~

continued from page 872.

The incidences and geographical distributions of infectious diseases are intimately tied to climate, which affects the life cycles of disease vectors. Cold temperatures often kill vectors, whereas unusually warm temperatures promote growth and reproduction of some vectors. Warming trends cause certain infections to reach altitudes and latitudes where cold temperatures would normally keep them at bay.

Global warming has raised concern that infectious diseases will spread to new regions. Based on evidence that the global average surface air temperature has increased 0.9°F (0.5°C) over the past century, and that sea level has increased 1 to 2 centimeters per decade, some researchers predict that average temperatures in the year 2050 will be 3.6°F (2°C) higher than they are today. Specific predictions are:

- a 25% increase in population sizes of black flies that transmit river blindness;
- 5 million more cases of schistosomiasis per year;
- with warmer temperatures, mosquitoes that cause dengue will be smaller and therefore bite more often and transmit the infection more readily;
- extension of African sleeping sickness beyond sub-Saharan Africa;
- spread of malaria to Kenya and Nairobi, causing a million more deaths per year;
- more dinoflagellates in the oceans, increasing incidence of paralytic shellfish poisoning.

To address the possible pending problem of infectious disease spread in response to global warming, public health officials recommend a shift in research focus from human health to the environment, such as exploring ecosystem dynamics in response to climate change. A better understanding of ecology may help prevent deaths from infectious disease in the coming century.

SUMMARY

1. Biomes are major types of communities and ecosystems with characteristic species and are located in large geographic areas.

2. The **tropical rain forest** is hot and wet, with diverse life. Competition for light leads to **vertical stratification.** Nutrient cycling is rapid.

3. Temperate deciduous forests require a growing season of at least 4 months, are vertically stratified, and have less diverse life than tropical rain forests. Tree shapes maximize sun exposure. **Decomposers** form soil from leaf litter.

4. Temperate coniferous forests have poor soil and a colder climate. Periodic fires occur in these areas.

5. The **taiga** is a very cold northern coniferous forest. Adaptations of conifers include needle shapes, year-round leaf retention, and conical tree shape.

6. Grasslands are temperate areas with less water than deciduous forests and more water than desert. The more moisture, the taller the grasses.

7. The **tundra** has very cold and long winters. A layer of **permafrost** lies beneath the surface. During the spring and summer, meltwater forms rivers and pools. Lichens are common in the treeless tundra, and animals include caribou, reindeer, lemmings, and snowy owls. **Eskers** are sandy ridges that break the icy terrain and are remnants of rivers that ran beneath glaciers.

8. Deserts have less than 8 inches (20 centimeters) of rainfall a year. Desert plants are well adapted for obtaining and storing water, with rapid life cycles and succulent leaves. Animals also minimize water loss with tough integuments, and they are active at night.

9. Freshwater ecosystems include standing water **(lentic systems)** and running water **(lotic systems).** The **littoral zone** of a lake is the shallow, lit area; the **limnetic zone** is the lit upper layer; the **profundal zone** is the dark deeper layer. In the littoral zone, most producers are rooted plants. In the limnetic zone, phytoplankton predominate. Nutrients fall from the upper layers and support life in the profundal zone.

10. Deep lakes in the temperate zone rely on **fall turnover** and **spring turnover** to mix oxygen and nutrients. Young, deep, **oligotrophic** lakes are clear blue, with few nutrients to support algae. Nutrients gradually accumulate, and algae tints the water green. The lake becomes a productive, or **eutrophic,** lake. In rivers, organisms are adapted to local current conditions.

11. Shoreline ecosystems include **estuaries,** rocky **intertidal zones,** sandy beaches, and **mangrove swamps.** Estuaries occur where a river empties into the sea. The richest marine ecosystems are coral reefs of coastal tropical waters.

12. The region of ocean near the shore is the **neritic zone.** Open water is the **oceanic zone** and includes the **benthic zone** (the bottom), the **abyssal zone** (the bottom region where light does not reach), and the **pelagic zone** (open water above the ocean floor). The most productive areas are in the neritic zones where **upwelling** occurs.

13. El Niño is a weakening or reversal of Pacific easterly trade winds. Seismic activity beneath the ocean may trigger El Niños. An El Niño alters sea temperature, sea level, and weather patterns in many areas. Many populations migrate or decline in response to blocked upwelling which alters food webs.

TO REVIEW . . .

1. What is the basis for distinguishing among different terrestrial biomes? aquatic biomes?

2. What adaptations do organisms have to survive conditions in the following biomes?

- **a.** tropical rain forest
- **b.** estuary
- **c.** mangrove swamp
- **d.** intertidal zone
- **e.** tundra
- **f.** taiga
- **g.** esker
- **h.** desert

3. Distinguish between vertical stratification and thermal stratification.

4. What do fall and spring turnover and upwelling have in common? How do they differ?

5. How can the tropical rain forest support diverse and abundant life with such poor soil?

6. What is permafrost?

7. Give an example of a lentic system and a lotic system.

8. How does photosynthetic activity differ in the zones of a lake?

9. What are the zones of an ocean?

TO THINK ABOUT . . .

1. Why would cutting down the trees of a mangrove swamp drastically alter the area?

2. Researchers hypothesize that seismic activity beneath the Pacific Ocean contributes to, or causes, El Niños. Suggest another hypothesis for the cause of El Niños, and state the type of evidence you would collect or experiments you might conduct to address the hypothesis.

3. Is a coral reef in the neritic or oceanic zone of the ocean?

4. Tick-borne illnesses such as Lyme disease and Rocky Mountain spotted fever occur seasonally during cooler periods. How would global warming affect incidence and distribution of these illnesses, compared to the effects of global warming on the spread of tropical infections?

5. How can a devastating fire be a natural part of a biome's dynamics? In what biome do fires occur regularly?

6. Why are tundra and coral reefs very fragile ecosystems?

7. Would a highly polluted and stagnant lake be oligotrophic or eutrophic?

8. Climatologists can predict El Niños, and epidemiologists can predict infectious disease outbreaks related to climate—with uncertainty. How should nations use this information to prevent future problems, in the face of overwhelming problems that already exist?

TO LEARN MORE . . .

References and Resources*

Broad, William J. April 25, 1995. Hot vents in the sea floor may drive El Niño. *The New York Times.* A controversial theory that may explain how El Niños originate.

Cohn, Jeffrey P. February 1996. The Sonoran desert. *BioScience.* This desert was once a forest.

Johnson, Barry, et al. March 1995. Past, present, and future concepts in large river ecology. *BioScience.* Human activities influence river-based ecosystems.

Krajick, Kevin. May 1996. An esker runs through it. *Natural History.* Ridges of sand and gravel, once rivers, break the bleak tundra landscape.

Ostfeld, Richard S., et al. May 1996. Of mice and mast. *BioScience.* A look at an eastern deciduous forest.

Patz, Jonathan A. January 17, 1996. Global climate change and emerging infectious diseases. *The Journal of the American Medical Association,* vol. 275. Global warming may spread infectious disease from the tropics to areas that are usually cooler.

Rutzler, Klaus, and Ilka C. Feller. March 1996. Caribbean mangrove swamps. *Scientific American.* Mangrove swamps have a unique collection of residents.

Trenberth, Kevin E. March 13, 1997. The use and abuse of climate models. *Nature*, vol. 386. Are computer predictions of climate change accurate and useful?

**Life*'s website features frequently updated references.

Many species have had to adapt to sharing habitats with humans. This week-old tern finds shelter from the sun and predators in a paper cup.

44 Environmental Concerns

chapter outline

***The* Exxon Valdez *Oil Spill* 890**

***Environmentalism* 891**
The 1960s and the 1990s
Environmental Insults May Be Subtle

***The Air* 892**
Air Pollution
Acid Precipitation
Thinning of the Ozone Layer
Global Warming

***The Waters* 897**
Chemical Pollution
Altered Rivers
Lake Eutrophication
Endangered Estuaries
Polluted Oceans

***The Land* 904**
Deforestation
Desertification

***Biodiversity and Resilience* 906**
Biodiversity and Ecosystem Stability
The Resiliency of Life

Moose as Environmental Barometers

Swedish moose are sick. In Alvsborg County alone, more than 1,500 of the large herbivores have died since 1985. The moose malady begins with weight loss and diarrhea. The animals develop ulcers, osteoporosis, infections, convulsions, blindness, and heart failure. Suggested causes include a viral plague or simply reduction in numbers to counter overpopulation. But an astute chemist developed an alternate hypothesis based on another symptom—a sick moose's hair changes color.

Recognizing that altered hair color is a sign of copper deficiency, Adrian Frank, a chemist for the Center for Metal Biology in Uppsala, Sweden, studied 4,360 samples of liver and kidneys taken from moose shot by hunters in Sweden in 1982. He analyzed the tissues for 14 elements and did the same for moose remains sampled in 1994. Results were striking: in the 12-year period, cadmium levels in moose tissues dropped by more than 30%, copper by 50%, and chromium by 80%. Yet levels of another metal, molybdenum, were up 20 to 40%. Molybdenum levels control the others, and when molybdenum rises, the others fall.

Frank hypothesized that a series of connected events was responsible for the declining moose population. The story begins with acid precipitation, but the direct cause of the moose illness is human intervention.

to be continued . . .

The *Exxon Valdez* Oil Spill

When Captain James Cook ventured into Alaska's Prince William Sound in 1778, he discovered a lush wilderness. The thousands of square miles of open water, islands, rocky outcroppings, shoreline, and mainland were home to millions of birds of more than 200 species, 30 species of land mammals (black bears, lynx, deer, beaver, and porcupines among them), and 10 species of marine mammals (including sea lions, whales, seals, dolphins, and otters). More than 200 types of marine invertebrates lived in the oceans, and the 300 streams running into the sea swarmed with fish.

In 1968, another natural feature of Prince William Sound would be its undoing—its deep-water port that provided docking and loading facilities for shipments of oil from the Prudhoe Bay oilfields. By 1977, the Trans-Alaskan Pipeline was in place, and tankers became a common sight in the many inlets of the sound. On March 24, 1989, an oil tanker, the *Exxon Valdez,* ran aground on Bligh Reef, spilling 11 million gallons (more than 50 million liters) of oil. The spreading, sticky black stain would snuff out much life.

Years after the *Exxon Valdez* oil spill, scientists are still assessing its impact on the ecosystem. The U.S. National Oceanic and Atmospheric Administration has tracked the oil (table 44.1). Populations of some native species have yet to recover. Aerial surveys done in the days following the disaster attempted to record biodiversity and compare it to surveys conducted in 1972 and 1973. The surveys identified damage to populations of 11 species of vertebrates, including birds, harbor seals, and killer whales.

Signs of long-term damage to life in Prince William Sound are appearing as reproductive problems in survivors. About half of the population of harlequin ducks survived, but many of the birds cannot reproduce. Similarly, about half of the herring offspring are abnormal. Salmon of a certain age produce deformed hatchlings, which indicates damage to their gametes.

Attempts to clean Prince William Sound endangered some species that had survived the pollution. Of 14,000 sea otters, 2,800 perished directly from the oil. Of 1,200 others that were captured, tranquilized, and scrubbed, only 197 were alive 2 months later to be released into a still-polluted inlet (fig. 44.1*a*). Some of the released otters carried a deadly herpes infection to healthy wild animals.

Scrubbing oil-covered rocks with very hot water under high pressure also had unexpected repercussions. Although the treatment cleansed rocky habitats (fig. 44.1*b* and *c*), it harmed many organisms that had survived the oil pollution. This was the fate of *Fucus* (rockweed), a light brown alga that had accounted for 90% of plant biomass in some areas of the sound. *Fucus* has grown back sporadically in some places, but it has yet to cover many areas in the shaggy carpet that it did before the oil spill.

A successful approach to reclaim the Prince William Sound ecosystem, **bioremediation,** uses microorganisms to cleanse the toxic environment. Bioremediation here entailed applying fertilizers to stimulate the metabolism of bacteria, popularly called "oil-eaters," that degrade crude oil components. Ten weeks after the oil spill, researchers from the U.S. Environmental Protection Agency spread nitrogen and phosphorus fertilizer on 750 selected areas along 74 miles (119 kilometers) of blackened shore. The fer-

Table 44.1 Fate of Oil from the *Exxon Valdez* Spill

Percentage	Fate
50%	Degraded on beaches, water, in tidal sediments
20%	Evaporated
12%	Remains in deep sediments
8%	Removed from environment with water skimmers
6%	Removed from sand, sediments, or water by cleanup crews
3%	Remains in tarry deposits on intertidal shores
1%	Unaccounted for

Source: National Oceanic and Atmospheric Administration.

(a)

(c)

(b)

FIGURE 44.1

Ecosystem recovery. (*a*) An oil-drenched, dying otter was a common sight following the spilling of 11 million gallons (50 million liters) of oil by the *Exxon Valdez* in Prince William Sound in March 1989. Efforts were made to clean many otters and birds, but few survived. (*b*) On May 2, 1989, the beach on Smith Island in Prince William Sound was drenched in oil. Workers washed the shore with pressurized hot water. (*c*) By 1996, the area was considerably cleaner, but ecologists continue to assess biodiversity. They use devices called quadrats to count species in small areas.

tilizer stimulated growth of the natural population of aquatic microorganisms that metabolize **polycyclic aromatic hydrocarbons** (PAHs), which are toxic organic compounds in the oil. Over the next 2 years, biologists watched. The treated shores cleansed much faster than untreated neighboring areas. Researchers calculated that toxic oil components degraded five times faster in the treated areas by comparing the PAH levels in these areas to levels of an organic compound that the microorganisms do not consume.

Ecologists continue to monitor depleted populations in Prince William Sound, such as those of herring and sea otters. The *Exxon Valdez* oil spill taught wildlife biologists what they can and cannot do to help an ecosystem recover from a severe and unnatural disturbance.

Mastering Concepts

1. List some of the effects of the oil spill in Prince William Sound.
2. What efforts have people taken to help the ecosystem recover, and what have been the results?

Environmentalism

"Saving the environment" is a popular concern for many people, but it is also a part of life science.

The 1960s and the 1990s

The 1960s were a time of great public concern over the environment. With celebrities adopting pet environmental causes and headlines exaggerating the extent of certain problems, the public at times lost sight of the science behind the situation. Nevertheless, new institutions, such as the U.S. Environmental Protection Agency, led the way in reducing some of the most serious sources of environmental pollution. **Environmentalism,** the philosophy of living in harmony with the global ecosystem, became a strong political force.

In the decades since the 1960s, continued growth of human populations and activities have elevated new environmental stresses to critical levels. In the 1990s, information from global environmental monitoring networks, including remote-sensing data from satellites, has revealed that the environment is a more complex system than we had imagined, and that upheavals—both human-generated and natural—are triggering complex changes in the biosphere.

Discharges of greenhouse gases, depletion of stratospheric ozone, deposition of acidic air pollutants, and fouling of lakes and streams are interacting to produce several major patterns of global change. Global warming, desertification, and great loss of biodiversity are among these effects. The major message from studying human-caused and natural disasters is that no environmental problem can be considered alone. The principles of ecology provide a framework for placing these challenges in perspective. This chapter continues the discussion of ecology by highlighting selected environmental problems.

Environmental Insults May Be Subtle

An environmental disaster need not be as obvious as the tarry, oil-slicked waters and beaches of Prince William Sound. No one could see the radiation that burst from the Chernobyl nuclear reactor in the former Soviet Union on April 25, 1986, sending a radioactive cloud around the world. Like the oil spill in Prince William Sound, the Chernobyl disaster had immediate and delayed effects on life.

Thirty-one people died soon after the leak, and hundreds suffered acute radiation sickness. By 1990, people still living in the area showed signs of chronic radiation sickness, with symptoms including fatigue, anemia, and poor appetite and vision. Epidemics of common illnesses suggested widespread immune system impairment. Deformed animals born months after the explosion offered vivid testament to the lingering effects of radiation exposure. Calves and pigs without eyes (fig. 44.2) and two-headed colts were not unusual. By 1994, people living near the Chernobyl plant had increased rates of cancer and thyroid disease, illnesses that take longer to produce symptoms.

Biology in Action 16.2 discusses genetic changes that resulted from the Chernobyl disaster, assessed a decade after the event. Biology in Action 44.1 describes a related problem—low levels of radiation that persist in the environment from weapons manufactured in the 1950s.

Mastering Concepts

1. What is environmentalism?
2. What types of environmental problems do we face today?
3. What were some effects on life of the explosion at the Chernobyl nuclear power plant?

FIGURE 44.2

The aftermath of a nuclear power plant accident. This eyeless pig is one of many farm animals born with severe defects following the April 25, 1986, explosion at the Chernobyl nuclear reactor in the former Soviet Union.

The Air

The atmosphere is the source of several environmental problems. These include pollution, acid precipitation, ozone depletion, and global warming.

Air Pollution

People have been polluting the air since ancient times. Analysis of lake sediments in Sweden and peat deposits in England reveal local lead air pollution during the Roman Empire, from 500 B.C. to A.D. 300. Then, lead mining and metal smelting were as common as during the Industrial Revolution, a time previously credited with introducing air pollution. Analysis of Greenland ice reveals that the smelters of the Roman Empire polluted the air on a global scale.

December 5, 1952, stands out as a more recent time when we became particularly aware of air pollution. On this day, an event called a temperature inversion occurred over London. A high-level layer of warm air blanketed the city, trapping cold air at ground level and preventing combustion products from burning of coal and other fossil fuels from dispersing upward. Sulfur dioxide (SO_2), which reacts in fog moisture to form sulfuric acid, was the main pollutant. This "London fog" sent hundreds of people to hospitals with respiratory symptoms, and the death rate soared. London-type air pollution occurs in urban and industrial regions where power plants and industries burn sulfur-rich coal and oil.

Another form of air pollution is **photochemical smog,** which results from vehicle emissions and the chemical reactions they undergo in the air. Photochemical smog forms in warm, sunny areas where automobile use is heavy. The ultraviolet radiation in sunlight causes nitrogen oxides and incompletely burned hydrocarbons to react to form ozone (O_3) and other oxidants. These compounds poison plants and cause severe respiratory problems in humans.

44.1 biology in action

A Legacy of Cold War

With the arms race and the Cold War over for now, ecologists are considering how to dispose of dangerous materials used to build weapons and how to respond to human health problems that the weapons industry has already caused. Epidemiologists are analyzing recently uncovered medical records that document the medical problems associated with long-term exposure to radiation in a 296,400-acre (1,200-square-kilometer) area in the former East Germany.

From a distance, the lake near Oberrothenback, Germany, appears inviting, but looks are deceiving. The lake harbors enough toxins to kill thousands of people. The waters contain heavy metals, low-level radioactive chemical waste, and 22,500 tons of arsenic. Radon, a radioactive by-product of uranium, permeates the soil. Many farm animals and pets that have drunk from the lake have perished and incidences of cancer and respiratory disorders among humans living nearby are especially high.

The deadly lake in Oberrothenback was once a dump for a factory that produced "yellow cake," the processed uranium ore once used to build atomic bombs. In the early 1950s, nearly half a million workers labored there and in surrounding areas in factories and mines. Records released in 1989, after the reunification of Germany, reveal that workers were given perks such as alcoholic beverages and better wages to work in the more dangerous areas. They paid a heavy price—tens of thousands died of lung ailments.

Today, the meticulously kept records may help answer a long-standing health question—what are the effects of human exposure to long-term, low-level radiation? Until now, the risks of such exposure have been extrapolated from health statistics amassed from the victims, survivors, and descendants of the atomic blasts in Hiroshima and Nagasaki in World War II (fig. 44.A). But a single large exposure may not have the same effect as prolonged low-level exposure. Perhaps what we learn from studying the effects of yellow cake can help protect modern workers who dismantle nuclear weapons (fig. 44.B).

Not only did the Germans record many details about the health problems of thousands of workers, they even preserved pieces of the lungs of those who died of lung cancer. The data also include effects on women, which has enabled researchers to identify gender differences in response to radiation exposure. Epidemiologists plan to continue to assess the health of people still living in the area and their descendants, which makes this a large-scale, prospective as well as retrospective study.

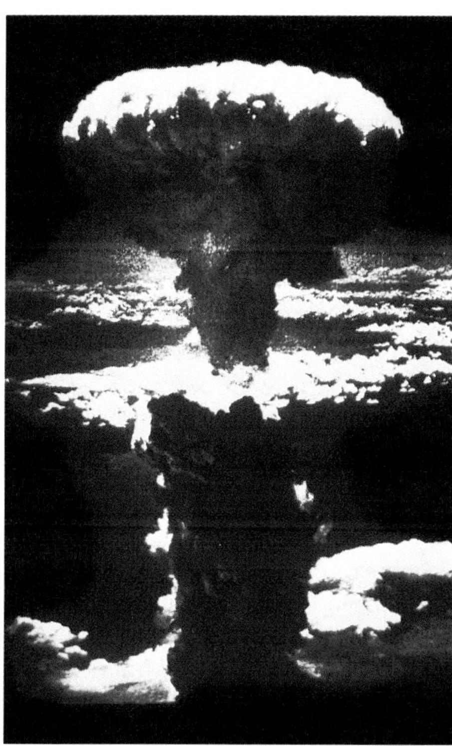

FIGURE 44.A
Much of what we know about the effect of radiation on the human body is extrapolated from data on people exposed to the bombs exploded over Japan in World War II. But this was a one-time, huge dose of radiation. Do long-term, lower exposures have different effects on health? Data from German workers exposed to uranium and other toxic substances used to manufacture bombs may provide the long-sought answer.

FIGURE 44.B
Plutonium is a human-made element used to manufacture bombs. A nuclear weapon contains a plutonium "pit," a sphere about the size of a softball, which is highly toxic. Such weapons are now being dismantled, and scientists are discussing how to dispose of the plutonium. The workers shown here are handling plutonium, in 1989. Note their protective suits and helmets.

The London fog incident began the age of active air pollution control. Today, such efforts focus not only on sulfur and nitrogen gases but also on the fine particulates that pervade the air. Particles smaller than 10 micrometers in diameter may be more dangerous than larger dust and smoke particles because they penetrate deeper into the lungs.

Acid Precipitation

Patterns of air circulation and geography can interact so that organisms far from the source of air pollution become ill. In 1852, British scientist Angus Smith coined the phrase "acid rain" to describe one effect of the Industrial Revolution on the clean British countryside. In the mid-1950s, studies on clouds in the northeastern United States, farmlands in Scandinavia, and lakes in England rediscovered the effects of **acid precipitation.**

Today, acid precipitation in the United States originates from smokestacks in the Midwest that burn coal and release sulfur and nitrogen oxides (SO_2 and NO_2) high into the atmosphere. Gasoline and diesel fuel burned in internal combustion engines and fumes from heavy metal smelters add to the problem. In the atmosphere, these oxides join water and fall over large areas as sulfuric and nitric acids (H_2SO_4 and HNO_3), forming acid rain, snow, fog, and dew.

Because the atmosphere contains carbon dioxide (CO_2) and water, all rainfall includes some carbonic acid and is therefore slightly acidic. Burned fossil fuels, however, have made the average rainfall in the eastern United States 25 to 60 times more acidic than normal. Acid precipitation is also a problem in the Pacific Northwest, the Rockies, Canada, Europe, the Orient, and the former Soviet Union.

Acid precipitation affects lake life dramatically. A typical small mountain lake is clear and blue, with a pH of around 8.0. Acid rain may increase the acidity by 1,000 (10^3) to 100,000 (10^5) times, dropping the pH range by 3.0 to 5.0 units. In this abnormally acidic environment, fish eggs die or hatch to yield deformed offspring. Amphibian eggs do not hatch at all. Crustacean shells do not harden. In very acidified lakes, most aquatic invertebrates and zooplankton die, and lake-clogging mosses, fungi, and algae replace aquatic flowering plants. Clinging algae outcompete phytoplankton and coat shallow nearshore areas, preventing trout from spawning. Organisms that feed on the doomed species must seek alternate food sources or starve, which disrupts or topples food chains. Over time, these conditions reduce lake life to a few species that can tolerate increasingly acidic conditions.

The effects of acid rain on lakes are difficult to study because of ecosystem complexity and lack of information on a lake's condition prior to acidification. In an experimental approach to studying effects of acid precipitation on lakes, researchers from the Department of Fisheries and Oceans in Winnipeg, Manitoba, Canada, have added sulfuric acid to a lake 200 miles (322 kilometers) southeast of the city since 1968. They have cataloged changes in the biotic, abiotic, and energy components of this experimental ecosystem. In that time, the lake's pH has dropped from 6.8 to 5.0.

Canada's experimentally acidified lake shows that life continues, but biodiversity plummets. Before acidification, the lake housed large populations of 80 species of midges, a type of insect. Today, 95% of the midge community is of one previously unknown species. A formerly rare midge variant able to thrive in low pH conditions survived as natural selection removed competing species.

The Canadian experimental lake shows that acid precipitation alters food webs and biogeochemical cycles, but it does not seem to affect primary production, nutrient levels, and decomposition rates. Natural events also counter acidification. Calcium compounds in the earth have a buffering effect, just as similar compounds do in stomach antacids. Bacteria on the needles of jack pine trees neutralize acid rain, and sphagnum peat mosses ringing lakes absorb and buffer acid rain. Bacteria deep within lakes break down nitrogen and sulfur compounds. Researchers at the Canadian lake hypothesize that if we stop polluting lakes, they would recover in just 5 to 10 years.

Acid precipitation alters forests too. In a coniferous forest high in the mountains of Fichtelgebirge, Germany, acid precipitation thins trees and yellows needles (fig. 44.3). Here, nitric and sulfuric acids trigger a cascade of effects. As soil pH drops, aluminum ions percolate toward tree roots, where they displace nutrients such as calcium ions (required for growth of twigs, leaves, stems, and trunks) and magnesium ions (part of chlorophyll). Loss of essential nutrients stunts tree growth. Yet the constant input of nitrogen, in nitric acid, stimulates trees to grow. These mixed environmental signals stress trees and make them less able to resist infection or to survive harsh weather. As a result, acid precipitation is thinning high-elevation forests throughout Europe and on the U.S. coast from New England to South Carolina.

Thinning of the Ozone Layer

Ultraviolet (UV) radiation can kill cells and induce mutations that may have delayed effects such as causing cancer, infertility, or birth defects. If the amount of UV light reaching the earth increases, it could kill plant cells, and this would disrupt food webs.

Fortunately, a layer of **ozone,** a bluish gas that forms high in the atmosphere when oxygen reacts with UV radiation, partially blocks UV. With each 1% decrease in ozone levels, the strength of UV radiation reaching the earth's surface increases by 1.28%. Chlorine, fluorine, and bromine—some of which enter the atmosphere as **chlorofluorocarbon (CFC) compounds**—can accelerate the breakdown of ozone. CFC compounds contain chlorine, fluorine, and carbon.

The colorless and odorless CFCs have found wide applications over the past 50 years because they are nontoxic and do not easily explode, burn, or corrode. They have been used in refrigerants, as propellants in aerosol cans, as fire retardants to clean electronic and computer parts, to soften pillows, and to produce foamed plastics. At fast-food restaurants, CFCs kept burgers hot and shakes cold.

Today, CFCs are no longer used because we recognize their danger, but they remain in the atmosphere. CFCs are incredibly stable, and once released into the environment—from a leaking refrigerator or a decaying container of foam—they persist for 60 to 120 years. Eventually, they rise in the upper atmosphere and react with UV light to release the chlorine compounds that break

FIGURE 44.3

Acid precipitation. (*a*) The first sign of damage from acidified soil may be browning conifer needles. (*b*) Acid fog, clouds, mist, and rain harm forests. Red spruce die as acid precipitation kills their needles. Other trees are harmed indirectly, as acidified soil releases aluminum ions, which block uptake of calcium and magnesium by root hairs. Nitrogen in pollutants stimulates plants to grow, yet they are also stressed by malnutrition. The result—increased susceptibility to other problems.

down ozone. Over the past 15 years, the ozone layer has thinned so much that a "hole" appears in satellite images over Antarctica. The ozone layer has also thinned over parts of Asia, Europe, and North America.

Natural events thin the ozone layer too. Although pollution contributed greatly to ozone depletion in the late 1970s, the 2 to 3% thinning the *Nimbus-7* satellite recorded in 1992 and 1993 was attributed more to Mount Pinatubo's June 1991 eruption. Chemical reactions that deplete ozone occur readily on the dust particles that the volcano sent into the atmosphere. Still, by 1995, when researchers had expected Pinatubo's influence to have waned, the ozone hole over Antarctica was still present (fig. 44.4). The persistence may be due to lingering effects of the volcano, to increased use of chlorine and bromine-containing chemicals, or to a cold wave in the stratosphere above Antarctica—or to a combination of these influences. Researchers do not know whether the ozone hole is temporary and localized or the beginning of global ozone depletion.

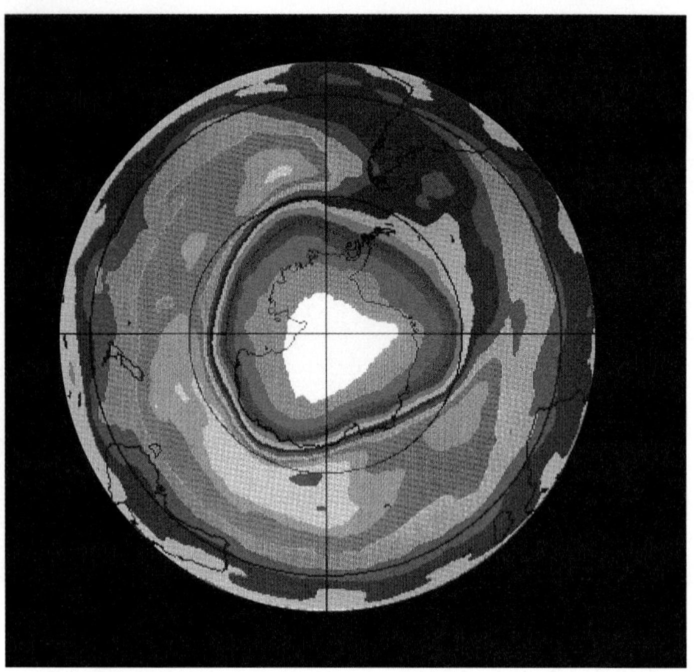

FIGURE 44.4

Climatologists do not know why the ozone hole over Antarctica persists.

Global Warming

Carbon dioxide is a colorless, odorless gas present in the atmosphere at a concentration of 350 parts per million. Although it is a minor atmospheric constituent, it has major effects on life by influencing temperature.

CO_2 that blankets the earth prevents radiation from exiting the atmosphere as quickly as it enters. CO_2 does not absorb shortwave solar radiation, including light waves. Thus, light reaches the earth's surface, where it is converted to heat. The surface radiates long-wave heat radiation outward, but CO_2 absorbs this radiation and reradiates some of it back toward the planet's surface. This traps heat near the earth's surface (fig. 44.5). The resulting increase in surface temperature is called the **greenhouse effect,** because CO_2 blocks heat escape as the glass panes of a greenhouse do. Other gases contribute to the greenhouse effect, including methane, nitrous oxide, and CFCs. These gases trap heat much more efficiently than CO_2 does, but because they are scarcer, they contribute only half as much to global warming.

The greenhouse effect was first noted in 1896, after industrialization came to the Western world. It was (and is) largely caused by burning fossil fuels such as coal, oil, and gas. Since then, total amounts of atmospheric CO_2 have increased by 15 to 30%. Today, burning fossil fuels, tropical deforestation, and other combustion activities release 5 million tons of CO_2 into the atmosphere each year (fig. 44.6).

Most scientists agree that the greenhouse effect has already changed the climate. Precipitation has increased at midlatitudes and decreased at low latitudes over the past 30 to 40 years much as computer models of global climate predict. A 1995 report by the U.S. Intergovernmental Panel on Climate Change states that the apparent global warming trend since 1900, and especially since 1980, is probably due to human activities. Data from weather stations around the world show that 1995 was the warmest year ever recorded (fig. 44.7).

Computer models predict that an average global warming of 4.5°F (2.5°C) may result from doubled CO_2 and might occur by A.D. 2040 to 2090. Warming will likely be much greater at high latitudes than in temperate and tropical regions. In the arctic, summer temperatures may average about 3.6°F (2°C) warmer and winter temperatures about 18°F (10°C) warmer. Precipitation and moisture will also change. Warmer temperatures will increase evaporation and thus precipitation. More moisture will probably reach high latitudes and increase precipitation there. At midlatitudes, continental interior regions may become drier in summer because snows will melt earlier in the spring, and summer temperatures will be higher.

Global warming will also affect the oceans. Sea level may rise 8 to 55 inches (20 to 140 centimeters), due to thermal expansion of warmer ocean water and increased melting of glaciers and ice caps. Ocean currents will change. In response, biodiversity will redistribute.

The basic predictions of models of climate change are not universally accepted. A few scientists argue that assumptions about how atmospheric moisture will amplify the influence of doubled CO_2 are not well supported. They believe that the influence of moisture, especially at high altitudes, opposes warming at lower levels in the atmosphere.

Mastering Concepts

1. What are major sources of air pollution?
2. How does acid precipitation form?
3. What effects do air pollution, acid precipitation, thinning of the ozone layer, and global warming have on life or are predicted to have on life?

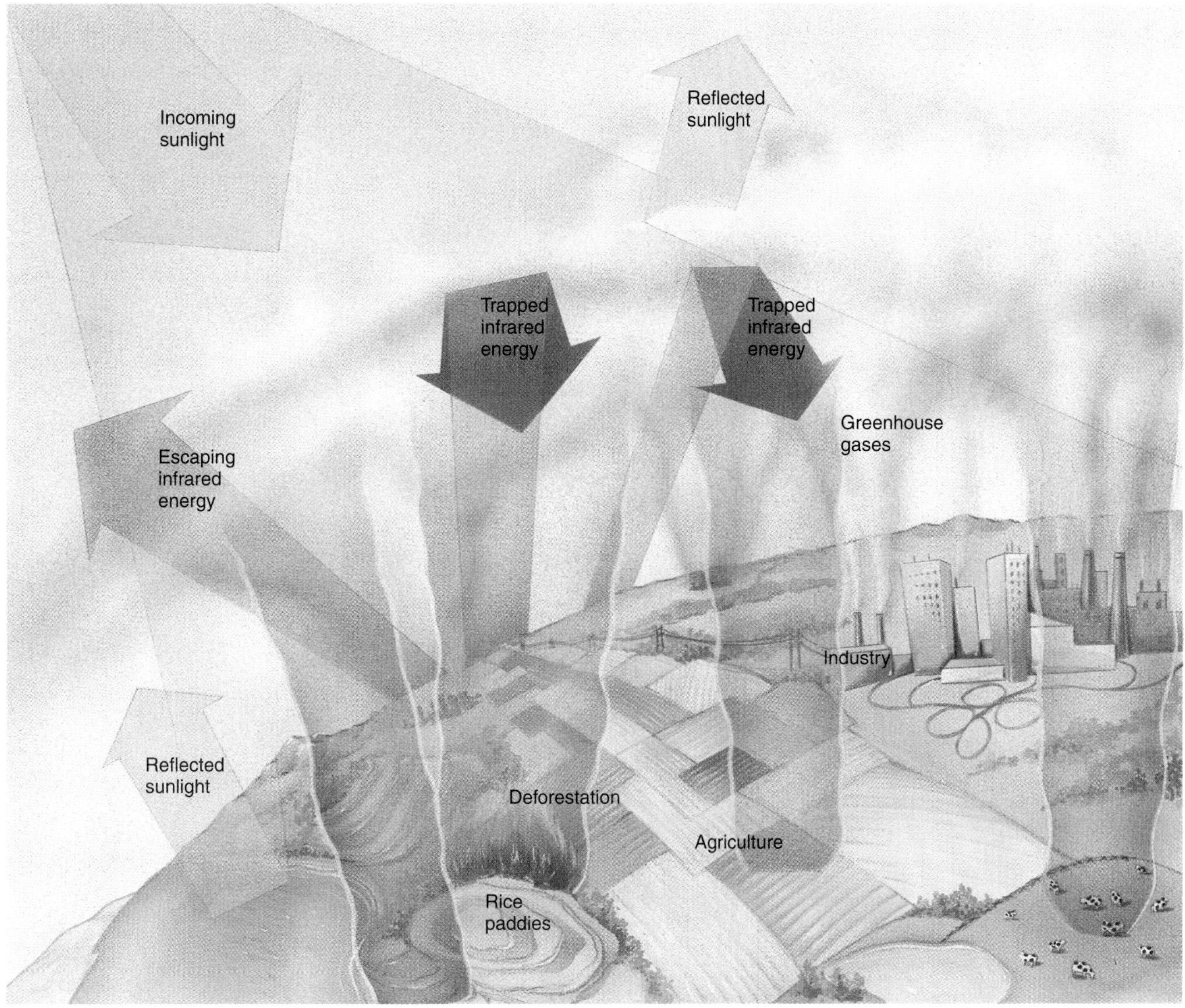

FIGURE 44.5

The greenhouse effect. Solar radiation heats the earth's surface. Some energy returns to the atmosphere as infrared radiation (heat), and some of this heat is trapped near the surface by increased levels of CO_2 and other gases. Industrial processes, farming, and deforestation produce these greenhouse gases.

The Waters

Like the air, the waters face human-induced challenges.

Chemical Pollution

Chapters 2 and 42 discussed DDT, a once widely used pesticide whose toxic effects are still felt. Other chemicals pollute aquatic ecosystems. **Polychlorinated biphenyls (PCBs)** are industrial chemicals used as insulation fluids in transformers and as plasticizers and fire retardants in various materials. They have entered aquatic environments through leakage and trash disposal. PCBs are biomagnified, sometimes reaching concentrations of hundreds to thousands of parts per million in birds and mammals at the tops of aquatic food chains. PCBs disturb reproduction in a number of bird and seal species.

Polycyclic aromatic hydrocarbons (PAHs) are another group of petroleum derivatives that are becoming major aquatic pollutants. Some PAHs are natural components of petroleum, and others form from combustion of fossil fuels. These chemicals enter aquatic ecosystems through oil pollution, urban runoff, and precipitation. Many PAHs cause liver cancers and fin deterioration of fish.

Mercury and cyanide are pollutants in gold-mining areas. Mercury is used to separate gold from lighter minerals. In the Amazon Basin, about 130 tons of mercury are discharged annually into rivers, and food chains in several rivers are heavily contaminated. In the western United States, cyanide is used to leach gold from old mine tailings and low-grade ore. Many birds and mammals have died after feeding or drinking at ponds containing water used in leaching for gold.

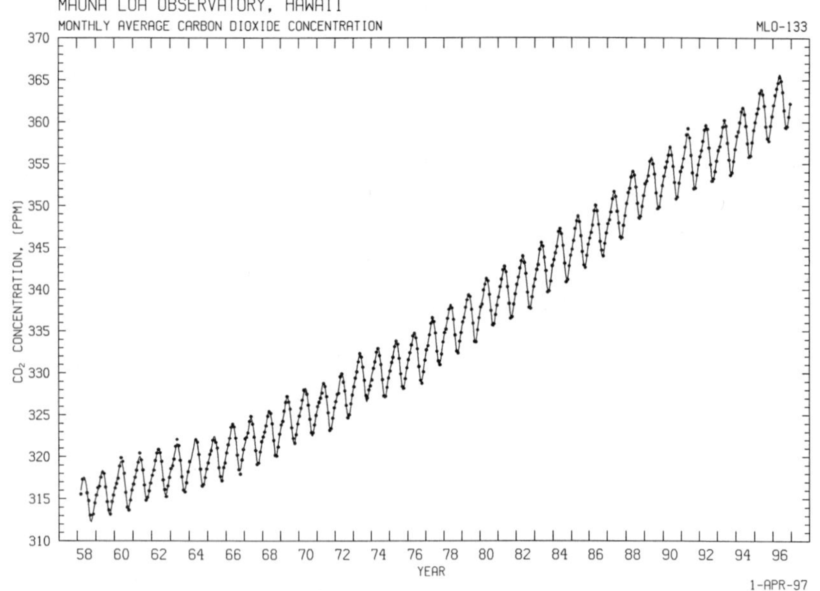

FIGURE 44.6

Atmospheric CO_2 levels continue to rise annually, as measured in an observatory at Mauna Loa, Hawaii.

Sources: Keeling, C.D., et al., in *Aspects of Climate Variability in the Pacific and the Western Americas,* Part 1, Appendix A, American Geophysical Union Monograph, vol. 55, pg. 165–263, (1989).
Keeling, C.D., Whorf, T.P., Whalen, M., van der Plicht, J. Interannual extremes in the rate of rise of atmospheric carbon dioxide since 1980, *Nature* 375, 666–70 (June 22, 1995).

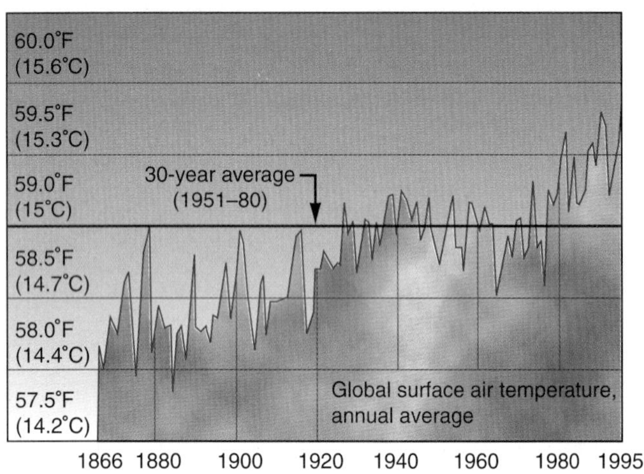

FIGURE 44.7

Whether global warming is a true warming trend or a normal part of long-term temperature fluctuation, data indicate that 1995 was a particularly warm year.

Source: NASA Goddard Institute for Space Studies.

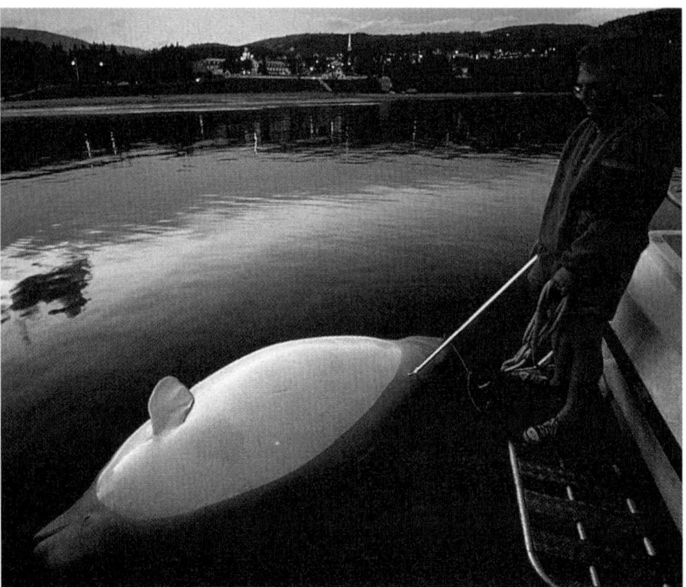

FIGURE 44.8

People aren't killing Arctic beluga whales in the St. Lawrence River directly—pollution is.

Selenium is another element that is toxic to animals in high concentrations. Selenium is abundant in some desert soils. Farming in desert regions leaches selenium from the soil and irrigation drainage waters carry it into aquatic ecosystems, where it becomes biomagnified. Kesterson National Wildlife Refuge in California, for example, received waste irrigation waters from San Joaquin Valley farms, which exposed plants and animals to high concentrations of selenium. Selenium poisoning has killed many water birds and caused many nesting birds to carry deformed embryos or become infertile.

Organisms living in polluted areas are often exposed to several toxins. Consider the 500 or so Arctic beluga whales that, following ancient instincts, swim from the Atlantic Ocean into the St. Lawrence River. Hunting brought their numbers down from 10,000 at the turn of the last century, but today, pollution is killing these whales. The St. Lawrence River contains a deadly mix of toxins that includes DDT and other organic pesticides, mercury, and PCBs.

Whales in the St. Lawrence River are dying from ailments rarely seen in other whale populations—tumors, stomach ulcers, thyroid and adrenal abnormalities, and impaired immunity (fig. 44.8). The scant milk that females produce is rich in PCBs and DDT, which are soluble in lipid. These toxins bioaccumulate in milk, and nursing offspring receive even higher doses than their mothers swim in. Biology in Action 44.2 describes another polluted river, the Rhine.

Altered Rivers

Because a river provides diverse habitats and can cover a large area, human intervention can have profound effects on life. For example, along the banks of the Mississippi River, levees built to prevent damage from flooding alter the pattern of sediment deposition. This confines nutrients and sediment to the river channel, which carries them to the Gulf of Mexico. There, the nutrients feed plankton, causing their populations to bloom. Red and brown tides, the result of such blooms, may kill fish, manatees, and other sea life.

Damming alters a river ecosystem too. Changes in water temperatures and nutrient levels alter populations of plankton, aquatic insects, and fishes, which creates great gaps in food webs. Channelization, or straightening the course of a meandering river, increases the water's flow rate, which erodes channel sediments and carries them downstream, where they can choke out stream communities.

Even a single, well-intentioned intervention can have unpredictable effects. When fishery managers introduced opossum shrimp into Flathead Lake, Montana, they meant to feed kokanee salmon. The shrimp, however, consumed the plankton that the salmon normally eat. Unable or unwilling to alter their diets as the planners had intended, the salmon starved. Soon, populations of bald eagles and grizzly bears declined as their food—salmon—became scarce. Another predator also disappeared—the tourists whose fondness for salmon had originally prompted the ill-fated introduction of shrimp.

Lake Eutrophication

Lakes age, and their populations change, as happens in any dynamic ecosystem. Human activity along a lake's shores can hasten these changes. Lake Tahoe is in the early stages of human-induced eutrophication.

Lake Tahoe lies along the California-Nevada state line and is the 10th deepest lake in the world. The lake formed about 11,000 years ago, making it young by geological standards. It was known only to the Paiute Indians before white settlers discovered it in 1844. Since then, human interference has taken a stiff toll.

Between 1870 and 1900, miners cut down forests surrounding the lake to build mine shafts. Since the late 1950s, the shores began to be transformed into a resort. As roads, houses, hotels, and casinos rose, nutrients from piles of fill flowed into the lake, causing the algae and weeds to bloom into a murky tangle. Pollution poured in from sewage, from runoff from fertilized golf courses and lawns, and from the air. Blackwood Creek was mined for gravel, sending debris to the lake's bottom. In the North End region, 100 acres of marsh vanished to make room for housing developments. Altogether, 75% of the marshes and 50% of the meadows surrounding Lake Tahoe were disturbed.

What lies ahead for Tahoe and other damaged lakes? If polluting continues, algal populations will bloom, and bacteria will decompose them, depleting the oxygen supply in deep waters. Bacterial activity may eventually cause the lake to smell like rotten eggs, and dead fish may litter its shores.

Natural protections may help Lake Tahoe recover. Wetlands around the lake block sediments and nutrients from reaching the water, which may retard eutrophication. Artificial marshes are being built to protect the lake from the condominiums on its shores. If construction companies, waste-dumping industries, and visitors leave the lake alone, its natural communities of organisms may be able to survive.

Yet the fate of the Great Lakes suggests that Lake Tahoe's pollution may linger. Dumping DDT and PCBs into the Great Lakes stopped years ago, but certain fishes still contain high levels of these chemicals. The reason for the persistent pollution is two newcomers to the ecosystem—the zebra mussel and a type of zooplankton. These organisms occupy low positions in the food web, and they concentrate DDT and PCBs remaining in the sediment. When animals eat the mussels and zooplankton, the toxins are biomagnified, eventually tainting trout, Coho salmon, and walleye.

Endangered Estuaries

Estuaries, as links between fresh water and salt water, are pivotal in ecology. Yet humans have drained and filled estuaries to build houses and dumped garbage and other pollutants in their waters. Laws now protect estuaries in some parts of the nation, but in other areas, damage has already gone too far.

The 64,000-square-mile (166,000-square-kilometer) Chesapeake Bay, on the eastern coast of the United States, touches six states and houses more than 200 species of fishes and 75 species of birds. The ecological changes of the past few decades here are ominous. The food web base is in upheaval, in response to influx of sewage, agricultural fertilizers, heavy metals, pesticides, and oil. Each day millions of gallons of waste flow into the bay.

Phytoplankton populations have bloomed and the composition of their communities has changed. Dinoflagellate and green algae populations are growing, while the numbers of diatoms are falling. Reduced light penetration has killed eelgrass. Eutrophication has sharply reduced the oxygen content of the deep water. As a result of these changes in food sources, the fish and crustacean populations are declining.

The Chesapeake Bay oyster is one victim of pollution. In colonial times, oysters lived up to 10 years and grew to a foot or more in length. Today, they barely live past 3 or 4 years and are about the size of a golf ball. In the 1890s, 15 to 18 million bushels of oysters were harvested a year; today the number is under a million. Blue crabs were once so plentiful that at certain times of the year a wader could not avoid stepping on them. The crabs are fast losing their feeding and molting grounds.

Cleanup efforts may help restore some of the Chesapeake's lost diversity. New sewage facilities are being built and old ones repaired. Floating plants introduced in some areas use the excess nitrogen. Ecologists hope to reduce the phosphorus and nitrogen flowing into the estuary by 40%.

Polluted Oceans

Like other ecological stresses, ocean fouling is a cascade of events. Pollutants pour in from animal droppings, human sewage and medical waste, motor oil, fertilizer from large farms and millions of lawns, plus industrial waste, oil spills, and garbage (fig. 44.9). Rivers deliver pollutants containing nitrogen and phosphorus to

44.2 biology in action

The Rhine—A River in Recovery

The Rhine River illustrates the types of chemicals that pollute rivers and how efforts to clean up the area can work. The Rhine borders or runs through Germany, France, Belgium, the Netherlands, and Switzerland. Pollutants poured in as Europe recovered from World War II, as agriculture and industry grew (fig. 44.C).

Organic compounds deplete a river's oxygen. They account for only 1% of pollutants but are difficult to study because there are thousands of them. Organic pollutants in the Rhine, however, fall into four broad categories:

- Detergents are surfactants, which break lipids into particles and disperse them. Since 1964, detergents have been biodegradable.
- Organic halogens include pesticides and chlorine used to bleach wood products in paper manufacturing.
- Hydrocarbons are petroleum products.
- Humic acids are harmless products of bioremediation. Bacteria cannot metabolize many organic compounds completely to CO_2. The remaining compounds are humic acids. They form naturally in the river and in sewage treatment plants.

Inorganic pollutants tend to leach from soil and sediment into water and also enter waters from human activities. Major inorganic pollutants in the Rhine include:

- Chloride ions (Cl^-) enter from natural sources at 33 to 165 pounds (15 to 75 kilograms) per second and from industry, sewage, and agriculture at 700 pounds (318 kilograms) per second. Much of this pollution came from a potash mine in Alsace, which released NaCl as a by-product for many years.
- Heavy metals in large amounts disrupt growth, development, and reproduction and interfere with metabolism and nervous system functioning. In the Rhine, industrial plants and corroded pipes have added chromium, zinc, and nickel, and chemical plants have added chrome, copper, cadmium, and mercury. Lead was a pollutant before it was removed from gasoline. The Rhine's water today is relatively low in heavy metal contamination because these pollutants have settled into the sediment.
- Nitrogen enters the water from fertilizer used on crops bordering the river, mostly as nitrates.
- Phosphorus enters the water from sewage. Phosphorus and nitrogen boost algal populations, which depletes oxygen and kills other species.

Multinational efforts to clean the Rhine began in the wake of a fire in a pesticide warehouse in 1986. Water used to extinguish the flames sent the toxic chemicals into the river, which affected populations of many types of invertebrates and fishes. The Rhine Action Program begun in 1989 identified 30 pollutants and then developed regulations and sewage treatments that had, by 1995, lowered levels of discharge of all of the pollutants by at least 50%. The Rhine may be on its way to recovery.

FIGURE 44.C
Sources of pollution of the Rhine River valley.

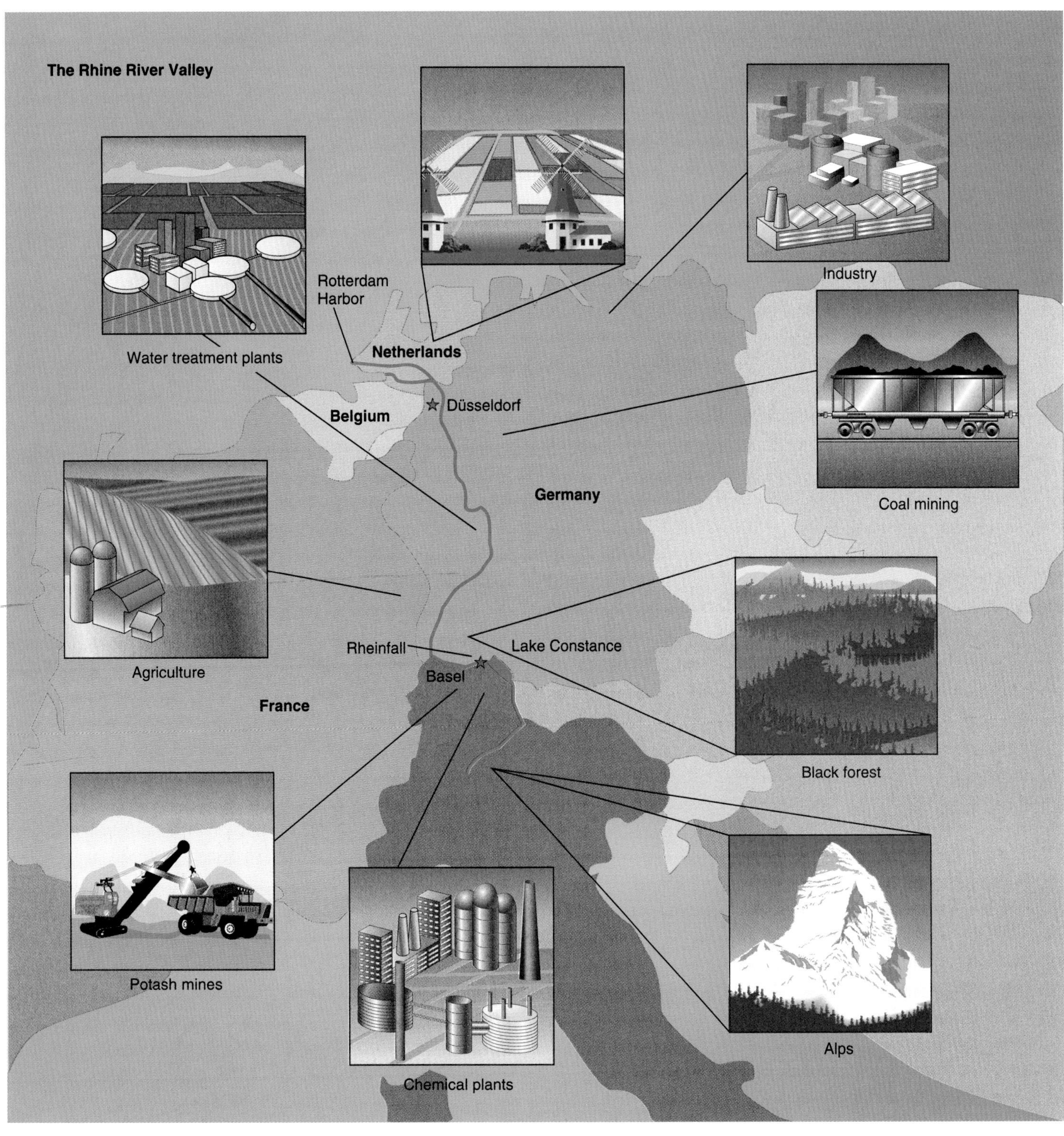
The Rhine River Valley
Water treatment plants
Rotterdam
Harbor
Netherlands
Industry
Coal mining
Düsseldorf
Belgium
Germany
Agriculture
Rheinfall
Lake Constance
Basel
France
Black forest
Potash mines
Chemical plants
Alps

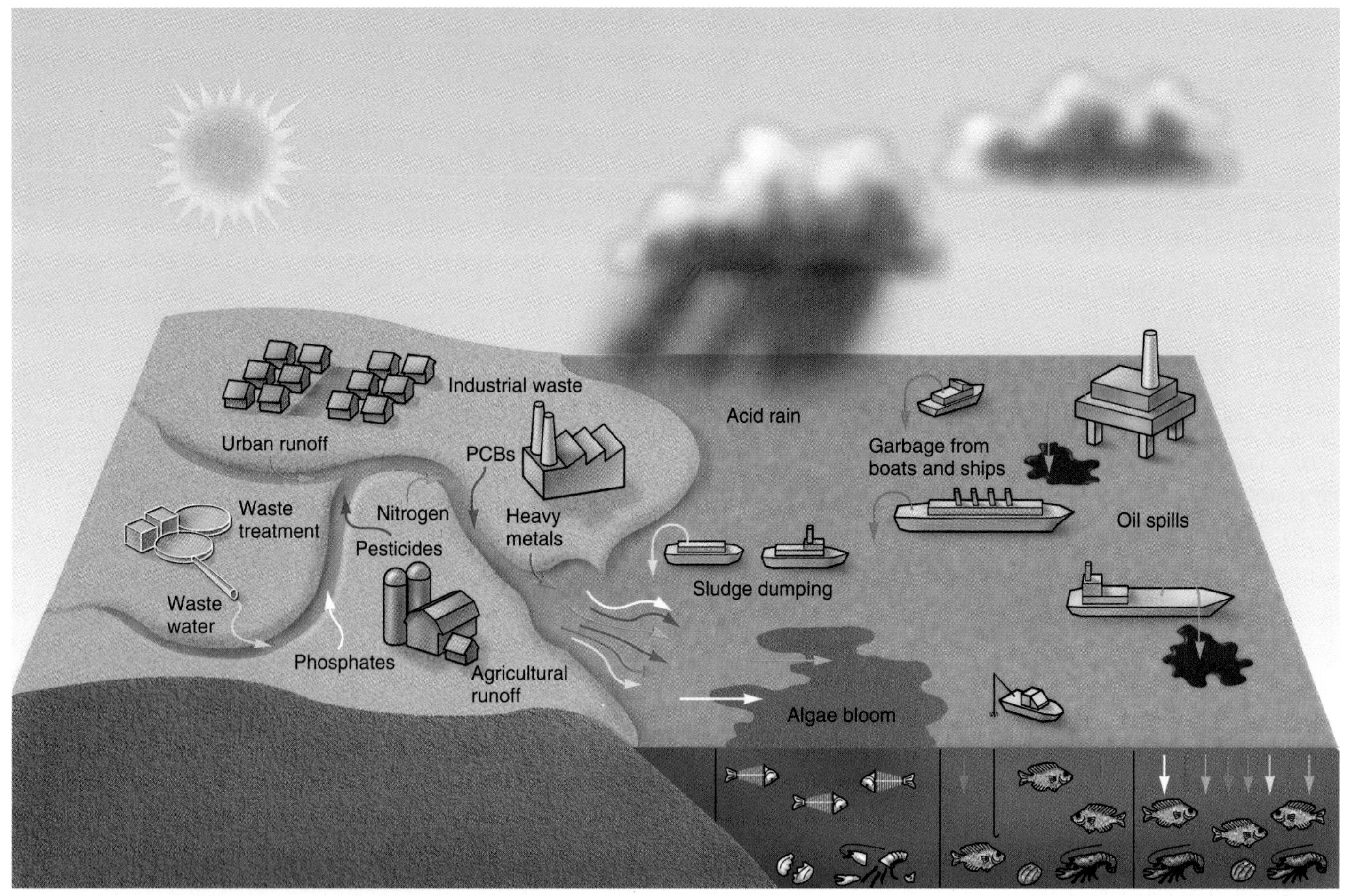
Industrial waste
Acid rain
Garbage from boats and ships
Urban runoff
PCBs
Waste treatment
Nitrogen
Heavy metals
Pesticides
Oil spills
Waste water
Sludge dumping
Phosphates
Agricultural runoff
Algae bloom
Many ocean pollution problems begin on the land.
Algal bloom blocks sunlight, removes oxygen, and produces toxins.
Fish that have eaten heavy metals, PCBs, and algal toxins are not fit for human consumption.
Fish and crustaceans develop sores and other symptoms from multiple environmental stresses.

FIGURE 44.9
Ocean pollution.

Table 44.2 Are the Oceans Trying to Tell Us Something?

Location	Event	Effect
Long Island Sound, Chesapeake Bay	Brown algal bloom, lack of oxygen	Many scallops dead
Raritan Bay, N.J.	"Dead zones" lack oxygen	One million fluke and flounder dead
N.J., N.Y., Mass., R.I.	Medical waste in water	Many beaches closed
East and West coasts	Toxins	Dolphins, fish, and crustaceans with holes and sores
North Sea	Polluted waters	Many dead seals
Baltic Sea	Polluted waters	Deformed seals
Oregon coast	Polluted waters	Salmon with tumors
Boston Harbor	0.5 billion gallons of sewage dumped daily	Shellfish with high levels of toxins
Florida Gulf coast	Toxic red tide	Florida manatee die-off ashore

44.3 biology in action

Adding Iron to the Ocean

Ecological research often takes investigators beyond the laboratory. This was the case for an experiment that probed lifeless patches of ocean.

The quest began with *observations:* Parts of the oceans have small amounts of phytoplankton, which shrinks food webs. Yet these areas are rich in major nutrients, such as nitrates and phosphates. Could too little of a trace element limit phytoplankton growth? Iron is a trace element required for photosynthesis and metabolism.

Next researchers formulated a *hypothesis:* If deficient iron limits phytoplankton population size, then adding iron to their environment will stimulate population growth.

Experiments took place in the laboratory and in the field.

Experiment #1: In a "shipboard nutrient bioassay," researchers grew phytoplankton in bottles and controlled the nutrients present. When they added iron, the phytoplankton population bloomed. But how closely does this experiment model the situation in the open ocean? More accurate experiments would take place in the ocean.

Experiment #2: In a project called "Iron Ex I," conducted in 1993, researchers added iron to a patch of ocean 800 miles (1,288 kilometers) west of the Galápagos Islands in the equatorial eastern Pacific. The experimental area covered 15,808 square acres (64 square kilometers). Unfortunately, the experiment ended early when the iron-enriched patch sunk.

Experiment #3: In 1996, "Iron Ex II" was more successful, because researchers added iron to a slightly larger patch of ocean in three doses. Biological changes were great and immediate—the growth rate of phytoplankton doubled, their numbers swelled 20-fold, and the nitrate level fell by 50% as photosynthetic rate soared. The diversity of the phytoplankton also shifted, with diatoms predominating. When the iron feeding stopped, the swell of life deflated.

In addition to explaining how lifeless ocean patches might form, Iron Ex II suggested a strategy to counter global warming. As the phytoplankton absorbed atmospheric CO_2 for photosynthesis, they released oxygen and an organic sulfur-containing compound (dimethyl sulfide) that helps them tolerate a salty environment (fig. 44.D). In the atmosphere, clouds form around dimethyl sulfide particles, and clouds lower temperature. Feeding the phytoplankton, then, simultaneously lowers CO_2 and temperature. But should we so drastically alter the environment?

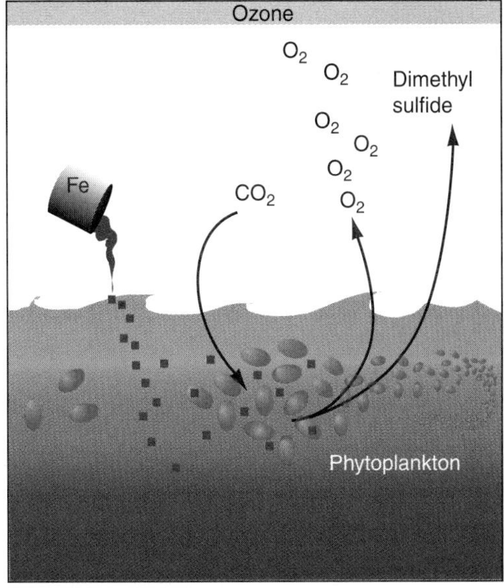

FIGURE 44.D
Adding iron to parts of the ocean transiently increases phytoplankton population size, which removes CO_2 from the atmosphere for photosynthesis, and releases oxygen (O_2) and dimethyl sulfide. Should we interfere with ocean ecosystems in this way to try to counter global warming?

marshes, wetlands, and estuaries. In salt water, chemical reactions encapsulate pollutants in particles, making them heavy enough to sink. Here they remain, unless disturbed by human action or extreme weather.

Once contaminated sediments are disturbed, a deadly chain reaction ensues. Nitrogen and phosphorus trigger algal bloom, which rapidly displaces other plants. Decomposing algae deplete oxygen and may produce toxins that cause paralytic shellfish poisoning, which ascends food webs and kills sea mammals. Table 44.2 identifies some geographic areas where ocean pollution has affected wildlife.

The top layer of the ocean supports vast webs of marine life, because this is where sunlight is strongest. This is also where petroleum-based pollutants concentrate, choking out life from above just as disturbed sediments kill from below. Biology in Action 44.3 describes another source of low biodiversity in some parts of the ocean—lack of iron.

Other forms of pollution with devastating effects on marine life are fishing nets and plastic. More than 2 million seabirds and 100,000 marine mammals, including 50,000 Alaskan fur seals, have been entrapped in net fragments or plastics (fig. 44.10). Some birds build their nests with plastic and feed their nestlings plastic bits. Sea turtles mistake floating blobs of plastic for their natural food, jellyfish, and swallow them. Plastic lodges in their intestines and kills them. The sea turtle, a very ancient animal that has weathered many extremes of oceanic existence, cannot adapt to plastic.

Like other biomes, oceans have natural defenses. Many toxins remain sequestered if undisturbed. The sheer volume of the

FIGURE 44.10

Animals can strangle when they become caught in discarded plastic straps and fishing net fragments.

oceans dilutes some pollutants. Microorganisms naturally degrade some pollutants. People are attempting to aid this natural healing. Cardboard devices are replacing plastic beverage carriers, ships are no longer supposed to dump waste at sea, cattle are being kept from rivers, and nearly everyone is more aware of what we discard and pour down the drain.

Mastering Concepts

1. What are some pollutants of aquatic ecosystems?
2. What types of changes adversely alter river ecosystems?
3. What changes occur with lake eutrophication?
4. Why is pollution in estuaries particularly dangerous?
5. How does pollution affect oceans?

The Land

Habitat destruction is perhaps most obvious on the land.

Deforestation

Tropical and temperate forests are disappearing. Tropical rain forests are shrinking in South America, Africa, and Southeast Asia. Once these areas are cleared to make room to plant crops, to grow trees for timber, or for animals to graze, nutrients in the ground wash away. Intense sunlight hardens some rain forest soils into a cementlike crust that cannot support plants. It takes at least 60 years for a forest to begin to regrow.

When trees die, food webs topple. Destroying the *Casearia corymbosa* tree in the tropical rain forest of Costa Rica, for ex-

FIGURE 44.11

Chimpanzees must forage farther for food in areas where "slash-and-burn" deforestation has destroyed the fruits it must eat to obtain enough water.

FIGURE 44.12

The light yellowish and pink areas in this satellite image indicate areas of deforestation in Brazil.

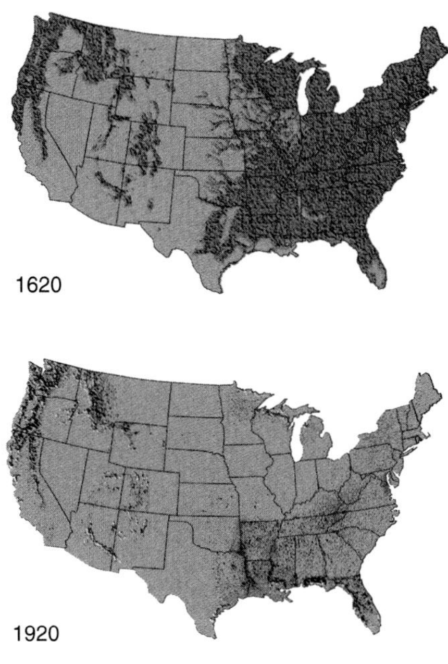

FIGURE 44.13

Deforestation in the United States. Dark areas represent forest.

ample, starves many of the 22 bird species that eat its fruits. Other trees whose seeds these birds disperse are then also threatened, as are other animals that eat the trees. Figure 44.11 shows a primate that depends upon juicy rain forest fruits to get enough water.

The current rate of forest destruction approaches 1% annually. Nearly half of the world's moist tropical forests have already been cleared. Satellite data indicate that from 1981 through 1990, tropical deforestation affected more than 7 million hectares (17.4 million acres) per year (fig. 44.12).

The shrinking tropical rain forests cap a long history of human interference. Settlement of the United States disrupted many natural ecosystems. In the first half of the eighteenth century, European settlers began clearing the land from east to west to create farmland and obtain fuel. By 1800, the lush Tidewater region of Virginia and the Carolinas had been cleared for rapidly expanding farms and towns. By 1840, the path of destruction had reached Louisiana and Arkansas.

The Industrial Revolution brought new machinery to cut down trees. Forests were sacrificed to build railroads and plantations, and soon cotton fields replaced the majestic magnolia and beech trees of the Old South. By 1880, virgin forest remained in less than 35% of the South. Still, a decade later, pioneers cut down trees in the southern Appalachians and Gulf coastal plain for timber and turpentine. Meanwhile, farms ruined in the Civil War showed new successional growth—nature's attempt to replace the forest. In the next century, the tanning industry in the Smoky Mountains claimed hemlock trees, and soybean fields replaced cypress trees and gum trees in what was left of the Mississippi valley. Today, although vast areas of managed pine forests and plantations occupy the region, less than 1% of the original temperate forests of the southeastern United States survives (fig. 44.13).

Desertification

Africa's Sahara is the driest and largest desert in the world. Its dust travels as far away as Florida. Along the Sahara's southern edge and in countries farther south, natural drought and human activities are reducing the productivity of the desert borderlands, a trend known as **desertification.** The most severely affected nations are those immediately south of the Sahara in an area called the Sahel.

The soil in the Sahel is so dry that seeds cannot germinate. Wells are dry. Starving cattle and goats have eaten the leaves on the few remaining trees, and horses have consumed the bark. Starving humans eat whatever they can find, including seeds that would in moister times produce the next year's crop. Birds, basking reptiles, and desert shrubs are rare, and some parts of the Sahel appear to be lifeless. These bleak conditions also prevail in Senegal in west Africa and in Ethiopia, Somalia, and Djibouti in east

(a)

(b)

FIGURE 44.14

An area of grassland with many species (*a*) better withstands a severe drought than a species-poor area (*b*).

Africa. Countries farther south, such as Kenya, Uganda, and Tanzania, are threatened as well. Botswana has had a drought for years, Zimbabwe has crop failures, and the people of Mozambique are malnourished.

The African drought began in 1968, following 10 years of plentiful rain. As grasses shriveled and trees died in the Sahel, farmers let their cattle browse farther to find food. The hills of Ethiopia were planted with crops, but they were not terraced, so whatever rain did come washed the plants away. Wars destroyed more life. Natural bodies of water were diverted to supply water to cities. Despite relief efforts, millions of people have starved to death in the African desert.

Drought is not unusual in this part of the world. The present drought is the third this century, and it has not endured as long as the drought from 1820 to 1840. Some ecologists fear that the present African drought is different because of its intensity and size. Although the current drought probably started naturally, short-sighted agricultural practices may be sustaining it, transforming what was once semidesert to the most lifeless land on the planet.

Desertification is not limited to Africa. Throughout the world, overgrazing, improper irrigation, unregulated vehicle activity, and other human impacts are degrading arid lands. In the United States, the Navajo-Hopi Indian Reservation in the Southwest is one of the world's largest areas of seriously desertified land.

Mastering Concepts

1. What events have led to tropical and temperate deforestation?
2. What factors promote desertification?

Biodiversity and Resilience

Genetic diversity in a population provides an insurance policy of sorts against disaster—should the environment change, individuals with certain genotypes are likely to survive. In ecosystems, species diversity provides some assurance of survival.

Biodiversity and Ecosystem Stability

The **diversity-stability hypothesis** holds that the more diverse and the greater the number of species in an ecosystem, the more it will be able to withstand a threat. Experiments conducted over an 11-year period in a Minnesota grassland support the hypothesis, but not all ecologists agree that the experiment accurately mirrors a natural ecosystem.

Grassland threats include fire, frost, hail, drought, insects, herbivores, and extreme temperatures. To assess an ecosystem's response to disaster, researchers measure the biomass of plants and catalog the number of species of a particular type of organism. Starting in 1982, researchers monitored 207 plots of land in the Minnesota grassland and determined plant biomass to measure ecosystem stability. They controlled the number of species by supplying certain amounts of nitrogen in fertilizer—the more nitrogen, the more plant species a particular plot supports.

After a severe drought in 1987–88, the researchers measured the plant biomass in the plots (fig. 44.14). Plots with the most species lost half their biomass, but plots with the fewest species lost seven-eighths of their biomass! The conclusion: In threatened situations, biodiversity provides stability. Under good environmental conditions, however, a species-poor area will also produce a large biomass.

The conclusion that species number alone ensures ecosystem survival is an oversimplification. Some ecologists suggest that an ecosystem remains stable not because of a large number of species but because of their diverse roles. A forest of plants of very different heights, for example, might withstand fire better than a forest of trees of similar height, even if both areas include the same number of species.

Another oversimplification of equating biodiversity with ecosystem stability is that species do not equally influence composition of an ecosystem. For example, moose on Isle Royale in Michigan control the vegetation (see fig. 42.7). In the same ecosystem, removing a beetle or fern species might not have as great an effect.

The Resiliency of Life

Life on earth has had many millions of years to adapt, diversify, and occupy nearly every part of the planet's surface. It would be very difficult to halt life on earth completely, short of a global catastrophe such as a meteor collision or a nuclear holocaust. The biosphere has survived mass extinctions in the past, and it probably will in the future.

Life has prevailed through many localized challenges—from natural events such as hurricanes, volcanic eruptions, and forest fires to human-caused garbage heaps, oil spills, and pollution. Organisms may perish—many of them—but the plasticity of genetic material ensures that, in most cases, some individuals will inherit what in one environment is a quirk but in another is the key to survival. Even when biodiversity is severely challenged, some species usually persist. Life does not completely end; it changes. Our own species will very likely someday be extinct.

Can we continue to rely on adaptation and diversity to maintain life on earth? Life may be resilient in an overall sense, yet it is fragile in its interrelationships. Intricate food webs easily collapse if a single member species declines or vanishes—such as the Antarctic krill or the ocean's phytoplankton that feed so many other species.

Maybe we have just been lucky, so far, that no single event has decimated so many key players in the game of life that other species could not replace them. But could multiple stresses combine to make the earth too inhospitable for life as we know it to continue? Will iron-poor, lifeless patches of oceans grow and coalesce, as ozone-depleted areas over the poles and stretches of dead desert do? A single eutrophied lake, an acid precipitation–ravaged forest, an oil-soaked shore, a cloud of radiation—alone they are terrible; together they could begin to unravel the tangled threads that tie all life together.

The answers to these compelling questions may lie within you and how you choose to live. This book has shown you the wonder of life, from its constituent chemicals to its cells, tissues, and organs, all the way up to the biosphere. Do nothing to harm life—and do whatever you can to preserve its precious diversity. For in diversity lies resiliency and the future of life on earth.

Mastering Concepts

1. How does biodiversity protect an ecosystem?
2. Why is the idea that biodiversity ensures ecosystem stability an oversimplification?
3. How does life persist despite environmental upheavals?

Moose as Environmental Barometers~

continued from page 890.

Moose eat a variety of plants—blueberries are a favorite. When factories and power plants throughout Europe began releasing large amounts of sulfur dioxide in the early 1980s, acid rain and snow fell on Sweden and killed the blueberry plants. Moose adapted their diets to the changing vegetation by frequenting farmers' fields.

The acid precipitation also altered the soil. As soil pH fell, certain heavy metals, such as zinc, manganese, and cadmium, became more available to animals grazing on the upper soil layers. At the same time, molybdenum became less available. But what harmed the moose, according to Adrian Frank's hypothesis, is not the condition set up by the acid precipitation but human response to the problem.

In the early 1980s, people began spreading lime on pastures, fields, lakes, and wetlands to raise soil pH. This, in turn, makes molybdenum less available and heavy metals more available to grazers. The liming, Frank contends, has added too much molybdenum to the mooses' diet and depleted copper, cadmium, manganese, zinc, and magnesium.

Frank, whose hypothesis is not accepted by all, urges that governments planning liming projects to counter acid precipitation first investigate effects on soil and the bioavailability of elements in the affected ecosystems. He also suggests yearly monitoring of the metal content of moose kidneys to track the problem. He concludes, "Because the moose is found in many areas of the Northern Hemisphere, it has the potential to serve as a monitor for toxic elements and environmental changes."

SUMMARY

1. Effects of human activities on the environment are difficult to predict because many complex factors interact.

2. Environmentalism is the philosophy of living in harmony with the global ecosystem. Environmental problems can affect the entire biosphere.

3. Air pollutants include heavy metals, fossil fuel emissions, **photochemical smog,** and fine particulates.

4. Acid precipitation forms when sulfur and nitrogen oxides from pollution react with water in the upper atmosphere to form hydrochloric and sulfuric acids. Acidification of lakes harms many organisms and changes aquatic communities. In coniferous forests, acid precipitation kills leaves and acidified soil releases aluminum, which robs roots of essential calcium and magnesium and stunts growth, as excess nitrogen stimulates trees to grow. The conflicting signals stress the trees.

5. Use of **chlorofluorocarbon compounds** has thinned the **ozone** layer, which protects life from UV radiation.

6. The **greenhouse effect** results from CO_2 and other gases trapping infrared radiation near the earth's surface. Agriculture, industry, burning fossil fuels, and destruction of tropical rain forests generate greenhouse gases. The greenhouse effect may contribute to global warming.

7. Pollutants of aquatic ecosystems include **polychlorinated biphenyls,** polycyclic aromatic hydrocarbons, and heavy metals. Erecting dams and levees and introducing nonnative species alters river ecosystems. Deposition of sediments and nutrients from development is eutrophying Lake Tahoe. Preserving estuaries is important because they are breeding grounds for many species. Heavy metals and other industrial wastes contaminate ocean sediments and foul beaches. Nutrient discharges into coastal waters stimulate phytoplankton blooms that produce toxins. Plastic and oil contaminate surface waters.

8. Poor agricultural practices are destroying tropical and temperate forests. **Desertification** reflects natural drought as well as poor agricultural practices.

9. The **diversity-stability hypothesis** states that an ecosystem with many diverse species is better able to survive a threat. The role of species in the ecosystem is important in ecosystem stability.

TO REVIEW . . .

1. What effects can each of these events have on an ecosystem?
- **a.** a major oil spill
- **b.** acid precipitation
- **c.** the greenhouse effect
- **d.** thinning of the ozone layer
- **e.** damming a river
- **f.** polluting an estuary

2. What natural protections do ecosystems have against the destructive effects of each of the following?
- **a.** acid precipitation
- **b.** eutrophication of lakes
- **c.** global warming from the greenhouse effect
- **d.** ocean pollution

3. Discuss two ways that pollutants can persist in an ecosystem even after their use ceases.

4. Describe an experiment mentioned in the chapter that provided information about an environmental problem.

5. How can too much of a nutrient alter an ecosystem?

6. Give an example of an environmental problem that had an immediate effect and one that had a delayed effect.

7. What are the sources of the following problems?
- **a.** acid precipitation
- **b.** the greenhouse effect
- **c.** pollution in the Rhine River

8. Why is the idea that all an ecosystem requires to survive environmental change is many species an oversimplification?

TO THINK ABOUT . . .

1. An oil spill pollutes a mile-long stretch of golden beach and clear water. What approaches could environmental scientists take to clean the area, based on what we learned from the *Exxon Valdez* experience?

2. Should we attempt to save populations threatened by an environmental disaster? Give an example in which a well-meaning attempt to save lives had unexpected results.

3. What is a limitation of trying to extrapolate results from experimental ecosystems to natural ecosystems?

4. What might the risks and benefits be of adding iron to apparently lifeless patches of ocean in order to counter global warming?

5. How have human interventions and/or natural events contributed to the following?
- **a.** depletion of the ozone layer
- **b.** global warming
- **c.** desertification
- **d.** deforestation

TO LEARN MORE . . .

References and Resources*

Baskin, Yvonne. February 1997. The work of nature. *Natural History.* The importance of number of species in an ecosystem is not well understood.

Beland, Pierre. May 1996. The beluga whales of the St. Lawrence River. *Scientific American.* Beluga whales store toxins in their fat, and mothers pass biomagnified toxins to their nursing young.

Frost, Bruce W. October 10, 1996. Phytoplankton bloom on iron rations. *Nature,* vol. 383. Adding iron to the ocean may counter global warming—but should we do it?

Hedin, Lars O., and Gene E. Likens. December 1996. Atmospheric dust and acid rain. *Scientific American.* Dust neutralizes acid rain.

Holloway, Marguerite. October 1996. Sounding out science. *Scientific American.* Legal battles over the *Exxon Valdez* oil spill hampered scientific efforts to clean up and learn from the disaster.

Kahn, Patricia. January 22, 1993. A grisly archive of key cancer data. *Science,* vol. 259. A zoologist from the former East Germany released data on exposure of weapons workers to uranium.

Malle, Karl-Geert. January 1996. Cleaning up the river Rhine. *Scientific American.* The river Rhine snakes along five European nations, and picks up many types of pollutants.

**Life*'s website features frequently updated references.

Units of Measurement Metric/English Conversions

Appendix A

Length

1 meter = 39.4 inches = 3.28 feet = 1.09 yards

1 foot = 0.305 meters = 12 inches = 0.33 yard

1 inch = 2.54 centimeters

1 centimeter = 10 millimeters = 0.394 inch

1 millimeter = 0.001 meter = 0.01 centimeter = 0.039 inch

1 fathom = 6 feet = 1.83 meters

1 rod = 16.5 feet = 5 meters

1 chain = 4 rods = 66 feet = 20 meters

1 furlong = 10 chains = 40 rods = 660 feet = 200 meters

1 kilometer = 1,000 meters = 0.621 mile = 0.54 nautical mile

1 mile = 5,280 feet = 8 furlongs = 1.61 kilometers

1 nautical mile = 1.15 miles

Area

1 square centimeter = 0.155 square inch

1 square foot = 144 square inches = 929 square centimeters

1 square yard = 9 square feet = 0.836 square meter

1 square meter = 10.76 square feet = 1.196 square yards = 1 million square millimeters

1 hectare = 10,000 square meters = 0.01 square kilometer = 2.47 acres

1 acre = 43,560 square feet = 0.405 hectare

1 square kilometer = 100 hectares = 1 million square meters = 0.386 square mile = 247 acres

1 square mile = 640 acres = 2.59 square kilometers

Volume

1 cubic centimeter = 1 milliliter = 0.001 liter

1 cubic meter = 1 million cubic centimeters = 1,000 liters

1 cubic meter = 35.3 cubic feet = 1.307 cubic yards = 264 U.S. gallons

1 cubic yard = 27 cubic feet = 0.765 cubic meter = 202 U.S. gallons

1 cubic kilometer = 1 million cubic meters = 0.24 cubic mile = 264 billion gallons

1 cubic mile = 4.166 cubic kilometers

1 liter = 1,000 milliliters = 1.06 quarts = 0.265 U.S. gallon = 0.035 cubic foot

1 U.S. gallon = 4 quarts = 3.79 liters = 231 cubic inches = 0.83 imperial (British) gallon

1 quart = 2 pints = 4 cups = 0.94 liter

1 acre foot = 325,851 U.S. gallons = 1,234,975 liters = 1,234 cubic meters

1 barrel (of oil) = 42 U.S. gallons = 159 liters

Mass

1 microgram = 0.001 milligram = 0.000001 gram

1 gram = 1,000 milligrams = 0.035 ounce

1 kilogram = 1,000 grams = 2.205 pounds

1 pound = 16 ounces = 454 grams

1 short ton = 2,000 pounds = 909 kilograms

1 metric ton = 1,000 kilograms = 2,200 pounds

Temperature

Celsius to Fahrenheit
$°F = (°C \times 1.8) + 32$

Fahrenheit to Celsius
$°C = (°F - 32) \div 1.8$

Energy and Power

1 erg = 1 dyne per square centimeter

1 joule = 10 million ergs

1 calorie = 4.184 joules

1 kilojoule = 1,000 joules = 0.949 British Thermal Units (BTU)

1 kilocalorie = 1,000 calories = 3.97 BTU = 0.00116 kilowatt-hour

1 BTU = 0.293 watt-hour

1 kilowatt-hour = 1,000 watt-hour = 860 kilocalories = 3,400 BTU

1 horsepower = 640 kilocalories

1 quad = 1 quadrillion kilojoules = 2.93 trillion kilowatt-hours

Metric Conversion

Appendix B

	Metric Quantities	Metric to English Conversion	English to Metric Conversion
Length	1 kilometer (km) = 1,000 (10^3) meters 1 meter (m) = 100 centimeters 1 centimeter (cm) = 0.01 (10^{-2}) meter 1 millimeter (mm) = 0.001(10^{-3}) meter 1 micrometer*(μm) = 0.000001 (10^{-6}) meter 1 nanometer (nm) = 0.000000001 (10^{-9}) meter *formerly called micron	1 km = 0.62 mile 1 m = 1.09 yards = 39.37 inches 1 cm = 0.394 inch 1 mm = 0.039 inch	1 mile = 1.609 km 1 yard = 0.914 m 1 foot = 0.305 m = 30.5 cm 1 inch = 2.54 cm
Area	1 square kilometer (km^2) = 100 hectares 1 hectare (ha) = 10,000 square meters 1 square meter (m^2) = 10,000 square centimeters 1 square centimeter (cm^2) = 100 square millimeters	1 km^2 = 0.3861 square mile 1 ha = 2.471 acres 1 m^2 = 1.1960 square yards = 10.764 square feet 1 cm^2 = 0.155 square inch	1 square mile = 2.590 km^2 1 acre = 0.4047 ha 1 square yard = 0.8361 m^2 1 square foot = 0.0929 m^2 1 square inch = 6.4516 cm^2
Mass	1 metric ton (t) = 1,000 kilograms 1 metric ton (t) = 1,000,000 grams 1 kilogram (kg) = 1,000 grams 1 gram (g) = 1,000 milligrams 1 milligram (mg) = 0.001 gram 1 microgram (μg) = 0.000001 gram	1 t = 1.1025 ton (U.S.) 1 kg = 2.205 pounds 1 g = 0.0353 ounce	1 ton (U.S.) = 0.907 t 1 pound = 0.4536 kg 1 ounce = 28.35 g
Volume (Solids)	1 cubic meter (m^3) = 1,000,000 cubic centimeters 1 cubic centimeter (cm^3) = 1,000 cubic millimeters	1 m^3 = 1.3080 cubic yards = 35.315 cubic feet 1 cm^3 = 0.0610 cubic inch	1 cubic yard = 0.7646 m^3 1 cubic foot = 0.0283 m^3 1 cubic inch = 16.387 cm^3
Volume (Liquids)	1 liter (l) = 1,000 milliliters 1 milliliter (ml) = 0.001 liter 1 microliter (μl) = 0.000001 liter	1 l = 1.06 quarts (U.S.) 1 ml = 0.034 fluid ounce	1 quart (U.S.) = 0.94 l 1 pint (U.S.) = 0.47 l 1 fluid ounce = 29.57 ml
Time	1 second (sec) = 1,000 milliseconds 1 millisecond (msec) = 0.001 second 1 microsecond (μsec) = 0.000001 second		

Periodic Table of Elements

Appendix C

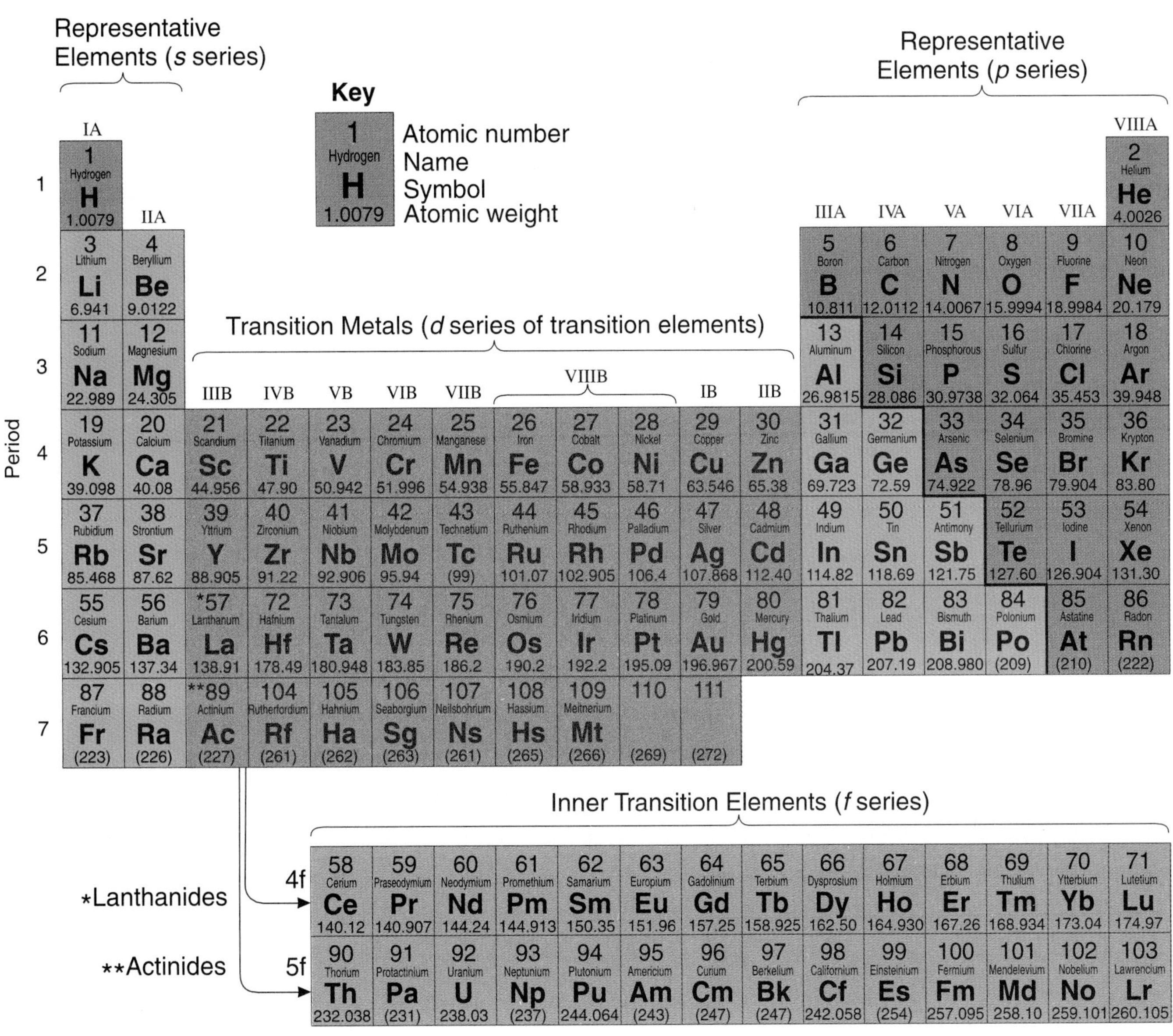

Period	IA	IIA	IIIB	IVB	VB	VIB	VIIB	VIIIB	VIIIB	VIIIB	IB	IIB	IIIA	IVA	VA	VIA	VIIA	VIIIA
1	1 Hydrogen H 1.0079																	2 Helium He 4.0026
2	3 Lithium Li 6.941	4 Beryllium Be 9.0122											5 Boron B 10.811	6 Carbon C 12.0112	7 Nitrogen N 14.0067	8 Oxygen O 15.9994	9 Fluorine F 18.9984	10 Neon Ne 20.179
3	11 Sodium Na 22.989	12 Magnesium Mg 24.305											13 Aluminum Al 26.9815	14 Silicon Si 28.086	15 Phosphorous P 30.9738	16 Sulfur S 32.064	17 Chlorine Cl 35.453	18 Argon Ar 39.948
4	19 Potassium K 39.098	20 Calcium Ca 40.08	21 Scandium Sc 44.956	22 Titanium Ti 47.90	23 Vanadium V 50.942	24 Chromium Cr 51.996	25 Manganese Mn 54.938	26 Iron Fe 55.847	27 Cobalt Co 58.933	28 Nickel Ni 58.71	29 Copper Cu 63.546	30 Zinc Zn 65.38	31 Gallium Ga 69.723	32 Germanium Ge 72.59	33 Arsenic As 74.922	34 Selenium Se 78.96	35 Bromine Br 79.904	36 Krypton Kr 83.80
5	37 Rubidium Rb 85.468	38 Strontium Sr 87.62	39 Yttrium Y 88.905	40 Zirconium Zr 91.22	41 Niobium Nb 92.906	42 Molybdenum Mo 95.94	43 Technetium Tc (99)	44 Ruthenium Ru 101.07	45 Rhodium Rh 102.905	46 Palladium Pd 106.4	47 Silver Ag 107.868	48 Cadmium Cd 112.40	49 Indium In 114.82	50 Tin Sn 118.69	51 Antimony Sb 121.75	52 Tellurium Te 127.60	53 Iodine I 126.904	54 Xenon Xe 131.30
6	55 Cesium Cs 132.905	56 Barium Ba 137.34	*57 Lanthanum La 138.91	72 Hafnium Hf 178.49	73 Tantalum Ta 180.948	74 Tungsten W 183.85	75 Rhenium Re 186.2	76 Osmium Os 190.2	77 Iridium Ir 192.2	78 Platinum Pt 195.09	79 Gold Au 196.967	80 Mercury Hg 200.59	81 Thallium Tl 204.37	82 Lead Pb 207.19	83 Bismuth Bi 208.980	84 Polonium Po (209)	85 Astatine At (210)	86 Radon Rn (222)
7	87 Francium Fr (223)	88 Radium Ra (226)	**89 Actinium Ac (227)	104 Rutherfordium Rf (261)	105 Hahnium Ha (262)	106 Seaborgium Sg (263)	107 Neilsbohrium Ns (261)	108 Hassium Hs (265)	109 Meitnerium Mt (266)	110 (269)	111 (272)							

*Lanthanides 4f	58 Cerium Ce 140.12	59 Praseodymium Pr 140.907	60 Neodymium Nd 144.24	61 Promethium Pm 144.913	62 Samarium Sm 150.35	63 Europium Eu 151.96	64 Gadolinium Gd 157.25	65 Terbium Tb 158.925	66 Dysprosium Dy 162.50	67 Holmium Ho 164.930	68 Erbium Er 167.26	69 Thulium Tm 168.934	70 Ytterbium Yb 173.04	71 Lutetium Lu 174.97
**Actinides 5f	90 Thorium Th 232.038	91 Protactinium Pa (231)	92 Uranium U 238.03	93 Neptunium Np (237)	94 Plutonium Pu 244.064	95 Americium Am (243)	96 Curium Cm (247)	97 Berkelium Bk (247)	98 Californium Cf 242.058	99 Einsteinium Es (254)	100 Fermium Fm 257.095	101 Mendelevium Md 258.10	102 Nobelium No 259.101	103 Lawrencium Lr 260.105

Tree of Life

Appendix D

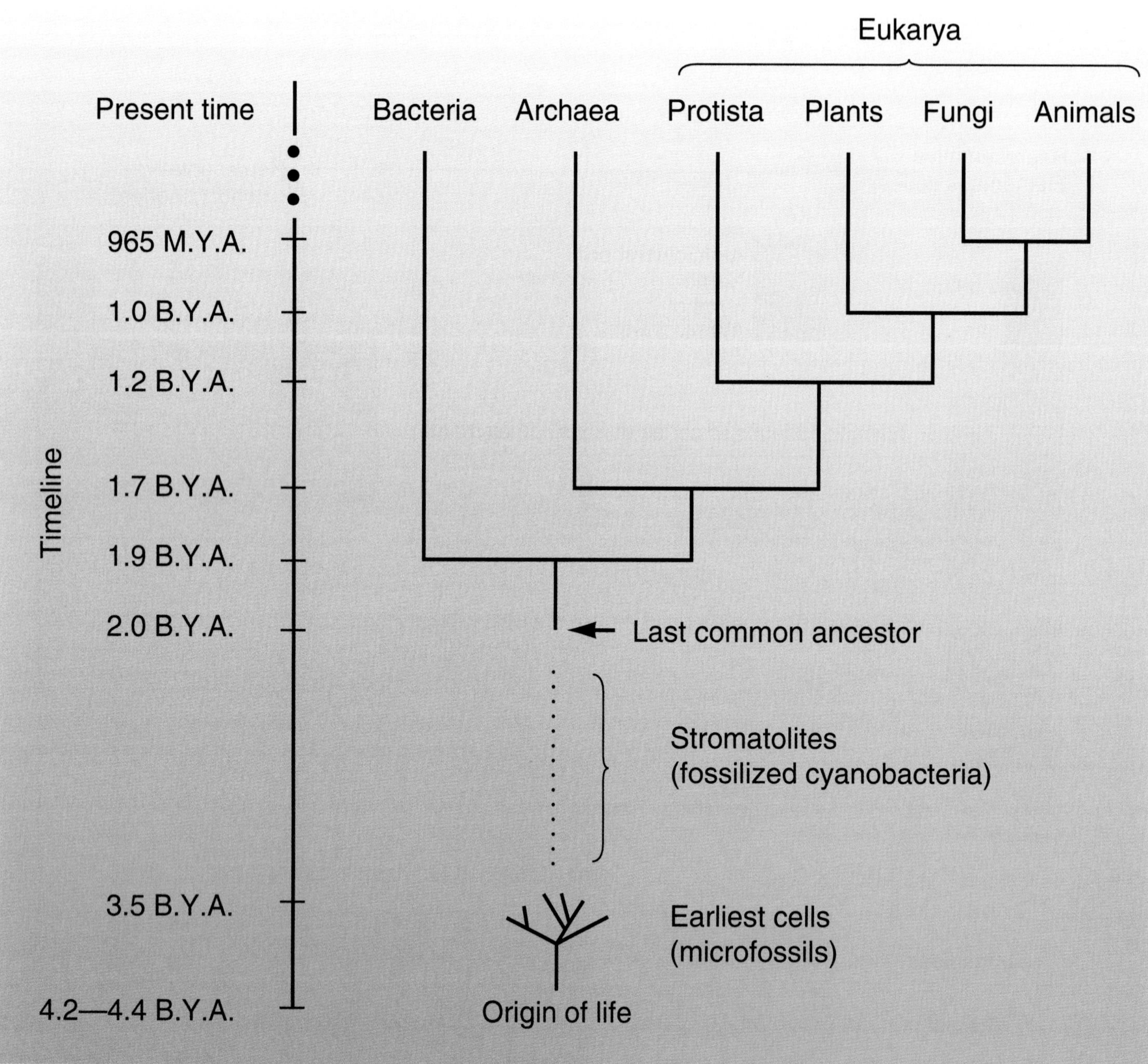

A tree of life showing possible evolutionary relationships among the major groups of organisms. Life scientists recognize three domains (Bacteria, Archaea, and Eukarya) and either five or six kingdoms. The six-kingdom scheme counts Archaea as a kingdom. The time scale is based on fossil evidence, and on evidence from protein amino acid sequence differences among modern species. The precise evolutionary relationship between Bacteria, Archaea, and Eukarya is still unclear. Based on molecular evidence, the Archaea and Eukarya are more closely related to each other than either is to the Bacteria.

Taxonomy

Appendix E

Kingdom Animalia

Kingdom Plantae

Kingdom Fungi

Kingdom Monera

Kingdom Protista

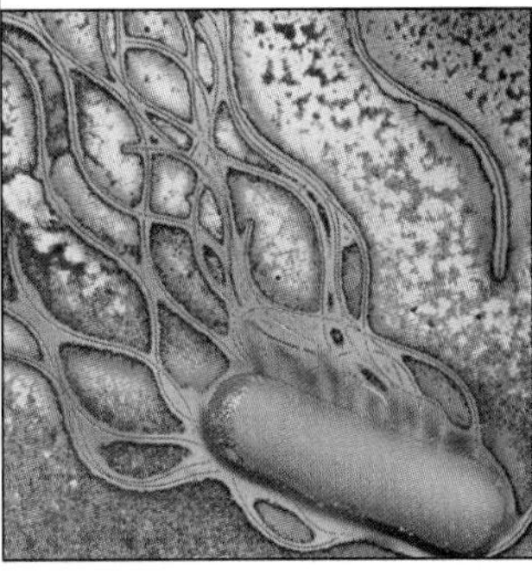

We can group the millions of living and extinct species that have dwelled on earth according to many schemes. Taxonomists group organisms to reflect both anatomical similarities and descent from a common ancestor. Two-, three-, four-, five-, and most recently, six-kingdom classifications have been proposed, as well as a system of three "domains" that supercede kingdoms. The five-kingdom scheme is outlined here with all phyla briefly described. Short statements explaining the rationale behind phyla groupings are given wherever possible, and indented subheadings reflect these groupings. Scientific names are followed by more familiar names of organisms. Figures from the text accompany the listing to help you visualize the wide diversity of life. Unit 6 visits the kingdoms in detail.

Kingdom Monera The monera are unicellular prokaryotes that obtain nutrients by direct absorption, by photosynthesis, or by chemosynthesis. Most monera reproduce asexually, but some can exchange genetic material in a primitive form of sexual reproduction. The six-kingdom classification system divides the monera into two kingdoms, bacteria and archaea. They differ in certain key molecular structures and environmental requirements, though neither has nuclei or organelles.

Phylum Bacteria The bacteria and cyanobacteria.

Phylum Archaea Unicellular organisms lacking nuclei and organelles that share biochemical characteristics with some eukaryotes.

Kingdom Protista Protista are the structurally simplest eukaryotes, and they can be unicellular or multicellular. They can absorb or ingest nutrients or photosynthesize. Reproduction is asexual or sexual. Some forms move by ciliary or flagellar motion, and others are nonmotile. The protists' early development differs from that of the fungi, plants, and animals. The kingdom includes the protozoa, algae, and the water molds and slime molds.

Protozoa Unicellular, nonphotosynthetic, lack cell walls.

Phylum Sarcomastigophora Locomote by flagella and/or pseudopoda and include the familiar *Amoeba proteus.*

Phylum Labyrinthomorpha Aquatic, live on algae.

Phylum Apicomplexa Parasitic, with characteristic twisted structure on anterior end at some point in life cycle.

Phylum Myxozoa Parasitic on fish and invertebrates.

Phylum Microspora Parasitic on invertebrates and primitive vertebrates.

Phylum Ciliophora The ciliates have cilia at some stage of the life cycle.

Algae Unicellular or multicellular, photosynthetic, some have cell walls. Distinguished by pigments.

Phylum Euglenophyta Unicellular and photosynthetic, with a single flagellum and contractile vacuole.

Phylum Chrysophyta Diatoms, golden-brown algae, and yellow-green algae. Unicellular and photosynthetic.

Phylum Pyrrophyta Dinoflagellates. Unicellular and photosynthetic.

Phylum Chlorophyta Green algae. Unicellular or multicellular, photosynthetic.

Phylum Phaeophyta Brown algae (kelps). Multicellular and photosynthetic.

Phylum Rhodophyta Red algae. Multicellular and photosynthetic.

Water and Slime Molds

Phylum Oomycota The water molds. Unicellular or multinucleate, with cellulose cell walls. Live in fresh water.

Phylum Chytridiomycota The chytrids. Multicellular, with chitinous cell walls. Aquatic.

Phylum Myxomycota Multinucleated, "acellular" slime molds.

Phylum Acrasiomycota Multicellular "cellular" slime molds.

Kingdom Fungi

With the exception of the yeasts, fungi are multicellular eukaryotes that decompose organisms to obtain nourishment. Fungal cells have chitinous cell walls. Phyla are distinguished by their mode of reproduction.

Division Zygomycota Reproduce with sexual resting spores.

Division Ascomycota Yeasts, morels, truffles, molds, lichens. Reproduce with sexual spores carried in asci.

Division Basidiomycota Mushrooms, toadstools, puffballs, stinkhorns, shelf fungi, rusts, and smuts. Reproduce by spore-containing basidia.

Division Deuteromycetes Mechanism of reproduction unknown.

Kingdom Plantae

Plants are multicellular, land dwelling, photosynthetic, and reproduce both asexually and sexually in an alternation of generations. Their cells have cellulose cell walls. Plants have specialized tissues and organs but lack nervous and muscular systems.

Nonvascular Plants (Bryophytes) Lack specialized conducting tissues and true roots, stems, and leaves. The gamete-producing reproductive phase predominates.

Division Bryophyta Liverworts, hornworts, mosses.

Vascular Plants (Tracheophytes) Xylem and phloem transport water and nutrients, respectively, throughout the plant body of roots, stems and leaves. The spore-producing reproductive phase predominates.

Primitive Plants Sperm cells travel in water to meet egg cells.

Division Pterophyta Ferns.

Division Psilophyta Whisk ferns.

Division Lycophyta Club mosses and others.

Division Spenophyta Horsetails.

Seed Plants Sperm cells and egg cells enclosed in protective structures.

Gymnosperms (naked seed plants) Male and female cones produce pollen grains and ovules.

Division Coniferophyta Conifers.

Division Cycadophyta Cycads.

Division Ginkgophyta Ginkgos.

Division Gnetophyta Gnetophytes.

Angiosperms (seeds in a vessel) The flowering plants.

Division Anthophyta Flowering plants.

Kingdom Animalia

The animals are multicellular with specialized tissues and organs, including nervous and locomotive systems. Cells lack cell walls. Animals obtain nutrients from food. Phyla are distinguished largely on the basis of body form and symmetry, characteristics generally established in the early embryo.

Mesozoa Simplest animals.

Phylum Mesozoa Very simple, wormlike parasites of marine invertebrates. Consist of only 20 to 30 cells.

Parazoa A separate branch from the evolution of protozoa to metazoa.

Phylum Placozoa A single species, *Trichoplax adhaerens,* characterized by two cell layers with fluid in between.

Phylum Porifera The sponges. Specialized cell types organized into canal system to transport nutrients in and wastes out.

Eumetazoa Animal phyla descended from protozoa.

Radiata Radially symmetric body plan. Sedentary, saclike bodies with two or three cell layers and a diffuse nerve net.

Phylum Cnidaria Hydroids, sea anemones, jellyfish, horny corals, hard corals.

Phylum Ctenophora Sea walnuts, comb jellies.

Bilateria Bilaterally symmetric body plan.

Protostomia (first mouth) Embryonic characteristics:

1. Mouth forms close to area of initial inward folding in very early embryo.
2. Spiral cleavage: At third cell division, second group of four cells sits atop first group of four cells, rotated 45°.
3. Determinate cleavage: Cell fates determined very early in development. If a cell from a four-celled embryo is isolated, it will divide and differentiate to form only one-quarter of an embryo.
4. The protostomes are further grouped by the way in which the body cavity (coelom) forms. A true coelom is a body cavity that develops within mesoderm, the middle layer of the embryo.

Acoelomates No coelom.

Phylum Platyhelminthes Flatworms.

Phylum Nemertina Ribbonworms.

Phylum Gnathostomulida Jawworms.

Pseudocoelomates Body cavity derived from a space in the embryo between the mesoderm and endoderm. The body cavity is called "pseudo" because it does not form within mesoderm. In the adult, the pseudocoelom is a cavity but it is not lined with mesoderm-derived peritoneum (seen in more complex forms).

Phylum Rotifera The rotifers. Small (40 μm–3 mm), intricately shaped organisms that have a structure on their anterior ends that resembles rotating wheels. The rotifers occupy a variety of habitats.

Phylum Gastrotricha Aquatic, microscopic, flattened organisms with a scaly outer covering.

Phylum Kinorhyncha Marine worms less than 1 mm long.

Phylum Nematoda Roundworms. Found everywhere, many parasitic.

Phylum Nematomorpha Horsehair worms. Juveniles are parasitic in arthropods; adults are free-living.

Phylum Acanthocephala Spiny-headed worms. Spiny projection from anterior end used to attach to intestine of host vertebrate. Range in size from 2 mm to more than a meter.

Phylum Entoprocta Nonmotile, sessile, mostly marine animals that look like stalks anchored to rocks, shells, algae, or vegetation on one end, with tufted growth on the other.

Eucoelomates Coelom forms in a schizocoelous fashion, in which the body cavity forms when mesodermal cells invade the space between ectoderm and endoderm, and then proliferate so that a cavity forms within the mesoderm.

Major Eucoelomate Protostomes Three phyla, with many species.

Phylum Mollusca Snails, clams, oysters, squids, octopuses.

Phylum Annelida Segmented worms.

Phylum Arthropoda Spiders, scorpions, ticks, mites, crustaceans, millipedes, centipedes, insects.

The Lesser Protostomes Seven phyla, including many extinct species. Little-understood offshoots of annelid-arthropod line.

Phylum Pripulida Bottom-dwelling marine worms.

Phylum Echiurida Marine worms.

Phylum Sipunculida Bottom-dwelling marine worms.

Phylum Tardigrada "Water bears." Less than 1 mm, live in water film on mosses and lichens.

Phylum Pentastomida Tongue worms. Parasitic on respiratory system of vertebrates, mostly reptiles.

Phylum Onychophora Velvet worms. Live in tropical rain forest and resemble caterpillars, with 14 to 43 pairs of unjointed legs and a velvety skin.

Phylum Pogonophora Beard worms. Live in mud on ocean bottom.

Lophophorates Three phyla distinguished by a ciliary feeding structure called a lophophore.

Phylum Phoronida Small, wormlike bottom-dwellers of shallow, coastal temperate seas. Live in a tube that they secrete.

Phylum Ectoprocta Bryozoa, or "moss animals." Aquatic, less than 1/2 mm long, live in colonies but each individual lives within a chamber secreted by the epidermis. Bryozoa look like crust on rocks, shells, and seaweeds.

Phylum Brachiopoda Lampshells. Attached, bottom-dwelling marine animals that have two shells and resemble mollusks, about 5 to 8 mm long.

Deuterostomia (second mouth)
Embryonic characteristics:

1. Mouth forms far from the area of initial inward folding in very early embryo.
2. Radial cleavage: At third cell division, second group of four cells sits directly atop first group.
3. Indeterminate cleavage: Cell fates of very early embryo not determined. If a cell from a four-cell embryo is isolated, it will develop into a complete embryo.
4. Coelom formation is enterocoelous. Body cavity forms from outpouchings of endoderm that become lined with mesoderm.

Phylum Echinodermata Sea stars, brittle stars, sea urchins, sea cucumbers, sea lilies. Radial symmetry in adult but larvae are bilaterally symmetric. Complex organ systems, but no distinct head region.

Phylum Chaetognatha Arrow worms. Marine-dwelling with bristles surrounding mouth.

Phylum Hemichordata Acorn worms and others. Aquatic, bottom-dwelling, nonmotile, wormlike animals.

Phylum Chordata Tunicates, lancelets, hagfishes, lampreys, sharks, bony fishes, amphibians, reptiles, birds, mammals. Chordates have a notochord, dorsal nerve cord, gill slits, and a tail. Some of these characteristics may only be present in embryos.

Glossary

A

aboral end *AB-or-al end* The end of an animal's body opposite its mouth. 511

abscisic acid *ab-SIS-ik AS-id* A plant hormone that inhibits growth. 578

abscission zone *ab-SCISZ-on ZONE* A region at the base of the petiole from which leaves shed. 546

abyssal zone *a-BIS-al ZONE* Part of the ocean that light does not reach. 883

accessory pigment Plant pigment other than chlorophyll that extends the range of light wavelengths useful in photosynthesis. 150

acetyl CoA formation *AS-eh-til FOR-MAY-shun* The first step in aerobic respiration. In the mitochondrion, pyruvic acid loses a carbon dioxide and bonds to coenzyme A to form acetyl CoA. 127

acid A molecule that releases hydrogen ions into water. 40

acid precipitation Low pH precipitation caused by sulfur and nitrogen oxides that form when pollution reacts with water in the upper atmosphere, forming hydrochloric and sulfuric acids. 894

acoelomate *a-SEE-lo-mayt* Animals that lack a coelom (body cavity) 511.

acrocentric *ak-ro-SEN-trik* A chromosome whose centromere divides the chromosome into a long arm and a short arm. 296

acrosome *AK-ro-som* A protrusion on the anterior end of a sperm cell containing digestive enzymes that enable the sperm to penetrate layers around the oocyte. 204

actin *AK-tin* A type of protein in the thin myofilaments of skeletal muscle cells. Also part of cytoskeletons. 100, 597

action potential An electrochemical change that occurs across the cell membrane of a neuron. 611

active immunity Immunity generated by an organism's production of antibodies and cytokines. 798

active site The part of an enzyme that provides catalysis. 50

active transport Movement of a molecule through a membrane against its concentration gradient, using a carrier protein and energy from ATP. 92

adaptation An inherited trait that enables an organism to survive a particular environmental challenge. 10

adaptive radiation *ah-DAP-tiv RAID-ee-AY-shun* The divergence of several new types of organisms from a single ancestral type. 381

adenine *AD-eh-neen* One of two purine nitrogenous bases in DNA and RNA. 51

adenosine triphosphate *ah-DEN-o-seen tri-FOS-fate* (ATP) A molecule whose three high-energy phosphate bonds power many biological processes. 92, 117

adhesion *ad-HE-jhun* The tendency of water to hydrogen bond to other compounds. 40

adipocyte *a-DIP-o-site* Fat cell. 594

adipose tissue *AD-eh-pose TISH-ew* Tissue consisting of cells laden with lipid. 44, 594

adrenal glands *ad-REE-nal GLANZ* Paired two-part glands that sit atop the kidneys and produce catecholamines, mineralocorticoids, glucocorticoids, and sex hormones. 667

adrenocorticotropic hormone *ah-DREEN-o-KOR-tah-ko-TROP-ik HOR-moan* (ACTH) A hormone of the anterior pituitary that stimulates the adrenal cortex to secrete hormones. 667

adventitious root *ad-ven-TISH-shus ROOT* Root that forms on stems or leaves. 546

aerial root *AIR-ee-al ROOT* Adventitious root that forms and grows up in the air. 550

afferent arteriole *AF-fer-ent are-TEAR-ee-olz* Branches of the renal artery that approach the proximal portion of a nephron. 781

age structure The distribution of ages of individuals of a population. 838

alcoholic fermentation *AL-ko-HALL-ik FER-men-TAY-shun* An anaerobic step that yeast use after glycolysis that breaks down pyruvic acid to ethanol and carbon dioxide. 137

algae *AL-gee* Unicellular or multicellular photosynthetic eukaryotes that generally lack roots, stems, leaves, conducting vessels, and complex sex organs. 458

allele *ah-LEEL* An alternate form of a gene. 200, 265

allergy mediator *AL-er-gee MEED-ee-a-ter* Biochemical, such as histamine and heparin, that mast cells release when contacting an allergen, causing allergy symptoms. 800

allopolyploid *AL-lo-POL-ee-ploid* An organism with multiple chromosome sets resulting from mating of individuals of different species. 414

alternation of generations The plant life cycle, which alternates between a diploid sporophyte stage and a haploid gametophyte stage. 192

alveolar duct *al-VEE-o-ler DUCT* In lungs, the narrowed ending of bronchioles that opens into a cluster of alveoli. 739

alveoli *al-VEE-o-li* Microscopic air sacs in the lungs. 739

amino acid *ah-MEEN-o AS-id* An organic molecule consisting of a central carbon atom bonded to a hydrogen atom, an amino group, a carboxylic acid, and an R group. 46

amino group *a-MEEN-o GROOP* A nitrogen atom single-bonded to two hydrogen atoms. 47

ammonia *ah-MOAN-ee-ah* A nitrogenous waste generated by deamination of amino acids (NH_3). 776

amoebocyte *a-MEE-bo-site* Mobile cell in a sponge. 515

ampullae *am-PULL-ee* The enlarged bases of the semicircular canals in the inner ear, lined with hair cells that detect fluid movement and convert it into action potentials. 655

anabolism *an-AB-o-liz-um* Metabolic reactions that use energy to synthesize compounds. 114

anaerobe *AN-air-robe* Organism that can live in an environment lacking oxygen. 127

anaerobic respiration *an-air-RO-bic res-per-A-shun* Cellular respiration in the absence of oxygen. 140

analogous structure *ah-NAL-eh-ges struc-chore* Body part in different species that is similar in function but not in structure that evolved in response to a similar environmental challenge. 394

anaphase *AN-ah-faze* The stage of mitosis when centromeres split and two sets of chromosomes part. 171

anaphase I *AN-ah-faze I* Anaphase of meiosis I; homologs separate. 201

anaphase II *AN-ah-faze II* Anaphase of meiosis II, when chromatids separate. 201

androecium *an-dro-EE-shee-um* The innermost whorl of a flower's corolla, consisting of male reproductive structures. 563

aneuploid *AN-you-ploid* A cell with one or more extra or missing chromosomes. 298

angiosperm *AN-gee-o-sperm* Flowering plant. 492

Animalia *an-i-MAIL-ya* The kingdom that includes the animals. 14, 429
annelid *AN-eh-lid* A segmented worm. 513
annuli *AN-yew-li* Demarcations between the segments of leeches. 521
antagonistic muscles *antag-o-NIS-tik MUS-uls* Two muscles or muscle groups that flank a bone and move it in opposite directions. 700
anther *AN-ther* Pollen-producing oval body at the tip of a stamen. 563
antheridia *an-ther-ID-ee-a* In algae, structures that contain sperm. In plants, multicellular male structures where gametes form. 464, 493
anthocyanin *an-tho-SI-ah-nin* Pigment produced in senescent plant cells. 587
antibody *an-tee-BOD-ee* Protein that B cells secrete that recognizes and binds to foreign antigens, disabling them or signaling other cells to do so. 790
anticodon *AN-ti-ko-don* A three-base sequence on one loop of a transfer RNA molecule that is complementary to an mRNA codon and joins an amino acid and its mRNA instructions. 329
antidiuretic hormone *an-ti-di-yur-ET-ik HOR-moan* (ADH) A hypothalamic hormone released from the posterior pituitary that acts on kidneys and smooth muscle cells of blood vessels to maintain the composition of body fluids. 665
antigen *AN-tee-gen* The specific parts of molecules that elicit an immune response. 85
antigen binding site *AN-tee-gen BIND-ing SITE* Specialized ends of antibodies that bind specific antigens. 796
antiparallelism *AN-ti-PAR-a-lel-izm* The head-to-tail relationship of the two rails of the DNA double helix. 313
anus *A-nus* The opening to the anal canal. 764
aorta *a-OR-tah* The largest artery; it leaves the heart. 720
apical meristems *A-pik-el MER-ee-stemz* Unspecialized plant cells that divide, near the tips of roots and shoots. 534
apolipoprotein *APER-o-LIP-o-PRO-teen* The protein parts of cholesterol-carrying molecules. 94
apoptosis *ape-o-TOE-sis* Programmed cell death. 177
appendicular skeleton *AP-en-DIK-u-lar SKEL-eh-ten* In vertebrates, the limb bones and the bones that support them. 685
appendix *ap-PEN-diks* A thin tube extending from the cecum in the human digestive system. 763
aqueous humor *AY-kwee-us HU-mer* In the eye, a nutritive, watery fluid between the cornea and the lens that focuses incoming light rays and maintains the shape of the eyeball. 649
aqueous solution *AWK-kwee-us so-LEW-shun* A solution in which water is the solvent. 40, 87
Archaea *ar-KEE-a* One of the three domains of life. Archaea are unicellular and have a combination of characteristics unlike those of the other two domains. 430
archegonia *arch-eh-GO-nee-a* Multicellular structures in plants where female gametes form. 493, 560
arteriole *are-TEER-ee-ol* Small, elastic blood vessel that arises from arteries and leads into capillaries. 708
artery *AR-teh-ree* Large, elastic blood vessel that leaves the heart and branches into arterioles. 708
arthropod *AR-throw-pod* Animals with three segmented body regions, a jointed exoskeleton, blood in body cavities, and a complex nervous system. Includes spiders and insects. 513
artificial selection Selective breeding. 381
ascending limb *ay-SEN-ding LIM* The distal portion of the loop of the nephron, which ascends from the kidney's medulla. 783
asci *ASS-key* Sexual cells of fungi. 481
ascomycete *ass-ko-mi-SEET* Fungus that has saclike asci that house ascospores. 478
asexual reproduction Reproduction in which a cell doubles its contents and then divides into two identical cells. 10, 192
atom *AT-um* A chemical unit, composed of protons, neutrons, and electrons, that cannot further break down by chemical means. 32
atomic number *a-TOM-ic NUM-ber* The number in an element's box in the periodic table that indicates number of protons or electrons. 33
ATP synthase *ATP SIN-thaze* An enzyme that allows protons to move through the mitochondrial membrane and trigger phosphorylation of ADP to ATP. 129
atriopore *AYT-ree-o-pore* The hole in a lancelet through which water exits the body. 526
atrioventricular node *A-tree-o-ven-TRIK-yu-lar NODE* (AV node) Specialized muscle cells that branch into a network of Purkinje fibers, which conduct electrical stimulation six times faster than other parts of the heart. 722
atrioventricular valve *A-tree-o-ven-TRIK-yu-lar VALV* (AV valve) Flap of tissue between each atrium and ventricle that moves in response to the pressure changes of ventricular contraction. 720
atrium *ATE-tree-um* An upper heart chamber. 710
atrophy *AT-tre-fee* Muscle degeneration resulting from lack of use or immobilization. 702
auditory nerve *AWD-eh-tore-ee NERV* Nerve fibers from the cochlea in the inner ear to the cerebral cortex. 654
autoantibody *AW-toe-AN-tee-bod-ee* Antibody that attacks the body's tissues, causing autoimmune disease. 800
autoimmunity *AW-toe-imm-YUN-i-tee* An organism's immune system attacking its own body. 800
autonomic nervous system *AW-toe-NOM-ik NER-ves sis-tum* Motor pathways that lead to smooth muscle, cardiac muscle, and glands. 622
autopolyploid *aw-toe-POL-ee-ploid* An organism with multiple chromosome sets. 414
autosomal dominant *AW-toe-soe-mal DOM-i-nent* An allele that masks the expression of another allele. 267
autosomal recessive *AW-toe-soe-mal re-SESS-ive* An allele whose expression is masked by another allele. 267
autosome *AW-toe-soam* A nonsex chromosome. 198
autotroph(ic) *AW-toe-trof* An organism that manufactures nutrients using energy from the environment. 14, 112, 146
auxin *AWK-zin* A type of plant hormone that elongates cells in seedlings, shoot tips, embryos, and leaves. 569
axial skeleton *AX-ee-al SKEL-eh-ten* In a vertebrate skeleton, the skull, vertebral column, ribs, and sternum. 685
axil *AX-el* The regions between a leaf stalk and stem. 541
axon *AKS-on* A thick branch of a neuron that sends messages. 596, 611

B

bacilli *ba-SILL-i* Rod-shaped bacteria. 443
Bacteria *bac-TEAR-e-a* One of the three domains of life, prokaryotes. 430
balanced polymorphism *BAL-anced POL-ee-MORF-iz-um* Stabilizing selection that maintains a genetic disease in a population because heterozygotes resist an infectious disease. 411
bark Tissues outside the vascular cambium. 551
Barr body *BAR BOD-ee* The dark-staining body in the nucleus of a female mammal's cell, corresponding to the inactivated X chromosome. 294
base A molecule that releases hydroxide ions into water. 40
basement membrane A thin layer in human skin that separates the epidermis from the dermis. 599
basidia *bass-ID-ee-a* Spore-bearing structures in basidiomycetes. 482
basidiomycete *bass-ID-eo-mi-SEET* Fungus that has club-shaped hyphae tips that contain spore-bearing basidia. 478
basilar membrane *BA-sill-ar MEM-brane* The membrane beneath hair cells in the cochlea of the inner ear that vibrates in response to sound. 653
B cells Lymphocytes that produce antibodies. 716, 791
benign tumor *bee-NINE TOO-mer* A noncancerous tumor. 180
benthic zone *BEN-thick ZONE* The bottom of the ocean. 883
bilaterally symmetrical *bi-LAT-er-a-lee sim-MET-rik-al* An animal body form in which only one plane divides the animal into mirror image halves. 512
bile A digestive biochemical that emulsifies fats. 757

binary fission *BI-nair-ee FISH-en* A type of asexual reproduction in which a cell divides into two identical cells. 192, 448
bioaccumulation *bi-o-a-cume-u-LAY-shun* Higher concentration of a chemical in cells than in the nonliving surroundings. 866
biodiversity *bi-o-di-VER-city* The spectrum of different life-forms. 12
bioenergetics *bi-o-n-er-JET-ix* The study of energy in life. 110
biogeochemical cycles *bi-o-gee-o-KEM-i-kal SI-kulls* Geological and biological processes that recycle chemicals vital to life. 864
biogeography *bi-o-gee-OG-grah-fee* The physical distribution of organisms. 374
biological evolution The process by which the genetic structure of populations changes over time. 374
bioluminescence *bi-o-loom-in-ES-ents* A chemical reaction that causes an organism to glow. 118
biomagnification *bi-o-mag-nif-i-KAY-shun* Increasing concentrations of a chemical with higher trophic levels. 866
biomes *BI-ohms* Major types of communities and ecosystems that are recognizable in large geographic areas. 859, 872
bioremediation *bi-o-ree-meed-e-AY-shun* Using organisms that metabolize toxins to clean the environment. 454, 890
biosphere *BI-o-sfere* The ecosystem of the entire planet. 9, 858
biotic community *bi-OT-ik cum-MUNE-it-ee* All the organisms in an area. 854
biotic succession *bi-OT-ik suk-SESH-in* Change among the populations in a biotic community. 856
bipedalism *by-PEED-a-liz-m* Walking upright on two limbs. 368
biramous *bi-RAYM-us* A double-branched arthropod appendage. 522
bivalve *BI-valv* A mollusk that has a two-part shell. 519
blade The flattened region of a leaf. 541
blastocyst *BLAS-toe-syst* Stage of human prenatal development that is a hollow, fluid-filled ball of cells. 218
blastomere *BLAS-toe-mere* A cell in a preembryonic organism resulting from cleavage divisions. 218
blastopore *BLAS-toe-pour* An indentation in a protostome embryo that develops into a mouth. 513
blood A complex mixture of cells suspended in a liquid matrix that delivers nutrients to cells and removes wastes. 594
blood-brain barrier Close knit cells of blood vessels in the brain, which limit substances that can enter the brain. 635
blood pressure The force blood exerts against blood vessel walls. 725
bolus *BO-lus* Food rolled into a lump by the tongue. 760
bond energy The energy required to form a particular chemical bond. 115
bone A connective tissue consisting of bone-building osteoblasts, stationary osteocytes, and bone-destroying osteoclasts, embedded in a mineralized matrix infused with spaces and canals. 594
boom and bust cycle A population that repeatedly and regularly increases and decreases in size. 844
braced framework A skeleton built of solid structural components strong enough to resist collapsing. 683
bracts *BRAKS* Floral leaves that protect developing flowers. 544
brain stem Part of the human brain closest to the spinal cord; controls vital functions. 625
bronchi *BRON-ki* Two tubules that branch from the trachea as it reaches the lungs. 737
bronchiole *BRON-ki-ol* Microscopic branch of the bronchi within the lungs. 737
bryophyte *BRI-o-fite* Plant that lacks vessels, including the mosses, liverworts, and hornworts. 492
budding Formation of a small progeny cell from a protozoan cell. 459
buffer system Pairs of weak acids and bases that maintain body fluid pH. 41
bulbourethral glands *BUL-bo-u-REE-thral GLANZ* Small glands near the male urethra that secrete mucus. 197
bulk element An element that an animal requires in large amounts. 32
bundle-sheath cell *BUN-dull SHEETH SEL* Thick-walled plant cell surrounding veins that functions in C_4 photosynthesis. 160
bursae *BURR-say* Small packets in joints that secrete lubricating fluid. 700

C

C_3 plant Plant that uses the Calvin cycle only to fix carbon dioxide. 159
C_4 photosynthesis *see-4 foto-SIN-the-sis* In plants, a biochemical pathway that prevents photorespiration. 160
calcitonin *KAL-sah-TOE-nin* A thyroid hormone that decreases blood calcium levels. 669
calyx *KA-liks* One of two outermost whorls of a flower. 563
canaliculi *can-al-IK-u-li* Passageways in bone that connect lacunae. 595
capacitation *cah-PASS-eh-TAY-shun* Activation of sperm cells in the human female reproductive tract. 217
carbohydrase *KAR-bo-HI-drase* Enzyme that breaks down certain disaccharides into monosaccharides. 762
carbohydrate *CAR-bo-HI-drate* Compound containing carbon, hydrogen, and oxygen, with twice as many hydrogens as oxygens; sugar or starch. 43
carboxyl group *kar-BOX-ill GROOP* A carbon atom double-bonded to an oxygen and single-bonded to a hydroxyl group (OH). 47
cardiac cycle The sequence of contraction and relaxation that makes up the heartbeat. 720
cardiac muscle cell Striated, involuntary, single-nucleated contractile cell in the mammalian heart. 598
cardiac output The volume of blood the heart ejects in 1 minute. 727
carotenoids *kare-OT-in-oydz* Yellow and orange plant pigments. 150, 587
carotid rete *care-OT-id REE-tee* A configuration of blood vessels in the brain that cools the brain. 770
carpel *KAR-pel* Leaflike structure in a flower that encloses ovules. 563
carrying capacity The theoretical maximum number of individuals an environment can support. 842
cartilage *CAR-teh-lij* A supportive connective tissue consisting of chondrocytes embedded in collagen and proteoglycans. 594
Casparian strip *kas-PAHR-ee-an STRIP* A single layer of tightly packed cells that forms the endodermis of a plant. 548
catabolism *cah-TAB-o-liz-um* Metabolic degradation reactions, which release energy. 114
catalysis *kat-AL-i-sis* Speeding a chemical reaction. 49
catecholamines *kat-ah-KOLE-ah-meens* Hormones of the adrenal medulla. 670
cell The structural and functional unit of life. 5, 60
cell body The enlarged portion of a neuron containing most of the organelles. 596, 611
cell cycle The life of a cell, in terms of whether it is dividing or in interphase. 168
cell membrane Proteins embedded in a lipid bilayer, which forms the boundary of cells. 66
cell population A group of cells with characteristic proportions in particular stages of the cell cycle. 179
cell theory The ideas that all living matter consists of cells, cells are the structural and functional units of life, and all cells come from preexisting cells. 61
cellular adhesion molecule *SEL-u-ler ad-HEE-jhun MOL-e-kuel* (CAM) A protein that enables cells to interact with each other. 85
cellular immune response The actions of T and B cells in the immune system. 797
cellular respiration Biochemical reactions that extract energy in mitochondria. 114, 126
cell wall A rigid boundary consisting of peptidoglycans in prokaryotic cells and cellulose in plant cells. 14, 64
central nervous system (CNS) The brain and the spinal cord. 611
centrioles *SEN-tre-olz* Paired, oblong structures consisting of microtubules in animal cells that organize the mitotic spindle. 74, 171
centromere *SEN-tro-mere* A characteristically located constriction in a chromosome. 171, 296
centrosome *SEN-tro-soam* A region near the cell nucleus that contains the centrioles. 171
cephalized *CEF-a-lized* An animal body form with a head end. 512
cerebellum *ser-eh-BELL-um* An area of the brain that coordinates muscular responses. 625
cerebrospinal fluid *sah-ree-bro-SPI-nal FLU-id* Fluid similar to blood plasma that bathes and cushions the central nervous system. 635

cerebrum *seh-REE-brum* The higher region of the brain that controls intelligence, learning, perception, and emotion. 625
cervix *SER-viks* In the female human, the opening to the uterus. 197
chaperone protein *shap-ER-one PRO-teen* A protein that stabilizes a growing amino acid chain. 336
chemical bond Attachments atoms form by sharing or exchanging electrons. 35
chemical equilibrium *KEM-e-cal e-kwil-IB-ree-um* When a chemical reaction proceeds in both directions at the same rate. 116
chemical reaction Interactions in which atoms exchange or share electrons, forming new chemicals. 35
chemiosmosis *KEM-ee-oss-MOE-sis* Phosphorylation of ADP to ATP occurring when protons that are following a concentration gradient contact ATP synthase. 135
chemoautotroph *KEEM-o-awt-o-trofe* An organism that derives energy from chemicals in nonliving surroundings. 112
chemoreception *KEEM-o-ree-SEP-shun* Smell and taste. 644
chemotroph *KEEM-o-trofe* An organism that obtains energy by oxidizing chemicals. 445
chlorenchyma *klor-REN-kah-mah* Chloroplast-containing parenchyma cells. 536
chlorophyll a *KLOR-eh-fill A* A green pigment plants use to harness the energy in sunlight. 75, 149
chloroplast *KLOR-o-plast* A plant cell organelle housing the reactions of photosynthesis. 74, 151
choanocytes *cho-AN-o-sites* Collar cells in sponges that produce a current that draws food and water into the central cavity. 515
chondrocyte *KON-dro-site* A cartilage cell. 595
chordate *KOR-dayt* Animal with a notochord, hollow nerve cord, and gill slits. 513
chorion *KOR-ee-on* A membrane that develops into the placenta. 218
chorionic villi *KOR-ee-ON-ik VIL-i* Fingerlike projections extending from the chorion to the uterine lining. 223
chromatid *KRO-mah-tid* A continuous strand of DNA comprising an unreplicated chromosome or one-half of a replicated chromosome. 171
chromosome *KRO-mo-soam* A dark-staining, rod-shaped structure in the nucleus of a eukaryotic cell consisting of a continuous molecule of DNA wrapped in protein. 168, 285
chyme *KIME* Semisolid food in the stomach. 760
chymotrypsin *KI-mo-TRIP-sin* A pancreatic enzyme that digests protein in the small intestine. 762
chytrid *KI-trid* Unicellular eukaryote with single flagella and chitinous cell walls. 458
cilia *SIL-ee-ah* Protein projections from cells that beat coordinately, moving cells and substances. 99
ciliary body *SIL-ee-air-ee BOD-ee* A highly folded, specialized structure in the center of the choroid coat of the human eye that houses the ciliary muscle, which alters the shape of the lens. 648
circadian rhythm *sir-KA-dee-en RITH-um* Regular, daily rhythm of a biological function. 587
classical conditioning A form of learning in which an animal responds in a familiar way to a new stimulus. 815
cleavage *KLEEV-ij* A period of rapid cell division following fertilization but before embryogenesis. 218
cleavage furrow *KLEEV-ij FUR-o* The initial indentation between two progeny cells in mitosis. 173
climax community The species in a community that persist when the environment doesn't change. 856
clitellum *cli-TELL-um* Part of the outside of earthworms that secretes mucus during copulation and a cocoon early in development. 521
clitoris *CLIT-or-is* A small, highly innervated bit of tissue at the juncture of the labia that is the female anatomical equivalent of the penis. 198
closed circulatory system A circulatory system that contains blood in vessels. 708
club mosses A type of seedless vascular plant. 497
cluster-of-differentiation (CD4) antigen *CLUSS-ter uv diff-er-en-she-AY-shun AN-teh-jen* A protein on a helper T cell that recognizes a nonself antigen on a macrophage, initiating an immune response. 797
Cnidaria *SNID-air-ee-a* An animal phylum whose members have a hollow two-layered body, with a jellylike interior and radial symmetry. 510, 516
cnidocytes *SNID-o-sites* Structures in cnidaria that contain stinging cells. 516
cocci *KOK-si* Round bacteria. 443
cochlea *COKE-lee-ah* The spiral-shaped, hindmost part of the inner ear, where vibrations are translated into nerve impulses. 653
codominant *ko-DOM-eh-nent* Alleles that are both expressed in the heterozygote. 273
codon *KO-don* A continuous triplet of mRNA bases that specifies a particular amino acid. 329
coefficient of relationship *co-eff-FISH-ent uv re-LAY-shun-ship* A measurement of how closely related two individuals are, based upon the proportion of genes they share. 278
coelom *SEE-loam* A central body cavity in an animal. 511
coenocytic *seen-o-SIT-ik* Fungal hyphae in which protoplasm streams continuously 477.
coenzyme *coe-EN-zime* An organic cofactor necessary for the function of certain enzymes. 119
coevolution *ko-ev-eh-LU-shun* The interdependence of two types of organisms for survival. 374
cofactor *COE-fac-tor* A biochemical necessary for an enzyme to function. 119
cohesion *co-HEE-jhun* The strong attraction of water molecules to each other. 40
coleoptile *KOL-ee-OP-tile* A sheathlike structure that covers the plumule in monocots. 568
collagen *COLL-a-jen* A connective tissue protein. 594
collecting duct A structure in the kidney into which nephrons drain urine. 783
collenchyma *kol-LEN-kah-mah* Elongated, living cells that differentiate from parenchyma and support growing regions of shoots. 536
colon *Kol-en* The large intestine. 763
colony-stimulating factors Secreted growth factors that stimulate blood cells to specialize. 717
columnar *col-UM-nar* Tall cells, as in epithelium. 594
commensalism *co-MEN-sal-izm* A symbiotic relationship in which one member benefits without affecting the other member. 434
compact bone A layer of solid, hard bone that covers spongy bone. 689
companion cells Cells that transfer carbohydrates to and from sieve tube members. 539
complementary base pairs Bonding of adenine to thymine and guanine to cytosine in the DNA double helix. 313
complement system A group of proteins that assist other immune defenses. 791
complex trait A trait caused by genes and the environment. 276
compound A molecule including different atoms. 34
concentration gradient Passive diffusion of ions from an area of high to low concentration. 88
concordance *kon-KOR-dance* A measure of the inherited component of a trait. The number of pairs of either monozygotic or dizygotic twins in which both members express a trait, divided by the number of pairs in which at least one twin expresses the trait. 279
conditioned stimulus A new stimulus coupled to a familiar or unconditioned stimulus, so that an animal associates the two. 815
cones Specialized receptor cells in the center of the retina that detect colors. 560, 648
conformation *KON-for-MAY-shun* The three-dimensional shape of a protein. 49
conifers *CON-i-ferz* One of four divisions of gymnosperms; tall, ancient trees. 498
conjugation *con-ju-GAY-shun* A form of gene transfer in bacteria. 000 A form of sexual reproduction in protozoa. 194, 449, 459
connective tissue Tissue consisting of cells embedded or suspended in a matrix, including loose and fibrous connective tissues, cartilage, bone, and blood. 594
constant regions The sequence of amino acids in the lower parts of heavy and light antibody chains, which vary in different antibody types. 795

contest competition When members of a population compete for a resource indirectly. 843
contractile vacuole *KON-tract-till VAK-u-ol* An organelle in a paramecium that pumps water out of the cell. 90, 459
convergent evolution *kon-VER-gent ev-o-LU-shun* Organisms that have similar adaptations to a similar environmental challenge but that are not related by descent. 379
cork cambium *KORK KAM-bee-um* The lateral meristem that produces the periderm, the covering on mature stems and roots. 550
cork cells Waxy cells on mature stems and roots. 552
cornea *KOR-nee-ah* A modified portion of the human eye's sclera that forms a transparent curved window that admits light. 648
corolla *kah-ROLE-ah* One of two outermost whorls of a flower, with no direct role in sexual reproduction. 563
corona radiata *kah-RONE-ah ray-dee-A-tah* Layer of cells around the oocyte. 217
corpus luteum *KOR-pis LU-te-um* A gland formed from an ovarian follicle that has recently released an oocyte; produces estrogen and progesterone. 674
cortex *KOR-teks* In plants, the ground tissue in between the epidermis and vascular tissue in stems. 541
cotyledons *KOT-ah-LEE-donz* Embryonic leaves in flowering plants that store energy for germination. 544
countercurrent exchange system Passage of heat from one blood vessel to another. 770
countercurrent flow Fluid flow in different parts of a continuous tubule in opposite directions, which maximizes the amount of a particular substance that diffuses out of the tubule. 734
coupled reactions *CUP-uld re-AC-shuns* Two chemical reactions that occur simultaneously and have a common intermediate. 119
covalent bond *KO-va-lent BOND* Atoms sharing electrons. 35
creatine phosphate *KRE-ah-tin FOS-fate* A molecule stored in muscle fibers that can donate its high-energy phosphate to ADP to regenerate ATP. 697
cristae *KRIS-ty* The folds of the inner mitochondrial membrane along which many of the reactions of cellular respiration occur. 71, 128
critical period The time during prenatal development when a structure is vulnerable to damage. 246
crop Food-storing structure in birds. 754
crossing-over Exchange of genetic material between homologous chromosomes during prophase of meiosis I. 200
cuticle *KEW-tah-kal* A tissue covering all of a plant except the roots. 536
cutin *KEW-tin* A fatty material that a plant's epidermal cells produce that forms a cuticle. 536
cyanobacteria *si-an-o-bak-TEAR-ee-ah* Unicellular algae that have bluish-green pigments. 64
cycads *SI-kadz* One of four divisions of gymnosperms; tall, ancient trees. 498
cyclic adenosine monophosphate *SI-klik a-DEN-o-seen mon-o-FOS-fate* (cAMP) A second messenger. 661
cyclin *SI-klin* A type of protein that controls the cell cycle. 177
cyst *SIST* A dormant form of a protozoan that has a cell wall and lowered metabolism. 459
cytochrome *SI-to-krome* An iron-containing molecule that transfers electrons in metabolic pathways. 121
cytogenetics *si-to-jen-ET-ix* Correlation of an inherited trait to a chromosomal anomaly. 286
cytokines *SI-toe-kines* Chemicals that B cells secrete. 791
cytokinesis *SI-toe-kin-E-sis* Distribution of cytoplasm, organelles, and macromolecules into two progeny cells in cell division. 168
cytokinins *SI-toe-KI-ninz* A class of plant hormones that promote division of the cell following nuclear divisions in seeds, roots, young leaves, and fruits. 577
cytoplasm *SI-toe-PLAZ-um* The jellylike fluid in which organelles are suspended in eukaryotic cells. 61
cytosine *SI-toe-seen* One of the two pyrimidine nitrogenous bases in DNA and RNA. 51
cytoskeleton *SI-toe-SKEL-eh-ten* A framework of arrays of protein rods and tubules in animal cells. 67
cytotoxic T cell *si-toe-TOCKS-ik T SEL* Immune system cell that kills nonself cells by binding them and releasing chemicals. 797

D

day-neutral plants Plants that do not rely on photoperiod to flower. 583
dead space The pharynx, the trachea, and the upper third of the lungs, which contain air not used in gas exchange. 742
deciduous tree *dah-SID-u-us TREE* Tree that sheds leaves at the end of a growing season. 546
decomposers *dee-kom-POZ-erz* Organisms that consume feces and dead organisms. 8, 862, 873
degenerate codons *de-JEN-er-at KO-donz* Different codons that specify the same amino acid. 344
dehydration synthesis *de-hi-DRA-shun SYN-theh-sis* Formation of a covalent bond between two molecules by loss of water. 43
deletion A chromosome missing genetic material. 301
dendrite *DEN-drite* Thin neuron branches that receive messages. 596, 611
denitrifying bacteria *de-NI-tri-fy-ing bak-TEAR-e-a* Bacteria that convert nitrites and nitrates to nitrogen gas. 865.
dense connective tissue Connective tissue that has dense collagen tracts. 594
density-dependent factor Event that kills when populations are large. 843
density-independent factor Event that kills irrespective of population size. 842
deoxyhemoglobin *de-OX-ee-HEEM-o-GLO-bin* Hemoglobin that is deep red after releasing its oxygen to tissues. 715
deoxyribonucleic acid *de-OX-ee-RI-bo-nu-KLAY-ic AS-id* (DNA) A double-stranded nucleic acid composed of nucleotides containing a phosphate group, a nitrogenous base (A, T, G, or C), and deoxyribose. 5, 51
deoxyribose *de-OX-ee-RI-bose* A five-carbon sugar that is part of DNA. 311
deposit feeder Animal that eats soil and strains out nutrients. 521
dermal tissue *DER-mal TISH-ew* Tissue that covers a plant. 534
dermis *DER-miss* In humans, the middle skin layer. 599
descending limb *de-SEN-ding LIM* The proximal portion of the loop of the nephron, which descends into the kidney's medulla. 783
determinate cleavage *de-TER-min-et CLEEV-edge* An early embryo in which cell fate is determined. 514
deuteromycete *doo-ter-o-mi-SEET* Fungus that lacks a distinctive sexual phase. 478
deuterostome *DOO-ter-o-stome* One of two coelomate lineages, characterized by radial cleavage, an anus, and indeterminate cleavage. 512
developmental pathway Series of events that forms a particular anatomical structure. 211
diaphragm A broad sheet of muscle that separates the thoracic cavity from the abdominal cavity. 742
diastole *di-ASS-tole-ee* Relaxation of the heart. 720
diatom *DIE-a-tom* An alga with distinctive silica walls. 467
dicot *DI-kot* A plant that has two seed leaves. 541
differentiate *diff-er-EN-shee-ate* Specialize. 211
diffusion *de-FUZE-jhun* Movement of a substance from a region where it is highly concentrated to an area where it is less concentrated without energy input. 88
dihybrid cross *DI-HI-brid KROS* Mating between individuals heterozygous for two particular genes. 269
dilution effect *di-LEW-shun e-FEKT* A behavior that decreases the chance of an event occurring. 822
dimorphic *di-MOR-fik* A fungus with unicellular and multicellular phases to the life cycle. 477
dinoflagellate *di-no-FLADJ-el-et* An alga with characteristics of two flagella. 466
diploblastic *dip-lo-BLAS-tik* An animal whose adult tissues arise from two germ layers in the embryo. 509
diploid cell *DIP-loid SEL* A cell with two copies of each chromosome. Also known as $2n$. 192
direct development Animals that hatch from eggs or are born resembling adults of the species. 512

directional selection Changes in the prevalence of a characteristic that reflects differential survival of individuals better adapted to a particular environment. 410

disaccharide *di-SAK-eh-ride* A sugar that consists of two bonded monosaccharides. 43

disruptive selection A population in which two extreme expressions of a trait are equally adaptive. 411

distal convoluted tubule *DIS-tel KON-vo-lu-tid TU-bule* The region of the kidney distal to the loop of the nephron and proximal to a collecting duct. 783

disulfide bond *di-SUL-fide bond* Attraction between two sulfur atoms within a protein molecule. 49

dizygotic twins *di-zi-GOT-ik TWINZ* Fraternal twins. 279

DNA polymerase *DNA po-LIM-er-ase* An enzyme that inserts new bases and corrects mismatched base pairs in DNA replication. 317

domain A taxonomic designation that supercedes kingdom. 12, 430

dominance hierarchy *DOM-eh-nance HI-er-ar-kee* A social ranking of members of a group of the same sex, which distributes resources with minimal aggression. 828

dominant An allele that masks the expression of another allele. 265

dormancy *DOR-man-see* A state of lowered metabolism that enables a plant to survive harsh climatic conditions. 587

dorsal *DOOR-sal* The back side. 512

double-blind An experimental protocol where neither participants nor researchers know which subjects received a placebo and which received the treatment being evaluated. 25

duodenum *do-o-DEE-num* The first section of the small intestine. 761

duplication A repeated portion of DNA in a chromosome. 301

E

echinoderm *ee-KINE-o-derm* Animal with a five-part body plan, radial symmetry, and a spiny outer covering. 513

ecology The study of relationships between organisms and their environments. 836

ecosystem A unit of interaction among organisms and their surroundings, including all life in a defined area. 9, 858

ectoderm *EK-TOE-derm* The outermost embryonic germ layer, whose cells become part of the nervous system, sense organs, outer skin layer, and its specializations. 220

ectomycorrhizae *ek-toe-mi-kor-IZ-ee* Parts of a fungus that ensheath a plant's roots but do not greatly penetrate the plant. 485

ectopic pregnancy *ek-TOP-ik PREG-nan-see* The implantation of a zygote in the wall of a fallopian tube rather than in the uterus. 239

ectoplasm *EK-toe-PLAZ-m* In an amoeba, an outer layer of thick, gel-like material. 692

ectotherm *EK-toe-therm* Animal that loses or gains heat to its surroundings by moving to where the temperature is suitable. 771

edema *eh-DEEM-ah* Swelling of a body part from fluid buildup. 727

effector *e-FEK-ter* A muscle or gland that receives input from a neuron. 611

efferent arteriole *EF-fer-ent are-TEAR-ee-ol* Branch of the renal artery that leaves the proximal portion of a nephron. 781

ejaculation *e-JAK-u-LAY-shun* Discharge of sperm through the penis. 197

elastin *e-LAS-tin* A type of connective tissue protein. 594

electrolyte *e-LEK-tro-lite* Solution containing ions. 40

electromagnetic spectrum *e-LEK-tro-mag-NET-ik SPECK-trum* A spectrum of naturally occurring radiation. 147

electron *e-LEK-tron* A negatively charged subatomic particle with negligible mass that orbits the atomic nucleus. 32

electronegativity *e-LEK-tro-neg-a-TIV-it-ee* The tendency of an atom to attract electrons. 38

electron orbital *e-LEK-tron OR-bit-al* The volume of space where a particular electron is 90% of the time. 35

element A pure substance consisting of atoms containing a characteristic number of protons. 32

El Niño *l NEEN-yo* A periodic slack in the trade winds that prevents upwelling along Peru's coast and changes global weather patterns. 883

embryonic induction *EM-bree-ON-ik in-DUK-shun* In an embryo, the ability of a group of specialized cells to stimulate neighboring cells to specialize. 226

embryonic stage *em-bree-ON-ik STAJE* Stage of prenatal development when organs form from a three-layered structure. 216

emergent property A quality that appears as biological complexity increases. 8

empiric risk *em-PEER-ik RISK* Risk calculation based on prevalence. 277

endergonic reaction *en-der-GONE-ik re-AK-shun* An energy-requiring chemical reaction. 116

endocrine system *EN-do-crin SIS-tum* Glands and cells that secrete hormones. 660

endocytosis *EN-doe-si-TOE-sis* The cell membrane's engulfing extracellular material. 93

endoderm *EN-doe-derm* The innermost embryonic germ layer, whose cells become the organs and linings of the digestive, respiratory, and urinary systems. 220

endodermis *en-do-DER-mis* The innermost part of a root's cortex. 548

endometrium *EN-doe-MEE-tree-um* The inner uterine lining. 197

endomycorrhizae *en-doe-mi-kor-IZ-ee* Fungal threads that penetrate a plant through the roots. 485

endoplasm *EN-doe-plaz-m* In an amoeba, an inner layer of sol-like cytoplasm. 692

endoplasmic reticulum *EN-doe-PLAZ-mik reh-TIK-u-lum* Interconnected membranous tubules and sacs that wind from the nuclear envelope to the cell membrane, along which proteins are synthesized (in rough ER) and lipids synthesized (in smooth ER). 70

endoskeleton *en-do-SKEL-eh-ten* An internal scaffolding type of skeleton in vertebrates. 684

endosperm *EN-do-sperm* A triploid tissue that nourishes the embryo in a seed. 568

endospore *EN-doe-spor* A walled structure that forms around a bacterial cell's nucleus, enabling the organism to survive harsh environmental conditions. 447

endosymbiont theory *EN-doe-SYM-bee-ont THER-ee* The idea that eukaryotic cells evolved from large prokaryotic cells that engulfed once free-living bacteria. 79

endothelium *end-o-THEEL-e-um* Layer of single cells that forms a capillary wall. 722

endotherm *EN-doe-THERM* Animal that uses metabolic heat to regulate its temperature. 771

energy The ability to do work. 110

energy nutrients Fats, proteins, and carbohydrates. 755

energy of activation Energy required for a chemical reaction to begin. 121

energy pyramid A diagram depicting trophic levels of energy stored in living tissues per unit time. 863

energy shell Levels of energy in an atom formed by electron orbitals. 35

entrainment *en-TRANE-ment* The resynchronization of a biological clock by the environment. 588

entropy *EN-tro-pee* Randomness or disorder. 8, 113

environmentalism The philosophy of living in harmony with the global ecosystem. 892

environmental resistance Mortality and emigration, which limit population growth. 842

enzyme *EN-zime* A protein that catalyzes a specific type of chemical reaction. 46

enzyme-substrate complex *EN-zime SUB-strate COM-plex* A transient structure that forms when a substrate binds an enzyme's active site. 50

epidemiology *EP-eh-dee-mee-OL-o-gee* The analysis of data derived from real-life, nonexperimental situations. 26, 384

epidermis *ep-ee-DERM-is* The outer integumentary layer in several types of organisms. 517, 536, 598

epididymis *EP-eh-DID-eh-mis* In the human male, a tightly coiled tube leading from each testis, where sperm mature and are stored. 197

epigenesis *ep-ee-GEN-i-sis* The idea that specialized tissue arises from unspecialized tissue in a fertilized ovum. 210

epiglottis *ep-ee-GLOT-is* Cartilage that covers the glottis, routing food to the digestive tract and air to the respiratory tract. 736

epinephrine *ep-ee-NEF-rin* (adrenaline) A hormone produced in the adrenal medulla that raises blood pressure and slows digestion. 670

epistasis *eh-PIS-tah-sis* A gene masking another gene's expression. 274

epithelial tissue *EP-eh-THEEL-e-al TISH-ew* Tightly packed cells that form linings and coverings. 594

epoch *Eh-poc* Time within a period. 359

equational division *ee-QUAY-shun-el deh-VISZ-un* The second meiotic division, when four haploid cells form from two haploid cells that are the products of meiosis I. 198

eras Very long periods of time of biological or geological activity. 359

erythropoietin *eh-rith-ro-po-EE-tin* A hormone produced in the kidneys that stimulates red blood cell production when oxygen is lacking. 715

esker *ES-kur* Extensive ridge of gravel and sand deposited by a river that ran under a glacier. 878

esophagus *ee-SOF-eh-gus* A muscular tube that leads from the pharynx to the stomach. 757

essential nutrient Nutrient that must come from food because the body cannot synthesize it. 755

estuary *ES-tu-air-ee* An area where fresh water in a river meets salty water of an ocean. 881

ethology *eth-OL-o-gee* Study of how natural selection shapes adaptive behavior. 812

ethylene *ETH-eh-leen* A simple organic molecule that is a plant hormone that hastens fruit ripening. 569

etiolated seedling *E-ti-o-LAY-tid SEED-ling* Seedling that has abnormally elongated stems, small roots and leaves, and a pale color, because it was grown in the dark. 586

euchromatin *u-KROME-a-tin* Light-staining genetic material. 296

euglenoid *YEW-glen-oyd* A type of algae. 466

Eukarya *yoo-KAR-ee-a* One of the three domains of life, including organisms that have eukaryotic cells. 430

eukaryotic cell *u-CARE-ee-OT-ik SEL* A complex cell containing organelles. 14, 63

euphotic zone *u-FOT-ik ZONE* Part of the ocean where photosynthesis occurs. 883

euploid *U-ployd* A normal chromosome number. 298

eusocial *yu-SO-shal* A population of animals that communicate, cooperate in caring for young, have overlapping generations, and divide labor. 819

eutrophic *yu-TRO-fik* An aging lake, containing many nutrients and decaying organisms, often tinted green with algae. 880

evaporative cooling Loss of body heat by evaporation of fluid from the body's surface. 773

evolution Changing gene frequencies in a population over time. 8

excision repair *ex-SIZ-jhun ree-PARE* Cutting pyrimidine dimers out of DNA. 319

excurrent siphon *ex-CUR-ent SI-fon* A structure that shoots water from a tunicate when it is disturbed. 526

exergonic reaction *ex-er-GONE-ik re-AK-shun* An energy-releasing chemical reaction. 116

exocrine glands *EX-o-crine GLANDZ* Structures that secrete substances through ducts. 660

exocytosis *EX-o-si-TOE-sis* Fusing of secretion-containing organelles with the cell membrane. 93

exon *EX-on* The bases of a gene that code for amino acids. 330

exoskeleton *ex-o-SKEL-eh-ten* A braced framework skeleton on the outside of an organism. 683

experiment A test to disprove a hypothesis. 21

experimental control An extra test that can rule out causes other than the one being investigated. 25

extinction *ex-TINK-shun* Disappearance of a type of organism. 815

extracellular digestion *ex-tra-SEL-yu-lar di-JEST-shun* Dismantling of food by hydrolytic enzymes in a cavity within an organism's body. 753

extraembryonic membranes *EX-tra-EM-bree-on-ik MEM-branz* Structures that support and nourish the mammalian embryo and fetus, including the yolk sac, allantois, and amnion. 216

F

facilitated diffusion *fah-SIL-eh-tay-tid dif-FU-shun* Movement of a substance down its concentration gradient with the aid of a carrier protein. 92

fallopian tubes *fah-LO-pee-an TUBES* In the human female, paired tubes leading from near the ovaries to the uterus. 197

fascicles *FAS-i-kulls* Groups of pine needles. 499

fatty acid A hydrocarbon chain that is a part of a triglyceride. 45

feather follicle *FETH-er FOLL-i-kull* Extensions of the epidermis from which a feather extends. 602

fecundity *fee-KUN-dit-ee* The number of offspring an individual produces. 842

fermentation *fur-men-TAY-shun* An energy pathway in the cytoplasm that extracts limited energy from the bonds of glucose. 137

fetal period *FEE-tal PEER-e-od* The final stage of prenatal development, when structures grow and elaborate. 216

fiber An elongated plant cell that forms strands. 536

fibrin *FI-brin* A threadlike protein that forms blood clots. 717

fibrinogen *fi-BRIN-o-jen* A protein that is the precursor of fibrin. 714

fibroblast *FI-bro-blast* A connective tissue cell that secretes collagen and elastin. 594

fibrous root system *FI-bres ROOT SIS-tum* A plant with short-lived first roots. 546

fiddlehead *FID-ul-hed* A leaf formation in a true fern. 498

filter feeder Animal that eats small organisms in an aquatic environment. 751

first filial generation *FURST FILL-e-al gen-er-A-shun* (F_1) The second generation in a genetics problem. 265

fixed action pattern An innate, stereotyped behavior. 813

flagellum *fla-GEL-um* A tail-like appendage on a prokaryotic cell that rotates, providing locomotion. 66, 99, 446

flame cell system A simple nephridia excretory system in flatworms. 778

flavin adenine dinucleotide *FLAY-vin AD-e-neen di-NUKE-lee-o-tide* (FAD) An electron carrier molecule that functions in certain metabolic pathways. 120

floridean starch *flor-ID-ee-in STARCH* A carbohydrate in red seaweeds. 469

follicle cell *FOL-ik-kel SEL* Nourishing cell surrounding an oocyte. 197

follicle-stimulating hormone *FOL-eh-kel STIM-u-la-ting HOR-moan* (FSH) A hormone made in the anterior pituitary that controls oocyte maturation, development of ovarian follicles, and their release of estrogen. 666

food chain The linear relationship of species that eat each other. 859

food vacuole *FOOD VAK-you-ole* A membrane-bounded sac that simple animals use for digestion. 752

food web Connections between species that eat each other. 862

foot An organ that mollusks use to locomote. 519

forebrain *FOR-brane* The front part of the vertebrate brain. 620

formed elements Blood cells and platelets. 711

founder effect Genetic drift when small groups found new settlements, partitioning a subset of the original population's genes. 408

fountain effect Splitting and regrouping of a school of fish, which confuses a predator. 822

fovea centralis *FO-ve-ah cenTRAL-is* An indentation in the retina opposite the lens that has only cones and provides visual acuity. 648

frameshift mutation *FRAME-shift mew-TAY-shun* A mutation that adds or deletes one or two DNA bases, altering the reading frame. 341

free energy The usable energy in the bonds of a molecule. 116

free radical Highly reactive by-product of metabolism that can damage tissue. 233

fronds *FRONDZ* Leaves of true ferns. 498

frustules *FRUS-tools* Two-part silica walls of diatoms that provide complex shapes. 467

fucoxanthin *foo-ko-ZAN-thin* A pigment that gives golden-brown algae their distinctive color. 467

fundamental niche *fun-da-MEN-tal NEESH* All the resources that a species could possibly use in its environment. 838

G

G_1 phase The gap stage of interphase when proteins, lipids, and carbohydrates are synthesized. 168

G_2 phase The gap stage of interphase when membrane components are synthesized and stored. 170

gallbladder *GALL-blad-er* A structure leading from the liver and toward the small intestine that stores bile. 757

gamete *GAM-eet* A sex cell. The sperm and ovum. 192

gametogenesis *ga-MEET-o-gen-i-sis* Meiosis and maturation; making gametes. 196

gametophyte *gam-EET-o-fite* Haploid stage of the plant life cycle. 192

gastric juice *GAS-trik JUICE* The fluid that stomach cells secrete that carries out chemical digestion. 760

gastrin *GAS-trin* A hormone that stomach cells secrete that stimulates secretion of more gastric juice. 760

gastrodermis *gas-tro-DERM-is* The inner tissue layer of a cnidarian. 517

gastrointestinal tract *GAS-tro-in-TES-ti-nal TRAKT* A continuous tubule along which food is physically and chemically digested. 757

gastropods *GAS-tro-podz* Snails and slugs. 519

gastrovascular cavity *gas-tro-VASC-yu-lar CAV-it-ee* The site of digestion in a cnidarian. 517

gastrula *GAS-troo-la* A three-layered embryo. 220

gemmae *GEM-ay* Structures that form asexually from liverworts that give rise to progeny. 494

gemmules *GEM-yools* Asexual structures in sponges. 516

gene A sequence of DNA that specifies the sequence of amino acids of a particular polypeptide. 51, 262

gene pool All the genes in a population. 404

generative nucleus *GEN-er-ah-tiv NU-klee-us* A haploid cell resulting from the mitotic division of a microspore in male plant reproduction. 563

genetic code Correspondence between specific DNA base triplets and amino acids. 51, 331

genetic drift Changes in gene frequencies caused by separation of a small group from a larger population. 408

genetic load Collection of deleterious alleles in a population. 410

genetic marker A detectable piece of DNA closely linked to a gene of interest whose precise location is unknown. 289

genotype *JEAN-o-type* Genetic constitution of an individual. 265

genotypic ratio *jean-o-TIP-ik RAY-shee-o* Proportions of genotypes among offspring of a genetic cross. 266

geological time scale A division of time into major eras of biological and geological activity, then periods within eras, and epochs within some periods. 359

germ cell Gamete or sex cell. 196

germinal mutation *GER-min-al mew-TAY-shun* A mutation in a sperm or oocyte. 341

germ layer A layer of the primordial embryo. 220

gibberellins *GIB-ah-REL-inz* A class of plant hormones that promote cell elongation and division in seeds, roots, shoots, and young leaves. 577

gill A highly folded respiratory surface for gas exchange in aquatic animals. 734

Ginkgo *GEENG-ko* A division of the gymnosperms; tree type. 499

gizzard The second part of a bird's stomach. 754

glomerular capsule *glo-MER-u-lar KAP-sool* The cup-shaped proximal end of the renal tubule that surrounds the glomerulus. 781

glomerular filtrate *glo-MER-u-ler FIL-trate* In a nephron in the kidney, the material that diffuses from the glomerulus to the glomerular capsule. 783

glomerulus *glo-MER-u-lus* A ball of capillaries between the afferent arterioles and efferent arterioles in the proximal part of a nephron. 780

glottis *GLOT-is* Opening from the pharynx to the larynx. 736

glucagon *GLU-ka-gon* A pancreatic hormone that breaks down glycogen into glucose, raising blood sugar levels. 664, 672

glucocorticoid *GLU-ko-KOR-tah-koid* Hormone that the adrenal cortex secretes that enables the body to survive prolonged stress. 667, 670

glycerol *GLI-sir-all* A three-carbon alcohol that forms the backbone of triglyceride fats. 45

glycocalyx *gli-ko-CAY-lix* A sticky layer of a bacterial cell wall that consists of proteins and/or polysaccharides. 446

glycolysis *gli-KOL-eh-sis* A catabolic pathway occurring in the cytoplasm of all cells. One molecule of glucose splits and rearranges into two molecules of pyruvic acid. 127

Gnetophyta *NEE-to-fit-a* A division of gymnosperms; tree type. 499

Golgi body *GOL-gee bod-ee* A system of flat, stacked, membrane-bounded sacs where sugars are polymerized to starches or bonded to proteins or lipids. 71

gonadotropic hormone *go-NAD-o-TRO-pik HOR-moan* A hormone made in the anterior pituitary that affects the ovaries or testes. 667

gradualism Slow evolutionary change. 415

granum *GRAN-um* A stack of flattened thylakoid discs that forms the inner membrane of a chloroplast. 74, 151

gravitropism *grav-eh-TROP-izm* A plant's growth response toward or away from gravity. 578

greenhouse effect Elevation in surface temperature caused by carbon dioxide buildup. 896

gross primary production Total amount of solar or geothermal energy converted to chemical energy in a certain time period by organisms in a given region. 863

ground tissue The tissue that makes up most of the primary body of a plant. 534

growth hormone (GH) A hormone produced in the pituitary gland that promotes growth and development of all tissues by increasing protein synthesis and cell division rates. 667

guanine *GWAN-een* One of the two purine nitrogenous bases in DNA and RNA. 51

guard cells Cells that control opening and closing of stomata in plants. 536

gymnosperm *JIM-no-sperm* Naked seed plant. 492

gynoecium *guy-no-EE-see-um* The second innermost whorl of a flower, consisting of female reproductive structures. 563

H

habitat *HAB-eh-tat* The physical and living surroundings of a species. 830

habituation *ha-BIT-ju-AY-shun* The simplest form of learning, in which an animal learns not to respond to certain irrelevant stimuli. 815

hair follicle *HAYR FOL-i-kul* An epidermal structure anchored in the dermis from which a hair grows. 599

half-life The time it takes for half the isotopes in a sample of an element to decay into a second isotope. 393

halophiles *HAL-o-files* Archaea that tolerate extremely salty surroundings. 451

haploid cell *HAP-loid SEL* A cell with one copy of each chromosome. Also called 1*n*. 192

hardwood Wood in dicots, such as oak, maple, and ash. 551

Hardy-Weinberg equilibrium *HAR-dee WINE-berg EE-kwah-LIB-ree-um* Maintenance of the proportion of genotypes in a population from one generation to the next. 405

heart valves Flaps in the heart that keep blood flow in one direction. 720

heartwood Wood in the center of a tree, where wastes collect. 551

heat capacity The amount of heat necessary to raise the temperature of a substance. 42

heat of vaporization *HEET of VA-por-i-ZA-shun* The amount of heat required to convert a liquid to vapor (gas). 42

heavy chains The two larger polypeptides of an antibody subunit. 795

helper T cell Lymphocyte that produces cytokines and stimulates activities of other immune system cells. 794

heme group *HEEM GROOP* An iron-containing complex that forms the oxygen-binding part of hemoglobin. 715

hemizygous *HEM-ee-ZY-gus* A gene on the Y chromosome in humans. 291

hemocoel *HEEM-o-seal* The body cavity of an arthropod. 522

hemocyanin *heem-o-SI-a-nin* A respiratory pigment in mollusks that carries oxygen. 710

hemoglobin *HEEM-o-GLO-bin* The iron-binding protein that carries oxygen in mammals. 710

hemolymph *HEEM-o-limf* The blood in arthropods. 522

hepatic portal system *heh-PAH-tik POR-tel SIS-tum* A division of the circulatory system that enables the liver to rapidly harness chemical energy in digested food. 725

heritability *herr-it-a-BILL-it-ee* The proportion of a trait attributable to heredity. 278

heterochromatin *het-er-o-KROME-a-tin* Dark-staining genetic material. 296

heterogametic sex *HET-er-o-gah-MEE-tik SEX* The sex with two different sex chromosomes. 290

heterotherm *HET-er-o-therm* An animal with variable body temperature. 771

heterotroph(ic) *HET-er-o-TROFE* An organism that obtains nourishment from another organism. 14, 146, 751

heterozygous *HET-er-o-ZI-gus* Possessing two different alleles for a particular gene. 265

highly conserved A protein or nucleic acid sequence that is very similar in different species. 395

hindbrain *HIND-brayne* The lower portion of the vertebrate brain, which includes the brain stem and controls vital functions. 620

hirudinea *hear-EW-din-ay* A class of annelids that includes the leeches. 521

histamine *HIS-tah-meen* An allergy mediator that dilates blood vessels and causes allergy symptoms. 791

holozoic *HOLE-o-ZO-ik* A protozoan that takes in food through phagocytic vacuoles. 459

homeostasis *HOME-ee-o-STA-sis* The ability of an organism to maintain constant body temperature, fluid balance, and chemistry. 9

homeotherm *HOME-e-o-therm* An animal with constant body temperature. 771

homeotic *home-ee-OT-ik* A mutant in which body parts form in the wrong places. 215

hominid *HOM-eh-nid* Animals ancestral to humans only. 367

hominoid *HOM-eh-noid* Animals ancestral to apes and humans. 367

homogametic sex *HO-mo-gah-MEE-tik SEX* The sex with two identical sex chromosomes. 290

homologous pairs *ho-MOL-eh-gus PAIRZ* Chromosome pairs that have the same sequence of genes. 198

homologous structures *ho-MOL-eh-gus STRUK-churs* Similar structures in different species that have the same general function, indicating descent from a common ancestor. 393

homozygous *HO-mo-ZI-gus* Possessing two identical alleles for a particular gene. 265

hornwort *HORN-wart* A type of bryophyte. 493

horsetail Seedless vascular plant of genus *Equisetum*. 497

human chorionic gonadotropin *YU-man KOR-ee-on-ik go-NAD-o-TRO-pin* (hCG) A hormone secreted by the preembryo and embryo that prevents menstruation. 218, 674

humoral immune response *HUME-er-al IM-mune ree-SPONZ* Secretion of antibodies by B cells in response to detecting a foreign antigen. 795

hybridoma *hi-bri-DOME-ah* An artificial cell created by fusing a B cell with a cancer cell that secretes a particular antibody. It mass-produces the antibody. 804

hydrocarbon *HI-dro-kar-bon* A molecule containing carbon and hydrogen. 38

hydrogen bond *HI-dro-gen BOND* A weak chemical bond between negatively charged portions of molecules and hydrogen ions. 39

hydrolysis *hi-DROL-eh-sis* Splitting a molecule by adding water. 43

hydrophilic *HI-dro-FILL-ik* Attracted to water. 39

hydrophobic *HI-dro-FOBE-ik* Repelled by water. 39

hydrostatic skeleton *Hi-dro-STAT-ik SKEL-eh-ten* The simplest type of skeleton, consisting of flexible tissue surrounding a constrained liquid. 683

hydroxide ion *hi-DROX-ide I-on* (OH^-) A molecule consisting of one oxygen and one hydrogen. 40

hyperthermophiles *hi-per-THERM-o-files* Archaea that use sulfur and tolerate extreme heat. 451

hypertonic *hi-per-TON-ik* The solution on one side of a membrane where the solute concentration is greater than on the other side. 89

hypertrophy *hi-PER-tro-fee* Increase in muscle mass, usually due to exercise. 701

hypha *HI-fa* A fungal thread, the basic structural unit of a multicellular fungus. 477

hypocotyl *hi-po-COT-ul* Stemlike region beneath seed leaves. 568

hypodermis *hi-po-DERM-is* An outer layer of a roundworm that secretes a protective cuticle. 519 The outermost protective layer of the cortex of a plant. 547

hypothalamus *Hi-po-THAL-eh-mus* A small structure beneath the thalamus that controls homeostasis and links the nervous and endocrine systems. 627

hypothesis *hy-POTH-eh-sis* An educated guess based on prior knowledge. 21

hypotonic *hi-po-TON-ic* The solution on one side of a membrane where the solute concentration is less than on the other side. 89

I

ileum *IL-ee-um* The last section of the small intestine. 761

imbibition *IM-bah-BISH-un* The absorption of water by a seed. 40

implantation *im-plan-TAY-shun* Nestling of the blastocyst into the uterine lining. 218

imprinting A type of learning that usually occurs early in life and is performed without obvious reinforcement. 816

inclusive fitness Fitness defined by personal reproductive success as well as by relatives sharing an individual's genes. 826

incomplete dominance A heterozygote whose phenotype is intermediate between the phenotypes of the two homozygotes. 273

incomplete penetrance A genotype that does not always produce a phenotype. 275

incus *INK-us* A small bone in the middle ear. 653

independent assortment The random organization of homologs during metaphase of meiosis I. 201, 269

indeterminate cleavage *in-dee-TER-min-et CLEEV-edge* A cleavage embryo in which the fate of the cells is not yet determined. 514

industrial melanism *in-DUS-tree-al MEL-an-iz-um* Coloration that is adaptive in a polluted area. 410

inferior vena cava *in-FEAR-ee-er VE-nah KA-vah* The lower branch of the largest vein that leads to the heart. 720

inflammation *in-flam-AY-shun* Increased blood flow and accumulation of fluid and phagocytes at the site of an injury, rendering it inhospitable to bacteria. 791

initiation complex *in-ish-e-A-shun COM-plex* A small ribosomal subunit, mRNA, and an initiator tRNA, joined. 333

initiation site *in-ish-e-A-shun SITE* The site on a chromosome where DNA replication begins. 317

innate *in-AYT* Instinctive. 812

inner cell mass The cells in the blastocyst that develop into the embryo. 218

insertion *in-SER-shun* The end of a muscle on a movable bone. 700

insight learning Ability to apply prior learning to a new situation without trial-and-error activity. 816

insulin *IN-sel-in* A pancreatic hormone that lowers blood sugar level by stimulating body cells to take up glucose from the blood and to metabolize or store it. 663, 672

integrin *in-TEG-rin* A cell adhesion molecule that helps direct a white blood cell through a capillary wall to the site of an injury. 103

integument *in-TEG-u-ment* Outer covering of an animal's body. 594

intercalary meristem *in-TER-kah-LARE-ee MER-eh-stem* Dividing tissue between mature regions of grass stem. 535

intercalated disks *in-TER-kah-LAY-tid DISKS* Tight foldings in cardiac muscle cell membranes that join adjacent cells. 692

intercellular junctions Structures that join cells, including tight junctions, gap junctions, desmosomes, and plasmodesmata. 96

interferon *in-ter-FEAR-on* A polypeptide produced by a T cell infected with a virus that diffuses to surrounding cells and stimulates them to manufacture biochemicals that halt viral replication. 791

interleukin *in-ter-LUKE-in* A class of immune system biochemicals. 792

intermediate-day plant Plant that flowers only when exposed to days of intermediate length, growing vegetatively at other times. 584

intermediate filament Cytoskeletal element intermediate in size between a microtubule and a microfilament. 100

intermembrane compartment *in-ter-MEM-brain cum-PART-ment* The space between a mitochondrion's two membranes. 128

interneuron *in-ter-NURE-on* A neuron that connects one neuron to another to integrate information from many sources and to coordinate responses. 611

internode *IN-ter-node* Part of stem between nodes. 541

interphase *IN-ter-faze* The period when the cell synthesizes proteins, lipids, carbohydrates, and nucleic acids. 168

intertidal zone *IN-ter-TI-dal ZONE* The region bordering an estuary where the tide recedes and returns. 881

intestinal ceca *in-TES-tin-al SEE-ka* The site of digestion and absorption in flatworms. 517

intracellular digestion *in-tra-SEL-yu-lar di-JEST-shun* Digestion within food vacuoles in cells. 753

intrinsic rate of increase Population growth rate if survival and reproduction are unrestricted. 840

intrinsic rate of natural increase Population growth rate under ideal conditions. 840

intron *IN-tron* Bases of a gene that are transcribed but are excised from the mRNA before translation into protein. 330

inversion *in-VER-shun* An introverted gene sequence. 301

invertebrate *in-VERT-e-brayt* An animal that does not have a backbone. 508

ion *I-on* An atom that has lost or gained electrons, giving it an electrical charge. 38

ionic bond *i-ON-ik bond* Attraction between oppositely charged ions. 38

ionizing radiation *I-o-nize-ing rade-e-A-shun* Radiation that ejects electrons from atoms. 147

irritability *IR-eh-tah-BIL-eh-tee* An immediate response to a stimulus. 10

islets of Langerhans *I-letz uv LANG-er-hanz* Clusters of cells in the pancreas that secrete hormones that control nutrient use. 672

isotonic *ice-o-TON-ik* When solute concentration is the same on both sides of a membrane. 88

isotope *I-so-tope* A differently weighted form of an element. 33

J

jejunum *jah-JU-num* The middle section of the small intestine. 761

J-shaped curve The shape of a curve depicting exponential population growth. 840

juvenile hormone *JU-vin-iyl HOR-moan* An insect hormone produced in larvae that controls metamorphosis. 665

K

karyokinesis *KAR-ee-o-kah-NEE-sus* Division of the genetic material. 168

karyotype *KAR-ee-o-type* A size-order chart of chromosomes. 286

keratin *KERR-a-tin* A hard protein that accumulates in the integument of many animals and forms specialized structures. 594

kidneys Paired organs consisting of millions of tubules that excrete nitrogenous waste and regulate water levels. 780

kilocalorie *KIL-o-KAL-o-ree* (kcal) The energy required to raise 1 kilogram of water 1 degree Celsius. 755

kinase *KI-nase* A type of enzyme that activates other proteins by adding a phosphate. 177

kinetic energy *kin-ET-ik EN-er-gee* The energy of motion. 112

Krebs cycle *KREBS SI-kle* The stage in cellular respiration that completely metabolizes the products of glycolysis. 127

K-selection A form of natural selection in which individuals are long-lived, late-maturing, and have few offspring that they intensively care for. Density-dependent factors regulate these populations. 844

L

labor Uterine contractions that expel a fetus. 216

lacteal *LAK-tee-ul* A lymph vessel that absorbs fat in the small intestine. 762

lactic acid fermentation *LAK-tik AS-id fer-men-TAY-shun* Conversion of pyruvic acid into lactic acid, occurring in some anaerobic bacteria and tired mammalian muscle cells. 139

lacunae *la-KEW-nay* Spaces in cartilage and bone tissue. 595

laminin *LAM-i-nin* A protein that binds skin layers. 102, 599

lancelets *LAN-si-letz* One of the three groups of chordates, including primitive invertebrates. 525

larynx *LAR-inks* The "voice box" and a conduit for air. 736

latent learning *LA-tent LERN-ing* Learning without reward or punishment; not apparent until after the learning experience. 816

lateral meristems *LAT-er-al MER-ee-stemz* Actively dividing plant cells that grow outward, thickening the plant. 534

law of segregation Allele separation during meiosis. 263

leader sequence A short sequence at the start of each mRNA that enables the mRNA to hydrogen bond with part of the rRNA in a ribosome. 332

leaf abscission *LEAF ab-SCISZ-on* Shedding of a tree's leaves as part of its life cycle. 546

lens The structure in the eye through which light passes and is focused. 648

lentic system *LEN-tik SIS-tum* Freshwater biome that has standing water, such as a lake or pond. 879

life expectancy Prediction of how long an individual will live, based on current age and epidemiology. 234

life span The longest a member of a species can live. 234

ligaments *LIG-ah-mentz* Tough bands of fibrous connective tissue that form a joint capsule. 700

ligand *LIG-and* A messenger molecule that binds to a cell surface protein. 94

ligase *LIG-aze* An enzyme that catalyzes formation of covalent bonds in the DNA sugar-phosphate backbone. 317

light chains The two smaller polypeptide chains in an antibody subunit. 795

lignin *LIG-nin* A carbohydrate in plants that adds support. 493

limnetic zone *lim-NET-ik ZONE* The layer of open water in a lake or pond that light penetrates. 879

linkage map *LINK-ege MAP* Diagram of gene order on a chromosome based on crossover frequencies. 288

linked genes Genes on the same chromosome. 286

lipase *LI-payse* Enzyme that chemically digests fats. 762

lipids *LIP-idz* Compounds containing carbon, hydrogen, and oxygen, but with less oxygen than carbohydrates. 44

lithotroph *LITH-o-trofe* Organism that obtains electrons from reduced inorganic compounds. 445

littoral zone *LIT-or-al ZONE* The shallow region along the shore of a lake or pond where sufficient light reaches to the bottom for photosynthesis. 879

liver The largest solid organ in the body. It detoxifies blood, stores glycogen and fat-soluble vitamins, synthesizes blood proteins, and monitors blood glucose level. 757

liverwort *LIV-er-wart* A type of bryophyte. 493

long-day plant Plant that requires light periods longer than some critical length to flower. 584

loose connective tissue Connective tissue with widely spaced fibroblasts and a few fat cells. 594

lotic system *LO-tik SIS-tum* Freshwater biome with running water, such as a river or stream. 879

low-density lipoprotein *LO DEN-sit-ee LIP-o-PRO-teen* A molecule that carries cholesterol. 94

lungs Paired structures that house the bronchial tree and the alveoli; the sites of gas exchange. 735, 739

luteinizing hormone *LU-ten-I-zing HOR-moan* (LH) A hormone made in the anterior pituitary that promotes ovulation. 666

lymph *LIMF* Blood plasma minus some large proteins, which flows through lymph capillaries and lymph vessels. 727

lymphatic system *lim-FAH-tik SIS-tum* A circulatory system that consists of lymph capillaries and lymph vessels that transport lymph. 727

lymph capillaries *LIMF CAP-eh-lair-eez* Dead-end, microscopic vessels that transport lymph. 727

lymph nodes *LIMF NODES* Structures in the lymphatic system that contain white blood cells and fight infection. 728

lymph vessels *LIMF VES-selz* Vessels that transport lymph and eventually empty into veins. 727

lysosome *LI-so-soam* A sac in a eukaryotic cell in which molecules and worn-out organelles are enzymatically dismantled. 71

M

macroevolution *MAK-ro-ev-eh-LU-shun* Large-scale evolutionary changes, such as speciation and extinction. 375

macromolecule *MAK-ro-MOL-e-kuel* A very large molecule. 35

macronucleus *MAK-ro-NEW-klee-us* A large nucleus in a protozoan cell. 459

macronutrients *MAK-ro-NU-tree-entz* Carbohydrates, fats, and proteins obtained from food in large amounts. 755

macrophage *MAK-ro-FAYGE* Very large, wandering phagocyte. 790

malignant tumor *mal-IG-nant TOO-mer* A cancerous tumor. 180

malleus *MAL-e-us* A small bone in the middle ear. 653

Malpighian tubule *mal-PIG-he-in TOOB-yul* An insect excretory structure. 778

mammary gland *MAM-a-ree GLAND* Milk-producing sweat gland derivative in mammals. 603

mangrove swamp *MAN-growv SWAMP* Tropical area that connects the sea and forests. 881

mantle A dorsal fold of tissue that secretes a shell in mollusks. 519

marsupial *mar-SOOP-e-al* A pouched mammal. 366

mast cell Immune system cell that releases allergy mediators when stimulated. 791, 800

matrix *MAY-trix* The inner compartment of a mitochondrion. The nonliving part of connective tissue. 128, 594

mechanoreception *mek-AN-o-ree-SEP-shun* Detecting touch. 644

medulla oblongata *mah-DULE-ah ob-long-AH-tah* The part of the brain stem nearest the spinal cord; regulates breathing, heartbeat, blood pressure, and reflexes. 627

medusa *med-EW-sa* The free-swimming form of a cnidarian. 517

megagametophyte *MEG-ah-gah-MEE-toe-fight* Large, egg-producing structure in a plant. 557

megakaryocyte *MEG-ah-KAR-ee-o-site* A huge bone marrow cell that breaks apart to yield platelets. 717

megasporangia *MEG-ah-spor-AN-gee-ah* Structures in which megaspores form. 557

megaspore *MEG-ah-spor* Structure in plants that yields female gametophyte. 557

megaspore mother cell *MEG-ah-spor MOTH-er SEL* A cell within an ovule in plants that divides meiotically to produce four haploid cells, three of which degenerate and one of which becomes the egg. 563

meiosis *mi-O-sis* Cell division that halves the genetic material. 192

melanocyte *mel-AN-o-site* Cell that produces melanin pigment. 598

melanocyte-stimulating hormone *mel-AN-o-site STIM-u-lat-ing HOR-moan* (MSH) A hormone produced between the anterior and posterior pituitary in some vertebrate species that controls skin pigmentation. 669

melatonin *mel-ah-TOE-nin* A hormone produced in the pineal gland that may control other hormones by sensing light and dark cycles. 675

memory cell Mature B cell, specific to an antigen already met, that responds quickly by secreting antibodies when that antigen is encountered again. 795

mesoderm *MEZ-o-derm* The middle embryonic germ layer, whose cells become bone, muscle, blood, dermis, and reproductive organs. 220

mesoglia *mez-o-GLEE-a* A jellylike substance in the middle of a cnidarian body. 517

mesohyl *MEZ-o-hile* A part of a sponge that contains amoebocytes. 515

mesophyll cell *MEZ-o-fill SEL* Thin-walled plant cell that takes part in C_4 photosynthesis. 160

messenger RNA *MESS-en-ger RNA* (mRNA) A molecule of ribonucleic acid that is complementary in sequence to the sense strand of a gene. 329

meta-analysis *ME-tah a-NAL-i-sis* A study that combines results of several studies. 27

metabolic pathway *met-a-BOL-ik PATH-way* A series of connected, enzymatically catalyzed reactions in a cell. 114

metabolism *meh-TAB-o-liz-um* The biochemical reactions that acquire and use energy. 8, 114

metacentric *met-a-SEN-trik* A chromosome whose centromere divides it into two similarly sized arms. 296

metamerism *MET-a-mer-izm* A body consisting of repeating segments. 521

metamorphosis *met-a-MORE-foe-sis* A developmental process in which an animal changes drastically in body form between the juvenile and the adult. 512

metaphase *MET-ah-faze* The second stage of cell division, when chromosomes align down the center of a cell. In mitosis, the chromosomes form a single line. 171

metaphase I *MET-ah-faze I* In meiosis I, when homologs align. 200

metaphase II *MET-a-faze II* When replicated chromosomes align down the center of the cell in meiosis II. 201

metapopulation *met-a-pop-eu-LAY-shun* Local populations that frequently exchange members. 838

metastasis *meh-TAH-stah-sis* Spreading of cancer. 181

methanogen *meth-AN-o-gen* Archaean that produces methane as a by-product of its metabolism. 451

microevolution *MIKE-ro-ev-eh-LU-shun* Subtle, incremental single-trait changes that underlie speciation. 375

microfilaments *MI-kro-FILL-ah-ments* Tiny actin rods in cells, especially contractile cells. 97

microgametophyte *MIK-ro-gah-MEE-toe-fight* Small sperm-producing structure in a plant. 557

micronucleus *mi-kro-NUKE-lee-us* A small nucleus in a protozoan cell. 459

micronutrients *MIKE-ro-NU-tree-entz* Vitamins and minerals, which are required in small amounts. 755

microsporangia *MIKE-ro-spor-AN-gee-ah* Structures in which microspores form. 557

microspore *MIKE-ro-spor* Structure in plants that gives rise to male gametophyte. 557

microspore mother cell *MIKE-ro-spor MOTH-er SEL* Cell in pollen sacs that divides meiotically to produce four haploid microspores. 563

microtubules *MI-kro-TU-bules* Long, hollow tubules of tubulin protein that move cells. 74, 97

microvilli *MIKE-ro-VIL-i* Tiny projections on the surfaces of epithelial cells that comprise intestinal villi. 762

midbrain *MID-brane* Part of the brain stem above the pons where white matter connects with higher brain structures and gray matter contributes to sight and hearing. 620

mineralocorticoids *MIN-er-ral-o-KOR-tah-KOIDZ* Hormones produced in the adrenal cortex that maintain blood volume and electrolyte balance by stimulating the kidneys to return sodium ions and water to the blood and to excrete potassium ions. 670

missense mutation *MISS-sentz mu-TAY-shun* A mutation that changes a codon specifying a certain amino acid into a codon specifying a different amino acid. 341

mitochondria *MI-toe-KON-dree-ah* Organelles within which the reactions of cellular metabolism occur. 71

mitosis *mi-TOE-sis* A form of cell division in which two genetically identical cells form from one. 168

mitotic spindle *mi-TOT-ik SPIN-del* A structure of microtubules that aligns and separates chromosomes in mitosis. 170

mixotroph *MIX-o-trof* A bacterium that uses inorganic molecules for energy and organic molecules for carbon. 445

mode of inheritance Whether a trait is autosomal or sex-linked, and dominant or recessive. 267

molecular evolution Estimating degrees of evolutionary relatedness by comparing informational polymer sequences among species. 395

molecular tree diagram A depiction of hypothesized relationships between species based on molecular sequence data. 397

molecular weight The sum of the weights of the atoms that make up a molecule. 35

molecule *MOL-eh-kuel* Combined atoms. 34

mollusk *MOLL-usk* Animal with a soft body, hard shell, three body regions, muscular feet, and circulatory, excretory, and nervous systems. 512

molting hormone An insect hormone produced in the late larva that triggers metamorphosis. 665

Monera *mone-AIR-ah* The kingdom that includes bacteria and cyanobacteria. 14, 430

monoclonal antibodies *MON-o-KLON-al AN-tee-bod-eez* Identical antibodies provided by a single B cell. 804

monocots *MON-o-kotz* Flowering plants that have one seed leaf. 541
monogamy *mono-OG-a-mee* Formation of a permanent male-female pair. 830
monohybrid cross *MON-o-HI-brid CROSS* Mating of individuals for a particular gene. 266
monomer *MON-o-mer* A single link in a polymeric molecule. 43
monosaccharide *MON-o-SAK-eh-ride* A sugar that is one five- or six-carbon unit. 43
monosomy *MON-o-SOAM-ee* Absence of one chromosome. 298
monotreme *MON-o-treem* An egg-laying mammal. 366, 527
monounsaturated (fat) *MON-o-un-SAT-yer-a-tid* A fatty acid with one double bond. 45
monozygotic twins *mon-o-zi-GOT-ik TWINZ* Identical twins resulting from the splitting of a fertilized ovum. 279
morphogens *MORF-o-genz* Proteins that form gradients that influence development. 215
morula *MORE-u-lah* The preembryonic stage of a solid ball of cells. 218
mosaic catastrophism *mo-ZAY-ik cat-ASS-trof-izm* The idea that a series of great floods molded the earth's features and caused extinctions and speciations. 376
moss The most common type of bryophyte. 493
motif *MOE-teef* A common part of a transcription factor. 327
motor neuron *MO-ter NEUR-on* A neuron that transmits a message from the central nervous system toward a muscle or gland, with a long axon and short dendrites. 611
motor unit A neuron and all the muscle fibers it contacts. 699
mucigel *MYUS-eh-gel* A slimy substance that root cap cells produce. 546
multicellular *mull-tee-SEL-u-lar* Consisting of many cells. 10, 60
multifactorial *mull-tee-fac-TORE-e-al* Traits molded by one or more genes and the environment. 276
multiple fission *MULL-ti-pull FISH-in* A form of asexual reproduction in protozoa in which a parent cell undergoes several nuclear divisions and then partitions off several progeny at one time. 459
muscle fasciculi *MUS-sel fah-SIK-u-li* Bundles of muscle fibers. 694
muscle fiber Skeletal muscle cell. 694
muscle spindles Receptors in skeletal muscle that monitor the degree of muscle tone or how many fibers contract at a time. 701
muscular tissue Tissue consisting of contractile cells that provide motion. 594
mutagen *MUTE-a-jen* An agent that causes a mutation. 339
mutant *MU-tent* A phenotype or allele that is not the most common for a certain gene in a population or that has been altered from the "normal" condition. 265
mutation *mu-TAY-shun* A change in a gene or chromosome. 265
mutualism *MU-chu-a-lism* Symbiosis that benefits both partners. 434
mycelium *mi-SELL-ee-um* An assemblage of hyphae that forms an individual fungus. 477
mycology *mi-KOL-o-gee* The study of fungi. 476
mycorrhizae *mi-kor-IZ-ee* Parts of a fungus that wrap around plant roots. 485
myelin sheath *MI-eh-lin SHEATH* A fatty material that insulates some nerve fibers in vertebrates, allowing rapid nerve impulse transmission. 614
myoepithelial cell *mi-o-ep-ee-THEEL-ee-al SEL* Cell that can contract and be part of a layer. 598
myofibrils *mi-o-FI-brilz* Cylindrical subunits of a muscle fiber. 694
myofilaments *mi-o-FILL-eh-mentz* Acin or myosin "strings" that make up myofibrils. 695
myosin *MI-o-sin* The protein that forms thick filaments in muscle tissue. 100, 597
myriapods *mi-REE-a-podz* Centipedes and millipedes. 522

N

nastic movement *NAS-tic MOVE-ment* Plant growth or movement that is not oriented toward the stimulus. 581
natural selection The differential survival and reproduction of organisms whose genetic traits better adapt them to a particular environment. 11
necrosis *neck-RO-sis* A form of cell death. 178
negative feedback Turning off of an enzyme's synthesis or activity when the product of the reaction that the enzyme catalyzes accumulates. 122
nematocyst *ne-MAT-o-cist* Stinging cell of cnidaria. 516
neonatology *NE-o-nah-TOL-eh-gee* Study of the newborn. 249
nephridia *nef-RID-ee-a* Networks of tubules in some invertebrates that have an excretory function. 778
nephron *NEF-ron* A microscopic tubular subunit of a kidney, consisting of a renal tubule and peritubular capillaries. 780
nephron loop *NEF-ron LOOP* A tubular section of nephron (kidney subunit) that lies between the proximal and distal convoluted tubules, where water is conserved and urine concentrates by a countercurrent exchange system. 783
Neptunism *NEP-tune-iz-um* The idea that a single great flood organized the features of the earth's surface present today. 376
neritic zone *ner-IT-ik ZONE* The coastal region of an ocean. 883
nervous tissue A tissue whose cells (neurons and neuroglia) form a communication network. 594
net primary production Chemical energy in plants for growth and reproduction after cellular respiration. 863
neural tube *NEUR-el TOOB* Embryonic precursor of the central nervous system. 226, 620
neuroglia *neur-o-GLEE-a* Cells associated with neurons in nervous tissue. 596
neuron *NEUR-on* A nerve cell, consisting of a cell body, a long "sending" projection called an axon, and numerous "receiving" projections called dendrites. 596
neurosecretory cell *NUR-o-SEK-rah-tore-ee SEL* Cell in the hypothalamus that functions as a neuron at one end but as an endocrine cell at the other by receiving neural messages and secreting the hormones ADH and oxytocin. 661
neurotransmitter *NEUR-o-TRANZ-mit-er* A chemical passed from a neuron to receptors on another neuron or on a muscle or gland cell. 597, 616
neurulation *NEUR-u-LAY-shun* Physical contact between the notochord and nearby ectoderm that triggers formation of the nervous system. 226, 620
neustonic *new-STON-ik* Algae that live where the water meets the air. 463
neutron *NEW-tron* A particle in an atom's nucleus that is electrically neutral and has one mass unit. 32
nitrifying bacteria *NI-tri-fy-ing bac-TEAR-e-a* Bacteria that convert ammonia in dead organisms into nitrites and nitrates, which other organisms can use. 865
nitrogen fixation *NI-tro-gen fix-A-shun* A microbial process that reduces atmospheric nitrogen gas to ammonia, which other organisms can use. 452
nitrogenous base *ni-TRODGE-eh-nus BASE* A nitrogen-containing compound that forms part of a nucleotide, giving it individuality. 51
node Area of leaf attachment. 541
node of Ranvier *NODE uv RON-vee-ay* A short region of exposed axon between Schwann cells on neurons of the vertebrate peripheral nervous system. 614
nondisjunction *NON-dis-JUNK-shun* Unequal partition of chromosomes into gametes during meiosis. 298
nonpolar covalent bond *non-POE-lar co-VAY-lent BOND* A covalent bond in which atoms share electrons equally. 38
nonsense mutation *NON-sents mu-TAY-shun* A point mutation that alters a codon that encodes an amino acid to one that encodes a stop codon. 341
nonshivering thermogenesis *non-SHIV-er-ing ther-mo-GEN-i-sis* Hormone-directed internal heat. 772
norepinephrine *NOR-ep-ee-NEF-rin* A hormone produced in the adrenal medulla that raises blood pressure, constricts blood vessels, and slows digestion, in response to a threat. 670
notochord *NO-toe-kord* A semirigid rod running down the length of an animal's body. 226, 524
nuclear envelope *NEW-klee-ar EN-vel-ope* A two-layered structure bounding a cell's nucleus. 70
nuclear pore *NEW-klee-ar POOR* A hole in the nuclear envelope. 70
nucleic acid *new-CLAY-ic AS-id* A biochemical that encodes an amino acid sequence. 50

nucleoid *NEW-klee-oid* The part of a prokaryotic cell where the DNA is located. 66
nucleolus *new-KLEE-o-lis* A structure within the nucleus where RNA nucleotides are stored. 71
nucleosome *NEW-klee-o-some* DNA wrapped around eight histone proteins as part of chromosome structure. 313
nucleotide *NEW-klee-o-tide* The building block of a nucleic acid, consisting of a phosphate group, a nitrogenous base, and a five-carbon sugar. 51
nucleus *NEW-klee-is* The central region of an atom, consisting of protons and neutrons. A membrane-bounded sac in a eukaryotic cell that contains the genetic material. 32, 61

O

obligate photoperiodism *OB-lah-get fo-toe-PER-ee-o-diz-um* A plant's requirement of a particular photoperiod in order to flower. 584
oceanic zone *O-shee-AN-ik ZONE* The open sea. 883
ocelli *o-SELL-ee* Eyespots; visual organs in flatworms. 644
octet rule *OC-tet ROOL* The tendency of an atom to fill its outermost shell. 35
olfactory epithelium *ol-FAK-tore-ee ep-e-THEEL-e-um* Neurons specialized to detect odors, in a small patch of tissue high in the nostrils. 646
oligochaetes *OL-eh-go-keets* A class of annelids that includes the earthworms. 521
oligotrophic *OL-ah-go-TRO-fik* A lake with few nutrients, usually very blue. 880
ommatidia *o-mah-TID-ee-ah* The visual units of a compound eye. 644
oncogene *ON-ko-jean* A gene that normally controls cell division but when overexpressed leads to cancer. 183
oocyte *OO-site* The female sex cell before it is fertilized. 196
oogenesis *oo-GEN-eh-sis* The differentiation of an egg cell from a diploid oogonium, to a primary oocyte, to two haploid secondary oocytes, to ootids, and finally, after fertilization, to a mature ovum. 204
oogonium *oo-GO-nee-um* The diploid cell where egg formation begins. 204, 464
open circulatory system A circulatory system in which blood is not always in vessels. 708
operant conditioning Trial-and-error learning, in which an animal voluntarily repeats any behavior that brings success. 816
operculum *o-PER-cu-lum* A flap of tissue that protects gills. 734
operon *OP-er-on* A series of genes with related functions and their controls. 327
optic nerve Nerve fibers that connect the retina to the visual cortex. 649
oral end The mouth. 511
organ A structure of two or more tissues that functions as an integrated unit. 5
organelles *OR-gan-NELLZ* Specialized structures in eukaryotic cells that carry out specific functions. 5, 60
organogenesis *or-GAN-o-gen-eh-sis* Development of organs in an embryo. 226
organotroph *or-GAN-o-trofe* Organism that obtains hydrogen or electrons from organic compounds. 445
orgasm *OR-gazz-m* A pleasurable sensation associated with sexual activity. 197
origin The end of a muscle on an immobile bone. 700
osculum *OS-kew-lum* A large hole at the top of a sponge that expels water and wastes. 515
osmoconformer *OZ-mo-con-FORM-er* An invertebrate whose ion concentrations match those in the ocean. 776
osmoregulation *OZ-mo-reg-u-LAY-shun* An organism's control of water and ion balance. 776
osmoregulator *OZ-mo-REG-u-lay-tor* An animal that controls its body fluid composition. 776
osmosis *oz-MO-sis* Passive diffusion of water. 88
ossicles *OSS-i-kulls* Calcium-containing structures in echinoderms' coverings. 523
osteoblast *OSS-tee-o-blast* A bone cell that secretes matrix. 595
osteoclast *OSS-tee-o-clast* A bone cell that degrades matrix. 595
osteocyte *OSS-tee-o-site* A mature bone cell in a lacuna. 595
osteon *OSS-tee-on* Concentric circles of osteocytes in bone. 688
osteonic canal *oss-tee-ON-ik ca-NAL* Portals that house blood vessels in bone. 688
osteoprogenitor cell *OSS-tee-o-pro-GEN-i-tor SEL* A cell lining a bone passageway that can differentiate into an osteoblast in the event of injury or growth. 595
ostia *OSS-tee-a* Surface pores on a sponge. 515
otolith *O-toe-lith* Calcium carbonate granules in the vestibule of the inner ear whose movements provide information on changes in velocity. 655
oval window A membrane between the middle ear and the inner ear. 653
ovary *O-var-ee* One of the paired female gonads that house developing oocytes. In a flowering plant, the carpels and the ovules they enclose. 197, 563
ovulation *OV-u-LAY-shun* The release of an oocyte from the largest ovarian follicle just after luteinizing hormone peaks in the blood in the middle of a menstrual cycle. 674
ovule *OV-yul* In flowering plants, a megasporangium that contains a megaspore mother cell. 563
oxidation *OX-e-DAY-shun* A chemical reaction that loses electrons. 117
oxidative phosphorylation *ox-i-DAY-tiv fos-for-e-LAY-shun* Phosphorylation of ADP to ATP coupled to protons moving down their concentration gradient across the inner mitochondrial membrane. 129
oxyhemoglobin *ox-ee-HEEM-o-GLO-bin* Bright red hemoglobin that has just picked up oxygen in the lungs. 715
oxytocin *ox-ee-TOE-sin* A hormone made in the hypothalamus and released from the posterior pituitary that stimulates muscle contraction in the mammary glands and the uterus. 664
ozone *O-zone* A bluish gas that forms high in the atmosphere when oxygen reacts with UV radiation. Consists of three oxygen atoms. 894

P

pacemaker Specialized cells in the wall of the right atrium that set the pace of the heartbeat. 720
Pacinian corpuscle *pah-SIN-ee-en KOR-pus-el* A receptor in the skin that senses touch. 656
paleontology *PAY-lee-on-TOL-ah-gee* Study of evidence of past life. 390
palisade mesophyll cells *PAL-eh-sade MEZ-o-fil SELZ* Columnar cells along the upper side of a leaf specialized for light absorption. 543
pancreas *PAN-kree-ass* A structure that has an endocrine part that produces somatostatin, insulin, and glucagon and a digestive part that produces pancreatic juice. 672, 764
parapodia *par-a-PODE-ee-a* Pairs of fleshy appendages that polychaete worms use to locomote. 521
parasitism *PAR-a-si-tizm* A symbiotic relationship in which one member harms the other. 435
parasympathetic nervous system *PAR-ah-SIM-pah-THET-ik NER-vus SIS-tum* Part of the autonomic nervous system that controls vital functions such as respiration and heart rate when at rest. 622
parathyroid glands *par-ah-THY-roid GLANZ* Four small groups of cells behind the thyroid gland that regulate calcium level in blood and tissue fluid, by secreting parathyroid hormone. 663
parenchyma *pah-REN-kah-mah* Abundant, unspecialized plant cells that can divide. 535
parental generation (P_1) The first generation in a genetic cross. 265
parsimony analysis *PAR-si-mone-ee eh-NAL-eh-sis* A statistical method to identify the most realistic evolutionary tree based on molecular data. 399
parthenogenesis *par-tho-GEN-eh-sis* Female reproduction without fertilization. 192
parturition *par-ter-I-shun* The birth of a baby. 216
passive immunity Immunity generated when an organism receives antibodies from another organism. 798
pedicellariae *ped-ee-SELL-ar-ay* Echinoderm pincers. 524
pedigree *PED-eh-gree* A chart showing relationships of relatives and which ones have a particular trait. 268
pelagic zone *pah-LA-gik ZONE* Water above the ocean floor. 883

penis *PEE-nis* Male sex organ. 197
pentaradial symmetry *pent-a-RAY-de-al SIM-it-ree* A body form with five arms, characteristic of echinoderms. 524
pepsin *PEP-sin* A stomach enzyme that chemically digests protein. 760
peptide bond *PEP-tide BOND* A chemical bond between two amino acids resulting from dehydration synthesis. 48
peptide hormone *PEP-tide HOR-moan* A hormone composed of amino acids that is water soluble but fat insoluble so that it cannot traverse a cell membrane. 662
pericardium *per-ee-KAR-de-um* Connective tissue sac that houses the heart. 718
pericycle *PEAR-ee-si-kel* A ring of parenchyma cells in a root's cortex that produces branch roots that burst through the cortex and epidermis into the soil. 548
periderm *PEAR-ee-derm* Outer covering of mature stems and roots. 541
period Time period within an era. 359
periodic table *peer-ee-OD-ic TA-ble* Chart that lists naturally occurring elements according to their properties. 32
peripheral nervous system *per-RIF-er-al NER-vous SIS-tum* (PNS) Neurons that transmit information to and from the central nervous system. 618
periplast *PER-ee-plast* A supportive protein structure on the inside face of the cell membrane of euglenoids. 466
peristalsis *pear-ee-STAL-sis* Waves of muscle contraction that propel food along the digestive tract. 759
peritubular capillaries *pear-ee-TOOB-yu-lar CAP-eh-LAIR-eez* Capillaries that surround renal tubules in kidney nephrons. 780
permafrost *PER-mah-frost* Permanently frozen part of the ground in the tundra. 878
peroxisome *per-OX-eh-soam* A membrane-bounded sac that buds from the smooth ER and that houses enzymes important in oxygen use. 72
petal *PET-al* Large and often vibrant flower part that may lure pollinators. 563
petiole *PET-ee-ol* The stalklike part of a leaf. 541
phagocytic vacuole *fag-o-SIT-ik VAK-u-ole* A space that forms within a protozoan cell when it engulfs food particles. 459
pharyngeal gill slits *farr-in-GEE-al GILL SLITZ* Chordate structures that function in feeding or respiration. 525
pharynx *FAHR-inx* The throat. 736
phelloderm *FEL-ah-derm* Living parenchyma cells in secondary growth in plants. 552
phenocopy *FEEN-o-kop-ee* An environmentally caused trait that resembles an inherited trait. 276
phenotype *FEEN-o-type* Observable expression of a genotype. 265
phenotypic ratio *feen-o-TIP-ik RAY-shee-o* Proportions of different phenotypes among offspring of a genetic cross. 266
pheromones *FER-eh-moanz* Biochemicals an organism secretes that elicit a response in another member of the species. 660
phloem *FLOW-m* Plant vessels that transport photosynthetic products. 496, 534
phospholipid *FOS-fo-LIP-id* A molecule consisting of a lipid and a phosphate that is hydrophobic at one end and hydrophilic at the other end. 85
phosphorylation *foss-for-eh-LAY-shun* Adding a phosphate (PO_4) group to a molecule. 119
photochemical smog *fo-to-KEM-i-kal SMOG* A form of air pollution caused by vehicle emissions. 892
photon *FOE-ton* A packet of light energy. 111, 147
photonasty *FO-toe-nas-tee* A nastic response in plants to light and dark. 582
photoperiodism *fo-toe-PER-ee-o-diz-um* A plant's ability to measure seasonal changes by the length of day and night. 583
photophosphorylation *FO-toe-FOS-for-eh-LAY-shun* A photosynthetic reaction in which energy released by the electron transport chain linking the two photosystems is stored in the high-energy phosphate bonds of ATP. 155
photoreactivation *fo-to-re-ac-ti-VAY-shun* A type of DNA repair in which an enzyme uses light energy to break pyrimidine dimers. 319
photoreception *fo-toe-ree-CEP-shun* Detecting light by means of pigment molecules in contact with sensitive membranes. 644
photorespiration *fo-to-res-per-A-shun* A process that counters photosynthesis. 160
photosynthesis *FO-toe-SIN-the-sis* The series of biochemical reactions that enable plants to harness sunlight energy to manufacture nutrient molecules. 73, 146
photosystem *FO-toe-SIS-tum* A cluster of pigment molecules that enables green plants to absorb, transport, and harness solar energy. 153, 154
phototroph *FO-toe-trofe* An organism that derives energy from the sun. 445
phototropism *fo-toe-TROP-iz-um* A plant's growth towards unidirectional light. 578
pH scale A measurement of how acidic or basic a solution is. 41
phycobilins *fi-ko-BILL-ins* Pigments of red algae. 469
phylogenies *fi-LODJ-ah-nees* Evolutionary relationships between species. 390
phylum *FI-lum* A major taxonomic group, just beneath kingdoms. 12
phytochrome *FI-toe-krome* A pale blue plant pigment that promotes flowering of long-day plants and inhibits flowering of short-day plants. 585
phytoplankton *FITE-o-PLANK-tun* Microscopic algae. 458
pili *pill-I* Short projections on bacterial cells used to attach to objects. 446
pinacocytes *pine-AK-o-sites* Cells in the body wall of a sponge. 515
pineal gland *pie-NE-al GLAND* A small oval structure in the brain that produces melatonin, a hormone that senses light and dark periods. 675
pinocytosis *pine-o-si-TOE-sis* A protozoan cell pinches inward and brings fluid into the cell. 459
pioneer species *pie-o-NEER SPEE-sheez* The first species to appear in an area devoid of life. 856
pistil *PIS-til* The female reproductive structures and their coverings in a flower. 563
pith *PITH* Ground storage tissue in the center of the stem in plants that have concentric cylinders of xylem and phloem. 541
pituitary gland *pah-TU-eh-tare-ee GLAND* A pea-sized gland in the human head that releases several types of hormones. 666
placebo *pla-SEE-bo* An inert substance used as an experimental control. 25
placenta *pla-CEN-tah* A specialized organ that connects a pregnant female placental mammal to unborn offspring. 216
plankton *PLANK-tun* Microscopic organisms that drift in large populations in water. 458
Plantae *PLAN-tay* The plant kingdom. 14, 429
plasma *PLAZ-ma* A watery, protein-rich fluid that forms the matrix of blood. 594, 711
plasma cells *PLAZ-mah SELZ* Mature B cells that secrete large quantities of a single antibody type. 795
plasmodium *plaz-MO-dee-um* A multinucleated mass of an acellular slime mold. 470
plastid *PLAS-tid* A plant organelle that encapsulates photosynthetic membranes. 151
platelet *PLATE-let* A cell fragment that is part of the blood and orchestrates clotting. 595, 714
Platyhelminthes *plat-ee-hel-MEN-theez* Flatworms. 510, 517
pleiotropic *PLY-o-TRO-pik* A genotype with multiple expressions. 276
plumule *PLU-mule* The epicotyl plus the first leaves of a young plantlet. 568
pluripotency *plur-e-POE-ten-see* A cell that retains the potential to specialize in any way. 210
pneumatophore *new-MAT-o-for* The floating portion of a cnidarian. 517, 550
point mutation *POYNT mu-TAY-shun* A change in a single DNA base. 341
polar body A small cell generated during female meiosis, enabling cytoplasm to be partitioned into just one of the four meiotic products, the ovum. 204
polar covalent bond *PO-lar co-VAY-lent BOND* A covalent bond in which electrons are attracted more toward one atom's nucleus than to the other. 38
polar nuclei *PO-lar NU-klee-i* The two nuclei in a cell of a plant's megagametophyte. 567
pollen grains Male microgametophytes. 560
pollen sacs The four microsporangia in an anther, the male part of a flower. 563
pollination Transfer of pollen from an anther to a receptive stigma. 567
polyandry *pol-ee-AN-dree* A mating system in which one female mates with several males. 830

polychaetes *POL-ee-ketes* Marine segmented worms. 521
polygamy *pol-IG-ah-mee* A mating system in which a member of one sex associates with several members of the opposite sex. 830
polygenic *pol-ee-JEAN-ik* A trait caused by more than one gene. 276
polygynandry *POL-ee-gine-AN-dree* Communal mating displays, involving several males and several females. 830
polygyny *pol-IJ-ah-nee* A mating system in which one male mates with several females. 830
polymer *POL-eh-mer* A long molecule composed of similar subunits. 43
polyp *POL-ip* The sessile form of a cnidarian. 517
polypeptide chain *pol-ee-PEP-tide CHAYN* A long polymer of amino acids. 46
polyploidy *POL-ee-PLOID-ee* A cell with extra chromosome sets. 297, 414
polyunsaturated (fat) *POL-ee-un-SAT-yer-a-tid FAT* A fatty acid with more than one double bond. 45
pons *PONZ* An oval mass in the brain stem where white matter connects the medulla to higher brain structures and gray matter helps control respiration. 627
population A group of interbreeding organisms living in the same area. 9, 404
population bottleneck A type of genetic drift. An event kills many members of a population, and a small number of individuals restore its numbers, restricting the gene pool. 409
population density The abundance of species per unit area or volume of habitat. 838
population dispersion The pattern in which individuals are scattered throughout the habitat. 838
Porifera *pore-IF-er-a* Sponges. 510, 515
positive feedback A biochemical pathway in which accumulation of a product stimulates its further production. 123, 664
postsynaptic cell *post-sin-AP-tik SEL* One of two adjacent neurons that receives a message. 617
potential energy The energy stored in the position of matter. 112
preembryonic stage *pre-em-bree-ON-ik STAYJ* Prenatal development before the organism folds into layers. 216
preformation *pre-for-MAY-shun* The idea that a gamete or fertilized ovum contains an entire preformed organism. 210
presynaptic cell *PRE-sin-AP-tik SEL* One of two adjacent neurons that transmits a message. 616
primary consumers Animals that eat plants and form the second trophic level. 862
primary growth Lengthening of a plant due to cell division in the apical meristems. 534
primary immune response The immune system's response to its first encounter with a foreign antigen. 795
primary nutrient deficiency Too little of a particular nutrient due to inadequate diet. 757
primary oocyte *PRI-mare-ee OO-site* An intermediate in ovum formation. 204
primary producers The species forming the base of a food web or the first link in a food chain. 859
primary spermatocyte *PRI-mare-ee spur-MAT-o-site* An intermediate in sperm formation. 204
primary (1°) structure The amino acid sequence of a protein. 49
primary succession Appearance of life in an area previously devoid of life. 856
primitive streak *PRIM-eh-tiv STREEK* The pigmented band along the back of a 3-week embryo that develops into the notochord. 226
principle of competitive exclusion The idea that two or more species that compete for the same limited resources cannot indefinitely occupy the same habitat. 845
principle of superposition The idea that lower rock layers are older than those above them. 376
prions *PRI-onz* Infectious protein particles. 77
product rule The chance of two events occurring equals the product of the chances of either event occurring. 271
profundal zone *pro-FUN-dal ZONE* The deep region of a lake or pond where light does not penetrate. 879
progenotes *pro-JEAN-notes* Collections of nucleic acid and protein that were forerunners to cells. 357
progesterone *pro-JES-ter-own* A hormone produced by the ovaries that controls secretion patterns of other reproductive hormones. 674
proglottids *pro-GLOT-ids* Structures in tapeworms that contain reproductive structures. 518
prokaryotic cell *pro-CARE-ee-OT-ik SEL* A cell that lacks organelles. Bacteria and cyanobacteria. 14, 63
prolactin *pro-LAK-tin* A hormone made in the anterior pituitary gland that stimulates milk production. 667
promoter *pro-MOW-ter* A control sequence near the start of a gene that attracts RNA polymerase and transcription factors. 327
pronuclei *pro-NU-kle-eye* The genetic packages of gametes. 218
prophase *PRO-faze* The first stage of cell division, when chromosomes condense and become visible. 171
prophase I *PRO-faze I* Prophase of meiosis I, when synapsis and crossing-over occur. 198
prophase II *PRO-faze II* Prophase of the second meiotic division. 201
prostaglandins *pros-tah-GLAN-dinz* Lipids released locally and transiently at the site of a cellular disturbance. 660
prostate gland *PROS-tate GLAND* A small gland that produces a milky, alkaline fluid that activates sperm. 197, 675
protein *PRO-teen* A polymer of amino acids. 46
Protista *pro-TEES-ta* The kingdom that includes unicellular, eukaryotic organisms, the protista. 14, 429
proton *PRO-ton* A particle in an atom's nucleus carrying a positive charge and having one mass unit. 32
protonephridia *pro-toe-nef-RID-ee-a* Structures in flatworms that help maintain internal water balance. 518
protoplasm *PRO-tow-plaz-m* Living matter. 60
protostomes *PRO-toe-stomes* One of two coelomate lineages, characterized by spiral cleavage, a blastopore that develops into a mouth, and determinate cleavage. 512
protozoa *PRO-toe-ZO-a* Single-celled eukaryotes often classified by their mode of movement, including amoebae, euglenae, and paramecia. 458
proximal convoluted tubule *PROX-eh-mel kon-vo-lu-tid TUBE-yule* Region of the nephron, proximal to the glomerular capsule, where selective reabsorption of useful components of the glomerular filtrate occurs. 783
pseudocoel *su-doe-SEEL* The fluid-filled hydrostatic skeleton of a roundworm. 519
pseudocoelomate *su-doe-SEEL-o-mayt* An animal that retains a body cavity from an early embryonic stage, but one that is not lined with mesoderm. 511
pseudoplasmodium *su-doe-plaz-MO-dee-um* The multicellular, mobile stage of the life cycle of a cellular slime mold. 471
pseudopod *su-doe-pod* The portion of an amoeba that extends outward, moving the organism. 459, 692
pseudostratified epithelium *su-doe-STRAT-if-eyed ep-eh-THEEL-e-um* A single layer of epithelium whose nuclei are at different levels, giving the appearance of more than one layer. 594
pulmonary artery *PULL-mo-nair-ee AR-ter-ee* The artery that leads from the right ventricle to the lungs. 720
pulmonary semilunar valve *PULL-mo-nair-ee SEM-ee-LOON-er VALVE* The valve that leads from the right ventricle to the pulmonary artery. 720
pulmonary veins *PULL-mo-nair-ee VAYNZ* Four veins that lead from the lungs to the left atrium. 720
pulp The soft inner part of a tooth, consisting of connective tissue, blood vessels, and nerves. 760
punctuated equilibrium *PUNK-chew-ate-ed ee-kwa-LIB-ree-um* The view that life's history has had periods of little change interrupted by bursts of rapid change. 415
pupil The opening in the iris that admits light into the human eye. 266, 648
purine *PURE-een* A type of organic molecule with a double ring structure, including the nitrogenous bases adenine and guanine. 313
Purkinje fibers *per-KIN-gee FI-berz* Muscle fibers that branch from the atrioventricular node and transmit electrical stimulation rapidly in the heart. 722

pyramid of biomass An ecological pyramid depicting weight of organisms at each trophic level. 863

pyramid of numbers An ecological pyramid depicting number of organisms at each trophic level. 863

pyrenoid *PIE-re-noyd* A starch-containing large chloroplast in a green alga. 465

pyrimidine *pie-RIM-eh-deen* A type of organic molecule with a single ring structure, including the nitrogenous bases cytosine, thymine, and uracil. 313

pyruvic acid *pi-ROO-vic AS-id* The products of glycolysis. 131

Q

quaternary (4°) structure *QUAT-eh-nair-ee STRUK-sure* The organization of polypeptide chains of a protein. 49

quiescent center *kwee-ES-cent CEN-ter* A reservoir of cells behind the root cap that can replace damaged cells in adjacent meristem. 547

R

radial cleavage *RAY-dee-al CLEE-vedge* The pattern of directly aligned blastomeres in the early deuterostome embryo. 513

radially symmetrical *RAY-dee-a-lee sim-ET-rik-el* An animal body form in which any plane passing from one end to the other divides the body into mirror images. 511

radicle *RAD-eh-kil* The first root to emerge from a seed. 546

radioles *RAY-de-oles* Ciliated structures on the anterior end of polychaete worms that filter small organisms and organic particles from seawater. 521

radiometric dating *RAD-ee-o-MET-rik DAY-ting* Using measurements of natural radioactive decay as a clock to date fossils. 393

radula *RAD-yew-la* A chitinous, tonguelike structure that mollusks use to eat. 519

reabsorption *re-ab-SORP-shun* The kidney's return of useful substances to the blood. 778

reactant *re-AK-tant* A starting material in a chemical reaction. 35

reaction center Clusters of chlorophyll and proteins that receive photon energy in photosynthesis. 153

reaction chain A sequence of releasers that joins several behaviors. 814

reading frame The DNA nucleotide corresponding to the first codon position in mRNA. 332

realized niche *RE-a-lized NEESH* The resources in a species' environment that it can actually use, considering competition and other limitations. 838

receptor-mediated endocytosis *re-CEP-ter ME-dee-a-ted en-do-ci-TOE-sis* Binding of a ligand by a cell surface protein that draws the ligand into the cell in a vesicle. 94

receptor potential A change in membrane potential in a neuron specialized as a sensory receptor, caused by redistribution of ions in response to the strength of the stimulus. 643

recessive *re-SESS-ive* An allele whose expression is masked by the activity of another allele. 265

reciprocal translocation *re-SIP-ro-kal tranz-lo-CAY-shun* Two nonhomologous chromosomes exchanging parts. 301

rectum *REK-tum* A storage region leading from the large intestine to the anus. 764

red blood cell (erythrocyte) A disc-shaped cell, lacking a nucleus, that contains hemoglobin. 594, 714

red marrow Immature blood cells and platelets in cavities in spongy bone. 689

reduction A chemical reaction in which electrons are gained. 117

reduction division Meiosis I, when the diploid chromosome number is halved. 198

reflex arc A neural pathway that links a sensory receptor and an effector. 623

regulatory enzyme *REG-u-la-tor-ee EN-zime* An enzyme that controls the activity of a biochemical pathway because it can be activated or deactivated by binding a compound other than its substrate under certain conditions. 122

relative dating An estimate of the time an organism lived based on the location of its fossils in rock layers. 392

releaser The specific factor that triggers a fixed action pattern. 814

releasing hormone A hormone produced by the hypothalamus that influences another gland. 666

renal cortex *RE-nal COR-tex* The outer portion of a kidney. 780

renal medulla *RE-nal med-U-la* The middle part of a kidney. 780

renal pelvis *RE-nal PEL-vis* The inner part of a kidney, where urine collects. 780

renal tubule *RE-nal TU-bule* The tubule part of a nephron, where toxins are added and nutrients recycled to the blood, forming urine. 780

replication fork A locally unwound portion of DNA where replication occurs. 317

respiratory chain A series of electron-accepting enzymes embedded in the inner mitochondrial membrane. 127

resting potential The electrical potential (−65 millivolts) on the inside of a neuron not conducting a nerve impulse. 612

reticular activating system *rah-TIK-u-lar AK-tah-vay-ting SIS-tem* (RAS) A diffuse network of cell bodies and nerve tracts that extends through the brain stem and into the thalamus; screens sensory input to the cerebrum. 627

retina *RET-eh-na* A sheet of photoreceptors at the back of the human eye. 648

reverse transcriptase *re-VERS tran-SCRIPT-aze* An enzyme that constructs a DNA molecule from an RNA molecule. 355, 798

R group *R GROOP* An amino acid side chain. 47

rhizoids *RI-zoydz* Hairlike extensions along the lower surface of a bryophyte that anchors the plant and absorbs water and minerals. 493

rhizomes *RI-zomes* Fleshy, underground stems that run horizontally. 497

rhodopsin *ro-DOP-sin* A pigment molecule stored in rod cells of the retina. Light splits rhodopsin, which depolarizes the rod cell and provokes an action potential. 644

ribonucleic acid *RI-bo-nu-KLAY-ik AS-id* (RNA) A single-stranded nucleic acid consisting of nucleotides containing a phosphate, ribose, and nitrogenous bases adenine, guanine, cytosine, and uracil. 51

ribose *RI-bose* The five-carbon sugar that is a structural component of RNA. 311

ribosomal RNA *RI-bo-SOAM-el RNA*(rRNA) RNA that, along with proteins, forms a ribosome. 329

ribosome *RI-bo-soam* A structure built of RNA and protein upon which mRNA anchors during protein synthesis. 66

ribozymes *RI-bo-zimes* Small RNAs that function as enzymes. 329

RNA polymerase *RNA poe-LIM-er-ase* An enzyme that takes part in DNA replication and RNA transcription. 317, 327

RNA primer *RNA PRI-mer* A small piece of RNA, inserted at the start of a piece of DNA to be replicated, which attracts RNA polymerase and is later removed. 317

rods Specialized receptor cells in the retina that provide black-and-white vision. 648

root Part of plant underground. 534

root cap A thimble-shaped protective structure on a root tip. 546

rosettes *ro-ZETTZ* Nonelongated stems. 541

r-selection A form of natural selection in which individuals are short-lived, reproduce early and have many offspring that get little care. Density-independent factors regulate these populations. 845

rugae *RU-guy* Folds in the mucosa of the stomach. 760

S

saccule *SAK-yul* A pouch in the vestibule of the inner ear, containing a jellylike fluid and lined with hair cells that contain calcium carbonate granules that move in response to changes in velocity, firing action potentials. 655

salivary amylase *SAL-eh-vare-ee AM-eh-lase* An enzyme produced in the mouth that begins chemical digestion of starch. 760

salt A molecule composed of cations and anions. 38

saltatory conduction *SAL-teh-tore-ee kon-DUK-shun* Jumping of an action potential between nodes of Ranvier in myelinated axons. 614

sample size The number of individuals in an experiment. 24

saprobes *SA-probez* and **saprophytes** *sa-pro-FITES* Organisms that obtain nutrients from dead plants and animals through absorption. 445, 476
sapwood *SAP-wood* Wood nearest the vascular cambium that transports water and dissolved nutrients within a plant. 551
sarcomere *SAR-ko-mere* A pattern of repeated bands in skeletal muscle. 695
sarcoplasmic reticulum *SAR-ko-PLASZ-mik reh-TIK-u-lum* The endoplasmic reticulum of a skeletal muscle cell. 694
satellite The end of a chromosome attached to the rest of the chromosome by a bridge. 296
saturated (fat) A triglyceride with single bonds between the carbons of its fatty acid tails. 45
savanna *sa-VAN-ah* A grassland. 877
scale A pinecone subunit that bears two ovules. 560
scavenger *SKAV-en-jer* Animal that eats the remains of other animals' meals. 862
scent glands Sweat gland derivatives that occur in different body regions in different species and produce fluids that evaporate to yield a distinctive odor. 603
Schwann cell *SHWAN sel* A type of neuroglia that forms a sheath around certain neurons. 597, 612
scientific method A systematic approach to interpreting observations, involving reasoning, predicting, testing, and drawing conclusions, and then putting them into perspective with existing knowledge. 18
sclera *SKLER-ah* The outermost white layer of the human eye. 648
sclereids *SKLER-ridz* Plant cells with a gritty texture found in pears and peanut hulls. 536
sclerenchyma *sklah-REN-kah-mah* Elongated supportive plant cells with thick, nonstretchable secondary cell walls. 536
scolex *SKO-lex* The holdfast organ of a tapeworm. 518
scramble competition When members of a population compete directly for a resource. 843
scrotum *SKRO-tum* The sac of skin containing the testes. 197
sebaceous glands *se-BAY-shis GLANDS* Glands in human skin that secrete a mixture of oils that soften hair and skin. 603
secondary consumer Carnivores, which form trophic levels beyond the second level. 862
secondary growth Thickening of a plant due to cell division in lateral meristems. 534
secondary immune response The immune system's response to subsequent encounters with a foreign antigen. 795
secondary nutrient deficiency Too little of a particular nutrient due to an inborn metabolic condition. 757
secondary oocyte *SEC-un-derry OO-site* A haploid cell that is an intermediate in ovum formation. 204
secondary spermatocyte *SEC-un-derry sper-MAT-o-site* A haploid cell that is an intermediate in sperm formation. 204
secondary (2°) structure The shape a protein assumes when amino acids close together in the primary structure chemically attract. 49
secondary succession Return of life to a devastated area. 856
second filial generation *SEK-und FIL-e-al jen-er-A-shun* (F_2) The third generation in a genetics problem. 265
second messenger A biochemical activated by an extracellular signal that transmits a message inside a cell. 103
secretion *seh-KREE-shun* A cell's release of a biochemical. Adding substances to the material in a kidney tubule. 778
secretory vacuole *SEK-re-tore-ee VAK-u-ole* An enzyme-containing sac in a protozoan. 459
seed A temporarily dormant sporophyte within a protective coat. 561
seed coat A tough outer layer that protects a dormant plant embryo and its food supply in a seed. 569
segmentation *SEG-men-TAY-shun* Localized muscle contractions in the small intestine that provide mechanical digestion. 761
selectively permeable *sah-LEK-tive-lee PERM-ee-ah-bul* A biological membrane that admits only some substances. 612
semicircular canal *SEM-ee-SIR-ku-ler kah-NAL* Fluid-filled structure in the inner ear that provides information on the position of the head. 653
semiconservative replication *sem-ee-con-SERV-a-tive rep-li-KAY-shun* Mode of DNA replication. Each double helix has one parental and one new strand. 317
semilunar valve *SEM-ee-LOON-er VALV* Tissue flaps in the artery just outside each ventricle that maintain one-way blood flow. 720
seminal fluid *SEM-in-el FLEW-id* Secretions that carry sperm. 197
seminal vesicles *SEM-in-el VES-eh-kels* In the human male, the paired structures that add fructose and prostaglandins to sperm. 197
seminiferous tubule *sem-i-NIF-er-us TUBE-yule* Tubule within the testis where sperm form and mature. 196
sensory adaptation Lessening of sensation with prolonged exposure to the stimulus. 644
sensory neuron A neuron that brings information toward the central nervous system, with a short axon and long dendrites that transmit the message from the stimulated body part to a cell body near the spinal cord. 611
sensory receptors Specialized neurons in the skin and sense organs that transmit messages in response to environmental stimuli. 597, 640
sepals *SEE-pelz* Leaflike structures that enclose and protect inner floral parts. 563
septate *SEP-tate* Structures that partition fungal hyphae. 478
setae *SEE-tay* Bristles on the sides of earthworms. 521, 645
sex chromosome A chromosome that carries genes that determine sex. 198
sex hormone Hormone that controls development of secondary sexual characteristics and prepares an animal for sexual reproduction. 670
sex-linked A gene on the X chromosome or a trait that results from activity of such a gene. 291
sex pilus *SEX PILL-us* A bacterial cell outgrowth that exchanges genetic material. 194, 449
sex ratio The proportion of males to females in a population. 838
sexual dimorphism *SEX-u-al di-MOR-fiz-um* The difference in appearance between males and females of the same species. 830
sexual reproduction The combination of genetic material from two individuals to create a third individual. 10, 192
sexual selection Natural selection of traits that increase an individual's reproductive success. 382
shoot Part of seedling aboveground. 534
short-day seedling Plant that requires light periods shorter than some critical length to flower. 583
sieve cells *SIV SELZ* Less specialized conducting cells in phloem. 539
sieve plate *SIV PLATE* End walls of aligned sieve tubes in phloem. 539
sieve tube members *SIV TOOB MEM-berz* More complex and specialized conducting cells in phloem that form long sieve tubes. 539
signal transduction The biochemical transmission of a message from outside the cell to inside. 85, 103
signature sequence DNA sequence unique to the members of specific taxonomic groups. 442
simple epithelium *SIM-pel ep-eh-THEEL-ee-um* A single layer of epithelium. 594
sinoatrial (SA) node *SI-no-A-tree-al NODE* Specialized cells in the wall of the right atrium that set the pace of the heartbeat. The pacemaker. 720
skeletal muscle cell Single, multinucleated cells that contract when actin and myosin filaments slide. They make up voluntary, striated muscle. 598
sliding filament model Sliding of protein myofilaments past each other to shorten skeletal muscle cells, leading to contraction. 695
slime molds Organisms that form acellular or cellular mobile masses that enable cells to collectively migrate to locate food. 458
smooth muscle cells Cells that make up involuntary, nonstriated contractile tissue that lines the digestive tract and other organs. 598
sodium-potassium pump A mechanism that uses energy released from splitting ATP to transport Na^+ out of cells and K^+ into cells. 92, 612

softwood Wood of gymnosperms. 551
solute *SOL-yoot* A chemical that dissolves in another, forming a solution. 40, 87
solution A homogenous mixture of a substance (the solute) dissolved in water (the solvent). 40
solvent *SOL-vent* A chemical in which others dissolve, forming a solution. 40, 87
somatic mutation *so-MAT-ik mew-TAY-shun* A mutation in a body cell. 341
somatic nervous system *so-MAT-ik NER-ves SIS-tum* Part of the motor pathways of the peripheral nervous system that leads to skeletal muscles. 622
somatostatin *so-MAT-owe-STAT-in* A pancreatic hormone that controls the rate of nutrient absorption into the bloodstream. 672
sori *SOR-ee* Dark spots on the undersides of most ferns that are collections of spore-producing structures. 498
speciation *SPE-she-AY-shun* Appearance of a new type of organism. 375
species A group of similar individuals that breed in nature only among themselves. 9
specific rate of population growth The number of births and deaths divided by the number of individuals in a population during a specific time period. 840
sperm The male sex cell. 196
spermatid *sper-ma-TID* An intermediate stage in sperm development. 204
spermatogenesis *sper-MAT-o-JEN-eh-sis* The differentiation of a sperm cell from a diploid spermatogonium, to primary spermatocyte, to two haploid secondary spermatocytes, to spermatids, and finally to mature spermatozoa. 203
spermatogonium *sper-mat-o-GOWN-e-um* A diploid cell that divides, yielding progeny cells that become sperm cells. 203
spermatozoa *sper-mat-o-ZO-a* Mature sperm. 204
S phase The synthesis phase of interphase, when DNA replicates and microtubules assemble from tubulin. 170
sphincter *SFINK-ter* Muscular ring that controls passage of a substance from one part of the body to another. 760
spicules *SPIK-yools* Glassy or limy material that make up sponge skeletons. 516
spines Modified leaves that protect plants from predators and excess sunlight. 544
spiracles *SPEER-a-kulls* Openings in the body wall of arthropods used for breathing. 522
spiral cleavage *SPY-ral CLEEV-edge* Pattern of early cleavage cells in protostomes, which resembles a spiral. 513
spleen Abdominal organ that produces and stores lymphocytes and red blood cells. 728
spongin *SPUNJ-in* Organic material in sponge skeletons. 516
spongocoel *SPUNJ-o-seel* The central cavity of a sponge. 515
spongy bone Flat bones and tips of long bones that have many large spaces between a web of bony struts. 689
spongy mesophyll cells *SPUN-gee MEZ-o-fil SELZ* Irregularly shaped chlorenchyma cells separated by large spaces that are beneath the palisade layer in leaves. 543
spontaneous generation The idea, proven untrue, that life can arise from nonliving matter. 353
spontaneous mutation A change in DNA sequence not caused by a mutagen, usually resulting from a DNA replication error. 339
sporophyte *SPOR-o-fite* Diploid stage of the plant life cycle. 192
spring turnover The rising of nutrient-rich lower layers of a lake and sinking of oxygen-rich layers from the top, often causing algal blooms. 880
squamous *SKWAY-mus* Flat, as in epithelium. 594
S-shaped curve A leveling off of a population growth curve. 843
stabilizing selection When extreme phenotypes are less adaptive than an intermediate phenotype. 411
stable isotope tracing *STA-bull ICE-o-tope TRAYS-ing* A method to chemically trace which species eat which species. 862
stamens *STA-menz* Male reproductive structures in flowers that consist of stalklike filaments that bear pollen-producing anthers at their tips. 563
stapes *STAY-peez* A small bone in the middle ear. 653
statocyst *STAT-o-sist* A fluid-filled cavity that contains minerals and sensory hairs that control balance and equilibrium. 645
statolith *STAT-o-lith* A mineral that functions as a mechanoreceptor. 645
stem cell An undifferentiated cell that divides to give rise to specialized cells. 178
steroid hormone *STAIR-oid HOR-moan* A lipid hormone that can cross a target cell membrane and enter the cell's nucleus. 662
stigma *STIG-mah* A pollen receptacle at the tip of a style in a flower. 563
stipe *STYP* Stemlike region of a brown alga. 467
stolon *STOL-on* Stem that grows along the soil surface. 541
stoma *STO-mah* Pore in a plant's cuticle through which water and gases exchange between the plant and the atmosphere. 159
storage leaves Fleshy leaves that store nutrients. 544
stratified epithelium *STRAT-eh-fyed ep-eh-THEEL-ee-um* Layered epithelium. 594
stroma *STRO-ma* The nonmembranous inner region of the chloroplast. 151
style A stalk that forms from an ovary in a flower. 563
subapical region *sub-APE-eh-kel REE-jen* The region behind the root cap, which is divided into zones of cellular division, cellular elongation, and cellular maturation. 547
subcutaneous layer *sub-kew-TAYN-ee-us LAY-r* In human skin, the layer of fat beneath the dermis. 599
suberin *SU-ber-in* A waxy, waterproof biochemical in the interior of a root's cortex. 548
substrate *SUB-strate* A reactant an enzyme acts upon. 50
substrate-level phosphorylation *SUB-strate LEV-ell fos-for-i-LAY-shun* ATP formation from transferring a phosphate group to ADP. 129
succulent stems *SUK-ku-lent STEMZ* Fleshy plant tissues that store water. 541
superior vena cava *su-PEER-ee-er VEE-nah KAY-vah* The upper branch of the largest vein that leads to the heart. 720
supernormal releaser A model that exaggerates a releaser and elicits a stronger response than does the natural object. 814
survivorship The fraction of a group of individuals that survive to a particular age. 842
suspension feeders Animals that capture food particles suspended in the water. 515
sweat glands Epidermal invaginations that produce sweat. 602
symbiosis *sim-bi-O-sis* A biological relationship of one type of organism living in or on another. 434
sympathetic nervous system Part of the autonomic nervous system that mobilizes the body to respond to environmental stimuli. 622
synapse, synaptic cleft *sin-APSE, sin-AP-tik CLEFT* A space between two neurons. 616
synapsis *sin-AP-sis* The gene-by-gene alignment of homologous chromosomes during prophase of meiosis I. 198
synaptic integration *sin-AP-tik in-teg-RAY-shun* A cell's overall response to many incoming neural messages. 617
syncytium *sin-SIK-tee-um* A mass of multinucleated cells. 175
synovial joint *sin-OV-ee-el JOYNT* A capsule of fluid-filled fibrous connective tissue between freely movable bones. 700
synovial membrane *sin-OV-ee-el MEM-brane* Lining of the interior of a joint capsule, which secretes lubricating synovial fluid. 700
synteny *SIN-ten-ee* Comparison of gene order on chromosomes between species. 396
systematics *sis-te-MAT-ix* Study of the evolutionary relationships among species. 427
systole *SIS-toll-ee* The heart's contraction. 720

T

taiga *TI-e-gah* The northern coniferous forest, north of the temperate zone. 876
taproot system A plant in which the first root enlarges to form a major root that persists through life. 546
target cell A cell that a hormone binds and directly affects. 660
taste receptors Specialized neurons that detect taste. 647

taxonomy *tax-ON-o-mee* Classification of organisms on the basis of evolutionary relationships. Taxonomic levels include, in order: domain, kingdom, phylum (or division), class, order, family, genus, and species. 12

T cell receptor A cell surface molecule that binds a T cell. 797

T cells T lymphocytes; a component of the immune system 716, 790

tectorial membrane *tek-TOR-ee-al MEM-brane* The membrane above hair cells in the cochlea of the inner ear that is pressed by hair cells responding to the basilar membrane's vibration in the presence of sound waves. 654

tegument *TEG-yew-ment* The protective body covering of certain flatworms. 517

telocentric *tell-o-SEN-trik* A chromosome with the centromere at the tip. 296

telomerase *tell-OM-er-ase* An enzyme that extends chromosome tips using RNA as a template. 175

telomere *TELL-o-meer* A chromosome tip. 175

telophase *TELL-o-faze* The final stage of cell division, when two cells form from one and the spindle is disassembled. 171

telophase I *TELL-o-faze I* Arrival of homologs at opposite poles in meiosis I. 201

telophase II *TELL-o-faze II* Nuclear envelope formation around meiotic products. 201

temperate deciduous forest *TEMP-er-et de-SID-u-us FOR-est* A terrestrial biome with a temperate climate where one or two species of deciduous trees predominate. 873

tendon *TEN-din* A heavy band of fibrous connective tissue that attaches a muscle to a bone. 694

tendril *TEN-dril* Shoot or modified leaf that supports plants by coiling around objects. 541

teratogen *teh-RAT-eh-jen* Something that causes a birth defect. 246

territoriality *ter-eh-tor-ee-AL-it-ee* An animal's marking and defending of a piece of land. 828

tertiary consumers *TER-she-air-ee con-SOOM-erz* Carnivores that eat other carnivores. 862

tertiary (3°) structure *TER-she-air-ee STRUK-sure* The shape a protein assumes when amino acids far apart in the primary structure chemically attract one another. 49

test Calcium-containing shell of protozoa. 458, 523

test cross Breeding an individual of unknown genotype to a homozygous recessive individual to reveal the unknown genotype. 266

testes *TES-teez* Paired male gonads containing seminiferous tubules, where sperm are manufactured. 196

testosterone *tes-TOS-tur-own* A sex hormone. 663

thalamus *THAL-eh-mus* A tight gray package of nerve cell bodies and glia beneath the cerebrum that relays sensory input to the appropriate part of the cerebrum. 627

thalli *THALL-i* Nonsexual parts of algae and fungi. 464, 477

theory An explanation for observations and evidence of natural phenomena. 18

thermal stratification *THER-mal STRAH-tah-fah-KAY-shun* Layers within lakes that have different temperatures. 879

thermocline *THER-mo-kline* A middle layer of a lake where water temperature changes rapidly and drastically. 879

thermodynamics *THERM-o-di-NAM-ix* Study of energy transformations in nature. 112

thermogenesis *therm-o-JEN-i-sis* Generating heat metabolically. 772

thermophiles *THERM-o-files* Archaea of genus *Thermoplasma,* which live under conditions of high heat and acid. 451

thermoregulation *THERM-o-reg-u-LAY-shun* Ability of an animal to balance heat loss and gain with the environment. 771

thigmomorphogenesis *THIG-mo-MOR-fo-GEN-ah-sis* A plant's responses to mechanical disturbances, including inhibition of cellular elongation and production of thick-walled supportive tissue. 582

thigmonasty *THIG-mo-NAS-tee* A nastic response to touch. 581

thigmotropism *THIG-mo-TRO-piz-um* A plant's growth response toward touch. 579

thorns Stems modified for protection. 541

thrombin *THROM-bin* A blood-clotting protein. 717

thromboplastin *THROM-bo-PLAS-tin* A protein released from blood vessel walls following injury that converts prothrombin to thrombin. 717

thrombus *THROM-bus* A blood clot that blocks a blood vessel. 717

thylakoids *THI-lah-koydz* Disclike structures that make up the inner membrane of a chloroplast. 75

thymine *THI-meen* One of the two pyrimidine bases in DNA. 51

thymus *THY-mis* A lymphatic organ in the upper chest where T cells learn to distinguish foreign from self antigens. 728

thyroid gland *THI-roid GLAND* In humans, a gland in the neck that manufactures thyroxine, a hormone that increases energy expenditure. 669

thyroid-stimulating hormone *THY-roid STIM-u-lay-ting HOR-moan* (TSH) A hormone made in the anterior pituitary gland in humans that stimulates the thyroid gland to release two hormones. 667

thyroxine *thy-ROX-in* A thyroid hormone that increases the rate of cellular metabolism. 665

tidal volume Volume of air inhaled or exhaled during a normal breath. 742

tissue Groups of cells with related functions. 5

tonsils Collections of lymphatic tissue in the throat. 728

trace element An element an animal requires in small amounts. 32

trachea *TRAY-kee-ah* The respiratory tube just beneath the layrnx. The windpipe. 736

tracheae *TRAY-kee-i* Branching system of tubules that brings the outside environment in close contact with an organism's cells so that gas exchange can occur. 733

tracheids *TRAY-kee-idz* Less specialized conducting cells in plants that are elongated, dead at maturity, and have thick walls. 538

transcription *tranz-SKRIP-shun* Manufacturing RNA from DNA. 326

transcription factor *tranz-SCRIPT-shun FAC-tor* A protein that turns on and off different genes in a particular cell. 327

transfer RNA (tRNA) A small RNA molecule that binds an amino acid at one site and an mRNA codon at another site. 329

translation *tranz-LAY-shun* Assembly of an amino acid chain according to the sequence of base triplets in a molecule of mRNA. 326

translocation *TRANZ-lo-KAY-shun* Exchange of genetic material between nonhomologous chromosomes. 301

transplantation *tranz-plan-TAY-shun* Replacing a diseased or damaged organ with one from a donor. 603

transposable element *tranz-POSE-a-bull EL-e-ment* Jumping gene. 343

transverse (T) tubule *TRANZ-verse TU-bule* Part of the sarcolemma that juts from the sarcoplasmic reticulum of a skeletal muscle cell. 694

trichome *TRI-koam* Outgrowth of a plant's epidermis that protects. 537

triglyceride *tri-GLI-sir-ide* A type of fat that consists of one glycerol and three fatty acids. 45

triiodothyronine *tri-i-ode-o-THY-ro-neen* A thyroid hormone that increases the rate of cellular metabolism. 669

trilobite *TRI-low-bite* Extinct arthropod that was once very abundant. 522

triploblastic *trip-lo-BLAS-tik* An animal whose adult tissues arise from three germ layers in the embryo. 511

trisomy *TRI-som-mee* A cell with one extra chromosome. 298

trocophore *TRO-ko-for* A mollusk larva. 519

trophic level *TRO-pik LEV-l A* feeding level in an ecological community. 859

trophoblast *TRO-fo-blast* A layer of cells in the preembryo that develops into the chorion and then the placenta. 218

trophozoites *tro-fo-ZO-ites* Form of the protozoan *Giardia lamblia* in a person's small intestine. 461

tropical rain forest A warm, moist terrestrial biome with a tree canopy. 872

tropic hormone *TRO-pik HOR-moan* A hormone that effects another hormone's secretion. 666

tropism *TRO-piz-um* Plant growth toward or away from an environmental stimulus. 578

tropomyosin *TRO-po-MI-o-sin* A type of protein in thin myofilaments of skeletal muscle cells. 695

troponin *tro-PO-nin* A type of protein in thin myofilaments of skeletal muscle cells. 695

true ferns The largest group of seedless vascular plants. 497

tube feet Cuplike structures on echinoderms that draw water in. 524

tube nucleus *TOOB NEW-klee-us* A haploid cell resulting from the mitotic division of a microspore in male plant reproduction. 563

tuber *TU-ber* Swollen region of stem that stores nutrients. 541

tubulin *TOOB-u-lin* Protein that makes up microtubules. 99

tumor necrosis factor *TOOM-er nek-RO-sis FAC-tor* (TNF) An immune system biochemical with varied functions in cancer, infection, and inflammation. 797

tumor suppressor gene *TOO-mer sup-PRESS-er JEAN* A gene which, when inactivated or suppressed, causes cancer. 183

tundra *TUN-drah* A band of land across the northern parts of Asia, Europe, and North America, where the climate is harsh and life is sparse. 878

tunicates *TOON-i-catz* One of three groups of chordates, including the sea squirts. 525

turgor pressure *TER-ger PRESH-er* Rigidity of a plant cell caused by water pressing against the cell wall. 90

twitch Rapid contraction and relaxation of a muscle cell following a single stimulation. 699

tympanal organ *tim-PAN-al OR-gan* A thin part of an insect's cuticle that detects vibrations and therefore sound. 645

tympanic membrane *tim-PAN-ik MEM-brane* The eardrum, a structure upon which sound waves impinge. 653

U

ultratrace element An element vital to an animal in very small amounts. 32

umbilical cord *um-BIL-ik-kel KORD* A ropelike structure that contains one vein and two arteries that connects a pregnant female placental mammal to unborn offspring. 216

unconditioned stimulus A stimulus that normally triggers a particular response. 815

uniramous *uni-RAY-muss* A single-lobed arthropod appendage. 522

unsaturated *un-SAT-yur-RAY-tid* (fat) A triglyceride with double bonds between some of its carbons. 45

upwelling Upward movement of nutrient-rich lower layers of a body of water. 883

uracil *YUR-eh-sil* One of two pyrimidine bases in RNA. 51

urea *u-REE-ah* A nitrogenous waste derived from ammonia. 776

ureter *YEW-re-ter* A muscular tube that transports urine from the kidney to the bladder. 780

urethra *u-RETH-rah* Tube that transports urine from the bladder out of the body. 197, 780

uric acid *YOUR-ik AS-id* A nitrogenous waste derived from ammonia. 776

urinary bladder *YOUR-en-air-ee BLAD-er* A muscular sac where urine collects. 780

uterus *U-ter-us* The muscular, saclike organ in a female placental mammal where the embryo and fetus develop. 197

utricle *U-trah-kel* A pouch in the vestibule of the inner ear filled with a jellylike fluid and lined with hair cells that contain calcium carbonate granules that move in response to changes in velocity, firing action potentials. 655

V

vaccine *vak-SEEN* A killed or weakened form of, or part of, an infectious agent that initiates an immune response so that when the real agent is encountered, antibodies are already available to deactivate it. 803

vacuole *VAK-yew-ole* A storage sac in a cell. 75

vagina *va-GINE-a* The birth canal in a female in many species. 197

valence electron *VAY-lense e-LEC-tron* Electron in the outermost shell. 35

van der Waals attractions *VAN dur walls a-TRAC-shuns* Dynamic attractions within or between molecules when oppositely charged regions attract. 39

variable A changeable factor in an experiment. 21, 25

variable regions Amino acid sequences that form the upper parts of heavy and light antibody chains, which vary greatly in different antibody types. 795

variably expressive *VAIR-ee-ah-blee ex-PRESS-ive* A phenotype that varies in intensity in different individuals. 275

vascular bundle *VAS-ku-ler BUN-del* Organized group of vascular tissues in stem. 541

vascular cambium *VAS-ku-ler KAM-bee-um* A thin cylinder of meristematic tissue in roots and stems that produces most of the secondary plant body. 550

vascular tissue *VAS-ku-ler TISH-ew* Specialized conducting tissue that forms leaf veins. 534

vas deferens *VAS DEF-er-enz* In the human male, a tube from the epididymis that joins the urethra in the penis. 197

vein In plants, strand of vascular tissue in leaves. In some animals, a large blood vessel arising from venules that returns blood to the heart. 708

venous valves *VEEN-is VALVES* Flaplike structures in veins that keep blood flowing in one direction. 725

ventral *VEN-tral* The belly or underside. 512

ventricles *VEN-treh-kelz* Spaces in the brain into which cerebrospinal fluid is secreted. Also, the two muscular heart chambers beneath the atria. 710

venule *VEN-yule* Vessel that arises from capillaries and drains into veins. 708

vertebral column *ver-TE-bral COL-um* Bones that protect the spinal cord. 685

vertebrate *VER-te-brayt* An animal that has a backbone. 508

vertical stratification *VER-ti-kal strat-if-i-KAY-shun* Layering of different plant species in the forest canopy. 872

vesicle *VES-i-cal* A membrane-bounded sac in a cell. 71

vessel element *VES-el EL-eh-ment* More specialized conducting cell in plants that is elongated, dead at maturity, and has thick walls. 538

vestibule *VES-teh-bule* A structure in the inner ear that provides information on the position of the head with respect to gravity and changes in velocity. 653

vestigial organ *ves-TEEG-e-el OR-gan* A structure that seems not to have a function in an organism but that resembles a functional organ in another species. 395

villi *VILL-i* Tiny projections on the inner lining of the small intestine that greatly increase surface area for nutrient absorption. 762

viroid *VIE-royd* Infectious genetic material. 77

virus An infectious particle consisting of a nucleic acid (DNA or RNA) wrapped in protein. 75

visceral mass *vis-er-al MASS* Part of the molluscan body that contains the digestive and reproductive systems. 519

vitreous humor *VIT-ree-us HU-mer* A jellylike substance behind the lens that makes up most of the volume of the eye. 649

vocal cords Two elastic tissue bands that cover the glottis. They vibrate as air passes, producing sounds. 736

W

water molds A diverse group of eukaryotes with cellulose cell walls that can aggregate into multinucleated forms. 458

water vascular system *WA-ter VAS-ku-ler SIS-tem* System of canals in echinoderms that admits seawater. 524

wavelength The distance a photon moves during a complete vibration. 147

whisk fern The simplest type of seedless vascular plant. 497

white blood cell A cell that helps fight infection. 594, 714

wild-type The most common phenotype or allele for a certain gene in a population. 265

X

X inactivation *X IN-ak-tah-VA-shun* Turning off one X chromosome in each cell of a female mammal at a certain point in prenatal development. 293

X-inactivation center *X IN-ak-tah-VA-shun SEN-ter* The part of the X chromosome that inactivates the X. 293

xylem *ZI-lem* Plant vessels that transport dissolved ions and water. 496, 534

Y

yeast A unicellular fungus. 477

Z

zona pellucida *ZO-nah pel-LU-seh-dah* A thin, clear layer of proteins and sugars that surrounds a secondary oocyte. 217

zone of cell differentiation The hindmost region of the subapical region of a root, where tiny root hairs protrude from epidermal cells. 547

zone of cell division The meristematic part of the subapical region in a plant's root. 547

zone of cell elongation The middle part of the subapical region of a root, where rapid cellular elongation lengthens the root. 547

zooplankton *ZO-O-PLANK-tun* Abundant protozoa and microscopic animals in bodies of water. 458

zygomycetes *zi-go-mi-SEETS* Fungi that have thick, black zygospores. 478

zygospore *zi-go-SPOOR* Sexual structure of zygomycetes that consists of zygotes within thick black coats. 481

zygote *ZI-goat* In prenatal humans, the organism during the first 2 weeks of development. Also called a preembryo. In plants, the fused egg and sperm that grows into a sporophyte. 192

Credits

Photographs

Table of Contents

Pages vii–xx: Digital Stock

Unit Openers

One: © Y. Arthus-Bertrand/Peter Arnold, Inc; **Two:** Courtesy of Dr. Markus Missler, University of Texas Southwestern Medical Center, Dallas; **Three:** Courtesy, National Zoological Park/photo: Jessie Cohen; **Four:** © Lawrence Migdale/Photo Researchers, Inc.; **Five:** © Jacky Beckett/American Museum of Natural History; **Six:** © Ron Sefton/ Bruce Coleman; **Seven:** © David Muench/Tony Stone Images; **Eight:** © Tom McHugh/Photo Researchers, Inc.; **Nine:** © Superstock

Chapter 1

Opener: © C.W. Perkins/Animals Animals; **1.1A:** © Adriana Rovers/Cornell University; **1.1B:** © Nelson Hairston, Jr.; **1.1C:** © Colleen Kearns; **BOX 1.1A:** © Amy E. Powers/Gamma Liaison; **1.2B:** © Gunter Ziesler/Peter Arnold, Inc.; **1.4A:** © David M. Phillips/Photo Researchers, Inc.; **1.4B:** © Runk/Schoenberger/Grant Heilman Photography; **1.4C:** © Mark Newman/Tom Stack & Assoc.; **1.5A,B:** © Michael and Patricia Fogden; **1.6(all):** © Steven Vogel; **1.7A:** © Kenneth Murray/Photo Researchers, Inc.; **1.7B:** © James L. Amos/Photo Researchers, Inc.; **1.B:** © John Kieffer/Peter Arnold, Inc.; **1.8A:** © Science Source/Photo Researchers, Inc.; **1.8B:** © D.P. Wilson/Science Source/Photo Researchers, Inc.; **1.8C:** © Joe McDonald/Visuals Unlimited; **1.8D:** © Ray Coleman/Photo Researchers, Inc.; **1.8E:** © James Balog/Tony Stone Images

Chapter 2

Opener: © Douglas Faulkner/Photo Researchers, Inc.; **2.1A:** © Inga Spence/Tom Stack & Associates; **2.1B:** Dr. James G. Sikarskie, DVM/Michigan State University; **2.1C:** Courtesy of World Wildlife Fund. Photo by Glen Fox; **2.2:** From V. Neman-Simha and F. Megraud, "In Vitro Model for *Campylobacter pylori* Adherence Properties," *Infection and Immunity,* 56(12):3329–3333, Dec. 1988. American Society for Microbiology; **2.4:** © Dr. Tony Brain/ SPL/Photo Researchers, Inc.; **BOX 2.A:** © Katy Payne; **2.7A:** © 1992 by: People for the Ethical Treatment of Animals; **2.7B:** © Michael Schwarz/ The Image Works; **2.8:** © 1995 Amgen, Inc.; **2.9:** © NIH/Science Source/Photo Researchers, Inc.; **2.10:** © Dr. Dennis Kunkel/Phototake; **2.11A:** © Philippe Plailly/SPL/Photo Researchers, Inc.; **2.11B:** Institut fur Mathematik und Datenverabeitung in der Medizin; **2.13:** Courtesy National Library of Medicine

Chapter 3

Opener: Todd Stevens, Pacific Northwest National Laboratory; **p. 35:** Courtesy National Institutes of Health Clinical Center; **BOX 3.A:** Johns Hopkins University Photography; **p. 41:** Maria Eisner, Cornell University; **3.10A:** © Tony Craddock/ Photo Researchers, Inc.; **3.10B:** © William E. Ferguson; **p. 44:** © John A. Baross, University of Washington; **3.12:** © Stephen J. Krasemann/Peter Arnold, Inc.; **3.13:** Courtesy of Diane R. Nelson, Ph.D.; **p. 46** : © A.B. Sheldon; **p. 47:** © The McGraw-Hill Companies, Inc./Bob Coyle, photographer; **3.15:** © Van Bucher/Photo Researchers, Inc.; **3.18:** © Darlyne A. Murawski/ Peter Arnold, Inc.; **p. 50:** From Lynn Jorde, John Carey, Ray White, *Medical Genetics,* fig. 10.6, p. 198. © Mosby Year Book, Inc.; **3.22:** © Stephen J. Krasemann/DRK Photo; **3.25A:** © Stephen J. Krasemann/Photo Researchers, Inc.; **3.25B:** © Richard R. Hansen/Photo Researchers, Inc.; **3.25C:** © Gerald Lacz/Peter Arnold, Inc.; **3.25D:** © John Cancelosi/Peter Arnold, Inc.; **3.26A:** © Ron Davis/Shooting Star; **3.26B:** © Frank Trapper/ Sygma; p. 51: © Robert Carr/Bruce Coleman

Chapter 4

Opener: © Jennifer Kramer and Sam Wells, Molecular Probes, Inc.; **4.1A,B:** © Manfred Kage/Peter Arnold, Inc.; **4.1C:** © Edwin A. Reschke/Peter Arnold, Inc.; **4.2A–D:** © Dr. Tony Brain/David Parker/SPL/Photo Researchers, Inc.; **4.4A:** © Fred Hossler/Visuals Unlimited; **4.4B:** © Boehringer Ingelheim International Gmbtt, photo by Lennart Nilsson, "The Body Victorious" Dell Publishing; **4.4C:** © Aaron H. Swihart and Robert C. MacDonald; **4.10:** From J.T. Staley, M.P. Bryant, N. Pfennig, and J.G. Holt, (Eds.), *Bergey's Manual of Systematic Bacteriology,* Vol. 3./p. 2224, fig. 25.37 © 1989 Williams and Wilkins Co., Baltimore.; **4.6:** © Manfred Kage/Peter Arnold, Inc.; **4.7B:** © Dr. S. Kim/Peter Arnold, Inc.; **4.8A:** © David M. Phillips/Visuals Unlimited; **4.8B:** © George J. Wilder/Visuals Unlimited; **4.8C:** © T.E. Adams/Visuals Unlimited; **4.9A1:** © SPL/Custom Medical Stock Photos; **4.9A2:** Courtesy of Dr. Eldon Newcomb, University of Wisconsin, Madison; **4.11B:** © Philip & Karen Smith/Tony Stone Images; **4.12A:** © Gordon Leedale/Biophoto Associates; **4.13A:** © Keith Porter/Photo Researchers, Inc.; **4.14B:** © K.G. Murti/Visuals Unlimited; **4.15A:** © D. Friend-D. Fawcett/Visuals Unlimited; **4.15B:** From S.E. Frederick and E.H. Newcomb, *Journal of Cell Biology,* 4:343–53, 1969; **4.16A:** © Barry Dowsett/SPL/Photo Researchers, Inc.; **4.17:** Theodore Diener/USDA/Plant Virology Laboratory; **4.18A:** © Reuters/Ian Hodgson/Archive Photos; **4.18B:** From D.C. Carlton, "Unconventional Viruses: The Ongoing and Disappearances of Kuru", *Science,* 197: 1976 Photo © The Nobel Foundation

Chapter 5

Opener: © Prof. Motta, Correr & Nottola/University "La Sapienza", Rome/SPL/Photo Researchers, Inc.; **5.1:** © NIBSC/SPL/Photo Researchers, Inc.; **5.2C:** © Gordon Leedale/ Biophoto Associates; **5.7A–C:** © David M. Phillips/Visuals Unlimited; **5.14B:** Olaf Mundigl and Pietro De Camilli, Yale University School of Medicine, Department of Cell Biology, New Haven, CT.; **5.18A:** © Biophoto Assoc./Photo Researchers, Inc.; **5.18B1:** © Ray Evert; **5.20:** © K.G. Murti/ Visuals Unlimited; **5.22A:** © CNRI/SPL/Photo Researchers, Inc.; **5.22B:** © Manfred Kage/Peter Arnold, Inc.; **5.23B:** © Bob Thomas/Tony Stone Images; **5.24B:** Courtesy Dr. Gary Floyd; **5.25:** © CNRI/SPL/Photo Researchers, Inc.; **5.26A:** © CNRI/SPL/Photo Researchers, Inc.; **5.27A:** © Dwight Kuhn; **BOX 5.A:** © The Kobal Collection; **BOX 5.B:** © American Medical Association, 1944

Chapter 6

Opener: © Gunter Ziesler/Gunter Ziesler; **6.1:** © Stephen Dalton/Photo Researchers, Inc.; **6.2A:** © Jeremy Walker/Tony Stone Images; **6.2B:** © Ed Degginger/Bruce Coleman; **6.2C:** © Greg Vaughn/Tony Stone Images; **6.2D:** © Robert Frerck/Odyssey Productions; **6.2E:** © Keith Wood/Tony Stone Images; **6.2F:** © Bob Pool/ Tom Stack & Associates; **6.3A:** © J. Edmond/ Visuals Unlimited; **6.3B:** © D. Foster/Visuals Unlimited; **6.4A:** © Bryan Yablosky/Duomo; **6.4B:** © Stephen Dalton/NHPA; **6.6:** © Manoj Shah/Tony Stone Images; **6.7(both):** © The McGraw-Hill Companies, Inc./Jim Shaffer, photographer; **6.9B:** © Corbis/Bettmann; **6.10B:** © Gamma Liaison; **BOX 6.A:** © Keith Kent/SPL/Photo Researchers, Inc.

Chapter 7

Opener: © Biophoto Associates/Science Source/Photo Researchers, Inc.; **7.1A:** © Lori Adamski Peek/Tony Stone Images; **7.2:** © Jeff Lepore/Photo Researchers, Inc.; **7.12:** © John Eastcott/Yva Momatiuk/Photo Researchers, Inc.; **7.13:** © David M. Phillips/Visuals Unlimited; **BOX 7.A:** © Angelina Lax/Photo Researchers, Inc.; **7.15:** Courtesy of Peter Hinkle, Cornell University; **BOX 7.B:** © Avril Ramage/Oxford Scientific Films/ Animals Animals/Earth Scenes; **7.17B:** © Vance Henry/Nelson Henry; **7.18A:** © Les Stone/Sygma; **7.18B:** © Network Productions/The Image Works

Chapter 8

Opener: © Tim Thompson/Tony Stone Images; **8.1:** © UPI/Corbis-Bettmann; **8.5:** © Galen Rowell/ Peter Arnold, Inc.; **8.6B:** © Dr. Jeremy Burgess/ SPL/Photo Researchers, Inc.; **8.6C:** © Grant Heilman Photography; **8.7A:** © David M. Phillips; **8.7C:** © Philip Sze/Visuals Unlimited; **BOX 8.B:** © 1995 Discover Magazine; **8.11:** © BioPhot; **8.14B:** Courtesy of J. D. Mauseth, University of Texas, Austin

Chapter 9

Opener: © CNRI/SPL/Photo Researchers, Inc.; **9.2A:** © Larry Lefever/Grant Heilman Photography; **9.2B:** © Barry L. Runk/Grant Heilman Photography; **9.4C:** From Dr. A.T. Sumner, "Mammalian Chromosomes from Prophase to Telophase," *Chromosoma,* 100:410–418, 1991. © Springer Verlag; **9.6, 9.7A–E, 9.9B:** © Ed Reschke; **9.12:** © R. Lyons/Visuals Unlimited; **9.15:** Taken from Schwartz & Osborne, Immunology Today, etc.. EM by Lucy Yin, University Massachusetts Microscope Facility.; **9.17A:** National Library of Medicine; **9.17B:** © Cecil H. Fox/Science Source/ Photo Researchers, Inc; **9.19A,B:** © Custom Medical Stock Photo; **9.20:** © Nancy Kedersha/ Immunogen/SPL/Photo Researchers, Inc.; **9.21:** © The Granger Collection

Chapter 10

Opener: © Francis LeRoy/Biocosmos/SPL/Photo Researchers, Inc.; **10.1A:** © Carolina Biological Supply Company/Phototake; **10.1B:** © Francis Gohier/Photo Researchers, Inc.; **BOX 10.A,B:** © Dwight Kuhn; **10.2B:** © William Ferguson; **10.3:** Courtesy of E.L. Wollman; **10.4B:** © R. Kessel and G. Shih/Visuals Unlimited; **10.12A:** © Professor P. Motta/Dept. of Anatomy/University of La Sapienza, Rome/SPL/Photo Researchers, Inc.; **10.14C:** Larry Johnson, Dept. of Veterinary Anatomy and Public Health, Texas A&M University. Originally appeared in Biology of Reproduction, 47:1091–1098, 1992.; **10.16:** Courtesy of Dr. Y. Verlinsky

Chapter 11

Opener: © Kathy Bushue/Tony Stone Images; **11.1A:** © Margot Granitsas/The Image Works; **11.1B(first 2 photos):** © John D. Cunningham/ Visuals Unlimited; **11.1B(right):** © Jerome Wexler/ Photo Researchers, Inc.; **11.3:** Furnished through the courtesy of C.L. Markert; **11.6A:** Courtesy of A. Villeneuve and B. Meyer; **11.7(top):** Courtesy of Christine Nusslein-Volhard; **11.7(bottom three):** Courtesy of Jim Langeland, Steve Paddock, Sean Carroll/ Howard Hughes Medical Institute, University of Wisconsin; **11.8(both):** Courtesy of F.R. Turner, Indiana University, Bloomington, IN; **11.10:** © David Scharf/Peter Arnold, Inc.; **11.13A:** © Petit Format/Nestle/Science Source/Photo Researchers, Inc.; **11.13B:** © Professors P.M. Motta & J. Van Blerkom/SPL/Photo Researchers, Inc.; **11.13C:** © Petit Format/Nestle/Science Source/ Photo Researchers, Inc.; **BOX 11.A:** © Integra Med; **BOX 11.B:** © Roger Pedersen, Ph.D.; **11.19E:** © Petit Format/Nestle/Science Source/Photo Researchers, Inc.; **BOX 11.EA and BOX 11.EB:** Courtesy of Dr. Brett Casey; **11.21:** © James Stevenson/SPL/Photo Researchers, Inc.; **11.23A:** © Gary Mulburn/Tom Stack & Associates; **11.23B:** © Jean-Paul Ferrero/Ardea London; **11.24(newborn, infant,child):** © Digitial Stock/Babies and Children; **11.24(adolescent and adult):** PhotoDisc/Family & Lifestyles; **11.25(top 2):** The Bettmann Archives; **11.25(third one down):** AP/Wide World Photos; **11.25(rest):** The Bettmann Archives; **11.26:** © J.L. Bulcao/Gamma Liaison

Chapter 12

Opener: © David Young Wolff/Tony Stone Images; **12.1A:** AP/Wide World Photos; **12.1B:** Courtesy of Cornell University; **12.3:** © CNRI/Phototake; **12.4:** People Weekly © Taro Yamasaki; **12.5:** © Luciano Amendola/Imago Press; **12.6C:** © Integra Med; **12.7:** © Jim Sulley/The Image Works; **12.9:** © Topham/The Image Works; **12.10B:** From Dr. A. Pytkowicz Streissguth, *Science* 209:353–361, July 18, 1980. © 1980 by the AAAS; **12.11:** © Philippe Plailly/Eurelios/SPL/Photo Researchers, Inc.; **12.14, 12.15:** © CDC/Science Source/Photo Researchers, Inc.; **12.16:** © Zeva Oelbaum/Peter Arnold, Inc.; **12.17:** Courtesy of Ward F. Odenwald, National Institute of Neurological Disease and Stroke

Chapter 13

Opener: © M. Kitada/Orange County Register; **13.1A,B:** © Archive Photos; **13.2:** Courtesy of W. Dorsey Stuart, University of Hawaii; **13.7A:** Courtesy of Dr. Myron Neuffer, University of Missouri; **13.7B:** © Nancy Hamilton/Photo Researchers, Inc.; **Table 13.4, first,** p. 273: © Norvia Behling/Animals Animals/Earth Scenes; **Table 13.4, second,** p. 273: © Grant Heilman/Grant Heilman Photography; **Table 13.4, third,** p. 273: © Robert Maier/Animals Animals/Earth Scenes; **Table 13.4, fourth,** p. 273: © Hans Burton/Bruce Coleman; **Table 13.4, fifth,** p. 273: © Jane Burton/Bruce Coleman; **13.19:** © Lester Bergman; **13.20A–C:** Courtesy Thomas Kaufman. Photo by Phil Randazzo and Rudi Turner.; **13.21:** Library Of Congress; **13.A:** © Redneck/Gamma Liaison

Chapter 14

Opener: From L. Chong, "A Human Telomeric Protein", *Science,* 270: 1663–1667, 1995. © American Association for the Advancement of Science. Photo courtesy, Dr. Titia DeLange.; **14.A:** © P.J. Devries; **14.B:** © Zig Leszcynski/Animals Animals; **14.6:** © BioPhoto Associates/Photo Researchers, Inc.; **14.10A:** © Horst Schaefer/Peter Arnold, Inc.; **14.10B:** © William E. Ferguson; **14.11A,B:** From Wilson and Foster, *Williams Textbook of Endocrinology,* 7/e © W.B. Saunders 1985; **14.11C,D:** Courtesy of The National Jewish Hospital and Research Center; **14.14B:** From P.C. Nowell and P.A. Hungerford. Chromosome studies normal and leukemic leukocytes. *Journal of the National Cancer Institute,* 25:1960, pp. 85–109.; **14.14C:** Courtesy of Dr. Frederick Elder; **14.14D:** Courtesy David Ward, Department of Genetics, Yale University School of Medicine; **14.16:** Courtesy Colleen Weisz; **14.18:** Courtesy of Lawrence Livermore National Laboratory; **14.19B:** From R. Simensen, R. Curtis Rogers, "Fragile X Syndrome," *American Family Physician,* 39:186 May 1989. © American Academy of Family Physicians.; **p. 305:** Courtesy of Integrated Genetics

Chapter 15

Opener: © Ken Eward/BioGrafx/Photo Researchers, Inc.; **15.1:** © Dr. Gopal Murti/SPL/ Photo Researchers, Inc.; **15.5:** © A.C. Barrington Brown/Photo Researchers, Inc.; **15.9A:** © 1948 M-C Escher Foundation-Baarn-Holland, All Rights Reserved; **15.15:** © Kenneth Greer/Visuals Unlimited

Chapter 16

Opener: Courtesy Beckman Instruments; **16.16A:** Courtesy O.L. Miller, Jr.; **16.19A,B:** © Bill Longcore/Photo Researchers, Inc.; **16.C:** © Sovfoto/Eastfoto; **16.25A:** © Science Photo Library/Photo Researchers, Inc.; **16.25B:** © Susan McCartney/Photo Researchers, Inc.

Chapter 17

Opener: © Alfred Pasieka/SPL/Photo Researchers, Inc.; **17.1:** © William E. Ferguson; **17.2:** © Peter J. Bryant/Biological Photo Service; **17.3:** © Transparency #3316, Courtesy Department of Library Services, American Museum of Natural History; **17.7B:** © Dr. Tony Brain/SPL/Photo Researchers, Inc.; **17.11A:** © Dr. Malcolm Walter; **17.11B,C:** Dr. Andrew Knoll/Harvard University; **17.12A:** © Ed Reschke/Peter Arnold, Inc.; **17.14A:** Field Museum of Natural History, Chicago (Geo 75400C); **17.14B:** © Martin Land/Science Photo Library/Photo Researchers, Inc.; **17.18:** © John Reader/SPL/Photo Researchers, Inc.; **17.20A:** © G. Hinter Leitner/Gamma Liaison; **17.20B:** © Burt Silverman/Silverman Studios, Inc.

Chapter 18

Opener: © Frans Lanting/Minden Pictures; **18.2:** © Frans Lanting/Photo Researchers, Inc.; **18.4:** © Costa Manos/Magnum Photos; **18.5B:** © William E. Ferguson; **18.7A(ostrich):** © Barbara von Hoffmann/Tom Stack & Associates; **18.7A(emu):** © Mark Newman/Photo Researchers, Inc.; **18.7A(rhea):** © Toni Angermayer/Photo Researchers, Inc.; **18.7(N. American wolf):** © Ralph Eagle/Photo Researchers, Inc.; **18.7B(tasmanian wolf):** © Jeff Rotman/Peter Arnold, Inc.; **18.7C(reptile):** © A.J. Copley/Visuals Unlimited; **18.7C(fish):** © Runk/ Schoenberger/Grant Heilman Photography; **18.7C(dolphin):** © D. Holden Bailey/Tom Stack & Associates; **18.8A–C:** © Peter Grant/Princeton University; **18.9A,B:** © Tom McHugh/Photo Researchers, Inc.; **18.9C:** © Robert and Linda Mitchell; **18.9D:** Courtesy of John Alcock; **18.A:** © Science VU/Visuals Unlimited; **18.10A:** © A.B. Dowsett/SPL/Photo Researchers, Inc.; **18.10B:** © David M. Phillips/Visuals Unlimited; **18.10C:** © Stanley F. Hayes, National Institutes of Health/SPL Photo Researchers, Inc.; **18.10C(inset):** © Cabisco/Phototake; **18.11:** © Stuart Franklin/ Magnum Photos; **18.12:** The Bettmann Archives

Chapter 19

Opener: © Transparency #K17089, Courtesy Department of Library Services, American Museum of Natural History; **19.3A:** © Bob Gossington/ Bruce Coleman; **19.3B:** © Jim Steinberg/Photo Researchers, Inc.; **19.3C:** © John Reader/SPL/Photo Researchers, Inc.; **19.3D:** Courtesy Department Library Services - American Museum of Natural History, Neg. No. 320496; **19.3E:** © David

Grimaldi/American Museum of Natural History; **19.7A:** Courtesy Dr. H. Hameister, Universitatulm Ulm, Abteilung Medizinische Genetik; **19.9A:** From James H. Asher and Thomas B. Friedman, *Journal of Medical Genetics* 27:618–626, 1990; **19.9B:** © Vickie Jackson; **19.9C:** From James H. Asher and Thomas B. Friedman, *Journal of Medical Genetics* 27:618–626, 1990; **19.12:** © Zoological Society of London; **19.A:** © Bob Taylor

Chapter 20

Opener: © Mike Severns; **20.1:** © Sandy Macys/Gamma Liaison; **20.A(multitoed cat):** © Scott Camazine/Photo Researchers, Inc.; **20.A(curl cat):** © Barb Zurawski; **20.3:** Courtesy of Dr. McKusick/Johns Hopkins Hospital; **20.6A,B:** © Michael Tweedie/Photo Researchers, Inc.; **20.8:** © Porterfield/Chickering/Photo Researchers, Inc.; **20.9A:** © Karl Ammann/Bruce Coleman; **20.9B:** © Norman Owen Tomalin/Bruce Coleman; **20.B:** Photo by L. Rieseberg and G. Seiler, Biology Department, Indiana University; **20.10A(top):** © E.R. Degginger/Photo Researchers, Inc.; **20.10A(bottom):** © Kjell B. Sandved/Visuals Unlimited; **20.10B(both):** © J. Wm. Schopf/UCLA; **20.11:** © Discover Magazine

Chapter 21

Opener: © Suzanne & Nick Gear/Tony Stone Images; **21.1:** © Alan D. Carey/Photo Researchers, Inc.; **Table 21.1, third,** p. 426: © Mary Clay/Tom Stack & Associates; **21.A(both):** © Barbara Mensch; **21.6:** Courtesy of Dr. JoAnn Burkholder; **21.B:** © Heather Angel; **21.C:** © The Granger Collection; **21.D:** © Gregory S. Paulson

Chapter 22

Opener: © P. Canumette/Visuals Unlimited; **22.1A:** © Arthur M. Siegelman/Visuals Unlimited; **22.1B:** © T.E. Adams/Visuals Unlimited; **22.1C:** From J.T. Staley, M.P. Bryant, N. Pfennig, and J.G. Holt, (Eds.), *Bergey's Manual of Systematic Bacteriology,* Vol. 3. © 1989 Williams and Wilkins Co., Baltimore. Prepared by G. Bentzen; photographed by the Laboratory of Clinical Electron Microscopy, U. of Bergen; **22.3:** Esther R. Angert & Norman R. Pace; **22.4A:** © Arthur M. Siegelman/Visuals Unlimited; **22.4B:** © George J. Wilder/Visuals Unlimited; **22.4C:** © Thomas Tottleben/Tottleben Scientific Company; **22.5A:** © Arthur M. Siegelman/Visuals Unlimited; **22.5B:** © E.S. Anderson/Photo Researchers, Inc.; **22.7A,B:** © T.J. Beveridge/Biological Photo Service; **22.A(top):** © Ifremer; **22.A(bottom 2):** © Jean Vacelet, Centre d'Oceanologie de Marseille; **22.9A:** Courtesy of Charles C. Brinton, Jr. and Judith Carnahan; **22.9B:** © Fred Hossler/Visuals Unlimited; **22.9C:** © George Musil/Visuals Unlimited; **22.9D:** From J. Hoeniger and C. Headley, *Journal of Bacteriology* 96:1844, 1968. American Society for Microbiology; **22.9E:** © G. Murti/SPL/Photo Researchers, Inc.; **22.12A,B:** © Arthur M. Siegelman/Visuals Unlimited; **22.13:** © Fred Lyon/Photo Researchers, Inc.; **22.14A:** © Science VU–D. Foster, Woods Hole Oceanographic Institution/Visuals Unlimited; **22.14B:** © WHOI/Visuals Unlimited; **22.15:** © Sylvan Wittwer/Visuals Unlimited; **22.16:** © Joe Munroe/Photo Researchers

Chapter 23

Opener: © Eric V. Grave/Photo Researchers, Inc.; **23.1:** © Roland Birke/Peter Arnold, Inc.; **Table 23.1, first,** p. 458: © Biophoto Associates/Science Source/Photo Researchers, Inc.; **Table 23.1, second,** p. 458: © Gary R. Robinson/Visuals Unlimited; **Table 23.1, third,** p. 458: © J. Robert Waaland/Biological Photo Service; **Table 23.1, fourth,** p. 458: © M.I. Walker/Photo Researchers, Inc.; **Table 23.1, fifth,** p. 458: © Carolina Biological Supply/Phototake; **23.2:** © Biophoto Associates/Science Source/Photo Researchers, Inc.; **23.A:** © Lennart Nilsson, *The Body Victorious,* Dell Publishing Company; **23.5A,B:** From M. Schaechter, G. Medoff, and D. Schlessinger (Eds.), *Mechanism of Microbial Disease,* 1989, Williams and Wilkins.; **23.6:** © Kenneth W. Fink/Bruce Coleman; **23.9A:** © M.I. Walker/Photo Researchers, Inc.; **23.9B:** © M. Kage/Peter Arnold, Inc.; **23.9C:** © L.L. Sims/Visuals Unlimited; **23.10:** © John D. Cunningham/Visuals Unlimited; **23.11:** © R. Kessel-G. Shin/Visuals Unlimited; **23.12:** © Heather Angel/Biofotos; **23.13:** © D.P. Wilson/Science Source/Photo Researchers, Inc.; **23.14A,B:** © John D. Cunningham/Visuals Unlimited; **23.15A:** © Randy Morse/Tom Stack & Associates; **23.16:** © Gary R. Robinson/Visuals Unlimited; **23.17:** © J. Robert Waaland/Biological Photo Service; **23.19:** © B. Beatty/Visuals Unlimited; **23.20A:** © Carolina Biological Supply/Phototake; **23.20B:** © David Scharf/Peter Arnold; **23.20C:** © Carolina Biological Supply/Phototake

Chapter 24

Opener: © E. Guiho/CNRI/SPL/Photo Researchers, Inc.; **24.1:** Courtesy of N. Allin and G.L. Barron, University of Gue; **24.2A,B:** Courtesy of G.L. Barron, The University of Guelph; **24.3A:** © David M. Phillips/Visuals Unlimited; **24.3B:** © William J. Weber/Visuals Unlimited; **24.4C:** Courtesy of Dr. Garry T. Cole/The Univ. of Texas at Austin; **Table 24.1, first,** p. 479: © Dwight Kuhn; **Table 24.1, second,** p. 479: © Doug Sherman/Geofile; **Table 24.1, third,** p. 479: Courtesy of G.L. Barron, The University of Guelph; **Table 24.1, fourth,** p. 479: © Andrew McClenaghan/Photo Researchers, Inc.; **24.5A:** © Biophoto Associates/Science Source/ Photo Researchers, Inc.; **24.5B:** © Bruce Iverson; **24.5C:** © Stanley Flegler/Visuals Unlimited; **24.6:** Photograph by B.A. Roy, reprinted by permission from *Nature* Vol. 362, Cover and page 57.; **24.7:** © R.S. Hussey/Visuals Unlimited; **24.8:** © Dwight Kuhn; **24.9A:** © John D. Cunningham/Visuals Unlimited; **24.9B:** © Doug Sherman/Geofile; **24.A:** © Holt Studios, Ltd./Animals Animals/Earth Scenes; **24.10B:** © Stanley L. Flegler/Visuals Unlimited; **24.11A:** © Bill Keogh/Visuals Unlimited; **24.11B:** © Hans Reinhard/Bruce Coleman, Inc.; **24.12A:** © Andrew McClenaghan/SPL/Photo Researchers, Inc.; **24.12B:** © E. Gueho/CNRI/SPL/Photo Researchers, Inc.; **24.13A:** © V. Ahmadjian/Visuals Unlimited; **24.13B:** © Steve & Sylvia Sharnoff/Visuals Unlimited; **24.13C:** © S.J. Kraseman/Peter Arnold, Inc.; **24.14:** © John D. Cunningham/Visuals Unlimited; **24.15A:** © Science VU-USDA/Visuals Unlimited; **24.15B, 24.16A,B:** © Everett S. Beneke/Visuals Unlimited; **24.16C:** © Dick Thomas/Visuals Unlimited; **24.16D:** © Cecil Fox/Photo Researchers, Inc.; **24.26E:** Courtesy of Pfizer Roerig, Inc.

Chapter 25

Opener: © Rod Planck/Tom Stack & Associates; **25.1:** © W. H. Hodge/Peter Arnold, Inc.; **Table 25.1, first,** p. 493: © Heather Angel/Biofotos; **Table 25.1, second,** p. 493: © Rod Planck/Tom Stack & Assoc.; **Table 25.1, third,** p. 493: © Alan G. Nelson/Animals Animals/Earth Scenes; **Table 25.1, fourth,** p. 493: © John Gerlach/Visuals Unlimited; **25.3A:** © Heather Angel/Biofotos; **25.3B:** © William E. Ferguson; **25.3C:** © Kingsley Stern; **25.4A and 25.4B:** Courtesy of Howard Crum, University of Michigan Herbarium; **25.4C:** © John D. Cunningham/Visuals Unlimited; **25.5:** © William E. Ferguson; **25.A:** © Ira Block; **25.6A:** © W. Ormerod/Visuals Unlimited; **25.6B:** © Bud Lehnhausen/Photo Researchers, Inc.; **25.6C:** © Joe McDonald/Visuals Unlimited; **25.6D:** © Rod Planck/Tom Stack & Associates; **25.7A,B:** © Robert & Linda Mitchell; **25.8:** © Cabisco/Visuals Unlimited; **25.9A:** © W.H. Hodge/Peter Arnold, Inc.; **25.9B:** © William E. Ferguson; **25.10A:** © Kjell B. Sandved/Butterfly Alphabet; **25.10B:** © David M. Dennis/Tom Stack & Associates; **25.11A:** © Biophoto Associates/Photo Researchers, Inc.; **25.11B:** © Alan G. Nelson/Animals Animals/Earth Scenes; **25.11C:** © Jeff Gnass; **25.11D:** © R.D. Estes/Photo Researchers, Inc.; **25.12A:** © Scott Camazine/Photo Researchers, Inc.; **25.12B:** © W.H. Hodge/Peter Arnold, Inc.; **25.12C:** © Noron Thomas/Photo Researchers, Inc.; **25.12D:** © W. H. Hodge/Peter Arnold, Inc.; **25.B1, B2:** © William E. Ferguson; **25.B3:** © Stephen J. Krasemann/Peter Arnold, Inc.; **25.14:** © Gilbert Grant/Photo Researchers, Inc.; **25.15:** © Renee Lynn/Photo Researchers, Inc.

Chapter 26

Opener: © Craig Young/Harbor Branch Oceanographic Institution; **26.1:** © C.P. Hickman/Visuals Unlimited; **26.6A:** © Bios/Peter Arnold, Inc.; **26.6B(caterpillar):** © Rod Planck/ Tom Stack & Associates; **26.6B(butterfly):** © John Shaw/Tom Stack & Associates; **26.8A:** © C.P. Hammond; **26.8B:** © Marty Snyderman/Visuals Unlimited; **26.10A:** © Tom Branch/Photo Researchers, Inc; **26.12:** © CNRI/SPL/Photo Researchers, Inc.; **26.13:** © A.M. Siegelman/Visuals Unlimited; **26.15A:** © Fred McConnaughey/Photo Researchers, Inc.; **26.15B:** © Chesher/Photo Researchers, Inc.; **26.15C:** © Fred Bavendam/Peter Arnold, Inc.; **26.16B:** © Michael DiSpezio; **26.17:** © William Ferguson; **26.18A:** © James C. Amos/ Photo Researchers, Inc.; **26.18B:** © Stephen Krasemann/Photo Researchers, Inc.; **26.18C:** © Zigmund Leszczynski/Animals Animals; **26.18D:** © David Scharf/Peter Arnold, Inc.; **26.19A:** © Rick M. Harbo; **26.19B:** © Andrew Martinez/Photo Researchers, Inc.; **26.20A:** © Michael DiSpezio; **26.20B:** © Runk/Schoenberger/Grant Heilman Photography; **26.20C:** © Russ Kinne/Photo Researchers, Inc.; **26.20D:** © Daniel W. Gotshall/ Visuals Unlimited; **26.20E:** © R. DeGoursey/ Visuals Unlimited; **26.20F:** © Joel Arrington/ Visuals Unlimited; **26.20G:** © E.R. Degginger/

Photo Researchers, Inc.; **26.20H:** © Karl Maslowski/Visuals Unlimited; **26.20I:** © Zigmund Leszczynski/Animals Animals; **26.21A:** © Runk/Schoenberger/Grant Heilman Photography; **26.22:** © Michael DiSpezio; **26.23A:** © Tom McHugh/Photo Researchers, Inc.; **26.23B:** © John D. Cunningham/Visuals Unlimited

Chapter 27

Opener: © Nicholas Parfitt/Tony Stone Images; **27.2:** © BioPhot; **27.3A:** Courtesy Long Ashton Research Station; **27.3B:** © Randy Moore/BioPhot; **27.4A:** © BioPhot; **27.4B:** © Bruno P. Zehnder/Peter Arnold, Inc.; **27.5:** © BioPhot; **27.6:** © David M. Phillips/Visuals Unlimited; **27.7:** From Yolanda Heslop-Harrison, "SEM of Fresh Leaves of Pinguicula", *Science* 667:173. © 1970 by the AAAS; **27.11A:** © Charles Gurche; **27.11B:** © Kenneth W. Fink/Photo Researchers, Inc.; **27.11C:** © Kjell Sandved/Butterfly Alphabet; **27.11D:** © Franz Krenn/Photo Researchers, Inc.; **27.11E:** © Dwight R. Kuhn; **27.12A–C, 27.13A:** © Kjell Sandved/Butterfly Alphabet; **27.13B:** © Alford W. Cooper/Photo Researchers, Inc.; **27.14A,B:** © Dwight R. Kuhn; **27.16A:** © Michael Ederegger/Peter Arnold, Inc.; **27.16B:** © Kingsley Stern; **27.17B:** © Ed Reschke; **27.19:** © John D. Cunningham/Visuals Unlimited; **27.20B:** © Runk/Schoenberger/Grant Heilman Photography; **27.21:** © P. Dayanandar/Photo Researchers, Inc.; **27.23A:** © A. W. Ambler/Photo Researchers, Inc.; **27.23B,C:** © W. H. Hodge/Peter Arnold, Inc.; **27.23D:** © William Ferguson; **27.25A:** © Manfred Kage/Peter Arnold, Inc.; **27.25B:** © BioPhot

Chapter 28

Opener: © Joyce Photographics/Photo Researchers, Inc.; **28.1:** © J. Alcock/Visuals Unlimited; **28.2A:** © Runk/Schoenberger/Grant Heilman; **28.2B:** © W.H. Hodge/Peter Arnold, Inc.; **28.2C:** © Runk/Schoenberger/Grant Heilman; **28.3A:** © William E. Ferguson; **28.4A:** © W.H. Hodge/Peter Arnold, Inc.; **28.6:** © R.J. Erwin/Photo Researchers, Inc.; **28.C,D:** © David Littschwager, Susan Middleton/Discover Magazine; **28.9A:** © William E. Ferguson; **28.9B:** © Francois Gohier/Photo Researchers, Inc.; **28.9C:** © Joe McDonald/McDonald Wildlife Photography; **28.10A,B:** © Thomas Eisner; **28.13A:** © R.J. Erwin/Photo Researchers, Inc.; **28.13B:** © Toni Michaels; **28.13C:** © James Welgos/Photo Researchers, Inc.; **28.14A:** © Adam Hart-Davis/SPL/Photo Researchers, Inc.; **28.14B:** Agricultural Research Service/USDA; **28.14C:** © Rod Planck/Photo Researchers, Inc.; **28.14D:** © W.H. Hodge/Peter Arnold, Inc.; **28.14E:** © William Ferguson; **28.15:** © Michael P. Gadomski/Photo Researchers, Inc.

Chapter 29

Opener: © Tim Thompson/Tony Stone Images; **29.1:** © Fred Bavendam; **29.2:** © David Newman/Visuals Unlimited; **29.3:** © R. Lyons/Visuals Unlimited; **29.4:** © W.H. Hodge/Peter Arnold, Inc.; **29.5:** © Jerome Wexler/Photo Researchers, Inc.; **29.A(both):** © Dr. Randy Moore; **29.6:** © John D. Cunningham/Visuals Unlimited; **29.7:** © William E. Ferguson; **29.8A:** © John Kaprielian/Photo Researchers, Inc.; **29.8B:** © Richard H. Gross; **29.8C:** © John Kaprielian/Photo Researchers, Inc.; **29.9A,B:** © Tom McHugh/Photo Researchers, Inc.; **29.14:** © Stephenie Ferguson/William Ferguson; **29.15:** © Runk/Schoenberger/Grant Heilman; **29.16:** © Galen Rowell/Peter Arnold, Inc.; **29.17:** © Dr. Steve Kay, Scripps Research Insitute; **29.B:** © Joanne Chory; **29.C:** © Philip Becraft, Department of Zoology & Genetics, Iowa State University

Chapter 30

Opener: © Max Aguilera-Hellweg 4:AM Productions; **Table 30.1, first,** p. 597: © J. & L. Weber/Peter Arnold, Inc.; **Table 30.1, second,** p. 597: © Edwin A. Reschke/Peter Arnold, Inc.; **Table 30.1, third,** p. 597: © Edwin A. Reschke/Peter Arnold, Inc.; **Table 30.1, fourth,** p. 597: © Edwin A. Reschke/Peter Arnold, Inc.; **Table 30.1, fifth,** p. 597: © Edwin A. Reschke/Peter Arnold, Inc.; **Table 30.2, first,** p. 597: © Ed Resechke; **Table 30.2, second,** p. 597: © OMIKRON/Science Source/Photo Researchers, Inc.; **Table 30.2, third,** p. 597: © Ed Reschke; **Table 30.2, fouth,** p. 597: Carolina Biological Supply Company; **Table 30.2, fifth,** p. 597: © Chuck Brown/Photo Researchers, Inc.; **Table 30.2, sixth,** p. 597: © J. & L. Weber/Peter Arnold, Inc.; **30.7:** AP/Wide World Photos; **30.8:** © CNRI/SPL/Photo Researchers, Inc.; **30.10A:** © Renee Lynn/Tony Stone Images; **30.10B:** © John Warden/Tony Stone Images; **30.10C:** © Art Wolfe/Tony Stone Images; **30.A:** © National Library of Medicine/Visible Human Project

Chapter 31

Opener: © CNRI/SPL/Science Source/Photo Reseachers, Inc.; **31.1:** © Leonard Lee Rue III; **31.2B:** © Secchi-League/Roussel-UCLAF/CNRI/SPL/Photo Researchers, Inc.; **31.6D:** Courtesy National Multiple Sclerosis Society; **31.8C:** © Don Fawcett/Photo Researchers, Inc.; **31.10:** © Howard Hall; **31.12B:** Courtesy Kathryn Tosney; **31.17:** © Dan McCoy/Rainbow; **31.18A:** © Science VU/Visuals Unlimited; **31.18B:** Courtesy of Dr. Marcus Raichle, Washington University Medical School; **31.B:** National Institute on Drug Abuse

Chapter 32

Opener: © Dave Watts/Tom Stack & Associates; **32.1A and 32.1B:** © Thomas Eisner; **32.A:** © Larry Miller/Photo Researchers, Inc.; **32.B:** © Denise Tackett/Tom Stack & Associates; **32.C:** © David R. Frazier/Photo Researchers, Inc.; **32.4A:** © Thomas Eisner; **32.4B:** © Fred Bavendam/Peter Arnold, Inc.; **32.5B:** © Veronika Burmeister/Visuals Unlimited; **32.10A:** © Frank S. Werblin; **32.15B:** © CNRI/SPL/Photo Researchers, Inc.; **32.18C:** © Lennart Nilsson, *Behold Man,* Little Brown and Company

Chapter 33

Opener: © Jeanne Drake/Tony Stone Images; **33.2:** © Fred Hossler/Visuals Unlimited; **33.4:** © Thomas Eisner; **33.8B:** © Dan Kline/Visuals Unlimited; **33.11A–D:** Clinical Pathological Conference on Acromegaly, Diabetes, Hypermetabolism, Protein Use and Heart Failure, *American Journal of Medicine* 20: 133, (1956). © Cahners Publishing Co., Inc.; **33.13B:** © Lester Bergman; **33.16B:** © Manfred Kage/Peter Arnold, Inc.; **33.17A,B:** Courtesy of Eli Lilly & Company; **33.A:** © G. Gransanti/Sygma

Chapter 34

Opener: NASA; **34.1A:** © Bill Bachman/Photo Researchers, Inc.; **34.1B:** © Dwight Kuhn; **34.3B:** © Nancy Stefton/Photo Researchers, Inc.; **34.5:** © Simon Brown/SPL/Photo Researchers, Inc.; **34.7:** From Kenneth R. Gordon, Ph.D., "Adaptive Nature of Skeletal Design," *BioScience* 39(11):784–790, Dec. 1989.; **34.8:** © Biophoto Associates/Photo Researchers, Inc.; **34.9:** © Chuck Brown/Photo Researchers, Inc.; **34.10:** © J. & L. Weber/Peter Arnold, Inc.; **34.13A:** © Ed Reschke; **34.13B:** © Manfred Kage/Peter Arnold, Inc.; **34.13C:** © Ed Reschke; **34.14:** © David M. Phillips/Visuals Unlimited; **34.14B:** © K. Jeon/Visuals Unlimited; **34.20A:** © James Dennis/Phototake; **34.21A,B:** © The McGraw-Hill Companies, Inc./Bob Coyle, photographer; **34.26A:** © Custom Medical Stock Photos; **34.26B:** © B.S.I.P./Custom Medical Stock Photos; **34.C:** © G. W. Willis, Ochsner Medical Institution/Biological Photo Service

Chapter 35

Opener: © P.M. Motta & S. Correr/SPL/Photo Researchers, Inc.; **35.4A:** Courtesy of Dr. Leland Clark; **35.4B:** Courtesy of Organogenesis Inc.; **35.6:** © Boehringer Ingelheim International GmbH, Photo: Lennart Nilsson, *The Body Victorious,* Pell Publishing Co.; **35.9A:** © NFBSC/SPL/Photo Researchers, Inc.; **35.9B:** © Prof. P. Motta/Dept. of Anatomy/University of "La Sapienza," Rome/SPL/Photo Researchers, Inc.; **35.B:** National Library of Medicine; **35.C:** © C.P. Hickman/Visuals Unlimited; **35.12A:** © H. & F. Michler/Peter Arnold, Inc.; **35.17:** © Lennart Nilsson, *Behold Man,* Little Brown & Co.; **35.21A:** © SIU/Photo Researchers, Inc.

Chapter 36

Opener: © CNRI/SPL/Photo Researchers, Inc.; **36.1:** © Douglas M. Munnecke/Biological Photo Service; **36.2A:** © Gregory Ochocki/Photo Researchers, Inc.; **36.2B:** © Dwight Kuhn; **36.2C:** © E.R. Degginger/Photo Researchers, Inc.; **36.2D:** © Soames Summerhay/Photo Researchers, Inc.; **36.2E:** © Tim Davis/Photo Researchers, Inc.; **36.4A:** © F. Gohier/Photo Researchers, Inc.; **36.4B:** © Andrew Wood/Photo Researchers, Inc.; **36.9B:** From Kessel and Kardon: *A Text Atlas of Scanning Electron Microscopy,* 1979 W.H. Freeman and Company; **36.B:** © Moredun Animal Health/SPL/Photo Researchers, Inc.; **36.C(both):** Courtesy of American Cancer Society; **36.15:** © Weinstein/Custom Medical Stock Photo; **36.16A:** © Scott Camazine/Photo Researchers, Inc.; **36.16B:** © CNRI/SPL/Photo Researchers, Inc.

Chapter 37

Opener: © Gerry Ellis; **37.1:** © Dwight Kuhn; **37.3A:** © James D. Watt/Planet Earth Pictures; **37.3B:** © John D. Cunningham/Visuals Unlimited; **37.4:** © Tom McHugh/Photo Researchers, Inc.; **37.7A:** © Arthus Bertrand/Photo Researchers, Inc.; **37.A:** © James Nachtwey/Magnum; **37.13B:** Courtesy Dr. Jean-Luc C. Urbain, Temple University

Hospital & Fels Institute; **37.17A:** © David Scharf/Peter Arnold, Inc.; **37.17B:** Courtesy of David H. Alpers, M.D.

Chapter 38

Opener: © Roger Eriksson; **38.1:** © Richard Nowitz; **38.A:** © Fred Bavendam/Peter Arnold; **38.3:** Courtesy of Eric Toolson; **38.6:** © Dwight Kuhn; **38.10:** © H.W. Silvester/Rapho/Photo Researchers, Inc.

Chapter 39

Opener: © R. Ellis/Sygma; **39.1:** Courtesy of Dr. Christopher J. Bayne; **39.3:** © Biology Media/Photo Researchers, Inc.; **39.4B:** © NIBSC/SPL/Photo Researchers, Inc.; **39.6B:** © Manfred Kage/Peter Arnold, Inc.; **39.10:** © McGraw-Hill Higher Education Group, Inc./Bob Coyle, photographer; **39.12:** © Dr. A. Liepins/SPL/Photo Researchers, Inc.; **39.14:** From Ivan Roitt, *Essential Immunology*, fig. 8.7, p. 130. Blackwell Scientific Publications Ltd. Photo courtesy of Prof. I. Brent; **39.15:** © Zeva Oelbaum/Peter Arnold, Inc.; **39.17:** © Sygma; **39.A(leopard):** © Tim Davis/Photo Researchers, Inc.; **39.A(cat):** © Renee Lynn/Photo Researchers, Inc.; **39.18B:** © Lennart Nilsson/Boehringer Ingelheim International Gmbh; **39.19A:** © David Scharf/Peter Arnold, Inc.; **39.19B:** © Phil Harrington/Peter Arnold, Inc.

Chapter 40

Opener: © J. Emmett Duffy, College of William & Mary, Virginia Institute of Marine Science; **40.1:** © J.A.L. Cooke; **40.2:** © K & K Ammann/Bruce Coleman; **40.4(all):** © Ernest Cooper; **40.6:** Nina Leen/Life Magazine © Time Warner, Inc.; **40.7:** Courtesy of Gary Meszaros; **40.A:** © Marty Snyderman; **40.8:** © Lynn Rogers; **40.9:** © Runk/Schoenberger/Grant Heilman; **40.C:** © Raymond A. Mendez; **40.D:** © Stephen Dalton/Photo Researchers, Inc.; **40.E:** © Louis Quitt/M.A.S./Photo Researchers, Inc.; **40.10:** © Horst Schafer/Peter Arnold, Inc.; **40.11:** © Laura Riley/Bruce Coleman, Inc.; **40.12A,B:** © Charles H. Janson; **40.15A:** Courtesy of David W. Pfennig; **40.16:** © Diana and P. Sullivan/Burce Coleman, Inc.; **40.17:** © Michael & Patricia Fogden; **40.18A:** © Lady Philippa/Scott/NHPA; **40.18B:** © Ernest Cooper; **40.19A–C:** Courtesy of Lowell L. Getz and Lisa L. Davis; **40.20:** © Paul Watson, Biology Department, University of New Mexico; **40.22A:** © Bildorchiv Okapia/Photo Researchers, Inc.; **40.22B:** © Mimi Forsyth/Bruce Coleman

Chapter 41

Opener: © Nicholas DeVore/Tony Stone Images; **41.A:** © C. Ray Cable; **41.B:** Courtesy of Montana Dept. of Fish, Wildlife, and Parks; **41.C:** © Tim Clutton-Brock; **41.9:** © Dr. Edward S. Ross; **41.10A:** © Laurence Gould/OSF/Animals Animals; **41.10B:** © Herb Segars; **41.12A,B:** © Norm Smith and Jim Dawson

Chapter 42

Opener: © Christine & Michel Denis-Huot; **42.2:** © Fritz Polking/Peter Arnold, Inc.; **42.4:** © Jeff Henry/Peter Arnold, Inc.; **42.6A–F:** Courtesy, Dr. Stuart Fisher, Department of Zoology, Arizona State University; **42.A:** © Brian Parker/Tom Stack & Associates; **42.B:** © M.P. Kahl, Jr./Photo Researchers, Inc.; **42.C:** © Uta Matthes-Sears/Univ. of Guelph; **42.D:** © Dr. D.W. Larson/Univ. of Guelph; **42.E:** Courtesy of Sandra Miller; **42.F:** © Tom Stack/Tom Stack & Associates; **42.7:** © Charlie Ott/Photo Researchers, Inc.; **42.8B:** © Steven C. Amstrup/U. S. Fish and Wildlife; **42.16:** © R.T. Smith/Ardea

Chapter 43

Opener: © Glenn Oakley; **43.2A:** © Jim Zipp/Photo Researchers, Inc.; **43.2B:** © Luiz C. Marigo/Peter Arnold, Inc.; **43.2C:** © L. West/Photo Researchers, Inc.; **43.2D:** © Toni Michaels; **43.2E:** © John Bova/Photo Researchers, Inc.; **43.2F:** © Stephen J. Krasemann/Peter Arnold, Inc.; **43.3:** © Roy Toft/Tom Stack & Associates; **43.4A:** © Carl R. Sams/Peter Arnold, Inc.; **43.4B:** © Stephen J. Krasemann/Photo Researchers, Inc.; **43.5:** © Thomas Kitchin/Photo Researchers, Inc.; **43.6:** © John Shaw/Bruce Coleman; **43.7:** © Fred Bruemmer/Peter Arnold, Inc.; **43.8A:** © Jeff Rotman; **43.8B:** © Jen & Des Bartlett/Bruce Coleman; **43.B:** © Nancy Sefton/Photo Researchers, Inc.; **43.10:** © Nigel Dennis/Photo Researchers, Inc.; **43.11A:** © David Hall/Photo Researchers, Inc.; **43.11B:** © Peter Parks/Oxford Scientific Films/Animals, Animals

Chapter 44

Opener: Dr. Jerome A. Jackson, Department of Biological Sciences, Mississippi State University; **44.1A,B:** © Vanessa Vick/Photo Researchers, Inc.; **44.1C:** © Stephen Ferry; **44.2:** © TASS/Sovfoto; **44.A:** © Scott Camazine/Photo Researchers, Inc.; **44.B:** Courtesy E.G. & G. Rocky Flats; **44.3A:** © Ray Pfortner/Peter Arnold, Inc.; **44.4:** © NASA/SPL/Photo Researchers, Inc.; **44.8:** © Flip Nicklin/Minden Pictures; **44.10:** © George Antonelis, National Marine Fisheries Serv.; **44.11A:** © Michael J. Balick/Peter Arnold, Inc.; **44.11B:** © Tom McHugh/Photo Researchers, Inc.; **44.12:** © NRSC/SPL/Photo Researchers, Inc.; 4**4.14A,B:** Courtesy of David Tilman

Appendix D

D–1, p. 913: © USDA/Photo Researchers, Inc.; **D–2,** p. 913: © William Ferguson; **D–3,** p. 913: © Michael Viard/Peter Arnold, Inc; **D–4,** p. 913: © Dorthy D. Beatty/Photo Researchers, Inc; **D–5,** p. 913: © Stephen E. Sutton/Duomo

Text/Line Art

Chapter 3

3.8: From Stuart Ira Fox, *Human Physiology,* 5th ed. Copyright © 1996 The McGraw-Hill Companies, Inc. All Rights Reserved. Reprinted by permission. **3.20:** From Stuart Ira Fox, *Human Physiology,* 5th ed. Copyright © 1996 The McGraw-Hill Companies, Inc. All Rights Reserved. Reprinted by permission.

Chapter 4

4.12B: From David Shier, et al., *Hole's Human Anatomy and Physiology,* 7th ed. Copyright © 1996 The McGraw-Hill Companies, Inc. Reprinted by permission. **4.13B:** From David Shier, et al., *Hole's Human Anatomy and Physiology,* 7th ed. Copyright © 1996 The McGraw-Hill Companies, Inc. Reprinted by permission. **4.16B:** © George V. Kelvin / Scientific American.

Chapter 5

5.2B: From David Shier, et al., *Hole's Human Anatomy and Physiology,* 7th ed. Copyright © 1996 The McGraw-Hill Companies, Inc. Reprinted by permission. **5.5:** From Randy Moore, et al., *Botany.* Copyright © 1995 The McGraw-Hill Companies, Inc. All Rights Reserved. Reprinted by permission. **5.13A,B:** From Bruce Alberts, et al., *Molecular Biology of the Cell.* Copyright © 1983 Garland Publishing Co., New York. Used by permission. **5.15:** From David Shier, et al., *Hole's Human Anatomy and Physiology,* 7th ed. Copyright © 1996 The McGraw-Hill Companies, Inc. Reprinted by permission. **5.17:** From David Shier, et al., *Hole's Human Anatomy and Physiology,* 7th ed. Copyright © 1996 The McGraw-Hill Companies, Inc. Reprinted by permission. **5.19:** From David Shier, et al., *Hole's Human Anatomy and Physiology,* 7th ed. Copyright © 1996 The McGraw-Hill Companies, Inc. Reprinted by permission. **5.24A:** From Randy Moore, et al., *Botany.* Copyright © 1995 The McGraw-Hill Companies, Inc. All Rights Reserved. Reprinted by permission.

Chapter 6

6.12 and **6.14:** From Purves, Orians, and Heller, *LIFE: The Science of Biology,* Third Edition. Copyright © 1992 Sinauer Associates, Inc., Sunderland, MA. Used by permission. **6.16:** From *Biology: The Science of Life,* 3rd Edition by Robert A. Wallace, Gerald P. Sanders and Robert J. Ferl. Copyright © 1991 by HarperCollins Publishers, Inc. Reprinted by permission of Addison-Wesley Educational Publishers Inc. **6.18:** From Peter H. Raven and George B. Johnson, *Understanding Biology,* 3d ed. Copyright © 1995 The McGraw-Hill Companies, Inc. All Rights Reserved. Reprinted by permission.

Chapter 7

7.8: From Randy Moore, et al., *Botany.* Copyright © 1995 The McGraw-Hill Companies, Inc. All Rights Reserved. Reprinted by permission. **7.14:** Adapted from Randy Moore, et al., *Botany.* Copyright © 1995 The McGraw-Hill Companies, Inc. All Rights Reserved. Reprinted by permission.

Chapter 8

8.6: From Randy Moore, et al., *Botany.* Copyright © 1995 The McGraw-Hill Companies, Inc. All Rights Reserved. Reprinted by permission. **8.8:** From Randy Moore, et al., *Botany,* 2d ed. Copyright © 1998 The McGraw-Hill Companies, Inc. All Rights Reserved. Reprinted by permission.

Chapter 10

10.4A: Redrawn from Scott F. Gilbert, *Developmental Biology,* 2d ed. Copyright © Sinauer Associates, Inc., Sunderland MA. Used by permission. **10.14A:** From Kent M. Van De Graaff, *Human Anatomy,* 4th ed. Copyright © 1995 The

McGraw-Hill Companies, Inc. All Rights Reserved. Reprinted by permission. **10.14B:** From David Shier, et al., *Hole's Human Anatomy and Physiology,* 7th ed. Copyright © 1995 The McGraw-Hill Companies, Inc. All Rights Reserved. Reprinted by permission.

Chapter 11

11.16: From Kent M. Van De Graaff and Stuart Ira Fox, *Concepts of Human Anatomy and Physiology,* 4th ed. Copyright © 1995 The McGraw-Hill Companies, Inc. All Rights Reserved. Reprinted by permission.

Chapter 15

15.11: Reproduced with permission of Quadrant Holdings Cambridge Ltd.

Chapter 16

16.3: Redrawn from A. Papavassiliou, *The New England Journal of Medicine,* vol. 332, no. 1, Jan. 6, 1995.

Chapter 17

17.4: From Cleveland P. Hickman and Larry S. Roberts, *Biology of Animals,* 6th ed. Copyright © 1994 The McGraw-Hill Companies, Inc. All Rights Reserved. Reprinted by permission.

Chapter 22

22.6: From Lansing M. Prescott, et al., *Microbiology,* 3d ed. Copyright © 1996 The McGraw-Hill Companies, Inc. All Rights Reserved. Reprinted by permission. **22.7:** From Lansing M. Prescott, et al., *Microbiology,* 3d ed. Copyright © 1996 The McGraw-Hill Companies, Inc. All Rights Reserved. Reprinted by permission.

Chapter 23

23.7: From Lansing M. Prescott, et al., *Microbiology,* 3d ed. Copyright © 1996 The McGraw-Hill Companies, Inc. All Rights Reserved. Reprinted by permission. **23.8:** From Lansing M. Prescott, et al., *Microbiology,* 3d ed. Copyright © 1996 The McGraw-Hill Companies, Inc. All Rights Reserved. Reprinted by permission. **23.15B:** From Lansing M. Prescott, et al., *Microbiology,* 2d ed. Copyright © 1993 The McGraw-Hill Companies, Inc. All Rights Reserved. Reprinted by permission. **23.15C:** From Randy Moore, et al., *Botany.* Copyright © 1998 The McGraw-Hill Companies, Inc. All Rights Reserved. Reprinted by permission.

Chapter 24

24.4: From Lansing M. Prescott, et al., *Microbiology,* 3d ed. Copyright © 1996 The McGraw-Hill Companies, Inc. All Rights Reserved. Reprinted by permission. **24.10A:** From Randy Moore, et al., *Botany,* 2d ed. Copyright © 1998 The McGraw-Hill Companies, Inc. All Rights Reserved. Reprinted by permission.

Chapter 26

26.2: From Stephen A. Miller and John P. Harley, *Zoology,* 3d ed. Copyright © 1996 The McGraw-Hill Companies, Inc. All Rights Reserved. Reprinted by permission. **26.11:** From Stephen A. Miller and John P. Harley, *Zoology,* 3d ed. Copyright © 1996 The McGraw-Hill Companies, Inc. All Rights Reserved. Reprinted by permission. **26.16A:** From Stephen A. Miller and John P. Harley, *Zoology,* 3d ed. Copyright © 1996 The McGraw-Hill Companies, Inc. All Rights Reserved. Reprinted by permission. **26.21B:** From Stephen A. Miller and John P. Harley, *Zoology,* 3d ed. Copyright © 1996 The McGraw-Hill Companies, Inc. All Rights Reserved. Reprinted by permission.

Chapter 27

27.17A: From Randy Moore, et al., *Botany.* Copyright © 1995 The McGraw-Hill Companies, Inc. All Rights Reserved. Reprinted by permission. **27.22:** From Randy Moore, et al., *Botany.* Copyright © 1995 The McGraw-Hill Companies, Inc. All Rights Reserved. Reprinted by permission.

Chapter 28

28.7: From Randy Moore, et al., *Botany.* Copyright © 1995 The McGraw-Hill Companies, Inc. All Rights Reserved. Reprinted by permission.

Chapter 29

29.10: From Randy Moore, et al., *Botany.* Copyright © 1995 The McGraw-Hill Companies, Inc. All Rights Reserved. Reprinted by permission. **29.11:** From Randy Moore, et al., *Botany.* Copyright © 1995 The McGraw-Hill Companies, Inc. All Rights Reserved. Reprinted by permission. **29.13:** From Raven, Evert, Eichhorn, *Biology of Plants* 4/e. Worth Publishers, New York, 1986. Used by permission.

Chapter 30

30.4: From David Shier, et al., *Hole's Human Anatomy and Physiology,* 7th ed. Copyright © 1996 The McGraw-Hill Companies, Inc. Reprinted by permission. **30.5:** From Kent M. Van De Graaff and Stuart Ira Fox, *Concepts of Human Anatomy and Physiology,* 4th ed. Copyright © 1995 The McGraw-Hill Companies, Inc. All Rights Reserved. Reprinted by permission. **30.9:** From Kenneth V. Kardong, *The Vertebrates.* Copyright © 1995 The McGraw-Hill Companies, Inc. All Rights Reserved. Reprinted by permission. **30.10:** From Kenneth V. Kardong, *The Vertebrates.* Copyright © 1995 The McGraw-Hill Companies, Inc. All Rights Reserved. Reprinted by permission. **30.11:** From Kenneth V. Kardong, *The Vertebrates.* Copyright © 1995 The McGraw-Hill Companies, Inc. All Rights Reserved. Reprinted by permission.

Chapter 31

31.8AB: From David Shier, et al., *Hole's Human Anatomy and Physiology,* 7th ed. Copyright © 1995 The McGraw-Hill Companies, Inc. All Rights Reserved. Reprinted by permission.

Chapter 32

32.17: From Stuart Ira Fox, *Human Physiology,* 5th ed. Copyright © 1996 The McGraw-Hill Companies, Inc. All Rights Reserved. Reprinted by permission. **32.19:** From David Shier, et al., *Hole's Human Anatomy and Physiology,* 7th ed. Copyright © 1996 The McGraw-Hill Companies, Inc. Reprinted by permission.

Chapter 33

33.6: From David Shier, et al., *Hole's Human Anatomy and Physiology,* 7th ed. Copyright © 1996 The McGraw-Hill Companies, Inc. Reprinted by permission. **33.8:** Adapted from Cleveland Hickman, et al., *Integrated Principles of Zoology,* 7th ed. Copyright © 1984 The McGraw-Hill Companies, Inc. Used by permission.

Chapter 34

34.11: From David Shier, et al., *Hole's Human Anatomy and Physiology,* 7th ed. Copyright © 1996 The McGraw-Hill Companies, Inc. Reprinted by permission. **34.13A:** From David Shier, et al., *Hole's Human Anatomy and Physiology,* 7th ed. Copyright © 1996 The McGraw-Hill Companies, Inc. Reprinted by permission. **34.13B:** From David Shier, et al., *Hole's Human Anatomy and Physiology,* 7th ed. Copyright © 1996 The McGraw-Hill Companies, Inc. Reprinted by permission. **34.13C:** From David Shier, et al., *Hole's Human Anatomy and Physiology,* 7th ed. Copyright © 1996 The McGraw-Hill Companies, Inc. Reprinted by permission. **34.17:** From David Shier, et al., *Hole's Human Anatomy and Physiology,* 7th ed. Copyright © 1996 The McGraw-Hill Companies, Inc. Reprinted by permission.

Chapter 35

35.10: From David Shier, et al., *Hole's Human Anatomy and Physiology,* 7th ed. Copyright © 1996 The McGraw-Hill Companies, Inc. Reprinted by permission.

Chapter 36

36.12: From David Shier, et al., *Hole's Human Anatomy and Physiology,* 7th ed. Copyright © 1996 The McGraw-Hill Companies, Inc. Reprinted by permission. **36.13:** From David Shier, et al., *Hole's Human Anatomy and Physiology,* 7th ed. Copyright © 1996 The McGraw-Hill Companies, Inc. Reprinted by permission.

Chapter 38

38.8: Adapted from Cleveland Hickman, et al., *Integrated Principles of Zoology,* 7th ed. Copyright © 1984 The McGraw-Hill Companies, Inc. Used by permission. **38.14:** Adapted from Cleveland Hickman, et al., *Integrated Principles of Zoology,* 7th ed. Copyright © 1984 The McGraw-Hill Companies, Inc. Used by permission. **38.16:** From David Shier, et al., *Hole's Human Anatomy and Physiology,* 7th ed. Copyright © 1996 The McGraw-Hill Companies, Inc. Reprinted by permission.

Chapter 39

39.2: From David Shier, et al., *Hole's Human Anatomy and Physiology,* 7th ed. Copyright © 1996 The McGraw-Hill Companies, Inc. Reprinted by permission. **39.6A:** From David Shier, et al., *Hole's Human Anatomy and Physiology,* 7th ed. Copyright © 1996 The McGraw-Hill Companies, Inc. Reprinted by permission. **39.7:** From David Shier, et al., *Hole's Human Anatomy and Physiology,* 7th ed. Copyright © 1996 The McGraw-Hill Companies, Inc. Reprinted by permission. **39.18A:** From David Shier, et al., *Hole's Human Anatomy and Physiology,* 7th ed. Copyright © 1996 The McGraw-Hill Companies, Inc. Reprinted by permission.

Chapter 40

40.14A: Reproduced with permission from Goodenough, Judith, *Animal Communication;* Carolina Biology Reader Series, No. 143. Copyright 1984, Carolina Biological Supply Co., Burlington, North Carolina, U.S.A.

Chapter 41

41.11: Adapted from *Natural History Magazine,* July 1989. Used by permission.

Chapter 42

42.3: Based on Eldon D. Enger, et al., *Concepts in Biology,* 6th ed., © 1991 The McGraw-Hill Companies, Inc. **42.8A:** From "Stable Carbon Isotopes in the Study of Food Chains," by Stephen Hart, *Biology Digest,* February 1990. Used by permission.

Chapter 43

43.1: Reprinted by permission of D.C. Heath & Co. from *Heath Biology* by McLaren, et al., 1991. **43.2:** Reprinted by permission of D.C. Heath & Co. from *Heath Biology* by McLaren, et al., 1991.

Chapter 44

44.6: C.D. Wheeling, et al., in *Aspects of Climate Variability in the Pacific and the Western Americas,* American Geophysical Union Monograph, vol. 55, pp. 165–236, (1989). Updated monthly from author. **44.13:** Redrawn from Andrew P. Dobson, *Conservation and Biodiversity,* Scientific American Books. Used by permission of W.H. Freeman and Company.

Index

A

A band, 697
ABO blood type, 273-75, 408, 796
Abomasum, 441
Aboral end, 511-13
Abortion
  elective, 252-53
  spontaneous, 204, 245, 249, 272, 298, 412
Abscisic acid, 576, 578
Abscission zone, 546, 678
Absorption, nutrient, 123, 762-64
Abyssal zone, 883-84
Acacia tree, 533
Acanthamoeba, 461
Acanthostega, 362, 364
Accelerated aging disorder, 233
Accessory pigment, 150, 153
Acclimation, 587
Accommodation, 649
Acellular slime mold, 470-72
Acetabularia, 465
Acetone, 139-40
Acetylcholine, 105, 617-18, 696
Acetylcholinesterase, 123
Acetyl CoA, 127-28, 131-33, 140-43
Acetylene, 37
Achlya, 470
Achondroplasia, 261, 269, 339
Acid, 40-42
Acid environment, 42
Acidosis, 41
Acid precipitation, 41, 890, 892, 894-95, 907
Acoelomate, 511-12, 528
Acquired characteristics, inheritance, 378
Acquired immune deficiency syndrome. *See* AIDS
Acrasiomycota, 470-72
Acrocanthosaurus, 391
Acrocentric chromosome, 296
Acromegaly, 667-68
Acrosomal reaction, 251-52
Acrosome, 204, 217-19
ACTH, 667-69, 679
Actin, 48, 96, 100, 102, 488, 597-98, 692, 694-99
Action potential, 611-12, 614
Activation energy, 121
Active immunity, 798
Active site, 50, 53, 121
Active transport, 91-93, 95
Acupuncture, 634
Acute mountain sickness, 745
Acyclovir, 253
Adam's apple. *See* Larynx
Adaptation, 10-12, 374, 640-41
"Adaptive episode," 801
Adaptive radiation, 381
Adder snake, 10, 112
Addiction, 104-5, 634
Addison's disease, 672
Adenine, 51, 311, 313
Adenosine deaminase deficiency, 326, 345
Adenosine diphosphate. *See* ADP
Adenosine triphosphate. *See* ATP
Adenylate cyclase, 103, 663
ADH. *See* Antidiuretic hormone
Adhesion receptor protein, 103
Adhesiveness, water, 40-41, 540
Adipocyte, 594, 597
Adipose tissue, 44, 594, 596-97
ADL. *See* Adrenoleukodystrophy
Adolescence, 231-32
Adoption study, 279-80
Adorable Louise (donkey), 238-39
ADP, 117
Adrenal cortex, 670-71, 678-79
Adrenal gland, 661, 667, 670-72
Adrenaline. *See* Epinephrine
Adrenal medulla, 632, 670-71, 678-79
Adrenocorticotropic hormone. *See* ACTH
Adrenoleukodystrophy (ADL), 72-74
Adulthood, 231-32
Adventitious bud, 556
Adventitious root, 546, 561, 569
Aegyptopithecus, 367
Aerial root, 550, 872, 882
Aerobe, 127, 445
Aerobic respiration. *See* Cellular respiration
Aerosol inhaler, 739
Afferent arteriole, 781
Afferent neuron. *See* Sensory neuron
Aflatoxin, 341, 475, 483
African bush baby, 665
African clawed frog, 211
African green monkey, 396
Africanized bee, 573
African pompany, 428
African rock python, 109-10, 123
African sleeping sickness, 461, 872, 886
Agar, 469
Agave sisalana, 536
Age of Amphibians, 362
Age of Cycads, 498
Age of Fishes, 362
Age structure, population, 838, 841-42
Aggregate fruit, 569
Aggression, 815, 824, 827-29
Aging, 232-33
  accelerated aging disorders, 233
Agnatha, 525
Agonist, 634
Agriculture, 369, 385, 462, 492
AIDS, 253-54, 384-85, 435, 459, 461, 463, 488, 716, 746, 789, 798-801
AIDS dementia, 254
Airborne disease, 449-51
Air pollution, 892-94
Air sac, 735
Albinism, 267, 270, 406
Albumin, 714, 785
Alcohol, 97, 185, 761, 785
  teratogenicity, 246-48
Alcoholic beverage, 137-38
Alcoholic fermentation, 137-39, 141
Alcohol-related neurodevelopmental disorder, 509
Alder, 873
Aldosterone, 672, 784-85
Alfalfa, 543
Algae, 458, 463-69, 883
  classification, 463-64
  lichens, 484
  types, 465-69
Alkaloid, 504, 648
Alkalosis, 41
Alkylating agent, 248, 340
Allantois, 223, 226
Allard, H.A., 583
Allele, 200, 265
  lethal, 272, 277
  multiple, 272-73, 277
  temperature-sensitive, 275-76
Allele frequency, 375, 404-12
Allergen, 800-803
Allergy, 800-802
Allergy mediator, 800-803
Alligator, 18, 22, 828-29
Allison, Anthony, 411
Allograft, 604
Allomyces, 469
Allopolyploid, 414
Allosaur, 365
Allosaurus, 391
Alopecia, 268, 271
Alpha-1-antitrypsin, 335
Alpha globin, 715
Alpha globin gene, 339, 341
Alpha helix, 49, 52
Alphaketoglutaric acid, 132
Alpine tundra, 858
Alternate leaf pattern, 542
Alternation of generations, 192-93, 492, 557, 560, 562
Altitude, 872-73

Altitude sickness, 745
Altruism, 826
Aluminum, 894-95
Alveolar bone, 760
Alveolar duct, 739-40
Alveolar pore, 740
Alveoli, 739-40
Alzheimer disease, 298, 335-36, 338, 618, 633
Amber, 392
Amboseli baboon, 828
Amelogenesis imperfecta, 293
American elm, 541
American robin, 750
American wolf, 381
Amino acid, 46-49, 358-59, 751, 763
Amino group, 47-48
2-Amino 5-nitrophenol, 341
Aminopterin, 248
Amish people, 408
Ammonia, 139-40, 142, 452, 776-77, 865, 867
Amnesia, 633
Amniocentesis, 287
Amnion, 220, 223
Amniote egg, 226, 363-65, 526
Amniotic cavity, 220-21, 226
Amniotic fluid, 226, 229
Amoeba, 460
Amoeba proteus, 692
Amoebic dysentery, 460-61
Amoebocyte, 515-16
Amoeboid movement, 459-60, 692-93
Amphetamine, 634
Amphibian, 510, 525-26, 733
    population declines, 836, 850
Amphioxus, 526
Amphistegina lobifera, 473
Amphotericin, 97, 488
Ampullae, semicircular canals, 655
Amylase
    pancreatic, 762-64
    salivary, 760, 763
Amyloid, 336
Amyotrophic lateral sclerosis, 633
Anabaena, 152
Anabolic steroid, 676
Anabolism, 114-15
Anaerobe, 127, 445
Anaerobic respiration, 140
Anal canal, 764
Analogous structures, 394
Anal sphincter, 759, 764
Anaphase
    meiosis I, 199, 201
    meiosis II, 201-2
    mitotic, 171, 173-74
Anaphylactic shock, 802
Anatomy, comparative, 393-94
Andrews, Tommie Lee, 314
Androecium, 563
Anemia, 102, 708, 715-16, 755-56
Anencephaly, 605, 621
Aneuploidy, 298-300
Angina at rest, 712
Angina of exertion, 712
Angina pectoris, 712
Angiosperm, 498, 501-4
    gymnosperm versus, 572
    life cycle, 563-72
    origins, 502
    tissue types, 534
Angstrom, 63
Animal
    body cavity, 511-12
    characteristics, 508
    development, 209-35, 509-14
    levels of organization, 509
    phyla, 514-28
    phylogenetic tree, 511
    symmetry, 511-12
Animal breeding, 406-7
Animal cell, 66-67
Animalia, 14, 429-30, 507-28, 914-15
Animal pollination, 567-68
Animal research, 24-26, 508-9
Animal rights movement, 25, 508
Animal society, 812, 832
Animal tissue, 593-607
Ankylosing spondylitis, 85
Ankyrin, 101-2
Annelid, 510-11, 513, 521, 619
Anopheles, 462
Anopheles gambiae, 411
Anorexia nervosa, 758
Ant, 207, 523, 820, 847
Antagonist, 634
Antagonistic muscles, 700-702
Antarctic ecosystem, 854-55, 858
Anteater, 527
Antenna, 522, 656
Antenna complex, 153
Antennapedia, 215-16
"Anterior," 512-13
Anterior pituitary, 666-69
Anther, 563, 566
Antheridia, 464, 466, 493, 497
Anther-smut fungus, 479
Anthoceros, 495
Anthocyanin, 587
Anthrax, 450-51
Anthropoidea, 527
Antibiotic, 764
Antibiotic resistance, 343, 386-87, 416, 420
Antibody, 44, 48, 716, 790, 795-98, 803
Anticoagulant, 718
Anticodon, 329-30, 333, 336
Antidiuretic hormone (ADH), 90-91, 397, 665-69, 679, 784-85
Antifreeze, 42, 771
Antigen, 48, 85, 790
Antigen binding site, 796
Antihistamine, 800
Antimicrobial substances, 791
Antiparallelism, 313, 316
Antisense technology, 334
Antisperm secretions, 239-40
Antler, 383, 603-4, 828
Anus, 196, 753, 764
Aorta, 711, 720-21, 725, 744-45
Aortic semilunar valve, 720-21
Apatosaur, 365-66
Apatosaurus, 503
Ape, 367, 395, 434, 527
Aphid, 194, 207
Apical complex, 461
Apical meristem, 534-35, 568, 572
Apicomplexa, 461
Apiculture, 573
Apis mellifera, 556, 572
Aplastic anemia, 716
Apolipoprotein, 94
Apopka alligator, 18, 22
Apoptosis, 168-70, 177-79, 233
Aporosaura anchietae, 10
Appendicular skeleton, 685-86
Appendix, 759, 763-64, 793
Apple, 556, 569, 587
Apple scab, 481, 486
Aquaporin, 90
Aquaspirillum serpens, 444
Aquatic ecosystem, 858-59
Aquatic organism, gas exchange, 733-34
Aqueous humor, 649
Aqueous solution, 40, 87-88
Aquifer, 865
Arabidopsis thaliana, 564-65, 580, 588-89
Arachnid, 521-23
Archaea, 13-14, 63, 351, 430-33, 440-41
    bacteria versus, 444-45
    characteristics, 450-52
    classification, 451-52
Archaean cell, 67-69
Archaeoglobus fulgifus, 351
Archaeology, 495
Archaeopteryx, 365
Archean era, 359
Archegonia, 493, 497, 562
Arctic fox, 878
Arctic hare, 878
Arctic tern, 817
Area, 909-10
Aristotle, 378
Armadillo, 25
Armillaria luteobubalina, 480
Aromatic hydrocarbon, 21
Arousal, 627-28
Arrhenius, Svante, 353
Arrhythmia, 712, 722-23
Arsenic, 893
Arteriole, 708, 722, 724
Artery, 708, 722-23
Arthritis, 700-701
Arthrobotrys anchonia, 476, 483
Arthropod, 510-11, 513, 521-23, 619-20, 708
    medically important, 523
Arthropod-borne disease, 450-51, 461
Artificial insemination, 241-42
Artificial seed, 559
Artificial selection, 381, 407
Artiodactyla, 527
Asbestos, 184
Ascaris lumbricoides, 518-19
Ascending limb, nephron loop, 782-83
Ascetospora, 461
Ascomycetes, 478-81, 485
Asconoid sponge, 515
Ascorbic acid. *See* Vitamin C
Ascospore, 479, 481
Ascus, 192, 479, 481
Asexual reproduction, 10, 14, 192-93, 195, 556-57
Ash tree, 551
Aspartame, 28
Aspartic acid, 161
Aspen, 857, 862
Aspergillosis, 488

Aspergillus, 475, 483
Aspergillus flavus, 483
Aspergillus fumigatus, 483, 488
Aspergillus niger, 133
Aspergillus parasiticus, 483
Aspirin, 677, 761
Assisted reproductive technologies, 241-45
  disasters, 242
Assisted suicide, 6
Association area, 628-29
Aster, 173
Asterias, 524
Asteroid shower, 353, 417
Asthma, 737-39, 802, 804
Astigmatism, 651-52
Astral rays, 171-73
Ataxia telangiectasia, 269, 322
Atherosclerosis, 712-13, 717
Athlete
  muscle fiber types and athletic ability, 703
  steroid abuse, 676
Athlete's foot, 476, 481
Atlantic salmon, 817
Atom, 7, 32-35
Atomic bomb, 893
Atomic number, 33
Atomic weight, 33
ATP, 117-21
  assay, 118
  formation, 129
    cellular respiration, 126-27
    electron transport system, 133-36
    fermentation, 137-40
    glycolysis, 129-31, 136
    Krebs cycle, 132-33, 136
    photosynthesis, 154-55
    respiratory chain, 127
    using energy from nutrients, 128-29
  lack of, 126, 143
  structure, 117, 119
  use, 128
    active transport, 92-93, 613
    bioluminescence, 118
    Calvin cycle, 158-59
    closure of Venus's flytrap, 581
    glycolysis, 130
    locomotor organelles, 99-100
    muscle contraction, 695-99
    photorespiration, 160
    signal transduction, 103
    translation, 333
ATP synthase, 129, 133-35, 155
Atrazine, 21
Atrial natriuretic peptide, 661, 786
Atriopore, 526
Atrioventricular (AV) node, 722
Atrioventricular (AV) valve, 720
Atrium, 710-11
Atrophy, 702
Auditory canal, 653
Auditory messages, 824
Auditory nerve, 653-54
Auditory tube, 653
Aurelia, 517
Auricle, planarian, 518
Australian hopping mouse, 783
Australopithecus, 368
Australopithecus afarensis, 368
Australopithecus anamensis, 368
Australopithecus boisei, 368
Australopithecus ramidus, 368
Australopithecus robustus, 368
Autoantibody, 800
Autograft, 604
Autoimmune disorders, 802
Autoimmunity, 233, 800
Autonomic nervous system, 621-22, 630-32
Autopolyploid, 414
Autosomal aneuploid, 298
Autosomal trait
  dominant, 267-70
  recessive, 267-70
Autosome, 198
  X-to-autosome ratio, 290
Autotroph, 14, 112, 146, 445, 459
Autumn leaves, 150-51, 587
Auxin, 569, 576-79, 586
Avery, Oswald, 309-11
Aves, 525-26
AV node. *See* Atrioventricular node
AV valve. *See* Atrioventricular valve
Axial skeleton, 685-86
Axillary bud, 541
Axon, 596, 598, 611-12
Azolla, 498

B

Baboon, 822, 828
Baby boom, 844
Baby F., 4, 15
Bacilli, 66, 443
Bacillus, 140
Bacillus anthracis, 450
Bacillus licheniformis, 444
Bacillus megaterium, 443
Background information, 20-21, 23
Bacteria, 14, 64, 439-54
  archaea versus, 444-45
  distinguishing characteristics, 446-48
  human pathogens, 449-50
  photosynthetic, 156-57
  reproduction, 448-49
  sponge-dwelling, 446
Bacteria (domain), 430-33
Bacterial vaginosis, 245-46
Bacteriochlorophyll, 156
Bacteriophage, 454, 786
Bacteriorhodopsin, 150
Bacteroides amylophilus, 441
Bacteroides succinogenes, 441
Badger, 828
Baking, 453
Balance, 645, 654-56
Balanced polymorphism, 411-12
Bald eagle, 899
Baleen, 602, 854
Baleen plate, 602, 752
Baleen whale, 230, 751-52, 854, 862
Ball-and-stick model, 37
Bamboo, 557, 585, 752
Banana, 501, 541, 577
Banded stilt, 860
Bank swallow, 817
B27 antigen, 85
Banting, Frederick, 660, 678
Banyan tree, 550
Bark, 551
Bark beetle, 854
Barley, 547
Barley blotch, 485
Barley powdery mildew, 485
Barnacle, 522, 838, 882
Barn swallow, 382
Barr, Murray, 294
Barr body, 294-95
Barrel cactus, 878
Basal ganglia, 605
Basal metabolic rate (BMR), 138
Base, 40-42
Basement membrane, 599, 601
Basidia, 478-79, 482
Basidiomycetes, 478-80, 482-83, 485
Basilar membrane, 653-54
Basking, 772
Basophil, 714, 716, 719
Bat, 230, 398-99, 527, 568, 640, 645
Bateson, William, 286
Bath sponge, 515-16
Bay-breasted warbler, 837
B cells, 719, 792-97, 803
Beach strawberry, 542
Beagle (Darwin's ship), 379-82
Beak (bird), 50, 754
Bean, 582
Bear, 9, 527, 839
Beaver, 783
Becker muscular dystrophy, 293, 335
Bedwetting, 785
Bee, 207, 290, 523, 567, 820
Beech, 377, 856-57
Beech fern, 491, 496
Beech-maple forest, 873
Bee dances, 824-26
Beehive, 820-21
Beet, 548
Beetle, 290, 522, 567
Behavior, 811-32
  effects of genes and experience, 812
  studying, 812-13
Behavioral genetics, 281
Behavioral isolation, 413
Beluga whale, 898
Benign tumor, 180
Benthic algae, 463
Benthic zone, 883-84
Beriberi, 755
Bering land bridge, 367
Bernard, Christiaan, 804
Berry, 569-70
Best, Charles, 660, 678
Beta globin, 708, 715
Beta globin gene, 339, 344
Beta-pleated sheet, 49, 52
Beta vulgaris, 314
Bewick's swan, 829-30
Bicarbonate, 743, 762
Biceps, 700, 702
Bichat, Marie Francois Xavier, 5
Bichir, 735
Bicoid protein, 215
Bicuspid valve, 720-21
Bighorn sheep, 382-83
Bilateral symmetry, 512-13, 611

Bile, 757, 762-63, 765
Bile duct, 765
Bile pigment, 765
Bile salts, 765
Binary fission, 192-93, 448, 451, 459-60, 464
Bindweed, 579
Binge-and-purge cycle, 758
Binomial name, 14, 426
Bioaccumulation, 866-68
Biochemical, 5
Biodiversity, 12-15, 427-28, 854, 892, 894, 906-7
  estimating, 436
Bioenergetics, 110
Bioethics, 4, 6
Biofeedback, 630
Biogenetic law, 210
Biogeochemical cycles, 864-66
Biogeography, 374, 379, 400
Biological clock, 588-89
Biological evolution, 374
Biological movement, 689-700
Biological society, 819-21
Biological species, 436
Bioluminescence, firefly, 118
Bioluminescent synchrony, 118
Biomagnification, 866-69, 897
Biomass, 863, 907
Biome, 859, 871-86
Bioremediation, 454, 890-91
Biosphere, 7, 9, 858-59
Biot, Jean Baptiste, 358
Biotic community, 854-55
Biotic potential, 842
Biotic succession, 856-58
Biotin, 755
Bipedalism, 368, 774
Bipolar neuron, 650
Biramous appendage, 522
Birch, 862
Bird, 510, 525-26
Bird of paradise, 383
Bird's nest fungi, 477, 482
Birds of prey, 648
Bird's song, 812-13
Birth, 229-30
Birth control, 250-53, 848-50
Birth control pill, 248, 250-51
Birth defect, 20, 246-49
Birth weight, 411
Bismuth sulfate, 766
1,3-Bisphosphoglyceric acid, 130-31
Bitis peringueti, 10
Bitter taste, 647-48
Bivalve, 519-20, 524, 751
Black bear, 818
Blackberry, 569
Blackburnian warbler, 837
Black Death. *See* Bubonic plague
Black-eyed Susan, 568
Black fly, 872, 886
Black locust, 556
Black mustard, 857
Black piedra, 487
Black smoker, 453
Black spruce, 856
Black-throated green warbler, 837
Black walnut, 856-57
Black widow spider, 523
Bladder (brown algae), 467-68
Bladder cancer, 184
Blade
  brown algae, 467-68
  leaf, 541-42, 545
Blastocyst, 218-20
Blastocyst cavity, 218, 221
Blastocystis hominis, 459
Blastomere, 218, 514
Blastomere separation, 225
Blastomyces dermatitidis, 477
Blastopore, 513-14
Blaustein, Andrew, 850
Blechnum, 499
Bleeding disorder, 717
Blighted ovum, 204
Blood, 594, 596-97
  composition, 711-18
  functions, 714
  pH, 41, 784
Blood-brain barrier, 96-97, 635
Blood cancer, 184
Blood cells, 684, 717-19
Blood clotting, 717
Blood group, ABO, 273-75, 408, 796
Bloodhound, 646
Blood poisoning, 449
Blood pressure, 713, 725-26, 781, 784-86
Blood substitute, 711
Blood vessel(s), 722-25
  circulation and, 725
  fatty plaques in, 47
Blood vessel equivalent, 711
Bloom, 452, 466-67, 883, 902-3
Bloom syndrome, 322
Blubber, 772
Blue crab, 899
Bluefin tuna, 771
Blue whale, 854
BMI. *See* Body mass index
BMR. *See* Basal metabolic rate
Bobcat, 846
Body, stomach, 760-61
Body cavity, 511-12
Body composition, 138
Body heat, sources, 771
Body mass index (BMI), 766
Body surface, gas exchange, 733
Body temperature. *See* Temperature regulation
Body weight, 765
Bog, 42, 495
Bohr model, 35
Bolus, 760
Bombay phenotype, 274
Bomb calorimeter, 755-56
Bond. *See* Chemical bond
Bond energy, 115-16
Bone, 594-97, 684, 688-89
  fracture repair, 690
  growth, 690-91
  remodeling, 595
  working with muscle, 700-704
Bonellia viridis, 291
Bone marrow, 793
Bone marrow transplant, 226, 486, 708, 717
Bone mass, 690
Bonner, James, 584
Bonnet, Charles, 207, 378
Bonobo, 831-32
Bony fish, 525
Boom and bust cycle, 844
Boreal forest, 858
Borna virus, 76
Borrelia burgdorferi, 384, 386, 864
Borrelia hermsii, 449
Bossiella, 469
Botanical garden, 427
Botulinum toxin, 94
Bouchard, Thomas, 280
Boveri, Theodor, 286
Bovine immunodeficiency virus, 801
Bovine spongiform encephalopathy. *See* Mad cow disease
Bowerbird, 383
Braced framework, 683
Brachiopod, 362, 861
Brachiosaurus, 503
Bract, 544
Brain, 621, 623-29
Brain damage, 624
Brain death, 4, 15
Brain injury, 610, 636
Brain scan, 624, 626
Brain stem, 625, 627
Brain wave, 628
Brassinolide, 588
Breadmaking, 477, 481, 488
Breakbone fever, 872
Breast cancer, 18, 20, 27, 29
Breast cancer gene, 302-3
Breathing, 732, 742
  control, 744-45
  high altitude, 745
"Breathing" respiration, 126-27
Brewing, 453, 481, 488
Briggs, Winslow, 578
Brine shrimp, 860
Bristle, 683
Bristlecone pine, 499-500, 553
British soldier lichen, 484
Brittle star, 523-24
Brittlewort, 466
Broccoli, 589
Brock, Thomas, 13
Bronchi, 736-39
Bronchial tree, 737
Bronchiole, 736-40
Bronchitis, 746
Brown, Louise Joy, 238-39
Brown adipose tissue, 44, 135, 772
Brown algae, 458, 463, 467-68
Brown-headed cowbird, 847
Brown recluse spider, 523
Brown rot, 481
Brucellosis, 449
Bryophyte, 492-96
Bryoria lichen, 484
Bryozoa, 415-16
Bubble boy, 799-800
Bubo, 435
Bubonic plague, 434-35, 450-51
Budding
  archaeal, 451
  protozoa, 459-60
  yeast, 477, 479
Buffalo, 754
Buffer system, 41, 894
Buffon, Georges, 352
Bufo americanus, 413

Bufo fowleri, 413
Bulbourethral gland, 196-97
Bulimia, 758
Bulk element, 32, 34
Bulk mineral, 756
Bull's-horn acacia, 847
Bullwinkle J. Moose, 404
Bumblebee, 567
Bundle-sheath cell, 158, 160-61
Burdock, 570-71
Burgess shale, 363, 392
Burn patient, 599, 606
Bursa, 700
Bursitis, 700
Butler, Reginald, 479
Butterfly, 110, 434, 513, 640, 664-65, 845
  photoreceptors on genitals, 642-43
Butterwort, 538
Buttress root, 550

C

Cactus, 162, 541, 544, 573, 878
Cadmium, 376, 890, 907
Caenorhabditis elegans, 25, 177, 213-15, 227, 518
Caesarean section, 229-30, 249
Caffeine, 341, 501, 785
Cai Lun, 492
Calcitonin, 669, 679
Calcium, 756
  blood, 670-71, 685
  bone, 685
  muscle contraction, 696-98
Calico cat, 294-95
California poppy, 177
Calloselasma rhodostoma, 374
Callus
  fracture repair, 690
  plant, 558-59
Calment, Jeanne, 210, 234-35
Calorie, 110
Calvaria, 374
Calvert family, 243
Calvin, Melvin, 159
Calvin cycle, 158-61
Calypogeia, 494
Calyx, 563
CAM. *See* Cellular adhesion molecule
Cambrian period, 360-62, 417, 508
Camel, 770
Camellia sinensis, 501
Camouflage, 10, 403, 410-11, 669, 878
cAMP. *See* Cyclic AMP
CAM plant, 162
Canadian red-sided garter snake, 814
Canaliculi, 595, 688-89
Cancer, 60, 80, 180-82, 234, 766
  causes, 182-85
  characteristics of cells, 180-83
  control of cell division, 175-76
  doctor-patient relationship, 185
  lactic acid fermentation in cancer cells, 139
  metastasis, 181
Cancer drugs, 99
Candida albicans, 487-88
Candidiasis, 488
Canine distemper, 806
Cannabis, 537
Cannibalistic behavior, 826-27, 829
Canola stem rot, 485
Canopy, 872
Cap, messenger RNA, 330
Capacitation, 217
Cape May warbler, 837
Capillary, 8, 708, 723-25
Capillary bed, 723-24
Capsid, viral, 75-76
Capsule, bacterial, 446
Capuchin monkey, 823-24
Carbaminohemoglobin, 743-44
Carbohydrase, 48, 762-63
Carbohydrate, 43-44, 54
  digestion, 751, 755, 762-63
  energy from, 142
Carbon, 35
  covalent bond formation, 37
Carbon-14, 393
Carbon cycle, 132, 865-66
Carbon dioxide, 35, 117, 865-66
  atmospheric, 896, 898
  blood, regulation of breathing rate, 744-45
  cellular respiration, 126, 132
  electron acceptor, 140
  from fermentation, 137-39
  transport in blood, 743-44
  use in photosynthesis, 146
Carbon dioxide fixation, 158-59
Carbonic acid, 41, 743-44
Carbonic anhydrase, 121, 743
Carboniferous period, 359, 362-63, 365
Carbon monoxide, 35-36, 247, 715, 747
Carbon signature, 862-63
Carboxyhemoglobin, 747
Carboxyl group, 47-48
Carcharodontosaurus saharicus, 390-91, 401
Carcinogen, 184
Cardia, 760-61
Cardiac cycle, 720-22
Cardiac muscle, 596, 598, 718, 720
Cardiac muscle cell, 689-93
Cardiac output, 727
Cardiomyopathy, familial hypertrophic, 269-70
Cardiovascular system. *See* Circulatory system
Caribou, 871, 877-78
Carnivora, 527
Carnivore, 750, 862, 864
Carnivorous fungus, 476
Carnivorous plant, 537, 544-45, 576, 581
Carotenoid, 150-51, 156, 464-66, 492, 587
Carotid artery, 725, 744-45
Carotid rete, 770, 773-74
Carpal, 686
Carpel, 563-65
Carr, Archie, 818
Carrageenan, 469
Carrier, 281, 292, 294
Carrier protein, 85, 91
Carrion flower, 568
Carrot, 548
Carrying capacity, 840, 842, 844-45
Cartilage, 594-97, 684, 687-88, 690, 700-701
Case-controlled study, 27
Casein, 48
Casparian strip, 548-49
Cast (fossil), 390, 502
Cat, 230, 527, 648, 846
  breeding, 407
  chromosomes, 396
  coat color, 273, 275, 294-95
  curl, 407
  feline immunodeficiency virus, 801
Catabolism, 114-15
Catalase, 72-74
Catalysis, 49
Cataract, 652
Catecholamine, 670, 679
Caterpillar, 434, 512, 664-65
Catfish, 645, 822
Cattail, 879
Cauliflower, 589
Cave art, 369
Cavia porcellus, 426
CCK. *See* Cholecystokinin
CD antigen, 797
CD4 receptor, 798-99
Ceboidea, 527
Cecum, 759, 763-64
Cedar waxwing, 571
Cell, 5, 7-8, 77
  complexity, 63-69
  differentiated, 211
  microscopy, 60-61
  specialized, 61, 100-102, 170, 594-98
  unit of life, 60-63
Cell body, neuron, 596, 598, 611-12
Cell connections, 96
Cell culture, plant, 558-59
Cell cycle, 168-75
  cancer cells, 180-82
  time to complete, 181
Cell death, 168-69. *See also* Apoptosis
Cell division, 167-69, 488
  control, 175-77
  in microgravity, 580
  tissue, 178-80
Cell division signal, 176-77
Cell fusion, 194-96
Cell implant, 606
Cell membrane, 65-68, 85-87
  archaean, 68
  fluid mosaic, 85, 87
  large-scale transport across, 93-95
  movement across, 87-93
Cell plate, 173
Cell populations, 179-80
Cell shape, 89-90, 101-2, 443
Cell size, 63-65, 69
Cell surface, 84
Cell surface protein, 85
Cell theory, 61-62
Cell-to-cell interaction, 103-6
Cellular adhesion molecule (CAM), 85, 103-6, 631-32
Cellular architecture, 84-106
Cellular clock, 175-76
Cellular immune response, 797
Cellular respiration, 114, 123-28, 132, 136, 162-64, 732, 865-66
  control, 136-37
Cellular slime mold, 470-72
Cellulase, 28
Cellulose, 14, 44, 46, 754
Cell volume, 64-65
Cell wall, 75
  algal, 464
  archaeal, 445
  bacterial, 444-45

fungal, 476
plant, 14, 68, 90, 492
prokaryotic, 64-65
Cementum, 760
Cenozoic era, 357, 359, 361, 366-67
Centimeter, 63
Centipede, 522-23
Central nervous system (CNS), 611, 618, 621
human, 622-23
Central vacuole, 72
Centriole, 67, 74, 171-72, 174-75
Centromere, 171-74
Centromere position, 296
Centrosome, 171-72
Century plant, 585
Cephalization, 512, 611
Cephalochordata, 525-26
Cephalopod, 519-20
Cercopithecoidea, 527
Cerebellum, 625-27
Cerebral cortex, 609, 629
Cerebral edema, 745
Cerebral hemisphere, 629
Cerebral palsy, 633
Cerebral specialization, 628
Cerebrospinal fluid, 635, 744-45
Cerebrovascular accident, 633
Cerebrum, 625-26, 629
Cereus, 573, 588
Cervical cancer, 180, 799
Cervical cap, 250-51
Cervical nerve, 625
Cervical vertebrae, 624, 685, 687
Cervix, 197
dilation, 229-30
incompetent, 245
Cetacea, 527
CFC. *See* Chlorofluorocarbon
CFTR. *See* Cystic fibrosis transmembrane conductance regulator
"Chain of being," 378
Chambered nautilus, 520
Chameleon, 113
Channelization, 899
Chaperone protein, 49, 336-37
Chara, 466
Chargaff, Erwin, 311
Charophyta, 466
Cheesemaking, 453, 483
Cheetah, 409-10
Chelicerata, 522-23
Chemical bond, 35-40, 115-16
Chemical digestion, 751, 760, 762
Chemical equation, 35
Chemical equilibrium, 116
Chemical evolution, 353-56
Chemical messages, 824-25
Chemical reaction, 35
energy changes, 115-16
Chemiosmosis, 134-36
Chemistry basics, 32-40
Chemoautotroph, 112
Chemonasty, 581
Chemoreception, 644, 646-48
Chemoreceptor, 744
Chemotroph, 445
Chemotropism, 581
Chernobyl disaster, 342, 410, 892
Cherry, 556, 587
Chesapeake Bay, 899
Chestnut blight, 481, 485
Chewing, 760
Chicken, 211
Chicken-to-duck experiment, 416-17
Chief cell, 760-61
Chiggers, 523
Childbirth, 634
Childhood, 231
Chimney sweep, 184
Chimpanzee, 395-97, 527, 813, 822, 832, 904
Chipmunk, 126
Chirality, 358-59
Chiroptera, 527
Chitin, 14, 44, 46, 476, 683
Chlamydia, 253
Chlamydia psittaci, 449
Chlamydomonas, 194-96, 465, 644
Chlordane, 21
Chlorella, 159, 465
Chlorenchyma, 536
Chloride, 756, 900
Chlorofluorocarbon (CFC), 894, 896
Chlorophyll, 75, 149-51, 464, 587
Chlorophyll a, 149-50, 154-55, 465-66, 492
Chlorophyll b, 150, 465-66, 492
Chlorophyll c, 466
Chlorophyta. *See* Green algae
Chloroplast, 61, 68, 74-75, 123, 125, 151-53, 536, 580
DNA, 79-81
Choanocyte, 515
Choanocyte chamber, 515
Choking, 738
Cholecalciferol, 45, 755
Cholecystokinin (CCK), 762
Cholera, 384-85, 440, 450-51, 454
cystic fibrosis and, 411-12
Cholera toxin, 411, 440, 454
Cholesterol, 45, 47-48, 94, 712
Chondrichthyes, 525
Chondrocyte, 595-97, 688
Chordata, 510-11, 513, 524-28
Chorion, 218, 221, 223
Chorionic villi, 221, 223, 226, 228
Chorionic villus sampling, 287
Choroid coat, 648-49
Christmas cactus, 584
Chromatid, 171-72, 174
Chromatin, 313
Chromium, 184, 756, 890
Chromosome, 171, 285-306, 317
comparing to determine species relatedness, 395-96
fetal, 286-87
health and, 302
structure, 296-97
studying, 286-87
Chromosome abnormality, smaller-scale, 300-301
Chromosome banding pattern, 395-96
Chromosome incompatibility, 413-14
Chromosome number, 286, 395-96
abnormal, 245, 297-300
Chromosome rearrangement, 300-301
Chronic bronchitis, 741
Chronic fatigue syndrome, 800
Chronic granulomatous disease, 293
Chronic respiratory disease, 234
Chronobiology, 576
Chroococcus turgidus, 66
Chrysanthemum, 586-87
Chrysophyta, 467
Chyme, 760
Chymotrypsin, 762-64
Chytrid, 458, 469-70
Chytridiomycota, 470
Cicada, 773
Cichlid fish, 290, 400
Cilia, 83, 99-100, 459-61, 646-47, 655
respiratory tract, 731-32, 736-37, 741, 790
Ciliary body, 648-49
Ciliary muscle, 648
Ciliated columnar epithelium, 596
Ciliated epithelium, 735-36
Ciliophora, 461
Cinchona bark, 462
Circadian rhythm, 576, 587-89, 677
Circulatory system, 599-600, 707-28
adaptations to heat tolerance, 773-75
control of circulatory functions, 726
disorders, 712-13
exercise and, 727
human, 710-26
invertebrate, 708-9
in pregnancy, 226
replacement parts, 710-11
types, 708
vertebrate, 710
Cithaerias menander, 845
Citric acid, 132-33, 137
Clade, 390
Cladistics, 390
Cladogram, 391
Cladonia cristatella, 484
Cladorhizidae, 446
Clam, 510, 513, 519-20, 751
Clamp connection, 482-83
Clarkia franciscana, 414
Clarkia rubicunda, 413-14
Class, 12, 14, 426, 433
Classical conditioning, 815
Classification, 425-36. *See also* Taxonomy
Claviceps purpurea, 482
Clavicle, 686
Clay, polymerization template on prebiotic earth, 355
Cleavage, 216, 218-20, 513-14
Cleavage furrow, 173-74, 227
Climate, 534, 553, 872-73, 885-86, 896
Climax community, 856, 858
Cline, 408
Clitellum, 521
Clitoris, 197-98
Cloaca, 754
Clone, 10, 556
Cloning, 225
Closed circulatory system, 708, 710
Closed system, 112, 114
Clostridium, 139
Clostridium acetobutylicum, 140
Clostridium botulinum, 449
Clostridium pectinovovum, 447
Clostridium perfringens, 449
Clostridium tetani, 449
Clot-busting drug, 712-13
Cloud, 865
Clover, 584

Club fungi. *See* Basidiomycetes
Club moss, 493, 496-97
CML. *See* Leukemia, chronic myeloid
Cnidaria, 510-11, 516-17, 619, 733
Cnidocyte, 516
CNS. *See* Central nervous system
Coal, 502
Coal Age, 363, 365
Coastal redwood, 499
Coastal region, 881-83
Coat color
  cat, 273, 275, 294-95
  rabbit, 273, 275
Cobalt, 756
Cocaine, 97, 204, 247, 618, 634
Cocci, 66, 443
Coccidioides immitis, 476, 487, 489
Coccidioidomycosis, 476, 489
Coccyx, 685, 687
Cochlea, 653-55
Cochlear canal, 653
Cochlear nerve, 654-55
Cochliobolus heterostrophus, 485
Cocklebur, 584
Cockroach, 662, 683
Coconut, 501, 571
Cocoon, 521, 664-65
Codeine, 634
Coding strand, 326, 328-29
Codominance, 273-74, 277
Codon, 329
  degenerate, 344
Coefficient of relationship, 278-79
Coelom, 511, 709
Coelomate, 511-12
Coenocytic hypha, 477-78
Coenzyme, 119-20
Coevolution, 374-75
Cofactor, 119
Coffee, 501
Cohesiveness, water, 40-41, 43, 540
Colchicine, 414
Cold (respiratory infection), 746
Cold temperature, coping with, 771-72
Cold-water drowning, 725
Coleoptile, 568-69, 572, 578
Coley, William, 804
Colias cesonia, 640
Collagen, 48, 344, 594, 597, 599, 601-3, 688
Collateral circulation, 712-13, 727
Collecting duct, 781-84
Collenchyma, 536, 539
Colloid, 661
Colon. *See* Large intestine
Colon cancer, 322
Colony-stimulating factor, 48, 717, 791, 806
Coloration, 410
Color blindness, 292-93, 652
Color vision, 648, 651
Colostrum, 798
Columnar epithelium, 594, 596-97
Coma, 633
Comet, 353, 359
Commensalism, 434-35
Common ancestor, 210, 427
Communicating canal, 688-89
Communication, 824-26
Community, 7, 9, 854-55
Compact bone, 689-92
Companion cell, 539
Comparative anatomy, 393-94
Comparative embryology, 395
Compass sense, 817
Competition, 837-38, 843, 845
Competitive exclusion, principle of, 845
Competitive inhibition, 122-23
Complementary base pairing, 53, 311-13, 317, 319, 328
Complement system, 791, 796-97
Complex carbohydrate, 44, 46
Complex trait, 276-80
Compound, 34-35
Compound eye, 522, 644-45
Compound leaf, 542
Compound microscope, 60
Compression fossil, 502
Computer experimentation, 26-27
Computerized database, taxonomy, 427
Concentration, 88
Concentration gradient, 88, 91, 613
Conclusion, 23
Concordance, 279
Concussion, 633
Condensation, 865
Conditional mutation, 275-77
Conditioned stimulus, 815
Condom, 250-51
Conductive deafness, 654
Cone, pine, 560-63
Cones (eye), 648-52
Confidentiality, doctor-patient relationship, 6
Conformation, protein, 49, 52
Confusion effect, 822
Congestive heart failure, 712-13
Conifer, 493, 498-500
Coniferous forest, 894
  temperate, 873-76
Conjoined twins, 238, 256
Conjugation, 192-95
  bacterial, 447-49
  protozoan, 459
Connective tissue, 594-97
Consanguinity, 410
Consciousness, 628
Constant region, 795
Consumer, 8-9, 14, 859-62, 864
Contact inhibition, 176-77, 181
Contest competition, 843
Continental drift, 417-18
Contour feather, 602
Contraception, 250-53, 848-50
Contractile ring, 175
Contractile vacuole, 90, 459, 778
Contractility, 689
Control, experimental, 20, 24-25
Controlled substance, 482
Convergent evolution, 379-81, 567
Copeland, Herbert F., 430
Copepod, 4-5, 522, 883
Copper, 33, 756
  deficiency, 890, 907
Coprinus, 485
Coprinus dominicana, 485
Coral, 362, 516-17, 881-83
Coria, Rodolfo, 401
Cork cambium, 550-52
Cork cell, 552
Cork oak tree, 552
Corn, 161, 536, 538, 541, 548-50, 569, 589
  kernel color, 267
  taxonomy, 432
Cornea, 648-49, 651, 804
Corn leaf blight, 485
Corolla, 563
Corona radiata, 217-19
Coronary artery, 712
Coronary bypass operation, 712-13
Coronary heart disease, 712
Coronary sinus, 720
Corpus callosum, 626
Corpus luteum, 674
Corte, Rosanna Della, 244
Cortex
  root, 549
  stem, 541
Cortisol, 672
Cortisone, 45
Cosmic rays, 341
Costus, 544
Cotton, 414, 537
Cotyledon, 544, 568-70
Cotylosaur, 365
Coumadin derivative, 248
Countercurrent exchange system, 770-73
Countercurrent flow, 734
Countercurrent heat exchanger, 772
Countercurrent multiplier system, 783
Coupled reactions, 119
Coupling of alleles, 287-89
Courtship behavior, 382, 822, 829
Covalent bond, 35-38, 40
Cowpox, 29, 805
C_3 plant, 159, 162
C_4 plant, 160-62
Crab, 510, 692, 695, 776, 882
Crabeater seal, 838
Cranial nerve, 622, 630
Cranial radiator, 774-75
Crassulacea, 162
Crayfish, 513, 522, 644
Creatine phosphate, 697-98
Creatinine, 784
Creationism, 370
Creativity, 629
Crenarchaeota, 451
Crepidula fornicata, 291
Cretaceous period, 359-61, 366, 417, 503
Cretaceous/Tertiary boundary, 417
Cretinism, 669
Creutzfeldt-Jakob syndrome, 77-78
Crick, Francis, 311, 331
Cricket, 645
Cricket frog, 640
Cri du chat syndrome, 301
Crinoid, 362, 861
Cristae, 71, 73, 128
Critical period, 246-47, 816
Cro-Magnons, 368-69
Crop (digestive), 753-54
Crop plant, 492
  transgenic, 335
Cross bridge, 696-97, 699
Crossing-over, 200, 286, 288-89
Crowd diseases, 385
Crowding, 829, 843
  contact inhibition, 176-77
Crown, tooth, 760

Crustacean, 521-23
Cryphonectria parasitica, 485
Cryptobiosis, 42-43
Cryptococcus neoformans, 483
Cryptoendolithic ecosystem, 860-61
Cryptoendolithic fungi, 480
Cryptomonas, 464
Cryptosporidiosis, 461
Cryptosporidium parvum, 461
Cuboidal epithelium, 594, 596-97
Cud, 441, 754
Culture, bacteria, 442
Cup fungus, 478
Curare, 618
Curl cat, 407
Cushing's syndrome, 672
Cutaneous fungal infection, 486
Cuticle, 48, 497, 536-37
Cutin, 536
Cutshall, Cynthia, 326
Cutting (plant), 192, 557
Cuttlefish, 645
Cyanide, 135, 613, 897
Cyanobacteria, 14, 64, 66, 80-81, 152, 352, 415-16, 440, 445, 463, 484, 493
Cyanocobalamin, 716, 755
Cycad, 493, 498-500, 503
Cycadeoid, 498
Cyclic adenosine monophosphate. *See* Cyclic AMP
Cyclic AMP (cAMP), 103, 661, 663
Cyclic ovulator, 665
Cyclin, 177-78
Cycliophora, 528
Cyclosporin, 804
Cyst
  algal, 433
  protozoan, 459, 461
Cystic fibrosis, 262, 268-69, 273, 281, 335-36, 338, 343, 746
  cholera and, 411-12
Cystic fibrosis transmembrane conductance regulator (CFTR), 87, 335-36, 411
Cytochrome, 119-21
Cytochrome c, 397
Cytogenetics, 286
Cytokine, 48, 728, 791-93, 797, 799, 805-6
Cytokinesis, 168, 171, 173-75, 201-2
Cytokinin, 576-77
Cytoplasm, 61, 70-71
Cytosine, 51, 311, 313
Cytoskeleton, 67, 69, 84, 96-103
Cytotoxic T cell, 797, 800

D

2,4-D, 577
Damming, 899
Danazol, 248
Dandelion, 547, 571, 857
Danish Adoption Register, 279
Daphnia, 522
Dark meat, 703
Darwin, Charles, 210, 376, 379-83, 577, 579
Darwin, Erasmus, 379
Darwin, Francis, 577, 579
Darwinian evolution, 373-87
Database homology searching, 336-37
Day-neutral plant, 583
DDE, 21-24, 27
DDT, 18-19, 21, 866, 868-69, 898-99
Dead space, 742
Deafness, 654
Death
  causes, 234
  definition, 4-5, 15
Death row inmate, 606
Decibel, 654
Deciduous forest, temperate, 873-76
Deciduous tree, 546
Decomposer, 8-9, 14, 441, 862, 865-67, 873
Decorator crab, 516
Dedifferentiation, 181
Deep-sea hydrothermal vent, 112, 452-53, 508
Deer, 527, 754, 817, 839
Deerfly, 523
Deer tick, 384, 386, 450
Defibrillator, implantable, 723
Defoliation, 146
Deforestation, 896, 904-5
Degenerate codon, 344
Degradation. *See* Catabolism
Dehydration synthesis, 43-44, 48
Deinococcus radiodurans, 439, 442
Deletion, 298, 300-301, 343
Delivery, 229-30
Deltadromeus agilis, 390
Delta globin, 728
Delta land, 881
DeLuca, Marlene, 118
Dendrite, 596-98, 611-13
Dendrochronology, 534, 552-53
Dengue, 384, 872, 886
Dengue hemorrhagic fever, 872
Denitrifying bacteria, 865, 867
Dense connective tissue, 594, 596-97
Density-dependent factors, 842-43
Density-independent factors, 842-43
Dentine, 760
Deoxyhemoglobin, 715
Deoxyribonucleic acid. *See* DNA
Deoxyribose, 51, 311, 326, 358-59
Dephlogisticated air, 149
Depo-Provera, 250-51
Deposit feeder, 521
Depression, 92-93, 618
Depth perception, 652
Dermal tissue, plant, 534-37
Dermis, 599, 601-3
Dermocentor, 523
Dermoptera, 527
DES. *See* Diethylstilbestrol
Descending limb, nephron loop, 782-83
Desensitization, 802
Desert, 858, 872, 874-75, 878-79
Desertification, 892, 905-6
DeSilva, Ashanti, 326, 345
Desmosome, 96-97
Detergent, 900
Determinate cleavage, 514
Deuteromycetes, 478-80, 483
Deuterostome, 511-14
Development, 9-10
  animal, 209-35, 509-14
  Caenorhabditis elegans, 213-15
  Drosophila melanogaster, 215-16
  nervous system, 620-21
  postnatal, 230-33
  prenatal. *See* Prenatal development
Developmental pathway, 211, 213-16
Devonian period, 359, 362, 364, 417
Diabetes insipidus, 90, 293, 784
Diabetes mellitus, 335, 604-5, 660, 672-73, 678, 766, 785
  type I (insulin-dependent, juvenile), 660, 672-73, 802
Diagnostic kit, monoclonal antibody-based, 805
Dialysis, 786
Diaminoanisole, 341
Diaminotoluene, 341
Diapause, 4-5, 873
Diaphragm (birth control), 250-51
Diaphragm (muscle), 742
Diaphragmatic hernia, 249
Diaptomus sanquineus, 5
Diastole, 720
Diastolic pressure, 726
Diatom, 458, 467
Diatomaceous earth, 467
Dicot, 541, 544, 550, 568, 570, 572
Dictyostelium discoideum, 472
Didinium, 459
Diencephalon, 627
Diet, cancer and, 185
Dietary fiber, 185
Diethylstilbestrol (DES), 18, 184
  DES daughters, 18
  DES sons, 18
Diet plan, 766
Differentiated cell, 211
Differentiation, 211-12, 214
Diffusion, 88, 91, 95, 708-9, 733
DiGeorge syndrome, 216
Digestion, 123, 750
Digestive enzyme, 72, 123
Digestive system, 599-600, 750, 753
  adaptations, 754
  human, 757-65
  types, 752-54
Dihybrid, 265
Dihybrid cross, 269, 271, 286-88
Dihydroxyacetone phosphate, 130
Dilution effect, 822
Dimethyl sulfide, 903
Dimorphic fungi, 477
2,4-Dinitrophenol, 135
Dinofilaria immitis, 519
Dinoflagellate, 152, 433, 458, 466-67, 886
Dinosaur, 365-66, 389-91, 417, 526
Dinosaur eggs, 389
Diphenylhydantoin (Dilantin), 248
Diphtheria, 384, 449-51
Diploblastic embryo, 509-11
Diplococcus pneumoniae, 309
Diploid, 192
Direct-contact infection, 450-51
Direct development, 512-13
Directional selection, 410-11
Disaccharide, 43-44
Discrimination tank, 816
Displacement behavior, 828
Disruptive selection, 411
Distal convoluted tubule, 782-84
Disulfide bond, 49, 51-52
Diuretic, 785
Diversion behavior, 828
Diversity-stability hypothesis, 906-7
Diverticulum, 753

Division (taxonomy), 12, 433
Dizygotic twins, 218, 223, 279-80
DNA, 5, 8, 51, 69
base sequence, 312-13
chloroplast, 79-81
comparing sequences to determine evolutionary relatedness, 397-98
functions, 308
hybridization technique, 397-98
mitochondrial, 79-81, 322, 342, 399, 426, 436, 818
prebiotic earth, 356
prokaryotic, 66
proof that it is genetic material, 309-11
repair, 233, 319-22, 850
repair disorders, 319, 322
repeated sequences, 339-41
replication, 53, 307, 313-15, 319
replication disorders, 322
RNA versus, 326
structure, 39, 53, 311-13, 316
transcription, 326-29
DNA fingerprinting, 314-15
DNA polymerase, 48, 317-18
DNA probe, 296-97
Doctor-patient relationship, 6
disclosure of prognosis to cancer patient, 185
Dodo bird, 374-75, 419
Dog, 230, 527, 806
breeding, 407
Dogface butterfly, 640
Dog tapeworm, 518
Dog tick, 523
Dolphin, 381, 643
Domain (taxonomy), 12, 14, 69, 430-33
Dominance hierarchy, 828-29
Dominant allele, 265
Dominant trait
autosomal, 267-70
dominance relationships, 273-74, 277
sex-linked, 292-93
Dopamine, 105, 281, 605, 618, 634
Dormancy, 4, 587
"Dorsal," 512-13
Dorsal lip, 210, 213
Double-blind study, 25
Double bond, 37, 45
Double fertilization, 568
Double helix, 51, 53, 311-12, 316
Down feather, 602
Down syndrome, 298, 300-302
Dragline silk, 49
Dragonfly, 522
Drone, 820
Dropping detection, 839
Drosophila melanogaster. See Fruit fly
Drought, 417, 878, 906-7
Drug(s)
crossing blood-brain barrier, 97
that alter neurotransmitters, 618
Drug addiction, 634
Drupe, 569-70
Duchenne muscular dystrophy, 293, 335, 339, 695
Dunker community, 408-9
Duodenum, 759, 761
Duplication, 298, 300-301, 343
Dust mite, 804
Dutch elm disease, 480-81, 485
Dwarfism, pituitary, 335, 667
Dynamic equilibrium, 88
Dynein, 99-100, 227
Dyslexia, 633
Dystrophin, 48, 102, 335, 695

E

Ear, 652-54
Eardrum. *See* Tympanic membrane
Earthworm, 521, 683-84, 733, 778-79
Ear wax, 790
Eastern white cedar, 860
Eating, evolution, 751-52
Eating disorder, 758
Ebola virus, 75-76, 386-87
ECG. *See* Electrocardiogram
Echinacea purpurea, 572
Echinodermata, 510-11, 513, 523-24
Echolocation, 645
Ecological isolation, 413
Ecological niche, 838
Ecological pyramid, 863
Ecology, 836
Ecosystem, 7, 9, 853, 858-63
energy flow through, 859-62, 864
stability, 906-7
Ectoderm, 220-22, 511, 620
Ectomycorrhizae, 485
Ectopic pregnancy, 239, 245
Ectoplasm, 692, 694
Ectotherm, 771
Edema, 727, 785
Edentata, 527
EEG. *See* Electroencephalogram
Effector, 611, 682
Efferent arteriole, 781
Efferent neuron. *See* Motor neuron
EGF. *See* Epidermal growth factor
Egg Mountain, 366
Egg nucleus, 568
Eggplant, 569
Egg-rolling response, 813-14
Eggshell, 868
Einstein, Albert, 147
Ejaculation, 197, 204, 217
Ejaculatory duct, 196
Eland, 774-75
Elastin, 48, 594-95, 597, 688
Eldredge, Niles, 415
Electrical gradient, 613
Electrocardiogram (ECG), 712, 722-23, 727
Electroencephalogram (EEG), 627-28
Electrolyte, 40
Electromagnetic spectrum, 147
Electron, 32, 34-38
Electron dot diagram, 37
Electronegativity, 38
Electron microscope, 62-63
Electron orbital, 35
Electron transport system, 121, 133-36
anaerobic, 127, 140
photosynthetic, 154-55
respiration, 114, 141
Electroperception, 657
Electrophoresis, 339
Electroreception, 639-40
Element, 32-34
Elementen, 262-63
Elephant, 112-13, 230, 512, 527, 754, 824
"language," 22-23
Elephantiasis, 519, 727
Elephant seal, 735, 838
Ellis-van Creveld syndrome, 408
El Niño, 883-86
Elodea, 61
Embolus, 717
Embryo, 226-28
early experiments with, 210-11
Embryology, 210
comparative, 395
Embryonic disc, 220
Embryonic induction, 226
Embryonic resemblances, 212
Embryonic stage, 216, 223-28, 246, 512
Embryo sac, 566-68
Embryo transfer, 238
Emergent property, 8
Emerging infectious disease, 384
Emigration, 838
Emotion, 627
Empedocles, 378
Emperor moth, 646
Emphysema, 335, 741, 746
Empiric risk, 277-78
Emu, 380-81
Enamel, 760
Encyrtid wasp, 435
Endangered species, 427-28
Endergonic reaction, 116
Endocarp, 570
Endochondral bone, 687
Endocrine gland, 660-62
Endocrine system, 599-600, 659-79
Endocytosis, 93-95, 105
Endoderm, 220-22, 511, 620
Endodermis, 548-49
Endometriosis, 239-40
Endometrium, 197
Endomycorrhizae, 485
Endoplasm, 692, 694
Endoplasmic reticulum, 70-71, 75
Endorphin, 618, 634
Endoskeleton, 684
Endosome, 93
Endosperm, 175, 566, 568-69, 571
Endospore, 447
Endosymbiont theory, 79-81, 164
Endothelial cell, 8
Endothelium, 722-23
Endotherm, 526, 771
Energy. *See also* ATP
in chemical reactions, 115-16
definition, 110-12
enzymes and, 121-23
flow through ecosystem, 859-62, 864
law of conservation of, 113
metabolism, 162-64
origins of energy pathways, 163-64
sources, 31, 54, 110-12
sources on primitive earth, 354
transfer from nutrients to ATP, 128-29
transformation, 114, 117-19, 123-24
units, 909
ways that cells release, 127-29
Energy content, 117
Energy nutrient, 755
Energy of activation, 121
Energy pyramid, 863-64

Energy shell, 35
ENOD40, 588-89
Enrichment culture, 442
Entamoeba histolytica, 461
Enterobius vermicularis, 519
Enterococcus faecalis, 443
Entomologist, 428
Entomophthora, 481
Entrainment, 588
Entropy, 8, 113-14
Envelope, viral, 75
Environment
  effect on gene expression, 275-76
  multifactorial traits, 276-80
  response to stimuli, 682
  in sex determination, 291
Environmental barometer, 890, 907
Environmental concerns, 889-907
Environmental decline, 849-50
Environmental disaster, 842
Environmentalism, 891-92
Environmental resistance, 842
Enzyme, 46, 49-50, 53, 114-15
  energy and, 121-23
  regulatory, 122-23
  temperature dependence, 13, 121-22
Enzyme-substrate complex, 50, 53
Eocene period, 361
Eosinophil, 714, 716, 719
Ephedra, 500
Epicotyl, 568, 570
Epidemiological study, 26-27
Epidemiologist, 18
Epidemiology, 26, 384
  evolution and, 384-87
Epidermal growth factor (EGF), 177
Epidermis, 178-80, 516-17, 598-602
  plant, 536
Epidermolysis bullosa, 599, 601
Epididymis, 196-97, 204-5
Epigenesis, 210
Epiglottis, 736-37
Epilepsy, 618, 633
Epinephrine, 618, 632, 670, 739
Epiphyseal plate, 690-91
Epiphyte, 872
Epistasis, 274-75, 277
Epithelial tissue, 594, 596-97
Epoch, 359
Epulopiscium fishelsoni, 64, 442
Equational division. *See* Meiosis II
Equilibrium (balance), 645, 654-56
Equisetum, 497-98
Equisetum fluviatile, 496
Era, 359
Ergot, 482
Ergotism, 482
Ermine, 878
Erysiphe graminis, 485
Erythrocytes. *See* Red blood cells
Erythropoietin, 715
Escape behavior, 812
Escherichia coli, 9, 66, 122, 195, 386, 447, 450, 770, 786
Escherichia coli O157:H7, 786
Esker, 871, 878
Esophagus, 736, 757, 759-61, 765
Essential nutrient, 755
Estrogen, 45, 176-77, 663, 665, 670-75, 679, 690
Estrogen mimic theory, 18-23
Estrogen receptor, 20-21, 23, 29
Estrus, 665
Estuary, 776, 881-82, 899
Ethane, 37
Ether, 28
Ethology, 812
Ethylene, 37, 569, 576-79
Etiolated seedling, 586
Euchromatin, 296
Eugenics, 281
Euglena, 460, 466
Euglenoid, 466
Euglenophyta. *See* Euglenoid
Eukarya (domain), 430-33
Eukaryote, 14, 431
Eukaryotic cell, 14, 63, 66-67
  evolution, 79-81
  prokaryotic cell versus, 69
Euphausia superba, 854
Euphotic zone, 883-84
Euploidy, 298
European beech, 377
Euryarchaeota, 451
Eurypterid, 362
Eusocial group, 819
Eutheria, 527
Eutrophication, 880-81, 899
Evaporation, 865
Evaporative cooling, 773
Evergreen tree, 876
Eversible pharynx, 518
Evolution, 8, 11, 15
  biological, 374
  chemical, 353-56
  eating, 751-52
  energy pathways, 163-64
  epidemiology and, 384-87
  eukaryotic cells, 79-81
  evidence for, 389-401
  evolution/creationism debate, 370
  forces of change, 403-20
  fungi, 485
  human, 367-70, 399
  metabolism, 357-58
  molecular, 395-400
  nervous system, 610-11
  pace, 415-16
  photosynthesis, 146-47
  plant, 493, 503-4
  theories through history, 378
  virus, 77
Excision repair, 319
Excitatory synapse, 617
Excretion, 865
Excurrent siphon, 526
Exercise
  body weight and, 766
  circulatory system and, 727
  effect on muscle, 701-3
  volume of air inhaled during, 743
Exergonic reaction, 116
Exocrine gland, 660, 662
Exocytosis, 93-95, 617
Exon, 330
Exoskeleton, 683, 685, 695, 733
Expanding cell population, 179
Expanding mutation, 343
Experiment, 21-23, 903
Experimental control, 20, 24-25
Experimental design, 24-28
Experimental results, 23, 28
Expiration, 742
Expiratory center, 744
Exponential growth, 840-42
Expressivity, 275, 277
External gills, 733-34
External respiration. *See* Breathing
Extinction (loss of conditioned reflex), 815
Extinction (species disappearance), 375, 416-19. *See also* Mass extinction
Extracellular digestion, 753
Extracorporeal shock wave lithotripsy, 786
Extraembryonic membrane, 216, 223, 226
Extraterrestrial life, 353
Extraterrestrial object, 417
Extraversion, 278
Extremophile, 42
Exxon Valdez oil spill, 890-91
Eye. *See also* Visual system
  invertebrate, 644-45
Eyeglasses, 651
Eyespot. *See* Ocelli

F

Facilitated diffusion, 91-92, 95
Facultative anaerobe, 140, 445
Facultative photoperiodism, 584
FAD, 119-20, 132-33, 136
Fainting, 633, 701
Falco peregrinus, 854, 868
Fallopian tube, 197
  blocked, 239-40
Fall turnover, 880
Familial hypercholesterolemia, 48, 269, 335
Familial hypertrophic cardiomyopathy, 269-70
Family (taxonomy), 12, 426, 433
Fanconi anemia, 322
Farsightedness, 651-52
FAS. *See* Fetal alcohol syndrome
Fascicle, 499
Fast-twitch fiber, 703
Fatal familial insomnia, 79
Fate map, 213-14
Fats, 44
Fat-soluble hormone, 663-64
Fat-soluble vitamin, 755
Fatty acid, 45, 142, 751
Feather, 382, 602, 771
Feather follicle, 602
Feather star, 524
Feces, 750, 754, 762-64
Fecundity, 842
Fecundity schedule, 842
Feedback inhibition. *See* Negative feedback
Feeders, 750
Feline immunodeficiency virus, 801
Female infertility, 239-40
Female reproductive system, human, 197-98
Femur, 395, 686
Fermentation, 127, 137-40, 163
  alcoholic, 137-39, 141
  lactic acid, 137, 139-41
Fern, 497-99, 557, 561
Ferritin, 48-49
Fertilin, 252
Fertility awareness methods, 250

Fertility drug, 239
Fertilization, 206, 213, 216-19
plants, 568
Fetal alcohol syndrome (FAS), 246-48, 509
Fetal cell sorting, 287
Fetal chromosomes, 286-87
Fetal circulation, 226
Fetal hemoglobin, 728
Fetal period, 216, 229-30, 246
Fetal skeleton, 687, 690-91
Fetal surgery, 249
Fetal tissue transplant, 605
Fever, 774, 792
F_1 generation. *See* First filial generation
F_2 generation. *See* Second filial generation
FGF. *See* Fibroblast growth factor
Fiber, plant, 536-37
Fibrillation, 712
Fibrin, 48, 123
Fibrinogen, 714
Fibroblast, 175, 594, 596-97
Fibroblast growth factor (FGF), 177
Fibroid, 239-40, 245
Fibronectin, 246
Fibrous root system, 546-47
Fibula, 686
Fiddlehead, 498-99, 561
Fiddler crab, 830
Field guide, 427
Fight-or-flight response, 630, 632, 659
Fig tree, 550
Fig Tree sediments, 359
Filarial worm, 519
Filoplume, 602
Filter feeder, 751
Finasteride, 675
Finch, Galápagos, 380-83
Fiord, 861
Fir, 499, 551, 560, 862, 876
Fire ant, 819
Fire coral, 516
Firefly, 118, 576
First breath, 732, 739
First filial (F_1) generation, 265
First law of thermodynamics, 113
First messenger, 103
Fish, 381, 510, 734
saltwater, 776-77
FISH. *See* Fluorescence in situ hybridization
Fishing net, 903-4
Five-kingdom classification, 14, 430-31
Fixed action pattern, 813-14
Flagella, 66, 99-100, 446-47, 459-60, 464, 466-67
Flame cell, 518, 778
Flashbacks, 482
Flatworm. *See* Platyhelminthes
Flavin, 579
Flavin adenine dinucleotide. *See* FAD
FlavrSavr tomato, 334
Flea, 523, 683
Fleshy root, 550
Fletcher, Marilyn, 704
Flight feather, 602
Flightless bird, 379-81
Flightless cormorant, 828
Flight muscle, 692
Flippers, 381
Flooding, 843, 881
Floodplain, 881
"Floras," 427
Floridean starch, 469
Flower
color, 273-74, 568, 824
formation, 564-65
fragrance, 556, 567-68
shape, 567
structure, 563
Flowering, 582-86
Flowering plant. *See* Angiosperm
Flower mimicry, 479-80
Fluidity, 40
Fluid mosaic, 85, 87
Fluke, 517-18
Fluorescence, 153
Fluorescence in situ hybridization (FISH), 296-97, 301
Fluorescent stain, 443
Fluorine, 756
Fluoxetine (Prozac), 92-93
Fly, 479
Flyger, Vagn, 839
Flying fox, 398-99
Flying lemur, 527
Fogging of canopy, 428
Folic acid, 755
deficiency, 185
Follicle cell, 197
Follicle-stimulating hormone (FSH), 666-69, 674-75, 679
Food-borne disease, 450-51
Food chain, 859, 863-64
Food inhalation, 738
Food label, 33
Food location, 640
Food poisoning, 66, 450
Food pyramid, 756
Food spoilage, 481
Food supplement, 677
Food vacuole, 752-53
Food web, 860, 862-63
Foolish seedling disease, 577
Foot, mollusk, 520
Foraging behavior, 837
Foraging efficiency, 822-23
Foramen magnum, 685
Foraminifera, 457-58, 460, 473
Forebrain, 620, 625-28
Forensic science, 314
Forewing, 522
Formed elements, blood, 711, 714-16
Fossil fuel, 502, 865-66, 892, 894, 896
Fossil record, 352, 359-70, 378-79, 389-90, 400-401, 415-16, 485
animal, 508
dating of fossils, 392-93
formation of fossils, 390-92
plants, 502
Fossombronia cristula, 494
Founder effect, 408
Foundry Cove, 375-77
Fountain effect, 822
Four-chambered heart, 710-11
Four-kingdom classification, 429-30
Four-o'clock, 587
Fovea centralis, 648-50
Fracture repair, 690
Fragavia chilensis, 542
Fragile X syndrome, 300, 304, 343
Fragmentation, 451, 464, 493
Frameshift mutation, 341, 343
Frank, Adrian, 890
Franklin, Rosalind, 311
Fraternal twins. *See* Dizygotic twins
Free energy, 116
Free nerve ending, 656
Free radical, 72, 233, 631, 633
Freezing, 42, 44
Freshwater biome, 879-81
Frog, 665
Frond, 498-99, 561
Frontal bone, 686
Frontal lobe, 628
Fructose, 43, 762
Fructose-1,6-bisphosphate, 130-31, 137
Fructose-6-phosphate, 130-31
Frugivore, 750
Fruit, 566
abscission, 578
dispersal, 570-71
formation, 569-70
ripening, 577-78
Fruit fly, 25, 215-16, 255, 821, 843
linked genes, 286-88
sex determination, 290
temperature-sensitive alleles, 275-76
X chromosome, 289
Fruiting body, 472
Frustule, 467
FSH. *See* Follicle-stimulating hormone
Fuchsia plant, 567
Fucoxanthin, 150, 467
Fucus, 890
Fulcrum, 700, 702
Fullerene, 417
Fumaric acid, 132
Fundamental niche, 838
Fundatrix, 194
Fundus, 760
Fungal body, 477-78
Fungi, 14, 430, 475-89, 914
characteristics, 476-80
classification, 480-84
compared to plants and animals, 476-77
diversity, 480
ecology, 485
evolution, 485
flower mimicry, 479-80
lichens, 484
pathogenic, 476, 483, 485-88
reproduction, 478-80
Fungus ball, 488
Fur, 774
Fur seal, 838
Furylfuramide, 341
Fusiform bacteria, 66

G

GABA, 618, 634
Gage, Phineas, 610
Gait, 692
Galactose, 43-44
Galago, 828
Galápagos finch, 380-83
Galápagos Islands, 373, 375, 380, 809
Galápagos mockingbird, 749
Galápagos tortoise, 375-76, 380
Gallbladder, 757, 759, 764-65

Gallstone, 765
Gamete, 191-92
  human, 196-98
  maturation, 201-6
Gamete intrafallopian transfer (GIFT), 243-44
Gametogenesis, 196, 216
  human, 196-98
Gametophyte, 192-93, 492-93, 497-98, 557, 560-62, 566
Ganglion cell, 650
Gap junction, 96-97
Garbanzo bean, 571
Garner, W.W., 583
Garrod, Archibald, 308
Gas exchange, 126, 732
Gas exchange system, 732-35
Gas transport, 742-43
Gastric juice, 760-61
Gastric pit, 760-61
Gastric ulcer, 750, 766
Gastrin, 760
Gastritis, 750, 766
Gastrodermis, 516-17
Gastroenteritis, 443
Gastrointestinal tract, 757
Gastropod, 519-20
Gastrovascular cavity, 516-17
Gastrula, 219-20, 514
Gastrulation, 216, 220-21
Gaucher disease, 269
Gemmae, 494
Gemmule, 516
GenBank, 337
Gene, 51, 262
  jumping, 28, 343
  linked, 286-90
  relationship to protein, 312-13
  selective activation, 211-12, 214
Gene expression, 326
  environmental influences, 275-76
Gene flow, 404
Gene pool, 404
Generation time, 416
Generative nucleus, 563, 566
Gene targeting, 334-35
Gene therapy, 226, 281, 326, 335, 345, 800
Genetic code, 51, 331-33, 344
  deciphering, 332-33
Genetic discrimination, 302-3
Genetic drift, 408-10, 412
Genetic load, 410
Genetic marker, 289-90
Genetic material, identification and description, 308-11
Genetic parent, 242
Genetics
  history, 262
  population, 404-15
  terminology, 265
  transmission, 261-81
Genetic variability, 192, 201, 344
Genital warts, 806
Genotype, 265
Genotype frequency, 405
Genotypic ratio, 266, 268
Genus, 12, 426, 433
Geographic barrier, 408, 413
Geological time scale, 359
Geology, 376-78
George III, King of England, 276
Geothermal energy, 110, 112, 858
German measles. *See* Rubella
Germ cell, 196
Germinal mutation, 341
Germination, seed, 40-41, 569-72, 586
Germ layer, 220-22
Gerontology, 234-35
Gerstmann-Straussler-Scheinker syndrome, 79
Gestational parent, 242
Gestation time, placental mammals, 230
Giant intestinal roundworm, 518-19
Giant sequoia, 550
Giardia lamblia, 461, 463
Giardiasis, 461
Gibberellin, 176, 576-77
Gibbon, 367, 527, 831
GIFT. *See* Gamete intrafallopian transfer
Giganotosaurus carolinii, 390-91, 401
Gigaspora margarita, 480
Gill(s), 710, 733-34, 776-77
Gill arch, 734
Gill raker, 734
Gingiva, 760
Ginkgo, 493, 499-500
Gizzard, 754
Glaucoma, 652
Glioma, 335
Glires, 436
Global climate, 418-19
Global warming, 419, 886, 892, 896-98, 903
Globin gene, 338-39, 341, 344
Globulin, 714
Glomerular capsule, 781-84
Glomerular filtrate, 783
Glomerulonephritis, 802
Glomerulus, 780-82, 786
Glottis, 736-37
Glowworm, 118
Glucagon, 48, 664-65, 672, 679
Glucocorticoid, 667, 670-72, 679
Glucose, 43-44, 751, 762
  blood, 663-65, 672-74
  energy from, 127. *See also* Glycolysis
  phosphorylation, 129
  production in photosynthesis, 146
  urine, 785
Glucose-6-phosphate, 130
Glucose-6-phosphate dehydrogenase deficiency, 293, 412
Gluphisia septentrionis, 40
Glycerol, 45, 142
Glycocalyx, 446
Glycogen, 44, 46, 72, 476
Glycolipid, 45, 87
Glycolysis, 128-31, 136-37, 141-42, 162-63
  evolution, 163-64
Glycoprotein, 85, 87
Gnathostomata, 525
Gnetophyte, 493, 499-500
Gnetum, 500
GnRH. *See* Gonadotropin-releasing hormone
Goat, 816
Gobea, Andrew, 345
Goblet cell, 764
Goiter, 669-70
Gold, 184
Golden-brown algae, 467
Golden orb weaver spider, 49
Goldenrod, 857
Golden-rumped elephant shrew, 830
Golgi body, 67-68, 70-71, 75
Gonad, 196
Gonadotropic hormone, 667
Gonadotropin-releasing hormone (GnRH), 674-75
Gondwana, 401, 417-18
Gonium, 464
Gonorrhea, 385, 450-51
Gonyaulax, 466
Gonyaulax tamarensis, 152
Goodall, Jane, 813
Gorilla, 395-97, 414, 527
Gould, Stephen Jay, 415
Gout, 293, 778
G_1 phase, 168-69, 171
G_2 phase, 170-71
G protein, 663
Gradient, 88
Gradualism, 415
Graft rejection, 84, 804
Gram-negative bacteria, 66, 444
Gram-positive bacteria, 66, 444
Gram staining procedure, 444
Grana, 74-75, 151, 159
Grape, 314
Grape downy mildew, 470
Grapefruit, 541
Grapsus grapsus, 769
Graptolite, 362
Grass, 856-57, 877
Grasshopper, 290, 522, 645
Grassland, 874-75, 877-78, 906-7
Grave's disease, 802
Gravitropism, 578-81
Gray fox, 845
Gray matter, 615, 622-23
Gray squirrel, 426
Gray tree frog, 683
Great white shark, 771
Green algae, 458, 465-66, 493
Green bean, 570
Greenbrier, 541
Greenhouse effect, 896-97
Greenhouse gas, 892
Green sea turtle, 735
Green turtle, 818
Greylag goose, 813-14
Griffith, Frederick, 309, 311
Grimmia, 494
Grizzly bear, 816, 839, 899
Grooming, 824
Gross primary production, 863
Ground squirrel, 813
Ground tissue, 534-36
Groundwater, 865
Group cohesion, 824-26
Group living, 812, 819-23, 832
Growth, 9-10, 230-33
Growth factor, 48
  cell division signals, 176-77
Growth hormone, 667-68, 679
Growth rings. *See* Tree rings
Growth spurt, adolescent, 667
Grypania spiralis, 362
Guanine, 51, 311, 313
Guano, 777
Guard cell, 536, 538, 545
Guinea pig, 426, 436

Gurdon, J.B., 211
Gut-associated lymphoid tissue, 793
Gymnodinium, 466
Gymnosperm, 492-93, 498-501, 550
angiosperm versus, 572
diversity, 499-500
life cycle, 560-62
uses, 500-501
Gynoecium, 563

H

Habitat, 836-38
Habitat destruction, 882, 904-6
Habituation, 104, 815
Haeckel, Ernst, 210, 395, 429
Hair, 50-51, 232, 599-602
Hair cell, 654-56
Hair color, 344
Hair follicle, 599, 601
Hairy ears, 293
Haldane, J.B.S., 354
Hales, Stephen, 148
Half-life, 393
Hallucinogen, 482-83
Halobacterium, 451-52
Halobacterium halobium, 644
Halobacterium salinarium, 440
Halophile, 451-52
Hamner, Karl, 584
Handedness in biomolecules, 358-59
Hangover, 785
Hantavirus, 75, 385
Haploid, 192
Hard palate, 736-37
Hardwood, 551
Hardy, H.H., 405
Hardy-Weinberg equation, 405
Hardy-Weinberg equilibrium, 405-6, 412
Harris' hawk, 846-47
Hartsoeker, Niklass, 203
Hashish, 537
Hawaiian Islands, 375
Hay fever, 802, 804
Hayflick, Leonard, 175
Hayflick limit, 175
Hazelnut, 873
hCG. *See* Human chorionic gonadotropin
HDL. *See* High-density lipoprotein
Headache, 633
Health, chromosomes and, 302
"Healthy ill," 302
Hearing, 642, 645, 652-54
Hearing loss, 654
Heart, 708
blood flow, 720
disease, 234, 766
human, 718-23
structural diversity, 710-11
structure, 718-19
Heart attack, 712-13
Heartbeat, 720-22
Heart murmur, 720
Heart rate, 727
Heart sounds, 720
Heart transplant, 605, 804
Heart valve, 720
replacement, 720
Heartwood, 551
Heartworm, 519
Heat capacity, 42-43
Heat of vaporization, 42-43
Heat tolerance, 772-75
Heavy chain, 795-96
Heavy metals, 376, 900
Hedge sparrow, 830-31
Height, 277-78
Heimlich maneuver, 738
HeLa cells, 180
Helianthus annuus, 415
Helianthus petriolaris, 415
Helicase, 317-18
Helicobacter pylori, 18, 20, 750, 766
Helix pomatia, 790
Helix-turn-helix motif, 327
Helper T cell, 794, 797, 799
Heme group, 715
Hemizygote, 291
Hemlock, 876
Hemochromatosis, 269
Hemocoel, 522
Hemocyanin, 710
Hemocyte, 790
Hemoglobin, 48-49, 339, 344, 708, 710, 715, 728, 735, 742-43
Hemolymph, 522, 708
Hemolytic anemia, 716, 802
Hemolytic uremic syndrome, 770, 786
Hemophilia, 335, 717
Hemophilia A, 292-94
Hemophilia B, 293
Hemorrhagic fever, 386
Henbane, 584
Heparin, 800
Hepatic artery, 725
Hepatic cecum, 526
Hepatic portal system, 725
Hepatic portal vein, 725
Hepatic vein, 725
Hepatitis B, 249
Heptachlor, 21
Herb(s), 856-57
Herbal ecstasy, 500
Herbal medicine, 504, 572
Herbicide, 21, 184, 573, 577, 635
Herbivore, 750, 862, 864
Hercules beetle, 383
Hereditary nonpolyposis colon cancer, 322
Hereditary persistence of fetal hemoglobin, 728
Heritability, 278
Hermaphrodite, 521
Hermit crab, 434
Heroin, 97, 634
Heron's bill, 587
Herpes simplex, 249, 253
Herring, 751
Herring gull, 814
Hershey and Chase experiment, 310-11
Heterochromatin, 296
Heterogametic sex, 290
Heterotaxy, 227
Heterotherm, 771
Heterotroph, 14, 146, 445, 459, 508, 751-52
Heterotrophy, 751-52
Heterozygote, 265
Hibernation, 9, 135, 772, 873
Hickory, 856-57
High altitude, breathing at, 745
High-density lipoprotein (HDL), 48, 712, 727
High-energy phosphate bond, 117-19
Highly conserved sequences, 395-96
Hindbrain, 620, 625-27
Hindgut, 753
Hippocampus, 630
Hippocrates, 785
Hippopotamus, 591, 828, 853
hearing, 642-43
Hiruda medicinalis, 718
Hirudin, 718
Hirundinea, 521
Histamine, 791, 800, 803
Histidine, 122-23
Histone, 313, 317
Histoplasma capsulatum, 486
Histoplasmosis, 486
Hives, 802
HIV infection, 75-76, 253-54, 385, 420, 798-801. *See also* AIDS
detection of infection, 320
fetal, 249
long-term nonprogressors, 799
transmission, 254-55, 799
HLA. *See* Human leukocyte antigen
Holdfast organ, 467-68
Holozoic heterotroph, 459
Homarus americanus, 523
Homeostasis, 9, 610-11, 660-62, 682
Homeotherm, 771
Homeotic gene, 215-16
Homeotic mutation, 215-16
Homing, 817-19
Hominid, 367-70
Hominidae, 527
Hominoid, 367
Hominoidea, 527
Homo erectus, 368-69
Homogametic sex, 290
Homo habilis, 368, 392
Homologous pairs, 198-99
Homologous structures, 393
Homo sapiens, 368-69, 527
Homosexuality, 255-56
Homozygote, 265
Homunculus, 203
Honeybee, 461, 556, 572, 734, 773, 820-21, 823-26
Honey-crop, 773
Honey locust, 542
Hooke, Robert, 60
Hookworm, 716
Hopi Indians, 406
Hops, 139
Hormone, 103, 660-62
cell division signals, 176-77
controlling levels, 663-64
invertebrate, 664-65
mechanism of action, 662-63
plant, 177, 569, 576-78
vertebrate, 665-66
Horn, 50, 382-83, 603-4
Horn coral, 390, 392
Hornwort, 493, 495
Horologium florae, 582
Horse, 527, 692-93, 695
Horsefly, 523
Horseshoe crab, 522-23, 710
Horsetail, 493, 496-98
Hot springs, 12-13, 451-52

House cat, 846
Housefly, 138
Howler monkey, 527
Human central nervous system, 622-23
Human chorionic gonadotropin (hCG), 218, 221, 252, 674, 679
Human circulatory system, 710-26
Human digestive system, 757-65
Human evolution, 367-70, 399
Human Genome Project, 302-3, 311
Human heart, 718-23
Human immune system, 790-806
Human leukocyte antigen (HLA), 85, 793-94
Human life span, 234-35
Human peripheral nervous system, 630
Human population growth, 847-50
Human prenatal development, 216-30
Human respiratory system, 735-42
Human sensory systems, 645-56
Human skeleton, 685-86
Human taxonomy, 426, 431-32
Human urinary system, 780-85
Humerus, 686-87
Humic acid, 900
Hummingbird, 567, 771
Humoral immune response, 793, 795-97
Hunger, 758
Hungerford, David, 286
Huntington disease, 269, 335, 343, 618, 633
Hurler disease, 72
Hutchinson-Gilford syndrome, 233
Hutton, James, 376-78
Hybrid, 382, 413, 415, 558
Hybridoma, 804-6
Hydra, 510, 513, 516-19, 661, 683, 708-9, 753
Hydrocarbon, 38, 900
Hydrocephalus, 293
Hydrochloric acid, 123, 760-61
Hydrogen, 35, 127-28
  from fermentation, 139
Hydrogenation (food processing), 45
Hydrogen bond, 39-40
Hydrogen peroxide, 72
Hydrogen sulfide, 112, 139, 164, 452
Hydrolysis, 43, 355
Hydronasty, 581
Hydronium ion, 40-41
Hydrophilic molecule, 39
Hydrophobic molecule, 39
Hydrostatic skeleton, 517, 521, 683-84
Hydrotropism, 581
Hydroxide ion, 40-41
Hydroxyapatite, 595
Hydroxyurea, 708, 728
Hyena, 845
Hyla chrysoscelis, 414
Hyla versicolor, 414
Hyles lineata, 567
Hypercholesterolemia, familial, 48, 269, 335
Hyperkalemia, 38
Hypernatremia, 38
Hypersaline environment, 452
Hypersomnia, 618
Hypertension, 677, 713, 726, 766
Hyperthermophile, 451-52
Hyperthyroidism, 670
Hypertonic solution, 89
Hypertrophic cardiomyopathy, familial, 269-70
Hypertrophy, 701
Hypha, 477-78, 480, 485
Hypocotyl, 568, 570, 572
Hypodermis, 519, 547
Hypoglycemia, 673-74
Hypohidrotic ectodermal dysplasia, 293
Hypokalemia, 38
Hyponatremia, 38
Hypotension, 726
Hypothalamus, 255, 626-27, 632, 666-67, 669, 771, 784
Hypothesis, 20-21, 23, 903
Hypothyroidism, 669
Hypotonic solution, 89
Hyracotherium, 695

I

I band, 697
Ice, 42-43
Ice age, 417-18
"Ice Man," 369-70
Ichthyologist, 428
Ichthyosaur, 365, 381
Ichthyosis, 293
Ichthyostega, 362, 364
ICSH. *See* Interstitial-cell-stimulating hormone
Identical twins. *See* Monozygotic twins
Idiotype, 795-96
Ig. *See* Immunoglobulin
Iguana, 373
Ileocecal valve, 764
Ileum, 759, 761
Iliac artery, 725
Iliac vein, 725
Ilium, 686
Imbibition, 40-41, 43, 571
Immigration, 838, 840
Immune response, 793-97
Immune system, 84, 178, 599-600, 789-806
  altering immune function, 802-6
  characteristics, 792-93
  development, 798
  disorders, 798-800
  human, 790-806
  neuroendocrine-immune interactions, 677
Immunodeficiency viruses, 801
Immunoglobulin (Ig), 795-97
Immunoglobulin A (IgA), 796
Immunoglobulin D (IgD), 796
Immunoglobulin E (IgE), 796, 803
Immunoglobulin G (IgG), 796, 802
Immunoglobulin M (IgM), 796
Immunosuppressant drug, 804
Immunotherapy, 804, 806
Impact theory, 417
Imperfect fungi. *See* Deuteromycetes
Implantation, 218, 221, 245
Impression fossil, 502
Imprinting, 815-16, 818
INAH 3, 255
Inborn errors of metabolism, 33, 308, 335
Inbreeding, 407
Inclusive fitness, 826
Incomplete dominance, 273-74, 277
Incomplete penetrance, 275
Incontinentia pigmenti, 292-93
Incus, 653
Independent assortment, 200-201, 265, 269-72
Indeterminate cleavage, 514
India ink, 443
Indoleacetic acid, 577
Induced mutation, 340-41
Induced ovulator, 666
Industrial melanism, 410
Industrial microbiology, 452-54
Infancy, 231
Infantile Refsums disease, 74
Infection, during pregnancy, 249
Infectious disease
  crowd diseases, 385
  emerging, 384
  mass extinctions and, 419
  in next century, 872, 886
Inferior vena cava, 720-21, 725
Infertility, 239
  female, 239-40
  male, 239
Infertility test, 240
Inflammation, 103, 677, 716, 791-92
Influenza, 234, 385, 746
Informed consent, 606
Infrared radiation, 147-48
Infrasound, 22-23
Ingenhousz, Jan, 149
Ingram, V.M., 339
Inheritance, mode of, 267-68
Inheritance of acquired characteristics, 378
Inhibiting hormone, 666
Inhibitory synapse, 617
Initiation codon, 336
Initiation complex, 333
Initiation site, replication, 317-18
Innate behavior, 812-14
Inner cell mass, 218, 220-21
Inner ear, 653
Insect, 510, 521-23, 644, 683, 692, 779
Insecticide, 21, 184, 617
Insectivora, 527
Insectivore, 750
Insect sting, 802
Insect-trapping leaf, 544
Insertion (mutation), 343
Insertion of muscle, 700, 702
Insight learning, 815-16
Insomnia, 618
Inspiration, 742
Inspiratory center, 744
Inspiratory muscles, 742
Insulin, 48, 97, 334, 336, 605, 660, 663, 665, 672-73, 678-79
Integrin, 48, 103, 106, 580
Integument, 594, 598-603
Intelligence, 278
Intelligent design theory, 370
Intercalary meristem, 535
Intercalated disk, 692-93, 720
Intercellular junction, 96-97
Interferon, 791, 797, 800, 806
Interleukin, 791-92, 797
Interleukin-1, 794
Interleukin-2, 800, 806
Intermediate-day plant, 583-84
Intermediate fiber, 703
Intermediate filament, 100
Intermembrane compartment, 128
Internal respiration, 732
Interneuron, 611, 613, 623
Internode, 541

Interphase, 168, 171, 174, 201
Interspecific competition, 845
Interspecific interaction, 845-47
Interstellar dust, 353
Interstitial cell, 205
Interstitial-cell-stimulating hormone (ICSH), 667, 669, 675, 679
Intertidal zone, 881
Intestinal ceca, 517
Intestinal flora, 763-64
Intestinal gas, 764
Intracellular digestion, 753
Intracytoplasmic sperm injection, 242
Intramembranous bone, 687
Intrauterine device, 250-51
Intrinsic rate of increase, 840-42
Intrinsic rate of natural increase, 840
Intron, 330-31
Inversion, 298, 300-301
Invertebrate, 508
In vitro fertilization, 204, 206, 242-43
Iodine, 669, 756
Ion, 38
Ion balance, 776
Ion channel, 612, 662
Ionic bond, 38-40
Ionizing radiation, 147-48, 439, 442
Iridium, 417-18
Iris, 648-49
Irish moss, 493
Irish potato famine, 470
Iron, 121, 715, 756
  in ocean, 903
Iron deficiency anemia, 716
"Iron Ex" experiments, 903
Iron lung, 682
Irreversibility, 114
Irrigation, 898
Irritability, 10-11
Ischium, 686
Island, 374-75, 409
Islets of Langerhans, 672
Isocitric acid, 132
Isoetes, 497
Isograft, 604
Isopod, 522
Isotonic solution, 88
Isotope, 33
Isotretinoin (Accutane), 248

J

Jackdaw bird, 830
Jack pine, 856, 894
Jacob, François, 326-27
Janssen, Johann, 60
Janssen, Zacharius, 60
Jaundice, 765
Jawed fish, 525
Jaw hearing, 643
Jawless fish, 362, 525
J chain, 796
Jejunum, 759, 761
Jellyfish, 510-11, 516-17, 619, 683, 860, 882
Jeningan, Joseph Paul, 606
Jenner, Edward, 29, 78, 805
Johanson, Donald, 368
Johnson, Anna, 243
Johnson, Ben, 676
Joint, 700-701
Jojoba oil, 46
J-shaped curve, 840, 847
Jumping gene, 28, 343
Jumping leg, 522
Juniper berry, 501
Jurassic period, 359, 365, 391, 401, 417, 498
Juvenile hormone, 666

K

Kallmann syndrome, 633
Kandler, Otto, 430
Kangaroo, 231, 527
Kangaroo rat, 773, 776, 878
Kaposi's sarcoma, 254, 798-99
Kartagener syndrome, 227
Karyokinesis. *See* Mitosis
Karyotype, 286-87, 296-97
Katydid, 645
Kay, Steven, 576
Kcal. *See* Kilocalorie
Kelp, 463
Kepone, 21
Keratin, 48-51, 594, 598-99
Keratitis, 461
Kernel color, 267
Kevorkian, Jack, 6
Kidney, 780-85
  control of kidney function, 784-85
  osmosis and, 90-91
Kidney cancer, 302
Kidney failure, 786
Kidney stone, 786
Kidney transplant, 604, 786, 804
Kilocalorie, 755-56
Kinase, 177-78
Kinase receptor, 589
Kinetic energy, 112-13
Kinetochore, 171-72
Kingdom, 12, 14, 426
  history of kingdom concept, 429-30
Kinky hair syndrome. *See* Menkes disease
Klinefelter syndrome, 299-300
Kluver-Bucy syndrome, 256
Knockout organism, 335
Köhler, Georges, 804
Kokanee salmon, 899
Komodo dragon, 189
Krebs cycle, 127-28, 131-33, 136, 140, 142
Krill, 602, 751-52, 854-55
K-selection, 844-45
Kuru, 79
Kwashiorkor, 758

L

Labium majora, 197
Labium minora, 197
Labor, 216, 229-30
  premature, 246
Laboratory culture, bacteria, 442
La Brea tar pits, 392
Labyrinthomorpha, 460
Labyrinthula, 460
Lachnospira multiparus, 441-42
Lacks, Henrietta, 180
lac operon, 326-27
Lactase, 762
Lacteal, 762, 764
Lactic acid, 137
Lactic acid fermentation, 137, 139-41
Lactobacillus bulgaris, 66
Lactococcus, 453
Lactose, 44, 326-27
Lactose intolerance, 269, 762
Lacunae, 595, 688-89
Lake, 858, 879-81, 899
Lake Tahoe, 899
Lake Torrens, 860
Lamarck, Jean Baptiste, 378
Lamellibrachia, 507
Laminarin, 467
Laminin, 48, 102, 599
Lancelet, 525-26
Landon, Michael, 185
Land snail, 790
Language, 627
Lanugo, 229
Large intestine, 757, 759, 763-65
Larva, 512, 514, 664-65, 843
Larynx, 736-37
Latent learning, 815-17
Lateral geniculate nucleus, 650
Lateral meristem, 534-35
Latitude, 872-73
Latrodectus, 523
Laurasia, 401
Laveran, Charles Louis Alphonse, 462
Law of segregation, 262-67
Law of use and disuse, 378
Laws of thermodynamics, 112-14
LDL. *See* Low-density lipoprotein
Leader sequence, 332
Leaf, 536, 538, 541-46
  adaptation to wind, 11
  autumn colors, 150-51, 587
Leaf abscission, 546
Leaf axil, 541
Leaf venation, 543-44
Learning, 626-27, 629-30, 812, 814-17, 823
Least tern, 843
Leatherback turtle, 735
Lecanora esculenta, 484
Leech, 508, 521
  medicinal, 717-18
Leeuwenhoek, Antonie van, 60-61, 203, 440
Left atrium, 711, 721
Left ventricle, 711, 721
Legionella pneumophila, 447, 449
Legionellosis, 384, 449, 451
Legionnaires' disease. *See* Legionellosis
Legume, 452-53, 550, 582
Leishmaniasis, 461
Lemon juice, 504
Lemur, 527
Length, 909-10
Lens, 648-49, 651
Lentic system, 879
Leprosy, 25
Leptin, 26
Lesch-Nyhan syndrome, 293
L'Esperance family, 243
Lethal allele, 272, 277
Lettuce, 586
Leucine zipper, 327
Leukemia, 716
  acute myeloblastic, 181
  chronic myeloid (CML), 181, 286, 304

Leukocyte adhesion deficiency, 106
Leukocytes. *See* White blood cells
Levee, 899
Levene, Phoebus, 311
Lever system, 682-84, 700-701
LH. *See* Luteinizing hormone
Lichen, 463, 478, 484, 856, 877, 879
Life
characteristics, 4-5
energy use and metabolism, 8-9
internal consistency, 9
irritability and adaptation, 10-11
organization, 5-8
reproduction, growth, and development, 9-10
history of life on earth, 359-70
origin of, 352-55
from outer space, 353
resiliency, 907
Life cycle
angiosperm, 563-72
gymnosperm, 560-62
moss, 193
plant, 555-74
Life expectancy, 234
Life span, 138, 210, 235
human, 234-35
Life support, withdrawal, 6
Ligament, 594, 700
Ligand, 94
Ligase, 317-18
Liger, 413
Light, 147-49
Light chain, 795
Light-dependent reactions, photosynthesis, 153-59
Light-harvesting complex, 156-57
Light-independent reactions, photosynthesis, 153, 158-59
Light microscope, 62-64
Lignin, 493, 497, 548
Limbic system, 646-47
Limestone, 458, 865-66
Liming, 907
Limnetic zone, 879-80
Lindow man, 495
Linen, 536
Linkage map, 288-90, 302-3
Linked genes, 286-90
Linnaeus, Carolus, 429, 582
Linum usitatissimum, 536
Lion, 1, 790, 806
Lipase, 48, 762-64
Lipid, 44-46, 54
digestion, 751, 762-63
energy from, 140-42
Lipofuscin, 233
Lipopolysaccharide, 444
Lipoprotein, 48
Liposome, 85, 335
Lissencephaly, 633
Lithium therapy, 248
Lithops, 542
Lithotroph, 445
Littoral zone, 879-80
Liver, 757, 764-65
Liver transplant, 605
Liverwort, 493-94
"Living sands," 458, 473
Lizard, 771
Lobe-finned fish, 362, 364
Lobster, 510, 522-23, 683, 824-25
Logistic growth curve, 843-44
Long bone, 690, 692
Long-day plant, 583-84
Long-eared owl, 817
Longevity, 210, 234-35
Long-horned wood-boring beetle, 382-83
Long-term memory, 630
Long-term synaptic potentiation, 630
Loose connective tissue, 594, 596-97
Lorenz, Konrad, 812, 816
Lorenzo's Oil, 72-73
Loris, 527
Lotic system, 879
Lou Gehrig disease. *See* Amyotrophic lateral sclerosis
Low-density lipoprotein (LDL), 48, 94, 712
Low-density lipoprotein (LDL) receptor, 335
Lower esophageal sphincter, 761
Lowland gorilla, 414
Loxosceles reclusa, 523
LSD, 481-82, 634
Luciferase, 118-19, 588
Luciferin, 118
Luciferyl adenylate, 118
Lucy (fossil), 368
Lumbar nerve, 625
Lumbar vertebrae, 685, 687
Lumbricus terrestris, 521
Lumpers, 433
Lung cancer, 180-82, 184, 741, 746, 893
Lungfish, 362
Lungs, 733, 735, 737, 739
Lung surfactant, human, 246, 739-40
Lunularia, 494
Luteinizing hormone (LH), 666-69, 674, 679
Lycoperdon, 477, 483
Lycoperdon perlatum, 487
Lycopod. *See* Club moss
Lycopodium, 497-98
Lycopodium selage, 496
Lyell, Charles, 378-79
Lyme disease, 384, 386, 450-51, 523, 864
Lymph, 727
Lymphatic duct, 727
Lymphatic system, 727-28
Lymph capillary, 727
Lymph node, 728, 793
Lymphocytes, 714, 716, 728, 792-94. *See also* B cells; T cell(s)
Lymph vessel, 727
Lyngbya, 416
Lynx, 610
Lyon, Gabrielle, 390
Lysergic acid diethylamide. *See* LSD
Lysosomal storage disease, 72
Lysosome, 67, 71-73, 75, 94, 473, 790
Lysozyme, 122, 790

M

McCarty, Maclyn, 309-11
McClintock, Barbara, 28
McElroy, William, 118
MacLeod, Colin, 309-11
Macroevolution, 375-76
Macromolecule, 7, 35, 43-54
Macronucleus, 459
Macronutrient, 755
Macrophage, 61, 65, 179, 790-94
Mad cow disease, 77-79
Maggot, 353
Magnaporthe grisea, 485
Magnesium, 756
Magnesium sulfate, 777
Magnetic resonance imaging (MRI), brain, 624, 626
Magnetism, navigational cue, 817
Magnolia, 567
Maiasaur, 366
Maidenhair tree, 499-500
Major groove, DNA, 312
Malaria, 461, 716, 872, 886
eradication, 462
history, 462
sickle cell disease and, 411-12
symptoms and course of infection, 462
Malayan pit viper, 374, 387
Male infertility, 239
Male reproductive system, human, 196-97
Malic acid, 132, 160, 162
Malignant tumor, 180
Malleus, 653
Malnutrition, 249
Malpighian tubule, 753, 778-79
Malthus, Thomas, 380-81
Maltose, 43, 751
Mammal, 510
Mammalia, 525-28
Mammary gland, 527, 603
Mammoth, 392, 419
Manatee, 527, 882
Mandible, 686
Mandrill, 527
Manganese, 756, 907
Mangrove swamp, 146, 881-82
Mangrove tree, 550
Manila hemp, 536-37
Manna lichen, 484
Mannitol infusion, 97
Mantle, 519-20
MAO. *See* Monoamine oxidase
Maple, 551, 571, 857
Maple syrup urine disease, 269
Map sense, 817-19
Maranta, 582-83
Marasmus, 758
Marattia, 498
Marfan syndrome, 269, 276
Margulis, Lynn, 79
Marijuana, 21, 584
Marine biome, 776, 881-84
Marine iguana, 771-72
Marine saltern, 452
Marrow cavity, 689
Marshall, Barry, 20, 750, 766
Marsupial, 231, 293, 366-67, 381, 527
Martin, 828
Mascignia macroptera, 538
Mash, 139
Mass, 909-10
Mass extinction
causes, 417
through geologic time, 418-19

Mast cell, 791, 800, 802-3
"Master control gene," 212
Maternity plant, 557
Mating, nonrandom. *See* Nonrandom mating
Mating behavior, 814-15, 824-25, 829-32
  primate, 831-32
Mating type, 195-96, 291, 662
Matrix
  connective tissue, 594
  mitochondrial, 128
Maturation, 196
Mauritius, 374-75
Maxilla, 686
Maxwell, James, 147
Meadow tree frog, 829
Measles, 384-85
Mechanical digestion, 760
Mechanoreception, 644-45, 652-56
Medicinal plant, 504
Medulla oblongata, 626-27, 649
Medusa, 517
Mefloquine, 462
Megabat, 398-99
Megagametophyte, 557, 561, 567
Megakaryocyte, 717, 719
Megasporangia, 557, 560, 562
Megaspore, 557, 562-67
Megaspore mother cell, 563-67
Meiosis, 192, 196, 198-201, 270
  compared to mitosis, 199
  Mendel's laws and, 265-66
Meiosis I, 198-201, 206, 266, 272, 298-99
Meiosis II, 198, 201-2, 206, 266, 272, 298-99
Meissner's corpuscle, 656
Melanin, 267, 577
Melanocyte, 598, 601
Melanocyte-stimulating hormone (MSH), 668-69, 679
Melanoma, 183, 335
Melatonin, 675, 677, 679
Membrane, precursors on prebiotic earth, 357
Membrane protein, 85-87
Memory, 278, 627, 630
Memory cell, 794-95, 797, 803
Mendel, Gregor, 262-66, 269-71
Mendelian trait
  altered Mendelian ratios, 272-77
  humans, 267-69
  unusual, 267
Mendel's first law, 262-67
Mendel's second law, 269-72
Meninges, 622
Meningitis, 633
Menkes disease, 33, 293
Menopause, 232, 670
Menstrual cycle, 674-75
  irregular, 239
Menstruation, 177, 197
Mental retardation, 246-48
Menthol, 537
Mercury, 868, 897-98
Meristem, 534-35, 547
Merozoite, 462
Merychippus, 695
Mesenteric artery, 725
Mesentery, 757
Mesoderm, 220-22, 511, 620
Mesoglea, 516-17
Mesohyl, 515
Mesophyll, 543-45
Mesophyll cell, 151, 158, 160-61
Mesozoic era, 357, 359, 365-66
Messenger RNA (mRNA), 70, 329
  processing, 330-31
  translation, 331-37
Meta-analysis, 27
Metabolic cycle, 115
Metabolic pathway, 114-15
  activation, 123
  inhibition, 122-23
Metabolic potential, 138
Metabolic rate, 770-71
Metabolism, 8-9, 114-17
  control, 122-23, 669-74
  origin, 357-58
  whole-body, 138
Metacarpal, 686
Metacentric chromosome, 296
Metamerism, 521
Metamorphosis, 512-13, 664-66
Metaphase
  meiosis I, 199-201, 266, 272
  meiosis II, 201-2, 266, 272
  mitotic, 171, 173-74
Metapopulation, 838
Metastasis, 181
Metatarsal, 686
Metatheria, 527
Metazoa, 400
Metchnikov, Ilya, 234
Meteor impact, 417-19
Meteorite, 353, 359
Methane, 34-38, 54, 117, 140, 446, 451, 896
Methanobrevibacter ruminantium, 451
Methanogen, 54, 140, 451
Methanosarcina, 451
Methanotrix, 451
Methotrexate, 248
Methoxychlor, 21
Methylation, of genes, 589
Methylene blue, 443
Methylprednisolone, 624
Metric conversion, 910
Metric/English conversions, 909
Meyerowitz, Elliot, 564
Microbat, 398-99
Microbial ecology, 441-42
Microbiologist, 428
Microcystis aeruginoda, 440
Microdeletion, 301
Microevolution, 375-76, 398
Microfilament, 97-98, 100-101
Microgametophyte, 557, 560-61
Microgravity, 580
Micrometer, 63
Micronucleus, 459
Micronutrient, 755-56
Microsatellite, 319
Microscope, 60-63
Microspora, 461
Microsporangia, 557, 560, 562
Microspore, 557, 562-63
Microspore mother cell, 563, 566
Microsporum canis, 486
Microsurgery, 718
Microtubule, 74, 97-100, 170-73, 175
Microvilli, 83, 762-64
Midbrain, 620, 625-27
Middle ear, 653
Midge, 894
Midgut, 753
Midrib, 542, 545
Miescher, Friedrich, 308, 311
Migration, 408, 412, 754, 817-19, 873
Mildew, 476
Milk, 69-71, 527, 667, 798
Milkweed, 847
Millar, Andrew, 576
Miller, Stanley, 353-54
Millet, 161
Millimeter, 63
Millipede, 522
Milstein, Cesar, 804
Mimicry, 479-80
Mimosa, 543, 581-82
Mineral, 756
  polymerization template on prebiotic earth, 355
Mineralocorticoid, 670-72, 679
Mining, 897
Minipill, 251
Minisatellite DNA, 342
Mink, 666
Minor groove, DNA, 312
Minute chromosome, 304
Miocene period, 361
Miohippus, 695
Miscarriage, 245
Mismatch repair, 319
Missense mutation, 341, 343
Mistletoe, 550
Mitchell, Peter, 133
Mite, 522-23
Mitochondria, 67-68, 71, 73, 75, 125
  DNA, 79-81, 322, 342, 399, 426, 436, 818
Mitochondrial membrane, 128, 133-36
Mitosis, 168-74
  compared to meiosis, 199
Mitotic aneuploidy, 298
Mitotic spindle, 170-75, 227
Mitral valve. *See* Bicuspid valve
Mitral valve prolapse, 720
Mixotroph, 445
Mobbing behavior, 822
Mockingbird, 641
Mode of inheritance, 267-68
Mold (fossil), 502
Mole, 527
Mole crab, 881
Molecular clock, 398-99
Molecular evolution, 395-400
Molecular formula, 37
Molecular methods, studying Moneran diversity, 442-43
Molecular tree diagram, 397-99
Molecular weight, 35
Molecule, 7, 34-35
Mollusca, 510-11, 519-20, 683, 708
Molting, 664, 683
Molting hormone, 666
Molybdenum, 890, 907

Monarch butterfly, 666, 847
Monera, 14, 430, 439-54, 913
  classification, 443
  distinguishing bacteria from archaea, 444-45
  energy acquisition, 445
  importance, 452-54
  studying diversity among, 442-43
Mongoose, 8, 828
Monilinia, 479
Monkey, 527
Monoamine oxidase (MAO), 281
Monoamine oxidase (MAO) inhibitor, 618
Monoclonal antibody technology, 804-6
Monocot, 541, 544, 549, 568-69, 572
Monocyte, 714, 716, 719
Monod, Jacques, 326-27
Monogamy, 830-31
Monohybrid, 265
Monohybrid cross, 266-68
Monomer, 43
Mononucleosis, 716
Monosaccharide, 43
Monosomy, 298-99
Monostroma, 464
Monotreme, 366, 527
Monounsaturated fatty acid, 45
Monozygotic twins, 10, 223, 255, 279-80
Monster beech, 377
Moose, 862, 876, 890, 907
Morbillivirus, 806
Morchella esculenta, 481
Morel, 481
Morgan, Thomas Hunt, 288
Mormon tea, 500
Morning glory, 579
Morphine, 504, 555, 634
Morphogen, 215
Mortality, 838, 840
Morula, 218-20
Mosaic, 298
Mosaic catastrophism, 376
Mosquito, 462, 523, 753, 814
Moss, 193, 493-94, 560, 856, 877
Moth, 567-68, 645
  salt-acquiring, 40
Motif, 327
Motion sickness, 655-56
Motivation, 627
Motor molecule, 97-98
Motor neuron, 611, 613, 623, 698
Motor response, 610
Motor unit, 699-700
Mountain gorilla, 414, 832
Mountain sickness, 745
Mount Pinatubo, 896
Mount St. Helens, 842-43, 857-58
Mourning gecko, 403
Mouse, 230, 666
  tricolored, 213
Mouth, 736, 757, 759-60, 765
Mouthpart, 522
Movement
  biological, 689-700
  musculoskeletal system and, 682-83
M phase, 171
MPTP, 635
MRI. *See* Magnetic resonance imaging
mRNA. *See* Messenger RNA
MSH. *See* Melanocyte-stimulating hormone
Mucigel, 546
Mucous membrane, 736, 790
Mucus, 741, 760-62
Mud puppy, 733
Mule, 413
Mullis, Kary, 320
Multicellular organism, 7, 10, 14, 60, 361-62, 466, 508
  origin, 510
Multifactorial trait, 276-80
Multiple alleles, 272-73, 277
Multiple fission, 459-60
Multiple sclerosis, 86, 806
Multiple vaccine, 803
Mumps, 301
Murchison meteorite, 353, 359
Musa textilis, 537
Muscle
  contraction, 695-99
  digestive tract, 759
  effect of exercise, 701-3
  invertebrate, 692, 695
  lactic acid fermentation, 139-40
Muscle cell, 102
  types, 689-93
Muscle fasciculi, 694, 696
Muscle fiber, 596, 694, 697-98
  types, 703
Muscle spindle, 701-2
Muscle tissue, 594, 596-98
Muscle tone, 701-2
Muscular dystrophy, 102, 293, 335, 343, 695
Muscular system, 599-600
  biological movement, 689-700
  working with bone, 700-704
Musculoskeletal plasticity, 682
Musculoskeletal system, 681-704
  functions, 684-85
  movement and, 682-83
  response to stimuli, 682
  vertebrate, 684-87
Mushroom, 476, 482-83, 485
Musk oxen, 822
Mussel, 520, 751, 881
Mustard weed, 564-65, 580, 588
Mutagen, 341
Mutant, 265
Mutation, 8, 265, 338-44, 412
  allele frequencies and, 410
  causes, 339-41
  effects, 338
  molecular clocks, 398-99
  natural protection against, 344
  types, 341-44
Mutation rate, 339, 342, 398-99
Mutualism, 434
Myasthenia gravis, 618, 633, 802
Mycelium, 477
Mycology, 476
Mycoplasma, 442
Mycorrhizae, 485, 550
Mycosis, human, 486-88
Mycosphaerella graminicola, 485
Myelin, 86, 597
Myelin sheath, 598, 612, 614-15, 698
Myoepithelial cell, 598-99
Myoepithelium, 596
Myofibril, 694-98
Myofilament, 695-96
Myoglobin, 48, 703, 735
Myosin, 48, 100, 102, 488, 597-98, 692, 694, 696-99
Myotonic dystrophy, 269, 343
Myriapod, 521-23
Myrtle warbler, 837
Myxedema, 669
Myxomycota, 470-72
Myxozoa, 461

N

NAD$^+$, 119-20, 136
  from alcoholic fermentation, 137-39
  use in glycolysis, 129-31
  use in Krebs cycle, 132-33
NADH, regulation of phosphofructokinase, 137
NADP$^+$, 119-20, 154-55
NADPH, 158-60
Nails, 603
Naked mole rat, 820
Naked seed, 498-501
Namib ant, 774-75
Nanochlorum eukaryotum, 64
Nanometer, 63
Nasal bones, 736
Nasal cancer, 184
Nasal cavity, 737
Nastic movement, 581-82
Natural birth control, 250
Natural disaster, 842
Natural history museum, 427
A Naturalist's Voyage on the Beagle, 380
Natural selection, 11-12, 374, 379-83, 387, 403, 410-12
  types, 410-11
Nature-nurture controversy, 279-81
Nautiloid, 362
Naval stores, 501
Navigation, 817-19
Neanderthals, 368-69
Nearsightedness, 651-52
Necator americanus, 519
Necrosis, 178
Nectar, 567, 773
Needham, John, 352, 371
Neem tree, 504
Negative feedback, 122-23, 137, 663, 670
Negative reinforcement, 816
Neisseria gonorrhoeae, 450
Nematocyst, 516
Nematoda, 510, 518-19
Nemertena, 510
Neonatology, 249
Nephila clavipes, 49
Nephridia, 520, 778-79
Nephridiopore, 778-79
Nephron, 780-84
Nephron loop, 782-84
Nephrostome, 778-79
Neptunism, 376
Nereis virens, 521
Neritic zone, 883-84
Nerve cell, 57, 61, 93-94
Nerve collar, 520
Nerve cord, 525, 620
Nerve fiber. *See* Axon

Nerve gas, 123, 617
Nerve growth factor (NGF), 178
Nerve impulse, 611-16
Nerve net, 518, 619
Nervous system, 599-600, 609-36, 661-62
  development, 620-21
  disorders, 631-35
  evolution, 610-11
  invertebrate, 619-20
  organization, 621-22
  protection, 635
  vertebrate, 620-22
Nervous tissue, 594, 596-97
Net primary production, 863
Netted leaf veins, 543-44
Neural groove, 228
Neural integration. *See* Synaptic integration
Neural tube, 226, 228, 620-21
Neural tube defect, 228, 278, 621
Neuroendocrine control, 666-67
Neuroendocrine-immune interactions, 677
Neurofibromatosis I, 269, 339
Neuroglia, 596
Neuromuscular junction, 696, 698
Neuron, 596-98, 609-11
Neurosecretory cell, 661
Neurospora crassa, 291, 481
Neuroticism, 278
Neurotoxin, 466
Neurotransmitter, 93-94, 597, 616-17, 631
  carbon monoxide and nitric oxide, 36
  disposal, 617
  imbalances, 618, 633
  types, 618
Neurotransmitter receptor, 634
Neurulation, 226, 620
Neustonic algae, 463
Neutron, 32-34
Neutrophil, 714, 716, 719, 790-91
Newborn, 230-31
Newt, 210, 213
Newton, Isaac, 147
New World hookworm, 519
New World monkey, 527
NGF. *See* Nerve growth factor
Niacin, 120, 755
Niagara Escarpment, 860-61
Niche, 836-38
Nicholas II, Tsar of Russia, 294, 308, 322
Nickel, 184
Nicotinamide adenine dinucleotide. *See* NAD^+
Nicotinamide adenine dinucleotide phosphate. *See* $NADP^+$
Nicotine, 97, 618
  addiction, 104-5
Nicotinic receptor, 105
Night blindness, 755
9 + 2 microtubule array, 99-100
Nitrate, 140, 184, 865, 867
Nitric oxide, 36
Nitrifying bacteria, 865, 867
Nitrite, 140, 184, 340-41, 865, 867
Nitrogen, 35
Nitrogenase, 865
Nitrogen cycle, 865, 867
Nitrogen fixation, 452-53, 498, 550, 865, 867
Nitrogenous base, 51, 311, 339, 356
Nitrogenous waste, 776-79
Nitrogen oxides, 894, 896
Noctiluca, 466
Noctiluca scintillans, 467
Nocturnal enuresis, 785
Node, 535, 541
Node of Ranvier, 612, 614-15
Nomadic people, 408
Noncoding strand, 328-29
Noncompetitive inhibition, 122-23
Nondisjunction, 298-99
Nonessential nutrient, 755
Nonoxynol-9, 250
Nonpolar covalent bond, 38
Nonrandom mating, 406-8, 412
NonREM sleep, 627-28
Nonsense mutation, 341, 343
Nonshivering thermogenesis, 772
Nonspecific defense, 790-92
Norepinephrine, 618, 632, 634, 670
Norplant, 250-51
Nose, 646-47, 735-36
Nosema apis, 461
Nosema bombicus, 461
Nostril, 736
Notochord, 226, 524-25, 620
Novelty seeking, 281
Nowell, Peter, 286
Nuclear envelope, 67-68, 70
Nuclear matrix protein, 327
Nuclear membrane, 174
Nuclear pore, 67, 70
Nuclear reactor, 342, 442, 892
Nuclear transplant, 211, 214
Nuclear weapon, 893
Nuclear winter, 146
Nuclease, 764
Nucleic acid, 50-54
  viral, 75
Nucleoid, 65-66, 447
Nucleolus, 67-68, 71, 173
Nucleoside analog, 799
Nucleosome, 313, 317
Nucleotide, 311-12
Nucleus
  atomic, 32
  cell, 61, 66-68, 70, 75
Nude mouse, 25-26
Nuptial flight, 820
Nutrient, 751
  deficiency, 756
  energy transfer to ATP, 128-29
  teratogenic, 248-49
Nutrition, human, 755-57
Nutritional label, 33

O

Oak, 9, 551, 856-57
Oak ape, 367
Oak-hickory forest, 873
Oberrothenback, Germany, 893
Obesity, 26, 138, 185, 281, 712-13, 766
ob gene, 26
Obligate aerobe, 445
Obligate anaerobe, 445
Obligate photoperiodism, 584
O'Brien, Chloe, 244
Observation, 19-20, 23, 28-29, 903
Occipital lobe, 628, 686
Occupational teratogens, 249
Ocean, 883-84
  iron content, 903
  pollution, 899-904
Oceanic zone, 883-84
Ocelli, 518, 644
Octet rule, 35
Octopus, 510, 513, 519-20, 645
Ocular albinism, 293
Odorant molecule, 646-47
Oenothera claviformis, 160
Oil of wintergreen, 588
Oil reserves, 452
Oils, 44
Oil spill, 454, 842, 890-91
Okazaki fragment, 319
"Oldest old," 210, 235
Old field succession, 856-57
Old World monkey, 527
Olfactory bulb, 646, 649
Olfactory cortex, 647
Olfactory epithelium, 646-47, 649
Olfactory nerve, 647
Olfactory receptor cell, 646
Olfactory sense cortex, 649
Olfactory tract, 649
Oligocene period, 361
Oligochaete, 521
Oligodendrocyte, 614
Oligosaccharide, 43-45
Oligotrophic lake, 880
Omasum, 441
Omega-3 fatty acid, 45, 47
Omega-6 fatty acid, 45
Ommatidia, 644-45
Omnivore, 750
Oncogene, 183-84
Oncologist, 185
One-child program, 849
One-opening digestive system, 753
Onion bulb, 544
Onion root tip, 173
On the Origin of Species, 382
Onychophora, 400
Oocyte, 191, 196-97, 201-6, 217-18
Oocyte banking, 244
Oocyte donation, 244
Oogenesis, 204-6
Oogonia, 204, 464, 466
Oomycota, 470
Oparin, Alexander I., 354
Open circulatory system, 708, 710
Open system, 112, 858
Operant conditioning, 815-16
Operculum, 734
Operon, 327
Ophiostoma novo-ulmi, 480, 485
Ophrys, 556
Opiate, 634
Opium, 634
Opium poppy, 504, 555
Opossum, 527
Opossum shrimp, 899
Opportunistic infection, 254, 435, 486-88, 798
Opposite leaf pattern, 542, 544
Opsin, 650, 652
Optic nerve, 649-50

Oral cancer, 184
Oral end, 511-13
Oral vaccine, 805
Orangutan, 395-97, 513, 527, 832
Orb weaving spider, 829
Orchid, 162, 550, 556, 567
Order (taxonomy), 12, 14, 426, 433
Ordovician period, 359, 362, 417
Organ, 5, 7-8, 509, 595
Organelle, 5, 7-8, 60, 62, 66, 69-75
Organic halide, 900
Organic molecule, 35, 43-54
handedness, 358-59
Organism, 5
Organization, 5-8
Organogenesis, 216, 226
Organotroph, 445
Organ system, 5, 7, 509, 594-95, 599-600
Organ transplant, 84
Orgasm, 197-98
Orientation, 817-19
Origin of muscle, 700, 702
Ornithologist, 428
Orthostatic hypotension, 724
Osculum, 515
Osmoconformer, 776
Osmoreceptor, 90-91, 669, 784
Osmoregulation, 776-77
Osmoregulator, 776
Osmosis, 88-91, 95
Osmotic pressure, 784
Ossification center, 691
Osteichthyes, 525
Osteoarthritis, 700
Osteoblast, 595, 597
Osteoclast, 595, 597
Osteocyte, 595, 597, 688-89
Osteon, 688-89
Osteonic canal, 688-89
Osteoporosis, 670, 690-91
Osteoprogenitor cell, 595
Ostia, 515
Ostrich, 379-81, 822-23, 835
Otolith, 655-56
Otter, 48
Outcrossing species, 567
Outer ear, 653
Oval window, 653
Ovarian follicle, 674
Ovary, 197, 661, 678-79
plant, 563, 566
Overpopulation, 836
Overweight, 766
Ovulation, 204-6, 239, 665-66, 674
Ovulation predictor test, 239
Ovule, 563, 566, 568
Ovum, 204-6, 217-18
Owl, 686-87, 754, 822
Oxaloacetic acid, 132-33, 161
Oxidation, 117
Oxidative phosphorylation, 129-30, 133-36
Oxygen, 35
cellular respiration, 126
from photosynthesis, 146-47, 164
transport in blood, 715, 742-43
Oxygen debt, 139
Oxygen-free radical, 72
Oxyhemoglobin, 715
Oxyluciferin, 118
Oxytocin, 660, 664, 666, 668-69, 679
Oyster, 510, 882, 899
Oystercatcher, 814
Ozone/ozone layer, 148, 164, 850, 892, 894-96

P

p53 protein, 168, 186
P680, 153
Pacemaker, 720-22
electronic, 712
Pacific yew, 499
Pacinian corpuscle, 656
Pack rat, 828
Pain perception, 634, 656
Palaeolyngbya, 416
Paleobotany, 502
Paleocene period, 361
Paleontology, 390-91
Paleozoic era, 357, 359, 361-65
Palisade mesophyll, 543, 545
Palmately compound leaf, 542
Palmellopsis, 464
Palolo worm, 521
Palp, 522
Pancreas, 661, 672-73, 678-79, 757, 764-65
transplant of cells from, 604-5
Pancreatic amylase, 762-64
Pancreatic cancer, 185
Pancreatic duct, 765
Panda, 752
Pando, 557
Pangaea, 391, 417-18
Panspermia, 353
Panting, 774
Pantothenic acid, 755
Papermaking, 492, 500-501, 505
Papilio xuthus, 642
Papillary muscle, 721
Parallel leaf veins, 543-44
Paralysis, 624-25
Paralytic shellfish poisoning, 467, 886, 903
Paramecium, 89-90, 194, 459-61, 753
Paramecium caudatum, 459
Paramylon, 466
Parapodia, 521
Parasite, 845-47
Parasitism, 435
Parasitoid, 435
Parasympathetic nervous system, 621-22, 630-31
Parathyroid gland, 661, 663, 670, 678-79
Parathyroid hormone (PTH), 663, 670-71, 679
Parenchyma, 535-36, 539
Parental (P) generation, 265
Parental investment, 830-31
Parental type, 287-89
Parent-infant interaction, 814
Parietal bone, 686
Parietal cell, 760-61
Parietal lobe, 628
Parkinson disease, 605-6, 618, 635
Parrot fever, 449
Parsimony analysis, 399
Parthenogenesis, 192, 194, 207, 290
Partial parthenogenote, 192, 207
Parturition, 216
Passenger pigeon, 419
Passion vine, 581
Passive aging, 233
Passive immunity, 798
Passive transport, 91
Pasteur, Louis, 139, 359, 371
Patella, 686
Paternal behavior, 823
Paternalism, doctor-patient relationship, 6
Pauling, Linus, 338-39
PCB. *See* Polychlorinated biphenyl
PCP, 634
PCR. *See* Polymerase chain reaction
Peach, 569
Pea plant, 544, 577
linked genes, 286-88
Mendel's experiments with, 262-66, 269-71
Pear, 537, 556
Peat, 495, 894
Pebrine, 461
Pecking order, 828
Pectin, 98
Pectoral girdle, 685-86
Pedicellariae, 524
Pedigree, 268, 270-71
Pelagic zone, 883
Pelletier, Pierre Joseph, 462
Pelvic cavity, 686
Pelvic girdle, 685-86
Pelvic inflammatory disease (PID), 253
Pelycosaur, 365
Penetrance, 275, 277
Penguin, 771
Penicillamine, 248
Penicillin, 28, 123, 386, 483, 784, 802
Penicillium, 483
Penicillium camembertii, 483
Penicillium notatum, 483
Penicillium roquefortii, 483
Penis, 196-97, 205, 780
Pentaradial symmetry, 524
PEP. *See* Phosphoenolpyruvate
Pepper, 569
Peppered moth, 410
Peppermint, 537
Pepsin, 760
Pepsinogen, 760-61
Peptic ulcer, 18, 20, 440-41
Peptidase, 763
Peptide bond, 48
Peptide hormone, 588-89, 662-63
Peptidoglycan, 64, 66, 444
Perception, 641, 646
Peregrine falcon, 854, 868
Pereissodactyla, 527
Pericardium, 718
Pericarp, 569
Pericycle, 548-49
Periderm, 551
Period, geological, 359
Periodic table, 32-34, 911
Periodontal ligament, 760
Periosteum, 689, 691
Peripheral nervous system, 618, 621
human, 630
Periplast layer, 466
Peristalsis, 759-60, 762
Peritubular capillary, 780, 782
Periwinkle, 504
Permafrost, 878
Permanent wave, 51
Permian period, 359, 363, 365, 417-19

Pernicious anemia, 716, 755, 802
Peronospora hyoscyami, 470
Peroxisomal disorder, 74
Peroxisome, 67-68, 72-75
Persistent vegetative state, 6
Personality, 610, 636
Pertussis, 385, 449-51, 716
Pesticide, 23, 898
PET. *See* Positron emission tomography
Petal, 563-65
Petiole, 541-42, 545
Petrification, 502
Petrified wood, 390, 392
Peziza, 478
Pfiesteria piscimorte, 433-34
PGAL. *See* Phosphoglyceraldehyde
pH, 41-42
  blood, 41, 784
Phaeophyta. *See* Brown algae
Phage. *See* Bacteriophage
Phagocyte, 790
Phagocytic vacuole, 459
Phagocytosis, 790-91
Phalanges, 686
Phallus impudicus, 483
Pharyngeal gill slits, 525
Pharyngitis, 746
Pharynx, 736-37, 753, 757, 759-60, 765
Phelloderm, 552
Phenocopy, 276-77
Phenotype, 265
Phenotypic ratio, 266
p-Phenylenediamine, 341
Phenylketonuria (PKU), 269, 412
Pheromone, 660, 662, 666, 814, 821, 831
Philadelphia chromosome, 286, 297, 304
Phloem, 497, 534, 539
Phocomelia, 246, 276
Phosphate bond, high-energy, 117-19
Phosphate rock, 866-67
Phosphoenolpyruvate (PEP), 129-31, 161-62
Phosphofructokinase, 130, 137
Phosphoglyceraldehyde (PGAL), 130-31, 158-59
Phosphoglycerate, 159-60
2-Phosphoglycerate, 130-31
3-Phosphoglycerate, 130-31
Phosphoglycolic acid, 160
Phospholipid, 45, 85-86
Phospholipid barrier, 88
Phospholipid bilayer, 85-87
Phosphorus, 35, 756, 900
Phosphorus cycle, 866-67
Phosphorylation, 119
Photochemical smog, 892-93
Photon, 111, 147-49
Photonasty, 581-83
Photoperiodism, 582-86
Photophosphorylation, 155
Photoreactivation, 319
Photoreception, 644-45, 648-52
Photoreceptor, 642-43
Photorespiration, 160-61
Photosynthate, 160, 162
Photosynthesis, 14, 73-75, 123-24, 145-65, 463, 865-66
  bacterial, 156-57
  early thoughts about, 148-49
  efficiency, 160
  evolution, 146-47, 163-64
  factors controlling, 163
  importance, 146-47
  light-dependent reactions, 153-59
  light-independent reactions, 153, 158-59
Photosystem, 151-53
Photosystem I, 153-58
Photosystem II, 153-58
Phototroph, 445
Phototropism, 578-79, 581, 586
Photuris versicolor, 118
Phragmoplast, 74, 173
Phycobilin, 469
Phycocyanin, 150, 469
Phycoerythrin, 150, 469
Phycologist, 428
Phyla, 433
Phylogenetic species, 436
Phylogenetic tree, animals, 511
Phylogeny, 390, 397-99
Phylum, 12, 14, 426
Physalia, 517
Physarum, 471
Phytochrome, 584-86
Phytoflagellate, 460
Phytoplankton, 458, 463, 467, 862-63, 879, 884, 903
Phytopthora infestans, 470
Picea glauca, 501
PID. *See* Pelvic inflammatory disease
Piebald cat, 273
Piedraia hortae, 487
Pig, 527, 605, 821-22
Pigeon, 817
Pigment, 149-51, 464, 579, 708-10
Pili, 440, 446-47, 454
Pillbug, 522
Pilot experiment, 24
Pinacocyte, 515
Pin cherry, 857
Pine, 499, 551, 560-62, 856-57, 876
Pineal gland, 627, 661, 675-76, 678-79
Pineapple, 162
Pine needle, 499
Pingelapese blindness, 409
Pinguicula grandiflora, 538
Pinnately compound leaf, 542-43
Pinocytosis, 459
Pinus longaeva, 499-500, 553
Pinworm, 519
Pioneer species, 856-58
Pirarucu fish, 873
Pistil, 563
Pitch, 876
Pitcher plant, 545
Pith, 541
Pituitary dwarfism, 667
Pituitary giant, 667
Pituitary gland, 627, 661, 666-69, 678-79
Pituitary tumor, 240
Pit viper, 374, 387
PKU. *See* Phenylketonuria
Placebo, 25, 29
Placenta, 216, 223, 226, 228, 230
Placental mammal, 367, 527
Planarian, 517-18, 778
Plankton, 458, 466, 860
Plant
  adaptation, 11
  characteristics, 492
  development, 572, 580
  evolution, 493, 503-4
  form and function, 533-53
  growth, 580
  life cycles, 555-74
  medicinal, 504
  parts of plant body, 540-50
  preserving diversity, 504-5
  primary growth, 534-40
  response to stimuli, 575-90
  secondary growth, 534-35, 550-52
  space travel, 580
  value of diversity, 492
Plantae, 14, 425, 429, 492-505, 914
Plant breeding, 314, 414-15, 558-59
Plant cell, 66, 68
Plant clock, 576
Plantlet, 557-59
Plant pathogen, 485
Plant pioneers, 484
Plasma, 594, 711-12, 715
Plasma cell, 67, 794-95, 797, 803
Plasmid, 447-49
Plasmodesmata, 96-98
Plasmodium, 461-62
Plasmodium (slime mold), 470-71
Plasmopara viticola, 470
Plastic, ocean pollution, 903-4
Plastid, 151-52
Platelets, 595, 714, 717, 719
Plate tectonics, 417-18
Platyhelminthes, 510-11, 517-18, 619-20, 644, 706-9, 733, 753
Platypus, 639-40, 657
Pleiotropy, 276-77
Pleistocene epoch, 361, 367
Pleurisy, 746
Pliny the Elder, 504
Plumule, 568-70
Pluripotency, 210
Pluripotent stem cell, 717-19
Plutonium, 893
Pneumatophore, 517, 550
Pneumocystis carinii, 459, 798
Pneumonia, 234, 746, 798-99
Poinsettia, 544, 584
Point mutation, 341
Polar bear, 209, 464, 646, 862, 878
Polar body, 204
  first, 206, 219
  second, 206, 219
Polar covalent bond, 38
Polarized membrane, 612
Polar nucleus, 566-68
Polio, 633, 682, 704
Polio vaccine, 682, 805
Pollen, 499, 563
Pollen grain, 560, 562, 566
Pollen sac, 566
Pollen tube, 561-62, 566, 568
Pollination, 560, 562, 566-68, 588
Pollinator, 556, 567-68, 573
Pollution, lichens as indicators, 484
Polyandry, 830-31
Poly A tail, 330
Polychaete, 521, 882
Polychlorinated biphenyl (PCB), 21, 454, 897-99
Polyclonal antibody response, 795
Polycyclic aromatic hydrocarbon, 891, 897

Polycystic kidney disease, 269
Polydactyly, 269, 275
Polygamy, 830
Polygenic trait, 276-77
Polygynandry, 830-31
Polygyny, 830-31
Polymer, 43
Polymerase chain reaction (PCR), 320-21, 453-54
Polymerization, prebiotic earth, 355
Polyp (cnidarian), 517
Polypeptide, 46, 52
Polyploidy, 297-98, 414
Polysaccharide, 44, 46
Polytrichum, 493
Polyunsaturated fatty acid, 45
Pome, 569-70
Pompe disease, 72
Pond, 879-81
Pongidae, 527
Pons, 626-27
Population(s), 7, 9, 404, 835-50
Population bottleneck, 409-10
Population decline, amphibians, 836, 850
Population density, 838
Population dispersion, 838
Population genetics, 404-15
Population growth
  human, 847-50
  potential, 840-42
Population size, regulation, 842-45
Porifera, 510-11, 515-16
Porocyte, 515
Porphyria, 269, 276
Porpoise, 643
Porter, Amy, 594, 607
Portuguese man-of-war, 516-17
Positive feedback, 123, 664
Positive reinforcement, 816
Positron emission tomography (PET), brain, 624, 626, 646
Postcoital test, 240
"Posterior," 512-13
Posterior pituitary, 666-69
Postmating reproductive isolation, 413-14
Postnatal growth, 230-33
Post-polio syndrome, 704, 805
Postreproductive individuals, 841-42
Postsynaptic neuron, 616-17
Potassium, 38, 756
  deficiency or excess, 38
  glomerular filtrate, 784
  nerve impulse transmission, 612-13
Potassium-40, 393
Potato, 10, 536, 541-42, 552
Potato blight, 470
Potato spindle-tuber disease, 77
Potential energy, 112-13, 540
Potts, Sir Percival, 184
Poverty, 849-50
Powdery mildew, 481
Power, 909
Prairie, 877
Prairie dog, 828
Prairie vole, 665, 830
Prayer plant, 582-83, 587
Prebiotic soup, 353-55
Precambrian time, 357, 359-61, 508
Precipitation (water cycle), 865, 896
Predation, 842
Predator, 610, 812, 822, 845-47
Preembryonic stage, 216-21
  preembryo as research subject, 224-25
Preformation, 210
Pregnancy
  age at, 298
  problems, 245-50
  termination. *See* Abortion
"Pregnancy block," 666
Pregnancy test, 221, 805
Preimplantation genetic diagnosis, 244-45, 514
Premating reproductive isolation, 413
Premature infant, 245-46
Prenatal development, 168
  birth defects, 246-49
  human, 216-30
Prenatal test, 226, 286-87, 298, 849
Prereproductive individuals, 841-42
Presynaptic neuron, 616
Prey, 822, 845-47
Prickles, 541
Prickly pear, 531
Priestley, Joseph, 148
Primary consumer, 862, 864, 869
Primary growth, plant, 534-40
Primary hyperoxaluria type I, 74
Primary immune response, 795
Primary motor area, 628-29
Primary nutrient deficiency, 756
Primary oocyte, 204, 206
Primary producer, 752, 859, 862-63, 869
Primary sensory cortex, 628-29
Primary spermatocyte, 204-5
Primary structure, protein, 49, 52
Primary succession, 856-58
Primary visual cortex, 650
Primate, 367, 428-29, 527
Primate mating behavior, 831-32
Primitive streak, 226, 228
Primordial soup, 353-55
Principle of competitive exclusion, 845
Principle of superposition, 376
Prion, 77-79
Probability, 271
Proboscida, 527
Proboscipedia gene, 275-76
Procambium, 548
Producer, 8-9, 14, 859-62, 864
Product rule, 271
Proflavine, 341
Profundal zone, 879-80
Progenote, 357-58
Progeria, 233
Progesterone, 248, 672, 674-75, 679
Proglottid, 518
Proinsulin, 336
Project Life Lion, 806
Prokaryote, 14, 430-31
Prokaryotic cell, 14, 63-66
  eukaryotic cell versus, 69
Prolactin, 239, 667-68, 679
Promoter, 327
Pronucleus, 219
Prophase
  meiosis I, 198-99, 201
  meiosis II, 201-2
  mitotic, 171-74
Prop root, 550, 569
Prosimii, 527
Prostaglandin, 660, 664, 676-77
Prostate gland, 196-97, 675
  enlargement, 675
Prostate-specific antigen (PSA), 675
Protease, 48
Protease inhibitor, 799
Protective behavior, 813, 832
Protein, 46-50, 52, 54
  comparing sequences to determine evolutionary relatedness, 396-97
  conformation, 49, 52
  digestion, 751, 762-63
  energy from, 140-42
  folding, 49, 336-37
  functions, 46, 48
  relationship to gene, 312-13
  structure, 47-49
  translation, 331-37
Protein gradient, 215-16
Protein kinase, 488
Protein-phospholipid bilayer, 85-87
Protein starvation, 758
Proteoglycan, 594
Proterozoic era, 359
Prothrombin, 717
Protista, 14, 429-30, 457-73, 913-14
Protoctista, 430
Proton, 32-34
Protonephridia, 518
Proton gradient, 133-36, 142, 155
Protoplasm, 60
Protoplast fusion, 558
Protostome, 511-14
Prototheria, 527
Protozoa, 458-63, 708-9, 752-53
  classification, 459
  human pathogens, 461-63
  types, 460-61
Proventriculus, 754
Proximal convoluted tubule, 783-84
"Prune belly," 298
PSA. *See* Prostate-specific antigen
Pseudoautosomal region, Y chromosome, 291
Pseudocoel, 519
Pseudocoelomate, 511-12
Pseudomonas, 140
Pseudoplasmodium, 471
Pseudopod, 459-60, 692, 694, 790
Pseudostratified epithelium, 594, 597
Psilocybe mexicana, 478, 483
Psilophyte, 362
Psilotum, 497
Psilotum nudum, 496
Psylla, 435
Ptarmigan, 878
PTH. *See* Parathyroid hormone
Puberty, 660
Pubis, 686
Puccinia graminis, 485
Puccinia monoica, 479-80
Puddling, 40
Puffball, 477, 482-83, 486-87
Pulmonary artery, 720, 725
Pulmonary circulation, 725
Pulmonary edema, high-altitude, 745
Pulmonary embolism, 717

Pulmonary semilunar valve, 720-21
Pulmonary vein, 721, 725, 740
Pulp, tooth, 760
Pumpkin, 544
Punctuated equilibrium, 415-16
Punnett, R.C., 286
Punnett square, 266, 271, 273
Pupation, 666
Pupfish, 413
Pupil, 648-49
Purine, 313
Purkinje fiber, 722
Pus, 785, 791
Pussy willow, 557
Pyloric sphincter, 761
Pylorus, 760-61
Pyramid of biomass, 863
Pyramid of numbers, 863-64
Pyrenoid, 464-65
Pyridoxine, 755
Pyrimidine, 313
Pyrimidine dimer, 319
Pyrodinium, 466
Pyrrhophyta. *See* Dinoflagellate
Pyrus communis, 537
Pyruvic acid, 129-31, 137, 141-42

Q

Quadrat, 891
Quagga, 398-99, 419
Quaternary epoch, 361
Quaternary structure, protein, 49, 52
Queen, 290, 819-21
Queen cell, 821
Queen mandibular pheromone, 821
Quiescent center, 547
Quillwort, 497
Quinine, 462
Quinlan, Karen Ann, 6

R

Rabbit, 436, 754
  coat color, 273, 275
Raccoon, 876
Radial cleavage, 513-14
Radial symmetry, 511-13, 611
Radiation exposure, 892-93
Radicle, 546, 568-70, 572
Radioactive isotope, 33, 342, 393
Radioactive waste, 893
Radiograstrogram, 761
Radiolarian, 458, 460
Radiole, 521
Radiometric dating, 393
Radius, 686
Radon, 184, 893
Radula, 519-20
Ragweed pollen, 804
Rain forest, tropical, 872-76, 904-5
Rape suspect, 314
Raspberry, 569
Ray (fish), 525
Reabsorption, 778, 783-84
Reactant, 35
Reaction center bacteriochlorophyll, 156
Reaction center chlorophyll, 151-53
Reaction chain, 814-15
Reading frame, 332
Realized niche, 838
Reasoning ability, 278, 816
Receptor, 85, 94-95, 662
Receptor-mediated endocytosis, 94-95, 97
Receptor potential, 643
Recessive allele, 265
Recessive trait
  autosomal, 267-70
  sex-linked, 291-92, 294
Reciprocal translocation, 301
Recombinant DNA technology, 66, 332, 334, 447-48, 453, 678
Recombinant type, 287-89
Rectum, 197, 759, 764-65
Red algae, 361-62, 458, 463, 469, 882
Red blood cells, 64, 89, 100-102, 594-97, 707, 710, 714-16, 742
Red deer, 844
Red hair, 344
Redi, Francesco, 353
Red kangaroo, 879
Red marrow, 689, 692, 714-15, 719
Red snapper, 428
Red tide, 152, 466-67, 902
Reduction, 117
Reduction division. *See* Meiosis I
Redwing blackbird, 822
Redwood, 560
Reed, 879
Reeve, Christopher, 624-25
Reflex, 623
Reflex arc, 623
Reflex center, 627
Regeneration, 558
Regulatory enzyme, 122-23
Reindeer, 836
Reindeer moss, 493, 878
Reinforcement, 816
Relapsing fever, 449
Relative dating, 392
Releaser, 814
Releasing hormone, 666-67, 679
Religion, 370
REM sleep, 627-28
Renal artery, 725, 782
Renal cortex, 780-81, 784
Renal medulla, 780-81, 784
Renal pelvis, 780-81
Renal tubule, 780-81
Renal vein, 725
Renewal cell population, 179
Renin, 785
Replication. *See* DNA, replication
Replication fork, 317-18
Reproduction, 9-10, 192, 840
  asexual, 10, 14, 192-93, 195, 556-57
  endocrine control, 674-75
  sexual, 10, 14, 192-93
Reproductive behavior, 665-66
Reproductive diversity, 192-96
Reproductive individuals, 841-42
Reproductive isolation, 413-14
Reproductive success, 382-83, 822
Reproductive system, 599-600
  female, 197-98
  male, 196-97
Reptilia, 510, 525-26, 771
Repulsion of alleles, 287-89
Research embryo, 224-25
Research process, 27
Reserpine, 618
Residual air, 742
Resiliency of life, 907
Resin, 501
Resources, 838
Respiration, 732, 865-66
  control, 744-45
Respiratory chain, 127-28, 133-34, 142
  disruption, 135
Respiratory cycle, 744
Respiratory distress syndrome, infantile, 246, 739, 746
Respiratory pigment, 708-10
Respiratory system, 599-600, 731-47
  disorders, 741, 746
  human, 735-42
Respiratory tree, 524
Resting potential, 612-14
Restriction enzyme, 314, 334
Restriction point, 168, 171
Resurrection plant, 497
Rete mirabile. *See* Carotid rete
Reticular activating system, 627
Reticulum, 441
Retina, 648-50
Retinal, 650, 652
Retinitis pigmentosa, 293
Retinol. *See* Vitamin A
Retrovirus, 253, 326, 798
Reverse transcriptase, 253, 355-57, 798-99
Reversible reaction, 116
R group, 47-48
Rhacomitrium, 494
Rhea, 380-81
Rhesus monkey, 527
Rheumatic fever, 802
Rheumatoid arthritis, 700-701, 802
Rhine River, 900-901
Rhizobium, 452-53, 550, 865
Rhizoid, 493, 560-61
Rhizome, 497-98, 561
Rhizopus stolonifer, 478, 481
Rhodophyta. *See* Red algae
Rhodopseudomonas acidophila, 156-57
Rhodopsin, 644, 650, 652
Rhodospirillum rubrum, 443
Rhynchosporium secalis, 485
Rhythmicity center, 744-45
Rhythm method, 250-51
Ribbonworm, 510
Rib cage, 685-86, 742
Riboflavin, 120, 185, 755
Ribonuclease, 52
Ribonucleic acid. *See* RNA
Ribose, 51, 311, 326, 358-59
Ribosomal RNA (rRNA), 329-30, 442-43, 459
Ribosome, 65-71, 75, 329, 337, 447, 786
  translation, 333-35
Ribozyme, 329
Ribulose bisphosphate (RuBP), 158-59
Rice, 492, 589
Rice blast, 485
Richtersveld coloureds, 408
Right atrium, 711, 721
Right ventricle, 711, 721

Rigor mortis, 697
Rinderpest virus, 806
Ringworm, 476, 486
Ripening, 577-78
Ritalin, 618
River, 881, 899-901
River blindness, 872, 886
RMP-7, 97
RNA, 51
  DNA versus, 326
  messenger. *See* Messenger RNA
  processing, 330-31
  ribosomal. *See* Ribosomal RNA
  transfer. *See* Transfer RNA
RNA polymerase, 317-18, 327-28, 445
RNA primer, 317-18
RNA world, 355-57
Robertsonian translocation, 301
Rock layers, 376-79
Rockweed, 890
Rocky Mountain spotted fever, 523
Rod(s), 648-52
Rodentia, 426, 436, 527
Roe deer, 828
Romanov family, 308, 322
Root
  plant, 534-35, 546-50, 572
  tooth, 760
Root apical meristem, 535
Root canal, 760
Root cap, 535, 546, 548, 579-80
Root hair, 535, 548
Root nodule, 452-53, 550, 865
Root tip, 535, 580
Rosebush, 541
Rosette, 541
Rosin, 501
Ross, Sir Ronald, 462
Rough endoplasmic reticulum, 67-68, 70-71, 73
Round dance, 824-26
Round window, 653
Roundworm, 25, 177, 510, 518-19
Roux, Wilhelm, 210
Royal jelly, 820
rRNA. *See* Ribosomal RNA
r-selection, 844-45
RU 486, 252
Rubella, 249
Rubisco, 158-60
RuBP. *See* Ribulose bisphosphate
RuBP carboxylase/oxygenase. *See* Rubisco
Rugae, 760-61
Rule of tens, 863-64
Rumen, 441-42, 451, 754
Ruminococcus albus, 441
Runner (plant), 541
Runner's high, 634
Rush, 879
Rust fungus, 479-83

S

Sabin, Albert, 805
Saccharin, 184
Saccharomyces cerevisiae, 192, 477, 481, 488
Saccule, 655
Sac fungi. *See* Ascomycetes
Sacral nerve, 625
Sacrum, 685, 687
Saguaro cactus, 878
Sahel, 905-6
St. Anthony's fire, 482
St. Matthew Island, 836
Salamander, 734, 822
Salgado, Leonardo, 401
Saliva, 760
Salivary amylase, 760, 763
Salivary gland, 757
Salk, Jonas, 805
Sally Lightfoot crab, 769
Salmonella enteritidis, 443, 449-50
Salmonella typhi, 443
Salmonellosis, 450-51
Salt, 38
  as preservative, 451-52
Saltatory conduction, 614-15
Salt gland, 772
Salty taste, 647-48
Sample size, 24
Sand dollar, 510, 523-24, 845-46
Sand lizard, 10
Sandworm, 521
Sandy beach, 881
SA node. *See* Sinoatrial node
Sap, 539
Saprobe, 445
Saprolegnia, 470
Saprophyte, 476-77, 485
Saprozoic heterotroph, 459
Sapwood, 551
Sarcodina, 460
Sarcolemma, 694, 696, 698
Sarcomastigophora, 460
Sarcomere, 695, 697
Sarcoplasmic reticulum, 694, 696-97
Sarracenia, 545
Satellite (chromosome), 296
Saturated fatty acid, 45
Savanna, 877-78
Sawfly, 773
Scale
  pine cone, 560
  snake, 50
Scale-leaf spike moss, 497
Scallop, 692, 884
Scanning electron microscope, 62-64
Scanning probe microscope, 62, 64
Scapula, 686
Scavenger, 862
Scent gland, 603
Scent marking, 828
Scent trail, 642
Schierenberg, Einhard, 213
Schistosomiasis, 872, 886
Schizophrenia, 618
Schleiden, Matthias J., 61-62
Scholastic achievement, 278
Schwann, Theodor, 61-62, 371
Schwann cell, 597-98, 614-15, 635, 698
Schweitzer, Arlette, 242
SCID. *See* Severe combined immune deficiency
Scientific method, 18-23, 28-29
Sclera, 648-49
Sclereid, 536-37
Sclerenchyma, 536, 539
Scleroderma, 802
Sclerotinia sclerotiorum, 485
Scolex, 518
Scorpion, 522-23, 878
Scouring rush, 498
Scramble competition, 843
Scrapie, 77
Scrotum, 196-97, 205, 239
Scurvy, 755
Sea anemone, 10, 434, 510, 516-17, 611, 619, 683, 733, 882
Sea cucumber, 523-24
Sea grass, 882
Seal, 686-87, 772, 838
Sea level, 896
Sea lily, 510, 523-24, 861
Sea lion, 809
Sea nettle, 516
Sea otter, 890-91
Sea snake, 771
Seasonal affective disorder, 675-76
Seasonal change, 873
Seasonal forest, 872
Seasonal frugivore, 750
Seasonal ovulator, 665
Seasonal response, plants, 582-87
Sea squirt, 525
Sea star, 523-24, 861
Sea surface monolayer, 860
Sea turtle, 771, 845, 903
Sea urchin, 510, 523-24, 845-46
Sea wasp, 516
Seaweed, 469, 881
Sebaceous gland, 601, 603
Secondary consumer, 862, 864, 869
Secondary growth, plant, 534-35, 550-52
Secondary immune response, 795
Secondary nutrient deficiency, 756
Secondary oocyte, 204, 206, 217-18
Secondary spermatocyte, 204-5
Secondary structure, protein, 49, 52
Secondary succession, 856-58
Second filial (F_2) generation, 265
Second law of thermodynamics, 113-14
Second messenger, 36, 103, 662-63
Second trimester, 229
Secretin, 762
Secretion, 69-71, 778
Secretory vacuole, 459
Secretory vesicle, 71
Sedge, 878
Seed, 498-504, 561, 566
  artificial, 559
  development, 568-70
  dispersal, 570-71
  dormancy, 569
  germination, 40-41, 569-72, 586
  naked, 499-501
  structure, 569-70
Seed bank, 504-5
Seed coat, 561, 566, 569-70
Seed fern, 498
Seedless vascular plant, 493, 496-98
Seedling, 586
Seed-producing vascular plant, 493, 498-504. *See also* Angiosperm; Gymnosperm
Segmentation (intestinal motility), 761-62
Segmented worm. *See* Annelid
Segregation, law of, 262-67
Selaginella lepidophylla, 497
Selectin, 103, 106
Selection. *See* Artificial selection; Natural selection

Selectively permeable membrane, 612
Selenium, 185, 756, 898
"Self," 84, 790, 800
Self-assembly, microtubules, 99
Self-pollination, 567
Semen, 204
Semen analysis, 240
Semicircular canal, 653-56
Semiconservative replication, 313-14
Semidesert, 872
Semilunar valve, 720, 722
Seminal fluid, 197, 675
Seminal vesicle, 196-97
Seminiferous tubule, 196, 204-5
Semisynthetic skin, 606
Senescence, plant, 578, 587
Sense(s), 639-57
Sense RNA, 328-29
Sensory adaptation, 644
Sensory deafness, 654
Sensory input, 610
Sensory integration, 610
Sensory neuron, 611, 613, 623, 641
Sensory receptor, 597, 623, 640-43
Sensory system, 640-41
  human, 645-56
  invertebrate, 644-45
  principles of sensory reception, 641-44
  vertebrate, 645
Sepal, 563-65
Septate hypha, 478
Sereno, Paul, 390
Serotonin, 92-93, 255, 618
Setae, 521, 645, 683-84
Severe combined immune deficiency (SCID), 799-800
Sex attractant, 814
Sex chromosome, 198, 290
Sex chromosome aneuploid, 298-300
Sex determination, 290-91
  humans, 290-92
Sex factor, 449
Sex hormone, 670-72
Sex-linked trait, 265, 293
  dominant, 292-93
  recessive, 291-92, 294
Sex pilus, 194-95, 447, 449
Sex ratio, 838
Sex ratio at birth, 849
Sexual dimorphism, 830
Sexually transmitted disease (STD), 253-55, 450
Sexual orientation, 255-56
Sexual reproduction, 10, 14, 192-93
Sexual seasons, 665-66
Sexual selection, 382-83, 831
Shark, 525, 602-3, 646
Sheep, 754
Shelf fungi, 482
Shell, mollusk, 519, 683
Shiga toxin, 786
Shivering thermogenesis, 772
Shoot, 534, 572
Short-day plant, 583-84
Short-tailed weasel, 845
Short-term memory, 630
Shrimp, 522, 884
Shrub, 856
Sickle cell disease, 269, 338-39, 708, 716, 728
  malaria and, 411-12
SIDS. *See* Sudden infant death syndrome
Siemens, Hermann, 280
Sierra dome spider, 830-31
Sieve cell, 539
Sieve plate, 539
Sieve tube member, 539
Sight. *See* Vision
Signal transduction, 85, 103, 337, 631, 633
Signature sequence, 442-43
Silent mutation, 344
Silk, spider, 49
Silkworm, 461
Silurian period, 359, 362
Silver, 184
Simian immunodeficiency virus, 801
Simple carbohydrate, 44
Simple diffusion, 88-89, 91
Simple epithelium, 594
Simple eye, 522
Simple leaf, 542-43
Simpson, George Gaylord, 415
Sinoatrial (SA) node, 720-22
Sinus, 685, 736
Sirenia, 527
Sisal, 536
Six-kingdom classification, 430-31
Skeletal muscle, 102, 596, 598, 703
  circulation and, 724-25
  contraction, 695-99
  macroscopic structure and function, 698-99
  organization, 694-96
  vertebrate, 692-700
Skeletal muscle cell, 692-93
Skeletal system, 599-600
  diversity, 683-87
Skeleton, 394
  cells and tissues of, 687-89
  child, 690-91
  fetal, 687, 690-91
  human, 685-86
  variations on vertebrate theme, 686-87
Skin, 598-603, 790
  semisynthetic, 606
Skin cancer, 168, 184, 186
Skin color, 598, 601
Skin graft, 604
Skinner, B.F., 816
Skinner box, 816
Skull, 635, 685-86
Skunk cabbage, 135
Slash-and-burn deforestation, 904
Sleep, 627-28
"Sleep movement," 582
Sliding filament model, 695-99
SLiME bacteria, 31-32, 54
Slime layer, 446
Slime mold, 458, 470-72
Slipper limpet, 291
Slow-twitch fiber, 703
Slug, 513, 683
Slug (slime mold), 471-72
"Small for gestational age" infant, 246
Small intestine, 757, 759, 761-65
Smallpox, 29, 77, 385
  destruction of stored virus, 78
  eradication, 805
Smallpox vaccine, 805
Smell, 642, 644, 646-47, 649
Smelting, 892, 894
Smith, John Maynard, 812
Smith, William, 378
Smoker's cough, 741
Smoking, 36, 182, 184, 712-13, 731-32, 736, 741
  nicotine addiction, 104-5
  during pregnancy, 247-48
Smooth endoplasmic reticulum, 67-68, 71
Smooth muscle, 596, 598
Smooth muscle cell, 689, 693
Smut, 482-83
Snail, 510, 513, 519-20, 872, 879
Snake, 395, 771
  tongue, 642
Snake venom, 8, 374, 387
Snapdragon, flower color, 273-74
Snapping shrimp, 811-12, 832
Sneeze, 736
Snout, 773-74
Snow, space beneath, 860-61
Snowshoe hare, 610, 844, 878
Snuff, 184
Soay sheep, 844
Social insect, 819-20
Social structure, 823-24
Society, biological, 819-21
Sodium, 38, 756
  deficiency or excess, 38
  in kidney, 783-84
  nerve impulse transmission, 612-14, 617
  salt-acquiring moths, 40
Sodium bicarbonate, 41
Sodium chloride, 34, 38-39
Sodium nitrite, 340-41
Sodium-potassium pump, 92, 612-14
Soft palate, 737
Soft-shelled crab, 683
Softwood, 551
Soil, 872, 907
Solar energy, 111, 858
Solute, 40, 87-88
Solution, 40
Solvent, 40, 87-88
Somaclonal variant, 559
Somatic embryo, 558-59
Somatic mutation, 341
Somatic nervous system, 621-22, 630
Somatostatin, 672, 679
Songbird, 812-13
Soot wart, 184
Sorghum, 161
Sori, 498-99
Sorrel, 582
Sour taste, 647-48
Southern aeshna dragonfly, 645
Soybean, 845
Soy sauce, 475, 481, 483
Space, life from, 353
Space-filling model, 37
Space travel
  circulatory system and, 724
  plants and, 580
Spallanzani, Lazzaro, 352, 371
Spanish moss, 162, 493
Sparrow, 830-31
Specialized cells, 61, 100-102, 170, 594-98
Speciation, 375, 404, 413-16
Species, 9, 12, 426, 433
  definition, 436
  number on earth, 435-36
Specific defense, 792-97

Specific rate of population growth, 840
Spectrin, 101-2
Spemann, Hans, 210
Spemann's organizer, 210, 213
Sperm, 20, 99, 191, 196, 203-5, 217-18
  maturation, 201-6
Spermatid, 204-5
Spermatogenesis, 203-4
Spermatogenous cell, 562
Spermatogonium, 203-5
Sperm bank, 241
Sperm count, 239-40
Spermicide, 250-51
Sperm whale, 824
Sphagnum moss, 495
S phase, 168, 170-71
Sphincter, 760
Sphygmomanometer, 726
Spicule, 515-16, 683
Spider, 510, 522-23, 830-31
Spider monkey, 527
Spider silk, 49
Spiderweb, 823
Spike moss, 497
Spina bifida, 621
Spinal cord, 621-23
  injury, 624-25
Spinal nerve, 622-23, 630
Spine, plant, 544
Spiracle, 522
Spiral cleavage, 513-14
Spiral leaf pattern, 544
Spirilla, 66, 443
Spirogyra, 465
Spiroplasma, 442
Spirotaenia, 464
Splachnum luteum, 494
Spleen, 727-28
Splitters, 433
Sponge, 10, 510, 513-16, 683, 685, 778, 882
  monera within, 446
Spongin, 516
Spongocoel, 515-16
Spongy bone, 689, 691-92
Spongy mesophyll, 543-45
Spontaneous generation, 352-53, 371
Spontaneous mutation, 339-40
Spontaneous process, 114
Sporangium, 498, 560-61
Spore
  algal, 464
  fern, 498
  fungal, 476, 479, 486-87, 489
  plant, 557
Spore mother cell, 560-61
Sporophyte, 192-93, 492-93, 497, 557, 560-62, 566
Sporothrix schenckii, 487
Sporotrichosis, 487
Sporozoa, 461
Sporozoite, 462
Spring turnover, 880
Spruce, 499, 551, 560, 862, 876
Squamous epithelium, 594, 596-97
Squid, 510, 519-20, 645, 683, 708
Squirrel monkey, 876
SRY gene, 229, 291-92
S-shaped curve, 843-44
Stabilizing selection, 411
Stable isotope tracing, 862-63
Stain, chromosomal, 296
Staining characteristics, Monera, 443-44
Stamen, 563-65
Stance, 774-75
Stanozolol, 676
Stapelia, 568
Stapes, 653
Staphylococcus, 387
Staphylococcus aureus, 384, 386
Starch, 44, 46
Starch granule, 580
Starfish, 510, 644, 708-9
Starling, 845
Starvation, 758
Stasis, 415-16
Static cell population, 179-80
Statocyst, 645
Statolith, 645
STD. *See* Sexually transmitted disease
Stegodyphus, 823
Stegosaur, 365
Stem, 541-42
Stem cell, 178-80, 204, 345
Sterilization (birth control), 250-51
Sternum, 685-86
Steroid, plant, 588
Steroid abuse, 676
Steroid hormone, 662
Sterol, 45
Stevens-Johnson syndrome, 594, 607
Stickleback fish, 814-15, 829
Stigeoclonium, 464
Stigma, 563
Stimulus, 10, 640-41, 815
Stimulus intensity, 643
Stinging nettle, 537
Stinkhorn, 482-83
Stipe, 467-68
Stolon, 541-42
Stomach, 757, 759-61, 765
Stomachbeat, 761
Stomach cancer, 181, 184
Stomata, 159-60, 162, 497, 536-38, 542-43, 545
Stone plant, 542
Stonewort, 466
Stop codon, 336, 341
Storage leaf, 544
Stork, 754
Stratified epithelium, 594, 597
Strawberry, 569
Strawberry stem rot, 482
Stream, 881
Strep throat, 746
Streptococcus thermophilus, 66
Streptokinase, 712
Stress, 139-40
Stress test, 727
Stretch receptor, 780
Striped morning sphinx moth, 567
Stroke, 234, 633, 713
Stroma, 151, 158-59
Stromatolite, 352, 390
Structural formula, 37
Sturtevant, Alfred, 288-89
Style, 563
Styracosaur, 392
Subapical region, 547-48
Subatomic particle, 32
Subclavian vein, 727
Subcutaneous fungal infection, 486-87
Subcutaneous layer, 599, 601
Suberin, 548
Submetacentric chromosome, 296
Subnivean space, 860-61
Substrate, 50
Substrate-level phosphorylation, 129-31, 136
Succession, biotic, 856-58
Succinic acid, 132
Succinomonas amylolytica, 441
Succulent, 878
Succulent stem, 541
Sucker (plant), 556-57
Suckerfish, 434-35
Sucking reflex, 246
Suckling, 70-71
Sucrose, 43-45
Sudden infant death syndrome (SIDS), 744-46
Sugar beet, 314
Sugarcane, 44, 160
Sulfanilamide, 123
Sulfate, 140
Sulfur, 35, 756
Sulfur oxides, 892, 894
Sulston, John, 213
Sunburn, 168, 186
Sundew, 576, 581
Sunflower, 415, 541, 575, 578
Superficial fungal infection, 486-87
Superior vena cava, 720-21, 725
Supernormal releaser, 814
Superovulation, 239
Superoxide dismutase, 233
Suprachiasmatic nucleus, 255
Surface area, 64-65
Surface tension, 739
Surface-to-volume ratio, 138
Surfactant. *See* Lung surfactant
Surrogate motherhood, 242-43
Surroundings, 112-13
Survivorship, 841-42
Suspension feeder, 515, 520-21, 524
Sutton, Walter, 286
Suture, 686, 690
Swainson's hawk, 817
Swallowing, 737, 760
Swallowtail butterfly, 642
Swamp, 42
Sweat gland, 601-2, 662
Sweating, 42, 773
Sweat test, 262
Sweetness, 45
Sweet potato, 548
Sweet taste, 647-48
"Swollen glands," 728
Symbion pandora, 508, 528
Symbiosis, 434-35, 441, 458, 473
Symmetry, animal body, 511-12
Symniosome, 473
Sympathetic nervous system, 621-22, 630-31
Symplocarpus foetidus, 135
Synalpheus regalis, 811-12, 832
Synapse, 612, 616
Synapsis, 198
Synaptic cleft, 616-17
Synaptic integration, 617

Synaptic knob, 616
Synaptic transmission, 616-18
Synaptic vesicle, 616
Syncytium, 175
Synesthesia, 646
Synovial fluid, 701
Synovial joint, 700-702
Synovial membrane, 700-701
Synteny, 396
Synthesis. *See* Anabolism
Syphilis, 6, 440
System, 112-13
Systematics, 427
Systemic circulation, 720, 725
Systemic fungal infection, 486-87
Systemic lupus erythematosus, 802
Systemin, 588
Systole, 720
Systolic pressure, 726

T

Tactile messages, 824
Tadpole, 512, 665, 829
Taenia pisiformis, 518
Taiga, 874-77
Tamandua anteater, 427
Tambopata, 428
Tamoxifen, 18, 29, 184
Tampon, 384, 386
Tannin, 581
Tapeworm, 517-18, 716
Taproot system, 546-47
Taq1 DNA polymerase, 321
Tardigrade, 42-43
Target cell, 660
Tarsal, 686
Tarsier, 527
Tasmanian wolf, 381
Taste, 642, 644-45, 647-49
Taste bud, 647-49
Taste sense cortex, 649
TATA box, 327
Tautomer, 339
Taxol, 99, 499
Taxon, 12
Taxonomy, 12, 14, 425-36, 913-15
    challenges, 433-35
    characteristics to consider, 428
    evolution, 429-35
    historical aspects, 427-28
    new organisms, 428
Tay-Sachs disease, 72, 86, 269, 410, 412, 633
T cell(s), 84, 178-79, 254, 345, 719, 728, 790-94, 797-98
T cell receptor, 797
Tea, 501
Tears, 790
Tectonic plates, 417-18
Tectorial membrane, 654
Tectum, 626-27
Tegument, 517-18
Telencephalon, 627
Telocentric chromosome, 296
Telomerase, 175-76
Telomere, 175-76, 296
Telophase
    meiosis I, 199, 201
    meiosis II, 201-2
    mitotic, 171, 173-74
Temperate coniferous forest, 873-76
Temperate deciduous forest, 873-76
Temperate forest, 905
Temperature
    enzyme function and, 13, 121-22
    water and, 42-43
Temperature inversion, 892
Temperature regulation, 770-75, 821-22
Temperature-sensitive allele, 275-76
Temperature units, 909
Temporal bone, 686
Temporal caste, 820
Temporal isolation, 413
Temporal lobe, 256, 628
Tendon, 694, 696
Tendril, 541-42, 544, 579
Tennis elbow, 700
Tent caterpillar, 821
Teratogen, 246-49
Teratoma, 207
Termite, 434, 458, 662, 820, 873, 877-78
Tern, 889
Terrestrial biome, 872-79
Terrestrial ecosystem, 858-59
Territoriality, 827-28
Territorial marker, 828
Tertiary consumer, 862, 864, 869
Tertiary epoch, 361
Tertiary structure, protein, 49, 52
Test (shell), 457-58, 523
Test cross, 266
Testicular cancer, 20
Testis, 196-97, 205, 661, 675, 678-79
Testosterone, 45, 663, 667, 672, 675-76, 679
Test tube baby, 238-39
Tetanus, 449, 699-700
Tetanus toxin, 94
Tetracycline, 248
Tetrahydrocannabinol, 21
Tetrapod, 362
Thalamus, 626-27, 649
Thalassemia, 716
Thalidomide, 246-47, 276
Thallus, 464, 477
Thecodont, 365-66
Theory, 18
Therapsid, 365
Thermal cycler, 321
Thermal stratification, 879-80
Thermocline, 879-80, 885
Thermodynamics, 112-14
Thermogenesis, 135, 772
Thermogenin, 135
Thermonasty, 581
Thermophile, 13, 451-52
Thermoplasma, 69, 451
Thermoregulation. *See* Temperature regulation
Thermotropism, 581
Thermus aquaticus, 13, 321, 453
Theropod, 389, 401
Thiamine, 755
Thick myofilament, 695, 697
Thigmomorphogenesis, 582
Thigmonasty, 581-82
Thigmotropism, 579, 581
Thin myofilament, 695, 697
Third trimester, 229
Thoracic cavity, 742
Thoracic duct, 727
Thoracic nerve, 625
Thoracic vertebrae, 685, 687
Thorn, 541-42
Thoroughfare channel, 724
Thottea, 543
Threat display, 828, 830
Three-chambered heart, 710-11
Three-kingdom classification, 429-30
Throat cancer, 184
Thrombin, 48, 717
Thromboplastin, 717
Thrombus, 717
Thrush, 481, 488
Thuja occidentalis, 860-61
Thylakoid, 75, 151-52, 155
Thylakoid space, 151
Thymine, 51, 311, 313, 326
Thymine dimer, 319
Thymus gland, 178, 232, 661, 727-28, 793, 798
Thyroid cancer, 60, 80, 184, 342
Thyroid follicle, 661
Thyroid gland, 138, 661, 665, 669-70, 678-79
Thyroid-stimulating hormone (TSH), 667-70, 679
Thyroxine, 138, 661, 665, 669-70, 679, 772, 829
Tibia, 686
Tick, 522
Tidal volume, 742
Tide pool, 423, 451
Tiger, 828
Tiger salamander, 826-27
Tight junction, 96-97
Tilapia, 605
Time, 910
Tinbergen, Niko, 812
Tinnitus, 654
Tissue, 5, 7-8, 509, 595
    cell division in, 178-80
    replacement of damaged tissue, 603-6
Tissue culture, 26
Tissue engineering, 603, 606
Tissue plasminogen activator, 712-13
Tmesipteris, 497
Toad, 413
Toadfish, 828
Toadstool, 482
Tobacco, 583
Tocopherol, 755
Tofu, 481
Tomato, 569, 577, 588
    FlavrSavr, 334
Tomato bushy stunt, 77
Tongue, 647, 649, 736, 760
    snake, 642
Tonicity, 88-89
Tonsils, 728
Tooth, 760
Tortoiseshell butterfly, 110
Touch, 656
Touch receptor, 601
Toxic shock syndrome, 384, 386
Toxin, 386
Toxoplasma gondii, 461-63
Toxoplasmosis, 461-63
Trace element, 32, 34
Trace mineral, 756

Trachea, 736-39
Tracheae, 522
Tracheal mite, 573, 823
Tracheal system, 733-34
Tracheid, 538-39
Tracheole, 522
Trade winds, 885
Transcription, 326-29
Transcription factor, 327, 565
Transferrin, 97
Transfer RNA (tRNA), 329-30
  translation, 333-35
"Transformer" mutation, 291
Transforming principle, 309-10
Transgenic technology, 334-35
Transitional form, 415
Transitional zone, 860
Translation, 326, 331-37
Translocation, 298, 301-2
Translocation carrier, 301
Transmission electron microscope, 62
Transmission genetics, 261-81
Transpiration, 537-38, 865
Transpiration-cohesion hypothesis, 540
Transplantation, 603, 804
  controversial transplants, 605
  nonhuman parts into humans, 605
  transplant types, 604
Transposable element, 343
Transverse tubules, 694
Tree of life, 912
Tree rings, 534, 550-53
Trekboers, 408
Treponema pallidum, 440
Triassic period, 359, 365, 417
Triceps, 700, 702
Triceratops, 366
Trichina worm, 519
Trichinella spiralis, 519
Trichome, 537-38, 541
Trichomonas vaginalis, 461
Trichomoniasis, 461
Trichonympha, 458
Tricolored mouse, 213
Tricuspid valve, 720-21
Tricyclic antidepressant, 618
Triglyceride, 45, 47
Triiodothyronine, 669-70, 679
Trilobite, 362-63, 417, 523
Triose, 43
Triple bond, 37
Triplet repeat, 631, 633
Triploblastic embryo, 511
Triplo-X, 299
Tris, 341
Trisomy, 298-99
Trisomy 13, 298
Trisomy 18, 298
Trisomy 21, 298, 300
tRNA. *See* Transfer RNA
Trochophore larva, 519-21
Trombiculid mite, 523
Trophic level, 859, 863-64, 866
Trophoblast, 218, 220
Trophozoite, 461
Tropical rain forest, 428, 436, 872-76, 904-5
Tropic hormone, 666
Tropism, 578-81
Tropomyosin, 695-96
Troponin, 695-96
True-breeding plant, 263-64, 286
True bug, 436
True fern, 493, 497-98
Truffle, 481
Trypanosoma brucei gambiense, 461
Trypanosoma brucei rhodiense, 461
Trypanosomiasis, 872
Trypsin, 762-64
Tryptophan, 618
Tsetse fly, 872
TSH. *See* Thyroid-stimulating hormone
T tubules. *See* Transverse tubules
Tubal ligation, 250-51
Tube feet, 524, 709
Tube nucleus, 562-63, 566
Tuber, 541-42
Tuberculosis, 234, 384, 404, 412, 420, 716, 746, 799
Tuber melanosporum, 481
Tube worm, 452-53, 507-8
Tubulin, 48, 96, 99, 170, 175
Tule tree, 550
Tulip, 425
Tulip poplar, 856-57
Tumbleweed, 571
Tumor necrosis factor, 791, 797
Tumor suppressor, 48
Tumor suppressor gene, 183-84
Tundra, 874-75, 878
Tunic, 526
Tunicate, 525-26
Turgor pressure, 90
Turkey tail bracket fungus, 477
Turner, Henry, 298
Turner syndrome, 298-99
Turpentine, 501
Turtle, 291, 434-35, 774
  frozen, 44
Tuskegee syphilis study, 6
Twins, 245
  conjoined, 238, 256
  dizygotic (fraternal), 218, 223, 279-80
  monozygotic (identical), 10, 223, 255, 279-80
Twin study, 279-80
Twitch, 699
Two-chambered heart, 710-11
Two-opening digestive system, 753-54
Tympanal organ, 645
Tympanic canal, 653
Tympanic membrane, 653
Typhoid fever, 443, 716
Typhus, 385
Tyrannosaurus rex, 391

U

Ulcerative colitis, 802
Ulna, 686
Ultratrace element, 32
Ultraviolet radiation, 147-48, 168, 184, 186, 319, 341, 850, 894
Ulvaria, 100
Umbilical artery, 226, 228
Umbilical cord, 216, 221, 223, 226
Umbilical cord blood, 226, 345, 718
Umbilical vein, 226, 228
Unconditioned stimulus, 815
Uncoupler, 135
Underwater sound, 643
Underweight, 138
Unicellular organism, 10, 60
Uniformitarianism, 376-78
Uniramia, 522
Uniramous appendage, 522
Unsaturated fatty acid, 45, 47
Unstable angina, 712
Upwelling, 883, 885
Uracil, 51, 326
Uranium, 893
Urbanization, 385
Urea, 776-77
Ureter, 780
Urethra, 196-97, 780
Urey, Harold C., 354
Uric acid, 776-79
Urinalysis, 676, 785
Urinary bladder, 184, 196-97, 780
Urinary sphincter, 780
Urinary system, 599-600
  disorders, 785-86
  human, 780-85
Urinary tract cancer, 184
Urinary tract infection, 785
Urine, 90-91, 780-85
Urochordata, 525-26
Usnea, 484
Uterine cancer, 184
Uterine contraction, 229-30, 664
Uterus, 197, 218, 221
Utricle, 655
Uvula, 736

V

Vaccine, 29, 77-78, 385, 454, 803, 805
Vacuole, 67-68, 75
Vagina, 197
Vaginal orifice, 197
Valence electron, 35-38
Valium, 97, 618, 634
Valley fever, 476, 487, 489
van der Waals attractions, 39-40
van Helmont, Jan, 148
Variable, 21, 25
Variable expressivity, 275
Variable region, 795
Varicose vein, 239
Variola. See Smallpox
Varroa mite, 573
Vasa recta, 783
Vascular bundle, 158, 541
Vascular cambium, 550-51
Vascular plant
  seedless, 493, 497-98
  seed-producing, 493, 498-504
Vascular tissue, plant, 534-35, 537-59
Vas deferens, 196-97, 204-5
Vasectomy, 250-51
Vasomotor center, 726
Vector, 461

"Vegetative force," 352, 371
Vein, 708, 723-25
  leaf, 543
Vein ending, 543
Velociraptor, 389
Velvet worm, 400
Venom. *See* Snake venom
Venous valve, 724-25
"Ventral," 512-13
Ventricle
  brain, 621
  heart, 710-11
Ventricular fibrillation, 712, 722-23
Venule, 708, 724
Venus's flytrap, 537, 544-45, 578, 581
Vernix caseosa, 229
Vertebral column, 622-23, 635, 685-87
Vertebrata, 508, 525-28
Vertical stratification, 872-73, 882
Vervet monkey, 822
Vesicle, 75
Vesicle trafficking, 94-95
Vessel element, 538-39
Vestibular canal, 653
Vestibule, 653-56
Vestigial organ, 395
Vetter, David, 799-800
Vibrio, 66
Vibrio cholerae, 411, 440, 449, 454
Vibrio vulnificus, 449
Viceroy butterfly, 641
Victoria, Queen of England, 294
Villi, 762-64
Violent behavior, 281
Violin spider, 523
Virchow, Rudolph, 62
Viroid, 77
Virulence, 420
Virus, 75-77
Visceral mass, 519-20
Visible Human Project, 606
Visible light, 147-48
Vision, 642, 644
Visual cortex, 652
Visual messages, 824-26
Visual system, 648-52
  converting light energy to neural messages, 650-52
  disorders, 651-52
  focusing the light, 649-50
Vital capacity, 742
Vitamin, 119-20
  fallacies and facts, 756
  fat-soluble, 755
  produced by intestinal flora, 763
  water-soluble, 755
Vitamin A, 248, 650, 755
  deficiency, 185
Vitamin B_1. *See* Thiamine
Vitamin B_2. *See* Riboflavin
Vitamin B_3. *See* Niacin
Vitamin B_6. *See* Pyridoxine
Vitamin B_{12}. *See* Cyanocobalamin
Vitamin C, 248-49, 717, 755
  deficiency, 185, 249
Vitamin D. *See* Cholecalciferol
Vitamin E. *See* Tocopherol
Vitamin K, 717, 755
Vitis vinifera, 314
Vitreous humor, 649-50
Vocal cords, 736
Volcano, 42, 356, 375, 418, 446, 842-43, 857-58, 885-86, 896
Vole, 342
Volume, 909-10
Volvox, 465-66
Vomeronasal organ, 642
Vomiting, 758
von Baer, Karl Ernst, 210, 212
von Frisch, Karl, 812
Vulture, 754, 862

W

Waardenburg syndrome, 397-98
Waggle dance, 825-26
Walking fern, 557
Walkingstick, 522
Wallace, Alfred Russel, 382
Warble, 854
Warbler, 837
Warren, J. Robin, 20, 750
Wasp, 207, 291, 523, 556, 820
Wasting syndrome, 254
Water, 40-43
  adhesiveness, 40-41, 540
  cohesiveness, 40-41, 43, 540
  conservation, 770, 776, 783-84
  movement in plants, 40-41, 540
  osmosis, 88-91
  requirement in humans, 755
  structure, 38
  temperature and, 42-43
Water balance, 776
Waterbear. *See* Tardigrade
Water-borne disease, 450-51
"Water breaking," 229
Water cycle, 865
Water flea, 522, 821
Water lily, 879
Water mold, 458, 470
Water pollution, 897-98, 900-901
Water potential, 540
Water-soluble hormone, 662-63
Water-soluble vitamin, 755
Water table, 865
Water vascular system, 524
Watson, James, 311
Wavelength, 147-48
Wax, 45-46, 48
Wax palm, 536
Wax plant, 162
Weather patterns, 885
Weaver ant, 828
Weddell seal, 838
Weight lifting, 702
Weight loss program, 766
Weinberg, W., 405
Weizmann, Chaim, 140
Welwitschia, 500
Welwitschia mirabilis, 541
Werner syndrome, 233
Wetlands, 899
Whale, 395, 527, 602, 772, 854-55
Wheat, 298, 547
Wheat leaf blotch, 485
Wheat rust, 485
Wheelis, Mark, 430
Whiptail lizard, 290
Whirling disease of fish, 461
Whisker, 656
Whisk fern, 493, 496-97
White, Timothy, 368
White blood cells, 84, 103, 106, 594, 596-97, 714, 716-17
White Cliffs of Dover, 458
White-crowned sparrow, 812-13
White-fronted bee-eater, 826
White matter, 615, 622-23
White meat, 703
White spruce, 501
Whittaker, Robert, 429-30
Whole-body metabolism, 138
Whooping cough. *See* Pertussis
Whorled leaf pattern, 542
Wiggle-matching, 553
Wild carrot, 857
Wild dog, 845
Wildlife census, 839
Wild type, 265, 343
Wilkins, Maurice, 311
Wilms tumor, 183
Wilson disease, 33
Wilting, 90
Wind-dispersed fruit, 571
Windpipe. *See* Trachea
Wind pollination, 568
Winfrey, Oprah, 766
Wing
  bird, 686-87
  insect, 692
Winterberry, 571
Withdrawal symptoms, 105, 634
Woese, Carl, 430
Wolf, 381, 771, 817, 824, 828
Wolff, Kaspar Friedrich, 210
Wolffia, 541
Wolf moss lichen, 484
Wolf spider, 829-30
Wood, 505, 550-51
Wood products, 501
Wood sorrel, 587
Wood tick, 523
Worker, 820-21
Wound healing, 177
Wrinkle, 602
Wuchereria brancrofti, 519

X

Xanthoma, 48
Xanthophyll, 150, 466
X chromosome, 289-92
  inactivation, 293-95
  X-to-autosome ratio, 290
Xenograft, 604
Xenopus laevis, 211, 227
Xeroderma pigmentosum, 319, 322
X-inactivation center, 293-95

X ray(s), 184, 340
X-ray diffraction, DNA, 311
XX male, 192, 207, 291
XY female, 291
Xylem, 497, 534, 538-40
XYY male, 300

Yam, 550
Y chromosome, 290-92
Yeast, 14, 139, 192, 477, 481, 487-88, 662
Yeast genome project, 488
Yeast infection, 476, 488
Yellow cake, 893
Yellow-green algae, 467
Yellow marrow, 689, 692
Yellow nutsedge, 845
Yersinia pestis, 435, 450
Y linkage, 293
Yogurt, 66
Yolk sac, 221, 223, 226

Z

Zamia, 499-500
Zebra mussel, 899
Zellweger syndrome, 74
ZIFT. *See* Zygote intrafallopian transfer
Zimmerman, Deborah, 509
Zinc, 756, 907
Zinc finger, 327
Z line, 697
Zona pellucida, 217-19, 251-52
Zone of cell differentiation, 547-48
Zone of cell division, 547-48
Zone of cell elongation, 547-48
Zoo, 427
Zooflagellate, 460-61
Zoonosis, 385
Zooplankton, 458, 863, 879, 883-84, 899
Zooxanthellae, 883
Z-scheme, 154-55
Zygomatic bone, 686
Zygomycetes, 478-81, 485
Zygospore, 479, 481
Zygote intrafallopian transfer (ZIFT), 244